D1134687

NEUROBIOLOGY OF BRAIN DISORDERS

NEUROBIOLOGY OF BRAIN DISORDERS

BIOLOGICAL BASIS OF NEUROLOGICAL AND PSYCHIATRIC DISORDERS

Edited by

MICHAEL J. ZIGMOND
Departments of Neurology, Neurobiology, and Psychiatry, University of Pittsburgh, Pittsburgh, Pennsylvania, USA

LEWIS P. ROWLAND
Neurological Institute, Columbia University Medical Center, New York, USA

JOSEPH T. COYLE
Harvard Medical School, McLean Hospital, Belmont, Massachusetts, USA

AMSTERDAM • BOSTON • HEIDELBERG • LONDON
NEW YORK • OXFORD • PARIS • SAN DIEGO
SAN FRANCISCO • SINGAPORE • SYDNEY • TOKYO
Academic Press is an imprint of Elsevier

Academic Press is an imprint of Elsevier
32 Jamestown Road, London NW1 7BY, UK
225 Wyman Street, Waltham, MA 02451, USA
525 B Street, Suite 1800, San Diego, CA 92101-4495, USA
The Boulevard, Langford Lane, Kidlington, Oxford OX5 1GB, UK

ISBN: 978-0-12-398270-4

British Library Cataloguing-in-Publication Data
A catalogue record for this book is available from the British Library

Library of Congress Cataloging-in-Publication Data
A catalog record for this book is available from the Library of Congress

Typeset by TNQ Books and Journals
www.tnq.co.in

Printed and bound in China

Dedication

To our students and patients who, over the years, have motivated us to produce this book

and

To Nancy Wexler, whose commitment to research and education about brain disorders has been an inspiration to each of us.

Contents

7. Rett Syndrome: From the Involved Gene(s) to Treatment

CHARLOTTE KILSTRUP-NIELSEN, NICOLETTA LANDSBERGER

8. Fragile X-Associated Disorders

SCOTT M. SUMMERS, RANDI HAGERMAN

II

DISEASES OF THE PERIPHERAL NERVOUS SYSTEM

9. Introduction

HENRY J. KAMINSKI

10. Myasthenia Gravis

LINDA L. KUSNER, HENRY J. KAMINSKI

11. Muscular Dystrophy

SAŠA A. ŽIVKOVIĆ, PAULA R. CLEMENS

12. Peripheral Neuropathies

MARIO A. SAPORTA, MICHAEL E. SHY

13. Diabetes and Cognitive Dysfunction

CATRINA SIMS-ROBINSON, BHUMSOO KIM, EVA L. FELDMAN

III

DISEASES OF THE CENTRAL NERVOUS SYSTEM AND NEURODEGENERATION

14. Introduction

ELENA CATTANEO, ALESSANDRO VERCELLI

15. Spinal Cord Injury

ALESSANDRO VERCELLI, MARINA BOIDO

IV

INFECTIOUS AND IMMUNE-MEDIATED DISEASES AFFECTING THE NERVOUS SYSTEM

V

DISEASES OF HIGHER FUNCTION

33. Disorders of Frontal Lobe Function

PETER PRESSMAN, HOWARD J. ROSEN

34. Stress

BRUCE S. McEWEN

35. Addictions

EDUARDO R. BUTELMAN, ROBERTO PICETTI, BRIAN REED, VADIM YUFEROV, MARY JEANNE KREEK

36. Sleep Disorders

BIRGITTE RAHBEK KORNUM, EMMANUEL MIGNOT

37. Fear-Related Anxiety Disorders and Post-Traumatic Stress Disorder

ARSHYA VAHABZADEH, CHARLES F. GILLESPIE, KERRY J. RESSLER

38. Obsessive–Compulsive Disorder

NASTASSJA KOEN, DAN J. STEIN

39. Schizophrenia

GLENN T. KONOPASKE, JOSEPH T. COYLE

40. Bipolar Disorder

HEINZ GRUNZE

41. Pain: From Neurobiology to Disease

MICHAEL S. GOLD, MIROSLAV "MISHA" BACKONJA

42. Migraine

DAVID BORSOOK, NASIM MALEKI, RAMI BURSTEIN

43. Depression and Suicide

MAURA BOLDRINI, J. JOHN MANN

VI

DISEASES OF THE NERVOUS SYSTEM AND SOCIETY

44. Introduction

MICHAEL J. ZIGMOND

45. Advances in Ethics for the Neuroscience Agenda

JUDY ILLES, PETER B. REINER

46. Burden of Neurological Disease

MITCHELL T. WALLIN, JOHN F. KURTZKE

47. Stress, Health, and Disparities

ZINZI D. BAILEY, DAVID R. WILLIAMS

Preface

Interest in understanding the basis of neurological and psychiatric disorders is thousands of years old. People of China and India, as well as the Egyptians and Greeks, all had ideas about how the brain worked and what caused the occasional functional abnormalities that they observed. Moreover, they often developed interventions to relieve symptoms, if not treat the disease. Indeed, the origins of neuroscience probably go back even farther. For example, trephination of the skull is thought to have been practiced as long as 7000 years ago and may have been designed to release evil spirits believed to be the cause of brain disorders. Since then, some of the ancient treatments have been found to be quite effective and have even served as the basis for much more recent interventions. However, the modern era of inquiry into the neurobiological basis of brain disorders did not begin until the nineteenth century. Several milestones along the path of that inquiry can be identified; here we mention just a few.

Rauwolfia serpentina is a shrub from which the people of India have been making a medicinal tea for thousands of years.[1,2] Among the many conditions for which it was used was "moon disease", which we now recognize as psychosis. In the early 1950s it was determined that most of the tranquilizing effects of the plant extracts resulted from a compound that was named reserpine. Over the next decade, Arvid Carlsson and colleagues, working first at the US National Institutes of Health, then at the University of Lund, and finally at the University of Göteborg, Sweden, demonstrated that the effects of this natural product were due to its depletion of the neurotransmitter dopamine from the striatum, as described in the Nobel Lecture by Arvid Carlsson.[3] This led to several key observations, including the discovery by Oleh Hornykiewicz in Vienna that Parkinson disease (PD) was associated with a loss of striatal dopamine and that many of the motor symptoms of PD could be reversed by administration of the dopamine precursor, L-dopa (see Chapter 19).[3,4]

The use of reserpine as a treatment for psychosis, together with the discovery of chlorpromazine for the treatment of schizophrenia and the realization in 1963 that it, too, acted by reducing dopaminergic transmission,[5] led to the focus on reducing dopaminergic transmission to treat schizophrenia (see Chapter 39). Likewise, the observation that a loss of dopamine was associated with PD, and that the behavior of reserpinized animals and patients with PD could both be improved by L-dopa, resulted in the use of drugs that activate dopamine receptors in the treatment of PD. This sequence of events, conducted over a period of less than 10 years, is a landmark in the use of behavioral and neurochemical approaches for studying the nervous system, and was largely responsible for initiating the twin fields of neuropharmacology and biological psychiatry.

There have been many other such moments in the emergence of biological approaches to neurological and psychiatric disorders. For example, Ernst Wilhelm von Brücke and colleagues, as well as their students (e.g. Sigmund Freud), working in Austria during the latter half of the nineteenth century, were among the first to apply laboratory methods to the study of the nervous system and to suggest that behavior could be understood through an understanding of biological events. The introduction of electrophysiology into neuroscience can be traced as far back as the seventeenth century to the work of Jan Swammerdam in Holland, although it is Luigi Galvani, working in Italy in the nineteenth century using nerve–muscle preparations, who is usually credited with initiating electrophysiology as an approach for understanding how the nervous system functions.[6] Neuropathology was introduced by Paul Oscar Blocq and Georges Marinesco in the late nineteenth century in Paris. During a postmortem examination, they found a tumor in the contralateral substantia nigra of a patient who had exhibited the symptoms of PD, as reviewed by Catala and Poirier.[7] In short, many of the principal tools for understanding the neurobiology of brain disorders – neuropathology, histochemistry, electrophysiology, biochemistry, and behavior – gradually emerged over the past 250 years as a result of investigators working in many different areas of the world. In the 1970s, two more approaches were added, molecular neurobiology and brain imaging. (For an excellent treatise on the history of neuroscience, see *Origins of Neuroscience: A History of Explorations into Brain Function*, by Stanley Finger,[8] and excellent articles in *The Journal of the History of Neuroscience*. For a timeline and an extensive bibliography of the history of neuroscience, see also the website of Eric Chudler at the University of Washington.[9] Additional material can be found on the website of the Society for Neuroscience.[10])

Our decision to assist in the teaching of the neuroscience of brain disorders by preparing this textbook began to take shape over three decades ago. The Marine Biological Laboratory (MBL) in Woods Hole, Massachusetts (USA) twice played a role in the origins of the project, as it has in the development of neuroscience more generally.[11,12] The first event occurred on a rainy weekend afternoon in 1979, when Edward Kravitz invited two individuals to speak on the neurobiology course that he was co-teaching there. They were Nancy Wexler, then a program officer at the US National Institute of Neurological Diseases and Stroke, and Marjorie Guthrie, the widow of Woody Guthrie. Marjorie spoke movingly about how Woody's Huntington disease affected him and their entire family; Nancy also commented on the disease. After the presentations, Marjorie, Nancy (who was to become the president of the Huntington's Disease Foundation and whose family has also suffered from that condition), Ed, Michael Zigmond, and several others on the course went to "The Captain Kidd", a popular hangout in Woods Hole, to continue the discussion. The group immediately began to talk about how moving the presentations by Marjorie and Nancy had been and how valuable it would be to expose others in the field to such experiences. Ed took this idea and ran with it, obtaining funding from the National Institutes of Health to underwrite the "Neurobiology of Disease" workshop now held each year just before the annual meeting of the Society for Neuroscience.

The second event was a six-day workshop for faculty on teaching about the neurobiology of disease in which the three editors of the present textbook (and many others) taught during August 2011. The objective was to provide the participants with information and instructional methods that would allow them to go back to their home institutions and mount, or substantially improve, a course on the neurobiology of disorders. Much of the impetus for moving from courses to a textbook – and a few of the book's authors (Ann McKee, Robert Brown) and consultants (Gerald Fischbach, Donald Price) – arose from that workshop. The hope was – and remains – that through this book still others will be able to develop courses on the neurobiology of disease. This textbook is not complete; there are separate chapters on the role of inflammation but not mitochondrial dysfunction, on PD but not Tourette syndrome, on depression but not anxiety, on traumatic brain injury, but not brain tumors. These and several other topics must await a second edition.

But this raises the question: Why this abiding interest in helping to stimulate training in the neurobiology of disease? It is not because we believe that basic research in this field is less important than research that more directly confronts disease. On the contrary, virtually all of our current understanding of the biological basis of brain disorders stems from discoveries made in basic science laboratories, as the examples given at the beginning of this Preface indicate (see also the excellent series of pamphlets produced by the Society for Neuroscience, "Research and Discoveries"[13]). However, knowing more about disorders of the nervous system can motivate researchers to work even harder, and who among us does not want their work to eventually make a difference in the lives of others? Moreover, we firmly believe in the aphorism of Louis Pasteur that "chance favors the prepared mind". We hope this textbook will aid in that preparation.

Michael J. Zigmond, PhD
Lewis P. Rowland, MD
Joseph T. Coyle, MD

References

1. Sen G, Bose K. *Rauwolfia serpentina*, a new Indian drug for insanity and hypertension. *Indian Med World*. 1931;21:194–201.
2. Lele RD. Beyond reverse pharmacology: mechanism-based screening of Ayurvedic drugs. *Ayurveda Integr Med*. 2010;1:257–65.
3. Carlsson A. A half-century of neurotransmitter research: impact on neurology and psychiatry (Nobel Lecture). *Chembiochem*. 2001;2:484–93. For a video of this lecture, see http://www.nobelprize.org/nobel_prizes/medicine/laureates/2000/carlsson-lecture.html.
4. Hornykiewicz O. The discovery of dopamine deficiency in the parkinsonian brain. *J Neural Transm Suppl*. 2006;70:9–15.
5. Baumeister AA. The chlorpromazine enigma. *J Hist Neurosci*. 2013;22:14–29.
6. Verkhratsky A, Krishtal OA, Petersen OH. From Galvani to patch clamp: the development of electrophysiology. *Pflugers Arch*. 2006;453:233–47.
7. Catala M, Poirier J. Georges Marinesco (1863–1938): neurologist, neurohistologist and neuropathologist. *Rom J Morphol Embryol*. 2012;53:869–77.
8. Finger S. *Origins of Neuroscience: A History of Explorations into Brain Function*. Oxford: Oxford University Press; 2001.
9. Chudler E. Milestones in neuroscience research. http://faculty.washington.edu/chudler/hist.html
10. Society for Neuroscience. History of neuroscience. http://www.sfn.org/About/History-of-Neuroscience/History-Resources.
11. Zottoli SJ. How the early voltage clamp studies of José del Castillo inform "modern" neuroscience. *Neuroscientist*. 2012;18:415–21.
12. Maienschein J. Neurobiology a century ago at the Marine Biological Laboratory, Woods Hole. *Trends Neurosci*. 1990;13:399–403.
13. Society for Neuroscience. The research and discoveries series. www.brainfacts.org.

Acknowledgments

This textbook has been a long time in gestation. Some time around 2005, Michael Zigmond was approached by Johannes Menzel, then at the Academic Press division of Elsevier, with a proposal to organize a textbook such as this one. Donald Price was soon brought into the conversations and over the next few years played a major role in shaping the project, providing suggestions for both topics and authors. Gerald Fischbach was also a source of excellent advice. In the end, we three agreed to carry the project through to its conclusion.

Although our initial editor at Elsevier, Susan Lee, helped to get the project started, it was Mica Haley and editorial project manager April Farr who made it happen – being amazingly patient with us and the authors as one deadline after another came and went. And, in addition to being patient, Mica provided invaluable suggestions at virtually every step along the way.

Working with Michael at the University of Pittsburgh, Susan Giegel and later Beth Fischer provided essential administrative assistance. Finally, we greatly appreciate the help of all those involved in the production of this textbook, including our copy editor, Charlotte Pover and project manager Chris Wortley.

The royalties generated from this book will be used primarily to support the purchase and distribution of this textbook to trainees in developing countries.

No grant support was specifically obtained for this project. However, our institutions, the University of Pittsburgh, Harvard Medical School/McLean Hospital, and Columbia University provided us with the facilities within which to carry out the work and, in some cases, support for our salaries.

List of Contributors

Stanley H. Appel Department of Neurology, Methodist Neurological Institute, Houston, Texas, USA

Miroslav "Misha" Backonja Departments of Neurology, Anesthesiology, and Rehabilitation Medicine, University of Wisconsin School of Medicine and Public Health, Madison, Wisconsin, USA; CRILifetree, Salt Lake City, Utah, USA

Zinzi D. Bailey Department of Social and Behavioral Sciences, Harvard School of Public Health, Boston, Massachusetts, USA

David R. Beers Department of Neurology, Methodist Neurological Institute, Houston, Texas, USA

Tommy K. Begay Norton School of Family and Consumer Sciences, College of Agriculture and Life Sciences, University of Arizona, Tucson, Arizona, USA

Anna Berti Psychology Department, University of Turin, Turin, Italy; Neuroscience Institute of Turin (NIT), University of Turin, Turin, Italy

Marina Boido Department of Neuroscience, Neuroscience Institute Cavalieri Ottolenghi, Turin, Italy

Maura Boldrini Columbia University, New York State Psychiatric Institute, New York, USA

David Borsook PAIN Group, Department of Anesthesia, Boston Children's Hospital and Harvard Medical School, Boston, Massachusetts, USA

R.H. Brown Jr. Department of Neurology, University of Massachusetts Medical School, Worcester, Massachusetts, USA

Rami Burstein Department of Anesthesia, Beth Israel Deaconess Hospital, Harvard Medical School, Boston, Massachusetts, USA

Vera Joanna Burton The Johns Hopkins University/ Kennedy Krieger Institute Residency in Neurodevelopmental Disabilities, Baltimore, Maryland, USA

Eduardo R. Butelman Laboratory on the Biology of Addictive Diseases, The Rockefeller University, New York, USA

Louis R. Caplan Beth Israel Deaconess Medical Center, Boston, Massachusetts, USA

F. Xavier Castellanos NYU Child Study Center, NYU Langone Medical Center, New York, USA; Nathan Kline Institute for Psychiatric Research, Orangeburg, New York, USA

Elena Cattaneo Department of Biosciences, University of Milan, Italy

Paula R. Clemens Neurology Service, Department of Veterans Affairs, University of Pittsburgh, Pennsylvania, USA, and Department of Neurology, University of Pittsburgh School of Medicine, Pittsburgh, Pennsylvania, USA

Samuele Cortese Child Neuropsychiatry Unit, Life and Reproduction Sciences Department, Verona University, Verona, Italy; NYU Child Study Center, NYU Langone Medical Center, New York, USA

Chiara Cossetti Department of Clinical Neurosciences, John van Geest Centre for Brain Repair, Wellcome Trust–MRC Stem Cell Institute and NIHR Biomedical Research Centre, University of Cambridge, Cambridge, UK

Joseph T. Coyle Harvard Medical School, McLean Hospital, Belmont, Massachusetts, USA

Daniel H. Daneshvar Center for the Study of Traumatic Encephalopathy, Alzheimer's Disease Center, Department of Neurology

Mahlon R. DeLong Department of Neurology, School of Medicine, Emory University, Atlanta, Georgia, USA; Udall Center of Excellence in Parkinson's Disease Research, Emory University, Atlanta, Georgia, USA

Eva L. Feldman Department of Neurology, University of Michigan, Ann Arbor, Michigan, USA

Francesca Garbarini Psychology Department, University of Turin, Turin, Italy

Thomas Gasser Department for Neurodegenerative Diseases, Hertie Institute for Clinical Brain Research, German Center for Neurodegenerative Diseases, University of Tübingen, Tübingen, Germany

Charles F. Gillespie Department of Psychiatry and Behavioral Sciences, Emory University School of Medicine, Atlanta, Georgia, USA

Michael S. Gold Pittsburgh Center for Pain Research, Department of Anesthesiology; Center of Neuroscience; Departments of Neurobiology and Medicine; Division of Gastroenterology Hepatology and Nutrition, University of Pittsburgh

Mary Lee Gregory The Johns Hopkins University/Kennedy Krieger Institute Residency in Neurodevelopmental Disabilities, Baltimore, Maryland, USA

Heinz Grunze Institute of Neuroscience, Newcastle University, Newcastle upon Tyne, UK

Randi Hagerman Medical Investigation of Neurodevelopmental Disorders (MIND) Institute, Department of Pediatrics, University of California – Davis, Sacramento, California, USA

James C. Harris Psychiatry and Behavioral Sciences, Pediatrics, Mental Health, and History of Medicine, The Johns Hopkins University School of Medicine, Baltimore, Maryland, USA

Norman J. Haughey Department of Neurology, Division of Neuroimmunology and Neurological Infections, The Johns Hopkins University School of Medicine, Baltimore, Maryland, USA

Judy Illes National Core for Neuroethics, Division of Neurology, Department of Medicine, The University of British Columbia, Vancouver, British Columbia, Canada

Raffaele Iorio Department of Laboratory Medicine and Pathology, Mayo Clinic, College of Medicine, Rochester, Minnesota, USA

James W. Ironside National CJD Research & Surveillance Unit, Western General Hospital, Edinburgh, UK

Henry J. Kaminski Department of Neurology, George Washington University, Washington, DC, USA

Charlotte Kilstrup-Nielsen Laboratory of Genetic and Epigenetic Control of Gene Expression, Department of Theoretical and Applied Sciences, Division of Biomedical Research, University of Insubria, Busto Arsizio, Italy

Bhumsoo Kim Department of Neurology, University of Michigan, Ann Arbor, Michigan, USA

Nastassja Koen Department of Psychiatry, University of Cape Town, Groote Schuur Hospital, Cape Town, South Africa

Glenn T. Konopaske Harvard Medical School, McLean Hospital, Belmont, Massachusetts, USA

Birgitte Rahbek Kornum Molecular Sleep Laboratory, Department of Diagnostics and Danish Center for Sleep Medicine, Copenhagen University Hospital Glostrup, Glostrup, Denmark

Mary Jeanne Kreek Laboratory on the Biology of Addictive Diseases, The Rockefeller University, New York, USA

Krister Kristensson Department of Neuroscience, Karolinska Institutet, Stockholm, Sweden

John F. Kurtzke VA Multiple Sclerosis Center of Excellence–East, Georgetown University School of Medicine, Department of Veterans Affairs Medical Center Neurology Service, Washington, DC, USA

Linda L. Kusner Department of Pharmacology and Physiology, George Washington University, Washington, DC, USA

Nicoletta Landsberger Laboratory of Genetic and Epigenetic Control of Gene Expression, Department of Theoretical and Applied Sciences, Division of Biomedical Research, University of Insubria, Busto Arsizio, Italy; San Raffaele Rett Research Center, Division of Neuroscience, San Raffaele Scientific Institute, Milan, Italy

Tong Li Department of Pathology, The Johns Hopkins University School of Medicine, Baltimore, Maryland, USA

Paweł P. Liberski Department of Molecular Pathology and Neuropathology, Medical University of Lodz, Lodz, Poland

Jennifer L. Lyons Department of Neurology, Division of Neurological Infections, Brigham and Women's Hospital and Harvard Medical School, Boston, Massachusetts, USA

Nasim Maleki PAIN Group, Department of Anesthesia, Boston Children's Hospital and Harvard Medical School, Boston, Massachusetts, USA

Giulia Mallucci Department of Clinical Neurosciences, John van Geest Centre for Brain Repair, Wellcome Trust–MRC Stem Cell Institute and NIHR Biomedical Research Centre, University of Cambridge, Cambridge, UK

J. John Mann Columbia University, New York State Psychiatric Institute, New York, USA

Justin C. McArthur Department of Neurology, Division of Neuroimmunology and Neurological Infections, The Johns Hopkins University School of Medicine, Baltimore, Maryland, USA

Bruce S. McEwen Laboratory of Neuroendocrinology, The Rockefeller University, New York, USA

Ann C. McKee Center for the Study of Traumatic Encephalopathy, Alzheimer's Disease Center, Department of Neurology, Department of Pathology, Boston University School of Medicine, Boston, Massachusetts, USA, VA Boston HealthCare System, Boston, Massachusetts, USA

Guy McKhann Mind/Brain Institute, Johns Hopkins University, Baltimore, Maryland, USA

Tatiana Melnikova Department of Pathology, The Johns Hopkins University School of Medicine, Baltimore, Maryland, USA

Emmanuel Mignot Stanford Center for Sleep Sciences, Stanford University School of Medicine, Palo Alto, California, USA

Andrew H. Miller Department of Psychiatry and Behavioral Sciences, Emory University School of Medicine, Atlanta, Georgia, USA

William C. Mobley Department of Neurosciences, University of California – San Diego, La Jolla, California, USA

Marco Neppi-Modona Psychology Department, University of Turin, Turin, Italy; Neuroscience Institute of Turin (NIT), University of Turin, Turin, Italy

Orna O'Toole Department of Laboratory Medicine and Pathology, Mayo Clinic, College of Medicine, Rochester, Minnesota, USA; Department of Neurology, Mayo Clinic, College of Medicine, Rochester, Minnesota, USA

Matthew P. Parsons Department of Psychiatry, Brain Research Centre, University of British Columbia, Canada

O.M. Peters Department of Neurology, University of Massachusetts Medical School, Worcester, Massachusetts, USA

Roberto Picetti Laboratory on the Biology of Addictive Diseases, The Rockefeller University, New York, USA

Sean J. Pittock Department of Laboratory Medicine and Pathology, Mayo Clinic, College of Medicine, Rochester, Minnesota, USA; Department of Neurology, Mayo Clinic, College of Medicine, Rochester, Minnesota, USA

Stefano Pluchino Department of Clinical Neurosciences, John van Geest Centre for Brain Repair, Wellcome Trust–MRC Stem Cell Institute and NIHR Biomedical Research Centre, University of Cambridge, Cambridge, UK

Peter Pressman Memory and Aging Center, University of California, San Francisco, California, USA

Donald L. Price Department of Pathology, The Johns Hopkins University School of Medicine, Baltimore, Maryland, USA; Department of Neurology, The Johns Hopkins University School of Medicine, Baltimore, Maryland, USA; Department of Neuroscience, The Johns Hopkins University School of Medicine, Baltimore, Maryland, USA

Charles L. Raison Department of Psychiatry, College of Medicine, University of Arizona, Tucson, Arizona, USA; Norton School of Family and Consumer Sciences, College of Agriculture and Life Sciences, University of Arizona, Tucson, Arizona, USA

Lynn A. Raymond Department of Psychiatry, Brain Research Centre, University of British Columbia, Canada

Brian Reed Laboratory on the Biology of Addictive Diseases, The Rockefeller University, New York, USA

Peter B. Reiner National Core for Neuroethics, Department of Psychiatry, The University of British Columbia, Vancouver, British Columbia, Canada

Kerry J. Ressler Department of Psychiatry and Behavioral Sciences, Emory University School of Medicine, Atlanta, Georgia, USA; Yerkes National Primate Research Center, Atlanta, Georgia, USA; Howard Hughes Medical Institute, Chevy Chase, Maryland, USA

Graham W. Rook Centre for Clinical Microbiology, Department of Infection, University College London, London, UK

Howard J. Rosen Memory and Aging Center, University of California, San Francisco, California, USA

Lewis P. Rowland Neurological Institute, Columbia University Medical Center, New York, USA

Aarti Ruparelia Department of Neurosciences, University of California – San Diego, La Jolla, California, USA

Mario A. Saporta Department of Neurology, Universidade Federal Fluminense, Rio de Janeiro, Brazil

Alena V. Savonenko Department of Pathology, The Johns Hopkins University School of Medicine, Baltimore, Maryland, USA; Department of Neurology, The Johns Hopkins University School of Medicine, Baltimore, Maryland, USA

Julia Schaeffer Department of Clinical Neurosciences, John van Geest Centre for Brain Repair, Wellcome Trust–MRC Stem Cell Institute and NIHR Biomedical Research Centre, University of Cambridge, Cambridge, UK

Helen E. Scharfman Departments of Child & Adolescent Psychiatry, Physiology & Neuroscience, and Psychiatry, New York University Langone Medical Center, New York, USA; The Nathan Kline Institute, Dementia Research, Orangeburg, New York, USA

Bruce K. Shapiro The Johns Hopkins University/Kennedy Krieger Institute Residency in Neurodevelopmental Disabilities, Baltimore, Maryland, USA

Michael E. Shy Department of Neurology, University of Iowa, Iowa City, Iowa, USA

Roger P. Simon The Neuroscience Institute, Morehouse School of Medicine, Atlanta, Georgia, USA

Catrina Sims-Robinson Department of Neurology, University of Michigan, Ann Arbor, Michigan, USA

Dan J. Stein Department of Psychiatry, University of Cape Town, Groote Schuur Hospital, Cape Town, South Africa

Scott M. Summers Medical Investigation of Neurodevelopmental Disorders (MIND) Institute, Department of Psychiatry and Behavioral Sciences, University of California-Davis, Sacramento, California, USA

Kiran T. Thakur Department of Neurology, Division of Neuroimmunology and Neurological Infections, The Johns Hopkins University School of Medicine, Baltimore, Maryland, USA

Luis B. Tovar-y-Romo Instituto de Fisiología Celular, Universidad Nacional Autónoma de Mexico, Mexico

Arshya Vahabzadeh Department of Psychiatry and Behavioral Sciences, Emory University School of Medicine, Atlanta, Georgia, USA

Alessandro Vercelli Department of Neuroscience, Neuroscience Institute Cavalieri Ottolenghi, Turin, Italy

Mitchell T. Wallin VA Multiple Sclerosis Center of Excellence–East, Georgetown University School of Medicine, Department of Veterans Affairs Medical Center Neurology Service, Washington, DC, USA

Thomas Wichmann Department of Neurology, School of Medicine, Emory University, Atlanta, Georgia, USA; Udall Center of Excellence in Parkinson's Disease Research, Emory University, Atlanta, Georgia, USA; Yerkes National Primate Research Center, Emory University, Atlanta, Georgia, USA

Clayton A. Wiley Division of Neuropathology, UPMC Presbyterian Hospital, Pittsburgh, Pennsylvania, USA

David R. Williams Department of Social and Behavioral Sciences, Harvard School of Public Health, Boston, Massachusetts, USA

Philip C. Wong Department of Pathology, The Johns Hopkins University School of Medicine, Baltimore, Maryland, USA; Department of Neuroscience, The Johns Hopkins University School of Medicine, Baltimore, Maryland, USA

Vadim Yuferov Laboratory on the Biology of Addictive Diseases, The Rockefeller University, New York, USA

Weihua Zhao Department of Neurology, Methodist Neurological Institute, Houston, Texas, USA

Michael J. Zigmond Departments of Neurology, Neurobiology, and Psychiatry, University of Pittsburgh, Pittsburgh, Pennsylvania, USA

Saša A. Živković Neurology Service, Department of Veterans Affairs, University of Pittsburgh, Pennsylvania, USA, and Department of Neurology, University of Pittsburgh School of Medicine, Pittsburgh, Pennsylvania, USA

CHAPTER

1

An Introduction: A Clinical Neuroscientist and Disorders of the Brain

Guy McKhann

Mind/Brain Institute, Johns Hopkins University, Baltimore, Maryland, USA

OUTLINE

INTRODUCTION

I will start by saying what this article is not about. This is not a commentary about how to perform and interpret an examination of the nervous system. Thus, I will not comment on the importance of obtaining a proper history, or the nuances of eliciting clinical responses as part of the examination. In other words, I will downplay the clinical aspects of my title. Those seeking such information might consider referring to one of many excellent textbooks on the subject.[1–3] Rather, I will focus on the application of neuroscience to current questions about diseases of the nervous system. I will emphasize those areas in which my background as a neuroscientist has helped me. Further, I will try to highlight areas where those in neuroscience can, and I hope will, help me as a clinician. This is not an extensive review of many topics; rather, I have focused on a few of the areas with which I am familiar. I am a neurologist, so much of my emphasis is on disorders that fall under the province of neurology. However, I have also included some comments about psychiatry.

As a clinician, I am interested in disease: What can go wrong with the nervous system? The word "disease" or "disorder", the somewhat softer term we have chosen to use in this textbook, implies a change from a person's normal state. That change can vary markedly from person to person. For example, I once was asked to evaluate a 70-year-old patient who had been one of the youngest graduates from the Massachusetts Institute of Technology (MIT), at age 14. His family and business associates reported that he was starting to have cognitive problems,

Neurobiology of Brain Disorders
http://dx.doi.org/10.1016/B978-0-12-398270-4.00001-X

no longer making complex business decisions or remembering details of business transactions. Routine cognitive testing was useless; he was off the positive end of the scales in everything. When we devised particularly difficult tests of memory and executive function (more about that later), we could pick up some deficits. However, the family proved to be correct and over the next 2 years, he developed clear symptoms of memory decline, and ultimately was diagnosed with Alzheimer disease (AD). In contrast is a 40-year-old woman with Down syndrome, who had been functioning with an IQ about 80. Her mother reported that her daughter was starting to have increasing problems with memory, and like the first patient, over the next 2 years she too had the symptoms of AD, as do many subjects with Down syndrome in later life. Clearly, these two people were starting at different baselines, and the change from that baseline was unique for each individual. (Further discussion of AD is covered in Chapter 21.)

LOCALIZATION OF LESIONS

In classical clinical neurology, the emphasis was on the location of a place in the brain associated with a specific clinical picture. This stems from the pioneering observations of Broca and Wernicke regarding language. Broca, for example, made two observations: people speak with the left side of the brain; and an anterior lesion of cerebral cortex, subsequently delineated as "Broca's area", leads to an expressive aphasia. The concept of specific regions of brain having very specific functions, as happens in the motor system, held until the application of imaging, both computed tomography (CT) and magnetic resonance imaging (MRI), for the correlation of lesions with clinical symptoms. These studies showed that subjects with similar language deficits, including a classical "Broca's aphasia", may have markedly different lesions. These observations have enhanced the concept of localization of function so that we now believe that rather than there being specific localization of functions in the brain, there exist networks of function that interrelate not only areas of cerebral cortex, but also deeper brain structures such as thalamus, basal ganglia, and cerebellum. Alterations of these networks are part of the pathophysiology underlying disorders of language, motor function, and consciousness, as well as psychiatric disorders.

The networks of the brain have been more fully established in the motor system, where not only cortical but subcortical structures are involved. These motor networks, originally delineated in the non-human primate, have been the basis for understanding not only normal and abnormal motor control but also the development of deep brain stimulation (DBS) for a variety of motor disorders, particularly Parkinson disease (PD), which will be discussed at length.[4]

IMAGING

If you were to ask what has been the primary advance that has changed our approach to the brain, both its normal and abnormal functioning, I would point to the continuing development of imaging of the brain. The major impact of imaging is that we can now study the living brain. The techniques are non-invasive; the subject just lies there and lets the machine do its thing. Whether one is determining the structure of the brain, abnormal growths, such as a tumor, how the brain is functioning, or whether there are abnormal accumulations, as occurs in AD, modern techniques are fast, accurate, safe, and relatively easy to interpret. Further, most results are interchangeable. Studies done in London or Singapore can be analyzed in Baltimore and *vice versa*.

Structural imaging by CT scanning began in the 1970s, followed by MRI in the 1980s. These techniques made it possible to obtain immediate information about structural abnormalities of the brain and revolutionized our diagnostic approach to stroke, head trauma, and brain tumors. Thus, we could immediately diagnose a patient with a head injury, seizure, or persistent headache. Many emergency rooms now have CT or MRI scanning available. Further advances in MRI have yielded techniques such as diffusion weighted imaging,[5] which indicates recent ischemic events in the brain, and diffusion tensor MRI,[6] which delineates white matter pathways in the brain.

Functional imaging demonstrates which part of the brain is being used as a subject performs various tasks. Functional imaging studies originally evaluated the use of glucose by positron emission tomography (PET), but now evaluate blood flow by functional magnetic resonance imaging (fMRI). PET still has a role in diagnosis, using labeled ligands to bind to substances in the brain and to receptors, as in the demonstration of Aβ mentioned below.[7]

SELECTIVE VULNERABILITY OF NEURONAL POPULATIONS

Are the changes in brain function related to changes in the basic abilities of brain systems, or has some external factor disrupted brain processes? In the first category are disease processes in which intrinsic functions of brain cells and systems are altered, such as occurs in neurodegenerative diseases, multiple sclerosis (MS), seizures, or disorders of consciousness and sleep. In the latter category, brain functions are altered by insults that originate outside the brain such as trauma, infections, effects of radiation, and responses to systemic diseases such as a cardiac arrest or exposure to toxins.

Regardless of type, we seldom encounter a generalized dysfunction of the brain, or "brain failure", as we do with other organs such as the liver or kidneys. The reaction of the nervous system to both external and internal insults can be surprisingly focused, involving only specific neuronal populations. For example, with a cardiac arrest in which the entire neuronal population has been deprived of oxygen and sustenance, only certain neuronal populations exhibit degeneration, specifically the Purkinje cells in the cerebellum, layers 2, 3, and 5 of the cerebral cortex, or the C-1 portion of the hippocampus, and lack of sustenance from hypoglycemia can have similar specificity (see Chapter 22). Certain viral infections can also lead to specific neuropathology, as in the case of poliomyelitis (see Chapter 27), which affects predominantly anterior horn cells. In genetically determined diseases, in which every cell in the body presumably carries the same gene mutation, only specific sets of neurons may be affected; for example, in a retinal degeneration, only photoreceptor cells; in a hereditary cerebellar ataxia, only Purkinje cells; in Huntington chorea, only neurons in the caudate and putamen. And the specificity need not be with respect to brain *region* but can also be with regard to neuronal *type*. For example, although for many years the focus of the neuropathology underlying PD was on the dopamine neurons of the substantia nigra, it now appears that many long, unmyelinated axons are affected both caudal and rostral to the nigra (see Chapter 19).

The neurodegenerative diseases all start by involving specific neuronal populations. Why this selective vulnerability occurs is not known, but it is not limited to humans. There is a reverse side to this question: What protects the non-involved neurons?

RECOVERY AFTER INJURY

After acute injuries to the brain there is, in most people, substantial recovery. This recovery is greater in younger people: children recover from an injury that would be devastating to an adult. Recovery is aided by therapy that is specific to the deficit, when started early, but some degree of recovery typically occurs spontaneously.

One of the mysteries in clinical neurology is: How does the brain recover from injury? And why does it sometimes not recover? Two people may have seemingly similar strokes involving the motor area of the right cerebral hemisphere, resulting in the inability to use the left arm and hand. Over the next 3 months subject A makes a slow but consistent recovery, and by 6 months he is using his left hand almost normally. The other, subject B, has made very little recovery, and is still quite disabled at 6 months. What occurred in subject A that did not occur in subject B? There are several possibilities. Perhaps the damage to the motor area was not as great in subject A and the part of the brain that controls the arm and hand was only slightly damaged, and recovered. Another possibility is that some other area of the brain takes over at least some of the functions of the motor area. Perhaps the motor area in the other side of the brain takes over. Or perhaps some area on the same side of the brain that we do not normally associate with motor function steps up.

Dr Steven Zeiler and his colleagues at Johns Hopkins[8] have taken a step towards solving this mystery. In their experiments, they first trained mice to perform the complicated task of reaching for and grasping a food pellet through a narrow slit using one paw. That training took nine to 10 sessions. The next step was to create an artificial stroke, by damaging the motor area controlling the trained paw. As expected, the mice lost their ability to get the food pellets. They next gave the mice "physical therapy", retraining them to get the pellets starting 2 days after the "stroke". That worked, and after a similar period of training the mice were back to where they had started, just like human subject A mentioned above.

In the human, one can use imaging to try to see which part of the brain is active, and gain some clue as to how the brain might have reorganized. Such studies have been conducted in people with strokes, but not with consistent results. In the mouse one can look at the brain and see what may be going on. Zeiler first found that the damaged neurons in the motor area had not recovered, so that was not the mechanism. However, another, nearby area of the brain, the medial premotor cortex, an area in front of the usual motor area, had become active. In healthy mice, damaging this medial premotor cortex did not cause any paralysis, so normally this is not a motor-control area, but apparently becomes one as a response to injury. Zeiler and colleagues then damaged the medial premotor cortex in mice with their motor area damaged, and saw the original motor deficit return.

But that is only part of the story. What is going on in the medial premotor cortex? This area contains inhibitory neurons that keep other premotor neurons inactive. After the injury to the motor cortex, these inhibitory neurons become quiet, allowing the other neurons to be more active and to take over motor functions. But why did human subject B not recover? Perhaps his lesion was larger, knocking out both the primary motor area and the equivalent of the premotor area. Or maybe the inhibitory neurons in the premotor area did not get turned off. Or maybe the subject did not receive adequate physical therapy. Addressing the timing, dose, and use of medicines are the next steps.

This mouse model has obvious human implications, but many questions remain. In the human, what is the equivalent of the medial premotor area? Can this area be focused on in studies of recovery? Are there similar inhibitory factors that have to be shut off or diminished to allow this human area to take over motor functions? Could this area in the human be suppressed by applying transcranial magnetic stimulation (TMS) to provide inhibition of the inhibitory cells to aid recovery?

Attempts to go from the results of therapy in a mouse model of stroke to an application in humans have a dismal record, as outlined in the discussion of stroke, below. However, most previous studies of animal models have been different. In the most commonly used model, a large stroke is induced in the mouse, and some agents are given to the mouse to try to make the stroke smaller. Many agents achieve that in mice, but not a single one has had any positive effect in humans. The studies I report here are different: they are targeting the *cellular mechanisms of recovery*. Thus, these observations by Zeiler and his colleagues move the enquiry in a different direction. For example, they ask: How does the brain injury remove the inhibitory influences that normally keep the medial premotor neurons less active? Does this area of the human brain that is the equivalent of the medial premotor area of the mouse become active during recovery? If it could be identified, would suppressing inhibition in the area by applying TMS to provide inhibition to the inhibitory cells in this area aid recovery?

STEM CELLS IN RECOVERY

In most of the brain, with the exception of specific areas such as the cerebellum or hippocampus, regeneration of new neurons does not occur and is not a mechanism of recovery. Thus, the possibility of introducing a source of new neurons from stem cells has raised great expectations. Dipping into the stem cell field is not easy, in part because the terminology is confusing and keeps changing. In general, stem cells have different properties depending on their source or origin[9]:

- *Embryonic stem cells* are derived from the embryo, usually mouse or human. These cells are pluripotential; that is, they are able to differentiate into any organ in the body. There is considerable information about the differential factors that will drive embryonic stem cells to form cells of a specific organ such as heart, skin, or brain. The study of embryonic stem cells is where the field first started, and this almost immediately raised ethical issues relating to the use of human embryonic tissue. There has been a period of federal restriction on this research in which the US National Institutes of Health would, at various times, (1) not support this research, (2) allow research but only with certain cell lines, and, currently, (3) allow some research but with regulations. This is a politically charged subject in the USA, with Democrats in the US Congress generally in favor of this research, and Republicans generally opposed.
- *Adult stem cells* are derived from a specific organ and lead to the formation of cells in that organ, which in the brain are neurons, oligodendrocytes, and astrocytes. In the brain, specific cells that can differentiate into more mature neurons and oligodendrocytes reside in two sites: in the wall of the lateral ventricle in the subventricular zone, with migration to the olfactory bulb and beyond; and in the subgranular zone, from which cells migrate to the dentate area of the hippocampus.
- *Induced pluripotent stem cells (iPSCs)* are derived from an adult tissue such as skin, and then "dedifferentiated" to a pluripotential stem cell. These cells have several potential advantages. First, being derived from adult tissue, their use avoids raising the ethical issues associated with embryo-derived cells. Second, they can be derived from the subject who is to receive the stem cells, thus avoiding the immunological mismatches associated with a foreign donor. This approach, like much in the stem cell field, is rapidly evolving. For example, the techniques involved in dedifferentiation that were originally feared to be potentially associated with tumor formation have been modified so this concern has been minimized. However, it is not clear that iPSCs have the same potential as embryonic stem cells. Another potential problem is that a cell from an individual with a disease may have the biological characteristics that were involved with the appearance of the disease in the first place, with the possible result that the iPSCs will reproduce the disease you are trying to treat.
- *Mesenchymal stem cells*, also called multipotent stromal cells, are derived from diverse tissue including bone marrow, adipose tissue, and amniotic fluid. They are pluripotential in that they may be limited to certain cell outcomes. As will be discussed in several of the disease-specific chapters that follow, stem cells, usually mesenchymal stem cells or neural stem cells, have been proposed to promote recovery in a variety of neurological conditions including stroke, MS, spinal cord, and brain injury, PD, and amyotrophic lateral sclerosis (ALS). In addition, they have been proposed in specific roles such as memory improvement in AD, and return of vision in hereditary blindness. Possible benefit may occur when these cells

differentiate into specific cell types such as neurons or oligodendrocytes. However, more commonly, these cells migrate to the site of the disease and release trophic factors that aid recovery.

The stem cells may be introduced by intravenous injection or directly into the brain tissue, as in the spinal cord in ALS. Not only in brain or spinal cord, but also in other tissues such as the heart, stem cells migrate to the site of injury. The mechanisms behind this migration are not known. For example, in animal models of stroke, stem cells given intravenously migrate to the site of injury, whereas in non-stroke animals, they stay in the vascular system. At the injury site, stem cells can differentiate into neurons or oligodendrocytes, release trophic factors which promote recovery, or suppress inflammation.[10]

Oligodendrocytes have been specifically sought in MS or metabolic diseases affecting the function of oligodendrocytes, such as Pelizaeus–Merzbacher disease. One of the primary lesions in MS, discussed in detail in Chapter 30, is demyelination. In many patients with MS clinical recovery occurs to some degree when remyelination occurs. In animal models of the disease, this remyelination, as well as a decrease in inflammation, is aided by the introduction of human mesenchymal stem cells. These stem cells do not differentiate into oligodendrocytes; rather, they produce hepatocyte growth factor, which is important in promoting both a decrease in inflammation and remyelination. Mesenchymal stem cells and hepatic growth factor are both being considered as therapy in MS in clinical trials.[10]

There have been several human trials involving injection of stem cells into the spinal cord in patients with ALS. Investigators at Emory University are currently carrying out a phase I safety trial involving 12 patients. No untoward problems have been seen. Thus, these investigators are moving on to a phase II efficacy trial and, at the time of writing, have included an additional six patients.[11] Similarly, a few safety trials in spinal cord injury have recently been instigated.

The actual use of stem cells as therapeutic agents in brain diseases is in its infancy. To the author's knowledge, no trials have yet shown clear-cut efficacy for stem cells in any brain disease.

Stem cells are an excellent example of the interaction between clinical scientists and more basic neuroscientists. Clinically oriented scientists know, and have access to, the human diseases. They may even have experience in developing animal models of these diseases. However, the basic cell biology of what is going on, as represented by such issues as dedifferentiation, factors involved in differentiation to a specific cell type, migration to an injury site, and mechanism of aiding recovery, are all problems awaiting attention by researchers in basic neuroscience.

BRAIN TRANSPLANTS

This is a different issue from stem cell implants. Here, brain tissue is transplanted to the brain as a therapy for a disease. The most striking findings have been in PD, where human fetal substantia nigra has been transplanted into the striatum of patients with PD. In controlled trials, which involved sham surgery in the control groups, a therapeutic response was found in some of the transplanted group. Indeed, the response was too good: some patients had symptoms compatible with overdose of L-dopa, with adventitious movements.[12] Unfortunately, two double brain placebo-controlled trials failed to see any efficacy as a result of fetal tissue transplants. However, questions have been raised about the experimental design and thus it remains unclear as to whether the transplantation of fetal brain tissue will be an effective intervention in this – or any – condition.[13] In addition, many of the ethical issues raised regarding the use of embryonic stem cells have also been raised regarding the use of fetal tissue.

In any event, these studies raise an important basic science question: Can exogenous cells containing transmitter be transplanted, survive, and provide the needed transmitter to other neurons? Some of these studies suggest that they can but that the exogenous cells are not integrated and are outside the usual control systems, and that this loss of control can be a limiting factor. Therapy with stem cells, differentiated as dopamine-producing neurons, if that becomes possible, may well have the same problem.[14]

NEUROLOGY AS A THERAPEUTIC FIELD

When I first decided to become a neurologist, many years ago, colleagues asked me why I wanted to do that, when all I could do was diagnose, with no therapy to provide. For many disease entities therapeutic possibilities have changed dramatically. New therapeutic approaches have changed our management of MS, stroke, epilepsy, PD, migraine, and human immunodeficiency virus (HIV) disease. In addition, understanding of the underlying disease mechanisms has led to clinical trials in many disorders, including AD, ALS, depression, epilepsy, Huntington disease, PD, schizophrenia, stroke, traumatic brain injury, and genetically determined disorders, to mention just a few. It is true that many of these attempts have shown relatively little efficacy and we still have a long way to go, but progress is being made.

The neurodegenerative diseases have several things in common. First, most are sporadic diseases. There are families with a specific, often dominantly inherited disease as occurs with AD, PD, and ALS, but they are quite rare. Second, they are usually late-onset diseases, with increasing occurrence after ages 60–70. One hypothesis for neurodegenerative diseases is an extension of the prion hypothesis, first applied to a group of diseases called transmissible spongiform encephalopathies, which includes Creutzfeldt–Jakob disease, kuru, and bovine spongiform encephalopathy (mad cow disease). This hypothesis was put forth by Stanley Prusiner, for which he won the Nobel Prize in 1997. As discussed in Chapter 23, in the prion hypothesis, a normal cellular protein, a prion (from "proteinaceous" and "infection") self-replicates into an abnormal, misfolded form that is toxic to cells. An important part of the hypothesis is the ability of the abnormal, misfolded protein to induce misfolding in normal prion protein in other cells. This self-replication spreads to other cells, so that a cascade of toxic cells damages the brain.[15] In neurodegenerative diseases there are accumulations of specific proteins such as β-amyloid (Aβ) and tau in AD, and tau in frontotemporal dementia and chronic traumatic encephalopathy after head injury. In PD there is the accumulation of α-synuclein in the Lewy bodies of specific neurons. Recent studies have suggested that all of these proteins can lead to transmissible disease after long incubation periods (years) following intracerebral injection. For example, human brain homogenate from AD patients injected into marmosets resulted in the appearance of Aβ plaques after 3–4 years.[16] This phenomenon was shown to be specific to the Aβ plaques, even synthetic plaques.[17]

ANIMAL MODELS OF HUMAN DISEASE

Of importance to basic neuroscientists is the development of *in vitro* and *in vivo* models of human disease. These are important not only in defining basic mechanisms but also in evaluating potential therapies. So far, going from the mouse, even the genetically altered mouse, to the human has been only marginally productive. However, successes have included the development of the 1-methyl-4-phenyl-1,2,3,6-tetrahydropyridine (MPTP) model of PD in the non-human primate, which provided a model for evaluating the affects of lesions on PD. Ultimately, these studies resulted in the development of DBS for the human parkinsonian patient.[4]

One of the outstanding examples of failure in going from an animal to human effectiveness has been in the area of stroke. It has been estimated that there have been over 500 trials of "neuroprotective compounds" in models of stroke, mostly in mice and rats. However, only antiplatelet compounds [aspirin and Plavix® (clopidogrel bisulfate)] and recombinant tissue plasminogen activator (tPA) are accepted therapies in humans. There are several reasons for these failures, but as outlined by van der Worp and colleagues,[18] the healthy, younger, male animals usually used do not reflect the aging human with comorbidities, particularly hypertension (in 75%) and diabetes (in >60%). Further, most animal models have built-in differences from what is possible in the human, such as a short time from the inducement of the stroke to onset of therapy (10 minutes in animals), change in stroke volume as an outcome rather than a functional recovery, and short follow-up (1–3 days in animals versus 3 months in humans).[18] Unfortunately, many pharmaceutical companies have abandoned stroke as a therapeutic goal because of previous failures of clinical trials. However, as will be discussed below, there may be ways to improve both the animal model and the clinical application.

DEVELOPMENT OF NEW DRUGS

The development of a new pharmacological approach to a disease is a long and extremely expensive process. Estimates suggest that it takes 12 years and costs around 1.2 billion USD from the first research on a drug to its final clinical acceptance. That cost figure can be sharply revised upwards if the pharmaceutical company includes the cost of their drug failures as part of the calculation. Even for a large company like Merck or GlaxoSmithKline, only a few drugs can be supported for expensive clinical trials. Brain drugs must compete with drugs for other diseases such as cancer or heart disease. Not surprisingly, decisions are influenced by recent successes or failures. Thus, if a drug for AD or stroke has failed in a recent trial, the company may be hesitant to conduct further trials of drugs for neurological disorders. That is exactly what is happening, as large pharmaceutical companies ("big pharma") have decreased their neuroscience operations and clinical trials of drugs for brain diseases.

A challenge to those in clinical neuroscience is to devise ways to simplify this process so that clinical trials can involve smaller numbers of subjects, have defined predictable outcome measures, be of shorter duration, and therefore cost less. For example, the population being studied can be simplified by being more homogeneous, with a better defined clinical course and defined outcomes. We can also learn from investigators in other fields. For example, in oncology various forms of cancer are being defined by their molecular defects. In breast cancer the presence or absence of a hormonal receptor (estrogen or herceptin) defines the population.

In general, in the development of a drug there is a jump from animal results to large trials in the human. As mentioned above, the results in animals may hint at clinical efficacy, but the animal model is often so far from the human disease, or the therapeutic approach in animals so impractical in the human, that using that data as the basis of an extremely expensive human trial is hazardous. There is a crucial step, or question, missing: Does the stuff do anything in the human?

CLINICAL TRIALS

Eventually, a potential treatment for patients has to be evaluated in a clinical trial. In this trial the projected therapy is compared with a control group, which may consist of those with no therapy, standard therapy, or a successful therapy. Ideally, subjects are randomized into the treatment group or control group. This randomization and study evaluation is carried out in a double-blind fashion, with neither the subjects nor the investigators knowing in which group a subject resides. The randomized control group is included to be a measure of the natural progression of the disease, to equalize the effects of gender, age, and local factors, and to minimize the bias of the investigators.

One of the factors that complicates these studies is the "placebo effect": that is, the control group responding in a positive way, as if they are receiving the actual therapy. These placebo responders appear to be using two mechanisms. The psychological one is related to the power of suggestion. If I tell you that what I am giving you will help your problem, you are prepared to respond positively. If I am quite positive, you will also be positive. Your response may have a biological component, such as when a change in neurotransmitter level or the release of endorphins results in a higher pain threshold. Not all subjects respond in the same way. One of my colleagues, Tony Ho, now at AstraZeneca and previously at Merck, ran into this placebo problem while he was directing a trial of a new compound for the treatment of pain. He noticed that some, but not all, subjects in the placebo group were responding almost as well as those receiving the new agent. In his next trial he started by giving both groups only the placebo. After a period he removed those who responded to placebo and randomized only the non-placebo responders in the next stage of clinical trial to treatment or placebo.[19] In other words, he eliminated the placebo responders.

A continuing problem is to devise clinical trials that are applicable to changing approaches to a disease. This arises in the comparison of two approaches to coronary artery disease: surgery with coronary artery bypass grafting (CABG) versus stents. (A stent is a mesh tube that keeps a previously occluded coronary vessel open.) The problem with such studies is that both groups, particularly those receiving stents, have made technical advances, so that a trial that requires 5 years to finish is out of date by the time it is completed. With stents, cardiologists have gone from bare metal stents to drug-eluting stents to prevent rethrombosis. So now there is a need for another study to determine whether the drug-eluting stents are actually better than the old, bare metal stents. Since some of these studies use survival as an outcome, they take time to be completed. The final answer is not in yet, but the results of several studies indicate that the death rate over 6 years is higher in the stent group than in the CABG group, particularly in patients with diabetes, and that the type of stent may not make a difference.[20]

There is no easy answer to the problem of progress in the diagnosis or treatment of a disease while a trial is in progress. This is particularly true in some trials that involve imaging techniques as part of the evaluation of subjects. What do you do about a new imaging technique, such as going from a 1.5 T MRI scanner to a 3.0 T MRI scanner? A possible solution is to evaluate a sample of subjects using both the older and the new imaging techniques for comparison purposes.

TRIALS IN ALZHEIMER DISEASE

Our concept of AD has changed over the years. When I was first training in neurology in the late 1950s, AD was thought of as a rare form of dementia, called a "presenile dementia" because it affected people in their forties and fifties. In the 1970s it became apparent that older people with dementia at that time called "senile dementia" or "hardening of the arteries" had the same pathology as those younger people first described by Alois Alzheimer. In recent years our concepts have changed even more. First, the disease process starts many years before we make the diagnosis based on the appearance of clinical symptoms such as impaired memory. Thus, criteria have recently been established for three phases of the disease: (1) a preclinical phase in which a subject is cognitively normal by history and usual cognitive testing, but has biomarker and pathological evidence of the disease; (2) an intermediate phase in which there is cognitive impairment, but not enough to make a diagnosis of dementia: this phase, referred to as minimal clinical impairment (MCI), is defined by cognitive change and biomarker evidence of underlying disease; and (3) finally, the dementia phase, in which people become increasingly impaired and have biomarkers of the disease.[21] The brain pathology of the dementia phase shows marked neuronal loss, particularly in the regions of the brain where the disease starts, such as the entorhinal cortex and hippocampus. From these early involved sites, the disease spreads to other areas, such as the frontal cortex. Unless some way of providing new, healthy neurons to an involved region is found, reversal of this advanced disease is unlikely. I wonder if one is doing any favors to the patient or their family by treating a severely demented patient when, at best, one might slow the progression to even more severe dementia. However, until recently it is exactly these demented patients who have been the focus of clinical trials. It would be preferable to treat the phase I, or early phase II subject with the goal of preventing or slowing the disease process in a relatively normal person. To accomplish this, we need to know whether it is possible to predict in a normal or minimally affected population who will go on to progress to MCI or dementia. There are several studies in progress of subjects, some with positive family histories or genetic risk factors, who are evaluated at baseline and then followed for years. In a study being carried out at Johns Hopkins by Marilyn Albert, some of the subjects have been followed for 16 years. Her study, and others, indicates that the combination of cognitive testing, changes on MRI, and cerebrospinal fluid (CSF) findings of biomarkers (see "Biomarkers of Disease", below) can identify a population that is at risk for progression.[22] However, these studies are applied to a population, and it is not clear whether they can be used with an individual subject.

BIOMARKERS OF DISEASE

In the preceding paragraphs I have referred to biomarkers of AD. Biomarkers are biological markers of a disease process. Some are related specifically to the disease and the disease mechanism, others reflect the effects of the disease, and some are markers of disease presence, but not related to disease mechanism or effects (e.g. prostate-specific antigen in prostate cancer). Currently, five biomarkers in two major categories are used in diagnosing and following disease progression in AD: (1) those related to brain amyloid Aβ deposition, specifically measures of Aβ in CSF and demonstration of amyloid on brain imaging by PET; and (2) those related to neuronal degeneration, including measures of atrophy on MRI structural imaging, measures of hypometabolism on fluorodeoxyglucose positron emission tomography (FDG-PET) and increased levels of tau and phosphorylated tau in CSF. These markers have been used to develop dynamic models of the three phases of AD.[23]

One advance in the area of AD is the ability to demonstrate the presence of amyloid in the living human by PET scanning. After years of preliminary work, two investigators, William Klunk and Chester Mathis at the University of Pittsburgh, USA, working with clinicians in Sweden, reported in 2002 and 2004 the demonstration of amyloid in AD subjects, and not in non-AD subjects.[7] Since then, other teams have used the Pittsburgh compound to evaluate subjects with possible AD. What originally looked like a clear-cut marker of the disease has become complicated. The major problem is that about 20–30% of those considered cognitively normal are positive for amyloid on PET scanning. It is not clear what these positives represent. If one waits long enough will they progress to show clinical indications of MCI or AD? In other words, are they representatives of the first, preclinical group mentioned above? Or is amyloid on PET scanning an imperfect marker? There is also an important ethical question raised by such results: Does one tell the individual of the finding, given that it may be a false positive, and even if it is not, we do not currently have a treatment to offer to patients beyond cognitive and physical exercise? This and other ethical dimensions of research are discussed in Chapter 45.

The second problem is that there appears to be a ceiling effect with this compound. In other words, amyloid is detected early in the disease process, but then there is a maximal detection by scanning, even though the disease continues to progress clinically. Thus, it is unlikely that amyloid imaging will be a good marker of attempts to alter the later stages of the disease.

A third problem that limits the widespread use of the Pittsburgh compound is its use of carbon-11, an isotope with a 20 minute half-life. Therefore, these studies require a cyclotron on site to make the labeled compound and then its immediate use in the scanner. Only a few medical centers have this combination of equipment available. Several companies have focused on a different labeled compound using fluorine-18, which has a half-life of 2 hours, enough time for a compound to be made at a central site and then be made available to those who cannot make their own compound. The first to appear was made by Avid Pharmaceuticals, and has been approved by the US Food and Drug Administration (FDA).[24] There are other fluorine-18 compounds in the pipeline.

A welcome advance would be the demonstration of tau by PET imaging. This would be helpful not only in AD, but also in other tauopathies such as frontotemporal dementia and in the chronic traumatic encephalopathy that occurs in athletes and military personnel after head injuries. In the latter situation tau accumulates in a different distribution from that seen in AD, and without the accompanying amyloid plaques. With the increasing concern about head injuries not only in professional athletes such as football players and boxers but also in younger amateur participants in contact sports, the ability to detect tau in brain quickly and reliably would be a significant breakthrough (see Chapter 16). There is one report of tau imaging in head injury from the University of California – Los Angeles, involving a small number of subjects,[25] and similar studies may be expected in the future. With successful tau imaging, a whole new set of questions would emerge. For example, are there genetic factors that determine a person's response to head injury in terms of accumulation of tau? If so, are there young people who could be identified as being at higher risk for subsequent behavioral and cognitive problems? Once there, is the accumulation of tau reversible? Are there medications that could modify the accumulation of tau?

There has been considerable interest by the pharmaceutical industry in AD because the potential market for a successful therapeutic compound is enormous. In 2013, it was estimated that 5.2 million people in the USA had AD; worldwide, the number jumps to more than 35 million. Unfortunately, clinical trials, almost all aimed at Aβ as the target, have been negative, and big pharma is getting restless.[26] Thus, a model of interaction has been developed between the US government, in this case the National Institute on Aging, an advocacy group (the Alzheimer's Association), various pharmaceutical companies, and academia. This group, known as the Alzheimer's Disease Neuroimaging Initiative, is developing and evaluating clinical, neuropsychological, biomarker, and imaging parameters in subjects with various stages of memory loss, from normal to MCI to AD. Of importance is that tests are carried out in various locations, but analyzed in central sites with standardized methods. The results are available to all researchers and provide background for future clinical trials of specific therapies. In its design, this group has been flexible, trying to take advantage of the rapidly changing techniques in imaging and biomarkers.[27] This model of collaboration could and should be applied to other diseases as well.

PSYCHIATRIC DISEASE

As I mentioned at the outset of this chapter, this is not an all-inclusive review of diseases of the nervous system, but focused on areas where I have first hand experience, particularly neurological diseases. However, if you combine those with depression (Chapters 40 and 43), schizophrenia (Chapter 39), anxiety (Chapter 37), obsessive–compulsive disorder (Chapter 38), addiction (Chapter 35), and suicide (Chapter 43) you are considering a large number of subjects. It is estimated that in any given year, 25% of Americans have a diagnosable mental illness, a total that probably exceeds all those with neurological disease. (I put AD in both camps.) These groups with psychiatric diseases have several things in common that set them apart from most neurological diseases at present:

- These entities are defined by their clinical symptoms, obtained by clinical history from the patient or an informant. This leads to controversy about the criteria for diagnosis, particularly at the ends of the age spectrum, in children and elderly people.
- These diseases do not yet have unequivocal biomarkers (with the exception of disorders lying at the interface between neurology and psychiatry, such as AD). Thus, there are no biological aids for diagnosis or measurement of disease course.
- Most have no definitive pathology. One cannot examine the brain at autopsy and determine that the patient suffered from depression or schizophrenia.
- Imaging studies have been helpful in demonstrating structural and functional abnormalities associated with psychiatric disorders, but have not yet yielded clear diagnostic differences. For example, the hippocampus can exhibit atrophy in schizophrenia, major depressive disorder, and post-traumatic stress disorder.
- Whereas many of these disorders have increased prevalence within a family, the genetics is complex at best. This is similar to neurological diseases such as AD, MS, and PD.

Despite these problems, therapy in psychiatric diseases, based on their neuropharmacology, has been more successful than for most neurological diseases. This has been particularly true in schizophrenia and the mood disorders. The treatments for these disorders were originally developed over 50 years ago by serendipity in clinical testing of novel agents, including chlorpromazine, imipramine, and lithium, with unknown mechanisms of action. Subsequent preclinical investigations identified potential mechanisms of action. For example, Julius Axelrod demonstrated that the antidepressant imipramine inhibited the synaptic uptake/inactivation of biogenic amine neurotransmitters, and Arvid Carlsson proposed that the mechanism of action of antipsychotic drugs was blockade of dopamine receptors. Axelrod and Carlsson received Nobel Prizes in Medicine or Physiology for these insights. The pharmaceutical industry capitalized on this knowledge to develop second generation drugs that were associated with greater specificity and fewer side effects.

In depression, the currently used medications act by influencing the transmitters serotonin, norepinephrine, and dopamine. The goal of therapy is to selectively increase the level of these transmitters at the synapse. For example, selective serotonin reuptake inhibitors increase the level of serotonin by inhibiting its reuptake by the presynaptic cell, making more serotonin available to the postsynaptic cell. Commonly used antidepressants such as sertraline, paroxetine, and fluoxetine work by this mechanism. Newer agents are more broadly based in their action, often affecting more than one transmitter.

GENETICS OF NEUROLOGICAL AND PSYCHIATRIC DISORDERS

It is hoped that understanding the genetics underlying a disorder will aid diagnosis, elucidating the underlying mechanism, and approaches to therapy. In some diseases, such as the metabolic diseases Tay–Sachs disease and metachromatic leukodystrophy, this has been the case. However, unlike many neurological disorders that follow Mendelian genetics, psychiatric disorders, AD, and MS are disorders of complex genetics in which multiple risk genes of modest effect interact with the environment to produce the phenotype. The principles of epigenetics are only just being applied to diseases such as schizophrenia. Newer genetic approaches such as the application of Genome-Wide Association Study (GWAS) to specific populations may prove to be helpful. This approach has been useful in establishing a location on chromosome 9p21 in ALS.[28] The usual application of GWAS techniques to a large population of subjects with ALS and controls was unrevealing. However, when the study was focused on a Finnish population, genes localized on chromosome 9 were indicated. Eventually, a hexanucleotide (CPORF72) was found.[29] It is anticipated that advances in genetics will lead to the identification of novel targets for the development of treatments that address the pathophysiology and will be more effective than current therapies.

TEMPERAMENT AND DISEASE

There is an area that spans the overlap between psychiatry and neurology (as well as other areas of medicine) and that is the effect of temperament, mood, or both on the clinical course of a disease. In other words, does a patient's emotional reaction to a disease affect their response to that disease? One striking example is the effect of depression on outcome after a heart attack. If a patient is depressed 1 month after a heart attack, his or her mortality is three to four times higher over the next year or two. Similarly, depression is associated with poor outcomes after CABG. The mechanisms underlying these associations are not clear. Nor is it established that antidepressive therapy will alter cardiac outcomes.[30] In other instances, depression is associated with poorer outcomes of disease, but it is not clear whether depression is causative as part of the disease process or the relationship in an association. This situation comes up when evaluating the outcomes in MS, AD, and PD, among others.

CONCLUSION

In looking back over my half-century of research and practice in clinical neuroscience, I see successes, failures, and lessons for the future. The major lesson I have learned is that clinical neuroscience has changed from being a descriptive field to being an active intervention field. For example, in stroke the role of a neurologist has gone from arranging physical therapy to considering immediate therapy with tPA, anticoagulants, and possible intra-arterial or surgical therapy, to preventive therapy keeping hypertension and cholesterol under control (see Chapter 30 for further discussion).

Even in an area where clinical progress has been slow, there is hope. In the most severe form of brain tumor, glioblastoma multiforme, despite treatment with both radiation and chemotherapy, life expectancy is not much better now than it was 10 years ago. With no therapy, survival is 4–6 months. With the currently accepted therapy of radiation followed by chemotherapy, survival has been extended to 14–16 months. Some, mainly younger, people live longer, up to 3–5 years. Currently, much attention is being paid to the cells of origin in this tumor, particularly cancer stem cells. The tumor is quite heterogeneous, with both tumor and non-tumor cells. Within the tumor cell population are cells that have the properties of stem cells, referred to as cancer stem cells. These cells account for the uncontrolled growth, resistance to therapy, and migration of tumor cells, and therapy directed at these cells is being tried.[31] This field desperately needs new ideas, and this is one of them.

I have already mentioned our inability to get at the underlying mechanism in neurodegenerative diseases. I do not know whether Prusiner is correct in his prion hypothesis as applied to these diseases, but at least it is a start. The question now is whether the prion hypothesis can lead to a treatment. This is not the only area where basic sciences are contributing approaches to therapy. Other examples are the use of trophic factors, such as nerve growth factor or brain-derived neurotrophic factor, as therapeutic agents, or attempts at gene replacement or modification, particularly in muscular dystrophies.

I will close by commenting on the development of new approaches to the treatment of brain disorders. The dependence on enormously expensive, prolonged, clinical trials is killing us. We need a change in philosophy to include smaller, focused trials that will provide indications of efficacy, and less "me-tooism" in considering new approaches. We must avoid acting like lemmings, following each other off the cliff. At least some participants should be sitting quietly with their feet on their desks, considering what should be the new approach, and then bravely going into the laboratory, followed by carrying out small, focused clinical trials. It is not clear to me that big pharma can adapt. The onus may have to shift back to academia, working in conjunction with smaller, more flexible pharmaceutical entities.

In conclusion, there are reasons to rejoice – we have come a long way; but also reasons for disappointment – there is so much farther to travel. I hope that the readers of this textbook will find the inspiration to join the struggle.

References

1. Larsen PD, Stansaas SS. Neurologic exam: anatomical approach. Library.med.utah.edu/neurologicexam/html/home_exam.html.
2. Naisman R. The neurologic exam. Neuro.wustl.edu/education/student-education/neurology-clerkship/history/the-neurological-exam/.
3. Ropper A, Samuels M. *Adams and Victor's Principles of Neurology*. 8th ed. New York: McGraw-Hill Professional; 2005.
4. DeLong M, Wichmann T. Deep brain stimulation for movement and other neurologic disorders. *Ann N Y Acad Sci*. 2012;1265:1–8.
5. Schaefer PW, Copen WA, Lev MH, Gonzalez RG. Diffusion weighted imaging in acute stroke. *Neuroimaging Clin N Am*. 2005;15:503–530.
6. Mori S. *Introduction to Diffusion Imaging*. ScienceDirect. Elsevier; 2007. www.ScienceDirect.com/science/book/978044428285.
7. Mathis CA, Mason NS, Lopresti BJ, Klunk WE. Development of positron emission tomography β-amyloid plaque imaging agents. *Semin Nucl Med*. 2012;42(6):423–432.
8. Zeiler SR, Gibson EM, Hoesch RE, et al. Medial premotor cortex shows a reduction in inhibitory markers and mediates recovery in a mouse model of focal stroke. *Stroke*. 2013;44(2):483–489.
9. Banerjee S, Williamson DA, Habibm N, Chataway J. The potential benefit of stem cell therapy after stroke: an update. *Vasc Health Risk Manag*. 2012;8:569–580.
10. Bai L, Lennon DP, Caplan AI, et al. Hepatocyte growth factor mediates mesenchymal stem cell-induced recovery in multiple sclerosis models. *Nat Neurosci*. 2012;15(6):862–870.
11. Glass JD, Boulis NM, Johe K, et al. Lumbar intraspinal injection of neural stem cells in patients with amyotrophic lateral sclerosis: results of a phase I trial in 12 patients. *Stem Cells*. 2012;30(6):1144–1151.
12. Evans JR, Mason SL, Barker RA. Current status of clinical trials of neural transplantation in Parkinson's disease. *Prog Brain Res*. 2012;200:169–198.
13. Bjorklund A, Kordower JH. Cell therapy for PD: what next? *Mov Disord*. 2013;28:110–115.
14. Hayashi T, Wakao S, Kitada M, et al. Autologous mesenchymal stem cell-derived dopaminergic neurons function in Parkinsonian macaques. *J Clin Invest*. 2013;123(1):272–284.
15. Prusiner SB. Cell biology. A unifying role for prions in neurodegenerative diseases. *Science*. 2012;336(6088):1511–1513.
16. Ridley RM, Baker HF, Windle CP, Cummings RM. Very long term studies of the seeding of beta-amyloidosis in primates. *J Neural Transm*. 2006;113(9):1243–1251.
17. Stöhr J, Watts JC, Mensinger ZL, et al. Purified and synthetic Alzheimer's amyloid beta (Aβ) prions. *Proc Natl Acad Sci U S A*. 2012;109(27):11025–11030.
18. van der Worp HB, Howells DW, Sena ES, et al. Can animal models of disease reliably inform human studies? *PLoS Med*. 2010;7(3):e1000245.
19. Ho TW, Backonja M, Ma J, Leibensperger H, Froman S, Polydefkis M. Efficient assessment of neuropathic pain drugs in patients with small fiber sensory neuropathies. *Pain*. 2009;141(1–2):19–24.
20. Kappetein AP, Head SJ, Morice MC, et al. on behalf of the SYNTAX Investigators. Treatment of complex coronary artery disease in patients with diabetes: 5-year results comparing outcomes of bypass surgery and percutaneous coronary intervention in the SYNTAX trial. *Circulation*. 2013;128(2):141–151.
21. McKhann GM. Changing concepts of Alzheimer disease. *JAMA*. 2011;305(23):2458–2459.
22. Moghekar A, Li S, Lu Y, et al. CSF biomarker changes precede symptom onset of mild cognitive impairment. *Neurology*. 2013;81(20):1753–1758.
23. Jack Jr CR, Knopman DS, Jagust WJ, et al. Tracking pathophysiological processes in Alzheimer's disease: an updated hypothetical model of dynamic biomarkers. *Lancet Neurol*. 2013;12(2):207–216.
24. Clark CM, Pontecorvo MJ, Beach TG, et al. AV-45-A16 Study Group. Cerebral PET with florbetapir compared with neuropathology at autopsy for detection of neuritic amyloid-β plaques: a prospective cohort study. *Lancet Neurol*. 2012;11(8):669–678.
25. Small GW, Kepe V, Siddarth P, et al. PET scanning of brain tau in retired national football league players: preliminary findings. *Am J Geriatr Psychiatry*. 2013;21(2):138–144.
26. Callaway E. Alzheimer's drugs take a new tack. *Nature*. 2012;489(7414):13–14.
27. Traynor BJ. A roadmap for genetic testing in ALS. *J Neurol Neurosurg Psychiatry*. 2014;May 85(5):476. http://dx.doi.org/10.1136/jnnp-2013-305726. [Epub ahead of print.]
28. Laaksovirta H, Peuralinna T, Schymick JC, et al. Chromosome 9p21 in amyotrophic lateral sclerosis in Finland: a genome-wide association study. *Lancet Neurol*. 2010;9(10):978–985.
29. Turner MR, Hardiman O, Benatar M, et al. Controversies and priorities in amyotrophic lateral sclerosis. *Lancet Neurol*. 2013;12(3):310–322.
30. Meijer A, Conradi HJ, Bos EH, et al. Adjusted prognostic association of depression following myocardial infarction with mortality and cardiovascular events: individual patient data meta-analysis. *Br J Psychiatry*. 2013;203:90–102.
31. Maugeri-Saccà M, Di Martino S, De Maria R. Biological and clinical implications of cancer stem cells in primary brain tumors. *Front Oncol*. 2013;3. 6.1.

SECTION I

DEVELOPMENTAL DISORDERS

CHAPTER

2

Introduction

Aarti Ruparelia, William C. Mobley

Department of Neurosciences, University of California – San Diego, La Jolla, California, USA

The nervous system functions through a beautiful but complex network of interacting genes and molecules, individual neurons that form synapses with other neurons, all of which act in concert with other neural circuits and networks to create the biological context in which behavior and cognition emerge. Neuroscientists are faced with the incredible challenge of deciphering the regulation of individual and societal behavior by dissecting, examining, and then defining events that occur at every level of the nervous system. This is compounded by the intrinsic complexities at each level and when linking between the levels – from genes to molecules, from individual neurons to circuits and networks, and from each of these levels to a thinking and behaving organism. Furthermore, neuroscientists are confronted with the scales that define the nervous system. Consider the length scale, which entails events that occur on the scale of nanometers to those that extend over meters, across a dimensional difference in scale of 10^9. Now consider the temporal scale, which could range from 10 microseconds to tens of years. Finally, consider the differences in other parameters, such as the status of nervous system development or the expression of disease pathogenesis, and additional complexity is introduced. Understanding the brain requires that we consider these scales and carry out the studies needed to explore and define events at each level followed by studies to model and link each of the levels.

The chapters that follow make the case that deciphering the biology of neuropsychiatric disorders is quintessentially a search for answers across the scales. Whatever the insights into pathogenesis and treatment, to address the impact of the disease in individuals one must integrate across levels of analysis to understand the disorder well enough to define experimental strategies, diagnostic protocols, therapeutic interventions, and the possible means by which to monitor the outcome of treatment interventions. Each disorder will be unique in the particulars with respect to genetic factors, their impact on neuronal and circuit function, the resulting manifestations as clinical phenotypes, the methods most appropriate for ongoing discovery efforts, and the strategies for defining disease targets. Common to all is the need to adequately address each of these important steps.

To fully benefit from the material presented in this section, the reader is encouraged to read each of the chapters. In so doing, it will be evident that approaching developmental disorders is a process that benefits from the contrasts in exploring disease biology as well as the similarities. Each disorder builds upon an increasingly rich body of clinical information as well as insights from the basic biology of neurological and psychiatric disorders, and yet each has its own story to tell.

In Chapter 3, Gregory, Burton, and Shapiro introduce developmental disabilities and speak at length about the underlying concepts, clinical insights, and tools used to understand them. They introduce the concept of development of the nervous system and how this informs and defines functional milestones as an infant matures; delays and aberrations in achieving these milestones in children are categorized as developmental disabilities. The authors point to etiologies of primary injury such as genetic, vascular, environmental, and inflammatory that may help in the understanding of neurodevelopmental disability. Elucidating these etiologies can provide prognosis, identification, and management of associated or comorbid conditions, and can lead to prevention and improved outcomes. Basic and advanced techniques that enable accurate and timely diagnosis and that can monitor the progression in a child are also addressed, to provide insight into the mechanisms underlying the dysfunction that can better inform management of the disorder.

The succeeding chapters chart a course that characterizes disorders that vary from common to quite uncommon. In so doing, they demonstrate the very different approaches that can be brought to bear in the laboratory and the clinic. Consider, for example, the

Neurobiology of Brain Disorders
http://dx.doi.org/10.1016/B978-0-12-398270-4.00002-1

contribution from Cortese and Castellanos (Chapter 4), whose review of attention deficit/hyperactivity disorder (ADHD) points to clinical evaluations and, in particular, to the use of neuroimaging tools in defining the brain regions and networks linked to clinical manifestations, as most productive. The large and expanding neuroimaging literature, and the use of tools that provide different types of information, allow the authors to define findings that are common across studies. They examine changes in children and adults, offer insights into the extent to which treatments do or do not affect the findings, and suggest new approaches for further exploring the underlying pathophysiology. ADHD is clearly a neurodevelopmental disorder whose impact on the structure and function of the brain is evident in many adults diagnosed during childhood. One can look forward to better and cheaper methods for examining brain structure and function and for defining the impact of behavioral and pharmacological interventions. Although animal models of ADHD have proved useful, the extent to which existing models can be used to advance the field is not clear.

The next three chapters report on disorders that are much less frequent but for which insights into the underlying physiology are being contributed with increasing frequency. A large part of this progress is due to the ability to define a specific genetic factor(s) responsible for pathogenesis. The specific genetic factor distinguishes the disorders discussed. The contribution from Ruparelia and Mobley reviews Down syndrome (DS) (Chapter 5), where the presence of an extra chromosome harboring 350–500 normal genes leads to perturbations in multiple systems. Scaling the levels of analysis in DS thus requires linking phenotypes to genes. Once considered too difficult to attempt, the completion of the human genome, the availability of mouse models that genetically replicate clusters of genes on human chromosome 21, and the use of increasingly sophisticated tools for neuroscience research have yielded important new insights into pathogenesis and pointed to potential treatment targets. The authors focus on cognitive impairment and decline from neurodevelopment to neurodegeneration in DS. They address the role of key candidate genes that may further elucidate the genotype–phenotype correlations and point to cognitive insights gained from mouse models of DS that have led to the emergence and development of pharmacotherapeutic targets.

A priori, it may be assumed that a disorder to mutations in a single gene may be less complex and present fewer obstacles for understanding pathogenesis and defining treatments. However, Summers and Hagerman, examining fragile X-associated disorders (Chapter 8), convincingly dispel the idea. Indeed, it is clear that while mutation of the *FMR1* gene on the X chromosome, through expansion of a trinucleotide repeat in the 5′ end next to the promoter, results in a reduction in the level of the *FMR1*-encoded protein called FMRP, the clinical consequences and approaches to treatment are far from simple. Remarkably, fragile X-related disorders can be neurodevelopmental or neurodegenerative. Complexity can be traced to a number of factors, including differences in the size of the expanded repeat, both loss-of-function and gain-of-function mechanisms, FMRP's role in regulating the transport and translation of hundreds of neuronally important RNAs, and interacting genetic factors. Owing to the defined genetic basis for fragile X, cellular and mouse models of the disorder have provided important insights.

Nielsen and Landsberger (Chapter 7) contribute a thorough review of Rett syndrome. As in fragile X, there is considerable complexity in the biology and clinical manifestations linked to changes in the gene for MeCP2, also located on the X chromosome. The richness of genetic and other clinical data reveals the scope of complexity of the disorders that result from *MeCP2* mutation. Indeed, the many mutations now documented, as well as specific mutations, result in a variety of clinical presentations in females. Typical Rett syndrome is also seen in males. In part, complexity is due to the type of *MeCP2* mutation, the extent to which the X chromosome bearing the mutation in females is inactivated, and other genetic factors, including mutations in other genes whose effect is to phenocopy *MeCP2* mutations. Moreover, it is now apparent that increased dose of a normal *MeCP2* gene is responsible for disease manifestations. The authors document clearly the important contributions made by studies in mouse models to inform understanding of pathogenesis and help to guide attempts at genetic and other treatments.

Now consider a spectrum of disorders that have multiple behavioral and biological phenotypes and which do not have a defined genetic fingerprint but, instead, have several genes that are identified as risk factors. In Chapter 6, Harris introduces the complexity of autism spectrum disorder (ASD), which is a heterogeneous grouping of neurodevelopmental disorders that results in deficits in social communication and social interaction and in restricted, repetitive patterns of behavior, interests, and activities. The increasing prevalence of ASD diagnoses urges a deeper understanding of the various components that lead to its clinical manifestation. To this extent, Harris describes the neurobiological features of brain growth and atypical neural connectivity that commonly occur in ASD. The author then proceeds to comment on neuroimaging studies, neurochemical investigations, and identification of genetic risk factors that have led to remarkable discoveries about the pathology of ASD. Based on these neurobiological developments, the chapter concludes with recommended future directions and classification refinements to treat ASD.

In summary, in documenting progress in a number of neurodevelopmental disorders, the chapters in this section show that studies on the neurobiology of such diseases are making discoveries and revealing new insights into the genetics, cell biology, and circuitry important for both the normal and abnormal nervous system. The progress documented herein allows the prediction that understanding pathogenesis will lead to effective treatments for reversing and possibly preventing these and other developmental disorders.

CHAPTER

3

Developmental Disabilities and Metabolic Disorders

Mary Lee Gregory, Vera Joanna Burton, Bruce K. Shapiro

The Johns Hopkins University/Kennedy Krieger Institute Residency in Neurodevelopmental Disabilities, Baltimore, Maryland, USA

OUTLINE

Neurobiology of Brain Disorders
http://dx.doi.org/10.1016/B978-0-12-398270-4.00003-3

INTRODUCTION

Neurodevelopmental disabilities are the functional expressions of disordered brain development. These disorders may arise from many underlying etiologies. The current explosion of knowledge in genetics and neuroscience has increased our understanding of these disorders and offers the potential for cure. This chapter will provide a framework for understanding neurodevelopment, explore the detection and diagnosis of developmental disabilities, describe techniques for investigation of the mechanisms of these disorders, and present principles of management.

BRAIN DEVELOPMENT

What is Development?

Development is the growth and maturation of an organism over time. Developmental medicine focuses on the neurobiology by which a child matures from conception to adulthood, evaluating, diagnosing, and treating children with functional consequences of disruptions to the typical pathways.

Development progresses in expected ways across individuals, unless something interferes with the normal developmental process. Genetic factors are the foundation of development. The different stages of development are dependent on the proper activation and deactivation of genes and the correct production of gene products. As our knowledge of genetics increases, we are discovering that genetics can place individuals at higher risk for developmental abnormalities when exposed to certain agents. Therefore, the role of genetics in development is becoming increasingly nuanced. Even in the absence of genetic anomalies, there can be abnormal development. Further, as our knowledge increases, it is likely that we will recognize more substances and exposures that interfere with developmental processes in more subtle ways.

Embryonic Development

Embryology is a vast field, which would require its own textbook to cover in even minimal detail. For the purposes of this chapter, the reader should understand that the maturational process from fertilized egg to term infant is ordered, proceeding according to a set schedule (Fig. 3.1). The organism is first one cell (the zygote), which then divides, and this process is repeated over and over again. Initially these cells are undifferentiated; they have the potential to form any part of the developing body (i.e. pluripotent stem cells). However, over time, cells become progressively more differentiated. They acquire particular characteristics of the mature cell type that they will become and lose the potential to form other types of cells. As these initial pluripotent stem cells are differentiating into more specialized cells, the organism that is being formed by these cells is also progressively differentiating. The growing mass of cells develops an axis and gradually begins to form the major structures and organ systems of the human body.

Development of the Nervous System

The development of the nervous system is especially complicated but can be viewed as four overlapping epochs: neurogenesis and cell differentiation; neuronal migration; neural circuitry and synaptogenesis; and synaptic pruning and myelination.[2]

Neurogenesis and Cell Differentiation

The precursor cells are located along the innermost layer of cells lining the neural tube in a region called the periventricular zone. These cells are rapidly dividing and may form either new stem cells or neuroblasts, which have the ability to form any of the many types of neurons or astrocytes and oligodendrocytes. The generation of these different cell types and the maintenance of an appropriate balance of cell types are crucial to brain development, and disruption of this balance can also have neurological consequences. As the brain develops into distinct regions, neural cell division and differentiation continue to produce the numerous and diverse neurons and glial cells of the nervous system. Most of the neurons of the adult brain are produced before birth. Neural generation in the adult brain is possible but only in specific regions and circumstances. Destruction of neurons that occurs after precursor cells have differentiated is therefore irreversible. Cell differentiation pathways are controlled by the activation or antagonism of several genes including *bHLH*, *Notch*, *Olig1*, *Olig2*, and *Nkx2.2*. Most differentiation appears to be controlled primarily by interactions between cells.

In vertebrates, the nervous system develops from ectoderm (one of the three initial germ cell layers). The dorsal-most (top) portion of the ectoderm forms the neural plate. The cells of the neural plate enfold to form the neural tube, which is the precursor for the CNS. As the pluripotent neural stem cells begin differentiating, the basic anatomic structures of the brain are simultaneously being formed. The neural tube contorts, bending and forming bulbous outpouches and constrictions creating the prosencephalon, mesencephalon, and rhombencephalon. Further division of these areas creates the telencephalon (which forms the cortex and hippocampi, and basal ganglia) and the diencephalon (which forms the thalamus and hypothalamus). The mesencephalon forms the midbrain, and the rhombencephalon forms the cerebellum, pons, and medulla.

Neuronal Migration

Neurons generated in the periventricular zone must migrate to distant sites, differentiate, and form functional neural circuits. The migrational process is complex and governed by genes, transcription factors, peptide hormones, and signaling molecules (including retinoic acid, sonic hedgehog, and bone morphogenetic proteins). There are many forms of disruption to this process, including teratogens and metabolic dysfunction.[3–5]

The most common neuronal migration system is travel along radial glia. Radial glial cell bodies are found along the ventricular zone but have long appendages that reach to the outer surface of the growing brain (Fig. 3.2). The appendages are used as scaffolding by the neuroblasts as they travel to their destinations in the neuronal layers of the cerebral cortex, hippocampus, and cerebellum. This migration requires the proper functioning of many components including extracellular matrices, cell adhesion molecules, and signal transduction molecules. As these cells migrate outward the cortex thickens and develops enfolding of the surface, forming sulci and gyri. Other forms of migration include migration along pre-existing axons. Those cells that are left lining the ventricles become ependymal cells, although there may be some remaining neural stem cells.

Humans have a substantially larger cortex than other primates. This increased cortical size is felt to serve the higher cognitive functions seen in humans. Any disruption in the migration pattern of neurons can result

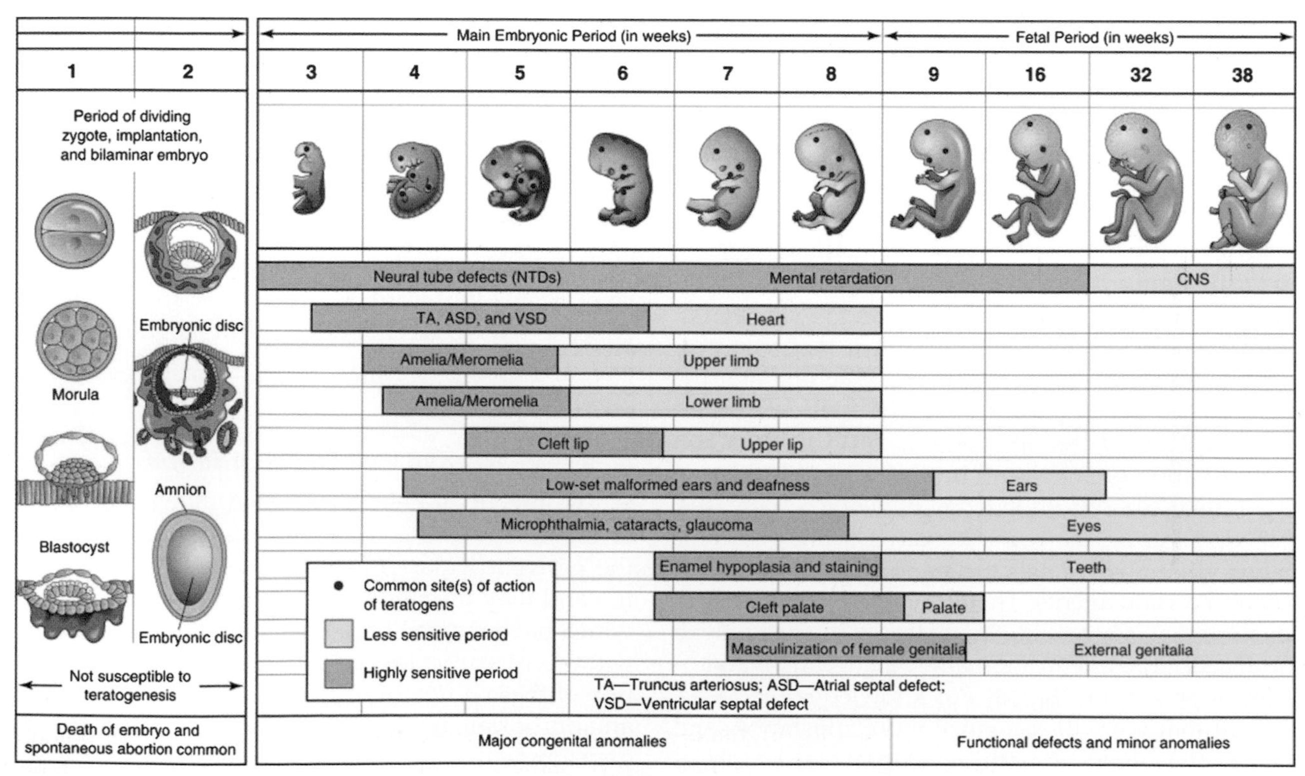

FIGURE 3.1 Embryological development and periods of susceptibility. *Source: Reprinted from Moore and Persaud.* Before We Are Born: Essentials of Embryology and Birth Defects. *Saunders/Elsevier; 2008,[1] with permission.*

in anomalous cortex development and functioning. Examples of failure of migration include heterotopias, which are areas of inappropriately placed gray matter, holoprosencephaly, which is inappropriate division of the cortical hemispheres, and schizencephaly, which is an inappropriate track of gray matter to the ventricles.

Neural Circuitries and Synaptogenesis

Once neurons and glia have reached their final destination, they remain part of an immature system. Neurons must link with one another via axonal growth. Axons must grow to their target locations, in a process dependent on the polymerization and depolymerization of cytoskeletal elements and guided by extracellular matrix cell adhesion molecules, growth factors, and chemorepellants. Once axons are in place, synapses (connections between neurons) must be formed selectively. The synapse is dependent on presynaptic and postsynaptic elements. The initial stages of synapse formation are dependent on many of the same molecules involved in axon guidance, including cadherins. Active zones in the presynaptic neuron are formed by the clustering of synaptic vesicles containing neurotransmitters. In the postsynaptic neuron, the postsynaptic density is formed and localizes receptors, channels, and other associated signaling molecules.

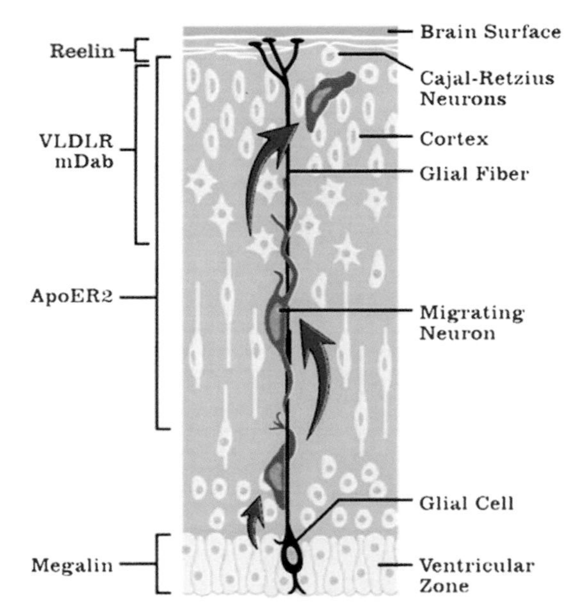

FIGURE 3.2 **Cell migration.** A schematic section through the cortex is shown. A radial glial cell is anchored in the ventricular zone and sends its guidance fiber to the surface of the brain, where it is anchored through several foot processes. After their birth in the ventricular zone, postmitotic neurons migrate along the guidance fiber to the surface then migrate horizontally to their final position. This process is guided by multiple signaling molecules. *Source: Reprinted from Trommsdorff* et al. Cell. *1999;97(6):689–701,[6] with permission. © 1999 Cell Press.*

Surplus neurons are created. The least optimal are destroyed by programmed cell death (apoptosis) to provide a final population of neurons that are ideally located and connected. Surplus synapses are formed and are pruned over time. The least optimal are destroyed by programmed cell death (apoptosis) to provide a final population of neurons that are ideally located and connected. Unlike the other processes discussed here, the branching and pruning of synapses continues well after birth.

Once synapses are established, they require continued support through neurotrophins, otherwise axons and dendrites will atrophy. The strength of the relative synaptic connection continues to be modulated throughout adult life and underpins learning and memory.[7,8]

Postnatal Brain Development

Although most neurons and glia have been formed by birth, the brain continues to mature extensively postnatally. As a result of the branching of dendrites and formation of new synapses, the amount of gray matter in the brain increases until early puberty. Then, extensive dendritic pruning takes over and continues throughout early adulthood (Fig. 3.3).

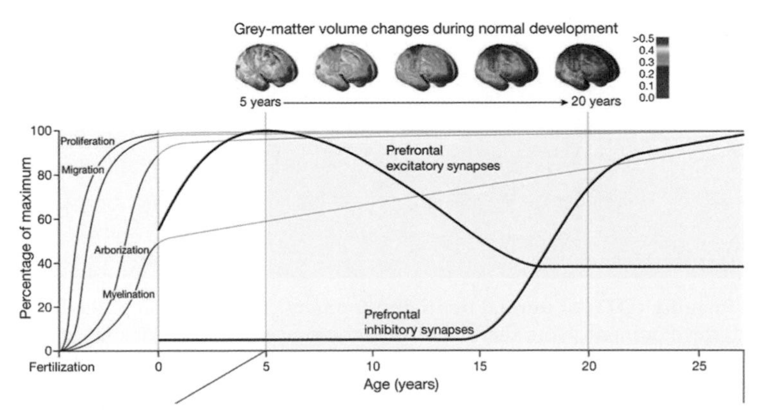

FIGURE 3.3 **Gray matter volume changes with development.** Normal cortical development involves proliferation, migration, arborization (circuit formation) and myelination, with the first two processes occurring mostly during prenatal life and the latter two continuing through the first two postnatal decades. The combined effects of pruning of the neuronal arbor and myelin deposition are thought to account for the progressive reduction of gray matter volume observed with longitudinal neuroimaging. *Source: Reprinted from Insel* Nature. *2010;468(7321):187–193,[9] with permission. © 2010 Nature Publishing Group.*

From birth until middle age, the brain undergoes a process of increasing myelination. The majority of myelination occurs in the early years of life; however, some continues well into adulthood (Fig. 3.4). Not all areas of the brain develop at the same speed. The cortex, particularly the prefrontal cortex, is the last area of the brain to structurally and functionally mature. This is an area of the brain felt to underlie many of our higher cognitive functions such as response inhibition, working memory, and motivation. The prefrontal cortex continues to form synapses later than other brain structures, and undergoes pruning of these synapses until much later in life.[11,12]

FUNCTIONAL DEVELOPMENT

Neural development underlies functional achievements in motor, language, cognitive, academic, and social domains. These achievements, or milestones, occur in an ordered fashion in children, similar to the maturational processes of brain development. A typically developing child sits, then crawls, then stands, then walks. Normal ages for early functional milestone acquisition are largely "preprogrammed" and independent of cultural influences. Milestone guidelines for multiple streams of development have been established and verified for the US population.[13]

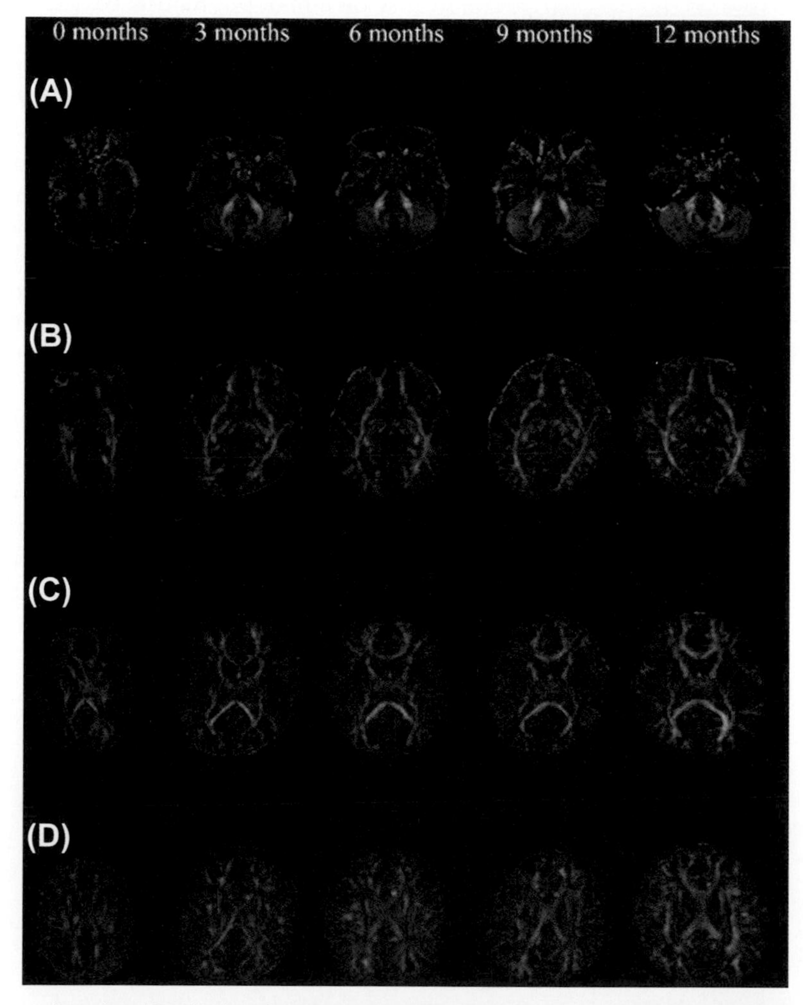

FIGURE 3.4 **Diffusion tensor imaging (DTI) of normal brain development.** The evolution of the DTI color maps shows the changes in white matter tracts during normal development. Note that the most significant changes occur during the first year, but that white matter continues to develop into adulthood. Representative axial color maps are shown at 0, 3, 6, 9, 12, 24, 36, and 48 months. The axial images are arranged from ventral to dorsal with A and E being the most ventral images and D and H being the most dorsal. A color map at adult age and the corresponding T1-weighted anatomical images are represented in the last two columns. alic: anterior limb of internal capsule; cbt: corticobulbar tract; cg: cingulum; cst: corticospinal tract; fx: fornix; icp: inferior cerebellar peduncle; ifo: inferior fronto-occipital fasciculus; ilf: inferior longitudinal fasciculus; mcp: middle cerebellar peduncle; ml: medial lemniscus; plic: posterior limb of internal capsule; slf: superior longitudinal fasciculus; unc: uncinate fasciculus. *Source: Reprinted from Hermoye* et al. Neuroimage. *2006;29(2):493–504,*[10] *with permission. © 2006 Elsevier Inc.*

Development is the end result of multiple interactions that may occur at the genetic, cellular, structural, or physiological level and may be influenced by the organism's environment. Arnold Gesell defined a system of assessing the developmental function of young children. He noted that: (1) development was an ordered process; (2) it encompassed several domains (streams): motor, language, problems solving, and personal–social; and (3) children with developmental disabilities generally achieved the same functional milestones as typically developing children, but did so at a slower rate. He felt that development could be quantified by calculating a ratio or developmental quotient (DQ).

Identifying developmental disability early is crucial to the well-being of the child and family. Identification is the first step in establishing a diagnosis and management program. A diagnosis allows the family to receive information regarding prognosis and guides the etiological investigation.

Children with developmental disabilities present to the practitioner by a variety of pathways; most common are failure to meet the developmental expectations, screening of high-risk populations, or because of other neurological concerns. Children usually come to attention because of failure to meet an age-appropriate expectation. Parental concern is generally a quite sensitive test for developmental delays. Children may also be referred for care because of identification of developmental disability by a non-relative, frequently through the school system. Children with developmental disabilities may also present to their health-care provider because of other health concerns, such as epilepsy or prematurity.

Parental concerns are closely linked to the child's chronological age. A neonate has very few functional expectations. Barring major disruptions of consciousness, major dysmorphisms, or organ failure, there will not be developmental concerns. In early infancy, children are expected to begin recognizing family and responding

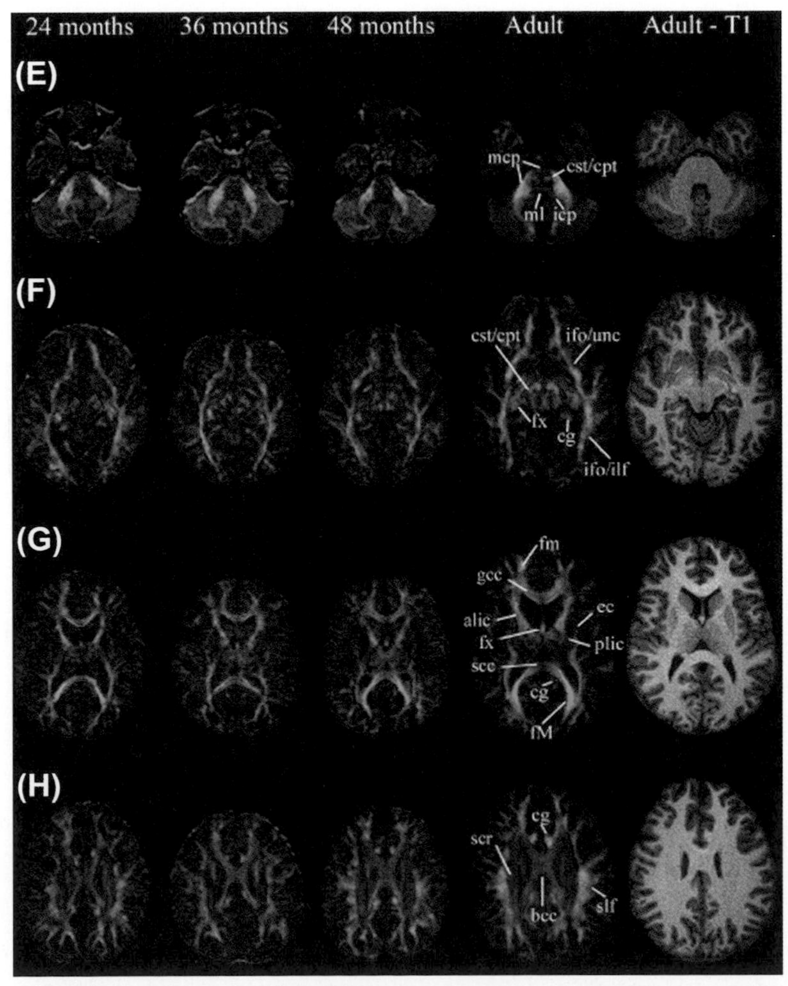

FIGURE 3.4 *(cont'd)*

to the world around them. Developmental concerns in the first few months of life generally regard sensory perception: "Can my baby hear and see?" The latter part of the first year of life through the second year is dominated by huge gross motor gains, and a delay in motor milestones is the most likely developmental concern bringing children with cerebral palsy or other infantile onset movement disorders to attention. Language rapidly increases after 2 years of age in typically developing children, and parents are most likely to seek care at this time for a child with limited vocabulary. When a child reaches school age, socialization, behavior, and school performance become the major markers of development and the major source for developmental referral (Table 3.1).

Delay connotes a lag in acquisition of milestones (either globally or in more specific streams of development). This is the most commonly understood and appreciated form of abnormal development. There is no strict definition of delay. In general, the delay necessary to qualify for services and interventions is less than 70–75% (two standard deviations + standard error), but this is determined by public policy, not scientific data. A lower cut-off results in fewer people qualifying for services. A seemingly small change in numerical criteria can have large implications (e.g. increasing the cut-off to qualify for intellectual disability from 70 to 75 results in doubling the number of people who meet the criteria).

TABLE 3.1 Developmental Disabilities

Disorder	Impairments	Prevalence
Attention deficit/hyperactivity disorder	Attention and/or hyperactivity/impulsivity	6–10%
Autism	Social communication	
	Restricted, repetitive interests	1%
Blindness	Vision (acuity or field of vision)	0.4%
Cerebral palsy	Movement or posture	0.1–0.3%
Deafness	Hearing	0.1%
Developmental coordination disorder	Movement	6%
Epilepsy		1%
Intellectual disability	General cognition; adaptive behavior	1–3%
Mixed receptive–expressive language disorder	Language	6% of preschool children
Specific learning disability	Academic achievement	4–7%

A child's development cannot be described by a single number. Asynchronous development or dissociation is observed when one or more streams of development progress at a different rate from others. The pattern of dissociation (streams that are developing aberrantly versus typically) and the rate of development of each of these streams lead to the early diagnosis of developmental disabilities. A child with cerebral palsy may show intact language and problem-solving abilities but impaired motor development. A child with intellectual disability presents with delays in language, problem solving, and self-help abilities, but may have a grossly normal rate of motor development (Table 3.2).

The description of individual streams should not be interpreted as meaning that these skills develop in isolation from each other. Many of the skills required for development in motor, language, cognition, behavior, and adaptive function overlap, and therefore the same phenotype can have a number of underlying causes. A child with a written language disability may have difficulty with the physical act of writing (fine motor), the ability to express ideas (higher order language difficulties), the ability to organize thoughts coherently (executive function and attention), or a combination of all three of these domains.[14]

Sometimes the developmental sequence is violated. Deviance describes acquisition of milestones in a non-sequential or uneven fashion in one or more streams of development. Examples include failing to achieve a specific milestone and "skipping ahead" (e.g. a child who learns to walk without ever learning to crawl) or uncoupling skills that are usually linked (e.g. a 30-month-old child with a large vocabulary but no two-word phrases). Deviant development is often underappreciated by parents, but often signals an underlying pathology or an overestimation of milestones (Table 3.3).

TABLE 3.2 Dissociation: Gateway to Early Diagnosis

Disorder	Gross Motor	Fine Motor/ Problem Solving	Language	Self-Help
Cerebral palsy	Decreased	± to Decreased	±	±
Intellectual disability	±	Decreased	Decreased	Decreased
Receptive–expressive language disorder	±	±	Decreased	±
Autism spectrum disorder	±	±	Decreased	Decreased

Quantifying Development

Development is measured functionally. By documenting the temporal pattern of a child's attainment of developmental milestones and comparing it with accepted norms, development can be quantified by determining the child's developmental age and dividing it by the chronological age; therefore, a typical child has a DQ of 100:

$$\text{Developmental Quotient (DQ)} = \frac{\text{Developmental Age}}{\text{Chronological Age}} \times 100$$

The DQ at a given point is a useful, if rough, estimate of a child's current function compared with standard peers. However, when measured at multiple time-points, the DQ provides a more precise measure of a child's rate of development. DQs are much more useful for indicating disability than precociousness. A DQ of 75 requires close monitoring; a DQ of 125 is largely meaningless for long-term prognostication. The lower the DQ, the more stable the score over time. A child who scores a DQ of 75 at one time-point may catch up, but a child with a DQ of 40 is unlikely to do so.

Developmental Disability: A Definition

Developmental disability (also called developmental disorder) is defined as a severe, chronic disability of an individual that:

- is attributable to a mental or physical impairment or a combination of mental and physical impairment
- is manifested before the individual attains the age of 22 years
- is likely to continue indefinitely
- results in substantial functional limitations in three or more of the following areas of major life activity: receptive and expressive language, learning, mobility, self-care, self-direction, capacity for independent living, economic self-sufficiency
- reflects the individual's need for a combination and sequence of special, interdisciplinary, or generic

TABLE 3.3 Developmental Milestones

Age	Gross Motor	Language	Fine Motor/Problem Solving	Adaptive
1 month	Chin off table in prone	Alerts to sound	Visually fixes	Looks cute
2 months	Chest off table	Social smile		
3 months	Neck control	Cooing	Visually follows ring in a circle; blinks to threat	
4 months	Supports on wrists	Orients to voice; laughs aloud	Hands open; manipulates fingers	
5 months	Transfers objects	"Ah-goo"	Transfers; reaches	
6 months	Sits without support	Babbling	Inspects object	
9 months	Crawls; pulls to stand	Mama/Dada non-specific; gesture language	Immature pincer grasp; object permanence; uncovers objects	Finger feeds
1 year	Walks	Says two words; follows one-step commands with gestures	Mature pincer grasp; makes crayon marks	Cooperates with dressing
15 months	Runs	Immature jargoning; one-step command with gesture	Imitates scribbles	
18 months		Mature jargoning 20-word vocabulary	Obtains object with stick; spontaneously scribbles; stacks three-cube tower	Uses spoon with yogurt
2 years		50-word vocabulary, 2-word sentences, 2-step commands	Stacks four cubes horizontally	Drinks out of cup
30 months	Walks without watching feet	Concept of "one", pronouns	Makes horizontal and vertical marks	Undresses self; helps with dressing; spears with fork
3 years	Walks up steps alternating feet	250-word vocabulary, 3–4-word sentences; names one color	Draws a circle; draws a person with head and one body part	Begins cooperative play
4 years	Hops on one foot	5–6-word sentences, 4000–6000-word vocabulary	Draws a square	Dresses unassisted; uses spoon with soup
5 years	Skips; climbs	Complex sentences; tells stories	Draws basic letters; prints name	Takes turns; ties shoes, zips, buttons; knows right vs left

services, individualized supports, or other forms of assistance that are of lifelong or of extended duration and are individually planned and coordinated.[15]

Developmental disabilities are defined by functional limitations, not etiology or mechanism. A developmental disability is "chronic" and is therefore a lifelong impairment to functioning. A single discrete injury or incident that results in brief physical and/or mental dysfunction does not qualify as a developmental disability if complete recovery is anticipated. Impairment in physical or mental ability may qualify an individual. The disability must result in functional impairments in three or more areas of a person's life. Impairments in activities of daily living are not regularly addressed in medical settings, but are often the most limiting aspects of a disability.

The disability must occur before adulthood. This age distinction is important owing to differences between injury to the juvenile and mature adult brain.

Why Limit Developmental Disabilities to Children?

First, the developing nervous system is not set. Injury to the juvenile brain is not the same as injury to the adult brain. This is in part due to the fact that synaptic plasticity continues to take place throughout childhood. Pruning continues well into young adulthood. Because both processes are still occurring in the immature brain, injury results in disruption of future development. In addition to any neuron death, there is the potential loss of cell connectivity because of axonal shearing and gliosis resulting from the injury. Neuronal connections that otherwise would have formed are impeded. Downstream effects of this interference may be substantial and focal injury may result in more global impairments as well as an injury whose manifestations may change over time.

The actual age of full brain development varies depending on the brain structure specified. If one were to focus on myelination of the prefrontal cortex, the age of maturity would be approximately 25 years. If the focus were on brain size, 90% would be age 3 and 95% would be age 5. While all of the definitions of the developmental disabilities are restricted to children, the variation is large: onset of symptoms for intellectual disability is before age 18, for attention deficit/hyperactivity disorder (ADHD) is before age 12, and for cerebral palsy is before age 3.

Second, the etiologies of brain dysfunction in childhood have a wide-ranging impact and result in many areas of dysfunction with high rates of comorbidity. Multiple disabilities are the rule. Even developmental disabilities that are defined as very specific functional deficits (e.g. epilepsy or specific learning disabilities) generally have impairments in other areas.

Third, the response of the immature nervous system may differ from that of adults. The child may be more resistant to insults because of plasticity or recruitment of compensatory functions. Alternatively, the immature nervous system may be more sensitive to insult because of selective vulnerability, secondary effects, decreased functional reserve, or arrest of developmental processes. The response to therapy and the modalities used in children differ from those of adults.

Fourth, the timing of the injury to the brain can result in different manifestations. Selective vulnerability refers to the windows of vulnerability that are evidenced during certain periods of development. This may be due to physiological maturation (e.g. control of blood flow to certain areas of the brain), cellular activity (cells that are rapidly dividing or metabolically active are more sensitive to injury), or compensatory mechanisms.

The predictable course of development allows certain insults to the developing fetus to be timed. The structures that are developing at the time of injury are those that may be affected (exposure to a teratogen at week 6–7 may affect heart development and major limb development, whereas later exposure to the same injurious agent at 10 weeks may preferentially affect facial feature development and intestinal rotation; see Fig. 3.1). In addition, injury may occur to a specific subpopulation of developing cells that are particularly susceptible to the causative agent. As the embryo/fetus develops, all cell types that develop from the injured cell population may be affected. This is why reliable clusters of physical findings are often seen in certain disorders (anomalads). The timing of injury during development and the type of cells preferentially susceptible to this injury determine the ultimate cluster of findings that will be apparent in a given individual.

Infants born prematurely demonstrate this specific cellular vulnerability. Between 24 and 40 weeks of gestation, premyelinating oligodendrocytes are in an active phase of development and are in a unique position for injury. The germinal matrix is particularly susceptible to hemorrhage in the 24–28 week period. There are gradations in extension and involvement. However, the resultant characteristic lesion, referred to as periventricular leukomalacia, and the accompanying neuronal and axonal disease can affect the cerebral white matter, thalamus, basal ganglia, cerebral cortex, brainstem, and cerebellum. It can place the infant at risk for cerebral palsy and learning difficulties depending on the extent of the injury. The same injury occurring after 32 weeks of gestation will not cause periventricular leukomalacia.[16]

Finally, in contrast to an adult brain, the consequences of injury to a developing brain are not solely the result of initial and secondary injury. Damaged brain regions can also suffer arrest of development. This arrest is not specific to particular etiologies, further complicating the overlapping phenotypes.

ETIOLOGY

The neurodevelopmental disabilities are defined functionally and describe the final pathway of many injury types occurring at many different stages of development. They do not speak to the mechanism of injury or severity. At this time, much of what is understood about neurodevelopmental disability is based on associations rather than causality. For example, Down syndrome is associated with intellectual disability but the mechanisms by which the additional genes provided by that extra chromosome 21 affect cognition are only just being determined.[17]

It is important to recognize that understanding the etiology of a particular neurodevelopmental disability often does not lead to a specific treatment option that improves a specific child's outcome. There are, however, other benefits to a family of a child with special needs. Accurate etiological diagnosis can provide prognosis, identification, and management of associated or comorbid conditions, assessment of recurrent rates, and counseling for families. On a societal level, understanding etiology can lead to prevention, improved treatments, and better outcomes. The etiology of primary injury can be classified broadly into the following categories: genetic; vascular; hypoxic–ischemic; traumatic; environmental, including toxins/exposures and nutritional; infectious, immunological, and inflammatory, metabolic; endocrine; and oncological.

Genetic

Genetic disorders affect the underlying material that guides development and can result in neurodevelopmental disabilities from a number of mechanisms, including the wrong amount of genetic material, missing genetic material, disorders of a single gene, and mitochondrial disorders that affect gene function. Each cell nucleus contains chromosomes, which comprise the DNA and genes. At the cellular level, chromosomal abnormalities such as trisomies, having an entire extra chromosome (e.g. Down syndrome), or duplications, rearrangements, and deletions of part of a chromosome (e.g. Williams syndrome) can have significant effects on development. Single gene defects (e.g. neurofibromatosis type I) can also disrupt the typical progression of development by providing too much, too little, or none of a protein, or by disrupting regulation of expression of other genes. Finally, mitochondrial disorders have a unique maternal inheritance and are associated with neurodegenerative disease with crises during illness or other times of stress (see Chapter 5).

Vascular

Vascular lesions in adults result primarily in stroke. While focal strokes can occur in the developing brain, they are less common than hypoxic and anoxic ischemic injuries. Vascular strokes may also be seen in association with vascular malformations, genetic disorders that increase the likelihood of clotting (e.g. deficiencies in factor V Leiden, protein C, or methyltetrahydrofolate), disorders of fatty acid metabolism [e.g. mitochondrial myopathy, encephalopathy, lactic acidosis, and stroke (MELAS)], or congenital heart defects.

Hypoxic–Ischemic

Hypoxia (lack of oxygen) and ischemia (cell death due to lack of blood flow) often co-occur. Recently, much attention has been paid to perinatal hypoxic ischemia as a cause for neonatal encephalopathy in term infants, which is associated with long-term neurological morbidity. Hypoxia can occur over a prolonged time, such as during a lengthy maternal infection, or as an acute severe event, such as during a placental separation. The consequences of different types of injury are still being delineated, and the etiology for a hypoxic event is not always discernible. The consequences vary, but severe neonatal encephalopathy is associated with death, intellectual disability, or cerebral palsy. Neonates who suffered moderate neonatal encephalopathy have lower scores on cognitive, motor, and language assessments in preschool and learning difficulties in school.[18]

Traumatic

Media exposure has raised awareness about sports-related injury, including the deleterious effects of repeated concussion and head trauma. Head trauma has differential effects over time as the brain develops. It is hypothesized that skull thickness, head size in relation to body size, and the stage of myelination all play a role in determining the type of injury caused by brain trauma. Depending on these structural differences, injury type can be different. In early infancy, diffuse axonal injury is much more likely owing to the shearing forces. Later, a coup–contracoup injury is more frequently seen. Long-term sequelae of severe head trauma can include seizures, hydrocephalus, cerebral palsy, blindness, deafness, learning, and behavioral problems. Children younger than the age of 2 years have poorer outcomes from traumatic brain injuries than do older children and adolescents. Repeated concussive injury is associated with difficulties in learning, behavior, and executive function.

Environmental

Exposure to Toxins

Exposure to toxins can be from internal exposure such as metabolite buildup from inborn errors of metabolism or from external exposures. Known agents that interfere with development (especially prenatally) are described as teratogens. For example, exposure to exogenous retinoids [such as the acne medication Accutane® (isotretinoin)] during pregnancy disrupts the signaling at retinoid receptors, causing abnormal activation of transcription factors and gene expression in the developing brain, resulting in neurological, cardiovascular, and craniofacial defects. Knowledge of teratogens is based largely on gross structural abnormalities resulting from exposures; more subtle developmental abnormalities linked to as yet unidentified teratogens are likely to be present.

Probably one of the most studied toxins in the prenatal period is alcohol.[19] Effects of alcohol on neurodevelopment have been known since the nineteenth century. A spectrum of dysfunction is now characterized that extends from alcohol-related neurodevelopmental disorders to fetal alcohol syndrome. Children with prenatal exposure to alcohol can have low cognitive function and maladaptive social functioning. Birth defects, growth retardation, a characteristic pattern of facial anomalies, and neurological abnormalities define the fetal alcohol syndrome. There is no known safe dose of alcohol in pregnancy and the American Academy of Pediatrics recommends abstinence from alcohol for women who are pregnant or who are planning a pregnancy.

Intrauterine exposure to illicit drugs is also a source of concern. Although the children of women who use drugs while pregnant show higher rates of neurodevelopmental disability, specific associations have not been defined. This is because exposures are rarely single; women who use drugs during pregnancy generally have other risk factors including multiple drug use, smoking, and alcohol use, inadequate prenatal nutrition, and delayed and infrequent prenatal care. Socioeconomic factors frequently confound studies of intrauterine exposures.

Postnatal toxic exposure can also have developmental consequences. Lead is one toxin that can cause developmental disability after the postnatal period. Lead was previously used in a number of household items and as an ingredient in house paint. Pica behaviors, eating non-food items, can lead to elevated levels of lead. Pica may be a normal developmental stage, is prolonged in children with developmental delays, or may be associated with iron deficiency. Lead is highly neurotoxic in acute large exposures and may result in encephalopathy, neuropathy, anemia, or death. Chronic low-level lead exposures are associated with intellectual disabilities, language disabilities, and difficulty with executive function. No safe threshold of lead level has been defined.[20]

Nutritional

In addition to overexposures, such as teratogens, deficiencies in necessary metabolic factors can have developmental consequences. Most of the adverse neurodevelopmental outcomes that are associated with nutritional factors are due to deficiencies. Throughout the world the leading cause of intellectual disability is iodine deficiency, which leads to thyroid underactivity. Vitamin A deficiency is a leading cause of blindness. Folic acid supplementation has been shown to decrease the incidence of spina bifida.

Infectious, Immunological, and Inflammatory

Intrauterine infection or exposure to maternal infection may be associated with premature birth, structural deformity, or intrauterine death. Prenatal care now involves testing for a number of diseases that have neurodevelopmental consequences. For example, rubella is a virus that typically causes a transient and relatively benign elevation temperature and a skin rash after exposure. However, if a woman has a first exposure to this virus during pregnancy it can lead to an intrauterine exposure for the fetus with resultant birth defects such as cataracts and other eye anomalies, congenital heart defects, deafness, and intellectual disability. The patterns of abnormalities seen are specific to the gestational age at exposure.

Further, mothers are routinely screened before delivery for group B streptococcus, bacteria that can colonize the vaginal canal in asymptomatic women but can cause serious infection including meningitis that can have devastating neurodevelopmental consequences if not treated early. Postnatal infection of the meninges or brain (encephalitis or meningitis) can also have significant morbidity and even mortality. The routine immunizations that are provided at well-child examinations have decreased the frequency and therefore the morbidity associated with such childhood infections (see the Centers for Disease Control and Prevention website for immunization schedules[21]).

Any insult can result in secondary inflammation that can sometimes be more damaging to the brain than the initial insult.[22] For example, trauma, hypoxic–ischemic injury, and infection can all trigger inflammation from excitotoxicity and free-radical development. Inflammation can also be the primary insult. Two major inflammatory diseases that can occur in childhood are acute disseminating encephalomyelitis and transverse myelitis. Chronic inflammation may be seen in some neurometabolic disorders.

Metabolic

Metabolic disorders should be suspected whenever a child shows unexplained regression (loss of milestones), lethargy, and mental status change without infection, multiorgan involvement during an illness that is not typical for the illness, or a pattern of greater than expected levels of illness in the setting of known mild infections. Some metabolic diseases present acutely, where others have a more indolent course. Neurological symptoms are common but generally non-specific, and include developmental delay or regression of development, epilepsy, movement disorders, and vision or hearing loss. Other organs that may be involved depending on the metabolic defect include eye, skin, muscle, liver, spleen, heart, and kidney.

Metabolism is the collection of biochemical reactions that take place in the cells of the body to produce the energy and molecular building blocks needed to sustain life. These chemical reactions make up the metabolic pathways of the body and are either catabolic (involved in breaking down molecules of protein, carbohydrates, or fat for the production of energy) or anabolic (using energy to make proteins, nucleic acids, etc., that are needed by the cells). Each metabolic pathway requires a series of enzymes to transform one molecule into another. If any enzyme in the pathway is not working, the final molecule of the pathway cannot be produced and there will be a collection of upstream precursor molecules that cannot be broken down further. Symptoms of a metabolic disease may be caused either by not having enough of the final molecule of the affected pathway or by the buildup of the precursor molecules. The severity of a metabolic illness is usually determined by the level of function of the abnormal enzyme.[23] Approaches to the treatment of inborn errors of metabolism are summarized in Fig. 3.5.

Endocrine

Stable endocrine function is critical to brain development. In the prenatal environment two important maternal hormones that affect fetal development are the thyroid hormones and insulin.

Thyroid hormones

Thyroxine is critical to brain development, especially during the second trimester, and hypothyroidism is associated with pre-eclampsia, low birth-weight infants, abruptio placentae, spontaneous miscarriage, and intellectual disability. Long-term deficits affect memory, visuospatial, and motor abilities. Fetal hypothyroidism, worldwide, is most commonly due to iodine deficiency. The most severely affected infants have neurological "cretinism", manifested by severe intellectual disability and impaired motor function. In areas with sufficient iodine, such as North America, maternal hypothyroidism is mainly due to autoimmune thyroid disease. The severity, timing of onset, and duration, as well as postnatal management, all influence fetal and neonatal brain development.

Glucose Control

Poor control of blood glucose can expose the fetus to increased risk owing to hypoglycemia (low blood sugar) and hyperglycemia (high blood sugar). Gestational diabetes with insulin dysregulation can lead to hyperglycemia. Hyperglycemia has been associated with large for gestation birth weight (macrosomia), fetal and neonatal mortality, pre-eclampsia, congenital–structural malformations, and perinatal mortality. Macrosomia can result in problems during labor, increased neonatal morbidity, and a long-term increased risk of obesity and metabolic disorders for exposed children. Maternal hyperglycemia has also been associated with learning and language disorders.

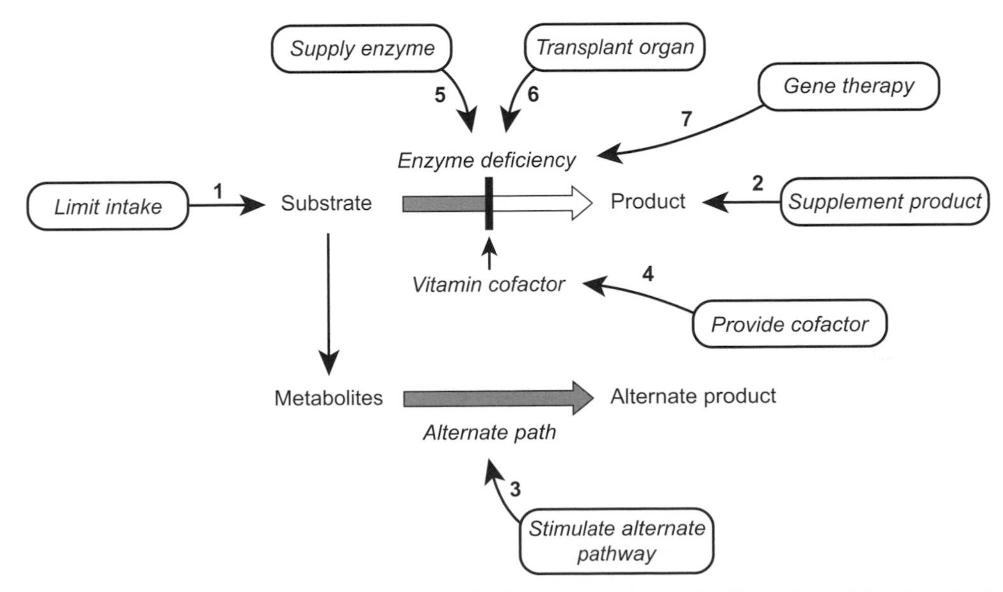

FIGURE 3.5 **Approaches to treatment of inborn errors of metabolism.** Treatment can be directed at (1) limiting the intake of a potentially toxic compound, (2) supplementing the deficient product, (3) stimulating an alternative metabolic pathway, (4) providing a vitamin cofactor to activate residual enzyme activity, (5) supplying the enzyme itself, (6) transplanting a body organ containing the deficient enzyme, and (7) gene therapy. *Source: Reprinted from Batshaw* et al. Children with Disabilities. *Brooks; 2013:325,*[24] *with permission.*

Oncological

Both cancer and the treatments used to cure it have neurocognitive effects.[25] Brain tumors are space-occupying lesions that have neurological consequences that are specific to where they are located in the brain as well as generalized cognitive effects. Resultant tumor infiltration, hydrocephalus, and surrounding swelling can also cause neurodevelopmental consequences. Some neurodevelopmental syndromes have specific associations with tumors. For example, tuberous sclerosis complex (TSC) is a neurodevelopmental disorder that can be defined as a clinical syndrome or genetically by mutations in *TSC1* or *TSC2*. These mutations result in overactivation of the mammalian target of rapamycin (mTOR) pathway. The major neurological features include subependymal nodules, subependymal giant cell astrocytomas, and cortical tubers. There are also overgrowth tumors in other organ systems including cardiac, renal, and ophthalmological. Comorbid conditions include epilepsy, autism, intellectual disability, and self-injury.[26]

Current treatments of childhood cancers include combinations of surgery, radiotherapy, and chemotherapy. Surgical consequences for brain tumors can include neurological deficits such as visual field cuts or motor function difficulties. Trauma from surgery can also result in more generalized deficits in overall cognition, language, memory, attention, and movement. Targeted radiotherapy therapy is still used in combination with chemotherapy for some brain tumors, despite known cognitive effects including significant drops in intelligence quotient (IQ). Therapies for non-neurological cancers may also have neurological consequences. Intrathecal CNS prophylaxis has replaced radiotherapy in treatment of children with leukemia such as acute lymphocytic leukemia (ALL) because of these effects. Chemotherapy has also been associated with hearing and balance problems as well as neurocognitive side effects ranging from attention and memory difficulties to cognitive decline. Younger age and female gender have been associated with worse effects. Increasing survival rates in childhood cancer have resulted in the reduction of treatment-related side effects as a main emphasis of current treatment goals.

TECHNIQUES

This is an exciting time to work in the field of neurodevelopment as there is a plethora of research tools available to identify the neuromolecular mechanisms of injury underlying specific phenotypes. In addition to understanding etiology for diagnostic purposes, identifying the underlying mechanisms opens the door for molecular treatment and even cure for a wide spectrum of neurodevelopmental disabilities. However, current understanding is still in the early stages and is limited to a few selected, rare disorders. A basic understanding of diagnostic techniques allows for an understanding of how an etiological diagnosis proceeds and gives insight into the mechanism of dysfunction.

Behavioral Testing and Therapy

Neuropsychological Tests

A variety of neuropsychological tests is available to elucidate the neurocognitive profiles of individuals.[27] Standard IQ tests (e.g. Stanford–Binet and Wechsler series) assess global intellectual functioning by testing performance on multiple tasks. Other tests are available that focus on and further detail specific areas of ability (e.g. language, memory, processing, and adaptive behavior). Multiple variations of these tests are available, depending on the age group and cognitive level of the subjects being tested. Each test has its own strengths and weaknesses. When evaluating a subject it is important to choose the correct test. The test should be well researched for validity and reliability. The test should be appropriate for the test subject's developmental age to avoid floor and ceiling effects. It is crucial that the test also specifically address the cognitive area of concern (if paragraph reading is the child's primary difficulty, then testing picture and single word identification will not properly address this concern).

Rating Scales

Rating scales are important tools in neurodevelopmental medicine. The most commonly used rating scales are for ADHD, but other rating scales are available for autism, mood disorders, anxiety, and adaptive behavior. Some scales focus on symptoms of only one diagnosis (narrow band) while others focus on symptoms for a variety of diagnoses (broad band). The rating scales document the frequency or severity of symptoms and can be filled out by multiple parents, caretakers, and teachers. This allows the practitioner to have a glimpse into the behavior of a patient in multiple environments and as assessed by multiple observers. The results of rating scales need to be interpreted with care as there is poor inter-reporter reliability and results can be skewed based on the reporter's time with the child, tolerance for behaviors, and understanding of questions asked.[28,29]

Behavioral Analysis

Behavioral analysis is a tool that requires substantial time and resources to be effective. The patient is monitored, and non-adaptive (e.g. "socially inappropriate", disruptive, self-injurious, or violent) behaviors are recorded, as are the situations in which these behaviors occur, any potential associated triggers, and any reinforcing or rewarding agents (e.g. attention for inappropriate behavior). A behavioral plan is created and applied in which unintentional rewards for the unwanted behavior are avoided, triggers are reduced, and more adaptive responses are taught and rewarded. Behavioral analysis is most commonly associated with autism treatment programs, but also has a long history of use in habilitating individuals with intellectual disability.

Other Behavioral Techniques

Many other behavioral techniques are used in evaluating and treating individuals with neurodevelopmental disabilities. One of the most commonly used (generally with one or more of the above methods) is observation of the subject. This can take place either in a controlled environment or in the subject's natural environment (e.g. school) and can be invaluable for collecting information about a child's level of functioning. Another frequently used behavioral technique is monitoring a subject's social behavior, specifically in scripted situations (such as with the Autism Diagnostic Observation Schedule) to determine social skills and adaptive behaviors. Social skills groups can then be used to teach and practice more adaptive ways of interacting. Comprehensive, individualized management programs can be developed by monitoring a child's behavior and measuring the change associated with an intervention (e.g. medication).

Structural Imaging

Ultrasound

Ultrasound uses sound waves to image structures within the body. Although it has limited applicability for neuroimaging in older children and adults because it does not penetrate the bone of the skull, it provides good imaging of the central structures of the brain for infants who still have open fontanelles. Ultrasound is the most widely used neuroimaging procedure in infants because it is fast, non-invasive, and inexpensive, and can be performed at the bedside, making it useful in severely ill infants. Ultrasound is an excellent tool for use in the emergent and intensive care setting, but it does not provide as much detail as magnetic resonance imaging (MRI).[30] Ultrasound may also allow measurement of blood flow to the brain in disorders that put children at risk for stroke (e.g. sickle cell disease).

Computed Tomography

Computed tomography (CT) is a method of obtaining multiple thin X-ray images throughout the brain (or other body parts) and, using a computer, essentially stacking the images to obtain images of the whole brain. CT provides an excellent view of the skull, major structural lesions, and acute bleeding, but does not have good resolution for viewing subtle details of brain structure, small lesions, and very acute ischemic changes. CT exposes the patient to substantial radiation and because of concerns about long-term risks the use of CT in children is limited to emergent situations. In most non-emergent situations, MRI is the preferred neuroimaging modality.[31]

Magnetic Resonance Imaging

MRI is the current state-of-the-art method of imaging the brain. It requires a high-powered magnet, very specialized equipment, and intricate computer analysis of data to develop high-resolution images. MRI uses electromagnetic pulses to align protons and measure their movement. This information is used to create two- or three-dimensional images of body parts. Manipulating various parameters provides a variety of images; the basic types are T1 weighted and T2 weighted. Some forms of MRI are best for viewing the basic structure of the brain (generally T1 class), while others are best for viewing lesions (generally T2 class). See the last column of Fig. 3.4 for examples of MRI images. This technology is constantly being improved and many specialized sequences are available for viewing specific aspects of brain pathology.

MRI provides excellent contrast of soft tissues and is therefore ideal for imaging the brain and spinal cord. Compared with CT, it detects smaller and more subtle lesions and can detect injuries such as ischemic strokes much earlier. MRI can be used to image the vasculature of the brain without the use of dyes; however, particularly for the neck and spine, vascular imaging is often improved with the use of contrast. Contrast agents such as gadolinium are useful for timing and diagnosing injuries. Because there is no radiation exposure, MRI scans can be performed repeatedly and used to track changes over time (e.g. measuring cortical growth in ADHD or change in myelin in leukodystrophies). Since MRI takes much longer to perform than CT scans, some children and adults with neurodevelopmental disorders may have difficulty remaining still long enough for MRI scans to be completed and general anesthesia may be required.[32]

Diffusion Tensor Imaging

Diffusion tensor imaging (DTI) is a subset of MRI that has become increasingly useful in examining connectivity between different regions of the brain by providing representations of the structural integrity of the fiber tracts. DTI tracks the diffusion of water molecules along white matter tracts of the nervous system.[33] See Fig. 3.4 for multiple examples of DTI.

Magnetic Resonance Spectroscopy

Magnetic resonance spectroscopy (MRS), also known as nuclear magnetic resonance spectroscopy, is an additional derivation of MRI that deserves special mention, particularly in developmental medicine. Unlike other MRI technologies, MRS allows investigation of the biochemical composition of a given area of the brain. For a chosen area of the brain (generally an area of apparent injury), the relative concentration of certain metabolites can be measured as peaks. MRS can assist with diagnosis of causes of brain injury, including tumors and metabolic diseases.

Muscle, Nerve, and Brain Biopsy

Muscle and nerve biopsies are crucial in the diagnosis of neuropathies and myopathies, which are disorders of the peripheral nervous system. Muscle and nerve biopsies are not routine tests in developmental disabilities (since the source of developmental disabilities is CNS dysfunction). Some diseases that are primarily considered disorders of the peripheral nervous system or muscle have a central component (e.g. some myopathies have comorbid intellectual or learning disability). Therefore, muscle or nerve biopsy may be critical in determining an etiology, but is only used if a disease of the peripheral nerves or muscle (respectively) is suspected. [34,35]

The use of tissue biopsies from the brain is not particularly common outside the realm of oncology, but pathological analysis is sometimes necessary to rule out diagnoses such as some infectious, immunology, and degenerative processes, particularly if such a diagnosis would warrant a change in management or offer a treatment.[36]

Physiological/Functional Tests

Physiological tests of the nervous system are designed to study the function of the nervous system rather than the structure.

Single Cell Recordings

Electrical recording of the currents in single neurons (techniques including whole cell and patch clamp recording) have produced invaluable information regarding how neurons function, including how they transmit information (e.g. action potentials, electrolyte currents, and receptor functions), but are largely limited to basic science research because they require invasive techniques. In clinical practice, less invasive procedures have been developed to determine the function and integrity of the nervous system.

Electromyography and Nerve Conduction Studies

Electromyography (EMG) is the measurement of electrical activity within the muscle. This test involves inserting a needle electrode into the muscle being studied and can determine abnormal activity at rest (which can be seen in disorders such as denervation syndromes and dystonias) and if the muscle does not electrically recruit additional fibers when it is required to activate (when the patient is told to contract that muscle). Nerve conduction studies (also called nerve conduction velocities) use electrical stimulation of the peripheral nerves to measure the speed of transmission of the signal. They can therefore diagnose demyelination syndromes, diseases of the neuromuscular junction, and nerve entrapment (pressure damage to nerves such as carpal tunnel syndrome). Nerve conduction studies use electrode discs placed on the skin over the nerve of interest and a recording electrode is placed over the muscles controlled by that nerve.[37]

Transcranial Magnetic Stimulation and Transcranial Direct Current Stimulation

Transcranial magnetic stimulation (TMS) and transcranial direct current stimulation (tDCS) are non-invasive ways of affecting the excitability of the cortex. TMS and tDCS are primarily clinical research tools for studying neuroconnectivity and the differences in connectivity, relative activation, and responses in various diseases. TMS uses a magnetic field to induce electrical current in the neurons of the cortex, whereas tDCS uses mild electrical stimulation applied by electrodes to the scalp (Fig. 3.6). Both TMS and tDCS can be used to either increase or decrease the excitability of an area of brain cortex. The effect of stimulation in this area of the brain on other connected areas can then be determined. Alternatively, the effect of stimulation of a brain area on performance can be measured.[38] Current applications include motor learning after stroke and as a potential treatment in other developmental paradigms.

Electroencephalography

Electroencephalography (EEG) is the use of multiple electrodes placed along the scalp to measure the inherent activity of the underlying brain areas. Unlike whole

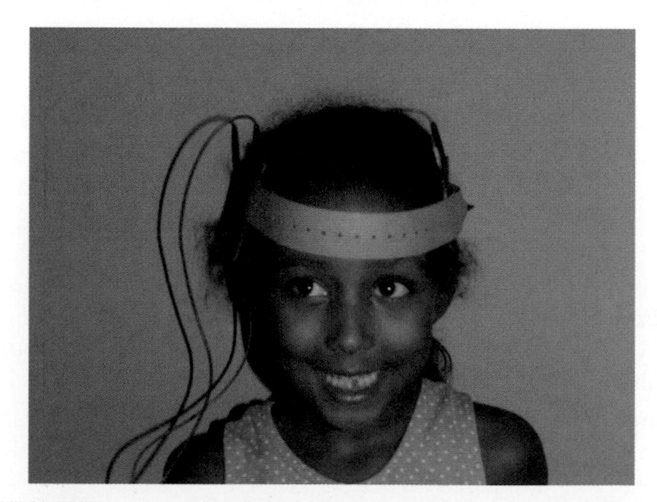

FIGURE 3.6 **A child receiving transcranial direct current stimulation to bilateral temporoparietal cortices in a right lateralization procedure.** The cathode (on left in blue) produces temporary mild inhibition of neuron firing. The anode (on right in pink) produces temporary mild increases in neuron firing. *Source: Courtesy of V. Joanna Burton, MD, PhD.*

cell clamping, each EEG electrode measures the activity of many neurons at once, giving a composite measure of neuron activity in a given area over time. This summation of postsynaptic potentials looks like waveforms of various frequencies. Clinically, EEG is primarily used to study the presence, location, and type of seizure activity. Many neurons firing in synchrony, as happens in a seizure, affects the overall activity of the surrounding brain region. EEG is also used to study brain state and in making diagnoses and outcome predictions for patients in altered mental states. EEG is also used in some research paradigms, where the relative bandwidth of wave frequencies is examined.[39]

Evoked Response Potentials

Evoked response potentials (ERPs) are a subset of EEG recording. ERPs measure the response of the brain to specific stimuli. By recording multiple ERPs resulting from a stimulus and averaging them, then subtracting baseline activity, the response of areas of the brain to a stimulus can be determined. Thus, the neuromechanics of perception are being pieced together. In addition, by examining ERPs of typical subjects and people with developmental disabilities, the differences in brain functioning in various developmental disabilities are being explored.[40]

Magnetoencephalography

Magnetoencephalography (MEG) is a method of mapping brain activity by measuring magnetic fields produced by the electrical activity of neurons. It is similar to EEG but measures magnetic instead of electrical activity resulting from postsynaptic evoked potentials. Magnetic fields are less distorted by the skull and scalp than electrical signals, resulting in better spatial resolution. MEG only measures the activity of neurons in a specific orientation, so signals have less interference due to the position of the neuron compared with the recording device. However, MEG is much more expensive than EEG and more easily distorted by ambient signals in the environment. MEG requires not only specialized equipment but also shielded areas.[41]

Functional Imaging

Positron Emission Tomography

The development of the positron emission tomography (PET) scan allowed three-dimensional functional imaging. A positron-emitting radionuclide (tracer) is attached to a biologically active molecule and administered. This tracer emits positrons that collide with electrons (generally within millimeters from the source) and create two gamma photons that are emitted in opposite directions. The gamma photons are detected and can be localized within the body and quantitatively measured. Three-dimensional images of tracer concentration within the body can be constructed. PET produces high-resolution images through the use of "coincidence detection" of the paired gamma photons. The concentration of the radioactive tracer and, therefore, the release of gamma photons is determined by where the biologically active molecule travels and is metabolized. The most common biologically active molecules chosen for PET are glucose analogues. Because glucose is used in metabolism, radiolabeled glucose analogues will be highest in areas of the body with the highest metabolic activity. Within the brain, this technology has been used to examine which areas of the brain are most active in performing a given task. PET is very good at localizing functions, but the temporal resolution is limited.[42] PET was used extensively in early studies of reading. PET can also be used to visualize radiolabeled ligands, which bind to receptors or transporters, to visualize their distribution or concentration in the brain. For example, the dystonia associated with Lesch–Nyhan syndrome has been linked to reduced dopamine D_2 receptor level in the striatum by PET scanning.[43]

Single-Photon Emission Computed Tomography

Single-photon emission computed tomography (SPECT) uses similar logistics and technology to PET. The SPECT tracer emits gamma radiation directly, as opposed to PET where the gamma radiation is emitted indirectly. Because SPECT does not use coincidence detection there is decreased resolution. SPECT is much less expensive than PET and uses longer lived and more variable radiolabeled tracers. Some radiolabeled tracers are able to measure cerebral blood flow, while others (similar to PET) are attached to glucose analogues and measure metabolism. The DaTscan has recently been developed, which measures dopamine transporters in the brain. A downside of both PET and SPECT is that they require the use of radioactive tracers.

Functional Magnetic Resonance Imaging

Functional magnetic resonance imaging (fMRI) is another form of functional brain imaging. It uses similar technology to standard MRI but looks at structural and temporal aspects of the image to determine the areas of highest metabolic activity. fMRI measures metabolic activity indirectly by examining the blood flow in the brain, which correlates with metabolic activity in the brain (hemodynamic response, i.e. greater blood flow = greater metabolic activity). Specifically, it uses the blood oxygen level-dependent contrast parameters, which take advantage of the difference in magnetization between oxygen-rich and oxygen-poor blood. Because fMRI does not require any artificial tracers, but measures the movement of intrinsic materials, there is no need for injections and no exposure to radiation. Therefore, it has largely supplanted PET in general research when examining areas of brain activation in different behavioral paradigms or examining changes in brain activity in illness. The equipment remains quite expensive. fMRI is currently used mainly in research to define potential mechanisms of autism, ADHD, and receptive–expressive language disorders, among others.[44]

Genetics

The field of genetics is rapidly advancing in both knowledge and testing techniques. Because of the speed of development of new techniques for evaluating genetics and the decreased time between test development and introduction into clinical practice, any text on genetics quickly becomes outdated.

Pedigree

The first diagnostic tool available to geneticists was the pedigree (a family tree which focuses on relatives of subjects and the presence or absence of traits), which allowed them to tease apart methods of inheritance. The pedigree is still a vital part of any genetic evaluation, but can only give clues regarding method of transmission, can be clouded by decreased penetrance and variability in expression, may confuse environmental causes with genetic causes, and provides little information regarding the location of the genetic mutation (except with mutations on the sex chromosomes).

Historical Genetic Tests

The first tool that looked specifically at genotype (genetic makeup), rather than phenotype (physical attributes), was the karyotype. In karyotyping, cells from a subject are induced to divide and are then arrested during cell division and chromatin stained, allowing researchers to view and identify the chromosomes by microscopy. Initially, karyotyping was limited to the diagnosis of aneuploidy (having an abnormal number of chromosomes, either too many or too few). By 1959, Down syndrome was recognized to result from three copies of chromosome 21 (i.e. trisomy 21). In the 1960s, geneticists developed methods of staining parts of the chromosome (banding), which allowed large translocations or deletions to be detected with karyotyping. Fluorescent *in situ* hybridization (FISH) was developed in the 1980s, which further increased the usefulness of karyotyping. By using specific probes to label certain base sequences on the chromosome (which essentially color coded chromosomes by their genes), much smaller translocations and deletions could be determined. It also allowed viewing and testing of areas of the chromosome that had previously been unobservable (e.g. subtelomeric abnormalities). Although karyotyping is still occasionally used to detect certain genetic anomalies, it has largely been supplanted by newer techniques.[45]

Polymerase Chain Reaction

Genetic sequencing became practical when the polymerase chain reaction (PCR) was developed in the 1980s. PCR denatures the DNA and makes multiple copies, therefore amplifying a single copy of a piece of DNA into thousands or millions of copies. DNA sequencing can determine unknown genetic sequences in which an amplification primer may be used in Sanger sequencing. Once the genetic sequence of interest is known, PCR can be used to produce hybridization probes to test other individuals in the family.

DNA Microarray

DNA microarray uses microchips with a collection of microscopic DNA spots and hybridization between two DNA strands (technology borrowed from Southern blot procedures) to simultaneously screen multiple regions of a genome and measure the levels of expression. It can detect deletions, duplications, and amplifications of any of the sites represented on the array. Microarray was first developed in the 1990s and has continued to increase in sophistication, allowing the detection of smaller genetic variants and larger numbers of genes. Prior microarray assays included comparative genome hybridization. Single nucleotide polymorphism (SNP) is currently the most sensitive microarray available. Microarray has rapidly become the standard in evaluating developmental disabilities and has largely displaced karyotyping owing to its increased resolution (although missing rare balanced translocations). Because microarrays determine the copy number, they are a useful tool for detecting large deletions, since the loss of a large area will show up as half the normal expression of that area on the genome. When standard microarrays are unrevealing, further specific testing can be ordered based on clinical suspicion (such as the X-linked intellectual disability panel, methylation studies for Angelman or Prader–Willi, or specific gene sequencing[46,47]).

Whole Exome and Whole Genome Sequencing

Exons are the regions of genes that code for proteins. The exome refers collectively to all of the exons in an individual. The genome represents all chromosomal genetic data including coding and non-coding (intron) regions, approximately 3 billion DNA letters. Exons represent only 1.5% of the genome but result in around 85% of monogenic diseases. Whole exome sequencing (WES) therefore focuses on the areas of the chromosome most likely to result in disease and significantly decreases the amount of data that must be examined. WES will not detect mutations in the introns. WES is now available for clinical use, but it is expensive.

Whole genome sequencing (WGS) examines both exons and introns. It necessarily involves large amounts of information and therefore is likely to result in many findings of unknown significance, particularly as so much of the intron data is poorly understood. WGS is not in clinical use currently, but is likely to be available soon. Of note, WES and WGS analyze gene sequence and are therefore very useful for detecting small sequence changes but do not look at copy variants. A large deletion or duplication of genetic material may not be detected on WES and WGS. Therefore, it is always recommended to perform a standard microarray before WES (and WGS when it is available). As arrays and gene sequencing use different technologies and gather different information, they should be considered complementary tests.[48]

Variants of Unknown Significance

There is no single "normal" genome. As every person's genome is distinct, with the exception of monozygotic twins, each will have a unique result on these tests. In addition, each gene may have multiple different functional codes. Just because the sequence of a gene found in an individual has not been encountered before, this does not mean that it is pathological. With the advent of microarrays and exome/genome sequencing, it is now possible to simultaneously obtain a great amount of genetic data. While some genetic findings are well known and closely associated with specific syndromes and many others are recognized to be normal variants, others are completely unknown. Therefore, it is possible to receive a report from these tests detailing known mutations associated with disease, describing the findings as "within normal limits" or, quite commonly, describing variants of unknown significance. As databases are being built linking phenotype and genetic variants, gradually many of these variants of "unknown significance" are being classified as pathological or as normal variants. Testing both biological parents may help to clarify the variants of unknown significance. A variant is considered likely to be pathological or causative if present in the affected patient but not present in either non-affected parent (or is present in an affected parent or another family member). Unfortunately, this assumes complete penetrance of gene and also presupposes that the link between genotype and phenotype in these few individuals is not coincidental.

Endophenotypes

Endophenotype is an epidemiological term used to connect behavioral symptoms with more well-understood structural phenotypes associated with known genetic causes or with abnormal genetic testing. It is being increasingly used in developmental disabilities, particularly looking at highly heritable polygenetic conditions such as ADHD, autism, and many psychiatric disorders. To be considered an endophenotype, a biomarker must fulfill four criteria: (1) it is associated with illness in the population; (2) it is heritable; (3) it is largely state independent (manifests in the individual whether or not the illness is active); and (4) within families, endophenotype and illness cosegregate. These strict criteria may not work for some endophenotypes of diseases that have more complex inheritance patterns and reduced penetrance.

Animal Models

Animal models provide valuable tools for studying neurodevelopmental diseases. Species with simple and relatively unvarying nervous system structures such as earthworms and sea snails can be used to study *in vivo* synaptic function. Species as varied as zebrafish, rats, and primates are used for behavioral studies. Of increasing importance is the transgenic mouse model. Mice can be bred to have specialized genetic codes. A gene can be deleted (knockout) or duplicated (knockin) to mimic known human genetic syndromes. In addition, rodents can be bred specifically for certain phenotypes (body structures and behavioral syndromes) that show similarities to the human phenotype of a neurodevelopmental disease.

These animal models can be used to perform studies that would be impossible in humans; however, all animal models must be interpreted with care. It is difficult to determine animal correlates for higher level cognitive functions such as language and executive function that are affected in neurodevelopmental disabilities, making it impossible to test their presence or absence in an animal model. Some human brain structures have no true correlates in animal models, limiting comparison. Increases or decreases in gene expression in animal models may not cause the same functional deficits seen in humans. Further, manipulation of genomes (knockin or knockout mice) often results in multiple changes in genetic expression aside from the direct gene target.

Metabolics

When there is concern over a metabolic disorder, a thorough workup is indicated. The pattern of involvement can help to guide the diagnostic workup. A thorough physical examination is always recommended and may help to determine the involvement of other organ systems. A few metabolic disorders present with dysmorphic features at birth, whereas other diseases may be associated with gradually progressive dysmorphism. Basic metabolic tests such as pH, comprehensive metabolic profile (including electrolytes, glucose level, liver and renal function tests, and serum protein levels), lactate, and ammonia should usually be sent for immediate analysis. The results of these tests can be obtained very quickly and may indicate one class of metabolic disorders. Other recommended tests (which often take a few days to return, but should be sent during the acute phase of illness) include serum amino acids, urine organic acids, urine mucopolysaccharides and oligosaccharides, serum very long-chain fatty acids, acyl carnitine profile, and transferrin electrophoresis. Often, in cases of unexplained encephalitis, cerebral spinal fluid may be sent to determine glucose, lactate, and neurotransmitter levels. Other metabolic tests may be indicated if a specific disorder is suspected.

PRINCIPLES OF MANAGEMENT

Management of neurodevelopmental disabilities can take place on multiple levels. Primary prevention seeks to prevent the disorder. Examples include childhood immunizations and folate supplementation during pregnancy. Secondary prevention refers to early identification in order to prevent or limit injury. Universal newborn screening programs for hearing and genetic disorders are examples of such screens (Fig. 3.7). Tertiary prevention is the treatment of disorders with the goal of correcting impairment, diminishing functional limitations, and increasing participation in society. Quaternary prevention attempts to change the way in which society functions to make it more accommodating to individuals with neurodevelopmental disabilities. This typically manifests as public policy and is a purpose of advocacy groups.

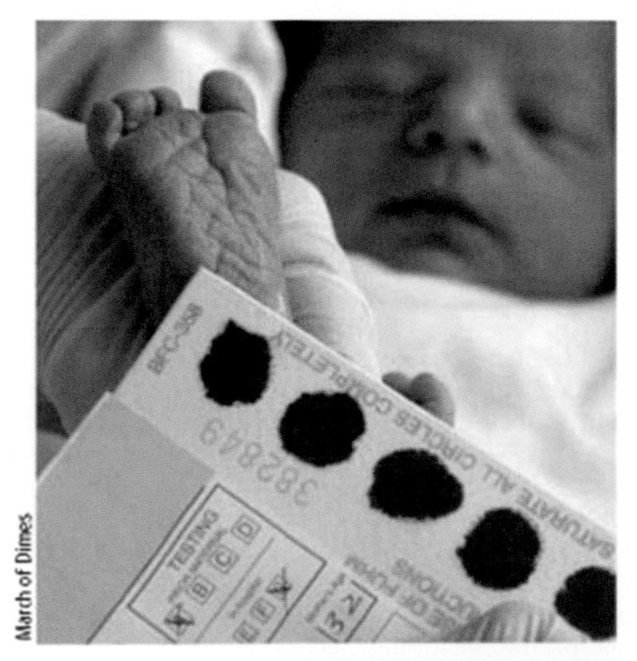

FIGURE 3.7 **Blood samples obtained by heel stick as part of the universal newborn screening program.** This public health program screens infants in the newborn period for severe but treatable illnesses that may not be clinically evident at birth. The specific panel of tests performed in the USA is determined by each state government and varies widely. *Source: Reprinted with permission from the March of Dimes Foundation. © 2013 Infant Health Research.*

Health management is traditionally thought of in terms of medication or surgery, and a wide range of medications is used in children with neurodevelopmental disabilities. However, it is a manifold process that begins with accurate diagnosis. Determining a child's current level of functioning in multiple domains allows the practitioner to determine which steps need to be put in place to help the child to reach their full potential. The domains that must be addressed include physiological stability and nursing needs, communication, activities of daily living, mobility and transport, work/school, leisure, behavior, and living environment. Not every child has needs in each of these domains, but considering the child's function in each of these areas it should result in a complete management program.[49]

A comprehensive management plan will include parental education, advocacy, treatment, and education issues. It also requires an understanding of the common comorbidities that occur with higher frequency in neurodevelopmental disabilities. Failure to recognize and address comorbid conditions is the most common reason for a failed treatment program.

Parental Education

Identifying a neurodevelopmental disability and counseling parents about their child will provide the best treatment in some cases. A diagnosis can also assist the family in goal-setting and long-term planning, identify sources of support, and assist in obtaining counseling. Having a diagnosis can help a family to understand that it is not their "fault". Disease-specific information can also be provided about additional medical concerns and likely developmental outcomes. Many disorder-specific support groups can be identified. Appropriate education of parents can minimize the expectation and ability mismatch that may exist. Although parents are generally quite accurate in estimating their child's functional abilities, this does not always result in expectations that are appropriate for the child's functional age. For example, if a child is 6 years old, but functioning like a 2- or 3-year-old, it is unreasonable to expect the child to independently complete a bedtime routine including using the bathroom, changing into pajamas, and brushing teeth. Guidance regarding appropriate expectations can help to avoid unnecessary stressors associated with placing a child in too demanding an environment, and potentially reduce mental health diagnoses related to adjustment disorder, maladaptive behaviors, and parent and child anxiety.

Advocacy

Advocacy for individuals with neurodevelopmental disabilities takes place both on an individual and family level and on a societal level. This means finding appropriate funding sources such as Supplemental Security Income and Medicaid to cover treatment costs. It also means helping the family to transition a child with neurodevelopmental disabilities into adulthood with consideration given to health insurance, guardianship, and medical power of attorney. Finally, advocacy provides information regarding activities for the child such as the Special Olympics, and for the family with local and national support groups such as The Arc, the Tuberous Sclerosis Alliance, and Children and Adults with Attention Deficit/Hyperactivity Disorder (CHADD).

Practitioners are in a unique position to understand the societal challenges for children with neurodevelopmental disabilities. Through participation with professional groups, they can raise awareness and help to institute systemic changes that cannot be brought about individually. Practitioners can also work on a legislative front to design appropriate health care and educational reform to improve the quality of life of their patients with neurodevelopmental disabilities. It is through the advocacy of various organizations that all 50 US states have adopted some form of newborn screening including universal hearing screening.

Specific Interventions

Treatment requires a layered approach including pharmacological, behavioral, therapeutic, and educational interventions.

Medication

Currently, most drug treatment in neurodevelopmental disabilities is symptomatic and, therefore, most of these medications are not specific. A few general principles can be applied to managing medications. Whenever a medication is prescribed it is important to weigh its risk–benefit ratio. Medications have both desired effects and unintended side effects. Understanding the side effects and monitoring closely for them can help to make the benefit worth the risk. There are additional special considerations for use in children and in populations with special needs.[50] Much less pharmacological research is carried out in pediatric and special needs populations, which means that many medications are used "off-label" in the pediatric population. Children may also metabolize medications differently from adults. Monitoring may require serum testing, and drawing blood in pediatric and special needs populations can be both mechanically (smaller veins in infants, contractures in patients with cerebral palsy) and behaviorally difficult (even adults do not generally enjoy needle sticks and children rarely sit still for such procedures).

An additional complication of medication management in pediatric and special needs populations is that many children with neurodevelopmental disabilities do not know how to swallow pills. Other formulations are available for most medications to try to address this issue: oral (capsules, liquid, chewable), transdermal (patches), intravenous, rectal, and intrathecal. Finding the right formulation requires understanding the patient well. A capsule, which can be opened and sprinkled into a tablespoon of apple sauce, may be considered for a child who cannot swallow a pill. However, if that child has sensitivities to different textures, he or she may not tolerate this either. Although many liquid formulations are not very palatable, some can be flavored.

Advances have been made in the treatment of some disorders as we understand more about their etiology, allowing for targeting of specific pathways and in some cases reversing molecular deficits. In some metabolic and genetic diseases, disease-modifying and curative agents are in use or under investigation.[51]

In metabolic disease, identifying the missing enzyme can allow for supplementation of that enzyme or a vitamin cofactor and a halt to the accumulation of toxic precursors. Other strategies include the use of metabolic inhibitors, depletion of stored substances, environmental manipulation, induction of metabolizing enzymes, supplying the missing enzyme, organ transplantation, limiting the frequency of undesirable genes, and introduction of missing genetic material. In most cases, if the impaired enzyme can have improved function or can be replaced, the consequences of the disease can be minimized. It is important to determine the presence of metabolic defects as soon as possible, because treatment works best before symptoms occur or become severe.

In phenylketonuria (PKU), a rare autosomal recessive inborn error of metabolism, patients lack an enzyme, phenylalanine hydroxylase, which normally converts phenylalanine to tyrosine. Without phenylalanine hydroxylase, these patients have a lack of tyrosine and a buildup of phenylalanine. If this buildup occurs during development, it causes intellectual disability and other neurological complications. Previously, the standard of care has been through dietary control: maintaining low levels of phenylalanine and supplementing synthetic proteins to prevent intellectual disability and ensure normal growth. Although avoidance of dietary phenylalanine remains integral to treatment, several additional targeted treatments exist today, including the use of sapropterin and phenylalanine ammonia lyase to reduce phenylalanine load. Sapropterin is a synthetic form of tetrahydrobipterin or BH_4, which is a cofactor for phenylalanine hydroxylase and has been shown to increase phenylalanine hydroxylase activity in some patients, thereby converting phenylalanine to tyrosine in some patients. The interplay of genetics and neurodevelopment again becomes important as researchers attempt to find genotype–phenotype correlations between patients who respond to sapropterin and those who do not. Current research has extended to gene therapy to replace the missing enzyme.[52] Owing to the severity of PKU if untreated and the ability to treat it through dietary modification, this was the first disease to be included in the newborn screening program (see Fig. 3.7).

In genetic syndromes such as fragile X (see Chapter 8) or TSC, understanding of the neurobiology has identified targets for treatment. Since the mutation in TSC results in an overexpression of mTOR, rapamycin, an mTOR inhibitor, is approved by the US Food and Drug Administration (FDA) for treatment of subependymal giant cell astrocytomas. It is also being studied for treatment of some of the other complications.

Behavior

Various behavior modification plans exist for working with maladaptive or unwanted behaviors. These range from simple sticker charts and counting to three to more involved token economies and external reward systems. Behaviors that are extreme may require functional behavioral analyses in order to develop applied behavior analysis therapies. The complexity of the behavior modification plan will depend on the frequency and severity of the behavior. For example, calling out in class every minute requires a more intensive intervention than occasionally forgetting to hand in homework. Similarly, intense behaviors generally require more intensive intervention. For example, self-injurious or aggressive behaviors generally require more structured functional analysis to identify the purpose of the behavior and then more intensive interventions.

Regardless of the complexity of the behaviors or the behavior modification plan, in each situation, identifying the antecedent (trigger) for the behavior as well as the consequences provides insight into the function of the maladaptive behavior. Approaching behavior occurring because a child does not understand a task will require a different approach from the same behavior in a child who does not like deviations from routine. Once an understanding of the function of the behaviors is understood, therapies can be designed and implemented to decrease maladaptive and increase appropriate behaviors.

Habilitative Therapy

Treatment plans require the clinician to identify and sometimes arrange participation in therapies with skilled professionals who provide high-intensity rehabilitation that targets functional impairment. Children with motor impairment are likely to benefit from the stretching and strengthening provided by physical therapy. Physical therapists also assist with bracing. Occupational therapists work with children with motor impairment to improve access to everyday activities and may assist with seating and other adaptive equipment. Children with articulation, fluency, and language disability may benefit from services provided by speech–language pathologists; these professionals may also assist with providing augmentative communication devices that circumvent the child's limited communication. Ideally, all therapists work with parents as part of a coordinated team to develop a comprehensive management program that prioritizes objectives and utilizes strategies to rehabilitate disability, circumvent problems, and provide accommodations for the disability.

Education

Education is the most important discipline for the management of children with neurodevelopmental disabilities because they are at risk for academic underachievement as a result of their diagnoses as well as comorbid learning, language, and attention difficulties. The goal of any educational plan developed for any child with special needs is to remediate areas of difficulty and provide accommodations to allow ongoing learning. For example, a student who has difficulty reading should be instructed at the appropriate reading level during language and arts lessons. This same child, however, should not be excluded from the science and social studies lessons that are appropriate for the child's cognitive or thinking level but may not be accessible because they require reading skills that the child does not yet possess. Appropriate accommodations could include having the reading portions provided on a tape, working with a "buddy", or having the child present their knowledge orally or pictorially rather than through writing.

Current US law entitles all children to a free and appropriate education in the least restrictive environment possible. Often, a diagnosis is required in order for a child to receive appropriate therapeutic services and special educational supports. This is achieved through the development of an individualized educational plan.[53] For students with disabilities who require accommodations, but do not meet criteria under educational law, schools also have services in place under article 504 of the Rehabilitation Act: a 504 plan can provide medical services (e.g. for a child with diabetes who requires nursing monitoring and insulin during the school day) or academic accommodations (e.g. for a child with ADHD).

Other Treatment Options

Nutritional and dietary modifications take advantage of the metabolic pathways that the human body uses to store and access energy to provide targeted treatments. Some neurodevelopmental disabilities respond well to particular diets. Vitamin B_{12} may correct the metabolic abnormalities seen in some forms of methylmalonic acidemia. Pyridoxine may be useful for treating some types of epilepsy. Higher than normal amounts of vitamin D may be required for children with cerebral palsy and with epilepsy because of limited sun exposure and the effects of medication. Some mitochondrial disorders and disorders of fatty acid metabolism may be responsive to nutritional therapies. The ketogenic diet is used in a number of specific epilepsy syndromes and for epilepsy refractory to multiple medications. Research into nutritional therapy in the absence of identified metabolic disorders has been inconclusive.

Sometimes routine amounts of nutritional factors may be toxic to some children with neurodevelopmental disorders. Galactosemia, PKU, branched-chain ketoaciduria, and urea cycle defects are examples of disorders treated by dietary restriction because their metabolic defects cause the accumulation of toxic products.

Research has also made advances in other treatment modalities including hypothermia, bone marrow and stem cell therapies, and even fetal surgeries. Advances in the field of obstetrics have created opportunities for a variety of *in vitro* fertilization techniques and raised the possibility of *in vitro* selection. Finally, the list of complementary and alternative medications in widespread use continues to expand.[54]

PRACTICE GUIDELINES

Several medical organizations have developed practice parameters to guide the assessment and treatment of different neurodevelopmental disabilities and specific syndromes. The American Academy of Neurology and the Child Neurology Society have worked together to create a number of such guidelines, as have the American Academy of Pediatrics, American College of Medical Genetics, and American Academy of Child and Adolescent Psychiatry. Often these guidelines overlap; however, when they are created independently there are some important differences between them. It is important to evaluate the purpose of the guidelines as well as the date they were created when incorporating them into practice. Guidelines or practice parameters are developed to provide a review of the evidence available either for identification and diagnosis or for treatment and management for the neurodevelopmental disability, comorbidities, and symptoms (Table 3.4). The guidelines provide a review of literature with class I evidence (statistical, population based) to class IV evidence (expert opinion, case reports). Recommendations are then provided based on a literature search and ranked by level of evidence. The goal is to standardize appropriate care.

TABLE 3.4 Sample of Existing Practice Guidelines

Examples of Practice Parameters	Example of Health-Care Supervision Guidelines
Practice Parameter: Screening and diagnosis of autism (2000)	Health-care supervision for children with William syndrome (2001)
Practice Parameter: Evaluation of the child with global developmental delay (2003)	Health supervision for children with Down syndrome (2011)
Practice Parameter: Diagnostic assessment of the child with cerebral palsy (2004)	Neurofibromatosis type 1 in genetic counseling practice (2007)
Evidence Report: Genetic and metabolic testing on children with global developmental delay (2012)	Neurodiagnostic evaluation of the child with a simple febrile seizure (2011)

CONCLUSION

Neurodevelopmental disabilities are a group of disorders that result from injury to the developing nervous system causing functional impairment. Multiple diagnoses are the rule because brain dysfunctions in childhood are the result of etiologies that have wide-ranging effects. The functional limitations may result directly from a primary injury, from secondary effects related to that injury, or from arrested brain development. While the functional limitations have been the historical focus of these disorders, new advances in genetics and related neurosciences offer the potential for amelioration and cure.

QUESTIONS FOR FURTHER RESEARCH

Clearly, much remains to be learned about developmental disorders. Among the questions that we ponder are the following. First, some neurodevelopmental diseases appear to result in impaired synaptic functioning due to defects on the molecular level: How might we improve synaptic functioning to develop treatments for these diseases? Second, many neurodevelopmental diseases are the result of impaired embryonic and fetal development: What effect does this have on our ability to treat these diseases in children and adults who already suffer from these diseases? And how much of the neural cause of these illnesses is reversible? Third, given that the etiology of many neurodevelopmental diseases is poorly understood, how valid are current animal models of these diseases? For instance, without knowing a specific cause for autism, how does one develop a useful animal model for research? Fourth, if a disease is caused by impairment in basic neuron functioning, embryogenesis, or another global factor, how is it that a person with that disease can survive after birth and develop (albeit on his or her own trajectory)?

When something basic goes wrong, how does so much "go right"? And finally, with advances in the treatment of neurodevelopmental diseases, people with previously degenerative and fatal diseases are surviving into adulthood and thriving, and therefore often having children of their own. What potential difficulties may result for their offspring? What methods can be used to intervene before conception and during pregnancy to limit morbidity and mortality?

References

1. Moore KL, Persaud TVN, eds. *Before We Are Born: Essentials of Embryology and Birth Defects*. 7th ed. Philadelphia, PA: Saunders/Elsevier; 2008.
2. Moore K, Persaud T, Torchia M. *The Developing Human, Clinically Oriented Embryology*. 9th ed. Philadelphia, PA: Elsevier; 2013.
3. Komada M. Sonichedgehog signaling coordinates the proliferation and differentiation of neural stem/progenitor cells by regulating cell cycle kinetics during development of the neocortex. *Congenit Anom (Kyoto)*. 2012;52(2):72–77.
4. Bronner ME, LeDouarin NM. Development and evolution of the neural crest: an overview. *Dev Biol*. 2012;366(1):2–9.
5. Alsina FC, Ledda F, Paratcha G. New insights into the control of neurotrophic growth factor receptor signaling: implications for nervous system development and repair. *J Neurochem*. 2012;123(5):652–661.
6. Trommsdorff M, Gotthardt M, Thomas Hiesberger T, et al. Disabled-like disruption of neuronal migration in knockout mice lacking the VLDL receptor and ApoE receptor 2. *Cell*. 1999;97(6): 689–701.
7. Kostović I, Jovanov-Milosević N. The development of cerebral connections during the first 20–45 weeks' gestation. *Semin Fetal Neonatal Med*. 2006;11(6):415–422.
8. Fields RD. Myelination: an overlooked mechanism of synaptic plasticity? *Neuroscientist*. 2005;11(6):528–531.
9. Insel TR. Rethinking schizophrenia. *Nature*. 2010;468(7321):187–193.
10. Hermoye L, Saint-Martin C, Cosnard G, et al. Pediatric diffusion tensor imaging: normal database and observation of the white matter maturation in early childhood. *NeuroImage*. 2006;29(2):493–504.
11. Fuster JM. Frontal lobe and cognitive development. *J Neurocytol*. 2002;31(3–5):373–385.
12. Lenroot RK, Giedd JN. Brain development in children and adolescents: insights from anatomical magnetic resonance imaging. *Neurosci Biobehav Rev*. 2006;30(6):718–729.
13. Shapiro BK, Gwynn H. Neurodevelopmental assessment of infants and young children. In: Accardo P, ed. *Capute & Accardo's neurodevelopmental disabilities in infancy and childhood*. 3rd ed. *Volume I: Neurodevelopmental diagnosis and treatment*; Baltimore, MD: Paul H. Brooks; 2008:367–382.
14. Shapiro BK. Academic underachievement: a neurodevelopmental perspective. *Rev Med Clin Condes*. 2011;22(2):211–217.
15. US Department of Health and Human Services. Developmental Disabilities Assistance and Bill of Rights Act of. *Public Law*. 2000:106–402. http://www.acl.gov/Programs/AIDD/DDA_BOR_ACT_2000/docs/dd_act.pdf.
16. Volpe J. Brain injury in premature infants: a complex amalgam of destructive and developmental disturbances. *Lancet Neurol*. 2009;8:110–124.
17. Ruparelia A, Pearn ML, Mobley WC. Cognitive and pharmacological insights from the Ts65Dn mouse model of Down syndrome. *Curr Opin Neurobiol*. 2012;22(5):880–886.
18. Jacobs S, Hunt R, Tarnow-Mordi W, Indur T, Davis P. Cooling for newborns with hypoxic ischaemic encephalopathy. *Cochrane Database Syst Rev*. 2007;(4):CD003311.
19. Paintner A, Williams AD, Burd L. Fetal alcohol spectrum disorders – implications for child neurology, Part I: Prenatal exposure and dosimetry. *J Child Neurol*. 2012;27(2):258–263.
20. Winneke G. Developmental aspects of environmental neurotoxicology: lessons from lead and polychlorinated biphenyls. *J Neurol Sci*. 2011;308:9–15.
21. Centers for Disease Control and Prevention. Immunization schedules. http://www.cdc.gov/vaccines/schedules. Accessed September 10, 2012.
22. Hagberg H, Gressens P, Mallard C. Inflammation during fetal and neonatal life: implications for neurologic and neuropsychiatric disease in children and adults. *Ann Neurol*. 2012;71:444–457.
23. Fong C. Principles of inborn errors of metabolism: an exercise. *Pediatr Rev*. 1995;16(10):390–395.
24. Batshaw ML, Roizen NJ, Lotrecchiano GR. *Children with Disabilities*. 7th ed. Baltimore, MD: Paul H. Brooks; 2013:325.
25. Butler RW, Haser JK. Neurocognitive effects of treatment for childhood cancer. *Ment Retard Dev Disabil Res Rev*. 2006;12:184–191.
26. Dyment DA, Sawyer SL, Chardon JW, Boycott KM. Recent advances in the genetic etiology of brain malformations. *Curr Neurol Neurosci Rep*. 2013;13(8):364.
27. Pennington BF. How neuropsychology informs our understanding of developmental disorders. *J Child Psychol Psychiatry*. 2009; 50(1–2):72–78.
28. Biederman J, Monuteaux MC, Kendrick E, Klein KL, Faraone SV. The CBCL as a screen for psychiatric comorbidity in paediatric patients with ADHD. *Arch Dis Child*. 2005;90(10): 1010–1015.
29. Langberg JM, Vaughn AJ, Brinkman WB, Froehlich T, Epstein JN. Clinical utility of the Vanderbilt ADHD Rating Scale for ruling out comorbid learning disorders. *Pediatrics*. 2010;126(5): e1033–e1038.
30. van Wezel-Meijler G, Steggerda SJ, Leijser LM. Cranial ultrasonography in neonates: role and limitations. *Semin Perinatol*. 2010;34(1):28–38.
31. Macias CG, Sahouria JJ. The appropriate use of CT: quality improvement and clinical decision-making in pediatric emergency medicine. *Pediatr Radiol*. 2011;41(suppl 2):498–504.
32. Dahmoush HM, Vossough A, Roberts TP. Pediatric high-field magnetic resonance imaging. *Neuroimaging Clin N Am*. 2012;22 (2):297–313, xi.
33. Nucifora PG, Verma R, Lee SK, Melhem ER. Diffusion-tensor MR imaging and tractography: exploring brain microstructure and connectivity. *Radiology*. 2007;245(2):367–384.
34. Menezes MP, North KN. Inherited neuromuscular disorders: pathway to diagnosis. *J Paediatr Child Health*. 2012;48(6): 458–465.
35. Mellgren SI, Lindal S. Nerve biopsy – some comments on procedures and indications. *Acta Neurol Scand Suppl*. 2011;191:64–70.
36. Schott JM, Reiniger L, Thom M, et al. Brain biopsy in dementia: clinical indications and diagnostic approach. *Acta Neuropathol*. 2010;120(3):327–341.
37. Weiss L, Silver JK, Weiss J. *Easy EMG*. Philadelphia, PA: Elsevier; 2004.
38. Wassermann EM, Epstein CM, Ziemann U, Walsh V, Paus T, Lisanby SH, eds. *Oxford Handbook of Transcranial Stimulation*. Oxford: Oxford University Press; 2008.
39. Abou-Khalil B, Misulis KE. *Atlas of EEG & Seizure Semiology*. Philadelphia, PA: Elsevier; 2005.
40. Luck SJ. *An Introduction to the Event-Related Potential Technique*. Cambridge, MA: MIT Press; 2005.
41. Ramantani G, Boor R, Paetau R, et al. MEG versus EEG: influence of background activity on interictal spike detection. *J Clin Neurophysiol*. 2006;23(6):498–508.
42. Kim S, Salamon N, Jackson HA, Bluml S, Panigrahy A. PET imaging in pediatric neuroradiology: current and future applications. *Pediatr Radiol*. 2010;40(1):82–96.

43. Wong DF, Harris JC, Naidu S, et al. Dopamine transporters are markedly reduced in Lesch–Nyhan disease. *in vivo Proc Natl Acad Sci U S A.* 1996;93:5539–5543.
44. Huettel SA, Song AW, Gregory M. *Functional Magnetic Resonance Imaging*. 2nd ed. Sunderland, MA: Sinauer; 2008.
45. Hogan AJ. Visualizing carrier status: fragile X syndrome and genetic diagnosis since the 1940s. *Endeavour*. 2012;36(2): 77–84.
46. Dave BJ, Sanger WG. Role of cytogenetics and molecular cytogenetics in the diagnosis of genetic imbalances. *Semin Pediatr Neurol.* 2007;14(1):2–6.
47. Coughlin II CR, Scharer GH, Shaikh TH. Clinical impact of copy number variation analysis using high-resolution microarray technologies: advantages, limitations and concerns. *Genome Med.* 2012;30(4):80.
48. Bras J, Guerreiro R, Hardy J. Use of next-generation sequencing and other whole-genome strategies to dissect neurological disease. *Nat Rev Neurosci*. 2012;13(7):453–464.
49. Accardo J, Shapiro BK. Neurodevelopmental disabilities: beyond the diagnosis. *Semin Pediatr Neurol*. 2006;12:242–249.
50. Risen SR, Accardo PJ, Shapiro BK. A clinical approach to the pharmacological management of behavioral disturbance in intellectual disability. In: Shapiro BK, Accardo PJ, eds. *Neurogenetic Syndromes: Behavioral Issues and Their Treatment*. Baltimore, MD: Paul H. Brookes; 2010:185–216.
51. Johnston MV, Gross RA, eds. *Principles of Drug Therapy in Neurology*. 2nd ed. New York: Oxford University Press; 2008.
52. Vernon H, Koerner C, Johnson M, Bergner A, Hamosh A. Introduction of sapropterindihydrochloride as standard of care in patients with phenylketonuria. *Mol Genet Metab*. 2010;100: 229–233.
53. Gartin BC, Murdick NL. IDEA 2014: the IEP. *Remedial Spec Educ. 2005*;26(6):327–331.
54. Hyman SL, Levy SE. Introduction: Novel therapies in developmental disabilities – hope, reason and evidence. *Ment Retard Dev Disabil Res Rev.* 2005;11:107–109.

CHAPTER

4

Attention Deficit/Hyperactivity Disorder

Samuele Cortese[*,†], *F. Xavier Castellanos*[†,‡]

*Child Neuropsychiatry Unit, Life and Reproduction Sciences Department, Verona University, Verona, Italy; †NYU Child Study Center, NYU Langone Medical Center, New York, USA; ‡Nathan Kline Institute for Psychiatric Research, Orangeburg, New York, USA

OUTLINE

INTRODUCTION

Attention deficit/hyperactivity disorder (ADHD) is the most common childhood neurobehavioral disorder, with an estimated worldwide prevalence of at least 5% in school-age children.[1] Impairing symptoms of ADHD persist in adulthood in as many as 65% of individuals with a childhood diagnosis.[2]

ADHD imposes an enormous burden on society in terms of psychological dysfunction, adverse vocational outcomes, stress on families, and societal financial costs. The US annual incremental costs of ADHD have been estimated at 143–266 billion USD.[3]

Currently, the mainstay of treatment, at least in the USA, is pharmacological, with psychostimulant medications (methylphenidate and amphetamines) as the first line, and non-stimulants[4] as secondary options. However, clinical normalization following pharmacological treatment only occurs in about half of treated children and the long-term effectiveness and safety of psychotropic treatments can no longer be definitively established for ethical reasons. In addition, some parents and clinicians have reservations about chronic medication use for a behavioral condition. Non-pharmacological treatments, such as dietary regimens, cognitive training, and neurofeedback, are also available, although the empirical evidence for

Neurobiology of Brain Disorders
http://dx.doi.org/10.1016/B978-0-12-398270-4.00004-5

their efficacy on ADHD core symptoms is generally weak to non-existent.[5] Behavioral therapies, delivered through parents and teachers, are effective for ADHD symptoms in childhood, but their availability is typically limited.

Despite being the most studied disorder in child psychiatry, the pathophysiology of ADHD remains elusive. Gaining insight into the neurobiology of ADHD would allow clinical investigators to: (1) complement the subjective diagnostic process with objective measures; (2) develop new therapeutic strategies based on pathophysiology; and (3) assess the effects of available treatments with objective instruments. In turn, this should decrease the high costs associated with ADHD.

This chapter reviews the clinical definition and characteristics of ADHD, and then focuses on the neuroimaging literature and on animal models of ADHD. The authors consider how these two research areas are informing the understanding of the pathophysiology of ADHD, and how research findings may be translated into diagnostic and therapeutic applications, and ask some key questions.

CLINICAL DESCRIPTION

According to the *Diagnostic and Statistical Manual of Mental Disorders*, 5th edition (DSM-5),[6] ADHD is defined by a pervasive and age-inappropriate pattern of inattention, hyperactivity–impulsivity, or both. Onset of several symptoms before the age of 12 years and their presence in two or more settings are required for the DSM-5 diagnosis. According to the DSM-5,[6] three presentations of ADHD can be described: combined, predominantly inattentive, and predominantly hyperactive/impulsive. DSM-5 criteria for the diagnosis of ADHD and ADHD types are reported in Table 4.1.

Besides the core behavioral symptoms, deficits in executive functions (defined as the set of cognitive skills that are necessary to plan, monitor, and execute a sequence of goal-directed complex actions, including inhibition, working memory, planning, and sustained attention) are commonly, although not universally, associated with ADHD.[4]

Clinical and epidemiological studies show that ADHD is frequently comorbid with other psychiatric and neurodevelopmental disorders, including oppositional defiant disorder, conduct disorder, substance use disorder, learning or language problems, anxiety disorders, and mood disorders.[4]

NEUROIMAGING STUDIES

Some of the earliest neuroimaging studies of ADHD were based on positron emission tomography (PET). However, starting in the 1990s, the lack of ionizing radiation and greater spatial resolution resulted in magnetic resonance imaging (MRI) becoming the most frequently used imaging modality. The voluminous MRI literature has been condensed by several comprehensive meta-analyses that are summarized here, starting with structural and functional imaging studies.

Structural Neuroimaging

Morphological Studies

A voxel-based morphometry (VBM) meta-analysis[7] of 14 datasets comprising 378 individuals with ADHD (children or adults) and 344 healthy comparisons showed significantly greater ADHD-related reductions in gray matter volume in the right lentiform nucleus (putamen plus globus pallidus) and the right caudate nucleus (Fig. 4.1). Increasing age was found to be associated with basal ganglia volumes that approached normal values. These results are in line with the alteration to the frontostriatal circuitry that was posited as the main pathophysiological alteration in ADHD.[8] However, no significant differences were found in the prefrontal cortex, despite the fact that six of the included studies reported significantly lower ADHD prefrontal volumes. This lack of consistency reflects the spatial heterogeneity of the abnormal prefrontal clusters in the six studies. The source of such heterogeneity remains unclear, but there are many possibilities, including the syndromic heterogeneity of the disorder, different recruitment patterns, and the generally low statistical power of most studies, which increases the probability of false-positive results.

Another meta-analysis of 11 (overlapping) VBM studies also included papers in which quantification was performed through manual tracing.[9] This report included 320 patients with ADHD and 288 healthy comparisons. It yielded similar findings to those of the previous meta-analysis,[7] and provided additional information on age-related differences in brain morphology. Analyzing VBM studies, the authors found significantly decreased volumes in the right putamen and globus pallidus in children, and in the anterior cingulate cortex, bilaterally, in adults. The meta-analysis of manual tracing studies ($n=7$, all in children) revealed significantly lower volumes of the caudate, bilaterally, in ADHD.

An earlier meta-analysis[10] of 21 structural imaging studies showed a significant reduction in the volume of cerebellar regions, in keeping with the extension of the frontostriatal model to a frontostriatal–cerebellar one. However, this meta-analysis was limited to regions of interest (ROIs), which constrains the generalizability of its findings.

Beside this meta-analytical evidence, more and more studies are revealing morphological alterations beyond frontostriatal regions.[11] In particular, decreased gray matter density in the occipital lobes in adults with ADHD is thought to reflect alterations in regions subserving early-stage subexecutive attentional mechanisms.[11]

TABLE 4.1 Diagnostic and Statistical Manual of Mental Disorders, 5th Edition (DSM-5) criteria for the diagnosis of attention deficit/hyperactivity disorder (ADHD)

A. A persistent pattern of inattention and/or hyperactivity–impulsivity that interferes with functioning or development, as characterized by (1) and/or (2):
  1. *Inattention*: Six (or more) of the following symptoms have persisted for at least 6 months to a degree that is inconsistent with developmental level and that negatively impacts directly on social and academic/occupational activities:
     Note: The symptoms are not solely a manifestation of oppositional behavior, defiance, hostility, or failure to understand tasks or instructions. For older adolescents and adults (age 17 and older), at least five symptoms are required.
     a. Often fails to give close attention to details or makes careless mistakes in schoolwork, at work, or during other activities (e.g. overlooks or misses details, work is inaccurate).
     b. Often has difficulty sustaining attention in tasks or play activities (e.g. has difficulty remaining focused during lectures, conversations, or lengthy reading).
     c. Often does not seem to listen when spoken to directly (e.g. mind seems elsewhere, even in the absence of any obvious distraction).
     d. Often does not follow through on instructions and fails to finish schoolwork, chores, or duties in the workplace (e.g. starts tasks but quickly loses focus and is easily sidetracked).
     e. Often has difficulty organizing tasks and activities (e.g. difficulty managing sequential tasks; difficulty keeping materials and belongings in order; messy, disorganized work; has poor time management; fails to meet deadlines).
     f. Often avoids, dislikes, or is reluctant to engage in tasks that require sustained mental effort (e.g. schoolwork or homework; for older adolescents and adults, preparing reports, completing forms, reviewing lengthy papers).
     g. Often loses things needed for tasks and activities (e.g. school materials, pencils, books, tools, wallets, keys, paperwork, eyeglasses, mobile telephones).
     h. Is often easily distracted by extraneous stimuli (for older adolescents and adults, may include unrelated thoughts).
     i. Is often forgetful in daily activities (e.g. doing chores, running errands; for older adolescents and adults, returning calls, paying bills, keeping appointments).
  2. *Hyperactivity and impulsivity*: Six (or more) of the following symptoms have persisted for at least 6 months to a degree that is inconsistent with developmental level and that negatively impacts directly on social and academic/occupational activities:
     Note: The symptoms are not solely a manifestation of oppositional behavior, defiance, hostility, or failure to understand tasks or instructions. For older adolescents and adults (age 17 and older), at least five symptoms are required.
     a. Often fidgets with or taps hands or feet or squirms in seat.
     b. Often leaves seat in situations when remaining seated is expected (e.g. leaves his or her place in the classroom, in the office or other workplace, or in other situations that require remaining in place).
     c. Often runs about or climbs in situations where it is inappropriate. (Note: In adolescents or adults, may be limited to feeling restless).
     d. Often unable to play or engage in leisure activities quietly.
     e. Is often "on the go", acting as if "driven by a motor" (e.g. is unable to be or uncomfortable being still for extended time, as in restaurants, meetings; may be experienced by others as being restless or difficult to keep up with).
     f. Often talks excessively.
     g. Often blurts out an answer before questions have been completed (e.g. completes people's sentences; cannot wait for turn in conversation).
     h. Often has difficulty waiting his or her turn (e.g. while waiting in line).
     i. Often interrupts or intrudes on others (e.g. butts into conversations, games, or activities; may start using other people's things without asking or receiving permission; for adolescents or adults, may intrude into or take over what others are doing).

B. Several inattentive or hyperactive–impulsive symptoms were present prior to age 12 years.

C. Several inattentive or hyperactive–impulsive symptoms are present in two or more settings (e.g. at home, school or work; with friends or relatives; in other activities).

D. There is clear evidence that the symptoms interfere with, or reduce the quality of, social, academic, or occupational functioning.

E. The symptoms do not occur exclusively during the course of schizophrenia or another psychotic disorder and are not better explained for by another mental disorder (e.g. mood disorder, anxiety disorder, dissociative disorder, personality disorder, substance intoxication or withdrawal).

Specify whether:

Combined presentation: If both Criterion A1 (inattention) and Criterion A2 (hyperactivity–impulsivity) are met for the past 6 months.
Predominantly inattentive presentation: If Criterion A1 (inattention) is met but Criterion A2 (hyperactivity–impulsivity) is not met for the past 6 months.
Predominantly hyperactive/impulsive presentation: If Criterion A2 (hyperactivity–impulsivity) is met and Criterion A1 (inattention) is not met for the past 6 months.

Specify if:

In partial remission: When full criteria were previously met, fewer than the full criteria have been met for the past 6 months, and the symptoms still result in impairment in social, academic, or occupational functioning.

Specify current severity:

Mild: Few, if any, symptoms in excess of those required to make the diagnosis are present, and symptoms result in no more than minor impairment in social or occupational functioning.
Moderate: Symptoms or functional impairment between "mild" and "severe" are present.
Severe: Many symptoms in excess of those required to make the diagnosis, or several symptoms that are particularly severe, are present, or the symptoms result in marked impairment in social or occupational functioning.

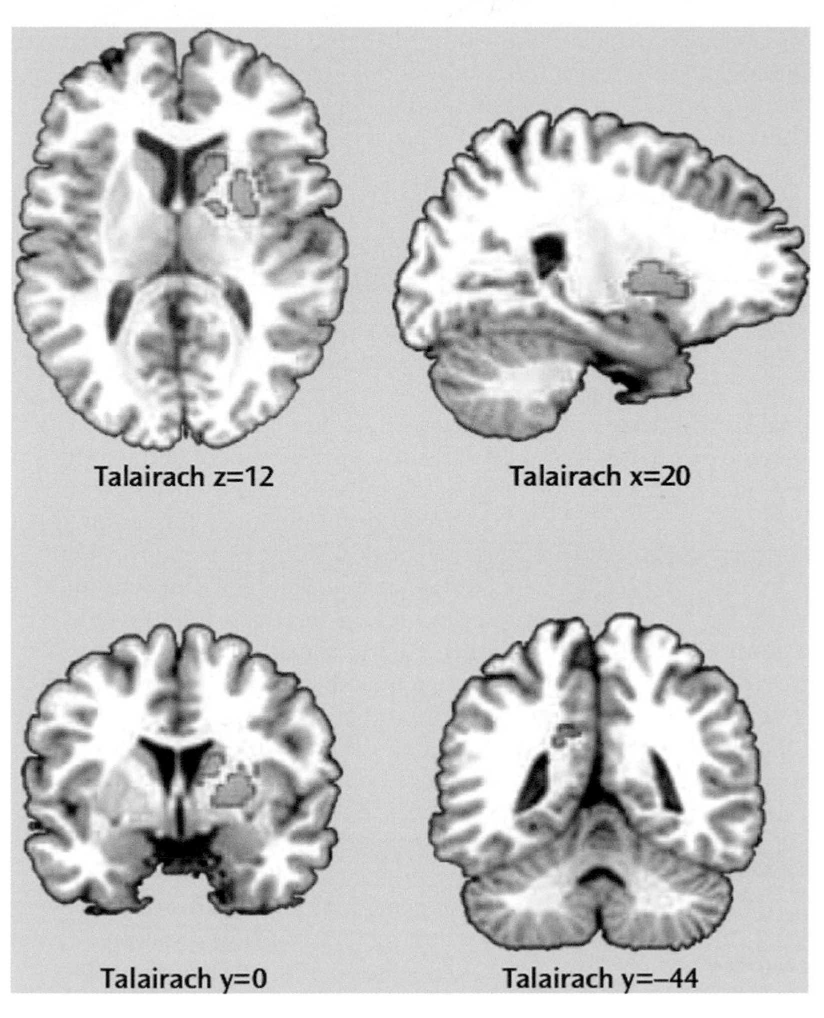

FIGURE 4.1 **Brain regions of smaller and larger gray matter volume in individuals with attention deficit/hyperactivity disorder (ADHD) in relation to healthy comparisons from a meta-analysis of 14 datasets.** Smaller volumes in individuals with ADHD in relation to healthy comparisons are indicated in green, larger volumes in orange. *Source: Reproduced from Nakao* et al. Am J Psychiatry. *2011;168:1154–1163,*[7] *with permission.*

Whereas most of the available studies have focused on volumetric parameters, other morphological aspects, such as cortical thickness, cortical surface area, and gyrification (e.g. Shaw *et al.*[12]) have also been explored. Although meta-analytical evidence from such studies is currently lacking, most tend to converge on reduced widespread cortical thinning, which is more evident in the middle and medial prefrontal cortex and posterior cingulate cortex.

The majority of the available morphological studies in adults with ADHD relied on retrospective recall of childhood symptoms, which may be problematic. An exception examined VBM and cortical thickness in a sample of adults with ADHD diagnosed in childhood (probands), followed up on average 33 years after initial diagnosis, and prospectively enrolled comparisons without childhood ADHD.[13] Proal and co-workers found significantly thinner cortex in probands (regardless of current diagnostic status) than in comparisons at mean age 41 in regions corresponding to the dorsal attentional network and limbic areas.[13] They also found significantly decreased gray matter density in probands in the right caudate, right thalamus, and bilateral cerebellar hemispheres. Of note, probands with persistent ADHD did not differ significantly from those with remitting ADHD when correcting for multiple comparisons. Adults with remitting ADHD had thicker cortex relative to those with persistent ADHD in largely medial ventral regions involved in emotion processing, including the orbitofrontal cortex, parahippocampus, and insula (uncorrected $p<0.05$). These tentative differences (since they only emerged with uncorrected statistical analyses) are consistent with compensatory changes in those regions as well as in visual occipital cortex. These results demonstrate that morphological differences are detectable in adulthood, regardless of current diagnostic status. Yet, despite being conducted within a longitudinal follow-up, the imaging in the Proal[13] study was cross-sectional, so that developmental hypotheses could not be tested directly.

A large sample of children with ADHD and comparisons has been followed prospectively at the US National Institute of Mental Health.[12] This cohort study began in 1991 and accrued more than 200 individuals with ADHD as well as a comparable number of healthy controls, scanned at approximately 2 year intervals on the same 1.5 T magnet until it was decommissioned in about

2010. Multiple seminal results have emerged from this landmark study (see papers by F. Castellanos, J. Giedd, J. Rapoport, or P. Shaw), including smaller global cerebral (including lobar) and cerebellar volume size in children with ADHD compared with healthy comparisons, and parallel developmental trajectories in most brain regions except for the caudate, which normalized in volume by mid-adolescence.[12] Given the non-linear inverted U shape of most cortical regional volumes, an index of cortical maturation was defined as the age at which peak cortical thickness was reached. Shaw and colleagues[2] found that cortical maturation lagged behind that of typically developing children in children with ADHD by 3–5 years. The delay was greatest in the lateral prefrontal cortex. The sequence in the development of regional development in typically developing children, with primary sensory and motor regions maturing earlier than higher order associative areas, was also found in the ADHD group; this supports the notion that trajectories of cortical development are delayed rather than abnormal. The only exception was the primary motor cortex, which matured earlier in ADHD than in typically developing children. These results suggest that a combination of early maturation in the motor area, not paralleled by a concurrent maturation of higher order motor control regions, may underlie the motor restlessness that is a prominent symptom of the disorder. Analysis of the trajectories of cortical thickness showed another important finding: when followed longitudinally in early adulthood, those ADHD patients who clinically remitted showed a normalization of initial delays, while those with persistent ADHD showed either fixed non-progressive cortical deficits or a tendency to diverge from the trajectories observed in typically developing subjects.

Finally, the morphometric literature has raised the question of whether persistent pharmacological treatment may account for the observed changes in brain structure. The earliest study to examine this question found abnormalities in medication-naïve children with ADHD that were more prominent than those found in treated children.[14] In addition, both VBM meta-analyses[7,9] found that stimulant treatment was associated with normalization of brain volumes. This evidence suggests that differences in brain morphology between individuals with ADHD and typically developing children are not ascribable to the effect of psychostimulants.

In summary, most of the available studies, using a cross-sectional design, have addressed the question: Do individuals with ADHD present with significant brain morphological differences compared with typically developing subjects? The convergent answer across studies has been affirmative and significant differences have been consistently reported in frontostriatal–(cerebellar) networks, although recent studies are extending the regions possibly involved in the pathophysiology of ADHD. While the bulk of the studies is cross-sectional, important longitudinal data are also available, pointing to ADHD as a neurodevelopmental disorder characterized by abnormal trajectories in cortical development. These longitudinal data have also started to elucidate the relationship between cortical trajectories and clinical outcome.[12]

Diffusion Tensor Imaging Studies

In keeping with the conceptualization of ADHD as a disorder of altered connectivity of structural and functional networks,[11] diffusion tensor imaging (DTI) is increasingly being used to assess abnormalities in white matter tracts. DTI relies on the assessment of the diffusion of water molecules. The most commonly used DTI index in studies of ADHD has been fractional anisotropy (FA). FA values can range from 0 (fully isotropic) in regions with free water diffusion to a theoretical maximum of 1 (fully anisotropic) in regions in which the movement of water would be restricted to one direction, such as within tracts of heavily myelinated axons. This index, which reflects a complex mixture of tissue properties, including axonal ordering, axonal density, and myelination, does not provide a direct measure of white matter integrity, but reflects physical differences in white matter structure and can thus be used as a starting point for exploring structural connectivity.[15] Interpretation of FA remains challenging, particularly considering its decrease in crossing fibers. Other diffusion metrics, such as axial and radial diffusivity, are underexplored in the ADHD literature.

A systematic review[16] of DTI studies of ADHD identified seven ROI studies and nine reports that assessed DTI indices across the whole brain [voxel-based analysis (VBA)]. ROI studies were based on the assessment of white matter regions and tracts chosen based on the theoretical conceptualization of ADHD, mostly reflecting the frontostriatal–cerebellar model. Collectively, ROI studies showed white matter structural alterations in several white matter regions and tracts, including the inferior and superior longitudinal fasciculus (SLF), anterior corona radiata, corticospinal tract, cingulum, corpus callosum, internal capsule, caudate nucleus, and cerebellum. However, ROI studies were characterized by high heterogeneity in terms of *a priori* chosen regions and methodology. Importantly, ROI studies cannot be pooled in meta-analyses, so it is difficult to distill the central findings from this initial group of studies. By contrast, the nine VBA DTI studies, encompassing a total of 173 individuals with ADHD and 169 healthy comparisons, could be pooled.[16] A meta-analysis revealed: (1) clusters of altered FA in the right anterior corona radiata, containing fibers from the SLF, a pathway connecting dorsolateral prefrontal regions and caudal–inferior parietal lobe, which allows prefrontal cortex to regulate focusing attention in space; the SLF (more precisely, the SLF II) also subserves spatial working

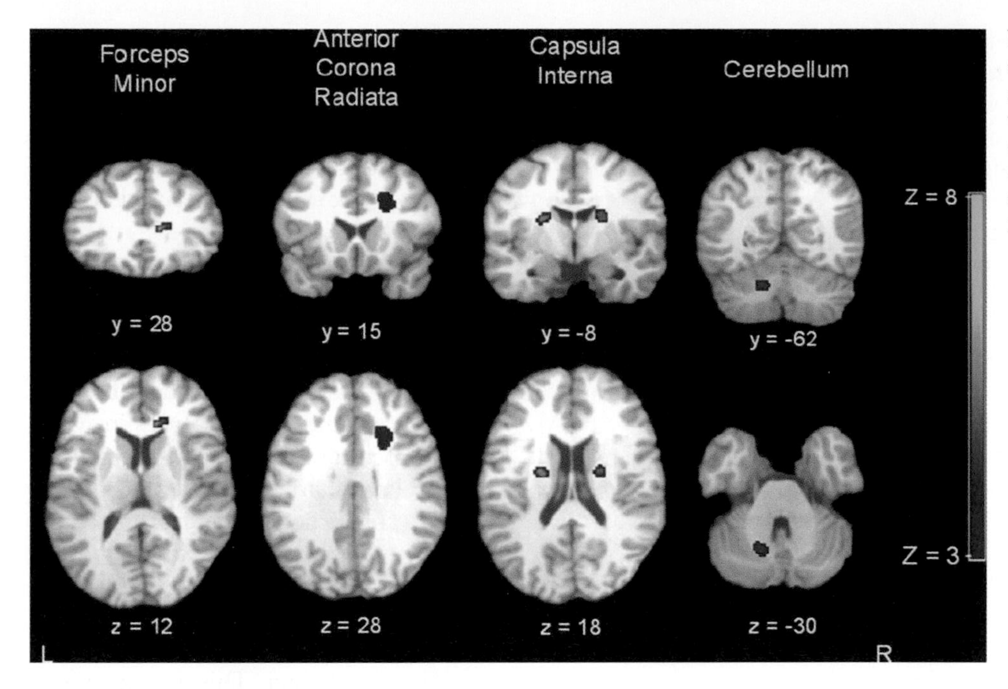

FIGURE 4.2 Clusters of white matter where fractional anisotropy was significantly different in individuals with attention deficit/hyperactivity disorder (ADHD) in relation to healthy comparisons. *Source: Reproduced from van Ewijk* et al. Neurosci Biobehav Rev. *2012;36:1093–1106,*[16] *with permission.*

memory, a consistent executive function deficit in individuals with ADHD; (2) a cluster of altered FA in the left cerebellar white matter; (3) two clusters in the internal capsule, at the level of the genu, containing motor fibers from the motor cortex to the cerebral peduncle; and (4) a small cluster in the forceps minor, close to the genu of the corpus callosum (Fig. 4.2). This evidence also supports the frontostriatal–cerebellar model of ADHD. However, the results of this meta-analysis remain preliminary given the small number of studies, which required pooling data from adults and children. In addition, although all nine included studies relied on a VBA approach, they used different techniques, such as tract-based spatial statistics or less advanced VBA approaches. Other important sources of across-study heterogeneity should also be taken into account, such as the effect of intelligence quotient (IQ), medication status, and inconsistencies in correction for multiple comparisons and in addressing the deleterious effects of head motion.

As for the morphological literature, DTI has also started to be approached from a developmental perspective. Although prospective longitudinal follow-up with DTI measures have not yet been published, Nagel and colleagues[17] were the first to assess DTI measures in a sample of prepubertal medication-naïve children. In addition to alteration of white matter structure in tracts already found to be abnormal in adolescents and in adults, the authors found evidence for altered white matter structure in the frontolimbic tract. Since this is among the last tracts to mature, it was hypothesized that the frontolimbic tract shows an altered developmental trajectory that cannot be appreciated in adolescence or in adulthood. Given that the frontolimbic tract is implicated in emotional regulation, this finding is in line with the tendency to extend the pathophysiology of ADHD beyond the classical executive dysfunction model to include impairments and deficits in other functions, which are also observed clinically.[4]

Although also cross-sectional, the study by Cortese and co-workers reported DTI analyses[18] from the same sample examined as part of a 33 year follow-up of childhood ADHD. As in the morphometric analyses, the authors found that probands exhibited significant changes in white matter structure relative to comparisons regardless of diagnostic status in adulthood. In particular, probands showed significantly lower FA than comparisons in the right superior and posterior corona radiata, right SLF, and in a left cluster including the posterior thalamic radiation, the retrolenticular part of the internal capsule, and the sagittal stratum ($p<0.05$, corrected). FA was significantly decreased relative to comparisons in several tracts in both probands with current and remitted ADHD, who did not differ significantly from each other. In summary, converging preliminary evidence supports white matter alterations in frontostriatal–cerebellar circuits as well as in parietotemporal–occipital regions that have been considered less central in ADHD.

Functional Imaging

Task-Based Functional Magnetic Resonance Imaging

Task-based functional magnetic resonance imaging (fMRI) studies represent the largest body of imaging research in ADHD. Results from different task-based fMRI ADHD imaging studies do not always overlap; this is due, in part, to analytical issues (e.g. different statistical software and approaches), often small sample sizes, and heterogeneity in sample characteristics (e.g. gender ratio, age, comorbidities). Recent meta-analyses address this issue and highlight brain regions that are consistently abnormal, in terms of activation, in individuals with ADHD in relation to healthy comparisons.

Hart and colleagues pooled 21 datasets of fMRI studies focusing on inhibition tasks, including 287 individuals with ADHD and 320 healthy comparisons.[19] The regions that showed significantly lower activation across studies in ADHD compared with comparisons when considering all inhibition tasks were the right inferior frontal cortex, supplementary motor area, anterior cingulate cortex, and striatothalamic areas. When analyses were limited to motor inhibition tasks, decreased activation was found in the right inferior frontal cortex and insula, right supplemental motor area, anterior cingulate cortex, right thalamus, left caudate, and right occipital lobe. Considering interference inhibition tasks, ADHD-related hypoactivated regions were found in the left anterior cingulate cortex, right inferior frontal cortex and insula, right caudate head, and left posterior insula/parietal lobe. Examining age effects by means of metaregression analysis, the authors found that the supplementary motor area and basal ganglia were underactivated exclusively in children with ADHD relative to comparisons, whereas the inferior frontal cortex and the thalamus were underactivated exclusively in adults with ADHD versus comparisons. The meta-analysis restricted to attentional tasks indicated a different specific network that was consistently hypoactivated in ADHD across studies: right dorsolateral prefrontal cortex, posterior basal ganglia, and thalamic and parietal regions. ADHD-related hyperactivation was found in the right cerebellum and left cuneus, interpreted by the authors as possible compensatory mechanisms. Given power issues, age effects could not be analyzed for attention-related tasks.

In another report focusing on the meta-analysis of 11 fMRI studies assessing timing-related tasks, comprising 150 patients and 145 healthy controls, Hart and co-workers reported reduced ADHD-related activation in the left inferior prefrontal cortex/insula, cerebellum, and left inferior parietal lobe, thus confirming the implication of these structures in timing tasks.[20] They interpreted the hypoactivation in the left frontoparietal–cerebellar deficits, in contrast with the right frontostriatal dysfunction, as an expression of the cognitive domain-specific neuronal deficits in ADHD. Although cerebellar areas were hyperactivated in ADHD during attention tasks, they were hypoactivated in timing tasks, suggesting that the abnormality in neuronal activation is task dependent, with different networks being deficient in ADHD depending on the specific cognitive domain being probed.

Cortese and colleagues conducted a meta-analysis of 55 fMRI ADHD studies using a different approach.[21] They combined task-based fMRI studies that relied on the most commonly used tasks in ADHD (i.e. working memory, inhibition, and tests of attention), for a total of 741 patients with ADHD and 801 comparison subjects. Besides performing task-specific meta-analyses, these researchers pooled all studies, regardless of task, to find possible common alterations. In children, significant ADHD-related hypoactivation was found in frontal regions and putamen bilaterally and in right parietal and right temporal regions. The authors also found that children with ADHD significantly hyperactivated the right angular gyrus, middle occipital gyrus, posterior cingulate cortex, and midcingulate cortex relative to children without ADHD. Restricting the analysis to adults, a different pattern emerged. Significant hypoactivation in ADHD was detected in the right central sulcus, precentral gyrus, and middle frontal gyrus. ADHD-related hyperactivation was observed in a region with a peak in the right angular and middle occipital gyri. In keeping with the perspective of ADHD as a disorder of large-scale networks, Cortese and co-workers[21] related the ADHD-related hypoactivated or hyperactivated regions to seven large-scale networks defined based on resting state functional magnetic resonance imaging (R-fMRI) patterns.[22] These networks, limited to cortical regions, are: (1) the *frontoparietal*, which comprises the lateral frontal pole, dorsal anterior cingulate, dorsolateral anterior prefrontal cortex, lateral cerebellum, anterior insula, and inferior parietal lobe, and is implicated in goal-directed executive processes; (2) the *ventral attentional*, which includes the temporoparietal junction, supramarginal gyrus, frontal operculum, and anterior insula, and is involved in orienting attention to salient stimuli; (3) the *dorsal attentional system*, which encompasses the intraparietal sulcus and frontal eye fields, and is implicated in top–down attention orienting according to internal goals; (4) the *default network*, including the posterior cingulate cortex, retrosplenial cortex, lateral parietal cortex/angular gyrus, medial prefrontal cortex, superior frontal gyrus, temporal lobe, and parahippocampal gyrus, and underpins self-referential processes typically suppressed during the execution of active externally oriented tasks; (5) the *somatomotor* network; (6) the *visual* network; and (7) the *limbic* network.

When abnormalities were mapped to these seven subnetworks in children (Fig. 4.3), ADHD-related hypoactivation was present mostly in systems involved in executive function (frontoparietal network) and attention (ventral attentional network). Significant hyperactivation in ADHD relative to comparison subjects was observed predominantly in the default, ventral attention, and somatomotor networks. The hypoactivation of the frontoparietal network is in line with executive dysfunction

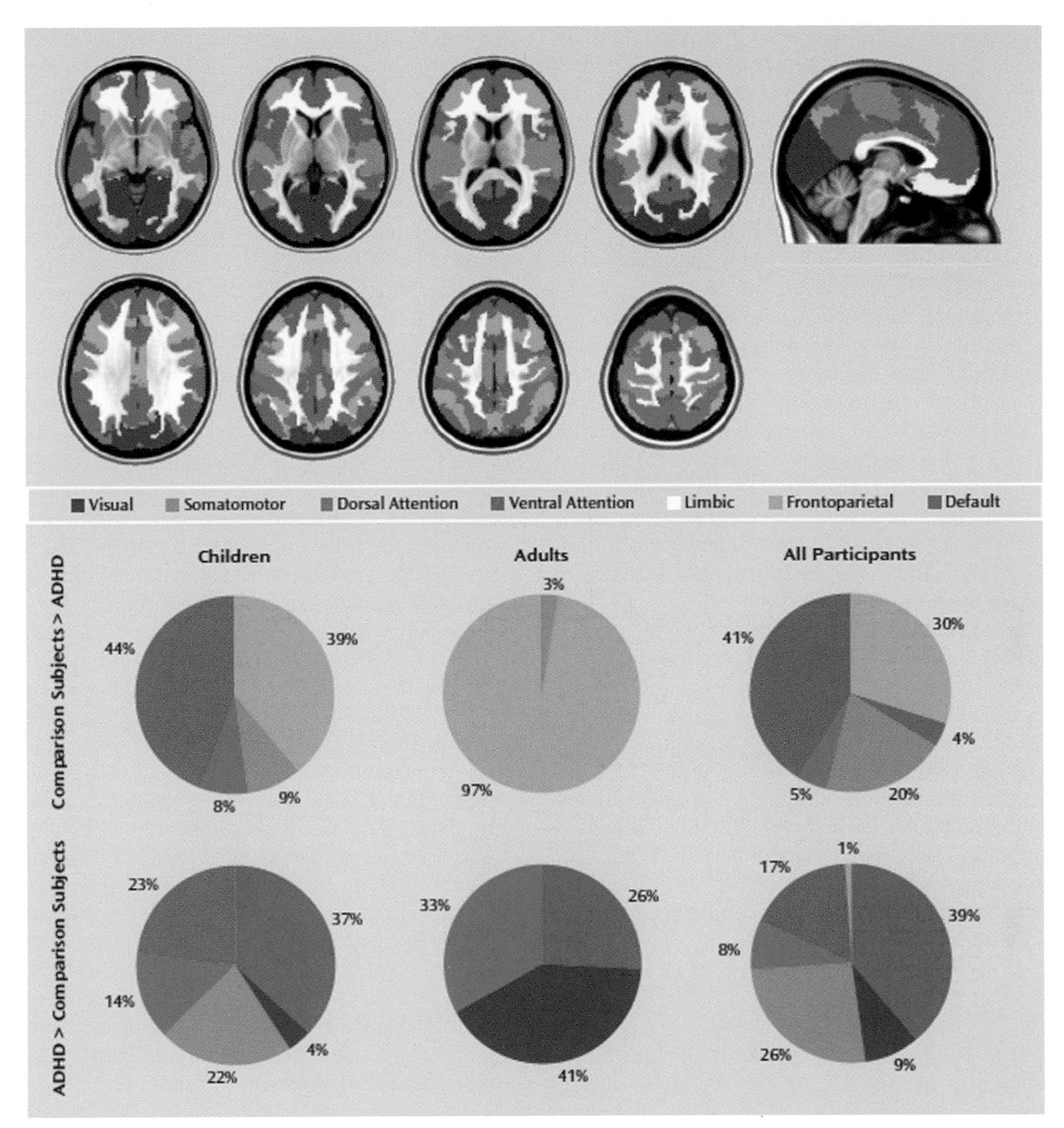

FIGURE 4.3 **Proportion of attention deficit/hyperactivity disorder (ADHD)-related hypoactivation or hyperactivation in meta-analyses of 55 functional magnetic resonance imaging studies.** The upper panel shows the seven canonical networks from Yao *et al.*[22] The pie charts in the lower panel represent the percentage of voxels from the contrasts attention deficit/hyperactivity disorder (ADHD) > Comparison and Comparison > ADHD in each of the seven canonical networks. *Source: Reproduced from Cortese* et al. Am J Psychiatry. *2012;169:1038–1055,*[21] *with permission.*

reported in children with ADHD. The hypoactivation of the ventral attentional network is in keeping with ADHD-related difficulties in detecting changes in the environment. However, hyperactivation was also observed in the ventral attentional system, which may underlie distractibility, which is commonly observed in children with ADHD. The hypoactivation of the somatomotor system may underpin abnormal function of the pyramidal system, which would be consistent with motor symptoms in ADHD that have been overlooked in most recent research (but see Gilbert *et al.*[23] for an exception).

With regard to brain regions that were more activated in ADHD than in comparisons, the hyperactivation of the default network, coupled with the hypoactivation of frontostriatal circuits, provides empirical support for the *default mode hypothesis of ADHD*. This hypothesis posits that the default network may be inadequately regulated by other task-active systems, and may consequently intrude on or disrupt ongoing cognitive performance, contributing to the spontaneous fluctuations in attention that appear to characterize ADHD.[24] Cortese and colleagues interpreted the ADHD-related hyperactivation in the somatomotor and visual systems as a potentially compensatory mechanism.[21] In other words, children with ADHD may try to compensate for deficits in higher order goal-directed executive functions using regions involved in sensory modalities.

Mapping meta-analytical abnormalities in adults to the seven canonical networks (Fig. 4.3), ADHD-related hypoactivation was found predominantly in the fronto-parietal system, while ADHD-related hyperactivation was mostly present in the visual, dorsal attention, and default networks. Two aspects deserve to be mentioned: (1) hypoactivation in the somatomotor system was less evident in adults than in children, possibly reflecting clinical observations that motoric hyperactivity decreases with age; and (2) enhanced hyperactivation was seen in the visual and dorsal systems, compared with children, which may suggest that adults rely on these systems to compensate. By revealing possible compensatory mechanisms, neuroimaging may provide useful information that cannot be detected through clinical assessment if the compensatory mechanisms are effective.

Concern that study results may be driven by medication status motivated further subanalyses in both Hart's and Cortese's meta-analyses. By means of metaregression, Hart and colleagues found that long-term stimulant medication was associated with more normal activation in the right caudate during attention tasks.[19] Limiting analysis to stimulant-naïve subjects, Cortese and colleagues confirmed the presence of significant differences in brain activation between individuals with ADHD and typically developing subjects,[21] casting doubts on the effect of medication as the factor accounting for functional brain alterations in ADHD.

Resting State Functional Magnetic Resonance Imaging

Although task-based fMRI represents the most used imaging modality in ADHD, an increasing number of investigations relies on R-fMRI. R-fMRI allows the detection of the coherence in patterns of spontaneous low frequency fluctuations in the blood oxygen level-dependent (BOLD) signals across brain regions. The absence of an explicit active task, as well as the ease of acquiring R-fMRI datasets (the participant is asked to lie still in the scanner for about 6 minutes for this part of the imaging study), has attracted researchers aiming to elucidate possible abnormalities of brain network interactions in ADHD. However, concerns about artifacts produced by motion make the use of R-fMRI challenging in ADHD and call for appropriate motion-correction procedures that are still being elaborated and debated in the field (for a discussion of this complex issue, see Yan *et al.*[25]).

The first R-fMRI studies of ADHD were published in 2006. Since then, an increasing number of investigators has proposed a plethora of metrics and analytical approaches to describe the alteration in coherence of the patterns of spontaneous BOLD fluctuations at rest across brain regions. Here, R-fMRI studies are highlighted that have explored the default mode network hypothesis of ADHD.[24] This hypothesis can be tested by contrasting indices of resting state functional connectivity between components of the default network and task-positive networks. In a preliminary study of this type, in a sample of 20 adults with ADHD and 20 healthy comparisons, Castellanos and colleagues found significantly less negatively correlated activity between an ROI "seed" located in the precuneus/posterior cingulate cortex (one of the principal components of the default network) and a seed in the dorsal anterior cingulate cortex (within the frontoparietal system, a task-positive network).[26] The authors also found ADHD-related decreases in functional connectivity among components of the default-mode network, in particular between the precuneus/posterior cingulate cortex and ventromedial prefrontal cortex.[26] Abnormal resting state connectivity within the components of the default network was also reported in children with ADHD. In a study of voxelwise functional connectivity in 23 children with ADHD and 23 typically developing comparisons, Fair and colleagues[27] found that the posterior cingulate cortex, a core default network ROI, was significantly correlated with a frontal default region (medial prefrontal cortex) in the comparison group, whereas this correlation was weak in the ADHD sample. A secondary analysis using graph theory confirmed stronger integration among default network regions in the comparison group.[27] Since correlated spontaneous intrinsic activity within the default network strengthens with age, the authors concluded that these results are consistent with the hypothesis of ADHD as a disorder of delayed neuronal development.

Another group[28] confirmed significantly decreased functional connectivity of the posterior cingulate cortex to the dorsal anterior cingulate cortex in a sample of medication-naïve boys with ADHD, thus addressing possible concerns that previous similar findings may have been ascribable to a history of stimulant treatment. In addition, in the healthy comparison group, but not in the ADHD sample, the anticorrelation between the dorsal anterior cingulate cortex and the posterior cingulate cortex increased with age. The study also showed a reduction in functional connectivity between the dorsal anterior cingulate cortex and the dorsomedial prefrontal cortex, which was not found in Castellanos' study in adults,[26] although the sample sizes were too small to make such contrasts meaningful. A study in medication-naïve adults with ADHD[29] found that the antiphase coherence between the dorsolateral prefrontal cortex (a component of the task-positive network) and default network components was correlated with measures of selective attention.

In summary, R-fMRI has allowed the testing of novel pathophysiological hypotheses of ADHD and is shedding light on dysfunctional circuits that could not be easily examined by previous neuroimaging approaches. Continued focus on the interplay among the default network, frontoparietal executive control networks, and the dorsal and ventral attentional networks is likely to be useful. At the same time, primary cortex such as visual or sensorimotor regions should not continue to be neglected.

EFFECTS OF ATTENTION DEFICIT/ HYPERACTIVITY DISORDER TREATMENTS ON BRAIN FUNCTION

Metaregression analyses report that stimulant use is associated with a tendency towards normalization of the structural[7] and functional[19] brain abnormalities associated with ADHD. Besides this aggregated evidence, double-blind, randomized, placebo-controlled pharmacological fMRI studies, conducted mostly by Rubia and colleagues, show, collectively, that a single clinical dose of stimulant medication leads to normalization in the pattern of brain activation in children with ADHD compared with controls during the execution of several tasks (e.g. error processing[30]). Although some studies have considered longer term stimulant treatment effects on brain structure or function, long-term randomized placebo-controlled studies assessing the effects of stimulants cannot be conducted for ethical reasons. Specifically, given the compelling evidence for the short-term efficacy of psychostimulants, assigning individuals to long-term double-blind placebo cannot be justified ethically. Therefore, information from metaregression analyses of structural and functional studies represents the best available evidence on the long-term effects of ADHD medications on the brain in humans.

In addition to the assessment of medication effects, neuroimaging studies have been used to explore the effects of non-pharmacological treatments. In a trial including 18 children with ADHD assigned either to cognitive training or a control training program, Hoekzema and colleagues found a focal increase in gray matter volume after a 2 week cognitive training program in the right posterior–inferior cerebellar lobule and bilateral dorsolateral (middle frontal) prefrontal cortex, in relation to the control group, suggesting that cognitive training may normalize some brain structural alterations in ADHD.[31] The authors speculated that the observed changes reflected alterations in neuronal morphology, such as sprouting of dendritic and axonal arbors and synaptic remodeling. The same research group also reported the effects on brain activity of a 10 day cognitive training program in 19 children with ADHD treated with either cognitive training or sham training.[32] The authors found increased activation in the cognitive training group compared with the sham group in the left orbitofrontal, right middle temporal, and bilateral dorsolateral prefrontal cortex, specifically the left superior frontal and right inferior frontal cortex, during a task of response inhibition. They also found increased activity after cognitive training in the right superior posterior cerebellum on a task of selective attention. The regions where enhanced activity was detected were also targeted by psychostimulant effects, and therefore both pharmacological and non-pharmacological treatments may act on similar circuits. Although groundbreaking, these studies are limited by the short duration of the training (2 weeks), the small sample sizes, and the lack of randomized design. Thus, the pioneering studies by Hoekzema and co-workers lay the groundwork for more rigorous investigation of the long-term effects of cognitive training and other non-pharmacological treatments. At present, these studies are too small to be given much weight.

SINGLE-PHOTON EMISSION COMPUTED TOMOGRAPHY AND POSITRON EMISSION TOMOGRAPHY STUDIES

Although single-photon emission computed tomography (SPECT) and PET studies have been largely superseded by MRI as the main neuroimaging modality used in ADHD, they still provide unique molecular information *in vivo*. In the case of ADHD, SPECT and PET have been mostly used to assess possible quantitative alterations of the dopamine transporter, which is thought to be involved in the mechanism of action of psychostimulants. A meta-analysis[33] of nine PET or SPECT studies, encompassing 169 patients with ADHD and 173 healthy comparisons, showed 14% higher average striatal dopamine transporter density in the ADHD group compared with the control sample. However, heterogeneity across studies was high and, considering only medication-naïve subjects, individuals with ADHD had lower, rather than higher, striatal dopamine density compared with controls. Indeed, PET studies have shown that methylphenidate, one of the most commonly used psychostimulants for ADHD, blocks dopamine transporters in the striatum. Therefore, the findings of this meta-analysis may seem counterintuitive. However, the authors speculated that increased striatal dopamine transporter density in patients receiving long-term treatment could represent an adaptive mechanism to chronic striatal dopamine transporter blockade. They also noted that their findings and hypothesis are in keeping with the clinical observation that methylphenidate is effective in improving ADHD symptoms in the short term but larger doses are sometimes required to maintain clinical effectiveness. By contrast, in a non-human primate study in which treatment parameters could be controlled exactly, PET examination of D_2/D_3 receptor and dopamine transporter binding availability failed to find any significant differences or any deleterious effects on growth, or vulnerability to abuse cocaine, following 1 year of treatment with doses producing expected blood levels during adolescence.[34]

CURRENT CLINICAL APPLICATIONS OF NEUROIMAGING STUDIES

Despite a burgeoning literature increasingly documenting differences between groups of individuals and comparison subjects, one could ask: And so what? How or when will this body of work become useful in the clinical management and, ultimately, the outcome of patients? It has to be clearly stated that, as for other psychiatric disorders, neuroimaging does not yet have direct clinical implications in the day-to-day clinical handling of patients with ADHD, except rarely to rule out neurological conditions in the differential diagnosis. Thus, clinicians and families should be advised against premature commercial misuse of imaging methods and techniques to establish the diagnosis and predict the prognosis of ADHD.

The lack of progress in developing objective diagnostic methods has not deterred investigators from continuing to attempt to do so. But perhaps it should raise the fundamental question of whether this goal is at all meaningful, even when it becomes attainable. After all, current psychiatric diagnoses are all made on clinical grounds. To design an objective tool that would converge perfectly on the subjective diagnosis would not advance our knowledge, it would just make the process more expensive. Instead, the reason for continuing to perform neuroimaging and neurobiological studies is to attain fundamental insights into pathophysiology, each of which may be relevant to only a proportion of individuals. The hope is that the combination of exploration, hypothesis generation, and hypothesis testing will yield mechanistic models of the key control processes that are likely to underlie psychiatric disorders.

The targets of our investigations are unlikely to be single or simple, and the introduction of multivariate analytical methods based on supervised machine learning[35] may help to accelerate the pace of exploration and discovery. Among supervised machine learning, support vector machine approaches are gaining popularity in clinical neuroscience for examining the diagnostic and prognostic potential of neuroimaging data at the individual (patient) level. Supervised machine learning includes a training phase in which a subset of the neuroimaging data is used to develop an algorithm aimed at discriminating between groups previously defined by the researcher (e.g. patients versus controls), and a testing phase in which the algorithm is used to predict the group to which a new observation belongs, without depending on clinical or demographic data.

The application of support vector machine and similar approaches to predict information from large datasets is still in its infancy in ADHD. This research needs to be encouraged, but at the same time, enthusiasm should be tempered, as shown by the results of the ADHD-200 Global Competition.[36] This was an initiative promoted by the ADHD-200 Consortium, an international group of investigators in the field of ADHD who, in March 2011, made available 491 datasets obtained in typically developing subjects and 285 children and adolescents (aged 7–21 years) with ADHD. The dataset included R-fMRI and structural data, as well as related phenotypic information including age, gender, diagnostic status, dimensional ADHD symptom measures, IQ, and lifetime medication status. In July 2011, the Consortium launched a competition: 197 datasets, from six different sites, were publicly released, without diagnostic labels, and investigators were invited to find the best analytical approach to predict the diagnostic status of the subjects. Fifty teams participated, attesting to the great interest in the scientific community. The top ranked team, from Johns Hopkins University, was able to correctly classify 94% of typically developing subjects using several approaches, including support vector machine. However, they were able to correctly classify only 21% of the individuals with ADHD, which highlights the much lower sensitivity of their method (for more detailed information, see ADHD-200 Consortium[36] and related articles in the online special issue). Therefore, although this competition showed the great international interest in open data sharing to support translational application of neuroimaging research, it underscored that the field still has a long way to go to attain clinically useful predictions.

FUTURE PERSPECTIVES IN THE NEUROIMAGING OF ATTENTION DEFICIT/HYPERACTIVITY DISORDER

The neuroimaging literature of ADHD has grown considerably in the past two decades, not only from a quantitative standpoint but also in the quality and rigor of most studies. Nevertheless, the high cost of any type of neuroimaging scan continues to make it difficult for any individual group to obtain large-scale samples. The demonstrated ability to aggregate some types of neuroimaging data across multiple sites offers the possibility of achieving the sort of scale commensurate with the complexity of the challenges of the human brain.

From a technical standpoint, one can expect continued development of existing methods and the emergence of nascent approaches. For example, near-infrared spectroscopy is beginning to be used in research and, owing to its lower costs, could facilitate translational applications of imaging. With regard to structural imaging, various diffusion-weighted methods are being developed, such as diffusional kurtosis imaging,[37] which address the many limitations of DTI metrics. Another challenge worth tackling is finding ways to integrate multiple images collected from the same individual using different approaches.

From a conceptual point of view, categorical studies of ADHD need to be supplemented by dimensional approaches, as encouraged by the US National Institute of Mental Health through its Research Domains Criteria project.[38] The combination of dimensional and categorical approaches in brain imaging of ADHD has already yielded interesting results.[39,40] The latter study was the first reasonably powered analysis to differentiate neuronal correlates of combined type ADHD from those of inattentive type ADHD. The ADHD-200 project was carried out to demonstrate the feasibility of this approach and was conducted without prior coordination or harmonization of phenotypic measures. The partial success of this approach should encourage funding agencies to catalyze improved versions of this type of collaborative effort.

With regard to clinical applications, the present authors make the following recommendations. First, neuroimaging has the potential to guide the implementation of novel pathophysiologically based therapeutic strategies. Despite a substantial therapeutic armamentarium for ADHD, there is substantial room for improvement, and many patients resist adhering to available treatments over the long term, particularly during adolescence. The implementation of therapeutic strategies based on the pathophysiology of the disorder could represent an important advance. Transcranial direct current stimulation (tDCS)[11] may offer an example. This technique delivers low-amplitude electric current to brain ROIs. Possible applications of tDCS in ADHD are supported by reports showing enhanced anticorrelated network coherence following real, but not sham, tDCS to the left dorsolateral prefrontal cortex (summarized in Castellanos and Proal[11]). Thus, R-fMRI and other neuroimaging approaches could be used to guide the placement of electrodes in future transcranial electrical stimulation studies of ADHD, to determine whether this method has a place in clinical settings.

Second, beyond diagnostic status, neuroimaging may be used to complement information related to the prediction of clinical outcome and treatment response. While likely to be the most complex challenge, the greatest reward from neuroimaging would come from being able to predict those likely to have the most negative outcomes, with elucidation of the specific involved mechanisms. For example, the long-term outcome of childhood ADHD is much worse for those who go on to develop conduct disorder and antisocial personality disorder.[41] The application of machine learning methods may be useful to predict responses to ADHD treatments and outcomes.

Third, another useful prediction concerns the outcome of preschoolers with ADHD-like symptoms. Currently, there are no predictors to identify those children with early evidence of ADHD symptoms who will develop the full syndrome during the school years or who will remit. Obtaining such predictive information, particularly if couched in terms of pathophysiologically relevant constructs such as imbalances among large-scale neural networks, could be used to target the most vulnerable children for potentially preventive interventions, while minimizing unnecessary risks and costs for those likely to remit with maturation.

EXPERIMENTAL ANIMAL MODELS

Introduction

In contrast to other psychiatric disorders such as bipolar disorder or schizophrenia, brain tissues are not generally available from individuals with ADHD. Thus, animal models may provide a source of tissue for genetics, neuroimaging, and neurophysiological studies. Animal models can also be used to test the effects of novel drugs, as well as the role of environmental factors, such as toxins, in increasing the risk of ADHD.

Overall, research on animal models in ADHD represents a small niche. Except for the study of attentional deficits and medication effects in monkeys,[34] current animal research on ADHD relies exclusively on rodent models, which are genetically more homogeneous and less expensive to maintain.[42] In addition, investigators can more easily manipulate environmental factors (e.g. diet or environment) with rodents.[42] Before presenting the main available rodent models, some methodological issues in the field are discussed.

Methodological Issues

Validation criteria have been proposed for an animal model to be meaningful in the translational perspective. First, the model needs to show face validity; that is, phenomenological similarities with the disorder of interest in humans.[43] With regard to ADHD, given its phenotypic heterogeneity, a first question is whether the model should mimic all three symptoms dimensions of the disorder (i.e. inattention, hyperactivity, and impulsivity), as endorsed by some researchers (e.g. Sontag *et al.*[43]), or whether it suffices that the model simulates a specific component, as suggested by others.[44] In addition, some proposed animal models mimic specific neuropsychological deficits (e.g. executive dysfunctions). However, executive dysfunction is not a defining feature of ADHD; although children or adults with ADHD may present with deficits in executive function, they are not uniformly manifest in ADHD. As such, face validity does not necessarily require including deficits in executive function.

Face validity is also not satisfied simply if a model simulates ADHD symptoms; the intensity and the quality of symptoms must also be considered. Regarding intensity, the DSM criteria underscore that the symptom must be "excessive" or "inappropriate". Since excessive and inappropriate are relative concepts, it is pivotal not only to select the animal model that mimics specific symptoms, but also to carefully choose the comparison model. For example, if the comparison strain displays hypoactivity, any conclusion on the face validity of an ADHD model in terms of hyperactivity will be fallacious. With regard to qualitative aspects, a fundamental feature of ADHD symptoms is that they are not constant; rather, their expression and intensity are modulated by the context. Clinicians know well, for example, that children with ADHD are not constantly inattentive; they are inattentive mostly in usual and repetitive situations, while they can pay attention in novel and rewarding contexts. ADHD is better conceived as a disorder of behavioral regulation rather than as a disorder entailing *fixed* deficits. Similarly, hyperactivity is not constant. Actimeters show that hyperactivity is modulated by context. For example, differences in actigraphic measures between children with ADHD and typically developing controls were minimal while the children watched television, but significant differences were observed during classes at school (as reviewed by Wickens *et al.*[44]). Therefore, an animal model displaying constant levels of hyperactivity would not qualify as a valid model of ADHD.

Given the challenge of reproducing the complex behavioral patterns of ADHD, the operational definition of "symptom" used by basic science researchers in animal models of ADHD differs from the clinical definition and features of ADHD in day-to-day practice. In particular, the animal literature necessarily relies on objective measures. For example, one of the most frequently used tests of sustained attention in rodents is the five-choice serial reaction time task (5-CSRTT). This test requires animals to monitor a horizontal array of five apertures and to withhold from responding until the onset of a stimulus, a brief flash of light presented pseudorandomly in one of the five holes. Then, the animal has to make a nose-poke response in the spatial location where the stimulus was presented to receive a food reward. The 5-CSRTT was modeled on the continuous performance test (CPT), used to assess human attentional processes. Given all the challenges in linking rodents to humans, the homology between the 5-CSRTT and the CPT is reasonable.

The second criterion for validity relates to construct validity,[43] meaning that models need to conform to an established or hypothesized pathophysiology. However, we lack precisely those sorts of pathophysiological models.[43] Therefore, criterion validity is still a distant goal.

The third criterion pertains to predictive validity,[43] namely to display reduction of symptoms following treatment that is effective in treating the human disorder, and to predict biological and behavioral aspects of the disorder that have not been observed in clinical evaluations, and predict novel treatment strategies. As discussed by Wickens and colleagues,[44] it has been argued that predictive validity is good only when the model responds to both amphetamines and methylphenidate (the two main types of stimulant medication used for ADHD). However, among individuals with ADHD, some respond either to amphetamines or to methylphenidate. Therefore, requiring a response to both drugs as a predictive validity criterion may not always be appropriate.

In summary, many methodological issues need to be addressed when considering the translational aspects of current animal models of ADHD. As correctly observed by Wickens and colleagues[44]: "Future progress requires close interaction of clinical and basic science perspectives to prevent the core behavioural characteristics of ADHD from being lost in translation to animal models". Therefore, there is still a long way to go in order to fully translate into clinical practice the findings from animal models of ADHD. With this in mind, the most common models currently available are reviewed below, along with their characteristics and limitations.

Current Rodent Models of Attention Deficit/Hyperactivity Disorder

Genetic Models

SPONTANEOUSLY HYPERTENSIVE RAT

Established by Sagvolden and Johansen,[45] the spontaneously hypertensive rat (SHR) is by far the most commonly studied animal model of ADHD. SHR was initially developed as a model of hypertension by breeding rats of the Wistar–Kyoto strain (WKY). The SHR strain is hyperactive in open field tests. However, the open field may not be an appropriate context in which to assess ADHD-related hyperactivity (i.e. situation-dependent hyperactivity). The SHR also displays impulsive behavior with characteristics similar to those reported in individuals with ADHD, including abnormal response to reward. As in individuals with ADHD, a high sensitivity to immediate reinforcement is noted in the SHR strain. The SHR also shows learning and memory deficits, although it is not impaired in tasks of sustained attention.[46] Therefore, the face validity of this model is considered to be good. Deficits observed in the SHR map to the frontostriatal system. Dopamine release in the prefrontal cortex, nucleus accumbens, and caudate–putamen is impaired in the SHR. It also shows reduced expression of D_4 receptors, which have been implicated by candidate gene studies of ADHD.[43] In addition to dopaminergic abnormalities, SHR shows elevated concentrations of norepinephrine (noradrenaline) in the locus ceruleus, substantia nigra, and prefrontal cortex.[43] Drug treatment studies have produced inconsistent results. Tests of stimulant effects using open field trials are prone to the same bias mentioned previously. Methylphenidate has not consistently reduced the impulsive response of SHRs in tests of reward sensitivity.[44] Therefore, although face validity is among the best available in animal models of ADHD, current empirical evidence does not support the SHR as a strain to test the efficacy of drugs for ADHD. Two additional issues to keep in mind are: (1) concern that hypertension may contribute to cognitive deficits; however, young SHRs are not hypertensive, suggesting that hypertension is not a confounder in young animals[43]; and (2) the choice of the control strain. In particular, WKY rats tend to show decreased activity levels; thus, it is not surprising that SHRs appear relatively hyperactive. For this reason, Sagvolden and Johansen[45] reported that the Wistar Kyoto substrain (WKY/NHsd) obtained from Harlan, UK, is likely to be the best control strain.

DOPAMINE TRANSPORTER KNOCKOUT MOUSE

The absence of the dopamine transporter (DAT) in this strain is associated with a marked decrease in dopamine clearance, which in turn induces compensatory changes such as reduction in dopamine release from nerve terminals, so that extracellular dopamine is increased about five-fold.[43] This strain has been proposed as a model of ADHD because of its high level of spontaneous locomotor activity; DAT knockout mice engage almost exclusively in repetitive straight movements around the perimeter of the enclosure.[44] They also show deficits in spatial memory. High doses of methylphenidate decrease locomotion in DAT knockout mice and this effect is partly mediated by increased serotonin (5-hydroxytryptamine), leading to a suggestion that changing serotonin levels may be beneficial in ADHD.[42] However, this translational insight has not been validated by clinical use of serotonin-selective reuptake inhibitors, and the role of DAT in ADHD is still unclear. The meta-analysis of SPECT and PET studies mentioned earlier in this chapter suggested that psychostimulant-naïve individuals with ADHD have decreased, rather than increased, DAT levels.[33] However, given that only nine studies were included in the meta-analysis, its conclusions need to be considered tentatively. Until the role of DAT in ADHD is better elucidated, the translational validity of the DAT knockout mouse model should be considered preliminary.

COLOBOMA MUTANT MOUSE

The coloboma mutant mouse strain was developed using neutron irradiation. This mouse has a mutation on the *SNAP-25* (synaptosomal-associated protein) gene. The associated protein is necessary for the fusion of the neurotransmitter vesicle with the presynaptic membrane to facilitate neurotransmitter release.[43] The coloboma mouse shows neurodevelopmental delays and behavioral alterations encompassing motor hyperactivity and impulsivity, including impaired inhibition in reward sensitivity tasks.[43] The deficit in the fusion of the vesicle could explain the reduction of dopamine release in the dorsal striatum found in this mouse. In addition to dopaminergic deficits, an increase in the concentration of norepinephrine in the striatum, locus ceruleus, and nucleus accumbens has been found. With regard to predictive validity, the mouse is responsive to amphetamines but not methylphenidate.[43] This is a potentially interesting model, although the role of *SNAP-25* in human ADHD is not clear.

NAPLES HIGH-EXCITABILITY RAT

The Naples high-excitability rat presents with increased activity depending on the environment; thus, it is not hyperactive in its home cage but displays increased motor activity with increasing environmental complexity compared with the Naples low-excitability rat.[42] As such, its hyperactivity patterns are the reverse of those exhibited by children with ADHD, who show increased hyperactivity in familiar contexts and reduced motor activity in novel situations. This rat exhibits hyperexpression of the DAT and D_1 receptor messenger RNA in the prefrontal cortex. Given the lack of studies on impulsivity and on the effects of psychostimulants, the validity of this model is uncertain pending further examination.

THYROID HORMONE RECEPTOR-β_1 TRANSGENIC MOUSE

This is a relatively novel model. The strain presents with a mutation of the thyroid hormone receptor-β_1 (*TR*β_1) gene, derived from patients diagnosed with resistance to thyroid hormone. In comparison with the wild type, the TRβ_1 transgenic mouse is hyperactive but is generally not impulsive and shows normal attentional functioning.[43] It responds to methylphenidate treatment. Although thyroid hormone influences several brain systems associated with the regulation of attention, impulsive behavior, and motor activity,[42] the majority of children with ADHD do not have abnormal thyroid hormone levels. However, based on observations, in the model, of elevated levels of thyroid stimulating hormone only at the age of 33 days (corresponding to childhood) but not in adulthood, it has been proposed that a short period of thyroid abnormalities during brain development may contribute to the phenotypic alterations that persist in childhood and eventually in adulthood in the absence of thyroid abnormalities.[43] This model offers interesting hypotheses for future studies.

Pharmacological Animal Models of Attention Deficit/Hyperactivity Disorder

6-HYDROXYDOPAMINE-INDUCED BRAIN LESION

This is a well-established model of Parkinson disease. In rats, it has been noted that 6-hydroxydopamine (6-OHDA)-induced brain lesions lead to spontaneous motor hyperactivity and impaired spatial learning before puberty. It is thought[43] that this is due to adaptive mechanisms; in particular, the remaining dopaminergic neurons release more dopamine, and both presynaptic D_2 autoreceptors and dopamine transporters are reduced. There is an increase in D_4 receptors, which correlates with the level of hyperactivity. As would be expected, lesioning dopaminergic neurons also affects other neurotransmitter systems. For example, a serotonergic hyperinnervation of the striatum has been reported.[43] Behavioral deficits improve after methylphenidate or D-amphetamine treatment. Some observations suggest that stimulants reduce hyperactivity of 6-OHDA-lesioned rats by inhibiting norepinephrine and serotonin transporters.[42] Future studies on this model need to include impulsive and inattentive dimensions.

NEONATAL HYPOXIA

Hypoxia induced in rats by nitrogen soon after birth leads to hyperactivity and cognitive deficits later on. Anoxia is a risk factor for ADHD.[42] At the brain level, there is an acute decrease in norepinephrine in the cortex and of dopamine in the striatum. Afterwards, at postnatal day 21, norepinephrine increases in the hippocampus.[43] It has been postulated that the cognitive changes are related to morphological changes in the hippocampus. Hyperactivity decreases following administration of D-amphetamine. Currently, there are no studies focusing on attention deficits or impulsivity in this model.

In addition to these models, rats reared in social isolation present with behavioral alterations including hyperactivity, impulsivity, aggression, and memory deficits.[42] However, they also present with other disturbances, such as anxiety, and therefore it is not clear how closely they model ADHD *per se*.

Remarks and Future Perspectives on Animal Models

Despite the methodological issues pointed out in this text, the body of research from animal models has expanded our knowledge on the pathophysiology of ADHD. A notion that appears evident from the previous models is that neurotransmitter alterations in ADHD go beyond the dopaminergic system, which has classically been the focus of attention in this disorder. At the very least, noradrenergic and serotonergic systems are involved. On the basis of animal model results, some researchers[42] have hypothesized that neurotransmitter changes reflect a more general impairment of neurotransmission, such as altered calcium signaling. An analysis[47] of genome-wide single-nucleotide polymorphism (SNP) data for five psychiatric disorders (autism spectrum disorder, ADHD, bipolar disorder, major depressive disorder, and schizophrenia) concurs with this hypothesis, showing that some of the SNPs surpassing the cut-off of genome-wide significance were related to voltage-gated calcium channel subunits, suggesting that calcium-channel activity genes may have pleiotropic effects on psychopathology.

So far, animal models have not been able to predict novel drugs that may be potentially useful for ADHD. Bearing in mind that intensity and manifestation of ADHD symptoms are not constant but rather context dependent, a fruitful research avenue may be to focus on the effects of drugs that modulate the sensitivity of the motor and attentional systems to contingencies. This route may avoid future efforts with animal models becoming lost in translation.[44]

CONCLUSION

Like all psychiatric syndromes, ADHD is heterogeneous. Despite many challenges, aspects of the neurobiology of ADHD are emerging from the neuroimaging literature. Implicated regions at the level of the cortex include frontoparietal networks handling executive control and top–down regulation of attention, as well as the interacting default network. Subcortical implicated regions include the striatum, thalamus, and cerebellum. Studies in adults further highlight the role of regions subserving transmission of emotional information to cognitive processing regions, and there is evidence implicating visual and sensorimotor cortex as well. It is unlikely that all these systems, which encompass nearly the entire brain, are involved in each and every person with ADHD. What is hoped is that we will soon be able to formulate a limited number of brain anatomic and functional dimensions that can be related to a limited number of differentiable quantitative phenotypes. Perhaps some of these will also be more closely related to specific genetic markers, which will then motivate in-depth analysis of their biological and cellular networks in conditional knockout/knockin models, etc. This strategy of incorporating categorical and dimensional perspectives is central to the Research Domains Criteria project,[38] and probably represents the most reasonable way forward.

QUESTIONS FOR FURTHER RESEARCH

- As genes associated with the risk of ADHD are increasingly identified, how can animal models be best utilized to illuminate the pathophysiology of ADHD?
- Pharmacological models of ADHD have focused predominantly on the catecholamines, dopamine and norepinephrine. These neuromodulators are unquestionably involved in mediating the therapeutic effects of medication treatments. However, the roles of the glutamatergic and GABAergic systems have been largely ignored. How might their functions be interrogated in model systems relevant to ADHD?
- In line with dimensional models of psychopathology, what are the dimensions of animal behavior that can be best linked to human homologous conditions? For example, locomotor hyperactivity in ADHD is environmentally modulated, rather than indiscriminate. How can that be best modeled in various species?
- What are the various ways in which impulsivity can be modeled in laboratory animals that are likely to be most relevant to ADHD?

References

1. Polanczyk G, de Lima MS, Horta BL, Biederman J, Rohde LA. The worldwide prevalence of ADHD: a systematic review and metaregression analysis. *Am J Psychiatry*. 2007;164:942–948.
2. Mannuzza S, Klein RG, Moulton III JL. Persistence of attention-deficit/hyperactivity disorder into adulthood: what have we learned from the prospective follow-up studies? *J Atten Disord*. 2003; 7:93–100.
3. Doshi JA, Hodgkins P, Kahle J, et al. Economic impact of childhood and adult attention-deficit/hyperactivity disorder in the United States. *J Am Acad Child Adolesc Psychiatry*. 2012;51:990–1002.
4. Pliszka S. Practice parameter for the assessment and treatment of children and adolescents with attention-deficit/hyperactivity disorder. *J Am Acad Child Adolesc Psychiatry*. 2007;46:894–921.
5. Sonuga-Barke E, Brandeis D, Cortese S, et al. Nonpharmacological interventions for ADHD: systematic review and meta-analyses of randomized controlled trials of dietary and psychological treatments. *Am J Psychiatry*. 2013;170:275–289.
6. American Psychiatric Association. *Diagnostic and Statistical Manual of Mental Disorders*. 5th ed. Washington, DC: APA; 2013.
7. Nakao T, Radua J, Rubia K, Mataix-Cols D. Gray matter volume abnormalities in ADHD: voxel-based meta-analysis exploring the effects of age and stimulant medication. *Am J Psychiatry*. 2011;168:1154–1163.
8. Castellanos FX. Toward a pathophysiology of attention-deficit/hyperactivity disorder. *Clin Pediatr (Phila)*. 1997;36:381–393.
9. Frodl T, Skokauskas N. Meta-analysis of structural MRI studies in children and adults with attention deficit hyperactivity disorder indicates treatment effects. *Acta Psychiatr Scand*. 2012;125:114–126.
10. Valera EM, Faraone SV, Murray KE, Seidman LJ. Meta-analysis of structural imaging findings in attention-deficit/hyperactivity disorder. *Biol Psychiatry*. 2007;61:1361–1369.
11. Castellanos FX, Proal E. Large-scale brain systems in ADHD: beyond the prefrontal–striatal model. *Trends Cogn Sci*. 2012;16:17–26.
12. Shaw P, Gogtay N, Rapoport J. Childhood psychiatric disorders as anomalies in neurodevelopmental trajectories. *Hum Brain Mapp*. 2010;31:917–925.
13. Proal E, Reiss PT, Klein RG, et al. Brain gray matter deficits at 33-year follow-up in adults with attention-deficit/hyperactivity disorder established in childhood. *Arch Gen Psychiatry*. 2011;68:1122–1134.
14. Castellanos FX, Lee PP, Sharp W, et al. Developmental trajectories of brain volume abnormalities in children and adolescents with attention-deficit/hyperactivity disorder. *JAMA*. 2002;288:1740–1748.
15. Jones DK, Knosche TR, Turner R. White matter integrity, fiber count, and other fallacies: the do's and don'ts of diffusion MRI. *Neuroimage*. 2013;73:239–254.
16. van Ewijk H, Heslenfeld DJ, Zwiers MP, Buitelaar JK, Oosterlaan J. Diffusion tensor imaging in attention deficit/hyperactivity disorder: a systematic review and meta-analysis. *Neurosci Biobehav Rev*. 2012;36:1093–1106.
17. Nagel BJ, Bathula D, Herting M, et al. Altered white matter microstructure in children with attention-deficit/hyperactivity disorder. *J Am Acad Child Adolesc Psychiatry*. 2011;50:283–292.
18. Cortese S, Imperati D, Zhou J, et al. White matter alterations at 33-year follow-up in adults with childhood attention-deficit/hyperactivity disorder. *Biol Psychiatry*. 2013;74:591–598.
19. Hart H, Radua J, Nakao T, Mataix-Cols D, Rubia K. Meta-analysis of functional magnetic resonance imaging studies of inhibition and attention in attention-deficit/hyperactivity disorder: exploring task-specific, stimulant medication, and age effects. *JAMA Psychiatry*. 2013;70:185–198.
20. Hart H, Radua J, Mataix-Cols D, Rubia K. Meta-analysis of fMRI studies of timing in attention-deficit hyperactivity disorder (ADHD). *Neurosci Biobehav Rev*. 2012;36:2248–2256.
21. Cortese S, Kelly C, Chabernaud C, et al. Toward systems neuroscience of ADHD: a meta-analysis of 55 fMRI studies. *Am J Psychiatry*. 2012;169:1038–1055.
22. Yeo BT, Krienen FM, Sepulcre J, et al. The organization of the human cerebral cortex estimated by intrinsic functional connectivity. *J Neurophysiol*. 2011;106:1125–1165.
23. Gilbert DL, Isaacs KM, Augusta M, Macneil LK, Mostofsky SH. Motor cortex inhibition: a marker of ADHD behavior and motor development in children. *Neurology*. 2011;76:615–621.
24. Sonuga-Barke EJ, Castellanos FX. Spontaneous attentional fluctuations in impaired states and pathological conditions: a neurobiological hypothesis. *Neurosci Biobehav Rev.* 2007;31:977–986.
25. Yan CH, Cheung B, Kelly C, et al. A comprehensive assessment of regional variation in the impact of micromovement head motion on functional connectomics. *Neuroimage*. 2013;76:183–201. http://www.ncbi.nlm.nih.gov/pubmed/23499792.
26. Castellanos FX, Margulies DS, Kelly C, et al. Cingulate–precuneus interactions: a new locus of dysfunction in adult attention-deficit/hyperactivity disorder. *Biol Psychiatry*. 2008;63:332–337.
27. Fair DA, Posner J, Nagel BJ, et al. Atypical default network connectivity in youth with attention-deficit/hyperactivity disorder. *Biol Psychiatry*. 2010;68:1084–1091.
28. Sun L, Cao Q, Long X, et al. Abnormal functional connectivity between the anterior cingulate and the default mode network in drug-naive boys with attention deficit hyperactivity disorder. *Psychiatry Res*. 2012;201:120–127.

29. Hoekzema E, Carmona S, Ramos-Quiroga JA, et al. An independent components and functional connectivity analysis of resting state fMRI data points to neural network dysregulation in adult ADHD. *Hum Brain Mapp*. 2013; Feb 18. http://dx.doi.org/10.1002/hbm.22250. [Epub ahead of print.] 2014;35:1261-1272.
30. Rubia K, Halari R, Mohammad AM, Taylor E, Brammer M. Methylphenidate normalizes frontocingulate underactivation during error processing in attention-deficit/hyperactivity disorder. *Biol Psychiatry*. 2011;70:255–262.
31. Hoekzema E, Carmona S, Ramos-Quiroga JA, et al. Training-induced neuroanatomical plasticity in ADHD: a tensor-based morphometric study. *Hum Brain Mapp*. 2011;32:1741–1749.
32. Hoekzema E, Carmona S, Tremols V, et al. Enhanced neural activity in frontal and cerebellar circuits after cognitive training in children with attention-deficit/hyperactivity disorder. *Hum Brain Mapp*. 2010;31:1942–1950.
33. Fusar-Poli P, Rubia K, Rossi G, Sartori G, Balottin U. Striatal dopamine transporter alterations in ADHD: pathophysiology or adaptation to psychostimulants? A meta-analysis. *Am J Psychiatry*. 2012;169:264–272.
34. Gill KE, Pierre PJ, Daunais J, et al. Chronic treatment with extended release methylphenidate does not alter dopamine systems or increase vulnerability for cocaine self-administration: a study in nonhuman primates. *Neuropsychopharmacology*. 2012;37:2555–2565.
35. Orru G, Pettersson-Yeo W, Marquand AF, Sartori G, Mechelli A. Using support vector machine to identify imaging biomarkers of neurological and psychiatric disease: a critical review. *Neurosci Biobehav Rev*. 2012;36:1140–1152.
36. ADHD-200 Consortium. The ADHD-200 Consortium: a model to advance the translational potential of neuroimaging in clinical neuroscience. *Front Syst Neurosci*. 2012;6:62.
37. Helpern JA, Adisetiyo V, Falangola MF, et al. Preliminary evidence of altered gray and white matter microstructural development in the frontal lobe of adolescents with attention-deficit hyperactivity disorder: a diffusional kurtosis imaging study. *J Magn Reson Imaging*. 2011;33:17–23.
38. Insel T, Cuthbert B, Garvey M, et al. Research domain criteria (RDoC): toward a new classification framework for research on mental disorders. *Am J Psychiatry*. 2010;167:748–751.
39. Chabernaud C, Mennes M, Kelly C, et al. Dimensional brain–behavior relationships in children with attention-deficit/hyperactivity disorder. *Biol Psychiatry*. 2012;71:434–442.
40. Fair DA, Nigg JT, Iyer S, et al. Distinct neural signatures detected for ADHD subtypes after controlling for micro-movements in resting state functional connectivity MRI data. *Front Syst Neurosci*. 2012;6:80.
41. Klein RG, Mannuzza S, Olazagasti MA, et al. Clinical and functional outcome of childhood attention-deficit/hyperactivity disorder 33 years later. *Arch Gen Psychiatry*. 2012;69:1295–1303.
42. Russell VA. Overview of animal models of attention deficit hyperactivity disorder (ADHD). *Curr Protoc Neurosci*. 2011. Chapter 9:Unit 9.35.
43. Sontag TA, Tucha O, Walitza S, Lange KW. Animal models of attention deficit/hyperactivity disorder (ADHD): a critical review. *Atten Defic Hyperact Disord*. 2010;2:1–20.
44. Wickens JR, Hyland BI, Tripp G. Animal models to guide clinical drug development in ADHD: lost in translation? *Br J Pharmacol*. 2011;164:1107–1128.
45. Sagvolden T, Johansen EB. Rat models of ADHD. *Curr Top Behav Neurosci*. 2012;9:301–315.
46. Bari A, Robbins TW. Animal models of ADHD. *Curr Top Behav Neurosci*. 2011;7:149–185.
47. Cross-Disorder Group of the Psychiatric Genomics Consortium. Identification of risk loci with shared effects on five major psychiatric disorders: a genome-wide analysis. *Lancet*. 2013;381(9875):1371–1379.

CHAPTER

5

Down Syndrome: A Model for Chromosome Abnormalities

Aarti Ruparelia, William C. Mobley

Department of Neurosciences, University of California – San Diego, La Jolla, California, USA

OUTLINE

INTRODUCTION

Chromosomal Disorders

Disruption of brain development, with a concomitant negative impact on intellectual function, is in some cases due to a chromosomal disorder. These disorders arise from a change in the amount of DNA in the genome or an abnormal arrangement of genes. The change in DNA content may involve loss or gain of an entire chromosome or a part of a chromosome. Distinct from chromosomal disorders are Mendelian or monogenic disorders and those in which complex traits are due to interactions among multiple genes. It is noteworthy, however, that the line between chromosomal disorders and monogenic disorders is bridged under conditions in which a chromosomal deletion includes several genes whose decreased dose creates what appears to be a constellation of Mendelian or monogenic disorders. These cases are referred to as contiguous gene syndromes.

While several mechanisms can be identified as contributing to the changes induced by adding or deleting DNA to the genome, a prominent one is that the change in the dose of one or more genes, i.e. copy number variation (CNV), directly or indirectly impacts brain structure or function. The detection of chromosomal disorders and other disorders of the genome is enabled by a number of routine and increasingly advanced technologies. Karyotype analyses with G-banding and fluorescent *in situ* hybridization (FISH) analysis are commonly used. Higher resolution is achieved through comparative genome hybridization (CGH) and genomic microarrays.

Neurobiology of Brain Disorders
http://dx.doi.org/10.1016/B978-0-12-398270-4.00005-7

We are fast approaching an era in which even very subtle changes in DNA content and sequence can be detected using tools that can be widely applied clinically. Using the latest generation of genomic microarrays, CNVs at the kilobase level can be detected. For example, in one study to identify chromosomal abnormalities using microarray analysis, pathogenic rearrangements were detected in subtelomeric regions in around 4% of patients, including approximately 3% of those in whom karyotype analysis was normal. Even more striking is the finding from genome studies in normal individuals that each human genome may contain 1000 large CNVs.

There are several distinct types of chromosomal disorders:

- *Aneuploidy*: due to gain or loss of a chromosome. The most common aneuploidy is trisomy 21, also called Down syndrome (DS; see below). Other aneuploidies featuring an extra copy of a chromosome include trisomy 13, trisomy 18, and those involving the sex chromosomes, i.e. 47,XXY (Kleinfelter syndrome) and 47,XYY in phenotypic males, and 47,XXX in phenotypic females. In a monosomy there is loss of a chromosome; these are most often linked to the sex chromosomes, e.g. phenotypic females with Turner syndrome are often 45,X.
- *Deletion, duplication, and inversion*: deletions are due to breakage and loss of a portion of a chromosome. They may be large or small. The former are often lethal. Microdeletions may be terminal, located at the end of a chromosome, or interstitial, located in the body of the chromosome. Miller–Dieker lissencephaly is an example of the former and Williams–Beuren syndrome and Prader–Willi syndrome are examples of interstitial deletions. Duplications represent an increase in copy number of one or more genes and are interstitial; they may be the reciprocal product of a duplication. Examples include Russell–Silver syndrome and Pelizeus–Merzbacher disease. Inversions are intrachromosomal rearrangements due to "flipping" the orientation of a genetic segment. Because there is often no net reduction in genes, there may be no phenotypic manifestations.
- *Translocation*: an exchange of chromosomal segments or whole chromosomal arms between different chromosomes. Robertsonian translocations involve the exchange of a chromosomal arm; chromosomes frequently affected include 13, 14, 15, 21, and 22. Reciprocal translocations are segmental exchanges between chromosomes. These may be balanced, with conservation of gene number, or unbalanced, in which cases there is loss of DNA.
- *Isochromosome formation*: in this case, there is creation of a chromosome consisting of two copies of either the short arm or the long arm. Abnormal splitting of the chromosome at mitosis may be responsible. In one example, isochromosome 21 is a rare cause of DS.

Intellectual Disability and Chromosomal Disorders

Intellectual disability is defined as emerging before the age of 18 years and where significant limitations are present in both intellectual functioning and adaptive behavior. The full spectrum of genetic changes in chromosomal structure reviewed above has been shown to underlie changes in intellect and behavior. Genetic changes are the most common cause of intellectual disability: the use of high-resolution analyses shows that CNV is an important source. Indeed, genomic rearrangements appear to cause 12–18% of developmental delay and intellectual disability. The rapid pace of genetic discovery suggests that genomic changes will ultimately be linked to an even greater percentage of those with behavioral difficulties and intellectual deficits.

Important for future studies of intellectual deficits will be efforts to: (1) define in detail the phenotypes linked to intellectual changes using quantitative methods that allow for careful comparison of individuals; (2) carry out longitudinal evaluation of the phenotypes; (3) define the genes implicated, the genetic mechanisms responsible, and the pattern of their expression; (4) clarify the underlying molecular basis for changes in brain function; and (5) define rational targets for treatment specific either to the genetic change or to the molecular mechanism that results. Although such studies will be challenging, the increasing armamentarium of genetic tools and methods to explore neurobiology promises success in elucidating mechanisms and defining treatments. Recent progress in DS encourages optimism.

DOWN SYNDROME

DS is a complex genetic condition caused by the trisomy of chromosome 21 (Hsa21). The triplication of Hsa21 leads to the overexpression of more than 500 genes and their encoded proteins that are involved in the regulation of neuronal cell number, transcription, translation, and synaptic plasticity, resulting in structural and functional defects in multiple systems. DS is a multifaceted condition characterized by invariable clinical phenotypes such as intellectual disability, hypotonia, craniofacial abnormalities, and the presence of Alzheimer disease (AD)-related pathology in the brains of subjects aged 40 years and above; in addition, at least 80 other variable phenotypes affect a subpopulation of DS, including acute megakaryoblastic leukemia, atrioventricular septal heart defects, and other neurological complications such as seizures and obstructive sleep apnea.

The incidence of DS occurs at a frequency of approximately one in 650–1000 live births worldwide. It is the most common genetic cause of intellectual disability and accounts for 15% of the population with intellectual disability. Owing to its high prevalence and multisystem dysfunction, DS is one of the most intensively studied aneuploidy conditions. Remarkable progress in medical treatment and social intervention in developed countries has increased the life expectancy of people with DS from an average of 12 years of age in the 1940s to 60 years of age at present, with an average increase of 1.7 years per year. Significantly, even with a prenatal diagnosis, the incidence of DS in not diminishing, possibly as a result of relaxed societal pressures of advanced maternal age and the reduced stigma of having a child with DS. Indeed, the past 25 years have shown an increase of almost 15% of women above the age of 35 years giving birth. Males with DS have an increased life expectancy of more than 3.3 years relative to females with DS. This comparative survival advantage over females may be ascribed to a greater incidence of congenital heart defects in females with DS, leading to reduced life expectancy. The gender-specific difference could also be explained by menopause occurring at an average age of 46 years in women with DS, which is approximately 4–6 years earlier than women in the general population. It is well established that precocious menopause and the accompanying drop in estrogen levels is a high risk factor for premature mortality through heart disease, stroke, breast cancer, and/or dementia. Another risk factor for premature mortality is the apolipoprotein E (ApoE) genotype; in DS and the general population, the presence of the ApoE ε_2 allele is associated with longevity and preserved cognitive functioning, whereas the ApoE ε_4 allele is associated with increased susceptibility to dementia and early mortality.

The identification and development of neuronal targets for pharmacotherapies to treat DS-related complications remains a crucial goal. The combination of remarkable advances in scientific research and medical technology has led to extraordinary results in increasing longevity. However, increased longevity is possibly linked to increased prevalence of the disorder and we are now faced with new challenges to tackle a larger population of adults with DS, who display premature aging accompanied by various age-related problems. Thus, in addition to the neurodevelopmental and neurodegeneration research and therapies, there is a growing importance to facilitate healthy adult functioning by providing specialized health care and investing in scientific research for the aging DS population. The contribution of studies from mouse models of DS has provided a pivotal framework through which the dissociation of genotype–phenotype correlations has begun to be elucidated and in which the pathological molecular and neurobiological processes underlying DS are being unraveled. However, much is still unknown, and further efforts to understand the pathogenesis that leads to the clinical manifestation of DS and its associated phenotypes are needed.

A comprehensive review of DS and its multiple system complexity is beyond the scope of this chapter. Instead, the focus is on cognitive impairments that characterize DS and specifically the learning and memory impairments that are marked by perturbed neurodevelopment, altered neuronal structure, synaptic plasticity deficits, and early-onset neurodegeneration. Insights from widely used mouse models of DS are incorporated to provide a deeper understanding of the clinical characterization of DS and the breakthroughs in pharmacological interventions that have emerged from these findings.

History of Down Syndrome

The first observations of DS were made during the time when genetics was being defined. In 1859, Charles Darwin published his pioneering theory of evolution in *The Origin of Species*. Soon afterwards, in 1865, Gregor Mendel published his laws of genetics based on meticulous experiments in plant hybridization. It was against this backdrop that in 1866 John Langdon Down published a seminal paper on his description of DS: *Observations on an Ethnic Classification of Idiots*. His observations were heavily influenced by Blumenbach's anthropology, which classified people into five races: Ethiopians, Malayans, Caucasians, Aztecs, and Mongolians. Based on this classification of race, and while working with an ardent ethnologist, Down attempted to ethnically categorize patients with mental retardation using skull contour measurements, including the diameter of the head, and from photographs he had taken, to identify specific facial features.

Although Down's ethnic classification was never widely accepted, his use of the term "Mongolism" to characterize people with DS was in keeping with these very limited observations (Box 5.1). It was only in 1961 that a group of 20 prestigious biomedical researchers signed a petition in *The Lancet*, asking for the term "Mongolism" to be replaced as its connotations were misleading and embarrassing. Following this, the Mongolian People's Republic delegation approached the Director General of the World Health Organization (WHO) at the 1965 World Health Assembly, requesting the term "Mongolian idiot" be removed as it was derogatory. The use of these objectionable terms was consequently diminished and superseded by "Down syndrome".[1]

Down believed that the cause of DS was not due to postnatal instances. Again, given limited insights into

BOX 5.1

DOWN'S DISCOVERY OF "MONGOLISM"

As eloquently captured in a biographic account of John Langdon Down,[1] Down described a typical member of this group as having features where: "The hair is not black, as in the real Mongol, but of a brownish colour, straight and scanty. The face is flat and broad, and destitute of prominence. The cheeks are roundish and extended laterally. The eyes are obliquely placed and the internal canthi more than normally distant from one another. The palpebral fissure is very narrow. The forehead is wrinkled transversely from the constant assistance which the levatores palpebrarum derive from the occipito-frontalis muscle in the opening of the eyes. The lips are large and thick with transverse fissures. The tongue is long, thick and much roughened. The nose is small. The skin has a slight dirty yellowish tinge, and is deficient in elasticity, giving the appearance of being too large for the body ... They are usually able to speak; the speech is thick and indistinct, but may be improved very greatly by a well directed scheme of tongue gymnastics. The co-ordinating faculty is abnormal, but not so defective that it cannot be strengthened".

Embedded with philosophical questions about the origin of the human race, Down wrote about the Mongolian type: "A very large number of congenital idiots are typical Mongols. So marked is this that, when placed side by side, it is difficult to believe the specimens compared are not children of the same parents ... The Mongolian type of idiocy occurs in more then ten per cent of the cases which are presented to me. They are always congenital idiots and never result from accidents after uterine life. They are, for the most part instances of degeneracy arising from tuberculosis in the parents ... The tendency in the present day is to reject the opinion that the various races are merely varieties if the human family having a common origin ... I cannot but think that the observations which I have recorded are indications that the differences in the race are not specific but variable. These examples of the results of degeneracy among mankind appear to me to furnish some arguments in favour of the unity of the human species".[1]

Through his practical observations, Down became philosophically astute and described Down syndrome as being a racial "retrogression" that broke down barriers of racial divisions by being able "to stimulate so closely the features of the members of another division".[1]

the nature of the disorder, he believed that it derived from degeneracy that occurred in parents with tuberculosis. Almost a century after the first observed cases of DS, the development of cytogenetics by pioneers such as Charles B. Davenport and Petrus J. Waardenburg led to the understanding that chromosomal irregularities may be the cause of intellectual disability and the speculation that non-disjunction could result in trisomy or monosomy and may be a cause of DS. It was only in 1959 that Jerome Lejeune confirmed that DS is, in fact, caused by the trisomy of chromosome 21.[2]

Aneuploidy in Down Syndrome

Chromosomal aneuploidies account for approximately 50% of spontaneous abortions before 15 weeks of gestation, with trisomies being responsible for 50% of spontaneous abortions. Trisomy 21 is caused by the non-disjunction of chromosome 21 during meiosis in either oogenesis or spermatogenesis, resulting in reduced recombination. Using DNA polymorphism analysis, it has been determined that parental origin of non-disjunction in trisomy 21 is maternal in almost 95% of cases[3] and from a meiosis I error in approximately 77% of maternal non-disjunction cases. The only well-established risk factor for non-disjunction appears to be advanced maternal age. Even though the rate of recombination remains constant with advancing maternal age, chromosome 21 chiasmata in aged oocytes are more susceptible to non-disjunction than young oocytes. Therefore, the aging of the oocyte is responsible for the maternal age effect in DS.

Trisomies fall into four categories dependent on the size of the genomic region: complete or whole-chromosome trisomies, partial trisomies (usually larger than 5 Mb), microtrisomies (shorter than 3–5 Mb), and triplication of single genes. An intermixed population of aneuploid and euploid cells defines mosaic trisomy 21. Complete trisomy of Hsa21 accounts for 95% of the population with DS, with the remaining 5% constituting mosaic or partial trisomies. The additional copy of Hsa21 results in increased expression of many genes encoded on this chromosome, but with some variations in expression levels of individual genes and between different tissues. This increase in gene dosage in Hsa21 genes has been proposed to cause the plethora of phenotypic alterations that characterize DS. Nevertheless, allelic differences in Hsa21 genes and an intricate interplay with other non-Hsa21 genes, as well as epigenetic influences and environmental factors, can all be envisioned to influence the phenotype seen in DS.[4]

Two differing, but not mutually exclusive, hypotheses emerged to explain the phenotypic variation observed in DS compared with the euploid population. The *gene dosage effect hypothesis* proposed that a subset of triplicated genes in the gene-rich distal part of Hsa21 (band 21q22), identified as the Down syndrome critical region (DSCR), was sufficient to cause specific DS phenotypic traits and that the severity of the syndrome would depend on the extent of trisomy of this genomic region. The *amplified developmental instability hypothesis* postulated that the majority of DS-associated phenotypes result from elevated activity of segments of genes that results in an overall genetic instability, rather than the direct contribution of specific genes in causing specific phenotypes. Thus, a greater number of triplicated genes would lead to greater susceptibility to developmental abnormalities.

Evidence from human studies of DS with triplicated regions outside the DSCR, and from mouse model studies demonstrating that homologous DSCR genomic regions in mice do not always recapitulate DS phenotypes, accumulatively argues against a DSCR in being sufficient to cause all DS-associated phenotypes. Nevertheless, a fundamental role for the gene dosage hypothesis is now apparent. Indeed, it is increasingly clear that increased doses of specific genes contribute significantly to particular phenotypes. It is also noteworthy that trisomy 21 results in the overexpression of Hsa21-encoded microRNAs, which could lead to decreased expression of certain target proteins and also contribute to the variance in DS-related phenotypes. Current evidence makes a compelling argument for a number of genes and susceptibility regions encoded on Hsa21; these, in turn, may be modulated by other loci on Hsa21 and elsewhere in the genome, as well as by other epigenetic processes, to collectively create a complex network that defines the impact of gene dosage expression on DS-associated phenotypes.

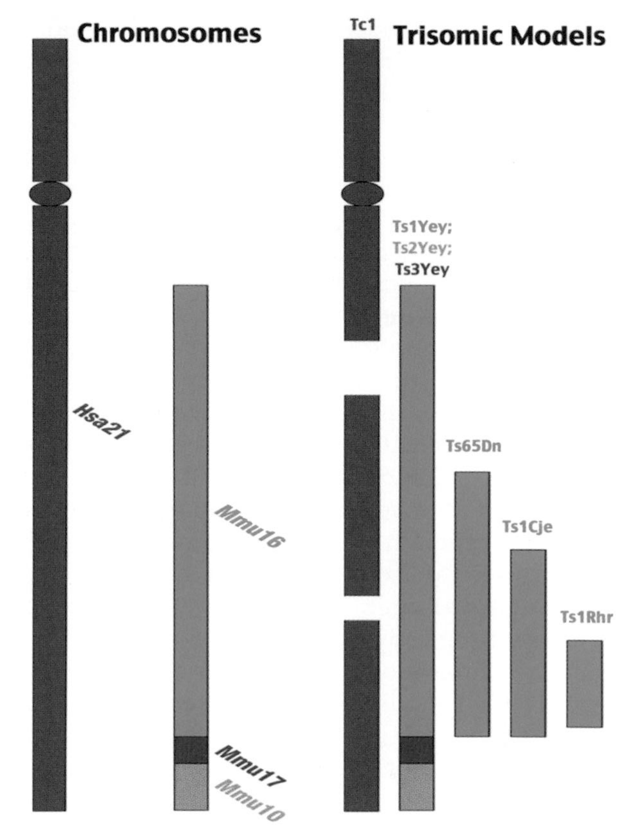

FIGURE 5.1 **Mouse models of down syndrome (DS).** Hsa21 is syntenic to three regions of the mouse genome. Approximately 150 of the homologous *Hsa21* genes are encoded on mouse chromosome 16 (Mmu16), and nearly 100 *Hsa21* homologous genes have synteny on Mmu10 and Mmu17. The majority of mouse models used for DS research are segmentally trisomic for regions of Mmu16. The Ts65Dn mouse model is the most widely used mouse model of DS and harbors 104 genes. The Ts1Cje mouse is trisomic for approximately 81 of the genes encoded on the Ts65Dn translocation. The Ts1Rhr mouse is trisomic for 33 Mmu16 genes that have conserved synteny with the Hsa21 DSCR. The Tc1 mouse is a transchromosomic mouse model that carries a freely segregating, almost complete copy of Hsa21 in addition to the normal complement of the mouse chromosome. The Ts1Yey;Ts2Yey;Ts3Yey model has been developed that is trisomic to all mouse regions syntenic to Hsa21. *Source: Adapted from Ruparelia* et al. Curr Opin Neurobiol. *2012;22:880–886.*[28]

MOUSE MODELS OF DOWN SYNDROME

Deciphering the correlation between the effects of trisomic Hsa21 genes and clinical aspects of the syndrome is an important goal in DS research. By thoroughly elucidating the underlying cellular, molecular, and neurobiological mechanisms that are responsible for DS-associated phenotypes, it should be possible to develop and validate potential therapeutic targets for treatment of these phenotypes.[5]

Examination of people with DS who have partial trisomy 21 resulting from translocations or duplications of various genomic segments of Hsa21 has led to crucial discoveries in mapping genes to specific phenotypes. However, the small number of partial trisomy cases, the substantial variation in clinical phenotypes within this population, and the restriction of intrusive studies limit further analyses in this population. The contribution of mouse models of DS has thus been instrumental in progressing our understanding of the clinical manifestations of DS, particularly that of cognitive deficits, and unraveling the pathogenesis through which these phenotypes arise. These mouse models, which recapitulate several features of DS including neuroanatomical, neurobiological, and behavioral phenotypes, have provided invaluable insight to identifying genes, mechanisms, and therapeutic targets.

Hsa21 harbors approximately 550 genes and is syntenic to three mouse genomic regions with high conservation of many genes between human and mouse, although important differences are noted.[6] Approximately 150 of the Hsa21 homogeneous genes are encoded on mouse chromosome 16 (Mmu16), and nearly 100 homologous Hsa21 genes are encoded on two smaller regions, Mmu17 and Mmu10 (Fig. 5.1).

Most DS mouse models are segmentally trisomic for Mmu16. The most widely studied model is the Ts65Dn mouse, which carries a reciprocal translocation that creates trisomy for 13.4 Mb of the 22.9 Mb Hsa21 syntenic region on Mmu16 and harbors approximately 104 genes that are orthologues of Hsa21 genes. The Ts65Dn mouse recapitulates several DS-associated phenotypes including learning and memory deficits. However, this model also contains 19 triplicated genes on Mmu17 not orthologous to Hsa21, raising the possibility that some phenotypes may not be specific to DS. The Ts1Cje mouse model is trisomic for approximately 81 of the genes encoded on the Ts65Dn translocation. The Ts1Rhr mouse model is trisomic for 33 Mmu16 genes that have conserved synteny with the Hsa21 DSCR. Although Ts1Cje and Ts1Rhr models exhibit DS-associated learning and memory deficits, they are generally manifest to a reduced extent and severity than in the Ts65Dn mouse model.

A transchromosomic mouse model that carries a freely segregating copy of Hsa21 in addition to the normal mouse chromosome complement was generated in 2005. The Tc1 mouse model is trisomic for approximately 80% of Hsa21 genes; however, it is mosaic for Hsa21, such that not every cell contains the human chromosome. Using chromosome-engineering technology, which allows duplication or deletions of large chromosomal segments, a Ts1Yey model that carries a duplication spanning the entire Hsa21 region syntenic with Mmu16 was recently developed. Soon afterwards, models that carried duplications for Hsa21 regions syntenic with Mmu10 (Ts2Yey) and with Mmu17 (Ts3Yey) were developed by the same researchers. The breeding of these mice enabled the generation of the most complete mouse model of DS currently available, the Ts1Yey;Ts2Yey;Ts3Yey model, which is trisomic for all the mouse orthologues of Hsa21 genes.

CLINICAL CHARACTERIZATION OF DOWN SYNDROME

Language

Intellectual difficulties are normally characterized by anomalies in language, learning, and memory. Difficulties in language development and communication are of particular concern as they unfortunately restrict societal acceptance. The type of cognitive function affected and the severity of the impairment vary among individuals with DS, making it difficult to identify a prototypical language profile. In DS, these language difficulties are commonly presented as impairments in morphosyntax, verbal short-term memory, and explicit long-term memory, combined with deficiency in acquiring new skills. Associative learning, visuospatial short-term memory, and implicit long-term memory, however, are relatively preserved. A recent meta-analysis examined language and memory skills in children with DS, comparing them with typically developing children matched on non-verbal mental age. The study concluded that children with DS exhibit deficits in expressive vocabulary, grammar, and verbal short-term memory; however, receptive vocabulary skills were similar to those of controls. This suggests a broad language deficit in children with DS that is prevalent in all language domains except for receptive vocabulary. Receptive vocabulary may be spared in DS because it is dependent on non-verbal responses. In addition, it may be less cognitively demanding than expressive vocabulary and grammatical skills.

Hearing losses occur in two-thirds of children with DS, and may contribute to their difficulties in language development. Otitis media is characterized by effusion in the middle ear and hearing loss and is prevalent in up to 78% of people with DS. Over 70% of Ts65Dn mice present with otitis media; however, no such impairments were found in Tc1 mice. This suggests that trisomic genes in the Ts65Dn mouse, which are disomic in the Tc1 mouse model, may predispose to otitis media.

Brain Anatomy

Altered brain morphology and perturbed neural connectivity may underlie the cognitive deficits seen in DS. Brain structures that mature later in development, such as the hippocampus, cerebellum, and frontal cortex, display disproportionate impairments.[7] Postmortem analysis and magnetic resonance imaging (MRI) studies document reduced brain volume and brachycephaly in individuals with DS compared with controls. These reductions are not universal and are most prominent in the cerebellum and hippocampus. The average cerebellum in DS is less than 75% the volume of controls. The parahippocampal gyrus is reported to be larger in people with DS, but the subcortical areas including the lenticular nuclei and posterior cortical gray matter are relatively preserved.

Mouse models of DS recapitulate these anatomical phenotypes. Morphometric analyses in Ts65Dn mice reveal altered brain shape and reduced cerebellar volume. Ts1Cje mice also demonstrate DS-associated craniofacial dysmorphology including smaller brains, hypoplasia of the cerebellum, and enlarged ventricles. Ts1Rhr mice only exhibit altered brain shape, raising the possibility that triplication of genes in the DSCR does not cause reduced cerebellar volume.

Neurodevelopment and Synaptic Plasticity

In DS, prenatal brains display delayed development and disorganized cortical lamination, with reduced dendritic arborization in cerebral cortical pyramidal cells and fewer synapses. Postnatal brains show degeneration of the cortical pyramidal neurons, severe dendritic and synaptic impairments, and a reduction in the number of hippocampal neurons and the number of granule cells in the cerebellar cortex. Abnormal spinogenesis is apparent within the first 2 years of life in DS, characterized by a failure of dendritic spines to achieve mature morphology and enlarged atrophic spine heads.[8] The development of the cerebellum is also delayed and precociously terminates.[9] Neuronal apoptosis, as measured by oxidative stress levels, is also increased in DS fibroblasts. It has been proposed that Hsa21-encoded genes Pbx regulating protein-1 (PREP1), a transcription factor involved in the regulation of organism size, and tetratricopeptide repeat domain (TTC3), an E3 ubiquitin ligase that targets AKT, contribute to the observed increase in apoptosis and increased proliferation.

Studies from mouse models show that excitatory glutamatergic projection neurons comprise approximately 90% of the neocortex; these neurons are generated from precursors in the dorsal telencephalic ventricular and subventricular zones. The remaining neurons are inhibitory γ-aminobutyric acid (GABA) interneurons, which are generated in the ventral ventricular zone of the ganglionic eminence. During development, the evolution of a normally functioning cortex involves, first, the neurogenesis of specific neurotransmission of excitatory and inhibitory neurons from their distinct origins of the brain, followed by the migration and differentiation of these neurons within the neocortex. Neuronal development and cognitive functioning are dependent on a balanced ratio of excitatory and inhibitory neurons; alterations in neuronal function and neurotransmission have been proposed to be a mechanism through which impairments in synaptic plasticity and long-term potentiation (LTP), a cellular biological correlate for learning and memory, occur.

In DS, the mechanisms and processes involved in controlling embryonic production and the allocation of neurons are perturbed. Reduced numbers of excitatory neurons are produced by the dorsal ventricular zone in DS fetal brains and in Ts65Dn mice. The combination of reduced proliferation of these cells and a lengthened cell cycle results in a reduced expansion of the maturing cortical layers, eventuating in delayed development and possibly contributing to later defective synapse formation. A gene encoded on Hsa21, dual-specificity tyrosine phosphorylation-regulated kinase (*DYRK1A*), is strongly expressed during embryonic neurogenesis, particularly in neural precursor cells, and is implicated in dorsal telencephalic ventricular zone proliferation. The overexpression of *DYRK1A* has been proposed to lead to premature neuronal differentiation, depletion of neural precursor cells available during neurogenesis, and inhibition of cell proliferation, presumably through deregulated NOTCH signaling.[10] DYRK1A also interacts with growth factors, transcription factors and cell cycle regulatory proteins that are also involved in neural cell proliferation and specification. Overexpression of *DYRK1A* has been implicated in perturbing a chromatin-remodeling complex, leading to impaired dendritic growth and deregulated pluripotency and embryonic stem cell fate in DS.

Along with the underproduction of cortical excitatory neurons, the excitatory–inhibitory ratio is further imbalanced by excessive production of a subset of forebrain inhibitory interneurons in the Ts65Dn mouse, particularly in the dorsal neocortex and hippocampus. This increase in the number of interneurons in precursors enhances the rate of cell production, which promotes neurogenesis specifically in the medial ganglionic eminence (MGE), resulting in overinhibition. During MGE neurogenesis, transcription factors can induce MGE precursor cells to divide to generate either inhibitory interneurons or oligodendrocytes. Two transcription factors, oligodendrocyte transcription factor-1 (*Olig1*) and lineage factor-2 (*Olig2*), are heavily expressed during this process. Both genes are triplicated in DS and have been implicated in the ventral telencephalon inhibitory neuronal phenotype in Ts65Dn mice. Normalizing expression of these genes to disomic levels in Ts65Dn mice resulted in the restoration of MGE neurogenesis and inhibitory neuron production to normal levels.

Ts65Dn mice display enlarged boutons and dendritic spine heads in cortical and hippocampal neurons. They also demonstrate excessive inhibition in the hippocampus and fascia dentate, which results in failed induction of LTP. The overinhibition phenotype displayed in Ts65Dn pyramidal neurons was rescued in Ts65Dn mice that had disomic copies of the *Olig1* and *Olig2* genes, implicating an additional role for the triplication of these genes in causing the excitatory–inhibitory imbalance. However, these genes may not be directly involved in causing synaptic plasticity deficits as Ts1Rhr mice demonstrate synaptic deficits but are not trisomic for *Olig1* and *Olig2*. Increased inhibitory input is a consequence of altered efficiency of the GABAergic system in the DG of Ts65Dn mice, as opposed to a decrease in inhibitory synapse density. In support of larger inhibitory synapses found in Ts65Dn mice, electrophysiological data have revealed enhanced $GABA_A$ and $GABA_B$ receptor-mediated neurotransmission with a concomitant reduction in paired-pulse ratios of evoked inhibitory postsynaptic currents (IPSCs), suggesting an increased presynaptic release of GABA. Impaired synaptic plasticity was also seen in Ts65Dn striatal cholinergic interneurons, highlighting a potentially novel and important role for the intrastriatal cholinergic system in the pathophysiology of DS-associated cognitive defects.

Enhanced postsynaptic $GABA_B$ signaling can be attributed to the triplication of *KCNJ6/Girk2* (G-protein-coupled inward rectifying potassium channel) that encodes the protein Girk2, which is a subunit of a channel that modulates postsynaptic $GABA_B$ receptors. In Ts65Dn mice, Girk2 is expressed throughout the brain, particularly in the hippocampus, leading to an almost two-fold increase in $GABA_B$-mediated Girk2 current. The overexpression of *Kcnj6* is associated with increased channel density, increased current, and increased inhibitory $GABA_B$ signaling, suggesting a functional consequence in the excitatory–inhibitory balance of neuronal transmission. The overexpression of *Kcnj6* also leads to an imbalance between $GABA_A$ and $GABA_B$ inhibition of CA1 pydramidal neurons, thereby perturbing hippocampal circuitry functioning. Enhanced $GABA_B$/Girk2 signaling has also been documented in Ts65Dn granule cells. Girk2 channels reduce membrane potential and consequently reduce neuronal excitability; lower excitability is proposed to impede *N*-methyl-D-aspartate (NMDA)-dependent plasticity, resulting in learning and memory impairments. Intriguingly, regulator of calcineurin-1 (*RCAN1*), another Hsa21-encoded gene, is a negative regulator of calcineurin that modulates the activation kinetics of the NMDA receptor by decreasing the opening times of the NMDA receptor channel. The triplication of RCAN1 can thus be conceived to also negatively impact NMDA-dependent plasticity.

Additional Hsa21 genes may play roles in brain structure and function (Table 5.1). Down syndrome cell adhesion molecule (DSCAM) plays a critical role in facilitating dendritic morphology and neuronal wiring during neurodevelopment, and contributing to efficient synaptic plasticity in adulthood. Overexpression of DSCAM in Ts1Cje hippocampal neurons was found to inhibit dendritic branching. Overexpression of DSCAM may lead to a loss of NMDA-mediated regulation of DSCAM local messenger RNA translation, resulting in the observed dendritic impairments in neurodevelopment and synaptic plasticity. Synaptojanin-1 (*SYNJ1*) encodes a presynaptic polyphosphoinositide phosphatase important for membrane trafficking and normal synaptic vesicle recycling, and also plays a significant role in maintaining the stability of GABAergic neurotransmission. The overexpression of superoxide dismutase-1 (*SOD1*) also upregulates GABAergic neurotransmission, and plays additional roles in reducing the number of hippocampal neuronal progenitor cells, reducing LTP, and enhancing sensitivity to degeneration and apoptosis. The triplication of the transcriptional repressor single-minded homolog-2 (*SIM2*) dramatically reduces expression of drebrin-1 (*DBN1*) by directly binding to its promoter. *DBN1* is a neuronally expressed gene that affects dendritic spine structure and neuritogenesis and is involved in modulating dendritic spine–cytoskeleton dynamics at postsynaptic terminals; decreased levels of DBN1 have been observed in cortices of people with AD and DS. Reduced DBN1 levels could contribute to changes in structure and function in DS. Overexpression of amyloid-β precursor protein (APP) is heavily linked to the neurodegeneration phenotype; nevertheless, a role for APP in neurodevelopment has been suggested. In this study, β-amyloid (Aβ) levels were lowered using a γ-secretase inhibitor, resulting in improved learning and memory conditions in young Ts65Dn mice. Therapies targeting the reduction of Aβ levels may thus improve cognitive function in young DS patients.

Similar to the reduced cerebellum phenotype in DS, Ts65Dn mice have a reduced cerebellar volume, with noteworthy reductions in the molecular layer and the internal granule layer and almost a 20% reduction in Purkinje cell and granule cell density. Sonic hedgehog (SHH) growth factor response regulates cell cycle length and neurogenesis in cells; decreased SHH is proposed to elongate the cell cycle length, resulting in increased proliferation rates and impaired neurogenesis. Support for this comes from evidence in the Ts65Dn mouse that a deficient mitotic response to the SHH growth factor leads to altered proliferation in cerebellar granule and neural crest progenitor cells, which could also explain the craniofacial dysmorphology prevalent in people with DS. Altered proliferation in cerebellar granule cells and neural progenitor cells along with increased cell death rates have also been reported in the Ts1Cje and Tc1 mouse models. Defective SHH signaling in neuronal precursor cells and the resulting impairments in cerebellar neurogenesis may be caused by overexpression of APP, which leads to the enhanced expression of SHH receptor patched-1 (Ptch1), an inhibitor of the SHH signaling pathway. These cerebellar alterations may also underlie the muscle hypotonia and fine motor control deficits present in DS.

Cognition and Behavioral Tests in Down Syndrome Mouse Models

Intellectual disability is apparent even in the very young person with DS; although this can range from a mild to moderate impairment, the intelligence quotient (IQ) for people with DS ranges from 20 to 80, with the mean IQ being approximately 50.[11] Age-related cognitive decline and dementia can further worsen cognition. Individuals with DS have been found to display age-related reductions in volume of frontal, temporal, and parietal lobes, accompanied by an age-related increase in cerebrospinal fluid volume. This suggests accelerated aging in DS, with certain brain regions predisposed to premature age-related cognitive decline and AD.[12]

TABLE 5.1 Contribution of Hsa21-Encoded Genes to Cognition

Hsa21 Gene	Physiological Role	Pathogenic Role
APP	Cell surface receptor and transmembrane precursor protein that promotes transcriptional activation	Forms the protein basis of Aβ plaques prevalent in AD and DS Leads to the overexpression of SHH receptor Ptch1, which inhibits SHH signaling pathway leading to impaired cerebellar neurogenesis Leads to enlarged early endosomes and a deficit in NGF retrograde axonal transport
DSCAM	Cell adhesion molecule with a crucial role in dendrite morphology and neuronal wiring	Inhibits dendritic branching and causes perturbed synaptic plasticity May lead to aberrant NMDA-mediated regulation of DSCAM local translation
DYRK1A	Kinase involved in regulating several signaling pathways and cell proliferation Involved in neurogenesis, particularly of neural precursor cells	Leads to premature neuronal differentiation, depletion of neural precursor cells and inhibition of cell proliferation Deregulates genes implicated in dendritic growth, cell pluripotency and embryonic stem cell fate Phosphorylates APP and may contribute to extracellular Aβ phenotype Phosphorylates tau at a key priming site and therefore can lead to NFTs
ITSN1	Multidomain adaptor protein implicated in membrane trafficking and synaptic transmission	Perturbed retrieval of synaptic vesicle proteins, including SYNJ1 and dynamin, during clathrin-dependent endocytosis
KCNJ6/GIRK2	Effector protein for $GABA_B$ receptors Modulates potassium channel current and density	Causes GABAergic excitatory–inhibitory imbalance through increased channel density, current and $GABA_B$ signaling Reduces membrane potential and neuronal excitability, thereby impeding NMDA-dependent plasticity
OLIG1 and *OLIG2*	Transcription factors implicated in oligodendrogenesis and neurogenesis, especially during MGE neurogenesis	Causes GABAergic excitatory–inhibitory imbalance through an overinhibition phenotype
PCBP3	Splicing factor important for post-transcription activities	Misregulates splicing of exon 10 of tau, resulting in abnormal ratios of tau isoforms leading to tauopathies and NFTs
PREP1	Transcription factor involved in the regulation of organism size	Increases neuronal apoptosis and enhanced proliferation in DS fibroblasts
RCAN1	Inhibits calcineurin-dependent signaling pathways affecting development Modulates NMDAR activation	Decreases opening probability of NMDAR channel; upregulates GSK3β to cause enhanced tau phosphorylation Interaction with DYRK1A could aggravate perturbed endocytosis and synaptic vesicle pore kinetics Aberrant dephosphorylation of SYNJ1 during depolarization of synaptic nerve terminals
SIM2	Transcriptional repressor implicated in synaptic plasticity and morphology	Reduces DBN1 levels, causing morphological cytoskeletal changes at postsynaptic terminals in dendritic spines
SOD1	Cytoplasmic protein involved in oxidative stress	Decreases hippocampal neuronal progenitors and LTP Increases sensitivity to degeneration and apoptosis Upregulates GABAergic neurotransmission
SYNJ1	Nerve terminal protein implicated in membrane trafficking and synaptic transmission	Inability to maintain stable GABAergic neurotransmission; perturbed dephosphorylation of phospholipids during clathrin-dependent endocytosis Leads to enlarged early endosomes
TTC3	An E3 ubiquitin ligase that targets the AKT kinase	Increases neuronal apoptosis and enhanced proliferation in DS fibroblasts

GABA: γ-aminobutyric acid; MGE: medial ganglionic eminence; NMDAR: *N*-methyl-D-aspartate receptor; AD: Alzheimer disease; DS: Down syndrome; SHH: sonic hedgehog; NGF: nerve growth factor; NMDA: *N*-methyl-D-aspartate; DSCAM: Down syndrome cell adhesion molecule; APP: amyloid-β precursor protein; NFT: neurofibrillary tangle; LTP: long-term potentiation.

Mouse models recapitulate DS-associated cognitive impairments, with learning and memory deficits assessed through batteries of behavioral tests. An ideal behavioral test would demonstrate the functionality of a particular brain region; however, most such tests report on specific involvement of linked brain regions. Regardless of this, behavioral tests in mouse models have provided unequivocal insights into brain regions involved in specific deficits, which have translated into the development and validation of tests created to assess cognitive levels in people with DS. One such example is the Arizona Cognitive Test Battery (ACTB), which includes neuropsychological tests of non-verbal general cognition (so as not to confound with language demands) and assessment of specific brain regions such as prefrontal, hippocampal, and cerebellar functioning. Importantly, it also considers genetic modifiers of variation and is applicable to a wide age range.[13]

A widely used navigation task used to assess spatial learning and memory is the Morris water maze (MWM). This task is heavily dependent on the involvement of the hippocampus as well as other brain regions including the entorhinal cortex, striatum, fimbria, and fornix. In this task, a transparent platform is submerged in a circular water tank that is filled with opaque water. The cued (visible) platform task involves mice finding the platform, which contains a visible cue such as a flag. This is normally initially carried out as a control experiment. In the hidden platform task, the platform is not visibly flagged and involves a complex spatial mapping strategy and visuospatial integration, including cues both inside and outside the maze to identify the fixed platform location, while the starting position of the mice is changed. In these tasks, learning is evaluated upon the latency to find the platform and the path trajectory undertaken. The probe trial is normally conducted after the hidden platform task is completed; the platform is removed, and the time spent in the quadrant that formerly contained the platform is assessed. Mouse models of DS appear to show similar performance in the cued MWM task; however, Ts65Dn and Ts1Cje mice demonstrate hippocampal-dependent learning deficits in the hidden platform variations of this task. The Ts1Yey mouse model and the triple trisomy (Ts1Yey;Ts2Yey;Ts3Yey) model also show these impairments; however, the Ts2Yey and Ts3Yey models do not display such deficits, suggesting a strong contribution of orthologous genes on Mmu16 playing a significant role in spatial learning. The Ts1Rhr mouse model also failed to demonstrate any deficits in this task, thereby further narrowing down on a segment on Mmu16 that is responsible for this phenotype. The Tc1 mouse model retained intact spatial recognition memory but demonstrated an impaired spatial working memory in other variations of the MWM.

The novel object recognition test (NORT) includes the involvement of the dentate gyrus and CA1–CA3 regions of the hippocampus as well as the perirhinal, insular, and medial prefrontal cortices. This task is based on the tendency for rodents to explore novel objects over familiar ones. However, impairments in recognition memory lead to the inability to discriminate between the two objects. Impairments in this task have been documented in Ts65Dn mice and in Ts1Rhr mice, suggesting the crucial contribution of the 33 trisomic orthologous genes on the DSCR for this phenotype. However, despite the presence of all these 33 genes, Ts1Cje mice displayed normal performance in the NORT, suggesting a more complex interaction of Hsa21 genes. Short-term, but not long-term, deficits were also apparent in Tc1 mice. In a slight variation of this task, the object-in-place test, mice are tested on remembering whether a familiar object has been moved to a new position over the course of a few hours, as opposed to a prolonged period of 24 hours in the NORT. Ts65Dn mice perform normally in this task, suggesting particular difficulties in declarative tasks that require long-term memory.

Short-term working memory is assessed by spontaneous alteration in a T-shaped or Y-shaped maze. This task is also based on the innate tendency of rodents to explore novelty by visiting different arms of the maze. Performance in this task is measured by the number of times a mouse enters an arm that is different from the one it recently visited (alterations), relative to the possible number of alterations it could have made. Ts65Dn, Ts1Cje, and Ts1Rhr mice demonstrate approximately a 20% lower spontaneous alteration rate compared with euploid mice, revealing short-term working memory deficits in these models. Tc1 mice did not differ from controls in spontaneous alteration rates.

Fear conditioning is administered to assess associations and induce fearful behavior, which is measured by a freezing response, i.e. a sudden cessation in movement. In this task, a conditioned stimulus, normally a tone and context, is paired with an unconditioned stimulus, usually a mild electric shock to the foot. A day later, the mice are placed in a novel chamber and exposed to the tone without receiving the foot shock. The tone and context both form the conditioned stimulus; however, distinct brain regions are differentially recruited. Acoustic tone fear conditioning is heavily dependent on the amygdala and contributions from the perirhinal and entorhinal cortices, and the thalamus; contextual fear conditioning is predominantly dependent on the hippocampus with crucial inputs from the cerebellum. Ts65Dn, Ts1Yey, and the triple trisomy mice all demonstrate reduced freezing when reintroduced in the contextual conditioned chamber. The Ts2Yey and Ts3Yey models show normal responses, suggesting that Hsa21 homologous genes on Mmu16, and not the other two mouse genomic regions, may be responsible for this phenotype. Ts65Dn demonstrate normal freezing response to acoustic conditioning, suggesting intact amygdala functioning in these mice.

Alzheimer Disease and Neurodegeneration

People with DS above the age of 45 years are highly susceptible to early-onset AD, with the prevalence of dementia increasing exponentially such that by the age of 65 years, over 75% have a clinical diagnosis of dementia. AD pathology impacts the entorhinal cortex followed by the hippocampus, parietal cortex, and prefrontal cortex.[14] The neuropathology of DS is essentially identical to that for AD in the changes detected, their regional distribution, relative time of appearance, and progression over time. Despite the full-blown clinical manifestation occurring later in life, the histopathological hallmark features of AD are seen earlier in people with DS; brain atrophy, deposits of extracellular Aβ and the accumulation of hyperphosphorylated tau that results in neurofibrillary tangles (NFTs) are all present in people with DS by the age of 40.[15–17] The dysfunction of the frontal lobe and the hippocampus in DS may predispose to AD, as these are the brain regions in which Aβ first accumulates during the early stages of AD. Under physiological conditions, APP produces healthy amounts of Aβ peptides; however, the triplication of *APP* leads to increases in Aβ peptide production and plaque deposits. Increased *APP* gene dose is believed to play a necessary role. Evidence for this is documented in case studies of five rare families with early-onset AD that have small internal duplications of Hsa21 and that include the triplication of *APP*.[18] Moreover, in cases of partial trisomy 21, which do not include triplication of *APP*, no neuropathological or neuropsychological presence of AD is evident.[19]

Mouse models of DS also encapsulate certain DS-associated neurodegenerative phenotypes. Ts65Dn mice do not produce Aβ plaque deposits or NFTs, unlike in DS and AD, but they do demonstrate a loss of basal forebrain cholinergic neurons (BFCNs), which is correlated with cognitive decline in these mice. The BFCNs are particularly vulnerable to neurodegeneration in DS and AD.[20] Ts1Cje mice show no evidence of BFCN neurodegeneration, thus providing a compelling argument that the contribution of one or more of the approximately 23 extra trisomic genes in the Ts65Dn mouse model are necessary for BFCN neurodegeneration. Degeneration of BFCNs is also related to a deficit in retrograde axonal transport of nerve growth factor (NGF), which is a neurotrophic factor that is important for the survival and maintenance of specific neurons. Compared with control mice, Ts65Dn mice display a severe impairment of NGF retrograde transport, which is six times worse than that in Ts1Cje mice.[21] In an attempt to understand the contribution of specific genes in causing this phenotype, *App* was identified as a triplicated gene that is present in the Ts65Dn mouse, but not in the Ts1Cje model. The third copy of *App* in Ts65Dn mice was knocked down, thus rendering its expression comparable to disomic levels. In these Ts65Dn mice, which harbor only two copies of *App*, NGF retrograde transport was restored to that of Ts1Cje mice. Significantly, normalizing App protein levels prevented the loss of BFCN in Ts65Dn mice. Further support for the role of overexpression of APP in causing abnormal axonal transport of NGF was demonstrated by deficits in retrograde NGF transport in a mouse carrying a human wild-type APP transgene and in another mouse harboring a human mutant APP transgene. The gene dosage imbalance of *App* has also been linked to the degeneration of locus ceruleus neurons, which occurs before the BFCN neuronal loss.

Endocytosis is a mechanism through which cells regulate their developmental and functional programs and communicate across the plasma membrane. In clathrin-mediated endocytosis, after internalization of the neurotrophin–receptor complex, the clathrin-coated vesicle becomes uncoated in the intracellular cytoplasm and fuses with early endosomes. The internalized molecules can be recycled back to the plasma membrane from the early endosome, transported to a late endosome or lysosomes, or transported retrogradely to the cell body (Fig. 5.2).

The BFCN terminal ends in Ts65Dn mice display enlarged early endosomes that contain markers for both NGF and APP, suggesting that the overexpression of *App* causes enlarged early endosomes, leading to disrupted retrograde NGF transport and, consequently, neurodegeneration[21] (Fig. 5.3). Ts1Cje mice do not display enlarged early endosomes and normalizing *App* gene dosage in Ts65Dn mice rescued the enlarged early endosome phenotype.[22] Collectively, these studies point to a gene dosage response of *App* in causing enlarged early endosomes. Endosomal abnormalities develop in neurons of brain regions most severely affected in AD and DS, including the hippocampus and neocortex as well as the basal forebrain. Early endosomes were documented to be 32-fold larger in volume in DS pyramidal neurons in lamina III of the prefrontal cortex compared with controls. Enlarged endosomes and a subsequent aberrant endocytic pathway have been observed as early as 2 months of age in DS brains and in DS fibroblasts.

The role of *App* in causing endocytic deficits was further examined by studying key APP proteolytic enzymes [β-site APP cleaving enzyme-1 (BACE-1) and γ-secretase] and various APP proteolytic fragments [Aβ and C-terminal APP fragment (βCTF)] in DS fibroblasts. Morphological and functional endocytic abnormalities in DS fibroblasts were reversed when the expression of APP or BACE-1 was lowered using short hairpin RNA

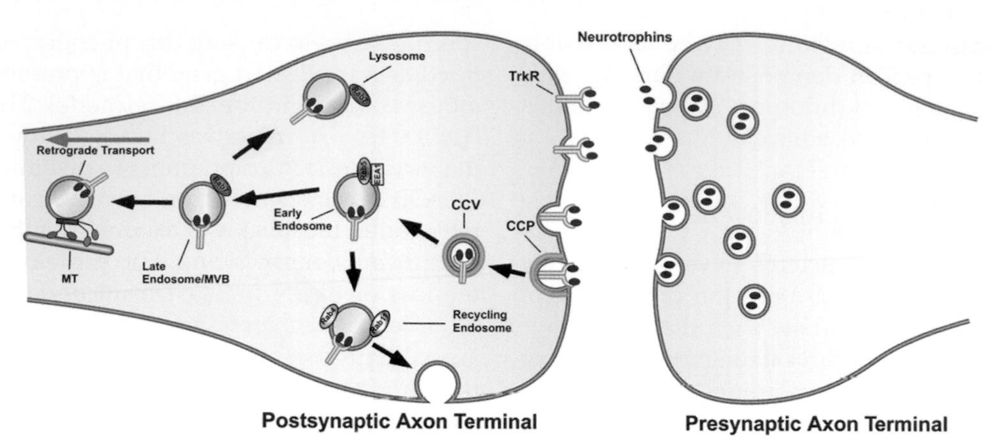

FIGURE 5.2 **Clathrin-mediated endocytosis at the axon terminal.** Neurotrophins bind to their Trk receptors (TrkR) and recruit clathrin at the plasma membrane to facilitate internalization into clathrin-coated pits (CCP), which then bud and pinch off to form clathrin-coated vesicles (CCV). The CCV becomes uncoated and fuses with an early endosome, Rab5, and its effector protein, EEA1. The cargo can be transported to the cell body by dynein motor proteins along the microtubules (MT). It can also be recycled back to the plasma membrane through Rab4- and Rab11-regulated endosomes. If further proteolysis is required, the cargo can be trafficked to Rab7-regulated late endosomes, or multivesicular bodies (MVB). Subsequently, the cargo can be transported to lysosomes for proteolysis.

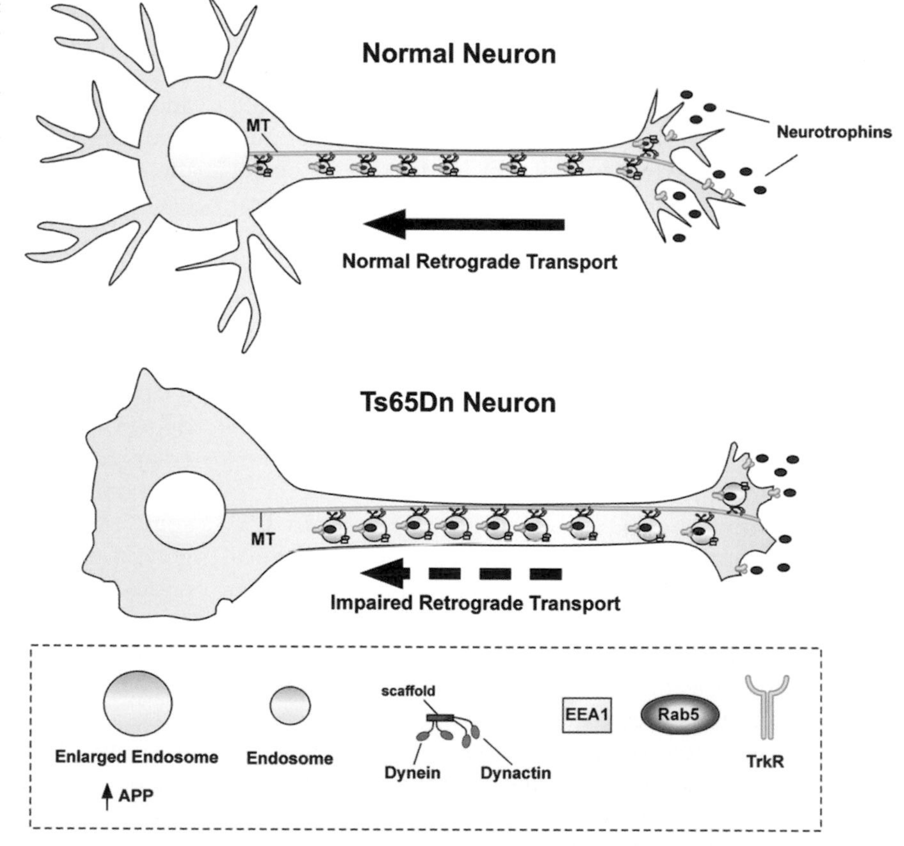

FIGURE 5.3 **Retrograde axonal transport of neurotrophins in normal and Ts65Dn neurons.** Healthy neurons (top) demonstrate normal retrograde transport of neurotrophins in early endosomes along the axon. Transport along microtubules (MT) is facilitated by regulatory and motor proteins including Rab5, EEA1, and the dynein–dynactin complex. Ts65Dn neurons (below) demonstrate impaired axonal transport. Overexpression of amyloid-β precursor protein (APP) in these mice is postulated to cause enlarged early endosomes, which impedes axonal transport of essential neurotrophins, leading to neurodegeneration.

constructs. Overexpression of wild-type APP was sufficient to induce endosomal pathology in normal disomic fibroblasts. Control fibroblasts were unaltered when transfected with a mutant form of APP, which lacked the amino acid sequence required for the β-site cleavage, establishing the importance of βCTF generated by BACE-1. A γ-secretase inhibitor, which pharmacologically reduced Aβ production but raised βCTF levels, also produced endosome pathology in control fibroblasts and worsened the pathology in DS fibroblasts. To further elucidate the role of βCTF, control fibroblasts that were transfected with a construct that elevated only the βCTF levels also displayed an increase in endosome size similar to that of increased APP expression. Reduced

BACE-1 expression, and thus lowered βCTF production, rescued endosomal pathology in DS fibroblasts. This provides strong evidence that endocytic pathway dysfunction is dependent on APP gene dose and processing and elevated levels of βCTF and that this phenotype may be independent of Aβ. It is conceivable that abnormal morphology and dysfunction of the endocytic pathway in DS and AD may be the mechanism through which neurodegeneration develops in several brain regions, including BFCNs.

Aside from *APP*, other Hsa21 genes have also been implicated in contributing to DS-associated neurodegeneration, aberrant endocytic pathway, and abnormal synaptic circuits. DYRK1A has been shown to functionally interact with and phosphorylate APP, and may contribute to the extracellular Aβ plaques seen in AD and DS. As mentioned earlier, another early histopathological feature of AD and DS is the formation of NFTs. None of the available DS mouse models produces NFTs; however, by the age of 3 months the Ts1Cje mouse displays hyperphosphorylation of tau, a microtubule-associated protein. This mouse model is not trisomic for APP. DYRK1A also encodes a kinase that phosphorylates tau at a key priming site. The triplication of DYRK1A is thus proposed to facilitate the hyperphosphorylation of tau in an APP-independent manner, leading to NFTs and the development of AD in people with DS. NFTs can also form through misregulated splicing of tau. Splicing misregulation of exon 10 of tau has been shown to result in abnormal ratios of tau isoforms, leading to tauopathies including AD and early-onset dementia in DS. A splicing factor on Hsa21, poly(rC)-binding protein-3 (PCBP3), was found to activate splicing of exon 10, suggesting a pathogenic role for this gene in forming NFTs and providing another mechanism through which people with DS may be susceptible to early-onset AD.

Rat cortical neurons incubated with exogenous Aβ have been found to induce oxidative stress, thereby mediating the upregulation of RCAN1 expression. Calcineurin is a serine–threonine phosphatase that dephosphorylates tau, and RCAN1 inhibits the phosphatase activity of calcineurin, resulting in increased phosphorylation of tau. Increased RCAN1 also causes the upregulation of glycogen synthase kinase-3β (GSK3β), a tau kinase, which also results in the hyperphosphorylation of tau. In addition, increased RCAN1 and phospho-tau levels were found in lymphocytes of people whose ApoE genotype contained the ε_4 allele, which has been linked to a higher risk of developing AD compared with other allelic variations.

Intersectin (*ITSN1*) and synaptojanin (*SYNJ1*) are also involved in synaptic vesicle endocytosis. ITSN1 is a multidomain adaptor protein, which functions with SYNJ1 and dynamin, a GTPase involved in the scission of newly formed vesicles, to facilitate the retrieval of synaptic vesicle proteins during clathrin-dependent endocytosis. SYNJ1 functions to dephosphorylate phospholipids during this process; this activity is regulated by DYRK1A and includes the phosphorylation of SYNJ1, dynamin-1, and amphiphysin. An additional role for RCAN1 is implicated in coordinating neurotrophin receptor endocytosis and axonal growth, by calcineurin-mediated dephosphorylation of an endocytic GTPase, dynamin-1. RCAN1 also interacts with DYRK1A and controls endocytosis and synaptic vesicle fusion pore kinetics. It dephosphorylates SYNJ1 during depolarization of synaptic nerve terminals. The overexpression of *SYNJ1* in a neuroblastoma cell line as well as in transgenic mice has been associated with the enlarged early endosome phenotype. This phenotype was reduced by silencing *SYNJ1* expression with RNA interference in DS fibroblasts, suggesting a therapeutic target other than APP for addressing endosomal morphology and trafficking abnormalities. Collectively, the increased gene dosage of one or more of these genes may make important contributions to perturbed endocytic pathways and synaptic functioning in DS, in part, by altering synaptic vesicle morphology and release probability and the size of available vesicle pools during synapses. In studies using *Drosophila*, overexpression of the gene homologues ITSN1, SYNJ1, and RCAN1 was found to cause abnormal synaptic morphology and impaired vesicle recycling.

Other Neurological Complications

Seizures in childhood and adulthood are linked to structural brain abnormalities and can also impair cognitive functioning, including language, and lead to a greater risk of injury and a higher mortality rate. Children with DS are more susceptible to infantile spasms, although little is known about the mechanisms underlying this condition. Six percent of children and adolescents with DS experience epileptic seizures. Seizures in DS tend to occur in a bimodal manner, with half the individuals presenting seizure onset before 1 year of age and the other half in the third decade or later. Significantly, people with DS over the age of 45 years who experience late-onset seizures are more likely to develop AD, and the seizure may even cause the onset of dementia. A 2012 study documented a marked decline in cognition in people with DS and intellectual disabilities who experience seizures compared with those who have not had a seizure.[23]

It has been demonstrated that Ts65Dn mice are highly susceptible to audiogenic-induced seizures, and that blockade of the metabotropic glutamate receptor-5 ($mGluR_5$) or passive immunization with anti-Aβ antibodies reduces the incidence of these seizures and the number of associated deaths. These data provide evidence that increased *APP* gene dose may result in a lowered seizure threshold. Therapeutic interventions targeted at lowering *APP* expression, modifying APP processing, or increasing the clearance of Aβ may be beneficial in treating or preventing seizures. In another study of human APP transgenic mice, epileptic activity induced by overexpression of Aβ was associated with the sprouting of inhibitory axons in the molecular layer of the dentate gyrus.

Moyamoya disease is a cerebrovascular condition characterized by reduced blood flow leading to greater susceptibility to ischemia, recurrent transient ischemic attacks, sensorimotor paralysis, and convulsions. Moyamoya disease has been reported to occur with a higher frequency in people with DS relative to the general population. A study observing over 500 patients found a 26-fold greater prevalence of DS in patients with coexisting moyamoya disease that presented with ischemic stroke, especially within Caucasian and Hispanic ethnicity groups. These data suggest that Hsa21 genes may be associated with the pathogenesis of moyamoya disease.

As a result of craniofacial dysmorphology including a small upper airway, midfacial hypoplasia, and muscular hypotonia, children with DS are more prone to developing obstructive sleep apnea. In 57% of 2–4-year-old children with DS, obstructive sleep apnea was linked with low IQ, behavioral disorders, visuospatial impairments, respiratory complications, and a reduced quality of life. Obstructive sleep apnea is due to a compromised cardiorespiratory circuit and a dampened sympathetic response to sleep-disordered breathing, which makes this population more susceptible to pulmonary hypertension and other cardiovascular complications compared with typically developing children. Even though adults with DS are predisposed to obstructive sleep apnea, little is known about this population. In one study of adults with DS, 94% displayed abnormal polysomnogram readings and this was significantly correlated with obesity. This study emphasizes the need to address obstructive sleep apnea in adults with DS, as the consequences of untreated obstructive sleep apnea may overlap with the manifestation of DS, leading to accumulated but potentially treatable symptoms.

Circadian rhythm behavioral patterns have been studied in Ts65Dn mice, in which the sleep–wake cycle was assessed within a light–dark photocycle; however, the current literature provides conflicting evidence. One study reported that adult Ts65Dn mice were more hyperactive and daily patterns of total and specific behavioral activity varied significantly compared with wild-type controls. In another study, Ts65Dn mice demonstrated a robust circadian rhythm behavioral pattern under a standard light–dark cycle, and also when moved from constant darkness to constant light. Further studies of this phenotype in other mouse models of DS may help to clarify the contributions of dysfunctional circadian timing and sleep disturbances to cognitive deficits.

Neuropsychiatric conditions are also more prevalent in people with DS than in the general population. Behavioral and psychiatric comorbidities are apparent in 20–40% of children and adolescents with DS, including disruptive behavior, attention deficit/hyperactivity disorder, conduct/oppositional disorder, and aggressive behavior. By adulthood, the externalizing symptoms of attention problems and aggressiveness tend to be replaced by internalizing symptoms of depression and loneliness, rendering approximately a quarter of all adults with DS with a comorbid psychiatric disorder, usually major depressive disorder or aggressive behavior.[24] It has been suggested that these age-related changes of loneliness and depression may also contribute to the greater susceptibility to AD. A high comorbidity between DS and autism is prevalent; symptoms of autism are 10 times more likely to occur in children with DS than in the general population. This subpopulation is also more likely to present a greater extent of impaired brain function, including increased propensity to seizures, than children with DS who do not have autism.[25]

DEVELOPMENT OF PHARMACOTHERAPY IN DOWN SYNDROME

The identification of morphological, neurobiological, and behavioral alterations in mouse models has been instrumental in defining underlying genetic and mechanistic changes.[26] This has enabled the development of treatment targets for pharmacological interventions to restore cognition in DS.[27,28] Potential therapeutic targets can be tested on mouse models of DS to assess the physiological, biochemical, and molecular changes that are caused by the treatment. In this regard, the Ts65Dn mouse has been particularly useful as treatment successes have led to the development of clinical trials for candidate drugs in humans with DS (Table 5.2).

Neuroprotective Peptides

Enhanced neurodevelopment has been reported by pharmacological intervention with neuroprotective peptides. In DS, vasoactive intestinal peptide (VIP) levels are altered, with reduced responsiveness to VIP stimulation demonstrated in cortical astrocytes of neonatal Ts65Dn mice. The VIP stimulation of astrocytes results in the release of peptides derived from naturally occurring glial proteins that are neuroprotective neurotrophic factors, including activity-dependent neuroprotective protein (ADNP) and activity-dependent neurotrophic factor (ADNF). Treatment of DS cortical neurons with active fragments of ADNP and ADNF, NAPVSIPQ (NAP), and

TABLE 5.2 Development of Pharmacological Compounds to Improve Cognition in Down Syndrome (DS)

Pharmacological Intervention	Neurobiological Pathway	Cognitive Benefits	Treatment Evidence
NAPVSIPQ+SALLRSIPA	Neuroprotective peptides	Increases neuronal survival, reduces degenerative morphological alterations, and enhances protection from oxidative damage and apoptosis	Prenatal treatment in Ts65Dn prevents developmental delays, glial deficits, and aberrant ADNF expression, and facilitates achievement of motor and sensory milestones at same time as euploid controls
Fluoxetine	SSRIs	Increases neurogenesis by enhancing proliferation and survival of neurons; restores 5-HT receptor and BDNF levels	Chronic treatment for 3 weeks in Ts65Dn mice increased neuronal proliferation and survival in DG; early pharmacotherapy also rescued proliferation deficits and restored 5-HT receptor and BDNF levels. A month after treatment ended, Ts65Dn mice displayed greater number of surviving cells, proliferating precursors, and granule cells. Behavioral deficits in contextual fear-conditioning task were reversed
Lithium	Mood stabilizers	Increases neurogenesis by restoring cell proliferation	Treatment with lithium increased cellular proliferation in the subventricular zone in Ts65Dn and euploid controls
Pentylenetetrazol	$GABA_A$ antagonist	Enhances LTP and reduces overinhibition	Chronic treatment in Ts65Dn mice reversed deficits in NORT and spontaneous alternation tasks. Improved cognition and LTP was sustained for up to 2 months; chronic treatment for 8 weeks corrected spatial learning and memory deficits in the MWM
Picrotoxin	$GABA_A$ antagonist	Enhances LTP and normalizes NMDAR-mediated currents	Treatment reversed deficits in NORT and normalized NMDAR-mediated currents, with sustained improvements lasting for 2 weeks
α_{5IA}	$GABA_A$ inverse agonist	Reduces overinhibition of GABAergic neurotransmission and restores normal synaptic plasticity	Chronic treatment improved cognitive deficits in MWM, normalized SOD1 overexpression, and enhanced learning-evoked immediate early genes expression levels; Roche is currently commencing a clinical trial using this compound
Memantine	NMDAR antagonist	Reduces excessive glutamate neurotransmission and increases BDNF levels; previously tested in AD to decrease rate of clinical deterioration	Treatment in Ts65Dn alleviated spatial learning deficits in MWM and improved performance in fear-conditioning tasks; long-term treatment improved spatial reference learning in MWM and recovered object discrimination ability in NORT; increased BDNF expression levels were displayed in hippocampus and frontal cortex; a clinical trial found no improvements in cognitive decline or dementia in people with DS above the age of 40
L-DOPS	Norepinephrine-mediated	Corrects norepinephrine deficiency and improves contextual memory	Treatment in Ts65Dn corrected norepinephrine production deficiency through norepinephrine-modulated adrenergic activation in hippocampus and locus ceruleus to restore learning and memory deficits in a contextual fear-conditioning task
Xamoterol	β_1-Adrenergic receptor partial antagonist	Corrects norepinephrine deficiency and improves contextual memory	Treatment in Ts65Dn corrected norepinephrine production deficiency and restored learning and memory deficits in a contextual fear-conditioning task

L-DOPS: L-threo-dihydroxyphenylserine; SSRI: selective serotonin reuptake inhibitor; GABA: γ-aminobutyric acid; NMDAR: *N*-methyl-D-aspartate receptor; 5-HT: 5-hydroxytryptamine; BDNF: brain-derived neurotrophic factor; LTP: long-term potentiation; ADNF: activity-dependent neurotrophic factor; NORT: novel object recognition test; MWM: Morris water maze; SOD1: superoxide dismutase-1.

SALLRSIPA (SAL), respectively, resulted in increased neuronal survival, reduced degenerative morphological alterations, and enhanced protection from oxidative damage and apoptosis.[29] The results of this study suggest that neuropeptides that innately possess neuroprotective properties against oxidative damage may be beneficial in preventing neuronal damage and preserving neuronal function. The efficacy of NAP+SAL neuroprotective peptides in preventing glial deficits and developmental delays through prenatal treatment was examined in Ts65Dn mice. Ts65Dn mice demonstrate delayed development in achieving motor and sensory milestones, downregulated ADNF expression, and glial deficits. Prenatal treatment with NAP+SAL in Ts65Dn mice resulted in achievement of motor and sensory milestones at the same time as euploid controls, restored ADNF expression, prevented glial deficits, and restored VIP levels. Prenatal treatment of NAP+SAL in euploid pups led to a significantly earlier achievement of developmental milestones compared with euploid pups treated with placebo.[30] This study identifies a potential treatment target that could be administered during pregnancy to improve developmental delays and glial deficits in DS. Additional studies to define the impact of treatment on brain structure and function will be important.

Selective Serotonin Reuptake Inhibitors and Mood Stabilizers

In an attempt to understand the mechanism underlying reduced neurogenesis and the behavioral phenotype in DS, Ts65Dn mice were exposed to chronic antidepressant treatment for over 3 weeks with a selective serotonin [5-hydroxytryptamine (5-HT)] reuptake inhibitor (SSRI), fluoxetine, which increased neurogenesis by enhancing proliferation and survival of neurons in the DG.[31] Based on this evidence, a study published in 2010 examined whether early pharmacotherapeutic intervention could improve neurogenesis and cognitive behavior.[32] In this study, neonatal Ts65Dn mice were treated with fluoxetine from postnatal day 3 to 15. By postnatal day 15, untreated Ts65Dn demonstrated defective proliferation in the DG, subventricular zone, striatum, and neocortex. These mice displayed normal 5-HT levels, but a lower expression of 5-HT_{1A} receptors and brain-derived neurotrophic factor (BDNF). Treatment with fluoxetine not only rescued the proliferation deficits but also restored the expression of 5-HT_{1A} receptors and BDNF levels to those of euploid controls. Significantly, a month after treatment ended, an increased number of surviving cells was found in the DG of Ts65Dn mice, with a greater number of cells with a neuronal phenotype, increased number of proliferating precursors, and more granule cells. Behavioral deficits were also reversed, as demonstrated by corrected memory performance in the contextual fear-conditioning task, a heavily hippocampal dependent task.

The administration of mood stabilizers, such as lithium, to increase neurogenesis was also examined in Ts65Dn mice.[33] Neurogenesis was assessed in the subventricular zone, a region that retains neurogenic potential across life. In untreated Ts65Dn, approximately 40% fewer cells were found in this region compared with euploid controls, suggesting severe impairments in proliferation. Treatment with lithium restored cell proliferation in the subventricular zone in Ts65Dn mice to levels similar to those in untreated euploid mice. Treatment with lithium in euploid mice also significantly increased the number of cells in this region. These studies provide further encouragement for the development of readily available and usable drug candidates, such as antidepressants and other mood stabilizers, that can be administered as early as prenatally to correct for the documented neurogenesis and cognitive impairments in DS. As with other potential therapeutic interventions, additional studies will be needed to guide future clinical trials and ensure that the treatment is safe and effective in people with DS.

γ-Aminobutyric Acid-A Antagonists and Agonists

Several pharmacological interventions have aimed to lower the excitatory–inhibitory imbalance by decreasing the excessive inhibition of GABAergic neurotransmission prevalent in Ts65Dn mice.[34] This imbalance is restored to normal by drugs that inhibit the $GABA_A$ receptors in Ts65Dn and Ts1Cje mice.[35,36] To improve cognition, Ts65Dn mice have been treated with the noncompetitive $GABA_A$ antagonists pentylenetetrazol (PTZ) and picrotoxin (PTX). Chronic treatment in Ts65Dn mice with PTZ reversed the deficits seen in the NORT and spontaneous alteration task; this improvement in cognition and LTP was sustained for up to 2 months, suggesting a lasting effect of treatments aimed at reducing overinhibition.[37] Subsequently, it was discovered that treating Ts65Dn chronically for 8 weeks did not alter sensorimotor abilities or locomotor activity in home cages; however, it did correct spatial learning and memory deficits in the MWM.[38] Treating Ts65Dn mice with PTX also reversed deficits in the NORT, with sustained improvements lasting for 2 weeks. Treatment with both these compounds suppressed excessive inhibition, resulting in improved induction of LTP and normalized NMDA receptor-mediated currents.

It was discovered in 2008 that cognitive impairments in Ts65Dn mice were restored by chronic treatment with an inverse agonist that is selective for the α_5 subunit of the $GABA_A$ benzodiazepine receptor (α_{5IA}). This compound not only improved cognitive deficits in the MWM, but also normalized SOD1 overexpression and enhanced the learning-evoked immediate-early gene expression levels.[39] On the basis of this development and with the ultimate aim of reducing GABAergic neurotransmission in the hippocampus and restoring normal synaptic plasticity, Roche announced the commencement of a phase I clinical trial to assess the safety of a drug selective for the α_5 subunit of $GABA_A$ receptors.[40] At the time of writing, this trial is still in its early stages, and it will be exciting to follow the results of the study.

N-Methyl-D-Aspartate Receptor Antagonists

Memantine is a non-competitive open-channel antagonist of NMDA receptors, which reduces abnormal activation of glutamate neurotransmission to improve learning and memory. Memantine has previously been tested in the context of AD to decrease the rate of clinical deterioration.[41] Administration of memantine in Ts65Dn alleviated spatial learning deficits in the MWM, and improved performance in a fear-conditioning task.[42,43] Long-term treatment with memantine improved spatial reference memory in the MWM and recovered object discrimination ability in NORT; however, activity in spontaneous alteration was still impaired.[44] Although memantine is known to be neuroprotective, histopathological analyses from this study revealed no morphological modifications indicative of protection from neurodegeneration in the BFCN or in the locus ceruleus. However, the hippocampus and frontal cortex displayed increased BDNF expression levels. Acute treatment with memantine even 30 minutes before testing was sufficient to enhance performance on the NORT. Despite mouse studies exhibiting promising benefits of memantine as a drug candidate, a well-conducted clinical trial reported memantine to be an ineffective pharmacological intervention for cognitive decline or dementia in people with DS above the age of 40.[45] The results from this trial suggest that effective therapies in AD may not necessarily confer the same benefits in DS, and that even though commonalities in the pathogenesis of disease may exist, the mechanism may vary. They also suggest that a drug may be more effective if administered earlier in the progression of the disease or syndrome.

Norepinephrine-Mediated Neurotransmission

Other pathways that modulate learning and memory have also been examined for pharmacological intervention. The contextual memory deficits prevalent in DS are hippocampal dependent, mediated by several innervating neurons including those originating from the locus ceruleus. These afferents use norepinephrine as a neurotransmitter. Norepinephrine signaling in the hippocampus has been suggested to be impaired in Ts65Dn mice because of neurodegeneration in the locus ceruleus; however, postsynaptic targets of innervation appear to be responsive to noradrenergic receptor agonists.[46] A norepinephrine prodrug, L-threo-dihydroxyphenylserine (L-DOPS), corrects the norepinephrine production deficiency through norepinephrine-modulated adrenergic activation in the hippocampus and restores learning and memory deficits in contextual fear-conditioning tasks in Ts65Dn mice. Treating Ts65Dn mice with xamoterol, a β_1-adrenergic receptor partial agonist, also targets norepinephrine deficiency and improves learning and memory deficits in the contextual fear-conditioning task.[46] These results implicate a correction of cognitive dysfunction in DS through restoration of norepinephrine-mediated neurotransmission.

CONCLUSION AND REMAINING ISSUES

The triplication of an extra chromosome and the resulting overexpression of over 500 genes affecting multiple organ systems suggested that DS would prove difficult to understand and treat. However, remarkable advances in medical intervention, progression in scientific research, and increased social care have deepened our understanding of DS, facilitated better health care, and pointed to potential therapeutic targets. Human studies assessing individuals with partial trisomy 21 have provided valuable insight into the contribution of various segments of the chromosome in causing specific DS-associated phenotypes. These findings have been extended and enhanced by studies in mouse models of DS, which have provided not only insights into genotype–phenotype correlations but also a deeper knowledge of the neurobiological pathways underlying pathogenic mechanisms. Collectively, this has led to considerable efforts to develop and validate neuronal targets for improving learning and memory deficits and addressing early-onset AD in DS. Consequently, several compounds are currently being developed for DS clinical trials. Importantly, however, no treatment has yet been approved and none of those reviewed above can be considered safe and effective at this time.

Despite incredible progress in understanding DS over the past half-decade, little is known about the contribution of specific genes and the neurobiological and molecular pathways they engage to cause clinical manifestation in DS. The development of mouse models that more closely recapitulate the gene dosage imbalance in DS and genome-wide association studies of people with DS will be pivotal in identifying dosage-sensitive trisomic genes and pathogenic mechanisms. An increased life expectancy in people with DS, which results in an increased prevalence of DS-associated symptoms, invites a marked increase in investment to tackle the emerging challenge of an aging DS population.

Acknowledgments

We acknowledge our grant sponsors: the National Institutes of Health (NS06672, NS24054, PN2EY016525), Down Syndrome Research and Treatment Foundation, Alzheimer's Association, Thrasher Research Fund, and Larry L. Hillblom Foundation.

References

1. Ward OC. John Langdon Down: the man and the message. *Downs Syndr Res Pract*. 1999;6(1):19–24.
2. Lejeune J, Gautier M, Turpin R. [Study of somatic chromosomes from 9 mongoloid children]. *C R Hebd Seances Acad Sci*. 1959;248(11):1721–1722.
3. Antonarakis SE. Parental origin of the extra chromosome in trisomy 21 as indicated by analysis of DNA polymorphisms. Down Syndrome Collaborative Group. *N Engl J Med*. 1991;324(13):872–876.
4. Dierssen M, Herault Y, Estivill X. Aneuploidy: from a physiological mechanism of variance to Down syndrome. *Physiol Rev*. 2009;89(3):887–920.
5. de la Torre R, Dierssen M. Chapter 1 – Therapeutic approaches in the improvement of cognitive performance in Down syndrome: past, present, and future. In: Mara D, Rafael De La T, eds. *Progress in Brain Research*; vol. 197. Oxford: Elsevier; 2012:1–14.
6. Hattori M, Fujiyama A, Taylor TD, et al. The DNA sequence of human chromosome 21. *Nature*. 2000;405(6784):311–319.
7. Nadel L. Down's syndrome: a genetic disorder in biobehavioral perspective. *Genes Brain Behav*. 2003;2(3):156–166.
8. Marin-Padilla M. Pyramidal cell abnormalities in the motor cortex of a child with Down's syndrome. A Golgi study. *J Comp Neurol*. 1976;167(1):63–81.
9. Haydar TF, Reeves RH. Trisomy 21 and early brain development. *Trends Neurosci*. 2012;35(2):81–91.
10. Tejedor FJ, Hammerle B. MNB/DYRK1A as a multiple regulator of neuronal development. *FEBS J*. 2011;278(2):223–235.
11. Bittles AH, Bower C, Hussain R, Glasson EJ. The four ages of Down syndrome. *Eur J Public Health*. 2007;17(2):221–225.
12. Lott IT. Chapter 6 – Neurological phenotypes for Down syndrome across the life span. In: Mara D, Rafael De La T, eds. *Progress in Brain Research*; vol. 197. Oxford: Elsevier; 2012:101–121.
13. Edgin JO, Mason GM, Allman MJ, et al. Development and validation of the Arizona Cognitive Test Battery for Down syndrome. *J Neurodev Disord*. 2010;2(3):149–164.
14. Braak H, Braak E. Neuropathological staging of Alzheimer-related changes. *Acta Neuropathol*. 1991;82(4):239–259.
15. Mann DM. The pathological association between Down syndrome and Alzheimer disease. *Mech Ageing Dev*. 1988;43(2):99–136.
16. Mann DM, Esiri MM. The pattern of acquisition of plaques and tangles in the brains of patients under 50 years of age with Down's syndrome. *J Neurol Sci*. 1989;89(2–3):169–179.
17. Wisniewski KE, Wisniewski HM, Wen GY. Occurrence of neuropathological changes and dementia of Alzheimer's disease in Down's syndrome. *Ann Neurol*. 1985;17(3):278–282.
18. Cabrejo L, Guyant-Marechal L, Laquerriere A, et al. Phenotype associated with APP duplication in five families. *Brain*. 2006; 129(Pt 11):2966–2976.
19. Prasher VP, Farrer MJ, Kessling AM, et al. Molecular mapping of Alzheimer-type dementia in Down's syndrome. *Ann Neurol*. 1998;43(3):380–383.
20. Holtzman DM, Santucci D, Kilbridge J, et al. Developmental abnormalities and age-related neurodegeneration in a mouse model of Down syndrome. *Proc Natl Acad Sci U S A*. 1996;93(23):13333–13338.
21. Salehi A, Delcroix JD, Belichenko PV, et al. Increased App expression in a mouse model of Down's syndrome disrupts NGF transport and causes cholinergic neuron degeneration. *Neuron*. 2006;51(1):29–42.
22. Cataldo AM, Petanceska S, Peterhoff CM, et al. App gene dosage modulates endosomal abnormalities of Alzheimer's disease in a segmental trisomy 16 mouse model of down syndrome. *J Neurosci*. 2003;23(17):6788–6792.
23. Lott IT, Doran E, Nguyen VQ, Tournay A, Movsesyan N, Gillen DL. Down syndrome and dementia: seizures and cognitive decline. *J Alzheimers Dis*. 2012;29(1):177–185.
24. Visootsak J, Sherman S. Neuropsychiatric and behavioral aspects of trisomy 21. *Curr Psychiatry Rep*. 2007;9(2):135–140.
25. Lott IT, Dierssen M. Cognitive deficits and associated neurological complications in individuals with Down's syndrome. *Lancet Neurol*. 2010;9(6):623–633.
26. Das I, Reeves RH. The use of mouse models to understand and improve cognitive deficits in Down syndrome. *Dis Model Mech*. 2011;4(5):596–606.
27. Ruparelia A, Wiseman F, Sheppard O, Tybulewicz VLJ, Fisher EMC. Down syndrome and the molecular pathogenesis resulting from trisomy of human chromosome 21. *J Biomed Res*. 2010;24(2):87–99.
28. Ruparelia A, Pearn ML, Mobley WC. Cognitive and pharmacological insights from the Ts65Dn mouse model of Down syndrome. *Curr Opin Neurobiol*. 2012;22(5):880–886.
29. Busciglio J, Pelsman A, Helguera P, et al. NAP and ADNF-9 protect normal and Down's syndrome cortical neurons from oxidative damage and apoptosis. *Curr Pharm Des*. 2007;13(11):1091–1098.
30. Toso L, Cameroni I, Roberson R, Abebe D, Bissell S, Spong CY. Prevention of developmental delays in a Down syndrome mouse model. *Obstet Gynecol*. 2008;112(6):1242–1251.
31. Clark S, Schwalbe J, Stasko MR, Yarowsky PJ, Costa AC. Fluoxetine rescues deficient neurogenesis in hippocampus of the Ts65Dn mouse model for Down syndrome. *Exp Neurol*. 2006;200(1):256–261.
32. Bianchi P, Ciani E, Guidi S, et al. Early pharmacotherapy restores neurogenesis and cognitive performance in the Ts65Dn mouse model for Down syndrome. *J Neurosci*. 2010;30(26):8769–8779.
33. Bianchi P, Ciani E, Contestabile A, Guidi S, Bartesaghi R. Lithium restores neurogenesis in the subventricular zone of the Ts65Dn mouse, a model for Down syndrome. *Brain Pathol*. 2010;20(1):106–118.
34. Kleschevnikov AM, Belichenko PV, Gall J, et al. Increased efficiency of the $GABA_A$ and $GABA_B$ receptor-mediated neurotransmission in the Ts65Dn mouse model of Down syndrome. *Neurobiol Dis*. 2012;45(2):683–691.
35. Belichenko PV, Kleschevnikov AM, Salehi A, Epstein CJ, Mobley WC. Synaptic and cognitive abnormalities in mouse models of Down syndrome: exploring genotype–phenotype relationships. *J Comp Neurol*. 2007;504(4):329–345.
36. Kleschevnikov AM, Belichenko PV, Villar AJ, Epstein CJ, Malenka RC, Mobley WC. Hippocampal long-term potentiation suppressed by increased inhibition in the Ts65Dn mouse, a genetic model of Down syndrome. *J Neurosci*. 2004;24(37):8153–8160.
37. Fernandez F, Morishita W, Zuniga E, et al. Pharmacotherapy for cognitive impairment in a mouse model of Down syndrome. *Nat Neurosci*. 2007;10(4):411–413.
38. Rueda N, Florez J, Martinez-Cue C. Chronic pentylenetetrazole but not donepezil treatment rescues spatial cognition in Ts65Dn mice, a model for Down syndrome. *Neurosci Lett*. 2008;433(1):22–27.
39. Braudeau J, Dauphinot L, Duchon A, et al. Chronic treatment with a promnesiant GABA-A alpha5-selective inverse agonist increases immediate early genes expression during memory processing in mice and rectifies their expression levels in a Down syndrome mouse model. *Adv Pharmacol Sci*. 2011;2011:153218.
40. Roche. Roche starts early stage clinical trial in Down syndrome. http://www.rocheusa.com/portal/usa/press_releases_nutley?siteUuid=re7180004&paf_gear_id=38400020&pageId=re7-425113&synergyaction=show&paf_dm=full&nodeId=1415-fbfa4d-37db2611e0953b3d6bec9c2782¤tPage=0. September 9, 2011.
41. Ferris SH. Evaluation of memantine for the treatment of Alzheimer's disease. *Expert Opin Pharmacother*. 2003;4(12):2305–2313.

42. Rueda N, Llorens-Martin M, Florez J, et al. Memantine normalizes several phenotypic features in the Ts65Dn mouse model of Down syndrome. *J Alzheimers Dis*. 2010;21(1):277–290.
43. Costa AC, Scott-McKean JJ, Stasko MR. Acute injections of the NMDA receptor antagonist memantine rescue performance deficits of the Ts65Dn mouse model of Down syndrome on a fear conditioning test. *Neuropsychopharmacology*. 2008;33(7):1624–1632.
44. Lockrow J, Boger H, Bimonte-Nelson H, Granholm AC. Effects of long-term memantine on memory and neuropathology in Ts65Dn mice, a model for Down syndrome. *Behav Brain Res*. 2011;221(2): 610–622.
45. Hanney M, Prasher V, Williams N, et al. Memantine for dementia in adults older than 40 years with Down's syndrome (MEADOWS): a randomised, double-blind, placebo-controlled trial. *Lancet*. 2012;379(9815):528–536.
46. Salehi A, Faizi M, Colas D, et al. Restoration of norepinephrine-modulated contextual memory in a mouse model of Down syndrome. *Sci Transl Med*. 2009;1(7):7ra17.

CHAPTER

6

Autism Spectrum Disorder

James C. Harris

Psychiatry and Behavioral Sciences, Pediatrics, Mental Health, and History of Medicine, The Johns Hopkins University School of Medicine, Baltimore, Maryland, USA

OUTLINE

INTRODUCTION

Leo Kanner first described autism in 1943 in his classic paper "Autistic disturbances of affective contact". Kanner wrote that we must "assume these children have come into the world with the innate inability to form the usual, biologically provided contact with people, just as other children come into the world with innate physical or intellectual handicaps".[1] Lack of interest in social contact and other characteristics that came to define the syndrome, such as delayed and deviant language development, restricted interest in activities, and stereotypical and repetitive patterns of behavior, were described in the first case report. Thus, from its first description autism was proposed as a neurobiological disorder.

Neurobiology of Brain Disorders
http://dx.doi.org/10.1016/B978-0-12-398270-4.00006-9

In the following year Kanner proposed the diagnostic category "early infantile autism", terminology that was subsequently used in the third version of the *Diagnostic and Statistical Manual of Mental Disorders* (DSM-III) in 1980.[2] Kanner excluded cases with known brain dysfunction and/or severe intellectual disability in making this diagnosis. Subsequently, the diagnosis was broadened to include infants, children, and adults at all cognitive levels as well as those with neurogenetic syndromes if they met behavioral diagnostic criteria. Thus, the current diagnosis is a broadly heterogeneous grouping. It includes individuals ranging from a severely intellectually disabled and non-verbal child with motor stereotypies and self-injury to a computer engineer with high-functioning autism and fluent language who is self-centered, has difficulty in gauging others' emotions, and exhibits ideational perservation about his obscure interests.

The neurobiology of affective contact and stereotyped/repetitive behavior in affected children remains a major theme in autism research. Autism may provide cues to better understanding how social cognition and affective engagement emerge in development and the functioning of social neuronal networks in the brain.

DSM-5, the current classification,[3] introduces a new term, autism spectrum disorder (ASD), which replaces the earlier DSM diagnostic category of pervasive developmental disorders. Moreover, it collapses the subgroups listed in DSM-IV (autistic disorder, Asperger disorder, Rett disorder, childhood disintegrative disorder, and pervasive developmental disorder not otherwise specified) into this one broad category. ASD is entirely defined by behaviors in DSM-5. Moreover, DSM-5 collapses the three core diagnostic domains in DSM-IV into two domains. In DSM-5 these are (1) deficits in social communication and social interaction, and (2) restricted and repetitive patterns of behavior, interests, and activities.

Ultimately, the identification of biomarkers – genetic, biochemical, physiological, or anatomical – that are specific to one or more of the features of ASD is expected to resolve controversies about clinical classification. Studies of identical twins make it clear that there is high heritability for ASD. However, ASD is not inherited in a simple Mendelian fashion. Many genes have been identified that may contribute to risk. Current neurobiological research focuses on social and affective neuroscience. Studies of affective development, social cognition, interpersonal reciprocity, and repetitive behaviors using neurobiological measures are guiding themes in research.

This chapter reviews the history, clinical features, classification, epidemiology, course, differential diagnosis, assessment, diagnostic instruments, etiologies, models, developmental issues, neurobiology, and treatment of ASD. Although most of the research summarized here comes from studies of typical autism, consideration is given to the full range of severity. The main emphasis is on understanding ASD from a neurobiological perspective as a grouping of disorders of early brain development that begin *in utero* but dynamically change over time and continue throughout life.[4]

HISTORY

The descriptive term "autism" was chosen by Kanner to emphasize the sense of social aloneness apparent to those who observed children with this condition. "Autism" refers to paucity of social self-awareness in relationship to others and in the use of the imagination. Kanner[1] described case histories of 11 children, 2–8 years of age, who presented with previously unreported behavior. He described them as socially remote, insistent on maintaining sameness in their environment, and with stereotypies and echolalia, the repeating of speech sounds made by others. These children failed to initiate socially meaningful anticipatory gestures. They did not reach to be picked up, ignored animate people in the environment, and appeared to be "in a world of their own". Kanner's initial report documented similarities in their behavior. His follow-up of these cases 28 years later documented the differences among them.[5]

Despite Kanner's early description of its features, autism was classified as a childhood form of schizophrenia in DSM-I. It was not until 1980 with the publication of DSM-III that specific diagnostic criteria for autism were introduced in a DSM. Before DSM-III, "childhood schizophrenia" or "childhood psychosis" was broadly used to diagnose children with severe psychiatric disturbances beginning in early life. Research in the 1970s made clear that autism was distinct from childhood schizophrenia based on the age of onset, symptom presentation, and clinical course.[6,7] Thus, 37 years after the original description of infantile autism (with onset before age 30 months) as a diagnostic category, it entered the official classification system. In DSM-III the term pervasive developmental disorder was introduced to describe deviant development in multiple developmental lines involving social skills, language, attention, and perception. The DSM-III definition specified an age of onset before 30 months of age, pervasive lack of responsiveness to others, deviant language development, unusual responses to the environment, and the absence of hallucinations and delusions as found in schizophrenia (see Chapter 39). Further revisions were made in DSM-III-R in 1987 because the original criteria applied best to younger and more severely impaired individuals and were considered too restrictive.

The DSM-III-R recognized the importance of changes in syndrome expression during development and included more developmentally focused criteria, leading to a change in the name of the category from "infantile autism" to "autistic disorder". Furthermore, its differentiation from schizophrenia was further clarified so that an individual with an autistic disorder might have both diagnoses if the additional diagnostic criteria for schizophrenia, such as the presence of hallucinations and delusions, were met. The major change in DSM-III-R was the introduction of the developmentally focused criteria. Yet DSM-III-R broadened the concept of autistic disorder substantially so that more false-positive cases were reported. DSM-III-R identification of more atypical cases complicated its use for both clinical and research purposes, leading to additional modifications in DSM-IV.

The changes introduced in DSM-IV in 1994 were developed to provide greater simplicity in diagnosis while maintaining compatibility with the 10th revision of the International Statistical Classification of Diseases and Related Health Problems (ICD-10) of the World Health Organization, but with greater emphasis on clinical judgment. Moreover, additional categories were added in DSM-IV under the "pervasive developmental disorder" terminology to include Rett disorder, childhood disintegrative disorder, and Asperger disorder.

"Asperger disorder" was introduced in DSM-IV based on renewed interest in people with high functioing autism. Hans Asperger, in 1944, described the clinical presentation of four children whose intelligence was in the normal range, with good grammar and vocabulary, but who were odd socially, and had poor non-verbal communication, limited interests, and poor social communication.[8] Asperger's paper (in German) drew little attention until 1981, when Lorna Wing brought it to general attention.[9] In DSM-5 Asperger disorder was incorporated under the umbrella term ASD.

CLINICAL FEATURES

Deficits in Social Communication and Interaction

ASD is characterized by persistent impairment in reciprocal social interaction and communication. The severity and nature of the social deficit vary with the child's age and developmental level but the deficit is present from very early childhood and impairs functioning at home, in school, and in the community. Because there is a range of presentations depending on developmental level, severity, and chronological age, the term "spectrum" is used. The impairment is sustained throughout the lifetime, however, compensation for some clinical features may occur across development thus the condition is dynamic, not static.

In infancy, children with ASD may resist cuddling and not mold to the parent when held. As toddlers and during their preschool years, they often ignore others, even bumping into them or walking over them as if they were unaware of their existence. They may not turn in recognition of being called by name or look at or towards someone seeking to engage them in conversation. Gaze avoidance may continue into school age and even into adulthood in a less striking form. Lower functioning individuals may be mute. Others who speak are one-sided in conversation and do not engage in reciprocal social exchange. Thus, they ignore the social conventions of taking turns in conversation and waiting for a reply before speaking again. As adults their deficits in social emotional reciprocity may be most apparent when having to respond to complex social cues, not knowing when to initiate a conversation or how to sustain one with others, and not appreciating what is socially intuitive in typical development.

Higher functioning children, adolescents, and adults may have learned many social skills but continue to have social deficits. These are recognized through socially intrusive behaviors, an inappropriate lack of awareness of others' feelings, and general misunderstanding of the negative impact that their behavior can have on others. This may result from limited ability to interpret the tone of voice or facial expression of another person. These higher functioning individuals have difficulty making friends when they wish to do so and in engaging others in play. Those who seek friends may be ostracized for their social awkwardness when they attempt to socialize. Thus, people with an ASD are rigid and often stereotyped in their social responses and need to be taught simple social rules and patterns of proper interacting, such as greeting another person. The verbal individual with an ASD may learn social rules, often by rote, but not apply learned rules appropriately in a social context.

The failure to acquire language at the chronologically expected age is the most frequent presenting concern by parents of preschool children with ASD. Early in life, children who are verbal may be echolalic, that is, they repeat a question back rather than responding to it (immediate echolalia). Such echolalia, a repetitive behavior, may be associated with a reversal of pronouns, that is, the child refers to himself as "you" or by name, rather than using the word "I" appropriately in conversation.

Most affected preschool children have some type of developmental language difficulty. These difficulties are not simply in language expression but often involve impaired comprehension and pragmatic use of language. Some may be mute, and others may have problems understanding conversation directed towards them. Others do develop language but speak unintelligibly or do not use appropriate sentence structure. Those who speak late may use jargon that does not have communicative intent. This includes phrases they have heard or memorized information from cartoons or television commercials. Verbal children may speak in a monotone, too softly or loudly, or in a singsong manner. Typically, there is a deficit in the use of speech rhythm and intonation (prosody) and failure to question to clarify the meaning of another's speech.

An abnormality in inner language development is also a characteristic feature and is most often demonstrated in observations of play routines that reveal the lack of flexibility in inner language use and imagination that is characteristic in ASD.

Acquired speech fluency can be misleading because many children with ASD lack comprehension of what is said to them. This is especially evident when questions are addressed to them about their personal life experiences. Others show relatively normal language development and speak more appropriately; however, they are often preoccupied with a narrow range of their own favorite topics. They pursue these topics in conversation, showing little regard for the interests or responses of the person with whom they are speaking. Moreover, they may perseverate by asking the same questions over and over again, even though they have already heard the answers. In some instances, they may repeat and recite phrases they have heard. In doing so they may exactly imitate the tone of voice and rhythm of the original speaker.

Deficits in pragmatic language use, a form of non-verbal communication, are most evident when the affected child does not use gestures in initiating social contact or conversation. Young children with ASD generally do not initiate anticipatory gestures, for example, to be picked up when approached or when they approach others. They demonstrate limited or no joint attention when engaged. This is shown through not using eye contact to engage another person when pointing to an object. They do not bring items to show to others and fail to follow the gaze of an adult looking towards an object.

Although children typically begin to point to things they like with one finger at about 9 or 10 months and begin shaking their head "no" by 1 year of age, affected children are limited in developing such non-verbal behaviors. Instead of pointing at a desired object, they may seek out objects for themselves or move the parent's hand towards the preferred object. Lacking the use of gesture for their communicative intent, an affected child may become distressed and cry or have a tantrum until an adult has guessed, often by trial and error, what the child seeks or needs. They must be taught how to participate in a person-to-person conversation: to look at the conversational partner; to interpret tone of voice, facial expression, and body language; to maintain the topic in conversation; to take turns.[10]

Restricted, Repetitive Patterns of Behavior, Interests, or Activities

Children with diagnoses of ASD routinely show restricted patterns of behaviors, interests (ideological perseveration), and activities. The presentation varies with age, cognitive ability, and the extent of environmental support. Behaviors may be simple repetitive patterns of movement (stereotypies and mannerisms) that include hand flapping (especially when excited), twirling, rocking, head banging, finger posturing, and sensory preoccupations. More complex behaviors can include repetitive use of objects (lining up toys), repetitive speech (echolalia, stereotyped use of phrases), and restricted interests (preoccupation with train schedules) and activities (insistence in following the same route when traveling).

Commonly, affected children may resist changes in routines in their everyday environment, preferring to maintain sameness. For example, they may be quite distressed when a familiar object is moved to a new location. Repetitive use of objects is common. This includes flicking a string, turning light switches on and off, repetitively tearing paper into shreds, or turning a toy car over and spinning the wheels rather than rolling it along. Some children may become preoccupied with letters and sound out words (hyperlexia) without understanding their meaning. Others may repetitively turn the pages in the telephone book. Some have verbal stereotypies, repeating the same statements over and over. When efforts are made to change or break their routines the child may resist and have a tantrum.

Some affected children and adults show hypersensitivity or hyposensitivity to sensory input. Thus, some inappropriately stroke silk stockings, repetitively smell objects, have odd food preferences, react negatively to changes in lighting, hold their hands over their ears to block out certain sounds (a vacuum cleaner, sirens), and show increased or decreased sensitivity to pain.

A lack of imagination in play is apparent in ASD from the preschool years. Play figures may be manipulated, used for self-stimulation, lined up, and used in repetitive ways rather than with imagination. Recognition that figurines, used in play, represent people is delayed. Higher functioning children can show forms of pretend play, such as feeding a stuffed animal or putting it to sleep. However, such pretend play tends to be repetitious and lacking in flexibility.

Verbal children, adolescents, and adults often become preoccupied with and become an expert through practice in very limited topics, such as making maps or repetitively copying timetables. Once a preoccupation is mastered they may verbally perseverate on a preferred theme and speak about their interest continually and inappropriately. As they grow older, many adolescents and adults who are higher functioning [those with higher intelligence quotient (IQ) and better language] learn to suppress their repetitive behaviors in public but pursue them in private. Such interests may continue to be a source of pleasure for them and may be used as motivators in behavioral treatment and education.[10]

DEFINITION AND CLASSIFICATION

In DSM-5 clinical features of ASD are summarized as persistent deficits in social communication and social interaction across multiple settings – home, school, or community – and restricted, repetitive patterns of behavior, interests, or activities. These include deficits in social reciprocity, non-verbal communicative behaviors used for social interaction, and skills in developing, maintaining, and understanding relationships. Because behaviors change with development, diagnostic criteria may be met based on a history of having met the criteria in the past.

Specifiers and moderators replace DSM-IV subgroups. Thus, within the diagnosis of ASD individual clinical characteristics are recorded using specifiers that include (1) with or without accompanying intellectual impairment, (2) with or without accompanying structural language impairment, (3) associated with a known medical/genetic or environmental/acquired condition, and (4) associated with another neurodevelopmental, mental, or behavioral disorder. Other specifiers focus on the autistic symptoms themselves (age at first recognition; onset with or without loss of established skills) and severity. Severity is based on the level of support needed for each of the two domains, social communication impairments, and restrictive, repetitive patterns of behavior. Table 2 in the DSM-5 manual gives examples for three levels of support for each domain: level 1, requiring support; level 2, requiring substantial support; and level 3, requiring very substantial support. The severity of each of the domains is recorded separately.

Specifiers provide clinicians with an opportunity to individualize the diagnosis and communicate a more complete clinical description of an affected person to others. For example, many individuals previously diagnosed with Asperger disorder in DSM-5 would be diagnosed as ASD without language or intellectual impairment. Specifiers are needed for a full clinical characterization. Specifiers can also be used to facilitate identifying subgroups for research case identification. It is expected that new research studies will require additional specifiers when developing research protocols.[11] Those wanting more detail about the diagnostic criteria for ASD should consult the DSM-5.[3]

EPIDEMIOLOGY

The prevalence of individuals diagnosed with ASD has risen substantially since the 1980s.[12] The rate for classic autism based on community samples in the 1960s was 2–4 per 10,000. Currently, the rate of the ASD (including classic autism) is 30–100 per 10,000. For classic autism it is 13–30 per 10,000.[13] The major rise in prevalence is not fully accounted for by identifying ASD associated with severe intellectual disability. It is more likely that more individuals are being identified with non-verbal IQ in the normal range help to account for the difference.[14]

This substantial increase in rates is documented in the USA, Japan, Scandinavia, and several other European countries. Because the increase is found in so many different countries it seems unlikely that the cause is environmental because an environmental factor would have to act simultaneously across diverse settings. The increase in rate started in the 1980s when the diagnostic term "pervasive developmental disorder" was introduced, broadening the diagnosis. With the change in definition, more high-functioning cases with better language skills and higher IQ were diagnosed. The focus in diagnosis shifted from a focus on a severe, highly deviant, discontinuous grouping to a continuum of deficits that were less severe. This new dimensional view reduces boundaries and makes differences from other developmental disorders and psychiatric syndromes more difficult to establish.[15]

A substantial proportion of the increase is believed to be the result of changes in diagnostic practices. Because of changes in criteria over time, trend analyses of datasets or of national registries are not sufficient to determine differences. Exploration of environmental risk factors should be based on prospective cohort or population-based case–control studies.[15] Thus, increased awareness of the diagnosis, inclusion of subthreshold cases, and changes in study methodology using systematic standardized instruments are important factors in understanding the increasing rates.[14]

The claim that increased prevalence is linked to the measles–mumps–rubella (MMR) vaccine or the mercury-containing preservative thimerosal in the vaccines has no empirical support.[16] The original 1988 paper making this claim was retracted by the journal *The Lancet* as scientifically flawed, and well-designed trials and meta-analyses have found no evidence of any association. In Japan, removal of MMR vaccine from use was not followed by any fall in the rate of autism, or even by a reduction in the rate of rise. Similarly, the discontinuation of use of the vaccine preservative thimerosal in Scandinavia was not followed by any change in the rising rate of autism. Although there may be a link to other prenatal or postnatal factors or other toxins, so far there has been no confirmation of the involvement of any environmental factor in the rising rates of ASD. One element that could contribute to a rise in rates, if one truly has occurred, may be the rising age of parenthood. There is evidence that increased parental age is associated with ASD; thus, older parenting may contribute to an increased rate of autism. Older parental age, particularly paternal age, is correlated with an increased risk of copy number variation (CNV) (submicroscopic chromosomal duplications or deletions) and of *de novo* single nucleotide variation

(SNV) (or mutation).[14,17] An increase in both CNV and deleterious *de novo* variants is reported in autism.[18]

ASD is diagnosed four times more often in males than in females. The reason for male vulnerability for ASD and other neurodevelopmental disorders is an area of ongoing interest. Females with an ASD diagnosis are more likely to have a co-occurring intellectual disability. Girls without intellectual impairments or language delays may go unrecognized, potentially because of subtler social or communication problems.[14]

Intellectual deficits co-occurs in a substantial number of ASD cases. The prevalence of intellectual deficits in classic autism is approximately 60%, with rates of severe and profound ID deficits ranging from 8% to as high as 40%. This raises issues about when to diagnose and how to apply the criteria, because the social deficit may lack specificity in low-functioning children. In contrast, when the full autism spectrum is considered, the rate of intellectual deficits is approximately 30 to 40%.

Approximately 5–10% of ASD cases are associated with a variety of neurogenetic disorders. When large community samples of cases of ASD are surveyed, neurogenetic syndromes are rare. Certain syndromes such as tuberous sclerosis and fragile X syndrome (see Chapter 8) include significant numbers (30%) who meet ASD criteria.

Finally, epilepsy (see Chapter 17) is associated with autism, with estimates of the occurrence of seizure that vary from 5 to 44%. Onset of epilepsy follows a bimodal distribution, with some children presenting with epilepsy in the first few years of life, often preceding the diagnosis of autism, and more commonly, others developing epilepsy in the teenage years. Signs of epilepsy may be present on electroencephalograms in children with ASD who have no clinical evidence of seizures. It is unclear whether these children are at higher risk of developing epilepsy later in life.

NATURAL HISTORY

There is considerable heterogeneity in the age of recognition of an ASD. Most children with an ASD are recognized in the second year of life. However, parents may report non-specific problems earlier with feeding, settling, and sleep. There may be limited responsiveness to others, reduced anticipatory gestures, and excessive quietness during infancy, or, in contrast, excessive irritability and screaming. If there is another child in the family with an ASD diagnosis, parents may be more sensitive in recognizing social deficits. Home videos from 12–18 months or earlier in life may document subtle abnormalities of development.

In most children with ASD, early language development is deviant, with difficulties in both the comprehension of speech and the expression of ideas. Although markedly impaired in verbal and non-vocal symbolic processes, affected children may be good at non-symbolic matching and assembly tasks.

Approximately one-quarter to one-third of all children with ASD in studies carried out in the USA, the UK, and Japan lose previously acquired language skills, most often between 18 and 24 months. Attention deficits and new stereotypical movements sometimes accompany loss of language use. These children may initially show near-normal development until as late as 18–24 months of age, when they regress. However, their clinical picture is generally indistinguishable from those with early onset.

Academic achievement and social adaptability may improve with special education. Verbal skills are the best predictor of social adaptive functioning. Academic achievement is related to intellectual functioning, and early non-verbal IQ shows a positive relationship to outcome. However, academic achievement declines when task demands for abstract reasoning exceed rote memory skills. Academically high-functioning children may be returned to special education classes because of deficits in interpersonal skills. For example, high-functioning affected children who have been mainstreamed in grade school sometimes return to special education during the high-school years.

Approximately 15–20% of children who show some autistic behavior in their early years gradually emerge from autistic social withdrawal. Some make a relatively good social adjustment although they may continue to have unusual and eccentric behaviors as adults. Adult adjustment is judged based on the capability for independent living and employability. Moderately impaired individuals may work successfully in areas where their careful attention to detail and preoccupations can be channeled into jobs requiring completion of repetitive tasks. They may be most successful in job settings that require the least interaction with others. Other employees who are aware of an obvious disability may help and support them. Achievement by the group who are most successful academically may be limited by their social deficits, particularly by difficulty in language comprehension and poor judgment in social situations. Employers may not appreciate the extent of their limitations in social problem solving and social adjustment. Successful social adjustment requires self-awareness of difference from others on the part of the person with ASD as well as special education programs specifically focused on his or her social, language, and cognitive impairment.

In one study,[19] the outcome of a supported employment program over a period of 8 years was examined for adults with autism or Asperger syndrome (with an IQ of 60 or above). Approximately two-thirds found employment. Of the 192 jobs obtained, most were permanent contracts and involved administrative, technical, or computing work. IQ, language skills, and educational attainments predicted success. Supportive employment is also beneficial for those with lower abilities. Job coaches are used to provide support for them at the job site.

Affective Development

Kanner's initial report focused on deficits in affective contact with others.[1] Learning to recognize and respond to another's emotional state is a developmental milestone. When affective development is monitored in children with ASD, aggression towards others, sadness when frustrated, and apparent joy when pursuing interests are reported. Although fears, for example of animals, may be expressed, general awareness of risk and danger and an understanding of dangerous situations are lacking. The transition from lack of social awareness to social engagement and perplexity about social situations may be heralded by exaggerated emotions and behavioral difficulties, including intrusive behavior.

Studies of affective development address the ability to comprehend affect in social relationships. Overall, people with an ASD do not attend to faces or use information from faces, as do typically developing children. In the preschool years (ages 2–4), children with ASD show fewer observed intervals of affective response, positive or negative, when interacting with familiar adults than age-matched controls. They may perform better than matched control subjects in recognizing faces that are shown to them upside down, suggesting that the lower portion of the face, rather than the upper portion, is used in facial recognition of another person. Thus, they may focus more on the mouth than on the eyes to identify facial expressions in others. They have more difficulty in matching videotaped segments or pictures of gestures or vocalizations, and more trouble understanding the situational context of photographed or drawn pictures of facial expressions.[20,21] If given a choice on which cue to use to identify others, affected children may use items of dress, such as hats, rather than facial expression to identify. They are less able to imitate an emotion when asked to do so, or to imitate an affect demonstrated by another person. Affective responses are generally reported to be flat and not contingent on the particular situation observed. Familiar teachers and caretakers suggest that children with ASD show all facial expressions except for surprise. When they look at themselves in the mirror, children with ASD show less positive affect and less self-consciousness than control children.

Psychiatric Disorders and Forensic Issues

As children with ASD reach adulthood, they often develop co-occurring psychiatric disorders. Depression and anxiety (discussed in Chapters 43 and 37, respectively) are particularly common. Depressive symptoms lead to poorer global functioning.[22] In one follow-up study of 135 children to age 21 years, 16% developed a definite new psychiatric disorder.[23] Five were diagnosed with an obsessive–compulsive disorder and/or catatonia. Eight were diagnosed with an affective disorder with marked obsessional features and three with complex affective disorder. One was diagnosed with bipolar disorder (discussed in Chapter 40) and one an acute anxiety state complicated by alcohol excess. There were no cases of schizophrenia, consistent with earlier studies that these are distinct conditions. However, there are reports of psychotic symptoms with hallucinations and delusions that tend to be diagnosed as brief reactive psychosis.

Because of their inappropriate social behavior, people with ASD may have legal problems. These range from misunderstandings caused by socially inappropriate behavior with strangers to apparent criminal activity linked to their obsessional interests and perseverations (see Chapter 38), and violent outbursts. Their deficits are important in forensic evaluations of culpability and determination of ASD as a mitigating factor when determining criminal responsibility.[24] This is particularly important when considering whether incidents of criminal or violent behavior are intended or unintended. Three characteristic deficits that are pertinent in forensic settings are theory of mind (understanding another person's perspective), regulation of emotions, and moral reasoning (understanding the consequences of one's actions). Published studies suggest rates of about 2% in a special hospital for criminal offenders for high-functioning ASD, although in one study violent behavior was rare.[24]

Predictors of Outcome

In children with an ASD, language skills are the best predictors of social outcome. Because of the frequent wide discrepancy in verbal and non-verbal IQ, non-verbal IQ is frequently used in determining outcome. Those with non-verbal IQ in the normal range have the best outcomes. Those with non-verbal IQ less than 50 in the preschool years are far less likely to acquire useful spoken language and have an increased risk of poor social functioning during adolescence and adulthood.[4]

Higher IQ and language skills alone are not sufficient to ensure good social outcomes. Targeted psychoeducational experiences beginning early in life are needed, and care must be directed towards ensuring appropriate transitions from primary to secondary school and into adult life. Even higher functioning people with ASD may find gaining employment and living independently challenging, and the support of the family and social agencies may be needed to maintain life in the community.

Key to good outcomes is appropriate early intervention. This entails understanding the autistic nervous system. Early interventions target social engagement. In early life, focused interventions may begin with facilitating joint attention,[25] verbal imitation, and social communicative aspects of adaptive skills.

DIFFERENTIAL DIAGNOSIS

Selective Mutism

In selective mutism, children speak normally in the home environment but do not speak in one or more other settings. In these children, early development is normal and includes appropriate social engagement and communication skills. Even when mute, the child demonstrates social reciprocity and does not show restricted or repetitive patterns of behavior and stereotypies.

Language Disorders

With language disorders, difficulty in communication may be associated with secondary social problems. However, specific language disorders (receptive and expressive) ordinarily are not associated with deficits in non-verbal communication, nor are they associated with restricted, repetitive patterns of behavior, interests, or activities.

Social (Pragmatic) Communication Disorder

Social (pragmatic) communication disorder[3] is a new diagnosis in DSM-5. As a disorder of pragmatic language it involves persistent deficits in social use of language and communication (e.g. greeting others, sharing information socially), impaired flexibly in changing context to meet the needs of the listener in a conversation, and impaired ability to follow the rules of social reciprocity in conversation (taking turns in speaking and listening, using verbal and non-verbal signals to regulate social conversation, making inferences, and understanding metaphor and humor in social exchange). An individual with a diagnosis of social communication disorder differs from a person with an ASD because he or she does not show restricted and repetitive behavior or interests. An ASD diagnosis supersedes that of social communication disorder if the criteria for ASD are met. When there are deficits in social communication it is essential to ask about past or current restricted or repetitive behavior to rule out ASD.

Intellectual Disability (Intellectual Developmental Disorder)

Behavior in individuals with intellectual disabilities may be difficult to differentiate from ASD in very young children and those with severe and/or profound levels of intellectual disability because of a failure to have developed language or symbolic skills and because simple repetitive behavior can occur in cases of intellectual disability. The diagnosis may be made when social communication and interaction are significantly impaired relative to the developmental level of the individual's non-verbal skills (e.g. fine motor skills, non-verbal problem solving). Conversely, intellectual disability is the appropriate diagnosis if there is no apparent discrepancy between the level of social communicative skills and other cognitive and intellectual skills.

Stereotypic Movement Disorder

Because motor stereotypies are diagnostic features of ASD, the diagnosis of stereotypic movement disorder is not made when repetitive behaviors are better explained by ASD. Moreover, certain repetitive behavior such as lining up objects and patterns of hand flapping are more consistently found in ASD. However, when stereotypies causing self-injury are a focus of treatment, both diagnoses may be given.

Attention Deficit/Hyperactivity Disorder

Deficits in executive functioning, sustained attention, overly focused attention, easy distractibility, and hyperactivity are common in individuals with an ASD [see Chapter 4 for a presentation of attention deficit/hyperactivity disorder (ADHD)]. When criteria are met in DSM-5, both diagnoses may be given when attention dysregulation, impulsiveness, or hyperactivity exceeds that typical for children of comparable mental age.

Schizophrenia

In early-onset child schizophrenia (see Chapter 39), a prodromal state with social impairment and atypical interests and beliefs may occur, which may be confused with the social deficits identified in ASD. However, hallucinations and delusions, which are defining features of schizophrenia, are not characteristic of an ASD. In differential diagnosis clinicians must recognize that thinking in individuals with an ASD is generally concrete and associational. For example, when responding to questions such as "Do you hear voices when no one is there?" a person may respond concretely (e.g. "Yes [on the radio]"), or when asked what you do when you cut your finger may make an associational response, "San Diego Clippers" and proceed to list statistics for each team member. Such responses are the result of a language disorder in ASD and do not indicate schizophrenia.

Rett Syndrome

Rett syndrome is a rare genetic disorder of known etiology (disruption of the X-linked gene *MECP2*, a translational repressor) that was categorized in DSM-IV as a pervasive developmental disorder (see Chapter 7). Unlike ASD, it occurs almost entirely in girls (rather than boys) and is phenotypically distinct from ASD. Moreover, unlike in ASD, where there may be a period of accelerated brain growth and macrocephaly, in Rett syndrome there is microcephaly and slowing of brain grown. In Rett syndrome, hand stereotypies are simple midline hand clasping (with loss of pincer grasp), whereas in ASD hand stereotypies are peripheral and complex, often hand flapping. Those with Rett syndrome test in the severe/profound range of intellectual disability, have seizure onset in early childhood, and show a distinct difference in postmortem neuropathology. In Rett syndrome there may be an encephalopathic regressive phase of social withdrawal (typically between 1 and 4 years of age) that differs from characteristic social deficits in ASD. After this period, a substantial proportion of affected young girls improve in their social relatedness. Because of these differences Rett syndrome is no longer classified as an ASD.

Severe Environmental Deprivation

When institutionalized children are both psychosocially deprived of interpersonal care and environmentally deprived of sensory stimulation from the beginning of life beyond 6 months of age, the term quasi-autism has been used. About one in six Romanian orphans who experienced this degree of severe deprivation have ongoing social deficits. Such deficits result from failure of environmental provision and lead to disturbed attachment behavior. However, these environmentally deprived children do not show the typical features of ASD.[26]

ASSESSMENT

Confirmation of the diagnosis of ASD is based on the clinical history, neuropsychiatric interview, and observational assessment. An interdisciplinary team of professionals who meet after the assessment period to develop a comprehensive treatment plan conducts the assessment. This interdisciplinary assessment includes a standardized intelligence test and other psychological tests, speech and language testing, and assessment by occupational and physical therapists and social workers as appropriate. Hearing testing may be indicated. When the child cannot cooperate in standard behavioral audiometry, brainstem auditory evoked response measures may be carried out.

Several psychometric assessment instruments are available for the assessment of ASD symptoms and behaviors. Structured instruments and rating scales are used in conjunction with diagnostic information drawn from the child's developmental history and reports from informants about behavior at home, in school, and in the community. The most comprehensive interview and observation scales are the Autism Diagnostic Interview™, Revised (ADI-R) and the Autism Diagnostic Observation Schedule™ (ADOS). These instruments were designed to evaluate children with a diagnosis of idiopathic autism. The validity of the ADI-R and ADOS in evaluating children with intellectual disability has poor to moderate agreement between the ADI-R and clinical judgment. The sensitivity and specificity of both the ADI-R and the ADOS are diminished in very young children and individuals with lower developmental ages.[27,28]

Although family members may inquire about blood and urine tests, genetic assessment, electrophysiological studies, and neuroimaging to confirm a diagnosis of an ASD, there are no specific biomarkers. However, testing is carried out to assure that the condition is not progressive and to rule out known metabolic disorders, neurological conditions, or neurogenetic syndromes that may be associated with the diagnosis.

The clinical history emphasizes the development of sociability, language development, imaginative play, the presence of stereotypies, and abnormal responses to sensory stimuli. Although autistic symptoms ordinarily are not related to birth events, birth history and history of intrauterine infections and postnatal infections and accidents that may involve the brain are included in the assessment history. Because of potential heritability, a family history of autism and/or other developmental disorders, specific psychiatric disorders, such as mood disorders, and conditions involving the brain are assessed.

The physical examination evaluates for signs of specific disorders that have been associated with autistic-like behavior, such as tuberous sclerosis, congenital rubella, and fragile X syndrome. The mental status examination is primarily observational for younger children. It includes efforts to engage the child in meaningful social interactions and, for verbal children, in conversation. Imaginative play is assessed using toys in younger children. For those with less severe involvement, subtle difficulties in the child's relatedness and imaginative play must be assessed. Observations are carried out to evaluate gaze avoidance, difficulties in initiating social communication, and problems with joint attention, stereotypies, and repetitive behaviors and interests.

NEUROPSYCHOLOGICAL PROFILE/ COGNITIVE FUNCTIONING

Cognitive impairment is a result of the neurodevelopmental disorder. The neuropsychological phenotype includes attention/arousal, long-term episodic memory, executive function, and social cognitive deficits. People with ASD generally have uneven profiles on subtests of versions of the Wechsler Adult Intelligence Scale (WAIS) and the Wechsler Intelligence Scale for

Children (WISC), in contrast to IQ-matched controls. The major differences are on subtests dealing with verbal abstraction, sequencing, visuospatial skills, and rote memory. These deficits are thought to impair normal language acquisition and social functioning.

The "theory of mind" paradigm[29] may be assessed. Theory of mind refers to the ability of normal children to attribute mental states, that is, beliefs, desires, and intentions, to themselves and to other people as a way of predicting and making sense of the mental states of others. Metarepresentational deficits are thought to impair an autistic person's comprehension of the mental states of others' behavior. Individuals with an ASD may show significantly poorer performance on tests of their understanding of others' beliefs and knowledge. However, an autistic person's social problems are not fully accounted for by conceptual impairment in interpersonal understanding, although this may be an essential feature. As Kanner proposed,[1,20,30] children with ASD lack a capacity to form affective contact with others and to develop intimate friendships as they grow older, despite their wish to do so. Their lack of understanding of others' beliefs and desires may not be an adequate explanation for the quality of their non-verbal communication disorder and relationship difficulties. Although executive dysfunction may be present, it is not a core neuropsychological deficit in ASD.

NEUROBIOLOGY

Although autism was first described in 1943, the first studies of possible neurobiological bases for ASD did not begin to appear until the late 1980s, almost half a century later. Many of these are neuroimaging studies of the brains of people with ASD diagnoses that seek to correlate the behavioral features and core impairments with differences in brain anatomy. Figure 6.1 shows brain regions that are proposed to be linked to social and communicative impairments and repetitive behaviors.[31]

The sections that follow review other neurobiological findings.

Trajectory of Brain Growth

ASD is a heterogeneous disorder with multiple behavioral and biological phenotypes. Accelerated brain growth during early childhood is a well-established biological feature of autism. Macrocephaly occurs in approximately 20% of affected individuals and was recognized in Kanner's original publication. It is usually due to megalencephaly (abnormal enlargement of the brain). At birth, head circumference is essentially in the typical range. Overgrowth is recognized during the first 18 months of life, when head growth typically accelerates. By 3–4 years of age there is an average increase in brain size by about 10% in those affected. Accelerated brain growth begins before most clinical features. Brain changes have been demonstrated using magnetic resonance imaging (MRI) and other imaging methods. These studies document the overall mean volume of the brain. Increased white matter and gray matter lead to increased volume of the cerebral cortex. These findings are not accounted for by the intellectual disability quotient, psychotropic medication use, or comorbid psychopathology.[32]

Cortical thickness, like brain volumes, may follow a period of early overgrowth followed by early arrested growth. One study evaluated 330 head circumference measures collected longitudinally between birth and 18 months of age from 35 male children with autism and a comparison group of 22 typically developing control subjects. Analyses revealed significantly thinner cortex in the ASD group with findings located predominantly in the left temporal and parietal lobes. Participants with ASD in another study had thinner cortex in the left fusiform/inferior temporal cortex compared with typically developing individuals. Thus, there may be a second period of abnormal cortical growth when greater thinning may be involved.[33]

Various mechanisms have been proposed with regard to whether the changes reflect an excess of neurons and/or reduced synaptic pruning during development. Neurogenesis is complete during uterine development in the prefrontal cortex and throughout the entire cerebral cortex. Developmental programmed cell death (apoptosis) occurs before and soon after birth. Such processes affect the net number of neurons in childhood. Thus, an increase in neurons would be consistent with prenatal origin. Postmortem cortical gray matter in the prefrontal cortex was examined in an autopsy study.[34] The investigators found 79% more neurons in the dorsolateral prefrontal cortex and 29% more in the mesial prefrontal cortex when seven brains of boys with autism were compared with matched controls.[34] These authors proposed that increased neuron number in the prefrontal cortex is correlated with accelerated postnatal brain growth and macrocephaly in early childhood. However, because the relationship is complex, research on neuron numbers in the prefrontal cortex is needed in both children and adolescents. Studies are required for those who do not have brain overgrowth as well as for non-autistic children with benign megalencephaly to clarify whether increased prefrontal neuron count in autism is associated only with autism. Cortical neurons are generated in prenatal life; therefore, a pathological overabundance of neurons indicates early developmental disturbances in critical brain regions.

The pattern of symptom onset has been closely studied in ASD but little is known about how it may be related to brain growth. Failure of developmental progression with loss of acquired skills is documented in

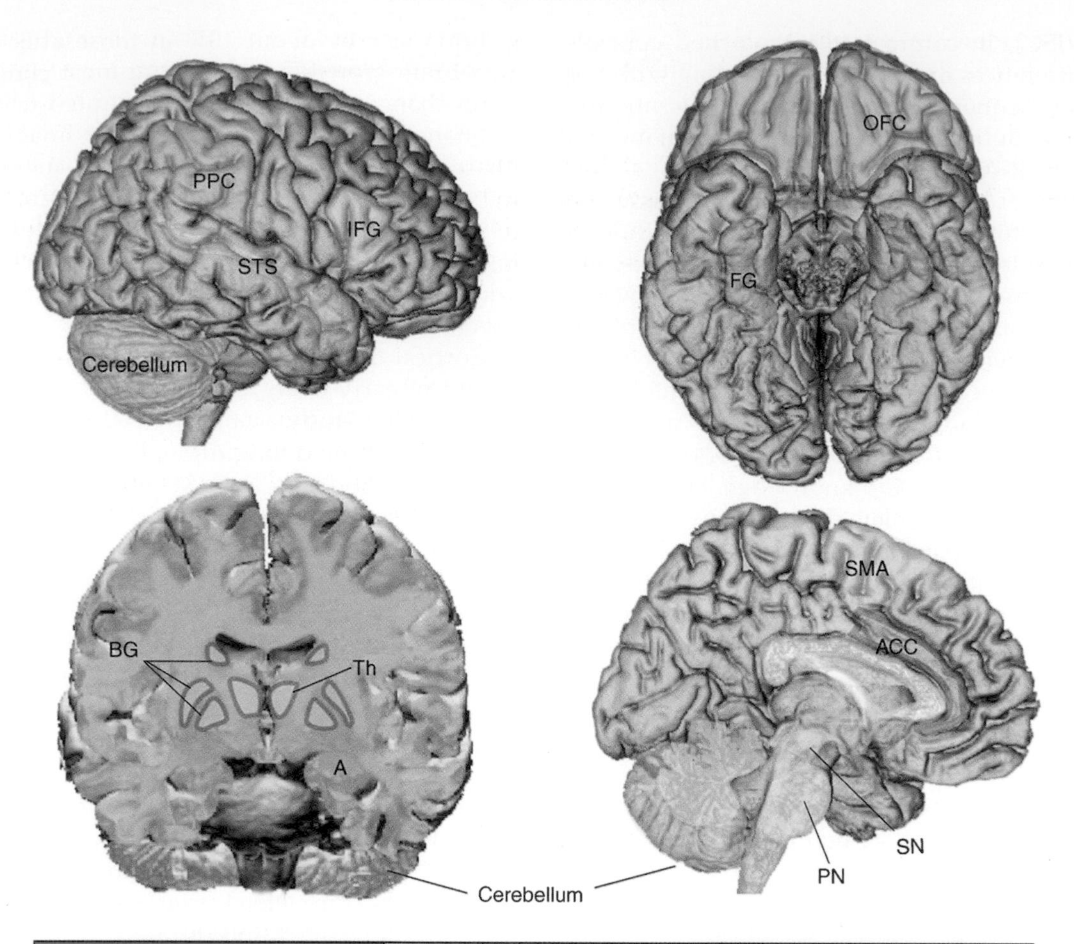

Social impairment	Communication deficits	Repetitive behaviors
OFC – Orbitofrontal cortex	IFG- Inferior frontal gyrus (Broca's area)	OFC – Orbitofrontal cortex
ACC – Anterior cingulate cortex	STS – Superior temporal sulcus	ACC – Anterior cingulate cortex
FG – Fusiform gyrus	SMA – Supplementary motor area	BG – Basal ganglia
STS – Superior temporal sulcus	BG – Basal ganglia	Th – Thalamus
A – Amygdala mirror neuron regions	SN – Substantia nigra	
IFG – Inferior frontal gyrus	Th – Thalamus	
PPC – Posterior parietal cortex	PN – Pontine nuclei cerebellum	

FIGURE 6.1 **Major brain regions that may be relevant to the core features of autism spectrum disorder.** These brain regions are linked with social behavior in animal studies and in lesion studies in human patients, or identified in functional imaging studies. They include regions of the frontal lobe, the superior temporal cortex, the parietal cortex, and the amygdala. Expressive language function is linked to Broca's area in the inferior frontal gyrus and portions of the supplementary motor cortex. Wernicke's area is essential for receptive language function and the superior temporal sulcus plays a role in both language and social attention. Repetitive or stereotyped behaviors of autism may involve the orbitofrontal cortex and caudate nucleus.[31]

25–35% of affected children in epidemiological studies. Typically, in the second year of life in affected children attention become diffuse, acquired word use is lost, and motor stereotypes become apparent. The relationship between total brain volume and onset of ASD symptoms was examined in affected 2–4-year-old boys and girls and comparisons were made between 53 cases with no regression and 61 who regressed, along with a comparison group.[35] When head circumference measurements from birth to 18 months of age were examined, abnormal brain enlargement was found most often in boys with behavioral regression. Evidence was found that brain enlargement is associated with ASD in preschool-age boys but not girls. In boys without regression the brain did not differ from controls. Thus, rapid head growth may be a risk factor in boys with onset of regression in the second year of life.

ASD is typically diagnosed by around 18 months of age. Little information is available on brain development at 6 and 12 months of age. A prospective infant sibling study completed longitudinal MRI scans at three time-points along with behavioral assessments. Fifty-five infants were examined: 33 "high-risk" (with an affected sibling) and 22 "low-risk" infants were imaged at 6–9 months; 43 of these (27 high-risk and 16 low-risk) were imaged at three time-points (6–9, 12–15, and 18–24 months of age). The 10 infants who developed ASD had significantly greater extra-axial fluid at 6–9 months, which persisted and remained elevated at 12–15 and 18–24 months, characterized by excessive cerebrospinal fluid in the subarachnoid space, particularly over the frontal lobes. The amount of extra-axial fluid detected as early as 6 months was predictive of more severe ASD symptoms at the time of outcome. Infants who developed ASD also had significantly larger total cerebral volumes at both 12–15 and 18–24 months of age. This is the first MRI evidence of brain enlargement in autism before 2 years of age.[36]

Studies are needed in children with ASD who do not have brain overgrowth, and other regions of interest such as the temporal lobe and amygdala should be examined. Early amygdala enlargement is reported in ASD on neuroimaging studies. The age at which abnormal amygdala enlargement begins was studied in 45 boys with ASD and 25 typical controls, and growth trajectories of the amygdala were examined longitudinally 1 year later. The amygdala was larger in children with ASD at baseline and 1 year later. Amygdala enlargement was present by 37 months of age in ASD, although substantial heterogeneity exists in amygdala and total cortical growth patterns. Clinical characterization of different amygdala growth patterns may have implications for treatment.[37]

NEUROIMAGING

Structural and functional MRI has been used extensively to examine the neuroanatomy of the brain in people with ASD. Current studies benefit from careful identification of cases using ADI-R research criteria for childhood autism. These studies have been carried out longitudinally and cross-sectionally as described above, in both children and adults. Adult studies allow the examination of regional brain changes that persist into adult life. However, our understanding of the relationship between ASD and the anatomy of specific brain regions is complicated by the non-replication of findings and small sample sizes. To clarify these relationships a large-scale multicenter MRI study was conducted in the UK. Comparisons were made between 89 men with ASD and 89 age-matched male controls. Subjects were high functioning, with full-scale IQs of 110 in those with ASD and 113 in controls.[38] Although the men with ASD were not significantly different from those in the control group on global volume measures they had regionally specific differences in gray and white matter volume. Increased gray matter was found in the anterior temporal and dorsolateral prefrontal regions, but decreased gray matter in the occipital and medial parietal regions in the ASD subjects. When gray matter brain systems were examined, adults with ASD showed changes in the cingulate gyrus, supplementary motor area, basal ganglia, amygdala, inferior parietal lobule, and cerebellum. Additional regional differences were found in the dorsolateral prefrontal, lateral orbitofrontal, and dorsal and ventral medial prefrontal cortices. These were accompanied by spatially distributed reductions in regional white matter volume.[38]

These findings are consistent with regional neuroanatomical abnormalities in ASD persisting into adulthood. Moreover, regional differences in neuroanatomy in this study were correlated with the severity of specific autistic symptoms based on the ADI-R. Adult males with an ASD diagnosis have differences in brain anatomy and connectivity associated with specific autistic features and traits consistent with ASD, a syndrome involving atypical neural connectivity.[39] Thus, structural brain mapping may be used to study morphological connectivity in ASD. Differences between genders must be considered as one source of heterogeneity, as studies suggest that ASD manifests differently in males and females. Males with autism have been disproportionately represented in research.

Neuroanatomical differences are reported between high-functioning males and females with ASD whose intelligence is in the average to above average range. Lai and colleagues[40] used neuroimaging to study the brains of 30 right-handed males and 30 right-handed females and matched controls. They found differences in males and females in gray matter and white matter regions of interest. In the females there was overlapping with brain structures that showed evidence of sexual dimorphism in matched controls. These findings suggesting gender-dependent neuroanatomy in ASD require replication but indicate the importance of stratifying by biological sex in neuroimaging studies. Future studies are needed to clarify whether these differences between males and females are linked to cognition and generalize to males and females whose intelligence is in the mild to severe range of intellectual disability (intellectual developmental disorder).

Understanding brain connectivity is important because structures are tightly coupled in development; they grow at the same time to establish networks. Correlations between frontal lobe gray matter volume and temporal lobe, parietal lobe, and subcortical gray matter are disrupted in ASD. Despite agreement that ASD is associated with altered brain connectivity, the nature of the deficit is poorly understood. There is evidence to suggest a complex functional phenotype characterized by both hypoconnectivity and hyperconnectivity

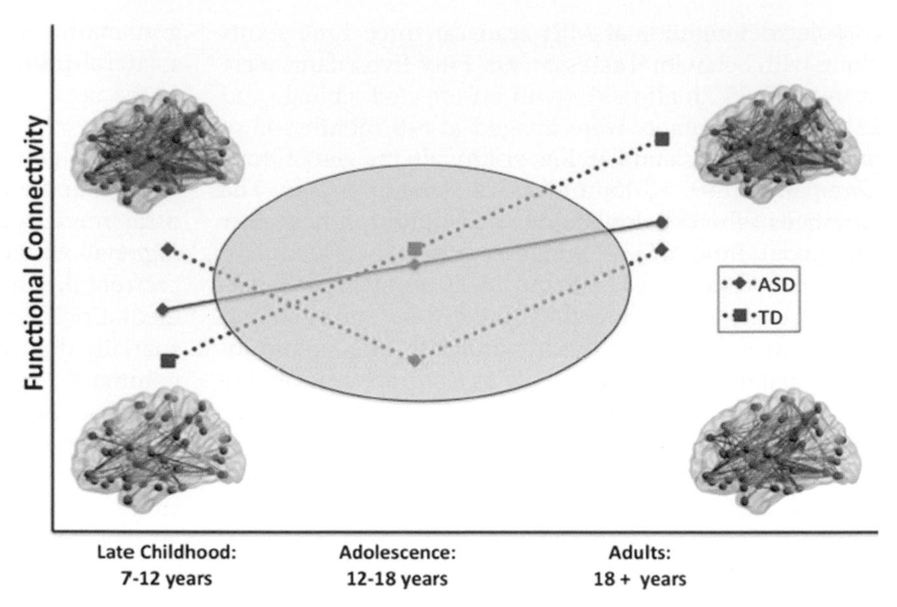

FIGURE 6.2 **Schematic model of two scenarios to account for developmental shift from intrinsic hyperconnectivity to hypoconnectivity in autism spectrum disorder (ASD).** In the first scenario the group with ASD (solid red line) shows a less steep developmental increase over the three age spans compared with a control group (TD: typical development). In the second scenario (dashed red line) there are anomalous patterns of connectivity across the pubertal period.[41]

involving large-scale brain systems. Studies involve task-based functional connectivity (synchronization of activation of a brain region to a cognitive challenge task) and resting-state functional connectivity in the absence of a task. Discrepant findings may be reconciled using a developmental perspective. A review of functional MRI studies of functional connectivity in children, adolescents, and adults suggests dynamic changes with aging. While in adolescents and adults with autism connectivity seems reduced compared with age-matched controls, in younger children functional connectivity seems to be increased. Thus, a developmental framework that considers prepubertal, adolescent, and adult subjects may resolve the conflicting results on hypoconnectivity and hyperconnectivity, and lead to a better understanding of the neurobiology of ASD.[41] As shown in the resting state, functional connectivity MRI studies are consistent with widespread hyperconnectivity in children, in contrast to hypoconnectivity in adolescents and adults (Fig. 6.2).

NEUROPHYSIOLOGY

Cognitive deficits in ASD result in strategies that excessively engage sensory systems, to the detriment of the more integrative processing needed for a person to be aware of contextual subtleties required for prediction. Thus, people with ASD manifest unusual processing when faced with unpredictable events. They show deficits in orienting to changing, novel sensory stimuli.[42] Studies of event-related potentials illustrate the psychophysiological mechanisms and neural bases underlying these deficits in ASD. Such dysfunction in building flexible prediction in ASD may result from impaired top–down control over several sensory and higher level information processing systems, consistent with underconnectivity.[41]

Other neurophysiological studies found changes with power spectrum analyses in the analysis of brain regions. Studies using event-related potentials and magnetoencephography to document face processing found decreased sensitivity to whether a face is upright or inverted, reduced responsiveness to repeated face presentation, abnormal eye-to-eye gaze, and abnormal hemispheric lateralization of processing in the cortex. Auditory processing and involuntary orienting to sound, in a wide network that involves the auditory cortex and multimodal sensory areas in the parietal lobe and dorsolateral prefrontal cortex, have been studied using these methods.[4]

NEUROPATHOLOGY

Autism is a disorder of neural development. The typical brain develops in several stages. These involve neuronal proliferation and migration, the establishment of dendritic arbors and synaptic connections, and later dendritic pruning and programmed cell death. Disruption in one or more of these stages could result in detrimental downstream effects. MRI studies of people with autism demonstrate aberrant brain development during early childhood involving the whole brain and more specifically in some regions such as the amygdala. However, given the limited resolution in MRI studies, postmortem human brain research is required to determine the neurobiological basis of MRI results. Studies of postmortem tissue may use MRI findings to target specific brain regions for study. Both of these approaches facilitate our understanding of the neuropathology of ASD.

Neuroanatomical studies have demonstrated abnormalities in the emotional or limbic brain and cerebellum.[43] Approximately 60 brains have been studied. In nine of 14 brains examined postmortem, increased cell packing density and smaller neuronal size were found. Twenty-one of 29 brains studied showed a decreased number of Purkinje cells in the cerebellum; in five cases changes were found in cerebellar nuclei and the inferior olive. More than half of the brains studied showed features of cortical dysgenesis in the cerebral cortex.

Unfortunately, most of the cases evaluated in earlier postmortem studies involved brains from individuals with a diagnosis of severe intellectual disability or who had comorbid seizure disorders. Epilepsy is associated with pathology of the cerebral cortex, amygdala, cerebellum, and hippocampal formation, regions implicated in autism. Therefore, comorbid disorders associated with ASD are of concern in interpreting the neuroanatomy of ASD. More neuropathological studies are needed, using new technologies with larger samples; it is essential to include and younger subjects free of comorbidities such as severe intellectual disability and epilepsy.

Most postmortem studies have not targeted regions of interest potentially linked to clinical features of ASD (Fig. 6.1). Abnormalities in face perception, a core feature of social disability in ASD, have been studied using functional MRI. The fusiform gyrus and other cortical regions supporting face processing in controls are hypoactive in patients with autism. One study, on seven postmortem brains of ASD subjects and 10 controls, examined this brain region for alterations in neuron density, total neuron number, and mean perikaryal volume with high-precision design-based stereology. Separate analysis of layers II, III, IV, V, and VI of the fusiform gyrus in patients with ASD showed significant reductions in neuron densities in layer III, total neuron numbers in layers III, V, and VI, and mean perikaryal volumes of neurons in layers V and VI, providing important insights into the cellular basis of abnormalities in face perception in autism. This hypothesis-based approach to neuropathology is to be encouraged.[44]

NEUROCHEMISTRY

Neurochemical investigations in ASD have focused on measuring neurotransmitter levels in blood and urine, dietary depletion of tryptophan [the dietary precursor of serotonin (5-hydroxytryptamine, 5-HT)], functional positron emission tomography (PET) and single-photon emission computed tomography (SPECT), and measurement of brain metabolites, especially *N*-acetyl aspartate (NAA), using proton spectroscopy. The most consistent finding is that of hyperserotoninemia in blood platelets in 30–50% of autistic people. Whole blood 5-HT levels are in the upper 5% of the normal range. Increases have also been noted in the broader ASD phenotype.[45] More important is 5-HT function in the brain rather than the periphery because, in addition to being a neurotransmitter, 5-HT serves as a growth factor and regulator of early neuronal development in the developing brain. Studies involving dietary tryptophan depletion and PET studies support central 5-HT deficits. Acute depletion of dietary tryptophan reduces 5-HT in the brain, and a worsening of symptoms has been reported with such tryptophan depletion.[4] PET studies using a serotonin precursor showed reduced synthesis. These findings are consistent with developmental dysregulation of 5-HT synthesis. There is evidence of significant reductions in 5-HT$_{1A}$ receptor binding density in superficial and deep layers of the posterior cingulate cortex and fusiform gyrus, and in the density of 5-HT$_{2A}$ receptors in superficial layers of the posterior cingulate cortex and fusiform gyrus.[46]

In contrast to serotonin, evidence for dopamine and norepinephrine is less compelling. Levels of homovanillic acid, the primary dopamine metabolite, in blood, urine, and cerebrospinal fluid, were not significantly different between people with ASD and controls. However, a PET study suggested low medial prefrontal dopamine activity. There are no consistent findings of deficits in the norepinephrine system.[4]

Proton MRI is a non-invasive way to study brain neurochemistry and perform *in vivo* quantification of biochemical and metabolite concentrations.[47] Both glutamate and γ-aminobutyric acid (GABA) have been studied. Glutamate is the primary excitatory neurotransmitter in the brain and is important in neuronal plasticity and higher cortical functioning. GABA is the primary inhibitory neurotransmitter in the brain. Peripheral measurements of these two neurotransmitters in the blood have shown conflicting findings. Postmortem studies have shown changes in the receptors for both glutamate and GABA in the hippocampus.

ASD is associated with widespread reductions in NAA, creatine and phosphocreatine, choline-containing compounds, *myo*-inositol, glutamate, glutamine, and GABA. These reductions suggest impaired neuronal function and/or metabolism. However, findings vary depending on the study and region of interest. Studies should control for variability in subjects' age and level of functioning to address neurodevelopmental levels and processes associated with ASD.[4] A meta-analysis identified 22 articles satisfying the criteria with measures of NAA, creatine, choline-containing compounds, *myo*-inositol, glutamate, and glutamine in the frontal, temporal, and parietal regions, amygdala–hippocampus complex, thalamus, and cerebellum. Random effect analyses showed significantly lower NAA levels in all the examined brain regions except the cerebellum; however, there was no significant difference in metabolite levels in adulthood. There were clear changes in the extent of abnormality in NAA levels, especially in the frontal lobes, in ASD.[48] These findings suggest that early transient brain expansion in ASD may be caused by an increase in non-neuronal tissues, such as glial cell proliferation.

Other investigators have correlated proton spectroscopy changes with social and cognitive functioning. In one study involving 77 3–6-year-old children with an ASD diagnosis (23 boys and eight girls), concentrations of NAA in the left amygdala and the bilateral orbitofrontal cortex were determined. Reductions in NAA were found in the left amygdala and in the orbitofrontal cortex bilaterally compared with those in a control group. NAA levels were correlated with ratings of the social quotient in the children with ASD, suggesting neuronal dysfunction in these brain regions.[49]

Oxytocin

ASD is proposed to result from many rare genetic variants. These involve common neurotransmitter or neurodevelopmental pathways. Moreover, for ASD multiple common polymorphisms confer risks for the disorder. Genetic associations with ASDs and oxytocin, vasopressin, and related proteins pertinent to both social or repetitive behavior disorders. Oxytocin is integral to social sensitivity throughout the life cycle. It is involved in social cognition, interpersonal bonding, trust, and stress management. Its release is highly sensitive to emotional and social context and it plays an important role in emotional regulation through parent–child attachment. Polymorphisms of the oxytocin receptor have been implicated in sensitivity to social cues. Knowledge of polymorphisms in the oxytocin genetic pathways will become important as more is learned about the epigenetic effects of social interactions. Moreover, oxytocin is being considered as an adjunctive treatment to facilitate social cognition and emotion regulation in ASD. Oxytocin modulates emotions and social judgments in part through actions on the brainstem and autonomic nervous system. It may alter perceptions of the social environment as safe or threatening. An essential issue is to determine whether oxytocin administration could be a drug treatment for an identified disorder or disorders such as ASD or might better be used as an adjunctive treatment, taking into account the context dependence of its effects.

Any theory regarding the risk for development of ASDs must consider the significant male bias in risk for developing this constellation of symptoms. An awareness of gender differences in the central regulation and expression of oxytocin and vasopressin may help in understanding aspects of ASDs.[50] Oxytocin is estrogen dependent, and in some cases levels of the peptide and its receptor are higher in females. Possibly relevant to ASDs is that levels of vasopressin in the extended amygdala–lateral septal axis of the nervous system are sexually dimorphic and higher in males. Moreover, males seem to be more sensitive than females to the actions of vasopressin, especially during development. Insensitivity to vasopressin or a lack of dependence on this peptide could be protective against ASDs in females. Oxytocin also could protect females, either directly or indirectly. A reduction in fear and an increased sense of safety or trust are expected to be protective in disorders such as ASDs that are associated with high levels of anxiety. Differences in coping mechanisms between males and females, especially to downregulate anxiety in social interactions, could be pertinent to ASDs. Disruptions in systems that rely on vasopressin may also increase the vulnerability to ASDs.

GENETIC AND ENVIRONMENTAL RISK FACTORS

Genetic research in ASD has accelerated in the past 10 years, led by advances in two areas. First, the development of standardized autism diagnostic tools such as the ADI-R and ADOS allowed international and cross-institutional collaboration and the collection of large datasets in the late 1990s and early 2000s. Second, genetic technology has advanced rapidly, as summarized by Geschwind[51] (Fig. 6.3).

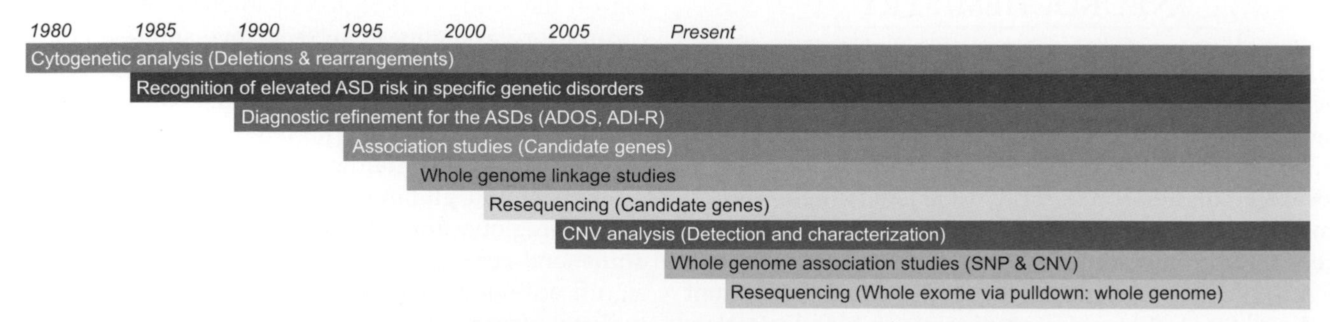

FIGURE 6.3 **Methodological changes that have accelerated progress in autism spectrum disorder (ASD) genetics.** Cytogenetic studies in the 1980s were followed by whole genome linkage studies, whole genome association studies [single-nucleotide polymorphism (SNP) and copy number variation (CNV)] and resequencing studies.[51] ADOS: Autism Diagnostic Observation Schedule; ADI-R: Autism Diagnostic Interview, Revised.

Cytogenetic studies of macroscopic chromosomal anomalies were described in the 1980s and into the 1990s. Candidate gene association studies began in the mid-1990s. The first large-scale genome linkage studies were published in the early 2000s. Chromosomal microarray identification of CNVs followed in the mid-2000s, with genome-wide association studies (GWAS) soon afterwards. The current wave of next generation sequencing technology allows characterization of all variation in the exome or even the full genome.

With each advance in genetic methodology, new risk variants have been identified in ASD. In almost all cases, however, studies have identified uncommon genetic variants that individually lead to a substantial increase in ASD risk. No single one of these uncommon risk variants is present in more than about 2% of children with ASD. Other than syndromal disorders (discussed below), the most common contributors to substantial ASD risk have been CNVs, such as chromosome 16p deletion or duplication, maternal chromosome 15q11 duplication, or 7q11 duplication, each present in 0.5–2% of children with ASD. Collectively, CNVs that are likely to contribute to ASD risk are present in about 9–10% of children with ASD. Sequencing has identified more uncommon variants that, when added to CNVs and syndromal cases, may yield a substantial genetic risk variant in up to 20% of children with ASD. These variants disrupt a large number of brain-expressed genes, including synaptic cell adhesion molecules such as neurexin, neuroligin, and shank family members.[51] Advances in next generation sequencing have enabled the discovery of a vast number of *de novo* mutations that confer a risk for ASD that include a number of chromatin remodeling genes.[52] Both *de novo* CNVs and *de novo* SNVs are more likely with advancing paternal age, which has been shown to be a robust risk factor for ASD.

Unlike most other behaviorally defined disorders, genetic testing is recommended in ASD. The American College of Medical Genetics recommends that every person with ASD should receive a chromosomal microarray (CMA) as the first line test to identify CNVs. However, whereas identified CNVs may indicate substantial risk, most are not specific for ASD itself. Many, such as chromosome 16p11 deletion, are also associated with intellectual disability and could potentially confer risk of ASD as a result of a more general cognitive "hit" that coincides with other ASD risk factors. Others, such as chromosome 16p11 duplication, appear to lead to risk that extends beyond ASD to ADHD and schizophrenia, but may also be observed in the absence of a neuropsychiatric phenotype. The clinical impact associated with most CNVs (and probably SNVs) is therefore likely to be probabilistic and not deterministic. Few can be understood as "causes" of ASD, even when the associated risk is substantial. Even when inherited within a family, the same CNV or SNV may be associated with multiple different neuropsychiatric phenotypes.

The other recommended genetic test for children with ASD is for fragile X syndrome (see Chapter 8), the most commonly observed genetic syndrome in ASD. A number of other syndromes, including tuberous sclerosis, also confer increased risk of ASD. The pattern of behavioral symptoms within these syndromes can be quite variable, with none leading to ASD in every case, and often in the minority of individuals. In contrast, most genetic syndromes associated with ASD risk confer intellectual disability in most or all cases. Even when ADI-R and ADOS results are consistent with ASD, the clinical presentation of individuals with these syndromes may differ from the overall group of people with ASD. For example, individuals with fragile X syndrome often show a characteristic pattern of aversion to eye contact and sensory sensitivity, yet may be quite socially engaged and motivated. These syndromes offer an opportunity to understand co-occurring risk factors that may result in a more autism-like picture in a subset of individuals. They may also offer a window into understanding the underlying neurobiology conferring risk of neurodevelopmental disorders in general. For example, mouse models of fragile X syndrome demonstrate increased signaling downstream of the glutamate mGlu5 receptor. Drugs that decrease mGlu5 receptor signaling can rescue many brain and behavioral phenotypes in the mouse and are now in clinical trials in fragile X syndrome. Although it must not be assumed that these drugs would benefit the broader group of individuals with ASD, an understanding of this specific syndromal disorder may yield new ideas about the neurobiology of neurodevelopmental disorders in general.

Heritability estimates and resulting models of ASD risk predict that common variants would also be identified that would have a smaller individual effect on risk but in a much broader portion of the ASD population. Unfortunately, however, GWAS have not identified replicable common gene variants in ASD, perhaps because of insufficient power. Encouraging GWAS results in schizophrenia have suggested that a larger sample size may yield these common variants in ASD as well. The heterogeneity that is characteristic in autism could potentially make finding common risk genes more challenging than in other neuropsychiatric disorders. Gene–gene and gene–environment interactions may also complicate the identification of risk factors. ASD could potentially involve epigenetic risk factors, or inherited or acquired changes in gene expression that do not result from change in the primary DNA sequence. Although there is limited evidence that epigenetic mechanisms are involved in ASD, one study reported a difference in epigenetic markers at the oxytocin receptor gene in ASD.[53]

With the exponential rise of genetic technology, risk genes appear to be much easier to detect than environmental susceptibility factors. Ongoing work suggests, however, that environmental risk factors, particularly prenatal and perinatal factors, play an important role. Emerging data suggest that extremely low birth weight is a robust risk factor for ASD, which is more common in survivors of the neonatal intensive care unit. Other risk factors, such as short interpregnancy interval and infection during pregnancy, have been reported in more than one study. Advanced parental age, particularly paternal age, also appears to be a robust risk factor and is increasing over time, particularly with the advent of reproductive technology. In the environmental arena, rare risk factors may also be easier to identify and study, such as fetal exposure to valproate, which has been identified as an environmental risk factor.

TREATMENT

Treatment programs for children with ASD are developmentally based, affectively oriented, and tailored to specific known deficits in an individual child.[10] Early childhood behavioral interventions have been shown to be the most likely to be successful.[54] Treatment programs must be sensitive to the needs and perceptions of the autistic child and provide guidance to parents. Even though our understanding of the neurobiological basis of ASD is growing and anatomical features are apparent, intervention can still be effective in helping children to compensate for their developmental deficits. Four general aims in the treatment of the autistic person are (1) to promote cognitive development, (2) to promote language development, (3) to promote social development, and (4) to promote overall learning. Besides these, behavior reduction and behavior enhancement strategies are needed, as is appropriate use of medications to treat for co-occurring conditions. The role of the parent is crucial for intervention in a child with ASD. The parent functions as cotherapist and plays an integral role in treatment. Parental counseling begins with clarification of the diagnosis and an explanation of the characteristics of an ASD.

Planned periods of interaction must be scheduled to promote social development. How intense they need to be is an area of ongoing study.[54] Cognitive development strategies focus on facilitation of active, meaningful experiences with planned periods of interaction. Because there is reduced cognitive capacity, direct teaching at the appropriate developmental level is required. Language development is facilitated by planned social interactions and interpersonal conversational exchanges to deal with social isolation and the lack of social reciprocity. Social reciprocity problems[25] result from deficits in joint attention that require structured reciprocal language between the person with an ASD and the therapist.

Direct instruction is necessary to teach the social use of language. It is carried out with differential reinforcement of language use rather than focusing only on speech. Since language development is limited, direct teaching must be targeted at the child's level of language comprehension. Alternative means, such as the use of signing, are needed for non-verbal children. The promotion of social development involves positive personal interaction that is pleasurable to the child. The lack of social approach and lack of social reciprocity in social interaction require structured settings. The lack of social awareness characteristic of ASD is addressed through direct teaching of skills that lead to social competence as well as early interpersonal programs.[54]

The basic principles of behavior modification are used in treatment based on a clear understanding of the deficits and types of abnormal behavior associated with ASD. The nature of the autistic deficit makes adaptability and generalization of targeted behaviors across settings difficult. A stable environmental context is necessary for treatment, and once this environment is established it should not be changed without careful consideration. Behavior management strategies are used to eliminate non-specific maladaptive behaviors. The behavioral approach is based on a functional analysis of behavior and application of learning theory. When using a behavioral technique, it is essential to determine which environmental features influence a behavior, not in children in general but in this particular child.

The types of behavior most commonly targeted in behavior management programs are aggression and self-injurious behavior. For these and other disruptive behaviors, avoidance of precipitants, help to establish coping skills, and differential reinforcement may be used as interventions. There is good evidence from randomized controlled trials that non-intensive interventions, especially those focused on communication and joint social interaction, may have a significant and positive impact on functioning in children with ASD.[54] Future challenges include assessing which treatments work for which children and identifying the individual characteristics that predict responsiveness to specific programs and approaches.

FUTURE DIRECTIONS

ASD is a highly heterogeneous and complex disorder. Refinements are needed in classification with continued efforts at subtyping. Longitudinal follow-up studies will improve our knowledge of developmental trajectories to validate subgroups and allow examination of their neurobiology. The following recommendations build on current developments in neurobiology.

- *Complex genetics*: Although ASD is highly genetic, its genetics are complex. Rare variants of substantial effect have been identified in ASD, but more common risk variants have been elusive. Parallel work in schizophrenia suggests that detection of common risk variants will be possible in larger sample sizes in GWAS. Next generation sequencing technology that allows every exon of every gene (the whole exome) to be examined simultaneously may now be used to identify risk variants underlying linkage peaks that were identified in families with multiple affected individuals. Subgrouping by biomarkers or a well-defined clinical profile may be necessary to deal with heterogeneity.
- *Refinements in case identification*: ASD is a multifactorial disorder. For research purposes, separating the genetics of social deficits from those of repetitive behaviors and investigating the genetics of specific behaviors such as joint referencing deficits and dysregulated sensory modulation may be beneficial. The use of DSM-5 specifiers is one approach to assist in subtyping.
- *Longitudinal studies with large cohorts*: Developmentally focused, prospective, longitudinal MRI studies, both morphological and functional, are needed. A few such prospective studies have been conducted, but over only a short time-frame. Cross-sectional findings, such as enlarged amygdala size in young children and smaller amygdala size in adolescents and adults, will require longitudinal studies to evaluate whether the amygdala is truly shrinking within some individuals with ASD. Neuroimaging studies can be linked with longitudinal studies of ASD symptoms to better understand the relationship between brain connectivity and circuits and the resulting pattern of behavior and development. Results from MRI studies should guide the choice of regions of interest for postmortem studies.
- *Proton spectroscopy*: The continued use of *in vivo* proton spectroscopy may be able to identify abnormalities in brain metabolites or neurotransmitters that could subgroup individuals within the spectrum. Correlation of proton spectroscopy findings with brain regions of interest for brain connectivity and clinical findings may clarify the systems and circuitry critical for social function and repetitive behavior.
- *Age of onset*: From a developmental perspective, age of onset or recognition is an important element. Progress is being made in refining brain changes in children with the regressive type of ASD who show rapid deterioration between 18 and 24 months of age. Studies in high-risk populations, such as younger siblings of affected children or survivors of extreme premature birth with low birth weight, already point to early markers of risk, but more work is needed for earlier identification.
- *Neuropathological studies*: Neuropathological studies have been confounded by the co-occurrence of seizure disorder, severe intellectual disability, comorbid diagnoses, and broad age groups studied. The Autism Brain Bank can be used to expand the database. Neuroimaging studies can be used to identify regions of interest and test hypotheses linked to brain and behavior. There is a need to focus on developmental cohorts and compare subjects who did and did not show accelerated early brain growth.
- *Animal studies*: Since we lack an understanding of common risk factors for ASD, animal models best target rare genetic or environmental risk factors that contribute a substantial degree of risk. Models of environmental risk are more challenging, since drugs such as valproate increase autism risk within an ambiguous time-window that may be difficult to model in an animal. Rather than focusing on developing animal models of ASD, these models can be used to understand how ASD risk factors affect the brain, potentially yielding an understanding of pathophysiology.
- *Neuroimaging studies*: With the possible exception of findings in the amygdala and the striatum, neuroimaging studies suggest that ASD is a distributed disorder. Functional connectivity and structural connectivity studies implicate aberrant long-distance communication within the brain in ASD, although it remains difficult to assess whether this is causal or a result of abnormal neuronal function. Ongoing work should focus on functional imaging and evoked electrocortical response techniques to examine functional coherence in brain circuits in relation to specific symptom patterns and behavior.
- *Treatment interventions*: The dramatic heterogeneity in ASD calls for individualized treatment with a developmental focus. Early, intensive behavioral intervention clearly helps some children, but treatment studies have largely clustered all children with ASD together, rather than clarifying which ones benefit. Randomized and large-scale studies are needed to understand the required intensity of treatment for different children, as well as which specific symptom domains can be expected to improve. Studies in targeted subpopulations will be especially important when genetic findings or biomarkers can be connected to potential avenues for treatment. In addition to core ASD symptoms, careful management of

co-occurring symptoms is essential, especially because some "recovered" children were initially characterized as having significant co-occurring disorders such as ADHD or anxiety symptoms. Parental and professional advocacy for multifaceted, individual treatment plans will continue to be important as more evidence-based treatments emerge.

Acknowledgment

Thanks to Jeremy Veenstra-VanderWeele, MD, Director, Division of Child and Adolescent Psychiatry, and Medical Director, Treatment and Research Institute for Autism Spectrum Disorders at Vanderbilt University, for his helpful suggestions for the sections on genetics and future directions in this chapter.

References

1. Kanner L. Autistic disturbances of affective contact. *The Nervous Child*. 1943;2:217–250.
2. American Psychiatric Association. *Committee on Nomenclature and Statistics. Diagnostic and Statistical Manual of Mental Disorders*. 1st ed. Washington, DC: APA; 1990.
3. American Psychiatric Association. *Committee on Nomenclature and Statistics. Diagnostic and Statistical Manual of Mental Disorders*. 5th ed. Washington, DC: APA; 2013.
4. van Engeland H, Buitelar JK. Autism spectrum disorders. In: Rutter M, Bishop DVM, Pine DS, et al., eds. *Rutter's Child and Adolescent Psychiatry*. 5th ed. Oxford: Blackwell; 2008:759–781.
5. Kanner L. Follow-up study of eleven autistic children originally reported in 1943. *J Autism Child Schizophr*. 1973;1:119–145.
6. Rutter M. Childhood schizophrenia reconsidered. *J Autism Child Schizophr*. 1972;2:315–338.
7. Rutter M. Autism: diagnosis and definition. In: Rutter M, Schopler E, eds. *Autism: A Reappraisal of Concepts and Treatment*. New York: Plenum Press; 1978:1–25.
8. Asperger H. "Autistic psychopathy" in childhood. In: Frith U, ed. *Autism and Asperger Syndrome*. New York: Cambridge University Press; 1991:37–92.
9. Wing L. Asperger syndrome: a clinical account. *Psychol Med*. 1981;11:115–129.
10. Harris JC. Autistic disorder. In: Harris JC, ed. *Developmental Neuropsychiatry: Assessment, Diagnosis, and Treatment of Developmental Disorders*; Vol. 2. New York: Oxford University Press; 1998:187–221.
11. Lai MC, Lombardo MV, Chakrabarti B, Baron-Cohen S. Subgrouping the autism "spectrum": reflections on DSM-5. *PLoS Biol*. 2013;11:e1001544.
12. Autism and Developmental Disabilities Monitoring Network Surveillance Year 2008 Principal Investigators; Centers for Disease Control and Prevention. Prevalence of autism spectrum disorders – Autism and Developmental Disabilities Monitoring Network, 14 sites, United States. *MMWR Surveill Summ*. 2008;2012(61):1–19.
13. Brugha TS, McManus S, Bankart J, et al. Epidemiology of autism spectrum disorders in adults in the community in England. *Arch Gen Psychiatry*. 2011;68:459–465.
14. Rutter M. Commentary: Fact and artefact in the secular increase in the rate of autism. *Int J Epidemiol*. 2009;38(5):1238–1239. author reply 1243–1244.
15. Fombonne E. Commentary: On King and Bearman. *Int J Epidemiol*. 2009;38(5):1241–1242. author reply 1243–1244.
16. Dales L, Hammer SJ, Smith NJ. Time trends in autism and in MMR immunization coverage in California. *JAMA*. 2001;285:1183–1185.
17. Harris JC. Autism risk factors: moving from epidemiology to translational epidemiology. *J Am Acad Child Adolesc Psychiatry*. 2012;51:461–463.
18. Marshall CR, Scherer SW. Detection and characterization of copy number variation in autism spectrum disorder. *Methods Mol Biol*. 2012;838:115–135.
19. Howlin P, Alcock J, Burkin C. An 8 year follow-up of a specialist supported employment service for high-ability adults with autism or Asperger syndrome. *Autism*. 2005;9(5):533–549.
20. Hobson PR. The autistic child's appraisal of emotion. *J Child Psychol Psychiatry*. 1986;28:321–342.
21. Hobson PR, Chidambi G, Lee A, Meyer J. Foundations for self-awareness: an exploration through autism. *Monogr Soc Res Child Dev*. 2006;71(2):vii–166.
22. Mazzone L, Postorino V, De Peppo L, et al. Mood symptoms in children and adolescents with autism spectrum disorders. *Res Dev Disabil*. 2013;34(11):3699–3708.
23. Hutton J, Goode S, Murphy M, Le Couteur A, Rutter M. New-onset psychiatric disorders in individuals with autism. *Autism*. 2008;12:373–390.
24. Lerner MD, Haque OS, Northrup EC, Lawer L, Bursztajn HJ. Emerging perspectives on adolescents and young adults with high-functioning autism spectrum disorders, violence, and criminal law. *J Am Acad Psychiatry Law*. 2012;40:177–190.
25. Kasari C, Freeman S, Paparella T. Joint attention and symbolic play in young children with autism: a randomized controlled intervention study. *J Child Psychol Psychiatry*. 2006;47:611–620.
26. Rutter M, Sonuga-Barke EJ, Castle J. Investigating the impact of early institutional deprivation on development: background and research strategy of the English and Romanian Adoptees (ERA) study. *Monographs of the Society for Research in Child Development*. 2010;75:1–20.
27. Ventola P, Kleinman J, Pandey J, et al. Differentiating between autism spectrum disorders and other developmental disabilities in children who failed a screening instrument for ASD. *J Autism Dev Disord*. 2007;37(3):425–436.
28. Gray KM, Tonge BJ, Sweeney DJ. Using the Autism Diagnostic Interview–Revised and the Autism Diagnostic Observation Schedule with young children with developmental delay: evaluating diagnostic validity. *J Autism Dev Disord*. 2008;38:657–667.
29. Colle L, Baron-Cohen S, Hill J. Do children with autism have a theory of mind? A non-verbal test of autism vs. specific language impairment. *J Autism Dev Disord*. 2007;37(4):716–723.
30. Chevallier C, Kohls G, Troiani V, Brodkin ES, Schultz RT. The social motivation theory of autism. *Trends Cogn Sci*. 2012;16:231–239.
31. Amaral DG, Schumann CM, Nordahl CW. Neuroanatomy of autism. *Trends Neurosci*. 2008;31:137–145.
32. Lainhart JE, Lange N. Increased neuron number and head size in autism. *JAMA*. 2011;306:2031–2032.
33. Wallace GL, Dankner N, Kenworthy L, Giedd JN, Martin A. Age-related temporal and parietal cortical thinning in autism spectrum disorders. *Brain*. 2010;133(Pt 12):3745–3754.
34. Courchesne E, Mouton PR, Calhoun ME, et al. Neuron number and size in prefrontal cortex of children with autism. *JAMA*. 2011;306:2001–2010.
35. Nordahl CW, Lange N, Li DD, et al. Brain enlargement is associated with regression in preschool-age boys with autism spectrum disorders. *Proc Natl Acad Sci U S A*. 2011;108:20195–20200.
36. Shen MD, Nordahl CW, Young GS, et al. Early brain enlargement and elevated extra-axial fluid in infants who develop autism spectrum disorder. *Brain*. 2013;136(Pt 9):2825–2835.
37. Nordahl CW, Scholz R, Yang X, et al. Increased rate of amygdala growth in children aged 2 to 4 years with autism spectrum disorders: a longitudinal study. *Arch Gen Psychiatry*. 2012;69:53–61.

38. Ecker C, Suckling J, Deoni SC, et al. Brain anatomy and its relationship to behavior in adults with autism spectrum disorder: a multicenter magnetic resonance imaging study. *Arch Gen Psychiatry*. 2012;69(2):195–209.
39. Ecker C, Ronan L, Feng Y, et al. Intrinsic gray-matter connectivity of the brain in adults with autism spectrum disorder. *Proc Natl Acad Sci U S A*. 2013;110:13222–13227.
40. Lai MC, Lombardo MV, Suckling J, et al. Biological sex affects the neurobiology of autism. *Brain*. 2013;136(Pt 9):2799–2815.
41. Uddin LQ, Supekar K, Menon V. Reconceptualizing functional brain connectivity in autism from a developmental perspective. *Front Hum Neurosci*. 2013;7:458.
42. Gomot M, Wicker B. A challenging, unpredictable world for people with autism spectrum disorder. *Int J Psychophysiol*. 2012;83:240–247.
43. Palmen SJ, van Engeland H, Hof PR, Schmitz C. Neuropathological findings in autism. *Brain*. 2004;127(Pt 12):2572–2583.
44. van Kooten IA, Palmen SJ, von Cappeln P, et al. Neurons in the fusiform gyrus are fewer and smaller in autism. *Brain*. 2008;131 (Pt 4):987–999.
45. Mulder EJ, Anderson GM, Kema IP, et al. Platelet serotonin levels in pervasive developmental disorders and mental retardation: diagnostic group differences, within-group distribution, and behavioral correlates. *J Am Acad Child Adolesc Psychiatry*. 2004;43:491–499.
46. Oblak A, Gibbs TT, Blatt GJ. Reduced serotonin receptor subtypes in a limbic and a neocortical region in autism. *Autism Res*. 2013;6(6):571–583.
47. Baruth JM, Wall CA, Patterson MC, Port JD. Proton magnetic resonance spectroscopy as a probe into the pathophysiology of autism spectrum disorders (ASD): a review. *Autism Res*. 2013;6:119–133.
48. Aoki Y, Kasai K, Yamasue H. Age-related change in brain metabolite abnormalities in autism: a meta-analysis of proton magnetic resonance spectroscopy studies. *Transl Psychiatry*. 2012 Jan 17;2:e69.
49. Mori K, Toda Y, Ito H, et al. A proton magnetic resonance spectroscopic study in autism spectrum disorders: amygdala and orbitofrontal cortex. *Brain Dev*. 2013;35:139–145.
50. Carter CS. Sex differences in oxytocin and vasopressin: implications for autism spectrum disorders? *Behav Brain Res*. 2007;176(1):170–186.
51. Geshwind D. Autism genetics and genomics: a brief overview and synthesis. In: Amaral D, Dawson G, Geshwind D, eds. *Autism Spectrum Disorders*. New York: Oxford University Press; 2011:812–826.
52. Sanders SJ, Murtha MT, Gupta AR, et al. De novo mutations revealed by whole-exome sequencing are strongly associated with autism. *Nature*. 2012;485(7397):237–241.
53. Gregory SG, Connelly JJ, Towers AJ, et al. Genomic and epigenetic evidence for oxytocin receptor deficiency in autism. *BMC Med*. 2009;7:62.
54. Howlin P, Magiati I, Charman T. Systematic review of early intensive behavioral interventions for children with autism. *Am J Intellect Dev Disabil*. 2009;114(1):23–41.

CHAPTER

7

Rett Syndrome: From the Involved Gene(s) to Treatment

Charlotte Kilstrup-Nielsen, Nicoletta Landsberger*, †*

*Laboratory of Genetic and Epigenetic Control of Gene Expression, Department of Theoretical and Applied Sciences, Division of Biomedical Research, University of Insubria, Busto Arsizio, Italy, †San Raffaele Rett Research Center, Division of Neuroscience, San Raffaele Scientific Institute, Milan, Italy

OUTLINE

INTRODUCTION

Imagine Anna, a wonderful 8-month-old girl sitting in her high chair and turning the pages of a book while looking at it. Imagine Anna's mother showing you other pictures of her daughter, smiling to her siblings or grasping objects. Everything seems normal, but then, a few months later, the pictures are different. Anna is not smiling any more, the expression on her face is different, the brightness has disappeared, and in many pictures Anna has protruding jaws (Fig. 7.1). Anna's mother tells me, "this is when I realized that something was changing .... At that time Anna's progress stopped, the ability to hold the book and turn its pages was lost, overcome by continuous stereotyped hand-wringing movements. Rett syndrome and its regression phase was taking Anna away, locking her in her body for good."

Anna is now 15. She is wheelchair bound, unable to talk and to play; like most girls affected by Rett syndrome (RTT) she suffers from seizures, hypotonia, constipation, scoliosis, osteopenia, and breathing irregularities. Like most girls affected by typical RTT she has a mutation in the X-linked *MECP2* gene.

Today, almost 30 years after RTT was internationally recognized as a unique disorder mainly affecting girls, we know that RTT is a rare genetic disease which,

Neurobiology of Brain Disorders
http://dx.doi.org/10.1016/B978-0-12-398270-4.00007-0

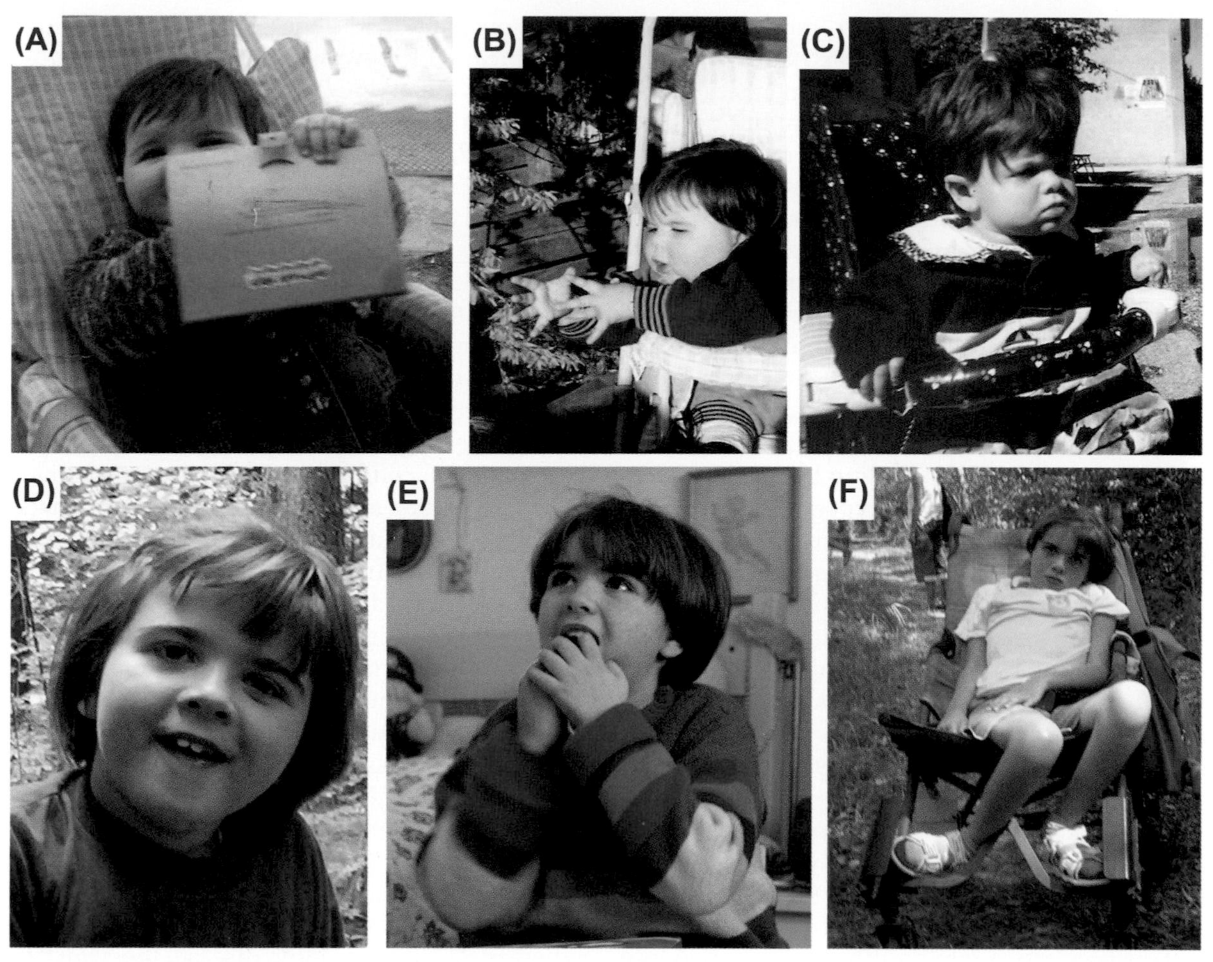

FIGURE 7.1 **Photographs showing Anna from when she was an apparently healthy little girl to an adolescent with fully developed Rett syndrome.** (A, B) Anna at 8 and 9 months of age was still using her hands. (C) At 15 months Anna was socially withdrawing. (D,E) Anna at 7 and 8 years, respectively: the typical hand stereotypies are evident in (E); in both images the intense eye gaze appears. (F) At 10 years of age Anna is wheelchair bound with a prominent scoliosis.

because of its prevalence (roughly 1 in 10,000 born girls), can be considered one of the most frequent causes of intellectual disability in females worldwide. Hundreds of different loss-of-function *MECP2* mutations have been associated with RTT or, less frequently, with other forms of intellectual disability, such as autism, schizophrenia, mental retardation, and Angelman-like syndrome; duplication and triplication of the *MECP2* gene have been identified as the genetic cause of the *MECP2* duplication syndrome in males. Altogether, these molecular data suggest that MeCP2 is a key protein in the brain, the level and functions of which cannot be altered without severe consequences.

This chapter will describe the clinical manifestations of RTT and other *MECP2*-related disorders, current knowledge of the role of the transcription factor MeCP2 and its expression, and perceptions of how defects in this protein can lead to neuronal and non-neuronal dysfunctions. The relevance of animal models in the study of RTT and MeCP2 functions will be discussed, along with data demonstrating that, at least in mice, RTT is not an irreversible condition. Finally, the chapter will focus on the most recent translational studies that may lead to the development of therapies for RTT.

CLINICAL FEATURES OF RETT SYNDROME AND OTHER *MECP2*-RELATED DISORDERS

RTT (MIM 312750) is traditionally considered a pediatric neurological disorder with a delayed onset of symptoms, which has to be clinically diagnosed relying on specific criteria.[1] Girls affected by typical RTT are born apparently healthy after a normal pregnancy and uneventful delivery, and appear to develop normally usually through the first 6–18 months of life. Then, their neurological development appears to arrest and, as the syndrome progresses, a regression phase occurs that leads to a documented loss of early acquired developmental skills, such as purposeful hand use, learned single words and babble, and motor skills. During the regression phase, patients develop gait abnormalities and almost continuous stereotypic hand wringing, washing, clapping, and mouthing movements that constitute the hallmark of the disease. Often, but not necessarily, typical RTT girls show a deceleration of head growth, leading to acquired microcephaly. During regression, autistic features often manifest, such as social withdrawal, irritability, expressionless face, hypersensitivity to sound, and indifference to the surrounding environment and social cues. It is worth noting that old observations and more recent videos have reported that soon after birth many RTT girls manifest subtle alterations of tone, feeding, crying, and spontaneous neonatal movements. Many other severe clinical features are associated with typical RTT, including breathing abnormalities (breath-holding, apnea, hyperventilation, and forced expulsion of air and saliva), seizures (which vary in severity and tractability), hypotonia and weak posture, scoliosis, weight loss, bruxism, hypotrophic cold-blue feet, severe constipation, and cardiac abnormalities, often presenting prolonged QT intervals.[2] RTT patients often live into adulthood, although a slight increase in the mortality rate is observed, which is often caused by sudden deaths, probably triggered by breathing dysfunctions and cardiac alterations.

As RTT girls grow up, they may improve their social behavior and communication, mostly by means of increased eye-to-eye contact and eye pointing; furthermore, seizures tend to improve after the teenage years. In contrast, as they move into adulthood they often lose weight, and acquire rigidity, parkinsonian features and severe motor deterioration, together with worsening of scoliosis. Thus, RTT represents a severely debilitating physical condition that requires constant assistance throughout life.

In addition to typical RTT, clinical variants or atypical RTT patients have been described. These individuals have only some of the clinical RTT features, which may be either milder or more severe. According to their clinical features, these patients are generally classified in four main distinct groups.[1,2] The *forme fruste* or *worn-down form* and the *preserved speech variant* represent the milder forms of RTT. The first has a later age of onset (between 1 and 3 years of age) and hand usage is preserved to some extent, whereas the second is mainly characterized by the capability of patients to speak few words. The *early-onset seizure variant* and the *congenital variant* are among the most severe forms; the former is distinguishable by the onset of severe, and often intractable, seizures within the first 5 months of life; the latter is characterized by a grossly abnormal initial development and the lack of the typical intense RTT eye gaze.

The genetic background of RTT patients is far from homogeneous. Most cases can be regarded as sporadic, in that they arise from *de novo* mutations. *MECP2* is the most commonly mutated gene (90–97%) in patients affected by typical RTT; however, only 50–70% of atypical cases are positive on *MECP2* testing. Mutations in other loci have been found in some atypical RTT patients; mutations in *CDKL5* are often identified in patients affected by the early-onset seizure variant,[3] whereas *FOXG1* mutations can be found in the congenital form.[4]

RTT was initially considered an exclusively female disease. Boys with *MECP2* mutations that cause typical RTT in girls usually show very severe postnatal encephalopathy, death within the first years of life, and lack of distinctive RTT features. However, it is now well accepted that boys with typical RTT exist; these patients may either carry the pathogenic mutation in a Klinefelter chromosomal condition (47,XXY) or as a somatic mosaicism, or harbor still unknown modifier gene(s), which suppress the infantile lethality generally observed in these patients. An *MECP2* duplication syndrome has also been defined; it is associated with duplications of Xq28, which includes the *MECP2* gene.[5] Cardinal features of the affected boys are early infantile hypotonia, delayed psychomotor development, severe intellectual disability, epileptic seizures, progressive spasticity and, in most cases, susceptibility to recurrent respiratory infections that may reduce survival. Importantly, even though this disorder remains to be further characterized, male patients often present common dysmorphic facial features and brain imaging studies report structural and progressive defects. It is worth observing that whereas RTT is a sporadic disease, the *MECP2* duplication syndrome is inherited with 100% penetrance from duplicated carrier mothers. In general, these mothers are asymptomatic because of extreme or complete skewed inactivation of the duplicated X chromosome; however, neuropsychiatric features, such as depression, anxiety, and autistic behaviors, or intellectual disability have been described in *MECP2* duplicated females.

To conclude, although the best defined *MECP2*-related condition is RTT, it is now recognized that mutations in this gene can cause a broad spectrum of neuropsychiatric disorders, the full spectrum of which needs to be fully defined. *MECP2* mutations can also be found in females displaying Angelman-like phenotypes, mild to severe mental retardation, learning disabilities and attention-deficit disorders, and occasionally general autism (Fig. 7.2).

DIAGNOSIS AND CLINICAL MANAGEMENT OF RETT SYNDROME

Since 1999, when the *MECP2* gene was discovered as the main genetic cause of RTT, diagnosis has relied on molecular genetic testing. Complete sequence analysis of *MECP2* coding regions, large insertion and deletion

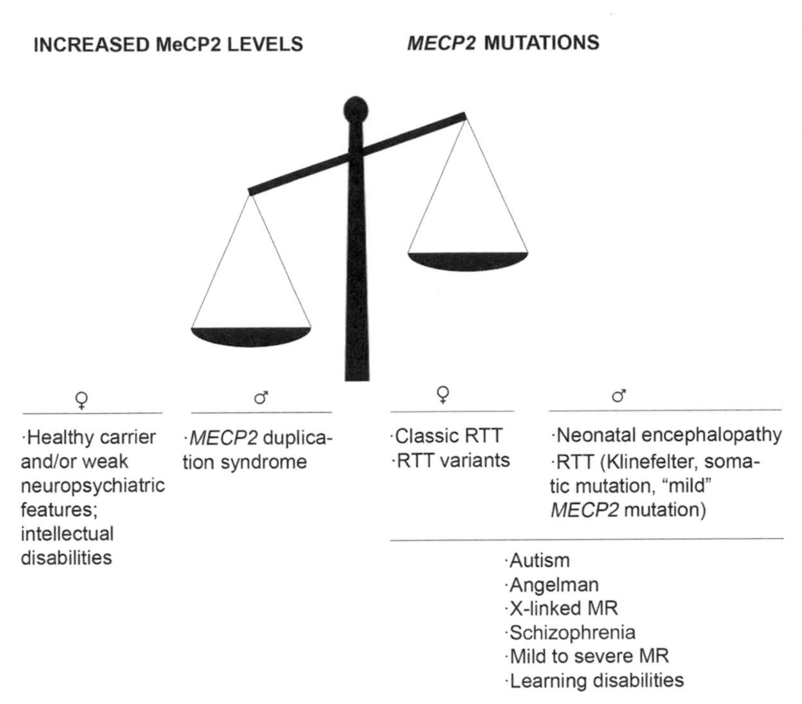

FIGURE 7.2 **Different pathological conditions are associated with mutations or duplications of *MECP2*.** An increase in *MECP2* copies is mostly associated with the *MECP2* duplication syndrome in males but can cause neuropsychiatric features or intellectual disability in females (left). *MECP2* mutations cause distinct neurological disorders in females and males depending on the mutation type and other known and unknown genetic modifiers.

analyses, and testing for exonic, multiexonic, and whole-gene deletions and duplications are available on a clinical basis. Whenever negative, genetic tests may include additional causative genes such as *FOXG1*, *CDKL5*, and *MEF2C*. However, considering that, first, even when using the best methodologies, almost 5% of typical RTT patients and 30–40% of atypical RTT individuals do not show any mutation in *MECP2*, and, second, *MECP2* mutations can cause several neuropsychiatric disorders, it appears that genetic alterations of this gene are neither absolutely required nor sufficient for a diagnosis of RTT. Thus, the diagnosis of RTT still remains clinical. Because of that, and to limit confusion, the diagnostic criteria of typical and atypical RTT have recently been revisited (Table 7.1).[1] Importantly, a diagnosis of RTT requires a documented period of regression that should occur within the first 5 years of life and should not be determined by any other primary cause of neurological dysfunction (such as brain injury, neurometabolic disease, or severe infection). In this perspective, the diagnosis of the congenital variant of RTT may not be fully appropriate, because these patients lack a clear history of regression.

Although *MECP2* mutations have profound effects on the brain, neurodegeneration has not been observed either in mice mimicking *MECP2* disorders (see "*Mecp2* Mouse Models", below) or in RTT patients. (As already observed, the duplication syndrome may constitute an exception.) Consistent with the lack of neuronal loss, several studies have demonstrated that RTT is not an irreversible condition in mice and phenotypic rescue is possible. However, no cure for RTT has been identified so far, and patients are only given symptomatic and supportive treatments. Clinical management requires lifelong multidisciplinary approaches. Physiotherapy, hippotherapy, music therapy, rehabilitation programs, and several other treatments aim to facilitate emerging skills, improve communication and social interactions, and control posture and movements, thereby limiting muscle waste. Pharmacological treatments may be used to control seizures, respiration, constipation, and other specific dysfunctions. Bracing or surgical intervention is often required to correct scoliosis. Increased tone in the Achilles tendons generally signals the onset of rigidity; ankle–foot orthoses together with physiotherapy are used to maintain foot alignment and reduce cord shortening, whereas various arm restraints can help to decrease stereotyped hand movements. Finally, psychosocial support for the family is often recommended.

GENETICS OF RETT SYNDROME: *MECP2* GENE, PATHOGENIC MUTATIONS, AND PHENOTYPIC OUTCOME

Even though the genetic origin of RTT was suggested by monozygotic twin concordance and few cases of vertical transmission of the disorder had been reported, the sporadic occurrence of the disease made the identification of the involved gene particularly challenging.

Considering that most of the affected patients with RTT were females, in 1983 Bengt Hagberg hypothesized that the causative gene was located on the X chromosome. In an X-linked dominant model, hemizygous males would not survive, whereas females, heterozygous

TABLE 7.1 Revised Diagnostic Criteria for Rett Syndrome (RTT)

RTT diagnostic criteria 2010	
	Consider diagnosis when postnatal deceleration of head growth is observed
Required for typical or classic RTT	
	A period of regression followed by recovery or stabilization[a]
	All main criteria and all exclusion criteria
	Supportive criteria are not required, although often present in typical RTT
Required for atypical or variant RTT	
	A period of regression followed by recovery or stabilization[a]
	At least two of the four main criteria
	Five out of 11 supportive criteria
Main criteria	
	Partial or complete loss of acquired purposeful hand skills
	Partial or complete loss of acquired spoken language[b]
	Gait abnormalities: impaired (dyspraxic) or absence of ability
	Sterotypic hand movements such as hand wringing/squeezing, clapping/tapping, mouthing and washing/rubbing automatisms
Exclusion criteria for typical RTT	
	Brain injury secondary to trauma (perinatally or postnatally), neurometabolic disease, or severe infection that causes neurological problems[c]
	Grossly abnormal psychomotor development in first 6 months of life[d]
Supportive criteria for atypical RTT[e]	
	Breathing disturbances when awake
	Bruxism when awake
	Impaired sleep pattern
	Abnormal muscle tone
	Peripheral vasomotor disturbances
	Scoliosis/kyphosis
	Growth retardation
	Small cold hands and feet
	Inappropriate laughing/screaming spells
	Diminished response to pain
	Intense eye communication: "eye pointing"

TABLE 7.1 Revised Diagnostic Criteria for Rett Syndrome (RTT)—cont'd

[a]*Because* MECP2 *mutations are now identified in some individuals before any clear evidence of regression, the diagnosis of "possible" RTT should be given to those individuals under 3 years old who have not lost any skills but otherwise have clinical features suggestive of RTT. These individuals should be reassessed every 6–12 months for evidence of regression. If regression manifests, the diagnosis should then be changed to definite RTT. However, if the child does not show any evidence of regression by 5 years, the diagnosis of RTT should be questioned.*

[b]*Loss of acquired language is based on best acquired spoken language skill, not strictly on the acquisition of distinct words or higher language skills. Thus, an individual who had learned to babble but then loses this ability is considered to have a loss of acquired language.*

[c]*There should be clear evidence (neurological or ophthalmological examination and magnetic resonance imaging/computed tomography) that the presumed insult directly resulted in neurological dysfunction.*

[d]*Grossly abnormal to the point that normal milestones (acquiring head control, swallowing, developing social smile) are not met. Mild generalized hypotonia or other previously reported subtle developmental alterations during the first 6 months of life are common in RTT and do not constitute exclusionary criteria.*

[e]*If an individual has, or has ever had, a clinical feature listed it is counted as a supportive criterion. Many of these features have an age dependency, manifesting and becoming more predominant at certain ages. Therefore, the diagnosis of atypical RTT may be easier for older individuals than for younger. In the case of a younger individual (<5 years old) who has a period of regression and at least two main criteria but does not fulfill the requirement of five out of 11 supportive criteria, the diagnosis of "probably atypical RTT" may be given. Individuals who fall into this category should be reassessed as they age and the diagnosis revised accordingly.*

Source: From Neul et al. Ann Neurol. *2010;68:944–950.*[1]

for the defective gene, would manifest the symptoms. Through linkage analysis performed on rare multiplex familial cases, in 1998 the region of interest was mapped to Xq28. This discovery led the laboratory of Huda Zoghbi to report, a year later, six different alterations in the *MECP2* gene, encoding the methyl-CpG-binding protein-2 (MeCP2), as disease-causing mutations. Thus, for the first time a deficient epigenetic regulation of gene expression was linked to the pathogenesis of RTT. Indeed, MeCP2 had already been discovered at that time and partially characterized by Adrian Bird and his collaborators; it was recognized as an abundant and ubiquitously expressed nuclear protein that selectively binds methylated DNA with no sequence preference and represses transcription *in vitro*. Accordingly, the protein was preferentially localized on methylated pericentric heterochromatin in mouse cells. Furthermore, in 1998, MeCP2 was demonstrated to participate in a complex containing histone deacetylases, thus providing the first link between two epigenetic mechanisms: DNA methylation and chromatin structure/post-translational modifications (PTMs).[6]

Structurally, the human MeCP2 protein is 486 residues long and contains four main functional domains: an 85-amino acid amino-terminal methyl-CpG binding domain (MBD), a 104-amino acid transcriptional repression domain (TRD), a nuclear localization signal (NLS) located within the TRD, and a carboxy-terminal domain that facilitates MeCP2 binding to both naked and nucleosomal DNA (Fig. 7.3A).

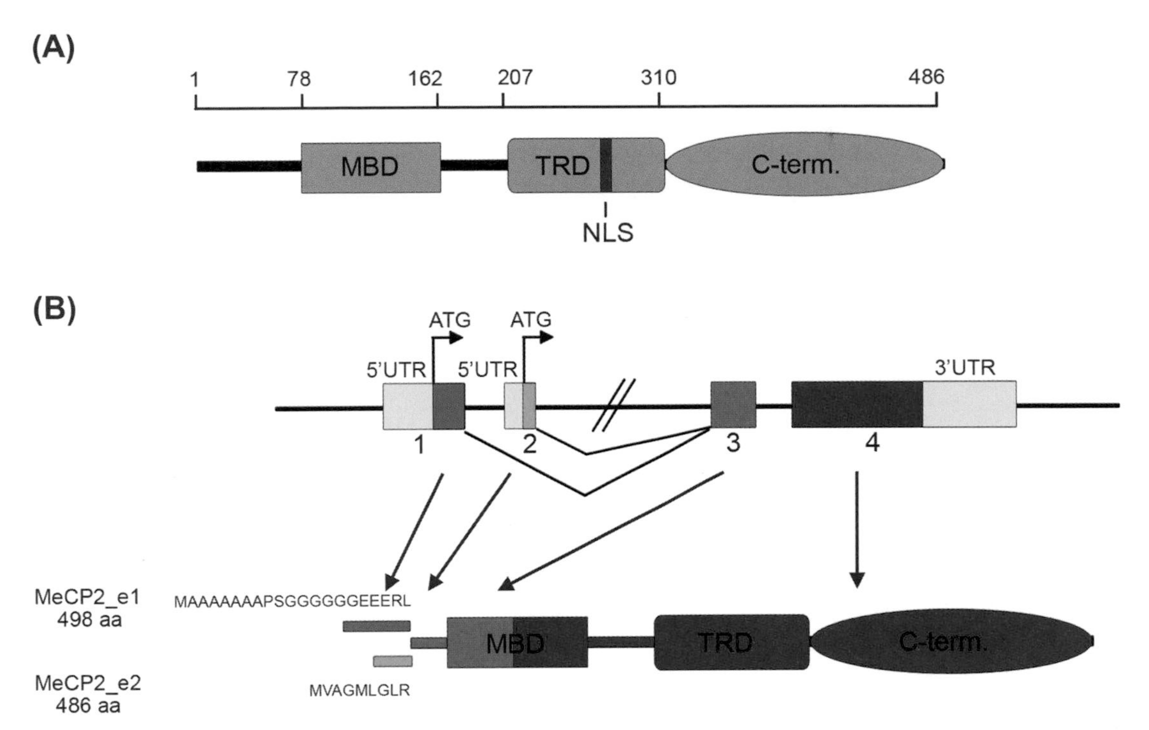

FIGURE 7.3 **Schematic illustration of MeCP2.** (A) MeCP2 contains within its N-terminus the methyl-DNA-binding domain (MBD), whereas the transcriptional repression domain (TRD) containing a nuclear localization signal (NLS) is located in the central portion of the protein. The C-terminus (C-term) consists of 176 amino acids. The scale bar above refers to the human MeCP2_e2 isoform. (B) MeCP2 is encoded by four exons, of which exons 1 and 2 are alternatively spliced to exon 3 (upper panel). MeCP2_e1 and MeCP2_e2 contain distinct extreme N-terminal regions as depicted in the lower panel. The color code indicates the correspondence between the distinct exons and various regions of the protein. The light gray regions in exons 1, 2, and 4 correspond to untranslated regions (UTR). ATG indicates the translational start sites in exons 1 and 2.

The *MECP2* gene spans almost 76 kb and consists of four exons (Fig. 7.3B). The gene is present in all vertebrates, but no *MECP2* orthologues have been found in invertebrates or plants. Among mammals, the MeCP2 protein is highly conserved. Because of alternative splicing, two different MeCP2 proteins are generated: the MeCP2_e1 isoform contains 21 unique N-terminal residues and is encoded by exons 1, 3, and 4; MeCP2_e2 is encoded by exons 2, 3, and 4 and has nine unique residues; all the remaining amino acids are identical between the two isoforms.[2] Although the two isoforms have distinct expression patterns, with MeCP2_e2 being 10 times less abundant than the e1 isoform in postnatal brain, it has generally been assumed that the two isoforms are functionally equivalent. A study in 2012 suggested that Mecp2_e2, but not e1, can promote neuronal cell death. The neurotoxic activity of MeCP2_e2 appears to be constrained by its preferential interaction with FOXG1, a transcription factor that, as previously mentioned, is involved in RTT.[7]

In addition to the coding regions, exons 1 and 2 of the *MECP2* gene express a short 5′ untranslated region (UTR) (167 nucleotides long), whereas exon 4 contains a large 3′ UTR that because of distinct polyadenylation sites can vary in length. A long form of over 8.5 kb appears to be preferentially expressed in brain, suggesting that the 3′ UTR may contain regulatory functions.

Over 500 RTT-causing *MECP2* mutations have been identified so far and are catalogued in web-based mutation databases, including http://mecp2.chw.edu.au/, http://www.mecp2.org.uk/, and http://aussierett.org.au/our-research/search-our-databases.aspx. Missense, nonsense, and frameshift mutations can be found in all domains of MeCP2, meaning that they are all required for the proper function of the protein. However, although missense mutations can be found throughout the entire open reading frame, a significant amount falls within the MBD; deletion/insertion mutations, in contrast, are more frequent in the C-terminal portion. Furthermore, there are eight hotspots of mutations, which account for almost 65% of all mutations found in classical RTT (Fig. 7.4). All of these hotspot mutations are C to T transitions, probably caused by unrepaired spontaneous deamination of methylated cytosines. Despite this being common to many other point mutations causing human genetic diseases, these findings are in good accordance with RTT being mainly a sporadic disease.

Several studies have addressed whether a relationship between genotype and phenotype exists; in other words, whether different types of mutation in *MECP2* can be associated with the severity of specific symptoms. These studies have generally not reported consistent results. Furthermore, it is well known that most

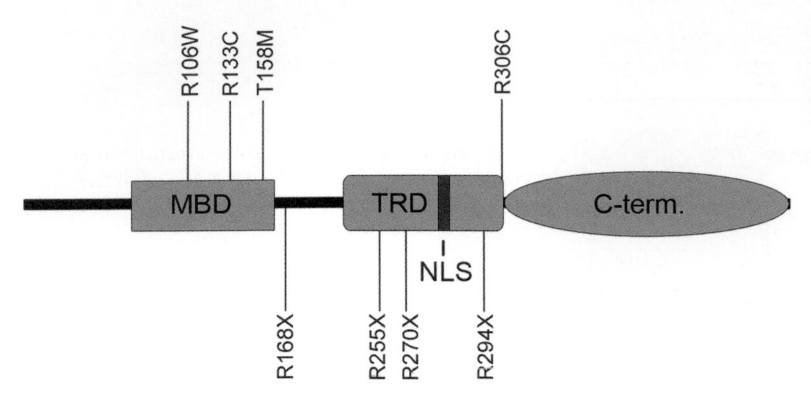

FIGURE 7.4 **Eight hot-spot mutations account for 65% of all *MECP2* mutations.** The four most common missense mutations are indicated above the protein and truncating mutations are indicated below. X indicates a truncating mutation.

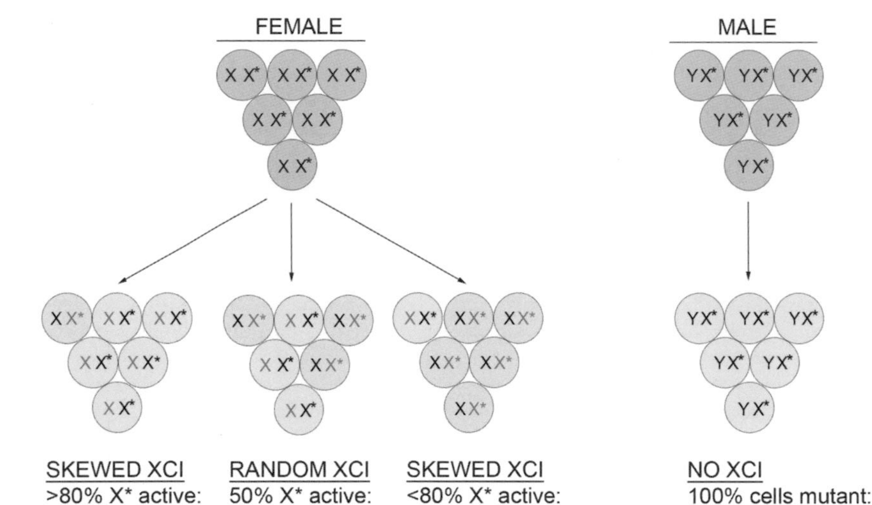

FIGURE 7.5 **X-chromosome inactivation.** All female cells contain two X chromosomes (left panel), one of which carries an *MECP2* mutation (X*). Random X-chromosome inactivation renders the tissue mosaic, with 50% of cells expressing the mutated allele (blue balls). Skewed inactivation, causing most cells to express the mutated allele, will lead to an aggravation of the phenotype (far left); in healthy carriers favorably skewing of X-inactivation results in silencing of the mutant allele in most cells (red balls, far right). In males, the X chromosome is not inactivated and an *MECP2* mutation will be expressed in all cells (right panel).

MECP2 mutations can lead both to typical RTT and one of its variant forms. Therefore, it is well accepted that the clinical severity of RTT cannot be predicted solely depending on the type of mutation: other factors play a major role in the phenotypic outcome and among these is the X-inactivation pattern. Because of X-inactivation, each cell in female mammals contains only one active sexual chromosome. Thus, since X-inactivation in general is random, a female RTT patient is usually a mosaic, characterized by having half of her cells expressing the normal *MECP2* allele and the other half the mutated one (Fig. 7.5). However, if this process is not random (skewed), the clinical phenotype is significantly affected. In favorable skewing, the mutated X chromosome is inactivated in most cells and no clinical manifestations are overt (as seen in silent carriers); alternatively, the mutated allele can be preferentially expressed, causing an aggravation of the phenotype that depends on the skewness of X-inactivation. It is generally accepted that genetic modifiers of *MECP2*, which remain to be identified, also affect the severity of the disease. Indeed, a few reports show the existence of girls who do not have RTT symptoms despite having common mutations in *MECP2* and a random X-inactivation pattern. It is likely that these girls have mutations in other genes that confer protection from damaged *MECP2*.[8]

To summarize, the extreme variability in the clinical phenotypes associated with *MECP2* mutations is the result of complex interactions among the type of mutation, X-chromosome inactivation, and the presence of genetic or epigenetic modifiers, or both.

MeCP2 MOUSE MODELS RECAPITULATING HUMAN MECP2-RELATED PATHOLOGIES

Several mouse models carrying different *Mecp2* alterations were generated during the early twenty-first century and are being continuously produced to improve the comprehension of the functions of MeCP2 and the mechanisms underlying RTT and other *MECP2*-related conditions (summarized in Table 7.2).[2,24,25] The Cre-Lox technology was used to generate the first two *Mecp2*-null mice that recapitulate very well many RTT features, thus providing a formal genetic proof of the involvement of

TABLE 7.2 Principal Mouse Models Useful for Understanding the Role of MeCP2 in Rett Syndrome (RTT) Pathophysiology

	Name	Description	Phenotype	Importance	Take-home message
Constitutive knockout	*Mecp2*-Bird [9]		Null allele	Neurological symptoms from 3–8 weeks including ataxic gait, hindlimb clasping, hypoactivity, tremor, breathing problems, piloerection, seizures. Death at 6–10 weeks	These mouse models have been instrumental in establishing that the loss of Mecp2 is sufficient to cause an RTT-like phenotype. These models remain those mainly used to study RTT pathophysiology
	Mecp2-Jae [10]		C-terminal peptide present	As Mecp2-Bird model	
Overexpression	*Mecp2*-Tg1 [11]		Two-fold ubiquitous overexpression	Symptoms at 10 weeks: forepaw clasping, aggressiveness, hypoactivity, seizures; 30% of mice die within 1 year	These mice showed that increased Mecp2 levels cause neurological symptoms in accordance with *MECP2* duplication syndrome in humans
Conditional knockout	Nestin-cKO [10]		Deletion of *Mecp2* in neurons and glia in embryogenesis	Similar to *Mecp2*-null mice with respect to symptoms and onset	Loss of Mecp2 in neurons is sufficient to cause a large array of RTT symptoms; Mecp2 critical in mature neurons
	CamKII-cKO [10]		Deletion of *Mecp2* in postmitotic neurons	Onset of symptoms at 3 months that recapitulate most symptoms in *Mecp2*-null mice	
	GFAP-cKO [12]		Postnatal deletion of *Mecp2* in astrocytes	Respiration deficits and some hindlimb clasping, smaller body size	Loss of Mecp2 in glia contributes to some RTT symptoms
Human pathogenic mutations	$Mecp2^{308/y}$ [13]; $Mecp2^{T158A/y}$ [14]; $Mecp2^{A140V/y}$ [15]; $Mecp2^{R168X/y}$ [16]		Ubiquitous expression of *Mecp2* with RTT causing missense or nonsense mutations	Milder and/or partial phenotypes with prolonged lifespan	Useful for understanding the pathophysiology of RTT
Adult inactivation	$Mecp2^{lox/}y$, CreESR^{T} [17–19]		TAM-induced inactivation of *Mecp2* at juvenile and adult stages	Male mice develop RTT symptoms including hypoactivity, tremors, hindlimb clasping, breathing difficulties, locomotor and learning deficits after Mecp2 removal and have shortened lifespan	Mecp2 functions required also in adult brain
Rescue models	*Mecp2*-Stop-Cre+TAM [20]		Reactivation of endogenous *Mecp2* in 80% of all cells	Reactivation in fully symptomatic male and female mice reverts most symptoms	Demonstrates that RTT is not an irreversible condition. Has been confirmed with other models
	$Mecp2^{Stop/y}$-hGFAPcreT2+TAM [12]		Expression of *Mecp2* in astrocytes of otherwise *Mecp2*-null mice	Partial rescue of lifespan, locomotion, respiration deficits	These mice confirm the contribution of glia in the RTT phenotype
Phospho-defective	$Mecp2^{S80A/y}$ [21]; $Mecp2^{S421A}$ [22]; $Mecp2^{S421,424A}$ [21,23]		*Mecp2* derivatives carrying substitutions of specific serines with the non-phosphorylatable alanine	Specific effects on locomotor activity and behavior	Useful for understanding *in vivo* the role of specific phosphorylation events

The color of the mice indicates the presence of normal Mecp2 levels (light gray), absence of Mecp2 (white), or increased levels (dark gray).
cKO: conditional knockout; GFAP: glial fibrillary acidic protein; TAM: tamoxifen; S: serine; A: alanine.

the *MECP2* gene in RTT. In particular, *Mecp2*-null males (*Mecp2*$^{-/y}$) have no apparent phenotype until 3–8 weeks of age, when they develop gross abnormalities, such as hindlimb clasping, stiff and uncoordinated gait, hypotonia, reduced spontaneous movements, tremors, breathing irregularities, and often seizures. Symptoms worsen over time and the animals lose weight and die by the age of 6–10 weeks. Compared with controls, these mice have smaller brains, reduced cortical thickness, and smaller and more densely packed neurons, showing immature synapses. Heterozygous female mice (*Mecp2*$^{-/+}$) are viable and fertile, and appear normal up to 4–6 months of life, when they start manifesting RTT-like symptoms, such as inertia, ataxic gait, hindlimb clasping, and often irregular breathing. In contrast to the null males, females usually survive longer than 10 months and become overweight.

It is important to observe that, because of the long time required for symptoms to appear in *Mecp2*$^{-/+}$ mice and the associated phenotypic variability, probably caused by differential X inactivation, most studies have focused on the *Mecp2*-null male model, which develops a condition similar to that observed in *MECP2*-mutated boys. Considering that RTT predominantly affects girls, it is reasonable to assume that the appropriate genetic mouse model of RTT should be the heterozygous female; on the other hand, *Mecp2*-null mice are considered a suitable model to address the function of the *Mecp2* gene and its role in the development of RTT-like phenotypes.[24]

Other models with less severe genetic lesions, often mimicking human mutations, have subsequently been generated. In 2002, Shahbazian and collaborators[13] generated an RTT mouse model (*Mecp2*$^{308/y}$) (Table 7.2) expressing a hypomorphic truncated Mecp2 derivative that, similarly to the case in many RTT patients, lacks the C-terminal domain but spares the MBD and TRD. Even though the overall phenotype is milder, it again recapitulates many features of RTT. The hemizygous males appear normal up to 6 weeks of age, when they start showing progressive, neurological phenotypes, such as tremors, hypoactivity, seizures, kyphosis, anxiety, forepaw stereotypies, and learning and memory deficits. These mice are fertile, with a normal body weight, and survive up to 1 year of age. Once again, the heterozygous females are less affected and display more variable symptoms. Although these three mouse models are the most commonly used, many others have been generated, including one carrying a truncating mutation, *Mecp2 R168X*, and two with different substitutions of single amino acids: *Mecp2 T158A* and *Mecp2 A140V*. The last three mutations represent RTT-causing mutations and are therefore useful in directly analyzing the pathophysiology involved in the disease.

As already mentioned, an *MECP2* duplication syndrome has recently been classified; therefore, a mouse line overexpressing Mecp2 has been generated (*Mecp2-TG1*). This model is still little uncharacterized[25] but the mice show hindlimb clasping, aggressiveness, hypoactivity, and seizures, and die between 20 weeks and 1 year of age, reinforcing the importance of finely tuned Mecp2 expression.

To better understand the etiology of RTT and the role of *Mecp2* in discrete brain regions or cell types, conditional knockout mice have been developed and characterized. The gene was initially inactivated in neurons and glia using a mouse line expressing a *Nestin*-driven Cre recombinase that causes reduced *Mecp2* expression starting from embryonic day 12 onwards. Overall, the obtained animals developed similarly to the null strain, suggesting that Mecp2 is required in postnatal phases of development and may have a role in neuronal maturation and/or maintenance. Accordingly, the inactivation of *Mecp2* in forebrain neurons during early postnatal development (mediated by a *CamK-Cre93*-expressing mouse line) led to similar, albeit milder and postponed symptoms. Subsequent studies, inactivating *Mecp2* in single brain areas or neuronal subtypes, have reproduced only a subset of the typical features of RTT. A lack of Mecp2 in dopaminergic neurons causes dysfunction of motor coordination, loss in serotoninergic neurons produces increased aggression, whereas *Mecp2* inactivation in the basolateral amygdala results in anxiety and learning deficits. Finally, the specific deletion of *Mecp2* in hypothalamic Sim1-expressing neurons reveals a role of Mecp2 in feeding behavior, aggression, and response to stress, whereas dysfunction in GABAergic neurons recapitulates many features of RTT.[25] Importantly, as already alluded to, despite all *Mecp2*-inactivating mutations investigated so far having profound effects on the brain, they are not associated with neuronal loss. Consistently, a breakthrough in the field came in 2007, when Bird and his collaborators demonstrated that *Mecp2* reactivation in symptomatic adult mice (either *Mecp2*$^{-/y}$ or *Mecp2*$^{-/+}$) results in an almost complete phenotypic rescue, demonstrating that RTT, at least in mice, is not an irreversible condition and MeCP2-related disorders can be treated even at late stages of disease progression.[20] These results have been confirmed by several subsequent studies.

Considering all the above, it appears that the presence of MeCP2 is not crucial during early development; rather, the effects of its deficiency ensue after the development of the brain is complete or, at least, when it has reached a certain stage of maturation. It is therefore plausible that RTT is not a neurodevelopmental disease and that MeCP2 functions are essential to maintain a fully functional mature neuron. This hypothesis has been tested by addressing the neurological consequences of an adult inactivation of the *Mecp2* gene. Three studies, published in 2011 and 2012, showed that depletion of *Mecp2* at different postnatal stages (from late juvenile animals, 3 weeks old, to adults aged 20 weeks, in both males and females) always causes the appearance of RTT-like phenotypes and premature death.[17–19] Several clinical and cellular features observed in the mouse models of RTT are reproduced upon the late deletion of *Mecp2*. Neuronal cells pack more densely and the brain shrinks even when *Mecp2* is deleted at the adult stage and in both hemizygous males and heterozygous females, which are characterized by a 50% mosaic depletion.[18] The authors speculate that RTT girls, in addition to the phase of developmental stagnation, may move through a period in which the mature brain shrinks while dendritic complexities are reduced. If this were the case, it would clarify why the hallmark of RTT is the presence of a regression phase.

In summary, mouse models of RTT have indicated that the lack of functional Mecp2 always causes severe and overlapping neurological symptoms, independently of the developmental stage in which Mecp2 is lost. Therefore, it appears that the mature brain continuously depends on MeCP2 functions. Considering all the findings, it can be postulated that RTT may be treatable even at late stages of the disease, but therapy would have to be continuously maintained.

MeCP2 EXPRESSION DURING BRAIN DEVELOPMENT: ROLE IN NEURONAL MATURATION AND/OR MAINTENANCE OF THE MATURE STATE

Once it was recognized that several, if not all of the symptoms manifested by the *Mecp2*-mutated mice are caused by the lack of the methyl-binding protein in the CNS, it became imperative to define the associated neuropathological abnormalities, as well as the expression pattern of the MeCP2 protein during brain development. Therefore, several studies have used immunoblots and immunohistochemistry to describe the spatial and temporal distribution of MeCP2, mainly in rodent and human brains. Even though some discrepancies exist among the various reports, possibly due to different sensitivity and specificity of the antibodies used against MeCP2, most of the obtained results converge on some common facts.

In general, MeCP2 has been described as a ubiquitously expressed factor, the abundance of which can vary significantly depending on the tissue and developmental stage. In particular, Mecp2 protein expression appears particularly high in brain, high in lung and spleen, moderate in kidney and heart, and quite low in stomach, small intestine, and liver. Most of the early studies agreed that the protein was absent in glial cells, a concept that has recently been completely revised (see "Rett Syndrome: Not Solely a Neuronal Disease", below). *Mecp2* messenger RNA (mRNA) levels do not always overlap with protein levels, suggesting that MeCP2 abundance is post-transcriptionally regulated.

All the reports agree that MeCP2 expression in brain mirrors neuronal maturation; in more details, its expression rises when neurons develop dendritic arbors and project axons, and when connectivity is established.[26,27] In an experimental model that prevents synapse formation, MeCP2 was shown to be present, albeit at a lower level than normal, suggesting that the presence of MeCP2 does not require synapse formation but its abundance depends on it. Furthermore, olfactory biopsies from RTT girls have been compared with those from healthy controls. Even though the overall number of neurons was comparable between the two samples, the RTT samples were characterized by an excess of immature neurons or abnormally structured neurons with retracted dendritic arbors and dispersed axons. Although, in principle, this may reflect a primary defect in neuronal maturation, these results are consistent with a need for MeCP2 to maintain the mature state, in agreement with the more recent data obtained by inactivating Mecp2 in adults (see "*Mecp2* Mouse Models", above). It is important to mention that MeCP2 levels remain elevated throughout adulthood, supporting the relevance of the protein for the functions of differentiated neurons.

The expression pattern of Mecp2 has been finely analyzed in embryonic and postnatal neocortex, including neuronal precursors.[27] The obtained results are consistent with previous data and show increasing expression of MeCP2 from the intermediate zone to the cortical plate correlating with the ontogeny of the CNS. In contrast, less solid data exist regarding Mecp2 expression in undifferentiated cells and through early cortical embryogenesis. Two reports describe the presence of Mecp2 in only a few undifferentiated nestin-positive cells and it is not clear whether the protein is already present in the neocortex at embryonic day 12.[28,29] The few experiments reported so far suggest that MeCP2 is not involved in cell proliferation or differentiation of neuronal precursors, therefore reinforcing its involvement in neuronal maturation and maintenance of the mature status.

NEUROMORPHOLOGICAL AND NEUROPHYSIOLOGICAL CONSEQUENCES OF MeCP2 DYSFUNCTION

The ability to connect clinical manifestations with underlying changes in morphological and physiological neurobiology is often a prerequisite for understanding the pathogenesis of a disease. However, despite fundamental and fast progresses in modeling RTT in animals, we still lack a clear comprehension of the functional role of MeCP2 in brain and the anatomical and physiological consequences of its dysfunction in the CNS of RTT children or *Mecp2*-mutated mice.

Because of space constraints, this section will focus on some results that, to the authors' knowledge, appear to be the most solid data and the most relevant to nonspecialized readers.

It is generally well accepted that the gross structure of RTT brains appears to be preserved and, in accordance with the experiments on phenotypic rescue, there are no obvious signs of degeneration, atrophy, gliosis, or inflammation.[2,26] Furthermore, as already mentioned, data collected so far suggest that the loss of MeCP2 does not affect early stages of CNS development: proliferation, neuronal specification, and migration do not seem to be influenced by the loss of Mecp2 in mice.[27]

The most conspicuous morphological abnormality reported in RTT patients is, as already described, reduced brain size and weight (a 12–34% reduction has been reported), associated with more subtle alterations, such as a reduced dendritic arborization, defects in spine density and morphology, and an increase in neuronal packing, leading to augmented cellular density.[30,31] This evidence has been strengthened by several studies performed on different *Mecp2* mouse models that show perfectly overlapping phenotypes.[27,32] Kishi and Macklis[27] speculated that the neuronal alterations that lead to a significant reduction in cortical thickness may be partly caused by deregulated brain-derived neurotrophic factor (BDNF)–tropomyosin receptor kinase B (TrkB) signaling in the RTT mice. *TrkB* and *Bdnf* conditional mice show morphological alterations that phenocopy those of RTT mice, but are characterized by much less severe symptoms. In addition to these results, a study published in 2009 used embryonic stem cell-derived neurons obtained from *Mecp2*-null mice to monitor the earliest steps of neuronal differentiation.[33] The authors demonstrate that, in the absence of Mecp2, neuronal nuclei (but not glial ones) fail to grow in size and remain significantly smaller; concomitantly, BDNF levels in neurons remain lower with respect to the control cells. In line with previous experiments, reexpression of *Mecp2* in mutant neurons rescues both alterations, thus confirming the direct involvement of Mecp2 in regulating nuclear size.

Dendritic arborization is an important maturation process during the formation of neuronal circuitry; dendritic spines and axons are the structures devoted to conducting and receiving electrical inputs, essential for neurons to communicate. Therefore, several reports have analyzed in detail dendritic and axonal morphology in neurons lacking Mecp2. In particular, an elegant study from Belichenko and colleagues[32] reported that most (33 out of 41) parameters analyzed were altered in *Mecp2*-mutant dendrites. Most strikingly, dendrites were swollen, spine density was altered (generally reduced, but increased in a few brain areas), and spines were characterized by a smaller head and a longer neck (Fig. 7.6). In accordance with previous studies in humans, ultrastructural studies showed that mutant dendrites contained larger mitochondria, with defects in the morphology of cristae. These changes were widespread, involving the hippocampus and motor cortex. Axonal organization was disrupted in the motor cortex in line with complementary, earlier studies suggesting that MeCP2 deficiency may lead to abnormal axonal projections. Importantly, similar results were obtained when observing hippocampal neurons of female RTT patients.[33]

Altogether, these findings suggest that Mecp2 dysfunction may have a strong impact on neural circuits. Accordingly, several studies have shown that *Mecp2*-null hippocampal slices are characterized by significantly reduced spontaneous excitatory synaptic transmission,

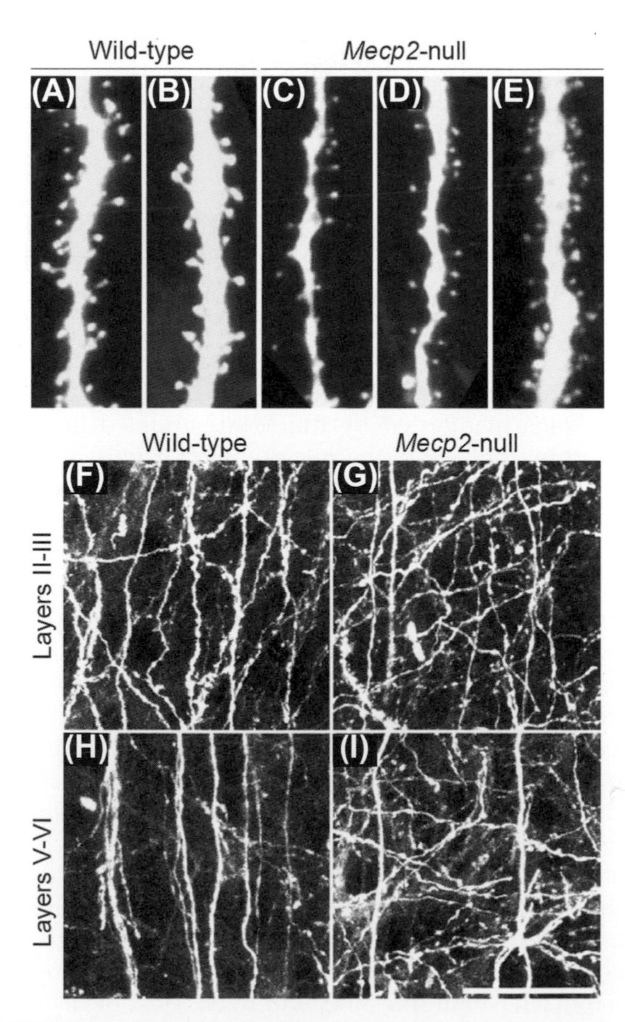

FIGURE 7.6 **Loss of Mecp2 alters neuronal morphology and organization.** (A–E) Dendritic spine density changes in the fascia dentata of *Mecp2* mutant mice (C–E) compared with wild-type mice (A,B). Both reduced and increased spine number can be observed. (F–I) Axonal distribution in motor cortex of mutant mice (G,I) compared with wild-type mice (F,H). *Source: Adapted from Belichenko* et al. J Comp Neurol. *2009;514:240–258.*[32]

deficits in long-term potentiation and long-term depression. Two-photon time-lapse imaging has shown that, at the onset of the disease, *Mecp2*-null somatosensory cortices display remarkable alterations in the dynamics of dendritic spines; in contrast, when maturation of the connectivity is complete, no differences in spine dynamics are evident in *Mecp2*-mutated mice compared with their wild-type controls.[34]

Finally, overexpression of MeCP2 also affects neuronal plasticity. Decreased dendritic branching and spine density, enhanced excitatory synaptic transmission, and altered glutamatergic transmission have been associated with mouse models of the *MECP2* duplication syndrome.

To conclude, it appears that MeCP2 dysfunctions compromise neuronal plasticity, a process of fundamental importance for learning and memory.

RETT SYNDROME: NOT SOLELY A NEURONAL DISEASE

Despite the presence of MeCP2 in many different non-neuronal cell types, several studies supported the idea that RTT was exclusively caused by the lack of MeCP2 in neurons. The two main findings in support of this hypothesis were, first, that MeCP2 was undetectable in glia, and, second, that conditional knockouts specific for neural stem/progenitor cells or postmitotic neurons led to phenotypic manifestations similar to those presented by the ubiquitous *Mecp2*-null mouse.

However, in 2008 and 2009, Western blot and immunostaining analyses clearly demonstrated that Mecp2 is present in glia, although at lower levels than in neurons.[35–37] Moreover, previous reports had demonstrated increased expression of glial genes, involved in neuropathogenic mechanisms, in postmortem female RTT brains, as well as altered glial metabolism in mouse models of RTT. Altogether, these results prompted the study of a possible role of glia in RTT. It is well established that glial cells (astrocytes, microglia, and oligodendrocytes) not only provide trophic and structural support to neurons, but also directly modulate synapse formation and neuronal plasticity. Accordingly, glial cells have been implicated in many neurological disorders; in particular, multiple sclerosis, amyotrophic lateral sclerosis, spinocerebellar ataxia, Huntington disease, and Parkinson disease have a well-documented astrocytic component, whereas microglia are receiving increasing attention in several neurological diseases, such as multiple sclerosis and Alzheimer disease.

Strikingly, Ballas and colleagues demonstrated that *Mecp2*-null astrocytes are unable to support normal growth of neighboring wild-type neurons.[36] By coculturing astrocytes from *Mecp2*$^{-/y}$ mice with wild-type and RTT neurons, or using their conditioned medium, they obtained neurons with stunted dendrites resembling those present in the RTT mouse models. Therefore, they hypothesized that, in RTT female patients, neurons may be affected in a non-cell autonomous fashion by astrocytes expressing the mutated *MECP2* allele and probably secreting a toxic causal factor. A subsequent publication proposed that Mecp2-deficient astrocytes not only are significantly abnormal, revealing morphological and molecular alterations, but also may propagate the Mecp2-deficient state to other astrocytes through gap junction communications, a possibility that is waiting for further support.[37] This publication questioned the effects seen using conditioned medium from *Mecp2*-null astrocytes and, thus, the hypothesis that they secrete a toxic agent.

The *in vitro* evidence has been expanded and supported *in vivo*, again using mouse models. Gail Mandel and her collaborators used inducible Cre mouse lines to address the consequences of removing or reactivating *Mecp2* selectively in astrocytes.[12] Importantly, they found that reintroduction of Mecp2 selectively in astrocytes in an otherwise null background significantly improved several clinical symptoms of the RTT mice, such as longevity, locomotor abilities, and respiratory patterns. Morphologically, neuronal dimensions and their dendritic arbors returned to a wild-type phenotype. In contrast, the selective removal of Mecp2 from astrocytes in an otherwise wild-type background did not affect longevity and neuronal morphology, but led to irregular breathing and minor motor defects. The authors interpreted these results by postulating that neurons play a major role in controlling the initiation of most RTT-like phenotypes, but not breathing, whereas astrocytes largely control the progression of the disease. This interpretation parallels similar results and the proposed roles of neurons and glia in familial amyotrophic lateral sclerosis.

The importance of glia in RTT has been further demonstrated by a 2012 study addressing whether microglia, constituted by brain-resident macrophage-like cells of hematopoietic origin, may play a role in this devastating disease.[38] Transplantation of wild-type bone marrow into irradiated *Mecp2*-null mice (thus permitting microglia engraftment of brain parenchyma) at postnatal day 28, when symptoms normally start appearing, arrested the progression of the disease. In other words, the presence of wild-type microglia in an *Mecp2*-null male mouse increased the lifespan, normalized the breathing patterns, improved locomotor activity, and restored normal body weight, whereas the neuronal dimension was unaffected. These benefits were impeded when phagocytic activity of the engrafting wild-type microglia was inhibited. Altogether, these results suggest that microglia have a major role in the pathophysiology of RTT and suggest that the inability of MeCP2-deficient microglia to clear debris caused by normal neuronal death and membrane shedding may lead to a suboptimal milieu for neurons that are already malfunctioning because of the *MECP2* mutation. However, it is important to mention that wild-type bone marrow transplantation in a 40-day-old *Mecp2*-null mouse was not able to ameliorate the symptoms, suggesting that this treatment can halt the progression of the disease but cannot rescue the phenotype.

These data prove the relevance of glia to RTT. Although future studies may demonstrate the contribution of other tissues to the clinical symptoms of RTT, it is important to recall that a role of MeCP2 in development of the heart and skeleton has been proposed.[39] Overexpression of Mecp2 in mice may result in fetal loss, caused by cardiac septum hypertrophy. The overexpressing animals are also characterized by skeletal abnormalities, further suggesting that Mecp2 plays key roles in non-neuronal tissues.

MeCP2: A MULTIFUNCTIONAL PROTEIN WHOSE PATHOGENIC MECHANISMS REMAIN UNSOLVED

Based on the observations that RTT is a not an irreversible disease, there is an urgent need to understand the functional roles of MeCP2 and their relevance to the pathobiology of RTT, in order to develop clinical applications.

As already mentioned, MeCP2 was originally isolated as a nuclear factor capable of binding *in vitro* a DNA probe containing at least one symmetrically methylated CpG-dinucleotide; *in vivo*, the protein was found to accumulate in mouse cells at pericentric heterochromatin, which contains highly methylated satellite DNA. The heterochromatic localization was lost in mouse cells devoid of methylated DNA, further supporting its specificity for methylated DNA.

Considering the well-known role of DNA methylation in transcriptional silencing, the ability of MeCP2 to repress transcription selectively from a methylated template was demonstrated both in cells and *in vitro*. A central domain of almost 100 residues (Fig. 7.3), the TRD, was found to be responsible for the transcriptional repression. The repressive activity could act at a long distance; indeed, silencing could be observed when the protein was tethered to DNA up to 2000 base pairs upstream of the transcription initiation site. It was then demonstrated that the TRD binds to corepressor complexes (Sin3A and NCoR) containing histone deacetylase activity, thereby linking the transcriptional repressive activity of MeCP2 to chromatin compaction (Fig. 7.7A).[6] Several other factors can interact with MeCP2, including the chromatin remodeling complexes Brahma and ATRX, the corepressors c-Ski, CoREST, and LANA, and the epigenetic factors DNA methyltransferase I and H3K9 histone methyltransferase, further reinforcing the link between MeCP2 and chromatin structure.[2,40] Even if *MECP2* mutations hit the entire length of the coding region it is not surprising that a significant part falls within the two principal domains. Only in a few cases have the functional consequences of RTT-causing mutations been clarified; in particular, the R106W mutation, which is one of the most frequent, abolishes the binding of MeCP2 to methylated DNA. Although this has not been demonstrated so far, the capacity of the TRD to recruit corepressors to chromatin is very likely to be impaired by some of the mutations falling within this domain.

The ability of MeCP2 to work as an architectural chromatin protein was supported by a report showing that MeCP2 is a potent chromatin-condensing factor, functioning directly without other corepressor or enzymic activities.[41] The authors demonstrated that depending on its molar ratio to nucleosomes, MeCP2 assembles novel secondary and tertiary chromatin structures independent of DNA methylation; these effects on large-scale chromatin organization represent a good explanation for the already known histone deacetylation-independent repressive activity exerted by MeCP2 and for its ability to repress transcription at a distance *in vivo* (Fig. 7.7B). These studies are particularly relevant since they suggest for the first time that MeCP2 is a multifunctional protein, whose functions (binding to methylated or non-methylated DNA; mechanism of chromatin compaction; long- or short-range effects) are strictly dependent not only on the presence of MeCP2 but also on its abundance.

In addition to its proposed roles in gene silencing and chromatin architecture, other functional roles have been proposed for MeCP2. A study linked the functions of MeCP2 to mRNA splicing because of its capacity to interact with YB1, a protein that has been implicated in regulation of alternative splicing (Fig. 7.7D). However, more studies are required to determine whether the absence of MeCP2 leads to defects in RNA splicing in RTT mouse models and human patients, and the possible involvement of such defects in the phenotypic outcome. Another report showed that MeCP2 may silence gene expression through the formation of chromatin-associated loops. Furthermore, based on transcriptional profiling studies of RNA purified from hypothalami and cerebella of RTT mice, it has been suggested that MeCP2, contrary to expectation, may also act as a transcriptional activator; an interaction between the methyl-binding protein and the transcriptional activator cAMP response element-binding protein (CREB) was identified (Fig. 7.7C).[42] Aberrant neuronal protein synthesis is often considered a cause of the clinical features of autism spectrum disorders, and the mammalian target of rapamycin (mTOR) pathway is a crucial regulator of neuronal cell soma size, dendrite arborization, synaptic function, structure, and

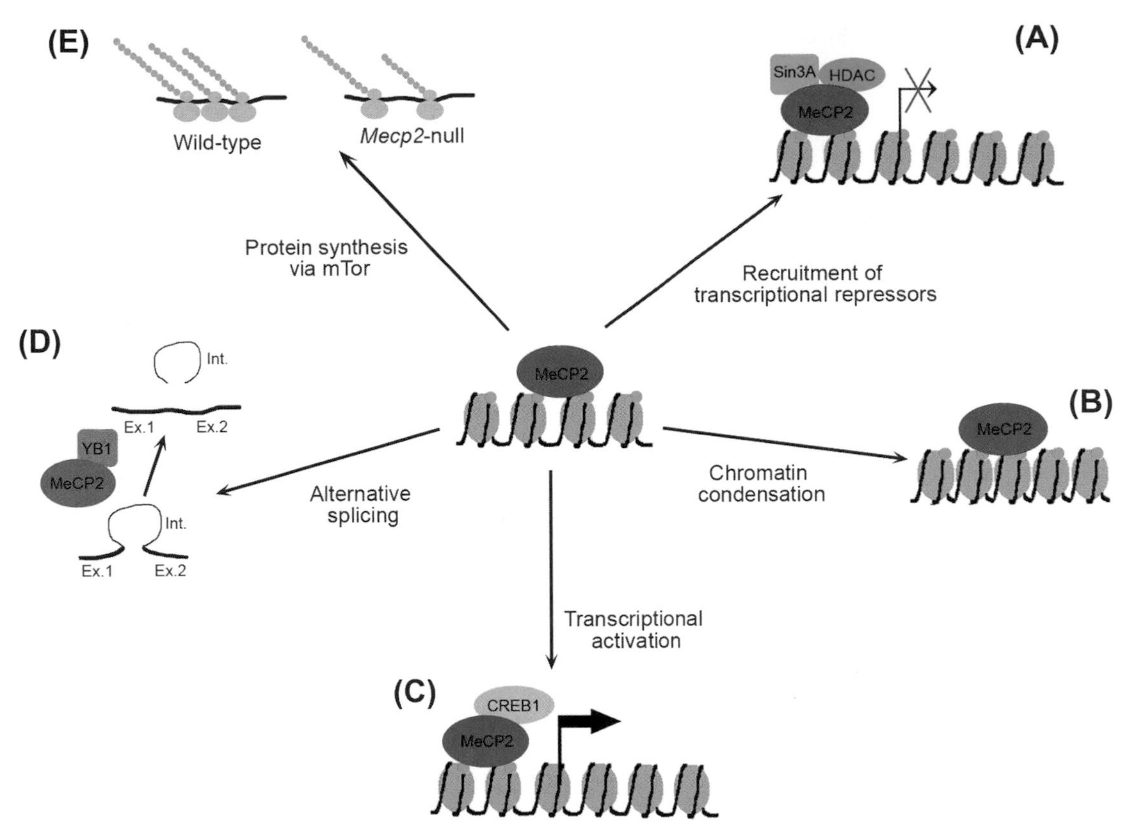

FIGURE 7.7 **MeCP2 is a multifunctional protein associating with methylated DNA (central part; light blue dots indicate methyl-CpGs).** (A) MeCP2 silences gene expression through the recruitment of transcriptional corepressors such as the Sin3A–HDAC complex. (B) MeCP2 can condense chromatin in the absence of other enzymic activities. (C) Through its interaction with CREB1, MeCP2 may act as a transcriptional activator. (D) A role in alternative splicing is suggested by the interaction of MeCP2 with YB1. (E) The absence of Mecp2 in knockout mice causes defects in protein synthesis in an mTOR-dependent manner.

plasticity. In line with this, a 2011 publication showed that AKT/mTOR signaling is reduced in both *Mecp2*-null mice and heterozygous females, and that protein synthesis is significantly impaired in these mice.[43] Although these studies did not address whether this translational defect is a direct cause of the disease or an early indirect effect, or whether it contributes to its progression, they suggest yet another function of MeCP2, that is, a direct or an indirect role in regulating protein synthesis and, therefore, cell homeostasis (Fig. 7.7E).

In an effort to shed light on the role played by MeCP2 in the brain, Skene and colleagues finely quantified MeCP2 levels in isolated neuronal nuclei.[44] Importantly, they found that a single nucleus of mature neurons contains 1.6×10^7 molecules of MeCP2, a number that corresponds roughly to one molecule every second nucleosome (3×10^7/nucleus). MeCP2 levels in glia and other tissues are 10 and 30 times less abundant, respectively. Since methylated CpG dinucleotides are about 5×10^7 per nucleus, in neurons, MeCP2 is sufficiently abundant to saturate a major fraction of the methylated sites of the genome. Accordingly, the authors demonstrated by ChIP-seq analysis that Mecp2 associates preferentially with methylated DNA and that it is genome-wide bound, tracking methylated CpG moieties. In line with a structural role of MeCP2, its deficiency leads to global changes in chromatin structure, such as an increase in histone acetylation and histone H1 levels. Indeed, MeCP2 can substitute for histone H1 in methylated chromatin. Elevated transcriptional noise has been observed from repetitive elements and L1 retrotransposons in neurons, but not in glia, in agreement with a global action of MeCP2 in neurons.[45] Altogether, these data suggest that the protein may use different molecular functions depending on its abundance.[41]

Considering all the above, it would probably be incorrect to describe MeCP2 as a transcription factor targeted to specific genes. Even though several laboratories have used genome-wide expression profile studies or candidate approaches, few reproducible changes have been reported. There may be several causes of this inconsistency. In addition to a global neuronal effect leading to subtle alterations in the expression levels of thousands of genes, which are difficult to reveal, specific expression changes may occur in the *Mecp2*-null brain. These may have a causative role, occurring early, before the onset of overt symptoms, or be secondary

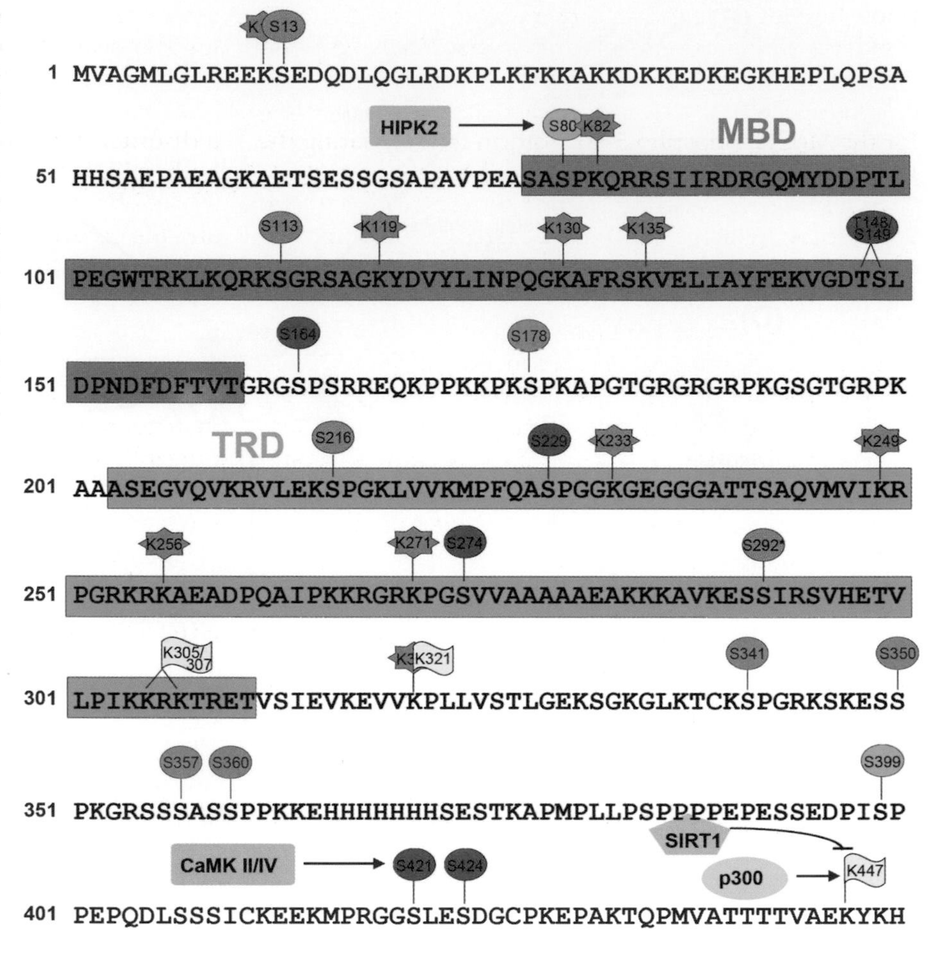

FIGURE 7.8 **The MeCP2 protein is subject to various post-translational modifications.** The sequence of the murine Mecp2_e2 protein is indicated with the methyl-DNA-binding domain (MBD) shown in blue and the transcriptional repression domain (TRD) in green. Amino acids that are subject to phosphorylation are represented by lollipops, of which the green and red ones represent phosphorylation in resting and activated neurons, respectively. The protein kinases HIPK2 and CamKII/IV are associated with phosphorylation of S80 and S421, respectively.[47] Yellow flags indicate acetylated amino acids. P300 is capable of acetylating K447 while SIRT1 causes the reciprocal process.[48] Orange stars represent ubiquitination of distinct lysine residues.[49]

effects, caused by the compromised clinical conditions of the animals or by compensatory mechanisms; therefore, different timings in the various analyses may have influenced the results. Furthermore, MeCP2 may regulate different subset of targets, depending on the brain region; if this were the case, studies applied to whole brains or to complex areas might have diluted away any significant transcriptional changes. Future analyses performed with separate brain regions or with distinct subtypes of neurons will be important to identify genes whose expression is regulated by MeCP2.

One potentially crucial aspect is that the actions of MeCP2 may be modulated. The possibility must be considered that MeCP2 acquires the ability to regulate the expression of specific genes through its PTMs, affecting its activity in response to extracellular cues. Indeed, several reports have testified that differential phosphorylation of MeCP2 in response to neuronal activity is a key mechanism by which the methyl-binding protein modulates gene expression. In particular, in rodent brain the two major phosphorylation sites under resting conditions are serine 80 (S80) and S399, whereas S424 and S421 show specific activity-dependent phosphorylation.[21,46] Several additional residues have been found to be phosphorylated in brain. These data, together with evidence that PTMs other than phosphorylation occur on MeCP2, further support the idea that a complex pattern of PTMs transforms the protein into a regulatory platform whose activities respond to various signaling pathways (Fig. 7.8).

In particular, two studies performed in cultured primary neurons suggested that activity-induced S421 phosphorylation precedes the release of Mecp2 from the promoter of BDNF, therefore releasing its repression.[46,50] These studies led to the hypothesis that Mecp2 may be able to regulate specific neuronal target genes; however, it did not exclude more diffuse effects. This interpretation has been challenged not only by Skene and colleagues,[44] but also by a report demonstrating that, *in vivo*, the phospho-S421 isoform of MeCP2 is globally bound to the genome.[22] Thus, although these studies clearly demonstrated that in mice Mecp2 S421 phosphorylation is induced by neuronal activation and regulates subtle aspects of cortical neuronal morphology, synaptic function, and behavior, they argue against a locus-specific role of MeCP2 and favor a global role for the MeCP2 phospho-S421 isoform in modulating the response of neuronal chromatin to activity.

Summing up, MeCP2 may have both global and local roles, which may be mediated by a combination of different PTMs affecting the affinity of MeCP2 for chromatin and/or specific protein interactors. It is apparent that MeCP2 phosphorylation is required for proper brain development and function *in vivo*. Accordingly, phosphorylation of S80 increases the affinity of MeCP2 for euchromatin in resting neurons and loss of this modification in mice leads to weight gain and decreased locomotor activity, two symptoms already observed in mice modeling the complete or partial loss of Mecp2.[21]

MeCP2 RESEARCH: FROM BENCH TO BEDSIDE

As mentioned already, several reports have demonstrated that phenotypic rescue is possible in *Mecp2*-deficient mice; although these studies did not employ clinically applicable therapeutic approaches, they suggest that *MECP2*-related disorders can be treated, even at late stages of disease progression. This proof of principle validates the next phase of research: finding strategies to cure RTT and related disorders. Gene therapy, targeting the primary underlying cause of the disease, may appear to be an optimal therapeutic strategy; however, it does not represent a valid approach for the near future because of several associated technical challenges (see "*MECP2* Gene Therapy", below); in particular, severe implications for this approach arise from the fact that modest perturbations in MeCP2 levels are deleterious for brain functioning. Therefore, RTT research is seeking processes downstream of MeCP2 that may be responsible for a subset of the symptoms and that can be therapeutically modulated. Sadly, as already discussed, at present we cannot say which are the most relevant functions of MeCP2 in RTT. The possibility that MeCP2 is a global structural protein affecting the chromatin state and expression of thousands of loci further complicates the identification of pharmacologically treatable "MeCP2 target genes". In general, a single deregulated molecular pathway may have a primary role in the pathophysiology of the disease, therefore representing an optimal therapeutic target; however, it may be a secondary effect caused by general compensatory mechanisms or by the compromised health of the subject, and thus be of minor interest for therapy. Several reports have proposed that successful therapeutic interventions should begin early, in the first few months of life, before severe clinical symptoms appear and during early critical periods in cortical development. It should be kept in mind that it is still not known whether RTT is a neurodevelopmental disease or a disorder of the maintenance of neuronal functions, or both; still, it is important to underline that the reversal of the phenotype upon *Mecp2* activation in adult mice, and the occurrence of an RTT-like disease when *Mecp2* is inactivated in adult brains, suggest that a specific time-window should not be mandatory for the treatment of RTT. It is reasonable that at least those treatments that are directed towards secondary effects (such as the possible accumulation of cellular debris in brain due to a deficiency of the *MECP2*-null macrophages) may be significantly more effective when administered early. Similarly, some RTT features may have to be treated before the nervous system is fully mature, whereas others can be ameliorated or fixed at any time.

MECP2 Gene Therapy

The discovery that postnatal activation of the *Mecp2* gene rescues most of the RTT-like features in mice, together with previous publications suggesting that RTT was exclusively a postmitotic neuronal disease, have led the scientific community to consider gene therapy as the most straightforward therapeutic strategy to cure the disease. The strategy would be to deliver a functional and long-term expressing copy of the *MECP2* gene to as many affected cells as possible. However, several challenges and lack of knowledge undermine the possibility of a successful approach. First of all, it has generally been assumed that exogenous *MECP2* should be introduced in brain neurons. However, as discussed, recent data suggest that astrocytes and microglia also participate in the RTT phenotype, implying that a complete reversal of the clinical symptoms may require *MECP2* delivery not only to neurons but possibly even to peripheral tissues. Depending on the targeted cells, different promoter and regulatory sequences may be chosen for expressing *MECP2*. Considering the necessity of expressing physiological levels of MeCP2, the identification of the best regulatory sequences represents an additional entanglement. Furthermore, even considering a cure limited to the CNS, obtaining a widespread transduction of neurons *in vivo* is far from trivial and requires the identification of the most appropriate vector. Currently, lentiviral and adeno-associated viral (AAV) vectors appear as the preferred choices and both have advantages and limitations. In brief, lentiviral vectors infect very efficiently postmitotic cells, such as neurons, and lead to a stable expression of the gene that becomes integrated within the genome. However, these viruses are unable to pass the blood–brain barrier and, therefore, have to be delivered by direct injection into the brain parenchyma; furthermore, they have very limited capacity to spread away from the injection site. Finally, their integration into the genome raises safety concerns linked to a possible insertional mutagenesis or tumorigenesis. As an alternative, the AAV9 serotype appears particularly promising for the therapy of RTT since the recombinant viruses can transduce non-proliferating cells and CNS by systemic administration, and confer long-term stable expression without inflammation and integration into the genome. However, since high viral titers are required to transduce many cells, the connected multiplicity of infection will be in conflict with the necessity of not overexpressing *MECP2*, which, as mentioned, is a dosage-sensitive gene. The dosage concern shows up even in cells infected by a single viral particle; indeed, since RTT female patients are mosaics with half of their cells expressing the wild-type *MECP2* allele, it is generally assumed that introduction of the therapeutic gene will be beneficial only when acquired by the cells expressing the mutated allele, but detrimental in cells expressing the functional protein. Furthermore, it is still not known whether some of the almost 600 different *MECP2* pathogenic mutations may also work as dominant negative, therefore abolishing the efficacy of a possible gene therapy.

A study published in 2013 used AAV9/*MECP2* gene transfer into the CNS of *Mecp2*-null mice to assay the transduction efficiency and the associated benefits or toxicity.[51] When the virus was injected in the CNS of neonatal transgenic mice, transduction efficiencies varied depending on the brain areas, reaching a maximum of almost 40% of cells in hypothalamus and a minimum of approximately 7% in the striatum. Analysis suggested a long-lasting expression of the exogenous MeCP2 at 1–1.25 times native levels. Importantly, in this condition the gene therapy improved some of the RTT-like symptoms and prolonged the lifespan of *Mecp2*-null mice. Furthermore, the authors generated an AAV9 vector containing the *MECP2* gene driven by a portion of the *Mecp2* promoter that should preferentially lead to neuronal MeCP2 expression. When the vector was administered by tail vein injection into juvenile *Mecp2*-null mice, only a few neurons (2.5% on average) were transduced, even when using high viral titers. However, despite the low transduction efficiency, a benefit in survival of the treated animals was observed, although it was accompanied by liver damage, possibly due to overloading with vector copies. The conclusion is that a therapeutic outcome is possible, provided that the appropriate vector, promoter system, and route of administration (direct brain injection, peripheral intravascular injection, or injection into the cerebrospinal fluid) are carefully selected.

Readthrough of *MECP2* Nonsense Codon Mutations

Up to 50% of typical RTT cases are caused by nonsense *MECP2* mutations associated with premature stop codons (PSCs). Aminoglycoside antibiotics, such as gentamicin, can induce suppression of nonsense codons and expression of full-length proteins by permitting ribosomal readthrough of PSCs. Mechanistically, gentamycin binds the ribosome and impedes codon/anticodon recognition, leading to the incorporation of a different amino acid at the PSC level, and thus causing a missense mutation (an issue that may be of some concern with MeCP2, which is particularly sensitive to residue alterations in several positions). This concept has already found clinical application in human diseases, such as Duchenne muscular dystrophy and cystic fibrosis. However, these studies have demonstrated significant toxicity at the required high doses and a limited ability to pass the blood–brain barrier. Because of these limitations a new generation of related compounds (including PTC124 and NB54) has been developed and is under testing. Considering all the above, several common nonsense *MECP2* mutations, either overexpressed in cultured cells or carried by fibroblasts obtained from patients, have been read through by aminoglycoside treatments, leading to low-level expression of full-length MeCP2 that was found to be correctly localized into the cell nucleus. The efficiency, which so far remains very limited, appears to be highly dependent on the compound and the nucleotide context of the nonsense mutation.[52] Therefore, even though these studies constitute a proof of concept of the possibility to treat some RTT patients with readthrough therapy, the several drawbacks (low efficiency, toxicity, and, possibly, the production of proteins containing a missense residue) make this approach unsuitable in the near future.

Pharmacological Modulation of MeCP2 Downstream Targets

Considering that even slightly unmatched levels of MeCP2 can be detrimental, which makes the task of restoring proper MeCP2 levels in cells particularly challenging, alternative promising interventions for RTT may be based on pharmacological approaches meant to modulate the molecular pathways altered by impaired MeCP2 activity.

Although convincing and reproducible data on the direct or indirect targets of MeCP2 are still lacking, the few pathways listed below are generally believed to be potentially relevant to therapeutic approaches.

- *Brain-derived neurotrophic factor*: BDNF is the molecular target of MeCP2 that has so far received most attention for therapy, owing to significant experimental evidence. Besides the already mentioned publication demonstrating the interaction of Mecp2 with the promoter of *Bdnf*,[46,50] several authors have reported a reduction in BDNF levels in RTT mouse brains. Furthermore, manipulation of the levels of this neurotrophic factor by breeding *Mecp2*-knockout mice with lines either overexpressing *Bdnf* or lacking one *Bdnf* allele has demonstrated that levels of this neurotrophin affect RTT symptoms in mice. Some phenotypic overlaps have been observed between mice lacking Mecp2 or BDNF in neurons. Therefore, various strategies have been used and are still being tested to increase BDNF levels in mouse models of RTT and to address the extent of phenotypic reversal. In particular, the levels of the neurotrophic factor have been elevated by administering to mice ampakines (positive modulators of AMPA receptors) or agonists of TrkB or, more recently, fingolimod, a modulator of sphingosine-1 phosphate receptor. In all cases, BDNF levels were partially rescued in mice lacking Mecp2 and an amelioration of some symptoms, such as respiratory functions, lifespan, locomotor activity, and neuronal size, was observed.
- *Insulin-like growth factor-1*: IGF-1 is another growth factor that plays a role in the CNS by promoting neuronal survival, synaptogenesis, and maturation of neuronal plasticity. Importantly, defects in IGF-1 signaling have been associated with autism spectrum disorders, and *Mecp2*-null mice and RTT patients express unusually high levels of IGF binding protein-3 (IGFBP-3) which, by limiting IGF-1, can depress its signaling. Importantly, and differently from BDNF, IGF-1 can cross the blood–brain barrier. Considering all this, Tropea and colleauges systemically treated a mouse model of RTT with an active peptide of IGF-1 and ameliorated lifespan, locomotor functions, breathing patterns, and heart rate, together with brain weight.[53] Since IGF-1 is already an authorized treatment, even in childhood, its tolerance was tested in six RTT patients in an open-label trial.[54] This study demonstrated that the drug is well tolerated in patients, paving the way to a double-blind placebo study involving a larger cohort of patients. (For an updated list of the ongoing and completed trials the reader should visit http://clinicaltrials.gov/ct2/results?term=rett+syndrome.)
- *Improving respiratory abnormalities*: Respiratory disorders are considered among the most devastating features of RTT patients and are generally characterized by erratic respiratory rhythm, breath-holding, and life-threatening apneas. Therefore, many current research efforts are aimed at understanding the pathophysiology of these disturbances and identifying pharmacological interventions. Of relevance to these studies, RTT mouse models faithfully mimic the breathing abnormalities of RTT. Summarizing most of the studies presented in the literature, it appears that at least five different neurochemical signaling systems – norepinephrine, serotonin, glutamate, γ-aminobutyric acid (GABA), and BDNF – are altered in mouse models of RTT and appear to contribute to their breathing phenotype.[55] Therefore, more than one system will have to be pharmacologically treated to restore normal respiratory activity in RTT patients. Ampakine treatment of *Mecp2*-null mice has already been mentioned to increase BDNF levels with a concomitant restoration of normal breathing. Other studies have found their rationale in several reports that demonstrated a reduction in bioamine levels (mainly norepinephrine, serotonin, and dopamine) in postmortem RTT brains and mouse models. Therefore, desipramine, an antidepressant drug, which enhances norepinephrine signaling by blocking its uptake, has been administered to RTT mouse models, demonstrating a positive evolution of breathing symptoms and apneas and a lengthened lifespan. The possibility of improving RTT breathing symptoms via augmented GABA inhibition has also been tested in animal models either using a serotonin agonist or by combining it with a blocker of GABA reuptake. A significant amelioration of apneas, reaching a wild-type phenotype, has been observed. Importantly, a 6-year-old RTT patient received an important respiratory benefit from benzodiazepine treatment. Alterations in acetylcholine and, more specifically, in cholinergic functions have also been observed in the RTT brain. In view of that, a diet enriched in choline was given to *Mecp2*-null mice; however, no significant improvement in the RTT phenotype was observed.[52]

CONCLUSION AND FUTURE CHALLENGES

Dramatic progress has been made since 1999, when the *MECP2* gene was discovered as the RTT-causing gene, providing geneticists with the appropriate molecular diagnostic tools and researchers with a plethora of animal models and molecular pathways that may be relevant for treatments. More importantly, and unexpectedly, several laboratories have in recent years been able to demonstrate that RTT is not an irreversible condition, at least in mice, which has boosted research on the pathophysiology of the disease as well as on the identification of the biological roles of MeCP2 and its targets. Despite this enormous acceleration of research in the RTT field, several questions remain unanswered and novel interrogations have emerged.

Diagnostically, we still have to identify the molecular causes of approximately 5–10% of typical RTT patients and 30–40% of atypical cases that do not show any defects in *MECP2*. So far, very few studies have addressed the possibility that *MECP2* mutations occur outside the coding region of the gene, such as exon–intron boundaries or intronic sequences that affect splicing, 5′ and 3′ UTRs or upstream regulatory regions. In particular, few data have been reported about sequence variants in the very long 3′ UTR, which probably influence mRNA stability, translation, and/or nuclear export. Mutations of the 3′ UTR may also alter microRNA target sites, thereby compromising the regulation of *MECP2* expression. In order to improve the diagnostic tools it will be necessary to understand whether the molecular analysis needs to be extended to include these *MECP2* regions as well. Almost all typical RTT patients will probably harbor a defect in the *MECP2* region but other loci, yet to be identified, may become associated with the variant forms of RTT in the near future. In fact, a high percentage of patients affected by these clinical conditions are negative for *MECP2* mutations and at least two novel genes (*CDKL5* and *FOXG1*) are associated with two specific RTT variants. Since the pattern of X-chromosome inactivation does not fully explain the phenotypic discordance in female patients with identical *MECP2* mutations, the presence of unknown disease modifiers has been postulated. Accordingly, a small number of reports shows the existence of girls free of RTT symptoms despite having common *MECP2* mutations; these girls may carry alterations in modifier genes that confer protection from dysfunctional *MECP2*.[8] The identification and characterization of such hypothesized "protective" modifier genes of MECP2 may lead to novel and powerful therapeutic approaches.

Back in 1992, MeCP2 was isolated as a methyl-binding protein highly enriched in heterochromatin, and several subsequent papers have confirmed this function and classified the protein as a transcriptional repressor; however, novel reports have challenged such a limited view of its biological role. Several other functions have been attributed to MeCP2, including a role in transcriptional activation and binding to non-methylated DNA. Most of the available data designate the function originally proposed by Bird as the main function of MeCP2; however, considering the ability of MeCP2 to form several labile protein–protein interactions and its diverse PTMs, the authors consider it relevant to investigate whether the diverse modified isoforms of MeCP2 have different networking capabilities and whether MeCP2, by changing its protein partners, may switch from a transcriptional repressor to an activator. A central point to be clarified is whether the protein works exclusively as a global architectural factor, tracking methylated DNA, which would make research into direct target genes futile, as suggested by Bird and coauthors.[44] The present authors suggest that the protein may be globally bound to methylated DNA in mature neurons, where its concentration is maximal, but in other tissues, or in immature neurons and glia, where its abundance is significantly lower, MeCP2 may behave as a specific transcriptional regulator. Furthermore, as already discussed, it is reasonable to conceive that out of the several PTMs, some of these may influence either MeCP2 binding to DNA or its interaction with specific coregulators, or both, therefore specifically affecting the expression of the bound genes.

The modulation of MeCP2 by PTMs may constitute a crucial aspect in its being needed to produce and maintain a mature neuronal phenotype. Contrary to a static picture of neuronal circuitry, the continuous remodeling of synaptic contacts and of the proteome of single neurons is required for proper connectivity. Although one tends to be more struck by the ability of nerve circuits to accumulate "memory", it should be kept in mind that equally important are the processes that allow neuronal circuitries to "forget", by appropriately canceling and/or remodeling previously established contacts. This suggests that not only proper establishment of neuronal connectivity, but also the dynamic maintenance of such connectivity constitute essential chores that mature neurons must continually accomplish. In this view, a crucial role would be played by proteins that can influence the overall status of neuronal chromatin, but can also be extensively modulated to finely regulate the expression of batteries of genes involved in plastic modifications in response to neuronal activity.

Whatever the modality used by MeCP2 to associate with the genome, a high priority for therapeutic intervention is the identification of the molecular pathways that are commonly altered when MeCP2 is malfunctioning and their association with specific clinical symptoms. Studies are needed to ascertain which of these pathways are deregulated because of the *MECP2* gene defect and which are secondary effects of a primary deregulation. Whereas primary effects will certainly have to be cured, secondary effects may not even appear in the event of effective causal or symptomatic therapy. The timing of intervention also has to be defined.

MeCP2 has also been suggested in some reports (not discussed here because of space constraints) to affect transcription by regulating the expression of non-coding RNAs, which are deregulated in RTT. Therefore, the question needs to be addressed of which and how non-coding RNAs influence the RTT phenotypic outcome and whether they represent a convenient target for therapies.

Furthermore, considering that MeCP2 abundance and PTMs may change depending on the area of the brain and the maturation stage of the included cells, it is of relevance to determine the target genes and pathways of MeCP2 during embryogenesis, and neuronal progenitors in adults, and to ascertain whether specific RTT symptoms are exclusively associated with a specific brain area. In this regard, it should be noted that although RTT has always been considered a neurodevelopmental disorder (a definition that, as already discussed, has been challenged by the discovery of the need to maintain MeCP2 expression in adult brain[17–19]), most attention has been paid to the role of Mecp2 in mature neurons, when its levels are at their highest, leaving largely incomplete the studies during embryogenesis, when the expression of the protein and its levels are still debated. As mentioned above, a body of evidence points to an early function of MeCP2, including family videos of the perinatal age of young patients, ultrasound recordings of few-day-old RTT mouse pups, and human hemizygous male patients who show the disease immediately after birth. Concerning the role of MeCP2 in different brain districts or tissues, several studies on conditional mice in which *Mecp2* has been selectively removed from specific areas have sought an association between specific symptoms and discrete brain areas. These studies appear to be highly relevant to therapy, given that viral administration of a therapeutic *MECP2* gene is severely limited by the low tropism and limited spatial spread of most viruses. By discovering which districts of the brain most urgently need intervention and in which areas unbalanced levels of MeCP2 have a more detrimental role, a localized therapeutic intervention may be implemented. Last, but not least, several studies have unequivocally demonstrated that RTT is not exclusively a brain disorder[12,36–39]; therefore, it is imperative to understand which other systems participate in the severe syndromic scenario that may be relevant targets for therapies.

Although gene therapy appears to be the most promising definitive approach, several issues make intervention at the level of the gene difficult. The development of a polycistronic vector, including a sequence expressing a small interfering RNA that will suppress the expression of the endogenous *MECP2* but not of the therapeutic gene, may be an interesting approach to be combined with the appropriate regulatory sequences and the best viral vector. A similar method has already been used by Zhou and colleagues in primary neurons.[46] Another challenging approach that is being pursued by some groups consists of reactivating the inactive X chromosome that carries the wild-type *MECP2* allele. Because of the X-linked gene dosage problem, it is conceivable that reactivation should be targeted only to the *MECP2* locus or the immediate vicinity. At present, no resources exist for such a targeted reactivation or for affecting only those cells that have inactivated the normal allele; thus, such a reactivation approach is not likely to be available in the near future.

Research into therapeutic approaches to RTT raises a novel challenge, namely the development of optimal cellular and animal models for drug screening and testing. The best cellular models are likely to come from neurons derived from patients' induced pluripotent stem cells (iPSCs). So far, few publications have demonstrated that MeCP2-deficient iPSCs that have differentiated into neurons recapitulate the deficits previously observed in primary neurons. Thus, considering that these cells offer an unlimited source of RTT cells with various genetic backgrounds, iPSCs hold enormous promise for the development of screens for drug discovery, provided that optimal molecular and phenotypic readouts are identified. Although several animal models have already been generated for RTT, it remains to be seen which are the best to be considered for preclinical trials, which gender should be treated, and whether more models, recapitulating human pathogenic mutations, need to be developed.

Acknowledgments

We would like to dedicate this chapter to Rita and Fausto, who taught us the importance of courage, strength, and love, and to Anna, who inspires our work every day. We would also like to thank ProRETT Ricerca for their continuous support, and all present and former members of the laboratory: your continuous work and efforts are essential to us. Furthermore, we deeply recognize the help of Riccardo Fesce in carefully reading the manuscript. Finally, we apologize to all colleagues whose work could not be cited here owing to space constraints. This work was supported by Fondazione Cariplo (grant 2010-0724 to NL), FP7-PEOPLE-ITN-2008 (CKN), Jerome Lejeune Foundation (CKN and NL), Telethon (grant GGP10032 to NL), and Ministero della Salute (Ricerca finalizzata 2008–Bando Malattie Rare to NL).

References

1. Neul JL, Kaufmann WE, Glaze DG, et al. Rett syndrome: revised diagnostic criteria and nomenclature. *Ann Neurol*. 2010;68:944–950.
2. Chahrour M, Zoghbi HY. The story of Rett syndrome: from clinic to neurobiology. *Neuron*. 2007;56:422–437.
3. Kilstrup-Nielsen C, Rusconi L, La Montanara P, et al. What we know and would like to know about CDKL5 and its involvement in epileptic encephalopathy. *Neural Plast*. 2012;2012:728267.
4. Florian C, Bahi-Buisson N, Bienvenu T. FOXG1-related disorders: from clinical description to molecular genetics. *Mol Syndromol*. 2012;2:153–163.
5. Van Esch H. MECP2 duplication syndrome. *Mol Syndromol*. 2012;2:128–136.
6. Jones PL, Veenstra GJ, Wade PA, et al. Methylated DNA and MeCP2 recruit histone deacetylase to repress transcription. *Nat Genet*. 1998;19:187–191.
7. Dastidar SG, Bardai FH, Ma C, et al. Isoform-specific toxicity of Mecp2 in postmitotic neurons: suppression of neurotoxicity by FoxG1. *J Neurosci*. 2012;32:2846–2855.
8. Takahashi S, Ohinata J, Makita Y, et al. Skewed X chromosome inactivation failed to explain the normal phenotype of a carrier female with MECP2 mutation resulting in Rett syndrome. *Clin Genet*. 2008;73:257–261.
9. Guy J, Hendrich B, Holmes M, Martin JE, Bird A. A mouse Mecp2-null mutation causes neurological symptoms that mimic Rett syndrome. *Nat Genet*. 2001;27:322–326.

10. Chen RZ, Akbarian S, Tudor M, Jaenisch R. Deficiency of methyl-CpG binding protein-2 in CNS neurons results in a Rett-like phenotype in mice. *Nat Genet*. 2001;27:327–331.
11. Collins AL, Levenson JM, Vilaythong AP, Richman R, Armstrong DL, Noebels JL. David Sweatt J, Zoghbi HY. Mild overexpression of MeCP2 causes a progressive neurological disorder in mice. *Hum Mol Genet*. 2004;13:2679–2689.
12. Lioy DT, Garg SK, Monaghan CE, et al. A role for glia in the progression of Rett's syndrome. *Nature*. 2011;475:497–500.
13. Shahbazian MD, Young JI, Yuva–Paylor LA, Spencer CM, Antalffy BA, Noebels JL, Armstrong DL, Paylor R, Zoghbi HY. Mice with truncated MeCP2 recapitulate many Rett syndrome features and display hyperacetylation of histone H3. *Neuron*. 2002;35:243–254.
14. Goffin D, Allen M, Xhang L, et al. Rett syndrome mutation MeCP2 T158A disrupts DNA binding, protein stability and ERP responses. *Nat Neurosci*. 2011;15:274–283.
15. Jentarra GM, Olfers SL, Rice SG, et al. Abnormalities of cell packing density and dendritic complexity in the MeCP2 A140V mouse model of Rett syndrome/X-linked mental retardation. *BMC Neurosci*. 2010;11:19.
16. Brendel C, Belakhov V, Werner H, et al. Readthrough of nonsense mutations in Rett syndrome: evaluation of novel aminoglycosides and generation of a new mouse model. *J Mol Med*. 2011;89: 389–398.
17. McGraw CM, Samaco RC, Zoghbi HY. Adult neural function requires MeCP2. *Science*. 2011;333:186.
18. Nguyen MV, Du F, Felice CA, et al. MeCP2 is critical for maintaining mature neuronal networks and global brain anatomy during late stages of postnatal brain development and in the mature adult brain. *J Neurosci*. 2012;32:10021–10034.
19. Cheval H, Guy J, Merusi C, De Sousa D, Selfridge J, Bird A. Postnatal inactivation reveals enhanced requirement for MeCP2 at distinct age windows. *Hum Mol Genet*. 2012;21:3806–3814.
20. Guy J, Gan J, Selfridge J, Cobb S, Bird A. Reversal of neurological defects in a mouse model of Rett syndrome. *Science*. 2007;315: 1143–1147.
21. Tao J, Hu K, Chang Q, et al. Phosphorylation of MeCP2 at Serine 80 regulates its chromatin association and neurological function. *Proc Natl Acad Sci USA*. 2009;106:4882–4887.
22. Cohen S, Gabel HW, Hemberg M, et al. Genome-wide activity-dependent MeCP2 phosphorylation regulates nervous system development and function. *Neuron*. 2011;72:72–85.
23. Li H, Zhong X, Chau KF, Willims EC, Chang Q. Loss of activity-induced phosphorylation of MeCP2 enhances synaptogenesis, LTP and spatial memory. *Nat Neurosci*. 2011;14:1001–1008.
24. Ricceri L, De Filippis B, Laviola G. Mouse models of Rett syndrome: from behavioural phenotyping to preclinical evaluation of new therapeutic approaches. *Behav Pharmacol*. 2008;19:501–517.
25. Na ES, Nelson ED, Kavalali ET, Monteggia LM. The impact of MeCP2 loss- or gain-of-function on synaptic plasticity. *Neuropsychopharmacology*. 2013;38:212–219.
26. Neul JL, Zoghbi HY. Rett syndrome: a prototypical neurodevelopmental disorder. *Neuroscientist*. 2004;10:118–128.
27. Kishi N, Macklis JD. MECP2 is progressively expressed in post-migratory neurons and is involved in neuronal maturation rather than cell fate decisions. *Mol Cell Neurosci*. 2004;27:306–321.
28. Jung BP, Jugloff DG, Zhang G, Logan R, Brown S, Eubanks JH. The expression of methyl CpG binding factor MeCP2 correlates with cellular differentiation in the developing rat brain and in cultured cells. *J Neurobiol*. 2003;55:86–96.
29. Schmid RS, Tsujimoto N, Qu Q, et al. A methyl-CpG-binding protein 2-enhanced green fluorescent protein reporter mouse model provides a new tool for studying the neuronal basis of Rett syndrome. *Neuroreport*. 2008;19:393–398.
30. Armstrong D, Dunn JK, Antalffy B, Trivedi R. Selective dendritic alterations in the cortex of Rett syndrome. *J Neuropathol Exp Neurol*. 1995;54:195–201.
31. Bauman ML, Kemper TL, Arin DM. Pervasive neuroanatomic abnormalities of the brain in three cases of Rett's syndrome. *Neurology*. 1995;45:1581–1586.
32. Belichenko PV, Wright EE, Belichenko NP, et al. Widespread changes in dendritic and axonal morphology in Mecp2-mutant mouse models of Rett syndrome: evidence for disruption of neuronal networks. *J Comp Neurol*. 2009;514:240–258.
33. Chapleau CA, Calfa GD, Lane MC, et al. Dendritic spine pathologies in hippocampal pyramidal neurons from Rett syndrome brain and after expression of Rett-associated MECP2 mutations. *Neurobiol Dis*. 2009;35:219–233.
34. Landi S, Putignano E, Boggio EM, Giustetto M, Pizzorusso T, Ratto GM. The short-time structural plasticity of dendritic spines is altered in a model of Rett syndrome. *Sci Rep*. 2011;1:45–51.
35. Rusconi L, Salvatoni L, Giudici L, et al. CDKL5 expression is modulated during neuronal development and its subcellular distribution is tightly regulated by the C-terminal tail. *J Biol Chem*. 2008;283:30101–30111.
36. Ballas N, Lioy DT, Grunseich C, Mandel G. Non-cell autonomous influence of MeCP2-deficient glia on neuronal dendritic morphology. *Nat Neurosci*. 2009;12:311–317.
37. Maezawa I, Swanberg S, Harvey D, LaSalle JM, Jin LW. Rett syndrome astrocytes are abnormal and spread MeCP2 deficiency through gap junctions. *J Neurosci*. 2009;29:5051–5061.
38. Derecki NC, Cronk JC, Lu Z, et al. Wild-type microglia arrest pathology in a mouse model of Rett syndrome. *Nature*. 2012;484: 105–109.
39. Alvarez-Saavedra M, Carrasco L, Sura-Trueba S. Demarchi Aiello V, Walz K, Neto JX, Young JI. Elevated expression of MeCP2 in cardiac and skeletal tissues is detrimental for normal development. *Hum Mol Genet*. 2010;19:2177–2190.
40. Guy J, Cheval H, Selfridge J, Bird A. The role of MeCP2 in the brain. *Annu Rev Cell Dev Biol*. 2011;27:631–652.
41. Georgel PT, Horowitz-Schere RA, Adkins N, Woodcock CL, Wade PA, Hansen JC. Chromatin compaction by human MeCP2. Assembly of novel secondary chromatin structures in the absence of DNA methylation. *J Biol Chem*. 2003;278:32181–32188.
42. Chahrour M, Jung SY, Shaw C, et al. MeCP2, a key contributor to neurological disease, activates and represses transcription. *Science*. 2008;320:1224–1229.
43. Ricciardi S, Boggio EM, Grosso S, et al. Reduced AKT/mTOR signaling and protein synthesis dysregulation in a Rett syndrome animal model. *Hum Mol Genet*. 2011;20:1182–1196.
44. Skene PJ, Illingworth RS, Webb S, et al. Neuronal MeCP2 is expressed at near histone-octamer levels and globally alters the chromatin state. *Mol cell*. 2010;37:457–468.
45. Moutri AR, Marchetto MC, Coufal NG, et al. L1 retrotransposition in neurons is modulated by MeCP2. *Nature*. 2010;468:443–446.
46. Zhou Z, Hong EJ, Cohen S, et al. Hu, L, Steen JA, Weitz CJ, Greenberg ME. Brain-specific phosphorylation of MeCP2 regulates activity-dependent Bdnf transcription, dendritic growth, and spine maturation. *Neuron*. 2006;52:255–269.
47. Bracaglia G, Conca B, Bergo A, et al. Methyl-CpG-binding protein 2 is phosphorylated by homeodomain-interacting kinase 2 and contributes to apoptosis. *EMBO Rep*. 2009;10:1327–1333.
48. Zocchi L, Sassone-Corsi P. SIRT1-mediated acetylation of MeCP2 contributes to BDNF expression. *Epigenetics*. 2012;7:695–700.
49. Gonzales ML, Adams S, Dunaway KW, LaSalle JM. Phosphorylation of distinct sites in MeCP2 modifies cofactor association and the dynamics of transcriptional regulation. *Mol Cell Biol*. 2012;32:2894–2903.
50. Chen WG, Chang Q, Lin Y, et al. Depression of BDNF transcription involves calcium-dependent phosphorylation of MeCP2. *Science*. 2003;31:885–889.
51. Gadalla KK, Bailey ME, Spike RC, et al. Improved survival and reduced phenotypic severity following AAV9/MECP2 gene transfer to neonatal and juvenile male Mecp2 knockout mice. *Mol Ther*. 2013;21(1):18–30.

52. Gadalla KK, Bailey ME, Cobb SR. MeCP2 and Rett syndrome: reversibility and potential avenues for therapy. *Biochem J*. 2011; 439:1–14.
53. Tropea D, Giacometti E, Wilson NR, et al. Partial reversal of Rett syndrome-like symptoms in MeCP2 mutant mice. *Proc Natl Acad Sci USA*. 2009;106:2029–2034.
54. Pini G, Scusa MF, Congiu L, et al. IGF1 as a potential treatment for Rett syndrome: safety assessment in six Rett patients. *Autism Res Treat*. 2012;2012:679801.
55. Katz DM, Dutschmann M, Ramirez JM, Hilaire G. Breathing disorders in Rett syndrome: progressive neurochemical dysfunction in the respiratory network after birth. *Respir Physiol Neurobiol*. 2009;168:101–108.

CHAPTER

8

Fragile X-Associated Disorders

Scott M. Summers[*,†], Randi Hagerman[*,‡]*

*Medical Investigation of Neurodevelopmental Disorders (MIND) Institute, †Department of Psychiatry and Behavioral Sciences, University of California-Davis, Sacramento, California, USA, ‡Department of Pediatrics, University of California – Davis, Sacramento, California, USA

OUTLINE

INTRODUCTION

This chapter will cover the whole spectrum of fragile X-associated disorders caused by mutations in the fragile X mental retardation 1 gene (*FMR1*). This gene is located on the bottom end of the X chromosome at Xq27.3, where a fragile site appears in those with fragile X syndrome (FXS) when cells are grown in folate-deficient tissue culture media (Fig. 8.1). The gene was sequenced in 1991 and it is characterized by a trinucleotide repeat expansion (CGGn) in the 5′ end next to the promoter region. Most individuals in the general population have 29 or 30 CGG repeats in *FMR1* and the protein produced by this gene (FMRP) is an RNA transporter protein that regulates translation of hundreds of messages that are important for synaptic plasticity and intellectual development.[1] FMRP regulates the translation of approximately 30% of all of the known genes associated with autism, including neuroligins, neurorexins, *PSD95*, *Arc*, *Shank*, *PTEN*, *NOC1*, and hundreds of others.[2] FXS is the most well known fragile X-associated disorder because it is the most common cause of inherited intellectual disability and the most common single gene cause of autism. Approximately 2–6% of those with an autism spectrum disorder (ASD) and 2–3% of those with intellectual disability will have FXS. FXS is caused by a full mutation, meaning over 200 CGG repeats, and this large expansion usually undergoes methylation, which shuts down transcription of the gene. Therefore, little or no *FMR1* messenger RNA (mRNA) is produced, leading to little or no FMRP (Fig. 8.2). It is the lack of FMRP that causes FXS.

A patient with the full mutation and FXS always receives the mutation from their mother, who is a carrier of a premutation (55–200 CGG repeats) or a full mutation herself. Those with a premutation are very common in the general population, 1 in 200 females and 1 in 400 males, whereas the full mutation occurs in approximately

Neurobiology of Brain Disorders
http://dx.doi.org/10.1016/B978-0-12-398270-4.00008-2

1 in 5000.[3] Male carriers of the premutation will always pass to all of their daughters only the premutation because only the premutation (not the full mutation) is able to survive in the sperm. The expansion to the full mutation only occurs when passed on by a female, so that these daughters will often go on to have children with the full mutation and FXS. Whether a mother with a premutation will pass on a premutation or a full mutation depends on the length of her premutation and the presence of AGG anchors that typically occur after every 10 CGG repeats.[4] Mothers with over 100 CGG repeats and no AGG anchors will always pass on a full mutation from this mutated X and the child who receives this will have FXS.

In the past those with the premutation were considered unaffected clinically and most have normal intellectual abilities with normal levels of FMRP. In 1991, it was initially reported that women with the premutation were at higher risk of having early ovarian failure (menopause before age 40); this is now termed fragile X-associated primary ovarian insufficiency (FXPOI).[5] Then, in 2001, five male premutation carriers were found to have a neurodegenerative disorder that was subsequently named the fragile X-associated tremor ataxia syndrome (FXTAS).[6,7] At the time of the discovery of FXTAS, the molecular underpinning of these problems was found to be elevated levels of *FMR1* mRNA that ranged from two to eight times normal.[8] This is the opposite of what occurs in FXS, where there is little or no *FMR1* mRNA (Fig. 8.2). The higher the CGG repeat within the premutation range, the higher the level of mRNA and the earlier the onset of FXTAS.

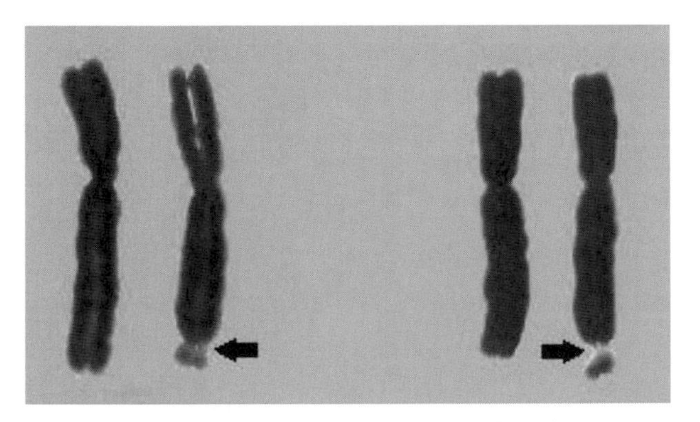

FIGURE 8.1 **Two pairs of X chromosomes from karyotypes of women with fragile X syndrome.** Note the fragile sites indicated by the black arrows.

RNA TOXICITY IN PREMUTATION CARRIERS

The elevated levels of *FMR1* mRNA present in carriers can lead to a gain-of-function RNA toxicity that is similar to what is seen in myotonic dystrophy.[9] The elevated levels of mRNA also have the expanded premutation CGG repeats on each strand (Fig. 8.2), which form secondary structures, such as hairpin formations. These structures bind proteins that are important for cellular function, such as Sam 68, DROSHA, and DGCR8.[10] These last two proteins are critical for maturing pri-microRNA so that in premutation carriers there is dysregulation of some microRNAs[10] in addition to oxidative stress. In culture of premutation neurons, there is oxidative stress, mitochondrial dysfunction, a decrease in the number of dendritic branches, and neuronal excitability with excessive spikes.[11,12] The cause of the clinical problems that occur in premutation carriers is primarily secondary to the RNA toxicity and protein dysregulation in the cell, although some individuals at the upper end of the premutation range have mild deficits of FMRP, which can also add to the dysfunction of the neuron. Some premutation carriers can present with intellectual disability and these individuals often have a significant deficit of FMRP or a second genetic defect such as a significant copy number variant (CNV) elsewhere in the genome.

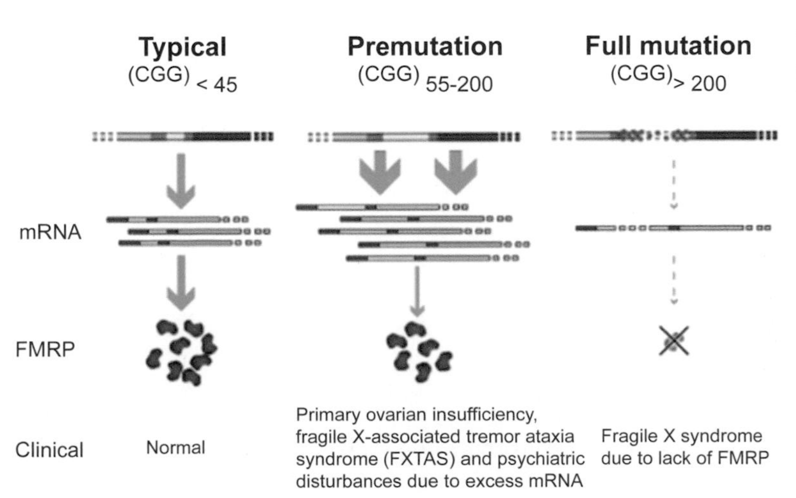

FIGURE 8.2 **Relationship between CGG repeat number, mRNA levels, FMRP, and clinical involvement among the various *FMR1* gene states.**

FRAGILE X SYNDROME

Most children with FXS are identified between 2 and 3 years of age because of language delay, which initiates a medical workup that includes fragile X DNA testing to document the CGG repeat expansion in *FMR1*. Any child with developmental delay or ASD should have fragile X DNA testing. These children often have poor eye contact, hand flapping, hand biting, hyperextensible finger joints, and prominent ears, although most look very normal physically (Fig. 8.3).

Females with FXS often present as less affected cognitively, but shyness, social anxiety, and language delays are common. Children with FXS are often tactilely defensive and hypersensitive to sound and other stimuli, and are usually labeled as having sensory integration problems with hyperactivity. Neurochemically, the lack of FMRP leads to enhanced presynaptic release of neurotransmitters[13] so that the synapse is flooded, the action potential is widened, and subtle differences in neuronal activation are not perceived. The inhibitory γ-aminobutyric acid (GABA) system is also downregulated in the absence of FMRP[14] and clinically disinhibition or impulsivity occurs. Often parents are overwhelmed by the hyperactivity and tantrums of children with FXS. Anxiety is often the underlying symptom for many of the behavioral problems; therefore, the use of a selective serotonin reuptake inhibitor, such as sertraline, is helpful early on.[15] Guanfacine is helpful for calming hyperarousal and hyperactivity, particularly for children under 5; after the age of 5 stimulants are useful for the majority of children.[16]

FIGURE 8.3 **Young adult male with fragile X syndrome (FXS) and unusual finger movements typical for the disorder.** Note the lack of any other obvious physical features associated with FXS.

As children with FXS age the intellectual disability becomes more apparent, and most boys have an intelligence quotient (IQ) less than 70, but 15% are high functioning with an IQ above 70. These boys are usually mosaics, meaning that some cells have the premutation and other cells have the full mutation. Mosaicism can also occur with methylation differences in the *FMR1* gene; therefore, some cells with the full mutation have a lack of methylation so that mRNA and FMRP are produced to some extent, whereas other cells have the full mutation that is fully methylated. The amount of FMRP that is produced correlates with IQ. Females with FXS always have one normal X chromosome while the other X has a full mutation; thus, they are typically producing FMRP from their normal X chromosome if it is on the active X. The activation ratio can be measured in blood: it represents the percentage of cells that have the normal X as the active X. A high activation ratio such as 0.70 means that 70% of cells have the normal X as the active X; this favorable activation ratio will be associated with a higher IQ than a ratio of 0.20. Approximately 30% of girls with FXS will have an IQ less than 70, 40% will have a borderline IQ, and 30% will have an IQ in the normal range, but problems with executive function deficits and attention deficit/hyperactivity disorder (ADHD) are common in girls with a normal IQ.

ASD occurs in 60% of boys with FXS and about 20–30% of girls with FXS. A workup of ASD is always indicated when a diagnosis of FXS is obtained, so that behavioral interventions for autism can be started as early as possible. The presence of autism also alerts the physician to a more careful workup for seizures, which occur in 20% of patients with FXS and are associated with the autistic phenotype.[17] Individuals with FXS plus autism typically have a lower IQ than those with FXS alone. Because their language and behavior are more dysfunctional, sometimes with aggression, the use of an atypical antipsychotic such as aripiprazole is more common in this subgroup.[16]

Targeted Treatments for Fragile X Syndrome

This is a new age of targeted treatments for FXS because research in animal models, such as the knockout mouse where the *fmr1* gene is eliminated, has shown upregulation of the metabotropic glutamate receptor-5 (mGluR5) pathway, downregulation of the $GABA_A$ pathway, and upregulation of matrix metalloproteinase-9 (MMP9) levels in FXS.[18] The results of human FXS trials of mGluR5 antagonists are promising[19] and a subgroup of children, particularly those with ASD or social

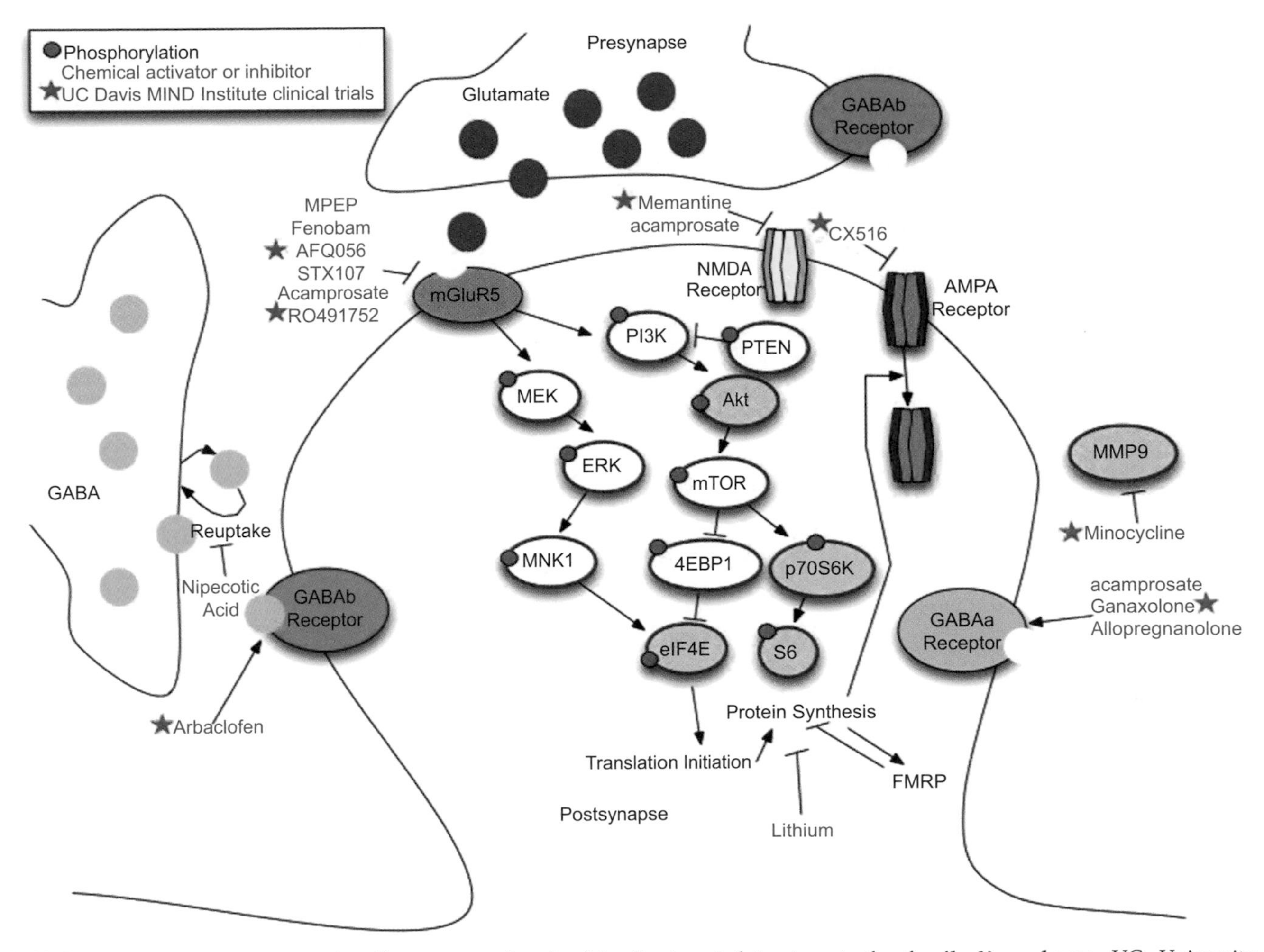

FIGURE 8.4 **Second messenger signaling systems involved in the targeted treatments for fragile X syndrome.** UC: University of California; MIND: Medical Investigation of Neurodevelopmental Disorders; MPEP: 2-methyl-6-(phenylethynyl)pyridine; GABA: γ-aminobutyric acid; NMDA: *N*-methyl-D-aspartate; MMP9: matrix metalloproteinase-9; FMRP: fragile X mental retardation protein. *Source: Adapted from Levenga* et al. Trends Mol Med. *2010;16(11):516–527.*[21]

deficits, responds well to arbaclofen (a $GABA_B$ agonist), which lowers glutamate at the synapse (Figs 8.4 and 8.5).[20] Minocycline, an antibiotic that lowers MMP9 levels, was shown to have a beneficial effect in children with FXS in a placebo-controlled trial (Fig. 8.5).[22] Minocycline can be prescribed clinically for children with FXS but it requires careful follow-up for side effects such as darkening of the skin or gums, graying of the teeth in young children, headache, rash, or blurred vision. On rare occasions, autoimmune problems including a lupus-like condition can develop, so an antinuclear antibody (ANA) titer should be followed every 6 months. In FXS the absence of FMRP leads to upregulation of many proteins since the translational suppression is removed. Minocycline lowers the elevated MMP9 levels that are present in FXS and improves synaptic connections. Minocycline can also stall translation in general, and because it is neuroprotective the beneficial effects in FXS may go beyond the improvement in MMP9 levels.[22] Ganaxolone, a $GABA_A$ agonist, is also being studied in FXS because there is significant downregulation in the GABA system in FXS (Fig. 8.4). The use of targeted treatments in FXS is emerging and as synaptic connections are strengthened it is important to use the best educational interventions, including apps for the iPad or similar tablets, which can enhance language and social development in addition to academic progress in reading and mathematics.

CLINICAL MANIFESTATIONS OF THE FRAGILE X PREMUTATION

Fragile X-Associated Premature Ovarian Insufficiency

The higher prevalence of menopause beginning before the age of 40 in premutation carriers was one of the first pathologies found in this population. It is now recognized as being one of the most common known

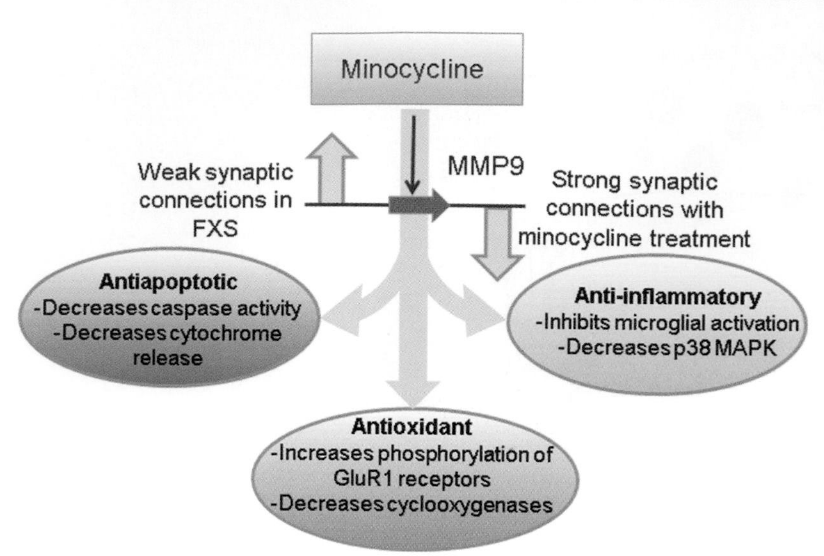

FIGURE 8.5 **The multiple therapeutic pathways of minocycline in treatment of fragile X syndrome (FXS).** MMP9: matrix metalloproteinase-9; MAPK: mitogen-activated protein kinase; GluR1: glutamate receptor-1.

causes of premature loss of fertility and affects roughly 20% of premutation carriers.[23] On average, carriers begin menopause 5 years earlier than the general population,[23] but this varies based on CGG repeat size, with midsized (80–99 repeat) carriers demonstrating an earlier menopause than those with higher or lower numbers in the premutation range.[24] This intermediate range of CGG repeat size may correlate with the highest, or toxic, levels of mRNA that are hypothesized to mediate the disorder. However, the age of onset of menopause is polygenic in nature, and the *FMR1* expansion is only one factor to consider. Even among premutation carriers, the age of menopause is best predicted by that of close family members.[25]

Fragile X-Associated Tremor Ataxia Syndrome

FXTAS represents the most severe end of the spectrum of premutation involvement. It is defined through the diagnostic clinical criteria outlined in Table 8.1. These criteria include the presence of tremor, which is typically an action or intention tremor, ataxia, cognitive changes including memory and executive function deficits, and white matter disease in the middle cerebellar peduncles and periventricular regions. Overall, only a subgroup of premutation carriers develops FXTAS as they age. In males in their fifties, only 17% have tremor and ataxia, but by their eighties 75% have these symptoms and presumably FXTAS.[26] Females are less affected by FXTAS, and overall only 16% develop these symptoms, presumably because of the protective effects of the normal X chromosome (Fig. 8.6).[27] Additional symptoms are common in FXTAS, including neuropathy, particularly in the legs, with symptoms of numbness and tingling; autonomic problems, such as hypertension; erectile dysfunction, orthostatic hypotension, cardiac arrhythmias, muscle weakness, and fatigue. As the ataxia progresses, patients eventually lose ambulation, and urinary and bowel incontinence occurs. Further cognitive decline to dementia is seen in approximately 50% of cases. Swallowing problems, choking, and eventually aspiration pneumonia often lead to death in elderly patients.

TABLE 8.1 Diagnosis of Fragile X-Associated Tremor and Ataxia Syndrome

All subjects must have a CGG expansion of *FMR1* between 55 and 200 repeats inclusive	
Diagnostic features	
Major clinical	Intention tremor or gait ataxia
Minor clinical	Executive functioning, short-term memory deficits, or Parkinsonism
Major radiological	White matter lesions of the middle cerebellar peduncles
Minor radiological	White matter lesions within the cerebrum
Histological	Intranuclear inclusions in neurons and astrocytes
Diagnostic category requirements	
Possible	One major clinical and one minor radiological
Probable	Two major clinical, or one minor clinical and one major radiological
Definite	One major clinical, one major radiological, and one major histological

The neuropathology of FXTAS includes eosinophilic intranuclear inclusions in neurons and astrocytes throughout the brain (Fig. 8.7). They are tau and synuclean negative but contain the excess *FMR1* mRNA

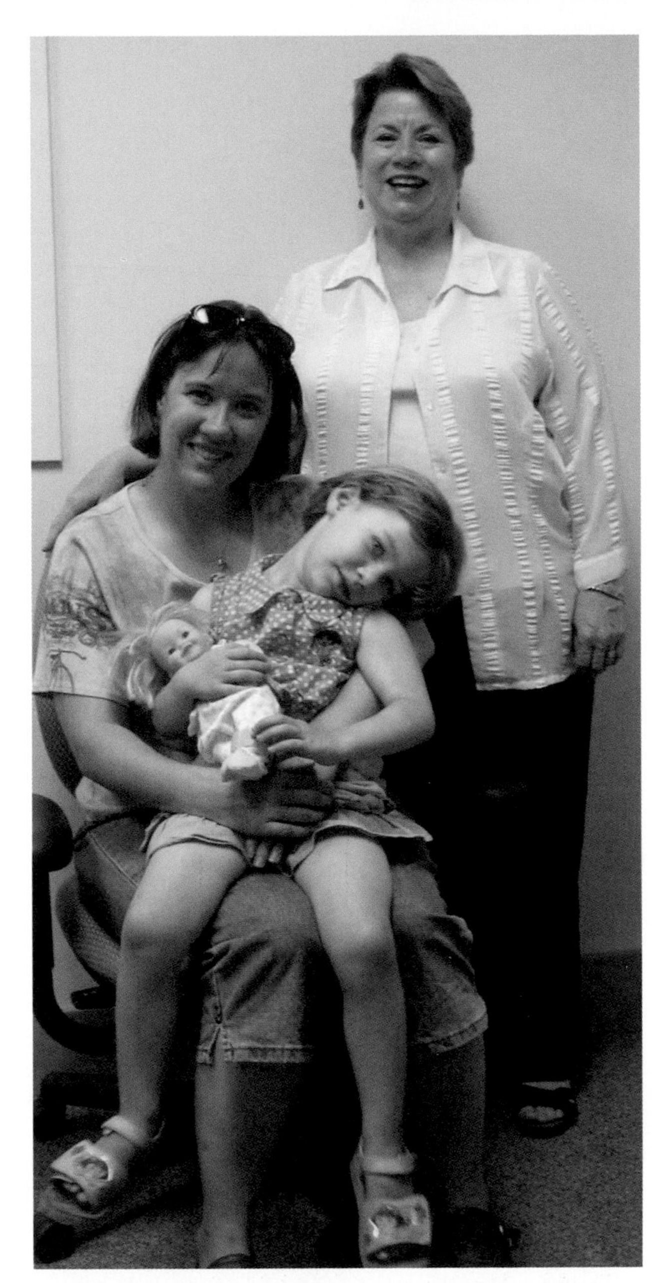

FIGURE 8.6 **Three generations of involvement in a fragile X family.** The grandmother has the premutation and neurological problems including vertigo and neuropathy related to the premutation. Seated is her daughter with the full mutation and mood disorders (treated with a selective serotonin reuptake inhibitor and lamotrigine), along with learning disabilities (treated with minocycline). On her mother's lap is the granddaughter, who has the full mutation and significant cognitive deficits, anxiety, and attention deficit/hyperactivity disorder (treated with minocycline and an mGluR5 antagonist).

and sequestered proteins mentioned above, including DROSHA and DGCR8.[10] The inclusions were added to the diagnostic criteria for FXTAS in 2004. Areas of high inclusion numbers include the hippocampus and amygdala, with higher numbers of astrocytes than neurons containing these inclusions. FXTAS inclusions are also seen in the peripheral nervous system, including the myenteric plexus in the intestines, and in organs including the Leydig cells of the testes, the adrenals, pituitary, thyroid, and heart.[28]

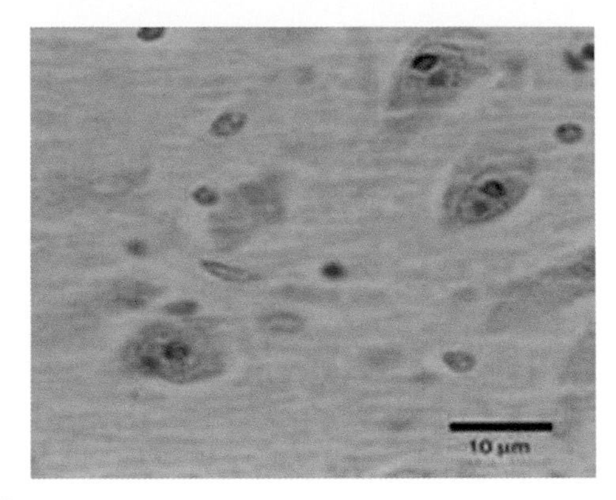

FIGURE 8.7 **Intranuclear inclusions found in the neurons of a patient with fragile X-associated tremor and ataxia syndrome (FXTAS).** The inclusions are stained red and are diagnostic of FXTAS.

PSYCHIATRIC MORBIDITY OF THE FRAGILE X PREMUTATION

While the more severe aspects of the carrier phenotype such as FXPOI and FXTAS were the first to be accepted, it was not until recently that a distinct psychopathology of the premutation carrier was generally acknowledged. The psychiatric features of the carrier state continue to be a widely researched area. The following is a brief overview of what is known about the prevalence and nature of various psychiatric conditions in individuals with the premutation.

Developmental Disorders

Although not at the levels seen in FXS, developmental disorders have generally shown higher rates in premutation carriers than in controls in the few studies conducted. Exact rates have been difficult to ascertain because of biases in children with more severe phenotypic expressions presenting to clinics, but the levels of ASD may be as high as 14% of boys and 5% of girls in girls carrying a premutation allele.[29] Seizure disorders, which occur in 13% of premutation carriers, can predispose children to the development of autism.[30] Compared with FXS, autistic symptoms in the premutation population are often mild and usually do not reach criteria for ASD. A different developmental disorder, ADHD, is much more common, with up to 38% of carriers afflicted in childhood, roughly three times the rate of age-matched controls.[31] Problems with ADHD often appear to persist into adulthood and severity has been directly linked to CGG repeat number.[32] Given the relatively high prevalence of the premutation, it is important to consider the premutation in individuals who present with ADHD and anxiety or ASD.

Cognition

Given that intellectual impairment is the most immediately apparent symptom of FXS and that premutation carriers can have some FMRP deficit, cognitive abilities in premutation carriers have long been a focus of research. There have been mixed results on this topic, with some studies showing no difference and others a subtle decrease of 6 points in full-scale IQ. It has become increasingly clear that these differences are much more apparent in those with higher CGG repeat numbers (generally above 100) and lower levels of FMRP. People with FXTAS also reliably score lower on IQ testing than age-matched controls; this is related to the dementing process that can accompany the syndrome. Impulse control and executive dysfunction are some of the most prominent problems with cognition in the FXTAS population.[33,34] The mouse model of FXTAS demonstrated a decline in visuospatial cognition that mirrors deficits seen in human premutation carriers.[35]

Depression

Research into depression in the premutation has produced a wide variety of results. Although depression was often described anecdotally in patients with the premutation, initial controlled studies failed to demonstrate higher prevalence rates. This may have been due to several factors including a lack of consideration for repeat size, a focus on female carriers protected through heterozygosity, small patient populations, poorly standardized instruments and, probably most importantly, a lack of understanding of FXTAS. More recent work, notably using the Symptom Checklist-90 questionnaire, has shown a higher prevalence of depressive symptoms in women with 100 or more CGG repeats than in carriers with fewer than 100 repeats.[36] Patients with FXTAS demonstrated significantly worse depressive symptoms on the Neuropsychiatric Inventory than age-matched controls. Literature reviews found the lifetime prevalence of major depression in premutation carriers to vary from 29 to 78%, generally higher than that found in large population studies.[37]

Anxiety

Evidence for anxiety, most notably social anxiety, has been demonstrated even in the mouse model of the premutation.[38] In humans, similar to depression, prevalence rates can vary based on FXTAS status in men, but the direction of the correlation varies. In a study directly comparing subjects with and without FXTAS with age-matched control data from the National Comorbidity Survey, premutation carriers without FXTAS showed a roughly two-fold higher prevalence of social phobia compared with controls, whereas those with FXTAS did not differ significantly.[39] The lifetime prevalence of panic disorder and specific phobias showed the reverse, with historical prevalences higher in people with FXTAS. Generalized anxiety and obsessive–compulsive disorder were not found to differ among the groups. However, the prevalence of all anxiety disorders in people with FXTAS is roughly three-fold greater than in age-matched controls. Non-FXTAS premutation carriers still showed a two-fold greater prevalence. These increased rates have been demonstrated to be independent of the presence of children with FXS.

Psychosis

Some of the first work carried out on neuropsychiatric aspects of the premutation demonstrated a higher rate of schizotypal personality traits.[40] Multiple case reports of a psychotic illness comorbid with the *FMR1* premutation have since been published, but linkage analysis has failed to demonstrate a connection with illnesses such as schizophrenia. Population studies have also failed to find higher rates of psychosis. However, a 2013 study showed that lower levels of FMRP correlate with a lower IQ and earlier onset of psychosis in those with schizophrenia but without a fragile X mutation.[41]

Substance Abuse and Dependence

Although substance abuse has been hypothesized as an exacerbating factor in FXTAS, as well as being related to executive dysfunction and white matter disease in FXTAS, substance abuse itself has been the topic of only a few studies involving the premutation. The largest study on the topic found that roughly one-third of premutation carriers had a history of substance abuse, significantly greater than age-matched controls but similar to non-premutation carrier family members (Fig. 8.8).[42] As the CGG repeat increases in the premutation range, the level of FMRP is lowered and the activity of the mGluR5 pathway increases; this biomarker is linked to substance abuse.

Other Pathology Found in the Fragile X Premutation

Beyond the psychopathology and the more clearly defined conditions of FXTAS and FXPOI, certain other correlations have been made with the presence of the premutation and FXTAS. Carriers with FXTAS have also been found to have higher prevalences of sleep apnea and hypertension.[27] Both of these may contribute to worsening or even development of the FXTAS condition through hypoxemia. Olfactory and visual disturbances[27] have also been described in this population. Premutation carriers with FXTAS have also demonstrated significantly higher rates of immune-mediated conditions such as autoimmune thyroid disease, irritable bowel syndrome, and fibromyalgia (Fig. 8.8).[43] There is also evidence that restless legs syndrome[44] and fatigue may be significantly higher among premutation carriers. It is not yet clear what role immune factors may play in the development of the neurodegeneration.

CONCLUSION

The many disorders associated with *FMR1* mutations make it imperative that clinicians are aware of both the neurodevelopmental and neurodegenerative aspects of the phenotype, and that they order fragile X DNA testing to find the genotype leading to these clinical problems. Psychiatric, endocrine, and neurological problems intertwine to lead to the clinical presentation, and identification of both the premutation and the full mutation will facilitate the development and clinical use of targeted treatments.

AREAS FOR FUTURE RESEARCH

Fragile X-associated disorders are a fertile area for research because FMRP is a ubiquitous protein that is a master regulator of synaptic plasticity. It is likely that deficits of FMRP will be seen in the general population without an *FMR1* mutation that will correlate with cognitive deficits and psychiatric problems. This has been reported in people with schizophrenia and there is evidence for similar deficits of FMRP in ASD, depression, bipolar disorder, and other conditions. The premutation

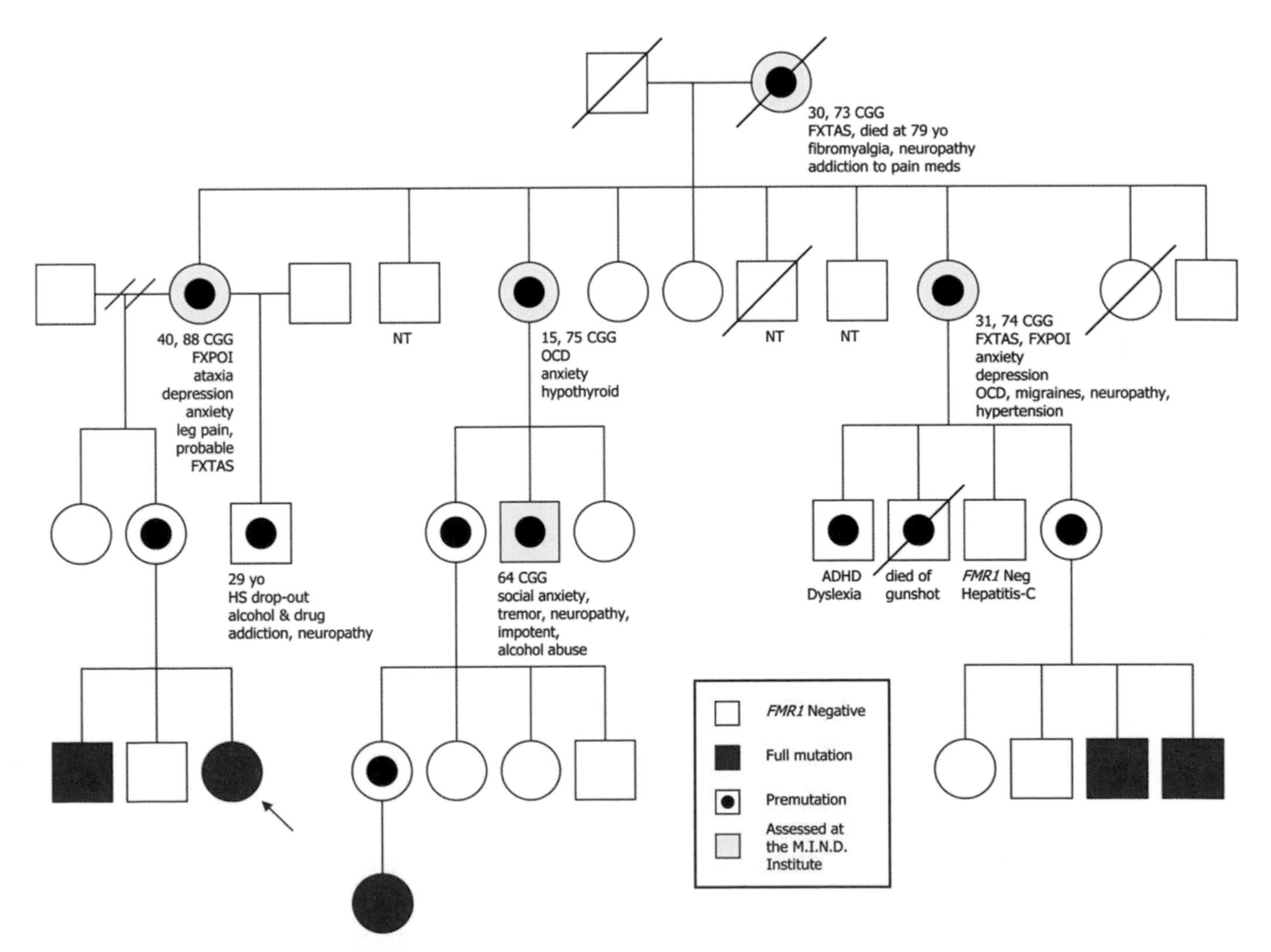

FIGURE 8.8 **A pedigree depicting five generations of fragile X-associated disorders.** FXTAS: fragile X-associated tremor and ataxia syndrome; FXPOI: fragile X-associated primary ovarian insufficiency; HS: high school; OCD: obsessive–compulsive disorder; ADHD: attention deficit/hyperactivity disorder; MIND: Medical Investigation of Neurodevelopmental Disorders.

neuron is particularly vulnerable to environmental toxins because of the oxidative stress that the elevated mRNA causes. The impact of environmental toxins on cognitive and psychiatric development can be studied easily in those with the premutation because they are common in the general population. The impact of multiple genetic hits such as CNVs in addition to the premutation is fertile ground for understanding ASD and learning problems coming from additive genetic and environmental factors. Research into targeted treatments has advanced significantly in patients with FXS, but to reverse intellectual disabilities will take a combination of targeted treatments in combination with educational endeavors. This will require further research but it is a worthy cause. Once the control of FMRP expression has been worked out, upregulating and downregulating FMRP may be extremely helpful not only for those with a fragile X mutation but also for many in the general population who have a deficit in FMRP.

Conflicts of Interest

Randi Hagerman has received funding from Roche, Novartis, Seaside Therapeutics, Forest, Curemark, and the National Fragile X Foundation for clinical trials in fragile X syndrome and/or autism. She has also consulted with Novartis, Genentech, and Roche regarding treatment in fragile X syndrome. Scott Summers has participated in clinical trials funded by Roche, Novartis, Seaside Therapeutics, and Marinus Pharmaceuticals.

Acknowledgments

This work was supported by National Institute of Health grants (HD02274, HD036071) and by the MIND Institute at the University of California at Davis Medical Center. Dr Scott Summers received support from the Department of Psychiatry and Behavioral Sciences at the University of California at Davis.

References

1. Darnell JC, Van Driesche SJ, Zhang C, et al. FMRP stalls ribosomal translocation on mRNAs linked to synaptic function and autism. *Cell*. 2011;146(2):247–261.
2. Iossifov I, Ronemus M, Levy D, et al. De novo gene disruptions in children on the autistic spectrum. *Neuron*. 2012;74(2):285–299.
3. Tassone F, Iong KP, Tong TH, et al. FMR1 CGG allele size and prevalence ascertained through newborn screening in the United States. *Genome Med*. 2012;4(12):100.
4. Yrigollen CM, Tassone F, Durbin-Johnson B, Tassone F. The role of AGG interruptions in the transcription of FMR1 premutation alleles. *PLoS ONE*. 2011;6(7):e21728.
5. Cronister A, Schreiner R, Wittenberger M, Amiri K, Harris K, Hagerman RJ. Heterozygous fragile X female: historical, physical, cognitive, and cytogenetic features. *Am J Med Genet*. 1991;38 (2–3):269–274.
6. Hagerman RJ, Leehey M, Heinrichs W, et al. Intention tremor, parkinsonism, and generalized brain atrophy in male carriers of fragile X. *Neurology*. 2001;57(1):127–130.
7. Jacquemont S, Hagerman RJ, Leehey M, et al. Fragile X premutation tremor/ataxia syndrome: molecular, clinical, and neuroimaging correlates. *Am J Hum Genet*. 2003;72(4):869–878.
8. Tassone F, Hagerman RJ, Taylor AK, Gane LW, Godfrey TE, Hagerman PJ. Elevated levels of FMR1 mRNA in carrier males: a new mechanism of involvement in the fragile-X syndrome. *Am J Hum Genet*. 2000;66(1):6–15.
9. Hagerman P. Fragile X-associated tremor/ataxia syndrome (FXTAS): pathology and mechanisms. *Acta Neuropathol*. 2013; 126(1):1–19.
10. Sellier C, Freyermuth F, Tabet R, et al. Sequestration of DROSHA and DGCR8 by expanded CGG RNA repeats alters microRNA processing in fragile X-associated tremor/ataxia syndrome. *Cell Rep*. 2013;3(3):869–880.
11. Cao Z, Hulsizer S, Cui Y, et al. Enhanced asynchronous Ca(2+) oscillations associated with impaired glutamate transport in cortical astrocytes expressing Fmr1 gene premutation expansion. *J Biol Chem*. 2013;288(19):13831–13841.
12. Chen Y, Tassone F, Berman RF, et al. Murine hippocampal neurons expressing Fmr1 gene premutations show early developmental deficits and late degeneration. *Hum Mol Genet*. 2010;19(1):196–208.
13. Deng PY, Rotman Z, Blundon JA, et al. FMRP regulates neurotransmitter release and synaptic information transmission by modulating action potential duration via BK channels. *Neuron*. 2013;77(4):696–711.
14. Heulens I, D'Hulst C, Van Dam D, De Deyn PP, Kooy RF. Pharmacological treatment of fragile X syndrome with GABAergic drugs in a knockout mouse model. *Behav Brain Res*. 2012;229(1):244–249.
15. Indah Winarni T, Chonchaiya W, Adams E, et al. Sertraline may improve language developmental trajectory in young children with fragile X syndrome: a retrospective chart review. *Autism Res Treat*. 2012;2012:104317.
16. Hagerman RJ, Berry-Kravis E, Kaufmann WE, et al. Advances in the treatment of fragile X syndrome. *Pediatrics*. 2009;123(1):378–390.
17. Berry-Kravis E, Raspa M, Loggin-Hester L, Bishop E, Holiday D, Bailey DB. Seizures in fragile X syndrome: characteristics and comorbid diagnoses. *Am J Intellect Dev Disabil*. 2010;115(6):461–472.
18. Gurkan CK, Hagerman RJ. Targeted treatments in autism and fragile X syndrome. *Res Autism Spectr Disord*. 2012;6(4):1311–1320.
19. Jacquemont S, Curie A, des Portes V, et al. Epigenetic modification of the FMR1 gene in fragile X syndrome is associated with differential response to the mGluR5 antagonist AFQ056. *Sci Transl Med*. 2011;3(64):64ra61.
20. Berry-Kravis EM, Hessl D, Rathmell B, et al. Effects of STX209 (arbaclofen) on neurobehavioral function in children and adults with fragile X syndrome: a randomized, controlled, phase 2 trial. *Sci Transl Med*. 2012;4(152):152ra127.
21. Levenga J, de Vrij FM, Oostra BA, Willemsen R. Potential therapeutic interventions for fragile X syndrome. *Trends Mol Med*. 2010;16(11):516–527.
22. Leigh MJ, Nguyen DV, Mu Y, et al. A randomized double-blind, placebo-controlled trial of minocycline in children and adolescents with fragile X syndrome. *J Dev Behav Pediatr*. 2013;34(3):147–155.
23. Sullivan AK, Marcus M, Epstein MP, et al. Association of FMR1 repeat size with ovarian dysfunction. *Hum Reprod*. 2005; 20(2):402–412.
24. Allen EG, Sullivan AK, Marcus M, et al. Examination of reproductive aging milestones among women who carry the FMR1 premutation. *Hum Reprod*. 2007;22(8):2142–2152.
25. Hunter JE, Epstein MP, Tinker SW, Charen KH, Sherman SL. Fragile X-associated primary ovarian insufficiency: evidence for additional genetic contributions to severity. *Genet Epidemiol*. 2008;32(6):553–559.
26. Jacquemont S, Hagerman RJ, Leehey MA, et al. Penetrance of the fragile X-associated tremor/ataxia syndrome in a premutation carrier population. *JAMA*. 2004;291(4):460–469.

27. Hagerman RJ, Hagerman PJ. Advances in clinical and molecular understanding of the FMR1 premutation and fragile X-associated tremor/ataxia syndrome. *Lancet Neurol*. 2013;12:786–798.
28. Hunsaker MR, Greco CM, Spath MA, et al. Widespread non-central nervous system organ pathology in fragile X premutation carriers with fragile X-associated tremor/ataxia syndrome and CGG knock-in mice. *Acta Neuropathol*. 2011;122(4):467–479.
29. Clifford S, Dissanayake C, Bui QM, Huggins R, Taylor AK, Loesch DZ. Autism spectrum phenotype in males and females with fragile X full mutation and premutation. *J Autism Dev Disord*. 2007;37(4):738–747.
30. Chonchaiya W, Au J, Schneider A, et al. Increased prevalence of seizures in boys who were probands with the FMR1 premutation and co-morbid autism spectrum disorder. *Hum Genet*. 2012;131(4):581–589.
31. Farzin F, Perry H, Hessl D,L, et al. Autism spectrum disorders and attention-deficit/hyperactivity disorder in boys with the fragile X premutation. *J Dev Behav Pediatr*. 2006;27(Suppl 2):S137–S144 .
32. Hunter JE, Epstein MP, Tinker SW, Abramowitz A, Sherman SL. The FMR1 premutation and attention-deficit hyperactivity disorder (ADHD): evidence for a complex inheritance. *Behav Genet*. 2012;42(3):415–422.
33. Kogan CS, Cornish KM. Mapping self-reports of working memory deficits to executive dysfunction in fragile X mental retardation 1 (FMR1) gene premutation carriers asymptomatic for FXTAS. *Brain Cogn*. 2010;73(3):236–243.
34. Yang JC, Simon C, Niu YQ, et al. Phenotypes of hypofrontality in older female fragile X premutation carriers. *Ann Neurol*. 2013;74:275–283. http://dx.doi.org/10.1002/ana.23933. [Epub ahead of print.]
35. Hocking DR, Kogan CS, Cornish KM. Selective spatial processing deficits in an at-risk subgroup of the fragile X premutation. *Brain Cogn*. 2012;79(1):39–44.
36. Johnston C, Eliez S, Dyer-Friedman J, et al. Neurobehavioral phenotype in carriers of the fragile X premutation. *Am J Med Genet*. 2001;103(4):314–319.
37. Bourgeois JA, Coffey SM, Rivera SM, et al. A review of fragile X premutation disorders: expanding the psychiatric perspective. *J Clin Psychiatry*. 2009;70(6):852–862.
38. Qin M, Entezam A, Usdin K, et al. A mouse model of the fragile X premutation: effects on behavior, dendrite morphology, and regional rates of cerebral protein synthesis. *Neurobiol Dis*. 2011;42(1):85–98.
39. Bourgeois JA, Seritan AL, Casillas EM, et al. Lifetime prevalence of mood and anxiety disorders in fragile X premutation carriers. *J Clin Psychiatry*. 2011;72(2):175–182.
40. Sobesky WE, Hull CE, Hagerman RJ. Symptoms of schizotypal personality disorder in fragile X women. *J Am Acad Child Adolesc Psychiatry*. 1994;33(2):247–255.
41. Kovacs T, Kelemen O, Keri S. Decreased fragile X mental retardation protein (FMRP) is associated with lower IQ and earlier illness onset in patients with schizophrenia. *Psychiatry Res*. 2013;210(3):690–693.
42. Kogan CS, Turk J, Hagerman RJ, Cornish KM. Impact of the fragile X mental retardation 1 (FMR1) gene premutation on neuropsychiatric functioning in adult males without fragile X-associated tremor/ataxia syndrome: a controlled study. *Am J Med Genet B Neuropsychiatr Genet*. 2008;147b(6):859–872.
43. Winarni TI, Chonchaiya W, Sumekar TA, et al. Immune-mediated disorders among women carriers of fragile X premutation alleles. *Am J Med Genet A*. 2012;158A(10):2473–2481.
44. Summers SM, Cogswell J, Goodrich JE, et al. Fatigue and body mass index in the fragile X premutation carrier. *Fatigue: Biomed, Health & Behav*. 2014;1–9.

SECTION II

DISEASES OF THE PERIPHERAL NERVOUS SYSTEM

CHAPTER

9

Introduction

Henry J. Kaminski

Department of Neurology, George Washington University, Washington, DC, USA

The peripheral nervous system is composed of autonomic, sensory, and motor neurons, skeletal muscles, and neuromuscular junctions. Each of these areas and their supporting cell components and basal lamina may be affected by disease. The spectrum of disorders that compromise peripheral nervous system function is vast. The four chapters in this section describe in detail diseases that serve as hallmarks for peripheral nervous system disease and illustrate basic principles. The disorders demonstrate the diversity of pathology that leads to dysfunction of the peripheral nervous system.

In Chapter 10, Kusner and Kaminski discuss myasthenia gravis, a disorder of neuromuscular junction function produced by a derangement of immune tolerance. Autoantibodies damage the postsynaptic membrane, leading to a compromise of neuromuscular transmission. Myasthenia gravis is arguably the best understood autoimmune disorder despite a lack of understanding regarding the etiology of disease induction. Diseases of the neuromuscular junction are the rarest of neuromuscular disorders, but their importance lies in their serving as the simplest model of synaptic pathology. The appreciation in the 1970s that antibodies against the acetylcholine receptor cause myasthenia gravis led to investigations that ultimately identified a spectrum of autoimmune channelopathies, including CNS pathologies such as limbic encephalitis caused by antibodies against potassium and *N*-methyl-D-aspartate receptors. The concept of a role of autoimmunity as an etiology of epilepsy and neurodegenerative disease continues to expand.

In Chapter 11, Živković and Clemens consider muscular dystrophies, genetic disorders with the defining feature of producing progressive skeletal muscle dysfunction. The pathogenic mutations affect genes that control muscle development, fiber structure, and intracellular signaling. Some muscular dystrophies involve other tissues, particularly the heart, while myotonic dystrophy compromises the function of nearly every tissue. In addition to the potentially broad organ involvement, the disorders vary considerably in their age of onset and pattern of skeletal muscle involvement. For example, boys with Duchenne muscular dystrophy caused by a mutation in the dystrophin gene develop definite weakness around age 4 and, until recent improvements in care, died in early adulthood, whereas boys with Becker muscular dystrophy and also a mutation in dystrophin develop weakness much later in life and can have a normal lifespan. Duchenne patients have weakness initially of shoulder and pelvic girdle muscles, whereas patients with fascioscapulohumeral dystrophies have highly selective muscle group weakness, as indicated by the name. The biological basis for the differential involvement of muscle groups is not known, but is a common feature of not only dystrophies but all neuromuscular diseases. Since the identification of the mutation of dystrophin as the cause of Duchenne muscular dystrophy in the 1980s, great advances have been made in understanding its pathogenesis, with early-phase gene therapy and disease-modifying treatments underway.

Saporta and Shy discuss peripheral neuropathies and Guillain–Barré syndrome in Chapter 12. Disorders of the peripheral nerves are highly diverse in pathogenesis and clinical presentation. The manifestation of neuropathies is dictated by the relative involvement of sensory, motor, or autonomic nerves, or the mixture of nerves, as well as whether the pathology produces chronic injury or rapid, fulminate damage. The etiology of peripheral neuropathy is broad, ranging from acquired neuropathies caused by toxins (heavy metal intoxication and pharmaceuticals) to metabolic pathology (endocrine disorders) and infections or genetic disease. Neuropathies can be divided into two basic categories based on whether injury is focused on the axon or the myelin. From a clinical perspective the distinction is critical. Diseases that damage myelin are often autoimmune in nature, such as acute or chronic inflammatory demyelinating polyneuropathies, which may be treated with

Neurobiology of Brain Disorders
http://dx.doi.org/10.1016/B978-0-12-398270-4.00009-4

immunomodulatory therapies. In contrast, pathology of the axon often is not amenable to specific treatment. For example, alcohol injures the axon and, although abstinence may arrest alcohol-induced neuropathy, the damage may not be reversed. Because of its generally slowly progressive nature and heterogeneity, the global burden on health of peripheral neuropathy is underappreciated and understudied.

The final chapter in this section, by Sims-Robinson, Kim, and Feldman (Chapter 13), focuses on diabetes mellitus, which is of worldwide importance and an increasing cause of morbidity and financial cost to society. Diabetes produces various types of peripheral nerve injury, which may affect motor, sensory, or autonomic nerves to varying degrees. The metabolic dysfunction produced by diabetes most commonly produces a symmetric polyneuropathy, which slowly and progressively compromises nerve function in a distal to proximal fashion. Patients may lose sensation slowly and not appreciate local injury to their feet, leading to abrasions and cuts to skin, which are susceptible to infection. Acute nerve injury also occurs, producing focal mononeuropathies, which are likely to be ischemic in etiology. Diabetes also causes a painful, lumbosacral plexopathy, which leads to unilateral leg paralysis. The condition is relatively uncommon and has recently been hypothesized to be related to an inflammatory injury. As noted in this chapter, diabetes can also have CNS consequences. Through acceleration of atherosclerosis, diabetes has the indirect effect of producing cerebral infarcts as well as compromised cellular function through hyperglycemia and hypoglycemia induced by treatment. Diabetes can also directly alter cognition through deranged blood brain function. Despite improvements in blood sugar control therapies and focused patient education for disease management, there has been little impact on moderating the severity of the neuromuscular complications produced by diabetes. This raises the general problem of lifestyle and health, as well as the determinants of lifestyle choices that are highly relevant to neuroscience but beyond the scope of this book.

A cursory review of each of these four disorders gives the impression that little overlap exists in their pathology, but this is clearly not the case. Muscular dystrophies are influenced by the inflammatory response and the metabolic derangement of diabetes may also lead to an inflammatory-mediated injury of peripheral nerves. In contrast, the autoimmune condition myasthenia gravis demonstrates a differential involvement of muscle groups, which can only be understood through improvements in basic skeletal muscle biology. The future for patients can only be improved by coordinated collaborations of clinical and basic scientists.

CHAPTER

10

Myasthenia Gravis

Linda L. Kusner, Henry J. Kaminski*†

*Department of Pharmacology and Physiology, George Washington University, Washington, DC, USA;
†Department of Neurology, George Washington University, Washington, DC, USA

OUTLINE

INTRODUCTION

Myasthenia gravis (MG) is an immune-mediated disorder of neuromuscular transmission with antibodies directed towards proteins of the neuromuscular junction, primarily the nicotinic acetylcholine receptor (AChR).[1,2] The autoimmune attack leads to skeletal muscle weakness with a characteristic of worsening with repetitive activity. To demonstrate the link in advances in the basic understanding of the disease to clinical medicine, this chapter begins with a brief historical overview of MG (Fig. 10.1).

As is the case for all diseases, an appreciation of a common set of symptoms and signs must come first and then a name is applied that "defines" the clinical phenotype. Physicians and scientists can then focus their study on pathogenesis and develop treatment approaches within the context of biological understanding of the time. Improvements in patient outcome may develop because of advances in overall medical care, rather than

Neurobiology of Brain Disorders
http://dx.doi.org/10.1016/B978-0-12-398270-4.00010-0

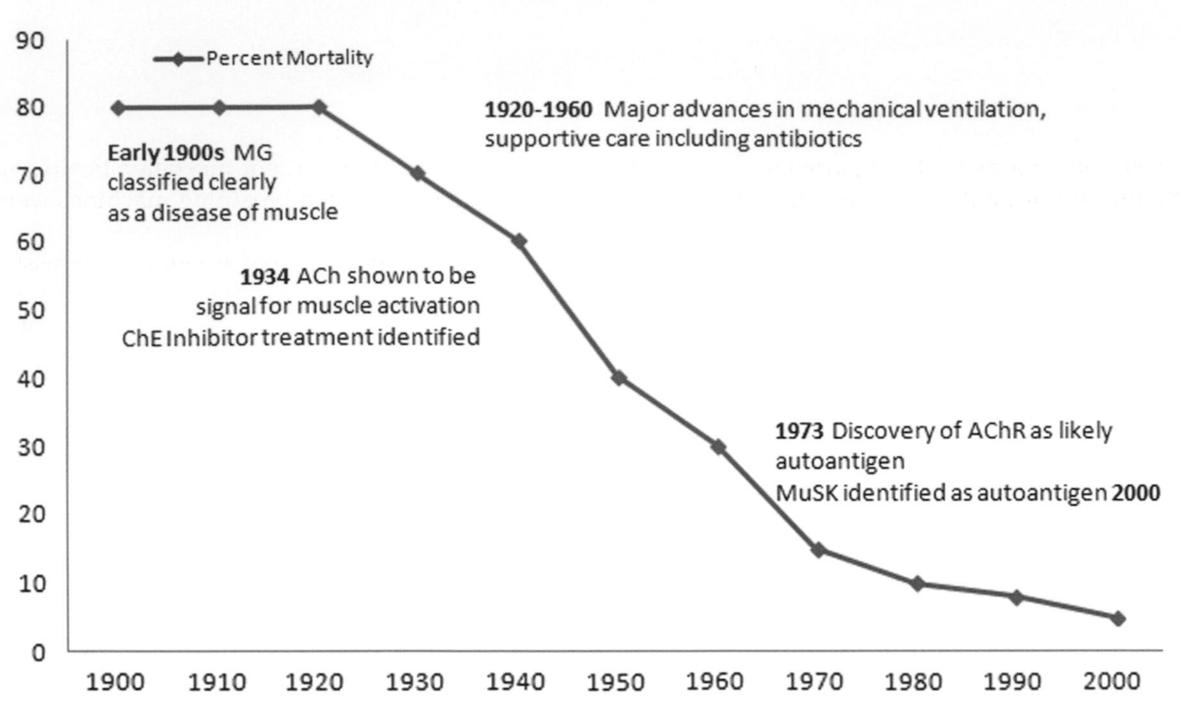

FIGURE 10.1 **History and mortality of myasthenia gravis (MG).** The graph shows estimates of mortality improvement over time. Key discoveries and therapeutic advance are noted. ACh: acetylcholine; AChR: acetylcholine receptor; MuSK: muscle-specific kinase.

specific to understanding of disease pathogenesis. For MG, a decrease in mortality (Fig. 10.1) occurred primarily because of the development of mechanical ventilation and antibiotics, and can one argue that it is only since the 1970s that the clinician and the research scientist have begun to synergistically improve patient care.

The Native American Chief Opechancanough, known for his leadership of the Powhattan attack on settlers in Virginia, was probably the first person to be described with MG. He suffered from excessive fatigue, with eyelids that were "so heavy that he could not see unless they were lifted up by his attendants", and he had to be carried on a litter by members of his tribe. The presence of drooping eyelids and general weakness with fatigue is a classic picture of MG. The first physician to describe patients with apparent MG was Thomas Willis, the great seventeenth century anatomist and physician, who wrote about several individuals with a chronic disorder marked by fluctuating strength and fatigue worsened by activity and improved with rest. Two-hundred years later, several case descriptions began to appear from throughout Europe, and in 1895 Friedrich Jolly labeled the disease "myasthenia gravis", combining Greek and Latin terms for grave muscle weakness.[3]

Before the 1930s, MG treatment was treated by the usual pharmacopeia of the day, strychnine, arsenic, and iodine. An Arizona physician with MG, Harriet Edgeworth, described in the *Journal of the American Medical Association* the benefit of ephedrine, which she had taken for menstrual symptoms, on manifestations of MG. Ephedrine has modest effects on neuromuscular transmission and is primarily a CNS stimulant. Ephedrine was used by some patients as adjunctive therapy until the early twenty-first century, but its use has been curtailed because of its cardiovascular complications and potential for abuse.

The year 1934 marks the development of what one may consider rational therapy for MG. Mary Walker was a house physician at St. Alfege's Hospital in Greenwich, England. She appreciated the similarity of the manifestations of MG to curare poisoning, which was treated with physostigmine; when she administered the drug subcutaneously to patients with MG she noted a dramatic improvement in weakness. Her reports of successful treatment of MG with physostigmine and prostigmine led to the widespread therapeutic use of cholinesterase (ChE) inhibition.[3] In addition, diagnostic tests were developed based on the administration of a ChE inhibitor to produce rapid improvement in strength, confirming the presence of MG. These pharmacological tests allowed diagnosis of milder forms of MG and identified patients at earlier stages of the disease. At the time of Walker's discovery, acetylcholine (ACh) had been identified by Otto Loewi as being involved in peripheral nerve and heart muscle signaling. In 1934 Henry Dale showed that ACh was a signal in skeletal muscle activation by nerve stimulation and that pyridostigmine and similar chemicals block the destruction of Ach, prolonging and enhancing the action of ACh.

Appreciation of organ pathology among patients with MG dates to the early decades of the twentieth century, when a connection between thymic pathology and MG was recognized with reports of either thymic hyperplasia or benign tumors in a majority of patients. The association of thymic abnormalities spurred surgical intervention. Ernst Sauerbruch performed a thymectomy for treatment of severe hyperthyroidism in a young woman and observed coincident improvement in myasthenic weakness. However, it was Alfred Blalock who removed the thymus gland specifically for treatment with MG in one patient with a thymic cyst. He performed a few additional operations with good outcomes and this meager evidence base led to the surgery's popularization throughout the world. Many decades followed before the role of the thymus in the immune system was appreciated.

Simpson and Nastuk in 1959–1960 independently proposed that MG is an autoimmune disorder. Nastuk observed that MG patients' sera applied to isolated nerve–muscle preparations compromised muscle contraction and the level of serum complement components correlated inversely with the severity of MG manifestations. Simpson supported an autoimmune origin for MG through observations of transient myasthenic weakness ("neonatal MG") in infants of mothers with MG, inflammatory infiltrates in muscles, and common pathological changes in the thymus of MG patients, and the association of MG with other putative autoimmune disorders. The discovery by Patrick and Lindstrom[4] that immunization of rabbits with AChR recapitulated key clinical and electrophysiological features of myasthenia confirmed the autoimmune etiology of MG. In 2000, Hoch and colleagues[5] discovered among patients without AChR antibodies a new autoantigen, the muscle-specific kinase (MuSK), which led to the definition of a new clinical phenotype and disease mechanism. Several groups in 2011 and 2012 identified low-density lipoprotein receptor-related protein-4 (LRP-4) as a new likely autoantigen in a subgroup of patients with MG.

Before the 1930s, it is estimated that less than one-tenth of patients with MG were diagnosed and they were the ones with the most severe weakness. Between 1915 and 1934, prevalence estimates of MG were about one per 200,000 of the population, and 70% of these patients died of respiratory failure and pneumonia, even though improvements in the care of severely weak patients had been made with the development of tracheostomy and endotracheal intubation and of negative pressure-assisted ventilation (the "iron lung") (Fig. 10.1). The greatest improvement in mortality occurred with better diagnostic accuracy of MG through the application of anti-acetylcholinesterase (anti-AChE) medications, including the edrophonium test, and the improved diagnostic accuracy led to an increase in prevalence estimates of MG to about 1 per 20,000. Treatment with AChE inhibitors, antibiotics, and perhaps thymectomy in combination with better ancillary therapies led to further decreases in patient mortality. Positive-pressure and volume-controlled breathing machines were associated with further improvements in death rates. In the era before immune-focused treatment, remissions were described in 20% of patients. This fact is complicated by evaluations of therapies in small numbers of patients, which at times can identify "remarkable" improvement. By the late 1970s into the 1980s, corticosteroids, azathioprine, plasmapheresis, and intravenous immunoglobulin were associated with further reductions in mortality. The discovery that the removal of pathogenic antibody by plasma exchange improved outcomes in MG offered additional support for the antibody-mediated basis. The common, clinical use of AChR antibody testing in the mid-1970s led to enhanced disease detection and raised prevalence estimates to about 1 per 17,000. Because of the diagnosis of patients with milder MG and improvement in therapeutics, mortality fell to 5%. With proper recognition and treatment, patients with MG now have a normal lifespan.[1,2]

IMMUNOPATHOGENESIS

MG is one of the few disorders to fulfill strict criteria for an autoimmune disease[6,7]: (1) the AChR and MuSK antigens can be administered to an animal and induce a disorder with several distinctive features of the human disorder; (2) administration of human autoantibodies or those from animals with experimental autoimmune myasthenia gravis (EAMG) reproduce the phenotype of the patients with MG; (3) immunoglobulin is bound to the neuromuscular junctions of patients and can be determined to bind the autoantigen; (4) AChR and MuSK antibodies can be detected in upwards of 95% of patients, and even the few patients without detectable autoantibodies by conventional radioimmunoassay have immunoglobulin bound to the neuromuscular junction[8]; and (5) treatments that reduce the serum concentration of anti-AChR antibody, such as plasma exchange or antibody deposition at the junction, improve weakness.

AChR antibodies are heterogeneous, with a variation in immunoglobulin G (IgG) subclasses, which include different subclasses of the IgG heavy chain and different types of light chain, and bind to various sites on the AChR.[9] AChR antibodies, which are the final effectors of pathology, produce the neuromuscular transmission defect by three mechanisms: compromise of the AChR, antigenic modulation, and complement mechanisms (Fig. 10.2).[9–11]

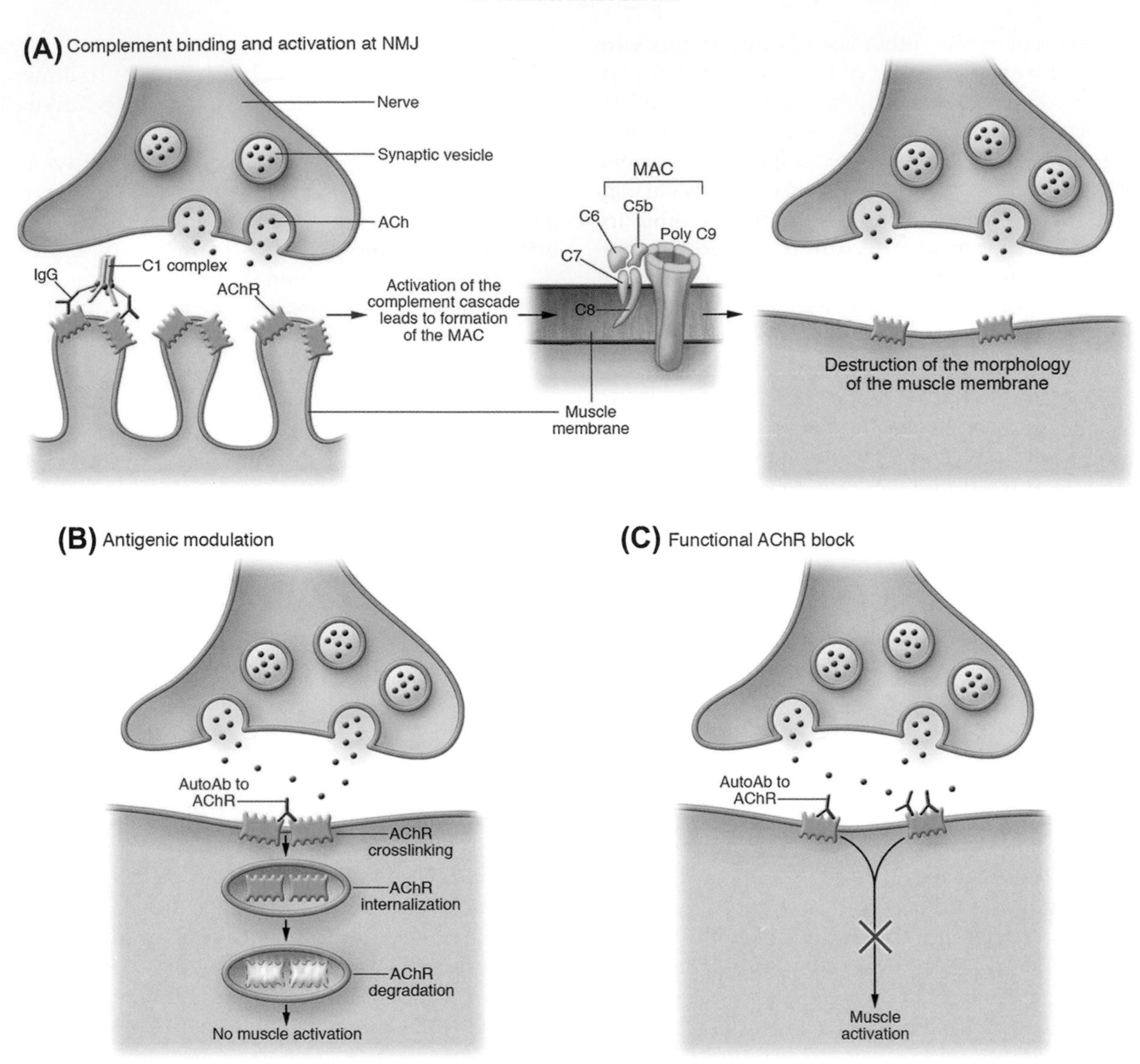

FIGURE 10.2 **Effector mechanisms of acetylcholine receptor (AChR) antibodies.** (A) Antibody binding to the AChR activates the complement cascade, resulting in the formation of the membrane attack complex (MAC) and localized destruction of the postsynaptic membrane. This leads to a simplified, postsynaptic membrane, which lacks the normal deep folds and has a relatively flat surface. NMJ: neuromuscular junction; ACh: acetylcholine; IgG: immunoglobulin G. (B) Antibodies may cross-link AChR molecules on the postsynaptic membrane, causing endocytosis of the cross-linked AChR molecules and their degradation (antigenic modulation). This ultimately leads to a reduced number of AChR molecules on the postsynaptic membrane. (C) Antibody binding the ACh-binding sites of the AChR causes functional block of the AChR by interfering with ACh-induced activation of the AChR. This results in failure of neuromuscular transmission. *Source: Conti-Fine* et al. J Clin Invest *2006;116: 2843–2854,*[7] *with permission.*

Compromise of the Acetylcholine Receptor

Antibodies may compromise AChR activity through blockade of the ACh binding site or inhibition of ion channel function, both of which would be expected to produce particularly severe weakness. However, such antibodies appear to be only a small fraction of the AChR antibodies among patients and not a major contributor to pathology. Serum from patients has been found to contain small amounts of AChR antibodies that recognize the cholinergic binding site. These antibodies can cause acute, severe weakness in animals, which suggests that antibodies may cause a functional block of the AChRs and produce profound weakness at low concentrations. Monoclonal antibodies that bind to the ACh binding site produce weakness in rodents within hours of injection by a direct blockade of ACh-induced opening of the ion channel.

Antigenic Modulation

Antigenic modulation is the ability of an antibody to cross-link two antigen molecules, which triggers cellular signals that cause accelerated endocytosis and degradation of those molecules. Immunoglobulin from patients with MG accelerates the degradation rate of the AChR *in vivo* and in cultured muscle cells. Not all AChR antibodies cause antigenic modulation, even though AChR antibodies are divalent with two antigen binding sites, which could be expected to link adjacent receptors. Therefore, the epitope location on the AChR surface may not allow for cross-linking to occur. In addition, disruption of the postsynaptic architecture produced by complement injury could interfere with cross-linking.

The contribution of antigenic modulation to human disease is not clear. Studies of antigenic modulation involve the application of patient sera to cultured cells expressing radioactively labeled AChR. A reduction in cell surface receptors was observed with addition of MG sera compared with sera from healthy individuals. Passive transfer of human MG sera may induce EAMG, but this requires large quantities of human antibodies.

Complement Mechanisms

Complement-mediated destruction of the neuromuscular junction is the most important mechanism by which antibodies produce weakness, as evidenced by several observations.[10,12] The postsynaptic surface in patients and EAMG animals contains complement component activation fragments and the membrane attack complex (MAC). Depletion of complement activity by various means protects animals from EAMG. Mice deficient in complement components are resistant to the development of weakness induced by EAMG. Mice with genetic deletion of cell surface modulators that protect cells from the activation of autologous complement, called intrinsic complement regulators, develop severe EAMG produced by administration of AChR antibodies. In addition, mice deficient in interleukin-12 (IL-12), a cytokine responsible for the Th1 response, develop minimal weakness after AChR immunization despite AChR antibody production. Analysis of the AChR-specific IgG content demonstrates a reduction in complement-fixing antibodies. Their resistance to EAMG indicates that without activation of complement, AChR antibodies cannot compromise neuromuscular transmission significantly.

Epitopes for Acetylcholine Receptor Antibody Binding

The AChR antibodies are polyclonal, binding to a heterogeneous set of epitopes on all subunits and representing all IgG subclasses. The majority of autoantibodies bind to the main immunogenic region (MIR), which is located on the extracellular loop of the α-subunit of the AChR (Fig. 10.3).[13] Why this region is particularly targeted is not known. The epitope and the type of heavy chain used by the AChR antibodies influence their pathogenic potential. Pathogenic antibodies must recognize

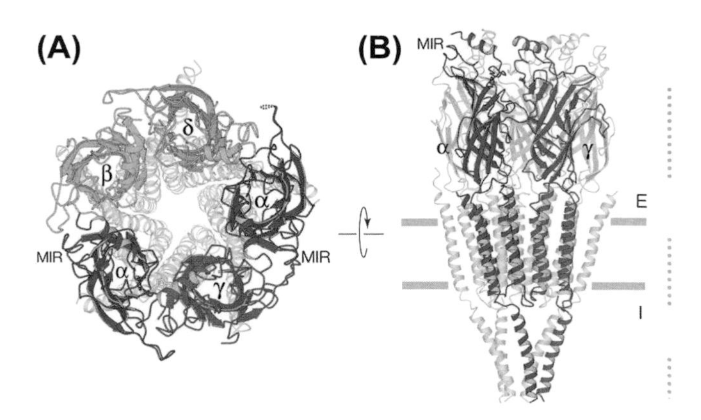

FIGURE 10.3 **Ribbon diagrams of the *Torpedo* acetylcholine receptor (AChR) as viewed (A) from the synaptic cleft and (B) parallel with the membrane plane.** The acetylcholine-binding domain is highlighted in (A) and only the front two subunits are highlighted in (B) (α, red; β, green; γ, blue; δ, light blue). Also shown are the locations of the main immunogenic region (MIR) and the membrane (horizontal bars; E: extracellular; I: intracellular). The dotted lines on the right denote the three main zones of subunit–subunit contacts. *Source: Unwin.* J Mol Biol. *2005;346:967–989,*[13] *with permission.*

regions of the AChR that are accessible in the intact postsynaptic membrane of the neuromuscular junction. Certain surface areas of the AChR, such as the MIR, may be particularly well situated to permit cross-linking of AChR or activation of complement. Other autoimmune disorders with antibodies directed at a protein antigen also possess particular epitopes, which dominate the immune response. Not all AChR antibodies are pathogenic. They may bind without cross-linking AChR and may be of a subtype that does not activate complement. Such factors appear to explain why serum AChR antibodies correlate poorly with disease severity.

T-Cell Influence and Loss of Tolerance

Autoantibody production in MG is a T-cell dependent process and a breakdown in tolerance towards self-antigens appears to be the primary abnormality of MG (Fig. 10.4). Several lines of evidence indicate that T cells have a critical role in MG pathogenesis. Treatment of patients with CD4 antibody leads to improvement and disappearance of AChR-induced T-cell responses *in vitro*. A decrease in the anti-AChR activity of circulating T cells is observed after thymectomy. Studies using cell transfer models *in vivo* indicate that AChR-specific $CD4^+$ T cells and B cells are required to produce EAMG in rats, whereas the transfer of B cells alone does not result in EAMG.

The thymus plays a key role in the responsiveness of lymphocytes to foreign antigens and in tolerance induction to self antigens. The immature T cells pass through the thymic cortex and those that recognize self major histocompatibility complex (MHC) antigens pass through to the medulla. During this stage in the thymic

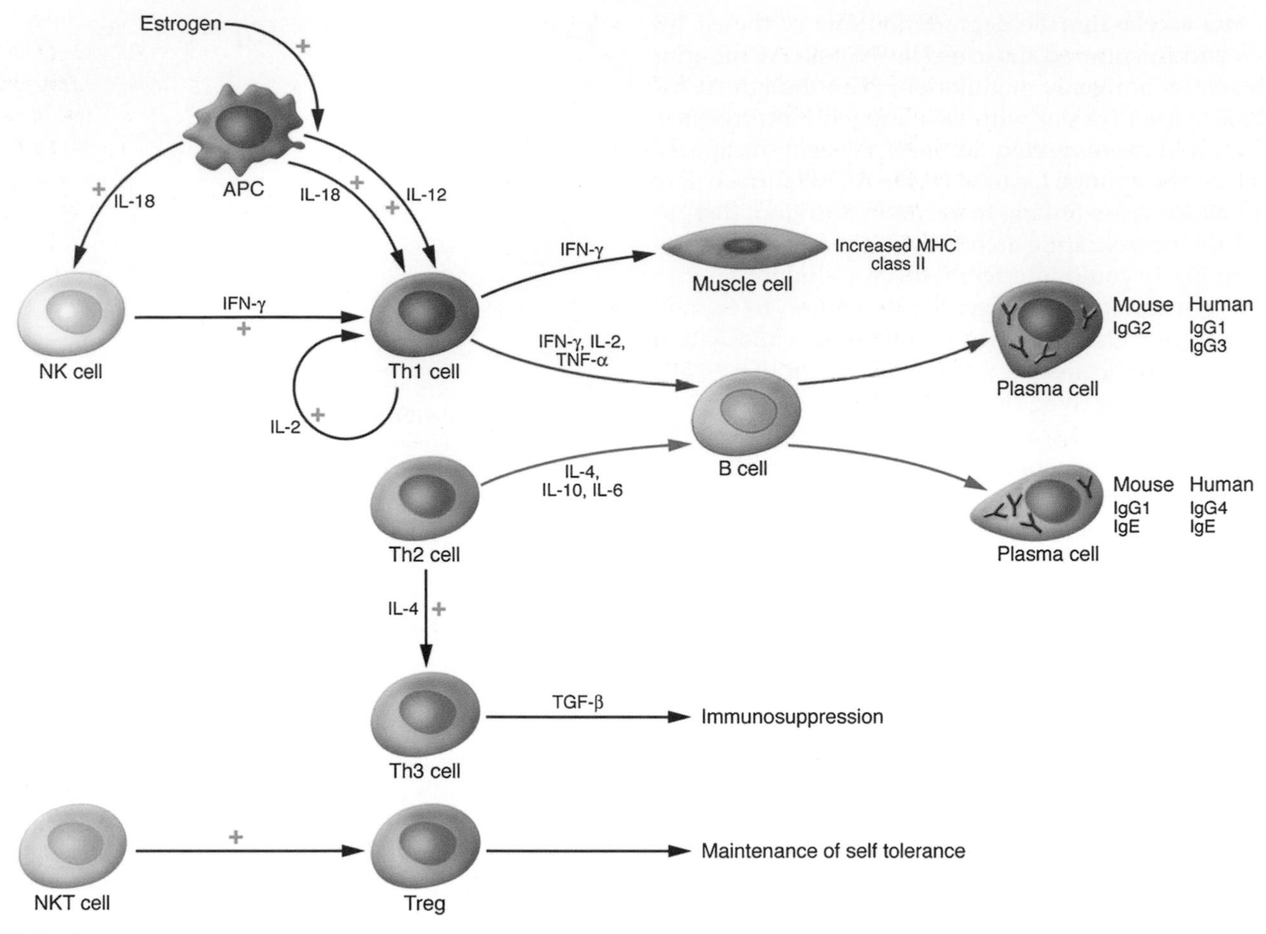

FIGURE 10.4 **Cytokine network and cells involved in the pathogenesis and immunoregulation of myasthenia gravis.** T-helper type 1 (Th1) cytokines stimulate production of immunoglobulin G (IgG) subclasses that bind and activate complement effectively, whereas Th2 cytokines stimulate the production of Ig classes and IgG subclasses that do not. The Th2 cytokine interleukin-4 (IL-4) is also a differentiation factor for Th3 cells, immunosuppressive cells that secrete transforming growth factor-β (TGF-β). The Th1 cytokine interferon-γ (IFN-γ) stimulates expression of major histocompatibility complex (MHC) class II molecules on the muscle cell membrane, thus facilitating presentation of muscle AChR. The IL-18 secreted by antigen-presenting cells (APCs) favors the differentiation of Th1 cells both directly and indirectly through the action of natural killer (NK) cells. CD1-d-restricted NKT cells can activate T regulatory cells (Tregs), thereby inhibiting autoimmune processes. TNF-α: tumor necrosis factor-α. *Source: Conti-Fine* et al. J Clin Invest. *2006;116:2843–2854,*[7] *with permission.*

cortex, T cells that would react towards self antigens are removed. Sequestration of antigens from the immune system may also be important in preventing autoimmune attack. Once in the medulla, the T cells differentiate into helper and suppressor cells and eventually are released to the periphery. Healthy individuals usually have T cells specific for autoantigens, including the muscle AChR, but their presence rarely leads to autoimmune diseases. Autoreactive T cells may survive clonal deletion during their maturation in the thymus because their T-cell receptors bind the self epitope/MHC with low affinity. These low-affinity autoreactive T cells may never become activated during life. However, the frequency of autoreactive T cells in healthy individuals indicates that mechanisms of peripheral tolerance must involve keeping in check the activity of self-reactive T cells. Failure of those mechanisms is a likely cause of MG and autoimmune diseases in general. T cells that express the transcription factor Foxp-3, referred to as T regulatory cells, play a crucial role in the maintenance of peripheral tolerance towards self antigens,[14] and impairment of their function has been demonstrated in patients with MG.

How a loss of tolerance develops in MG is not understood, but thymic abnormalities appear to be important in the pathogenesis of MG.[1,2,15] (1) Eighty percent of patients demonstrate hyperplasia of the thymus. The perivascular spaces are enlarged and filled with lymphoid tissue. Similar to the secondary lymph follicles of peripheral lymph nodes, active germinal centers and high numbers of mature T cells ($CD4^+$, $CD8^-$, or $CD4^-$/$CD8^+$) are present. T and B cells may be cultured from the MG thymus that are able to mount a productive AChR antibody response. (2) Differences in thymic histology correlate with certain clinical features indicating that the thymic cellular environment affects the disease manifestation. (3) About 10% of patients have a thymoma, a neoplasm of the thymus. Such patients have no sex or human leukocyte antigen (HLA) associations, but have high titers of AChR antibody and a low frequency of other autoimmune diseases. Thymoma appears to lead to a loss of tolerance through localized reduction in expression of the autoimmune regulator (AIRE). AIRE is a transcriptional regulator that controls immunological tolerance by regulation of promiscuous expression of tissue-specific self antigens in the thymus. (4) The thymus contains proteins that are antigenically similar, or identical, to the AChR. The expression of antigenically similar proteins to the AChR in the thymus would provide a source of antigen for autosensitization of T cells towards the skeletal muscle AChR. There also exists an active complement-mediated attack against thymic cells and the thymus from MG patients has reductions of intrinsic complement inhibitors. (5) Finally, removal of the thymus may improve the clinical course of MG.[16]

Similar to antibody heterogeneity, the epitope repertoire of anti-AChR T cells in patients is complex and specific for an individual patient. With increasing disease duration, T-cell sensitization spreads to larger parts of the AChR, with T cells recognizing epitopes formed by a number of sequence regions on each AChR subunit. A few sequence regions are recognized in most or all MG patients, and by large numbers of AChR-specific T cells. They form epitopes that are both universal and immunodominant. These universal epitopes sensitize pathogenic T cells and stimulate the synthesis of AChR antibodies. Stimulation of the autoimmune reaction could be by molecular mimicry or in response to superantigen. Superantigens, found on viruses or bacteria, stimulate particularly powerful proliferative responses, which could stimulate anergic T cells directed towards self antigens like the AChR.[17]

Muscle-Specific Kinase and Low-Density Lipoprotein Receptor-Related Protein-4 Myasthenia Gravis

In contrast to the decades of research that characterized AChR antibody-mediated MG, studies on MuSK- and LRP-4-related MG have been limited. Clinical observation and laboratory investigation in the early twenty-first century have led to rapid appreciation of the pathogenic differences between AChR and MuSK MG. MuSK is a receptor tyrosine kinase located on the postsynaptic surface of the muscle and the activation of MuSK leads to clustering of AChR in the postsynaptic membrane. Antibodies directed against MuSK[5] appear to reduce the concentration of AChR membranes of cultured cells. Animal models have confirmed that EAMG can be induced by immunization with purified antigen or the administration of MuSK antibodies.[18–20] In humans there is a reduction in AChRs in MuSK MG but damage to the postsynaptic surface is more limited and complement deposition less prominent. This is consistent with the activity of the IgG_4 subclass, which does not fix complement but is the prime driver of pathology in patients with MuSK MG.

LRP-4 serves as a receptor for agrin, which becomes a complex with MuSK, and thereby is similarly involved in AChR cluster formation on the postsynaptic surface.[21] In 2011, antibodies directed at LRP-4, but not MuSK or the AChR, were identified in patients with MG.[22] Thus far, the mechanisms by which LRP-4 antibodies produce disease have not been characterized in detail. The antibodies are of the IgG_1 subtype, which does fix complement, and cell-based assays indicate the potential to inhibit clustering. Unlike for AChR and MuSK MG, an animal model has not been produced and human muscle histopathology has not been assessed. Therefore, LRP-4 remains to be proven as a pathogenic autoantigen.

DEFECT IN NEUROMUSCULAR TRANSMISSION

The defining characteristic of MG is a reduction in the safety factor for neuromuscular transmission. The safety factor is defined as the difference between the membrane potential and the threshold potential for initiation of an action potential. Quantal release, AChR conduction and density, postsynaptic architecture, sodium channel density, and AChE activity all contribute to the endplate potential and influence the safety factor.[23] AChRs are normally activated only once in response to ACh released from the nerve terminal. Inactivation of AChE, which is a therapeutic strategy, prolongs the duration of action of ACh and slows the decay of the endplate current. Postsynaptic folds form a high-resistance pathway that focuses endplate current flow on voltage-gated sodium channels concentrated in the depths of the folds, which serves to reduce the action potential threshold at the postsynaptic surface and therefore increases the safety factor. In addition, the destruction of the postsynaptic membrane by complement disrupts the secondary synaptic folds and reduces sodium channel density, each of which lowers the safety factor. Reduced temperature enhances ACh release, which offers a biological basis for the Ice Pack test, which is used in diagnosis (see "Diagnosis", below). MG is characterized by a reduction in the endplate potential necessary for action potential generation secondary to the reduction in functional skeletal muscle AChR. With sustained activity, the amount of ACh is relatively reduced and at a damaged junction is insufficient to activate the remaining AChR to achieve an endplate potential above the threshold required to generate an action potential.

Differential Muscle Involvement by Myasthenia Gravis

Patients with MG demonstrate a broad and variable spectrum of muscle group involvement (see below). The basis for this differential involvement is likely to be due to a combination of inherent differences in physiological properties among muscles, a given muscle's capability to adapt to injury, and the specific nature of the antibody-mediated damage. The only muscle group that has been examined in detail for its preferential involvement by MG is the group of extraocular muscles (EOMs), which move the eyes. The EOMs are highly specialized muscles that differ from limb muscles in basic fiber morphology, gene expression characteristics, functional requirements, and neuromuscular junction architecture. (For a thorough discussion see Spencer and Porter.[24])

There are four major explanations for the different susceptibility to MG of skeletal and ocular MG. (1) The simplest explanation lies in the precision required to maintain alignment of the visual axes, so that any weakness of the EOMs produces symptoms of double or blurred vision, bringing the individual to prompt medical attention. In contrast, a small reduction in force generation of a limb muscle may not be readily appreciated or may be ignored more easily by a patient. (2) The EOM junctions are subject to high rates of neuronal stimulation, at times reaching 500 Hz. Repetitive stimulation leads to a relative reduction in ACh release from the nerve terminal; however, at a normal junction, this is inconsequential because of a high safety factor. At a junction with reduced AChR the safety factor would be compromised and perhaps the particularly high firing frequencies of the ocular motor neurons place EOM junctions at risk for transmission failure. (3) EOM junctions demonstrate fundamental differences from those of other skeletal muscle. Most dramatically, about 20% of the muscle fibers are multiply innervated with small *en grappe* neuromuscular junctions. These fibers have tonic contractile characteristics and do not generate action potentials. Therefore, a safety factor does not exist and any reduction in endplate potential induced by loss of AChR will decrease contractile force. The *en plaque* junctions have a paucity, sometimes a complete absence, of junctional folds, and the EOM singly innervated fiber junctions have less prominent synaptic folds. (4) Finally, EOMs express lower levels of intrinsic complement inhibitory proteins, making them more susceptible to antibody-mediated complement injury.[25]

Patients with antibodies directed at MuSK have a differential involvement of muscle groups.[2] Clinical studies have demonstrated a propensity for involvement of cranial nerve innervated muscles. In addition, these patients show distinct muscle atrophy that is rarely evident among AChR antibody-positive patients. The biological basis for these observations has not been determined, although it has been appreciated that levels of MuSK differ among muscle groups.

ANIMAL MODELS OF MYASTHENIA GRAVIS

Understanding of MG has benefited greatly from robust models of the human disease established primarily in rodents, but also other animals, including primates.[26–28] Two basic approaches are used to produce AChR antibody-mediated EAMG. Passive transfer EAMG involves intraperitoneal or intravenous administration of antibodies against the AChR. Monoclonal antibodies produced through hybridomas have been extensively used to induce EAMG because of high reproducibility. Less exact in practice are polyclonal antibodies from patients' sera or from other animals that have EAMG induced by immunization with AChR. When produced in Lewis rats, the animals become weak, quickly reaching a peak severity in 48–72 hours, and have some of the typical features of human MG, including fatigue, improvement with ChE inhibitor treatment, and decremental responses to repetitive nerve stimulation. Histological analysis demonstrates complement deposition at the junction but also an inflammatory response, which is not observed in humans. The value of this model is that it reproduces the final common pathway of antibody-mediated injury. Passive transfer EAMG has been used to evaluate potential therapeutics, which focus on effector mechanisms,[11] such as complement inhibition or immunoglobulin Fab arm exchange. However, the model does not produce a breakdown in tolerance and, therefore, cannot reveal potential immune regulatory mechanisms.

Immunization with peptide fragments or whole AChR purified from mammals or the electric organ of eels or rays produces EAMG, which more closely mimics the human disease.[26] Active EAMG is characterized by an acute phase in the first 2 weeks of disease induction, at which time rodents become weak with muscle inflammation with a prominent macrophage infiltration and neuromuscular junction necrosis, but with low levels of circulating AChR antibody, which then increase over time. Rodents recover from this acute phase, but over the following weeks will develop high levels of AChR antibody, complement deposition, and reduced AChR density at the neuromuscular junction. Electron microscopy reveals simplification of the junctional folds and sloughed postsynaptic membrane (Fig. 10.5). Lewis rats are the most commonly used strain. Rats have an advantage over mice in that they develop weakness within 4–6 weeks of AChR immunization. Rats also demonstrate

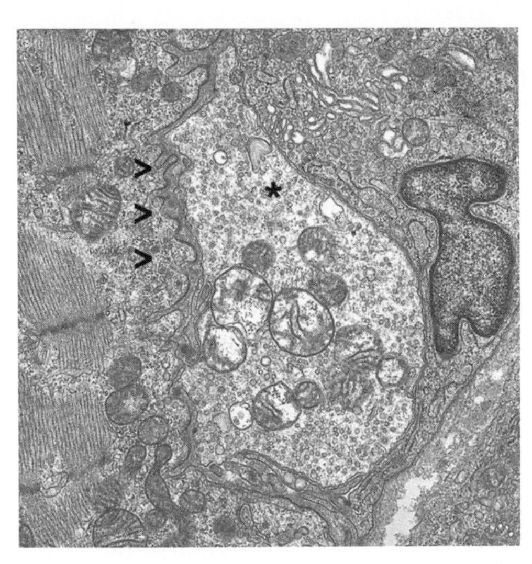

FIGURE 10.5 **Electron micrograph of a neuromuscular junction from a mouse with experimental autoimmune myasthenia gravis.** The arrowheads point to simplified postsynaptic membrane. The synaptic cleft has electron-dense material, which is probably sloughed postsynaptic membrane. The asterisk marks the nerve terminal.

more severe symptoms than mice. Wild-type C57b mice require multiple immunizations by the methods typically used and develop less severe weakness because of a high safety for neuromuscular transmission. Mice offer the advantage of the availability of numerous transgenic strains to dissect the pathogenic mechanisms. The limitations that distinguish the active EAMG model from human MG are that exogenous antigenic stimulation is needed to produce disease and stimulation with antigens must be repeated to perpetuate the autoimmune response. In addition, the fluctuations of weakness and autoimmune activity that occur over time among humans are not observed in animals. Despite common features of mammalian immune systems, there are differences in the basic regulatory pathways of human, mouse, and rat. Therefore, caution is required when extrapolating conclusions based on animal investigation to humans.

Spontaneous Myasthenia Gravis in Animals

Dogs, cats, and goats develop AChR-antibody related MG. Canine MG shares the same dominant epitopes and has similar clinical features to humans. Because a large portion of a dog's esophagus is composed of skeletal muscle, the dog develops "mega-esophagus", a profound dilatation of the esophagus that impairs eating. The dog demonstrates fatigue with improved activity level with rest. Although of potential interest for evaluation of therapeutics, the natural history of canine MG is that of spontaneous remission within the first 6 months of disease onset, which limits its utility in preclinical trials.[29]

EPIDEMIOLOGY AND GENETICS

The annual incidence of about 3 per million and prevalence estimate of MG of 200 per million qualifies MG for orphan disease status.[1] Studies consistently indicate that MG occurs more commonly among young women and older men, with an overall female predominance of about 3 to 2. Two distinct age groups appear to exist, with onset before or after 50 years. In the early-onset group, men develop MG approximately 10 years later than women, while in the later onset group the peak age of onset is between 70 and 80 years for both genders. The age of onset appears relevant to predisposing genetic factors. Women with early onset are more likely to be HLA-A1, B8, or DRw3-positive, whereas men with late onset have HLA-A3, B7, or DRw2.1 associations. The prevalence of late-onset MG may be on the increase for reasons not thought to be related to improved diagnostic methods. The prevalence of MG in children below the age of 15 years is significantly lower than in adults.

All ethnic groups are affected, with a suggestion that the frequency of MG is higher among African Americans than whites, with perhaps a greater percentage of patients with purely ocular myasthenia among African Americans. Ocular myasthenia has a higher frequency of onset before puberty and appears to be more common among Asian populations. A study from South Africa indicates that the black population has higher rates of a treatment-resistant ocular myasthenia than white South Africans. Further, analysis found a single-nucleotide polymorphism within the regulatory region of the decay-accelerating factor gene (*DAF*), which encodes an intrinsic complement inhibitor associated with the severe ocular manifestations.[30] DAF protein is concentrated at neuromuscular junctions and protects against complement-mediated endplate injury in mice with EAMG. MuSK antibody-positive MG also appears to be more common among African Americans than among whites. Populations closer to the equator have higher rates of MuSK MG than those in northern regions of Europe, in which the rates are low.[1,2,31]

A genetic predisposition for MG is based on several lines of evidence. While the risk of developing MG can be estimated to be about 0.01% in the general population, MG occurs in 4% of family members of patients with MG. Studies of dizygotic and monozygotic twins can be used to calculate a heritability index of 0.65 for MG. Such a level places MG in the range of Alzheimer disease and above multiple sclerosis for genetic predisposition. As mentioned above, HLA associations are evident in patients with MG. The *MYSA1* locus is associated with MG and thymic hyperplasia. Linkage dysequilibrium analysis identified a marker in the *CHRNA1* gene as being closely associated with MG. Finally, as mentioned already, a polymorphism has been identified in the promoter region of the *DAF* gene in a subgroup of patients with MG. In contrast to genetic factors, little is understood about the environmental triggers of MG.[1,2]

CLINICAL PHENOTYPE

MG produces either transient or persistent weakness with a hallmark of abnormal fatigability affecting any or all skeletal muscles.[1,2] Fatigue is appreciated by the patient by an increase in weakness with continuous activity. For example, an individual may have no manifestations upon awakening, but when washing or combing their hair the arm position cannot be maintained and rest is required for the shoulder girdle muscles. In addition to the activity-dependent variations in weakness, the severity of the underlying disease varies so that patients may have exacerbations for several weeks followed by spontaneous improvement. Infections or use of drugs that alter neuromuscular transmission may also worsen weakness or lead to subclinical MG being unmasked. Because of the broad range of presentations, which can involve a variety of muscle groups, coupled with its relative rarity and the variable severity of weakness, MG can be difficult for the clinician to identify. The sections below discuss particular clinical phenotypes.

Ocular Myasthenia

In 15% of MG patients, weakness is restricted to the ocular muscles, which are the levator palpebrae and the EOMs. Levator weakness produces droopy eyelids, while the most common symptom of EOM weakness is double vision, although some patients do not appreciate two images but rather complain of dizziness or unsteadiness when walking because of impaired gaze shifting. In nearly half of patients with MG, ocular manifestations will be the first symptoms of MG.

Almost all patients during the course of their illness develop double vision or ptosis. Hence, the diagnosis of MG may need to be questioned if ocular manifestations do not occur. Within 6 months of presentation with visual disturbance about half of patients develop weakness beyond the ocular muscles. Three years after onset, only 6% of those with ocular myasthenia develop generalized disease. Rarely, decades after onset a patient with purely ocular myasthenia may develop general weakness.

Predominantly Bulbar Myasthenia Gravis

About 20% of patients with MG initially develop prominent weakness of cranial nerve innervated muscles, which has led some clinical classifications of MG to place such "bulbar" patients into a set category. Such patients have a predominance of symptoms localized to the head and neck muscles with relative sparing of limb manifestations. Because of palatal muscle weakness, speech becomes nasal and regurgitation of liquids through the nose may occur. Weakness of jaw muscles leads to "jaw drop", which compromises speaking and eating and causes some patients to hold their jaw closed with their hands. Facial muscle weakness produces a smooth face with a lack of emotion. The reason for the targeting of this muscle group is not known.

Generalized Myasthenia Gravis

Most patients will have weakness that involves all skeletal muscles to varying degrees. In addition to the clinical manifestations described above, patients have compromised limb strength, which leads to difficulties with daily activities of walking, opening jars, shaving, or climbing stairs. Within the category of generalized disease, patients may have remarkably isolated weakness. Weakness of neck extensors or distal limb muscles may occur, producing foot drop or wrist drop, which mimic peripheral nerve involvement. Life-threatening respiratory muscle weakness is the most grave consequence of MG and nearly always occurs in the context of generalized weakness, but rarely may occur as an isolated manifestation. Once breathing difficulty becomes so severe that mechanical ventilation is required, the patient is deemed to be in a "myasthenic crisis".

Paraneoplastic Myasthenia Gravis

About 10% of patients with MG have a tumor of the thymus gland, a thymoma. The MG arises as result of an immune response directed towards the tumor. Thymomas have been found to be deficient in expression of the AIRE, which is a transcriptional regulator that controls tolerance. The deficiency leads to promiscuous expression of tissue-specific antigens within the tumor, which in the case of MG would be the AChR, and this is hypothesized to induce AChR antibody formation. MG is the most common autoimmune disorder associated with thymomas, but others are neuromyotonia (autoantigen voltage-gated calcium channel), stiff person syndrome (glutamic acid decarboxylase), red cell aplasia (unknown autoantigen), and limbic encephalitis (neuronal antigens).[32]

Paraneoplastic MG is a more severe disease. Patients have lower rates of remission and higher mortality than patients without thymoma. Removal of the tumor does not eliminate the disease, indicating that a fundamental loss of immune tolerance has occurred. Thymoma-associated MG patients are uniformly positive for AChR antibodies and also have other antibodies, including striational, titin, and ryanodine receptor antibodies. Thymomas grow slowly over many years causing compression of blood vessels, heart, and lungs. In a large series, three-quarters were fully encapsulated at operation, with no evidence of local tumor invasion; the remainder invaded adjacent structures: the pleura, lung, and pericardium. Less than 1% had more distant metastases. Most patients whose tumor extended outside a capsule that defines the margin of the tumor died within 5 years after surgery. More recent surveys reveal a better prognosis.

Neonatal Myasthenia Gravis

Of infants born to mothers with MG, about one-fifth are weak at birth, owing to placental transmission of AChR antibodies. The development of neonatal MG does not correlate with the severity or duration of the mother's illness or with her serum AChR antibody concentration, and infants born to seronegative mothers may also have neonatal MG. A more frequent occurrence of neonatal MG does occur with higher ratios of fetal AChR to adult AChR antibodies in the mother. Optimal treatment of mothers with MG may limit the development of neonatal MG; however, this has not been studied extensively. Mothers with a history of a child with neonatal MG are more likely to have subsequent infants affected by neonatal MG.

In two-thirds of infants, the weakness is detected within hours of birth, in three-quarters within the first day, and in the remainder on the second or third day. The most common signs are feeble cry and facial weakness, which occur in 95% of affected infants. Difficulty in feeding, with weakness of sucking and swallowing, generalized weakness, respiratory weakness and inability to swallow pharyngeal secretions, may cause airway obstruction and cyanosis. Only 15% of infants may have ptosis or ocular misalignment, a much lower incidence than in adults, although this may relate to the difficulty in identifying ocular muscle weakness in newborns. An improvement with ChE inhibitors confirms the diagnosis of neonatal MG.

ChE inhibitors and supportive care are typically all the treatment necessary. Plasma exchange should be reserved for the most severe cases. In one series, 11% of infants with neonatal MG died 1–21 days after birth, mainly because of inadequate treatment. In infants who recover, the duration of disease ranges from 5 to 57 days, and recovery is almost invariably complete, with no recurrence. Recovery is probably related to the clearance of pathogenic antibodies from the mother.

Arthrogryposis is a severe congenital malformation produced by weakness of skeletal muscle *in utero*, which causes skeletal muscle deformity. Arthrogryposis rarely occurs in infants of mothers with MG, and these women may not have evidence of MG, with the exception of serum AChR antibodies. The identification of an antibody-mediated injury of an infant from an asymptomatic mother led to the hypothesis that subclinical, maternal autoimmune disease may produce pathology in children, including CNS diseases such as autism and schizophrenia.

Muscle-Specific Kinase Antibody Myasthenia Gravis

Patients with MuSK MG tend to have a distinct syndrome, although there is a great overlap with AChR antibody-mediated MG. Patients tend to be women and have more prominent bulbar and neck weakness and a greater frequency of myasthenic crisis. In contrast to patients with AChR MG, muscle atrophy may be evident. ChE inhibitors are less effective in treating MuSK MG than in patients with AChR antibodies. A retrospective evaluation concluded that plasmapheresis is better therapy for severe weakness. Overall, there is also a suggestion that MuSK patients do not respond as well to treatment and have lower rates of remission compared with AChR antibody MG.

Drug-Induced Myasthenia Gravis

Drugs may precipitate a worsening of weakness or unmask subclinical weakness in patients with neuromuscular transmission disorders, and do so by several mechanisms. Many drugs interfere directly with neuromuscular transmission, including magnesium, antiepileptics, cardiovascular agents, and antibiotics, especially aminoglycosides and macrolides. Tandutinib is a tyrosine kinase inhibitor that was under evaluation for treatment of malignancies. Within days to a few months of tandutinib treatment, weakness and electrophysiological abnormalities developed, consistent with abnormal neuromuscular transmission. The hypothesis is that the drug interfered with MuSK signaling. Some individuals develop an autoimmune reaction when exposed to a drug. Penicillamine is primarily used as a treatment for rheumatoid arthritis, and is capable of reducing disulfide bonds, which may lead to exposure to antigenic targets. The AChR has a disulfide bond in proximity to the ACh binding site on the extracellular surface of the AChR, which offers a potential site for modification. Some patients exposed to penicillamine develop mild manifestations of MG with elevations of AChR antibody and electrophysiological abnormalities of MG. With discontinuation of penicillamine, the disease resolves after several months. MG is among several autoimmune disorders that may occur with exposure to penicillamine. Interferon-α_{2b}, used in the treatment of cancer and chronic hepatitis, may induce severe MG, but by unknown mechanisms. Mice with an engineered overexpression of interferon-γ at the neuromuscular junction develop manifestations similar to MG, but over time develop a necrotizing myopathy. A class of "statin" medications, which are commonly used for the treatment of elevated cholesterol, may in rare cases induce an apparent neuromuscular transmission disorder. Again, the mechanism is not clear but may be related to mitochondrial dysfunction at the neuromuscular junction. The presynaptic nerve terminal has a high concentration of mitochondria and the statin-induced disorder may be related to focal energy failure, which impairs neuromuscular transmission. This points to the possibility of mitochondrial dysfunction at the neuromuscular junction contributing to the neuromuscular transmission defect of MG. For an extensive discussion of drug- or toxin-induced neuromuscular transmission disorders see Kaminski.[1]

DIAGNOSIS

The clinical diagnosis of MG is confirmed by bedside evaluations, electrodiagnostic studies, and serology for autoantibodies. All of these evaluations are compromised in some way and their use requires appreciation of their limitations.[1,2]

Since Walker's observation, evaluation of strength after administration of a ChE inhibitor has been used for the diagnosis of MG. In adults, intravenous edrophonium is administered, whereas in children intramuscular neostigmine is easier to use. If unequivocal improvement occurs in a weakened muscle, the test is considered positive. Given the potential for a placebo-based improvement or bias of the evaluator, the test is most useful if improvement in ptosis or the strength of an EOM is demonstrable, because of the objective nature of this response. The physician should not talk themselves into a positive response and should set clear standards for positive results. Unequivocal interpretation of the response is more difficult when assessment is based on improvement in limb strength or bulbar function. Patients with bulbar muscle weakness may complain of worsened swallowing because of excess secretions produced by ChE inhibition. Muscarinic adverse effects, including tearing, sweating, gastrointestinal cramps, and diarrhea, may occur, as well as bradycardia with hypotension. Patients with asthma may develop bronchospasm.

False-positive edrophonium tests are described with motor neuron disease, Lambert–Eaton syndrome, and intracranial mass lesions. False-negative tests are relatively common. Within the proper clinical context, a single positive edrophonium test is strong support for the diagnosis of MG. Despite there being no evidence that ChE inhibitors are unsafe, edrophonium testing has become less popular in the USA.

Non-pharmacologically based evaluations have been developed to replace edrophonium. The Rest test involves the patient lying down for 30 minutes with eyes closed and ptosis being evaluated before and after. The Ice Pack test is performed in a patient with ptosis by placing a cooling pack placed over their orbits for 5 minutes. Improvement in ptosis or eye movement is assessed. Some patients may have difficulty tolerating the ice pack. These tests lack the decades of clinical experience of the edrophonium test and have uncertain sensitivity and specificity.

Three AChR antibody tests are used for serological diagnosis. The only one used in clinical trials for diagnosis is the AChR binding antibody, which utilizes a standard radioimmunoassay methodology with human AChR as the antigen. About 85% of patients with generalized MG have elevated titers of binding antibodies, while at best 50% of those with ocular myasthenia will be positive. Seropositivity increases with time and therefore repeat testing is beneficial to confirm the diagnosis. AChR antibodies are detected at low levels in other conditions (patients with thymoma, amyotrophic lateral sclerosis, and rheumatoid arthritis, as well as family members of patients with MG) when there is no clinical or electrophysiological evidence of MG. With these rare exceptions, the AChR binding test is highly sensitive.

α-Bungarotoxin almost irreversibly binds to the AChR and blocks the ACh binding site. This reagent is used to determine the content of AChR-blocking antibodies in patient sera. To determine the ability of patient serum to block the binding of ACh, serum is preincubated with antigen and then incubated with ^{125}I-labeled bungarotoxin. A reduced level of binding compared with control serum indicates the presence of blocking antibodies. However, less than 1% of patients have blocking antibodies without detectable binding antibodies, and therefore the blocking antibody test has no clinical value. AChR-modulating antibodies are detected using cultured human cells. The patient serum, when applied to the cell cultures, increases degradation of AChR by cross-linking receptors and the reduction in AChR is determined by a decrease in radioactive bungarotoxin. The modulating assay has a similar sensitivity but is technically more difficult than the binding assay. False-positive results are common and technical errors should be considered, especially in the setting of normal binding, which again limits the test's clinical usefulness.

About 3% of patients with generalized MG and a very small number of those with ocular myasthenia have antibodies against MuSK. These patients, except in exceedingly rare situations, do not have antibodies against the AChR. About 5–7% of patients are seronegative for MuSK and AChR, and confirmation of the diagnosis relies on other diagnostic testing. Patients seronegative for autoantibodies may represent a special subset of acquired autoimmune MG. Some of these patients are likely to harbor other autoantibodies. In 2011, antibodies were detected to LRP-4 in sera of double-negative patients; however, commercial tests for these autoantibodies are not yet available, and evaluation of the pathogenic nature of LRP-4 antibodies is ongoing. The determination of antibody status is clinically important not only for diagnosis, but also to guide treatment, as discussed below. It is likely that some seronegative patients have high-affinity antibodies to neuromuscular junction proteins that are at such low levels in the serum that they cannot be identified by conventional assays.

Striational antibodies were the first autoantibodies identified among patients with MG; however, they have limited diagnostic value and may be observed in other autoimmune diseases. They are found with greater frequency in patients with thymoma, including those without MG, and in normal elderly people. The presence of antistriational antibodies is predictive of a thymoma in men between the ages of 40 and 60. However, since the gold standard for diagnosis of thymoma is chest imaging, assessment of striational antibodies is of limited use even in thymoma diagnosis.

Electrodiagnostic studies are an important aspect of the evaluation of possible MG. With 2–3 Hz repetitive nerve stimulation, a decremental response of the compound muscle action potential is identified in about three-quarters of patients and despite the absence of generalized weakness, patients with ocular myasthenia also demonstrate a decremental response, but in a lower percentage than generalized MG patients. Single-fiber electromyography to evaluate the consistency of neuromuscular transmission is the most sensitive test for detecting abnormalities consistent with MG.

In addition to diagnostic testing, patients confirmed to have MG require ancillary evaluations. Coincident autoimmune thyroid disease, which may contribute to weakness, is present in up to 10% of patients. Computed tomography or magnetic resonance imaging of the chest is needed to test for thymoma. In patients with respiratory symptoms, pulmonary function tests quantify the severity of weakness and the potential need for more emergency therapy. MG occurs more frequently among patients with rheumatoid arthritis and systemic lupus erythematosus. Therefore, the clinician should evaluate patients for other autoimmune diseases.

Limitations of Diagnostic Evaluations as Biomarkers

The diagnostic tests described above are biomarkers for MG; however, none can serve as surrogates for clinical endpoints in practice or in clinical trials.[33] There is a great need in the field of MG for biomarker discovery to be integrated coherently into both preclinical and clinical efforts. The long-term benefit that would result from validated markers of efficacy in humans would be enormous. If these discoveries have parallels in animal studies the potential to translate novel treatments from preclinical studies to the bedside would be greatly enhanced. There is also benefit in the early termination of preclinical efforts that have little chance of success, in order to save the hundreds of millions of dollars that are expended on early-phase trials in humans.

TREATMENT

There is no cure for MG.[1,2,34] All therapies are designed to moderate the severity of the disease and the physician needs to attempt to minimize the adverse effects of the treatments. Among the therapies for MG, only ChE inhibitors and thymectomy can be considered to have been applied to MG specifically, while all others were used for MG based first on application to other disorders. Given the ongoing dissection of the mechanisms of MG pathogenesis, there is an opportunity to develop disease-specific treatments. This section will review conventional therapies for MG.

Cholinesterase Inhibitors

Pyridostigmine bromide is typically the first treatment for MG. It is begun with 30–60 mg individual doses given every 3–4 hours, with doses of greater than 120 mg rarely being effective. Dosing intervals of pyridostigmine range from 3 to 6 hours. As described above, muscarinic side effects may limit their use, although muscarinic blockers, such as glycopyrrollate (Robinul®) may be effective in limiting these complications. Excess and thick saliva may worsen swallowing and it may appear as if a patient is worsening. Such apparent paradoxical responses should be appreciated and lead to a reduction in ChE inhibitor dose. In patients with myasthenic crisis, discontinuation of ChE inhibitors is recommended to limit secretions while on artificial ventilation. Muscle twitches, fasciculations, and cramps also occur.

Only rare patients will have effective resolution of symptoms with ChE inhibitor treatment. Their lack of efficacy arises from the variability in patient activity level and involvement of muscle groups as well as the inability of the drug to establish effective steady-state levels in the blood. Patients are often aware that the effectiveness of ChE inhibition diminishes over time. This may be explained by the expression of a soluble splice variant of the ChE gene, which is not effectively inhibited by pyridostigmine.

Thymectomy

Since the popularization of thymectomy in the first half of the twentieth century, the surgery has become an accepted treatment, in particular at the time of initial diagnosis. However, an evidence-based review that evaluated all reliable retrospective studies of thymectomy concluded that the surgery may improve the chance of remission.[16] Comparison among clinical studies is complicated because of variations in surgical approach (trans-cervical, trans-sternal, robotic, or videoscopic), study-specific methods and outcome definitions, and characteristics of patients. A clinical trial is ongoing, comparing patients randomized to thymectomy and prednisone versus prednisone alone. With improvements in surgical technique, anesthesia, and respiratory care, operative morbidity and mortality approach zero. Thymectomy is thought to be best performed within the first few years of diagnosis and is usually recommended for patients with no significant contraindications to surgery and under the age of 65. With advances in surgical methods, it has become unclear which surgical approach should be used, although it is agreed that the greater the extent of thymic tissue removal the better.

Glucocorticoids

Corticosteroids are consistently the most effective treatment for MG despite having never been subjected to a randomized, controlled trial. Their benefit is so clear that such a trial would be unethical. However, many patients have contraindications to therapy, such as diabetes, severe osteoporosis, and morbid obesity. Prednisone is the usual glucocorticoid treatment used and is recommended for patients with ocular and generalized weakness that does not improve significantly with ChE inhibitor treatment. There is no consensus on the dose and manner of initiation of corticosteroid therapy, but usually 60 mg of prednisone daily is instituted until clinical improvement occurs, followed by alternate-day steroid treatment with gradual tapering as tolerated. Some clinicians believe that worsening of strength may occur within the first days of treatment initiation and advocate gradual initiation of corticosteroids. Eighty percent of patients may expect significant improvement or resolution of weakness on such a regimen during the first months of therapy.

Treatment must be maintained for months with a gradual tapering dose dependent on adverse effects and return of weakness. All patients receiving prednisone have complications. These may be mild, such as irritability, insomnia, acne, and rounding (moon) facial appearance. About 30% of patients have severe complications, which include diabetes, compression fractures from osteoporosis, and hypertension. Rarely, patients develop severe depression or mania, which will require specific treatment. It is beyond the scope of this chapter to detail the multitude of complications produced by chronic prednisone treatment. The adverse effects of prednisone motivated the need for steroid-sparing agents, described below. The importance of reducing the need for steroids is demonstrated in the common use of steroid reduction as a primary outcome measure for clinical trials.[33]

Azathioprine (Imuran®)

Azathioprine interferes with T- and B-cell proliferation by inhibition of nucleic acid synthesis as a purine analogue. Azathioprine is usually used in combination with corticosteroids in patients who have or are likely to develop complications from steroids, although it has been used alone. The efficacy of azathioprine in reducing the daily dose of corticosteroid was demonstrated in a randomized trial, but only after 18 months of treatment. Azathioprine treatment requires monitoring for liver and hematological toxicity. Of great concern is the possibly increased rate of lymphoma in patients treated with azathioprine, and this risk needs to be explained to the patient. About 10% of patients develop fever and influenza-like symptoms upon initiation of therapy, necessitating discontinuation of the medication. Adverse reactions appear to be related to activity of the thiopurine *S*-methyltransferase enzyme, which metabolizes azathioprine.

Mycophenolate Mofetil (Cellcept®)

Mycophenolate mofetil affects primarily T- and B-cell proliferation through inhibition of guanosine nucleotide synthesis, which is expected to have a slow onset of action. Based on retrospective evaluations with brief follow-up indicating its efficacy in reducing corticosteroid dose and reduced AChR antibody levels, two randomized, controlled trials were performed, each of which was less than 6 months in duration and failed to demonstrate efficacy for generalized MG. The trials have been criticized for several reasons, including their short duration, and myophenolate mofetil remains a commonly used medication for MG in the USA. Mycophenolate is generally well tolerated, except for initial minor gastrointestinal intolerance. Rarely, anemia, thrombocytopenia, and leukopenia may occur. Mycophenolate may increase the risk of malignancy and occasional cases of progressive multifocal leukoencephalopathy have been reported.

Tacrolimus (Prograf®)

Tacrolimus is a macrolide antibiotic that inhibits calcineurin with specific inhibitory effects on T cells. Tacrolimus has been evaluated in a randomized trial and large retrospective studies. The drug may produce kidney and hypertension as well as tremor and nausea. The agent is provided in two divided doses of 0.075–0.1 mg/kg with an intention to achieve a trough level of 7–10 ng/ml.

Cyclosporine

Cyclosporine has similar activity to tacrolimus, but is only rarely used because of its poor adverse effect profile compared with other agents. Cyclosporine has been used with benefit in the therapy of steroid-resistant MG patients. Corticosteroids were discontinued or decreased in nearly all patients in clinical trials of the agent; however, 5% of patients could not tolerate the drug. Adverse effects of hypertension and renal insufficiency limit its use. The drug, derived from a fungus, is a cyclic undecapeptide that blocks T-helper cell synthesis of cytokines, in particular through IL-2 and IL-2 surface receptors. The agent also interferes with the transcription of genes critical to T-cell function.

Plasma Exchange

Although never subjected to a randomized control trial, plasma exchange is well established as treatment for severe exacerbations of MG and myasthenic crisis. Typically, exchanges are carried out to remove one to two plasma volumes three times per week, up to six exchanges. Some patients show improvement within days, whereas others may take up to 2 weeks. Treatment reduces immunoglobulin levels rapidly. Rebound weakness may occur in weeks if concomitant immunosuppressive treatment is not begun. Because of the need for specialized equipment and support staff, plasma exchange is limited to use by major medical centers. Complications of therapy are related to the frequent need for large-bore intravenous catheters, which lead to infections. Patients may experience muscle cramps, nausea and vomiting, and hypotension. A severe reaction to replacement fluid could result in anaphylaxis.

Intravenous Immunoglobulin

Intravenous immunoglobulin (IVIg) is similar to plasma exchange in its rapid onset of action and use for treatment of severe weakness. The primary mechanism of action is through anti-idiotype networks. The typical treatment dose is 2 g/kg divided over 2–5 days. A response to treatment may occur within days but studies suggest 3 weeks as the expected time for a significant benefit. IVIg may produce life-threatening adverse effects of myocardial infarction, stroke, and pulmonary embolism. To date, no clear risk factors have been identified in patients at risk for these complications. Among patients with congestive heart failure or renal insufficiency, IVIg may induce worsening of the underlying conditions. IVIg is generally well tolerated, with common adverse effects of headache and nausea. Overall, this may be a relatively inexpensive treatment for hospitalized patients. IVIg is used as chronic therapy in patients resistant to standard immunosuppressives to reduce the need for corticosteroids; however, this practice has never undergone formal evaluation.

Other Therapies

Cyclophosphamide is used to treat MG resistant to corticosteroids and one study found that half of patients were asymptomatic after 1 year. Some centers have used high-dose cyclophosphamide coupled with bone marrow reconstitution in an attempt to induce tolerance to the AChR, but MG has recurred in some of the few patients treated thus far. Rituximab, an anti-CD20 antibody, has been used to treat MG, despite the lack of a significant database to support its efficacy in AChR antibody-positive MG.

Treatment is designed to produce a resolution of symptoms (remission) with limited adverse effects. Treatment is individualized based on the severity of the disease, the patient's lifestyle and career, associated medical conditions, and the patient's assessment of risk and benefit of various therapies. For example, a young woman with an active career may require prednisone and immunosuppressive treatment, but be educated about the potential risk of medications for a developing fetus. In contrast, an older individual with a sedentary lifestyle and mild general weakness with diabetes mellitus may be best treated with ChE inhibitors only. Often a discussion of costs of medication is necessary. In the USA, a typical course of IVIg costs 12,000–14,000 USD, while mycophenolate costs upwards of 1000 USD per month. In contrast, the cost of prednisone for 1 month is about 25 USD, but the cost of complications is not factored into the price of the medication.[35] Depending on the patient's insurance, medication costs may not be covered, thus limiting their treatment options solely for economic reasons.

CONCLUSION

MG is often considered the best understood autoimmune disorder and is one of the few disorders that fulfill strict criteria for autoimmunity. Therefore, MG has long served as a model for the study of autoimmunity and its evaluation has led to discoveries of other diseases that focus the immune attack on ion channels, from the Lambert–Eaton syndrome (presynaptic calcium channel) to neuromyotonia (potassium channels) to limbic encephalitis (potassium channels, *N*-methyl-D-aspartate receptors) and neuromyelitis optica (aquaporin-4 channel).[36,37] Patients with MG generally respond well to treatment and as a group have a normal lifespan; however, mortality of patients who require hospitalization is upwards of 10%. In addition, as the adverse effect profiles of immune therapy are poor, there is a great need to improve treatment.

The most critical question in MG is what triggers and maintains the autoimmune reaction against the neuromuscular junction. In all likelihood, the mechanisms involved will vary among patient groups: early versus late onset, ocular myasthenia seronegative patients versus ocular myasthenia with AChR antibodies, and thymoma patients, for example. Genetic susceptibility and environmental exposures will assist in the understanding of MG pathogenesis, but their identification will require large, comprehensive studies, which are challenging and expensive to perform. The precise characterization of underlying disease mechanisms will allow specific interventions to be developed, but even without this precise knowledge therapeutics need to be enhanced. As basic science investigations continue to find effective biomarkers for early identification of disease progression, the prediction of adverse effects and determination of ultimate outcome are necessary to guide more precisely preclinical and clinical trials. For the field to advance, collaborations between basic and clinical scientists will be critical.

References

1. Kaminski HJ, ed. *Myasthenia Gravis and Related Disorders*. 2nd ed. Totowa, NJ: Springer (Humana Press); 2009.
2. Engel AG. *Myasthenia Gravis and Myasthenic Disorders*. Oxford: Oxford University Press; 2012.
3. Keesey JC. A history of treatments for myasthenia gravis. *Semin Neurol*. 2004;24:5–16.
4. Patrick J, Lindstrom J. Autoimmune response to acetylcholine receptor. *Science*. 1973;180:871–872.
5. Hoch W, McConville J, Helms S, Newsom-Davis J, Melms A, Vincent A. Auto-antibodies to the receptor tyrosine kinase MuSK in patients with myasthenia gravis without acetylcholine receptor antibodies. *Nat Med*. 2001;7:365–368.
6. Cavalcante P, Bernasconi P, Mantegazza R. Autoimmune mechanisms in myasthenia gravis. *Curr Opin Neurol*. 2012;25:621–629.
7. Conti-Fine BM, Milani M, Kaminski HJ. Myasthenia gravis: past, present, and future. *J Clin Invest*. 2006;116:2843–2854.
8. Leite MI, Jacob S, Viegas S, et al. IgG1 antibodies to acetylcholine receptors in "seronegative" myasthenia gravis. *Brain*. 2008; 131:1940–1952.
9. Lindstrom J, Luo J, Kuryatov A. Myasthenia gravis and the tops and bottoms of AChRs: antigenic structure of the MIR and specific immunosuppression of EAMG using AChR cytoplasmic domains. *Ann N Y Acad Sci*. 2008;1132:29–41.
10. Kusner LL, Kaminski HJ, Soltys J. Effect of complement and its regulation on myasthenia gravis pathogenesis. *Exp Rev Clin Immunol*. 2008;4:43–52.
11. Gomez AM, Van Den Broeck J, Vrolix K, et al. Antibody effector mechanisms in myasthenia gravis – pathogenesis at the neuromuscular junction. *Autoimmunity*. 2010;43:353–370.
12. Tuzun E, Huda R, Christadoss P. Complement and cytokine based therapeutic strategies in myasthenia gravis. *J Autoimmun*. 2011; 37:136–143.
13. Unwin N. Refined structure of the nicotinic acetylcholine receptor at 4A resolution. *J Mol Biol*. 2005;346:967–989.
14. Kornete M, Piccirillo CA. Functional crosstalk between dendritic cells and Foxp3(+) regulatory T cells in the maintenance of immune tolerance. *Front Immunol*. 2012;3:165.
15. Cavalcante P, Le Panse R, Berrih-Aknin S, et al. The thymus in myasthenia gravis: site of "innate autoimmunity"? *Muscle Nerve*. 2011;44:467–484.
16. Gronseth GS, Barohn RJ. Practice parameter: Thymectomy for autoimmune myasthenia gravis (an evidence-based review): report of the Quality Standards Subcommittee of the American Academy of Neurology. *Neurology*. 2000;55:7–15.
17. Conti-Fine BM, Diethelm-Okita B, Ostlie N, Wang W, Milani M. Immunopathogenesis of myasthenia gravis. In: Kaminski HJ, ed. *Myasthenia and Related Disorders*. New York: Humana Press; 2009:43–70.
18. Jha S, Xu K, Maruta T, et al. Myasthenia gravis induced in mice by immunization with the recombinant extracellular domain of rat muscle-specific kinase (MuSK). *J Neuroimmunol*. 2006;175:107–117.
19. Shigemoto K, Kubo S, Maruyama N, et al. Induction of myasthenia by immunization against muscle-specific kinase. *J Clin Invest*. 2006;116:1016–1024.
20. Cole RN, Ghazanfari N, Ngo ST, Gervasio OL, Reddel SW, Phillips WD. Patient autoantibodies deplete postsynaptic muscle-specific kinase leading to disassembly of the ACh receptor scaffold and myasthenia gravis in mice. *J Physiol*. 2010;588:3217–3229.
21. Kim N, Stiegler AL, Cameron TO, et al. Lrp4 is a receptor for Agrin and forms a complex with MuSK. *Cell*. 2008;135:334–342.
22. Higuchi O, Hamuro J, Motomura M, Yamanashi Y. Autoantibodies to low-density lipoprotein receptor-related protein 4 in myasthenia gravis. *Ann Neurol*. 2011;69:418–422.
23. Slater CR. Structural factors influencing the efficacy of neuromuscular transmission. *Ann N Y Acad Sci*. 2008;1132:1–12.
24. Spencer RF, Porter JD. Biological organization of the extraocular muscles. *Prog Brain Res*. 2005;151:43–80.
25. Kaminski H, Zhou Y, Soltys J, Kusner L. The complement hypothesis to explain preferential involvement of extraocular muscle by myasthenia gravis. In: Leigh R, Devereaux M, eds. *Advances in Understanding Mechanisms and Treatment of Infantile Forms of Nystagmus*. New York: Oxford University Press; 2008.
26. Christadoss P, Poussin M, Deng C. Animal models of myasthenia gravis. *Clin Immunol*. 2000;94:75–87.
27. Baggi F, Antozzi C, Toscani C, Cordiglieri C. Acetylcholine receptor-induced experimental myasthenia gravis: what have we learned from animal models after three decades? *Arch Immunol Ther Exp (Warsz)*. 2012;60:19–30.
28. Wu B, Goluszko E, Huda R, Tuzun E, Christadoss P. Experimental autoimmune myasthenia gravis in the mouse. *Curr Protoc Immunol*. 2011. Chapter 15:Unit 15.23.
29. Shelton GD, Lindstrom JM. Spontaneous remission in canine myasthenia gravis: implications for assessing human MG therapies. *Neurology*. 2001;57:2139–2141.
30. Heckmann JM, Uwimpuhwe H, Ballo R, Kaur M, Bajic VB, Prince S. A functional SNP in the regulatory region of the decay-accelerating factor gene associates with extraocular muscle pareses in myasthenia gravis. *Genes Immun*. 2010;11:1–10.
31. Tsiamalos P, Kordas G, Kokla A, Poulas K, Tzartos SJ. Epidemiological and immunological profile of muscle-specific kinase myasthenia gravis in Greece. *Eur J Neurol*. 2009;16:925–930.
32. Shelly S, Agmon-Levin N, Altman A, Shoenfeld Y. Thymoma and autoimmunity. *Cell Mol Immunol*. 2011;8:199–202.
33. Benatar M, Sanders DB, Burns TM, et al. Recommendations for myasthenia gravis clinical trials. *Muscle Nerve*. 2012;45:909–917.
34. Gilhus NE, Owe JF, Hoff JM, Romi F, Skeie GO, Aarli JA. Myasthenia gravis: a review of available treatment approaches. *Autoimmune Dis*. 2011;2011:847393.
35. Guptill JT, Marano A, Krueger A, Sanders DB. Cost analysis of myasthenia gravis from a large US insurance database. *Muscle Nerve*. 2011;44:907–911.
36. Vincent A. Autoimmune channelopathies: well-established and emerging immunotherapy-responsive diseases of the peripheral and central nervous systems. *J Clin Immunol*. 2010;30(suppl 1):S97–S102.
37. Vincent A. Autoimmune channelopathies: John Newsom-Davis's work and legacy. A summary of the Newsom-Davis Memorial Lecture 2008. *J Neuroimmunol*. 2008;201–202:245–249.

CHAPTER

11

Muscular Dystrophy

Saša A. Živković, Paula R. Clemens

Neurology Service, Department of Veterans Affairs, University of Pittsburgh, Pennsylvania, USA, and Department of Neurology, University of Pittsburgh School of Medicine, Pittsburgh, Pennsylvania, USA

OUTLINE

INTRODUCTION

The term "muscular dystrophy" describes a heterogeneous group of inherited progressive disorders of muscle characterized by destruction of muscle and its replacement by fatty and fibrous tissue (Fig. 11.1).[1] Muscular dystrophies often have an early onset, but initial symptoms may also be reported in early or even late adulthood. While most patients report muscle weakness as the predominant symptom, the quality of life may also be greatly impacted by the involvement of one or more other organ systems, including the cardiovascular system, gastrointestinal tract, and brain. Genes that are mutated causing muscular dystrophies play major roles in muscle development and maintenance of muscle function including reinforcing and repairing the plasma membrane. The physiology of muscle fibers is significantly affected by exposure to considerable mechanical forces involved in contractility and specific regeneration and energy demands. Clinical manifestations of muscular dystrophies are closely related to a multitude of developmental, transcriptional, and metabolic pathways that are affected by mutations of genes causing particular types of muscular dystrophy. Furthermore, phenotypic heterogeneity is observed when mutations in the same gene manifest with different distinct phenotypes even within the same family (e.g. distal or proximal myopathy, isolated cardiomyopathy) (Table 11.1).

The muscular dystrophies described in this chapter (see Table 11.1) reflect the broad spectrum of clinical presentations, genetic causes, pathological mechanisms, and approaches to treatment. The most common muscular dystrophy is Duchenne muscular dystrophy (DMD), which is named after Dr Guillaume Duchenne

Neurobiology of Brain Disorders
http://dx.doi.org/10.1016/B978-0-12-398270-4.00011-2

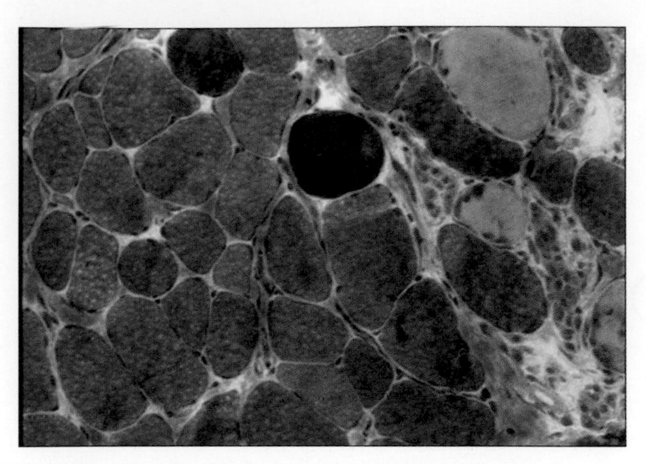

FIGURE 11.1 **Modified Gomori trichrome staining of a muscle tissue section from a young boy with Duchenne muscular dystrophy.** The pathological changes observed are common to all causes of dystrophic muscle and include variation in muscle fiber size, muscle fiber necrosis and regeneration, hypercontracted muscle fibers, inflammation, and increases in connective tissue.

de Boulogne, who gave the initial patient description in 1868. DMD is caused by mutations in the dystrophin gene, and striated muscle histopathology in DMD shows a complete loss of dystrophin from the sarcolemma of muscle fibers and cardiomyocytes. A second disorder is clinically similar but not identical and is caused by mutations in the same gene as DMD. This second disorder is allelic to DMD, and is called Becker muscular dystrophy (BMD). BMD is characterized by a partial loss of dystrophin and a less severe phenotype than DMD. Facioscapulohumeral dystrophy (FSHD) was first described by Drs Landouzy and Dejerine in 1884 as a "progressive atrophic myopathy". FSHD is usually manifest by a pattern of weakness that involves the face and limb muscles asymmetrically.[2] Yet another type of dystrophy is myotonic dystrophy (DM), initially described by Dr Hans Steinert and by Drs Fred Gibb and H.P. Batten in 1909. Principal skeletal muscle features of DM include delayed muscle relaxation after a forceful but painless contraction, resulting in muscle stiffness and weakness. Prominent systemic features include cardiac arrhythmias, daytime hypersomnolence, and early development of cataracts. Two distinct types of DM, type 1 (DM1) and type 2 (DM2), are determined by different genes, demonstrating the principle of genetic heterogeneity.[3]

Oculopharyngeal muscular dystrophy (OPMD) is an adult-onset disorder that presents later in life than most other muscular dystrophies and is characterized by ptosis (eyelid muscle weakness), difficulty swallowing, and proximal limb muscle weakness.[4]

Limb-girdle muscular dystrophies (LGMD) comprise a heterogeneous group of diseases with predominantly proximal weakness, a dystrophic pattern on muscle biopsy, and mutations of various genes crucial for muscle function.[5] Depending on the type of LGMD, clinical symptoms may have an early or a later onset with variable pace of progression and multisystemic involvement (Table 11.1).

Emery–Dreifuss muscular dystrophy (EDMD) is characterized by early joint contractures with slowly progressive weakness and cardiomyopathy.[6] Muscle weakness in patients with EDMD typically involves proximal upper (scapulohumeral) and distal lower (peroneal) extremity muscles, and motor function is further impaired by the development of contractures.

PATHOPHYSIOLOGY AND GENETICS

Muscular dystrophies are caused by mutations of genes that play major roles in diverse physiological processes such as muscle contraction, gene transcription, or organ development. Many of the abnormal or absent proteins associated with the development of the muscular dystrophies have important roles in membrane repair, muscle contractility, or structure of the cytoskeleton. Depending on the type of dystrophy and the gene involved, heredity may follow autosomal dominant (DM, FSHD, OPMD, LGMD1, EDMD2/4/5), autosomal recessive (OPMD, LGMD2, EDMD3), or X-linked (DMD, BMD, EDMD1) inheritance. Spontaneous new mutations also occur. The variety of genetic disruptions that cause muscular dystrophy includes point mutations (DMD, BMD, LGMD, EDMD), multiple exon deletions and duplications (DMD, BMD), chromosomal fragment deletions (FSHD1, DMD/BMD), and expansion of nucleotide repeats (DM1, DM2, OPMD) (Table 11.2). These mutations can also cause a variety of allelic disorders. In recessive disorders, carriers are typically asymptomatic, although some may eventually develop some symptoms that are typically not as severe as in affected patients with homozygous (autosomal) or hemizygous (X-linked) mutations.

Duchenne and Becker Muscular Dystrophies

DMD and BMD are caused by the disruption of function of the largest known gene, dystrophin, owing to gene deletions, duplications, or point mutations. The majority of mutations in dystrophin are large, intragenic exon deletions (65%) or duplications (5%). Deletions or duplications can preserve the reading frame of the transcript (in-frame), which usually results in the milder BMD phenotype, or disrupt the reading frame of the transcript (out-of-frame), which results in the more severe DMD phenotype. The remaining mutations are point mutations that disrupt messenger RNA (mRNA) splicing or cause premature chain termination (nonsense mutations). DMD and BMD are X-linked disorders that

TABLE 11.1 Genetics of Muscular Dystrophies

Syndrome	Gene Product	Gene Function	Chromosomal Location	Heredity	Typical Age of Onset	Allelic Syndromes
DMD/BMD	Dystrophin	Sarcomere protein	Xp21.2	X-recessive	DMD before age 5 years; BMD after age 7 years	Isolated cardiomyopathy
DM1	DMPK	Threonine kinase	19q	AD	Infancy to 40 years	–
DM2	ZNF9	RNA binding protein	3q21.3	AD	8–60 years	–
FSHD	D4Z4[a]	Derepression of germline transcription factor[a]	4q35	AD	Infancy to 25 years	–
OPMD	PABPN1	Role in nuclear protein aggregation	14q11.2-q13	AD/AR	After age 40 years	–
LGMD 1A	Myotilin	Sarcomere protein	5q31.2	AD	Variable	Myofibrillar myopathy
LGMD 1B	Lamin A/C	Nuclear envelope protein	1q22	AD	Before age 20 years	EDMD2/3; familial partial lypodystrophy Dunnigan type (FPLD); AR axonal neuropathy (CMT2B1); mandibuloacral dysplasia (MAD), progeria syndromes
LGMD 1C	Caveolin	Transmembrane protein; vesicular traffic and scaffolding	3p25.3	AD	Age 5 years to adulthood	Rippling muscle disease (RMD); hyperCKemia; distal Tateyama myopathy (MDT); hypertrophic cardiomyopathy
LGMD 1D	DNAJB6	Chaperone protein	6q22	AD	20–60 years	Dilated cardiomyopathy (CMD1F)
LGMD 2A	Calpain-3	Muscle-specific protease	15q15	AR	Infancy to 20 years	–
LGMD 2B	Dysferlin	Transmembrane sarcolemmal protein	2p13.2	AR	10–30 years	Miyoshi myopathy (MMD1); distal myopathy (DMAT)
LGMD 2C	γ-Sarcoglycan	Dystrophin-associated glycoprotein	13q12	AR	Infancy to 10 years	
LGMD 2D	α-Sarcoglycan	Dystrophin-associated glycoprotein	17q21.33	AR	2–15 years	
LGMD 2E	β-Sarcoglycan	Dystrophin-associated glycoprotein	4q12	AR	3–16 years	
LGMD 2F	δ-Sarcoglycan	Dystrophin-associated glycoprotein	5q33	AR	2–10 years	AD familial dilated cardiomyopathy (CMD1L)
LGMD 2G	Telethonin	Sarcomeric Z-disc protein	17q12	AR	9–15 years	
LGMD 2H	Trim32	E3 ubiquitin ligase	9q33.1	AR	8–27 years	
LGMD 2I	Fukutin-related protein	Sarcoplasmic glycosylating enzyme	19q13.3	AR	Before age 5 years	Congenital muscular dystrophy (MDC1C/MDDGA5)

Continued

TABLE 11.1 Genetics of Muscular Dystrophies—cont'd

Syndrome	Gene Product	Gene Function	Chromosomal Location	Heredity	Typical Age of Onset	Allelic Syndromes
LGMD 2J	Titin	Sarcomeric protein with serine kinase activity	2q24.3	AR	0–30 years	Tibial muscular dystrophy (TMD); familial hypertrophic cardiomyopathy (CMH9)
LGMD 2K	POMT-1[b]	O-Mannosylation of proteins	9q34.1	AR	Infancy to 10 years	Congenital muscular dystrophy (MDDGA1)
LGMD 2L	Anoctamin 5	Chloride channel	11p13-p12	AR	30–40 years	Distal myopathy (MMD3)
LGMD 2M	Fukutin	Sarcoplasmic glycosylating enzyme	9q31	AR	Infancy	Congenital muscular dystrophy (MDDGA4/B4/C4); cardiomyopathy, dilated (CMD1X)
LGMD 2N	POMT-2[c]	O-Mannosylation of proteins	14q24	AR	–	Congenital muscular dystrophy (MDDGA2)
LGMD 2O	POMGnT1[d]	O-Mannosylation of proteins	1p34.1	AR	6–40 years	Congenital muscular dystrophy (MDDGA3)
EDMD1	Emerin	Nuclear envelope, interacts with lamin	Xq28	X-recessive	10–20 years	
EDMD2/3	Lamin A/C	Nuclear envelope protein	1q22	AD (EDMD2); AR (EDMD3)	Missense –2 years; frameshift –30 years	LGMD1B; familial partial lipodystrophy Dunnigan type (FPLD); autosomal recessive axonal neuropathy (CMT2B1); mandibuloacral dysplasia (MAD); progeria syndromes
EDMD4	Nesprin-1/SYNE1	Nuclear envelope protein	6q25.1-q25.2	AD	11 years	Cerebellar ataxia (SCAR8)
EDMD5	Nesprin-2/SYNE2	Nuclear envelope protein	14q23.2	AD	Childhood to 40 years	
EDMD6	FHL1	RNA binding protein	Xq26.3	X-recessive	–	X-linked reducing body myopathy; X-linked myopathy with postural muscle atrophy (XMPMA)

DMD/BMD: Duchenne/Becker muscular dystrophy; DM1/2: myotonic dystrophy type 1/2; FSHD: facioscapulohumeral muscular dystrophy; OPMD: oculopharyngeal muscular dystrophy; LGMD: limb-girdle muscular dystrophy; EDMD: Emery–Dreifuss muscular dystrophy; AD: autosomal dominant; AR: autosomal recessive.

[a]*Deletion of non-coding D4Z4 repeats at 4q35 causes inappropriate expression of DUX4*

[b]*protein-O-mannosyltransferase-1*

[c]*protein-O-mannosyltransferase-2*

[d]*protein-O-mannosyl-β1,2N-acetylglucoamintransferase-1.*

primarily affect boys, but different clinical manifestations may affect up to 10% of female carriers. Clinical manifestations in most symptomatic carriers are related to skewed X inactivation, although rarely chromosomal rearrangements may occur.[7]

The dystrophin protein is a sarcomeric cytoskeletal protein that plays a critical role in the dystrophin–glycoprotein complex, a protein complex that links intracellular actin filaments to laminin in the extracellular matrix of the muscle fiber and maintains the stability of the plasma membrane, which is exposed to the mechanical stress of contraction (Fig. 11.2). The amino-terminus of dystrophin binds to the intracellular actin molecule and the carboxy-terminus of dystrophin links it to a complex of glycoproteins in the sarcolemma. Loss of dystrophin protein because of out-of-frame mutations is accompanied by a secondary loss of other proteins in the dystrophin–glycoprotein complex. The breach of integrity of the dystrophin–glycoprotein complex at the sarcolemma results in muscle plasma membrane fragility such that normal muscle contraction causes an influx of extracellular calcium that triggers protease-mediated muscle fiber degradation. Despite attempts by diseased muscle to regenerate, ultimately, muscle fiber regeneration fails and muscle fibers are replaced with fibrous and fatty connective tissue.

TABLE 11.2 Genetic Alterations in Muscular Dystrophies

Point mutations
Missense mutations
Nonsense mutations
Deletions and duplications
In-frame
Out-of-frame
Deletions of non-coding repeats (e.g. D4Z4 in FSHD1)
Expansion of triplet/quadruplet repeats

FSHD: facioscapulohumeral dystrophy.

Facioscapulohumeral Dystrophy

FSHD is caused by a toxic gain of function and inappropriate expression of a germline transcription factor, DUX4.[8] The most common genetic cause of FSHD, which is found in 95% of patients, is a deletion of a variable number of copies of the repeated 3.3 kb macrosatellite D4Z4 at chromosome location 4q35.2, and this is also known as the FSHD1 variant. This is an example of molecular disease pathology caused by a variation in copy number of a macrosatellite repeat. Individuals with 11–150 D4Z4 repeats or deletion of the entire array do not have FSHD. Individuals with 1–10 D4Z4 repeats manifest clinical FSHD. The D4Z4 repeat deletion leads to DNA hypomethylation of the contracted D4Z4 allele and histone modification. These epigenetic changes in chromatin configuration result in upregulation of expression of the downstream *DUX4* gene. It is thought that this toxic gain of function of the *DUX4* gene, which codes for a putative double homeobox protein, causes transcriptional overexpression of other 4q35 genes, thus manifesting disease.[8,9] Less frequently, the FSHD phenotype is associated with hypomethylated D4Z4 DNA despite a normal number of D4Z4 repeats and this type is referred to as the FSHD2

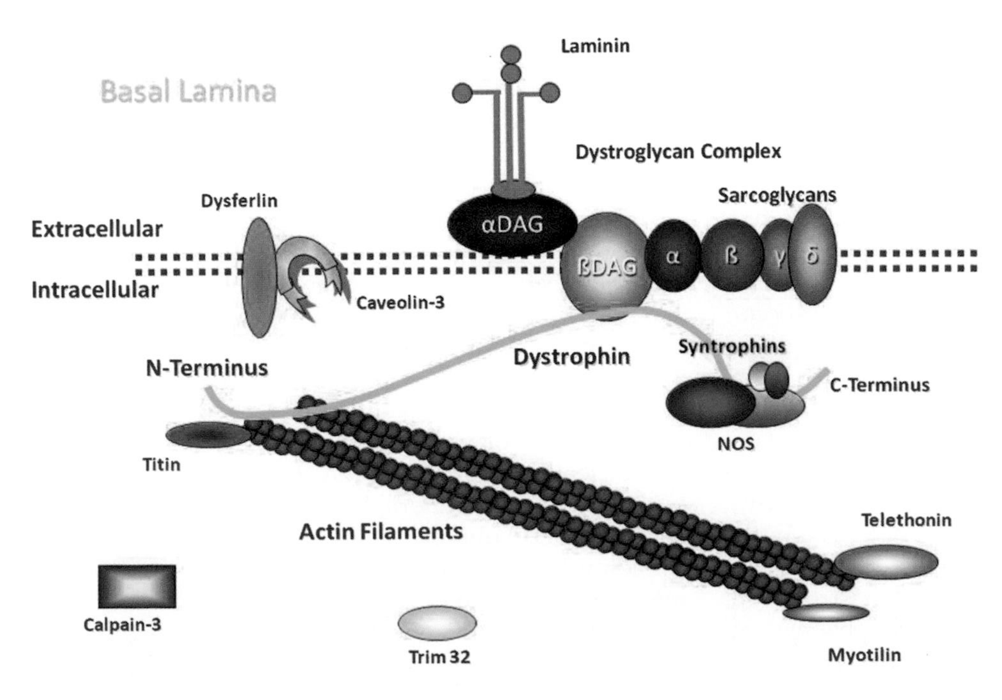

FIGURE 11.2 **Schematic diagram of muscle proteins whose dysfunction can cause muscular dystrophy.** DAG: dystrophin-associated glycoprotein; NOS: nitric oxide synthase.

variant.[10] Experimental studies have suggested a role of pLAM polymorphism on chromosome 10 which stabilizes the DUX4 transcript, leading to development of the FSHD phenotype even in the absence of a D4Z4 deletion.[8]

Myotonic Dystrophy Types 1 and 2

The myotonic dystrophies, DM1 and DM2, share a common genetic toxic gain of function and autosomal dominant inheritance. Unstable DNA triplet or quadruplet repeats expand pathologically; repeat expansions interfere with the export of mRNA molecules from the nucleus to the cytoplasm. DM1 is caused by a CTG repeat expansion in the gene encoding DMPK, and DM2 is caused by a CCTG quadruplet repeat expansion in the *ZNF9* gene. The accumulation of aberrant RNA molecules also adversely affects normal mRNA splicing, which, in turn, affects the expression of multiple downstream targets (>200 genes in a mouse model of DM1).[11] These events precipitate depletion of RNA binding proteins and activation of protein kinase C in DM1 and depletion of RNA proteins in DM2. The function of RNA binding *muscleblind* (Mbnl; Mbnl 1/2/3) proteins, which play a significant role in normal muscle and brain development, is altered in DM1/2, but the pathophysiology is not fully understood. Typical features of DM1/2 are found in a mouse model that has knockout of *Mbnl1*, including myotonia, cataracts, and alternative splicing of the muscle chloride channel.[12] Experimental studies also showed developmental abnormalities in the brain of *Mbnl2* knockout mice, which correlated with reported clinical abnormalities in DM1, including daytime hypersomnia, rapid eye movement sleep abnormalities, and altered spatial memory.[13] DMPK also acts as a protein kinase that has an inhibitory effect on myosin phosphatase target subunit (MYPT1) and protein phospholemman, which play a role in maintaining muscle contractility and relaxation.[14]

Genetic anticipation (younger age at onset in succeeding generations) has been reported in DM1, but this phenomenon, which is thought to occur as a result of further expansions of the unstable triplet or quadruplet repeat with successive generations, may not occur in DM2. In DM1, the size of the CTG repeat expansion may dramatically increase in offspring (especially with maternal transmission), leading to an earlier onset and greater severity of clinical manifestations, However, taking each patient individually, the severity of DM1 is not proportionate to the number of CTG repeats. Similarly, there are also rare reports of contractions of CTG repeats in the *DMPK* gene associated with paternal transmission in DM1.[3] The size of CTG repeats is also unstable within different tissues and may lead to different repeat expansion sizes in different tissues, a phenomenon referred to as somatic mosaicism. There seems to be an increased frequency of recessive mutations of chloride channel type 1 (CLCN1) in families with DM2, which has the potential to amplify muscle myotonia, but the clinical importance of this finding remains uncertain.[15] Expression of CLCN1 is also altered in DM1 and DM2 owing to aberrant alternative splicing, resulting in impaired chloride channel function and membrane hyperexcitability in myotonic dystrophies.[16] In contrast to DM, myotonia congenita is a non-dystrophic cause of myotonia that is attributed to CLCN1 mutations and can be autosomal dominant (Thomsen) or recessive (Becker).[17] Small premutation expansions of the triplet repeat in the *DMPK* gene have been reported in patients with isolated cataracts without musculoskeletal symptoms. It is possible that descendants of these patients may develop DM1 with a further expansion of CTG repeats in the *DMPK* gene.[18]

Oculopharyngeal Muscular Dystrophy

OPMD is caused by a polyalanine expansion in the gene polyadenylate-binding protein nuclear 1 gene (*PABPN1*; previously also known as *PABP2*).[4] Transmission of OPMD is typically autosomal dominant, but it may also follow an autosomal recessive pattern. The most typical histological finding is the presence of intranuclear inclusions consisting of unbranched tubular filaments. These inclusions are present only in muscle fibers, and not in other cell types. Polyalanine-expanded PABPN1 triggers the formation of insoluble intranuclear aggregates of toxic proteins containing PABPN1, poly(T)-RNA, components of the ubiquitin–proteasome pathway, and transcription factors. A possible mechanism that could lead to cell death involves sequestration of RNA binding proteins and mRNA, but the exact cascade of events remains largely obscure. The cellular role of PABPN1 is still unclear but it seems to participate in the control of expression of muscle-specific genes at the transcriptional level.

Limb-Girdle Muscular Dystrophies

LGMDs are a heterogeneous group of disorders caused by altered expression of various genes that encode transmembrane proteins, ubiquitin ligase, ion channels, and sarcomeric proteins, and result in a slowly progressive form of weakness starting in proximal upper and lower extremity (limb-girdle) muscles with other clinical features depending on the LGMD type. Based on the inheritance pattern, LGMDs have been classified as the LGMD1 subtype, which demonstrates autosomal dominant heredity, and the LGMD2 subtype, which is autosomal recessive.

Mutations of sarcomeric proteins myotilin, dysferlin, telethonin, and titin cause dominant and recessive

LGMD, types 1A and 2B/G/J, respectively (Table 11.1). Myotilin is a sarcomeric protein that binds F-actin, cross-links actin filaments, and contributes to stabilization and anchorage of thin filaments. Mutations of myotilin may cause LGMD1A or distal myofibrillar myopathy.[19] LGMD2B is caused by mutations of the transmembrane sarcolemmal protein dysferlin. Dysferlin plays a role in caveolar signaling, and interacts with the proteins calpain-3 and caveolin-3, which are associated with LGMD1C and LGMD2A, respectively. Experimental studies have demonstrated a role for dysferlin in muscle regeneration, membrane repair, and myoblast differentiation.[20] Demonstrating phenotypic heterogeneity, mutations of the dysferlin gene can cause either LGMD2B or distal myopathies. Telethonin (LGMD2G) is another sarcomeric protein found exclusively in skeletal and cardiac muscle and serves as an anchor for titin, which gives the sarcomere an increased capacity for mechanical resistance. LGMD2J is caused by mutations of the gene encoding titin, which is the largest protein known. Titin maintains the integrity of the sarcomere and also has serine kinase properties. It plays an important role in muscle fiber assembly, force transmission, and maintenance of resting muscle tension.

Another group of dominant and recessive muscular dystrophies is caused by mutations of genes that encode proteins playing a role in membrane transport and repair. LGMD1C is caused by mutations of the gene encoding caveolin-3, which regulates vesicular trafficking within the muscle fiber. Caveolin-3 is a muscle-specific member of the protein family that contributes important components of the caveolar subcompartments of the plasma membrane. Caveolin-3 is associated with the dystrophin–glycoprotein complex and plays a major role in muscle development, energy metabolism, and signaling. Mutations of caveolin-3 were also reported in patients with rippling muscle disease, which is a rare autosomal dominant disorder characterized by mechanical hyperirritability of muscle, electrically silent muscle contractions triggered by mechanical stimulation, and distal myopathy. Mutations in caveolin-3 can also cause idiopathic hyperCKemia, characterized by high serum levels of creatine kinase without symptoms of myopathy.

LGMD1D is caused by mutations of the gene encoding chaperone protein DNAJB6, which is found in the Z-disc structure of the sarcomere. DNAJB6 interacts with heat-shock proteins, stimulates ATP hydrolysis, and inhibits activation of protease caspase-3.

Recessive LGMD2A is caused by mutations of the gene encoding a muscle-specific calcium-dependent protease, calpain-3, which plays an important role in muscle adaptation to physical stress.[21] Calpain-3 belongs to a family of intracellular, non-lysosomal, cysteine proteases, and participates in muscle membrane repair and remodeling.

LGMD2C–F are caused by mutations of four different sarcoglycans, which are proteins in the dystrophin–glycoprotein complex that play an important role in the maintenance of myofiber integrity. The sarcoglycan complex functions as a unit affected by dysfunction or loss of any of its members, and mutations of one of the sarcoglycans may result in secondary loss of the other sarcoglycans.[22] Sarcoglycans appear to play a major role in maintaining sarcolemmal integrity, although many questions remain.

LGMD2H is caused by mutations of tripartite motif-containing protein-32 (Trim32), which is a ubiquitin ligase that localizes to the Z-line in skeletal muscle and plays a role in protein degradation. Reflecting a lack of complete consensus on LGMD nomenclature, LGMD2H is sometimes referred to as LGMD2L. Mutations of Trim32 result in impairment of substrate ubiquitination that thus interferes with protein degradation.

Another group of autosomal recessive muscular dystrophies, LGMD2I/K/M/N, is caused by the genes encoding enzymes that modify proteins by glycosylation of α-dystroglycan. These enzymes play a significant role in organ development. In addition to mutations in these genes causing LGMD, allelic disorders caused by mutations in the same genes include congenital muscular dystrophies, which may be accompanied by developmental anomalies of the eye and brain. Histopathological muscle studies in LGMD2I/K/M/N show altered glycosylation of dystroglycans, which are major components of the dystrophin–glycoprotein complex. Both α-dystroglycan and β-dystroglycan are translated from a single mRNA encoded by the gene *DAG1*. α-Dystroglycan is a membrane-associated protein that binds to laminin in the extracellular matrix and to an extracellular domain of β-dystroglycan, which is an integral membrane protein.

Mutations of the gene encoding fukutin-related protein (FKRP) can cause recessive LGMD2I or congenital muscular dystrophy with or without eye and brain abnormalities.[23] The exact role of FKRP is still largely uncertain, but experimental studies have suggested that it plays an important role in maintaining adequate glycosylation of dystroglycans.

Mutations in fukutin (LGMD2M) were initially identified as a cause of Fukuyama congenital muscular dystrophy, which is a congenital muscle disorder associated with brain development abnormalities. Histopathological studies demonstrated decreased glycosylation of dystroglycan in muscle from patients with LGMD2M.

Protein O-mannosylation is an essential protein modification in organ development, but the exact metabolic pathways have not been fully elucidated. The only well-characterized substrate of O-mannosylation in human tissue is α-dystroglycan, and the loss of O-mannosylation may affect the interaction of dystroglycan with the extracellular matrix. Mutations of genes encoding enzymes POMT1 and POMT2 cause recessive LGMD2K or LGMD2N, or congenital muscular dystrophy with developmental anomalies of the eye and brain.

Mutations of putative chloride channel anoctamin-5 are associated with proximal myopathy (LGMD2L) and with distal myopathy (MMD3).[24] Anoctamin-5 is an integral membrane glycoprotein with high expression levels in skeletal and cardiac muscle, growth-plate chondrocytes and osteoblasts. Anoctamin-5 expression also follows a specific pattern during embryonic development, but its role remains largely unclear.

Emery–Dreifuss Muscular Dystrophy

EDMD is characterized by early contractures, muscle weakness, and cardiac arrhythmias. Depending on the subtype, EDMD is associated with an X-linked (EDMD1/6), an autosomal dominant (EDMD2/4/5), or an autosomal recessive (EDMD3) pattern of inheritance. Different types of EDMD are caused by mutations of nuclear envelope proteins that play a role in nuclear trafficking, DNA replication, chromatin segregation, gene transcription, RNA processing, and cell cycle progression. Gene mutations that cause EDMD affect the ability of the entire cell to respond to stress. EDMD1 is an X-linked dystrophy caused by mutations in the gene that encodes emerin, which localizes to the inner nuclear membrane and plays a significant role in the maintenance of the integrity and stability of the nucleus. Autosomal dominant EDMD2 and autosomal recessive EDMD3 are caused by mutations of the gene encoding nuclear envelope protein lamin A/C. Lamin A/C mutations interfere with nuclear trafficking. Mutations in the lamin A/C gene can also cause one of the autosomal dominant forms of LGMD, LGMD1B. Experimental studies demonstrate an accumulation of inner nuclear membrane protein Sun1 in disorders caused by mutations in the lamin A/C gene, and the phenotype is alleviated by blocking accumulation of Sun1 in mutant fibroblast cells and in transgenic animals.[25] However, the exact role of this protein remains unclear. Transcription of the lamin A/C gene facilitates the production of two isoforms of lamin, lamin A and lamin C, by alternative splicing. These two proteins play a significant role in maintaining nuclear integrity and activation of gene transcription. Mutations of lamin may also accelerate processes associated with aging.

Two other variants of EDMD, which are caused by mutations of nesprin1 and nesprin2, have been described. The nesprin proteins are also expressed in the nuclear envelope, where they play a role linking the nucleoskeleton with the cytoskeleton. Another X-linked variant of EDMD is caused by mutation of the gene encoding four-and-a-half LIM domains protein-1 (FHL1), which is another RNA binding protein that serves as a transcriptional coactivator.

EPIDEMIOLOGY

Our understanding of the epidemiology of the muscular dystrophies continues to evolve owing to greater use of improved diagnostic methods. Advances in molecular diagnostics have also allowed the diagnosis of milder and atypical clinical cases that may not have been properly classified previously.

The most common muscular dystrophy is caused by dystrophin deficiency. The more severe phenotype of dystrophin deficiency is DMD, which affects 1 in 3500 male births, while the milder dystrophin deficiency phenotype, BMD, affects 1 in 18,000 male births. Although the inheritance of these disorders is X linked, up to one-third of cases are caused by new mutations of the dystrophin gene.

FSHD is the third most common muscular dystrophy, with an estimated prevalence of 1 in 15,000 to 1 in 20,000.[26] FSHD has a high penetrance, approximately 95%, and a worldwide distribution.

DM1 is the second most common muscular dystrophy overall and the most common muscular dystrophy in adults, with a prevalence ranging from 1 in 8000 (Caucasian populations) to 1 in 20,000 (Asian and African populations). Epidemiological population studies of DM2 are sparse, but its prevalence may be significantly underestimated as limited studies of Italian and Finnish populations suggest a prevalence of 1 in 1800.[27]

OPMD has a worldwide distribution with a prevalence of 1 in 100,000, with much higher frequency concentrated in several clusters including Bukhara Jews (1 in 600) and French Canadians (1 in 1000).

The epidemiology of LGMD is highly variable as some subtypes of LGMD are rare and some subtypes have higher endemic occurrence in a particular ethnic group (e.g. LGMD2L in Manitoba Hutterites) or are described in only a few families (e.g. LGMD2M)[1] in the USA. The most common LGMDs have been LGMD2B (dysferlin; 18% of all LGMDs), LGMD2C–F (sarcoglycans; 15% of all LGMDs), LGMD2I/K/M/N (dystroglycan glycosylation; 15% of all LGMDs), LGMD2A (calpain-3; 12% of all LGMDs), and LGMD1C (caveolin; 1.5% of all LGMDs).[28] Deficiency of telethonin (LGMD2G) is very rare and has been reported mostly in patients from Brazil, with considerable phenotypic variability. Mutations of anoctamin-5, which can cause LGMD2L or non-dysferlin Miyoshi-type myopathy, have been described in 1 in 400,000 in northern England.

Less is known about worldwide epidemiology of different forms of EDMD. The prevalence of EDMD2 (lamin A/C) has been estimated in different populations at 1–3 in 100,000.

CLINICAL MANIFESTATIONS

Muscle Involvement

Progressive muscle weakness is a cardinal feature of muscular dystrophies, and it may affect various groups of muscles at different ages. Weakness usually involves proximal muscles, but early involvement of distal muscles may also occur. Weakness of axial muscles may lead to the development of scoliosis. Respiratory muscle weakness ultimately can cause respiratory failure. Most patients report the insidious onset of weakness or limitations of physical activity or even motor development delay before a fully developed clinical presentation and diagnosis of muscular dystrophy. Paradoxically, mutations of dysferlin (LGMD2B) can be associated with a superior athletic ability that permits participation in sports before the onset of symptoms.

The clinical symptoms of muscular dystrophy depend on the extent and complement of individual muscles affected. The pattern of muscle involvement is predominantly or exclusively proximal or distal, with acute or chronic progression, muscle pain and stiffness, fatigue or muscle atrophy (e.g. calf wasting). It is not fully understood why certain muscles are more affected than others in patients with muscular dystrophy.

Mutations of the same gene can cause different distinct phenotypes, which demonstrates the principle of phenotypic heterogeneity (e.g. lamin A/C–LGMD1B or EDMD2/3; dysferlin–LGMD2B or Miyoshi myopathy), or similar phenotypes of different severity (e.g. dystrophin deficiency causing DMD or BMD).

DMD and BMD are X-linked recessive disorders affecting males.[7] In DMD, selective involvement of limb-girdle and proximal limb muscles results in the early development of exaggerated lumbar lordosis. Other cardinal findings include toe-walking (caused by ankle contractures) and calf hypertrophy. Most patients are diagnosed by the age of 5 years and progressive proximal weakness leads to the loss of independent ambulation before the age of 13 years.[7] Respiratory and cardiac insufficiency are usually apparent in the second decade. By the end of the second decade, there are prominent contractures, with weakness and loss of muscle mass that preclude almost all independent mobility. Patients ultimately require total care for activities of daily living and respiratory support. Death usually occurs during the third decade owing to respiratory or cardiac failure.

The phenotype of BMD is broader than DMD, and includes any patient with dystrophin deficiency who does not meet the criteria of onset and loss of ambulation described above for DMD. BMD can present with similar symptoms to DMD, but have a later onset and slower progression. Alternatively, BMD can present with isolated cardiomyopathy, isolated cognitive deficits, or a syndrome of myalgia and myoglobinuria. Female carriers of dystrophin mutations are most often asymptomatic; however, in the setting of skewed X inactivation, manifesting carrier females may have proximal muscle weakness or cardiomyopathy. Up to 10% of female carriers of one dystrophin mutation have symptoms of muscle weakness.

FSHD is characterized by weakness of facial, scapular, and proximal upper extremity muscles. Clinical features of the two variants of FSHD (FSHD1 and FSHD2) are indistinguishable.[2] Weakness is often asymmetric. Affected family members may show a large spectrum of severity from very mild to severe. Diffuse body pain is a typical feature of FSHD and can be difficult to treat. The progression of muscle weakness in FSHD is often slow and life expectancy is typically not affected. Only approximately 20% of FSHD patients require a wheelchair for mobility by the age of 50 years.[2]

Myotonic dystrophies affect predominantly distal (DM1) or proximal (DM2) muscles and predominant symptoms include muscle weakness and stiffness attributable to delayed muscle relaxation, which is called myotonia (Fig. 11.3). DM1 typically shows distal muscle wasting and weakness. A more severe form of congenital DM1 has been described, predominantly in children of DM1-affected mothers. Severely affected babies with congenital DM1 often require respiratory support, and the baby is typically hypotonic or in severe cases may have arthrogryposis (multiple congenital joint contractures), with mortality of up to 16%. Congenital DM1 is caused by a very large expansion of the triplet repeat in the *DMPK* gene. In contrast to DM1, DM2 predominantly affects proximal muscles. A severe congenital form of DM2 has not been reported.

OPMD typically manifests with ptosis, diplopia, and progressive dysphagia, usually not until the fifth or sixth decade. Patients also later develop proximal limb-girdle weakness of variable severity. Life expectancy is normal.[4]

Limb-girdle muscular dystrophies share a predilection for proximal muscle involvement and have other clinical features that distinguish individual variants of LGMD depending on subtype. Many LGMDs share the causative genes with allelic disorders, which include distal myopathies, isolated cardiomyopathies, and even axonal neuropathy (Table 11.1). The onset of muscle weakness in LGMDs is variable depending on LGMD subtype. As in DMD and BMD, LGMD2C–F (sarcoglycan disorders) may show calf pseudohypertrophy, but

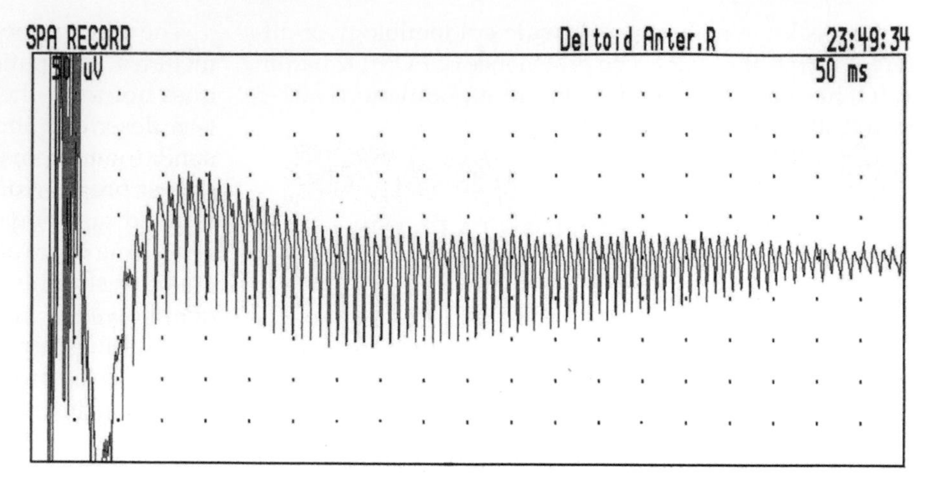

FIGURE 11.3 **Electrical myotonia on needle electromyography in a patient with myotonic dystrophy type 1.** Electrical myotonia can be observed in myotonic dystrophies, non-dystrophic myotonic muscle disorders, and some other muscle disorders, including acid maltase deficiency.

muscle histopathology shows normal dystrophin levels and inheritance is autosomal recessive.

EDMD is characterized by early muscle contractures at the elbows, Achilles tendons, and paravertebral cervical area. The prominent contractures can further exacerbate the decreased function due to muscle weakness. The distribution of muscle weakness in EDMD is predominantly proximal in the upper extremities and distal in the lower extremities.

Multisystemic Involvement

Multisystemic involvement in muscular dystrophies may reflect the widespread expression of mutated genes or may occur as a secondary process related to muscle weakness (e.g. respiratory failure caused by diaphragm weakness) (Table 11.3). Asymmetric axial weakness may precipitate the development of scoliosis. As scoliosis becomes more severe, it may also exacerbate respiratory insufficiency. Similarities between the two types of striated muscle, skeletal and cardiac, may explain the occurrence of cardiomyopathy in some types of muscular dystrophy, and clinical manifestations of cardiac involvement include a decrease in cardiac systolic and diastolic function that may progress to heart failure or arrhythmias.[29] Cardiac involvement does not depend on the extent of skeletal muscle disease, and heart failure or arrhythmias may show minimal or no signs of skeletal myopathy. Progressive heart failure and sudden death due to arrhythmias are major contributors to mortality in some types of muscular dystrophy, especially DMD, BMD, EDMD, DM1, and LGMD1B/E and LGMD2D–F/I.[29] Less frequently, cardiac involvement is also present in FSHD. Overall, there is a great variability in the severity of cardiac involvement, but some patients with muscular dystrophy may develop cardiac disease that requires the therapeutic intervention of pacemakers, defibrillators, or even heart transplantation.[30]

TABLE 11.3 Multisystem Manifestations of Muscular Dystrophies

Type of Dystrophy	Frequent Extraskeletal Muscle Manifestations	Uncommon Extraskeletal Muscle Manifestations
DMD/BMD	Cardiac, pulmonary, CNS	
FSHD	Hearing loss, retinal teleangiectasias, CNS	Respiratory, cardiac
DM1/DM2	Cataracts, endocrine, cardiac, gastrointestinal, CNS, skin	Respiratory (frequent with congenital DM1), neuropathy
OPMD	Swallowing difficulties	CNS
LGMD	Cardiac, respiratory	Scoliosis
EDMD	Cardiac, joint contractures	

DMD/BMD: Duchenne/Becker muscular dystrophy; FSHD: facioscapulohumeral muscular dystrophy; DM1/2: myotonic dystrophy type 1/2; OPMD: oculopharyngeal muscular dystrophy; LGMD: limb-girdle muscular dystrophy; EDMD: Emery–Dreifuss muscular dystrophy; CNS: central nervous system complications.

In otherwise asymptomatic female carriers of dystrophin mutations, cardiac involvement may develop.

In EDMD, frequent arrhythmias are a major medical problem and may cause sudden death. Arrhythmias may even precipitate cardioembolic brain infarcts. Ventricular dysfunction is usually less severe in EDMD, and heart failure is relatively rare. In recessive EDMD, there are only rare reports of arrhythmias in otherwise asymptomatic carriers.

Cognitive difficulties of variable severity have been reported in patients with DMD, DM1/2 (especially with congenital DM1), and LGMDs (Fig. 11.4).[3,7]

FSHD is typically not associated with cardiomyopathy and respiratory involvement is very rare. Ocular involvement includes the development of retinal

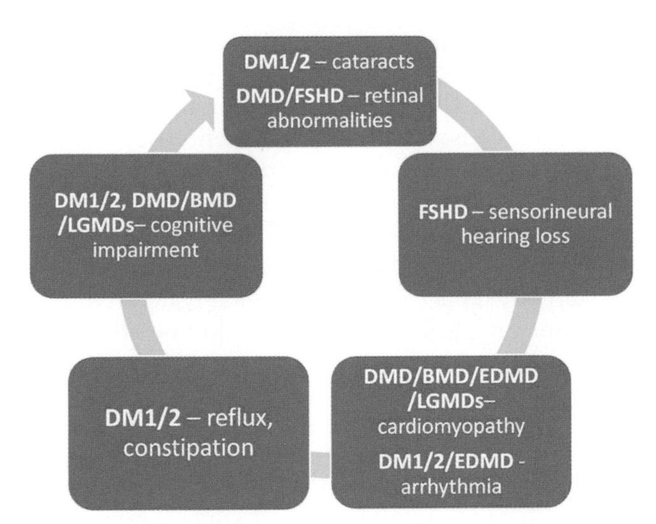

FIGURE 11.4 **Multisystemic involvement demonstrates the spectrum of clinical manifestations of various muscular dystrophies.** BMD/DMD: Becker/Duchenne muscular dystrophy; DM1/2: myotonic dystrophy type 1/2; EDMD: Emery–Dreifuss muscular dystrophy; FSHD: facioscapulohumeral muscular dystrophy; LGMD: limb-girdle muscular dystrophy.

telengiectasias that may eventually lead to exudative retinopathy and retinal detachment (Coates syndrome). Early-onset severe FSHD may be associated with severe hearing loss, which may be misinterpreted as cognitive dysfunction (Fig. 11.4).[2]

Cataracts are a major hallmark of DM1 and DM2. An early onset of cataracts, before the age of 50 years, may be an important diagnostic clue. Multisystemic involvement in DM1 has been well characterized and also includes respiratory insufficiency and frequent sleep abnormalities, cognitive impairment, and gastrointestinal dysmotility. Endocrine disturbances, which include insulin resistance, hypothyroidism, and male hypogonadism and infertility, are also frequent complaints in DM1. Multisystemic involvement in DM2 is similar to DM1, but DM2 patients are more likely to experience body pain than DM1 patients.[3] Cognitive dysfunction and structural brain abnormalities are often found in DM1 and to a lesser extent in DM2.

DIAGNOSIS

Diagnostic workup of patients with suspected muscular dystrophy is typically pursued as part of an evaluation of progressive muscle weakness in symptomatic individuals, and also of asymptomatic relatives of affected patients. Diagnosis is based on clinical features, family history, and laboratory testing, which may include muscle tissue histology and genetic studies.[1] Clinical findings alone may not be sufficient to provide a reliable diagnosis as there is often overlap of symptoms and signs (e.g. proximal weakness) in different forms of muscular dystrophy. Muscle biopsy and genetic testing are often needed for definitive diagnosis. However, clinical findings are helpful in narrowing the differential diagnosis and facilitating the diagnostic workup. Magnetic resonance imaging of muscle may also be helpful to demonstrate characteristic involvement of individual muscles or groups of muscles.[31] Electrodiagnostic testing (electromyography and nerve conduction studies) may facilitate the diagnostic process by showing the electrical features of myopathy or myotonia, but these tests may not be necessary in most cases of suspected muscular dystrophy because the clinical features are so characteristic.

General histology and specific protein expression can be analyzed in muscle biopsy tissue. The muscle sample is snap-frozen or paraffin-embedded and then sectioned. Staining techniques applied to muscle tissue sections can reveal morphological changes in myofibers or signs of inflammation. Immunohistochemistry is used to detect the presence and localization of specific proteins.

Genetic testing is required to determine the causative mutation for muscular dystrophies. However, genetic testing may be unrevealing if the correct form of muscular dystrophy has not been suspected and, as a result, genetic testing may not be performed for the correct gene. Furthermore, there are technological limitations that may prevent identification of a gene mutation. One of the pitfalls of genetic testing is that clinically meaningful mutations must be distinguished from DNA polymorphisms. Therefore, careful analysis of the clinical importance of newly described mutations is always needed. Technical advances in genetic testing have facilitated tremendous gains in the ability to diagnose different types of hereditary disorder, including muscular dystrophies, in clinically affected patients and carriers. Some of the modern technologies, such as next generation sequencing, may decrease the cost of genetic testing, which can improve access to specific mutation detection.[32] In those cases where genetic testing is not possible or not informative, muscle biopsy may be needed to establish the diagnosis.

Diagnostic algorithms depend on individual type of muscular dystrophy. For example, DMD may be suspected because of a delayed onset of ambulation or clumsiness and falls early in life. The first testing to be done is to measure the serum level of creatine kinase (CK). CK is an enzyme that is present in muscle; its release into serum in large amounts is a sensitive, but non-specific indicator of muscle fiber degeneration. A markedly elevated level of serum CK is consistent with, but not diagnostic of, DMD. The next level of testing is to search for single or multiple exon deletions or duplications in the dystrophin gene, because deletions are found in 65% of cases and duplications in 5% of cases of DMD. If this testing is negative, then typically, the next step is to sequence the dystrophin gene, looking for smaller mutations observed in the remaining 30% of cases.

For DM, FSHD, EDMD, and OPMD, the clinical phenotype generally provides sufficient guidance to determine the genetic testing to perform. If genetic testing is not revealing, then muscle biopsy can confirm the presence of a muscular dystrophy at a minimum, usually by the presence of variation in muscle fiber size, small fibers with centrally placed nuclei suggesting muscle regeneration, muscle fiber necrosis, and other morphological changes. A variable amount of inflammation may be detected in muscle tissue depending on the type of muscular dystrophy.

The LGMD disorders present the greatest challenge to achieve a specific genetic diagnosis. The multiple genetic subtypes and the current limitations of mutation testing for these disorders compound the diagnostic dilemma. Often, it is not possible to identify the specific subtype of LGMD. However, the limitations of genetic mutation testing for LGMD are likely to improve with advances in genetic diagnostic technologies.

GENETIC COUNSELING

The value of genetic counseling for families affected by muscular dystrophies cannot be overestimated. A precise and reliable diagnosis is a prerequisite to useful counseling. The complexity of genetic testing and its potential uses have multiplied over the past several years. However, limitations are still present and an individual phenotype cannot be always predicted based on the genotype.[33] While genetic information may provide the odds, this is still short of predicting whether (or when) a medical condition will develop in an individual. In addition, the number of potential legal issues has also expanded. These include disclosure and misuse of private health information, discrimination based on genetic information, and misdiagnosis. Disclosure of private health information is regulated by law, but other potential legal and medical consequences must be considered. When an individual is identified as an asymptomatic carrier of a clinically important allele, there is an increased risk of carrying the same allele and of sharing the consequent risk of disease in offspring. If there were an intervention that could modify the risk of progression to disease, then there would be a clear advantage for other family members to be informed. However, disclosure may be prevented by some affected individuals denying release of this information to family members for any of a variety of reasons, including personal disagreements and concern for the potential for discrimination.

Newborn Screening

Newborn screening (NBS) was established as a public health initiative in the 1960s to identify infants at a higher risk for certain disorders, who would benefit from early diagnosis and management. Most disorders identified by NBS, although not apparent at birth, would cause death or permanent disability if left untreated. Therefore, early treatment depends on early detection.

In the mid-1970s, it was demonstrated that a screening test for DMD could be performed on the dried blood spots already collected by the established NBS program.[34] Serum CK levels are elevated in newborns with DMD, but can also be elevated in other muscle conditions, including traumatic muscle damage during the birth process. For this reason, testing the newborn CK level detects all infants with DMD, but also generates many false-positive results.[35]

There have been several trials of NBS for DMD, and it is currently offered for newborn boys with parental consent in some countries, including Wales, Canada, and Belgium. The German infant screening program ended in November 2011 (G. Scheuerbrandt, personal communication, December 7, 2011).

There are currently four hospitals in the USA that offer optional NBS for DMD, through a research study funded by the Centers for Disease Control and Prevention (CDC) and conducted by investigators at the Columbus Children's Research Institute in Ohio.[36] In the past, NBS was offered in other parts of the USA, including parts of Pennsylvania (Pittsburgh), New York, Oregon, Iowa, and Texas, as well as parts of Brazil, France, New Zealand, and Puerto Rico.

As treatments for muscular dystrophies continue to be developed, the need for earlier treatment may arise, which would support NBS to identify individuals with muscular dystrophy presymptomatically. Even now, there is strong support for NBS for DMD among families affected by DMD, mainly because of the potential to help them to prepare for the future care of their child.

TREATMENT AND OUTCOMES

Treatment of Muscular Dystrophies

At the time of writing, treatment of muscular dystrophies is largely limited to experimental treatment studies exploring gene replacement therapy, gene function modulation, cell-based treatments, and various more conventional pharmacological treatments. Advances in gene therapy for muscular dystrophies reveal the emergence of a variety of strategies targeting different types of gene dysfunction.[37] These strategies will probably require tailoring according to the mutation profile of individual patients. These advantages have led to the concepts of personalized medicine, pharmacogenomics, and mutation-specific therapies.[38]

Gene replacement therapies have been largely based on delivery of a functional recombinant gene using replication-deficient viral vectors. Additional strategies include upregulation of expression from complementary (substitute) genes and readthrough of nonsense mutations.

Most viral vector-mediated gene transfer preclinical studies have been conducted in animal models of DMD. The *mdx* mouse has a point mutation in exon 23 of the murine dystrophin gene that causes premature chain termination of the dystrophin protein. The *mdx* mouse, therefore, is a biochemical and genetic animal model of DMD that has provided great utility for preclinical studies. Viral vector-mediated gene delivery studies have used adenoviral, herpes simplex viral (HSV), lentiviral, and adeno-associated viral (AAV) vectors. The 11 kb coding sequence of the dystrophin complementary DNA (cDNA) could only be accommodated by high-capacity adenovirus, from which all viral coding sequences were removed, and HSV. Neither adenoviral nor HSV vectors have proved efficient for transduction of skeletal muscle. Therefore, much of the more recent focus of viral vector-mediated gene delivery of the dystrophin cDNA has been on AAV vectors. Because of the transgene-carrying capacity of AAV vectors being limited to 4–5 kb, the dystrophin cDNA must be truncated. Based on the milder phenotypes of some BMD patients, the truncated dystrophin transgene has been constructed with a large in-frame deletion of the central rod domain of the protein. In transgenic mouse and gene transfer studies, these truncated dystrophin proteins have been demonstrated to be functional and to provide therapeutic benefit, although not to the degree of full-length constructs.

Multiple serotypes of AAV have been identified and characterized. Each serotype has a unique panel of tissue transduction efficiencies. AAV6, AAV8, and AAV9 efficiently transduce skeletal and cardiac muscle from a systemic vector delivery. Therefore, these AAV serotypes have been most extensively used for muscle transduction.

Based on the strength of preclinical studies of AAV vector-mediated gene delivery, pilot human studies have begun. The ability to transduce human skeletal muscle by direct injection without significant safety concerns has been demonstrated. Future studies are needed to explore systemic delivery and to understand the immune consequences.

A second promising future therapy for DMD is the use of exon skipping to convert an out-of-frame deletion mutation to an in-frame, larger deletion by manipulating splicing of the mRNA using antisense oligonucleotides. Two different oligonucleotide chemistries, 2-*O*-methyl and morpholino, have shown promise in preclinical and early clinical studies. By this therapeutic strategy, the hope is to convert a DMD phenotype to a milder BMD phenotype.

For the up to 15% of DMD patients who have nonsense mutations (a single-point mutation that changes an amino acid codon to a stop codon), therapies that promote readthrough of nonsense mutations are of potential benefit. Initial nonsense-mediated suppression studies tested aminoglycoside antibiotics, which demonstrated therapeutic potential but also toxicity that limited this potential. Researchers identified small molecules that promote nonsense-mediated suppression for study *in vitro* and *in vivo*. One compound, called Ataluren (previously known as PTC124), has been subjected to phase IIa and IIb human clinical studies.[39] The results to date of therapeutic benefit have not been completely convincing, such that further study will be required.

RNA interference has emerged as a potentially powerful technology to address the therapy of different forms of muscular dystrophy. Its use in converting out-of-frame to in-frame transcripts in DMD is described above. Use of this technology is also being applied in preclinical studies of mutation-specific therapy for DM and FSHD. Animal studies have shown potential benefits of targeting expanded trinucleotide repeats in a model of DM1.[40] Preclinical studies have also shown potential benefits of therapeutic RNA interference that may inhibit toxic gain-of-function and DUX4 muscle toxicity in an animal model of FSHD.[41] More conventional pharmacological treatments are based on the modulation of downstream events in the pathological process of muscular dystrophy.[42] Treatment of DMD with corticosteroids slows the progression of the disease and prolongs ambulation by 2–3 years.[7] The physiology of the beneficial effects of corticosteroids in DMD is not well understood and proposed mechanisms include stabilization of the muscle membrane, modulation of gene expression, improvement in the balance of muscle protein synthesis or degradation, and reduction of inflammation. However, long-term use of corticosteroids is complicated by frequent side effects, especially weight gain and behavioral symptoms.

At this time, treatment options in OPMD are still mostly supportive but experimental preclinical studies show potential benefits of the antiprion medications 6-aminophenanthridine and guanabenz acetate, which have demonstrated their ability to limit protein aggregation and toxicity of PABPN1 in preclinical studies.[43]

Treatment of LGMDs also remains mostly supportive but a few small studies have shown potential benefits of corticosteroids in LGMD2I. Initial human clinical gene replacement studies showed sustained expression of the α-sarcoglycan gene in LGMD2D after AAV gene transfer.[44]

Treatment of EDMD is largely limited to attempts to delay the development of limb contractures, and treatment of arrhythmias and cardiomyopathy. Preclinical studies are exploring the possible use of inhibition of extracellular signal-regulated and c-Jun N-terminal kinases to prevent the development of cardiomyopathy associated with lamin A/C mutations.[45]

Supportive Treatment

Long-term care of patients with chronic muscular dystrophies requires a multidisciplinary collaborative effort from a team that includes neurologists, internists, physical medicine and rehabilitation specialists, physical and occupational therapists, pulmonologists, cardiologists, and speech pathologists.

Physical and Occupational Therapy

Cautious exercise and stretching are generally beneficial for patients with neuromuscular disorders including muscular dystrophies, but patients should be advised to avoid intense workouts leading to exhaustion and exercise-induced muscle injury. Maximal muscle contraction may damage the cytoskeletal framework, with myofibrillar disruption and subsequent worsening of weakness and muscle soreness.

Joint contractures may be a significant clinical issue in patients with progressive weakness and reduced mobility, and may also occur early, especially with EDMD and DMD. While orthoses are the mainstay of prevention of contractures, some patients may need surgical interventions including Achilles tenotomy for correction of foot deformity. Provision of a prolonged passive stretch of the Achilles tendon by the night-time use of ankle–foot orthotics can prevent or delay the need for surgical intervention.

Scoliosis is usually a late complication of axial weakness and may be treated conservatively with bracing. More severe progression may require spinal fusion.

Pulmonary Evaluation and Management

Hypoventilation in muscular dystrophies is typically a manifestation of respiratory muscle weakness and is treated with non-invasive and invasive ventilatory support. Often the first intervention is non-invasive positive-pressure ventilation, which provides adequate support for many patients, especially those with less severe respiratory insufficiency. With progression to severe respiratory insufficiency, invasive breathing support may be required. Daily use of a cough assist device is an effective means to promote good pulmonary hygiene and decrease the likelihood of developing pneumonia. Regular vaccinations for influenza and pneumococcus are recommended.

Sleep-disordered breathing associated with muscular dystrophy is typically associated with hypoventilation, but central sleep apnea and abnormal sleep architecture are especially common with DM1.[3]

Cardiac Evaluation and Management

Cardiac involvement in muscular dystrophies usually affects cardiac muscle function causing cardiomyopathy (affecting contractility and cardiac output) or the cardiac conduction system, which may provoke potentially fatal arrhythmias (even in the absence of significant ventricular dysfunction).[40] Impaired cardiac output due to ventricular dysfunction is treated with standard heart failure treatment, which often includes angiotensin-converting enzyme inhibitors and β-blockers. Preclinical studies showed potential benefits of pharmacological inhibition of mitogen-activated protein kinase in an animal model of laminopathy (lamin A/C; EDMD2/3, LGMD1B).[46] Rarely, end-stage heart failure may require cardiac transplantation.[30] The most common indication for cardiac transplantation for patients with muscular dystrophies is heart failure due to BMD. Cardiac arrhythmias can be managed by implementing a pacemaker or defibrillator.

Other Management Issues

Swallowing impairment and weakness of jaw muscles may interfere with adequate nutrition and cause weight loss. When a patient reaches the point at which oral nutrition cannot meet physiological requirements and weight loss ensues, supplemental feeding using a gastric tube is required.

Chronic disorders, including muscular dystrophies, are frequently associated with comorbid mood disorders in patients and their caregivers.

Patients with DM1 or DM2 should be regularly monitored for development of cataracts as early cataracts are one of the hallmarks of the clinical presentation of both DM1 and DM2. The symptoms of delayed muscle relaxation and myotonia in myotonic dystrophy or myotonia congenita may improve with mexiletine treatment.

Neuromuscular weakness and lack of activity increase the risk of osteoporosis in patients with muscular dystrophies and this is even more pronounced in patients treated with long-term corticosteroids (e.g. in DMD). Therefore, attention to vitamin D nutrition is important to reduce the morbidity associated with osteoporosis. Additional studies are needed to characterize bone health in the muscular dystrophies.[47]

CONCLUSION

So far, the twenty-first century has witnessed great progress in understanding the physiology of normal and diseased muscle. A complex cascade of events causes the muscular dystrophies with the involvement of diverse proteins. Many of these proteins play important roles in muscle development, repair, and regeneration. We are now entering the realm of personalized medicine, which should enable clinicians to diagnose patients early and accurately and to eventually develop treatments suited to individual patients' needs and underlying disorders. Further development of the field and ongoing lines of inquiry will lead in the direction of novel diagnostic tests and treatment protocols for muscular dystrophies.

Scientists should focus not only on an expanded understanding of individual types of muscular dystrophy and mutation-specific strategies of treatment, but also on shared pathways of muscle injury and regeneration.

Acknowledgments

We appreciate the contributions of Saman Eghtesad, who originally designed Fig. 11.3, and Molly Wood, who helped us to develop our concept of newborn screening for muscular dystrophy. The authors take full responsibility for the contents of this chapter, which does not represent the views of the Department of Veterans Affairs or the United States Government.

References

1. Manzur AY, Muntoni F. Diagnosis and new treatments in muscular dystrophies. *Postgrad Med J*. 2009;85(1009):622–630.
2. Tawil R, Van Der Maarel SM. Facioscapulohumeral muscular dystrophy. *Muscle Nerve*. 2006;34(1):1–15.
3. Udd B, Krahe R. The myotonic dystrophies: molecular, clinical, and therapeutic challenges. *Lancet Neurol*. 2012;11(10):891–905.
4. Brais B. Oculopharyngeal muscular dystrophy: a polyalanine myopathy. *Curr Neurol Neurosci Rep*. 2009;9(1):76–82.
5. Rocha CT, Hoffman EP. Limb-girdle and congenital muscular dystrophies: current diagnostics, management, and emerging technologies. *Curr Neurol Neurosci Rep*. 2010;10(4):267–276.
6. Muchir A, Worman HJ. Emery–Dreifuss muscular dystrophy. *Curr Neurol Neurosci Rep*. 2007;7(1):78–83.
7. Bushby K, Finkel R, Birnkrant DJ, et al. Diagnosis and management of Duchenne muscular dystrophy, Part 1: Diagnosis, and pharmacological and psychosocial management. *Lancet Neurol*. 2010;9(1):77–93.
8. Lemmers RJ, van der Vliet PJ, Klooster R, et al. A unifying genetic model for facioscapulohumeral muscular dystrophy. *Science*. 2010;329(5999):1650–1653.
9. Geng LN, Yao Z, Snider L, et al. DUX4 activates germline genes, retroelements, and immune mediators: implications for facioscapulohumeral dystrophy. *Dev Cell*. 2012;22(1):38–51.
10. de Greef JC, Lemmers RJ, van Engelen BG, et al. Common epigenetic changes of D4Z4 in contraction-dependent and contraction-independent FSHD. *Hum Mutat*. 2009;30(10):1449–1459.
11. Du H, Cline MS, Osborne RJ, et al. Aberrant alternative splicing and extracellular matrix gene expression in mouse models of myotonic dystrophy. *Nat Struct Mol Biol*. 2010;17(2):187–193.
12. Kanadia RN, Johnstone KA, Mankodi A, et al. A muscleblind knockout model for myotonic dystrophy. *Science*. 2003;302(5652):1978–1980.
13. Charizanis K, Lee KY, Batra R, et al. Muscleblind-like 2-mediated alternative splicing in the developing brain and dysregulation in myotonic dystrophy. *Neuron*. 2012;75(3):437–450.
14. Kaliman P, Llagostera E. Myotonic dystrophy protein kinase (DMPK) and its role in the pathogenesis of myotonic dystrophy 1. *Cell Signal*. 2008;20(11):1935–1941.
15. Suominen T, Schoser B, Raheem O, et al. High frequency of co-segregating CLCN1 mutations among myotonic dystrophy type 2 patients from Finland and Germany. *J Neurol*. 2008;255(11):1731–1736.
16. Mankodi A, Takahashi MP, Jiang H, et al. Expanded CUG repeats trigger aberrant splicing of ClC-1 chloride channel pre-mRNA and hyperexcitability of skeletal muscle in myotonic dystrophy. *Mol Cell*. 2002;10(1):35–44.
17. Miller TM. Differential diagnosis of myotonic disorders. *Muscle Nerve*. 2008;37(3):293–299.
18. Medica I, Teran N, Volk M, Pfeifer V, Ladavac E, Peterlin B. Patients with primary cataract as a genetic pool of DMPK protomutation. *J Hum Genet*. 2007;52(2):123–128.
19. Olive M, Goldfarb LG, Shatunov A, Fischer D, Ferrer I. Myotilinopathy: refining the clinical and myopathological phenotype. *Brain*. 2005;128(Pt 10):2315–2326.
20. Glover L, Brown Jr RH. Dysferlin in membrane trafficking and patch repair. *Traffic*. 2007;8(7):785–794.
21. Ojima K, Kawabata Y, Nakao H, et al. Dynamic distribution of muscle-specific calpain in mice has a key role in physical-stress adaptation and is impaired in muscular dystrophy. *J Clin Invest*. 2010;120(8):2672–2683.
22. Duggan DJ, Gorospe JR, Fanin M, Hoffman EP, Angelini C. Mutations in the sarcoglycan genes in patients with myopathy. *N Engl J Med*. 1997;336(9):618–624.
23. Boito CA, Melacini P, Vianello A, et al. Clinical and molecular characterization of patients with limb-girdle muscular dystrophy type 2I. *Arch Neurol*. 2005;62(12):1894–1899.
24. Bolduc V, Marlow G, Boycott KM, et al. Recessive mutations in the putative calcium-activated chloride channel Anoctamin 5 cause proximal LGMD2L and distal MMD3 muscular dystrophies. *Am J Hum Genet*. 2010;86(2):213–221.
25. Chen CY, Chi YH, Mutalif RA, et al. Accumulation of the inner nuclear envelope protein Sun1 is pathogenic in progeric and dystrophic laminopathies. *Cell*. 2012;149(3):565–577.
26. Flanigan KM, Coffeen CM, Sexton L, Stauffer D, Brunner S, Leppert MF. Genetic characterization of a large, historically significant Utah kindred with facioscapulohumeral dystrophy. *Neuromuscul Disord*. 2001;11(6–7):525–529.
27. Suominen T, Bachinski LL, Auvinen S, et al. Population frequency of myotonic dystrophy: higher than expected frequency of myotonic dystrophy type 2 (DM2) mutation in Finland. *Eur J Hum Genet*. 2011;19(7):776–782.
28. Moore SA, Shilling CJ, Westra S, et al. Limb-girdle muscular dystrophy in the United States. *J Neuropathol Exp Neurol*. 2006;65(10):995–1003.
29. Hermans MC, Pinto YM, Merkies IS, de Die-Smulders CE, Crijns HJ, Faber CG. Hereditary muscular dystrophies and the heart. *Neuromuscul Disord*. 2010;20(8):479–492.
30. Wu RS, Gupta S, Brown RN, et al. Clinical outcomes after cardiac transplantation in muscular dystrophy patients. *J Heart Lung Transplant*. 2010;29(4):432–438.
31. Mercuri E, Pichiecchio A, Allsop J, Messina S, Pane M, Muntoni F. Muscle MRI in inherited neuromuscular disorders: past, present, and future. *J Magn Reson Imaging*. 2007;25(2):433–440.
32. Bell CJ, Dinwiddie DL, Miller NA, et al. Carrier testing for severe childhood recessive diseases by next-generation sequencing. *Sci Transl Med*. 2011;3(65):65ra64.
33. Burga A, Lehner B. Beyond genotype to phenotype: why the phenotype of an individual cannot always be predicted from their genome sequence and the environment that they experience. *FEBS J*. 2012;279(20):3765–3775.
34. Zellweger H, Antonik A. Newborn screening for Duchenne muscular dystrophy. *Pediatrics*. 1975;55(1):30–34.
35. Drousiotou A, Ioannou P, Georgiou T, et al. Neonatal screening for Duchenne muscular dystrophy: a novel semiquantitative application of the bioluminescence test for creatine kinase in a pilot national program in Cyprus. *Genet Test*. 1998;2(1):55–60.
36. Mendell JR, Shilling C, Leslie ND, et al. Evidence-based path to newborn screening for Duchenne muscular dystrophy. *Ann Neurol*. 2012;71(3):304–313.
37. Mendell JR, Rodino-Klapac L, Sahenk Z, et al. Gene therapy for muscular dystrophy: lessons learned and path forward. *Neurosci Lett*. 2012;527(2):90–99.
38. Tremblay J, Hamet P. Role of genomics on the path to personalized medicine. *Metabolism*. 2013;62(suppl 1):S2–S5.

39. Finkel RS. Read-through strategies for suppression of nonsense mutations in Duchenne/Becker muscular dystrophy: aminoglycosides and ataluren (PTC124). *J Child Neurol*. 2010;25(9):1158–1164.
40. Wheeler TM, Leger AJ, Pandey SK, et al. Targeting nuclear RNA for in vivo correction of myotonic dystrophy. *Nature*. 2012;488(7409):111–115.
41. Wallace LM, Liu J, Domire JS, et al. RNA interference inhibits DUX4-induced muscle toxicity in vivo: implications for a targeted FSHD therapy. *Mol Ther*. 2012;20(7):1417–1423.
42. Abdel-Hamid H, Clemens PR. Pharmacological therapies for muscular dystrophies. *Curr Opin Neurol*. 2012;25(5):604–608.
43. Barbezier N, Chartier A, Bidet Y, et al. Antiprion drugs 6-aminophenanthridine and guanabenz reduce PABPN1 toxicity and aggregation in oculopharyngeal muscular dystrophy. *EMBO Mol Med*. Jan 2011;3(1):35–49.
44. Mendell JR, Rodino-Klapac LR, Rosales XQ, et al. Sustained alpha-sarcoglycan gene expression after gene transfer in limb-girdle muscular dystrophy, type 2D. *Ann Neurol*. 2010;68(5): 629–638.
45. Bushby K, Finkel R, Birnkrant DJ, et al. Diagnosis and management of Duchenne muscular dystrophy, Part 2: Implementation of multidisciplinary care. *Lancet Neurol*. 2010;9(1):177–189.
46. Wu W, Muchir A, Shan J, Bonne G, Worman HJ. Mitogen-activated protein kinase inhibitors improve heart function and prevent fibrosis in cardiomyopathy caused by mutation in lamin A/C gene. *Circulation*. 2011;123(1):53–61.
47. Morgenroth VH, Hache LP, Clemens PR. Insights into bone health in Duchenne muscular dystrophy. BoneKEy Reports 2012;(1):9. Accessed November 1, 2012.

CHAPTER

12

Peripheral Neuropathies

Mario A. Saporta, Michael E. Shy†*

*Department of Neurology, Universidade Federal Fluminense, Rio de Janeiro, Brazil, †Department of Neurology, University of Iowa, Iowa City, Iowa, USA

OUTLINE

PERIPHERAL NERVOUS SYSTEM BIOLOGY

Peripheral Nerve Structure

Peripheral nerves are composed of axons from ventral spinal cord motor neurons, dorsal root ganglion neurons, autonomic neurons, and their respective supportive cells, including Schwann cells and ganglionic satellite cells (Fig. 12.1A). Mixed sensorimotor nerves are formed by fibers from the ventral and dorsal roots at each spinal level. In the cervical, brachial, and lumbosacral areas, the mixed spinal nerves form plexuses from which arise the major anatomically defined limb nerves. Each mixed nerve is composed of large numbers of myelinated and non-myelinated nerve fibers of varying diameter. The large myelinated axons originate from motor neurons and large-fiber sensory nerves that mediate proprioception and vibratory senses and originate from the dorsal root ganglion. Small, thinly myelinated and non-myelinated axons primarily convey nociception and autonomic functions.[1] Myelinated fiber density in the human sural nerve varies from 7000 to 9000 fibers/mm^2, while unmyelinated fiber density varies from 30,000 to 40,000 fibers/mm^2.

Preganglionic sympathetic autonomic fibers begin in the intermediolateral column of the spinal cord and

Neurobiology of Brain Disorders
http://dx.doi.org/10.1016/B978-0-12-398270-4.00012-4

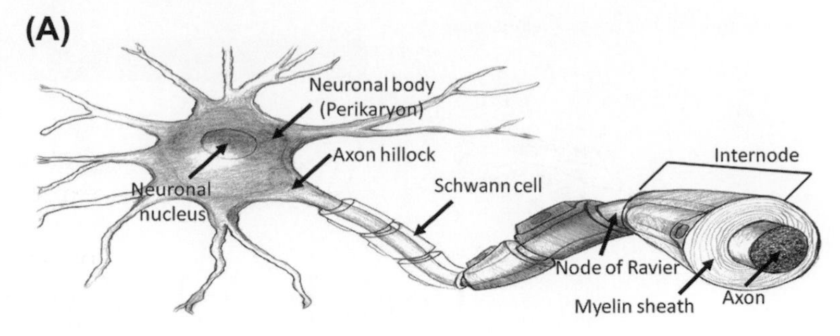

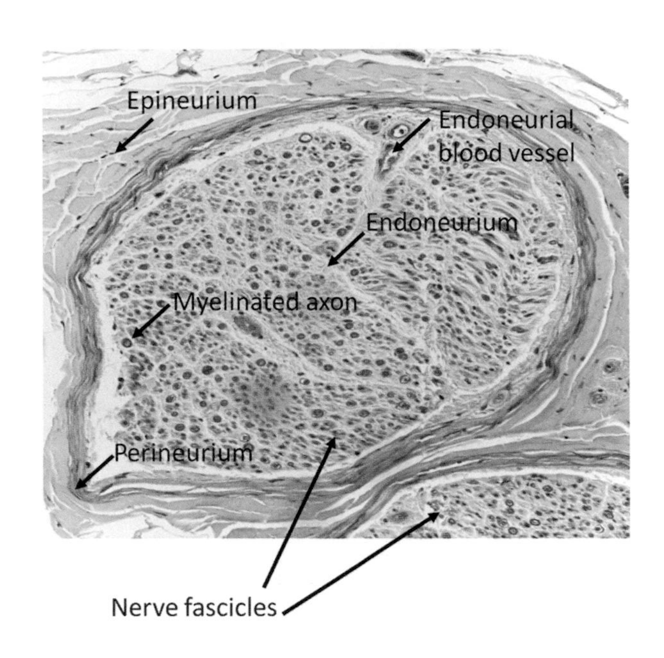

FIGURE 12.1 **Structure of a peripheral nerve.** (A) A peripheral neuron, its axon, and associated Schwann cells. (B) Cross-section of a sural nerve biopsy illustrating the main connective tissue sheaths of the peripheral nerve. Gomori trichrome stain, 20×.

synapse at sympathetic trunk ganglia. Preganglionic parasympathetic fibers travel long distances from their cell bodies in the brainstem or sacral spinal cord to reach terminal ganglia near the organs that the parasympathetic fibers innervate.

Peripheral nerves are organized in bundles or fascicles of nerve fibers (axons and their associated Schwann cells). A sheath of squamous cells and connective tissue limits each fascicle, forming the perineurium. A collagenous matrix (epineurium) further ensheathes the fascicles and also encloses the vascular bundles that travel along with peripheral nerves and give rise to the nerve's vascular supply. Each individual nerve fiber is also embedded in an intrafascicular connective tissue (endoneurium) (Fig. 12.1B). The perineurium is a metabolically active structure that forms a barrier to diffusion of small molecules between the epineurium and endoneurium, maintaining a constant microenvironment in the endoneurial space.[2]

Cellular Components of the Peripheral Nerve

Approximately 50% of the volume within the perineurium is occupied by axons and Schwann cells, while up to 30% is composed of endoneurial fluid and collagen. Of all the nuclei present within the perineurium, 10% are from fibroblasts, up to 9% belong to endogenous macrophages, and the remainder is Schwann cells. CD4 and CD8 lymphocytes are not normally found within the endoneurial space of nerves.

Schwann cells are the main cellular component of peripheral nerves. They are mostly derived from the neural crest, although some Schwann cells originate from the ventral portion of the neural tube and migrate to the extremities. Schwann cells mature by specific developmental steps that will eventually determine their phenotype as either myelinating or non-myelinating cells (Fig. 12.2). This process depends on interactions between the developing Schwann cell and its neighboring axons. In spinal nerves, Schwann cell development involves two embryonic transitional stages: (1) differentiation of Schwann cell precursors (SCPs) from neural crest cells; and (2) maturation of SCPs into immature Schwann cells. The presence of SCPs is not required for the initial development of peripheral nerves. Even in the absence of SCPs, the initial outgrowth and guidance of axons to their target areas in the limbs occur normally.

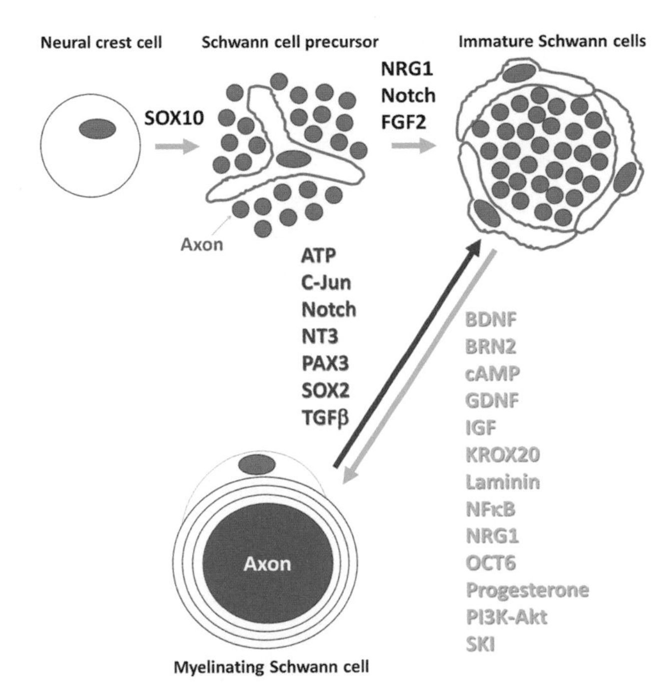

FIGURE 12.2 **Schematic representation of the main steps in Schwann cell development.** The main positive and negative red regulators of myelination are listed in color. SOX10: SRY (sex determining region Y) box 10; NRG1: neuregulin-1; FGF2: fibroblast growth factor-2; ATP: adenosine triphosphate; NT3: neurotrophin-3; PAX3: paired box gene 3; TGFβ: transforming growth factor-β; BDNF: brain-derived neurotrophic factor; BRN2: brain 2 class III POU-domain protein; cAMP: cyclic adenosine monophosphate; GDNF: glial cell-line derived neurotrophic factor; IGF: insulin-like growth factor; KROX20: early growth response 2 (EGR2); NFκB: nuclear factor-κB; OCT6: octamer-binding transcription factor-6; PI3K: phosphatidylinositol-3-kinase; SKI: v-ski sarcoma viral oncogene homologue. *Source: Adapted from Jessen and Mirsky.* Nat Rev Neurosci. *2005;6(9):671–862.*[3]

However, nerves lacking SCPs have abnormal development of fascicles and project their axons abnormally within their final target tissues, including to neuromuscular junctions. SCPs also provide important trophic support to the developing axons and neurons, and loss of SCPs leads to neuronal degeneration. SCPs are also dependent on axon-derived factors for survival, mainly neuregulin-1 (NRG1) and its effect on ErbB2/B3 receptors in the SCPs. NRG1 also plays a central role in the transition of SCPs to immature Schwann cells. Other regulators of this developmental transition include endothelin, the notch pathway, and the transcription factor AP2α. Immature Schwann cells undergo profound phenotypic changes as they differentiate into myelinating and non-myelinating Schwann cells. Large-diameter axons (>1 μm in diameter) undergo radial sorting, a process in which they establish a one-to-one relationship with individual Schwann cells, in preparation for myelination. Radial sorting is dependent on interactions between β_1-integrin receptors in the Schwann cell membrane and laminins in their basal lamina, and is mediated by the GTPases Rac1 and Cdc2. At this point, Schwann cells exit the cell cycle and commit to a myelinating phenotype, as promyelinating Schwann cells. Smaller diameter axons are surrounded by non-myelinating Schwann cells, some of which establish contact with multiple axons (Remak bundle). Myelinating and non-myelinating Schwann cells also differ in their expression of transcription factors and cell surface markers. While myelinating Schwann cells express Krox20, periaxin, myelin protein zero, myelin basic protein, myelin-associated glycoprotein (MAG), and Connexin 32, non-myelinating Schwann cells express p75, nCAM, L1, glial fibrillary acidic protein (GFAP), growth-associated protein-43, and Ran-2. Comprehensive reviews on the development of Schwann cells can be found in Woodhoo and Sommer[4] and Jessen and Mirsky.[3]

Myelination

After radial sorting has taken place and a one-to-one relationship between axon and promyelinating Schwann cell has been established, myelination can commence. The myelin sheath has a dual role of providing support for mature axons, while electrically insulating them for faster (saltatory) action potential propagation. Axonal signals, mainly in the form of neuregulin 1 type III, act to induce a set of changes in myelin gene expression in Schwann cells via receptors ErbB2 and ErbB3, which will eventually culminate in the formation of the myelin sheath. The main signaling pathways established to date include the JNK/c-jun, ERK1/2, and PI3/AKT pathways, and it is believed that the balance between them will determine whether a Schwann cell will remain myelinating or will dedifferentiate and re-enter the cell cycle, thus proliferating. Well-established transcriptional positive regulators of this process include Krox-20 (early growth response gene-2, or Egr-2) and its associated proteins, nucleic acid-binding protein (NAB) 1 and 2, Sox-10, Oct-6, and Brn1 and Brn2. Calcium signaling also is believed to play a role in myelination. Calcineurin, a phosphatase activated by calcium, dephosphorylates nuclear factor of activated T cells (NFAT) c3 and c4, resulting in translocation to the nucleus. NFAT c4 complexes with Sox-10, a transcription factor required for myelination, to activate the promoter and myelin-specific enhancer of the Krox-20 gene, which globally regulates peripheral myelination by promoting cell cycle exit and resistance to apoptotic death, by upregulating expression of the main myelin genes and by suppressing markers of the immature Schwann cell stage. Other factors involved in normal myelination include cyclic AMP and the transcription factor nuclear factor-κB (NF-κB).

Schwann cells can readily dedifferentiate, unlike most differentiated cells, especially when deprived of axonal contact (in injured nerves, for example). By doing so, Schwann cells alter their program of gene expression and provide an environment through which axons can regrow after injury and subsequently remyelinate. Negative regulators of myelination are activated in Schwann cells that lose contact with their axons and drive this phenotypic change towards what is characterized as the "denervated state", which is similar to immature Schwann cells. These negative regulators include c-Jun, Notch, Sox-2, Pax-3, Krox-24, and Egr-3. Therefore, myelination is determined by a balance between two opposing transcriptional programs that allow Schwann cells to perform their highly specialized role of myelinating axons, while maintaining enough phenotypic flexibility to dedifferentiate when necessary and help in the process of axonal regeneration (Fig. 12.2).

The Myelinated Axon

The interaction between myelinated Schwann cells and axons promotes molecular changes in the axonal membrane (axolemma) forming specialized domains with very specific functions. These include the nodes of Ranvier, the paranodes, the juxtaparanodes, and the internodes (Fig. 12.3A). The paranode–node–paranode (PNP) regions are the most complex part of a myelinated fiber. They are responsible for the propagation of action potentials as saltatory impulses. The PNP regions are also where most axonal–Schwann cell interactions vital for the survival of both structures take place. The node of Ranvier presents a uniquely high concentration of sodium channels, mainly of the Sca8/PN4 family, and Na^+/K^+-ATPases (Fig. 12.3B), in keeping with its main physiological function in action potential propagation. A set of membrane proteins localized to the nodal axolemma, including ankyrinG, neurofascin, n-CAM, and tenascin, probably plays an important role in localizing and anchoring Na^+ channels to the nodal regions.

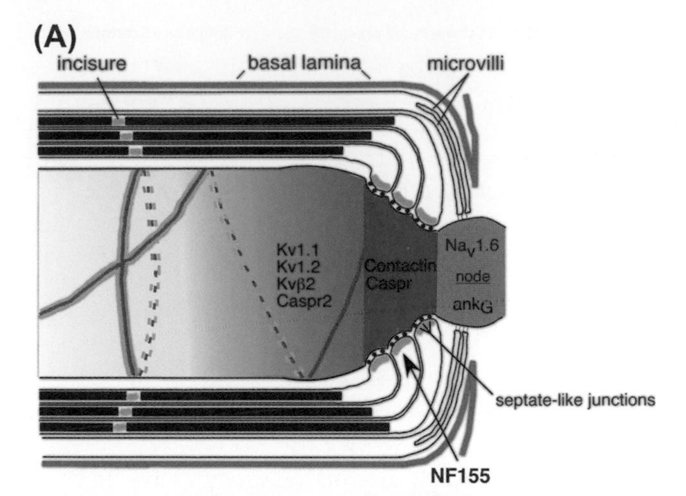

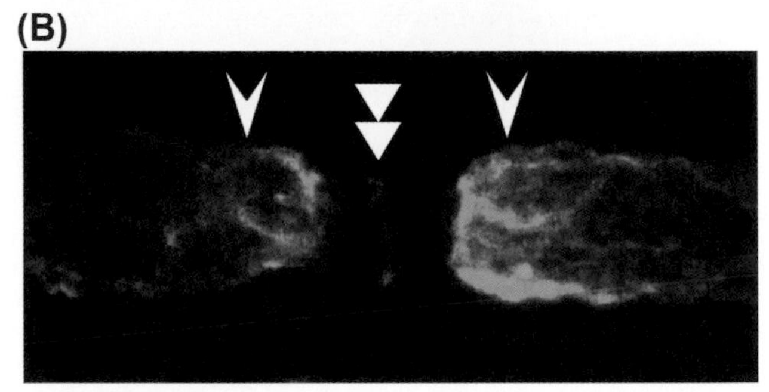

FIGURE 12.3 **Myelinated axon.** (A) Schematic depiction of the node, paranode, and juxtaparanode. Kv1.1, Kv1.2, Kvβ2: voltage-gated potassium channels; Na_V: voltage-gated sodium channel; ank_G: ankyrin G; NF155: neurofascin 155. (B) Laser scanning confocal micrograph of Na^+ and K^+ channels. A myelinated fiber teased from rat sciatic nerve was labeled with a rabbit antiserum against voltage-gated Na^+ channels (rhodamine) and a monoclonal antibody against Kv1.2 (fluorescein). Na^+ channels are restricted to the node (double arrowheads), whereas Kv1.2 channels are found in the juxtaparanodal region (arrowheads). *Source: Courtesy of Dr Steven Scherer, University of Pennsylvania.*

The main function of paranodes is to act as an insulant between the nodal and juxtaparanodal areas, preventing direct exchange of ions between these two regions. This is accomplished by connections between the myelin sheath and the axolemma, the axoglial junctions, which act similarly to tight junctions by preventing the free diffusion of small molecules. The main membrane proteins involved in this structural interaction are from the Caspr, contactin, and neuroligin families. The juxtaparanode, a region extending 10–15 μm from the paranodes, is rich in delayed rectifying K^+ channels (Kv1.1 and Kv1.2); its main physiological function is dampening the excitability of myelinating fibers, thus avoiding ectopic impulse generators within the peripheral nerve (Fig. 12.3B). The microscopic organization of the internode is relatively simple compared with the other regions described previously. The internode corresponds to the length of an entire Schwann cell and varies from 100 to 2000 μm.

The myelin sheath itself can be divided into two domains, compact and non-compact myelin, each of which contains specific sets of proteins. Compact myelin forms the bulk of the myelin sheath; non-compact myelin is found in paranodes and in Schmidt–Lanterman incisures. The Schwann cell basal/abaxonal surface apposes the basal lamina. The Schwann cell basal lamina contains laminin-2 (comprised of α_2/merosin, β_1, and γ_1-laminin chains), type IV collagens, entactin/nidogen, fibronectin, N-syndecan, and glypican. The basal/adaxonal Schwann cell membrane contains the integrin $\alpha_6\beta_4$ and dystroglycan, both of which probably bind to laminin-2. The apical/adaxonal surface apposes the axon and is highly enriched in MAG, which may bind to molecules on the axonal surface. The inner and outer edges of the Schwann cell membrane, which contact the adjacent layer of the myelin sheath, are called the inner and outer mesaxon, respectively. Gap junctions are present in paranodes, incisures, and inner and outer mesaxons, and they mediate a radial pathway across the myelin sheath that acts as a shortcut for diffusion of molecules across the Schwann cell. Radial pathway diffusion is estimated to be 3 million times faster than diffusion through the cytoplasm of a myelinated Schwann cell wrapped around an axon, owing to the geometry of the myelin sheath. Compact myelin in the peripheral nervous system (PNS) is largely composed of lipids, mainly cholesterol and sphingolipids, including galactocerebroside and sulfatide. The main myelin proteins are myelin protein zero (MPZ or P0), peripheral myelin protein 22 kDa (PMP22), and myelin basic protein (MBP). P0 is by far the most abundant protein and is the main adhesive molecule in compact myelin. It has an immunoglobulin G (IgG)-like extracellular domain that forms tetramers in *cis*, which interact with each other in *trans*. PMP22 is a hydrophobic intrinsic membrane protein of unknown function. The stoichiometry of PMP22 is critical for the integrity of compact myelin, as demonstrated by the occurrence of human diseases associated with abnormal copy number variation of the *PMP22* gene (CMT1A and HNPP, described in "Pathobiology", below). A comprehensive description of the molecular architecture of myelinated nerves can be found in Arroyo and Scherer.[5]

Mechanisms of Nerve Injury

Pathological processes affecting the peripheral nerves usually involve two types of injury mechanism: axonal degeneration and demyelination. Combinations of these two mechanisms are usually seen, underscoring the role of Schwann cell–axonal interaction in the homeostasis of peripheral nerves.

Axonal degeneration can occur following at least four different mechanisms. Wallerian degeneration occurs when an axon is transected (axotomy) and consists of the breakdown of the axonal segment distal to the transection and its associated myelin sheath with a subsequent reparative stage. Regeneration (sprouting) of axons from the proximal stump of the transected nerve begins almost immediately after axotomy and is concomitant to Schwann cell proliferation within the tube formed by their original basal lamina. Nerve regeneration progresses at a rate of 1–3 mm/day, resulting in the appearance of clusters of small, thinly myelinated fibers (regenerating clusters). Distal axonopathy (or dying-back neuropathy) is observed in symmetrical subacute or chronic peripheral neuropathies that affect preferentially long and large axons, leading to a length-dependent neuropathy, with symptoms progressing for the distal lower extremities to more proximal segments. The pathogenesis of this process is still poorly understood, but there is growing evidence that this length dependency may be related to abnormalities in fast axonal transport. Loss of myelinated large fibers is observed, with some regeneration occurring to different degrees, depending on the precipitating factor. Neuronopathy, the lesion of the neuronal cell body, usually leads to synchronous injury to the axon, rendering regeneration impossible. Sensory neuronopathies are usually associated with a restricted group of pathologies, including paraneoplastic syndromes, pyridoxine intoxication, and Sjögren disease. The exact mechanisms underlying selective loss of dorsal root ganglion sensory neurons are still unknown. The lack of vascular barriers in dorsal root ganglion and free sensory terminals (with subsequent retrograde axonal transport of toxins to the neuronal body) may render these cells more susceptible to circulating toxins or antibodies. Abnormalities in axonal caliber, including axonal atrophy and axonal swelling, are associated with specific chronic neuropathies. Axonal atrophy can be observed in uremic, diabetic, toxic, paraproteinemic, and inherited peripheral neuropathies. Axonal swelling has been identified in patients with a specific form of inherited neuropathy (giant axonal neuropathy) and some toxic neuropathies.

Disease processes affecting the Schwann cells and myelin sheath typically cause primary segmental demyelination. This process begins at the paranodes and involves the myelin internodes discontinuously. Denuded axons send signals that drive Schwann cell proliferation. These newly proliferated Schwann cells will remyelinate the axons, forming small, intercalated internodes. These can be easily appreciated on teased-nerve preparations as internodes of unequal size, shorter than normal and with thinner myelin sheaths, compared with unaffected internodes. Repeated episodes of segmental demyelination and remyelination lead to proliferation of Schwann cells and of basement membrane deposits arranged concentrically around a relatively intact axon, which can be completely denuded of myelin or may have a thin myelin sheath. This pathological hallmark of repetitive demyelination and remyelination is named onion-bulb formation and can be found in inherited and acquired demyelinating peripheral neuropathies. Widening of myelin lamellae may be observed in specific demyelinating neuropathies, including inherited neuropathies associated with mutation in the *MPZ* gene (CMT1B) and paraproteinemic neuropathies with anti-MAG antibodies. Focal hypermyelination (tomaculous neuropathy) can be seen in patients with hereditary neuropathy with liability to pressure palsies, some other inherited neuropathies, alcoholic neuropathy, and paraproteinemic neuropathies. Demyelinating neuropathies ultimately lead to length-dependent axonal degeneration that usually correlates more with clinical impairment than does the demyelination itself. How demyelination disrupts axoglial signaling is not well understood but is the subject of active investigation.

CLINICAL MANIFESTATION AND DIAGNOSTIC MODALITIES IN PERIPHERAL NEUROPATHIES

The evaluation of a patient with suspected peripheral neuropathy begins with the history and physical examination to demonstrate peripheral nerve disease and proceeds to neurophysiological testing to characterize whether the process is demyelinating or axonal. Other specific tests can then be ordered to look for possible specific causes of the neuropathy.

Symptoms of peripheral neuropathy are determined by the type of fiber involved (large or small caliber, motor, sensory, and/or autonomic) and not by the basic pathological process (demyelination or axonal degeneration). Therefore, the main symptoms of a peripheral neuropathy include sensory loss, weakness, abnormal balance, and autonomic dysfunction, and they may happen whether the neuropathy is primarily demyelinating or axonal. Because of the length-dependent nature of most peripheral neuropathies, symptoms usually begin in the most distal segments of the limbs, especially in the lower extremities, and progress in an ascending pattern; as a general rule, sensory symptoms begin in the hands when sensory symptoms in the legs have progressed up to the knees. Sensory symptoms of peripheral neuropathies may reflect disease of small, thinly myelinated or nonmyelinated fibers subserving pain and temperature, as well as large myelinated fibers subserving position sense. Common symptoms of small-fiber sensory neuropathy include intense pain (pins and needles) or dysesthesias (feeling as though the feet are ice cold) and difficulty determining whether bath water is hot or cold with the foot. Large-fiber sensory loss usually impairs balance, which may be worse in the dark when vision cannot overcome the loss of proprioception. Loss of proprioception is also frequently length dependent, so a patient may improve balance by lightly touching a wall with the hand to improve proprioceptive input to the brain.

Deep and superficial muscles that are innervated by the peroneal nerve, such as the tibialis anterior and peroneus brevis and longus muscles, often cause more symptoms than do the plantar flexion muscles innervated by the tibial nerve, such as the gastrocnemius. As a result, tripping on a carpet or curb and spraining one's ankle are frequent symptoms. In the hands, symptoms typically involve difficulty with fine movements, such as using buttons or zippers and inserting and turning keys in locks. Muscle cramps frequently occur with motor or sensorimotor neuropathies.

Autonomic symptoms are frequent in neuropathies associated with diabetes or amyloidosis and include urinary retention or incontinence, abnormalities of sweating, constipation alternating with diarrhea, and light-headedness when standing. Impotence is frequent. Peripheral neuropathies usually affect both motor and sensory nerves, thereby causing both weakness and sensory loss. However, certain neuropathies are predominantly sensory, such as diabetes, or motor, such as multifocal motor neuropathy (MMN). Most neuropathies are symmetrical and length dependent. Pronounced asymmetries in symptoms suggest specific disorders, such as mononeuritis multiplex or hereditary neuropathy with liability to pressure palsies. It is also useful to know whether symptoms are acute (<1 month), subacute (<6 months), or chronic (>6 months). For example, Guillain–Barré syndrome develops over a period of days to weeks, whereas chronic inflammatory demyelinating polyradioneuropathy (CIDP) evolves over months and inherited neuropathies may develop over years.

Neurological examination will help to confirm the diagnosis of a peripheral neuropathy and determine whether the neuropathy is symmetrical or asymmetrical, length dependent or not, sensorimotor or predominantly sensory or motor, and these features will help to narrow down the differential diagnosis of a specific patient.

Sensory loss is usually in a stocking–glove distribution in both large- and small-fiber neuropathy, especially in length-dependent processes. Cold, erythematous, or bluish discolored feet suggest loss of small-fiber function. Large-fiber sensory loss, or "sensory ataxia", can often be detected by an abnormal gait (tabetic or stomping gait), characterized by a wide-base, abnormal elevation of the legs and slamming of the foot hard on to the ground, often exacerbated by darkness. Patients may also have trouble tandem walking and accurately locating their thumb with the opposite index finger with their eyes closed. Pseudoathetosis, a characteristic irregular tremor of the fingers, can also be seen in more severe cases. In length-dependent neuropathies, weakness is often most pronounced in foot dorsiflexion and eversion and usually progresses to the muscles of plantar flexion before more proximal muscles become involved. In the upper extremities, the intrinsic hand muscles are usually the first affected. Wasting of muscle is prominent in many neuropathies, regardless of whether they are primary axonal or primary demyelinating disorders, because even demyelinating neuropathies are associated with secondary axonal degeneration. Atrophy frequently occurs in muscles of dorsiflexion, such as the tibialis anterior, and in intrinsic hand muscles, such as the first dorsal interosseus. Fasciculations, which appear as small twitches of the muscle, are sometimes present, particularly in axonal neuropathies.

The complete absence of reflexes early in the course of a neuropathy suggests a demyelinating neuropathy; for example, the absence of reflexes in early childhood is often the first detectable abnormality in children with inherited demyelinating neuropathies. Alternatively, the absence of ankle reflexes but the presence of normal patellar or upper extremity reflexes is common in length-dependent axonal neuropathies, both acquired and inherited. Reflexes may be present in small-fiber neuropathies.

After the diagnosis of a peripheral neuropathy is established by the history and neurological examination, nerve conduction studies and electromyography (EMG) are usually performed to confirm the diagnosis and determine whether the neuropathy is primarily demyelinating or axonal and symmetrical or asymmetrical. The main parameters used to determine the basic pathological process of a neuropathy on electrophysiology are the size (amplitude) of a nerve response and the velocity of action potential propagation. Motor nerve conduction velocities measure conduction over the main body of nerves but not their proximal or distal portion. Distal motor latencies and F-wave latencies measure velocities over the distal and proximal portions of the nerves, respectively. When slowing is roughly the same over the proximal, distal, and main portion of the nerve, the slowing is said to be uniform. When the slowing is multifocal or asymmetrical, either along the same nerve or between different nerves, the slowing is said to be non-uniform. Slowed conduction velocities (to less than 70% of normal) suggest that the neuropathy is primarily demyelinating. On the other hand, in axonal neuropathies, amplitudes of the compound muscle action potential or sensory nerve action potential are reduced, while conduction velocities remain unchanged, unless the loss of fast-conducting large fibers leads to relative slowing of conduction velocity. The presence of spontaneous activity on EMG, such as fibrillations or positive sharp waves, suggests that an acute or active process is damaging axons and denervating muscle. The presence of large, polyphasic motor units suggests partial reinnervation of muscle by regenerating axons (i.e. a more chronic process). Recruitment of motor units is also reduced in patients with demyelinating and axonal neuropathies.

Other less frequently used methods of studying the PNS of patients with suspected peripheral neuropathies include quantitative sensory testing, sensory evoked potentials, autonomic function studies, and skin and nerve biopsy. Nerve biopsy is occasionally indicated to address specific questions, such as whether vasculitis, tumor, or another infiltrative or metabolic disorder is present. Biopsy of sural nerves is performed just above the ankle. After biopsy, patients lose sensation over the region on the lateral aspect of the foot that is innervated by the sural nerve, and transient painful dysesthesias may develop around the biopsy site. Teased sural nerve fiber analysis can demonstrate segmental demyelination or remyelination, and electron microscopy can demonstrate features of nerve regeneration and identify specific pathological processes. Epidermal skin biopsies with quantification of the loss of small epidermal nerve fibers may aid in the diagnosis of sensory neuropathies, particularly in neuropathies that involve a loss of small fibers, such as diabetes mellitus, human immunodeficiency virus (HIV) infection, or chemotherapeutic drugs. Skin biopsies also are increasingly being used to evaluate dermal myelinated sensory nerves.

Evaluation of all patients with suspected neuropathy should include blood glucose (and oral glucose tolerance test, if deemed necessary), vitamin B_{12} and metabolites (methylmalonic acid and homocysteine), and serum protein immunofixation electrophoresis.[6] If the history and EMG are consistent with exposure to a toxin or a vitamin deficiency state, specific testing is indicated. In unexplained sensory neuropathies, HIV testing should be considered. In selected patients, electrodiagnostic studies will suggest the need to test for specific antibodies, such as antibodies reacting to ganglioside GM_1 or MAG. Genetic testing will be discussed in the next section of this chapter.

INHERITED NEUROPATHIES

Inherited neuropathies, collectively known as Charcot–Marie–Tooth disease (CMT), are a group of genetically and phenotypically heterogeneous peripheral neuropathies associated with mutations or copy number variations in over 70 distinct genes[7] (Table 12.1). They can manifest as motor and sensory neuropathies (HMSN), pure motor neuropathies (dHMN), or sensory and autonomic neuropathies (HSAN), depending on the specific gene affected. Autosomal dominant forms are subdivided into demyelinating (CMT1) and axonal (CMT2) forms based on electrophysiological and neuropathological criteria. X-linked (CMTX) and autosomal recessive (CMT4) forms are also seen. Each type of CMT is subdivided according to the specific genetic cause of the neuropathy. For example, the most common form of CMT1, termed CMT1A, is caused by a duplication of a fragment of chromosome 17 containing the *PMP22* gene (see below).

Epidemiology

The prevalence of CMT is about 1 in 2500, with a global distribution and no ethnic predisposition. CMT1A (17p11.2 duplication) accounts for 60–70% of demyelinating CMT patients (around 50% of all CMT cases), CMT1X for approximately 10–20% of CMT cases, CMT1B for less than 5% of patients, and CMT2 for about 20% of cases. The prevalence of hereditary neuropathy with liability to pressure palsies is not known, but about 85% of patients with clinical evidence of this syndrome have a chromosome 17p11.2 deletion.

TABLE 12.1 Classification of Charcot–Marie–Tooth Disease

Type	Gene/locus	Specific phenotype
AUTOSOMAL DOMINANT (AD) CMT1		
CMT1A	*Dup 17p* (*PMP22*)	Classic CMT1
	PMP22 (point mutation)	Classic CMT1/DSS/CHN/HNPPs
CMT1B	*MPZ*	CMT1/DSS/CHN/intermediate/CMT2
CMT1C	*LITAF*	Classic CMT1
CMT1D	*EGR2*	Classic CMT1/DSS/CHN
CMT1E	*NEFL*	CMT2 but can have slow MNCVs in CMT1 range ± early-onset severe disease
HNPP	*Del 17p* (*PMP22*)	Typical HNPP
HNPP	*PMP22* (point mutation)	Typical HNPP
X-LINKED CMT1 (CMT1X)		
CMT1X	*GJB1*	Intermediate ± patchy MNCVs/male MNCVs less than female MNCVs
AUTOSOMAL RECESSIVE (AR) DEMYELINATING CMT (CMT4)		
CMT4A	*GDAP1*	Demyelinating or axonal, usually early onset and severe/vocal cord and diaphragm paralysis described/rare AD
		CMT2 families described
CMT4B1	*MTMR2*	Severe CMT1/facial/bulbar/focally folded myelin
CMT4B2	*SBF2*	Severe CMT1/glaucoma/focally folded myelin
CMT4C	*SH3TC2*	Severe CMT1/scoliosis/cytoplasmic expansions
CMT4D (HMSNL)	*NDRG1*	Severe CMT1/gypsy/deafness/tongue atrophy
CMT4E	*EGR2*	Classic CMT1/DSS/CHN
CMT4F	*PRX*	CMT1/more sensory/focally folded myelin
CMT4H	*FGD4*	CMT1
CMT4J	*FIG4*	CMT1
CCFDN	*CTDP1*	CMT1/gypsy/cataracts/dysmorphic features
HMSN-Russe	*10q22-q23*	CMT1
CMT1	*PMP22* (point mutation)	Classic CMT1/DSS/CHN/HNPPs
CMT1	*MPZ*	CMT1/DSS/CHN/intermediate/CMT2
AUTOSOMAL DOMINANT (AD) CMT 2		
CMT2A	*MFN2*	CMT2/usually severe/optic atrophy
CMT2B	*RAB7A*	CMT2 with predominant sensory involvement and sensory complications
CMT2C	*12q23-q24*	CMT2 with vocal cord and respiratory involvement
CMT2D	*GARS*	CMT2 with predominant hand wasting/weakness or dHMN V
CMT2E	*NEFNEFL*	CMT2 but can have slow MNCVs in CMT1 range ± early-onset severe disease
CMT2F	*HSPB1* (*HSP27*)	Classic CMT2 or dHMN II
CMT2G	*12q12-q13.3*	Classic CMT2
CMT2L	*HSPB8* (*HSP22*)	Classic CMT2 or dHMN II

TABLE 12.1 Classification of Charcot–Marie–Tooth Disease—cont'd

Type	Gene/locus	Specific phenotype
CMT2	*MPZ*	CMT1/DSS/CHN/intermediate/CMT2
CMT2 (HMSNP)	*3q13.1*	CMT2 with proximal involvement
AUTOSOMAL RECESSIVE (AR) CMT2 (ALSO CALLED CMT4)		
AR CMT2A	*LMNA*	CMT2 proximal involvement and rapid progression described/also causes muscular dystrophy/cardiomyopathy/lipodystrophy
AR CMT2B	*19q13.1-13.3*	Typical CMT2
AR CMT2	*GDAP1*	CMT1 or CMT2 usually early onset and severe/vocal cord and diaphragm paralysis described/rare AD CMT2 families described
DOMINANT INTERMEDIATE CMT (DI-CMT)		
DI-CMTA	*10q24.1-25.1*	Typical CMT
DI-CMTB	*DNM2*	Typical CMT
DI-CMTC	*YARS*	Typical CMT
HEREDITARY NEURALGIC AMYOTROPHY (HNA)		
HNA	*SEPT9*	Recurrent neuralgic amyotrophy

AD: autosomal dominant; AR: autosomal recessive; CHN: congenital hypomyelinating neuropathy; CMT: Charcot–Marie–Tooth; CTDP1: CTD phosphatase subunit 1; Del: deletion; DMN2: dynamin 2; DSN: Dejerine Sottas neuropathy; Dup: duplication; EGR2: early growth response 2; FGD4: FYVE: RhoGEF and PH domain containing 4; FIG4: FIG 4 homologue; GARS: glycyl tRNA synthetase; GDAP1: ganglioside-induced differentiation associated protein 1; GJB1: gap junction protein β1; HNPP: hereditary neuropathy with liability to pressure palsies; HSP22: heat shock 22 kDa protein 8; HSP27: heat shock 27 kDa protein 1; KIF1Bb: kinesin family member 1B-b; LITAF: lipopolysaccharide-induced tumor necrosis factor; LMNA: lamin A/C; MCV: motor conduction velocity ; MFN2: mitofusin 2; MPZ: myelin protein zero; MTMR2: myotubularin-related protein 2; MTMR13: myotubularin-related protein 13; NDRG1: Nmyc downstream regulated gene 1; NEFL: neurofilament: light polypeptide 68 kDa; PMP22: peripheral myelin protein 22; PRX: periaxin; RAB7: member RAS oncogene family; SEPT9: septin 9; SH3TC2: SH3 domain and tetratricopeptide repeats 2; YARS: tyrosyl tRNA synthetase.

Source: Reprinted from Reilly and Shy. J Neurol Neurosurg Psychiatry. *2009;80(12):1304–1314,*[8] *with permission.*

Pathobiology

Most genes identified in patients with inherited neuropathies can be mapped to specific Schwann cell or axonal functions (Fig. 12.4). Most types of CMT1 are associated with mutations in myelin protein genes, including *PMP22* (CMT1A), *MPZ* (CMT1B), and *GJB1* (CMT1X), or other genes associated with Schwann cell function, including transcriptional control of myelination (*EGR2*, CMT1D) and intracellular trafficking (*LITAF/SIMPLE*, CMT1C; *MTMR2* and *MTMR13*, CMT4B1 and 4B2, among others)[9] (Fig. 12.4A). Impaired interactions between axons and mutant Schwann cells seem to be the common process linking all forms of demyelinating CMT, which lead to segmental demyelination, remyelination, and secondary axonal degeneration, all features observed in nerve biopsy from these patients. Repetitive cycles of demyelination and remyelination cause the concentric Schwann cell lamellae around axons (onion-bulb formations) usually present on nerve biopsies (Fig. 12.5A, B), with loss of both small- and large-diameter myelinated fibers. Focal, sausage-like thickenings of the myelin sheath (tomacula) are characteristic of hereditary neuropathy with liability to pressure palsies (HNPP), but may also be found in other forms of CMT1, particularly CMT1B. In CMT1, disability typically correlates better with secondary axonal degeneration than with demyelination itself, thereby demonstrating the importance of Schwann cell–axonal interactions in demyelinating disease.

Mutations in genes associated with axonal structure and function are associated with CMT2 (Fig. 12.4B, C). Some examples include mutations in proteins of the neuronal cytoskeleton (*NEFL*, CMT2E; *GIGAXONIN*, giant axonal neuropathy), proteins associated with axonal mitochondrial dynamics (*MFN2*, CMT2A; *GDAP1*, CMT2K) and proteins associated with regulation of membrane and intracellular trafficking (*RAB7*, CMT2B, for example).[10] Other CMT2-related genes code for small heat shock proteins (HSPB1, HSPB3, HSPB8), amino-acyl-tRNA synthetases (GARS, YARS, AARS, HARS, KARS), and proteins involved in lipid metabolism (e.g. SPTLC1 and SPTLC2), and are ubiquitously expressed. Why mutations in these genes specifically affect axons of peripheral neurons remains to be determined. The pathological hallmark of CMT2 is axonal degeneration with loss of all types of nerve fibers in the absence of onion-bulb formation.

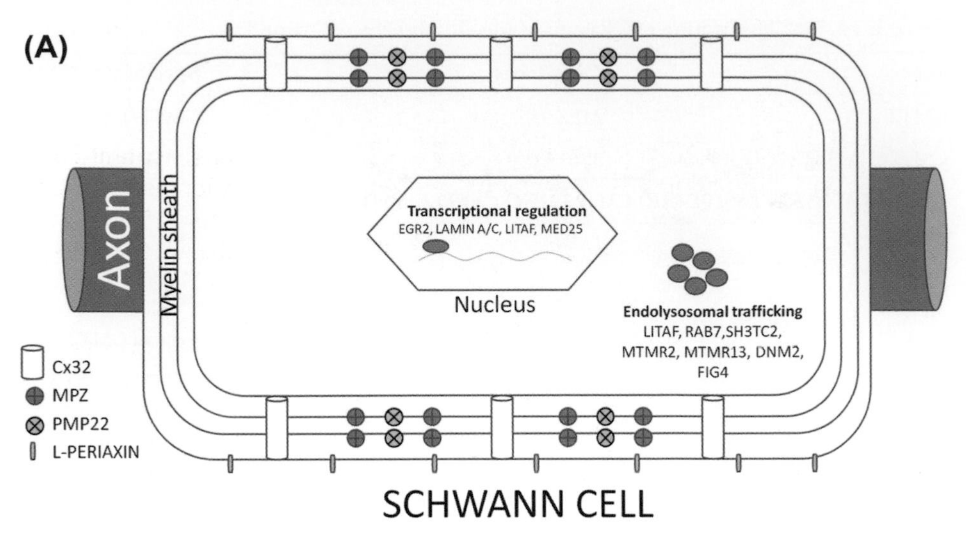

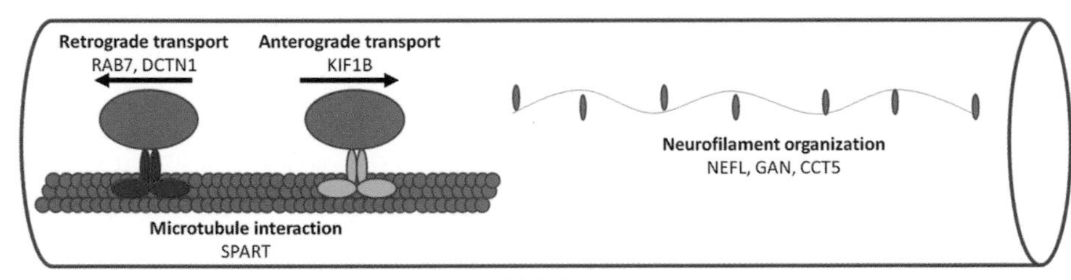

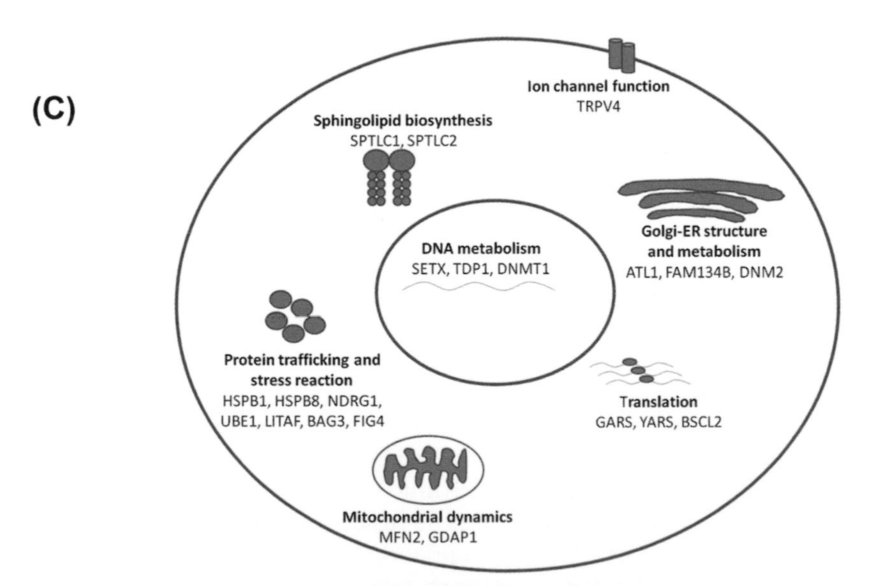

FIGURE 12.4 **Main cellular compartments and functions affected by genes associated with Charcot–Marie–Tooth disease: (A) Schwann cell; (B) axon; (C) neuronal body.** ER: endoplasmic reticulum. For other abbreviations, see footnote to Table 12.1.

Models of Inherited Neuropathies

Pathophysiological studies in inherited neuropathies have relied mostly on transgenic animal models of specific forms of CMT. Knockout and knockin mice have been used to study different types of demyelinating CMT. The addition of extra copies of the *pmp22* gene in rodents has generated models to study CMT1A, reproducing the increased PMP22 messenger RNA (mRNA) expression in peripheral nerves and the progressive demyelination and axonal loss similar to what is observed in CMT1A patients. Different strategies to reduce the toxic effect of PMP22 overexpression have been tested using this model, including ascorbic acid and a progesterone

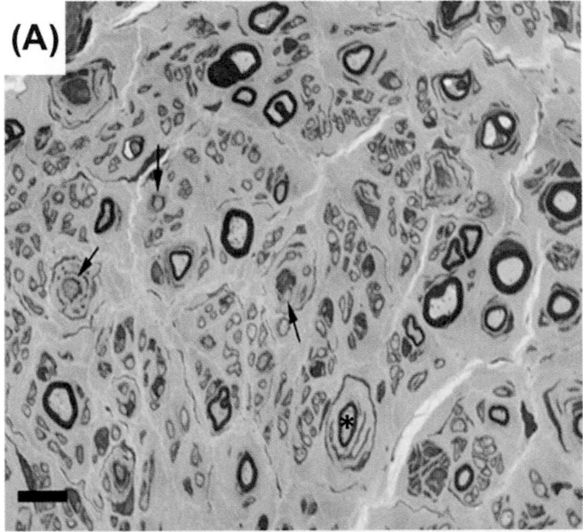

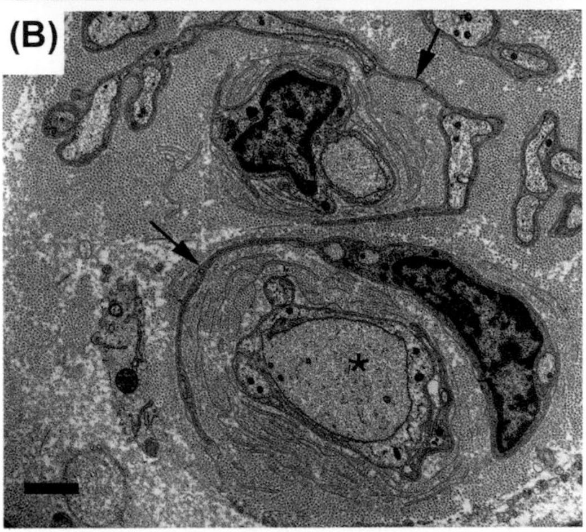

FIGURE 12.5 **Pathological findings in demyelinating Charcot–Marie–Tooth (CMT) disease.** (A) Cross-section of a sural nerve biopsy of a patient with demyelinating CMT showing numerous demyelinated axons (arrows) and occasional classical onion bulbs (asterisk). Thionine and acridine orange stain. Scale bar = 10 μm. (B) Electron micrograph showing a demyelinated axon (asterisk) surrounded by some excess basal lamina. The Schwann cells associated with small, unmyelinated axons are abnormally long and attenuated (arrows). Scale bar = 1 μm.

antagonist (onapristone). Despite demonstrating promising results in the rodent model, ascorbic acid failed to demonstrate any therapeutic effect in several clinical trials with CMT1A patients. Point mutations in the *PMP22* gene can also cause demyelinating neuropathy (CMT1E) in humans. Two mouse strains, Trembler (Tr) and Trembler-J (TrJ), were identified to carry spontaneous disease-associated mutations (L16P and G150D) in the *pmp22* gene. Studies using these animals have demonstrated accumulation of misfolded PMP22 in the endoplasmic reticulum of Schwann cells, and different strategies aiming to reduce this toxic accumulation, including curcumin, have been shown to mitigate the phenotype of the mice, suggesting that this approach could benefit patients with CMT1E. Similar findings were observed in mouse models of early-onset CMT1B, associated with specific mutations in the *mpz* gene (S63del and R98C). In these models, severe demyelination (Fig. 12.6A) and reduced nerve conduction velocity (NCV) are observed, similar to patients with these two point mutations. Mutant S63del and R98C MPZ accumulate in the endoplasmic reticulum of Schwann cells, triggering an unfolded protein response and dedifferentiation of these cells, with consequent demyelination.

The heterozygous *pmp22* knockout mouse ($pmp22^{+/-}$) is an authentic model for hereditary neuropathy with liability to pressure palsies, a human disease associated with haploinsufficiency of *PMP22* and characterized by focal areas of myelin thickening (tomacula) (Fig. 12.6B). This model has been used to study mechanisms associated with conduction block, an electrophysiological feature of this disease.

Models for CMT1X include the mice knockdown for *Gjb1* ($Gjb1^{-/-}$) and the R142W knockin mice. Their phenotypes suggest that this type of CMT is associated with a loss of function of the Connexin32 protein. Also, studies using these models suggest a role of the immune system in the pathophysiology of the disease, with some neurological improvement being seen in $Gjb1^{-/-}$ mice after downregulation of T cells or macrophages.

The two animal models of CMT2 that have been developed are the mutant mouse for the neurofilament light gene (NEFL, CMT2E) and the mutant mouse for the heat shock protein B1 (HSPB1, CMT2F). Transgenic mouse models expressing mutant forms of HSPB1 (S135F and R127W) exclusively in neurons present phenotypes resembling CMT2, including progressive decline of motor functions starting at age 6 months, axonal degeneration, and denervation of the neuromuscular junction. HSPB1 mutants show an increased affinity for microtubules, resulting in their overstabilization, at a presymptomatic stage. This sustained overstabilization seems to activate regulatory mechanisms that induce deacetylation, which may be pathogenic. This process and the associated mouse phenotype can be rescued using histone deacetylase inhibitors.

Mouse models of human diseases have several limitations and are time consuming to develop. Transgenic technology has to be used carefully as the mouse strain and its genetic background cannot interfere with the phenotype, and promoter of the transgene should allow moderate expression of the target protein in order to avoid artifacts linked to overexpression. Nevertheless, several mouse models of inherited neuropathies have allowed significant insights into the pathomechanisms of these conditions and assessment of therapeutic strategies.

Other animal models developed to study inherited neuropathies include *Drosophila melanogaster* and *Caenorhabditis elegans* models. The known genome and short reproductive cycle of *Drosophila* make this an attractive

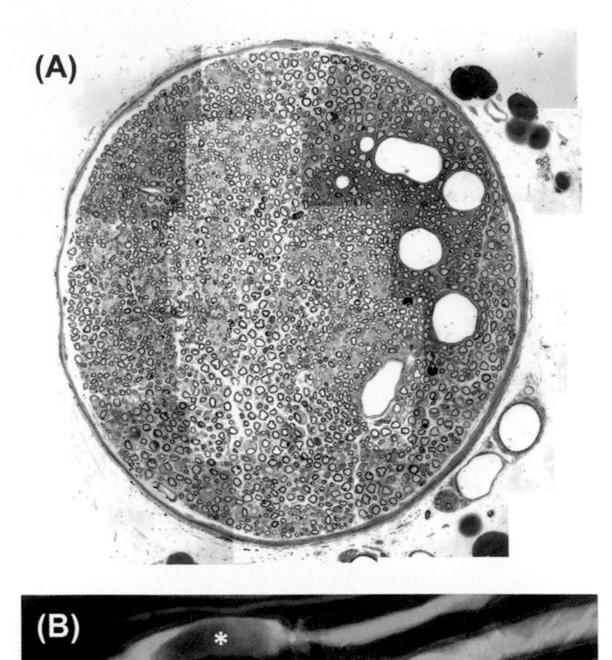

FIGURE 12.6 **Animal models of inherited neuropathies.** (A) Cross-section of a fascicle from a sciatic nerve of an mpzR98C$^{+/-}$ mouse (a model of early-onset Charcot–Marie–Tooth 1B) demonstrating abnormally thin myelin sheaths, even around large-diameter axons, in the absence of classical onion bulbs. (B) Reconstruction of a confocal microscopy image series of a sciatic nerve teased-fiber preparation from a pmp22$^{+/-}$ mouse showing a tomaculous formation (arrow) compressing the enclosed segment of axon (asterisk). This pathological feature is commonly seen in hereditary neuropathy with liability to pressure palsies (HNPP).

and less expensive model. A *Drosophila* model of a dominant intermediate form of CMT (DI-CMTC) associated with a mutation in the *YARS* gene, coding for the enzyme tyrosyl-tRNA synthetase (TyrRS), has been reported. Specific expression of YARS mutants in *Drosophila* neurons induces a progressive deficit of motor function associated with specific degeneration of the associated axons. A *C. elegans* model of CMT was also reported. In this model, a sporadic mutation in HARS (hars-1 Arg137Gln) was overexpressed in γ-aminobutyric acid (GABA)ergic neurons of the nematode and resulted in gross morphological defects in commissural axons, denoted by failure to reach the dorsal nerve cord, axonal beading, defasciculation, and breaks in the visualized GABAergic dorsal nerve cord. As a consequence, the worms developed a progressive loss of motor neuron function and coordination resembling the human CMT phenotype. Biological differences between mammals and invertebrates regarding the PNS and peripheral glial cells may impact the mechanisms underlying disease phenotype and the pharmacokinetic and pharmacodynamic of potential therapeutics, limiting the translation of findings using these models into therapies for patients. A review of animal models used in the study of inherited neuropathies can be found in Bouhy and Timmerman.[11]

Diagnosis

The typical clinical course of most inherited neuropathies includes normal initial development followed by gradual weakness and sensory loss appearing within the first two decades of life. Affected children are often slow runners and have difficulty with activities that require balance (e.g. skating, walking along a log). Ankle–foot orthoses are frequently required by the third decade. Fine movements of the hands for activities such as turning a key or using buttons and zippers may be impaired, but the hands are rarely as affected as the feet (an exception is CMT2D, caused by *GARS* mutations). Most patients remain ambulatory throughout life and have a normal lifespan.

A minority of CMT patients has a more severe phenotype with delayed motor milestones and onset in infancy, termed Dejerine–Sottas neuropathy. Particularly severe cases are classified as congenital hypomyelination if myelination appears to be disrupted during embryological development. Many patients have *de novo* autosomal dominant disorders, and the term Dejerine–Sottas neuropathy is currently used primarily to denote severe early-onset clinical phenotypes regardless of the inheritance pattern.

The diagnosis of a patient with CMT relies on the combined analysis of family history/inheritance pattern, neurophysiological study, and developmental milestones, as previously described.[12] Four genes (*PMP22, GJB1, MPZ,* and *MFN2*) are responsible for most cases of CMT. NCV testing can distinguish between demyelinating and axonal neuropathies. Most CMT1 patients, particularly those with CMT1A, have a uniformly slow NCV of about 20 m/s. Asymmetrical slowing, which is characteristic of hereditary neuropathy with liability to pressure palsies, may be found in patients with missense mutations in *PMP22, MPZ, EGR2,* and *GJB1*. Although axonal loss and reduced compound muscle action potential and sensory nerve action potential amplitudes are characteristic of CMT2, virtually all forms of CMT1 have axonal loss as well as demyelination. Specific clinical manifestations can help to direct genetic testing further, especially in rare cases. Whole genome sequencing has been used to determine the genetic abnormality in families with unknown CMT and will probably become the standard method for investigating undetermined cases of CMT, as it will soon be less expensive than sequencing multiple single genes using direct sequencing techniques.

Treatment

There is no specific therapy for the inherited neuropathies, but clinical and genetic counseling and symptomatic and rehabilitative treatment are important. A detailed family history and often examination of family members are required for prognosis and genetic counseling.

Ankle–foot orthoses to correct footdrop may return gait and balance to normal for years. Foot surgery is sometimes offered to correct inverted feet, pes cavus, and hammer toes. Surgery may improve walking, alleviate pain over pressure points, and prevent plantar ulcers. However, foot surgery is generally unnecessary and does not improve weakness and sensory loss.

Ascorbic acid, antagonists to progesterone, and subcutaneous injections of neurotrophin-3 have demonstrated improvement in animal models of CMTlA, but failed to demonstrate any therapeutic effect in patients.

IMMUNE-MEDIATED NEUROPATHIES

Peripheral neuropathies associated with abnormal humoral and cellular immunological responses against self antigens in the PNS (autoimmunity) include a variety of disease entities, such as inflammatory (Guillain–Barré syndrome, CIDP and variants), vasculitic, paraproteinemic, and paraneoplastic neuropathies. Other neuropathies in which an inflammatory component can be identified include some infectious and metabolic neuropathies, including diabetic radiculoplexopathy. This section will review some of the most common forms of immune-mediated neuropathies.

Guillain–Barré Syndrome

Guillain–Barré syndrome is an acute, acquired, inflammatory peripheral neuropathy characterized by elevated cerebrospinal fluid (CSF) protein levels with low CSF cell counts (cytoalbumological dissociation) and a monophasic progression with at least partial recovery. Guillain–Barré syndrome is subdivided into acute inflammatory demyelinating polyneuropathy, acute motor and sensory axonal neuropathy, acute motor axonal neuropathy, and the Miller–Fisher syndrome, according to patients' phenotypes. Acute inflammatory demyelinating polyneuropathy accounts for up to 97% of cases of Guillain–Barré syndrome in North America and Europe. It is a sporadic disorder with an incidence of 0.6–1.9 cases per 100,000 in North America and Europe. Men are more likely to be affected than women (1.4 to 1 ratio). In 60% of cases, acute inflammatory demyelinating polyneuropathy is preceded by a respiratory tract infection (e.g. cytomegalovirus, Epstein–Barr virus) or gastroenteritis (*Campylobacter jejuni*). Acute motor axonal neuropathy

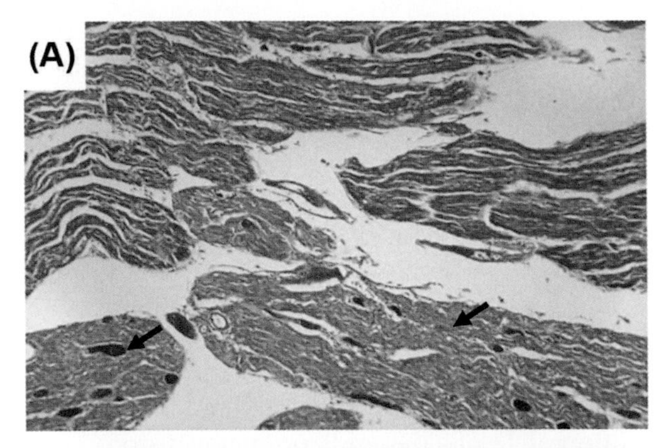

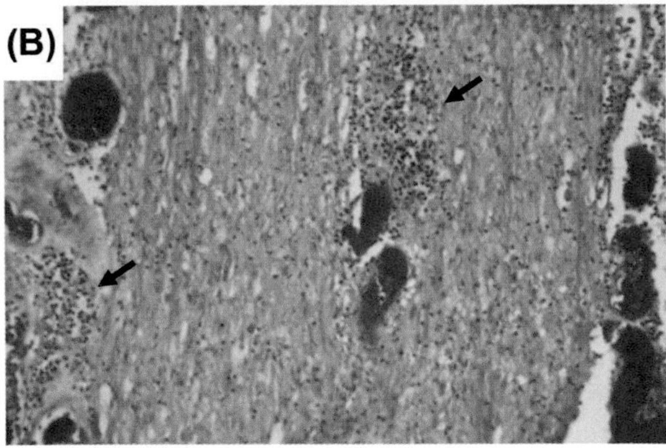

FIGURE 12.7 **Autopsy specimen of a patient with Guillain–Barré syndrome.** The figure demonstrates the involvement of the proximal peripheral nervous system (spinal rootlets) by demyelination (arrows in A) and lymphocytic inflammation (arrows in B). Luxol fast blue stain. *Source: Courtesy of Dr Steven A. Moore, University of Iowa.*

and acute motor and sensory axonal neuropathy are rare in North America and Europe but more frequent in China, Japan, Mexico, Korea, and India.

Both cellular and humoral immunity seem to be involved in Guillain–Barré syndrome. The triggering mechanisms are probably the result of postinfectious molecular mimicry in which nerve antigens (mostly PNS gangliosides) are attacked by the immune system because they resemble antigens present in microbes. Assays with antiganglioside antibodies, bacterial toxins, and lectins have characterized potential immunogenic regions of diarrhea-associated *C. jejuni* strains, for example. In the acute phase of the disease, disseminated foci of segmental demyelination can be observed in perivenular spaces, associated with endoneurial edema and mononuclear cellular infiltrates. The pathological hallmark of typical Guillain–Barré syndrome is macrophage-mediated attack on the Schwann cell or the axolemma. Macrophages cross the basal lamina, displace the Schwann cell cytoplasm, and insert themselves between the outer myelin lamellae, progressively destroying the myelin and exposing the associated axon. Inflammatory lesions tend to predominate in the proximal regions of the PNS (Fig. 12.7).

Experimental autoimmune neuritis has been used to model inflammatory neuropathies. It can be induced in rats, mice, rabbits, and guinea pigs by immunization with peripheral nerve tissue or specific myelin proteins, including MPZ, P2, PMP22, MBP, and MAG. Around 2 weeks after inoculation, animals develop a phenotype that resembles the human disease, including weakness and ataxia. Peripheral nerve infiltration by lymphocytes and macrophages with segmental demyelination and secondary axonal damage is observed. $CD4^+$ T cells reactive against myelin proteins play a crucial role in the pathogenesis of immune-mediated neuropathies. Activation of T cells depends on major histocompatibility complex (MHC) class II-bound peptide antigen presentation by antigen-presenting cells and costimulatory signals provided by B7-1/B7-2:CD28/CTLA-4 interactions and CD40L:CD40 pathways. Adhesion molecules, including selectins, integrins, and matrix metalloproteinases, regulate migration of activated lymphocytes across the blood–nerve barrier. Activated T cells cross the blood–nerve barrier and encounter specific or cross-reactive antigens, amplifying the immune response via proinflammatory cytokines and chemokines, recruiting macrophages and B lymphocytes to the peripheral nerve. Peripheral neuropathy ensues as a result of different mechanisms, including direct phagocytic attack by macrophages, damage from Th1 cytokines and oxygen free radicals, T-cell mediated cytotoxicity, complement-dependent attack, and antibody-mediated functional impairment. Despite being a useful model to study immunological disease mechanisms, experimental autoimmune neuritis has failed to identify the antigenic targets of autoreactive T cells in Guillain–Barré syndrome and CIDP. In contrast, several antibodies against PNS antigens have been implicated in the pathogenesis of axonal subtypes of Guillain–Barré syndrome and some forms of chronic demyelinating neuropathies, including antibodies against MAG and antibodies against the gangliosides GD_{1b}, GM_1, and GQ_{1B}. Aside from axonal injury, antiganglioside antibodies can also exert synaptopathic actions at the neuromuscular junction, characterized by an initial dramatic increase in the frequency of miniature endplate potentials followed by calpain-mediated synaptic necrosis and failure of neuromuscular transmission.[13]

Weakness, which is the most common initial symptom in both acute inflammatory demyelinating polyneuropathy and acute motor and sensory axonal neuropathy, can vary in severity, from mild cases, with only some difficulty in walking, to severe forms, with total quadriplegia and respiratory failure. The most common manifestation is leg weakness that subsequently "ascends" into the arms. Bilateral weakness of facial muscles (facial diplegia) occurs in about 50% of cases. Although Guillain–Barré syndrome has been described as an "ascending paralysis", proximal weakness is common, and 5% of cases have isolated cranial nerve involvement that subsequently descends into the limbs. Sensory loss occurs in the majority of patients. The autonomic nervous system is involved in about 65% of patients. Length-dependent weakness without sensory loss develops in patients with acute motor axonal neuropathy, including cranial nerve involvement in about 25% of cases. Miller–Fisher syndrome consists of the triad of ophthalmoplegia, ataxia, and areflexia. Facial weakness, ptosis, and pupillary abnormalities may be present.

The diagnosis of acute inflammatory demyelinating polyneuropathy and acute motor and sensory axonal neuropathy is based on the history, physical examination, and CSF evaluation. Deep tendon reflexes are decreased or absent, weakness is usually symmetrical, and the CSF is abnormal with cytoalbuminological dissociation. In both acute inflammatory demyelinating polyneuropathy and acute motor and sensory axonal neuropathy, the CSF should have fewer than five white blood cells (WBCs)/ml. If the CSF cell count is greater than 50 WBC/ml, another diagnosis, such as in HIV infection or Lyme disease, should be considered. Elevated CSF protein may not be apparent in the first 7–10 days of the illness; in up to 10% of cases, CSF protein levels remain normal. Acute inflammatory demyelinating polyneuropathy is distinguished from acute motor and sensory axonal neuropathy by nerve conduction studies. NCVs in Miller–Fisher syndrome are generally normal. Approximately 5% of patients with Guillain–Barré syndrome have Miller–Fisher syndrome, and more than 85% of these patients have polyclonal antibodies that react with the ganglioside GQ_{1b}.

The differential diagnosis varies in different parts of the world. In North America, poliomyelitis has been eradicated, but other viral illnesses may induce polio-like syndromes which can be mistaken for Guillain–Barré syndrome, including ECHO 70, coxsackievirus, West Nile virus, and rarely, rabies. Because these diseases are not demyelinating disorders, they can be distinguished from acute inflammatory demyelinating polyneuropathy by their normal NCVs. However, the results of electrodiagnostic studies are similar in both acute motor axonal neuropathy and the polio-like syndromes, thus making distinction between acute motor axonal neuropathy and these syndromes difficult. Tick paralysis is caused by a toxin within the tick and can mimic Guillain–Barré syndrome, particularly in children. Usually, removal of the tick is associated with improvement within hours, although progression can occur. Botulism produces a rapid flaccid paralysis. Patients have ophthalmoplegia, bulbar weakness, dry mouth, constipation, and orthostatic hypotension, but sensory symptoms do not develop. Other entities that can mimic Guillain–Barré syndrome are acute spinal cord compression, acute transverse myelitis, and vascular myelopathies, all of which are characterized by decreased reflexes before the development of upper motor neuron signs such as increased reflexes. Carcinomatous or lymphomatous meningitis can also cause a rapidly developing quadriparesis, but both are associated with elevated CSF cell counts.

Patients with Guillain–Barré syndrome require hospitalization because of the potential for respiratory failure. Pulmonary function tests should be performed frequently; a vital capacity of less than 1 liter or a negative inspiratory force of less than −70 suggests that ventilator support may be needed in an intensive care unit setting. Autonomic instability and difficulty swallowing also need to be monitored.

Therapies directed at modulating the immune system are effective in Guillain–Barré syndrome. Intravenous immunoglobulin (IVIg) therapy, which is the preferred treatment, is typically given as 2 g/kg divided over 2–5 days within the first 2 weeks. Although plasma exchange and IVIg are equally effective, at least in the first 2 weeks, IVIg is often preferred for its convenience unless there are contraindications, such as low serum IgA levels, renal failure, or severe hypertension. Plasma exchange, usually in the form of four exchanges of 1.5 liters of plasma spread over a period of 10 days, is also effective. Two plasma exchanges may be sufficient in mild cases, and six exchanges are not superior to four in severely affected patients. This therapy should ideally be administered within the first 2 weeks and not later than 4 weeks from clinical onset. Ten percent of patients with Guillain–Barré syndrome relapse after initially responding to plasma exchange or IVIg; they usually respond to a second cycle of the previously effective treatment. The combined use of plasma exchange followed by IVIg does not improve the prognosis.

Corticosteroids in different forms (intravenous methylprednisolone, oral prednisolone or prednisone, intramuscular adrenocorticotropic hormone) are not beneficial. A trial of intravenous methylprednisolone (500 mg/day for 5 days) in association with IVIg versus IVIg alone showed a slight initial advantage over IVIg alone for the combined treatment but no benefit in terms of long-term disability.

Fifty percent of patients progress to their nadir, or maximum disability, within 2 weeks, 75% within 3 weeks, and more than 90% within 4 weeks of the onset of symptoms. With modern supportive care, acute mortality is about 2%. After a brief period of stabilization, slow spontaneous recovery occurs over a period of weeks or months. Most patients either undergo complete recovery or are left with minor sequelae; about 20% have a persistent disability. The long-term prognosis depends at least in part on the extent of axonal loss. Patients with low compound muscle action potential amplitudes in the upper extremities are more likely to have a poor prognosis. Patients with acute motor axonal neuropathy will recover after approximately 2 months, but the extent of recovery may be less than in Guillain–Barré syndrome. In general, the prognosis for recovering from Miller–Fisher syndrome is excellent. A complete review can be found in Hughes and Cornblath.[14]

Chronic Inflammatory Demyelinating Polyradiculoneuropathy and Variants

CIDP is a chronic acquired demyelinating sensorimotor neuropathy that may be monophasic, relapsing, or progressive, developing over a period of at least 2 months. CIDP occurs in all age groups with a mean age range of 30–50 years. Women are slightly more likely to be affected than men. Antecedent events are less common than in Guillain–Barré syndrome; they occur in around 30% of patients and include upper respiratory infections, gastrointestinal infections, vaccinations, surgery, and trauma. The pathophysiology of CIDP is very similar to Guillain–Barré syndrome. Nerve biopsy shows macrophage-mediated segmental demyelination, occasional endoneurial lymphocytic T-cell infiltrates, and endoneurial edema. The MHC class I and II antigens are upregulated, and there are often deposits of immunoglobulins and complement split products on the outer Schwann cell membranes or myelin sheaths. CIDP can be passively transferred to animals by patient sera, but no clear autoantigen has been identified.

Weakness and sensory loss begin insidiously and progress over a period of months to years. Weakness is commonly proximal as well as distal and patients can become bedridden. Loss of proprioception from damage to large-diameter sensory nerves may affect balance and result in an action tremor. Deep tendon reflexes are usually absent or markedly decreased. Facial weakness (15%), ptosis or ophthalmoparesis (5%), and papilledema occur occasionally. Variant forms include pure motor, pure sensory, and multifocal disease. The multifocal variant of CIDP is also known as Lewis–Sumner syndrome and is characterized by asymmetrical conduction blocks in both motor and sensory nerves. Lewis–Sumner syndrome responds to the same treatment options as CIDP.

Diagnosis is based on clinical symptoms and signs, electrodiagnostic studies, and CSF examination. Nonuniform, asymmetrical slowing on nerve conduction studies is characteristic. One portion of a nerve may have different conduction from another. For example, if damage is primarily in the spinal roots, proximal conduction velocities and F-wave latencies may be most affected. Compound muscle action potentials are generally reduced because of the concomitant axonal degeneration that occurs with demyelinating neuropathies. However, temporal dispersion and conduction block may also reduce the amplitude of muscle action potential in any demyelinating neuropathy. Sensory nerve conduction is also slow in CIDP, but because sensory nerve action potentials are often not detectable, sensory conduction velocity may be unmeasurable.

CSF results resemble those of acute inflammatory demyelinating polyneuropathy: WBC counts are usually less than 10 cells/mm^3, and protein levels are higher than 60 mg/dl. CSF cell counts higher than 50/mm^3 suggest another diagnosis, such as HIV infection or hematological malignancy. CSF protein levels may be normal early in the course of CIDP.

CIDP is distinguished from acute inflammatory demyelinating polyneuropathy by its time-course. A similar manifestation can also occur in diabetes, lymphoma, monoclonal gammopathies, and asymmetrical inherited neuropathies. Corticosteroids are effective in more than two-thirds of patients with CIDP. A standard approach is to use oral prednisone (1 mg/kg/day) for 6–8 weeks, followed by slow tapering over 3–12 months to a maintenance level of about 0.1 mg/kg/day. A response to prednisone may take months to occur, and occasionally patients may worsen before they respond. As a result, plasma exchange or IVIg is often used as initial treatment.

Because of the side effects of long-term corticosteroids or the lack of response to them, azathioprine, cyclosporine, cyclophosphamide, methotrexate, mycophenolate mofetil, rituximab, and interferon-α or -β have been used with variable success in uncontrolled reports.

Multifocal motor neuropathy (MMN) is characterized by progressive, predominantly distal and asymmetrical limb weakness, mostly affecting the upper limbs with minimal or no sensory impairment. The prevalence of MMN is estimated at 2 per 100,000. Men are more frequently affected than women (2.6 to 1 ratio). Initial symptoms develop in 80% between the ages of 20 and 50 years, with a mean age at onset of 40. The diagnosis is established by the presence of multifocal, persistent partial conduction blocks on motor but not sensory nerve conduction studies. MMN is considered to be an autoimmune neuropathy based on its clinical improvement with immunologically based therapies and because of a frequent association with antiglycolipid antibodies. Patients with MMN often have serum antibodies that react with ganglioside GM_1, and these titers decrease during effective treatment. GM_1 is highly represented in neural membranes at the nodes of Ranvier, compact myelin, and the motor endplate at the neuromuscular junction. A blocking effect on mouse distal motor nerve conduction has been induced *in vitro* by sera from MMN patients with and without high anti-GM_1 antibody titers. These data support the presence of serum factors responsible for conduction block in the sera of patients with MMN, although these factors are not invariably related to anti-GM_1 antibodies. Owing to this unique pathophysiology, MMN is now considered a separate entity from CIDP.

IVIg (2 g/kg) is the initial treatment for MMN, and almost 80% of patients respond within a week. However, improvement is typically brief (3–6 weeks), so repeated treatments are required indefinitely. Patients may eventually become refractory to IVIg, and another agent may be needed, such as rituximab (e.g. 375 mg/m^2 weekly for 4 weeks) or azathioprine (2–3 mg/kg/day). Plasma exchange and corticosteroids are generally ineffective and have been associated with worsening neuropathy in some patients.[15] A review can be found in Köller *et al.*[16]

Paraproteinemic Neuropathies

A paraprotein consists of immunoglobulins of the same isotype, all produced by a single clone of abnormally proliferating lymphocyte/plasma cells. In some cases the M protein is part of a malignant lymphoproliferative disease such as multiple myeloma, solitary plasmacytoma (IgG and IgA), Waldenström macroglobulinemia (IgM), lymphoma, chronic lymphocytic leukemia (IgM), primary amyloidosis, or cryoglobulinemia. In most instances, however, monoclonal gammopathy is not initially associated with any of these disorders and is classified as a monoclonal gammopathy of uncertain significance (MGUS). Monoclonal gammopathies occur in up to 8% of patients with peripheral neuropathy of unknown etiology. However, MGUS is frequent, being found in 1% of the population older than 50 years and in 3% older than 70 years, and most subjects with MGUS do not have neuropathy. In some cases the co-occurrence of neuropathy and M protein may be a coincidence, but in others the M protein is clearly related to the neuropathy.

The prevalence of neuropathy is higher in patients with IgM compared with IgG or IgA M proteins. The prevalence of symptomatic neuropathy associated with IgM monoclonal gammopathy in patients older than 50 years is approximately 20 per 100,000. In half of such patients, the M protein reacts with either the HNK1 carbohydrate moiety of MAG or with other glycoproteins (MPZ, PMP22) and glycolipids (sulfoglucuronylparagloboside and lactosaminylparagloboside). IgM M proteins associated with neuropathy may also bind to other neural antigens.

In patients with IgG monoclonal gammopathy and neuropathy, the relationship is less clear than with IgM. Although about 10% of patients with multiple myeloma have neuropathy, in most cases the M protein does not react with a neural antigen, and patients do not improve with immunotherapy. Conversely, approximately 50% of patients with the osteosclerotic form of myeloma have neuropathy, often associated with the non-neurological manifestations of organomegaly, endocrine abnormalities, and brown, tannish discoloration of the skin, known as POEMS syndrome. Similarly, about 50% of patients with light-chain amyloidosis have neuropathy.

In patients with IgM M proteins that immunoreact with MAG, nerve biopsies demonstrate segmental demyelination with deposits of M protein and complement. There is often a widening of myelin lamellae on sural nerve biopsies, which is not necessary for diagnosis. High titers (>1:10,000) of anti-MAG IgM antibodies are associated with neuropathy, and intraneural or systemic injection of anti-MAG IgM M proteins causes complement-mediated demyelination of nerves in animals. Most patients with anti-MAG neuropathies are initially seen in their sixth to seventh decade of life with dysesthesias and paresthesias in their legs and unsteadiness while walking because of loss of proprioception. Physical examination shows a length-dependent large-fiber sensory neuropathy. Weakness may develop later.

NCVs are slow (around 25 m/s), with pronounced delays in distal motor latencies, thus prompting the designation distal acquired demyelinating symmetrical neuropathy to distinguish the disorder from CIDP. Treatment of neuropathies associated with monoclonal gammopathy is similar to that for CIDP. However, patients with anti-MAG-related neuropathies do not respond as well to treatment as do patients with CIDP. Anecdotal data support the benefit of rituximab in some patients. Progression of the neuropathy of monoclonal gammopathy disables about 25% of patients after 10 years and 50% after 15 years. The course of patients with osteosclerotic myeloma and neuropathy depends on the response to treatment of the myeloma. In patients whose myeloma responds to treatment, more than 50% have improvement in neuropathy. In other patients with sensorimotor neuropathies associated with plasma cell dyscrasias, the course is variable and the M protein may not be related to their neuropathy.[1]

Vasculitic Neuropathies

Neuropathy can occur in systemic vasculitis associated with other organ systems, as well as in non-systemic vasculitis affecting just nerve and muscle. Vasculitic neuropathies typically present as painful acute or semiacute axonal mononeuritis multiplex. There is acute motor and sensory loss in multiple nerve territories. The number of nerves involved may be extensive enough to make the distinction between a multifocal and diffuse neuropathy difficult. Occasionally, vasculitic neuropathy can present as sensory neuropathy, trigeminal neuropathy, compressive neuropathy, or autonomic neuropathy.

Systemic vasculitic neuropathy is more common than non-systemic vasculitic neuropathy. Peak ages at onset of both are the fifth to eighth decades, but vasculitis can occur at any age. Neuropathy, particularly mononeuritis multiplex, is common in several forms of systemic vasculitis. More than 50% of patients with Churg–Strauss syndrome, 40–50% with Wegener granulomatosis, 35–75% with polyarteritis nodosa, and a majority of patients with mixed cryoglobulinemia develop neuropathy. Patients with Sjögren syndrome are often initially found to have sensory neuropathies. Rheumatoid arthritis evolves into systemic rheumatoid vasculitis in 5–15% of patients, and vasculitic neuropathy will develop in about 50% of these patients. Neuropathies are uncommon in systemic lupus erythematosus.

In patients with mononeuritis multiplex, axonal degeneration develops as a result of nerve ischemia caused by the vasculitic process. Immune-mediated inflammation and necrosis of blood vessel walls occlude the vessel's lumen, thereby resulting in ischemic damage. Small arteries or arterioles (50–300 μm) are most commonly affected, particularly those that occur in watershed areas between the distribution of the major nutrient arteries of proximal nerves. True nerve infarcts are rare. The immune-mediated inflammation is associated with antibody–antigen complexes that are deposited in the wall of the blood vessel. Antibodies also bind directly to endothelial cell antigens. In both circumstances, complement is activated, as evidenced by deposition of membrane attack complex. Chemotactic factors then recruit neutrophils, which release proteolytic enzymes and generate toxic oxygen free radicals. The sensory neuropathy of Sjögren syndrome probably results from the infiltration of dorsal root ganglia by cytotoxic T cells.

Patients typically have a relatively sudden onset of painful, focal or multifocal weakness or sensory loss. These symptoms reflect ischemia anywhere along the length of the nerves, generally in the lower extremities. Onset usually occurs rapidly (hours to days) as a result of the abrupt, ischemic etiology.

Nerve biopsy of clinically affected sensory nerves (sural, superficial peroneal, or superficial radial) is often necessary and is justified because therapy may be aggressive and long term. Superficial peroneal nerve biopsy may be combined with muscle biopsy from the same incision. Pathological features diagnostic of vasculitis occur in 60% of patients, and less specific features such as multifocal loss of fibers occur in others. Findings diagnostic of vasculitis include destruction of the vessel and inflammation within the vessel wall. Fibrinoid necrosis, vessel wall scarring, recanalization, neovascularization, and hemosiderin are common, but not essential histopathological features of vasculitis. Although nerve biopsy is the gold standard for diagnosis, clinical, serological, and electrophysiological findings can suggest the diagnosis. For example, EMG and NCV studies can distinguish between mononeuritis multiplex and a symmetrical neuropathy. It is essential to confirm NCV abnormalities in a nerve before biopsy, which is always required before treatment. An acute or a subacute onset of asymmetrical weakness or sensory loss in the distribution of individual nerves suggests mononeuritis multiplex, particularly in the setting of a known connective tissue disorder. Systemic symptoms, such as unexplained weight loss and purpura, or constitutional symptoms, such as fever, myalgias, arthralgias, pulmonary disease, abdominal complaints, rashes, or night sweats, suggest systemic vasculitis in a patient with mononeuritis multiplex.

The erythrocyte sedimentation rate is usually elevated in the systemic vasculitides but is normal in non-systemic cases. Perinuclear and cytoplasmic antineutrophil cytoplasmic antibody suggests Wegener granulomatosis or Churg–Strauss syndrome. Hepatitis C is usually associated with the presence of cryoglobulins. Serum complement levels, extractable nuclear antigen, angiotensin-converting enzyme levels, serum protein electrophoresis, and HIV serology are generally indicated. CSF analysis is not usually helpful in cases of vasculitic neuropathy but may be needed to exclude infectious (e.g. Lyme disease) or other inflammatory causes.

Acute or subacute mononeuritis multiplex may also result from diabetes, sarcoidosis, Lyme disease, and malignant infiltration of nerves. MMN with conduction block and Lewis–Sumner syndrome can resemble vasculitic mononeuritis multiplex. Sensory neuropathies similar to those in Sjögren syndrome may occur in patients with diabetes, paraneoplastic syndromes associated with anti-Hu antibodies, and pyridoxine intoxication.

Corticosteroid therapy is used for most vasculitic neuropathies. Oral prednisone (1 mg/kg) is appropriate for relatively mild cases, but intravenous methylprednisolone (1000 mg/day for 3–5 days) may be indicated as initial treatment in severe cases. Daily dosing is commonly used for the first 2 months or longer if the disease remains active. Subsequently, the dose is gradually tapered, with a transition to alternate-day dosing and discontinuation depending on the clinical picture and associated systemic features. Corticosteroid treatment may be adequate for Churg–Strauss syndrome, but additional medication is generally needed in other forms of systemic vasculitic neuropathy. In most cases of Wegener granulomatosis and microscopic polyangiitis, combined therapy with glucocorticoids and oral cyclophosphamide (2 mg/kg/day) or weekly oral methotrexate (7.5 mg/week) is used. Azathioprine (2–3 mg/kg/day) is also used. Because patients with non-systemic vasculitic neuropathy may recover spontaneously or have a relatively benign course, low-dose or alternate-day oral prednisone (60–80 mg/day) is often adequate therapy. Azathioprine or weekly methotrexate can be used as a glucocorticoid-sparing agent. Doses such as 60 mg of prednisone on alternate days, 2–3 mg/kg/day of azathioprine, or 7.5–15 mg/week of methotrexate, are reasonable starting doses that can ultimately be tapered if the treatments prove effective. Most systemic and non-systemic vasculitis responds to treatment, and patients make at least partial recovery with gradual return of function after a static period. The prognosis of patients with non-systemic vasculitis is better than that of patients with systemic vasculitis, with fewer episodes of nerve damage; the disease may be monophasic or relapsing/remitting over a period of years, and most patients recover the ability to walk. A review of the vasculitic neuropathies can be found in Pagnoux and Guillevin.[17]

Paraneoplastic Neuropathies

Paraneoplastic neuropathies are hypothesized to be the result of host immune responses to a tumor antigen or antigens that are also present in neural tissues. They are not caused by metastatic invasion of neural tissue, radiation therapy or chemotherapy, metabolic, vascular, or hormonal disturbances, or opportunistic infections. Paraneoplastic syndromes occur in less than 1% of patients with cancer; peripheral neuropathy is only one of the paraneoplastic syndromes. Although more than 25% of patients with cancer have evident neuropathy on neurological examination, the relationship to malignancy is unclear in most. Paraneoplastic neuropathy may develop before, during, or after the tumor is diagnosed. In certain tumors, neuropathies are distinctive and should prompt a thorough investigation for cancer. Small cell carcinoma of the lung is by far the most common underlying neoplasm, followed by carcinoma of the stomach, breast, colon, rectum, ovary, and prostate.

Subacute sensory neuronopathy, the most characteristic paraneoplastic neuropathy, results from an immune-mediated ganglionitis that destroys sensory neurons in the dorsal root ganglia. Mononuclear inflammatory infiltrates composed of $CD4^+$ and prominent $CD8^+$ T cells, along with plasma cells, are found in the stroma surrounding the dorsal root ganglion neurons. Other findings include atrophy of the dorsal roots, loss of sensory neurons, which appear to be replaced by a proliferation of satellite cells (Nageotte nodule), axonal degeneration, and secondary degeneration of the dorsal column of the spinal cord. Inflammatory infiltrates can also be found in peripheral nerves or muscle. Sural nerve biopsies typically reveal only loss of myelinated nerve fibers and are not useful for diagnosis.

Subacute sensory neuronopathy is characterized by subacute, progressive impairment of all sensory modalities and is associated with severe sensory ataxia and areflexia. It may precede the diagnosis of tumor by months or even years. At onset, patients may have shooting pain and burning sensations. Other symptoms include numbness, tingling, and a progressive sensory loss that may be asymmetrical. Symptoms usually progress rapidly to involve all four limbs, the trunk, and the face. Findings may then stabilize, although by this time the patient is often totally disabled. Occasional patients have an indolent course. Neurological examination reveals loss of deep tendon reflexes and involvement of all modalities of sensation; large-fiber modalities such as vibration and joint position sense are most severely affected. The loss of position sense may lead to severe sensory ataxia with pseudoathetoid movements of the hands and an inability to walk despite normal strength. Cranial nerve involvement may cause sensorineural deafness, loss of taste, and facial numbness. The frequent asymmetrical pattern of symptoms sometimes suggests a radiculopathy or plexopathy. A paraneoplastic encephalomyelitis characterized by patchy, multifocal neuronal loss in regions of the cerebral hemispheres, limbic system, cerebellum, brainstem, spinal cord, and autonomic ganglia often develops in patients with subacute sensory neuronopathy. Autonomic symptoms include impotence, dry mouth, and constipation.

The diagnosis is based on recognizing the typical neuropathy in the setting of malignancy. The results of routine laboratory studies are generally normal. The diagnosis is supported by finding serum polyclonal IgG anti-Hu antibodies, also called antineuronal antibodies type 1, or by indirect immunofluorescence or immunohistochemistry, and confirmed by Western blot analysis.

Subacute painful, asymmetrical neuropathy or neuronopathy in an elderly patient should prompt a search for carcinoma of the lung because small cell lung cancer accounts for more than 80% of the associated tumors. Subacute sensory neuronopathy has also been reported in patients with adenocarcinoma of the lung, breast, ovary, stomach, colon, rectum, and prostate, as well as Hodgkin and non-Hodgkin lymphoma. In patients with no evidence of cancer, detection of anti-Hu antibodies should prompt a computed tomography study of the chest with special attention to the mediastinal lymph nodes. The use of whole body positron emission tomography with fluorodeoxyglucose has been advocated for early diagnosis in patients with anti-Hu antibodies or clinical suspicion of subacute sensory neuronopathy because it may reveal neoplastic adenopathy months before computed tomography or magnetic resonance imaging can detect them.

Subacute sensory neuronopathy responds poorly to plasma exchange, IVIg, or various immunosuppressant medications, even when such treatment is started early in the course of the disease, possibly before the loss of sensory neurons. Even successful treatment of the tumor rarely induces remission of subacute sensory neuronopathy, but the symptoms may stabilize.

Sensorimotor neuropathy occurs in approximately 25% of patients with all types of tumors. The neuropathy can have an acute or subacute onset, with a progressive or relapsing/remitting course. Because no antineuronal antibody has been specifically associated with these neuropathies, their paraneoplastic nature is not established. Severe or relapsing neuropathies often precede the diagnosis of cancer, but the search for malignancy is generally limited to a chest radiograph, stool samples for blood, and routine blood tests. There are no specific treatments for these neuropathies and their progression does not necessarily correlate with that of the malignancy. A nonsystemic vasculitic neuropathy, which may also involve muscle, occurs with various types of tumor, including small cell lung cancer, lymphoma, and carcinoma of the kidney, stomach, and prostate. Neurological symptoms may develop either before or after the tumor is diagnosed. The neuropathy is subacute and progressive and usually affects older men. Like many paraneoplastic disorders, these neuropathies often do not respond well to treatment, which is similar to that for the vasculitic neuropathies. A comprehensive review on paraneoplastic neuropathies can be found in Rudnicki and Dalmau.[18]

OTHER NEUROPATHIES

Infectious Neuropathies

Infectious diseases can harm the PNS by distinct mechanisms, including the direct invasion of the peripheral nerve by pathogens or the action of a specific neurotoxin, or by eliciting a crossed immune reaction against peripheral nerve antigens, as seen in acquired immune neuropathies.

Neuropathy Associated with Leprosy

Leprosy is the most common cause of peripheral neuropathy in developing countries, although it is infrequent in the Western world. Leprosy may be manifested in different forms, depending on the host's immune system. Patients with normal cell-mediated immunity are more likely to have a tuberculoid form characterized by hypopigmented skin lesions associated with decreased sensation. In patients with abnormal cell-mediated immunity, the more severe lepromatous form with large disfiguring lesions may develop. A mononeuritis multiplex pattern with prominent superficial sensory loss is the most typical clinical manifestation of leprosy.

Mycobacterium leprae infects nerves from the epineurial lymphatics and blood vessels, enters the endoneurial compartment through its blood supply, and invades Schwann cells, ultimately causing demyelination. Recent studies have identified several binding targets for *M. leprae* in the Schwann cell membrane or basal lamina. These include α-dystroglycan and α_2-laminins in the Schwann cell surface, and laminin-2 in the basal lamina. Other studies have also demonstrated the ability of myelin P0 to bind *M. leprae* and a histone-like, laminin-binding protein (Hlp/LBP) expressed by *M. leprae* has been identified that plays a role in binding to Schwann cells and other cells. *Mycobacterium leprae* also binds to ErbB2, a Schwann cell receptor for neuriguiin-1 which is a critical mediator of Schwann cell–axon interaction. After *M. leprae* adheres to the Schwann cell surface, the bacterium is slowly ingested. Schwann cells apparently provide an environment suitable for preservation and proliferation of *M. leprae*, consistent with the longstanding histopathological observations that *M. leprae* appears to persist and grow within Schwann cells in human nerves. Binding of an *M. leprae*-derived lipoprotein to toll-like receptor-2 (TLR2) in human Schwann cells, both *in vitro* and *in vivo*, has been reported to result in apoptosis. *Mycobacterium leprae* binding also alters Schwann cell expression of a small number of genes (GFAP, TGFb1, NCAM, ICAM, N-cadherin, and L1) and activation of the transcription factor NF-κB. Rapid demyelination follows adherence of *M. leprae* to Schwann cells, even in the absence of immune cells, suggesting a contact-dependent mechanism. However, an immune response may also be directed at *M. leprae*-infected Schwann cells. Human Schwann cells express MHC II molecules after infection with *M. leprae*. These cells process and present *M. leprae*, and some of its protein and peptide antigens to MHC class II-restricted $CD4^+$ T cells, and are efficiently killed by these activated T cells.[19]

The final downstream consequence of these multiple mechanisms is segmental demyelination, which has been documented in teased-nerve fiber studies of leprosy patients. However, paranodal demyelination and axonal atrophy have also been identified and linked to abnormal phosphorylation of neurofilament proteins in the axon. The exact mechanisms behind this process are still unknown.

If treated early, neuropathies in leprosy can improve significantly. The World Health Organization recommends combination therapy that includes dapsone (50–100 mg/day or 200–250 mg/week), rifampicin (600 mg/month), and clofazimine (100 mg/day).

Neuropathies Associated with Human Immunodeficiency Virus Infection

The PNS may be involved in all phases of HIV infection. The most common peripheral neuropathy is a distal, painful, sensory axonal polyneuropathy that is very similar to the toxic neuropathy caused by nucleoside reverse transcriptase inhibitors (NRTIs), including zidovudine, zalcitabine, didanosine, stavudine, and lamivudine. When an iatrogenic neuropathy is suspected, discontinuation of NRTIs may improve symptoms. Conversely, a neuropathy caused by HIV is likely to stabilize or improve with antiretroviral treatment.

Inflammatory neuropathies such as chronic or acute inflammatory demyelinating polyneuropathy can also occur in the early stages of HIV infection; the CSF cytoalbumin dissociation usually seen with these conditions may not be evident in these patients because of a mild CSF mononuclear pleocytosis. The response of these neuropathies to plasma exchange or IVIg is generally good. In later stages of HIV infection, cytomegalovirus may cause either an acute lumbosacral polyradiculopathy as a result of direct invasion of nerve roots or a mononeuritis multiplex through a vasculitic mechanism.

Neuropathies Associated with Herpes Zoster

Varicella-zoster virus usually remains latent in cranial or spinal ganglia after resolution of a systemic infection. Reactivation, which tends to occur in elderly people or immunocompromised patients, causes a vesicular skin eruption accompanied by pruritus and dysesthesias. Herpes zoster normally undergoes spontaneous resolution but is frequently followed by a severe postherpetic neuralgia, which is defined as pain persisting for more than 6 weeks after the rash appears. Early treatment with oral acyclovir (800 mg, five times daily for 7 days) may reduce both the duration of the acute phase and the chance of postherpetic neuralgia developing, which is usually treated with symptomatic drugs for neuropathic pain.

Neuropathy Associated with Lyme Disease

Borrelia burgdorferi causes a disease with three stages. In the first stage, shortly after and in the area of a tick bite, a non-pruritic rash (erythema migrans) appears and spontaneously disappears after a few weeks. The second stage is frequently associated with neurological complications such as lymphocytic meningitis and focal and multifocal peripheral and cranial neuropathies; characteristic manifestations are unilateral or bilateral facial palsy and radiculitis. The third stage is associated with severe neurological complications, including encephalopathy, encephalomyelitis, and a predominantly sensory axonal polyneuropathy. A lymphocytic pleocytosis in CSF and demonstration of *B. burgdorferi* infection in serum or CSF are the main laboratory findings.

Neuropathy Associated with Diphtheria

Although vaccination has made diphtheria rare in developed countries, it is still an important cause of subacute neuropathy in developing countries. Some strains of *Corynebacterium diphtheriae* produce a potent neurotoxin that causes palatal weakness, accommodation deficits, and extraocular palsies. This acute manifestation is followed by an ascending paralysis secondary to a demyelinating neuropathy that shares many clinical features with acute inflammatory demyelinating polyneuropathy.

The neuropathy caused by the neurotoxin usually resolves with resolution of the infection. The diphtheria organism can be eradicated by therapy with antibiotics such as erythromycin (2 g/day intravenously divided twice daily for adults) or penicillin (procaine penicillin G, 1.2 million units/day intramuscularly divided twice daily for 14 days). However, the neuropathy, as with other manifestations of the disease, generally requires treatment with diphtheria antitoxin, a hyperimmune antiserum produced in horses. Depending on the severity of the disease, antitoxin is administered intramuscularly or intravenously (80,000–120,000 units for extensive disease with a duration of 3 or more days).

Toxic and Deficiency Syndromes

In Western countries, toxic neuropathies are frequently the side effects of medications rather than a result of environmental exposure (Table 12.2). In most cases, iatrogenic neuropathy is manifested as a length-dependent or dying-back axonal neuropathy. Treatment requires a correct diagnosis and discontinuation of the drug, but improvement is often slow and may take several months.[20]

Diabetic and Other Metabolic Neuropathies

Diabetes is the most common cause of neuropathy in the Western world. Diabetic neuropathy occurs in 8–70% of patients with diabetes, depending on the criteria used to diagnose neuropathy, and patients with retinopathy or overt albuminuria are over twice as likely to have neuropathy. Distal symmetrical polyneuropathies are the most common diabetic neuropathy, but distal autonomic neuropathy is also common. For example, impotence develops in 20–60% of diabetic men, but widespread autonomic dysfunction develops in less than 5% of diabetic patients. Diabetic peripheral neuropathies can be separated into two large groups: (1) symmetrical, predominantly sensory or autonomic neuropathies (or both); and (2) asymmetrical mononeuropathies or plexopathies. A complete account of the pathophysiology and clinical manifestations of diabetic neuropathy can be found in Chapter 13.

TABLE 12.2 Substances Associated with Toxic Neuropathies

Antineoplastic agents	Vincristine, paclitaxel (Taxol®), cisplatin, suramin, thalidomide
Antimicrobials	Chloroquine, dapsone, isoniazid, metronidazole, nitrofurantoin
Cardiac medications	Amiodarone, perhexiline, hydralazine
Other medications	Colchicine, tacrolimus, gold salts, phenytoin, disulfiram (Antabuse®), pyridoxine (vitamin B_6)
Heavy metals	Lead, arsenic, mercury, thallium
Chemical compounds	Acrylamide, carbon disulfide, ethylene glycol, hexacarbons, organophosphate esters, vacor

Source: Adapted from Shy ME. Peripheral neuropathies. In: Goldman L, Schafer AI, eds. Goldman's Cecil Medicine. *24th ed. Philadelphia, PA: Elsevier; 2012:2396–2409.*[1]

CONCLUSION

To summarize, peripheral nerves are elegant dynamic structures in which glial cells and the neurons they ensheathe interact with each other in a true symbiotic fashion. PNS neurons instruct Schwann cells to myelinate. Axonal integrity depends on Schwann cells and myelin. Schwann cells can alter their phenotype depending on the axon they contact. PNS neurons can extend processes more than 1 m in length and maintain these axons through the life of an individual. PNS axons can regenerate, in part because of signals from Schwann cells. The molecular signals that regulate these interactions are at the heart of how glia interact with axons. Understanding the molecular basis of these interactions is necessary to develop rational therapies for the many genetic and acquired disorders that affect peripheral nerves.

QUESTIONS FOR FURTHER RESEARCH

- What are the internal and external signals that activate positive or negative regulation of myelination, and how can they be manipulated to treat demyelinating neuropathies?
- Axons can regenerate for up to approximately 18 months after being transected. What are the mechanisms responsible for this limited timeframe and how can they be altered to prolong and promote axonal regeneration?
- Why do mutations in ubiquitously expressed genes, such as *MFN2*, *GDAP1*, or *GARS*, specifically affect axons of peripheral neurons?
- What are the environmental and genetic modifiers that can account for the phenotypic variability observed in some forms of CMT (e.g. CMT1A)?
- What are the immunological and biological mechanisms behind the phenotypic variability observed in distinct types of immune-mediated neuropathy? Specifically, what are the mechanisms behind the different therapeutic responses observed (e.g. CIDP responds well to glucocorticoids, whereas Guillain–Barré syndrome and MMN do not)?

References

1. Shy ME. Peripheral neuropathies. In: Goldman L, Schafer AI, eds. *Goldman's Cecil Medicine*. 24th ed. Philadelphia, PA: Elsevier; 2012:2396–2409.
2. Berthold C-H, Fraher JP, King RHM, Rydmark M. Microscopic anatomy of the peripheral nervous system. In: Dyck PJ, Thomas PK, eds. *Peripheral Neuropathy*. 4th ed. Philadelphia, PA: Elsevier; 2005:35–92.
3. Jessen KR, Mirsky R. The origin and development of glial cells in peripheral nerves. *Nat Rev Neurosci*. 2005;6(9):671–862.
4. Woodhoo A, Sommer L. Development of the Schwann cell lineage: from the neural crest to the myelinated nerve. *Glia*. 2008;56(14):1481–1490.
5. Arroyo EJ, Scherer SS. On the molecular architecture of myelinated fibers. *Histochem Cell Biol*. 2000;113(1):1–18.
6. England JD, Gronseth GS, Franklin G, Carter GT, Kinsella LJ, Cohen JA, et al. Practice Parameter: evaluation of distal symmetric polyneuropathy: role of laboratory and genetic testing (an evidence-based review). Report of the American Academy of Neurology, American Association of Neuromuscular and Electrodiagnostic Medicine, and American Academy of Physical Medicine and Rehabilitation. *Neurology*. 2009;72(2):185–192.
7. Rossor AM, Polke JM, Houlden H, Reilly MM. Clinical implications of genetic advances in Charcot–Marie–Tooth disease. *Nat Rev Neurol*. 2013;9(10):562–571.
8. Reilly MM, Shy ME. Diagnosis and new treatments in genetic neuropathies. *J Neurol Neurosurg Psychiatr*. 2009;80(12):1304–1314.
9. Berger P, Niemann A, Suter U. Schwann cells and the pathogenesis of inherited motor and sensory neuropathies (Charcot–Marie–Tooth disease). *Glia*. 2006;54(4):243–257.
10. Bucci C, Bakke O, Progida C. Charcot–Marie–Tooth disease and intracellular traffic. *Prog Neurobiol*. 2012;99(3):191–225.
11. Bouhy D, Timmerman V. Animal models and therapeutic prospects for Charcot–Marie–Tooth disease. *Ann Neurol*. 2013;74(3):391–396.
12. Saporta AS, Sottile SL, Miller LJ, Feely SM, Siskind CE, Shy ME. Charcot–Marie–Tooth disease subtypes and genetic testing strategies. *Ann Neurol*. 2011;69(1):22–33.
13. Soliven B. Autoimmune neuropathies: insights from animal models. *J Peripher Nerv Syst*. 2012;17(suppl 2):28–33.
14. Hughes RA, Cornblath DR. Guillain–Barré syndrome. *Lancet*. 2005;366(9497):1653–1666.
15. Nobile-Orazio E, Gallia F, Tuccillo F, Terenghi F. Chronic inflammatory demyelinating polyradiculoneuropathy and multifocal motor neuropathy: treatment update. *Curr Opin Neurol*. 2010;23(5):519–523.
16. Köller H, Kieseier BC, Jander S, Hartung HP. Chronic inflammatory demyelinating polyneuropathy. *N Engl J Med*. 2005;352(13):1343–1356.
17. Pagnoux C, Guillevin L. Peripheral neuropathy in systemic vasculitides. *Curr Opin Rheumatol*. 2005;17(1):41–48.
18. Rudnicki SA, Dalmau J. Paraneoplastic syndromes of the peripheral nerves. *Curr Opin Neurol*. 2005;18(5):598–603.
19. Scollard DM. The biology of nerve injury in leprosy. *Lepr Rev*. 2008;79(3):24253.
20. Morrison B, Chaudhry V. Medication, toxic, and vitamin-related neuropathies. *Continuum (Minneap Minn)*. 2012;18(1):139–160.

CHAPTER

13

Diabetes and Cognitive Dysfunction

Catrina Sims-Robinson, Bhumsoo Kim, Eva L. Feldman

Department of Neurology, University of Michigan, Ann Arbor, Michigan, USA

OUTLINE

DIABETES MELLITUS

Diabetes mellitus is a complex metabolic disorder primarily due to either a lack of insulin production or an inability of cells to respond to insulin, leading to impaired fasting glucose levels (hyperglycemia) and abnormally high hemoglobin A_{1C} levels. According to the American Diabetes Association, approximately 25.8 million people in the USA are affected by diabetes. Furthermore, it is predicted that nearly 79 million adults over the age of 20 have prediabetes, which is associated with an increased risk of developing diabetes. Diabetes is an epidemic not only in the USA but also worldwide, as it is estimated that nearly 347 million adults in the world have diabetes. Diabetes may be the result of genetic abnormalities, drugs, diseases affecting the pancreas, or pregnancy (gestational); however, the exact mechanisms behind most adult cases are not fully understood. The two most common types of diabetes are called type 1 and type 2 diabetes.

Type 1 Diabetes

Type 1 diabetes, which is also known as insulin-dependent diabetes, is associated with hyperglycemia and insulin deficiency. This insulin deficiency is due to the destruction of the β cells of the pancreatic islets of Langerhans, which are responsible for homeostatic insulin production. Thus, patients with type 1 diabetes require exogenous insulin delivered by either regular injections or an implanted pump. Type 1 diabetes accounts for only 5–10% of all the diagnosed cases of diabetes. Although

Neurobiology of Brain Disorders
http://dx.doi.org/10.1016/B978-0-12-398270-4.00013-6

type 1 diabetes can occur at any age, this form of diabetes primarily affects children and adolescents. The cause of type 1 diabetes is not fully understood; however, autoimmunity, genetics, and environmental factors are thought to play a role in the onset of the disease. There is currently no way to prevent type 1 diabetes. Curing type 1 diabetes would ultimately require a means to restore β-cell function; thus, most research in this area focuses on promoting β-cell regeneration or transplantation. Until approaches to prevent or cure the disease become available, management of complications that arise from type 1 diabetes is the key clinical goal and an intense area of research.

Type 2 Diabetes

Type 2 diabetes, also known as insulin-independent diabetes, is the most common form of diabetes. It accounts for approximately 90–95% of all cases of diabetes. Unlike type 1 diabetes, type 2 diabetes is preventable in most cases with proper diet and exercise, and its epidemic growth is largely the result of societal abandonment of healthy living practices. Genetics may also increase an individual's risk for the disease. Individuals with type 2 diabetes have hyperglycemia in addition to insulin resistance, which is defined as a state of reduced responsiveness of target tissue to normal circulating levels of insulin. Besides restoration of proper diet and exercise, pharmacological agents can reduce hyperglycemia and these are standard components of therapy for type 2 diabetes.

An additional risk factor for type 2 diabetes as well as cardiovascular disease is the group of physiological dysfunctions called the "metabolic syndrome". The National Cholesterol Education Program's Adult Treatment Panel III (NCEP/ATP III) has identified six components of the metabolic syndrome: abdominal obesity, atherogenic dyslipidemia, elevated blood pressure, insulin resistance with or without glucose intolerance, proinflammatory state, and prothrombotic state. The diagnosis of metabolic syndrome consists of the presence of abdominal obesity plus any two of the four following additional factors: elevated serum triglycerides, reduced high-density lipoprotein (HDL)-cholesterol, elevated blood pressure, or elevated fasting plasma glucose level. The National Health and Nutrition Examination Survey study reports that metabolic syndrome affects 34% of adults in the USA. The incidence of metabolic syndrome increases with age and obesity.

Obesity is the major contributing factor in prediabetes. Obesity is characterized by having an excess amount of adipose tissue, with a body mass index of 30 or higher. Obesity typically results from the consumption of more calories than the number of calories the body expends; however, other factors that may contribute to obesity are genetics, environment, behavior, and socioeconomic status. Obesity is an epidemic in the USA, with approximately two-thirds of adults and one-third of children and adolescents being classified as either overweight or obese. The economic burden of obesity is estimated to be more than 150 billion USD annually. According to the Diabetes Prevention Program, a patient with prediabetes is likely to develop type 2 diabetes within 10 years unless significant weight loss occurs. Abdominal fat accumulation closely correlates with abnormal insulin signaling and increases the risk of diabetes.

COMPLICATIONS ASSOCIATED WITH DIABETES

The long-term dysregulation of glycemic homeostasis that occurs with both type 1 and type 2 diabetes leads to a variety of complications that can affect the circulatory system, nervous system, and other organ systems, and diabetes even has profound impacts on the CNS. Although there are multiple physiological and biochemical problems that lead to diabetic complications, one unifying mechanism that affects multiple organ systems is damage to the vasculature. Vasculature complications are grouped as either microvascular or macrovascular. Microvascular complications are defined as damage to the small blood vessels including the capillaries, arterioles, and venules, and are associated with retinopathy and neuropathy. Macrovascular complications include damage to large blood vessels such as the arteries, and include atherosclerosis.

Retinopathy

Retinopathy is primarily due to damage of the blood vessels in the retina and is the leading cause of blindness. It is one of the most common microvascular complications associated with diabetes. Both type 1 and type 2 diabetes are associated with retinopathy. Patients with type 1 diabetes may develop evidence of retinopathy within 20 years of diagnosis; however, retinopathy may develop several years before a diagnosis of type 2 diabetes. The mechanisms that underlie diabetic retinopathy have been the subject of extensive research and great strides have been made to reduce complications in treating diabetic patients with retinopathy. A role for the CNS is implemented in the pathogenesis of diabetic retinopathy. An early sign of retinopathy is the loss of pericytes, which are cells of the CNS that associate with capillaries to provide vascular stability. Furthermore, intact cell-to-cell communication among neuronal, glial, microglial, vascular, and pigmented epithelial cells of the retina is vital to normal vision. Thus, one extensive area of research involves providing support to further strengthen or redefine the cellular environment using stem cells. More research is needed to determine both the short-term and long-term benefit of such therapies.

Neuropathy

Neuropathy is probably due to a combination of damage in the blood vessels that support the nerve and metabolic dysfunctions of diabetes such as oxidative stress (discussed below in the context of cognitive impairment). Neuropathy is the leading cause of recurrent infection, non-healing ulcers, and amputations in the toes and feet among diabetic patients; thus, diabetic neuropathy is the most morbid and costly of all diabetic complications. Neuropathy affects nearly 60% of all patients with diabetes, making it one of the most common of all diabetic complications. The onset of diabetic neuropathy correlates with the duration of diabetes. According to the Diabetes Control and Complications Trial research group, chronic hyperglycemia correlates well with the development of neuropathy; thus, strict glycemic control is one of the few currently available treatment options. The involvement of the CNS in neuropathy is now recognized and is the subject of extensive research. Neuropathy is discussed in greater detail in Chapter 12.

Atherosclerosis

Atherosclerosis is a condition that develops when plaque accumulates within the arterial walls and constricts blood flow. Atherosclerosis is the most common cause of cardiovascular (heart) disease and stroke. According to the American Heart Association, adults with diabetes are up to four times more likely to have cardiovascular disease or a stroke than non-diabetic individuals. Furthermore, nearly 65% of diabetic patients die from either cardiovascular-associated disease or stroke. The coexistence of other metabolic disorders also contributes to the severity and effects of cardiovascular disease and stroke associated with diabetes. Although lifestyle changes are essential to reducing the effects of atherosclerosis, β-adrenergic receptor blockers, which reduce the effects of the sympathetic nervous system, are one of the most widely prescribed treatments for cardiovascular disease. The sympathetic nervous system, a division of the autonomic nervous system, utilizes the CNS to maintain homeostasis in the heart by accelerating heart rate, increasing cardiac contractility, reducing venous capacitance, and constricting vessel resistance. The sympathetic nervous system plays an important role in contributing to heart failure in patients with cardiovascular disease.

Atherosclerosis may also cause stroke, which occurs when tissue within the brain is damaged owing to a disruption of blood flow such as blockage in a blood vessel. A stroke may lead to life-changing debilitating complications such as paralysis, cognitive dysfunction, and emotional instability. Stroke is discussed in further detail in Chapter 22.

Cognitive Impairment

The correlation between diabetes and cognition has been gaining more attention over the past decade. Cognition involves a group of mental functions including attention, memory, learning, verbal recall, concentration, reasoning, problem solving, and decision making. Deficits in aspects of cognition, such as memory and attention, have been documented in both type 1 and type 2 diabetes (Table 13.1).[1] The age of onset, degree of glycemic control, and duration of diabetes affect the severity of cognitive deficits in patients with type 1 diabetes.[2] Mild cognitive impairment, a disorder that involves a decline in cognition, usually does not significantly impede daily independent function. Mild cognitive impairment does not meet diagnostic criteria for dementia but is a risk factor for dementia. Mild cognitive impairment is thought to either be an early indicator of dementia or an increased risk factor for developing dementia later in life.

Dementia is a broad term that describes an array of diseases and conditions as a consequence of neuronal death or damage. Several disorders or conditions fall under the term dementia and many forms are unrelated to diabetes. Diabetes is, however, associated with the prevalence of vascular dementia and Alzheimer disease (AD). Vascular dementia is the second most common cause of dementia, accounting for nearly 20% of all cases. Vascular dementia is caused by brain damage from cerebrovascular disease (discussed in Chapter 22) or cardiovascular disease. Although stroke is the most common cause, vascular dementia may also be attributed to amyloid protein accumulation in cerebral blood vessels and in many cases coexists with AD. AD, which is discussed in further detail in Chapter 21, is the most

TABLE 13.1 Cognitive Deficits Associated with Diabetes

Type 1 Diabetes Cognitive Deficits	Type 2 Diabetes Cognitive Deficits
Slowing of information processing	Psychomotor speed
Psychomotor efficiency	Frontal lobe/executive function
Attention	Attention
Visuoconstruction	Verbal fluency
Memory	Memory
Visual–motor skills	Processing speed
Visual–spatial skills	Complex psychomotor function
Motor speed	
Vocabulary	
General intelligence	
Motor strength	

Source: Adapted from Sims-Robinson et al. Nat Rev Neurol *2010;6(10):551–559.*[1]

TABLE 13.2 Diabetes and Risk of Developing Alzheimer Disease

Reference	Patients with Diabetes/ Total Number of Patients	Risk Ratio (95% CI)
Ott *et al.* (1999)[3]	692/6370	1.9 (1.2–3.1)
Brayne *et al.* (1998)[4]	25/376[a]	OR 1.4 (1.1–17.0)
Yoshitake *et al.* (1995)[5]	70/828	2.2 (1.0–4.9)
Peila *et al.* (2002)[6]	900/2574[a]	1.7 (1.0–2.8)
MacKnight *et al.* (2002)[7]	503/5574[a]	1.2 (0.8–1.8)
Xu *et al.* (2004)[8]	114/1301	HR 1.3 (0.8–1.9)
Leibson *et al.* (1997)[9]	1455/75,000[a]	SMR 1.6 (1.3–2.0)
Luchsinger *et al.* (2005)[10]	231/1138[a]	HR 2.4 (1.8–3.2)
Arvanitakis *et al.* (2004)[11]	27/824[a]	HR 1.7 (1.1–2.5)

Patients with type 2 diabetes have a greater than two-fold increased risk of developing Alzheimer disease compared with individuals without this type of diabetes. Data represent number of patients at baseline or [a]at follow-up.
CI: confidence interval; HR: hazard ratio; OR: odds ratio; SMR: standardized morbidity ratio.

common form of dementia, accounting for 50–70% of all cases. AD is characterized by cognitive deficits, and by neuropathological markers including extracellular β-amyloid (Aβ)-containing senile plaques and intracellular neurofibrillary tangles.

Although some studies have failed to support a link between diabetes and AD, multiple population-based studies have reported that patients with diabetes have an increased risk of developing AD compared with age- and gender-matched non-diabetic people.[1] Analysis of nine high-quality studies demonstrates that individuals with type 2 diabetes have a more than two-fold increased risk of developing AD compared with individuals without type 2 diabetes (Table 13.2).[12] In all of the studies that indicate a positive association between the two diseases, obesity, dyslipidemia, and high blood pressure have all been identified as potential risk factors for both diabetes and AD.

UNDERLYING MECHANISMS LINKING DIABETES AND ALZHEIMER DISEASE

The potential mechanisms underlying diabetic complications include impaired insulin signaling, abnormal glucose metabolism, mitochondrial dysfunction, and impaired cholesterol metabolism. According to the Centers for Disease Control and Prevention, cognitive impairment is a growing challenge to the USA because of the economic burden and the efforts required to care for patients. Given the impact of both diabetes and cognitive impairment on the economy and the expected rise in diabetes over the next decade, it is imperative to understand the mechanisms that link these two conditions. Thus, the remainder of this chapter will focus on the potential mechanisms linking diabetes and cognitive impairment, with a focus on AD.

Normal Insulin Signaling

Insulin is a major anabolic hormone and is responsible for maintaining glucose homeostasis. Insulin facilitates the uptake and transport of glucose into tissues for use in energy metabolism. The major sites of glucose transport in the periphery in humans are skeletal muscle and adipocytes. Insulin facilitates this transport by binding to the insulin receptor. The insulin receptor is a tetrameric glycoprotein composed of two subunits, α and β. Insulin binds to the extracellular α-subunit of the insulin receptor, which results in the autophosphorylation and subsequent activation of the intracellular β-subunit (Fig. 13.1).[13] Once activated, the insulin receptor phosphorylates several intracellular substrates, including insulin receptor substrate family members (IRS1–4) and Shc. Phosphorylated tyrosine residues on IRS then serve as docking sites for the p85 subunit of phosphatidylinositol 3-kinase (PI3-K). Activated PI3-K phosphorylates phosphatidylinositol 4,5-bisphosphate (PIP2) to generate phosphatidylinositol 3,4,5-trisphosphate (PIP3). Increased PIP3 stimulates phosphoinositide-dependent protein kinase (PDK), resulting in the activation of Akt (also known as protein kinase B) and atypical protein kinase C. Akt plays important roles in glycogen, lipid, and protein synthesis, cell survival, and the anti-inflammatory response.[14,15] Phosphorylation of Shc by the insulin receptor activates another signaling pathway, the mitogen-activated protein kinase (MAPK) branch of insulin signaling.[14,16] Phosphorylation of Shc attracts growth factor receptor-binding protein-2 (Grb-2), which recruits and activates the GTP-exchange factor SOS and small GTP protein Ras. Activated Ras triggers the activation of a sequence of kinases, ultimately leading to the full activation of MAPKs. MAPK pathway activation by insulin signaling is responsible for gene expression, cell growth, and mitogenesis.

Insulin Signaling in the CNS

In the periphery, insulin plays a key role in regulating glucose metabolism. Unlike peripheral muscle and adipose tissue, CNS neurons are not dependent on insulin for the uptake of glucose, but they do respond in other ways to insulin. The source of insulin in the CNS has

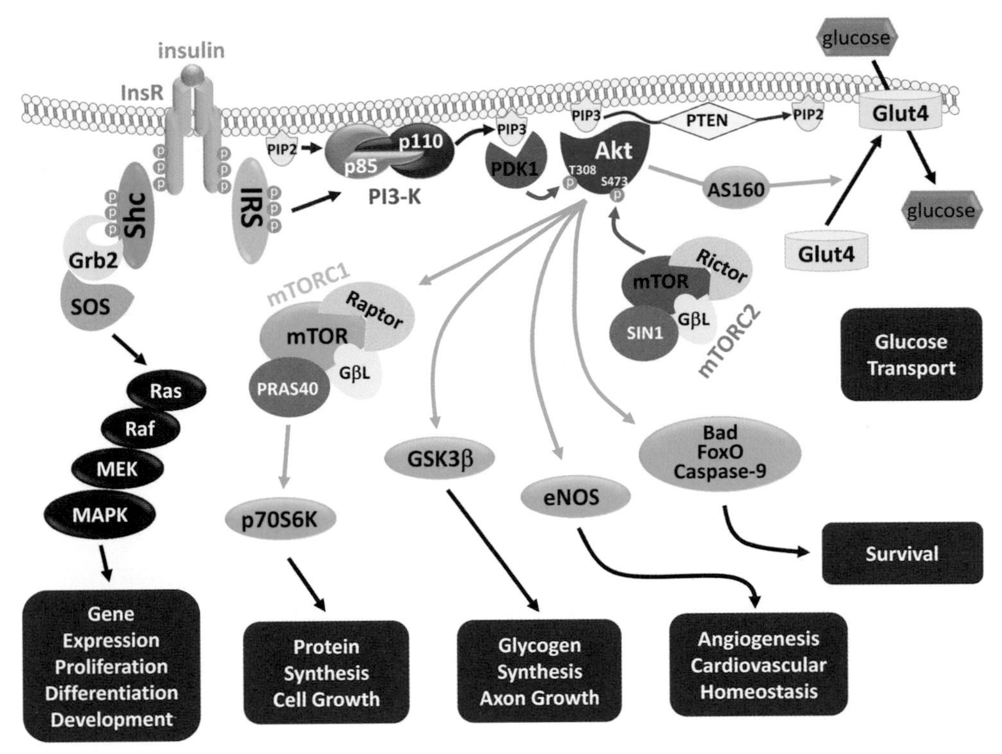

FIGURE 13.1 **Insulin signaling pathways.** The binding of insulin to the insulin receptor (InsR) stimulates the autophosphorylation of the receptor and recruits docking proteins Shc and insulin receptor substrate (IRS). IRS phosphorylation stimulates phosphoinositide 3-kinase (PI3-K) activation, which leads to the generation of phosphatidylinositol-3,4,5-triphosphate (PIP3) by phosphorylating phosphatidylinositol-4,5-bisphosphate (PIP2). PDK1 and Akt are recruited to the plasma membrane by binding to PIP3 through their PH domains. PDK1 phosphorylates a threonine residue in the catalytic domain of Akt. Full activation of Akt requires a second phosphorylation of a serine residue at the hydrophobic motif by the mTORC2 complex. Alternatively, phosphorylated Shc recruits Grb2/SOS, which stimulates the mitogen-activated protein kinase (MAPK) signaling pathway by sequential activation of Ras, Raf, MEK, and MAPK. Substrates activated by MAPK and PI3-K/Akt pathways mediate various downstream biological responses of insulin including cell survival and glucose metabolism. *Source: Adapted from Kim and Feldman.* Trends Endocrinol Metab *2012;23(3):133–141.*[13]

been a topic of debate. Although some evidence supports the idea of *de novo* insulin synthesis within the CNS, the majority of insulin in the adult CNS is produced by pancreatic β cells and transported across the blood–brain barrier. Thus, the amount of insulin in the CNS is a direct reflection of the amount of insulin circulating in the periphery. Since insulin deficiency occurs during type 1 diabetes, it is likely that the level of insulin in the CNS is also deficient. However, type 2 diabetes is associated with an initial acute increase in insulin in the periphery (hyperinsulinemia) and a subsequent chronic hyperinsulinemia. Acute hyperinsulinemia leads to an increase in insulin levels in the CNS; however, chronic hyperinsulinemia is associated with a decrease in insulin levels since insulin crosses the blood–brain barrier by a saturable transport process.[17] Further research is warranted to understand the relationship between diabetes and insulin signaling in the periphery and in the CNS.

As in the periphery, insulin in the CNS rapidly binds to the insulin receptor to promote signaling. The insulin receptor and its downstream signaling molecules have been detected throughout CNS neurons (Fig. 13.2) including the olfactory bulb, cerebral cortex, hippocampus, hypothalamus, and amygdala. The insulin receptor is concentrated in the cell body and synapses of neurons. The presence of the insulin receptor at the synapse suggests a role for insulin in the regulation of synaptic activity and in learning and memory. The lack of insulin in the CNS associated with insulin deficiency in type 1 diabetes, and chronic hyperinsulinemia in type 2 diabetes leads to a decrease in insulin signaling, which impairs learning and memory.[18] The next section will explore the abnormal insulin signaling that occurs during diabetes and how this may contribute to AD pathology.

Abnormal Insulin Signaling and Resistance

Insulin signaling within the CNS is impaired in both type 1 and type 2 diabetes. Insulin deficiency is associated with the cognitive deficits observed in type 1 diabetes. For example, insulin could prevent changes in cognition, including spatial learning and long-term potentiation, in rats induced with type 1 diabetes (Fig. 13.3).[2,19] In contrast, chronic hyperinsulinemia and insulin resistance in type 2 diabetes lead to impaired insulin signaling and contribute to cognitive impairment associated with type 2 diabetes.

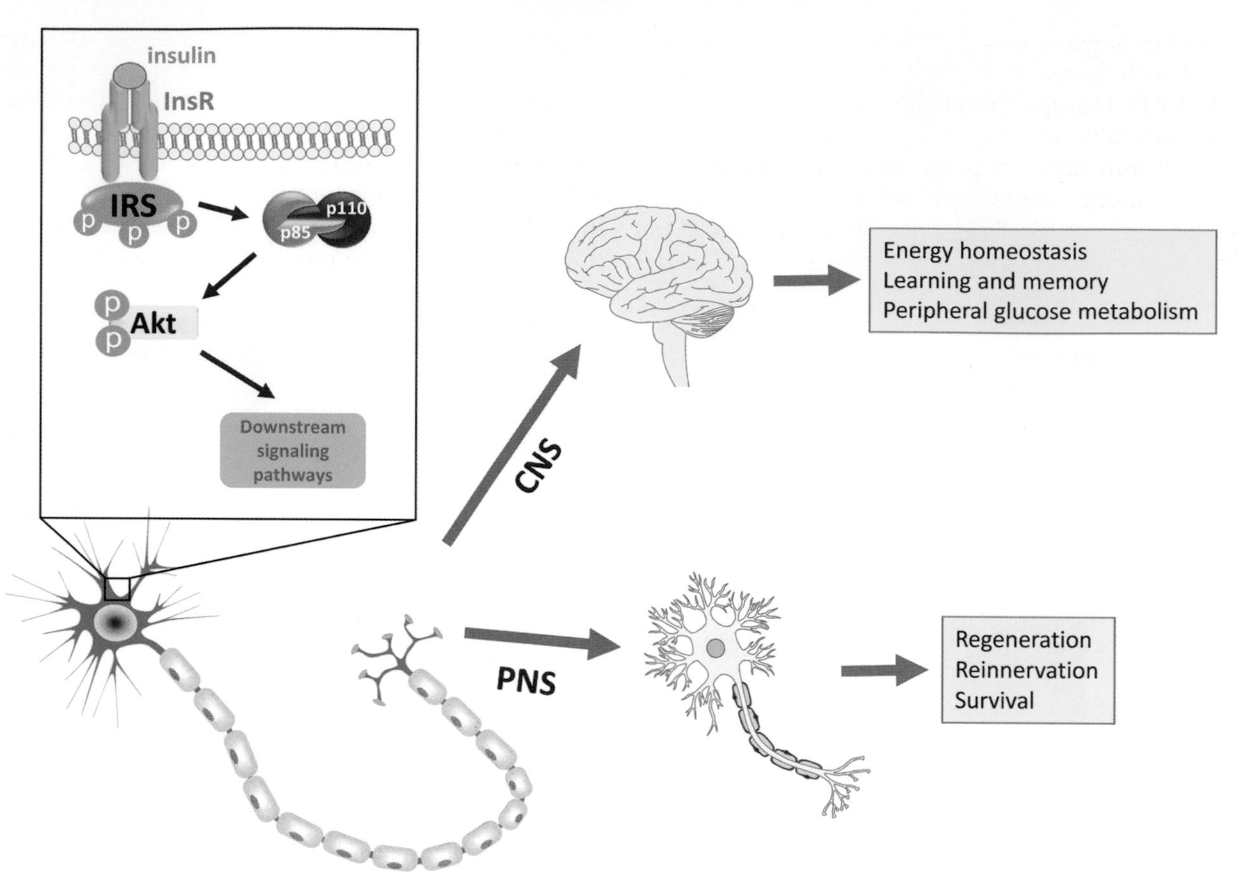

FIGURE 13.2 **Actions of insulin on the central and peripheral nervous system.** Even though neurons are generally not considered insulin dependent, they are responsive to insulin. Insulin receptors (InR) are widely distributed in both the central and peripheral nervous system. Insulin signaling plays important roles in various aspects of neuronal functions, as indicated in the figure. *Source: Adapted from Kim and Feldman.* Trends Endocrinol Metab *2012;23(3):133–141.*[13]

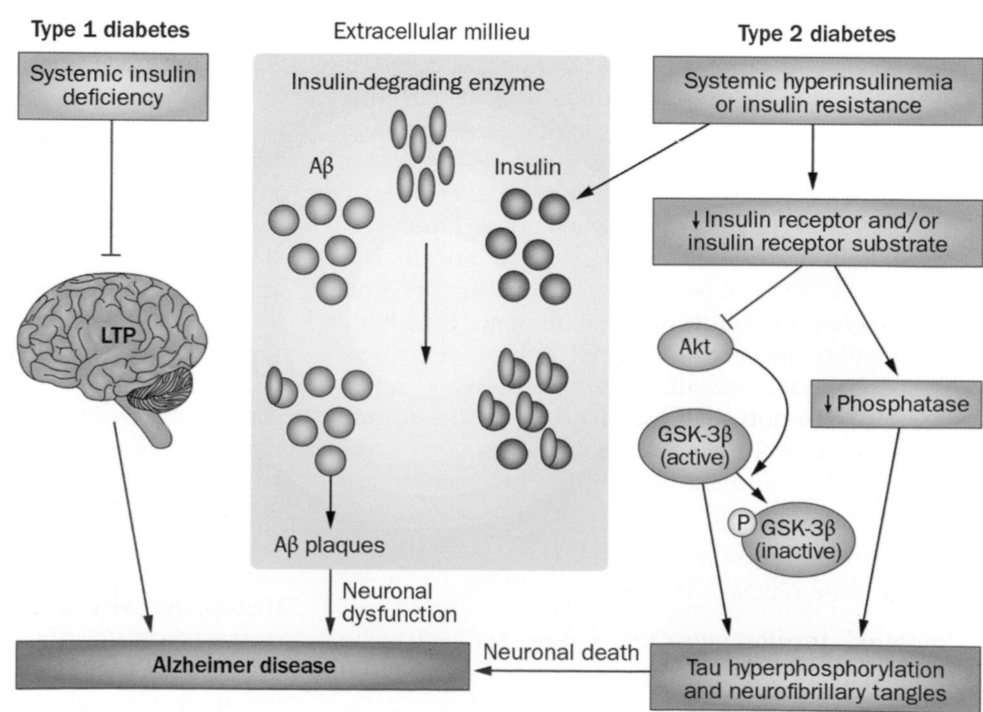

FIGURE 13.3 **Altered insulin signaling in diabetes may contribute to Alzheimer disease pathophysiology.** In type 1 diabetes, insulin deficiency attenuates long-term potentiation and may lead to deficits in spatial learning and memory. In type 2 diabetes, insulin resistance leads to both amyloid-β (Aβ) plaque formation and tau hyperphosphorylation. During hyperinsulinemia, insulin and Aβ compete for insulin-degrading enzyme, leading to Aβ accumulation and plaque formation. A decrease in insulin receptor signaling leads to inhibition of Akt and dephosphorylation (activation) of glycogen synthase kinase-3β (GSK-3β), and results in tau hyperphosphorylation. P: phosphate. *Source: Adapted from Sims-Robinson* et al. Nat Rev Neurol *2010;6(10):551–559.*[1]

These studies suggest that defective insulin signaling is associated with decreased cognitive ability and the development of AD. Disruption of insulin signaling makes the neurons vulnerable to metabolic stress, thus accelerating neuronal dysfunction. Intranasal insulin administration improved working memory in both human and animal studies (see below). Furthermore, the intrahippocampal delivery of insulin, but not insulin-like growth factor-1 (IGF-1), improved hippocampal-dependent spatial working memory. Insulin receptor messenger RNA (mRNA) and protein levels are upregulated in the hippocampus in association with short-term memory formation after a spatial learning experience. The poor cognitive performance associated with diabetes or AD may be due to insulin resistance or decreased insulin receptor signaling. On the other hand, neuronal insulin receptor knockout mice, despite complete loss of neuronal insulin signaling, demonstrate little alteration in spatial learning. Brain-specific IRS-2 knockout mice have enhanced hippocampal spatial memory, suggesting that IRS-2 acts as a negative regulator of memory formation. Therefore, impairment in insulin signaling in the brain may have different effects on cognition depending on other factors that are not yet understood.

In both type 1 and type 2 diabetes, the lack of insulin also affects downstream insulin signaling. The phosphorylation of IRS proteins on tyrosine residues activates insulin signaling and stimulates glucose transport through the downstream activation of PI3-K. The PI3-K-dependent signaling pathway, which is critical for the metabolic effects of insulin, is usually affected in people with diabetes. IRS proteins are abnormally phosphorylated at serine residues during diabetes. In general, the serine phosphorylation of the insulin receptor inhibits insulin signaling (even though it can act as a positive regulator in certain situations). Increased serine phosphorylation of IRS prevents its interaction with the insulin receptor and downstream signaling molecules, induces mislocalization of IRS, and enhances insulin receptor degradation by the ubiquitin–proteasome-mediated pathway. These changes prevent IRS–PI3-K association and impair insulin signaling. A reduction in the expression of IRS and activation of PI3-K are evident in genetically obese and high-fat fed animals, both in peripheral neurons (dorsal root ganglion cells) and in the brain. PI3-K activates Akt, which affects insulin signaling and metabolism. Alterations in Akt activity are evident during diabetes and insulin resistance. Phosphorylated MAPK, which is increased in patients with type 2 diabetes, can phosphorylate IRS1 at specific serine residues.[13]

Contribution of Impaired Insulin Signaling to Alzheimer Disease Pathology

Impaired insulin signaling affects the expression and metabolism of both Aβ and tau. The insulin deficiency associated with type 1 diabetes and the chronic hyperinsulinemia associated with type 2 diabetes lead to a decrease in insulin transport and impair insulin signaling. Insulin signaling impairment correlates with Aβ accumulation.[20] Intranasal administration of insulin reverses Aβ accumulation. Although the underlying mechanism is not completely understood, it is likely that this effect is due to the ability of insulin to reduce the expression of the amyloid precursor protein (APP), presenilin-1 (PS1), and presenilin-2 (PS2), all of which participate in the amyloidogenic pathway (see Chapter 21 for more information on amyloid processing). Intranasal insulin, which bypasses the blood–brain barrier, is a promising therapy for AD and perhaps for other neurodegenerative diseases. Intranasal insulin does not alter the levels of glucose and insulin in the periphery. While clinical trials thus far have been promising, more research is necessary to reveal the long-term effects of this therapy *in vivo*. Furthermore, additional studies are required to determine whether the delivery of insulin to the CNS in people with diabetes will delay or prevent the progression of AD, particularly with regard to Aβ metabolism.

In one study intranasal insulin was effective in improving memory only at a low dose of insulin, while a high dose of insulin did not have this effect.[21] The levels of insulin administered in the CNS are important, because hyperinsulinemia can lead to the accumulation of Aβ (Fig. 13.3). Insulin and Aβ compete for the insulin-degrading enzyme in neurons and microglia. The higher affinity of insulin for the insulin-degrading enzyme ultimately leads to Aβ accumulation. Thus, acute hyperinsulinemia, which also leads to an increase in insulin transport into the brain, leads to Aβ accumulation. Furthermore, treating mice with AD-like symptoms with rosiglitazone, an insulin sensitizer, reduces Aβ levels and improves memory[22]; however, a larger study is needed to evaluate the potential of insulin sensitizer treatment in humans to improve cognition and reduce Aβ accumulation. Thus, insulin sensitizers may provide a beneficial effect by increasing insulin signaling and decreasing the levels of insulin available to compete with Aβ for degradation by the insulin-degrading enzyme.

Insulin also regulates tau phosphorylation and may play a role in the development of neurofibrillary tangles.[1] Insulin receptor signaling leads to the activation of two major signaling pathways: MAPK and AKT signal transduction pathways. MAPK signaling is a required component of cell differentiation, cell proliferation, and cell death, whereas AKT signaling is involved in the regulation of cell growth, cell proliferation, protein synthesis, and cell survival. MAPK is increased in AD and localizes with Aβ plaques and neurofibrillary tangles. Specifically, MAPK colocalizes with aggregated tau in mouse models of AD.

One of the key signaling molecules activated downstream of Akt is glycogen synthase kinase-3 (GSK-3) αβ.

GSK-3α increases Aβ production by stimulating amyloid precursor protein (APP) γ-secretase activity, while GSK-3β is the major tau kinase responsible for its hyperphosphorylation and formation of neurofibrillary tangles.[23,24] Impaired insulin signaling activates GSK3αβ by reducing its phosphorylation, which affects both Aβ and tau neuropathology. By contrast, inhibition of GSK-3β attenuates APP processing and inhibits hyperphosphorylated tau-associated neurodegeneration (Fig. 13.3). The regulation of GSK-3β in the hippocampus and cortex changes in response to changes in glucose and insulin concentrations. In type 2 diabetes increased GSK-3β activity reduces glucose clearance and may contribute to insulin resistance.

Increased tau phosphorylation, the major component of neurofibrillary tangles, is evident in mouse models of either type 1 or type 2 diabetes. This change in the phosphorylation status of tau in mice with type 1 diabetes positively correlates with impaired learning and memory deficits.[25,26] Tau phosphorylation and cleavage is considerably more pronounced in mice with type 2 diabetes than in mice with type 1 diabetes.[25] Considering that increased tau phosphorylation is also observed in postmortem brain samples from patients with type 2 diabetes, patients with this form of the disease may also experience cognitive deficits as a result of dysfunctional glucose metabolism.

Consequences of Abnormal Glucose Metabolism and Oxidative Stress

Since glucose is the sole source of energy for the brain, its precise regulation and metabolism is required for proper CNS function; thus, hyperglycemia can lead to a number of pathophysiological processes, including oxidative stress and the formation of advanced glycation endproducts (AGEs), which can cause brain damage. Insulin signaling and normal glucose metabolism are vital for normal brain function. Reduced glucose metabolism has been observed in the cortex and hippocampus of patients with AD. Patients may also have increased fasting plasma insulin levels, or a decreased ratio of cerebrospinal fluid (CSF) to plasma insulin. Intravenous administration of either insulin while maintaining normal blood glucose levels or glucose alone facilitates cognitive functions in patients with AD as well as in healthy older adults.[27] Thus, normal glucose metabolism is required for the performance of cognitive functions and data suggest that impairments in glucose metabolism may contribute to cognitive dysfunction. The adverse impact of impaired glucose metabolism on cognitive functioning may be caused, in part, by the formation of AGE, an increase in oxidative stress, and subsequent local inflammation in the brain.[1]

Abnormal glucose metabolism increases the production of free radicals such as reactive oxygen species (ROS) and reactive nitrogen species (RNS). This overproduction of free radicals exhausts the antioxidant capacity of the cell and results in oxidative stress, which is a hallmark of both type 1 and type 2 diabetes. Oxidative stress is increased not only in patients with diabetes but also in AD patients compared with healthy controls. Furthermore, animal models of AD also exhibit a decrease in antioxidant capacity and an elevation in the levels of oxidative stress before the development of Aβ plaques and neurofibrillary tangles. The increase in the levels of Aβ associated with oxidative stress may be due to an increase in APP levels or by modulating APP processing. On the other hand, studies in postmortem brain samples from patients with AD and in transgenic mice suggest that Aβ production contributes to the increased levels of oxidative stress.

Abnormal glucose metabolism and oxidative stress both contribute to the formation of AGEs. AGEs are a heterogeneous group of molecules formed by irreversible, non-enzymic reactions between sugars and the free amino groups of proteins, lipids, and nucleic acids. Auto-oxidation of glucose leads to the formation of hydroxyl radicals. Hydroxyl radicals are intermediates in the AGE pathway and the main source of endogenous AGE (Fig. 13.4). Although the formation and accumulation of AGEs occur during normal aging, the process is exacerbated in patients with diabetes, and the binding of AGE to its receptor induces a series of biological processes that cause further diabetic complications. Both Aβ plaques and neurofibrillary tangles are positive for AGE immunoreactivity. Furthermore, hippocampal neurons from patients with this neurodegenerative disease have also been shown to contain Aβ, AGEs, and receptors for AGE-positive granules.[28,29] Whether AGE modification of Aβ and tau protein is a primary or secondary event in AD is controversial; however, the participation of AGEs in the progression of AD is recognized and accepted. AGE-induced glycation of Aβ and tau protein leads to Aβ aggregation and the formation of neurofibrillary tangles, respectively.[30] An increase in receptors for AGE expression in neurons and glia is exhibited in diabetic mice with cognitive impairments compared with control mice.[31] Although an increase in AGE staining in postmortem brain slices was evident in patients with AD and diabetes,[32] another study failed to detect a difference in AGE staining in neurofibrillary tangles and senile plaques between patients with AD and patients with AD and diabetes.[33] Thus, more work is necessary to understand the role of AGEs in AD pathology.

Oxidative stress is also responsible for eliciting an inflammatory response (Fig. 13.4). The inflammatory response is characterized by a local cytokine-mediated acute phase, activation of the complement cascade, and

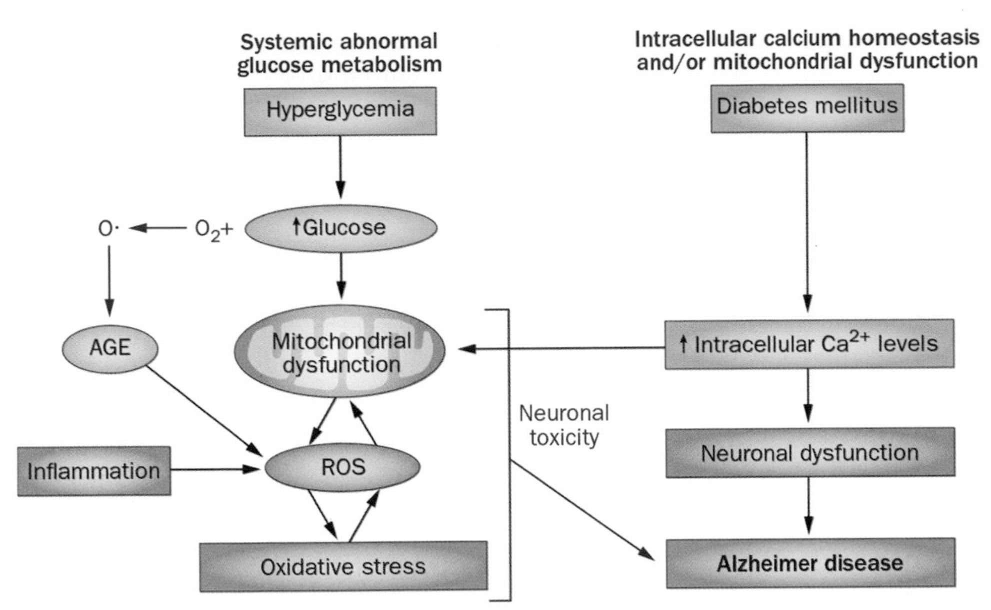

FIGURE 13.4 **Pathological mechanisms associated with diabetes may cause Alzheimer disease (AD).** Mitochondrial dysfunction, oxidative stress, and dysregulated calcium homeostasis are all associated with diabetes and may be contributory factors to the development of AD. Glucose auto-oxidation can lead to advanced glycation endproduct (AGE) formation and, as a result, oxidative stress, which is associated with mitochondrial dysfunction. Oxidative stress combined with an increase in intracellular calcium result in a feedforward cycle of continued mitochondrial damage that can cause neuronal death and hence contribute to AD pathology. ROS: reactive oxygen species. *Source: Adapted from Sims-Robinson* et al. Nat Rev Neurol *2010;6(10):551–559.*[1]

subsequent cell damage. This response is initiated by activated microglia and reactive astrocytes surrounding extracellular Aβ deposits. Cytokines and chemokines, including interleukins, tumor necrosis factor, and macrophage inflammatory protein-1α, are all upregulated in patients with AD and animal models of AD.[34] Although many epidemiological studies suggest that anti-inflammatory drugs would be beneficial in decreasing the risk of AD, experimental treatment trials in patients with AD failed to provide a beneficial effect. Perhaps the negative results of these trials were due to the low number of subjects or the subjects being too far into the course of their disease to detect any improvements. More studies are required in animal models to determine whether anti-inflammatory drugs can prevent or delay the onset of AD.

Contribution of Lipid Metabolism to Alzheimer Disease

Diabetes is associated with both dyslipidemia and hypercholesterolemia. The lack of glucose utilization for energy metabolism in diabetes leads to an increase in lipolysis, the breakdown of lipids for energy. Typically, there is an increase in the levels of both cholesterol and triglycerides in the periphery due to lipolysis. In the periphery, lipids require carrier molecules known as lipoproteins, such as very low-density lipoprotein (VLDL), low-density lipoprotein (LDL), and HDL. The levels of triglycerides and LDL-cholesterol are elevated in type 1 diabetes, especially without tight glycemic control. Type 2 diabetes is associated with an increase in the levels of triglycerides, VLDL- and LDL-cholesterol, and a decrease in HDL-cholesterol levels.

Hypercholesterolemia is associated with an increased risk of type 2 diabetes. In addition, patients with both hypercholesterolemia and type 2 diabetes have an increased risk of cognitive decline compared with healthy individuals.[35] Hypercholesterolemia is present in 70% of patients diagnosed with diabetes.[36] Patients with type 2 diabetes typically have an accumulation of cholesterol within pancreatic β cells, which leads to a decrease in insulin secretion and β-cell mass, and a worsening of diabetes symptoms.[37] Hypercholesterolemia increases Aβ production by increasing the expression of β-secretase and receptors for AGE, which transport Aβ from the circulation to the brain, and decreasing the expression of insulin-degrading enzyme and LDL receptor-related protein, which play a role in the clearance of Aβ from the brain to circulation.[38] Furthermore, cholesterol, cholesterol oxidase, and apolipoprotein E colocalize with Aβ in fibrillar plaques in transgenic AD mouse models,[39] and cholesterol and oxidized cholesterol (oxysterols) accumulate in the dense core of Aβ plaques,[40] suggesting that cholesterol may play a role in the formation of senile plaques. In the brain, cholesterol is oxidized to 24-hydroxycholesterol (24-HC) by 24-hydroxylase, and 24-HC levels in plasma and CSF reflect neuronal cholesterol synthesis. In addition, 27-HC is one of the main oxysterols found in circulation, and the ratio of these two oxysterols is altered in neurodegenerative disease.[41] The level of 27-HC in the brain is elevated in AD, and may subsequently increase the production of Aβ.[42] Therefore, it has been proposed that drugs that reduce cholesterol and can cross the blood–brain barrier, such as statins, may reduce the risk of AD.[43] However, although initial studies reported that statins may be useful in the prevention of AD, recent studies demonstrated that statins are ineffective in preventing AD. Additional studies are currently underway to determine whether statins are a viable treatment in AD, and more work is needed to fully elucidate the role of cholesterol in AD pathology and progression.

Mitochondrial Dysfunction

Both diabetes and AD are associated with deficits in mitochondrial activity.[1] In animal models of type 1 diabetes, neuronal mitochondria exhibit a decrease in antioxidant capacity due to an increase in oxidative stress. Furthermore, inherited defects in mitochondrial DNA cause an insulin-deficient form of diabetes that resembles type 1 diabetes. The activity of oxidative enzymes within mitochondria is lower in patients with type 2 diabetes. Obesity is associated with smaller mitochondria and reduced antioxidant capacity. Mitochondrial dysfunction triggers neuronal degeneration and cell death; therefore, it is thought to also contribute to AD pathophysiology. Exactly how mitochondrial dysfunction contributes to AD pathology is not fully understood. The APP associates with the outer mitochondrial membrane, and insertion of Aβ protein into the mitochondrial membrane disrupts the mitochondrial electron transport chain. The consequences of this disruption are attenuation of cellular energy production and increased formation of ROS. The electron transport chain may also be negatively affected by the presence of β-secretase and Aβ, which inhibit cytochrome oxidase in the presence of copper, within mitochondria. Cytochrome oxidase is an iron-containing enzyme within mitochondria that plays an important role in cell respiration. Both dysfunction of mitochondrial electron transport proteins and a decrease in cytochrome oxidase activity are associated with AD. Furthermore, neurons affected by AD pathology exhibit an overall decrease in mitochondrial mass, an increase in cytoplasmic mitochondrial DNA, and an increase in cytochrome oxidase.

Mitochondria are essential for ATP synthesis and maintaining calcium homeostasis. The maintenance of calcium homeostasis is necessary for normal neuronal function. The calcium hypothesis states that dysregulation of calcium homeostasis is a central process in brain aging and in age-related diseases. Both type 1 and type 2 diabetes are associated with an increase in intracellular calcium levels (Fig. 13.4).[1] Abnormal calcium homeostasis is common in patients with diabetes and animal models of diabetes. Insulin deficiency is associated with calcium overload. Calcium overload affects metabolic function by activating calcium-dependent protein kinases, phosphatases, proteases, phospholipases, and lysosomal enzymes. The high levels of intracellular calcium in the pancreatic β cells may impair insulin secretion in type 2 diabetes. Dysregulation of calcium homeostasis seems to occur as a result of decreased mitochondrial function, as the concentration of calcium is increased in the brain tissue of patients with AD. The levels of intracellular calcium levels are greater in AD neurons containing neurofibrillary tangles than in neurons from healthy control patients. Furthermore, levels of calcium-dependent proteases are increased in neurons containing neurofibrillary tangles. Transglutaminase, a calcium-activated enzyme that induces cross-linking of tau protein, is also increased in AD.[44] Increased intracellular calcium may also play a role in Aβ metabolism by enhancing APP processing. More research is needed to understand the role of abnormal calcium homeostasis in other cellular organelles, especially the organelles that may play a role in abnormal protein processing such as the endoplasmic reticulum.

ANIMAL MODELS OF DIABETES AND ALZHEIMER DISEASE

Diabetes Models

Although there are numerous animal models of diabetes, none of them fully recapitulates the effect of diabetes in humans. Nonetheless, experimental animal models are valuable tools to study diabetes. Most of the studies are carried out in rodent models. The benefits of using mice or rats are: a relatively short lifespan enabling the study of aging effects; a short gestation period; small size, allowing easy and less costly maintenance; and the large range of resources available for genetic manipulation.

The most widely used model of type 1 diabetes is induction of diabetes with a streptozotocin (STZ) injection. STZ is a cytotoxic agent that destroys the pancreatic β cells. STZ-injected mice or rats display the hallmarks of type 1 diabetes including hyperglycemia, elevated fasting blood glucose, glycated hemoglobin, and weight loss. The Ins.D^{d}1 and $Ins2^{Akita}$ (C57BL/6J X $Ins2^{C96Y}$ and C57BL/6-*Ins2*^{Akita}/J) mouse models of spontaneous type 1 diabetes have a mutation in the insulin 1 and 2 genes, respectively. $Ins2^{Akita}$ mice are hyperglycemic and hypoinsulinemic, and display polydipsia and polyuria as early as 3–4 weeks of age. The diabetic phenotype is more severe and progressive in the male than in the female. Non-obese diabetes (NOD) mice and Bio Breeding (BB) rats develop type 1 diabetes owing to the autoimmune destruction of pancreatic β cells, and display severe hyperglycemia, hypoinsulinemia, and ketosis.

The db/db mice (BKS.Cg-*m*+/+*Lepr*^{db}/J and B6.Cg-*m Lepr*^{db}/++/J) are spontaneous type 2 diabetes models resulting from the autosomal recessive mutation in the leptin receptor. These mice display hyperphagia, severe obesity, hyperinsulinemia, and hyperglycemia as early as 4 weeks of age. A similar diabetic phenotype is seen in B6-ob/ob (B6.V-*Lep*^{ob}/J) mice, which results from the lack of leptin. B6-ob/ob mice display hyperphagia, hyperglycemia, glucose intolerance, and elevated plasma insulin.

Both db/db and ob/ob mice become rapidly obese, reaching up to three times the normal weight of their wild-type littermate controls. Feeding wild-type animals a high-fat diet (45–60% of calories derived from fat) is often used to induce type 2 diabetes. Under these conditions, animals develop similar physiology to human obesity including hyperglycemia, insulin insensitivity, hypercholesterolemia, hypertriglyceridemia, and elevated plasma leptin. There is much variability in the response to a high-fat diet, depending on animal strains and the choice of foods.

Alzheimer Disease Models

The first animal models of AD were developed around the turn of the twenty-first century. Initially, the animal models of AD exhibited only one of the classic pathological features of AD, either Aβ-containing plaques or neurofibrillary tangles. However, models are now available that display both plaques and neurofibrillary tangles. Several reviews are available detailing some of the commonly used mouse models of AD.[45–47] Despite over a decade of intense research, all the currently used mouse models are based on the rare familial forms of AD, and mouse models for sporadic (late-onset) AD are still lacking. Nevertheless, the available mouse models have provided tremendous advances in our knowledge of the pathophysiology of AD. This section will highlight a few of the currently available animal models of AD.

There are several mouse models that display the Aβ pathology. The transgenic mouse models often differ by the type of mutation, including the APP_{V717F} (APP Indiana mutation); Tg2576, APP23, and C3-3 mouse lines expressing $APP_{K670N/M671L}$ (Swedish double mutation) with different promoters; TgCRND8 and J20 mice ($APP_{K670N/M671L}+APP_{V717F}$); and hAPP-Arc mice ($APP_{K670N/M671L}+APP_{V717F}+APP_{E22G}$). These mice invariably develop age-dependent amyloid plaques due to increased production of Aβ peptides, especially Aβ42, and also display cognitive impairment. Even though expression of the wild-type or mutant (M146L or M146V) presenelin gene (*PSEN1*) does not cause detectable amyloid pathology or memory deficit, when expressed in a mutant APP background, it accelerates amyloid deposition. Thus, the APP/PS1 mice were developed by crossing Tg2576 mice ($APP_{K670N/M671L}$) with mice expressing human $PS1_{M46L}$. In the APP/PS1 mouse model, Aβ plaques are visible at 6 months of age compared with 9 months in Tg2576 mice. Another mouse model, the 5XFAD mouse ($APP_{K670N/M671L+I716V+V717I}+PS1_{M146L+L286V}$), displays the Aβ pathology as early as 1.5 months of age. Despite extensive amyloid deposition in most of the transgenic mice with APP and PS1/2 mutations, these mice do not display the global neuronal loss observed in human AD brains. Several lines of these transgenic mice, however, display progressive loss of synapses and dendrites around amyloid plaques. Another prominent difference between AD transgenic mouse models and human AD is the almost complete absence of neurofibrillary tangles.

Neurofibrillary tangles containing hyperphosphorylated tau are one of the key neuropathological hallmarks of AD. Tau dysfunction can cause neurodegeneration and AD-like dementia in the absence of Aβ pathology. Furthermore, reducing endogenous tau levels by crossing human APP mice with Tau knockout mice ($Tau^{-/-}$) prevents cognitive impairment despite the presence of high levels of Aβ.[48] The first tau transgenic models were generated by expressing wild-type four-repeat (4R) and three-repeat (3R) human tau. These mice display tau hyperphosphorylation present in dystrophic neurites, neurodegeneration in the brainstem and spinal cord, astrocyte activation and astrogliosis in the cortex, and progressive motor weakness. However, wild-type tau transgenic mice do not develop true neurofibrillary tangles (only pretangle formation is evident) or noticeable neuronal loss despite high tau expression, and do not display cognitive impairment. Even though no tau mutation is identified in AD patients, the discovery of tau mutations in frontotemporal degeneration with Parkinsonism linked to chromosome 17 prompted scientists to develop transgenic mice expressing tau mutants. The models include JPNL3, TauV337M, TauR406W, TauP301S, and TauVLM (G272V + P301L + R406W). These mice develop neurofibrillary tangle accumulation at various ages along with neurodegeneration. However, because of the moderate to severe motor weakness and paralysis that occur in these models, comprehensive cognitive testing has not been possible in most of the strains. The rTg (tau_{P301L}) 4510 mice expressing P301L tau display age-dependent neurofibrillary tangle formation starting as early as 2.5 months of age and cognitive impairment as early as 4 months of age.

Multigenic AD models have been created by crossing two or three transgenic mouse models. TAPP mice were generated by crossing Tg2576 (expressing $APP_{K670N/M671L}$) with JPNL3 ($MAPT_{P301L}$) mice. TAPP mice display similar Aβ pathology to the parental Tg2576 mice; however, they also develop enhanced neurofibrillary tangle pathology along with neuronal loss. Similarly, injection of Aβ42 into the brain of P301L tau mutant mice, but not wild-type tau transgenic mice, resulted in a five-fold increase in neurofibrillary tangles. Furthermore, triple transgenic (3xTg-AD) mice harboring APPswe, $PSEN1_{M146V}$, and $MAPT_{P301L}$ display both amyloid and tau pathology. These mice develop age-dependent synaptic dysfunction and cognitive impairment that correlates with the accumulation of intraneuronal Aβ before the appearance of plaque or tangle formation.

Concomitant Diabetes and Alzheimer Disease in Animal Models

Increased tau phosphorylation, the major component of neurofibrillary tangles, is evident in mouse models of either type 1 or type 2 diabetes compared with control animals.[25] This change in the phosphorylation status of tau in mice with type 1 diabetes positively correlates with impaired learning and memory deficits. Tau phosphorylation and cleavage are considerably more pronounced in mice with type 2 diabetes than in mice with type 1 diabetes. Several studies demonstrate that the induction of diabetes in mouse models of AD leads to an acceleration of AD neuropathology. For example, the induction of type 1 diabetes by means of an intraperitoneal injection of streptozotocin in transgenic mice prone to tau pathology is associated with an increase in hyperphosphorylated tau and elevated levels of insoluble tau, compared with diabetic non-transgenic mice.[49] Results from a study conducted by Jolivalt and colleagues indicate that insulin deficiency enhances AD pathology in transgenic mice with type 1 diabetes. Increases in $A\beta_{42}$, immunoreactive Aβ plaques, GSK-3β activation, and tau phosphorylation, which all contribute to neurodegeneration and neuronal loss, were evident in the mice after the induction of experimental diabetes.[50] By contrast, induction of type 2 diabetes does not increase levels of Aβ in the brains of transgenic mice that develop AD pathology, but it does cause early onset of cognitive impairment.[51] More studies are needed to fully elucidate the link between diabetes and AD.

CONCLUSION

Diabetes is a complex disorder that leads to a vast array of complications including retinopathy, neuropathy, atherosclerosis, and cognitive impairment. This chapter has provided an overview of CNS involvement in diabetes-related complications. Impaired insulin signaling, abnormal glucose metabolism, mitochondrial dysfunction, and impaired cholesterol metabolism are the major mechanisms under investigation for tying diabetes to cognitive impairment, and these mechanisms are likely to be involved in other diabetic complications. Improving understanding of the contributions of these mechanisms to cognitive impairment and other diabetic complications is a key focus of diabetes research and a prerequisite for the development of targeted therapies. Model systems to study diabetic complications have dramatically improved and provide an excellent platform for exploring the role of the mechanisms outlined in this chapter in diabetic complications.

Acknowledgments

The authors acknowledge Judith Bentley for editing and formatting this chapter. This work was supported by grants from the National Institute of Diabetes and Digestive and Kidney Diseases (R24-DK082841), National Institute of Aging (NIA T32-AG000114), Animal Models of Diabetic Complications Consortium (NIH U01-DK076160), the A. Alfred Taubman Medical Research Institute, and the Program for Neurology Research and Discovery.

References

1. Sims-Robinson C, Kim B, Rosko A, Feldman EL. How does diabetes accelerate Alzheimer disease pathology? *Nat Rev Neurol*. 2010;6(10):551–559.
2. Brands AM, Biessels GJ, de Haan EH, Kappelle LJ, Kessels RP. The effects of type 1 diabetes on cognitive performance: a meta-analysis. *Diabetes Care*. 2005;28(3):726–735.
3. Ott A, Stolk RP, van Harskamp F, Pols HA, Hofman A, Breteler MM. Diabetes mellitus and the risk of dementia: the Rotterdam Study. *Neurology*. 1999;53:1937–1942.
4. Brayne C, Gill C, Huppert FA, et al. Vascular risks and incident dementia: results from a cohort study of the very old. *Dement Geriatr Cogn Disord*. 1998;9:175–180.
5. Yoshitake T, Kiyohara Y, Kato I, et al. Incidence and risk factors of vascular dementia and Alzheimer's disease in a defined elderly Japanese population: the Hisayama Study. *Neurology*. 1995;45:1161–1168.
6. Peila R, Rodriguez BL, Launer LJ. Type 2 diabetes, APOE gene, and the risk for dementia and related pathologies: the Honolulu-Asia Aging Study. *Diabetes*. 2002;51:1256–1262.
7. MacKnight C, Rockwood K, Awalt E, McDowell I. Diabetes mellitus and the risk of dementia, Alzheimer's disease and vascular cognitive impairment in the Canadian Study of Health and Aging. *Dement Geriatr Cogn Disord*. 2002;14:77–83.
8. Xu WL, Qiu CX, Wahlin A, Winblad B, Fratiglioni L. Diabetes mellitus and risk of dementia in the Kungsholmen project: a 6-year follow-up study. *Neurology*. 2004;63:1181–1186.
9. Leibson CL, Rocca WA, Hanson VA, et al. Risk of dementia among persons with diabetes mellitus: a population-based cohort study. *Am J Epidemiol*. 1997;145:301–308.
10. Luchsinger JA, Reitz C, Honig LS, Tang MX, Shea S, Mayeux R. Aggregation of vascular risk factors and risk of incident Alzheimer disease. *Neurology*. 2005;65:545–551.
11. Arvanitakis Z, Wilson RS, Bienias JL, Evans DA, Bennett DA. Diabetes mellitus and risk of Alzheimer disease and decline in cognitive function. *Arch Neurol*. 2004;61:661–666.
12. Biessels GJ, Staekenborg S, Brunner E, Brayne C, Scheltens P. Risk of dementia in diabetes mellitus: a systematic review. *Lancet Neurol*. 2006;5(1):64–74.
13. Kim B, Feldman EL. Insulin resistance in the nervous system. *Trends Endocrinol Metab*. 2012;23(3):133–141.
14. Duarte FH, Jallad RS, Amaro AC, Drager LF, Lorenzi-Filho G, Bronstein MD. The impact of sleep apnea treatment on carbohydrate metabolism in patients with acromegaly. *Pituitary*. 2013;16(3):341–350.
15. Chaves RN, Duarte AB, Rodrigues GQ, et al. The effects of insulin and follicle-simulating hormone (FSH) during in vitro development of ovarian goat preantral follicles and the relative mRNA expression for insulin and FSH receptors and cytochrome P450 aromatase in cultured follicles. *Biol Reprod*. 2012;87(3):69.
16. de Oliveira Baraldi C, Moises EC, de Jesus Ponte Carvalho TM, et al. Effect of type 2 diabetes mellitus on the pharmacokinetics of metformin in obese pregnant women. *Clin Pharmacokinet*. 2012;51(11):743–749.

17. Neumann KF, Rojo L, Navarrete LP, Farias G, Reyes P, Maccioni RB. Insulin resistance and Alzheimer's disease: molecular links and clinical implications. *Curr Alzheimer Res*. 2008;5(5):438–447.
18. Duarte AI, Moreira PI, Oliveira CR. Insulin in central nervous system: more than just a peripheral hormone. *J Aging Res*. 2012;2012:384017.
19. Biessels GJ, Kamal A, Urban IJ, Spruijt BM, Erkelens DW, Gispen WH. Water maze learning and hippocampal synaptic plasticity in streptozotocin-diabetic rats: effects of insulin treatment. *Brain Res*. 1998;800(1):125–135.
20. Devi L, Alldred MJ, Ginsberg SD, et al. Mechanisms underlying insulin deficiency-induced acceleration of beta-amyloidosis in a mouse model of Alzheimer's disease. *PLoS ONE*. 2012;7(3):e32792.
21. Craft S, Baker LD, Montine TJ, et al. Intranasal insulin therapy for Alzheimer disease and amnestic mild cognitive impairment: a pilot clinical trial. *Arch Neurol*. 2012;69(1):29–38.
22. Pedersen WA, McMillan PJ, Kulstad JJ, Leverenz JB, Craft S, Haynatzki GR. Rosiglitazone attenuates learning and memory deficits in Tg2576 Alzheimer mice. *Exp Neurol*. 2006;199(2):265–273.
23. Muyllaert D, Kremer A, Jaworski T, et al. Glycogen synthase kinase-3beta, or a link between amyloid and tau pathology? *Genes Brain Behav*. 2008;7(suppl 1):57–66.
24. Takashima A. GSK-3 is essential in the pathogenesis of Alzheimer's disease. *J Alzheimers Dis*. 2006;9:309–317 (3 Suppl).
25. Kim B, Backus C, Oh S, Hayes JM, Feldman EL. Increased tau phosphorylation and cleavage in mouse models of type 1 and type 2 diabetes. *Endocrinology*. 2009;150(12):5294–5301.
26. Jolivalt CG, Lee CA, Beiswenger KK, et al. Defective insulin signaling pathway and increased glycogen synthase kinase-3 activity in the brain of diabetic mice: parallels with Alzheimer's disease and correction by insulin. *J Neurosci Res*. 2008;86(15):3265–3274.
27. Watson GS, Craft S. Modulation of memory by insulin and glucose: neuropsychological observations in Alzheimer's disease. *Eur J Pharmacol*. 2004;490(1–3):97–113.
28. Sasaki N, Fukatsu R, Tsuzuki K, et al. Advanced glycation end products in Alzheimer's disease and other neurodegenerative diseases. *Am J Pathol*. 1998;153(4):1149–1155.
29. Sasaki N, Toki S, Chowei H, et al. Immunohistochemical distribution of the receptor for advanced glycation end products in neurons and astrocytes in Alzheimer's disease. *Brain Res*. 2001; 888(2):256–262.
30. Ledesma MD, Bonay P, Colaco C, Avila J. Analysis of microtubule-associated protein tau glycation in paired helical filaments. *J Biol Chem*. 1994;269(34):21614–21619.
31. Toth C, Schmidt AM, Tuor UI, et al. Diabetes, leukoencephalopathy and rage. *Neurobiol Dis*. 2006;23(2):445–461.
32. Girones X, Guimera A, Cruz-Sanchez CZ, et al. N epsilon-carboxymethyllysine in brain aging, diabetes mellitus, and Alzheimer's disease. *Free Radic Biol Med*. 2004;36(10):1241–1247.
33. Heitner J, Dickson D. Diabetics do not have increased Alzheimer-type pathology compared with age-matched control subjects. A retrospective postmortem immunocytochemical and histofluorescent study. *Neurology*. 1997;49(5):1306–1311.
34. Akiyama H, Barger S, Barnum S, et al. Inflammation and Alzheimer's disease. *Neurobiol Aging*. 2000;21(3):383–421.
35. Yaffe K. Metabolic syndrome and cognitive decline. *Curr Alzheimer Res*. 2007;4(2):123–126.
36. Harris MI. Hypercholesterolemia in diabetes and glucose intolerance in the US population. *Diabetes Care*. 1991;14(5):366–374.
37. Ishikawa M, Iwasaki Y, Yatoh S, et al. Cholesterol accumulation and diabetes in pancreatic beta-cell-specific SREBP-2 transgenic mice: a new model for lipotoxicity. *J Lipid Res*. 2008;49(12): 2524–2534.
38. Sharma S, Prasanthi RPJ, Schommer E, Feist G, Ghribi O. Hypercholesterolemia-induced Abeta accumulation in rabbit brain is associated with alteration in IGF-1 signaling. *Neurobiol Dis*. 2008;32(3):426–432.
39. Burns MP, Noble WJ, Olm V, et al. Co-localization of cholesterol, apolipoprotein E and fibrillar Abeta in amyloid plaques. *Brain Res*. 2003;110(1):119–125.
40. Mori T, Paris D, Town T, et al. Cholesterol accumulates in senile plaques of Alzheimer disease patients and in transgenic APP(SW) mice. *J Neuropathol Exp Neurol*. 2001;60(8):778–785.
41. Bjorkhem I, Heverin M, Leoni V, Meaney S, Diczfalusy U. Oxysterols and Alzheimer's disease. *Acta Neurol Scand*. 2006;185:43–49.
42. Prasanthi JR, Huls A, Thomasson S, Thompson A, Schommer E, Ghribi O. Differential effects of 24-hydroxycholesterol and 27-hydroxycholesterol on beta-amyloid precursor protein levels and processing in human neuroblastoma SH-SY5Y cells. *Mol Neurodegen*. 2009;4:1.
43. Jick H, Zornberg GL, Jick SS, Seshadri S, Drachman DA. Statins and the risk of dementia. *Lancet*. 2000;356(9242):1627–1631.
44. Johnson GV, Cox TM, Lockhart JP, Zinnerman MD, Miller ML, Powers RE. Transglutaminase activity is increased in Alzheimer's disease brain. *Brain Res*. 1997;751(2):323–329.
45. McGowan E, Eriksen J, Hutton M. A decade of modeling Alzheimer's disease in transgenic mice. *Trends Genet*. 2006;22(5):281–289.
46. Morrissette DA, Parachikova A, Green KN, LaFerla FM. Relevance of transgenic mouse models to human Alzheimer disease. *J Biol Chem*. 2009;284(10):6033–6037.
47. Gotz J, Ittner LM. Animal models of Alzheimer's disease and frontotemporal dementia. *Nat Rev Neurosci*. 2008;9(7):532–544.
48. Roberson ED, Scearce-Levie K, Palop JJ, et al. Reducing endogenous tau ameliorates amyloid beta-induced deficits in an Alzheimer's disease mouse model. *Science*. 2007;316(5825):750–754.
49. Ke YD, Delerue F, Gladbach A, Gotz J, Ittner LM. Experimental diabetes mellitus exacerbates tau pathology in a transgenic mouse model of Alzheimer's disease. *PLoS ONE*. 2009;4(11):e7917.
50. Jolivalt CG, Hurford R, Lee CA, Dumaop W, Rockenstein E, Masliah E. Type 1 diabetes exaggerates features of Alzheimer's disease in APP transgenic mice. *Exp Neurol*. 2010;223(2):422–431.
51. Takeda S, Sato N, Uchio-Yamada K, et al. Diabetes-accelerated memory dysfunction via cerebrovascular inflammation and Aβ deposition in an Alzheimer mouse model with diabetes. *Proc Natl Acad Sci U S A*. 2010;107(15):7036–7041.

SECTION III

DISEASES OF THE CENTRAL NERVOUS SYSTEM AND NEURODEGENERATION

CHAPTER

14

Introduction

Elena Cattaneo, Alessandro Vercelli†*

***Department of Biosciences, University of Milan, Italy, †Department of Neuroscience, Neuroscience Institute Cavalieri Ottolenghi, Turin, Italy**

Brain disease is placing an increasing economic and social burden on developed countries. The increase in the mean age of the population and improved emergency and chronic care allow many patients to survive for years or even decades with a genetic disease or after a sudden insult. The chapters in this section highlight the relevance, incidence, and prevalence of this class of diseases while focusing on the major ones. A detailed taxonomy is provided for epilepsy (Chapter 17), amyotrophic lateral sclerosis (ALS, Chapter 18), Parkinson disease (PD, Chapter 19), Alzheimer disease (AD, Chapter 21), and prion disease (Chapter 23), covering their various causes and clinical features.

Brain diseases result from a mixture of genetic, environmental, and accidental factors, which may affect the brain at the molecular, cellular, and circuit levels. Whereas spinal cord injury (SCI, Chapter 15), traumatic brain injury (TBI, Chapter 16) and, to a certain extent, stroke (Chapter 22) are mostly caused by unpredictable accidents, they may be facilitated by environmental agents and behavior. Some neurodegenerative diseases, such as AD, PD, and ALS, are frequently associated with specific genes. In cases such as Huntington disease (Chapter 20), there is a strict genotype–phenotype correlation.

Many brain diseases share pathological hallmarks and pathogenic mechanisms. Edema, ischemia, impaired axonal transport, hemorrhage, inflammation, and rupture of the blood–brain barrier are involved in the early phases of TBI, SCI, stroke, and epilepsy. In many brain diseases the generation of free radicals damages neurons and triggers a molecular cascade leading to cell death. Excitotoxicity occurring in stroke, epilepsy, AD, PD, HD, and ALS can result in apoptosis and autophagic cell death, and may also be involved in TBI and SCI. In some neurodegenerative diseases, cytoplasmic aggregates are found, such as synuclein-positive inclusions (Lewy bodies) in PD, mutant huntingtin aggregates in HD, neurofibrillary tangles and β-amyloid aggregates in AD, misfolded proteins in ALS, and misfolded prion proteins in prion diseases.

Although several brain diseases were initially considered cell autonomous, such as PD (affecting dopaminergic neurons), HD (affecting inhibitory spiny stellate projection neurons), and ALS (affecting upper and lower motoneurons), there is increasing evidence of a non-secondary role for glial cells. Glial cells and the molecules they release or do not release may have a detrimental effect on disease progression. Astroglia and microglia therefore play a role in diseases, and may be detrimental or protective depending on their state of activation, either modulating or contributing to neuroinflammatory mechanisms.

In some chapters in this section, neurodegenerative diseases are also considered in terms of anatomical circuits that are affected and modified by neuronal loss or axonal interruption, leading to the specific symptoms that characterize the disease. The regenerative potential of the adult nervous system is limited by growth-inhibiting factors, and intrinsic cell growth programs that are turned off following lesion cannot be switched on to restore lost connections. Adult neurogenesis, which seems to be enhanced in several diseases, such as in stroke, is not sufficient to replace the lost neurons. Oligodendrocytes are major players in brain repair, since on one hand they die and are lost in axonal degeneration, while on the other hand they can either prevent or promote axonal regeneration, depending on their developmental stage.

These chapters give an exhaustive overview of the major animal models of brain disease. Although these models cannot reproduce all of the pathological and behavioral features and symptoms of human disease, they are useful tools to study mechanisms and to test new therapies. Treatments in use and under experimentation are also described.

Neurobiology of Brain Disorders
http://dx.doi.org/10.1016/B978-0-12-398270-4.00014-8

Many questions remain unanswered, as brain disorders represent both an opportunity and a challenge to researchers. How does a gene mutation that affects a ubiquitously expressed molecule lead to damage of specific cell types, such as striatal spiny projection neurons in HD or motoneurons in ALS? What are the molecular and cellular mechanisms of cell death, and how can we interfere with them to prevent neuronal loss? What is the role of neuroinflammation in disease, and how does it develop at a molecular and cellular level? What happens when neurons are lost, and how many neurons must die before neural circuits and networks are significantly altered and the pathological and structural changes induce behavioral and cognitive deficits? Is cell replacement an option, and are there other possibilities? The understanding of the neurobiology of disease can take advantage of experimental studies on cellular and animal models of diseases and on molecular pathways and functional analysis in the normal brain. On the other hand, the observation of the loss of function due to disease has always been a useful cue to understand brain function and to study brain connectivity and circuits.

CHAPTER

15

Spinal Cord Injury

Alessandro Vercelli, Marina Boido

Department of Neuroscience, Neuroscience Institute Cavalieri Ottolenghi, Turin, Italy

OUTLINE

INTRODUCTION

Spinal cord injury (SCI) affects approximately 6 million people worldwide. The incidence of SCI in the USA alone varies from 25 to 59 new cases per million population per year, with an average of 40 per million. Incidence data are comparable only for regions in North America, Western Europe, and Australia,[1] although the incidence of SCI in the rest of the world is much lower than in the USA owing to an overall lower number of acts of violence and motor vehicle crashes in the other countries.[2] The highest incidence occurs in adolescents and young adults (half aged 16–30 years), but there is a growing prevalence of SCI in elderly people, primarily because of falls.[3] The male-to-female ratio for SCI is approximately 4 to 1. Working and sports SCIs are most commonly caused by vehicular crashes, violence, falls, and accidents.[1]

SCI often results in significant dysfunction and disability, determining long-lasting and irreversible deficits, such as partial or complete paralysis and loss of sensation below the level of the injury, with little chance of recovery.[4] The level of the injury refers to the neuromer of the spinal cord, which usually causes sensorimotor impairment below the corresponding vertebral body due to the cranial shift of the spinal cord during development. Tetraplegia, with sensorimotor loss below the head, results from an SCI at a level between C1 and T1. Injury between T2 and S5 results in paraplegia, which is disability in the lower body and legs.

SCI physically and psychologically affects not only the individual, but also the family and the whole society, affecting quality of life and life expectancy, with a huge economic burden because of the considerable costs of post-traumatic care.[4,5]

Neurobiology of Brain Disorders
http://dx.doi.org/10.1016/B978-0-12-398270-4.00015-X

This chapter discusses the anatomopathological, cellular, and molecular events occurring in the spinal cord following injury, together with perspectives for therapy to promote axonal regrowth and to restore connectivity.

TYPES OF INJURY AND GLIAL SCAR FORMATION

The pathophysiology of SCI can be subdivided in two phases: primary and secondary.[6] The initial mechanical trauma, due to compression, distraction (misalignment of fragments), laceration or transection, damage to blood vessels, and disruption of axonal fibers and local neurons, interrupts the anatomical continuity of the cord; in a few minutes, the spinal cord swells, occupying the entire diameter of the spinal canal at the level of injury, and exceeds the blood pressure, which results in secondary ischemia. Finally, the release of toxic molecules leads to secondary damage.[5]

The wave of secondary cell death, which mainly affects neurons and oligodendrocytes, spreads both rostrally and caudally from the site of impact, leading to structural and functional damage. Key secondary injury mechanisms include damage of spinal cord blood vessels and ischemia, glutamatergic excitotoxicity, oxidative cell stress, lipid peroxidation, and inflammation, all of which alone or in concert can stimulate apoptosis and cell death.

Several cytokines produced by leukocytes, macrophages, and lymphocytes can induce an inflammatory reaction in the injured spinal cord, resulting in the formation of a cystic cavity and activation of glial cells around the site of the lesion. Two types of scar tissue are formed in the lesion site: one is a glial scar consisting of reactive astrocytes, activated microglia, and glial precursor cells; the second is a fibrotic scar, formed by fibroblasts invading the lesion site. Both glial and fibrotic scars can become an obstacle for axonal regeneration.

The pathological events following injury are strictly dependent on the kind of lesion. The most frequent is compression injury, either persistent or transient, mostly due to bone fractures, dislocations, and disc ruptures. Spinal distraction can result in flexion, extension, rotation, or dislocation, with consequent shearing or stretching of the spinal cord and/or its blood supply. Finally, laceration and rupture can determine partial or complete transection.[7]

Experimental models can hardly mimic the accidental human pathology, and no single animal model can reproduce all the different physiopathological mechanisms. The two most common rodent models consist of compression and transection. Specific devices can reproduce a standard compression lesion, such as a clip, a weight-drop impactor, an electromagnetic injury device, or an epidural balloon to be inflated. Non-impact models include photochemical injury by endovenous administration of the photosensitive Rose Bengal dye followed by irradiation, to reproduce vascular damage. These models mimic the tissue destruction occurring in humans, and allow study of the inflammatory process and glial cyst formation. The other model, monolateral or bilateral transection,[8] which is mostly restricted to white matter, allows study of axonal degeneration and regeneration and the amount of spontaneous axonal sprouting and regrowth (Fig. 15.1).

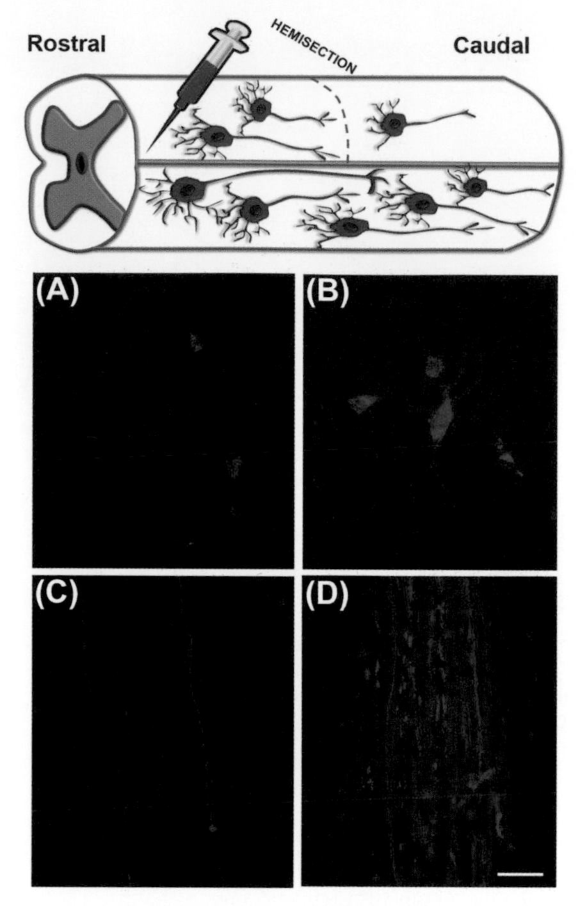

FIGURE 15.1 **Descending and ascending projections in the spinal cord below injury in the hemisection model.** Fluoro-Ruby, a fluorescent dextran, used as both anterograde and retrograde neuronal tracer, was injected in the cervical spinal cord after lumbar hemisection. A strong reduction was observed ipsilaterally to hemisection in the labeling of cell bodies [(A) ipsilateral vs (B) contralateral, ascending projections] and of axons [(C) ipsilateral vs (D) contralateral, descending projections]. Scale bar = 50 μm (A,B), 30 μm (C,D).

TIME-COURSE OF POSTINJURY CHANGES

The microenvironment in the spinal cord changes rapidly after injury. Several consecutive phases, the immediate, acute, intermediate, and chronic stages of SCI, can be identified (Fig. 15.2)[6]:

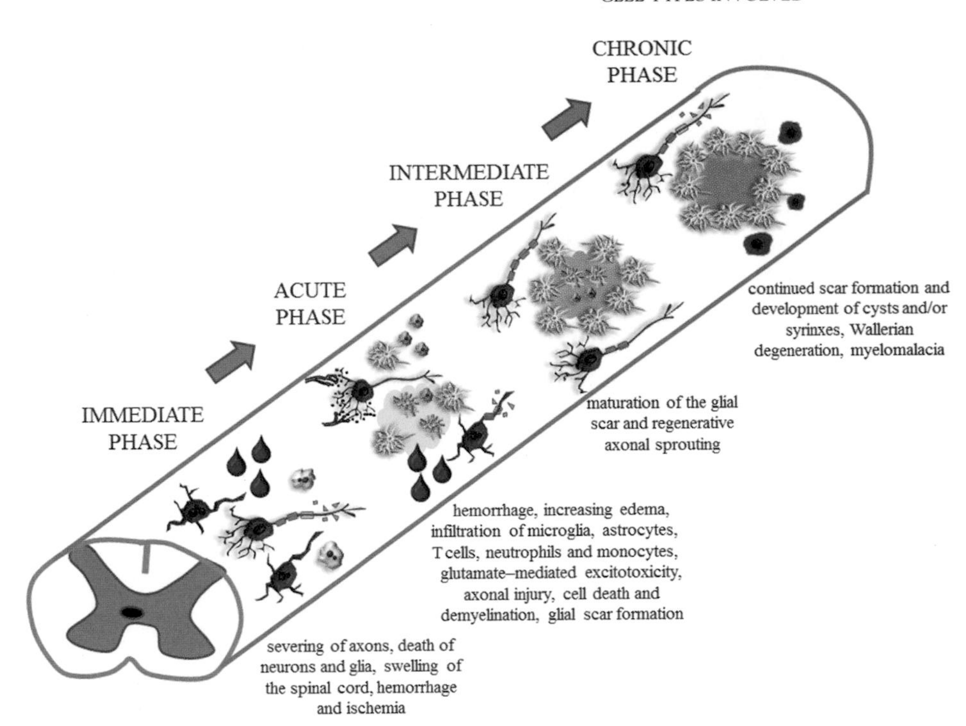

FIGURE 15.2 **Time-course of a spinal cord injury.**

- The *immediate phase* (0–2 hours) is characterized by traumatic severing of axons, death of neurons and glia, spinal shock, swelling of the spinal cord, mechanical disruption of cell membranes, and hemorrhage and ischemia resulting from vascular disruption, both rostral and caudal to the injury.
- The *acute phase* (2 hours to 2 weeks) is characterized by continuing hemorrhage, increasing edema, inflammation consisting of infiltration by microglia, astrocytes, T cells, neutrophils and invading monocytes, free radical production, ionic dysregulation, glutamate-mediated excitotoxicity, rupture of the blood–brain barrier (BBB), with immune-associated neurotoxicity leading to axonal injury, cell death, demyelination, and glial scar formation.
- The *intermediate phase* (2 weeks to 6 months) is characterized by continued maturation of the glial scar and regenerative axonal sprouting, although not sufficient for functional recovery.
- The *chronic phase* (>6 months) is characterized by maturation and stabilization of the lesion: continued scar formation and development of cysts or syrinxes, ongoing Wallerian degeneration, with axonal fragmentation and degradation after injury and myelomalacia, the final stage of necrotic death after SCI, and delayed neurological dysfunction leading to ascending paralysis, brainstem symptoms, and neuropathic pain.

Inflammation occurs immediately after SCI, with a rapid increase in inflammatory cytokines [interleukin-1α (IL-1α), IL-1β, IL-6, and tumor necrosis factor-α (TNF-α)] and a consequent decrease in the next 24 hours.[9] Therefore, treatments in the acute phase of SCI can modulate the early phases of inflammation, microglia activation, and astrogliosis; however, the outcome of an injury is unpredictable in this phase, and the lesion needs stabilizing before effective treatment can be designed.

The weeks before the complete formation of the glial scar may be the optimal time to maximize benefit from any treatment, as this is when maximum repair in the injured CNS can be achieved; for example, grafted Schwann cells survive best immediately or 10 days after injury, when post-traumatic microcavitation and macrocavitation are strongly reduced.

The chronic phase is characterized by the formation of the glial scar and the cyst, both of which strongly inhibit axonal regeneration. Treatment in this phase could give new hope to people who have been functionally impaired for a long time.

CELL TYPES INVOLVED

Neurons, astrocytes, microglial cells, oligodendrocytes, and endothelial cells are all involved in SCI pathophysiology (Fig. 15.3).

Neurons

The pathophysiological processes triggered by SCI lead to cell death,[6] at first by necrosis and later by apoptosis.[10] These events are induced by a loss of ionic homeostasis, namely dysregulation of Ca^{2+} concentration, leading to the activation of calpains, mitochondrial

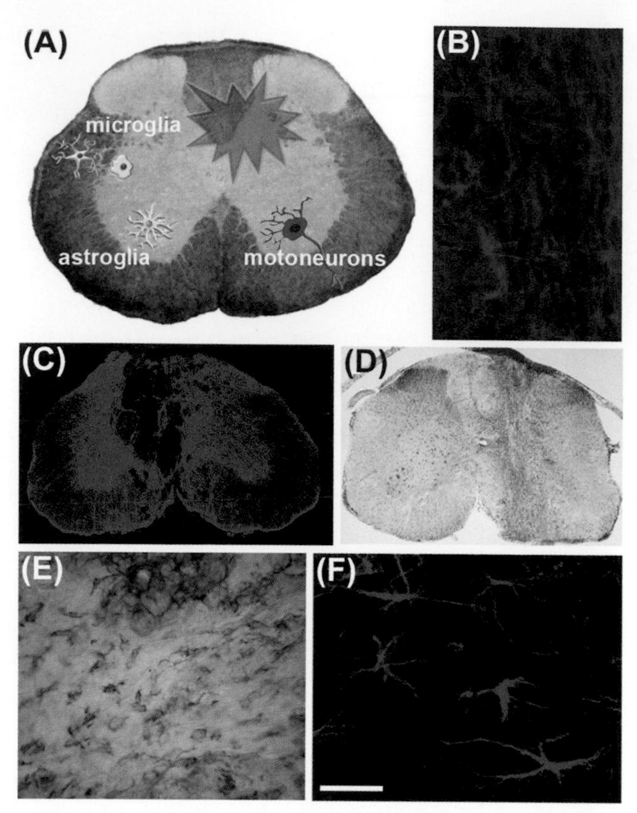

FIGURE 15.3 **Histological features of spinal cord injury (SCI).** (A) Spinal cord injury involves neurons, glia, and axons. (B) Serotonin-positive raphespinal axons: these fibers degenerate after SCI, undergoing demyelination and Wallerian degeneration. (C,D) The major obstacle to axonal regeneration is represented by the glial cyst [C: glial fibrillary acidic protein (GFAP)-labeled astrocytes; D: Nissl staining]. (E,F) The lesion site is characterized by a central core, mainly filled with activated microglia (E: CD11b-positive), surrounded by reactive astrocytes (F: GFAP-positive), forming the glial scar. Scale bar = 100 μm (B), 400 μm (C,D), 60 μm (E), 20 μm (F).

dysfunction, and production of free radicals.[6] Excitotoxicity is a major event inducing both neuronal and glial cell death, by the excessive activation of glutamate receptors and leading to the influx of Na^+ and Ca^{2+}, particularly through *N*-methyl-D-aspartate (NMDA) receptors.[11] The production of reactive oxygen and nitrogen species (ROS and RNS) can be extremely toxic to neurons: this oxidative damage elicits a self-sustaining production of ROS as a feedback response, leading to massive neuronal death. Finally, oligodendrocyte loss, which results in axonal demyelination, can indirectly cause neuronal death owing to atrophy of the soma. Spinal locomotor circuits undergo a dramatic rearrangement following SCI; nevertheless, sparing of 10–15% of descending pathways may allow some locomotor recovery.[12] Stepping movements can be elicited in primates after complete transection of the spinal cord. As soon as the acute phase of SCI resolves, the polysynaptic spinal reflex reappears together with locomotor electromyographic activity seen following assisted locomotion. This activity results in exaggerated reflexes, muscle tone, and spasms for a few months, probably due to rearrangement of polysynaptic reflexes and synaptic connections. The spinal pattern generator, comprising the neural circuits underlying stepping movements, must be stimulated by peripheral afferents such as proprioceptive inputs. Locomotor activity is lost around 1 year after SCI, due in part to the loss of afferent activity from descending pathways and in part to changes in muscle fibers. This leads to decreased electromyographic activity, which is called EMG exhaustion and is probably due to interneuronal mechanisms. One possible explanation is that the loss of descending projections and of appropriate peripheral afferents causes a predominantly inhibitory activity.

Oligodendrocytes

Demyelination of axons is a common feature in SCI, consisting of vesicular myelin degeneration and enlargement of the interaxonal spaces.[13,14] Infiltrating macrophages promote the recruitment of oligodendrocyte precursors; activated spinal microglia can induce either oligodendrocyte cell precursor proliferation and oligodendrocyte neogenesis or death. Beginning 2 weeks postinjury axon remyelination occurs, as a result of the proliferation and differentiation of oligodendrocyte precursors. Mature oligodendrocytes do not participate in remyelination. Rather, remyelinating oligodendrocytes derive from a nerve/glial antigen-2 (NG2)-positive subpopulation of oligodendrocyte precursors. Proliferating oligodendrocyte precursor cells are found mainly around the lesion, possibly because of a differential distribution of growth factors: fibroblast growth factor-2 (FGF-2), a mitogen for oligodendrocyte precursor cells, is upregulated in the 4 weeks after injury. Other trophic factors implicated in oligodendrocyte precursor proliferation and maturation, such as ciliary neurotrophic factor (CNTF) and insulin-like growth factor (IGF), and their receptors, are also upregulated following SCI. SCI also results in increased levels of neurotrophins, such as neurotrophin-3 (NT-3) and brain-derived neurotrophic factor (BDNF). The reactivation of notch signaling in axons after SCI and the upregulation of the polysialylated form of the neural cell adhesion molecule (PSA-NCAM) in the site of the lesion prevent maturation of olygodendrocyte precursors and remyelination. New insights into the mechanisms underlying remyelination could be provided by the study of microRNAs (miRNAs), as SCI has been shown to alter the expression of around 100 miRNAs.[15]

Astroglia

Usually, astrocytes provide both structural and physiological support to neurons. They are involved in neurotransmitter regulation, ion homeostasis, maintenance of the BBB, and production of extracellular matrix molecules.[16] Astrocytes are induced after injury, generating the glial scar, which is a physical and molecular barrier to axonal regeneration. Thereafter, astrocytes become hypertrophic, giving rise to reactive gliosis, characterized by extended processes and increased production of glial fibrillary acidic protein (GFAP) and vimentin.[17] The complete molecular cascade leading to astrogliosis remains to be elucidated, but transforming growth factor-β (TGF-β), IL-1 and two inflammatory cytokines, interferon-γ and basic FGF-2, seem to be involved.[18] Although astrogliosis is detrimental to functional recovery after SCI, it is conserved throughout vertebrate evolution,[19] suggesting that this process could also provide an advantage to survival. Glial scar may stabilize fragile CNS tissue after injury by repairing the BBB, preventing neuronal degeneration, and limiting the spread of damage, thus protecting the healthy tissue from uncontrolled damage.[18]

Microglia and Macrophages

Macrophages protect the local environment from pathogens, maintain tissue homeostasis, and phagocytose dead and dying cells. Resident macrophages in the CNS (i.e. microglia) participate in development by removing apoptotic cells and inappropriate neural connections. In the normal, healthy brain these cells represent the resting microglia, characterized by numerous cytoplasmic processes. In response to any disturbance of nervous system homeostasis, microglia rapidly change their phenotype and become activated. Fully activated microglia retract their cytoplasmic processes and become amoeboid. Following an injury, macrophages from the peripheral circulation and those derived from resident microglia are attracted to the injury site and participate in the inflammatory response.[20] Similarly to astrocytes, microglia can exert both detrimental and beneficial effects, depending on environmental signals, on the cell surface receptors they express, and on the intracellular signaling pathways that are activated. On the one hand, microglia produce proinflammatory cytokines such as IL-1β and TNF-α and cytotoxic molecules such as ROS and RNS. On the other hand, microglia can exert neuroprotective effects by releasing TGF-β and macrophage–colony-stimulating factor receptor. Moreover, TNF-α signaling (through TNF receptor superfamily member 1B) can promote remyelination by stimulating the proliferation of oligodendrocyte progenitors. Finally, microglia cells can phagocytose dead and damaged cells and clear tissue debris. Neutrophils enter the spinal cord 3–24 hours, monocytes 2–3 days, and dendritic cells 3–7 days after SCI, and they remain there for months. Since they have short half-lives, it is reasonable to believe that there is sustained recruitment of myeloid cells into the injured spinal cord.[21]

Endothelial Cells

The blood–brain/spinal cord barrier is organized as a specialized system of non-fenestrated endothelial cells and accessory structures, including the basement membrane, pericytes, and astrocytic endfeet processes. The barrier protects the CNS and regulates the transport of nutrients, hormones, metabolites, and other blood constituents, providing the stable microenvironment that is fundamental for physiological neuronal function. The restrictive nature of the barrier is mainly attributed to a complex negatively charged glycocalyx, able to repulse circulating proteins.[22] SCI triggers microvascular damage involving endothelial cell death and breakdown of the blood–brain/spinal cord barrier. The primary insult causes hemorrhage starting from the center of the injury and spreading to both the gray and white matter. During the first 24 hours postinjury endothelial cells die by necrosis, and in the following days by apoptosis. The barrier is disrupted and its permeability altered, exposing the injured spinal cord to inflammatory and vasoactive molecules, which lead to edema, mediated by matrix metalloproteinase activity.[23]

ROLE OF THE EXTRACELLULAR MATRIX AND GROWTH INHIBITORS

Many molecular players are involved in limiting spontaneous regeneration of the CNS following trauma, including chondroitin sulfate proteoglycans (CSPGs), myelin-associated growth inhibitors, and chemorepulsive axon guidance molecules.[24]

Chondroitin Sulfate Proteoglycans

CSPGs, such as versican, neurocan, and brevican, represent an important class of inhibitory extracellular matrix molecules, consisting of proteins and glycosaminoglycans. They are involved in CNS development, cell migration, maturation and differentiation, cell survival, and tissue homeostasis. They are crucial components of perineuronal nets, the highly condensed matrix that surrounds cell bodies and proximal dendrites of some classes of neuron. However, after trauma CSPGs become one of the main component of the fibrotic core of the scar, contributing to the inhibition of axonal regeneration and limiting neuronal plasticity.[25] They bind two classes of growth-inhibitory receptor expressed on the surface of axons, thus activating specific growth-inhibitory pathways. Therefore, the enzymic degradation of glycosaminoglycans, using chondroitinase ABC (ChABC, for the three isoforms of glycosaminoglycans), could represent a therapeutic approach. ChABC treatment can be combined with other regenerative strategies, including the use of biomaterials, neurotrophic factors, or cell transplantation.[24] Unfortunately, ChABC treatment is limited by its short period of enzymic activity at 37°C (body temperature), its inability to cross the BBB, and its high immunogenicity.

Myelin-Associated Growth Inhibitors

This class includes Nogo and myelin-associated glycoprotein (MAG). Nogo is expressed on the surface of myelin sheath and can reduce axonal growth after CNS injury. Three differentially spliced isoforms (Nogo A, B, and C) are known; the main isoform of CNS Nogo, Nogo A, triggers growth cone collapse, inhibits neurite outgrowth, and limits corticospinal tract regeneration after SCI.[26] MAG is a transmembrane glycoprotein expressed by oligodendrocytes and Schwann cells in the periaxonal layer of myelin, which is involved in the maintenance of myelinated axons in the adult nervous system. On the one hand, it can promote neurite outgrowth from newborn dorsal root ganglion neurons, while on the other hand it plays a growth inhibitory role in the adult.

Chemorepulsive Axon Guidance Molecules

Several axon guidance molecules are expressed in the adult CNS, where they are implicated in network stabilization by limiting neuronal growth, and their expression is regulated following injury.[27] Among them, semaphorins, a large family of secreted and membrane-associated proteins involved in axon repulsion and/or attraction, are expressed following injury, mainly in the glial scar, where they regulate its formation. In addition, meningeal fibroblasts reaching the lesion site express Sema3A, Sema3B, Sema3C, Sema3E, and Sema3F. Semaphorins can also contribute to the inhibitory properties of oligodendrocyte-lineage cells. Similarly, other chemorepulsive axon guidance molecules, such as ephrin, netrin, and Wnt, limit axonal elongation in the injured CNS, in some cases also inducing the formation of the glial scar and apoptosis.

CELL DEATH FOLLOWING SPINAL CORD INJURY

Cell loss occurs at the lesion site and in the surrounding regions, owing to trauma, ischemia, and inflammatory reactions. Cell death mechanisms affect mostly neurons and oligodendrocytes, and include necrosis, in the first hours following trauma, apoptosis, which may occur early or may be delayed, and autophagy, which has been considered less in the past. All these mechanisms of cell death underscore the role of mitochondria, which represent a promising target for therapy.[28] In the first hours and days following SCI, the molecular pathways involved in cell death are activated; for example, c-Jun-N-terminal kinase (JNK) and its target c-Jun are upregulated starting from 1 hour to 3 days after SCI. The effects of JNK activation are multifaceted: JNK is expressed in three isoforms, of which one, JNK3, is more involved than the others in cell death in the nervous system. JNK3 activation is involved in oligodendrocyte apoptosis. JNK blockade leads to decreased c-Jun phosphorylation and caspase-3 cleavage and myelin sparing.[29]

In addition to neuronal death in the area of injury (discussed above), motoneurons may die from transynaptic cell death. The loss of supraspinal afferents can induce atrophy and apoptosis of motoneurons.

As early as 15 minutes following experimental contusion of the rat spinal cord, oligodendrocytes die owing to necrosis and apoptosis. Oligodendrocyte cell death occurs for at least 3 weeks after a lesion, spreading cranially and caudally from the site of injury. Autophagy has been implicated in oligodendrocyte death in the first 3 weeks after injury. Proteolytic enzymes, neuroinflammatory mechanisms, ROS, and excitotoxicity contribute to oligodendrocyte cell death. Oligodendrocytes are specifically vulnerable to oxidative stress and excitotoxicity. They express glutamate receptors, such as 2-amino-3-(5-methyl-3-oxo-1,2-oxazol-4-yl)propanoic acid (AMPA), kainate receptors, and NMDA receptors. In addition, elevated ATP, released from several cell types after SCI, binds to P2X purinoceptor-7 (P2X7) in oligodendrocytes, leading to Ca^{2+} overload.

GENETIC AND EPIGENETIC CONTROL OF AXONAL GROWTH

Axonal growth during development consists of outgrowth, elongation, fasciculation, branching, and pruning. These aspects of axonal growth in the embryo are under the influence of both the environment and an intrinsic program. In the adult CNS, the potential to grow axons is both intrinsically and extrinsically decreased, whereas recapitulation of ontogenetic axonal growth is fundamental for the regenerative process. When the spinal cord is lesioned, both ascending and descending pathways are fully or partially transected, the corticospinal tract is retracted, axonal regeneration is almost absent, and sprouting is limited.[30] Nonetheless, some descending tracts, such as the raphespinal and the rubrospinal tracts, seem to be more plastic in the adult than the corticospinal tract. Growth-related genes (*GAP-43*, *c-JUN*, *galectin-1*, and *βII-tubulin*) are upregulated in raphe and red nuclei, but not in upper motoneurons.[31]

The sensory tract originating from the dorsal root ganglia is a unique system for the study of both peripheral and central nerve regeneration. The peripheral branch regenerates after injury, whereas regeneration of the lesioned central branch is inhibited by the environment. Nevertheless, when the two lesions are combined, regeneration of the central branch is enhanced by the conditioning peripheral injury. Peripheral axon injury triggers protein kinase A activity, which increases cyclic AMP (cAMP) levels through the phosphorylation of cAMP response element-binding protein (CREB).

The level of axotomy, that is, the distance from the cell body, also seems to be relevant to regeneration. Several growth-associated genes, such as the oncogene *c-jun/ap-1*, *l1cam/ncaml1*, *atf3*, and *krox2-4/egr1*, are increased in upper motoneurons after proximal intracortical but not distal spinal axotomy.

A growing list of regeneration-associated genes has been identified. They include transcription factors such as c-JUN, CREB, signal transducer and activator of transcription-3 (STAT3), activating transcription factor-3 (ATF3), retinoic acid receptor-β (RARβ), and p53; cytoskeleton and growth cone-associated proteins such as α-tubulin, Coronin-1β, Ras-related protein in brain (Rab13), microtubule associated protein-1 (MAP-1), growth-associated protein-43 (GAP-43), and cytoskeleton-associated protein-23 (CAP-23); cell adhesion molecules such as NCAM, cell adhesion molecule L1 (L1cam), and transient axonal glycoprotein-1 (TAG1 or contactin-2); and cytokines and extracellular matrix components, including synaptosomal-associated protein-25 (SNAP-25), candidate plasticity-related gene-15 (CPG15/neuritin), Galectin-1, Galanin, and small proline-rich protein-1 (SPRR1).

Jaerve and colleagues[32] provided a detailed analysis of genes expressed in sensorimotor cortex following SCI, and showed time-dependent changes in the expression of genes. Some genes, involved in wounding responses, cytoskeletal reorganization, and cell survival, are related to the immediate response; those involved in oxidative stress responses relate to the early response; and those involved in protein biosynthesis, apoptotic processes, and synaptic reorganization are mostly related to the late response. Antiscar treatment strongly modifies the expression of growth-related genes such as galanin (*GAL*), glial cell-derived neurotrophic factor (*GDNF*), kalirin (*KALRN*), LIM domain only-4 (*LMO4*), galectin-1 (*LGALS1*), paternally expressed-3 (*PEG3*), serum/glucocorticoid regulated kinase (*SGK*), mitogen-activated protein kinase 8 interacting protein (*MAPK8IP*), actin-β (*ACTB*), IL-4, IL-6, vitamin D_3 receptor (*VDR*), enabled homologue (*ENAH*), and S100 calcium binding protein-β (*S100B*); axonal guidance genes, such as roundabout homolog1 (*robo1*), reticulon4receptor (*rtn4r/nogor*), unc-5 homolog A,B (*unc5A, 5B*), dihydropyrimidinase-like 5 (*dpysl5/crmp5*), paired-Ig-like receptorB (*pirb*), neuropilin 1 (nrp1), receptor-like tyrosine kinase (*ryk*), cyclin-dependent kinase 5 (*cdk5*), semaphorin 6B (*sema6b/semaZ*), contactin2 (*cntn2/tax1*), neural cell adhesion molecule L1 (*l1cam/ncaml1*), fasciculation and elongation protein zeta1 (*fez1*), and activated leukocyte cell adhesion molecule (*alcam*); and also specifically regulates other genes. (See Jaerve *et al.*[32] for details.)

The response in gene expression to SCI is age dependent, since aging reduces spontaneous axon sprouting of corticospinal, serotonergic, raphespinal, and catecholaminergic ceruleospinal tracts. One day following injury, genes associated with ubiquitination are upregulated in young animals, whereas those associated with complement were upregulated in aged animals. In the subacute stage (1 week after SCI), cAMP and chemokine signaling were upregulated in young animals, and the complement system and Notch signaling in aged animals. In the chronic stage (35 days after SCI), *RhoA*, *Erk5*, and *PI3K/Akt* were upregulated in young animals, and *CNTF* and *Oct-4* in aged animals. Therefore, it may be hypothesized that the aged spinal cord, which expresses lower levels of *necdin*, *neurofascin*, and *plasticity related gene-1*, is more susceptible to injury. Nevertheless, antiscar treatment elicits a similar pattern of gene expression in both young and aged animals in the chronic phase. Some of these genes are growth and transcription factors, such as CNTF, IGF-1, BDNF, Krüppel-like growth factor-7 (KLF7), doublecortin (DCX), CREB1, and ATF2, which are usually downregulated after lesion. This suggests that, although the response to SCI in the older animal is less efficient than in the young, both may respond positively to therapy.

A molecular pathway regulating axonal growth has been identified in phosphatase and tensin homologue (PTEN), on chromosome 10. Compensatory sprouting of corticospinal tract axons after SCI is markedly enhanced by *Pten* deletion, which activates mammalian target of rapamycin (mTOR) and Akt signaling and inhibits other signaling molecules such as glycogen synthesis kinases (GSK-3) and phosphatidylinositol 3,4,5-trisphosphate (PIP3). mTOR activation enhances protein synthesis, to provide the building blocks for axonal growth, and GSK-3 inhibition may enhance axonal transport.

The expression of seven miRNAs is significantly altered following SCI. Among these, three miRNAs exhibit spatial dysregulation, since MiR129-1 and 129-2 are suppressed cranially to the injury site, whereas miR146a is increased caudally. Although miRNAs can regulate several genes simultaneously, they play a role in the maladaptive changes occurring after SCI; research on miRNAs could provide insights for therapeutic strategies.

INFLAMMATORY AND MALADAPTIVE IMMUNE RESPONSES AND THE BLOOD–BRAIN BARRIER

The role of inflammation in brain damage is discussed at length in Chapter 25. Although much of the focus of that chapter is on neurodegenerative disease, there are several reports indicating astrogliosis and microglial activation in several different models of SCI.[33] This includes the influences of these events on somatic hypertrophy and thickened branches. Most studies highlight the activation both in the early and chronic phases of SCI (time) and at different levels, above, below, and at the lesion site (space). This leads to the idea of gliopathy. Glutamate, ATP, ROS, neurotrophic factors, and proinflammatory cytokines, whose receptors are expressed by astrocytes and microglia, are released following SCI. Dysregulation of glutamate transporters can lead to alterations in K^+ and Ca^{2+} homeostasis at the glial–neuronal interface, which causes hyperexcitability of the dorsal horn neurons, thus representing the substrate for neuropathic pain.

In both the early and chronic stages, glia and neurons release proinflammatory cytokines, such as IL-1β, TNF-α, IL-6, and leukemia inhibiting factor (LIF), which are also produced by infiltrating macrophages and T lymphocytes in the chronic phase. SCI induces the release of cytokines such as scaffold protein MEK partner-1 (MP-1) and monocyte chemoattractant protein-1 (MCP-1). Many cytokines are transported through the BBB by a saturable, receptor-mediated transport system. For example, the transport system of TNF-α is enhanced at different time intervals following SCI, depending on the lesion. The transport of LIF is saturable and is increased in SCI.

The BBB is a physical barrier formed by endothelial cells that in normal conditions prevents the free exchange of molecules between the blood compartment and the CNS. The main constituents are therefore non-fenestrated endothelial cells and their tight junctions, together with pericytes, astroglial processes, and the extracellular matrix. Both astroglia and neurons interact with endothelial cells, influencing not only blood flow but also permeability, thus leading to the definition of the neurovascular unit.[34]

The BBB in the spinal cord has some specificities, allowing a blood–spinal cord barrier (BSCB) to be defined in terms of glycogen deposits (present only in BSCB microvessels), decreased expression of transporter molecules and of tight junction and adhering junction proteins, and, finally, increased permeability.[35] Breakdown of the BSCB occurs 5 minutes after trauma, whereas its restoration requires weeks: alterations in the basement membranes, widening of tight junctions, oxidative stress, generation of free radicals, and nitric oxide release lead to secondary inflammation. Moreover, disruption of microvessels results in vessel regression and neovascularization.

Restoration of the BSCB, which occurs after the second week from injury, correlates with improved motor function in rats. In the chronic phase of SCI, some patients develop syringomyelia, that is, the formation of fluid-filled cysts, owing to post-traumatic arachnoiditis and obstruction of the flow of cerebrospinal fluid and also to the alterations in the BSCB.

NEUROPATHIC PAIN AND AUTONOMIC DYSREFLEXIA

Unilateral hemisection or spinal contusion leading to incomplete SCI can result in allodynia (pain induced by a usually non-painful stimulus) and hyperalgesia (exaggerated response to a painful stimulus). The mechanisms underlying these symptoms are first related to hyperexcitability of dorsal horn neurons due to changes in the expression of ion receptors, loss of descending serotonergic fibers, reduced γ-aminobutyric acidergic (GABAergic) inhibition, and changes in the expression of metabotropic glutamate receptors. In addition, rewiring of local dorsal horn circuits, and reorganization of dorsal afferents and nociceptive networks may be responsible for neuropathic pain in SCI.[36] Microglia and astroglia participate in neuropathic pain owing to maladaptive changes in the production of cytokines and neurotrophic factors such as BDNF and prostaglandin GE_2. Positive results and pain relief have been obtained with deep brain stimulation, following electrode implantation in the thalamus, in the periaqueductal or periventricular gray matter or in the internal capsule.

Autonomic dysreflexia is a well-known complication of SCI, and consists of changes in blood pressure and cardiac rhythm, detrusor muscle dyssynergia and hyperreflexia, and anal continence.[37]

THERAPEUTIC TOOLS IN SPINAL CORD INJURY

SCIs represent a major therapeutic challenge because of the interruption of descending and ascending pathways, the formation of glial scars or cysts, and the presence of inhibitory molecules that prevent axonal growth in the adult nervous system. Nevertheless, the spinal cord is often incompletely severed, and sparing of 10–15% of fibers allows some recovery of function. Different therapeutic approaches have been tested, including preventing glial scar formation, removing inhibitory molecules, and stimulating axonal growth by administering trophic factors. The organization of neural circuits has been enhanced through physical exercise. More invasive approaches have also been suggested, such as transplanting different stem cell types with the aim of replacing and restoring neural circuits that have been lost or, more simply, modulating the environment in which axons could regrow. Different types of scaffolds could facilitate axonal growth and the engraftment of stem cells. Finally, gene therapy could enhance the growing potential of severed axons.

Exercise

After an SCI, a patient may undergo rehabilitative training, including exercising on a treadmill, bicycling, and swimming, to train the spinal cord to regain function, especially locomotion. There is increasing evidence that physical exercise can exert a strong modulatory effect on neuritic growth and plasticity, for example by forming or strengthening synapses. In one study, treadmill training induced recovery in injured rats by activating the local circuitry, which retained some ascending and descending fibers.[38] However, other authors report limited and modestly effective results.

Physical activity can upregulate the expression and release of several growth factors, especially BDNF, FGF-2, and IGF-1, which play a role in regulating synaptic efficacy, growth cone extension, and progenitor cell proliferation and differentiation. The effect of exercise is already evident in the intact rat spinal cord 3 days after voluntary wheel running, when expression of BDNF, its receptor trkB, and downstream effectors on synaptic plasticity (synapsin-I, CREB, and GAP-43) increase strongly. Thus, the level of BDNF is reduced after SCI, but rapidly restored by exercise. The restorative effect of running is often reported only when exercise is associated with an enriched environment. However, running is the critical element enhancing neurogenesis and BDNF levels, whereas enrichment by itself does not exert these effects. Exercise can also modulate hypersensitivity and cellular inflammation after spinal cord contusion; indeed, exercise strongly reduces allodynia or attenuates hyperalgesia, probably by normalizing BDNF and NT-3 messenger RNA levels in the cord and at the periphery. Increased BDNF, induced by exercise, enhances growth-promoting signals and reduce growth-inhibitory signals.

Exercise influences the activity of inhibitory transmitters after SCI. Glycine and GABA levels are significantly increased, but can be restored to normal levels by treadmill step training, especially in interneurons that make synapses with motoneurons directly involved in the motor task.

An important aspect in relation to the effects exerted by training concerns the type of exercise, that is, whether the activity is passive or forced, such as in treadmill training, swimming, or automated running wheels, or voluntary, when patients or animals spontaneously participate in exercise. Voluntary wheel running is a stress-free activity, simply based on the motivation of the subject. It is well known that the motivational state, linked to the hypothalamic–pituitary axis, is crucial to the therapeutic effectiveness of exercise. However, it is difficult to quantify the time spent by an animal in physical activity, as this depends on its sedentary or active nature and is therefore hard to standardize.

A combination of epidural electrical stimulation and a cocktail of serotonergic and dopamine receptor agonists, together with treadmill-based training and voluntary exercise, has been shown to induce an extensive remodeling of cortical projections, including the formation of brainstem and intraspinal relays in the spared circuitries.[39]

Cell Therapy

In the early twenty-first century, cell therapy emerged as a new tool for several diseases of the nervous system, since stem cells can provide trophic and immunomodulatory factors to enhance axonal growth, to modulate the environment, and to reduce neuroinflammation. Cell-based approaches for spinal cord functional repair center on two fundamental directions that are not mutually exclusive: restitution of white matter long tracts (regenerative approaches) and cell (i.e. neuronal or oligodendrocyte) replacement.[40]

Different types of stem cells have been tested, according to their properties and the therapeutic aims (Fig. 15.4). They differ from each other in origin, developmental stage, stage of differentiation, and fate lineage.

Embryonic stem cells (ESCs) are isolated from the inner cell mass of the blastocyst. ESCs can replicate indefinitely without aging, are pluripotent, give rise to genetically normal cells, and can be easily manipulated genetically. They can differentiate into functional neurons. In addition, they are able to differentiate into glia, such as oligodendrocytes, capable of remyelination and promoting regeneration in the injured axons in SCI and functional recovery. However, ESCs have an intrinsic potential to cause teratomas after transplantation.

Neural stem cells and neural precursor cells (NSCs and NPCs) can be isolated from many regions of the CNS of embryonic and adult mammals and propagated in culture in the presence of epidermal growth factor and/or basic FGF as neurospheres. They have the capacity for self-renewal and multipotency. After delayed transplantation of embryonic derived NPCs into the injured rat spinal cord, differentiation into glial and neuronal lineages as well as modest functional improvement have been reported. NSCs display a lower tumorigenic risk than ESCs.

Adult bone marrow is easily accessible, containing both hematopoietic stem cells (HSCs) and mesenchymal stem cells (MSCs). The use of these cells does not raise the ethical concerns associated with ESCs. They can be collected from the patient or from donors following informed consent. MSCs have an anti-inflammatory potential, by decreasing microglia and astroglia activation, and their use in diseases of the nervous system has a positive functional outcome. Moreover, they can

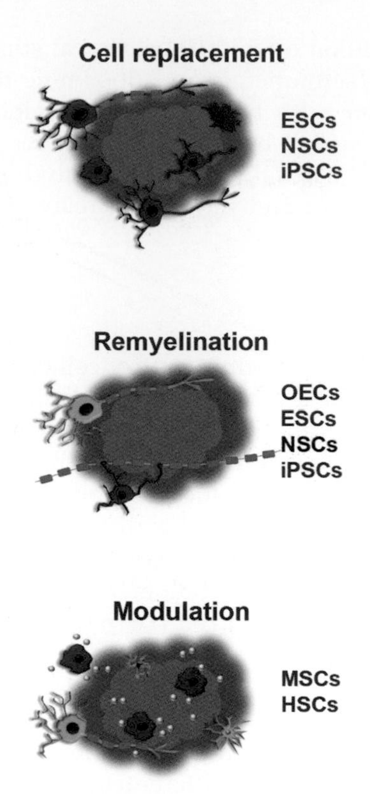

FIGURE 15.4 **Cell therapy.** The figure summarizes the main mechanisms of action of transplanted stem cells (red). Embryonic stem cells (ESCs), neural stem cells (NSCs), and induced pluripotent stem cells (iPSCs) are able to differentiate into both functional neurons and glia (cell replacement); in addition, stem cells, such as olfactory ensheathing cells (OECs), can promote remyelination and regeneration of the host (green) injured axons. Mesenchymal stem cells (MSCs) and hematopoietic stem cells (HSCs), as well as ESCs, NSCs, and iPSCs, display an anti-inflammatory potential, modulate microglia and astroglia activation, and can secrete neurotrophic factors (yellow dots), supporting a positive functional outcome.

be transplanted in association with neurotrophic factors such as IL-6, BDNF, nerve growth factor (NGF), vascular endothelial growth factor (VEGF), and NT-3.

Olfactory ensheathing cells (OECs) surround olfactory axons and facilitate their lifelong regeneration. They are attractive for their plasticity and allow axons to cross glial scars as well as the boundary between the peripheral and central nervous systems. OECs can stimulate tissue repair and neuroprotection, enhance axonal regeneration and remyelination, activate angiogenesis, and influence the endogenous glia after lesion. However, there are contrasting reports, probably owing to the changes in their biological properties with increasing age and/or passage number.

An innovative approach in regenerative medicine is the *ex vivo* use of transdifferentiated cells for reprogramming somatic cells (e.g. fibroblasts) by inserting some transcription factors (OCT4, SOX2, KLF4, and MYC) to obtain induced pluripotent stem cells (iPSCs). When transplanted into injured murine spinal cords, iPSCs allowed consistent functional recovery attributed to remyelination and serotonergic fiber regrowth.

Finally, all cell types tested for cell therapy can be genetically modified to enhance therapeutic potential and to express high levels of neurotrophic factors (NT-3, NT-4/5, BDNF, GDNF, NGF). Gene therapy can provide injured axons and neurons with a local source of trophic molecules, to stimulate neuronal survival and possibly axonal growth.

Scaffolds

Since the glial scar acts as a physical barrier and the lesion area is characterized by many inhibitory factors, axonal regeneration is very limited. To overcome these factors, implantable scaffolds can bridge the gap.

The biomaterials used for SCI treatment must be biocompatible with the host tissue, display an adjustable rate of degradation with non-toxic discards, and have specific mechanical properties, especially elasticity, associated with anti-inflammatory and regenerative properties. These scaffolds are viscoelastic injectable gels, which can conform to the shape of the lesion *in situ* and perfectly fill the gap. They can be associated with cells and other substances such as microparticles and nanoparticles, and growth factors. Therefore, scaffolds not only support regeneration but also can enhance cell survival after transplantation and promote cell differentiation into desired phenotypes.

Scaffolds can be made of both natural and synthetic materials. Natural materials include fibrin, hyaluronic acid, alginate, collagen, agarose, chitosan, matrigel, and methylcellulose hydrogels. Although they contain intrinsic amino acid useful for cell adhesion and are easily degraded by the organism, they can trigger an immune response. Fibrin is the most commonly used, because of its biocompatibility, biodegradability, flexibility, and plasticity. When used alone, it can promote regeneration and delay accumulation of reactive astrocytes at the lesion site. These positive results can be enhanced by the association with stem cells, such as bone marrow stem cells, or growth factors such as NT-3, NGF, or platelet-derived growth factor, determining an increase in axon density and in stem cell vitality and differentiation.

Collagen and hyaluronic acid are widely used and their implantation creates a favorable environment for nerve regeneration, significantly improving the recovery of locomotor and sensory functions. Moreover, hyaluronic acid can modulate the immune response by inhibiting lymphocyte migration and proliferation, granulocyte phagocytosis and degranulation, and macrophage motility, positively influencing scar formation. Collagen can promote regeneration, but it is a component of the glial scar and its role is still debated. However, grafted collagen filaments in adult rats with SCI enhance neurite outgrowth and promote partial functional recovery, when the collagen filaments are aligned rostrocaudally.

Synthetic scaffolds such as biodegradable [polyester of lactic acid (PLA), polyester of glycolic acid (PGA), and polyethylene glycol (PEG)] or non-biodegradable hydrogels, methacrylate-based, allow better control of the chemical and physical characteristics of the material. The best synthetic materials are PGA and their copolymers, for their biodegradability and bioresorbability *in vivo*. NSCs and Schwann cells can be associated with a PGA scaffold to enhance their growth-promoting properties on axons across the transected spinal cord, to reduce the loss of tissue and the glial scar, and to improve motor behavior. Other scaffolds, such as poly(*N*-isopropylacrylamide)-co-poly(ethylene glycol) (PNIPAAm-PEG) and HPMA-RGD hydrogels [*N*-(2-hydroxypropyl)-methacrylamide with attached amino acid sequences Arg–Gly–Asp], seeded with MSCs, have achieved positive results in SCI as well.

Natural or synthetic materials with different properties can enhance the physical and biological properties of both types. For example, the association of collagen and PGA fibers has been used in human patients with peripheral nerve injuries.

Gene Therapy

It has been reported that the adult spinal cord fails to activate the intrinsic growth program during injury, thus limiting axon regeneration. To boost regrowth, a gene transfer strategy has been proposed, either by gene transfer into cells to be engrafted or by direct injection *in vivo* (Fig. 15.5). Gene transfer of transcription factors such as CREB or STAT3 enhances axonal growth but *in vivo* effects are limited to short-distance sprouting around the lesion site.

Neuronal overexpression of proteins playing a major role in CNS development, such as RARβ and neuronal calcium sensor-1 (NCS-1), has shown promise in augmenting the growth of injured axons. Increased PI3K/Akt signaling by viral overexpression of NCS-1 in corticospinal neurons induces collateral sprouting of uninjured axons and short-distance regeneration of injured axons in the spared gray matter. The latter study used lentiviral vectors to express the gene of interest together with green fluorescent protein to specifically label axons of transduced neurons. A similar approach for the rapid cloning and identification of genes with growth-promoting properties has also been described. Genes that are developmentally downregulated and promote neurite growth may also provide targets for future gene transfer studies of injured neurons. Activating neurotrophin signaling in injured neurons, by viral BDNF gene delivery or protein infusions to rubrospinal and corticospinal neurons, can also increase the expression of genes associated with regeneration such as GAP-43 and βIII-tubulin, thereby enhancing axonal sprouting. However, the activation of intrinsic receptors for BDNF is insufficient to induce corticospinal axon growth into a BDNF-expressing graft. Only if intrinsic neuronal growth mechanisms are enhanced by overexpression of trkB in adult corticospinal motor neurons can corticospinal axons regenerate into cellular grafts expressing BDNF. While these studies demonstrated, for the first time, the regeneration of injured corticospinal axons into a cellular graft, trkB trafficking was limited to subcortical axons; and only if BDNF-expressing grafts were placed in a subcortical lesion site was axon growth observed. In contrast, BDNF delivery in the spinal cord failed to induce axon regeneration, probably owing to the lack of trkB transport to more distal spinal cord axons. Eliciting corticospinal regeneration into spinal cord lesions will therefore require a means to enhance trkB trafficking along injured axons.

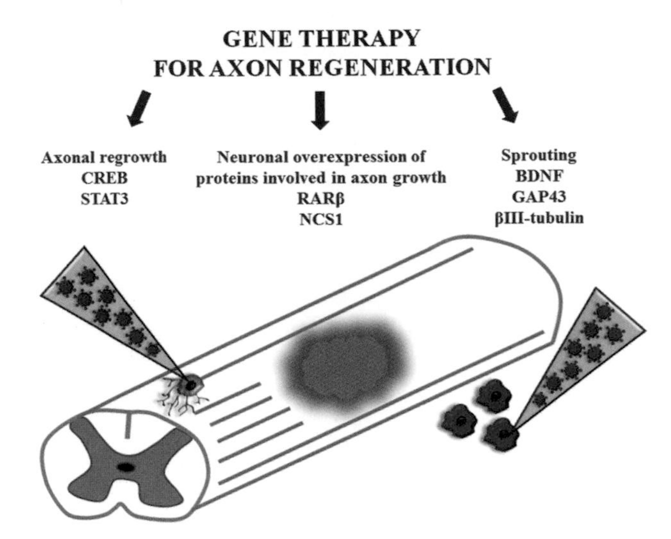

FIGURE 15.5 **Gene therapy.** This therapeutic approach boosts the limited intrinsic regenerative potential of injured spinal cord: both transplanted cells (red) and host cells (blue) can be modified by gene transfer (pipette tip). Depending on the transfected genes, it is possible to induce axonal regrowth (*CREB*, *STAT3*), to overexpress proteins involved in fiber growth (*RARβ*, *NCS1*), and to enhance sprouting (*BDNF*, *GAP43*, *βIII-tubulin*).

QUESTIONS FOR FURTHER RESEARCH

There is still much work to be done in this field and it will require active involvement from both basic and translational neuroscientists. Many of these needs will have become obvious from the material presented in this chapter. For example, we desperately need therapies, including those aimed specifically at the early and late phases of SCI. We need to learn how to prevent glial scar formation, dissolve glial scar when it does form, promote the intrinsic growth program of motor neurons, and prevent neuronal and oligodendroglial cell death. We should explore ways to create a permissive and supportive environment to axonal growth by manipulating the extracellular matrix, creating bridges by transplanting stem cells and implanting scaffolds, supplying trophic factors, reinforcing intrinsic spared neural networks, and stimulating axonal growth by specific rehabilitation exercises. Six million people with SCI is a large number. Add to this the number of caregivers and the social and economic impacts and one begins to see the huge dimensions of the worldwide problem. Only through a concerted effort by many people in many corners of neuroscience can we be assured of the progress that is so urgently required.

References

1. Devivo MJ. Epidemiology of traumatic spinal cord injury: trends and future implications. *Spinal Cord*. 2012;50:365–372.
2. Cripps RA, Lee BB, Wing P, Weerts E, Mackay J, Brown D. A global map for traumatic spinal cord injury epidemiology: towards a living data repository for injury prevention. *Spinal Cord*. 2011;49:493–501.
3. Mothe AJ, Tator CH. Advances in stem cell therapy for spinal cord injury. *J Clin Invest*. 2012;122:3824–3834.
4. Thuret S, Moon LDF, Gage FH. Therapeutic interventions after spinal cord injury. *Nat Rev Neurosci*. 2006;7:628–643.
5. McDonald JW, Sadowsky C. Spinal-cord injury. *Lancet*. 2002;359:417–425.
6. Rowland JW, Hawryluk GW, Kwon B, Fehlings MG. Current status of acute spinal cord injury pathophysiology and emerging therapies: promise on the horizon. *Neurosurg Focus*. 2008;25:E2.
7. Dumont RJ, Okonkwo DO, Verma S, et al. Acute spinal cord injury, part I: Pathophysiologic mechanisms. *Clin Neuropharmacol*. 2001;24:254–264.
8. Boido M, Rupa R, Garbossa D, Fontanella M, Ducati A, Vercelli A. Embryonic and adult stem cells promote raphespinal axon outgrowth and improve functional outcome following spinal hemisection in mice. *Eur J Neurosci*. 2009;30:833–846.
9. Nakamura M, Houghtling RA, MacArthur L, Bayer BM, Bregman BS. Differences in cytokine gene expression profile between acute and secondary injury in adult rat spinal cord. *Exp Neurol*. 2003;184:313–325.
10. Beattie MS, Hermann GE, Rogers RC, Bresnahan JC. Cell death in models of spinal cord injury. *Prog Brain Res*. 2002;137:37–47.
11. Park E, Velumian AA, Fehlings MG. The role of excitotoxicity in secondary mechanisms of spinal cord injury: a review with an emphasis on the implications for white matter degeneration. *J Neurotrauma*. 2004;21:754–774.
12. Dietz V. Neuronal plasticity after a human spinal cord injury: positive and negative effects. *Exp Neurol*. 2012;235:110–115.
13. Balentine JD. Pathology of experimental spinal cord trauma. II. Ultrastructure of axons and myelin. *Lab Invest*. 1978;39:254–266.
14. Banik NL, Powers JM, Hogan EL. The effects of spinal cord trauma on myelin. *J Neuropathol Exp Neurol*. 1980;39:232–244.
15. Liu NK, Wang XF, Lu QB, Xu XM. Altered microRNA expression following traumatic spinal cord injury. *Exp Neurol*. 2009;219:424–429.
16. Fitch MT, Silver J. CNS injury, glial scars, and inflammation: inhibitory extracellular matrices and regeneration failure. *Exp Neurol*. 2008;209:294–301.
17. Renault-Mihara F, Okada S, Shibata S, Nakamura M, Toyama Y, Okano H. Spinal cord injury: emerging beneficial role of reactive astrocytes' migration. *Int J Biochem Cell Biol*. 2008;40:1649–1653.
18. Silver J, Miller JH. Regeneration beyond the glial scar. *Nat Rev Neurosci*. 2004;5:146–156.
19. Larner AJ, Johnson AR, Keynes RJ. Regeneration in the vertebrate central nervous system: phylogeny, ontogeny, and mechanisms. *Biol Rev Camb Philos Soc*. 1995;70:597–619.
20. David S, Kroner A. Repertoire of microglial and macrophage responses after spinal cord injury. *Nat Rev Neurosci*. 2011;12:388–399.
21. Hawthorne AL, Popovich PG. Emerging concepts in myeloid cell biology after spinal cord injury. *Neurotherapeutics*. 2011;8:252–261.
22. Noble LJ, Mautes AE, Hall JJ. Characterization of the microvascular glycocalyx in normal and injured spinal cord in the rat. *J Comp Neurol*. 1996;376:542–556.
23. Mautes AE, Weinzierl MR, Donovan F, Noble LJ. Vascular events after spinal cord injury: contribution to secondary pathogenesis. *Phys Ther*. 2000;80:673–687.
24. Sharma K, Selzer ME, Li S. Scar-mediated inhibition and CSPG receptors in the CNS. *Exp Neurol*. 2012;237:370–378.
25. Galtrey CM, Fawcett JW. The role of chondroitin sulfate proteoglycans in regeneration and plasticity in the central nervous system. *Brain Res Rev*. 2007;54:1–18.
26. Akbik F, Cafferty WB, Strittmatter SM. Myelin associated inhibitors: a link between injury-induced and experience-dependent plasticity. *Exp Neurol*. 2012;235:43–52.
27. Giger RJ, Hollis II ER, Tuszynski MH. Guidance molecules in axon regeneration. *Cold Spring Harb Perspect Biol*. 2010;2. a001867.
28. McEwen ML, Sullivan PG, Rabchevsky AG, Springer JE. Targeting mitochondrial function for the treatment of acute spinal cord injury. *Neurotherapeutics*. 2011;8:168–179.
29. Repici M, Chen X, Morel MP, et al. Specific inhibition of the JNK pathway promotes locomotor recovery and neuroprotection after mouse spinal cord injury. *Neurobiol Dis*. 2012;46:710–721.
30. Tetzlaff W, Kobayashi NR, Giehl KM, Tsui BJ, Cassar SL, Bedard AM. Response of rubrospinal and corticospinal neurons to injury and neurotrophins. *Prog Brain Res*. 1994;103:271–286.
31. Di Giovanni S. Molecular targets for axon regeneration: focus on the intrinsic pathways. *Expert Opin Ther Targets*. 2009;13:1387–1398.
32. Jaerve A, Schiwy N, Schmitz C, Mueller HW. Differential effect of aging on axon sprouting and regenerative growth in spinal cord injury. *Exp Neurol*. 2011;231:284–294.
33. Gwak YS, Kang J, Unabia GC, Hulsebosch CE. Spatial and temporal activation of spinal glial cells: role of gliopathy in central neuropathic pain following spinal cord injury in rats. *Exp Neurol*. 2012;234:362–372.
34. Hawkins BT, Davis TP. The blood–brain barrier/neurovascular unit in health and disease. *Pharmacol Rev*. 2005;57:173–185.
35. Bartanusz V, Jezova D, Alajajian B, Digicaylioglu M. The blood–spinal cord barrier: morphology and clinical implications. *Ann Neurol*. 2011;70:194–206.
36. Deumens R, Joosten EA, Waxman SG, Hains BC. Locomotor dysfunction and pain: the scylla and charybdis of fiber sprouting after spinal cord injury. *Mol Neurobiol*. 2008;37:52–63.
37. Krassioukov A, Warburton DE, Teasell R, Eng JJ. Spinal Cord Injury Rehabilitation Evidence Research Team. A systematic review of the management of autonomic dysreflexia after spinal cord injury. *Arch Phys Med Rehabil*. 2009;90:682–695.
38. Edgerton VR, Roy RR. Robotic training and spinal cord plasticity. *Brain Res Bull*. 2009;78:4–12.
39. van den Brand R, Heutschi J, Barraud Q, et al. Restoring voluntary control of locomotion after paralyzing spinal cord injury. *Science*. 2012;336:1182–1185.
40. Garbossa D, Boido M, Fontanella M, Fronda C, Ducati A, Vercelli A. Recent therapeutic strategies for spinal cord injury treatment: possible role of stem cells. *Neurosurg Rev*. 2012;35:293–311.

CHAPTER

16

Traumatic Brain Injury

Daniel H. Daneshvar[*,†,‡], Ann C. McKee[*,†,‡,§,¶]*

[*]Center for the Study of Traumatic Encephalopathy, [†]Alzheimer's Disease Center, [‡]Department of Neurology, and, [§]Department of Pathology, Boston University School of Medicine, Boston, Massachusetts, USA, [¶]VA Boston HealthCare System, Boston, Massachusetts, USA

OUTLINE

INTRODUCTION: EPIDEMIOLOGY AND CLASSIFICATIONS

Traumatic brain injury (TBI) occurs when a force transmitted to the head or body results in neurological dysfunction. TBI is a major public health problem worldwide; in the USA alone, an estimated 1.7 million people sustain a TBI requiring hospital evaluation annually.[1] These episodes result in 1.365 million emergency room visits and 275,000 hospital admissions annually, with associated direct and indirect costs estimated to have been 60 billion USD in the USA in 2000.[1,2] However, if people evaluated

Neurobiology of Brain Disorders
http://dx.doi.org/10.1016/B978-0-12-398270-4.00016-1

for TBI in outpatient settings or in military facilities were considered, the actual number would be considerably higher.[3]

TBI is the leading cause of death and disability for people between ages 1 and 44 years, and an estimated 5.3 million Americans, almost 2% of the population, live with long-term disabilities due to a prior TBI.[3,4] The age groups most likely to sustain a TBI are children aged 0–4 years, adolescents aged 15–19, and adults aged 65 years or older (Figs 16.1 and 16.2).[2] Between ages 5 and 25, rates of TBI emergency department visits are approximately two times higher for males compared to females (Fig. 16.3).[2] Motor vehicles are a major cause of death and disability from TBI, particularly in young people, in children younger than 4 years, and people older than 65 years.

Falls account for most TBIs (Fig. 16.4).[2] Among active military service members, blasts from improvised explosive devices (IEDs), mortar rounds, and rocket-propelled grenades are a leading cause of TBI, and many other activities experienced during military training, recreation, and combat put military personnel at high risk for TBI.[3]

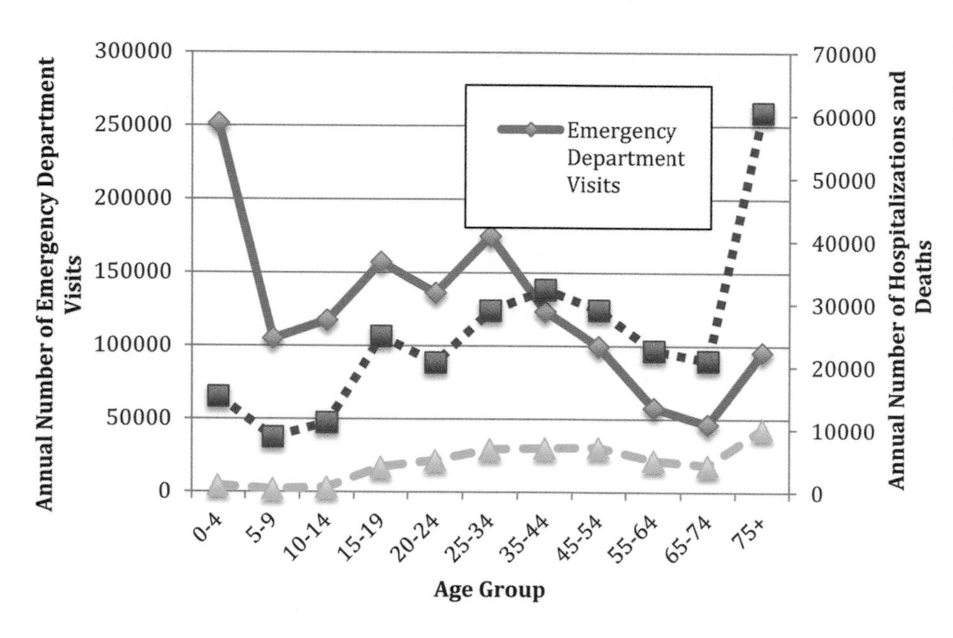

FIGURE 16.1 **Estimated number of emergency room visits, hospitalizations, and deaths, in the USA, by age group, 2002–2006.** *Source: Data from Faul* et al. *US Department of Health and Human Services; 2010.*[2]

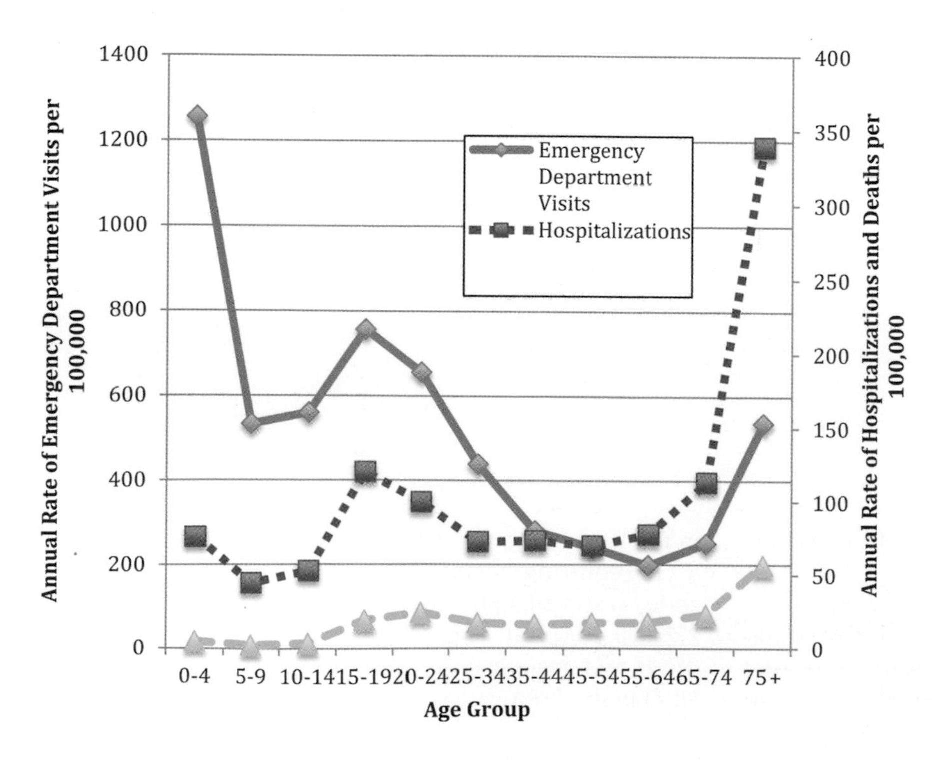

FIGURE 16.2 **Estimated rate of emergency room visits, hospitalizations, and deaths per 100,000, in the USA, by age group, 2002–2006.** *Source: Data from Faul* et al. *US Department of Health and Human Services; 2010.*[2]

Severity of Injury

Severity of TBI can be quantified using a variety of measures. The two most commonly used assessment scales, the Glasgow Coma Scale (GCS) score and the duration of loss of consciousness (LOC) or post-traumatic amnesia (PTA), base TBI severity on clinical symptoms. Although patient recovery following TBI is often projected based on measured injury severity, the extent to which each of these severity assessments correlates with outcome is unclear.

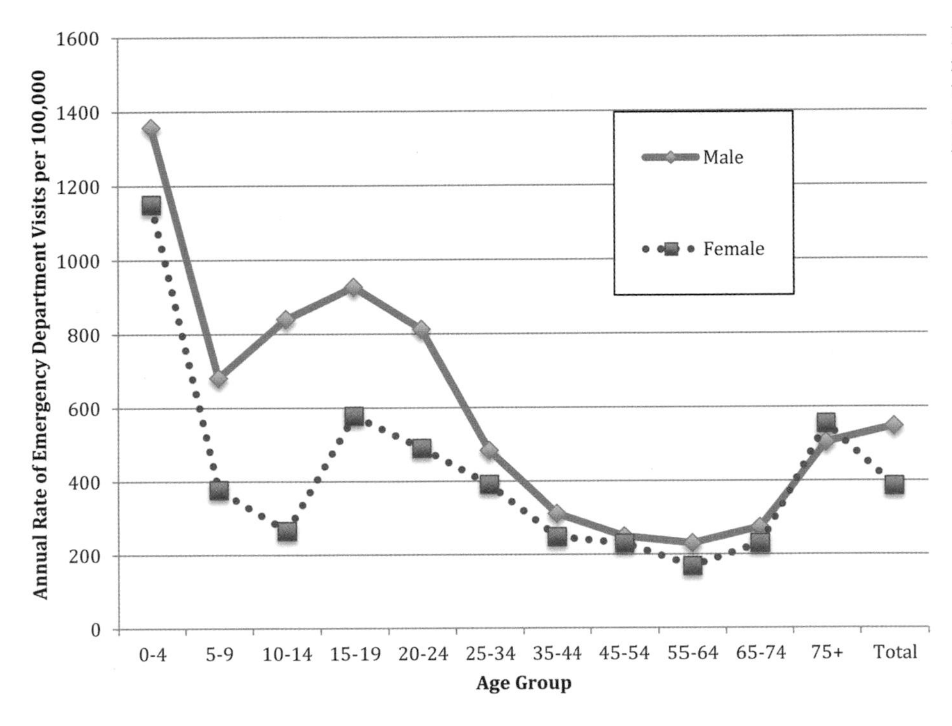

FIGURE 16.3 **Gender differences in estimated rate of emergency room visits per 100,000, in the USA, by age group, 2002–2006.** *Source: Data from Faul* et al. *US Department of Health and Human Services; 2010.*[2]

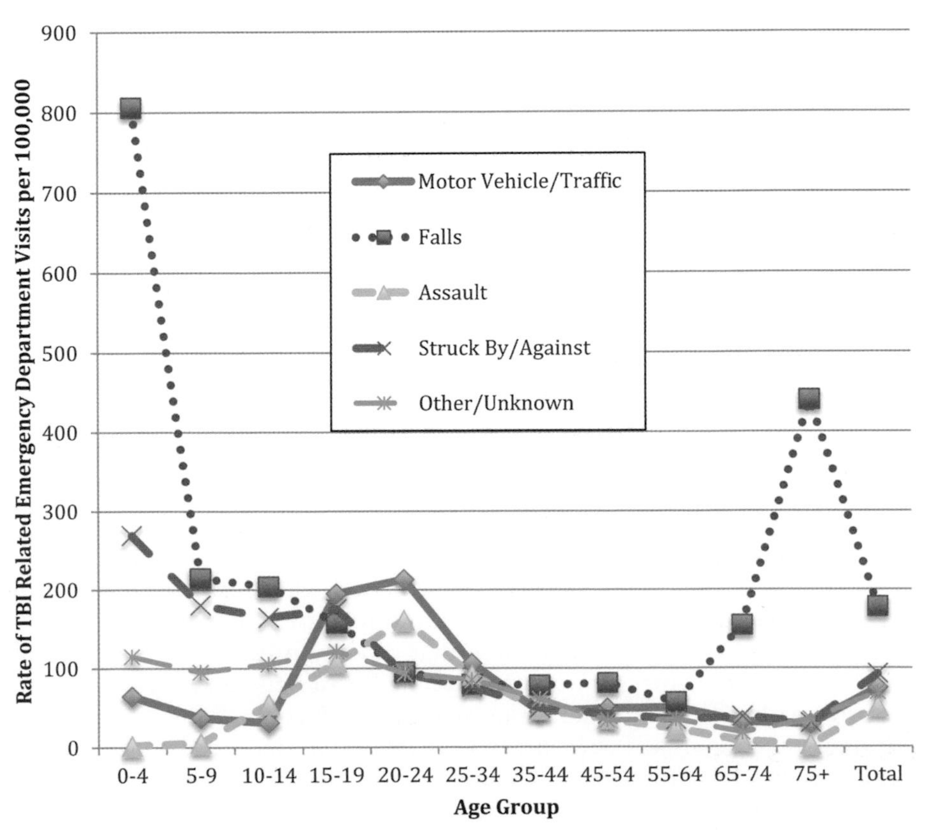

FIGURE 16.4 **Estimated rate of traumatic brain injury (TBI)-related emergency room visits per 100,000, in the USA, by external cause and age group, 2002–2006.** *Source: Data from Faul* et al. *US Department of Health and Human Services; 2010.*[2]

TABLE 16.1 Standard Glasgow Coma Scale Scores and Severity

Score	1	2	3	4	5	6
Eye opening	none	to pain	to speech	spontaneous	N/A	N/A
Verbal response	none	incomprehensible sounds	inappropriate words	confused	oriented	N/A
Motor response	none	extension	abnormal flexion	withdraws	localizes	obeys commands
Sum score	Severe TBI		Moderate TBI		Mild TBI	
	3≤8		8≤12		12≤15	

N/A: not applicable; TBI: traumatic brain injury.

The GCS can be used to grade TBI as mild, moderate, or severe (Table 16.1).[5] The main advantage of the GCS is its simplicity and, as a result, it can be used to compare outcomes across a series of patients.[5] An estimated 75–85% of all TBIs are categorized as mild TBI, with a GCS score of 13–15.[6] Mild TBI includes concussion as well as some subconcussive impacts and some blast injuries from IEDs. There is often full neurological recovery after mild TBI, although 15–30% of subjects develop persistent neurocognitive and behavioral changes.[7] Furthermore, all grades of TBI, including mild, can produce long-term physical, emotional, behavioral, and cognitive consequences that permanently affect an individual's ability to return to work and perform routine activities.[3,7]

Sports-related concussive mild TBI is especially common, with an estimated 1.6–3.8 million concussions occurring annually in the USA.[3] Concussion is particularly frequent in American football, where 4.5% of high-school, 6.3% of collegiate, and 6.6% of professional players are diagnosed with at least one concussion per season.[1,8] However, these numbers greatly underestimate the frequency of concussion because the injuries often resolve spontaneously and many are not reported.[1] When athletes are asked about concussion symptoms, 47% of high school and 70.4% of college football players endorse concussion symptomology.[1] Mild TBI from closed head injury occurs in a wide variety of other sports, including boxing, cheerleading, hockey, lacrosse, rugby, soccer, and wrestling.[1] In general, athletes are at higher risk of concussion during competitions than in practice sessions and, within a given sport, females report more concussions than males.[1] Although the rate of concussion has steadily increased over recent decades, this trend in part reflects improved concussion detection.[1]

In moderate TBI (GCS 9–13), the patient is usually lethargic or stuporous, while in severe TBI (GCS 3–8), the patient is comatose, unable to open the eyes or follow commands. Patients with severe TBI are at high risk for secondary brain injury including hypotension, hypoxemia, and brain swelling.[9] In these lower ranges of GCS score,[3–9] there is a direct linear relation to a poor outcome (death, vegetative state, or severe neurological disability). Advancing age, particularly over the age of 60 years, is also associated with an increased risk of a poor outcome.[9]

The severity of TBI may also be categorized by the duration of LOC and PTA; a mild TBI is defined as LOC of less than 1 hour and PTA for less than 24 hours. A moderate TBI occurs when there is LOC between 1 and 24 hours or PTA for 1–7 days. TBI is considered severe when there is LOC for more than 24 hours or PTA for more than 1 week.[10] Several studies have reported that these measures may have better correlation with patient outcome.[10]

Focal Versus Diffuse Injury

TBIs are typically classified as focal or diffuse based on the presence or absence of focal lesions. Although injuries are considered predominantly focal or diffuse based on this criterion, there is increasing recognition that there is overlap between these pathologies, especially in cases of severe TBI.[11] Mass lesions, such as contusion, subdural hematoma, and epidural hematoma, are considered focal injuries, whereas diffuse injury encompasses widely distributed damage, such as those caused by diffuse axonal injury (DAI), hypoxic–ischemic injury, and microvascular injury. The mortality rate for focal injuries has been reported to be approximately 40% and, for diffuse injuries, approximately 25%.[12] The classifications for common injury types are listed in Table 16.2.

Primary Versus Secondary Injury

Neuronal injuries resulting from TBI can also be characterized based on whether they result directly from the initial trauma, or indirectly. Primary injuries following TBI are those that directly result from the external mechanical force leading to brain tissue deformation and disruption of normal brain function. Types of mechanical forces include acceleration, deceleration, rotational forces, blast winds, blunt impact, and penetration by a projectile. These forces directly damage the blood

TABLE 16.2 Categories of Neurological Damage following Traumatic Brain Injury

Injury mechanism	Focal	Diffuse
Primary	Skull fracture	Diffuse axonal injury
	Cortical contusion	Petechial hemorrhage
	Focal hemorrhage	Blast injury
	Intracranial hematoma	Excitotoxicity
	Focal axonal injury	
Secondary	Microvascular injury	
	Hypoxic–ischemic injury	
	Neuroinflammation	
	Hypometabolism	
	Edema and herniation	
	Excitotoxicity	

vessels, neurons, axons, dendrites, and glia in a focal, multifocal, or diffuse pattern, and initiate a dynamic series of complex cellular, inflammatory, mitochondrial, neurochemical, and metabolic alterations.[11] The magnitude of the primary injury resulting from an impact can typically only be modified by the use of preventive measures, such as protective equipment.

Although the immediate neurological damage produced by primary traumatic forces is usually not alterable, TBI sets in motion a progressive cascade of secondary events that are potentially reversible. Whereas specific primary injury mechanisms are typically tied to specific injuries, secondary injury mechanisms can result from either focal or diffuse injuries (Table 16.2). Secondary brain injury occurs as a complication of the different types of primary brain damage and includes ischemic and hypoxic damage and cerebral swelling, the consequences of raised intracranial pressure, hydrocephalus, and infection.[11] Within hours of the trauma, cellular and vasogenic fluid accumulates in the brain with resulting cerebral edema, elevated intracranial pressure, and cerebral ischemia. Brain dysfunction and morbidity are further increased by a reduction in cerebral blood flow or oxygen content below a threshold level or by cerebral herniation.[11]

PRIMARY EFFECTS OF TRAUMATIC BRAIN INJURY

Specific injuries typically have consistent focal and diffuse primary effects. Common injury mechanisms can therefore be distinguished based on whether they are associated primarily with focal or diffuse pathology. Those with focal pathology tend to have more disparate symptoms, based on the specific foci of injury, compared with diffuse injuries.

Focal Injuries

Skull Fractures

The presence of a skull fracture indicates that the impact has had considerable force and is a significant risk factor for mortality and morbidity.[13] However, many fatal head injuries are closed head injuries without evidence of skull fracture[13] and many patients with a skull fracture do not have evidence of serious brain injury. A patient with a skull fracture following TBI is significantly more likely to have subarachnoid, subdural, or epidural hemorrhage.[13]

Cortical Contusion

Contusions are considered hallmarks of TBI, and their presence confirms that a head injury has occurred. They most often involve the inferior aspect of the frontal lobes, pole, and inferolateral part of the temporal lobes where brain tissue comes in contact with irregular bony protuberances in the skull base. Contusions are focal injuries that result from damage to small blood vessels and other components of the brain parenchyma producing hemorrhages at right angles to the cortical surface. Contusions typically are most severe at the gyral crests but may extend through the cortex into the subcortical white matter as a wedge-shaped necrotic area. They may occur directly beneath the impact site, as "coup" contusions, or opposite the site of impact, as "contrecoup" contusions. A herniation contusion secondary to raised intracranial pressure may occur at the margins of brain herniation sites. Laceration results when there is a physical disruption of the parenchyma of the brain, and laceration–contusions are sometimes found together at the surface of the brain.[11]

Hemorrhage and Hematoma

Traumatic brain hemorrhage results from tearing of blood vessels at the moment of head impact. Gradually expanding, delayed post-traumatic hematomas may not be apparent clinically until hours or days after the initial injury, when they cause elevated intracranial pressure and herniation.[11] These injuries are quite common; in fact, traumatic subdural hemorrhage and traumatic subarachnoid hemorrhage are the most common nonconcussive injuries seen in individuals hospitalized following TBI.[14] The resulting hemorrhage, in some cases, may extend over an entire hemisphere.

Diffuse Injuries

Diffuse Vascular Injury

Petechial hemorrhages are common findings in fatal cases of severe TBI. The main collisions responsible for these injuries involve rapid accelerations and decelerations, causing capillary shearing. These hemorrhages are not typically visible using most current neuroimaging techniques; however, they may coalesce into larger lesions with progressive secondary hemorrhage.

Diffuse Axonal Injury

Although axonal injury may be focal, axonal injuries are most often diffuse or multifocal.[11] The elasticity of the neural tissue allows it to deform in response to normal head movement, and when the relatively large human brain is exposed to rapid acceleration and deceleration, the force may exceed the maximum elasticity of the tissue, resulting in DAI.[15] Severe traumatic injury may result in primary axotomy, whereas less severe injuries produce focal pathological abnormalities resulting in delayed secondary axotomy.[11] The microscopic injury of DAI is generally poorly detected with conventional structural imaging, although diffusion tensor imaging (DTI) has provided experimental evidence of DAI and may prove a useful clinical tool in the future.[16]

Excitotoxicity and Oxidative Stress

In addition to the traumatic stretch injury of axons and other cellular compartments after TBI, neurotransmitters, including glutamate, are abruptly released with massive increases in intracellular calcium, glucose hypermetabolism, kinase activation, and diminished cerebral blood flow. Functional magnetic resonance imaging (fMRI) studies have detected alterations in brain activation patterns in individuals with persistent symptoms after mild TBI.[17] These abnormal brain activation patterns can persist for months after injury, despite normal neurocognitive task performance.[17] The discrepancy between fMRI and neurocognitive testing may be the result of functional reallocation of neurocognitive resources as a compensatory mechanism, followed by a more prolonged period of microstructural recovery.[18]

Blast Injury

Blast injury is particularly relevant given the increased use of IEDs in current combat spheres; it is currently considered the signature TBI for active military service personnel. A blast injury results from the rapid transmission of an acoustic wave through the brain tissue, along with accompanying blast winds.[19] Although blast injuries can be heterogeneous and quite different from traumatic impact injury,[19] kinematic analysis of experimental blast injury in mice reveals that blast wind-induced head oscillations at angular accelerations are the primary pathway by which blast exposure initiates acute brain injury, and that the resultant cerebral injury is similar to the effects of repetitive concussive impacts. Blast injuries can also cause diffuse and focal hemorrhage and edema as vessels and tissue rapidly contract and expand, several times within a fraction of a second, and even small blast injuries produce considerable microvascular pathology.

SECONDARY EFFECTS OF TRAUMATIC BRAIN INJURY

Both diffuse and focal primary injuries can initiate pathological cascades resulting in similar secondary injuries. Unlike primary injuries, the extent of secondary injuries can be modified via treatment. Prompt and proper treatment is all the more important as these injuries are the leading cause of in-hospital mortality in TBI patients.[20] Secondary injuries may be the natural extension of primary injuries, such as damage due to reactive oxygen species generated from damaged tissue or insufficient oxygenation due to vascular damage. Alternatively, secondary injuries may occur when otherwise beneficial functions, such as clearing tissue debris, cause damage.[20]

Vascular Injury

Progressive secondary hemorrhage is one of the most damaging secondary injuries resulting from TBI.[20] This secondary injury occurs within hours of the TBI and results in expansion of tissue damage, due not only to increased intracranial pressure, but also to ischemia, hypoxia, free radical formation, and induction of inflammation.

Hypoxic–Ischemic Injury

Hypotension, hypoxia, and ischemia are common secondary events following TBI and comprise the majority of prehospitalization secondary injuries in severe TBI.[20] This injury is an important target for research because secondary hypoxic episodes have been established as causes of death. However, as excessive hyperoxemia is also associated with tissue damage, any potential therapeutic would need to be carefully monitored.

Neuroinflammation

The endogenous and exogenous neuroinflammatory response to TBI compounds the primary injury. TBI causes rapid microglial activation, resulting in the release of proinflammatory cytokines and other neurotoxic products that generate free radicals.[21] In addition, neutrophils are recruited to phagocytose and clear cellular debris after TBI and, in the process, more free radicals are released.[21] These free radicals cause tissue damage and harm otherwise healthy cells, thereby propagating tissue injury.[21]

Hypometabolism

Fluorodeoxyglucose positron emission tomography (FDG-PET) and fMRI, as well as animal models, have shown evidence of hypometabolism in military veterans and athletes exposed to TBI. Although the precise cause of this finding is unknown, it is thought that either DAI or impaired mitochondrial function is responsible.[22] Regardless of etiology, studies suggest that hypometabolism following TBI is associated with depression and worse prognosis.[22]

Edema and Herniation

Secondary brain swelling encompasses both edema and congestion and, together with hematomas, is the major contributor to increased intracranial pressure. The edema is caused because TBI alters the permeability of the blood–brain barrier, resulting in altered fluid homeostasis. It may eventually lead to distortion, shift, and herniation of the brain. Edematous swelling often occurs around contusions and intracerebral hemorrhages, whereas swelling of one hemisphere may result from a combination of congestion and edema. In addition to the damage caused by increased intracranial pressure, the alteration of solute concentration can impair neuronal function.[23] Diffuse swelling of the entire brain owing to hyperemia occurs in young children, even after an apparently trivial injury, but is uncommon in adults. Generalized brain swelling produces gyral flattening, sulcal narrowing, and ventricular collapse.[23]

CHRONIC EFFECTS OF TRAUMATIC BRAIN INJURY

The Institute of Medicine has examined several epidemiological studies of dementia in individuals exposed to TBI with loss of consciousness, and concluded that there is sufficient evidence to link the two.[4,24] Additional neuropathological analysis is necessary to determine the pathology underlying the dementia, with some studies suggesting a link between brain trauma and Alzheimer disease (AD), Parkinson disease (PD), amyotrophic lateral sclerosis (ALS), and chronic traumatic encephalopathy (CTE).[7,17,25,26]

Alzheimer Disease

TBI has been implicated in the pathogenesis of AD. After age, family history, and ApoE4 genotype, the risk factor with the strongest linkage to AD is a history of TBI. Some studies have also suggested that TBI is associated with an earlier onset of AD. β-Amyloid (Aβ) plaques and intra-axonal Aβ deposits have been found in approximately one-third of TBI patients who died after TBI, even in young people.[27] Murine models show transient elevation of β-amyloid precursor protein (APP) and intra-axonal Aβ deposits[28] after acute TBI; the cleavage of APP results in the Aβ plaques characteristic of AD.[28]

The bulk of human data supporting this relationship comes from epidemiological studies that found a relationship between TBI and dementia in later life.[4,24] However, without neuropathological confirmation, it is unclear what specific neurodegeneration underlies the clinical symptoms of dementia.[7] More research is warranted to elucidate the relationship between AD and TBI, and to ensure neuropathological confirmation of disease in cases of clinically diagnosed probable and possible AD.

Parkinson Disease

In addition to AD, a link has been suggested between TBI and PD. While this relationship has been less explored than that between TBI and AD, animal models have shown that TBI may result in α-synuclein deposition, which results in the Lewy bodies characteristic of PD and Lewy body disease.[29] However, whether the link between PD and TBI is due to the accumulation of α-synuclein inclusions in neurons of the substantia nigra, as has been suggested by some experimental studies,[29] or to neurofibrillary degeneration of the substantia nigra, such as occurs in CTE,[30] or both, is unclear.

Chronic Traumatic Encephalopathy

To date, all pathologically diagnosed cases of CTE have come from individuals with a history of minor brain trauma, and that trauma has usually been repetitive.[25,30] Not all individuals diagnosed with CTE experienced symptoms after acute mild TBI, suggesting that neurological symptoms at the time of injury are not necessary for the development of CTE.[25,30] CTE typically affects athletes or military personnel involved in high-risk activities, such as collision or contact sports or military service, and other individuals who are exposed to considerable repetitive brain injury, such as head-banging.[25,30,31] Initially, CTE was referred to as "punch drunk" owing to its strong association with boxers, particularly slugging boxers who had experienced prolonged punishment to the head.[25,31] Later, the terms "dementia pugilistica" and CTE were used to indicate a neurodegenerative disease that developed after traumatic exposure from varying sources.[25,30] In addition to boxers, CTE has been identified in American football players, ice hockey players, professional wrestlers, military veterans, physically abused individuals, and people with poorly or uncontrolled epilepsy. One dwarf with CTE had worked for 15 years as a circus clown; he had participated in "dwarf-throwing events" and had been knocked unconscious "a dozen times".[25]

Neuropathology of Chronic Traumatic Encephalopathy

Like AD and most other neurodegenerative diseases, a definite diagnosis of CTE can only be made at postmortem neuropathological examination.[30,31] Currently, there are no widely used clinical diagnostic criteria or biomarkers of disease that can be used to aid in the clinical diagnosis of CTE,[31] and as a result, CTE is defined principally by its neuropathological features (Table 16.3). The earliest neuropathological descriptions of CTE were based on relatively isolated cases of boxers.[25] Gross neuropathological alterations in these early cases of CTE described cerebral atrophy, ventricular enlargement, cavum septum pellucidum, and pallor of the substantia nigra. General histological stains such as cresyl violet and silver stains were used to describe neurofibrillary tangles (NFTs) and neuronal loss in the cerebral cortex, diencephalon, and brainstem, usually in the absence of senile plaques.

The earliest extensive description of neuropathological findings and clinical symptoms was by Corsellis, Bruton, and Freeman-Browne in their landmark report detailing the neuropathological findings in 15 former boxers (reviewed in McKee *et al.*[25]). They summarized the most common neuropathological findings as: (1) a reduction in brain weight, (2) enlargement of the lateral and third ventricles, (3) thinning of the corpus callosum, (4) cavum septum pellucidum with fenestrations, (5) scarring and neuronal loss of the cerebellar tonsils, (6) neurofibrillary degeneration of the substantia nigra, and (7) neurofibrillary degeneration of the cerebral cortex. Only four of the 15 cases had senile plaques. Later, the use of immunocytochemical techniques on 14 of these 15 cases and six additional boxers determined that 19 of the 20 cases had widespread diffuse Aβ deposits.[25] The neuropathological findings of three boxers were reported by Hof and colleagues (reviewed in McKee *et al.*[25]), who noted that the NFTs in CTE tended to be superficially distributed in layers II and III in the neocortical areas and were generally denser than in AD, in the absence of Aβ deposition. Hof also reported high densities of NFTs in the temporal cortex, especially perirhinal and inferior temporal cortex, and amygdala.

TABLE 16.3 Pathological Criteria for the Diagnosis of Chronic Traumatic Encephalopathy

Perivascular foci of p-tau immunoreactive ATs and NFTs
Irregular cortical distribution of p-tau immunoreactive NFTs and ATs with a predilection for the depth of cerebral sulci
Clusters of subpial and periventricular ATs in the cerebral cortex, diencephalon, and brainstem
NFTs in the cerebral cortex located preferentially in the superficial layers.

AT: astrocytic tangle; NFT: neurofibrillary tangle.
Source: Adapted from McKee et al. Brain *2013;136(Pt 1):43–64.*[30]

Neuropathological alterations in five young men, age 23–28 years, were reported by Geddes (reviewed in McKee *et al.*[25]). Two were boxers, one was a soccer player, one was "mentally subnormal" with a long history of head-banging, and another had epilepsy and frequently struck his head during seizures. The brains of the two young boxers showed the most damage; microscopically, there were argyrophilic, tau-positive neocortical NFTs, strikingly arranged in groups around small intracortical blood vessels, usually associated with neuropil threads and granular tau positive neurons; there was no Aβ deposition. A neuropathological analysis of several postmortem human brains from US military veterans exposed to blast mild TBI, young adult athletes with history of concussive mild TBI, and age-matched controls with no history of blast exposure, mild TBI, or concussion, found similar perivascular foci of tau pathology in the brains of athletes with a history of concussive or subconcussive injury as well as in blast-injured veterans.

The largest study of pathologically confirmed CTE[30] delineated the neuropathological alterations of 68 men, ranging in age from 17 to 98 years (mean 59.5 years), including 64 athletes, 21 military veterans (86% of whom were also athletes), and one individual who engaged in self-injurious head-banging behavior. Of those 68 cases of CTE, CTE was the sole diagnosis in 43 cases (63%); eight (12%) were also diagnosed with motor neuron disease (MND), seven (11%) with AD, 11 (16%) with Lewy body disease, and four (6%) with frontotemporal lobar degeneration. The spectrum of hyperphosphorylated tau (p-tau) pathology in CTE ranged in severity from focal perivascular clusters of NFTs in the frontal neocortex (similar to the p-tau pathology found by Geddes in young athletes; reviewed by McKee *et al.*[25]) to severe tauopathy affecting widespread brain regions, including the medial temporal lobe. The ordered and predictable progression of p-tau pathology among the 43 cases of pure CTE allowed a staging scheme of progressive pathology to be proposed, as CTE stages I–IV.

GROSS PATHOLOGICAL FEATURES OF CHRONIC TRAUMATIC ENCEPHALOPATHY

The gross pathological features of CTE are generalized cerebral atrophy, with a predilection for the frontal, temporal, and medial temporal lobes; enlarged ventricles; thinning of the corpus callosum; cavum septum pellucidum, often with fenestrations; thinning of the hypothalamic floor; and shrinkage of the mammillary bodies.[25,26] These changes are not apparent in early-stage CTE (stage I–II), begin to appear in stage III disease, and are fully evident in stage IV. In most cases, there is also pallor of the locus ceruleus and substantia nigra, which can be severe.

MICROSCOPIC PATHOLOGY OF CHRONIC TRAUMATIC ENCEPHALOPATHY

DISTRIBUTION OF HYPERPHOSPHORYLATED TAU Microscopically, CTE is characterized by the aggregations of p-tau protein as NFTs and astrocytic tangles throughout the brain.[7,25,26,30,31] Perivascular multifocal tau pathology localized to the depths of cortical sulci is a unique feature of CTE not found in any other tauopathy or neurodegeneration, including AD (Table 16.4). The progression of CTE has been described in four stages (Fig. 16.5).

STAGES OF CHRONIC TRAUMATIC ENCEPHALOPATHY

- *Stage I*: Tau is found in focal, perivascular clusters as NFTs. Astrocytic tangles are most prominent at the sulcal depths of cerebral cortex, typically involving superior and dorsolateral superior frontal cortex. The surrounding cortex is typically unremarkable and rare NFTs may be found in the deep nuclei, such as the locus ceruleus.[30,31]
- *Stage II*: Multiple discrete clusters of perivascular p-tau NFTs and astrocytic tangles are found in the sulcal depths, most commonly in frontal, temporal, parietal, insular, and septal cortices. NFTs are typically found in the superficial layers of adjacent

TABLE 16.4 Distinctions in Hyperphosphorylated tau (p-tau) Pathology between Alzheimer Disease (AD) and Chronic Traumatic Encephalopathy (CTE)

	Pathological features	AD	CTE
Tau protein			
	6 isoforms	All 6 isoforms present	All 6 isoforms present
	3 or 4 repeat tau	3 repeat and 4 repeat tau present	3 repeat and 4 repeat present
Cell type			
	Neuronal	NFTs and pretangles	NFTs and pretangles
	Astrocytic	Not present[a]	Prominent ATs
Neuronal domain			
	Cell body	Prominent	Prominent
	Dendrite	Prominent	Prominent
	Axon	Sparse	Prominent
Cell origin			
	Perivascular	Not present	Prominent NFTs and ATs
	Foci at depths of cerebral sulci	Not present	Prominent NFTs and ATs
	Irregular, patchy cortical distribution	Not present	Prominent
	Cortical laminae	Predominantly laminae III and V	Predominantly laminae I–III
	Subpial ATs	Not present	Prominent
	Periventricular ATs	Not present	Present
Distribution			
	Mild pathology	Braak stages I–III: NFTs in entorhinal cortex, amygdala, and hippocampus	CTE stages I–II: NFTs in focal epicenters in cerebral cortex, usually frontal lobe
	Advanced pathology	Braak stages IV–VI: High density of NFTs in widespread cortical areas and medial temporal lobe, uniform distribution. Low densities of NFTs in basal ganglia and brainstem; none in mammillary bodies. White matter tracts relatively uninvolved	CTE stages III–IV: High density of NFTs in widespread cortical areas and medial temporal lobe, patchy irregular distribution. High densities of NFTs in thalamus, hypothalamus, mammillary bodies, and brainstem. Moderate densities of NFTs in basal ganglia, especially nucleus accumbens. Prominent p-tau pathology in white matter tracts

AT: astrocytic tangle; NFT: neurofibrillary tangle.

[a]*Low densities of 4R immunoreactive "thorn-shaped astrocytes" are found in the temporal lobe of some older subjects and older subjects with AD.*

Source: Adapted from McKee et al. Brain *2013;136(Pt 1):43–64.*[30]

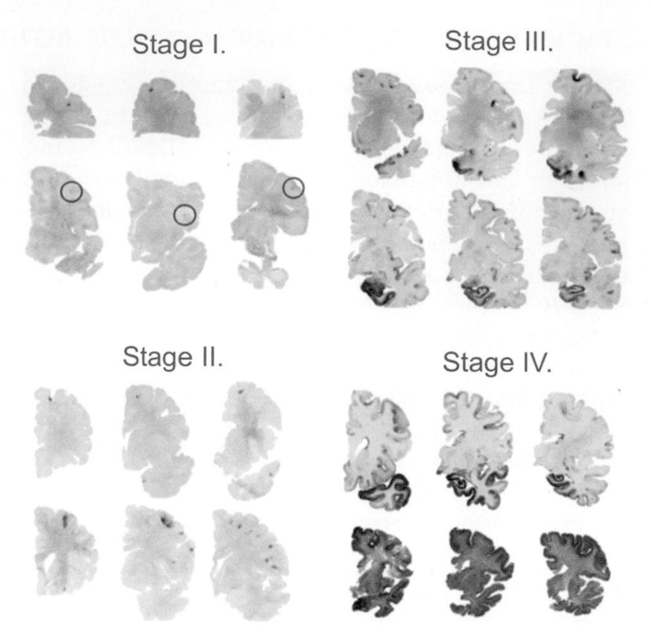

FIGURE 16.5 **Neuropathological stages of chronic traumatic encephalopathy (CTE).** Stage I CTE is marked by hyperphosphorylated tau (p-tau) pathology restricted to discrete foci in the cerebral cortex, commonly in the superior, dorsolateral, or lateral frontal cortices, as well as perivascularly and at the sulcal depths (black circles). Stage II CTE is defined by increased p-tau epicenters at the depths of the cerebral sulci and localized spread of neurofibrillary pathology from these epicenters superficially to the adjacent cortex, with little p-tau found in the medial temporal lobe. In stage III CTE, p-tau pathology is widespread; the frontal, insular, temporal, and parietal cortices show neurofibrillary degeneration. The greatest severity in now in the frontal and temporal lobe, concentrated at the depths of the sulci. Also by stage III CTE, the amygdala, hippocampus, and entorhinal cortex show neurofibrillary pathology. In stage IV CTE, severe p-tau pathology affects most regions of the cerebral cortex and the medial temporal lobe, typically sparing calcarine cortex in most cases. All images are CP-13 immunostained 50 mm tissue sections. *Source: Adapted from McKee* et al. Brain *2013;136(Pt 1):43–64.*[30]

cortex surrounding the epicenters. NFTs are also found in the nucleus basalis of Meynert and locus ceruleus. Rare NFTs may be found in entorhinal cortex, amygdala, hippocampus, thalamus, substantia nigra, and dorsal and median raphe nuclei of the midbrain.[30,31]

- *Stage III*: Medial temporal lobe structures including the hippocampus, entorhinal cortex, and amygdala show dense NFTs, and there is widespread involvement of the frontal, temporal parietal, insula, and septal cortices. NFTs are densely found in olfactory bulbs, hypothalamus, thalamus, mammillary bodies, substantia nigra, and dorsal and median raphe nuclei.
- *Stage IV*: There are dense NFTs in widespread regions of the CNS, with prominent neuronal loss and gliosis of the neocortex and an increasing predominance of tau pathology in astrocytes compared with NFTs. There is prominent neuronal loss in CA1 of the hippocampus, subiculum, and cerebral cortices, and increasing myelinated fiber and axonal loss in the white matter. Severe p-tau pathology is found in the cerebral cortex, diencephalon, basal ganglia, and brainstem, and also affects white matter tracts and spinal cord. Primary visual cortex is generally spared.[30,31]

Tau in CTE is similar biochemically to tau in AD, being composed of both 3- and 4-microtubule binding repeat tau, with a ratio of 4 to 3 microtubule binding repeat tau of about 1, indistinguishable soluble and insoluble tau, and six abnormally phosphorylated tau isoforms with the same electrophoretic mobility pattern.[30,31]

TAR DNA-BINDING PROTEIN-43 PATHOLOGY The majority of CTE cases also have Tar DNA-binding protein-43 (TDP-43) abnormalities, with TDP-43 pathology as intraneuronal and intraglial inclusions and neurites found in more than 80% of CTE cases, and increasing TDP-43 pathology a characteristic of advancing CTE severity.[26]

AXONAL DAMAGE In addition to p-tau pathology, axonal pathology is present at all stages of CTE and appears to progress with stage of CTE.[25,30,32] In the earliest stages of disease, phosphorylated neurofilament immunohistochemistry shows distorted axonal varicosities in cortex, subcortical white matter, and deep white matter tracts of the diencephalon. By stage III, severe axonal loss and pathological profiles are found in the subcortical white matter, and are most severe in the frontal and temporal lobes. In advanced CTE, there is widespread axonal loss, with frequent severely distorted axonal profiles widely distributed in the subcortical white matter.[30]

INFLAMMATION Neuroinflammation, seen as astrocytosis and activated microgliosis of the white matter, is a consistent feature of both CTE and TBI.[32] Clusters of microglia are found in the subcortical white matter, and there is typically a robust astrocytosis.[32]

AMYLOID-β PEPTIDE Aβ peptide deposits are found in 40–50% of CTE cases, are significantly associated with age at death, and are not found in the early stages of CTE (stages I and II).[25,26,30] Aβ plaques in CTE are generally less dense and predominantly of the diffuse type in CTE,[25,26,30] whereas in AD Aβ plaques are more widely distributed and consist of both neuritic and diffuse plaques, with neuritic plaques a requirement for a diagnosis of AD.[30]

LEWY BODIES α-Synuclein-positive Lewy bodies are found in approximately 20% of CTE cases, and are found in subjects who are significantly older than those without Lewy bodies.[30]

Chronic Traumatic Encephalopathy and Comorbid Disease

CTE is associated with the development of other neurodegenerations, notably Lewy body disease, MND, AD, and frontotemporal lobar degeneration.[25,30] Among 68 cases of pathologically confirmed CTE, coexistent Lewy body disease was found in 16% of cases, MND in 12%, AD in 11%, and frontotemporal lobar degeneration in 4%, suggesting that either repetitive trauma or the accumulation of tau pathology in CTE provokes the deposition of other abnormal proteins involved in neurodegeneration.[30]

Pathologically, CTE can be distinguished from other neurodegenerative diseases, including AD. In CTE, there is an irregular distribution of tau pathology at the depths of the cerebral sulci, with a prominent perivascular distribution, subpial tau in astrocytes, and NFTs distributed primarily in superficial cortical layers II and III, in the relative absence of Aβ deposition (Table 16.4). Periventricular regions show intense ependymal immunostaining for tau. Axonal varicosities and neuropil threads in the subcortical and deep white matter are also tau immunopositive. In AD, the cortical distribution of NFTs is diffuse, preferentially involves laminae III and V, and there is no accentuation at depths of sulci, around small blood vessels, in the subpial or periventricular regions. Neuritic Aβ plaques are a necessary feature of all cases of AD, whereas sparse Aβ plaques are found in fewer than half of cases of CTE, and diffuse plaques predominate over neuritic plaques.

Clinical Symptoms of Chronic Traumatic Encephalopathy

CTE is clinically characterized by a progressive decline in memory and executive functioning; mood and behavioral disturbances that include depression, apathy, impulsivity, anger, irritability, suicidal behavior, and aggressiveness; gait changes that resemble parkinsonism; and, eventually, progression to dementia.[16,25,30,31] The clinical diagnosis of early stages of CTE can be complicated by an overlap between the symptoms of early CTE and prolonged postconcussion syndrome.[7] The cognitive symptoms of CTE primarily include episodic memory impairment and executive dysfunction.[25,30,31] Although cognitive symptoms affect the majority of individuals with CTE, these symptoms typically present later in the disease course.[30,31] Mood and behavioral disturbances associated with CTE include depression, apathy, impulsivity, anger, irritability, suicidal behavior, and aggressiveness.[25,30,31] These issues, which are often disconcerting to family members, are often the first reported.[31] However, it is unclear whether these symptoms are the first to appear or are recognized by sensitive observers.[16]

CLINICAL PROGRESSION OF CHRONIC TRAUMATIC ENCEPHALOPATHY

Whereas early-stage CTE may be asymptomatic, advanced CTE stages are associated with increased clinical impairment.[25,31] Although tau burden may be associated with the severity of cognitive impairment in CTE, this relationship has not been fully explored.[25,30,31] In stage I CTE, headache and loss of attention and concentration are found in most cases, although short-term memory difficulties, aggressive tendencies, depression, difficulties with planning and organization, and explosivity may also be reported[30,31] and are common in stage II disease. In stage III CTE, symptoms are memory loss, decreased attention and concentration, executive dysfunction, depression or mood swings, explosivity, suicidal tendencies, visuospatial difficulties, aggression, and cognitive impairment.[30,31] By stage IV CTE, executive dysfunction and memory loss predominate and dementia is common. Most individuals also exhibit a profound loss of attention and concentration, executive dysfunction, language difficulties, explosivity, aggressive tendencies, paranoia, depression, and gait and visuospatial difficulties. Less common symptoms include impulsivity, dysarthria, and parkinsonism.[16,30,31]

A case series of athletes with neuropathologically confirmed CTE suggested that there may be two distinct manifestations of CTE: one that tends to becomes evident at a younger age with symptoms of behavioral (e.g. impulsivity, violence) or mood changes (e.g. depression, hopelessness), and another that typically manifests later in life with initial symptoms of cognitive impairment (e.g. episodic memory deficits, executive dysfunction).[31]

DEMENTIA IN CHRONIC TRAUMATIC ENCEPHALOPATHY

Advanced stage CTE is associated with cognitive impairment, and dementia is likely by stage IV.[30,31] Most subjects diagnosed with stage IV CTE, in the absence of other comorbidities, were clinically diagnosed with dementia during life and would have met the core diagnostic criteria for AD dementia of both the National Institute on Aging–Alzheimer's Association (NIA-AA) Workgroup and the National Institute of Neurological and Communicative Disorders and Stroke and the Alzheimer's Disease and Related Disorders Association (NINCDS-ADRDA).[31] Specifically, individuals with CTE dementia exhibit functional impairment and cognitive or behavioral impairment in at least two domains. In addition, those with CTE dementia meet specific criteria for AD dementia, including insidious onset, progressive course, and initial deficits in either episodic memory or a non-amnestic domain including language, visuospatial, or executive functioning,[31] suggesting that dementia in the two disorders is difficult to distinguish clinically. Additional work to identify differences between the clinical presentation of CTE and AD dementia is needed to improve the specificity of the clinical diagnosis of CTE.

Potential Mechanisms

Several animal models of mild TBI may serve to elucidate the relationship between traumatic injury and development of CTE. In one study, wild-type mice exposed to a single controlled sublethal blast developed neuropathological changes similar to those in humans exposed to blast-related or concussive CTE, including phosphorylated tau immunoreactivity, microvascular pathology, axonopathy, widespread astrocytosis, and microgliosis.[32] These findings suggest that a single blast injury or multiple concussive head impacts initiate similar neuropathological cascades that result in CTE. Although NFTs were not found, this discordance may be due to the fact that wild-type murine tau is aggregated differently from human tau, or that 2 weeks is too soon to produce fibrillar intraneuronal tau inclusions. Further investigation using a mouse line that is transgenic for human tau (e.g. hTau), a longer postinjury period, or multiple blast exposures, is warranted. Additional factors such as other genetic contributions, inflammatory response, age, gender, and other factors with potential to affect CTE pathology could also be explored.[32]

Another study examined the differences in cognitive impairment and pathology between mice exposed to single and repetitive mild TBI. Using a closed head impact model, the authors found short-term behavioral abnormalities in mice exposed to a single mild TBI in the form of transient deficits in motor function and spatial memory. Pathologically, these mice exhibited reactive astrocytosis and sparse amyloid precursor protein-immunoreactive axonal pathology in the corpus callosum.[33] However, in mice exposed to five mild TBIs, administered at 48 hour intervals over 8 days, mores cognitive impairment, microglial activation, reactive astrocytosis, and multifocal axonal pathology were found.[33] However, that model induced more severe traumatic injury than is usually associated with CTE; specifically, there were focal contusion injuries in a subset of both exposure groups.[33]

A third study used a repeated closed head impact system to examine the development of tau pathology in 18-month-old hTau mice, mice that express wild-type human tau isoforms on a null murine tau background.[34] Although hTau mice unexposed to head impacts have some tau pathology at 18 months, there was a significant increase in p-tau immunoreactivity in mice exposed to repetitive mild TBI, but not in mice exposed to single mild TBI. There was also an increase in inflammatory markers, including reactive gliosis and microglial activation, more in the repetitive mild TBI group than in the single mild TBI or control groups,[34] although no perivascular or subpial tau pathology was observed.[34] These and future studies are warranted to continue to advance our understanding of CTE pathogenesis.

Although the precise mechanisms that tie repetitive mild TBI and single TBI to abnormal protein accumulation and neurodegeneration remain to be discovered, axonal injury produced by the initial trauma and aggravated by subsequent traumatic injuries is likely to play a fundamental role. In acute TBI, the brain undergoes shear deformation that elongates and injures axons, small blood vessels, and astrocytes.[35] Unmyelinated axons may be particularly vulnerable given their length and high ratio of axolemma to cytoplasm. Traumatic axonal injury results in alterations in axolemmal permeability, ionic shifts including massive influx of calcium, and release of caspases and calpains that trigger the misfolding, truncation, phosphorylation, and aggregation of many proteins, including tau and TDP-43, the breakdown of the microtubules and neurofilaments, and disrupted axonal transport.[35,36] With increasing axonal injury from repetitive mild TBI or moderate to severe axonal injury from single TBI, toxic p-tau aggregates accumulate in cytoplasm, where they are misrouted into the somatodendritic compartment.[22] Aggregates of misfolded p-tau eventually overwhelm normal clearance mechanisms, allowing p-tau to spread transynaptically and interneuronally, probably involving protein templating mechanisms and extracellular cerebrospinal fluid (CSF) clearance pathways.[37] In addition, acute TBI disrupts the microvasculature, damaging the blood–brain barrier, inducing an inflammatory cascade and microhemorrhages that lead to iron deposition and generation of free radicals, which exacerbates the axonopathy, microvascular disruption, and spread of toxic p-tau aggregates.[36,37] In addition, p-tau aggregates provoke the deposition of other toxic proteins, including Aβ, and TDP-43, and ultimately result in further neuronal and axonal loss and neurodegeneration.[26,35] Furthermore, the mechanism by which an initial multifocal axonal neurodegeneration develops around small blood vessels or in the depths of cortical sulci is probably explained by the physics of shear deformation of the brain: stress and resultant axonal injury are greatest at the interface of two tissues with differing viscoelastic properties (such as seen between blood vessels and brain, or gray and white matter), and the depths of the sulci and immediate perivascular region are areas of stress concentration.[22] In addition, the local distribution of traumatic axonal injury to the subcortical white matter at the sulcal depths directly correlates with the distribution of p-tau pathology in the overlying cortex.[30]

Motor Neuron Disease

TDP-43 proteinopathy is a feature of 85% of CTE cases, and is universally present in stage IV CTE. In addition, approximately 10% of individuals with CTE develop progressive MND that appears clinically indistinguishable from sporadic ALS (Table 16.5).[26] Typically, individuals with CTE and MND present with motor weakness, atrophy, and fasciculations,[25,30] and develop mild cognitive and behavioral symptoms several years after the onset of their motor disorder. Although the co-occurrence of both TDP-43 and tau proteinopathies may be coincidental, a study conducted by the Centers for Disease Control and Prevention indicated that individuals who play more than 5 years in the National Football League have a 4.31% higher risk of developing ALS and a 3.86% higher risk of developing dementia compared with controls matched for age and gender, supporting the concept that playing football increases the likelihood of developing MND and CTE.

Differential Diagnosis of Chronic Traumatic Encephalopathy

The major differential diagnosis in early-stage CTE is prolonged postconcussion syndrome. However, the onset of disease in CTE is insidious, often decades after the initiating injuries, whereas postconcussion syndrome develops immediately after the injury and is generally non-progressive.[7] More advanced CTE with cognitive impairment must be distinguished from AD or frontotemporal dementia (FTD).[7] Symptoms of CTE usually begin decades before the usual onset of AD and have slower progression.[30,31] CTE usually begins with behavior and personality changes at midlife, similar to FTD, but the clinical course of CTE is generally slower than that typically associated with FTD. In addition, all cases of CTE have had a history of brain trauma, whereas individuals with FTD do not typically have such a history.[7,31]

In the setting of exposure to repetitive mild TBI, motor weakness, atrophy, and fasciculations need to be distinguished from sporadic ALS, especially if there is evidence of concurrent impairment in cognition, mood, or behavior.

TABLE 16.5 Criteria for the Diagnosis of Chronic Traumatic Encephalopathy–Motor Neuron Disease (CTE-MND)

Clinical diagnosis of definite amyotrophic lateral sclerosis using the revised El Escorial criteria for the diagnosis of amyotrophic lateral sclerosis
Pathological diagnosis of CTE
Degeneration of lateral and ventral corticospinal tracts of the spinal cord
Marked loss of anterior horn cells from cervical, thoracic, and lumbar spinal cord with gliosis
TDP-43- or phosphorylated TDP-43-positive neuronal, glial, neuritic, or intranuclear inclusions in anterior horn cells and white matter tracts of spinal cord

TDP-43: Tar DNA-binding protein-43.
Source: Adapted from McKee et al. Brain *2013;136(Pt 1):43–64.*[30]

Future Areas for Research

Beyond repetitive, minor brain trauma, the risk factors for CTE remain unknown but determining how genetics, gender, low cognitive reserve, and other environmental exposures, such as drug or steroid use, may contribute to the development of CTE is clearly indicated. Moreover, there is currently no way to diagnose CTE during life; developing biomarkers to enable the identification of CTE is urgently needed, not only to monitor potential therapies and rehabilitative strategies, but also to counsel individuals and families with suspected CTE and to guide support services.

Potential Risk Factors for Chronic Traumatic Encephalopathy

APOLIPOPROTEIN E

Apolipoprotein E (ApoE) is the dominant apolipoprotein in the brain, primarily synthesized by astrocytes, although neurons and microglia may also contribute to its production.[22] The *APOE* gene has three alleles: the *APOE* ε_3 allele is the most common, *APOE* ε_4 is the strongest susceptibility gene for AD, and *APOE* ε_2 is protective against AD.[38] After TBI, membrane lipids and ApoE-immunoreactive plaques are often found in the brain, with an allele dose-dependent relationship between lipid density and *APOE* genotype.[27] Allelic variation in *APOE* in boxers and professional football players following a single TBI has also been implicated in prolonged recovery time and lower cognitive performance.[39] As a result, there is sufficient evidence to warrant examination of ApoE as a potential risk factor for the development of CTE after head impacts.

Several studies have reported higher than expected prevalence of *APOE* ε_4 carriers and homozygotes in individuals with neuropathologically confirmed CTE, providing further evidence that ApoE may be associated with CTE development.[25,30,31] Specifically, in a large cohort of neuropathologically confirmed CTE with and without other neurodegenerative disease there were significantly more *APOE* ε_4 homozygotes.[30,31] In individuals with CTE and comorbid neurodegenerative disease, the prevalence is akin to that found in AD.[31] Similarly, in a clinical study of 30 professional boxers, Jordan and colleagues found a significant association between severity of neurological impairment, high boxing exposure, and the *APOE* ε_4 allele.[39]

Although no genes have been established as risk factors for CTE, it is very likely that there is a genetic component that influences the pathology of CTE. Other genes of interest include *MAPT*, for its role in p-tau in both AD and FTD, and *TARDBP*, for its role in TDP-43.[40,41]

GENDER

Gender differences may play a role in the development of CTE. Most neuropathologically confirmed cases have been identified in men; however, this may simply reflect the demographics of mild TBI, including contact sports and military service. Women appear to be at greater risk for mild TBI and postconcussion syndrome, which show a higher incidence in women than in men within a single activity.[1] These differences may be based on hormonal differences, neck and muscle strength, or simply more reporting of somatic symptoms.[1] Any potential increased risk of mild TBI in women may influence the risk of developing CTE.

COGNITIVE RESERVE

Cognitive reserve may also play a role in the development and clinical course of CTE. Specifically, in neurodegenerative diseases such as AD, cognitive reserve has been suggested as protective against clinical manifestations,[31] in that it has been argued that occupational and educational attainment can increase one's "reserve" and confer relative resistance to neuropathological changes.[31] In diseases such as AD, cognitive reserve provides an explanation for individual differences in measured cognitive deficits despite similar degrees of neuropathology.[31]

Potential Biomarkers

The lack of biomarkers for clinical diagnosis hampers the ability to detect and monitor the clinical course, assess the efficacy of therapies and rehabilitation techniques, and gauge prognosis. Furthermore, without a method for clinical diagnosis, the true incidence and prevalence of CTE cannot be easily determined.

CEREBROSPINAL FLUID

One of the most promising potential biomarkers of CTE is analysis of CSF protein concentrations. Because CSF is in direct contact with neural tissue, it may provide an early indicator of CTE pathology, specifically in the form of tau and p-tau, but potentially earlier indicators of axonal damage as well. Specifically, tau, p-tau, and the p-tau/amyloid-β1–42 ratio have been effective in the clinical and preclinical diagnosis of neurodegenerative diseases, including AD and frontotemporal lobar degeneration.[42,43] However, CSF tau may not be an early marker of CTE, and another biomarker may be needed to detect CTE preclinically.

BLOOD

Plasma provides another potential biomarker for CTE. Specifically, blood–brain barrier disruption subsequent to head injury may result in abnormal levels of plasma proteins. For example, astrocytic protein S100B increases in blood acutely in response to head impacts.[44] Although S100B is typically used as a marker for acute injury, the perivascular deposition of p-tau characteristic of CTE may suggest a blood–brain barrier disruption, suggesting that markers of such disruption could provide an early diagnostic signature of CTE.

MAGNETIC RESONANCE IMAGING

Macroscopic neuropathological changes are often observed in late-stage CTE. As such, structural MRI may be useful in identifying these changes. Structural MRI can identify early pathological changes in other diseases, including AD, FTD, and dementia with Lewy bodies.[45] Structural MRI should be helpful in identifying general brain atrophy, as well as atrophy of specific structures, such as the amygdala, cavum septum pellucidum, and septal perforations.[25,30] Structural MRI can detect atrophy in former athletes with CTE-like symptoms.[16,31]

FUNCTIONAL MAGNETIC RESONANCE IMAGING

Neuropathological changes associated with CTE may result in observable functional deficits, perhaps even before clinical symptoms are manifest. Specifically, fMRI has been used in the diagnosis of multiple neurodegenerative diseases including AD, FTD, and dementia with Lewy bodies, and may be useful in the diagnosis of CTE.[16,31] Repetitive head impacts sustained by high-school football players are strongly associated with changes in an n-back fMRI task.[46] Although these findings are acute changes associated with subconcussive injury, they may provide assistance in identifying CTE as well.

DIFFUSION TENSOR IMAGING

DTI is sensitive to DAI, an indicator of TBI, which has been implicated in the pathogenesis of CTE[22]; DTI may therefore serve as an early marker of CTE. Preliminary evidence has indicated changes in fractional anisotropy in former professional athletes subjected to repetitive brain trauma, and compared with normal controls.[16] However, more rigorous analysis is needed to draw further conclusions from these data, especially because these findings may be a result of structural changes unrelated to CTE.

MAGNETIC RESONANCE SPECTROSCOPY

Magnetic resonance spectroscopy uses clinical magnetic resonance scanners to non-invasively measure *in vivo* brain biochemical metabolites, including *N*-acetyl aspartate, creatine, choline, glutamate, glutamine, and *myo*-inositol.[47] Changes in these metabolites are seen in response to single and repeated mild TBI, and correlate with clinical symptoms.[47] A pilot study found significant increases in choline, glutamate, and glutamine in former professional athletes with a history of repetitive brain trauma compared with age-matched controls.[48]

SUSCEPTIBILITY WEIGHTED IMAGING

Susceptibility weighted imaging can reveal breakdown of the blood–brain barrier by detecting heme iron and can image small microhemorrhages after TBI.[16] As microvascular pathology is found experimentally in animal models of blast injury, and vessels surrounded by hemosiderin-laden macrophages are commonly found in the white matter of individuals with CTE, susceptibility weighted imaging may be useful in the diagnosis of CTE.[16]

POSITRON EMISSION TOMOGRAPHY

PET is a non-invasive diagnostic imaging modality using isotope-labeled molecular probes that bind to biomolecules with high specificity and affinity. Several investigative PET ligands designed to target Aβ plaques in AD have advanced to various stages of clinical evaluation, including [^{18}F]FDDNP, [^{11}C]PIB, and [^{18}F]Florbetapir.[49–51] However, most cases of CTE do not show substantial Aβ pathology; indeed, all cases of early-stage CTE (stage I–II) to date have been Aβ negative.[30] PET ligands that specifically target p-tau are currently in development. A pilot study of five former professional football players suggested that [^{18}F]FDDNP may be useful in the detection of CTE[52]; however, it is not clear that [^{18}F]FDDNP provides specific identification of tau. Two recently identified ligands, [^{18}F]T807 and [^{18}F]T808, have shown high binding affinity and good selectivity for p-tau aggregates *in vitro* and *ex vivo* in rodent brains.[53] Pilot PET scans of humans with mild cognitive impairment or AD showed patterns of radiotracer accumulation correlated with cognitive impairment, mimicking the progression of p-tau pathology in AD.[53] Given that p-tau in CTE is biochemically similar to p-tau in AD, and CTE p-tau and AD p-tau consist of 3- and 4-repeat tau, [^{18}F]T807 and [^{18}F]T808 represent promising potential biomarkers of CTE.[25,30]

CONCLUSION

The effects of moderate to severe TBI are related to the severity of the initial and secondary injuries and are a leading cause of disability. Mild TBI, by far the most common form of TBI, has been largely overlooked as a major health concern until very recently. Although most individuals recover from mild TBI in a few days or weeks, mild TBI can have long-term cognitive, behavioral, and psychological consequences, including depression. The factors involved in the long-lasting effects of mild TBI are the focus of intense research. Of specific interest is the relationship between mild TBI and neurodegenerative disease, including CTE.

CTE develops after repetitive mild TBI including concussion and subconcussion. The exact pathogenic relationship between acute mild TBI and the development of CTE is not clear, but undoubtedly involves acute axonal injury, disruption of the axolemma, persistent axonal transport failure, microvascular injury, cytoskeletal disruption, and tau hyperphosphorylation and aggregation. The clinical symptoms of CTE usually develop many years after exposure to repetitive but minor brain trauma. The most common early symptoms of CTE are depression, headache, attention and concentration difficulties, and short-term memory loss, which typically appear in midlife. Neuropathologically, CTE is associated with diminished brain weight; marked atrophy of the medial temporal lobe, cerebral cortex, subcortical white matter, thalamus, hypothalamus, and mammillary bodies; enlargement of the lateral and third ventricles; septal abnormalities; and depigmentation of the locus ceruleus and substantia nigra. The progression of p-tau pathology in CTE follows a predictable sequence that can be divided into four stages, I–IV. Stage I is characterized by focal, perivascular, deposits of p-tau at the base of the cerebral sulci; later stages involve progressively more expansive regions of neocortex, medial temporal lobe, diencephalon, basal ganglia, brainstem, and spinal cord. The early, focal cortical p-tau pathology of CTE is distinctive from other tauopathies, and unlike the early limbic degeneration and later neocortical involvement typical of AD.[25,26,30] TDP-43 abnormalities are also found in most CTE cases; in advanced CTE, TDP-43 pathology is often severe and widespread. As tau and TDP-43 deposition increases, there is a parallel increase in axonal pathology and loss. CTE may be associated with other neurodegenerative diseases, including AD, PD, Lewy body disease, frontotemporal lobar degeneration, and MND. Most individuals with CTE and MND have symptoms of MND at a young age and develop subtle cognitive and behavioral symptoms a few years later. While the pathogenic mechanisms of CTE remain to be fully elucidated, progressive axonal degeneration is followed by the accumulation and transneuronal propagation of toxic aggregated proteins. CTE can be diagnosed only at autopsy. However, promising efforts to develop DTI, fMRI, magnetic resonance spectroscopy, CSF and blood p-tau and neuroinflammatory markers, and p-tau PET ligands as biomarkers are underway to diagnose and monitor the course of disease in living subjects. Future therapeutic research in mild TBI will need to address acute mild TBI as well as the long-term progressive neurodegeneration that follows. Currently, the best therapy is prevention of the initial traumatic injury and continued public education about detection and management.

DIRECTIONS FOR FUTURE RESEARCH

Although the neurological effects of TBI have been studied for decades, many important questions remain and warrant investigation. One major avenue for future research involves exploring the specific mechanisms underlying neurological injury. In the case of moderate and severe TBI, improved understanding of disease mechanisms would lead to more effective acute interventions and long-term recovery strategies. As for neurodegenerative disease associated with TBI, understanding the precise mechanisms responsible for pathogenesis would result in improved preventive measures. Specific mechanistic questions involve how one's risk of disease is changed by exposure to distinct types of impact (e.g. collisions resulting in rotational versus linear acceleration), duration between successive impacts, or impacts at different magnitudes sustained at different ages. Improved understanding would allow for changes in protective equipment, rule changes, and recovery guidelines, which would ultimately decrease disease burden. In addition, understanding the disease mechanism would result in understanding the genetic and environmental risks associated with CTE.

Although there are several promising biomarkers, there is currently no proven way to diagnose CTE *in vivo*. The discovery of new potential biomarkers, in conjunction with animal models and human studies with pathological confirmation of disease, will allow for the identification of a cohort of individuals with CTE. This step is necessary to testing potential treatments and eventual disease cures.

Acknowledgments

We gratefully acknowledge the use of resources and facilities at the Edith Nourse Rogers Memorial Veterans Hospital (Bedford, MA), the help of all members of the Center for the Study of Traumatic Encephalopathy at Boston University and the Boston VA, as well as the individuals and families whose participation and contributions made this work possible. This work was supported by The Department of Veterans Affairs, Veterans Affairs Biorepository (CSP 501), Translational Research Center for Traumatic Brain Injury and Stress Disorders (TRACTS) Veterans Affairs Rehabilitation Research and Development Traumatic Brain Injury Center of Excellence (B6796-C), National Institute of Aging Boston University Alzheimer's Disease Center [P30AG13846, supplement 0572063345-5], National Institute of Aging Boston University Framingham Heart Study R01 [AG1649], Sports Legacy Institute, and the National Operating Committee on Standards for Athletic Equipment. This work was also supported by an unrestricted gift from the National Football League, the Andlinger Foundation, and Worldwide Wrestling Entertainment.

References

1. Daneshvar DH, Nowinski CJ, McKee AC, Cantu RC. The epidemiology of sport-related concussion. *Clin Sports Med*. 2011;30(1):1–17. vii.
2. Faul M, Xu L, Wald MM, Coronado VG. *Traumatic Brain Injury in the United States: Emergency Department Visits, Hospitalizations and Deaths 2002–2006*. Washington, DC: US Department of Health and Human Services; March 2010.
3. Langlois JA, Rutland-Brown W, Wald MM. The epidemiology and impact of traumatic brain injury: a brief overview. *J Head Trauma Rehabil*. 2006;21(5):375–378.
4. Shively S, Scher AI, Perl DP, Diaz-Arrastia R. Dementia resulting from traumatic brain injury: what is the pathology? *Arch Neurol*. 2012;69(10):1245–1251.
5. Teasdale G, Jennett B. Assessment of coma and impaired consciousness: a practical scale. *Lancet*. 1974;ii:81–84.
6. Centers for Disease Control and Prevention. *Report to Congress on Mild Traumatic Brain Injury in the United States: Steps to Prevent a Serious Public Health Problem*. Atlanta, GA: National Center for Injury Prevention and Control; 2003.
7. Daneshvar DH, Riley DO, Nowinski CJ, McKee AC, Stern RA, Cantu RC. Long-term consequences: effects on normal development profile after concussion. *Phys Med Rehabil Clin N Am*. 2011;22(4):683–700. ix.
8. Pellman EJ, Powell JW, Viano DC, et al. Concussion in professional football: epidemiological features of game injuries and review of the literature – part 3. *Neurosurgery*. 2004;54(1):81–94. discussion 94–96.
9. Hukkelhoven CW, Rampen AJ, Maas AI, et al. Some prognostic models for traumatic brain injury were not valid. *J Clin Epidemiol*. 2006;59(2):132–143.
10. Forde CT, Karri SK, Young AM, Ogilvy CS. Predictive markers in traumatic brain injury: opportunities for a serum biosignature. *Br J Neurosurg*. 2014;28(1):8–15.
11. Graham DI, Gennarelli TA, McIntosh TK. Trauma. In: Graham DI, Lantos PL, eds. *Greenfield's Neuropathology*. London: Arnold; 2002:823–898.
12. Marshall LF, Gautille T, Klauber MR, et al. The outcome of severe closed head injury. *J Neurosurg*. 1991;75:S28–S36.
13. Tseng WC, Shih HM, Su YC, Chen HW, Hsiao KY, Chen IC. The association between skull bone fractures and outcomes in patients with severe traumatic brain injury. *J Trauma*. 2011;71(6):1611–1614. discussion 1614.
14. Harvey LA, Close JC. Traumatic brain injury in older adults: characteristics, causes and consequences. *Injury*. 2012;43(11):1821–1826.
15. Smith DH, Wolf JA, Lusardi TA, Lee VM, Meaney DF. High tolerance and delayed elastic response of cultured axons to dynamic stretch injury. *J Neurosci*. 1999;19(11):4263–4269.
16. Baugh CM, Stamm JM, Riley DO, et al. Chronic traumatic encephalopathy: neurodegeneration following repetitive concussive and subconcussive brain trauma. *Brain Imag Behav*. 2012;6(2):244–254.
17. Chen JK, Johnston KM, Petrides M, Ptito A. Recovery from mild head injury in sports: evidence from serial functional magnetic resonance imaging studies in male athletes. *Clin J Sports Med*. 2008;18(3):241–247.
18. Cubon VA, Putukian M, Boyer C, Dettwiler A. A diffusion tensor imaging study on the white matter skeleton in individuals with sports-related concussion. *J Neurotrauma*. 2011;28(2):189–201.
19. Nakagawa A, Manley GT, Gean AD, et al. Mechanisms of primary blast-induced traumatic brain injury: insights from shock-wave research. *J Neurotrauma*. 2011;28(6):1101–1119.
20. Ghajar J. Traumatic brain injury. *Lancet*. 2000;356(9233):923–929.
21. Morganti-Kossmann MC, Satgunaseelan L, Bye N, Kossmann T. Modulation of immune response by head injury. *Injury*. 2007;38(12):1392–1400.
22. Blennow K, Hardy J, Zetterberg H. The neuropathology and neurobiology of traumatic brain injury. *Neuron*. 2012;76(5):886–899.

23. Marmarou A. A review of progress in understanding the pathophysiology and treatment of brain edema. *Neurosurg Focus*. 2007;22(5):E1.
24. Committee on Gulf War and Health. Gulf War and Health. *Long-Term Consequences of Traumatic Brain Injury*. Volume 7. Washington, DC: National Academies Press; 2009.
25. McKee AC, Cantu RC, Nowinski CJ, et al. Chronic traumatic encephalopathy in athletes: progressive tauopathy after repetitive head injury. *J Neuropathol Exp Neurol*. 2009;68(7):709–735.
26. McKee AC, Gavett BE, Stern RA, et al. TDP-43 proteinopathy and motor neuron disease in chronic traumatic encephalopathy. *J Neuropathol Exp Neurol*. 2010;69(9):918–929.
27. Horsburgh K, Cole GM, Yang F, et al. Beta-amyloid (Abeta)42(43), abeta42, abeta40 and apoE immunostaining of plaques in fatal head injury. *Neuropathol Appl Neurobiol*. 2000;26(2):124–132.
28. Chen XH, Johnson VE, Uryu K, Trojanowski JQ, Smith DH. A lack of amyloid beta plaques despite persistent accumulation of amyloid beta in axons of long-term survivors of traumatic brain injury. *Brain Pathol*. 2009;19(2):214–223.
29. Uryu K, Richter-Landsberg C, Welch W, et al. Convergence of heat shock protein 90 with ubiquitin in filamentous alpha-synuclein inclusions of alpha-synucleinopathies. *Am J Pathol*. 2006;168(3):947–961.
30. McKee AC, Stein TD, Nowinski CJ, et al. The spectrum of disease in chronic traumatic encephalopathy. *Brain*. 2013;136(Pt 1):43–64.
31. Stern RA, Daneshvar DH, Baugh CM, et al. Clinical presentation of neuropathologically-confirmed chronic traumatic encephalopathy in athletes. *Neurology*. 2013;81(13):1122–1129.
32. Goldstein LE, Fisher AM, Tagge CA, et al. Chronic traumatic encephalopathy in blast-exposed military veterans and a blast neurotrauma mouse model. *Sci Transl Med*. 2012;4(134). 134ra60.
33. Mouzon B, Chaytow H, Crynen G, et al. Repetitive mild traumatic brain injury in a mouse model produces learning and memory deficits accompanied by histological changes. *J Neurotrauma*. 2012;29(18):2761–2773.
34. Ojo JO, Mouzon B, Greenberg MB, Bachmeier C, Mullan M, Crawford F. Repetitive mild traumatic brain injury augments tau pathology and glial activation in aged hTau mice. *J Neuropathol Exp Neurol*. 2013;72(2):137–151.
35. Johnson VE, Stewart W, Smith DH. Axonal pathology in traumatic brain injury. *Exp Neurol*. 2013;246:35–43.
36. Prins ML, Alexander D, Giza CC, Hovda DA. Repeated mild traumatic brain injury: mechanisms of cerebral vulnerability. *J Neurotrauma*. 2013;30(1):30–38.
37. Frost B, Jacks RL, Diamond MI. Propagation of tau misfolding from the outside to the inside of a cell. *J Biol Chem*. 2009;284(19):12845–12852.
38. Farrer LA. Genetics and the dementia patient. *Neurologist*. 1997;3:13–30.
39. Jordan BD, Relkin NR, Ravdin LD, Jacobs AR, Bennett A, Gandy S. Apolipoprotein E epsilon4 associated with chronic traumatic brain injury in boxing. *JAMA*. 1997;278(2):136–140.
40. Lee SE, Tartaglia MC, Yener G, et al. Neurodegenerative disease phenotypes in carriers of MAPT p.A152T, a risk factor for frontotemporal dementia spectrum disorders and Alzheimer disease. *Alzheimer Dis Assoc Disord*. 2013;27(4):302–309.
41. Czell D, Andersen PM, Morita M, Neuwirth C, Perren F, Weber M. Phenotypes in Swiss patients with familial ALS carrying TARDBP mutations. *Neurodegener Dis*. 2013;12(3):150–155.
42. Cruchaga C, Kauwe JS, Harari O, et al. GWAS of cerebrospinal fluid tau levels identifies risk variants for Alzheimer's disease. *Neuron*. 2013;78(2):256–268.
43. Irwin DJ, McMillan CT, Toledo JB, et al. Comparison of cerebrospinal fluid levels of tau and Abeta 1-42 in Alzheimer disease and frontotemporal degeneration using 2 analytical platforms. *Arch Neurol*. 2012;69(8):1018–1025.
44. Marchi N, Bazarian JJ, Puvenna V, et al. Consequences of repeated blood–brain barrier disruption in football players. *PLoS ONE*. 2013;83:e56805.
45. O'Brien JT. Role of imaging techniques in the diagnosis of dementia. *Br J Radiol*. 2007;80(Spec No 2):S71–S77.
46. Breedlove EL, Robinson M, Talavage TM, et al. Biomechanical correlates of symptomatic and asymptomatic neurophysiological impairment in high school football. *J Biomech*. 2012;45(7):1265–1272.
47. Lin AP, Liao HJ, Merugumala SK, Prabhu SP, Meehan III WP, Ross BD. Metabolic imaging of mild traumatic brain injury. *Brain Imag Behav*. 2012;6(2):208–223.
48. Lin AP, Ramadan S, Box H, Stanwell P, Stern R, Mountford C. *Neurochemical changes in athletes with chronic traumatic encephalopathy*. Chicago, IL: Paper presented at the Annual Meeting of the Radiological Society of North America; December 1, 2010.
49. Tauber C, Beaufils E, Hommet C, et al. Brain [18F]FDDNP binding and glucose metabolism in advanced elderly healthy subjects and Alzheimer's disease patients. *J Alzheimers Dis*. 2013;36(2):311–320.
50. Klunk WE, Engler H, Nordberg A, et al. Imaging brain amyloid in Alzheimer's disease with Pittsburgh Compound-B. *Ann Neurol*. 2004;55(3):306–319.
51. Clark CM, Schneider JA, Bedell BJ, et al. Use of florbetapir-PET for imaging beta-amyloid pathology. *JAMA*. 2011;305(3):275–283.
52. Small GW, Kepe V, Siddarth P, et al. PET scanning of brain tau in retired national football league players: preliminary findings. *Am J Geriatr Psychiatry*. 2013;21(2):138–144.
53. Chien DT, Bahri S, Szardenings AK, et al. Early clinical PET imaging results with the novel PHF-tau radioligand [F-18]-T807. *J Alzheimers Dis*. 2013;34(2):457–468.

CHAPTER

17

Epilepsy

Helen E. Scharfman[*,†]

[*]Departments of Child & Adolescent Psychiatry, Physiology & Neuroscience, and Psychiatry, New York University Langone Medical Center, New York, USA; [†]The Nathan Kline Institute, Dementia Research, Orangeburg, New York, USA

OUTLINE

INTRODUCTION

History

Epilepsy is a disease with an extremely long history, suggesting that the causes of epilepsy have been present for centuries and are not related to the industrialized world. The first descriptions of seizures are found in texts from ancient civilizations. For example, in a document from Mesopotamia, an individual having a seizure was described with "... a neck that turns left, hands and feet are tense and eyes wide open, froth flowing from the mouth and consciousness being lost". Without the understanding

Neurobiology of Brain Disorders
http://dx.doi.org/10.1016/B978-0-12-398270-4.00017-3

of the nervous system that we have today, most of these behaviors were attributed to demons "seizing" a person. The word "epilepsy" has its origin in the Greek word *epilambanein*, which is a combination of *epi* (upon) and *lambanein* (to seize or take hold of). In Rome, epilepsy was known as a curse from the gods, and in Greece epilepsy was referred to as the "sacred disease". During the height of Christianity, over 40 patron saints were established for epilepsy, a number exceeded only by catastrophic illnesses like the plague. St Valentine, who is famous as the patron saint of lovers, is also a patron saint of epilepsy. St Valentine may have been a doctor who died on February 14, AD 273, and an association was made with epilepsy for a reason that is currently unclear. Another suggestion is that St Valentine "cured" epilepsy in his fiancée, most likely because he witnessed a remission of seizures, which occurs in some people with epilepsy.

Hippocrates was the first to suggest that epilepsy was not a disease related to demons but a biological problem in the brain. However, the association of epilepsy with evil spirits or demonic influence persisted. Remarkably, a negative conception of people with epilepsy, religious or not, still exists in some parts of the world today. Even in industrialized nations, patients with epilepsy often describe a stigma associated with the words epilepsy and epileptic. This view has led to suggestions that alternative terms be used for epilepsy, such as seizure disorder, and antiseizure drug (ASD) instead of antiepileptic drug. A great deal of effort has been made to dispel negative views about epilepsy.

Clinical Overview

In general, epilepsy is defined by spontaneous seizures that occur intermittently and recurrently. The word "spontaneous" means that the seizure is unprovoked. However, in some epilepsy syndromes a person experiences a seizure whenever a specific stimulus occurs, such as a flashing light, a particular sound, or excessive alcohol intake. This type of epilepsy is called provoked or reflex epilepsy.[1,2]

There is some disagreement about the exact number of seizures necessary for the diagnosis of epilepsy. Until recently, epilepsy was defined by any number greater than one, because even two seizures that were unprovoked would discriminate epilepsy from other disorders. This definition has been debated because seizures may occur in many conditions besides epilepsy, such as head injury, gastrointestinal illness accompanied by electrolyte depletion, or drug abuse. For this reason and others, it was suggested in 2005 that only one spontaneous seizure be required for a diagnosis of epilepsy as long as there were signs that more seizures were likely to occur. For example, an individual who had their first seizure could be diagnosed with epilepsy if neuroimaging showed a structural abnormality consistent with a type of epilepsy, such as a cortical tuber, which is the defining pathology in tuberous sclerosis.

Terminology

One of the long-standing debates in the field of clinical epileptology, as well as basic research in epilepsy, relates to terminology. Not only are there debates about the exact definition for the term "epilepsy" but there also is some difficulty in defining the term "seizure". One definition of seizure is "hypersynchronous neuronal activity", but the term hypersynchronous is subjective because various types of synchronous neuronal activity are characteristic of the normal brain, such as alpha, theta, or delta rhythms. Some definitions include a long duration of hypersynchronous neuronal activity, such as greater than 10 seconds, because some normal individuals may exhibit brief periods of electroencephalographic (EEG) activity that are seizure-like, but long periods are extremely rare. Nevertheless, EEG manifestations of seizures are sometimes hard to differentiate from normal EEG activity. Therefore, seizures that are accompanied by convulsive movements are the easiest to define as seizures. These events may be called convulsions or convulsive seizures or, in the UK, fits. Movements can be deceptive, however. Individuals who faint, for example, are similar to those with atonic seizures; in each case, the person suddenly loses postural tone and falls. Narcolepsy is an example of a condition where sudden falls are common, but in this case there is no EEG activity that resembles a seizure. Conditions where movements occur that are similar to a type of epilepsy (such as atonic seizures), but are not accompanied by EEG manifestations of seizures, are called non-epileptic seizures or psychogenic non-epileptic seizures (Table 17.1).

Other terms are also debated. For example, "anticonvulsant" was commonly used to refer to drugs used to treat epilepsy, until it was recognized that these medications can also reduce non-convulsive seizures. The term antiepileptic drug is widely used today but is not ideal because these drugs are also used in some psychiatric disorders in which there may be no evidence of seizures by EEG. In these people it may be that specific brain areas are overly active, and therefore may benefit from a drug that reduces neuronal activity even though the excessive activity is never sufficiently robust to be detected by EEG. As a result of these observations, and the stigma associated with the term epileptic (described above), the term antiseizure drug (ASD) has been suggested.

TABLE 17.1 Terminology

Term (Synonyms)	Definition
Absence seizure (petit mal)	A seizure that is accompanied by staring and unresponsiveness, with 2 Hz spike-and-wave discharges recorded from multiple EEG leads, bilaterally
Antiseizure drug	A drug that reduces seizures
Anticonvulsant drug	A drug that reduces convulsions
Antiepileptic drug	A drug that reduces the severity or frequency of seizures in epilepsy
Atonic	A sudden fall or loss of posture, often called a "drop" or "drop attack"
Benign epilepsy	Epilepsy that resolves with time
Clonic	Rhythmic muscular contractions and relaxations
Convulsions (behavioral seizures)	Movements that accompany epileptic seizures
Complex partial seizure (psychomotor)	A focal seizure that is accompanied by a loss of awareness or responsiveness
Epilepsy (seizure disorder)	Recurrent spontaneous (unprovoked) seizures
Epileptic	A subject (person or animal) with epilepsy
Epileptic seizures	Seizures that are accompanied by hypersynchronous neuronal activity
Epileptogenesis	The process whereby the normal adult brain becomes epileptic
Focal (localized, partial)	A seizure that occurs in one area of the brain
Generalized (diffuse, distributed)	A seizure that involves multiple brain areas in both hemispheres
High-frequency oscillations	Rhythmic potentials that are between 100 and 500 Hz; sometimes called ripples (usually ripples refer to oscillations < 200 Hz) or fast ripples (typically > 200 Hz)
Ictal	The period during which a seizure occurs
Interictal	The period between seizures
Interictal spikes	A transient voltage deflection on the EEG that occurs between seizures
Myoclonic	Involuntary twitching of a small subset of muscles
Non-epileptic (pseudo-, psychogenic) seizures	Behavior that is similar to epileptic seizures but the EEG is normal
Paroxysmal depolarization shift	A sudden, large depolarization of a neuron, usually with action potentials at the peak of the depolarization, and typically lasting < 500 ms
Preictal	The period immediately before a seizure
Ripples or fast ripples	*See* High-frequency oscillations
Secondarily generalized	A partial seizure that changes to one that involves both hemispheres
Seizure	Hypersynchronous neural activity with an EEG pattern that is abnormal, typically lasting > 10 seconds, except for absence seizures (which can be brief)
Simple partial seizure	A seizure that occurs primarily in one brain region, and awareness is unaffected
Spike-and-wave (spike–wave)	A rhythmic EEG pattern that is typically 3 Hz in humans, and >6 Hz in lower mammals, with a short voltage deflection (spike) and slow component (wave)
Tonic	Stiffness or rigidity reflecting persistent activity of extensor muscles
Tonic–clonic seizure (grand mal)	A seizure that is accompanied by tonic–clonic movements and both hemispheres in the EEG

EEG: electroencephalogram.

Incidence and Prevalence

Epilepsy is extremely common and pervasive, affecting over 50 million people worldwide. In the USA, between 1.3 and 2.8 million people have epilepsy, a number comparable to that for diseases that typically are considered much more prevalent, such as Alzheimer's disease (2.5 million) and schizophrenia (2.4 million). Likewise, whereas it is estimated that 1 in 88 children will develop autism, 1 in 26 will develop epilepsy.[2]

In the USA and other industrialized nations, the incidence, or the number of new cases of epilepsy per year, is between 40 and 70 per 100,000. In developing countries the incidence is higher, at 100–190 per 100,000 people. Low socioeconomic status has been related to an increased risk of epilepsy. Indeed, it has been estimated that 80% of the patients with epilepsy in the world live in underdeveloped countries. Age is also a factor, with epilepsy mainly occurring in children and elderly people.[3]

The prevalence, or total number of individuals who have epilepsy out of the total number of people who were studied, is 5–10 per 1000. The lifetime prevalence, or fraction of individuals who at some point in their life will have epilepsy, has been estimated to be 3% in the USA.[4]

Mortality and Sudden Unexpected Death in Epilepsy

Although epilepsy is defined by seizures, mortality is a common outcome and often underappreciated. Many epilepsy-related deaths occur suddenly and unexpectedly, a phenomenon termed sudden unexpected death in epilepsy (SUDEP). It is estimated that 1 of every 10,000 new diagnoses of epilepsy will die from SUDEP; the number is higher for patients who are refractory to medication and are eligible for surgery: approximately 9 of every 1000 of these patients have SUDEP.[1,2]

Cost Burden

Not surprisingly, epilepsy is a disease associated with huge costs. There is not only a financial burden on patients, who typically face a lifetime coping with their illness, but also a reduced quality of life, which is hard to define in dollars. There is also a cost to families, who may need to care for the patient and support them if they cannot drive or work. Furthermore, there is a cost to society related to lost productivity in the workforce. The annual direct medical care cost of epilepsy in the USA is more than 9.6 billion USD.[2] The total direct and indirect costs, comprising medical care costs plus lost earnings and productivity, are approximately 15.5 billion USD.

CLASSIFICATION OF THE SEIZURES AND THE EPILEPSIES

Seizures Versus Epilepsy

Describing epilepsy is difficult because it is not a single disorder but many, reflected in the term "the epilepsies". Historically, characteristics of seizures have been used to define the type of epilepsy, with terms such as "petit mal" referring to staring spells or literally absences, which led to the term "absence epilepsy" for patients with this type of seizure (Fig. 17.1A). In contrast to petit mal, "grand mal" refers to major movements of the limbs associated with repetitive contraction and relaxation. Both petit mal and grand mal seizures involve many sites in both hemispheres; thus, the seizures are generalized, in contrast to the term focal, which refers to seizures that are confined to one portion of the brain (Table 17.2). In grand mal seizures, the rhythmic contractions and relaxations of the muscles are called tonic–clonic seizures or generalized tonic–clonic seizures (GTCS). The EEG during a generalized tonic seizure is shown in Fig. 17.1(B). Consciousness (often called responsiveness or awareness) is lost in both absence seizures and GTCS, but the outcome to an observer is vastly different. In absence epilepsy, one cannot be sure that someone is having a seizure without EEG because they usually sit quietly, staring into space. GTCS, which is the type of seizure most people associate with epilepsy, involves major movements of the limbs. Therefore, before the advent of the EEG in the first part of the twentieth century, many people with absence epilepsy were undiagnosed. With EEG, categorization and diagnosis of the epilepsies became much more advanced (Tables 17.3 and 17.4).

Characteristics of seizures remain central to classification of the epilepsies. Seizures are divided into two categories: localized or focal (also called simple or partial) versus generalized (also called diffuse or distributed) (Table 17.2). Focal seizures begin in one brain region whereas generalized seizures occur in many brain areas in both hemispheres at the same time. However, some seizures start focally and then become generalized; this is called secondary generalization.

The EEG activity during seizures is extremely variable, but within a given epilepsy syndrome it is similar. Therefore, the EEG activity plays a critical role in defining the type of epilepsy of any given patient. Figure 17.1 shows an example of a seizure from an individual with absence epilepsy beside an example from a person with another type of epilepsy who had a tonic seizure. The absence seizure is distinguished by a very specific EEG pattern composed of a transient (spike) followed by a slow wave, or "spike and wave" (Fig. 17.1A). There is an abrupt onset, at which time responsiveness is lost,

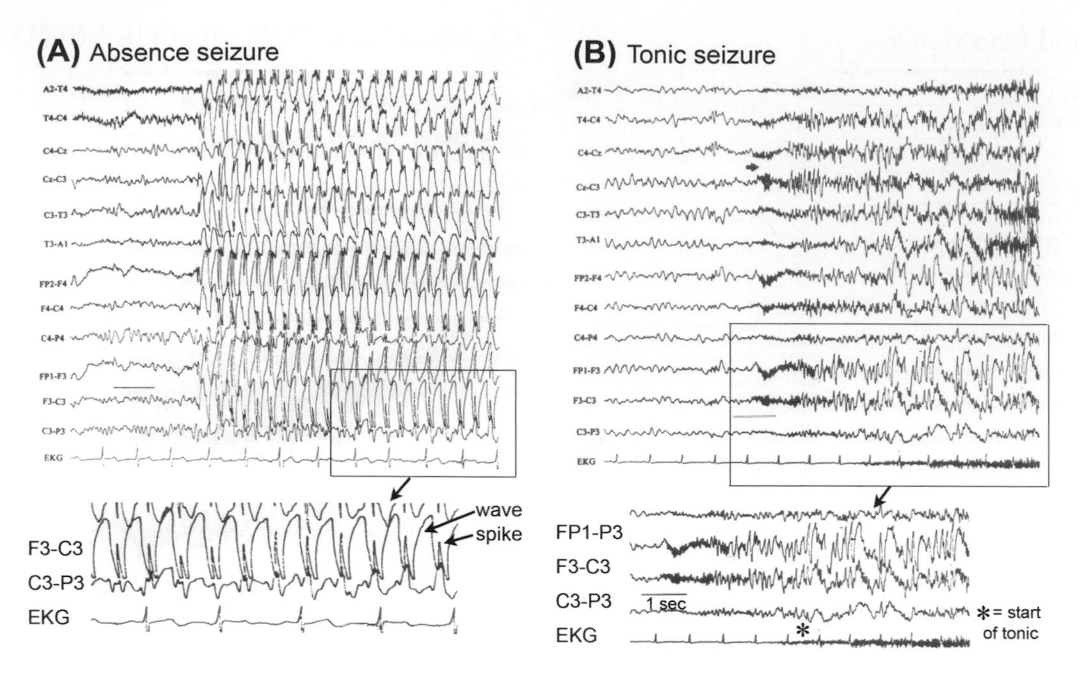

FIGURE 17.1 **The use of electroencephalography (EEG) to distinguish types of epilepsy.** (A) Example of a seizure in absence epilepsy. There is a repetitive spike-and-wave pattern, with approximately three spike–wave cycles per second. At the bottom, the area of the EEG that is boxed is enlarged. (B) Example of a tonic seizure in a patient with a different kind of epilepsy than absence epilepsy. The EEG illustrates many differences from the absence seizure, such as the lack of a 3 Hz spike–wave pattern, and rhythmic activity that is faster than 3 Hz. At the bottom, the electromyogram (EMG) is shown to illustrate that muscular activity began during the onset of the tonic seizure. *Source: Adapted from Sperling and Clancy. In:* Epilepsy: A Comprehensive Textbook. *Philadelphia, PA: Lippincott-Raven, 1997:849-885.*[5]

and an abrupt termination, when awareness resumes. The spikes and slow waves occur repetitively at 3 Hz (Fig. 17.1A). The tonic seizure shown in Fig. 17.1(B) is very different: it does not begin simultaneously in all areas of the brain, but eventually all brain areas are generating fast rhythmic activity. A spike-and-wave cycle never occurs.

Until very recently, epilepsy was divided into three major categories, and often still is: idiopathic, symptomatic, and cryptogenic. Provoked, or reflex epilepsies are a minor, fourth category (Table 17.3). Focal or generalized seizures are subdivisions within each of the three major categories (Table 17.3). Idiopathic refers to an epilepsy syndrome with an unknown cause, but presumably a genetic basis, with the gene(s) not yet identified. Symptomatic epilepsies, in contrast, are related to a known structural cause, such as a cortical malformation or a head injury. The word cryptogenic reflects a cause that is unclear. In 2010, the International League Against Epilepsy (ILAE), the largest international group dedicated to epilepsy, proposed a new set of guidelines for the definitions of seizure types and the epilepsies (Tables 17.2 and 17.3).[6] Regarding seizures, it was suggested that words such as "partial" seizure be de-emphasized in favor of "focal". With respect to the classification of the epilepsies, it was suggested that the term "idiopathic" be replaced by "genetic", and instead of "symptomatic", "structural/metabolic" be adopted. "Unknown" was offered as a replacement for "cryptogenic", to be more direct and transparent with terms. Although these guidelines are not completely settled, new ways to classify the epilepsies have already been suggested that take into account the new guidelines. One proposal is shown in Table 17.3.[7] In this classification, some of the older terms are still used, but there is a new emphasis that is consistent with the proposed guidelines. The ILAE guidelines also suggested a classification where epilepsies are divided into constellations of symptoms called "electroclinical syndromes". An electroclinical syndrome is defined as "a complex of clinical features, signs and symptoms that together define a distinctive, recognizable clinical disorder". Divisions into constellations have been suggested before (Fig. 17.2). In this figure, the age of onset, severity, and focal versus generalized seizure types are distinguished in a way that establishes roughly four constellations.

The current classifications of the epilepsies do not specify the relative proportions of the different types of epilepsies, and how many of the epilepsies are treatable or still lack a therapeutic approach that stops seizures. The most common epilepsy syndrome is temporal lobe epilepsy. These individuals have focal seizures that start in the temporal lobe and are accompanied by a loss of awareness; secondarily generalized seizures can occur. There are characteristics that suggest that a symptomatic categorization is best for this

TABLE 17.2 Classification of Seizures

Before 2010[a]	After 2010[b]
I. Focal (partial)	**I. Focal**
Simple partial	With or without aura
Complex partial	With or without awareness
Secondary generalized	With or without motor/autonomic manifestations
	With or without evolution to both hemispheres
II. Generalized	**II. Generalized**
Absence	Absence
Myoclonic	– Typical
Clonic	– Atypical
Tonic	– Absence with special features (myoclonic, eyelid myoclonia)
Tonic–clonic	Myoclonic
Atonic	– Myoclonic
	– Myoclonic atonic
	– Myoclonic tonic
	Tonic
	Tonic–clonic
	Atonic
	III. Unknown or unclear
	Epileptic spasms

[a]*Standard classification.*
[b]*Classification suggested by the International League Against Epilepsy (ILAE) in 2010.*[6]

TABLE 17.3 Classification of the Epilepsies: (a) Major Categories; (b) Detailed Classification

(a) Before 2010[a]		After 2010[b]
Idiopathic	→	Genetic
Symptomatic	→	Structural/metabolic
Cryptogenic	→	Unknown
(b) Engel (2001)[c]		**Shorvon (2011)[d]**
I. Localization-related		**I. Idiopathic/genetic**
A. Idiopathic		A. Pure epilepsies due to single gene disorders
B. Symptomatic		B. Pure epilepsies with complex inheritance
C. Cryptogenic		
II. Generalized		**II. Symptomatic/structural–metabolic**
A. Idiopathic		A. Predominantly genetic or developmental
B. Symptomatic or cryptogenic		B. Predominantly acquired
C. Symptomatic		
III. Undetermined whether focal or generalized		**III. Cryptogenic/unknown cause**
A. Both		
B. Equivocal		
IV. Special syndromes		**IV. Provoked**
Situation-related		A. Provoking factors
		B. Reflex epilepsies

[a]*Standard categories used before 2010.*
[b]*2010 guidelines from the International League Against Epilepsy (ILAE).*
[c]*Detailed classification of standard categories and subcategories.*[7]
[d]*A recent view that takes into account 2010 suggestions from the ILAE.*[8]

type of epilepsy, because there is usually a loss of neurons in a particular pattern in hippocampus, originally called Ammon's horn or hippocampal sclerosis.[10] After it was shown that more areas than hippocampus were typically damaged, a term with a broader scope was invoked: mesial temporal sclerosis (MTS).[10] The damage is highly variable, however, and some patients even lack an identifiable lesion; in these individuals, the cause may be genetic.[11,12] Therefore, there may be types of temporal lobe epilepsy that are genetic, and types that are structural/metabolic and are associated with MTS. Many of these patients are successfully treated by ASDs, but approximately 20–30% continue to have seizures. Individuals who have seizures that are not controlled by ASDs are called drug-resistant or refractory, and are a major clinical concern.[13]

Comorbidities

Seizures are the major symptom in epilepsy, but not the only symptom. Additional symptoms may be dysfunctions associated with interruption of normal activity in the part of the brain where seizures occur. For example, in temporal lobe epilepsy, memory loss can occur because the temporal lobe is critical to memory. Although the seizures could be causally related to these comorbidities, it is often hard to prove cause and effect. In some cases, the seizures do seem causally related to the comorbid condition, such as in malnourished patients. In these cases, the comorbidity (malnutrition) may be the cause of the seizures. Symptoms in patients on medications are especially hard to interpret because of interactions between seizures, medications, and the comorbidity. For example, some medications increase appetite, which may contribute to a comorbid condition: obesity. According to the 2013 Institute of Medicine report,[2] a comorbidity is "the co-occurrence of two

TABLE 17.4 Classification of the Epilepsies: Expanded

Engel (2001)	Shorvon (2011)
I. Localization-related	**I. Idiopathic/genetic**
A. Idiopathic	A. Pure epilepsies due to single gene disorders
Benign childhood epilepsy with centrotemporal spikes	Benign neonatal febrile convulsions
Childhood epilepsy with occipital paroxysms	Severe myoclonic epilepsy in infancy
Primary reading epilepsy	Autosomal dominant frontal lobe epilepsy
B. Symptomatic	B. Pure epilepsies with complex inheritance
Temporal lobe epilepsy, frontal lobe epilepsy, parietal lobe epilepsy, occipital lobe epilepsy	Idiopathic generalized epilepsy
C. Cryptogenic	
II. Generalized	**II. Symptomatic/structural–metabolic**
A. Idiopathic	A. Predominantly genetic or developmental
Benign neonatal febrile convulsions	Childhood epilepsy syndromes (West syndrome, Lennox–Gastaut syndrome)
Benign myoclonic epilepsy in infancy	Progressive myoclonic (Unverricht–Lundbord disease, Lafora)
Juvenile absence epilepsy	Neurocutaneous (tuberous sclerosis, Sturge–Weber syndrome)
Juvenile myoclonic epilepsy	Other neurological single-gene disorders (Angelman syndrome, Rett syndrome)
Epilepsy with generalized tonic–clonic seizures or myoclonic absences	Disorders of chromosome function (Down syndrome, fragile X syndrome)
B. Symptomatic or cryptogenic	Developmental anomalies of cerebral development (focal cortical dysplasia)
West syndrome	B. Predominantly acquired
Lennox–Gastaut syndrome	Mesial temporal lobe sclerosis (acquired temporal lobe epilepsy)
Epilepsy with myoclonic–astatic seizures or myoclonic absences	Perinatal or infantile causes (neonatal seizures, cerebral palsy)
C. Symptomatic	Cerebral trauma (head injury)
Non-specific etiology (early myoclonic encephalopathy with suppression burst)	Cerebral infection (meningitis, encephalitis)
Specific syndromes	Cerebral immunological disorders
	Degenerative or other neurological conditions (Alzheimer disease)
III. Undetermined whether focal or generalized	**III. Cryptogenic**
A. Both	
Neonatal	
Severe myoclonic epilepsy in infancy	
Epilepsy with continuous spike–waves during slow-wave sleep	
Landau–Kleffner syndrome	
B. Equivocal	
IV. Special syndromes (situation-related)	**IV. Provoked/unknown cause**
Febrile convulsions	A. Provoking factors
Isolated status epilepticus	Fever, sleep–wake cycle
	B. Reflex epilepsies
	Photostimulation, auditory

Note that not all epilepsy syndromes are listed.

supposedly separate conditions at above chance levels" and comorbidities can be distinguished into categories (Table 17.5). Guidelines proposed by the ILAE and American Epilepsy Society suggest the following four types of comorbidity:

- *Essential comorbidity*: Two conditions share an underlying causal factor (e.g. a cortical malformation may impair development and also cause seizures).
- *Secondary comorbidity*: One condition directly contributes to mechanisms involved in producing another condition (e.g. temporal lobe seizures may lead to functional impairments related to the temporal lobe, such as memory impairment).
- *Iatrogenic comorbidity*: The treatment of epilepsy leads to or exacerbates another condition (e.g. side effects of an ASD, such as sedation).
- *Situational or contextual comorbidity*: Social and environmental factors, as a consequence of epilepsy, may lead to a comorbid condition (e.g. loss of the ability to drive may lead to social isolation and depression).

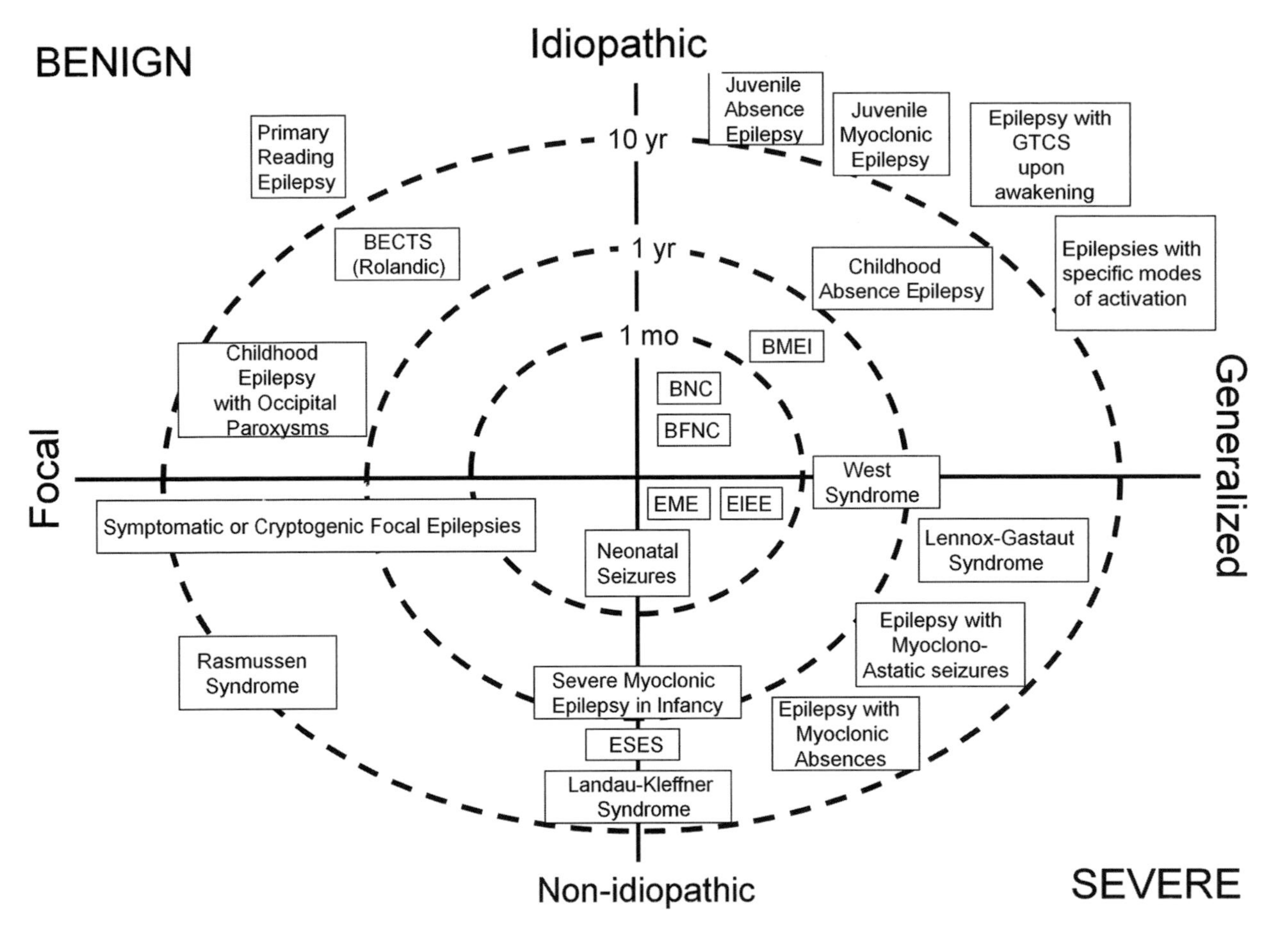

FIGURE 17.2 **The epilepsy solar system.** A perspective on epilepsy classification is shown. This depiction of four quadrants, each including several epilepsy syndromes that are similar (e.g. upper right: idiopathic and generalized) is similar to the idea of the ILAE (2010) that the epilepsies should be considered as constellations of syndromes with common characteristics. Typical age of onset is denoted by the concentric circles, with neonatal onset near the center and older ages farther out. GTCS: generalized tonic–clonic seizure; BECTS: benign epilepsy with centrotemporal spikes; BMEI: benign myoclonic epilepsy in infancy; BNC: benign neonatal convulsions; BFNC: benign familial neonatal convulsions; EME: early myoclonic epilepsy; EIEE: early infantile epileptic encephalopathy; ESES: electrical status epilepticus in sleep. *Source: Adapted from Nguyen The Tich and Pereon.* Epilepsia. *1999;40:531–532.*[9]

TABLE 17.5 Comorbidities in Epilepsy

Somatic	Neurological	Psychiatric	Intellectual/Cognitive	Infectious/Immune	Nutritional/Dietary
Anemia	Cerebral palsy	ADHD	Down syndrome	Encephalitis	Malnutrition
Asthma	Chronic pain	Alzheimer disease	Fragile X syndrome	Glioma	Obesity
Diabetes	Migraine	Autism	Intellectual disability (MR)	Meningitis	
Fibromyalgia	Stroke	Depression	Memory loss	Neurocysticerosis	
Gastrointestinal	Rett syndrome	Schizophrenia			
	Hearing loss	Substance abuse			
	Vision loss	Suicidality			
	Sleep disorders				

ADHD: attention deficit/hyperactivity disorder; MR: mental retardation.

MECHANISMS UNDERLYING SEIZURES

Understanding Seizures by Their Electroencephalographic Correlates

Interictal Spike and Ictal Events

Before the 1960s, mechanisms underlying seizures were poorly defined. Various hypotheses for epilepsy had been proposed, but until the foundations of modern neurophysiology were in place, most hypotheses were vague. Two methodological advances made a great difference: the ability to record from single neurons to clarify the neurophysiological correlates of EEG seizures, and the first methods for experimental induction of seizures in laboratory animals. Using topical application of penicillin to the cortical surface, Ayala and colleagues made some of the first intracellular recordings of cortical neurons during a seizure (Fig. 17.3).[14] They recorded activity of cortical neurons during interictal spikes as well as the ictal period. Interictal spikes are large transient voltage deflections, around 50–100 milliseconds long, that occur rhythmically in the period between seizures; the seizure itself is the ictus or ictal period (Figs 17.3 and 17.4). Ayala and colleagues learned that during the interictal spike, there was a sudden, large paroxysmal depolarization shift (PDS) in cortical principal cells. The depolarization triggered a high-frequency burst of action potentials on its peak, and then action potentials decayed, with resting potential ultimately being restored (Fig. 17.3). During ictal periods, a PDS initially occurred and the depolarization was greatly prolonged (Fig. 17.3). These observations led to the view that the PDS was central to seizure generation. It was hypothesized that either the neurons that exhibited the PDS were altered intrinsically (i.e. cortical principal cells became epileptic neurons) or the interactions between cortical neurons were altered, creating an epileptic aggregate. Ultimately, it was suggested that the PDS was caused by a giant synaptic potential; this is larger than a normal synaptic potential, but with similar underlying mechanisms, namely glutamate release from afferent fibers acting at ionotropic glutamate receptors [2-amino-3-(3-hydroxy-5-methyl-isoxazol-4-yl) propanoic acid (AMPA)/kainate and *N*-methyl-D-aspartate (NMDA) receptors[17]]. Like normal glutamatergic transmission, the initial depolarization of the PDS was mediated by AMPA receptors and the slower component of the depolarization was attributed to NMDA receptors. However, the idea of epileptic neurons continues to be raised.

The research that followed the initial studies of the PDS pursued the mechanisms underlying seizures in more detail, using recordings from both cortical or hippocampal principal cells and the γ-aminobutyric acidergic (GABAergic) interneurons located nearby, which are primarily responsible for synaptic inhibition of the principal cells. A major question was the neurobiological basis of the transition from the interictal to ictal state; that is, the transition to seizure. From the recordings, three states were described: (1) the interictal period; (2) the preictal period, immediately before the onset of a seizure; and (3) the ictal period (Figs 17.3 and 17.4). The idea of a silent preictal period changed after recordings from GABAergic neurons showed that during the preictal period there is an increase in GABAergic neuron firing, causing $GABA_A$ receptor-mediated inhibition of principal cells (Figs 17.3 and 17.4). There also is a gradual increase in excitation of principal cells by glutamatergic inputs (Fig. 17.4).[16] Therefore, there is no silent period necessarily. Another important event during the preictal period is a gradual increase in the extracellular concentration of potassium ($[K^+]_o$). The increase in $[K^+]_o$, and possibly other factors, leads to a shift in chloride flux through $GABA_A$ receptors, which normally is inward, and normally causes hyperpolarization of the cell. Increased $[K^+]_o$ affects K^+/Cl^- cotransporters (e.g. KCC2[18]) so that chloride flux is outward instead of inward, and a depolarization occurs in principal cells in response to GABA. In addition, there may be a gradual failure of GABA release during the preictal period because of persistent GABAergic neuron firing, causing depolarization block. However, postsynaptic $GABA_A$ receptors do not appear to change.[16]

The onset of the seizure is heralded by the depolarizing effects of GABA, reduced GABA release, and persistent glutamatergic activity.[16] Elevated $[K^+]_o$ could also play a role by depolarizing principal neurons, which can change their firing behavior from a tonic to a bursting mode.[19] A bursting mode would be consistent with the type of firing at the onset of the seizure. As the onset of the ictal period begins, the failure of local GABAergic inhibition and presence of recurrent collaterals between principal cells could be sufficient to cause hypersynchrony[20] and potentially explain how seizures spread from one brain area to adjacent regions.[21] Other mechanisms may also be important, such as dendritic potentials in principal cells, related to GABAergic innervation of dendrites or dendritic ion channels governing backpropagation of action potentials.[22]

Seizure Termination

One of the fundamental aspects of epilepsy is that most patients have seizures for only a small fraction of the time. When seizures occur, they last for only a brief time (seconds to minutes), because there are effective inherent mechanisms for seizures to self-terminate, meaning that they stop on their own. Therefore, many investigators have attempted to understand the endogenous mechanisms that stop seizures, because this knowledge may lead to new drugs to treat epilepsy. However, although our understanding of potential mechanisms for seizure

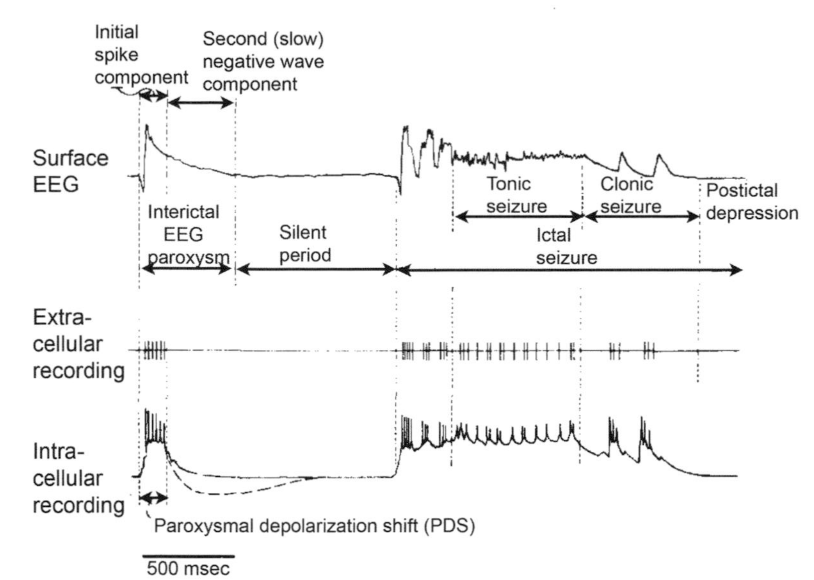

FIGURE 17.3 **The cellular correlates of interictal spikes and seizures.** A recording of the electroencephalogram (EEG) from the surface of the brain (top), near cortical neurons (center, extracellular recording), and inside a cortical neuron (bottom, intracellular recording). Recordings were made from a cat where the convulsant penicillin had been applied to the cortical surface to create an area where spontaneous seizures occurred periodically. During a spike that occurred between seizures (an interictal spike), cortical neurons fired action potentials on a large, sudden depolarization called a paroxysmal depolarization shift (PDS). During the seizure (the ictal event), a prolonged depolarization with action potentials occurred. Together, these recordings explained the activity of cortical neurons during spikes and seizures that were recorded at the cortical surface. *Source: Adapted from Ayala* et al. J Neurophysiol *1970;33:73–85.*[14]

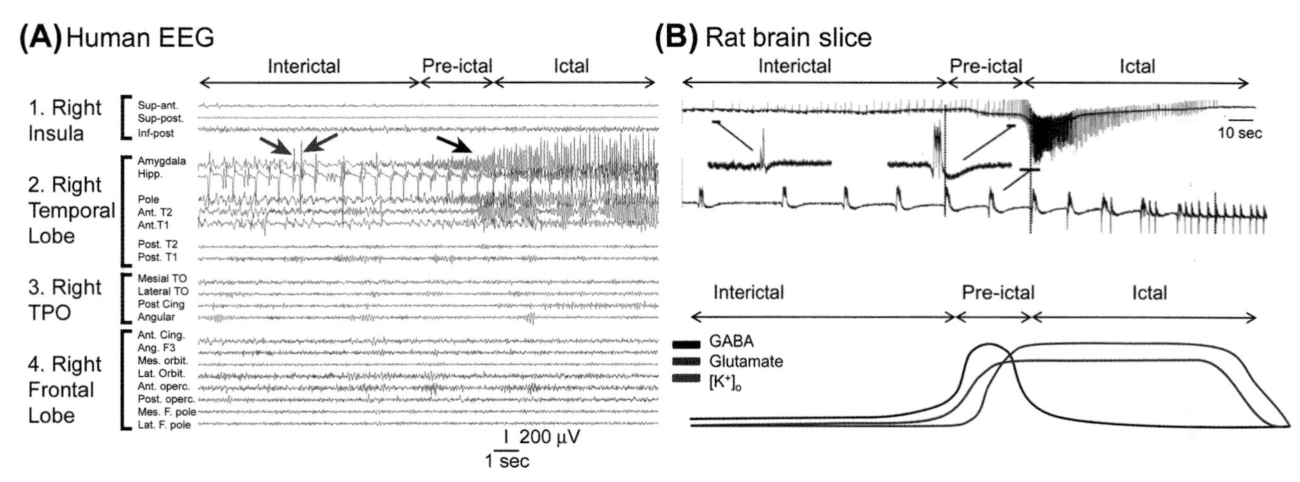

FIGURE 17.4 **Mechanisms underlying the transition from normal brain activity to seizures.** (A) An example of a focal seizure that started in the temporal lobe. The black arrow on the electroencephalogram (EEG) points to the start of the seizure, called the ictal period. The red arrows point to interictal spikes, which change frequency before the seizure begins, during the preictal period. (B) Top: Example of a seizure-like event in a slice of hippocampus of a rat, exposed to the convulsant 4-aminopyridine. Parts of the seizure (designated by the horizontal bars) are expanded below. Bottom: Schematic illustrating the changes in the activity of γ-aminobutyric acidergic (GABAergic) neurons (black), glutamatergic principal cells (blue), and the concentration of extracellular K^+ ($[K^+]_o$; red) during the transition between interictal, preictal, and ictal periods. During the preictal period, GABAergic neurons increase their firing rate. $[K^+]_o$ rises and glutamatergic neurons become depolarized. Glutamatergic neurons increase their activity as GABAergic inhibition becomes depolarizing instead of hyperpolarizing, and GABA release fails (because of GABA depletion, depolarization block, or other mechanisms). At this point the seizure begins. At the end of the seizure, mechanisms that normally control excitability of glutamatergic and GABAergic neurons are restored, such as the normal $[K^+]_o$ by Na^+/K^+-ATPase. *Source: (A) Adapted from Ryvlin* Epileptic Dis *2006;8 (Suppl 2):S37–S56.*[15]*; (B) adapted from Zhang* et al. Epilepsy Res *2011;97:290–299.*[16]

termination is extensive, the endogenous mechanisms that stop an ictal event, and ways to enhance them, are still debated. Below are four mechanisms that are considered to contribute to seizure termination.

- *Action potential repolarization and voltage-dependent K^+ channels*: Mechanisms that repolarize neurons after action potential generation are a logical place to look for mechanisms that stop seizures. The vast majority of repolarization after action potentials is attributed to voltage-gated K^+ channels. The A-type K^+ channel and delayed rectifier type of K^+ channels are primarily responsible for repolarization after a single action potential. During a persistent depolarizing input, M-type K^+ channels also contribute by slowing action potential firing frequency, which is called spike frequency adaptation. The M current appears to decrease its expression in some neurons in animal models of epilepsy, and the ASD retigabine exerts its effect by opening these channels.[23] Ca^{2+}-dependent K^+ currents are also important in regulating the hyperpolarizations that follow action potentials (afterhyperpolarizations).

- *The Na^+/K^+ pump and other mechanisms that regulate $[K^+]_o$*: After prolonged depolarizations, such as those that occur during ictal discharges, extracellular K^+ rises and the Na^+/K^+ pump is a primary mechanism for restoring $[K^+]_o$. Cellular energy depletion inhibits the pump because it depends on ATP. Inhibitors of the Na^+/K^+ pump, such as ouabain, promote seizure activity in hippocampal slices, particularly at young ages. The susceptibility of young ages suggests that a vulnerability of the Na^+/K^+ pump may be one of the reasons that children are susceptible to seizures. Astrocytes are a major regulator of $[K^+]_o$, as discussed below.
- *GABAergic inhibition*: If release of GABA becomes impaired during the onset of a seizure, restoring GABA by either enhanced GABA reuptake into the nerve terminal or increasing GABA synthesis could help to restore GABAergic function. In addition, the mechanisms that control the preferred direction of chloride flux through $GABA_A$ receptors are important. Carbonic anhydrase plays an important role as the enzyme that catalyzes the hydration of CO_2. There are five families of carbonic anhydrase (α–ε), with the α family primarily responsible for mammalian forms. Of the more than 15 types of α-carbonic anhydrases, many are expressed in the brain and implicated in seizures. Inhibition of carbonic anhydrase in brain slices or deletion in transgenic mice has proconvulsant effects; these are likely to be mediated by acidification of the extracellular milieu, which can facilitate actions of glutamate at NMDA receptors. In addition, water balance is disrupted by carbonic anhydrase inhibition, which is potentially important because neuronal swelling increases excitability. However, carbonic anhydrase may also be an indirect regulator of KCC2 because of its effects on the concentration of HCO_3^-, which influences the direction of chloride flux through $GABA_A$ receptors. The restoration of the normal direction of this chloride flux, into the cell, has been suggested to be responsible for termination of seizures. For this reason, carbonic anhydrase inhibition is one potentially important mechanism of action of several ASDs, acetazolamide being the most well known (see Table 17.9).
- *Brain gates*: There are several sites in the brain where seizure activity is normally difficult to induce, and it has been proposed that these areas act as barriers or gates to seizure propagation. The gates can stop a seizure from developing, truncate seizures if they have begun, or prevent them from reaching parts of the brain that control movement. As a result, seizures do not trigger convulsive behaviors. The locations of these gates are diverse, and include the substantia nigra, subthalamic nucleus, superior colliculus, reticular activating system, thalamus,[24] and dentate gyrus.[25] What causes the gating behavior is not clear, but in the case of the reticular activating system, it is suggested that the widespread innervation of cortical areas is the reason for its robust effects on seizures. In the dentate gyrus, it has been proposed that there are several inherent mechanisms in principal cells of the region, the granule cells, which make the dentate gyrus operate like a gate to seizures. These include intrinsic properties of granule cells such as a very hyperpolarized resting potential, and strong spike frequency adaptation, as well as robust GABAergic inhibition of granule cells.[25] Because the granule cells are situated between cortex and area CA3 of hippocampus, if the granule cells do not discharge action potentials, area CA3 may not either, stopping cortical seizure activity from affecting area CA3.[25]

Neuromodulators

The brain synthesizes many potent anticonvulsant molecules, often packaged in dense core vesicles in GABAergic neurons, preferentially released during high-frequency activity (the type of activity that would occur during a seizure). One example is neuropeptide Y (NPY), which acts on five types of NPY receptor, located presynaptically and postsynaptically. Actions at Y2 and Y5 receptors inhibit glutamate release from hippocampal pathways involved in seizure propagation, and infusion or overexpression of NPY leads to a reduced susceptibility or fewer seizures. NPY synthesis normally occurs in a subset of GABAergic neurons, and increases after seizures in both the GABAergic neurons that normally synthesize NPY and other cells that normally do not produce NPY. Therefore, it has been proposed that upregulation of NPY is an endogenous anticonvulsant response of the CNS to seizures. As a result, gene therapy to increase synthesis of NPY has been proposed as a therapeutic strategy for epilepsy.[26] Other endogenous molecules that have anticonvulsant effects include somatostatin, adenosine, and endocannabinoids.

The brain also produces neuromodulators that have excitatory effects and promote neuronal outgrowth. In the context of the normal brain, these neuromodulators may be beneficial by supporting neuronal activity and plasticity. In the context of epilepsy, blocking their receptors may be therapeutic. One example is the neurotrophin brain-derived neurotrophic factor (BDNF), which acts at TrkB receptors to support synaptic plasticity (long-term potentiation) and structural plasticity (increased dendritic spine density, axon sprouting). BDNF infusion into the hippocampus of an adult rat induces seizures, suggesting that blockade of TrkB could be therapeutic in epilepsy.[27]

Three-Hertz Spike-and-Wave Rhythms

The EEG of an individual with absence epilepsy distinguishes it from virtually all other types of epilepsy, as described above. Understanding the causes of this rhythm could lead to better treatment. Experiments in animals first established that the 3 Hz spike-and-wave rhythm was produced by oscillations in thalamocortical circuitry, where cortical principal cells excite thalamic GABAergic neurons in the reticular nucleus of the thalamus, which inhibit the thalamic relay cells that project back to cortex (Fig. 17.5). Neurophysiological recordings from the primary cell types in the thalamocortical circuit established that the 3 Hz rhythm is caused by action potentials of the corticothalamic and thalamocortical neurons during the spike of the spike-and-wave discharge; the wave is associated with a hyperpolarization of the neurons (Fig. 17.5). A critical first step in producing the spike-and-wave rhythm appears to be increased cortical excitability (Fig. 17.5).[30] The increased activity of cortical neurons causes the GABAergic neurons of the reticular nucleus to fire more intensely, releasing more GABA on to thalamic relay cells (Fig. 17.5). The increase in GABA release activates not only $GABA_A$ receptors but also $GABA_B$ receptors, which leads to a longer period of inhibition of the thalamic relay cells (Fig. 17.4). As inhibition wanes, the unique ion channels of the relay cell, including a T-type calcium channel, cause the relay cell to exhibit a rebound burst discharge. The T-type channel is normally inactivated at resting potential but deinactivates with hyperpolarization; the consequence of activation is calcium entry into the relay cell, leading to a depolarization. Because of the particular combination of ion channels expressed by the relay cell, including T-type calcium channels and also hyperpolarization-activated cyclic nucleotide-gated cation channels (I_h), the relay cell oscillates at about 3 Hz. This rhythmic firing of relay cells entrains the cortical cells, so that the entire thalamocortical circuitry begins to oscillate at a 3 Hz spike-and-wave

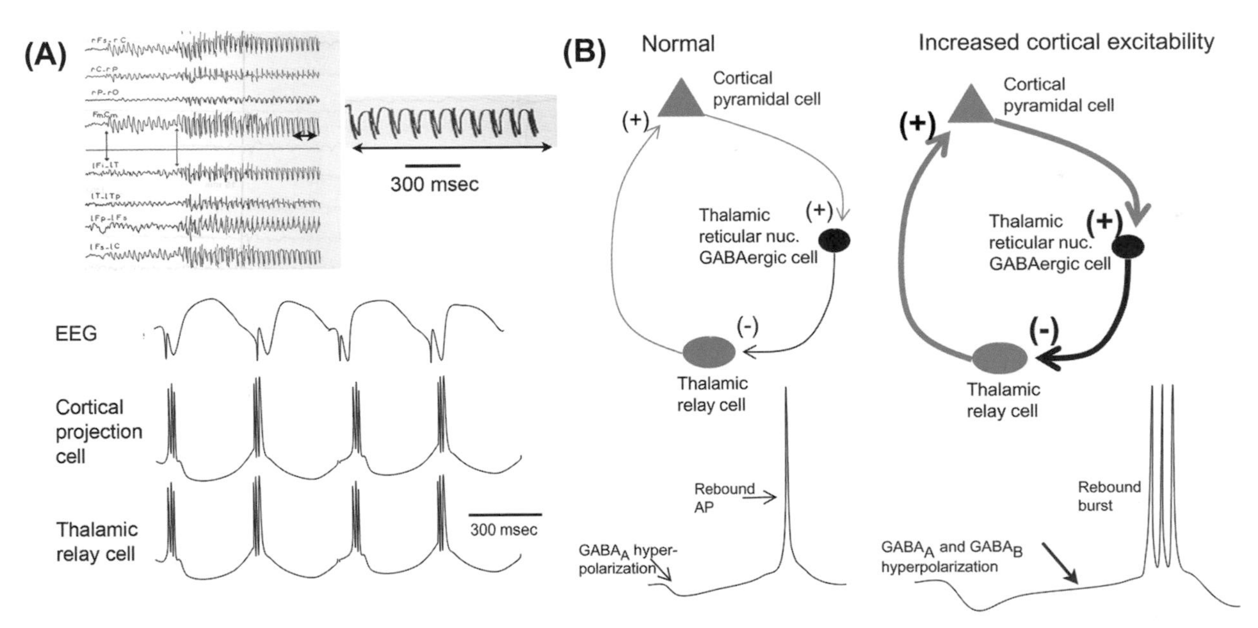

FIGURE 17.5 **The neuronal correlates of 3 Hz spike-and-wave discharge.** (A) Top: Example of an absence seizure with 3 Hz spike-and-wave discharge. The part of the recording indicated by the double-sided arrow is expanded on the right. Bottom: Schematic of the neuronal activity in corticothalamic principal cells and thalamocortical relay cells during a spike-and-wave discharge. During the spike there is neuronal discharge and during the wave there are hyperpolarizations. EEG: electroencephalogram. (B) Schematic of some of the circuitry involved in spike-and-wave discharge. The corticothalamic neuron, typically a cortical pyramidal cell, innervates γ-aminobutyric acidergic (GABAergic) neurons of the thalamic reticular nucleus, which innervate glutamatergic thalamic relay cells. The relay cells project back to cortex. When there is increased cortical excitability, the corticothalamic cell excites GABAergic neurons more, leading to greater release of GABA on thalamic relay cells. In contrast to the normal situation, where GABA release primarily activates $GABA_A$ receptors on the relay cells, there is activation of $GABA_A$ and $GABA_B$ receptors on the relay cell, causing a larger and more prolonged hyperpolarization in the relay cell (red arrows). When the hyperpolarization increases, there is a more effective deinactivation of T-type calcium channels of the relay cell. As a result, there is a greater T-type calcium current when the hyperpolarization decays, leading to enhanced firing of the relay cell. The relay cell then activates the corticothalamic cell, leading to a cycle of hyperpolarization and burst firing at approximately 3 Hz.[28] AP: action potential. *Source: (A) Adapted from Gastaut.* Epilepsy – The Electroclinical Correlates. *Oxford: Blackwell; 1954.*[29]

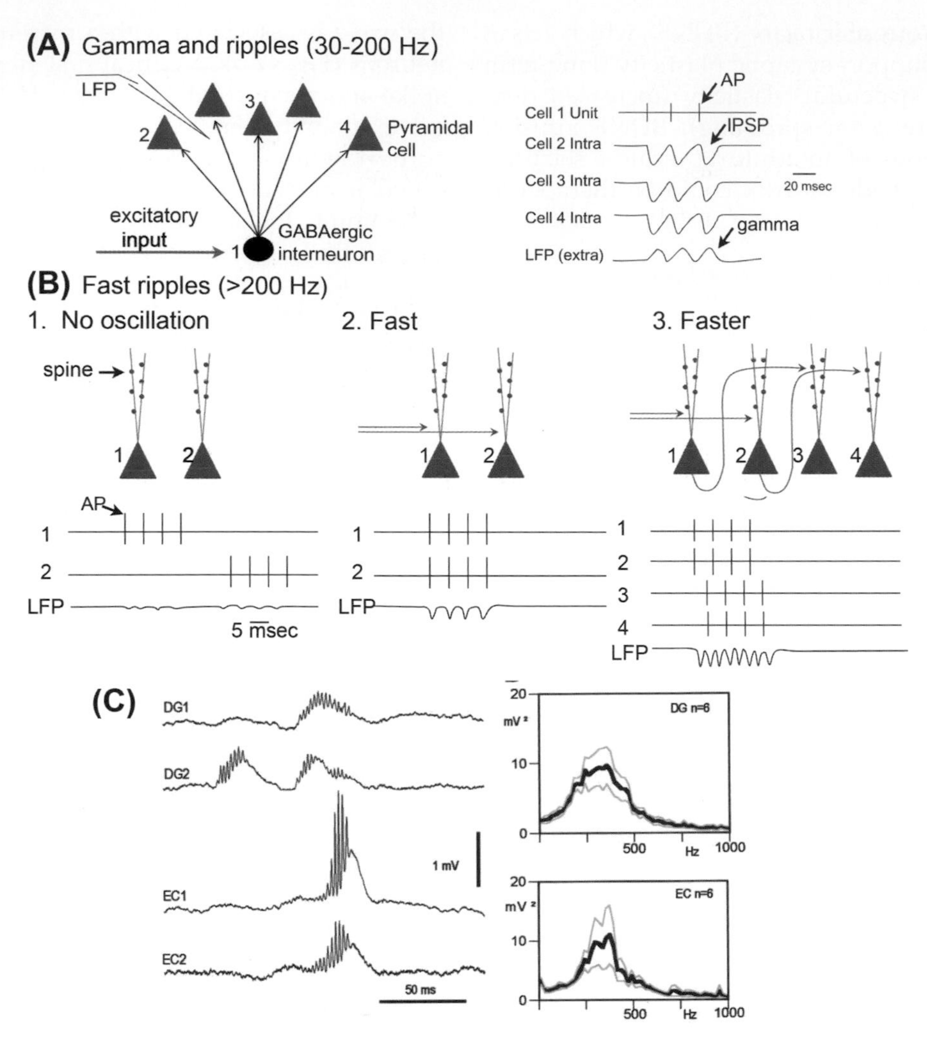

FIGURE 17.6 **High-frequency oscillations and their underlying mechanisms.** (A) High-frequency oscillations refer to several types of oscillations, typically over 100 Hz. Gamma oscillations are between 30 and 100 Hz. Ripples typically refer to oscillations between 100 and 200 Hz. The underlying mechanisms are generally attributed to the effects of synchronized γ-aminobutyric acidergic (GABAergic) neurons on principal cells. A shared glutamatergic input is one way in which the GABAergic neurons are synchronized. Because GABAergic neurons innervate multiple principal cells, they can silence principal cell activity synchronously. Effects that are detected extracellularly [local field potential (LFP)] are shown on the right. When the GABAergic neurons fire [indicated by extracellular recordings of action potentials (APs)], synchronized inhibitory postsynaptic potentials (IPSPs) occur in the principal cells (cells 2–4). As chloride ions enter $GABA_A$ receptors on the principal cells, positive waves occur in the LFP. (B) Fast ripples are higher frequency oscillations, typically 200–500 Hz, and are considered to be a hallmark of epileptic tissue. Mechanisms for high-frequency oscillations are typically non-synaptic because synapses would slow the oscillation to a frequency below 200 Hz. Instead, glutamatergic afferents that are divergent, innervating many principal cells, may synchronize them (compare B1 to B2). The influx of Na^+ during APs would be detected as negative oscillations (ripples) in the LFP, and the frequency of the ripples would reflect the frequency of synchronous firing of the principal cells. Faster frequencies could theoretically occur by a mechanism shown schematically in B3, where glutamatergic principal cells sprout on to principal cells nearby. (C) High-frequency oscillation from a recording from the dentate gyrus (DG) and entorhinal cortex (EC) with power spectral analyses on the right, showing peak frequencies between 200 and 500 Hz. *Source: (B) From Kohling and Staley* Epilepsy Res. *2011;97:318–323.*[36] *(C) From Bragin* et al. J Neurosci *2002;22:2012–2021.*[31]

rhythm. For this reason, absence seizures are exacerbated by drugs that enhance GABAergic inhibition. In contrast, drugs that block the T-type calcium channel, such as ethosuximide, are highly effective.

High-Frequency Oscillations

The normal brain exhibits spontaneous EEG rhythms at many frequencies, from relatively slow oscillations (<1 Hz) to higher frequencies (>100 Hz). In normal adults, these rhythms are well characterized, including the well-known alpha, beta, and delta rhythms. In the past few decades, the mechanisms underlying faster oscillations, such as the cellular mechanisms for gamma oscillations (30–120 Hz), have been characterized. Synchronous discharges of GABAergic interneurons appear to be responsible for gamma oscillations, because they synchronously hyperpolarize principal cells (Fig. 17.6A); extracellular recordings near the principal cell bodies show an

TABLE 17.6 Experimental Models of Seizures

In Vitro Models of Seizure-Like Activity[a]		*In Vivo* Models of Seizures	
Preparations	**Treatments**	**Preparations**	**Treatments**
Tissue culture	Convulsants	Lower species	Convulsants
- Primary cultures	Other chemicals	- Zebrafish	Other chemicals
- Organotypic cultures	Altered extracellular	- Invertebrates	Altered CSF/plasma
Brain slices	- Ion concentrations	Rodents (mice, rats, other)	- pH
Isolated brain preparations	- pH	- Anesthetized	- Osmolarity
- Perfused guinea pig brain	- Osmolarity	- Unanesthetized	- Immune stimulus
- Isolated hippocampus	Electrical stimulus	Other lower mammals	- Change to blood–brain barrier
	Optogenetics	Non-human primates	Electrical stimulus
	Transgenic tissue	- Monkey	Optogenetics
	Viral injection[b]	- Photosensitive baboon	Transgenic tissue
			Viral injection

[a]In vitro *preparations do not exhibit seizures in the clinical sense, so the phrase "seizure-like activity" is often used instead.*
[b]*Use of viral vectors to alter gene expression.*

oscillation at approximately gamma rhythm, corresponding to the repetitive inward flux of chloride during each GABAergic hyperpolarization (also called inhibitory postsynaptic potentials or currents) (Fig. 17.6A). Higher frequency ripples (100–200 Hz) are also considered to be a result of GABAergic mechanisms, but other mechanisms contribute, such as gap junctions.[32]

Higher frequency oscillations (200–500 Hz), often called fast ripples, have captured attention in epilepsy because they are recorded in patients with epilepsy as well as animal models,[33–35] but are normally rare. These oscillations are too fast to be explained by synaptic mechanisms. Using normal hippocampal slices exposed to elevated $[K^+]_o$, It has been suggested that glutamatergic afferents initiate the fast ripple, but intrinsic mechanisms that allow the principal cells to fire at 200–500 Hz are ultimately responsible for it (Fig. 17.6).[36] When action potentials from adjacent pyramidal cells are synchronous, a fast ripple would be generated extracellularly at a frequency proportional to the interspike interval, and reflect the repetitive inward Na^+ currents during the action potentials (Fig. 17.6).[36] There are several potential reasons why fast ripples may be found preferentially in epileptic tissue, such as an increase in the number of recurrent excitatory connections between principal cells, as a result of axon outgrowth or sprouting (Fig. 17.6) (sprouting is discussed further below, under "Lessons Learned from Status Epilepticus Models"). In this case, the action potentials in the synchronized population of neurons could trigger action potentials in other neurons that are not innervated by the original glutamatergic afferents. Extremely fast oscillations would be possible. Other mechanisms may also contribute to fast ripples, such as increased variability in the onset of action potentials activated by glutamatergic inputs.

A great deal of interest has been generated by the finding that the epileptic brain may be characterized by a type of fast oscillation that is rarely detected under normal conditions. Fast oscillations could be a useful biomarker of the disease, which would be helpful in diagnosis and treatment. Furthermore, the mechanisms that underlie fast oscillations may provide targets for new ASDs. These ideas have led investigators to study various mechanisms related to fast ripples, and it has been suggested that drugs that block gap junctions may be therapeutic because they would block fast ripples.

Understanding Seizures by Principles of Neuronal Excitability: The Balance of Excitation and Inhibition

A common perspective on seizures is that they represent an imbalance between excitation and inhibition. This general idea is useful because it allows one to appreciate the diverse ways in which the brain can produce seizures, leading to many ideas for new drugs to stop seizures. Table 17.6 provides a summary of the current methods to induce seizures in laboratory animals, and Table 17.7 lists the tests that are commonly used to assess the efficacy of new ASDs.

Although the concept that a seizure arises when there is an imbalance between excitation and inhibition has been useful in epilepsy research, one could argue that this is an oversimplification, because seizures in people with epilepsy are often rare, occurring once a day, once a week, or even once a month. If there

TABLE 17.7 Standardized Tests for Anticonvulsant Drug Screening

Seizure Model[a]	Clinical Seizure Type[b]
Pentylenetetrazol (metrazol) administration	Myoclonic/clonic seizures
Maximal electroshock test	Generalized tonic–clonic seizures
Rodents with absence epilepsy (genetic)	Absence seizures
γ-Hydroxybutyrate administration	Absence seizures
Kindling model	Focal seizure with secondary generalization
Kainic acid- or pilocarpine-induced epilepsy	Focal seizure with secondary generalization (drug-resistant)

[a]Models of experimental seizures used in tests of potential anticonvulsant drugs at the National Institutes of Health.[1]
[b]For each seizure model, the response to a drug is considered predictive of the effect that drug would have on the type of clinical seizure it simulates.

were simply an imbalance, for example excessive excitation, one would expect seizures to occur most of the time. Furthermore, some types of epilepsy are associated with many complex changes in the brain, including changes in glia and vasculature. Indeed, most ASDs have multiple sites of action (see below). This has been one reason for developing animal models to study epileptogenesis and epilepsy that are different from the methods used to study seizures (Table 17.8).

Neuronal Properties: Ions, Ion Channels, and Channelopathies

Neurons in the CNS have diverse intrinsic properties, but some generalizations can be made. For example, the major ions that pass through ligand- or voltage-gated ion channels are distributed so that Na^+, Ca^{2+}, and Cl^- are concentrated extracellularly and K^+ is concentrated intracellularly. The consequence is an approximately −60 mV resting membrane potential. Therefore, conditions where electrolyte balance becomes impaired (such as severe diarrhea after cholera) may result in seizures. Mg^{2+} is an important example of a critical mineral, because it normally maintains a voltage-dependent block of NMDA receptors. There is also a potential consequence of perturbations in pH and osmolarity: acidic pH influences NMDA receptors and can exacerbate seizures, and a disruption of osmolarity can lead to neuronal swelling and non-synaptic interactions that facilitate seizures.

It is also relevant to consider the voltage-gated ion channels that are required for action potential generation. Mutations in the subunits of the voltage-gated sodium channels which are responsible for fast depolarization during the rising phase of the action potential are the basis for several genetic forms of epilepsy in childhood, including severe myoclonic epilepsy in infancy (also called Dravet syndrome) (see Table 17.8 and Fig. 17.7A). This type of epilepsy is often called a channelopathy because that is considered to be the only defect in the disease, and the defect is a loss-of-function mutation in a gene that normally gives rise to an ion channel. Other channelopathies also give rise to epilepsy. For example, mutations in the *KCNQ* gene family, encoding Kv7 channels, lead to abnormal K^+ channels and disruption of the M current. *KCNQ* mutations are responsible for a type of childhood epilepsy called generalized epilepsy with febrile seizures plus (GEFS+) (Table 17.8 and Fig. 17.7B).[37] Voltage-gated ion channels are not the only examples of channelopathies; a mutation in the genes responsible for subunits of the nicotinic acetylcholine receptor leads to an epilepsy syndrome called autosomal dominant frontal lobe epilepsy (ADFLE) (Table 17.8).[38]

Synaptic Transmission: Glutamate, γ-Aminobutyric Acid, and Others

Many aspects of synaptic transmission are targets for ASDs or are related to causes of epilepsy. Virtually every component of synaptic transmission is a target, including trafficking of neurotransmitters and their receptors to the nerve terminal or dendrites, transmitter release and reuptake, postsynaptic receptor expression and function, and postsynaptic signaling pathways and their regulation. This applies to the major neurotransmitters in the CNS, glutamate and GABA, as well as other neurotransmitters such as acetylcholine and norepinephrine.

Neurodevelopment

From the discussion above, there already would seem to be sufficient vulnerabilities in the nervous system to explain the diversity of seizures and epilepsy, but there is more to consider in the circuit organization of the CNS. Many forms of epilepsy arise from defects in the essential structure of the brain, such as inability to form the normal pattern of lamination in the neocortex. Some of these developmental disorders are caused by single gene mutations, such as lissencephaly, where *Lis1* is mutated and the normal gyri and sulci of the neocortex are absent.[39] Other developmental disorders give rise to malformations such as tubers in tuberous sclerosis, where the mammalian target of rapamycin (mTOR) pathway is disrupted.[40] Additional examples are often discovered by investigators who introduce a novel mutation in a transgenic mouse to study a specific question in neurobiology, and find that seizures occur; these mice may provide insight into the many types of epilepsy that have been considered "idiopathic" in the past (Table 17.8).

TABLE 17.8 Experimental Models of Epileptogenesis and Epilepsy

Type of Epilepsy	Animal Model
I. Genetic	
BNC, BFNC	*KCNQ* mutations
SMEI, GEFS+	*SCN1A/B* mutations
ADFLE	*CHRNA/B* mutations
Absence epilepsy	Spontaneous mutations in mice (e.g. *tottering*)
Absence epilepsy	Inbred rat strains (GAERS, Wag/Rij)
Neocortical epilepsy	EL mouse
II. Structural/metabolic	
A. Predominantly genetic or developmental	
1. West syndrome (infantile spasms)	CRH injection, betamethasone/NMDA, doxorubin/LPS/chlorophenyalanine, neonatal tetrodotoxin, *ARX* knockout mice
2. Progressive myoclonic	
Unverricht–Lundborg disease	*CSTB* (Cystatin B) knockout mice
Lafora disease	*EPM2A* (Laforin) knockout mice
3. Neurocutaneous	
Tuberous sclerosis	*TSC1/2* knockout mice, Eker rat
4. Other single-gene disorders	
Angelman syndrome	*Ube3a* or *GABRB3* knockout mice
Rett syndrome	*MECP2* knockout mice
5. Disorders of chromosome function	
Down syndrome	Ts65Dn transgenic mice
Fragile X syndrome	*Fmr1* knockout mice
6. Developmental anomalies of cerebral development	
Focal cortical dysplasia	Prenatal irradiation of MAM, p35 knockout mice
Polymicrogyria	Neonatal freeze lesion
Heterotopias	*Lis1* knockout mouse
B. Predominantly acquired	
1. Mesial temporal sclerosis/TLE	Experimental status epilepticus
2. Perinatal or infantile causes	Neonatal febrile seizures or hypoxia
3. Cerebral trauma	Fluid percussion injury, cortical undercut, focal alumina gel, penicillin or tetanus toxin
4. Cerebral infection	Lipopolysaccharide infection, viral encephalitis
5. Cerebral immunological disorders	
Rasmussen encephalitis	Antibodies to GluR3
6. Other degenerative or neurological conditions	
Stroke	Carotid occlusion, vascular manipulations
Alzheimer disease	APP, tau, PS1 mutations/overexpression
III. Cryptogenic	Transgenic mice with epilepsy but unclear etiology
IV. Provoked	Kindling (electrical. chemical)
Auditory stimulation	Audiogenic seizure-susceptible (DBA) mice
Photic stimulation	Photosensitive baboon (*Papio papio*)
Drug-induced	Alcohol withdrawal
Hormonal	Mild status epilepticus, female rats

Experimental models of epileptogenesis and epilepsy are listed with the same classification scheme as the categorization of the epilepsies in Table 17.4. Not all types of epilepsy or all animal models are listed.
BNC: benign neonatal convulsions; BFNC: benign familial neonatal convulsions; SMEI: severe myoclonic epilepsy in infancy; GEFS+: generalized epilepsy with febrile seizures plus; ADFLE: autosomal dominant frontal lobe epilepsy; TLE: temporal lobe epilepsy; *KCNQ*: gene encoding the Kv7 potassium channel family; *SCN1A/B*: gene encoding the α- and β-subunits of the Nav1.1 voltage-dependent sodium channel; *CHRNA/B*: gene encoding the α- and β-subunits of the nicotinic cholinergic receptor; GAERS: genetic absence epilepsy rats from Strasbourg; WAG/Rij: Wistar–albino–Glaxo from Rijswijk; EL: EL/Suz; CRH: corticotropin-releasing hormone; NMDA: *N*-methyl-D-aspartate; TSC: tuberous sclerosis complex; MAM: methoxymethanol; APP: amyloid precursor protein; PS1: presenilin 1; DBA: Dilute Brown Non-Agouti.

Non-Neuronal Mechanisms of Seizures

Another perspective on seizure generation de-emphasizes the typical neuron-centric view of the CNS and considers other cells and processes that are essential to normal brain function, such as astrocytes, the vasculature, the immune system, and subcellular energy metabolism.

Astrocytes contribute to the regulation of seizures in many ways. Astrocytes are considered essential because they can buffer $[K^+]_o$ and therefore keep the extracellular concentration of $[K^+]_o$ relatively low. Astrocytes express membrane-bound transporters that are critical to the synthesis and uptake of transmitters such as glutamate and GABA. Their presence at synapses limits the spatial and temporal exposure of synapses to neurotransmitter. As a result, astrocytes regulate the duration of synaptic potentials and

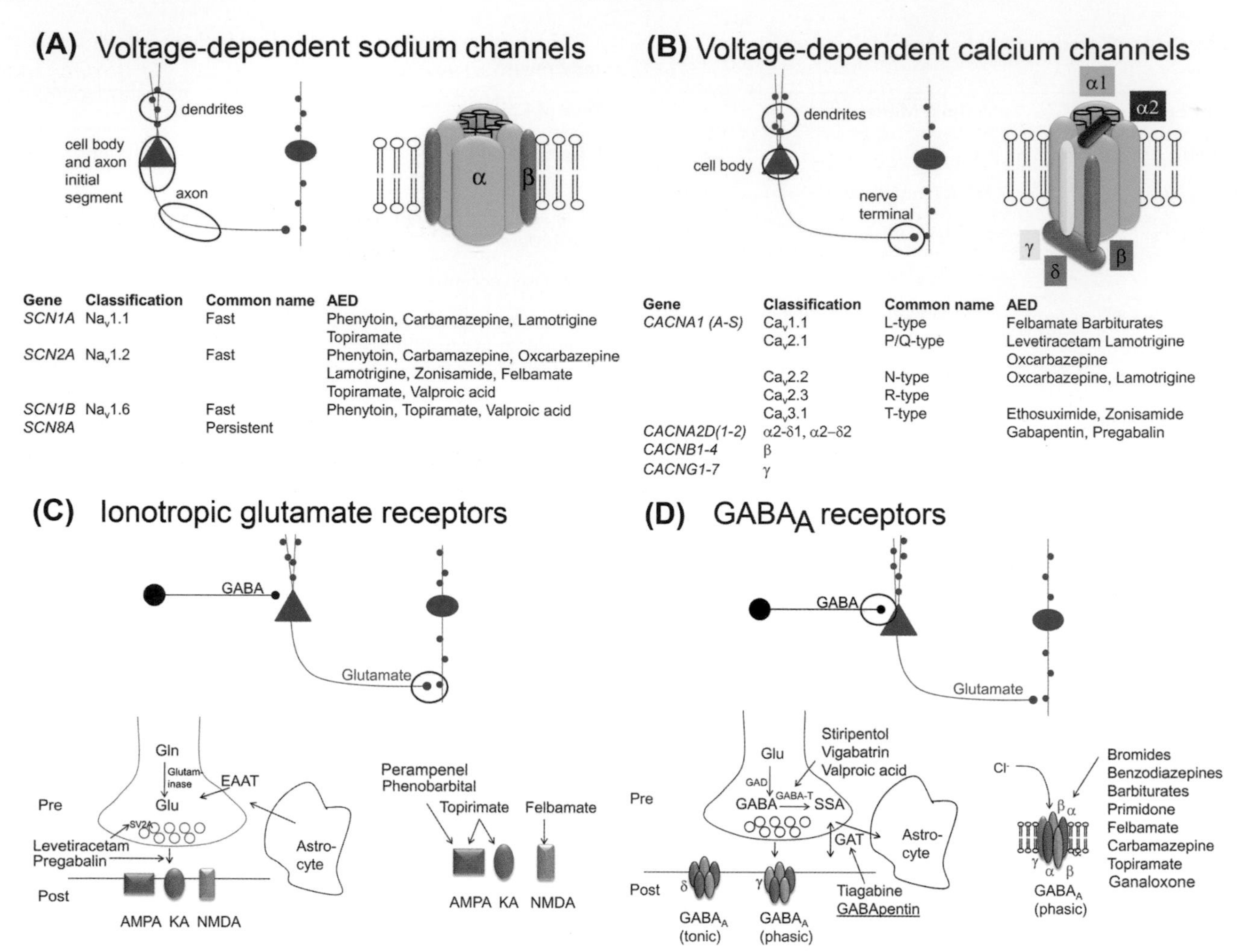

Gene	Classification	Common name	AED
SCN1A	$Na_v1.1$	Fast	Phenytoin, Carbamazepine, Lamotrigine Topiramate
SCN2A	$Na_v1.2$	Fast	Phenytoin, Carbamazepine, Oxcarbazepine Lamotrigine, Zonisamide, Felbamate Topiramate, Valproic acid
SCN1B	$Na_v1.6$	Fast	Phenytoin, Topiramate, Valproic acid
SCN8A		Persistent	

Gene	Classification	Common name	AED
CACNA1 (A-S)	$Ca_v1.1$	L-type	Felbamate Barbiturates
	$Ca_v2.1$	P/Q-type	Levetiracetam Lamotrigine Oxcarbazepine
	$Ca_v2.2$	N-type	Oxcarbazepine, Lamotrigine
	$Ca_v2.3$	R-type	
	$Ca_v3.1$	T-type	Ethosuximide, Zonisamide
CACNA2D(1-2)	α2-δ1, α2–δ2		Gabapentin, Pregabalin
CACNB1-4	β		
CACNG1-7	γ		

FIGURE 17.7 **Mechanisms of action of antiseizure drugs (ASDs).** (A) Inhibition of voltage-dependent sodium channels by ASDs. Top left: Location of channels in the plasma membrane of a typical cortical neurons Channels are primarily axonal, but some are also present on some dendrites. Top right: The structure of the sodium channel consists of multiple pore-forming α-subunits (gold) and accessory β-subunits (green). Bottom: Sodium channels that are targets of ASDs, with gene, numerical designation, and common name. Antiepileptic drugs (AEDs) that are known to act on these channels are listed on the right. (B) Inhibition of voltage-dependent calcium channels by ASDs. Top left: Voltage-dependent calcium channels are located in many parts of cortical neurons, including dendrites, somata, and the nerve terminal. Top right: The structure of voltage-dependent calcium channels consists of pore-forming subunits (α_1, green) and several accessory subunits (α_2, purple; β, blue; γ, yellow; δ, purple). Bottom: Nomenclature for calcium channels and examples of drugs that act on the channels. (C) ASDs that act on ionotropic glutamate receptors. Top: Ionotropic glutamate receptors are located mainly at synapses. Some receptors are also present immediately outside the synapse (extrasynaptically) and on astrocytes (not shown). GABA: γ-aminobutyric acid. Bottom left: Drawing of a glutamatergic synapse, with sites of action of ASDs in red. Bottom right: Subtypes of ionotropic glutamate receptors are shown schematically, with drugs that act on the receptors shown in red. Gln: glutamine; Glu: glutamate; EAAT: excitatory amino acid transporter; SV2A: synaptic vesicle glycoprotein type 2A; AMPA: 2-amino-3-(3-hydroxy-5-methyl-isoxazol-4-yl) propanoic acid; KA: kainic acid; NMDA: *N*-methyl-D-aspartate. (D) ASDs that act on GABA receptors. Top: GABA receptors (primarily $GABA_A$ or $GABA_B$) are found at the synapses of GABAergic neurons and can be extrasynaptic. Bottom left: Schematic of a GABAergic synapse, with actions of ASDs in red. $GABA_A$ receptors are composed of multiple subunits (α–δ) and mediate phasic (synaptic) inhibition or tonic (non-synaptic) inhibition depending on the presence of the δ-subunit. $GABA_B$ receptors are not shown because antiseizure drugs do not influence them, although agonists facilitate absence seizures (e.g. γ-hydroxybutyrate). Bottom right: $GABA_A$ receptor, with drugs that act on the receptor in red. Barbiturates and benzodiazepines have specific binding sites. GAD: glutamic acid decarboxylase; GABA-T: GABA transaminase; GAT: GABA transporter; SSA: succinic acid semialdehyde. *Source: Adapted from Meldrum and Rogawski MA.* Neurotherapeutics *2007;4:18–61.*[23]

exposure of extrasynaptic receptors (receptors outside the synapse, often adjacent to synaptic receptors) to transmitters. Astrocytes also have endfeet, which are located at the blood–brain barrier (discussed further below). Recent research has shown that astrocytes play a role in synaptic plasticity and behavior, functions previously attributed to neurons. In the type of epilepsy that occurs after brain injury, astrocytes invade the area of injury and transform the local environment, often creating a scar-like region (gliosis) that impedes regrowth in the local area and prevents the entry of ASDs. Thus, oral administration of ASDs may be unsuccessful because drug concentration in the area of gliosis is too low.

The vasculature in the area of gliosis is also a potential problem in drug delivery. In resected tissue from patients with drug-resistant epilepsy, the new blood vessels that grow in the epileptic brain are often contorted or have abnormal characteristics.[41] Related to this point is the critical role of the blood–brain barrier and its changes in epilepsy. During acute seizures, the blood–brain barrier is weakened, and proteins from the periphery can enter the brain. Some of these proteins can exert adverse effects. For example, when serum albumin enters the brain, it can be removed by astrocytes, which decrease K^+ uptake as a result, increasing excitability.[42,43] Normally, transporters such as *p*-glycoprotein and multidrug-resistant proteins are present to prevent the entry of toxic substances into the brain. In epilepsy, these transporters may be upregulated at the blood–brain barrier, leading to drug resistance.[44]

Another important factor in epilepsy that is not regulated directly by neurons is inflammation and the immune response of the brain.[45,46] A series of studies in animal models of epilepsy, later corroborated by tissue specimens from patients with epilepsy, showed that seizures cause a dramatic proinflammatory signal mediated by microglia secreting cytokines such as interleukins. Interleukin-1β has been shown to mediate several effects that are proconvulsant, either indirectly or directly acting on neurons.[46] Antagonism of the interleukin-1β receptor has an anticonvulsant effect and has been suggested as a new target for ASDs. Many types of epilepsy appear to be associated with chronic inflammation, suggesting that anti-inflammatory drugs could be therapeutic. One example is Rasmussen encephalitis, where antibodies against glutamate receptors are present in the brain.

Subcellular metabolism is another component of CNS function that is involved in epilepsy. This has been discussed since the time of Galen, who promoted the idea that decreased food intake would cure patients of epilepsy.[47] It is still unclear what causes alterations in energy metabolism to be anticonvulsant, but several hypotheses have been proposed (see "Ketogenic Diet", below). Other aspects of cellular metabolism may be important in epilepsy, besides those that govern the production of energy substrates. For example, some epilepsy syndromes are caused by single gene defects in enzymes responsible for lysosomal metabolism of macromolecules, leading to their accumulation. In neuronal ceroid lipofucsinoses, autosomal recessive mutations in *CLN* genes cause lipofucsin to accumulate inside cells. Mutation in *CLN3*, which produces the lysosomal enzyme battenin, causes a form of neuronal ceroid lipofucsinosis called Batten disease. One of the progressive myoclonic epilepsies is due to a mutation in *CLN8*. Neimann–Pick disease, type C, where there are mutations in the *NPC1* or *NPC2* gene, is often associated with seizures; *NPC1* and *NPC2* normally contribute to lysosomal transport of cholesterol and other lipids.

MECHANISMS OF EPILEPTOGENESIS AND EPILEPSY

Epileptogenesis

Although research into the basic mechanisms underlying seizures has provided important information, other scientists have focused on the question: How does a normal brain become capable of producing repeated seizures (i.e. epileptogenesis and epilepsy)? People with epilepsy may have a genetic predisposition for epilepsy from birth, and only when there are sufficient additional changes or "second hits" will a seizure occur. A second hit could be a brain injury, or a time of life when changes in the body (hormones, aging) increase the likelihood of a seizure. Indeed, many people are diagnosed with epilepsy late in life, after steroid hormones such as estrogen, progesterone, and testosterone decline. Metabolites of progesterone and testosterone, so-called neurosteroids suggest why: they bind to $GABA_A$ receptors and facilitate the actions of GABA.[48]

Kindling: An Animal Model of Epileptogenesis

In 1967, one of the first methods to study epileptogenesis was discovered by Graham Goddard while attempting to understand the process in the brain that underlies memory.[49] In stimulating the brain repeatedly, intending to simulate what might occur during learning, he found that animals exhibited a gradual increase in response to the stimulus that ultimately was manifested by a convulsive seizure.[49] This process, called "kindling" because of the resemblance to kindling a fire, became a major area of research in epilepsy.

Studies of kindling revealed that it could occur by many mechanisms, for example by stimulating various parts of the brain on a daily basis with an indwelling electrode (electrical kindling), or infusing a chemical into the brain on a regular basis using an implanted cannula (chemical kindling). These studies suggested that repeated excitation in the brain could initiate changes (plasticity) and that neural plasticity was a key element of epileptogenesis.[50,51] Chemicals were identified that were extremely effective in kindling, and areas of the brain were defined where kindling was easier to elicit. Identifying these chemicals and areas was useful because they indicated particular neurochemical vulnerabilities and neuroanatomical sites of vulnerability.

Animal Models of Epilepsy Induced by Excitotoxicity

Kindling was criticized as a model of epilepsy because stimulation was required to elicit a seizure. In epilepsy, it has been argued, seizures are spontaneous; or at least that is true for the vast majority of patients. Initially, investigators had few methods to model epilepsy, except for topical application of convulsants to neocortex, creating a localized focus. A major advance came with the discovery that kainic acid, injected systemically or directly into the brain, could cause a lesion to the brain in laboratory rats that was similar to what is observed in temporal lobe epilepsy. Kainic acid caused severe seizures and neuronal death in areas CA1, CA3, and the dentate gyrus hilus, with sparing of the dentate gyrus granule cell layer. This pattern of neuronal loss was similar to MTS in temporal lobe epilepsy. The animals developed spontaneous recurrent seizures after several weeks, and the seizures persisted for the rest of their lives. This model of temporal lobe epilepsy has provided many opportunities to investigate how animals can become epileptic.[52]

Before discussing the benefits of this animal model and the lessons learned from the experiments, it should be noted that the model is not without criticism. For example, kainic acid induces a period of continuous seizures, called status epilepticus (SE), and areas of the brain that are not always affected in temporal lobe epilepsy are damaged in the animals. In addition, adult rodents are typically used; however, temporal lobe epilepsy is typically considered to be a disease that begins in response to an injury or insult early in life, and SE may not occur. One counter-argument is that the induction of SE is merely a tool to cause a syndrome similar to temporal lobe epilepsy. Although some investigators use anticonvulsants shortly after the onset of SE, to decrease the severity of SE and reduce the extent of brain damage, the debate continues. Notably, Turski and colleagues used pilocarpine, a muscarinic cholinergic agonist, as an alternative to kainic acid.[53] The pilocarpine model produces a slightly different pattern of damage from kainic acid, but epilepsy still emerges, making it a useful alternative. Other investigators have produced additional animal models, using methods that simulate risk factors for temporal lobe epilepsy, such as complex febrile seizures, neonatal hypoxia/ischemia, and traumatic brain injury.[54] Remarkably, it has been difficult to induce epilepsy in aged rodents, although it often develops late in life in humans. Therefore, species differences are an important consideration in epilepsy research using laboratory animals.

Lessons Learned from Status Epilepticus Models

Once it had been established that a pattern of hippocampal neuronal loss similar to patients with temporal lobe epilepsy could be induced in a rat by SE, a valuable animal model was established to understand underlying mechanisms of epileptogenesis and epilepsy. It was assumed that the neuronal loss in the rats caused the epilepsy, since rats that had SE exhibited neuronal damage within days, but spontaneous recurrent seizures (i.e. the epilepsy) did not develop until afterwards. For this reason, many investigators tried to clarify why a particular pattern of neuronal loss occurred. First, the pattern of damage in the rat was better defined, and it was agreed that certain neurons in hippocampus were vulnerable (selective vulnerability) whereas others were relatively resistant. Several hypotheses for neuronal vulnerability and resistance were developed, including the idea that neurons were vulnerable if they had a weak capacity to buffer intracellular calcium, which was later disputed. Certain molecules such as STAT3 appear to be weakly expressed in vulnerable neurons, suggesting other molecular mechanisms. Possible causes of selective vulnerability are still debated.

At the same time, the assumption that neuronal loss caused epilepsy was questioned. Many animal models of temporal lobe epilepsy were developed where neuronal loss was negligible. For example, animal models are now established where young rats are subjected to elevated temperature to simulate febrile seizures, or a period of hypoxia, which are common precipitating factors for temporal lobe epilepsy in children. The extent of neuronal loss is modest, but the animals develop seizures later in life. On the other hand, the seizures in these animals are not associated with robust convulsive movements, and are infrequent compared with those in rats that experience SE in adulthood. When neonatal insults are induced to produce more neuronal injury, a more robust convulsive epilepsy is induced in the long term.

A more detailed examination of the hippocampus after SE has led to some great insights. For example, it was noted that one hallmark of neuronal loss in hippocampus was a pattern of axonal outgrowth of granule cells in the dentate gyrus called mossy fiber sprouting. Mossy fiber sprouting has now been identified both in the SE models and in tissue resected from patients with drug-resistant temporal lobe epilepsy.[55] Anatomical and physiological analyses of the sprouted axons showed that mossy fiber sprouting reflected a novel excitatory circuit where granule cells axons developed synaptic connections with other granule cells (recurrent excitatory circuitry). The idea that epilepsy was a product of increased recurrent excitatory circuitry became popular, because recurrent excitation appears to increase in other brain regions besides the dentate gyrus, and in other animal models of epilepsy such as kindling. Increased recurrent excitation provides an explanation for the emergence of fast ripples in epilepsy (Fig. 17.6). However, other studies suggest that mossy fiber sprouting may not be as important as once thought, or at least it is more complicated that initially suggested.

For example, strategies to reduce mossy fiber sprouting do not necessarily stop epilepsy, although drugs with ideal specificity have not yet been tested. In addition, some investigators have found that mossy fiber sprouting is associated with increased inhibition, not increased excitation, because the abnormal sprouted axons innervate GABAergic neurons. As more work has been done, the complexity of the changes in GABAergic inhibition of the granule cells in the dentate gyrus after SE has increased, with alterations in the presynaptic GABAergic neurons and the postsynaptic GABA receptors.

Some advances in our understanding of epileptogenesis after SE have come with improved technology: video-recordings can be used to monitor rats in their home cage continuously, for weeks or even months. From detailed studies of seizures over time, it has become clear that the brain is not silent between SE and the emergence of epilepsy. Instead, there is a gradual increase in abnormal neuronal activity and it continues to rise even after epilepsy is established. These findings have led to the idea that there is a progressive change in the brain during epileptogenesis that starts with an initial insult and continues potentially for life. The implications are depicted in Fig. 17.8A. At the start of the timeline is a person exposed to one of many potential risk factors for temporal lobe epilepsy, such as a birth injury, an infection, febrile seizures, or perinatal hypoxia. Then, a series of rapid and slower changes occurs, with the initial effects triggering the later alterations. The process may be accelerated or prolonged depending on genetic predisposition, the presence of a second injury, and the environment (e.g. stressful living conditions). Essentially, the plasticity inherent in the brain leads to a process that ultimately lowers seizure threshold sufficiently for a spontaneous seizure to occur. As shown in Fig. 17.8(B1-2), this process may not stop. Instead, any time a seizure occurs, it can initiate another period of changes. One of the underlying reasons is that many genes are regulated by neuronal activity, with an increase in neuronal activity either increasing or decreasing transcription and translation.[56,57]

The specific changes that occur on the fastest and increasingly slower timescales are shown in Fig. 17.8(C1-3). After the initial period of SE, animals

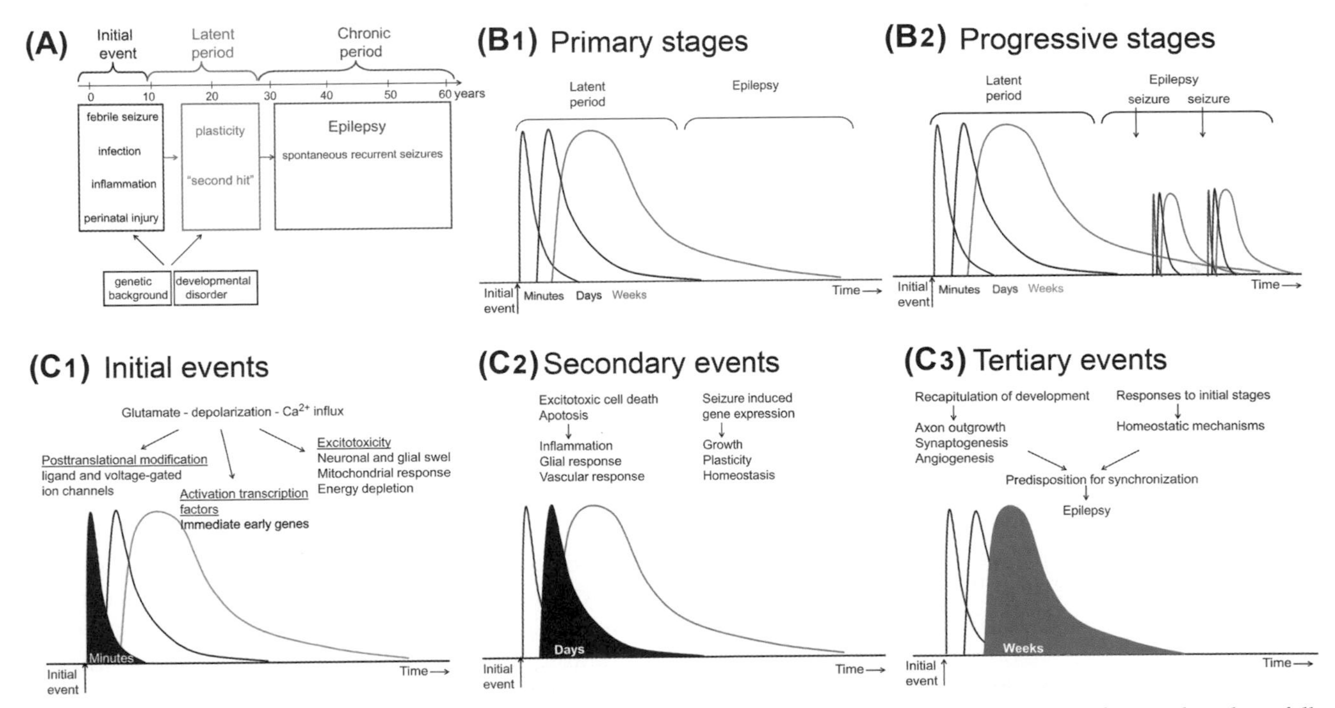

FIGURE 17.8 **Stages of epileptogenesis after an insult or injury.** (A) Schematic of the stages of epileptogenesis that are thought to follow a brain insult or injury early in life, ultimately causing temporal lobe epilepsy. On the *x*-axis is age in years. Below the timeline, from left to right, are three stages: blue, the initial stage, where an precipitating event occurs; green, a latent period when responses of the CNS to an early life injury occur, and may include a second "hit" (second insult/injury, not necessarily the same as the first); and red, a chronic stage where spontaneous recurrent seizures occur (epilepsy). These stages are influenced by genes and development, indicated at the bottom. Genetic predisposition may lead to a more rapid and severe progression; abnormal development (biological or environmental) may also lead to a more rapid onset of epilepsy. (B) 1. The effects of the initial precipitating insult or injury in (A) are shown in more detail. There are rapid responses and slower responses to the initial precipitating insult or injury, with short timescales (minutes, days, weeks) in experimental animals but in humans the stages are more protracted, requiring decades in temporal lobe epilepsy. 2. The chronic period (epilepsy) is shown. Notably, there is potential for chronic seizures to initiate additional changes in the brain other than those that occurred earlier. (C) Specific changes in the brain caused by the initial insult or injury shown in (A) and (B). 1. Rapid changes, occurring within minutes in the rodent brain. 2. Delayed effects in response to the rapid changes, occurring within days in the rodent brain. 3. Effects that occur in the weeks that follow the initial insult, which may be compensatory.

undergo dramatic changes caused by the large rise in intracellular calcium that accompanies high-frequency neuronal activity. Alterations in some cells are excitotoxic and lead to swelling and neuronal death. The surviving neurons exhibit a number of changes in gene expression and are influenced by the parallel changes in the extracellular milieu, astrocytes, and openings in the blood–brain barrier that allow molecules to enter the neuropil from the peripheral circulation. Secondary changes ensue, including the alteration of astrocytes in the area of the injured neurons, becoming reactive glia. The changes in gene expression lead to new proteins being synthesized in surviving neurons and redistribution of proteins along the dendrites and axons. Over the course of days to weeks, new circuits form from existing neurons that grow new collaterals, refine dendrites and dendritic spines, and reorganize circuitry. Then there are responses to these new circuits, in an attempt to re-establish homeostasis. In some individuals who develop epilepsy after a precipitating insult in childhood, there may be a malfunction in these homeostatic mechanisms, which differentiates them from other children who may have a similar injury, but for some reason never develop epilepsy. Even in rodents, where genetic and environmental factors are less variable, there can be variability between animals.

TREATMENT OF EPILEPSY

Antiseizure Drugs

One might think that after a seizure almost anyone would be prescribed a medication to stop it from occurring again. However, many seizures are not accompanied by long-term risk for more seizures and are simply a result of a specific situation, such as hypoglycemia. Therefore, drugs are the first line treatment of epilepsy, but there are established guidelines when to administer ASDs. According to guidelines of the American Academy of Neurology, the decision to treat should be made on the basis of the risk of recurrence, which is established after diagnostics such as neuroimaging. A high risk is usually suggested by a structural abnormality that is consistent, for example, with temporal lobe epilepsy.

ASDs have been in use since Charles Locock discovered potassium bromide in 1857. This chemical acts by simulating chloride, passing through the chloride channel coupled to the $GABA_A$ receptor. Therefore, potassium bromide enhances the effects of GABA at $GABA_A$ receptors. Phenobarbital, introduced in 1912, also acts by facilitating actions of GABA at $GABA_A$ receptors, but by another mechanism. The barbiturates and benzodiazepines bind to particular subunits of the $GABA_A$ receptor, and when bound, GABA has greater effects. Importantly, severe seizures can cause these subunits to change, and the subunit where benzodiazepines bind is replaced by another. Therefore, at the onset of SE, benzodiazepines are often the drug of choice, but have diminishing efficacy with time.[58]

Phenobarbital was the standard for epilepsy care until Tracy Putnam and H. Houston Merritt discovered the beneficial effects of phenytoin in 1939. This was a major advance because phenytoin caused less sedation than phenobarbital. The next decades brought a wave of new medications for epilepsy: the benzodiazepines, other GABAergic drugs such as primidone (1952), drugs that acted primarily on calcium channels (ethosuximide, 1955) or drugs that had multiple effects (carbamazepine, 1963; valproic acid, 1967) and novel effects, such as inhibition of carbonic anhydrase (acetazolamide, 1953). After the advent of the anticonvulsant screening program in the USA in 1971 at the National Institutes of Health (NIH), and growth at academic centers and pharmaceutical companies throughout the world, another generation of ASDs became available: oxcarbazepine (1990), felbamate (1993), gabapentin (1993), lamotrigine (1990), topiramate (1995), tiagabine (1996), levetiracetam (1999), zonisamide (2000), pregabalin (2004), and eslicarbazepine acetate (2009).

At the present time, valproic acid is the most widely prescribed ASD and the first choice for GTCS; carbamazepine is the first line drug for partial seizures, with levetiracetam and lamotrigine also commonly used in partial epilepsy (Table 17.9).[59] Recommendations of the American Academy of Neurology suggest that in newly diagnosed patients, the following ASDs are useful: carbamazepine, phenytoin, valproic acid, phenobarbital, GABApentin, lamotrigine, oxcarbazepine, and topiramate. If the newly diagnosed patient has partial epilepsy or mixed seizures, gabapentin, lamotrigine, oxcarbazepine, and topiramate are recommended. In childhood, lamotrigine is suggested. In cases that are refractory to medication, seven ASDs are suggested: gabapentin, lamotrigine, topiramate, tiagabine, oxcarbazepine, levetiracetam, and zonisamide.

Regrettably, 20–30% of newly diagnosed patients are not seizure free after being treated with ASDs.[59] Statistically, patients who fail to have seizures controlled by the first two ASDs have a high risk of failure with subsequent ASDs, leading some patients to be evaluated for surgical treatment or alternatives to pharmacotherapy. For patients who are pharmacologically refractory, meaning that they do not have seizure control after trying several ASDs, a localized focus with a structural lesion such as MTS indicates that they are candidates for surgical resection of the hippocampus (see "Surgery", below).

TABLE 17.9 Antiseizure Drugs (ASDs)

		Voltage-Dependent Sodium Channels	Voltage-Dependent Calcium Channels	Glutamate Receptors	GABA Receptor	Other Mechanism	Use
ACTH						Suppresses CRH	Infantile spasms
Aromatic allylic alcohols	Stiripentol				X		Sever myoclonic epilepsy in infancy
Barbiturates	Phenobarbital			X	X		Many types
Benzodiazepines	Clobazam				X		Many types
	Diazepam, Lorazepam				X		Status epilepticus
Benzoxazoles	Zonisamide	X	X	?	?	Carbonic anhydrase inhibitor?	Many types
Bromides	Potassium bromide				X		
Carbamates	Felbamate	?	X	NR2B	X		Focal seizures, Lennox–Gastaut
Carboxamides	Carbamazepine	X			X		Many types
Fatty acids	Valproic acid	X	X		X	HDAC inhibitor	Many types
Deoxybarbiturates	Primidone	X			X		Many types
GABA analogues	GABApentin		X		X		Many types
	Pregabalin		X			Inhibits transmitter release	Focal-onset seizures
	Vigabatrin				X		Many types, Lennox–Gastaut
Hydantoins	Phenytoin	X					Many types
Oxazolidinones	Trimethadione		X				Absence seizures
Piperidines	Tiagabine				X		Focal-onset seizures
Succinimides	Ethosuximide		X				Absence seizures
Sulfamate-substituted polysaccharides	Topiramate	X		X	X	Carbonic anhydrase inhibitor	Many types, Lennox–Gastaut
Sulfonamides	Acetazolamide					Carbonic anhydrase inhibitor	Absence and myoclonic seizures
Triazines	Lamotrigine	X	?	X		H current	Many types
New	Retigabine					Opens KCNQ K^+ channels	Focal-onset seizures
	Perampanel			AMPA			Focal-onset seizures
	Ganaxalone				X		Many types

The major classes of ASD are listed with examples, mechanistic information, and clinical use. X: Actions at voltage-dependent sodium channels, calcium channels, ionotropic glutamate receptors, γ-aminobutyric acid-A ($GABA_A$) receptors, or other mechanisms of action; ?: data not clear, because of conflicting reports, results that are not easily interpreted, or generalization from experimental preparations to humans is not clear.
ACTH: adrenocorticotropic hormone; NR2B: NR2B subunit of the *N*-methyl-D-aspartate receptor; AMPA: 2-amino-3-(3-hydroxy-5-methyl-isoxazol-4-yl) propanoic acid; CRH: corticotropin-releasing hormone; HDAC: histone deacetylase.

As shown in Table 17.9, ASDs comprise a chemically diverse group of drugs, but most of them, where the mechanism is known, appear to target four general neurobiological mechanisms: voltage-dependent sodium channels, voltage-dependent calcium channels, glutamatergic transmission, or GABAergic transmission. These four mechanisms are depicted schematically in Fig. 17.7. Phenytoin is the prototypical antagonist of voltage-dependent sodium channels, and ethosuximide is the prototypic antagonist of calcium channels, blocking the T-type calcium channel. Although many drugs affect glutamatergic transmission, few only do so; in contrast, several drugs appear to act only at GABAergic synapses, interfering with GABA synthesis, GABA reuptake, or the $GABA_A$ receptor.

Surgery

For most patients, surgery is recommended only if ASDs fail. However, in many cases, surgery is an excellent option. In patients with MTS and drug-resistant epilepsy, removal of the hippocampus is suggested, with much better outcome if the seizure onset zone is in the area where pathology (i.e. MTS) exists, and secondary sites of pathology are not detected by neuroimaging. In a retrospective review of outcomes, patients with a localized lesion before surgery had the best long-term outcome. However, some reports are less favorable with respect to long-term seizure control. In patients with other types of epilepsy, such as tuberous sclerosis, removal of the tuber is associated with seizure control, and even when there is more than one tuber a multistaged approach has been successful. To standardize the outcomes from surgery, Jerome Engel Jr suggested a classification, now known as the Engel classification. There are four classes: class I is "freedom from disabling seizures", class II is "rare disabling seizures", class III is "worthwhile improvement", and class IV is "no worthwhile improvement".[7]

For specific epilepsy syndromes, such as hemi-megencephaly, chronic encephalitis, congenital hemiplegia, or Sturge–Weber syndrome, other surgical options exist, including hemispherectomy (removal of the majority of one hemisphere). Outcome is excellent, with over 85% of patients improving and 60% becoming seizure free; cognitive function also improves in many patients. Another approach is to sever neuronal connections within a cortical area. This option is attractive when the seizure onset zone is located in an area where preservation of critical functions would be lost if the area was removed. Frank Morrell pioneered the approach in which multiple subpial transections are made to sever connectivity as much as possible while preserving function. Corpus callosotomy is another method that attempts to block the spread of seizures.

Other Therapies

Ketogenic Diet

The ketogenic diet (KD) was first tested in people with epilepsy in 1921 by Russell Wilder,[47] but was used less often once phenobarbital and other ASDs became available. In the 1990s, there was a resurgence of the KD after the son of a well-known movie producer was successfully treated and several programs about the KD were televised. The KD is typically used in children with drug-resistant epilepsy, and is successful in diverse types of epilepsy, such as severe myoclonic epilepsy in infancy, tuberous sclerosis, and infantile spasms.

The diet is a high-fat, low-carbohydrate, and low-protein diet, where the ratio of fat to carbohydrate and protein (in grams) is approximately 4 to 1. The response to the diet can be rapid in some patients and may take several weeks in others. The diet may be continued if seizures persist; in those patients where seizures stop, recurrence of epilepsy occurs in only 20%.

The reason for the efficacy of the KD is not entirely clear. It had been assumed that caloric restriction was an important component, because fasting can decrease seizures in patients with epilepsy[47] and laboratory animals. Therefore, it had been recommended that the KD be administered with caloric restriction to 75% of the recommended dietary allowance (RDA). In addition, caloric restriction increases the production of ketones by the KD. Some studies suggest that caloric restriction is essential to the ability of the KD to stop seizures, but others have shown that it is not essential.

There are many hypotheses for the mechanism of action of the KD. The correlation of ketone levels with seizure control has led many to believe that ketones are directly responsible. In addition, the major ketone bodies, β-hydroxybutyrate, acetyl coenzyme A and acetone, all have actions that reduce seizures in laboratory animals. Another hypothesis is that the KD leads to greater production of mitochondria in neurons, and more energy. One of the reasons for increased mitochondria may be the effect of the KD to shift ATP production from glycolysis in the cytoplasm to mitochondria. Increased mitochondria could support ATPases like the Na^+/K^+ pump and help neurons to repolarize more quickly during seizures. As a consequence, seizures would be shortened.

A shift in cellular ATP production to mitochondria could prevent seizures in another way, based on the idea that ATP produced by glycolysis is near ATP-dependent K^+ channels at the plasma membrane, but mitochondrial ATP production is not. Therefore, if glycolysis were reduced, ATP-mediated inhibition of the K^+ channel would be reduced. The consequence would be K^+ channel opening, causing hyperpolarization of the cell and reducing excitability.

Another explanation for the effects of the KD is based on the increased production of GABA and reduced release of glutamate-caused excess of the ketone body acetyl coenzyme A, which causes a shunt in the Krebs cycle towards production of α-ketoglutarate, which produces glutamate. One would think that an increase in glutamate levels would increase the likelihood of seizures, instead of having a protective effect, but there are two reasons why a reduction in seizures may occur instead: (1) glutamate is the precursor to GABA, so more GABA is produced; and (2) there may be a defect in the transport of glutamate into synaptic vesicles in the presence of the ketone body acetoacetate, because acetoacetate inhibits the vesicular glutamate transporter vGLUT-2. The enhanced concentration of adenosine caused by the KD has also been suggested to mediate the effects of the diet on seizures, because adenosine is an endogenous compound with anticonvulsant actions.

Brain Stimulation

In 1938, Bailey and Bremer found that the electrocardiogram of cats was altered upon stimulation of the vagus nerve.[60] That observation was overlooked for some time because it was assumed that the effect was related to hypotension, which ordinarily follows vagal nerve stimulation (VNS). However, Zanchetti and colleagues showed that the effect was not likely to be related to hypotension.[61] It was subsequently shown that stimulation of the vagus nerve could desynchronize the EEG, and stop seizures in dogs treated with the convulsant strychnine. These findings led to the idea that stimulating the vagus nerve may prevent seizures in patients with epilepsy. In 1997, the US Food and Drug Administration approved an implantable device for focal epilepsy. For practical reasons related to preservation of the battery, and safety reasons, a low frequency of stimulation was chosen. Despite concerns that cardiac and other peripheral side effects of vagal nerve stimulation would develop, the implantable stimulator has not been associated with a high frequency of side effects and has been a successful adjunct to ASD treatment of epilepsy.

The mechanism of action of VNS is still unclear. The vagus nerve is the 10th cranial nerve, releases acetylcholine, and is responsible for involuntary actions associated with the autonomic nervous system. It regulates the heart, gastrointestinal tract, and other organs. The vagus nerve also regulates sensory functions related to the ears and tongue. In the brain, the cell bodies of the vagus nerve are located in the brainstem. The mechanism of VNS stimulation may be related to the fact that seizure control often takes weeks or months, and typically continues to improve over years. The idea that the anticonvulsant effect takes time suggests that reorganization of neural circuits, either in the brainstem or in the forebrain where the brainstem systems project, is responsible for the efficacy of VNS. Other hypotheses include an increase in GABA levels, and decreased inflammation. Other types of stimulation have also been studied in epilepsy. Responsive stimulation is a procedure that attempts to abort a seizure by stimulating the area where the seizure starts, using an implanted electrode, shortly after it begins. In the first published results of patients with drug-resistant focal epilepsy, there was a 38% reduction in seizures in the patients who were stimulated, compared with 17% in the control group.[62]

Deep brain stimulation involves electrodes in structures such as the thalamus. A study in which the anterior nucleus of the thalamus was stimulated showed efficacy for drug-resistant focal seizures with or without secondary generalization. Similarly to VNS, efficacy appeared to increase over time: after 3 months, there was a 40% median reduction in seizures; after 2 years, it was 56%, compared with approximately 15% of controls.

Cooling

Baldwin and colleagues were the first to describe the anticonvulsant effects of cooling.[63] These reports were followed by many studies in diverse preparations showing that hot temperatures exacerbated seizures while cooling decreased them. Devices to cool the brain have repeatedly shown efficacy,[64] but so only one case report, in a patient with tumor-related epilepsy, shows promise.

Stem Cells

The use of transplanted cells to improve brain function has been proposed for many diseases. In epilepsy, the idea that transplantation would be effective was also tested, but met with little success initially. Improved methods to implant GABAergic neurons into neocortical sites have been made possible by selection of neurons from the medial ganglionic eminence, where they normally develop. Initial studies from several laboratories all show promise. Experiments in animals with neocortical transplantation show that the transplanted GABAergic neurons survive, are able to release GABA, and hyperpolarize adjacent neurons. When the hippocampus was targeted, transplantation also gave rise to GABAergic neurons and was able to stop seizures. However, other mechanisms may also be relevant besides those mediated by GABA, such as the ability to restore the expression of glial-derived neurotrophic factor (GDNF) to astrocytes. Shetty and colleagues showed that GDNF is an anticonvulsant and is expressed widely by transplanted cells, but the expression of GABA was relatively weak.[65]

SUMMARY

Epilepsy is one of the most complex neurological disorders. It requires a broad knowledge of neurology and neuroscience, not only to understand the patient but also to advance research so that better diagnostics and therapeutics can be developed. Here, the current conceptions about clinical epilepsy have been summarized, starting with basic terminology, epidemiology, and categorization of the epilepsies. How seizures arise at the level of single neurons in cortical circuits has been summarized with a historical perspective, starting with the first observations of seizures using EEG and subsequent recordings from single neurons in anesthetized preparations or brain slices of experimental animals. From these studies and many others, it is now known that seizures arise by various mechanisms, leading to the diversity of the epilepsies.

With the tools provided by modern neuroscience and molecular biology, many of the initial conceptions about mechanisms have been expanded, or new explanations have been suggested. For example, many of the "idiopathic" epilepsies that were previously considered to have no clear cause are now recognized to have a genetic contribution or arise from mutations in single genes. New animal models have been developed and have changed many of the prevailing views, such as the neurocentric view of seizures and epilepsy. Now it is well accepted that astrocytes, microglia, the vasculature, and the blood–brain barrier play critical roles in seizures and epilepsy.

ASDs are the primary therapeutic approach for epilepsy. Surgery can be an effective treatment, however, and there is increasing interest in alternatives such as the KD and brain stimulation. While this is an impressive armamentarium, much more research is needed because many epilepsy syndromes lack effective treatments or lead to debilitating side effects. In some of the epilepsies, where medication controls seizures in some individuals, there are others whose seizures are not controlled by available ASDs; this is called drug resistance. Addressing these limitations of current clinical treatments is a priority for future research.

Acknowledgments

The author acknowledges the NIH, Alzheimer's Association, and the New York State Office of Mental Health.

References

1. National Institute of Neurological Disorders and Stroke. *Seizures and epilepsy: hope through research;* 2010. Accessed April 1, 2013. www.ninds.nih.gov/disorders/epilepsy/detail_epilepsy.htm.
2. Institute of Medicine. *Epilepsy Across the Spectrum: Promoting Health and Understanding*. Washington, DC: National Academies Press; 2012.
3. Hauser WA. Seizure disorders: the changes with age. *Epilepsia*. 1992;33(suppl 4):S6–S14.
4. Hauser WA, Kurland LT. The epidemiology of epilepsy in Rochester, Minnesota, 1935 through 1967. *Epilepsia*. 1975;16:1–66.
5. Sperling MR, Clancy RR, Ictal EEG. In: *Epilepsy: A Comprehensive Textbook*. Philadelphia, PA: Lippincott-Raven, 1997:849–885.
6. Berg AT, Berkovic SF, Brodie MJ, et al. Revised terminology and concepts for organization of seizures and epilepsies: report of the ILAE commission on classification and terminology, 2005–2009. *Epilepsia*. 2010;51:676–685.
7. Engel J. *Surgical Treatment of the Epilepsies*. New York: Raven Press; 1987.
8. Shorvon SD. The etiologic classification of epilepsy. *Epilepsia*. 2011;52:1052–1057.
9. Nguyen The Tich S, Pereon Y. Letter to the Editor. *Epilepsia*. 1999;40:531–532.
10. Scharfman HE, Pedley TA. Temporal lobe epilepsy. In: Gilman S, ed. *The Neurobiology of Disease*. New York: Elsevier; 2006:349–369.
11. Lowenstein DH. Seizures and epilepsy. In: Fauci AS, Braunwald E, Kasper DL, et al., eds. *Harrison's Principles of Internal Medicine*. 17th ed. New York: McGraw-Hill; 2008:2498–2512.
12. Noebels JL. The biology of epilepsy genes. *Annu Rev Neurosci*. 2003;26:599–625.
13. Kwan P, Schachter SC, Brodie MJ. Drug-resistant epilepsy. *N Engl J Med*. 2011;365:919–926.
14. Ayala GF, Matsumoto H, Gumnit RJ. Excitability changes and inhibitory mechanisms in neocortical neurons during seizures. *J Neurophysiol*. 1970;33:73–85.
15. Ryvlin P. Avoiding falling into the depths of the insular trap. *Epileptic Dis*. 2006;8(Suppl 2):S37–S56.
16. Zhang ZJ, Valiante TA, Carlen PL. Transition to seizure: from "macro"- to "micro"-mysteries. *Epilepsy Res*. 2011;97:290–299.
17. Johnston D, Brown TH. Giant synaptic potential hypothesis for epileptiform activity. *Science*. 1981;211:294–297.
18. Kahle KT, Staley KJ, Nahed BV, et al. Roles of the cation-chloride cotransporters in neurological disease. *Nat Clin Pract Neurol*. 2008;4:490–503.
19. Beck H, Yaari Y. Plasticity of intrinsic neuronal properties in CNS disorders. *Nat Rev Neurosci*. 2008;9:357–369.
20. Traub RD, Wong RK. Cellular mechanism of neuronal synchronization in epilepsy. *Science*. 1982;216:745–747.
21. Trevelyan AJ, Schevon CA. How inhibition influences seizure propagation. *Neuropharmacology*. 2013;69:45–54.
22. Yaari Y, Yue C, Su H. Recruitment of apical dendritic T-type Ca^{2+} channels by backpropagating spikes underlies de novo intrinsic bursting in hippocampal epileptogenesis. *J Physiol*. 2007;580:435–450.
23. Meldrum BS, Rogawski MA. Molecular targets for antiepileptic drug development. *Neurotherapeutics*. 2007;4:18–61.
24. Lado FA, Moshe SL. How do seizures stop? *Epilepsia*. 2008;49:1651–1664.
25. Hsu D. The dentate gyrus as a filter or gate: a look back and a look ahead. *Prog Brain Res*. 2007;163:601–613.
26. Noe F, Nissinen J, Pitkanen A, et al. Gene therapy in epilepsy: the focus on NPY. *Peptides*. 2007;28:377–383.
27. McNamara JO, Scharfman HE. Temporal lobe epilepsy and the BDNF receptor, trkb. In: Noebels JL, Avoli M, Rogawski MA, Olsen RW, Delgado-Escueta AV, eds. *Jasper's Basic Mechanisms of the Epilepsies [internet]*. 4th ed. Bethesda, MD: National Center for Biotechnology Information; 2012.
28. Destexhe A, Contreras D, Steriade M. Mechanisms underlying the synchronizing action of corticothalamic feedback through inhibition of thalamic relay cells. *J Neurophysiol*. 1998;79:999–1016.
29. Gastaut H. *Epilepsy – The Electroclinical Correlates*. Oxford: Blackwell; 1954.
30. McCormick DA, Pape HC. Properties of a hyperpolarization-activated cation current and its role in rhythmic oscillation in thalamic relay neurones. *J Physiol*. 1990;431:291–318.

31. Bragin A, Mody I, Wilson CL, Engel Jr J. Local generation of fast ripples in epileptic brain. *J Neurosci*. 2002;22:2012–2021.
32. Traub RD, Draguhn A, Whittington MA, et al. Axonal gap junctions between principal neurons: a novel source of network oscillations, and perhaps epileptogenesis. *Rev Neurosci*. 2002;13:1–30.
33. Bragin A, Engel Jr J, Wilson CL, Fried I, Mathern GW. Hippocampal and entorhinal cortex high-frequency oscillations (100–500 Hz) in human epileptic brain and in kainic acid-treated rats with chronic seizures. *Epilepsia*. 1999;40:127–137.
34. Fisher RS, Webber WR, Lesser RP, Arroyo S, Uematsu S. High-frequency EEG activity at the start of seizures. *J Clin Neurophysiol*. 1992;9:441–448.
35. Traub RD, Whittington MA, Buhl EH, et al. A possible role for gap junctions in generation of very fast EEG oscillations preceding the onset of, and perhaps initiating, seizures. *Epilepsia*. 2001;42:153–170.
36. Kohling R, Staley K. Network mechanisms for fast ripple activity in epileptic tissue. *Epilepsy Res*. 2011;97:318–323.
37 Burgess DL. Neonatal epilepsy syndromes and GEFS+: mechanistic considerations. *Epilepsia*. 2005;46(suppl 10):51–58.
38. Lerche H, Shah M, Beck H, Noebels JL, Johnston D, Vincent A. Ion channels in genetic and acquired forms of epilepsy. *J Physiol*. 2012;591:753–764.
39. Schwartzkroin PA, Walsh CA. Cortical malformations and epilepsy. *Ment Retard Dev Disabil Res Rev*. 2000;6:268–280.
40. Wong M. A critical review of mTOR inhibitors and epilepsy: from basic science to clinical trials. *Expert Rev Neurother*. 2013;13:657–669.
41. Marchi N, Lerner-Natoli M. Cerebrovascular remodeling and epilepsy. *Neuroscientist*. 2012.
42. van Vliet EA, da Costa Araujo S, Redeker S, van Schaik R, Aronica E, Gorter JA. Blood–brain barrier leakage may lead to progression of temporal lobe epilepsy. *Brain*. 2007;130:521–534.
43. Ivens S, Kaufer D, Flores LP, et al. TGF-beta receptor-mediated albumin uptake into astrocytes is involved in neocortical epileptogenesis. *Brain*. 2007;130:535–547.
44. Sills GJ. The multidrug transporter hypothesis of refractory epilepsy: corroboration and contradiction in equal measure. *Epilepsy Curr*. 2006;6:51–54.
45. Devinsky O, Vezzani A, Najjar S, De Lanerolle NC, Rogawski MA. Glia and epilepsy: excitability and inflammation. *Trends Neurosci*. 2013;36:174–184.
46. Vezzani A, French J, Bartfai T, Baram TZ. The role of inflammation in epilepsy. *Nat Rev Neurol*. 2010;7:31–40.
47. Wheless JW. History and origins of the ketogenic diet. In: Stafstrom CE, Rho JM, eds. *Epilepsy and the Ketogenic Diet*. New York: Humana; 2004:31–50.
48. Reddy DS. Role of anticonvulsant and antiepileptogenic neurosteroids in the pathophysiology and treatment of epilepsy. *Front Endocrinol (Lausanne)*. 2011;2:38.
49. Goddard GV. Development of epileptic seizures through brain stimulation at low intensity. *Nature*. 1967;214:1020–1021.
50. Sutula TP. Mechanisms of epilepsy progression: current theories and perspectives from neuroplasticity in adulthood and development. *Epilepsy Res*. 2004;60:161–171.
51. Scharfman HE. Epilepsy as an example of neural plasticity. *Neuroscientist*. 2002;8:154–173.
52. Bertram E. The relevance of kindling for human epilepsy. *Epilepsia*. 2007;48(suppl 2):65–74.
53. Turski WA, Cavalheiro EA, Schwarz M, Czuczwar SJ, Kleinrok Z, Turski L. Limbic seizures produced by pilocarpine in rats: behavioural, electroencephalographic and neuropathological study. *Behav Brain Res*. 1983;9:315–335.
54. Pitkanen A, Moshe SL, Schwartzkroin PA. *Models of seizures and epilepsy*. New York: Elsevier; 2006.
55. Buckmaster PS. Mossy fiber sprouting in the dentate gyrus. In: Noebels JL Avoli M, Rogawski MA, Olsen RW, Delgado-Escueta AV, eds. *Jasper's Basic Mechanisms of the Epilepsies [internet]*. 4th ed. Bethesda, MD: National Center for Biotechnology Information; 2012.
56. Scharfman HE. Seizure-induced neurogenesis in the dentate gyrus and its dependence on growth factors and cytokines. In: Binder DK, Scharfman HE, eds. *Growth Factors and Epilepsy*. Hauppauge: Novasciences; 2006:1–40.
57. Lukasiuk K, Dabrowski M, Adach A, Pitkanen A. Epileptogenesis-related genes revisited. *Prog Brain Res*. 2006;158:223–241.
58. Goodkin HR, Joshi S, Kozhemyakin M, Kapur J. Impact of receptor changes on treatment of status epilepticus. *Epilepsia*. 2007;48 (suppl 8):14–15.
59. Goldenberg M. Overview of drugs used for epilepsy and seizures. *Pharmacol Ther*. 2010;35:393–415.
60. Bailey P, Bremer F. A sensory cortical representation of the vagus nerve (with a note on the effects of low blood pressure on the cortical electrograms). *J Neurophysiol*. 1938;1:405–412.
61. Zanchetti A, Wang SC, Moruzzi G. The effect of vagal afferent stimulation on the EEG pattern of the cat. *Electroencephalogr Clin Neurophysiol*. 1952;4:357–361.
62. Morrell MJ. Responsive cortical stimulation for the treatment of medically intractable partial epilepsy. *Neurology*. 2011;77: 1295–1304.
63. Baldwin M, Frost LL. Effect of hypothermia on epileptiform activity in the primate temporal lobe. *Science*. 1956;124(3228): 931–932.
64. Smyth MD, Rothman SM. Focal cooling devices for the surgical treatment of epilepsy. *Neurosurg Clin N Am*. 2011;22:533–546.
65. Waldau B, Hattiangady B, Kuruba R, Shetty AK. Medial ganglionic eminence-derived neural stem cell grafts ease spontaneous seizures and restore GDNF expression in a rat model of chronic temporal lobe epilepsy. *Stem Cells*. 2010;28:1153–1164.

CHAPTER

18

Amyotrophic Lateral Sclerosis

O.M. Peters, R.H. Brown Jr

Department of Neurology, University of Massachusetts Medical School, Worcester, Massachusetts, USA

OUTLINE

INTRODUCTION

Amyotrophic lateral sclerosis (ALS) is a neurodegenerative disease of the motor system characterized by focal and then generalized weakness leading to paralysis and death from respiratory failure. Symptoms arise from the loss of corticospinal, brainstem, and spinal motor neurons. ALS may be sporadic, with an unknown cause, or familial, in which genetic mutations predispose to ALS.

DIAGNOSIS OF AMYOTROPHIC LATERAL SCLEROSIS

The cornerstone for the diagnosis of ALS is evidence of progressive deterioration of function of motor neurons in the brain, brainstem, and spinal cord. The diagnosis of ALS requires dysfunction of two populations of motor neurons: those that innervate skeletal muscle (lower motor neurons, which in the spinal cord designates

Neurobiology of Brain Disorders
http://dx.doi.org/10.1016/B978-0-12-398270-4.00018-5

anterior horn cells), and those that reside in the motor cortex of the brain (corticospinal motor neurons) and send axons into the spinal cord to innervate lower motor neurons either directly or indirectly via spinal interneurons. Rarely one encounters slowly progressive weakness arising primarily from selective pathology in the upper motor neurons (designated primary lateral sclerosis) or the lower motor neurons (progressive muscular atrophy); in some instances, these progress to more typical ALS with involvement of both populations of motor neurons.

Loss of the corticospinal neurons causes weakness with stiffness, ascribed to heightened, reflex-stimulated activity of the lower motor neurons. Degeneration of cortical motor neurons that innervate brainstem motor nuclei (corticopontine and corticomesencephalic motor neurons) is correlated with a distinctive clinical state in which emotional reflexes are exaggerated. Patients with this feature, designated pseudobulbar affect, are unable to avoid an excessive expression of emotion (laughing or crying) in response to emotional stimuli that are only mild (e.g. minimally amusing or upsetting).

Loss of motor neurons causes weakness with diminished, slack motor tone. Conclusive diagnosis of ALS depends on both the presence of these features and the absence of other neurological findings (sensory loss, impaired control of bowel and bladder). Establishing the diagnosis typically also entails excluding a number of defined causes of neuromuscular weakness such as Lyme disease, diabetes, vitamin B_{12} deficiency, or antibody-mediated motor neuron dysfunction (e.g. motor neuropathy with conduction block). Electrophysiological studies are usually central to the diagnostic process, because they may detect evidence of denervation in limbs even before weakness is evident; moreover, they can distinguish between recently denervated and chronically reinnervated muscles. Such studies are also important in excluding alternative diagnoses such as generalized peripheral neuropathies or motor neuropathies arising from focal block of motor nerve conduction.

CLINICAL CHARACTERISTICS OF AMYOTROPHIC LATERAL SCLEROSIS

Several cardinal features are core clinical characteristics of ALS. It typically begins in midlife, with a mean onset of about 55 years (although there are many exceptions). It almost always begins focally and spreads, in a pattern that suggests progressive involvement of contiguous sets of motor neurons within the spinal cord. In the majority of cases, the onset is in a distal limb (Fig. 18.1). Less frequently, there may be initial involvement of bulbar motor neurons with resulting difficulty with speech, chewing, and swallowing. Very infrequently, the earliest manifestation is respiratory failure; without limb or bulbar involvement, the diagnosis may be elusive. As noted, ALS spares most non-motor functions. Moreover, some motor neurons, including those that innervate the oculomotor muscles and the urinary and bowel sphincters, are not affected until very late into the course of the illness.

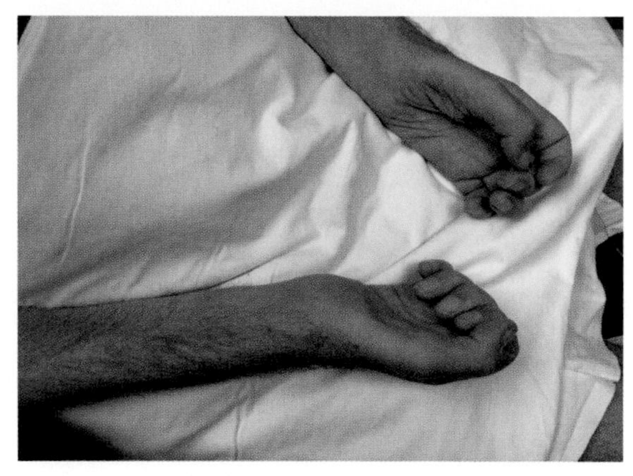

FIGURE 18.1 **Clinical signs of amyotrophic lateral sclerosis (ALS).** In late-stage ALS, denervation of the hands and forearms leads to extensive muscle wasting. Because there is preferential denervation of extensor more than flexor muscles, the atrophied hands show finger flexion at rest. There is profound wasting in the entire hand musculature, which is particularly apparent in the base of the thumb (thenar eminence). By the time the denervation is this pronounced in the upper extremities, there usually is extensive involvement of all other limb muscles, often associated with bulbar weakness.

In a subset of cases, the weakness evolves concurrently with (or is even preceded by) a distinctive type of dementia, variously designated frontotemporal dementia (FTD) or frontotemporal lobe dementia. FTD is characterized by frontal lobe behavioral features, including loss of executive function, impulsivity, disinhibition, and apathy, with relative preservation of memory. FTD may entail language disturbances such as progressive loss of the ability to comprehend words and speak (primary progressive aphasia) or loss of meaningful speech content (semantic aphasia). Rarely, FTD may be associated with some types of movement disorders. FTD is distinguished from Alzheimer disease by several clinical features, including relative preservation of memory, somewhat earlier onset (e.g. around the same age as the onset of ALS), frontal behavioral changes and, in some cases, the aphasia syndromes, which are unusual in Alzheimer disease.

Most cases of ALS are sporadic (sALS). However, about 10–15% are transmitted through families as Mendelian dominant traits (fALS). This has led to an intensive effort to identify ALS genes. Since the first ALS gene (cytosolic Cu/Zn superoxide dismutase or *SOD1*) was identified in 1993,[1] more than 30 ALS genes have been reported. Indeed, most of the current hypotheses about the pathogenesis of ALS, and most of the strategies for developing therapies, have been suggested through investigations of ALS genetics. With the advent of remarkable new technologies for the analysis of exomes and genomes, it is now possible to test the hypothesis that apparently sALS is caused by combinations of genetic variants. It is conceivable that polygenic factors may be critical in more than the 10% of cases that are Mendelian.

NATURAL HISTORY OF AMYOTROPHIC LATERAL SCLEROSIS

Survival in ALS dictated ultimately by respiratory and, to an important degree, nutritional status. Bulbar-onset cases often progress to respiratory paralysis earlier than in limb-onset cases. Without ventilator support, survival in bulbar disease is 3–4 years, in contrast with limb-onset cases whose survival to ventilation may be 5 or more years. With ventilation, survival may be many years or indeed even decades. Individuals with bulbar onset and early impairment of swallowing may experience substantial weight loss. This correlates with shortened survival, a feature reversed by early placement of feeding tubes and maintenance of appropriate nutrition.

AVAILABLE TREATMENTS FOR AMYOTROPHIC LATERAL SCLEROSIS

Like almost all neurodegenerative diseases, ALS has been refractory to virtually all trials of therapy. A single compound, riluzole, was approved for ALS by the US Food and Drug Administration for ALS in 1995. Riluzole is thought to block calcium-mediated activation of postsynaptic glutamate receptors by blocking sodium channels. Riluzole may also block postsynaptic *N*-methyl-D-aspartate (NMDA) receptors. This drug extends survival in ALS by about 10–15%, a subtle effect that has been evident in both prospective and retrospective studies.

Interventions in ALS other than riluzole, whose benefit is modest at best, are indirect and symptomatic. Maintenance of nutritional status is critical; as bulbar symptoms worsen, a gastric feeding tube is often necessary to maintain caloric intake, and to avoid the risk of aspiration pneumonia that accompanies ingestion of food when the laryngeal musculature cannot adequately protect the airway. As diaphragmatic force weakens, it is customary to augment the volume of inspired air using positive airway pressure, typically applied with a nose or face mask. Over time, as respiratory failure advances further, the positive airway pressure usually does not suffice for adequate ventilation. The next level of support, which many people with ALS elect not to pursue, is to place a small hole in the trachea (tracheostomy) through which full ventilation can be achieved using a mechanical positive pressure respirator. Other measures such as oropharyngeal suctioning are important to avoid aspiration.

NEUROBIOLOGICAL BASIS OF AMYOTROPHIC LATERAL SCLEROSIS

Advances in high-throughput genetic screening have led to the discovery of ALS-associated mutations in more than 30 genes (Table 18.1). The identities of these genes and the RNA and proteins they produce have defined several often overlapping biological processes that underlie the pathogenesis of ALS (Figs 18.2 and 18.3).

Generation of Reactive Oxygen Species

Copper/zinc superoxide dismutase-1 (*SOD1*) was the first gene associated with hereditary ALS (Fig. 18.4). More than 160 mutations have now been identified in the *SOD1* gene, accounting for approximately 20% of fALS cases and a further 1% of sALS. The SOD1 protein normally functions in detoxifying (dismutating) the superoxide radicals (O_2^-) that are a byproduct of normal metabolic processes. One hypothesis for the role of SOD1 in motor neuron degeneration postulated that ALS-related mutations disrupt its normal function and impede its ability to detoxify reactive oxygen species, for example by altering the confirmation of its copper/zinc binding domains. According to this view, subsequent oxidative damage might then ensue, leading to motor neuron cytotoxicity. Several factors weigh against this loss-of-function model. That mutant SOD1-related ALS migrates heterozygously in pedigrees as a dominant trait argues strongly against simple loss of function; unless there is haploinsufficiency, the presence of one normal allele and the corresponding normal protein should suffice to prevent oxidative cytotoxicity. Equally compelling was the observation in 1994 that mice bearing transgenes that express high levels of mutant SOD1 protein develop an ALS-like phenotype. If anything, those initial lines of transgenic ALS mice had supranormal levels of dismutation activity. Moreover, it was reported in 1996 that SOD1 null mice were viable and did not exhibit motor neuron loss, although they did reveal modest electrophysiological evidence of a low-grade, subclinical axonopathy.[2] Taken together, these points argued convincingly that the fundamental molecular defect in SOD1-mediated ALS was one or more adverse properties conferred by the mutant SOD1 protein.

These early experiments highlighted the importance of assessing the biological significance of pathological genes and proteins rather than *in vitro*. Transgenic expression of mutant SOD1 protein in mice produced an ALS phenotype; at least in initial models, the mutant protein

TABLE 18.1 Functional Characterization of Selected Amyotrophic Lateral Sclerosis (ALS) Genes

Gene	Phenotype	Protein Stability/Metabolism	Prion Domains	DNA/RNA Biology	Cytoskeletal Function	Environmental Toxin Exposure	Other
FALS GENES THAT PREDOMINANTLY INVOLVE PROTEIN PATHOLOGY AND TRANSPORT FUNCTIONS							
SOD1	fALS	Inclusions			Impairs axonal transport		Metabolizes superoxide anion
VAPB	fALS	Inclusions			Vesicle function/axonal transport		
Dynactin	fALS				Retrograde transport		
OPTN	fALS	Colocalizes with intrusions					
VCP	ALS/FTD/IBM/Paget disease	Inclusions/altered protein trafficking					Multisystem pathology
SQSTM1/p62	fALS	Colocalizes with intrusions					
Ubiquitin 1	fALS	Protein metabolism					
Ubiquitin 2	fALS	Protein metabolism					
Alsin	Childhood ALS, HSP				Endosome trafficking		
FALS GENES THAT PREDOMINANTLY INVOLVE RNA BIOLOGY							
TDP43	Adult and childhood ALS	Most frequent component of inclusions in fALS/sALS	Prion domain	Contains DNA/RNA binding domains			Participates in multiple aspects of RNA biology
FUS	Adult and childhood ALS	Inclusions	Prion domain	Contains DNA/RNA binding domains			Participates in multiple aspects of RNA biology
TAF15	fALS		Prion domain	DNA/RNA binding			
C9orf72	fALS/sALS/FTD/PD	Inclusions		RNA foci/toxicity			
Senataxin	fALS			DNA/RNA helicase			
ANG	fALS/sALS			RNase			
HNRNPA1	fALS	Inclusions	Prion domain				
HNRNPA2B1	fALS	Inclusions	Prion domain				
FALS GENES IMPLICATED IN OTHER PATHOLOGICAL PATHWAYS							
PFN1	fALS				Actin polymerization		
PON1–3	ALS					Organophosphate exposure	
NTE	Childhood HSP					Organophosphate exposure	

Continued

TABLE 18.1 Functional Characterization of Selected Amyotrophic Lateral Sclerosis (ALS) Genes—cont'd

Gene	Phenotype	Protein Stability/Metabolism	Prion Domains	DNA/RNA Biology	Cytoskeletal Function	Environmental Toxin Exposure	Other
Sigma-R1	ALS						Calcium channel regulation
DAO	fALS						D-amino acid production
SELECTED sALS GENES							
Ataxin-2	sALS						CAG expansions increase risk
KIFAP3	sALS						Influences duration
CHGB	sALS				Kinesin-associated protein		Influences onset/interacts with SOD1
UNC13A	sALS						Regulates transmitter release/influences risk
EphA4	sALS				Axon outgrowth		Influences duration
CREST	sALS		Prion domain				Influences risk

Although the pathogenic roles for the majority of known genes associated with fALS are not well understood, it is useful to group these into two broad functional categories that entail perturbations in: (1) protein stability and quality control; and (2) functional properties of RNA/DNA binding proteins.
fALS: familial amyotrophic lateral sclerosis; sALS: sporadic amyotrophic lateral sclerosis; FTD: frontotemporal dementia; IBM: inclusion body myopathy; HSP: hereditary spastic paraparesis; PD: Parkinson disease.

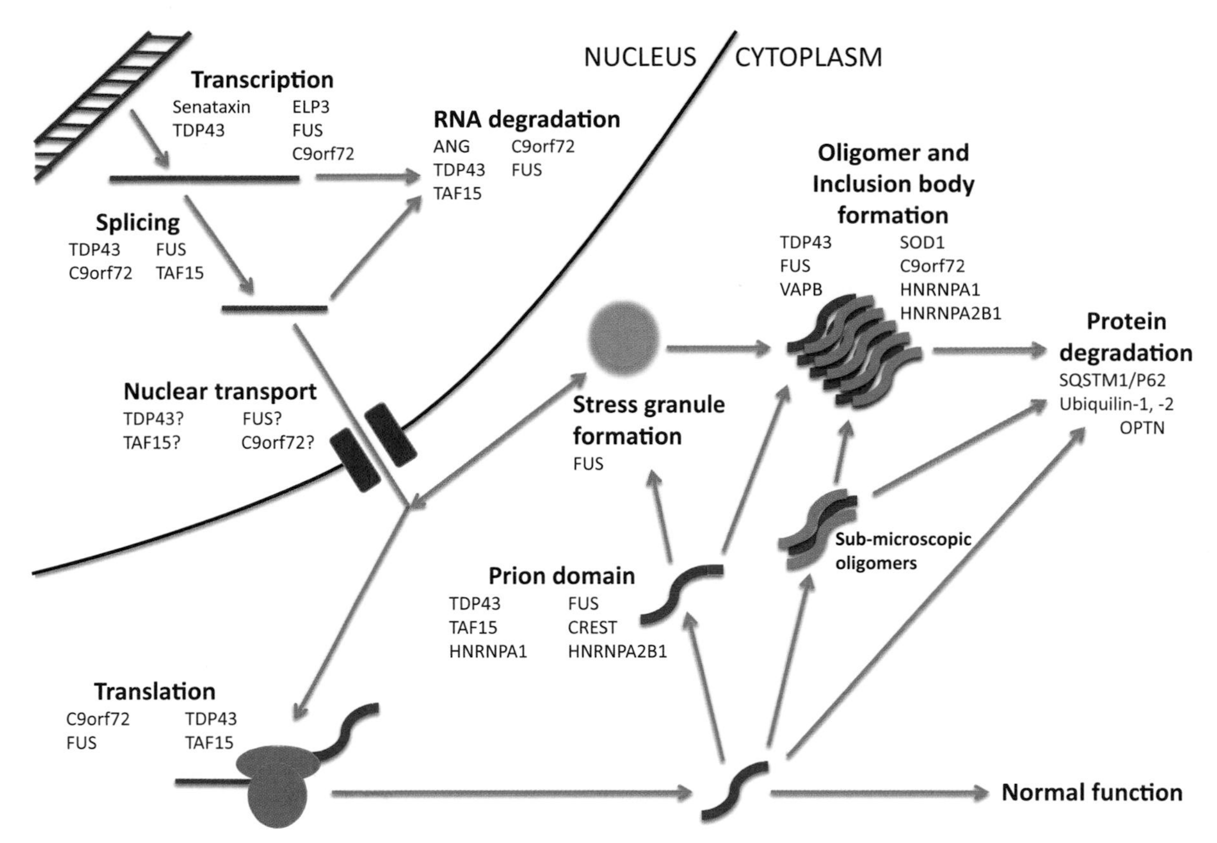

FIGURE 18.2 **Overview of pathogenic molecules and pathways implicated by amyotrophic lateral sclerosis (ALS) genes.** One may envision two broad functional categories of ALS genes: those that alter protein stability, quality control, and function, often entailing protein misfolding and aggregation; and those that impinge on the processing and function of RNA, affecting processes such as transcription, splicing and translation. The presence of prion domains in several of the implicated ALS proteins suggests that they may propagate misfolding and aggregation in adjacent cells, causing spread of disease throughout the CNS.

was surprisingly well tolerated in motor neurons in culture. Reciprocally, although the SOD1-null mice did not develop overt motor neuron disease, the SOD1-null neurons were difficult to culture *in vitro*, suggesting that the culture milieu entailed oxidative stressors (e.g. supraphysiological ambient oxygen pressures) that challenged the viability of neurons in the highly artificial *in vitro* setting. Furthermore, embryonic fibroblasts derived from SOD1-null mice were substantially less viable and more sensitive to oxidative stress than control cells.

Protein Toxicity

Postmortem analysis of CNS tissue from ALS patients has robustly demonstrated that motor neurons loss is accompanied by the presence, in both neurons and non-neuronal cells, of inclusion bodies composed of densely packed protein. Whether these proteinacious inclusions cause neurodegeneration or are simply markers of the underlying death process remains unclear. During the early to mid-1990s it was observed that analogous inclusions were present in many other neurodegenerative disorders (e.g. Alzheimer disease, Parkinson disease, dementia with Lewy bodies, Huntington disease, spinocerebellar ataxia). As in ALS, in these other disorders there is debate about the primacy of the various aggregates: Are they directly toxic or are they an indirect manifestation of some other pathway to cell death? One view is that the critical mutant proteins produce submicroscopic oligomers that mediate the primary pathology.

The following types of protein accumulation are prominent in fALS, although others have been detected.

SOD1

Immunohistochemical staining of postmortem tissues from fALS patients demonstrated insoluble SOD1 in spheroidal, ubiquitinated, inclusion bodies in the cytoplasm of spinal motor neurons.[3] Possible toxicity of these inclusions has been modeled both *in vitro* and in animal model systems, demonstrating that mutant forms of SOD1 peptide produce misfolding tertiary structures with an increased propensity to aggregate spontaneously. *In vitro* studies have also demonstrated that, when post-translationally oxidized, wild-type SOD1 is also able to misfold into aggregate-prone conformations comparable to those formed by mutant SOD1. This species of oxidized wild-type SOD1 is detected in the spinal cord of sALS cases.[4]

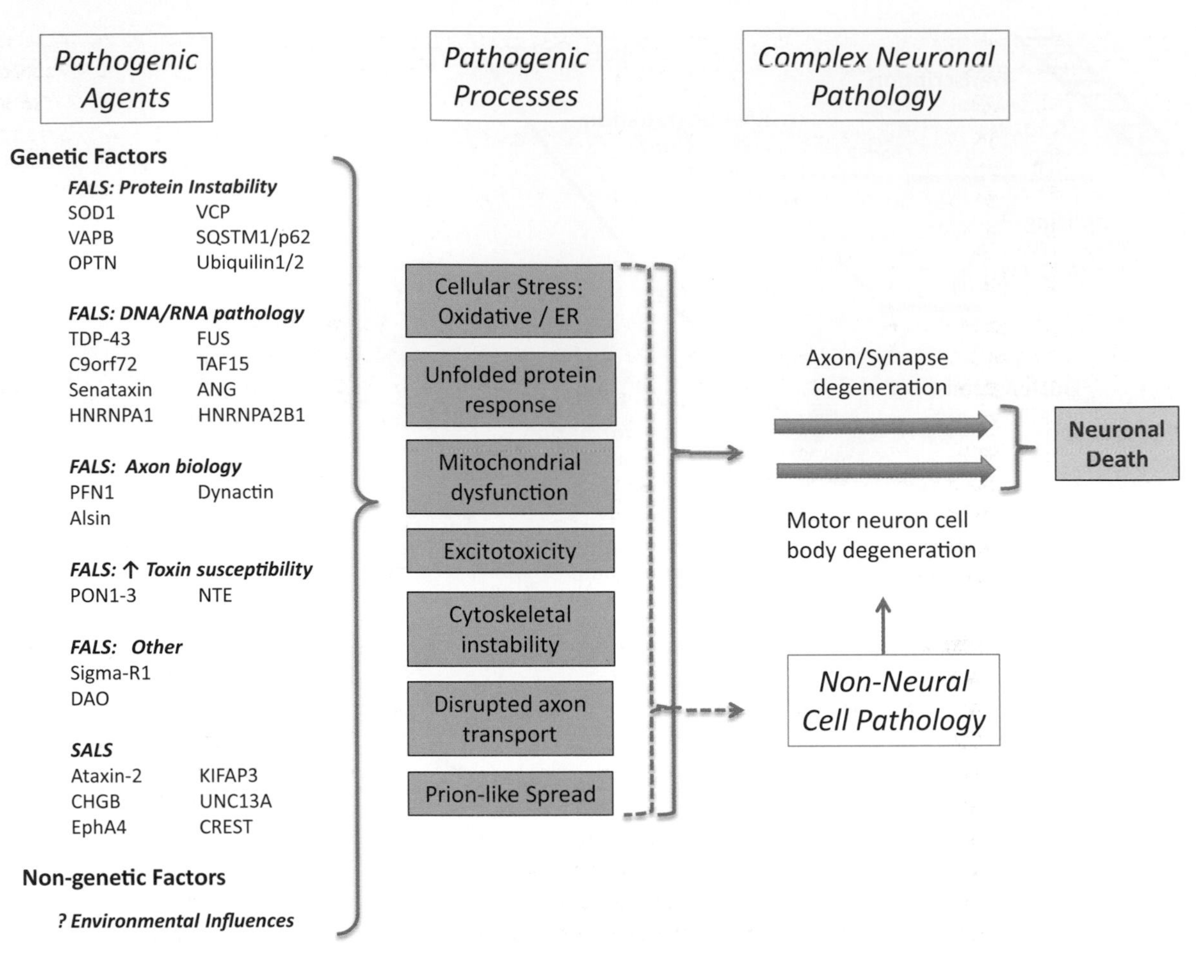

FIGURE 18.3 **Schematic diagram of parallel cell death pathways in amyotrophic lateral sclerosis (ALS).** In this scheme, genetic and non-genetic factors initiate pathogenic processes that impair diverse functions in neurons which, acting in parallel, culminate in neuronal death. Non-neural cells modify these processes and thereby affect clinical features such as disease duration. One implication of this framework is that successful treatment of this disease will probably require interventions in multiple processes affecting diverse neuronal compartments (e.g. dendrites, soma, axon, synapse).

TDP-43

The most commonly detected accumulating protein in ALS is the 43 kilodalton ribonucleoprotein transactive response DNA binding protein 43 (TDP-43) (Fig. 18.5). TDP-43 was first implicated in sALS and FTD; subsequently, multiple reports established that germline mutations in the *TDP-43* gene cause about 5% of dominantly transmitted fALS, 1% of sALS, and rare instances of hereditary frontotemporal lobar degeneration (FTLD).[5] TDP-43 contains two RNA recognition motifs, predicted nuclear localization and export sequences, and a glycine rich C-terminus to which the majority of known mutations localize.[6] Hyperphosphorylated TDP-43 is a component of ubiquitin and p62 immunoreactive inclusion bodies in sALS and fALS.[7] Another finding in ALS is the presence of 25 kDa C-terminal fragments of TDP-43, indicating that this protein undergoes post-translational modification in ALS. When misfolded, TDP-43 forms intranuclear and cytoplasmic inclusion bodies in neurons and glia, usually appearing as dense, ubiquitinated spheroidal or skein-like structures. A staging system has been composed to describe the time-course of TDP-43 inclusion burden throughout the CNS in ALS patients; this highlights a pattern of progressively spreading misfolded TDP-43 throughout the brain, beginning in discrete regions of the motor cortex and brainstem motor nuclei.[8] The pathobiology of TDP-43 is not confined to ALS; TDP-43 inclusions have been detected in several other neurodegenerative diseases including FTLD, Parkinson disease, Alzheimer disease, Huntington disease, and spinocerebellar ataxia. By contrast, they are notably absent in mutant SOD1- and FUS-associated fALS.

FUS

Fused in sarcoma/translocated in liposarcoma (FUS/TLS or FUS) is an RNA/DNA binding ribonucleoprotein with a structure that is generally homologous to that of TDP-43, encompassing an RNA recognition motif, predicted nuclear import and export sequences, various glycine-rich regions, and a zinc finger domain[6] (Fig. 18.5).

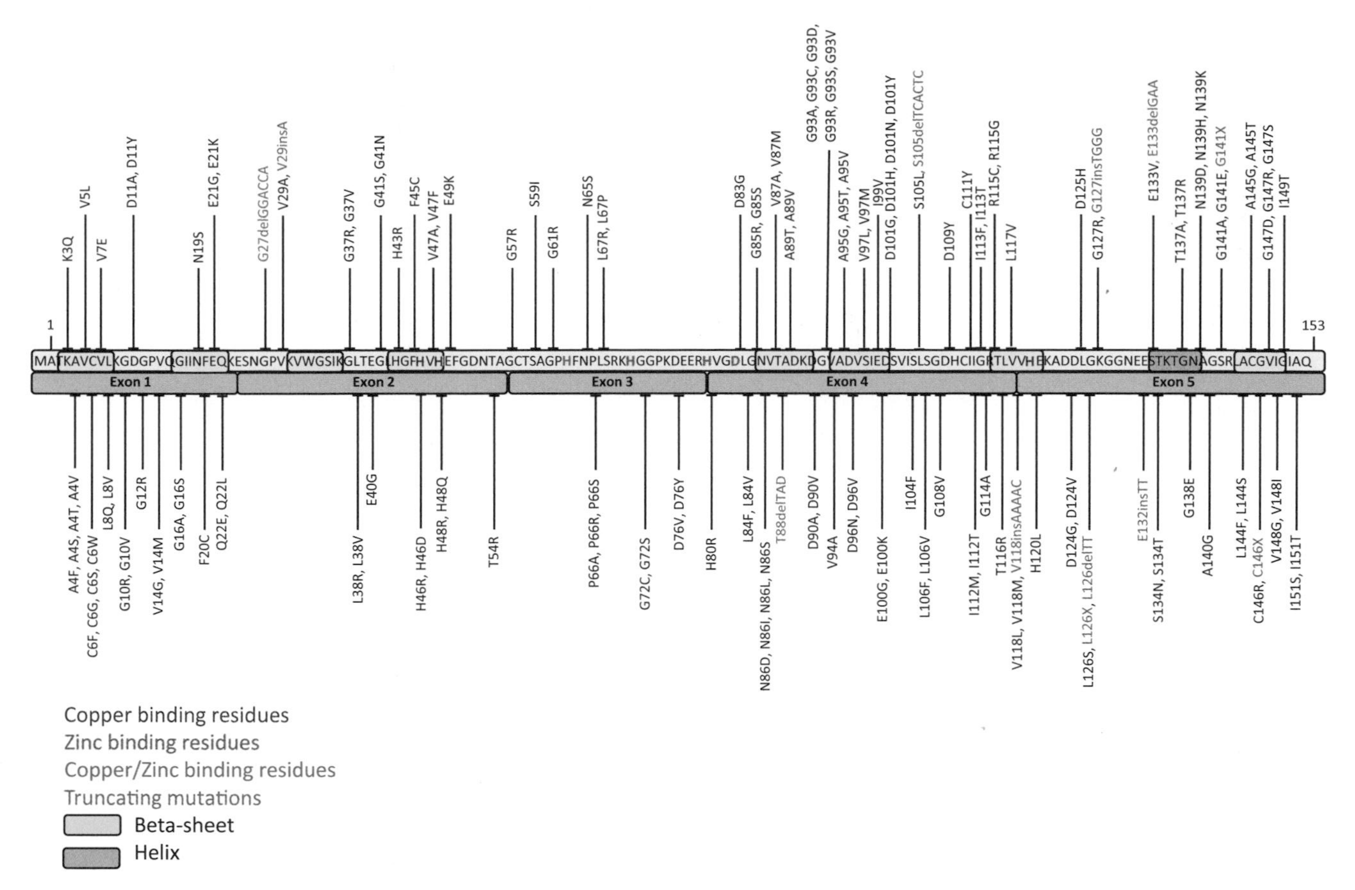

FIGURE 18.4 **Structure of superoxide dismutase-1 (SOD1) and the position of familial amyotrophic lateral sclerosis-associated SOD1 mutations.** More than 160 mutations have been identified across the peptide sequence of SOD1. These are overwhelming missense mutations; there is a small set of truncation mutations (highlighted above) that eliminate a few amino acids at the carboxy-terminus. No mutations predict absence of the SOD1 protein. To varying degrees, these mutations are thought to alter tertiary structure and conformational stability, sometimes imparting a propensity to aggregate. *Source: Mutations mapped using the Amyotrophic Lateral Sclerosis Online Genetics Database,* http://alsod.iop.kcl.ac.uk.

Mutations in the *FUS* gene account for approximately 5% of fALS and 1% of sALS.[9,10] FUS-positive cytoplasmic inclusion bodies have been reported within neurons and glia in the brain and spinal cord of fALS and hereditary FTD patients harboring *FUS* mutations, although they are extremely rare in sALS. FUS-positive inclusion bodies are generally ubiquitin and p62 positive and not immunoreactive for TDP-43.

C9orf72

The most recently identified ALS gene, *C9orf72*, encodes an open reading frame on chromosome 9. In up to approximately 30% of fALS, 8–10% of sALS, and 0.5% of controls, the C9orf72 gene harbors an expansion of a GGGGCC intronic repeat sequence.[11,12] Because this hexanucleotide expansion occurs in a non-coding region, it is predicted not to encode a protein but to generate a lengthy RNA transcript from the expanded DNA. fALS patients with a C9orf72 mutation generally develop typical TDP-43 aggregate pathology; in addition, a subset of neurons within the hippocampus and cerebellum contains a unique form of star-shaped or small dot-like inclusion bodies that are TDP-43 negative. These structures contain a variety of proteins, including (unexpectedly) dipeptides corresponding to three possible open reading frames of the intronic GGGGCC expansion, Gly–Ala, Gly–Pro, and Gly–Arg.[13,14] Translation of this apparently non-coding sequence was attributed to the rare phenomenon described as repeat associated non-ATG-initiated (RAN) translation, a mechanism that has been documented in other repeat-associated neurodegenerative disorders. The relevance and causative role of RAN-translated dipeptides in ALS are currently unexplored.

It is therefore clear that disturbances of protein stability and formation of protein inclusions and aggregates are hallmarks of all forms of ALS. As noted, it remains to be determined whether these are directly toxic or indirectly denote some other pathogenic event (such as toxicity caused by oligomeric forms of the same protein). Also unknown is whether the host cells can metabolize the proteins through normal protein degradation. The presence of aggregates may reflect not only an abundance of mutant, conformationally unstable protein but also a diminished

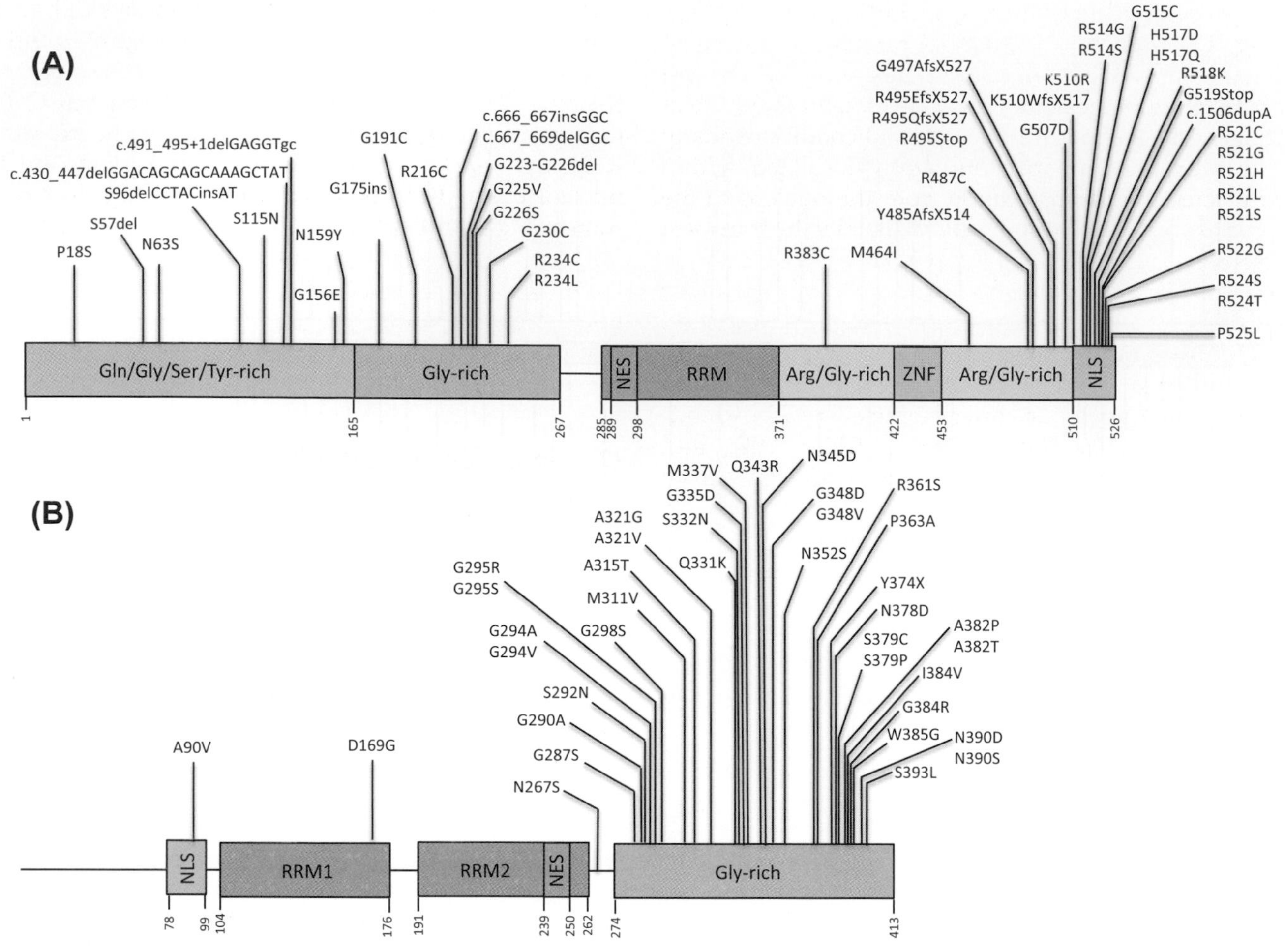

FIGURE 18.5 **Doman structure of FUS and TDP-43 and position of familial amyotrophic lateral sclerosis-associated mutations.** (A) Several mutations, including truncations, are found within the nuclear localization sequence of FUS, suggesting a probable cause for the cytosolic mislocalization of this protein in fALS. Additionally within FUS, and more abundantly in TDP-43 (B), many mutations are located within the protein-interacting glycine rich domain. RRM: RNA recognition motif; NES: nuclear export sequence; NLS: nuclear localization sequence; ZNF: zinc finger domain. *Source: Functional domains predicted using UniProt, NLS predicted using* http://nls-mapper.iab.keio.ac.jp/cgi-bin/NLS_Mapper_form.cgi, *NES predicted using* http://www.cbs.dtu.dk/services/NetNES, *the Amyotrophic Lateral Sclerosis Online Genetics Database (ALSoD),* http://alsod.iop.kcl.ac.uk. *Source: Adapted from Lagier-Tourenne et al 2010.*[6]

capacity for protein degradation. Genetic screening has identified several ALS genes that implicated defective protein clearance as an important component of fALS. For example, X-linked ALS is associated with dominantly inherited mutations in the gene encoding ubiquilin-2,[15] a member of the ubiquilin family of proteins that regulate the degradation of ubiquitinated proteins. *In vitro* experiments demonstrated that the ALS-associated mutations in ubiquilin-2 impair proteasome-mediated protein degradation. The ubiquilin-2 protein has been detected in inclusion bodies in the brains and spinal cord of fALS cases carrying UBQLN2 mutations, and also in sporadic cases of ALS and ALS-FTD. The importance of the ubiquilins in this disease has been highlighted by the finding that mutations in the *ubiquilin-1* gene initiate a childhood form of motor neuron disease. fALS-associated mutations have also been detected in *SQSTM1*, the gene that encodes the ubiquitin-binding scaffold protein, sequestome-1/p62. p62 normally binds ubiquitin, prominently decorating inclusion bodies. These data convincingly support the hypothesis that disruption of protein clearance and quality control mechanisms contribute to the pathogenesis of ALS.

RNA Processing Deficits and RNA Toxicity

The understanding of ALS pathology has undergone a vast change following the identification of fALS-associated mutations in genes that encode DNA/RNA interacting proteins, including *TDP-43* and *FUS*, and less frequently *TAF15*, *EWSR1*, *ANG*, *SETX*, *ELP3*, *Ataxin-1* and *-2*, *hnRNP-A1* and *-A2B1*, and *CREST*. The most common of these DNA/RNA binding genes, *TDP-43* and *FUS*, have been implicated in multiple DNA and

RNA processing mechanisms, including transcription, slicing, RNA transport, microRNA processing and translation (reviewed by Lagier-Tourenne *et al.*[6]). Several characteristics of these proteins may explain their role in pathogenesis. Although under normal conditions many of these proteins are predominantly localized within the nucleus, their translocation from the nucleus to the cytoplasm occurs physiologically, usually to transport messenger RNA (mRNA) for translation but also for degradation in processing bodies (or p-bodies) or under conditions of cellular stress. In the latter circumstance, mRNAs are exported and sequestered in translationally quiescent stress granules until translation is required. In ALS, neuronal nuclei are often depleted of RNA binding proteins such as TDP-43, FUS, TAF15, EWSR1, and ataxin-2, with accompanying mislocalization from nucleus to cytoplasm in the same neurons. Notably, in the *FUS* gene, multiple mutations have been detected that mutate or eliminate the FUS nuclear localization sequence. These observations point to both loss- and gain-of-function models of ALS pathology. Because many of these proteins play important roles in DNA/RNA processing and signaling in the nucleus, mislocalization from nucleus to cytoplasm probably impairs these intranuclear functions, disrupting activities such as splicing. Alternatively, mislocalization may confer a pathological gain of function: DNA/RNA binding proteins TDP-43, FUS, TAF15, EWSR1, hnRNP-A1, and A2B1 carry prion-like domains that normally assist in orderly accumulation in stress granules. In the gain-of-function model, the mislocalization of these proteins to the cytosol permits their irreversible accumulation in stress granules, release from which is hindered by the abnormally strong interaction of their prion domains. Fibril formation of the prion domains of multiple proteins then results in large inclusion body formation. The prion domains are also able to interact without entering stress granules. The presence of prion domains may explain the spread of disease throughout the brain. Through secretion or leakage from atrophied neurons, extracellular prion peptides may be taken up by adjacent neurons, propagating further aggregation and spread of pathology to regions distant from the original focus of disease.[15] Such a mechanism has been suggested in the staging of TDP-43 pathology; TDP-43 inclusions were found to initiate in the motor cortex and brainstem, and subsequently spread throughout the brain.[8]

The hexanucleotide expansion within an apparently non-coding region of *C9orf72* introduces other mechanisms whereby a pathologically expansion in an RNA transcript can contribute directly to disease (reviewed by Todd and Paulson[16]). First, as noted above, although the pathologically expanded GGGGCC hexanucleotide repeat is located in a 5′ non-coding, regulatory region of *C9orf72*, it can be RAN transcribed to form multiple minipeptide repeats. Second, fluorescent *in situ* hybridization experiments confirm the presence of multiple GGGGCC-rich RNA foci within the nucleus of affected neurons. By analogy with other intronic repeat disorders (e.g. myotonic dystrophy), these deposits of expanded RNA may serve as sinks that effectively deplete the nucleus of key RNA binding factors, thereby disrupting transcription and splicing. Third, these expanded RNA tracts may generate illegitimate antisense RNA fragments that may interact unfavorably with as yet unidentified targets. Analysis of the C9orf72 expansion should elucidate these and other mechanisms by which this unusual mutation contributes to disease.

Disrupted Cytoskeletal Function

Motor neurons are uniquely large cells, with axons reaching lengths upwards of 1 m and a volume thousands of times greater than most other cells. Several lines of evidence substantiate that disruption of normal cytoskeletal stability and function can contribute to ALS pathogenesis. Rare mutations have been identified in genes encoding cytoskeleton filament subunits, the neurofilament heavy subunit and the intermediate filament subunit, peripherin. Large-caliber motor neurons, those richest in neurofilament proteins, are the earliest to degenerate in both human disease and models of ALS. Neurofilaments have been detected within cytoplasmic inclusion bodies in ALS and several other neurodegenerative diseases. Misregulation of cytoskeleton assembly or disassembly may also contribute to disease. fALS-associated mutations have been identified in profilin-1, a protein that regulates actin polymerization.[17] *In vitro*, fALS-associated mutations decrease polymeric γ-actin formation and cause growth cone abnormalities or impeded normal axon outgrowth.

Disturbances in other neuronal processes critical in maintaining an intact cytoskeleton have also been associated with ALS. Axonal transport is essential for efficient conveyance of vesicles and organelles between cell body and distal synapses. In rodent models of ALS, reduced velocity of slow axonal transport is one of the earliest, prodromal abnormalities. Direct application of mutant SOD1 protein to squid axoplasm preparations slows anterograde transport.[18,19] Mutations have been detected in several ALS genes that encode proteins involved in axon transport. The first identified, *DCTN1*, encodes the p150 subunit of the dynactin protein complex that links vesicle and organelle cargos with the retrograde transport motor dynein.[20] In some populations, polymorphisms that reduce expression of kinesin-associated protein-3 (*KIFAP3*) gene are associated with an extended lifespan in sALS.[21] KIFAP3 forms a trimeric motor complex with the kinesis proteins KIF3A and KIF3B, which function in anterograde transport and chromosome cytokinesis. Finally, a familial mutation has been identified in vesicle-associated membrane protein/synaptobrevin-associated membrane protein B (VAPB), an essential component of the vesicle and membrane trafficking machinery; this is particularly important in trafficking of vesicles from the endoplasmic reticulum.[22] Expression in *Drosophila* of VAPB bearing an ALS-associated mutation (VAPB^{P58S}) triggered accumulation of VAPB protein in the endoplasmic reticulum, with corresponding neurodegeneration and motor deficits.

Contribution of Non-Neuronal Cell Types

A seminal finding in the early twenty-first century was that cellular pathology in ALS is not restricted to neurons. Postmortem tissues from patients demonstrate prominent infiltration and activation of microglia and astrocytes in regions of degeneration. Protein inclusion bodies are frequently detected within glia. This raises the important possibility that there may be non-cell autonomous influences on the cell death process in motor neurons. Several *in vitro* model systems have been used to determine which non-neuronal cell types contribute to ALS pathogenesis.

The concept that pathology in multiple cell types is necessary to induce motor neuron death in transgenic mutant SOD1 ALS is supported by reports that when transgenic expression of human SOD1^{G37R}, SOD1^{G93A}, and SOD1^{G85R} is restricted to neurons, motor neuron degeneration is not observed, despite high intracellular levels of the mutant proteins. This is striking, given that when expressed ubiquitously in mice, these alleles robustly induce motor neuron death with accumulation of SOD1 in spinal motor neurons.[23,24] That is, these experiments imply that the full expression of the disease requires the presence of the mutant protein in both motor neurons and neighboring non-neuronal cells. A parallel experiment in chimeric ALS mice was consistent with this concept. Survival in ALS mice was dramatically prolonged when motor neurons with mutant SOD1 were surrounded by non-neuronal cells that lack the mutant protein; these studies unequivocally document the fact that motor neuron death from mutant SOD1 can be rescued non-cell autonomously.[25] The converse, that motor neuron degeneration can be induced in wild-type motor neurons by exposure to neighboring cells with mutant SOD1, has not been shown. Indeed, at this point, a reasonable conclusion is that in the transgenic SOD1 ALS mice, the presence of mutant SOD1 in the motor neuron is the *sine qua non* for the occurrence of motor neuron disease. Non-neuronal cells can powerfully modify the resulting phenotype, but have not been shown to induce chronic progressive motor neuron disease.

Microglia

Pronounced infiltration of microglia is typical in tissue surrounding degenerating neurons in postmortem human tissue and rodent models. Selective silencing of mutant SOD1 in microglia in transgenic SOD1^{G37R} mice that otherwise ubiquitously express mutant SOD1^{G37R} significantly prolonged survival.[26] In other experiments, transplantation of a donor population of wild-type glia into SOD1^{G93A} mice also significantly slowed disease progression.[27] The reciprocal transplantation of microglia bearing mutant SOD1^{G93A} into wild-type mice did not cause motor neuron pathology. These findings are consistent with the view that although not a trigger of pathogenesis, the presence of mutant SOD1 in microglia contributes to the outcome of this disease.

Astrocytes

Experimental data for the role of astrocytes have been difficult to interpret. As in experiments with microglia, selective, nearly complete silencing of expression of SOD1^{G37R} or SOD1^{G85R} in astrocytes extends the lifespan of the corresponding ALS mice.[28] Again by analogy with microglial studies, in the converse experiment, overexpression of mutant SOD1^{G86R} selectively in astrocytes did not induce neurodegeneration or motor dysfunction. These findings contrast with tissue culture experiments.[29] When cultured on a monolayer of astrocytes derived from mutant transgenic SOD1, wild-type motor neurons undergo degeneration and death. Furthermore, exposing both wild-type and SOD1 mutant primary motor neurons to astrocyte-conditioned media was lethal to the motor neurons, suggesting that astrocytes secrete one or more undefined factors that are toxic to neurons.

Myelinating Glia in the Central and Peripheral Nervous System

The role of SOD1 in myelinating cells has also been assessed. Excision of SOD1^{G85R} expression from Schwann cells, the myelinating glia of the peripheral nervous system, resulted in a shortened lifespan in transgenic mice, despite no change in the progression of motor axon loss.[30] As the SOD1^{G85R} mutant protein retains SOD activity, it was suggested that the accelerated degeneration is a loss of superoxide detoxification in Schwann cells.

The role of dysfunction of oligodendrocytes, a class of CNS myelinating glia, has also been investigated in a rodent model of ALS. In the degenerating spinal cord of SOD1^{G93A} mice, loss of early-born oligodendrocytes during disease progression is matched by maturation of oligodendrocytes from their progenitor NG2^{+} cells; thus, although there is no net change in the number of oligodendrocytes, the population in mutant mice is more immature.[31] These adult-born, immature oligodendrocytes, were functionally and morphologically abnormal, and inefficient in remyelination. Furthermore, in experiments analogous to those described above in microglia and astrocytes, partial silencing of expression of $SOD1^{G85R}$ in $NG2^{+}$ cells significantly prolonged disease duration and increased the lifespan of transgenic mice. These data suggest that maintenance of an appropriate pool of oligodendroglial cells is beneficial to motor neurons harboring mutant SOD1 protein. The basis for this remarkable finding is not clear. One explanation is that correct apposition of oligodendroglial cells to axons and cell bodies is essential for normal metabolic and trophic support of motor neurons. Oligodendrocytes provide axons with the essential metabolite lactate via monocarboxylate transporter-1 (MCT1); reduced levels of MCT1 correlate with motor neuron degeneration in both the $SOD1^{G93A}$ mouse and sALS patients.[32] Another explanation for the role of oligodendroglial cells is in the more conventional role played by this cell type in ensuring normal myelination and electrophysiological function.

Environmental Contribution to Pathogenesis

Exposure to environmental toxins is thought to contribute to an increased risk of developing ALS, although firm proof that any one environmental neurotoxin directly causes ALS has been elusive.

Pesticides and Herbicides

First noted in an Italian cohort, an elevated incidence of ALS in agricultural workers suggested that exposure to insecticide and herbicides may contribute to the pathogenesis of the disease. ALS is not unique in this regard; exposure to paraquat notably increases the risk of Parkinson disease. However, many population studies have explored the association between chronic exposure to pesticides and ALS incidence, without conclusive findings.

β-N-Methylamino-L-Alanine

The high incidence of an ALS–Parkinsonian dementia complex disease in the South Pacific island of Guam and the Japanese Kii Peninsula has been associated with exposure to the non-proteinogenic amino acid β-*N*-methylamino-L-alanine (BMAA). Produced by cyanobacteria, BMAA was ingested through a traditional diet rich in flour produced from seeds of the cycad plant (which has a symbiotic relationship with cyanobacteria), and in much greater concentrations from the flesh of animals that consume the seeds. Although the mechanism by which BMAA might contribute ALS is unclear, unusually high levels of BMAA have been detected within the brains of ALS patients from Guam and regions of Florida where BMAA contamination of seafood is apparent. Furthermore, the incidence of ALS is high in regions surrounding lakes rich in cyanobacteria blooms.

Organophosphates

Organophosphates (OPs) irreversibly phosphorylate acetyl-cholinesterase, leading to an accumulation of inactivated forms of the enzyme in cholinergic synapses. In both animals and humans, a striking axonal neuropathy afflicting both the central and peripheral nervous system is induced after a delay by exposure to OPs. This OP-induced axonopathy can cause irreversible, disabling sensorimotor dysfunction. Two proteins associated with OP metabolism and toxicity have been associated with an increased risk of sALS. Paraoxonase (PON) proteins are a class of esterase that metabolize oxidized lipids and in some instances OPs. Several studies have identified a link between the PON locus and increased risk of sALS. In addition, in the three PON proteins encoded by this locus, rare coding mutations have been reported in ALS cases.[33,34] However, a large meta-analysis of all association studies of the PON locus failed to confirm significant genetic linkage with risk of sALS. Therefore, the role of PON in ALS awaits further definition. A second target of OP toxicity, neuropathy target esterase (NTE), has also been associated with lower motor neuron disease. In rare families, recessive loss-of-function mutations in the NTE gene cause childhood-onset, corticospinal degenerations (hereditary spastic paraparesis). Mutational analyses have not disclosed seemingly causal NTE mutations in ALS, however.

Epistasis in Amyotrophic Lateral Sclerosis Susceptibility and Phenotype

Virtually all of the fALS genes identified to date have appeared to migrate as single-gene Mendelian defects through fALS pedigrees. Indeed, virtually all of the major discoveries in the genetics of human diseases since around the year 2000 have involved monogenic disorders. The possibility that ALS susceptibility, or aspects of the ALS phenotype, may reflect the presence of an adverse combination of genetic variants (a phenomenon termed *epistasis*) that, taken by themselves, are not cytotoxic, has been posed but not investigated. That this model is plausible is suggested by the observation that mutations in C9orf72, TDP-43, or FUS can lead to ALS or FTLD or a combination of these diseases. That is, it seems likely that one or more factors determine whether the primary lesion in a gene such as *C9orf72* causes predominant motor neuron or frontal lobe degeneration. Moreover, studies consistent with this oligogenic model of ALS have reported that some patients carry both the *TARDBP* N352S mutation, which is autosomal dominant but only partially penetrant, and a second mutation in other ALS-associated genes (e.g. *C9orf72* or *angiogenin*).[35]

Factors that modify phenotype in ALS have also been reported. One example was generated in a screen of genes that modify the penetrance of SOD1 pathology in zebrafish. Knockdown of the *Rtk2* gene, the zebrafish homologue of the EphA4 ephrin receptor tyrosine kinase, rescued motor axon abnormalities in zebrafish with mutant-SOD1 induced motor pathology.[36] In SOD1^{G93A} transgenic mice, heterozygous ablation of *EphA4* significantly increased lifespan. In a Dutch ALS cohort, *EPHA4* expression levels correlated inversely with onset and survival. Because EphA4 normally functions to inhibit or curtail axon outgrowth, it is plausible that reduced EphA4 levels are associated with enhanced axonal capacity to sprout in response to injury, presumably enhancing maintenance of innervation at the neuromuscular junction.

MODEL SYSTEMS OF AMYOTROPHIC LATERAL SCLEROSIS TOXICITY

Animal models have become an essential tool in understanding the pathobiology of ALS and are critical in the race to produce new therapeutic strategies to treat the disease. Genetically modified rats, mice, fish, flies, and worms have all been produced with the intent of modeling specific pathological features of ALS, with varying degrees of accuracy and usefulness. When designing or choosing to use these models in experiments, many factors must be taken into consideration. Two main types of phenotype are most frequently produced: gain of function, in which genetic manipulation results in an excess of a normal function or an entirely novel adverse function for a protein; and loss of function, in which normal function of a protein is impaired, either through the deletion of a gene or through mutations that result in a functionally impaired product. It is possible for both loss- and gain-of-function phenotypes to exist in parallel; for example, in TDP-43 and FUS, gained functions such as increased cytoplasmic protein aggregation may accompany the loss of nuclear RNA/DNA interactions.

The choice of model organism is also critical to experimental design, each species having its own merits and limitations. Although the CNS of a rodent has much more in common with that of a human than that of a fruit fly, the genome of the fly can be modified to an extent impossible to match in a rodent. The following section will present highlights of model organisms, focusing on the extent to which their phenotypes resemble ALS in humans, what they have told us about the pathogenic mechanisms involved in the disease, and how they have been used in the development of therapeutics.

Rodent Models

Transgenic rodents, particularly mice, comprise the most widespread class of model organism in ALS research, and offer many advantages. The nervous systems of mice and rats bear close anatomic resemblances to that of humans, particularly with regard to organization of the motor system. The rodent lifespan is short enough to allow testing for changes in survival over a reasonable timescale but long enough for age-associated phenotypes to develop. The genomes of both mice and rats have been fully sequenced, allowing genetic manipulation including knocking-out of genes, expression of transgenes, targeted knockin of mutations, and inducible models. Numerous robust tests of motor, sensory, and cognitive behavior are available, allowing for behavioral phenotyping to complement molecular, biochemical, and histological characterization.

SOD1

SOD1 was the first mutated gene to be linked to fALS, and accordingly has been the focus of the vast majority of ALS model organism studies. A diverse range of rodents has been produced, either null for SOD1 or expressing wild-type or mutant forms of the human gene under the control of various types of promoters. The discussion here will be limited to the most common rodent models and lessons learned from them about the pathology of ALS.

Overexpression of wild-type forms of SOD1 has been repeatedly demonstrated not to cause disease in rodents.[37] In contrast, expression of several fALS-associated mutant forms of this gene causes neurodegeneration, with a phenotype that recapitulates many features in human motor neuron disease. Currently, the most commonly used and best characterized rodent model of ALS is mice that express the G93A SOD1 mutant protein from a 12 kb genomic construct in which the human promoter drives transcription and splicing of the human SOD1 RNA (Fig. 18.6). The natural history of disease progression of the SOD1^{G93A} mouse has been well defined through numerous experiments. Like most transgenic SOD1 mice, the SOD1^{G93A} mice develop motor deficits reminiscent of ALS. These manifest initially as weakness of the hindlimbs at 3–4 months of age and then progress to involuntary clasping and loss of leg splay response when the animal is lifted by the tail. Thereafter, first the hindlimbs and then the forelimbs weaken and waste. As weakness advances, feeding becomes difficult leading, in conjunction with denervational atrophy, to significant loss of body mass. Death follows, typically at about 5 months of age. These mice are typically killed during the fifth to sixth month, when they lose the ability to right themselves.

As in the human counterpart, the underlying cause of motor deficits in SOD1^{G93A} mice is the selective loss of motor neuron populations; spinal motor neurons and their axons are progressively lost during disease progression, as are those within orofacial brainstem nuclei. In a pattern reminiscent of human ALS, occulomotor neurons are resistant to the motor neuron death

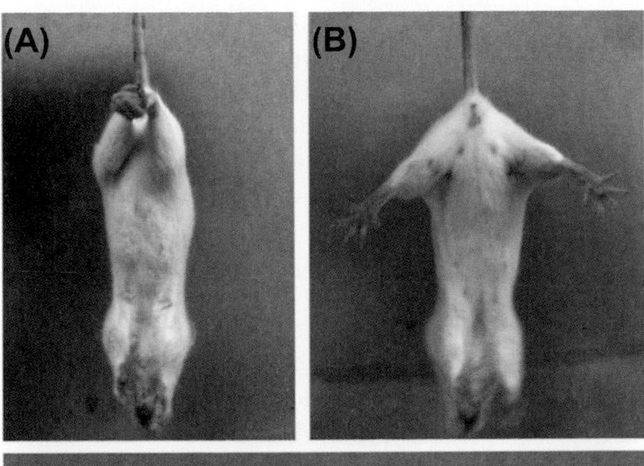

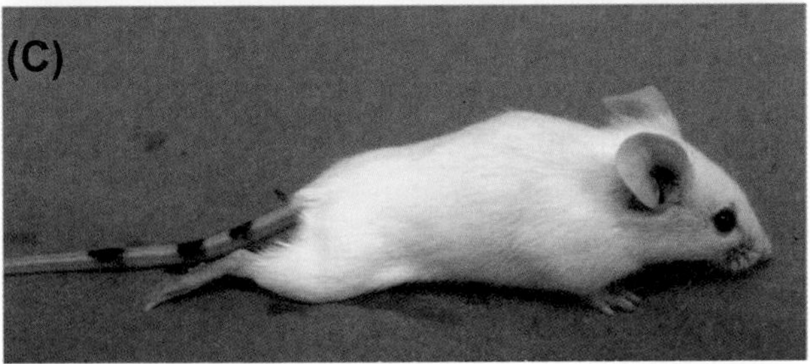

FIGURE 18.6 **The SOD1^{G93A} transgenic mouse develops a phenotype recapitulating amyotrophic lateral sclerosis.** End-stage SOD1-G93A mice on a mixed SJL/Bl6 background showing hindlimb paresis/paralysis. When elevated by the tail, late-stage SOD1^{G93A} mice (A) show clear loss of the hindlimb splay reflex normally seen in littermate wild-type animals (B). Weakness of hindlimbs progresses to full paralysis, generally observed as dragging of the legs (C), accompanied by loss of body mass through a combination of diminished muscle tone and reduced feeding.

process. Loss of motor neurons is accompanied by steadily increasing levels of both microgliosis and astrogliosis. Denervation of neuromuscular junctions is seen at early stages in the disease, with fastest firing motor units being most sensitive. Similarities with the pathology seen in human ALS are further demonstrated by assessment of the peripheral nerves of SOD1^{G93A} mice, where a selective pattern of motor axon loss is seen. Large-caliber fibers appear to be considerably more vulnerable to SOD1^{G93A}-mediated damage than smaller caliber fibers. Sensory deficits in SOD1^{G93A} mice are limited, although slight morphological changes in sensory neurons have been reported. Motor neuron loss in the SOD1^{G93A} mouse is associated with the accumulation of cytoplasmic inclusion bodies containing SOD1 or ubiquitin. These are first observed as small inclusions in the synapses and axonal cytoplasm of affected neurons, followed by the formation of large spherical and amorphic inclusions in the perikaryal cytoplasm. The SOD1^{G93A} mouse thus presents several pathological characteristics that are typically seen in humans carrying SOD1 mutations, most notably motor behavior deficits, selective loss of motor neurons, and accumulation of SOD1 in inclusion bodies.

Several lines of mutant SOD1 rats have also been generated. Although these have not greatly furthered the understanding of the disease mechanism of ALS beyond what is known from mice, their larger size is beneficial in some studies related to therapy development. ALS-like motor system neurodegeneration has been demonstrated in rats expressing G93A or H46R mutant SOD1[38,39] but not in those expressing wild-type human SOD1. SOD1^{G93A} rats develop progressive motor dysfunction and die prematurely, associated with substantial loss of spinal motor neurons, atrophy of axons, and accumulation of SOD1 into inclusion bodies. Glial pathology is also evident, with SOD1-positive inclusions in astrocytes and progressive loss of the astrocyte glutamate transporter EAAT2 in the spinal cord ventral horn.

Because the phenotype of the SOD1 transgenic rodents shares many clinical and pathological features with ALS patients, the animals are frequently used in the preclinical development of novel ALS therapeutics. Therapy trials have included countless drug trials, cell replacement strategies, and genetic manipulations. Notable recent studies have attempted to silence expression of the offending mutant SOD1 protein using inhibitory RNA directed against SOD1. Several delivery strategies have been tested. Small interfering RNA has been tested in ALS rats using intrathecal delivery of antisense oligonucleotides targeting SOD1 G93A; this intervention significantly decreased SOD1 RNA and protein levels in the brain, spinal cord, and cerebrospinal fluid of the rats.[40] These positive findings were followed by phase I pilot safety testing of intrathecally delivered anti-SOD1 antisense oligonucleotides in individuals with ALS. The therapy was well tolerated, but the study design and limited cohort size precluded meaningful evaluation of the outcome. It is reasonable to predict that as more ALS-associated genes are identified and the next generation of transgenic rodents is produced (see below), the focus of research and therapeutic development will naturally shift from the SOD1 mice to the newer models.

TDP-43

Since the identification of TDP-43 as a principal feature of ALS pathology, numerous transgenic and knockout rodent models have been generated. These can be split into two groups that address gain of function through transgenic expression of wild-type and mutant forms of human *TARDBP*, and loss of function through deletion of the mouse homologue.

The majority of transgenic mouse lines now available express high neuronal levels of either wild-type or mutant human *TARDBP* under the transcriptional control of neuron-specific promoters. These animals have provided insight into the gain-of-function pathogenic mechanisms that TDP-43 invokes. A prominent initial observation is that the pathogenic mechanisms may not be specific to the disease alleles, as forced expression of wild-type and mutant *TDP43* transgenes in mice yields nearly indistinguishable phenotypes. Motor abnormalities have been reported in all models, manifested as changes in gait progressing to paralysis of the limbs, and are associated with evidence of motor neuron cell body and neurite degeneration. As with the SOD1 transgenic mice, the onset of motor deficits is associated with abnormalities in motor neurons, their neurites and neuromuscular junctions, and premature death. Although 25 kDa and 35 kDa C-terminal fragments of TDP-43 are detectable in both wild-type and mutant *TARDBP* transgenic mice, TDP-43 inclusion bodies are notably absent. Without this latter hallmark pathological feature it is difficult to conclude that any one of the above models truly recapitulates TDP43 pathology. A more interesting recapitulation of disease is seen in mice ubiquitously overexpression of wild-type human TDP-43 genomic fragments, for which the endogenous promoter region regulates transcription.[41] Wild-type, G348C, and A315T human *TARDBP* transgenic mice also develop motor abnormalities and degeneration of motor neurons and, unlike in the neuron-specific transgenic mice, ubiquitinated TDP-43 inclusions were present. Rat transgenic models of TDP-43 overexpression have also been produced. Developmental overexpression of a minimal M337V mutant *TARDBP* gene resulted in paralysis and premature death of founder rats 30 days after birth, whereas rats expressing the wild-type gene were unaffected.[42] Phosphorylated mutant TDP-43 was mislocalized to the cytoplasm, but inclusion bodies were not seen. Expression of TDP43-M337V selectively in astrocytes was also sufficient to cause degeneration of motor neurons, motor dysfunction, and premature death.[43] The authors attributed the death of neurons to loss of protective activity by astrocytes through the depletion of astrocyte glutamate transporters EAAT1 and EAAT2, which are essential components in buffering glutamate levels at dendritic synaptic clefts. Given the diverse roles of TDP-43 in DNA/RNA processing (see above), is not surprising that TARDBP null mice are non-viable, with embryos unable to develop beyond the blastocyst stage. Heterozygous loss of TARDBP is insufficient for lethality, owing to a compensatory upregulation of the gene's expression. TDP-43 also appears to be essential to long-term maintenance of healthy neurons, with mice developing motor system atrophy and dying prematurely after its conditional deletion from adult motor neurons.

It is clear that while many features of ALS are reproduced in currently available rodent models of TDP-43 pathology, no one system recapitulates both the gain-of-function and loss-of-function hypotheses. The transgenic genomic fragment model produced by Swarup and colleagues[41] probably represents the most useful current model, in which there is ubiquitious expression of wild-type human *TARDBP* controlled by it endogenous promoter, resulting in motor neuron degeneration and inclusion body formation. Future model rodents should ideally address the question of whether the protein mediates toxicity through function gained, such as aberrant oligomerization, or lost, such as disrupted DNA/RNA interaction.

FUS

Because of interest in FUS as a component in the liposarcoma-associated fusion oncogene TLS-CHOP, *FUS/TLS* null mice were developed more than a decade ago, well before discovery of FUS/TLS mutations in ALS. FUS null mice are viable but, at least in some strains, display severe abnormalities in the mechanism of homologous recombination, and have defective spermatogenesis and hypersensitivity to irradiation-induced chromosomal damage. Because FUS mutations are a small subset of hereditary ALS cases, it is not surprising that only a few mouse transgenic models of wild-type FUS or mutant FUS-mediated pathology have been generated to date.

Both rats and mice overexpressing human *FUS* are viable. Some deficits in spatial learning and memory have been observed in aged human FUS transgenic rats, associated with loss of hippocampal and cortical neurons.[44] Expression of R521C mutant human FUS in adult rats generates a severe phenotype, characterized by progressive motor deficits and premature death, with denervation of muscles and atrophy of skeletal muscle without loss of spinal motor neurons or their axons. FUS transgenic mice develop more pronounced motor phenotypes than their transgenic rat counterparts. Mice expressing HA-tagged human wild-type FUS[45] or a truncated human FUS[46] were viable and initially indistinguishable from non-transgenic littermates. However, both developed pronounced motor deficits and degeneration of motor neurons, and died prematurely. Overexpression of human FUS in mice has suggested that a negative feedback mechanism controls expression of FUS, as endogenous murine *FUS* levels are reduced in response to increased levels of transgenic human *FUS*.

Non-Rodent Models

Although rodents are the most commonly used model organisms of ALS there are several limitations to their use. Manipulation of the rodent genome is generally time consuming and sometimes difficult to achieve. The lifespan of rodents, although in some aspects beneficial, makes assessment of slowly progressing, age-dependent phenotypes a lengthy process. A further consideration is the substantial cost of rodent colony maintenance. In light of these drawbacks, many non-mammalian model organisms of ALS have been produced, and have been useful in furthering our understanding of the disease.

Zebrafish

Zebrafish (*Danio rerio*) is a useful small-vertebrate model organism with a simplified nervous system with strong homology to the mammalian CNS. The zebrafish genome has been fully sequenced. The accessibility of motor neurons in zebrafish larvae and the translucence of their flesh make *in vivo* imaging and electrophysiological recording from motor neurons easily achievable. Several zebrafish models of ALS have been generated. Expression of mutant zebrafish SOD1[47] or human SOD1 either transiently[48] or stably[49] produces motor function deficits and dose-dependent motor axon abnormalities, including aberrant axonal branching, decreases in axon length, and denervation of the neuromuscular junction. Both knockout and overexpression of TDP-43 in zebrafish produce motor deficits and abnormalities in motor axon morphology. A second TDP-43-like homologue exists in zebrafish, expressed by the gene *Tardbpl* as a compensatory splice variant for the reduced Tardbp. For effective knockout of TDP-43, both genes must be inactivated. Similarly, both the knockdown and overexpression of mutant FUS in zebrafish cause abnormalities in motor neuron morphology and impede efficient synaptic transmission at the neuromuscular junction.[50] A further benefit of zebrafish as a model organism is the ability to rapidly generate new model systems in response to developments in understanding of a disease. Following the identification of C9orf72 in late 2011, a knockdown model of zebrafish homologue C9orf72 was quickly generated and characterized.[51] Data from these animals suggest that loss of C9orf72 may contribute to the onset of motor dysfunction, with null animals showing reduced movement, associated with motor axon abnormalities similar to those seen is the above fish models of ALS. This motor dysfunction could be rescued by expression of human C9orf72 mRNA. However, since these data were generated through morpholino-induced knockdown of zC9orf72, they should be treated with some caution as off-target effects may contribute to the observed phenotype.

Drosophila melanogaster

Drosophila melanogaster is another well-characterized research organism that has substantial advantages in producing novel genetic models of disease. The *Drosophila* genome is relatively simple compared with that of mammals, allowing highly complex genetic manipulations to be made with relative ease. A large number of cell-type specific promoters, including those active in motor neurons, have been defined in *Drosophila*, allowing the expression or ablation of genes of interest in a tissue-specific manner. Although the CNS of *Drosophila* is primitive in comparison to mammals, it is nonetheless a highly sophisticated system complete with a multitude of different neuron and glia cell type, which generate complex motor and cognitive behaviors. The life cycle of *Drosophila* also permits assessment of the impact of a given gene mutation on development and function at both the larval and adulthood stages, each with differing analytical uses. With regard to ALS, a significant biochemical difference between *Drosophila* and higher organisms exists in the type of neurotransmitters used in vertebrate motor neurons (acetylcholine in vertebrate animals but glutamatergic in *Drosophila*), although the overall structure of the neuromuscular junction is similar. To date, *Drosophila* models of ALS have not offered a great deal of insight into the pathogenic mechanism of disease beyond what has been learned from rodent modeling. Loss of SOD1 in *Drosophila* shortens the lifespan, with overexpression conferring protection from oxidative stress. Models of mutant SOD1 overexpression in *Drosophila* are few. In one report, restricted overexpression of wild-type, A4V, and G85R proteins produced deficits in motor function, presenting as poor performance in a climbing assay, combined with age-dependent neuronal electrophysiological dysfunction and cytoplasmic accumulation of SOD1 protein in motor neurons.[52] Knockout and overexpression of various forms of TDP-43 and FUS invariably result in lethality in *Drosophila*, with larvae rarely successfully reaching adulthood. Early death is associated with reduced locomotion in larvae and the few surviving adult flies. *Drosophila* will continue to offer a useful tool for rapidly modeling newly identified mutations in ALS. Furthermore, as more disease-relevant *Drosophila* models of ALS are produced, for example with knockin mutation rather than overexpression and ablation, they will become a useful tool in screening for genes that could modify the progression of ALS pathology.

Caenorhabditis elegans

Although the nematode *Caenorhabditis elegans* also has a very simple nervous system, basic neuronal mechanism are conserved across *C. elegans*, flies, and rodents. As a model organism, *C. elegans* has many benefits, most notably an anatomy that is defined beyond that of any other multicellular organism, a readily modified genome, and optical transparency that allows high-resolution imaging. Pan-neuronal expression of human G85R mutant SOD1 is sufficient to induce motor deficits that are not seen *C. elegans* wild-type SOD1.[53] As in the rodent models, progressive motor deficits are accompanied by cytoplasmic accumulation of SOD1. Lines of *C. elegans* expressing mutant forms of TDP-43 and FUS have also been generated. Knockout experiments suggest that the *C. elegans Tardbp* homologue *TDP-1* plays an essential role in response to stress induced by oxidation and protein misfolding, and subsequently contributes to lifespan. In accordance with vertebrate models, overexpression of wild-type or mutant human TDP-43 resulted in severe motor dysfunction; however, only the mutant forms of TDP-43 caused loss of motor neurons and accumulation of TDP-43 into insoluble aggregates. In *C. elegans*, neuron-restricted expression of wild-type, missense mutants or C-terminal truncated mutant forms have given support to a gain-of-function mechanism of disease, with motor dysfunction and cytoplasmic mislocalization of FUS only seen in lines expressing mutated FUS. *Caenorhabditis elegans* has also proved useful in testing model disease mechanisms generated from observations in postmortem tissues from ALS patients. Both downregulation of the retrograde axonal transport motor protein dynactin-1 and the accumulation of autophagosomes have been described in patients with sALS. RNA interference-mediated knockdown of the *C. elegans* dynactin-1 homologue dmc-1 in motor neurons caused significant motor dysfunction, with impaired axonal transport.[54] In validation of the hypothesized mechanism, ultrastructural analysis detected the accumulation of autophagosomes in motor neurons of affected animals.

Patient-Derived Induced Pluripotent Stem Cells

The ability to reprogram patient-derived fibroblasts into induced pluripotent stems cell (iPSCs) has allowed considerable advances to be made in modeling diseases *in vitro*. iPSCs potentially allow the effects of disease-associated mutations to be tested in any conceivable cell type. Although this technique is still in the relatively early stages of development, protocols have been established to differentiate iPSCs into functioning motor neurons that appear morphologically and physiologically comparable to those *in vivo*. To generate motor neurons, human fibroblasts are extracted from a skin biopsy and reprogrammed into iPSCs using a cocktail of reprogramming transcription factors. iPSCs are then differentiated into motor neurons through treatment with transcription factors such as *Ascl1*, *Brn2*, *Myt1l*, *Lhx3*, *Hb9*, *Isl1*, and *Ngn2*.[55] iPSC-derived neurons and glia have so far been generated from patients harboring mutations in the VAPB, C9orf72, and TDP-43-encoding genes. iPSCs will almost certainly become a valuable tool for future ALS research, both in modeling and delineating the pathogenic mechanisms associated with disease and in preclinical testing of therapeutic modalities.

Conclusion to Model Systems

As more genes become associated with ALS, researchers will need to become less reliant on the current gold standard model, the SOD1^{G93A} transgenic mouse, and to develop more diverse and relevant models of the disease. This will mean surmounting technical challenges such as those encountered in cloning of the massive hexanucleotide repeat contained in C9orf72 and in generating more relevant models of TDP-43 and FUS/TLS mutations. Fortunately, continuing advances in methodology for the manipulation of genomic DNA sequences, such as zinc finger endonucleases, transcription activator-like effector nucleases (TALENs), and clustered regulatory interspaced short palindromic repeat (CRISPR), permit more efficient production of complex forms of model rodents. This new technology should enable the generation of polygenic models of neurodegeneration. It is also hoped that these advances, coupled with new insights into the biology and epidemiology of ALS, will allow the creation of informative rodent models of sALS.

FUTURE DIRECTIONS

Many unresolved questions will need continuing research into ALS in the coming years, and should enhance our capacity to understand and treat this devastating disease. Which aspects of the neurobiology of motor neurons render them susceptible to this disease, and why are some motor neurons spared? In inherited forms of ALS, in which the mutant proteins are expressed from conception onwards, why does a lethal gene mutation not ignite fulminant motor neuron death until after several decades? Can polygenic, epistatic, genetic factors be defined in both familial and sporadic ALS? What is the role of epigenetic modification of the genome in ALS susceptibility? How does the same mutation in the same family lead to ALS in some individuals while others develop ALS and FTD? Can environmental toxins be identified that augment the risk of sporadic forms of neurodegenerative diseases? Most importantly, from the cumulative insights garnered in these investigations, can innovative and effective therapies for ALS be devised?

Certain dimensions of this disease may be particularly fruitful for study. Next generation sequencing is already transforming the analysis of ALS genetics. Several consortia have generated thousands of ALS exomes, with genomes almost certainly to follow. The same technologies permit high-volume RNA transcript analysis (RNAseq), providing an unprecedented opportunity to study profiles of gene expression and gene splicing events. Such studies will almost certainly illuminate large sets of candidate ALS genes and pathways. Indeed, as a corollary, it will be essential to develop the appropriate methods to validate or discard critical pathways. This roadmap from candidate molecules to validated pathways will probably be indispensible in the effort to understand the susceptibility of motor neurons to ALS and define targets for therapy development. Moreover, it is likely to enhance efforts to define biomarkers for this disease. Biomarkers are devoutly sought, as they should offer additional insights into pathogenesis, accelerate diagnosis (in the USA, about 1 year lapses from the first symptom to the diagnosis of ALS), and provide surrogate indicators of disease activity that improve the cost-effectiveness of ALS treatment trials.

Acknowledgments

RHB receives support from the NINDS, the ALS Therapy Alliance, the ALS Association, the Angel Fund, the Al-Athel Foundation, the Pierre L. de Bourgknecht ALS Research Foundation, Project ALS, and P2ALS. OMP is supported by the ALS Therapy Alliance, the Angel Fund, and Project ALS.

References

1. Rosen DR, Siddique T, Patterson D, et al. Mutations in Cu/Zn superoxide dismutase gene are associated with familial amyotrophic lateral sclerosis. *Nature*. 1993;362(6415):59–62.
2. Reaume AG, Elliott JL, Hoffman EK, et al. Motor neurons in Cu/Zn superoxide dismutase-deficient mice develop normally but exhibit enhanced cell death after axonal injury. *Nat Genet*. 1996;13(1):43–47.
3. Shibata N, Asayama K, Hirano A, Kobayashi M. Immunohistochemical study on superoxide dismutases in spinal cords from autopsied patients with amyotrophic lateral sclerosis. *Dev Neurosci*. 1996;18(5–6):492–498.
4. Bosco DA, Morfini G, Karabacak NM, et al. Wild-type and mutant SOD1 share an aberrant conformation and a common pathogenic pathway in ALS. *Nat Neurosci*. 2010;13(11):1396–1403.
5. Sreedharan J, Blair IP, Tripathi VB, et al. TDP-43 mutations in familial and sporadic amyotrophic lateral sclerosis. *Science*. 2008; 319(5870):1668–1672.
6. Lagier-Tourenne C, Polymenidou M, Cleveland DW. TDP-43 and FUS/TLS: emerging roles in RNA processing and neurodegeneration. *Hum Mol Genet*. 2010;19(R1):R46–R64.
7. Neumann M, Sampathu DM, Kwong LK, et al. Ubiquitinated TDP-43 in frontotemporal lobar degeneration and amyotrophic lateral sclerosis. *Science*. 2006;314(5796):130–133.
8. Brettschneider J, Del Tredici K, Toledo JB, et al. Stages of pTDP-43 pathology in amyotrophic lateral sclerosis. *Ann Neurol*. 2013; 74(1):20–38.
9. Kwiatkowski Jr TJ, Bosco DA, Leclerc AL, et al. Mutations in the FUS/TLS gene on chromosome 16 cause familial amyotrophic lateral sclerosis. *Science*. 2009;323(5918):1205–1208.
10. Vance C, Rogelj B, Hortobagyi T, et al. Mutations in FUS, an RNA processing protein, cause familial amyotrophic lateral sclerosis type 6. *Science*. 2009;323(5918):1208–1211.
11. Dejesus-Hernandez M, Mackenzie IR, Boeve BF, et al. Expanded GGGGCC hexanucleotide repeat in noncoding region of C9ORF72 causes chromosome 9p-linked FTD and ALS. *Neuron*. 2011;72(2):245–256.
12. Renton AE, Majounie E, Waite A, et al. A hexanucleotide repeat expansion in c9orf72 is the cause of chromosome 9p21-linked ALS-FTD. *Neuron*. 2011;72(2):257–268.
13. Mori K, Weng SM, Arzberger T, et al. The C9orf72 GGGGCC repeat is translated into aggregating dipeptide-repeat proteins in FTLD/ALS. *Science*. 2013;339(6125):1335–1338.
14. Ash PE, Bieniek KF, Gendron TF, et al. Unconventional translation of C9ORF72 GGGGCC expansion generates insoluble polypeptides specific to c9FTD/ALS. *Neuron*. 2013;77(4):639–646.
15. Deng HX, Chen W, Hong ST, et al. Mutations in UBQLN2 cause dominant X-linked juvenile and adult-onset ALS and ALS/dementia. *Nature*. 2011;477(7363):211–215.
16. Todd PK, Paulson HL. RNA-mediated neurodegeneration in repeat expansion disorders. *Ann Neurol*. 2010;67(3):291–300.
17. Wu CH, Fallini C, Ticozzi N, et al. Mutations in the profilin 1 gene cause familial amyotrophic lateral sclerosis. *Nature*. 2012; 488(7412):499–503.
18. Kanaan NM, Pigino GF, Brady ST, Lazarov O, Binder LI, Morfini GA. Axonal degeneration in Alzheimer's disease: when signaling abnormalities meet the axonal transport system. *Exp Neurol*. 2013;246:44–53.
19. Morfini GA, Bosco DA, Brown H, et al. Inhibition of fast axonal transport by pathogenic SOD1 involves activation of p38 MAP kinase. *PLoS ONE*. 2013;8(6):e65235.
20. Munch C, Sedlmeier R, Meyer T, et al. Point mutations of the p150 subunit of dynactin (DCTN1) gene in ALS. *Neurology*. 2004; 63(4):724–726.
21. Landers JE, Melki J, Meininger V, et al. Reduced expression of the kinesin-associated protein 3 (KIFAP3) gene increases survival in sporadic amyotrophic lateral sclerosis. *Proc Natl Acad Sci U S A*. 2009;106(22):9004–9009.
22. Nishimura AL, Mitne-Neto M, Silva HC, et al. A mutation in the vesicle-trafficking protein VAPB causes late-onset spinal muscular atrophy and amyotrophic lateral sclerosis. *Am J Hum Genet*. 2004;75(5):822–831.
23. Lino MM, Schneider C, Caroni P. Accumulation of SOD1 mutants in postnatal motoneurons does not cause motoneuron pathology or motoneuron disease. *J Neurosci*. 2002;22(12):4825–4832.
24. Pramatarova A, Laganiere J, Roussel J, Brisebois K, Rouleau GA. Neuron-specific expression of mutant superoxide dismutase 1 in transgenic mice does not lead to motor impairment. *J Neurosci*. 2001;21(10):3369–3374.
25. Clement AM, Nguyen MD, Roberts EA, et al. Wild-type nonneuronal cells extend survival of SOD1 mutant motor neurons in ALS mice. *Science*. 2003;302(5642):113–117.
26. Boillee S, Yamanaka K, Lobsiger CS, et al. Onset and progression in inherited ALS determined by motor neurons and microglia. *Science*. 2006;312(5778):1389–1392.
27. Beers DR, Henkel JS, Xiao Q, et al. Wild-type microglia extend survival in PU.1 knockout mice with familial amyotrophic lateral sclerosis. *Proc Natl Acad Sci U S A*. 2006;103(43):16021–16026.
28. Yamanaka K, Boillee S, Roberts EA, et al. Mutant SOD1 in cell types other than motor neurons and oligodendrocytes accelerates onset of disease in ALS mice. *Proc Natl Acad Sci U S A*. 2008;105(21):7594–7599.
29. Nagai M, Re DB, Nagata T, et al. Astrocytes expressing ALS-linked mutated SOD1 release factors selectively toxic to motor neurons. *Nat Neurosci*. 2007;10(5):615–622.
30. Lobsiger CS, Boillee S, McAlonis-Downes M, et al. Schwann cells expressing dismutase active mutant SOD1 unexpectedly slow disease progression in ALS mice. *Proc Natl Acad Sci U S A*. 2009;106(11):4465–4470.
31. Kang SH, Li Y, Fukaya M, et al. Degeneration and impaired regeneration of gray matter oligodendrocytes in amyotrophic lateral sclerosis. *Nat Neurosci*. 2013;16(5):571–579.
32. Lee Y, Morrison BM, Li Y, et al. Oligodendroglia metabolically support axons and contribute to neurodegeneration. *Nature*. 2012;487(7408):443–448.
33. Landers JE, Shi L, Cho TJ, et al. A common haplotype within the PON1 promoter region is associated with sporadic ALS. *Amyotroph Lateral Scler*. 2008;9(5):306–314.
34. Ticozzi N, LeClerc AL, Keagle PJ, et al. Paraoxonase gene mutations in amyotrophic lateral sclerosis. *Ann Neurol*. 2010;68(1): 102–107.
35. van Blitterswijk M, van Es MA, Hennekam EA, et al. Evidence for an oligogenic basis of amyotrophic lateral sclerosis. *Hum Mol Genet*. 2012;21(17):3776–3784.
36. Van Hoecke A, Schoonaert L, Lemmens R, et al. EPHA4 is a disease modifier of amyotrophic lateral sclerosis in animal models and in humans. *Nat Med*. 2012;18(9):1418–1422.

37. Gurney ME, Pu H, Chiu AY, et al. Motor neuron degeneration in mice that express a human Cu,Zn superoxide dismutase mutation. *Science*. 1994;264(5166):1772–1775.
38. Nagai M, Aoki M, Miyoshi I, et al. Rats expressing human cytosolic copper–zinc superoxide dismutase transgenes with amyotrophic lateral sclerosis: associated mutations develop motor neuron disease. *J Neurosci*. 2001;21(23):9246–9254.
39. Howland DS, Liu J, She Y, et al. Focal loss of the glutamate transporter EAAT2 in a transgenic rat model of SOD1 mutant-mediated amyotrophic lateral sclerosis (ALS). *Proc Natl Acad Sci U S A*. 2002; 99(3):1604–1609.
40. Winer L, Srinivasan D, Chun S, et al. SOD1 in cerebral spinal fluid as a pharmacodynamic marker for antisense oligonucleotide therapy. *JAMA Neurol*. 2013;70(2):201–207.
41. Swarup V, Phaneuf D, Bareil C, et al. Pathological hallmarks of amyotrophic lateral sclerosis/frontotemporal lobar degeneration in transgenic mice produced with TDP-43 genomic fragments. *Brain*. 2011;134(Pt 9):2610–2626.
42. Zhou H, Huang C, Chen H, et al. Transgenic rat model of neurodegeneration caused by mutation in the TDP gene. *PLoS Genet*. 2010;6(3):e1000887.
43. Tong J, Huang C, Bi F, et al. Expression of ALS-linked TDP-43 mutant in astrocytes causes non-cell-autonomous motor neuron death in rats. *EMBO J*. 2013;32(13):1917–1926.
44. Huang C, Zhou H, Tong J, et al. FUS transgenic rats develop the phenotypes of amyotrophic lateral sclerosis and frontotemporal lobar degeneration. *PLoS Genet*. 2011;7(3):e1002011.
45. Mitchell JC, McGoldrick P, Vance C, et al. Overexpression of human wild-type FUS causes progressive motor neuron degeneration in an age- and dose-dependent fashion. *Acta Neuropathol*. 2013;125(2):273–288.
46. Shelkovnikova TA, Peters OM, Deykin AV, et al. Fused in sarcoma (FUS) protein lacking nuclear localization signal (NLS) and major RNA binding motifs triggers proteinopathy and severe motor phenotype in transgenic mice. *J Biol Chem*. 2013;288(35):25266–25274.
47. Ramesh T, Lyon AN, Pineda RH, et al. A genetic model of amyotrophic lateral sclerosis in zebrafish displays phenotypic hallmarks of motoneuron disease. *Dis Model Mech*. 2010;3(9–10):652–662.
48. Lemmens R, Van Hoecke A, Hersmus N, et al. Overexpression of mutant superoxide dismutase 1 causes a motor axonopathy in the zebrafish. *Hum Mol Genet*. 2007;16(19):2359–2365.
49. Sakowski SA, Lunn JS, Busta AS, et al. Neuromuscular effects of G93A-SOD1 expression in zebrafish. *Mol Neurodegener*. 2012;7:44.
50. Armstrong GA, Drapeau P. Loss and gain of FUS function impair neuromuscular synaptic transmission in a genetic model of ALS. *Hum Mol Genet*. 2013;22(21):4282–4292.
51. Ciura S, Lattante S, Le Ber I, et al. Loss of function of C9orf72 causes motor deficits in a zebrafish model of amyotrophic lateral sclerosis. *Ann Neurol*. May 30, 2013. [Epub ahead of print.]
52. Watson MR, Lagow RD, Xu K, Zhang B, Bonini NM. A drosophila model for amyotrophic lateral sclerosis reveals motor neuron damage by human SOD1. *J Biol Chem*. 2008;283(36):24972–24981.
53. Wang J, Farr GW, Hall DH, et al. An ALS-linked mutant SOD1 produces a locomotor defect associated with aggregation and synaptic dysfunction when expressed in neurons of. *Caenorhabditis elegans*. *PLoS Genet*. 2009;5(1):e1000350.
54. Ikenaka K, Kawai K, Katsuno M, et al. dnc-1/dynactin 1 knockdown disrupts transport of autophagosomes and induces motor neuron degeneration. *PLoS ONE*. 2013;8(2):e54511.
55. Son EY, Ichida JK, Wainger BJ, et al. Conversion of mouse and human fibroblasts into functional spinal motor neurons. *Cell Stem Cell*. 2011;9(3):205–218.

CHAPTER

19

Parkinson Disease and Other Synucleinopathies

Thomas Gasser, Thomas Wichmann†, ‡, §, Mahlon R. DeLong†, ‡*

*Department for Neurodegenerative Diseases, Hertie Institute for Clinical Brain Research, German Center for Neurodegenerative Diseases, University of Tübingen, Tübingen, Germany; †Department of Neurology, School of Medicine, Emory University, Atlanta, Georgia, USA; ‡Udall Center of Excellence in Parkinson's Disease Research, Emory University, Atlanta, Georgia, USA; §Yerkes National Primate Research Center, Emory University, Atlanta, Georgia, USA

OUTLINE

Neurobiology of Brain Disorders
http://dx.doi.org/10.1016/B978-0-12-398270-4.00019-7

INTRODUCTION

Parkinson disease (PD) is the second most common neurodegenerative disorder after Alzheimer disease (see Chapter 21). PD is characterized by the constellation of motor features known as "parkinsonism", which includes the triad of tremor, slowness of movement, and muscular rigidity. James Parkinson, in his 1817 publication "An essay on the shaking palsy", gave the classic description of this disorder.

The discovery of dopamine in the brain and its depletion in the basal ganglia in patients with PD in the 1950s and 1960s was followed by the demonstration that the symptoms of parkinsonism respond to the oral administration of levodopa (L-dopa), the immediate precursor of dopamine. The degeneration of dopaminergic neurons in the substantia nigra pars compacta (SNc), together with eosinophilic intracellular inclusions (so-called Lewy bodies) in degenerating neurons, remains central to concepts of the pathology of PD, and is thought to account for many of the motor signs and symptoms of the disease. The pathological process, however, is now known to precede and to extend beyond that in the SNc to other areas of the central and even the peripheral nervous system.[1] The widespread pathology accounts for many of the non-motor features of the disease, including autonomic, sleep, cognitive, and psychiatric disturbances, which constitute a significant portion of the clinical burden, particularly as the disease progresses.

Most cases of PD are termed "sporadic" or "idiopathic", reflecting the absence of an identified cause for the disorder. Although only a small proportion of cases of parkinsonism are hereditary, the mapping and cloning of gene mutations in these rare cases, as well as advances in molecular biology, have led to major insights into the neurobiology of this disorder, in particular the central role of α-synuclein, in both genetic and idiopathic PD. In most patients, both genetic and environmental factors appear to play a role. This chapter discusses the clinical features of PD and related synucleinopathies, our understanding of the etiology, pathogenesis, and pathophysiology of the disease, as well as current and potential future approaches to treatment.

CLINICAL FEATURES OF PARKINSON DISEASE

According to the widely accepted UK Brain Bank criteria, a diagnosis of PD is likely in a patient with slowness of movement (bradykinesia) and either tremor at rest or muscular rigidity. A unilateral onset of the symptoms, the presence of tremor at rest, and a positive response to levodopa strongly favor the diagnosis. There is no reliable diagnostic test for PD, so that the diagnosis is based primarily on history and clinical features.

Motor Features

Whereas PD is progressive, its clinical course varies greatly in different patients. "Benign" forms of PD with predominant tremor and slow progression are recognized at one end of the spectrum, and more rapidly progressive cases with prominent disturbances of gait and balance and cognitive decline at the other.

Bradykinesia

Bradykinesia is perhaps the most fundamental and most disabling aspect of PD, interfering with many aspects of daily living, including speech (hypophonia), handwriting (micrographia), rising from a chair, walking, and dressing. Whole body movements such as walking, turning, and sitting down are sometimes described as occurring *en bloc*, that is, like a statue and lacking fluidity. The term bradykinesia strictly refers to the slowness of movement but is also often used to include the overall paucity of movement and difficulty with movement initiation (akinesia).

Bradykinesia is associated with a slow buildup of muscle activation in agonist muscles and has been thought to be due to a vaguely defined "failure of energy". Patients generally do not perceive themselves as moving slowly or speaking softly, indicating that bradykinesia has perceptual as well as purely motor components.

Another aspect of impaired motility seen in varying degrees in patients with PD that may contribute to slowness is the loss of automaticity of movement that translates thought or intent into action. Patients must, thus, switch from a learned or "habit" mode to a more conscious, volitional "goal-directed" mode of action. This is reflected in their difficulty in executing simultaneous movements, such as walking and reaching for objects in a pocket.

Tremor at Rest

Tremor at rest, typically at a frequency of 4–6 Hz, is seen in up to 80% of patients. Tremor most often appears unilaterally in the fingers and hand, and then gradually ascends to more proximal portions of the arm. In most cases, it spreads to the opposite side within a year from onset of the clinical signs and symptoms. The lower extremities may be involved, but less so than the arms. Head and voice tremor are distinctly uncommon, although lip, tongue, or jaw tremor may be present. Parkinsonian tremor is called a "rest" tremor because it is present with the patient in repose, and typically diminishes or disappears with voluntary movement. This tremor pattern is unlike that seen in essential or cerebellar outflow tremor, which is manifest with movement.

FIGURE 19.1 **Historical depiction of a patient with Parkinson disease.** *Source: Reproduced from Gowers.* A Manual of Diseases of the Nervous System. *London: J. & A. Churchill; 1886.*[2]

Tremor, like most movement disorders, is made worse by arousal and stress.

Rigidity

Rigidity is felt by the examining physician with the patient fully relaxed, as a uniform resistance to passive movement of the joints throughout the full range of motion, giving rise to a characteristic "plastic" quality. Brief interruptions of resistance, representing subclinical tremor, especially in the arms, may give rise to a sensation of "cog wheeling".

Posture and Balance

Most patients exhibit a degree of flexed posture of the upper body and upper limbs (see the classic depiction in Fig. 19.1). Patients may later develop balance problems owing to impaired postural reflexes. Postural instability with loss of balance is one of the most disabling components of advanced PD, leading to falls and injuries. Loss of postural reflexes may account for involuntary backward falls (retropulsion) when off balance, or festination (progressing involuntarily from walking to running), resulting from a tendency to be propulsed forward with short, accelerating steps in an effort to retain balance.

Freezing of Gait

This is a sign of advanced disease and manifests as hesitation or arrest of movement, such as at the initiation of gait, when turning, upon entering smaller spaces such as a doorway or crowded room, or when approaching a chair. Episodes of freezing of gait are often preceded by a series of accelerating steps of decreasing length. Freezing may also affect other activities, such as speech.

Abnormal Eye Movements

Slow, hypometric saccades, complex changes in reflexive saccades, and impairments in smooth pursuit movements are often seen in PD.

Motor Features in Advanced Parkinson Disease

As the disease progresses, additional movement abnormalities may develop, including difficulties with swallowing and speaking, abnormal postures of the trunk and limbs, and worsening of postural instability. Unlike the cardinal motor manifestations of PD, most of these late-stage motor manifestations of the disease do not respond to levodopa replacement.

Non-Motor Features

Numerous non-motor features of PD have long been associated with the disease.[3] Those that may precede the development of motor symptoms and signs include a loss of olfaction (anosmia), constipation, sleep disturbances, and depression. These "premotor" features are now being explored as a means to diagnose the disease at the earliest possible stage.

Sleep Disorders

Disorders of sleep–wakefulness regulation are common in PD. While the presence of significant parkinsonism or side effects of antiparkinsonian medications may disrupt sleep, many individuals with PD have *bona fide* sleep disorders, such as rapid eye movement (REM) sleep behavioral disorder (RBD), whose main feature is physically "acting out" of dreams during episodes of REM sleep. RBD may precede the onset of motor signs by several years. Other sleep problems include reduced sleep efficiency due to sleep fragmentation, and excessive daytime sleepiness.

Disturbances of Autonomic Function

Orthostatic hypotension, constipation, urinary urgency and frequency, excessive sweating, and seborrhea are prominent features of the disease. Orthostatic hypotension may also be drug induced, but, in severe cases, is more likely to result from sympathetic denervation of the heart. Constipation resulting from involvement of peripheral nerves and direct involvement of intestinal motility may precede the onset of motor features by a decade or more.

Disturbances of Sensation

Paresthesias and pain also occur in PD. Approximately 60–80% of parkinsonian patients exhibit various forms of pain and discomfort. The pain often responds to dopaminergic therapy.

Mood Changes

Changes in mood accompany the motor features of PD in most patients as the disease progresses. Very common among these is depression, which affects up to 50% of patients with PD and can occur at any phase of the illness. In many patients, depression may be difficult to differentiate from apathy. Anxiety is also very common in PD patients. In advanced cases of PD, changes in mood and anxiety may accompany the "wearing off" of motor benefit of dopaminergic medications, due to decreasing dopamine levels. These symptoms are responsive to medication adjustments that reduce "off times".

Cognitive Impairment

In the original description of the disease, James Parkinson did not include intellectual impairment as part of the syndrome. Neuropsychological studies, however, have clearly demonstrated cognitive deficits as an integral part of the parkinsonian syndrome. The impairment is most apparent in tasks that require speed, simultaneous processing, and shifts in cognitive patterns, resulting in an overall appearance of an executive "frontal–striatal" deficit. It is now believed that about one-third of PD patients have signs of cognitive impairment, with longer duration of disease, advancing age, and later age of onset being the most important risk factors. The cumulative prevalence of dementia in PD may be as high as 75% in patients who survive for more than 10 years beyond the onset of the motor symptoms. The most established risk factors for early dementia in PD are old age, severity of the parkinsonian motor signs and symptoms (with the exception of tremor), severity of postural and gait disturbances, and visual hallucinations.

The pathology observations of Braak and co-workers (see e.g. Goedert *et al.*[1]), described in more detail below, offer a parsimonious explanation for the development of cognitive impairment in later stages of PD, as they describe a sequential spread of pathology, ascending from the lower brainstem via the midbrain to entorhinal and frontoparietal cortical areas. However, in some patients dementia develops early, either simultaneously or even preceding motor impairment. This form of dementia is usually called dementia with Lewy bodies (DLB). The term "Lewy body" refers to characteristic pathological features that are seen in PD and related disorders (see "Pathology and Pathogenesis", below). DLB was originally considered a clinicopathological entity with a specific combination of clinical features, including onset of dementia within 1 year or even preceding the onset of parkinsonism, fluctuations of attention, frequent visual hallucinations, and severe autonomic disturbances. Today, most investigators agree that DLB is a part of a disease spectrum of Lewy body diseases, which range from pure parkinsonism without major cognitive impairment to severe dementia with abundant cortical pathology with or without the parkinsonian motor syndrome.[1] In most cases with DLB, a variable degree of Alzheimer disease pathology is also present. The precise relationship between clinical and pathological features of PD and Alzheimer disease in DLB is unclear. At least part of the interaction may be acting on the genetic level, as it has been shown that the major genetic risk factor for Alzheimer disease, the $Apo\varepsilon_4$ allele, is also a risk factor for dementia in PD and DLB.

DIAGNOSIS OF PARKINSON DISEASE

The frequency of misdiagnosis of PD is still 10–20%, even by well-trained specialists. As discussed, parkinsonism accompanies conditions other than PD. Multiple systems atrophy (MSA) and progressive supranuclear palsy (PSP) are examples of "atypical" parkinsonism. Patients with these disorders have features of parkinsonism, but have differing etiology and pathology, and additional neurological findings.[4] Whereas MSA will be discussed below the broader differential diagnosis of atypical and secondary forms of parkinsonism (e.g. drug exposure, structural brain lesions, metabolic disorders such as Wilson disease) is beyond the scope of this chapter. These conditions are sometimes difficult to distinguish from PD early in the course of the disease, but these forms of parkinsonism can usually be differentiated from sporadic PD within the first 3–4 years of symptom onset. For further information on these conditions, the reader is directed to one of the sources referenced in Chapter 1.

From a clinical standpoint, it is important to rule out drug-induced parkinsonism, since it is generally reversible by removal of the offending agent. This is most often seen with drugs that block dopamine receptors (e.g. neuroleptic agents, or drugs used in gastrointestinal medicine, such as metoclopramide), or with agents that interfere with catecholamine release (e.g. reserpine or tetrabenazine). Drug-induced parkinsonism closely resembles PD, but the tremor is often less prominent.

Although there is no laboratory test that unequivocally confirms the diagnosis of PD, single-photon emission computed tomography (SPECT) can be used to detect the high-affinity dopamine transporter protein present on dopamine terminals and estimate the loss of those terminals, and is sometimes justified (Fig. 19.2). These scans show a characteristic reduction of putamenal dopamine transporter in cases of PD and can differentiate PD from drug-induced parkinsonism as well as other causes of tremor. Unfortunately, these scans cannot reliably separate PD from the most common atypical neurodegenerative forms of parkinsonism that show a similar loss of dopaminergic terminals.

ETIOLOGY OF PARKINSON DISEASE: CLUES FROM EPIDEMIOLOGY AND GENETICS

Diseases without an obvious causative agent, such as most cases of PD, are usually thought to arise from an interplay between genetic and environmental factors. Researchers have conducted a large number of epidemiological and genetic studies in order to untangle and define these different components. The strongest risk factor for PD and other neurodegenerative diseases, and one of the least well understood, is age. (For an in-depth discussion of epidemiology, see Chapter 46.)

Epidemiology and Environmental Risk Factors

Prevalence estimates of PD vary considerably between studies, depending on methodology and diagnostic criteria, with figures ranging between 100 and 300 per 100,000 individuals.[5] As in other neurodegenerative diseases, prevalence is age dependent. Most estimates assume an age-adjusted prevalence of about 1% in people aged over 60 and 3–4% in people aged over 80. There is a consistently higher prevalence in men than in women, with an odds ratio of approximately 1.5 to 1.8. PD occurs worldwide, but some studies show lower prevalence rates in Asian or African countries. It is unclear whether this reflects different methodologies, different age structures of the surveyed populations, or true ethnic differences.[5]

Twin Studies

Although it is clear now that some cases of PD are caused by mutations in single genes and that genetic risk factors contribute to disease risk in sporadic cases (see below), earlier epidemiological studies provided little evidence for a major genetic contribution. Twin studies showed low concordance rates for PD, in both monozygotic and dizygotic twins. Even large, twin-registry based studies could not detect a difference in prevalence between monozygotic and dizygotic twins, indicating that strong non-genetic modifying factors must be operative.

A limitation of twin studies, however, is that they are usually cross-sectional. When asymptomatic co-twins of affected individuals were studied for the integrity of the dopaminergic system using positron emission tomography (PET) (Fig. 19.2), subclinical changes suggestive of

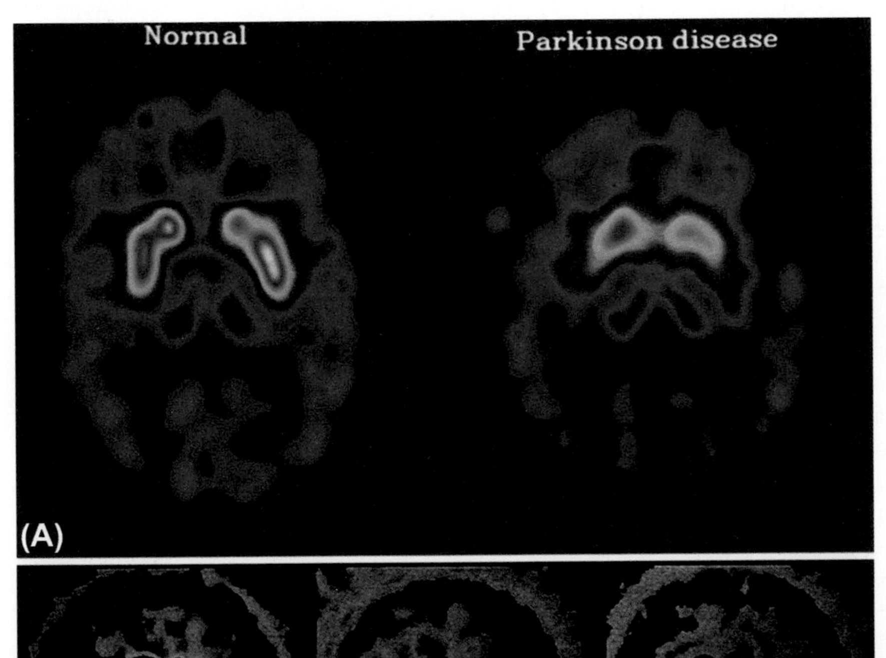

FIGURE 19.2 **The progression of Parkinson disease (PD) can be followed in humans through *in vivo* imaging techniques.** (A) [^{18}F]Fluorodopa positron emission tomography (PET) scans in a normal and parkinsonian patient. This PET label provides a measure of dopamine metabolism. Note the diminished uptake, particularly in the posterior putamen in the PD patient. (B) Result of β-CIT-SPECT [2β-carbomethoxy-3β-(4-iodophenyl) tropane single-photon emission computed tomography] scanning in patients at various stages (rated according to the Hoehn and Yahr system) of PD. This imaging label helps to visualize the distribution of the dopamine transporter, corresponding to the distribution of dopamine terminals. *Source: Reproduced from Olanow* et al. Neurology. *2001;56(Suppl 5): S1–S88,*[6] *with permission.*

dopamine loss were identified in 45% of monozygotic and 29% of dizygotic twins. When twin cohorts were followed for several years, concordance rates for nigrostriatal dysfunction increased from 55% at baseline to 75% in monozygotic twins and from 18 to 22% in dizygotic twins, suggesting a significant genetic component (reviewed by Wirdefeldt *et al.*[5]).

Family Studies

Families have been described in which PD follows a clear Mendelian inheritance pattern of autosomal dominant transmission. In some of these families, single gene defects have been identified (see below). The rarity of such cases, however, does not necessarily reflect the genetic contribution to the PD population as a whole. In fact, when familial aggregation of the disease was studied on a population level, first degree relatives of patients with PD were found to have a higher risk of developing PD, although odds ratios vary considerably between studies, ranging from about 1.5 to 7.5 or even higher. The discrepancies are probably mostly related to patient selection (e.g. population based versus referral center based). Familial aggregation was usually stronger in early-onset than in late-onset disease.

Environmental Risk Factors

Epidemiological studies have provided only limited evidence for specific environmental exposures as risk factors for PD, with two notable exceptions: smoking and coffee drinking. Many studies have indicated that both act as protective factors in a dose-dependent fashion. Case–control studies have provided strong evidence that the relative risk of developing PD in smokers compared with those who had never smoked was reduced to 0.6–0.3. That is, PD was 40–70% less likely to occur in smokers than in non-smokers. For any given duration of smoking, the number of cigarettes per day was not related to PD risk, whereas increasing duration was found to be associated with decreased PD risk for a given smoking intensity. Similarly, many studies have shown a protective effect of coffee drinking on PD risk. In a meta-analysis of eight case–control and five cohort studies, the pooled relative risk for coffee drinkers was 0.69, with a dosage effect of 0.75 per three additional cups of coffee per day.[7]

A possible relationship between environmental toxins, particularly herbicides and pesticides, and PD was identified in the 1980s, when it was discovered that exposure to 1-methyl-4-phenyl-1,2,3,6,-tetrahydropyridine (MPTP), a mitochondrial toxin structurally related to the herbicide paraquat, resulted in degeneration of dopaminergic neurons in humans and experimental animals, causing chronic parkinsonism.[8] The suspicion was further fueled by studies showing a higher prevalence of PD in agricultural regions and/or associated with well water drinking, suggesting that exposure to pesticides, herbicides, insecticides, and other potential environmental toxins may increase the risk of developing PD. A meta-analysis of 19 case–control studies published between 1998 and 1999 reported that the pooled risk of PD related to exposure to pesticides was increased twofold compared with non-exposed individuals (reviewed by Wirdefeldt *et al.*[5]), and a more recent study of pesticide use in agricultural areas in central California found that use of pesticides raises the PD risk by a factor of three. Evidence for an influence of other lifestyle factors on PD risk, including specific diets, physical activity, body mass index, tea drinking, or alcohol use, is limited and controversial.[5]

Other Associations

Several lines of evidence point to a role of inflammation in the development of PD (see Chapter 25). These include the presence of activated microglial cells in autopsy specimens from patients with PD, increased levels of inflammatory markers in the serum and/or cerebrospinal fluid of PD patients, and genetic evidence of an association with polymorphisms in genes related to inflammatory pathways. The influence of the use of non-steroidal anti-inflammatory agents on the occurrence of PD has also been studied. Although some of the studies are contradictory, most show a decreased risk of PD in patients who regularly use non-aspirin, non-steroidal anti-inflammatory drugs.

Several reports described an increased incidence of a history of head trauma in patients with PD, although other studies have failed to confirm this observation. Both positive and negative associations with cancer have also been described. There is fairly consistent evidence for a lower risk of at least some cancers among PD patients. with a notable exception of a positive association between PD and skin cancers, including melanomas.

Genetic Causation of Parkinson Disease

Genetic research in the past several years has advanced our knowledge of the disease etiology and the associated molecular pathways. These findings have also provided insight into the molecular events leading to neurodegeneration in general, with evidence emerging that at least some of the pathways identified in the rare genetic variants of PD are also important in the common sporadic form of the disease.

Autosomal Dominant Forms of Parkinson Disease

The discovery of the first PD-causing mutation in 1997[9] initiated a new era in PD research. The mutation was identified in a large family (referred to as the Contursi kindred) with an autosomal dominantly inherited form of PD in a gene called α-*synuclein* (the gene is abbreviated as *SNCA*, the protein as αSYN), with a single base-pair change from alanine (A) to threonine (T) at position 53 of αSYN (A53T). This discovery soon led to the recognition that αSYN is the main constituent of the characteristic intracytoplasmic inclusions, long considered to be a hallmark of PD, the Lewy body.[1] This finding has put αSYN at the center of the current understanding of the molecular pathogenesis of PD, although only very few further pathogenic point mutations in the *SNCA* gene have been detected so far. Further insight into the link between *SNCA* and PD came from the discovery that duplications and triplications of the (otherwise intact) entire *SNCA* gene[10] can also lead to dominantly inherited PD (with a dose-dependent effect: patients with triplications have an earlier onset than those with duplications), showing that even the wild-type protein can be pathogenic, if present in excessive concentrations. The currently favored hypothesis states that either amino acid changes in αSYN with a subsequent alteration of its physicochemical properties or overexpression of the gene may lead to an increased tendency of the protein to assume a β-pleated sheet instead of an α-helical conformation, which may then lead to the formation of oligomeric and fibrillar aggregates, eventually resulting in neuronal dysfunction and cell death (Fig. 19.3). This sequence of events has been recapitulated in a number of animal models. However, the precise relationship between mutations, aggregate formation, and their toxicity remains poorly understood.

While familial PD caused by *SNCA* mutations is exceedingly rare, mutations in another gene, the gene for the leucine-rich repeat kinase-2 (*LRRK2*),[11] were found to be a much more common cause of PD, accounting for 3–15% of familial cases of the disease, and for about 1–2% of "sporadic" cases in individuals of European descent. An even higher prevalence of a specific mutation, G2019S, was reported to occur in up to 20% in Ashkenazi Jewish and 40% of North African Arab populations, in both sporadic and familial cases.[12] Cases with *LRRK2* mutations generally have a somewhat later onset than most other Mendelian forms of PD (average about 59 years, similar to the sporadic disease). Both share, at least in the majority of cases, the typical αSYN pathology with Lewy bodies and Lewy neurites, although a few LRRK2 cases have been published with nigral degeneration without Lewy bodies or even tau pathology. The function(s) of LRRK2 are not known, but it has been suggested that it associates with membranes and that it may be involved in presynaptic vesicle trafficking, cytoskeletal dynamics, autophagy, or the maintenance of the nuclear architecture. The way in which *LRRK2* mutations cause PD is also not clear. LRRK2 is a large multidomain protein that contains several functionally important regions, including – in addition to several protein–protein interaction domains – a "Ras of complex" protein (ROC) GTPase, and a kinase domain. The known pathogenic mutations in *LRRK2* are all located in the ROC, C-terminal of ROC (COR), and kinase domains. There is evidence that increased kinase activity of LRRK2 or the GTPase activity of LRRK2 may contribute to the toxicity of pathogenic LRRK2 mutants.

Mutations in *SNCA* and *LRRK2* explain only a fraction of the familial PD cases with apparent dominant inheritance. Although large families suitable for a classic linkage analysis/positional cloning approach are very rare in any late-onset disorders such as PD, it is likely that new exome sequencing technologies will reveal a number of additional genes in the yet unclassified families.

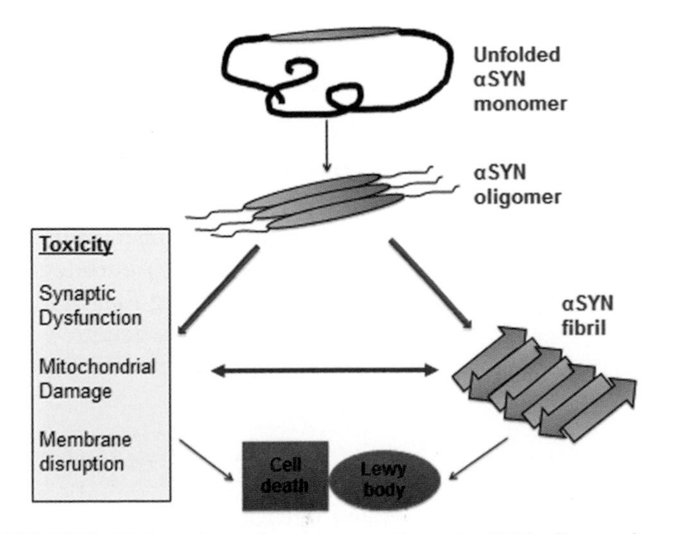

FIGURE 19.3 **Schematic representation of αSYN oligomerization, fibril formation, and Lewy body formation.** It is thought that the toxic moieties are αSYN oligomers, rather than mature aggregates as found in Lewy bodies.

Autosomal Recessive Forms of Parkinson Disease

In 1998, the identification of a gene mutated in an autosomal recessive form of PD, autosomal recessive juvenile Parkinson disease (ARJP), was reported.[13] The age at onset in ARJP cases is usually below 40 years, but the disease sometimes starts even in the second decade of life, with a clinical picture similar to sporadic PD. The first ARJP gene was named *parkin*, and its function was identified as that of an E_3 ubiquitin ligase. Later, two other genes causing similar early-onset recessive forms of PD were identified: *PINK1* and *DJ-1*.[14] Flies deficient in *parkin* and *PINK1* were later found to have mitochondrial dysfunction, with PINK1 apparently acting upstream of parkin in the same pathway. A currently favored mechanistic model suggests that cytoplasmic parkin is recruited to damaged mitochondria, where it is phosphorylated by the mitochondrial kinase PINK1, and subsequently ubiquitinates other mitochondrial membrane proteins, thereby promoting the elimination of dysfunctional mitochondria through a specific form of autophagy, called mitophagy. As damaged mitochondria produce increased amounts of toxic oxygen radicals, a failure of their clearance will lead to increasing cellular damage. These findings are particularly interesting in light of earlier findings identifying mitochondrial dysfunction as a possible etiological factor in sporadic PD. A single autopsy of a case with PINK1-related ARJP showed extensive Lewy pathology, whereas the majority of the few cases with *parkin* mutations did not. This discrepancy still needs to be resolved.

Genetic Risk Factors for Sporadic Parkinson Disease

Several genome-wide association studies (GWAS) in large cohorts of patients with sporadic PD and controls have been carried out to identify the contribution of common genetic variants to disease risk. More than 20 risk loci have been identified, generally acting with an odds ratio of 1.2 to 1.5.[15] However, even if the strongest risk alleles are combined, the relative risk of developing the disease is increased by no more than two- to four-fold. The two strongest risk factors for sporadic PD turned out to be variants in *SNCA* and microtubule-associated protein tau (MAPT). Thus, the GWAS findings also provided evidence that the close resemblance between familial and sporadic forms on the clinical and pathological level also has its correspondence in their genetic underpinning. The mechanism by which these variants act is not understood.

A considerably stronger genetic risk factor for PD has been found by astute clinical observation. It had previously been observed that some patients with Gaucher disease had neurological sequelae and Lewy pathology on postmortem examination. In addition, relatives of patients with Gaucher disease had PD more frequently than would be expected. A mutation in one allele of β-glucocerebrosidase (*GBA*), the gene that causes Gaucher disease when mutated in a homozygous or compound heterozygous fashion, increases the risk of PD about five-fold.[16] It is possible that point mutations in the *GBA* gene lead to misprocessing of the protein in the endoplasmic reticulum and result in a disturbance of protein homeostasis. This, in turn, may affect αSYN processing, thereby initiating the pathogenic cascade leading to PD.

Genetic Models for Parkinson Disease

The identification of genes causing PD has provided researchers with the opportunity to generate animal models for the disease by introducing mutated versions or eliminating (knocking out) the respective genes in experimental organisms. These models may ultimately provide a window into the molecular pathogenesis of PD on the cellular level. Many different model organisms have been reported, from yeast and *Caenorhabditis elegans* to *Drosophila*, mice, rats, and even primates. However, although in some cases interesting aspects of involved pathogenic mechanisms have been uncovered, most of these models have been disappointing as they failed to recapitulate some of the specific aspects of the human disease. For example, a number of transgenic mouse lines expressing mutated or wild-type human αSYN under different promoters have been generated. These models usually show somatodendritic accumulation of transgenic αSYN and an age-dependent motor phenotype, but fail to display the predilection for the dopaminergic system characteristic of the human disease. Similarly, a bacterial artificial chromosome transgenic model for LRRK2 displayed minimal neurodegeneration at best, making this a weak model of PD.

With progress in stem cell biology, the generation of human induced pluripotent stem cells (iPSCs) will provide a further, complementary system to study the molecular consequences of disease-causing mutations. It has already been shown that transfection of differentiated patient-derived fibroblasts with just four transcription factors initiates a process that leads to a reprogramming of these cells to a cell type that resembles, in many ways, embryonic stem cells. Using a cocktail of factors, these iPSCs can then be differentiated into a host of differentiated cell types including dopaminergic neurons (Fig. 19.4). This technology opens the possibility of generating *ex vivo* human cellular models for neurodegenerative diseases, by culturing, reprogramming, and transdifferentiating fibroblasts of patients with genetically defined forms of the disease. Although much remains to be learned about the proper use of these novel models, initial results are promising.[17]

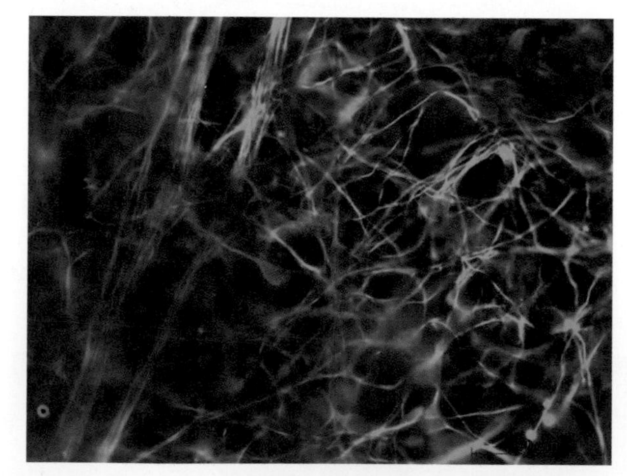

FIGURE 19.4 **Neurons *in vitro*, differentiated from induced pluripotent stem cells from a patient with Parkinson disease.** Neuron-specific tubulin III is stained in green, tyrosine hydroxylase (a marker of dopaminergic cells) in red. Only a minority of neurons are positive for tyrosine hydroxylase. Cell nuclei are stained blue.

Selective Neuronal Degeneration in Parkinson Disease

Most PD-associated genes, such as *SNCA*, *LRRK2*, and *parkin*, are expressed throughout the nervous system and even in other tissues. Protein aggregates are generally found widely distributed in the nervous system. A consistent feature in PD is the fact that dopamine neurons appear to be particularly vulnerable.[1] This phenomenon has been widely studied, and the genes involved in monogenetic PD, as described above, as well as the study of the properties of dopaminergic neurons, both shed light on this interesting question. The discovery of loss-of-function mutations in *parkin* and *PINK1* as causes for ARJP and the elucidation of their role in mitochondrial quality control have refocused interest on mitochondrial dysfunction as one of the final common steps to neurodegeneration in PD. A mitochondrial complex I deficiency selective to the substantia nigra has been long recognized. Disturbed homeostasis of PD-relevant proteins may be a contributing factor, as mitochondrial abnormalities have been observed in transgenic mice overexpressing mutant αSYN. Deficiencies of protein degradation, either by an impaired ubiquitin–proteasomal system or by decreased chaperone-mediated autophagy, have been observed in models of PD, potentially further increasing cellular stress.

As these pathways are not restricted to the substantia nigra, the specific vulnerability of dopaminergic neurons may be due, in part, to the intrinsic properties of these cells.[18] Dopamine neurons show a number of features that may explain their increased sensitivity: these neurons exhibit spontaneous bursting activity (also discussed below), which leads to unusually high energy demand and also to high calcium influx and consequently mitochondrial oxidative stress. Energy demand is further increased by the anatomic properties of dopamine neurons. It has been estimated that each of these highly arborized cells projecting to the striatum gives rise to more than 1,000,000 synapses.[18]

Pathology and Pathogenesis

Degeneration of melanized dopaminergic neurons in the SNc and intracytoplasmic αSYN-positive aggregates in the form of Lewy bodies and Lewy neurites are the pathological hallmarks of PD (Fig. 19.5). However, Lewy pathology is widespread in end-stage PD, with αSYN aggregates being found in basal forebrain cholinergic and brainstem norepinephrine and serotonin neurons, as well as neurons of the olfactory system, the cerebral cortex, and even the spinal cord and the peripheral autonomic nervous system, and does not necessarily correlate with cell death. Braak and co-workers have shown that neurons are likely to be affected in a sequential manner, starting in the olfactory bulb and intestine and spreading to some of the lower brainstem nuclei, such as the dorsal nucleus of the vagal nerve, and subsequently ascending via the midbrain to thalamic and cortical areas (Fig. 19.6).[1] This pattern of spreading of pathology may underlie the observation that some non-motor symptoms may precede the typical motor manifestations by many years. It is only when this ascending pathology has reached the midbrain (Braak stage 3) that dopamine neurons are affected, eventually leading to dopamine deficiency in their main projection area, the striatum, which in turn is the direct cause of most of the motor symptoms of the disease (Fig. 19.6).

There is no simple one-to-one relationship between Lewy body pathology and PD. Lewy pathology occurs in about 10% of elderly individuals without signs and symptoms of PD during life, a condition called incidental Lewy bodies (ILBs). Whereas ILBs may denote a preclinical stage of PD, ILBs with a wide range of distribution and density can occur in clinically healthy individuals. Another possibility is that Lewy body formation itself is not part of the disease pathogenesis, but represents a protective mechanism that helps resilient neurons to deal with misfolded αSYN.[20] Furthermore, αSYN pathology is not found in most patients with parkinsonism caused by *parkin* mutations and also not in some with PD due to *LRRK2* mutations,[11] suggesting that the disease process that leads to nigral degeneration is not invariably linked to the formation of mature Lewy bodies. In fact, there is considerable evidence that lower molecular weight αSYN oligomers rather than large aggregates are crucial in the pathogenesis of PD. If this is the case, the formation of αSYN oligomers and of Lewy bodies could both be secondary to an initial problem with αSYN homeostasis and indicate excessive and/or mislocalized

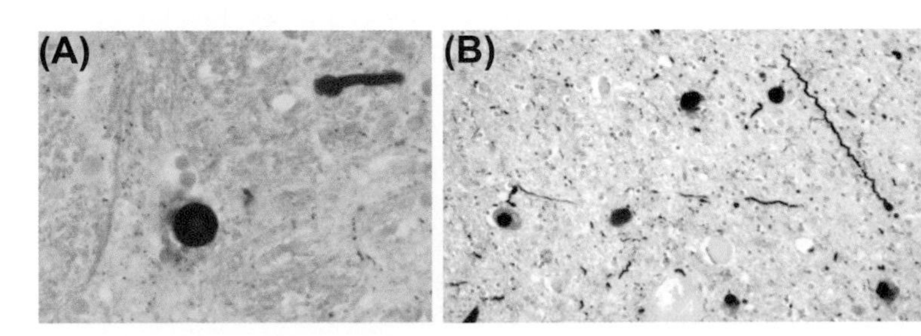

FIGURE 19.5 **Classical histopathological features of Parkinson disease.** (A) Micrograph of a hematoxylin and eosin stained section of the midbrain, showing the typical ring-like structure of a brainstem Lewy body. (B) Micrograph of an αSYN staining of a midbrain section showing Lewy bodies and Lewy neurites, i.e. proteinaceous material found in degenerating axons. *Source: Images courtesy of Prof. Dr Manuela Neumann, Tübingen, Germany.*

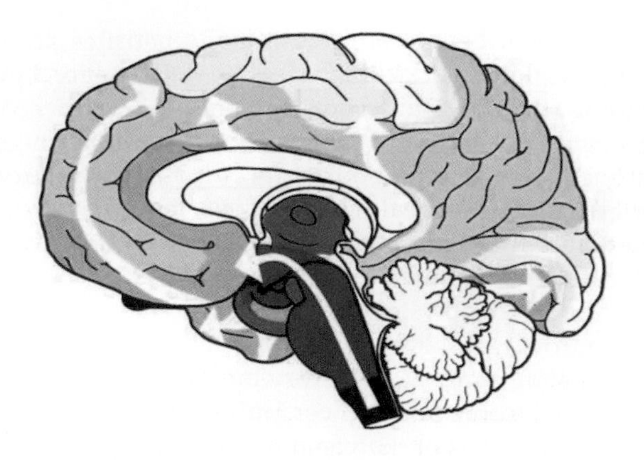

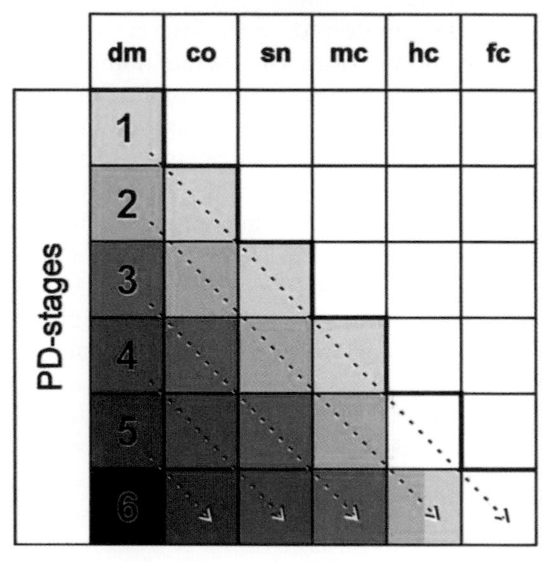

FIGURE 19.6 **Progression of Parkinson disease (PD)-related intraneuronal pathology.** Top: The pathological process targets specific subcortical and cortical induction sites. Lesions initially occur in the dorsal IX/X motor nucleus and frequently in the anterior olfactory nucleus. Thereafter, less susceptible brain structures gradually become involved (white arrows). Bottom: Topographic expansion of the lesions and growing severity on the part of the overall pathology. co: ceruleus–subceruleus complex; dm: dorsal motor nucleus of the glossopharyngeal and vagal nerves; fc: first order sensory association areas, premotor areas, as well as primary sensory and motor fields; hc: high order sensory association areas and prefrontal fields; mc: anteromedial temporal mesocortex; sn: substantia nigra. *Source: Reproduced from Braak* et al. Neurobiol Aging. *2003;24:197–211,*[19] *with permission.*

accumulation of abnormally folded αSYN. Subsequent events may be governed by different modifying factors and proceed at different speed and intensity, explaining the apparent discrepancies between Lewy bodies and neuronal dysfunction and damage in a proportion of cases.

The seminal findings of Braak and colleagues, demonstrating an apparent "ascending" evolution of Lewy pathology, also sparked the concept of a cell-to-cell transmission of αSYN pathology, similar to processes in prion diseases such as kuru or Creutzfeldt–Jakob disease (CJD) (see Chapter 23). Prion diseases are an example of the astonishing fact that a disease can be both inherited (approximately 10% of cases of CJD are familial and are caused by mutations in the prion protein gene) and transmissible. The hypothesis that prion-like transmission may be operative in PD has received support from autopsy findings in two cases of PD patients who had received heterologous fetal grafts into the striatum years before death (see Chapter 23). Surviving embryo-derived neuronal cells in these patients showed Lewy body-like pathology, staining positively with αSYN antibodies at autopsy. Since then, evidence has accumulated that the spreading of protein aggregates may be a relevant mechanism in the development of PD pathology. αSYN monomers and oligomers have been shown to be released from neurons in a calcium-dependent manner, have been found in cerebrospinal fluid, and could be shown to be taken up by endocytosis and seed αSYN aggregation in neurons.

A coherent picture of the molecular pathogenesis of PD is beginning to emerge. Overexpression of αSYN, due to mutations or genetic polymorphisms of the gene, abnormal intracellular trafficking on the translational or post-translational level, or environmental influences, leads to a local (probably presynaptic) increase in αSYN concentration. This in itself may not reach a critical threshold. However, protein homeostatic mechanisms, such as the action of molecular chaperones, proteasomal protein degradation, or autophagy, deteriorate with advancing age, which is well recognized as the strongest risk factor for neurodegeneration. Also strongly correlated with age, mitochondrial integrity is increasingly impaired, leading to additional oxidative stress. Under these multiple unfavorable influences, the aggregation-prone αSYN protein then assumes a β-sheet structure and forms oligomers. Once a seed of these oligomers is formed, it will recruit other αSYN molecules, owing to their templating prion-like properties, both within the cell and to neighboring neurons.[21] Neuronal demise follows, but this again is probably a multifactorial process, involving cell-autonomous (e.g. stress from unfolded proteins, energy failure, and reduction of antioxidants) and non-cell-autonomous (e.g. inflammatory) processes.

Relationship between Parkinson Disease and Multiple Systems Atrophy

MSA is a sporadic neurodegenerative disease that shares several clinical and neuropathological features with PD. Clinically, two main forms are distinguished: MSA of the parkinsonian type (MSA-P) and MSA of the cerebellar type (MSA-C). Whereas MSA-P is more common than MSA-C in European populations, this ratio seems to be reversed in Japan. Although motor symptoms can thus be variable, the cardinal symptom of MSA recognized in recent diagnostic criteria is that of autonomic failure, manifesting with orthostatic blood pressure dysregulation, urinary retention or incontinence, or disturbances of sweating. In contrast to PD, response to levodopa treatment is usually poor, although some limited therapeutic benefit can be seen in up to 30% of cases.[22] The pathological picture is dominated by neuronal and glial pathology, with αSYN-positive aggregates mainly in oligodendroglial cells, called glial cytoplasmic inclusions.

Patients with several mutations of the *SNCA* gene, such as the A53T point mutation or the SNCA locus triplication, generally considered to be forms of hereditary PD, share clinical and pathological features with MSA, including prominent autonomic disturbances and also oligodendroglial αSYN pathology, suggesting that both diseases may share not only features but also disease mechanisms.

Pathophysiology

Research into the pathophysiology of PD has been greatly facilitated by animal models. The earliest models relied on systemic injections of reversible blockers of dopamine synthesis (e.g. the tyrosine hydroxylase blocker α-methyl-*p*-tyrosine), release (e.g. the vesicular monoamine-transporter blocker reserpine) or dopamine receptor blockers (e.g. haloperidol). Studies in rodents treated with these agents were essential for the discovery of the dopaminergic basis of parkinsonism in the 1950s, but have the disadvantage of pharmacological and spatial non-selectivity. In subsequent models, permanent lesions of dopaminergic neurons were generated, using toxins such as the locally injected 6-hydroxydopamine (6-OHDA) or the systemically administered compound MPTP. The toxicity of MPTP was first discovered in humans in the early 1980s.[8] Although not toxic itself, MPTP is metabolized in the brain into the toxin MPP^+ by the enzyme monoamine oxidase type B (MAO-B). MPP^+ is then selectively taken up by dopaminergic cells of the SNc. MPP^+ interferes with complex I of the electron transport chain, which leads to cell death. Since its discovery, MPTP has become an essential research tool in non-human primates, in which it produces a very convincing parkinsonian phenotype.[8] Lesions produced with MPTP in monkeys were initially thought to be highly specific for the dopaminergic system, but there is now compelling evidence that other brain areas, such as the serotonergic dorsal raphe, the noradrenergic locus ceruleus, and the central median–parafascicular (CM-Pf) of the thalamus, degenerate in MPTP-treated animals just as they do in PD, further extending the reliability and usefulness of this model. Toxin-induced models of PD have been criticized as not mimicking the actual etiology of the disease. However, 6-OHDA acts by increasing oxidative stress and MPTP by inhibiting mitochondrial complex I, both of which are likely to play a role in the pathophysiology of PD. Although the onset of degeneration caused by such toxins is much more rapid than is the slow, progressive degeneration associated with PD, if one can reduce oxidative stress or mitochondrial dysfunction caused by a toxin, a comparable strategy is worth exploring in treating the disease itself. Finally, it is undeniable that toxin-based models have been invaluable in the development of symptomatic treatments.

Relevant Anatomical Considerations

The basal ganglia are components of massive parallel and largely closed cortical–subcortical circuits (Fig. 19.7), in which information is sent from different cortical areas to spatially separate domains of the basal ganglia, processed, and then returned to the frontal cortical area of origin via the thalamus.[24] These domains include, broadly, motor, oculomotor, prefrontal (or associative), and limbic circuits. Each circuit is, in turn, thought to be comprised of multiple functionally and anatomically distinct subcircuits, taking origin from separate cortical areas and traversing largely parallel pathways through the basal ganglia and thalamus, and then returning to the cortical area of origin. The pattern of functional connectivity between the cerebral cortex and the striatum that was identified in animal studies has been confirmed in PET and functional magnetic resonance imaging studies in humans.

Although the evidence for parallel processing within the broad functional domains is strong, there is also evidence for some integration or "funneling" within the individual domains. In addition, limbic information appears to cross circuit boundaries.[25]

The motor aspects of PD are believed to result mostly from disturbances in the basal ganglia thalamocortical "motor circuit".[26] The motor circuit (Figs 19.7 and 19.8, left side) originates from precentral and postcentral sensorimotor areas, including the primary motor cortex, supplementary motor area, premotor cortex, cingulate motor area, and related postcentral sensory cortical areas. The circuit then engages specific somatotopically organized motor portions of the basal ganglia, including the putamen, the sensorimotor area of the striatum, postcommissural putamen, dorsolateral subthalamic nucleus (STN), and ventral posterolateral internal segment of the globus pallidus (GPi). The "motor" projections from GPi reach the ventrolateral nucleus in the thalamus (VL), which then projects back to the primary motor cortex and the various premotor areas. It is noteworthy that although the striatum receives input from both precentral and postcentral cortex, output is directed almost entirely to specific areas in the frontal lobe.

The striatum and STN are the main entry points for cortical and thalamic inputs to the basal ganglia, and the GPi and the substantia nigra pars reticulata (SNr) provide basal ganglia output to the thalamus and brainstem. The

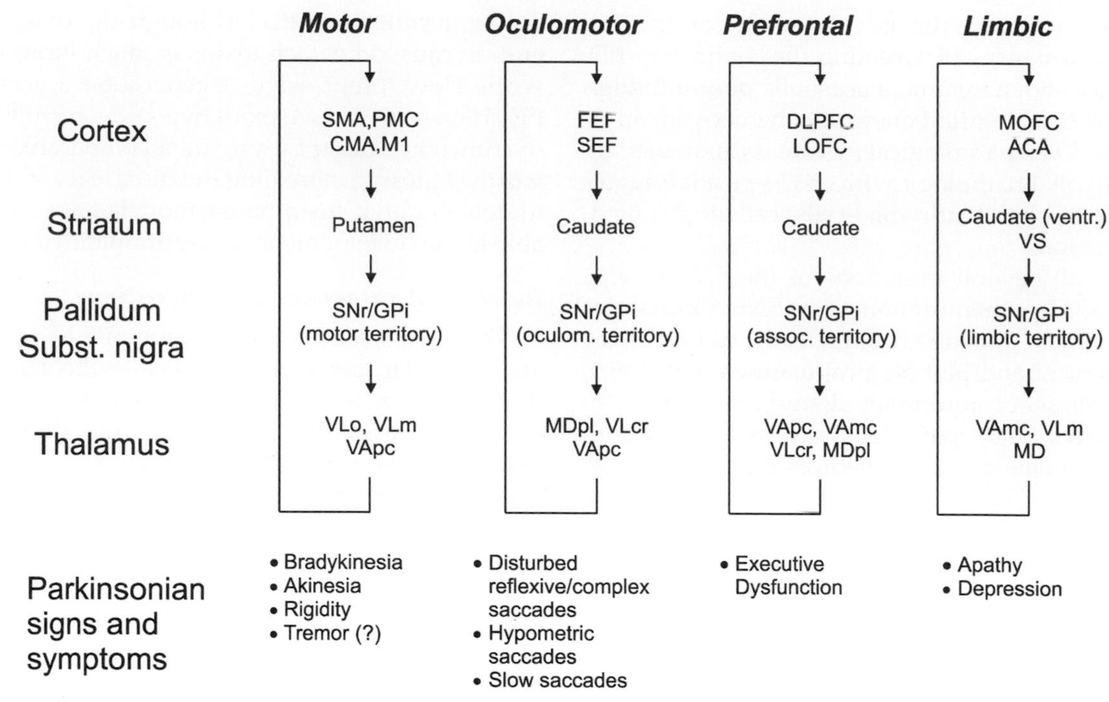

FIGURE 19.7 **Intrinsic anatomy of the parallel basal ganglia–thalamocortical circuits and the proposed relationship to parkinsonian motor signs and symptoms.** ACA: anterior cingulate area; CMA: cingulate motor area; DLPFC: dorsolateral prefrontal cortex; FEF: frontal eye fields; GPi: internal globus pallidus; LOFC: lateral orbitofrontal cortex; M1: primary motor cortex; MDpl: mediodorsal nucleus of thalamus, pars lateralis; MOFC: medial orbitofrontal cortex; PMC: premotor cortex; SEF: supplementary eye field; SMA: supplementary motor area; SNr: substantia nigra pars reticulata; VAmc: ventral anterior nucleus of thalamus, pars magnocellularis; VApc: ventral anterior nucleus of thalamus, pars parvocellularis; VLcr: ventrolateral nucleus of thalamus, pars caudalis, rostral division; VLm: ventrolateral nucleus of thalamus, pars medialis; VLo: ventrolateral nucleus of thalamus, pars oralis; VS: ventral striatum. *Source: Reproduced from Wichmann and Delong* Neuron. *2006;52:197–204,*[23] *with permission.*

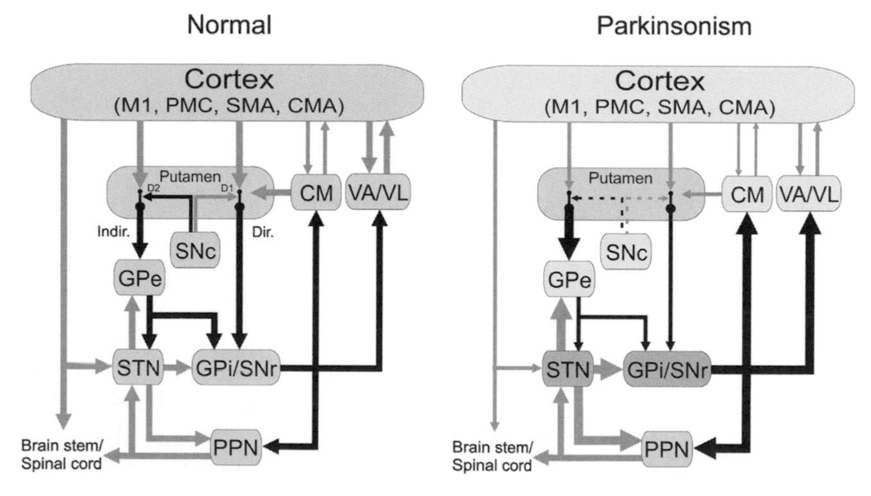

FIGURE 19.8 **Parkinsonism-related changes in overall activity (rate model) in the basal ganglia–thalamocortical motor circuit.** Black arrows indicate inhibitory connections and gray arrows indicate excitatory connections. The thickness of the arrows corresponds to their presumed activity. CM: centromedian nucleus of thalamus; CMA: cingulate motor area; Dir.: direct pathway; D1 and D2: dopamine receptor subtypes; GPe: external globus pallidus; GPi: internal globus pallidus; Indir.: indirect pathway; M1: primary motor cortex; PMC: premotor cortex; PPN: pedunculopontine nucleus; SMA: supplementary motor area; SNc: substantia nigra pars compacta; SNr: substantia nigra pars reticulata; STN: subthalamic nucleus; VA: ventral anterior nucleus of thalamus; VL: ventrolateral nucleus of thalamus. *Source: Reproduced from Galvan and Wichmann.* Clin Neurophysiol. *2008;119(7):1459–1474,*[26] *with permission.*

striatum and GPi/SNr are linked via two anatomically distinct pathways (Fig. 19.8). The direct pathway arises from striatal medium spiny neurons that project monosynaptically to neurons in GPi and SNr, while the indirect pathway arises from other striatal neurons that project to the external pallidal segment (GPe). The GPe then projects to GPi and SNr either directly or via the STN.

Modulation of the synaptic efficacy of corticostriatal projections is seen as the primary function of dopamine released from terminals of the nigrostriatal tract. In the healthy state, dopamine acts (via D_2 receptors) to inhibit information transfer at cortical synapses on striatal medium spiny neurons that give rise to the indirect pathway, and facilitates (via D_1 receptors) corticostriatal transmission on medium spiny neurons of the direct pathway. At least as a first approximation, the two pathways are thought to have opposing actions on the output of the basal ganglia (Fig. 19.8). Thus, activation of the direct pathway increases inhibition of the (tonically active) neurons in GPi and SNr. Because GPi/SNr projections to the thalamus and other areas are inhibitory, the reduced GPi/SNr activity will translate into disinhibition of the thalamic and other targets. Activation of the indirect pathway will have the opposite effect, resulting in a net increase in GPi/SNr activity, and greater inhibition of thalamic and brainstem targets of GPi/SNr output.

The STN also receives projections from cortical motor and non-motor areas. The cortical–STN projection and the subsequent STN–GPi/SNr projections are together referred to as the "hyperdirect" pathway, emphasizing that this pathway may provide a faster route for cortical input to modulate the activity of the output nuclei of the basal ganglia. There is evidence that the corticosubthalamic pathway may play a role in the inhibition or cancellation of ongoing movements.

In addition to the outputs to the thalamus, the basal ganglia output nuclei send projections to brainstem areas, such as the superior colliculus and the pedunculopontine nucleus (PPN). These projections are phylogenetically older than those to the thalamus and are found at the earliest stages of evolution, in birds and the earliest vertebrates such as the lamprey.[27] Moreover, evidence exists for corresponding parallel loop systems in these animals which resemble those that later developed with the expansion of the cerebral cortex.[28] Descending PPN projections reach other brainstem systems, and ascending PPN projections are returned to the basal ganglia, the thalamus, and the basal forebrain. The PPN is implicated in the regulation of gait and balance. The superior colliculus projection is involved with the control of saccadic and head and neck movements.

Other projection systems may also be important for our understanding of parkinsonism. For instance, a massive system of collaterals from the pallidothalamic and nigrothalamic projections reaches the thalamic caudal intralaminar nuclei (i.e. the centromedian and parafascicular nuclei, CM/Pf). These projections are not components of the transthalamic corticocortical re-entrant loops mentioned above, but function as a thalamostriatal positive-feedback system.[29] CM/Pf projections to the striatum are also implicated in the regulation of procedural learning and attention processes. These thalamostriatal projections degenerate very early in the course of PD, potentially contributing to the early executive dysfunction and procedural learning deficits in PD patients.

Finally, there is evolving evidence that the basal ganglia and cerebellum are more closely interconnected than previously thought.[30] Thus, the STN outflow appears to reach the cerebellum via projections to the pontine nuclei, and cerebellar output reaches the striatum via the thalamus. Anatomical, neuroimaging, and functional neurosurgical studies are providing increasing evidence that the cerebellum and the basal ganglia are involved in the pathophysiology of tremor at rest.

Pattern of Dopamine Loss in the Striatum

Pathology and imaging studies indicate that dopaminergic SNc neurons and their projections to the striatum degenerate slowly (over decades), with a caudal-to-rostral progression, starting in the caudal putamen, and progressing to involve the rostral putamen and caudate nucleus (see Fig. 19.2). Early loss in the putamen, the sensorimotor component of the motor loop, leads to the early features of PD. Recognizable motor or non-motor signs appear only after degeneration of the nigrostriatal neurons affecting at least 50% of these cells, corresponding to approximately a 70% loss of the striatal dopamine supply. The loss of the nigrostriatal projection not only leads to reduced dopamine levels in the striatum, but triggers secondary changes, such as alterations in the density and sensitivity of dopamine receptors, as well as a reduction in the density of dendritic spines on medium spiny neurons, which may alter corticostriatal transmission.

Discharge Rate Changes in the Basal Ganglia

An early circuit model of PD, the "rate" model (Fig. 19.8, right side), was based on the idea that dopamine loss would disturb the balance of activity in the direct and indirect pathways within the motor circuit of the basal ganglia (discussed by Galvan and Wichmann[26]). Loss of dopamine in the striatum was thought to lead to reduced discharge rates in striatal projection neurons of the direct pathway and to increased activity of striatal projection neurons of the indirect pathway, thus leading to increased net (inhibitory) output of GPi neurons upon recipient thalamic neurons, resulting in the hypokinetic features of PD.

The rate model was supported by single unit recording studies in MPTP-treated monkeys, studies in human patients undergoing electrophysiological studies as part of neurosurgical treatments, and optogenetic studies in rodents.[31] Together, these studies found increased firing rates in the GPi and STN and reduced rates in GPe, thalamus, and motor cortex (see examples in Fig. 19.9 and Galvan and Wichmann[26]). Animal studies using metabolic markers of neuronal activity also reported changes in the basal ganglia, thalamus, and motor cortices consistent with the rate model. One of the strongest lines of evidence in favor of the important role of overactivity of neurons in the STN and GPi was the fact that lesions of the STN as well as pallidotomies (GPi lesions involving the motor area) have immediate and dramatic anti-parkinsonian effects.[32] The rate model was highly important because it provided a strong rationale for the use of ablative procedures that targeted the motor loop in the STN and GPi in cases of PD (see below).

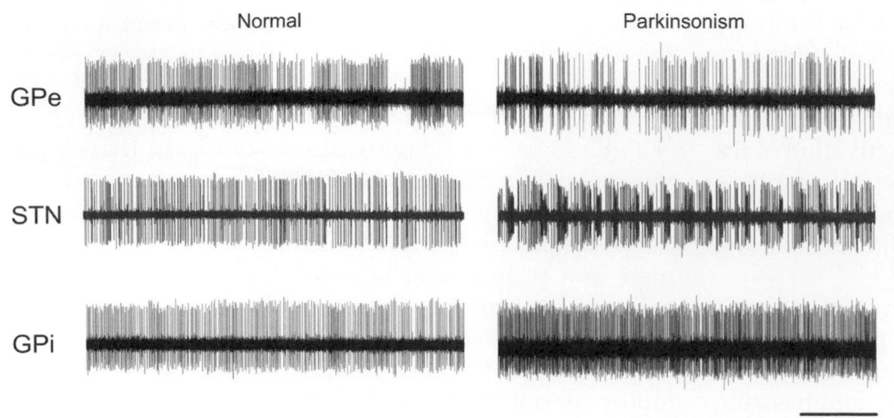

FIGURE 19.9 **Changes in the activity of single cells in external globus pallidus (GPe), subthalamic nucleus (STN), and internal globus pallidus (GPi) of 1-methyl-4-phenyl-1,2,3,6,-tetrahydropyridine (MPTP)-treated monkeys.** Shown are examples of separate neurons, recorded with standard extracellular electrophysiological recording methods in normal and parkinsonian animals. Each data segment is 5 s in duration. *Source: Reproduced from Galvan and Wichmann.* Clin Neurophysiol. *2008;119(7):1459–1474,*[26] *with permission.*

Changes in Activity Patterns of Basal Ganglia Networks

Although it was initially positively received, the rate model quickly fell out of favor since it was shown to be inconsistent with the finding that lesions of the motor thalamus did not result in akinesia, and that lesions of GPi did not lead to involuntary movements (dyskinesias). Since then greater emphasis has been now placed on abnormalities in firing patterns of output neurons, including an increased tendency of these neurons to discharge in burst, to exhibit oscillatory firing patterns (Fig. 19.9), and to fire synchronously with neighboring neurons.[26]

Increases in burst firing have been reported to occur in neurons in the GPi, GPe, and STN in neurotoxin models of PD, and in recordings from patients with PD. Increased burst firing has also been described for neurons in GPi-recipient regions of the thalamus and in motor cortical regions in parkinsonian primates. Burst discharges are often found combined with oscillatory firing patterns. Neural network simulations and tissue culture experiments suggest that oscillatory fluctuations in firing rates may arise as network phenomena. Oscillations are also observed in local field potential (LFP) recordings with macroelectrodes, which reflect synchronous synaptic membrane potential fluctuations of groups of neurons. LFPs can be recorded from deep brain stimulation (DBS) electrodes implanted in the basal ganglia of patients with PD. Analysis of oscillations in such LFP records has revealed the occurrence of oscillatory activity in the 10–35 Hz range (beta range) throughout the extrastriatal basal ganglia (specifically in the STN), which can be suppressed by dopaminergic replacement therapies (Fig. 19.10).[34] In parallel with the presence of beta-band oscillatory activities, gamma-band oscillatory activities (frequencies > 35 Hz) are found to be reduced in the basal ganglia and cortex of parkinsonian patients and in animal models of the disease. At least part of the motor disorder in parkinsonian patients may, thus, arise from an imbalance between the (prokinetic) gamma-band oscillations and the (antikinetic) beta-band oscillations in the basal ganglia–thalamocortical motor circuit.[35] A related abnormality throughout the basal ganglia, thalamus, and cortex is the presence of increased synchrony of neuronal discharge.

Many of the described abnormalities can be at least partially restored by treatment with dopaminergic agents, suggesting that they are the result of dopamine deficiency. However, damage to other neurotransmitter systems in PD, such as the loss of norepinephrinergic cells in the locus ceruleus and other catecholaminergic brain stem regions, or the aforementioned structural changes in the basal ganglia and associated areas, may also contribute to the pathophysiology.

Relationship of Rate and Pattern Changes to Parkinsonism

The relationship of the different electrophysiological findings to the parkinsonian motor signs remains uncertain. The fact that neurosurgical interventions that interrupt or alter the activity of the basal ganglia thalamocortical motor circuit are effective in treating the symptoms of the disease suggests that abnormalities in the basal ganglia are important for the expression of parkinsonian motor signs. Early studies suggested that some of the electrophysiological abnormalities occur before symptom onset, giving rise to the idea that parkinsonism manifests at a time when the remaining healthy brain systems are no longer able to compensate for abnormal basal ganglia activity patterns that are the result of dopamine loss. A link between abnormal basal ganglia activity patterns and parkinsonism is also supported by findings of experiments in patients with implanted DBS electrodes, in whom parkinsonism can be worsened by DBS at low frequencies. However, these results contrast with those of studies that found that abnormalities such as oscillatory synchronous discharge or LFP changes in

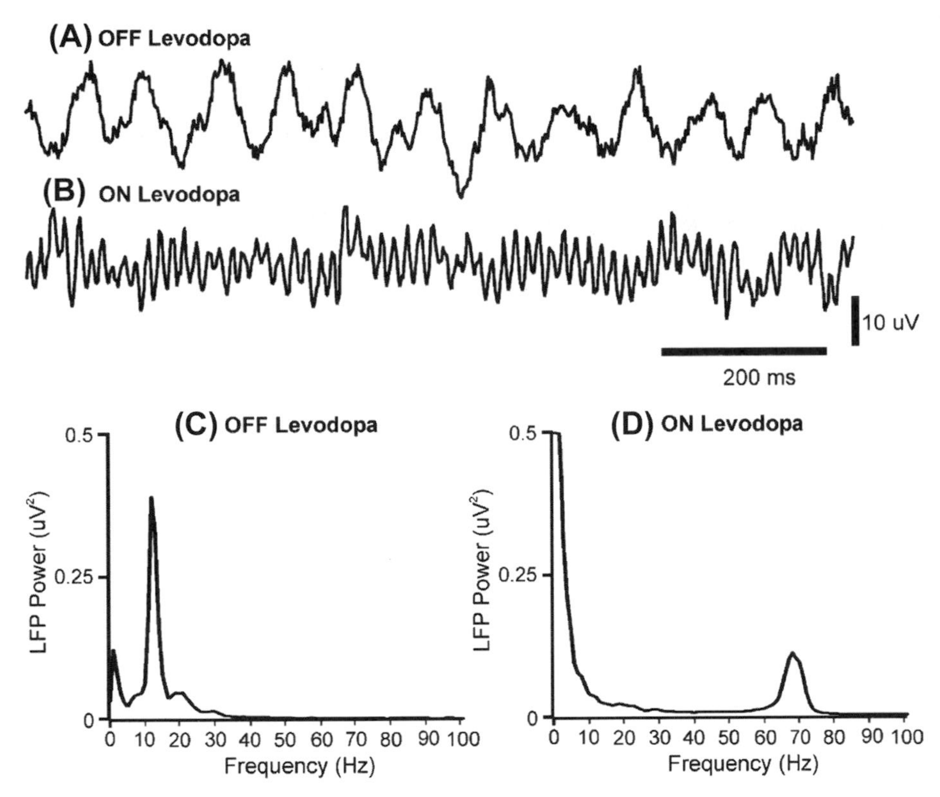

FIGURE 19.10 **Field potential signals recorded in a patient with Parkinson disease (PD) with a macroelectrode positioned in the subthalamic area.** (A) Field potential signals recorded after overnight withdrawal of medication. (B) Field potential signals recorded after subsequent levodopa challenge. (C) Power spectrum of field potentials recorded after overnight withdrawal of medication (140s recording). (D) Power spectrum of field potential signals recorded after subsequent levodopa challenge (140s recording). There was a spectral peak at around 13Hz off medication, and at around 70Hz after levodopa treatment. *Source: Reproduced from Brown and Williams.* Clin Neurophysiol. *2005;116:2510–2519,*[33] *with permission.*

the basal ganglia develop only after the onset of overt parkinsonism (discussed in Galvan and Wichmann[26]).

Role of Non-Motor Basal Ganglia Circuits in Parkinson Disease

Thus far, this discussion has considered largely the pathophysiological changes resulting from loss of dopamine in the motor circuit of the basal ganglia. However, it is likely that dopamine loss in other basal ganglia circuits also affects motor function as well as other aspects of behavior (Fig. 19.7). While dopamine probably modulates synaptic activity uniformly across the striatum, its loss in the different circuits would lead to different behavioral outcomes because of the different functions of these circuits.

For instance, one of the critical functions of dopamine that may be altered in PD is its role in learning and habit formation, requiring both associative and motor circuits. The motor circuit is thought to be involved prominently in learned or habitual behavior, while the associative circuit appears to be involved preferentially in motor learning and goal-directed behavior. During the active process of learning a motor task, increased activity occurs primarily in portions of the associative (prefrontal) circuit, while the later execution of previously learned behaviors is associated with activity in the motor circuit (Fig. 19.7). As mentioned earlier, the motor deficits in patients with PD are correlated with the prominent loss of dopamine in the posterior putamen, the striatal portion of the motor circuit, while the dopamine content of the caudate nucleus, the striatal portion of the associative (prefrontal) circuit, remains relatively intact until later stages of the disease (Fig. 19.2). The differential loss of dopamine in the two circuits may account for the progressive disruption of automatic or habitual movements in PD, and the greater reliance on the volitional, goal-directed mode of action control.[36]

The parallel loop organization also provides a substrate for understanding the oculomotor and non-motor behavioral changes that accompany PD (Fig. 19.7). One of the functions of the associative (prefrontal) circuit appears to be the organization of information in order to facilitate a behavioral response, while the anterior cingulate circuit (a component of the limbic circuit) is critical for motivated behavior, and the orbitofrontal circuit (another component of the limbic circuit) integrates relevant limbic and emotional information needed for an appropriate behavioral response to a given situation. The corresponding abnormalities, impaired executive functions, decreased motivation (apathy, depression), and impulsivity are all common manifestations of frontal–subcortical circuit dysfunction.

It should be noted that activity changes in these circuits in PD may also result from alterations in transmitter systems other than the dopaminergic system, such as the serotonergic and the norepinephrine systems, the effects of antiparkinsonian medications, or pathological abnormalities affecting the function of these circuits at locations outside the basal ganglia.

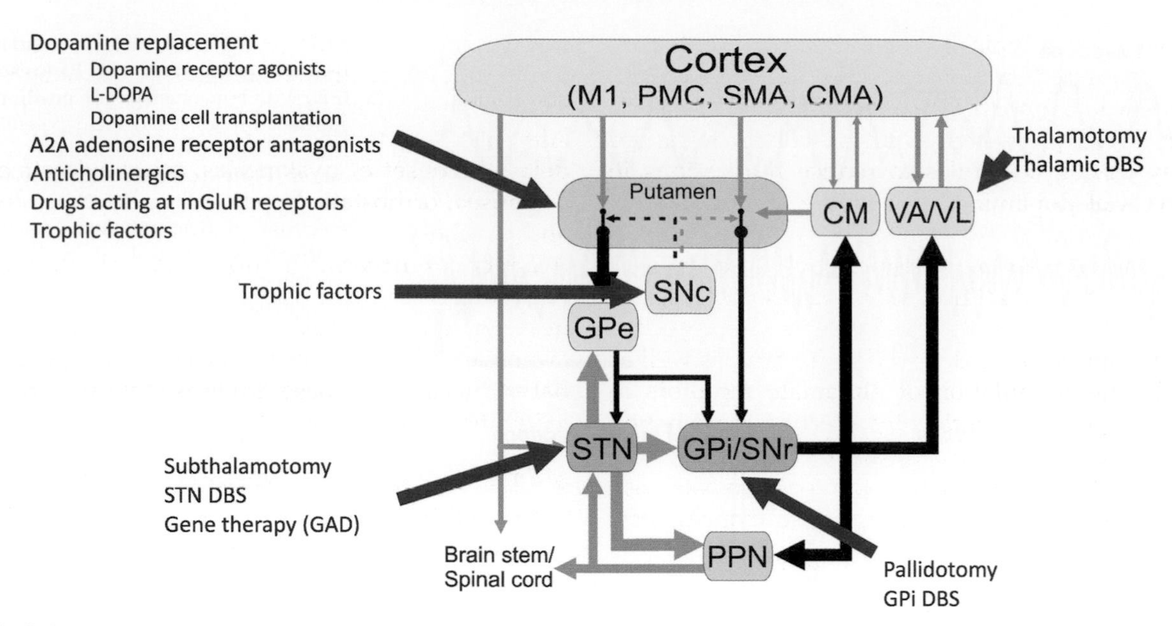

FIGURE 19.11 **Major areas of symptomatic therapies in Parkinson disease (PD).** CM: centromedian nucleus of thalamus; CMA: cingulate motor area; DBS: deep brain stimulation; GPe: external globus pallidus; GPi: internal globus pallidus; M1: primary motor cortex; mGluR: metabotropic glutamate receptor; PMC: premotor cortex; PPN: pedunculopontine nucleus; SMA: supplementary motor area; SNc: substantia nigra pars compacta; SNr: substantia nigra pars reticulata; STN: subthalamic nucleus; VA: ventral anterior nucleus of thalamus; VL: ventrolateral nucleus of thalamus.

TREATMENT OF PARKINSON DISEASE

An important goal of therapy in PD is to maintain function and quality of life, and to preserve independence while avoiding drug-induced complications.[37] The major approaches to the treatment of PD are shown in Fig. 19.11.

At least early in the course of the disease, the cardinal features of PD can be treated successfully with dopaminergic agents with relatively few side effects. Patients should be treated with these and other treatments as soon as symptoms begin to interfere with function, so that the maximal level of physical and mental activity can be maintained. Another obvious treatment goal should be to slow disease progression. However, thus far, the search for such disease-modifying treatments has not proven successful.

Physical Exercise

Before considering drug and surgical therapies, emphasis should be placed on the importance of physical exercise for PD.[38] All patients should be strongly encouraged to participate in regular exercise, for its general cardiovascular, musculoskeletal, metabolic, and neuropsychological effects. This goal should be maintained throughout the course of disease. Animal studies have shown that aerobic exercise increases cerebral catecholamine levels, alters the level of the dopamine transporter, enhances resistance to oxidative stress and neurotoxins, and increases levels of growth factors, such as brain-derived neurotrophic factor (BDNF) and glial cell-line derived neurotrophic factor (GDNF), which may be relevant to maintaining the health of dopaminergic and other neurons in the brain.

Drug Treatment

Levodopa

Levodopa, introduced in the 1960s, remains the single most effective agent for treating the symptoms of PD and can help patients remain active for years.[37] Levodopa is converted in the brain and in the periphery into dopamine. It is usually administered together with a peripherally acting blocker of the levodopa-converting enzyme, dopa decarboxylase, such as carbidopa or benserazide, to reduce peripheral dopamine-related side effects such as nausea or vomiting, and to optimize brain delivery.

While providing remarkable restoration of function, the long-term use of levodopa is frequently associated with a number of complications, including motor fluctuations, a term used to describe the appearance of parkinsonian signs experienced by many patients between doses. Fluctuations are due to the "wearing off" of benefit as well as the unpredictable and poorly understood sudden switches between "on" and "off" states.

Equally troublesome is the development of involuntary movements (dyskinesias) associated particularly with the peak effect of levodopa and, in some cases, the wearing off of the drug effect (diphasic dyskinesia).

Dyskinesias, which typically arise after 5 years or more of treatment, may become disruptive to normal motor function. The incidence of drug-induced dyskinesias has decreased greatly, however, as physicians have learned to modify dosing schedules to reduce fluctuations in blood levels of dopamine, to decrease the cumulative dose of levodopa, and to use other antiparkinsonian agents, such as dopamine receptor agonists (see below). The details of the induction and expression of dyskinesias remain elusive, but there is evidence that dyskinesias result from pulsatile delivery of the drug as well as non-continuous stimulation of glutamate receptors in the striatum, perhaps specifically affecting the function of direct-pathway medium spiny neurons.

Levodopa (and dopamine receptor agonists; see below) can also induce behavioral disturbances, particularly in young-onset PD patients.[37] These impulse control disorders include the dopamine dysregulation syndrome, characterized by an addictive overuse of medication, drug hoarding, and neglect of social interactions, work, and personal relations, as well as "punding" (obsessive, repetitive behaviors, often involving the handling or manipulation of objects related to the patient's prior interests or hobbies). Impulse control disorders also include a variety of behaviors that may significantly interfere with functioning, including pathological gambling, compulsive shopping, hypersexuality, and internet overuse.

Another side effect of dopaminergic agents, particularly in patients with pre-existing cognitive decline, is the development of psychotic symptoms, ranging from mild hallucinations to paranoid ideation. These symptoms can become very distressing to patients and their caregivers and are a frequent cause for hospitalization.

Although there has been a long-standing controversy regarding possible toxic effects of levodopa on dopamine neurons resulting from its oxidative metabolism, which generates free radicals, there is no clear evidence from clinical studies that levodopa adversely affects the progression of PD. Recent trials of levodopa/carbidopa suggest that there is no advantage in delaying treatment with levodopa and that early treatment may even be beneficial in terms of disease progression.

Dopamine Receptor Agonists

Dopamine receptor agonists[37] have been widely used for decades, as monotherapy or adjuvants to levodopa. The most commonly used agonists for treatment of PD act primarily on dopamine D_2-like receptors (D_2, D_3, and D_4 receptors), and include the orally available drugs pramipexole and ropinirole, the transdermally delivered drug rotigotine, and apomorphine, an injectable agent.

Compared with levodopa, dopamine receptor agonists offer the advantages of reliable absorption and transport into the brain, no requirement for further metabolism, and a longer half-life. These agents effectively treat the cardinal features of PD, although benefits do not reach those achievable by levodopa treatment. Although their use in newly diagnosed PD patients may delay the onset of dyskinesias, other side effects, such as nausea, orthostatic hypotension, daytime sleepiness, and psychosis, are more common than with levodopa treatment. Furthermore, dopamine receptor agonists, even more so than levodopa, are implicated in the development of impulse control disorders. The growing awareness of these complications has tempered the initial enthusiasm for these drugs as monotherapy and has led to the use of lower doses.

Monoamine Oxidase Type B Inhibitors

MAO-B inhibitors,[37] which reduce the breakdown of dopamine in the brain, are used primarily for symptomatic therapy but have also been investigated for potential neuroprotective benefit. The first trial of an MAO-B inhibitor as a neuroprotective agent was the Deprenyl and Tocopherol Antioxidative Therapy of Parkinsonism (DATATOP) trial in the 1980s. This placebo-controlled study showed that the MAO-B inhibitor deprenyl (selegiline) delayed the need for levodopa treatment in newly diagnosed patients. While this was initially thought to reflect a neuroprotective effect, it is now believed to relate to a mild symptomatic effect of the drug. Nonetheless, the DATATOP study highlighted the need to refine clinical methodology for assessing disease progression and led to subsequent studies of drugs for their potential to modify the course of PD.

The more recently released MAO-B inhibitor rasagiline has also been examined as a neuroprotective agent, but studies have produced controversial results. Rasagiline is approved for use as a symptomatic agent in early and advanced disease. Although much less effective than dopamine receptor agonists or levodopa, rasagiline is now frequently used in early PD as monotherapy and has been shown to modestly decrease symptoms. Both rasagiline and selegiline also reduce motor fluctuations.

Inhibitors of Catecholamine Ortho-Methyltransferase

The catecholamine *ortho*-methyltransferase (COMT) inhibitors entacapone and tolcapone augment the effects of levodopa by blocking the enzymatic degradation of levodopa in the periphery. When used in conjunction with carbidopa/levodopa, COMT inhibitors help to alleviate wearing-off symptoms and increase on-time by increasing the half-life of levodopa. Common side effects include gastrointestinal and typical dopaminergic side effects, including sleep disturbances and increased dyskinesias. Tolcapone is more potent than entacapone, but its use requires the monitoring of liver enzymes because of potential toxicity.

Amantadine

Introduced as an antiviral agent for the treatment of influenza in the 1960s, amantadine was serendipitously discovered to have antiparkinsonian effects (reviewed by Smith *et al.*[37]). Amantadine acts on multiple receptors in the brain. Its mild to moderate antiparkinsonian effects may relate to the fact that it has multiple sites of action including dopaminergic effects, and the blockade of acetylcholine and glutamate [*N*-methyl-D-aspartate (NMDA)] receptors. Amantadine is the only available drug with antidyskinetic effects, which is attributed to its non-competitive NMDA-receptor blocking properties. However, side effects of amantadine include anticholinergic effects, livedo reticularis (purplish discoloration of the skin), worsening of cognitive function, and even psychosis, particularly in patients with advanced disease.

Anticholinergic Drugs

Anticholinergic drugs are the oldest antiparkinsonian agents, dating back to the late nineteenth century, when the naturally occurring alkaloid atropine was first used in PD patients. In the 1950s, synthetic formulations of muscarinic receptor blockers became available, including benztropine, trihexyphenidyl, and ethopromazine. These agents have not been as rigorously tested as more recently introduced drugs, but, in general, their effects are modest compared with those of levodopa, with the exception of the treatment of tremor. Because the currently available anticholinergics block all subtypes of muscarinic receptor non-selectively, they have numerous side effects that limit their usefulness, particularly in older patients. The recent development of allosteric modulators of specific muscarinic receptor subtypes may allow selective targeting of muscarinic receptors in the basal ganglia associated with movement disorders (particularly M_4 and M_5), while sparing those associated with cognitive (M_1) and peripheral autonomic side effects (M_2 and M_3).

Drugs Undergoing Evaluation

Although a great deal of attention has been focused on developing drugs that act on the dopamine system, the side effects of many of these drugs are considerable. It is therefore important to explore agents that act on other transmitter systems free from such side effects. The following are examples of promising classes of such agents.

A_{2A} ADENOSINE RECEPTOR ANTAGONISTS

Adenosine is a ubiquitous purine, acting on G-protein coupled adenosine receptors. Of these, A_{2A} receptors are the most common subtype in the basal ganglia, and are expressed primarily in the dorsal striatum, nucleus accumbens, olfactory tubercle, and GPe.[37] In the striatum, A_{2A} receptors are colocalized with D_2 dopamine receptors and other constituents of the indirect pathway of the basal ganglia. The idea that A_{2A} receptor blockers might be a novel form of treatment had its origin in epidemiological data (see "Environmental Risk Factors", above) indicating that caffeine, a non-selective adenosine receptor antagonist, may have therapeutic and neuroprotective effects in PD. It is postulated that blockade of A_{2A} receptors at the striatopallidal γ-aminobutyric acidergic (GABAergic) synapse or at corticostriatal glutamatergic synapses may help to restore the balance between the direct and indirect pathways and, thereby, alleviate the motor symptoms of PD and, possibly, as found in animal studies, reduce levodopa-induced dyskinesias. Several A_{2A} receptor antagonists have been developed. Despite promising preclinical evidence, the testing of the first generation of these agents showed only modest benefits in human trials. Studies assessing the efficacy of later generation compounds are underway.

METABOTROPIC GLUTAMATE RECEPTOR LIGANDS

The G-protein coupled metabotropic glutamate receptors (mGluRs) are expressed throughout the basal ganglia, where they play a major role in regulating excitability and synaptic transmission. These receptors affect neuronal activity via a variety of mechanisms including the modulation of ion channels, release of intracellular calcium, and functional interactions with ionotropic glutamate receptors and other G-protein coupled receptors (D_2 dopamine receptors, A_{2A} adenosine receptors). Because of their varied effects, different pharmacological properties and specificity, the mGluRs are promising potential drug targets.[37] These drugs are better tolerated than drugs acting at ionotropic receptors because they have less drastic effects on synaptic transmission, and have a more circumscribed distribution in the brain.

Among the mGluR subtypes, $mGluR_5$ and $mGluR_4$ are currently being investigated as targets for PD therapy. The recent development of specific mGluR active drugs with favorable pharmacokinetic profile (including allosteric modulators), toxicity, and brain permeability may allow the use of these agents, particularly as symptomatic treatment and in treatment of drug-induced dyskinesias. Blockade of $mGluR_5$ has antiparkinsonian and antidyskinetic effects in a variety of animal models as well as early trials in patients. There is also preclinical evidence that the combination therapy of A_{2A} receptor and $mGluR_5$ receptor antagonists has synergistic antiparkinsonian efficacy. It will be of interest to study the benefits of smaller, subtherapeutic doses of levodopa in combination with A_{2A} and $mGluR_5$ antagonists.

AGENTS ACTING ON NOREPINEPHRINE TRANSMISSION

L-Threo-dihydroxyphenylserine (L-DOPS) is a precursor of norepinephrine, which was recently FDA approved for use in patients with autonomic failure and orthostatic hypotension. The drug may also have beneficial effects on freezing of gait.

TREATMENT OF NON-MOTOR ASPECTS OF PARKINSON DISEASE

Despite the significant impact of non-motor features of PD on the quality of life being well recognized, only a small number of well-controlled clinical trials has specifically addressed these aspects of the disease. It is important to realize that few of these are modified by levodopa or other antiparkinsonian agents. This topic will not be discussed here, but is covered in recent reviews.[39]

Surgery for Parkinson Disease

Functional Neurosurgery

Although ablative procedures such as thalamotomy and pallidotomy were used widely in the 1950s and 1960s, particularly for intractable tremor, they were largely abandoned following the introduction of levodopa. Since the 1990s, however, there has been a renaissance in functional neurosurgery for PD. A significant factor that encouraged the reintroduction of neurosurgical procedures (initially pallidotomy) for PD was the progress made in understanding the pathophysiological mechanisms underlying the disease, particularly the demonstration of abnormal and excessive activity in the STN and GPi and the success of subthalatomotomy in reversing parkinsonism in the primate MPTP model of PD.[32] These studies provided a clear rationale for surgical therapy. Subsequently, surgical approaches gained popularity after a novel surgical approach, DBS, which comprises chronic high-frequency electrical stimulation via implanted electrodes into the same targets used for ablation, was shown to have results similar to the ablative procedures.

DBS has largely replaced ablative procedures as an antiparkinsonian treatment because of its reversibility, adjustability, and less invasive nature.[40] It is an effective treatment for all of the cardinal motor features of PD, including bradykinesia, tremor, drug-induced motor fluctuations, and dyskinesias. It is generally agreed that a prerequisite for DBS also surgery for PD is the demonstration of a clear clinical response to levodopa treatment (with the exception of drug-resistant tremor). Success is dependent on patient selection, lead placement, and postoperative programming of the device. Contraindications to surgery include signs of atypical PD, dementia, and psychiatric disorders. DBS is a purely symptomatic treatment that does not appear to alter the progression of the disease. DBS also does not alter levodopa-refractory motor aspects of PD such as problems with speech and handwriting, freezing of gait, or postural instability.

The most common DBS targets are the motor portions of the GPi and STN, major nodes of the motor circuit. The procedure is usually performed bilaterally, although unilateral DBS is effective for asymmetrical symptoms and can even have significant bilateral effects. Most of the recent trials comparing DBS in STN or GPi have shown that both procedures similarly reduce parkinsonian motor signs, off time, drug-induced dyskinesias, and motor fluctuations. In patients with advanced PD, DBS at either target improves the patient's quality of life and has been shown to be more effective than medical management. In contrast to patients with GPi DBS, those with STN DBS are often able to substantially reduce their medications. While DBS is currently used as a treatment in patients with advanced disease, it is also being explored as treatment in early phases of the disease to see whether it can alter the clinical course of the disorder, by intervening before adverse effects of abnormal circuit activity alter system function.

The risk of significant surgical complications with DBS, such as intracerebral hemorrhage, is small (1–2%). Reduced verbal fluency and declines in executive functions, as well as postoperative depression (even resulting in suicide), anxiety, and apathy, have been reported in a small number of patients. Such inadvertent behavioral effects appear to be less frequent with GPi DBS than with STN DBS.[40]

DBS at other targets may be useful in patients with PD with specific clinical problems. For instance, patients with severe tremor, but no other parkinsonian signs, can be effectively treated with thalamic DBS. However, even in such cases, the STN is targeted more often than the thalamus, because of the expectation that such patients will eventually develop additional parkinsonian signs that do not respond to thalamic DBS. Because of the failure of STN and GPi DBS to ameliorate postural instability and gait difficulties, such as freezing, other targets, including the PPN, are being explored in a number of small pilot studies, with conflicting outcomes. Benefits in the treatment of freezing of gait have also been reported with combined DBS of STN and SNr. Other targets under study for treatment of PD include the caudal zona incerta, the CM nucleus of the thalamus GPe, and motor cortex.

Although DBS has been in use for PD treatment since the 1990s, the simple approach of continuous high-frequency stimulation has changed little. In part, this is due to its overall success in relieving the motor symptoms without the complications of antiparkinsonian medication. However, recent studies have explored the use of novel electrode design and stimulation paradigms. The use of closed-loop systems triggered by changes in oscillatory activity, measured at the cortical level or at the implantation site of the DBS electrodes in the basal ganglia, has shown favorable effects in studies in human patients and experimental animals.[41]

The mechanisms of action of DBS in PD remain controversial. Although DBS was initially thought to act like a lesion because of the similarity in outcomes with ablation, the mechanism of action of DBS is now viewed as more complex. DBS clearly influences neurons in the target region as well as passing fibers within the range of stimulation. Stimulation in the STN evokes complex excitatory and inhibitory effects in the GPi and alters oscillatory resonance characteristics of the STN-GPi network. Moreover, STN probably modulates the activity of nearby pallidothalamic fibers. PET studies in humans and optogenetic and electrophysiological studies in animals provide evidence that some of the effects of STN DBS also involve antidromically mediated effects on the cerebral cortex (discussed in Devergnas and Wichmann[42]).

A prominent effect of DBS may be to block or override the effects of abnormal basal ganglia output on thalamocortical and brainstem networks, replacing it with patterns of activity that are less disruptive. Using imaging and other methods in humans, STN DBS has been shown to normalize intracortical inhibitory mechanisms and to induce widespread changes (including an at least partial normalization of activity patterns) in frontal motor areas at rest and during the execution of movement tasks.[42]

Other Surgical Approaches

CELL TRANSPLANTATION

Cell transplantation for PD has been carried out by transferring dopamine-producing cells into the patient's striatum. Several sources of dopamine-producing cells have been used, including adrenal cells, dopaminergic glomus cells from the carotid body, and retinal pigmented epithelium cells. Controlled clinical trials of these agents, however, have not proven successful.

Animal studies have demonstrated that grafted fetal mesencephalic dopamine neurons form synaptic contacts with host neurons and release dopamine. In a small number of cases, fetal cell transplantation in patients with PD induced significant, long-lasting clinical improvements. However, fetal cell transplantation is now essentially on hold, because two large blinded randomized controlled clinical trials have shown that the benefit from fetal tissue transfer is at best modest. Moreover, several patients have developed symptomatic graft-induced dyskinesias that required, in some cases, subsequent pallidotomy or pallidal DBS. Besides obvious ethical problems, a critical shortcoming of these transplantation approaches is that they do not address the widespread non-dopaminergic pathology and symptomatology of PD.

The implantation of stem cell-derived dopaminergic neurons as a source of dopamine has been explored. However, transplantation of such cell populations, which are highly capable of proliferation, into the brain raises major safety concerns, since unforeseen proliferation of transplanted cells may result in the formation of brain tumors, as has been seen in other transplantation paradigms.

GENE TRANSFER

The delivery of genes to brain neurons in living animals through transfection with neurotropic viruses is another approach under study. Gene delivery is being used to transfer various enzymes needed for dopamine synthesis (e.g. amino acid decarboxylase or tyrosine hydroxylase) to the striatum. Another approach is to convert the (excitatory) glutamatergic cells in the STN to (inhibitory) GABA cells, in order to decrease the discharge of GPi and SNr neurons and thus reduce symptoms, similar to a subthalamotomy. Positive results of a blinded pilot study of this latter approach were published in 2011, showing that the viral transfection of the synthesizing enzyme glutamic acid decarboxylase is safe and potentially beneficial.[43] The documented effectiveness was, however, inferior to that of STN DBS or ablation. Another form of gene transfer therapy, induction of neurturin (NTN) expression, is discussed in the next section. The safe use of gene therapy for PD remains to be proven.

TROPHIC FACTORS

The transfer or stimulation of trophic factors to the brain may become a key therapeutic tool for neurorestoration in PD and other neurodegenerative conditions. Growth factors include members of the neurotrophin family, such as nerve growth factor (NGF), BDNF, neurotrophin-3, and neurotrophin-4/5, as well the GDNF family, such as GDNF and NTN. GDNF has been most studied because of its powerful effects on growth, survival, and protection of midbrain dopaminergic neurons against toxic insults. Since neither GDNF nor NTN crosses the blood–brain barrier, direct injection into the brain or ventricles is required. Despite promising preclinical data, two double-blinded clinical trials with GDNF administration into the lateral ventricles or the striatum were disappointing because of the development of significant side effects and lack of benefit. Interest has shifted towards the use of NTN, which, in preclinical studies, was shown to powerfully affect midbrain dopaminergic cell growth. As in the GDNF trials, however, a trial of intrastriatal NTN administration proved unsuccessful.[44] BDNF, mesencephalic astrocyte-derived neurotrophic factor (MANF), and the related cerebral dopamine neurotrophic factor (CDNF) are three other targets of interest for the possible development of growth factor-based therapies. These agents have not been systematically studied in human patients with PD.

Neuroprotective Strategies and Biomarker Development

The prospect of potentially being able to halt or at least slow the progression of PD has always been a goal of physicians. Numerous therapeutic approaches based on pathophysiological concepts and preclinical trials with various toxin models of PD have been tried, but none has shown evidence of a disease-modifying effect in patients with PD. This lack of effective neuroprotective approaches has many explanations, including the possibility that our understanding of the pathogenic and pathophysiological steps that trigger and cause the progression of PD is incomplete. One important aspect of the problem is the lack of disease models that capture the progressive nature of the disease, especially the widespread pathology beyond dopamine depletion. In this regard, transgenic models offer a clear advantage over the other animal models discussed in this chapter, but these suffer from a lack of behavioral effect. Another highly important challenge is the absence of reliable biomarkers that would predict future disease or reflect the progression of PD in patients. The realization that PD motor symptoms develop relatively late in the course of disease progression makes the need for such early predictive and "trait" biomarkers obvious. Several clinical, serological, and imaging biomarkers for PD are now being studied. Another approach is to combine the presence of early non-motor symptoms, such as anosmia or RBD, with SPECT, PET, or magnetic resonance imaging findings indicative of dopamine deficiency, to identify patients at risk of PD before the appearance of overt motor symptoms. An extensive clinical test of this approach is underway, which attempts to detect at-risk individuals in a cohort of first degree relatives of PD patients by using changes in olfaction, sleep, and dopamine transporter SPECT imaging as biomarkers.

The identification of genes associated with PD provides another approach to identify individuals at risk of developing PD. Although most monogenic mutations account for only a very small proportion of the global PD population, the mutation of specific genes such as *LRRK2* may be associated with as much as 40% of certain PD patients in some populations (see "Autosomal Dominant Forms of Parkinson Disease", above). Imaging studies have demonstrated abnormal dopaminergic function in the striatum in small groups of presymptomatic individuals with *LRRK2* or other mutations. Various large-scale consortium efforts are currently underway to characterize the status of striatal dopamine imaging in larger cohorts of *LRRK2* family relatives.

Brain imaging can also be used to examine non-dopaminergic markers that are potentially indicative of PD pathology, such as inflammation, mitochondrial dysfunction, αSYN deposition, or protein misfolding. The usefulness of these efforts to identify patients with PD relies on the development of specific ligands with high sensitivity and suitable longitudinal time-course assessment.

Finally, it may be possible to develop methods to identify biochemical pathways associated with the progressive development of PD in presymptomatic stages of PD, by identifying changes in gene expression, protein levels, or metabolites in cerebrospinal fluid or blood.

CONCLUSION

PD is a complex neurodegenerative disease that has traditionally been diagnosed clinically by its motor features, but is actually far more protean in nature and associated with a number of complex and debilitating non-motor features, related to the widespread progressive pathology. Some of the non-motor features may develop before the diagnostic motor features. Key discoveries in the past two decades in molecular biology and genetics as well as basal ganglia circuitry, function, and pathophysiology have moved the field forward enormously.

Despite significant advances in symptomatic treatments, however, PD remains a progressive disorder resulting in worsening of the clinical burden of motor, cognitive, and psychiatric impairments. Much remains to be learned about the etiology and pathogenesis of PD, but the exploration of novel molecular targets in genetic animal models and the start of large-scale clinical studies of PD biomarkers may lead to the successful development of neuroprotective therapies. The development of novel drugs and surgical strategies aimed at optimizing therapeutic outcome while minimizing treatment side effects must continue in parallel with the efforts to reduce disease progression. The major research topics in this field include the search for better disease models, clarification of the cause or effect nature of aspects of pathology and pathophysiology, the development of biomarkers for preclinical PD, and the continued search for better symptomatic and disease-modifying therapies.

References

1. Goedert M, Spillantini MG, Del Tredici K, Braak H. 100 years of Lewy pathology. *Nature Rev Neurol*. 2012;9(1):13–24.
2. Gowers WR. *A Manual of Diseases of the Nervous System*. London: J. & A. Churchill; 1886.
3. Lim SY, Lang AE. The nonmotor symptoms of Parkinson's disease – an overview. *Mov Disord*. 2010;25(suppl 1):S123–S130.
4. Wenning GK, Litvan I, Tolosa E. Milestones in atypical and secondary parkinsonisms. *Mov Disord*. 2011;26(6):1083–1095.
5. Wirdefeldt K, Adami HO, Cole P, Trichopoulos D, Mandel J. Epidemiology and etiology of Parkinson's disease: a review of the evidence. *Eur J Epidemiol*. 2011;26(suppl 1):S1–S58.
6. Olanow CW, Watts RL, Koller WC. An algorithm (decision tree) for the management of Parkinson's disease (2001): treatment guidelines. *Neurology*. 2001;56(11 Suppl 5):S1–S88.

7. Hernan MA, Takkouche B, Caamano-Isorna F, Gestal-Otero JJ. A meta-analysis of coffee drinking, cigarette smoking, and the risk of Parkinson's disease. *Ann Neurol*. 2002;52(3):276–284.
8. Langston JW, Irwin IMPTP. current concepts and controversies. *Clin Neuropharmacol*. 1986;9(6):485–507.
9. Polymeropoulos MH, Lavedan C, Leroy E, et al. Mutation in the alpha-synuclein gene identified in families with Parkinson's disease. *Science*. 1997;276(5321):2045–2047.
10. Singleton AB, Farrer M, Johnson J, et al. α-Synuclein locus triplication causes Parkinson's disease. *Science*. 2003;302(5646):841.
11. Zimprich A, Biskup S, Leitner P, et al. Mutations in LRRK2 cause autosomal-dominant parkinsonism with pleomorphic pathology. *Neuron*. 2004;44(4):601–607.
12. Ozelius LJ, Senthil G, Saunders-Pullman R, et al. LRRK2 G2019S as a cause of Parkinson's disease in Ashkenazi Jews. *N Engl J Med*. 2006;354(4):424–425.
13. Kitada T, Asakawa S, Hattori N, et al. Mutations in the parkin gene cause autosomal recessive juvenile parkinsonism. *Nature*. 1998;392(6676):605–608.
14. Gasser T, Hardy J, Mizuno Y. Milestones in PD genetics. *Mov Disord*. 2011;26(6):1042–1048.
15. Lill CM, Roehr JT, McQueen MB, et al. Comprehensive research synopsis and systematic meta-analyses in Parkinson's disease genetics: the PDGene database. *PLoS Genet*. 2012;8(3):e1002548.
16. Sidransky E, Nalls MA, Aasly JO, et al. Multicenter analysis of glucocerebrosidase mutations in Parkinson's disease. *N Engl J Med*. 2009;361(17):1651–1661.
17. Reinhardt P, Schmid B, Burbulla LF, et al. Genetic correction of a LRRK2 mutation in human iPSCs links parkinsonian neurodegeneration to ERK-dependent changes in gene expression. *Cell Stem Cell*. 2013;12(3):354–367.
18. Bolam JP, Pissadaki EK. Living on the edge with too many mouths to feed: why dopamine neurons die. *Mov Disord*. 2012;27(12):1478–1483.
19. Braak H, Del Tredici K, Rüb U, de Vos RA, Jansen Steur EN, Braak E. Staging of brain pathology related to sporadic Parkinson's disease. *Neurobiol Aging*. 2003;24(2):197–211.
20. Adler CH, Connor DJ, Hentz JG, et al. Incidental Lewy body disease: clinical comparison to a control cohort. *Mov Disord*. 2010;25(5):642–646.
21. Olanow CW, Prusiner SB. Is Parkinson's disease a prion disorder? *Proc Natl Acad Sci U S A*. 2009;106(31):12571–12572.
22. Wenning GK, Geser F, Krismer F, et al. The natural history of multiple system atrophy: a prospective European cohort study. *Lancet Neurol*. 2013;12(3):264–274.
23. Wichmann T, DeLong MR. Deep brain stimulation for neurologic and neuropsychiatric disorders. *Neuron*. 2006;52:197–204.
24. Alexander GE, DeLong MR, Strick PL. Parallel organization of functionally segregated circuits linking basal ganglia and cortex. *Annu Rev Neurosci*. 1986;9:357–381.
25. Haber SN, Fudge JL, McFarland NR. Striatonigrostriatal pathways in primates form an ascending spiral from the shell to the dorsolateral striatum. *J Neurosci*. 2000;20(6):2369–2382.
26. Galvan A, Wichmann T. Pathophysiology of parkinsonism. *Clin Neurophysiol*. 2008;119(7):1459–1474.
27. Grillner S, Robertson B, Stephenson-Jones M. The evolutionary origin of the vertebrate basal ganglia and its role in action-selection. *J Physiol*. 2013;591(Pt 22):5425–3541.
28. Reiner A, Hart NM, Lei W, Deng Y. Corticostriatal projection neurons – dichotomous types and dichotomous functions. *Front Neuroanat*. 2010;4:142.
29. Galvan A, Smith Y. The primate thalamostriatal systems: anatomical organization, functional roles and possible involvement in Parkinson's disease. *Basal Ganglia*. 2011;1(4):179–189.
30. Bostan AC, Strick PL. The cerebellum and basal ganglia are interconnected. *Neuropsychol Rev*. 2010;20(3):261–270.
31. Kravitz AV, Freeze BS, Parker PR, et al. Regulation of parkinsonian motor behaviours by optogenetic control of basal ganglia circuitry. *Nature*. 2010;466(7306):622–626.
32. Bergman H, Wichmann T, DeLong MR. Reversal of experimental parkinsonism by lesions of the subthalamic nucleus. *Science*. 1990;249(4975):1436–1438.
33. Brown P, Williams D. Basal ganglia local field potential activity: character and functional significance in the human. *Clin Neurophysiol*. 2005;116:2510–2519.
34. Brown P, Oliviero A, Mazzone P, Insola A, Tonali P, Di Lazzaro V. Dopamine dependency of oscillations between subthalamic nucleus and pallidum in Parkinson's disease. *J Neurosci*. 2001;21(3):1033–1038.
35. Brown P. Bad oscillations in Parkinson's disease. *J Neural Transm Suppl*. 2006;70:27–30.
36. Redgrave P, Rodriguez M, Smith Y, et al. Goal-directed and habitual control in the basal ganglia: implications for Parkinson's disease. *Nat Rev Neurosci*. 2010;11(11):760–772.
37. Smith Y, Wichmann T, Factor SA, Delong MR. Parkinson's disease therapeutics: new developments and challenges since the introduction of levodopa. *Neuropsychopharm*. 2012;37(1):213–246.
38. Zigmond MJ, Cameron JL, Hoffer BJ, Smeyne RJ. Neurorestoration by physical exercise: moving forward. *Parkinsonism Relat Disord*. 2012;18(suppl 1):S147–S150.
39. Meissner WG, Frasier M, Gasser T, et al. Priorities in Parkinson's disease research. *Nat Rev Drug Discov*. 2011;10(5):377–393.
40. Bronstein JM, Tagliati M, Alterman RL, et al. Deep brain stimulation for Parkinson disease: an expert consensus and review of key issues. *Arch Neurol*. 2010;68(2):165–171.
41. Rosin B, Slovik M, Mitelman R, et al. Closed-loop deep brain stimulation is superior in ameliorating parkinsonism. *Neuron*. 2011;72(2):370–384.
42. Devergnas A, Wichmann T. Cortical potentials evoked by deep brain stimulation in the subthalamic area. *Front Syst Neurosci*. 2011;5:30.
43. Lewitt PA, Rezai AR, Leehey MA, et al. AAV2-GAD gene therapy for advanced Parkinson's disease: a double-blind, sham-surgery controlled, randomised trial. *Lancet Neurol*. 2011;10(4):309–319.
44. Marks Jr WJ, Bartus RT, Siffert J, et al. Gene delivery of AAV2-neurturin for Parkinson's disease: a double-blind, randomised, controlled trial. *Lancet Neurol*. 2010;9(12):1164–1172.

CHAPTER

20

Huntington Disease

Matthew P. Parsons, Lynn A. Raymond

Department of Psychiatry, Brain Research Centre, University of British Columbia, Canada

OUTLINE

OVERVIEW OF HUNTINGTON DISEASE

The first in-depth description of what we have come to know as Huntington disease (HD) was written in 1872 by the American physician George Huntington.[1] In this classic manuscript, Dr Huntington describes hereditary, motor and cognitive aspects of the disease, which appear as if due to "... some hidden power, something that is playing tricks". Through observation, he noted three "peculiarities" of the disease: (1) its hereditary nature; (2) a tendency towards insanity and suicide in the family; and (3) that it manifested as a grave disease only in adult life. He also realized the true severity of the disease, particularly during the later stages, explaining that "the effect is ridiculous in its extreme" and that the motor deficits associated with the disease are "anything but pleasing to witness". He closes by saying, "I have drawn your attention to this form of chorea gentlemen, not that I consider it of any great practical importance to you, but merely as a medical curiosity, and as such it may have some interest".

Today, over 140 years later, the HD literature is filled with thousands of peer-reviewed, original research articles from the efforts of scientists around the globe who, as a whole, have made extraordinary progress in advancing understanding of this disease. During this time, the genetic mutation that causes HD has been identified, along with a plethora of neurobiological consequences of this mutation that will be highlighted later in this chapter. The chapter also presents the symptoms and progression of HD as well as many of the strategies being employed today for the future treatment of HD.

What is Huntington Disease?

HD is a fatal, autosomal dominant neurodegenerative disorder that results from a CAG repeat expansion in the gene encoding the huntingtin protein. If this repeat

Neurobiology of Brain Disorders
http://dx.doi.org/10.1016/B978-0-12-398270-4.00020-3

occurs fewer than 36 times, the "normal" huntingtin protein (commonly containing 15–25 CAG repeats) is expressed. However, 36 or more repeats result in a huntingtin protein with an expanded polyglutamine (polyQ) tract, and this relatively simple modification to a single protein is all that is required to produce the entire gamut of cognitive, psychiatric, and motor impairments that characterize HD. In general, glutamine repeat lengths of 36–40 result in incomplete penetrance, whereas lengths greater than 40 repeats result in full penetrance, with the age of onset and disease severity being negatively and positively correlated with repeat length, respectively. A particularly alarming aspect of HD is that it is dominantly inherited and therefore each child of an affected parent has a 50% chance of inheriting the mutated gene and subsequently developing the disease. For this reason, current prevalence estimates are just the tip of the iceberg. For simplicity, for the remainder of this chapter, "htt" will be used to refer to the non-expanded "normal" huntingtin protein, while "mhtt" (for mutant huntingtin) will refer to the protein with a pathological number of glutamine repeats that results in full penetrance.

Despite being born with mhtt, an HD gene mutation carrier will generally not show any significant, clinically detectable symptoms until adulthood. Symptoms may first appear only as slight changes in personality or mild cognitive deficits that may go unnoticed for many years. Clinical diagnosis occurs at an average age of 39, although in cases of extremely long repeat lengths (60 or more), juvenile-onset HD may be observed. Following diagnosis, there is a progressive worsening of symptoms over a 15–30 year period until death, which is commonly due to inanition or choking as a result of severe progressive dysphagia. HD has sometimes been referred to as Huntington chorea, a term that describes the most obvious feature of the disease: the jerky, involuntary dance-like movements that result from mhtt-induced damage to the brain's motor circuitry. Although chorea is certainly a major factor contributing to diagnosis, the term "Huntington disease" is preferred to "Huntington chorea" as the motor symptoms represent only a single aspect of the disease and chorea is a poor marker of disease severity and disability.

Where is Huntington Disease?

Today, North American prevalence estimates range from 5–10 individuals per 100,000 of the general population, affecting males and females equally. Although this number may seem somewhat small, it translates to roughly 30,000 Americans, each of whom has a 50/50 chance of passing the disease on to his or her offspring. Prevalence rates in North America, Europe, and Australia are generally similar (with some reports suggesting a higher prevalence in Europe), whereas rates in Asia are much lower, affecting less than 1 person in 100,000 (Fig. 20.1).[2] Many believe the current figures to be an underestimation as improved diagnostic tools and increased lifespans are resulting in more HD diagnoses later in life in carriers with repeat lengths of 36–39. Furthermore, as HD commonly manifests in midlife or later, many currently unaffected individuals who carry the HD gene mutation have already had children, and it is therefore likely that a significant population of young mhtt carriers exists that has not been genetically tested and is currently symptom free. Regardless of its precise prevalence, this is a devastating disease that has a tremendous impact on many families from all over the world.

Ethical Issues Regarding Huntington Disease

A known genetic mutation leads to HD. Each child of an affected parent has a 50% chance of developing the disease. A relatively simple genetic test can detect the presence of the *mhtt* gene early in life even though

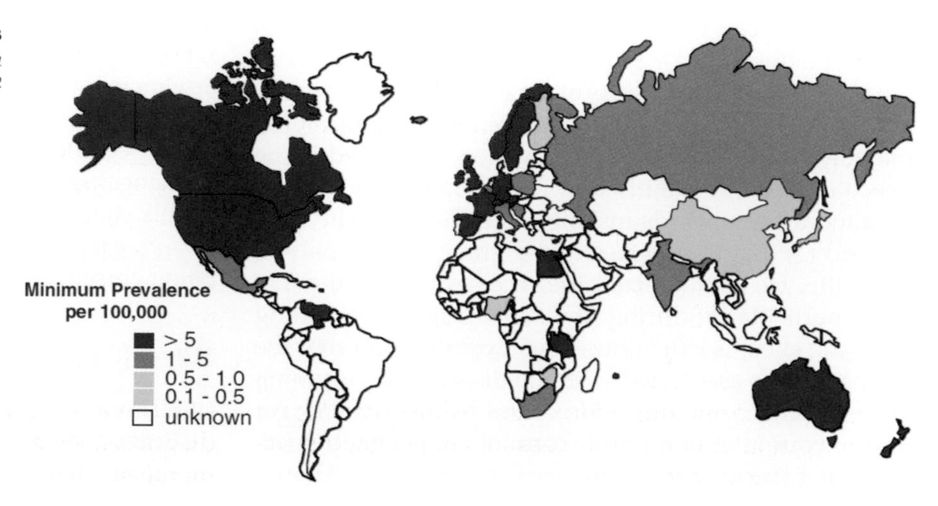

FIGURE 20.1 **Global prevalence estimates of Huntington disease.** *Source: Reprinted from Warby* et al. Eur J Hum Genet. *2011;19:561–566,*[2] *with permission.*

the disease will not manifest until later in life. The symptoms of HD are severe, long-lasting, debilitating, and eventually fatal. There is currently no cure. With the advent of predictive testing, at-risk adolescents are faced with a life-altering choice. The ability to provide a definitive test to determine whether an individual will develop an incurable disease undoubtedly raises numerous ethical issues and for these reasons, predictive testing in HD is arguably one of the most ethically challenging aspects of any neurodegenerative disorder. Who, other than the individual undergoing testing, has the right to be informed of a positive result? How should a positive result be disclosed? How will the individual and his or her family members and friends react to a positive result? On the one hand, one may wonder why such information should be made available to someone who, regardless of CAG repeat status, will remain symptom free for many years. Will this information lead to depression, anxiety, or even suicide during a period that would otherwise be symptom free? On the other hand, it is likely that an individual's career path, family planning, and other life decisions will be heavily influenced by the test results. For some at-risk individuals, a definitive answer either way is simply better than not knowing. When dealing with such an issue, it is clear that a person's reaction to the results of the test will depend on a variety of factors, such as their personal coping strategy and beliefs, as well as the level of support they have from friends and family.

What are the benefits of predicting a currently incurable disease? As mentioned, the relief of uncertainty and the ability to plan accordingly for the future are clear reasons that can drive certain individuals to seek testing. In addition, predictive testing and clinical research studies in healthy HD gene mutation carriers are critical to advancing the search for therapies to delay HD onset. Large cohorts of these identified HD gene mutation carriers are participating in research studies in an effort to identify accurate biomarkers of disease progression. Without such biomarkers, clinical trials evaluating the effectiveness of novel treatment agents would not be feasible.

What are some of the potential harms of predictive genetic testing for HD? For many individuals, receiving the news that they carry the HD gene mutation affects their relationships with family and friends, who may treat them differently because of expectations about how the mutation may alter the individual's behavior and future health. Even if the HD gene mutation carrier does not divulge this information, they may distance themselves from their inner circle to protect their loved ones from the perceived burden of future disease. Moreover, for many, divulging this information publicly can trigger discrimination at work, as well as loss of health, disability, and life insurance.

Whether the benefits outweigh the potential risks is central to the ethical dilemma of predictive testing. Regardless of one's point of view, it is widely agreed upon that the decision to undergo genetic testing is an individual's decision. Physicians do not have the right to reveal test results to anyone other than the at-risk individual, unless given written consent. Quality programs are in place to ensure that individuals undergoing predictive testing receive proper pretest and post-test counseling and are prepared as much as possible for the receipt of such news. Perhaps a surprising statistic to some is that well under half (estimates range from 5 to 20%) of those at risk for HD actually decide to undergo genetic testing.

A related and perhaps even more ethically challenging issue concerns the predictive testing of minors. International guidelines state that predictive testing should not be done for minors unless there is an obvious medical benefit. Therefore, most at-risk individuals must reach the age of majority before they are able to take the test. Allowing the testing of minors would neglect rules of autonomy and confidentiality. However, some argue that predictive testing should be available to minors at the request of their parents. Nonetheless, the benefits and harms of the testing of minors have yet to be adequately studied in a large population and, therefore, it is likely to remain an uncommon practice in the years to come. The fact that, when old enough to consent, fewer than one in five people actually seek predictive testing provides evidence against the testing of minors.

Progression of Huntington Disease

Despite the common genetic mutation underlying all HD cases, there is a great variability in symptoms between patients. It is well established that CAG repeat length plays a major role in determining the age of symptom onset, accounting for roughly 70% of the variance. Nonetheless, there is still a considerable diversity in symptomatology that is independent of repeat length and is likely to be due to other genetic and environmental factors. To describe a typical case of the progression of HD is difficult as there is so much individual variation, particularly in the non-motor aspects of the disease. In this section, the reader should bear in mind that not all of the mentioned symptoms apply equally to all affected individuals. The term "presymptomatic" has been used to describe the stage before the "symptomatic" stage, with the latter referring to the clinical detection of significant motor dysfunction leading to the diagnosis of HD. However, it is apparent that this "presymptomatic" stage often includes a host of subtle yet detectable symptoms and signs occurring up to 15 years before HD diagnosis. As a result, the term "prodromal" (or "premanifest") has been adopted to describe this stage, while "manifest" HD refers to the stages of disease progression following a clinical diagnosis that is based on overt motor abnormalities.

Prodromal Huntington Disease

The prodromal stage of HD refers to the period of months or years before clinical diagnosis in which subtle cognitive, psychiatric, and even motor alterations progress as the HD mutation carrier nears diagnosis. Striatal atrophy as well as a generalized reduction in white matter volume are evident during this stage and correlate well with estimates of time to diagnosis.[3] It is particularly important to understand the scope, progression, and variability of changes occurring in the prodromal stage so that the effectiveness of treatments aimed at delaying or preventing disease onset can be properly evaluated.

Researchers have used a wide variety of tests to measure cognitive function in individuals with prodromal HD. Age and CAG repeat length can be used to divide the prodromal HD group into subgroups defined by their estimated time to diagnosis. A study published in 2011 extensively examined a cohort of 738 prodromal HD gene mutation carriers and 168 control participants to assess cognitive function.[4] All participants had a family history of HD and had chosen (before the study) to undergo predictive genetic testing. The HD group was then divided into "Near", "Mid", and "Far" groups, defined by an estimated less than 9 years, 9–15 years, or more than 15 years until diagnosis, respectively. The cognitive battery consisted of 19 tests assessing a broad range of cognitive functions that have been implicated in the early stages of HD, including planning and reasoning, serial response time, emotion recognition, working memory, episodic memory, non-verbal reasoning, verbal fluency, and executive function. The Near group performed significantly worse than controls on nearly all tests, and the Mid group also showed impaired performance on most tests. In contrast, the performance in the Far group was similar to that of controls on all tests except for the emotion recognition task. The authors concluded that measureable changes in cognitive function occur in HD gene mutation carriers years before diagnosis. The fact that the extent of performance impairment was dependent on the estimated time to diagnosis suggests that the cognitive dysfunction in prodromal HD develops over time rather than being present throughout an HD mutation carrier's lifespan.

Psychiatric issues are also common in both prodromal and manifest HD. High prevalences of psychological stress, anxiety, irritability, depression, and apathy have all been associated with prodromal HD. Suicide rates soar to five to 10 times higher than the general population. Although less frequent, obsessive–compulsive symptoms and psychosis have also been noted in prodromal HD. It could be argued that these psychiatric changes reflect a reaction to the knowledge the HD mutation carrier has of his or her condition, rather than an organic cause. High rates of such psychiatric conditions are also observed in individuals with non-pathological repeat lengths in families with a history of HD, and these may arise in part from the physical and emotional burden of caring for close family members with the disease. It is also possible that mhtt itself contributes directly to psychiatric illness, independently of a patient's concerns of being at risk. Unlike other prodromal changes, the severity of psychiatric manifestations does not seem to progress with proximity to diagnosis.

Although the gross motor abnormalities determine conversion to manifest HD, subtle motor alterations have been detected in the prodromal stage. Impaired balance, finger-tapping speed, and oculomotor saccades have all been described before diagnosis. Although certain studies have failed to find gross motor impairments in prodromal HD, this is likely to be a reflection of both symptom variability among carriers and the sensitivity of the tests.

Lastly, energy metabolism and sleep–wake regulation are commonly affected in prodromal HD. The average body mass index for HD gene mutation carriers before diagnosis is significantly lower than for the general population and progressively decreases as the disease manifests. Weight loss continues despite increased caloric intake, suggesting early and progressive dysfunction of hypothalamic neural circuits or efficiency of mitochondrial energy coupling in brain and muscle. Sleep disturbances, although variable, are also regularly reported to begin in prodromal HD, and their severity and progression appear to be independent of CAG repeat length.

Manifest Huntington Disease

Virtually all of the prodromal symptoms described above gradually worsen as the patient enters the manifest stage of HD. Mood disorders, personality changes, and cognitive decline are all highly associated with manifest HD. The clinical diagnosis of HD requires overt motor manifestations that may or may not be associated with cognitive or psychiatric symptoms. Presentation of typical motor abnormalities together with a family history of HD is sufficient for diagnosis. Once manifest, the hallmark of HD is chorea: irregular and unintended spasmodic movements, particularly of the trunk and limbs. Chorea, fine motor incoordination, and impaired postural reflexes dominate early manifest HD and gradually worsen over time until akinesia and rigidity dominate later in the disease. The lack of motor control also affects the mouth and tongue as well as the larynx, pharynx, and respiratory system. As a result, dysarthria and dysphagia occur and can become so severe that the death of an HD patient is typically due to aspiration, choking, or inanition.

ANIMAL MODELS OF HUNTINGTON DISEASE

The severity of HD described in the previous section highlights the necessity of finding a cure. However, studying the mechanisms underlying mhtt-induced neurodegeneration is difficult in human tissue. The quest to find a cure for HD has therefore prompted the advent of a number of animal models of HD in which such mechanistic work can be performed much more readily. The availability of animal models (particularly genetic mouse models) since the late 1990s has greatly facilitated and enhanced knowledge of the neurobiology of HD. Many features of HD in humans are recapitulated in these models, which may not be too surprising considering the genetic nature of HD, and many new discoveries made in animal models seem to extend to the human condition.

Before the discovery of the HD gene, the early animal models of HD relied on intrastriatal injections of glutamate receptor agonists such as kainic acid or quinolinic acid to produce an HD-like phenotype. More recently, genetically modified mice with a varying number of CAG repeat expansions have become available. Massive amounts of clinically relevant data have been drawn from these mice in a relatively short period regarding the mechanisms of mhtt toxicity and potential therapeutic strategies. Many of the findings presented in the sections to follow arise from a combination of both human and animal studies. First, though, it is worthwhile to briefly mention some of the different mouse models that have led to dramatic changes in the HD research field. These mouse models can be separated into three main categories:

- those that express only an N-terminal fragment of human mhtt in addition to endogenous wild-type murine htt
- those that express full-length human mhtt in addition to endogenous wild-type murine htt
- those that express a pathological CAG repeat within the endogenous murine htt (also known as knockin models).

The R6/2 mouse was the first genetic mouse model of HD and is still widely used today. Only the N-terminal fragment of mhtt (exon 1) is expressed in these mice and contains approximately 150 CAG repeats. This mhtt is expressed in addition to full-length endogenous murine htt. The expression of exon 1 of mhtt proved sufficient to result in rapidly progressive motor symptoms, weight loss, and a dramatically shortened lifespan. Symptoms are detectable quite early and death generally occurs by about 12–18 weeks of age in these mice. Although striatal volume loss is evident, there is a lack of neuronal loss in these mice.

YAC and BAC models are the most common HD mice that incorporate full-length mhtt into their genome; they are so called because they utilize a yeast artificial chromosome (YAC) or bacterial artificial chromosome (BAC) to carry the large genomic human HD DNA. Again, this mhtt is expressed in addition to endogenous murine htt. Currently, the most common YAC model has 128 CAG repeats (YAC128) whereas the BAC model has 97 repeats. Both models develop progressive motor abnormalities later than R6/2 mice and also show progressive volume and cell loss in the striatum, as well as in extrastriatal regions including cortex, hypothalamus, and white matter.

Lastly, knockin models introduce CAG repeats of varying lengths into the murine HD gene. This is sufficient to result in HD-like motor abnormalities and neuropathology in a repeat length-dependent manner, but the phenotype is relatively subtle and slowly progressive.

The differences between these models (e.g. age of onset, severity of symptoms/neuropathology, lifespan) make it difficult to declare a single most appropriate model to use, and therefore all are still commonly used in research throughout the world. One criticism common to all HD mouse models regards the requirement for abnormally long repeat CAG lengths to be able to reliably observe measureable effects within the rodent's lifespan. Nonetheless, it is undeniable that data from the various HD mouse models have helped and continue to help our understanding of human HD in ways that were unimaginable before their introduction. The past two decades have also seen the development of a variety of other HD animal models, ranging from flies, to worms, sheep, and non-human primates, all of which have contributed in unique ways to a deeper understanding of the disease. Together, these animal models will be instrumental to the development of any future therapeutics that delay or prevent HD onset.

NEUROPATHOLOGY OF HUNTINGTON DISEASE

Despite the ubiquitous expression of mhtt in HD patients, the striatum (caudate and putamen) is subject to the most substantial volume and cell loss throughout the course of the disease. Even the naked eye can easily distinguish between a postmortem brain slice through the level of the striatum of a control subject and that of an HD patient (Fig. 20.2). Upon closer inspection of the heterogeneous cell populations within the striatum, it becomes apparent that not all cell types are affected equally. The cell composition within the striatum is dominated by γ-aminobutyric acidergic (GABAergic) spiny projection neurons (SPNs) while cholinergic and

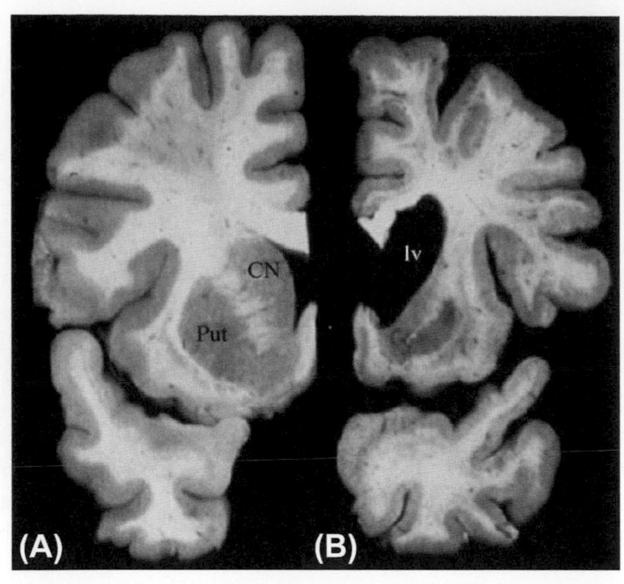

FIGURE 20.2 **Coronal section through the brain of (A) a healthy control subject and (B) a 62-year-old female with Huntington disease (HD) (B).** Note the stark atrophy of the caudate nucleus (CN) and putamen (Put) as well as the enlarged lateral ventricle (lv) in the HD brain. *Source: Reprinted from Ryu* et al. Pharmacol Ther. *2005;108:193–207,*[5] *with permission.*

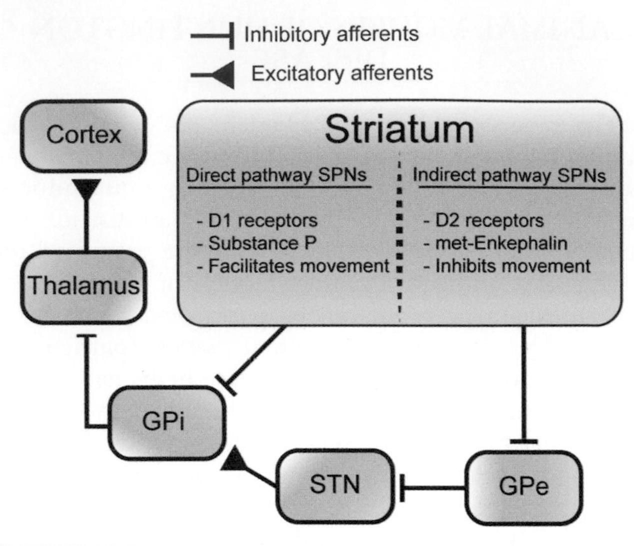

FIGURE 20.3 **Simplified schematic summary of the output of the striatal spiny projection neurons (SPNs), highlighting the direct and indirect pathways.** GPe: external globus pallidus; GPi: internal globus pallidus; STN: subthalamic nucleus.

GABAergic interneurons make up a much smaller percentage of the overall neuronal count. Of the SPNs, which account for roughly 95% of neurons in the striatum, two chemically and functionally distinct types exist:

- direct pathway spiny projection neurons (dSPNs) that express dopamine D_1 receptors and the neuropeptide substance P; these neurons send projections directly to the internal segment of the globus pallidus
- indirect pathway spiny projection neurons (iSPNs) that express dopamine D_2 receptors and the neuropeptide met-enkephalin; these neurons send projections to the external segment of the globus pallidus.

The anatomical differences between SPN subtypes results in a functional dichotomy in which dSPN activation facilitates movement whereas iSPN activation inhibits movement (Fig. 20.3). In HD, these SPN subpopulations differ in their temporal order of degeneration with a preferential loss of iSPNs earlier in the disease followed by the loss of dSPNs. The relatively early loss of iSPNs appears to account for the diminished ability to inhibit unwanted movements, giving rise to chorea. These dance-like movements gradually worsen and then generally abate in later stages, when rigidity and akinesia dominate. This switch is correlated with the eventual death of dSPNs. Striatal interneurons, particularly the cholinergic interneurons, are largely spared in the disease.

While the most striking changes are observed in the striatum, the effects of the HD mutation are far from exclusive to this region. As mentioned, alterations in cognitive function are among the earliest changes to occur in prodromal HD and can be observed decades before the onset of motor symptoms. Cognitive deficits that precede clear neuropathology and overt motor symptoms are also evident in many mouse models of HD, facilitating the study of disease mechanisms underlying cognitive dysfunction. For example, YAC128 mice exhibit a progressive deficit in motor learning beginning at about 2 months of age, well before the documented changes in brain volume or cell counts. These mice also perform poorly on tests of spatial memory and tests that require a rapid change in strategy to solve the task (reversal learning).[6] The non-motor symptoms of HD are likely to be due to the dysfunction and/or degeneration of extrastriatal regions. Cortical atrophy has been consistently observed from imaging studies in prodromal and manifest HD patients, but varies in regional extent. In autopsy brain tissue from HD patients, the group in which motor signs and symptoms were the dominant feature showed significantly more atrophy in the primary motor cortex than those who were afflicted with more severe mood symptoms, which showed more neuronal loss in the anterior cingulate cortex.[7] Similar to the striatum, cortical atrophy occurs more readily in certain cell populations, with pyramidal neurons of layers III, V, and VI being the most susceptible. Subcortical white matter loss is also evident in presymptomatic mhtt carriers. These extrastriatal alterations are so common in HD that morphometric analyses of striatal as well as white matter volume by magnetic resonance imaging may provide an accurate biomarker of disease progression.[3]

Additional areas such as the hippocampus, hypothalamus, and thalamus have also been shown to be affected in HD.

The neuropathological hallmark of HD is the presence of mhtt aggregates. While htt is a soluble, largely cytoplasmic protein, a polyQ expansion within its structure results in dense insoluble aggregates found in both the cytoplasm and nucleus of neurons. Although these aggregates are indeed indicative of the presence of the pathological polyQ-expanded htt protein, they do not appear to be the toxic species that gives rise to the disease. These aggregates will be discussed in more detail later (see "Intranuclear Inclusions", below).

As mentioned, the effects of mhtt are not restricted to the striatum. Nonetheless, the most striking neurodegenerative changes associated with HD occur in this region. Why, then, is the striatum so vulnerable to the HD mutation if the mutated protein is expressed ubiquitously? This is an important question that, so far, has no definitive answer. To understand the striatal vulnerability in HD, it must first be appreciated that HD appears not to be the result of a single physiological effect induced by the mutated protein. Rather, countless lines of evidence from laboratories around the world indicate that mhtt expression results in a host of cellular and molecular alterations which together appear to be responsible for progressive neurodegeneration over the course of many years. These changes, and potential reasons why the striatum may be more heavily influenced by them, will be discussed throughout the following section on the neurobiology of HD. Although much emphasis has been placed on the striatal SPNs owing to their susceptibility in the disease, it is important to keep in mind that HD is now recognized as a whole-brain disease and that many cellular consequences of mhtt identified for the striatum are probably not restricted to this region.

NEUROBIOLOGY OF HUNTINGTON DISEASE

Structure and Function of the Huntingtin Protein

Huntingtin (htt) is a soluble 3144 amino acid (348 kDa) protein, with the highest levels of expression being found in the CNS and testes. The N-terminal 17 amino acids, or N17 region, has been identified as a critical region that plays a role in htt localization, aggregation, and toxicity. Immediately following the N17 region is the polyQ tract, which contains as many as 35 CAG repeats in those individuals not afflicted by HD. Following the polyQ stretch is a polyproline-rich region, which is then followed by clusters of HEAT repeats (α-helix-loop-α-helix motifs), which are important for various interactions between proteins. A cytoplasmic retention signal and nuclear export signal have been identified on the N- and C-terminals of htt, respectively. The amino acid sequence of htt contains important sites of post-translational modification including sites of ubiquitination/SUMOylation, phosphorylation, palmitoylation, acetylation, and proteolysis. Many of the post-translational modification sites are found within or adjacent to four separate regions enriched in proline, glutamic acid, serine, threonine, and predicted cleavage sites termed PEST domains. These sites are of great research interest because, as we will see later in this chapter, slight post-translational modifications of mhtt can dramatically influence its cytotoxicity. A schematic of the htt structure highlights some regions of interest (Fig. 20.4).

Before describing the neurobiological effects of mhtt, it is important to understand the function of the htt protein with 35 glutamine repeats or fewer (i.e. "normal" htt). While much evidence suggests that the majority of HD pathogenesis is the result of a toxic gain of function

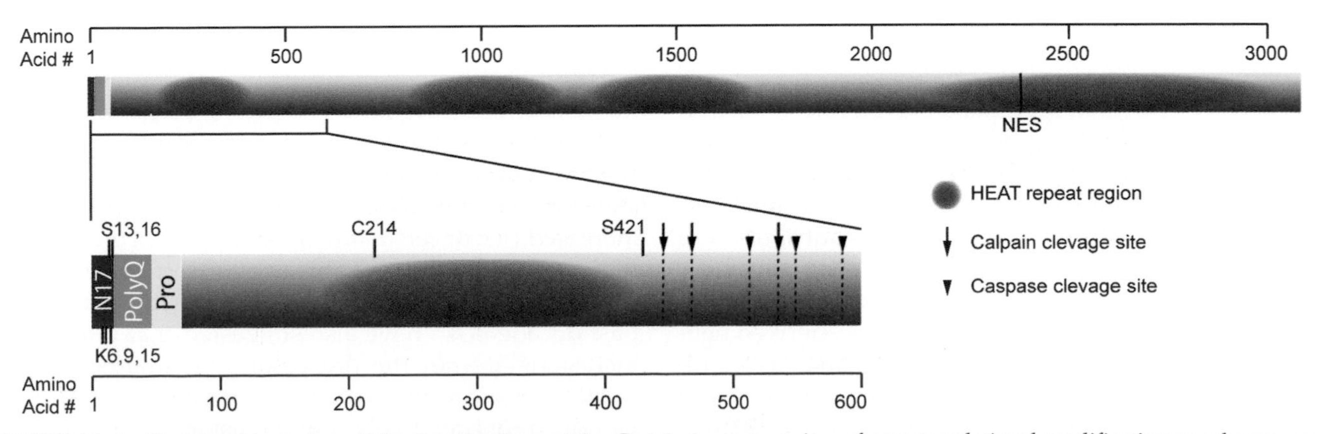

FIGURE 20.4 **Structural overview of the huntingtin protein.** Certain important sites of post-translational modification are shown, as are sites of calpain and caspase cleavage. Clusters of HEAT repeats are denoted in red. Note the particularly high abundance of post-translational modification sites within the N17 region. NES: nuclear export signal; Pro: proline-rich region. *Source: Adapted from Ehrnhoefer* et al. Neuroscientist. *2011;17:475–492*[8] *and Cattaneo* et al. Nat Rev Neurosci. *2005;6:919–930.*[9]

in mhtt due to the polyQ expansion, there is also considerable evidence suggesting that at least some of the disease progression can be attributed to a loss of function of the normal htt protein. It is well established that htt can increase cell survival and resistance to toxicity when overexpressed in its unexpanded form.[9] Whereas full deletion of the *htt* gene is lethal to the embryo (suggesting a critical role for htt function in development), a 50% reduction of htt expression in animals can result in enhanced cell death, partial mimicry of HD, and exacerbation of behavioral symptoms in the presence of mhtt. Unfortunately, this prosurvival function of htt complicates the targeting of the *htt* gene as a cure for HD and, as a precaution, gene therapies should be geared towards the specific knockdown of *mhtt*. However, the study of the natural functions of htt may spawn pharmacological interventions aimed at enhancing its prosurvival features in an effort to restore its normal function in HD or to counteract the toxicity of mhtt, or both.

Surprisingly, the precise cellular role for htt is rather ill defined. Ascribing a role to this protein is complicated by its ubiquitous presence throughout the brain and body as well as its subcellular localization. In neurons, htt has been found to associate with the nucleus, endoplasmic reticulum, and Golgi, and can also be found associated with vesicular structures in neurites. Such a widespread localization of htt suggests broad roles for this protein in cellular function, roles that are just beginning to be uncovered.

Biochemical approaches such as yeast two-hybrid screens, affinity chromatography, and immunoprecipitation studies have identified a wide variety of proteins that interact directly with htt,[10] consistent with the multiple HEAT repeat regions within its amino acid sequence. Such a large interactome suggests that htt may be an important scaffold protein within the cell. Furthermore, htt interacts with many proteins that are involved in intracellular transport. It has been postulated that "htt is able to coordinate the binding of multiple types of motor proteins to vesicular cargo, most probably by acting as a scaffold that can differentially bind to many proteins".[11] It has become clear that htt indeed plays a role in the trafficking of vesicles and organelles. Reducing htt levels or expanding the polyQ tract to a pathological length can obstruct htt's facilitatory role in intracellular trafficking.

Perhaps the most widely recognized contribution of normal htt to cellular function, and one that is likely to contribute to its prosurvival properties, is its involvement in the transcription and trafficking of brain-derived neurotrophic factor (BDNF). Although BDNF itself is a critical factor for cell growth and survival, it is not produced in significant quantities by striatal neurons. Rather, the large majority of BDNF support to the striatum is derived from cortical afferent fibers. This anatomical arrangement requires the transcription of BDNF within cortical neurons and successful BDNF trafficking along the axon to presynaptic release sites within the striatum. It appears that htt can promote both BDNF transcription and its transport along microtubules.[12,13] The finding that this effect is lost in the presence of polyQ-expanded mhtt, resulting in a striatum that is deprived of BDNF, is an excellent example of how the loss of function of normal htt in HD may contribute to disease pathology.

Intranuclear Inclusions

Immunohistochemical analysis of early mouse models of HD revealed the presence of visible mutant htt nuclear aggregates (termed "intranuclear inclusions").[14] Soon thereafter, these aggregates were observed in the postmortem human HD brain and are now considered a neuropathological hallmark of HD. Such aggregates can only be stained with an antibody directed towards N-terminal and not C-terminal epitopes of mutant htt, suggesting that the mutant protein is cleaved and it is the N-terminal fragment that accumulates. Initially, there were many proponents of the "aggregation theory", which posited that these aggregates represent the initial critical step that triggers downstream effects which eventually lead to neurodegeneration. Thus, according to such a hypothesis, preventing intranuclear inclusions would essentially cure the disease. It turns out to be not quite that straightforward.

As the consequences of mhtt aggregation began to be studied in more detail, researchers were divided into different camps. There were many who still held on to the notion that these inclusions were the toxic trigger, although numerous lines of evidence began suggesting that the presence of inclusions did not correlate well with cell death and, in fact, may actually protect neurons from death. Thus, there were some who believed the aggregates to be toxic, some who believed them to be protective, and still others who believed them to be irrelevant to disease progression. One of the main arguments against their role in toxicity came from a study using longitudinal survival analysis, in which individual living cells were imaged over time to determine the fate of mhtt-expressing cells with aggregates versus those without aggregates.[15] The study found no evidence to suggest that the presence of aggregates predicted an increased risk of cell death. In fact, it found the opposite: those cells that lacked visible aggregates were more vulnerable to cell death. Further lines of evidence over the past decade or so have also suggested that intranuclear inclusions are not the precursor to neurodegeneration in HD. For example, postmortem analysis of HD brains demonstrated relatively few inclusions in the striatum, the area affected first in HD. Within the striatum, most of the aggregates were found in the cholinergic interneurons, which are largely spared in the disease, rather than in the SPNs. There is also a poor correlation between the presence of aggregates and cell death in mouse models of HD.

The progression of mutant htt from a soluble protein to an insoluble aggregated intranuclear or cytoplasmic inclusion represents two extremes of its conformation. It is now recognized that there are intermediate steps along this route and it is thought that the toxic species is one that is made before the "macroaggregates". Smaller "microaggregates" are visible at the electron microscopy level, and a monoclonal antibody has been developed that recognizes only low molecular weight conformations of mutant htt. The expression of these monomeric or small oligomeric forms of mhtt is highly predictive of cell death.[16] The macroaggregates are now postulated to act as a coping response, as they can sequester some of the toxic forms of mutant htt.

Huntington Disease: A Cell-Autonomous Disease?

A major topic in the HD field has been to address why striatal SPNs are particularly affected by the disease while certain other cell types appear to be relatively spared or die later than SPNs. As a consequence of this investigation, it has become important to address whether or not the disease is cell autonomous; that is, whether mhtt expression in SPNs alone can account for their dysfunction and death or whether there is a requirement for mhtt to be present in other components of the SPN's local and network environments. In light of the fact that HD is a brain-wide disease and is by no means restricted to the striatum, it should be clarified that the study of cell-autonomous and non-autonomous mechanisms in HD is concerned with the effects of mhtt when expressed solely in the postsynaptic cell of interest compared with its effect on that cell when mhtt is expressed entirely within the context of its specific neuronal network.

Based on rodent studies, the neurodegeneration in HD is likely to be due to a combination of cell-autonomous and non-autonomous effects, with pathological interactions between cells being central to disease pathogenesis. The best way to address the question of cell autonomy in HD is to selectively express mhtt in a particular cell type and observe the resultant biochemical, behavioral, and neuropathological effects. In 2005, the laboratory of William Yang generated mice that expressed exon 1 of mhtt either exclusively in the cortex or widely throughout the brain. Despite the observation of progressive accumulation of nuclear aggregates in the mhtt-expressing cortical neurons, there were no clear motor deficits or neuropathology when mhtt was restricted to the cortex. However, when the same mhtt fragment was expressed widely throughout the brain, clear mhtt aggregates as well as progressive neurodegeneration, reactive gliosis, and locomotor deficits were observed.[17] Two years later, the same group generated mice that expressed exon 1 of mhtt exclusively in the striatum. These mice displayed mhtt aggregation and an increased sensitivity of *N*-methyl-D-aspartate (NMDA) receptors in the striatum, but again lacked significant neuropathology or motor deficits.[18] These results demonstrate that the exclusive expression of mhtt in the cortex or striatum is sufficient to induce certain changes typical of HD pathogenesis, but argue against a purely cell-autonomous mechanism of neurodegeneration.

There is also considerable evidence to suggest that astrocytes are a critical player in HD. One role of astrocytes is to rapidly clear extracellular glutamate that is released by neural transmission. This is accomplished by glutamate transporters such as GLT-1. The restriction of mutant htt expression to just astrocytes results in an age-dependent gliosis, neurological symptoms, and reduced GLT-1 expression.[19] The reduction of GLT-1 in mutant htt-expressing astrocytes is thought to result in excessive extracellular glutamate, which contributes to excitotoxic cell death. This excessive glutamate tone occurring in tandem with enhanced SPN NMDA receptor sensitivity is a potentially lethal combination of alterations for the striatal SPNs. Thus, mhtt expression probably contributes to cell death through both cell-autonomous and non-autonomous mechanisms.

Excitotoxicity and N-Methyl-D-Aspartate Receptors

Considerable evidence suggests that excitotoxicity is a major contributing factor to the neurodegeneration observed in HD. Excitotoxicity generally arises from prolonged neuronal stimulation by the excitatory neurotransmitter glutamate. More specifically, calcium influx through activated NMDA receptors mediates the toxic effect. One of the earliest animal models of HD was based on injections of quinolinic acid, an endogenous NMDA receptor agonist, into the striatum. This resulted in the selective degeneration of SPNs and recapitulated certain behavioral and biochemical aspects of HD.[20] Striatal interneurons, which are spared in HD, survive these intrastriatal injections of quinolinic acid. Enhanced NMDA-mediated currents and sensitivity to NMDA-induced excitotoxicity have been observed in various animal models of HD. Support for the excitotoxicity hypothesis has also come from studies of postmortem brains from human HD patients. Relative to controls, brain tissue from early symptomatic and even presymptomatic HD patients is characterized by a loss of glutamate binding sites and a reduction in NMDA receptor subunit messenger RNA (mRNA), consistent with a loss of cells that contain NMDA receptors.

Calcium influx through NMDA receptors is not always toxic. In fact, it is necessary for normal development and plays an important role in cell survival as well as synaptic plasticity. Eliminating NMDA receptor activity can also result in apoptosis, making the treatment of HD with general NMDA receptor blockers a poor option. This highlights the importance of basic scientific investigation into how NMDA receptors can have both positive and negative effects on cell health. It was initially thought that the level of calcium influx through NMDA receptors determined cell fate. According to that view, low to moderate levels of calcium were related to cell survival while larger amounts of calcium influx were associated with cell death. While this idea should not necessarily be completely discounted, more recent experiments in cell culture suggest that the precise location of the activated NMDA receptors determines whether the subsequent calcium influx will be beneficial or harmful to the cell. Under certain experimental conditions, NMDA receptors found within the postsynaptic density (synaptic NMDA receptors) are preferentially associated with downstream signaling pathways that promote cell survival and inhibit cell death. In contrast, NMDA receptors found outside the synapse (extrasynaptic NMDA receptors) can be preferentially associated with signaling pathways promoting cell death and inhibiting cell survival.[21]

Excitotoxicity is a common process in the brain and is certainly not specific to striatal SPNs. Yet, SPNs are the first cell type to die in HD. So, if HD is a result of excitotoxic cell death, then why are SPNs particularly vulnerable? A number of reasons, depicted in Fig. 20.5, may explain why[22]:

- Striatal SPNs are the target for large amounts of glutamate released from dense glutamatergic afferent fibers. The glutamatergic innervation of the striatum arises from the cortex and the thalamus.
- Aside from the large amounts of glutamate released into the striatum, there is considerable evidence to suggest impaired uptake of glutamate in HD. Normally, much of the released glutamate is rapidly cleared by transporter-mediated uptake mechanisms. These mechanisms are impaired in HD, resulting in abnormally high levels of ambient glutamate that can contribute to excitotoxicity.[23]
- HD is associated with an increase in the extrasynaptic localization of NMDA receptors in striatal SPNs.[24,25] Thus, the high levels of glutamate released as a result of cortical/thalamic activity and a lack of uptake within the striatum are likely to enhance the activation of these NMDA receptors at extrasynaptic sites. Memantine, an NMDA receptor antagonist that can preferentially block extrasynaptic receptors, improves cell survival as well as skilled motor learning and motor performance in the YAC mouse model of HD.
- While NMDA receptors throughout the developed brain are dominated by $GluN_{2A}$-containing subunits, the striatum contains a relatively high

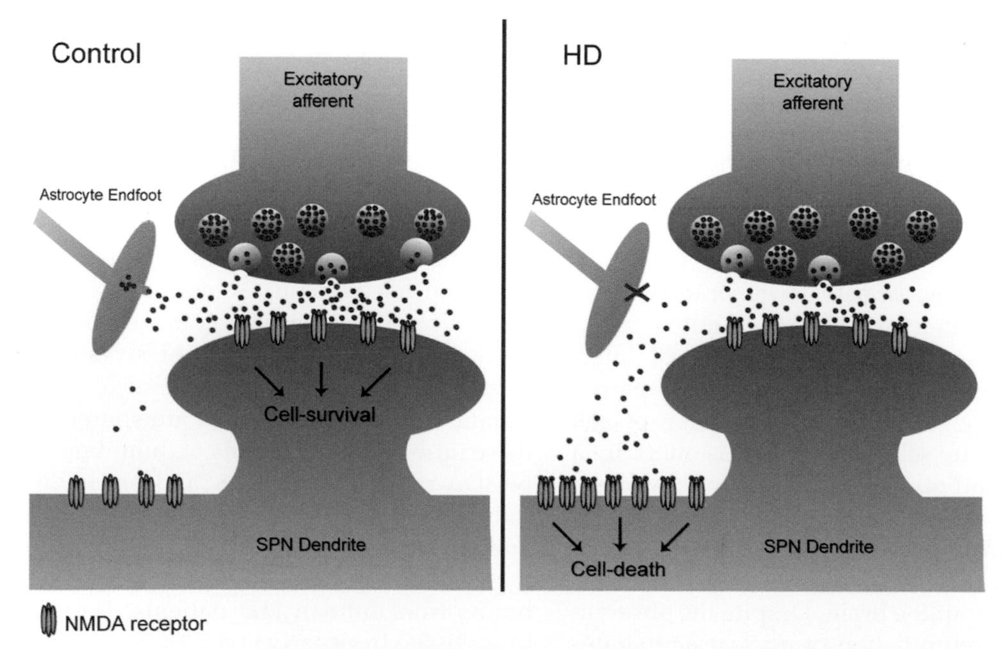

FIGURE 20.5 **Excitotoxicity in Huntington disease (HD).** Striatal spiny projection neurons (SPNs) receive large amounts of glutamate released from excitatory afferents, namely from the cortex and thalamus. Under normal conditions, much of the released glutamate is taken up by glial glutamate transporters before it can spill over to extrasynaptic sites. However, in HD, the dysfunction of glial glutamate transporters results in more glutamate diffusing outside the synapse, where it can activate cell death-associated extrasynaptic *N*-methyl-D-aspartate (NMDA) receptors, which are enhanced in early HD.

number of $GluN_{2B}$ subunits, and surface expression of these subunits is specifically increased in the presence of mutant htt. Signaling via $GluN_{2B}$-containing receptors is more closely associated with excitotoxicity and cell death compared with signaling via $GluN_{2A}$-containing receptors. The cell death induced by NMDA receptor agonists can be blocked by $GluN_{2B}$ subtype-specific antagonists.

In all, striatal SPNs are already the direct target of large quantities of glutamate release. The enhanced $GluN_{2B}$ expression, extrasynaptic NMDA receptor localization, and reduced glutamate uptake as a result of mutant htt undoubtedly contribute to the pathogenesis of HD.

Transcriptional Dysregulation

Upon close examination of the subcellular localization of both htt and mhtt, it was found that normal htt is chiefly a soluble cytoplasmic protein while mhtt (particularly N-terminal fragments of mhtt) associates much more prominently with the nucleus. This nuclear translocation of mhtt in HD provided the first clue to suggest a potential effect of mhtt on gene expression. From studies of postmortem brain tissues from HD patients, it is known that there are specific patterns of loss of genes encoding various neuropeptides such as substance P and preproenkephalin, as well as of neurotransmitter receptors such as dopamine receptors that show region specificity and cannot be entirely explained solely by the cell loss associated with the disease. The use of HD animal models has confirmed many of these findings and has advanced our understanding of the transcriptional dysfunction in HD and the mechanisms by which these occur.

The transcriptional dysregulation in HD influences genes that are critical to many different facets of neuronal function, including neurotransmitter synthesis, neurotransmitter receptors, intracellular signaling pathways, and synaptic proteins. Microarray studies have been valuable to our understanding of the types of genes altered in the disease state. Perhaps not surprisingly, given the severity of the disease progression, the general consensus is that genes involved in cell growth and survival, calcium homeostasis, and synaptic transmission are typically downregulated, whereas those with a role in cell stress and apoptosis are generally upregulated. Progressive abnormalities in gene expression can be observed relatively early in HD, again supporting transcriptional dysfunction independent of cell loss.

Although mhtt nuclear aggregates provided major insight into the possibility of a transcriptional dysregulation in HD, the presence of the aggregates themselves does not predict subsequent gene changes within a given cell. Thus, the nuclear inclusions observed in HD brains are not the cause of transcriptional alterations. So, how can mhtt influence the expression of so many different genes? As mentioned, htt has numerous binding partners within the cell. A commonly observed feature of the htt protein is the way in which the presence of a pathogenic polyQ repeat can alter its interactions with other proteins. With regard to gene expression, htt can interact (either directly or indirectly) with many transcription factors and coactivators, such as repressor element-1 silencing transcription factor (REST), CREB binding protein (CBP), and peroxisome proliferator-activated receptor-γ coactivator-1α (PGC-1α). The altered association of mhtt with these proteins and other transcription-regulating proteins appears to be the main mechanism by which mhtt can influence gene expression. For example, htt interacts indirectly with REST, a well-established repressor of neuronal gene expression. When associated with htt, REST is sequestered in the cytoplasm, where it fails to regulate target gene expression. However, in the presence of a pathogenic polyQ expansion within htt, REST is translocated to the nucleus, where it can repress the expression of a number of target genes. The BDNF deficit observed in Huntington disease is a result of mutant htt-mediated translocation of REST to the nucleus and resultant suppression of the *Bdnf* gene (Fig. 20.6). REST can also translocate to the nucleus in the absence of the htt protein, suggesting a role for the non-expanded htt

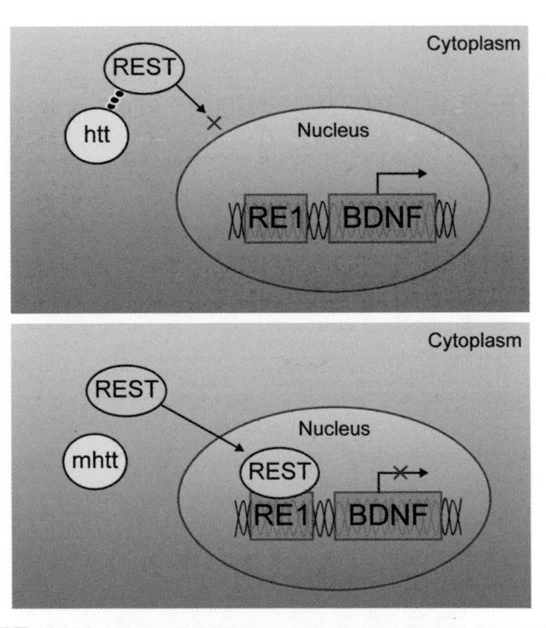

FIGURE 20.6 **Brain-derived neurotrophic factor (BDNF) transcription is regulated by huntingtin (htt).** Repressor element 1-silencing transcription factor (REST) is sequestered in the cytoplasm through an indirect association with htt. This prevents its access to the nucleus, thereby permitting transcription of the BDNF gene (top). In Huntington disease (HD), the association between mhtt and REST is weaker and REST is able to enter the nucleus, where it represses BDNF transcription (bottom). RE-1: repressor element-1. *Source: Adapted from Zuccato and Cattaneo.* Nat Rev Neurol. *2009;5:311–322.*[26]

protein in gene regulation via retention of REST in the cytoplasm. Similarly, many other transcriptional regulatory proteins are altered in HD as a result of an enhanced or decreased association with htt/mhtt. Another example is CBP, which interacts more strongly with mhtt than with htt, resulting in a suppression of neuroprotective cAMP response element-binding protein (CREB)-mediated gene transcription in HD. As a result of mhtt altering the function of numerous transcription factors and coactivators, many genes are dysregulated in HD (Fig. 20.7).

Mitochondrial Dysfunction

HD (including prodromal HD) is associated with cellular energy deficits as a result of mitochondrial dysfunction. HD patients show a decrease in glucose metabolism and weight loss despite an equal or even increased caloric intake. A decrease in the activity of mitochondrial respiratory complexes II, III, and IV has been reported in HD from rodents to humans, as has a reduction in mitochondrial membrane potential, calcium buffering capacity, and expression of oxidative phosphorylation enzymes. Furthermore, administration of the mitochondrial toxin 3-nitropropionic acid mimics many aspects of HD. N-terminal fragments of mhtt interact with mitochondria, suggesting that direct interactions between mhtt and mitochondria may underlie some of these metabolic changes in HD. On the other hand, it has become clear that the transcriptional dysregulation induced by mhtt described above impacts cell energetics in a major way. Many of the genes altered by mhtt encode proteins involved in mitochondrial function; mhtt decreases PGC-1α expression by about 30% in the striatum, with no changes observed in the cerebellum and hippocampus.[28] PGC-1α is a transcriptional coactivator involved in many vital neuronal functions such as mitochondrial biogenesis and oxidative phosphorylation; mhtt (but not htt) associates with the endogenous PGC-1α promoter and interferes with the transcriptional pathway responsible for PGC-1α expression. When mice lacking PGC-1α are crossed with HD mice, the degree of striatal degeneration and motor dysfunction is enhanced. In contrast, direct lentiviral delivery of PGC-1α to the striatum of R6/2 mice completely blocks striatal atrophy, providing good evidence that PGC-1α can be neuroprotective in HD and that mitochondrial function in HD plays a key role in its pathogenesis.

Post-Translational Modifications

As previously mentioned, the htt protein contains regions that are subject to a variety of post-translational modifications, particularly in its N17 domain. Some important examples that will be discussed below include phosphorylation, SUMOylation, acetylation, palmitoylation, and proteolysis. Since these post-translational modifications can significantly influence the toxicity of mhtt they may be targeted as potential therapeutic strategies to combat HD.

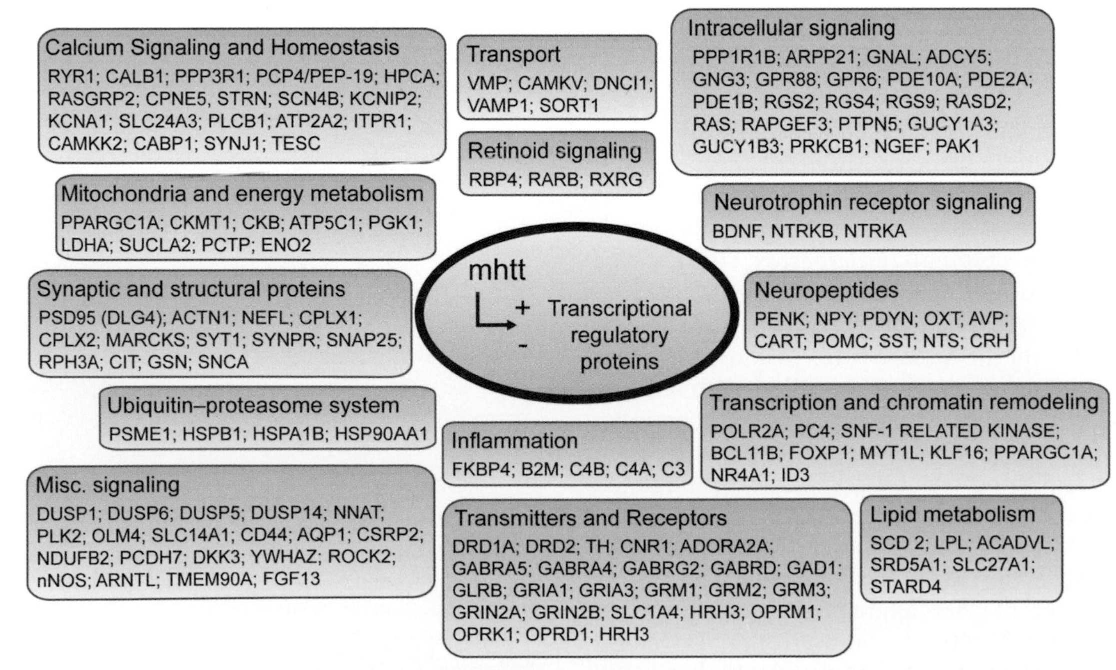

FIGURE 20.7 **Transcriptional dysregulation in Huntington disease (HD).** The mutant huntingtin protein (mhtt) can increase or decrease the function of numerous transcriptional regulatory proteins, with the result that a vast array of genes becomes dysregulated in HD. Genes are grouped according to function. *Source: Adapted from Seredenina and Luthi-Carter.* Neurobiol Dis. *2012;45:83–98.*[27]

Phosphorylation is a common post-translational modification that involves the addition of a phosphate group to certain amino acids (most commonly serine and threonine). Phosphorylation is achieved by protein kinases, whereas the removal of a phosphate group, or dephosphorylation, is mediated by protein phosphatases. The htt protein has various sites of phosphorylation and, in general, normal htt is more heavily phosphorylated than its mutant form. This suggests that increasing the phosphorylation state of mutant htt may have protective effects. In one study, insulin-like growth factor exerted a neuroprotective effect in mutant htt-expressing striatal neurons. The effect depended on the activation of protein kinase Akt and was significantly attenuated when serine 421 on the htt protein was mutated to a phosphoresistant alanine. Thus, insulin-like growth factor may be neuroprotective in HD by increasing htt phosphorylation on serine 421. Notably, processing of Akt is impaired in human HD brains, which may result in reduced kinase activity in the disease state.[29]

SUMOylation, another post-translational modification implicated in HD pathogenesis, refers to the attachment of small ubiquitin-like modifier (SUMO) proteins to lysine residues. SUMOylation competes with ubiquitination (the post-translational attachment of ubiquitin molecules) for the same lysine residues on the htt protein. With regard to the polyQ-expanded protein, ubiquitination has been implicated in a reduction of toxicity and SUMOylation with an enhancement of toxicity. Ubiquitination generally targets proteins for degradation and therefore an increased clearance of mhtt is likely to underlie its protective effect. However, SUMOylation appears to stabilize the toxic form of htt (and actually decreases the amount of macroaggregates). Work on htt SUMOylation has shed additional light on the vulnerability of striatal SPNs in HD. Rhes, a small guanine nucleotide binding protein, binds to mutant htt with greater affinity than it does wild-type htt. When bound, Rhes increases the SUMOylation of mutant htt, thereby stabilizing its toxic conformation and subsequently increasing cell death.[30] The fact that Rhes is highly expressed in the striatum and not in extrastriatal regions suggests that Rhes-dependent SUMOylation of mutant htt may be another key contributor to SPN vulnerability in HD.

Acetylation was first identified for histone proteins. High histone acetylation is generally associated with an open, relaxed chromatin state that facilitates gene transcription. Supporting a transcriptional dysregulation, histone acetylation is decreased in HD. Histone deacetylase inhibitors, which promote increased acetylation, are therefore potential therapeutic targets for HD and have shown some promising results in mouse models. In addition, mhtt can be acetylated at lysine 444 and this promotes its trafficking to autophagosomes for degradation. Thus, increasing the acetylation of mhtt itself may help to rid cells of the toxic protein and ameliorate neuronal dysfunction and death.

Palmitoylation, the post-translational addition of the 16-carbon chain fatty acid palmitate to cysteine residues, is also impaired in the presence of a polyQ expansion. Some of the known palmitoyl acyltransferases, the enzymes responsible for palmitoylation, are htt interacting proteins. Specifically, htt interacting proteins 14 (DHHC17) and 14L (DHHC13) are palmitoyl acyltransferases that palmitoylate htt as well as numerous synaptic proteins. Many of the substrates of DHHC17 and 13 show reduced palmitoylation in the presence of mhtt, implicating hypopalmitoylation as a possible contributor to disease pathogenesis.[31] As the palmitoylation of synaptic proteins plays a major role in their subcellular localization and adherence to the plasma membrane, hypopalmitoylation of synaptic proteins in HD may result in a host of synaptic deficits and impair cell-to-cell communication.

One of the hallmarks of HD is the presence of N-terminal fragments that aggregate intracellularly. Although it now appears that these aggregates do not cause cell death, their presence is indicative of the abnormal processing of mhtt in HD. Proteolysis, the last modification to be discussed here, is the process by which mhtt can be broken down into these smaller fragments that aggregate in the disease. What happens if we prevent the breakdown of mutant htt? Can we prevent multimerization of mhtt and inhibit aggregate formation? Can we reverse the toxicity associated with the mutant protein? To answer these questions, researchers have looked at the role of proteases in the pathogenesis of HD. Members of both the caspase and calpain families of proteases can cleave htt, and there is evidence suggesting that the activity levels of these proteases are enhanced in the HD brain and that mhtt cleavage is required to produce toxic N-terminal fragments. One study demonstrated that caspase-1 activity was increased in the brains of HD mice and human HD patients.[32] The authors also investigated whether a decrease in caspase-1 activity in HD mice could influence disease progression. Expressing a dominant-negative caspase-1 mutant in R6/2 mice resulted in improvement of motor performance, delayed formation of aggregates, and increased survival.[32] Similarly, expression of mutant htt resistant to caspase-6 cleavage also resulted in delayed nuclear translocation of htt, prevented neurodegeneration, and improved some of the motor deficits normally seen in the YAC model of HD.[33] Calpain cleavage also contributes to mhtt toxicity, as mutation of calpain cleavage sites in the mhtt sequence decreases toxicity. In sum, targeting the caspase or calpain family of proteins to reduce the amount of mhtt cleavage is a promising therapeutic strategy for HD.

Neuroinflammation in Huntington Disease

There is evidence of neuroinflammation in HD. However, it is difficult to determine the extent to which neuroinflammation contributes to or is a reaction to HD pathology. As reviewed in Section IV of this textbook, microglia are the resident immune cells of the CNS and become activated in response to infectious agents, CNS injury, and cell stress. When active, they can release a variety of factors, including cytotoxic and cytoprotective agents. Acute neuroinflammation is generally beneficial, helping the brain to rid itself of infectious agents and clearing necrotic cellular debris. However, chronic neuroinflammation can damage cells through the prolonged release of proinflammatory cytokines and the increased production of superoxide and nitric oxide. In HD brains, mhtt is found in glial cells as well as neurons, and many regions (including the striatum) display an increase in the overall numbers of microglia and the number of activated microglia. The extent of microglia activation correlates well with the degree of neuronal loss in HD. *In vivo* positron emission tomography (PET) imaging studies have determined that microglial activation is evident early in prodromal HD and is closely associated with neuronal dysfunction, suggesting that neuroinflammation in HD is chronic and that imaging microglial activity may be an important biomarker to track disease progression. While neuroinflammation in HD may not be the initiating factor of disease pathogenesis, it is highly likely that it can exacerbate cell death and disease progression.

TREATMENT OF HUNTINGTON DISEASE

Current Treatments

It should be evident by now that there are many neurobiological consequences to the huntingtin mutation. Despite our vast understanding of the genetics and neurobiology of HD, there is no cure at present and effective therapeutic options specific to the disease are limited. The medications currently prescribed to HD patients are aimed at alleviating individual symptoms rather than ameliorating the underlying neuropathology. Chorea is one clinical feature amenable to amelioration with medical therapy, although it is not the most debilitating motor abnormality of HD. The challenge in treating chorea is finding the proper balance between inhibiting unwanted movements and preserving postural balance and voluntary movement. Tetrabenazine was the first, and remains the only drug approved by the US Food and Drug Administration (FDA) for the treatment of chorea in HD. This drug, a vesicular monoamine transporter inhibitor that depletes presynaptic dopamine to a greater extent than other monoamines, gained FDA approval in 2008 after a double-blind, randomized, placebo-controlled clinical trial demonstrated a significant beneficial effect on chorea in HD patients.[34] Unfortunately, the drug has no beneficial effect on the various non-motor symptoms in HD, and may even increase depression and anxiety as dose-dependent side effects. Tetrabenazine is also relatively expensive. Therefore, it is commonly used as a second choice treatment for HD, falling behind antipsychotics (neuroleptics).

Neuroleptics, in particular second generation atypical neuroleptics such as risperidone or olanzapine, are commonly prescribed in HD as multipurpose agents with the potential benefit of alleviating chorea, certain psychiatric symptoms, and even the dramatic weight loss associated with HD. These drugs are much less expensive than tetrabenazine but can also have adverse effects on voluntary movement and cognitive processing speed, both of which are much more common with typical neuroleptics such as haloperidol. With regard to psychiatric symptoms, antidepressants such as selective serotonin reuptake inhibitors (SSRIs) or mood stabilizers such as valproic acid can be beneficial to the depression or irritability and impulsivity commonly associated with HD. However, as the non-motor symptoms of HD are particularly variable, there is no single treatment that works for all patients. Rather, the precise medication prescribed to patients is based on the range and severity of symptoms and responses to initial treatments. Similarly, therapeutic interventions to alleviate any cognitive symptoms would also have to be based on the individual. Unfortunately, no effective treatment has been cited for the cognitive deterioration in HD. Donepezil, an acetylcholinesterase inhibitor used to treat Alzheimer disease, and latrepirdine, an antihistamine drug with reported cholinesterase-inhibiting actions, both appear to offer no beneficial effect on cognitive dysfunction. Similarly, atomoxetine, a selective norepinephrine reuptake inhibitor approved for Alzheimer disease treatment, failed to show any cognitive benefits in a clinical trial. Early, open trials of memantine, also used to treat cognitive and behavioral symptoms in Alzheimer disease, have shown promising results but a phase III trial has not yet been undertaken.

Patients with HD may also benefit from various types of non-pharmacological strategies to cope with particular aspects of the disease. For example, physical therapy may have beneficial effects on certain motor skills and counseling may help to alleviate psychiatric aspects of HD. Speech therapy may also be of importance, especially since dysphagia is quite common later in the disease and is often the cause of death. Similarly, a nutritionist may help with the dramatic weight loss observed in HD by designing meal plans that help a patient to gain weight in a healthy manner without resorting to simply overeating energy-dense foods that are high in saturated fats and simple sugars. In all, current therapeutic strategies to alleviate the burden of HD are multidimensional and unique to the individual.

Future Treatment

Although many drugs are still being tested for their effect on the symptoms of HD, most experts in the field believe that the best treatments will prove to be those that target the disease as a whole rather than the individual symptoms. One way to do this is to use our broad knowledge base of the neurobiology of HD to prevent or minimize cell dysfunction and death from occurring in the first place. Some of these strategies, such as inhibiting enzymes that slice mhtt into toxic fragments or decrease excitotoxicity by blocking NMDA receptors (preferably extrasynaptic), have already been mentioned. Approaches that have proven effective in animal models must first be thoroughly tested in clinical trials.

Another strategy to target the disease in its entirety, and perhaps the most obvious approach, is to eliminate the mhtt protein itself. The entire spectrum of symptoms – from chorea to depression to dysphagia – is a result of mhtt protein expression. Thus, a great deal of research effort in the HD field has turned to gene-silencing approaches to eliminate mhtt expression completely. This approach benefits from preventing the initial toxic effects of mhtt rather than attempting to treat downstream pathophysiology as a result of mhtt expression. While this may seem like the obvious target with regard to a therapeutic strategy – especially considering the broad range of neurobiological consequences of the mutation – it is certainly not without its challenges. First, htt is critical to development and normal cellular function and therefore the ideal treatment should aim to eliminate mhtt while leaving normal htt intact. Although studies in non-human primates suggest that reducing htt expression in adulthood is well tolerated, this idea lacks evidence based on long-term follow-up and in-depth behavioral, cognitive, and physiological testing. Thus, based on our knowledge of htt's function, eliminating htt and mhtt together in the same cells may be a dangerous therapeutic approach. HD is a brain-wide disease, and this presents a second major challenge to eliminating mhtt. Therapies targeting mhtt that cannot cross the blood–brain barrier would require direct central administration. Not only is this invasive and costly, but also the therapeutic effect is limited by an agent's diffusion radius following administration and therefore multiple injections may be required. Lastly, any unwanted side effects of these treatments may prove difficult to reverse. Side effects from medications can be sidestepped by switching medications until finding the one that is right for the individual. Unfortunately, with gene-interfering strategies, the effect is long lasting, which may work to its advantage in HD treatment but may also produce sustained unwanted side effects. This highlights the need for highly selective gene-interfering agents and well-controlled preclinical and clinical trials.

OTHER CAG REPEAT DISORDERS

Although space constraints have required that the authors focus on HD in this chapter, it is important to remember that HD is not the only CAG repeat disorder. Mutations resulting in polyQ-containing proteins are known to give rise to nine different disorders. These proteins, like mhtt, become improperly folded and processed and generally acquire a toxic gain of function. As with HD, the other CAG repeat disorders are associated with a relatively late onset (typically 30–40 years, with the age of onset being inversely proportional to the repeat length), autosomal dominant inheritance (with the exception of one X-linked recessive disorder), and repeat instability such that expansions can increase from one generation to the next. The nine CAG repeat disorders are HD, spinal and bulbar muscular atrophy (SBMA; also known as Kennedy disease), dentatorubropallidoluysian atrophy (DRPLA), and various spinocerebellar ataxias [SCAs; type 1, 2, 3 (Machado–Joseph disease), 6, 7, and 17].

Kennedy disease was the first disease to be recognized as a CAG repeat disorder and is at present the only known X-linked CAG repeat disorder. The expansion occurs in exon 1 of the gene encoding the androgen receptor and affects an estimated 1–2 individuals per 100,000. Non-pathogenic repeat lengths range from nine to 37, with 38 or more repeats being pathogenic. As in HD, N-terminal fragments of the mutated protein form intranuclear inclusions which may reflect a cellular coping mechanism, although some argue that aggregation is a critical step towards neurodegeneration as these inclusion bodies are seen in both postmortem tissue and animal models of Kennedy disease. Regardless of this, nuclear aggregation of the mutated androgen receptor is a key hallmark of disease presence and occurs both within and outside the CNS. The neuronal populations most vulnerable to neurodegeneration in Kennedy disease are the ventral horn motor neurons of the spinal cord and cranial nerve motor nuclei in the brainstem, with the exception of the extraocular motor nuclei. As a result, patients develop fasciculations, weakness, and atrophy of the facial, bulbar, and limb musculature, often in midlife and progressing to the point where a wheelchair is a common requirement roughly 15 years after diagnosis. Other symptoms include mild feminization, impotence, testicular atrophy, and reduced fertility. Kennedy disease is androgen dependent in that the mutated androgen receptor requires ligand binding for its nuclear translocation and subsequent toxic effects. For this reason, female mutation carriers are generally asymptomatic and the effectiveness of androgen deprivation therapies in males is being tested in clinical trials. The underlying neurobiology of Kennedy disease shows considerable overlap with that of HD, with transcriptional dysregulation, mitochondrial dysfunction, and impaired axonal transport all playing key roles in mutant androgen receptor-induced toxicity.

DRPLA is another CAG repeat disorder, resulting from a CAG expansion in the gene encoding a transcriptional coregulator called atrophin-1. The resulting polyQ-expanded protein results in widespread neurodegeneration, with the most notable atrophy occurring in the globus pallidus, cerebellum, and pons. Significant white matter atrophy has also been observed. Clinically similar to HD, the toxic effects of this mutation generally result in cerebellar ataxia, myoclonus, chorea, epilepsy, and dementia. The average age of onset is roughly 30 years and symptom severity progresses with time. While less is known of the underlying neurobiology of DRPLA, there appear to be similarities with that of HD, including altered interactions between proteins, transcriptional dysregulation, metabolic dysfunction, and excitotoxicity.

Currently, there are 31 known SCAs, six of which make up the remainder of the nine known CAG repeat disorders (Fig. 20.8). CAG expansions in the ataxin-1, 2, 3, and 7 genes give rise to SCA1, 2, 3, and 7, respectively, while

SCA#	Affected gene	CAG length	Symptoms in addition to ataxia, dysarthria and oculomotor symptoms
1	Ataxin-1	44+	Dysphagia, pyramidal and extrapyramidal signs, sensory deficits, cognitive decline
2	Ataxin-2	34+	Dysphagia, rigidity, bradykinesia, peripheral neuropathy, executive dysfunctions, cognitive decline
3	Ataxin-3	52+	Dysphagia, pyramidal and extrapyramidal signs, sensory deficits, peripheral neuropathy, amyotrophy and rarely Parkinsonism
6	CACNA1A	21+	Dysphagia, tremor, somatosensory deficits
7	Ataxin-7	36+	Progressive loss of vision, pyramidal signs
17	TBP	47+	Chorea, dystonia, rigidity, pyramidal signs, psychiatric symptoms, cognitive decline and epilepsy

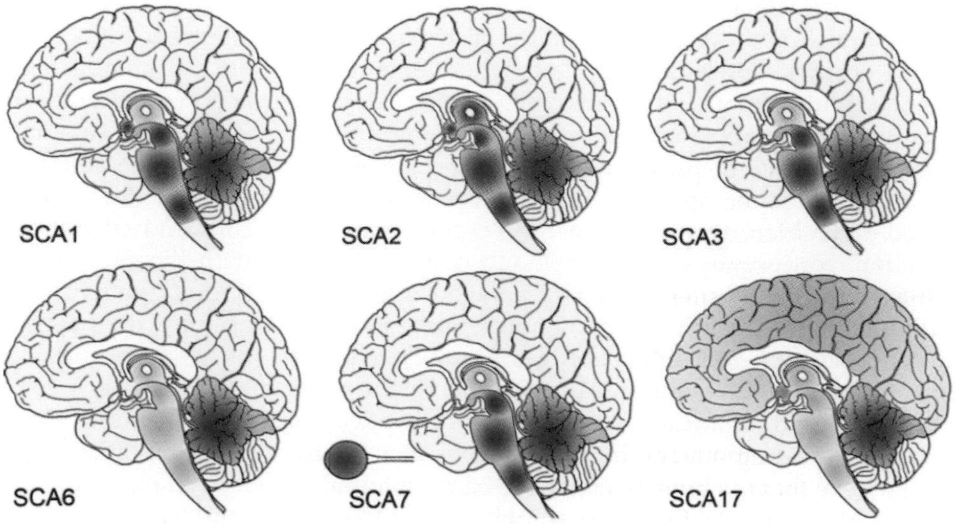

SCA1 SCA2 SCA3

SCA6 SCA7 SCA17

FIGURE 20.8 **The CAG repeat spinocerebellar ataxias (SCAs).** Summary of the affected genes, pathogenic CAG repeat lengths and associated symptoms of the six SCAs caused by CAG expansions. The schematic midsagittal sections highlight the regions with the most notable atrophy in each disorder (marked atrophy shown in light red, severe atrophy in dark red). Note the retinal atrophy unique to SCA7. *Source: Table adapted and figure reprinted from Seidel* et al. Acta Neuropathol. *2012;124:1–21,[35] with permission.*

CAG expansion in the *CACNA1A* gene generates SCA6 and CAG expansion in the *TBP* gene generates SCA17. All are highlighted by widespread neurodegeneration, with the most pronounced atrophy generally occurring in the cerebellum and brainstem. Severe retinal ganglion cell degeneration is associated with SCA7. As with the other CAG repeat disorders, the pathological hallmark of the SCAs is a nuclear accumulation of the mutated protein, and whether these represent a toxic or protective form of the protein remains an issue for debate. Symptoms of SCAs generally include ataxia, oculomotor dysfunction, dysarthria, and progressive cognitive deterioration. Vision is severely affected in SCA7 as this disorder is associated with the degeneration of retinal ganglia cells. Gaining a better understanding of the normal function of the proteins encoded by these SCA genes and their alteration in disease should aid the development of future therapeutic interventions.

QUESTIONS FOR FURTHER RESEARCH

A wealth of knowledge regarding the underlying neurobiology of HD caused by mhtt has accumulated over the two decades since the HD gene mutation was identified. Therapeutic strategies have been proposed, some of which are currently being tested in controlled clinical trials. The ultimate goal of any therapeutic strategy is to delay or completely prevent the disease. The importance of studying prodromal HD cannot be underestimated, because achieving a cure relies on the identification of accurate biomarkers of subclinical progression. If the disease cannot be tracked in the premanifest stage, then the effectiveness of medications and treatments over time cannot be accurately determined during clinical trials. Answering questions such as how early and how aggressively HD patients should be treated also requires accurate biomarker monitoring. Thus, continuing the search for HD biomarkers is a prerequisite for the future treatment of the disease. Other researchers will use animal models to focus on designing gene-interfering agents that eliminate mhtt. Challenges inherent in this approach include minimizing side effects, specifically targeting mhtt while preserving htt, and achieving adequate access of the agent throughout the brain. Additional basic science research can help to determine the importance of some of these key issues. For example, while specific gene silencing of mhtt is clearly superior to general htt knockdown in theory, a gene therapy that is 100% specific to mhtt may not be feasible. It is therefore important to know how well varying degrees of normal htt knockdown during adulthood are tolerated (if at all), thereby providing a standard of specificity that a particular gene therapy must reach before being considered as a viable treatment. Similarly, studies focusing on the contributions of cell-autonomous and non-autonomous processes to HD pathogenesis can help to determine which brain regions to specifically target with such treatments if whole-brain access proves troublesome.

Further development of our understanding of the cellular mechanisms of HD is likely to lead to novel drugs that delay neuronal dysfunction and cell death; therefore, basic science research should be expected to continue at a rapid pace. Key to the pharmacological prevention of HD is the identification of the early mhtt-induced cellular events that occur well in advance of cell death. Reversal of these early changes may delay or completely prevent manifest HD from occurring. A good example is the recent identification of elevated extrasynaptic NMDA receptors in early HD. Is elevated extrasynaptic NMDA receptor expression specific to the striatum or does it occur throughout the brain? Can the presence of these receptors alone account for enhanced vulnerability to excitotoxic insult or is there a requirement for the coincident dysfunction of glutamate uptake by glial cells, providing a means by which these receptors can become activated? Is the elevated extrasynaptic NMDA expression also true for the human condition and, if so, at which stage of disease progression does it occur? Other early events in HD include the post-translational modifications of mhtt, which play a critical role in its toxic effects. Which post-translational modifications are the most critical to mhtt toxicity? What accounts for the elevated expression and activity of enzymes responsible for the cleavage of mhtt into smaller, toxic fragments?

These are just some examples of the many questions that remain in the HD field. Some of these are likely to be addressed in the near future and it is evident that our overall understanding of this debilitating disease will continue to expand rapidly. With the dedicated efforts and collaborations of researchers around the world – from identifying the smallest mhtt-induced cellular changes to conducting large-scale clinical trials – the possibility of a definitive therapy for HD, and ultimately for all of the CAG repeat disorders, is being viewed with much optimism and excitement.

References

1. Huntington G, On chorea, George Huntington MD. *J Neuropsychiatry Clin Neurosci*. 2003;15:109–112.
2. Warby SC, Visscher H, Collins JA, et al. HTT haplotypes contribute to differences in Huntington disease prevalence between Europe and East Asia. *Eur J Hum Genet*. 2011;19:561–566.
3. Paulsen JS, Langbehn DR, Stout JC, et al. Detection of Huntington's disease decades before diagnosis: the Predict-HD study. *J Neurol Neurosurg Psychiatry*. 2008;79:874–880.
4. Stout J, Paulsen J, Queller S, et al. Neurocognitive signs in prodromal Huntington disease. *Neuropsychology*. 2011;25:1–14.
5. Ryu H, Rosas HD, Hersch SM, Ferrante RJ. The therapeutic role of creatine in Huntington's disease. *Pharmacol Ther*. 2005;108:193–207.

6. Van Raamsdonk J, Pearson J, Slow E, Hossain S, Leavitt B, Hayden M. Cognitive dysfunction precedes neuropathology and motor abnormalities in the YAC128 mouse model of Huntington's disease. *J Neurosci*. 2005;25:4169–4249.
7. Thu DC, Oorschot DE, Tippett LJ, et al. Cell loss in the motor and cingulate cortex correlates with symptomatology in Huntington's disease. *Brain*. 2010;133:1094–1110.
8. Ehrnhoefer DE, Sutton L, Hayden MR. Small changes, big impact: posttranslational modifications and function of huntingtin in Huntington disease. *Neuroscientist*. 2011;17:475–492.
9. Cattaneo E, Zuccato C, Tartari M. Normal huntingtin function: an alternative approach to Huntington's disease. *Nat Rev Neurosci*. 2005;6:919–930.
10. Li SH, Li XJ. Huntingtin–protein interactions and the pathogenesis of Huntington's disease. *Trends Genet*. 2004;20:146–154.
11. Caviston J, Holzbaur E. Huntingtin as an essential integrator of intracellular vesicular trafficking. *Trends Cell Biol*. 2009;19:147–202.
12. Gauthier LR, Charrin BC, Borrell-Pages M, et al. Huntingtin controls neurotrophic support and survival of neurons by enhancing BDNF vesicular transport along microtubules. *Cell*. 2004;118:127–138.
13. Zuccato C, Ciammola A, Rigamonti D, et al. Loss of huntingtin-mediated BDNF gene transcription in Huntington's disease. *Science*. 2001;293:493–498.
14. Davies SW, Turmaine M, Cozens BA, et al. Formation of neuronal intranuclear inclusions underlies the neurological dysfunction in mice transgenic for the HD mutation. *Cell*. 1997;90:537–548.
15. Arrasate M, Mitra S, Schweitzer ES, Segal MR, Finkbeiner S. Inclusion body formation reduces levels of mutant huntingtin and the risk of neuronal death. *Nature*. 2004;431:805–810.
16. Miller J, Arrasate M, Brooks E, et al. Identifying polyglutamine protein species in situ that best predict neurodegeneration. *Nat Chem Biol*. 2011;7:925–934.
17. Gu X, Li C, Wei W, et al. Pathological cell–cell interactions elicited by a neuropathogenic form of mutant Huntingtin contribute to cortical pathogenesis in HD mice. *Neuron*. 2005;46:433–477.
18. Gu X, Andre VM, Cepeda C, et al. Pathological cell–cell interactions are necessary for striatal pathogenesis in a conditional mouse model of Huntington's disease. *Mol Neurodegen*. 2007;2:8.
19. Bradford J, Shin JY, Roberts M, Wang CE, Li XJ, Li S. Expression of mutant huntingtin in mouse brain astrocytes causes age-dependent neurological symptoms. *Proc Natl Acad Sci U S A*. 2009;106:22480–22485.
20. Beal MF, Kowall NW, Ellison DW, Mazurek MF, Swartz KJ, Martin JB. Replication of the neurochemical characteristics of Huntington's disease by quinolinic acid. *Nature*. 1986;321:168–171.
21. Hardingham GE, Bading H. Synaptic versus extrasynaptic NMDA receptor signalling: implications for neurodegenerative disorders. *Nat Rev Neurosci*. 2010;11:682–696.
22. Fan MM, Raymond LA. N-methyl-D-aspartate (NMDA) receptor function and excitotoxicity in Huntington's disease. *Prog Neurobiol*. 2007;81:272–293.
23. Behrens PF, Franz P, Woodman B, Lindenberg KS, Landwehrmeyer GB. Impaired glutamate transport and glutamate–glutamine cycling: downstream effects of the Huntington mutation. *Brain*. 2002;125:1908–1922.
24. Milnerwood AJ, Gladding CM, Pouladi MA, et al. Early increase in extrasynaptic NMDA receptor signaling and expression contributes to phenotype onset in Huntington's disease mice. *Neuron*. 2010;65:178–190.
25. Okamoto S, Pouladi MA, Talantova M, et al. Balance between synaptic versus extrasynaptic NMDA receptor activity influences inclusions and neurotoxicity of mutant huntingtin. *Nat Med*. 2009;15:1407–1413.
26. Zuccato C, Cattaneo E. Brain-derived neurotrophic factor in neurodegenerative diseases. *Nat Rev Neurol*. 2009;5:311–322.
27. Seredenina T, Luthi-Carter R. What have we learned from gene expression profiles in Huntington's disease? *Neurobiol Dis*. 2012;45:83–98.
28. Cui L, Jeong H, Borovecki F, Parkhurst C, Tanese N, Krainc D. Transcriptional repression of PGC-1alpha by mutant huntingtin leads to mitochondrial dysfunction and neurodegeneration. *Cell*. 2006;127:59–128.
29. Humbert S, Bryson EA, Cordelieres FP, et al. The IGF-1/Akt pathway is neuroprotective in Huntington's disease and involves Huntingtin phosphorylation by Akt. *Dev Cell*. 2002;2:831–837.
30. Subramaniam S, Sixt KM, Barrow R, Snyder SH. Rhes, a striatal specific protein, mediates mutant-huntingtin cytotoxicity. *Science*. 2009;324:1327–1330.
31. Young FB, Butland SL, Sanders SS, Sutton LM, Hayden MR. Putting proteins in their place: palmitoylation in Huntington disease and other neuropsychiatric diseases. *Prog Neurobiol*. 2012;97:220–238.
32. Ona VO, Li M, Vonsattel JP, et al. Inhibition of caspase-1 slows disease progression in a mouse model of Huntington's disease. *Nature*. 1999;399:263–267.
33. Graham RK, Deng Y, Slow EJ, et al. Cleavage at the caspase-6 site is required for neuronal dysfunction and degeneration due to mutant huntingtin. *Cell*. 2006;125:1179–1191.
34. Huntington Study Group. Tetrabenazine as antichorea therapy in Huntington disease: a randomized controlled trial. *Neurology*. 2006;66:366–372.
35. Seidel K, Siswanto S, Brunt ER, den DW, Korf HW, Rub U. Brain pathology of spinocerebellar ataxias. *Acta Neuropathol*. 2012;124:1–21.

CHAPTER

21

Alzheimer Disease

Alena V. Savonenko[*,†], *Tatiana Melnikova*[*], *Tong Li*[*], *Donald L. Price*[*,†,‡], *Philip C. Wong*[*,‡]

[*]Department of Pathology, The Johns Hopkins University School of Medicine, Baltimore, Maryland, USA; [†]Department of Neurology, The Johns Hopkins University School of Medicine, Baltimore, Maryland, USA; [‡]Department of Neuroscience, The Johns Hopkins University School of Medicine, Baltimore, Maryland, USA

OUTLINE

INTRODUCTION

Alzheimer disease (AD) is characterized by progressive deficits in memory, and cognitive and behavioral impairments that ultimately lead to dementia. It affects more than 5.4 million individuals in the USA. The prevalence, cost of care, impact on individuals and caregivers, and lack of mechanism-based treatments make AD one of the most challenging diseases. The syndrome of AD results from dysfunction and death of neurons in specific regions and circuits, particularly those populations of nerve cells subserving memory and cognition.[1,2] Characteristics of AD neuropathology include accumulations of intracellular and extracellular protein aggregates. Abnormally phosphorylated tau assembles into paired helical filaments (PHFs) that aggregate into neurofibrillary tangles (NFTs) in the neuronal perikarya and dystrophic neurites. The second pathological hallmark is the extracellular deposition of β-pleated assemblies of Aβ peptide, forming diffuse and neuritic senile plaques.[2]

In the 1970s and 1980s, neurochemical examination of brain samples from cases of AD led to the demonstration

Neurobiology of Brain Disorders
http://dx.doi.org/10.1016/B978-0-12-398270-4.00021-5

of a dramatic loss of cortical cholinergic innervations, and subsequent neuropathological studies found degeneration of basal forebrain magnocellular neurons and cholinergic deficits in the cortex and hippocampus. These observations led to the introduction of cholinesterase inhibitors as a first treatment for AD. Evidence of the involvement of glutamatergic systems in hippocampal and cortical circuits in AD, coupled with information about glutamate excitotoxicity [mediated, in part, by *N*-methyl-D-aspartate receptors (NMDA-R)], led to a second therapy being approved by the US Food and Drug Administration, NMDA-R antagonists. Both of these therapeutic strategies have given modest and transient symptomatic benefit in some patients.[3]

Observations of autosomal dominant inheritance in families with early-onset, familial Alzheimer disease (fAD) in concert with the work of geneticists resulted in the discovery of mutations in genes encoding the amyloid precursor protein (APP) and the presenilins (PS1 and PS2). Although the exact mechanisms affected by each mutation differ, the general outcome of fAD-associated mutations is an increase in production of $A\beta_{1-40}$ and/or $A\beta_{1-42}$ peptides or the $A\beta_{1-40}/A\beta_{1-42}$ ratio. The presence of specific alleles of other genes such as apolipoprotein E (*ApoE*) has been found to be a risk factor for late-onset disease.[4,5] The mechanisms affected by the risk factors associated with late-onset AD are likely to include alterations in Aβ metabolism, Aβ aggregation and clearance, or cholesterol homeostasis.

Intensive study of the mechanisms of generation of Aβ peptides resulted in the discovery of sequential endoproteolytic cleavages of APP by two membrane-bound enzymes, termed β-site APP cleaving enzyme-1 (BACE1) and γ-secretase. Using mouse models with genetically altered activities of BACE1 or γ-secretase, both of these enzymes have been experimentally validated as high-priority therapeutic targets for AD therapy.[6–8] Based on preclinical studies, pharmacological inhibition of these activities has been predicted to decrease the generation of Aβ and to ameliorate cognitive decline in AD. However, when these novel mechanism-based experimental therapies were moved into clinical trials, they showed lower than expected efficacies in ameliorating functional deficits, and prominent adverse effects. These discrepancies between the outcomes of preclinical and clinical trials are forcing a re-evaluation of views of the disease, its models, and ways to resolve this translational dilemma.[9]

Alzheimer Disease and Mild Cognitive Impairment

AD is named after Dr Alois Alzheimer, who first observed the disease in 1901 in a 51-year-old woman named Augusta D, who experienced excessive feelings of jealously towards her husband as one of first signs. "Very soon she showed rapidly increasing memory impairments; she was disoriented, carrying objects to and fro in her flat and hid them. Sometimes she felt that someone wanted to kill her and began to scream loudly. ... After 4.5 years of sickness she died".[10] Postmortem investigation of the brain revealed a high concentration of protein plaques and NFTs. Alzheimer reported peculiar changes in the neurofibrils: "In the centre of an otherwise almost normal cell there stands out one or several fibrils due to their characteristic thickness and peculiar impregnability".[10] He also described the typical plaques: "Numerous small miliary foci are found in the superior layers. They are determined by the storage of a peculiar material in the cortex".[10]

In the decade after Alzheimer's case report dozens of articles described observations of senile plaques and NFTs in a variety of patients with dementing diseases, including post-traumatic stress dementia, amyotrophic lateral sclerosis, Down syndrome, toxic conditions, and dementia pugilistica.[11] Diagnostic classifications, however, were confusing because senile plaques and NFTs were also found in the brains of older individuals who had shown no signs of dementia.[11] Similar modern histopathological reports of "silent plaques" attracted much attention after years of failure of clinical trials with anti-amyloid therapy.

In the first version of the *Diagnostic and Statistical Manual of Mental Disorders* (DSM) in 1952, the term "Alzheimer's disease" was used for cases with presenile onset and "chronic organic brain syndrome" was used more frequently for senile brain disease. The diagnosis of AD re-emerged in the late 1960s because of the unprecedented increase in the number of people who were living to be 85 and older. Increased human lifespan in the twentieth century revealed a new "epidemic" of aging-related neurodegenerative diseases. The DSM-IV, published in 1994, referred to AD as the most common cause of dementia. In the twenty-first century, this trend has been further amplified by aging "baby-boomers", who make up the most rapidly growing part of the population.

In 2011, new criteria for AD and mild cognitive impairment (MCI) were developed, which emphasize that the pathophysiological processes underlying AD begin decades before clinical symptoms appear.[12] The recognition that there are many patients whose cognition is not normal for their age but who do not meet criteria for dementia led to the classification of MCI.

Whereas based on the previous guidelines, a physician could diagnose only AD dementia, the new criteria recognize three stages of AD progression (Fig. 21.1). The first (preclinical AD) stage occurs before changes in cognition are detected; the second stage (MCI due to AD) occurs when cognitive symptoms emerge but daily function is preserved; and the third stage (AD dementia) is diagnosed when deficits in multiple

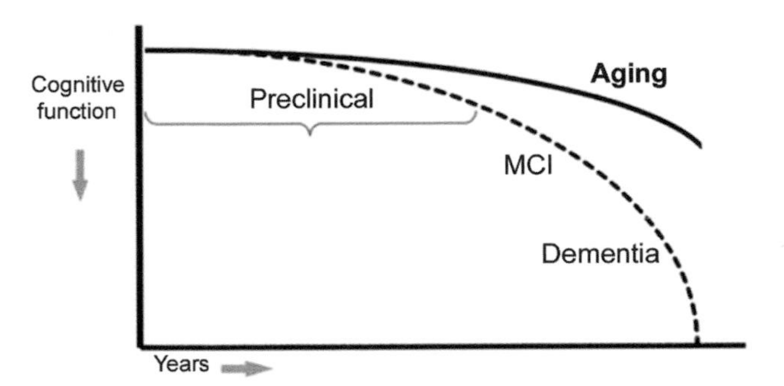

FIGURE 21.1 **Model of the clinical trajectory of Alzheimer disease (AD).** The stage of preclinical AD precedes mild cognitive impairment (MCI) and encompasses the spectrum of presymptomatic autosomal dominant mutation carriers, asymptomatic biomarker-positive older individuals at risk for progression to MCI due to AD and AD dementia, and biomarker-positive individuals who have demonstrated subtle decline from their own baseline that exceeds that expected in typical aging, but would not yet meet criteria for MCI. This diagram represents a hypothetical model for the pathological–clinical continuum of AD but does not imply that all individuals with biomarker evidence of AD-pathophysiological process will progress to the clinical phases of the illness. *Source: Modified from Sperling* et al. Alzheimers Dement. *2011;7(3):280–292,*[11] *with permission.*

cognitive domains coincide with deterioration in day-to-day function.[12] This new set of criteria relies heavily on the development and incorporation of biomarkers into clinical practice.[13,14]

Early Diagnosis of Alzheimer Disease

Well-validated biomarkers of AD, MCI, and possibly preclinical stages are urgently needed to develop and test therapeutics and to take advantage of the opportunity for early preventive treatments (Fig. 21.2). In cases of AD, the levels of Aβ peptides, $A\beta_{42}$ in particular, in the cerebrospinal fluid (CSF) are often low, and levels of phosphorylated tau may be higher than in controls. Magnetic resonance imaging (MRI) often discloses regional brain atrophy in AD patients, particularly involving the hippocampus and entorhinal cortex. Moreover, rates of atrophy, which correlate with changes in clinical status, may have predictive value for diagnosis. Positron emission tomography (PET) using [^{18}F]deoxyglucose or single-photon emission computed tomography (SPECT) commonly demonstrates decreased glucose utilization and the early reduction of regional blood flow in the parietal and temporal lobes.

The most exciting advance in imaging relevant to AD is the use of PET with antiamyloid radiolabeled tracers: Pittsburgh compound B (^{11}C-PIB), ^{18}F-Florbetapir, and others. These brain-penetrant tracers bind to Aβ with high affinity and can visualize the burden of Aβ amyloid *in vivo*. In comparison with controls, subjects with AD show marked retention of the tracer

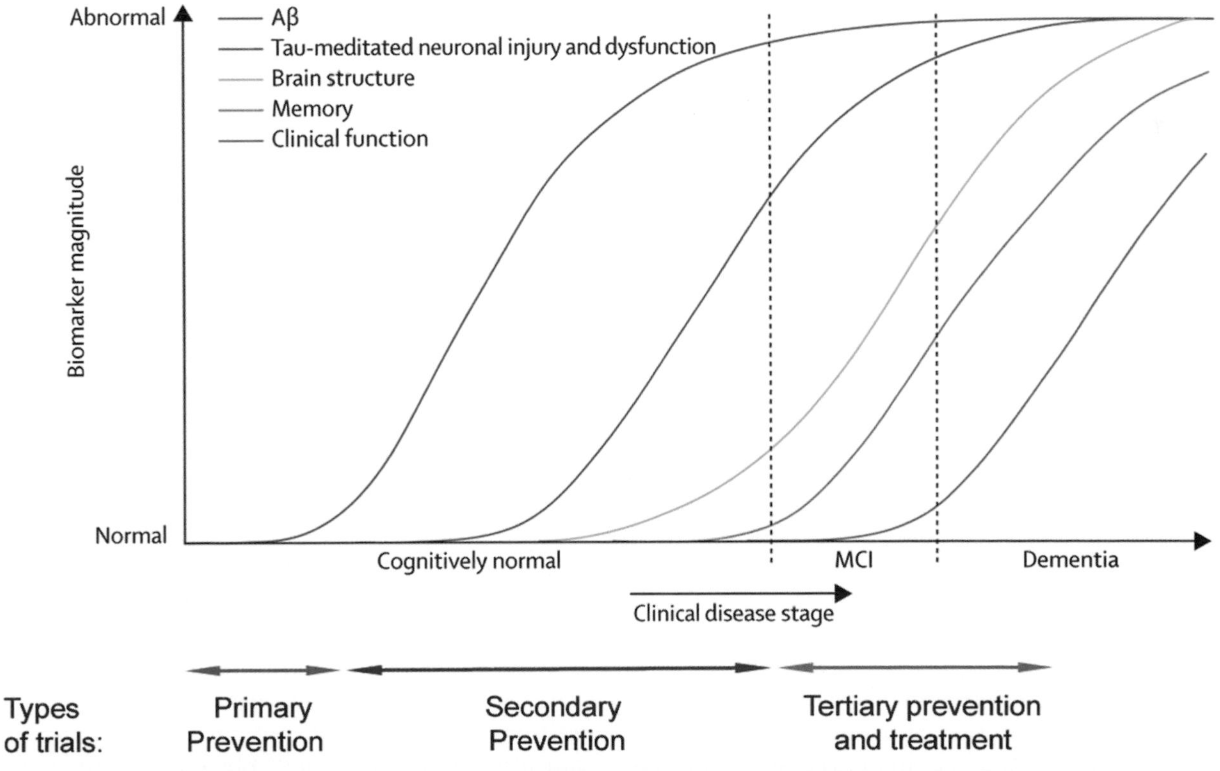

FIGURE 21.2 **Dynamics of biomarkers in the Alzheimer pathological cascade.** β-Amyloid (Aβ) is identified by cerebrospinal fluid (CSF) $A\beta_{42}$ or positron emission tomography (PET) amyloid imaging. Tau-mediated neuronal injury and dysfunction is identified by CSF tau or fluorodeoxyglucose-PET. Brain structure is measured by structural magnetic resonance imaging. MCI: mild cognitive impairment. *Source: Modified from Jack* et al. Lancet Neurol. *2010;9(1):119–128,*[13] *with permission.*

in several areas of brain that commonly accumulate amyloid. This amyloid imaging approach should eventually prove useful for enhancing the accuracy of diagnosis and assessing the efficacy of new anti-amyloid therapeutics. However, some instances of unexpectedly high uptake in cognitively normal individuals or poor tracer retention in clinically diagnosed cases of AD have implied individual variability in levels of the burden of Aβ and may decrease the predictive value of this measure.

Prospective, longitudinal study of individuals with autosomal dominant inherited mutations that result in AD supports the theoretical concept of long-term pathophysiological changes, at least in these familial cases of AD.[15] Biomarkers of brain amyloidosis ($A\beta_{42}$) (see Figs 21.2 and 21.4) in the CSF were estimated to decline 25 years before expected onset of symptoms. Aβ amyloidosis measured by PIB-PET and increased in levels of tau in the CSF were detected 15 years before the expected onset of symptoms. These pathophysiological changes in biomarkers were followed by cerebral hypometabolism and impaired episodic memory 10 years and global cognitive impairment 5 years before expected symptom onset.[15] Although the results of this study require confirmation with the use of longitudinal data on onsets of symptoms (expected onset of symptoms was estimated based on a history of AD in previous generations or relatives of patients), these data provide information on the rate and order of pathophysiological processes and the types of biomarker that should be considered at different stages of AD.

Further development of non-invasive imaging tools for the visualization of soluble Aβ would aid greatly in early diagnosis and the assessment of therapeutic efficacy. Longitudinal studies of blood and CSF laboratory measurements, imaging of amyloid and tau, and parallel documentation of cognitive status will help in finding appropriate markers that not only distinguish populations of patients from populations of controls, but also can be used as predictive or classification markers for a particular individual.[13]

Autosomal Dominant Mutations in Familial Alzheimer Disease

The majority of cases with sporadic AD exhibit the first clinical signs in the seventh decade. However, some individuals develop disease in midlife; in these cases, a family history of the illness is more likely. The genetics of AD are complex and heterogeneous, and exhibit age-related patterns: rare, early-onset familial AD mutations in APP and PS genes are transmitted in an autosomal dominant fashion, whereas late-onset AD without clear familial segregation is believed to reflect the influences of several risk factors. Early-onset disease (<65 years) represents less than 2% of the cases of AD.[1,20,21] These AD-causing mutations occur in three different genes located on three different chromosomes, but influence a common biochemical pathway: the production of Aβ species leading to a relative overabundance of $A\beta_{42}$. To date, more than 160 mutations in three genes (*APP*, *PS1*, and *PS2*) have been reported to cause this type of illness. The most frequently mutated gene, *PS1*, accounts for the majority of AD cases with onset before age 50. An overview of disease-causing mutations is available at the Alzheimer Disease & Frontotemporal Dementia Mutation Database (http://www.molgen.ua.ac.be/ADMutations) and Alzheimer Research Forum website (http://www.alzforum.org/mutations).

The majority of mutations in *APP*, *PS1*, and *PS2* affect BACE1 and γ-secretase proamyloidogenic cleavages of APP to increase the levels of all Aβ species or the relative amounts of toxic $A\beta_{42}$ (discussed below). Individuals with trisomy 21 or Down syndrome have an extra copy of APP (and other genes) in the putative obligate Down syndrome region; these individuals develop AD pathology relatively early in life.

Increased Risk Associated with the Apolipoprotein E_4 Allele

Late-onset AD, representing the great majority of all cases of AD, is defined as disease appearing after 65 years of age. To date, only the ε_4 allele of the apolipoprotein E gene (chromosome 19q13) has been consistently replicated in a large number of studies across many ethnic groups. While *ApoE*-ε_4 is a susceptibility allele, ε_2, a low-frequency allele, exhibits a weak protective effect. *ApoE*-ε_4 is neither necessary nor sufficient to cause AD, but operates as a genetic risk modifier by decreasing the age of onset in a dose-dependent manner. The biochemical consequences of *ApoE*-ε_4 in pathogenesis of AD are not yet fully understood, but this variant has been hypothesized to influence Aβ metabolism, Aβ aggregation and clearance, and cholesterol homeostasis. It is likely that additional late-onset AD loci or their combinations remain to be identified, since *APP*, *PS1*, *PS2*, and *ApoE* appear to account for less than 50% of the genetic variance in AD. Research has identified gene variants encoding ubiquilin-1 (*UBQLN1*) and sortilin-1 (*SORL1*) as risk factors, possibly by influencing splicing and trafficking, respectively. It is unclear how many of the newly recognized susceptibility loci will prove to be significant risk factors as opposed to causative variants. It should be emphasized that, in 500 independent association studies, no single gene has been demonstrated to contribute a risk approaching the same degree of consistency as *ApoE*-ε_4.

NEUROPATHOLOGY OF ALZHEIMER DISEASE

Neuritic Plaques

Neuritic Aβ plaques are composed of Aβ deposits of β-pleated sheet peptides surrounded by swollen neurites (nerve terminals) (Fig. 21.3). $A\beta_{1-40,42}$ and $A\beta_{11-40,42}$ amyloid peptides are derived by β- and γ-secretase cleavage of APP, as described below (Fig. 21.4). Aβ multimers assemble into β-pleated sheets, into protofilaments, and finally into fibrils; these aggregates are birefringent when stained with Congo Red or thioflavin dyes and viewed in polarized light or fluorescence illumination, respectively. Considerable debate exists concerning the Aβ species and conformational state exhibiting the greatest toxicity, with plaques, fibrils, protofibrils, or oligomers proposed as principal offenders. It is now believed that multimers, sometimes termed Aβ-derived diffusable ligands (ADDLs), are the main toxic entities.

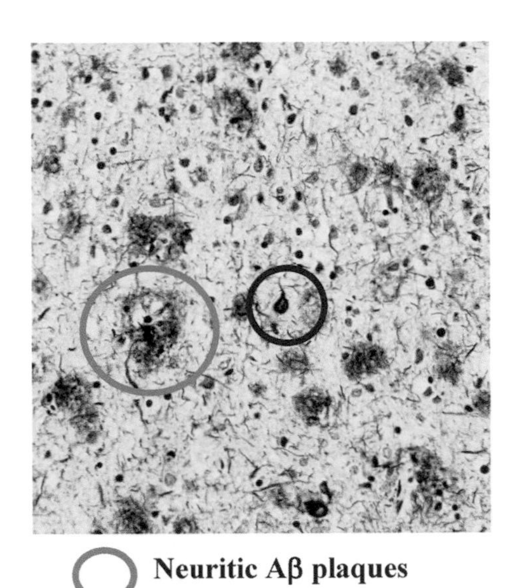

FIGURE 21.3 **Plaques and tangles.** Photomicrograph showing a neuritic plaque and a neurofibrillary tangle in an individual with autopsy-verified Alzheimer disease (Bielschowsky's silver stain.)

In one model, APP and proamyloidogenic secretases are transported to terminals and thus neurons are thought to be the major source of APP that gives rise to Aβ. At synapses, BACE1 cleaves APP to form amyloidogenic C-terminal derivatives, which are then cleaved by γ-secretase to generate $A\beta_{40}$, $A\beta_{42}$, and $A\beta_{43}$ peptides. Released normally at terminals, Aβ may influence synaptic functions, perhaps behaving as a modulator depressing activity at excitatory, glutamatergic synapses via the NMDA receptor.[16] With increasing accumulations of $A\beta_{42}$ multimers at terminals, synaptic functions, including long-term potentiation, are disrupted. In this scenario, neuritic amyloid plaques, a classical feature of AD, are complex structures, representing sites of Aβ-mediated damage to synapses associated with degeneration of neurites and disconnection of terminals from targets. Surrounding plaques are astrocytes and microglia, which produce cytokines, chemokines, and other factors (including complement components) involved in inflammatory processes that may increase the damage to neuronal circuits.

Neurofibrillary Tangles

Tangles are fibrillar intracytoplasmic inclusions in the cell bodies and proximal dendrites of affected neurons, whereas neuropil threads and neurites are predominantly swollen, filament-containing dendrites

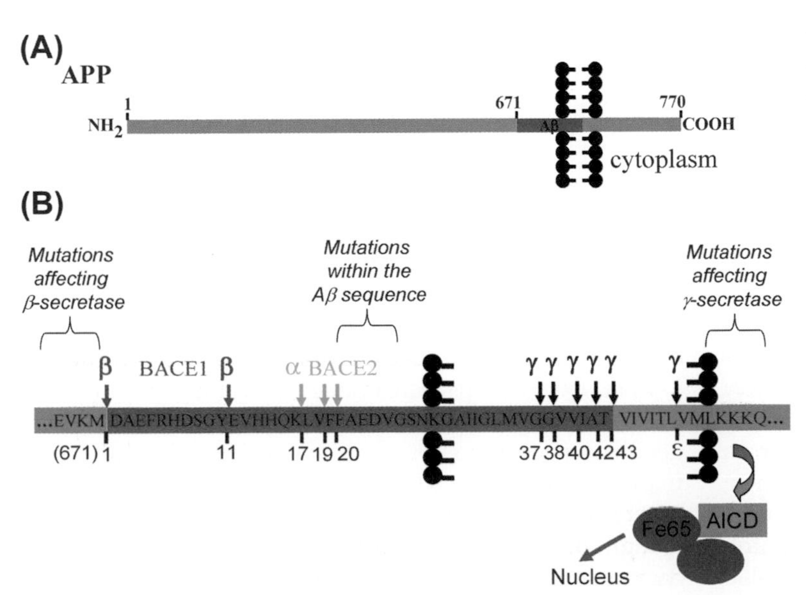

FIGURE 21.4 **Structure of (A) amyloid precursor protein (APP) and (B) Aβ domain of APP.** α, β, and γ relate to α, β, and γ-secretase activities, respectively. ε cut has been shown to depend on γ-secretase activity. AICD: APP intracellular domain; Fe65: a nuclear adaptor protein.

and distal axons and terminals, respectively (Fig. 21.3). These intracellular lesions are rich in PHFs comprised of poorly soluble hyperphosphorylated isoforms of tau, a low molecular weight microtubule-associated protein.[17] In human brain, alternative splicing from a single tau gene leads to the formation of six tau isoforms, consisting of three isoforms of three-repeat tau (3-R tau) and three isoforms of four-repeat (4-R) tau, the latter derived by inclusion of exon 10 in the transcript. Normally, tau, synthesized in neuronal cell bodies, is transported anterograde in axons, where it acts, via repeat regions that interact with tubulin, to stabilize tubulin polymers critical for microtubule assembly and stability.[17] The post-translationally modified tau, which exhibits abnormal conformations, differs somewhat in the different tauopathies: in cases of AD, the PHFs are comprised of six isoforms of tau; in contrast, the inclusions occurring in cases of progressive supranuclear palsy and cortical basal degeneration are characterized by 4-R tau, while the inclusions seen in individuals with Pick disease are enriched in 3-R tau.[17]

In one hypothetical model linking Aβ and phosphorylated tau, $A\beta_{42}$ damages terminals leading to synaptic disconnections which, perhaps preferentially in primates with six isoforms of tau, leads to a retrograde signal that ultimately triggers the activation of kinases (or the suppression of phosphatases) whose activities lead to excessive phosphorylation of tau at certain residues. Subsequently, conformational changes in the protein cause the formation of PHFs. Since the cytoskeleton is essential for maintaining cell geometry and for the intracellular transport of proteins and organelles, disturbances of the cytoskeleton can lead to alterations in axonal transport which, in turn, can compromise the functions and viability of neurons. Eventually, affected nerve cells die (possibly by apoptosis) and extracellular tangles remain as "tombstones" of the nerve cells destroyed by disease (Fig. 21.3).

GENETICS AND MOLECULAR BIOLOGY OF ALZHEIMER DISEASE

The theories on mechanisms of AD emerged from studying the neuropathology and biochemistry of plaques and tangles. In 1984, George Glenner and colleagues isolated the principal protein that constitutes Aβ-amyloid, first from AD brains and then from the brains of patients with Down syndrome. This approximately 4 kDa protein is now known as Aβ but was originally named "β-protein" to reflect the earlier findings from studies of amyloid proteins from other organs (liver, spleen). Fibrils extracted from amyloid proteins usually have a cross-β X-ray diffraction pattern. Determination of amino acid residues and their sequence in Aβ enabled the cloning of the gene from which Aβ is derived. This gene encodes a large transmembrane protein, APP.

Cleavage of Amyloid Precursor Protein by β- and γ-Secretase

APP is processed by β- and γ-secretase enzymes, resulting in the release of the ectodomain of APPs, the production of a cytosolic fragment termed APP intracellular domain (AICD), and the generation of several Aβ peptides (Fig. 21.4). In the CNS, Aβ peptides are generated by sequential endoproteolytic cleavages of neuronal APP by two membrane-bound enzymes: BACE1 cleaves APP at the Aβ +1 and +11 sites to generate APP-β carboxyl-terminal fragments (APP-βCTFs)[6,8]; γ-secretase complex, via regulated intramembranous proteolysis, cleaves APP-βCTFs at several sites including $A\beta_{40}$, $A\beta_{42}$, and $A\beta_{43}$ to form these peptides. The γ-secretase cleavage of APP-βCTF or αCTF releases the AICD, which forms a multimeric complex with Fe65, a nuclear adaptor protein (reviewed in De Strooper *et al.*[18]). It has been suggested that the complex of Fe65 and Aβ, or Fe65 alone (in a new conformation), enters the nucleus and binds to the histone acetyltransferase, Tip60, to influence gene transcription. This signaling mechanism is analogous to that occurring in the Notch1 pathway following the S3 cleavage of NEXT to produce NICD. In other cells in other organs, APP can also be cleaved endoproteolytically within the Aβ sequence through alternative, non-amyloidogenic pathways involving α-secretase [tumor necrosis factor-α converting enzyme (TACE)] or BACE2. The α-secretase and BACE2 cleavages, which occur in non-neural tissues, preclude the formation of Aβ peptides and thus are thought to protect these organs from Aβ amyloidosis.

BACE1 and BACE2, encoded by genes on chromosomes 11 and 21, respectively, are transmembrane aspartyl proteases that are directly involved in the cleavage of APP.[6,8] BACE1 preferentially cleaves APP at the +11 to +1 sites of Aβ in APP and this enzyme is essential for the generation of Aβ (Fig. 21.4). Significantly, the APP Swedish mutation (APPswe; see "Amyloid Precursor Protein and Mutations", below) is cleaved perhaps 100-fold more efficiently at the +1 site than is wild-type APP. Thus, this mutation greatly increases BACE1 cleavage and accounts for the elevation of Aβ species in the presence of this mutation. The expression of BACE1 is increased in certain regions of the brain in some cases of sporadic AD. Thus, BACE1 is the main neuronal β-secretase and is responsible for the critical penultimate proamyloidogenic cleavages. Although BACE1 messenger RNA (mRNA) is present in a variety of tissues (particularly the pancreas), levels of this protein are low in most non-neural tissues. In the pancreas, BACE1 mRNA is high, but the transcript is alternatively spliced to produce a smaller protein incapable of cleaving APP.

BACE2 mRNA, which is present in a variety of systemic organs, is very low in neural tissues, except for scattered nuclei in the hypothalamus and brainstem. BACE2 activity appears to be virtually undetectable in brain regions involved in AD. BACE2 is responsible for the generation of antiamyloidogenic cleavages at +19/+20 of Aβ (Fig. 21.4). Thus, BACE2, an antiamyloidogenic enzyme, acts like α-secretase or TACE, which cleave between residues 16 and 17 of the Aβ peptide.

γ-Secretase is essential for the regulated intramembraneous proteolysis of a variety of transmembrane proteins. It is a multiprotein catalytic complex, which includes: PS; Nicastrin (Nct), a type I transmembrane glycoprotein; and Aph-1 and Pen-2, two multipass transmembrane proteins.[18] PS1 is isolated with γ-secretase under specific detergent-soluble conditions; it is selectively cross-linked or photoaffinity labeled by transition state inhibitors. Substitutions of aspartate residues at D257 in transmembrane segment 6 (TM6) and at D385 in TM7 have been reported to reduce secretion of Aβ and cleavage of Notch1 *in vitro*; $PS1^{-/-}$ cells show decreased levels of secretion of Aβ. Aph-1 and Pen-2 are transmembrane proteins with seven and two transmembrane regions, respectively. The functions of these proteins and their interactions with each other in the complex and in γ-secretase activity are not yet fully defined. PS1 may act as an aspartyl protease, function as a cofactor critical for the activity of γ-secretase, and play an additional role in trafficking of APP or proteins critical for enzymic activity to the proper compartment for γ-secretase cleavage. In one model, Aph-1 and Nct form a precomplex that interacts with PS. Subsequently, Pen-2 influences the cleavage of PS into two fragments; thus, all proteins are critical for the activities of the γ-secretase complex. Significantly, γ-secretase cleaves both Notch-1 and APP, generating intracellular peptides termed NICD and AICD, which influence transcription. As described above, the AICD interacts with FE65, a cytosolic adapter, and this interaction leads a signal which influences transcription. Results of targeting of PS1, Nct, and Aph-1 in mice are consistent with the concept that these are critical components of the γ-secretase complex. The phenotypes of targeted PS1, Nct, and APH-1 mice are the result of impaired Notch1 signaling.

Amyloid Precursor Protein and Mutations

APP, a type I transmembrane protein existing as several isoforms, is abundant in the nervous system, rich in neurons, and transported anterograde in axons to terminals. The three most abundant isoforms of APP are APP770, APP751, and the predominantly neuronal APP695. The specific functions of APP are not yet fully understood. From the N- to C-terminal the APP domains include a heparin binding and growth factor-like domain (HBD1/GFLD), a copper binding domain (CuBD), a zinc binding domain (ZnBD), an acidic region (DE), a Kunitz-type protease inhibitor domain (KPI; not present in APP695), a second heparin binding domain (HBD2), a random coiled region (RC), the amyloid-beta domain (Aβ), and an intracellular C-terminal domain (AICD). APP production is developmentally regulated with expression increasing during neuronal differentiation, maximal during synaptogenesis, and declining when mature connections are established, implying that APP expression is important during development and aging. Some studies suggest a role of APP in dendritic spine formation, synaptic transmission, learning and memory, neurite outgrowth and motility, and neurogenesis. APP is shown to homodimerize as well as interact with soluble APP fragments and other receptors to influence downstream signaling events. For example, soluble APP fragments created by cleavage of α-secretase (sAPPα) (Fig. 21.4) may serve a neuroprotective function, whereas soluble APP fragments created by cleavage of β-secretase may be toxic and trigger axon degeneration.

The numerous mutations in *APP* linked to AD can be roughly divided into three classes based on their locations relative to the Aβ sequence. *APP* mutations located immediately before the Aβ sequence (Fig. 21.4) usually modify the efficacy of β-secretase cleavage. One of the best known examples is a double mutation involving codons 670 and 671 named after a Swedish family with fAD (APPswe mutation). This double mutation enhances many-fold the BACE1 cleavage at the N-terminus of Aβ; the result is substantial elevation in levels of full-length Aβ peptides. *APP* mutations located immediately after the Aβ sequence (Fig. 21.4) are thought to affect the enzymic activity of γ-secretase or modify cleavage to increase the relative content of the longer form of Aβ, $A\beta_{42}$. For example, with APP717 mutations, γ-secretase cleavage is altered, leading to an increased ratio of $A\beta_{42}/A\beta_{40}$. Finally, a third type of *APP* mutation is located within the Aβ sequence. Although the mechanism of these mutations is not fully understood, they appear to influence the biology of Aβ by promoting local oligomer/fibril formation or changing the propensity of Aβ to bind to other proteins such as ApoE (see below) and affecting Aβ clearance.

Presenilin-1 and -2 and Mutations

PS1 and PS2, two highly homologous and conserved 43–50 kDa multipass transmembrane proteins, are involved in Notch 1 signaling pathways critical for cell fate decisions. They are endoproteolytically cleaved to form an N-terminal fragment (≈28 kDa) and a C-terminal fragment (≈18 kDa), both of which (along with several other proteins described below) are critical components of the γ-secretase complex. Nearly 50% of cases of early-onset fAD are linked to over 100 different mutations in *PS1*. A small number of *PS2* mutations cause autosomal dominant fAD in several pedigrees. The majority of abnormalities in *PS* genes are single amino acid missense mutations that change the γ-secretase activities and increase the relative levels of the toxic $A\beta_{42}$ peptides. In addition to changes in Aβ production, *PS1* mutations can play roles in cell-signaling events (some modulated by interaction with APP) in the initiation of apoptosis and development of neurodegeneration.

Apolipoprotein E Alleles

ApoE carries cholesterol and other lipids in the blood. In humans, three alleles exist: $ApoE_2$, $ApoE_3$, and $ApoE_4$. The $ApoE_3$ allele is most common in the general population (frequency of 0.78), whereas the allelic frequency of $ApoE_4$ is 0.14. However, in clinic-based studies, the $ApoE_4$ allelic frequency in patients with late-onset disease (>65 years of age) is 0.50; thus, the presence of $ApoE_4$ increases the risk of AD. Significant differences exist in the abilities of *ApoE* isoforms to bind Aβ and these features of the individual protein are hypothesized to differentially influence aggregation, deposition and/or clearance of Aβ by different ApoE isoforms.

Modeling Aβ Amyloidosis and Tau Pathologies

Animal models that recapitulate critical aspects of disease progression are essential for understanding the disease mechanism and developing successful AD therapy. Early discoveries of mutations in *APP* and presenilins (*PS1* and *PS2*) in cases of fAD[19] set the stage to create multiple transgenic mouse models of Aβ amyloidosis using a variety of strategies.[20,21] These animal models range from mice transgenic for a single gene to more complex double and triple transgenic animals, which reproduce important features of AD, including elevated levels of Aβ (particularly more amyloidogenic $A\beta_{1-42}$ peptide), amyloid plaques, reductions in neurotransmitter markers; age-related cognitive impairments, tau-immunoreactive PHF, and less commonly (in case of double or triple transgene), death of some neuronal populations. There is remarkable consistency among different APP transgenic mice in terms of the age-dependent cellular abnormalities characteristic of AD: Aβ amyloid deposits, neuritic plaques, and glial responses.[20] These histopathological profiles have been identified in mice that express different isoforms of mutant human APP and with several different transgene constructs. A key factor is that the production of Aβ peptide is elevated sufficiently to induce plaque-related pathology.

Despite the success of transgenic approaches in reproducing some of the features of Alzheimer-type cerebral amyloidosis, the modeling of another cardinal feature of AD, tau-related pathology, has proved more complicated. Originally, some researchers hypothesized that robust deposition of Aβ amyloid in mouse models would also result in the development of intracellular tau aggregates analogous to NFTs and neuropil threads. However, tau pathology was rarely observed in APP transgenic models and mainly was represented by increased tau phosphorylation. It has been suggested that the paucity of tau abnormalities in various lines of mutant mice with Aβ amyloidosis may be related to differences in tau isoforms expressed in these species as compared to humans. To model tau pathology in mice, researchers used genetic approaches to overexpress human wild-type or mutated tau.[17,22–25] Transgenic mice that expressed full-length human tau showed abnormal distribution of hyperphosphorylated tau in some neurons, but overall neuropathology in these mice was limited. On the other hand, transgenic mice expressing tau mutations linked to autosomal dominant frontotemporal dementia with parkinsonism (FTDP), such as tau_{P301L}, form abnormal neuronal tau-containing filaments that have striking similarities with the NFTs observed in human cases of AD or FTDP.[22,23]

Since Aβ appears to be a trigger for AD, researchers have used animal models to test whether the development of tau pathology, as well as the other neurodegeneration occurring in the AD brain, are downstream events of the Aβ pathology. Some evidence has accumulated to confirm dynamic interactions between Aβ- and tau-related pathologies. The tau filaments in the brains of tau transgenic mice are considerably less numerous than in AD brains; however, an injection of $A\beta_{42}$ fibrils into the brains of tau_{P301L} mice dramatically increases the number of tangles in neurons projecting to the sites of Aβ injection.[22] Interactions between Aβ- and tau-related pathologies were also demonstrated in mice that coexpress APPswe and tau_{P301L} and exhibit enhanced tangle-like pathology in the limbic system and olfactory cortex.[23] These observations are consistent with the hypothesis that Aβ, if present in proximity to axon terminals, can facilitate the formation of tangles in neuronal cell bodies. Further attempts to create a mouse model that combines amyloidosis and tau pathology led to development of a triple transgenic mouse (3xTg-AD), which was generated by microinjecting APP_{swe} and tau_{P301L} into single cells derived from monozygous $PS1_{M146V}$ knockin mice.[25] These mice develop age-related plaques and tangles and, alongside other models,[26] have been a valuable tool to investigate functional outcomes of Aβ and tau pathology. When Aβ levels were reduced by an immunization approach in the 3xTg-AD mice, NFT pathology was significantly reduced. In addition, tau appears to function downstream of at least some of the effects of Aβ-related toxicity. Knocking out endogenous mouse tau attenuated the toxic effects of Aβ amyloidosis and increased resistance to the epileptogenic agent pentylenetetrazole in transgenic mice expressing human APP. Taken together, these data suggest that the presence of Aβ amyloidosis in brain plays a key role in the development of tau pathology and further downstream events associated with neuronal dysfunction. However, more recent studies suggest that tau aggregation may develop independently of the presence of Aβ peptides. When the *BACE1* gene is knocked out in 3xTg-AD mice, Aβ amyloidosis is completely suppressed, while the development of tau pathology is unchanged. It should be kept in mind that all these data were obtained from studies using frontotemporal dementia-related tau transgenic mice. In the future, animal models that more faithfully represent Aβ and tau pathogenesis in cases of AD are needed to understand the dynamic interactions between these pathologies.

It is important to note that no transgenic or mutant mouse model can provide an all-encompassing view of the biology of a human disease, and particularly a disease involving changes in cognitive capacities such as AD. Only a consensus about the most common and reproducible features from different AD models can ensure appropriate translation of preclinical findings into realistic expectations for effective experimental therapies in the clinic.[27]

Revisiting the Amyloid Cascade Hypothesis

The use of transgenic models of AD in the early twenty-first century has significantly furthered our understanding of the pathogenesis of the disease. The original amyloid cascade hypothesis proposed that the cause of neurodegeneration in AD is deposition of Aβ into plaques, a process by which an initial early Aβ insult leads to a series of downstream events ranging from inflammation to synaptic loss triggering hyperphosphorylation of tau and, finally, to the death of susceptible neurons. Strong correlations between levels of Aβ plaques and cognitive deficits have been reported in different mouse models of amyloidosis, supporting the roles of Aβ plaques in memory decline. A lack of significant correlations between plaque load and dementia in AD patients[24] may be explained by the notion that Aβ plaques that began to accumulate at the very early stages of disease could have less predictive power for dementia scores than events occurring much later in the cascade of pathologies, such as accumulation of NFTs. The idea of the critical role of Aβ plaques in cognitive decline survived a more stringent test for causality when a newly discovered anti-Aβ active immunization approach was used in APP transgenic models leading to amelioration of Aβ deposits and to rescue of memory deficits. However, results of the passive immunization approach demonstrate that cognitive deficits in mouse models of amyloidosis can be acutely rescued by systemic treatment with anti-Aβ antibodies without significant changes in levels of amyloid plaques.[28,29] These findings prompted a revisiting of the amyloid cascade hypothesis to suggest that total amounts of Aβ accumulated during aging in the form of plaques may represent a surrogate marker for small "non-plaque" Aβ assemblies that play a primary role in memory impairment.[29] The amyloid hypothesis was revised to include multiple Aβ assemblies as possible toxic entities: fibrils, protofibrils, dimers, trimers, dodecamers, and ADDLs. Considerable debate still exists concerning which of the Aβ species and conformational states are the principal toxic entity; however, it is likely that multiple Aβ species and assemblies are balanced and represent a spectrum of toxicities dominated by various Aβ assemblies at different stages of disease.

Developments in our understanding of tau-related mechanisms in AD have many analogies to the amyloid story. The idea that NFTs are the main offenders mediating neuronal death and cognitive deficits has been revised to view NFTs, like amyloid plaques, as the final pathological "tombstones" rather than the main neurotoxic agents. Neurotoxicity has been attributed to tau species that are intermediate between normally phosphorylated tau and the hyperphosphorylated fibrils.[17] As in mice modeling Aβ amyloidosis, a dissociation between cognitive recovery and the continuous presence of aggregates (in this case NFTs) has been demonstrated in mice that conditionally overexpress mutated tau (Tet-off:Tau_{P301} mice).[30] The same study demonstrated that in addition to amelioration of cognitive deficits, the inhibition of Tau_{P301} production stopped progression of neuronal loss but was ineffective in halting further accumulation of NFTs. The data from this and other studies served as a basis for further refinement of the amyloid cascade hypothesis to incorporate the idea that some facets of the cascade may become self-propagating and independent from the initial trigger.[9,31]

It is clear that the amyloid cascade hypothesis must be refined to include non-amyloid factors, including some functions of AD-related genes, that may contribute significantly to AD. Potential mechanisms that could be operative in the pathogenesis of AD include defective endolysosomal trafficking, altered intracellular signaling cascades, and impaired neurotransmitter release.[32] An integrated view of the amyloid-dependent and -independent mechanisms could promote molecular understanding of the pathogenesis and help to reconcile the findings that cannot be explained solely by the amyloid hypothesis.

A discrepancy between the results obtained in preclinical models and the results of recent clinical trials serves as a call to revisit and adjust our theoretical views. When such discrepancies happen too often and in large trials this calls for a paradigm shift that questions every stage of translation: from how we model the disease to how we run clinical trials.

In the following sections, some examples of AD therapeutics are presented to discuss whether preclinical studies in transgenic models would have been useful in predicting the limitations in efficacy and side effects observed in clinical trials.

CURRENT AND FUTURE THERAPIES FOR ALZHEIMER DISEASE

Cholinergic Hypothesis of Alzheimer Disease

Despite substantial progress in understanding the molecular mechanisms and neurobiology of AD, recent therapies have been based on early observations of the pathology and biochemistry of AD. Initial studies showed a severe loss of cholinergic markers in the cerebral cortex that was correlated with senile plaques and dementia scores in AD. Later discoveries revealed that brains of patients with advanced AD are characterized by severe loss of cholinergic neurons providing major inputs to the cortex and hippocampus: the nucleus basalis and septal nuclei.[1] These studies led to a cholinergic hypothesis of AD[33] that served as a rationale for the development of acetylcholinesterase inhibitors (AChEIs) as a treatment. These agents prolong the action of acetylcholine at the postsynaptic cholinergic receptors and enhance cholinergic functions. More recently, AChEIs have been shown to have a number of additional effects that potentially have disease-modifying qualities such as neuroprotection and modulation of the β-amyloid pathway through activation of nicotinic acetylcholine receptors. Activation of the M_1 muscarinic receptors (M_1 mAChR) also has disease-modifying potential, as M_1 selective muscarinic agonists have been shown to decrease Aβ levels and tau hyperphosphorylation *in vitro* and to rescue cognitive deficits and decrease $A\beta_{42}$ and tau pathologies in relevant *in vivo* models.

Treatment with AChEIs, however, results in a modest therapeutic effect, only temporarily halting disease progression.[3] The rather mild effect of AChEIs on memory deficits is not surprising considering the outcomes of early studies in aged monkeys which, like humans, develop neuritic plaques. These studies showed a very narrow range of effective concentrations of AChEIs that moderately improved memory performance. The determination of appropriate doses can be further complicated by a dramatic interindividual variability in the optimal dose effective in aged subjects (monkeys and humans). When translated to the clinic, this narrow dose–response characteristic of AChEIs results in a mild average response in a population of AD patients, with only some patients showing cognitive improvement (responders), presumably due to a particular stage of cholinergic decline or other individual characteristics.

The benefits of treatment for dementia, and of AChEI treatment in particular, are more complex than an improvement on a cognitive measure. In a situation where mechanism-based treatments are not available, any treatment, even with relatively low benefits, is highly valuable for patients and caregivers and can make an important difference to their quality of life. In addition, AChEIs have beneficial effects on some behavioral symptoms of AD (physical aggression, screaming, restlessness, anxiety, depression, apathy, agitation, hallucinations, delusions, and sleep disturbances),[34] although none of these treatment effects is large.[3] These behaviors have serious consequences for patients and caregivers, worsening their quality of life and resulting in earlier institutionalization. AChEI-induced amelioration of psychiatric symptoms in AD patients can be related to the well-known role of the cholinergic system in attention and the emergent link between attentional deficits and development of at least some neuropsychiatric symptoms.[34]

Alzheimer Disease as a Failure of Multiple Neurotransmitter Circuits

In addition to the degeneration of cholinergic neurons, AD is associated with the early and progressive degeneration of monoaminergic neurons: serotonergic (5-hydroxytryptamine) neurons in the raphe and the noradrenergic neurons in the locus ceruleus. Mouse models of amyloidosis (without obvious neurofibrillary pathology) demonstrate that mechanisms related to Aβ production and accumulation are necessary and sufficient for degeneration of monoaminergic neurons. Degeneration of these neurons in APPswe/PS1ΔE9 mice appears first at axon terminals in the cortical/hippocampal areas with Aβ pathology and progresses to cell bodies in brain areas spared Aβ pathology. This overall pattern of neurodegeneration is consistent with findings in AD, in which monoaminergic neuronal loss occurs without local deposition of Aβ. The progression of monoaminergic neurodegeneration and cholinergic deficits in APPswe/PS1ΔE9 mice coincides with the onset and progression of cognitive abnormalities, with episodic-like memory being the most sensitive to these insults.

The nature of AD as a failure of multiple neurotransmitter systems was recognized as the main reason for the low efficacy of AChEIs even at the time when the cholinergic hypothesis was first formulated: "It may ... be necessary to simultaneously improve ... the balance between the cholinergic and other neurotransmitter systems in order to substantially reduce behavioral impairments".[33] This statement still outlines directions for future research. Extensive observations from rodent models indicate that deficits in a single neuromediator system (reproduced by a pharmacological blockade or lesion) may be necessary but not sufficient to reproduce cognitive impairment (reviewed in Kenton *et al.*[35]). Simultaneous pharmacological blockade of at least two neuromediator systems results in more dramatic and more easily detectable memory deficits.[35] These experimental data support the idea that when multiple neurotransmitter systems fail, amelioration of cognitive impairment requires treatments targeting multiple systems.

Development of Mechanism-Based Therapeutics

Several strategies have been investigated for possible disease-modifying effects in AD (Fig. 21.5). Some studies include drug targets focused on amyloid processing, including inhibition of Aβ production, facilitation of Aβ breakdown or clearance, and interference with Aβ oligomerization.[9,37,38] Tau-focused therapies are being developed, including drugs inhibiting tau phosphorylation or stabilizing microtubules.[17] There has also been an interest in developing anti-inflammatory and neurotrophic/neuroprotective agents or dietary vitamin supplementation. Despite numerous clinical trials, to date no experimental agent has survived the ultimate test of benefits in a phase III clinical trial. Failures include trials with AN1792 (active anti-Aβ immunization), Dimebon (an antihistamine with additional multiple mechanisms of action), the antioxidant *Gingco biloba*, the γ-secretase modulator tarenflurbil, and the γ-secretase inhibitor (GSI) semagacestat. Most of these therapeutics, particularly those focused on antiamyloid strategies, have been validated in mouse models of amyloidosis. The following sections discuss some of these experimental therapeutics, and whether limitations in their efficacy and side effects can be observed or predicted in AD mouse models.

Therapeutics Targeting Production of Aβ

INHIBITION OF β-SECRETASE

Since the discoveries of mutations associated with familial forms of AD, intensive studies have been

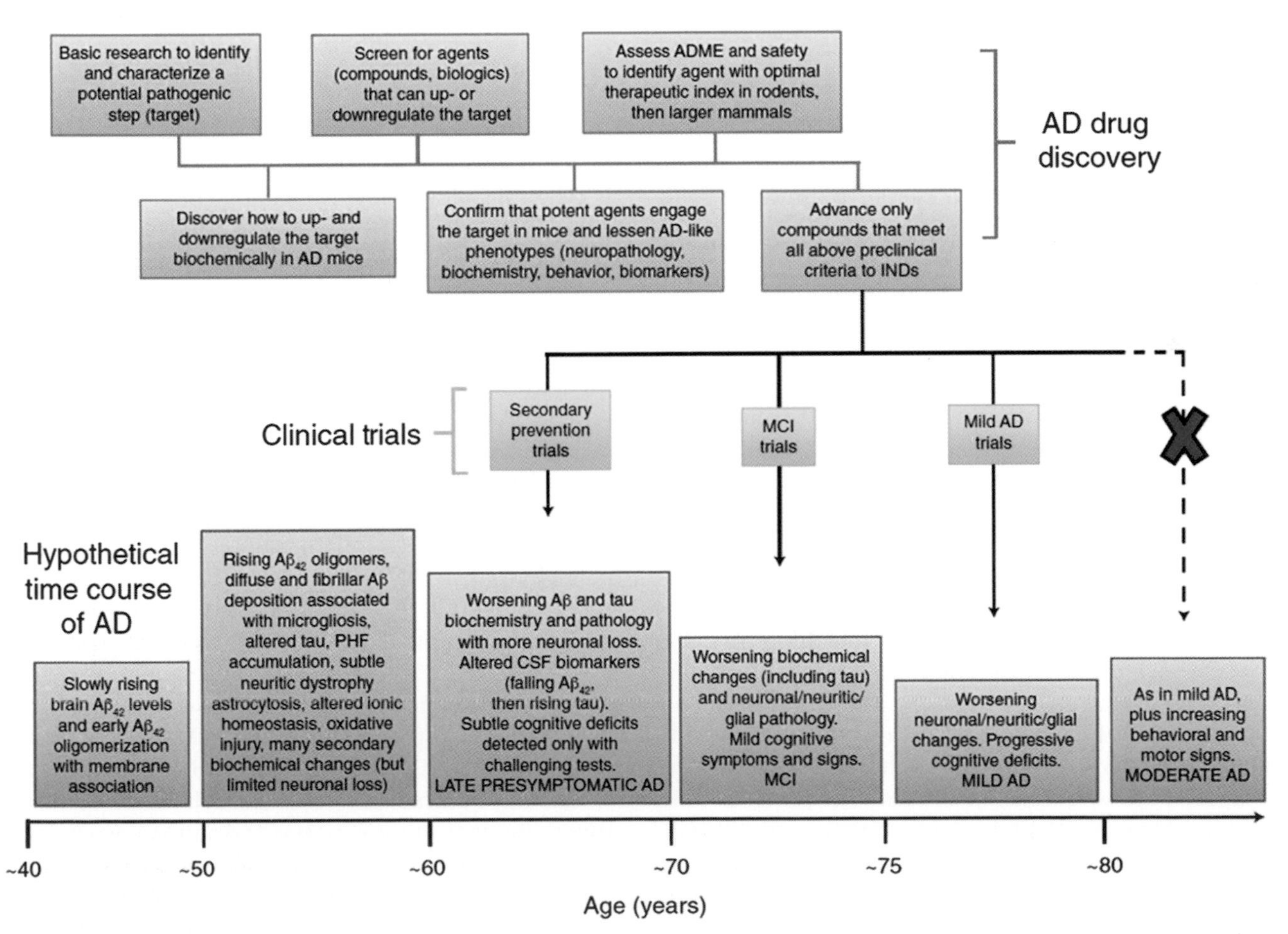

FIGURE 21.5 **Intersecting disease-modifying agents for Alzheimer disease (AD) with the course of the disease.** Blue boxes show steps in the discovery of compounds or biologics appropriate for investigational new drugs (INDs) in AD. Red boxes indicate speculative stages in the long presymptomatic and symptomatic phases of AD in a hypothetical individual who undergoes Aβ buildup for one of several possible reasons (e.g. presenilin or APP mutation, *ApoE$_4$* inheritance, or increased BACE activity) and develops MCI by around age 70. Green boxes indicate categories of clinical trial depending on the stage of AD. Red X show trials with anti-Aβ treatment in moderate AD that are unlikely to have any beneficial effects on AD. PHF: paired helical filaments; CSF: cerebrospinal fluid. *Source: Modified from Selkoe.* Nat Med. *2011;17(9):1060–1065,*[36] *with permission.*

initiated to understand the cell biology and biochemistry of Aβ production from its precursor, APP. Because BACE1 cleavage of APP is a critical rate-limiting step in Aβ amyloidosis, it has been suggested that inhibition of BACE1 would ameliorate Aβ deposition in AD.[8] Supporting this notion are studies demonstrating that deletion of BACE1 prevents Aβ secretion in cultured neurons and in the brain,[6] and that mutant APP mice lacking BACE1 do not develop Aβ plaques or Aβ-related memory deficits.[39,40]

Original optimism for the development of pharmacological BACE1 inhibitors was based on the successful precedent in drug development of an inhibitor of an HIV protease that is, like BACE1, an aspartyl protease. The discovery of such BACE1 inhibitors has proved to be difficult because the β-secretase active site for substrate recognition has a large cleft that is structurally incompatible with the requirement for small-molecule blood–brain barrier-penetrating inhibitors with high potency and selectivity. Significant advances have been made, however, and the first candidates (LY2886721, Eli Lilly and Co.; CTS21166, CoMentis; HPP854, TransTech Pharma) have progressed to clinical trials.

INHIBITION OF BACE1

In preclinical studies, some of the BACE1 inhibitors that were conjugated to a carrier peptide to facilitate penetration of the blood–brain barrier were able to reduce brain levels of Aβ after systemic injections in mutant APP transgenic mice. A study of a new generation of BACE1 inhibitors (without conjugated carriers) showed brain penetration sufficient to reduce the interstitial fluid concentration of Aβ in the brain that coincided with a reduction of Aβ in plasma.[41] In addition, treatment with this inhibitor ameliorated cognitive deficits in the Tg2576 model.[41] Although the degree of inhibition of new production of Aβ was significant (≈50% and 60% Aβ reduction in interstitial fluid and plasma, respectively), there were no acute effects on memory deficits. Treatment with the BACE1 inhibitor required at least 4 months for cognitive benefits to be detectable.[41] Another important finding from this study is a modulation of efficacy with age: transgenic mice older than 16 months did not benefit from the treatment as did younger mice. The decrease in efficacy with advanced age or disease stage is in agreement with earlier studies in which BACE1 inhibition was modeled by genetic ablation of a *BACE1* gene.[40] Full deletion of BACE1 in an aggressive model of amyloidosis, APPswe/PS1ΔE9 mice, prevented both Aβ deposition and age-associated cognitive abnormalities. However, functional outcomes of partial BACE1 deletion in BACE1 heterozygous knockout mice (≈50% of BACE1 activity) declined as aging progressed, possibly owing to compromised Aβ clearance mechanisms in aged animals[40] or aging-related increase in *BACE1* expression. In this study, despite a 50% reduction in BACE1 activity, 20–24-month-old mice had levels of Aβ deposition as high as in mice with full BACE1 activity.

Another group has published data on the effects of a structurally different non-competitive BACE1 inhibitor (TAK-070) that ameliorated Aβ pathology and behavioral deficits in Tg2576 mice.[42] Although the reduction in Aβ levels was modest, the authors proposed that the increase in sAPPα by non-competitive BACE1 inhibition may have an additional benefit, resulting in amelioration of cognitive deficits. In contrast to the previously discussed BACE1 inhibitor, TAK-070 resulted in cognitive benefits after short-term (2 week) treatment when tested in young Tg2576 mice (5 months of age). Acute efficacy in rescuing cognitive deficits has been documented in young mice of the same model using a passive immunization approach.[29] Although no testing of TAK-070 effects was presented for older mice, based on other preclinical studies the cognitive benefits of TAK-070 may decrease as aging and disease progress.

Detailed investigation into the efficacy and limitations of partial inhibition of BACE1 is particularly important because complete inhibition of this enzyme could be associated with problems. Aβ may normally play an important role in modulating activities of certain synapses[16] and strong BACE1 inhibition may result in Aβ deficiency below physiological levels. In addition, other putative substrates for BACE1 have been identified, indicating that BACE1 has multiple functions. β-Amyloid precursor protein-like proteins (APLP1 and APLP2) are processed by BACE1 to act via the same nuclear target (Tip60 in a complex with Fe65), suggesting that BACE1 cleavage regulates a common function of APP and APLPs in neurons. Other substrates of BACE1 include the low-density lipoprotein receptor-related protein (LRP), β-galactoside α2,6-sialyltransferase I (ST6Gal I), the adhesion protein P-selectin glycoprotein ligand-1 (PSGL-1), the β-subunit of voltage-gated sodium channels (VGSCβ), the neural cell adhesion molecule close homologue of L1 (CHL1) that is involved in neurite outgrowth, and vascular endothelial growth factor receptor-1 (VEGFR1).

The proteolytic role of BACE1 has also been confirmed in the processing of neuregulin-1 (NRG1), a ligand for members of the ErbB family of receptor tyrosine kinases, which have numerous roles in CNS development and functions, including synapse formation, plasticity, neuronal migration, myelination of central and peripheral axons, and the regulation of neurotransmitter expression. In addition to these physiological functions, *NRG1* has been linked to an increased risk of schizophrenia, and mice with complete genetic deletion of BACE1 demonstrate numerous behavioral traits consistent with schizophrenia-related endophenotypes.[27] Study of direct infusion of a β-secretase inhibitor into cerebral ventricles of adult mice demonstrated strongly reduced levels of Aβ but no changes in the processing of NRG1.[43] These data indicate that the role of BACE1, in relation to at least some of its substrates, may be developmentally regulated.

Additional preclinical studies are necessary to ascertain the efficacy and safety of new BACE1 inhibitors and to explore the limitations in their efficacy as a function of aging and stage of disease progression.

INHIBITION OF γ-SECRETASE

The γ-secretase complex catalyzes the final cleavage of APP and has been considered to be a significant target for therapy. As demonstrated by gene targeting strategies, this complex is critically dependent on the presence of PS1, PS2, and Pen-2, as well as Nct and Aph-1a.[18] Both genetic and pharmaceutical lowering of γ-secretase activity decreases production of Aβ peptides in cell-free and cell-based systems and reduces levels of Aβ in mutant mice with Aβ amyloidosis.[40] However, like β-secretase, γ-secretase cleaves multiple substrates.[18] The role of presenilins in the cleavage of the Notch receptor was identified before the discovery of its role in APP processing.[18] Interference with Notch signaling has been recognized as a basis for the most important potential side effects of inhibition of γ-secretase activity. These side effects include gastrointestinal bleeding, skin cancer, and autoimmune issues.

Owing to early recognition of interference with Notch signaling, research into the efficacy of new GSIs usually addresses the possibility of side effects developing due to concomitant changes in Notch signaling. Importantly, Notch inhibition is viewed as an advantage for the treatment of certain cancers that have excessive Notch signaling. For these antineoplastic agents, the strategy is to attain a therapeutic window for GSIs that would allow for benefit at doses and durations of treatment small enough to avoid side effects. In AD, initial development of GSIs also focused on finding a balance between therapeutic benefits and side effects. However, because drugs for AD therapeutics must enter the brain, blood–brain barrier permeability of GSIs is an additional hurdle that complicates finding a balance between beneficial and side effects. The latest research on GSIs for AD has focused on ways to dissociate the activities of γ-secretase on APP and Notch processing.

PRECLINICAL AND CLINICAL TRIALS OF γ-SECRETASE INHIBITORS

Several companies have attempted to identify and develop potent and selective GSIs. Although some of these GSIs have reached phase III clinical trials (reviewed in Henley *et al.*[7]), to date none has been successful. One of the most recent disappointments with GSIs was a phase III clinical trial comparing semagacestat (LY450139; Eli Lilly, www.lilly.com) with placebo in more than 2600 patients with mild-to-moderate AD. Trials were started in March to September 2008 and halted in August 2010 owing to worsening observed in cognitive assessments and activities of daily living compared with the placebo group. An increased risk of skin cancer was also observed, as expected from preclinical studies with genetic inhibition of γ-secretase.[44] In the phase I and II clinical trials, this side effect was not observed, probably owing to a shorter duration of treatment (< 14 weeks). Phase II studies showed no significant cognitive benefits in patients with mild-to-moderate AD and the lack of effect was attributed to the short duration of the treatment (< 14 weeks). If phase III trials were to confirm an absence of beneficial effect on functional outcomes, this could be explained by the late start of treatment relative to the onset of the disease. Aggregation of Aβ peptides into oligomers and ultimately into plaques, a process that represents an initial early event in disease progression, may lead to pathological mechanisms that are relatively independent from the initial trigger or become irreversible.[9,31] This assumption predicts that anti-Aβ therapies will be increasingly ineffective as disease progresses. This view is supported by findings from the AN1793 clinical trial, in which almost complete removal of Aβ plaque in some immunized patients resulted in no differences in time to severe dementia. The most alarming finding from the clinical trials with semagacestat is not a lack of cognitive benefits but what appears to be an aggravation of cognitive decline. It is unclear whether such negative outcomes could be expected from preclinical studies or early phases of clinical studies.

Semagacestat (LY450139) is a highly potent GSI that has been tested extensively in animal models and humans (reviewed in Henley *et al.*[7]) (Fig. 21.1). In an APP transgenic mouse model (PDAPP), LY450139 lowered brain, CSF, and plasma Aβ levels. Importantly, when LY450139 was administered to wild-type mice (not expressing mutated APP), a GSI-induced decrease in plasma Aβ concentration was followed by a significant increase at later time-points. Similar dynamics were observed in beagle dogs.[7] These data were interpreted as a possible effect of the GSI in the periphery, since in both wild-type mice and dogs increases in plasma Aβ were not associated with simultaneous elevations in the CSF or brain. The pharmacodynamics of LY450139 and its effects on CSF and plasma Aβ were extensively studied in healthy volunteers. A single dose of this compound led to biphasic changes in the levels of Aβ in CSF, particularly in levels of $A\beta_{1-42}$, with an initial decrease in levels of Aβ reaching a plateau between 6 and 15 hours after treatment, followed by an increase between 20 and 32 hours.[37] In contrast to previous interpretations (see above), these data strongly suggest an involvement of central effects of the GSI in mediating a biphasic Aβ response. Importantly, the biphasic dynamics of the CSF Aβ concentration were dose dependent and more pronounced for $A\beta_{1-42}$ than for $A\beta_{1-40}$.[37] At the dose used in the phase III clinical trials (100–140 mg), the GSI resulted in a predominant decrease in $A\beta_{1-40}$ followed by a second phase of increase in $A\beta_{1-42}$. The biphasic dynamics of Aβ in response to LY450139 were reported earlier in the plasma of healthy volunteers after a single oral dose of the drug.[45] In this case, an initial dose-dependent decrease in plasma Aβ was followed by a dramatic increase (>300%). The increase in plasma Aβ could be caused by rising concentrations of substrate for γ-secretase during the period of enzyme inhibition, which result in an increase in Aβ after LY450139 is no longer present.[45]

The "Aβ rise" effect of GSIs has been widely observed as a phenomenon when the same GSI increases Aβ at low concentration and decreases it at higher concentrations. The extent of GSI-induced rise in Aβ has been shown to be more pronounced for $A\beta_{1-42}$ than for $A\beta_{1-40}$ peptides.[46] These data are in agreement with findings in human volunteers after a single dose of LY450139.[37] Considering that $A\beta_{1-42}$ has rapid aggregation kinetics, even small increases in $A\beta_{1-42}$ or the $A\beta_{1-42}/A\beta_{1-40}$ ratio can significantly affect the neurotoxicity of total Aβ. Transient increases in $A\beta_{1-42}$ concentrations in the CSF and plasma of healthy volunteers after LY450139 administration probably reflect successful clearance of these peptides from the CNS. These GSI-induced increases in Aβ might not occur in a situation with impaired Aβ clearance and/or in the presence of Aβ plaques. Aβ plaques, owing to fast sequestration of soluble Aβ, change the balance between the interstitial fluid and peripheral compartments. Indeed, detection of changes in the Aβ levels in CSF proved to be more difficult in AD patients, with a majority of studies reporting no significant effects of GSIs, including LY450139, on $A\beta_{1-42}$ and $A\beta_{1-40}$.[47] This lack of observed changes may be misleading and rather indicates that $A\beta_{1-42}$ and $A\beta_{1-40}$ concentrations in CSF are not useful biomarkers of the drug activity in the AD brain. In contrast to unchanged levels of $A\beta_{1-42}$ and $A\beta_{1-40}$, significant increases were found in the CSF levels of $A\beta_{1-14}$, $A\beta_{1-15}$, and $A\beta_{1-16}$ peptides,[47] which may serve as better biomarkers of γ-secretase inhibition. Similar changes in CSF levels of short Aβ peptides were found after treatment with GSIs in mutant APP mouse models. This outcome could be explained by an increased amount of substrate (C99 APP-CTF) for α-secretase after inhibition of γ-secretase (reviewed in Portelius *et al.*[47]).

Another important nuance in understanding the discrepancies in the effects of GSIs between recent clinical studies and studies in APP mouse models comes from data showing that overexpression of APP, and APPswe in particular, precludes the Aβ rise effect of GSIs.[46] Although the exact mechanisms for the lack of GSI-induced rise in Aβ are not clear (high substrate availability and/or shift in GSI potency),[37] the APP transgenic mouse models that were widely used to study the effects of GSIs in the early twenty-first century (Tg2576, PDAPP, TgCRND8, and others) may have resulted in a misleading impression of a higher degree of inhibition in Aβ production. Even in experimental conditions that are favorable for detecting positive outcomes, GSIs reduced brain levels of Aβ significantly when tested in young APP mice but efficacy was reduced in old APP mice. Similar age-related limitations in the efficacy of γ-secretase inhibition were shown by genetic ablations of proteins comprising the γ-secretase complex.[44,48] Based on these studies in APP animal models, it may be expected that treatment with LY450139, when started in older AD patients with well-established Aβ amyloidosis, will not result in reduction of Aβ levels in the brain. Considering the Aβ rise effect of GSI and LY450139 in particular, LY450139 treatment could be expected to increase brain Aβ levels, including levels of oligomeric $A\beta_{1-42}$ peptides. The latter outcome would be in agreement with the observed worsening of cognitive symptoms.

Therapeutics Targeting Clearance of Aβ

ACTIVE IMMUNIZATION WITH Aβ

Pioneering studies by Schenk and colleagues[49] showed that active immunization with Aβ attenuates levels of Aβ peptides and plaques in the brain of an APP transgenic model of AD (PDAPP mice). Importantly, preclinical efficacy in reducing Aβ loads was demonstrated for both young and older animals. In young mice, immunization essentially prevented the formation of Aβ plaques. In older mice, treatment started after the onset of plaque deposition, but was effective in markedly reducing the extent and progression of AD-like neuropathologies.[49] Not long thereafter, the first human trial of AD immunotherapy with AN1792 was initiated, in which an $A\beta_{1-42}$ synthetic peptide with the QS21 adjuvant was administered to patients with mild-to-moderate AD. However, the trial was terminated because of the development of aseptic meningoencephalitis as a complication of the vaccine. Anti-Aβ antibodies raised in AD patients as a result of active immunization recognized β-amyloid plaques, diffuse Aβ deposits, and vascular β-amyloid in brain blood vessels. T-cell and microglial activation have been suspected as potential mechanisms of meningoencephalitis. Indeed, postmortem analysis of brain sections revealed decreased Aβ plaques in regions of the neocortex associated with activated microglia and T-cell infiltrates in the CNS, compared with unimmunized patients with AD.

PASSIVE ANTI-Aβ IMMUNIZATION

Passive immunotherapy has significant advantages over the active immunization approach, allowing for better control over the duration of treatment, overcoming problems with low responders, and allowing for careful selection of the antibodies to maximize efficacy and minimize adverse events.

To date, the most exciting findings regarding the effects of anti-Aβ immunotherapy in rescuing cognitive deficits come from a study in which acute systemic treatment with anti-Aβ antibody reversed memory deficits in an APP mouse model.[28] Importantly, since the duration of the treatment was short (days), the memory improvement was not associated with any detectable changes in the brain Aβ burden, suggesting that removal of Aβ plaques is not necessary for the beneficial effect of immunization. Instead, a dramatic increase in concentration of Aβ was observed in the blood,[28] leading to the hypothesis that a soluble pool of Aβ that can be easily removed from the brain is responsible for cognitive deficits. A significant correlation between an antibody-induced increase in Aβ plasma concentration and Aβ amyloid load in the brain[28] suggests that there is a dynamic equilibrium between a removable pool of Aβ species and an aggregated Aβ sequestered into plaques. This hypothesized equilibrium may explain correlations observed between Aβ plaque load and memory deficits in different APP transgenic models. The studies on acute passive immunization[28,29] seriously challenged the original role of Aβ plaques in mediating memory deficits and intensified investigations into which type of "soluble" non-plaque Aβ species is responsible for cognitive toxicity. A number of groups initiated the development of conformation-specific anti-Aβ antibodies with higher affinity to oligomeric species. Some of these antibodies have also been demonstrated to acutely improve learning and memory in APP transgenic mice.

EFFICACY AND SIDE EFFECTS: EXPECTATIONS FROM PRECLINICAL STUDIES OF PASSIVE ANTI-Aβ IMMUNIZATION

The discovery of acute memory improvement after anti-Aβ passive immunization brought hope that this approach might rapidly reduce cognitive impairment in AD patients apart from any effect on amyloid deposition. However, additional preclinical studies showed significant limitations in the efficacy of acute treatments with anti-Aβ antibodies. Immunization required longer duration of the treatment to be effective in mice of advanced age or with greater amyloid deposition.[50] Another factor limiting the efficacy of anti-Aβ passive immunization has been demonstrated in 3xTg-AD mice that combine Aβ plaque- and tau-related pathology. In these mice, anti-Aβ passive immunization led to memory benefits only if there was amelioration in tau-related pathology. The latter was resistant to change in the short time-window of an acute treatment.[25]

Data on the effects of a single administration of an anti-Aβ antibody (solanezumab; Eli Lilly) in AD patients have been made available.[51] These AD patients were treated with a humanized version of the murine antibody (m266.2) that was originally used to discover the acute reversal of memory deficits in mouse models.[28] As in preclinical studies, a significant dose-dependent increase in concentrations of Aβ in plasma and CSF was observed after a systemic injection of the antibody in AD patients.[51] However, in contrast to the preclinical findings, a single administration of the antibody did not coincide with significant changes in cognitive scores. This negative functional outcome is consistent with expectations from preclinical models showing low efficacy of passive immunization in the setting of advanced amyloidosis and the presence of tau pathology, as discussed above.

Preclinical studies elucidated several side effects that can be expected from the passive immunization approach. Based on the knowledge that meningoencephalitis is one of the major side effects of anti-Aβ vaccination, the potential for this outcome was investigated in APP mouse models after passive immunization. In the Tg2576 APP mouse model, meningoencephalitis was observed after a passive transfer of NAB61 antibody. Histologically, these observations were consistent with inflammation triggered by antibody binding to Aβ angiopathy. Another side effect discovered in APP mouse models after passive anti-Aβ immunotherapy is cerebral microhemorrhages associated with amyloid-laden vessels. Additional preclinical studies demonstrated that exacerbation of hemorrhages depends on antibody recognition of the deposited form of Aβ. Antibodies that are raised to different domains of Aβ but share high affinity to Aβ deposits increase vascular Aβ angiopathy and microhemorrhages. In contrast, antibody that does not bind deposited Aβ does not result in this complication. Reports from a phase II clinical trial with bapineuzumab, an anti-Aβ antibody that is raised against the N-terminal fragment of Aβ and binds to Aβ plaques, indicate a low incidence of vasogenic edema that could reflect cerebral amyloid angiopathy and antibody-induced changes in vascular permeability. This side effect was more prevalent with increasing dose of the antibody and in ApoE-ε_4 carriers, indicating that some subgroups of AD patients may be more prone to antibody-induced vascular damage and should be evaluated at a lower dose range in future studies. A study of anti-Aβ vaccination in APPswe/NOS2 knockout mice pointed to increased vascular expression of endothelial nitric oxide synthase as another factor that may increase susceptibility to microhemorrhages and possibly serve as a basis for interindividual variability in vascular complications. Recognition of microhemorrhages as a possible side effect of passive immunization approaches led to additional preclinical research in an attempt to find protocols and antibody modifications that would minimize this complication. For example, deglycosylation of antibodies has been shown to retain the memory-enhancing and amyloid-ameliorating properties of the immunotherapy while attenuating the increased vascular Aβ deposition and microhemorrhages observed with unmodified immunoglobulin G.

LACK OF EFFICACY OF ANTI-Aβ IMMUNIZATION IN CLINICAL STUDIES

Data are available from phase III clinical trials using long-term passive immunization (bapineuzumab, Janssen, Pfizer, and Johnson & Johnson; and solanezumab). In one of these trials (NCT00575055), bapineuzumab, an anti-Aβ antibody, or a placebo was administered intravenously to more than 1100 patients with mild-to-moderate AD who carry an $ApoE_4$ allele, the strongest genetic risk factor for AD. Unfortunately, after the planned 18 months of treatment bapineuzumab failed to have any meaningful beneficial effect in this cohort of patients. Shortly after this setback, other trials that used bapineuzumab in $ApoE_4$ non-carriers were halted owing to a lack of benefits in cognitive and functional outcome measures. Some of the trial participants underwent PET scans to quantify amyloid load in the brain, and another not fully overlapping subset of patients underwent spinal taps to measure levels of tau and phospho-tau in the CSF. MRI was also carried out to analyze changes in brain volume. These data, when fully analyzed, may provide some important information on changes in amyloid load and markers of neurodegeneration to explain the negative results of the trials and to indicate whether starting treatment earlier would be more efficacious. For example, the results of an earlier study with bapineuzumab demonstrated that the treatment lowered brain fibrillar Aβ by about 10%, but this effect may be insufficient, particularly in later stages of the disease.

The anti-Aβ antibody solanezumab was being tested in two phase III clinical trials in a population of AD patients similar to that in trials with bapineuzumab (mild-to-moderate AD). In comparison with bapineuzumab trials, the solanezumab treatment may have been slightly more aggressive and included monthly injections of the antibody for 80 weeks. (The regimen of treatment with bapineuzumab was once every 13 weeks for 72 weeks.) However, both trials with solanezumab have been halted because of a lack of benefits in cognitive and functional outcomes. More encouraging are combined results from two solanezumab studies which suggest modestly slower mental decline in patients with milder disease. The patients with mild AD showed a difference of nearly 2 points in the roughly 90-point score on thinking abilities, indicating that the solanezumab-induced improvement was very modest and fell short of clinically significant improvement. It is important to note, however, that patients with milder AD have fewer functional deficits. This may limit the window in which to see an improvement unless the outcome measures are adjusted to increase sensitivity to impairments characteristic of the early stages of AD.

Further detailed analyses of biomarkers accumulated during the bapineuzumab and solanezumab trials (CSF biomarkers, brain volumetric analysis by MRI, and PET scans for Aβ plaque burden) are important to understand whether and how these biomarkers relate to measurements of cognitive decline or benefits. Biomarker data reported so far for the bapineuzumab trial (1121 $ApoE_4$ carriers and 1331 non-carriers) indicate that the treatment prevents accumulation of Aβ in the brain of patients with mild-to-moderate AD and lowers phospho-tau (p-tau). These results confirm efficient target engagement and beneficial secondary effects on markers of neurodegeneration. Failure to observe expected functional and cognitive benefits of passive immunization in mild-to-moderate AD in these clinical trials indicates that antiamyloid therapy is not effective in the late stages of the disease.

How much earlier an antiamyloid treatment needs to be initiated to be functionally and cognitively beneficial is unclear. If AD is a disease with long preclinical and clinical stages of Aβ accumulation and oligomerization, then an antiamyloid treatment should be started as early as possible. However, accurate diagnosis at such early stages of sporadic AD is problematic. This raises an important question that will be critical for the proper design of antiamyloid treatments: What is the latest stage of the disease at which modifying the production or clearance of Aβ stops or at least delays the progression of AD? Results of recent anti-Aβ passive immunization trials (bapineuzumab) have showed that halting the progression of Aβ amyloidosis in patients with mild-to-moderate AD does not change the dynamics of their cognitive decline (see Dr Sperling's slides on the EFNS website http://www2.kenes.com/efns/info/Pages/WebcastingRecordedsessions.aspx).

CONCLUSION

The research described in this chapter has greatly enhanced our understanding of AD. However, disappointing results of clinical trials in AD have raised concerns about current hypotheses and therapeutic targets for AD-modifying drugs. The failures of clinical trials aimed at reducing production or enhancing clearance of Aβ can be explained, in part, by a number of reasons, including targeting the wrong pathophysiological mechanisms, low efficacies on drug targets, and testing the drugs at the wrong stage of the disease. Detailed analyses of the outcomes of these clinical trials, as well as a realization that the pathophysiological process of AD starts decades before the onset of clinical symptoms, indicate that the most likely reason for the failure of these clinical trials is testing the right target at the wrong stage of the disease.[12] If the window of opportunity for antiamyloid drugs is indeed in the early stages of AD (MCI or preclinical AD), prevention trials would be the right approach to test the amyloid cascade hypothesis. However, moving clinical trials to the early stages of AD requires the development of CSF and imaging biomarkers with the sensitivity to detect *in vivo* changes in Aβ production and accumulation, tau phosphorylation, and the degrees of neurodegeneration. Another hurdle on this road is the development of cognitive measures that can be used to track very early changes that are sensitive to early AD pathology.

We are on the threshold of implementing novel treatments based on an understanding of the neurobiology and progressive nature of AD. Future discoveries may lead to the design of mechanism-based therapies with multiple targets that can be tested in models of AD. Eventually, these approaches could be introduced into clinical settings at appropriate stages of disease progression for the benefit of patients with this devastating disease.

QUESTIONS FOR FURTHER RESEARCH

Although we have learned a great deal about AD, there is still much to discover. The following examples are some of the most important areas for future research. (1) Non-invasive imaging tools are needed for the visualization of soluble/oligomeric Aβ that would aid in early diagnosis and the assessment of therapeutic efficacy. (2) Biomarkers (e.g. amyloid, tau, neurodegeneration) must be established that not only distinguish a population of patients from a population of controls, but also can be used as predictive/classification markers on an individual level. (3) Cognitive measures need to be developed that can be easily implemented and used for tracking very early cognitive changes that are specific to AD pathology. (4) Research is needed to define the window of opportunity in the progression of the disease, during which interventions must be initiated to reduce the production or clearance of Aβ, to stop or delay disease progression. (5) Researchers need to develop mechanism-based therapies with multiple targets and find ways to confirm the molecular mechanisms of their additive efficacy. (6) There is a desperate need for new mouse and rat models of AD that reflect multiple characteristics of AD (amyloidosis, tau pathology, and neurodegeneration), which allow for better prediction of drug efficacy in human trials. Parallel research questions could be fashioned for virtually all the neurodegenerative diseases covered in this textbook. The hope is that these and other major issues will attract the attention of the next generation of researchers so that the progress made in the last several decades can continue and, ultimately, interventions that significantly modify the incidence and progression of AD and other neurodegenerative diseases can become a reality. Surely there can be few more urgent priorities.

Disclosure/Conflict of Interest

AVS, TM, TL, and PCW have nothing to report. DLP is on the scientific advisory boards of Biogen Idec and Satori Pharmaceuticals Inc. (Cambridge, MA, USA), but does not have current research support from these companies.

Acknowledgments

We wish to thank our many colleagues at JHMI and other institutions who have contributed to the original work cited in this review and have engaged us in helpful discussions. We are grateful to Dr A. Huberman for careful editing and discussion of the manuscript. Because of space constrains we were unable to cite all studies relevant to topics discussed. Aspects of this work were supported by grants from the US Public Health Service (NS45150, NS47308, NS41438, NS49088, NS10580, AG005416, AG14248, MH086881, NS047225, TW008019) as well as the Metropolitan Life Foundation, Adler Foundation, Alzheimer's Association, American Health Assistance Foundation, Rotary CART Fund, Wallace Foundation, Ellison Medical Foundation, and Bristol-Myers Squibb Foundation.

References

1. Whitehouse PJ, Price DL, Struble RG, Clark AW, Coyle JT, Delon MR. Alzheimer's disease and senile dementia: loss of neurons in the basal forebrain. *Science*. 1982;215(4537):1237–1239.
2. Braak H, Braak E. Neuropathological staging of Alzheimer-related changes. *Acta Neuropathol*. 1991;82(4):239–259.
3. Birks J. Cholinesterase inhibitors for Alzheimer's disease. *Cochrane Database Syst Rev*. 2006;(1):CD005593.
4. Bertram L, Tanzi RE. Genome-wide association studies in Alzheimer's disease. *Hum Mol Genet*. 2009;18(R2):R137–R145.
5. Kim S, Swaminathan S, Shen L, et al. Genome-wide association study of CSF biomarkers Abeta1-42, t-tau, and p-tau181p in the ADNI cohort. *Neurology*. 2011;76(1):69–79.
6. Cai H, Wang Y, McCarthy D, et al. BACE1 is the major beta-secretase for generation of Abeta peptides by neurons. *Nat Neurosci*. 2001;4(3):233–234.
7. Henley DB, May PC, Dean RA, Siemers ER. Development of semagacestat (LY450139), a functional gamma-secretase inhibitor, for the treatment of Alzheimer's disease. *Expert Opin Pharmacother*. 2009;10(10):1657–1664.
8. Vassar R, Kovacs DM, Yan R, Wong PC. The beta-secretase enzyme BACE in health and Alzheimer's disease: regulation, cell biology, function, and therapeutic potential. *J Neurosci*. 2009;29(41):12787–12794.
9. Golde TE, Schneider LS, Koo EH. Anti-abeta therapeutics in Alzheimer's disease: the need for a paradigm shift. *Neuron*. 2011;69(2):203–213.
10. Maurer K, Volk S, Gerbaldo H. Auguste D and Alzheimer's disease. *Lancet*. 1997;349(9064):1546–1549.
11. George D, Whitehouse PJ. The classification of Alzheimer's disease and mild cognitive impairment: enriching therapeutic models. In: Ballenger JF, Whitehouse PJ, Lyketsos CG, Rabins PV, Karlawish JHT, eds. *Treating Dementia: Do We Have a Pill for It?*. Baltimore, MD: Johns Hopkins University Press; 2009:5–25.

12. Sperling RA, Aisen PS, Beckett LA, et al. Toward defining the preclinical stages of Alzheimer's disease: recommendations from the National Institute on Aging–Alzheimer's Association workgroups on diagnostic guidelines for Alzheimer's disease. *Alzheimers Dement*. 2011;7(3):280–292.
13. Jack Jr CR, Knopman DS, Jagust WJ, et al. Hypothetical model of dynamic biomarkers of the Alzheimer's pathological cascade. *Lancet Neurol*. 2010;9(1):119–128.
14. Jack Jr CR, Knopman DS, Jagust WJ, et al. Tracking pathophysiological processes in Alzheimer's disease: an updated hypothetical model of dynamic biomarkers. *Lancet Neurol*. 2013;12(2):207–216.
15. Bateman RJ, Xiong C, Benzinger TL, et al. Clinical and biomarker changes in dominantly inherited Alzheimer's disease. *N Engl J Med*. 2012;367(9):795–804.
16. Kamenetz F, Tomita T, Hsieh H, et al. APP processing and synaptic function. *Neuron*. 2003;37(6):925–937.
17. Brunden KR, Trojanowski JQ, Lee VM. Advances in tau-focused drug discovery for Alzheimer's disease and related tauopathies. *Nat Rev Drug Discov*. 2009;8(10):783–793.
18. De Strooper B, Vassar R, Golde T. The secretases: enzymes with therapeutic potential in Alzheimer disease. *Nat Rev Neurol*. 2010;6(2):99–107.
19. Sherrington R, Rogaev EI, Liang Y, et al. Cloning of a gene bearing missense mutations in early-onset familial Alzheimer's disease. *Nature*. 1995;375(6534):754–760.
20. Price D, Martin L, Savonenko A, Li T, Laird F, Wong P. Selected genetically engineered models relevant to human neurodegenerative disease. In: Rosenberg RN, DiMauro S, Paulson HL, Ptacek L, Nestler EJ, eds. *The Molecular and Genetic Basis of Neurologic and Psychiatric Disease*. Philadelphia, PA: Lippincott Williams & Wilkins; 2007:35–62.
21. Savonenko AV, Borchelt DL. Transgenic mouse models of Alzheimer's disease and episodic memory. In: Dere E, Easton A, Nadel L, Huston JP, eds. *Handbook of Episodic Memory*. Amsterdam: Elsevier; 2008:553–573.
22. Gotz J, Chen F, van Dorpe J, Nitsch RM. Formation of neurofibrillary tangles in P301l tau transgenic mice induced by Abeta 42 fibrils. *Science*. 2001;293(5534):1491–1495.
23. Lewis J, Dickson DW, Lin WL, et al. Enhanced neurofibrillary degeneration in transgenic mice expressing mutant tau and APP. *Science*. 2001;293(5534):1487–1491.
24. Giannakopoulos P, Herrmann FR, Bussiere T, et al. Tangle and neuron numbers, but not amyloid load, predict cognitive status in Alzheimer's disease. *Neurology*. 2003;60(9):1495–1500.
25. Oddo S, Vasilevko V, Caccamo A, Kitazawa M, Cribbs DH, LaFerla FM. Reduction of soluble Abeta and tau, but not soluble Abeta alone, ameliorates cognitive decline in transgenic mice with plaques and tangles. *J Biol Chem*. 2006;281(51):39413–39423.
26. Roberson ED, Scearce-Levie K, Palop JJ, et al. Reducing endogenous tau ameliorates amyloid beta-induced deficits in an Alzheimer's disease mouse model. *Science*. 2007;316(5825):750–754.
27. Savonenko AV, Melnikova T, Laird FM, Stewart KA, Price DL, Wong PC. Alteration of BACE1-dependent NRG1/ErbB4 signaling and schizophrenia-like phenotypes in BACE1-null mice. *Proc Natl Acad Sci U S A*. 2008;105(14):5585–5590.
28. Dodart JC, Bales KR, Gannon KS, et al. Immunization reverses memory deficits without reducing brain Abeta burden in Alzheimer's disease model. *Nat Neurosci*. 2002;5(5):452–457.
29. Kotilinek LA, Bacskai B, Westerman M, et al. Reversible memory loss in a mouse transgenic model of Alzheimer's disease. *J Neurosci*. 2002;22(15):6331–6335.
30. Santacruz K, Lewis J, Spires T, et al. Tau suppression in a neurodegenerative mouse model improves memory function. *Science*. 2005;309(5733):476–481.
31. Herrup K. Reimagining Alzheimer's disease – an age-based hypothesis. *J Neurosci*. 2010;30(50):16755–16762.
32. Pimplikar SW, Nixon RA, Robakis NK, Shen J, Tsai LH. Amyloid-independent mechanisms in Alzheimer's disease pathogenesis. *J Neurosci*. 2010;30(45):14946–14954.
33. Bartus RT, Dean III RL, Beer B, Lippa AS. The cholinergic hypothesis of geriatric memory dysfunction. *Science*. 1982;217(4558):408–414.
34. Pinto T, Lanctot KL, Herrmann N. Revisiting the cholinergic hypothesis of behavioral and psychological symptoms in dementia of the Alzheimer's type. *Ageing Res Rev*. 2011.
35. Kenton L, Boon F, Cain DP. Combined but not individual administration of beta-adrenergic and serotonergic antagonists impairs water maze acquisition in the rat. *Neuropsychopharmacology*. 2008;33(6):1298–1311.
36. Selkoe DJ. Resolving controversies on the path to Alzheimer's therapeutics. *Nat Med*. 2011;17(9):1060–1065.
37. Bateman RJ, Siemers ER, Mawuenyega KG, et al. A gamma-secretase inhibitor decreases amyloid-beta production in the central nervous system. *Ann Neurol*. 2009;66(1):48–54.
38. Lemere CA, Masliah E. Can Alzheimer disease be prevented by amyloid-beta immunotherapy? *Nat Rev Neurol*. 2010;6(2):108–119.
39. Ohno M, Sametsky EA, Younkin LH, et al. BACE1 deficiency rescues memory deficits and cholinergic dysfunction in a mouse model of Alzheimer's disease. *Neuron*. 2004;41(1):27–33.
40. Laird FM, Cai H, Savonenko AV, et al. BACE1, a major determinant of selective vulnerability of the brain to amyloid-beta amyloidogenesis, is essential for cognitive, emotional, and synaptic functions. *J Neurosci*. 2005;25(50):11693–11709.
41. Chang WP, Huang X, Downs D, et al. Beta-secretase inhibitor GRL-8234 rescues age-related cognitive decline in APP transgenic mice. *FASEB J*. 2011;25(2):775–784.
42. Fukumoto H, Takahashi H, Tarui N, et al. A noncompetitive BACE1 inhibitor TAK-070 ameliorates Abeta pathology and behavioral deficits in a mouse model of Alzheimer's disease. *J Neurosci*. 2010;30(33):11157–11166.
43. Sankaranarayanan S, Price EA, Wu G, et al. In vivo beta-secretase 1 inhibition leads to brain Abeta lowering and increased alpha-secretase processing of amyloid precursor protein without effect on neuregulin-1. *J Pharmacol Exp Ther*. 2008;324(3):957–969.
44. Li T, Wen H, Brayton C, et al. Moderate reduction of gamma-secretase attenuates amyloid burden and limits mechanism-based liabilities. *J Neurosci*. 2007;27(40):10849–10859.
45. Siemers E, Skinner M, Dean RA, et al. Safety, tolerability, and changes in amyloid beta concentrations after administration of a gamma-secretase inhibitor in volunteers. *Clin Neuropharmacol*. 2005;28(3):126–132.
46. Burton CR, Meredith JE, Barten DM, et al. The amyloid-beta rise and gamma-secretase inhibitor potency depend on the level of substrate expression. *J Biol Chem*. 2008;283(34):22992–23003.
47. Portelius E, Dean RA, Gustavsson MK, et al. A novel Abeta isoform pattern in CSF reflects gamma-secretase inhibition in Alzheimer disease. *Alzheimers Res Ther*. 2010;2(2):7.
48. Chow VW, Savonenko AV, Melnikova T, et al. Modeling an anti-amyloid combination therapy for Alzheimer's disease. *Sci Transl Med*. 2010;2(13). 13ra11.
49. Schenk D, Barbour R, Dunn W, et al. Immunization with amyloid-beta attenuates Alzheimer-disease-like pathology in the PDAPP mouse. *Nature*. 1999;400(6740):173–177.
50. Wilcock DM, Rojiani A, Rosenthal A, et al. Passive immunotherapy against Abeta in aged APP-transgenic mice reverses cognitive deficits and depletes parenchymal amyloid deposits in spite of increased vascular amyloid and microhemorrhage. *J Neuroinflamm*. 2004;1(1):24.
51. Siemers ER, Friedrich S, Dean RA, et al. Safety and changes in plasma and cerebrospinal fluid amyloid beta after a single administration of an amyloid beta monoclonal antibody in subjects with Alzheimer disease. *Clin Neuropharmacol*. 2010;33(2):67–73.

CHAPTER

22

Cerebrovascular Disease – Stroke

*Louis R. Caplan**, *Roger P. Simon*†

*Beth Israel Deaconess Medical Center, Boston, Massachusetts, USA; †The Neuroscience Institute, Morehouse School of Medicine, Atlanta, Georgia, USA

OUTLINE

DEFINITION OF STROKE

Stroke is a term used to describe brain injury caused by an abnormality of the blood supply to the brain. The word is derived from the fact that most stroke patients are struck suddenly by the blood vessel problem and abnormalities of brain function begin quickly, sometimes in an instant.

Stroke is a broad term that describes a variety of different blood vessel diseases. Since treatment depends on the type of stroke and the location of the blood vessels involved, doctors need to determine what caused the vascular and brain injury and where the abnormalities are located.

BRAIN LESIONS CAUSED BY CEREBROVASCULAR DISEASE

Strokes can be divided into two very broad groups: hemorrhage and ischemia.[1] Hemorrhage refers to bleeding inside the skull, either into the brain or into the fluid surrounding the brain. The second major types of stroke are ischemic, a term that refers to a lack of blood. The brain depends on a constant supply of sugar and oxygen to function normally. When a part of the brain is not supplied with an adequate supply of blood, that part of the brain becomes unable to perform its normal functions. Hemorrhage and ischemia are polar opposites.

Neurobiology of Brain Disorders
http://dx.doi.org/10.1016/B978-0-12-398270-4.00022-7

Hemorrhage is characterized by too much blood inside the skull, whereas in ischemia there is insufficient blood supply. Brain ischemia is much more common than hemorrhage. About four strokes out of every five are ischemic.

Hemorrhage

Bleeding into the brain substance (called variously intracerebral or parenchymatous hemorrhage) tears and disconnects important gray and white matter connections and pathways. The bleeding is due to rupture of small blood vessels, arterioles, and capillaries in the brain parenchyma. The blood usually oozes into the brain under pressure and forms a localized, often round or elliptical blood collection referred to as a hematoma. The hematoma separates normal brain structures and interrupts brain white matter tracts. Parenchymatous hemorrhages occur in a localized region of the brain (Fig. 22.1A). The degree of damage depends on the location, rapidity, volume, and pressure of the bleeding.

Intracerebral hemorrhages are at first soft and dissect along white matter fiber tracts. When bleeding dissects into the ventricles or onto the surface of the brain, blood is introduced into the cerebrospinal fluid. The blood in the hematoma clots and solidifies, causing swelling of adjacent brain tissues. Later, blood is absorbed, and after macrophages have cleared the debris, a cavity or slit forms that may disconnect brain pathways.

Hematomas exert pressure on brain regions adjacent to the collection of blood and can injure these surrounding tissues. Edema often develops around the hematomas, further adding to the extra volume of abnormal tissue in the brain. Large hemorrhages are often fatal because they increase pressure within the skull, squeezing vital regions in the brainstem. The intracranial cavity is a closed system. The bony skull and dura mater act as a fortress protecting the brain from outside injury. In adverse situations, such as swelling or hemorrhage arising inside the fortress, these structures can function as a prison, restricting and strangulating their enclosed contents and forcing herniation of tissue from one compartment to another.

Bleeding into the fluid around the brain is referred to as subarachnoid hemorrhage, in which the blood collects and stays under the arachnoid membrane that lies over the pia mater. An artery carrying blood under systemic blood pressure leaks and spills blood rapidly into the spinal fluid that circulates around the brain and spinal cord. In most instances the leaking structure is an aneurysm (Fig. 22.1B). The sudden release of blood under high pressure increases the pressure inside the skull and causes severe, sudden-onset headache, often with vomiting. The sudden increase in pressure causes a lapse in

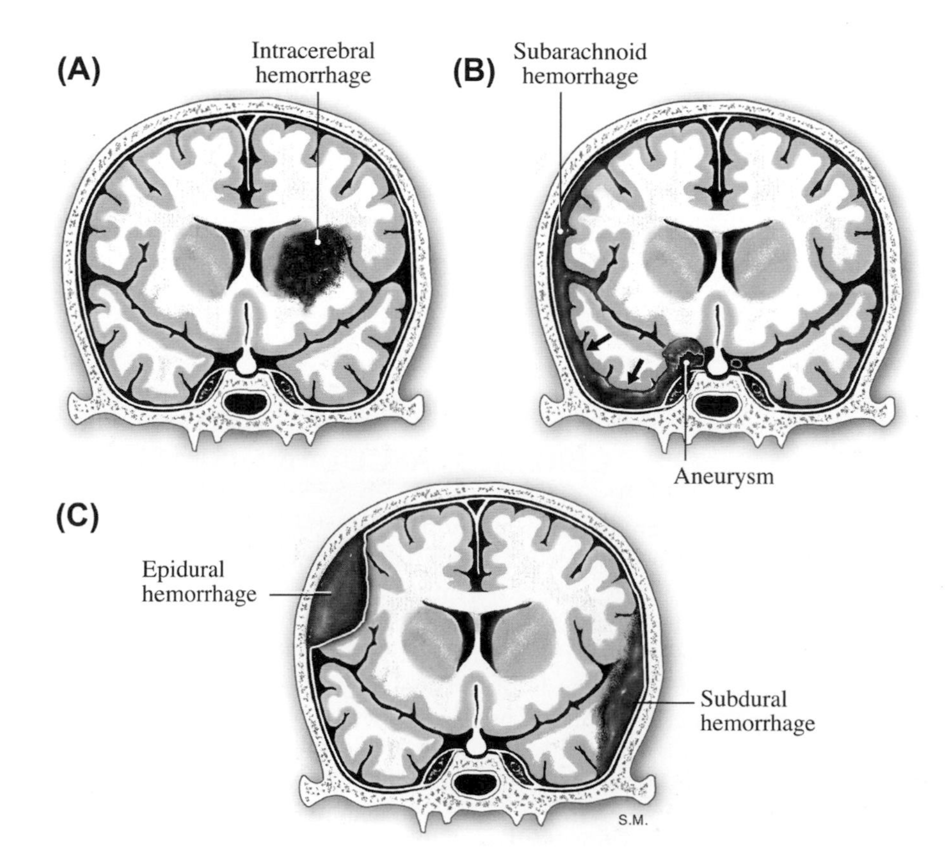

FIGURE 22.1 **Artist's drawings of hemorrhages within the skull at various locations.** (A) Intracerebral (intraparenchymatous) hemorrhage in the basal ganglion region on the right (white dot). (B) Leakage of blood around the brain (black arrows) from an aneurysm (white dot). (C) The lines point to collections of blood in the subdural space (between the arachnoid and dural membranes) and an epidural hemorrhage outside the dura. *Source: From Caplan.* Stroke. *Demos Medical Publishing; 2006,*[2] *with permission.*

brain function so that the patient may stare, drop to the knees, or become confused and unable to remember. Most often the symptoms in patients with subarachnoid hemorrhage relate to diffuse abnormalities of brain function since usually there is no bleeding into one part of the brain. At times, the leakage of blood into the subarachnoid space emits from a vascular malformation or an excessive bleeding tendency.

In contrast to the situation in subarachnoid hemorrhages, in patients with intracerebral hemorrhages, the hematoma is localized and causes loss of function related to the area damaged by the local blood collection. For example, if the bleeding is into the left cerebral hemisphere, the patient often has weakness and loss of feeling in the right limbs and loss of normal speech. A hemorrhage into the cerebellum will, instead, cause dizziness and a loss of balance.

Subdural and epidural hemorrhages are most often caused by traumatic head injuries to blood vessels (Fig. 22.1C). In subdural hemorrhages the bleeding is usually from veins that lie in the space between the arachnoid membrane and the dura mater. In epidural hemorrhages the bleeding usually results in tearing of meningeal arteries. Often there is an accompanying skull fracture that tears a meningeal artery. Blood accumulates much more rapidly when it issues from arteries than it does when the source is one or more small veins, so that symptoms usually develop soon after head injury in patients with epidural hemorrhages. In subdural hemorrhages the bleeding can be slow, so that symptoms of headache and brain dysfunction may be delayed for weeks after the head injury.

Ischemia and Infarction

There are three main categories of brain ischemia, referred to as thrombosis, embolism, and systemic hypoperfusion, each indicating a different mechanism of blood vessel injury or reason for decreased blood flow.

- *Thrombosis* refers to a local problem within a blood vessel that supplies a portion of the brain (Fig. 22.2A). The process is often a disease of the blood vessel, which narrows blood flow through that vessel. Often, a thrombus is superimposed on the localized vascular process. In some patients with blood hypercoagulability, blood vessels can become blocked by a primary clotting disorder without a major underlying vascular lesion. The accompanying decreased blood flow to the brain region supplied by the local vascular problem renders the brain ischemic. If the ischemia is severe or prolonged enough, then infarction develops.
- *Embolism* refers to a breaking loose of materials (often thrombi, but occasionally bacteria, cholesterol crystals, fat, foreign bodies, etc.) from a proximal source to block a distant vessel farther along the circulatory path (Fig. 22.2B). The vessel blocked downstream causes ischemia and infarction to a localized brain region in the same way that a primary vascular lesion ("thrombosis") does. Embolic instances are often characterized by the donor source (usually the heart, aorta, or a proximal artery or vein), by the location of the recipient artery, and by the substance that makes up the embolus.
- *Systemic hypoperfusion* is characterized by a global decrease in blood flow to the head rather than a localized decrease as occurs in thrombosis and embolism (Fig. 22.2C). Abnormal performance of the pump (heart) could lead to low pressure in the system. Abnormally slow or fast heart rhythms, cardiac arrest, and failure of the heart to pump blood adequately can all diminish blood flow to the head and brain. Another cause of diminished circulatory function is a lowering of blood pressure and blood flow due to an inadequate amount of blood and fluid in the vascular compartment of the body. Bleeding, dehydration, and loss of fluid into body tissues (shock) can all lead to inadequate brain perfusion. Hypotension from any cause can lead to global brain ischemia and syncope.

The decreased whole brain perfusion can cause different patterns of ischemic damage. In many patients those regions that are most vulnerable to a decrease in necessary nutrients (oxygen and sugar) are injured. This pattern is termed selective vulnerability. In other patients infarcts develop in areas described as border-zone regions. These regions lie between the major supply arteries. Figure 22.3 illustrates the concept of border zones.

Global ischemia produces non-focal cessation of blood flow to the entire brain, as occurs in cardiac arrest. Energy failure from impaired oxidative metabolism occurs rapidly as the substrates for ATP, oxygen, and glucose fall. Neuronal injury and cell death begin within 4 minutes, occurring predominantly in selectively vulnerable cell populations of the pyramidal neurons of the hippocampus, neurons of the middle lamina of the cortical mantel, and nuclei of the lateral thalamus. The mechanism of this selective vulnerability remains uncertain, as does the delayed cell death of the CA1 sector neurons of the hippocampus occurring 24–72 hours later. The morphological hallmark of neuronal injury is ischemic cell change characterized by microvacuolization in the cytoplasm (Fig. 22.4). At the ultrastructural level, the vacuoles mainly represent dilated mitochondria with Ca^{2+} loading (addressed below) (Fig. 22.5).

In patients with brain embolism and thrombosis, one artery is usually blocked, leading to dysfunction of the

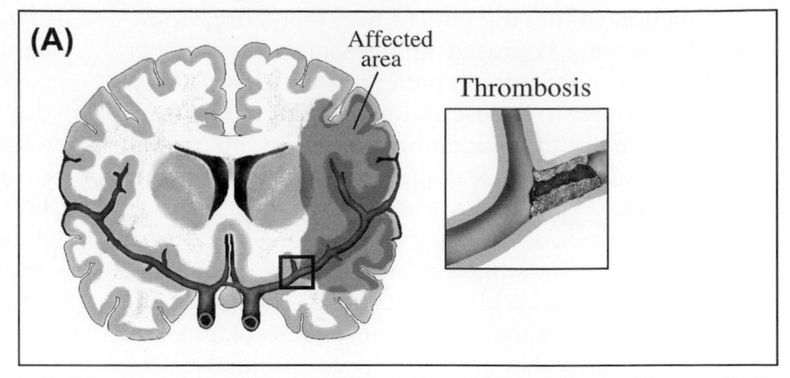

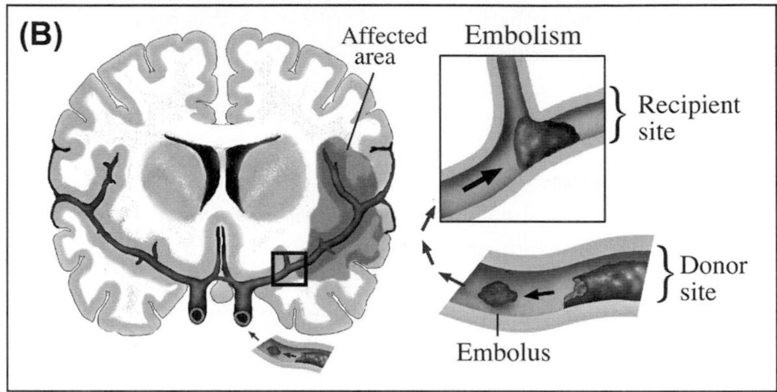

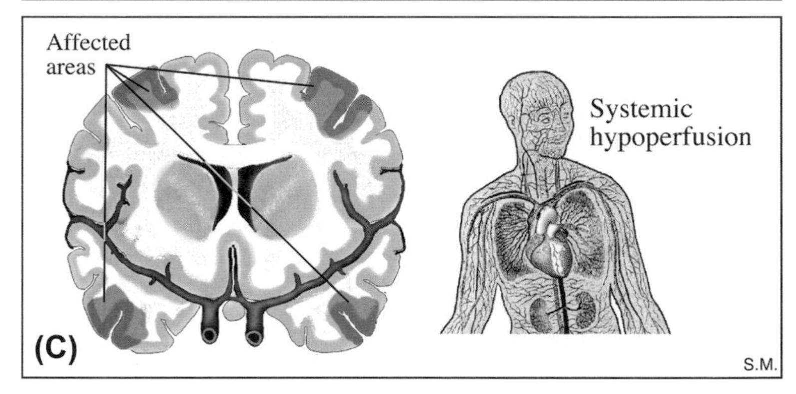

FIGURE 22.2 **Artist's drawings of various mechanisms of brain ischemia/infarction.** (A) An infarct in the right side of the cerebral hemisphere (gray shaded zone). The insert depicts an occluded thrombosed artery located in the middle cerebral artery indicated in the small box on the left drawing. (B) A similar area of infarction in the right cerebral hemisphere (gray shaded zone). The insert depicts an embolus that has traveled to the portion of the middle cerebral artery designated in the box in the left drawing. (C) A systemic circulatory problem with diminished blood flow to the whole brain. The gray shaded areas represent infarction in both cerebral hemispheres in areas located between major supply areas (the border zones are illustrated in Fig. 22.3). *Source: From Caplan.* Stroke. *Demos Medical Publishing; 2006,*[2] *with permission.*

part of the brain supplied by that blocked artery. This is reflected in focal abnormalities of brain function, such as weakness of the limbs on one half of the body. In this respect the abnormalities are similar to those found in patients with local brain hematomas. In contrast, systemic hypoperfusion leads to more diffuse abnormalities such as light-headedness, dizziness, confusion, dimming of vision, and hearing loss. Patients appear pale and generally weak. These symptoms are caused by a generalized reduction in blood flow and not by loss of function in one local region of the brain.

Focal ischemia produces impaired blood flow in the distribution of an artery supplying a portion of the brain such as occurs in thrombotic and embolic strokes. Here, there is a gradient of energy failure, maximum in the ischemic core and less so in the surrounding brain tissue, referred to as the penumbra. Na^+/K^+-ATPase, responsible for maintaining intracellular K^+ concentration, and requiring two-thirds of the total energy expenditure in the brain, fails, with resultant K^+ leakage and depolarization of adjacent cells. Cortical spreading depression follows, producing further neuronal and astrocytic depolarization and energy use, and infarct extension. Cellular depolarization causes intracellular Ca^{2+} entry through activation of voltage-gated Ca^{2+} channels and voltage-sensitive glutamate receptor-gated channels. Anaerobic metabolism results in lactate accumulation and acidosis in the ischemic brain regions. The presence of acidosis activates Ca^{2+}-permeable, acid-sensing ion channels, further exacerbating calcium dysregulation. Intracellular Ca^{2+} buffering and sequestration mechanisms in mitochondria and endoplasmic reticulum

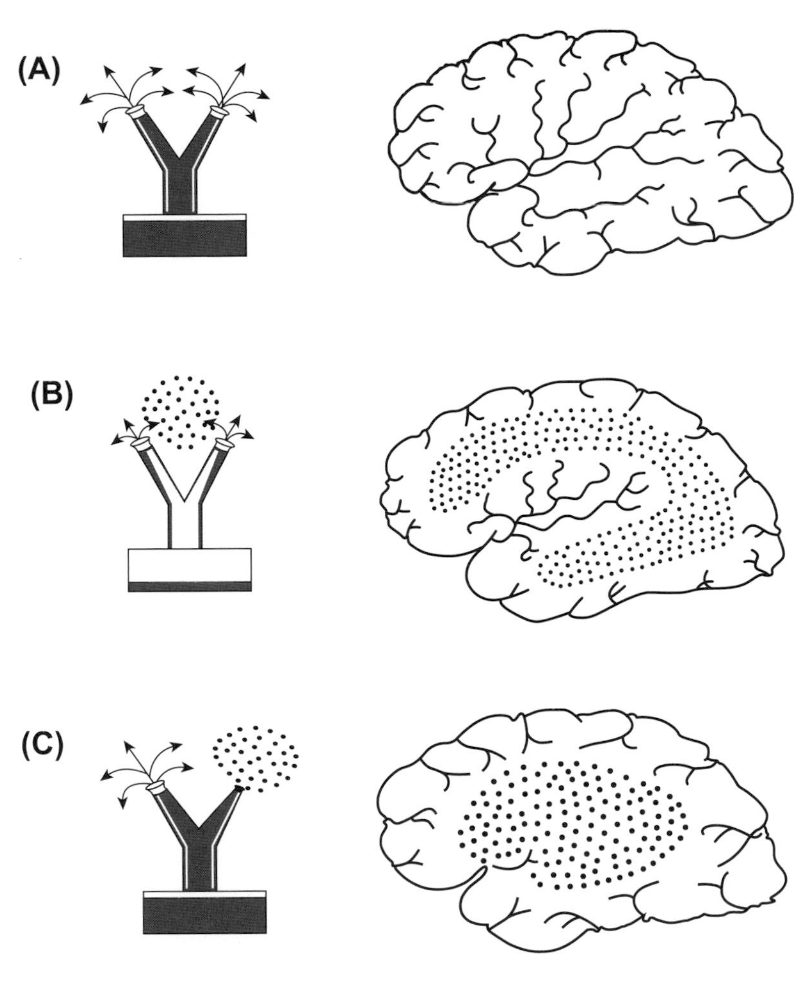

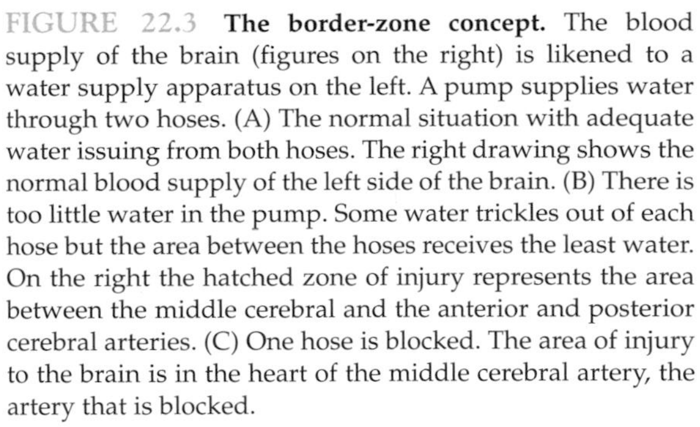
FIGURE 22.3 **The border-zone concept.** The blood supply of the brain (figures on the right) is likened to a water supply apparatus on the left. A pump supplies water through two hoses. (A) The normal situation with adequate water issuing from both hoses. The right drawing shows the normal blood supply of the left side of the brain. (B) There is too little water in the pump. Some water trickles out of each hose but the area between the hoses receives the least water. On the right the hatched zone of injury represents the area between the middle cerebral and the anterior and posterior cerebral arteries. (C) One hose is blocked. The area of injury to the brain is in the heart of the middle cerebral artery, the artery that is blocked.

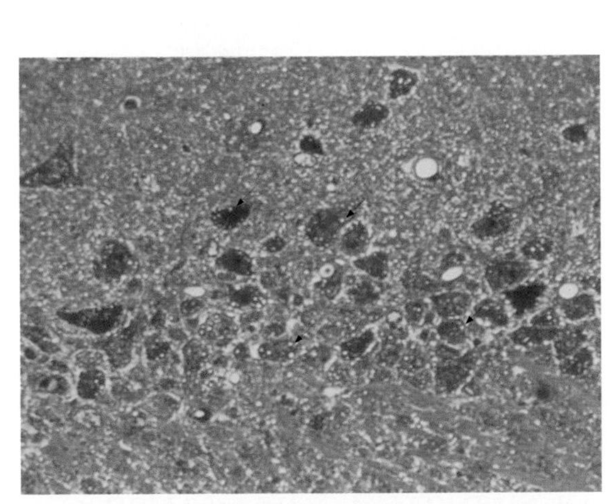
FIGURE 22.4 **Ischemic cell change showing microvacuolization of the cytoplasm (arrows).** Hippocampal pyramidal neurons in rat following 30 minutes of global ischemia followed by 30 minutes of reperfusion (Toluidine blue staining).

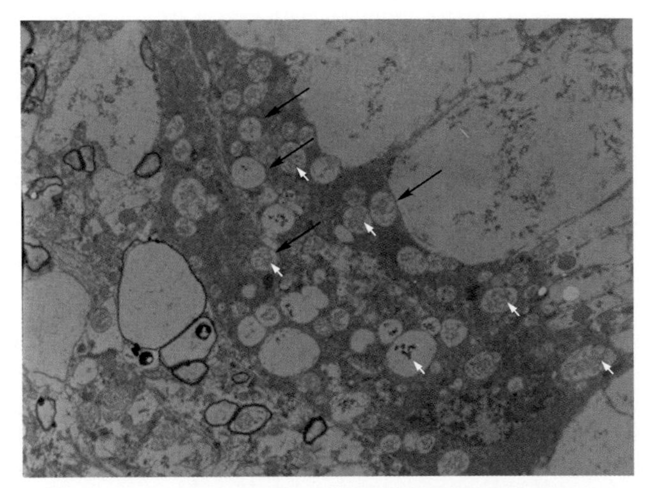
FIGURE 22.5 **Ultrastructure of microvacuolated hippocampal neuron.** Vacuoles (black arrows) are dilated mitochondria which show calcium loading (calcium pyroantimonate deposits; white arrows).

become overwhelmed (Fig. 22.6), catabolic processes are activated, mitochondrial function fails, and cell death pathways are mobilized.

Cell death takes place most rapidly in the infarct core and more slowly in the penumbra. Rapid cell death occurs by the process of necrosis, with cell and organelle swelling, membrane rupture, and spilling of cellular contents into the extracellular space with consequent inflammation. Organelle swelling results in cytoplasmic microvacuolization (Fig. 22.5). In the penumbra, cell death occurs

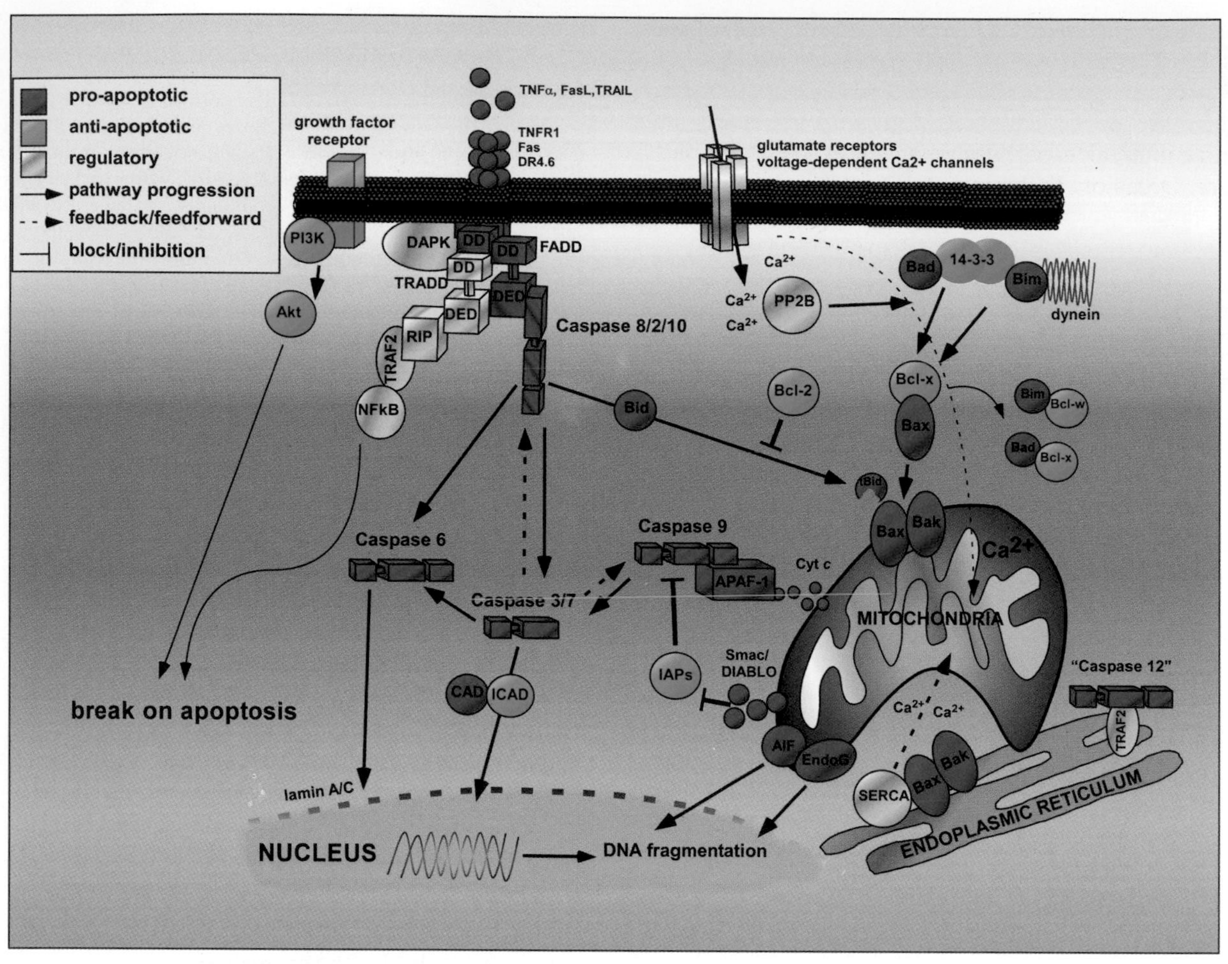

FIGURE 22.6 **Major apoptosis pathways.** Extrinsic pathway: tumor necrosis factor (TNF) and other death receptor agonists, caspase-8 and caspase-3 activation; intrinsic pathway: mitochondrial membrane compromise with resultant cytochrome *c* (Cyt *c*) release, caspase-9 and caspase-3 activation. AIF: apoptosis-inducing factor; APAF-1: apoptosis protease-activating factor-1; CAD: caspase-activated DNase; DAPK: death-associated protein kinase; DD: death domain; DED: death effector domain; DISC: death-inducing signaling complex; DR: death receptor; EndoG: endonuclease G; FADD: Fas-associated death domain; FasL: Fas ligand; IAP: inhibitor of apoptosis protein; ICAD: inhibitor of caspase-activated DNase; NFkB: nuclear factor-κB; PI3K: phosphatidylinositol 3-kinase; PP2B: protein phosphatase 2 B (calcineurin); RIP: receptor-interacting protein; SERCA: sarcoplasmic/endoplasmic Ca^{2+}-ATPase; Smac: second mitochondria-derived activator of caspase; TNFR1: tumor necrosis factor receptor-1; TRADD: TNF receptor-associated death domain; TRAF2: TNFR-activating factor-2; TRAIL: TNF receptor apoptosis-inducing ligand. *Source: From Henshall and Simon.* J Cereb Blood Flow Metab. *2005;25:1557–1572.*[3]

in a slower, energy-dependent, and more regulated way, characterized by programmed cell death. Here, in contrast to necrosis, affected cells demonstrate shrinkage, membrane blebbing without rupture, aggregation of chromatin about the nuclear membrane, preservation of intracellular organelle integrity with subsequent dispersal of cell contents in membrane-bound apoptotic bodies, and finally DNA fragmentation. Inflammation does not result.

The importance of differentiating among the types of ischemia should appear obvious. In patients with systemic hypoperfusion attention is directed to the heart, blood, and circulatory system. In patients with thrombosis and embolism, heart, aortic, and vascular imaging is used to clarify the nature of the occlusive lesion in the arteries that supply the region of the brain that is ischemic or infarcted. At times, the vascular problem leads to a temporary decrease in perfusion and a temporary abnormality of brain function. These attacks are referred to as transient ischemic attacks (TIAs).

VASCULAR PATHOLOGIES CAUSING BRAIN ISCHEMIA AND HEMORRHAGE

Hemorrhage

Hypertension

Hypertension is the most common and important condition that causes brain hemorrhage.[1,4] Sudden increases in blood pressure severely stress small arteries in the brain and can cause them to break with resultant hemorrhage into brain substance. Chronic poorly controlled hypertension also produces wearing down of the walls of small arteries so that small outpouchings (microaneurysms) develop. These outpouchings are not visible grossly but are seen only microscopically. The walls of these outpouchings are thin and can break especially when exposed to high blood pressure. Occasionally hypertension can cause rupture of small arteries on the surface of the brain. Bleeding in that case is subarachnoid into the fluid around the brain.

Aneurysms and Vascular Malformations

Bleeding can also ensue from abnormal blood vessels: aneurysms and vascular malformations. Aneurysms are outpouchings from arteries (see Fig. 22.1B). The angiograms in Fig. 22.7 show aneurysms of various sizes. They are usually located on large arteries along the base of the brain and are especially common at branchpoints where two arteries meet. The walls of aneurysms often contain weak spots that can break, especially if the blood pressure is high. Aneurysms usually rupture into the subarachnoid space, but occasionally into the brain, or into both the subarachnoid space and the brain. Although weak points in the artery may be present from birth, hypertension and other factors often lead to a gradual increase in the size of aneurysms. Large aneurysms (Fig. 22.7B) may contain thrombi that can potentially embolize to more distal branches of the parent artery.

Vascular malformations are congenital lesions that involve blood vessels. They usually arise from a failure of normal development of vascular networks that are present in the fetus, but some malformations are acquired during life. There are five types of vascular malformation:

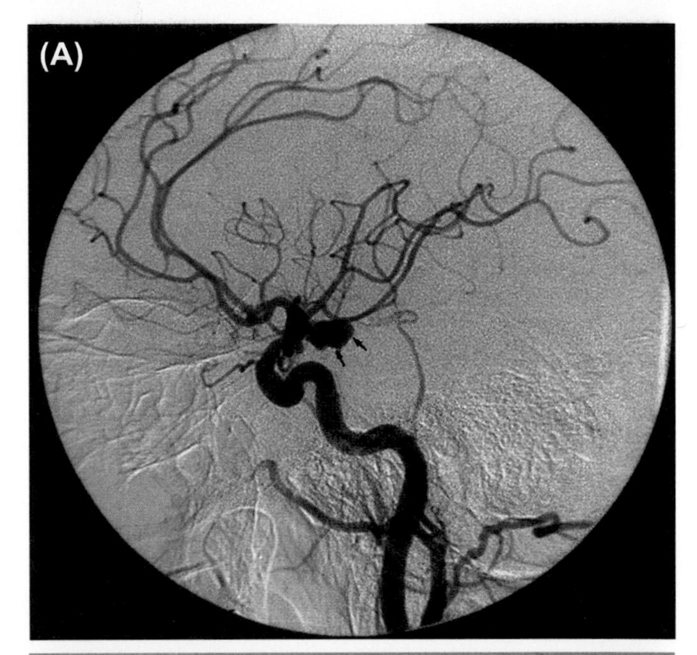

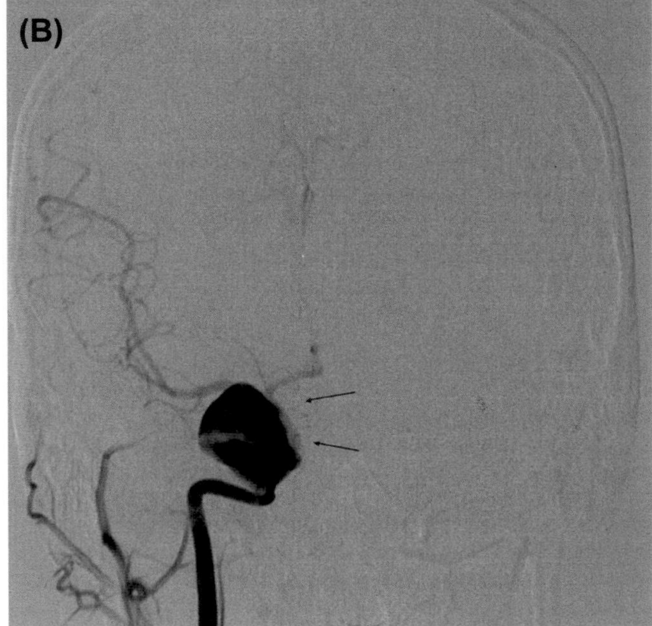

FIGURE 22.7 **Cerebral digital subtraction angiograms showing arterial aneurysms.** (A) Carotid angiogram, lateral view. An aneurysm (small black arrows) originating from the carotid artery is pointing posteriorly. (B) Carotid angiogram, anteroposterior view. A giant aneurysm (small black arrows) is shown. *Source: From Caplan.* Caplan's Stroke: A Clinical Approach. *4th ed. Saunders Elsevier; 2009,*[5] *with permission.*

- *Arteriovenous malformations (AVMs)* contain arteries that communicate directly with veins. Normally, blood goes from large to small arteries (arterioles) and then to a bed of tiny blood vessels called capillaries. The blood is then drained by veins. In an AVM, blood goes directly from arteries to veins without an intervening capillary network. Veins are thin walled and cannot withstand the pressure present in the arterial system. Veins within an AVM often rupture. Bleeding is usually into the brain, but can also be into the subarachnoid space if the AVM abuts on the brain surface. The size of the component vessels that make up AVMs vary greatly but the largest vessels are always veins. Figure 22.8(A) shows a drawing of an AVM and Fig. 22.8(B) and (C) show an MRI and an angiogram from a patient with an AVM.
- *Cavernous angiomas (cavernomas)* differ from AVMs in that they have no direct arterial supply. They are composed of a compact mass of tiny capillaries located close together in a capsule that separates the angioma from the rest of the brain. Most often they are located in the brain. When they bleed, the hemorrhage is usually contained by the capsule within the brain substance. Rebleeding is common and these lesions often cause seizures. Figure 22.9(A) shows a drawing of a cavernous malformation and Fig. 22.9(B) is an MRI from a patient with a cavernous malformation.
- *Developmental venous anomalies (DVAs)* are the most common type of vascular malformation found in

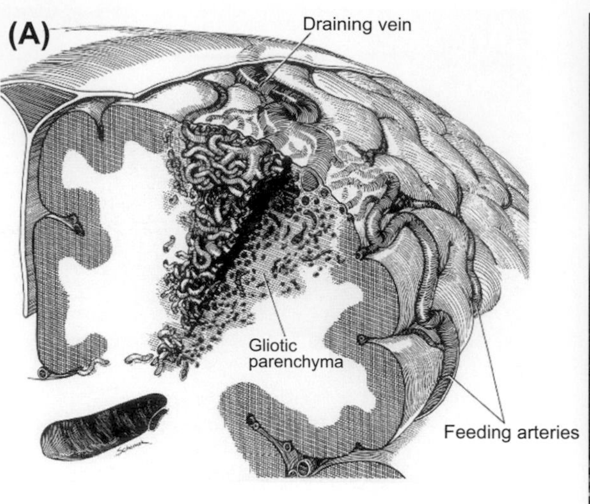

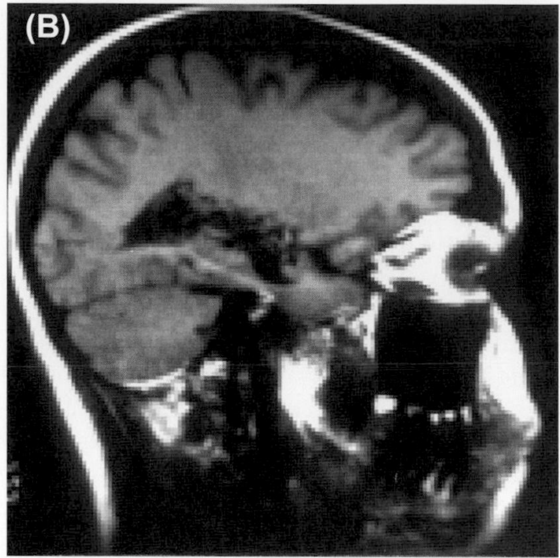

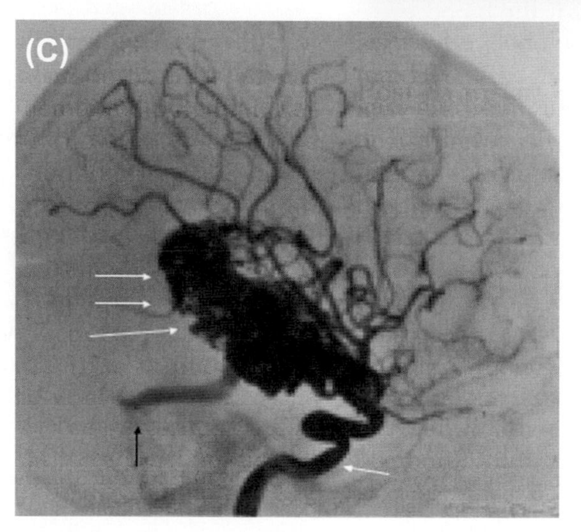

FIGURE 22.8 **Arteriovenous malformations (AVMs).** (A) Artist's drawing of an AVM, showing feeding arteries, parenchymal component, and draining vein. (B) Magnetic resonance imaging (MRI), sagittal view, showing the parenchymal hypervascularity as a black area in the temporal lobe of the brain. (C) Digital subtraction cerebral angiogram lateral view from the patient whose MRI is shown in (B). The middle cerebral artery branch of the internal carotid artery (single white arrow) is supplying a tangle of vessels in the brain parenchyma (three white arrows). The parenchymal lesion is draining into a vein (small black arrow). *Source: From Caplan.* Caplan's Stroke: A Clinical Approach. *4th ed. Saunders Elsevier; 2009,[5] with permission.*

the brain by modern brain imaging or after death. (They were formerly called venous angiomas.) In a DVA, at birth there is a deficiency of draining veins, so that some remaining veins must drain a larger portion of brain than is customary. DVAs rarely cause brain hemorrhages but can cause localized brain edema and infarction especially if draining veins become occluded. These lesions, like cavernomas, predispose to epileptic seizures. Figure 22.10(A) shows a drawing of a DVA and Fig. 22.10(B) shows a DVA in a brain specimen at necropsy.
- *Telangiectasias* (very small dilated capillaries admixed with brain tissue) and *venous varices* (very dilated draining veins) are other types of vascular malformation, but they rarely cause serious brain hemorrhage.

Hematological Abnormalities

Several hematological abnormalities can provoke cranial bleeding. The most common is iatrogenic prescription of anticoagulant agents: heparins, warfarin, dabigatran, or rivoraxaban. Thrombocytopenia and congenital or acquired deficiencies in constituents of the body's coagulation cascade also cause bleeding. The most well-known hereditary disorder is hemophilia, which is a deficiency in factor VIII (antihemophiliac globulin). In patients with bleeding tendencies, hemorrhage is usually into a number of different locations but the most devastating is bleeding into the brain.

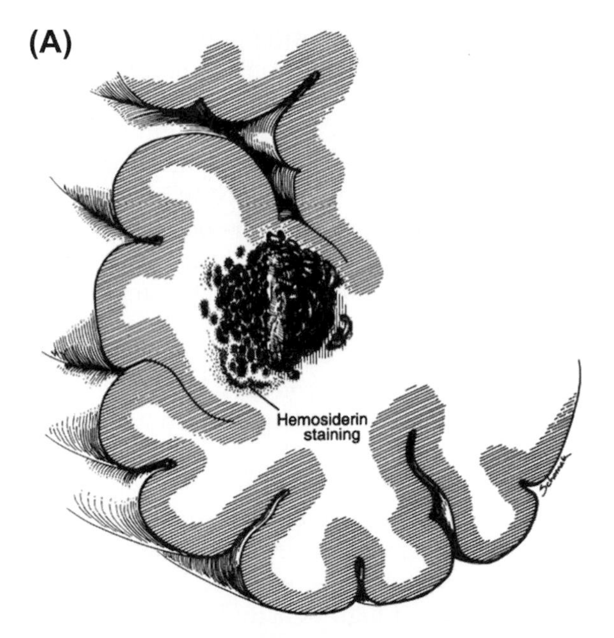

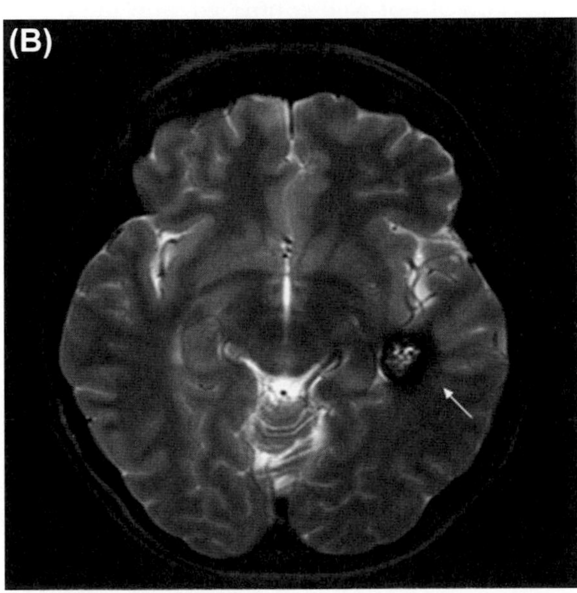

FIGURE 22.9 **Cavernous angiomas.** (A) Artist's drawing of a cavernous angioma. (B) Magnetic resonance imaging: gradient echo image showing a cavernous angioma (white arrow). *Source: From Caplan.* Caplan's Stroke: A Clinical Approach. *4th ed. Saunders Elsevier; 2009,*[5] *with permission.*

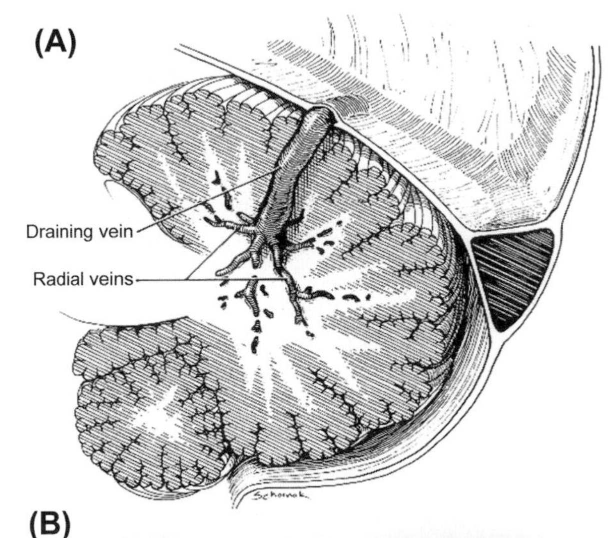

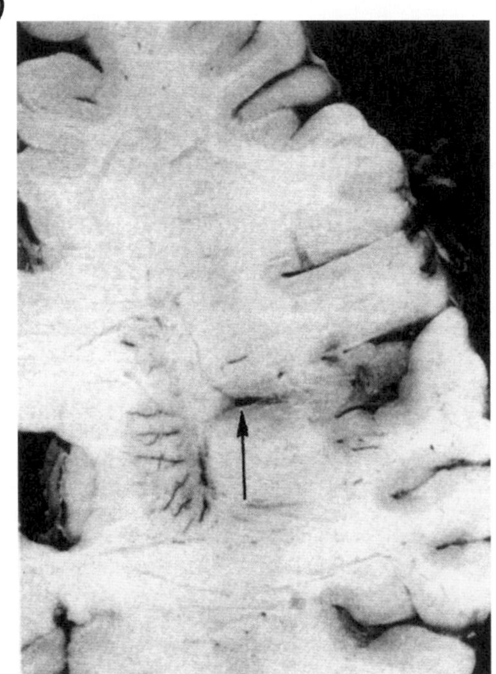

FIGURE 22.10 **Developmental venous anomalies (DVAs).** (A) Artist's drawing of a DVA. (B) Brain specimen showing a radial pattern of small veins draining into a larger vein (black arrow). *Source: From Caplan.* Caplan's Stroke: A Clinical Approach. *4th ed. Saunders Elsevier; 2009,*[5] *with permission.*

Ischemia and Infarction

Atherosclerosis

Atherosclerosis (*athero* refers to fatty accumulations) and arteriosclerosis (literally, hardening of arteries) are terms often used interchangeably for degeneration of the components of the arterial wall.

This degeneration is characterized by the development of plaques on the inside of arteries, and wear and tear affecting the wall of the arteries leading to decreased elasticity and stiffness of the arteries. Atherosclerotic plaques (atheromas) develop in the aorta and in the large arteries in the neck and head that supply the brain with blood. The earliest atherosclerotic abnormalities are fatty streaks that are visible as regions of yellowish discoloration of the intima of the aorta and large- and medium-sized cerebrovascular arteries. Fatty streaks develop in childhood. The fat comes from circulating blood lipids. Lipid accumulates in smooth muscle cells beneath the intima to form these fatty streaks. Later in life, firm fibrous plaques develop in the same regions as these fatty streaks. These plaques consist of lipid, smooth muscle, fibrous tissue, connective tissue, white blood cells, and crystals of cholesterol. Some plaques are soft while others are very firm and even calcified.

When atherosclerotic plaques enlarge they narrow arterial lumens, cause turbulence of flow, and diminish distal brain perfusion. The irregular surfaces of plaques attract platelets that stick to each other and form bonds with fibrin, a protein formed in the blood from fibrinogen. These platelet–fibrin clumps are white and so are often called white clots. Cracks in plaques activate clotting factors in the blood. Red blood cells form a mesh with fibrin; these erythrocyte–fibrin thrombi are termed red clots. Figure 22.11 shows necropsy specimens of an intracranial atherosclerotic lesion in which a plaque has ruptured, causing a luminal thrombus to form.

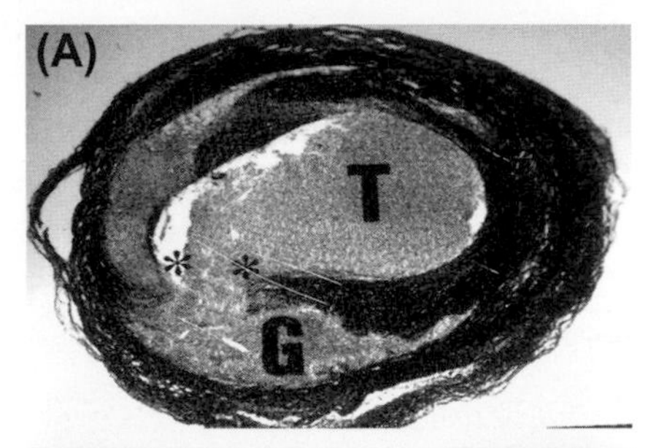

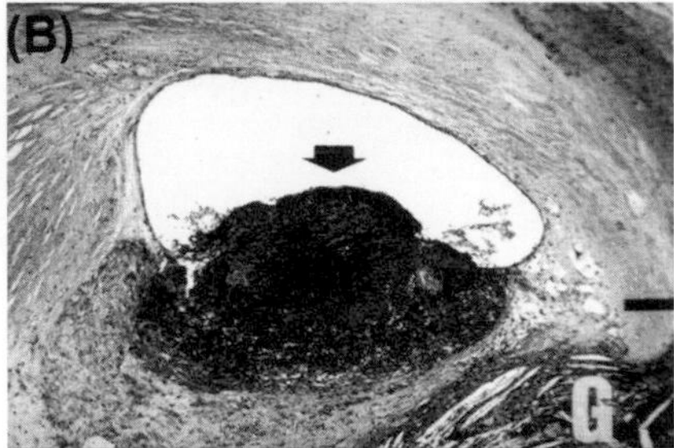

FIGURE 22.11 **An atherosclerotic intracranial artery at necropsy.** (A) An atherosclerotic plaque (G) has ruptured (area between the asterisks) and the contents of the plaque along with tissue factor have entered the arterial lumen (T). (B) A thrombus has formed in the arterial lumen. *Source: From Ogata* et al. J Neurol Neurosurg Psychiatry. *1994;57:17–21.*[6]

Atherosclerotic large artery abnormalities cause ischemia in three major ways: (1) severe luminal narrowing markedly decreases blood flow, leading to brain ischemia in the territory of the compromised artery (hypoperfusion); (2) plaques or occlusive thrombi mechanically block branches of the main arteries, leading to hypoperfusion in the distribution of these branches of the artery; and (3) propagation and embolization of thrombi cause occlusion of distal branches. Emboli can consist of red thrombi, white platelet–fibrin aggregates, or elements of plaques such as cholesterol crystals.

Some conditions promote and accelerate the development of atherosclerotic plaques. These include hypertension, especially if not well controlled, cigarette smoking, high blood cholesterol levels (elevated low-density lipoproteins and reduced high-density lipoproteins), diabetes, especially if not well controlled, and the metabolic syndrome.

Hypertension

High blood pressure (hypertension) leads to wear and tear on arteries. Picture a plumbing situation in which the water pressure is quite high. The pipes would rust and the walls of the pipes may thin with time. High blood pressure accelerates the development of atherosclerotic changes in the large arteries of the neck and head. Plaque development is more severe and occurs earlier in life than when blood pressure remains normal. Hypertension also leads to thickening of the walls of small arteries in the brain. This thickening narrows the lumen of the arteries and can lead to infarcts deep within the brain. Figure 22.12(A) shows a penetrating artery damaged by severe hypertension.

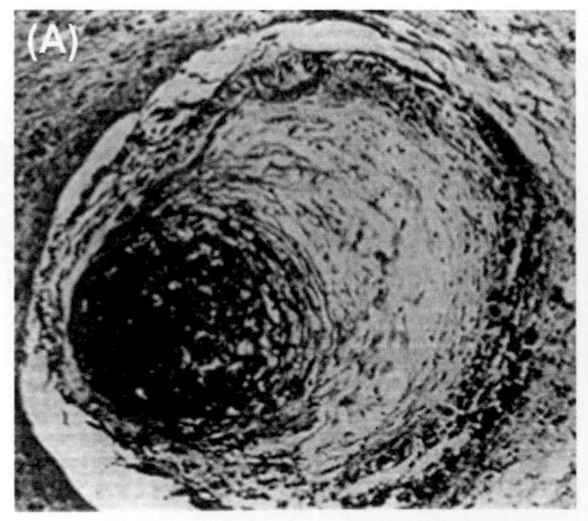

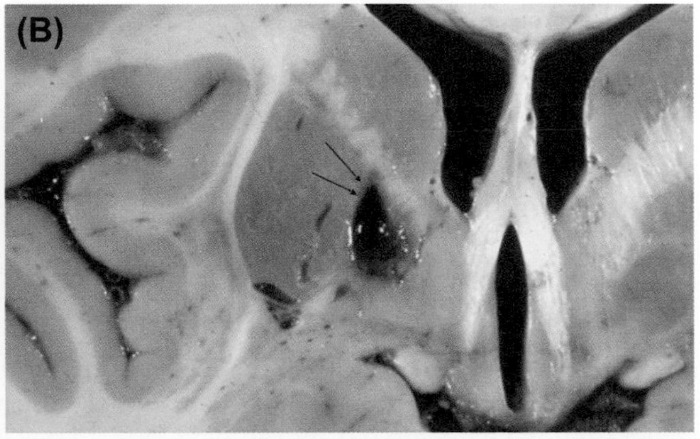

FIGURE 22.12 **Penetrating artery disease caused by hypertension.** (A) Photomicrograph of a small penetrating artery showing lipohyalinosis and fibrinoid necrosis in the arterial wall. The lumen is considerably compromised. (B) Necropsy specimen showing a cavity (black arrows) due to an old lacunar infarct that is located in the medial basal ganglia (mostly the globus pallidus) and extends through the internal capsule in a patient with a hemiplegia during life. *Source: (A) Courtesy of C. Miller Fisher MD. (B) From Caplan.* Caplan's Stroke: A Clinical Approach. *4th ed. Saunders Elsevier; 2009,*[5] *with permission.*

Fibrinoid material has accumulated in the lumen and has severely compromised the lumen. Figure 22.12(B) shows a small deep infarct caused by occlusion of a penetrating artery damaged by hypertension. These deep infarcts are often referred to as lacunes (meaning holes).

Hypertension can also lead to a rupture of small arteries in the brain. These hemorrhages most often develop in the deep structures in the brain parenchyma that are supplied by arteries that penetrate from the middle, anterior, and posterior cerebral arteries and the vertebral and basilar arteries. The basal ganglia, thalamus, pons, and cerebellum are the most common sites of these hypertensive hemorrhages. Hypertensive hemorrhages can also develop in the white matter of the cerebral lobes. Hypertension, when uncontrolled, is the single most important risk for brain ischemia and brain hemorrhage.

Arterial Dissection

The term dissection refers to a tear in an artery. Sudden movements and stretching as well as direct injury to an artery can cause the wall of an artery to tear. Portions of the neck arteries destined to supply the brain are anchored in the neck and where they pass into the skull. Other portions of these arteries are quite mobile and can be stretched and torn. When a tear occurs, it causes bleeding in the arterial wall. The wall swells and may block blood flow in the lumen of the artery. Figure 22.13 shows a drawing of an arterial dissection. The clot in the arterial wall can be discharged into the lumen of the artery and from there embolize into the brain. Arteries in the skull can also be torn, causing brain infarction or subarachnoid hemorrhage. Dissections are important causes of stroke in children and young adults. Dissections are also a complication of therapeutic neck manipulation.

Fibromuscular Dysplasia

This uncommon condition involves the wall of the artery. There is an excess amount of smooth muscle and connective tissue in the wall of the artery. The excess tissue can narrow the arterial lumen, and the excess smooth muscle can contract and narrow the lumen. This vasoconstriction can block blood flow to the brain, causing infarction. Fibromuscular dysplasia is more common in women than in men. The angiogram in Figure 22.14 shows the contractile portions of an artery that harbors fibromuscular dysplasia. This disorder can also involve the arteries to the kidneys (renal arteries), causing hypertension. Individuals with fibromuscular dysplasia have a higher than normal frequency of also harboring aneurysms in arteries of the skull.

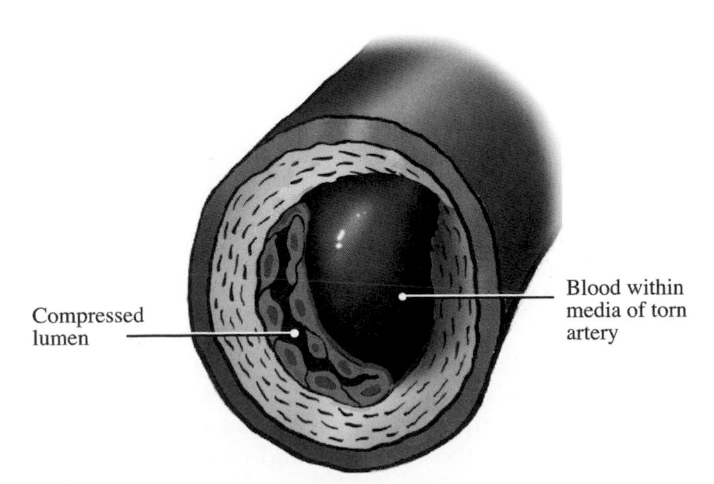

FIGURE 22.13 **Artist's drawing of a cross-section of an artery containing a dissection.** Blood is within the arterial wall and the lumen is severely compromised. *Source: From Caplan.* Stroke. *Demos Medical Publishing; 2006,*[2] *with permission.*

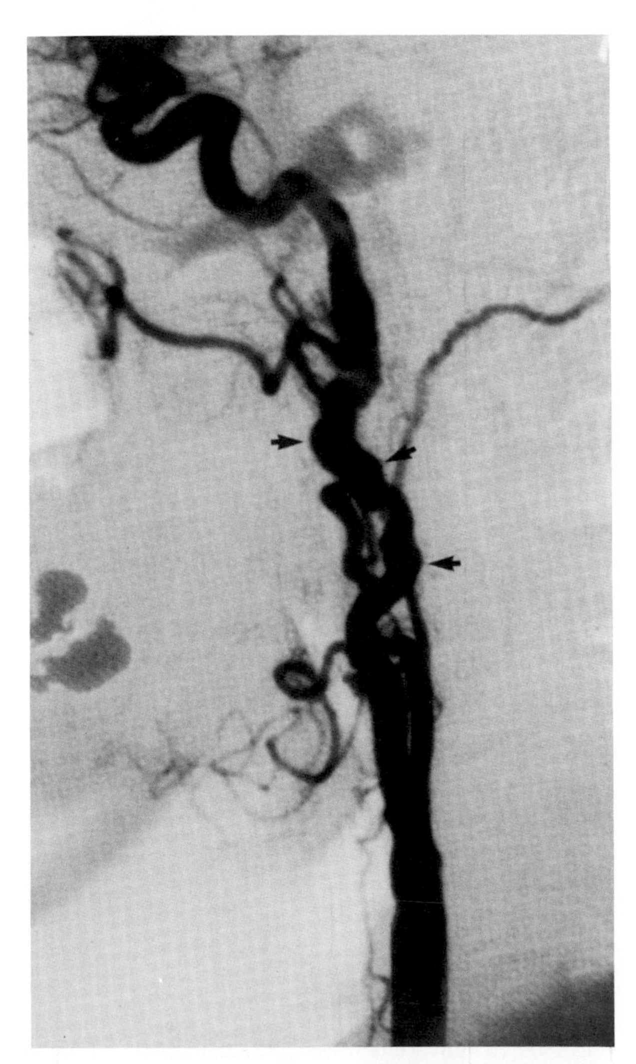

FIGURE 22.14 **Carotid angiogram, subtraction lateral view.** Contractile areas that represent regions of fibromuscular dysplasia are shown with black arrows. *Source: From Caplan.* Caplan's Stroke: A Clinical Approach. *4th ed. Saunders Elsevier; 2009,*[5] *with permission.*

Arterial Dolichoectasia (Dilatative Arteriopathy)

Some blood vessels that supply the brain can become quite elongated and dilated and follow a tortuous and windy course with frequent loops and curves. The most frequent medical term used for this type of abnormality is dolichoectasia (Greek *dolichos*, elongation, and *ectasia*, dilatation). The abnormality can involve the arteries in the neck or in the skull; often both are elongated and tortuous. The widening and lengthening of arteries can slow blood flow. In some patients blood even backs up and temporarily flows down the blood vessels back towards the heart. The elongation can distort the orifices of branches, and slowing of blood flow can stimulate blood clotting in the arteries. TIAs and minor strokes are common and occasionally the arteries break and lead to subarachnoid bleeding. Dolichoectasia is caused by abnormalities in the medial and elastic coats of the arteries. It is more common in the intracranial vertebral and basilar arteries but can affect any cervical or intracranial artery. Figure 22.15 is an angiogram showing a dilated ecstatic basilar artery that contains a thrombus.

Heart and Aortic Conditions that Cause Brain Ischemia and Infarction

Blood clots and other particles can be released into the bloodstream from the heart and aorta. These particles are carried in the flowing blood towards the brain and other organs, driven forward in the blood vessels by the force of the contractions of the heart and the blood pressure in the arteries. Arteries become smaller as they reach the organs that they supply, especially after branchings. Depending on their size and makeup, particles can become stuck at branch points or where the arterial passage in which they flow is smaller than the embolic material (Fig. 22.2B). Blockage of an artery deprives the brain of blood and nutrition. The blockage may be temporary, the embolic material breaking up and slipping through the block point. If the obstruction to blood flow lasts long enough, the brain becomes infarcted. Embolic materials include red and white thrombi, bacteria from valvular infected vegetations, fibrin, and calcium.

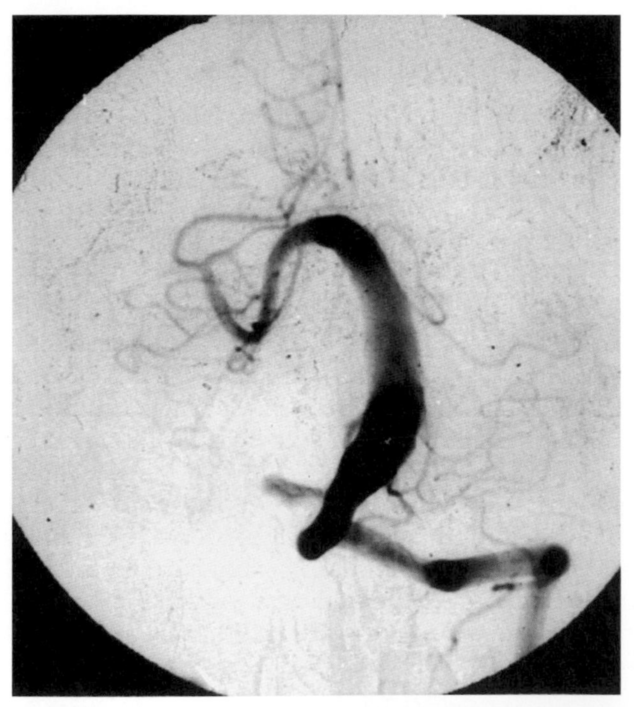

FIGURE 22.15 **Vertebrobasilar digital subtraction angiogram showing a very irregular dilated ectatic basilar artery with extensive atheromatous plaques.** There is a thrombus in the lumen of the dilated artery. The branches of the rostral basilar artery are not well opacified because of reduced antegrade blood flow. *Source: From Caplan.* Caplan's Stroke: A Clinical Approach. *4th ed. Saunders Elsevier; 2009,*[5] *with permission.*

Various heart conditions and diseases lead to brain embolism (Table 22.1). Thrombi can form in the atria or ventricles. Clots in the left atrium and left ventricle can be expelled to travel to the brain and other organs.

Atrial fibrillation is a very common condition that becomes even more common as people age. It is present in about 1 in every 200 individuals, and as many as 5% of people over age 60. Atrial fibrillation describes inefficient, irregular contractions of the atria. The atria become dilated and blood can pool within them because of the inefficient contractions. The pooling of blood leads to stagnation and the formation of red clots in the atria and the atrial appendage. These clots can then pass into the ventricles and from there into the aorta and into the arteries feeding the brain and other organs. Figure 22.16 shows a thrombus in the left atrium of a patient who had a brain embolism.

Myocardial infarcts (heart attacks) are another common source of brain embolism. Atherosclerosis, described above as a very important cause of brain ischemia and infarction, is the most important cause of heart ischemia and infarction. Atherosclerotic plaques in the coronary arteries that supply the muscle (myocardium) of the heart become blocked with plaques. Blockage of these arteries leads to infarction of portions of heart muscle. The damage can lead to poor heart muscle contractility (hypokinesis) and even formation of bulges and outpouches in the ventricles (ventricular aneurysms). This damage to the heart muscle leads to deposition of clots in the interior of the heart. These clots can then be pumped by heart contractions into the aorta and bloodstream.

Brain embolism is also a major risk in patients who develop congestive heart failure. In this condition the heart cannot pump out the blood that is brought into it. Blood pools in the ventricles and can clot. Coronary artery disease and hypertension are the most frequent causes of heart failure but other less common conditions that affect heart muscle (myocardium) can also lead to heart failure and brain embolism.

TABLE 22.1 Cardiac Sources of Emboli

1. Coronary artery disease and myocardial infarction
2. Arrhythmias
Atrial fibrillation
Sick-sinus syndrome
3. Heart failure
4. Valvular heart disease
Rheumatic, especially mitral stenosis and insufficiency
Mitral annulus calcification
Mitral valve prolapse
Calcific aortic stenosis
Bacterial endocarditis
Non-bacterial thrombotic endocarditis
5. Myocardiopathies
Alcoholic cardiomyopathy
Cocaine cardiomyopathy
Peripartum myocardiopathy
Myocarditis
Sarcoidosis
Fabry disease
Amyloidosis
6. Cardiac tumors
Myxomas
Fibroelastomas
7. Septal abnormalities (paradoxical embolism)
Atrial septal defects
Patent foramen ovale
Atrial septal aneurysms

Various inflammatory conditions (myocarditis) and other disorders (myocardopathies) that affect heart muscle can cause poor pumping function similar to myocardial infarction and so also predispose to clots that form in the heart and later embolize to the brain.

The heart valves can also be the site of disease that leads to brain embolism. The valves in the heart serve very important functions. They open to allow blood to flow in the desired direction and then close to prevent blood from flowing backwards. Valvular disease can cause hardening of the valves, which impairs mobility and narrows the space available for blood to flow. This is termed valvular stenosis. When a valve fails to close efficiently, allowing backflow, the condition is called valve insufficiency.

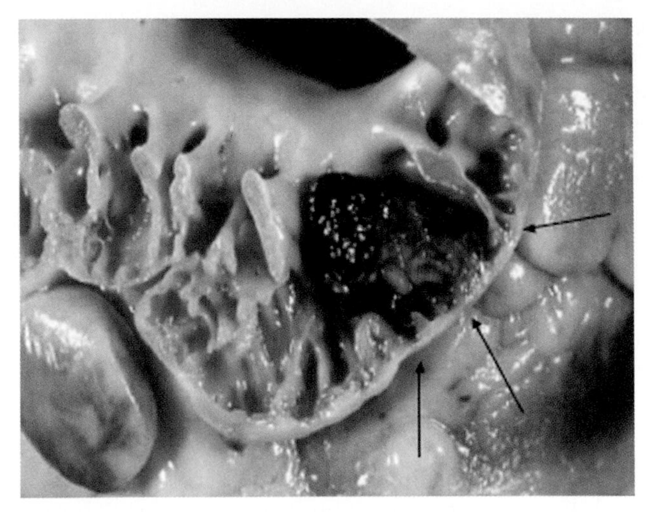

FIGURE 22.16 **A thrombus (three black arrows) in the left atrium at necropsy.** *Source: Ogata et al.* Ann Neurol. *2008;63(6):770-781,*[7] *with permission.*

Rheumatic fever was once the major cause of heart valve inflammation and disease. Although the frequency of rheumatic fever and rheumatic heart disease is decreasing it is still an important cause of heart disease. This condition often affects the mitral and aortic valves to cause mitral insufficiency and mitral stenosis.

Aortic stenosis and aortic insufficiency are also common. Some children are born with abnormal heart valves. Aging can also lead to degenerative changes in the valves. One relatively common valve condition, which is especially frequent in women, is mitral valve prolapse. This means that portions of the mitral valve go backwards into the atrium instead of going entirely into the left ventricle. Mucoid material can be deposited in the mitral valve to cause this abnormal functioning. Heart valves can also be damaged by infection. This is usually termed bacterial endocarditis. Some patients with cancer and other debilitating disease develop deposits of fibrin and fibrous vegetations on their heart valves. This condition is called non-bacterial thrombotic endocarditis. Valve diseases can lead to discharge of a number of different types of particles into the bloodstream, including blood white and red clots, pieces of calcium, bacteria, and fibers that collect along the valve.

Some heart problems are congenital, meaning that the defects are present at birth and remain. In some patients holes exist between the left and right atria or ventricles. These are referred to as atrial or ventricular septal defects. For many reasons, an individual can develop a blood clot (thrombus) in a vein somewhere in the body, especially in a leg vein. Sitting for a long time in one position, crossing the legs and so compressing the veins, abnormal leg veins, and abnormally increased tendency for blood clotting are just some of the reasons for leg vein thrombosis. When the clot first forms, it can break loose and go to the heart with the returning venous blood. These clots most often go into the right atrium, through the tricuspid valve into the right ventricle and then through the pulmonary arteries into the lungs; this condition is called pulmonary embolism and can be serious. When there is a hole between the two atria, a thrombus formed in the leg veins reaching the right atrium can pass through the hole (atrial septal defect) into the left atrium and then through the mitral valve into the left ventricle. From there, the clot goes through the aortic valve into the aorta and finally into one of its systemic branches. If it enters one of the arteries to the brain, the embolus can cause a brain infarct. When defects cause clots to pass between the two circulations because of defects, the stroke caused is termed paradoxical embolism.

Before birth, since the fetal lungs do not function and do not breathe air, a hole called the foramen ovale exists in the wall that separates the left and right atria. This allows blood to go through the mother's circulatory system for oxygenation. At birth the hole usually closes but, in about 30% of individuals, it does not fully close. When the hole remains open it is usually referred to as a patent foramen ovale.

The aorta is the largest artery in the body. The arteries destined to supply the brain arise from the beginning portion of the aorta in the chest. This part of the aorta and the aorta in the abdomen are regions in which atherosclerosis is often very severe. Atheromatous plaques in the aorta in the chest can block arteries to the head. Red and white clots often form on the surface of aortic plaques. These clots as well as calcium particles and pieces of cholesterol within plaques can break loose and be carried by the bloodstream into the arteries that feed the brain. Clamping of the aorta during heart surgery is an important cause of emboli to the arteries supplying the brain. Figure 22.17 shows a necropsy specimen of a severely atherosclerotic aorta.

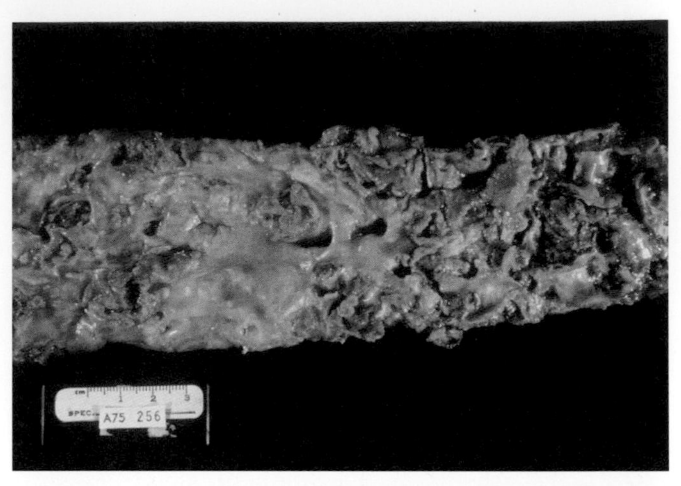

FIGURE 22.17 **Aorta at necropsy from a patient whose transesophageal echocardiogram before surgery showed severe disease of the ascending aorta and aortic arch with mobile protruding plaques.** This patient died after coronary artery bypass surgery, having never awoken after the procedure. *Source: Kindly submitted by Dr Denise Barbut. From Caplan.* Caplan's Stroke: A Clinical Approach. *4th ed. Saunders Elsevier; 2009,[5] with permission.*

Blood Abnormalities Contributing to Brain Ischemia and Infarction

The ability of the blood to clot is a very important defense mechanism. Physical injuries frequently cause breakage of small blood vessels. Components of the blood plug the regions of blood vessel injury to prevent excess bleeding. A deficiency of these blood factors leads to excess bleeding, whereas other conditions can lead to excess clotting.

The two most important parts of the clotting system are blood platelets and proteins in the blood and blood vessel walls that promote blood clotting. Platelets (also called thrombocytes, a term that literally means clotting cells) are tiny cells that circulate in the blood. When there is a blood vessel injury, platelets circulating in the blood are drawn to the injured region. They adhere to the point of injury and stick together to form a plug. When there are too many platelets (thrombocytosis) clotting is excessive. Excess bleeding can develop when there are too few platelets (thrombocytopenia). Some blood proteins (antithrombin III, protein C, and protein S) when present in normal amounts inhibit the blood from clotting. When there is a deficiency of one of these substances, usually from birth, then blood clotting is excessive. When other blood clotting proteins are excessive, for example, factor VII, VIII, or XII, there is an increased tendency for the blood to clot. Some medical conditions, including cancer, also lead to an increased tendency of the blood to clot. Acute infections and inflammatory conditions such as inflammatory bowel disease increase clotting factors and promote thrombosis. Thrombi can develop in blood vessels that have plaques and in apparently normal small blood vessels.

FACTORS AFFECTING TISSUE SURVIVAL IN PATIENTS WITH BRAIN ISCHEMIA AND INFARCTION

The survival of the brain regions at risk depends on a number of factors: the adequacy of collateral circulation, the state of the systemic circulation, serological factors, changes in the obstructing vascular lesion, resistance in the microcirculatory bed, and brain edema and intracranial pressure.

Adequacy of the Collateral Circulation

Congenital deficiencies in the circle of Willis and prior occlusion of potential collateral vessels decrease the available collateral supply. Hypertension or diabetes diminishes blood flow in smaller arteries and arterioles and thus reduces the potential of the vascular system to supply blood flow to the needy region.

State of the Systemic Circulation

Cardiac pump failure, hypovolemia, and increased blood viscosity all reduce cerebral blood flow. The two most important determinants of blood viscosity are the hematocrit and fibrinogen levels. In patients with hematocrits in the range of 47–53%, lowering of the hematocrit by phlebotomy to below 40% can increase cerebral blood flow by as much as 50%. Blood pressure is also important. Elevation of blood pressure, except in malignant ranges, increases cerebral blood flow. Surgeons take advantage of this fact by injecting catecholamines to raise blood pressure and flow during the clamping phase of carotid surgery. Low blood pressure significantly reduces cerebral blood flow. In some patients, the balance is so tenuous that simply sitting in bed or standing lowers collateral pressure enough to induce symptoms. Low blood and fluid volume also limits available blood flow in collateral channels. Many older individuals limit their fluid intake, especially in the evening, to avoid getting up at night to urinate. After a stroke there may not have been any fluid intake because of swallowing difficulty or lack of feeding during the trip to the hospital and hospital encounters.

Serological Factors

Blood functions as a carrier of oxygen and other nutrients. Hypoxia is clearly detrimental because each milliliter of blood delivers a less than normal oxygen supply. Low blood sugar similarly increases the risk of cell death, and elevated blood sugar can also be detrimental to the ischemic brain. Elevated serum calcium levels and high blood-alcohol content are other potential important detrimental variables.

Changes in the Obstructing Vascular Lesion

Embolic occlusive thrombi do not adhere to the vessel wall of the recipient artery and frequently move on. The moving embolus can block a more distal intracranial artery, causing added or new ischemia, or it may fragment and pass through the vascular bed. Clot formation activates an endogenous thrombolytic system that includes tissue plasminogen activator (tPA). Inhibitors of tPA are also present. Sudden obstruction of a vascular lumen can cause reactive vasoconstriction (spasm), which in turn causes further luminal compromise. Thrombolysis, passage of clots, and reversal of vasoconstriction all promote reperfusion of the ischemic zone. If reperfusion occurs quickly enough, the stunned, reversibly ischemic brain may recover quickly. The occlusive clot may propagate further proximally or distally along the vessel, blocking potential collateral channels. The distal end of the thrombus can also break loose and embolize to an intracranial receptive site. Hypercoagulable states promote such extension of thrombi.

Resistance in the Microcirculatory Bed

The vast majority of cerebral blood flow does not occur in the large macroscopic arteries at the base of the brain or along the surface. Most blood flow passes through microscopic-sized vessels: the arterioles, capillaries, and venules. Resistance to flow in these small vessels is affected by prior diseases, such as hypertension or diabetes, which often cause thickening of arterial and arteriolar walls. Experimental animals and patients who have been hypertensive or diabetic before an arterial occlusion fare worse than individuals who were previously normotensive, presumably because of these microcirculatory changes.

Both hyperviscosity and diffuse thromboses in the capillaries and microvessel bed greatly reduce flow through the microcirculation. Ischemic insults may produce biochemical changes that lead to platelet activation, clumping of erythrocytes, and plugging of the microcirculation. These changes are sometimes posited to induce a "no reflow" state in the microvascular bed, even when large arteries are reperfused. In general, studies of cerebral blood flow are sensitive to changes in resistance in the microcirculatory bed. Blood flow is inversely proportional to resistance in the vascular bed, the majority of which is microcirculatory.

Brain Edema and Increased Intracranial Pressure

Edema and pressure changes in the brain and cranial cavity are important influences on the survival of brain tissue and patient recovery after vascular occlusion or hemorrhages. There are two types of brain edema: water accumulation inside cells, termed cytotoxic or dry edema; and fluid in the extracellular space, termed vasogenic or wet edema because the cut surface of the brain oozes edema fluid. Cytotoxic edema is caused by energy failure, with movement of ions and water across the cell membranes into cells. Extracellular edema is influenced by hydrostatic pressure factors, especially increased blood pressure and blood flow, and by osmotic factors. When proteins and other macromolecules enter the brain extracellular space because of breakdown of the blood–brain barrier, they exert an osmotic gradient that pulls water into the extracellular space. This vasogenic edema accumulates most in the cerebral and cerebellar white matter because of the difference in compliance between gray and white matter.

Brain swelling caused by cytotoxic edema means a large volume of dead or dying brain cells, which portends a bad outcome. In contrast, edema in the extracellular space does not necessarily imply neuronal injury, and fluid in the extracellular compartment can potentially be mobilized and removed. In any case, severe edema may cause gross swelling of the brain; shifts in position of brain tissue, with potential pressure damage; and herniation of brain contents from one compartment to another. Vasogenic edema is especially common after venous occlusions. Because of the lack of drainage, fluid backs up into brain tissue.

Edema adds to the volume of tissue in the cranium. Substances that are usually not present in brain tissue, such as hematomas, tumors, and abscesses, also add to the volume of tissue. The increased volume can increase intracranial pressure leading to increased morbidity and decreased cerebral blood flow. When intracranial pressure is increased, the pressure in the venous sinuses and draining veins must also increase if blood is to be drained normally from the cranium. There must be a gradient between venous pressure and intracranial pressure for drainage to occur. For tissue perfusion to occur, arterial pressure must exceed venous pressure. Blood flow is compromised in the presence of arterial and venous occlusions. When the intracranial venous system contains occlusions, venous pressure is increased. The limited drainage often causes fluid to back up into the brain, causing vasogenic edema. Increased intracranial pressure places an additional stress on the system, forcing even higher the flow values required for tissue survival. Brain edema and increased intracranial pressure cause headache, decreased consciousness, and vomiting. Pressure shifts and herniation cause pressure-related damage to adjacent tissues and signs of dysfunction of the compressed structures. Pressure shifts and herniations are most common early in the course after intracerebral hemorrhage, owing to the additional presence in the brain of an extra mass of tissue (hematoma). Edema after brain infarction usually develops after days, so that pressure changes occur later than in patients with parenchymatous hemorrhages.

DEATH OF CELLS IN THE CNS, AND NEUROPROTECTIVE AND REPARATIVE MECHANISMS

Cell Death

Several final pathways contribute to and facilitate cell death.[8–10] Excitotoxicity refers to the pathological effect of excitatory neurotransmitters, particularly glutamate, which is present in high concentrations in the extracellular space of ischemic brain. The increased glutamate concentrations are the result of depolarization-induced synaptic release, reversal of astrocytic glutamate uptake, and activation of voltage-sensitive glutamate receptor-gated channels. Extracellular glutamate binds to the postsynaptic *N*-methyl-D-aspartate (NMDA)-preferring receptor, particularly the extrasynaptic receptors containing an NR2B subunit. The NMDA receptor gates a Ca^{2+}-permeable ion channel. Accordingly, its activation produces additional Ca^{2+} influx. Furthermore, Ca^{2+} entry by this route leads to activation of neuronal nitric oxide synthase, with resultant production of reactive oxygen species producing mitochondrial damage, DNA damage, ion channel activation, protein modification, and triggering of cell death pathways. Glutamate also binds to the 2-amino-3-(5-methyl-3-oxo-1,2-oxazol-4-yl) propanoic acid (AMPA) receptor, which gates Na^{2+} and K^+, but in some configurations, those lacking the glutamate receptor-5 subunit can also flux Ca^{2+}.

The cell death pathways that are most completely understood are those involved in the programmed cell death phenomena of apoptosis. Two major gene families regulate apoptosis: caspases (cysteine–aspartate proteases) and Bcl-2 (an oncogene family of 20 members which positively or negatively regulate apoptosis). These gene products can be activated by two pathways: extrinsic or intrinsic (the latter predominate in stroke). Extrinsic pathways are activated via ligand binding to cell surface death receptors, for example tumor necrosis factor-α. The intrinsic pathway is initiated from the movement of proapoptotic proteins from cytoplasm to mitochondria with resultant alteration of mitochondrial membrane permeability and cytochrome *c* release producing activation of the initiator caspase, caspase-9, which in turn activates the executioner caspase, caspase-3 (Fig. 22.6). In the extrinsic pathway, cleavage of the initiator caspase, caspase-8, activates the executioner caspase, caspase-3. Both pathways converge on inhibitor of caspase-activated DNase and lead to the release of caspase-activated DNase, which translates to the nucleus, where it produces DNA fragmentation (a hallmark of apoptosis). A caspase-independent form of programmed cell death occurs with apoptosis-inducing factor translocation from the mitochondria to nucleus with resultant DNA degradation.

Survival and Repair Mechanisms

Neurogenesis occurs in the normal adult brain, as proliferating precursors are found in the rostral subventricular zone and in the subgranular zone of the dentate gyrus. Subventricular zone progenitors migrate towards the olfactory bulb (in rodents) via the rostral migratory stream and subgranular neurons migrate to the adjacent granule cell layer in the dentate gyrus. After acute focal ischemic brain injury, proliferation, differentiation, and migration of these cells are enhanced. It is unlikely these new cells could actually repopulate infarcted brain, but they may promote survival of ischemic brain by way of growth factor release or inflammatory suppression, or via other neurotrophic roles. For example, progenitor cell proliferation seems to be required for ischemic tolerance. Angiogenesis is also important since new vessels can also develop.

Ischemic tolerance refers to endogenous neuroprotective mechanisms, which are widely evolutionarily conserved and active in mammalian brain. [11–14] In tolerance, a brief ischemic stress (preconditioning) can alter the brain's response to subsequent prolonged severe ischemia, resulting in attenuated ischemic injury (tolerance) such as protection against stroke. The preconditioned brain, when challenged by severe ischemic stress, reprograms its response from one of injury induction to one of neuroprotection. Tolerance is dependent on protein synthesis, requires 24 hours to develop, and lasts for about a week. The molecular basis for this protection has been investigated using genomic and proteomic techniques. These studies show that prior ischemic stress changes the brain's genomic and proteomic responses to subsequent ischemic injury to produce the protective phenotype. At the transcriptional level, the protected phenotype in brain is that of broad-based transcriptional suppression (gene silencing), which has features similar to the molecular and cellular functions seen in the brains of hibernating animals. At the translational level, newly synthesized proteins include polycomb protein complexes, which are epigenetic regulators of transcription and appear to drive the protective gene silencing. Cross-tolerance refers to non-ischemia preconditioning (e.g. seizures or endotoxin), which also induces tolerance to ischemia. Remote tolerance refers to ischemic in a peripheral organ (e.g. limb ischemia via blood pressure cuff) that also induces tolerance to brain ischemia. This last form of tolerance is under investigation in human clinical trials.

QUESTIONS FOR FURTHER RESEARCH

- Can stem cells be engineered for stroke neuroprotection or stroke recovery? How should the engineered cells be given: intravenously, intra-arterially, or directly into the ischemic brain? When should they be best given: hyperacutely, risking that they may not survive because of decreased blood supply, subacutely, or chronically? To which patients?
- Which factors regulate cortical remodeling and how can this biology be leveraged for stroke treatment?
- Should apoptotic or necrotic processes be the major therapeutic targets for stroke?
- How can epigenetics be harnessed for therapeutic approaches to acute stroke and to recovery?
- Which interventions and agents are useful in enhancing recovery and which can retard or worsen recovery?

References

1. Caplan LR. *Navigating the Complexities of Stroke*. New York: Oxford University Press; 2013.
2. Caplan LR. *Stroke*. New York: Demos Medical Publishing; 2006.
3. Henshall DC, Simon RP. Epilepsy and apoptosis pathways. *J Cereb Blood Flow Metab*. 2005;25:1557–1572.
4. Caplan LR, van Gijn J, eds. *Stroke Syndromes*. 3rd ed. Cambridge: Cambridge University Press; 2012.
5. Caplan LR. *Caplan's Stroke: A Clinical Approach*. 4th ed. Philadelphia, PA: Saunders Elsevier; 2009.
6. Ogata J, Masuda J, Yutani C, Yamaguchi T. Mechanisms of cerebral artery thrombosis: a histopathological analysis on eight necropsy cases. *J Neurol Neurosurg Psychiatry*. 1994;57:17–21.
7. Ogata J, Yutani C, Otsubo R, et al. Heart and vessel pathology underlying brain infarction in 142 stroke patients. *Ann Neurol*. 2008;63(6):770–781.
8. Moskowitz MA, Lo EH, Iadecola C, Moskowitz MA, Lo EH, Iadecola C. The science of stroke: mechanisms in search of treatment. *Neuron*. 2010;67(2):181–198.
9. Lo EH. A new penumbra: transitioning from injury into repair after stroke. *Nat Med*. 2008;14(5):497–500.
10. Lo EH, Dalkara T, Moskowitz MA. Mechanisms, challenges and opportunities in stroke. *Nat Rev Neurosci*. 2003;4(5):399–415.
11. Dirnagl U, Becker K, Meisel A. Preconditioning and tolerance against cerebral ischaemia: from experimental strategies to clinical use. *Lancet Neurol*. 2009;8(4):398–412.
12. Keep RF, Wang MM, Xiang J, Hua Y, Xi G. Is there a place for cerebral preconditioning in the clinic? *Transl Stroke Res*. 2010;1(1):4–18.
13. Stapels M, Piper C, Yang T, et al. Polycomb group proteins as epigenetic mediators of neuroprotection in ischemic tolerance. *Sci Signal*. 2010;3(111):ra15. http://dx.doi.org/10.1126/scisignal.2000502.
14. Stenzel-Poore MP, Stevens SL, Lessov NS, et al. Effect of ischaemic preconditioning on genomic response to cerebral ischaemia: similarity to neuroprotective strategies in hibernation and hypoxia-tolerant states. *Lancet*. 2003;362(9389):1028–1037.

CHAPTER

23

Prion Diseases

Paweł P. Liberski, James W. Ironside*†

*Department of Molecular Pathology and Neuropathology, Medical University of Lodz, Lodz, Poland; †National CJD Research & Surveillance Unit, Western General Hospital, Edinburgh, UK

OUTLINE

INTRODUCTION

Prion diseases or transmissible spongiform encephalopathies (TSEs) are neurodegenerative (i.e. non-inflammatory) diseases characterized by the deposition of a misfolded isoform of prion protein in the CNS and some other organs.

In humans, prion diseases include kuru; Creutzfeldt–Jakob disease (CJD), which occurs in four types (sporadic, familial or genetic, iatrogenic, and variant); Gerstmann–Sträussler–Scheinker (GSS) disease; fatal familial insomnia (FFI); and variably protease-sensitive prionopathy (VPSPr). Several prion diseases are known in animals. Natural scrapie occurs in sheep, goats, and mouflons (Fig. 23.1); bovine spongiform encephalopathy (BSE) in cattle; BSE passaged to lion, tiger, cheetah, puma, bison, and exotic antelopes (kudu, oryx, nyala); chronic wasting disease (CWD) in captive and free-range cervids; and transmissible mink encephalopathy (TME) in ranch-reared mink.

Neurobiology of Brain Disorders
http://dx.doi.org/10.1016/B978-0-12-398270-4.00023-9

FIGURE 23.1 **A sheep affected by scrapie.** Large areas of the animal's skin are devoid of wool because of scratching.

TABLE 23.1 Frequencies of the Codon 129 Allele

Population	Met Met (%)	Val Val (%)	Homozygotes (%)	Heterozygotes (%)
Healthy Caucasian population	40	10	50	50
European, sCJD	67	17	84	16
Healthy Japanese population	92	0	92	8
Japanese, sCJD	97	1	98	2

sCJD: sporadic Creutzfeldt–Jakob disease.
Source: Modified from Brown et al. Emerg Infect Dis. *2012;18(6):901–907.*[2]

CAUSES AND PATHOGENESIS OF PRION DISEASES

Nomenclature

The nomenclature of prion protein (PrP) species is not standardized. PrP^c is a normal cellular isoform, while PrP^d is the pathological misfolded protein (d, from disease). PrP^d is operationally defined as resistant to proteinase K (PK), hence PrP^{res}, and insoluble in denaturing detergent; however, some pathological isoforms of PrP^d have recently been found not to be PK resistant (hence VPSPr). Thus, the authors prefer to use the neutral term PrP^d, which denotes the misfolded species of PrP that is associated with disease, whether it is PK resistant or not. PrP 27–30 is the proteolytic cleavage product of PrP^d.

Codon 129

A gene encoding PrP (*PRNP*) is polymorphic at codon 129 (Val or Met).[1] In European populations, including in the UK, the dominant allele is 129^{Met}. This 129^{Met} dominance diminishes towards Africa and South-East Asia. In Japan, the frequency of the codon 129^{Val} is so low that it is regarded as a mutation, not a polymorphism. 129^{Val} is the dominant allele in Papua New Guinea and certain Native American populations. The frequencies of codon 129 alleles are shown in Table 23.1. The homozygosity at codon 129 predisposes to infection with prions; the most striking example is variant Creutzfeldt–Jakob disease (vCJD), where all clinical cases have been 129^{Met} homozygotes.

Prions, Prionoids, and Infectious Amyloid

Prion diseases are caused by an accumulation of the abnormal misfolded isoform (PrP^d) of normal cellular protein (PrP^c). Thus, the conversion of PrP^c into PrP^d underlies the disease pathogenesis. In familial prion diseases, mutation within the *PRNP* gene is responsible for this conversion. In sporadic prion diseases, the pathogenic mechanism is unclear and may be due to either a somatic mutation or a spontaneous conversion.

PrP^c is a highly conserved 253 amino acid sialoglycoprotein encoded by a cellular gene mapped to chromosome 20 in humans and chromosome 2 in mice.[3] The gene (*PRNP* in humans) is ubiquitous and highly conserved; it has been cloned in numerous mammalian species, including marsupials, and there are analogues of this gene in birds, reptiles, amphibians, and fish; the latter possess two PrP genes, *PrP1* and *PrP2*. A functional protein is proteolytically cleaved to remove 22 amino acids from the N-terminus and 23 amino acids from the C-terminus; the remaining protein is 209 amino acids long. PrP 27–30, a proteolytic cleavage product of PrP^d, was first discovered as a protein copurifying with scrapie infectivity in extracts derived from hamster brains infected with the 263K strain of scrapie, which led to the conclusion that PrP is a part or a whole of infectivity.

The prion hypothesis, which is deeply rooted in the association between PrP and infectivity, was formulated by Stanley B. Prusiner in 1982.[4] The hypothesis suggested that the scrapie agent was a proteinaceous infectious particle, because infectivity was dependent on protein but resistant to methods known to destroy nucleic acids. Ultrastructurally, PrP^d is visualized as amyloid fibrils (Fig. 23.2) ("prion rods", according to Prusiner) and PrP^d, at least in plaques, meets the criteria of an amyloid: the protein has cross-β-sheet conformation, its plaques are congophilic and birefringent in polarized light, and it binds thioflavin-based dyes.

Like many amyloid proteins, PrP 27–30 (27–30 kDa) is a proteolytic cleavage product of a precursor protein, PrP^d 33–35. However, PrP^d 33–35 is not the primary product of the cellular gene. It has an amino acid sequence

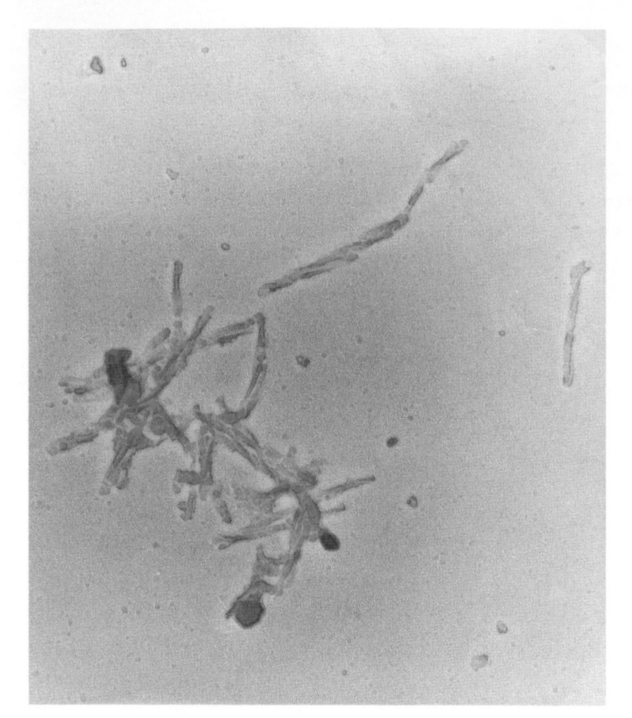

FIGURE 23.2 **Amyloid fibrils composed of pathological misfolded prion protein (PrP^d).** (Negative staining electron microscopy.)

and post-translational modifications (e.g. N-linked glycosylation sites at positions Asn-181 and Asn-197, a disulfide bond between Cys-179 and Cys-214, and the attachment of glycosylphosphatidylinositol to the cell membrane) identical to those of PrP^c 33–35, but strikingly different physicochemical features. In particular, PrP^c is completely degraded by digestion with PK but PrP^d is only partially degraded, yielding PrP^d 27–30. Since PrP is a glycoprotein with two Asn-glycosylation sites, it may exist as deglycosylated, monoglycosylated, and diglycosylated isoforms of different electrophoretic mobility.

As early as the mid-1980s, D. Carleton Gajdusek introduced the term "β-fibrilloses of the brain" or "transmissible amyloidoses" to stress the resemblance of prion rods to many different non-transmissible amyloids. This similarity to other amyloids, in the form of amyloid plaques, was appreciated even before the era of protein chemistry of PrP, in the 1950s, when similarities of kuru plaques to plaques in Alzheimer disease (AD) prompted Gajdusek to call kuru "a galloping senescence of the juvenile".

All amyloids, irrespective of amino acid sequence, are formed in a seeded nucleation mechanism in which a small aggregate (oligomers) of a few amyloid monomers (a seed or a nucleus) seeds or nucleates the assembly of a normal non-amyloid precursor into β-sheet amyloid fibrils. This self-perpetuating process leads to the formation of more and more amyloid. Aguzzi termed these self-aggregating proteins "prionoids".[5] "True" prions differ substantially from all other prionoids – prions are infectious in the microbiological term, spread laterally between individuals, and cause real epidemics such as kuru and vCJD in humans and BSE in animals, not to mention iatrogenic Creutzfeldt–Jakob disease (iCJD). It is also necessary to reflect on the term "infectious". According to the 28th edition of *Stedman's Medical Dictionary*, "infectious" "denotes a disease due to the action of a microorganism". Thus, peptides should not be described as infectious even if those structures replicate. Paul Brown, in his lecture at the Neuroprion meeting in Montreal in 2011, said that you may think of rust as a parallel; rust on a metal surface expands (i.e. replicates) but nobody would call it infectious. Both Gajdusek and Lansbury and Caughey[6] evoked the metaphor of *ice-nine* to embrace all amyloids. This stable isoform of ice, created by Kurt Vonnegut in his novel *Cat's Cradle*, catastrophically converts all the world's water into ice-nine. While there is no evidence of lateral case-to-case spreading of non-transmissible amyloidoses, there is one possible exception in the transmission of AA (reactive or secondary) amyloidosis in captured cheetahs, where it is responsible for some 70% of deaths.[7] Amyloid fibrils, detected in cheetah feces, serve as a vehicle of transmission.

All sporadic neurodegenerations are diseases of old age. This epidemiological phenomenon is consistent with a view that spontaneous conformational change from soluble, monomeric precursor protein into an insoluble amyloid aggregate is accomplished via a seeded nucleation process characterized by a long lag phase. Several predictions can be made on the basis of this assumption. First, an increase in the precursor monomer concentration may favor nucleation and thus shorten the lag phase because the availability of the substrate is increased. This is particularly observed when a mutation in a gene encoding for the precursor of amyloid occurs. Second, an increase in the number of seeds should lead to amplification of the nucleation reaction. This is also observed when the PrP^d is fragmented either by sonication [protein-misfolding cyclic amplification (PMCA)] or by shaking [quaking-induced conversion (QuIC)], and forms the basis of recent diagnostic procedures (see "Laboratory Tests", below).

There are several protein misfolding disorders, the most widely known of which include AD, Parkinson disease, and other α-synucleinopathies, frontotemporal dementias in which abnormally phosphorylated microtubule-associated protein-τ (MAP-τ) protein accumulates and, finally, polyglutamine expansion diseases such as Huntington disease and certain spinocerebellar ataxias. The proteins involved in each of these are different [β-amyloid (Aβ), α-synuclein, MAP-τ, and huntingtin] but the molecular mechanism is almost exactly the same, namely a seeding–nucleation mechanism.

β-Amyloid Peptide and Alzheimer Disease

AD is the most common human dementia, caused by accumulation of two amyloids: the Aβ peptide in the form of both diffuse and neuritic plaques and MAP-τ in the form of neurofibrillary tangles (NFTs) composed of paired helical filaments (PHFs). There is evidence that Aβ-containing extracts from AD brain can seed Aβ-amyloidosis in non-human primates or in TgAPP mice when injected intracerebrally or peripherally.[8,9] The latter suggests that Aβ seed can be translocated from the periphery to the brain exactly as prions travel from peripheral sites of inoculation using neuroanatomical pathways. However, the experiments in TgAPP mice could be interpreted as being due to preformed seeds barely accelerating a process of plaque formation that would eventually occur anyway. To bypass this objection, Prado and Baron[10] used transgenic rats that overexpress amyloid precursor protein (APP) with the Swedish double mutation (K670N-M671L) and the Indiana mutation (V642F) and which do not develop plaques spontaneously. Those rats also developed plaques following inoculation with Aβ-rich AD brain extracts. Furthermore, also analogous to prion-associated PrP, the Aβ in the seeds is PK resistant. As in PMCA, extensive sonication of the Aβ seeds into smaller oligomers enhances Aβ-plaque induction. The induced Aβ aggregates take the form of either diffuse or amyloid plaques. When TgAPP mice were seeded intraperitoneally, most of the Aβ aggregates developed around the blood vessels as a congophilic angiopathy, a characteristic trait of AD.[9] N-terminally truncated Aβ containing pyroglutamate is more toxic than Aβ42 and induces prion-like misfolding of Aβ42.[11] When TgAPP mice were injected with Aβ-containing extracts from old TgAPP mice, the march of aggregates was observed from the area proximal to the inoculation site into more distant regions. Analogous to findings with prion diseases,[12] stainless steel wires coated with Aβ-containing extracts induced cerebral Aβ aggregates in the TgAPP mice.[13] The latter finding may be of practical importance because of a possibility of inducing Alzheimer changes following neurosurgery on incipient AD brain.

Microtubule-Associated Protein-τ

MAP-τ is a microtubule binding protein that, in hyperphosphorylated isoforms, forms PHFs in NFTs. It also forms PHFs and straight filaments in many different frontotemporal dementias including Pick disease, progressive supranuclear palsy, corticobasal neurodegeneration and frontotemporal lobar degeneration, and parkinsonism linked to chromosome 17; the latter is caused by mutations in the *MAPT* gene encoding for MAP-τ. Neuroanatomical studies by Braak and Braak demonstrated that in AD brain MAP-τ marches from restricted areas of entorhinal cortex into the hippocampal formation and other cortical regions.[14] Modern molecular approaches in transgenic mice expressing MAP-τ in a restricted region of entorhinal cortex suggested that this progressive march is trans-synaptic and accomplished via a seeded polymerization mechanism that leads to the acquisition of the abnormal conformation of MAP-τ.[15] MAP-τ spreads from cell to cell separated by synapses and follows neuroanatomical pathways. The same phenomenon was described in prion disease in the early 1980s.[16] Analogous to the seeding of Aβ plaques by Aβ-rich extracts, the seeding of MAP-τ filamentous inclusions was initiated by injection of brain extracts from P301S transgenic mice into ALZ17 transgenic mice that normally do not exhibit MAP-τ inclusions.[17]

α-Synuclein

α-Synuclein is a highly soluble unfolded protein that accumulates in Lewy bodies and Lewy neurites in Parkinson disease and other synucleinopathies. Mutations in the gene encoding α-synuclein (*SNCA*) are linked to familial Parkinson disease. Like other amyloids, α-synuclein acquires a cross-β-sheet structure in the seeded nucleation process.[18] *In vitro*, cells transfected with preformed fibrils composed of α-synuclein form Lewy-body like intracellular inclusions. When intrastriatal neuronal grafting was performed to alleviate some signs and symptoms of Parkinson disease, Lewy bodies appeared in grafted neurons. Human α-synuclein spreads from neurons in Tg mice expressing human α-synuclein into grafted naïve neurons.[19] Furthermore, rat nigral neurons expressing human α-synuclein spread α-synuclein to transplanted embryonic ventral mesencephalic neurons. Small areas of human α-synuclein were surrounded by a larger ring of rat α-synuclein, suggesting a seeding mechanism.

KURU

Kuru was the first human neurodegenerative disease transmitted to chimpanzees. It was reported to Western medicine in 1957 by Gajdusek and Vincent Zigas,[20] but the first cases probably appeared at the end of the nineteenth century. The recognition of kuru as a neurodegenerative disease that is also transmissible (i.e. infectious) and subsequent evidence that CJD belongs to the same category led to the Nobel Prize being awarded to Gajdusek in 1976 and to Prusiner in 1997.

Kuru in the Fore language of Papua New Guinea means to shiver from fever or cold. Kuru was restricted to natives of the Fore linguistic group in Papua New Guinea's eastern highlands. The disease was spread by ritualistic endocannibalism (consumption of extended family members as part of a mourning ritual). The incidence of kuru increased in the 1940s and 1950s to reach a mortality rate in some hamlets of 35 per 1000 among a population of 12,000 Fore people, and greatly distorted the gender ratio. In particular, in the South Fore, the female to male ratio was 1 to 1.67, in contrast to the 1 to 1 ratio in unaffected Kamano people. This ratio increased to 1 to 2 and even 1 to 3 in certain South Fore settlements.

The practical absence of kuru cases in South Fore among children born after 1954 and the rising of age of kuru cases year by year suggested that transmission of kuru to children stopped in 1950s when cannibalism ceased to be practiced among the Fore people.

Kuru is an always fatal cerebellar ataxia accompanied by tremor and choreiform and athetoid movements. In contrast to the neuropathological picture, it is remarkably uniform in clinical signs, symptoms, and evolution. The progressive dementia is barely noticeable in patients with kuru, and then only late in the course of the illness. However, patients often display emotional changes, including inappropriate euphoria and compulsive laughter (hence the journalistic "laughing death" or "laughing disease"), or apprehension and depression. Kuru is divided into three clinical stages: ambulant, sedentary, and terminal. The duration of kuru, measured from the onset of prodromal signs and symptoms until death, is about 12 months (range 3–23 months).

There is an ill-defined prodromal period characterized by headache and limb pains, frequently in the joints: first in the knees and ankles, followed by elbows and wrists; sometimes, interphalangeal joints are first affected, with abdominal pains and loss of weight. This period lasts for a few months.

The prodromal period is followed by the ambulant stage, the end of which is defined when the patient is unable to walk without a stick. As patients were well aware that kuru heralded death in about a year, they became withdrawn and quiet. A fine shivering tremor, starting in the trunk, amplified by cold and associated with goose flesh, is often followed by titubation (staggering) and other abnormal movements. Attempts to maintain balance result in clawing of the toes and curling of the feet. Plantar reflex is always flexor while clonus, in particular ankle but also patellar clonus, are hallmarks of the clinical picture. Resting tremor is a cardinal sign of kuru. A horizontal convergent strabismus is a typical sign, especially in younger patients; nystagmus is common but the papillary responses are preserved.

The second sedentary stage begins when the patient is unable to walk without constant support and ends when he or she is unable to sit without it. Postural instability, severe ataxia, tremor, and dysarthria progress endlessly through this stage.

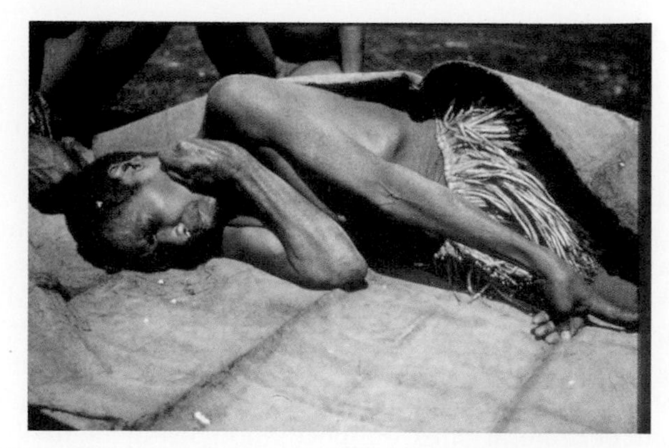

FIGURE 23.3 **A woman in terminal stage of kuru.** *Source: From the late D. Carleton Gajdusek.*

In the third stage (Fig. 23.3), the patient is bedridden and doubly incontinent, with dysphasia and primitive reflexes, and eventually succumbs in a state of advanced starvation. Extraocular movements are jerky or slow and rigid. A strong grasp reflex occurs, as well as fixed dystonic postures, athetosis, and chorea.

Neuropathology of Kuru

According to classical descriptions, neurons are shrunken and hyperchromatic or pale, with dispersion of Nissl substance or contained intracytoplasmic vacuoles similar to those described in scrapie.[21] Astroglial and microglial proliferation is widespread; the latter form rosettes and appear as rod or amoeboid types or as macrophages (gitter cells). The most striking neuropathological feature of kuru is the presence of numerous amyloid plaques (Fig. 23.4A–C). Kuru plaques are metachromatic and stain with periodic acid–Schiff, Alcian blue, and Congo red, and some are weakly argentophilic.

Renewed interest in kuru pathology was provoked by the appearance in 1997 of a variant form of CJD characterized by numerous amyloid plaques, including florid or daisy plaques – a kuru plaque surrounded by a corona of spongiform vacuoles. To this end, a few studies have re-evaluated the historic material.[21–23] In contrast to the classical studies described above, these papers stressed the presence of typical spongiform change present in deep layers (III–V) of the cingulate, occipital, entorhinal, and insular cortices, and in the subiculum. Spongiform change was also observed in the putamen and caudate, and some putaminal neurons contained intraneuronal vacuoles. Spongiform change was prominent in the molecular layer of the cerebellum, in the periaqueductal gray matter, basal pontis, central tegmental area, and inferior olivary nucleus. The spinal cord showed only minimal spongiform change.

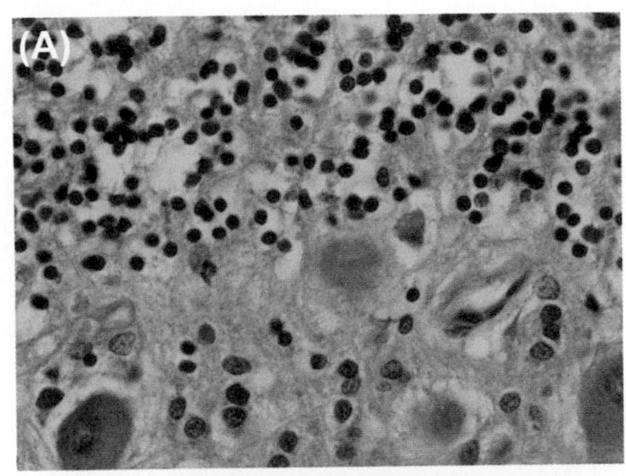

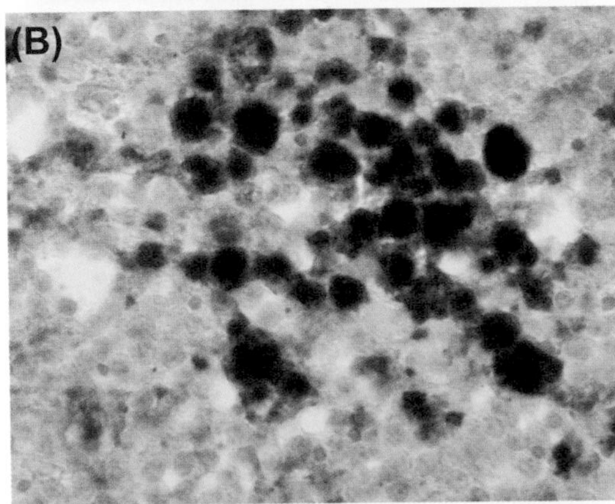

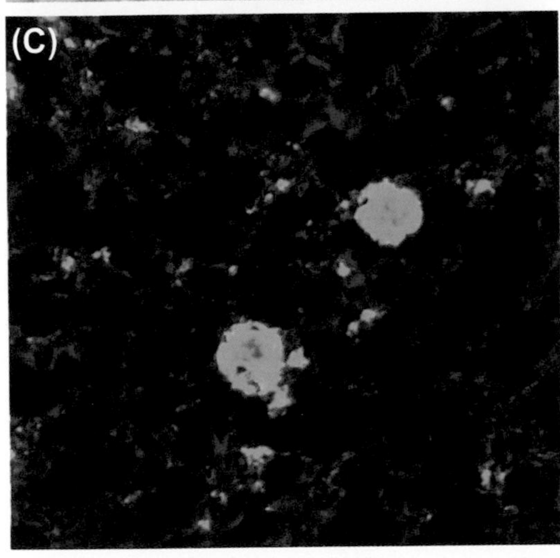

FIGURE 23.4 **Amyloid plaques in kuru.** (A) Typical amyloid plaque in the center (H&E stain); (B) numerous plaques stained immunohistochemically by antibodies raised against prion protein (PrP); (C) confocal laser microscopy image of amyloid plaques (green) against the background of glial fibrillary acidic protein (GFAP)-positive reactive astrocytes (red). *Source: Courtesy of Dr Beata Sikorska, Lodz, Poland.*

Genetics and Molecular Biology of Kuru

Individuals of $129^{\text{Val Val}}$ and $129^{\text{Met Val}}$ genotypes were susceptible to kuru, but those of $129^{\text{Met Met}}$ genotype were overrepresented in the younger age group and those of $129^{\text{Val Val}}$ $129^{\text{Met Val}}$ in much older people. People who survived the epidemic were characterized by almost the total absence of $129^{\text{Met Met}}$ homozygotes. A genome-wide study has confirmed a strong association of kuru with a single nucleotide polymorphism (SNP) localized within codon 129 but also with two other SNPs localized within genes *RARB* (the gene encoding retinoic acid receptor-β) and *STMN2* (encoding SCG10).[24]

The practice of endocannibalism underlying the kuru epidemic created a selective pressure on the PrP genotype. As in CJD, homozygosity at codon 129 ($129^{\text{Met Met}}$ or $129^{\text{Val Val}}$) is overrepresented in kuru. Furthermore, among Fore women over 50 years of age, there is a remarkable overrepresentation of heterozygosity ($129^{\text{Met Val}}$) at codon 129, which is consistent with the interpretation that $129^{\text{Met Val}}$ makes an individual resistant to TSE agents and that such a resistance was selected by cannibalistic rites. Because of this $129^{\text{Met Val}}$ heterozygote advantage, it has been suggested that the heterozygous genotype at codon 129 has been sustained by a widespread ancient practice of human cannibalism.

CREUTZFELDT–JAKOB DISEASE

Sporadic Creutzfeldt–Jakob Disease

According to the Centers for Disease Control and Prevention,[25] sporadic Creutzfeldt–Jakob Disease (sCJD) can be diagnosed by standard neuropathological techniques, immunocytochemically, by Western-blot confirmed protease-resistant PrP, and/or by the presence of scrapie-associated fibrils. sCJD is probably present if there is rapidly progressive dementia and at least two of the following four clinical features: myoclonus, visual or cerebellar signs, pyramidal/extrapyramidal signs, and akinetic mutism. In addition, sCJD is indicated by a positive result on at least one of the following laboratory tests: a typical electroencephalogram (EEG) (periodic sharp-wave complexes) (Fig. 23.5) during an illness of any duration, a positive 14-3-3 cerebrospinal fluid (CSF) assay in patients with a disease duration of less than 2 years, magnetic resonance imaging (MRI) high signal abnormalities in the caudate nucleus and/or putamen on diffusion-weighted (DW) imaging or fluid-attenuated inversion recovery (FLAIR). sCJD may also account for a more progressive dementia if the duration of the illness is less than 2 years and there are at least two out of the aforementioned clinical signs, and the absence of a positive result for any of the three laboratory tests noted above that would classify a case as "probable".

Iatrogenic Creutzfeldt–Jakob Disease

iCJD is a progressive cerebellar syndrome caused by inadvertent contamination with CJD, by infection with human growth hormone (hGH) and gonadotrophins, by

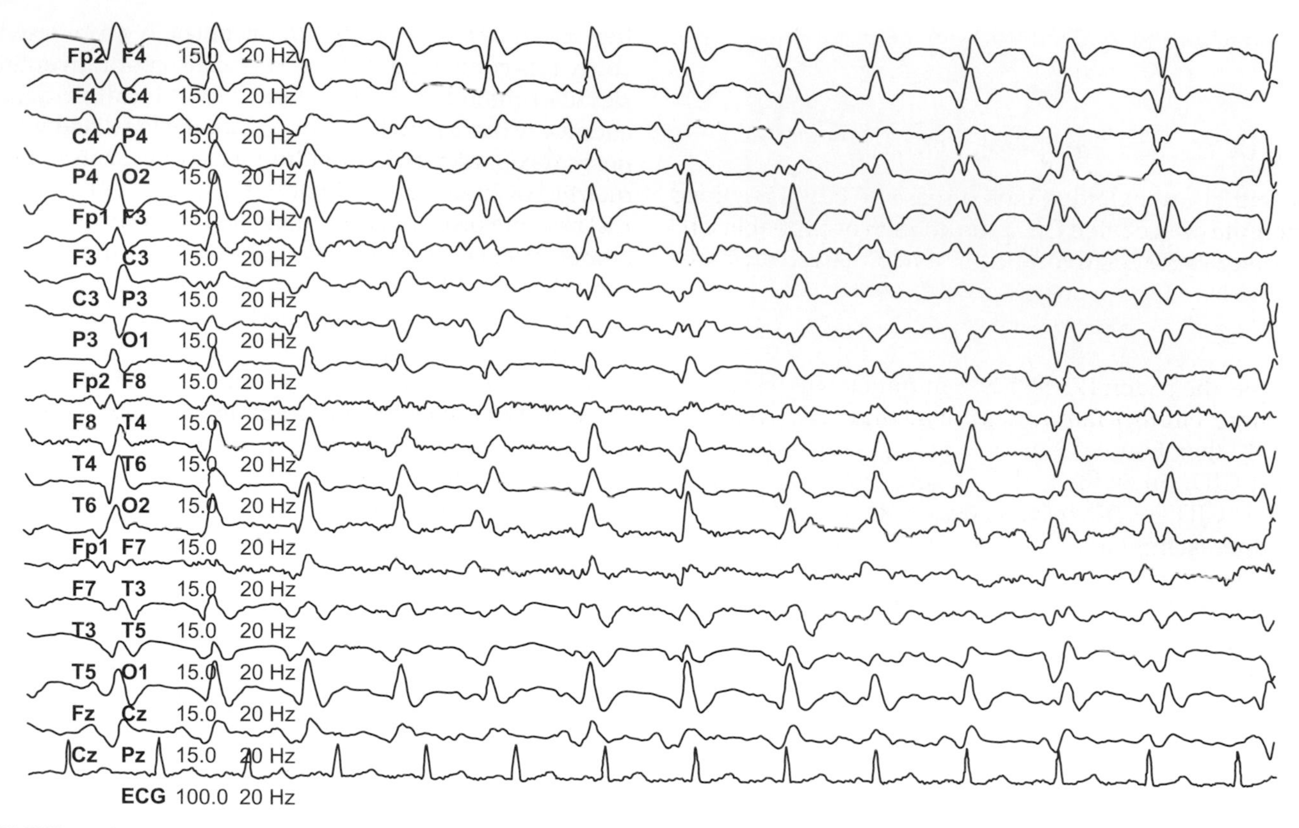

FIGURE 23.5 **A typical periodic pattern of sharp-wave complexes on electroencephalography (ECG) in sporadic Creutzfeldt–Jakob disease.**

using contaminated dura mater and cornea, or by using sterotactic EEG electrodes that had been previously used on a patient with CJD.[2] A summary of iCJD cases is shown in Table 23.2.

Human Growth Hormone

The total number of iCJD cases following hGH administration is 226.[2] Most of these occurred in France (119 cases out of 1880 recipients), the UK (65 cases out of 1800 recipients), and the USA (29 cases out of 7700 recipients). In the USA, no cases have been seen in those treated with hGH after 1977. The mean incubation period is 17 years, with a range of 5–42 years. The majority of those cases are $129^{\text{Met Met}}$ homozygotes. Although $129^{\text{Met Val}}$ heterozygotes have a longer incubation period than $129^{\text{Met Met}}$ homozygotes, some iCJD cases with $129^{\text{Met Met}}$ present an exceptionally long incubation period much in excess of 30 years. The clinical picture is characterized by the absence of dementia at the onset of disease, which is similar to kuru and vCJD; the latter two types of prion disease are thought to originate from peripheral inoculation.

Dura Mater

At the time of writing, the total number of cases from contaminated dura mater is 228 and new cases are occurring worldwide. The mean incubation period is 12 years (range 1.3–30 years). The clinical picture is akin to that of sCJD. Japanese cases may present additional signs and symptoms, such as slow progression and plaques in neuropathology. Some Japanese patients exhibit occasional florid plaques in the brain and the presence of a pulvinar

TABLE 23.2 Summary of Iatrogenic Creutzfeldt–Jakob Disease

Category	No. of Cases	Incubation Period: Range (mean) (years)	Clinical Syndrome
Dura mater	228	1.3–30 (12)	Cerebellar, visual, dementia
Neurosurgery	4	1–2.3 (1.4)	Cerebellar, visual, dementia
Stereotactic EEG needles	2	1.3, 1.7	Cerebellar, dementia
Corneal transplant	2	1.5, 2.7	Cerebellar, dementia
hGH	226	5–42 (17)	Cerebellar
Gonadotrophins	4	12–16 (13.5)	Cerebellar
Blood cells – vCJD	3	6.5; 7.8; 8.3	Psychiatric, sensory, dementia, cerebellar – typical vCJD

EEG: electroencephalography; hGH: human growth hormone; vCJD: variant Creutzfeldt–Jakob disease.
Source: Modified from Brown et al. Emerg Infect Dis. *2012;18(6):901–907.*[2]

sign on MRI, not unlike those signs regarded as typical of vCJD, but with no other features of vCJD.

Familial Creutzfeldt–Jakob Disease

Familial Creutzfeldt–Jakob disease (fCJD)[26] is defined as definite or probable CJD plus definite or probable CJD in a first degree relative, and/or a neuropsychiatric disorder with a disease-specific *PrP* gene mutation.

Mutations at several codons are found in fCJD; a detailed list is provided in Liberski and Surewicz.[27] For example, the codon 178^{Asp} 129^{Val} mutation was described in a large Finnish family with CJD and later found in several other families. The clinical symptoms are as in typical CJD, but with earlier onset, and an absence of periodic CJD and myoclonic jerks. Codon 178^{Asp} 129^{Met} mutation is linked to fatal familial insomnia (FFI),[28] which is characterized by sleep disorders, nocturnal hallucinations, behavioral disturbances, and autonomic dysfunction. Changes on neuropathology are confined to the thalamus and PrP^{d} accumulation may be limited. Parkinsonism and dementia are seen with the 183^{Val} 129^{Met} mutation, and neuropathology reveals spongiform change. The codon 200^{Lys} 129^{Met} mutation results in a Gly to Lys substitution.[29] Several healthy individuals from families at risk for CJD tested positive for the presence of the codon 200 mutation, suggesting that the mutation itself is not sufficient for development of disease or that penetration of this mutation is not complete.

Octarepeat expansions may occur in region 51–91 of the *PRNP* gene.[30] This region is not part of PrP 27–30, so the fact that its expansion leads to CJD or GSS is unexplained. Additional octapeptide repeats, as described in detail in Liberski and Surewicz,[27] may manifest as typical CJD or as a CJD-like disease with a younger age of onset than in typical CJD.

Creutzfeldt–Jakob Disease Subtypes

CJD is a disease of very diverse signs and symptoms, with an obvious need for subclassification.[31] The most widely used classification, by Parchi and Gambetti, has been validated against a large number of CJD cases. This classification is based on the presence of either PrP^{d} type 1 (21 kDa) or type 2 (19 kDa) and the status of polymorphic codon 129 (Met or Val). PrP^{d} types 1 and 2 are unglycosylated C-terminal core peptides of PrP^{d} truncated at the N-terminus; the primary cleavage site of PrP^{d} type 1 is at residue 82 while that of type 2 is at residue 97. PrP^{d} types 1 and 2 may coexist in the same brain. A detailed description of the various subtypes of CJD is provided by Parchi and Saverioni.[31]

Type 1 sCJD (MM1/MV1) is the most frequent subtype, found in nearly half of all sCJD cases. This type is observed in $129^{Met\ Met}$ homozygotes or $129^{Met\ Val}$ heterozygotes. The mean age at onset is 65 years and duration of the disease is typically short, on average 4 months. Signs and symptoms include cognitive decline leading to dementia, myoclonic jerks, cortical blindness, ataxia, and other involuntary movements. In one-quarter of cases, neurological signs are unilateral. In the vast majority of cases, a typical pattern of periodic sharp-wave complexes (PSWCs) is seen on the EEG and 14-3-3 is detected in CSF. DW-MRI demonstrates signal hyperintensity in the basal ganglia or cerebral ribbon. Neuropathologically, spongiform change is found, consisting of small vacuoles of 2–10 μm diameter in deeper cortical layers and the molecular layer of the cerebellum. The anterior part of the brain is more involved than the posterior. Immunohistochemically, the synaptic type of PrP^{d} accumulation is dominant.

Type 2 (VV2), the second most frequent subtype of sCJD (approximately 15%), is also called the ataxic type or Brownell–Oppenheimer syndrome. The mean age at onset is approximately 60 years and the duration of the disease about 6 years. The most frequent sign is cerebellar ataxia, and dementia may appear later in the course of the disease. Myoclonic jerks are absent in about 30% of cases. EEG demonstrates only non-specific changes. This subtype may be suspected in a patient with typical MRI changes but no periodic EEG. The 14-3-3 is detected in the CSF in some 80% of cases. DW-MRI shows hyperintense changes in the basal ganglia and thalamus, while involvement of cerebral cortex is less prominent and limited to limbic cortex. The neocortex is often spared. Neuropathological changes are typical and spongiform change frequently has a laminar distribution. The basal ganglia and thalamus may be more involved than the cerebral cortex. Immunohistochemically, PrP^{d} forms plaque-like deposits which are reminiscent of typical plaques but are not amyloid. The other types of PrP^{d} deposit are perineuronal and synaptic types. In cases of less than 5 months' duration, PrP^{d} is present in large amounts in the brain except for the neocortical areas; the latter become involved with a longer duration of the disease.

The kuru plaque subtype (MV2) is the third most common type of sCJD (approximately 8%). Phenotypically, it resembles type 2 sCJD, but the duration of the disease is longer, often over 2 years. The mean age at onset is 60 years. The typical signs and symptoms are gait ataxia and dementia. EEG and CSF studies give similar results to type 2 sCJD, and changes similar to those in type 2 are seen on DW-MRI, with hyperintensity in the basal ganglia and thalamus; the latter may be reminiscent of the pulvinar sign of vCJD. Neuropathological changes are similar to those encountered in type 2, but the hallmark is the presence of true kuru amyloid plaques in the granular and Purkinje cell layer of the cerebellum.

The MM2C cortical subtype is one of the least common subtypes (around 1%). The clinical phenotype comprises progressive dementia and aphasia. Myoclonic jerks and PSWCs on EEG are usually absent. DW-MRI shows hyperintense signals in the cortical ribbon. Neuropathologically, large confluent vacuoles, formerly called status spongiosus, are present. These changes are observed in the cerebral cortex, basal ganglia, and thalamus. Neuronal loss and astrocytic reaction are also seen. The cerebellum is largely spared. A perivacuolar (coarse) pattern of PrP^d and plaque-like deposits are observed.

VV1 is also very uncommon (around 1%). The mean age of onset is 39 years, which is the youngest among CJD patients, and the mean duration of disease is 15 months. The main symptom is frontotemporal dementia, usually without motor signs. DW-MRI shows involvement of the cerebellar cortex. Neuropathological changes are severe. PrP^d accumulation is mild and of the synaptic type. Frontal lobes are involved less than occipital lobes, and the involvement of the cerebellum and the thalamus is mild.

Occasional sporadic prion diseases have a phenotype almost identical to that of FFI. The mean age of onset of fatal sporadic insomnia (FSI, MM2T subtype) is 50 years and the duration of disease is 15 months. A typical symptom is insomnia, with difficulties in initiating and maintaining sleep, frequent arousals, acting out dreams, and motor signs including myoclonic jerks, diplopia, dysarthria, dysphagia, and pyramidal signs. Autonomic changes, as in FFI, are prominent. No PSWCs on the EEG and no signal hyperintensity on DW-MRI are observed. Neuropathologically, involvement of the thalamus is dominant with severe atrophy, neuronal loss, and astrogliosis. Amyloid plaques are not seen in the latter location but may be observed in the cerebral cortex. As in FFI, the PrP^d peptide is of type 2 but it is present in much lower amounts than in other sCJD subtypes. The main difference between FSI and FFI lies in the glycosylation profile: in FFI unglycosylated PrP^d predominates, whereas in FSI, as in sCJD, the distribution is 26% diglycosylated, 40% monoglycosylated, and 34% unglycosylated glycoforms.

Mixed forms of sCJD are also found. sCJD has thus been classified into pure subtypes [MM1/MV1, MV2K (with kuru plaques), MM/MV2C, MM2T (FFI and FSI, thalamic), and VV1] and mixed subtypes [MM/MV1+2C, MM/MV2C+1, VV2+1, MV2K+1, and MM2T+C]. For details on these mixed subtypes of sCJD, see Parchi and Saverioni.[31]

VPSPr is a novel sporadic prion disease described by Gambetti and coauthors in 2012.[32] VPSPr occurs on the background of every 129 codon genotype (MM, MV, and VV, with a respective distribution of 10%, 23%, and 67%). This distribution of the codon 129 alleles among VPSPr patients is in strict contrast to sCJD, where $129^{Met\ Met}$ is the most common. The clinical picture consists of psychiatric disturbances, disinhibition, euphoria, impulsiveness or apathy, semantic or nominal aphasia, dysarthria, and cognitive decline leading to frank dementia. The $129^{Met\ Val}$ VPSPr cases are characterized by parkinsonian features, ataxia, and myoclonic jerks; aphasia is rare. The mean age of onset is 72 years and disease duration is 45 months. The clinical picture in $129^{Met\ Met}$ cases consists of parkinsonism, ataxia, and dementia accompanied by myoclonic jerks. A case of $129^{Val\ Val}$ demonstrated DW-MRI changes similar to those of sCJD and a typical synchronous EEG pattern. Neuropathological examination demonstrates vacuoles of size in between those typical for sCJD MM1/MV1 subtype and large confluent vacuoles of sCJD MM2 subtype. PrP^d is seen as target-like deposits or microplaque deposits in $129^{Met\ Val}$ or $129^{Met\ Met}$ cases, particularly in the cerebellum. The hallmark of VPSPr is a banded ladder pattern of seven or more PrP^d species on Western blot; the PrP^d bands range from 7 to 27 kDa in size and the diglycosylated glycoform is lacking.

Neuropathology of Creutzfeldt–Jakob Disease

Macroscopically, non-specific brain atrophy is often but not always visible.[33] At the light microscopy level, a classical triad of spongiform change (Fig. 23.6), neuronal loss, and astrocytic gliosis is visible. PrP^d is detectable by immunohistochemical methods or paraffin-embedded tissue blot. Spongiform change consists of small neuropil vacuoles in contrast to intraneuronal vacuoles more typical for animal prion diseases, such as scrapie, BSE, or CWD. Neuronal loss and astrogliosis, accompanied by microglial proliferation, are variable but the hippocampal formation and dentate gyrus are spared in the majority of cases. Astrogliosis is most prominent in deeper cortical layers; in the cerebellum, the proliferation of Bergmann glia is observed.

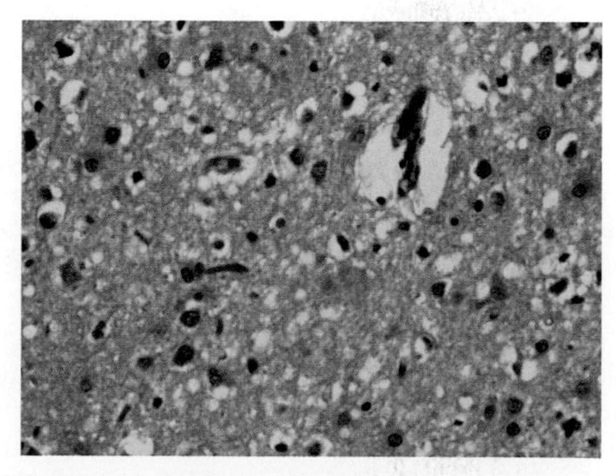

FIGURE 23.6 **Typical spongiform change in a case of sporadic Creutzfeldt–Jakob disease.** (H&E stain.)

Immunohistochemical detection of PrP remains the gold standard in diagnosis of prion disease. Most, if not all, available antibodies, recognize both PrP^{c} and PrP^{d}; thus, to visualize only PrP^{d}, PrP^{c} must be removed by hydrated or hydrolytic autoclaving or formic acid pretreatment, or both. Immunoexpression of PrP^{d} may be of several types: synaptic or diffuse, small punctuate depositions in the cerebral and the cerebellar cortex (Fig. 23.7A); perivacuolar deposits (Fig. 23.7B); perineuronal deposits (Fig. 23.7C); coarse deposits; plaque-like deposits, which are reminiscent of amyloid plaques but do not have tinctorial properties of amyloid; or kuru plaques.

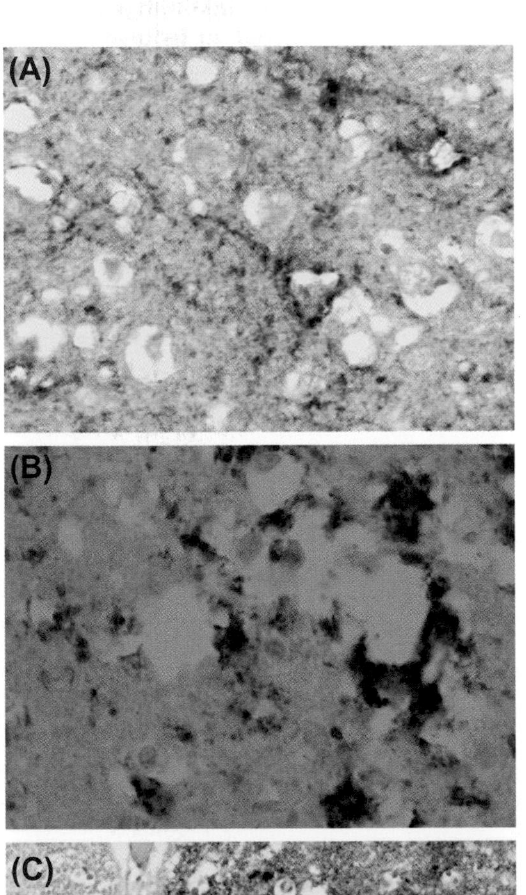

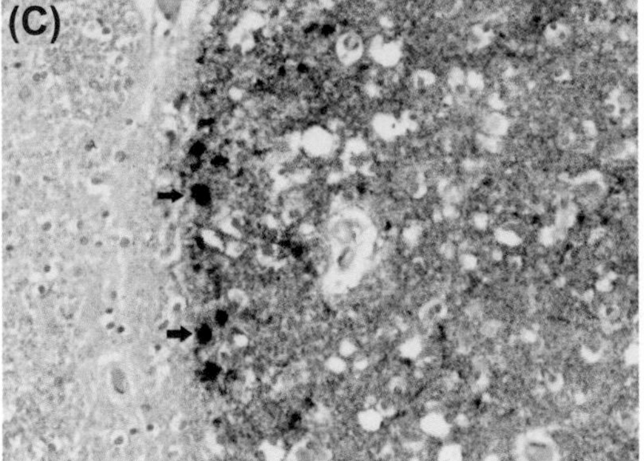

FIGURE 23.7 **Different patterns of pathological misfolded prion protein (PrP^{d}) immunoreactivity.** (A) In the affected cerebellum, the synaptic pattern is visible in the granular cell layer on the right, while numerous plaque-like structures are labeled with arrows; (B) perivacuolar pattern; (C) perineuronal pattern.

VARIANT CREUTZFELDT–JAKOB DISEASE

vCJD was identified in the UK in 1996 by the National Creutzfeldt–Jakob Disease Surveillance Unit at the University of Edinburgh.[34] This unit was established in 1990 to undertake surveillance of all types of CJD in light of the epidemic of BSE in UK cattle and the unknown implications of BSE for human health. The methodology used in the unit depends on the voluntary notification of cases of suspected CJD by health-care professionals across the UK. These cases are then investigated and subject to detailed clinical and neuropathological assessment, including genetic investigations wherever possible. Autopsy rates for suspected CJD cases have generally been high, but have recently reduced in line with the general decline in autopsy rates in the UK and other countries.

BSE was first reported in the UK in 1987,[35] but it is likely that cases occurred in the earlier years of that decade. Detailed epidemiological studies on BSE indicated that the most likely source of infection was consumption of contaminated meat and bonemeal animal feed. A series of bans on the use of this animal feed subsequently allowed the UK BSE epidemic to be controlled, and it has now been almost completely eradicated in the UK. BSE has occurred in many countries across the world, largely as a consequence of the import of contaminated animal feed and infected cattle from the UK, although the UK has had by far the highest number of cases. Around 180,000 cases of BSE were identified in the UK on the basis of clinical signs and symptoms in infected cattle, but the total number of infections is likely to have been much higher. Asymptomatic BSE infections were not detected until active surveillance with biochemical screening for BSE in cattle in abattoirs was instigated in the late 1990s and subsequently used more widely; it has been estimated that the total number of BSE infections in the UK may be as high as 2–3 million cases.[36] BSE has spread to other species in the UK by the consumption of contaminated animal feed (to antelopes in zoos and domestic cats) or contaminated cattle carcasses (to large wild cats in zoos). No recent cases have been identified, reflecting improved BSE control measures and the decline of BSE in UK cattle. BSE has also been transmitted experimentally to other species (including sheep), the results of which identified BSE as a novel single prion strain capable of infecting a wide range of mammals.

Given the large numbers of BSE-infected cattle in the UK, it is highly likely that contaminated bovine tissues (predominantly brain and spinal cord) entered the human food chain in the late 1980s and early 1990s, largely through the consumption of food products made from mechanically recovered meat. Other potential routes of BSE exposure have been proposed, including occupational exposure (e.g. in abattoir and farm workers) and via medical products such as vaccines, which had been prepared using bovine materials. No evidence to support the transmission of BSE to humans by these alternative routes has yet been identified in the UK or elsewhere, but there is epidemiological evidence to support the oral route of transmission of BSE via contaminated meat products.

Variant Creutzfeldt–Jakob Disease in Relation to Other Human Prion Diseases

Since the initial descriptions of Creutzfeldt–Jakob disease in the twentieth century, a widening spectrum of human prion diseases has been identified that includes sporadic, genetic, and acquired forms (Table 23.3). The most common human prion disease is sCJD, a worldwide disorder with a relatively uniform incidence of 1–2 per million of the population per annum. Studies on the naturally occurring polymorphism at position 129 in the *PRNP* gene in sCJD in the UK have revealed an excess of homozygosity in comparison with the normal population (Table 23.4). The clinical and neuropathological phenotype of sCJD is variable and depends on the *PRNP* codon 129 genotype and the isotype of abnormal PrP in the brain as determined by Western blotting studies. In order to establish vCJD as a novel prion disease, it was important to compare the first cases of vCJD with the full range of clinical and neuropathological phenotypes in sCJD.

Likewise, it was important to compare vCJD with the other acquired forms of human prion disease, particularly iCJD. Kuru is now virtually extinct, but there are still cases of iCJD occurring in the UK in hGH and dura mater graft recipients. Transmission of CJD by contaminated neurosurgical instruments has also occurred in the UK in the past century, although no recent cases have been identified. Until the emergence of BSE, there was no evidence that other prion diseases occurring in animals, particularly scrapie in sheep and goats, were pathogenic to humans. However, the evidence of BSE transmission by the oral route to other species and the recognition of BSE as a novel strain of prions distinct from scrapie renewed concerns that it might represent a hazard to human health.

TABLE 23.3 Classification of Human Prion Diseases

Idiopathic	
	Sporadic Creutzfeldt–Jakob disease
	Sporadic fatal insomnia
	Variably protease-sensitive prionopathy
Genetic	
	Familial Creutzfeldt–Jakob disease
	Gerstmann–Sträussler–Scheinker syndrome
	Fatal familial insomnia
Acquired	
Human origin	Iatrogenic Creutzfeldt–Jakob disease
	Kuru
Bovine origin	Variant Creutzfeldt–Jakob disease

Clinical Features of Variant Creutzfeldt–Jakob Disease

Up to the end of February 2014, 177 cases of definite or probable vCJD had been identified in the UK, with 51 additional cases in 11 other countries (Table 23.5). Of the UK patients, 57% were male and 43% female. The

TABLE 23.4 Codon 129 Prion Protein Gene (*PRNP*) Polymorphisms in Creutzfeldt–Jakob Disease (CJD) and Normal Caucasian Population

	Codon 129 Polymorphism		
	Met Met	**Met Val**	**Val Val**
Normal population	39%	50%	11%
Sporadic CJD	63%	19%	18%
Variant CJD	100%	–	–

Source: National CJD Research & Surveillance Unit (NCJDRSU).[37]

TABLE 23.5 Worldwide Incidence of Variant Creutzfeldt–Jakob Disease (vCJD) as at August 2012

Country	Total no. of Primary Cases (no. alive)	Total no. of Secondary Cases: Blood Transfusion (no. alive)	Cumulative Residence in UK >6 Months during Period 1980–1996
UK	173 (0)	3 (0)	176
France	27 (2)	–	1
Republic of Ireland	4 (0)	–	2
Italy	2 (0)	–	0
USA	3[a] (0)	–	2
Canada	2 (0)	–	1
Saudi Arabia	1 (0)	–	0
Japan	1[b] (0)	–	0
Netherlands	3 (0)	–	0
Portugal	2 (0)	–	0
Spain	5 (0)	–	0
Taiwan	1 (0)	–	1

[a]*The third US patient with vCJD was born and raised in Saudi Arabia and has lived permanently in the USA since late 2005. According to the US case report, the patient was probably infected as a child when living in Saudi Arabia.*
[b]*The case from Japan had resided in the UK for 24 days in the period 1980–1996.*
Source: National CJD Research & Surveillance Unit (NCJDRSU).[37]

median age at onset of disease was 29 years and the median age at death 30 years, in comparison with 67 years and 68 years, respectively, in sCJD.[37] The youngest vCJD case was aged 12 years at onset, while the oldest case was aged 74 years. The median duration of illness from the onset of first symptoms to death was 14 months in vCJD, compared with 4 months in sCJD.

The clinical features of vCJD are relatively stereotyped and much more uniform than in sCJD, where a wide range of clinical features is recognized. Behavioral and psychiatric features, including depression, anxiety, and agitation, usually dominate the initial presentation. These non-specific symptoms are often followed by unusual sensory symptoms including pain and dysesthesia in the face or limbs. Other neurological abnormalities occur around 6 months after the initial symptoms, with cerebellar ataxia, chorea, dystonia, visual abnormalities, and cognitive impairment. The final phase of the illness is usually characterized by akinetic mutism.

EEG studies in vCJD may be normal in the initial stages of the illness, but show non-specific abnormalities in the later stages. Two cases of vCJD with periodic triphasic discharges in the EEG (similar to those occurring in sCJD) have been reported in the final stages of the illness. Routine examination of the CSF in vCJD is unremarkable, as with sCJD. The CSF 14-3-3 test is less sensitive in vCJD than in sCJD. A positive result may provide some support for the diagnosis but a negative result by no means excludes the diagnosis.[37] However, increased levels of phospho-tau have been identified in the CSF in vCJD.

The MRI in vCJD shows a characteristic abnormality seen in the posterior thalamic region (the pulvinar sign), which has been found in over 90% of pathologically proven vCJD cases (Fig. 23.8). This abnormality is best seen on FLAIR sequencing and has been present on MRI within 3 months of symptom onset in a few patients. The finding of a positive MRI scan allows a categorization of "probable vCJD" according to the World Health Organization (WHO) diagnostic criteria (Table 23.6).[38]

Unlike other forms of CJD, vCJD shows accumulation of the disease-associated PrP in the lymph nodes, spleen, tonsil, and appendix. Tonsil biopsy has been used as a supportive diagnostic test in vCJD, particularly to provide support for the diagnosis in cases with negative MRI scans or who have atypical clinical features. A positive tonsil biopsy allows a diagnosis of probable vCJD (Table 23.6).

Neuropathology of Variant Creutzfeldt–Jakob Disease

In keeping with the stereotypic clinical features, the pathological phenotype of vCJD is relatively uniform compared with sCJD. However, variations in the severity of the pathological lesions occur, with the most severe

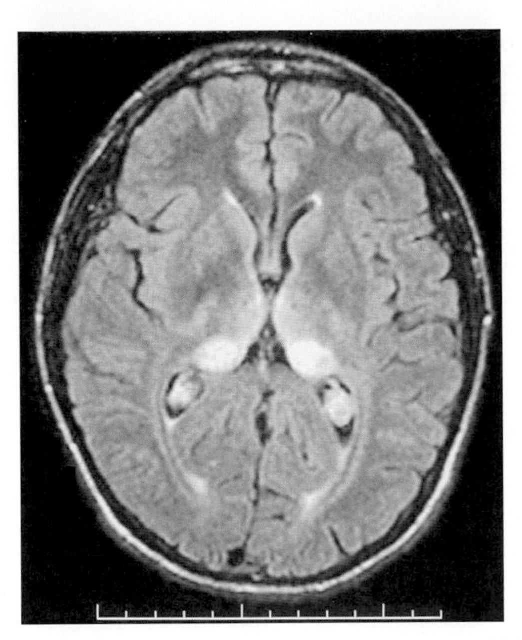

FIGURE 23.8 **The pulvinar sign of variant Creutzfeldt–Jakob disease.** Transverse fluid-attenuated inversion recovery magnetic resonance imaging sequence showing bilateral and symmetrical high signal in the pulvinar nuclei of the thalamus. *Source: Courtesy of Dr David Summers, Edinburgh, UK.*

TABLE 23.6 World Health Organization Criteria for the Clinical Diagnosis of Variant Creutzfeldt–Jakob Disease (vCJD)

I.	A. Progressive neuropsychiatric disorder
	B. Duration of illness >6 months
	C. Routine investigations do not suggest an alternative diagnosis
	D. No history of potential iatrogenic exposure
	E. No evidence of a familial form of TSE
II.	A. Early psychiatric symptoms
	B. Persistent painful sensory symptoms
	C. Ataxia
	D. Myoclonus or chorea or dystonia
	E. Dementia
III.	A. EEG does not show the typical appearance of sporadic CJD (or no EEG performed)
	B. MRI brain scan shows bilateral symmetrical pulvinar high signal
IV.	A. Positive tonsil biopsy
	Definite: IA and neuropathological confirmation of vCJD
	Probable: I and 4/5 of II and IIIA and III B, OR I and IVA
	Possible: I and 4/5 of II and IIIA

TSE: transmissible spongiform encephalopathy; EEG: electroencephalography; MRI: magnetic resonance imaging.
Source: World Health Organization; 2012.[38]

pathology occurring in cases with a lengthy clinical illness. Cerebral and cerebellar atrophy are most conspicuous in cases with a clinical history of over 18 months' duration. Cerebral atrophy is accompanied by ventricular dilatation, with a corresponding loss of white matter volume.

On microscopy, spongiform change is widespread within the cerebral cortex and most severe in the occipital cortex. All cortical layers can be involved in a patchy distribution, but there is accentuation of spongiform change around the amyloid plaques in the cerebral cortex, mostly in a microvacuolar pattern (Fig. 23.9).[39] Confluent spongiform change is rare in the cerebral cortex. Neuronal loss in the cerebral cortex is variable, but most severe in the primary visual cortex in the occipital lobe. In cases with a lengthy clinical history there is widespread and severe loss of neurons throughout the cerebral cortex, with accompanying astrocytosis. The neuronal populations in the hippocampus are generally preserved, even in cases with severe cerebral cortical neuronal loss.

In contrast, confluent spongiform change is prominent in the caudate nucleus and putamen. Neuronal loss in the basal ganglia is most evident in cases with severe spongiform change. Patchy spongiform change is present in most of the thalamic nuclei, the hypothalamus, and globus pallidus. Neuronal loss in the thalamus is most severe in the posterior nuclei, particularly in the pulvinar, which also shows marked astrocytosis. Mild spongiform change is present detected in the periaqueductal gray matter in the midbrain, accompanied by neuronal loss and astrocytosis. However, spongiform change is scanty in the pontine nuclei and medulla, and is generally absent in the gray matter of the spinal cord. In cerebellar cortex, spongiform change is conspicuous in the hemispheres and vermis. Confluent spongiform change is observed in the molecular layer of the cerebellum in a patchy distribution, often associated with amyloid plaques.

The amyloid plaques in vCJD plaques are easily identified on hematoxylin and eosin (H&E) stains (Fig. 23.9) as fibrillary structures with a dense core surrounded by pale radiating fibrils, and surrounded by spongiform change. These "florid" plaques occur most frequently at the bases of the gyri in the occipital cortex and cerebellar cortex. Electron microscopy of the amyloid plaques in vCJD has demonstrated masses of radiating fibrils at the periphery, with abnormal neurites similar to those seen at the periphery of the Aβ plaques in AD. Immunoelectron microscopy shows PrP accumulation both in the amyloid fibrils and in some of the abnormal cell membranes surrounding the plaques, but no NFTs have been identified on either light or electron microscopy.

Immunocytochemistry in Variant Creutzfeldt–Jakob Disease

The florid plaques in the cerebral and cerebellar cortex stain intensely on immunocytochemistry for PrP, which also labels numerous smaller plaques, often arranged in irregular clusters, that are not evident on routine stains (Fig. 23.10).[39] These cluster plaques are present in all vCJD cases. There is also widespread amorphous pericellular deposition of PrP around small neurons in the cerebral and cerebellar cortex. Occasional loose PrP deposits are also present outside the basement membrane of capillaries in the cerebral and cerebellar cortex, but not as an amyloid angiopathy.

A predominantly perineuronal pattern of PrP accumulation is characteristically present in the basal ganglia, often with linear decoration of dendrites and axons. A synaptic pattern of immunoreactivity with occasional

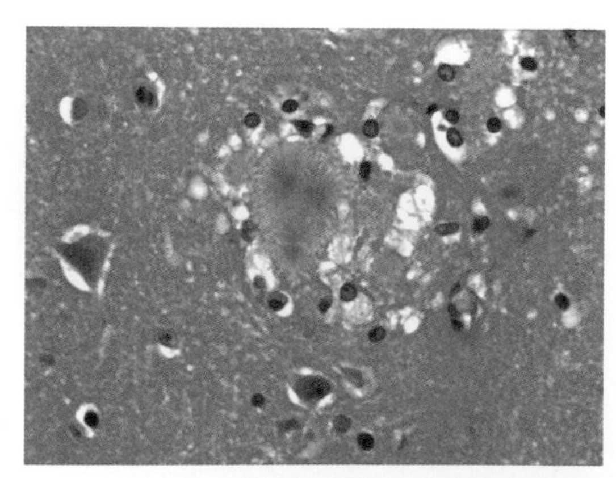

FIGURE 23.9 **A florid plaque in the frontal cortex in a patient with a 12-month clinical history of variant Creutzfeldt–Jakob disease.** The plaque has a dense eosinophilic core with a pale fibrillary periphery and is surrounded by spongiform change. (H&E stain.) *Source: Courtesy of Dr Diane Ritchie, Edinburgh, UK.*

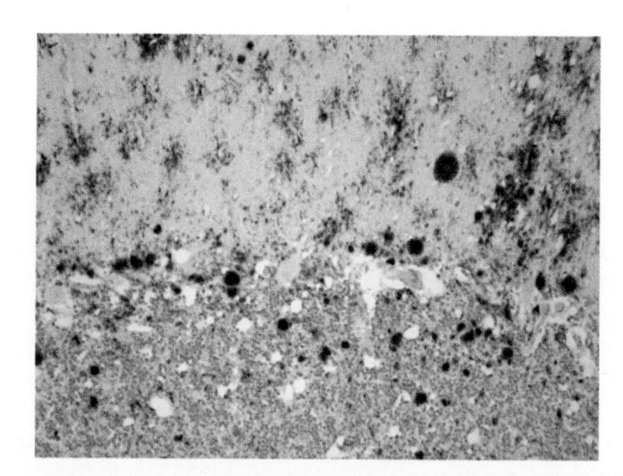

FIGURE 23.10 **Immunocytochemistry of prion protein (PrP) in the cerebellar cortex in variant Creutzfeldt–Jakob disease (same case as Fig. 23.9).** Intense labeling of the florid plaques is seen, but also multiple smaller plaques and amorphous accumulations of PrP that are not visible on routine staining. (12F10 anti-PrP antibody.) *Source: Courtesy of Dr Diane Ritchie, Edinburgh, UK.*

plaques is detected in the thalamus, hypothalamus, and globus pallidus, while synaptic and perineuronal neuronal PrP accumulation is also present in the brainstem and spinal cord. No PrP plaques are present in these structures, and staining for PrP in the leptomeninges is negative.

Immunohistochemistry for glial fibrillary acidic protein demonstrates clearly the severe astrocytosis in the posterior thalamus.[34] This technique also demonstrates astrocytosis in the inferior colliculi and periaqueductal gray matter in the midbrain, and in the cerebral and cerebellar cortex, particularly in relation to areas of severe neuronal loss and less frequently around the margins of amyloid plaques. Immunohistochemistry for ubiquitin and phosphorylated tau labels the neuritic processes around the amyloid plaques in the cerebral cortex and the cerebellum in vCJD, but no NFTs have been identified by these techniques.

PrP accumulation is readily detectable by immunohistochemistry identified in follicular dendritic cells and macrophages in many germinal centers in the pharyngeal, lingual, and palatine tonsil, and in germinal centers in the thymus, the appendix, Peyer's patches in the ileum, spleen (Fig. 23.11), and lymph nodes from many regions in the body.[39]

Western blotting to detect the protease-resistant form of PrP (PrP^{res}) in the brain in vCJD has shown a unique PrP^{res} isoform, in which the unglycosylated band migrates at around 19 kDa, similar to the type 2 PrP^{res} in sCJD (Fig. 23.12).[39] However, unlike in sCJD, the diglycosylated isoform of PrP^{res} predominates in vCJD, providing a molecular profile for vCJD that is distinct from other forms of human prion disease. This molecular profile is also present in lymphoid tissues, where the predominance of the diglycosylated band is often more pronounced than in the brain. A similar PrP^{res} molecular profile is present in the brain in BSE and in other BSE-related infections in felines and in experimental BSE transmissions to mice.

Current Concerns in Variant Creutzfeldt–Jakob Disease

vCJD is unique in human prion diseases because it results from an acquired infection from a non-human species, and also because of the widespread distribution of the infectious agent in the body. Abnormal PrP was identified by immunocytochemistry and Western blot examination in lymphoid tissues in vCJD, but not in sporadic or other forms of CJD, allowing the use of tonsil biopsy as an aid to diagnosis in some vCJD cases. Experimental transmission studies have confirmed that infectivity is present in lymphoid tissues in vCJD, although at levels that are around 2–3 logs lower than in the brain. These findings have given rise to concerns that vCJD may be transmitted accidentally, by surgical instruments used on lymphoid tissues (such as in tonsillectomy procedures), or by blood transfusion or blood products. Although no evidence for the transmission of vCJD by surgical instruments has so far emerged, concerns over potential infectivity in blood in vCJD have been reinforced by the experimental transmission of BSE by blood transfusion in a sheep model, at a preclinical stage in the infection. In the UK, steps have been implemented by the blood transfusion services to reduce these potential risks, particularly the implementation of leukodepletion to remove white blood cells from red blood cell concentrates.

To date, epidemiological studies have identified four instances of likely vCJD transmission by the transfusion of non-leukodepleted red blood cell concentrates donated by individuals who, although asymptomatic at the time of diagnosis, subsequently died from vCJD.[37] Three of the recipients developed vCJD after incubation periods of 6.5–8 years, and died with typical clinical features and neuropathology. These three recipients were all methionine homozygotes at codon 129 in the *PRNP* gene (i.e.

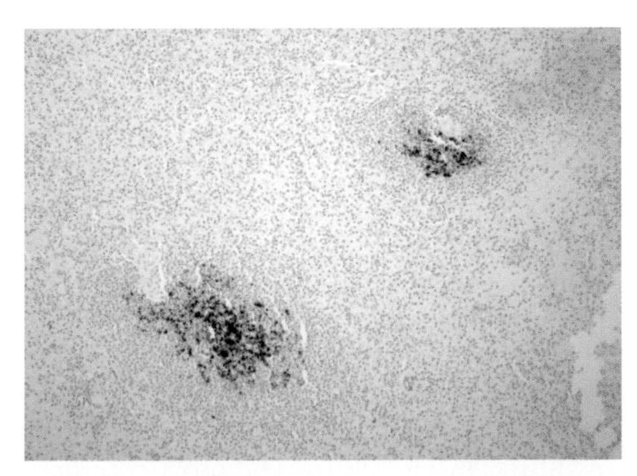

FIGURE 23.11 **Immunocytochemistry of prion protein (PrP) in the spleen in variant Creutzfeldt–Jakob disease.** Intense labeling of a germinal center is seen, with positivity appearing in follicular dendritic cells. (12F10 anti-PrP antibody.) *Source: Courtesy of Dr Diane Ritchie, Edinburgh, UK.*

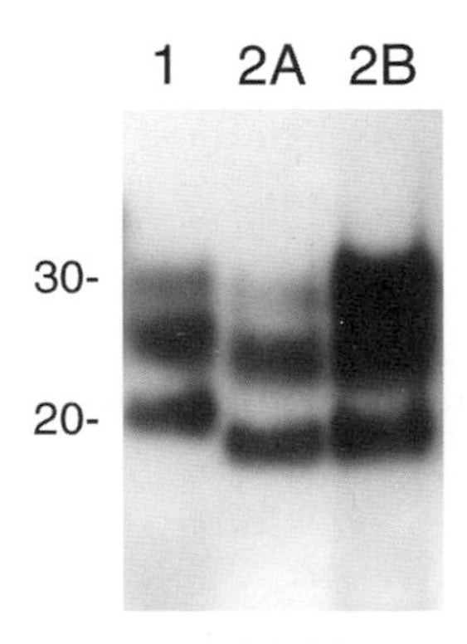

FIGURE 23.12 **Western blot analysis of PrP^{res} in the brain showing type 1 (1), type 2A (2A) and type 2B (2B), typical of sporadic Creutzfeldt–Jakob disease (types 1 and 2A) and variant Creutzfeldt–Jakob disease (type 2B).** The position of molecular weight markers is indicated, with their masses shown in kilodaltons. PrP^{res}: pathological misfolded prion protein, resistant to proteinase K. *Source: Courtesy of Dr Mark Head, Edinburgh, UK.*

$129^{Met\ Met}$). The fourth recipient died 5 years after transfusion with no neurological symptoms. Autopsy revealed no evidence of vCJD in the nervous system, and immunohistochemistry and Western blot analysis for abnormal PrP in the brain and spinal cord were negative. However, immunohistochemistry and Western blot analysis of the lymphoid tissues showed evidence of abnormal PrP accumulation in the spleen and lymph nodes. The molecular profile of the PrP^{res} in the spleen was characteristic of vCJD. However, this individual was a $129^{Met\ Val}$ heterozygote, indicating that this genotype is also susceptible to vCJD infection. A similar finding was made in a hemophiliac patient in the UK who had received large volumes of UK-sourced plasma; although this patient showed no neurological symptoms before death, autopsy studies found PrP^{res} accumulation in the spleen. This patient was also a heterozygote at codon 129 in the *PRNP* gene.

Considerable difficulties exist in attempting to predict the future numbers of vCJD cases in the UK and elsewhere. Although there was earlier evidence of an increase in the incidence of the disease in the UK, this has not been sustained and no new cases were identified in 2012. However, two new cases of vCJD were identified in France in 2012. A summary of the worldwide cases of vCJD is given in Table 23.5.[37] Whether these observations relate to differences in exposure or susceptibility to BSE infection (with a correspondingly variable incubation period) is unknown. Continued surveillance for all forms of CJD is required to answer these questions, and to determine whether individuals with $129^{Met\ Val}$ or $129^{Val\ Val}$ *PRNP* genotypes will also be susceptible to BSE infection. The identification of such cases would have an impact on the likely numbers of future vCJD patients, and neuropathology is a key part in the investigation and characterization of such cases.

Two retrospective immunocytochemical studies to detect abnormal PrP in surgically removed appendixes from otherwise normal individuals in the UK have been undertaken over the past 10 years to establish the prevalence of vCJD infection. The most recent reports indicate that the prevalence of vCJD infection may be as high as 1 in 2000 in the UK,[40] reinforcing the need for ongoing surveillance and detailed analysis of all cases of prion disease to establish whether vCJD will emerge in *PRNP* codon 129 valine homozygotes and heterozygotes in the UK, despite the lack of current new cases in *PRNP* codon 129 methionine homozygotes.

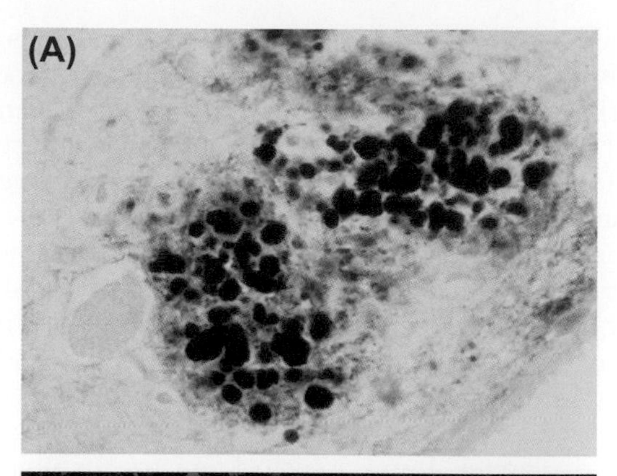

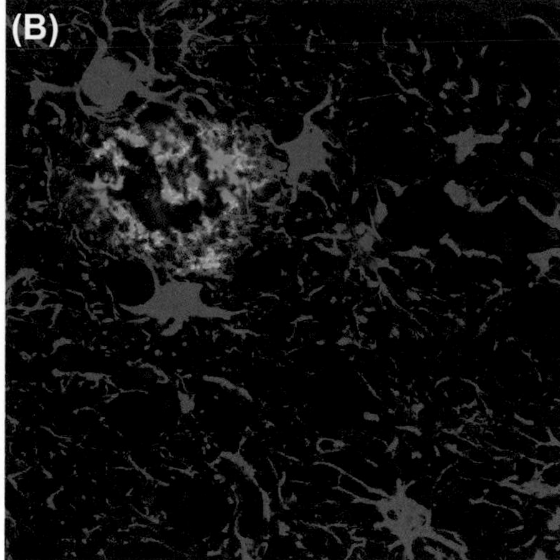

FIGURE 23.13 **Multicentric plaques of Gerstmann–Sträussler–Scheinker disease (GSS).** (A) Multicentric plaques of GSS. Immunohistochemistry with anti-prion protein (PrP) antibodies; (B) confocal laser microscopy of the amyloid plaque of GSS. Green: PrP; red: glial fibrillary acidic protein (GFAP)-immunoreactive astrocytes.

GERSTMANN–STRÄUSSLER–SCHEINKER DISEASE

GSS is a slowly progressive neurodegenerative disease, with an autosomal dominant mode of inheritance. This very rare disease has a prevalence of only 1–10 per 100 million and involves progressive ataxia, dementia, encephalopathy, and multicentric PrP plaques (Fig. 23.13A,B).

The first ("H") family with this disease, in Vienna, was reported by Gerstmann, Sträussler, and Scheinker in 1936. The transmissibility of GSS was reported by Masters and colleagues,[41] although inocula derived from only five brains with 102^{Leu} have been transmitted. In 1981, Masters described the "CG" family[41]; although phenotypically similar to GSS with regard to amyloid plaques, this family was later shown to represent familial British dementia, with a different genetic alteration from that in GSS.

Several codon mutations are involved in GSS. These are described in detail in Liberski and Surewicz,[27] and most of them are shown in Fig. 23.14. The most common mutations have been found on codons 102, 105, 117, and 198.

Codon 102 mutation (102^{Leu} 129^{Met}) has been found in families from Japan, Germany, Israel, Hungary, Poland, the UK, and Italy, and in the original H family. The

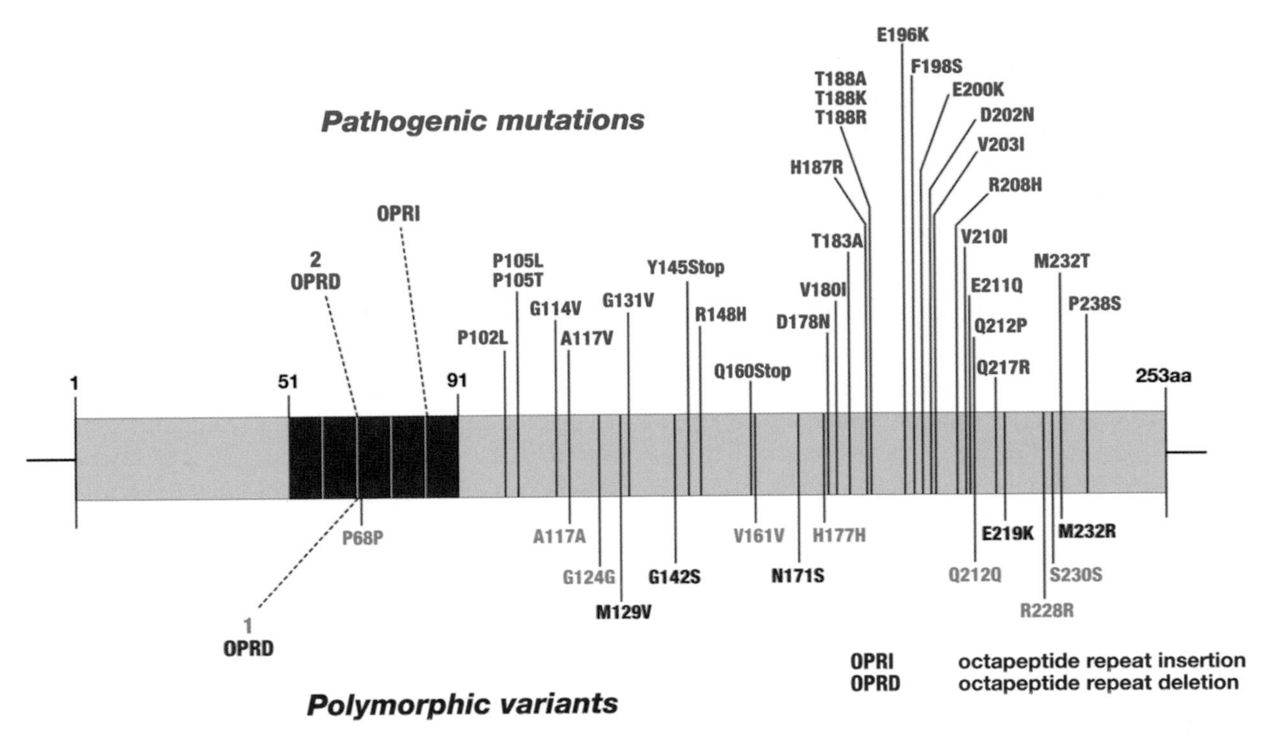

FIGURE 23.14 **Mutations of the open reading frame of the *PRNP* gene.**

disease course comprises slowly progressive cerebellar ataxia and later dementia. The last case of GSS from the original H family, however, had features of typical CJD, with early dementia and a periodic EEG.[42] Families with codon 102 mutation have varied neurological signs and symptoms. The classical ataxic type of GSS starts in the second to sixth decade and lasts for a few months to years. Symptoms include dysartria, alterations of saccadic eye movements, pyramidal and extrapyramidal signs, and cognitive changes leading to dementia. A CJD-like disease type with myoclonic jerks and a periodic EEG pattern is sometimes seen. MRI demonstrates mild atrophy of the cerebellum and the brain.

A codon 105 mutation (105^{Leu} 129^{Val}) has been found in five Japanese families. Early symptoms include spastic paraparesis with brisk reflexes; tetraplegia, dementia, and rigidity appear in the terminal stage. Illness starts around 40–50 years of age and lasts for 6–12 years. PrP deposits are found mainly in the cerebral cortex, less often in the striatum, and occasionally in the cerebellum. There may be sparse NFTs composed of PHFs.

Families with a mutation on codon 117 (117^{Val} 129^{Val}) (telencephalic GSS) have dementia but are otherwise atypical for GSS cerebellar ataxia. "Pure" dementia was seen in earlier generations of an Alsatian family, compared with a mixture of signs and symptoms, including dementia, in later generations.

A mutation on codon 198 (198^{Ser} 129^{Val}) was found in a family from Indiana and in another, unrelated, family. Patients are homozygous or heterozygous for Val at codon 129. The Indiana kindred is characterized by pyramidal and cerebellar signs, dementia, dysarthria, progressive clumsiness and difficulties in walking, parkinsonian features, optokinetic nystagmus, and sleep disturbances. An early symptom is changes in saccadic eye movements. Disease onset is between 40 and 70 years of age and it lasts for approximately 5 years, although the disease course may be accelerated to 1–2 years. Neuropathological examination reveals changes otherwise typical of GSS. Neurites around plaques contain NFTs composed, not unlike those of AD, of hyperphosphorylated MAP-τ. Spongiform change is occasionally seen around the plaques.

LABORATORY TESTS

Surrogate Markers

The 14-3-3 protein is released from damaged neurons and is detectable in higher amounts in the CSF in sCJD. However, since 14-3-3 is also increased in other CNS disorders involving acute neuronal damage, such as stroke and encephalitides, it is a non-specific marker. It is detected in approximately half of vCJD cases.

The hyperphosphorylated isoform of MAP-τ forms PHFs of NFTs in AD and frontotemporal dementia. Mutations in the MAP gene encoding it segregate with frontotemperal dementia linked to chromosome 17. The level of MAP-τ in the CSF in CJD is high (1533–12,215 pg/ml) compared with other dementive disorders (233–640 pg/ml) and controls (109–296 pg/ml). In patients with fCJD, the MAP-τ concentration ranges from 70 to 35,5000 pg/ml; its concentration in GSS ranges from 97 to 4120 pg/ml.

The CSF concentration of S-100, an acidic Ca^{2+} binding protein, may be increased in CJD (2–117 ng/ml) compared with healthy cases (1–19 ng/nl).

Detection of Pathological Prion Protein

To detect PrP^d, its propensity for self-propagation through seeded/template conformational conversion has been utilized in PMCA, the amyloid seeding assay, and the QuIC reaction.[42]

The basis of PMCA is conversion of PrP^c by PrP^d in a test sample in a repeated cycle of incubation and sonication. Although it is difficult to perform, serial automated PMCA has been developed. As PMCA can detect as little as 1.2 ag (10^{-18} g) of PrP^d, it is possible to detect PrP^d in the blood and in environmental samples, such as water. PMCA has been further improved by using recombinant PrP^c instead of brain homogenate (recombinant PMCA).

In the amyloid seeding assay, thioflavin T is used to detect PrP^d, instead of the Western blot as in PMCA. The source of PrP^c is a recombinant protein. This assay is capable of detecting PK-sensitive PrP^d.

QuIC is one of the new generation of tests used to detect PrP^d.[43] It differs from PMCA by replacing sonication with intermittent shaking. Analogously to PMCA, PrP^d seeds present in a tested sample convert $rPrP^c$ to PrP^d. A real-time (RT)-QuIC test has recently been developed, which detects newly formed PrP^d by thioflavin T and uses a high-throughput multiwall plate format. RT-QuIC can detect 10^{-10}-fold dilutions of sCJD or 1 fg of PrP^d, and demonstrates 100% specificity and more than 80% sensitivity in detecting PrP^d. While MM1/MV1, VV2 and MV2k brain homogenates seeded equally efficiently, MM2C and VV1 were less reproducible and efficient, while vCJD homogenates were less efficient than sCJD in RT-QuIC. In a 2012 study, 108 CSF samples taken from 56 patients with all subtypes of CJD and 52 controls were studied by RT-QuIC, and positive results were obtained for 50 CJD samples,[43] which translates to 91% sensitivity and 98% specificity. RT-QuIC has similar sensitivity and higher specificity compared with 14-3-3 detection.

CONCLUSION

The field of prion diseases has advanced a long way from obscure diseases of unknown etiology, through the discovery of transmissible agents, to the novel concept of misfolded proteins, which, by a mechanism of seeded polymerization, underlies their pathogenesis. It soon was discovered that many other neurodegenerative diseases, such as AD and Parkinson disease, are in reality prionoids – supposedly non-transmissible misfolded protein disorders. They are also brain amyloidoses because aggregated proteins meet the criteria for amyloid.

The history of these human disorders can be traced to the seminal discovery of kuru. Kuru, now a nearly extinct exotic disease of a cannibalistic tribe in the remote highlands of Papua New Guinea, still exerts an influence on many aspects of neurodegeneration research. First, it demonstrated that a non-inflammatory neurodegenerative disease in humans could result from an infectious agent, then called a "slow virus". This discovery heralded a new class of human diseases, including CJD, GSS, and FFI. Parenthetically, CJD was posited as a possible equivalent of kuru on the basis of non-specific neuropathological findings, and GSS was identified as linked because of the presence of numerous amyloid plaques, not unlike kuru plaques.

One may also speculate on what would have happened if kuru had not been discovered or had not existed. The infectious nature of CJD would probably not have been suspected until the identification of cases of iCJD in recipients of hGH or dura mater, or in the vCJD outbreak in the UK. For decades, CJD and GSS would have remained as obscure neurodegenerations of merely academic interest. The familial forms of CJD would not have benefited from *PRNP* gene analysis, but only later would have been studied by linkage analysis and possibly reverse genetics. The whole field would have probably remained of only arcane interest to veterinarians until the BSE epidemic began to exert its devastating effect. The discovery of vCJD would have been delayed, as no CJD surveillance would have been initiated. And, perhaps most importantly, the sea change in scientific inquiry that has led to the paradigm-shifting concept of protein-misfolding diseases, including not only the neurodegenerative but also an increasing number of non-neurological disorders, would have been delayed by decades.

QUESTIONS FOR FURTHER RESEARCH

Although a basic understanding of prion diseases has already been achieved, multiple unanswered questions and poorly understood issues remain. At the very basic level, we do not have any insight into the cause of sporadic diseases. Specifically, it is unclear what initiates the conversion of the normal into the abnormal isoform of PrP. We also do not understand why sporadic diseases are nearly always transmissible experimentally while many familial diseases are not. GSS is perhaps the best example, as only GSS caused by the 102 point mutation has been transmitted while other mutations have not. In this regard, novel animal models are highly desirable, and the discovery that bank voles (*Myodes glareolus*) are almost universally susceptible to prion disease may facilitate transmission studies.

In clinical neurobiology, the most important practical issue is to develop improved methods to diagnose prion diseases and to increase the sensitivity of the QuIC. This will enable the detection of PrP^d in bodily fluids long before the development of clinical signs and symptoms. Since no therapeutic agent of any usefulness has so far been discovered, innovative effective therapies would be very desirable.

A critically important caveat is CWD in mule deer, elk, and moose. As this is a highly infectious disease, its transmissibility potential to humans should be studied in humanized transgenic mice. Another important question concerns the relationship between synthetic prions and real ones, and whether they are really identical at the cellular and molecular levels.

Prionoids have become a hot topic in the prion field. As misfolded proteins are prion like in their nucleation/polymerization formation and intracellular and intercellular spread, the most important question is whether diseases such as AD and amyotrophic lateral sclerosis (ALS) could spread between individuals. The recent finding of three ALS cases among recipients of cadaveric hGH, provided within the framework of the National Hormone and Pituitary Program, is disturbing and warrants urgent inquiry.

References

1. Lloyd S, Mead S, Collinge J. Genetics of prion disease. *Top Curr Chem*. 2011;305:1–22.
2. Brown P, Brandel JP, Sato T, et al. Iatrogenic Creutzfeldt–Jakob disease, final assessment. *Emerg Infect Dis*. 2012;18(6):901–907.
3. Basler K, Oesch B, Scott M, et al. Scrapie and cellular PrP isoforms are encoded by the same chromosomal gene. *Cell*. 1986;46(3):417–428.
4. Prusiner SB. Novel proteinaceous infectious particles cause scrapie. *Science*. 1982;216(4542):136–144.
5. Aguzzi A, Rajendran L. The transcellular spread of cytosolic amyloids, prions, and prionoids. *Neuron*. 2009;64(6):783–790.
6. Lansbury Jr PT, Caughey B. The chemistry of scrapie infection: implications of the "ice 9" metaphor. *Chem Biol*. 1995;2(1):1–5.
7. Zhang B, Une Y, Fu X, et al. Fecal transmission of AA amyloidosis in the cheetah contributes to high incidence of disease. *Proc Natl Acad Sci U S A*. 2008;105(20):7263–7268.
8. Meyer-Luehmann M, Coomaraswamy J, Bolmont T, et al. Exogenous induction of cerebral beta-amyloidogenesis is governed by agent and host. *Science*. 2006;313(5794):1781–1784.
9. Eisele YS, Obermüller U, Heilbronner G, et al. Peripherally applied Aβ-containing inoculates induce cerebral beta-amyloidosis. *Science*. 2010;330(6006):980–982.
10. Prado MA, Baron G. Seeding plaques in Alzheimer's disease. *J Neurochem*. 2012;120(5):641–643.
11. Nussbaum JM, Schilling S, Cynis H, et al. Prion-like behaviour and τ-dependent cytotoxicity of pyroglutamylated amyloid-β. *Nature*. 2012;485(7400):651–655.
12. Zobeley E, Flechsig E, Cozzio A, Enari M, Weissmann C. Infectivity of scrapie prions bound to a stainless steel surface. *Mol Med*. 1999;5(4):240–243.
13. Eisele YS, Bolmont T, Heikenwalder M, et al. Induction of cerebral beta-amyloidosis: intracerebral versus systemic Aβ inoculation. *Proc Natl Acad Sci U S A*. 2009;106(31):12926–12931.
14. Braak H, Braak E. Neuropathological staging of Alzheimer-related changes. *Acta Neuropathol*. 1991;82(4):239–259.
15. Liu L, Drouet V, Wu JW, et al. Trans-synaptic spread of tau pathology in vivo. *PLoS ONE*. 2012;7(2):e31302.
16. Fraser H. Neuronal spread of scrapie agent and targeting of lesions within the retino-tectal pathway. *Nature*. 1982;295(5845):149–150.
17. Clavaguera F, Bolmont T, Crowther RA, et al. Transmission and spreading of tauopathy in transgenic mouse brain. *Nat Cell Biol*. 2009;11(7):909–913.
18. Luk KC, Song C, O'Brien P, et al. Exogenous alpha-synuclein fibrils seed the formation of Lewy body-like intracellular inclusions in cultured cells. *Proc Natl Acad Sci U S A*. 2009;106(47):20051–20056.
19. Angot E, Steiner JA, Lema Tomé CM, et al. Alpha-synuclein cell-to-cell transfer and seeding in grafted dopaminergic neurons in vivo. *PLoS ONE*. 2012;7(6):e39465.
20. Gajdusek DC. Kuru and its contribution to medicine. *Philos Trans R Soc B*. 2008;363:3697–3700.
21. Liberski PP, Sikorska B, Lindenbaum S, et al. Kuru: genes, cannibals and neuropathology. *J Neuropathol Exp Neurol*. 2012;71(2):92–103.
22. Hainfellner JA, Liberski PP, Guiroy DC, et al. Pathology and immunohistochemistry of a kuru brain. *Brain Pathol*. 1997;7:547–554.
23. Brandner S, Whitfield J, Boone K, et al. Central and peripheral pathology of kuru: pathological analysis of a recent case and comparison with other forms of human prion diseases. *Philos Trans R Soc B*. 2008;363:3755–3763.
24. Mead S, Uphill J, Beck J, et al. Genome-wide association study in multiple human prion diseases suggests genetic risk factors additional to PRNP. *Hum Mol Genet*. 2012;21(8):1897–1906.
25. Centers for Disease Control and Prevention. *CDC's Diagnostic Criteria for Creutzfeldt-Jakob Disease (CJD)*; 2010. Accessed 28.02. 14. http://www.cdc.gov/ncidod/dvrd/cjd/diagnostic_criteria.html.
26. Kong Q, Surewicz WK, Petersen RB, et al. Inherited prion diseases. In: Prusiner SB, ed. *Prion Biology and Diseases*. Cold Spring Harbor, NY: Cold Spring Harbor Laboratory Press; 2004. 673–775.
27. Liberski PW, Surewicz WK. Molecular genetics of Gerstmann–Sträussler–Scheinker disease and Creutzfeldt–Jakob disease. *Genetics*. 2013;2:2.
28. Budka H. Fatal familial insomnia around the world. Introduction. *Brain Pathol*. 1998;8(3):553.
29. Goldfarb LG, Mitrová E, Brown P, Toh BK, Gajdusek DC. Mutation in codon 200 of scrapie amyloid protein gene in two clusters of Creutzfeldt–Jakob disease in Slovakia. *Lancet*. 1990;336(8713):514–515.
30. Surewicz WK, Apostol MI. Prion protein and its conformational conversion: a structural perspective. *Top Curr Chem*. 2011;305:135–167.
31. Parchi P, Saverioni D. Molecular pathology, classification, and diagnosis of sporadic human prion disease variants. *Folia Neuropathol*. 2012;50(1):20–45.
32. Puoti G, Bizzi A, Forloni G, Safar JG, Tagliavini F, Gambetti P. Sporadic human prion diseases: molecular insights and diagnosis. *Lancet Neurol*. 2012;11(7):618–628.
33. Head MW, Ironside JW. Review: Creutzfeldt–Jakob disease: prion protein type, disease phenotype and agent strain. *Neuropathol Appl Neurobiol*. 2012;38(4):296–310.
34. Will RG, Ironside JW, Zeidler M, et al. A new variant of Creutzfeldt–Jakob disease in the UK. *Lancet*. 1996;347:921–925.

35. Wells GA, Scott AC, Johnson CT, et al. A novel progressive spongiform encephalopathy in cattle. *Vet Rec*. 1987;121:419–420.
36. Smith PG, Bradley R. Bovine spongiform encephalopathy (BSE) and its epidemiology. *Br Med Bull*. 2003;66:185–198.
37. National CJD. Research & Surveillance Unit (NCJDRSU). Accessed 28.02.14. www.cjd.ed.ac.uk.
38. World Health Organization. Variant Creutzfeldt–Jakob disease. Fact sheet no. 180; revised February 2012. http://www.who.int/mediacentre/factsheets/fs180/en. Accessed 28.02.14.
39. Ironside JW, McCardle L, Horsburgh A, Lim Z, Head MW. Pathological diagnosis of variant Creutzfeldt–Jakob disease. *APMIS*. 2002;11:79–87.
40. Gill ON, Spencer Y, Richard-Loendt A, et al. Prevalent abnormal prion protein in human appendixes after bovine spongiform encephalopathy epizootic: large scale survey. *BMJ*. 2013;347:f5675. http://dx.doi.org/10.1136/bmj.f5675.
41. Masters CL, Gajdusek DC, Gibbs Jr CJ. Creutzfeldt–Jakob disease virus isolations from the Gerstmann–Sträussler syndrome with an analysis of the various forms of amyloid plaque deposition in the virus-induced spongiform encephalopathies. *Brain*. 1981;104(3):559–588.
42. Hainfellner JA, Brantner-Inthaler S, Cervenakova L, et al. The original Gerstmann–Sträussler–Scheinker family of Austria: divergent clinicopathological phenotypes but constant PrP genotype. *Brain Pathol*. 1995;5:201–211.
43. McGuire LI, Peden AH, Orrú CD, et al. Real time quaking-induced conversion analysis of cerebrospinal fluid in sporadic Creutzfeldt–Jakob disease. *Ann Neurol*. 2012;72(2):278–285.

SECTION IV

INFECTIOUS AND IMMUNE-MEDIATED DISEASES AFFECTING THE NERVOUS SYSTEM

CHAPTER

24

Introduction

Clayton A. Wiley

Division of Neuropathology, UPMC Presbyterian Hospital, Pittsburgh, Pennsylvania, USA

Mammals have evolved a broad spectrum of defenses to combat the potential of a wide variety of infectious agents to invade the CNS. These defenses range from simple structural barriers like the skull and meninges, to highly intricate and interacting arms of the immune system. For the most part these defenses do a spectacular job of isolating the CNS from colonizing microbes, but occasionally the defense is circumvented or, worse, the immune reaction goes awry and attacks the CNS. The six chapters in this section outline infectious and immune-mediated diseases of the CNS. Like all tissues in the body, the CNS responds to any pathological insult with an innate immune response frequently referred to as inflammation. In the first two chapters (Chapters 25 and 26), Appel and Raison describe how inflammation affects neurological and psychiatric disease. In Chapter 27, Kristensson examines the spectrum of infectious organisms, and in Chapter 28, Lyons and colleagues drill down into a single infectious disease, AIDS, to demonstrate the complexity of host and pathogen interactions. The final two chapters (Chapters 29 and 30) look at the other side of the coin and assess how the CNS is damaged when immunological control is lost and attacks the host.

While many neurodegenerative diseases have strong inheritable components, of the common diseases discussed in Chapter 25 only around 10% have a transmissible genetic defect so the vast majority has some acquired defect. Appel and co-authors examine the link between neuroinflammation and neurodegeneration. Deciphering which is cause and which is effect is a true Gordian knot. Critical to understanding these diseases is the idea of non-cell autonomous neurodegeneration. That is, neuronal death is not the result of purely neuronal dysfunction but rather the sum of neuronal dysfunction amplified on the background of microglial activation. Many of the well-described neurotoxins [e.g. 1-methyl-4-phenyl-1,2,3,6-tetrahydropyridine (MPTP) in animal models of Parkinson's disease] have shown that for a compound to be neurotoxic *in vivo* requires the presence of microglia or T cells. This surprising observation has been extended to show that both microglia and T cells can exhibit neuroprotective and neurotoxic profiles. The authors use the most common neurodegenerative disease, Alzheimer disease, and its associated animal models to illustrate the hypothesis of an extracellular neurotoxin (Aβ peptide) eliciting a danger signal cascade that drives microglia away from a neuroprotective and into a neuroinflammatory phenotype. This leads to a spiraling cycle of further microglial activation and more neurotoxic inflammation. Appel and colleagues also use their extensive studies in amyotrophic lateral sclerosis (ALS) to illustrate the unexpected role of T cells in modulating neurodegenerative disease. Using transgenic mice with bone marrow transplantation, the authors also illustrate the potentially protective versus toxic role of T cells in modulating a neurodegenerative disease. Evaluations of ALS patients have supported a role of the adaptive immune system in modulating prognosis of this human disease.

In Chapter 26, Raison and colleagues examine bidirectional interactions between the immune system and psychiatric disease. Many immunological diseases are tightly associated with psychiatric symptoms and, *vice versa*, many psychiatric diseases are associated with immunological dysfunction. Which came first: the chicken or the egg? Perhaps the best described connection between the immune and nervous systems is the hypothalamic–pituitary–adrenal (HPA) axis. Steroid production after stimulation of the HPA axis is a powerful means of suppressing immune function. Depression profoundly affects the HPA axis and yet anyone who has been ill for an extended period can readily attest to the associated depression. The capacity for immune stimuli (e.g. lipopolysaccharide and polyinosinic:polycytidylic acid) to robustly and reproducibly induce depressive-like behavior is well documented in many animal models, while modern cytokine therapy in humans has

Neurobiology of Brain Disorders
http://dx.doi.org/10.1016/B978-0-12-398270-4.00024-0

confirmed the potent connection between exogenous cytokine administration and behavioral changes (e.g. interferon therapy and depression).

In Chapter 27, Kristensson provides a broad overview of infections of the CNS. From the wide variety of microbes in the environment, a modest number has developed the capacity to evade the host defenses and invade the nervous system. Rather than catalog all of these pathogens, Kristensson has selected the most important of these agents to illustrate their invasive mechanisms and the host's response. The brain is surrounded by impressive physical barriers on a gross and microscopic scale. Once within the CNS, however, the host–microbe battle is on a less even footing. The microenvironment of the CNS must be carefully and precisely maintained; thus, when it becomes the battleground on which the immune system attempts to vanquish the invader, there are great costs to homeostasis. The mammalian host has evolved two complementary systems of defense: the innate and adaptive immune systems. Through ubiquitously expressed pathogen recognition molecules the host immediately reacts to invading pathogens with an innate immune defense. While this is quite effective, the host takes no chances and uses the innate immune response to initiate the development of an adaptive immune response, such that if the host survives the first attack, a rapid, targeted, and amplified response can ensue on any future invasion by the same organism. For the most part we have evolved measures that protect us from environmental pathogens; however, as microbes mutate and ecosystems are disrupted, we face new emergent infections that we must adapt to.

Chapter 28 focuses on a single viral infection, HIV, and drills into the complexities and interlocking nature of viral, immune, and CNS factors that result in neurological disease. With the discovery of HIV in the 1980s, the severity of AIDS epidemic was quickly appreciated and elicited a concerted response on part of the scientific community. Every tool in the biomedical armamentarium to combat infectious disease has been brought to bear on abrogating the epidemic and associated neurological disease. In a relatively short time the molecular biology of HIV replication was fully elucidated. But it is a long way from understanding the molecular biology of a virus to understanding the pathogenesis of an infection. The intricate tango danced between virus and immune system almost defies imagination. The pathogenesis of CNS dysfunction has been examined from almost every angle, including disturbance of every cell component in the brain and aberrations of CNS metabolism. Like other microbes, HIV has discovered that hiding within host monocytes evades the immune system and provides the Trojan horse to traffic into the brain. Once there, it persists as if the brain and body compartments were completely separate. Indeed, the brain reservoir of HIV has become one of the most important challenges to eradicating the virus. While combination antiretroviral therapy has been nothing short of a miracle, we still do not fully understand the disease or have completely effective therapies. Unfortunately, highly penetrant antiretroviral drugs bring with them their own neurotoxic properties, thus complicating therapy.

Chapter 29 examines how autoimmune diseases offer insights into the functional complexity of the nervous system. Autoimmune diseases are tragic experiments of nature wherein the immune system targets self rather than non-self antigens. Through aberrant recognition of self-antigens, both humoral and cellular arms of the immune system become pathologically activated. While idiopathic autoimmune diseases develop, three-quarters of the time they are harbingers of even worse news when they antedate the diagnosis of cancer (i.e. paraneoplastic disorders). The complexity of nervous system structure is reflected in the complexity of the wide variety of different autoimmune neurological diseases. This autoimmune targeting of individual cells to individual surface molecules and then to individual neurotranscription factors is manifest as bizarre and sometimes quirky neurological disease. Neurological disease associated with the development of antibodies to neuron-specific cell surface proteins is perhaps the most readily understood and readily treated (e.g. with plasmapheresis or intravenous immunoglobulin). But autoimmune cell-mediated responses can also develop to intracellular proteins and these can also be manifest as highly selective neuronal dysfunction. Unfortunately, the cell-mediated diseases are more difficult to treat and even successful surgical removal of associated tumors is not associated with neurological improvement.

Chapter 30 uses the paradigmatic autoimmune neurological disease multiple sclerosis (MS) to describe the complexities of discovering the pathogenesis of and treating a devastating neurological disease. Despite over a century of study the etiology (or etiologies) of MS remains elusive. Genetic, environmental, and infectious factors all play an important role. Discordance in disease incidence among identical twins highlights the importance of environmental factors. Human migration studies clearly show a risk (or protective factor) that is acquired during the first 15 years of life. Despite a number of excellent animal models, it has not been possible to discern which factor is most influential in MS. Without a known etiology we are left treating neurological symptoms. Every type of immunosuppression has been tried, including steroids, myeloablative chemotherapy, plasmapheresis, novel monoclonal antibodies, and preventing immune cell trafficking, with only limited success for all of them. As for so many other neurological disorders, it would appear that much of the basic biology of the immune and nervous system remains to be determined so that creative new therapies can be designed.

But what are the mechanisms by which immune activation produces behavioral disturbances? Here lies a challenge for future scientists. Immune stress during critical neural development can also disrupt multiple levels of development (neurogenesis, neuronal migration, etc.). Novel hypotheses have emerged regarding the role of early inflammation (e.g. *in utero*) and the development of psychiatric diseases such as schizophrenia and autism. Studying these diseases is difficult for a variety of reasons. First of all, the temporal relationship between the immune stress (e.g. *in utero* infection) and the psychiatric disease (e.g. autism) could extend to years. Given the ubiquity of many infections, how can such temporally separated events be linked with a high degree of confidence? When inflammatory responses disrupt fundamental cell survival, how can one tease apart more subtle disruptions of functional brain neurocircuitry? Between the complexity of the immune system and the complexity of the nervous system, how can one create biologically significant animal models for testing hypotheses about connection of the systems?

CHAPTER

25

Role of Inflammation in Neurodegenerative Diseases

Stanley H. Appel, David R. Beers, Weihua Zhao

Department of Neurology, Methodist Neurological Institute, Houston, Texas, USA

OUTLINE

INTRODUCTION

An alternative title for this chapter could have been "Neuroinflammation Fans the Flames of Neurodegeneration", because it will demonstrate that neuroinflammation plays a major role in a wide variety of diverse neurological disorders. Indeed, this is true not only for neurological disease but also for psychiatric disease (see Chapter 26). Neuroinflammation is a prominent pathological feature of neurodegeneration in amyotrophic lateral sclerosis (ALS), Parkinson disease (PD), and Alzheimer disease (AD), and is characterized by activated microglia and infiltrating T-lymphocytes at sites of neuronal injury. (More details on these neurological conditions can be found in Chapters 18, 19, and 21, respectively.) The key question is whether this neuroinflammatory response contributes to the pathogenesis of neuronal injury or is a consequence of the neurodegenerative process. Widespread inflammation and immune activation are common reactions to a diverse array of infections resulting in marked CNS pathology. Inflammation is also present in diseases such as multiple sclerosis (MS) and CNS vasculitis, and has prompted the development of therapeutics that can limit the

Neurobiology of Brain Disorders
http://dx.doi.org/10.1016/B978-0-12-398270-4.00025-2

inflammatory cascade and resulting pathology. However, in these neurodegenerative diseases, significant therapy has been lacking, and only modest symptomatic treatment is the best that can be offered at present. The inflammatory response in these disorders is much more subtle than in CNS infections or MS, and the importance of activated microglia and T-lymphocytes has been disputed. Nevertheless, more recent data suggest that neuroinflammation makes a significant contribution to the neurodegenerative process.

PD, AD, and ALS present predominantly as sporadic, and to a lesser extent, inheritable disorders. The etiology of the sporadic forms of disease is largely unknown, but discovery of specific mutations responsible for inheritable forms of each of these disorders has dramatically increased our understanding of pathogenic pathways applicable to both inheritable and sporadic forms of disease, and has led to the development of transgenic animal models. Each of these neurodegenerative disorders is heterogeneous in onset and progression, and is expressed as multiple clinical phenotypes. Analysis of brain and spinal cord specimens of these disorders provides evidence that multiple molecular pathways contribute to neuronal injury. These include misfolded and aggregated proteins, dysfunction of autophagy and the ubiquitin–proteasome pathway, mitochondrial dysfunction, increased reactive oxygen species (ROS), alterations in axonal transport, and impaired RNA metabolism. All are key events that can initiate and provoke neurodegeneration. The heterogeneity of the clinical phenotypes may be attributable at least in part to the multiplicity of these compromised pathways. The question is whether compromise of any or all of these perturbations within neurons is sufficient to cause neuronal death. Data from the mutant Cu^{2+}/Zn^{2+} superoxide dismutase (mSOD1) mouse model of ALS provide a potential answer,[1] because expression of the genetic defect solely in motor neurons does not give rise to progressive disease. Furthermore, expression solely in microglia also does not cause motor neuron injury or give rise to disease, nor does expression solely in astrocytes. Expression of mSOD1 must be present in motor neurons as well as glia to amplify motor neuron injury, leading to progressive disease and shortened survival. Decreasing the expression of mSOD1 in microglia or astrocytes can slow the rate of disease progression and lengthens survival with no effect on onset, whereas decreasing the expression of mSOD1 in motor neurons delays the onset of disease, and thereby also lengthens survival. Thus, in the animal model of ALS, dysfunction of the multiple compromised pathways within neurons is not sufficient to cause neuronal cell death; motor neuron injury is non-cell autonomous and disease progression depends on a well-orchestrated involvement of motor neurons, glia, and T-lymphocytes.

Extrapolating from the mSOD1 mouse transgenic model of ALS, it is now hypothesized that neurons do not die alone in the neurodegenerative disorders and that cell death is non-cell autonomous, which means that signaling between neurons and glia is necessary to promote progressive neuronal injury and accelerated death. Regardless of whether misfolded and aggregated proteins leading to mitochondrial dysfunction and increased ROS, or altered RNA processing, are the earliest events in motor neurons, cell death depends on the well-orchestrated participation of non-neuronal cells including microglia and astrocytes. One thesis is that neuroinflammation is a necessary component of this non-cell autonomous process and depends on neuron–glial–T-lymphocyte signaling, fostered and initiated by the compromised intraneuronal pathways. Signals from injured neurons are initially communicated to microglia, members of the innate immune system, which following activation can induce activation of other glia such as astrocytes, and thereby coordinate a communal response to neuronal injury. In response to alterations in the activation states of the innate immune microglia and astrocytes, T-lymphocytes, as members of the adaptive immune system, join this process by infiltrating the CNS at sites of neuronal injury. Thus, the evidence favoring non-cell autonomy suggests that neuroinflammation is not just the consequence of neuronal injury, but an active participant in disease pathogenesis; both innate and adaptive immune systems respond to, as well as contribute to, the neurodegenerative disease pathology and tissue destruction.

MICROGLIA: CONVERGENCE POINT FOR PROMOTING OR COMPROMISING NEURONAL SURVIVAL

Microglia are the resident innate immunocompetent cells of the CNS, and exist in a variety of phenotypic states.[2] Their number, morphology, surface receptor expression, and production of growth factors and cytokines can readily change depending on environmental cues. In the resting state, microglia display a small cell soma and numerous branching processes (a ramified morphology). In healthy brain tissue, these processes are dynamic structures that extend and retract, sampling and monitoring their microenvironment.[3] In the presence of an activating stimulus, microglial cell-surface receptor expression is modified and the cells change from a monitoring role to one of protection and repair. The changes reflect altered activation states induced by signals that arise from the surrounding environment, most prominently the neurons. As with macrophages, microglia exhibit a functional continuum of phenotypic states from a protective anti-inflammatory state (M2 microglia) to a cytotoxic proinflammatory state (M1 microglia). Subtypes of M2 microglia have been described based on their gene expression profile, with each phenotype manifesting diverse functional states. The M2 alternatively activated (M2a) phenotype, induced by interleukin-4 (IL-4) or IL-13, expresses the proteins arginase-1, FIZZ1, and chitinase 3-like protein (Ym1). The M2a phenotype can be distinguished from the M2b phenotype induced by immune complexes and toll-like receptor (TLR) agonists, and the M2-deactivated (M2c) phenotype induced by IL-10 or transforming growth factor-β (TGF-β) with increased expression of chemokine receptor-2 (CCR2) and scavenger receptors. M2 microglia can enhance the release of neurotrophic factor such as insulin-like growth factor-1 (IGF-1), brain-derived neurotrophic factor (BDNF), and glial-derived neurotrophic factor (GDNF), secrete anti-inflammatory cytokines (IL-10, IL-4, IL-13, TGF-β), and assist in the resolution of inflammation (Fig. 25.1).

Neuroprotection Mediated by Neuronal–Microglial Signaling

Neurons can enhance neuroprotection by the CD200–CD200R pathway. CD200 is a glycoprotein expressed on neurons and can promote protection through binding the microglial CD200 receptor (CD200R). CD200 and CD200R belong to the immunoglobulin superfamily and CD200 transmits an immunoregulatory signal through CD200R to limit microglial proinflammatory activity.[4] Fractalkine–fractalkine receptor (CX3CL1–CX3CR1) signaling is another protective neuronal–microglial pathway.[5] CX3CL1 (a chemokine with the C-X3-C motif) is a large chemokine that is released from neurons and binds to CX3CR1 which is constitutively expressed on microglia, thereby enhancing the neuroprotective and repair milieu. As microglia assume this neuroprotective and phagocytic role, their morphology changes from ramified to amoeboid, and performs beneficial functions, such as scavenging neurotoxins, removing dying cells and cellular debris, and secreting anti-inflammatory cytokines and trophic factors that promote neuronal survival.

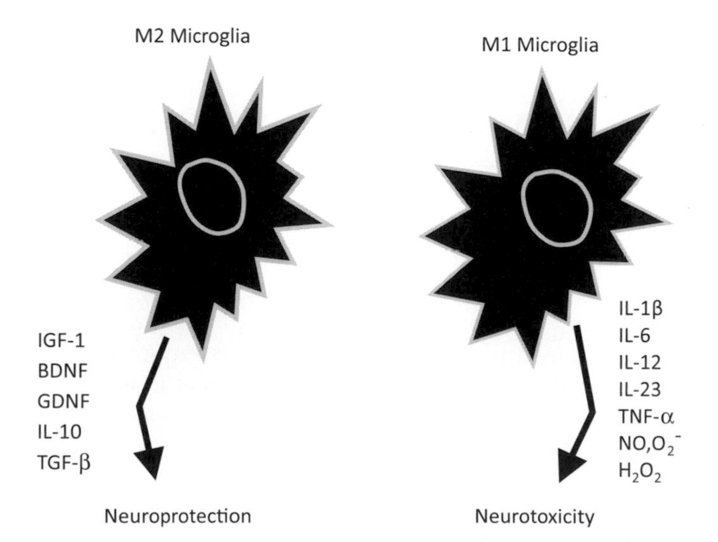

FIGURE 25.1 **Microglia: innate immune cells of the CNS that mediate the balance between neuroprotection and cytotoxicity.** M2 microglia mediate neuroprotection by the release of anti-inflammatory cytokines and neurotrophic factors. M1 microglia mediate neurotoxicity by the release of reactive oxygen species and proinflammatory cytokines. IGF: insulin-like growth factor; BDNF: brain-derived neurotrophic factor; GDNF: glial-derived neurotrophic factor; IL: interleukin; TGF: transforming growth factor; TNF: tumor necrosis factor; NO: nitric oxide; O_2^-: superoxide radical; H_2O_2: hydrogen peroxide.

Neurotoxicity Promoted by M1 Activated Microglia

In response to injury, microglia contribute to a proinflammatory response by shifting towards an M1 phenotype, producing and secreting potent ROS such as superoxide radicals ($O_2^{\bullet-}$) and nitric oxide (NO), proinflammatory cytokines [tumor necrosis factor-α (TNF-α) and IL-1β]; the M1 proinflammatory microglial response also includes the reduced release of neurotrophic factor. Other secreted inflammatory mediators include IL-6 and the chemokine macrophage inflammatory protein-1α (MIP-1α). Many cytokines, such as interferon-γ (IFN-γ), and compounds such as lipopolysaccharide (LPS) can directly promote proinflammatory M1 microglial activation *in vitro*, and may act directly on microglia *in vivo*. Extrapolating from the *in vitro* data, neurotoxin-mediated microglial activation offers a potential pathway for microglial contribution to neurodegeneration.

Several neurotoxins that activate microglia provide models of dopaminergic cell injury. Following exposure to neurotoxins such as the mitochondrial complex I inhibitor MPP^+ (derived from 1-methyl-4-phenyl-1,2,3, 6-tetrahydropyridine, MPTP), and the dopamine analogue 6-hydroxydopamine (6-OHDA), microglia become activated, enhancing release of proinflammatory mediators and promoting neuronal injury and a self-propagating and self-sustaining neurodegenerative process.[6] Mutations that give rise to oxidized, nitrated, and aggregated or truncated proteins can also promote neuronal injury. *In vitro*, the altered protein species released from neurons can activate microglia and induce a proinflammatory M1 state. Secretion of ROS and cytokines is increased, and neuronal injury is aggravated, leading to cell death. Oligomers of misfolded, oxidized proteins join neurotoxins, LPS, and IFN-γ in being able to convert M2 anti-inflammatory microglia towards M1 proinflammatory microglia. At least *in vitro*, the oxidized proteins directly cause inflammation, promoting neuronal injury and self-propagating neurodegeneration. A similar *in vivo* process can certainly explain how α-synuclein could initiate a self-propagating pathology in models of PD as well as PD itself.[7] Comparable neurodegenerative pathways could be implicated in AD and ALS, with β-amyloid (Aβ) and tau representing inflammation-promoting oligomerized proteins in AD, and mSOD1, transactive response DNA binding protein-43 (TDP-43), and fused in sarcoma (FUS) protein representing inflammation-producing aggregated and/or oxidized proteins in ALS. Regardless of the initiating factor, overproduction of these oligomers has the potential to trigger a self-perpetuating inflammatory response that, if left unresolved, may contribute to the death of vulnerable neuronal populations.

T-LYMPHOCYTES: NEUROPROTECTION AND NEUROTOXICITY

The CNS has traditionally been considered immunologically privileged, meaning that inflammatory cells are excluded under normal circumstances owing to a relatively impenetrable blood–brain barrier. However, it is now clear that whereas peripheral immune access to the CNS is restricted and tightly controlled, the CNS is capable of dynamic immune and inflammatory responses to a variety of insults. Even in the presence of an intact blood–brain barrier, the CNS can elicit immune responses, primarily triggered by the brain's resident innate immune microglia. Furthermore, signaling from the CNS can result in site-specific entry of adaptive immune cells (T-lymphocytes). In the presence of injury, the CNS can promote the entry of adaptive immune cells, and restrain or amplify an ongoing neuroinflammatory process.

CD4$^+$ T-Lymphocytes

CD4$^+$ T-helper (Th) lymphocytes are a major lymphocyte population that plays an important role in governing acquired immune responses to diverse antigens. Many CD4$^+$ lymphocyte subsets have been described, but here the focus will be on four distinct subsets [Th1, Th2, Th17, and natural/inducible regulatory T-lymphocytes (nTregs and iTregs)], each of which has specialized functions to control immune responses.[8] In neurodegenerative diseases, these four T-lymphocyte subsets fall into two main classes: those that are neuroprotective, Th2 lymphocytes and Tregs; and those that are proinflammatory and neurotoxic, Th1 and Th17 lymphocytes (Fig. 25.2). Clearly, such a designation is oversimplified and does not fully explain all the relevant immune interactions, but it does provide a way to model the involvement of the immune system in neuroprotection or cytotoxicity. Whether T-lymphocytes polarize towards a neuroprotective or a cytotoxic phenotype can be influenced by the cytokine milieu, which in turn is dependent on monocytes and dendritic cells in the periphery and microglia in the CNS; this phenomenon is often referred to as "functional plasticity". Dendritic cells, monocytes, and microglia can act as antigen-presenting cells (APC) that present cognate antigens on their surface in the context of major histocompatibility complex class II (MHC-II). The presentation of a cognate antigen within an MHC class II molecule on an APC is initially recognized by a T-lymphocyte through a selective T-cell receptor (TCR). Interaction of costimulatory molecules on the APC with CD28 on the T-lymphocyte prompts expansion of the T-lymphocyte population, and secretion from APCs of cytokines, which direct the differentiation into effector T-lymphocyte (Teffs) subtypes. Thus, the specific cytokine milieu created by activated APCs is critical for specifying immune responses.

CD4$^+$ T-Lymphocyte Cytokines

The capacity of APC-derived cytokines as well as those released from activated CD4$^+$ T-lymphocytes, dendritic cells, and other cell types helps to polarize T-lymphocytes to amplify or suppress the immune response.

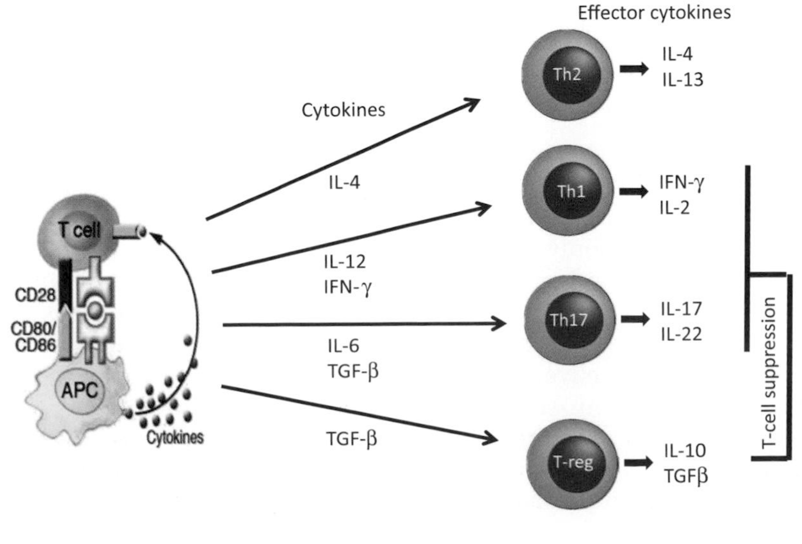

FIGURE 25.2 Differentiation of naïve T-lymphocytes into specific phenotypes by cytokines released from antigen-presenting cells. Each subtype of T-lymphocytes in turn releases effector cytokines which regulate immune responses. APC: antigen-presenting cell; IL: interleukin; IFN: interferon; Th: T-helper cell; T-reg: regulatory T cell; TGF: transforming growth factor.

The generation of Th1 lymphocytes depends on IFN-γ and IL-12 in synergy with IL-18. TGF-β secretion can polarize naïve cells towards a Treg phenotype. However, the presence of IL-6, in addition to TGF-β, polarizes the T-lymphocytes to a Th17 phenotype, which is maintained by the presence of IL-23. The absence of IL-12 and IL-18 induces the production of the Th2 cytokine IL-4 by T-lymphocytes, which acts in an autocrine fashion to polarize committed Th2 lymphocytes. Th2 lymphocytes release IL-4, IL-5, and IL-13. Tregs release IL-10, TGF-β, and IL-4. Th1 lymphocytes release IL-2 and IFN-γ; Th17 lymphocytes release IL-17, IL-22, and IL-26. The function of Tregs $CD4^+$/CD25high/FoxP3 lymphocytes is most pertinent for the discussion of neuroprotection since these inducible Tregs suppress the clonal expansion of the proinflammatory and cytotoxicity-promoting Th1 and Th17 lymphocytes. However, Tregs are not end-stage differentiated and the downregulation of FoxP3 by various cytokines (e.g. TNF-α) can abrogate the suppressive action of Tregs and permit the expansion of proinflammatory Teffs and promote neuronal injury.

Studies of facial nerve denervation in rodents provide cogent evidence for the neuroprotective role of both the innate and adaptive immune cells. Within hours of unilateral axotomy, microglia accumulate around the cell bodies of facial motor neurons. Within several days, T -lymphocytes are recruited to the same motor neuron somas in a site-specific manner. Mice that lack functional T and B lymphocytes have significantly reduced facial motor neuron survival, while reconstitution of these mice with wild-type splenocytes restores facial motor neuron survival to wild-type levels. Transplanting $CD4^+$ T-lymphocytes can rescue the axotomized facial motor neurons.[9] The overall effect is neuroprotective, with functional restoration of the facial motor neuron, its axon, and its synaptic connections. Although the mechanism of Th2-mediated neuroprotection in this axotomy model has not been clearly delineated, the ability of Th2 lymphocytes to release IL-4 and promote an M2 microglial-mediated repair and regenerative phenotype may well explain the beneficial effects.

Studies in models of ALS suggest an active dialogue between innate immune microglia and adaptive immune T-lymphocytes. Th2 lymphocytes and Tregs secrete neurotrophic factors, which directly protect motor neurons. IL-4, IL-10, and TGF-β secreted by Th2 lymphocytes and/or Tregs help to maintain M2 microglia, and M2 microglia in turn signal Th2 lymphocytes and Tregs. Tregs have been shown to differentiate macrophages directly toward the M2 state. M2 microglia can induce suppressive Tregs, suggesting that in the early stages of neurodegenerative disorders, neuroinflammation primarily mediated by M2 microglia and neuroprotective Tregs/Th2 lymphocytes may play a prominent role in promoting neuroprotection and repair, while in later more chronic stages the immune reactivity of proinflammatory M1 microglia and Th1 lymphocytes may fan the flames of neuronal cell injury and cell death.[2,10]

PARKINSON DISEASE

PD is a chronic neurodegenerative disease that presents clinically with tremor, rigidity, and bradykinesia, and pathologically with compromise of dopamine as well as non-dopamine pathways, with significant degeneration of the dopamine nigrostriatal pathway (see Chapter 19). The earliest symptoms of PD are loss of olfaction, the presence of marked constipation, and rapid eye movement behavior disorder. The motor manifestations of PD have been attributed to the loss of nigrostriatal dopamine neurons, especially those projecting to the putamen. The pathological hallmark of PD is the Lewy body, an α-synuclein-containing intraneuronal cytoplasmic inclusion.[11] Since the report that a mutation in α-synuclein gave rise to familial parkinsonism, there has been an explosion of evidence documenting the central role of α-synuclein in the pathophysiology of both motor and non-motor signs as well as dopaminergic and non-dopaminergic systems in PD, prompting the delineation of PD as an α-synucleinopathy. The central role of α-synuclein has been reinforced by reports that increased copy number of the α-synuclein gene caused familial parkinsonism, and increased α-synuclein expression in transgenic mice mimicked parkinsonism. In PD, the extent of dopamine terminal loss in the striatum appears to be more pronounced than the loss of substantia nigra (SN) dopamine neurons, suggesting that alterations of dopamine terminals and a "dying-back" process may represent one of the earliest manifestations of disease. Although the cause of such dying back is unclear, it is of considerable interest that oxidized, nitrated, and aggregated oligomers of α-synuclein are located at the presynaptic terminal and could alter synaptic vesicle trafficking.

Clinical Spectrum of Parkinson Disease

PD is predominantly a sporadic disease, but at least 5% of cases are familial, presenting with mutations in α-synuclein, Parkin (an E_3 ubiquitin ligase), and more commonly LRRK2 (leucine-rich repeat kinase), as well as other relatively uncommon mutations (UCHL1, PINK1, DJ-1). Familial cases have helped to confirm the molecular pathways that contribute to SN injury, just as familial cases have helped to define pertinent metabolic pathways in AD and ALS. The pathological progression of PD, which appears to ascend from the lower brainstem to the midbrain, has been linked by Braak staging to initial involvement of neurons with highest messenger

RNA (mRNA) expression of α-synuclein followed by compromise of neurons with lower expression of the protein.[12] Olfactory tubercle and dorsal motor nucleus of the vagus express high levels, followed by the locus ceruleus, amygdala, and SN pars compacta. This stereotypic pattern of pathology in PD has been attributed to a prion-like spread of misfolded α-synuclein from neuron to neuron (see Chapter 23).[13] *In vitro*, synthetic α-synuclein preformed fibrils seed the aggregation of soluble endogenous α-synuclein in primary wild-type mouse neuronal cultures. Intracerebral inoculation of these preformed α-synuclein fibrils *in vivo* accelerates neurological symptomatology and death of non-transgenic mice.[14] These experimental studies suggest that such prion-like spread could explain the development of Lewy bodies in embryonic mesencephalic neuronal grafts in PD patients many years after grafting.

Immune Responses in Parkinson Disease

Although high levels of α-synuclein have been proposed to initiate and propagate PD pathology, other factors must also be involved because patients with high α-synuclein levels without symptoms of PD have been reported, as well as patients with low α-synuclein levels and definite PD. These other factors prominently include the presence of immune–inflammatory alterations. A genome-wide association study (GWAS) documented a cogent association of PD with a highly polymorphic region of HLA-DR, previously associated with numerous inflammatory disorders such as rheumatoid arthritis.[7] HLA-DR immunoreactive microglia are prominent in the SN of PD patients, and are known to interact with T-lymphocytes. Degeneration of the ventral midbrain dopamine neurons in PD is accompanied by the presence of activated microglia with increased inducible nitric oxide synthase (iNOS), TNF-α, IL-1β, IFN-γ and nuclear translocation of nuclear factor-κB (NF-κB). In PD, activated microglia exhibit a classically activated M1 phenotype. Increased lipid peroxidation, as well as carbonyl- and nitrotyrosine-modified proteins, is also observed in CNS nigral tissue. Activated microglia surrounding dopamine neurons are noted in postmortem tissues of people who had developed symptoms and signs of parkinsonism following exposure to the neurotoxin MPTP. This discovery led to the development of mouse models of MPTP-induced parkinsonian, and provided a reproducible model to investigate the role of neuroinflammation and immunomodulation in the pathophysiology of dopamine neuronal injury. In the MPTP model, initiation of neuronal injury has been attributed to MPP^+-mediated blockade of mitochondrial complex I. In mouse striatal slices and isolated dopamine neurons, the blockade of complex I increases the production of ROS, and enhances oxidation and nitration of dopamine neuronal proteins including α-synuclein. However, MPP^+, the active constituent of MPTP, can also directly act on microglia, inducing the expression of iNOS mRNA and protein. MPP+ can also induce the mRNA expression of pre-IL-1β and TNF-α in microglia. Once activated, the proinflammatory microglial cytokine and chemokine secretions can directly induce dopamine neuron death. In the MPTP model, M1 microglia were associated topographically and temporally with dopamine neurodegeneration, with induction of microglial NOS and NADPH oxidase (NOX). Mice defective in microglial NOS or NOX secreted less NO or $O_2^{\bullet-}$ and exhibited less neuronal loss, with reduced levels of carbonyl- and nitrotyrosine-modified proteins despite MPTP treatment. Such data suggest that, in the MPTP model, activated microglia injure dopamine neurons, possibly by peroxynitrite generated from $O_2^{\bullet-}$ and NO. Monocyte chemoattractant protein-1 (CCL2) is upregulated in the striatum and the ventral midbrain. Astrocytes are the predominant source of CCL2 in the striatum and the SN, and dopamine neurons in the SN constitutively express CCL2. Deletion of the genes that encode CCL2 and its major receptor CCR2 does not affect MPTP-induced striatal dopamine depletion, thus mitigating a role for the recruitment of peripheral monocytes to the CNS in the pathogenesis of dopaminergic cell loss.

Microglial Activation by α-Synuclein

Are mitochondrial dysfunctions, increased ROS, and misfolded α-synuclein within dopamine neurons sufficient to cause dopamine neuron death in the MPTP model? More specifically, is the MPTP model an example of cell autonomous changes within dopamine neurons, or is the process non-cell autonomous, requiring the presence of microglia? Misfolded aggregates, including conformationally altered α-synuclein, are degraded poorly by chaperone-mediated autophagy and block the degradation of other substrates. Inhibition of autophagy further enhances misfolding of α-synuclein, leading to the selective degeneration of PD dopamine neurons. However, neuronal injury *per se* may not be sufficient to cause dopamine neuronal death. Recent data suggest that misfolded α-synuclein-induced cytotoxicity is non-cell autonomous and requires the participation of non-neuronal cells such as microglia. Normal monomeric and aggregated forms of α-synuclein are secreted from dopamine neurons, and in turn, the aggregated α-synuclein activated microglia upregulate NOX and ROS production (Fig. 25.3). Inhibitors of microglial production of NO and $O_2^{\bullet-}$ attenuate the *in vitro* dopamine neurotoxicity. Expression of the human α-synuclein protein in transgenic mice results in accumulation of insoluble α-synuclein aggregates

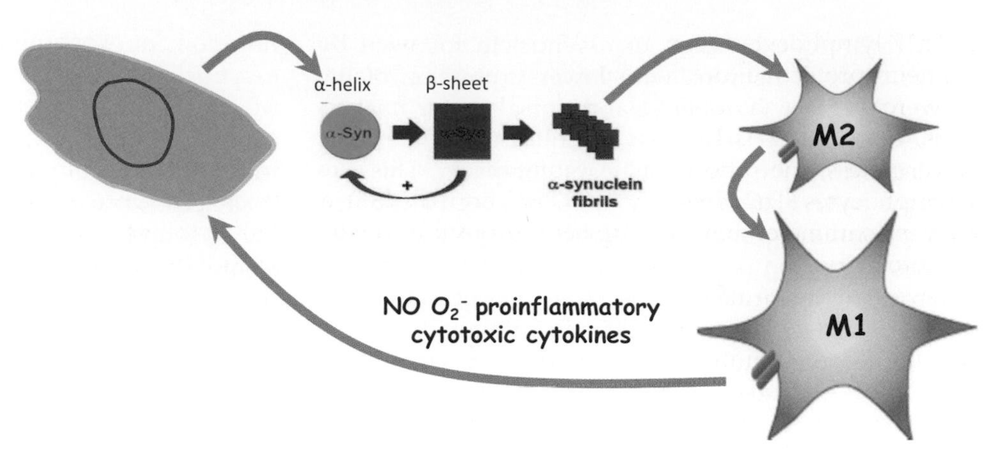

FIGURE 25.3 **Neuron–microglia signaling in Parkinson disease.** α-Synuclein (α-Syn) released from neurons forms β-sheets and fibrils that activate microglia to an M1 phenotype, releasing free radicals and cytokines that enhance neuronal injury and establish a self-propagating cascade. NO: nitric oxide; O_2^-: superoxide radical.

in nigral neurons, and makes dopamine neurons more vulnerable to LPS-induced microglia-mediated neurotoxicity. Stereotaxic injection of LPS into the SN of these transgenic mice causes not only microglial activation but also formation of cytoplasmic inclusions of aggregated oxidized/nitrated α-synuclein and subsequent death of dopaminergic nigral neurons. Thus, dopamine neurons exposed to MPP^+ have increased accumulation and secretion of misfolded and aggregated α-synuclein, but the MPP^+-mediated dopamine neuronal injury is not sufficient to cause neuronal death; microglia are necessary to deliver the *coup de grâce* by releasing free radicals and proinflammatory cytokines. The increased release of misfolded and aggregated α-synuclein from injured dopamine neurons thus has a self-propagating effect, causing further oxidation and further release of α-synuclein from dopamine neurons, and further amplifying dopamine neurodegeneration. In brief, α-synuclein released from injured dopaminergic neurons can activate microglia, enhancing the release of proinflammatory ROS and cytokines and promoting death of injured dopaminergic neurons and injuring otherwise unaffected neurons. This immune pathway involving microglia activated by misfolded α-synuclein represents an alternative to the neuron-to-neuron spread of the prion hypothesis of neurodegenerative diseases.[13]

T-Lymphocytes in MPTP Models

T-lymphocytes also participate in the ongoing inflammatory process. Both $CD4^+$ and $CD8^+$ T-lymphocytes are present in the SN of PD patients. Whether these infiltrating T-lymphocytes serve a destructive or a protective function in human PD cannot be readily inferred from end-stage autopsy tissue. Once again, experimental animal models have provided insights as to the potential role of immunomodulation in the pathogenesis of dopaminergic neuronal injury. In MPTP-treated mice, T-lymphocytes enhance cytotoxicity. In the absence of functioning T-lymphocytes, SCID, $Rag1^{-/-}$, or $TCR^{-/-}$ mice are relatively resistant to MPTP-induced SN dopaminergic cell degeneration.[15] This result suggests that the T-lymphocytes promote cytotoxicity, which would also be supported by the demonstration that in mice lacking $CD4^+$ T-lymphocytes, MPTP causes significantly less dopamine cell death. Cell death is also attenuated in mice transplanted with splenocytes with impaired ability to initiate Fas ligand-mediated cell killing. $CD8^+$ T-lymphocytes do not participate, since MPTP-induced dopamine cell death is not changed in mice that lack $CD8^+$ T-lymphocytes. Furthermore, transfer of whole T-lymphocyte populations from mice immunized with nitrated α-synuclein (i.e. misfolded) accelerates MPTP-induced dopamine cell loss.[16] Whether this cytotoxicity is promoted by Th1 or Th17 lymphocytes was not specifically defined in these experiments. Nevertheless, it is clear that MPTP-mediated dopamine cell killing requires microglia as well as CD4 T-lymphocytes.

T-lymphocytes also can be neuroprotective; transplantation of T-lymphocytes from mice immunized with the immunomodulatory drug glatiramer acetate (GA) attenuates the MPTP-induced SN cell loss. IL-4 provides neuroprotection from toxicity mediated by activated microglia,[17] and studies in ALS animal models indicate that Th2 lymphocytes as well as Tregs release IL-4 and modulate the proinflammatory M1 phenotype, suggesting that a similar mechanism may be relevant to PD.[18] In MS, GA has been documented to induce Th2-polarized immune responses that could mediate neuroprotection through the release of IL-4, suppressing the production and release of free radicals from activated microglia. Alternatively, GA could enhance Tregs and increase the production and release of the anti-inflammatory cytokines IL-4, IL-10, and TGF-β, and decrease microglial release of proinflammatory factors. GA may also modulate the differentiation and survival of Th17 lymphocytes through a mechanism involving signal transducer and activator of transcription (STAT3) and/or retinoic acid receptor-related orphan receptor-γ (RORγ), which are necessary for the differentiation and survival of Th17 lymphocytes. At present, it is not clear whether the neuroprotective functions of GA in the MPTP models were due to enhanced Th2 lymphocyte or Treg functions, or reduced Th17- or Th1-lymphocyte functions. Nevertheless, the ratio of neuroprotective to cytotoxic T-lymphocytes (Th2/Tregs to Th1/Th17) appears to be a relevant parameter in mediating neuroprotection versus neurotoxicity.

Several studies of the MPTP model suggest that increased Tregs mediate neuroprotection and Th17 lymphocytes promote neurotoxicity. Adoptive transfer of CD3-activated Tregs to MPTP-intoxicated mice provided greater than 90% protection of the nigrostriatal system. The response was dose dependent and paralleled modulation of microglial responses and upregulation of GDNF and TGF-β. Transplantation of Teffs provided no significant neuroprotective activities. Tregs were found to mediate neuroprotection, possibly through suppression of microglial responses to aggregated and nitrated α-synuclein. *In vitro*, Tregs suppressed microglial synthesis and the release of ROS that was induced by misfolded α-synuclein, whereas Teffs exacerbated microglial release of ROS. Thus, neuroprotection was achieved through modulation of microglial oxidative stress and inflammation. The neurotoxicity in this model was also mediated by Th17 lymphocytes, with Tregs dysfunction in nitrated-α-synuclein-induced T-lymphocytes.[19] Purified nTregs and iTregs induced by vasoactive intestinal peptide (VIP) reversed the nitrated α-synuclein nigrostriatal degeneration. Combinations of adoptively transferred nitrated α-synuclein and VIP immunocytes or nTregs administered to MPTP mice attenuated microglial inflammatory responses and led to robust nigrostriatal protection. Taken together, these results demonstrate that Tregs can suppress nitrated α-synuclein-induced neurodestructive immunity.

The functional state of $CD4^+$ T-lymphocytes is highly dependent on the diverse cytokine milieu, mediated by whether secretions are from anti-inflammatory M2 or proinflammatory M1 microglia. Both microglia and T-lymphocytes can serve protective as well as cytotoxic functions. The combination of M2 monocyte/microglia and Tregs/Th2 lymphocytes is neuroprotective and actively suppresses M1/Th1/Th17-mediated toxicity; the combination of M1 monocyte/microglia and Th1/Th17 lymphocytes is cytotoxic. The timing of protection and cytotoxicity remains to be determined. Is the milieu protective early in the course of experimental models of PD as well as in PD patients, and toxic later in disease, as noted in models of ALS? The answers are not yet available. Also critical for developing meaningful therapy is determining the specific signals that mediate the switch from a neuroprotective immune state to a cytotoxic immune state.

ALZHEIMER DISEASE

The etiology of sporadic AD in humans is unknown. In familial cases, mutations in amyloid precursor protein (APP) or components of its processing machinery (β-secretase and γ-secretase) result in increased levels of $A\beta_{1-40}$ and $A\beta_{1-42}$ peptides which comprise the senile plaque, a pathological hallmark of AD (see Chapter 21). Based on these familial cases as well as transgenic mice with mutations in APP, Aβ peptides have been implicated as initiating disease and causing the neuritic plaque pathology not only in familial disease, but also in sporadic disease.

Aβ accumulation is clearly caused by an imbalance between Aβ production and clearance. Metabolic labeling studies in older patients with sporadic AD permitted the measurements of $A\beta_{42}$ and $A\beta_{40}$ production and clearance rates in the CNS, and documented that clearance rates for both $A\beta_{42}$ and $A\beta_{40}$ are impaired in AD compared with controls.[20] No differences could be demonstrated in $A\beta_{40}$ or $A\beta_{42}$ production rates. Thus, impairment in Aβ clearance is one of the significant drivers of pathophysiology in sporadic AD.

Less than 1% of AD cases are familial forms, with autosomal dominant inheritance and onset of cognitive dysfunction before age 65.[21] Autosomal dominant familial AD can be attributed to mutations in one of three genes: APP and presenilins 1 and 2 (PS1 and PS2),[22] all of which increase the production of $A\beta_{42}$, the main component of senile plaques. Some of the mutations alter the ratio between $A\beta_{42}$ and the other major forms, for example, $A\beta_{40}$, without increasing $A\beta_{42}$ levels. This suggests that presenilin mutations can cause disease even if they lower the total amount of Aβ produced, and may point to the importance of other roles of presenilin or the greater significance of oligomers of $A\beta_{42}$ rather than the plaque content of $A\beta_{42}$. Most cases of AD are sporadic. Nevertheless, genetic differences may act as risk factors, with the most common being the inheritance of the ε_4 allele of the apolipoprotein E (*APOE* ε_4) gene.[23] Between 40 and 80% of people with AD possess at least one *APOE*ε_4 allele. The *APOE*ε_4 allele increases the risk of the disease by three times in heterozygotes and by 15 times in homozygotes.

Microglia in Alzheimer Disease: Role of β-Amyloid

The deposition of the Aβ peptide promotes cerebral neuroinflammation by activating microglia, which display an activated phenotype surrounding plaques. In human AD and mice transgenic for mutant APP, microglia are found in large numbers around neuritic amyloid plaques, but not diffuse plaques, suggesting that they could play a role in the conversion of diffuse to neuritic senile plaques but not in the origin of diffuse plaques. In autopsy specimens, the highest mean density of microglia is associated with neuritic plaques rather than with diffuse plaques. Microglia are more prevalent in diffuse plaques in patients who also have neuritic plaques than in individuals who only have diffuse plaques, suggesting that activated microglia may promote conversion of diffuse to neuritic plaques.

Neuritic plaques are likely to represent the gravestones of Aβ pathology, and the presence of activated microglia surrounding such plaques may enhance phagocytosis and lower Aβ plaque burden, preventing rather than exacerbating Aβ neuronal injury. The role of microglia surrounding plaques is presumably to clear the pathological deposits of Aβ through phagocytosis and degradation, and limiting microglial accumulation and phagocytosis increases Aβ deposition. As AD progresses, microglia adopt a chronically activated phenotype, secreting proinflammatory cytokines including IL-1β, which can actually impair microglial clearance functions. Thus, increased secretion of proinflammatory cytokines can increase Aβ deposition by impairing clearance by microglia. Clearance is also impaired by APOEε_4, which limits transport from the brain parenchyma into the peripheral circulation.[24]

Clearly, extrapolating from end-stage tissue obtained at autopsy has limitations and cannot define the dynamic process promoting disease pathology in AD patients or in any other neurodegenerative disease. Nevertheless, increasing evidence from experimental models suggests that an inflamed CNS environment may limit microglial phagocytic functions and thereby enhance plaque deposition rather than plaque removal. Consequently, the microenvironment of the brain can influence whether microglia perform beneficial or deleterious functions in pathophysiological states.

Microglia and TREM2 in Alzheimer Disease

Recent reports substantiate the role of microglia in AD; triggering receptor expressed on myeloid cells-2 (TREM2) receptor appears to function as a gateway for controlling microglial responses. TREM2 controls two signaling pathways which regulate the reactive phenotype.[25] The first of these pathways regulates phagocytosis. Increased expression of TREM2 on microglia is coupled to enhanced phagocytic pathways by potentially removing cell debris and Aβ in AD and to increased alternatively activated M2 protective microglia. The other signaling pathway suppresses inflammatory reactivity and cytokine production and secretion. TREM2 control of constitutive cytokine signaling may promote survival by prompting the secretion of TNF-α at levels that potentiate survival and repair pathways through the TNF receptor-2 (TNFR2). The expression of TREM2 rises in parallel with cortical levels of Aβ; compromised TREM2 function is likely to have ramifications for the clearance of cell debris and possibly the removal of Aβ in AD. At the same time, abrogation of the TREM2 suppression of cytokine secretion, potentially effected by TREM2 variants, may fuel inflammatory cascades, leading to a systemic inflammatory response and neuronal death. In an animal model of MS (see Chapter 30), blocking of TREM2 exacerbates disease, whereas boosting TREM2 signaling ameliorates disease. These effects of TREM2 suggest that the immune system is clearly involved in AD.

Despite their limitations, experimental models of AD do provide meaningful insights. Deposition of Aβ peptide drives cerebral neuroinflammation by activating microglia, and prolonged exposure to Aβ leads to persistent activation of microglia in these models. Studies in the APP/PS1 AD transgenic mouse model have documented increased production of IL-12 and IL-23 subunit p40 by microglia. This finding is of considerable interest since the concentration of p40 is increased in the cerebrospinal fluid of AD patients. Genetic ablation of the IL-12/IL-23 signaling molecule p40 in APP/PS1 mice results in a decreased cerebral Aβ load, and injections of p40 antibodies into the cerebral ventricles significantly reduces the concentration of soluble Aβ and reverses cognitive deficits in aged APP/PS1 mice. The microglia-mediated increases in IL-12 and IL-23 may have significant effects on the T-lymphocyte population; IL-12 can promote the differentiation of proinflammatory Th1 lymphocytes, while IL-23 facilitates conversion of Tregs into proinflammatory Th17 lymphocytes, thereby downregulating the Teff suppressive functions of Tregs and enhancing the overall proinflammatory milieu.

Microglial Inflammasomes in Alzheimer Disease

The pathway of microglial activation by Aβ appears to require both interaction at the cell surface through scavenger receptors or CD14/toll-like receptors (TLRs) and intracellular activation of the NOD-like receptor family, pyrin domain containing-3 (NLRP3) inflammasome, which can sense danger signals, such as inflammatory crystals and aggregated proteins, including Aβ. IL-1β is produced as a biologically inactive proform and requires caspase-1 for activation and secretion. Caspase-1 activity is controlled by inflammasomes, and was found to be increased in brain lysates from AD patients compared with controls, consistent with chronic inflammasome activation. Increased caspase-1 processing was also noted in aged APP/PS1 transgenic mice, which express a human/mouse chimeric APP and human presenilin-1, each carrying mutations associated with familial AD, leading to chronic deposition of Aβ, neuroinflammation, and cognitive impairment. In an elegant series of experiments, inflammasome NLRP3 knockout mice ($Nlrp3^{-/-}$) were crossed into APP/PS1 mice to assess the contribution of the NLRP3 inflammasome to the pathogenesis of AD.[26] In these APP/PS1/$Nlrp3^{-/-}$ mice, caspase-1 cleavage was absent, brain IL-1β amounts were reduced to wild-type levels, and the severe deficits in spatial memory of aged APP/PS1 mice were corrected. In addition, Aβ plaque volume was reduced, suggesting that phagocytosis was increased in APP/PS1/$Nlrp3^{-/-}$ mice. In addition to phagocytosis, microglia contribute to Aβ clearance through proteolytic enzymes, such as insulin-degrading enzyme (IDE), and IDE was increased in brain homogenates of APP/PS1/$Nlrp3^{-/-}$ mice. These data suggest that NLRP3 activation negatively affects the microglial clearance function in AD. Activation of NLRP3 in microglia is also associated with increased NOS, one of the hallmarks of the classically activated M1 microglia, while expression of markers of alternatively activated M2 microglia, such as FIZZ1, arginase, and IL-4, were significantly increased in APP/PS1/$Nlrp3^{-/-}$ mice. In contrast, cerebral nitric oxide synthase-2 (NOS2), a hallmark of M1 microglia, was reduced in inflammasome-deficient APP/PS1/$Nlrp3^{-/-}$ mice. NLRP3 deficiency results in skewing of activated microglial cells towards an M2 activated state. This M2 phenotype is also characterized by increased Aβ clearance and enhanced tissue remodeling. In AD, the upregulation of NOS2 results in tyrosine nitration of several proteins, including Aβ, thereby accelerating its aggregation and seeding of new plaques; APP/PS1/$Nlrp3^{-/-}$ mice had less nitrated Aβ and a reduced average plaque size as well as fewer nitrated plaque cores. Thus, the microglial NLRP3 inflammasome plays a key role in Aβ-mediated inflammatory responses; Aβ-induced activation of the NLRP3 inflammasome may enhance AD progression by promoting a harmful chronic inflammatory tissue, resulting in synaptic dysfunction, cognitive impairment, and restricted microglial clearance functions.

T-Lymphocytes in Alzheimer Disease

Although there has been considerable clarification of the diverse roles of microglia in Aβ-mediated inflammatory responses, a role for adaptive immune T-lymphocytes, and more specifically Aβ reactivity of $CD4^+$ and $CD8^+$ T-lymphocytes, remains unclear, largely owing to the conflicting data from many previous investigations. A recent study of $CD4^+CD28^+$ T-lymphocyte populations in the peripheral blood of AD and control patients showed that soluble Aβ alone was unable to stimulate significant or specific proliferation of these T-lymphocyte populations. However, exposure of *in vitro* prestimulated lymphocytes to soluble Aβ peptides significantly enhanced the proliferative response of $CD4^+CD28^+$ T-lymphocytes in both AD and aged control populations, leading to increased secretion of TNF-α, IL-10, and IL-6. Therefore, no evidence could be provided for specific reactivity to Aβ epitope(s) in AD compared with controls. However, it would be a mistake to assume that there is no specific adaptive immune reactivity to Aβ; a more reasonable conclusion, as suggested in the discussions on PD and ALS in this chapter, is that investigations are needed into subpopulations of T-lymphocytes (e.g. Tregs, Th2, Th1, Th17) during various stages of disease and in patients with differing rates of progression to definitively support or refute a role for an adaptive immune response.

Tau in Alzheimer Disease

Tau is another important potential contributor to the neurodegenerative and neuroinflammatory pathogenesis of AD. The normal function of tau is to promote tubulin assembly into microtubules and stabilize microtubules. Hyperphosphorylation of the tau protein results in the self-assembly of tangles of paired helical filaments and straight filaments, and when misfolded, otherwise soluble tau can form extremely insoluble aggregates. In AD, altered processing of tau results in hyperphosphorylation, aggregation and proteolytic cleavage of tau.

A key question is: What causes these alterations in tau? More specifically, does neuroinflammation play a role in promoting hyperphosphorylation, truncation, and aggregation of tau? Although these queries cannot be answered definitively in human AD, compelling evidence from transgenic mice suggests that neuroinflammation is present before evidence of significant modifications of tau, and the release of proinflammatory cytokines from activated microglia appears to initiate pathological modifications of neuronal tau. In primary cortical neurons, microglia activated by Aβ have increased IL-1β production and release, activated neuronal p38-MAPK, and increased tau phosphorylation (Fig. 25.4). *In vitro*, both NO and IL-6 can promote tau phosphorylation inside neuronal cells. Furthermore, injection of LPS into the brains of transgenic mice expressing human mutant tau P301L or a 3× transgenic (Tg)-AD model (PS1M146V + APPKM670/671NL + tauP301L) induced significant microglial activation and increased tau phosphorylation at Ser199/202 and Ser396; the increased phospho-tau levels were dependent on TLR-4 and IL-1β signaling. Tau phosphorylation was further enhanced in mice lacking the microglia-specific CX3CR1.[5] Gene delivery of the proinflammatory cytokine TNF-α into the brains of 3×Tg-AD mice induced microglial activation and enhanced intracellular levels of hyperphosphorylated tau. These results strongly suggest that microglia are activated before the appearance of misfolded truncated tau and release cytokines that induce significant tau phosphorylation.

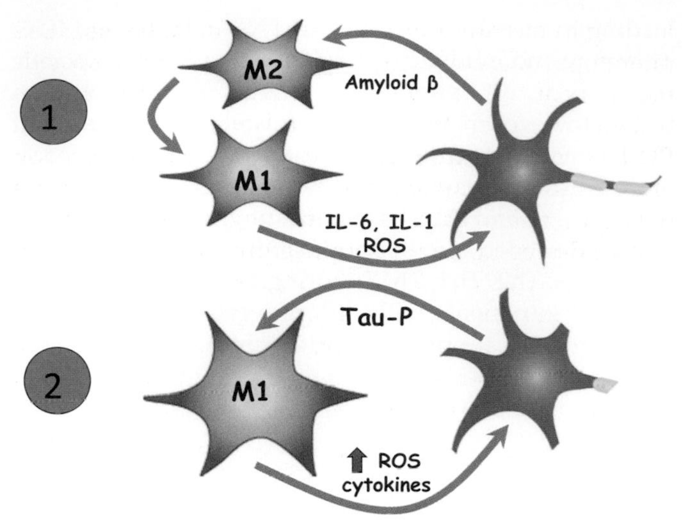

FIGURE 25.4 **Neuron–microglia signaling in Alzheimer disease.** Amyloid-β released from neurons promotes the conversion of microglia from an M2 to an M1 phenotype (1) .The M1 microglia release interleukin-6 (IL-6) and reactive oxygen species (ROS) that cause neuronal injury and foster the formation of phosphorylated and truncated tau, which is released from injured neurons and further promotes microglia-mediated neurotoxicity (2).

Misfolded truncated tau can, in turn, enhance neuroinflammatory responses. Under normal circumstances, tau is an intracellular cytoplasmic protein, but in AD misfolded, truncated, and phosphorylated tau can accumulate in neurofibrillary tangles as well as in the extracellular space and in the cerebrospinal fluid. Although this accumulation has been reported to result from leakage from dying and dead neurons, it also can result from extrusion from viable neurons in exosomes.[27] Thus, misfolded and truncated tau is available in the extracellular space to act as a potent inflammatory mediator. *In vitro*, purified recombinant truncated tau transformed microglia into reactive M1 phenotypes; mRNA levels of three MAPKs (JNK, ERK1, p38β) and transcription factors AP-1 and NF-κB were increased, resulting in enhanced mRNA expression of IL-1β, IL-6, TNF-α, and NO, and release of NO, proinflammatory cytokines (IL-1β, IL-6, TNF-α), and tissue inhibitor of metalloproteinase-1. In transgenic rats, expression of non-mutated human truncated tau (151–391, 4R) derived from sporadic AD patients induced activation of microglia and astrocytes, and neurofibrillary degeneration. In brief, the Aβ-promoted release of proinflammatory cytokines from activated microglia promotes alterations in the functional and structural properties of intraneuronal tau, and the structurally altered misfolded and truncated tau released from neurons further enhances microglial activation, thereby promoting a progressive self-propagating neuroinflammatory and neurodegenerative process.

It is not clear what initiates the increased Aβ oligomerization and subsequent microglial activation in AD. However, positron emission tomography (PET) studies with Aβ ligands suggest that amyloid plaques may occur at an early stage and possibly before either the appearance of misfolded phosphorylated tau or evidence of cognitive dysfunction.[28] This sequence of Aβ upregulation before misfolded tau would suggest that increased Aβ might activate microglia, enhancing release of proinflammatory cytokines, which in turn would downregulate protective M2 phagocytic function and induce significant neuronal tau phosphorylation and truncation. The released misfolded and truncated tau could then further enhance the proinflammatory microglial activation. Thus, both Aβ and tau promote and accelerate a self-propagating neurodegenerative process.

AMYOTROPHIC LATERAL SCLEROSIS

ALS, also known as Lou Gehrig disease, is a rapidly progressive disease of upper and lower motor neurons, resulting in atrophy and weakness of arms and legs, spasticity, dysarthria and dysphagia, and compromised breathing leading to death, typically within 4–6 years (see Chapter 18). Over 90% of cases have no family history of ALS, and are considered to be sporadic, while 10% of cases have a clear positive family history of disease. Mutations in SOD1 were long considered the most common cause of familial amyotrophic lateral sclerosis (fALS), and led to the development of mSOD1 transgenic animal models and a dramatic increase in our understanding of the pathogenesis of motor neuron disease. Mutations in SOD1 (now numbering more than 150) account for 20% of fALS cases. An exciting discovery from two different laboratories[29,30] indicates that the most common cause of fALS is due to increased hexanucleotide repeats in a non-coding region of the *C9orf72* gene, accounting for 40% of fALS cases and 7–10% of sporadic amyotrophic lateral sclerosis (sALS).[31] Although it is clear that the mutations in SOD1 cause disease by a toxic gain-of-function mechanism, the mechanism of C90rf72-mediated disease is not resolved; suggestions include haploinsufficency, toxic gain of function, a novel dipeptide repeat mechanism, or sequestration of vital proteins by RNA with multiple repeats analogous to the toxic mechanisms of RNA in myotoxic dystrophy.

Insights from the mSOD1 Transgenic Mouse

As in the case in virtually all other neurodegenerative disease, despite tremendous strides in basic science investigations, ALS therapy is at best symptomatic, and no medication is available to halt the inexorable progression of disease. Most advances in our understanding of the pathogenesis of motor neuron disease have come from investigations of the transgenic mSOD1 animal models. In these models, the earliest changes occur at the neuromuscular junction and appear before any loss of the motor neuron cell bodies can be detected in the spinal cord.[32] This suggests that motor neuron pathology may begin at the distal axon and proceed in a dying-back process, similar to that discussed above with respect to PD. As in patients with PD and AD, neuroinflammation is a prominent pathological feature of both fALS and sALS. The immune pathology is characterized by activated microglia and infiltrating T-lymphocytes at sites of spinal cord motor neuron injury.[33] In addition, increased monocytes/macrophages, dendritic cells, and CCL2 are present in ALS spinal cord tissue. Increased CCL2 is known to enhance the trafficking of peripheral cells such as monocytes and T-lymphocytes into the CNS. Depending on their phenotype and activation status, T-lymphocytes may cross-talk with neurons and microglia, and either damage or protect neurons from stressful stimuli. However, as in PD and AD patients, the presence of activated microglia or T-lymphocytes, or both, cannot define their neuroprotective or cytotoxic functions.

Microglia in Models of Amyotrophic Lateral Sclerosis

Transgenic models of ALS have provided clues to the important role of the immune system in motor neuron injury and cell death. T-lymphocytes, activated microglia, immunoglobulins, and dendritic cells are present in the spinal cords of mSOD1 mice.[34] Activated microglia are observed at an early age, even before signs of apparent weakness. During late stages of disease, immune activation of microglia towards an M1 phenotype is noted, with increased expression of iNOS, IL-1β, IL-6, and TNF-α. A protective function of innate immune microglia during the early stages of disease was demonstrated in the mSOD1 model employing PU.1 knockout mice, which lack an important hematopoietic transcription factor and require bone marrow transplantation (BMT) in the very early postnatal period to survive.[35] In this model, parenchymal microglia are absent at birth and the bone-marrow derived cells became the parenchymal "microglia"; the genotype of the transplanted "microglia" is determined by the genotype of the bone marrow donor. After transplantation into $mSOD1^{G93A}/PU.1^{-/-}$ mice, wild-type donor cells slowed motor neuron loss, prolonged disease duration, and increased survival time. On the other hand, $mSOD1^{G93A}$ donor cells accelerated motor neuron loss. To investigate possible neuroprotective microglial mechanisms, wild-type and $mSOD1^{G93A}$-expressing microglia were compared *in vitro*. Wild-type microglia were less activated and activatable than microglia expressing $mSOD1^{G93A}$. Furthermore, activated wild-type-derived microglia secreted significantly less NO as well as $O_2^{\bullet -}$, and induced less motor neuron injury compared with $mSOD1^{G93A}$-expressing microglia. In addition, wild-type microglia secreted more IGF-1 than $mSOD1^{G93A}$ microglia, thereby providing a more neuroprotective environment. The *in vivo* benefit observed after wild-type BMT of $mSOD1^{G93A}/PU.1^{-/-}$ mice is attributable to the decreased secretion of free radicals and increased secretion of IGF-1 from wild-type microglia. These data suggest that parenchymal wild-type microglia significantly contribute to neuroprotection when $mSOD1^{G93A}$ is expressed in motor neurons. Furthermore, the study confirmed the importance of microglia as a double-edged sword; microglia lacking mSOD1 enhanced motor neuron protection and slowed disease progression, whereas microglia with increased mSOD1 expression increased disease progression and motor neuron injury. mSOD1 microglia isolated from ALS mice at disease onset expressed higher levels of Ym1, CD163, and BDNF (markers of M2) mRNA and lower levels of NOX2 (a marker of M1) mRNA compared with mSOD1 microglia isolated from ALS mice at end-stage disease.[36] More importantly, when cocultured with motor neurons, the mSOD1 M2 microglia were neuroprotective and enhanced motor neuron survival compared with cocultured mSOD1 M1 microglia; end-stage mSOD1 M1 microglia were toxic to motor neurons. Thus, adult microglia isolated from ALS mice at disease onset have an M2 phenotype and protect motor neurons, whereas microglia isolated from end-stage disease ALS mice have an M1 phenotype and are neurotoxic. It is important to note that microglia can express a continuum of phenotypes from M2 to M1 and do not express just two phenotypes. Yet in the mSOD1 mouse, during the transition from the early slowly progressing stages to the later rapidly progressing stages, microglia appear to undergo a transformation exhibiting a preponderance of M1 markers.

The fact that mSOD1 expression in microglia alone cannot initiate disease, and expression is required in both motor neurons and microglia, suggests that proteins or other signals released from motor neurons may participate in the transformation of M2 to M1 microglia, and thereby contribute to neurotoxicity. This proposed paradigm is similar to suggested explanations of data from PD and AD models, where in the former misfolded and aggregated α-synuclein released from stressed dopamine neurons is postulated to activate microglia, and in the latter misfolded phosphorylated and truncated tau is postulated to enhance microglial activation, initiating a self-propagating amplification of neuronal cell injury. In the mSOD1 ALS model, the signal could be mSOD1 itself, because mSOD1 protein bound to chromogranin (a secretory granule protein) is secreted from neural cells and triggers microgliosis and neuronal death in mixed spinal cord cultures (Fig. 25.5). Although mSOD1 protein did not directly injure motor neurons, it activated microglia and triggered release of ROS, and induced

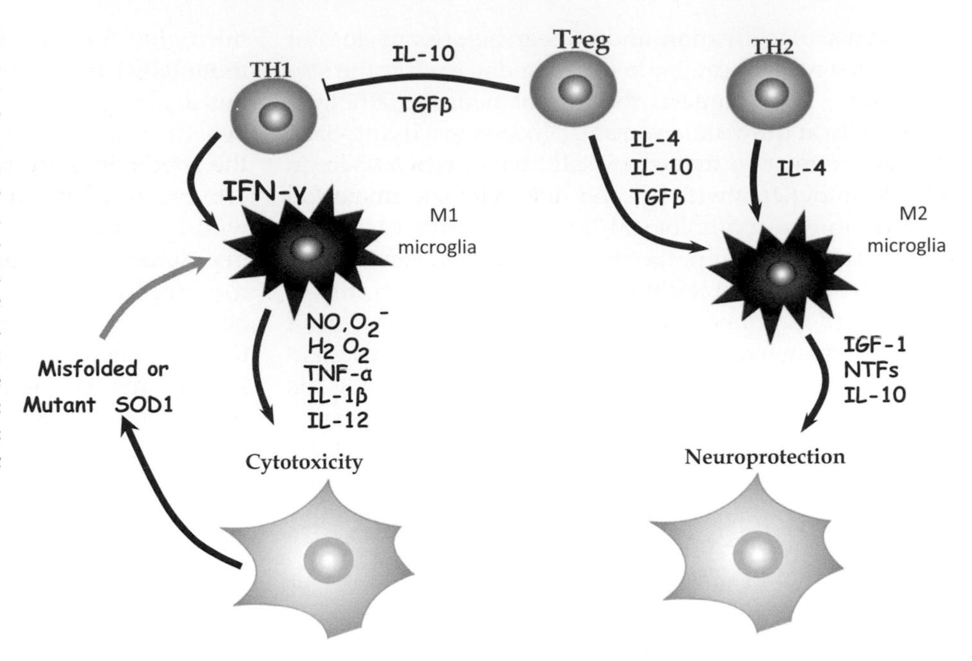

FIGURE 25.5 **Neuron–microglia–T-cell signaling in amyotrophic lateral sclerosis.** Motor neurons are maintained in a neuroprotective state by M2 microglia and T-helper and regulatory lymphocytes (Th2 and Treg). Misfolded superoxide dismutase-1 (SOD1) released from stressed and injured motor neurons converts microglia into a proinflammatory M1 phenotype, which is abetted by Th1 lymphocytes to release reactive oxygen species (ROS) and cytokines that aggravate and accelerate the ongoing motor neuron injury. IFN: interferon; IL: interleukin; TGF: transforming growth factor; IGF: insulin-like growth factor; TNF: tumor necrosis factor; NO: nitric oxide; O_2^-: superoxide radical; H_2O_2: hydrogen peroxide; NTF: neurotrophic factors.

motor neuron injury and cell death through a CD14/TLR pathway.[37] Of possible relevance to sALS, wild-type SOD1 acquires binding and toxic properties of mSOD1 through oxidative damage, and activates microglia and induces motor neuron death in spinal cord cultures.[38] Such an altered SOD1 species may also be present in spinal cord motor neurons of patients with sALS. An important trigger of the inflammatory reaction induced in microglia by misfolded mSOD1 is IL-1β. Cytoplasmic accumulation of mSOD1 activates the inflammasome NLRP3 and caspase 1 and proteolytically matures IL-1β. Knockout mice lacking caspase 1 or IL-1β had attenuated inflammatory pathology and extended lifespan.[39] A similar sensing of Aβ as a danger signal by the microglial NLRP3 inflammasome promoted the inflammatory pathway in models of AD.

Role of T-Lymphocytes in Models of Amyotrophic Lateral Sclerosis

T-lymphocytes also participate in the pathogenesis of motor neuron injury and death. In slowly progressing stages of disease in the mSOD1 model of ALS, endogenous T-lymphocytes, and more specifically Tregs, are increased, augmenting IL-4 expression and protective M2 microglia. In rapidly accelerating stages of disease, Tregs are decreased primarily through the downregulation of FoxP3 expression. During the stable disease phase, the total number of Tregs and their FoxP3 expression are increased in ALS mice compared with wild-type mice; during the rapidly progressing phase, the number of mSOD1 Tregs and FoxP3 expression decreases while the proliferation of mSOD1 Teffs increases. To determine whether T-lymphocytes modulate the microglial–motor neuron dialogue in mSOD1 mice, mSOD1 mice were bred with RAG2 knockout mice as well as with $CD4^+$ knockout mice that lack functional T-lymphocytes or $CD4^+$ T-lymphocytes.[34] The unexpected result was that motor neuron disease appeared at a much earlier age, suggesting that the absence of $CD4^+$ T-lymphocytes represents the absence of a neuroprotective T-lymphocyte population. This result is the opposite of what is found in the MPTP experiments in PD models, where the absence of $CD4^+$ T-lymphocytes significantly delays disease, suggesting that the $CD4^+$ T-lymphocytes are cytotoxic.[15] Furthermore, in the mSOD1/$CD4^{-/-}$ transgenic mice, the relatively slow phase of disease progression is eliminated. Proinflammatory and cytotoxic factors increase, anti-inflammatory and neurotrophic factors decrease, and survival decreases. Following BMT and restoration of $CD4^+$ T-lymphocytes, cytotoxicity is suppressed; the slowly progressive phase of disease is restored and survival prolonged. The $CD4^+$ neuroprotective T-lymphocytes promote the M2-activated microglial phenotype, enhancing release of anti-inflammatory and neurotrophic factors.

T-Regulatory Lymphocytes Mediate Neuroprotection in Models of Amyotrophic Lateral Sclerosis

To determine the specific population of cells responsible for the protective effects of BMT, wild-type $CD4^+$ T-lymphocytes were passively transferred into ALS mice lacking functional T-lymphocytes.[10] The transplants lengthened disease duration and prolonged survival. The passive transfer of endogenous Tregs from early disease stage donor mSOD1 mice into recipient ALS mice is even more beneficial in sustaining IL-4 levels and M2 microglia, and results in lengthened disease duration and prolonged survival; the stable disease phase is extended by 88% using mSOD1 Tregs. *In vitro* mSOD1 Tregs suppress the cytotoxic microglial factors NOX2 and iNOS through an IL-4-mediated mechanism, whereas Teffs are minimally effective; IL-4 inhibitory antibodies block the suppressive function of mSOD1 Tregs, and conditioned media from mSOD1 Tregs or the addition of IL-4 reduce microglial NOX2 expression.[18] The combination of IL-4, IL-10, and TGF-β is required to inhibit the proliferation of mSOD1 Teffs by mSOD1 Tregs isolated during the slow phase, while inhibition of mSOD1 Teffs by mSOD1 Tregs during the rapid phase, as well as wild-type Teffs, is not dependent on these factors. Thus, mSOD1 Tregs transplanted during the early slowly progressing phase of disease suppress microglial toxicity and mSOD1 Teff proliferation through different mechanisms; microglial activation is suppressed through IL-4 whereas mSOD1 Teffs are suppressed by IL-4, IL-10, and TGF-β. These data suggest that mSOD1 Tregs contribute to the slowly progressing phase in ALS mice through their interaction with microglia and sustain the M2 phenotype by secretion of IL-4. During the rapidly accelerating stages of disease Tregs/Th2 lymphocytes are markedly decreased and Th1 lymphocytes are the predominant phenotype. Therefore, during the early stages, neuroprotection is influenced by Tregs/Th2/M2 signaling, whereas during the rapidly progressing stages, neurodegeneration appears to be driven by Th1/M1 signaling.

T-Regulatory Cells in Patients with Amyotrophic Lateral Sclerosis

Clinical progression in the mSOD1 mouse is relatively predictable because of genetic homogeneity. However, this is far from the case in human ALS, where heterogeneity of disease is the rule. Can the mSOD1 transgenic mouse experiments provide insights into the heterogeneity of disease and the variable rates of progression in ALS patients? Onset of human sALS can be at any age from 20 to 88; symptoms and signs can start in arms, legs, speech, breathing, or even swallowing, and disease progression from onset to death can be as short as 6 months or as long as 40 years. Although the cloned mice undergo a relatively predictable disease course and differ from ALS patients, the mSOD1 mouse data permit the development of testable hypotheses. The mSOD1 mouse data would predict that rapid progression in ALS patients is associated with decreased Tregs and decreased FoxP3 per cell, while slow progression would be expected to be associated with increased Tregs and increased FoxP3 per cell. This prediction was confirmed. In patients' peripheral blood, both numbers of Tregs and their FoxP3 mRNA expression are reduced in rapidly progressing ALS patients and inversely correlate with disease progression; patients with low numbers of Tregs progress rapidly while patients with high FoxP3 progress slowly.[40] These results clearly support the predictions from transgenic mSOD1 disease; Tregs are decreased in rapidly progressing ALS patients and mSOD1 mice. In addition, the mRNA levels of FoxP3, TGF-β, IL-4, and Gata3 (a Th2 transcription factor) are reduced in rapidly progressing patients and inversely correlate with disease progression. No differences in IL-10, Tbx21 (a Th1 transcription factor), or IFN-γ expression are found between slowly and rapidly progressing patients. A 3.5 year prospective study with a second, larger cohort reveals that early reduced FoxP3 levels are indicative of progression rates at collection and predictive of future rapid progression and attenuated survival. After 3.5 years, 35% of ALS patients with low FoxP3 at the time of the initial blood draw had died, compared with only 13% of ALS patients with high FoxP3. Collectively, these data suggest that in ALS patients, as in the mSOD1 mice, Tregs and Th2 lymphocytes have a neuroprotective benefit and slow disease progression. Importantly, early reduced FoxP3 levels can be used to identify rapidly progressing patients. The cumulative mouse and human ALS data suggest that increasing the levels of Tregs in patients with ALS at early stages in the disease course may be of therapeutic value, and may slow the rate of disease progression and stabilize patients for longer periods.

CONCLUSION

In the neurodegenerative diseases PD, AD, and ALS, neurons do not die alone; neuronal injury is non-cell autonomous and depends on a well-orchestrated dialogue involving glia, T-lymphocytes, and neurons. The immune response is not merely a passive consequence of injury, but actively influences and significantly contributes to the balance of neuroprotection and neurotoxicity, and thereby mediates neuronal viability and neuronal injury. Regardless of whether mutations in a specific protein initiate a cascade of intraneuronal injury in familial disease, or whether the cause of the initial neuronal injurious process is undefined as in sporadic disease, a similar inflammatory response ensues. In both familial and sporadic cases of PD, AD, and ALS, differing temporal and mechanistic compromise of intraneuronal organelles may contribute to varying sites of onset as well as to varying rates of disease progression. The common theme is that immune system involvement may contribute to similar clinical expressions of disease, but variations in an individual patient's immune responsiveness may also contribute to the heterogeneity of clinical presentation and progression.

Although the molecular signals of intercellular communication have not been clearly defined and substantiated, data from experimental models of neurodegenerative disease suggest a general hypothetical and testable scheme of how neuroinflammation may enhance neuroprotection and neurotoxicity. In this scheme, communications from neurons inform microglia about the intraneuronal functional state; a neuronal signal for neuroprotection may activate M2 microglia to enhance release of anti-inflammatory neurotrophic factors and cytokines to sustain neuronal health, and intensify any needed repair and regenerative process. The specifically emitted neuronal signals for neuroprotection are not clearly defined, although CD200 and CX3CL1 are potential candidates for communicating a message of neuroprotection. Then the adaptive immune system becomes involved, with IL-4 released from Th2 lymphocytes and IL-4, IL-10, and TGF-β released from Tregs, which collectively can promote repair and regeneration. The Tregs with intact functional FoxP3 have the potential to enhance an anti-inflammatory microglia phenotype as well as suppress the proliferation of proinflammatory Teffs.

With deterioration in neuronal health, the severely stressed neurons release danger signals to convert microglia towards an M1 phenotype that secretes free radicals and proinflammatory cytokines and exacerbates neuronal injury, inducing a self-propagating cytotoxic cascade. What determines the threshold for this neuronal danger signaling is far from clear. Activated M1 microglia, in turn, release IL-1β and promote astroglial activation and a cytotoxic communal response. The adaptive immune system responds to this cytotoxic milieu with a downregulation of Treg protective functions, and a significant increase in Teffs and Th17 lymphocytes, the former secreting IFN-γ and the latter IL-17. It is unknown what neuronal signals communicate the intraneuronal danger to activated microglia *in vivo*. *In vitro*, it is clear that misfolded proteins or immunogenic peptide fragments in each of these disorders have the potential of mediating this response: α-synuclein in PD; Aβ and tau in AD; and mSOD1, TDP-43, and FUS in ALS. It is therefore possible that these same signals are operative *in vivo* when the threshold of intraneuronal sequestration or digestion is exceeded.

Confirming the specific molecular signals that mediate neuronal–microglial and T-lymphocyte–microglial interactions in PD, AD, and ALS is an exciting and fruitful area for future investigation. Are the misfolded proteins the neuronal–microglial signals or do peptide fragments of these proteins convey the danger signals? Do the microglia act as APCs and thereby activate immune pathways? The answers to these questions will help to identify novel therapeutic targets in these devastating neurodegenerative diseases. What is most critical is a greater understanding of how to limit populations of cells mediating cytotoxic immunomodulation and how to expand populations of cells mediating neuroprotective immunomodulation. Suppressing both neuroprotective and cytotoxic immune responses at the same time has not been effective clinically; suppressing both positive and negative reactivity does not change the balance of cytotoxicity and neuroprotection. To change the balance, the cytoxic arm must be downregulated and simultaneously the neuroprotective arm must be upregulated. Therapies that can achieve this goal offer the potential for sustaining quality of life and hope for the future.

References

1. Clement AM, Nguyen MD, Roberts EA, et al. Wild-type nonneuronal cells extend survival of SOD1 mutant motor neurons in ALS mice. *Science*. 2003;302:113–117.
2. Appel SH, Beers DR, Henkel JS. T cell–microglial dialogue in Parkinson's disease and amyotrophic lateral sclerosis: are we listening? *Trends Immunol*. 2010;31:7–17.
3. Raivich G. Like cops on the beat: the active role of resting microglia. *Trends Neurosci*. 2005;28:571–573.
4. Koning N, van Eijk M, Pouwels W, et al. Expression of the inhibitory CD200 receptor is associated with alternative macrophage activation. *J Innate Immun*. 2010;2:195–200.
5. Bhaskar K, Konerth M, Kokiko-Cochran ON, Cardona A, Ransohoff RM, Lamb BT. Regulation of tau pathology by the microglial fractalkine receptor. *Neuron*. 2010;68:19–31.
6. Liberatore GT, Jackson-Lewis V, Vukosavic S, et al. Inducible nitric oxide synthase stimulates dopaminergic neurodegeneration in the MPTP model of Parkinson disease. *Nat Med*. 1999;5:1403–1409.
7. Appel SH. Inflammation in Parkinson's disease: cause or consequence? *Mov Disord*. 2012;27:1075–1077.
8. Jiang S, Dong C. A complex issue on CD4(+) T-cell subsets. *Immunol Rev*. 2013;252:5–11.
9. Xin J, Mesnard NA, Beahrs T, et al. CD4+ T cell-mediated neuroprotection is independent of T cell-derived BDNF in a mouse facial nerve axotomy model. *Brain Behav Immun*. 2012;26:886–890.
10. Beers DR, Henkel JS, Zhao W, et al. Endogenous regulatory T-lymphocytes ameliorate amyotrophic lateral sclerosis in mice and correlate with disease progression in patients with amyotrophic lateral sclerosis. *Brain*. 2011;134:1293–1314.
11. Goedert M, Spillantini MG, Del Tredici K, Braak H. 100 years of Lewy pathology. *Nat Rev Neurol*. 2013;9:13–24.
12. Braak H, Del Tredici K, Rüb U, de Vos RA, Jansen Steur EN, Braak E. Staging of brain pathology related to sporadic Parkinson's disease. *Neurobiol Aging*. 2003;2003(24):197–211.
13. Olanow CW, Prusiner SB. Is Parkinson's disease a prion disorder? *Proc Natl Acad Sci U S A*. 2009;106:12571–12572.
14. Luk KC, Kehm V, Carroll J, et al. Pathological α-synuclein transmission initiates Parkinson-like neurodegeneration in nontransgenic mice. *Science*. 2012;338:949–953.
15. Brochard V, Combadière B, Prigent A, et al. Infiltration of CD4+ lymphocytes into the brain contributes to neurodegeneration in a mouse model of Parkinson disease. *J Clin Invest*. 2009;119:182–192.
16. Benner EJ, Banerjee R, Reynolds AD, et al. Nitrated alpha-synuclein immunity accelerates degeneration of nigral dopaminergic neurons. *PLoS ONE*. 2008;3:e1376.
17. Zhao W, Xie W, Xiao Q, Beers DR, Appel SH. Protective effects of an anti-inflammatory cytokine, interleukin-4, on motoneuron toxicity induced by activated microglia. *J Neurochem*. 2006;99:1176–1187.

18. Zhao W, Beers DR, Liao B, Henkel JS, Appel SH. Regulatory T-lymphocytes from ALS mice suppress microglia and effector T-lymphocytes through different cytokine-mediated mechanisms. *Neurobiol Dis*. 2012;48:418–428.
19. Reynolds AD, Stone DK, Hutter JA, Benner EJ, Mosley RL, Gendelman HE. Nitrated α-synuclein-induced alterations in microglial immunity are regulated by $CD4^{+}$ T cell subsets. *J Immunol*. 2009;182:4137–4149.
20. Mawuenyega KG, Sigurdson W, Ovod V, et al. Decreased clearance of CNS beta-amyloid in Alzheimer's disease. *Science*. 2010;330:1774.
21. Blennow K, de Leon MJ, Zetterberg H. Alzheimer's disease. *Lancet*. 2006;368:387–403.
22. Waring SC, Rosenberg RN. Genome-wide association studies in Alzheimer disease. *Arch Neurol*. 2008;65:329–334.
23. Strittmatter WJ. Apolipoprotein E: high-avidity binding to beta-amyloid and increased frequency of type 4 allele in late-onset familial Alzheimer disease. *Proc Natl Acad Sci U S A*. 1993;90:1977–1981.
24. Deane R, Sagare A, Hamm K, et al. apoE isoform-specific disruption of amyloid beta peptide clearance from mouse brain. *J Clin Invest*. 2008;118:4002–4013.
25. Jonsson T, Stefansson H, Steinberg S, et al. Variant of TREM2 associated with the risk of Alzheimer's disease. *N Engl J Med*. 2013;368:107–116.
26. Heneka MT, Kummer MP, Stutz A, et al. NLRP3 is activated in Alzheimer's disease and contributes to pathology in APP/PS1 mice. *Nature*. 2013;493:674–678.
27. Saman S, Kim W, Raya M, et al. Exosome-associated tau is secreted in tauopathy models and is selectively phosphorylated in cerebrospinal fluid in early Alzheimer disease. *J Biol Chem*. 2012;287:3842–3849.
28. Sperling RA, Karlawish J, Johnson KA. Preclinical Alzheimer disease – the challenges ahead. *Nat Rev Neurol*. 2013;9:54–58.
29. DeJesus-Hernandez M, Mackenzie IR, Boeve BF, et al. Expanded GGGGCC hexanucleotide repeat in noncoding region of C9ORF72 causes chromosome 9p-linked frontotemporal dementia and amyotrophic lateral sclerosis. *Neuron*. 2011;72:245–256.
30. Renton AE, Majounie E, Waite A, et al. A hexanucleotide repeat expansion in C9ORF72 is the cause of chromosome 9p21-linked ALS-FTD. *Neuron*. 2011;72:257–268.
31. Robberecht W, Philips T. The changing scene of amyotrophic lateral sclerosis. *Nat Rev Neurosci*. 2013;14:248–264.
32. Fischer LR, Culver DG, Tennant P, et al. Amyotrophic lateral sclerosis is a distal axonopathy: evidence in mice and man. *Exp Neurol*. 2004;185:232–240.
33. Henkel JS, Engelhardt JI, Siklós L, et al. Presence of dendritic cells, MCP-1, and activated microglia/macrophages in amyotrophic lateral sclerosis spinal cord tissue. *Ann Neurol*. 2004;55:221–235.
34. Beers DR, Henkel JS, Zhao W, Wang J, Appel SH. $CD4^{+}$ T cells support glial neuroprotection, slow disease progression, and modify glial morphology in an animal model of inherited ALS. *Proc Natl Acad Sci U S A*. 2008;105:15558–15563.
35. Beers DR, Henkel JS, Xiao Q, et al. Wild-type microglia extend survival in PU.1 knockout mice with familial amyotrophic lateral sclerosis. *Proc Natl Acad Sci U S A*. 2006;103:16021–16026.
36. Liao B, Zhao W, Beers DR, Henkel JS, Appel SH. Transformation from a neuroprotective to a neurotoxic microglial phenotype in a mouse model of ALS. *Exp Neurol*. 2012;237:147–152.
37. Zhao W, Beers DR, Henkel JS, et al. Extracellular mutant SOD1 induces microglial-mediated motoneuron injury. *Glia*. 2010;58:231–243.
38. Appel SH, Zhao W, Beers DR, Henkel JS. The microglial–motoneuron dialogue in ALS. *Acta Myol*. 2011;30:4–8.
39. Meissner F, Molawi K, Zychlinsky A. Mutant superoxide dismutase 1-induced IL-1beta accelerates ALS pathogenesis. *Proc Natl Acad Sci U S A*. 2010;107:13046–13050.
40. Henkel JS, Beers DR, Wen S, et al. Regulatory T-lymphocytes mediate amyotrophic lateral sclerosis progression and survival. *EMBO Mol Med*. 2013;5:64–79.

CHAPTER

26

Role of Inflammation in Psychiatric Disease

Charles L. Raison[*,†], *Graham W. Rook*[‡], *Andrew H. Miller*[§], *Tommy K. Begay*[†]

*Department of Psychiatry, College of Medicine, University of Arizona, Tucson, Arizona, USA; †Norton School of Family and Consumer Sciences, College of Agriculture and Life Sciences, University of Arizona, Tucson, Arizona, USA; ‡Centre for Clinical Microbiology, Department of Infection, University College London, London, UK; §Department of Psychiatry and Behavioral Sciences, Emory University School of Medicine, Atlanta, Georgia, USA

OUTLINE

INTRODUCTION

The best established findings linking mental illness with immunity may be summarized as follows: in aggregate, individuals diagnosed with a wide range of currently recognized psychiatric diseases show significant increases in peripheral inflammatory biomarkers and CNS markers of immune activation, while simultaneously evincing reduced effectiveness of immune elements important for protection against pathogens. (For reviews of material discussed in this chapter, see references 1–6; see also Chapter 25 for an introduction to inflammation and a discussion of neurological disorders.) But, as is often the case in psychiatry, even this overall pattern of findings is suggestive, not conclusive, and raises more questions than it answers. The goal in this chapter is to use the most important of these unanswered questions as a means of organizing

Neurobiology of Brain Disorders
http://dx.doi.org/10.1016/B978-0-12-398270-4.00026-4

the rapidly expanding field of psychoneuroimmunology into a coherent perspective (Table 26.1).

The most important question to be answered is whether associations between immune abnormalities and psychiatric disease reflect the fact that these abnormalities cause psychiatric illness, or are better explained by the fact that brain changes associated with these disease states secondarily affect immune functioning.[7] Or might it be the case that associations between immunity and mental illness result from no direct causative links whatsoever, but rather reflect the fact that conditions often found among people with mental illness, such as obesity, smoking, and poor sleep, also directly affect immune function? The importance of these issues to the field cannot be overstated. If the immune changes reported in psychiatric conditions are causative, then gaining an exact understanding of their nature will transform our understanding of disease pathogenesis and will point towards myriad novel immune-based treatment modalities. On the other hand, if these associations more often reflect the fact that altered CNS functioning secondarily affects the immune system, then scientific efforts would be better spent in trying to identify and reverse the CNS changes that impair immune functioning and its potential impact on physical health, rather than focusing on the immune system *per se*.

Increasing evidence supports each of these possibilities, highlighting the fact that associations between altered immune functioning and psychiatric disease cannot be subsumed under any simple causative paradigm. Behavioral symptoms relevant to psychiatric disease can be reliably produced when immune changes similar to those observed in association studies are induced in both animal and human model systems designed to explore the effect of immune alterations on brain function.[1] On the other hand, a rapidly expanding literature reports with high uniformity that brain-based states of mental functioning associated with psychiatric disease (e.g. stress, anxiety, despair, anger) affect CNS efferent pathways, such as the hypothalamic–pituitary–adrenal (HPA) axis and autonomic nervous system (ANS) in ways known to simultaneously activate an inflammatory response and suppress certain aspects of adaptive immune functioning, such as T-cell activity specific for antiviral and antineoplastic protection.[3] Finally, multiple lines of evidence suggest that much of the association between abnormal immune functioning and psychiatric illness is likely to be accounted for by the fact that people with mental illness engage in lifestyles and are burdened with demographic factors that are themselves associated with many of the immune changes seen in psychiatric disease.[4] However, before assuming that associations between psychiatric disease and altered immune function are potentially spurious, it is important to consider that many of the lifestyle and demographic factors associated with mental illness appear also to be risk factors for the development of these very same diseases.[8] This raises the intriguing possibility that altered immune functioning may be a causative mechanism by which these factors produce psychiatric disease. Viewed from this perspective, what once looked like confounders appear as another line of evidence in support of the notion that changes in immune functioning may contribute directly to the development of psychiatric disease.

TABLE 26.1 Immunity and Psychiatric Disease: Important Unanswered Questions

Do changes in immune functioning most often cause psychiatric disease, reflect the effects of brain changes associated with psychiatric disease, or result from factors associated with both psychiatric disease and these immune changes?
If alterations in immune functioning contribute to the pathogenesis of psychiatric disease, do the most relevant alterations occur in the periphery or in the CNS?
In the context of psychiatric illness, is it more often the case that peripheral immune changes secondarily drive alterations in CNS immune functioning, or vice versa?
Are there specific patterns of immune abnormality that preferentially promote the development of one psychiatric condition as opposed to another, or conversely are there specific changes in CNS functioning that preferentially produce psychiatric disease states associated with specific immune changes?
Are immune abnormalities best conceived of as trait or state markers for psychiatric disease?
To what degree are immune changes contributory to psychiatric disease states as a whole versus contributory to only the subset of individuals with any given disease that also show these changes?

This chapter addresses each of these three causative perspectives in turn before summarizing the current knowledge on the nature and relevance of immune abnormalities in specific conditions. In each of these sections, key unanswered questions, as enumerated in Table 26.1, will be highlighted. Before examining these key issues, a review of the immune system is presented, focused on points of particular relevance to brain–body interactions likely to contribute to the pathogenesis of psychiatric disease.

EVIDENCE THAT THE IMMUNE SYSTEM IS INVOLVED IN PSYCHIATRIC DISEASE PATHOGENESIS

If inflammatory processes are involved in the pathogenesis of mental illness, then activation of the inflammatory response might be expected to reproduce the symptoms that define the various psychiatric diseases associated with increased inflammation. Is this the case? As with many issues in psychiatry, the answer is yes and no. It is clear from animal and human studies that activation of the inflammatory response produces many symptoms of psychiatric disease, but not all. Inflammation produces many core symptoms of mood and anxiety disorders, but generally does not promote the types of psychotic and social/obsessive symptoms that define schizophrenia and autism, two disorders strongly associated in cross-sectional studies with increased inflammation. But before concluding that inflammatory processes do not contribute to disease pathogenesis in these conditions, we need to carefully consider what we mean by "cause". While inflammatory activation in adults does not routinely produce psychosis or autism, substantial animal data, as well as powerful circumstantial evidence in humans, suggest that increased inflammation very early in life, especially in the prenatal and perinatal periods, may promote changes in CNS development and function that significantly increase the risk of manifesting autism later in childhood and schizophrenia in adulthood.[6] Moreover, increasing evidence suggests that reversing immune changes associated with both autism and schizophrenia may promote symptomatic improvement,[9] again consistent with the notion that immune abnormalities – while perhaps not contributing as directly to disease expression as is the case in depression – play important, and incompletely understood, causative roles in these disorders. Finally, with the exception of certain genetic conditions, diseases always occur at the nexus of external forces and internal vulnerabilities. Circumstances that increase inflammation early in life increase the risk of developing schizophrenia or autism, and factors that increase inflammation later are associated with an increased risk of mood and anxiety disorders.

Behavioral Abnormalities Associated with Immune Activation

Multiple animal and human model systems have been developed to examine the effect of inflammatory activation on behavior. These models provide a remarkably consistent picture of the effect of inflammation on emotion, cognition, and behavior. Whether the stimulus is an infectious agent, isolated molecules from an infectious agent [e.g. lipopolysaccharide (LPS)] or cytokines produced by the body in response to infection, activation of the innate immune inflammatory response produces a motivated, stereotyped behavioral/physiological state that is known as sickness behavior, or sickness syndrome.[2] The core features of this syndrome are known to anyone who has had influenza, and include change in body temperature (typically fever), lethargy, fatigue, sleep and appetite disturbance, anhedonia (loss of interest and pleasure), social withdrawal, cognitive dulling, psychomotor retardation (moving slowly), and hyperalgesia (increased sensitivity to pain).

Towards the end of the twentieth century, it was recognized that these symptoms were similar to those observed in animals subjected to various psychological stressors and mirrored many symptoms seen in humans with major depressive disorder (MDD). For example, we think of fever as a paradigmatic sign of sickness, but are less often inclined to know that, as a group, medically healthy individuals with MDD also demonstrate increased body temperature in the mild febrile range. In the context of infection, it is widely accepted that sickness behavior enhances host survival. Fever has been shown in many studies to enhance pathogen-focused immune activity and to have direct antimicrobial effects. Following this line of reasoning, it has been suggested that depression (and the genes that promote it) may have been retained in humans precisely because, like sickness, it enhanced pathogen host defense in the face of the types of environmental adversity that promoted its development across human evolution (Fig. 26.1).[10]

Despite the similarities between sickness and depression, both animal and human studies strongly suggest that they are related but not identical. Again, anyone who has had a serious infection and who has struggled with clinical depression can attest to the fact that depression is characterized by a group of closely related negative emotions and cognitions that are not typically experienced in response to an acute infection. The validity of this perception is supported by research that employs treatment with the cytokine interferon-α (IFN-α) as a model for examining the effect of chronic inflammatory activation on behavior and its biological substrates.

Although an effective therapeutic strategy for some cancers and for chronic hepatitis C virus infection, IFN-α is notorious for inducing both sickness behavior and an array of depressive and anxiety symptoms. Indeed, while rates of depression vary widely between studies, taking the relevant literature as a whole, it appears that 20–50% of patients meet criteria for MDD during chronic IFN-α therapy, and the prevalence of individual depressive symptoms and subsyndromic depressive states is far higher.[1] Moreover, a comparison of symptoms in patients with IFN-α-induced depression versus medically healthy depressed patients revealed a large degree of overlap in both symptom expression and severity (with these groups being more symptomatically similar to each other than either was to people receiving IFN-α who did not develop depression). Further supporting the similarity between IFN-α-induced depression and depression in other populations is the fact that depression induced by IFN-α can be prevented and/or treated by conventional antidepressants. These findings highlight the importance of IFN-α treatment as a model system for understanding how chronic inflammation produces psychiatric disease. IFN-α also provides an invaluable perspective on how sickness and depression/anxiety are

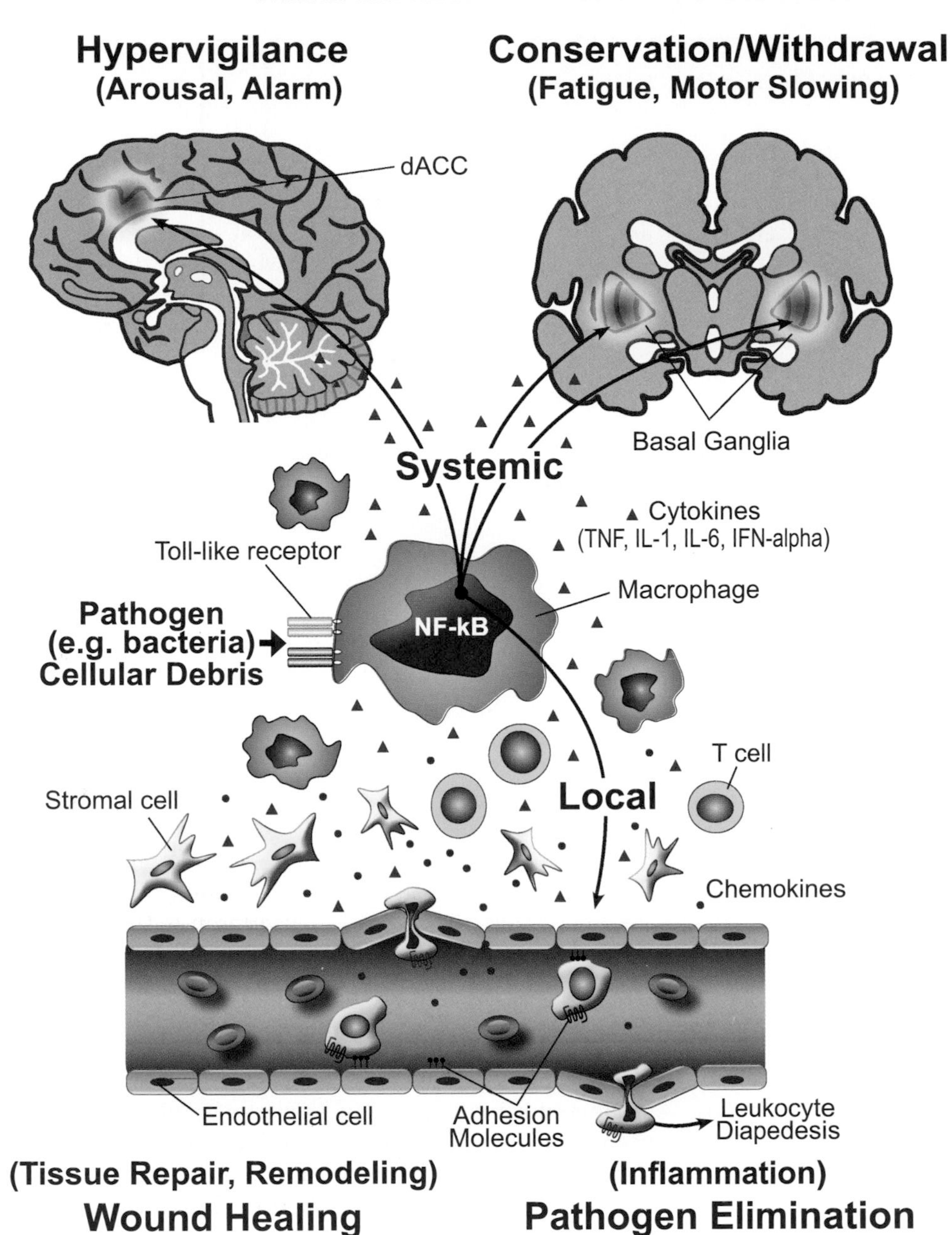

FIGURE 26.1 **Integrated suite of immunological and behavioral responses to infection and wounding that comprise pathogen host defense.** Upon encountering a pathogen or cellular debris from tissue damage, the body reacts with an orchestrated local and systemic response that recruits both immunological and nervous system elements. The response is initiated by interaction of pathogens and/or cellular debris with pattern recognition receptors such as toll-like receptors on relevant immune cells including macrophages that in turn are linked to inflammatory signaling pathways such as nuclear factor-κB (NF-κB), a lynchpin transcription factor in the host defense cascade. Release of cytokines [including tumor necrosis factor-α (TNF-α), interleukin-1 (IL-1), IL-6, and interferon-α (IFN-α)] and chemokines, as well as the induction of adhesion molecules, attracts and activates cells such as T cells at the site of infection/wounding, leading to the cardinal signs of inflammation (redness, heat, swelling, and pain) and ultimately promoting local pathogen elimination and wound healing. Cytokines and cells in the peripheral circulation mediate the systemic host response, which engages neural circuits in the brain that mediate hypervigilance [dorsal anterior cingulate cortex (dACC)] to avoid further wounding and pathogen exposure and conservation/withdrawal (basal ganglia), which promotes the shunting of energy resources to pathogen elimination and wound healing.

related, and this perspective closely replicates findings from animal models that use inflammatory stimuli. As in animal models, the acute administration of the inflammatory stimulus (in this case IFN-α) produces sickness symptoms in almost everyone. These symptoms emerge immediately following the first IFN-α injection and tend to resolve over the following few days, only to re-emerge following the next injection a week later. This pattern persists throughout treatment, although many patients find that sickness symptoms attenuate over time. Many sickness symptoms are identical, or closely related, to neurovegetative symptoms of depression (i.e. sleep and appetite disturbance), and so one might say that IFN-α immediately produces some depressive symptoms. Figure 26.2 describes findings from the use of IFN-α as a human model system.

What is striking is that the symptoms that most differentiate depression from sickness, such as sad and anxious mood, feelings of hopelessness and despair, and perceptions of diminished self-worth, only appear more gradually over the first few weeks of treatment. These differing time-courses have been taken as evidence that there are two syndromes caused by inflammation: one comprised of neurovegetative/sickness symptoms and a second that is more depression specific. Consistent with this notion, selective serotonin reuptake inhibitor (SSRI) antidepressants are more effective at preventing depression-specific (e.g. depressed mood, anxiety) than

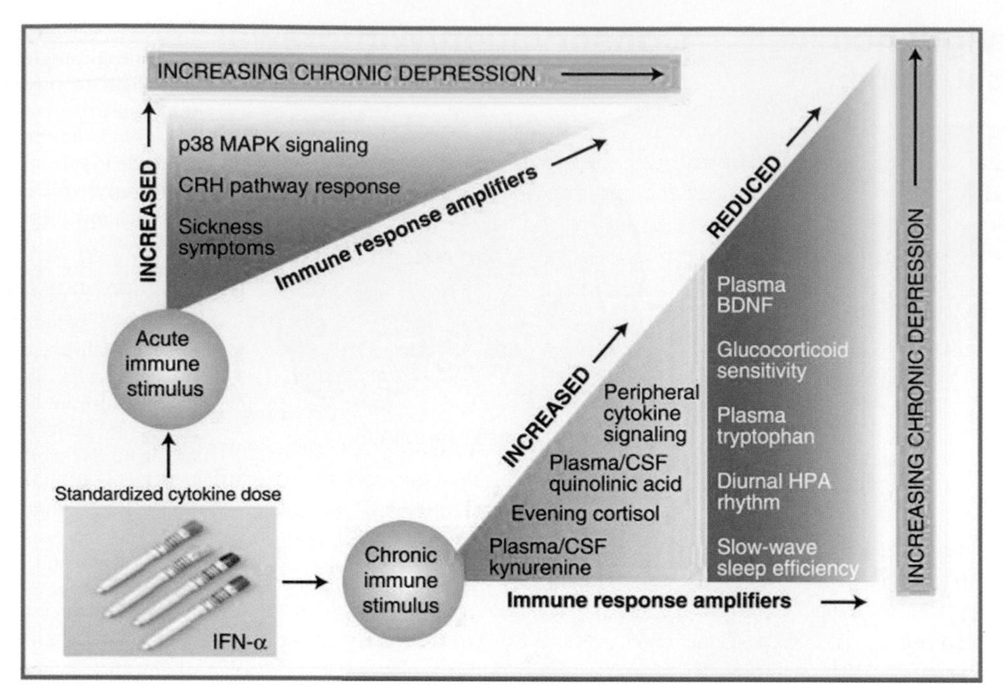

FIGURE 26.2 **Evidence supporting the importance of immune response amplifiers in the etiology of immune-based depression.** Studies using exposure to a constant dosage of inflammatory input have identified multiple pathways by which immune activation produces depressive symptoms. This figure illustrates patterns of functional vulnerability in such pathways, using treatment with interferon-α (IFN-α) as a model system for chronic cytokine exposure. Immune response amplifiers identified in response to an initial dose of IFN-α include increased activation of p38 mitogen-activated protein kinase (MAPK), an important intracellular proinflammatory signaling cascade, and increased response of corticotropin-releasing hormone (CRH) pathways. Although not a physiological pathway *per se*, increased sickness-type symptoms in response to a first dose of IFN-α predict the later development of full major depressive disorder (MDD). In terms of patterns of functional disability in response to chronic cytokine exposure, evidence suggest that people with any of the following changes are at significantly increased risk of developing significant depressive symptoms during IFN-α treatment: (1) increased peripheral proinflammatory [e.g. tumor necrosis factor-α (TNF-α) or interleukin-6 (IL-6)] cytokine concentrations; (2) flattening of the diurnal cortisol rhythm and increased evening cortisol; (3) increased peripheral and central nervous system concentrations of kynurenine and its metabolite quinolinic acid; and (4) disrupted sleep, reflected by reduced slow-wave sleep (SWS) and diminished sleep efficiency. Importantly, all of these changes have been observed in the context of idiopathic MDD in medically healthy individuals. CSF: cerebrospinal fluid; BDNF: brain-derived neurotrophic factor; HPA: hypothalamic–pituitary–adrenal axis.

sickness symptoms in the context of IFN-α treatment. However, it is also clear that sickness and depression are intimately related in the context of immune activation, over and above their phenomenological overlap.

Multiple converging lines of evidence from animal studies support the idea that sickness and depressive-type behavior are related but not identical. For example, immune stimuli that mimic either bacterial (i.e. LPS) or viral infection [i.e. polyinosinic:polycytidylic acid (poly I:C)] have been repeatedly shown to produce depressive-like behavior in standard model systems (e.g. forced swim test, sucrose preference, tail suspension) even at doses too low to produce overt sickness.[2] Similarly, doses of these inflammatory stimuli sufficient to produce acute sickness symptoms produce lingering depressive-like behavior long after sickness has resolved. This dissociation is most striking in rodents receiving bacille Calmette–Guérin (BCG), where depressive-like behavior remains apparent up to 3 weeks postinfection, long after the resolution of sickness. As in other rodent models, depressive-like behavior is associated with stimuli-induced increases in inflammatory cytokines, in this case IFN-γ and tumor necrosis factor-α (TNF-α). Conversely, knocking down the IFN-γ gene (i.e. use of IFN-$\gamma^{-/-}$ mice) or administration of the TNF-α antagonist etanercept ameliorates BCG-induced depressive-like behavior.

The notion that inflammatory processes are capable of initiating depression over and above their direct sickness-generating effects is supported by data from animal studies showing that sickness and depression-like effects of cytokine activation may be subserved by different physiological mechanisms. As with other inflammatory stimuli, LPS activates the tryptophan catabolizing enzyme indoleamine 2,3-dioxygenase (IDO), which results in tryptophan being shunted away from serotonin/melatonin towards the production of kynurenine and its downstream neurologically active metabolites kynurenic acid and quinolinic acid.[1] In rodents administered LPS, blockade of IDO activation, either indirectly with the anti-inflammatory agent minocycline or directly with the IDO antagonist 1-methyltryptophan (1-MT), prevents development of depressive-like behavior. However, in fractalkine receptor-deficient mice [CX(3)CR1($^{-/-}$)] that show profoundly increased and prolonged CNS microglial activation in response to LPS, administration of 1-MT blocks microglial activation and abrogates depressive-like behavior while having no effect on more sickness-specific symptoms of weight loss and reduced locomotor activity.[11] Similarly, administration of ketamine, which can block the effects of quinolinic acid on the *N*-methyl-D-aspartate (NMDA) receptor, has been shown to abrogate LPS-induced depressive-like behavior while leaving sickness behavior intact.

Long-Term Behavioral Abnormalities Associated with Immune Activation Prenatally or in Early Life

Although immune activation only very rarely produces psychotic phenomenon acutely in humans, substantial data suggest that exposure to increased inflammation either prenatally or early in life may produce long-term changes in CNS functioning that promote the development of two of the most serious, and chronic, mental disorders: schizophrenia and autism.[6] For example, multiple epidemiological studies have found associations between *in utero* exposure to a range of infectious organisms (e.g. influenza virus, *Toxoplasma gondii*, rubella virus, genital–reproductive infections) and the development of psychosis in general, and schizophrenia in particular in adult life. Fewer data are available for autism; however, a large epidemiological study from Denmark found that *in utero* exposure to maternal influenza infection associated with a two-fold increased risk and prolonged episodes of maternal fever caused a three-fold increased risk of infantile autism.[12] Although not the only potentially causative pathway, many other prenatal factors linked to an increased risk for the development of schizophrenia and autism have also been associated with increased levels of inflammatory biomarkers during pregnancy, including obstetric complications, gestational diabetes, severe physical (e.g. famine) or psychosocial stressors, and maternal depression. Directly supporting a role for increased inflammation in the pathogenesis of schizophrenia are several longitudinal studies showing that increased maternal inflammatory biomarkers predict the development of schizophrenia in adult offspring. Although not entirely consistent in terms of the specific cytokines involved, studies have implicated elevated TNF-α and interleukin-8 (IL-8) in this association. Increased maternal levels of IL-8 in the second and third trimesters have also been found to be associated with greater ventricular cerebrospinal fluid (CSF) volume in adult schizophrenia spectrum cases, as well as reduced volume of the left entorhinal cortex and right posterior cingulate.

These human findings are strongly reinforced by a compelling animal literature demonstrating that a variety of immune stimuli (i.e. stimuli that activate both antiviral and antibacterial immunity) administered prenatally (i.e. to the mother) affect CNS development and adult functioning in ways directly relevant to both schizophrenia and autism.[6] Various prenatal immune challenge models in rodents have been shown to produce an array of adult abnormalities relevant to severe mental illness, including impaired or deficient social interaction, repetitive and stereotyped behaviors, deficient working memory, impaired executive functioning (e.g. impaired discrimination in reversal learning and spatial/non-spatial information processing) and impaired sensorimotor gating as assessed by prepulse inhibition. Importantly, these abnormalities are accompanied by neurochemical and neuroanatomical abnormalities common to both autism and schizophrenia, including deficient reelin expression in prefrontal cortex and hippocampus, and altered serotonin (5-hydroxytryptamine) and dopamine metabolism, as well as morphological changes in multiple brain areas (e.g. basal ganglia, amygdala, hippocampus) relevant to psychiatric disease. Offspring exposed to certain highly pathogenic viruses *in utero* have reduced numbers of dopaminergic neurons and behavioral abnormalities in adulthood.[13] These strains show strong tropism for dopaminergic neurons *in vitro*, with resultant activation of the lynchpin intracellular inflammatory molecule nuclear factor-κB (NF-κB) and cellular apoptosis.

Consistent with these data from rodents, a study examining the effect of prenatal inflammation on adult behavior and physiology in primates found that exposure of pregnant macaques to low-dose LPS resulted in an array of behavioral abnormalities in the offspring, including deficient prepulse inhibition.[14] Prenatal exposure to low-dose LPS also resulted in significantly increased brain volume in the affected offspring, a finding repeatedly observed in human children with autism. Finally, both animal and human data suggest that the long-term impact of inflammatory activation on the developing CNS does not stop at birth but persists across early life. For example, large, population-based epidemiological studies have found that individuals with a history of an autoimmune condition early in life have a 29–45% increased risk of subsequently developing schizophrenia. A history of severe early life infection (as indexed by hospitalization) increased the risk of subsequent schizophrenia by 60%, and the effects of autoimmunity and infection were synergistic when combined in the same individuals, increasing the risk of subsequent schizophrenia by 125%, with these associations remaining significant after adjustment for other potentially confounding risk factors.[15] Animal studies are generally consistent with these human findings in suggesting that inflammatory stimulation very early in postnatal life has similar long-term effects to stimulation *in utero*.

The data linking various surrogates for immune activation (including measures of inflammatory biomarkers) remain circumstantial, but animal studies have been far more rigorous in demonstrating that inflammatory mediators drive the long-term effects of immune stimulation early in development on CNS morphology and function. For example, the long-term disease-relevant effects of administration of either IL-1β or IL-6 are abrogated if these cytokines are delivered in concert with strategies that block their biological effects (e.g. antibody against IL-6 or IL-1-receptor antagonist).[6] Similarly, mice genetically modified to overexpress the anti-inflammatory cytokine IL-10 demonstrated significant reductions in behavioral abnormalities induced by *in utero* exposure to poly I:C. Finally, rats exposed to LPS in the neonatal period had disrupted patterns of spontaneous synchronized cortical activity known to be essential for the formation of neural networks and in the regulation of neuronal survival and developmentally appropriate apoptosis (programmed cell death).[16] These effects, which are highly relevant to both schizophrenia and autism, were mediated by TNF-α derived from activated microglia cells, consistent with increasing evidence that immune molecules in general, and microglial activity in particular, is essential for many aspects of normal neural development, including neurogenesis, neuronal migration, axon guidance, synapse formation, and activity-dependent refinement of circuits. Importantly, blockade of TNF-α activity reversed the effects of neonatal LPS on neural firing patterns and apoptosis.

Perhaps the most striking example of the power of immune processes to both guide and disrupt neural development and thereby species-typical behavior comes from a study using a mouse model of maternal immune activation (MIA) that produces a variety of behaviors in offspring relevant to autism in humans. Prenatal immune stimulation reduced the number of circulating $CD4^+$ $TCR\beta^+$ Forkhead Box 3 $(Foxp3)^+$ $CD25^+$ T-regulatory cells (Tregs), while increasing IL-6 and IL-17 production by $CD4^+$ T cells. Transplantation of immunologically normal bone marrow into irradiated MIA offspring reversed the autism-relevant behavioral abnormalities observed in these animals, even though the heads of these animals were protected from any direct effects of the radiation. These findings are consistent with work demonstrating that peripheral T and monocytic cell functioning – specifically the homing of these cells to the meninges of the brain in response to psychosocial stress – is required for learning in rodents in a process that drives hippocampal brain-derived neurotrophic factor (BDNF) production via an IL-4-dependent mechanism.[17] Taken together, these findings raise the possibility that peripheral elements of both the innate and acquired immune system may serve as a type of "second brain" that play essential roles in brain development, behavioral plasticity, and stress resilience.

Mechanisms by which Immune Activation (Inflammation) Produces Behavioral Disturbance

The fact that exposure to inflammatory stimuli produces behavioral disturbances similar to those seen in psychiatric disorders is one piece of evidence suggesting that the immune system may contribute to disease pathogenesis. A second piece of evidence comes from the fact that inflammatory processes induce a number of changes in brain–body functioning that have been repeatedly implicated in the etiology of mood and anxiety disorders. Moreover, these changes have been shown to predict the development of behavioral symptoms in the context of inflammatory activation in both animal and human model systems, further strengthening the likelihood that the immune system may actually produce, rather than just reflect, psychiatric disease.

Effects of Inflammation on Neuroendocrine Mechanisms

Alterations in the hypothalamic–pituitary–adrenal (HPA) axis are among the best replicated findings in patients with a range of psychiatric conditions associated with increased inflammation, such as MDD, bipolar disorder, post-traumatic stress disorder (PTSD), and schizophrenia. Animal and human models suggest that activation of peripheral inflammation reproduces many of these alterations. For example, administration of inflammatory cytokines to laboratory animals has been shown to profoundly stimulate not only adrenocorticotropic hormone (ACTH) and cortisol release but also the expression and release of corticotropin-releasing hormone (CRH). Acute exposure to inflammatory stimuli also powerfully activates the HPA axis in humans, although this effect attenuates with repeated exposure. Strongly implicating HPA axis responses to inflammation in the development of behavioral disturbance are findings that the acute ACTH and cortisol response to a first injection of IFN-α correlates with the subsequent development of depressive symptoms during IFN-α therapy in patients with cancer.[1] In contrast to acute administration of IFN-α, chronic IFN-α administration is associated with flattening of the diurnal curve and increased evening cortisol concentrations, both of which correlate with the development of depression and fatigue, and which are commonly seen in the context of MDD.

Flattening of the diurnal cortisol rhythm is associated with glucocorticoid resistance, as indexed by non-suppression of cortisol by dexamethasone in the dexamethasone suppression test. Non-suppression of cortisol on this test and related *in vivo* and *in vitro* assays, including measures of immune cell glucocorticoid receptor (GR) sensitivity, is a common finding in MDD. It is generally believed that this glucocorticoid resistance results from decreased expression or function of the GR, or both. A rich literature indicates that inflammatory cytokines can disrupt GR function and decrease GR expression. For example, IFN-α has been shown to inhibit GR function by activating signal transducer and activator of transcription-5 (STAT5), which in turns binds to the activated GR in the nucleus, thus disrupting GR-DNA binding. A similar protein–protein interaction between the GR and NF-κB in the nucleus has also been described. Moreover, IL-1α and β have been shown *both in vitro* and *in vivo* to inhibit GR translocation from the cytoplasm to the nucleus through activation of p38 mitogen-activated protein kinase (MAPK). Studies using IL-1 knockout mice demonstrate that stress-induced alterations in GR translocation leading to glucocorticoid resistance are mediated by IL-1, given that these mice fail to exhibit impaired GR translocation following social disruption stress. Thus, the effects of cytokines on GR function may lead to a feed-forward cascade, such that increased inflammation through its effects on the GR undercuts the ability of glucocorticoids to restrain inflammatory responses, leading to further increases in inflammation and reduced GR function.

Effects of Inflammation on Brain Biochemical Pathways Relevant to Psychiatric Disease

Figure 26.3 provides a schematic for many of the CNS biochemical effects reported to be induced by peripheral

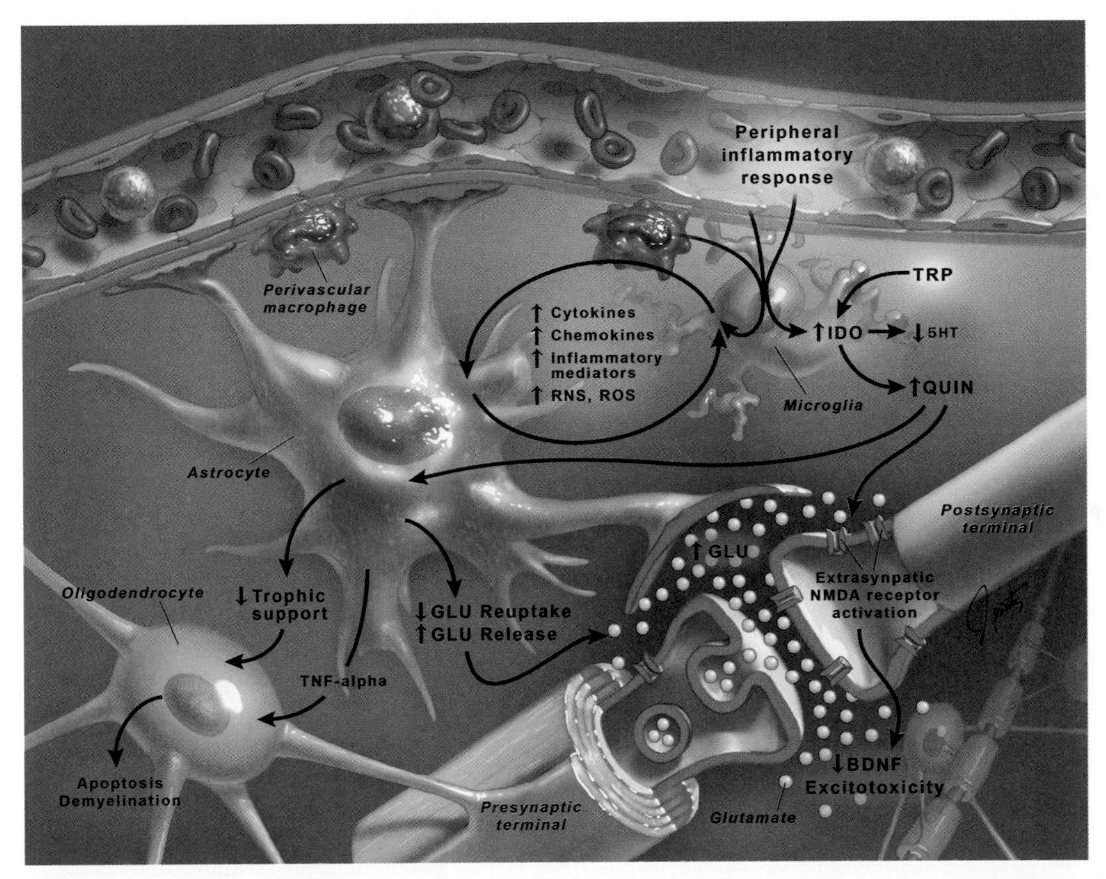

FIGURE 26.3 **Effects of the CNS inflammatory cascade on neural plasticity.** Microglia are primary recipients of peripheral inflammatory signals that reach the brain. Activated microglia, in turn, initiate an inflammatory cascade whereby release of relevant cytokines, chemokines, inflammatory mediators, and reactive nitrogen and oxygen species (RNS and ROS) induces mutual activation of astroglia, thereby amplifying inflammatory signals within the CNS. Cytokines including interleukin-1 (IL-1), IL-6, and tumor necrosis factor-α (TNF-α), as well as interferon-α (IFN-α) and IFN-γ (from T cells), induce the enzyme indoleamine 2,3-dioxygenase (IDO), which breaks down tryptophan (TRP), the primary precursor of serotonin [5-hydroxytryptamine (5-HT)] into quinolinic acid (QUIN), a potent *N*-methyl-D-aspartate (NMDA) agonist and stimulator of glutamate (GLU) release. Multiple astrocytic functions are compromised owing to excessive exposure to cytokines, QUIN, and RNS/ROS, ultimately leading to downregulation of glutamate transporters, impaired glutamate reuptake and increased glutamate release, as well as decreased production of neurotrophic factors. Of note, oligodendroglia are especially sensitive to the CNS inflammatory cascade and suffer damage due to overexposure to cytokines such as TNF-α, which has a direct toxic effect on these cells, potentially contributing to apoptosis and demyelination. The confluence of excessive astrocytic glutamate release, its inadequate reuptake by astrocytes and oligodendroglia, activation of NMDA receptors by QUIN, increased glutamate binding and activation of extrasynaptic NMDA receptors [accessible to glutamate released from glial elements and associated with inhibition of brain-derived neurotrophic factor (BDNF) expression], decline in neurotrophic support, and oxidative stress, ultimately disrupts neural plasticity through excitotoxicity and apoptosis.

immune activation. In addition to the biochemical and functional effects of cytokines discussed below, an emerging literature suggests that increased peripheral inflammatory markers correlate with reduced volume of brain areas implicated in psychiatric disease, in both normal and symptomatic individuals.[18]

MONOAMINE NEUROTRANSMITTER FUNCTION

The fact that SSRIs are able to prevent or treat depressive symptoms during chronic exposure to IFN-α provides strong evidence that serotonin pathways are involved in cytokine effects on behavior.[1] Genetic studies have complemented this work by demonstrating that polymorphisms in the promoter region of the serotonin transporter gene (5-HTTLPR) are associated with IFN-α-induced mood and anxiety symptoms in patients with hepatitis C. In addition, IFN-α-associated increases in CSF concentrations of IL-6 were found to be negatively correlated with the serotonin metabolite 5-hydroxyindoleacetic acid (5-HIAA), which was, in turn, negatively correlated with the severity of IFN-α-induced depression.

As with serotonin, dopamine metabolism in the CNS has been repeatedly shown to be affected by activation of both brain and bodily inflammatory pathways (Fig. 26.4).[1]

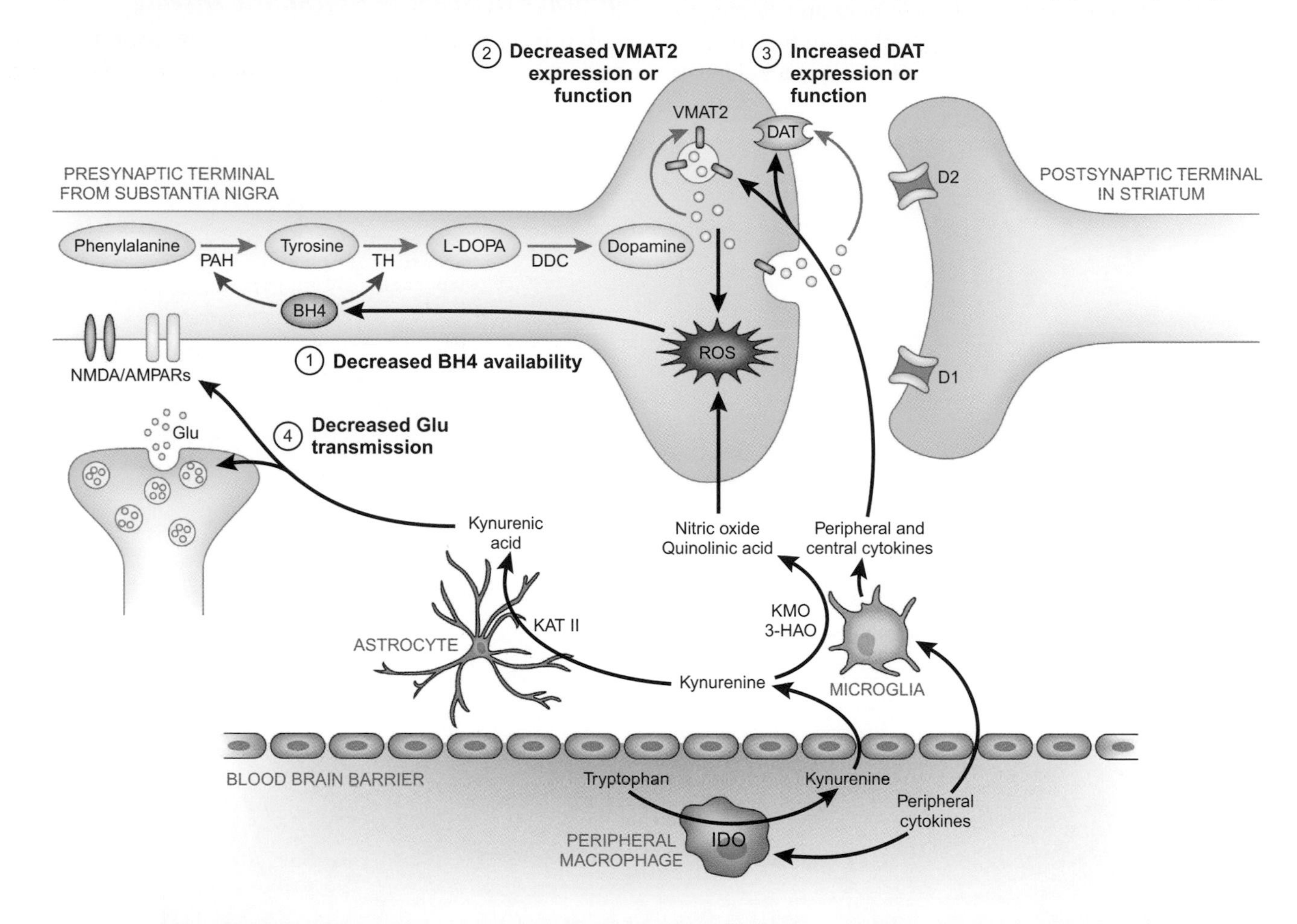

FIGURE 26.4 **Potential mechanisms of inflammatory cytokine effects on basal ganglia dopamine synthesis and release.** Evidence indicates that inflammatory cytokines from the periphery, or those produced locally by activated microglia or infiltrating macrophages, can produce nitric oxide, as well as quinolinic acid through indoleamine 2,3-dioxygenase (IDO) and kynurenine pathways, both of which contribute to oxidative stress and reactive oxygen species (ROS) generation. Increased ROS and inflammation-induced nitric oxide contribute to (1) oxidation of tetrahydrobiopterin (BH4), a cofactor required for the conversion of phenylalanine to tyrosine and tyrosine to L-3,4-dihydroxyphenylalanine (L-DOPA), which are necessary for the synthesis of dopamine. Furthermore, some evidence exists that inflammatory cytokines may (2) decrease the expression or function of the vesicular monoamine transporter-2 (VMAT2), and/or (3) increase expression or function of the dopamine transporter (DAT). Dysregulation of dopamine transport and vesicular packaging can increase cytosolic dopamine, leading to auto-oxidation and generation of ROS and neurotoxic quinones. Finally, cytokine activation of IDO in peripheral immune cells or microglia also produces kynurenic acid from kynurenine by kynurenine aminotransferase II (KAT-II) activity in astrocytes. Kynurenic acid can lead to (4) reduced glutamate (glu) neurotransmission by antagonism of glu receptors and release, consequently decreasing glu-evoked dopamine release in the striatum. Although not pictured, excessive cytokine-induced release of glutamate and quinolinic acid may also contribute to increased oxidative stress and excitotoxicity. 3-HAO: 3-hydroxyanthranilic acid oxygenase; AMPAR: 2-amino-3-(5-methyl-3-oxo-1,2-oxazol-4-yl) propanoic acid receptor; D1: dopamine receptor-1; D2: dopamine receptor-2; glu: glutamate; DDC: dopamine decarboxylase; KMO: kynurenine 3-monooxygenase; NMDAR: *N*-methyl-D-aspartic acid receptor; PAH: phenylalanine hydroxylase; TH: tyrosine hydroxylase.

For example, neuroimaging studies have revealed altered blood flow and metabolic activity in basal ganglia nuclei (brain areas heavily innervated and modulated by dopaminergic projections) during exposure to both acute and chronic inflammatory stimuli.

Consistent with these findings, studies in non-human primates have reported that reduced CSF concentrations of the dopamine metabolite homovanillic acid (HVA) were associated with depressive-like huddling behavior in response to chronic IFN-α administration, which is consistent with older studies showing similar huddling responses following chronic administration of the monoamine-depleting drug reserpine. Rodent studies have also indicated that cytokines target the basal ganglia and dopamine pathways. For example, acute administration of LPS to adult mice led to an approximately 50% long-term decrease in tyrosine hydroxylase-expressing neurons in the substantia nigra after 10 months, and chronic administration of LPS produced a 70% reduction in nigral dopaminergic neurons within 10 weeks. These effects appear to be mediated in part by LPS-induced production of TNF-α and the related activation of oxidative stress.

MECHANISMS OF INFLAMMATORY EFFECTS ON MONOAMINE METABOLISM

Multiple mechanisms have been explored by which inflammatory cytokines appear to affect monoamine neurotransmission. This section focuses on three of the most intensively studied: IDO, p38 MAPK, and tetrahydrobiopterin (BH4).

INDOLEAMINE 2,3-DIOXYGENASE IDO is an enzyme expressed in multiple cell types including macrophages, dendritic cells, microglia, astrocytes, and neurons when activated by a number of cytokines alone or in combination, including IFN-γ, TNF-α, IL-1, and IL-6.[5] IDO catabolizes tryptophan, the primary amino acid precursor of serotonin, away from serotonin and into kynurenine and, by extension, a number of downstream metabolites. This effect has been shown to contribute to both immune tolerance and enhanced ability to fight a number of intracellular infections (i.e. toxoplasmosis). As mentioned above, animal studies highlight the importance of IDO activation for the production of depressive-like behavior in response to inflammatory stimuli.[11] Similarly, IDO has been repeatedly shown to play an important role in cytokine-induced depression in response to IFN-α in humans. Several studies have reported correlations between IFN-α-induced depression and decreases in tryptophan and increases in kynurenine and/or the kynurenine to tryptophan ratio in plasma.[19–21]

Given the pivotal role played by serotonin in mood and anxiety disorders, it is not surprising that much of the initial attention focused on serotonin depletion as the likely culprit in IDO-associated depression. However, interest is shifting to the potential role of kynurenine and its downstream metabolites in mediating inflammatory effects relevant to mood, anxiety, and psychotic disorders (Fig. 26.5).[5] Data in laboratory animals demonstrate that intraperitoneal administration of kynurenine can induce depressive-like behavior, as manifested by increased immobility in the forced swim test and tail suspension test. In the CNS kynurenine is further catabolized into the neuroactive metabolites kynurenic acid and quinolinic acid, both of which are increased in the CSF of IFN-α-treated patients and correlate strongly with increases in depressive symptoms during treatment. Quinolinic acid, which is produced primarily in microglia and infiltrating macrophages, has multiple effects (including direct binding) that lead to activation of the NMDA receptor. Quinolinic acid also stimulates lipid peroxidation and other oxidative stress processes. In combination, these activities promote excitotoxicity in the brain. Consistent with this, excessive quinolinic acid has been implicated in a number of neurodegenerative disorders, including Huntington disease, amyotrophic lateral sclerosis, Alzheimer disease, and dementia secondary to infection with human immunodeficiency virus (HIV).[22–24] Kynurenic acid, which is produced primarily in astrocytes, inhibits the α_7-nicotinic acetylcholine receptor and by this mechanism reduces glutamate release, which, in turn, attenuates the release of dopamine. For example, intrastriatal administration of kynurenic acid to rodents leads to marked reductions in extracellular dopamine concentrations. Supporting the behavioral relevance of kynurenic acid, kynurenine aminotransferase II knockout mice (i.e. lacking the enzyme to produce kynurenic acid) exhibited significant increases in cognitive performance (object recognition, spatial discrimination, passive avoidance) compared with wild-type animals.[25,26]

MITOGEN-ACTIVATED PROTEIN KINASE Both *in vitro* and *in vivo* data have established that stimulation of p38 MAPK increases the expression and function of the serotonin transporter (SERT). For example, treatment of mouse midbrain and striatal synaptosomes with IL-1β and TNF-α increases SERT activity, an effect reversed by the p38 antagonist SB203580. The relevance of these effects has been demonstrated by studies showing that increases in SERT activity contribute to LPS-induced depressive-like behaviors. LPS-associated increases in SERT activity and depressive-like behaviors were reversed by pharmacological inhibition of p38 MAPK. Recent data also strongly support a role for p38 signaling in the development of cytokine-induced depression in humans. Specifically, increased phosphorylation (activation) of p38 MAPK following the first injection of IFN-α was associated with the development of depression and fatigue following 3 months of IFN-α treatment. MAPK pathways have also been found to influence the dopamine

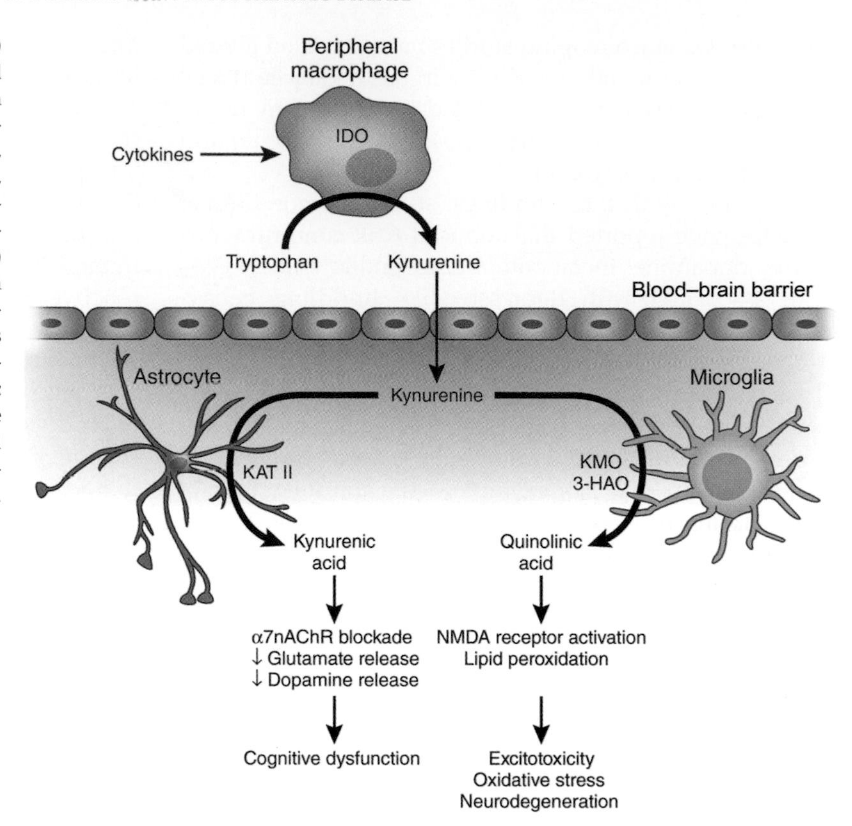

FIGURE 26.5 **Indoleamine 2,3-dioxygenase (IDO) and the kynurenine pathway in inflammation-induced CNS pathology.** Cytokine-induced activation of IDO in peripheral immune cells (e.g. macrophages and dendritic cells) or cells in the brain (e.g. microglia, astrocytes, and neurons) leads to the production of kynurenine, which is converted to kynurenic acid (KA) by kynurenine aminotransferase II (KAT-II) in astrocytes or quinolinic acid by kynurenine-3-monooxygenase (KMO) and 3-hydroxy-anthranilic acid oxygenase (3-HAO) in microglia or infiltrating macrophages. Through blockade of the a7nAChR, KA can reduce glutamate release as well as the release of dopamine, both of which can contribute to cognitive dysfunction. By contrast, quinolinic acid, through activation of the *N*-methyl-D-aspartate (NMDA) receptor, can increase glutamate release and lead to lipid peroxidation, thus contributing to excitotoxicity, oxidative stress, and ultimately neurodegeneration.

transporter (DAT). For example, hDAT-expressing cells transfected with a constitutively activate MAPK kinase (MEK) exhibit increased dopamine reuptake, whereas treatment of rat striatal synaptosomes with MEK inhibitors decreases dopamine reuptake in a concentration- and time-dependent manner.

TETRAHYDROBIOPTERIN BH4 is an essential enzyme cofactor for tryptophan hydroxylase and tyrosine hydroxylase, which are the rate-limiting enzymes for the synthesis of serotonin and of dopamine and norepinephrine, respectively. BH4 is also a cofactor for nitric oxide synthase (NOS), which converts arginine to nitric oxide (NO) and has downstream effects on glutamatergic function. BH4 is very labile and highly sensitive to nonenzymic oxidation driven by inflammation, which leads to an irreversible degradation of BH4 to dihydroxyanthopterin (XPH_2). This has been confirmed in both animal and human studies (Fig. 26.6).

For example, intramuscular injection of IFN-α to rats decreases CNS concentrations of BH_4 through stimulation of NO. Conversely, treatment with an inhibitor of NO reverses the inhibitory effects of IFN-α on brain concentrations of both BH_4 and dopamine. Chronic treatment with IFN-α in humans has been shown to significantly increase CSF concentrations of dihydrobiopterin (BH2), suggesting increased degradation of BH4. As predicted from reduced BH4 activity, the ratio of phenylalanine to tyrosine was also increased and this correlated with fatigue severity, as well as with reductions in CSF levels of dopamine and HVA. Finally, decreased CSF concentrations of BH4 correlated with increased levels of IL-6 in the CSF, supporting the notion that the induction of CNS inflammation by IFN-α, as indexed by IL-6, led to degradation of BH4, a decrease in monoamine bioavailability, and the development of symptoms.

GLUTAMATE AND γ-AMINOBUTYRIC ACID

Inflammatory processes profoundly affect glutamate metabolism, based on a rich literature demonstrating that inflammatory cytokines decrease the expression of glutamate transporters on relevant glial elements and increase the release of glutamate from astrocytes.[27] In contrast to neuronally released glutamate, which signals within the synaptic cleft, glutamate released by astrocytes preferentially accesses extrasynaptic NMDA receptors, which can mediate excitotoxicity and lead to decreased production of trophic factors including BDNF. As noted above, in addition to direct effects on glial functioning, cytokines induce quinolinic acid via stimulation of IDO, which leads to glutamate release through direct activation of the NMDA receptor.[5] Cytokines including TNF-α and IL-1 also induce both astrocytes and microglia to release reactive oxygen and nitrogen species. These amplify oxidative stress and thereby impair glutamate reuptake, while stimulating glutamate release and endangering

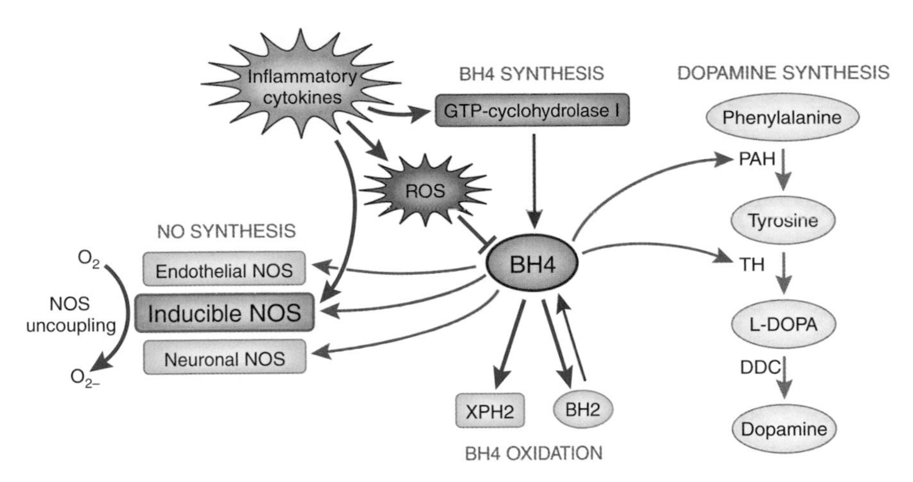

FIGURE 26.6 **Effects of inflammatory cytokines on BH4 synthesis and oxidation, and nitric oxide activity, which contribute to decreased BH4 availability for dopamine synthesis.** Dopamine synthesis relies on the conversion of tyrosine to L-3,4-dihydroxyphenylalanine (L-dopa) by tyrosine hydroxylase (TH), the rate-limiting enzyme for dopamine synthesis. Phenylalanine is converted to tyrosine by phenylalanine hydroxylase (PAH), and both TH and PAH require BH4 as a cofactor. BH4 is also a cofactor for the nitric oxide synthases (NOS), which convert arginine to nitric oxide (NO). BH4 is redox sensitive and readily oxidized reversibly to dihydrobiopterin (BH2) and irreversibly to dihydroxanthopterin (XPH_2). Although inflammatory cytokines induce expression of GTP-cyclohydrolase I, an enzyme necessary for BH4 synthesis, they also increase both reactive oxygen species (ROS) and inducible NOS activity. Increased oxidation of BH4 by ROS as well as increased inducible NOS activity result in NOS uncoupling and the preferential generation of free radicals from O_2 rather than NO. This production of free radicals further contributes to oxidative stress and reduction of BH4, ultimately leading to decreased BH4 availability for dopamine synthesis. Red arrows indicate effects of inflammatory cytokine on BH4 metabolism and activity.

relevant cell types including oligodendrocytes, which are especially vulnerable to oxidative damage. Loss of glial elements such as oligodendrocytes in several mood-relevant brain regions, including the subgenual prefrontal cortex and amygdala, has emerged as a fundamental morphological abnormality in MDD.

Effects of Inflammation on Functional Brain Neurocircuitry

Given that immune activation also produces behavioral disturbances reminiscent of many psychiatric disorders (depending on the model), it might be predicted that inflammatory processes are capable of altering brain activity in ways likely to promote behavioral disturbances. Many studies suggest this is the case, and implicate two brain areas in particular, the anterior cingulate cortex (ACC) and the basal ganglia, as being especially prone to changes in activity when impacted by cytokines.

BASAL GANGLIA

Using positron emission tomography (PET), early studies in patients undergoing IFN-α therapy revealed marked increases in glucose metabolic activity in the basal ganglia which correlated with symptoms of fatigue.[1] These increases are consistent with metabolic changes reported in patients with Parkinson disease and may reflect increased oscillatory burst activity in relevant basal ganglia nuclei secondary to dopamine depletion, which is a known biochemical effect of chronic immune activation. Psychomotor retardation is a prominent behavioral effect of cytokines that, when severe, has been reported to produce frank parkinsonism that can be reversed by treatment with the dopaminergic agent L-dopa. Using functional magnetic resonance imaging (fMRI), typhoid vaccination was also found to alter activation in the basal ganglia. Specifically, compared with controls, vaccinated volunteers exhibited increased evoked activity in the substantia nigra that was associated with both prolonged reaction times and increased peripheral blood concentrations of IL-6.[28]

ANTERIOR CINGULATE CORTEX

Another brain region reliably influenced by cytokine administration or inflammatory stimuli that promote cytokine activation is the dorsal anterior cingulate cortex (dACC). Compared with controls, patients chronically exposed to IFN-α demonstrated significantly greater activation of the dACC (Brodmann area 24) during an fMRI-based task of visuospatial attention.[29] Despite a low rate of error commission, a strong correlation was found between the degree of dACC activation and the number of errors made during the task in patients receiving IFN-α, but not in control subjects. This pattern of increased dACC activation in response to error is also observed in patients with neuroticism, high trait anxiety, and obsessive–compulsive disorder. Suggesting that an array of inflammatory stimuli targets the dACC, typhoid vaccination has been reported to produce increased dACC activity during a high-demand color word Stroop task.

The primary function of the dACC has traditionally been considered to be error detection and conflict monitoring. This concept has been expanded such that the dACC is now considered to be a type of "neural alarm system" that can both detect and respond to (with arousal and distress orchestrated by autonomic activation) threatening environmental stimuli in the social domain.[30] Because cytokines sensitize dACC responsivity, this may contribute to the anxiety and emotional arousal that often accompany chronic exposure to inflammatory stimuli such as IFN-α. Finally, it should be noted that the effects of inflammatory stimuli on the ACC appear to be context dependent. Although most studies have noted the strongest effects in the dACC, activation patterns in response to typhoid vaccination were more robust in subgenual ACC when the task used was explicitly emotional in nature, consistent with the centrality of the subgenual ACC in the pathophysiology of MDD.[31]

EVIDENCE THAT PATTERNS OF CNS ACTIVITY ASSOCIATED WITH PSYCHIATRIC DISEASE AFFECT IMMUNE FUNCTIONING IN HEALTH-RELEVANT WAYS

In contrast to the extensive literature demonstrating pathways whereby activation of the peripheral immune system affects CNS functioning, remarkably little is known about specific mechanisms within the CNS by which cognitive processes and appraisals affect immunity. The identification of these processes is one of the great tasks awaiting researchers working in the field of immune–brain interactions. More is known about the role of stress outflow pathways in modulating the effect of CNS processes on peripheral immunity (Fig. 26.7), but even here questions outnumber answers.

Psychosocial Stress and Promotion of the Proinflammatory Phenotype

Psychosocial stress is the best replicated risk factor for many psychiatric diseases, especially mood and anxiety disorders.[10] This is true whether stress is considered as the occurrence of actual negative events or as the perception of such events. This latter fact may explain why dispositional factors that promote either stressful events or perceptions that life is dangerous, unpredictable, or negative also powerfully predict depression, and have also been associated with immune changes. Conversely, traits such as dispositional optimism have been repeatedly observed to be associated with reduced levels of peripheral inflammatory biomarkers, especially in response to psychosocial stress. In addition to being a primary proximate risk factor for the development of mood, anxiety, and psychotic disorders, psychosocial stress is a powerful risk factor when considered from a developmental perspective. Specifically, early life adversity has been repeatedly shown to markedly increase the risk of developing a wide range of psychiatric conditions in adulthood and to be associated with changes in immune functioning linked to both mental and physical disease.[32]

Human studies examining the effect of psychosocial stress on immunity tend to use one of two methodologies: either naturalistic life stressors or exposure to any of a variety of standardized laboratory psychosocial stressors. Despite significant differences between these methodologies, results are fairly consistent and indicate that stress tends to promote general inflammatory activity while suppressing various measures of more specific immunity, including innate immune antigen presentation and B-cell functioning (often assessed as antibody responses to non-specific mitogens or in response to antigen-specific vaccines). Psychosocial stress may also impair acquired immune processes that normally regulate inflammation. For example, acute laboratory stressors have been reported to reduce plasma concentrations of the anti-inflammatory cytokine IL-10, as well as circulating $CD4^{+}CD25^{+}$ Treg cells, which are an important source of IL-10. Moreover, acute stress appears to downregulate Foxp3, an important T-cell immunoregulatory transcription factor. Chronic traumatic life stress has also been associated with reduced numbers and percentages of circulating $CD4^{+}CD25^{+}Foxp3^{+}$ Treg cells, a finding in line with many studies showing that PTSD, like MDD, is associated with increased circulating levels of innate immune proinflammatory cytokines.

Despite these fairly consistent findings, it is important to highlight the fact that the relationship between stress and immune–inflammatory activity is complex, and there is increasing evidence to suggest that acute and chronic stress may differentially affect immune functioning. The tendency for stress to promote chronic inflammation at the expense of more specific forms of immune responsivity is far more apparent in the context of chronic stress than in acute stress. In fact, many studies suggest that acute stressors may actually prime the immune response in ways that optimize antipathogen capacity.[33] Perhaps most strikingly in this regard, women exposed to a laboratory psychosocial stress test just before receiving an influenza vaccine increased subsequent antibody production, consistent with many studies observing similar immune-enhancing effects of acute stress in animal models. In these women, the increased antibody production was associated with evidence of heightened autonomic arousal and increased poststressor production of IL-6. These findings point to the complexity of stress effects on immune functioning, given that various indices of sympathetic activation have been repeatedly associated with various measures of impaired immune capacity, including reduced gene

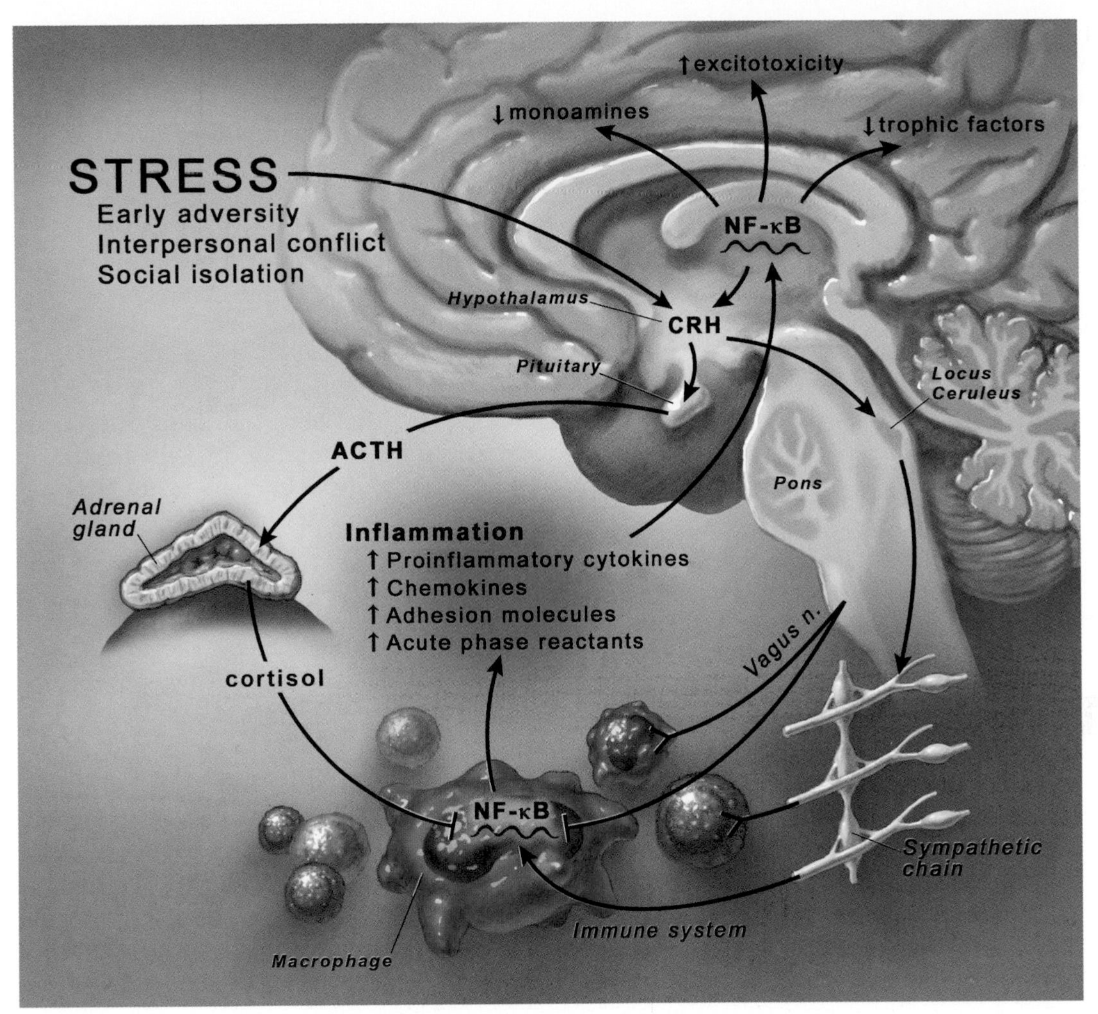

FIGURE 26.7 **Stress-induced activation of the inflammatory response.** Psychosocial stressors activate CNS stress circuitry, including corticotropin-releasing hormone (CRH) and ultimately sympathetic nervous system outflow pathways via the locus ceruleus. Acting through α- and β-adrenergic receptors, catecholamines released from sympathetic nerve endings can increase nuclear factor-κB (NF-κB) DNA binding in relevant immune cell types including macrophages, resulting in the release of inflammatory mediators that promote inflammation. Proinflammatory cytokines, in turn, can access the brain, induce inflammatory signaling pathways including NF-κB, and ultimately contribute to altered monoamine metabolism, increased excitotoxicity, and decreased production of relevant trophic factors. Cytokine-induced activation of CRH and the hypothalamic–pituitary–adrenal axis, in turn, leads to the release of cortisol, which, along with efferent parasympathetic nervous system pathways (e.g. the vagus nerve), serves to inhibit NF-κB activation and decrease the inflammatory response. In the context of chronic stress and the influence of cytokines on glucocorticoid receptor function, activation of inflammatory pathways may become less sensitive to the inhibitory effects of cortisol, and the relative balance between the proinflammatory and anti-inflammatory actions of the sympathetic and parasympathetic nervous systems, respectively, may play an increasingly important role in the neural regulation of inflammation. ACTH: adrenocorticotropic hormone.

expression in pathways linked to antigen presentation and attenuated activation of cytolytic CD8+ T cells, and given that increased IL-6 production in response to vaccination in chronically stressed older adults has been associated with reduced antibody responses.

Even the generalization that acute stress enhances, while chronic stress suppresses, adaptive immune function must be held loosely, given data over the last several years that chronic social stressors in rodents appear to enhance many aspects of innate and adaptive immune function.[34] Indeed, chronic social stress in mice has been shown to increase costimulatory molecules on dendritic cells, such as CD80 and major histocompatibility complex (MHC) type I, that are important for CD8+ T-cell activation and to increase – not diminish – the number of antigen-specific memory CD8+ T cells that are critical for establishing virus-specific immunological memory. This enhanced adaptive immune response was shown to result in more rapid influenza virus clearance from the lung, highlighting the fact that even chronic stress, in certain model systems (social disruption paradigm in this case) can benefit host defense against pathogens. Consistent with this idea, chronic social stress in animals has also been shown to prime activity in a variety of innate immune cell types, including macrophages and natural killer cells, with resultant increased capacity of these cells to kill pathogens in both *in vivo* and *in vitro* assays.

Stress Outflow Mechanisms by Which Psychosocial Stress Affects Immunity

Stressors of all sorts initiate profoundly complex cascades of biochemical activity in the brain and body that, while far from perfect, have evolved to help organisms to survive in the face of environmental changes, whether these changes signal danger, opportunity, or, as is usually the case, both.[10] Anything approaching a full explication of these complexities is far beyond the scope of this chapter. Accordingly, a brief summary is provided of the immune effects of the stress pathways best understood in terms of relevance to disease development: the HPA axis and the ANS, which includes the sympathetic nervous system (SNS) and parasympathetic nervous system.

Hypothalamic–Pituitary–Adrenal Axis

Cortisol, the body's primary stress hormone and end-product of HPA axis activation, is one of the most powerful anti-inflammatory chemicals known, which accounts for the widespread use of cortisol-like glucocorticoid compounds in the treatment of inflammatory and autoimmune conditions. Given the anti-inflammatory properties of glucocorticoids and the fact that so many psychiatric conditions linked to stress are associated with increased peripheral inflammatory biomarkers, it might be expected that these conditions would also be associated with reduced glucocorticoid signaling, and a great deal of evidence supports this. The simplest potential measure of reduced glucocorticoid signaling is reduced circulating concentrations of cortisol (or corticosterone in rodents). This has been observed in several psychiatric conditions, including PTSD and MDD, although the data are not fully consistent. However, psychiatric conditions, including MDD, bipolar disorder, and schizophrenia, have also been found to be characterized by increased plasma cortisol levels, raising the intriguing question of how, if cortisol is so profoundly anti-inflammatory, a condition could be simultaneously associated with inflammation and increased cortisol. Increasingly, the answer appears to revolve around the fact that conditions associated with increased peripheral cortisol are often also associated with resistance to cortisol, so that the signal received by relevant tissues may actually be decreased, not increased, despite the surfeit of circulating hormone.[1]

Many animal studies demonstrate that a variety of social stress paradigms that increase inflammatory activity also induce glucocorticoid resistance in various peripheral immune tissues and do so in concert with the risk of wounding any given animal faces in the stress paradigm. It is hypothesized that the development of glucocorticoid resistance in response to these types of stressors may serve the adaptive purpose of allowing an HPA axis response needed for the metabolic demands of the stressor, while at the same time releasing inflammatory processes needed for tissue healing from regulatory control. Despite these highly replicable animal data, however, confirming a link between glucocorticoid resistance and stress-induced increases in inflammation in humans has been more challenging. Although some data suggest that reduced cortisol responses to acute laboratory stressors in humans are associated with an increased inflammatory responses (e.g. increased NF-κB activity in peripheral leukocytes), many studies that have taken the straightforward approach of correlating changes in cortisol with changes in cytokines have produced null results. However, studies using more advanced *in vitro* or gene expression methodologies for examining the glucocorticoid signal (and not just circulating glucocorticoid levels) suggest that reduced glucocorticoid signaling may indeed contribute to the increased inflammation observed in response to stress, and by extension, in the context of psychiatric disease. For example, women with PTSD related to childhood abuse demonstrated increased activity of NF-κB in peripheral mononuclear cells that correlated both with symptom severity and with an *in vitro* measure of glucocorticoid resistance (concentration of dexamethasone needed to suppress LPS-induced TNF-α production by 50%).[35] Also consistent with a role for decreased glucocorticoid signaling in stress-related psychiatric conditions is the repeated finding that psychosocial stressors, as well as factors linked to stress (such as loneliness), are associated with significant increases in the expression of genes containing promoter response elements for NF-κB and significant decreases in genes containing promoter elements for the GR. Individuals undergoing chronic caregiver stress showed the expected increase in the balance of NF-κB/GR gene expression and had elevated plasma concentrations of several inflammatory biomarkers, but no differences from controls in cortisol secretion, again highlighting the fact that plasma concentrations of cortisol are not always a reliable marker for the underlying signaling capacity of the HPA axis.[35]

Autonomic Nervous System

Stressors of all sorts induce fairly stereotyped changes in ANS activity characterized by various degrees of SNS activation and parasympathetic withdrawal. As with the HPA axis, the effects of sympathetic and parasympathetic signaling on immunity are complex, but many studies suggest that, *in toto*, SNS activation and parasympathetic withdrawal drive increased inflammatory activity in the periphery and the CNS.[1,3] For example, catecholamines acting through α- and β-adrenergic receptors increase cytokine expression in both the brain and periphery of rats, and α-adrenergic antagonists attenuate increased peripheral blood concentrations of IL-6 induced by altitude stress in humans.[1] In addition, in animal models, α- and β-adrenergic agonists directly activate NF-κB *in vivo* and *in vitro*. As discussed above, chronic social stressors in animals (i.e. social disruption stress) have a variety of immune-activating effects that may enhance host defense against pathogens introduced in relation to the stressor as a result of wounding. Strikingly, in the social disruption stress paradigm many of these immune effects, including splenomegaly and increased plasma concentrations of IL-6, TNF-α, and monocyte chemoattractant protein-1 (MCP-1), were reversed by pretreatment with the β-receptor antagonist propranolol.[34] Furthermore, flow-cytometric analysis of cells from propranolol-pretreated mice indicated that β-adrenergic blockade reduced the social disruption stress-induced increase in the percentage of $CD11b^+$ splenic macrophages and significantly decreased expression of a range of markers associated with immune activation, including toll-like receptor-2 (TLR2), TLR4, and CD86 on the surface of these cells. Social disruption stress has also been shown to increase the trafficking of $CD11b^+/CD45^{high}/Ly6C^{high}$ macrophages to the brain and to produce a variety of changes in functional and morphological markers in microglial cells consistent with activation and hence increased inflammatory activity in the CNS. Importantly, all of these changes were abrogated by pretreatment with propranolol. Also implicating sympathetic signaling in the effect of stress on immunity is a study showing that multiple long-term immune changes driven by early life stress in rhesus macaques (i.e. enhanced expression of genes involved in inflammation, cytokine signaling, and T-lymphocyte activation, and suppression of genes involved in several innate antimicrobial defenses including type I IFN antiviral responses) were associated with increased activity of cAMP response element binding protein (CREB), an important intracellular transcription factor that is stimulated by catecholaminergic signaling.[36]

Given that stress so reliably leads to increased circulating concentrations of both norepinephrine and epinephrine, one might expect clear evidence that the SNS is a significant contributor to the increased levels of inflammatory biomarkers observed in response to psychosocial stress in humans. In fact, the situation is not unlike the one for cortisol, where it is surprising how few overall data exist to support this expectation. Although several studies have suggested that stress-induced increases in catecholamines correlate with increased inflammation, just as many studies have observed null results, again pointing to the fact that circulating levels of hormones or neurotransmitters are only one small part of the signaling capacity/efficiency of the far more complex physiological systems of which they are a part. This point is brought home by an *in vivo* study in humans showing that reduced β-adrenoreceptor sensitivity was associated with increased plasma concentrations of C-reactive protein (CRP), much as reduced glucocorticoid sensitivity has also been associated with increased inflammatory biomarkers. Consistent with animal data, *in vitro* studies of human monocytic cells have shown that α_1-adrenoreceptor stimulation increases LPS-induced production of IL-1β mediated by the p38 MAPK and protein kinase C signaling pathways, whereas α_1 blockade abrogates this effect.

As noted above, stressors not only activate the SNS, but also promote reductions in parasympathetic signaling to various bodily organs and compartments.[1] Given the generally opposite effects of SNS and parasympathetic signaling on bodily processes, it would be expected that if SNS activation generally increases inflammatory markers, parasympathetic signaling might demonstrate anti-inflammatory processes. And indeed an emerging literature indicates that the parasympathetic nervous system does generally produce anti-inflammatory effects. For example, stimulation of efferent parasympathetic nervous system fibers including the motor vagus has been shown to reduce mortality secondary to LPS administration in laboratory rats via reduction of sepsis, while also reducing LPS-induced activation of NF-κB as well as TNF-α. These inhibitory effects on the inflammatory response are mediated by vagal release of acetylcholine that, in turn, activates the α_7-subunit of the nicotinic acetylcholine receptor (nAChR), which regulates both cytokine transcription and translation (often referred to as the "cholinergic anti-inflammatory pathway"). In addition, recent data suggest that there may be a cellular component to this inhibitory cholinergic reflex. Indeed, adoptive transfer of Teff cells from vagotomized mice was shown to aggravate colitis in association with increased inflammatory scores and reduced Treg cells.

Because stress produces SNS activation and parasympathetic withdrawal, it has been difficult to determine which of these mechanisms is the primary driver of the various inflammatory processes that are enhanced by psychosocial stress, and which mechanism, therefore, is of most relevance to the link between psychiatric disease and altered immune functioning. This issue is one of the great unanswered questions in the field of psychoneuroimmunology. However, be this as it may, because the impact of stress on the SNS and parasympathetic branches of the ANS produces largely synergistic effects on immune functioning, it is perhaps not surprising that it has been easier to demonstrate that overall changes in ANS function are associated with inflammatory activity than it has been to demonstrate the relative importance of specific inflammatory effects of either SNS or parasympathetic activity in humans. Thus, several studies have found associations in humans between increased heart rate and decreased heart rate variability (HRV) (both of which to various degrees reflect summed ANS activity) and increased peripheral inflammatory biomarkers (mostly cytokines and CRP). These associations have been observed in the context of acute stressors and at rest, and have been observed in medically healthy individuals, as well as in people with MDD and/or a variety of medical conditions. These associations have been especially consistent for low-frequency HRV, which is the most physiologically enigmatic of HRV measures, with ongoing debates regarding the degree to which it reflects SNS versus vagal, parasympathetic, activity.

EVIDENCE THAT ENVIRONMENTAL FACTORS THAT PROMOTE PSYCHIATRIC MORBIDITY MAY DO SO BY ALTERING IMMUNE FUNCTION

An important line of circumstantial evidence linking immune abnormalities to the pathogenesis of psychiatric illness is the fact that most identified risk factors for these diseases are known to alter immune functioning, and generally to do so by increasing inflammatory activity.[4] This is the case for both proximate and developmental risk factors for these conditions. For example, proximal risk factors for the development of depression include medical illness, psychosocial stress, sedentary lifestyle, obesity, and dietary patterns built around processed foods, all of which have been shown to be associated with increased peripheral inflammatory biomarkers.[10] Similarly, many of the best replicated developmental, or early life, risk factors for psychiatric morbidity also affect the immune system in ways that are increasingly recognized as being likely to change adult CNS structure and function. Examples of such immunogenic risk factors include exposure to psychosocial stress *in utero*, prenatal infection, nutritional deficiency, obstetric complications including hypoxia, and early life stressors.[6] Finally, larger societal factors associated with the risk of both adult- and childhood-onset psychiatric disturbances may exert their effects in part via the immune system. There is increasing evidence that many features of modernity have disrupted ancient coevolved relationships with the microbial and parasitic world both outside and within the body, and especially within the gut, which hosts what has been widely considered to be the most complex microbial environment on Earth.[1,8] Examples of such features of modernity include increased hygiene, widespread use of antibiotics and other antimicrobials, and changes in dietary patterns.

Providing a comprehensive review of the association of psychiatric risk factors with altered immune function is far beyond the scope of this chapter. Therefore, this chapter focuses on two of the most important examples: psychosocial stress (see above) and altered relationships with the microbial/parasitic world as a result of modern hygienic lifestyles, as discussed below. Increasing evidence suggests that these two apparently disparate domains may be linked in surprising ways.

Role of Disrupted Human–Microbial and Human–Parasitic Relationships in the Development of Psychiatric Disease

The finding that psychosocial stressors that promote the development of many psychiatric diseases also activate inflammatory signaling pathways has become a cornerstone in our understanding of the ways in which immune changes may contribute to mental illness. A second assumption, not as often explicitly stated, but supported by a number of studies, is that individuals tend to "breed true" when it comes to immune functioning, meaning that levels of immune activity at rest and in response to stressors tend to be fairly stable within, and across, individuals. Again, although not typically made explicit, without this assumption a statement such as "early life adversity promotes increased inflammation in adulthood" would be senseless. A final, and almost never questioned, assumption is that the primary drivers of immune, neuroendocrine, and CNS interactions exist within individuals and not within larger collectivities, such as cultural or ecological systems. A pragmatic result of this foundational view has been that scientific results obtained from individuals in modern, industrialized societies (and from animal studies conducted in highly artificial environments) have been unquestioningly assumed to reflect invariant truths of human and mammalian biology. However, preliminary findings from anthropological and immunological studies issue a sharp critique to many of these assumptions and point to the possibility that the field's focus on immune activation and inflammation within specific individuals may miss the important point that immune disturbances of relevance to psychiatric pathology may be as much about larger societal and environmental factors as about individuals, and as much about what is missing as about what is present. And what is missing may be the wide world of immunoregulatory micro-organisms and parasites with which humans coevolved, but that are absent from modern environments.

Directly challenging the assumption that findings from individuals in developed countries are universally applicable are results from recent studies by Thomas McDade and colleagues. McDade found that in tribal and traditional societies in the Amazon, levels of CRP varied strikingly within individuals across time in ways not typically observed in the modern world. Levels of CRP spiked to very high levels in response to acute infections and then plummeted rapidly to very low levels upon recovery, pointing to the fact that in environments with high levels of microbial and pathogen exposure, interactions with these organisms may drive immune functioning to such a degree that other interindividual and intraindividual factors are of far less relevance than they are in the modern world, where levels of inflammation and immune activity show higher levels of within-individual stability.[37] Consistent with this possibility, McDade and colleagues reported that the link between psychosocial stress and inflammation may be less apparent in environments more like those within which humans evolved than in modern environments.[37] Specifically, they found that prospectively assessed exposure in childhood to environments in the Philippines rife with infectious, microbial, and parasitic agents were associated with reduced baseline CRP in adulthood and, even more strikingly, appeared to disconnect the link between psychosocial stressors and increased inflammation. In individuals raised in microbial-rich environments, neither loss of a parent in childhood nor current levels of stress were associated with increased CRP, whereas in individuals raised in more hygienic early environments in the Philippines the predicted associations between both childhood and current psychosocial adversity and increased inflammation, as indexed by CRP, were clearly apparent.[38] If confirmed in subsequent studies, these findings raise the transformative possibility that many of the interactions among immune reactions, the CNS, and behavior that have been assumed to be invariant components of human functioning based on studies conducted in the developed world may in fact reflect the far more local distorting effects of modern environments than any ultimate truths about the human condition.

These findings would be less striking if they existed in isolation, but they are part of an increasing body of evidence suggesting that modern environmental practices have disrupted human relationships with a range of micro-organisms in ways that may help to account for a wide range of immune-based pathologies that are far more common in industrialized societies than in environments more consistent with those in which humans evolved.[8] Indeed, overwhelming data demonstrate that the prevalence of T-helper type 1 (Th1)/T-helper type 17 (Th17)-mediated autoimmune and inflammatory bowel and T-helper type 2 (Th2)-mediated allergic/asthmatic conditions increased dramatically in the developed world during the twentieth century, with increases in the incidence of immune-mediated disease in the developing world during the same period closely paralleling the adoption of Western lifestyles. For example, the incidence of asthma, hay fever, type 1 diabetes, inflammatory bowel disease, and multiple sclerosis increased an average of two- to three-fold in industrialized nations between 1950 and the present, with increases occurring earlier in the most developed countries and later in countries in the process of adopting Western lifestyles.

Unlike the focus on factors that are patently proinflammatory that has characterized the study of immune factors relevant to psychiatry, more comprehensive theories of evolutionary mismatch, focused on the loss of microbially modulated immunoregulation, have been put forth to account for this increase in allergic, atopic, and autoimmune conditions. This broader perspective, which extends the systems approach applied to brain–body interactions to include human–microbial and human–parasitic interactions, may help to account for the striking increase in certain psychiatric conditions (e.g. autism and major depression) that has occurred over the past half-century or so and may provide insights into novel immune-based interventions for psychiatric conditions.

Initial theorizing regarding the role of human–microbial interactions in the explosion of immune disorders in the twentieth century focused on the possibility that reduced exposure to childhood infectious diseases in the modern world might be the culprit. As originally articulated in the "hygiene hypothesis", it was theorized that factors unique to industrialized societies, including improved sanitation, modern medicine, and smaller family size, reduce the prevalence and change the timing of childhood infections in a way that promotes the development of allergy and asthma. Because infections typically mobilize Th1-type inflammatory responses, a necessary correlate of this idea – supported by immunological understandings of the time – was that the loss of infection-induced Th1 activation early in life released Th2 processes from appropriate regulatory control, with resultant increases in allergy and asthma. However, it soon became clear that, rather than decreasing (as would be predicted from the Th1–Th2 balance idea), an array of Th1-mediated autoimmune conditions was also exploding in prevalence in exactly the same countries in which allergy was on the rise. Moreover, allergies are most common in inner cities where childhood infections are rife and least common in isolated rural communities.

A potential resolution to this dilemma, implicating other types of microbial and parasitic agents with powerful effects on regulatory dendritic cells and Tregs, began to emerge in the late 1990s. As eventually articulated in the "old friends" hypothesis, rather than resulting from increased childhood infectious morbidity, the link between modernity and increased inflammatory, atopic, and autoimmune disease might be better explained by disruptions of ancient associations with mostly non-pathogenic micro-organisms once ubiquitous in both the external and internal human environment, which have been largely (or completely) banished from our world of plastic and pavement and processed foods.

Rather than providing immune regulation and instruction via Th1 activation, these micro-organisms induce and maintain an adaptive level of immune suppression by stimulating T cells to differentiate along regulatory, rather than Th1 or Th2 lines, with a resultant increased production of anti-inflammatory, immunoregulatory cytokines, especially IL-10 and transforming growth factor-β (TGF-β) (Fig. 26.8). Many strands of evidence suggests that deficits in regulatory function in either innate (e.g. dendritic) or adaptive immunity (e.g. Treg) produce a wide range of disturbances that align remarkably well with the array of immune-related conditions increased in the modern world. For example, genetic defects of the gene encoding the transcription factor Foxp3 lead to the X-linked autoimmunity–allergic dysregulation syndrome (XLAAD), which includes aspects of allergy, autoimmunity, and enteropathy.[8]

Underlying the old friends perspective is an assumption that mammals coevolved with an array of organisms that, because they were always present and needed to be tolerated, took on a role as inducers of immunoregulatory circuits. Such organisms include various microbiotas and commensals inhabiting the gut, skin, and lungs; chronic infections picked up at birth, helminths that persist for life, and environmental organisms from animals, mud, and untreated water with which humans were in daily contact in the environments in which they evolved and lived until recently (often referred to as pseudocommensals). The profound alterations in our relationships with these organisms that has resulted from modern living conditions results in a society-wide tendency towards defective immunoregulation that, depending on the genetic background of any given individual, can manifest as a variety of chronic inflammatory disorders, including allergies, inflammatory bowel disease, autoimmunity, and psychiatric conditions associated with altered measures of immune function (Fig. 26.8).

While the data implicating alterations in human–microbial interactions with psychiatric disease are limited and largely circumstantial, significant animal data indicate that these interactions have immune and

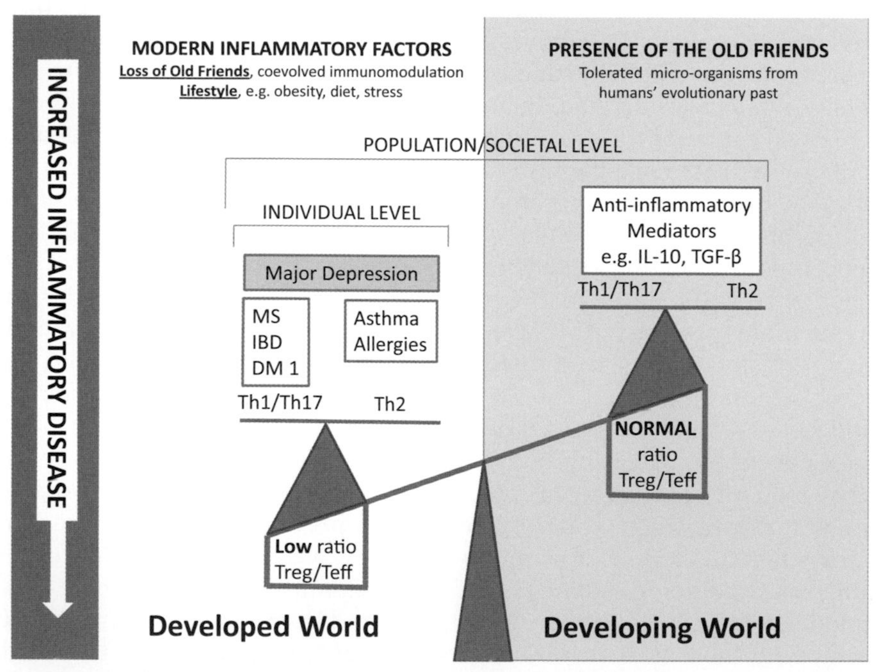

FIGURE 26.8 **Loss of contact with "old friends" and increased inflammatory conditions in the modern world.** In populations adequately exposed to old friends, such as many societies in the developing world, priming of regulatory T cells (Treg) is sufficient to maintain an appropriate balance of Treg to effector T cells (Teff), with the result that inappropriate inflammation is generally constrained. When contact with the old friends is disrupted as a result of modern cultural practices (e.g. sanitation, water and food treatment, modern medicines), priming of Treg is inadequate, with the result that the ratio of Treg to Teff is low. In this situation, the population as a whole is at risk for a variety of syndromes attributable to inadequate termination of inflammatory responses to any of a range of environmental stimuli. Consistent with this, although the prevalence of serious infections is significantly reduced in the industrialized world, rates of chronic inflammatory conditions (e.g. autoimmune diseases, allergies, asthma, and cardiovascular disease) have been shown in many studies to be much higher than in the developing world. Some individuals have a genetic background and/or immunological history that places them at risk for disorders, such as multiple sclerosis (MS), inflammatory bowel disease (IBD) and type 1 diabetes mellitus (DM), characterized by overactive/uncontrolled T helper (Th) type 1 and/or Th17 activity. In other individuals, Th2 responses are more liable to inadequate control, resulting in asthma and allergic disorders. While not developing gross immune-related pathology, a further group of individuals with inadequate termination of either Th1 or Th2 inflammatory responses is susceptible to CNS effects of cytokines, including major depressive disorder (MDD). It should be noted, however, that conditions associated with Th1/Th17 and Th2 dysregulation are highly comorbid with MDD. IL-10: interleukin-10; TGF-β: transforming growth factor-β.

behavioral effects of direct relevance to psychiatric functioning. Indeed, significant *in vitro* and *in vivo* preclinical data indicate that all three classes of old friends (microbiota, pseudocommensals, and parasites) have the potential to prevent or ameliorate pathology in animal models for autoimmune, allergic, and inflammatory bowel diseases, as well as neoplasms and certain infections, and do so via reductions in inflammatory activity. As a striking example of this, a single polysaccharide (polysaccharide A) from a *Bacteroides* species common in the microbiota largely corrected the subnormal and functionally distorted development of the immune system that occurs in germ-free mice (which lack a microbiota) and was protective against *Helicobacter hepaticus*-induced inflammatory colitis via induction of IL-10-producing Treg cells. Intragastric administration of the pseudocommensal *Mycobacterium vaccae* was shown to induce an immunoregulatory cytokine profile in mesenteric lymph nodes and the spleen, to reduce eosinophilic infiltrates in the lung following an intragastric allergen challenge, and, over several months of exposure, to reduce circulating levels of TNF-α and IL-4 in humans with tuberculosis. Prebiotics shown to increase *Bifidobacterium* species in the rodent gut markedly reduce plasma concentrations of IL-1β, TNF-α, IL-6, and MCP-1, and metabolic products from gut microbiota reduce inflammation in animal models of a variety of human autoimmune and allergic disorders, as well as in *in vitro* preparations of human peripheral blood mononuclear cells.

In addition to direct effects on immune functioning, there is evidence to suggest that the gut microbiome profoundly affects other whole body physiological processes, including peripheral pain sensitivity, sleep, and metabolism, that are affected by activity in inflammatory signaling pathways and that are abnormal in a variety of psychiatric conditions.[8] Moreover, an increasing number of animal studies suggests that host–microbial interactions early in development have profound effects on stress physiology and behavior by affecting the development and functioning of neuroendocrine and CNS processes relevant to the development of resilience or disease in response to stress.[39] For example, germ-free mice demonstrate abnormal responses to restraint stress, specifically increased ACTH and cortisol production, together with reduced expression of BDNF in the cortex and hippocampus, reduced GR in the cortex, and raised CRH in the hypothalamus, all of which have been repeatedly observed in the context of MDD. Oral reconstitution with a normal microbiota normalizes the HPA axis function if done at 6 weeks, but not if done later. Early mono-association with *Bifidobacterium infantis* also normalized HPA axis function, but mono-association with enteropathogenic *Escherichia coli* made the abnormalities more severe.

Taken together, the available data raise the intriguing possibility that disrupted human–microbial relationships might augment the immune–inflammatory and hence deleterious effects of psychosocial stress in the modern world, whether this stress occurs in adulthood, *in utero*, or in childhood. Again, the data are most compelling for the microbiota, which is known to be markedly different between individuals living in industrialized societies and those living more traditional lifestyles. Increasing evidence suggests that the microbiota can affect both brain development and functioning through a variety of signaling pathways. For example, germ-free mice have reduced levels of serotonin and norepinephrine in the cortex and hippocampus, and increased norepinephrine, dopamine, and serotonin turnover in the striatum, as well as diminished gene expression for BDNF in the hippocampus, amygdala, and cingulate cortex.[40] These animals also show alterations in long-term potentiation and patterns of synaptogenesis in the CNS. Perhaps the most compelling animal data demonstrating the ability of the microbiota to influence brain functioning and stress-relevant behavior come from studies showing that transplanting the microbiota from one strain of mouse to another causes the recipient mice to take on the behavioral repertoire of the donors. Specifically, when anxiety-prone germ-free BALB/c mice are colonized with microbiota from a far more gregarious strain (NIH Swiss mice), the transplanted animals demonstrate a striking increase in exploratory behavior. Conversely, germ-free NIH Swiss mice colonized with the microbiota from specific pathogen-free (SPF) BALB/c mice exhibit a reduction in exploratory behavior. That the microbiota institute these changes by affecting CNS functioning is supported by the fact that all the behavioral changes in the transplant experiments were associated with changes in BDNF levels in the hippocampus.

It is important to emphasize that the relationship between the microbiota and stress observed in animal models is bidirectional. Just as the composition of the microbiota appears to affect stress responses, stress changes the composition of the microbiota. For example, in mice the stress of maternal separation for 3 hours per day from postnatal days 2 to 12 has long-term effects on the subsequent 16S ribosomal RNA diversity of the microbiota that persists into adulthood, compared with control adults that were not exposed to maternal separation as pups. Similarly, prenatal stressors have been shown to alter the microbiome in rhesus monkeys by reducing the overall number of bifidobacteria and lactobacilli during adulthood. In addition to the ability of early life stressors to program baseline microbiota composition across the lifetime, animal studies demonstrate that acute stressors in adulthood can powerfully alter the composition of the microbiota. These alterations may then contribute to the behavioral changes observed in response to stress. For example, chronic exposure of mice to a grid flooring stressor produced significant changes in microbiota composition, which correlated with several circulating immune markers and with anxiety as assessed by closed arm entries and total time spent in the elevated plus-maze, time spent in a light/dark box, and time spent in the inner zone and total time spent in the open field test.

The data linking the microbiota and other environmental "old friends" organisms with psychiatric illness in humans lags far behind the more conclusive evidence derived from animal models. Nonetheless, a small double-blind, placebo-controlled trial found that treatment with *trans*-galacto-oligosaccharide (a prebiotic that increases gut *Bifidobacteria*) reduced anxiety in patients with irritable bowel syndrome, although the degree to which reduced anxiety resulted from improved bowel function is unclear. Similarly, 2 months of treatment with a *Lactobacillus* species reduced anxiety, but not depressive symptoms, in patients with chronic fatigue syndrome compared with placebo. Perhaps the most compelling data for any of the old friend organisms come from a large randomized trial of *M. vaccae* administration in patients with end-stage lung cancer, in which it was observed that the addition of *M. vaccae* to standard chemotherapy significantly improved quality of life in general, and depressive and anxiety symptoms in particular. The rapidly expanding understanding of the relevance of human–microbial and human–parasitic interactions to disease pathogenesis suggests that in addition to symptomatic treatment, organism-based interventions may show promise, if delivered early in life, as preventive or ameliorative strategies not just for depression and anxiety but perhaps for schizophrenia and autism as well.

EVIDENCE THAT PSYCHIATRIC CONDITIONS ARE ASSOCIATED WITH ALTERATIONS IN PERIPHERAL AND CNS IMMUNE ACTIVITY

This chapter concludes by addressing the most obvious question in relation to the association between the immune system and psychiatry: What is the evidence that psychiatric diseases are characterized by immune system abnormalities? Hundreds, and perhaps thousands, of studies have been conducted over the past 30 years examining cross-sectional associations between various psychiatric conditions and immune functioning in the periphery and CNS. This literature on this subject is vast: a summary is provided in Table 26.2, which lists many of the most commonly reported immune abnormalities associated with various psychiatric disorders,[41–45] with the strong caveat that negative results are almost always available to counter every positive finding that is reported. Moreover, as noted in Table 26.1, despite the surfeit of data, many key questions remain unanswered, including (1) whether specific immune changes are more strongly associated with some diagnoses than others (i.e. disease state specificity); (2) whether observed abnormalities reflect trait (and hence underlying vulnerability) versus state factors within any given disease state; (3) where the primary causal immune locus is located (brain versus body, lymphoid versus other cytokine-secreting tissues); and (4) whether immune abnormalities reported for any given disease state are relevant to the condition as a whole or only for individuals with the disease who actually manifest the abnormality.

Most researchers in the field would probably agree that elevations in proinflammatory mediators are the most commonly replicated finding of relevance to the association between psychiatric disease and the immune system. The source for this increase in inflammatory signaling, however, remains a matter of much debate, given that a range of tissues in the body can produce these molecules, including cells of the innate (i.e. monocytes and macrophages) and acquired (i.e. T cells) immune systems, as well as endothelial cells and adipocytes (which may help to account for the association of obesity with increased inflammation). As might be predicted from these findings, increasing data suggest that psychiatric diseases are characterized by hyperactivity of both peripheral monocytic cells and CNS microglia. Activation of peripheral T cells has also been observed, as have both increased and reduced numbers of Treg cells (reported to be reduced in depression, reduced and increased in bipolar disorder, but increased in schizophrenia). Strong associations between abdominal fat and increases in inflammatory biomarkers (especially CRP and IL-6) and between obesity and depression suggest that these cells may also represent causal foci for some of the link between MDD and increased inflammatory markers. The role of adipocytes in accounting for the increased inflammatory biomarkers observed in other conditions, such as schizophrenia and bipolar disorder, is less well studied. Finally, much interest has been paid over the years to the question of the degree to which various psychiatric conditions might be characterized by relative increases in Th1 versus Th2 immunity, with conflicting results. There is evidence to suggest that highly inflammatory Th17 type cells may also be overactive in certain contexts, such as recent-onset schizophrenia.

Despite the brain having been viewed for years as an "immune privileged organ", increasing evidence suggests that major psychiatric conditions are associated

TABLE 26.2 Immune Abnormalities Associated with Various Psychiatric Disorders

Disease State	Immune Abnormalities	Strength of Findings
Major depression	Increased blood concentrations of IL-6, TNF-α, CRP, sIL-2R, haptoglobin, PGE_2, increased WBC count, overall leukocytosis, reduced NK cell numbers, reduced percentage of T cells, increased CD4/CD8 ratio, reduced T and NK cell function, reduced lymphocyte proliferative response to mitogen	Meta-analyses
	Circulating plasma/serum/blood immune molecules	Single studies
	Increased plasma/serum levels of IL-1α, IL-1β, IL-1 receptor antagonist, the sCD8 molecule, neopterin, sTNFR1 and sTNFR2, MIP-1, eotaxin (CCL11), reduced serum/plasma concentrations of IL-10 (increases also reported), MCP-1 (CCL2) (increases also reported), and RANTES/CCL5	
	Stimulated production of immune molecules	
	Increased stimulated production of IFN-γ	
	Measures of cellular number/function	
	Reduced circulating Forkhead 2 (foxp3) positive $CD3^+CD25^+$ Tregs	
	Increased numbers and percentages of T cells bearing T-cell activation markers, such as $CD2^+CD25^+$, $CD3^+CD25^+$, and $HLA\text{-}DR^+$	
	Changes in intracellular signaling pathways (MAPK, NF-κB, etc.)	
	Gene expression profiling results	
	CNS findings	
	Increased CSF concentrations of IL-1β; activation of microglial cells	
Bipolar disorder	Changes in transcriptome profiles in multiple brain regions overlap with schizophrenia and major depressive disorder. Genes involved in energy metabolism and mitochondrial function are downregulated, genes involved in immune response and inflammation are upregulated, and genes expressed in oligodendrites are downregulated	?
Schizophrenia	Increased serum levels of IL-1α, IL-1β, IL-8, TNF-α, and IL-6, and increased leukocyte mRNA levels of IL-1α, IL-6, and TFN-α	Single study
	State markers: increased in acutely relapsed and first episode psychosis inpatients for IL-1β, IL-6, and TGF-β. In contrast, IL-12, IFN-γ, TNF-α, and sIL-2R appear to be trait markers. In CSF, IL-1β significantly decreased versus controls	Meta-analyses
	Increased levels of lymphocytes, CD3, CD4, and CD4/CD8 ratio, and decrease in CD3% in drug-naïve first episode psychosis. In longitudinal studies, CD4/CD8 ratio appears to be a state-related marker, as it decreases after antipsychotic treatment for acute exacerbations of psychosis. CD56 appears to be a trait marker, as levels significantly increase after antipsychotic treatment for relapse	Meta-analyses
	Significant association between maternal IL-8 levels during second trimester and risk of schizophrenia. Elevated third trimester homocysteine levels may elevate schizophrenia risk through developmental effects on brain structure and function and/or subtle damage to placental vasculature	Multiple studies
	Dysfunction in activation of type 1 immune response associated with decreased activity of tryptophan/kynurenine metabolism, 2,3-dioxygenase. Decreased activity associated with production of kynurenic acid (NMDA antagonist in CNS) and reduced glutamatergic neurotransmission in schizophrenia. Inflammatory state associated with increased PGE_2 production and increased COX-2 expression	Single study
	Dysregulation of kynurenine pathway results in hyperfunction or hypofunction of active metabolites, associated with neurodegenerative and other neurological disorders, depression and schizophrenia	Single study
Obsessive–compulsive disorder	Reduced plasma serum levels of IL-1β, with no significant difference in IL-6 and TNF-α levels. Stratified subgroup analysis revealed possible moderating effects of age and medication use on IL-6 levels. Studies including children on psychotropic medication had lower plasma IL-6 levels. Stratified subgroups analysis revealed a moderating effect of comorbid depression of TNF-α levels. Elevated TNF-α levels reported in studies including individuals with comorbid depression	Meta-analyses

Continued

TABLE 26.2 Immune Abnormalities Associated with Various Psychiatric Disorders—cont'd

Disease State	Immune Abnormalities	Strength of Findings
Autism	Maternal influenza infection associated with two-fold increased risk of infantile autism, prolonged episodes of fever caused three-fold increased risk of infantile autism, and use of antibiotics during pregnancy was a potent risk factor for autism spectrum disorder/infantile autism	Single study

IL: interleukin; TNF: tumor necrosis factor; CRP: C-reactive protein; sIL2R: soluble interleukin-2 receptor; PGE_2: prostaglandin E_2; WBC: white blood cell; NK: natural killer; sTNFR: soluble TNF-α receptor; MIP: macrophage inflammatory protein; MCP: monocyte chemoattractant protein; RANTES: regulated on activation, normally T-cell-expressed and T-cell-secreted; IFN: interferon; Treg: T-regulatory cell; HLA: human leukocyte antigen; MAPK: mitogen-activated protein kinase; NF-κB: nuclear factor-κB; CSF: cerebrospinal fluid; TGF-β: transforming growth factor-β; NMDA: *N*-methyl-D-aspartate; COX-2: cyclooxygenase-2;

with activation of microglial cells, which although not directly descended from monocytes, serve as the brain's mononuclear phagocytic cells, and are the primary source of inflammatory signaling in the CNS. For example, postmortem studies have reported increased microglial density and activity in schizophrenia, which has been correlated with IL-1β gene expression in dorsolateral prefrontal cortex. Increased activation of microglial cell surface markers has also been reported in MDD, as has increased microglial quinolinic acid production in the subgenual ACC in patients with severe MDD, and a trend for this association in bipolar disorder. Compared with healthy controls, deceased patients who had bipolar disorder were found to have significantly higher protein and messenger RNA levels of IL-1β, the IL-1 receptor (IL-1R), myeloid differentiation factor 88, NF-κB, and astroglial and microglial markers in the frontal cortex, all consistent with microglial activation. Similarly, widespread immune activation based on gene expression analysis has been reported in the frontal cortex (Brodmann area 10) in patients with MDD, with elevations observed in IL-1α, IL-2, IL-3, IL-5, IL-8, IL-9, IL-10, IL-12A, IL-13, IL-15, IL-18, IFN-γ, and lymphotoxin-α (TNF superfamily member-1). These postmortem findings have been supported by a PET study using (R)-[^{11}C]PK11195, a ligand that binds to the translocator protein (formerly known as the peripheral benzodiazepine receptor), which is concentrated in reactive microglia. These investigations found evidence of microglial activation in the brains of individuals with recent-onset schizophrenia.

Although individual studies have reported various differences in central and peripheral immune measures across the major psychiatric conditions, meta-analyses of peripheral immune biomarkers conducted for schizophrenia, bipolar disorder, and MDD suggest a common pattern of increased inflammatory biomarkers. All three conditions are associated with increased circulating TNF-α. MDD and schizophrenia are both associated with increased circulating IL-6, and schizophrenia and bipolar disorder are both associated with increased soluble receptor for the proinflammatory T-cell cytokine IL-2 (sIL-2R). Perhaps the most consistent difference between the conditions is the increase in the anti-inflammatory cytokine TGF-β that has been repeatedly observed in schizophrenia but not in the other major disorders, which have been more reliably associated with reduced measures of anti-inflammatory tone.

Psychiatric disorders are not static entities but rather wax and wane with time, and often progress. Given this, one might expect that many studies would have followed individuals longitudinally to evaluate how these time-associated changes affect the association between the disorder and immune functioning. In fact, very few data are available in this regard and most studies that have examined immune changes associated with different stages of any given illness have done so by cross-sectionally comparing different individuals at different disease stages (e.g. mania versus depression, early versus late in schizophrenia course) rather than following the same individuals over time. In general, findings suggest that immune abnormalities exist both between episodes and within episodes, with abnormalities typically being more intense and wide-ranging within episodes than when the disease is in relative remission. Similarly, many, but not all, studies suggest that effective treatment reduces levels of inflammatory biomarkers, and often these decreases are correlated with symptomatic improvement. Finally, some evidence suggests that the association between immune activation and disease presence may strengthen as the disease progresses, as reported in bipolar disorder. Consistent with this, the association between depression and inflammation increases with age, although the degree to which this reflects ongoing effects of the disease or the fact that older people are more likely to have conditions that drive inflammation and secondarily produce depression is an open question. On the other hand, anti-inflammatory interventions seem to be especially effective early in the disease course of schizophrenia. This suggests that immune processes may be more causally linked to disease pathogenesis early in disease than later in the illness, when the damage might be done and irreversible brain changes are no longer amendable to the attenuation of tissue-damaging immune processes.

A final question is whether psychiatric disorders are best understood as being associated with immune

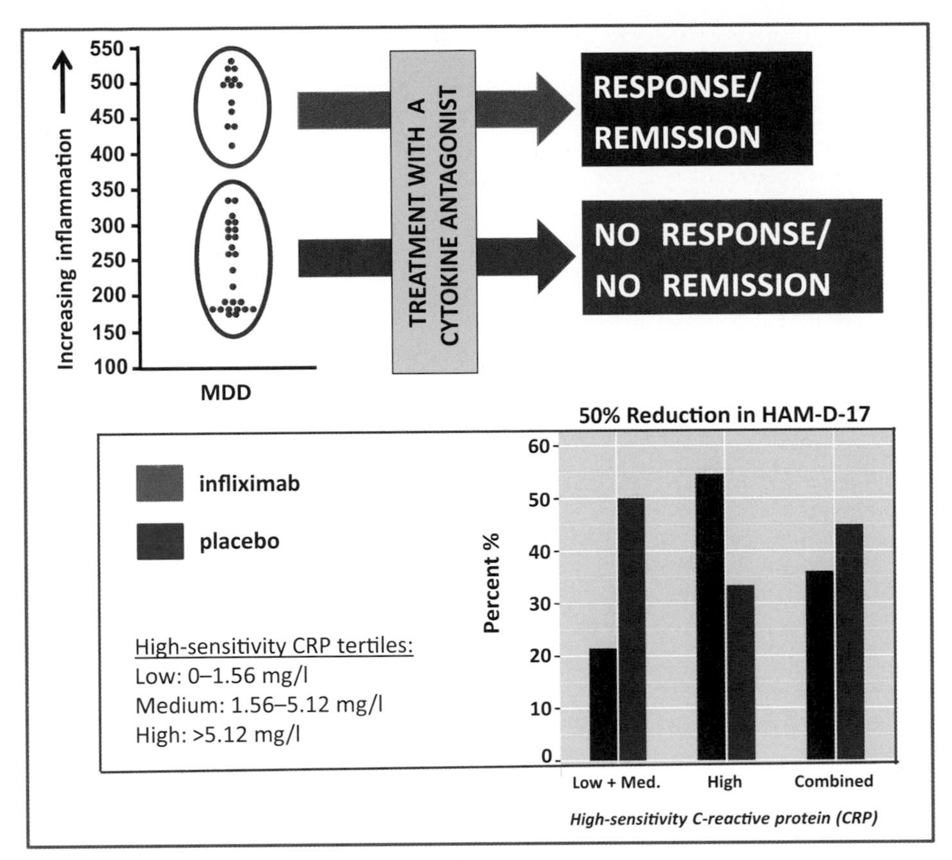

FIGURE 26.9 **Evidence that immune abnormalities may identify treatment-relevant subgroups within current psychiatric diagnostic categories.** If immune pathways contribute to disease pathogenesis in psychiatric conditions such as major depressive disorder (MDD), it might be predicted that these contributions would be most relevant for individuals with abnormalities in immune function. Patients with MDD demonstrate a wide range of values for any given measure of peripheral immune activity, with some patients having values higher than those typically seen in matched non-depressed populations and others having values as low as, or occasionally lower than, values observed in normal individuals. If the measure of immune activity in question contributes to depressive pathogenesis, it would be predicted that blocking this factor would produce a therapeutic response in those with elevated levels of the measure, while having no effect in individuals with normal or low levels of the factor. This idea was tested by randomizing 60 medically stable individuals with treatment-resistant depression (TRD) to receive either three infusions of the tumor necrosis factor-α (TNF-α) antagonist infliximab or a salt-water placebo. In the group as a whole, no effect was seen for infliximab compared with placebo. However, a linear relationship was observed between baseline plasma concentrations of both C-reactive protein (CRP) and TNF-α and antidepressant response to infliximab, illustrated in the figure by a graph showing differential antidepressant response rates to infliximab versus placebo based on a tertile split in pretreatment plasma concentrations of CRP. HAM-D-17: Hamilton Rating Scale for Depression.

abnormalities in general, or as being conditions in which only a subset of affected individuals (i.e. those with evidence of immune abnormalities) should be considered as having an immune system contribution to their particular pathology. This issue may be obscured by many of the statements that have been made in this chapter, suggesting that psychiatric diseases are associated with this or that immune change. In fact, for all psychiatric conditions immune values significantly overlap with normal comparison subjects. Is it possible that people with schizophrenia or depression who have immune values comparable to controls (or even lower in some cases) might still have an immune abnormality, or does it make more sense to assume that these individuals have come to their pathologies via non-immune mechanisms? One way to test this would be to block immune pathways and evaluate whether therapeutic effects were widespread within any given disease state, or were limited to those with evidence of abnormal immune functioning. A recent study has done this, as described in Fig. 26.9,[46] and the results strongly suggest that, for MDD at least, immune pathways may only be relevant to disease pathogenesis in those who demonstrate increased activity in these same pathways. Indeed, in a group of subjects with treatment-resistant depression, only those with high inflammation, as reflected by a CRP greater than 5 mg/l, exhibited a clinically significant response to a TNF antagonist. Whether this will hold true for the other psychiatric conditions associated with immune activation is an open question. Nevertheless, these early findings suggest that there may be subgroups of psychiatric patients with immune alterations that can be identified by relevant peripheral immunological biomarkers and treated with immune-targeted therapies.

References

1. Haroon E, Raison CL, Miller AH. Psychoneuroimmunology meets neuropsychopharmacology: translational implications of the impact of inflammation on behavior. *Neuropsychopharmacology*. 2012;37(1): 137–162.
2. Dantzer R, O'Connor JC, Freund GG, Johnson RW, Kelley KW. From inflammation to sickness and depression: when the immune system subjugates the brain. *Nat Rev Neurosci*. 2008;9(1):46–56.
3. Irwin MR, Cole SW. Reciprocal regulation of the neural and innate immune systems. *Nat Rev Immunol*. 2011;11(9):625–632.
4. O'Connor MF, Bower JE, Cho HJ, et al. To assess, to control, to exclude: effects of biobehavioral factors on circulating inflammatory markers. *Brain Behav Immun*. 2009;23(7):887–897.
5. Schwarcz R, Bruno JP, Muchowski PJ, Wu HQ. Kynurenines in the mammalian brain: when physiology meets pathology. *Nat Rev Neurosci*. 2012;13(7):465–477.
6. Miller BJ, Culpepper N, Rapaport MH, Buckley P. Prenatal inflammation and neurodevelopment in schizophrenia: a review of human studies. *Prog Neuropsychopharmacol Biol Psychiatry*. 2013;42:92–100.
7. Raison CL, Miller AH. Is depression an inflammatory disorder? *Curr Psychiatry Rep*. 2011;13(6):467–475.
8. Raison CL, Lowry CA, Rook GA. Inflammation, sanitation, and consternation: loss of contact with coevolved, tolerogenic microorganisms and the pathophysiology and treatment of major depression. *Arch Gen Psychiatry*. 2010;67(12):1211–1224.
9. Keller WR, Kum LM, Wehring HJ, Koola MM, Buchanan RW, Kelly DL. A review of anti-inflammatory agents for symptoms of schizophrenia. *J Psychopharmacol*. 2013;27(4):337–342.
10. Raison CL, Miller AH. The evolutionary significance of depression in Pathogen Host Defense (PATHOS-D). *Mol Psychiatry*. 2013;18(1): 15–37.
11. Corona AW, Norden DM, Skendelas JP, et al. Indoleamine 2,3-dioxygenase inhibition attenuates lipopolysaccharide induced persistent microglial activation and depressive-like complications in fractalkine receptor (CX(3)CR1)-deficient mice. *Brain Behav Immun*. 2013;31:134–142.
12. Atladottir HO, Henriksen TB, Schendel DE, Parner ET. Autism after infection, febrile episodes, and antibiotic use during pregnancy: an exploratory study. *Pediatrics*. 2012;130(6):e1447–e1454.
13. Landreau F, Galeano P, Caltana LR, et al. Effects of two commonly found strains of influenza A virus on developing dopaminergic neurons, in relation to the pathophysiology of schizophrenia. *PLoS ONE*. 2012;7(12):e51068.
14. Willette AA, Lubach GR, Knickmeyer RC, et al. Brain enlargement and increased behavioral and cytokine reactivity in infant monkeys following acute prenatal endotoxemia. *Behav Brain Res*. 2011;219(1):108–115.
15. Benros ME, Nielsen PR, Nordentoft M, Eaton WW, Dalton SO, Mortensen PB. Autoimmune diseases and severe infections as risk factors for schizophrenia: a 30-year population-based register study. *Am J Psychiatry*. 2011;168(12):1303–1310.
16. Nimmervoll B, White R, Yang JW, et al. LPS-induced microglial secretion of TNFα increases activity-dependent neuronal apoptosis in the neonatal cerebral cortex. *Cereb Cortex*. 2013;23(7):1742–1755.
17. Kipnis J, Gadani S, Derecki NC. Pro-cognitive properties of T cells. *Nat Rev Immunol*. 2012;12(9):663–669.
18. Frodl T, Amico F. Is there an association between peripheral immune markers and structural/functional neuroimaging findings? *Prog Neuropsychopharmacol Biol Psychiatry*. 2014;48:295–303.
19. Capuron L, Neurauter G, Musselman DL, et al. Interferon-alpha-induced changes in tryptophan metabolism. relationship to depression and paroxetine treatment. *Biol Psychiatry*. 2003;54(9):906–914.
20. Capuron L, Ravaud A, Neveu PJ, Miller AH, Maes M, Dantzer R. Association between decreased serum tryptophan concentrations and depressive symptoms in cancer patients undergoing cytokine therapy. *Mol Psychiatry*. 2002;7(5):468–473.
21. Bonaccorso S, Marino V, Puzella A, et al. Increased depressive ratings in patients with hepatitis C receiving interferon-alpha-based immunotherapy are related to interferon-alpha-induced changes in the serotonergic system. *J Clin Psychopharmacol*. 2002;22(1):86–90.
22. Guidetti P, Schwarcz R. 3-Hydroxykynurenine and quinolinate: pathogenic synergism in early grade Huntington's disease? *Adv Exp Med Biol*. 2003;527:137–145.
23. Guillemin GJ, Meininger V, Brew BJ. Implications for the kynurenine pathway and quinolinic acid in amyotrophic lateral sclerosis. *Neurodegener Dis*. 2005;2:166–176.
24. Guillemin GJ, Brew BJ, Noonan CE, Takikawa O, Cullen KM. Indoleamine 2,3 dioxygenase and quinolinic acid immunoreactivity in Alzheimer's disease hippocampus. *Neuropathol Appl Neurobiol*. 2005;31:395–404.
25. Schwarcz R, Pellicciari R. Manipulation of brain kynurenines: glial targets, neuronal effects, and clinical opportunities. *J Pharmacol Exp Ther*. 2002;303:1–10.
26. Wu HQ, Rassoulpour A, Schwarcz R. Kynurenic acid leads, dopamine follows: a new case of volume transmission in the brain? *J Neural Transm*. 2007;114:33–41.
27. Muller N. Inflammation and the glutamate system in schizophrenia: implications for therapeutic targets and drug development. *Expert Opin Ther Targets*. 2008;12(12):1497–1507.
28. Brydon L, Harrison NA, Walker C, Steptoe A, Critchley HD. Peripheral inflammation is associated with altered substantia nigra activity and psychomotor slowing in humans. *Biol Psychiatry*. 2008;63(11):1022–1029.
29. Capuron L, Pagnoni G, Demetrashvili M, et al. Anterior cingulate activation and error processing during interferon-α treatment. *Biol Psychiatry*. 2005;58(3):190–196.
30. Eisenberger NI, Cole SW. Social neuroscience and health: neurophysiological mechanisms linking social ties with physical health. *Nat Neurosci*. 2012;15(5):669–674.
31. Harrison NA, Brydon L, Walker C, Gray MA, Steptoe A, Critchley HD. Inflammation causes mood changes through alterations in subgenual cingulate activity and mesolimbic connectivity. *Biol Psychiatry*. 2009;66(5):407–414.
32. Danese A, McEwen BS. Adverse childhood experiences, allostasis, allostatic load, and age-related disease. *Physiol Behav*. 2012;106(1): 29–39.
33. Dhabhar FS. A hassle a day may keep the pathogens away: the fight-or-flight stress response and the augmentation of immune function. *Integr Comp Biol*. 2009;49(3):215–236.
34. Hanke ML, Powell ND, Stiner LM, Bailey MT, Sheridan JF. Beta adrenergic blockade decreases the immunomodulatory effects of social disruption stress. *Brain Behav Immun*. 2012;26(7): 1150–1159.
35. Miller GE, Chen E, Sze J, et al. A functional genomic fingerprint of chronic stress in humans: blunted glucocorticoid and increased NF-kappaB signaling. *Biol Psychiatry*. 2008;64(4):266–272.
36. Cole SW, Conti G, Arevalo JM, Ruggiero AM, Heckman JJ, Suomi SJ. Transcriptional modulation of the developing immune system by early life social adversity. *Proc Natl Acad Sci U S A*. 2012;109(50):20578–20583.
37. McDade TW. Early environments and the ecology of inflammation. *Proc Natl Acad Sci U S A*. 2012;109(suppl 2):17281–17288.
38. McDade TW, Hoke M, Borja JB, Adair LS, Kuzawa C. Do environments in infancy moderate the association between stress and inflammation in adulthood? Initial evidence from a birth cohort in the Philippines. *Brain Behav Immun*. 2013;31:23–30.
39. Rook GAW, Lowry CA, Raison CL. Microbial "old friends", immunoregulation and stress resilience. *Evol Med Public Health*. 2013;(1):46–64.
40. Cryan JF, Dinan TG. Mind-altering microorganisms: the impact of the gut microbiota on brain and behaviour. *Nat Rev Neurosci*. 2012;13(10):701–712.

41. Dowlati Y, Herrmann N, Swardfager W, et al. A meta-analysis of cytokines in major depression. *Biol Psychiatry*. 2010;67(5):446–457.
42. Howren MB, Lamkin DM, Suls J. Associations of depression with C-reactive protein, IL-1, and IL-6: a meta-analysis. *Psychosom Med*. 2009;71(2):171–186.
43. Miller BJ, Buckley P, Seabolt W, Mellor A, Kirkpatrick B. Meta-analysis of cytokine alterations in schizophrenia: clinical status and antipsychotic effects. *Biol Psychiatry*. 2011;70(7):663–671.
44. Miller BJ, Gassama B, Sebastian D, Buckley P, Mellor A. Meta-analysis of lymphocytes in schizophrenia: clinical status and antipsychotic effects. *Biol Psychiatry*. 2013;73(10):993–999.
45. Munkholm K, Vinberg M. Vedel Kessing L. Cytokines in bipolar disorder: a systematic review and meta-analysis. *J Affect Disord*. 2013;144(1–2):16–27.
46. Raison CL, Rutherford RE, Woolwine BJ, et al. A randomized controlled trial of the tumor necrosis factor antagonist infliximab for treatment-resistant depression: the role of baseline inflammatory biomarkers. *JAMA Psychiatry*. 2013;70(1):31–41.

CHAPTER

27

Infections and Nervous System Dysfunction

Krister Kristensson

Department of Neuroscience, Karolinska Institutet, Stockholm, Sweden

OUTLINE

INTRODUCTION

We are all surrounded by microorganisms such as bacteria, parasites, fungi, and viruses, collectively called microbes. The microbes can attack and interact with each other, with plants, and with cells and organisms of invertebrate or vertebrate host animals. A minor proportion of them can cause diseases after entering the body (i.e. infections: Latin *inficere*, in, inside; *facere*, do, make; to influence from inside), and are then called pathogens. The genetic codes of microbes are harbored in nucleic acids, whereas the replication of more recently defined pathogens, called prions, is encoded by a misfolded protein. In an infected multicellular host microbes can replicate inside or outside a cell; respectively, intracellular or extracellular microbes.

A very severe threat to a host is spread of pathogens to the nervous system, where they can cause various types of dysfunction, depending on the different regions under attack, their interactions with various cellular elements, and the different host immune responses that aim to combat the invaders. Infections limited to

Neurobiology of Brain Disorders
http://dx.doi.org/10.1016/B978-0-12-398270-4.00027-6

meninges cause meningitis, to the brain parenchyma cause encephalitis, and to the spinal cord cause myelitis; often, combinations such as meningoencephalitis or encephalomyelitis occur. Some microbes cause persistent infections in the nervous system to promote their own survival, provided that they do not kill the host or are eliminated by the host's defenses.

This chapter will examine how various pathogens cause a variety of CNS dysfunctions and the mechanisms of these changes. First, infections will be placed into a broader context, with a discussion on how the mechanisms by which microbes interact with a cell have evolved, and how cells and organisms defend themselves from microbial attacks.

MICROBE–HOST CELL INTERACTIONS

Microbes and Evolution

Microbes have been a driving force both in adaptation to the environment and in the evolution of host cells and animals. This is particularly evident for viruses, which depend on a host cell for replication and survival. These interactions have a long and intensive history and they may be viewed as an "arms race" giving advantages or losses to either the virus or the host cell. Sometimes an optimal interaction with a gain or added value for both may occur. Killing a host has no long-term survival value for a microbe, because that would deprive the microbe of its life source. There are innumerable virus–cell interactions; for instance, in apparently clean surface seawater about 10 million viral particles are found per milliliter, with an average of 10 viral particles per target bacterial or plankton cell to interact with and use for replication. The diversity of microbes is also extremely high: viral metagenomics have disclosed about 1 million viral genotypes in 1 kg of marine sediment.

Viruses depend on host cell mechanisms for their replication, assembly, and release or transmission. For this purpose they have learned to hijack the synthetic and transport machinery of the cell for their own advantage. They can also introduce new genes into the host genome in their favor, such as genes that enhance photosynthesis and thereby make more energy available for replication. A transfer of genes from viruses can more dramatically alter the function of a host cell by switching a bacterial cell from a non-pathogenic to a pathogenic one; for example, a viral gene encoding cholera toxin gives toxicity to otherwise non-toxic *Vibreo cholerae* bacteria.[1] Viruses may survive for extended periods not only by replicating large numbers of copies to be released for new infections, but also by injecting their genome into the host cell, as in some DNA viruses, and forming plasmids. The viral prophage elements are thereby maintained by transversal inheritance during unfavorable environmental conditions until these become favorable again, whereupon viruses are activated and released. Some viruses encoded by RNA can persist in eukaryotic cells at a very low level of replication, while retroviruses harbor reverse transcriptases that transcribe their genomic RNA into a complementary DNA molecule to be incorporated in the host cell genome.

Harboring viruses may sometimes be advantageous for the survival of the host. During evolution in the marine world, cellular functions were modified and shaped to cope with or prevent attacks by other, more pathogenic viruses. For instance, viruses may suppress unnecessary metabolic activities of the cell[2] or expression of functional surface molecules used by other pathogens for their entry. Viruses in the sea can also incorporate host cell DNA and transduce it to other cells. Since microbes in seawater constitute the largest gene pool on Earth, viruses play a major role in vertical transmission as well as horizontal gene transfer within this pool. Thus, virus–host cell interactions have participated profoundly in driving evolution and shaping cellular functions in marine organisms, and this also applies to life on land. A prominent example of the latter is the retroviral *env* genes, adopted by mammals, which encode an important component of the placenta.[3]

Entry of Microbes That Cause Disease (Pathogens) Into the Body

In their uncountable interactions with cells over millions of years, microbes have probably explored all possible metabolic pathways that can provide them with any advantage, but the host cells or organisms have also developed mechanisms, or immune responses, to protect themselves. These immune responses do not occur if an optimal condition has been established where survival of microbes may add a value to the life of a host. For example, about 1000 different viruses and a large number of bacteria thrive in the human gut. These organisms play an essential role in food digestion and provide essential nutrients. However, it has been estimated that about 1400 microbes in our environment can cause human diseases,[4] that is, be pathogenic, and a complexity of immune responses has evolved against these as defense mechanisms.

Several microbes can survive by establishing a long-term, persistent, dormant or latent, infection in a host, and this is probably the most ancient type of infection. The geographical spread of these microbes can be enhanced if they can interact with both a definite host and a different intermediate host with other habits and mobility, known as a vector. Pathogens that only cause acute infections are probably of more recent origin, because they need a high population density to survive, not to die out. It has been estimated that a population size of about 300,000–800,000 individuals is needed to maintain infections with measles, meaning that human infections with this virus did not appear before the rise of large cities, such as Babylon.[4] Several of the human pathogenic microbes originate from animals and transmissions from animal to humans are called zoonoses. For instance, retroviruses such as human immunodeficiency virus (HIV) and human T-cell leukemia virus (HTLV) probably originated from monkeys, measles and tuberculosis from cattle, and influenza virus from pigs and birds.

Pathogens that attack humans can spread via water, air, or contaminated food, or through direct skin contact. Microbes have developed various strategies to enter the organism.

The skin is covered by a layer of dead material, which provides the host with a barrier. This barrier can, however, be penetrated by protease-secreting larvae of some parasites, by bites of infected insects or animals, and by injections of pathogens. Microbes that need the metabolism of a cell for replication (intracellular microbes) may find epithelial cells in the respiratory, gastrointestinal, or urogenital tract to enter and multiply. Infection of a cell is facilitated if the microbe can attach to the cell surface; for this purpose, microbes use a large variety of attachment molecules or receptors expressed on cellular surfaces. They have thereby adapted to surface molecules that serve other cellular functions. Intracellular pathogens have the advantage that they can avoid antimicrobial molecules in the circulation (humoral immune response) by hiding inside cells. Extracellular microbes are instead directly exposed to these responses. The strategies that the host and the microbes have developed to detect and avoid each other, respectively, will now be highlighted.

IMMUNE RESPONSES TO INVADING PATHOGENS

Evolution of Immune Responses

As a defense against a microbial attack, various immune mechanisms have evolved at both cellular and systemic levels. Studies on the nematode *Caenorhabditis elegans* have revealed several molecular microbe–host interactions that are conserved across the animal kingdom.[5] Such interactions can result in production of antimicrobial peptides that form a component of the inducible innate immune response. Induction of some of these antimicrobial peptides in the epidermis and the intestine can also be regulated by neuronal activities; for example, epidermal immunity against fungi is induced by signals from the nervous system, while a sustained increased neuronal secretion in the intestines, similar to conditions under long-term psychological stress, can increase the sensitivity to infections with human pathogenic *Pseudomonas aeruginosa* bacteria. Host phagocytes can also produce reactive oxygen species (ROS) and nitric oxide (NO) that kill microbes, but at the same time be protected from protein or DNA damage by glutathione and thioredoxin, heat-shock transcription factors, and a branch of the unfolded protein response in the endoplasmic reticulum that plays a role in maintaining protein homeostasis (proteostasis).

Caenorhabditis elegans can distinguish bacteria that are pathogenic from those that provide nutrients by 32 chemosensory neurons located at each end of its body. In this way, the nematode's innate response can directly recognize peptides secreted by certain pathogenic bacteria (*Serratia*) strains and repel them, but with regard to other bacteria (Gram-negative bacteria), the nematodes have learned by association whether they are pathogenic or a source of food. This associative learning that elicits either an attraction or avoidance behavior is mediated by serotonergic neurons. An example of commensal or mutualistic interaction is the life cycle of a close relative to *C. elegans* called *Heterorhabditis bacteriphora*, which interacts with the enteric bacterium *Photorhabdus luminescens* to the benefit of both organisms.[5]

Innate Immune Responses

The microbe–host interactions in mammalian species build partly on the evolutionarily conserved pathways, but some novel mechanisms have been added to increase the complexity by which humans can interact with the microbes in the environment. The innate immune response employs both constitutively expressed antimicrobial molecules and molecules that are induced when encountering the pathogen. In both invertebrates and vertebrates a system of such inducible innate immune response molecules is released when pattern recognition receptors (PRRs) on the cells recognize pathogen-associated molecular patterns (PAMPs) released by the invading microbes. Different classes of PRR are expressed on cellular membranes, for example toll-like receptors (TLRs), and in the cytosol, for example retinoic acid-inducible-I (RIG-I)-like receptors (RLRs) or Nod-like receptors (NLRs). The TLRs were originally detected in gene-manipulated *Drosophila melanogaster* and play a role in their protection from fungal attacks. They encompass 12 members identified in mice and 10 in humans. Those expressed on the surface at the plasma membrane (TLR1, 2, 4, 5, 6, and 11) recognize various bacterial and protozoan components, while those expressed in endolysosomes (TLR3, 7, 8, and 9) recognize nucleic acids from RNA viruses as well as DNA sequences that become exposed when microbes are degraded in these organelles.[6] By this system, humans have learned to distinguish pathogenic *Salmonella* bacteria, which release

PAMPs, from nourishing salmon, which do not contain such molecules. Activation of the TLRs induces expression of various cytokines via the nuclear factor-κB (NF-κB) pathway in the cell, as well as interferon-α and -β (IFN-α/β). Such molecules, when released, activate mechanisms that probably prevent and eliminate most infections. The RLRs and NLRs recognize genomic RNA generated by replicating single-stranded RNA viruses as well as bacterial components (i.e. peptidoglycan components) in the cytoplasm of an infected cell, respectively, to inhibit further replication and spread of the pathogens. Natural killer (NK) cells are also an important component of the innate immune response. They can non-specifically kill microbe-infected cells[7] (Fig. 27.1).

The innate immune response can, thus, distinguish foreign molecules from species-specific molecules (it keeps the memory of the species) as well as pathogenic from non-pathogenic microbes. However, it keeps no memory of an invading pathogen, which implies that the microbial attack can be repeated over and over again without changes in the innate immune response elicited every time. When innate immune responses are insufficient, the adaptive immune response has to be called in to increase the chances for clearance or control of the pathogenic microbes.

Adaptive Immune Responses

Adaptive immune responses take several days to reach a threshold at which they are effective in controlling an infection. They are usually efficient owing to their restrictive specificity. An adaptive immune response keeps the memory of the specific invading microbe for years or decades.

One of the first cells that an invading pathogen encounters is the dendritic cell. These cells have extensive branches of cellular processes to provide a large surface area from which microbes can be incorporated by endocytosis or phagocytosis. The endocytosed particles are then cleaved into fragments by special proteases and the fragments are presented at the cell surface in context with major histocompatibility complex (MHC)

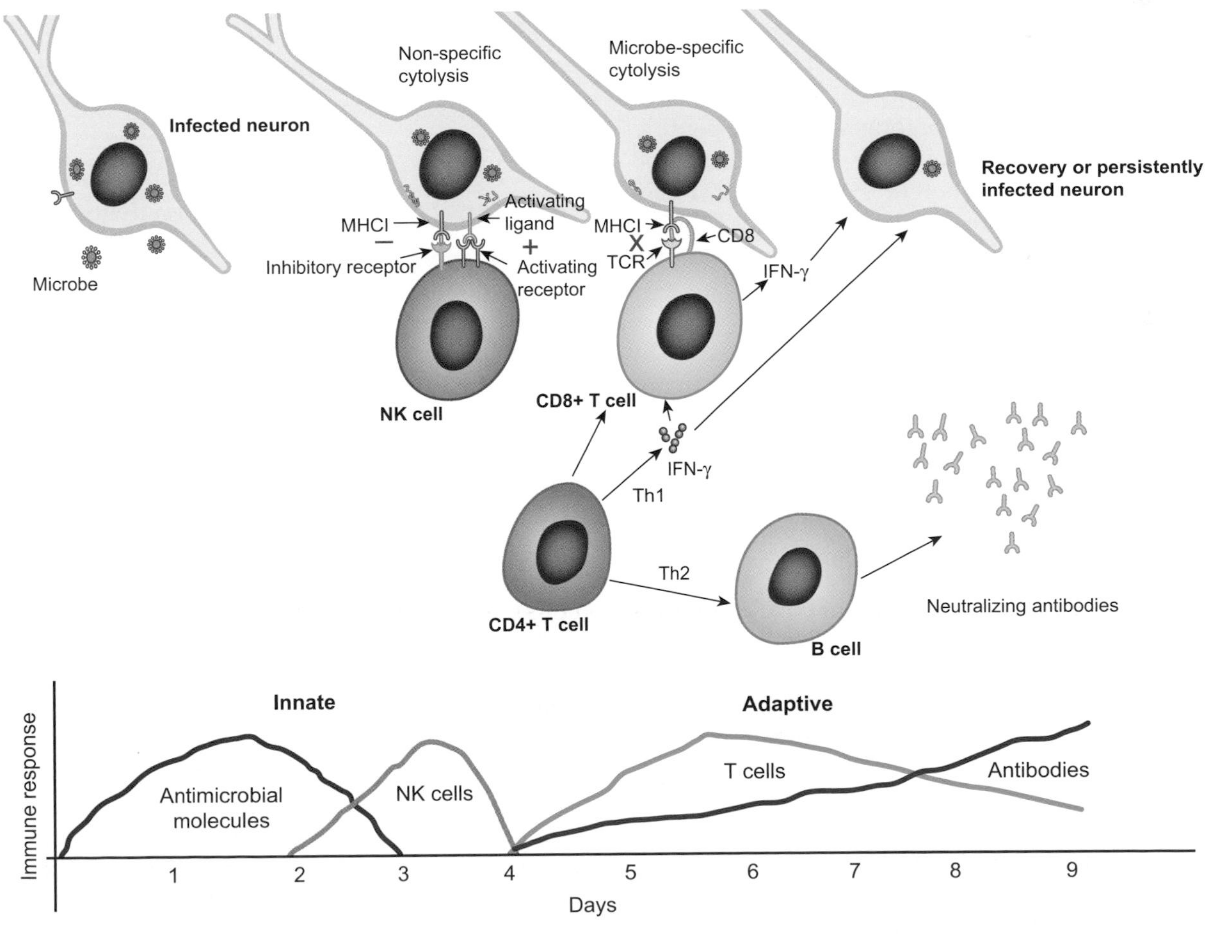

FIGURE 27.1 **Immune responses to an invading microbe.** Innate response: release of antimicrobial molecules and activation of natural killer (NK) cells. Adaptive response: T and B cells. Major histocompatibility complex class I (MHCI) molecules on the infected target cells play two opposite roles: in interaction with NK cells they inhibit killing, while in interaction with $CD8^+$ T cells they trigger killing or interferon-γ (IFN-γ) secretion. TCR: T-cell receptor; Th: T-helper cell.

class I or II molecules. The microbe-activated dendritic cells are detached from their surroundings and migrate by chemotaxis to the regional lymph nodes, where they present the fragments of the pathogens to T cells. The PAMPs are very important here: they induce dendritic cell migration and stimulate the expression of a subset of molecules necessary for T-cell activation, called costimulatory molecules.

T cells undergo their late developmental stage in the thymus (hence their name). After elimination of those that can recognize host-derived proteins (and therefore be deleterious to the organism) they are released as naïve cells, which search for foreign peptides in the secondary lymphoid organs. Upon detection of a foreign peptide presented by PAMP-activated dendritic cells, the T cells become peptide specific, proliferate and differentiate, and are ready to perform their protective duties. For this purpose, these sensitized T cells will circulate in the bloodstream and infiltrate the infected areas. In a virus-infected cell in such areas, viral proteins synthesized in the cytosol are cleaved by proteasomes, and fragments pass into the endoplasmic reticulum, where they are attached to pockets in the MHC-I molecules, which are expressed on most cells in the body. The sensitized T cells recognize and bind to infected cells via their specific receptors for the MHC-I molecules (now transported to the cell surface) containing a pathogen-derived peptide. Binding is enhanced by a cofactor in T cells called CD8. These cytotoxic T cells are therefore called $CD8^+$ T cells and they can either reduce replication of a pathogen inside a cell by secreting the cytokine IFN-γ or kill the infected cell by secreting perforin (Fig. 27.1).

The T-helper (Th) cells recognize peptides derived from foreign proteins incorporated and processed by macrophages, and expressed on their surfaces in the context of MHC class II molecules. Differently from MHC-I, MHC-II molecules are expressed only by macrophages, dendritic cells, and B cells. Binding to the T-cell receptor of these cells is enhanced by a cofactor called CD4, hence their name, $CD4^+$ T cells. This binding induces a release of a variety of cytokines. One set of such cytokines, which includes IFN-γ, is directed against intracellular microbes (Th1 response), while another set of cytokines, including interleukin-4 (IL-4), stimulates proliferation of B cells and tissue repair (Th2 response). The former response shifts the reaction towards control or killing the pathogen-containing cells to prevent further spread in the tissues, while the latter favors the neutralization of some extracellular pathogens and is especially important in the defense against worms. Thus, a Th1 response is important to detect and eliminate pathogens inside a cell that are inaccessible to antibodies. It has been estimated that for each given type of antigen there is only space in the body of a mouse for 100–1000 T cells, and a robust response therefore requires that the few stimulated T cells expand and generate hundreds of thousands of cells to be effective. The capacity of the immune system is further illustrated by the fact that more than 10^8 different antibodies can be produced in humans to recognize any possible pathogen that may invade the organism.

After elimination of the pathogen, an anti-inflammatory response starts to operate to downregulate the inflammation, remove the infiltrated T cells, and restore the tissue. The early-phase macrophage response with production of antimicrobial NO and ROS is now replaced by alternative macrophage activation (mediated by IL-4) with release of restorative molecules. During this contraction phase of the immune response most of the effector T cells die by apoptosis, but a few remain as memory T cells and these undergo a slow turnover. They contain high levels of RNA ready for protein synthesis, can be rapidly activated, and proliferate upon a new infection with the same pathogen. The quiescence of both naïve and memory T cells is tightly controlled by self versus non-self peptide presentation and by the cytokine IL-7, but the distinction between the two sets of T cells can be blurred as homeostatic cues fluctuate and there is more flexibility in the quiescent states than previously thought.[8]

What Makes the Nervous System Immune Responses Special: Immune Privilege?

Ideally, a balance is kept by the cell in its effort to kill the microbe and to protect itself from protein damage caused by antimicrobial molecules. This is most evident in the nervous system, in which harmful effects of immune responses could be particularly deleterious because most neurons cannot be renewed and proper connections in the networks not re-established. In fact, the nervous system has been considered to have "immune privilege", a term that was coined in the 1940s by Sir Peter Medawar following his observation that foreign tissues transplanted into the anterior chamber of the eye were not rejected as in other tissues.

There are several reasons for dampened immune responses in the CNS. The occurrence of dendritic cells is limited to the leptomeninges, which implies that materials implanted in the brain parenchyma may not be detected to initiate an adaptive immune response. In addition: (1) the brain parenchyma is separated from the bloodstream by the blood–brain barrier (BBB), which prevents diffusion of antibodies and limits infiltration of T cells; (2) neither neurons nor astrocytes normally express MHC-I molecules, which minimize T-cell recognition and killing by cytotoxic T cells; (3) a number of anti-inflammatory molecules are constitutively expressed, such as transforming growth factor-β

(TGF-β), which dampens inflammatory responses; and (4) macrophages are normally limited to perivascular sites, and microglial cells, although showing a scavenger function to remove dead cells, may not release as harmful molecules as macrophages upon activation. To avoid detection in a cell some viruses, such as adenoviruses and cytomegaloviruses, have developed mechanisms to downregulate MHC-I expression on the cell surface; in neurons they do not have to do this and several microbes can persist in the nervous system for extended periods, even for the lifespan of an individual. However, this immune privilege does not mean that infections cannot be controlled or pathogens eliminated from the nervous system. MHC-I molecules may be upregulated in neurons to make them susceptible to cytotoxic T-cell attacks; this has been shown in cultures of neurons treated with IFN-γ, which is known to be a strong inducer of MHC, and functionally silenced by tetrodotoxin.[9] In addition, a non-cytolytic clearance or control of pathogens in neurons has been described, whereby microglia can readily be induced to express MHC molecules and present incorporated viral materials to T cells. Upon activation, the sensitized T cells may release IFN-γ, which controls the growth of a number of pathogens in neuronal cells.[10]

Studies in the early twenty-first century have shown that innate immune responses can also be elicited in the nervous system. All TLRs are present on various cell elements, and inflammasomes, which are multisubunit complexes in the cytosol that process pro-IL-1β and pro-IL-18 into their active forms, can be induced. These cytokines play a crucial role in induction of inflammatory responses as well as adaptive immunity.[11] Consequently, "immune specialized" has been suggested as a better term to describe immune reactions in the nervous system. Representative examples of various modes for interplays between microbes and the immune responses in the nervous system will be given in the sections dealing with the individual infections.

Cross-Talk Between the Innate and Adaptive Immune Responses in the Brain

The innate and adaptive immune responses do not operate separately from each other. For many, if not most, infections of the innate immune response may eliminate the invading microbes, but at the same time alert the adaptive response. Several molecules released during the innate response thus activate the adaptive response. An example of such cross-talk in the brain is given by experimental infection in mice with lymphocytic choriomeningitis virus (LCMV), an RNA virus that belongs to the family Arenaviridae. It readily infects rodents but also occasionally infects humans to cause meningitis or encephalitis. After injection in adult mice, viral antigens appear in the choroid plexus and meninges, followed by inflammatory cells (hence the term choriomeningitis). Viral particles from the choroid plexus are released into the cerebrospinal fluid (CSF), invade the white matter surrounding the ventricles, and infect microglial cells. Viral RNA binds to and activates TLRs in microglia that produce IFN-α/β as an immune response. When secreted, this cytokine induces among other molecules a chemokine called CXCL10 in astrocytes and endothelial cells. CXCL10 attracts and retains invasion of a few sensitized T cells into the brain parenchyma to initiate the adaptive immune response. When such sensitized T cells recognize viral peptides in MHC-I molecules on surfaces of infected microglial cells, they become activated to secrete high levels of IFN-γ, which is a much stronger inducer of CXCL10 than IFN-α/β. Thereby, the inflammatory response is augmented; the innate response molecule IFN-α/β has served a role to call in the adaptive immune response into the brain, that is, a cross-talk between the two responses in the brain.[12] A similar chain of events occurs in connection with other infections in the brain (see below).

INVASION OF PATHOGENS IN THE NERVOUS SYSTEM

If an infection cannot be controlled at the site of microbe invasion (the skin, or the respiratory, gastrointestinal, or urinary tract), the pathogens may spread into the lymphatic system and into the bloodstream. Fortunately, subsequent invasion of the nervous system is a rare event because both the peripheral system and the CNS are equipped with barriers that, besides their original function to keep constancy in the internal environment, also prevent microbes from passing readily into the parenchyma. However, mechanisms have evolved by which a number of microbes under certain circumstances can cross or circumvent such barriers and enter the nervous system (Fig. 27.2).[13] Once inside the nervous system, microbes localize to various areas depending on their route of invasion and the presence of cells that may incorporate and support their replication.

Blood–Brain Barrier

The BBB is formed by cerebral endothelial cells that are linked with each other by tight junctions, which prevent the passage of macromolecules between the cells. These endothelial cells also show a paucity of transcellular traffic of endocytotic vesicles (transcytosis), which only allows transfer of selected molecules into the brain parenchyma. Rarely, microbes in the bloodstream infect and replicate in cerebral endothelial cells and, thus, gain access to the brain parenchyma. Nipah virus, which belongs to the family Paramyxoviridae and has caused epidemics in South-East Asia among pig farmers, is an example. The endothelial cells are destroyed, hemorrhages ensue, and the infection is associated with a high mortality rate. Another and less traumatic way for a microbe to enter the brain is to be transported by transcytosis without infecting the endothelial cells and be released from the abluminal surface of the vessel into the parenchyma. A number of microbes can also infect white blood cells (WBCs) that circulate in the bloodstream. Under normal circumstances, there is a low influx of both monocytes and T cells across the BBB and this may open the way for microbes concealed within them to enter, in a "Trojan horse" mechanism (Fig. 27.2). Examples of this are *Toxoplasma* parasites and *Listeria* bacteria in monocytes, measles virus in T cells, and HIV in both monocytes and T cells. It should be pointed out that infiltration of WBCs into the brain occurs not at the capillary, but at the postcapillary venule level. These vessels are surrounded by a basement membrane formed by astrocytes, in addition to a basement membrane formed by the endothelial cells, and this adds to the complexity of the barriers against WBCs and microbes, as will be described below.

Blood–Cerebrospinal Fluid Barrier

The endothelial cells of the leptomeningeal vessels are also linked by tight junctions, but they are more permeable to macromolecules than intracerebral vessels, and surrounded by basement membranes and cells distinct from those in the brain parenchyma. Some microbes may therefore cross these vessels more frequently than the cerebral vessels and reach the subarachnoid space; several bacterial and some viral infections are limited to the leptomeninges.

Other areas in the brain that can be the site for an early attack by microbes are the choroid plexus and the circumventricular organs (CVOs). In these areas the endothelial cells of the vessels are fenestrated and allow passage of macromolecules into the stroma. In experimental studies some pathogens can target these organs, but with the exception of HIV-infected monocytes, infiltration by pathogens in humans is not well described. Further spread into the CSF is hindered by the epithelial cells and tanycytes that outline the choroid plexus and CVOs, respectively.

Peripheral Nerve Barriers

The vessels in peripheral nerve root ganglia are also fenestrated and this may facilitate localization of extracellular pathogens such as trypanosomes and perhaps spirochetes to these sites.

A frequently used pathway to enter the nervous system is through peripheral nerve fibers. The interior of the peripheral nerve fibers (endoneurium) is secluded from the surrounding tissues by the perineurium and from the blood by the endothelial cells in the endoneural vessels; the perineurial cells as well as these endothelial cells are linked by tight junctions. However, several categories of axon terminals end openly with no surrounding perineurium and can therefore be the target of microbial attacks. Viruses can be incorporated by endocytosis and transferred within endocytotic vesicles in association with the retrograde axonal transport machinery to the cell bodies in the peripheral ganglia or spinal cord and brainstem (Fig. 27.3A). Viruses can only replicate after reaching the cell bodies, which contain the machinery required for synthesis of the various viral components. New viral particles may then spread to other cells in the CNS along neuroanatomical pathways.

Well-known examples of this route of invasion are *Herpes simplex* virus (HSV), rabies and poliomyelitis viruses, but certain arboviruses, neurovirulent influenza A virus strains, and the bacterium *Listeria monocytogenes* can also propagate along these pathways. An interesting route for neuroinvasion is provided by the olfactory nerve. The neurons in the olfactory epithelium are directly exposed to the external environment

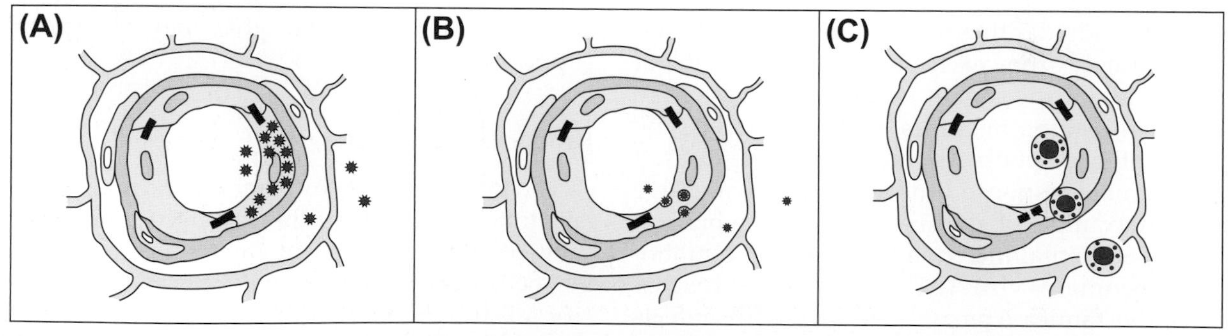

FIGURE 27.2 **Mechanisms for pathogenic invasion of the CNS across the blood–brain barrier: (A) Endothelial cell infections; (B) Transcytosis; (C) Trojan horse mechanism.**

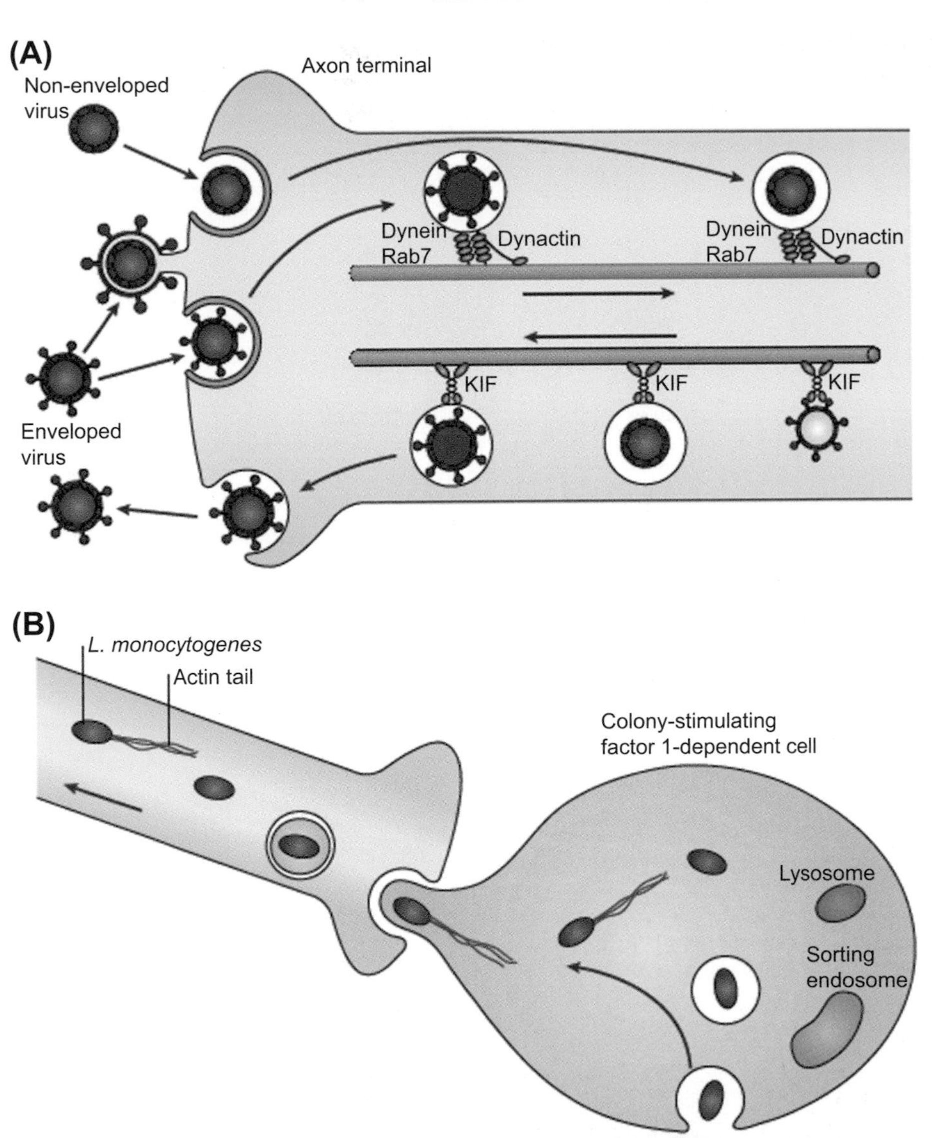

FIGURE 27.3 **Axon terminal uptake and transport of pathogens.** (A) An enveloped virus may be incorporated by endocytosis or fuse its envelope with the cell membrane to release the naked capsid into the axoplasm. A non-enveloped virus is incorporated by endocytosis. Endosomes marked by Rab7 become destined for retrograde transport to the nerve cell bodies. Such endosomes are transported in association with the microtubular transport system, including the mechanoenzyme dynein and the regulatory protein dynactin. Naked viral capsids may associate with the same transport system, but mechanisms for this are unclear. Newly formed viral particles or components in the nerve cell body can be carried by anterograde transport along microtubules associated with kinesin family members (KIFs) to be released by exocytosis. (B) *Listeria monocytogenes* may reach axon terminals through an initial cycle of replication in macrophages or dendritic cells. After uptake and lysis of the phagosome membrane, the bacteria multiply in the cytosol of the cell. The new bacteria can polymerize actin asymmetrically, producing actin tails, which propel them through the cytoplasm and into an axon terminal. Following *L. monocytogenes*-induced lysis of the enclosing two membranes, the bacteria acquire new actin tails to propel them towards the cell body. *Source: From Kristensson.* Nat Rev Neurosci. *2011;12:345–357.*[13]

in the nasal cavity. In experimental animals a number of viruses can attack these neurons and propagate to various areas in the brain connected to the olfactory bulbs, but whether this is a portal of entry to the human brain is not known (Fig. 27.4).

PATHOGENS CAUSING NERVOUS SYSTEM DYSFUNCTION

Pathogen–Nervous System Interactions

For the benefit of the host it is essential to prevent microbes from reaching the nervous system, whereas the microbe could gain an advantage by entering this specialized environment. The interactions between nervous tissues and microbes may be viewed as a game of chess with two players developing strategies of combined attack and defense in an extremely complex network, but with rules that are far from clear. In such interplays some microbes may have caused changes in the infected host that optimize the chances for their transmission to a vector or a predator host. Evolutionary pressure may thus have selected for microbes that alter odor, locomotor activity, as well as foraging and social behavior that increase attraction or make the host conspicuous for a vector or predator, which promotes dispersal of the parasites.[14] The parasitological literature is full of such examples and some are dramatic; for example, nematode-infected ants are less aggressive and their gasters, filled with parasites, turn into the shape of red fruits to be picked up by birds (Fig. 27.5).[15] Behavioral changes can also favor survival of the infected host. An evident example is the sickness response of the host to an infection that

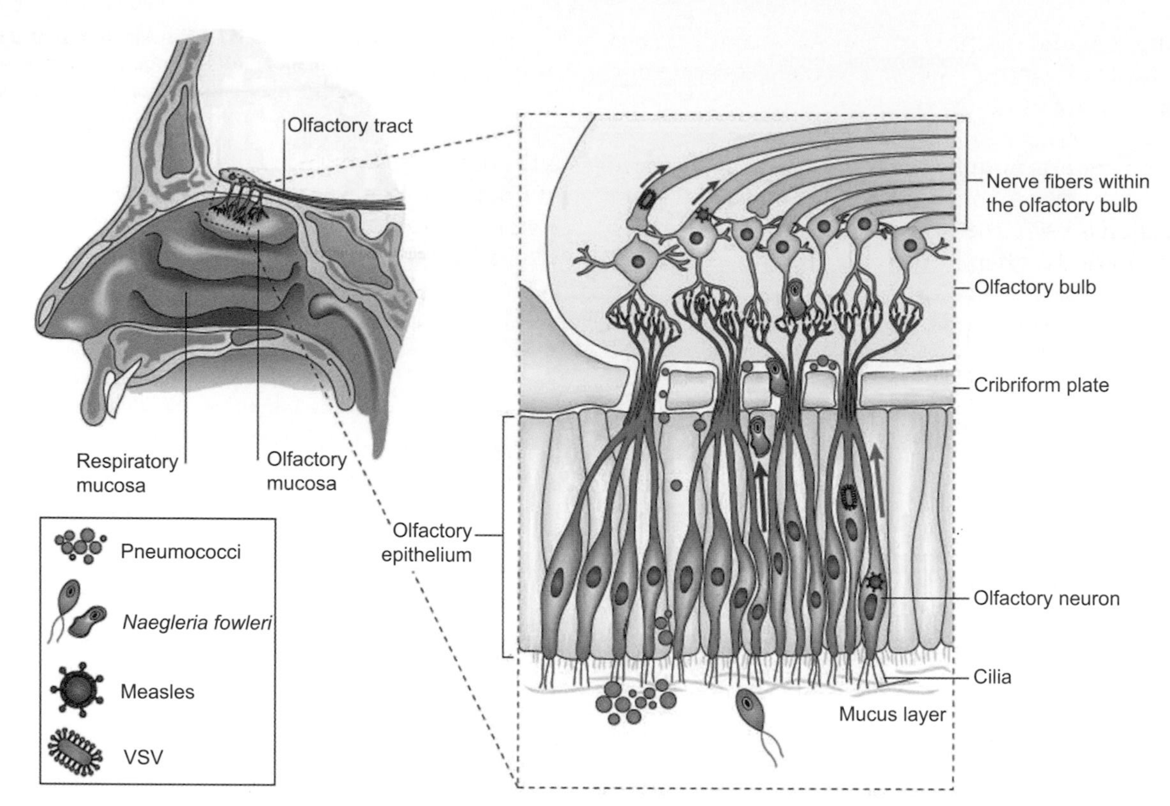

FIGURE 27.4 **Olfactory route of neuroinvasion.** Olfactory neurons are unique because their dendrites come into direct contact with the external environment and their axons are in direct contact with the brain (the olfactory bulb). Viral particles can infect olfactory neurons and be transported along axons to the olfactory bulb. From here, some viruses may spread via anterograde transport to reach the limbic system, whereas others may be taken up by axon terminals and spread via retrograde transport to neurons that project to the olfactory bulbs (e.g. serotonergic raphe neurons). Bacteria (e.g. pneumococci) often colonize the nasal cavity. In a similar way to the amoeba *Naegleria fowleri*, they may potentially pass the olfactory mucosa and, via the holes of the cribriform plate, reach the subarachnoid space and cause meningitis. VSV: vesicular stomatitis virus. *Source: From Kristensson.* Nat Rev Neurosci. *2011;12:345–357.*[13]

FIGURE 27.5 **Changes to the color, shape, and behavior of an ant infected with a nematode.** The body is transformed into a berry-like structure that can easily be picked up by a bird to disseminate the parasite. (A) Uninfected and (B) infected ant. *Source: From Hughes* et al. Curr Biol. *2008;18:R294–R295,*[15] *courtesy Stephen P. Yanoviakof with permission.*

includes increased time spent in sleep, social withdrawal, decreased sexual activity, reduced food and water intake, and hyperalgesia. These changes are advantageous, since they reduce motility which conserves energy needed for the generation of fever; fever inhibits microbe replication and enhances immune responses. They also limit the chances of spread of microbes between individuals in the society. Although it is always interesting to analyze the course of a disease from either the microbe's or the host's point of view, most microbe–host interactions cause brain dysfunctions, such as seizures and death, with no obvious advantage for either of them.

Infections can be acute (days to weeks), subacute (weeks to months), or chronic (months to years). The outcome can vary from full recovery or recovery with functional deficits (which sometimes appear after a long time interval) to persistent or latent infections with no or only minimal signs of disease. Chronic infections can show remission and relapses of clinical symptoms, and the outcome is often unpredictable: sometimes recovery, sometimes death, and often long-lasting disabilities. Chronic infections are distinguished from slow infections, a concept introduced from veterinary science based on observations on diseases in Icelandic sheep. Slow infection is used to denote infections with very long incubation periods (often years) and with a disease that follows a set pattern with slow deterioration (months to years) that, if untreated, inevitably leads to death.

Instead of a catalogue of the numerous pathogens that affect the nervous system, a selection of microbe infections will be presented with a focus on experimental work on relevant human diseases. These illustrate various interactions between pathogens and the immune responses, routes of neuroinvasion, and the variety of dysfunctions, ranging from behavior and sleep disturbances to late-onset unprovoked seizures (epilepsy) and neurodegeneration.

Parasitic Infections

This section will describe nervous system infections with traditional parasites, which are organisms with life cycles that employ an intermediate host in addition to their definite host, namely protozoa and helminths. Protozoa are unicellular organisms that can grow either outside or inside a cell, whereas the larger helminths, in general, grow extracellularly.

Malaria

In addition to anemia, cerebral malaria is the most dreaded complication of this infection. Malaria is most prevalent in sub-Saharan Africa, where it causes the deaths of about 2 million children each year. Four species of the malaria parasites are pathogenic in humans, but *Plasmodium falciparum*, which is prevalent on the African continent, causes most of the deaths; about 1–2% of infections with this species result in cerebral malaria.

After a bite by an infested mosquito, the parasites are transferred to the liver where they replicate and infect red blood cells (RBCs). In RBCs the parasites undergo a series of maturation steps and can be transmitted to a new blood-sucking mosquito that spreads the infection to other individuals. Children harboring the gametocytes of the parasites (transmissible to mosquitoes) show an increased attractiveness to mosquitoes, providing an example of a pathogen that causes changes in the odor of the host to promote its spread.[16]

The parasites inside the RBCs are well protected from both the innate and the adaptive immune responses, but they face a problem of being eliminated in the spleen, in which infected RBCs can be recognized and destroyed. To avoid this, a parasite-derived protein is expressed on the surface of infected RBCs. In humans this protein binds to intercellular adhesion molecule-1 (ICAM-1) expressed on the surface of endothelial cells, which is induced by cytokines such as tumor necrosis factor-α (TNF-α) and IFN-γ released by the immune response to the infection. Following attachment of infected RBCs to endothelial cells they are prevented from being killed in the spleen. Infested RBCs can also attach to other RBCs, infected or not, whereby they form rosettes. Sequestered RBCs, attached to vessel walls or forming rosettes, can inhibit blood flow through cerebral capillaries. One main theory behind cerebral malaria is that mechanical blockade of these capillaries results in ischemia. Since children with cerebral malaria may wake up from their coma suddenly, ischemic injuries may not be the whole explanation and an immunological theory has also been proposed. This implies disturbances related to the immune responses in cerebral endothelial cells that may transiently disturb the functions in the neurovascular units (Fig. 27.6B).[18]

Cerebral malaria is defined by the presence of unarousable coma with the exclusion of all other encephalopathies, and the presence of the asexual form of the parasite in the blood. By ophthalmoscopy, characteristic vascular changes can be seen in the retina that aid in diagnosis. The disease is the most common cause of acute seizures and status epilepticus in children in malarial endemic countries in sub-Saharan Africa. The mortality rate is high and in surviving children, who have recovered from the acute attack, late-onset unprovoked seizures (epilepsy, if recurrent) are frequent. These children also run an increased risk of developing cognitive deficits and behavioral disturbances. Malaria is therefore a leading cause of disability of the nervous system in tropical countries.[19]

Neurocysticercosis

The pork tapeworm, *Taenia solium*, is the most common parasitic infection in the world and gives rise to cysticercosis, a major cause of epilepsy in Latin America, Africa, Eastern Europe, India, and China. The larvae of this worm lodge in fluid-filled cysts (cysticerci) in skeletal muscle of the intermediate host, the pig. Humans are the definite hosts and following ingestion of poorly cooked infected meat, the larvae hatch in the intestine, and the tapeworm develops and sheds eggs (>100,000 daily) into the environment. After ingestion of these eggs from fecal contaminated food, embryos (onchospheres) attach to and cross the intestinal mucosa, reach the bloodstream, and disseminate throughout the body. Most of the embryos are killed by the immune response; however, a few may survive in the small vessels, particularly in skeletal muscle and brain, where they lodge and develop into the larval stage (cysticerci).

Clinical manifestations depend on the number of cysts and their location in the nervous system, but are usually associated with seizures or increased intracranial pressure. Single asymptomatic cysts are common. They can persist for 3–5 years without eliciting an immune response, since the larvae secrete molecules that are

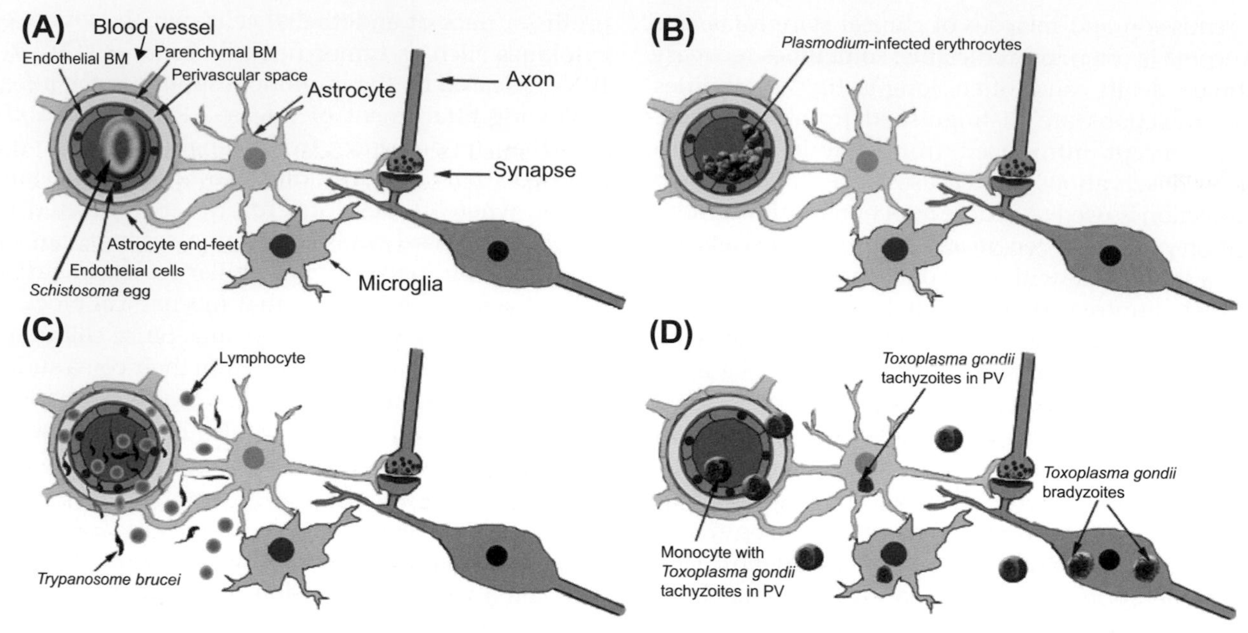

FIGURE 27.6 **Interactions between parasites and the neurovascular unit.** (A) *Schistosoma;* (B) *Plasmodium* (malaria); (C) African trypanosomes; and (D) *Toxoplasma.* BM: basement membrane; PV: parasitophorous vacuole. *Source: From Masocha and Kristensson.* Virulence. *2012;3(2): 202–212,*[17] *with permission.*

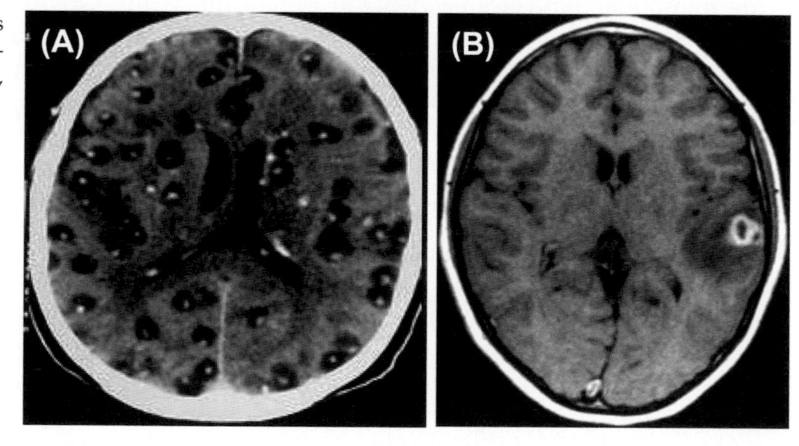

FIGURE 27.7 **Neurocysticercosis.** (A) Multiple cysts with scolex in the brain; (B) a solitary degenerating cyst surrounded by edema. *Source: Courtesy of Professor Arturo Carpio, Columbia University, New York.*

anti-inflammatory or mask their presence. When the larvae have reached their normal attrition and die or are killed by an antiparasitic drug, an immune reaction can appear around the dying cyst, and seizures are probably elicited by this localized release of inflammatory molecules.[20] An important decision whether or not to use antiparasitic drugs to kill the parasite has to be made in asymptomatic patients in whom the larvae are still alive. In some endemic countries measures are taken to eliminate the parasite from the environment by combined human and porcine treatments (Fig. 27.7).

Schistosoma

Neuroschistosomiasis is caused by subspecies of the helminth *Schistosoma* and is most prevalent in Brazil and sub-Saharan African countries; about 200 million people are infected worldwide and up to 10% of them develop severe disease. The larvae of the *Schistosoma* worms can secrete proteases that penetrate the skin and then spread via the bloodstream to the lungs and liver. They mature and reproduce in the mesenterial and portal veins. Eggs are seeded into the gut and urinary bladder for further spread to other individuals. In some infected people, the eggs shed into the bloodstream. The most common *Schistosoma* species that infect humans are *S. mansoni*, *S. haematobium*, and *S. japonicum*. The small round eggs of *S. japonicum* can travel to the brain, while the larger eggs of the other two species remain in the domain of the lower spinal cord, where they lodge in capillaries to induce a granulomatous inflammation (a granuloma is a nodule of tightly packed macrophages that surround and engulf microbes). Spinal cord schistosomiasis is the most well-known CNS complication, manifesting as progressive pain and weakness of the limbs. More localized deposition of eggs in the brain can follow worms that have migrated in vessels to the infested areas. Early treatment can be successful and is needed to prevent severe disabilities.[21]

Toxoplasma

Toxoplasma gondii parasites were first isolated from the Tunisian rodents gondi and the name toxoplasma is derived from their bow-like shape. Felidae are the definite hosts of the parasite and almost any warm-blooded animal, including rodents, birds, sheep, pigs, and humans, can serve as an intermediate host to promote the wide geographical spread of the parasites. *Toxoplasma* undergoes three development forms, namely the sporozoites produced in oocysts, the rapidly replicating tachyzoites, and the slowly multiplying, dormant bradyzoites (in tissue cysts). The parasite shows two distinct life cycles: one is the enteroepithelial cycle in the small intestines of the definite host, leading to the production of oocysts; the other is extraintestinal in the intermediate hosts, leading to the formation of cysts in the tissues.

Humans can be infected from soil contaminated with oocysts derived from cat feces. After being ingested, the oocysts release sporozoites, which infect the intestinal epithelia. From infected epithelial cells, tachyzoites are then released into the bloodstream and circulate by being carried within infected monocytes. The parasites can infect most tissues in an organism, but skeletal muscle and the brain are favored sites for persistent infection. Infected monocytes can penetrate the BBB by a Trojan horse mechanism and the parasites are released to infect astrocytes and neurons. Inside these cells the tachyzoites develop into slowly growing bradyzoites, which will become enclosed by a cyst wall (Fig. 27.6D). Under the control of the immune response, dormant *Toxoplasma* can persist protected in the cysts for the lifetime of the intermediate host. Under immunosuppressive conditions, such as in acquired immunodeficiency syndrome (AIDS), control may be weakened and the infection can flare up, with rapidly growing tachyzoites, and a severe necrotizing toxoplasmic encephalitis follows, which can be lethal if not treated. Seizures are common and toxoplasmic encephalitis, which responds readily to treatment, is an important differential diagnosis in HIV-positive patients with seizures. Toxoplasma parasites can also readily pass the placenta and cause congenital toxoplasmosis in the newborn. In the fetal brain the parasites grow in ependymal and subependymal cells to cause necrotizing periventricular lesions that can occlude the aqueduct and cause hydrocephalus.[22]

Although providing a safe haven for the parasite, the brain is at the same time a prison and the parasite runs the risk of being extinguished when the intermediate host dies (if not eaten before by the cat). Experimental studies have indicated that toxoplasma-infected rodents may more easily fall victim to predators than uninfected rodents, since they have lost their natural fear of the smell of cats. Rodents usually show aversion to this smell, but toxoplasma-infected rodents may even be attracted to it. Whether toxoplasma-induced changes in behavior favor parasite survival ("manipulation of the host's behavior by the parasite", as it has been called) or a non-specific sickness response to the infection is under debate (Fig. 27.8). Studies suggesting differences in personality traits between toxoplasma-seropositive and -seronegative individuals have to be verified, and epidemiological studies suggesting associations between toxoplasma and neuropsychiatric disorders remain elusive. For example, analysis of archived neonatal dried blood samples shows a correlation between maternal toxoplasma exposure and the development of non-affective psychosis that appears later in life.[23]

Trypanosoma

Trypanosoma brucei species or African trypanosomes cause human African trypanosomiasis (HAT) or sleeping sickness. Living parasites are stained with trypan dyes and were first detected as a cause of disease in wild game and cattle (called nagana). At the turn of the twentieth century a large outbreak of sleeping sickness in Uganda killed two-thirds of the population in the affected area. During that epidemic, trypanosomes were found in the CSF from patients and unsuccessful attempts were made

FIGURE 27.8 **Toxoplasma.** Infants should stay away from cats because of the risk of toxoplasma infections. *Source: Photograph by Karolina Kristensson.*

by Paul Ehrlich and collaborators to treat the disease with trypan dyes. The treatment failure was explained when the researchers discovered the BBB using these dyes.

Subspecies of *Trypanosoma brucei* (*T.b.*) replicate in the subcutaneous tissues following the bite of an infected intermediate host, the tsetse fly. Subspecies *T.b. gambiense* has humans as definite hosts, *T.b. rhodesiense* wild game and cattle, and *T.b. brucei* rodents; this last subspecies is therefore suitable for experimental studies in its natural hosts, mice and rats. After circulating in the bloodstream (Fig. 27.9), the parasites spread to most visceral organs and skeletal muscles. In rodent models they appear in the nervous system first in the leptomeninges and in areas with fenestrated vessels, namely the choroid plexus, CVO, and peripheral nerve root ganglia. Signs or symptoms of nervous system dysfunctions related to involvement of these areas may include pruritus as well as disturbances related to hypothalamic functions. A deep sensation of pain, called the Winterbottom sign after the colonial doctor who described it in himself, is characteristic. Later during the infection, parasites cross the BBB at the level of postcapillary venules. Entry into the nervous system therefore occurs in two phases.

Passage across the BBB in the second phase is a multistep process similar to that of T cells. T cells first attach to adhesion molecules (ICAM and vascular cell adhesion molecule) induced by cytokines such as TNF-α on the luminal side of cerebral endothelial cells. They then cross the endothelial cells either by a transient opening of the tight junctions (paracellular) or through a transcytosis-like process through the endothelial cells. The next step is regulated by the two basement membranes surrounding the vessels, the endothelial basement membrane and the parenchymal basement membrane. The laminin composition of the former determines whether it permits T-cell invasion, while the latter must be transiently opened by cleavage by matrix metalloproteases of a protein, dystroglycan, that anchors it to the astrocytic end-feet.[24] If this cleavage is inhibited, the T cells accumulate between the two basement membranes and stay like perivascular cuffs on hold waiting for further activation to infiltrate the brain parenchyma. Similarly, a perivascular accumulation of trypanosomes is seen in IFN-γ receptor knockout mice, indicating that this cytokine is involved in the process of invasion of both trypanosomes and T cells (Fig. 27.6C). For this purpose, several IFN-γ-inducible molecules are involved, and T-cell and trypanosome invasion of the brain follow a similar pattern to LCMV (described in "Cross-Talk Between the Innate and Adaptive Immune Responses in the Brain", above); both pathogens first target the choroid plexus, and then invade mostly the white matter bordering the cerebral ventricles. In HAT, the most prominent inflammatory cell infiltration is seen in the white matter and the infection is therefore called leukoencephalitis.[25]

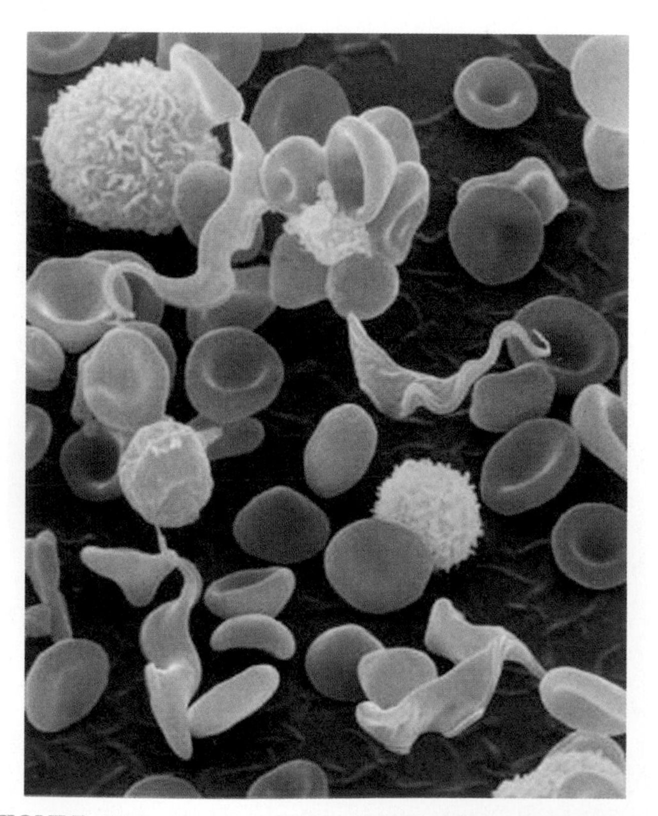

FIGURE 27.9 ***Trypanosoma brucei.*** Four African trypanosomes (light blue) are seen among red and white blood cells. *Source: Courtesy of Professor Michael Duszenko, University of Tuebingen.*

HAT patients as well as infected experimental rodents develop disturbances in the sleep pattern with narcolepsy-like sleep episodes during the day and wakeful episodes during the night. In contrast to classical narcolepsy, the sleep episodes are not accompanied by cataplexy. Such changes in sleep pattern and circadian rhythms may reflect the presence of trypanosomes and inflammatory responses in the CVOs. For instance, some of the hypothalamic neurons that regulate transitions between sleep and wakefulness as well as circadian rhythms project to the arcuate–median eminence complex, the latter component being a CVO. In addition, patients develop Parkinson-like changes and eventually dementia. The disease is invariably fatal if left untreated. At early stages of the disease relatively non-toxic drugs that can cure the infection are available, whereas at later stages treatment relies on drugs with an intolerable degree of toxicity. However, the diagnostic criteria to make this important clinical distinction are still under debate.[26] HAT declined in prevalence during the first part of the twentieth century only to re-emerge in hundreds of thousands of cases during the last decades. Through increased surveillance and early treatment the incidence is now decreasing.

African trypanosomes provide an example of how an extracellular microbe may evade the immune response. By secreting factors that reduce the innate immune response and by paradoxically triggering a Th1 immune response (which primarily aims to combat intracellular and not extracellular pathogens) that results in parasite neuroinvasion and protection in the immune-specialized nervous system, the survival of the trypanosomes is favored; the Gambian form of sleeping sickness can last for years. The trypanosomes are also programmed to switch surface coat protein to avoid being eliminated by antibodies; in this way, more than 1000 different coat proteins can be sequentially changed. Trypanosomes provide another example of manipulation of host behavior by the parasite, since their spread is promoted by alterations in the feeding behavior of infected tsetse flies.[27]

Acanthamoeba

Acanthamoeba are worldwide parasites living in warm water in lakes, spas, and non-disinfected swimming pools, as well as in soil. They can spread to the brain across the BBB, but also along the olfactory route after inhalation, and cause an almost invariably fatal infection whereby the parasites secrete proteases that digest the brain tissue. Fortunately, infection of humans with this brain-eating parasite is a rare event (about 200 cases reported) because treatment is not effective; only a few patients with severe disabilities have survived.[28] A similar number of people dying from infections with the related amoeba *Balamuthia mandrillaris*, originally found in baboons, has been reported.

Bacterial Infections

A number of bacteria can spread to the meninges and cause meningitis. Infections with *Escherichia coli* are serious threats to newborns after delivery under poor hygienic conditions. Later, the most common bacteria that cause meningitis are *Streptococcus pneumococcae*, *Neisseria meningitidis*, and *Haemophilus influenzae*. *Streptococcus pneumococcae* is very prevalent and colonizes the nasal cavity in otherwise healthy individuals who serve as carriers. Depending on both host and bacterial factors, the bacteria may occasionally spread into the bloodstream and cross the leptomeningeal vessels into the subarachnoid space.[29] In experimental models pneumococci may also reach the subarachnoid space along the olfactory route, passing across the olfactory epithelium and the cribriform plate. The CSF provides a good substrate for the growth of the bacteria and the inflammatory response to them is dominated by neutrophil leukocytes. After treatment a high proportion of recovered children develops late-onset unprovoked seizures even 5–6 years after the meningitis. Epilepsy is associated mainly with meningitis due to *S. pneumococcae* (up to 15% of patients) followed by *H. influenzae*, while meningitis due to *N. meningitides* does not significantly increase the risk. The mechanisms behind these late-onset disturbances are unclear.

Listeria

Listeria monocytogenes is one of the few bacteria, if not the only bacterium, that can infect neurons. It is widespread in the soil, but seldom infects people. However, the infections are important because they are often severe and lethal.[30] Infections *in utero* or during early life cause malformations or necrotic lesions in the brain after the pathogens have spread in the bloodstream within monocytes and crossed the BBB by a Trojan horse mechanism. Rarely in adult humans, but commonly in sheep, the bacteria can enter the brainstem along the trigeminal nerve. After passage through an epithelial lesion in the oral cavity, they infect monocytes or dendritic cells by binding to specific receptors. Once inside the cell, the endocytosed bacteria pass lysosomal membranes into the cytosol. Cellular actin polymerizes asymmetrically at one end of the bacterium, which contains a polymerizing enzyme. The actin tails propel the bacteria towards the plasma membrane, from which they bud into and are engulfed by a neighboring cell, in which they repeat the replication cycle after escaping into the cytosol. This cell-to-cell spread does not involve receptors. Axon terminals in contact with infected cells may therefore incorporate *Listeria*, which then propel retrogradely in axons to the trigeminal ganglia and into the brainstem (Fig. 27.3B). The ensuing brainstem encephalitis is often asymmetrical and associated with a high lethality, since antibiotics cannot easily reach the bacteria inside neurons located behind the BBB. Exposure of humans to *Listeria* occurs through contaminated food, such as cheese or salmon, and of sheep when they are cutting their second teeth. Death in sheep is often preceded by a circling behavior; so-called circling disease.

Leprosy

Although the number of patients with leprosy has markedly been reduced, new cases still appear. The disease is a leading cause of non-traumatic peripheral neuropathies in the world and is caused by *Mycobacterium lepra*. The bacteria are obligate intracellular organisms and replicate slowly in Schwann cells, but do not enter axons or spread to the CNS. They are well protected behind the peripheral nerve barriers, and the Schwann cells serve as a reservoir for further dissemination of the infection. Over time, an immune response- or bacteria-mediated injury of myelin and axons occurs to cause predominantly sensory disturbances; sensory nerve fibers are the early target of attack by the pathogen that invades though ulcerations in the skin. The bacteria may hijack intracellular signaling pathways in Schwann cells to their own advantage. Thus, more non-myelin-forming Schwann cells appear and this is crucial for survival of the bacteria, since they are more favorable for bacterial replication than the myelin-forming ones.[31] Another mycobacterium, *M. tuberculosis*, spreads from the lung via the blood to the meninges to cause a characteristic form of meningitis or into the brain parenchyma to form localized tuberculomas.

Spirochetes

The tick-borne spirochete *Borrelia burgdorferi*, distantly related to *Treponema pallidum* (which causes syphilis, a major cause of nervous system diseases before the antibiotic era: meningovascular and spinal syphilis, tabes dorsalis, and general paresis), causes Lyme disease, which can be associated with severe inflammation in peripheral nerve roots (radiculitis) leading to pain, and with meningitis. Whether the radiculitis reflects localization of the spirochetes to peripheral nerve root ganglia, in which the vessels are fenestrated, is unclear; the pathogenic mechanisms are not well understood because of a paucity of adequate animal models.

Bacterial Toxins and Molecular Mimicry

A number of bacteria can affect the nervous system at a distance by the release of toxins. Two classical examples are *Clostridium botulinus* and *tetani* toxins, which are metalloproteases that inhibit synaptic vesicle release. After ingestion, botulinum toxin spreads via the bloodstream to affect the open-ended axon terminals of motor neurons. After endocytotic uptake and escape into the cytosol, the toxin cleaves SNAREs (soluble *N*-ethylmaleimide-sensitive factor attachment protein receptors) to block synaptic vesicle release, followed by flaccid paralysis. Of concern is wound botulism in drug addicts, which results from injection of bacterial spores. Tetanus toxin, tetanospasmin, spreads from a wound in a bacterial-infected skeletal muscle to be incorporated into axon terminals at neuromuscular junctions. It is then transported retrogradely within carrier vesicles to motoneuron cell bodies, escapes degradation in lysosomes, and is released to block release of synaptic vesicles from axon terminals of inhibitory neurons; a process that is called synaptic transcytosis, with poorly defined mechanisms. This causes increased motoneuron activity that results in tetanic spasms.[32] *Bacillus anthracis* releases the anthrax toxin, which does not reach the neurons but disrupts the tight junctions between cerebral endothelial cells and causes lethal bleeding in the brain.

Bacteria, as well as other microbes, can also affect the nervous system from a distance by molecular mimicry, immune cross-reactivity between a microbe and the host tissue. The best example of a disease caused by mimicry is Sydenham chorea, which is characterized by involuntary movements and neuropsychiatric disturbances following infection with group A streptococci. The precise mechanisms are not clear, but streptococcus-specific antibodies react with neuronal antigens in basal ganglia to cause the dysfunctions.

Viral Infections

Viruses are small entities that originally were identified by their ability to pass bacteria-proof filters. Their genomic elements are nucleic acids, they are obligatory intracellular, and they depend on the host cell's synthetic machinery for their replication. Most animal viruses contain a genome of single-stranded RNA, some contain double-stranded RNA, and some contain DNA. RNA viruses have a high mutation frequency that facilitates adaption to changes in the environment, whereas DNA viruses are more stable and have proof-reading mechanisms. Retroviruses combine the flexibility of RNA viruses with the stability of DNA viruses, since they contain a reverse transcriptase that transcribes RNA into DNA. Some viruses are enveloped by a membrane that is relatively sensitive to injuries, whereas other viruses are naked and more resistant to environmental hazards. Incorporation of viruses into a cell is enhanced by binding to various molecules expressed on host cell surfaces that, in addition to their normal biological functions, act as viral receptors. To sustain an infection, the synthetic machinery of the cell must allow viral replication.

Some examples will be given that illustrate different ways by which viruses spread and interact with the CNS. RNA viruses, which can attack and sometimes persist in the nervous system, will be described first, followed by DNA viruses and retroviruses, which often establish latent or slow infections. For a comprehensive book on viral infections of the nervous system, see Johnson.[33]

RNA Viruses

Rabies virus is a classical neurotropic virus that belongs to the group of Rhabdoviruses, which are enveloped viruses that have a broad range of host organisms varying from plants to mammals. In nature the rabies virus infects mainly dogs, wolves, and foxes, but also bats. The virus is transmitted by a bite of a rabid animal. It replicates very slowly, for months or even years, in the skeletal muscles, from which newly formed viral particles are transmitted to nerve endings at the neuromuscular junctions or muscle spindles. The virus is transported in endocytotic vesicles to the spinal cord and dorsal root ganglia by retrograde axonal transport. After replication in the nerve cell bodies the virus spreads anterogradely in the peripheral nervous system, infects salivary glands, buds from the epithelial cells, and is released into saliva ready for transmission to the next individual. At the same time the virus has reached neurons in the brainstem, cerebellum, and thalamus. In dogs, this elicits aggressive and biting behavior, which favors spread of viruses in the saliva within populations (Fig. 27.10). Humans infected with rabies can, in addition to general signs of encephalitis, show agitated behavior and hyperexcitability. Half of the patients develop hydrophobia (pharyngeal spasms when seeing or drinking water). Although reduced γ-aminobutyric acid (GABA) levels have been described in rabies-infected brains in experimental studies, the mechanisms behind these behavioral changes are poorly understood. Some patients develop an ascending paralysis, for reasons that also are not clear.

Following entry of the rabies virus into the CNS, infections are almost invariably fatal, because the virus in neurons can evade the immune responses; sensitized T cells that approach infected neurons are killed and viral proteins prevent the neurons from undergoing apoptosis. Brains from patients who have died from rabies therefore often show only minimal signs of infiltration of inflammatory cells or nerve cell death. Conspicuous hallmarks of the infection are instead the Negri inclusion bodies in neurons, which are used as a diagnostic tool for rabies in aggressive dogs. The Negri bodies consist of viral nucleoproteins enclosing TLRs in the cytosol, which may hinder TLR activation and induction of innate immune responses.[35] Vaccination can be given as either pre-exposure prophylaxis or postexposure treatment. The latter is only effective before the virus has entered the nervous system; after that event, few patients survive even with aggressive treatment.

Poliomyelitis viruses are small non-enveloped enteroviruses belonging to the family Picornaviridae. They are disseminated in water and after an initial infection of the gastrointestinal tract they occasionally (about 1 in 150 infected individuals; the tip of the iceberg) spread via the bloodstream to skeletal muscles, and then within carrier vesicles they move retrogradely in axons to attack motoneurons in the spinal cord. The virus kills the neurons and causes flaccid paraparesis when more than two-thirds of the cells have been eliminated. Patients often survive with varying degrees of disability, but the paresis may start to progress decades later in the "postpoliomyelitis syndrome". In contrast to rabies, polioviruses do not attack the sensory system. Through extensive vaccination programs the virus is close to becoming eradicated, although these efforts have met unprecedented challenges.

Measles virus belongs to the family of Paramyxoviridae. It is one of the most infectious viruses in the world and extensive efforts are undertaken for its elimination. Measles is an example of a virus that has a high propensity to spread to the CNS via the bloodstream, as shown in experimental animals. In humans, up to 50% of children with ordinary measles may have changes in electroencephalographic recordings, but it is not known whether these reflect a viral invasion of the brain. However, measles virus can reach the brain in infants in the first 2 years of life. The virus persists and causes a disease called subacute sclerosing panencephalitis (SSPE) when children reach school age; they show behavioral

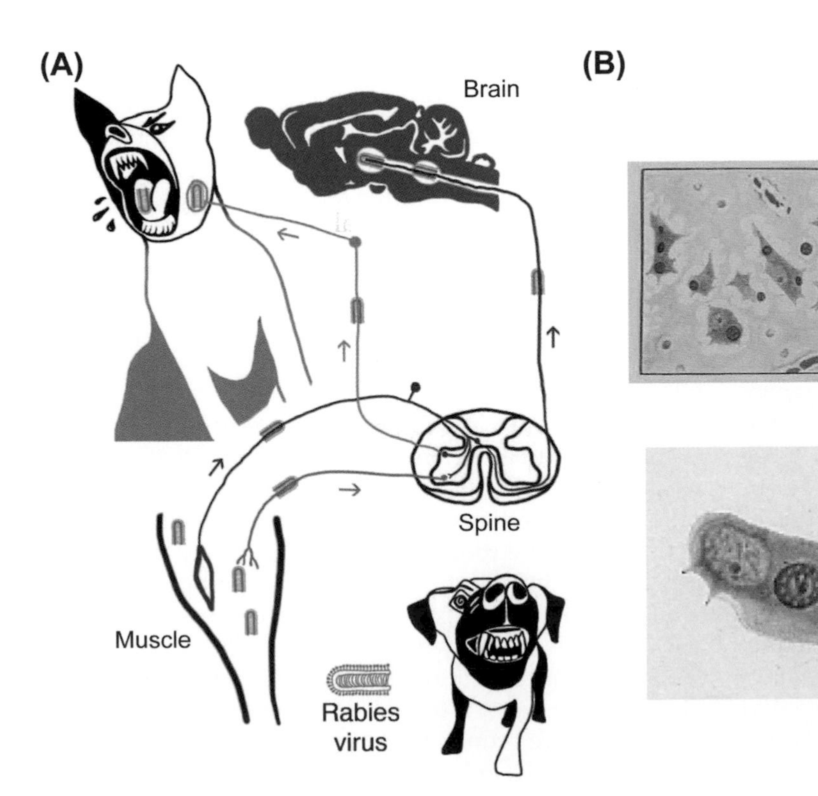

FIGURE 27.10 **(A) Spread of the rabies virus in the nervous system causes rabid, aggressive behavior in dogs; (B) Negri bodies in rabies-infected cells.** *Source: (A) Drawings by Karolina Kristensson; (B) drawings by Adelchi Negri, who detected these bodies, 1903; reproduced in C. Fermi.* La Rabbia. *Siena; 1950.*

changes and do not perform well academically. This is followed by severe neurological disturbances. The disease progresses for 6 months to 3 years and ends invariably in death.

Viral elements are most prominent in inclusion bodies seen in the nuclei of oligodendroglia, accompanied by demyelination and widespread inflammatory cell infiltration. Measles can also cause a persistent infection in the brain of immunodeficient adults, which months or years later progresses into a measles virus inclusion body encephalitis that also leads to death. *Mumps* virus belongs to the same family of viruses as measles virus and has a strong tendency to involve the CNS. Certain strains of mumps virus are more neurovirulent than others; infections of the nervous system may indeed be more common than infection of the salivary glands for these strains. Mumps infections are usually mild and transient, but in some epidemiological studies have been associated with non-affective psychosis later in life.

Rubella virus, which belongs to the family of Togaviruses, can cause a subacute lethal infection similar to SSPE in children. Infections with this virus during pregnancy can also cause malformations of the brain in offspring.

Several viruses can spread via infected ticks and mosquitoes, and are therefore called arthropod-borne viruses or arboviruses. An example of these is *tick-borne encephalitis* virus (TBEV, a member of the Flaviviridae family), which is common in Russia, Eastern Europe, and Scandinavian countries. After being inoculated through the bite of an infected tick, the virus can spread via the bloodstream and cause meningitis. It can also attack motor neurons in the spinal cord, probably after retrograde axonal transport from the skeletal muscles, to cause paralysis and, rarely, death, especially in elderly or immunosuppressed individuals. A Siberian variant of the virus causes severe encephalitis with a high fatality rate after epilepsia partialis continua, a term for recurrent focal motor seizures for extended periods, appearing with a long latency period after TBEV infections.

Two prominent examples of mosquito-borne viruses are the flaviviruses *Japanese encephalitis* and *West Nile encephalitis* virus. The former causes epidemics in Japan, which became prevalent in the 1920s. The infection then spread south-west and reached India. The virus can attack the nervous system and cause Parkinson-like disturbances. West Nile virus derives from northern Uganda, around the source of the West Nile river. It has spread to Egypt and Eastern Europe, and to the USA, where it is now the most common cause of viral meningitis. Most infected people remain asymptomatic, some develop signs of meningitis, and a few develop more severe encephalitis.[36] In mouse models, these two viruses spread to the CNS both via the bloodstream and along peripheral nerves to attack and destroy neurons.

Chikungunya (meaning "that which bends up" in the Bantu language Makonde) viruses belong to the family of Alphaviridae and are endemic in some African regions. Small epidemics sometimes flare up, but because the virus is highly contagious, a population rapidly develops immunity, which keeps the infection under control within the community. In recent years a more neurovirulent variant of the virus has appeared and spread among islands in the Indian Ocean; for example, in La Réunion more than half the population was infected and more than 80% of the infected population developed signs of disease, often manifesting as extreme fatigue. In children magnetic resonance imaging disclosed changes in periventricular white matter. The virus has spread to India and to Mediterranean countries, where infected mosquitoes have been identified.

Two viruses that have attracted experimental attention are *lymphocytic choreomeningitis* virus (LCMV, family Arenaviridae) and *Borna disease* virus (BDV, family Bornaviridae). These viruses can cause no direct cellular destruction; rather, they cause non-cytolytic viral infections. If mice are infected with LCMV during the first day of life, when the immune response is immature, the viruses can reach neurons in the cerebral cortex and persist there for the rest of the life of the mouse. Infections later, when the immune system has matured, cause a marked infiltration of T cells in the choroid plexus and meninges (choriomeningitis) followed by periventricular infiltration in the brain and death, as described above. LCMV has been extensively used as a prototype of an immune-mediated viral disease of the brain.[12] In BDV-infected adult mice there is also a marked T-cell infiltration in the brain parenchyma, causing death. However, as with LCMV, the virus can persist in neurons if the mice are infected early in life, interfering with neuronal signaling and synaptic activities, and causing behavioral disturbances.[37] Animals infected in nature may also show behavioral changes, most notably in horses in the cavalry town of Borna, Germany, where the disease was first described. Attempts to link BDV infections to human neuropsychiatric disorders have been equivocal.

DNA Viruses

Herpes simplex virus (HSV), a DNA virus with humans as the host animal, belongs to the large family of Herpesviridae. The viruses spread from their initial site of replication in the epidermis by retrograde axonal transport to peripheral nerve root ganglia, in which they may establish a latent infection. During the latency period, early viral transcripts are expressed. This expression triggers release of IFN-γ from sensitized T cells in ganglia, which blocks further replication of the viruses. During episodes of stress, trauma, or ultraviolet illumination, the viral replication is more pronounced, it cannot be blocked by IFN-γ, and the T cells are triggered to release perforin, which kills the infected neurons. In the meantime, new viral particles are formed and transported anterogradely in axons to infect the skin. New blisters arise and the virus can spread to another individual.

Rarely, HSV spreads to the brain, but in about 1 in 250,000 individuals it can attack the limbic system including the hippocampus. The virus spreads probably along the olfactory route to reach these structures and the lesions are usually asymmetrical. Some children who develop HSV encephalitis have defects in the TLR-3 signaling pathway, which impedes the innate immune response and cross-talk with the adaptive immune response. The early signs of HSV encephalitis include behavioral disturbances, after which the disease progresses to coma. If left untreated, the disease is often fatal, with hemorrhage and edema in the medial temporal lobe. Patients who recover may be left with severe memory and behavior disturbances, the Klüver–Bucy syndrome. Only occasionally does HSV attack the brainstem to cause a rapidly fatal infection. Such infections are probably the result of virus spreading from the trigeminal ganglia along the nerve roots, taking the wrong direction to the center instead of the periphery (the usual route after reactivation).

Another *herpesvirus* is varicella-zoster virus, which also remains latent in peripheral nerve root ganglia. After a childhood varicella infection, the virus can be harbored in the ganglia under immune control, only to become activated, usually after 65 years of age, when there is an age-dependent decrease in the adaptive immune response. The reactivated viruses spread anterogradely to cause blisters in the corresponding dermatome; it was by a careful mapping of the distribution of zoster blisters that the dermatomes were initially described at the turn of the twentieth century.

Urinary infections with *Papovaviruses* are very common in childhood. The viruses may spread to the brain to establish latent infections in the myelin-forming oligodendroglial cells. During states of immunodeficiency the viruses can be activated to cause myelin degeneration, resulting in a severe and mostly lethal disease called progressive multifocal leukoencephalitis (PML). HIV infections (see below) are one of the major causes of immunodeficiencies that induce PML, which is also a concern when aggressive immunosuppressive therapy is instituted. This virus is an example of an infectious agent that can cause demyelination in the human CNS. Experimentally, a number of viruses, such as mouse hepatitis virus and Theiler viruses, have been used to analyze mechanisms of virus-induced demyelination in both the central and peripheral nervous system.

Retroviruses

The human immunodeficiency virus-1 (HIV) causes AIDS. After spread to the brain it can be associated with behavioral disturbances that lead to dementia and death. The viral particles reach the brain predominantly via the bloodstream and pass the BBB in macrophages by a Trojan horse mechanism. From these invading monocytes the virus can be transmitted to perivascular macrophages and microglial cells, which can fuse to form giant multinuclear cells as a result of expression of viral fusion proteins on their surfaces.

Like trypanosomes, the infected monocytes can be found in the choroid plexus, and HIV predominantly infiltrates the white matter of the brain, at least in the early stages. Astrocytes can also be infected, but these cells do not provide the proper synthetic machinery for replication of the virus, although their functions may be disturbed. Neurons are not infected but their functions are altered and they may undergo degeneration. Molecules involved in neuronal dysfunction and degeneration include viral proteins such as *tat* and *nef*, immune response molecules, and excitotoxic amino acids, such as glutamate, when their uptake by infected astrocytes is impeded. Although the infection initially involves cerebral white matter, degenerative changes appear subsequently in the cortical neurons, leading to HIV-associated neurocognitive disorders (HAND) (Fig. 27.11). Infection used to lead invariably to death and was a slow infection; the disease visna ("wither away" in Icelandic, where the disease was first described) is its sheep counterpart. Since combination antiretroviral therapy has come into widespread use, severe HIV-associated dementia has become less common, although milder forms of HAND are still prevalent, but the pathogenesis is still uncertain.

HTLVs type 1 and 2 were first isolated from infected T cells in leukemia patients. HTLV-1 has infected humans for thousands of years and it is estimated that 10–20 million people worldwide are infected. The majority remains asymptomatic, and it is not clear why the lifetime risk of developing disease is only about 1% for the infected individuals. HTLV-1-infected $CD4^+$ T cells may pass the BBB into the thoracic segments of the spinal cord to cause HTLV-1-associated myelopathy, also called tropical spastic paraparesis. Similarly to HIV, the virus does not infect the neurons but indirectly causes functional disturbances manifesting as limb weakness and spasticity. Virus-specific $CD8^+$ T cells probably recognize the infected $CD4^+$ T cells in the spinal cord and release cytokines that can affect neuronal functions. Later, degenerative changes may appear. It is not known why the infected T cells preferentially invade the white matter of the thoracic spinal cord, and the mechanisms for late-appearing neurodegenerative changes remain elusive.

During evolution retrovirus sequences have been incorporated into the genome of animals as endogenous retroviral elements. It has been estimated that about 8% of the human genome consists of such sequences. In the past they have mainly been regarded as "junk", since most of them do not encode proteins. However, in mice whole endogenous retroviral particles can appear and cause a motor neuron disease in combination with infections with another virus, which by itself is non-pathogenic. Some studies indicate recombination between deficient endogenous retroviral elements and resurrection of infectious retrovirus in immunodeficient mice.[40] Although retroviral transcripts are induced in a number of human nervous system diseases, ranging from multiple sclerosis and amyotrophic lateral sclerosis to schizophrenia, it is not known whether they are selectively activated and play any pathogenic role in the diseases or merely serve as markers of leaky transcription of different genes.

Prion Infections

The genetic code for all the infectious microbes described so far is carried by nucleic acids. Prions are unusual in this respect, since these transmissible agents are devoid of nucleic acids. Instead, they consist of a misfolded isoform of normal protein, called the prion protein (PrP^c). The abnormal prion protein, called PrP^{Sc}, was discovered first and this is the reason why the normal isoform later received the name of PrP^c. The disease-associated PrP^{Sc} has the superscript Sc because it was first described in mouse brains infected with material from sheep with scrapie, which was later defined as a prion disease. Replication of prions occurs when one PrP^{Sc} molecule interacts with a PrP^c molecule, which then become misfolded to form another PrP^{Sc}. Synthetic PrP^{Sc} has been produced from recombinant PrP^c, but the process is facilitated by still undefined cellular factors. Both PrP isoforms are encoded by the same chromosomal gene, designated *PRNP* in humans and *Prnp* in mice. PrP^c is normally expressed on the cell surface linked to the plasma membrane by a glycolipid anchor.[41] It is expressed on several cells including WBCs in the body, but at particularly high concentrations in neurons. The function of PrP^c remains to be clarified, but since it is a copper binding protein a neuroprotective role against

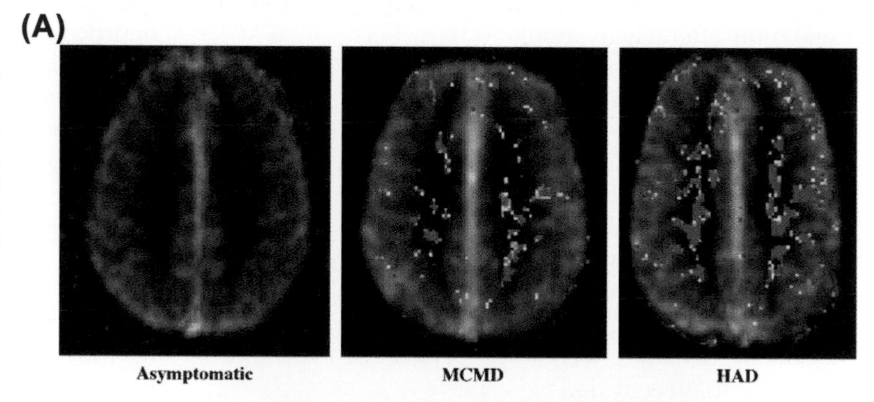

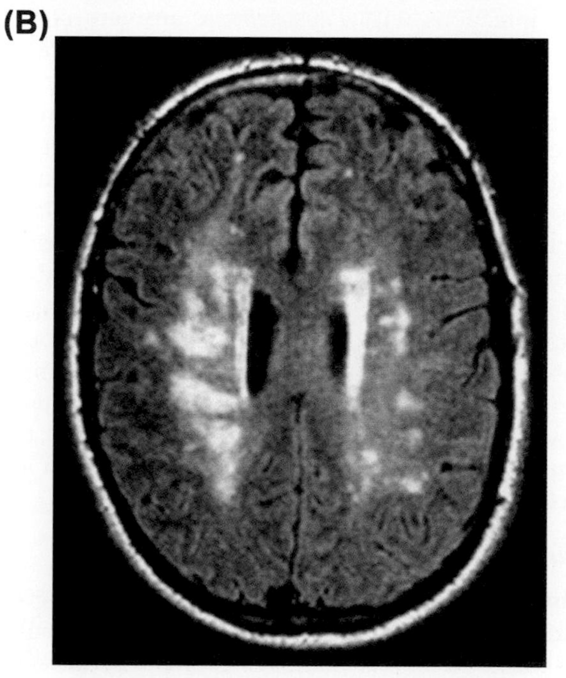

FIGURE 27.11 **(A) Perfusion magnetic resonance imaging maps showing regions of increasing cerebral blood volume with advancing human immunodeficiency virus (HIV) infection: asymptomatic, minor cognitive motor disorder (MCMD), and HIV-associated dementia (HAD); (B) computed tomography scans of HIV-infected brain.** *Source: (A) From Tucker* et al. J Neuroimmunol. *2004;157:153–162,*[38] *with permission; (B) from Offiah and Turnbull.* Clin Radiol. *2006;61:393–401,*[39] *with permission.*

oxidative stress has been suggested, among other potential physiological roles.

Prion infections cause slowly progressive neurodegenerative diseases that invariably lead to death; they are slow infections. In animals, they cause scrapie in sheep and goats, bovine spongiform encephalopathy (BSE) in cattle, chronic wasting disease in deer, and transmissible mink encephalopathy in mink. In humans, they cause Creutzfeldt–Jakob disease (CJD), of which there are four variants: sporadic, familial, iatrogenic, and a new variant (vCJD). All familial variants are associated with various mutations in the *PRPN*. vCJD has been observed in about 200 patients following the large epidemic of BSE in the 1980s. Prion diseases are unique because, in addition to being sporadic or hereditary, they are transmissible; that is, brain material from patients with either type of disease can transmit the disease to another individual. This is why they have caused the iatrogenic variant following transplantations of, for example, cornea and dura mater from infected individuals.

Prions cause slow infections, usually after a long incubation period that can last over decades. Prions do not induce an immune response since they are derived from a self protein. Paradoxically, immune cells seem to facilitate propagation of prions to the CNS following an inoculation in the periphery. Follicular dendritic cells serve as an important link in this spread to peripheral axon terminals and further propagation by retrograde axonal transport to the spinal cord or brainstem, as seen in mouse experiments. In the brain, prions cause neurodegeneration that starts at the synaptic level followed by spongiform degeneration of neurons, which means that large vacuoles appear in the neuronal cytoplasm, and a very strong astrocytic reaction. In contrast to most other infections, there is no inflammatory cell infiltration in the brain.

Various areas of the brain can be affected depending on the strain of the prion, host genetic factors, and the pathways for spread of the prions. Classical CJD targets mainly the cerebral cortex and striatum to cause dementia and myoclonic jerks. In the familial variant called fatal familial insomnia, the most severe neurodegenerative changes occur in the ventral and mediodorsal nuclei of the thalamus, and the patients experience rapidly progressive insomnia followed by hallucinations, coma, and death. A familial variant called Gerstmann–Sträussler–Scheinker syndrome affects mainly the cerebellum to cause ataxia and tremor, and this was also observed in a disease called kuru, which spread between people in Papua New Guinea by endocannibalism. Kuru was the first human prion disease that was shown to be transmissible; in this case to chimpanzees in the 1960s. With the prohibition of cannibalism, kuru more or less disappeared. Since some of the more common neurodegenerative diseases have a similar course of events that slowly progresses to an inevitable end in death and are characterized by accumulations of other misfolded proteins in the brain, research on the potential applications of the concept of prions is of current interest.

FUTURE DIRECTIONS

Studies on the interactions between infectious agents and the nervous system have the prospect to continue to provide new knowledge on a number of important pathobiological problems. For instance, analyses of molecules involved in combating an invading pathogen that at the same time do not harm neuronal cells and networks may lead to a better understanding of mechanisms that regulate the balance between neuroprotective and neurodegenerative signaling in diseases. Such studies may also reveal new host-derived molecules that could be of therapeutic use; the search for such molecules will be crucial with the increasing problem of resistance to antibiotics. Information on the neurobiology of prion infections may also provide novel concepts of the pathogenesis of neurodegenerative diseases, for example on how the misfolded proteins that characterize them are formed and propagated in the nervous system as well as their role in neurotoxicity.

Problems to be addressed include the pathophysiology of the long incubation periods of prion diseases and late-onset effects on the brain which can be manifested as epilepsy or behavioral disturbances, by several microbial infections described in this chapter. These mechanisms are of interest for understanding the pathogenesis of neuropsychiatric diseases, such as schizophrenia, which may be initiated by an early life infection or event, but give rise to symptoms in the second or third decade of life. Of fundamental interest to explore is whether the diseases reflect changes in the nervous tissues that evolve progressively and become clinically manifest only when they have reached a threshold of severity or extent, or disturbances in homeostatic cellular balances that at some point transit to an irreversible state leading to death of both cell and organism. Depending on this knowledge, treatment strategies may be devised to prevent the outbreak of a disease or the appearance of the long-term effects.

Although it is often useful to discuss microbe–host interactions from the point of view of favor to either one of them, the most successful interactions are symbiotic, providing added value for both organisms. Future studies on microbe–host interactions in simple systems will provide new knowledge on the evolution of molecular interplays that generate symbiotic behavior and the role that neurons may play in immunity.[42] Since microbes have a remarkable capacity to resist therapeutic interventions either by the evolution of resistance or by hiding from the immune responses owing to their coevolution, a stronger integration of concepts of evolutionary biology into pathobiology is also foreseen.[43] Such integration is particularly important for the development of new means to control vector-transmitted pathogens.

From the discovery point of view, the days of "microbe hunters",[44] which peaked in the mid-nineteenth century, are not over. Viral metagenomic analyses have estimated that only 1% of all viral diversity has been explored so far.[45] With the introduction of novel techniques, hitherto unknown microbes have already been detected in healthy blood donors, and we may see a new revolution in the discovery of infectious agents that play a role in some of the many diseases affecting the nervous system and whose etiology is still unknown. Increased understanding of the potential roles played by non-coding regions of retroviral elements incorporated in the human genome in transcriptional and translational regulations in nervous tissues is also foreseen. Furthermore, epigenetic mechanisms can control gene expression in microbes and there is an increasing literature indicating that microbes can influence epigenetic factors to regulate transcription in host cells. Studies of such mechanisms have the potential to provide new prospects for an understanding of how early life infections may cause nervous system dysfunctions later in life.

Finally, in a historical perspective, microbes not only are known as the cause of diseases in the nervous system, but also have been of help in investigating the nervous system. For instance, the map of dermatomes was established from clinical descriptions of the distribution of *Herpes zoster* blisters in the skin, the BBB was detected following attempts to treat trypanosome infections with trypan dyes, sleep–wakefulness-regulating areas in the brain were disclosed in neuropathological studies of patients who had died from encephalitis lethargica in the 1920s, axonal transport in the retrograde direction was predicted in studies of neurovirulent viruses, and synaptic functions were investigated using tools derived from bacterial toxins. Viruses are now widely used as transneuronal tracers and as vectors to deliver genes into neurons, and developments in the use of microbes to explore the development, connectivity, and function of the nervous system could therefore be predicted.

Acknowledgments

The studies have been supported by grants from the US NIH/Fogarty (1R21NS064888-01A1) and the Swedish Research Council (04480).

References

1. Rohwer F, Thurber RV. Viruses manipulate the marine environment. *Nature*. 2009;459:207–212.
2. Breitbart M. Marine viruses: truth or dare. *Annu Rev Mar Sci*. 2012;4:425–448.
3. Mallet F, Bouton O, Prudhomme S, et al. The endogenous retroviral locus ERVWE1 is a bona fide gene involved in hominoid placental physiology. *Proc Natl Acad Sci USA*. 2004;101:1731–1736.
4. Brüssow H. Europe, the bull and the minotaur: the biological legacy of a Neolithic love story. *Environ Microbiol*. 2009;11:2778–2788.
5. Tan M-W, Shapira M. Genetic and molecular analysis of nematode–microbe interactions. *Cell Microbiol*. 2011;13:497–507.
6. Takeucchi O, Akira S. Pattern recognition receptors and inflammation. *Cell*. 2010;140:805–820.
7. Ljunggren H-G, Kärre K. In search of the "missing self": MHC molecules and NK cell recognition. *Immunol Today*. 1990;11:237–244.
8. Hamilton SE, Jameson SC. CD8 T cell quiescence revisited. *Trends Immunol*. 2012;33:224–230.
9. Neumann H, Medana IM, Bauer J, Lassmann H. Cytotoxic T lymphocytes in autoimmune and degenerative CNS diseases. *Trends Neurosci*. 2002;25:313–319.
10. Griffin DE, Metcalf T. Clearance of virus infection from the CNS. *Curr Opin Virol*. 2011;1:216–221.
11. Hanamsagar R, Hanke ML, Kielian T. Toll-like receptor (TLR) and inflammasome actions in the central nervous system. *Trends Immunol*. 2012;33:333–342.
12. Thomsen AR. Lymphocytic choriomeningitis virus-induced central nervous system disease: a model for studying the role of chemokines in regulating the acute antiviral CD8+ T-cell response in an immune-privileged organ. *J Virol*. 2009;83:20–28.
13. Kristensson K. Microbes' roadmap to the neuron. *Nat Rev Neurosci*. 2011;12:345–357.
14. Poulin R. Manipulation of host behaviour by parasites: a weakening paradigm? *Proc R Soc Lond B*. 2000;267:787–792.
15. Hughes DP, Kronauer DJC, Boomsma JJ. Extended phenotype: nematodes turn ants into bird-dispersed fruits. *Curr Biol*. 2008;18:R294–R295.
16. Lacroix R, Mukabana WR, Gouagna LC, Koella JC. Malaria infection increases attractiveness of humans to mosquitoes. *PLoS Biol*. 2005;3:1590.
17. Masocha W, Kristensson K. Passage of parasites across the blood–brain barrier. *Virulence*. 2012;3(2):202–212.
18. Combes V, El-Assaad F, Faille D, Jambou R, Hunt NH, Grau GER. Microvesiculation and cell interactions at the brain–endothelial interface in cerebral malaria pathogenesis. *Prog Neurobiol*. 2010;91:140–151.
19. Kariuki SM, Ikumi M, Ojal J, et al. Acute seizures attributable to falciparum malaria in an endemic area on the Kenyan coast. *Brain*. 2011;134:1519–1528.
20. Mahanty S, Garcia HH. Cysticercosis and neurocysticercosis as pathogens affecting the nervous system. *Prog Neurobiol*. 2010;91:172–184.
21. Ferrari TC, Moreira PR. Neuroschistosomiasis: clinical symptoms and pathogenesis. *Lancet Neurol*. 2011;10:853–864.
22. Black MW, Boothroyd JC. Lytic cycle of Toxoplasma gondii. *Microbiol Mol Biol Rev*. 2000;64:607–623.
23. Blomström Å, Karlsson H, Wicks S, Yang S, Yolken RH, Dalman C. Maternal antibodies to infectious agents and risk for non-affective psychoses in the offspring – a matched case–control study. *Schizophr Res*. 2012;140:25–30.
24. Sorokin L. The impact of the extracellular matrix on inflammation. *Nat Rev Immunol*. 2010;10:712–723.
25. Kristensson K, Nygård M, Bertini G, Bentivoglio M. African trypanosome infections of the nervous system: parasite entry and effects on sleep and synaptic functions. *Prog Neurobiol*. 2010;91:152–171.
26. Kennedy PG. Human African trypanosomiasis of the CNS: current issues and challenges. *J Clin Invest*. 2004;113:496–504.
27. Van den Abbeele J, Caljon G, de Ridder K, de Baetsekier P, Coosemans M. *Trypanosoma bruzei* modifies the tsetse salivary composition, altering the fly feeding behavior that favors parasite transmission. *PLoS Pathogens*. 2010;6:1–9.

28. Heggie TW. Swimming with death: *Naegleria fowleri* infections in recreational waters. *Travel Med Inf Dis*. 2010;8:201–206.
29. Thornton JA, Durick-Eder K, Tuomanen EI. Pneumococcal pathogenesis: "innate invasion" yet organ-specific damage. *J Mol Med*. 2010;88:103–107.
30. Drevets DA, Bronze MS. *Listeria monocytogenes*: epidemiology, human disease, and mechanisms of brain invasion. *FEMS Immunol Med Microbiol*. 2008;53:151–165.
31. Tapinos N, Rambukkana A. Insights into regulation of human Schwann cell proliferation by Erk1/2 via a MEK-independent and p56Lck-dependent pathway from leprosy bacilli. *Proc Natl Acad Sci USA*. 2005;102:9188–9193.
32. Caleo M, Schiavo G. Central effects of tetanus and botulinum neurotoxins. *Toxicon*. 2009;54:593–599.
33. Johnson RT. *Viral Infections of the Nervous System*. 2nd ed. Philadelphia, PA: Lippincott-Raven; 1998.
34. Fermi C. *La Rabbia. Siena*. 1950.
35. Lafon M. Subversive neuroinvasive strategy of rabies virus. *Arch Virol*. 2004;(suppl 18):149–159.
36. Davis LE, DeBiasi R, Goade DE, et al. West Nile virus neuroinvasive disease. *Ann Neurol*. 2006;60:286–300.
37. Prat CM, Schmid S, Farrugia F, et al. Mutation of the protein kinase C site in Borna disease virus phosphoprotein abrogates viral interference with neuronal signaling and restores normal synaptic activity. *PLoS Pathog*. 2009;5:1–10.
38. Tucker KA, Robertson KR, Lin W, et al. Neuroimaging in human immunodeficiency virus infection. *J Neuroimmunol*. 2004;157(1):153–162.
39. Offiah CE, Turnbull IW. The imaging appearances of intracranial CNS infections in adult HIV and AIDS patients. *Clin Radiol*. 2006;61(5):393–401.
40. Young GR, Eksmond U, Salcedo R, Alexopoulou L, Stoye JR, Kassiotis G. Resurrection of endogenous retroviruses in antibody-deficient mice. *Nature*. 2012;491(7426):774–778.
41. Prusiner SB. *Prion Biology and Diseases*. 2nd ed. New York: Cold Spring Harbor Laboratory Press; 2004.
42. Kawli T, He F, Tan M-W. It takes nerves to fight infections: insights on neuro-immune interactions from. *C. elegans Dis Model Mech*. 2010;3:721–731.
43. Little TJ, Allen JE, Babayan SA, Matthews KR, Colegrave N. Harnessing evolutionary biology to combat infectious diseases. *Nat Med*. 2012;18(2):217–220.
44. De Kruif P. *Microbe hunters*. New York: Harcourt, Brace & Co; 1926.
45. Mokili JL, Rohwer F, Dutilh BE. Metagenomics and future perspectives in virus discovery. *Curr Opin Virol*. 2012;2:63–77.

CHAPTER

28

Pathobiology of CNS Human Immunodeficiency Virus Infection

Jennifer L. Lyons[†], *Luis B. Tovar-y-Romo*[‡], *Kiran T. Thakur*[*], *Justin C. McArthur*[*], *Norman J. Haughey*[*]

[†]Department of Neurology, Division of Neurological Infections, Brigham and Women's Hospital and Harvard Medical School, Boston, Massachusetts, USA; [‡]Instituto de Fisiología Celular, Universidad Nacional Autónoma de Mexico, Mexico; [*]Department of Neurology, Division of Neuroimmunology and Neurological Infections, The Johns Hopkins University School of Medicine, Baltimore, Maryland, USA

OUTLINE

Neurobiology of Brain Disorders
http://dx.doi.org/10.1016/B978-0-12-398270-4.00028-8

INTRODUCTION

CNS complications due to infection by the human immunodeficiency virus type 1 (HIV-1) were first described early in the epidemic, when severe neurocognitive disease attributable to the virus was common in individuals with profound immunodeficiency. The pathophysiology underlying this disease relates directly to infection of cells in the CNS and also indirectly to immune dysregulation induced by the virus in both the CNS and the periphery. All cell types in the CNS suffer damage mediated by HIV-1, although few are actually infected. HIV infection paradoxically triggers chronic immune activation in addition to immune suppression, which allows further infection. In the CNS, this cycle of infection and inflammation leads to parenchymal injury and varying degrees of the clinical phenotype, a subcortical dementing illness. Since the introduction of combination antiretroviral therapy (cART), the epidemiology of neurocognitive disease related to HIV-1 infection has shifted towards less severe forms, although the prevalence of these manifestations has not decreased and the exact pathophysiological mechanisms underlying this shift are unknown. This chapter details basic concepts of CNS HIV-1 infection and injury with additional attention to the consequences of HIV-mediated immune dysregulation on the CNS (reviewed by McArthur *et al.*[1]).

HUMAN IMMUNODEFICIENCY VIRUS GENETICS AND GENOMIC ORGANIZATION OF HIV-1

HIV is a human retrovirus belonging to the subfamily Lentivirinae. Two major types of HIV stem from the primates where they originated: HIV-1, which was originally a chimpanzee virus; and HIV-2, which infected sooty mangabeys. HIV-1 is seen throughout the world, whereas HIV-2 is restricted to West Africa, where it co-occurs with HIV-1. The neurovirulence of HIV-2 is largely unknown, although systemically HIV-2 is less pernicious. HIV-1 has been categorized into four groups: M (major), O (outlier), and two new groups, N and P. HIV-1 group M is responsible for the vast majority (>90%) of all HIV infection in the world. This group itself has subcategories known as clades or subtypes that are also regional. There are nine subtypes of HIV-1 (A, B, C, D, F, G, H, J, and K), and there are several recombinant forms among these subtypes, known as circulating recombinant forms (CRFs). Although the distribution of subtypes varies throughout the world, clade or subtype B is predominant in the USA, Europe, and Australia, where most studies on the neuropathology of HIV have been conducted.[1] Thus, as with HIV-2, the neuropathology of other subtypes of HIV-1 is largely uncharacterized. This chapter covers HIV-1 and presumptively clade B neuropathology (designated hereafter as "HIV", for simplicity).

The structural proteins of HIV are encoded in three conserved genes (reviewed by Sundquist and Krausslich[2]): group-specific antigen gene (*gag*), polymerase (*pol*), and envelope (*env*). Gag is a polyprotein precursor that includes matrix antigen (MA), capsid (CA), nucleocapsid (NC), and p6 (Fig. 28.1). MA (p17) guides Gag translocation to the plasma membrane by regulating myristoylation of a conserved glycine. This modification promotes interactions with phospholipid and phosphatidylinositol-(4,5)-bisphosphate components of the inner host cell membrane. CA (p24) interacts with cellular cyclophylin A to increase infectivity, possibly by blocking host antiviral factors. NC (p7) plays a role in the encapsidation of full-length, unspliced genomic RNA into virions and participates in reverse transcription and integration. p6 is the final segment of Gag encoding a 6 kDa proline-rich protein that binds the cellular endosomal sorting complex required for transport (ESCRT). This step is critical for the newly formed viral particles to split from the host cell membrane.

Pol is the polyprotein precursor of three different enzymes that function in genome reverse transcription and integration. Pol encodes a viral protease (PR), reverse transcriptase (RT), and integrase (IN). The third structural gene *env* encodes gp160, a precursor glycoprotein that is cleaved to produce the smaller glycoproteins surface (SU or gp120) and transmembrane (TM or gp41). gp120 and gp41 associate through non-covalent interactions to form trimeric complexes that localize to the surface of the viral particles. These proteins are critical for virion binding to host cell membranes.

HIV also codes for two regulatory and four accessory proteins (Fig. 28.1). The transactivator of transcription (Tat) is a regulatory factor that forms a ternary complex with the transcription factor pTEFb and the hairpin stem-loop secondary structure of the Tat response element. This interaction increases transcription of the HIV genome mediated by RNApol II.[2] The regulator for expression of viral proteins (Rev) binds to the Rev responsive element (RRE), which is present on the messenger RNAs (mRNAs) that code for structural genes and also for most of the accessory proteins. These interactions facilitate the nuclear export of the unspliced forms of their primary transcripts. Given that expression of these genes requires Rev function, they are referred to as late viral proteins. Rev-independent genes such as Tat, Nef, and Rev itself are considered to be early viral proteins.

Accessory HIV proteins include viral infectivity factor (Vif), viral protein R (Vpr), viral protein unique (Vpu), and negative factor (Nef). Vif blocks expression of host cell factors that interfere with HIV replication, like APOBEC3G. Vpr helps to import viral preintegration

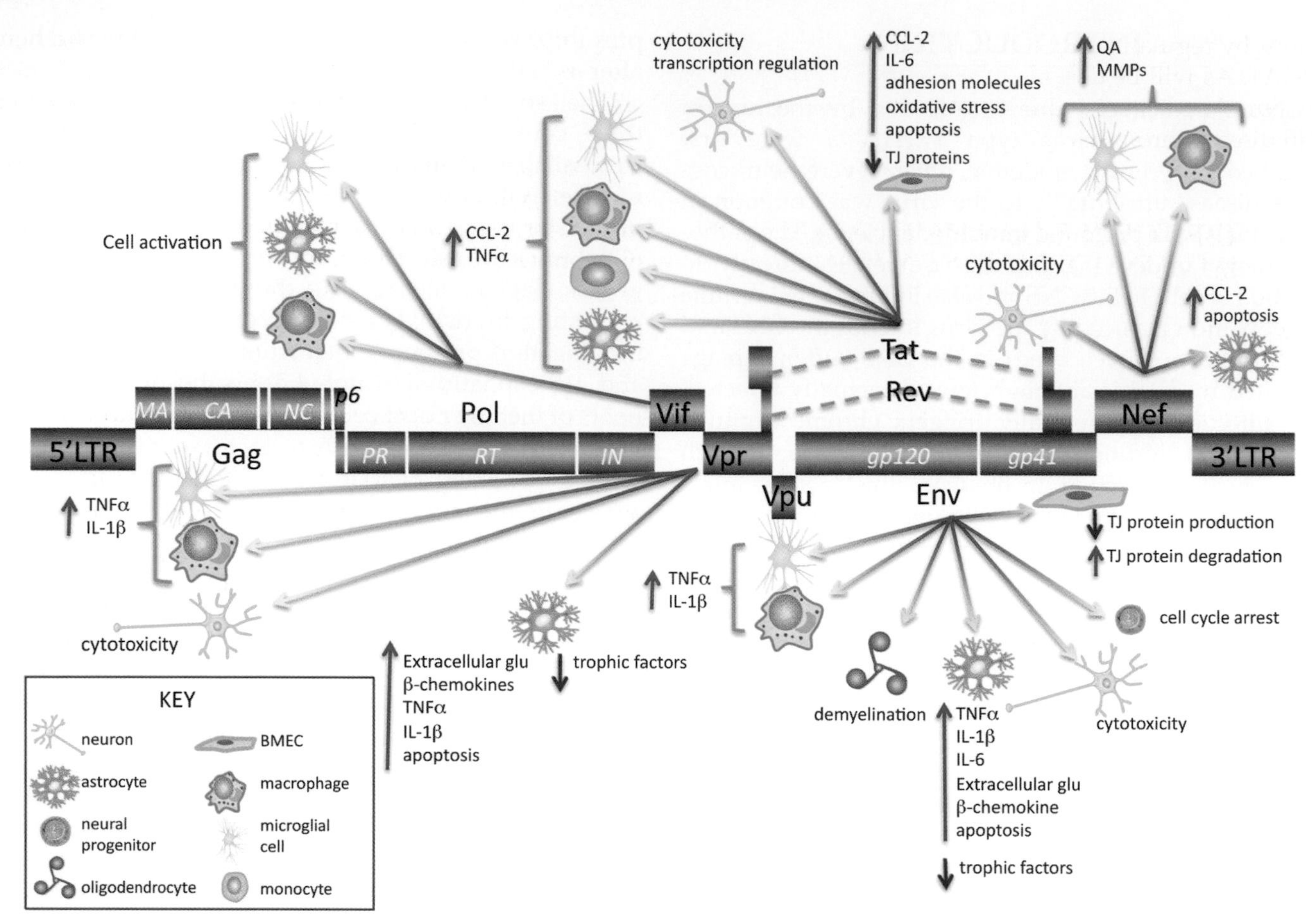

FIGURE 28.1 **HIV RNA genomic organization and potential CNS effects of proteins encoded.** *gag*, *pol*, and *env* are conserved genes that encode for structural proteins, and *vif*, *vpr*, *vpu*, *rev*, *tat*, and *nef* are genes encoding accessory proteins. Vif can be found in astrocyte and microglial cell lines and macrophages and is associated with activation of these cell types. Vpr is expressed in microglia and macrophages and increases production of the cytokines tumor necrosis factor-α (TNF-α) and interleukin-1β (IL-1β). Vpr is toxic to neurons by a caspase-dependent mechanism. It induces astrocyte dysfunction, increasing extracellular glutamate (glu), and in high concentrations is toxic to astrocytes. gp120 affects multiple cell types. It is shed by microglial cells and macrophages and activates these cells to produce proinflammatory cytokines including TNF-α and IL-1β. In astrocytes, it induces apoptosis, and/or activates astrocytes to decrease trophic factor production, increase β-chemokine production, increase production of TNF-α, IL-1β, and IL-6, and produce excess excitatory amino acids. gp120 acts on oligodendrocytes to effect demyelination. In neural progenitor cells, gp120 induces cell cycle arrest, and in neurons it is both directly and indirectly cytotoxic. It also interacts with brain microvascular endothelial cells (BMECs) to decrease zona occludin-1 and -2 production and increase their degradation, compromising the blood–brain barrier (BBB). gp41 activates microglial cells and results in TNF-α and IL-1β release from microglial cells and astrocytes. gp41 also demonstrates nitric oxide-mediated neurotoxicity. Tat is released by monocytes, macrophages, microglial cells, and astrocytes and can additionally cross the BBB independently. Tat enters neurons, glial cells, endothelial cells, and immune cells by the heparan sulfate and/or low-density lipoprotein receptors. It is directly excitotoxic to neurons and mediates neuronal death both directly via *N*-methyl-D-aspartate receptors and indirectly via immune activation. Tat also is transported to the neuronal nucleus, where it can regulate transcription. Tat activates astrocytes, macrophages, and microglial cells, increasing CCL-2 and TNF-α production. In endothelial cells, Tat can induce CCL-2 and IL-6 production, E-selectin expression, and downregulation of tight junction (TJ) proteins, exacerbating inflammation and infection, and it can also trigger oxidative stress and apoptosis. Nef is expressed in astrocytes, causing increased CCL-2 production and infiltration of macrophages. It also triggers apoptosis in astrocytes via c-kit and is neurotoxic. In macrophages and microglia, Nef upregulates quinolinic acid (QA) production and increases matrix metalloproteinases (MMPs). Rev is cytotoxic to neurons through cell membrane interactions, and defective Rev function in astrocytes contributes to restriction of infection in these cells.

complexes to the cell nucleus and induces a growth arrest of target cells at the G_2 phase of their cell cycle to optimize conditions for expression of viral genome. Vpu is a membrane protein that catalyzes the degradation of CD4 in the endoplasmic reticulum, blocking the interaction between nascent gp120 and the CD4 viral receptor that is necessary for Env assembly into virions. Vpu also contributes to virion release from the plasma membrane by blocking CD317, a cellular factor that prevents release of enveloped viruses from the plasma membrane. Nef is a multifunctional factor that also contributes to the downregulation and subsequent degradation of CD4 in infected cells. Nef also downregulates major histocompatibility complex class I (MHC-I) proteins to prevent cytotoxic T cells from recognizing infected cells and modulates the amount of cholesterol incorporated into virions by regulating the cellular cholesterol transporter ABCA1. As will be described in this chapter, several of these viral proteins have a central role in HIV-mediated CNS disease through a variety of mechanisms.

LIFE CYCLE OF THE HUMAN IMMUNODEFICIENCY VIRUS

Binding, Fusion, and Entry

HIV and other enveloped viruses gain entry into cells through interactions with cell surface receptors. The primary cell surface receptor for HIV binding is the CD4 glycoprotein receptor. The HIV coat protein gp120 forms trimeric complexes that interact with CD4 receptors expressed on the surface cells and involved in immunological functions, including T cells, monocytes, macrophages, microglial cells, and cells of dendritic origin. gp120 binding to CD4 induces a conformational shift in gp120, which exposes residues important for coreceptor binding. There are two primary coreceptors for HIV. The chemokine receptor CCR5 preferentially interacts with M-tropic strains of HIV, while CXCR4 preferentially interacts with T-tropic strains. Dual tropic strains of HIV are capable of using either CXCR4 or CCR5 as coreceptors (see "Tropism and Coreceptor Use", below). These initial binding events expose hydrophobic residues on the HIV surface protein gp41. Insertion of gp41 into the plasma membrane of the host cell triggers fusion of viral and host membranes. This fusion of virus and host cell membranes preferentially occurs in specialized membrane microdomains known as lipid rafts. These regions of cellular membranes are enriched in glycolipids that directly interact with gp120 to stabilize the fusion complex. In the CNS, these noncovalent interactions of HIV with glycolipids may be especially important for virus binding and entry into cells that do not express CD4.[2]

Integration, Assembly, and Release

Cells that undergo productive infection produce infectious virions in a multistep process that begins with integration of the viral genome into the host cell genome. This process involves a conversion of the viral RNA template to a DNA molecule by the viral reverse transcriptase. Integration of the proviral DNA is accomplished by viral integrase. Host DNA binding proteins may facilitate integration of proviral DNA at specific locations by inducing conformational changes in the DNA molecule.

Viral assembly begins with synthesis of the Gal–Pol precursor protein. The viral protease element coded in the Pol gene cleaves the Gag–Pol proprotein into several peptides including matrix, capsid, nucleocapsid, p6, reverse transcriptase, and integrase that are required for the assembly of infectious viral particles. Mature virions are assembled at the inner face of the cells plasma membrane. For this step, Gag is trafficked to the plasma membrane where it forms multimers that associate with other HIV proteins, such as Env. Several molecular cues direct Gag to lipid rafts in the plasma membrane where the final steps of viral assembly occur. These include post-translational modifications such as acylation, myristolyation, prenylation, and palmitoylation. HIV RNA genome transcripts are directed to the site of virion formation by interactions between RNA *cis*-acting elements and the NC portion of Gag. The HIV RNA genome is encapsidated in the form of dimers along with transfer RNAs (tRNAs) that function as primers for reverse transcription and a series of cellular RNA molecules.[2]

The final step in the virus life cycle is budding from the host plasma membrane. The viral envelope is formed from the cell plasma membrane and contains many elements consistent with budding from lipid rafts. Viral envelopes are enriched in the ganglioside GM1 and cholesterol, while phospholipid species that are not abundant within these membrane microdomains are generally absent.

Tropism and Coreceptor Use

Tropism is an important determinant of HIV pathogenesis and persistence of infection in the periphery and the CNS. Viral tropism refers to the type of cell in which infection is established. HIV strains are classified as macrophage (M) tropic, T-cell (T) tropic, or dual tropic. Viral coreceptor utilization affects but does not entirely determine tropism. In general, M-tropic HIV preferentially uses CCR5 as a coreceptor (R5 virus), while T-tropic HIV-1 preferentially uses CXCR4 (X4 virus). Dual tropic HIV-1 is capable of using either CXCR4 or CCR5.[3] However, since both macrophages and T cells are capable of expressing CCR5 and CXCR4, and each cell type is capable of supporting infection by either strain of virus, coreceptor use and tropism can be discordant. Often, there is a transition during infection from R5 to X4 tropism. R5 virus initially infects mucosal dendritic cells and macrophages, and memory CD4 T cells that express CCR5, especially in the gut-associated lymphoid tissue (GALT), are subsequently infected. Macrophages and T cells harboring R5 HIV disseminate and establish reservoirs in various compartments (see Gorry and Ancuta[3] for a review of viral reservoirs). X4 tropic HIV typically arises later in infection and is associated with a rapid decline in CD4 T cells, CD8 cell apoptosis, and an impaired humoral response, suggesting a link between the emergence of X4 tropic virus, clinical deterioration, and progression to AIDS. However, since HIV continues to mutate during the entire course of infection, individual infections can end up advancing to any permutation of X4, R5, or dual tropic virus.

ESTABLISHMENT OF HUMAN IMMUNODEFICIENCY VIRUS INFECTION

Peripheral Infection and Viral Dissemination

The major routes of HIV transmission are sexual contact, mother to child, injection drug use with contaminated needles, and exposure to infected blood or blood products, with sexual transmission being the most common route of exposure worldwide. The inaugural event is breach of a mucosal barrier, with traumatic abrasions or ulcerations linked to sexually transmitted diseases greatly facilitating viral entry. Dendritic cells in the genital and rectal mucosa are generally the initial cells infected, and HIV that uses CCR5 as its coreceptor (R5 virus) is chiefly the type that is transmitted.

Initial infective events produce a burst of viremia that drives infection and massive depletion of CD4 T cells, particularly in the GALT. The immune reaction to the virus causes activation of CD8 T cells and virus-specific antibody production, which in turn promotes rebound of CD4 T cells and slowed viral replication, but the mucosal lymph system does not recover to its preinfective condition. Host management of infection at this stage establishes a set-point of viral production during which clinical manifestations may be absent for years. If left untreated, however, the virus almost inevitably degrades the immune system at a later stage, rendering the individual vulnerable to opportunistic infections. In addition, the loss of GALT allows for ongoing microbial translocation that results in chronic immune activation throughout HIV infection[4] (see "Monocyte Trafficking", below).

Progression of HIV is caused in part by this constant immune activation that detrimentally affects the immune system through a combination of direct loss of infected CD4 T lymphocytes and bystander cell killing. This induces an insidious decline in numbers of CD4 T cells, which eventually leads to acquired immunodeficiency syndrome (AIDS), a concept supported by studies in both humans and animals. For example, in simian immunodeficiency virus (SIV) infection of sooty mangabeys, there is chronic high-level viremia but only low immune activation, and the animals do not progress to AIDS. Humans infected with HIV, however, have expansion of both the innate and adaptive immune response. First, HIV infection results in formation of memory B and T lymphocytes with distinct immunophenotypes. Second, dysfunction of B cells leads to hypergammaglobulinemia and T cells are activated, resulting in CCR5 expression. Next, monocytes become hyperactivated with increased expression of CD14. In addition, the innate immune system is activated as shown by increased inflammatory markers (e.g. C-reactive protein) and increased circulating levels of proinflammatory cytokines. Finally, immune activation is also associated with increased cardiovascular disease and mortality. This is evidenced by increased levels of soluble CD163, a marker of monocyte activation, in HIV-infected patients with non-calcified coronary plaques despite viremic control[5] and by an increased mortality rate in those with elevated levels of soluble CD14 (sCD14), a marker of immune activation.[6]

Despite the effects of cART in controlling HIV, immune activation remains prominent in HIV-infected individuals. Individuals with T cells that are persistently activated have lower CD4 T-cell gains during immune reconstitution with cART. Thus, there is a paradox of an activated immune system, which can be associated with disease progression.

These systemic events set the stage for CNS infection and damage, and they predispose to opportunistic infections of the nervous system. After initial infection, HIV rapidly makes its way to the CNS to establish infection, despite this being a heavily guarded location. Furthermore, as will be discussed later in this chapter, nervous system and systemic activation of chronic inflammation affect normal processes in the brain.

ENTRY OF HUMAN IMMUNODEFICIENCY VIRUS INTO THE CNS

Blood–Brain Barrier

The blood–brain barrier (BBB) serves as a partition between the CNS and the peripheral circulation. It is composed of brain microvascular endothelial cells (BMECs) lined by a basal lamina that is surrounded by astrocyte end-feet, supported by pericytes, and connected to neurons (an entity known as the neurovascular unit). HIV can initially cross an intact BBB, but infection results in breakdown of the selective permeability of the BBB through both immune dysregulation and effects of the virus itself. As described below, shed viral proteins trigger a variety of effects on cell signaling cascades, resulting in dysregulation of gene expression. Furthermore, cytokine release from activated immune cells has a variety of downstream effects that are deleterious to BBB integrity. These events have complex effects on propagation of CNS HIV infection, immune-mediated damage, opportunistic infection, and maintenance of BBB dysfunction.

BMECs are specialized endothelial cells that make up the cerebral microvasculature and maintain the structural integrity of the BBB. These endothelial cells, unlike their peripheral counterparts, lack fenestrations and possess an extensive network of tight junctions, all of which serve to strictly regulate transmigration into the brain. BMECs also contain cell surface efflux pumps that actively expel many compounds that otherwise might cross the abluminal surface into the parenchyma. In the setting of HIV, these cells play key roles both in CNS infection and in treatment response.

BMECs connect to one another by adherens junctions and tight junctions, and interactions between these proteins of neighboring cells serve to maintain the selective permeability of the BBB. Tight junctions consist of transmembrane claudin and occludin and intracellular zonula occludens (ZO) proteins. Claudins serve as the backbone for tight junctions, with occludins serving as regulatory proteins and ZOs offering structural support. Adherens junctions are composed of transmembrane cadherins and intracellular catenins, vinculin, and α-actinin. HIV compromises many of these proteins and interactions, as tight junction disruption triggered by viral proteins and/or immune activation contributes to the increased permeability of the BBB. Tat reduces expression of occludin, perhaps owing to alteration in cyclooxygenase-2 (COX-2) expression, but nonetheless this degrades tight junction integrity. gp120 treatment of cultured BMECs results in proteasomal degradation of ZO proteins ZO-1 and ZO-2, and transgenic mice that secrete gp120 have elevated levels of intercellular adhesion molecule-1 and vascular cell adhesion molecule-1 (ICAM-1 and VCAM-1, involved in transmigration of immune cells), with associated BBB dysfunction. Tumor necrosis factor-α (TNF-α) and interleukin-1β (IL-1β) released by activated monocytes and monocyte-derived macrophages (MDMs) also reduce tight junction protein expression (reviewed by Strazza *et al.*[7]).

The basal lamina is an extracellular matrix laid down by BMECs and maintained by astrocytes. As a network of proteins (collagens, laminin, and fibronectin) to which the BMECs are affixed and that participates in communication among the cells of the neurovascular unit, the basal lamina plays an important role in BBB upkeep that HIV infection can disrupt. For instance, Nef, Tat, and gp120 all upregulate expression of matrix metalloproteinases (MMPs). These enzymes are key regulators of neuronal health by a variety of mechanisms, but in the setting of infection and inflammation contribute to breakdown of basal lamina proteins and therefore compromise the integrity of the BBB.

Astrocytes are also instrumental in BBB integrity through their foot processes. Astrocyte end-feet contact the basal lamina on which the BMECs lie to provide support and signaling through trophic factors and to maintain selective permeability. They communicate directly with neighboring end-feet via gap junctions. Astrocytes are activated or damaged in HIV infection from a variety of sources, including non-productive infection, shed viral proteins, and chronic immune activation. Since astrocytes communicate with one another, the extracellular matrix, and BMECs, their dysfunction translates to a loss of several components critical in maintaining BBB integrity. Details on HIV-mediated damage to both astrocytes and BMECs (beyond effects on tight junction and extracellular proteins) can be found later in this chapter (see "Mechanisms of CNS Injury").

Drugs of abuse such as cocaine and methamphetamine, the use of which is prevalent in some HIV-infected populations, may additionally increase the permeability of the BBB.[7] First, use of these substances triggers the upregulation of molecules that ultimately facilitate transmigration of HIV across the BBB, such as MMPs, which degrade extracellular proteins and damage the BBB, and dendritic cell-specific ICAM-3, which functions in cellular adhesion and signaling. Cocaine binds to and enters BMECs, causing further disruption of tight junctions and promoting entry of cells such as monocytes and lymphocytes carrying the virus. Increased BBB permeability in cocaine and methamphetamine users is one mechanism by which psychostimulant substance use may lead to high viral burden in the CNS. Taken together, HIV-mediated and comorbid drug-use mediated insults all result in initiation and maintenance of BBB dysfunction, which not only propagates HIV-mediated infection and inflammation, but also may allow entrance of otherwise barred molecules and microbes.

Monocyte Trafficking

Peripheral blood monocytes, which express CD4 and CCR5 and as such are infected productively by HIV, can cross the BBB upon their activation (Fig. 28.2). A paramount consequence to the loss of the GALT is permissive microbial translocation from the gut mucosa into the circulation, the feature of which is substantially increased plasma levels of endotoxin [lipopolysaccharide (LPS)], a component of Gram-negative bacterial cell walls. LPS is bound by CD14, a constitutively expressed surface protein found on monocytes, resulting in monocyte activation and release of CD14 in soluble form (sCD14). This is facilitated by lipopolysaccharide binding protein (LBP), whose hepatic synthesis is upregulated in the setting of increased plasma concentrations of LPS. Cytokines released by activated monocytes in addition to sCD14 include TNF-α and IL-1β, which initiate a cascade that perpetuates chronic immune activation[4] and results in increased trafficking of cells that may harbor infection into the CNS. This ultimately contributes to neuronal damage, and increased plasma LBP, sCD14, and LPS

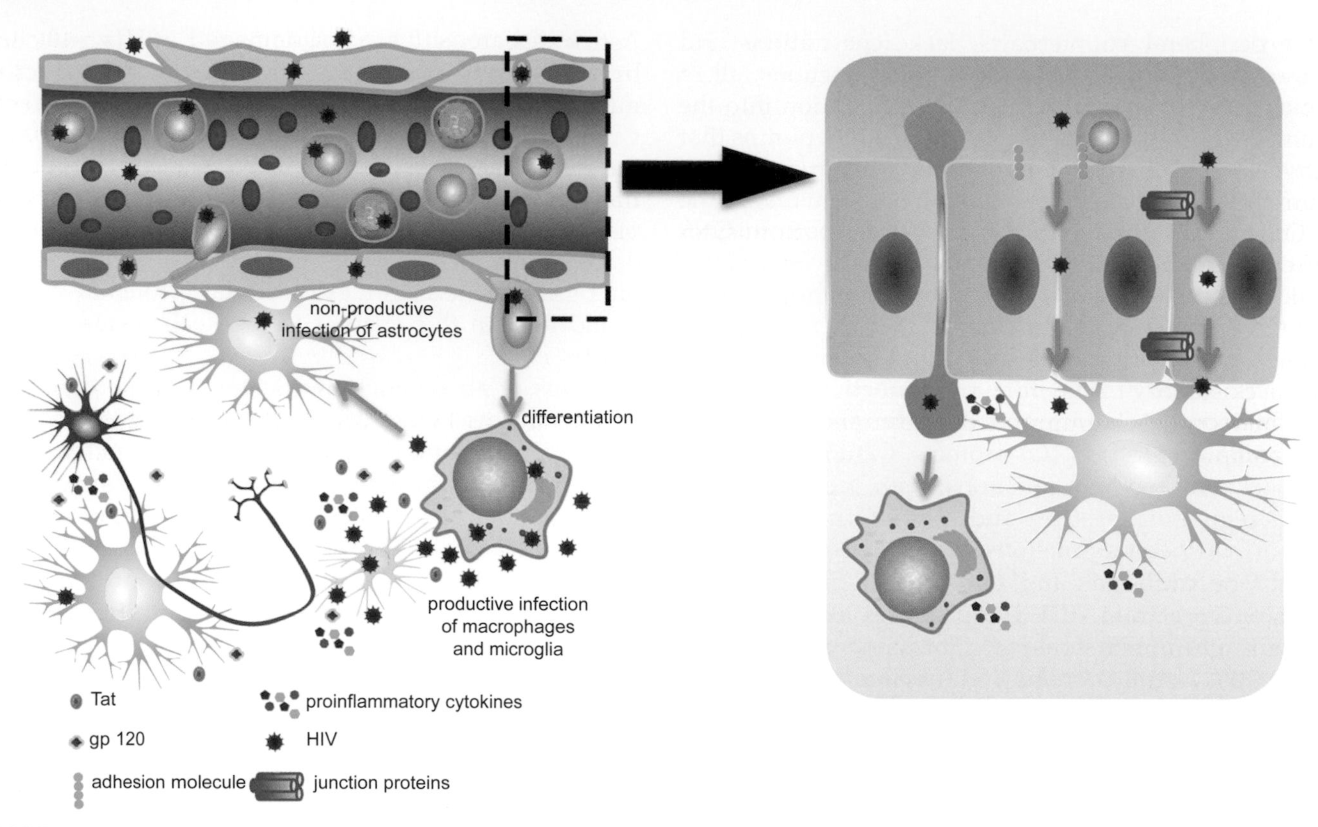

FIGURE 28.2 **HIV breach of the blood–brain barrier (BBB).** Ingress of infected, activated monocytes by means of diapedesis across intact BBB occurs initially. In the setting of ongoing infection, chronic immune activation, shed viral proteins, and infected and/or activated cells lead to upregulation of β-chemokines, vascular adhesion molecules, and loss of adhesion molecules, resulting in increased trafficking of infected and/or activated monocytes. Once across the BBB, monocyte differentiation to macrophages exacerbates infection and inflammation. Other reported mechanisms of breach include adsorptive transcytosis of cell-free virus through a CD4-independent mechanism and paracellular passage in the setting of a leaky BBB.

have all been implicated in the clinical manifestation of CNS HIV infection, HIV-associated neurocognitive disorders (HAND). Although cART does decrease plasma levels of LPS, chronic immune activation apparently continues to varying degrees despite treatment and seems to play a role in the epidemiology of post-cART HAND even in the setting of viral control.

Monocyte activation leads to the release of cytokines and inflammatory mediators and ultimately results in differentiation into macrophages. As mentioned above, TNF-α and IL-1β are released, and these two cytokines have a myriad of effects on many different cell types in the promotion of ongoing HIV infection and inflammation that is described throughout this chapter. At the BBB, they induce expression of selectins and cell adhesion molecules (VCAM-1) on BMECs. Interactions between BMEC selectins and activated monocytes initiate monocyte transmigration across the BBB, a complex series of events beginning with tethering and rolling of the monocyte on the vascular wall followed by the formation of focal adhesions to the vasculature and ultimately the migration of blood cells through capillary walls into the tissues, or diapedesis.[8] Diapedesis is facilitated by cell adhesion molecules and involves the opening of endothelial junctions and monocyte interactions with junctional proteins. The process is enhanced by chemokine gradients created by CCL-2, released by activated astrocytes. Monocytes hence successfully actively negotiate even an intact BBB, and if they are infected with HIV, virus infection can then be established within perivascular macrophages. This has been termed the "Trojan horse" mechanism of CNS HIV infection (although it occurs for other pathogens as well). It is thought to be the predominant route of HIV introduction into the CNS and can occur within days of initial infection.[9]

CD16, an immunoglobulin Fc receptor present on monocytes that is upregulated on their activation, also plays a role in the transmigration of monocytes across the BBB. As mentioned above, monocytes constitutively express CD14, but the mature $CD14^{+}CD16^{+}$ monocyte subpopulation ordinarily comprises about 5–10% of peripheral blood monocytes in non-infected individuals. However, these cells can increase to more than 40% in HIV-infected individuals, especially in advanced disease,[8] and levels increase in individuals with HAND.

The expansion of $CD14^{+}CD16^{+}$ monocytes in the CNS is due to increased trafficking of activated or infected monocytes into the CNS, especially in response to the β-chemokine CCL-2,[8] a finding supportive of increased monocyte trafficking and neuroinflammation in the neuropathogenesis of HIV.

Once inside the CNS, peripheral blood monocytes differentiate into perivascular macrophages. Since HIV-infected MDMs are harboring replication-competent virus, their ingress results in productive HIV infection being established in the brain. This is greatly facilitated by the existence of an incompetent BBB, and HIV infection and chronic immune activation mediate further BBB compromise.

T Cells

As T cells routinely traffic into the CNS and are the primary target of HIV infection, they ostensibly play a role in CNS breach by HIV. Studies in cerebrospinal fluid (CSF) lend evidence to T-cell derived virus in this compartment,[10] although the CSF is not necessarily representative of the parenchyma. In addition, most isolated virus from the CNS originates from monocytes.[11] Therefore, the higher rate of ingress of monocytes into the CNS and their subsequent residence in the parenchyma make them the likely main source of introduction of CNS infection, and macrophages and microglial cells constitute the productively infected populations inside the confines of the BBB.

Transcytosis Across Microvascular Endothelial Cells

Cell-free HIV experimentally breaches the BBB either transcellularly or paracellularly (Fig. 28.2). The latter occurs as a consequence of breakdown of tight junctions and adherens junctions that connect BMECs to one another, as described above. The former occurs when HIV traverses BMECs, a process known as adsorptive transcytosis. The interactions HIV has with BMECs and proteins both directly and indirectly result in loss of BBB integrity and ability of the virus to mediate injury to the brain. This cycle allows for ongoing BBB damage and infection propagation both by cell-free virus and by the major route of HIV entry to the brain via infected monocyte trafficking.

HIV gains access to BMECs experimentally,[11] and although this is not thought to be the principal mechanism of introduction of HIV into the CNS, it nevertheless does represent a potential means of entry into the brain (Table 28.1). It does not seem to result in productive infection of the endothelial cell, and as BMECs lack CD4 on their surfaces, viral entry must occur via a CD4-independent route. The interaction of mannose-6-phosphate receptor with HIV gp-120 has been shown to be involved in this. HIV binds, internalizes, and traverses the endothelial cell via adsorptive endocytosis to be released subsequently across the basement membrane into the brain parenchyma. Tat has been shown to facilitate endothelial transcytosis of HIV across the BBB.

Choroid Plexus

The choroid plexus is a frond-like epithelial layer situated atop a vascular stromal core lacking the tight junctions seen in the BBB and contains immune cells capable of productive HIV infection. There are several lines of evidence pointing to choroid plexus invasion by HIV (Table 28.1).[12] The lack of tight junctions between endothelium and stroma renders the latter relatively easily accessible by the virus, and both HIV-infected T lymphocytes and monocytes have been detected in choroid plexus. Further, HIV infection has also been demonstrated in choroid epithelial cells, and given their direct contact with the ventricles, this may be another entry mechanism of the virus into the brain.

TABLE 28.1 Type of Infection and Outcome of Cells within the CNS

CNS Cell Type	CD4 Expression	Type of HIV Infection	Outcome
Neurons	No	Not infected	Bystander damage, death
Astrocytes	No	Restricted infection	Activation, loss of homeostatic functions, damage to BBB
Microglial cells	Yes	Productive infection	Activation, syncytia formation, death
Resident macrophages	Yes	Productive infection	Activation, syncytia formation, death
Oligodendrocytes	No	Not infected	Bystander damage, death
Microvascular endothelial cells	No	Possible restricted infection vs transcytosis	Damage to BBB
Choroid plexus epithelial cells	No	Possible restricted infection	Damage to blood–CSF barrier
Pericytes	Possible low-level expression	Possible productive infection	Damage to BBB

BBB: blood–brain barrier; CSF: cerebrospinal fluid.

CNS HUMAN IMMUNODEFICIENCY VIRUS INFECTION BY CELL TYPE

Microglial Cells

Microglia are bone-marrow derived cells from myeloid precursors that migrate to the CNS early in development, forming a long-lived population involved in innate immunity and neuronal support. Microglial cells can be found in an active or relatively quiescent (ramified) state. A ramified microglial cell involves itself in immune surveillance, and an activated microglial cell upregulates production of cytokines, chemokines, and reactive oxygen species involved in the innate immune response. Secretory products of activated microglia such as IL-1, IL-6, IL-10, CCL-2, and CXCL8 contribute to immune defense but can also damage the infected brain and break down the BBB.[7] In addition to productive infection by HIV, activated microglia can gather around degenerating neurons and necrotic tissue to remove cellular debris. These aggregates are termed microglial nodules and constitute one feature of HIV encephalitis (HIVE), the neuropathological term for histological changes seen in the brain due to HIV infection, in addition to multinucleated giant cells (Fig. 28.3), monocytic perivascular cuffing, and astrocytosis.[1] This exemplifies how in HIV infection these cells both contribute to and are affected by ongoing damage in the CNS.

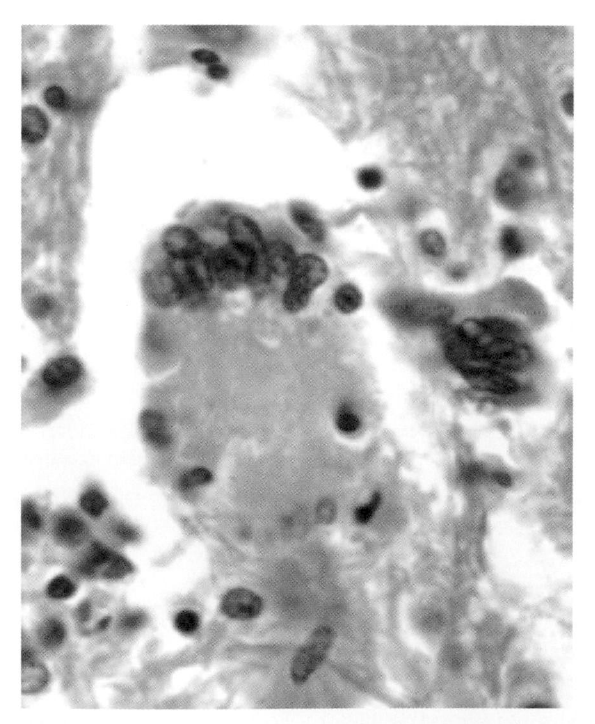

FIGURE 28.3 **Hematoxylin and eosin stain of HIV-infected brain demonstrating a multinucleated giant cell, one pathological hallmark of HIV encephalitis.** *Source: Photograph courtesy of Dr Carlos Pardo, Johns Hopkins Hospital.*

Given their common origins to monocytes and MDMs, the mechanisms and consequences of microglial HIV infection generally mirror those of these cells as described throughout this chapter. Microglia constitutively express CD4, CCR5, and CXCR4 and are productively infected by a CD4-dependent mechanism. Furthermore, they are primarily subject to infection by M-tropic virus. A key difference, however, is that the turnover rate of a microglial cell is in the order of years, as opposed to months for a macrophage, and so once infected, these cells may act as a barrier to eradication.

The consequences of microglial infection are complex and include effects of productive infection and cellular activation.[11] Microglial cells (and macrophages) express viral proteins including gp120, gp41, Tat, Vif, Vpr, and Nef, each of which has direct and indirect neurotoxic and neuroinflammatory effects (Fig. 28.1). Increased microglial number in HIV infection correlates with cognitive decline, and with close association between apoptotic neurons and markers of microglial activation within the deep gray structures. Further detail on damage resulting from microglial infection is given later in this chapter.

Macrophages

As mentioned above, macrophages are the result of monocyte differentiation in tissues and are key actors in both the innate and adaptive immune systems. They engulf and digest cellular debris and foreign pathogens and also trigger targeted responses by lymphocytes and other adaptive immune cells by antigen presentation. Macrophage infection by HIV (Figs 28.4 and 28.5) also results in their activation, which in turn mediates CNS damage through propagation of both infection and chronic immune activation if left unchecked, and

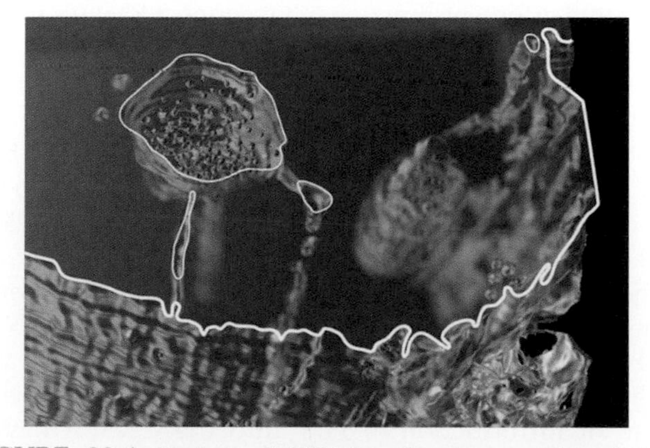

FIGURE 28.4 **Pockets of HIV.** Ion abrasion scanning electron microscopic image. Infected human monocyte-derived macrophages demonstrate HIV particles (red) in internalized endocytic vesicles, which are connected to the cell surface by long channels. *Source: Photograph courtesy of Drs Donald Bliss and Sriram Subramaniam, National Institutes of Health/National Cancer Institute.*

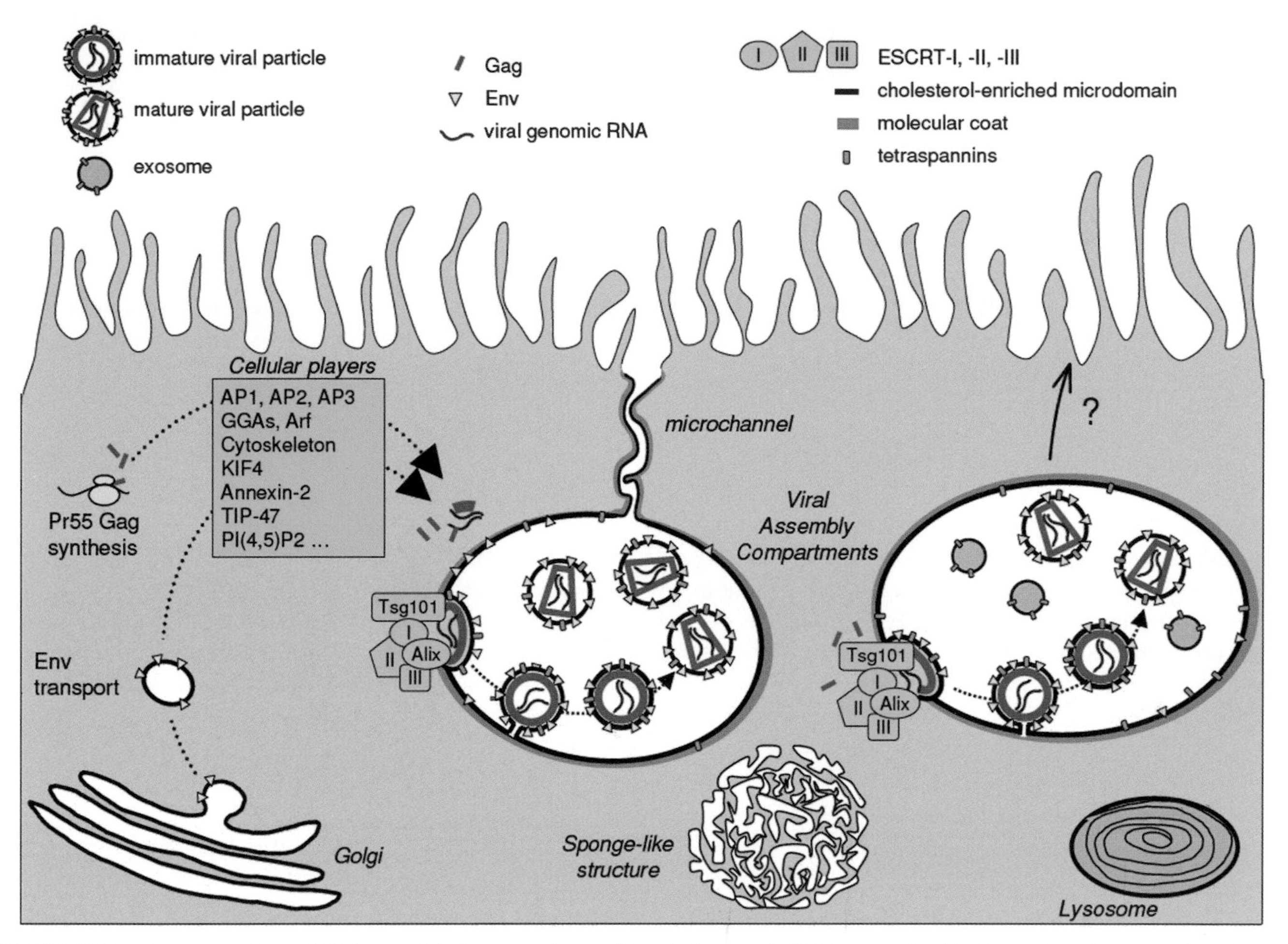

FIGURE 28.5 **A current view of HIV assembly in macrophages.** The viral genomic RNA transcribed in the nucleus is exported to the cytoplasm. The transmembrane envelope (Env) protein is produced in the endoplasmic reticulum and transits through the Golgi apparatus while Gag is synthesized on free cytosolic ribosomes. Both Env and the Gag precursors are targeted to the assembly site through unidentified pathways. The sites of Gag/Env interaction, Gag multimerization and binding to viral genomic RNA also remain elusive. The main cellular factors suspected to play a role in these trafficking events are indicated; nevertheless, most of the time their roles have still to be established in macrophages. The assembly process requires the hijacking of the cellular endosomal sorting complexes required for transport (ESCRT) machinery and occurs on cholesterol- and tetraspanin-enriched membrane microdomains. The assembly compartment can be connected at least transiently to the plasma membrane through thin microchannels that do not allow virion passage. The limiting membrane of the viral assembly compartment as well as the microchannels often exhibit thick molecular coats of obscure composition. *Source: Open access reprint from Benaroch* et al. Retrovirology. *2010;7:29.*[16]

to as yet not well understood extents even in the setting of cART.

Mechanisms of ingress into and egress from macrophages are identical regardless of the location of the macrophage in the body. Ingress utilizes cholesterol-rich lipid rafts present in the plasma membrane. It is controversial whether fusion occurs at the macrophage cell surface or through a vesicular mechanism, but several studies support the role of endocytosis. Macropinocytosis is a receptor-independent phagocytic vesicular pathway used by macrophages to sample the external microenvironment. It has been implicated in macrophage viral entry,[13] with fusion events occurring subsequently within the vesicles. A novel, cholesterol-dependent and caveolae-independent mechanism has been identified and is similar to macropinocytosis.[14] Although endocytosis risks lysosomal degradation, productive infection has been demonstrated through endocytic pathways.[13] β-Chemokines that bind CCR5 and inhibit viral entry in lymphocytes, such as regulated on activation, normally T-cell-expressed and T-cell-secreted (RANTES), macrophage inflammatory protein-1α(MIP-1α), and MIP-1β lack potency in macrophages and therefore support an endocytic pathway of entry in these cells (Fig. 28.4).[15]

HIV can utilize the cellular machinery for intracellular vesicle trafficking and microvesicular particle biogenesis for the assembly and budding of infectious particles (reviewed by Sundquist and Krausslich[2]). HIV budding and microvesicular particle biogenesis require the recruitment of their component proteins at the membrane site of assembly and different cellular mechanisms account for this. One of these mechanisms is ubiquitination, which provides an endocytic sorting cue for plasma membrane proteins to traffic to endosomes and microvesicular particle biogenesis. In the biogenesis of exosomes, ubiquitination is necessary for directing cargo proteins to exosomes by means of the ESCRT. Similarly, retrovirus budding depends on ubiquitination of Gag, which localizes this protein essential for viral assembly to the plasma membrane. This process is also mediated by ESCRT components such as Tsg101, a factor that functions in vacuolar protein sorting and is considered a marker of multivesicular bodies and microvesicles (Fig. 28.5).

Beyond productive infection and its downstream effects, macrophages experimentally have exhibited some noteworthy interactions with HIV including direct cell-to-cell transfer of virus and viral proteins, non-productive, CD4-independent entry, and macrophage-mediated cultivation of T cells. First, in the setting of endocytic entry into macrophages, HIV is capable of retrograde transport to the endoplasmic reticulum and Golgi apparatus where, following virus-induced cytoskeletal rearrangement, it can be shuttled to neighboring cells through cell-to-cell bridging conduits.[2] The viral protein Nef can also be transported from macrophages to B cells in a similar manner.[17] Both of these findings provide evidence of a possible evasion by the immune system. Second, HIV has the capacity to enter macrophages and microglia via gp120 interaction with surface mannose receptor in a CD4-independent fashion, although this mechanism of infection experimentally did not result in productive infection of these cells.[18] Finally, a subset of CD10-expressing macrophages may cultivate T cells and support productive HIV infection; these "nurse macrophages" demonstrate self-renewal capabilities and imply a source of ongoing infection.[19] All of these findings suggest a high level of complexity in the relationship between HIV and macrophages with implications for both the initial CNS infection and possible eradication of disease.

Astrocytes

Astrocytes are an additional population of glial cells that HIV can infect (reviewed by Gorry *et al.*[20]). Infection rates have been reported from less than 1 to 19%, but these are the most abundant cell types in the CNS. Given the large total number and extensive connectivity networks, astrocyte infection is thus potentially an important contributor to HIV neuropathology, even though the infection is usually restrictive rather than productive. Although astrocytes express CCR5 and CXCR4, they do not express CD4, and so their infection occurs via a CD4-independent mechanism (Table 28.1). This may be achieved by utilization of CCR5 or by a different receptor. Studies in human astrocyte cell lines have demonstrated viral exploitation of the human mannose receptor, a C-type lectin receptor, for CD4-independent astrocyte infection. The mannose receptor has also been implicated in non-productive infection of brain macrophages.[18] DC-SIGN, another C-type lectin receptor that is also expressed by astrocytes, has been shown to mediate viral infection. These CD4-independent pathways to intracellular delivery of virus involve endocytosis as opposed to virus–cell membrane fusion at the surface. Infection of astrocytes stimulates cell activation, but HIV readily binds astrocytes to effect activation regardless of viral internalization and infection.

Although infection of astrocytes is non-productive, genomic integration does occur. *In vitro* studies show that early viral proteins Nef, Rev, and Tat can be generated at low levels (Fig. 28.1), but production of viral particles and proteins is otherwise largely ineffective. There are many reasons for this, including inefficient entry into the cell and restricted HIV expression intracellularly.[20] However, despite the fact that this infection is non-productive, it is harmful and may be a barrier to eradication, as described below (see "Viral Latency", below).

Infection of Other CNS Cells

As noted above, productive infection only occurs in MDMs and microglial cells, with astrocytes non-productively infected and BMECs and choroid plexus infectable *in vitro*. However, productive infection of pericytes has been reported; these are support cells in the neurovascular unit and play a role in maintenance of BBB integrity.[21] Neurons, oligodendrocytes, and neural progenitor cells resist infection but nonetheless suffer damage in the setting of CNS disease caused by HIV, discussed later.

CNS ESCAPE AND VIRAL LATENCY

Viral Escape

Once HIV has invaded the CNS, viral replication and mutation can become disparate to those in the blood. This creates potential for resistance mutations that differ from those in the periphery. Although CNS viral variants can re-enter the peripheral circulation, which has implications in reservoir establishment, the more proximate possible outcome is inadequate control of replicating virus in the CNS by cART regimens based on peripheral blood genotypes. Perhaps at least partially attributable to inadequate CNS penetration of antiretroviral therapy, these escape variants can contribute to encephalopathy that may improve or resolve with resistance-based adjustment of antiretroviral therapy. Conceivably, however, if resistance mutations have been archived in long-lived and productively infected CNS microglial cells, a resistant reservoir of the virus may result. In addition to a similar scenario in the periphery, the added element of the BBB more or less walling off these latent variants could pose a special obstacle to complete viral elimination.

Viral Latency

In the periphery, there is establishment of a reservoir of latently infected T cells that serves as a barrier to eradication. cART is unable to eradicate infection in cells already harboring virus but instead diminishes viral burden and restores immune function by prevention of infection propagation into uninfected cells and by reliance on HIV-mediated and/or natural cell turnover for infected cells. Therefore, cells that harbor integrated virus maintain the potential to produce HIV virions in the appropriate circumstances and are therefore considered to be latently infected.

There are two mechanisms by which latency develops in the body. First, replication-competent virus can infect resting CD4 cells. However, the environment is inauspicious for viral integration into the host cell genome. This reservoir is thus short-lived and is more important in untreated patients than in those on suppressive cART, whose barrier to complete virus elimination results from a second type of latency, which comes from prior infection of activated T cells that permit viral genome integration into host DNA and that thereafter progress to a memory state. Because infection of activated cells predisposes to complete integration of the viral genome, reversion to quiescence renders infection unreachable by antiretroviral medication for the duration of the latent state, as mentioned above. This translates into long-lasting infection that can re-emerge with discontinuation of antiretroviral treatment even years later. With non-suppressive cART, resistance mutations can emerge and utilize this as an archiving mechanism in the body; inadequate control in the CNS could set up a similar situation.

Decay modeling of the CD4 T-cell latency in the periphery has projected a requisite treatment course of perhaps 60 years to fully eradicate this reservoir[22] by passive elimination of long-lived, latently infected cells. Contributing to the maintenance of viral latency are histone deacetylases (HDACs), enzymes that act to repress transcription. However, controversially, if such resting cells were purposefully stirred from dormancy, this timeline could be shortened substantially. This idea has been a target of therapeutic latency interruption by inhibitors of HDAC both *in vitro* and *in vivo*. Although trials with valproic acid failed to obliterate the viral reservoir, vorinostat, another HDAC inhibitor, is under investigation in clinical trials.[23] Disruption of viral latency in combination with suppressive antiretroviral therapy portends a cure to infection that may be more widely applicable than the only current "sterilizing" cure of bone marrow transplantation with R5 mutated cells that resist HIV infection, which is viable only for those without X4, dual, or mixed tropic virus.

An additional possibility that could significantly impede dormancy disruption strategies is similar establishment of latency in the cells infected in the CNS.[24] Since the productively infected cells in the CNS have very long half-lives – macrophages in the order of months and microglial cells in the order of years – these cells plausibly can harbor dormant virus that might re-emerge later. Indeed, *in vitro* studies have demonstrated the presence of latently infected MDMs retaining the ability to produce virus. One theory of latency accomplishment in these cells is proviral DNA mutations in Tat or the transactivation response element, which customarily initiate viral transcription, and with alternative capabilities for recovery of viral replication. One implication of this concept is that in the setting of segregated viral evolution in the CNS mentioned above, latent virus may contain unmeasured resistance that could institute reactivated infection in the future.

As the quest for eradication involves disruption of latency, there must be careful consideration of the brain as a potential viral sanctuary and of the consequences of viral reactivation in the vicinity of neurons and glial cells that suffer collateral damage from the virus, as this could produce irreversible clinical deficits. This also holds true for non-productively infected astrocytes, whose infection contributes to latency at the very least by its contribution to ongoing immune activation and low-level viral protein shedding. Infected astrocytes contribute to the cycle of ongoing immune activation but also must continue critical homeostatic functions in the CNS, albeit dysfunctionally. This poses a predicament for eradication tactics that target cell activation and destruction, as astrocytes may not be susceptible to such strategies. The converse is that continued restricted infection may also set the stage for ongoing CNS injury. These factors all argue for the need for further study into latency specifically in the CNS as a hurdle to eradication.

Because the major obstacle to the complete elimination of HIV is the latently infected cell, there is a need for strategies to address this issue both in the systemic compartment and in the CNS. Latent infection in CNS macrophages may prove a barrier to eradication of HIV infection given the long rate of decay of these cells, and thus alternative therapeutic strategies may need to be developed to disrupt latency in the CNS from those used in the periphery.

MECHANISMS OF CNS INJURY

Damage to the HIV-infected brain is induced both directly by the virus and indirectly by virally generated immune dysregulation and metabolic disarray, resulting in damaged neurons, glial cells, myelin, and the microvasculature. CNS infection can be established within days of initial infection,[9] and thus symptomatic disease can be preceded by years of silent or subclinical damage. As will be detailed below, there are many overlapping mechanisms of injury among the cell types in the CNS, and some effects of HIV on the brain parenchyma appear to be global. For instance, there is chronic aberrant clearance of misfolded proteins due to virus-mediated alterations in ubiquitin-proteasome function, which has been linked to cognitive dysfunction.[25]

HIV pathology can be found throughout the brain, but it has classically had a predilection for subcortical white matter and the basal ganglia from pre-cART era studies (Fig. 28.6) (reviewed by McArthur *et al.*[1]). Imaging studies in advanced disease frequently demonstrate global atrophy and abnormal signal in the subcortical white matter, but in well-controlled individuals volumetric differences have been demonstrated in more discrete brain areas, particularly the basal ganglia.

As will be detailed below, many soluble mediators of CNS damage are released in HIV infection. The "holy grail" of HIV neuropathology is to harness one or a group of these biomarkers to use in predicting and tracking disease, and for differentiating HIV-associated neuropathology from other co-occurring diseases. Although there have been many candidates in blood, CSF, and metabolites measured by imaging modalities (reviewed by Brew and Letendre[26]), a definitive marker has yet to be discovered.

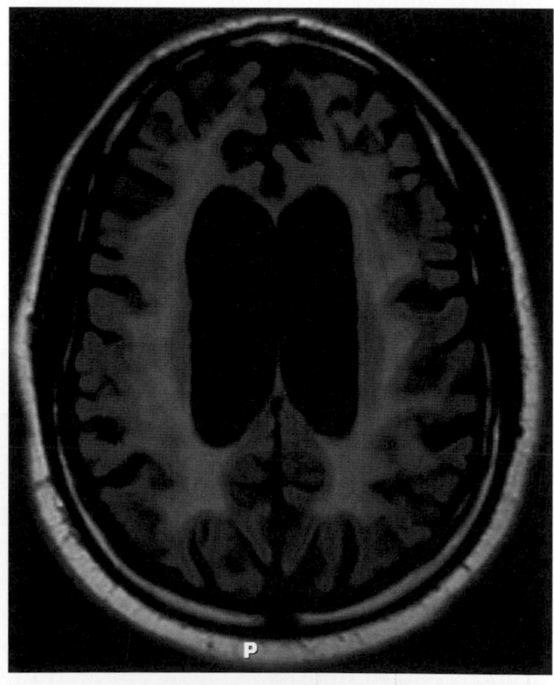

FIGURE 28.6 **Common neuroimaging findings attributable to chronic HIV infection.** Axial T2 fluid attenuated inversion recovery (FLAIR) image of the brain from a patient with HIV-associated dementia (HAD) demonstrating both marked sulcal widening and ventricular enlargement consistent with diffuse atrophy, and diffuse, confluent hyperintensity in the subcortical white matter.

Monocytes, Monocyte-Derived Macrophages, and Microglial Cells

Once across the BBB, activated monocytes harboring HIV differentiate into macrophages and produce viral particles that disseminate infection. These infected MDMs in the CNS also shed viral proteins, which independently are neurotoxic and immune activating, and express viral surface glycoprotein on their own cell membranes, creating an adhesive surface. These cells then fuse with neighboring macrophages and microglial cells, whether infected or uninfected, forming multinucleated giant cells (Fig. 28.3). These syncytial cells are the hallmark of HIVE and result at least in part from HIV-mediated cytoskeletal rearrangement to create contact points between cells.[17] Syncytial cells also contain replication-competent virus and thus propagate infection before their demise.

MDMs and microglial cells activated by infection trigger a cascade of injurious events to the brain parenchyma and suffer injury themselves. Phagocytosis, a primary function of a healthy macrophage, can malfunction owing to impaired membrane remodeling. Further, these cells emit proinflammatory cytokines such as TNF-α and IL-1β, which in turn stimulate astrocytosis, exacerbate immune activation, and induce apoptosis; TNF-α has also been implicated in BBB breach by cell-free virus *in vitro*, as mentioned above. Moreover, MDMs produce chemokines such as CCL-2, MIP-1α and MIP-1β, which attract monocytes (CCL-2) and lymphocytes (MIP-1α and MIP-1β), furthering the inflammatory state and infection.

In addition to releasing mediators of inflammation, infected MDMs and microglial cells generate a whole host of neurotoxins. The bulk of the neurotoxins released is described elsewhere in this chapter (see "CNS Metabolic Complications of HIV Infection", below), including quinolinic acid and excitatory amino acids such as glutamate, L-cysteine, and arachidonic acid; these products generate neurotoxic free radicals, causing dendritic thinning, synaptic dysfunction, and death. MDM infection has also been shown to lead to hyperphosphorylation of retinoblastoma protein, an important cell cycle regulator, which ultimately leads to neurodegeneration.

Notably, microglial cell response to infection may not to be entirely detrimental to the parenchymal microenvironment. In health, these cells produce neurotrophic factors such as nerve growth factor (NGF), brain-derived neurotrophic factor (BDNF), and neurotrophins. With HIV infection, these cells are stimulated to produce BDNF and neurotrophin-3 (NT-3). The trophic factors function via tyrosine kinase receptors on activated astrocytes and neurons and may promote neuronal well-being, although the overall result of their production remains largely unknown.

Finally, as MDMs and microglial cells contain an abundance of cell surface receptors that render them responsive to a variety of stimuli,[11] viral proteins, cytokines, and chemokines that infected cells produce, act on neighboring MDMs and microglial cells. With ongoing microbial translocation from the gut, immunogens such as LPS can exacerbate immune activation via toll-like receptor (TLR) signaling pathways. This cycle of infection, immune activation, and cellular damage is responsive to cART. Despite viral control in the periphery, CNS damage can continue to varying degrees. This is evident clinically by overt or subclinical neurocognitive impairment, and persistent CNS inflammation even in aviremic individuals. The reasons for this are not well understood but may relate to a delicate balance between viral control beyond the BBB (i.e. the effectiveness of antiretroviral therapy in the CNS), anti-inflammatory effects of the drugs, and their CNS toxicity.

Astrocytes

Despite restriction of virus production, infected astrocytes do become activated or damaged. This has widespread effects on the CNS, as these cells must contribute to BBB integrity, produce neurotrophic factors such as NGF and IL-6, facilitate neuronal interactions, detoxify excitatory amino acids (EAAs), and sustain the microglial ramified state.[27] These processes become deranged with astrocyte activation and proliferation, resulting in decreased neuronal and BBB support, neuronal demise from excess EAAs, and microglial activation. In addition, infected astrocytes can mediate apoptosis of neighboring, uninfected astrocytes via gap junctions,[28] which can directly result in loss of BBB integrity.

Activated, infected astrocytes produce proinflammatory cytokines and chemokines and shed viral proteins such as gp120 and Tat that interact with astrocytes regardless of infection status to elicit upregulation of cytokine and chemokine production (Fig. 28.1).[27,29] Nuclear factor-κB (NF-κB) is a transcription factor that is responsive to TNF-α, IL-1,[30] and gp120,[31] and it translocates to the cell nucleus to regulate gene expression. Furthermore, activated astrocytes are more susceptible to apoptosis, possibly mediated by caspase-3/Fas;[9] Nef mediates astrocyte apoptosis by c-kit. Of the cytokines and chemokines released by astrocytes, monocyte chemoattractant protein-1 (MCP-1 or CCL-2) is an important β-chemokine that becomes involved in trafficking of peripheral blood monocytes (which may be activated or HIV infected) across the BBB. Elevated levels of CCL-2 in the plasma and CSF have been demonstrated in severe forms of HAND, and this has become one of several potential biomarkers being studied for tracking disease.[26] Similarly, in SIV infection, monocyte chemotactic protein-3 (MCP-3 or CCL-7), another β-chemokine involved in monocyte trafficking, has increased expression in TNF-α-stimulated astrocytes with consequent neuroinvasion by infected peripheral monocytes.[32] Next, interferon-γ inducible protein-10 (IP-10 or CXCL-10), which is chemotactic for T cells in addition to having antiviral effects, is also expressed by activated astrocytes[31,33] and has been implicated in HAND. Both gp120 and Tat induce astrocyte production of interleukin-8 (IL-8 or CXCL-8), an α-chemokine involved in trafficking many types of leukocytes, primarily neutrophils.[31] RANTES or CCL-5 is a β-chemokine involved in monocyte and lymphocyte trafficking where expression in astrocytes is induced by gp120.[34] TNF-α, IL-1β, and IL-6 are all cytokines involved in perpetuation of immune activation; they are secreted primarily by activated mononuclear immune cells in HIV, but their production is also upregulated in astrocytes stimulated by gp120 (Fig. 28.1).[35,36]

In addition to the above, HIV-infected astrocytes mediate proteomic change of infected microglial cells, enhancing viral production in the microglia and magnifying neurotoxicity.[37] They also trigger TNF-α, IL-1β, and IL-6 production by monocytes, with consequences identical to the production of these cytokines by other cells.[38] Damage to astrocytes due to HIV-mediated metabolic pathway derangements is discussed later (see "CNS Metabolic Complications of HIV Infection").

Brain Microvascular Endothelial Cells

HIV-mediated CNS damage involving BMECs can be considered both in the context of effects on the BBB and through parenchymal effects (see "HIV Entry into the CNS", above). BMECs not only are damaged themselves by the virus, but also contribute to the inflammatory milieu of CNS HIV infection.[39,40]

HIV-induced immune dysregulation affects and is contributed to by BMECs. Evidence supporting this histologically is perivascular cuffing by infected or activated monocytes and MDMs, one hallmark of HIVE. In cocultures of BMECs and HIV-infected macrophages, soluble factors released from the former induce upregulation of the proinflammatory chemokines MIP-1α, MIP-1β, and RANTES expression from the latter. Further, Tat increases mRNA expression of CCL-2, another chemokine that is attractant for monocytes, in BMECs (Fig. 28.1). Attraction of inflammatory cells to the BBB exacerbates both infection and inflammation.

Viral proteins that have been shed from infected cells can cause BMEC damage. The viral product envelope glycoprotein gp120 and Tat have been shown in both *in vivo* and *in vitro* studies to be directly toxic to human endothelial cells (Fig. 28.1), and both probably work alongside vasoactive molecules such as prostaglandins, nitric oxide (NO), substance P, and MMPs to damage endothelial cell wall integrity. gp120 can bind to BMECs via CXCR4 or CCR5, leading to degradation of vascular basement membrane, downregulation of vascular tight junction proteins occludin and ZO-1 and ZO-2, impaired tight junction protein assembly, and proteasomal degradation of tight junction proteins.

The relationship between Tat and BMECs is quite complex. First, interactions between Tat and BMECs lead to oxidative stress and reduced intracellular levels of the antioxidant glutathione. Second, Tat causes changes in the cleavage of poly(ADP-ribose) polymerase, DNA laddering, and incorporation of fluorescein into the nicked chromosomal DNA of the BMECs, leading to apoptosis of BMECs and resulting in the increased permeability of the brain microvascular endothelial monolayer. Third, Tat causes increased endothelial permeability through activation of vascular endothelial growth factor receptors-1 and -2 and tyrosine kinase- and mitogen-activated protein kinase-dependent pathways. Fourth, BMEC expression of E-selectin, an adhesion molecule that serves in tethering and transmigration of monocytes across the BBB, increases with experimental exposure to Tat. These mechanisms collectively lead to BMEC demise or increased permeability of the BBB, again opening the door to further parenchymal infection and inflammation.

Oligodendrocytes

Oligodendrocytes, the myelin-producing glial cells in the brain, evade HIV infection. Nevertheless, myelin pallor has been reported in the context of HIVE.[41] Myelin damage appears to be due to both primary viral protein-mediated damage and the effects of virally induced innate immune dysregulation. TNF-α, released predominantly from activated macrophages but also from astrocytes and microglial cells, damages myelin both directly and indirectly via its effect on BBB integrity, allowing entry of myelinotoxins from the peripheral blood. In CNS HIV infection, increases in free radical nitric oxide (·NO) metabolites have also been demonstrated,[42] which in demyelinating disease have been implicated in oligodendrocyte damage and death. Moreover, gp120 affects oligodendrocyte function, also leading to demyelination (Fig. 28.1).[43] The most common HIV-associated pathology in the spinal cord is a vacuolar myelopathy, which is characterized histopathologically by vacuolation of the lateral corticospinal tracts and dorsal columns with a heavy burden of lipid-laden macrophages but no axonal loss.[44] Compared to those without vacuolar myelopathy or HIV-uninfected individuals, stained sections demonstrate significantly higher amounts of TNF-α, implicating this cytokine in pathogenesis.[45] Thus, although oligodendrocytes are not primary targets of HIV-mediated CNS damage, there is evidence that this occurs.

Neurons

Neurons escape infection by HIV, but the effects on neurons situated in a toxic milieu of cytokines, chemokines, toxins, and viral proteins are protean (reviewed by Kaul *et al.*[27]). Most neurotoxic substances mediate neuronal damage by activation of the *N*-methyl-D-aspartate receptor (NMDAR), which allows an influx of calcium and induction of apoptosis (see "Glutamate Metabolism", below). This is evident as atrophy, reduced synaptic density, and dendritic loss. Excitotoxic injury to neurons also leads to release of fractalkine, a cell surface protein that communicates with other neurons and inflammatory cells, and which has been associated with the severity of HAND.[1]

As neuronal damage and death can be topographically distinct from the location of productive infection, neuronal demise derives from soluble mediators of degeneration.[27] Neurons express CXCR4 and CCR5 and as such are susceptible to deleterious effects of shed gp120, which can interact with these receptors (Table 28.1, Fig. 28.1). These interactions can trigger caspase-mediated apoptotic signaling cascades. RANTES (CCL-5), MIP-1α, and MIP-1β, which are β-chemokines released by activated macrophages that bind CXCR4 and CCR5, are able to inhibit this. Additional mediators of neuronal injury include viral proteins Nef, Tat, and Vpr. Tat induces neuronal apoptosis and also is directly cytotoxic to the neuronal cell. Tat-mediated damage is seen particularly in sites of high NMDAR density, specifically the striatum, dentate gyrus, and CA3 region of the hippocampus. Nef promotes the production of quinolinic acid by macrophages and microglial cells, which in turn mediates excitotoxic neuronal damage. Vpr induces caspase-mediated apoptosis of neurons indirectly by inducing the release of inflammatory cytokines, such as IL-1β and TNF-α, from macrophages and microglial cells (Fig. 28.1).

In addition to viral protein-mediated damage, much of the inflammatory milieu and many products of aberrant metabolism in the CNS of HIV-infected individuals mediate neuronal damage. An excess of excitotoxic metabolites such as quinolinic acid, arachidonic acid, NO, and glutamate is present in the brains of HIV-infected individuals, as described below (see "CNS Metabolic Complications of HIV Infection"). Inflammatory mediators released by monocytes, MDMs, microglial cells, and astrocytes (detailed above) act on neurons as well. For instance, TNF-α and IL-1β released from activated microglia, macrophages, and astrocytes may damage neurons through free radical generation or other mechanisms. Hence, despite a lack of CD4 expression or infection by HIV, neuronal toxicity is a key component of HIV-mediated CNS damage and is achieved through bystander mechanisms.

Neural Progenitor Cells

Several lines of evidence indicate that adult neural progenitor cells also suffer damage inflicted by HIV. These cells are located in the dentate gyrus of the hippocampus and in the subependymal zone, and can proliferate and differentiate into neurons, astrocytes, and oligodendrocytes. Although the impetus for and outcome of these activities are unknown, they are thought to act on learning, memory, and cognition. This population could therefore play an important role in neurodegenerative disease. In animal models of HIV-associated dementia (HAD), HIV envelope protein gp120, which can be shed from the virus constitutively, elicits cell cycle arrest and decrement in adult neural progenitor cell numbers (Fig. 28.1).[46] These cells can be rescued by exercising the animal or administering the selective serotonin reuptake inhibitor paroxetine,[47] but the implications of these observations are as yet unknown, and no experimental evidence connects amnestic symptoms of HAND to these findings.

CNS METABOLIC COMPLICATIONS OF HUMAN IMMUNODEFICIENCY VIRUS INFECTION

In health, the brain maintains intricately intertwined metabolic pathways for proper function, and it is no surprise that HIV-triggered disruption at any point can elicit widespread downstream effects. Many viral proteins and cytokines admitted into or released inside the CNS as a result of HIV infection have potent interactions with lipids and amino acids, and HIV-mediated disturbance of physiological metabolism overall translates once again into parenchymal damage and neuronal quietus. That said, many of the relationships between HIV and metabolic pathways in the CNS are not well understood. CNS HIV infection as it relates to the kynurinine pathway, glutamate metabolism, cysteine, NO, arachidonic acid and eicosanoid metabolism, and also sphingosine and ceramide, is discussed here.

Tryptophan Metabolism

Products of tryptophan metabolism have been implicated in the neuropathogenesis of HIV.[48] Tryptophan (Trp) is an essential amino acid involved in the metabolic pathways for serotonin and subsequently melatonin and for nicotinamide adenine dinucleotide (NAD^+) (Fig. 28.7). In the latter pathway, also known as the kynurenine pathway, tryptophan is first catabolized by indoleamine 2,3-dioxygenase (IDO) to kynurenine. Effectors of innate immunity (e.g. macrophages) possess all the enzymes required for the kynurenine pathway, and IDO can be induced by cytokines such as IFN-γ, IFN-α, and to a lesser extent TNF-α. Other stimulants of the kynurenine pathway include platelet activating factor (a product of activated monocytes and macrophages, among other cell types) and HIV proteins Tat and Nef. A downside to the induction of this pathway is that quinolinic acid, an intermediary in the kynurenine pathway, is both activating and toxic to the CNS. Quinolinic acid stimulates astrocytosis, is excitotoxic to neurons via the NMDAR, and can induce apoptosis in astrocytes, neurons, and oligodendrocytes. Kynurenine also has vasoactive effects, which are likely to exacerbate CNS infection and inflammation. Furthermore, as pericytes are capable of kynurenine production, the integrity of the BBB can be compromised.

In HIV specifically, deranged kynurenine pathway metabolism has been tied to pathogenic mechanisms. Reduced CSF and serum tryptophan levels, increased ratios of kynurenine to tryptophan, and elevated quinolinic acid levels have all been reported. This has been linked to progression of clinical manifestations of the disease. Notably, quinolinic acid levels decline with effective cART.

Glutamate Metabolism

Glutamate is the principal excitatory amino acid neurotransmitter in the CNS and is found throughout the cerebral cortex. As it does not cross the BBB, glutamate is manufactured in the brain by glutaminase-mediated metabolism of glutamine, a freely available amino acid

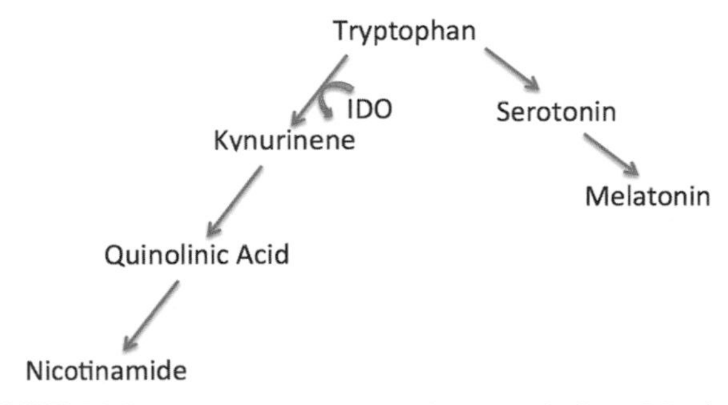

FIGURE 28.7 **Overview of tryptophan metabolism.** Metabolism of tryptophan results in the neurotransmitters serotonin and melatonin and, via the kynurenine pathway (KP), to nicotinamide adenine dinucleotide. Indoleamine 2,3-dioxygenase (IDO) catalyzes the first step in the KP and is upregulated in the setting of HIV infection. Subsequent generation of kynurenine and quinolinic acid, intermediaries in KP, contributes to HIV-mediated CNS damage.

in the extracellular space. Glutamate is released from presynaptic terminals upon stimulation and consequently depolarizes postsynaptic neurons by binding to high-affinity receptors. There are two types of postsynaptic receptor: ionotropic, which are ligand-gated ion channels, and metabotropic, which are coupled to G-protein mediated signaling pathways. The former type comprises *N*-methyl-D-aspartate (NMDA) and α-amino-3-hydroxy-5-isoxazolpropionate (AMPA)/kainate receptors. The latter is composed of three main subtypes with several subgroups that drive second messenger pathways. Glutamatergic neurotransmission is terminated by removal of neurotransmitter from the synaptic cleft, which is accomplished by transporters expressed on glial cells and neurons.

Extracellular concentrations of glutamate are kept low, primarily by astrocytes, which play a key role in glutamate reuptake, thereby preventing toxicity. The loss of glutamate extracellular concentration control results in the overactivation of its receptors, leading to deleterious effects on neurons in a series of processes collectively known as glutamate excitotoxicity. Excitotoxicity occurs when there is at least one of three scenarios: an elevated concentration of extracellular glutamate (possibly due to increased release from presynaptic terminals or to the spill of this metabolite from cells responding to stress or breakage), a reduced capacity or alterations in glutamate transport, or an increased sensitivity of glutamatergic receptors, leading to overactivation even in the presence of normal physiological concentrations of the neurotransmitter.

The most common mechanism of excitotoxicity is mediated by NMDARs capable of conducting large fluxes of Ca^{2+} that eventually trigger, in an uncontrolled manner, a series of degradation enzymes such as proteases, lipases, and nucleases that ultimately lead to neuronal death. This phenomenon occurs in acute CNS injuries including stroke, trauma, and ischemia, where there is a rapid elevation in the concentration of extracellular glutamate resulting from cellular breakage. The increased extracellular glutamate diffuses in the tissue surrounding the initial site of lesions, causing progressive excitotoxic death in the course of a few hours. In contrast, in chronic neurodegenerative diseases, for example amyotrophic lateral sclerosis, excitotoxicity occurs over an extended period, which is likely to be the scenario in the case of HAND.

In HIV infection, many abnormalities in glutamate metabolism have been documented.[11,27] Elevated levels have been found in the CSF of infected individuals.[49] Such high glutamate concentrations in the CNS could come from many different sources, including its release from infected macrophages. MDM and microglial infection by HIV also results in upregulation of glutaminase. In addition, infected macrophages can release other closely related molecules capable of acting on NMDARs, such as quinolinic acid (see "Tryptophan Metabolism", above) and NTox. Vpr mediates increased extracellular glutamate related to reduced custodial activities by astrocytes (Fig. 28.1). Glutamate may have a synergistic toxic effect along with HIV protein gp120, and gp120 itself contributes to the inhibition of glutamate uptake in astrocytes. It also promotes a reorganization of NMDARs within lipid rafts, interfering with their normal internalization after activation and generating a potentiation in calcium flux through these channels. Tat can bind NMDARs, inducing overactivation of glutamatergic transmission. NMDAR antagonists ameliorate the noxious effects of HIV in models of neurotoxicity *in vitro*,[50] *in vivo* in gp120 transgenic mice, and in mouse models of HIV encephalitis. Finally, there is a reduction in the expression of the highly selective glutamate transporter EAAT2 in brains of HIV-infected patients with or without HIV encephalitis, and the high levels of TNF-α released from activated microglia and infected macrophages, together with an increase in the synthesis and release of NO, disturb glutamate transport. These events can lead to neuronal demise through glutamate interaction with cell surface receptors either by means of sodium influx and hypertonic swelling or by calcium influx and induction of the apoptosis signaling cascade (reviewed by Erdmann *et al.*[51]).

Cysteine and Arginine Metabolism

Abnormal metabolism of L-cysteine and L-arginine is also seen in HIV. L-cysteine is a substrate for neuronal glutathione synthesis and contributes to cellular detoxification pathways, and as such is a necessary antioxidant at physiological concentrations. Excess L-cysteine functions as a neuronal excitotoxin via NMDARs (see "Glutamate Metabolism", above). MDMs release L-cysteine when stimulated by TNF-α or IL-1β and also by gp120 and gp41, although viral protein-triggered release may be mediated by the cytokines (Fig. 28.1).

Arginine is oxidized by a family of enzymes, the nitric oxide synthases (NOS), to the proinflammatory free radical nitric oxide (·NO). Further metabolism yields peroxynitrite, nitrate, and nitrite, and many cell types, including macrophages, endothelial cells, microglia, and astrocytes,[52] possess the machinery to carry out arginine metabolism. In health, NO has an important place in neuroprotection, neurotransmission, and synaptic plasticity, but in pathological scenarios, its excess contributes to neurotoxicity and neurodegeneration. Elevated levels of nitrite and nitrate have been shown in CSF and also in serum of HIV-positive patients with demonstrable BBB dysfunction.[42] ·NO and components of its metabolism also mediate HIV-related neurotoxicity and link to severity of clinical disease and neuropathology.[53] Finally, some evidence suggests that NO contributes to myelin damage in HIV infection.[54,55]

Arachidonic Acid Metabolism

Arachidonic acid is an unsaturated fatty acid found ubiquitously in plasma membranes, where it is bound to phospholipid. Its metabolism produces prostaglandins, prostacyclins, thromboxanes, and leukotrienes (eicosanoids). Eicosanoids serve as cell signaling molecules that are significant mediators of immune responses. In HIV-1 infection, gp120 and microglial activation induce cleavage of arachidonic acid from cell membrane phospholipids by phospholipase A_2, releasing it into the microenvironment.[56] Arachidonic acid itself is excitotoxic to neurons, and additionally it is metabolized via the COX or lipoxygenase pathway. Increased COX-2 expression has been shown *in vivo* and *in vitro* to be associated with the presence of gp120 and Tat. The COX pathway yields prostaglandins and thromboxanes, namely prostaglandin H_2 and subsequently prostaglandin E_2 or thromboxane A_2, which are associated with increased synthesis of proinflammatory TNF-α and IL-1β. Increased levels of these inflammatory mediators are found in HAD and parallel neurological severity. However, the story is a bit more convoluted, because prostaglandin E_2 also has suppressive effects on HIV replication.[57] In addition, both leukotriene B_4 and prostaglandin E_2 are associated with reduced CCR5 expression on the cell surface, ostensibly rendering the cell less infectable. However, leukotriene B_4 is a potent chemoattractant. In the systemic immune reconstitution inflammatory syndrome (IRIS), an aberrant inflammatory state in HIV during which an overzealous recovering immune system results in immunological mayhem, the drug montelukast, a leukotriene receptor antagonist, has been reported to be beneficial.[58] The presence of arachidonic acid metabolites in HIV infection imparts a complicated picture that is not fully understood.

Lipid Metabolism

Sphingolipids are sphingosine-containing compounds that comprise ceramides, sphingomyelins, and glycosphingolipids; they are critical fatty components in cell membranes and serve in cell signaling. Ceramides, the most rudimentary sphingolipids that constitute sphingomyelins and glycosphingolipids, are generated *de novo* or by sphingomyelinase-catalyzed cleavage of sphingomyelin to ceramide and phosphocholine. Ceramide produced from sphingomyelin hydrolysis undergoes further metabolism to GD3 ganglioside, which moves to the mitochondrion, interrupts the electron transport chain, and activates caspases, inducing apoptosis. This process is triggered by many stress signals, such as TNF-α and Fas/FasL, whose concentrations increase with immune activation and HIV infection.[59] In addition, gp120 and Tat can precipitate increased levels of sphingomyelin and ceramide in brain tissue and CSF of HIV-infected subjects with mild or moderate to severe dementia, as can homozygosity for *APOE4*.[60]

HIV-mediated alteration in CNS metabolism is complex and involves many pathways, which when imbalanced result in a buildup of neurotoxins and induction of apoptosis. As with other facets of CNS damage, both immune dysregulation and viral proteins impart these aberrances from healthy functioning. The interrelation of many pathways translates into widespread effects that frequently extend beyond any one specific pathway. This is yet another example of the far-reaching ramifications of HIV infection.

EXPERIMENTAL MODELS

Blood–Brain Barrier Models

Experimental models of the BBB have aided in expanding the understanding of CNS HIV infection and BBB damage mediated by the virus. Given the intimate interactions and cross-talk between cells comprising the BBB, *in vitro* models frequently employ cocultures of endothelial cells and astrocytes across a porous membrane, allowing for astrocyte foot processes to permeate the membrane and establish connections with endothelial cells. This construct has been used as an *in vitro* representation of the BBB to evaluate effects of HIV infection on junctional proteins, cell-to-cell interactions, and selective permeability.

Animal Models of HIV

The interactions between HIV and the brain are complex, intertwined, and evolving. This, coupled with the relative inaccessibility for intense study of the human CNS, poses a significant quandary for research in this area. However, several animal models have been developed. Rodent models using humanized mice, hybrid viruses, or transgenic mice capable of viral protein shedding, and non-human primate models have all been used to study CNS HIV infection and immune dysregulation.

In humans, the shedding of gp120 and tat proteins causes cellular injury in the CNS. Brain-tissue specific transgenic mouse models have been generated both for tat and for gp120 using a glial fibrillary acidic protein (GFAP) promoter. The animals have both clinical and neuropathological abnormalities (reviewed by Jaeger and Nath[61]) and have been used to probe the interactions of drugs of abuse with HIV-induced injury.[62]

Beyond rodent models transgenic for viral protein production, models have been developed to examine the neuropathological effects of infection by the virus. These include a humanized mouse model for HIVE whereby M tropic virus infected into macrophages is introduced into the brains of mice with severe combined immunodeficiency (SCID). These mice constitutively express p24 antigen and demonstrate neuropathology analogous to that seen in HIVE.[63] Another model has been created by infecting mice with chimerized virus using the envelope protein of a murine leukemia virus hybridized to HIV. This model reproduces several aspects of HIV neuropathogenesis and chronic inflammation but lacks gp120 specifically given the construct.[64]

Although rodent models of disease are logistically attractive, the most widely used HIV models are SIVs in non-human primates. Several species of primates can be infected with SIV strains, including sooty mangabeys, African green monkeys, pigtail macaques, and rhesus macaques, but only macaques are widely used as models for human disease. However, the "natural" viral infection in these animals more closely represents HIV-2 than HIV-1. One particularly useful model has been the rapid encephalitis model developed by Mankowski and colleagues.[65] This pigtail macaque model requires inoculation with a combination of a neurovirulent strain of SIV with an immunosuppressive strain. The animals reproducibly develop encephalitis and simian acquired immunodeficiency syndrome (SAIDS) within 3 months and have been used in a variety of experiments to probe neuropathogenesis and treatment effects. Another strategy has been the generation of the chimeric simian/human immunodeficiency virus (SHIV), which incorporates human *env*, *tat*, *vpu*, and *rev* into the SIV genome for a more representative infection of HIV-1. Both SIV and SHIV models can result in CNS infection and encephalitis, and these tools have been instrumental in understanding viral neuroinvasiveness and CNS effects of immune dysregulation (reviewed by Williams *et al.*[66]).

As mentioned above, concepts of the neuropathogenesis of HIV continue to evolve with improvement in viral control and prolongation of disease-free survival in those infected; ideally, the animal models should mirror the state of infection and immune activation seen in humans. With attempts at eradication of virus and viral reservoirs or vaccination against disease comes the necessity for surrogate disease models to study. Therefore, animal models of HIV neuropathology have been and will continue to be integral in understanding mechanisms of disease in the setting of infection, ongoing inflammation, and antiretroviral use.

CLINICAL MANIFESTATIONS OF CNS HUMAN IMMUNODEFICIENCY VIRUS INFECTION

In the era before the emergence of cART, severe neurocognitive disease attributable to HIV infection was common in those with profound immunodeficiency (CD4 <200). In the pre-cART era pathological examination demonstrated productive HIV replication, inflammatory infiltrates and gliosis in white matter and basal ganglia, along with diffuse neuronal loss, termed HIV encephalitis (reviewed by McArthur *et al.*[1]).

Despite the tremendous advances in antiretroviral therapies both at the level of drug discovery and in implementation strategies, neurological disease in the form of HAND remains prevalent and in many cases is disabling. However, the semiology of the neurocognitive dysfunction due to HIV has evolved over time. Early in the epidemic, the dementing illness was known as the AIDS dementia complex (ADC) was observed in up to 20% of those with late-stage disease. Since the widespread use of cART, however, less severe cognitive dysfunction is increasingly seen,[1] and the clinical phenotype has shifted from a severe subcortical dementia to a milder disease with mixed cortical and subcortical features.

In 2007, researchers published revised diagnostic criteria describing a spectrum of diseases, based on severity and collectively known as HAND. This umbrella term includes asymptomatic neurocognitive impairment, mild cognitive disorder, and the most severe form, HAD.[67] In the cART era, the diagnoses are not necessarily sequentially progressive. Symptoms can plateau, worsen, or even regress with reversion to less severe HAND diagnoses. The mechanisms underlying these shifts are not well understood but are likely to represent interplay among cART-mediated attenuation of viral effects and immune dysregulation, ongoing damage attributable to low-level viral replication or HIV-associated immune activation, and perhaps toxicities due to the long-term use of cART. The neuropathological features of the milder forms of HAND are much less well catalogued, since people with HIV infection live substantially longer and thus the death and autopsy rates have declined precipitously. Although highly productive HIV encephalitis is uncommon in cART-treated individuals, astrocytes may contribute to ongoing CNS injury in this population owing to non-productive infection and chronic immune activation. Synaptodendritic injury with sustained inflammation and less prominent neuronal loss is seen and the frequency of HIV encephalitis has dropped from 54% in the pre-cART era to 15% in the cART era. As earlier detection and institution of ever more potent cART regimens become standard, virus-associated CNS injury may continue to evolve.[1]

With the substantially increased survival of people with HIV treated with cART (with lifespan estimates in developed countries into the eighth decade, as in HIV-negative populations),[68] the effects of comorbidities such as accelerated cardiovascular and cerebrovascular disease, chronic hepatitis C coinfection, and age-related neurodegenerative disease have all come into sharper focus. First, HIV-associated vasculopathy is a term that has been used to describe a range of HIV-associated cerebrovascular changes, including arterial stenosis, aneurysm, vasculopathy, and accelerated atherosclerosis (reviewed by Benjamin *et al.*[69]). The etiology of the vasculopathy is unknown. It may be caused directly by HIV infection, although direct evidence of virus in vessels has not been seen; another possibility is that it is an indirect effect of sustained inflammation on the vasculature. Furthermore, some cART regimens provoke the metabolic syndrome and hypertriglyceridemia, which may increase stroke risk, although definitive data are lacking to make a strong connection. Next, chronic hepatitis C appears to contribute to radiological changes, to increased cognitive deficits, and, experimentally, to neurotoxicity.[70] Finally, HAND is more frequent or severe in older individuals. Amyloid deposition occurs with greater frequency in HIV-positive brains, but the pattern is distinct from that seen in Alzheimer disease. It is beyond the scope of this chapter to cover these topics comprehensively, but they comprise a few comorbid conditions that affect the neuropathology of HIV.

EFFECTS OF COMBINATION ANTIRETROVIRAL THERAPY ON CNS HUMAN IMMUNODEFICIENCY VIRUS PATHOLOGY

The neuropathogenesis of HIV is a complex and evolving field. Infection of supportive cells in the CNS is accompanied by immune activation and dysregulation and shed viral proteins to cause damage throughout the neuraxis. However, this has been modulated, or in some instances perhaps compounded, by cART. Although HIVE is still seen in cART-treated individuals, chronic forms have emerged that may relate to controlled viral replication and atypical aging. In addition, as mentioned above, clinically this is associated with more mild, prolonged neurocognitive deficits instead of overt dementia. As treatment trends shift towards offering treatment to all HIV-infected individuals regardless of immune status, and with some using boosted protease inhibitor monotherapy, the epidemiology and pathogenesis of CNS complications of HIV may continue to change.

The effects of prolonged cART on the neuropathogenesis of HIV are incompletely understood. For instance, antiretroviral effectiveness in the CNS seems to be a double-edged sword, as higher concentrations suppress viral replication but may also be neurotoxic. In addition, despite control of viral replication, viral protein shedding and immune activation are not fully controlled by antiretroviral medications, and as such smoldering CNS damage can result. Reservoirs of infection in the CNS (tissue macrophages, microglial cells, and potentially astrocytes) are not eradicated by antiretroviral medications, and these cells can periodically shed virus and/or viral proteins that also mediate damage by means mentioned above. Finally, devastating inflammatory complications (IRIS) during the weeks to months immediately following the initiation of antiretroviral therapy can cause permanent neurological damage or even death.

Antiretroviral medications have varying abilities to cross the BBB and concentrate in the CNS, and regimens with high penetration effectiveness may impact neurocognitive outcome and neurological sequelae. However, there are conflicting data as to the clinical importance of CNS penetration, and some have shown worse neurocognitive performance of patients on regimens with high CNS penetration effectiveness. In the setting of viral escape in the CNS, raising the penetration effectiveness score of a cART regimen has resulted in improved outcomes.[71] However, individual medications demonstrate differential likelihood for *in vitro* neurotoxicities such as dendritic damage and neuropil shrinkage. Taken together, these findings imply that a cART regimen must be optimized to effect immune system restoration and systemic and CNS viral suppression while minimizing neurotoxicity. This remains a controversial issue, and there are no clear guidelines for approaching CNS protection or restoration by antiretroviral regimens.

IRIS manifests as a paradoxical clinical worsening of HIV-infected patients after the start of treatment. It typically occurs systemically, and overt neurological involvement is much less common but does occur. Mechanistically, CNS IRIS begins with immune system recovery and is heralded by a rapid and dramatic expansion of memory CD4 T cells within weeks of starting cART. This expansion is replaced by naïve CD4 T-cell production in the thymus, which occurs in the first few months after initiation of cART. CD8 T cells, which destroy target cells identified by MHC-I antigens, also increase in the weeks following cART initiation. Expansion of CD8 T cells during immune system recovery can translate to an uncontrolled response to antigen recognition, sometimes despite very low levels of circulating antigen. In the CNS, this can lead to one of two subcategories of IRIS: paradoxical and unmasking. The former refers to a clinical worsening of a previously known disease (most commonly an opportunistic infection such as tuberculosis, toxoplasmosis, or progressive multifocal leukoencephalopathy) after starting cART and the latter to a new manifestation of previously undiagnosed disease.[72] Unfortunately, little is known about prevention or treatment of CNS IRIS, as modulation of a recovering immune system proves a difficult challenge.

CONCLUSION AND FUTURE CHALLENGES

The neuropathology of HIV in the brain is a complex latticework of productive infection, non-productive infection, immune activation, and bystander damage. It not only involves cells of CNS origin but also is intimately intertwined with systemic infection and its consequences. It contributes significantly to clinical morbidity and mortality, and poses a significant barrier to the eradication of infection given the potential reservoirs of viral genome within the brain.

Sustained inflammation despite virological control appears to be a critical element in producing CNS damage, with synaptodendritic injury as a key pathological outcome. Viral genetics may also play a role, and there may be clade differences driving differences in neurovirulence. In addition, long-term survival with chronic immune activation and aging in HIV infection is associated with an increased likelihood of abnormal protein deposition in the brain. There are synergistic or confounding effects of drugs of abuse, comorbid diseases, and coinfections, especially with hepatitis C. Furthermore, the relative CNS penetration of antiretroviral drugs may be important in determining HIV suppression within the CNS. As these seemingly disparate but interacting occurrences in HIV infection all affect neurological outcome, reliable means of predicting and tracking disease are and will continue to be in high demand but as yet do not exist. Antiretroviral treatment has brought changes to the pathophysiology and outcomes of infection, but much work lies ahead in understanding and abrogating the effects of CNS HIV infection in the era of cART.

Because of the dynamic nature of HIV neuropathology, and since many aspects are still unknown, there is great need for translational research that can be taken from bench to bedside. First and foremost is the need for development of practical biomarkers that are validated and specific for the detection and monitoring of HAND. Given that there are both viral and immunological components to HAND, this will probably require a combination of markers and better stratification of clinical disease, as to date there has not been a single marker of immune activation or of the virus that accurately predicts or follows HAND. Second, a critical need is for improved preventive strategies and therapeutic interventions for HAND. The delicate balance between viral destruction and virus-mediated dysregulation of the immune system makes immune modulation a tricky but likely necessary component of HIV management in order to combat or prevent neurocognitive disease. In addition, the effects of aging, and perhaps accelerated aging, of the brain in HIV-infected individuals on HAND are largely unknown but definitely an emerging issue given the extended life expectancies with cART. Important to achieving these goals will be a closer integration of research efforts between infectious disease specialists and neuroscientists. Third, the management and prevention of CNS IRIS is a major issue that can potentially disrupt the otherwise lifesaving measures of cART; better understanding of immune modulation and manipulation of the BBB in patients who develop this would be extremely useful. Fourth, the balance between antiretroviral benefit and toxicity to the CNS remains to be adequately delineated and correlated with outcomes. Finally, eradication of reservoirs in the brain needs careful consideration, as the nature of infection and secondary damage differ from that in the periphery. It is also worth mentioning that the neurovirulence of other strains and forms of HIV that are relatively common in developing countries remains an understudied area of research.

In conclusion, although there have been great strides in the understanding and management of HIV infection overall, its neurological aspects are ripe for investigation and may have a substantive clinical impact on infected individuals. The development of preventive and adjunctive therapies and clinical tools for tracking disease remain the most urgent need in this field.

Acknowledgments

Supported by RR 025005 from the National Center for Research Resources (NCRR), a component of the National Institutes of Health (NIH), and NIH Roadmap for Medical Research. The contents of this chapter are solely the responsibility of the authors and do not necessarily represent the official view of NCRR or NIH. Information on NCRR is available at http://www.ncrr.nih.gov/. This study was also supported by award 5P30MH075673-06 from the National Institute of Mental Health (NIMH) (PI JCM).

References

1. McArthur JC, Steiner J, Sacktor N, Nath A. Human immunodeficiency virus-associated neurocognitive disorders: mind the gap. *Ann Neurol*. 2010;67:699–714.
2. Sundquist WI, Krausslich HG. HIV-1 assembly, budding, and maturation. *Cold Spring Harb Perspect Med*. 2012;2:a006924.
3. Gorry PR, Ancuta P. Coreceptors and HIV-1 pathogenesis. *Curr HIV/AIDS Rep*. 2011;8:45–53.
4. Brenchley JM, Price DA, Schacker TW, et al. Microbial translocation is a cause of systemic immune activation in chronic HIV infection. *Nat Med*. 2006;12:1365–1371.
5. Burdo TH, Lo J, Abbara S, et al. Soluble CD163, a novel marker of activated macrophages, is elevated and associated with noncalcified coronary plaque in HIV-infected patients. *J Infect Dis*. 2011;204:1227–1236.
6. Sandler NG, Wand H, Roque A, et al. Plasma levels of soluble CD14 independently predict mortality in HIV infection. *J Infect Dis*. 2011;203:780–790.

7. Strazza M, Pirrone V, Wigdahl B, Nonnemacher MR. Breaking down the barrier: the effects of HIV-1 on the blood–brain barrier. *Brain Res*. 2011;1399:96–115.
8. Williams DW, Eugenin EA, Calderon TM, Berman JW. Monocyte maturation, HIV susceptibility, and transmigration across the blood brain barrier are critical in HIV neuropathogenesis. *J Leukoc Biol*. 2012;91:401–415.
9. Valcour V, Chalermchai T, Sailasuta N, et al. Central nervous system viral invasion and inflammation during acute HIV infection. *J Infect Dis*. 2012;206:275–282.
10. Neuenburg JK, Sinclair E, Nilsson A, et al. HIV-producing T cells in cerebrospinal fluid. *J Acquir Immune Defic Syndr*. 2004;37:1237–1244.
11. Gonzalez-Scarano F, Martin-Garcia J. The neuropathogenesis of AIDS. *Nat Rev Immunol*. 2005;5:69–81.
12. Burkala EJ, He J, West JT, Wood C, Petito CK. Compartmentalization of HIV-1 in the central nervous system: role of the choroid plexus. *AIDS*. 2005;19:675–684.
13. Marechal V, Prevost MC, Petit C, Perret E, Heard JM, Schwartz O. Human immunodeficiency virus type 1 entry into macrophages mediated by macropinocytosis. *J Virol*. 2001;75:11166–11177.
14. Carter GC, Bernstone L, Baskaran D, James W. HIV-1 infects macrophages by exploiting an endocytic route dependent on dynamin, Rac1 and Pak1. *Virology*. 2011;409:234–250.
15. Dragic T, Litwin V, Allaway GP, et al. HIV-1 entry into CD4+ cells is mediated by the chemokine receptor CC-CKR-5. *Nature*. 1996;381:667–673.
16. Benaroch P, Billard E, Gaudin R, Schindler M, Jouve M. HIV-1 assembly in macrophages. *Retrovirology*. 2010;7:29.
17. Xu W, Santini PA, Sullivan JS, et al. HIV-1 evades virus-specific IgG2 and IgA responses by targeting systemic and intestinal B cells via long-range intercellular conduits. *Nat Immunol*. 2009;10:1008–1017.
18. Trujillo JR, Rogers R, Molina RM, et al. Noninfectious entry of HIV-1 into peripheral and brain macrophages mediated by the mannose receptor. *Proc Natl Acad Sci U S A*. 2007;104:5097–5102.
19. Gartner S, Liu Y, Natesan S. De novo generation of cells within human nurse macrophages and consequences following HIV-1 infection. *PLoS ONE*. 2012;7:e40139.
20. Gorry PR, Ong C, Thorpe J, et al. Astrocyte infection by HIV-1: mechanisms of restricted virus replication, and role in the pathogenesis of HIV-1-associated dementia. *Curr HIV Res*. 2003;1:463–473.
21. Nakagawa S, Castro V, Toborek M. Infection of human pericytes by HIV-1 disrupts the integrity of the blood–brain barrier. *J Cell Mol Med*. 2012;16(12):2950–2957.
22. Finzi D, Blankson J, Siliciano JD, et al. Latent infection of CD4+ T cells provides a mechanism for lifelong persistence of HIV-1, even in patients on effective combination therapy. *Nat Med*. 1999;5:512–517.
23. Archin NM, Liberty AL, Kashuba AD, et al. Administration of vorinostat disrupts HIV-1 latency in patients on antiretroviral therapy. *Nature*. 2012;487:482–485.
24. Nath A, Clements JE. Eradication of HIV from the brain: reasons for pause. *AIDS*. 2011;25:577–580.
25. Nguyen TP, Soukup VM, Gelman BB. Persistent hijacking of brain proteasomes in HIV-associated dementia. *Am J Pathol*. 2010;176:893–902.
26. Brew BJ, Letendre SL. Biomarkers of HIV related central nervous system disease. *Int Rev Psychiatry*. 2008;20:73–88.
27. Kaul M, Garden GA, Lipton SA. Pathways to neuronal injury and apoptosis in HIV-associated dementia. *Nature*. 2001;410:988–994.
28. Eugenin EA, Berman JW. Gap junctions mediate human immunodeficiency virus-bystander killing in astrocytes. *J Neurosci*. 2007;27:12844–12850.
29. Roberts TK, Buckner CM, Berman JW. Leukocyte transmigration across the blood–brain barrier: perspectives on neuroAIDS. *Front Biosci*. 2010;15:478–536.
30. Beg AA, Baltimore D. An essential role for NF-kappaB in preventing TNF-alpha-induced cell death. *Science*. 1996;274:782–784.
31. Shah A, Kumar A. HIV-1 gp120-mediated increases in IL-8 production in astrocytes are mediated through the NF-kappaB pathway and can be silenced by gp120-specific siRNA. *J Neuroinflammation*. 2010;7:96.
32. Renner NA, Ivey NS, Redmann RK, Lackner AA, MacLean AG. MCP-3/CCL7 production by astrocytes: implications for SIV neuroinvasion and AIDS encephalitis. *J Neurovirol*. 2011;17:146–152.
33. Williams R, Yao H, Dhillon NK, Buch SJ. HIV-1 Tat co-operates with IFN-gamma and TNF-alpha to increase CXCL10 in human astrocytes. *PLoS ONE*. 2009;4:e5709.
34. Shah A, Singh DP, Buch S, Kumar A. HIV-1 envelope protein gp120 up regulates CCL5 production in astrocytes which can be circumvented by inhibitors of NF-kappaB pathway. *Biochem Biophys Res Commun*. 2011;414:112–117.
35. Shah A, Verma AS, Patel KH, et al. HIV-1 gp120 induces expression of IL-6 through a nuclear factor-kappa B-dependent mechanism: suppression by gp120 specific small interfering RNA. *PLoS One*. 2011;6:e21261.
36. Ronaldson PT, Bendayan R. HIV-1 viral envelope glycoprotein gp120 triggers an inflammatory response in cultured rat astrocytes and regulates the functional expression of P-glycoprotein. *Mol Pharmacol*. 2006;70:1087–1098.
37. Wang T, Gong N, Liu J, et al. HIV-1-infected astrocytes and the microglial proteome. *J Neuroimmune Pharmacol*. 2008;3:173–186.
38. Barber SA, Herbst DS, Bullock BT, Gama L, Clements JE. Innate immune responses and control of acute simian immunodeficiency virus replication in the central nervous system. *J Neurovirol*. 2004;10(suppl 1):15–20.
39. Annunziata P. Blood–brain barrier changes during invasion of the central nervous system by HIV-1. Old and new insights into the mechanism. *J Neurol*. 2003;250:901–906.
40. Kanmogne GD, Schall K, Leibhart J, Knipe B, Gendelman HE, Persidsky Y. HIV-1 gp120 compromises blood–brain barrier integrity and enhances monocyte migration across blood–brain barrier: implication for viral neuropathogenesis. *J Cereb Blood Flow Metab*. 2007;27:123–134.
41. Budka H. Human immunodeficiency virus (HIV)-induced disease of the central nervous system: pathology and implications for pathogenesis. *Acta Neuropathol*. 1989;77:225–236.
42. Giovannoni G, Miller RF, Heales SJ, Land JM, Harrison MJ, Thompson EJ. Elevated cerebrospinal fluid and serum nitrate and nitrite levels in patients with central nervous system complications of HIV-1 infection: a correlation with blood–brain-barrier dysfunction. *J Neurol Sci*. 1998;156:53–58.
43. Bernardo A, Agresti C, Levi G. HIV-gp120 affects the functional activity of oligodendrocytes and their susceptibility to complement. *J Neurosci Res*. 1997;50:946–957.
44. Sartoretti-Schefer S, Blattler T, Wichmann W. Spinal MRI in vacuolar myelopathy, and correlation with histopathological findings. *Neuroradiology*. 1997;39:865–869.
45. Tan SV, Guiloff RJ, Henderson DC, Gazzard BG, Miller R. AIDS-associated vacuolar myelopathy and tumor necrosis factor-alpha (TNF alpha). *J Neurol Sci*. 1996;138:134–144.
46. Okamoto S, Kang YJ, Brechtel CW, et al. HIV/gp120 decreases adult neural progenitor cell proliferation via checkpoint kinase-mediated cell-cycle withdrawal and G1 arrest. *Cell Stem Cell*. 2007;1:230–236.
47. Lee MH, Wang T, Jang MH, et al. Rescue of adult hippocampal neurogenesis in a mouse model of HIV neurologic disease. *Neurobiol Dis*. 2011;41:678–687.
48. Kandanearatchi A, Brew BJ. The kynurenine pathway and quinolinic acid: pivotal roles in HIV associated neurocognitive disorders. *FEBS J*. 2012;279:1366–1374.

49. Ferrarese C, Aliprandi A, Tremolizzo L, et al. Increased glutamate in CSF and plasma of patients with HIV dementia. *Neurology*. 2001;57:671–675.
50. Chen W, Sulcove J, Frank I, Jaffer S, Ozdener H, Kolson DL. Development of a human neuronal cell model for human immunodeficiency virus (HIV)-infected macrophage-induced neurotoxicity: apoptosis induced by HIV type 1 primary isolates and evidence for involvement of the Bcl-2/Bcl-xL-sensitive intrinsic apoptosis pathway. *J Virol*. 2002;76:9407–9419.
51. Erdmann NB, Whitney NP, Zheng J. Potentiation of excitotoxicity in HIV-1 associated dementia and the significance of glutaminase. *Clin Neurosci Res*. 2006;6:315–328.
52. Pietraforte D, Tritarelli E, Testa U, Minetti M. gp120 HIV envelope glycoprotein increases the production of nitric oxide in human monocyte-derived macrophages. *J Leukoc Biol*. 1994;55:175–182.
53. Zhao ML, Kim MO, Morgello S, Lee SC. Expression of inducible nitric oxide synthase, interleukin-1 and caspase–1 in HIV-1 encephalitis. *J Neuroimmunol*. 2001;115:182–191.
54. Merrill JE, Ignarro LJ, Sherman MP, Melinek J, Lane TE. Microglial cell cytotoxicity of oligodendrocytes is mediated through nitric oxide. *J Immunol*. 1993;151:2132–2141.
55. Mitrovic B, Ignarro LJ, Montestruque S, Smoll A, Merrill JE. Nitric oxide as a potential pathological mechanism in demyelination: its differential effects on primary glial cells in vitro. *Neuroscience*. 1994;61:575–585.
56. Bertin J, Barat C, Methot S, Tremblay MJ. Interactions between prostaglandins, leukotrienes and HIV-1: possible implications for the central nervous system. *Retrovirology*. 2012;9:4.
57. Hayes MM, Lane BR, King SR, Markovitz DM, Coffey MJ. Prostaglandin E(2) inhibits replication of HIV-1 in macrophages through activation of protein kinase A.. *Cell Immunol*. 2002;215:61–71.
58. Hardwick C, White D, Morris E, Monteiro EF, Breen RA, Lipman M. Montelukast in the treatment of HIV associated immune reconstitution disease. *Sex Transm Infect*. 2006;82:513–514.
59. Haughey NJ, Steiner J, Nath A, et al. Converging roles for sphingolipids and cell stress in the progression of neuro-AIDS. *Front Biosci*. 2008;13:5120–5130.
60. Cutler RG, Haughey NJ, Tammara A, et al. Dysregulation of sphingolipid and sterol metabolism by ApoE4 in HIV dementia. *Neurology*. 2004;63:626–630.
61. Jaeger LB, Nath A. Modeling HIV-associated neurocognitive disorders in mice: new approaches in the changing face of HIV neuropathogenesis. *Dis Model Mech*. 2012;5:313–322.
62. Theodore S, Cass WA, Maragos WF. Methamphetamine and human immunodeficiency virus protein Tat synergize to destroy dopaminergic terminals in the rat striatum. *Neuroscience*. 2006;137:925–935.
63. Persidsky Y, Limoges J, McComb R, et al. Human immunodeficiency virus encephalitis in SCID mice. *Am J Pathol*. 1996;149:1027–1053.
64. Potash MJ, Chao W, Bentsman G, et al. A mouse model for study of systemic HIV-1 infection, antiviral immune responses, and neuroinvasiveness. *Proc Natl Acad Sci U S A*. 2005;102:3760–3765.
65. Mankowski JL, Spelman JP, Ressetar HG, et al. Neurovirulent simian immunodeficiency virus replicates productively in endothelial cells of the central nervous system in vivo and in vitro. *J Virol*. 1994;68:8202–8208.
66. Williams R, Bokhari S, Silverstein P, Pinson D, Kumar A, Buch S. Nonhuman primate models of NeuroAIDS. *J Neurovirol*. 2008;14:292–300.
67. Antinori A, Arendt G, Becker JT, et al. Updated research nosology for HIV-associated neurocognitive disorders. *Neurology*. 2007;69:1789–1799.
68. Nakagawa F, May M, Phillips A. Life expectancy living with HIV: recent estimates and future implications. *Curr Opin Infect Dis*. 2013;26:17–25.
69. Benjamin LA, Bryer A, Emsley HC, Khoo S, Solomon T, Connor MD. HIV infection and stroke: current perspectives and future directions. *Lancet Neurol*. 2012;11:878–890.
70. Vivithanaporn P, Maingat F, Lin LT, et al. Hepatitis C virus core protein induces neuroimmune activation and potentiates human immunodeficiency virus-1 neurotoxicity. *PLoS One*. 2010;5:e12856.
71. Peluso MJ, Ferretti F, Peterson J, et al. Cerebrospinal fluid HIV escape associated with progressive neurologic dysfunction in patients on antiretroviral therapy with well-controlled plasma viral load. *AIDS*. 2012;26(14):1765–1774.
72. Johnson T, Nath A. Immune reconstitution inflammatory syndrome and the central nervous system. *Curr Opin Neurol*. 2011;24:284–290.

CHAPTER

29

Autoimmune and Paraneoplastic Neurological Disorders

Raffaele Iorio, Orna O'Toole*, †, Sean J. Pittock*, †*

*Department of Laboratory Medicine and Pathology, Mayo Clinic, College of Medicine, Rochester, Minnesota, USA;
†Department of Neurology, Mayo Clinic, College of Medicine, Rochester, Minnesota, USA

OUTLINE

Neurobiology of Brain Disorders
http://dx.doi.org/10.1016/B978-0-12-398270-4.00029-X

INTRODUCTION

The study of autoimmune neurological diseases, namely autoimmune neurology, is an emerging and rapidly evolving subspecialty. Neural-specific autoantibodies serve as helpful diagnostic biomarkers and immunotherapy may be beneficial. Autoimmune neurology intersects with many of the traditional neurological subspecialties including cognitive behavioral neurology, movement disorders, epilepsy, neuro-oncology, neuromuscular disorders, autonomic neurology, and demyelinating disorders. Antibodies targeting neural antigens have been documented in several autoimmune neurological diseases. Since most antigenic targets are expressed diffusely through the nervous system, neurological autoimmunity is often multifocal involving multiple levels of the neuraxis from cerebral cortex to spinal cord to the neuromuscular and autonomic synapses and muscle.

Target antigens include intracellular (nuclear and cytoplasmic enzymes, transcription factors, and RNA binding proteins) and plasma membrane (neurotransmitter receptors, ion channels, water channels, and channel-complex proteins) proteins. The molecular identification of these antigenic targets has provided insights into the pathogenic mechanisms underlying many autoimmune neurological disorders. Autoantibodies specific for neural antigens can arise idiopathically or in the context of neoplasm (i.e. paraneoplastic). Paraneoplastic disorders are often progressive, although early diagnosis and treatment may benefit some patients. The identification of paraneoplastic autoantibodies helps to direct the search for cancer and provides prognostic information. Unfortunately, many neurologists still suspect a paraneoplastic or autoimmune etiology only in the setting of certain syndromic manifestations, for example stiff man syndrome with amphiphysin autoantibody or opsoclonus–myoclonus associated with type 2 antineuronal nuclear autoantibody (ANNA-2/anti-Ri). This traditional view has now changed somewhat, as it is now recognized that most onconeural autoantibodies are more frequently associated with diverse, often multifocal neurological manifestations that extend beyond a unique syndrome.[1] Autoantibodies binding to extracellular epitopes of synaptic proteins have been identified in several forms of autoimmune neurological disorders. These discoveries have radically changed the diagnostic approach to several clinical problems such as epilepsy, cognitive disturbance, movement disorders, and psychiatric symptoms. Furthermore, comprehensive serological testing for these autoantibodies is now available to physicians through commercial clinical service laboratories worldwide.

This chapter, after a brief description of the immune system, focuses on the immunobiology, pathophysiology, and clinical presentations of autoimmune neurological disorders for which neural antigen-specific autoantibodies serve as diagnostic aids.

THE IMMUNE SYSTEM

The word immunity is derived from the Latin *immunitas*, which referred to the protection from legal prosecution offered to Roman senators during their appointment. In biological sciences the term immunity means protection of the organism against infectious agents. The cells and molecules responsible for immunity constitute the immune system, and their collective and organized reaction to microbes define the immune response. Other than defending the organism from infectious agents, the immune system performs tumor surveillance, promotes healing, and prevents damage mediated by dying cells. In physiological conditions the immune system normally does not react to its own (self) antigens, a state known as tolerance. In the setting of autoimmune diseases the immune system fails to recognize its own constituent parts as self and it mounts an aberrant immune response against its own cells and tissues. Thus, a delicate balance must be maintained between the protective effects of the immune response and potential deleterious effects. The immune system has two functional branches: the innate and the adaptive immune system.[2]

Innate Immune System

The innate immune system provides the early line of defense against microorganisms. It consists of non-specific cellular and biochemical defense mechanisms that are in place even before infection and are poised to respond rapidly to infections. These immune defenses react only to microbes (and to the products of injured cells) and do not confer long-lasting protective immunity to the host. The principal components of innate immunity are:

- *anatomical barriers*: physical (epithelial surfaces) and chemical (e.g. lysozyme and phospholipase in tears and saliva can damage the bacterial cell wall)
- *phagocytes* (macrophages, neutrophils) that are capable of phagocytosing foreign pathogens, and natural killer (NK) cells that kill microorganisms and cancer cells via cell-mediated cytoxicity
- *blood proteins*, such as the acute-phase proteins (e.g. C-reactive protein) and members of the complement system directly activated by components of the microbial cell wall, namely the alternative complement pathway
- *cytokines*: proteins that act as messengers facilitating intercellular communication within the immune system and between the immune system and other systems of the body.

The adaptive immune system recognizes specific pathogens and is able to develop and maintain long-lasting protection and a more rapid response in case of recurrent infection (immunological memory).

Adaptive Immune System

The adaptive immune system consists of the following components.

- *Antibodies*: Antibodies or immunoglobulins (Igs) are Y-shaped glycoproteins that can specifically bind to a variety of free antigens. Antibodies are produced by B cells, are present on their plasma membrane, and are abundantly secreted in the serum. Antibodies recognize specific microbial and other antigens through their antigen binding sites and bind to the Fc receptor expressed on the phagocyte's cell surface, facilitating antigen removal. Some subclasses of immunoglobulin are capable of activating complement via their Fc portion, thereby lysing their targets. The *humoral immune response* indicates the immunological response mediated by antibodies.
- *B lymphocytes*: B cells have the primary function of synthesizing antibodies. Antigen binding to B cells stimulates proliferation and maturation of that particular B cell, with subsequent increase in antigen-specific antibody production, resulting in the development of antibody-secreting plasma cells. Most B lymphocytes express major histocompatibility complex (MHC) class II antigens on the plasma membrane and can present the antigens to T lymphocytes.
- *T lymphocytes*: T cells can recognize specific antigens via their T-cell receptors. T cells may be classified into two main subsets: helper T cells expressing CD4 antigen on their cell surface and cytotoxic T cells expressing CD8 on their surface. $CD4^+$ T cells recognize antigen presented in association with MHC-II on the surface of antigen-presenting cells (APCs). CD4 T cells provide help to promote B-cell maturation and antibody production and secrete *cytokines* to modulate the immune response. CD8 T cells, also known as cytotoxic T cells, recognize antigen in association with MHC-I antigen on the surface of most cells and play a key role in the cytolytic immune response targeting viruses, intracellularly replicating bacteria, and cancerous cells. Cytotoxic T cells can damage target cells via the release of degrading enzymes, such as granzyme B and perforin, and cytokines such as interferon-γ (IFN-γ). The *cell-mediated immune response* denotes the immunological response in which T lymphocytes play a major role.
- *Antigen-presenting cells*: APCs present antigens to T cells. They are located in the skin, lymph nodes, spleen, and thymus. Unlike B cells that can recognize free antigen, T cells are capable only of recognizing antigen in the context of self MHC molecules. APCs process antigen intracellularly and peptide fragments are loaded on to MHC-II molecules to become displayed on the cell surface and accessible to CD4 helper T cells. The primary APCs are macrophages, dendritic cells, and Langerhans cells.

PATHOGENIC MECHANISMS OF NEURAL ANTIGEN-SPECIFIC AUTOIMMUNITY

Neural-specific autoantibodies are generally classified based on the location of the target antigen (Fig. 29.1): (1) intracellular (nuclear or cytoplasmic) autoantibodies, which are probably not pathogenic but serve as surrogate markers of a cell-mediated immune response, and are generally highly predictive of cancer (Table 29.1); or (2) plasma membrane antibodies, which bind to the extracellular domain of neural cell surface proteins and thus have pathogenic potential (Table 29.2). The two classes of antibody commonly coexist (Table 29.3).

Autoantibodies Specific for Intracellular Antigens

Autoantibodies specific for intracellular neural antigens serve as surrogate markers for a cytotoxic $CD8^+$ T-cell mediated injury (Fig. 29.2). Antigenic proteins inside intact cells are inaccessible to circulating antibodies. However, in a proinflammatory environment, the intracellular

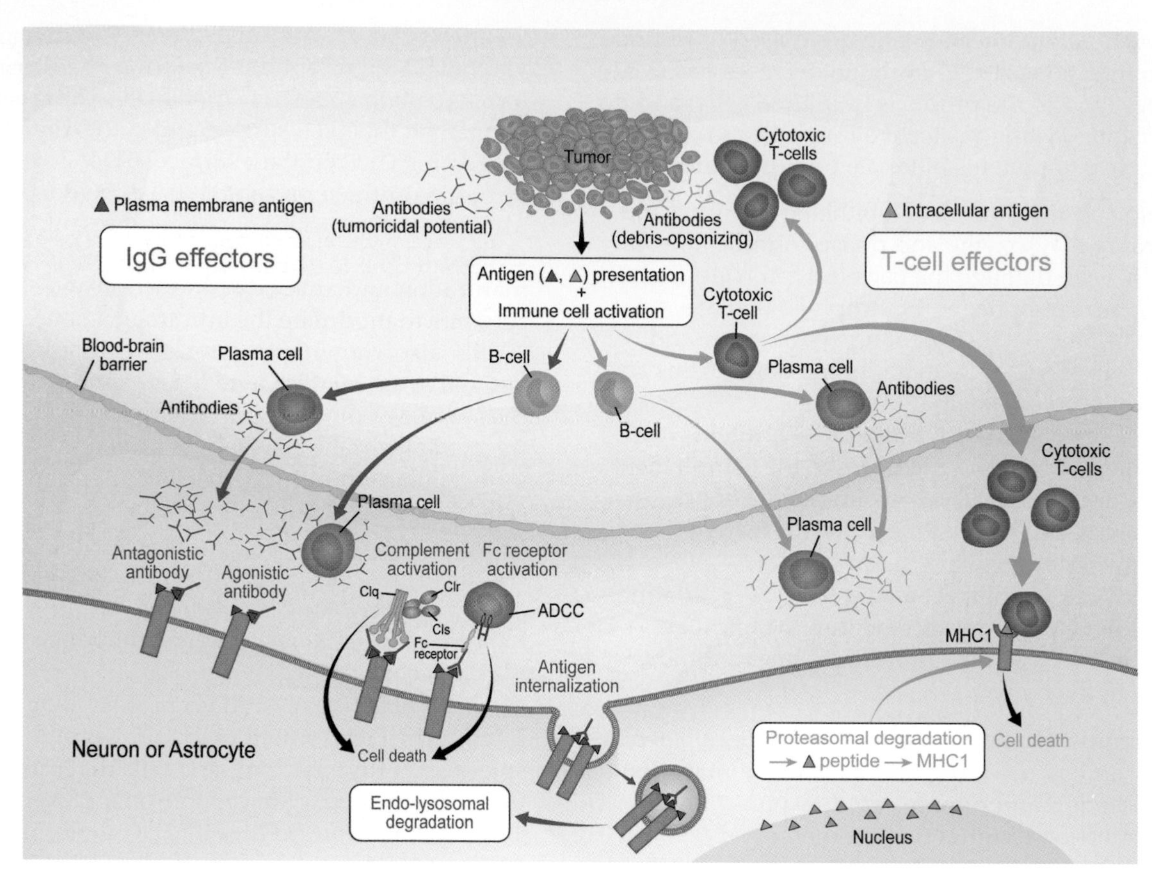

FIGURE 29.1 **Paraneoplastic neural autoantibodies and immunopathogenic mechanisms.** Tumor-targeted immune responses are initiated by onconeural proteins expressed in the plasma membrane (red triangle) or in the nucleus cytoplasm or nucleolus (green triangle) of certain tumors. These antigens are presented to the adaptive immune system and immune cell activation results. These antigens are also expressed in neural cells (neurons or glia) and thus are coincidental targets. Antibodies targeting plasma membrane antigens are effectors of injury (red): antibodies (red) directed at neural cell plasma membrane antigens [e.g. voltage-gated potassium channel (VGKC) complex, *N*-methyl-D-aspartate (NMDA), 2-amino-3-(5-methyl-3-oxo-1,2-oxazol-4-yl)propanic acid (AMPA), γ-aminobutyric acid-B ($GABA_B$) receptor, aquaporin-4 (AQP4)] are effectors of cellular dysfunction or injury through multiple effector mechanisms. These mechanisms include receptor agonist or antagonist effects, activation of the complement cascades, activation of Fc receptors [leading to antibody-dependent cell mediated cytotoxicity (ADCC)], and antigen internalization (antigenic modulation), thereby altering antigen density on the cell surface. Antibodies targeting nuclear or cytoplasmic antigens are serum markers of a T-cell effector mediated injury (green): intracellular antigens (green triangles) are not accessible to immune attack *in situ*, but peptides derived from intracellular proteins are displayed on upregulated major histocompatibility complex class I (MHC1) molecules in a proinflammatory cytokine milieu after proteasomal degradation, and are then accessible to peptide-specific cytotoxic T cells. Antibodies [green, e.g. antineuronal nuclear autoantibody type 1 (ANNA-1), Purkinje cell cytoplasmic autoantibody type 1 (PCA-1)] targeting these intracellular antigens (green) are detected in both serum and cerebrospinal fluid but are not pathogenic. In clinical practice, these antibodies serve as diagnostic markers of a T-cell predominant effector process. IgG: immunoglobulin G. *Source: Modified from Diamond* et al. Nat Rev Immunol. *2009;9:449–456,*[3] *with permission from the Nature Publishing Group.*

TABLE 29.1 Neuronal Nuclear Cytoplasmic Antibodies

Antibody	Oncological Association	Neurological Presentation
PCA-1 (anti-Yo)	Ovarian or other müllerian (gynecological tract) adenocarcinoma, breast adenocarcinoma	Cerebellar ataxia, brainstem encephalitis, myelopathy, radiculopathies, peripheral neuropathies
PCA-2	Small-cell carcinoma	Limbic encephalitis, ataxia, brainstem encephalitis, Lambert–Eaton syndrome, peripheral and autonomic neuropathies
PCA-Tr	Hodgkin lymphoma	Cerebellar ataxia, limbic encephalitis, autonomic dysfunction
ANNA-1 (anti-Hu)	Small-cell carcinoma; neuroblastoma (children)	Limbic encephalitis, brainstem encephalitis, chorea, autonomic neuropathies, sensory neuronopathy, other peripheral neuropathies
ANNA-2 (anti-Ri)	Small-cell carcinoma, breast adenocarcinoma	Brainstem encephalitis

TABLE 29.1 Neuronal Nuclear Cytoplasmic Antibodies—cont'd

Antibody	Oncological Association	Neurological Presentation
ANNA-3	Aerodigestive carcinomas (e.g. lung, esophagus)	Brainstem encephalitis, limbic encephalitis, myelopathy, peripheral neuropathy
AGNA	Small-cell carcinoma	Neuropathy, Lambert–Eaton syndrome, limbic encephalitis
CRMP-5-IgG (anti-CV2)	Small-cell carcinoma, thymoma	Subacute-onset dementia, personality change, aphasia, depression, chorea, ataxia, optic neuritis and retinitis, myelopathy, radiculopathy, neuropathy, Lambert–Eaton syndrome
Amphiphysin-IgG	Small-cell carcinoma, breast adenocarcinoma	Encephalitis, neuropathy, myelopathy, encephalopathy, stiff man syndrome, cerebellar degeneration
Ma1, Ma2	Testicular (Ma2) and breast, colon or testicular carcinoma (Ma1 and Ma2)	Limbic encephalitis, hypothalamic disorder, brainstem encephalitis
GAD65-IgG	Rare reports of thymoma, renal cell carcinoma, breast or colon adenocarcinoma (usually with other paraneoplastic antibodies)	Limbic encephalitis, seizures, stiff man syndrome and variants, brainstem encephalitis, ophthalmoplegia, cerebellar ataxia, parkinsonism, chorea, myelopathy

PCA: Purkinje cytoplasmic antibody; Tr: Trotter (named after John Trotter who first described this antibody in 1976); ANNA: antineuronal nuclear antibody; AGNA: antiglial neuronal nuclear antibody type 1; CRMP-5: collapsin response mediator protein-5; IgG: immunoglobulin G; GAD65: 65 kDa isoform of glutamic acid decarboxylase.

TABLE 29.2 Antibodies Targeting Ion Channels and Other Plasma Membrane Proteins

Antibody	Oncological Association	Neurological Presentation
VGKC complex	Various (20% of patients have small-cell lung carcinoma, thymoma or adenocarcinoma of breast or prostate)	Limbic encephalitis, frontotemporal dementia-like presentation, brainstem encephalitis, cerebellar ataxia, extrapyramidal disorders, psychiatric disturbances, sleep disorders, myoclonus, peripheral and autonomic neuropathy, chronic pain
NMDA receptor	Ovarian teratomas (50% of female patients)	Limbic encephalitis, psychiatric disturbances (anxiety, psychosis), dyskinesias, catatonia, seizures, central hypoventilation and autonomic instability, opsoclonus–myoclonus
AMPA receptor	Thymic tumors, lung carcinoma, breast adenocarcinoma	Limbic encephalitis, nystagmus, seizures
$GABA_B$ receptor	Small-cell lung carcinoma, other neuroendocrine neoplasia	Limbic encephalitis, orolingual dyskinesias
P/Q- and N-type VGCC	Small-cell carcinoma, breast or gynecological adenocarcinoma	Encephalopathies, myelopathies, neuropathies, Lambert–Eaton syndrome
NMO-IgG	Rare reports of thymoma and breast adenocarcinoma	Relapsing, severe optic neuritis, transverse myelitis, encephalopathies
Neuronal ganglionic AChR	Adenocarcinoma, thymoma, small-cell carcinoma (30% of patients)	Dysautonomic
Muscle AChR	Thymoma, thymic carcinoma, lung carcinoma	Myasthenia gravis
Glycine receptor	Thymoma, lymphoma (10% of patients)	Stiff man syndrome and variants
$GluR_1$ and $GluR_5$	Hodgkin lymphoma	Cerebellar ataxia and limbic encephalitis (Ophelia syndrome)
Muscle striational	Typically occurs with other paraneoplastic antibodies and is unlikely to predict cancer in isolation	Myasthenia gravis (with and without thymoma)

VGKC: voltage-gated potassium channel; NMDA: *N*-methyl-D-aspartate; AMPA: 2-amino-3-hydroxy-5-methyl-4-isoxazole-propionic acid; GABA: γ-aminobutyric acid; VGCC: voltage-gated calcium channel; NMO: neuromyelitis optica; IgG: immunoglobulin G; AChR: acetylcholine receptor; CASPR: contactin-associated protein; GluR: glutamate receptor.

TABLE 29.3 Frequency (%) of Autoantibodies Coexisting with Seven Defined Paraneoplastic Neuronal Nuclear and Neuronal Cytoplasmic Immunoglobulin G Encountered in Sera of 553 Patients (January 2000 to December 2003)

Coexisting Antibodies	ANNA-1 ($n=217$)[a]	CRMP-5 ($n=208$)	PCA-1 ($n=101$)	PCA-2 ($n=43$)	Amphiphysin ($n=26$)	ANNA-2 ($n=17$)	ANNA-3 ($n=10$)
NUCLEAR/CYTOPLASMIC							
ANNA-1	–	17	0	12	8	23	10
CRMP-5	17	–	0	44	19	12	20
PCA-1	0	0	–	0	0	0	0
PCA-2	2	9	0	–	8	0	10
Amphiphysin	1	2	0	5	–	6	0
ANNA-2	2	1	0	0	4	–	0
ANNA-3	1	1	0	2	0	0	–
Cumulative frequency of coexisting neuronal nuclear/cytoplasmic IgG (1 or more)	19	28	0	50	31	29	30
ION CHANNEL/STRIATIONAL							
Ca^{2+} channel, P/Q type	14	13	2	14	11	18	10
Ca^{2+} channel, N type	14	14	2	9	11	18	10
K^{+} channel	3	5	1	9	4	0	0
Ganglionic AChR	5	2	3	0	4	6	0
Muscle AChR	5	8	2	2	4	0	0
Striational	3	5	0	2	4	0	0
Cumulative frequency of coexisting ion channel/striational IgG/IgM (1 or more)	28	31	9	30	23	18	20
Overall frequency of coexisting antibodies (1 or more)	43	57	9	63	38	35	40

ANNA: antineuronal nuclear antibody; CRMP-5: collapsin response mediator protein-5; PCA: Purkinje cytoplasmic antibody; IgG: immunoglobulin G; AChR: acetylcholine receptor; IgM: immunoglobulin M.

[a]*Number of sera positive for each listed antibody; the sum of positive autoantibody markers exceeds 553 because 30% of patients had more than one autoantibody.*

Source: Pittock et al. Ann Neurol. *2004;56:715–719, with permission from John Wiley and Sons.*[19]

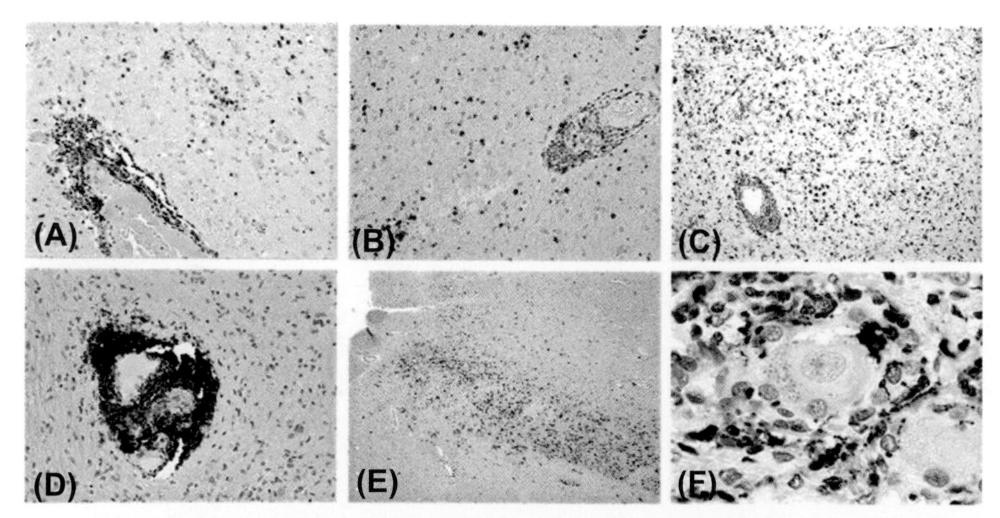

FIGURE 29.2 **Neuropathology of amphiphysin antibody associated paraneoplastic encephalitis.** (A) Immunophenotyping of perivascular and parenchymal mononuclear leukocyte infiltrates ($CD45^{+}$) in the medulla. Most of the parenchymal leukocytes were $CD8^{+}$ lymphocytes (B) or macrophages (C). B lymphocytes ($CD20^{+}$) were restricted to perivascular location (D). In the medial lemniscus of the pons, marked macrophage infiltration was seen (E). Scattered microglia and $CD8^{+}$ lymphocytes were prominent throughout the parenchyma of the spinal cord (not shown) and dorsal root ganglia (F). *Source: Pittock* et al. Ann Neurol. *2005; 58:96–107,* [21] *with permission from John Wiley and Sons.*

proteins are degraded into a variety of peptide epitopes and loaded on to upregulated MHC class I molecules to become displayed on the cell surface and accessible to peptide-specific cytotoxic T cells. Under the influence of IFN-γ, the constitutive proteolytic enzyme composition of the cellular proteasome switches to an immunoproteasome that cleaves proteins tagged for degradation into peptide fragments (typically eight to 11 amino acids long) different from the self peptides that establish and maintain immunological tolerance. These non-self autoantigenic peptides are predicted to resemble those produced by the distant tumor cells undergoing attack by the innate immune system initiating the paraneoplastic immune response. A single cell may have up to 250,000 molecules of MHC-I with bound epitope on its surface. During cell-mediated immunity, MHC-I molecule with bound peptide on the surface of neural cells can be recognized by a complementary-shaped T-cell receptor on the surface of a cytotoxic $CD8^+$ T lymphocyte to initiate destruction of the cell containing the endogenous antigen. The key effector molecules of the $CD8^+$ T-cell mediated attack are perforin and granzyme B. Perforin, a protein partially homologous to the terminal components of the membrane attack complex of complement, produces pores of up to 20 nm in diameter on target membranes leading to the cytolysis of the target cells. Granzyme B is a serine protease that is stored in cytoplasmic granules within cytotoxic T cells. Once released, granzyme B binds its receptor, the mannose-6-phosphate/insulin-like growth factor II receptor, and is endocytosed but remains arrested in endocytic vesicles until released by perforin. Once in the cytosol, granzyme B induces the apoptosis of the target cells through various pathways.

In vitro studies of T cells derived from patients with paraneoplastic cerebellar degeneration related to ovarian carcinoma persuasively support the role of onconeural peptide-specific $CD8^+$ T cells as effectors of the inflammatory cytotoxic neuropathology accompanying autoantibodies specific for neural intracellular antigens. Albert and colleagues demonstrated expanded populations of MHC class I-restricted $CD8^+$ cytotoxic T cells specific for a peptide derived from the cdr2 protein in a patient with acute-phase cerebellar ataxia.[4] $CD8^+$ cytotoxic T-cell responses were elicited in peripheral blood lymphocytes of two patients when presented with specific peptide antigen by human leukocyte antigen (HLA)-matched cells. Neuropathological study of autopsied cases of paraneoplastic neurological disorders also supports a role for $CD8^+$ cytotoxic T-cell response.[5]

ANNA-1 (Anti-Hu)

Antineuronal nuclear antibody type 1 (ANNA-1, also known as anti-Hu) binds to the Hu family of RNA binding proteins, which participate in post-transcriptional regulation of RNA in postmitotic neurons.[6] These antigens are restricted to neurons (Fig. 29.3C, D) and neuroendocrine malignancies exemplified by small-cell carcinoma. ANNA-1 is a biomarker of a tumor immune response against a pulmonary or less often an extrapulmonary small-cell carcinoma, childhood neuroblastoma or thymoma. Peptide-specific $CD8^+$ T cells are demonstrable in peripheral blood of seropositive patients using HLA-matched tetramer complexes. Classical $CD8^+$ T cells secreting IFN-γ (type I cytotoxic T cells) characterize the early immune response, while $CD8^+$ T cells secreting interleukin-5 (IL-5) and IL-13 (type II cytotoxic T cells) predominate during the chronic phase of the disease.[6] Neurological manifestations associated with ANNA-1 include, in decreasing order of frequency: peripheral neuropathy, limbic encephalitis. encephalomyelitis, and gastrointestinal dysmotility.[1]

ANNA-2 (Anti-Ri)

Antineuronal nuclear autoantibody type 2 (ANNA-2, also known as anti-Ri) binds the NOVA family of RNA binding proteins (NOVA-1, 55 kDa; and NOVA-2, 80 kDa) responsible for alternative splicing of neuronal transcripts that encode synaptic proteins (Fig. 29.3A, B).[7] Associated primarily with small-cell lung cancer and breast cancer, this antibody is much less common than ANNA-1.[8] Neurological manifestations associated with ANNA-2 include opsoclonus–myoclonus (about one-third of all patients), brainstem encephalitis, cerebellar degeneration, jaw dystonia and laryngospasm, and myelopathy.[8,9]

ANNA-3

Antineuronal nuclear autoantibody type 3 (ANNA-3) is identified using an indirect immunofluorescence assay based on its specific pattern of binding on a composite of mouse neural and non-neural tissues. Its target antigen, a protein of 170 kDa, remains unknown. ANNA-3 is a rare antibody associated with lung cancer of various types. Neurological manifestations associated with ANNA-3 include cerebellar degeneration, brainstem and limbic encephalitis, myelopathy, sensory neuronopathy, and peripheral and autonomic neuropathies.[10]

PCA-1 (Anti-Yo)

Purkinje cell cytoplasmic autoantibody (PCA-1, also known as anti-Yo) binds the 52 kDa cerebellar degeneration-related protein-2 (cdr2), which downregulates DNA transcription through inhibition of c-Myc. PCA-1 is found virtually exclusively in association with gynecological cancers: müllerian carcinoma of the uterus or ovary and breast carcinoma. Cerebellar degeneration is the most frequent neurological association (>90%), although other levels of the nervous system including the spinal cord and peripheral nerve may also be affected.[11] These diverse neurological manifestations are consistent with the distribution of the PCA-1 antigen, which is not confined to the cerebellum and is expressed in spinal motor neurons and Schwann cells of peripheral nerves (Fig. 29.4).[12,13]

PCA-2

Purkinje cell antibody type 2 (PCA-2) is identified using an indirect immunofluorescence assay based on its specific pattern of binding on a composite of mouse neural and non-neural tissues. Its target antigen is a 280 kDa protein of unknown molecular identity. It is most commonly associated with lung cancer. Neurological manifestations include encephalopathy, cerebellar degeneration, or involvement of both the peripheral and autonomic nervous systems.[14]

PCA-Tr

Purkinje cell antibody-Tr (PCA-Tr) is named after the physician (Dr John Trotter) who first described cerebellar degeneration in association with Hodgkin lymphoma, in which it is almost exclusively described.[15] PCA-Tr targets the delta-notch-like epidermal growth factor-related receptor.[16] Immunoreactivity is noted as a fine, sparkling appearance in Purkinje cell perikaryon and intracellular organelles of the molecular layer.[17] This autoantibody can also be occasionally detected in association with non-Hodgkin lymphoma and lung cancer. In addition to cerebellar degeneration, limbic encephalitis and autonomic dysfunction are rarely seen.

CRMP-5-IgG

Collapsin response mediator protein-5 (CRMP-5) antibodies bind to the 62 kDa protein of the same name involved in axonal development in the early nervous system.[18] Immunoreactivity is typical in mouse cerebellar molecular layer, midbrain, and myenteric plexus of gut smooth muscle. Small-cell lung carcinoma and thymoma are the most frequent tumor accompaniments.[19] Associated neurological disorders include cerebellar degeneration, chorea, optic neuropathy, retinopathy, myelopathy, radiculoneuropathies, and autonomic dysfunction.

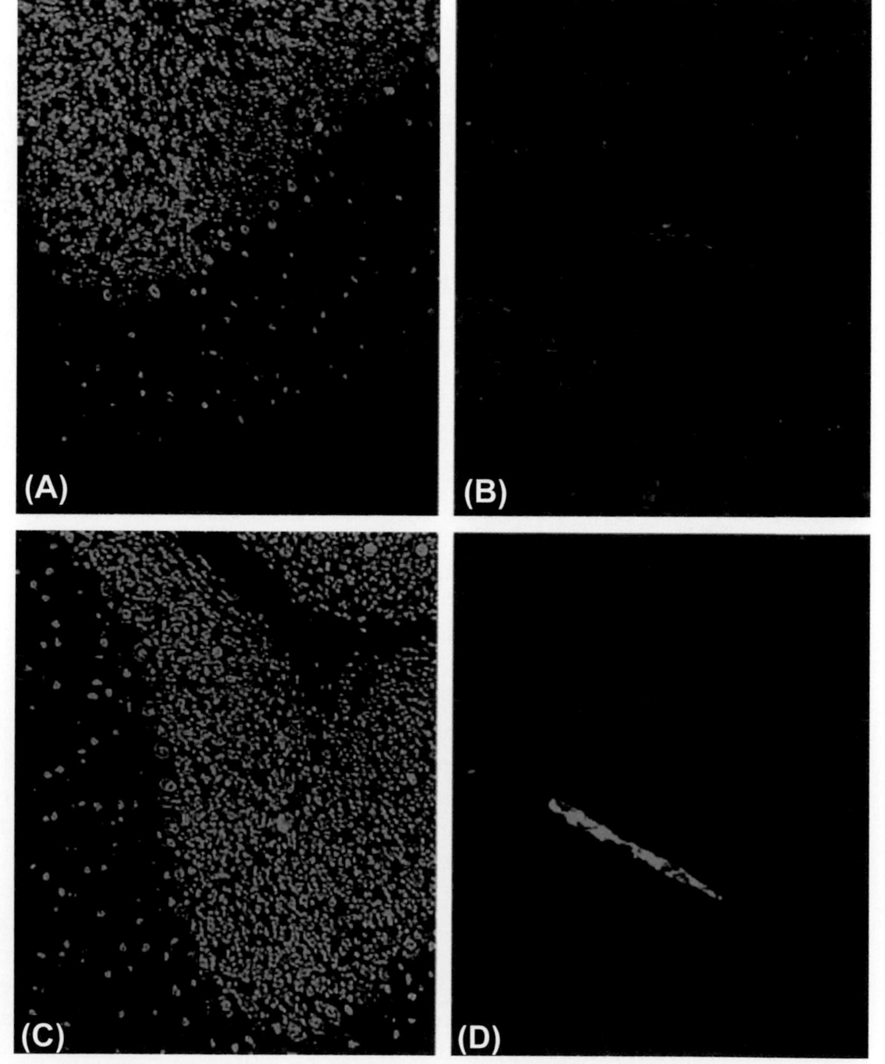

FIGURE 29.3 **Indirect immunofluorescence.** In cerebellar cortex, antineuronal nuclear autoantibody type 2 (ANNA-2) (A) and ANNA-1 (C) stain nuclei and cytoplasm of all neurons, whereas nucleoli are spared. In gut, ANNA-1 (D) stains nuclei and cytoplasm of enteric neurons, but not smooth muscle or mucosa (top) or renal tubules (kidney is at bottom). ANNA-2 (B) does not stain gut or renal tubules. (200× magnification.) *Source: Pittock* et al. Ann Neurol. *2003;53:580–587,* [8] *with permission from John Wiley and Sons.*

Amphiphysin

Amphiphysin antibodies (Fig. 29.2C) target a 128 kDa synaptic vesicle-bound protein that works with dynamin to retrieve membrane constituents after neurotransmitter exocytosis (Fig. 29.5). First described in relation to stiff man syndrome occurring in the setting of breast carcinoma,[20] its spectrum is now appreciated to be much wider. It is commonly associated with paraneoplastic encephalopathy, cerebellar degeneration, myelopathy, and neuropathy. Classic amphiphysin-associated stiff man syndrome is rare (< 10%). Both breast and small-cell lung carcinoma are seen frequently with this antibody.[21]

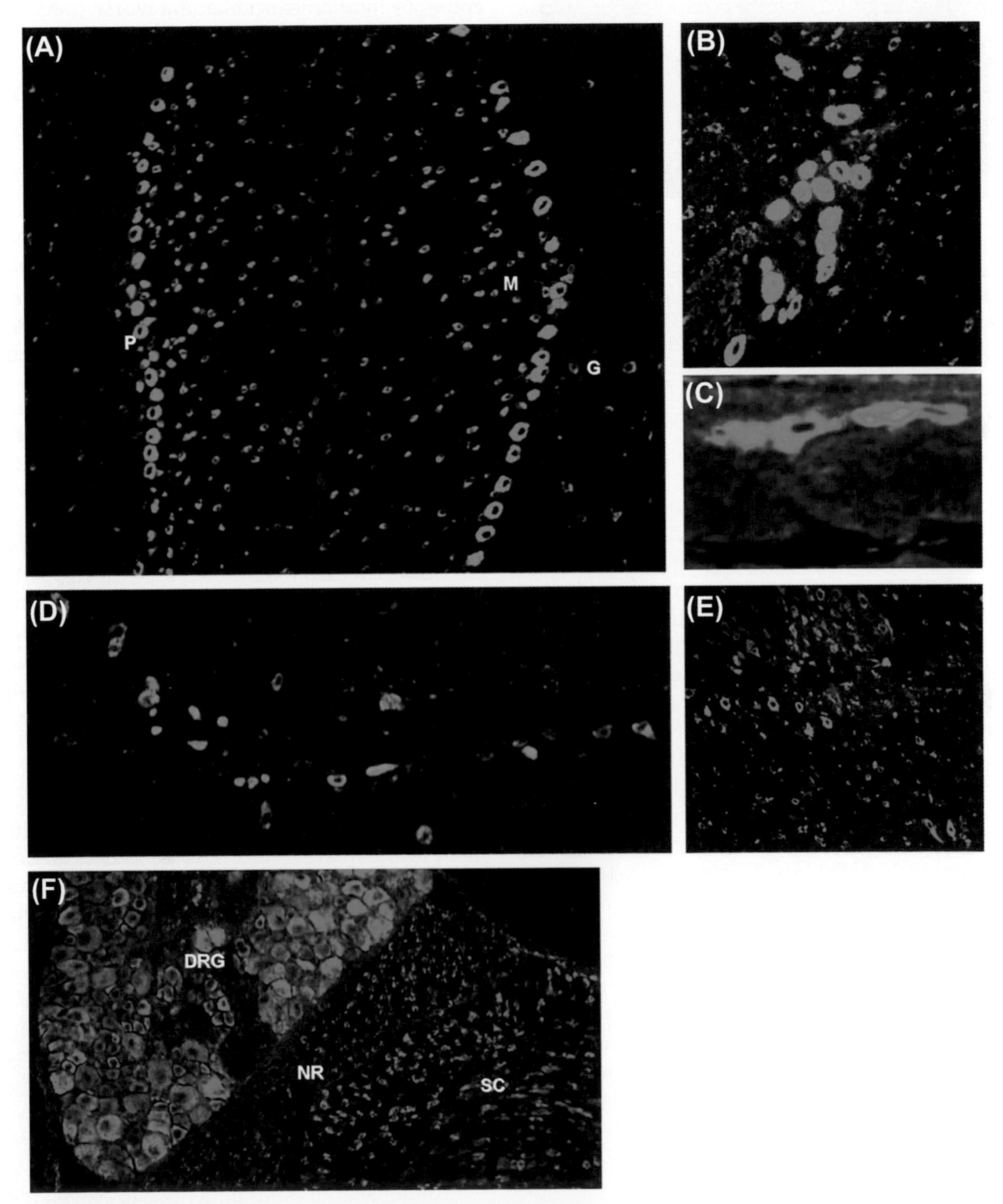

FIGURE 29.4 **Distribution of Purkinje cell cytoplasmic autoantibody type 1 (PCA-1) immunoreactivity in mouse neural tissues.** Neural-specific staining was assured by absorbing serum three times with bovine liver powder (at 1:240 or greater dilution). Patient immunoglobulin G (IgG) binds to discrete cytoplasmic elements: (A) Purkinje (P), molecular (M), and Golgi (G) neurons in cerebellar cortex; (B) cerebellar dentate neurons; (C) myenteric plexus, ganglionic neurons; (D) hippocampal neurons; (E) midbrain neurons; (F) spinal cord neurons (SC), nerve root (NR), dorsal root ganglion (DRG). Scale bars (A–F) = 40 μm. *Source: McKeon* et al. Arch Neurol. *2011;68(10):1282–1289,*[13] *with permission from the American Medical Association.*

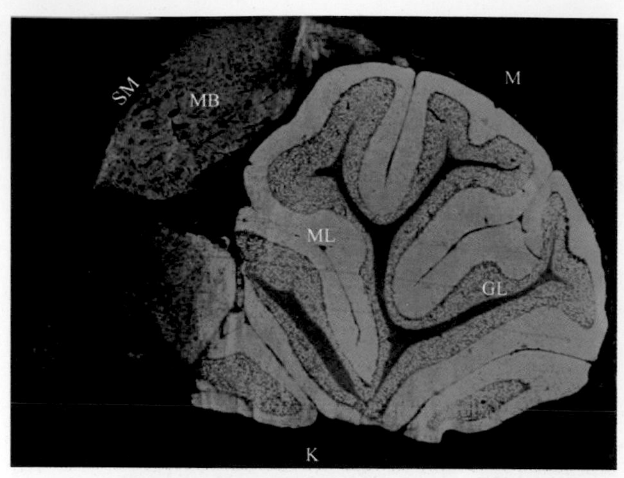

FIGURE 29.5 **Immunofluorescence showing binding of amphiphysin immunoglobulin G (IgG) in a patient's serum (at 1:120 dilution) to a composite substrate of mouse cerebellum, midbrain, gut mucosa and smooth muscle, and kidney.** There is intense synaptic staining of the cerebellar cortical molecular layer (ML), granular cell layer (GL, synaptic patches render a cobblestone pattern) and midbrain (MB). Purkinje neurons are unstained. Stained enteric neural elements are seen throughout the smooth muscle layer (SM). The kidney (K) and mucosa (M) are unstained. (50× magnification.) *Source: Pittock* et al. Ann Neurol. *2005;58:96–107,* [21] *with permission from John Wiley and Sons.*

A 2010 study reported a "stiffness phenotype" in rats injected intrathecally with patient serum-derived amphiphysin-IgG and described electrophysiological dysfunction when the serum was applied to cultured mouse neurons. However, this paper did not provide convincing evidence that amphiphysin antibodies were pathogenic and the authors failed to address the fact that neural autoantibodies (targeting intracellular and plasma membrane) commonly coexist.[22]

AGNA-1 (SOX-1)

Antiglial neuronal nuclear antibody type 1 (AGNA-1, also known as SOX-1) targets the transcription factor sex-determining region Y (SRY)-box 1 of the SOX-1 protein. Immunoreactivity in mouse brain shows staining restricted to the Bergmann glial cells and ependymal cells. AGNA is strongly associated with small-cell lung cancer. It commonly coexists serologically with calcium channel antibodies, and in the context of Lambert–Eaton myasthenic syndrome[23] indicates a paraneoplastic rather than an idiopathic etiology. It is important to note that it is not restricted to Lambert–Eaton syndrome and is commonly associated with a broad spectrum of neurological manifestations.

Ma-1 and Ma-2(Ta)

Ma-1 and Ma-2(Ta) are neuronal nuclear proteins weighing 40 and 42 kDa, respectively, with a speculated involvement in RNA transcription and regulation of apoptosis.[24,25] Autoantibodies binding to these antigens are detected in the setting of limbic and brainstem encephalitides, disorders of the diencephalon and cerebellar ataxia. Exclusive Ma-2 positivity (also known as anti-Ta) is associated with testicular germ-line cancers in males. Brainstem encephalitis is most common in this cohort and typically can have a good outcome. Dual Ma-1/Ma-2 positivity (also known as anti-Ma) is more commonly seen in females and is associated with breast, colon, or ovarian cancers and a worse outcome.

GAD65

Glutamic acid decarboxylase (GAD) (65 kDa) is the synaptic vesicle-associated antigen which catalyzes synthesis of γ-aminobutyric acid (GABA) from L-glutamate. It is associated with neurological and endocrinological autoimmunity. Identifiable on immunofluorescence by prominent synaptic staining of the granular molecular layers and midbrain in mouse cerebellum, it is frequently associated neurologically with stiff man syndrome, cerebellar ataxia, and type 1 (insulin-dependent) diabetes mellitus.[26] This antibody has also been reported with epilepsy, encephalomyelopathies, and extrapyramidal disorders.[27] A syndromic, predominantly brainstem disorder characterized by axial rigidity, postural instability, and ophthalmoplegia has been reported in African-American patients. This antibody, when occurring on its own, is rarely paraneoplastic. However, it is commonly seen in association with other paraneoplastic antibodies, which would increase suspicion for an underlying carcinoma. Because of its intracellular location, treatments targeting this antibody should not improve patients' conditions; however, response to immunotherapy has been reported in 50% of patients with stiff man syndrome, although most patients do require additional therapies such as benzodiazepines.

Peripherin

Antibodies to this type III neuronal intermediate filament protein are important in neuronal development and repair, particularly where the central and peripheral nervous systems link, but also in the peripheral and autonomic nervous systems. Mouse composite brain, gut, and kidney immunoreactivity demonstrates prominent cerebellar white matter, myenteric plexus and vascular sympathetic nerve staining. It is commonly associated with thyroid, pancreatic, or ovarian autoimmunity. Over 50% of patients typically present with limited or global dysautonomia.[28]

Striational Muscle Antibodies

Striational antibodies target heterogeneous components of skeletal muscle; the most frequent antigens are titin and the ryanodine receptor. They are often detected in patients with myasthenia gravis (MG) with or without thymoma and also in patients undergoing paraneoplastic evaluation for various neurological disorders. These antibodies frequently coexist with other paraneoplastic antibodies and when occurring alone are unlikely to predict cancer. Thymic carcinoma, sarcoma, bladder transitional cell carcinoma, renal clear cell carcinoma, chronic lymphocytic leukemia, and adenocarcinoma of various sites (lung, colon, ovarian) were all detected in a case series published in 2013.[29]

Autoantibodies Specific for Synaptic Plasma Membrane Antigens

Autoantibodies that bind to accessible cell surface epitopes have pathogenic potential (Fig. 29.1). Immunoglobulins specific for neural plasma membrane proteins involved in cell-to-cell communication cause synaptic autoimmune disorders and are not found in healthy subjects. Autoantibodies of IgG class are the major effectors of synaptic autoimmunity.

The concept of synaptic autoimmunity was first proposed in 1960 as the basis of the postsynaptic neuromuscular transmission defects in MG and was subsequently extended to plasma membrane receptors in the endocrine system, and the central and autonomic nervous systems. In the past few years the discovery of new antibodies targeting cell surface proteins expressed on neural cells of the CNS has broadened the spectrum of synaptic autoimmunity. Synaptic autoantibodies bind with high affinity to the ectodomain of channels and receptors for neurotransmitters or hormones, and impair their function directly or indirectly.

Autoantibodies targeting cell surface antigens can directly block the function of the plasma membrane protein, activating or preventing the allosteric transition. For example, acetylcholine receptor (AChR)-blocking antibodies may block the binding of acetylcholine to the receptor, leading to poor muscle contraction.

Indirect pathogenic mechanisms mediated by autoantibodies specific for cell surface antigens comprise the activation of the complement cascades resulting in the osmotic lysis of the target cells, activation of Fc receptors, leading to antibody-dependent cell-mediated cytotoxicity (ADCC), and the internalization (antigenic modulation) and degradation of the antigen, thereby altering antigen density on the cell surface (Fig. 29.1).[1,30]

Aquaporin-4-IgG

Aquaporin-4 (AQP4), the predominant water channel of the CNS, is the target antigen of an autoantibody biomarker (also known as NMO-IgG) that distinguishes a spectrum of inflammatory astrocytopathies, exemplified by neuromyelitis optica (NMO), from multiple sclerosis and other demyelinating diseases of the CNS.[31] Two AQP4 isoforms, M1 and M23, exist in the plasma membrane as homotetrameric and heterotetrameric intramembranous particles that have identical extracellular residues. Multiple molecular sequelae of NMO-IgG–AQP4 interaction plausibly account for the diversity of NMO pathology: edema, inflammation, demyelination, and necrosis.[1] The binding of NMO-IgG to astrocytes initiates: (1) partial endocytosis of AQP4 (Fig. 29.6) and the major CNS glutamate transporter [excitatory amino acid transporter-2 (EAAT2)] that is linked non-covalently to AQP4; (2) complement activation (Fig. 29.7A–D); (3) blood–brain barrier disruption; (4) inflammation and Fc-receptor-dependent cell-mediated astrocyte injury; and (5) impairment of water flux.[32–34] Four independent groups have reported animal model data supporting a role for AQP4-specific IgG as an effector of the inflammatory demyelinating pathology of NMO (Table 29.4). AQP4-IgG is rarely detected in paraneoplastic settings. Patients may present in the same manner as non-paraneoplastic cases with severe, recurrent optic neuritis or longitudinally extensive transverse myelitis, or both. The most commonly encountered neoplasms are breast carcinoma and thymoma.[1]

Muscle Acetylcholine Receptor Antibodies

Antibodies to skeletal muscle nicotinic acetylcholine receptors (mAChRs) are associated with the majority of cases of acquired MG.[40] These autoantibodies can target the adult mAChR (pentamers composed of α_1, α_2, β, δ, and ε subunits) or the fetal mAChR (in which a γ subunit replaces the ε subunit). Transplacental transfer of maternal mAChR-IgG can cause developmental musculoskeletal pathology as well as neonatal MG. AChR antibodies are usually of complement-activating classes IgG_1 and IgG_3. The fatigability and electrophysiological defect characteristic of MG reflects a loss of postsynaptic AChRs below the critical safety factor that ensures a muscle action potential. AChR loss is due to complement-mediated membrane lysis, AChR internalization (modulation) due to surface cross-linking by IgG, and direct impairment of AChR function by antibodies. These antibodies impair neuromuscular transmission when injected into animals.[41]

Weakness may be generalized or restricted to extraocular muscles. Approximately 15% of adult cases are paraneoplastic, and in 95% of those cases, the neoplasm identified is a thymic cortical epithelial tumor.[42] About 10% of patients with generalized MG do not have detectable AChR antibodies and do not harbor any thymic neoplasms, but respond favorably to immunotherapy. In these patients thymectomy confers no benefit. One-third of "seronegative" patients have autoantibodies specific for an alternative clinically pertinent postsynaptic antigen, identified by Hoch and colleagues as the muscle-specific kinase (MuSK).[43] The MuSK transmembrane protein is linked functionally to the agrin–lipoprotein receptor-related protein-4 (Lrp4) complex and is essential for clustering of

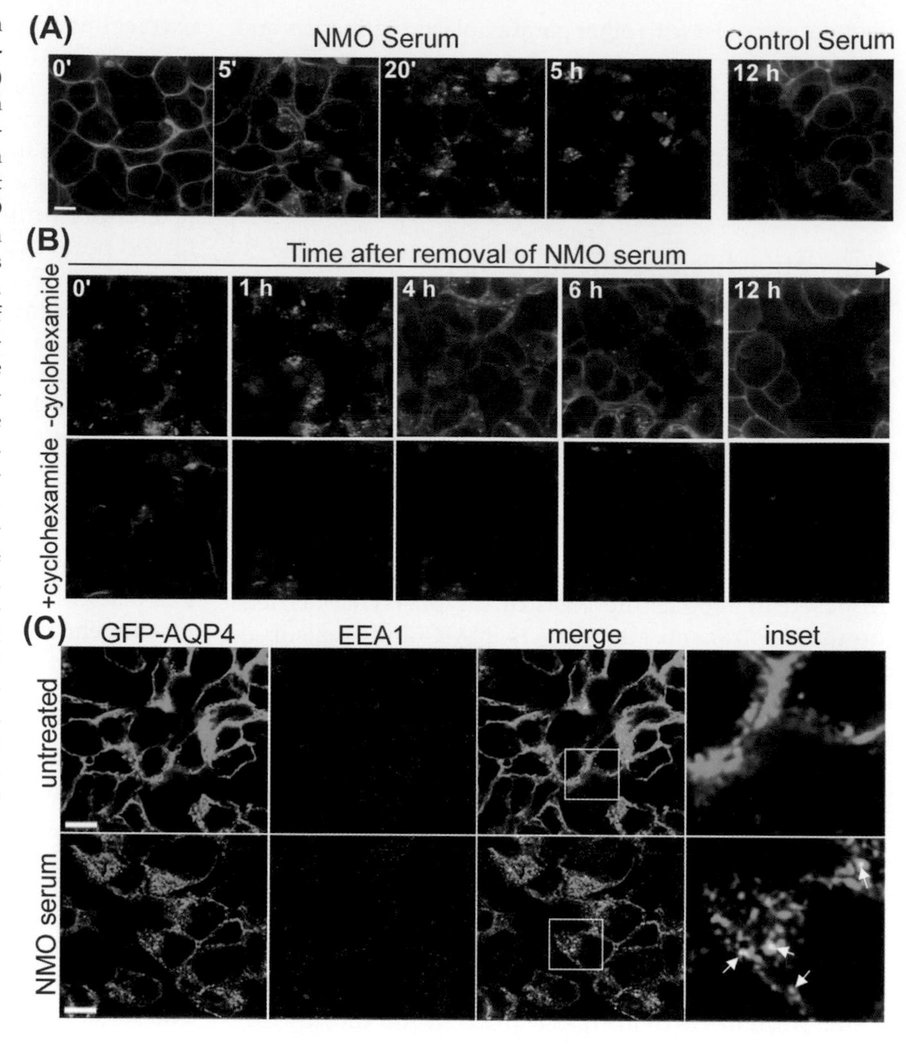

FIGURE 29.6 **Binding of immunoglobulin G (IgG) to surface aquaporin-4 (AQP4) initiates rapid and reversible downregulation.** (A) Time-lapse imaging of green fluorescent protein (GFP)-AQP4 in transfected HEK-293 cells during 12 hour exposure to serum of a patient with neuromyelitis optica (NMO) or a control patient (at 37°C). Surface AQP4 is lost rapidly in NMO patient serum, but not in control serum. With loss of surface AQP4, small vesicular structures appear in the cytoplasm and rapidly coalesce. (B) Restoration of surface AQP4 after removal of NMO patient serum requires new protein synthesis. The abundant small vesicles observed in the absence of cycloheximide reflect newly synthesized AQP4 in transit to the plasma membrane for insertion. In the presence of cycloheximide, residual foci of cytoplasmic fluorescence gradually disappear, reflecting continued degradation. (C) Binding of IgG in NMO patients' serum to surface AQP4 triggers endocytosis and targets AQP4 to the endolysosomal pathway. Early endosomal vesicles (identifiable by red EEA1 immunoreactivity) do not colocalize with AQP4 in untreated cells. Within 30 minutes of adding NMO patients' serum, several EEA1 foci coincide with cytoplasmic vesicles containing AQP4 (yellow in the merged panel; white arrows in the enlarged panel). Scale bars = 10 μm. *Source: Hinson* et al. Neurology *2007;69:2221–2231,*[32] *with permission from Wolters Kluwer Health.*

AChR in the postsynaptic membrane. The distribution of weakness in patients with autoimmune MuSK MG is more focal than in autoimmune AChR MG and prominently involves bulbar, neck, and respiratory muscles.[44]

Ganglionic Acetylcholine Receptor Antibodies

Nicotinic AChRs mediate fast synaptic transmission in sympathetic, parasympathetic, and enteric peripheral autonomic ganglia. These receptors are similar structurally to AChR pentamers found at the neuromuscular junction, but they contain α_3 subunits, associated with β_4 or β_2 subunits and sometimes an α_5 subunit. Up to 30% of patients with ganglionic AChR antibodies have cancer, most frequently adenocarcinoma of lung, gastrointestinal tract, and prostate, although small-cell lung cancer and thymoma have also been described.[45] Clinical presentations are diverse, including encephalopathy, dysautonomia, and peripheral neuropathy. Pandysautonomia is more frequently associated with extremely high titers of antibodies.[46] Ganglionic AChR antibodies may coexist with other paraneoplastic antibodies and have also been reported in association with muscle AChR autoantibodies in MG.[19,46] The pathogenicity of ganglionic AChR-IgG has been confirmed in animal models. For example, immunization with a recombinant extracellular segment of the α_3 AChR subunit induces clinical and electrophysiological signs of autoimmune ganglionopathy in rabbits with experimental autoimmune autonomic neuropathy (EAAN) (Fig. 29.8A–H),[47] and IgG isolated from serum of seropositive patients can transfer these signs to mice.[48]

P/Q-Type Voltage-Gated Calcium Channel Antibodies

Autoantibodies targeting the P/Q-type voltage-gated calcium channels (VGCCs) are detected in over 85% of cases of Lambert–Eaton myasthenic syndrome (LEMS).[49] Neuromuscular weakness is frequently accompanied by limited dysautonomia (sicca syndrome or erectile dysfunction). If more widespread dysautonomia is present (cardiovascular or bladder involvement), or calcium channel antibodies and/or AGNA antibodies coexist then suspicion for paraneoplastic disorder should be heightened. Small-cell lung

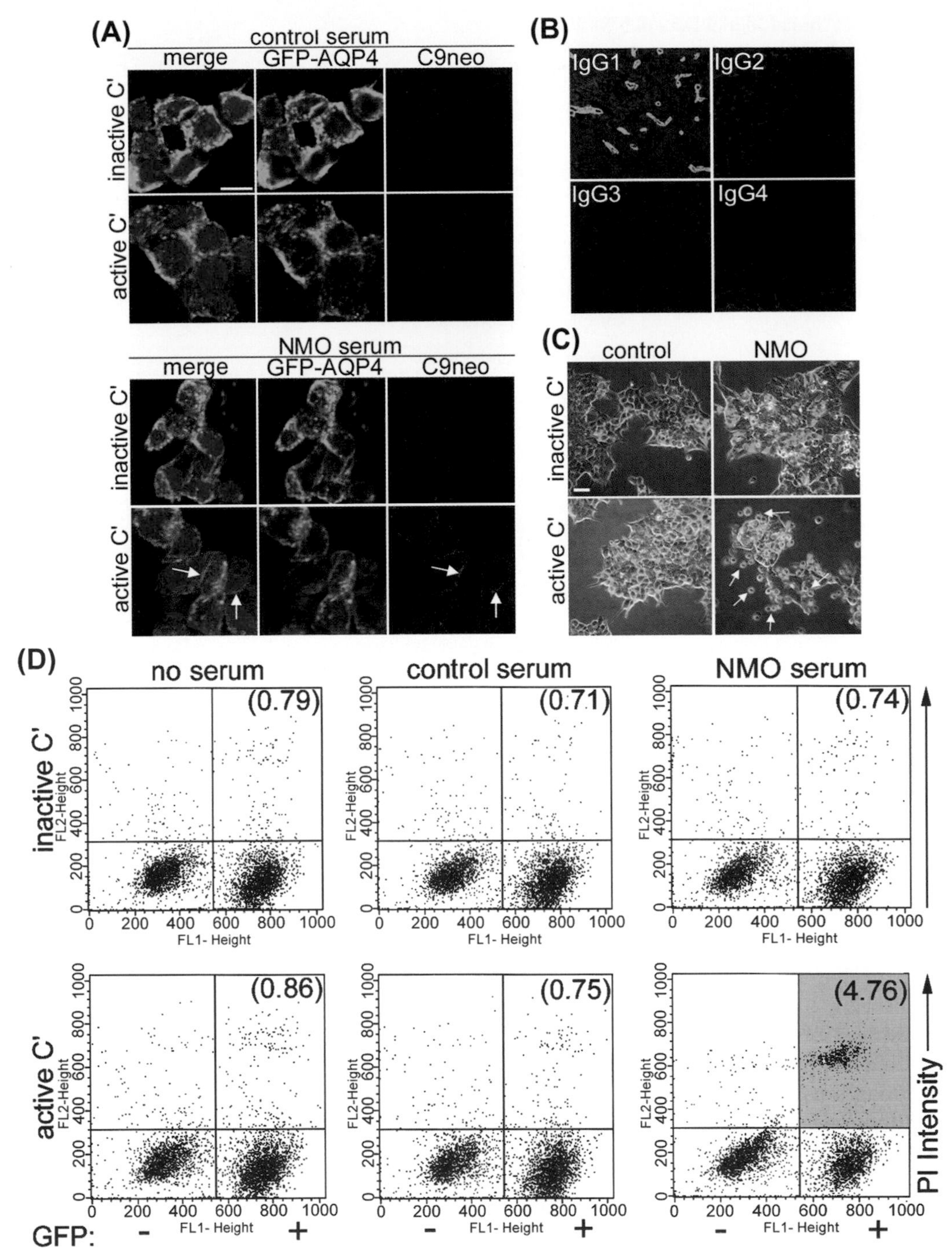

FIGURE 29.7 **The membranolytic attack complex of complement is assembled after immunoglobulin G (IgG) binds to surface aquaporin-4 (AQP4).** (A) Formalin-fixed aquaporin-4 (AQP4)-transfected HEK-293 cells were exposed to control human serum or neuromyelitis optica (NMO) patient serum in the presence of active or inactive complement (C′). Plasma membrane incorporation of the terminal C9neo component (arrows) requires NMO patient serum and active complement. (B) IgG subtypes detected using fluorescein-conjugated subtype-specific antihuman IgG probes (γ_1, γ_2, γ_3, and γ_4 heavy chain-specific, respectively) on mouse brain sections. Only IgG_1 was detected (five out of five individual NMO patients' sera and a serum pool representing 10 additional seropositive patients). Production of this major complement-activating subclass of human IgG reflects a proinflammatory helper T-cell immune response. Its vasculocentric binding pattern coincides with AQP4-rich sites. Specificities and optimal dilutions of the IgG subclass-specific reagents were confirmed using smears of human myeloma IgGs (kindly provided by Dr Robert Kyle, Mayo Clinic). One patient's serum contained IgG_3 antinuclear antibody (not shown). Scale bars = 10 μm. (C) Living cells exposed to control or NMO patient serum. In the presence of AQP4-specific IgG and active complement (1.5 hours at 37°C), cells expressing AQP4 (green) are selectively detached from the monolayer, indicated by arrows (transmitted bright-field image). (D) Comparison of fold increase in membrane permeability of HEK-293 cells expressing AQP4 (in parentheses) is in reference to the population not expressing AQP4. Flow cytometric analysis: *x*-axis, green fluorescent protein (GFP); *y*-axis, propidium iodide (PI) intensity. Membrane permeability is increased 4.76-fold (i.e. 476%) with an NMO patient's serum and active complement (shaded quadrant) and less than 1.0-fold in a control patient's serum or no human serum. *Source: Hinson* et al. Neurology. *2007;69:2221–2231,*[32] *with permission from Wolters Kluwer Health.*

TABLE 29.4 Summary of Animal Model Data in NMO

Study Methods	Post-IgG Outcome	Control Findings	Reference
NMO serum injected 4 times intraperitoneally into rats at EAE onset (MBP immunization)	*Day 4:*	No effect with IgG_{MS} or IgG_{normal}	Kinoshita *et al.*[35]
	EAE augmented	Minor effect with $IgG_{NMO\text{-}AQP4\text{-}absorbed}$	
	AQP4 and GFAP immunoreactivity loss, prominent around vessels		
	Deposition of human-IgG and rat complement products		
	Astrocyte damage/loss		
	Massive inflammation (macrophages, neutrophils, eosinophils) in gray matter		
AQP4-IgG$^+$ serum injected once intraperitoneally into rats at EAE onset (intravenous MBP T cells)	*Day 1:*	No effect with:	Bradl *et al.*[36]
	EAE augmented	– $IgG_{seronegative\text{-}NMO}$	
	Loss of AQP4, GFAP, and S100β around vessels	– IgG_{MS}	
	Deposition of human-IgG and rat complement	– $IgG_{other\ neurological\ disorders}$	
	Mature astrocyte damage/loss	– IgG_{NMO} plus non-pathogenic T cells	
	Lesional granulocytes and macrophages (not eosinophils) greatly increased		
$rAb_{AQP4\text{-}specific}$, injected once intravenously into rats at EAE onset (MBP peptide immunization)	*Day 1.25:*	No effect with:	Bennett *et al.*[37]
	No EAE augmentation recorded	– $rAb_{measles\text{-}specific}$	
	Perivascular deposition of human IgG and rat complement	– $rAb_{human\text{-}AQP4\text{-}specific}$	
	Astrocyte depletion, myelin vacuolization		
AQP4-IgG$^+$ serum plus human complement injected 1–3 times into the right cerebral hemisphere in mice	*Hour 12:*	No effect with:	Saadoun *et al.*[38]
	Ipsilateral AQP4 loss	– $IgG_{non\text{-}NMO\ control}$	
	Perivascular deposition of complement products	– IgG_{NMO} into AQP4-null mice	
	Myelin breakdown	Minor inflammation with IgG_{NMO} plus complement inhibitor	
	Glial swelling		
	Axonal injury		
	Polymorphonuclear cells in vessels		
	Day 7:		
	Contralateral visual field neglect		
	Massive macrophage infiltration, extensive demyelination, reactive astrocyte loss, neuronal death, and perilesional gliosis		

IgG: immunoglobulin G; NMO: neuromyelitis optica; MBP: myelin basic protein; EAE: experimental autoimmune encephalomyelitis; AQP4: aquaporin-4; GFAP: glial fibrillary acidic protein; S100β: astrocytic protein; $rAb_{AQP4\text{-}specific}$: recombinant monoclonal IgG derived from cerebrospinal fluid plasma cell in a patient with NMO.
Source: From Journal Watch Neurology *2010;12:52–53,*[39] *with permission from the Massachusetts Medical Society.*

cancer is identified in up to 50% of cases of Lambert–Eaton syndrome. Twelve percent of patients will also have either muscle AChR or striational coexisting antibodies.[1] Paraneoplastic encephalomyelopathy and cerebellar ataxia are also described with coexisting N- and PQ-type calcium channel antibodies in association with lung, breast, or ovarian carcinoma. Non-paraneoplastic autoimmune dysautonomia and cerebellar ataxia are also described. Polyclonal IgG isolated from LEMS patients can transfer clinical, electrophysiological, and ultrastructural nerve terminal lesions characteristic of LEMS to mice, in the absence of complement. However, to date no animal model of LEMS induced by active immunization has been confirmed, and no monoclonal or affinity-purified antibody of P/Q-type VGCC specificity has been shown to cause the neuromuscular or autonomic transmission defect of LEMS in experimental animals.[1]

N-Type Voltage-Gated Calcium Channel Antibodies

These antibodies frequently co-occur with PQ-type antibodies and Lambert–Eaton syndrome, although in isolation do not appear to predict the syndrome or cancer. They also commonly coexist with other paraneoplastic antibodies and associations with idiopathic autoimmune dysautonomia have been reported. The presence of N-type antibodies in association with other paraneoplastic antibodies should increase suspicion for underlying cancer.[1]

Voltage-Gated Potassium Channel Complex Antibodies

The voltage-gated potassium channels (VGKCs) modulate neuronal excitability, axonal conduction, and neurotransmitter release in the central, peripheral, and autonomic nervous systems. To date, only a small number of VGKCs are recognized as pertinent to neurological autoimmunity. Those are the Shaker type Kv1 channels that are sensitive to the snake venom α-dendrotoxin (Kv 1.1, Kv 1.2, and Kv 1.6). VGKCs form macromolecular complexes interacting with cell adhesion molecules and scaffolding proteins including contactin-2, contactin-associated protein-2 (CASPR2), membrane-associated guanylate kinases, disintegrin, metalloproteinase-22 (ADAM22), and a soluble binding partner of ADAM22, leucine-rich, glioma-inactivated-1 (LGI1) protein.[1] Serum autoantibodies that bind to solubilized macromolecular complexes containing VGKCs ligated with $[^{125}I]\alpha$-dendrotoxin are detected most sensitively by radioimmunoprecipitation.[1]

Traditional neurological accompaniments include limbic encephalitis, Isaac syndrome (peripheral nerve hyperexcitability syndrome), and Morvan syndrome (Isaac syndrome plus psychiatric and sleep disturbances with hallucinations). More recently, a wider spectrum of disorders has been appreciated, which includes limbic encephalitis, reversible dementia-like syndrome, psychiatric disturbances, autoimmune epilepsy, sleep disorders, autonomic and peripheral neuropathy, and pain syndromes.[50]

LGI1 and CASPR2 are two antigenic targets in the potassium channel complex that have been well characterized.[51,52] However, up to 54% of patients positive for potassium channel complex antibodies remain negative for these complex subunits, suggesting that there is at least one further target yet to be identified. It has been reported that LGI1 antibody-positive patients tend to have a CNS syndrome such as limbic encephalitis, whereas CASPR2 antibody-positive patients present with predominantly peripheral nervous system disorders. This is controversial, and Klein and colleagues have reported that both antibodies can affect all levels of the nervous system.[53] Cancer is present in the minority of VGKC complex patients (16%) and does not particularly associate with CASPR2 of LGI1 antibodies. Lung cancer, thymoma, and hematological malignancies are most frequently reported.[1]

Ionotropic Glutamate Receptor Antibodies

N-Methyl-D-aspartate receptors (NMDARs) are glutamate-gated cation channels involved in hippocampal synaptic transmission, plasticity, and long-term potentiation, a mechanism that underlies learning and memory development. The receptors are heterotetrameric complexes formed by subunits derived from three related families: NR1, NR2, and NR3. The NR1 subunit is the predominant target antigen of NMDAR-specific antibodies.[54] Binding of the antibody causes internalization of the NMDAR, decreasing receptor-mediated excitatory postsynaptic currents.[1] Typical features of this neuropsychiatric syndrome are a viral-like prodrome followed by psychiatric disturbances with associated orofacial and limb dyskinesias and seizures. Central hypoventilation and autonomic instability follow, requiring prolonged intensive care in most cases. This syndrome is more common in females and one-third of women over 18 years with this disorder have an undiagnosed ovarian teratoma. Teratomas are more common among African-Americans with this disorder.[55] It is important to remember that mature teratoma will not be demonstrated well on positron emission tomography–computed tomography (PET-CT), therefore transvaginal ultrasound is the preferred imaging modality. Cases where no underlying tumor is found have a worse prognosis and are more likely to relapse. Antibody-depleting therapies and tumor removal optimize recovery, which can be complete in approximately 75% of early-treated patients. The outcome where the diagnosis is missed can be devastating, with severe residual cognitive deficits and refractory epilepsy.[55] Although NMDAR encephalitis

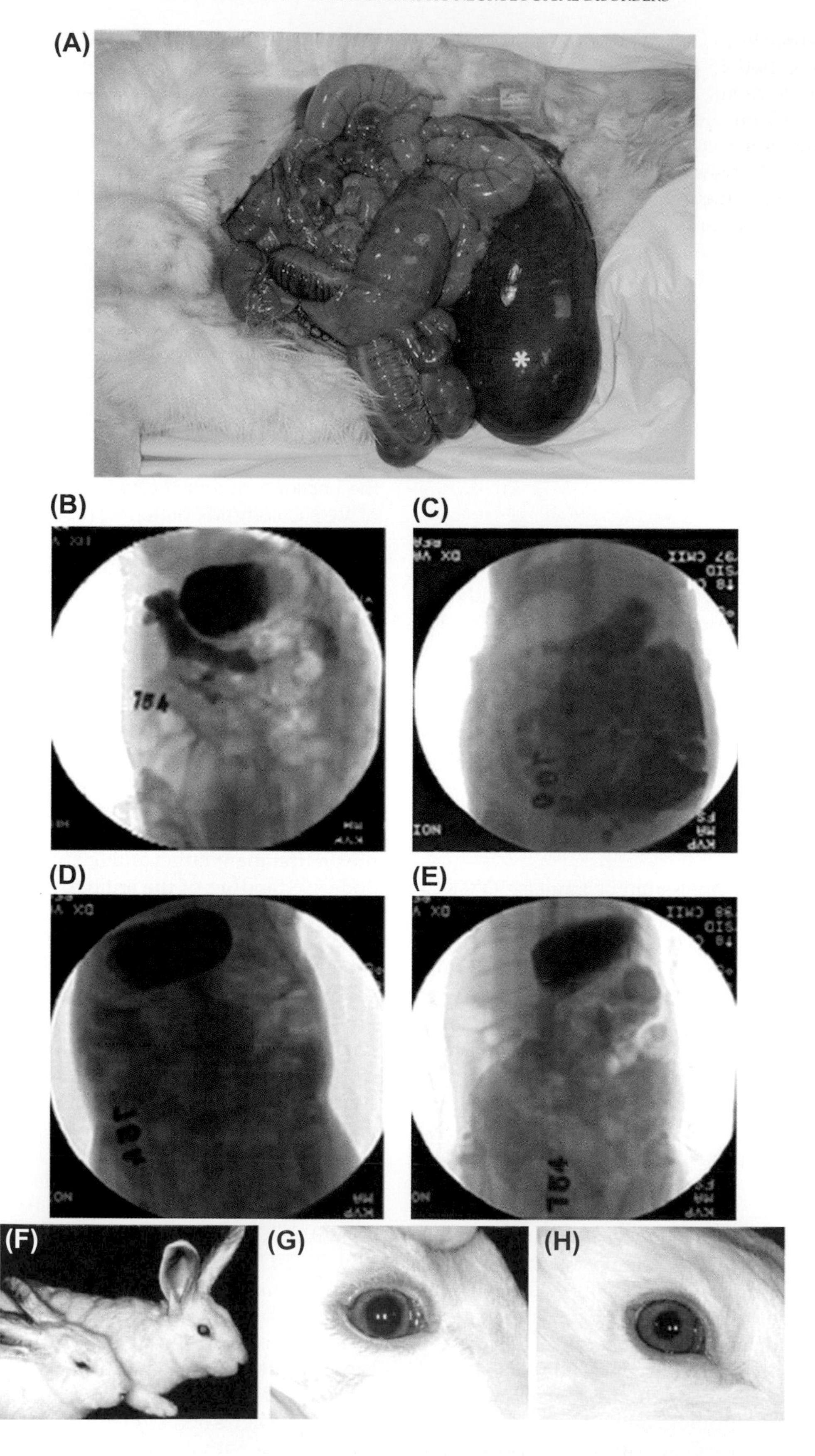
(A)
(B)
(C)
(D)
(E)
(F)
(G)
(H)

is classically associated with the IgG class of antibody, a 2012 report suggests that patients with IgA targeting the NMDAR also have a neurological syndrome distinct from NMDAR encephalitis. These patients all had a prolonged dementia-like syndrome and some cases responded at least transiently to immunotherapy. This raises the question of whether the spectrum of NMDAR autoimmunity will widen with time.[56]

Metabotropic Glutamate Receptors

Metabotropic glutamate receptors are G-protein coupled receptors that modulate neuronal activity by activating intracellular signaling pathways. Eight different receptor types are divided into three groups on the basis of structure and function: I ($mGluR_1$ and $mGluR_5$), II ($mGluR_2$ and $mGluR_3$), and III ($mGluR_4$, $mGluR_6$, and $mGluR_8$). Only $mGluR_1$ and $mGluR_5$ have been documented to be pertinent autoantigens. $mGluR_1$ is particularly prevalent in cerebellar Purkinje cells and $mGluR_5$ in the hippocampus. $mGluR_1$ antibodies have been described in a case series of patients with cerebellar ataxia, two of whom had Hodgkin lymphoma. Autoantibodies targeting $mGluR_5$ have been associated with two cases of limbic encephalitis and Hodgkin lymphoma (Ophelia syndrome).[57]

AMPA Receptor Antibodies

The 2-amino-3-(5-methyl-3-oxo-1,2-oxazol-4-yl)propanic acid (AMPA) receptor mediates most fast excitatory neurotransmission in the brain and is another antigen associated with limbic encephalitis. A series of 10 patients with antibodies to one or both of the $GluR_1$ and $GluR_2$ receptors and limbic encephalitis was published in 2009.[58] Similarly to NMDAR antibodies, AMPA receptors are internalized upon AMPA receptor antibody binding. Cancers (small-cell and non-small-cell lung cancer, thymoma, and breast cancer) were identified in seven out of the 10 patients.

$GABA_B$ Receptor Antibodies

$GABA_B$ receptors are G-protein coupled receptors that, functionally linked to potassium channels, elicit both presynaptic and slow postsynaptic inhibition. They are heterodimers of a $GABA_{B1}$ subunit (ligand binding) and a $GABA_{B2}$ subunit (responsible for signaling and membrane targeting). $GABA_B$-specific antibodies are detected in the setting of limbic encephalitis associated most often with small-cell lung carcinoma.[59] $GABA_B$-IgGs block receptor function, but do not cause receptor internalization. Breast cancer and small-cell lung cancer have both been described. Improvements reported with plasmapheresis support the likely pathogenicity of these antibodies, which frequently coexist with GAD65 and VGCC antibodies.[60]

Glycine Receptor

Antibodies to the glycine receptor α_1 subunit have been reported in patients with encephalopathy, myelopathy, myoclonus, and excessive startle. In the index case the term PERM was coined (progressive encephalopathy, rigidity, and myoclonus). A 2013 study reported glycine antibody positivity among patients with stiff man syndrome seronegative for GAD65 and amphiphysin antibodies.[61] Thymoma has been reported in one case.[62]

AUTOIMMUNE AND PARANEOPLASTIC NEUROLOGICAL DISEASES

Paraneoplastic and autoimmune disorders of the nervous system may be organized anatomically in a rostrocaudal order. Symptoms present in an acute or subacute manner and rapidly progress. Historically, these disorders were classified according to the syndromes they produced rather than by the antibodies associated with them. In practice it is more useful to recognize that both unifocal (pure limbic encephalitis in a patient with VGKC antibody) and multifocal (retinopathy and myelopathy in a patient with CRMP-5-IgG) syndromes may occur with most antibodies. Important clues to an underlying malignancy include recent weight loss, a personal or family history of cancer, a history of smoking and social or occupational exposure to carcinogens such as asbestos, and a personal or a family history of autoimmunity.

FIGURE 29.8 **Dysautonomia of gut and bladder in seropositive rabbits.** (A) Dilated loops of bowel (intestinal pseudo-obstruction) and enlarged bladder (megacystis, indicated by an asterisk) in a rabbit with severe experimental autoimmune autonomic neuropathy. (B–E) Radiological images of barium in transit through the gut of two rabbits (orientation, head at top). Images in (B), (D), and (E) are from a rabbit with experimental autoimmune autonomic neuropathy (EAAN) on days 74–77 after α_3-glutathione *S*-transferase (GST) immunization. The image in (C) is from a control rabbit on day 74 after adjuvant only. (B) Barium (black) initially leaving enlarged stomach enters abnormally dilated proximal duodenum approximately 20 minutes after 40 ml gavage. Distended gas-filled loops of surrounding small intestine appear white. (C) At 6 hours, barium is distributed throughout the bowel of the control rabbit. (D, E) In the EAAN rabbit, most barium remains in the stomach at 6 hours (D) and at 72 hours (E). All controls evacuated barium by day 3. Ocular signs of dysautonomia on day 28 after primary immunization. (F) Ptosis in an α_3-GST responder rabbit (left), a sign of sympathetic denervation, was not observed in the adjuvant-inoculated control rabbit (right) or in seronegative rabbits. (G) In the same EAAN rabbit, lack of pupillary response to light is a sign of parasympathetic denervation. (H) Light stimulus induced prompt pupillary constriction in the control rabbit shown in (A). *Source: Lennon* et al. J Clin Invest. *2003;111:907–913,*[47] *with permission from the American Society for Clinical Investigation.*

TABLE 29.5 Numbers (% Frequency) and Types of Cancer Detected in Seropositive Patients with Adequate Follow-up for the Period 2000–2003

			Carcinoma Detected							
			Lung							
IgG	Total Patients	Adequate clinical Information[a]	SCLC	NSCLC	Breast	Ovary	Fallopian Tube/Uterus	Thyoma	Other	Patients with Histologically Proven Cancer[b]
ANNA-1	217	142	93 (66)	6 (4)	0 (0)	0	0	3 (2)	12 (8)	114 (80)
CRMP-5	208	113	53 (47)	7 (6)	2 (2)	0	0	10 (9)	12 (11)	84 (74)
PCA-1	101	68	0 (0)	0 (0)	9 (13)	43 (63)	9 (13)	0	1 (1)	62 (91)
PCA-2	43	19	10 (53)	5 (26)	0 (0)	0	0	0	2 (10)	17 (89)
Amphiphysin	26	21	10 (48)	0 (0)	8 (38)	0	0	0	0	18 (86)
ANNA-2	17	14	3 (21)	2 (14)	3 (21)	0	0	0	0	8 (57)
ANNA-3	10	8	2 (25)	2 (25)	0 (0)	0	0	0	1 (12)	5 (62)

SCLC: small-cell lung carcinoma; NSCLC: non-SCLC.
[a]*Tumor identified or results of relevant imaging studies available.*
[b]*Apart from patients with histologically proven cancer, additional seropositive patients had a chest imaging abnormality identified that warranted further investigation: 27 (19%) for antineuronal nuclear antibody-1 (ANNA-1); 18 (6%) for collapsin response mediator protein-5 (CRMP-5); 0% for Purkinje cytoplasmic antibody-1 (PCA-1); two (10%) for PCA-2; one (5%) for amphiphysin-immunoglobulin G (IgG); six (43%) for ANNA-2; and five (25%) for ANNA-3. Continued cancer surveillance is recommended for patients without proven neoplasm. Positron emission tomography scanning is proving most sensitive for cases with otherwise normal imaging studies, or a serial abnormality that is deemed "stable".*
Source: Pittock et al. Ann Neurol. *2004;56:715–719,*[19] *with permission from John Wiley and Sons.*

In the diagnostic evaluation of a patient with a suspected autoimmune or paraneoplastic neurological disorder it is important to define the levels of the neuraxis involved. Ancillary diagnostic tools that measure the extent of nervous system involvement include nerve conduction studies for large fiber involvement, autonomic reflex testing and thermoregulatory sweat testing for dysautonomia, neuropsychometric studies for cognitive disorders, movement laboratory studies for myoclonus and hyperexcitability, and gastrointestinal transit studies for dysmotility (limited forms of dysautonomia). These objective measures provide a baseline measure of severity and can be compared with postimmunotherapy measures in assessing whether immunotherapy provides any objective benefit.

Examination of the cerebrospinal fluid (CSF) may provide supportive evidence of nervous system inflammation (e.g. lymphocytic pleocytosis, elevated protein, oligoclonal bands). Cytological evaluation, if positive, may indicate lymphoma or other malignancy.

Comprehensive testing for neural-specific autoantibodies is preferable to testing for individual antibodies, since the associated neurological manifestations are protean and often go beyond the scope of the reported literature. Furthermore, the detection of coexisting autoantibodies may further inform the prediction of cancer type (Table 29.5, Fig. 29.9). For example, amphiphysin antibodies in women may be associated with breast cancer or small-cell lung cancer, but if coexisting ANNA-1 is detected this narrows the search to small-cell cancer.[19] In 70% of patients with paraneoplastic neurological disorders, the neurological symptoms precede the diagnosis of tumor. Of those who undergo assessment for cancer after neural antibody detection, one will be identified in up to 80%. If the search reveals a neoplasm atypical for the paraneoplastic antibody found, then consideration should be given to the possibility of a second more typical occult cancer. CT is recommended as a screening tool in the first instance. If CT is negative PET-CT is the next most appropriate test. Often the only abnormality is in the mediastinal region, so particular attention should be paid to this area. All patients should be up to date with prostate, breast, cervical, and colon screening procedures appropriate to age. Transvaginal, testicular, and transrectal ultrasound are the preferred screening procedures for germ cell tumors and prostate cancer. It is important to note that PET-CT is not an appropriate test in the evaluation of a female with anti-NMDAR encephalitis as 99% of teratomas are mature and cystic and do not show increased fluorodeoxyglucose (FDG) uptake.[63] In patients with a known paraneoplastic antibody, if primary screening is negative then repeat screening after 3–6 months and every 6 months up until 4 years has been proposed. Where initial studies are negative, at least one additional PET-CT body scan or CT scan of the thorax, abdomen, and pelvis at 3–6 months is probably reasonable. In Lambert–Eaton syndrome screening for 2 years is felt to be adequate.

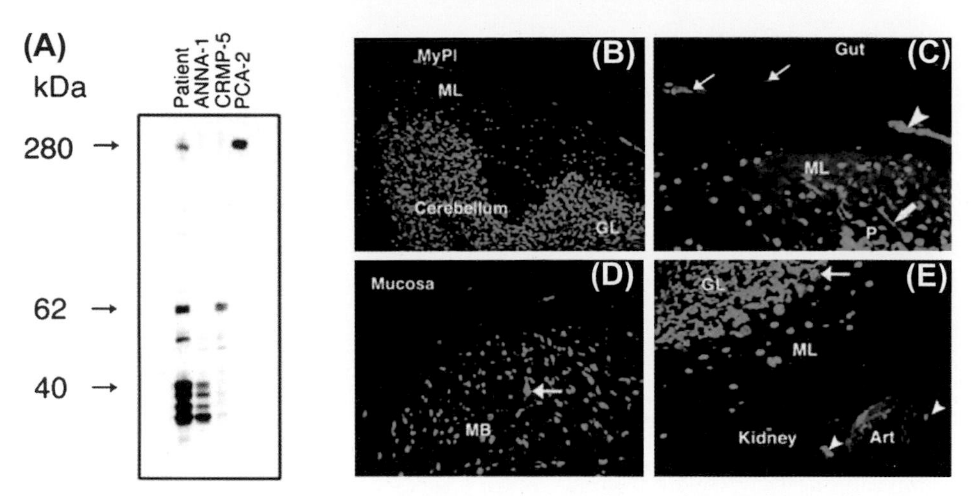

FIGURE 29.9 **Paraneoplastic autoantibodies commonly coexist.** Western blot of serum from a 78-year-old man with a subacute multifocal neurological presentation. Immunoreactive bands characteristic of three paraneoplastic autoantibodies were identified: antineuronal nuclear autoantibody type 1 (ANNA-1), Purkinje cell cytoplasmic autoantibody type 1 (PCA-2), and collapsin response mediator protein-5 (CRMP-5)-immunoglobulin G (IgG). Small-cell lung carcinoma was predicted and subsequently found. Control lanes show reference sera for each autoantibody. Native antigenic proteins in an aqueous extract of rat cerebellar cortex were reduced in 2-mercaptoethanol and denatured in 2% sodium dodecyl sulfate and separated by electrophoresis in 10% polyacrylamide gel. (B–E) Immunofluorescence showing binding of IgG in this patient's serum (at 1:120 serum dilution) to a composite substrate of mouse cerebellum, midbrain, gut mucosa and smooth muscle, and kidney; (B) cerebellum, gut mucosa and smooth muscle (100× magnification); (C) cerebellum, gut mucosa and smooth muscle (200× magnification); (D) midbrain and gut smooth muscle (200× magnification); (E) cerebellum and kidney (200× magnification). ANNA-1: immunoreactivity in neuronal nuclei (sparing nucleoli) and perikarya in the granular layer (GL; B, E), molecular layer (ML; B, E), and Purkinje cells (E, arrow) of cerebellum, midbrain (MB; D, arrow), and ganglionic neurons of myenteric plexus (B, MyPl; C, arrowhead); CRMP-5-IgG: diffuse immunoreactivity in synapse-rich regions of cerebellar ML (B, C) and enteric ganglionic cytoplasm and nerve fibers (C, thin arrows); PCA-2: immunoreactivity in Purkinje cell cytoplasm (P), extending into dendrites (C, thick arrow) in the ML, and nerve trunks innervating arterioles (Art) in the kidney (E, arrowheads). *Source: Pittock* et al. Ann Neurol. *2004;56:715–719,*[19] *with permission from John Wiley and Sons.*

LEVELS OF THE NEURAXIS AFFECTED BY PARANEOPLASTIC AND AUTOIMMUNE SYNDROMES

This section describes the autoimmune and neurological disorders according to the level of the neuraxis involved (Table 29.6): cerebral cortex and limbic system, basal ganglia and extrapyramidal system (chorea, opsoclonus–myoclonus), cerebellum, diencephalon, brainstem, spinal cord, peripheral nerves, autonomic nervous system, neuromuscular junction, or muscle.

Cerebral Cortex and Limbic System

In clinicopathological terms, the autoimmune disorders involving the cerebral cortex and the limbic system can be divided into three main categories: (1) limbic encephalitis, characterized by the triad of subacute memory loss, psychiatric disturbance and cognitive disorders, and seizures; (2) autoimmune epilepsy, a type of epilepsy associated with neural-specific autoantibodies that responds to immunotherapy; and (3) autoimmune dementia, characterized by immunotherapy-responsive cognitive impairment.

Limbic Encephalitis

Limbus is a Latin word meaning border. The limbic system is located at the border between the neocortex and the subcortical structures (diencephalon) and includes the hippocampus, amygdala, septal nuclei, cingulate cortex, entorhinal cortex, perirhinal cortex, and parahippocampal cortex. These last three cortical areas comprise different portions of the medial region of the temporal lobe. The limbic system has a critical role in emotional behavior and in the development of long-term memory. It also influences the endocrine and autonomic nervous systems.

Limbic encephalitis was originally described, in 1968, as a rare cliniconeuropathological entity, characterized by memory loss, seizures, and psychiatric disturbances, caused by an inflammatory process involving the mesial region of the temporal lobes and associated with an underlying cancer. Since then, several onconeural antibodies binding to intracellular proteins have been documented in the setting of limbic encephalitis. Furthermore, autoantibodies targeting neuronal cell surface antigens are being recognized. Some of these antibodies are associated with specific types of neoplasm.

Paraneoplastic limbic encephalitis is associated with small-cell lung cancer, germ cell tumors of the testes, breast cancer, Hodgkin lymphoma, and rarely thymoma or immature teratoma of the ovary. The most commonly reported associated antibodies include ANNA-1, Ma2 (with or without Ma1), CRMP-5-IgG, amphiphysin, VGKC complex, and NMDAR antibodies. GABAB-B, AMPA, and $mGluR_5$ antibodies are less common.[1,60]

TABLE 29.6 Neurological Manifestations According to the Nervous System Level Involved and Serological Associations

Level	Disorder	Neural Antibody Associations
Cerebral cortex	Limbic encephalitis	VGKC complex antibodies > CRMP-5-IgG > ANNA-1, NMDAR > VGCC-IgG > GABA$_B$R-IgG, Ma and Ta, AGNA, AMPAR-IgG > amphiphysin-IgG > muscle AChR, ganglionic AChR antibodies
	Autoimmune epilepsy	VGKC complex antibodies, GAD65-IgG, CRMP-5-IgG, ANNA-1 > NMDAR-IgG > GABA$_B$-IgG
	Autoimmune dementia	VGKC complex antibodies, ANNA-1, NMDAR-IgA, AMPAR, GABA$_B$R antibodies
Diencephalon	Hypothalamic dysfunction	Ma and Ta, LGI-1-IgG, AQP4-IgG > ANNA-1
Basal ganglia	Chorea	CRMP-5-IgG > GAD65-IgG, ANNA-1, ANNA-2, VGKC complex antibodies > amphiphysin-IgG
Cerebellum	Cerebellar ataxia	PCA-1, PCA-Tr, ANNA-1 > CRMP-5-IgG, VGCC (PQ type or N type), PCA-2 > mGluR-1
Brainstem	Brainstem encephalitis	VGCC (PQ and N type), CRMP-5-IgG, PCA-2, ANNA-1, muscle and ganglionic AChR, ANNA-2, amphiphysin-IgG, Ma and Ta
	Stiff man syndrome	Amphiphysin-IgG (39% of women; 12% of men), GAD65-IgG (rarely associated with cancer)
Cranial nerves	Olfactory, ocular, bulbar, and motor neuropathies	CRMP-5-IgG > ANNA-1 = VGCC (N type > PQ type) > muscle and ganglionic AChR and striational
Spinal cord	Myelopathy and myoclonus	VGCC = CRMP-5-IgG > amphiphysin > ganglionic > VGKC > ANNA-2 = ANNA-1, Ma and Ta
Peripheral nerves and ganglia	Sensory neuronopathy and sensorimotor neuropathies	ANNA-1 > CRMP-5-IgG > amphiphysin > ganglionic AChR, amphiphysin, muscle AChR, striational
Neuromuscular junction	Lambert–Eaton syndrome	VGCC (PQ type > N type) > muscle AChR or striational = ganglionic AChR
	Myasthenia gravis	Muscle AChR > > MuSK-IgG
Muscle	Polymyositis/dermatomyositis	Anti-tRNA synthetase (Jo-1-IgG and others)
	Necrotizing myopathy	SRP-54 and 72-IgGs, HMGCr-IgG
Autonomic and enteric nervous system	Dysautonomias, gastrointestinal dysmotilities	Ganglionic AChR, muscle AChR, VGCC (N type), VGKC, striational > ANNA-1 > CRMP-5-IgG > VGCC (N type > PQ type), peripherin-IgG

VGKC: voltage-gated potassium channel; CRMP-5: collapsin response mediator protein-5; IgG: immunoglobulin G; ANNA: antineuronal nuclear antibody; NMDA: *N*-methyl-D-aspartate; VGCC: voltage-gated calcium channel; GABA: γ-aminobutyric acid; AGNA: antiglial neuronal nuclear antibody type 1; AMPA: 2-amino-3-hydroxy-5-methyl-4-isoxazole-propionic acid; AChR: acetylcholine receptor; GAD65: 65 kDa isoform of glutamic acid decarboxylase; LGI: leucine-rich, glioma-inactivated protein; AQP4: aquaporin-4; PCA: Purkinje cytoplasmic antibody; Tr: Trotter; GluR: glutamate receptor; MuSK: muscle-specific kinase.

Anti-NMDAR encephalitis presents with a stereotypical variant of limbic encephalitis predominantly affecting females, with a viral-like prodrome, psychiatric manifestations, memory loss, seizures, reduced level of consciousness, central hypoventilation, autonomic instability, and movement disorder, and is frequently associated with ovarian teratoma in females.[55]

Magnetic resonance imaging (MRI) is abnormal in 60–80% of limbic encephalitis, best seen as abnormally high signal in both medial temporal lobes on coronal T2 and fluid-attenuated inversion recovery (FLAIR) sequences. CT-PET may also show increased FDG uptake in the corresponding regions. If the patient is seizing then epileptiform abnormalities or slow waves may be seen on electroencephalography emanating from one or both temporal lobes. CSF may show evidence of inflammation such as lymphocytic pleocytosis, raised protein, or CSF-specific oligoclonal bands. Important differential diagnoses to consider are infectious encephalitides, lupus cerebritis, and other forms of dementia with prominent behavioral disturbance or hallucinations such as frontotemporal dementia and dementia with Lewy body disease. Treatment of an underlying cancer itself may bring about recovery, although a period of prolonged immunosuppression is typically required. Corticosteroids, intravenous immunoglobulin, plasmapheresis, and cyclophosphamide have been used with variable success. Appropriate antiepileptic agents and psychiatric therapies may also be required. If therapy is delayed patients may be left with hippocampal atrophy, chronic seizure disorder, and memory impairment. Of those cases associated with classic intracellular onconeural antibodies, testicular cancer in association with Ma2 antibodies (without Ma1, also known as Ta) has the greatest likelihood of successful treatment. ANNA-1 and small-cell cancer-associated paraneoplastic limbic encephalitis do poorly. VGKC and *N*-methyl-D-aspartate (NMDA)-related cases respond in 70–80% of cases to immunotherapy and removal of tumor. NMDA cases where no tumor is identified are more likely to relapse at a later date.[55]

Autoimmune Epilepsy

An estimated 0.5–1% of the world's population is epileptic. Although structural, metabolic, and genetic causes of epilepsy have been identified, the majority of epilepsies are classified as having unknown cause. Epilepsy is usually controlled with antiepileptic drugs. However, around 30% of patients with epilepsy do not respond to antiepileptic drugs and are classified as having pharmacoresistant epilepsy. Evidence is mounting that some epilepsies may be immune mediated and where a specific autoantibody marker has been identified the term autoimmune epilepsy has been applied.[64] Autoantibodies commonly found in association with autoimmune epilepsy are VGKC complex antibodies, in particular LGI1 antibodies, GAD65-IgG, ANNA-1, and CRMP-5-IgG. When present, these antibodies predict a favorable response to the immunotherapy. A particular epileptic syndrome characterized by faciobrachial dystonic seizures has been described in association with LGI1 antibodies. This syndrome usually precedes the onset of full-blown limbic encephalitis and if recognized and treated properly may prevent the onset of amnesia and psychiatric disturbance.[65]

Autoimmune Dementia

Autoimmune dementia can be defined as an impairment of at least two different domains of cognitive function (where there is no delirium, but other features considered atypical for a neurodegenerative disorder are present) with the following accompaniments[66]: (1) clinical, radiological, or serological evidence supporting an autoimmune etiology; (2) exclusion of other causes of dementia, particularly reversible causes; and (3) an objectively documented favorable response to immunotherapy.

Patients with autoimmune dementia usually present with an acute or a subacute disorder of memory, thinking, or behavior. They have a fluctuating confusional state accompanied by one or more symptoms of memory loss, agitation, hallucinations, and focal seizures. Various autoantibody accompaniments have been described in the setting of autoimmune dementia (see Table 29.6). However, one or more of the full range of dementia symptoms may be encountered, including impairments in learning and retaining new information; disturbances in language, behavior, orientation, reasoning, and praxis; and difficulty handling complex tasks. Impairments in attention and consciousness (delirium) are common in autoimmune dementias but are not always present. Fluctuations in cognitive abilities from day to day are common in autoimmune dementias but may also be observed in patients with neurodegenerative disorders, particularly diffuse Lewy body disease. Although uncommon, a history of spontaneous remission suggests an autoimmune etiology, but remission also may occur in toxic, metabolic, and psychogenic disorders, and depression. Therefore, it is important to exclude other causes of reversible dementia to make the diagnosis of autoimmune dementia.[67] Predictors of response to immunotherapy reported by Flanagan and colleagues are summarized in Table 29.7.

Basal Ganglia and Extrapyramidal System

Chorea

Paraneoplastic movement disorders are exceedingly rare, accounting for less than 1% of all paraneoplastic neurological syndromes. Paraneoplastic chorea typically presents at 60–70 years of age. It must be distinguished from other choreiform disorders such as other autoimmune choreas and Huntington disease. Clues to a paraneoplastic etiology include severe chorea at onset, older age, male gender, reported recent significant weight loss and coexisting peripheral neuropathy. It is most commonly associated with CRMP-5-IgG and ANNA-1. The most commonly associated cancers are small-cell lung cancer, hematological disorders (Hodgkin and non-Hodgkin lymphoma, leukemia), and adenocarcinoma (breast, colon, pancreas) of various sites. It has also been reported in renal carcinoma, germinoma, and non-small-cell lung carcinoma. Imaging is typically normal, although Vernino and colleagues reported increased T2 and FLAIR signal in the caudate and putamen consistent with basal ganglionitis in five out of eight patients with chorea and CRMP-5-IgG antibodies.[68] Although some patients respond to cancer therapy alone, additional immunotherapy may

TABLE 29.7 Summary of Predictors of Immunotherapy Response by Univariate Logistic Regression[a]

	Total (n=72)	Responders (n=46)	Non-responders (n=26)	p^b	OR (95% CI)
Clinical					
Subacute onset	52 (72)	43 (93)	9 (35)	<0.001	27.1 (6.5–112.2)
Fluctuating course	47 (65)	42 (91)	5 (19)	<0.001	44.1 (10.7–181.6)
Headache	12 (17)	11 (24)	1 (4)	0.06	7.9 (0.9–64.8)
Tremor	21 (29)	19 (41)	2 (8)	0.007	8.4 (1.8–40.1)
Family history of dementia	26 (36)	10 (22)	16 (62)	<0.001	0.1 (0.04–0.4)
Neural autoantibodies	31/70 (44)	24/45 (53)	7/25 (28)	0.03	3.1 (1.1–8.8)
Cation channels	18/58 (31)	16/37 (43)	2/21 (10)	0.01	8.0 (1.6–39.2)
VGKC (nmol/l), median (range)[c]	11/58 (19)	10/37 (27)	1/21 (5)	0.05	8.1 (1.0–68.6)
		0.87 (0.13–4.22)	0.05		
TPO antibody	40/68 (59)	22/44 (50)	18/24 (75)	0.05	0.3 (0.1–1.0)
CSF analysis					
High protein (>100 mg/dl) or pleocytosis	19/67 (28)	17/43 (40)	2/24 (8)	0.02	6.9 (1.4–33.1)
Time to treatment (months), mean ± SD	16 ± 17.9	11 ± 12.1	25 ± 22.6	0.005	0.95 (0.91–0.98)

OR: odds ratio; CI: confidence interval; VGKC: voltage-gated potassium channel; TPO: thyroid peroxidase; CSF: cerebrospinal fluid.
[a]*Data are shown as number (percentage) of patients unless otherwise indicated*
[b]*p values from univariate logistic regression analysis*
[c]*reference range: 0.00–0.02 nmol/l.*
Source: Flanagan et al. Mayo Clin Proc. *2010;85:881–897,*[67] *with permission from Elsevier.*

be beneficial in up to 60% of cases. Survival is related to the underlying cancer and is reported as being generally poor (approximately 1.5 years) in a review by the PNS EuroNetwork.[69] Non-paraneoplastic autoimmune chorea is also recognized in association with lupus, antiphospholipid syndrome, and Syndenham chorea (a parainfectious disorder occurring after β-hemolytic streptococcus infection). These cases are younger, milder, and extremely responsive to immunotherapy.[70]

Opsoclonus–Myoclonus–Ataxia Syndrome

This rare disorder with chaotic involuntary multidirectional eye movements and myoclonus is a well-known paraneoplastic disorder, particularly of children. It may also have an autoimmune, postinfectious, toxic, or metabolic etiology. The behavior of the syndrome in children and adults appears to be distinct. In children, the most common tumor is neuroblastoma, found in over 50% of pediatric cases. The most common onconeural antibody identified in children is ANNA-2.[1] The syndrome may remit spontaneously after tumor resection but frequently relapses spontaneously or during intercurrent illness. Many cases are refractory to standard immunotherapies. Cytotoxic immunotherapy (corticosteroids, azathioprine, and cyclophosphamide) and antibody-depleting therapy (intravenous immunoglobulin and plasmapheresis) have all been used with variable success. Although eye movements may return to normal, children can be left with long-term cognitive, behavioral, and neurological sequelae. In adults, this disorder may also be idiopathic or paraneoplastic. The paraneoplastic group tends to be older, and to have more severe disease associated with encephalopathy. Imaging is typically normal, whereas CSF may show inflammatory changes. Lung and breast cancer are the most commonly identified carcinomas and ANNA-2 is the most common paraneoplastic antibody. A review of adult-onset disease by Klaas and colleagues suggests that adult-onset disease may be more responsive to immunotherapy than childhood disease, although the cohort described was predominantly idiopathic rather than paraneoplastic.[71]

Cerebellum

Cerebellar degeneration is one of the most common of the paraneoplastic neurological syndromic presentations. Patients may present in a similar fashion to vestibular neuronitis or posterior circulation stroke with nausea, vomiting, diplopia, vertigo, or dizziness. This is soon followed by ataxia, dysmetria, and dysarthria. Blurred vision with oscillopsia is common. Once onset has occurred rapid deterioration within weeks to months to maximum disability is typical. Clues to the etiology are persistent symptoms without relief, multidirectional or downbeat nystagmus or opsoclonus on examination, with normal brain and cerebral vasculature imaging. Commonly associated neoplasms include small-cell lung carcinoma (VGCC, ANNA-1), Hodgkin lymphoma (PCA-Tr), and gynecological and breast malignancies (PCA-1).[1] Other neural antibodies that may be associated less frequently with this syndrome include ANNA-2, ANNA-3, PCA-2, CRMP-5-IgG, GAD65-IgG, and $mGluR_1$. Important differential diagnoses to consider include alcohol or vitamin deficiency (thiamine, vitamin E) related cerebellar degeneration, toxic cerebellar disorder (antiepileptic agents), infectious or postinfectious cerebellitis, Creutzfeldt–Jakob disease, celiac disease related ataxia, Miller Fisher variant acute inflammatory demyelinating polyneuropathy, and whether the patient is known to have had cancer or treatment for cancer (cerebellar metastases or chemotherapy-associated toxicity). Pathology of the cerebellum demonstrates extensive Purkinje cell loss with or without visible inflammatory infiltrate throughout the cerebellum. There is no consensus on the best management of these patients. Initial treatment is aimed at eradicating the underlying cancer followed by immunotherapy. Some patients may stabilize. Corticosteroids, intravenous immunoglobulin, plasmapheresis, cyclophosphamide, and tacrolimus have all failed to show significant promise in treatment trials. There are, however, occasional case reports of successful treatment supporting use of immunotherapy, especially if a patient is still ambulatory. PCA-Tr- and ANNA-2-associated paraneoplastic cerebellar degeneration are more responsive than ANNA-1 or PCA-1 disease. In theory, those associated with intraneuronal antibodies are T-cell mediated and therefore should respond better to corticosteroids and/or cyclophosphamide. In a similar fashion, cell surface antibodies may be pathogenic in themselves and may respond to any of the immune therapies mentioned above. Those with most severe deficits are least likely to respond to treatment.

Diencephalon

Hypothalamic dysfunction with narcolepsy-like excessive daytime sleepiness, cataplexy, and hypnogogical hallucinations occurs frequently with Ma2 antibody (Ta, 30% of cases) and testicular carcinoma. Typically, patients will have a multifocal disorder, with accompanying encephalopathy, seizures, or endocrine deficits atypical in standard narcolepsy.[72] These patients also have low hypocretin levels similar to the non-paraneoplastic disorder. Diencephalic dysfunction resulting in euvolemic hyponatremia (due to the syndrome of inappropriate antidiuresis) and narcolepsy can be part of the clinical spectrum of NMO and encephalitis associated with VGKC complex antibodies.[1]

Brainstem

Disorders of eye movement and balance, and multiple cranial neuropathies are typical findings, occasionally in association with parkinsonism or sleep disorders. Men with Ma2 antibodies and testicular carcinoma who present with these symptoms have a favorable prognosis. Ma1 and Ma2 combined (also called anti-Ma) antibodies occur in patients of both genders, who may have breast, colon, or testicular carcinoma, and have a worse prognosis neurologically. Patients with ANNA-2 (anti-Ri) also present most commonly with brainstem encephalitis. Neurological manifestations include opsoclonus myoclonus, jaw dystonia, and paroxysmal laryngospasm. Small-cell lung carcinoma and breast adenocarcinoma have both been described with ANNA-2.[9]

CLIPPERS

Chronic lymphocytic inflammation with pontine perivascular enhancement responsive to steroids (CLIPPERS) is a rare, possibly (no autoantibody biomarker) autoimmune neurological disorder that is exquisitely responsive to steroids. Patients typically present with brainstem/cerebellar symptoms such as ataxia, cranial nerve palsies, pseudobulbar affect, and occasionally associated symptoms of myelopathy. Radiologically scattered protean regions of increased T2 and perivascular gadolinium enhancement are seen predominantly in the pons region (Fig. 29.10). Similar small areas of punctuate enhancement may be seen in the cervical or thoracic spinal cord. CSF may demonstrate mild increases in protein, white cells, and/or oligoclonal bands, which may be transient. The lymphocytic inflammation referred to in the name corresponds to the perivascular CD-3-positive T-lymphocytic infiltrate seen predominantly in the white matter of these patients. Brain biopsy is not always feasible and thorough exclusion of other inflammatory and infectious CNS disorders is required before the diagnosis should be made. Patients are extremely responsive to high-dose corticosteroids but typically relapse when these are withdrawn. Chronic steroid therapy with slow transition to another oral immunosuppressant is usually required. Importantly, most patients do not have lesions elsewhere in the brain and prognosis can be good when the syndrome is recognized early.[73]

Spinal Cord

Myelopathy

Paraneoplastic myelopathies are rare. The majority of cases are older (60–65 years) with a history of smoking. It occurs equally in males and females. The most frequently identified tumor is small-cell cancer of the lung although breast, renal clear cell, thyroid papillary carcinomas and Hodgkin lymphoma have also been described.[5] The most frequently cited antibodies associated are CRMPs, ANNA-1, amphiphysin and PCA-1. Imaging findings vary from normal to longitudinally extensive transverse myelitis. A useful clue to paraneoplastic etiology can be selective bilateral involvement of specific tracts such as the dorsal columns or corticospinal tracts. Optimum treatment is unknown but removal of primary tumor is recommended

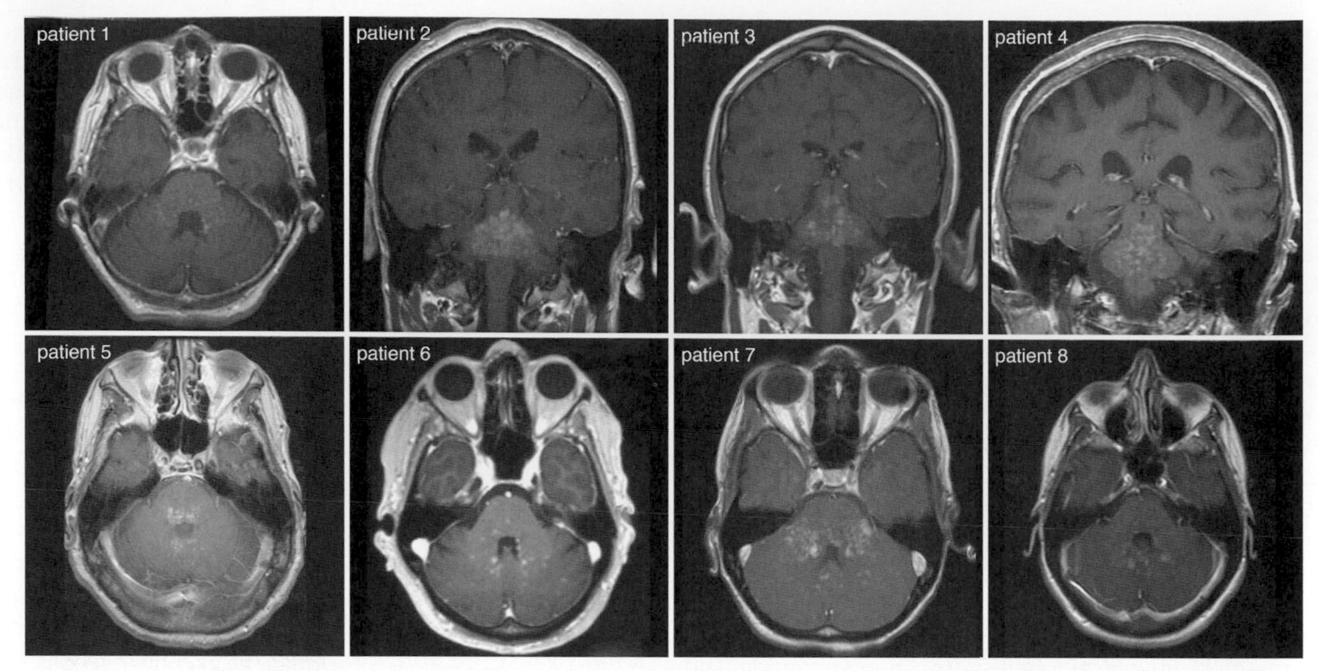

FIGURE 29.10 **Characteristic pontine-predominant magnetic resonance imaging (MRI) characteristics of chronic lymphocytic inflammation with pontine perivascular enhancement responsive to steroids (CLIPPERS) syndrome.** MRI of brain with gadolinium showing foci of pontine enhancement with a curvilinear pattern highly suggestive of a perivascular distribution in all eight patients. *Source: Pittock* et al. Brain. *2010;133:2626–2634,*[72] *with permission from Oxford University Press.*

TABLE 29.8 Sensitivity and Specificity of Six Aquaporin-4-Immunoglobulin G Assays[a]

	NMO (*n* = 35)	NMOSD (*n* = 25)	Total (*n* = 60)	Controls (*n* = 86)	Sensitivity	Specificity	ROC-AUC
IIF	17	12	29	0	48.3	100.0	0.742
FACS	25	21	46	0	76.7	100.0	0.883
CBA-O	24	20	44	0	73.3	100.0	0.867
ELISA-R (5.0)	18	18	36	0	60.0	100.0	0.800
FIPA-O	16	16	32	0	53.3	100.0	0.767
FIPA-M	16	16	32	2	53.3	97.7	0.755

NMO: neuromyelitis optica; NMOSD: neuromyelitis optica spectrum disorder; ROC: receiver operating characteristics curve; AUC: area under the curve; IIF: indirect immunofluorescence; FACS: fluorescence-activated cell sorting; CBA: cell-based assay; O: Oxford; ELISA: enzyme-linked immunosorbent assay; R: RSR/Kronus; FIPA: fluorescence immunoprecipitation assay; M: Mayo;
[a]*Results for blinded study of 146 samples on six assays with calculated sensitivities and specificities. The final column is a measure of assay accuracy.*
Source: Waters et al. Neurology. *2012;78:665–671, with permission from Wolters Kluwer Health.*[75]

and a trial of immunotherapy (high-dose steroids orally or intravenously, intravenous immunoglobulin and/or plasmapheresis) is reasonable. Outcomes are generally poor with over half becoming wheelchair dependent.[74]

Neuromyelitis Optica

NMO is a disease that mainly affects the spinal cord and the optic nerve. An autoantibody specific for AQP4 (NMO/AQP4-IgG) is a clinically validated serum biomarker that distinguishes a spectrum of astrocytopathies exemplified by NMO from multiple sclerosis and other demyelinating diseases of the CNS. Multiple sclerosis has no distinguishing biomarker and calls for different therapies. NMO is more common in females and in Asian and African-American ethnic groups. It is typically an idiopathic autoimmune disorder, although paraneoplastic cases are probably underestimated. Patients may have recurrent optic neuritis and/or longitudinally extensive transverse myelitis (defined as an inflammatory lesion involving at least three segments of spinal cord) in the setting of positive antibodies to AQP4. The most sensitive assay is able to detect AQP4-specific IgG in up to 80% of patients with NMO (Table 29.8). It remains to be determined whether other

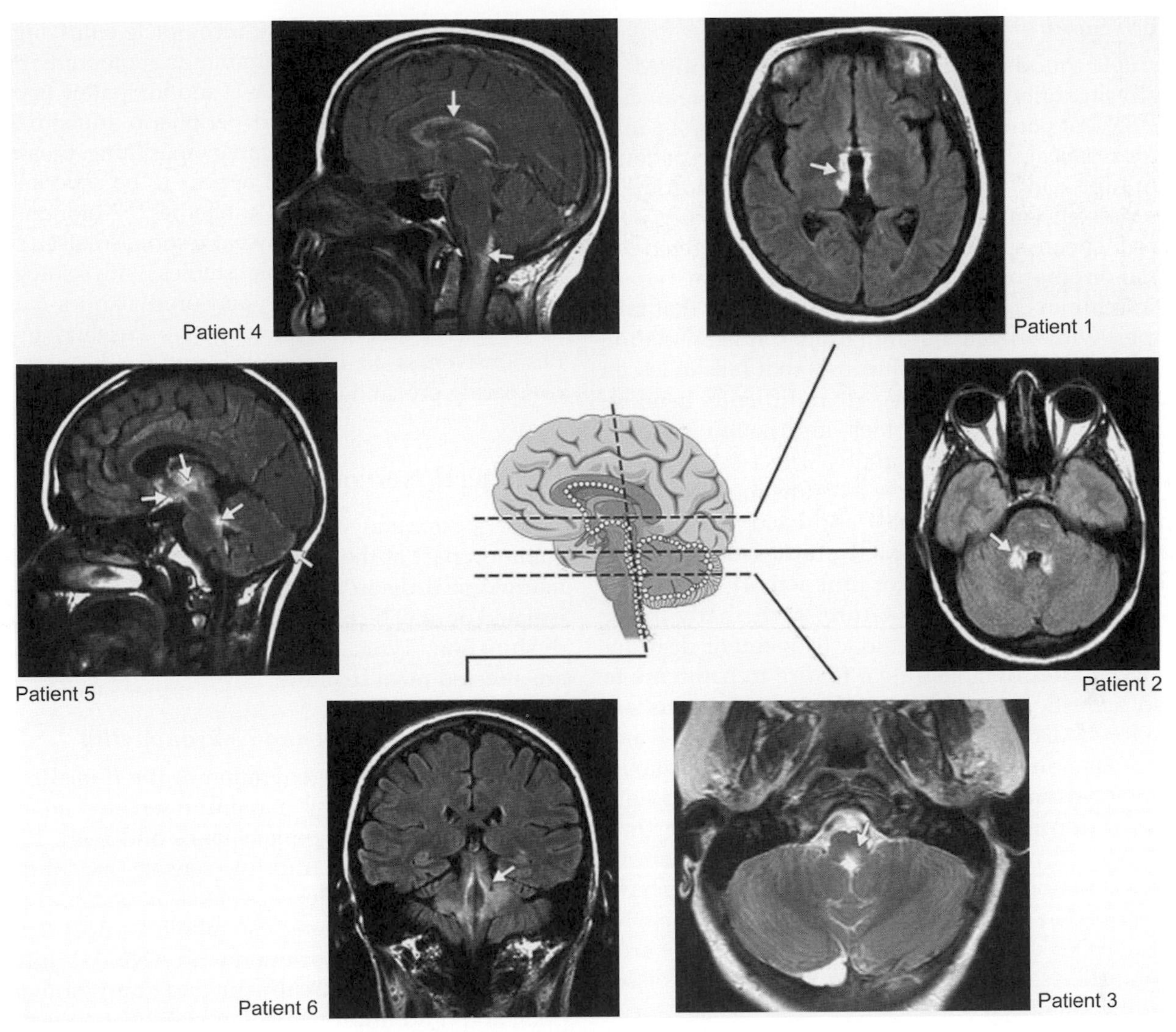

FIGURE 29.11 **Representative magnetic resonance imaging (MRI) showing localization of brain lesions in periependymal regions that are known to have high aquaporin-4 expression (white dots on midline sagittal section diagram).** Dashed black lines represent anatomical level [axial (patients 1–3) and coronal (patient 6)] of MRIs as they relate to the diagram. White arrows indicate abnormality of fluid-attenuated inversion recovery (FLAIR) or T2 signal. Patient 1 had FLAIR signal abnormality around the third ventricle with extension into the hypothalamus; patient 2 had FLAIR signal abnormality around the fourth ventricle; patient 3 had T2 signal abnormality in periaqueductal and peri-fourth ventricular distribution; patient 4 had FLAIR signal abnormality in periependymal regions surrounding the lateral ventricles (including the fornix and a longitudinal signal abnormality extending into the lower brainstem from a contiguous lesion in the upper cervical cord); patient 5 had FLAIR signal abnormalities in the thalamus, hypothalamus, and optic chiasm, extending into the superior cerebellar peduncle and tissue surrounding the fourth ventricle, and in a subpial location in the cerebellar hemispheres; and patient 6 had FLAIR signal abnormality in tissue surrounding the fourth ventricle with extension into cerebellar peduncles. *Source: Pittock* et al. Arch Neurol. *2006;63:964–968,*[76] *with permission from the American Medical Association.*

antigenic targets may account for the pathogenesis of NMO in patients who lack detectable AQP4-IgG. Despite historic dogma that NMO lesions are restricted to optic nerve and spinal cord, recent clinical, radiological, and pathological studies have documented brain involvement in NMO spectrum disorders, particularly in children (Fig. 29.11). The involvement of circumventricular organs (that are enriched in AQP4) accounts for the frequency of intractable nausea and vomiting and the syndrome of inappropriate antidiuresis as presenting or heralding symptoms of NMO.[1] Thus, progress in understanding the immunobiology of AQP4 autoimmunity necessitates continuing revision of the clinical diagnostic criteria for NMO spectrum disorders. NMO attacks may be severe, so early and lifelong immunosuppression is recommended to avoid significant disability once the disorder is recognized. Steroids or plasmapheresis, or both, are used in acute attacks, with azathioprine, mycophenylate, or rituximab used to maintain remission. Eculizumab has been reported to dramatically stop attacks in active and immunosuppressant-resistant NMO.[77]

Stiff Man Syndrome

Although this disorder was originally identified in amphiphysin antibody-positive patients with breast carcinoma, 85% of patients have neither amphiphysin antibodies nor cancer.[78] Furthermore, only 4% of patients with amphiphysin antibodies have stiff man syndrome.[21] Patients present with muscle stiffness and rigidity with associated severe painful spasms often triggered by emotional or environmental stimuli. Patients are hyperekplectic with an exaggerated startle response that fails to dampen with repeated stimulation. On examination patients have an exaggerated lordosis that fails to lessen on lying supine. They also have great difficulty bending forwards and touching their toes, and have extremely poor balance despite an apparently intact sensory system. Their gait is described as a "gunslinger's walk", with arms constantly partially abducted anticipating a fall. Electromyographic (EMG) studies demonstrate continuous simultaneous motor unit activity of affected agonist and antagonist musculature, particularly in the thoracic and lumbar spine region. Movement disorder laboratory studies demonstrate a failure to habituate to repeated acoustic startle. The majority of patients are positive for GAD65-IgG in their serum and/or CSF and may also have autoimmune diabetes, cerebellar ataxia, or epilepsy. Serum levels of GAD65-IgG associated with stiff man syndrome are in a much higher range than those associated with diabetes. Rare tumor associations have been described. α_1-Glycine subunit antibodies have been described in patients with stiff man syndrome who are both GAD65 antibody positive and negative.[78] True paraneoplastic-associated stiff man syndrome is associated with small-cell lung cancer and breast cancer. Patients tend to be older, with asymmetrical cervical and distal limb involvement. There may be associated spinal myoclonus and pruritis. Optimum management of this disorder involves the identification and treatment of any underlying tumor. Although response to steroids has been described in paraneoplastic cases, the benefits of intravenous immunoglobulin, plasmapheresis, and other immunotherapies are unknown. The majority of patients find some symptomatic benefit with very high-dose benzodiazepines and baclofen.

Autonomic Nervous System

Autoimmune Dysautonomia

Autonomic system disorders may present with symptoms of confined limited dysautonomia (i.e. isolated gastrointestinal dysmotility) or generalized pandysautonomia. Dry eyes, dry mouth, gastrointestinal dysmotility symptoms (early satiety, nausea, vomiting, abdominal cramping and constipation), orthostatic hypotension, heat intolerance (inability to sweat), and urological symptoms (bladder frequency, urgency, incomplete emptying, impotence) may all indicate an autoimmune autonomic disorder. Both paraneoplastic (ANNA-1) and idiopathic (ganglionic AChR, VGCC receptor, and peripherin antibodies) cases have been reported. Adenocarcinomas (lung, breast, colon, pancreas, uterine, prostate) appear to be especially common with ganglionic AChR antibodies.[45] Autonomic reflex testing, thermoregulatory sweat testing, small nerve fiber testing, gastrointestinal transit studies (with scintography), gastrointestinal manometry, and urodynamics may assist in diagnosis and provide objective baseline measures. Responsiveness to immunotherapy can be assessed by comparing pre- and post-therapy test results.

Peripheral Nervous System

The peripheral nervous system is the single most common part of the neuraxis affected by paraneoplastic neurological disorders. Peripheral neuropathy may be isolated or may be part of more widespread neurological dysfunction. Typically, the sensory nerves and ganglia are affected most severely.

Sensory Neuropathy and Neuronopathy

T-cell mediated degeneration of the dorsal root ganglia neurons produces a painful sensory neuropathy. Patients complain of poor balance and pain. Examination reveals a loss of multiple sensory modalities with reduced or absent reflexes, sensory ataxia, and pseudoathetosis. Small-cell lung carcinoma is most frequently cited as a cause in association with ANNA-1 antibodies. Response to cancer treatment and immunotherapy is poor, and mild improvement or stabilization is typically the best outcome. Non-paraneoplastic autoimmune sensory neuropathies may also occur in the setting of connective tissue disorders such as Sjögren disease, but will not be discussed further here.

Peripheral Nerve Hyperexcitability

Acquired neuromyotonia or Isaac syndrome is a disorder of peripheral nerve hyperexcitability associated with VGKC complex antibodies. Because the associated antibody targets the cell surface it is felt to be pathogenic. The patient complains of muscle cramping and twitching that can be extremely distressing. Muscles show delayed relaxation or pseudomyotonia after exercise. EMG studies should demonstrate fibrillations, fasciculations, and doublet or multiplet single unit discharges that have a high interburst frequency.[79] In some cases patients also experience confusion, mood alteration, sleep disturbances, hyperhidrosis, and hallucinations (Morvan syndrome). A minority of patients have an associated thymoma, small-cell lung carcinoma, or other malignancy. Although CASPR2 was felt to be the main antibody target within the VGKC complex in these disorders, there have been some cases associated with LGI1 and further autoantigens in this complex remain to be identified. Treatment of peripheral nerve hyperexcitability involves removal of any identified tumors, symptomatic therapy with anticonvulsant medications (phenytoin and carbamazepine), and consideration of immunotherapy such as intravenous immunoglobulin or plasmapheresis in refractory cases.

Neuromuscular Junction

Myasthenia Gravis and Lambert–Eaton Myasthenic Syndrome

MG and LEMS are the characteristic disorders described with fluctuating weakness and fatiguability of muscles. Eye, proximal limb, oropharyngeal, and respiratory musculature may be involved in myasthenia together or in isolation. Respiratory failure may be life threatening. If eye involvement alone is present for longer than 2 years it is unlikely to generalize thereafter. Spontaneous exacerbations and remissions are typical. Certain drugs may exacerbate symptoms. The clinical diagnosis may be made at the bedside using the edrophonium test (cholinesterase inhibitor) if patients have a clearly demonstrable weakness; rapid and dramatic improvement is seen after injection, but this is short lived. EMG shows a 10% decrement in the compound muscle action potential after supramaximal stimulation. mAChR antibodies are detected in 90% of cases. Approximately 15% of patients with MG and AChR antibodies harbor a thymoma. The presence of muscle striational antibody or CRMP-5 should increase suspicion for underlying thymoma or lung cancer. Any thymic tumors identified should be removed. Moreover, in patients younger than 60 years removal of thymus regardless of pathology may improve disease control.

Muscle-specific kinase antibodies are present in 40% of AChR seronegative patients. These patients usually experience a prevalent involvement of cranial and bulbar muscles and a high frequency of respiratory crises.[44] Long-acting acetylcholinesterase inhibitors are used as symptomatic therapy in addition to chronic immunosuppression with corticosteroids alone or in combination with other immunosuppressive agents such as azathiaprine and mycophenylate. Acetylcholinesterase therapy seems to be less efficient in MG with MuSK antibodies. Acute exacerbations of MG (myasthenic crises) can be treated with intravenous immunoglobulin or plasmapheresis.

Lambert–Eaton syndrome follows a clinically less dramatic course. Patients are often generally weak. Proximal limb weakness is typical and there may be eye involvement. Tendon reflexes may be absent but facilitated by exercise. Autonomic dysfunction is commonly encountered with complaints of dry mouth and impotence. Nerve conduction studies reveal a reduced compound muscle action potential amplitude at baseline, decrement with slow repetitive nerve stimulation but increments of over 100% after exercise or high-frequency stimulation. Antibodies targeting postsynaptic PQ- and N-type VGKCs are often found: the PQ type are more indicative of LEMS, whereas N type may heighten suspicion for cancer. Calcium channel antibodies frequently coexist with other antibodies and AGNA rarely occurs in the absence of calcium channel antibodies. Small-cell lung cancer is most frequently identified in this group. If a patient has LEMS but no cancer is identified, cancer screening should be continued for at least 2 years. Treatment of LEMS is aimed at tumor removal, symptomatic therapy with 3,4-diaminopyridine, and additional immunosuppression as required.

Muscle

Inflammatory Myopathies

Inflammatory myopathies are a heterogeneous group of diseases, of which polymyositis and dermatomyositis are the most frequent and well characterized. These two myopathies share several clinical features, such as a subacute proximal muscle weakness and evidence of inflammation on muscle biopsy. Characteristic cutaneous features often help the treating clinician to differentiate patients with dermatomyositis from those with other inflammatory myopathies. A violaceous eruption on the upper eyelids and in rare cases on the lower eyelids, known as "heliotrope rash", and an erythematous rash over the extensor surfaces of the metacarpophalangeal, proximal interphalangeal, and distal interphalangeal joints (known as Gottron papules) are characteristic features of dermatomyositis. Immune-mediated necrotizing myopathy (IMNM) is a distinct type of inflammatory myopathy with the same clinical features of polymyositis and dermatomyositis but with a different pathology on muscle biopsy. Characteristic findings are abundant degenerating, regenerating, and necrotic cells with few or no inflammatory cells. Another type of myopathy, inclusion body myositis, is sometimes included among inflammatory myopathies. However, because the primary role of inflammation in the pathogenesis of inclusion body myositis is still debated, this section will focus on polymyositis, dermatomyositis, and IMNMs.

The diagnostic evaluation of inflammatory myopathies includes serum creatine kinase concentrations, serum autoantibodies, electromyography, and muscle biopsy. Serum creatine kinase concentration should always be elevated in polymyositis and IMNMs, whereas dermatomyositis or inclusion body myositis can present with the creatine kinase concentration within the normal range.

The most common types of autoantibody found in the setting of polymyositis or dermatomyositis are antibodies targeting the aminoacyl-tRNA synthetase (Jo-1-IgG being the most common). Patients with these antibodies usually have a common constellation of clinical features, including autoimmune myopathy, interstitial lung disease, non-erosive arthritis, and fever, as well as hyperkeratotic lesions on the lateral and palmar aspects of the fingers (mechanic hands). Collectively, these diverse manifestations are referred to as the antisynthetase syndrome.

IMNMs are usually associated with autoantibodies targeting the signal recognition particle, a ribonucleoprotein that binds the signal sequences of newly synthesized proteins and facilitates their translocation into the endoplasmic reticulum. Autoantibodies binding to hydroxymethylglutaryl-coenzyme A reductase (HMGC-r) have been documented in patients who developed IMNM after long-term statin treatment. This type of myopathy differs from the more common toxic myopathy related to statin therapy, which improves after discontinuation of treatment. The lack of recovery after discontinuing the statin treatment should raise the suspicion of an IMNM with HMGC-r antibodies.[80]

Cytotoxic CD8 T cells that surround and invade non-necrotic muscle fibers are characteristic pathological findings on muscle biopsy from patients with polymyositis. Polymyositis is a cell-mediated disorder in which the immune attack is directed against some unknown antigen(s) displayed on the muscle fibers. The characteristic histological feature of dermatomyositis on muscle biopsy is perifascicular atrophy (small fibers surrounding a core of more normal-sized fibers deeper in the fascicle). The expression of membrane attack complex and immunoglobulins in capillaries and small blood vessels and MHC-I expression on perifascicular muscle fibers usually precedes the development of perifascicular atrophy. Dermatomyositis is thought to be a CD4 T-cell mediated disease in which the humoral immune response plays an important role.

Inflammatory myopathies, in particular dermatomyositis, can be associated with cancer. The frequency of neoplasia in patients with dermatomyositis, previously thought to be 50%, is more probably in the region of 10–15%. The most common detected neoplasms are breast, lung, ovary, and stomach cancers, as well as lymphoma.

CONCLUSION

The era of neural antibody discovery has arrived and this discovery is the major driver in the field of autoimmune neurology. Autoimmune neurological disorders, whether idiopathic or paraneoplastic, may respond to immunotherapy and some are potentially reversible, underscoring the importance of early diagnosis and treatment. A growing spectrum of clinical, radiological, and oncological associations is now being recognized for these autoantibodies. Recognition of novel biomarkers, advanced serological interpretive insights, increased understanding of the pathogenic impact of binding of neural antibodies to their plasma membrane neuronal and glial targets, and future identification of novel pathogenic mechanisms will lead to the formulation of individual patient-specific therapies.

FUTURE DIRECTIONS

Although we have learned a great deal, many central questions remain unanswered. These include: What is the seminal event (in the absence of cancer) that initiates an autoimmune response? Which events take place at a molecular level that allow access of both cellular and humoral components of the immune system to privileged areas (e.g. the CNS)? Why is there such diverse and regionally specific injury when target antigen is widely distributed? Which factors trigger relapsing disease?

Development and improvement of animal models representative of the disease in question are needed to further elucidate some of the immunopathogenic mechanisms of both antibody-mediated diseases (e.g. NMO and AQP4 autoimmunity, encephalitis and VGKC complex/NMDAR autoimmunity) and those that are predominantly mediated by cytotoxic T cells (e.g. limbic encephalitis and seropositivity for ANNA-1, CRMP5-IgG, and anti-Ma2). These models could use active immunization to investigate the initial immune response against the target antigen, the generation of neural antibodies, and the passage of antibodies and immune cells across the blood–brain barrier. Improved understanding of the specific components of disease pathogenesis will facilitate the development of new therapies.

Current strategies of broad-spectrum or selective B-cell immunosuppression carry significant long-term risk and often provide only small benefit. Unfortunately, there is an absence of randomized controlled trials for most paraneoplastic and idiopathic autoimmune neurological disorders, and standard of care approaches vary and are based on small retrospective case series or occasional open-label trials. Future therapeutic approaches to autoimmune neurological disorders may include: (1) repurposing of drugs currently being used in rheumatology and hematology, such as inhibition of complement activation and inhibition of glutamate toxicity or cytokine-mediated events (e.g. IL-6); (2) novel approaches, such as small molecules that bind neural antibodies or prevent their binding to their respective targets; and (3) improving drug pharmacokinetics, such as intrathecal delivery.

References

1. Iorio R, Lennon VA. Neural antigen-specific autoimmune disorders. *Immunol Rev*. 2012;248:104–121.
2. Delves PJ, Roitt IM. The immune system. *N Engl J Med*. 2000;343:37–49.
3. Diamond D, Huerta PT, Mina-Osorio P, Kowal C, Volpe BT. Losing your nerves? Maybe it's the antibodies. *Nat Rev Immunol*. 2009;9:449–456.
4. Albert ML, Darnell JC, Bender A, Francisco LM, Bhardwaj N, Darnell RB. Tumor-specific killer cells in paraneoplastic cerebellar degeneration. *Nat Med*. 1998;4:1321–1324.
5. McKeon A, Pittock SJ. Paraneoplastic encephalomyelopathies: pathology and mechanisms. *Acta Neuropathol*. 2011;122:381–400.
6. Roberts WK, Deluca IJ, Thimas A, et al. Patients with lung cancer and paraneoplastic hu syndrome harbor HuD-specific type 2 $CD8^+$ T cells. *J Clin Invest*. 2009;119:2042–2051.
7. Perrone-Bizzozero N, Bolognani F. Role of HuD and other RNA-binding proteins in neural development and plasticity. *J Neurosci Res*. 2002;68:121–126.
8. Pittock SJ, Lucchinetti CF, Lennon VA. Anti-neuronal nuclear autoantibody type 2: paraneoplastic accompaniments. *Ann Neurol*. 2003;53:580–587.
9. Pittock SJ, Parisi JE, McKeon A, et al. Paraneoplastic jaw dystonia and laryngospasm with antineuronal nuclear autoantibody type 2 (anti-Ri). *Arch Neurol*. 2010;67:1109–1115.
10. Chan KH, Vernino S, Lennon VA. ANNA-3 anti-neuronal nuclear antibody: marker of lung cancer-related autoimmunity. *Ann Neurol*. 2001;50:301–311.
11. Hetzel DJ, Stanhope CR, O'Neill BP, Lennon VA. Gynecologic cancer in patients with subacute cerebellar degeneration predicted by anti-Purkinje cell antibodies and limited in metastatic volume. *Mayo Clin Proc*. 1990;65:1558–1563.
12. Corradi JP, Yang C, Darnell JC, Dalmau J, Darnell RB. A post-transcriptional regulatory mechanism restricts expression of the paraneoplastic cerebellar degeneration antigen cdr2 to immune privileged tissues. *J Neurosci*. 1997;17:1406–1415.
13. McKeon A, Tracy JA, Pittock SJ, Parisi JE, Klein CJ, Lennon VA. Purkinje cell cytoplasmic autoantibody type 1 accompaniments: the cerebellum and beyond. *Arch Neurol*. 2011;68(10):1282–1289.
14. Vernino S, Lennon VA. New Purkinje cell antibody (PCA-2): marker of lung cancer-related neurological autoimmunity. *Ann Neurol*. 2000;47:297–305.
15. Trotter JL, Hendin BA, Osterland CK. Cerebellar degeneration with Hodgkin disease. An immunological study. *Arch Neurol*. 1976;33:660–661.
16. de Graaff E, Maat P, Hulsenboom E, et al. Identification of delta/notch-like epidermal growth factor-related receptor as the Tr antigen in paraneoplastic cerebellar degeneration. *Ann Neurol*. 2012;71:815–824.
17. Graus F, Gultekin SH, Ferrer I, Reiriz J, Alberch J, Dalmau J. Localization of the neuronal antigen recognized by anti-Tr antibodies from patients with paraneoplastic cerebellar degeneration and Hodgkin's disease in the rat nervous system. *Acta Neuropathol*. 1998;96:1–7.
18. Yu Z, Kryzer TJ, Griesmann GE, Kim K, Benarroch EE, Lennon VA. CRMP-5 neuronal autoantibody: marker of lung cancer and thymoma-related autoimmunity. *Ann Neurol*. 2001;49:146–154.
19. Pittock SJ, Kryzer TJ, Lennon VA. Paraneoplastic antibodies coexist and predict cancer, not neurological syndrome. *Ann Neurol*. 2004;56:715–719.
20. De Camilli P, Thomas A, Cofiell R, et al. The synaptic vesicle-associated protein amphiphysin is the 128-kD autoantigen of stiff-man syndrome with breast cancer. *J Exp Med*. 1993;178:2219–2223.
21. Pittock SJ, Lucchinetti CF, Parisi JE, et al. Amphiphysin autoimmunity: paraneoplastic accompaniments. *Ann Neurol*. 2005;58:96–107.
22. Geis C, Weishaupt A, Hallermann S, et al. Stiff person syndrome-associated autoantibodies to amphiphysin mediate reduced GABAergic inhibition. *Brain*. 2010;133:3166–3180.
23. Sabater L, Titulaer M, Saiz A, Verschuuren J, Gure AO, Graus F. SOX1 antibodies are markers of paraneoplastic Lambert–Eaton myasthenic syndrome. *Neurology*. 2008;70:924–928.
24. Schuller M, Jenne D, Voltz R. The human PNMA family: novel neuronal proteins implicated in paraneoplastic neurological disease. *J Neuroimmunol*. 2005;169:172–176.
25. Dalmau J, Gultekin SH, Voltz R, et al. Ma1, a novel neuron- and testis-specific protein, is recognized by the serum of patients with paraneoplastic neurological disorders. *Brain*. 1999;122(Pt 1):27–39.
26. Solimena M, Folli F, Denis-Donini S, et al. Autoantibodies to glutamic acid decarboxylase in a patient with stiff-man syndrome, epilepsy, and type I diabetes mellitus. *N Engl J Med*. 1988;318:1012–1020.
27. Pittock SJ, Yoshikawa H, Ahlskog JE, et al. Glutamic acid decarboxylase autoimmunity with brainstem, extrapyramidal, and spinal cord dysfunction. *Mayo Clin Proc*. 2006;81:1207–1214.
28. Chamberlain JL, Pittock SJ, Oprescu AM, et al. Peripherin-IgG association with neurologic and endocrine autoimmunity. *J Autoimmun*. 2010;34:469–477.
29. McKeon A, Lennon VA, Lachance DH, Klein CJ, Pittock SJ. A case–control study of striational antibodies in a paraneoplastic context. *Muscle Nerve*. 2013;47(4):585–587.
30. Howard Jr FM, Lennon VA, Finley J, Matsumoto J, Elveback LR. Clinical correlations of antibodies that bind, block, or modulate human acetylcholine receptors in myasthenia gravis. *Ann N Y Acad Sci*. 1987;505:526–538.
31. Lennon VA, Kryzer TJ, Pittock SJ, Verkman AS, Hinson SR. IgG marker of optic–spinal multiple sclerosis binds to the aquaporin-4 water channel. *J Exp Med*. 2005;202:473–477.
32. Hinson SR, Pittock SJ, Lucchinetti CF, et al. Pathogenic potential of IgG binding to water channel extracellular domain in neuromyelitis optica. *Neurology*. 2007;69:2221–2231.
33. Hinson SR, Roemer SF, Lucchinetti CF, et al. Aquaporin-4-binding autoantibodies in patients with neuromyelitis optica impair glutamate transport by down-regulating EAAT2. *J Exp Med*. 2008;205:2473–2481.
34. Hinson SR, Romero MF, Popescu BF, et al. Molecular outcomes of neuromyelitis optica (NMO)-IgG binding to aquaporin-4 in astrocytes. *Proc Natl Acad Sci U S A*. 2012;109:1245–1250.
35. Kinoshita M, Nakatsuji Y, Kimura T, et al. Anti-aquaporin-4 antibody induces astrocytic cytotoxicity in the absence of CNS antigen-specific T Cells. *Biochem Biophys Res Commun*. 2010;394(1):205–210.
36. Bradl M, Misu T, Takahashi T, et al. Neuromyelitis optica: pathogenicity of patient immunoglobulin in vivo. *Anna Neurol*. 2009;66(5):630–643.
37. Bennett JL, Lam C, Kalluri SR, et al. Intrathecal pathogenic anti-aquaportin-4 antibodies in early neuromyelitis optica. *Ann Neurol*. 2009;66(5):617–629.
38. Saadoun S, Waters P, Bell BA, Vincent A, Verkman AS, Papadopoulos MC. Intra-cerebral injection of neuromyelitis optica immunoglobulin G and human complement produces neuromyelitis optica lesions in mice. *Brain*. 2010;133(Pt 2):349–361.
39. Lennon VA. MiniReview Article: The role of immunoglobulin G in neuromyelitis optica: does IgG play a causal role in the pathology of neuromyelitis optica? *J Watch Neurol*. 2010;12(7):52–53.
40. Lindstrom JM, Seybold ME, Lennon VA, Whittingham S, Duane DD. Antibody to acetylcholine receptor in myasthenia gravis. Prevalence, clinical correlates, and diagnostic value. *Neurology*. 1976;26:1054–1059.

41. Lennon VA, Lindstrom JM, Seybold ME. Experimental autoimmune myasthenia gravis: cellular and humoral immune responses. *Ann N Y Acad Sci*. 1976;274:283–299.
42. Drachman DB. Myasthenia gravis. *N Engl J Med*. 1994;330: 1797–1810.
43. Hoch W, McConville J, Helms S, Newsom-Davis J, Melms A, Vincent A. Auto-antibodies to the receptor tyrosine kinase MuSK in patients with myasthenia gravis without acetylcholine receptor antibodies. *Nat Med*. 2001;7:365–368.
44. Evoli A, Tonali PA, Padua L, et al. Clinical correlates with anti-MuSK antibodies in generalized seronegative myasthenia gravis. *Brain*. 2003;126:2304–2311.
45. McKeon A, Lennon VA, Lachance DH, Fealey RD, Pittock SJ. Ganglionic acetylcholine receptor autoantibody: oncological, neurological, and serological accompaniments. *Arch Neurol*. 2009;66:735–741.
46. Vernino S, Low PA, Fealey RD, Stewart JD, Farrugia G, Lennon VA. Autoantibodies to ganglionic acetylcholine receptors in autoimmune autonomic neuropathies. *N Engl J Med*. 2000;343: 847–855.
47. Lennon VA, Ermilov LG, Szurszewski JH, Vernino S. Immunization with neuronal nicotinic acetylcholine receptor induces neurological autoimmune disease. *J Clin Invest*. 2003;111:907–913.
48. Vernino S, Ermilov LG, Sha L, Szurszewski JH, Low PA, Lennon VA. Passive transfer of autoimmune autonomic neuropathy to mice. *J Neurosci*. 2004;24:7037–7042.
49. Lennon VA, Kryzer TJ, Griesmann GE, et al. Calcium-channel antibodies in the Lambert–Eaton syndrome and other paraneoplastic syndromes. *N Engl J Med*. 1995;332:1467–1474.
50. Klein CJ, Lennon VA, Aston PA, McKeon A, Pittock SJ. Chronic pain as a manifestation of potassium channel-complex autoimmunity. *Neurology*. 2012;79:1136–1144.
51. Lai M, Huijbers MG, Lancaster E, et al. Investigation of LGI1 as the antigen in limbic encephalitis previously attributed to potassium channels: a case series. *Lancet Neurol*. 2010;9:776–785.
52. Irani SR, Alexander S, Waters P, et al. Antibodies to Kv1 potassium channel-complex proteins leucine-rich, glioma inactivated 1 protein and contactin-associated protein-2 in limbic encephalitis, Morvan's syndrome and acquired neuromyotonia. *Brain*. 2010;133:2734–2748.
53. Klein CJ, Lennon VA, Aston PA, et al. Insights from LGI1 and CASPR2 potassium channel complex autoantibody subtyping. *JAMA Neurol*. 2012;70(2):229–234.
54. Dalmau J, Tuzun E, Wu HY, et al. Paraneoplastic anti-N-methyl-D-aspartate receptor encephalitis associated with ovarian teratoma. *Ann Neurol*. 2007;61:25–36.
55. Dalmau J, Lancaster E, Martinez-Hernandez E, Rosenfeld MR, Balice-Gordon R. Clinical experience and laboratory investigations in patients with anti-NMDAR encephalitis. *Lancet Neurol*. 2011;10:63–74.
56. Pruss H, Holtje M, Maier N, et al. IgA NMDA receptor antibodies are markers of synaptic immunity in slow cognitive impairment. *Neurology*. 2012;78:1743–1753.
57. Lancaster E, Martinez-Hernandez E, Titulaer MJ, et al. Antibodies to metabotropic glutamate receptor 5 in the Ophelia syndrome. *Neurology*. 2011;77:1698–1701.
58. Lai M, Hughes EG, Peng X, et al. AMPA receptor antibodies in limbic encephalitis alter synaptic receptor location. *Ann Neurol*. 2009;65:424–434.
59. Lancaster E, Lai M, Peng X, et al. Antibodies to the GABA(B) receptor in limbic encephalitis with seizures: case series and characterisation of the antigen. *Lancet Neurol*. 2010;9:67–76.
60. Lancaster E, Martinez-Hernandez E, Dalmau J. Encephalitis and antibodies to synaptic and neuronal cell surface proteins. *Neurology*. 2011;77:179–189.
61. McKeon A, Martinez-Hernandez E, Lancaster E, et al. Glycine receptor autoimmune spectrum with stiff-man syndrome phenotype. *JAMA Neurol*. 2013;70(1):44–50.
62. Hutchinson M, Waters P, McHugh J, et al. Progressive encephalomyelitis, rigidity, and myoclonus: a novel glycine receptor antibody. *Neurology*. 2008;71:1291–1292.
63. McKeon A, Apiwattanakul M, Lachance DH, et al. Positron emission tomography-computed tomography in paraneoplastic neurologic disorders: systematic analysis and review. *Arch Neurol*. 2010;67:322–329.
64. Quek AM, Britton JW, McKeon A, et al. Autoimmune epilepsy: clinical characteristics and response to immunotherapy. *Arch Neurol*. 2012;69:582–593.
65. Irani SR, Michell AW, Lang B, et al. Faciobrachial dystonic seizures precede LGI1 antibody limbic encephalitis. *Ann Neurol*. 2011;69:892–900.
66. McKeon A, Lennon VA, Pittock SJ. Immunotherapy-responsive dementias and encephalopathies. *Continuum (Minneap Minn)*. 2010;16:80–101.
67. Flanagan EP, McKeon A, Lennon VA, et al. Autoimmune dementia: clinical course and predictors of immunotherapy response. *Mayo Clin Proc*. 2010;85:881–897.
68. Vernino S, Tuite P, Adler CH, et al. Paraneoplastic chorea associated with CRMP-5 neuronal antibody and lung carcinoma. *Ann Neurol*. 2002;51:625–630.
69. Vigliani MC, Honnorat J, Antoine JC, et al. Chorea and related movement disorders of paraneoplastic origin: the PNS EuroNetwork experience. *J Neurol*. 2011;258(11):2058–2068.
70. O'Toole O, Lennon VA, Ahlskog JE, et al. Autoimmune chorea in adults. *Neurology*. 2013;80(12):1133–1144.
71. Klaas JP, Ahlskog JE, Pittock SJ, et al. Adult-onset opsoclonus–myoclonus syndrome. *Arch Neurol*. 2012:1–10.
72. Dalmau J, Graus F, Villarejo A, et al. Clinical analysis of anti-Ma2-associated encephalitis. *Brain*. 2004;127:1831–1844.
73. Pittock SJ, Debruyne J, Krecke KN, et al. Chronic lymphocytic inflammation with pontine perivascular enhancement responsive to steroids (CLIPPERS). *Brain*. 2010;133:2626–2634.
74. Flanagan EP, McKeon A, Lennon VA, et al. Paraneoplastic isolated myelopathy: clinical course and neuroimaging clues. *Neurology*. 2011;76(24):2089–2095.
75. Waters PJ, McKeon A, Leite MI, et al. Serologic diagnosis of NMO: a multicenter comparison of aquaporin-4-IgG assays. *Neurology*. 2012;78:665–671.
76. Pittock SJ, Weinshenker BG, Lucchinetti CF, et al. Neuromyelitis optica brain lesions localized at sites of high aquaporin 4 expression. *Arch Neurol*. 2006;63:964–968.
77. Pittock SJ, McKeon A, Mandrekar JN, Weinshenker BG, Lucchinetti CF, Wingerchuk DM. Pilot clinical trial of eculizumab in AQP4-IgG-positive NMO. *Abstract T1826, American Neurological Association Annual Meeting*. October 2012.
78. McKeon A, Robinson MT, McEvoy KM, et al. Stiff-man syndrome and variants: clinical course, treatments, and outcomes. *Arch Neurol*. 2012;69:230–238.
79. Hart IK, Maddison P, Newsom-Davis J, Vincent A, Mills KR. Phenotypic variants of autoimmune peripheral nerve hyperexcitability. *Brain*. 2002;125:1887–1895.
80. Mammen AL. Autoimmune myopathies: autoantibodies, phenotypes and pathogenesis. *Nat Rev Neurol*. 2011;7:343–354.

CHAPTER

30

Multiple Sclerosis

Julia Schaeffer, Chiara Cossetti, Giulia Mallucci, Stefano Pluchino

Department of Clinical Neurosciences, John van Geest Centre for Brain Repair, Wellcome Trust–MRC Stem Cell Institute and NIHR Biomedical Research Centre, University of Cambridge, Cambridge, UK

OUTLINE

INTRODUCTION

General Definition

Multiple sclerosis (MS) is a chronic inflammatory demyelination disease of the human central nervous system (CNS) that affects young adults and over subsequent decades transforms into a progressive neurodegenerative disorder associated with major clinical disabilities.[1] The main pathological lesions of MS are multiple sclerotic plaques displaying demyelination of white and gray matter in the brain and spinal cord.[2] Within these focal lesions the myelin sheaths and the oligodendrocytes are destroyed after myelin protein components are recognized by cells of the immune system.[3] This inflammatory reaction includes the activation of myelin-specific $CD4^+$ autoreactive T cells and their differentiation into a pro-inflammatory T-helper-1 (Th1)-like phenotype, and also $CD8^+$ T cells, B cells secreting myelin-directed antibodies and cytokines, and other factors of the immune system.[4,5]

Neurobiology of Brain Disorders
http://dx.doi.org/10.1016/B978-0-12-398270-4.00030-6

During the initial state of the inflammatory response, activated lymphocytes recruited from the periphery to the CNS cross the blood–cerebrospinal fluid (CSF) barrier and the blood–brain barrier (BBB).[6,7] The presence of activated immune cells, as well as immunoglobulins and cytokines in the CSF of MS patients compared with control donors, suggests the participation of the immune system in MS pathogenesis.[6,7] The resulting pathological features in the CNS are the (chronic) demyelinated plaques, which consist of a well-demarcated hypocellular area characterized by the loss of myelin, relative preservation of axons, and formation of astrocytic scars. Lesions can be observed in the optic nerves, the periventricular white matter, brainstem, cerebellum, and spinal cord white matter, and can also involve gray matter.[1,2] Within MS lesions, myelin, oligodendrocytes, and axons are destroyed, and axonal dysfunction is likely to result from nerve conduction disorders following demyelination.

The concept of MS as an autoimmune disease is confirmed by experimental autoimmune encephalomyelitis (EAE), the most widely used animal model of MS. EAE is typically induced in animals either by injection of an emulsion containing a fragment of a myelin membrane protein or a homogenate of spinal cord tissue, or by injection of myelin antigen-specific T cells. Following disease induction, activated T cells cross the BBB and initiate a disease course similar to that observed in MS (inflammation, demyelination, and axonal degeneration).[4] Genetic susceptibility to MS is associated with several genes of the major histocompatibility complex (MHC) regions, for example the *HLA-DRB1* gene on chromosome 6p21; most of them are likely to have immune response functions.[6]

In MS pathogenesis, demyelination results in a progressive loss of structure and function of neurons, namely neurodegeneration, due to impaired propagation of action potentials across the demyelinated axons. No explanation has yet been given for the temporal relationship between inflammation, demyelination, and axonal degeneration, which could be a key point in understanding MS pathogenesis. It is still unclear whether inflammatory reactions cause neurodegeneration through demyelination, or whether MS is a primary neurodegenerative disease with secondary inflammation and demyelination. The accumulation of demyelinated lesions may lead to the neurological features of MS and explain the heterogeneity of the disease. During early stages of MS, spontaneous remyelination – the process by which oligodendrocyte progenitor cells (OPCs) re-ensheath demyelinated axons – helps to restore axonal conduction. However, over time, remyelination fails to compensate for the progression of inflammation-driven demyelination and consequent axonal damage/dysfunction leads to permanent, irreversible neurological decline.[2,8]

Different Forms of Multiple Sclerosis

The current classification of MS clinical forms distinguishes at least four clinical patterns.

- *Relapsing remitting multiple sclerosis (RRMS)* is the most common form (about 85% of cases) and is characterized by discrete attacks (relapses) over days or weeks, described as episodes of acute worsening of neurological functions. These are followed by complete or partial recovery periods (remissions) over months or years, supposedly without any disease progression.
- *Secondary progressive multiple sclerosis (SPMS)* is characterized by gradual neurological decline with occasional minor recoveries. Most cases of RRMS (50% within 10 years of the initial diagnosis) evolve into SPMS (Table 30.1 and Fig. 30.1).
- *Primary progressive multiple sclerosis (PPMS)* affects about 15% of patients with MS. This clinical pattern of disease is characterized by a steady functional worsening from the onset of symptoms, without identifiable relapses or remissions.
- *Progressive relapsing multiple sclerosis (PRMS)* is also characterized by steadily worsening symptoms from the onset, although with clear acute relapses with or without recovery observed. Heterogeneity of clinical features within these patterns increases the difficulty in understanding the pathogenic mechanisms responsible for the onset of the disease.[9]

Therapeutic agents that have been proposed so far primarily target the immune system and help to slow the progressive deteriorating effects of the disease, but do not cure it or reverse the damage.

Epidemiology

MS usually affects young adults, between 20 and 40 years old, with a later onset of disease for PPMS than for RRMS.[10] MS displays different incidence depending on gender, age, geographic distribution, and ethnic origin. Women are more susceptible to MS, the ratio women to men approaching 2:1 to 3:1, which suggests the possible involvement of sex hormones in susceptibility to MS.[11] However, the clinical pattern of PPMS does not show a female predominance.[10]

Pediatric MS accounts for 3–4% of cases and corresponds to symptom onset before age 18. The initial disease course of MS in children and adolescents is similar to that of adults affected by RRMS, with similar symptoms including cognitive deficits. Despite the variability in studies, it seems that pediatric MS patients reach the progressive stage of the disease later than patients with adult-onset MS (within 20 years from onset). Limited data are available regarding the immunopathogenesis

TABLE 30.1 Clinical Forms of Multiple Sclerosis

Clinical Form	Disease Course
Relapsing–remitting multiple sclerosis (RRMS)	85% of cases
	Age of onset between 20 and 30 years
	Characterized by discrete episodes of acute worsening of a given neurological function (relapses) over days to weeks, followed by complete or partial recovery periods (remissions) over months or years
	Female to male ratio between 2:1 and 3:1
	Early clinical symptoms: weakness, diminished coordination of limbs, optic neuritis, and sensory disturbance
Secondary progressive multiple sclerosis (SPMS)	Evolution of 50% of RRMS cases within 10 years of the initial diagnosis
	Characterized by gradual neurological decline with occasional minor recoveries
	Clinical features: inability to walk, progressive paralysis, brain and spinal cord atrophy
Primary progressive multiple sclerosis (PPMS)	About 15% of cases
	Later onset than RRMS (about 10 years)
	Characterized by steady functional worsening from the onset of the disease, without identifiable relapses or remissions
	Female to male ratio: 1:1
	Clinical features: starts with clinical disability, brain and spinal cord atrophy
Progressive relapsing multiple sclerosis (PRMS)	May be a variant of PPMS
	Characterized by steadily worsening disease from the onset, with clearly superimposed acute relapses with or without recovery

Source: Lublin and Reingold. Neurology. *1996;46(4):907–911.*[9]

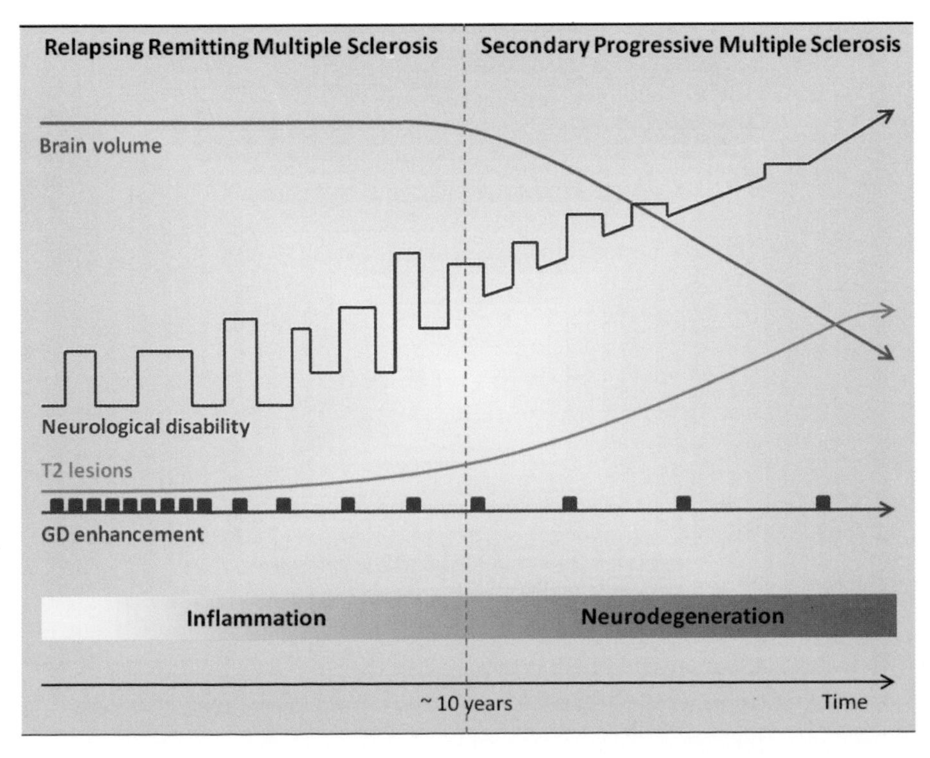

FIGURE 30.1 **Schematic description of the clinical evolution of relapsing–remitting into secondary progressive multiple sclerosis over the decades.** The graph shows neurological disability, brain atrophy, frequency of inflammatory events [T1 lesions with gadolinium (GD) contrast enhancement showing blood–brain barrier breakdown] and global level of tissue damage (T2 lesions).

of MS with childhood onset. However, pediatric MS patients display an increased activation of the innate immune system, at a higher level than observed in adult MS patients, whose pathophysiology is more dominated by the activation of the adaptive immune response. Magnetic resonance imaging (MRI) of pediatric MS cases shows fewer T2-weighted lesions and a higher frequency of large diffuse MS lesions than in adult MS cases. The first acute episode in MS at an early age has to be distinguished from a single neurological event, such as acute disseminated encephalomyelitis symptoms, which include those of optic neuritis.[12]

MS affects approximately 300,000 people in the USA and about 1 million worldwide,[13] raising critical socioeconomic, health, and care issues. Population- and family-based studies have observed a north to south gradient in disease prevalence in the northern hemisphere and the opposite in the southern (Fig. 30.2). The prevalence of MS per 100,000 population is higher than 60 in Europe and North America,[14] whereas the risk of developing MS in Asia and Africa is low. The geographic distribution of MS may be explained by environmental factors such as sunlight exposure or climate, as well as genetic susceptibility among populations. Migration studies have shown that the risk of developing MS is low if an individual migrated from a region with a high prevalence rate to one with a low prevalence rate before age 15, whereas it did not change after the age of 15, supporting the hypothesis that individuals born in low-risk areas benefit form long-lasting protection without transmission to their children. It has been suggested that ultraviolet light exposure may negatively influence disease development through its suppressive effects on the immune system, or through its involvement in the biosynthesis of vitamin D, one metabolite of which may have a role in inflammation.[11] Other behavioral or lifestyle factors such as industrialization, urban living, pollution, smoking habits, or diets may explain the disease distribution and the increasing worldwide prevalence over recent decades.[7]

ETIOLOGY OF MULTIPLE SCLEROSIS

Genetic Factors

MS is thought to emerge in genetically susceptible individuals who encounter external factors that trigger the inflammatory reaction against self antigens in the CNS. Different strategies – population-based association studies, family-based linkage methods, systematic genome screens – have been used to identify candidate genes that may have a role in the initiation of the disease.[15] Several family studies in different European regions have shown that first, second, and third degree relatives of affected

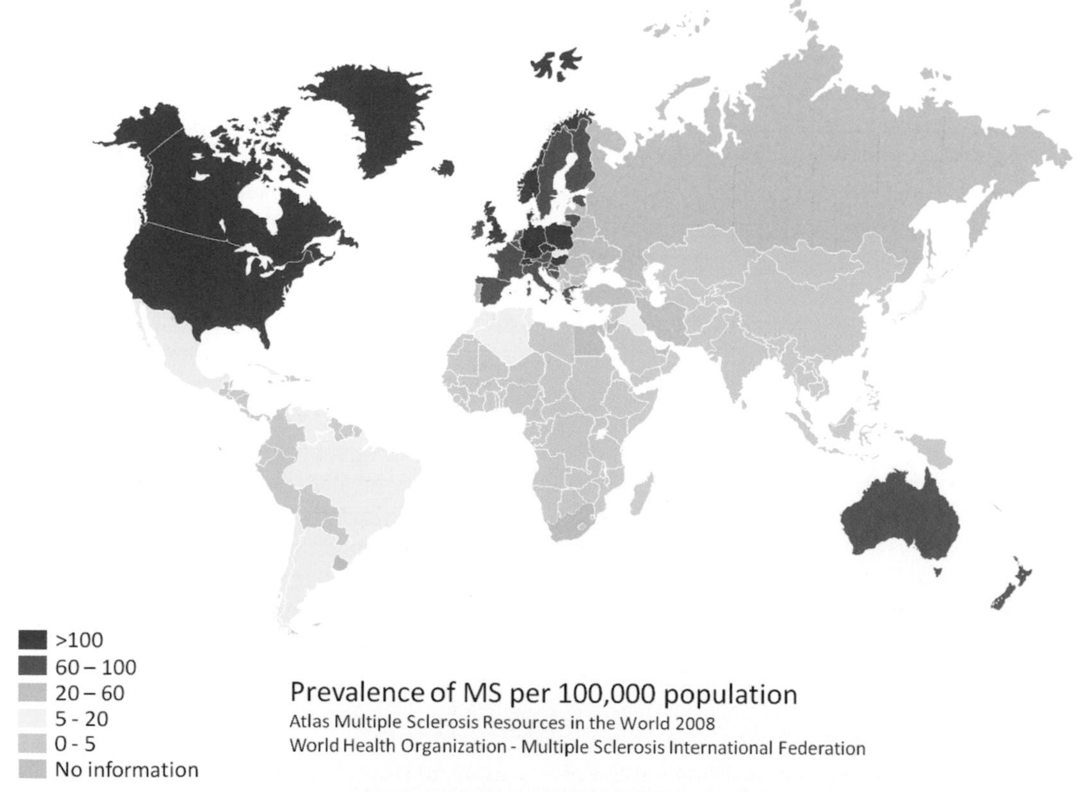

FIGURE 30.2 **Prevalence of multiple sclerosis (MS) per 100,000 population in 2008.** *Source: Data from World Health Organization; 2008. Atlas Multiple Sclerosis Resources in the World 2008.*[14] *Figure produced using Servier Medical Art.*

individuals display a higher risk of developing MS.[16] Studies of twins have shown an excess of monozygotic concordance compared with dizygotic concordance. Adoption studies and comparison of half-siblings and full-siblings reveal that genetic sharing is essential in familial aggregation of the disease, although discordance between most monozygotic twins also suggests a critical role of environmental factors in disease onset.[16]

The involvement of genes of the MHC region on chromosome 6p21 in disease susceptibility is one of the most consistent results of genetic association.[16] In particular, it is clear that specific haplotypes of the human leukocyte antigen (HLA)-DRB1*1501, -DQB1*0602, and -DQB1*0603 of MHC class II are strongly associated with higher susceptibility to MS. Higher frequencies of HLA class II allele DRB15 have been found in patients showing early-onset disease.[16] This increased susceptibility may be due to the *HLA-DRB1* and/or *HLA-DQB1* genes themselves, related to the function of these molecules in the immune response: antigen presentation or binding, T-cell recognition and selection of peptides, and intracellular signaling pathways. HLA-DR and -DQ molecules have been suggested to have binding characteristics that promote presentation of specific sets of self peptides, for example myelin peptides.[7] MS-associated HLA-DR and -DQ molecules could play a role in the determination of T-cell repertoire.[7] Although there is less information about genetic risk associated with HLA class I alleles, additional susceptibility alleles in class I loci, independent from class II alleles, have been identified. HLA-A*0301 increases the genetic risk of developing MS independently of DRB1*15, DQB1*06, and HLA-A*0201 decreases general susceptibility. HLA class I molecules may participate in the disease via the same mechanisms as HLA class II molecules, or via their interaction with cytotoxic $CD8^+$ T cells and natural killer (NK) cells.[17]

Polymorphisms of the T-cell receptor (TCR) have also been suggested to be associated with autoimmunity and with susceptibility to MS, with no significant evidence.[16] An allele of apolipoprotein E_4 ($ApoE_4$), which is a factor in lipid metabolism and has a significant role in remodeling and repair of nerve tissue, may be linked to faster progression of MS.[7,18] Cytotoxic T-lymphocyte antigen (CTLA)-4 is thought to be a candidate susceptibility gene for many autoimmune diseases, such as Graves disease, type 1 diabetes, and MS. CC-chemokine receptor 5 polymorphism has been shown to influence MS pathogenesis. Other candidates whose functions are related to immune response have been proposed, including interleukin-1 (IL-1) receptor gene, immunoglobulin loci, and tumor necrosis factor (TNF)-α and β alleles, but without conclusive evidence.[7]

Technological advances, including great progress in complete genome sequencing, have enabled the analysis of thousands of single-nucleotide polymorphisms (SNPs) in thousands of individuals, and the subsequent identification of linkage to different genomic regions.[17] Genome-wide association studies (GWAS) are based on the screening of statistical differences between cases and controls for the most frequent alleles.[17] GWAS confirmed the strong association of the *HLA-DRB1* risk locus on chromosome 6p21.3, but susceptibility genes outside the HLA region also came out with a genome-wide significance ($p < 10^{-8}$). These included *IL2Rα* (encoding interleukin receptor-2α) on chromosome 10p15 and *IL7R* (encoding interleukin receptor-7) on chromosome 5p13, both involved in cell survival, apoptosis, and immune response; and *CD58* (lymphocyte function-associated antigen-3) on chromosome 1p13, participating in cell-to-cell adhesion and immune response.[18]

The International Multiple Sclerosis Genetics Consortium (IMSGC) in collaboration with the Wellcome Trust Case Control Consortium (WTCCC2) has performed a large-scale GWAS, based on a dataset of 9772 MS cases versus 17,376 non-affected individuals.[19] This study highlighted more than 50 non-MHC regions containing SNPs with strong evidence of association. Among the identified MS risk loci, 29 have never been reported previously (Table 30.2). Some of the candidate genes identified by the study encode molecules involved in the immune response (*IL2Rα, IL7R, TNFRSF1A, TNFSF14*) or in costimulatory and signal transduction (*CD58, CD86, STAT3, TAGAP*). Another candidate gene, *CYP27B1*, is related to vitamin D metabolism, which may influence the autoimmune reaction in MS. The mechanisms involving the molecules coded by susceptibility genes *IL2Rα* and *VCAM-1* are targeted by therapies for MS.[19] Another GWAS identified three SNPs that have never been reported with MS, located at the *EOMES*, *MLANA*, and *THADA* loci, which show strong evidence of association ($p < 10^{-9}$).[20] Some additional SNPs display suggestive evidence of association with MS, and are known to be associated with autoimmune disorders such as Crohn disease, psoriasis, and rheumatoid arthritis.[19,20]

The role of SNP *TNFRSF1A*, encoding tumor necrosis factor receptor-1 (TNFR1), has been investigated through functional studies. The MS risk allele leads to expression of a soluble form of TNFR1, thus blocking TNF binding and subsequent nuclear factor-κB (NF-κB) signaling and TNFR1-mediated apoptosis. The MS-associated TNFR1 variant may act like TNF-blocking drugs, which have been suggested to promote or exacerbate the disease course.[21]

The list of MS susceptibility-associated genes is far from being complete and requires proper interpretation and further functional studies.

Environmental Factors

Based on the geographic prevalence of the disease decreasing with exposure to sunlight, some studies have suggested that solar radiation may be protective in MS.

TABLE 30.2 Non-Major Histocompatibility Complex Single-Nucleotide Polymorphisms (SNPs) with Genome-Wide Significant Association with Multiple Sclerosis

Chromosome	SNP	*p* Value	Risk Allele	Alleles	Gene of Interest	Biological Function
1	rs11581062	2.50×10^{-10}	G	A/G	*VCAM-1*[a]	Cell adhesion
1	rs1335532	2.00×10^{-9}	A	A/G	*CD58*	Cell-to-cell adhesion, immune response
3	rs11129295	1.20×10^{-9}	T	C/T	*EOMES*[a]	Developmental processes
3	rs9282641	1.00×10^{-11}	G	A/G	*CD86*	Cell activation and survival, immune response
5	rs6897932	2.60×10^{-6}	C	C/T	*IL7R*	Cell survival, immune response
6	rs1738074	6.80×10^{-15}	C	C/T	*TAGAP*[a]	Cell activation, immune response
9	rs2150702	3.28×10^{-8}	G	A/G	*MLANA*[a]	Melanosome biogenesis
10	rs3118470	2.00×10^{-9}	C	C/T	*IL2RA*	Apoptosis, immune response
11	rs630923	2.80×10^{-7}	C	A/C	*CXCR5*	Cell migration, immune response
12	rs1800693	1.80×10^{-10}	C	C/T	*TNFRSF1A*	Activation of NF-κB signaling, apoptosis, regulation of inflammation
12	rs12368653	2.00×10^{-7}	A	A/G	*CYP27B1*	Vitamin D metabolism
17	rs9891119	4.60×10^{-7}	C	A/C	*STAT3*	Cell differentiation, immune response
17	rs8070463a	9.55×10^{-8}	T	T/C	*TBX21*	Developmental processes, control of cytokine expression
19	rs1077667	9.40×10^{-14}	C	C/T	*TNFSF14*[a]	Costimulatory factor, cell proliferation, apoptosis

NF-κB: nuclear factor-κB.
[a]*Novel susceptibility genes identified by Sawcer* et al.[19]
Source: Gourraud et al.,[17] *Oksenberg* et al.,[18] *Sawcer* et al.,[19] *and Patsopoulos* et al.[20]

A first mechanism is based on the immunosuppressive effect of solar ultraviolet radiation: sunlight exposure may increase suppressor T-cell activity and decrease helper and autoreactive T-cell activity.[7,18] A second mechanism that has been proposed involves vitamin D biosynthesis, which is responsive to sunlight exposure. The prevalence of MS is low where vitamin D is abundant, as in sunny regions. It has been suggested that the hormone 1,25-dihydroxyvitamin D_3 [1,25-$(OH)_2D_3$], which is the most effective metabolite of vitamin D, may have a protective effect on genetically susceptible individuals. Receptors for this molecule are found in particular on activated T lymphocytes, and have been shown to be overrepresented in Japanese patients with MS. Strong evidence based on EAE models suggests that 1,25-$(OH)_2D_3$ is a natural inhibitor of the autoimmune mechanisms underlying MS.[11]

Behavioral and lifestyle influences are also thought to enhance the risk of MS. For example, smoking has been proposed to be associated with MS emergence in a large population of individuals living in Norway.[22] The risk of developing MS among smokers has been shown to be twice as high as among non-smokers, which may in part be explained by tobacco's toxic effects on the immune system and CNS.[22]

Epidemiological studies have highlighted a possible influence of parasitic infections on the MS disease course. MS patients infected with helminths, eukaryotic parasitic worms living and feeding in the host organism (e.g. in the digestive tract), show significantly fewer relapses, decreased activity on MRI, and a generally reduced disease severity compared with uninfected patients. There is evidence of a direct immunosuppressive function of helminthic infections in the inflammatory-driven response in MS.[23] Parasitic infection may be associated with the secretion of suppressive cytokines such as IL-10 and transforming growth factor-β (TGF-β) by regulatory T cells, and with the inhibition of T-cell proliferation and production of proinflammatory cytokines such as interferon-γ (IFN-γ) and IL-12. The reduction in parasitic infections may thus account for the increased risk of developing MS observed in developed countries.[23]

Infectious Factors

Several infectious agents have been proposed as a possible cause of MS, such as Epstein–Barr virus (EBV), rubella virus, measles virus, retroviruses, herpes virus, and zoster virus,[6,7] but such associations are unconfirmed. The most consistent evidence of involvement of a potential infectious etiology for MS is based on epidemiological studies.

EBV is a ubiquitous B-lymphotropic herpes virus that is responsible for infectious mononucleosis; it generally differs from other members of the herpes virus family because it may affect and transform antibody-producing B cells and plasma cells in the periphery as well as in the brain.[24] Epidemiological studies have reported a near-complete presence of anti-EBV antibodies in MS patients compared with healthy controls, who express a lower seropositivity.[7] Higher humoral responses to EBV has been found in the CSF of MS patients and antigen-specific immunoblotting has demonstrated that EBV proteins, EBV nuclear antigen-1 (EBNA-1) and the early protein BRRF2 may be targets for oligoclonal immunoglobulin G (IgG) bands in the CSF of MS patients.[24] Increased levels of EBV-specific $CD4^+$ and $CD8^+$ T-cell responses have been reported in the CSF and blood of MS patients,[24] supporting the hypothesis of a specific immune activation against the virus of viral proteins. Ectopic lymphoid structures in non-lymphoid tissues (e.g. the CNS) targeted by chronic inflammatory processes are thought to play a role in maintaining immune responses against persistent antigens.[25] The question of whether accumulation of EBV-infected cells in the CNS is a primary event in MS or a consequence of another disease mechanism is still unanswered.

Varicella zoster virus (VZV) DNA has been detected in CSF and peripheral blood mononuclear cells taken during MS relapse from patients with RRMS, which suggests an association between VZV infection and MS.[7]

Human endogenous retroviruses (HERVs) may play a role in autoimmune diseases and more particularly MS. HERV sequences have been isolated from cell culture of MS patients. Increased antibody reactivity to specific HERV epitopes is found in MS serum and CSF, and cell-mediated immune responses have also been reported. Although HERV-encoded proteins can have neuropathogenic effects, the causal role of HERVs in MS pathogenesis remains to be determined.[26]

Human herpes simplex virus-6 (HHV-6) is a β-herpes virus, which is T-lymphotropic and may also infect many different host cells such as monocytes. It has been proposed that RRMS patients experienced HHV-6 active infection, supporting the idea of a potential association between the viral active replication and relapse episodes. The link between SPMS and the potential role of HHV-6 is less well described, but it is thought that HHV-6 does not actively contribute to SPMS, whereas it may act as a trigger for MS attacks in patients with RRMS.[27]

Chlamydia pneumoniae is a bacterial pathogen of the respiratory tract that is responsible for community-acquired pneumonia. In spite of controversial results regarding its presence in the CSF of MS patients, *C. pneumoniae* is thought to be neurotropic and has been associated with CNS infections, since it has been shown to infect the CNS vasculature, initiating neurodegenerative disease.[7] Although MS patients are more likely to display *C. pneumoniae* in their CSF, evidence for the role of this bacterium in the etiology of MS is incomplete.

Other Factors

There is evidence supporting the influence of sex hormones in disease activity in both MS and EAE. Some studies have shown that pregnancy may be associated with a lower relapse rate in MS, with an increase in disease activity in the postpartum period. MRI studies support the idea of a correlation between estradiol and progesterone levels and gadolinium-enhancing lesions in women affected by MS; women with high levels of estradiol and low levels of progesterone have significantly higher levels of disease activity than women with low levels of both hormones. Experimental studies have suggested a protective effect of testosterone in EAE mice, and others have shown a decrease in disease severity correlated to high levels of estriol in EAE models, either after administration or during pregnancy. Sex hormones may influence MS onset and activity by affecting immune reactivity, including cytokine secretion, although the exact mechanism is not known.[11]

A positive strong correlation between dietary fat and MS prevalence has been found in some ecological studies. Linoleic acid, a polyunsaturated essential fatty acid, may significantly reduce the progression and severity of MS disability. Different mechanisms have been proposed to explain this influence on the etiology of MS. Myelin is composed of 75–80% lipids, of which polyunsaturated fatty acids are the major component. Polyunsaturated fatty acids may participate in membrane stability, susceptibility of the myelin sheath to demyelinating agents, and immune function.[11]

Investigating the environmental and non-genetic mechanisms underlying MS pathogenesis is difficult because of the large sample sizes required, and genetic causes may or may not be involved. However, new treatments for MS could be proposed based on such studies.

IMMUNE PATHOGENESIS OF MULTIPLE SCLEROSIS

Experimental animal models reinforce the concept of an autoimmune reaction in the CNS. It has been suggested that the immune system recognizes CNS myelin as foreign and is subsequently activated to attack it.[7] One hypothesis for this initiation is the possible activation of autoreactive cells by cross-reactivity between self antigens and foreign agents, a phenomenon known as molecular mimicry.

Molecular mimicry occurs when peptides of self antigens and infectious agents share sequences or have structural similarities, leading T and B cells to react with CNS antigens. Myelin basic protein (MBP), the major constituent of the myelin sheath of oligodendrocytes, is one autoantigen that shares similar MHC-II binding and TCR motifs with viruses such as herpes simplex virus (HSV).[7] When recognizing self antigens, cross-reactivity of self reactive T cells with foreign antigens may trigger activation of these cells, leading to crossing of the BBB and tissue damage upon recognition of antigens in the brain[28] (Fig. 30.3).

A second hypothesis for infectious induction of MS is bystander activation of autoreactive immune T cells, which occurs non-specifically during infections. Viral infections cause the activation of antigen-presenting cells (APCs) such as dendritic cells, which then trigger preprimed autoreactive T cells to initiate the autoimmune response.

Another mechanism of bystander activation involves inflammatory cytokines, superantigens, and toll-like receptor (TLR) activation, which subsequently activate autoreactive T cells in a TCR-independent manner. Virus-specific T cells can also initiate bystander activation depending on specific TCR recognition. Activated virus-specific T cells migrate to the infected CNS, where they kill infected cells presenting viral epitopes. This results in secretion of inflammatory cytokines including nitric oxide by dying cells, such as $CD8^+$ T cells and macrophages, leading to the destruction of self tissue and release of autoantigens. In the infectious context presentation of autoantigens may then activate autoreactive T cells.[7]

Pathophysiology

MS is considered an autoimmune disease, occurring when Th1 cells recognize components of the myelin sheath.[5,7] The pathological hallmark of MS is the demyelinated plaque, which consists of a defined hypocellular area characterized by loss of myelin, relative preservation of axons, and formation of an astrocytic scar. Lesions are found in optic nerves, periventricular white matter, brainstem, cerebellum, and spinal cord white matter, and they often surround medium-sized blood vessels.[1] Demyelination can also involve the gray matter, associated with meningeal inflammation. Acute highly inflamed white matter lesions are also characterized by axonal transection, probably due to a lack of myelin protection and oligodendrocyte support.[6] MRI diagnosis and CSF findings influence the classification of clinical syndromes associated with MS. High-field MRI highlights several distinct areas for MS lesions: infratentorial, callosal, juxtacortical, periventricular, and other white matter areas.[29] Variations in clinical disability during disease progression may be explained by structural differences among lesions, as assessed by the water diffusion coefficient and the degree of lesion hypointensities on T1-weighted MRI.[5]

Four pathological MS patterns have been identified based on the extent of myelin loss, localization and extension of plaques, oligodendrocyte destruction, and relative contribution of different immune cells.[30] Pattern I is dominated by infiltration of T cells and macrophages and may involve proinflammatory molecules such as TNF-α and IFN-γ; pattern II displays antibody and complement deposition and involves myelin oligodendrocyte glycoprotein (MOG)- and MBP-specific antibodies; pattern III is characterized by distal oligodendrogliopathy-associated demyelination, and may be induced by focal cerebral ischemia or toxic virus; and pattern IV results from a metabolic defect leading to non-apoptotic degeneration of primary oligodendroglia and occurs primarily in PPMS. All these patterns start from T-cell mediated inflammation and result in active demyelination and subsequent acute axonal injury. Lack of trophic support by oligodendrocytes then leads to chronic axonal injury in inactive demyelinated plaques.[30]

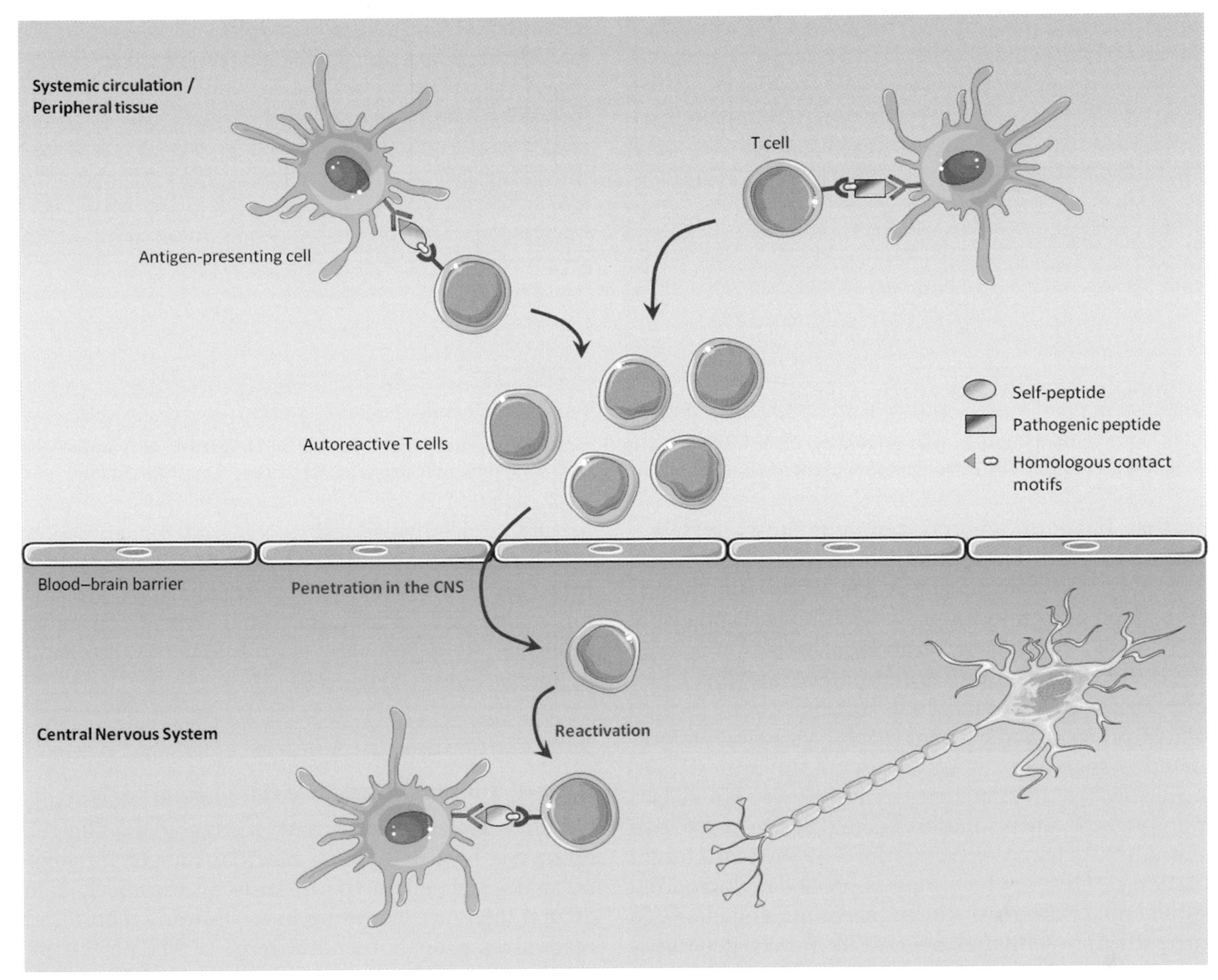

FIGURE 30.3 **Possible mechanism of cross-reactivity between self antigens (myelin basic protein, proteolipid protein, myelin oligodendrocyte glycoprotein) and infectious agents (Epstein–Barr virus, herpes simplex virus).** *Figure produced using Servier Medical Art.*

Autoimmunity-Related Events

During the initial state of the inflammatory response, activated T cells with encephalitogenic potential are recruited from the periphery to the CNS. Upon activation, T cells express integrins on the cell surface, which enable them to bind to specialized capillary endothelial cells of the BBB. Then, they migrate across the disrupted BBB through the endothelium and the endothelial basal lamina into the CNS parenchyma.[6] Following activation and recruitment of leukocytes, endothelial cells reorganize their membrane in multiple cup-shaped microdomains enriched with cellular adhesion molecules such as intracellular cell adhesion molecule-1 (ICAM-1) and vascular cell adhesion molecule-1 (VCAM-1). These molecules surround the migrating T cells and allow their passage across the endothelium through individual cells.[26] MRI clinical observations of acute and chronic active MS lesions reveal the breakdown of the BBB, which allows the infiltration of lymphocytes and leukocytes into the CNS. The release of proinflammatory cytokines such as IFN-γ and TNF-α activates cerebral endothelial cells and modulates the BBB phenotype by induction of several inflammatory genes. They affect BBB integrity through several mechanisms including inhibition of junctional protein expression. Activated T cells that migrate across the BBB express matrix metalloproteinases, which can disrupt the subendothelial basal lamina[31] (Fig. 30.4). This process is followed by a second amplification of the immune response within the CNS, as pathogenic T cells are reactivated by fragments of myelin antigens and other factors are involved in myelin degradation and damage to oligodendrocytes. Numerous non-specific, blood-derived and CNS inflammatory cells release myelin-toxic substances (e.g. lymphotoxin, nitric oxide, and perforins).[3] Activated macrophages and microglial cells secrete proinflammatory cytokines such as TNF-α and IFN-γ, toxic reactive oxygen species (ROS), or proteolytic and lipolytic enzymes. Cytotoxic activity of $CD8^{+}$ T cells may also participate in damaging the

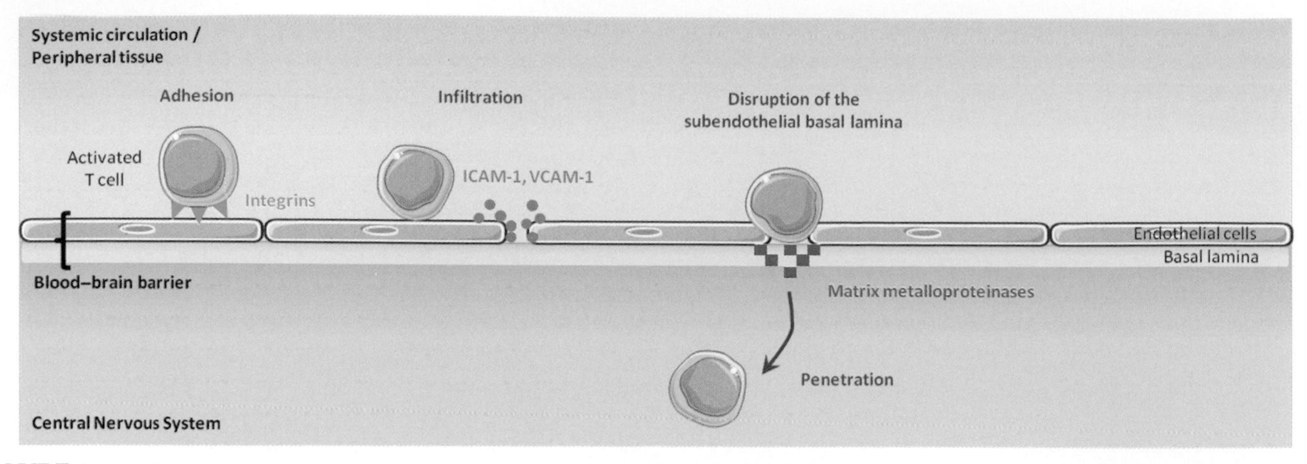

FIGURE 30.4 **Blood–brain barrier-related events leading to disruption of subendothelial basal lamina and migration of lymphocytes into the CNS.** ICAM: intracellular cell adhesion molecule; VCAM: vascular cell adhesion molecule. *Figure produced using Servier Medical Art.*

axons that have undergone demyelination.[3] Intrathecal synthesis of immunoglobulins is increased in MS patients. Once inflammation has started, myelin-specific B cells as well as antibodies infiltrate the disrupted BBB, enter the CNS, and release antibodies that contribute to myelin destruction through different mechanisms. These include complement-mediated opsonization, which facilitates phagocytosis by macrophages, complement-mediated cytolysis, or stimulation of antibody-dependent cell-mediated cytotoxicity by binding to NK cells.[6]

Spontaneous remyelination occurs in the context of inflammation-driven demyelination, via the recruitment and action of OPCs, which may express developmental genes, usually expressed during early myelogenesis, to re-ensheath demyelinated areas with newly generated myelin. Studies have highlighted the reactivation of oligodendrocyte transcription factor-1 (Olig-1) during remyelination in MS patients, which initiates the process of myelin regeneration. Other molecules including chemokine CXC and CC receptors are responsible for the recruitment of OPCs in the damaged areas.[5] Human and experimental models of CNS demyelination have reported that spontaneous remyelination occurs in response to myelin injury, and that modulation of the immune response may promote the remyelination process in patients with acute or early MS.[8] However, spontaneous remyelination fails to compensate for demyelination, resulting in chronic axonal injury and permanent neurological disability. Despite a large number of surviving OPCs in many MS lesions, failure of myelin repair may be due to an inadequate or insufficient recruitment of OPCs, or to a failure of recruited OPCs to differentiate into remyelinating oligodendrocytes.[8] Depletion of OPCs in some patients may be explained by the effect of MOG on OPCs: patients displaying MOG-directed antibodies may be deficient in OPCs.[8] Repeated remyelination and demyelination may also lead to progressive loss of function and depletion of OPCs.[8] Demyelinated axons may be unresponsive to remyelination because of intrinsic axolemmal alterations, for example expression of inhibitory cell surface molecules such as polysialylated neural cell adhesion molecule (PSA-NCAM), neurofilament fragmentation, or energy failure[6] (Fig. 30.5).

The Major Immune Players

Role of T-Lymphocyte Mediated Immunity

Autoreactive T cells may recognize myelin-related epitopes. T cells that are reactive to self antigens are normally deleted in the thymus by the mechanism of central tolerance, resulting in an immune repertoire that recognizes exogenous pathogens. In MS defects in self tolerance may lead to production of antigen-specific T cells with TCR alterations that are able to recognize self proteins and produce autoimmune reactions.[7] During inflammatory events, autoreactive T cells release cytokines that participate in migration and recruitment of cells to specific targets, initiating the inflammatory response.[6] Myelin-protein specific T cells that are retrieved from the peripheral circulation in healthy adults as well as in MS patients are controlled by peripheral tolerance that involves costimulatory signaling, transcriptional and epigenetic mechanisms, and regulatory T (Treg) cells.[7] Although not completely understood, dysfunction of Treg cells is thought to participate in MS pathogenesis.[32] Through a direct action on Treg cells, tolerogenic or immunogenic dendritic cells maintain immune reactions and thus contribute to the homeostasis of CNS immunity. Altered or dysfunctional dendritic cells have been described in MS patients.[32] Autoreactive T cells undergoing clonal expansion and clonotypes with unique TCR repertoires have been detected in MS lesions. $CD4^+$ T cells are found predominantly in the perivascular infiltrates and in the parenchyma, while $CD8^+$ T cells are identified in brain tissue as well as in

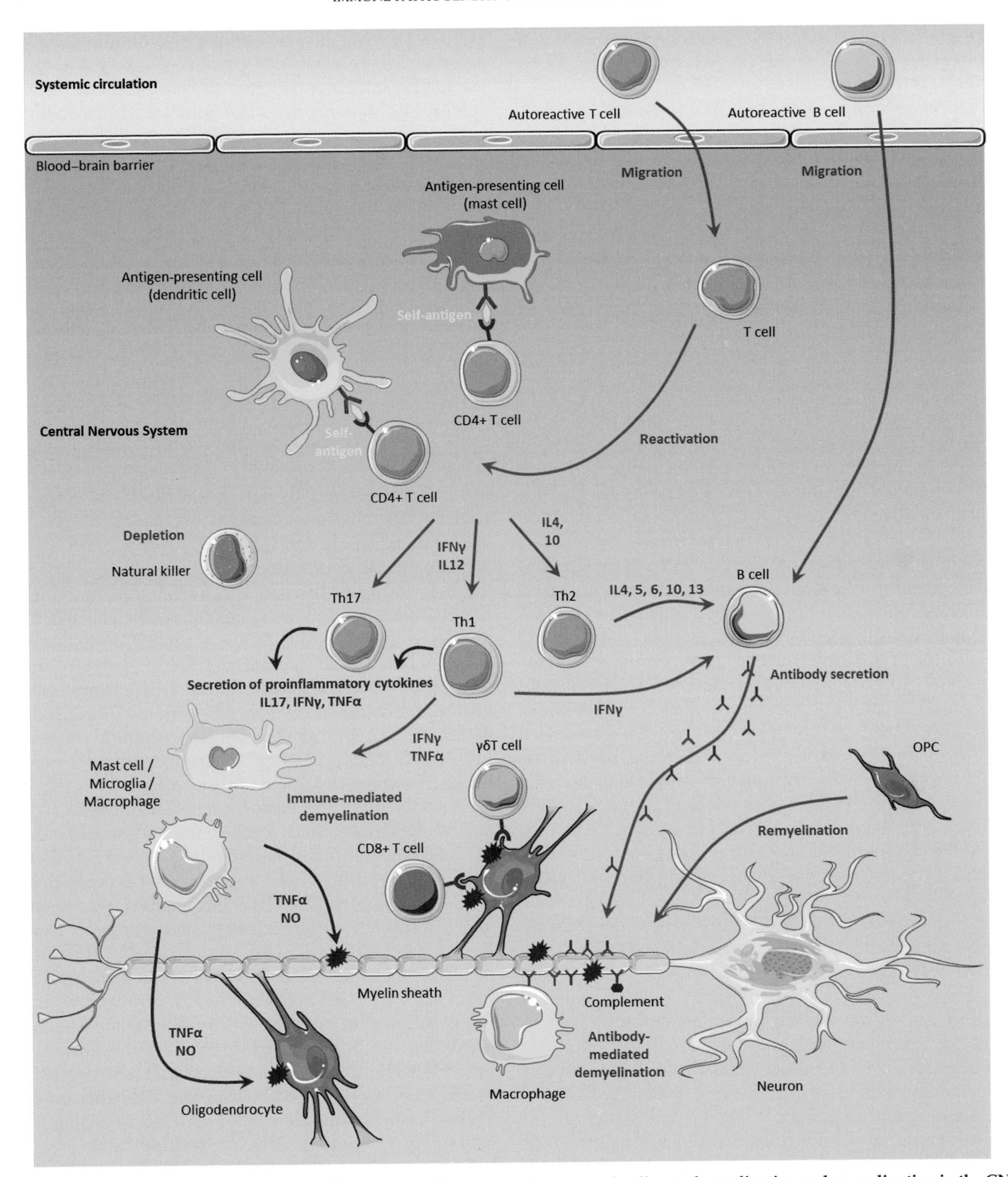

FIGURE 30.5 **Inflammation-driven demyelination: possible inflammatory events leading to demyelination and remyelination in the CNS.** IFN: interferon; IL: interleukin; TNF: tumor necrosis factor; NO: nitric oxide; Th: T-helper cell; OPC: oligodendrocyte progenitor cell. *Figure produced using Servier Medical Art.*

peripheral blood.[7] These cells may represent a pool of memory T cells, which could be recruited during acute attacks, but this mechanism remains unclear.

Current evidence exists that MS primarily involves $CD4^+$ autoreactive T cells as a central element for the autoimmune hypothesis of MS. $CD4^+$ T cells have been identified as CNS- and CSF-infiltrating cells in MS patients. In the EAE model, the injection of myelin protein components or spinal cord homogenate into animals induces a $CD4^+$ T-cell driven autoimmune disease course that is similar in many respects to that observed in MS.[7] Some HLA class II molecules are thought to increase the genetic risk of developing MS. Their role as antigen-presenting molecules to pathogenic $CD4^+$ T cells also supports the importance of $CD4^+$ T cells in the immunopathogenesis of MS. The frequency of high-avidity myelin-specific $CD4^+$ T cells that respond to low antigen concentrations to myelin proteins has been shown to be significantly higher in MS patients than in healthy controls, and they express mostly a proinflammatory phenotype.[7]

T-helper (Th) cells are $CD4^+$ effector cells that are differentiated in response to exposure to specific interleukins. Th cells may express different phenotypes based on their cytokine secretion profile and provide immune responses to mediate protection against pathogens.[33] When T cells are polarized to Th1 (proinflammatory phenotype), inflammation is promoted. Th1 cells secrete IFN-γ in large quantity, and also TNF-α, lymphotoxin-α (LT-α), and IL-2. IFN-γ produced by Th1 mediates protection against intracellular pathogens by activating macrophages. The number of mononuclear cells that secrete IFN-γ is increased in blood and CSF of MS patients, and IFN-γ is present in MS lesions.[34] In EAE, antibodies against LT-α and TNF-α have been shown to modulate the disease course, supporting the role of Th1 cells in EAE.[34]

Chemokines and their receptors play a central role in the inflammatory reaction. Among the different chemokine receptors, CXCR3 and CCR5 show preferential expression on Th1 cells. Chemokines induce leukocyte adhesion to the endothelium, and play an important role in the infiltration of inflammatory T cells through the BBB into the CNS.[7] Deregulated Th1 response to myelin components leads to an autoimmune reaction and brain tissue inflammation in EAE.

The damaging role of proinflammatory cytokines in MS is controversial. IFN-γ delivered intrathecally to EAE mice may increase the expression of the anti-inflammatory molecule TNFR1 on CNS-infiltrated lymphocytes. The subsequent binding of TNF-α on TNFR1 induces apoptosis of these cells. Mice treated with IFN-γ show a significant decrease in EAE inflammatory features (e.g. number of inflammatory infiltrates) and a general inhibition of the disease course.[35] These data challenge the exclusive detrimental function of proinflammatory cytokines such as IFN-γ in EAE and in MS, and the exclusive use of anti-inflammatory therapies in MS.

The Th2 phenotype is characterized by secretion of anti-inflammatory cytokines including IL-4, IL-5, IL-6, IL-10, and IL-13. Th2 cells induce B-cell proliferation and stimulate antibody production.[30] IL-4 and IL-13 provide positive feedback for Th2 cell differentiation, and IL-4 may inhibit the development of Th1 and Th17 proinflammatory phenotypes. Th2 responses can downregulate autoimmune disease. MS is associated with a parallel upregulation of proinflammatory Th1 cytokines and an immune response downregulation of anti-inflammatory Th2 cytokines such as IL-10 and TGF-β, a factor produced by activated T lymphocytes that has suppressive effects on both T- and B-cell related immunity.[34]

A distinct $CD4^+$ T-cell subpopulation has been identified as (IL-17-producing) Th17 cells. Proinflammatory cytokine IL-17 and regulated cytokine IL-23 are increased in CNS lesions and in peripheral blood mononuclear cells of MS patients. Increased peripheral blood Th17 cells may be associated with MS severity.[36] IL-17 receptors are seen in acute and chronic MS plaques.[33] In EAE models, induced disease is more severe after transfer of IL-17-producing $CD4^+$ T cells precultured with IL-23 than after transfer of Th1 cells, supporting the hypothesis of a critical role for IL-23 in EAE.[5] The Th17 population and IL-23-induced secretion of IL-17 are emerging as critical mediators of chronic autoimmune diseases. Both Th1 and Th17 cell populations may promote inflammation in MS, acting in parallel through different mechanisms of action[36]; however, the impact of Th17 cytokines remains unclear. Overexpression of T-cell specific IL-17A in EAE mice does not exacerbate the pathological course of the disease, and deficiency in IL-17A and IL-17F does not show a clear beneficial impact on the development of EAE. This suggests that Th17 cells may exert their pathogenicity through other factors or mechanisms of action. IL-17A produced by Th17 cells may disrupt the BBB by inducing the production of ROS, which are harmful to brain endothelium permeability (downregulation and reorganization of tight junction molecules), leading to breakdown of the BBB and infiltration of immune cells into the CNS. Conversely, IL-17A-deficient mice show lower levels of ROS production and BBB disruption.[36] These findings suggest a role for IL-17 in maintaining BBB integrity, but not in disease development.

Dysfunction of the suppressor function of Treg cells may be associated with MS. Treg cells can be classified into two subsets, natural and inducible Treg cells, according to the molecules expressed on their surface or their cytokine secretion profile.[32] Treg-cell mediated suppression plays a central role in the control of autoreactive T cells and the induction of peripheral tolerance. Although the frequency of $CD4^+/CD25^+$ Treg cells does not differ from healthy controls, CD4+/CD25+ Treg cells in MS patients appear to be functionally impaired or to have deficits in their maturation or in the thymic output. EAE studies further support this hypothesis, and the adoptive transfer of CD4+/CD25+ Treg cells from naïve mice prevents or reduces the severity of EAE. Inactivation or depletion of Treg cells using anti-CD25 antibody may increase susceptibility to EAE and prevent secondary remissions in myelin SJL mice immunized with proteolipid protein (PLP).[32]

CD8+ T cells have been detected in MS plaques and may have regulatory functions in the progression of the disease. Experimental studies have suggested that myelin-specific CD8+ T cells induce severe CNS autoimmunity, which has similarities to MS. Prominent oligoclonal expansions of CD8+ memory T cells have been found in the CSF and brain tissue of MS patients. The CD8+ cytotoxic T-cell response to MBP is increased in MS patients.[7] Cytotoxic CD8+ T cells may kill glial cells, leaving axons exposed; they mediate axonal transection in active MS lesions.[3,6]

Role of Humoral Immunity

The observation that immunoglobulin levels are increased in the CSF of MS patients is supporting evidence for a role of B cells and humoral immunity in MS immunopathogenesis.[7] It has also been observed that B cells, plasma cells, and myelin-specific antibodies are present in chronic and acute MS lesions with demyelination.[37] Mutant mice engineered to produce high levels of MOG-specific antibodies displayed higher EAE severity and more progressive disease, even after disease induction with myelin components other than MOG.[37] In progressive MS, organized ectopic lymphoid follicles have been observed close to inflammation-driven demyelinated MS lesions. They resemble secondary lymphoid organs in the CNS and contain proliferating B cells and follicular dendritic cells. The formation of ectopic (CNS) lymphoid tissue may play a role in maintaining immune responses against persistent antigens.[25]

Antibodies directed against proteins and lipids of the myelin sheath and against molecules in the CNS are secreted by B cells that have migrated into the CNS or from serum that has extravasated through the BBB.[13] Secretion of myelin-specific antibodies by B cells may lead to the destruction of myelin within plaques. B cells act as APCs for autoreactive T cells, and antibodies participate in the uptake and processing of antigen. B cells also provide costimulation to autoreactive T cells. Antibodies can opsonize myelin for macrophage phagocytosis and thus cause demyelination, and also target and address other cells to inflammatory sites.[37] B cells may exert a beneficial role in MS by influencing cytokine production to maintain a balance in the Th1/Th2 profile through their antigen-presenting function. Antibodies that normally recognize CNS components may also promote myelin repair and enhance remyelination, or have a regulatory function, for example by inducing T-cell anergy.[37]

Role of Innate Immune Mechanisms

Inflammation in MS may be initiated by Th1 cells that cross the BBB after activation in the periphery, recognize their target antigen, and activate non-specific effectors, for example macrophages and microglia, which are largely responsible for demyelination and axonal damage. Dendritic cells present in the CSF compartment and in MS lesions may contribute to the chronicity of MS and may play a role in PPMS.[32] TLRs present on dendritic cells may display inappropriate signaling in MS and EAE, leading to a potential inhibition of the immunosuppressive effects of Treg cells and thus contributing to the maintenance of MS.[7] Mature dendritic cells may be retained in the inflamed tissue and drive naïve T cells to a Th1 phenotype by secreting IL-12.[32]

Mast cells are increased in the CSF of MS patients, as well as in MS plaques and acute lesions. They have been suggested to act as APCs and to mediate a shift in the Th1/Th2 population.[7] NK cells have been associated with MS, based on their observed depletion in peripheral blood, plaques, and CSF of MS patients, as well as in EAE. NK could be involved in suppression of autoimmunity by producing cytokines or by inducting target lysis.[7] γδT cells have been found in chronic MS lesions and in the CSF of MS patients. These cells lyse oligodendrocytes via perforin, supporting a role in MS pathogenesis.[7]

Role of Self Antigens

The most studied candidate self antigens are the proteins that constitute the myelin sheath, such as MBP, MOG, myelin-associated glycoprotein (MAG), and PLP. MBP and PLP are the most abundant proteins that compose the myelin. PLP has been shown to induce autoreactive T- and B-cell responses in MS.[7] These proteins are expressed within the intracellular surface of myelin membranes of the sheath and are involved in maintaining the structure of compact myelin. MOG is less abundant and is specifically expressed on the surface of myelin, making it a target for demyelinating antibodies and cellular immune responses in MS. All of these self antigens can induce EAE in mice, rats, guinea pigs, and non-human primates.[7] Other myelin and non-myelin antigens targeted by CD4+ T cells include 2′,3′-cyclic nucleotide 3′-phosphodiesterase (CNPase), myelin-associated oligodendrocytic basic protein (MOBP), oligodendrocyte-specific glycoprotein (OSP), αB-crystalline, S100β protein, and myelin lipid components.[7]

Damage to myelin can be considered as an epiphenomenon following non-specific polyclonal activation of T and B cells by bacterial or viral antigens, or due to molecular mimicry, when proteins in the pathogen share molecular features with self antigens, such as specific myelin epitopes.[6]

CLINICAL FEATURES OF MULTIPLE SCLEROSIS

Clinical Features in Different Types of Multiple Sclerosis

There is much variability in the clinical features of MS, in the onset as well as in the progression of the disease. Symptoms result from the interruption of myelinated tracts in the CNS, whereas the peripheral nervous system is not disturbed.[6] The frequency of new symptoms depends on the age of onset. If the onset is before 20 years of age, the majority of symptoms relate to optic sensitivity, while motor deficits are prevalent with onset at later ages. The most common features of MS are paralysis, sensory disturbances, lack of coordination, and visual impairment.[13]

Current models for MS disease progression support the occurrence of two overlapping and connected stages, one driven by inflammation and one by neurodegeneration, with no clear understanding of their temporal relationship. Most MS cases start with attacks of neurological disturbance, which are probably mediated by autoimmune reactions leading to inflammatory lesions and demyelination. These attacks are followed by complete or partial recovery and thought to be free of disease progression. About half of these cases turn into a progressive phase characterized by steady worsening of the disease with superimposed relapse episodes, and often characterized by an inability to walk and progressive paralysis. Occasionally, clinical disability begins with the progressive phase, in the PPMS clinical pattern. The progressive stage is likely to be driven by degeneration of both the myelin sheath and the axons, and subsequent brain and spinal cord atrophy underlie motor symptoms[13] (Table 30.3).

TABLE 30.3 Clinical Features of Multiple Sclerosis

Lesion Site	Clinical Symptoms
Cerebrum	Hemifacial weakness and pain
	Motor impairments
	Cognitive deficits (memory and attention deficits, slowed information processing)
	Psychiatric features (dementia, bipolar disorder, depression)
Cortex	Motor, sensory, and cognitive disability
Optic nerve	Optic neuritis (often monocular acute blurring or loss of vision)
Cerebellum	Postural and action tremor
	Limb incoordination
	Gait instability
	Ataxia
Brainstem	Diplopia (double vision)
	Vertigo
	Impaired speech and swallowing
	Paroxysmal symptoms
Spinal cord	Weakness
	Diminished dexterity
	Spasticity (paraparesis, shaking, progressive ambulatory disability, stiffness)
	Autonomous disturbances (sexual impotence, bladder dysfunction, constipation)
	Pain (Lhermitte sign)
Other	Fatigue
	Temperature sensitivity

Source: Compston and Coles[1] and Hauser and Oksenberg.[6]

Clinical Symptoms

Clinical symptoms of MS often correlate with the functional localization of impaired conduction in MS lesion.[1] Lesions in the cerebrum result in hemifacial weakness, pain, and motor impairments, as well as the cognitive deficits that are common in advanced cases, including memory and attention problems and slowed information processing.[1,6] Later in the disease, patients may display psychiatric manifestations: dementia, bipolar disorders, pathological laughter and crying, and psychosis. Depression is experienced by 60% of patients.[6]

Cortical demyelination is believed to follow white matter pathology and to be rare in acute or early relapsing MS. Cortical plaques lead to focal cortical deficits underlying motor, sensory, and cognitive disability in MS.[1,6] Imaging in patients with high-strength magnets, functional MRI, and positron emission tomography, as well as cortical thickness and CNS atrophy measures, may provide useful information to characterize gray matter demyelination and better correlate cortical atrophy with specific disabilities, including cognitive dysfunction, fatigue, and depression.[2]

One of the most common initial symptoms is visual impairment, including reduced visual acuity and unilateral painful loss of vision. Demyelination commonly affects the visual system from the optic nerve to the cerebrum, since approximately 40% of the brain is involved in visual function. Among the neuro-ophthalmic symptoms, optic neuritis is the most clinically isolated demyelinating syndrome. Optic neuritis is caused by inflammation or demyelination of the optic nerve and results in acute blurring or loss of vision, usually monocular.[1]

Some common early symptoms are caused by MS lesions located in the brainstem. They include diplopia (double vision), vertigo, impaired speech and swallowing, and paroxysmal symptoms thought to originate from discharges along demyelinated axons.[1,6] Lesions in the cerebellum are responsible for postural and action tremor, and for limb incoordination, gait instability, and ataxia.[1,6]

The 2005 revision of the criteria for MS diagnosis focuses on characterizing lesions of the spinal cord and aims to simplify the diagnosis of PPMS.[38] Spinal cord lesions can be detected with high sensitivity using high-field MRI.[6] Lesions in the descending motor pathways of the spinal cord lead to weakness or diminished dexterity in the limbs, especially the legs. Spasticity, including spastic paraparesis, shaking, and difficulty walking, progressively turns into complete disability, and stiffness is also observed as the disease worsens. Lesions of motor and sensory fibers of the spinal cord and hypoactivity result in autonomic disturbances such as sexual impotence, bladder dysfunction, and constipation.[1,6] Ancillary symptoms include the Lhermitte sign, an electrical sensation running down the spine and into the limbs triggered by neck flexion, which is attributed to spontaneous discharges of demyelinated axons.[1,6] Fatigue and temperature sensitivity are other symptoms that occur in most patients[1,6] (Table 30.3).

Measures of Disease Progression

Both the clinical definition and therapeutic trials now require quantification of physical disability in MS. The Kurtzke Expanded Disability Status Scale (EDSS) is the standard measurement of neurological impairment in MS commonly used in clinical trials. A measure of overall neurological impairment was first given by the Disability Status Scale (DSS), which ranges from 0 (normal neurological examination) to 10 (death due to MS). The DSS evolved into the EDSS by dividing each step into two and defining functional system grades at each step. The EDSS quantifies disability in eight functional systems: pyramidal, cerebellar, brainstem, sensory, bowel and bladder, visual, cerebral, and other. It ranges from 1.0 to 4.5 for people affected with MS who are fully ambulatory, and from 5.0 to 9.5 for people with MS with impaired ambulation. The maximum grade, 10.0, represents death due to MS.[39] Although the Kurtzke EDSS is a standard method to evaluate disability, its use is limited by inconsistencies between evaluators and its insensitivity to clinical changes at a specific grade.

Cognitive dysfunction is not strongly correlated with the EDSS score, and can be rated through tests such as the Paced Auditory Serial Addition Test (PASAT) and the Symbol–Digit Modalities Test (SDT). Other clinical scales such as the Scripps Neurological Rating Scale (SNRS) and the Trojano Neurological Rating Scale are based only on the neurological examination. The Multiple Sclerosis Functional Composite (MSFC) includes quantitative functional measures of three key clinical issues of MS: leg functional ambulation, arm/hand function, and cognitive function. This measure of disability may be a consistent method by which to define the clinical status of MS patients.

Quantifying MRI techniques may be more sensitive, objective, and reliable than clinical methods in detecting and defining MS disease activity, although many MS lesions have no clinically detectable symptoms.[6]

PROGRESSIVE MULTIPLE SCLEROSIS AS AN UNMET NEED

About half of RRMS patients shift to a SPMS within 10 years of the initial diagnosis; this clinical form is characterized by a steady progression of neurological damage with or without relapses and the occasional minor remission or plateau.[9] Only about 15% of MS patients develop PPMS, with slow continuous worsening of neurological decline and no identifiable relapses or remissions.[9] PPMS displays a later (about 10 years) onset than RRMS. Available data on clinical phenotypes show differences in epidemiology of RRMS and progressive MS, raising the issue of a different pathogenic mechanism being responsible for the onset of the disease. The female to male ratio for PPMS is about 1:1, whereas it is between 2:1 and 3:1 for RRMS.[10] The genetic risk of developing MS associated with MHC-II genes is similar in RRMS and in PPMS, as well as other identified susceptibility genes.[40] Although the geographic distribution of progressive MS does not reflect the general distribution of MS, progressive MS is more frequent in relatively resistant populations (e.g. blacks and Asians).[10]

It is hypothesized that SPMS and PPMS emerge as the result of axonal and neuronal damage, consequent to accumulating multifocal inflammatory events, and leading to permanent and irreversible neurological disability. It is likely that the conversion of RRMS into SPMS occurs when the CNS fails to compensate for demyelination and subsequent axonal loss with spontaneous recovery mechanisms.

Comparison in Pathophysiology

Demyelinating lesions of white matter, identified by the prominence of macrophages containing myelin degradation products, are more frequent in RRMS than in SPMS or PPMS, whereas the inflammatory reaction is more pronounced in SPMS than in PPMS. There is also a clearer cortical demyelination in progressive MS than in RRMS. Acute axonal damage associated with inflammation and lack of trophic support from myelin sheaths seems to be more prominent in progressive MS than in RRMS.[10] Progressive MS is characterized by increased demyelination in white and gray matter, with less inflammation, which is more diffuse, within the lesions, and with acute axonal damage.

In progressive MS, inflammation is trapped behind a closed BBB. The severity of inflammation of the meninges and the formation of lymph follicle structures are associated with the activity of cortical lesions and with activation of microglial cells. Activated T cells release diffusible factors in the parenchyma, leading to CNS damage, either directly or indirectly via microglia activation. Soluble factors may include specific demyelinating antibodies, but they have not been identified.[25] In progressive MS, neurodegeneration may result in overreaction of microglial cells which produce neurotoxic factors upon stimulation by T cells.[25]

Another feature of demyelination and neurodegeneration (e.g. loss of structure and/or function of neurons) in progressive MS is mitochondrial injury. Damaged axons found in active lesions of chronic MS display profound mitochondrial alterations, leading to energy failure. Accumulation of Na^+ is observed in the axoplasm. As axonal Na+ concentrations increase, the Na^+/Ca^{2+} exchanger exchanges axoplasmic Na^+ for extracellular Ca^{2+}, leading to a toxic excess of Ca^{2+} in the axon and finally axonal degeneration. It is unclear what initiates mitochondrial impairment in MS, but it is likely that tissue injury is a result of inflammatory processes in the CNS.[25]

Comparison Based on Brain Imaging Techniques

MRI studies reveal smaller and less frequent gadolinium-enhanced lesions, indicating areas of active inflammation, in PPMS compared with SPMS patients. Lesions observed in PPMS may be associated with the progression and severity of disability. Gadolinium-enhanced brain inflammatory lesions in PPMS are more commonly observed in early disease.[40] Multifocal T2-hyperintense lesions of PPMS frequently affect the spinal cord, most commonly in the cervical region.[10] Proton magnetic resonance spectroscopy enables the quantification of disease evolution by measuring the white matter concentration of brain *N*-acetyl aspartate (NAA), a marker of neuronal integrity.[6,10] The concentration of NAA as assessed by proton magnetic resonance spectroscopy appears to be reduced more in PPMS and SPMS than in RRMS.[40] An increase in diffusion abnormalities on diffusion tensor imaging is seen in PPMS, attesting to an accumulation of pathology in gray matter over time.[40]

Magnetization transfer ratio imaging allows quantification of cerebral myelin content, and can be used to assess demyelination and remyelination. Patients with PPMS display a reduction in magnetization transfer ratio values in gray matter, which is associated with clinical disability and cognitive impairment. These values do not differ significantly in PPMS and RRMS cases.[10]

Quantitative MRI shows that cerebral atrophy of both gray matter and white matter occurs frequently in early PPMS, to a smaller extent in SPMS, and even less in RRMS.[10] Brain atrophy is an early clinical feature of PPMS and an indication of disease progression. Spinal cord atrophy is more pronounced in PPMS and SPMS than in RRMS. Cortical lesions using double-inversion-recovery imaging are observed in PPMS patients more than in RRMS, relative to disease duration,[10] again suggesting that gray matter abnormalities are correlated with clinical disability in PPMS.

TREATMENT OF MULTIPLE SCLEROSIS

MS therapy can be divided into specific and symptomatic approaches. MS-specific treatments approved by the US Food and Drug Administration (FDA) include injectable therapies (IFN-β and glatiramer acetate), oral therapies (teriflunomide, dimethyl fumarate, and fingolimod), and infusion therapies (mitoxantrone and natalizumab). Detailed effects and mechanisms of action of long-term MS drugs are summarized in Table 30.4. These treatments mainly target the ongoing inflammation and therefore are effective only in patients with RRMS, or those with SPMS who continue to have relapses. The short aims of specific therapy in MS are to reduce clinical relapses and to decrease the accumulation of new MRI lesions, as they both are linked with poor prognosis. The long aim of specific MS therapy is to delay the evolution from relapsing to progressive MS, as progressive MS is primarily dominated by neurodegenerative processes that are mainly indifferent to the available therapy.

To this end, a two-fold strategy is adopted, which consists of short-term treatment with steroids that aims to reduce the accumulation of disease burden after a relapse, and long-term treatment with licensed drugs that aim to steady the MS process.

As such, patients who have PPMS or SPMS without superimposed relapses only benefit from symptomatic therapy. Symptomatic treatments encompass a wide range of drugs directed to treat a very wide range of symptoms that includes walking impairment, spasticity, fatigue, bladder dysfunction, pain, and depression. Among the symptomatic drugs, dalfampridine is the sole therapy approved to treat walking impairment exclusively in MS patients.

TABLE 30.4 Treatments for Multiple Sclerosis: Effects and Potential Mechanisms of Action

Treatment	Dose	Route of Administration	Frequency of Dose	FDA Approval	Target	Mechanism of Action	Effects	Stage of the Study
FIRST LINE								
IFN-β[41]								
IFN-β_{1b}	250 μg	s.c.	Every other day	1993/2009	Immune system	Modulates cytokine expression profile to an anti-inflammatory phenotype	Reduced relapse rate, disability progression, and MRI activity vs placebo	Recombinant form currently studied for control of disease course and development of brain lesions vs placebo (ClinicalTrials.gov identifier: NCT01464905)
IFN-β_{1a}	30 μg	i.m.	Once a week	1996		Inhibits T-cell activation and proliferation		
IFN-β_{1a}	22 μg	s.c.	3 times a week	2002		Blocks T-cell migration across endothelium		
	44 μg	s.c.	3 times a week					
Glatiramer acetate[42]	20 mg	s.c.	Daily	1996	Immune system	Modulates cytokine expression profile to an anti-inflammatory phenotype	Reduced relapse rate and accumulation of disability vs placebo	
	40 mg	s.c.	3 times a week (PMID: 23686821)	2014		Induces antigen-specific suppressor T cells		
						Reduces antigen presentation		
						Corrects $CD8^+$ T-cell regulatory deficit		
Teriflunomide[43]	7 mg	Oral	Daily	2012	Inhibits mitochondrial dihydroorotate dehydrogenase	Modulates T-cell responses	Reduced relapse rate, disability progression, and MRI activity vs placebo	
	14 mg							

Continued

TABLE 30.4 Treatments for Multiple Sclerosis: Effects and Potential Mechanisms of Action—cont'd

Treatment	Dose	Route of Administration	Frequency of Dose	FDA Approval	Target	Mechanism of Action	Effects	Stage of the Study
Dimethyl fumarate[44,45]	240 mg	Oral	Twice daily	2013	Activates the antioxidant transcription factor Nrf2 pathway	Upregulates Th2 response	Reduced relapse rate and MRI activity vs glatiramer acetate	
						Reduces cell death	Reduced disability progression vs placebo	
SECOND LINE								
Natalizumab[46]	300 mg	i.v.	Every 4 weeks	2006	Targets α_4 subunit of VLA-4 receptor	Prevents migration of leukocytes across BBB into CNS	Reduced relapse rate, disability progression, and MRI activity vs placebo	
Fingolimod[47]	0.5 mg	Oral	Daily	2010	Modulates S1PR types 1, 2, 3, and 5	Prevents CCR7-positive lymphocytes, including naïve and central memory T cells, from exiting lymph nodes	Reduced relapse rate, disability progression, and MRI activity vs IFN-β_{1a}	
THIRD LINE								
Mitoxantrone[48]	12 mg/m^2 (lifetime cumulative limit 140 mg/m^2)	i.v.	Every 3 months	2000	Inhibitor of topoisomerase II	B- and T-cell suppression	Reduced relapse rate, disability progression, and MRI activity vs placebo and vs IFN-β_{1a}	
						Inhibits T-cell migration		

FDA: Food and Drug Administration; IFN: interferon; s.c.: subcutaneous; i.m.: intramuscular; i.v.: intravenous; Nfr2: nuclear factor (erythroid-derived 2)-related factor 2; VLA: very late antigen; S1PR: sphingosine-1-phosphate receptor; Th: T-helper; BBB: blood–brain barrier; CCR: chemokine receptor; MRI: magnetic resonance imaging.

Short-Term Treatments for Acute Relapse

Glucocorticoids have short-term effects on the speed of functional recovery in patients with acute attacks of MS and are widely used in the treatment of acute exacerbations in RRMS. They exert different immunomodulatory mechanisms of action, including inhibition of antigen presentation,[6] anti-inflammatory effects such as reduction of edema and arachidonic acid metabolites, decrease of proinflammatory cytokine expression,[7] and inhibition of lymphocyte proliferation.[33] Usually, they are administered intravenously at high doses during relapse periods of MS and steroid tapering is not suggested. A 2013 trial with a small number of patients reported that administration of bioequivalent doses of oral methylprednisolone is not inferior to intravenously methylprednisolone in terms of the safety and efficacy profile (measured as EDSS and MRI activity 4 weeks after MS relapse).[49] However, before any extension to clinical practice, these promising results have to be confirmed in larger clinical trials with longer follow-up.

Long-Term Treatments

The long-term relapsing MS treatment scenario is quickly growing and therapeutic decisions strictly depend on the course of MS. In general, two different strategies can be adopted.

The first is the *escalation strategy*, which is a therapeutic strategy based on a reasonable decision-making procedure in which drugs with the best risk–benefit ratio are first preferred and, if needed, drugs with increasing power or toxicity (but that are not necessarily more effective) are successively adopted.[50]

The second strategy is the *induction strategy*, which represents a more aggressive approach in which powerful immunosuppressant drugs are used right from the beginning to tackle the disease process hard and early with the aim of resetting the immune system. However, this clinical benefit is gained at the expense of the risk of clinically relevant side effects.[50] Patients who suffer from a non-aggressive form of RRMS are candidates for the escalation strategy, whereas those who suffer from an aggressive form of RRMS are more likely to be put through to an induction strategy.

In a simplified scenario, approved long-term MS therapy can be grouped into three categories based on their presumed target or mechanism of actions: (1) inhibitors of immune cell trafficking, such as natalizumab and fingolimod: by blocking very late antigen-4 (VLA-4) from binding to VCAM-1, natalizumab prevents adhesion of leukocytes to the endothelium and their subsequent transmigration into the CNS[33,51]; by targeting the receptors of sphingosine-1-phosphate (S1PR), fingolimod inhibits the egress of T cells from lymph nodes; (2) inhibitors of cell replication, such as teriflunomide and mitoxantrone: teriflunomide inhibits the mitochondrial dihydroorotate dehydrogenase, which reduces the *de novo* synthesis of pyrimidine nucleotides in proliferating cells, but does not inhibit the salvage pathway used by resting cells; mitoxantrone affects the DNA topoisomerase II, leading to an impairment of T- and B-cell proliferation[33]; (3) immunomodulators, such as dimethyl fumarate, interferons, and glatiramer acetate: dimethyl fumarate directly activates the antioxidant transcription factor nuclear factor (erythroid-derived 2)-related factor-2 (Nrf2) pathway, leading to a counteraction of oxidative stress; interferons promote antagonism of IFN-γ-mediated MHC upregulation on APCs, modulate apoptosis,[6] and inhibit T-cell activation and proliferation[33]; finally, glatiramer acetate promotes a shift in cytokine expression to an anti-inflammatory profile, induces antigen-specific suppressor T cells, and inhibits antigen presentation.[41]

All of the FDA-approved drugs to treat MS reduce annual relapse rate, stabilize or decrease the MRI lesion burden, and limit the accumulation of disability. Nevertheless, they vary significantly in magnitudes of benefits. Interferons, glatiramer acetate, and the recently approved teriflunomide are expected to be approximately 30% effective against relapses, 38% effective at slowing short-term progression, and 50% effective at reducing enhancing lesions and T2 lesions on MRI.[52] Conversely, fingolimod, natalizumab, mitoxantrone, and the recently licensed dimethyl fumarate reduce annual relapse rate by up to 68%, progression rate by up to 42%, and enhancing and Th2 lesions by up to 90% and 85%, respectively.[52–54] However, moving beyond efficacy, the first generation of MS therapies (interferons and glatiramer acetate) displays a well-defined reasonably safe profile, whereas the use of the more efficacious natalizumab and mitoxantrone is limited by safety concerns.

Natalizumab carries the risk of progressive multifocal leukoencephalopathy[55]; this risk is directly related to evidence of previous JC virus exposure, duration of natalizumab exposure, and previous use of immunosuppressants. Mitoxantrone carries the risk of leukemia (which approaches 1%) and of cardiotoxicity, limiting its use at a life cumulative dose of 140 mg/m^2.[56] The safety profile of the recently approved oral therapies (fingolimod, teriflunomide, and dimethyl fumarate) remains to be proven in a general MS population. None of these drugs has been shown to be safe during pregnancy or breastfeeding, and in women who need to continue therapy during pregnancy, only glatiramer acetate appears to be relatively safe (pregnancy category B).[57] Emerging treatments for RRMS are summarized in Table 30.5.

TABLE 30.5 Treatments for Multiple Sclerosis: Evidence from Clinical Trials

Treatment	Target/Mechanism of Action	Effects	Phase	Stage of the Study
CLINICAL TRIALS IN RRMS				
Alemtuzumab[58] (approved by EMA)	Monoclonal antibody, targets molecule CD52	Reduced relapse disability progression and MRI activity rate vs IFN-β_{1a}	III	Currently studied for long-term safety and efficacy (ClinicalTrials.gov identifier: NCT00930553)
Laquinimod	Modulates adaptive T-cell immune responses (by interfering with nuclear factor-κB)	Reduced relapse rate, disability progression, and MRI activity vs placebo	III	Currently studied for efficacy, safety, and tolerability vs IFN-β_{1a} (ClinicalTrials.gov identifier: NCT01975298)
Dalcizumab[59,60]	Monoclonal antibody, binds to CD25	Reduced relapse rate, disability progression, and MRI activity vs placebo	III	Currently studied for safety and efficacy vs IFN-β_{1a} (ClinicalTrials.gov identifier: NCT01064401)
Ocrelizumab[61]	Monoclonal antibody, binds to CD20 antigen	Reduced MRI activity vs placebo and IFN-β_{1a}	III	Currently studied for the potential reduction in disease activity vs IFN-β_{1a} (ClinicalTrials.gov identifier: NCT01247324; NCT01412333)
CLINICAL TRIALS IN SPMS				
Ocrelizumab	Monoclonal antibody, binds to CD20 antigen	NA	III	Currently studied for safety and efficacy in PPMS vs placebo (ClinicalTrials.gov identifier: NCT01194570)
Fingolimod	Modulates S1PR types 1, 2, 3, and 5	NA	III	Currently studied for safety and efficacy in PPMS vs placebo (ClinicalTrials.gov identifier: NCT00731692)
Siponimod	Modulates S1PR types 1 and 5	NA	III	Currently studied for safety and efficacy in SPMS vs placebo (ClinicalTrials.gov identifier: NCT01665144)
Mastinib	Tyrosine kinase inhibitor, primarily targets mast cells	NA	II/III	Currently studied for safety and efficacy in PPMS or relapse-free SPMS vs placebo (ClinicalTrials.gov identifier: NCT01450488)
Natalizumab	Targets the α_4 subunit of VLA-4 receptor	NA	II/III	Currently studied for safety and efficacy in PPMS and SPMS vs placebo (ClinicalTrials.gov identifier: NCT01077466; NCT01416181)

RRMS: relapsing–remitting multiple sclerosis; EMA: European Medicines Agency; MRI: magnetic resonance imaging; IFN: interferon; SPMS: secondary progressive multiple sclerosis; S1PR: sphingosine-1-phosphate receptor; VLA: very late antigen; NA: not applicable; PPMS: primary progressive multiple sclerosis;
Source: Loma and Heyman[33] and Lutterotti and Martin.[51]

Symptomatic Treatment: Aminopyridines

Potassium-blocking drugs such as aminopyridines (4-aminopyridine and 3,4-diaminopyridine) have been proposed as treatment for MS to improve nerve conduction in demyelinated axons. Potassium channels are found in high density in the internodal part of the axon covered by myelin, and they tend to decrease action potential amplitude and duration. Potassium channel blockers have the opposite effect. They can increase action potential amplitude and duration, improving nerve conduction in experimental demyelinating diseases.[62] However, their wide use in MS has always been hindered by the lack of evidence-based guidelines for aminopyridines as symptomatic treatment of MS, and by safety concerns about their side effects (e.g. seizures).[63] Dalfampridine, the sustained release version of 4-aminopyridine, overcomes safety concerns and improves walking impairment in about one-third of treated MS patients.[64,65]

Autologous Hematopoietic Stem Cell Transplantation

The transplantation of hematopoietic stem cells (HSCs) is a widely used therapy for patients with hematopoietic malignancies and solid tumors. Autologous HSC transplantation has also been successfully used to target autoimmune responses in many autoimmune diseases, including MS.[66]

A graft containing HSCs is obtained from the patient's bone marrow or mobilized from the marrow into the circulation. Non-myeloablative chemotherapy and immune-depleting biological agents are administered before the transplantation to destroy the autodestructive immune system. The immune system is regenerated from the infused autologous HSCs.[66] Following HSC transplantation in patients with severe progressive MS, improvement or stabilization of the neurological condition has been observed over a follow-up of almost 4 years, with a decrease or stabilization of EDSS and no clear correlation with toxicity of the conditioning regimen.[66] HSC transplantation is more likely to be effective for MS patients with active CNS inflammation and poor disability, by controlling the inflammatory response and reducing clinical relapses. In addition, it may slow or stop further neurological decline, with an improvement in disability over the years.[66]

Autologous HSC transplantation performed in MS patients since the late 1990s has proven to be a successful treatment; it is based on replacement of the immune system rather than suppression. However, HSC transplantation does not seem to be effective in all patients; in particular, SPMS patients experience poor outcomes compared with patients with aggressive highly active MS, but this therapy remains promising.[66] Phase I and II studies have shown encouraging results. Phase III randomized controlled trials in MS, and in autoimmune diseases such as rheumatoid arthritis, are currently being carried out in Europe and North America.[66]

Future Challenges for Treatment of Progressive Multiple Sclerosis

Current Prospects

At the time of writing, no treatments have been approved for both PPMS and SPMS without superimposed relapses. Moreover, the last phase III clinical trials with interferons, glatiramer acetate, and rituximab in progressive MS had negative results.[67] However, several new compounds are currently undergoing clinical development for progressive MS therapy, including immunomodulatory as well as non-selective and selective immunosuppressive drugs. Emerging therapies for progressive MS are summarized in Table 30.5.

Neuroprotective Strategies

Neuroprotective, restorative, and repair-promoting treatment strategies are slowly emerging. New therapeutic strategies have evolved that specifically target the neurodegenerative aspect of MS (e.g. anti-LINGO-1, riluzole ion channel blockers, statins). However, although some of these molecules displayed neuroprotective effects in animal models of MS, very few have been tested in clinical MS. Currently, the antibodies anti-LINGO-1 and riluzole are the most promising treatments for progressive MS.

LINGO-1 is a protein expressed on oligodendrocytes and neurons, which regulates OPC differentiation and remyelination. LINGO-1 is upregulated in MS plaques, and in animal models of MS LINGO-1 antagonism improves remyelination.[68] A phase II clinical trial with anti-LINGO-1 antibody (BIIB033) is ongoing in subjects with RRMS when used concurrently with interferons (ClinicalTrials.gov identifier: NCT01864148).

Riluzole (Rilutek®) inhibits glutamatergic activity by blocking the release of glutamate and aspartate and by altering the function of glutamate receptors and sodium channels. Riluzole decreases disease severity in EAE.[69] A phase II clinical trial on 15 patients with PPMS showed that riluzole seems to reduce the rate of cervical cord atrophy and the development of hypointense T1 brain lesions on MRI, although its effect on decreasing brain atrophy is poor.[70]

Stem and Progenitor-Cell Based Therapies

New therapeutic strategies for the progressive stage of MS are based on promoting reconstitution of functional myelin. Oligodendrocyte-mediated remyelination depends on the transcription factor Olig-1, which could be a target for MS therapy. Remyelination is also enhanced by cytokines such as IL-1β, and is affected by a chronic inflammatory or highly gliotic environment.[6] Growth factors have been shown to expand OPCs.[6] Therapies aimed at axonal protection based on Na^+ channel inhibitors may control the Na^+ influx in the axon and limit the entry of extracellular Ca^{2+}. Two sodium channel blockers, phenytoin (Dilantin®) and flecainide (Tambocor®), have neuroprotective effects in EAE models,[6] and may be a therapy for progressive MS.

Chronic demyelination and remyelination failure may be due to concurrent loss of OPCs and oligodendrocytes, following a decrease in recruitment of OPCs. OPC repopulation in areas of demyelination appears to be one potential therapeutic strategy to promote remyelination. OPCs have been used to promote remyelination in rodent models of focal CNS acute demyelination.[6] Human OPCs engrafted in a myelin-deficient mouse restore axonal myelination, with extensive myelin production and compaction at 12 weeks post-transplantation.[6] Yet, the environment inhibits or fails to promote differentiation or regeneration. In a context of demyelination, this could be overcome by a therapy based on engineered precursor cells, which may deliver therapeutic molecules to damaged areas to confer protection or to stimulate remyelination.[6]

The idea that stem cells could be a therapy in chronic inflammation-mediated demyelinating and neurodegenerative diseases is based on neurogenesis occurring at an adult age within specific areas of the brain. Stem cells, of either embryonic or adult origin, retain the ability to renew themselves for long periods and to differentiate into a wide range of specialized cell types. These functional properties may overcome the limitations of disease-modifying therapies in neurological disorders. In MS, stem-cell based therapeutic strategies focus on two issues that may have a neuroprotective effect on demyelinated areas: immunomodulation and remyelination. Transplantation of mesenchymal stem cells (MSCs) and neural stem/precursor cells (NPCs) shows protective effects in EAE, which are likely to be based on immune regulation and promotion of repair in the CNS.[71]

Intravenously infused MSCs have a beneficial effect on the severity and progression of EAE. MSCs can be isolated from some connective adult tissues such as bone marrow. They participate in the regulation of the immune system, for example by inducing T-cell anergy, altering B-cell proliferation, or migrating to inflammatory regions under specific conditions to attenuate the inflammatory response. MSCs may also promote repair in damaged tissues by secreting stimulating factors. Clinical trials are being carried out in patients with MS to assess the therapeutic relevance and safety of MSC injection.[71,72] These include a double-blind, randomized, cross-over phase I/II study that is evaluating the safety and the efficacy of the intravenous administration of autologous MSCs to patients with active MS resistant to currently available therapies [MEsenchymal StEm Cells for Multiple Sclerosis (MESEMS) trial; Clinicaltrials.gov identifier: NCT01854957].

In EAE, transplanting NPCs protects animals from paralysis.[71] The mechanism remains unclear, but it seems that NPCs exert neuroprotective function by a mechanism other than cell replacement. They improve remyelination, rescue degenerated axons, and modulate the immune response.[71] NPCs display immunomodulatory properties in the injured CNS. In both acute and chronic EAE models intracerebroventricularly injected NPCs migrate to inflamed areas and regulate inflammation, leading to a decrease in demyelination and axonal damage. A higher frequency of T-cell apoptosis has been observed upon transplantation of NPCs in EAE. Systemically injected NPCs can modulate peripheral immunomodulation in lymph nodes and attenuate severity in EAE.[73] NPCs directly injected in damaged areas of the brain also exert therapeutic effects by rescuing degenerated axons and improving the remyelination process in demyelinated areas. Intravascularly or intrathecally injected NPCs show only limited potentiation for remyelination, but they can improve the survival and regeneration of endogenous neural cells, including glial and neural progenitors. NPCs may also express and deliver neurotrophins and growth factors, which play a role in the neuroprotective effect. Clinical trials are being developed to investigate the safety and effects of NPC transplantation.[71,73]

Transplanted NPCs remain in the niche of endogenous neural stem cells in an undifferentiated state, or migrate out of the niche in a defined differentiated phenotype, depending on the inflammatory conditions. Yet the exact mechanisms underlying the NPC response to the microenvironment are unknown. The concept of therapeutic plasticity aims to describe the various therapeutic actions of stem cell transplants *in vivo*, particularly the ability to adapt functions to the specific microenvironment. Some functions of transplanted NPCs, including immune regulation, are based on intercellular communication. NPCs may communicate with host cells (e.g. cells of the immune system) via secretion of cytokines, chemokines, growth factors, and other mediator molecules, cell-to-cell interactions, or secretion of extracellular membrane vesicles.[73]

FUTURE DIRECTIONS

One of the major challenges in MS is to understand the mechanisms underlying the pathogenesis of the disease. MS is a complex disease that affects young adults who are genetically susceptible and are then exposed to a precipitating environmental factor or infectious agent. MS lesions are likely to originate from an autoimmune disorder, which involves various cell types and modulating molecules of the immune system. Autoimmunity may be the consequence of molecular mimicry, that is, the activation and recruitment of autoreactive immune cells due to structural homology between a self antigen and an exogenous pathogenic protein. Similarities in the disease observed in both human MS and EAE provide strong evidence for this hypothesis.[7] Susceptibility genes have been identified and genome-wide analysis is now available. Improving techniques of genomic investigation may lead to a better understanding of genetic associations with the disease.[20]

Many aspects remain unclear, including functional relations between the myelin sheath and the axon. White matter demyelinated lesions of the CNS in MS patients start with destruction of the myelin sheath of neurons by inflammatory reactions. The progressive stage of the disease is then dominated by axonal damage and neurodegeneration of chronically demyelinated neurons. However, it is still uncertain whether inflammation in MS is primary or secondary.[2] Less is known about MS cortical lesions, but it seems that inflammation does not occur in damaged areas in the gray matter. Brain imaging has suggested that the pathogenesis of MS includes a distinct and more global cortical pathology. In addition, as anti-inflammatory therapies appear to have no effect on the progressive stage, the question is raised whether inflammation initiates tissue damage in progressive MS.

Another major question is whether PPMS is the same disease as RRMS. The underlying mechanisms in both clinical patterns remain unknown. However, differences in disease course, epidemiology, and clinical pathology assessed by MRI suggest different forms of pathogenesis in RRMS and PPMS.[10]

Elucidating the cause of MS may provide key information in understanding the chronology of events leading to demyelination and axonal degeneration. Together with a consistent animal model, this could be helpful in further determining relevant targets for MS treatments. Therapeutic strategies aim at modulating the immune response, protecting the CNS, potentiating remyelination, and improving nerve conduction. Future prospects in MS therapy are based on stem cells; their systemic transplantation has been shown to protect the CNS from degeneration driven by chronic inflammation in EAE, through immunomodulatory or neuroprotective mechanisms.[47]

References

1. Compston A, Coles A. Multiple sclerosis. *Lancet*. 2002;359(9313): 1221–1231.
2. Trapp BD, Nave KA. Multiple sclerosis: an immune or neurodegenerative disorder? *Annu Rev Neurosci*. 2008;31:247–269.
3. Brück W. The pathology of multiple sclerosis is the result of focal inflammatory demyelination with axonal damage. *J Neurol*. 2005;252(suppl 5):v3–9.
4. Fletcher JM, Lalor SJ, Sweeney CM, Tubridy N, Mills KH. T cells in multiple sclerosis and experimental autoimmune encephalomyelitis. *Clin Exp Immunol*. 2010;162(1):1–11.
5. Frohman EM, Racke MK, Raine CS. Multiple sclerosis – the plaque and its pathogenesis. *N Engl J Med*. 2006;354(9):942–955.
6. Hauser SL, Oksenberg JR. The neurobiology of multiple sclerosis: genes, inflammation, and neurodegeneration. *Neuron*. 2006;52(1):61–76.
7. Sospedra M, Martin R. Immunology of multiple sclerosis. *Annu Rev Immunol*. 2005;23:683–747.
8. Franklin RJ. Why does remyelination fail in multiple sclerosis? *Nat Rev Neurosci*. 2002;3(9):705–714.
9. Lublin FD, Reingold SC. Defining the clinical course of multiple sclerosis: results of an international survey. National Multiple Sclerosis Society (USA) Advisory Committee on Clinical Trials of New Agents in Multiple Sclerosis. *Neurology*. 1996;46(4):907–911.
10. Antel J, Antel S, Caramanos Z, Arnold DL, Kuhlmann T. Primary progressive multiple sclerosis: part of the MS disease spectrum or separate disease entity? *Acta Neuropathol*. 2012;123(5):627–638.
11. Coo H, Aronson KJ. A systematic review of several potential non-genetic risk factors for multiple sclerosis. *Neuroepidemiology*. 2004;23(1–2):1–12.
12. Yeh EA, Chitnis T, Krupp L, et al. Pediatric multiple sclerosis. *Nat Rev Neurol*. 2009;5(11):621–631.
13. Steinman L. Multiple sclerosis: a two-stage disease. *Nat Immunol*. 2001;2(9):762–764.
14. World Health Organization. *Atlas: Multiple Sclerosis Resources in the World 2008*. Geneva: WHO; 2008.
15. Compston A. The genetic epidemiology of multiple sclerosis. *Philos Trans R Soc Lond B Biol Sci*. 1999;354(1390):1623–1634.
16. Dyment DA, Sadovnick AD, Ebers GC, Sadnovich AD. Genetics of multiple sclerosis. *Hum Mol Genet*. 1997;6(10):1693–1698.
17. Gourraud PA, Harbo HF, Hauser SL, Baranzini SE. The genetics of multiple sclerosis: an up-to-date review. *Immunol Rev*. 2012;248(1):87–103.
18. Oksenberg JR, Baranzini SE, Sawcer S, Hauser SL. The genetics of multiple sclerosis: SNPs to pathways to pathogenesis. *Nat Rev Genet*. 2008;9(7):516–526.
19. Sawcer S, Hellenthal G, Pirinen M, et al. Genetic risk and a primary role for cell-mediated immune mechanisms in multiple sclerosis. *Nature*. 2011;476(7359):214–219.
20. Patsopoulos NA, Esposito F, Reischl J, et al. Genome-wide meta-analysis identifies novel multiple sclerosis susceptibility loci. *Ann Neurol*. 2011;70(6):897–912.
21. Gregory AP, Dendrou CA, Attfield KE, et al. TNF receptor 1 genetic risk mirrors outcome of anti-TNF therapy in multiple sclerosis. *Nature*. 2012;488(7412):508–511.
22. Riise T, Nortvedt MW, Ascherio A. Smoking is a risk factor for multiple sclerosis. *Neurology*. 2003;61(8):1122–1124.
23. Correale J, Farez MF. The impact of parasite infections on the course of multiple sclerosis. *J Neuroimmunol*. 2011;233(1–2):6–11.
24. Owens GP, Gilden D, Burgoon MP, Yu X, Bennett JL. Viruses and multiple sclerosis. *Neuroscientist*. 2011;17(6):659–676.
25. Bradl M, Lassmann H. Progressive multiple sclerosis. *Semin Immunopathol*. 2009;31(4):455–465.
26. Prat A, Antel J. Pathogenesis of multiple sclerosis. *Curr Opin Neurol*. 2005;18(3):225–230.
27. Alvarez-Lafuente R. de las Heras V, García-Montojo M, Bartolomé M, Arroyo R. Human herpesvirus-6 and multiple sclerosis: relapsing–remitting versus secondary progressive. *Mult Scler*. 2007;13(5): 578–583.
28. Libbey JE, McCoy LL, Fujinami RS. Molecular mimicry in multiple sclerosis. *Int Rev Neurobiol*. 2007;79:127–147.
29. Wattjes MP, Harzheim M, Kuhl CK, et al. Does high-field MR imaging have an influence on the classification of patients with clinically isolated syndromes according to current diagnostic MR imaging criteria for multiple sclerosis?. *AJNR Am J Neuroradiol*. 2006;27(8):1794–1798. 4.
30. Lassmann H, Brück W, Lucchinetti C. Heterogeneity of multiple sclerosis pathogenesis: implications for diagnosis and therapy. *Trends Mol Med*. 2001;7(3):115–121.
31. Minagar A, Alexander JS. Blood–brain barrier disruption in multiple sclerosis. *Mult Scler*. 2003;9(6):540–549.
32. Zozulya AL, Wiendl H. The role of regulatory T cells in multiple sclerosis. *Nat Clin Pract Neurol*. 2008;4(7):384–398.
33. Loma I, Heyman R. Multiple sclerosis: pathogenesis and treatment. *Curr Neuropharmacol*. 2011;9(3):409–416.
34. Navikas V, Link H. Review: cytokines and the pathogenesis of multiple sclerosis. *J Neurosci Res*. 1996;45(4):322–333.
35. Furlan R, Brambilla E, Ruffini F, et al. Intrathecal delivery of IFN-gamma protects C57BL/6 mice from chronic–progressive experimental autoimmune encephalomyelitis by increasing apoptosis of central nervous system-infiltrating lymphocytes. *J Immunol*. 2001;167(3):1821–1829.
36. Jadidi-Niaragh F, Mirshafiey A. Th17 cell, the new player of neuroinflammatory process in multiple sclerosis. *Scand J Immunol*. 2011;74(1):1–13.
37. Cross AH, Trotter JL, Lyons J. B cells and antibodies in CNS demyelinating disease. *J Neuroimmunol*. 2001;112(1–2):1–14.

38. Polman CH, Reingold SC, Edan G, et al. Diagnostic criteria for multiple sclerosis: 2005 revisions to the "McDonald criteria". *Ann Neurol*. 2005;58(6):840–846.
39. Noseworthy JH. Clinical scoring methods for multiple sclerosis. *Ann Neurol*. 1994;36. Suppl:S80–S85.
40. Miller DH, Leary SM. Primary–progressive multiple sclerosis. *Lancet Neurol*. 2007;6(10):903–912.
41. Goodin DS, Frohman EM, Garmany GP, et al. Disease modifying therapies in multiple sclerosis: report of the Therapeutics and Technology Assessment Subcommittee of the American Academy of Neurology and the MS Council for Clinical Practice Guidelines. *Neurology*. 2002;58(2):169–178.
42. Johnson KP, Brooks BR, Cohen JA, et al. Copolymer 1 reduces relapse rate and improves disability in relapsing–remitting multiple sclerosis: results of a phase III multicenter, double-blind placebo-controlled trial. The Copolymer 1 Multiple Sclerosis Study Group. *Neurology*. 1995;45(7):1268–1276.
43. O'Connor P, Wolinsky JS, Confavreux C, et al. Randomized trial of oral teriflunomide for relapsing multiple sclerosis. *N Engl J Med*. 2011;365(14):1293–1303.
44. Gold R, Kappos L, Arnold DL, et al. Placebo-controlled phase 3 study of oral BG-12 for relapsing multiple sclerosis. *N Engl J Med*. 2012;367(12):1098–1107.
45. Fox RJ, Miller DH, Phillips JT, et al. Placebo-controlled phase 3 study of oral BG-12 or glatiramer in multiple sclerosis. *N Engl J Med*. 2012;367(12):1087–1097.
46. Polman CH, O'Connor PW, Havrdova E, et al. A randomized, placebo-controlled trial of natalizumab for relapsing multiple sclerosis. *N Engl J Med*. 2006;354(9):899–910.
47. Cohen JA, Barkhof F, Comi G, et al. Oral fingolimod or intramuscular interferon for relapsing multiple sclerosis. *N Engl J Med*. 2010;362(5):402–415.
48. Hartung HP, Gonsette R, Konig N, et al. Mitoxantrone in progressive multiple sclerosis: a placebo-controlled, double-blind, randomised, multicentre trial. *Lancet*. 2002;360(9350):2018–2025.
49. Ramo-Tello C, Grau-Lopez L, Tintore M, et al. A randomized clinical trial of oral versus intravenous methylprednisolone for relapse of MS. *Mult Scler*. 2013 Oct 21. [Epub ahead of print].
50. Rieckmann P. Concepts of induction and escalation therapy in multiple sclerosis. *J Neurol Sci*. 2009;277(suppl 1):S42–S45.
51. Lutterotti A, Martin R. Getting specific: monoclonal antibodies in multiple sclerosis. *Lancet Neurol*. 2008;7(6):538–547.
52. Hillert J. In the coming year we should abandon interferons and glatiramer acetate as first line therapy for MS: no. *Mult Scler*. 2013;19(1):26–28.
53. Methner A, Zipp F. Multiple sclerosis in 2012: novel therapeutic options and drug targets in MS. *Nat Rev Neurol*. 2013;9(2):72–73.
54. Damal K, Stoker E, Foley JF. Optimizing therapeutics in the management of patients with multiple sclerosis: a review of drug efficacy, dosing, and mechanisms of action. *Biologics*. 2013;7:247–258.
55. Langer-Gould A, Atlas SW, Green AJ, Bollen AW, Pelletier D. Progressive multifocal leukoencephalopathy in a patient treated with natalizumab. *N Engl J Med*. 2005;353(4):375–381.
56. Bruck W, Gold R, Lund BT, et al. Therapeutic decisions in multiple sclerosis: moving beyond efficacy. *JAMA Neurol*. 2013;70(10):1315–1324.
57. Cree BA. Update on reproductive safety of current and emerging disease-modifying therapies for multiple sclerosis. *Mult Scler*. 2013;19(7):835–843.
58. Cohen JA, Coles AJ, Arnold DL, et al. Alemtuzumab versus interferon beta 1a as first-line treatment for patients with relapsing–remitting multiple sclerosis: a randomised controlled phase 3 trial. *Lancet*. 2012;380(9856):1819–1828.
59. Gold R, Giovannoni G, Selmaj K, et al. Daclizumab high-yield process in relapsing–remitting multiple sclerosis (SELECT): a randomised, double-blind, placebo-controlled trial. *Lancet*. 2013;381(9884):2167–2175.
60. Comi G, Jeffery D, Kappos L, et al. Placebo-controlled trial of oral laquinimod for multiple sclerosis. *N Engl J Med*. 2012;366(11):1000–1009.
61. Kappos L, Li D, Calabresi PA, et al. Ocrelizumab in relapsing–remitting multiple sclerosis: a phase 2, randomised, placebo-controlled, multicentre trial. *Lancet*. 2011;378(9805):1779–1787.
62. Sheean GL, Murray NM, Rothwell JC, Miller DH, Thompson AJ. An open-labelled clinical and electrophysiological study of 3,4 diaminopyridine in the treatment of fatigue in multiple sclerosis. *Brain*. 1998;121(Pt 5):967–975.
63. Solari A, Uitdehaag B, Giuliani G, Pucci E, Taus C. Aminopyridines for symptomatic treatment in multiple sclerosis. *Cochrane Database of Systematic Reviews (Online)*. 2002;(4). CD001330.
64. Goodman AD, Brown TR, Krupp LB, et al. Sustained-release oral fampridine in multiple sclerosis: a randomised, double-blind, controlled trial. *Lancet*. 2009;373(9665):732–738.
65. Goodman AD, Brown TR, Edwards KR, et al. A phase 3 trial of extended release oral dalfampridine in multiple sclerosis. *Ann Neurol*. 2010;68(4):494–502.
66. Atkins HL, Freedman MS. Hematopoietic stem cell therapy for multiple sclerosis: top 10 lessons learned. *Neurotherapeutics*. 2013;10(1):68–76.
67. Comi G. Disease-modifying treatments for progressive multiple sclerosis. *Mult Scler*. 2013;19(11):1428–1436.
68. Mi S, Pepinsky RB, Cadavid D. Blocking LINGO-1 as a therapy to promote CNS repair: from concept to the clinic. *CNS Drugs*. 2013;27(7):493–503.
69. De Jager PL, Hafler DA. New therapeutic approaches for multiple sclerosis. *Annu Rev Med*. 2007;58:417–432.
70. Kalkers NF, Barkhof F, Bergers E, van Schijndel R, Polman CH. The effect of the neuroprotective agent riluzole on MRI parameters in primary progressive multiple sclerosis: a pilot study. *Mult Scler*. 2002;8(6):532–533.
71. Martino G, Franklin RJ. Baron Van Evercooren A, Kerr DA, Group SCiMSSC. Stem cell transplantation in multiple sclerosis: current status and future prospects. *Nat Rev Neurol*. 2010;6(5):247–255.
72. Connick P, Kolappan M, Crawley C, et al. Autologous mesenchymal stem cells for the treatment of secondary progressive multiple sclerosis: an open-label phase 2a proof-of-concept study. *Lancet Neurol*. 2012;11(2):150–156.
73. Cossetti C, Alfaro-Cervello C, Donegà M, Tyzack G, Pluchino S. New perspectives of tissue remodelling with neural stem and progenitor cell-based therapies. *Cell Tissue Res*. 2012;349(1):321–329.

SECTION V

DISEASES OF HIGHER FUNCTION

CHAPTER

31

Introduction

Joseph T. Coyle
Harvard Medical School, McLean Hospital, Belmont, Massachusetts, USA

OUTLINE

This section, "Diseases of Higher Function", covers an array of disorders that have in common impairments in cognition, executive function, affect, and motivation. Grouping them in this way avoids the outmoded distinction between "functional" conditions (e.g. psychiatric disorder) and conditions with obvious neuropathology (e.g. neurological disorders). Brain imaging studies carried out since the 1990s have made it clear that all of these disorders are associated with significant structural and functional neuropathology. In each chapter, the authors have provided a comprehensive review of the clinical manifestations, the neurobiological understanding of the disorders based on brain imaging and available postmortem data, genetics, and current treatment approaches.

Several common themes run through theses chapters. First, brain imaging, primarily magnetic resonance imaging (MRI), has transformed how we now think about these disorders as unequivocally due to brain dysfunction and not a consequence of unconscious conflicts. Specifically, functional magnetic resonance imaging (fMRI) conducted with challenges to "stress" the particular psychopathology has robustly distinguished patients from controls. For example, having subjects with schizophrenia perform cognitive tasks reveals the characteristic "hypofrontality" (see Konopaske and Coyle, Chapter 39), and the deviant emotional processing in anxiety disorders is associated with inappropriate activation of the amygdala in non-threatening conditions (Vahabzadeh, Gillespie, and Ressler, Chapter 37). These findings have been complemented by "default mode" imaging, which has disclosed aberrant neuronal connectivity that is congruent with the fMRI findings.

Second, it is now apparent that genetics play an important role in the risk for developing these disorders, even when environmental factors represent a dominant risk factor such as is the case for post-traumatic stress disorder (PTSD). In fact, 30% of the risk for PTSD is determined by heritable factors, and several candidate genes have been proposed.[1] Nevertheless, in contradistinction to many heritable neurological disorders such as Huntington disease or Tay–Sachs disease, psychiatric disorders and those affecting higher function such as Alzheimer disease involve complex genetics whereby multiple genes of modest effect interact with the environment to produce the phenotype. For example, recent meta-analyses of genome-wide association studies (GWAS) in schizophrenia comprising more than 80,000 subjects have revealed approximately 100 genes that achieve GWAS significance.[2]

Third, impairments in neuroplasticity characterize most of these disorders. The past two decades have seen a move from the view that the brain is hard-wired to one in which the brain is understood to be highly malleable and responsive to its environment. The precise molecular mechanisms that mediate functional plasticity, such as long-term potentiation and long-term depression, have been elucidated. These processes are linked to actual structural changes with a continuous process of synaptic remodeling driven by the secretion of growth and trophic factors such as brain-derived neurotrophic factor (BDNF). Cortical atrophy in schizophrenia, most severely affecting the frontal and temporal lobes, and hippocampal atrophy in PTSD and major depressive disorder are associated with decreased BDNF expression in relevant brain structures and reduced downstream

Neurobiology of Brain Disorders
http://dx.doi.org/10.1016/B978-0-12-398270-4.00031-8

intracellular mediators of synapse formation and retention (see Boldrini and Mann, Chapter 43). In the case of addiction, neuronal pathways comprising the reward systems and executive functions are rewired to mediate the addicted state (Butelman *et al.*, Chapter 35), and neuropathic pain is associated with reorganization of sensory pathways in the dorsal horn and in the thalamus (Gold and Backonja, Chapter 41).

Fourth, as a correlate of neuroplasticity, there has been a resurgence of interest in psychological interventions. In contrast to the past, when one type of psychological treatment (insight-oriented psychotherapy) was considered effective for most psychiatric disorders, increasing attention is being paid to the development of disorder-specific, evidence-based psychological interventions. In disorders such as depression and anxiety disorders, manual-based cognitive–behavioral therapy can be as effective as pharmacological interventions. As antidepressants augment BDNF expression in the hippocampus, it is not surprising that the combination of a selective serotonin reuptake inhibitor and cognitive–behavioral therapy is more persistently effective than either treatment alone.[3]

Given the salience of environmental factors and the complex genetics that characterize these disorders, it is not surprising that epigenetic processes have come forward as important variables affecting risk for disorders. This is a new, complex, and rapidly moving area of investigation. DNA and histone methylation and histone acetylation can produce structural modifications of DNA (but not its base sequence) that persistently alter gene expression, even affecting the next generation (see Boldrini and Mann, Chapter 43). MicroRNAs located in non-coding regions of the genome regulate the expression of a family of genes at the level of their messenger RNA. Allelic variants of a microRNA that would not be detected in the transcriptome could have a substantial impact on a family of genes important for brain function.[4] For example, *miR132*, which regulates the expression of several proteins involved in synapse formation and stability, has been implicated in schizophrenia.

Finally, with the publication of the fifth edition of *Diagnostic and Statistical Manual of Mental Disorders*, it is ironic that our advancing understanding of the genetics underlying psychiatric disorders has raised questions about the validity and relevance of a classification system constructed around syndromes and not on neuroscientific and genetic findings.[5] For example, the same copy number variant (i.e. deletion or reduplication of the genome) has been associated with autism, schizophrenia, epilepsy, and mental subnormality.[6] Family studies reveal elevated risk for bipolar disorder and autism in probands with schizophrenia. In the light of the complex genetics, shared pathological processes mediated by one set of risk genes may be modified by other risk genes to affect the syndromic manifestations. By identifying the risk genes that shape phenotype, it may be possible to develop drugs that address the primary pathophysiology, which could be more effective than existing symptomatic treatments.

It is hoped that these chapters leave the reader both knowledgeable about the clinical conditions affecting higher functions and impressed with the transformative effects that neuroscience has had in understanding the underlying pathophysiology. As we move away from "received wisdom" to agnostic scientific approaches, it is likely that we will witness a dramatic reordering of how we understand these conditions, which will be grounded in genetics and neuroscience.

References

1. Skelton K, Ressler KJ, Norrholm SD, Jovanovic T, Bradley-Davino B. PTSD and gene variants: new pathways and new thinking. *Neuropharmacology*. 2012;62(2):628–637.
2. Schwab SG, Wildenauer DB. Genetics of psychiatric disorders in the GWAS era: an update on schizophrenia. *Eur Arch Psychiatry Clin Neurosci*. 2013;263(suppl 2):S147–S154.
3. Hollon SD, Ponniah K. A review of empirically supported psychological therapies for mood disorders in adults. *Depress Anxiety*. 2010;27(10):891–932.
4. Coyle JT. MicroRNAs suggest a new mechanism for altered brain gene expression in schizophrenia. *Proc Natl Acad Sci U S A*. 2009;106(9):2975–2976.
5. Cuthbert BN, Insel TR. Toward the future of psychiatric diagnosis: the seven pillars of RDoC. *BMC Med*. 2013;11:126.
6. Hochstenbach R, Buizer-Voskamp JE, Vorstman JA, Ophoff RA. Genome arrays for the detection of copy number variations in idiopathic mental retardation, idiopathic generalized epilepsy and neuropsychiatric disorders: lessons for diagnostic workflow and research. *Cytogenet Genome Res*. 2011;135(3–4):174–202.

CHAPTER

32

Disorders of Higher Cortical Function

Anna Berti[*,†], *Francesca Garbarini*[*], *Marco Neppi-Modona*[*,†]

[*]Psychology Department, University of Turin, Turin, Italy; [†]Neuroscience Institute of Turin (NIT), University of Turin, Turin, Italy

OUTLINE

Neurobiology of Brain Disorders
http://dx.doi.org/10.1016/B978-0-12-398270-4.00032-X

INTRODUCTION: FROM NEUROPSYCHOLOGY TO MENTAL STRUCTURE

Focal brain lesions, that is, pathological events such as ischemic attack, tumors, and even infective diseases that affect relatively small parts of the cerebral cortex, can cause substantial but circumscribed impairments of either primary or cognitive nervous function, or both. The observed alterations depend on the areas and the cerebral circuits involved. If the lesion affects the primary motor areas, which are located in the frontal lobes of both hemispheres, and in particular in Brodmann area (BA) 4, and the axons that go from this cortical area to the spinal motoneurons, a paresis contralateral (most of the sensorimotor pathways are crossed) to the brain damage will be observed. Similarly, damage to primary sensory areas and pathways would cause contralateral sensory disturbances (e.g. anesthesia of contralateral limbs when sensory parietal cortices are damaged or blindness of one visual hemifield from damage to the primary visual areas in the occipital lobes). When the damage is localized outside primary motor and sensory cortices, complex cognitive deficits may be observed, involving what are called higher brain functions. In these latter cases the observed disorders may range from impairments of the cognitive analysis of input stimuli to impairments of the motor programming and selection of output signal (therefore, aspects that, although related to perception and movements, cannot be ascribed to the simple deficit in stimulus analysis or to the strength deficit related to the paralysis) or may involve functions such as language, memory, decisional processes, or domain-specific disturbance of conscious awareness (see below).

The discipline that studies the relationship between brain damage and higher cognitive functions is called neuropsychology. While its clinical goal is the description, diagnosis, and treatment of the disorders consequent upon the brain damage, the main experimental aim, as widely discussed in the seminal book by Tim Shallice, *From Neuropsychology to Mental Structure*,[1] is to draw inferences from the pathological conditions to normal functions, on the assumption that if the lesion to a particular brain area has provoked a specific deficit then that part of the brain is involved, underpins, or is necessary to sustain the normal function. Scientific neuropsychology dates back to the second half of the nineteenth century, when the observation that there was a strict relation between focal brain damage and disturbances of higher mental functions became a matter of debate among neurologists, who started to systematically study and report single cases of patients with domain-specific cognitive impairment to unveil the neural bases of cognition. The observation that cognitive impairment could derive from specific brain damage gave rise to many studies that not only described the relation between the symptoms and the lesion but also proposed real "models" for the altered function. Indeed, Broca in 1861[2] not only suggested a relationship between the presence of a brain lesion and a language problem, but indicated in the third frontal circonvolution the exact locus of linguistic functions, therefore establishing a precise anatomoclinical correlation.

The first neuropsychologists, on the bases of their observations, proposed simple models of cerebral functions in the form of diagrams in which gray matter structures, underlying higher cognitive functions, and white matter pathways, connecting different centers, were represented. These models could explain the disorders already observed and also tried to predict new syndromes on the basis of the acquired knowledge. The great merit of the "diagrammists" was that they realized that brain injuries can elucidate the organization of the cognitive systems. Although observations similar to those reported by Broca had been made previously, the scientific community was not ready to accept them. The scientific discoveries in neurology and neurophysiology, which identified in specific brain areas specialized regions for the control of sensorimotor functions, also created the cultural conditions for considering a cause–effect relationship between injury and impaired function plausible for cognitive disorders, thus initiating a localization–association approach for the study of higher brain functions. Although the acceptance of the relationship between brain injury and altered behavior paved the way for the study of the neurobiological bases of cognitive functions, these models had some limitations, mainly related to the lack of solid psychological theories, the idiosyncratic choice of the patients to be studied, and the lack of methodological constraints and quantitative analysis. Therefore, the neuropsychological models were adjusted and adapted over the course of the twentieth century depending on the theoretical paradigms that were dominant in the neuroscientific and psychological fields.

First, the classical method used by neuropsychology, the study of clinical cases, underwent some changes owing to the modified scientific conditions. The development of anatomical knowledge, on the one hand, and the need to apply a quantitative approach to the study of patients, which would allow a greater objectivity for neurological observations and a less idiosyncratic choice of the patients to be studied, on the other hand, imposed, in the second half of the twentieth century, the transition from the study of individual isolated cases to the study of groups of patients, who were selected according to strict criteria established *a priori*. For instance, patients could be selected on the basis of the symptom one wanted to study or according to the lesion site. Control groups of neurologically intact subjects were also considered as the normal reference system.

At the same time, the study of neuropsychological disorders began to be standardized through the use of validated tests, which permitted the gathering of data suitable for statistical evaluation. These advances led to the acquisition of reproducible and fundamental knowledge on various aspects of human cognition, mainly related to the differences in intrahemispheric and interhemispheric neurofunction. In the second half of the twentieth century, the introduction of the cognitive psychology approach and the model of human information processing, which proposed a multicomponent structure of the cognitive system, provided a useful theoretical framework for the study of the injured brain. Indeed, if it is plausible to assume the existence of functionally segregated cognitive centers, then it is possible that a center could be selectively damaged by a brain lesion, causing a very specific and isolated disorder of cognition. As Smith Churchland puts it: "So long as the brain functions normally, the inadequacies of common-sense framework can be hidden from view, but with a damaged brain the inadequacies of theory are unmasked".[3]

The general model of the functioning of the cognitive system was not very different from the diagrammists' models insofar as different brain centers were supposed to give rise to separated and anatomically segregated cognitive functions, connected by interhemispheric and intrahemispheric pathways, and the single case method was again considered the ideal way to unveil the structure of the cognitive system. However, in the modern version of the classical associations, clinical cases are evaluated with both sophisticated methodological and statistical criteria and detailed *in vivo* neuroanatomical investigations. It is worth noting that studies in patients with brain lesions, obtained with the methods of the modern neuropsychological research, have often disclosed the limits of the anatomoclinical inferences related to the idea that brain function is localized in a single area of cortical tissue. Often similar disorders can be observed in the presence of damage in different brain structures, which suggests that a given function is not localized in one single area of the brain, but that the damaged region is part of a circuit or system constituted by several cortical–subcortical centers. The idea that a function is distributed in the brain does not mean that all brain areas participate equally in that function, but that the function depends on the cooperation of specific, but different, parts of the brain. Therefore, although adjusted by the newly acquired knowledge on cerebral organization, the principle that focal damage to the brain can cause circumscribed and specific impairment of cognitive function still holds.

Lesion studies have significantly contributed to the definition and clarification of the cognitive prerogative of the human mind, falsifying the ultracognitive view that mental operations, made possible by our information processing systems, do not depend on the structural characteristics and organization of the biological substrate with which we are endowed.

LANGUAGE DISORDERS

As already mentioned, the birth of modern neuropsychology is usually identified as Paul Broca's presentation at the Société Anatomique, in 1861, of his study on the anatomical localization of the articulated language in the inferior frontal gyrus.[2] In Broca's pioneering findings, the tight link between cognitive functions and neuroanatomical structures became evident: if a specific structure (the "Broca area") is damaged, a specific function (speech production) is lost. This section will focus on the neuropsychological study of language deficits, from the "classical" aphasias to a more complex model of the linguistic functions inside the human brain. A new research field, aiming to explain "normal" linguistic processing within the theoretical context of the embodied semantic, is also investigated.

Aphasia: Definition and Treatments

Aphasia is a disturbance of the production and/or the comprehension of language caused by selective damage in specific brain areas, usually located in the left hemisphere (which is dominant for language in both right-handed and left-handed people). In aphasia, multiple aspects of language can be compromised, including the syntax (the grammatical structure of sentences), the lexicon (the collection of words that denote meanings), and the morphology (the combination of phonemes, single speech sounds, into morphemes, the smallest meaningful units of a word). There are different profiles of aphasia, depending on which of these linguistic aspects is most compromised in a specific patient. Most cases of aphasia are caused by trauma or stroke, cerebral tumors, or degenerative dementias. The correct diagnosis and effective treatment of aphasia have a great clinical importance because this deficit has a devastating impact on the patient's quality of life. Clinical interventions for aphasia aim to activate dysfunctional brain networks supporting linguistic processing and communicative intent. These interventions generally fall into three broad categories: speech–language rehabilitation treatments; pharmaceutical treatments; and direct brain-stimulation therapies, such as transcranial magnetic stimulation and transcranial direct current stimulation. After a brain injury there is a spontaneous recovery period, but there is converging evidence that language recovery may be enhanced by a program of rehabilitation therapy.

Classical Aphasia

Historically, on the basis of both clinical and postmortem anatomical observations, language has been localized in two major brain areas: Broca anterior frontal area (Brodmann area, BA44) for speech production and Wernicke posterior temporal area (BA22) for speech comprehension. The first functional model of language, proposed by Lichtheim in 1885,[4] was an attempt to explain linguistic processing within this left temporofrontal network, giving an account of a wide range of known aphasic symptoms. The novelty of Lichtheim's approach was to define language syndromes in terms of damage to the components of the model, in the same way as neuropsychologists nowadays do with the large spectrum of sensory, motor, and cognitive deficits affecting brain-damaged patients.

In the classical view of aphasia the following syndromes were described.[5]

Broca Aphasia

There are two variants of Broca aphasia. In classical Broca aphasia, lesions actually involve not only Broca area (BA44 and 45) but also the surrounding frontal regions (BA8, 9, 10, and 46) and the underlying white matter and basal ganglia. Patients have a dramatic loss of speech fluency and a specific form of agrammatism, characterized by the inability to organize words and sentences according to grammatical rules. Moreover, the patients' capacity to repeat sentences is compromised, as well as their ability to assemble phonemes correctly (they often show phonemic and phonetic paraphasias). When lesions are restricted to the Broca area, a milder and more transient form is observed. Traditionally, from Broca's original description,[2] Broca aphasia is considered a deficit of speech production, but some of the difficulties noted in language production have also been found to occur in language comprehension, in particular when the grammatical structure is complex as, for instance, in passive sentences.

Wernicke Aphasia

Wernicke aphasia is caused by damage to the posterior sector of the left auditory association cortex (BA22), often involving other surrounding areas (BA37, 39, and 40) and the underlying white matter. In patients with Wernicke aphasia, speech is fluent (effortless, melodic, and produced at normal rates) but the content is often unintelligible because of frequent errors in phoneme and word choice (patients often show phonemic and semantic paraphasias and neologisms). Patients with Wernicke aphasia have difficulties in comprehending sentences uttered by other people, showing a deficit at a semantic–lexical level. The deficit of comprehension can be extended to the written language.

Conduction Aphasia

Conduction aphasia shares with Broca and Wernicke aphasias the inability to repeat sentences, a defective assembly of phonemes, and an impaired naming ability, but it differs from them in the relatively preserved speech production and auditory comprehension. In classical descriptions (Wernicke's model[6] and Lichtheim's model[4]), conduction aphasia was uniquely ascribed to lesions of the arcuate fasciculus, a white matter pathway connecting the Wernicke and Broca areas. It is now known that, although the subcortical projections linking temporal, parietal, and frontal cortices are often damaged in patients affected by this kind of language problem there is no evidence that a pure white matter disconnection can cause conduction aphasia. In addition, the involvement of the left inferior parietal lobe (BA40), the left primary auditory cortices (BA41 and 42), and the insula seems to be necessary to cause the symptoms.

Global Aphasia

Patients with global aphasia have almost completely lost the ability to comprehend language and formulate speech, combining the features of both Broca and Wernicke aphasia. Global aphasia is usually caused by large anteroposterior damage, widely involving the language regions, the basal ganglia region, and the insula.

Transcortical Aphasias

These types of aphasia can be distinguished from all others by the fact that the ability to repeat sentences is normal. The motor variant usually occurs after left frontal lesions above and in front of the Broca area. The sensory variant is caused by lesions in temporal or parietal cortices, in the vicinity of the Wernicke area.

Anomic Aphasia

Damage to the left anterior temporal cortices (BA20, 21, and 38) severely impairs the ability to retrieve words, but is not accompanied by any grammatical, phonemic, or phonetic difficulty, causing a "pure" naming deficit. More specifically, when the damage is confined to the left temporal pole (BA38) patients have a deficit in the ability to retrieve proper nouns of places and people; when the lesion involves BA20 and 21 the defect encompasses the ability to retrieve both proper and common nouns.

Towards a More Complex Model of Language in the Human Brain

Although the anatomoclinical correlations between the different types of aphasia symptom and the damage to specific brain areas were thoroughly confirmed in the twentieth century by a large number of neuropsychological and neurostructural studies, in recent decades the complex networks underlying language have been explored using innovative techniques such as functional magnetic resonance imaging (fMRI) and diffusion tensor imaging. This approach led to the development of a more dynamic, connectionist approach to the study of the anatomical correlates of aphasias. On the one hand, areas originally thought to be specialized for language have been shown to be also involved in cognitive and perceptual functions not directly related to language; on the other hand, it has been demonstrated that language does not exclusively rely on the Wernicke–Broca language network.[7] Furthermore, recent neuropsychological studies on the consequences of acquired brain lesions in children and adults have suggested that the outcomes of aphasias have a greater variability than predicted by the classical models, depending on the patient's age, the etiology, site, and size of the lesion, and the initial severity of the impairment.[8] Whereas sudden brain lesions affecting specialized areas often result in severe aphasia, the clinical pattern is different in the case of slowly growing lesions such as low-grade gliomas. Slow tumor evolution allows for compensatory mechanisms to develop through the recruitment of intrahemispheric and interhemispheric neuronal networks (i.e. perilesional and/or contralateral homologous brain regions). As suggested by the case study of a "patient speaking without Broca's area",[7] in which the left inferior frontal gyrus resection (including Broca area), due to tumor infiltration, did not lead to severe language impairments, the efficiency of brain plasticity can, in some instances, compensate for the anatomical specialization of linguistic functions.

Embodied Semantics

Neuropsychological research on language functions is not confined to the study of aphasia, but also includes the study of "normal" linguistic processing in healthy subjects. An innovative and fecund research paradigm has developed, which aims to explain linguistic functions within the theoretical framework of embodied cognition.[9,10] From this perspective, not only perceptual representation but also high-level cognitive processes, such as concept formation and language, are essentially based on motor programs. Contrary to the classical approach in cognitive science, in which concepts are viewed as amodal and arbitrary symbols, the embodied hypothesis argues that concepts must be grounded on sensorimotor experiences to be meaningful. In line with this view, neuroimaging studies have shown a somatotopic pattern of activation along cortical motor areas for the observation of actions involving different body parts, as well as for action-related language comprehension (see Aziz Zadeh and Damasio[11] for a review). For example, the concept of "grasping" would be represented in sensorimotor areas that code grasping actions; the concept of "kicking" would be represented by sensorimotor areas that control kicking actions; and so forth. The signals related to the common goal for a variety of specific actions (i.e. grasping with the mouth or with the hand or toes, along with related affordances) would also converge on a particular group of neurons, thus providing a more general representation of "grasping". This would be the neurobiological substrate of generalized conceptual representations, which, in turn, could be related to language description.[11] This theory has been extended by Lakoff and colleagues to include metaphors. Thus, the phrase "kick off the year" would also involve the motor representations related to kicking, just as the phrase "grasp the explanation" would involve motor representations related to the control of the hand.[12] In a future perspective, exploring the link between these conceptual representations and metaphorical language will be especially important, as it has been proposed that much of abstract thinking is performed metaphorically.[12]

MEMORY DISORDERS: AMNESIA

Amnesia, a profound disorder of memory functions, is a clear example of how we can propose a hypothesis about the structure of a cognitive process and make inferences from the combination of association and dissociation of symptoms present in a particular kind of neuropsychological syndrome. Since the work of Hebb,[13] the idea that memory is equally distributed throughout different brain regions, being intimately related to perceptual and intellectual functions,[14] was strongly challenged by those scientists who believed that, although memory processes are indeed distributed in different cortical and subcortical structures, specific aspects are processed by different areas and relatively independently from other cognitive functions. This latter position was dramatically confirmed by the study of patients who, immediately after selective brain surgery for the treatment of intractable epilepsy, showed dramatic, specific, and dissociated memory impairments. The most enlightening observations were made by Brenda Milner, Susan Corkin, and their co-workers[15–17] on patient HM. This patient, after bilateral resection of the medial temporal lobes, was left with a complex syndrome characterized by severe memory impairments that affected some memory capacities but not others, and by a considerable sparing of other cognitive functions. First of all, it was noted that he had a severe anterograde amnesia, with which he could not acquire new memories, dating from the day of his brain damage. Moreover, he could not retrieve some of the memories that he acquired before surgery (retrograde amnesia), although access to facts and events from times remote from surgery were still possible. This latter finding implies that medial temporal lobe structures are not the final locus for storing old memories. After learning, memories that initially require the integrity of medial temporal lobe structures must be reorganized and stored somewhere else in the brain to become independent from these structures. According to many authors, once a memory has been fully consolidated, its storage and reactivation may depend on processes and structures located in neocortex.[18] The extent of retrograde amnesia may be taken as an index of how long the consolidation process lasts.

Despite his severe memory problems, HM's other cognitive and perceptual capacities were intact. For instance, his intelligence quotient (IQ) was higher after compared with before surgery and no main perceptual problems were detected. The memory disorders affected both verbal and non-verbal tasks, as well as stimuli presented in all sensory modalities. From these observations the inference was made that memory functions can be kept separated from other cognitive functions and from more primary perceptual and sensory processes.[18] It is also worth noting that the impairments described above involved what are normally indicated as long-term memory capacities. However, short-term memory was intact in HM. That means that he could repeat a sequence of digits immediately after their presentation and that he could follow a conversation providing that it did not become too long and based on data that were presented before the talk.[19] Therefore, the dissociation between long and short memory processes was another fundamental finding from the study of HM.

As already mentioned, within long-term memory processes, anterograde amnesia refers to the impossibility of learning new facts and HM was severely impaired in this domain. However, and somehow astonishingly considering his otherwise dramatic memory impairment, HM showed intact capacities of acquiring new motor and perceptual skills[20–22] (procedural memory), clearly demonstrating that amnesia associated with hippocampal damage affects what are indicated as declarative memory processes (the ability to recall facts and events related to specific personal experience, episodic memory), leaving intact other memory operations related to the functioning of separated brain structures and systems.[23] It is important to note that HM's profound memory disorder resulted from brain damage that was not confined to the hippocampus but also involved the amygdala and the adjacent parahippocamapal gyrus. Another patient, RB, who had a lesion limited to the hippocampus following an ischemic attack, had a similar, but less severe memory disorder.[24] Taken together, these results indicate that damage to the hippocampus is sufficient to cause a clinically significant amnesic disorder and that additional damage to adjacent structures aggravates the symptoms.

The study of HM revealed the most important aspects of how memory is structured and organized in the brain. As Squire and Wixted wrote in 2011: "These findings established the fundamental principle that memory is a distinct cerebral function, separable from other perceptual and cognitive abilities, and also identified the medial aspect of the temporal lobe as important for memory. The early descriptions of HM can be said to have inaugurated the modern era of memory research, and the findings from HM enormously influenced the direction of subsequent work".

It is worth remembering that there are, unfortunately, other pathological conditions that can induce severe amnesic disorders. Alcohol abuse, leading to thiamine deficit and to damage to diencephalic brain structures, including the mammillary body, can be the cause of Korsakoff syndrome, where a severe anterograde deficit is always observed (for a review see Fama *et al.*[25]). Infections affecting the brain, such as herpes simplex virus encephalitis, can cause disastrous amnesia that may extend beyond episodic problems and involve semantic knowledge.

DISORDERS OF MOVEMENT EXECUTION: APRAXIA

Apraxia has been defined as the inability to carry out learned, skilled motor acts despite preserved motor and sensory functions, coordination, and comprehension. The major types are described in this section.

Ideomotor Apraxia

Ideomotor apraxia (IMA) is the impaired ability to perform a skilled gesture with a limb upon verbal command and/or by imitation. It can be shown for both meaningful motor acts that do not imply objects and gestures that imply object use. In this latter case, patients may not be able to perform the pantomime (i.e. to show how to use an object without actually manipulating it) or may not be able to use the actual object. Many authors also consider the inability to imitate meaningless gestures as apraxia. Patients with IMA show errors in the temporal and spatial sequencing of movements, in their amplitude and configuration, and in limb position in space.

Various dissociations have been described. For instance, a voluntary–automatic dichotomy has been described according to which patients seem to be able to perform in their daily life activities the same acts that they are unable to perform when requested by the examiner. Differences in the performance related to the type of gesture to be produced have also been described. For instance, there may be differences in the performance between actions implying use of an object (transitive actions) and actions not implying use of an object (intransitive actions, e.g. waving goodbye). Finally, apraxia is often more severe for meaningless actions than for meaningful actions, although the opposite dissociation has also been described. Given these dissociations, it is essential that the diagnosis of IMA include a wide variety of tasks.

Ideational Apraxia

Patients with ideational apraxia (IA) have difficulties in executing a sequence of actions when performing a complex multistage task (e.g. making coffee). Some authors also distinguish between IA and conceptual apraxia, identifying with this latter condition an impairment of object or action knowledge. Patients with conceptual apraxia may misuse objects, can be impaired in matching objects with the corresponding actions, or may be unable to judge whether an action is correctly or ill formed. Patients with IA or conceptual apraxia tend to be seriously disabled in their everyday life. IA and IMA can be doubly dissociated.

Orofacial Apraxia

Orofacial apraxia is an impairment in the execution of skilled movements involving the face, mouth, tongue, larynx, and pharynx (e.g. blowing a kiss or whistling) when requested by the examiner. Similarly to IMA, automatic movements of the same muscles are often preserved. Orofacial and limb apraxias often coexist but can be dissociated.

Limb-Kinetic Apraxia

The term limb-kinetic apraxia describes inaccurate or clumsy distal movements of the limbs contralateral to the lesioned hemisphere. Limb-kinetic apraxia differs from classical ideomotor apraxia because it tends to be independent of modality (e.g. verbal command versus imitation) and there is typically no voluntary–automatic dissociation.

Constructional Apraxia

Constructional apraxia is a particular type of apraxia where patients are unable to spontaneously draw objects, copy figures and build blocks or patterns with sticks following damage to the left or right hemisphere. Hence, constructional apraxia seems to reflect the loss of the ability to integrate perceptual, categorical, and coordinate spatial relations with the motor actions necessary to complete a constructive task. Constructional apraxia can be dissociated from the other types of apraxia.

Anatomy of Apraxia

Interhemispheric Localization

The two major forms of apraxia (IMA and IA) are more frequent following left hemisphere lesions. Nevertheless, in a few instances apraxia has also been described after lesions of the right hemisphere, suggesting that right hemisphere structures can also support skilled movements to a certain extent. Specifically, right hemisphere "praxic" structures can be recruited following injuries to the left hemisphere and also as a result of training and execution of highly practiced familiar actions.

Intrahemispheric Localization

IMA and IA have been associated with lesions to the parietal and frontal cortices of the left hemisphere, as well as with white matter connections between these areas. fMRI studies during gesture execution in healthy subjects have confirmed the involvement of left parietal and frontal regions, including the dorsolateral frontal and intraparietal cortex. A particular form of IMA can be seen after callosal lesions; in this case, it only involves the left hand, being the consequence of the disconnection of the right hemisphere premotor and motor areas from the left hemisphere praxis centers.

The basal ganglia also play an important role in praxis via bilateral connections with frontal and parietal areas.

Finally, it is important to consider the relationship between limb apraxia and aphasia. Apraxia and aphasia frequently coexist following left hemisphere damage, raising the question of whether there is a common feature underlying both disturbances (some authors point to an asymbolic problem) or whether linguistic dysfunction is responsible for apraxia because patients do not understand the examiner's requests. However, dissociations between aphasia and apraxia have been reported, thereby falsifying both hypotheses. Moreover, aphasia cannot account for apraxia in cases of gesture imitation. So, rather than there being a causal relationship between aphasia and apraxia, these two symptoms are likely to be associated in the same patient owing to lesions affecting adjacent neural substrates for language and gestures.

Neuropsychological Models of Apraxia

Different models to understand the neurocognitive mechanisms underlying apraxic disorders have been proposed. Here, the most influential ones will be reviewed.

Liepmann's Model

In 1920, Hugo Liepmann proposed a disconnection model of praxic disorders according to which the representation of an action (space–time plan) is stored in the left parietal lobe.[26] In order to execute the action with the right hand, the space–time plan is retrieved and then reaches the primary motor areas through the left prefrontal cortex. Left-hand actions (ultimately controlled by the right hemisphere) are possible because the action plan reaches right premotor and motor areas through the corpus callosum. In ideomotor apraxia action representations and limb kinetics are intact, whereas frontoparietal connections are disrupted, causing an inability to execute normal actions. Ideational apraxia would result from direct lesion to the action representation area and limb-kinetic apraxia from disruption of "kinesthetic–innervatory engrams" in the left frontal lobe.

Geschwind's Model

Norman Geschwind proposed a disconnection model of apraxia routed on Liepmann's model, according to which the verbal command for the action of the right hand is comprehended in the Wernicke area and is transferred to the ipsilateral premotor and motor areas via the superior longitudinal fasciculus.[27] A lesion to this pathway would spare gesture comprehension but compromise action performance elicited by verbal command. To explain failure of action imitation and object use, Geschwind proposed that visual association and premotor areas are connected through the same pathways running through language and motor areas, but this notion remains controversial.

Heilman and Rothi's Model

Heilman and Rothi proposed an alternative representational model of apraxia in which left anterior premotor–motor regions are responsible for gesture production and left posterior parietal regions store the representation of learned movements and are responsible for gesture comprehension and discrimination.[28] Lesions to anterior regions would cause ideomotor apraxia, and lesions to posterior regions ideational apraxia as well as difficulties in movement production. Some neuropsychological data have supported this model.

To account for modality-specific dissociations in praxic disorders, in subsequent versions of their model Heilman and Rothi proposed separate processing routes for auditory and visual inputs, conveying information in a specific "action semantic system", dissociable from other semantics, which would activate an "action reception lexicon" connected with an "action production lexicon".[29] A separate "non-lexical route", which bypasses the action semantic system, would be responsible for the ability to imitate novel and meaningless gestures.

Key Questions

The discovery that particular neuronal populations (mirror neurons) are active both when an action is produced and when an individual observes that same action[30] poses a challenge for praxis models that hypothesize separate structures for action production and action recognition/comprehension. In particular, if the same representations subserve both action execution and recognition, the question arises of whether the perception/comprehension of a movement is constrained by its executional knowledge. The complexity of the praxis system suggests that it should be better studied through a multidisciplinary approach encompassing knowledge from lesion studies, fMRI data from healthy individuals, neurophysiological data from animal studies, and psychophysical and computational models of complex action execution (for reviews see references [31–33]). This could help to reach a more detailed mapping of large-scale neuronal networks underlying praxis, the assessment of alterations in these networks following injury, and their relationship with behavioral dysfunctions. A better understanding of these issues will lead to the development of more efficacious therapeutic and rehabilitative interventions.

DISORDERS OF VISUAL RECOGNITION: AGNOSIA

The neuropsychological disorder known as agnosia refers to the impairment of stimulus recognition in one modality in the absence of perceptual deficits, memory problems, and general intellectual impairment. This disorder is intriguing both scientifically and clinically and its study has contributed to shedding light on how the normal visual system functions. When the impaired recognition relates to objects in general, the condition is called object agnosia; when the unrecognized visual stimulus is a face, it is called prosopagnosia. An overview of the major types of agnosia is presented here.

Object Agnosia

In object agnosia, patients do not recognize objects in one specific input modality (visual, tactile, or auditory), whereas the same objects can be promptly recognized when presented through a different input channel. The perceptual nature of the disorder is testified by the fact that it cannot be ascribed to the co-occurrence of sensory elementary deficits, memory problems, naming difficulties (aphasia), and general intellectual impairment (patients are well aware of their predicament).

Visual Agnosia

Although it is a relatively rare neurological symptom, with some 100 cases published between 1890 and 1990, its study has greatly contributed to the understanding of how the process of visual recognition is organized in the human brain. There is no standard taxonomy of visual agnosias, but most neuropsychologists agree with Lissauer's original distinction between apperceptive and associative types,[34] depending on the lower or higher processing stage of visual information affected by the brain lesion. Because this account has continued to be used in the neuropsychological literature to the present day, it is used here as a general framework.

Apperceptive Agnosia

Apperceptive agnosia is evident when patients are unable to recognize objects because they cannot see them properly, in the absence of elementary visual deficits. It is thought to arise from a breakdown at relatively early stages of visual processing, where the elementary features of the stimulus are analyzed. Object recognition through verbal description by the examiner is, instead, preserved. In apperceptive agnosia, shape perception is abnormal in such a way that patients cannot recognize or copy pictures, letters, or even simple geometric shapes. In most cases of apperceptive agnosia, the brain damage is diffuse, often caused by carbon monoxide poisoning. In the rare cases with circumscribed brain lesions, damage primarily affected the ventral occipitotemporal cortex bilaterally.

According to a widely accepted interpretation, apperceptive agnosia can be considered a deficit of shape perception resulting from defective perceptual grouping of an object's local features into a global percept. However, clinical findings show that apperceptive agnosia covers a wide spectrum of disorders, some of which fall in between apperceptive and associative agnosia. Hence, two neuropsychologists, Riddoch and Humphreys,[35] proposed to differentiate between distinct subtypes of apperceptive agnosia, each corresponding to a defective processing stage along the hierarchically organized stream of visual information processing leading to conscious object perception (according to Marr's computational model of vision[36]). These subtypes are shape agnosia and integrative agnosia, which are closer to the apperceptive type; and transformational agnosia and agnosia due to impairment of internal object representation, which are closer to the associative type.[35]

- *Shape agnosia* results from a deficit of the initial processing stage of visual recognition and consists of the inability to organize the sensory input into a unified percept. Patients complain of blurred or unclear vision and are unable to discriminate stimulus boundaries from the background or other contiguous or overlapping shapes, as well as the orientation and size of the input.
- *Integrative agnosia* consists of the inability to integrate single object features into a global shape, in the presence of the ability to identify single object details. This deficit is more severe when the object shape is defined by high-frequency details or when overlapping figures must be identified, but is reduced when silhouettes of objects with reduced internal details are used for discrimination.
- *Transformational agnosia* is a deficit of perceptual categorization (first described in 1982 by Warrington[37]), which occurs when patients can recognize objects presented in a canonical view, but fail when they are presented in non-canonical views. What is lost is the ability to manipulate the mental representation of the object and to match it with its perceptual image.
- *Agnosia due to impairment of internal representations of objects* occurs when a structural description of the object is formed normally but its internal representation stored in presemantic memory cannot be accessed through a given route to match on-line descriptions encoded by the visual system.

Associative Agnosia

This type of agnosia occurs when patients can form a structural description of the visual object (object copy is preserved), yet are unable to recognize it. Associative agnosic patients cannot identify objects even by nonverbal means (e.g. by pantomiming their use or grouping together dissimilar objects from the same semantic category); however, recognition is preserved in the tactile modality (by touching the object) or from a spoken definition. Intrahemispheric location of the lesion is generally occipitotemporal, either unilateral (with a prevalence of left hemispheric lesions) or bilateral.

Associative agnosia has been explained as a deficit of the activation of the semantic associations related to the visual percept: patients, despite being able to form a normal visual representation of the stimulus, are unable to access the knowledge related to it and therefore recognition is prevented. Therefore, patients fail in semantic categorization and association tasks, as well as in the description of the semantic attributes of an object. In visual naming tasks, errors tend to be semantic (e.g. "knife" for "fork"), sometimes with the production of the superordinate ("flower" for "daisy"), although errors totally unrelated to the stimulus (e.g. "horse" for "chair") can be observed. Miming the use of a visually presented object is also impaired, whereas the task is carried out correctly if a verbal description of the object is offered. According to one view, associative agnosia is the consequence of a disconnection between visual areas and other brain centers responsible for language or memory. This hypothesis, however, does not account for the inability of agnosic patients to convey information non-verbally and to access old knowledge through vision. Another possibility is that semantic knowledge cannot be accessed through the visual modality because the lesion has damaged the connection between the areas that process the stimulus and semantic memory (semantic access agnosia). A different interpretation assumes that stored visual memory representations have been partially or totally damaged so that the newly formed visual percepts cannot be matched against any stored knowledge and, therefore, recognition is impossible. In this case, however, the disorder should be observed in all modalities and, therefore, cannot be considered a pure perceptual disorder but instead becomes a true disorder of semantic memory.

Tactile Agnosia

Tactile agnosia is the inability to recognize objects through touch, in the absence of elementary sensory deficits. Recognition in other modalities is preserved. The locus of the lesion involves the posterior–inferior portion of the parietal lobe and can be unilateral or bilateral. The deficit has been rarely studied and the interpretations are similar to those offered for visual agnosia.

Acoustic Agnosia

Acoustic agnosia (or aphasia) is the inability to name an object through sound (e.g. failure to name a bunch of keys given the sound the keys make when shaken) with preserved ability to recognize objects in other sensory domains.

Face Agnosia (Prosopagnosia)

The term prosopagnosia refers to the inability to recognize familiar faces. The deficit is confined to the identification of physiognomic traits, as shown by the fact that identification is preserved through non-physiognomic cues, such as voice, a particular item of clothing, a scar, or gait. Usually patients do not recognize friends, acquaintances, and famous people. In the most severe cases patients cannot even recognize their own face in the mirror. However, perceptual categorization of the stimulus is preserved (patients know that a face is a face) as well as the ability to differentiate faces by gender, race, age, and emotional expressions.

Psychophysical and neuropsychological studies on face recognition abilities in healthy and brain-damaged patients have revealed that familiar and unfamiliar face processing follows dedicated routes in the left and right hemisphere, respectively. This distinction led the neuropsychologist A.L. Benton to differentiate between two independent face processing deficits produced by brain damage: apperceptive prosopagnosia, which refers to a defective perceptual processing of face information and is brought out by unfamiliar face tasks; and associative prosopagnosia, which involves an additional mnestic component and is elicited by familiar face tasks.[38]

Apperceptive and Associative Prosopagnosia

According to the model of visual recognition proposed by Bruce and Young, the identification of a face is the final stage in a sequence of operations made by distinct, hierarchically organized, information processing modules distributed along the occipitotemporal ventral pathway of the brain.[39] At the earlier stages, perceptual face processing results in the construction of an object-centered, tridimensional structural description of the face. If this processing level is damaged, patients are unable to recognize familiar faces and to match different pictures of unfamiliar faces (apperceptive prosopagnosia). At later processing stages, the structural description of the face activates an abstract representation of it stored in recognition units responsible for the feeling of familiarity. Then, the information gains access to the semantic memory representation (identity nodes) containing the knowledge related to that particular known face. From the identity nodes, information finally accesses the modules containing the person's name. The anatomical–functional independence of the names module is confirmed by the existence of anomia for proper names following left brain damage (a rare occurrence) and by the frequent inability experienced by non-prosopagnosic people to retrieve the name of an otherwise well known person ("name on the tip of the tongue" phenomenon).

Associative prosopagnosia is normally consequent upon an impairment at the level of either recognition units or identity nodes, or both: patients are able to match unfamiliar faces, but fail to recognize familiar faces. Knowledge of the people to whom the faces belong can be accessed through other sensory modalities (e.g. sound) and is intact.

Unconscious Face Recognition

This phenomenon is based on the fact that some prosopagnosic patients may show, in indirect tasks, normal responses to famous faces, thus presenting with some degree of implicit processing of the unrecognized stimuli. This is particularly evident in psychophysiological and neurophysiological measures, such as skin conductance, where prosopagnosic patients may show an increase in electrodermal activity when presented with familiar faces, but not with unfamiliar faces (like healthy subjects). Perhaps the most convincing interpretation of this phenomenon is that conscious recognition requires a higher activation threshold from the visual input than implicit recognition. According to this view, if the lesion producing prosopagnosia completely impairs the function of recognition units, both implicit and explicit recognition will be impossible. If the impairment is only partial, instead, the output from recognition units will be sufficient for unconscious recognition but insufficient for overt recognition (for reviews on agnosia see references [40–42]).

DISORDERS OF SPATIAL REPRESENTATION: UNILATERAL NEGLECT

Unilateral neglect (UN) is a neurological syndrome first described around the turn of the twentieth century. Its core symptom is the loss of conscious awareness for the left side of the perceptual and mental space.

Clinical Manifestations

UN patients behave "… as if they were no longer able to perceive and conceive the existence of the left side of somatic and extrasomatic space".[43] For example, in the acute phase after stroke, more frequently affecting the right hemisphere, patients often show a more or less complete deviation of the eyes and head towards the right space, fail to respond to left-side visual and auditory stimulation, tend to underuse their left upper and lower limbs (in the absence of hemiplegia or severe motor impairment), do not explore the left side of their body or environment, and may forget to dress the left side of their body or to wash the left side of their face. UN patients are often unaware of their symptoms (anosognosia).

UN may be diagnosed by means of simple bedside paper-and-pencil tests. When asked to bisect a line segment, UN patients misbisect it to the right of the objective midpoint; when asked to cross out line segments printed on a sheet of paper (Albert's test) they omit to cross out a number of left-side segments; in drawing or recalling from memory simple objects (e.g. a daisy or a clock face) they omit, misarrange, or distort left-side details (Fig. 32.1).

Neglect symptoms may also occur outside the visual domain, in the haptic (somatic and extrasomatic) space (e.g. blindfolded patients asked to touch their left hand or shoulder with their right hand may fail to do it, or asked to collect objects spread over a table may not find them), the auditory space (patients asked to locate sound sources may mislocate them to the right side of space), and the representational space (patients asked to form the mental image of a familiar view – a town square, a room, a map of a country – and to describe its details from two opposite vantage points may omit left-side details of the mental images).

Clinicoanatomical Correlations

UN is more frequent following lesions to the right hemisphere, with an incidence varying from 30 to 43% after right hemisphere stroke.[43,44] UN may be localized within a hemisphere, and has been associated with both cortical and subcortical lesions. Brain areas whose lesion is most frequently associated with UN are the temporal, parietal, frontal, and occipital lobes, the basal ganglia, and the thalamus.[44]

Course of Unilateral Neglect

The majority of the symptoms of neglect recede more or less completely following the acute phase of the disease. However, depending on the severity of neglect, some symptoms may persist for weeks or years. The mechanisms sustaining the functional restoration of neglect symptoms are not yet fully understood and include plastic processes arising in surviving circuits in the damaged hemisphere as well as in structures of the undamaged hemisphere.

Interpretation of the Syndrome

Since the 1980s, several interpretations of UN have been proposed, which differ according to the processing level at which the causal mechanisms of the syndrome are thought to operate: the sensory, attentional, and representational level.

Sensory Interpretations

This group of interpretations considers the causal mechanisms of UN as relatively "peripheral": contralesional (left) sensory information would go undetected owing to (1) a sort of global extinction phenomenon (stimuli addressed to the damaged hemisphere are more or less correctly perceived if presented in isolation, but are suppressed if given in association with

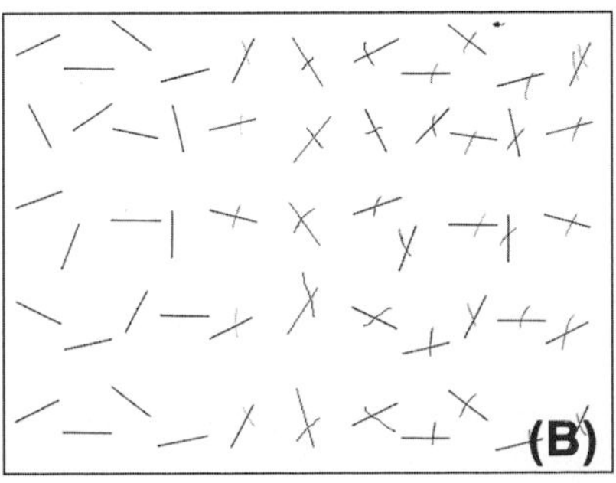

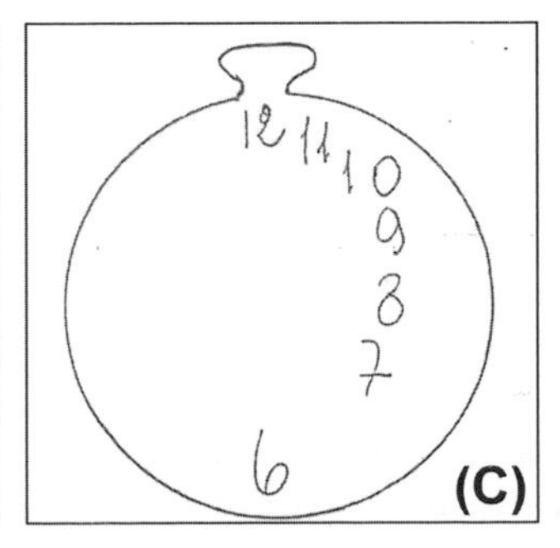

FIGURE 32.1 **Pencil-and-paper tests in unilateral neglect (UN).** (A) Copies of daisies from a model (upper part of the figure) by two patients with UN; (B) Albert's cancellation task: note that the patient has omitted to cancel the majority of left-side segments; (C) drawing from memory of a clock face, by a patient with UN.

stimuli presented to the undamaged hemisphere); (2) impairment of eye movements towards the left space; or (3) impairment of orienting reflex towards left stimuli. These interpretations refer to impaired processing of stimuli external to the CNS and lost most of their validity after it was ascertained that UN can be found at the level of mental imagery.[43]

Attentional Interpretations

According to these interpretations, lesions to the right hemisphere impair the ability of UN patients to efficaciously orient attention leftwards, with the consequence of becoming unaware of contralesional sensory events. The model that is often used to interpret UN symptoms was proposed by Kinsbourne.[45] This model posits the existence of two antagonist attentional vectors, each depending on one hemisphere and directing attention contralaterally. A further assumption of the model is that the left hemisphere vector normally predominates over the right hemisphere one, so that the right side of space is attentionally privileged even in healthy people. This model predicts that damage to one hemisphere would impair the corresponding attentional vector and therefore cause a pathological imbalance of attention towards the ipsilesional hemispace. If the left hemisphere is lesioned, the imbalance is less pronounced, and neglect unapparent, because it releases the weaker right hemisphere vector; by contrast, if the right hemisphere is lesioned, the predominant left hemisphere vector is released, biasing attention more strongly rightwards and causing neglect for the left side of space. According to this model, the attentional impairment caused by the lesion should be distributed along the left–right dimension of space, with a continuous gradient ranging from maximum to minimum severity in the left and right extreme spatial sectors, respectively. Behavioral and neurophysiological data confirm this prediction. In this respect, it is worth noting that single-neuron electrophysiology showed that in the postarcuate cortex of the monkey (the lesion of which gives rise to contralateral neglect) 29% of neurons have exclusively contralateral, 3% exclusively ipsilateral, and 68% bilateral receptive fields.[46] Such a distribution of space-coding neurons is compatible with the gradient of the attentional bias along the left–right dimension of space described for UN patients.

Representational Interpretations

As already mentioned, UN patients asked to remember and describe a familiar view (a town square, a map of their country, or the layout of their home) from a definite vantage point may neglect contralesional details, in the absence of long-term spatial memory deficits. Such pioneering clinical observations, first reported by Edoardo Bisiach and co-workers,[47] led them to conclude that UN is also a deficit of the endogenous mental representation of the contralesional extent of perceptual space. Representational accounts of UN do not entail that attentional interpretations are inadequate; they imply that the concept of attention should be extended to comprise "… the processes of generation and transformation of mental representations" and "… it would therefore refer to the dynamics of representational processes themselves, rather than to separate processes".[43] Bisiach and colleagues[48,49] updated the representational explanation in order to explain an apparently paradoxical behavior shown by some neglect patients. When asked to extend a segment towards the left in order to double its original length, some patients overextend it, so as to create a left half-segment much longer than the right half-segment. This behavior cannot be accounted for by sensory, attentional, and representational interpretations, all predicting absence of or minimal contralesionally directed behavior. Bisiach and co-workers proposed that "This phenomenon may be functionally interpreted as a left–right pathological anisometry of the medium in which within- and between-objects spatial relations are episodically represented. As a consequence of right brain damage, this medium becomes relatively more and more relaxed toward the contralesional and more and more compact toward the ipsilesional side".[49] Accordingly, any object confronting that scale would shrink on the contralesional side and stretch out on the ipsilesional side, so that UN patients erroneously displace rightwards the subjective midpoint of line segments to be bisected; similarly, when asked to duplicate leftwards a half-segment located ipsilesionally, they will overextend it leftwards in order to subjectively perceive it as identical to the right half-segment (Fig. 32.2).

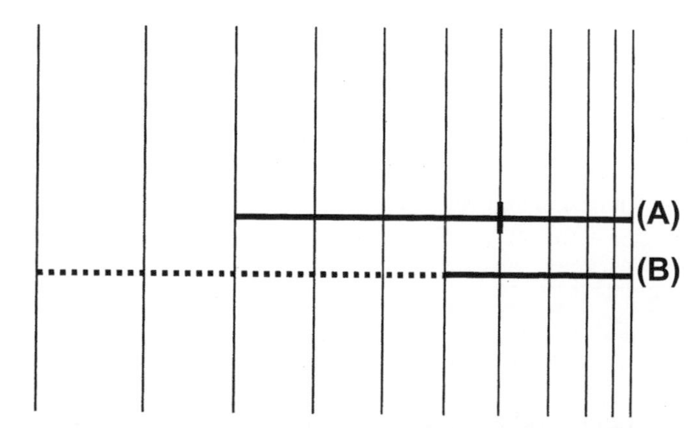

FIGURE 32.2 **Progressive leftward relaxation of the medium for space representation in unilateral neglect (UN).** (A) Horizontal extension of the left and right sides of a line segment are underestimated and overestimated, respectively, leading UN patients to misplace rightwards the subjective middle of the segment; (B) similarly, they overextend a segment leftwards to double its original length.

Implications for Cognitive Science

Data collected from UN patients have important theoretical implications for cognitive processes in general and for the understanding of the structure of mental representation in particular. It has long been debated whether the structure of mental contents is symbolic (linguistic) or analogic. The neuropsychological evidence from UN strongly supports the notion of an analogue structure of spatial cognition, similar to the topological relationship existing between external space and the surface of the retina and of the striate visual cortex: "… analogue relationships, such as those existing between perceived objects (and events) and the perceiving system, may also exist between those objects (and events) and the brain mechanisms capable of representing them even in their absence. … In the same way as lesion of part of the calcarine cortex gives rise to a sensory scotoma, so spatially circumscribed lesions of critical brain structures may give rise to topologically correspondent representational disorders".[43]

Another issue connected with UN is the "destiny" of neglected information: is it completely lost or is it processed at some level? A large body of neuropsychological evidence suggests that neglected information is processed up to a semantic level (see "Neglect: A Model for Spatial Awareness", below).

A final general remark concerns cognitive models of brain functions. In order to be considered fully reliable in their explanatory power, they need to be grounded on a solid neurobiological basis. This is also true for the representational model of UN, which is still lacking its neurobiological counterpart. Future research should seek an effective explanation of the phenomena of contralesional misrepresentation in the changes that are likely to be induced by unilateral brain lesions in the receptive fields of neurons involved in coding spatial relationships. Unilateral lesions may affect not only the metrics according to which spatial relations are processed in the brain, but also the degree to which, and the way in which, contents organized within a certain area of the pathologically uneven medium reach conscious processing levels. The study of changes in the responsiveness of single space-coding neurons following selective brain lesions may, therefore, shed light on the links between space representation and consciousness.

CONSCIOUS AWARENESS

Understanding the relationship between neural activity and subjective conscious experience is one of the most fascinating and challenging goals of modern neuroscience. In this section, this topic will be explored, starting from neuropsychological disorders in which different aspects of conscious awareness can be selectively impaired. In particular, the text will focus on three syndromes, namely blindsight, neglect, and anosognosia for hemiplegia (AHP), that can be used as models for the study of visual, spatial, and motor awareness, respectively. Finally, a modular model of conscious awareness will be proposed.

Blindsight: A Model for Visual Awareness

Awareness of a stimulus is usually defined as the subjective experience of the existence of a stimulus and the related recognition of its physical and semantic properties. Conscious identification of a visual stimulus can be reported verbally or can be inferred by the observer's overt behavior in response to stimulus presentation. Indeed, if an observer is aware of the presented stimuli, he or she can act upon them. A crucial issue in cognitive neuropsychology has been whether there is the possibility of stimulus elaboration without conscious awareness of it. A convincing example of this possibility is the blindsight syndrome.[50,51] The phenomenology of blindsight has two principal features. The first is the loss of visual awareness associated with damage to the primary visual cortex (V1; BA17). The second is the residual capacity of blind patients to use visual signals to guide behavioral responses. Pöppel and colleagues' pioneering study showed that these patients, when requested to look at the "unseen" targets, are able to direct their eyes towards these stimuli.[52] Weiskrantz and colleagues further investigated this phenomenon, showing that blindsight patients, although completely unaware of the presence of a visual stimulus in the blind field, are nevertheless able to point to it correctly and "guess" its orientation.[50] Since then, many studies have been conducted in many different laboratories around the world, which demonstrated residual processing of different attributes of the unseen stimuli by blindsight patients. For instance, Tamietto and de Gelder explored emotion detection in blindsight.[53] They demonstrated that patients could correctly guess the emotion expressed by faces presented in their blind field. Therefore, blindsight studies have shown that many stimulus attributes can be processed in the absence of conscious vision.

One possible explanation of this paradoxical phenomenon is that even though a large part of the visual cortex is damaged, tiny islands of healthy tissue are spared. The visual processing carried out in this spared tissue is not enough to provide conscious perception, but nevertheless is sufficient to sustain blindsight. In this view, conscious awareness would be a matter of threshold; that is, when, after damage to the brain, the cortical activity is too low and does not reach a sufficient level for conscious activation of sensory input, visual awareness is lost. However, at least in some patients, it has been demonstrated that when no residual island of healthy tissue remains, as in surgical resection of V1 or in hemispherectomy, blindsight can still be present.[54] An alternative hypothesis is that V1 activity has a direct role in generating visual awareness. If so, consciousness would not be a matter of threshold but would depend on the integrity of a specific brain circuit, while the possibility of blindsight would depend on the integrity of a different circuit. In this respect it is important to refer to the complex neuroanatomy of vision. First of all, it must be kept in mind that visual information uses many different pathways from the retina to the brain. The primary visual pathway projects from the retina to V1, in the occipital lobe, via an intermediate station in the lateral geniculate nucleus of the thalamus. From V1, visual information reaches the extrastriate cortex along the ventral (occipitotemporal) and the dorsal (occipitoparietal) stream. However, a minority of fibers originating from the retina takes a secondary route to the superior colliculus and the pulvinar. These two subcortical structures are interconnected and also send direct projections to the extrastriate visual cortex, bypassing V1. Considering these anatomical characteristics of the visual pathways, one possibility is that V1 is necessary for conscious recognition of visual stimuli. If it is damaged, visual consciousness is prevented. However, the integrity of alternative pathways, such as the retinal–collicular–extrastriate cortex pathway, can allow some visual processing without awareness, as in blindsight patients. Many studies have demonstrated the importance of the collicular nucleus for unconscious vision. Other studies have shown that V1 is necessary, but not sufficient, for full visual awareness. Indeed, other brain regions must be activated (such as prefrontal area 46) to gain full consciousness of visually presented stimuli.

Neglect: A Model for Spatial Awareness

A disorder that has shed light on spatial awareness is the neglect syndrome. Patients with unilateral spatial neglect, in contrast to blindsight patients, may not have any primary visual impairment. Nonetheless, they fail to respond to stimuli, objects, and even people located on their contralesional side, usually the left hemispace (see "Disorders of Spatial Representation: Unilateral Neglect", above).

Despite patients' behavior suggesting the opposite, it is possible to show that the neglected stimuli can in some cases be fully processed. Marshall and Halligan reported a case of a woman with a severe visual neglect who explicitly denied any difference between the drawing of an intact house and that of a burning house when the features relevant to the discrimination were on the neglected side.[55] However, when forced to choose the house she would prefer to live in she consistently indicated the non-burning one, showing an implicit knowledge of the information she was unable to consciously report. Further studies showed that the patient's choice was actually based on high-level visual processing. In Berti and Rizzolatti's study, patients were required to respond as rapidly as possible to target stimuli (pictures of animals or vegetables) presented to the normal field by pressing one of two keys according to the category of the targets.[56] The influence of priming stimuli, pictures of animals or vegetables, presented to the neglected field on target reaction times was studied. By combining different pairs of primes and targets, three different experimental conditions were obtained. In the first condition, "highly congruent", the target and the prime stimuli belonged to the same category and were physically identical; in the second condition, "congruent", the stimuli represented two elements of the same category but were physically dissimilar; in the third condition, "non-congruent", the stimuli represented one exemplar from each of the two categories of stimuli. The results showed that the responses were facilitated not only in the highly congruent condition, but also in the congruent condition, suggesting that patients with neglect are able to process stimuli presented to the neglected field up to a categorical level of representation even when they deny the presence of the stimulus in the neglected field.

The dual visual streams discussed above may be used to explain non-conscious perception in neglect patients. As already mentioned, visual information from V1 reaches the extrastriate cortex along the ventral (occipitotemporal) and the dorsal (occipitoparietal) stream. The ventral stream (also known as the "what pathway") travels to the temporal lobe and is involved with object identification and recognition. The dorsal stream ("where pathway") terminates in the parietal lobe and is involved with processing the object's spatial location relevant to the viewer.[57] Because the neglect syndrome is usually associated with parietal lesions, which spare occipitotemporal areas, the demonstration that neglect patients can process visual shapes presented to the neglected side (up to a semantic processing level), although counterintuitive psychologically, is not very surprising. Indeed, the capacity of shape analysis and categorization shown by neglect patients is exactly what one would expect from the functional–anatomical properties of the intact ventral (occipitotemporal) visual pathway. The really surprising aspect of Berti and Rizzolatti's findings is that, despite the presence of semantic elaboration, patients appeared unaware of the stimuli presented in the affected hemispace.[56] The authors proposed that the encoding of space is a necessary prerequisite for conscious perception. If spatial encoding is prevented or impaired, as it is in neglect, the presence of the stimulus does not reach the conscious level.

Anosognosia for Hemiplegia: A Model for Motor Awareness

Anosognosia for hemiplegia (AHP) is a clinical condition in which movement awareness is dramatically altered. The phenomenon was named anosognosia (from the Greek for "lack of knowledge for the illness") in 1914, by the French neurologist Joseph Babinski.[58] AHP is usually observed in patients with right-brain damage, who obstinately deny that there is something wrong with their contralesional limbs, despite the presence of severe left paralysis (for a review see Pia *et al.*[59]). If asked to produce an action with the paralyzed limb, some patients appear convinced that they are actually performing it, even though sensory and visual evidence from the affected motionless side should indicate that no movement has been performed.

Several authors have proposed that AHP may be conceptualized as a selective disorder of motor cognition,[60–62] on the basis of computational models of motor production and motor control[63–65] (Fig. 32.3). These models posit that, in the presence of a normal intentional attitude, once the appropriate motor commands have been selected and sent to the muscles for the execution of the desired movement, a prediction of the sensory consequences of the movement is formed and will be successively compared with the feedback associated with the actual execution of the intended movement. According to Blakemore and colleagues,[64] this prediction, based on the efference copy of the programmed motor act, constitutes the signal upon which motor awareness is constructed. A first consequence of the above-mentioned hypothesis is that if consciousness of a motor act precedes the sensory feedbacks related to a specific movement,[67] then one should expect to observe motor awareness for a certain movement even in the absence of any observable motor event.

This is exactly what happens in hemiplegic patients affected by AHP, who seem to be a perfect model to verify the relationship between motor awareness and motor intention. It has been proposed that the denial behavior in AHP patients may be due to direct damage to the comparator system, localized in the premotor (BA6) and insular area.[62,68] This may impair the motor monitoring process, preventing patients from distinguishing between movement and no-movement states. However, the evident feeling of movement that AHP patients (erroneously) report experiencing may arise from intact motor intentionality, due to normal activity in other areas (mostly involving the parietofrontal circuit) that implement intention-programming related processing, which are usually spared in AHP patients.[62] Therefore, although AHP patients may not be able to monitor the mismatch between motor prediction and actual execution, because of damage to the comparator, they may still be able to program movements and form predictions, with the consequence of constructing an illusory, but neurologically grounded, motor awareness.

This hypothesis has been confirmed by studies[69,70] showing that the subjective experience of movement reported by AHP patients has objective consequences on their motor behavior. Using bimanual motor tasks, in which AHP patients were asked to simultaneously perform movements with both hands, the authors found that the movements of the intact hand were influenced by the intended but not executed movements of the paralyzed hand. This "influence" produces both spatial[69] and temporal[70] coupling effects, comparable to those found in healthy subjects actually performing bimanual tasks. These findings in AHP patients clearly show that motor awareness can be constructed even in the absence of movement execution, solely on the basis of normal intentional processes.

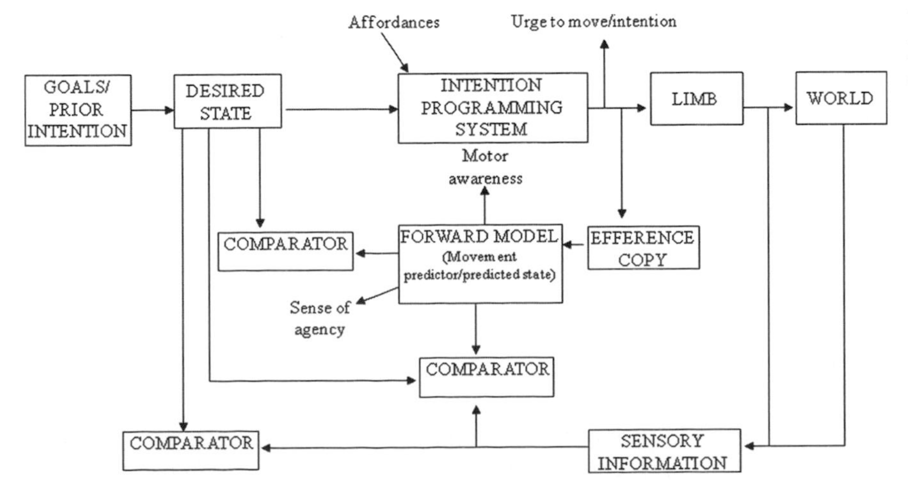

FIGURE 32.3 **Anosognosia for hemiplegia.** Modified version of the feed-forward model of action generation.[65,66]

A Modular Model for Conscious Awareness

The subjective experience that people have of themselves is reported to be, in normal conditions, a feeling of unity. The "illusion" of unity of the self, assumed by common-sense theories of consciousness, is evident in the normal experience of correspondence between the actual presence or absence of a stimulus and the presence or absence of a subjective experience of it. An alternative view would suggest that consciousness and self-consciousness do not have a unitary, monolithic structure, but instead have a composite nature, subserved by the activity of different brain mechanisms distributed in specialized brain areas. Such a view would predict that focal brain damage should not cause a generalized impairment of conscious experience or conscious self-monitoring, but should instead result in domain-specific disorders of awareness.[66] The discovery of selective disorders of conscious awareness (such as blindsight, neglect, and AHP) has provided evidence for a composite nature of conscious processes, as opposed to a unitary one. Although the construction of fully conscious states may need the co-occurrence of activity in different parts of the brain, initial modular activation in dedicated brain structures seems to be necessary for domain-specific awareness.

FUTURE DIRECTIONS

Throughout this chapter the authors have raised some of the many questions that remain to be answered as researchers drive to achieve a more complete understanding of the multitude of disorders of higher cortical function. To do so will require the combined effort of individuals with expertise in histology, electrophysiology, imaging, and behavioral analysis working both in animal models and in clinical populations. The hope is that some of the readers of this chapter will take up this critical challenge.

References

1. Shallice T. *From Neuropsychology to Mental Structure*. Cambridge, MA: Cambridge University Press; 1988.
2. Broca P. Perte de la parole, ramollissement chronique et destruction partielle du lobe antérieur gauche. *Bulletin de la Société d'Anthropologie*. 1861;2:235–238.
3. Smith Churchland P. *Neurophilosophy*. Cambridge, MA: MIT Press; 1986.
4. Lichtheim L. On aphasia. *Brain*. 1885;7:433–484.
5. Damasio AR. Aphasia. *N Engl J Med*. 1992;326:531–539.
6. Wernicke C. *Der aphasische Symptomencomplex*. Breslau: Max Cohn & Wiegert; 1874.
7. Plaza M, Gatignol P, Leroy M, Duffau H. Speaking without Broca's area after tumor resection. *Neurocase*. 2009;15:294–310.
8. Anderson VA, Morse SA, Catroppa C, Haritou F, Rosenfeld JV. Thirty month outcome from early childhood injury: a prospective analysis of neurobehavioral recovery. *Brain*. 2004;124:2608–2620.
9. Varela FJ, Thompson E, Rosch E. *The Embodied Mind. Cognitive Science and Human Experience*. Boston, MA: MIT Press; 1991.
10. Garbarini F, Adenzato M. At the root of embodied cognition: cognitive science meets neurophysiology. *Brain Cogn*. 2004;56:100–106.
11. Aziz Zadeh L, Damasio A. Embodied semantics for actions: findings from functional brain imaging. *J Physiol*. 2008;102:35–39.
12. Lakoff G, Johnson M. *Philosophy in the Flesh: The Embodied Mind and Its Challenge to Western Thought*. New York: Basic Books; 1999.
13. Hebb DO. *The Organization of Behavior*. New York: Wiley; 1949.
14. Lashley KS. *Brain Mechanisms and Intelligence: A Quantitative Study of Injuries to the Brain*. Chicago, IL: Chicago University Press; 1929.
15. Scoville WB, Milner B. Loss of recent memory after bilateral hippocampal lesions. *J Neurol Neurosurg Psychiatry*. 1957;20:11–21.
16. Squire LR. The legacy of patient HM for neuroscience. *Neuron*. 2009;61:6–9.
17. Corkin S. What's new with the amnesic patient HM? *Nat Rev Neurosci*. 2002;3:153–160.
18. Squire LR, Wixted JT. The cognitive neuroscience of human memory since HM. *Annu Rev Neurosci*. 2011;34:259–288.
19. Eichenbaum H. What HM taught us. *J Cogn Neurosci*. 2012;25:14–21.
20. Milner B. Les troubles de la memoire accompagnant des lesions hippocampiques bilaterales. In: Passouant P, ed. *Physiologie de l'hippocampe*. Paris: Centre National de la Recherche Scientifique; 1962:257–272.
21. Corkin S. Tactually-guided maze learning in man: effects of unilateral cortical excisions and bilateral hippocampal lesions. *Neuropsychologia*. 1965;3:339–351.
22. Corkin S. Acquisition of a motor skill after bilateral medial temporal lobe excision. *Neuropsychologia*. 1968;6:225–265.
23. Eichenbaum H, Cohen NJ. *From Conditioning to Conscious Recollection: Memory Systems of the Brain*. New York: Oxford University Press; 2001.
24. Zola-Morgan S, Squire LR, Amaral DG. Human amnesia and the medial temporal region: enduring memory impairment following a bilateral lesion limited to field CA1 of the hippocampus. *J Neurosci*. 1986;6:2950–2967.
25. Fama R, Pitel AL, Sullivan EV. Anterograde episodic memory in Korsakoff syndrome. *Neuropsychol Rev*. 2012;22:93–104.
26. Liepmann H. Apraxie. *Ergebn ges Med*. 1920;1:516–543.
27. Geschwind N. Disconnexion syndromes in animals and man. *Brain*. 1965;88:237–294.
28. Rothi LJG, Ochipa C, Heilman KM. A cognitive neuropsychological model of limb praxis. *Cogn Neuropsychol*. 1991;8(6):443–458.
29. Heilman KM, Rothi LJ. Apraxia. In: Heilman KM, Valenstein E, eds. *Clinical Neuropsychology*. 3rd ed. New York: Oxford University Press; 1993:141–163.
30. Rizzolatti G, Craighero L. The mirror-neuron system. *Annu Review Neurosci*. 2004;27:169–192.
31. De Renzi E, Faglioni P. Apraxia. In: Denes G, Pizzamiglio L, eds. *Handbook of Clinical and Experimental Neuropsychology*. Hove: Psychology Press; 1999:421–440.
32. Goldenberg G. Apraxia and the parietal lobes. *Neuropsychologia*. 2009;47:1449–1459.
33. Petreska B, Adriani M, Blanke O, Billard AG. Apraxia. A review. *Prog Brain Res*. 2007;164:61–83.
34. Lissauer H. Ein fall von seelenblindheit nebst einem beitrag zur theorie derselben. *Arch fur Psychiatrie*. 1890;21:222–270.
35. Riddoch MJ, Humphreys GW, eds. *Visual Object Processing: A Cognitive Neuropsychological Approach*. Hove: Laurence Erlbaum; 1987.
36. Marr D. *Vision: A Computational Investigation into the Human Representation and Processing of Visual Information*. San Francisco, CA: Freeman; 1980.

37. Warrington EK. *Agnosia: The Impairment of Object Recognition*. Amsterdam: Elsevier; 1985.
38. Benton AL. The neuropsychology of facial recognition. *Am Psychol*. 1980;35:176–186.
39. Bruce V, Young A. Understanding face recognition. *Br J Psychol*. 1986;77:305–327.
40. Behrmann M, Nishimura M. Agnosias. *WIREs Cogn Sci*. 2010;1: 203–213.
41. De Renzi E. Agnosia. In: Denes G, Pizzamiglio L, eds. *Handbook of Clinical and Experimental Neuropsychology*. Hove: Psychology Press; 1999:371–407.
42. Farah MJ. *Visual Agnosia. Disorders of Object Recognition and What They Tell Us About Normal Vision*. Cambridge, MA: MIT Press; 1990.
43. Bisiach E. Unilateral neglect. In: Denes G, Pizzamiglio L, eds. *Handbook of Clinical and Experimental Neuropsychology*. Hove: Psychology Press; 1999:479–495.
44. Buxbaum L. On the right (and left) track: twenty years of progress in studying hemispatial neglect. *Cogn Neuropsychol*. 2006;23: 184–201.
45. Kinsbourne M. Mechanisms of unilateral neglect. In: Jeannerod M, ed. *Neurophysiological and Neuropsychological Aspects of Spatial Neglect*. Amsterdam: Elsevier; 1987:69–86.
46. Rizzolatti G, Gentilucci M, Matelli M. Selective spatial attention: one center, one circuit or many circuits? In: Posner MI, Marin OSM, eds. *Attention and Performance XI*. Hillsdale, NJ: Erlbaum; 1985:251–265.
47. Bisiach E, Luzzatti C. Unilateral neglect of representational space. *Cortex*. 1978;14:129–133.
48. Bisiach E, Pizzamiglio L, Nico D, Antonucci G. Beyond unilateral neglect. *Brain*. 1996;119:851–857.
49. Bisiach E, Ricci R, Neppi-Modona M. Visual awareness and anisometry of space representation in unilateral neglect: a panoramic investigation by means of a line extension task. *Consc Cogn*. 1998;7:327–355.
50. Weiskrantz L, Warrington EK, Sanders MD, Marshall J. Visual capacity in the hemianopic field following a restricted occipital ablation. *Brain*. 1974;97:709–728.
51. Leopold DA. Primary visual cortex: awareness and blindsight. *Annu Rev Neurosci*. 2012;35:91–109.
52. Pöppel E, Held R, Frost D. Residual visual function after brain wounds involving the central visual pathways in man. *Nature*. 1973;243:295–296.
53. Tamietto M, de Gelder B. Neural bases of the non-conscious perception of emotional signals. *Nat Rev Neurosci*. 2010;11:697–709.
54. Tomaiuolo F, Ptito M, Marzi CA, Paus T, Ptito A. Blindsight in hemispherectomized patients as revealed by spatial summation across the vertical meridian. *Brain*. 1997;120:795–803.
55. Marshall JC, Halligan PW. Blindsight and insight in visuo-spatial neglect. *Nature*. 1988;336:766–767.
56. Berti A, Rizzolatti G. Visual processing without awareness: evidence from unilateral neglect. *J Cogn Neurosci*. 1992;4:345–351.
57. Goodale MA, Milner AD. Separate visual pathways for perception and action. *Trends Neurosci*. 1992;15:20–25.
58. Babinski J. Contribution à l'étude des troubles mentaux dans l'hémiplégie organique cérébrale (anosognosie). *Revue Neurologique*. 1914;27:845–848.
59. Pia L, Neppi-Mòdona M, Ricci R, Berti A. The anatomy of anosognosia for hemiplegia: a meta-analysis. *Cortex*. 2004;40:367–377.
60. Gold G, Adair JC, Jacobs DH, Heilman KM. Anosognosia for hemiplegia: an electrophysiologic investigation of the feed-forward hypothesis. *Neurology*. 1994;44:1804–1808.
61. Frith CD, Blakemore SJ, Wolpert DM. Abnormalities in the awareness and control of action. *Philos Trans R Soc Lond B Biol Sci*. 2000;355:1771–1788.
62. Berti A, Bottini G, Gandola M, et al. Shared cortical anatomy for motor awareness and motor control. *Science*. 2005;309:488–491.
63. Wolpert DM, Ghahramani Z, Jordan MI. An internal model for sensorimotor integration. *Science*. 1995;269:1880–1882.
64. Blakemore SJ, Wolpert DM, Frith CD. Abnormalities in the awareness of action. *Trends Cogn Sci*. 2002;6:237–242.
65. Haggard P. Conscious intention and motor cognition. *Trends Cogn Sci*. 2005;9:290–295.
66. Spinazzola L, Pia L, Folegatti A, Marchetti C, Berti A. Modular structure of awareness for sensorimotor disorders: evidence from anosognosia for hemiplegia and anosognosia for hemianaesthesia. *Neuropsychologia*. 2008;46:915–926.
67. Libet B, Gleason CA, Wright EW, Pearl DK. Time of conscious intention to act in relation to onset of cerebral activity (readiness-potential). The unconscious initiation of a freely voluntary act. *Brain*. 1983;106:623–642.
68. Karnath HO, Baier B, Nagle T. Awareness of the functioning of one's own limbs mediated by the insular cortex? *J Neurosci*. 2005;25:7134–7138.
69. Garbarini F, Rabuffetti M, Piedimonte A, et al. "Moving" a paralysed hand: bimanual coupling effect in patients with anosognosia for hemiplegia. *Brain*. 2012;135:1486–1497.
70. Pia L, Spinazzola L, Rabuffetti M, et al. Temporal coupling due to illusory movements in bimanual actions: evidence from anosognosia for hemiplegia. *Cortex*. 2013;49(6):1694–1703.

CHAPTER

33

Disorders of Frontal Lobe Function

Peter Pressman, Howard J. Rosen

Memory and Aging Center, University of California, San Francisco, California, USA

OUTLINE

INTRODUCTION

The human frontal lobes are a complex, multifaceted part of the brain. Among all regions of the brain, the frontal lobes have undergone the largest expansion in humans compared with other non-human primates (excluding the great apes). They occupy 41% of the human brain and are considered by many to be the seat of the higher faculties that have allowed humans to dominate their environment.[1] To many, frontal lobe function is synonymous with executive function, a term that is meant to encompass the capacity of the frontal lobes to control other cognitive processes and make them more efficient through attention, inhibition of basic responses in pursuit of a more complex goal, abstraction, and other mechanisms. Not all executive functions are primarily frontally mediated, however. More importantly, the frontal lobes are divided into a number of regions that participate in a variety of processes including motor function, memory, language, and social and emotional functions. Clinical disorders reflecting all these different functions can be seen with damage or dysfunction in the frontal lobe, the first example of which was Dr John Harlow's description of Phineas Gage who, after having a tamping iron forced through the medial portions of his frontal lobe, had a dramatic change in personality. The goal of this chapter is to review the functions of these various frontal systems in order to help readers to link specific types of disorders to the frontal systems involved.

FRONTAL TOPOGRAPHY

The frontal lobes contain a large number of regions, usually identified based on sulcal and gyral anatomy (Fig. 33.1). Histologically, most of these regions are six-layered neocortex, but the frontal lobes also contain regions of simple two-layered allocortex in the primary olfactory region, transitional mesocortex in the cingulate regions, and idiotypic cortex in the precentral gyrus. This histological variety reflects the involvement of the frontal lobes in functions ranging from basic, unimodal sensory and motor processing, monomodal premotor cortex, to multimodal integration in six-layered association cortex.

Neurobiology of Brain Disorders
http://dx.doi.org/10.1016/B978-0-12-398270-4.00033-1

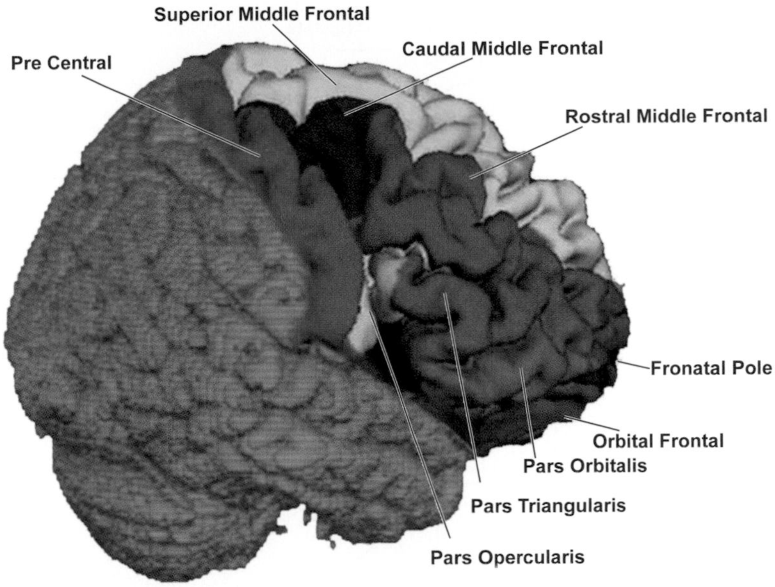

FIGURE 33.1 **The gyri of the frontal lobes.** *Source: Modified from Desikan* et al. Neuroimage. *2006;31(3):968–980.*[2]

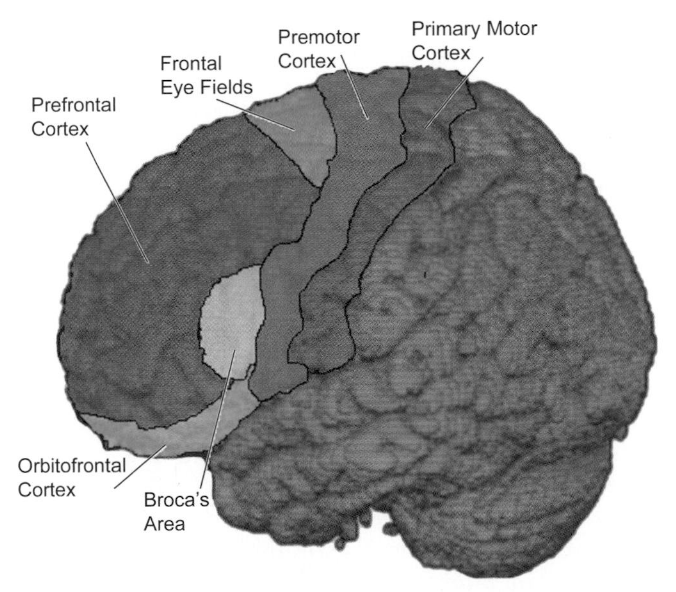

FIGURE 33.2 **Functional regions of the lateral frontal lobe.** Functional regions superimposed on a template brain from MRIcron. Because these regions are histologically and functionally defined, the borders are less distinct in reality. *Source: Rorden and Bret.* Behav Neurol. *2000;12:191–200.*[3]

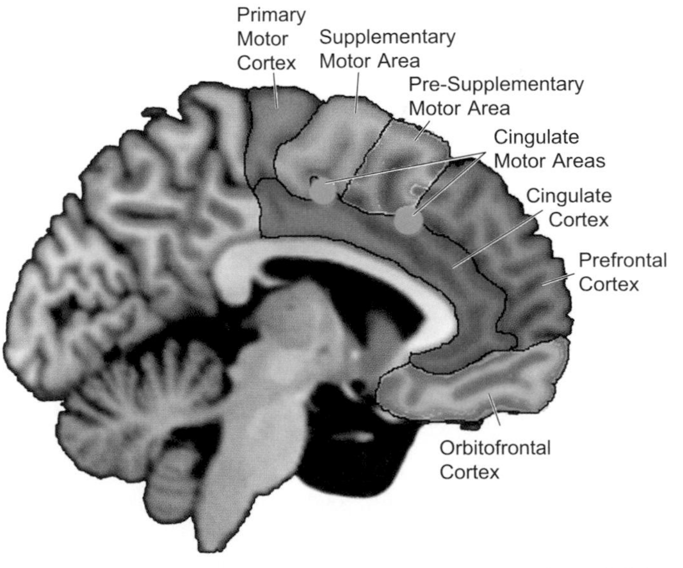

FIGURE 33.3 **Functional regions of the medial frontal lobe.** Functional regions superimposed on a template brain from MRIcron. Because these regions are histologically and functionally defined, the borders are less distinct in reality. *Source: Rorden and Bret.* Behav Neurol. *2000;12:191–200.*[3]

Based on their connectivity, functional activity, and symptoms that occur after injury, the gyri of the frontal regions are often grouped into larger regions that relate to specific functions (Fig. 33.2). The primary motor cortex (PMC), premotor, and supplementary motor regions predominantly consist of Brodmann areas (BA) 6 and 4, and are the more posterior aspects of the frontal lobe. The prefrontal cortex (PFC) consists of the more anterior surfaces, and is often divided into three major divisions: the dorsolateral frontal cortex, which includes the middle frontal gyrus and the adjacent portions of the superior and inferior frontal gyri; the orbitofrontal cortex (OFC), which includes the inferior portion of the inferior frontal gyrus and all the gyri on the orbital surface of the frontal lobe; and the medial frontal structures, including the anterior cingulate cortex (ACC) and medial portion of the superior frontal gyrus (Fig. 33.3). These anatomical groupings will be referred to frequently in this chapter. In addition, the insula is anatomically very close to the frontal lobes. Parts of the insula are histologically similar in particular to the orbital frontal regions and seem to mediate similar functions. Thus, some discussions of insular function will be included in this chapter.

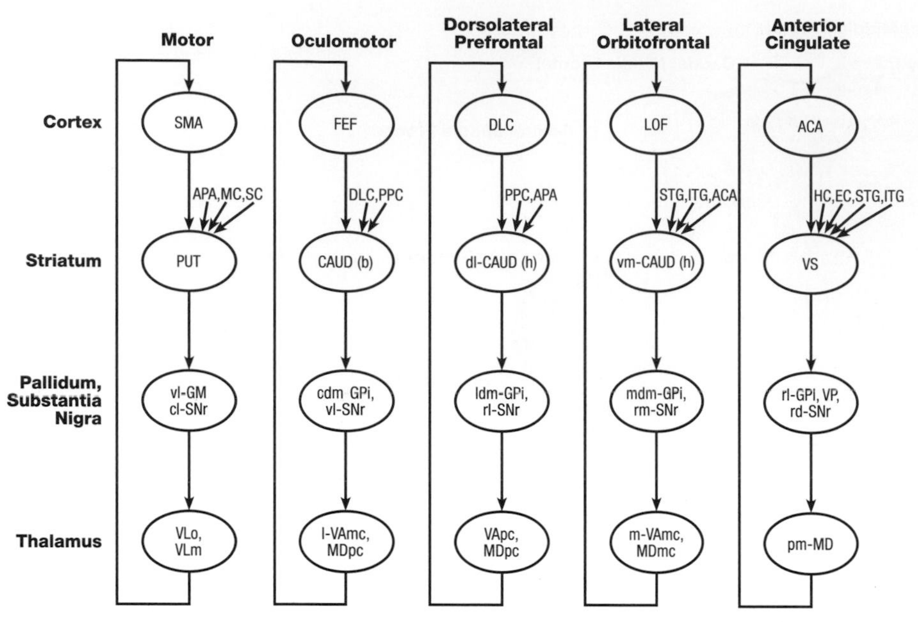

FIGURE 33.4 **Corticostriatothalamocortical loops.** ACA: anterior cingulate area; APA: arcuate premotor area; CAUD: caudate nucleus head (h) and body (b); DLC: dorsolateral prefrontal cortex; EC: entorhinal cortex; FEF: frontal eye fields; GPi: globus pallidus pars interna; HC: hippocampal cortex; ITG: inferior temporal gyrus; LOF: lateral orbitofrontal cortex; MC: motor cortex; MDpl: medialis dorsalis pars paralamellaris; MDmc: medialis dorsalis pars magnocellularis; MDpc: medialis dorsalis pars parvocellularis; PPC: posterior parietal cortex; PUT: putamen; SC: somatosensory cortex; SMA: supplementary motor area; SNr: substantia nigra pars reticulata; STG: superior temporal gyrus; VAmc: ventralis anterior pars magnocellularis; VApc: ventralis anterior pars parvocellularis; VLm: ventralis lateralis pars medialis; VLo: ventralis lateralis pars oralis; VP: ventral pallidum; VS: ventral striatum; cl: caudolateral; cdm: cudal dorsomedial; dl: dorsolateral; l: lateral; ldm: lateral dorsomedial; m: medial; mdm: medial dorsomedial; pm: posteromedial; rd: rostrodorsal; rl: rostrolateral; rm: rostromedial; vm: ventromedial; vl: ventrolateral. *Source: Poston KL, Eidelberg D. Functional brain networks and abnormal connectivity in the movement disorders. NeuroImage 2012;62(4):2261-70.*

While frontal cortical dysfunction is often the cause of symptoms referable to the frontal lobes, lesions involving the basal ganglia and thalamus can sometimes cause similar symptoms. This can be explained by the fact that each of the major cortical frontal systems is interconnected with specific subcortical regions in a functional loop. Prior anatomical and functional work has identified five such circuits (Fig. 33.4), called the corticostriatothalamocortical loops.[4]

In addition to these cortical–subcortical loops, other networks incorporating frontal and non-frontal cortical and subcortical regions have been described.[5] These networks include the prefrontal network for executive functions and comportment, the limbic network, the perisylvian network for language, and the dorsal parietofrontal network for spatial orientation. More recent work with functional neuroimaging has led to descriptions of similar networks, including the dorsal executive network and a salience network.[6] The existence of these networks underscores the fact that frontal lobe regions do not function in isolation, so that many disorders caused by frontal lobe lesions could be caused by lesions to other brain regions as well.

CORTICAL MOTOR SYSTEMS

Control of movement is implemented at many levels of the neuroaxis, including the spinal cord, cerebellum, and basal ganglia. This section will focus only on the functions of the cortical motor system, which is responsible for the initiation and planning of movements, ensuring that the movements are performed with precision, and modifying complex combinations of movements as appropriate to the environment. There are three main cortical cerebral regions: the primary motor cortex (PMC), consisting of BA4 in the precentral gyrus; the premotor cortex (preMC), consisting of lateral BA6, just anterior to the PMC; and the medial frontal lobe motor areas, including the presupplementary motor area (preSMA), supplementary motor area (SMA), and adjacent motor areas in the cingulate sulcus [cingulate motor area (CMA)]. The SMA and the CMA are collectively referred to as the supplementary motor complex (SMC). A summary of the relevant characteristics of each motor region is provided in Table 33.1.

Primary Motor Cortex

The PMC is comprised of unimodal idiotypical cortex occupying the posterior part of the precentral gyrus, the anterior bank of the central sulcus, and the anterior part of the paracentral lobule. The region is histologically distinguished by large neurons known as the gigantopyramidal cells of Vladamir A. Betz. BA4 receives projections from the preMC, the SMC, and multiple parietal regions.

TABLE 33.1 Characteristics of the Major Cerebral Motor Areas

Motor Region	Connectivity	Physiology	Lesion Syndrome
Primary motor cortex	Pyramidal tract	Fine motor control	Weakness: distal > proximal, extensor > flexor (arms), flexor > extensor weakness (legs)
	Other motor regions	Somatotopically organized	Hyperreflexia
	Contralateral BA4	Larger representation for more distal structures	Spasticity
	BA3, 2, 1, and 5		
	Thalamus: ventral anterior, ventral lateral nuclei		
	Brainstem		
PreSMA	Prefrontal cortex (BA46)	No somatotopic organization	Diminished motor learning
	Thalamus, ventral anterior, pars caudalis	More preparatory neuronal activity than SMA	Utilization behavior
		More complex movement than SMA	Alien limb
			Apraxia
SMA	Primary motor cortex	Somatotopically organized	Akinetic mutism
	Parietal lobe	More automatic movement than preSMA	Hypokinesia
			Utilization behavior
			Alien limb
			Apraxia
Premotor area	Pyramidal tract	Preparatory firing	Difficulty with retrieval of complex motor plans
	Thalamus area X	Firing in association with reaching movements	
	Parietal cortex	Firing in association with grasping movements	
		Mirror neurons	
		Increased activity during motor learning	

SMA: supplementary motor area; BA: Brodmann area.

Fibers from the PMC descend through the pyramidal tract (about one-third of pyramidal tract neurons originate in PMC) ultimately to synapse on neurons in the brainstem and the anterior horn of the spinal cord. The connectivity and functional activity in this region point to its role in initiation and control of fine movements. PMC neurons fire approximately 60 milliseconds before movement, and fire faster with low levels of muscular force.[7] While stimulation studies show that PMC neurons control only contralateral limbs, unilateral stimulation in regions controlling the upper face, soft palate, laryngeal muscles, masticatory muscles, and trunk usually cause bilateral movement. Landmark studies by Penfield in 1950 demonstrated an organization to the PMC representation of movement, with a progression through face, arm, trunk, and legs from inferolateral to medial superior portions of the frontal lobe.[8] These studies also showed that larger regions of the PMC are devoted to movement of the face, tongue, and distal musculature than to more proximal muscle groups.

Injury limited to the pyramidal tract and PMC results in a complex set of symptoms. Lesions that cause electrical excitability limited to these regions (seizures) cause focal clonic jerking, the location of which depends on the seizure focus. Seizures can sometimes represent the organization of PMC neurons by progressively involving face, then arm, then leg contralateral to the seizure focus (called a Jacksonian march after Hughlings Jackson, who described the phenomenon in 1868). Lesions such as stroke, trauma, or inflammation can cause severe weakness in early stages, which often evolves over time to become quite mild. The distal musculature tends to be weaker than the proximal, perhaps owing to a proportionally greater cortical representation. Even when weakness is absent, fine finger movements may be impaired, supporting theories that BA4 is particularly important for fine coordination. In addition to weakness, lesions of the corticospinal tract cause a classic constellation of "upper motor neuron" findings on neurological examination, which include spastic rigidity, meaning a velocity-dependent change in muscle resistance to passive stretch, as well as an increase in stretch reflexes. Abnormal reflexes may become apparent, such as the Babinski sign in which the toes fan up and apart when the bottom of the foot is scratched. The exact constellation of symptoms depends on the location and extent of the cortical lesion.

Premotor Cortex

The preMC lies directly anterior to the PMC, communicates with the other cortical motor areas, striatum, and thalamus, and contributes to the pyramidal tract. Physiological studies have shown that preMC regions become active during complex sensory-guided movements, and often fire 100 milliseconds or more before the beginning of an action.[9] The preMC can be subdivided into a number of regions with different properties. Movements such as climbing and reaching, which require truncal support, are more anterior, and simpler and more distal movements are posterior. Most researchers now generally divide the premotor cortex into four sections: a dorsal and caudal section (PMDc), a dorsal and rostral section (PMDr), a ventral and caudal section (PMVc), and a ventral and rostral section (PMVr). PMDc has been associated with reaching movements, PMDr with learning response rules and eye movement, PMVc with sensory guidance of movement and peripersonal space, and PMVr with guiding the hand to the mouth.[10] Some neurons in PMVr fire not only when an animal performs an action, but also when the subject observes a similar action being performed by another individual. These neurons have been called "mirror neurons", and are probably part of a network necessary for internal representation of movements.[11] Just anterior to BA6 is another premotor region, BA8, also called the supplementary eye field. Projections from this region guide bilateral eye movements in the midbrain. Stimulation of BA8 will drive the eyes towards the contralateral hemisphere.

Natural lesions of the preMC usually also involve the adjacent PMC. Strategic lesions in monkeys have suggested a loss of ability to generate specific responses to particular cues. There have been several reports of apraxia after premotor lesions (see below). Degeneration or lesions affecting BA8 commonly lead to difficulties making a voluntary eye movement to the opposite side, or may result in the eyes drifting towards the hemisphere ipsilateral to the lesion.

Medial Frontal Motor Area

Once entirely known as the supplementary motor area (SMA), the medial frontal motor areas include multiple regions dedicated to control of movement. These regions become active before movements occur. For example, scalp electroencephalographic (EEG) recordings from humans demonstrate a slowly increasing negative potential called the *Bereitschaftspotential* over this region before movements begin.[12] This signal may be diminished in Parkinson disease, in which voluntary action is disrupted. Electrophysiological and functional imaging studies have indicated that the medial frontal motor areas should be divided into the SMA, a more anterior preSMA, and a CMA. The SMA contributes to the pyramidal tract, connects to PMC and parietal cortex, and contains a somatotopically arranged map of the body. Unlike PMC, activity in the SMA can lead to movement of either hand. The preSMA has no pyramidal connections or connections to PMC or parietal lobes, but connects to other prefrontal regions such as dorsolateral PFC. Somatotopic organization is not found in the preSMA and electrical stimulation of the preSMA region rarely results in discrete movements.[13] Compared with SMA, the preSMA is involved in more complex motor tasks that are less automatic. Acquisition of a particular skill increases activity in SMA while continued practice on a task decreases activation in the preSMA. While less is known about the CMA, which is located along the banks of the cingulate sulcus (see Fig. 33.2), its organization appears similar to SMA in that the more posterior regions, ventral and dorsal CMA (CMAv and CMAd) are more intimately connected with PMC and parietal lobe than the more anterior region (CMAr).[13] Electrical stimulation of CMAv and CMAd produces movement, whereas stimulation of CMAr does not.

Few studies have differentiated between any of the medial frontal motor areas. Stimulation of the SMA leads to complex movements of body parts involving several different joints. A seizure in the medial frontal motor region classically produces vocalization, postural movements of the extremities (most on the side opposite the stimulation but often bilateral), and head rotation away from the side of stimulation. If the seizure remains focal and the patient does not lose consciousness despite bilateral motor involvement, this may lead to a faulty diagnosis of psychogenic seizure, especially if the electrical abnormalities are too deep to be detected by surface EEG. Unilateral tumor resection of medial frontal motor regions may result in loss of spontaneous movement of the contralateral limb. In more severe cases, the patient may suffer akinetic mutism, in which he or she is immobile and unresponsive. This is followed subacutely by hypokinesia, including reduced facial emotional expression and reduced spontaneous speech. When the patient has recovered, normal movements may be recovered except for slowing of rapid finger movement. Subtle deficits in the ability to perform bimanual tasks, such as tying a shoelace or transferring objects from one hand to the other, have also been described. Neurodegeneration affecting medial frontal motor areas can lead to a variety of disorders, including repetitive or compulsive movements, utilization behavior, in which the patient is driven to use a nearby object despite there being no reason to do so, "alien limb" phenomenon, in which an extremity performs semipurposeful movements outside the patient's control, and apraxia (see below). Both apraxia and alien limb phenomena are classic components of a corticobasal syndrome, which may result from histopathological protein deposits consistent with corticobasal degeneration, Alzheimer disease, or progressive supranuclear palsy.[14] Patients with a corticobasal syndrome have a characteristic dorsal pattern of atrophy that involves the PMC, preMC, and the medial frontal motor areas, along with other frontal and parietal regions.[15]

Apraxia

Apraxia is a motor disorder often associated with frontal lobe injury, including the PFC. Apraxia refers to an inability to perform a skilled motor activity despite intact strength, sensation, attention, memory, and drive, with no other movement disorders such as dystonia, tremor, or chorea to interfere. Patients with apraxia may describe a loss of ability to manipulate commonly used tools, or the loss of ability to perform something more complex, such as sewing or knitting. The most common type of apraxia assessed through neurological examination is ideomotor apraxia, meaning that patients cannot demonstrate skilled movements through pantomime. Several types of error may be made, such as incorrect finger configuration, improper timing, or moving the wrong joints. A common error is to substitute a body part as the tool, rather than pantomiming use of the tool as requested. For example, when asked to demonstrate how to use a pair of scissors, the subject may instead hold out the index and middle fingers and scissor them together as if playing rock–paper–scissors. Additional types of apraxia include dissociation apraxia, where pantomiming an action to command is impaired but the subject may still imitate the examiner and use the desired object; conduction apraxia, where imitation is also impaired; conceptual apraxia, where patients are unable to associate a use with a particular tool; and ideational apraxia, where a patient is unable to sequence a series of actions such as lighting a match and then blowing it out. Apraxia is not a well-localized disorder, and many different lesions can lead to apraxia, including parietal, frontal, and subcortical lesions. It is almost always associated with injury to the left hemisphere or the corpus callosum, and has been described specifically in the context of preMC and medial frontal motor lesions.[16] Apraxia is often seen in patients with neurodegenerative disorders. These may predominantly impact the parietal lobes, such as Alzheimer disease, or the frontal lobes. It is a common feature of corticobasal degeneration. The term apraxia has also been used to describe difficulty with complex, not necessarily routine tasks, such as an inability to dress properly (dressing apraxia) or draw (constructional apraxia). These uses of the term apraxia may not always indicate a motor disorder; for example, dressing or constructional apraxia may represent visuospatial disorders due to a right parietal lesion.

CORTICAL INFLUENCE ON THE AUTONOMIC NERVOUS SYSTEM

Portions of the frontal lobe including the ACC and ventromedial PFC, as well as the insula, communicate with the hypothalamus and with sympathetic and parasympathetic nuclei in the brainstem and spinal cord, allowing them to influence autonomic function. Thus, seizures in the left insula and frontal lobe have been associated with ictal bradycardia, asystole, and sudden death. Large acute lesions involving the insula have also been associated with cardiac autonomic dysfunction. Some suggest that stroke involving the right insula is more likely to cause cardiac arrhythmia than is stroke in the insula on the left side.[17] The lateralization of the sympathetic and parasympathetic functions, however, has not always been consistently found in the epilepsy literature.[18] Slowly worsening neuronal function, as seen in neurodegenerative disorders, can also lead to sympathetic dysfunction. William Seeley and colleagues (personal communication, 2013) have demonstrated cardiac changes in patients with degeneration of the right anterior insula.

Urethral and anal sphincters are represented in the medial inferior frontal cortex, inferior to the leg region of the PMC and anterior to the ACC. The superomedial precentral gyrus is active during voluntary contraction of the pelvic floor, while abdominal straining is associated with activity in the superior lateral portion of the precentral gyrus. The sensation of bladder fullness probably involves the medial frontal cortex. Positron emission tomography studies on micturition indicate involvement of the PFC as well as the cingulate gyrus, and forebrain lesions may cause urge incontinence. The bladder becomes overactive with injury to the right anterior cingulate, suggesting that this region may be particularly important for suppression of bladder activity. Fibers in this region may be damaged during ventricular extension in disorders such as normal pressure hydrocephalus, leading to the urinary incontinence that, along with gait abnormalities and dementia, classically defines the disorder.[19]

COGNITIVE FUNCTIONS OF THE FRONTAL CORTEX

Anatomically, the classic cognitive functions associated with the frontal lobes are usually attributed to the prefrontal regions. The PFC includes all of the cortex anterior to the premotor and primary motor regions. There are three other ways that this cortex has been defined. Electrophysiologically, stimulation of PFC does not usually produce movements. Histologically, these regions are characterized by the presence of a cortical granular layer IV, although some of the regions traditionally included among the prefrontal regions, such as the Broca area (BA44) are actually dysgranular or even agranular, as in caudal BA11 and orbital BA47. Another method of defining the PFC includes those regions that receive projects from the mediodorsal nucleus of the thalamus.[20] Cognitive functions of the frontal cortex with neuroanatomical correlates are outlined in Table 33.2.

Attention

Life presents us with more information than we can meaningfully process. We are therefore required to select, prioritize, and filter information, shifting focus to new stimuli as needed. The term attention has been used in many different ways, and consists of numerous components. Selective attention refers to the ability to attend to certain objects, stimuli, or concepts to the exclusion of competing internal and external stimuli. Sustained attention or vigilance refers to an individual's ability to maintain this focus over a period of time. Divided attention is an ability to multitask, responding to more than one task at a time. Alternating attention is the ability to shift focus as needed from one task to another. Attentional capacity varies not only between individuals, but within the same individual under different conditions such as fatigue, depression, or illness.

In order to assess attention, it is necessary to confirm a sufficient degree of arousal so that the subject is in an awake and alert state. Arousal depends on upper brainstem projection systems (the ascending reticular activating system), as well as thalamic systems including

TABLE 33.2 Characteristics of the Major Cerebral Cognitive Functions

Cognitive Domain	Associated Frontal Cortical Regions	Connectivity	Frontal Roles	Lesion Syndrome
Memory	DLPFC	Diffuse	Encoding strategies	Amnesia
	OFC		Working memory	
			Autobiographical memory	
Attention	Frontal eye field	Diffuse	Selecting stimuli	Distractibility
	DLPFC		Sustained attention	Diminished multitasking
	PreSMA		Shifting focus	Encephalopathy
	ACC			
	Anterior insula			
Executive function	DLPFC	Parietal cortices, other PFC regions, motor cortices	Organization	Disorganization
			Sequencing	
			Prioritization	
			Abstraction	
			Generation	
Language	BA44 (Broca area) and subcortical white matter	Frontal cortices	Language production	Broca-type aphasia
		Arcuate fasciculus to Wernicke area	Grammar	Aphemia

DLPFC: dorsolateral prefrontal cortex; OFC: orbitofrontal cortex; SMA: supplementary motor area; ACC: anterior cingulate cortex; BA: Brodmann area; PFC: prefrontal cortex.

the midline, ventral medial, and intralaminar thalamic nuclei. Hypothalamic histaminergic systems and basal forebrain cholinergic systems are also important to arousal. Lesions of the brainstem or bilateral thalamus, or diffuse damage to the cerebral hemispheres can disrupt these projection systems and diminish overall alertness.

Managing attention is a complex process that involves many regions outside the frontal lobe. The filtering of incoming stimuli begins even at the level of unimodal sensory cortex, with further filtering occurring in heteromodal regions of the frontal and parietal lobes.[21] This filtering is related to the task at hand, so that neurons in spatial attention networks give enhanced responses to targets likely to attract visual and manual grasp, rather than neutral stimuli during manual tasks. One of the important roles played by the frontal lobes in these processes is direction of attention to novel stimuli.[22] In addition, sustained attention has been associated with the PFC in BA8, 9, 44, and 46. BA46 has been also associated with divided attention tasks.[23] A ventral attention network has also been posited, which includes parts of the middle and inferior frontal gyrus and the temporoparietal cortex. This helps us to orient to something rapidly, even if it interrupts a goal, and lets us decide whether we should continue to focus on the new stimulus or go back to the task at hand. Prefrontal control of attention is partially lateralized, with relative right hemispheric specialization. Thus, functional tasks of either sustained or divided sensation demonstrate greater activation of the right hemisphere.

Lesions in the PFC can thus be associated with deficits on tasks requiring sustained attention, such as tasks requiring button presses in response to specific stimuli, as well as divided attention tasks, such as those requiring subjects to track multiple stimuli spontaneously. In daily life, such deficits may be seen in distractibility and difficulty with multitasking (e.g. not being able to accomplish a task if the radio or television is on). Failure to attend to novel stimuli may result in perseveration, in which a patient fixates on a task or stimulus despite its no longer being relevant to the immediate situation. Lesions of the right hemisphere are more likely to give rise to inattentive states than lesions of the left hemisphere. A particular phenomenon observed with right hemisphere lesions but rarely with left-sided lesions is neglect of the contralateral (usually left) half of space. In the most severe cases, such patients may not attend to the left side of space in any way, so that if someone approaches them or speaks to them from their left side, they will look for them on their right (see Chapter 32). Subtler, but more common manifestations of this phenomenon can be seen when patients cannot identify sensory stimuli (visual, somatosensory, auditory) on the left if they are simultaneously receiving similar stimulation on the right. This phenomenon can be observed with left frontal and/or parietal lesions and points to another aspect of the unique role of the right hemisphere in mediating spatial attention.

Memory

Although the structures typically associated with memory are the hippocampus and associated structures, the frontal lobe plays an important part is several aspects of memory. This is because there are other types of memory besides the episodic memory functions mediated by hippocampal systems and also because the frontal lobe plays a role in supporting the functions of the hippocampal systems in episodic memory.

One memory function strongly associated with the frontal lobes is working memory. This term refers to the ability to hold and manipulate information in mind for relatively brief durations (seconds to minutes). A classic example is holding a telephone number in mind long enough to dial the number. In clinical practice, this ability is often tested using tasks such as the Digit Span task, where patients must repeat strings of digits either forwards or in reverse order. Typical digit spans for normal adults are seven forwards and five backwards. An equivalent test of visual working memory involves pointing to a series of objects, then asking the subject to point to those objects in the same or reverse order. In her groundbreaking work with single-cell recording in 1988, Patricia Goldman-Rakic demonstrated that neurons in the dorsolateral frontal cortex function in a way that makes working memory possible, firing during periods when an animal must hold information in memory and then stopping their firing when the animal can act on this information. Functional brain imaging studies have demonstrated similar activities in humans.[24] Accordingly, patients with frontal lobe injury often complain of trouble remembering what they were going to do when they come into a room, and more astute patients may perceive specific trouble remembering short bits of information such as telephone numbers if they do not write them down.

Episodic memory refers to the ability to retain more complex memories involving events and the people involved in those events. Once established, such memories can last from a few minutes or hours to a lifetime. The process of creating such memories has been conceptualized as proceeding in phases including initial processing of information to be remembered (encoding), strengthening and retention of these memories over time (consolidation), and recall of the information when needed (retrieval). While a wealth of evidence has demonstrated that encoding and consolidation sufficient to produce lasting memory cannot occur without the hippocampus and associated systems, the frontal lobes participate in encoding. Functional imaging studies show activation of the PFC during both encoding and retrieval, and further suggest a lateralization of these functions, with the left hemisphere being specialized for encoding and the right for retrieval.[25] There are conflicting data as to whether this holds true for both verbal and non-verbal stimuli. Right hemispheric activity may be greater for non-verbal stimuli during both encoding and retrieval.[26]

Accordingly, patients with frontal lobe dysfunction may complain of memory problems that are similar in type and degree to those seen with hippocampal injury, specifically trouble retaining information over time, so that they must repeat questions and write things down to remember them. Sophisticated versions of list learning tasks used to assess memory have been developed that demonstrate how both frontal and hippocampal memory deficits can lead to encoding deficits. On some versions of these tasks, the list of words is divided into semantic categories such as types of furniture, fruit, and animals. Whereas normal individuals are able to use the semantic associations to facilitate organization of the learned material, patients with frontal lobe dysfunction may not be able to organize in this fashion, which limits their ability to remember the list items. Similarly, the frontal lobes appear to be involved in producing other mnemonic strategies, as patients with frontal lobe lesions do not spontaneously use such strategies when trying to memorize lists.[27] In short, poor learning strategies impair proper encoding in patients with frontal lobe damage. Because the information is never firmly fixed in the brain, a patient with deficient encoding will not recall information on later trials.

Patients with frontal lobe damage also show deficits in retrieval of information previously learned, because such retrieval depends on strategies that must be implemented by the individual. Thus, patients with frontal lobe injury are more likely than patients with hippocampal systems disease to have difficulty recalling older information from before the injury.[28] List learning tasks can also detect this type of problem by including trials where memory is cued (e.g. "tell me all the fruits on the list") or recognition memory trials where subjects shown previously studied word lists are asked to identify whether a newly presented word was on the list. A pattern of poor spontaneous recall with good cued recall and/or recognition would suggest a retrieval deficit and indicate frontal lobe dysfunction as the cause. However, when hippocampal systemic dysfunction is mild, recognition may be relatively spared, which limits the reliability of this pattern in many settings. During retrieval, patients with frontal lobe damage appear to be more susceptible to interference from competing information. For example, in one study participants learned related word pairs (e.g. fish–hook) and were then taught new associations for the first word in each pair (e.g. fish–ocean). When presented with the first word (fish), patients with frontal lobe damage were more prone to erroneously generate the first association (hook) despite being asked specifically to recall the second relationship.[29] This type of finding relates to a number of studies indicating that even if patients do remember information, they may misidentify where or when they heard it (referred to as a source memory deficit).

Lastly, the phenomenon of confabulation, where patients not only have difficulty recalling information but rather make up information in response to queries, is typically associated with damage to the ventromedial frontal region, PFC, and basal forebrain.[30] Studies have supported confabulation as being the result of impaired search processes during memory retrieval.

Executive Functions

The term "executive function" encompasses a wide range of cognitive activities. Examples include organizing a response to a complex problem, sequencing tasks, prioritizing external stimuli, and abstraction. These functions support and overlap other frontal processes such as activating complex motor patterns, organizing language, and selecting appropriate social behaviors. Fundamentally, these executive functions depend on the ability to hold and manipulate information not available in the surrounding environment.[31] Thus, most tasks attempting to tap into executive functions are sensitive to impairment in other frontally and non-frontally mediated processes. Although there has been a variety of approaches to encapsulate or unify these various aspects of frontal lobe function under a unifying organizational structure, none is completely satisfactory. For example, there is some evidence that the PFC of the left hemisphere is more devoted to particular elements within a sequence of events, and the right integrates information across events. Others have hypothesized that the right PFC is more involved with the generation of a plan, whereas the left hemisphere is more involved with its execution. Still others describe differences between task setting, associated with the left dorsolateral PFC, and error detection, which correlates with right dorsolateral prefrontal cortical activation. Manifestations of executive function are myriad with frontal lobe injuries, and neuropsychological tests can document impairments encompassing all of the executive functions discussed above.

One common aspect of impaired executive function is an inability to maintain focused effort on a given task. Clinically, this would be manifest in being unable to complete tasks that were started. Executive function also includes generation, including the ability to produce new words or designs. Certain tasks can be used to test both aspects simultaneously, for instance by asking the subject to generate as many words as they can in one minute using specific rules, such as all the words beginning with a particular letter (e.g. F) or falling into a specific category (e.g. animals). Owing to the reliance on language, these tests are more sensitive to left frontal dysfunction, and should only be interpreted as representing frontal lobe injury if other language functions are intact. A visual test of generation can be performed by asking the subject to generate as many designs as they can, again constrained by specific rules.

Another aspect of frontal lobe executive dysfunction is trouble with planning and executing complex tasks. In daily life, this could be manifest in a variety of settings, such as planning trips or events such as parties, and even in everyday tasks such as cooking that require timing and sequencing. There are many neuropsychological tasks aimed as assessing this type of ability. One example is sequencing tasks that are used to assess the patient's planning, as in the Luria hand sequence in which the subject must repeatedly touch the table with a fist in three alternating positions: a closed fist, an open palm, and the side of the hand. The Trails B test assesses a patient's ability to shift between established sets such as numbers and letters (e.g. 1→A→2→B→3).

Inhibition is another aspect of executive function. The term in and of itself is broad, but in the context of frontal lobe function it is used to refer to the ability to inhibit the tendency to respond to the environment in automatic ways and instead perform tasks in service of more complex goals. Inhibition of responses to cognitively or socially relevant stimuli is related to frontal lobe function. In this section, problems with inhibition for cognitive tasks will be described. Social disinhibition will be discussed below. In daily life, cognitive disinhibition may be manifest as impulsive decision making. When more severe, problems with inhibition may be reflected in utilization behavior and environmental dependency, terms that were originally coined by Lhermitte. In utilization behavior, patients carry out patterns of movement that are entirely dependent on external stimuli. For example, patients will attempt to drink from an empty cup set in front of them, or use a hammer when no nails are present. Environmental dependency describes a tendency to passively follow the gestures of others, even if inhibiting those cues would be more appropriate. For example, patients may passively ask for the same items a friend orders when they are in a restaurant together, repeat the last phrase stated by others (echolalia), and mimic the movements of others. This probably reflects a failure of conceptual generation as well as inhibition.[32] Bedside and formal neuropsychological testing for inhibition includes simple tasks such as asking a patient to tap twice whenever the examiner taps once, and tap once every time the examiner taps twice, which requires the patient to inhibit the tendency to mimic the examiner. Designed in 1935, the Stroop test is perhaps the best known test of inhibition. In this task, the patient is presented with a card with names of colors written in various conflicting colors. For example, the word "red" may be written in green. The patient is instructed not to read the word, but instead to name the color of the ink.

Abstraction is the process through which verbal or visual information is absorbed and meaning abstracted beyond superficial features. For example, the ability to interpret the meaning of a proverb such as "people in glass houses should not throw stones" requires an ability to think beyond the literal glass house to reflect on a metaphor for hypocrisy. To do so means integrating concepts of social constructs, fragility, and more. Similarly, the human ability to use tools requires the ability to look at something like a rock and see potential uses based on features of the rock such as weight, hardness, or shape. Multiple brain regions participate in the process of abstraction, and the ability varies between individuals, but the loss of the ability to abstract has often been associated with lesions in the PFC.

Insight and Self-Monitoring

Brain dysfunction from many causes is associated with complete or incomplete lack of awareness of the deficit induced by their injury – a form of anosognosia. Although there are several likely mechanisms leading to this impairment, many lesion studies have indicated a link to the frontal lobes, either with imaging or by correlating the presence or severity of anosognosia with impairment on tasks tapping executive function. Some have hypothesized the presence of a dedicated system for conscious monitoring of one's own abilities in the frontal lobes. EEG and magnetic resonance imaging (MRI) studies have demonstrated activity in the ACC associated with errors, with the degree of activity often correlating with corrective actions, further supporting the role of the frontal lobe, particularly the medial frontal regions, in self-monitoring.

Language

Beginning with the landmark observations of Broca in 1861, the frontal lobe has been recognized as having a central role in language processing. While Broca hypothesized that the left inferior frontal region was *the* center for language in the brain, subsequent observations by Wernicke and many others have demonstrated that language is processed in a distributed network of regions in the left frontal and parietal lobe around the sylvian fissure. The specific deficits associated with damage to the frontal portion of this network have been well delineated.

Language impairment caused by frontal lobe lesions is typically characterized by hesitant speech, with relatively short phrases and long pauses between words or phrases as the patient attempts to find the words to express their thoughts. The pattern of speech is often described as non-fluent, which is meant to refer to the low numbers of words produced in a given amount of time, and is often referred to as a Broca type of aphasia. Additional features include incorrect pronunciation of words, called paraphasias. In general with aphasia, two types of paraphasia are common: semantic paraphasias, where patients substitute semantically related words for the intended words (e.g. "couch" for "bed"); and phonemic paraphasias, where incorrect speech sounds (phonemes) are substituted into words (e.g. "spake" for "snake"). Broca aphasia is typically associated with phonemic paraphasias. It is also characterized by agrammatism, a term that describes the fact that these patients tend to drop grammatical function words such as prepositions, adjectives, and adverbs, using mostly nouns and some verbs. For instance, if asked to describe how they would make an omelette, they might say "Eggs ... ham ... well, I ham ... veggies ... heat ...". Speech in these settings can be limited to very few isolated words, sometimes referred to as telegraphic speech, and in the most severe forms of Broca aphasia patients may be mute. Language comprehension in this type of aphasia is generally good, although it has been observed that patients with Broca aphasia have difficulty with comprehension of grammatically complex sentences. For example, if provided a sentence such as "the lion was killed by the tiger", a patient with Broca aphasia often incorrectly believes that the tiger is the animal that has been killed. Lastly, patients with aphasia due to frontal lobe lesions tend to be well aware of their deficits, which can result in frustration, but probably contributes to a better functional outcome compared with aphasias from lesions in other locations. Written language generally parallels verbal language in all types of aphasia, including the Broca type, and patients will show the same deficits when asked to repeat phrases or sentences as they do when trying to speak spontaneously.

Broca aphasias can be seen with damage to the left frontal lobe due to a variety of lesions ranging from strokes and tumors through postictal aphasias to neurodegenerative diseases. In a classic study published in 1978, Mohr and colleagues demonstrated that a lesion affecting only the cortex of the left inferior frontal lobe does not cause a full Broca type of aphasia but rather causes an initial mutism that evolves into a limited difficulty with articulation that does not affect other aspects of language (e.g. writing).[33] The more typical Broca aphasia is associated with a more extensive injury involving the left inferior frontal gyrus (Broca area) along with anterior portions of the frontal operculum (bank of the sylvian fissure) and anterior insula. When neurodegenerative diseases affect language as the first and most significant symptom, this is referred to as primary progressive aphasia (PPA). There are several clinical presentations of PPA with the clinical presentation being closely associated with the brain regions affected, but when the degenerative process is centered in the left inferior frontal region, the clinical presentation is similar to Broca aphasia from other focal brain lesions, being principally characterized by slow effortful speech, agrammatism, and difficulty with articulation. This syndrome has been labeled the non-fluent variant of PPA.[34]

Although the region most commonly discussed with regard to aphasias affecting the frontal lobe is the Broca area, lesions in other regions cause similar or associated symptoms. For instance, a lesion higher in the left frontal lobe, around the anterior portion of the middle frontal gyrus, can be associated with a syndrome called transcortical motor aphasia. Such lesions can be caused by watershed-type infarctions from cerebral hypoperfusion. The syndrome is very similar to Broca aphasia except that repetition is relatively preserved, and it has been interpreted as a disconnection between the speech production regions in the left inferior frontal cortex and the motor initiation centers in the medial frontal lobes.[35]

EMOTION, MOTIVATION, AND SOCIAL BEHAVIOR

The initial rapid and reflexive aspects of emotional processing are implemented in phylogenetically ancient systems, often referred to as the limbic network and composed of subcortical and cortical structures including the hypothalamus, limbic cortex, paralimbic cortical belt, limbic striatum such as the nucleus accumbens, the limbic pallidum, the ventral tegmental area, the habenula, and the limbic and paralimbic thalamic nuclei.[5] Some of the paralimbic structures reside in the frontal lobes and adjacent anterior insula. Through the actions of these and other regions, the frontal lobe plays important roles in emotional processing (Table 33.3).

Emotions are complex, and considerable debate regarding how to define emotions continues. Here, emotions are conceptualized as the process of rapidly generating patterned psychological, physiological (autonomic, endocrine), and behavioral responses to environmental stimuli.[36] Generating these responses involves several processes beginning with evaluation of the provoking stimulus, which can be internal (e.g. a thought) or external. In emotions parlance, this is referred to as appraisal. While the emotional reactions to these stimuli are often thought of as automatic, stereotyped, and uncontrollable, in fact humans have a considerable ability to control these reactions based on situational demands. The processes of appraisal, reactivity, and regulation represent core aspects of emotional functioning, and the frontal lobe and adjacent insula have important roles in all of them.

TABLE 33.3 Characteristics of the Major Cerebral Emotion Areas

Frontal Cortical Region	Connectivity	Physiology	Lesion Syndrome
ACC	Rostral: amygdala, PFC, hypothalamus, brainstem	Motivation and drive	Abulia
	Caudal: intralaminar thalamic nuclei, parietal lobe, motor cortex, PFC	Directing attention	Akinetic mutism
		Mediation of autonomic responses,	Inattention
		Assigning emotional valence to stimuli	
Insula	PFC	Assigning emotional valence to stimuli,	Disinhibition, socially inappropriate behavior
	Thalamus	Mediation of disgust response	
Ventral tegmental area	Ventral striatum	Reward processing	Apathy
	Nucleus accumbens	Drive	Obsessive–compulsive behavior
	Paralimbic ACC		
	OFC		
	Amygdala		
	Hippocampus		
OFC	Insula	Reward/punishment	Impulsivity
	Hippocampus	Emotional regulation	Impaired social behavior
	Parahippocampus	Extinction of learned responses	Emotional changes
	Amygdala		
	Hypothalamus		
	Striatum		

ACC: anterior cingulate cortex; OFC: orbitofrontal cortex; PFC: prefrontal cortex.

Although not always couched in terms of these constructs of appraisal, reactivity, and regulation, a large body of literature related to emotion has delineated the network of brain regions that help us to identify rewards in the environment, the cues that predict them, how much we want them, and how to get them. This network involves a number of subcortical regions, including the ventral tegmental area in the brainstem, also known as the mesolimbic reward center, the septal nuclei, the ventral striatum (nucleus accumbens, ventral caudate, and putamen), as well as paralimbic areas of the PFC (including the ACC and the OFC). The nucleus accumbens is of particular importance, receiving projections from the ventral tegmental area, the paralimbic areas of the PFC, and the amygdala. It projects to the PFC via the ventral pallidum and the mediodorsal nucleus of the thalamus. The glutamateric projections from the PFC determine the intensity and balance of the behavioral response to a reward, and the amygdala assigns an affective tone to the stimulus (e.g. a pleasurable or noxious balance). Some disorders associated with frontal lobe injury clearly reflect its role in reward-related processing.

Appraisal

Appraisal is probably the process least associated with frontal lobe functions. Ample evidence indicates that appraisal begins very early in the processing of incoming stimuli, in subcortical regions and sensory cortex. The amygdala in the anteromedial temporal lobe plays a key role in these processes, but is beyond the scope of this discussion.

Physiological and lesion studies do provide evidence, however, of frontal involvement in appraisal. The medial frontal regions, especially the ACC, are active during viewing of emotional faces in functional MRI studies,[37] and during tasks where decisions are in part mediated by the emotional content of the stimulus.[38] Cytoarchitectonically, the ACC is mesocortical, and comprises part of the limbic system. The ACC can be divided into several divisions, all of which contribute to the integration of emotional and cognitive information to influence actions and attention. The rostral part of the ACC can be divided into an affective division and a visceromotor area. The caudal part can be divided into a cognitive division, a motor division, and a nociceptive area. The affective division is extensively connected with the amygdala and PFC. Through the visceromotor area, the affective ACC directs the autonomic nuclei of the hypothalamus and the brainstem. The nociceptive area of the ACC receives nociceptive information from intralaminar thalamic nuclei. As a whole, the anterior cingulate forms a network that motivates directed and sustained attention towards emotionally relevant stimuli. The posterior insula represents interoceptive information about the body's physiological status, allowing us to interpret our own somatic reactions to stimuli. The information is moved along a gradient towards the anterior insula, where information from the frontal lobes and anterior temporal lobes interact to assign emotional values to different stimuli.

Lesions to ventral and medial frontal regions impair recognition of emotions.[39] In some cases, damage to specific cortical regions is thought to cause relatively selective impairment in appreciating specific emotions. For example, fear has been linked closely with the amygdala, and lesions to the insula cause impairment in processing disgust.

An important aspect of appraisal is the ability to know what other people are feeling, a component of empathy that is sometimes referred to as emotional perspective taking. This ability is commonly lost in frontotemporal dementia, probably because of tissue loss in the medial PFC and adjacent paracingulate gyrus.[40]

Emotional Reactivity

The expression of complex visceromotor and somatic behavior associated with emotion can be generated by subcortical regions such as the hypothalamus with no further higher cortical input.[41–43] Cortical regions, in particular the insula and ACC, also play a role in generating these reactions.

The effects of excessive cingulate activity may be found in the case of ACC epilepsy, which is sometimes associated with an aura of fear, and some patients produce non-mirthful laughter. ACC seizures can lead to altered affective states and expression and influence visceromotor activity. Interictally, patients may display behavior reminiscent of psychopathology and some patients have reduced aggression or fear, decreased motivation, obsessive–compulsive behavior, and aberrant social behavior.[44] Consistent with the role of frontal cortex in reward processing, medial frontal lobe lesions can produce profound apathy or abulia. Patients with this degree of apathy may move only rarely, only eating when fed and only speaking in monosyllables, with a complete lack of interest in their environment. Acutely, abulia may result from ischemia of the anterior cerebral arteries or hemorrhage from an anterior communicating artery due to the proximity of this vasculature to the anterior cingulate and medial frontal cortices. More slowly progressive symptoms could represent a slowly growing mass such as a midline meningioma or glioblastoma, chronic infection such as human immunodeficiency virus, or a neurodegenerative disease such as frontotemporal dementia. Neoplasms of the third ventricle that create an obstructive hydrocephalus may also lead to apathy or akinetic mutism. In neurodegenerative illnesses, apathy may slowly progress, initially presenting only as social withdrawal, which may be perceived by others as depression or selfishness. Psychiatric disorders such as obsessive–compulsive disorder and compulsive gambling can certainly be thought of as disorders of reward processing, and functional and structural imaging studies have demonstrated abnormalities of frontal lobe function in these and many other psychiatric disorders.

In addition to its role in recognizing disgust, the insula is important in mediating expressions of disgust.[45] Lesions affecting insula or ACC can cause emotional blunting, such that noxious stimuli fail to evoke the appropriate emotional response. In addition to emotional perspective taking, it has been posited that empathy involves covert simulation of the behavior of those with whom we empathize. Our face smiles when a friend smiles at us, for example. The inferior frontal gyrus and inferior parietal lobule are both involved in simulating other's emotional expressions as well as other motor movements.[46] Similarly, the anterior insula is activated when seeing someone else experience disgust.[47] This overlap between regions activated by watching another's emotion or action and regions activated by direct experience of an emotion or action is sometimes referred to as shared representations.[47] Although the precise role for this process in empathy and other emotional functions is not well established, loss of this form of emotional reactivity may also underlie impaired empathy seen in diseases such as frontotemporal dementia.

Emotional Regulation

There are times when emotions must be regulated. These may be to meet social expectations, such as not laughing at a funeral, or to maintain attention on a task, as when a medical student must regulate feelings of disgust during their first cadaver dissection. There are different strategies for regulation of emotion. Two of the most commonly used are cognitive reappraisal, which refers to the ability to reinterpret the meaning of an emotional stimulus so that we feel less emotion (e.g. reminding yourself that a scary movie is only a movie);

and suppression, which refers to the ability to hide our emotional display even though we still feel strong emotion. Most research on emotion regulation has focused on reappraisal, and has indicated that dorsomedial and dorsolateral frontal regions play important roles. This is consistent with the idea that this type of regulation probably invokes processes such as directed attention, working memory, abstraction, and memory, depending on the situation. Specific studies of emotion regulation are uncommon in neurological disease, but some studies have identified regulation deficits in frontotemporal dementia, and a recent study of patients with pseudobulbar affect suggested that the disorder is related to impaired regulation of emotion, possibly due to frontal dysfunction.[48] Many neurological injuries that typically affect the frontal lobes, such as Alzheimer disease and traumatic brain injury, are associated with irritability and agitation characterized by quick emotional reactions, probably representing impaired emotional regulation, although this has not been formally studied. Certain psychiatric diseases, such as bipolar disorder and borderline personality disorder, have been characterized as disorders of emotional regulation, and while some have been linked to frontal dysfunction, formal studies of the links between frontal function and emotional regulation are relatively uncommon.

The OFC has been identified as playing a particular role in emotional regulation, specifically updating and reversing previously learned cue–reward or cue–punishment associations. This has been studied most in the setting of fear conditioning. While the amygdala is the key structure for linking a neutral cue (e.g. a sound like a bell) with an outcome (e.g. shock), the OFC is critical in abolishing the response once the association is no longer true (several bell sounds not followed by a shock), a process called extinction. Inability to extinguish these types of association has been hypothesized to contribute to post-traumatic stress disorder. Studies of the OFC have also linked this region with a broader role in updating the value of a stimulus. For instance, OFC neurons that fire in response to seeing a tasty food stop firing when people or animals are sated with that particular food, indicating that OFC accesses information about internal state (e.g. fullness, memory for foods recently eaten) in order to code the current value of the stimulus. The insula has been shown to alter its activity in a similar way. Failure of these mechanisms may lead to bizarre reward-related behavior. For instance, patients with frontotemporal dementia have been known to overeat, continuing to do so even after they are full, resulting in significant weight gain. Such patients are also often driven to hoard specific items (videos, toys, coins) in large amounts, filling rooms in their house. Failure of these mechanisms is likely to contribute to abnormalities in social behavior seen in this disorder. For instance, patients may touch strangers inappropriately because OFC neurons are not signaling that the context (a stranger, a public place) is inappropriate. Related research has examined emotional contributions to decision making in patients with OFC lesions using "gambling" tasks with several virtual decks of cards that require subjects to learn over many trials which decks are financially advantageous. Patients with OFC lesions are drawn towards the "bad" desks despite being able to voice their awareness that those decks will lose them money. This seems to be related to an inability to mount an emotional response when presented with a card from one of those decks.[49] This may be related to the role of the OFC in changing our perceptions about which environmental stimuli predict rewards and which do not. Because of these types of impairment, patients with diseases affecting the OFC make notoriously poor decisions that can be financially ruinous, despite being able to verbalize the reason these decisions are bad. Typical pathological causes for these types of injury include traumatic brain injury, tumors in the skull above the orbits, and frontotemporal dementia.

The Right Hemisphere and Emotion

Several theories of emotion posit differing roles for the right and left hemispheres in emotional processing. Some theories attribute most of emotional processing to the right hemisphere, including theories that suggest that the organization of right hemisphere emotion is similar to the organization of language in the left hemisphere, with expression mediated by the right frontal lobe and comprehension of emotion mediated by the right parietal lobe.[50] Still other theories, supported by EEG and MRI studies, propose that the right hemisphere is more involved in the processing of emotions characterized by withdrawal behavior, such as fear, sadness, and disgust, while the left hemisphere is more involved with emotions characterized by approach, such as happiness and perhaps anger. These theories have been proposed as justification for findings that left hemisphere lesions are often associated with depression. While the roles of the right and left hemispheres in emotion are still being delineated, evidence from neurodegenerative diseases such as frontotemporal dementia supports a unique role for right hemisphere structures in social and emotional functions.

CONCLUSION AND QUESTIONS FOR FURTHER RESEARCH

Far beyond being the seat of executive function, the frontal lobes mediate a variety of important processes, ranging from motor to language to emotional functions. If there is a unifying theme to the functions of the frontal lobe, it is probably monitoring, control, and modification of more basic functions. Although the variety of symptoms associated with frontal lobe injury is well documented, and many neuroscientific studies of animals and humans have elucidated some of the basic functions of the frontal lobes such as working memory and coding of rewards to some degree, there is still a vast amount to be learned about how basic functions of the frontal lobe are accomplished and how failure in these mechanisms leads to the cognitive, language, social, and emotional dysfunction seen with frontal lobe injury.

In the past century, the frontal lobes have gone from being sometimes regarded as a silent and relatively unimportant part of the brain, subject to such disregardful procedures as the bedside lobotomy, to being recognized as an important hub to much of what makes us uniquely human. Research into frontal lobar function continues, and innumerable important questions remain to be addressed. How does sensorimotor information become synthesized into the coordination of movement? Can electrical stimulation of particular frontal structures lead to improvements in memory or attention? Is there a more physiologically meaningful way to organize the multiple aspects of executive function? In addition to the well-studied lexical aspects of language, how do the frontal lobes mediate non-verbal communication? How do frontal networks interplay to create an accurate sense of the self and others? How can mutations in a DNA molecule lead to dramatic changes in social and emotional behavior? Which therapies, pharmacological and otherwise, can enhance the frontal lobe function in those who have been damaged by trauma, stroke, inflammation, or neurodegenerative illnesses? Are there molecular commonalities in diseases traditionally considered disparate but which share frontal lobar dysfunctions, such as schizophrenia or frontotemporal dementia? These and many other outstanding questions regarding frontal lobe function will provide a rich source of projects for neuroscientists studying brain function in many species for the foreseeable future.

References

1. Semendeferi K, Lu A, Schenker N, Damasio H. Humans and great apes share a large frontal cortex. *Nat Neurosci*. 2002;5(3):272–276.
2. Desikan RS, Segonne F, Fischl B, et al. An automated labeling system for subdividing the human cerebral cortex on MRI scans into gyral based regions of interest. *Neuroimage*. 2006;31(3):968–980.
3. Rorden C, Brett M. Stereotaxic display of brain lesions. *Behav Neurol*. 2000;12:191–200.
4. Alexander GE, DeLong MR, Strick PL. Parallel organization of functionally segregated circuits linking basal ganglia and cortex. *Annu Rev Neurosci*. 1986;9:357–381.
5. Mesulam M-M. Behavioral neuroanatomy. In: Mesulam M-M, ed. *Principles of Behavioral and Cognitive Neurology*. New York: Oxford; 2000:1–120.
6. Seeley WW, Menon V, Schatzberg AF, et al. Dissociable intrinsic connectivity networks for salience processing and executive control. *J Neurosci*. 2007;27(9):2349–2356.
7. Hepp-Raymond MC, Trouche E, Wiesendanger M. Effects of unilateral and bilateral pyramidotomy on a conditioned rapid precision grip in monkeys (*Macaca fascicularis*). *Exp Brain Res*. 1974;21:519–527.
8. Penfield W, Jasper H. *Epilepsy and the Functional Anatomy of the Human Brain*. 2nd ed. Boston, MA: Little, Brown and Co; 1954.
9. Wise SP. The primate premotor cortex: past, present and preparatory. *Annu Rev Neurosci*. 1985;1985(8):1–19.
10. Preuss TM, Stepniewska I, Kaas JH. Movement representation in the dorsal and ventral premotor areas of owl monkeys: a microstimulation study. *J Comp Neurol*. 1996;371(4):649–676.
11. Rizzolatti G, Luppino G, Matelli M. The organization of the cortical motor system: new concepts. *Electroencephalogr Clin Neurophysiol*. 1998;106:283–296.
12. Deecke L, Kornhuber HH. An electrical sign of participation of the mesial "supplementary" motor cortex in human voluntary finger movement. *Brain Res*. 1978;159:473–476.
13. Nachev P, Kennard C, Husain M. Functional role of the supplementary and pre-supplementary motor areas. *Nat Rev Neurosci*. 2008;9(11):856–869.
14. Lee SE, Rabinovici GD, Mayo MC, et al. Clinicopathological correlations in corticobasal degeneration. *Ann Neurol*. 2011;70(2):327–340.
15. Seeley WW, Crawford RK, Zhou J, Miller BL, Greicius MD. Neurodegenerative diseases target large-scale human brain networks. *Neuron*. 2009;62(1):42–52.
16. Heilman KM. Apraxia. *Continuum (Minneap Minn)*. 2010;16(4):86–98.
17. Tokgozoglu SL, Batur MK, Topuoglu MA, Saribas O, Kes S, Oto A. Effects of stroke localization on cardiac autonomic balance and sudden death. *Stroke*. 1999;30(7):1307–1311.
18. Britton J, Ghearing GR, Benarroch EE, Cascino GD. The ictal bradycardia syndrome: localization and lateralization. *Epilepsia*. 2006;47(4):737–744.
19. Tsakanikas D, Relkin N. Normal pressure hydrocephalus. *Semin Neurol*. 2007;27(1):58–65.
20. Markowitsch HJ, Pritzel M, Divac I. Cortical afferents to the prefrontal cortex of the cat: a study with the horseradish peroxidase technique. *Neurosci Lett*. 1979;11:115–120.
21. Mesulam MM. Attentional and confusional states. *Continuum (Minneap Minn)*. 2010;16(4):128–139.
22. Daffner K, Mesulam M, Scinto L, et al. The central role of the prefrontal cortex in directing attention to novel events. *Brain*. 2000;123(5):923–939.
23. Pardo J, Fox P, Raichle M. Localization of a human system for sustained attention by positron emission tomography. *Nature*. 1991;349(6304):61–64.
24. Grady CL, McIntosh AR, Bookstein F, Horwitz B, Rapoport SI, Haxby JV. Age-related changes in regional cerebral blood flow during working memory for faces. *Neuroimage*. 1998;8(4):409–425.
25. Nyberg L, Cabeza R, Tulving E. PET studies of encoding and retrieval: the HERA model. *Psychon Bull Rev*. 1996;3:135–148.
26. Lee A, Robbins T, Smith S, et al. Evidence for asymmetric frontal-lobe involvement in episodic memory from functional magnetic resonance imaging and patients with unilateral frontal-lobe excisions. *Neuropsychologia*. 2002;40:2420–2437.
27. Gershberg F, Shimamura A. Impaired use of organizational strategies in free recall following frontal lobe damage. *Neuropsychologia*. 1995;13:1305–1333.
28. McKinnon MC, Svoboda E, Levine B. The frontal lobes and autobiographical memory. In: Miller BL, Cummings JL, eds. *The Human Frontal Lobes*. New York: Guilford Press; 2007:227–248.
29. Shimamura A. Memory and frontal lobe function. In: Gazzaniga M, ed. *The Cognitive Neurosciences*. Cambridge, MA: MIT Press; 1995:803–813.

30. Gilboa A, Moscovitch M. The cognitive neuroscience of confabulation: a review and a model. In: Baddeley A, Kopelman M, Wilson B, eds. *The Handbook of Memory Disorders*. 2nd ed. Chichester: Wiley; 2002:315–342.
31. Goldman-Rakic P. The prefrontal landscape: implications of functional architecture for understanding human mentation and the central executive. *Philos Trans R Soc Lond*. 1996;351:1445–1453.
32. Lhermitte F, Pillon B, Serdaru M. Human autonomy and the frontal lobes. Part I: Imitation and utilization behavior: a neuropsychological study of 75 patients. *Ann Neurol*. 1986;19(4):326–334.
33. Mohr JP, Pessin MS, Finkelstein S, Funkenstein HH, Duncan GW, Davis KR. Broca's aphasia: pathologic and clinical. *Neurology*. 1978;28(4):311–324.
34. Gorno-Tempini ML, Hillis AE, Weintraub S, et al. Classification of primary progressive aphasia and its variants. *Neurology*. 2011;76(11):1006–1014.
35. Freedman M, Alexander M, Naeser M. Anatomic basis of transcortical motor aphasia. *Neurology*. 1984;34(4):409–417.
36. Rosen HJ, Levenson RW. The emotional brain: combining insights from patients and basic science. *Neurocase*. 2009;15(3):173–181.
37. Phan KL, Wager T, Taylor SF, Liberzon I. Functional neuroanatomy of emotion: a meta-analysis of emotion activation studies in PET and fMRI. *Neuroimage*. 2002;16(2):331–348.
38. Bush G, Luu P, Posner MI. Cognitive and emotional influences in anterior cingulate cortex. *Trends Cogn Sci*. 2000;4(6):215–222.
39. Hornak J, Bramham J, Rolls ET, et al. Changes in emotion after circumscribed surgical lesions of the orbitofrontal and cingulate cortices. *Brain*. 2003;126(Pt 7):1691–1712.
40. Fernandez-Duque D, Hodges SD, Baird JA, Black SE. Empathy in frontotemporal dementia and Alzheimer's disease. *J Clin Exp Neuropsychol*. 2010;32(3):289–298.
41. Bard P. A diencephalic mechanism for the expression of rage with special reference to the sympathetic nervous system. *Am J Physiol*. 1928;84:490–515.
42. Hess WR. *Diencephalon: Autonomic and Extrapyramidal Functions*. New York: Grune and Stratton; 1954.
43. Panksepp J. *Affective Neuroscience*. New York: Oxford; 1998.
44. Alkawadri R, So NK, Van Ness PC, Alexopoulos AV. Cingulate epilepsy: report of 3 electroclinical subtypes with surgical outcomes. *JAMA Neurol*. 2013:1–8.
45. Calder AJ, Lawrence AD, Young AW. Neuropsychology of fear and loathing. *Nat Rev Neurosci*. 2001;2(5):352–363.
46. Jabbi M, Swart M, Keysers C. Empathy for positive and negative emotions in the gustatory cortex. *Neuroimage*. 2007;34:1744–1753.
47. Decety J, Jackson PL. The functional architecture of human empathy. *Behav Cogn Neurosci Rev*. 2004;3(2):71–100.
48. Balakrishnan P, Rosen H. The causes and treatment of pseudobulbar affect in ischemic stroke. *Curr Treat Options Cardiovasc Med*. 2008;10:216–222.
49. Bechara A, Tranel D, Damasio H, Damasio AR. Failure to respond autonomically to anticipated future outcomes following damage to prefrontal cortex. *Cereb Cortex*. 1996;6(2):215–225.
50. Ross ED, Monnot M. Neurology of affective prosody and its functional–anatomic organization in right hemisphere. *Brain Lang*. 2008;104(1):51–74.

CHAPTER

34

Stress

Bruce S. McEwen

Laboratory of Neuroendocrinology, The Rockefeller University, New York, USA

OUTLINE

INTRODUCTION

"Stress" is a commonly used word that generally refers to experiences that cause feelings of anxiety and frustration because they push people beyond their ability to successfully cope. Besides time pressures and daily hassles at work and home, there are stressors related to economic insecurity, poor health, and interpersonal conflict. A feeling of lack of control is often a key element of being "stressed out", whether due to time pressures or a person's feeling of hopelessness about an aspect of his or her own life. There are also rare situations that are life-threatening (accidents, disasters, or violence) and these evoke the classical "fight or flight" response. In contrast to daily hassles, these stressors are acute and yet they also usually result in chronic stress and altered health behaviors in the aftermath of the tragic event.

The most common stressors are ones that operate chronically, often at a low level, and that cause people to behave in certain ways. For example, being stressed out may cause people to be anxious and or depressed, to lose sleep at night, to eat comfort foods and take in more calories than their bodies need, and to smoke or drink alcohol excessively. Being stressed out may also cause them to neglect seeing friends, or fail to take time off for relaxation, or to engage in regular physical activity as they, for example, sit at a computer and try to escape from under the burden of too much to do in too little time. People may take medication (e.g. anxiolytics and sleep-promoting agents) to help them to cope, and, with time, their bodies may increase in weight because of sleep deprivation, overeating, and too little physical activity.

The brain is the organ that decides what is stressful and determines the behavioral and physiological responses, whether health promoting or health damaging (Fig. 34.1).[1] The brain is a biological organ that changes under acute and chronic stress and directs, as well as responds to, many systems of the body (e.g. metabolic, cardiovascular, and immune) that are involved in

Neurobiology of Brain Disorders
http://dx.doi.org/10.1016/B978-0-12-398270-4.00034-3

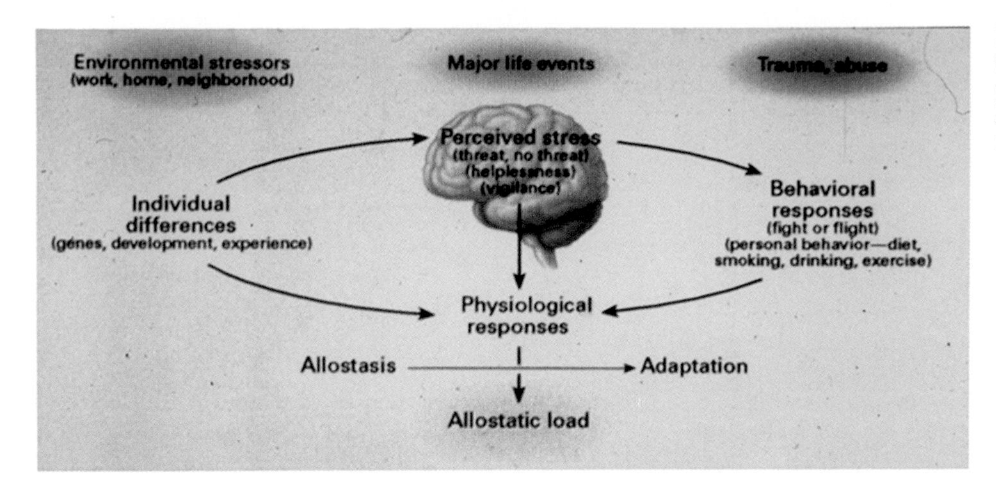

FIGURE 34.1 **Central role of the brain in allostasis and the behavioral and physiological response to stressors.** *Source: From McEwen.* N Engl J Med. *1998;338:171–179,*[1] *with permission.*

the short- and long-term consequences of being stressed out.[2]

What does chronic stress do to the body and brain? This chapter summarizes some of the current information, placing particular emphasis on how the stress hormones can play both protective and damaging roles in the brain and body, depending on how tightly their release is regulated, and discusses some of the approaches for dealing with stress in our complex world.

TYPES OF STRESS

This chapter will use the following classifications of types of stress: good stress, tolerable stress, and toxic stress (Table 34.1) (see http://developingchild.harvard.edu).

Good stress is a term used in popular language to refer to the experience of rising to a challenge, taking a risk, and feeling rewarded by an often positive outcome. Selye used a related term, "eustress".[3] Good self-esteem and good impulse control and decision-making capability, all functions of a healthy architecture of the brain, are important here. Even adverse outcomes can be growth experiences for individuals with such positive, adaptive characteristics, and toughening or inoculation to stressors is an aspect that is often discussed.

Tolerable stress refers to those situations where bad things happen, but the individual with healthy brain architecture is able to cope, often with the aid of family, friends, and other individuals who provide support. Here, distress refers to the uncomfortable feeling related to the nature of the stressor and the degree to which the individual feels a lack of ability to influence or control the stressor, according to Lazarus and Folkman.[4]

Finally, toxic stress refers to the situation in which bad things happen to an individual who has limited support, and who may also have brain architecture that reflects the effects of adverse early life events that have impaired the development of good impulse control and judgment and adequate self-esteem. Here, the degree or duration of distress, or both, may be greater. With toxic stress, the inability to cope is likely to have adverse effects on behavior and physiology, and this will result in a higher degree of allostatic overload, as will be explained in the next section.

DEFINITION OF STRESS, ALLOSTASIS, AND ALLOSTATIC LOAD

In spite of the further definitions of types of stress, "stress" is still an ambiguous term and has connotations that make it less useful in understanding how the body handles events that are stressful, both good and bad. Insight into these processes can lead to improved understanding of how best to intervene (see "Interventions that Change the Brain and Improve Health", below). There are two sides to this story: on the one hand, the body responds to almost any event or challenge by releasing chemical mediators (e.g. catecholamines that increase heart rate and blood pressure) that help an individual to cope with the situation; on the other hand, chronic elevation of these same mediators (e.g. chronically increased heart rate and blood pressure) produces chronic wear and tear on the cardiovascular system that can result, over time, in disorders such as strokes and heart attacks. For this reason, the term "allostasis" was introduced by Sterling and Eyer in 1988[5] to refer to the active process by which the body responds to daily events and maintains homeostasis (allostasis literally means "achieving stability through change").[6] Because chronically increased allostasis can lead to disease, the term allostatic load or overload has been introduced to refer to the wear and tear that results from either too much stress or inefficient management of allostasis, such as not turning off the response when it is no longer needed.[1] Other forms of allostatic load are summarized in Fig. 34.2 and involve

TABLE 34.1 Levels of Stressful Experiences

	Level of Stress		
	Good	**Tolerable**	**Toxic**
Cause	Any challenge (e.g. giving a speech)	Adverse life events buffered by supportive relationships	Unbuffered adverse events of greater duration and size
Result	Sense of mastery and control	Coping and recovery	Poor coping and compromised recovery; increased lifelong risk of physical and mental disorders
Brain architecture	Healthy	Healthy	Compromised
Attributes	Good self-esteem, judgment, and impulse control	Good self-esteem, judgment, and impulse control	Lack of self-esteem, poor judgment, and lack of impulse control

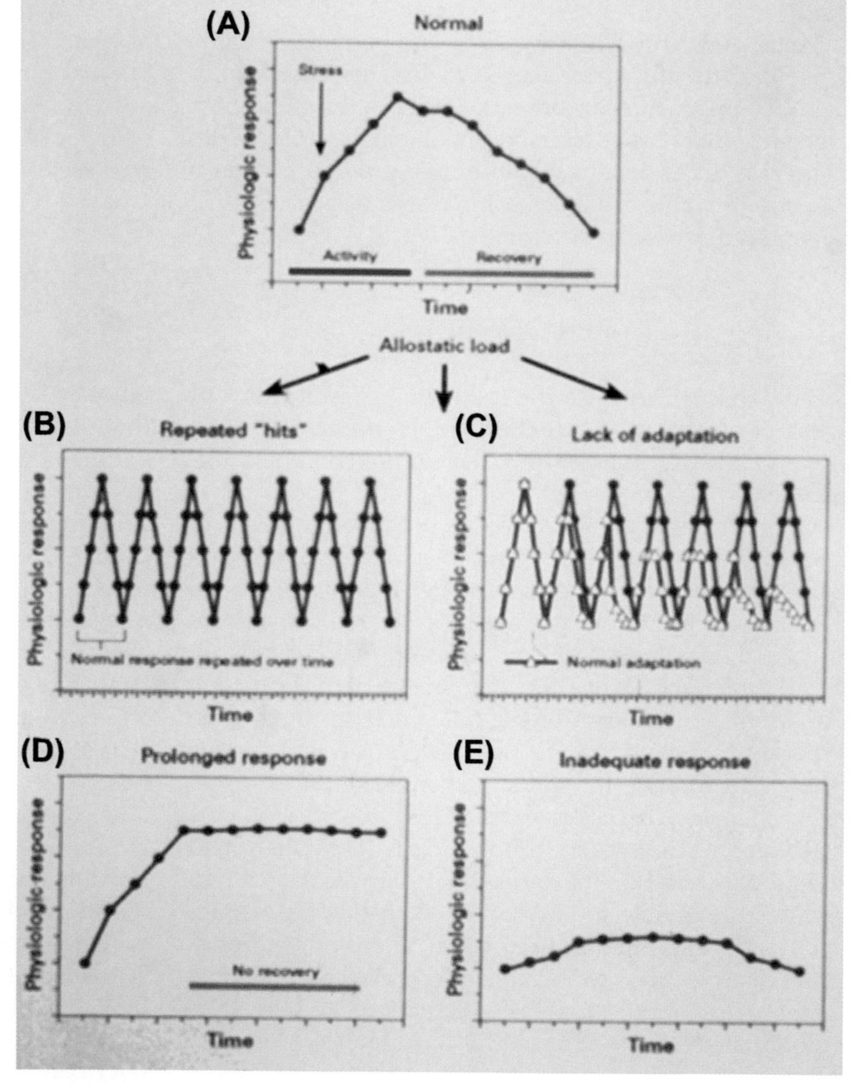

FIGURE 34.2 **Four types of allostatic load.** (A) Normal allostatic response, in which a response is initiated by a stressor, sustained for an appropriate interval, and then turned off. (B-E) Four conditions that lead to allostatic load: (B) repeated hits from multiple stressors; (C) lack of adaptation; (D) prolonged response due to delayed shutdown; and (E) inadequate response that leads to compensatory hyperactivity of other mediators (e.g. inadequate secretion of glucocorticoid, resulting in increased levels of cytokines that are normally counter-regulated by glucocorticoids). *Source: From McEwen.* N Engl J Med. *1998;338:171–179,*[1] *with permission.*

not turning on an adequate response in the first place or not habituating to the recurrence of the same stressor, and thus dampening the allostatic response and reducing the accumulation of allostatic load.[7,8]

RESPONSE TO STRESSORS: PROTECTION AND DAMAGE

Protection and damage are two contrasting sides of the physiology involved in defending the body against the challenges of daily life, whether or not we call them stressors. Besides epinephrine (adrenaline) and norepinephrine (noradrenaline), there are many mediators that participate in allostasis, and they are linked together in a network of regulation that is non-linear (Fig. 34.3), meaning that each mediator has the ability to regulate the activity of the other mediators, sometimes in a biphasic manner.[7]

Glucocorticoids produced by the adrenal cortex, in response to adrenocorticotropic hormone from the pituitary gland, are the other major stress hormones. Proinflammatory and anti-inflammatory cytokines are produced by many cells in the body, and they regulate each other and are, in turn, regulated by glucocorticoids and catecholamines. Whereas catecholamines can increase proinflammatory cytokine production, glucocorticoids are known to inhibit this production. However, there are exceptions, in proinflammatory effects of glucocorticoids that depend on dose and cell or tissue type. The parasympathetic nervous system also plays an important regulatory role in this non-linear network of allostasis, since it generally opposes the sympathetic nervous system and, for example, slows the heart and also has anti-inflammatory effects.

What this non-linearity means is that when any one mediator is increased or decreased, there are compensatory changes in the other mediators that depend on the time-course and level of change of each of the mediators. Unfortunately, one cannot measure all components of this system simultaneously and must rely on measurements of only a few of them in any one study. The non-linearity must be kept in mind when interpreting the results.[7]

A good example of the biphasic actions of stress (i.e. protection versus damage) is in the immune system, in which an acute stressor activates an acquired immune response via mediation by catecholamines, glucocorticoids, and locally produced immune mediators and, yet, a chronic exposure to the same stressor over several weeks has the opposite effect and results in immune suppression.[9] The acute immune enhancement is good for enhancing immunization, fighting an infection or repairing a wound, but is deleterious to health for an autoimmune condition, such as psoriasis or Crohn disease. On the other hand, the immune suppression is good in the case of an autoimmune disorder and deleterious for fighting an infection or repairing a wound. In an immune-sensitive skin cancer, acute stress is effective in inhibiting tumor progression, whereas chronic stress exacerbates progression.[10]

POSITIVE EFFECTS OF GLUCOCORTICOIDS ON NEURONAL FUNCTIONS AND STRUCTURE

Glucocorticoids and catecholamines mediate the immune-enhancing effects of acute stress.[9] They also mediate the memory-enhancing effects of acute stress in fear-learning situations and do this via actions in a

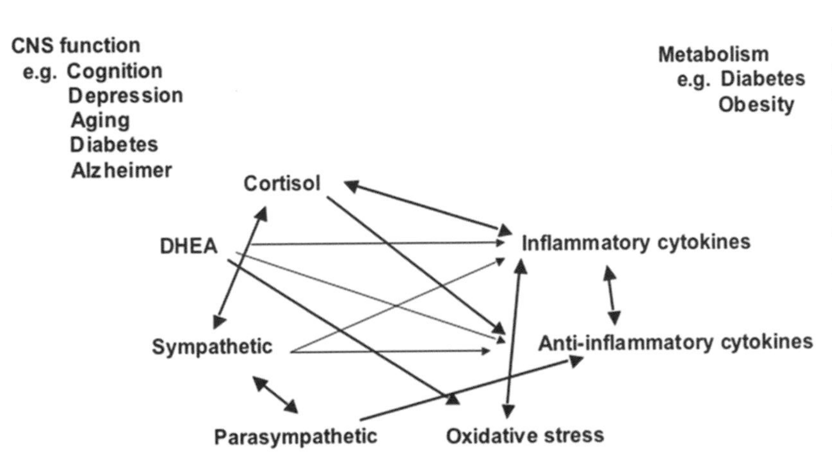

FIGURE 34.3 **Non-linear network of mediators of allostasis involved in the stress response.** Arrows indicate that each system regulates the others in a reciprocal manner, creating a non-linear network. Moreover, there are multiple pathways for regulation, e.g. inflammatory cytokine production is negatively regulated via anti-inflammatory cytokines, as well as via parasympathetic and glucocorticoid pathways, whereas sympathetic activity increases inflammatory cytokine production. Parasympathetic activity, in turn, contains sympathetic activity. *Source: Modified from McEwen.* Dialogues Clin Neurosci. *2006;8:367–381.*[7]

number of brain regions, including the hippocampus and amygdala.[2] Physiological levels of glucocorticoids also promote synapse turnover and dendrite growth, and activation of the trkB receptor for brain-derived neurotrophic factor (BDNF),[11] as well as promoting reversal of monocular deprivation of the visual system driven by visual stimulation.[12] Glucocorticoids also increase endocannabinoid levels in brain and contribute to the neural modulation of excitatory and inhibitory activity.[13] They promote the translocation of glucocorticoid receptors into mitochondria, where they have neuroprotective effects by enhancing calcium sequestration, thus reducing oxidative stress in the rest of the cell.[14,15] Finally, timed elevation of glucocorticoids blocks the delayed anxiety-inducing effect of a single traumatic stressor in an animal model of post-traumatic stress disorder (PTSD), and appears to have similar effects when given to human trauma victims during or immediately after the traumatic event.[16]

STRESS IN THE NATURAL WORLD

The operation of allostasis in the natural world provides some insight into how animals use this response to their own benefit or for the benefit of the species. As an example of allostasis, in spring, a sudden snowstorm causes stress to birds and disrupts mating, and stress hormones are pivotal in directing the birds to suspend reproduction, to find a source of food, and to relocate to a better mating site or at least to delay reproduction until the weather improves.[17] As an example of allostatic load, bears preparing to hibernate for the winter eat large quantities of food and put on body fat to act as an energy source during the winter. This accumulation of fat is used to survive the winter and provide food for gestation of young; in contrast, the fat accumulation occurs in bears that are captive in zoos and eating too much, partially out of boredom, while not exercising. The accumulation of fat under these latter conditions can be called allostatic overload, referring to a condition that is associated with pathophysiology,[17] as is all too common in humans. Yet, allostatic overload can also have a useful purpose for the preservation of the species, such as in migrating salmon or the marsupial mouse, which die of excessive stress hormone actions after mating; the stress, and allostatic load, are caused for salmon, in part, by the migration up the rapidly flowing rivers but also because of physiological changes that represent accelerated aging.[17] The result is freeing up food and other resources for the next generation. In the case of the marsupial mouse, it is only the males that die after mating, due apparently to a response to mating that reduces the binding protein, corticosteroid binding globulin, for glucocorticoids and renders them much more active throughout the body.

CIRCADIAN DISRUPTION

Because the brain is the master regulator of the neuroendocrine, autonomic, and immune systems, as well as behavior (see Fig. 34.1), alterations in brain function by chronic stress can have direct and indirect effects on the cumulative allostatic overload. One of the key systems in the brain and body that regulate homeostasis of these varied physiological and behavioral variables is the circadian system. Based in the suprachiasmatic nucleus of the hypothalamus, the brain's clock controls rhythms in the rest of the brain and body through both neural and diffusible signals. Biological clocks at the molecular level have been detected in almost every body organ and tissue so far examined, and these clocks are synchronized by the suprachiasmatic nucleus directly (by way of neural connections) or indirectly through hormonal signals (e.g. cortisol, melatonin) or behavioral outputs (e.g. feeding). The clock also regulates the timing of sleep, and as such sleep and circadian systems interact to regulate rest–activity cycles and keep an organism in synchrony with the external environment. Disruption of these key homeostatic systems could clearly contribute to allostatic overload.[18,19]

Reduced sleep duration has been reported to be associated with increased prediabetes, as well as body mass and obesity, in the US National Health and Nutrition Examination Survey (NHANES).[20] Sleep restriction to 4 hours of sleep per night increases blood pressure, decreases parasympathetic tone, increases evening cortisol and insulin levels, and promotes increased appetite, possibly through the elevation of ghrelin, a proappetitive hormone, along with decreased levels of leptin.[21] Moreover, increased levels of proinflammatory cytokines are seen with sleep deprivation, along with decreased performance in tests of psychomotor vigilance, and this has been reported to result even from a modest sleep restriction to 6 hours per night.[19]

Circadian disruption has sometimes been overlooked as a separate, yet related, phenomenon to sleep deprivation. In modern industrialized societies, circadian disruption can be induced in numerous ways, the most common of which are shift work and jet lag.[22] A longitudinal study in a cohort of nurses in night-shift work found that exposure to night work can contribute to weight gain and obesity. Moreover, alternating shift work was an independent risk factor for the development of obesity in a large longitudinal study of male Japanese shift workers. Numerous mouse models have also contributed to the understanding of the relationship between circadian disruption and metabolism, with

CLOCK mutant mice showing altered basal metabolism and a tendency towards obesity and metabolic dysregulation, while normal C57Bl/6 mice housed in a disrupted 10 hour light/10 hour dark cycle show accelerated weight gain and disruptions in metabolic hormones.[22] Behaviorally, circadian disruption can contribute to cognitive impairments. A study of long-recovery versus short-recovery flight crews found that short-recovery crews had impaired performance in a psychomotor task, reacting more slowly and with more errors compared with a long-recovery crew.[2] The same mouse model of circadian disruption shows cognitive inflexibility and shrinkage of dendrites in the medial prefrontal cortex.[22]

KEY ROLE OF THE BRAIN IN RESPONSE TO STRESS

The brain is the key organ of the stress response because it determines what is threatening and, therefore, stressful, and also controls the behavioral and physiological responses that were discussed earlier (see Fig. 34.1). There are enormous individual differences in the response to stress, based on people's experiences early in life and in adult life.[1] Positive or negative experiences in school, at work, or in romantic and family interpersonal relationships can bias an individual towards either a positive or a negative response in a new situation. For example, someone who has been treated badly in a job by a domineering and abusive supervisor or has been fired will approach a new job situation differently from someone who has had positive experiences in employment. Early life experiences perhaps carry an even greater weight in terms of how an individual reacts to new situations (see "Early Life Experiences", below).

THE BRAIN AS A TARGET OF STRESS

Hippocampus

One of the ways that stress hormones modulate function within the brain is by changing the structure of neurons. The hippocampus is one of the most sensitive and malleable regions of the brain and is also very important in cognitive function.[23] Within the hippocampus, the input from the entorhinal cortex to the dentate gyrus is ramified by the connections between the dentate gyrus and the CA3 pyramidal neurons. One granule neuron innervates, on average, 12 CA3 neurons, and each CA3 neuron innervates, on average, 50 other CA3 neurons via axon collaterals, as well as 25 inhibitory cells via other axon collaterals. The net result is a 600-fold amplification of excitation, as well as a 300-fold amplification of inhibition, that provides some degree of control of the system.

As to why this type of circuitry exists, the dentate gyrus–CA3 system is believed to play a role in the memory of sequences of events, although long-term storage of memory occurs in other brain regions. But, because the DG–CA3 system is so delicately balanced in its function and vulnerability to damage, there is also adaptive structural plasticity, in that new neurons continue to be produced in the dentate gyrus throughout adult life, and CA3 pyramidal cells undergo a reversible remodeling of their dendrites in conditions such as hibernation and chronic stress. The role of this plasticity may be to protect against permanent damage. As a result, the hippocampus undergoes a number of adaptive changes in response to acute and chronic stress.[23]

One type of change involves the replacement of neurons. The subgranular layer of the dentate gyrus contains cells that have some properties of astrocytes (e.g. expression of glial fibrillary acidic protein) and which give rise to granule neurons. After bromodeoxyuridine administration to label the DNA of dividing cells, these newly born cells appear as clusters in the inner part of the granule cell layer, where a substantial number of them will go on to differentiate into granule neurons within as little as 7 days. In the adult rat, 5000–9000 new neurons are born per day and survive with a half-life of 28 days.[24] There are many hormonal, neurochemical, and behavioral modulators of neurogenesis and cell survival in the dentate gyrus, including estradiol, insulin-like growth factor-1, antidepressants, voluntary exercise, and hippocampal dependent learning. With respect to stress, certain types of acute stress and many chronic stressors suppress neurogenesis or cell survival in the dentate gyrus, and the mediators of these inhibitory effects include excitatory amino acids acting via *N*-methyl-D-aspartate receptors and endogenous opioids.[25]

Another form of structural plasticity is the remodeling of dendrites in the hippocampus. Chronic restraint stress causes retraction and simplification of dendrites in the CA3 region of the hippocampus. Such dendritic reorganization is found in both dominant and subordinate rats undergoing adaptation of psychosocial stress in the visible burrow system, and is independent of adrenal size.[23] This result emphasizes that it is not adrenal size or presumed amount of physiological stress *per se* that determines dendritic remodeling, but a complex set of other factors that modulate neuronal structure. Indeed, as noted above, in species of mammals that hibernate, dendritic remodeling is a reversible process and occurs within hours of the onset of hibernation in European hamsters and ground squirrels, and is also reversible within hours of wakening of the animals from torpor.[26] This implies that reorganization of the cytoskeleton is taking place rapidly and reversibly and that changes in dendrite length and branching are not damage but a form of structural plasticity.

Regarding the mechanism of structural remodeling, adrenal steroids are important mediators of remodeling of hippocampal neurons during repeated stress, and exogenous adrenal steroids can also cause remodeling in the absence of an external stressor. The role of adrenal steroids involves many interactions with neurochemical systems in the hippocampus, including serotonin, γ-aminobutyric acid, and excitatory amino acids. Probably the most important interactions are those with excitatory amino acids, such as glutamate. Excitatory amino acids released by the mossy fiber pathway play a key role in the remodeling of the CA3 region of the hippocampus, and regulation of glutamate release by adrenal steroids may play an important role.[25]

Among the consequences of restraint stress is the elevation of extracellular glutamate levels, leading to induction of glial glutamate transporters, as well as increased activation of the nuclear transcription factor phosphoCREB. Moreover, 21 day chronic restraint stress causes depletion of clear vesicles from mossy fiber terminals and increased expression of presynaptic proteins involved in vesicle release. Taken together with the fact that vesicles that remain in the mossy fiber terminal are near active synaptic zones and that there are more mitochondria in the terminals of stressed rats, this suggests that chronic restraint stress increases the release of glutamate.[23]

Prefrontal Cortex and Amygdala

Repeated stress also causes changes in other brain regions, such as the prefrontal cortex and amygdala. Repeated stress causes dendritic shortening in the medial prefrontal cortex, but produces dendritic growth in neurons in the amygdala, as well as the orbitofrontal cortex.[27] Along with many other brain regions, the amygdala and prefrontal cortex also contain adrenal steroid receptors; however, the role of adrenal steroids, excitatory amino acids and other mediators has not yet been extensively studied in these brain regions. Nevertheless, in the amygdala, there is some evidence regarding mechanism, in that tissue plasminogen activator is required for acute stress not only to activate indices of structural plasticity but also to enhance anxiety. These effects occur in the medial and central amygdala, not in the basolateral amygdala, and the release of corticotropin-releasing hormone (CRH) acting via CRH1 receptors appears to be responsible.[25]

Acute stress induces spine synapses in the CA1 region of the hippocampus, and both acute and chronic stress also increases spine synapse formation in the amygdala, but chronic stress decreases it in the hippocampus. Moreover, chronic stress for 21 days or longer impairs hippocampal-dependent cognitive function and enhances amygdala-dependent unlearned fear and fear conditioning, which are consistent with the opposite effects of stress on hippocampal and amygdala structure. Chronic stress also increases aggression between animals living in the same cage, and this is likely to reflect another aspect of hyperactivity of the amygdala. Behavioral correlates of remodeling in the prefrontal cortex include impairment in attention set shifting, possibly reflecting structural remodeling in the medial prefrontal cortex.[28]

TRANSLATION TO THE HUMAN BRAIN

Much of the impetus for studying the effects of stress on the structure of the human brain has come from the animal studies summarized thus far. Although there is, so far, little evidence regarding the effects of ordinary life stressors on brain structure, there are indications from functional imaging of individuals undergoing ordinary stressors, such as counting backwards, that there are lasting changes in neural activity, and a 20 year history of chronic perceived stress has been linked to smaller hippocampal volume.[28] The study of depressive illness and anxiety disorders has also provided some insights. Life events are known to precipitate depressive illness in individuals with certain genetic predispositions. Moreover, brain regions such as the hippocampus, amygdala, and prefrontal cortex show altered patterns of activity on positron emission tomography and functional magnetic resonance imaging, and these structures decrease in volume with recurrent depression. The volume of the amygdala has been reported to increase in the first episode of depression, whereas hippocampal volume is not decreased.

It has been known for some time, for example from the classic work of the late Edward Sachar, that stress hormones, such as cortisol, are involved in psychopathology, reflecting emotional arousal and psychic disorganization rather than the specific disorder *per se*. As summarized in this chapter, it is now known that adrenocortical hormones enter the brain and produce a wide range of effects upon it.[2] In Cushing's disease, there are depressive symptoms that can be relieved by surgical correction of the hypercortisolemia. Both major depression and Cushing's disease are associated with chronic elevation of cortisol that results in gradual loss of minerals from bone and abdominal obesity. In major depressive illness, as well as in Cushing's disease, the duration of the illness and not the age of the subjects predicts a progressive reduction in volume of the hippocampus, determined by structural magnetic resonance imaging. Moreover, atrophy of the hippocampus has been reported in a variety of other anxiety-related disorders, such as PTSD and borderline personality disorder, suggesting that this is a common process reflecting chronic imbalance in the activity of adaptive systems, such as the hypothalamic–pituitary–adrenal axis, but also including endogenous neurotransmitters, such as glutamate.[28]

Another important factor in hippocampal volume and function is glucose regulation. Poor glucose regulation is associated with smaller hippocampal volume and poorer memory function in individuals in their sixties and seventies who have mild cognitive impairment (MCI), and both MCI and type 2 diabetes are recognized as risk factors for dementia.[29]

Inflammation is also a factor related to many disorders of modern life, and elevation of cytokines such as interleukin-6 has been linked to reduced hippocampal volume, as well as to poor sleep, and there is also more systemic inflammation associated with lower socioeconomic status, which, in turn, is also related to reduced white matter, which insulates brain circuits and keeps them working efficiently.[28] Last, but not least, chronic jet lag in humans is also linked to smaller temporal lobe volume, dysregulated cortisol and cognitive impairment, and circadian disruption in an animal model causes both obesity and neuronal remodeling in the prefrontal cortex, along with increased cognitive inflexibility.[2]

EARLY LIFE EXPERIENCES

Several studies have documented that early life adversity becomes embedded in brain architecture and in the body, with major implications for health throughout the lifespan. Chronic dysregulation of adaptive systems, referred to as allostatic overload, involving epigenetic mechanisms, shapes the developing brain and body's biological vulnerability to disease, as well as its responsiveness to potential interventions. Of particular relevance for children are experiences of abuse and neglect.[30] On the physiological level, adverse childhood experiences are associated with dysregulated cardiovascular, metabolic, and immunological function, which in turn feed into numerous systemic and behavioral disease conditions. Chaos in the home and inconsistent parenting impair brain development, and this can lead to disturbed cognitive function, unstable mood, low self-esteem, and numerous unhealthy activities, including overeating, substance abuse, sexual acting-out, and other forms of legal or illegal risk-taking.[31] For example, a 10 year history of a child growing up in a home where the mother has chronic depression is associated with the child having a larger amygdala.[32] Furthermore, low self-esteem and low locus of control are associated with a smaller hippocampus and a reduced ability to turn off the cortisol stress response.[33] Children growing up without adequate verbal stimulation and in chaotic, unstable home environments are vulnerable to impaired cognitive function, increased systemic inflammation, cardiovascular disease, substance abuse, antisocial behavior, and depression.[34,35]

Animal models can provide insights into brain mechanisms underlying the biological embedding of early experience.[36] Prenatal stress of the mother can impair features of normal brain development of the offspring; prolonged separation of infant from mother also impairs other aspects of brain development and function. Consistency, as well as quality of maternal care, plays a powerful role in lifelong patterns of reduced anxiety and efficient stress reactivity, as well as social and cognitive development.[37,38] Moreover, there are transgenerational effects that appear to be behaviorally transmitted by the mother to the female offspring and involve epigenetic modifications of glucocorticoid receptor genes, among other genes.[39] In contrast, inconsistent maternal care and maternal anxiety, for example, from food insecurity, produce anxiety in offspring and appear to contribute to metabolic syndrome and predisposition to diabetes, which itself has adverse effects on the brain. Thus, the behavioral and physiological consequences of early life abuse and neglect are profound, and the epigenetic concept of behavioral transmission of abuse and its effects on human brain function are being explored at the level of epigenetic regulation of gene expression.

So far, the important roles of the environment and experiences of individuals in the health outcomes have been emphasized, but genetic differences also play an important role. Different alleles of commonly occurring genes determine how individuals will respond to experiences. For example, work by many investigators, including Steve Suomi at the National Institutes of Health, and Avshalom Caspi and colleagues in a study in Dunedin, New Zealand, has shown that the short form of the serotonin transporter gene is associated with a number of conditions, such as alcoholism, and individuals who have this allele are more vulnerable to responding to stressful experiences by developing depressive illness.[40] This allelic variant results in lower expression of the transporter, compared with the long allele. In childhood, individuals with an allele of the monoamine oxidase A gene are more vulnerable to abuse in childhood and more likely themselves to become abusers and to show antisocial behaviors compared with individuals with another commonly occurring allele. Yet another example is the consequence of having the Val66Met allele of the BDNF gene on hippocampal volume, memory, and mood disorders. Nevertheless, in a positive, nurturing environment, as formulated by Suomi and by Bruce Ellis at the University of Arizona and Tom Boyce at the University of British Columbia, children with these same alleles may show better than average outcomes, which has led these alleles being called reactive alleles rather than bad genes (orchids versus dandelions for the less reactive allele).[41]

INTERVENTIONS THAT CHANGE THE BRAIN AND IMPROVE HEALTH

What can be done to remediate the effects of chronic stress, as well the biological embedding associated with early life adversity? Interventions may involve pharmaceutical, as well as behavioral, or top–down, interventions (i.e. interventions that involve integrated CNS activity, as opposed to pharmacological agents), including cognitive behavioral therapy, physical activity, and programs that promote social support, integration, meaning, and purpose in life. More targeted interventions for emotional and cognitive dysfunction may arise from fundamental studies of such developmental processes as the reversal of amblyopia and other conditions by releasing the brakes that retard structural and functional plasticity. It should be noted that many of these interventions that are intended to promote plasticity and slow decline with age, such as physical activity and positive social interactions that give meaning and purpose, are also useful for promoting positive health and eudamonia independently of any notable disorder and within the range of normal behavior and physiology.

Moreover, interventions towards changing physiology and brain function may be useful when adaptation to a particular environment has resulted in an individual who then chooses, or is forced, to adapt to a different (e.g. more or less threatening or nurturing) environment. Concerning biological embedding in neural architecture and the balance of neurochemical systems, in the case of adversity or shifting environments, one can hope at least to compensate for, even if one cannot reverse, those effects of early life adversity.[42] However, it is perhaps premature to draw that conclusion, since the ultimate limits of adult brain plasticity are still unknown, as will be discussed below.

A powerful top–down therapy (i.e. an activity, usually voluntary, involving activation of integrated nervous system activity, as opposed to pharmacological therapy, which has a more limited target) is regular physical activity, which has actions that improve prefrontal and parietal cortex blood flow and enhance executive function. Moreover, regular physical activity, consisting of walking for an hour a day, 5 out of 7 days a week, increases hippocampal volume in previously sedentary adults.[43] This finding complements work showing that fit individuals have larger hippocampal volumes than sedentary adults of the same age range. It is also well known that regular physical activity is an effective antidepressant and protects against cardiovascular disease, diabetes, and dementia. Intensive learning has also been shown to increase the volume of the human hippocampus.[44]

Social integration and support, and finding meaning and purpose in life, are known to be protective against allostatic load and dementia. Programs such as the Experience Corps, which promote these along with increased physical activity, have been shown to slow the decline of physical and mental health and to improve prefrontal cortical blood flow in a similar manner to regular physical activity.[45–47]

Depression and anxiety disorders are examples of a loss of resilience, in the sense that changes in brain circuitry and function, caused by the stressors that precipitate the disorder, become locked in a particular state and thus need external intervention. Indeed, prolonged depression is associated with shrinkage of the hippocampus and prefrontal cortex. While there appears to be no neuronal loss, there is evidence for glial cell loss and smaller neuronal cell nuclei, which is consistent with a shrinking of the dendritic tree after chronic stress. A few studies indicate that drug treatment may reverse the decreased hippocampal volume in unipolar and bipolar depression, but the possible influence of concurrent cognitive behavioral therapy is unclear.

Depression is more prevalent in individuals who have had adverse early life experiences. Neurotrophic factors, such as BDNF, appear to play a role in depression, and elevation of BDNF by diverse treatments, ranging from antidepressant drugs to regular physical activity, may be a key feature of successful treatment. Other potential applications include the reported ability of fluoxetine to enhance recovery from stroke.[48] Underlying this new view of treatment is that the drug, or some other intervention such as physical activity, is opening a window of opportunity that should also involve a targeted, positive behavioral intervention, such as cognitive behavioral therapy in the case of depression or intensive physiotherapy to counteract the effects of a stroke.[49] This is consistent with studies of animal models in which ocular dominance imbalance from early monocular deprivation can be reversed by patterned light exposure in adulthood while plasticity is being facilitated by fluoxetine, or by food restriction or intermittent elevation of glucocorticoid levels via drinking water. Investigations into the mechanisms underlying the re-establishment of a new window of plasticity are focusing on the balance between excitatory and inhibitory transmission and removing molecules that put the brakes on such plasticity.[50]

In this connection, it is important to reiterate that successful behavioral therapy, which is tailored to individual needs, can produce volumetric changes in both the prefrontal cortex in the case of chronic fatigue and the amygdala in the case of chronic anxiety.[51] This reinforces two important messages: first, that plasticity-facilitating treatments should be given within the framework of a positive behavioral or physical therapy intervention; and second, that negative experiences during the window may make matters worse. In that connection, it should be noted that BDNF also has the ability to promote pathology, as in seizures.

CONCLUSION

If being stressed out has such pervasive effects on the brain as well as the body, in what ways can an individual act to reduce the negative effects and enhance the ability of the body and brain to deal with stress with minimal consequences? At the same time, what programs and policies can business and policy makers in government enact to create an environment for reducing stress and promoting brain and body health, and thereby prevent or slow down disease? The answers may be simple and obvious, but are often difficult to achieve.

From the standpoint of the individual, a major goal should be to try to improve sleep quality and quantity, have good social support and a positive outlook on life, maintain a healthy diet, avoid smoking, and take regular moderate physical activity. Concerning physical activity, it is not necessary to become an extreme athlete, and any amount of moderate physical activity seems to help, as noted in relation to enlarging hippocampal volume.[51]

From the standpoint of policy, the goal should be to create incentives at home and in work situations, and to build community services and opportunities that encourage the development by the individual of beneficial lifestyle practices such as regular physical activity, adequate rest and time off, good diet, and good social interactions. In the case of early life adversity, policies and programs that promote better family and child health will save huge amounts of human suffering and money in the long run (see http://developingchild.harvard.edu).

As simple as the solutions seem to be, changing behavior and solving problems that cause stress at work and at home can be difficult and may require professional help on the personal level, or even a change of job or profession. Yet these are important goals because the prevention of later disease is very important for full enjoyment of life and also to reduce the financial burden on the individual and on society.

Nevertheless, many people often lack the proactive, long-term view of themselves, or feel that they must maintain a stressful lifestyle and, if they deal with these issues at all, they want to treat their problems with a pill. Sleeping pills, anxiolytics, β-blockers, and antidepressants are all drugs that are used to counteract some of the problems associated with being stressed out. Likewise, drugs that reduce oxidative stress or inflammation, block cholesterol synthesis or absorption, or treat insulin resistance or chronic pain can help to deal with the metabolic and neurological consequences of being stressed out. All are valuable to some degree, and yet each has its side effects and limitations that are based in part on the fact that all of the systems that are dysregulated in allostatic overload are also systems that interact with each other and perform normal functions when properly regulated. Because of the non-linearity of the systems of allostasis, the consequences of any drug treatment may be either to inhibit the beneficial effects of the system in question or to perturb other systems that interact with it in a direction that promotes an unwanted side effect. So, the best solution would seem to be not to rely solely on such medications, but to find ways to change one's lifestyle in a positive way.[2]

Being able to change lifestyle and associated behavior is not just an individual matter and may become easier with changes via another level of intervention, namely, policies in government and business. The Acheson Report from the UK in 1998 recognized that no public policy should be enacted without considering the implications for health of all citizens. Thus, basic education, housing, taxation, setting a minimum wage, and addressing regulations on occupational health and safety and environmental pollution are all likely to affect health via a myriad of mechanisms. Providing higher quality food and making it affordable and accessible in poor, as well as affluent, neighborhoods is necessary for people to eat more healthily, providing they also learn what types of food to eat. Likewise, making neighborhoods safer and more congenial and supportive, as shown in the Chicago neighborhoods study of Felton Earls, Steven Buka, and colleagues,[52] can improve opportunities for positive social interactions and increased recreational physical activity.

However, governmental policies are not the only way to reduce allostatic load. For example, businesses that encourage healthy lifestyle practices among their employees are likely to gain reduced health insurance costs and possibly a more loyal workforce. Above all, policy makers and business leaders need to be made aware of the broader issues of improving health and preventing disease, and the fact that they make economic sense, as well as being "the right thing to do".[2]

Finally, there are programs in existence that combine some of the key elements just described, namely, education, physical activity, and social support, with one other ingredient that is hard to quantify: namely, finding meaning and purpose in life. One such program is the Experience Corps, which takes elderly volunteers and trains them as teachers' assistants for younger children in neighborhood schools. This program not only improves the education of the children, but also benefits the elderly volunteers and improves their physical brain function and mental health.[46] We can only hope that politicians and business leaders will listen to and heed the advice of science, which often is reinforcing common sense, in helping to address the pervasive problems of stress in our world.

Acknowledgments

Research support for the author is from NIH grants MH41256, NS07080, and MH58911. The author also acknowledges organizations that have extended his understanding of stress into the human social domain, namely the MacArthur Foundation Research Network on Socioeconomic Status and Health (http://www.macses.ucsf.edu/) and the National Scientific Council on the Developing Child (see http://developingchild.harvard.edu/).

References

1. McEwen BS. Protective and damaging effects of stress mediators. *N Engl J Med*. 1998;338:171–179.
2. McEwen BS. Physiology and neurobiology of stress and adaptation: central role of the brain. *Physiol Rev*. 2007;87:873–904.
3. Selye H. The evolution of the stress concept. *Am Scient*. 1973;61: 692–699.
4. Lazarus RS, Folkman S, eds. *Stress, Appraisal and Coping*. New York: Springer; 1984.
5. Sterling P, Eyer J. Allostasis: a new paradigm to explain arousal pathology. In: Fisher S, Reason J, eds. *Handbook of Life Stress, Cognition and Health*. New York: John Wiley & Sons; 1988:629–649.
6. McEwen BS, Stellar E. Stress and the individual. Mechanisms leading to disease. *Arch Intern Med*. 1993;153:2093–2101.
7. McEwen BS. Protective and damaging effects of stress mediators: central role of the brain. *Dialogues Clin Neurosci*. 2006;8: 367–381.
8. Juster RP, McEwen BS, Lupien SJ. Allostatic load biomarkers of chronic stress and impact on health and cognition. *Neurosci Biobehav Rev*. 2010;35(1):2–16.
9. Dhabhar FS, Malarkey WB, Neri E, McEwen BS. Stress-induced redistribution of immune cells – from barracks to boulevards to battlefields: a tale of three hormones – Curt Richter Award Winner. *Psychoneuroendocrinology*. 2012;37(9):1345–1368.
10. Dhabhar FS, Saul AN, Daugherty C, Holmes TH, Bouley DM, Oberyszyn TM. Short-term stress enhances cellular immunity and increases early resistance to squamous cell carcinoma. *Brain Behav Immun*. 2010;24(1):127–137.
11. Jeanneteau F, Garabedian MJ, Chao MV. Activation of Trk neurotrophin receptors by glucocorticoids provides a neuroprotective effect. *Proc Natl Acad Sci USA*. 2008;105:4862–4867.
12. Spolidoro M, Baroncelli L, Putignano E, Maya-Vetencourt JF, Viegi A, Maffei L. Food restriction enhances visual cortex plasticity in adulthood. *Nat Commun*. 2011;2:320.
13. Hill MN, McEwen BS. Involvement of the endocannabinoid system in the neurobehavioural effects of stress and glucocorticoids. *Prog Neuropsychopharmacol Biol Psychiatry*. 2010;34(5):791–797.
14. Du J, Wang Y, Hunter R, et al. Dynamic regulation of mitochondrial function by glucocorticoids. *Proc Natl Acad Sci USA*. 2009;106: 3543–3548.
15. Du J, McEwen BS, Manji HK. Glucocorticoid receptors modulate mitochondrial function. *Commun Integr Biol*. 2009;2:1–3.
16. Rao RP, Anilkumar S, McEwen BS, Chattarji S. Glucocorticoids protect against the delayed behavioral and cellular effects of acute stress on the amygdala. *Biol Psychiatry*. 2012;72(6):466–475.
17. McEwen BS, Wingfield JC. The concept of allostasis in biology and biomedicine. *Horm Behav*. 2003;43:2–15.
18. Karatsoreos IN, McEwen BS. Psychobiological allostasis: resistance, resilience and vulnerability. *Trends Cogn Sci*. 2011;15(12): 576–584.
19. McEwen BS. Sleep deprivation as a neurobiologic and physiologic stressor: allostasis and allostatic load. *Metabolism*. 2006;55: S20–S23.
20. Engeda J, Mezuk B, Ratliff S, Ning Y. Association between duration and quality of sleep and the risk of pre-diabetes: evidence from NHANES. *Diabet Med*. 2013;30(6):676–680.
21. Spiegel K, Leproult R, Van Cauter E. Impact of sleep debt on metabolic and endocrine function. *Lancet*. 1999;354:1435–1439.
22. Karatsoreos IN, Bhagat S, Bloss EB, Morrison JH, McEwen BS. Disruption of circadian clocks has ramifications for metabolism, brain, and behavior. *Proc Natl Acad Sci USA*. 2011 Jan 25; 108(4):1657–1662.
23. McEwen BS. Stress and hippocampal plasticity. *Annu Rev Neurosci*. 1999;22:105–122.
24. Gould E. How widespread is adult neurogenesis in mammals? *Nat Rev Neurosci*. 2007;8:481–488.
25. McEwen BS. Stress, sex, and neural adaptation to a changing environment: mechanisms of neuronal remodeling. *Ann N Y Acad Sci*. 2010:1204. Suppl:E38–E59.
26. Magarinos AM, McEwen BS, Saboureau M, Pevet P. Rapid and reversible changes in intrahippocampal connectivity during the course of hibernation in European hamsters. *Proc Natl Acad Sci USA*. 2006;103:18775–18780.
27. McEwen BS, Morrison JH. The brain on stress: vulnerability and plasticity of the prefrontal cortex over the life course. *Neuron*. 2013;79(1):16–29.
28. McEwen BS, Gianaros PJ. Stress- and allostasis-induced brain plasticity. *Annu Rev Med*. 2011;62:431–445.
29. Gold SM, Dziobek I, Sweat V, et al. Hippocampal damage and memory impairments as possible early brain complications of type 2 diabetes. *Diabetologia*. 2007;50:711–719.
30. Anda RF, Butchart A, Felitti VJ, Brown DW. Building a framework for global surveillance of the public health implications of adverse childhood experiences. *Am J Prev Med*. 2010;39(1):93–98.
31. Evans GW, Gonnella C, Marcynyszyn LA, Gentile L, Salpekar N. The role of chaos in poverty and children's socioemotional adjustment. *Psychol Sci*. 2004;16:560–565.
32. Lupien SJ, Parent S, Evans AC, et al. Larger amygdala but no change in hippocampal volume in 10-year-old children exposed to maternal depressive symptomatology since birth. *Proc Natl Acad Sci USA*. 2011;108(34):14324–14329.
33. Pruessner JC, Baldwin MW, Dedovic K, et al. Self-esteem, locus of control, hippocampal volume, and cortisol regulation in young and old adulthood. *Neuroimage*. 2005;28:815–826.
34. Farah MJ, Shera DM, Savage JH, et al. Childhood poverty: specific associations with neurocognitive development. *Brain Res*. 2006;1110:166–174.
35. Danese A, McEwen BS. Adverse childhood experiences, allostasis, allostatic load, and age-related disease. *Physiol Behav*. 2012;106(1):29–39.
36. Lupien SJ, McEwen BS, Gunnar MR, Heim C. Effects of stress throughout the lifespan on the brain, behaviour and cognition. *Nat Rev Neurosci*. 2009;10:434–445.
37. Akers KG, Yang Z, DelVecchio DP, et al. Social competitiveness and plasticity of neuroendocrine function in old age: influence of neonatal novelty exposure and maternal care reliability. *PLoS ONE*. 2008;3(7):2840.
38. Parker KJ, Buckmaster CL, Sundlass K, Schatzberg AF, Lyons DM. Maternal mediation, stress inoculation, and the development of neuroendocrine stress resistance in primates. *Proc Natl Acad Sci USA*. 2006;103:3000–3005.

39. Francis D, Diorio J, Liu D, Meaney MJ. Nongenomic transmission across generations of maternal behavior and stress responses in the rat. *Science*. 1999;286:1155–1158.
40. Suomi SJ. Risk, resilience, and gene × environment interactions in rhesus monkeys. *Ann N Y Acad Sci*. 2006;1094:52–62.
41. Obradovic J, Bush NR, Stamperdahl J, Adler NE, Boyce WT. Biological sensitivity to context: the interactive effects of stress reactivity and family adversity on socioemotional behavior and school readiness. *Child Dev*. 2010;81(1):270–289.
42. McEwen BS. *Brain on stress: How the social environment gets under the skin. Proc Natl Acad Sci USA*. 2012;109(suppl 2):17180–17185.
43. Erickson KI, Voss MW, Prakash RS, et al. Exercise training increases size of hippocampus and improves memory. *Proc Natl Acad Sci USA*. 2011;108(7):3017–3022.
44. Draganski B, Gaser C, Kempermann G, et al. Temporal and spatial dynamics of brain structure changes during extensive learning. *J Neurosci*. 2006;26(23):6314–6317.
45. Boyle PA, Buchman AS, Barnes LL, Bennett DA. Effect of a purpose in life on risk of incident Alzheimer disease and mild cognitive impairment in community-dwelling older persons. *Arch Gen Psychiatry*. 2010;67(3):304–310.
46. Carlson MC, Erickson KI, Kramer AF, et al. Evidence for neurocognitive plasticity in at-risk older adults: the experience corps program. *J Gerontol A Biol Sci Med Sci*. 2009;64(12):1275–1278.
47. Seeman T, Epel E, Gruenewald T, Karlamangla A, McEwen BS. Socio-economic differentials in peripheral biology: cumulative allostatic load. *Ann N Y Acad Sci*. 2010;1186:223–239.
48. Chollet F, Tardy J, Albucher JF, et al. Fluoxetine for motor recovery after acute ischaemic stroke (FLAME): a randomised placebo-controlled trial. *Lancet Neurol*. 2011;10(2):123–130.
49. Castren E, Rantamaki T. The role of BDNF and its receptors in depression and antidepressant drug action: reactivation of developmental plasticity. *Dev Neurobiol*. 2010;70(5):289–297.
50. Bavelier D, Levi DM, Li RW, Dan Y, Hensch TK. Removing brakes on adult brain plasticity: from molecular to behavioral interventions. *J Neurosci*. 2010;30(45):14964–14971.
51. Davidson RJ, McEwen BS. Social influences on neuroplasticity: stress and interventions to promote well-being. *Nat Neurosci*. 2012;15(5):689–695.
52. Lochner KA, Kawachi I, Brennan RT, Buka SL. Social capital and neighborhood mortality rates in Chicago. *Soc Sc Med*. 2003;56:1797–1805.

CHAPTER

35

Addictions

Eduardo R. Butelman, Roberto Picetti, Brian Reed, Vadim Yuferov, Mary Jeanne Kreek

Laboratory on the Biology of Addictive Diseases, The Rockefeller University, New York, USA

OUTLINE

INTRODUCTION

The addictions are considered chronic relapsing brain diseases, with behavioral manifestations and wide-ranging morbidity.[1,2] Exposure to cocaine or amphetamines (psychostimulants), μ(mu)-opioid receptor (MOP-r) agonists (such as heroin and abused prescription opioids), alcohol, and nicotine can lead to addictions, which have the commonality that the patient continues their use in the face of medical complications (which can be severe and even fatal), social and financial problems, and legal issues. The cumulative public health, societal, and economic costs of the addictions are massive, and are compounded by comorbidity with other psychiatric disorders, primarily depression and anxiety. The onset of addictive diseases is often, but not exclusively, detected in adolescence or early adulthood, and this has maintained an active interest in age-dependent mechanisms in addiction neurobiology.

Neurobiology of Brain Disorders
http://dx.doi.org/10.1016/B978-0-12-398270-4.00035-5

Addiction Pharmacotherapy

A crucial approach to ameliorate the clinical impact of the addictions is pharmacotherapy, which targets neurobiological mechanisms involved in the trajectory of these diseases. The first successful development of a treatment for an addictive disease was methadone, as demonstrated in studies at the Rockefeller University starting in 1964, in patients addicted to the MOP-r agonist heroin.[3] The research team demonstrated that methadone, an orally bioavailable and longer lasting MOP-r agonist, prevented withdrawal symptoms and craving in heroin addicts.[3] They went on to demonstrate that, at sufficient doses, methadone attenuated heroin-induced subjective effects when heroin was administered while methadone was on board, and prevented relapse to heroin. At appropriate dose levels of methadone, 60–75% of patients enrolled in well-managed maintenance programs will successfully abstain from heroin for more than a year.[1,4] Methadone maintenance treatment normalizes some of the profound neuroendocrine changes observed in heroin-addicted patients.[1] Methadone has thus effectively served as a pharmacotherapeutic treatment for heroin addiction for half a century. A second, more recently introduced, maintenance treatment for heroin addiction is buprenorphine (see "μ-Opioid Receptor Agonist Addiction: Heroin and Prescription Opioids", below).

Although maintenance therapy has been successful in promoting adherence in the case of heroin addiction, a similar strategy for nicotine has proven to be less effective, with nicotine replacement therapy (NRT) promoting smoking cessation of over 1 year in only approximately 20% of patients; therapy duration was typically much shorter than 1 year in most studies on the effectiveness of NRT. Notably, there has been no successfully developed pharmacotherapeutic for psychostimulants. In the case of alcohol, disulfiram (which acts as a counterabuse agent, stimulating nausea when alcohol is taken on board) and naltrexone (an opioid receptor antagonist) have both been approved, as well as acamprosate. Novel approaches are being studied, based on increased knowledge of the neurobiology of alcoholism (see "Alcoholism", below).

Treatments for addiction can also involve various forms of psychotherapy, cognitive behavioral therapy, contingency management, and social support and rehabilitation. Perhaps the most well known are the self-help treatment groups, including Alcoholics Anonymous. Such approaches can be successful in some cases, with success rates between 10 and 30% depending on the class of substance and method of treatment. Such non-pharmacological approaches can be used in combination with pharmacotherapeutic treatment. Greater translational understanding in the behavioral neurobiology of the addictions may also eventually help to further improve behaviorally based therapeutic approaches.

Importance of Opioid Receptor and Neuropeptide Systems in Addiction Neurobiology

μ-Opioid receptors (MOP-r), κ-opioid receptors (KOP-r), δ-opioid receptors (DOP-r), and cognate opioid neuropeptide systems [pro-opiomelanocortin (*POMC*)/β-endorphin, dynorphins, and enkephalins] are involved in several aspects of addiction neurobiology. MOP-r are the main pharmacodynamic target of abused compounds such as heroin and prescription opioids.[5] However, more broadly, endogenous opioid receptor/neuropeptide systems modulate endogenous reward processes, have important regulatory roles in addiction-like trajectory, and undergo plasticity based on stage of exposure, withdrawal, or relapse. As a major example, this laboratory and others found that repeated exposure to cocaine resulted in an upregulation in prodynorphin (*Pdyn*) messenger RNA (mRNA).[6] The dynorphins are the main endogenous KOP-r agonists, and play important countermodulatory roles to dopaminergic neurotransmission, in brain areas involved in reward and addictive states.[6]

Much of the current state of knowledge in addiction neurobiology has been aided by the cloning of the three opioid receptors (DOP-r, MOP-r, and KOP-r). Among the main translational advances following from receptor cloning is the finding that a specific functional single-nucleotide polymorphism (SNP) in the gene coding for MOP-r (*OPRM1 A118G*) is clinically relevant in vulnerability to heroin addiction, and in addiction pharmacotherapy (see "The Genetics of Addiction", below).[7]

Public Health and Prevention

In terms of public health, nicotine addiction, with concomitant chronic exposure to tobacco-related chemicals, results in severe medical problems, including pulmonary disorders, cardiovascular disorders, and various cancers, especially lung cancer. Chronic, long-term alcohol exposure results in disorders of the liver and the nervous system, and is also responsible for fetal alcohol syndrome due to abuse by pregnant women (see "Alcoholism", below). Furthermore, alcohol intoxication, which is often accompanied by both an increase in risky behaviors and a decrease in motor coordination and attentional focus, is responsible for considerable morbidity and mortality, especially in motor vehicle and other accidents. In the case of MOP-r agonists such as heroin, a major public health consequence has been and remains the spread of bloodborne diseases such as HIV/AIDS and hepatitis C, through the sharing of needles and other risk behaviors. The respiratory depressant effects of MOP-r agonists, such as heroin and oxycodone, are the direct cause of many emergency visits and overdose deaths. Similarly, for psychostimulants such as cocaine and amphetamines (including methamphetamine), overdose leads to considerable morbidity and mortality. Additional medical complications related to psychostimulants are related to route of administration, with snorting leading to deterioration of the mucosal membrane, smoking to pulmonary problems, and injections to contagion with infectious diseases.

As with other medical conditions, the strategies to combat morbidity and mortality are based on disease prevention and treatment. Although there are some commonalities in the progression and characteristics of the addictions to the different substances, actual strategies for combating the addictions differ, given the differences in the mechanisms of biological action and resulting long-term biological changes underlying each specific addiction.

Historically, the use of nicotine-containing products (tobacco) and alcohol has been socially acceptable in many countries (with exceptions over time and place, and age restrictions), resulting in significant exposure by large numbers of people. Recognition of the public health costs of addiction to tobacco products and alcohol has led to increasing efforts to prevent addiction, largely through measures such as education, restrictions on access by young people, and taxation to increase costs and thus decrease consumption.

MOP-r agonists (morphine and other prescription opioids) continue to have a major place in clinical analgesia.[8] Thus, heroin (diacetyl-morphine, acting primarily as a prodrug for active metabolites such as morphine) is not used medically, other than in specific exceptions (e.g. as diamorphine in the UK). Prevention of heroin addiction has centered on interdiction and control efforts, aimed at limiting supply, and on education. Such measures, along with the scheduled status of heroin, have the goal of dissuading initial use of heroin in the population. In the case of prescription opioids, such as oxycodone and hydrocodone, prevention centers on regulation of prescription availability and on education. The considerable increase in prescription opioid addiction and overdoses underscores the need for greater understanding of addiction neurobiology in the trajectory of prescription opioid addiction, leading to improved interventions.

Psychostimulants such as cocaine, amphetamine, and methamphetamine are scheduled, because of their potential for abuse, although there are specific medical uses for certain compounds in this class. Prevention of addiction is largely based on limiting supply and availability, as well as education. There are no currently approved pharmacotherapies for psychostimulant addiction (see "Cocaine and Amphetamines; Pharmacotherapeutics", below).

Although the prevention strategies do result in some success, there continues to be a fairly stable segment of the population that expose themselves to these substances and may develop addictive diseases. The widespread availability of tobacco and alcohol, and the ability to obtain prescription opioids may not change dramatically in the foreseeable future.

Given the widespread and damaging nature of the addictive diseases, touched upon above, the development of additional pharmacotherapeutic agents is of paramount importance for public health. Improving our understanding of the neurobiology of the addictions may provide more flexibility and greater breadth in treatment approaches. Therefore, the following sections summarize the large literature on the trajectory of addictions, the methodologies used to study addiction neurobiology, and the neurobiology of addictions to opioids, cocaine, alcohol, and nicotine, including an understanding of the genetic basis of propensity to addiction.

TRAJECTORY OF ADDICTIONS AND UNDERLYING NEUROBIOLOGY

As mentioned above, addictions are chronic relapsing diseases, with plasticity at behavioral levels and at underlying neurobiological levels. The trajectory of addictions, in humans and in animal models, can be operationally defined in different ways, but can be conceptualized as having several stages, commencing with experimentation, escalation of self-exposure (reaching regular high-dose exposure), and in specific cases withdrawal and abstinence (of varying duration), followed by short- or long-term relapse (with re-escalation of self-exposure).[9] Studies have also found that short-term versus chronic administration of drugs of abuse, such as cocaine, results in differential and progressive neurobiological changes, including prodynorphin mRNA expression in more dorsal striatal areas, terminal fields of the nigrostriatal system.[10] Such progressive neurobiological changes may underlie in part the transition from early abuse to addiction-like behaviors, with associated dose escalation and high-dose regular exposure.[6,11]

The behavioral and neurobiological hallmarks during these stages in trajectory are not identical among major drugs of abuse such as short-acting MOP-r agonists (e.g. heroin or abused prescription opioids), cocaine and psychostimulants (including methamphetamine), and alcohol. These differences arise from the specific pattern of direct neuropharmacological effects of each drug, followed by differential longer term changes in mRNA and peptide, epigenetic, and neural network and learning events (see relevant sections below).

Patterns of self-exposure in addictions also differ among drugs, and can be studied with appropriate animal models. Thus, self-exposure to MOP-r agonists (after the initial experimentation stage) results in a relatively stable pattern of multiple daily exposure, with highly aversive neuroendocrine, autonomic, and subjective withdrawal signs emerging if such regular use is withheld.[12] Cocaine addiction tends to be characterized by repeated "binge" administration in rapid succession, with escalation in overall exposure over time.[6,11,13] Cocaine withdrawal is not characterized by the same somatic and autonomic signs as MOP-r agonist withdrawal, but results in anhedonia, dysphoria, and anxiety-like behaviors.[14] Alcohol abuse or addiction can also have binge-like episodes and eventually regular high-dose exposure. The alcohol withdrawal syndrome includes anhedonia, dysphoria, and anxiety-like behaviors,[15] and can be associated with high morbidity, including seizures that require medical management. The role of specific neurobiological consequences in cycles of withdrawal and relapse is an area of active research.

Relapse Prevention: A Major Challenge

As stated above, addictions tend to be relapsing diseases, with the patient's health being placed at risk by consecutive cycles of re-exposure. Using appropriate neuropharmacological agents to give short-term relief from withdrawal signs does not provide a long-term resolution of the relapsing nature of addictions. Studies in humans and in animal models have identified several neurobiological substrates that can mediate relapse-like behaviors, particularly stress-responsive systems (e.g. α_2-adrenergic receptors and corticotropin-releasing hormone), the KOP-r system, and vasopressin receptor systems.[16,17] Relapse (and relapse-like behavior in animal models; see below) can occur at prolonged intervals, well after classic withdrawal signs are observed. Several types of stimulus can precipitate or exacerbate relapse-like behavior. These include stressful events, with activation of stress-responsive systems in the brain and hypothalamic–pituitary–adrenal (HPA) axis, activation of KOP-r systems, and stimuli associated with prior drug-taking or withdrawal.

ANIMAL BEHAVIORAL MODELS TO STUDY ADDICTIONS

Animal models are invaluable tools to study addictions. Research in humans is limited by practical and ethical issues, and by the difficulty in assessing whether behavioral and neurobiological (e.g. neuroimaging) differences between addicted patients and healthy volunteers are pre-existing causes or the consequences of drug use. However, in animals, conditions leading to drug administration can be controlled, and relationships between a behavior (e.g. drug self-administration) and its causes or antecedents (e.g. impulsivity) and consequences (e.g. neurobiological changes) can be more directly investigated.

Several animal models are available to study specific aspects of addiction neurobiology, including trajectory. Scientists select appropriate models depending on the aspects and conditions under study. For instance, drug administrations can be passive, where the investigator administers the drug to the animal, acutely, chronically, or according to a binge pattern[18]; or active, where the animal self-administers the drug. Passive and active administration of drugs can differ to a degree in terms of their neurobiological effects, for example owing to associative changes.

A variety of behavioral tests is available to allow the dissection of different aspects of addiction. A few examples will be described in brief. One common model to study anxiety-like effects observed during drug exposure or withdrawal is the elevated plus maze (EPM). The apparatus for the EPM is a cross-shaped maze with two arms protected by high walls (closed arms) and two arms with very low walls (open arms). The maze is elevated above the floor. An animal that spends a relatively long amount of time in the closed arms is considered to show anxiety-like symptoms. For example, rats that receive a cocaine injection spend less time on the open arms of the EPM. A test that is used to quantify both locomotor and anxiety-like behaviors is the open field test. This apparatus is constituted by an open area, which rats or mice can freely explore. The three main parameters studied with this test are the distance traveled, which is a measure of the effect on locomotion, the time spent in the central part of the apparatus and the vertical behavior (i.e. rearing), which are measures of anxiety-like behavior. As an example, the injection of a psychostimulant increases the distance traveled in the open field (hyperlocomotion).

When studying addictions, it is important to assess rewarding and reinforcing properties of the addictive substance at specific stages in the trajectory. The terms "reward" and "reinforcement" are often used as synonyms. "Reward" is a more general term, used when discussing stimuli and neurobiological systems related to hedonic states (e.g. a subjectively pleasurable hedonic state, such as euphoria, can be postulated to be also a reward). A "reinforcer" is a more operationally defined term, which is useful in some experimental or clinical settings. Thus, a reinforcer is an event (e.g. administration of a drug) that when associated with a behavior or response increases the probability for a behavior or response to occur in the future, in the same organism. Subjects (e.g. experimental animals) will exhibit behaviors that increase the probability of obtaining positive reinforcers (e.g. food, or drugs such as MOP-r agonists or cocaine). Conversely, subjects will exhibit behaviors that will increase the probability of escape or avoidance of negative reinforcers (e.g. escaping anxiety or withdrawal states). Hence, broadly stated, classic rewarding/addictive drugs, such as cocaine or heroin, can act as positive reinforcers.

Behavioral models suited to study positive reinforcement or more generally rewarding events are conditioned place preference (CPP), self-administration, and intracranial self-stimulation (ICSS) assays. Negative reinforcement or aversion can be studied with conditioned place aversion (CPA), with ICSS, and with appropriately designed self-administration assays (e.g. responding to avoid or escape a stimulus).

CPP uses classical or Pavlovian conditioning, consisting of the temporal association between an unconditioned stimulus (US) and a conditioned stimulus (CS). A US, such as palatable foods, drugs of abuse, or aversive stimuli such as shock, results in certain approach or escape behaviors (or internal neurobiological events) even in the absence of training, called the unconditioned response (UR). A CS is a stimulus that when temporally paired with a US results in a conditioned behavioral or neurobiological response, called the conditioned response (CR). A classic example is represented by Pavlov's experiments with dogs.[19] These dogs salivated (UR) in the presence of food (US) and were trained to associate the presence of food with the sound of a bell (CS), so that eventually they salivated at the sound of the bell even without food. In the most basic version of CPP assays, the rodent is placed in an apparatus with two distinct chambers (e.g. with different textures and shading patterns on the floor). Subsequently, drug injections (e.g. the postulated rewarding drug, or US) are paired with the subjects confined to one chamber, whereas vehicle injections are paired with the subjects confined to the other chamber. After pairing, the chamber environment is associated with the rewarding drug (US), and thus acquires CS properties. In the test session, the subject is free to explore both chambers, in a drug-free state. If the subject spends more time in the drug-paired chamber, this is considered to be the expression of CPP, consistent with the conclusion that the drug US has rewarding effects. Conversely, withdrawal-induced aversive effects can be observed as CPA. In this case, the animal's drug withdrawal state is paired with one chamber. When the animal is free to explore the whole apparatus, the aversive effect of the withdrawal state (US) is discerned by the subject spending less time in the chamber paired with the "withdrawal" US.

The formation of a classical US–CS pairing does not depend on the subject exhibiting a particular behavior; the expression of the US–CS pairing is discerned by changes in behavior during the test (e.g. a preference for the drug-paired chamber). By contrast, operant or instrumental conditioning creates an association between a behavior (e.g. the animal's response) and the delivery of a consequence (e.g. a drug). Intravenous or oral self-administration and ICSS are both operant conditioning techniques.

Intravenous self-administration (IVSA) experiments are performed in chambers or boxes appropriate to the type of animal (e.g. rodent or non-human primate), containing levers, bars, or holes, speakers, light, and food magazine, and connected to computer-controlled equipment for responses and reinforcements.[20] Typically, the experimental animal has to perform a task (e.g. pressing a lever or poking their nose through a hole) to receive the reinforcer (e.g. a drug injection at a particular unit dose). The drug is delivered directly into the vein where a catheter has been surgically inserted previously. There is a strong relationship between the abuse potential of a drug (e.g. MOP-r agonists or psychostimulants) and the readiness with which it is self-administered. Several parameters of IVSA can be modified to model different aspects of drug-taking and abuse, the main ones being the unit dose of the drug delivered, the schedule of reinforcement, and the length of an IVSA session. The relationship between the unit dose of a reinforcing drug and the number of unit doses received follows a typical inverted U-shaped curve. Thus, the rate of drug self-administration behavior increases as the unit dose increases, up to a certain unit dose, and then it decreases as the unit dose is increased further. The ascending limb of the inverted U-shaped curve is thought to reflect more directly the reinforcing properties of a drug, whereas the descending limb may be the result of the non-specific disruption of self-administration due to various factors, such as toxicity, accumulation of the drug, satiety, or disruptive stereotypies.[21]

Schedules of reinforcement are manipulated to determine the time and the frequency of the reinforcement. These include: the *fixed ratio schedule*, where each reinforcer is delivered after a specific number of responses has been performed; the *variable ratio schedule*, where each reinforcer is delivered after an average number of responses has been performed; the *fixed interval schedule*, where each reinforcer is delivered after a specific amount of time has passed; the *variable interval schedule*, where each reinforcer is delivered after an average amount of time has passed; and the *progressive ratio schedule*, where the number of responses necessary to obtain a reinforcer is increased according to specific equations after a reinforcer has been delivered; eventually, the number of responses required to obtain a reinforcer will be so high that the animal will not complete the schedule. The last ratio completed in a progressive ration session is defined as the "breaking point" and can be considered as a biomarker for the maximal amount of work that a subject will expend for a particular drug reward.

Schedules of reinforcement for particular IVSA experiments can be manipulated and designed based on the goals, focus, and hypothesis of specific experiments, affording great insight into particular behavioral and underlying neurobiological changes in addiction models. The length of IVSA sessions also affects the drug intake pattern. Short sessions (e.g. up to 2 hours) typically result in a stable number of reinforcers taken over days. However, when the duration of IVSA sessions is prolonged, the number of reinforcers taken is not stable, but escalates across days.[13,22] This across-session escalation is thought to model certain stages in the addiction trajectory (see "Trajectory of Addictions and Underlying Neurobiology", above).

Typically, the unit dose of drug is kept constant throughout the experiment, so the animal escalates only the number of reinforcers obtained. However, when rats of vulnerable strains can select between different unit doses of cocaine[23] or heroin[24] in daily 18-hour long sessions, they escalate the unit dose as well. This self-selected unit dose escalation can also model other aspects of addiction trajectory, such as the transition between initial experimentation and addiction.

ICSS is another operant procedure used in rats and mice; its main conceptual difference from IVSA is that instead of an intravenous catheter, animals are implanted with intracranial electrodes in relevant brain areas (e.g. the medial forebrain bundle) and can self-administer electrical stimulations. Thus, subjects may directly stimulate the selected putative reinforcement-relevant circuits in the brain. Some advantages of ICSS include the possibility to target discrete brain regions, the rapid acquisition of this behavior due to the powerful reinforcing effects of ICSS, and the relative lack of satiating effects secondary to extensive exposure.[25] Among all the available ICSS procedures, two main procedures are used: the rate-frequency curve-shift procedure and the discrete-trial current-intensity threshold procedure (for a detailed description of these procedures, see Vlachou and Markou[25]). All ICSS procedures measure the reward threshold, which represents a quantitative assessment of the stimulation efficacy. The elevation of the threshold indicates a decrease in the reinforcing effect of the stimulation because higher current intensities or frequencies are required for the animal to perceive stimulation as rewarding. In contrast, lowering the threshold indicates an increase in the reinforcing effects of the stimulation because lower current intensities or frequencies are required to make the stimulation rewarding. For example, the administration of cocaine or MOP-r agonists lowers the ICSS threshold, whereas during withdrawal, the threshold is increased (mimicking dysphoria or anhedonia).

RESEARCH TECHNIQUES IN HUMANS

Several techniques can be used to study addictions as brain dysfunctions in humans. Information can thus be gleaned on pharmacodynamics of drugs of abuse or pharmacotherapeutic agents, or on the behavioral and neurobiological processes underlying addiction, after appropriate approval has been provided by the institutional review board, which evaluates all aspects of such studies in humans. Many of the techniques used in humans have methodological similarities to the *in vivo* techniques in experimental rodents and non-human primates, and are a crucial step in the bidirectional (bench to bedside and *vice versa*) translation of neurobiological research.[26]

Behavioral Pharmacology Experiments

Quantitative, laboratory-based pharmacology studies can be used to discern the effects of different drugs of abuse, and the effects of pharmacotherapeutic agents on these drugs. Behavioral effects of drugs of abuse, pharmacotherapies, and history of abuse or addiction can be quantified with appropriate scales[27] or with operant procedures (see "Animal Behavioral Models to Study Addictions", above). Notably, behavioral endpoints in humans can be of critical value in the interpretation and design of neuroimaging studies, using positron emission tomography (PET) or functional magnetic resonance imaging (fMRI), for example (see "Neuroimaging Studies", below).[28]

Neuroendocrinology Experiments

Addictive drugs have direct effects on major neuroendocrine systems, and also cause long-term adaptations in neuroendocrine control that can be observed after chronic exposure or withdrawal. Neuroendocrine experiments (and novel biomarker studies with cytokines or other circulating signal molecules) are experimentally attractive in humans, owing to their quantitative and relatively non-invasive nature. Neuroendocrine biomarkers, especially those related to the stress-responsive HPA axis, such as adrenocorticotropic hormone and cortisol, can be used to monitor aspects of addiction neurobiology at specific stages in a patient's trajectory[1] or genotype-dependent effects.[6] Other biomarkers, including prolactin, can also be used to monitor *in vivo* pharmacodynamics (e.g. MOP-r or KOP-r ligands) in healthy volunteers or patients.[29]

Neuroimaging Studies

Neuroimaging techniques (e.g. PET and fMRI; see also Chapter 3) in healthy human volunteers and patients with addictive diseases at different stages in their trajectory are now part of the armamentarium in the field of addiction neurobiology. Dopamine D_2-like receptors can be imaged with [^{11}C]raclopride and are of great importance, given the involvement of the dopaminergic system in reward and addiction processes. PET ligands are also available for MOP-r, and have been used to define levels of MOP-r occupancy for specific pharmacotherapeutic treatment, such as methadone or buprenorphine,[30] and of genotype-dependent responses in MOP-r occupancy.[31] PET studies have also shown that MOP-r function is relevant to clinical status in cocaine addiction, including craving and relapse.[28] This is consistent with animal studies indicating that the MOP-r system is involved in the downstream adaptations occurring in cocaine addiction models. The development of novel PET ligands for human use, for other molecular targets, will expand research into other neurobiological systems involved in addictions to specific agents (e.g. cocaine, MOP-r agonists, and alcohol). fMRI has also been used to delineate differences in brain function in patients at specific stages in their addiction trajectory, for different drugs of abuse, and these may be correlated with subjective or behavioral variables.

Postmortem Human Brain Studies

Although it is possible to model aspects of brain pathophysiology of addictions in animals, the examination of the human brain remains invaluable for these complex disorders.[32] Postmortem human brain studies are critical to the understanding of addiction neurobiology from a genetic, molecular, cellular, and neurochemical standpoint. The direct study of brain tissue allows detection of disease-related alterations at the cellular, synaptic, and molecular levels. Thus, studies of the postmortem human brain represent a critical and complementary approach to *in vivo* (e.g. human neuroimaging) studies, as well as a translational bridge between findings in humans and in animal models.

High-throughput microarray technology enables investigators to simultaneously examine changes in gene expression across hundreds of genes or even entire genomes.[33] Such studies are likely to increase our understanding of the consequences and complexities of addiction. For example, a significant decrease in the expression of numerous genes encoding proteins involved in presynaptic release of neurotransmitters was found in heroin abusers, a finding not seen in a cocaine-abusing cohort. These microarray studies showed differential alterations of prodynorphin (*PDYN*) expression in the nucleus accumbens; it increased in cocaine abusers and decreased in heroin abusers. Thus, the profiles of nucleus accumbens gene expression associated with chronic cocaine or heroin addiction are highly distinguishable, despite common downstream effects of these drugs on dopamine neurotransmission in this brain region. Other studies from postmortem brain (e.g. correlating neurochemical analysis with genotyping and clinical status) have led to novel mechanistic hypotheses that may be tested *in vitro* or in appropriate animal models.

Several studies have demonstrated changes in brain pathways subserving emotions and cognition induced by abused drugs and alcohol. Such changes may underlie structural reconfiguration of synaptic connections with exposure to cocaine. In particular, the upregulation of the *RECK* gene, an inhibitor of matrix metalloproteinases, and downregulation of the axon guidance gene ephrin type B receptor-4 (*EPHB4*) were observed in the hippocampi of cocaine abusers.[34] Extracellular matrix remodeling in the hippocampus may be a persisting effect of chronic abuse that contributes to the compulsive and relapsing nature of cocaine addiction.

The quality, and therefore the information value, of a postmortem human study depends on the extent to which it uses well-characterized brain specimens, a well-constructed experimental design, and appropriate controls for potential confounds. The validity of postmortem human brain research in the addictions also relies on accurate clinical and psychopathological diagnosis. A prospective recruitment approach, when feasible from a practical and ethical perspective, may be optimal in order to obtain information in a systematic and standardized manner and to minimize difficulties involved with retrospective selection of subject groups in this setting.

BASIC NEUROBIOLOGY OF SELECTED ADDICTIONS

Cocaine and Amphetamines

Cocaine is an alkaloid derived from the leaves of the coca plant and is available as either a hydrochloride salt or a free base. The salt form is water-soluble and can be self-administered intranasally or dissolved, and injected intravenously. The free base form is crystalline and is insoluble in water. This form, sometimes called "crack", needs to be volatized (smoked) to be consumed. The route of administration affects the speed of delivery to the brain and its rewarding effects. Intravenously administered or smoked cocaine produces a fast delivery of the drug to the brain and stronger rewarding effects than intranasal administration. Cocaine is then metabolized by liver and plasma cholinesterases, producing inactive water-soluble metabolites.

The main effect of cocaine is blocking of the dopamine reuptake transporter (DAT), although it also acts at norepinephrine and serotonin transporters (NET and SERT). Dopamine plays a fundamental role in reward, mood, and psychosis in several psychiatric and neurological disorders. Two main dopaminergic neural pathways are involved in the mechanisms responsible for reward, reinforcement, and addiction: the mesocorticolimbic pathway, projecting from the ventral tegmental area to the medial prefrontal cortex and the nucleus accumbens (or ventral striatum); and the nigrostriatal pathway, projecting from the substantia nigra pars compacta to the dorsal striatum. When dopamine is released in the synaptic cleft by dopaminergic neurons (i.e. neurons synthesizing dopamine), it is then removed by three mechanisms: dopamine is reabsorbed by the dopaminergic neurons through the DATs; it is metabolized by the monoamine oxidases; and it diffuses outside the synaptic cleft. Of these three, DAT plays the major role in removing dopamine. By blocking DAT, cocaine enhances synaptic dopamine levels. In addition to blocking DAT, cocaine blocks serotonin and norepinephrine transporters. Serotonin is another monoamine neurotransmitter that, together with dopamine, modulates certain forms of impulsivity. Norepinephrine is a monoamine neurotransmitter synthesized from dopamine, and modulates stress, arousal, and fight-or-flight (e.g. sympathetic) responses, of relevance to addiction neurobiology.

Experiments in knockout mice lacking DAT and SERT showed that DAT single knockout mice self-administer cocaine and exhibit cocaine-induced CPP. However, double knockout (DAT/SERT) mice show no cocaine-induced CPP. Experiments in rhesus monkeys revealed that blockade of DAT and SERT produces a stronger decrease in cocaine IVSA than the blockade of DAT alone. These results indicate that the dual manipulation of DAT and SERT may be an important approach in the development of treatments against cocaine abuse.

Amphetamines (including methamphetamine) increase extracellular levels of dopamine through several mechanisms. They reduce the amount of DAT present at the cell surface and reverse DAT transport function; that is, the transporter exchanges amphetamine molecules for dopamine molecules, which are effluxed from the cell into the extracellular space. In the cytoplasm, amphetamine stimulates the tyrosine hydroxylase to synthesize more dopamine and inhibits the monoamine hydroxylase. Moreover, in a normal situation, after dopamine undergoes reuptake by DAT and is released into the cytoplasm, it is transported into vesicles by the vesicular monoamine transporter (VMAT); amphetamine inhibits and/or reverses VMAT function.

Long-term cocaine use in humans has neurological and behavioral consequences, such as neurocognitive impairment, psychosis (e.g. paranoia and hallucinations), aggression, depression, insomnia, tremors, anorexia, repetitive movements (stereotypy), muscle rigidity, convulsions, anxiety, cardiovascular effects, and sudden death.[35]

Today, the main treatments for psychostimulant addiction are behavioral and psychosocial, and include cognitive behavioral therapy, contingency management, and community reinforcement, alone or in combination. Cognitive behavioral therapy aims at teaching how to change maladaptive behaviors and improve coping skills. Contingency management uses operant training methods, with the aim of achieving progressively desirable goals (e.g. abstinence). In this case, reinforcements, such as financial rewards or vouchers for goods or services, are given to promote or reach drug abstinence. With the community reinforcement approach, these reinforcements come from the social support of a group or the patient's family, or both.

Pharmacotherapeutics

Despite numerous studies, no pharmacological treatment for addiction to cocaine or other psychostimulants, including methamphetamine, has been approved by the US Food and Drug Administration (FDA). Among the modalities under study are vaccines that bind to the abused molecule (e.g. cocaine), thus decreasing pharmacodynamic effects after exposure. Preclinical and clinical research is focusing on compounds that act on dopamine, norepinephrine, and glutamate, such as modafinil, disulfiram, and methylphenidate, and also on opioid receptor antagonists, such as naltrexone and nalmefene.

μ-Opioid Receptor Agonist Addiction: Heroin and Prescription Opioids

MOP-r agonist-containing preparations such as opium (of which morphine is the main active constituent) have been used for millennia for their pharmacological properties, especially analgesia. Such formulations can also be rewarding and addictive when administered by ingestion, inhalation, or intranasally. MOP-r agonists continue to be of critical importance as analgesics for moderate to severe pain, and their appropriate use in the relief of such pain is highly supported and necessary. In recent years, abuse and addiction to prescription opioids has become a serious concern, in the USA and elsewhere. This phenomenon may include a segment of patients prescribed MOP-r agonists for pain indications (iatrogenic addiction), as well as those who commence exposure for non-medical reasons. Heroin, the widely abused MOP-r agonist, is a semi-synthetic morphine analogue (diacetyl morphine); prescription opioids that are abused include oxycodone, hydrocodone, and oxymorphone.[8]

The initial rewarding effects of these ligands, acting at $G_{i/o}$-coupled MOP-r (encoded by *OPRM1*), are thought to involve prominently, but not exclusively, inhibition of γ-aminobutyric acidergic (GABAergic) interneurons in mesolimbic and nigrostriatal dopaminergic pathways, resulting in disinhibition of dopamine neurons with concomitant increased dopamine release in terminal fields (e.g. nucleus accumbens and caudate putamen). At more advanced stages in the addiction trajectory, neurobiological adaptations in these pathways and other CNS areas (including noradrenergic sites in locus ceruleus) undergo functional plasticity, resulting in a syndrome of dependence and withdrawal, which includes dysphoria, autonomic activation, and activation of the HPA axis. Avoidance of such withdrawal can also become a negative reinforcer (motivator) for continued MOP-r agonist use.

Tolerance to MOP-r agonists is classically observed as a decrease in their pharmacodynamic effects (e.g. analgesia or rewarding effects) after chronic exposure. Mechanisms of tolerance have been an area of intensive study, involving specific adaptations at the molecular, neurobiological, and system levels. It may be postulated that MOP-r dose escalation during addiction trajectory is partially due to tolerance to rewarding effects, and that this dose escalation then results in continuing adaptations that further maintain the cyclical nature of dependence and withdrawal within addiction.[13] The presence of dysphoria and anhedonia (reward deficits) may further motivate continued MOP-r agonist use after repeated exposure.

Several studies show that the pharmacodynamics of MOP-r agonists differ among ligands (e.g. specific prescription opioids, some of which are targets of considerable abuse), in terms of signaling efficacy (e.g. in GTPγS assays), receptor cycling, desensitization, and internalization, as well as in certain *in vivo* models. The impact of differential (and "biased") pharmacodynamic and second messenger system activation by specific MOP-r ligands on addiction neurobiology is an area of active research.

Pharmacotherapy

There are several approved medications for MOP-r agonist addiction. The best known treatment involves chronic maintenance with the orally available MOP-r agonist methadone, or with the MOP-r partial agonist buprenorphine. Appropriate chronic treatment with methadone or buprenorphine results in a lack of a perceived "high" or reward, blockade of superimposed short-acting (i.e. abused) MOP-r agonists, and a lack of withdrawal signs between consecutive (typically daily) pharmacotherapy. Methadone is also a weak *N*-methyl-D-aspartate (NMDA) antagonist, which may decrease the propensity for tolerance induction caused by MOP-r agonists. Furthermore, buprenorphine, apart from having low efficacy (e.g. partial agonist effects) at MOP-r, has low efficacy effects at KOP-r.[36] The impact of this KOP-r activity of buprenorphine (and other ligands with limited efficacy at KOP-r) is an area of active research in addiction neurobiology and its comorbidity with psychiatric disorders such as depression and anxiety. Other medications, in particular a depot formulation of the opioid antagonist naltrexone, have also been recently approved.

Alcoholism

Alcoholism is a relapsing addictive disorder with massive public health consequences worldwide. These involve damage due to brain dysfunction *per se* (accidents, high-risk behaviors, neurobiological damage, withdrawal-induced seizures, psychiatric comorbidity), to alcohol's influence on other systems (liver cirrhosis), or to *in utero* exposure (fetal alcohol syndrome). Despite its use over millennia, the direct targets of alcohol's effects are elusive, although studies show that alcohol causes synaptic changes in various systems, including glutamatergic and GABAergic systems. Exposure to alcohol in experimental animal models, mimicking short- or long-term exposure in alcoholics, also results in characteristic molecular and epigenetic changes in specific brain areas, including striatal areas important for reward and motivated behaviors.

Alcohol, in common with major addictive agents and non-drug rewards, results in an acute increase in striatal dopamine release, in animal models and human studies.[37] This effect may underlie, at least in part, alcohol's rewarding effects, especially in early abuse. However, understanding later stages of abuse (e.g. heavy escalating exposure, withdrawal and relapse cycles) is also critical to more appropriately model and study the neurobiology of alcoholism at a basic level. These stages may be recapitulated, to some degree, in different experimental animal models. Alcohol self-administration, its neurobiological effects, and its pharmacotherapy, can be modeled in rodents or non-human primates. Other studies have focused on discrete lines of experimental animals (e.g. rats or mice) that exhibit high or low sensitivity to alcohol effects, of value in "gene by environment" studies, and on modeling the neurobiology of differential vulnerability to alcoholism.

Other major neuronal systems in addition to the dopaminergic system have been found to mediate alcohol reward, including the MOP-r system, for which β-endorphin and the enkephalins are considered the main endogenous ligands. Notably, naltrexone, an antagonist with relative selectivity for MOP-r, is approved for alcoholism pharmacotherapy. MOP-r genetic polymorphisms (e.g. the A118G *OPRM1*-r SNP) moderate vulnerability to alcoholism *per se* and to naltrexone pharmacotherapy, as tested in specific populations.

Other neuropeptidergic systems, primarily associated with stress-responsive and reward-related systems, including corticotropin-releasing factor, vasopressin, orexin, and dynorphins (the latter acting at KOP-r), also undergo neuroplastic changes during alcohol exposure, and affect alcohol's behavioral and rewarding effects, or the severity of dysphoria, anhedonia, and stress observed during alcohol withdrawal. Polymorphisms of genes encoding proteins expressed in some of these systems have been implicated in vulnerability to alcoholism, suggesting functional roles that may be further elucidated mechanistically.

Pharmacotherapy

The major pharmacological agents studied for alcoholism treatment are opioid receptor ligands, notably the opioid antagonist naltrexone (and its congener nalmefene), which have high affinity for MOP-r but also considerable affinity at KOP-r. The precise mechanism of naltrexone's therapeutic effectiveness in alcoholism may thus involve occupancy at opioid receptor sites in CNS regions involved in reward, or in other regions involved in mood and stress responsivity. Other approved agents against alcoholism include acamprosate, which may act as a modulator of glutamatergic function.[38]

Nicotine and Tobacco

Nicotine is the primary reinforcing substance in the tobacco plant, and the underlying cause of the addictive nature of tobacco smoking and chewing. Nicotine reinforcement leads to tobacco addiction, overcoming what is often an initial aversion in terms of taste and other effects from acute tobacco consumption. The addiction often continues in spite of minor to severe health problems, from shortness of breath to major pulmonary problems, emphysema, and cancer.

The widespread availability and marketing of tobacco, and its near universal legality, have led to its being the most highly damaging addictive substance throughout the world. The increased use of tobacco during the nineteenth and twentieth centuries was accompanied by increased health problems. The massive public health costs of tobacco led to attempts to curtail its use, attempts that, while marked by some measures of success, have had variable results in different parts of the world. This is in part due to continued efforts by those who manufacture, market, and sell tobacco products, especially cigarettes, who have a vested interest in resisting or circumventing regulatory efforts designed to reduce tobacco use, especially in countries where prevention efforts are less effective. The efforts of regulation and education have resulted in sharply decreased overall use in many countries, although in countries with a less well-developed regulatory infrastructure, tobacco use continues to climb.

In the case of cigarettes, normal use by addicted persons involves the smoking of several (10–80) cigarettes daily. Each dosing session, of 3–10 minutes of smoking, results in the vaporization of chemicals present in the tobacco leaves, which are taken in via inhalation. Following inhalation, there is rapid absorption through the lungs of a bolus of nicotine into the bloodstream. The nicotine molecules rapidly traverse the blood–brain barrier into the CNS, followed by binding and activation of nicotinic acetylcholine receptors.

One of the primary treatment modalities for tobacco addiction is NRT, in which nicotine is usually delivered buccally via gum or lozenges or transdermally via an adhesive patch. This method of nicotine delivery differs in crucial ways from the normal mode of delivery in cigarettes, in that the rate of nicotine onset and offset is considerably reduced in comparison to smoking. Of major health importance, the nicotine delivered in the context of NRT is not associated with tobacco smoking (or, in the case of snuff/chewing tobacco, prolonged oral tobacco exposure), which contains a number of additional chemicals that are largely responsible for the medical complications of tobacco addiction. Although NRT has been consistently demonstrated to reduce cravings for nicotine during abstinence from smoking, the number of smokers achieving long-term abstinence from smoking has been demonstrated to be two-fold better than with placebo or attempting self-motivated quitting alone, with up to 20% of patients achieving abstinence after 1 year. Thus, the success rate for treatment of smokers with NRT is considerably lower than a conceptually similar approach, methadone maintenance for heroin addiction. This may be due, in part, to differences in the typical duration of maintenance treatment, with methadone maintenance treatment typically being of indefinite duration, while most studies of NRT have durations of pharmacotherapeutic administration of 8–20 weeks.

Nicotine's primary site of action is the nicotinic acetylcholine receptor (nAChR), a ligand-gated ion channel, expressed on the neuronal membrane by neurons diffusely located throughout the nervous system, including at the neuromuscular junction.[39] Microdialysis studies in rodents have demonstrated that, similarly to other drugs of abuse, extracellular dopamine in the nucleus accumbens is acutely elevated in response to nicotine, as a result of specific nAChR activation of dopamine neurons in the ventral tegmental area projecting to this region. These findings have been confirmed in human PET dopamine receptor binding displacement studies. Moreover, during nicotine withdrawal, reductions in extracellular dopamine have been reported. Normalization of such reductions may be desirable for treatment. Bupropion, which inhibits transport of dopamine and norepinephrine, resulting in elevated extracellular levels of these neurotransmitters, has been proven to be effective in promoting long-term abstinence. Studies of buproprion, which was initially developed to treat depression, have demonstrated a similar measure of efficacy to NRT in promoting smoking cessation.

In rodent studies, chronic nicotine administration results in alterations in nAChR signaling and subunit composition in the basal ganglia. Gene manipulation studies in transgenic mice have greatly contributed to our understanding of the role of specific components of the complex nAChR system in mediating nicotine addiction. For instance, mice in which the gene for the β_2 subunit was deleted exhibited a lack of dopaminergic response following nicotine administration. These findings await confirmation in human studies, but point to potential therapeutic targets, if suitable nAChR-subunit discriminating compounds can be developed. Certain compounds with discrimination against different nAChR subunits have already been demonstrated. One approved ligand is varenicline, which is a partial agonist of the $\alpha_4\beta_2$ subtype, with full agonism of α_7-containing receptors.[40]

Pharmacotherapy

There are several approved pharmacotherapies for nicotine addiction, including NRT, chronic dopamine/norepinephrine reuptake blockade, and nAChR partial agonism. However, even in the optimal studies, at a timepoint of 1 year following cessation attempts, only one in five smokers remains abstinent. One reason for this may be genetic variation (see "The Genetics of Addiction", below). A study of genetic variations in response to varenicline and bupropion in smokers demonstrated that different genetic variants had effects on the odds of success, with genetic variants in nAChRs contributing to the likelihood of varenicline response, and genetic variants in the liver-metabolizing enzyme CYP2B6 contributing to responsivity to bupropion. Such findings portend the use of genetics to guide treatment decisions for addicted patients, through personalized medicine, which may increase overall treatment success rates.

Other Addictions and Related Conditions

Addictions can occur with other substances, including cannabinoids and solvents, and to novel emerging compounds ("designer drugs", many of which are pharmacologically related to psychostimulants). Non-drug reinforcers such as food and gambling can also result in behaviors and neurobiological alterations that share some mechanisms with the addictions.

Marijuana

Marijuana (*Cannabis sativa*) contains bioactive cannabinoids, with δ-9-tetrahydrocannabinol being the primary active ingredient responsible for the pleasurable subjective effects. The primary route of administration is smoking, although oral administration is common, with the inclusion of marijuana in baked goods or extraction in herbal tea preparations. Although the marijuana has been legalized or decriminalized in many locales, it is still predominantly illegal, and the ramifications of chronic use include incarceration. Marijuana addiction is a common condition, for which treatment may be sought for teenagers and others. Like other drugs of abuse, marijuana leads to acute alterations in dopamine release, with reductions in extracellular dopamine levels during withdrawal; recurrent chronic use can result in changes in components of the endogenous opioid system.

Gambling

Gambling addiction, also referred to as problem gambling or pathological gambling, is marked by intense urges to engage in gambling that is deleterious to the addict's financial and social well-being. Gambling typically takes place in the context of casino gambling or casino-like online gambling games. Of interest to basic neurobiology and modeling, such contexts for gambling behavior involve both classical and operant conditioning components. Understanding of the neurobiological and genetic basis of gambling addiction lags behind that of other addictions. However, it appears that the mediators of problem gambling also involve dopaminergic and endogenous opioid systems. Human PET imaging studies demonstrate that gambling is associated with dopamine release in the dorsal and ventral striatum.[41] Pharmacotherapy with opioid antagonists, including nalmefene, has been studied clinically.[42]

Food

There are functional and operational similarities between addictions and certain aspects of food consumption in obesity, which causes major current and projected morbidity in the industrialized world. In obese human subjects with binge-eating disorder, food stimulation leads to increased striatal dopamine release as measured by *in vivo* PET imaging, in a similar fashion to drugs of abuse.[43] In animal models of binge eating and excessive intake and in human experiments, the endogenous opioid system plays an important functional role in neurobiological systems also involved in addictive diseases.

Summary

Our understanding and the cumulative weight of evidence for altered brain function in cases of addiction to other substances or behaviors currently lags far behind those described in the earlier sections of this chapter (namely MOP-r agonists such as heroin, psychostimulants, nicotine, and alcohol). However, similarities in the brain systems affected, including the common substrates of striatal dopamine and the endogenous opioid system, are clear. It is possible that some common neurobiological dysfunctions among the addictive substances may allow for common classes of treatment agents.

THE GENETICS OF ADDICTION

Many individuals self-administer alcohol and drugs of abuse on an occasional basis. However, a subset of individuals develops specific addictions. Addictions are complex relapsing disorders, caused by genetic, epigenetic, exposure history and environment, and drug-induced factors. Identification of genetic and epigenetic factors should increase our understanding of the addictions, help in the development of new treatments, and advance personalized medicine. A large study of male twin pairs from the Vietnam Era Twin Registry showed that opiate abuse and dependence had a genetic risk of 54%, with 38% specific to heroin addiction and 16% shared across addictions to other drugs (heroin, marijuana, psychostimulants, sedatives, and psychedelics and hallucinogens).[44]

Studies aimed at identifying genes involved in addiction may use families or unrelated subjects. The use of families allows linkage of chromosomal regions with behavioral phenotypes, but suffers from low power and from difficulties in ascertaining family members of addicted individuals. A more powerful and practical approach is to use unrelated subjects in gene association studies. These subjects are classified as "cases" or controls. Association studies with a case–control design can be performed in two ways. The first approach is the hypothesis-driven candidate gene study, and the second is the unbiased genome-wide association study (GWAS). In the first approach, genes are selected based on knowledge of the biological functions of the product of that gene, the pharmacodynamic targets of the drug (e.g. MOP-r/*OPRM1*), or pharmacokinetic mechanisms (e.g. metabolic enzymes or transporter proteins). Most of these studies have used individual SNP analysis. In GWAS, up to a million genetic variants, both in genes and in intergenic regions, are interrogated simultaneously using high-density microarrays. The development of a high-throughput GWAS was facilitated by the Human Genome Project and the International Human Haplotype Map Project (HapMap), along with the development of high-throughput genotyping technologies. Difficulties associated with statistical analysis of GWAS (e.g. due to multiple comparison corrections) can be decreased by the use of hypothesis-driven arrays, for example focusing on clusters of mechanistically related genes.

Genetic Association Studies in the Opioid Receptor and Neuropeptide Gene Systems

The opioid receptor genes, and the genes encoding their cognate neuropeptides, are of great underlying mechanistic interest in the field of addiction neurobiology (see "Importance of Opioid Receptor and Neuropeptide Systems in Addiction Neurobiology", above). The gene encoding MOP-r (*OPRM1*) is a target in clinical genetic association studies, for two main reasons; first, MOP-r are the direct pharmacodynamic targets of heroin and prescription opioids in the context of both abuse and clinical analgesia[1,4,8]; and second and more broadly, the MOP-r system is involved in the rewarding and addictive effects of other drugs, including cocaine and alcohol. The most extensively studied *OPRM1* SNP is the functional *A118G* (rs1799971) polymorphism, with studies focusing on different phenotypes, patient populations, and stages of addiction.

Because of the involvement of the prodynorphin/KOP-r system (*PDYN/OPRK1*) (see "Importance of Opioid Receptor and Neuropeptide Systems in Addiction Neurobiology", above) in the regulation of basal and drug- and stress-induced release of dopamine in the dopaminergic nigrostriatal and mesolimbic–mesocortical systems, gene variants of *PDYN* and *OPRK1* have been the focus of several association and functional studies. One of the most studied polymorphisms in *PDYN* is a 68 base-pair tandem repeat polymorphism (rs35286281) located 1250 base pairs upstream of the 5′-untranslated region of exon 1. This polymorphic region, which contains a putative activator protein-1 transcription complex, is found in 1–5 copies. Two studies using stringent diagnostic criteria showed an increased risk for the development of cocaine dependence or codependence on cocaine and alcohol, but only in African-American subjects having three or four of the 68 base-pair tandem repeats, and with no effect in Caucasian subjects. In contrast, a significant association of three *PDYN* SNPs in another gene region, the 3′-untranslated region was found with both cocaine dependence and cocaine–alcohol codependence in Caucasians, but not in African Americans. Other aspects of the opioid system have also been investigated in association studies, including proenkephalin (*PENK*), and the κ- (*OPRK1*) and δ-opioid receptors (*OPRD1*).

Genetic Association Studies in Other Systems

A vast number of studies has reported the association of variants of genes and drug addiction-related phenotypes, but the results are not always consistent. Variants of the genes encoding alcohol-metabolizing enzymes *ADH1b*, *ALDH2*, and *ADH1c* have been found to have a protective influence on alcohol consumption. Variants in the chromosome 15 cluster of genes encoding subunits of the nicotinic acetylcholine receptor, including *CHRNA5/CHRNA3/CHRNB4*, are among the most robustly replicated association signals for nicotine addiction. Several other genes have been implicated in various aspects of cocaine addiction. These include dopaminergic SNPs in *DRD2/ANKK1*, dopamine β-hydroxylase, and catechol-*O*-methyltransferase; POMC; orthologues of genes regulating circadian rhythms (*CLOCK*, *PER1*, and *PER2*); and tryptophan hydroxylase-2.

Inconsistencies in genetic association findings across studies, or in meta-analyses, may be explained by several factors, including differences in phenotyping, severity, sample size, statistical approaches, ethnic heterogeneity, and population stratification, and large phenotype range. Case–control association studies of drug addiction have a complication in that the control group, for the most part, has never been self-exposed to an illicit drug, whereas the case group represents people with specific pre-existing behaviors related to the initial illicit drug-taking.

Confirmation of the findings from candidate gene and GWAS requires replication in similar, independent cohorts. The failure of some GWAS to replicate previous studies may be due to the architecture of the various microarrays used (such as poor coverage of alleles in a particular region or absence of the functional allele), low penetrance of the variant of interest, or low statistical power. The success of GWAS is dependent on many factors including the frequency of "risk" alleles, sample size, and individual effect sizes, as well as the representation of markers on the array. Large studies can suffer from a lack of accurate phenotyping, and an increase in the diversity of ethnic backgrounds and comorbidities.

The primary challenge in GWAS is the profound burden of multiple testing, which requires gene variants to exceed a threshold p value of 5×10^{-8} for statistical significance. Initial GWAS of smoking (e.g. $N=2000$), alcohol dependence ($N=1884$–3865), and cannabis dependence (e.g. $N=3054$) failed to find any statistically significant associations. Considerably larger samples and combined GWAS data are needed to produce large meta-analyses to identify all the variants involved in a disease.

However, smaller studies have proven valuable in addiction research. Several GWAS have used relatively modest sample sizes of subjects with carefully defined phenotypes and well-defined ethnicities, and have provided the bases for novel hypotheses.[45]

Both candidate gene studies and GWAS continue to provide important clues regarding the sources of genetic variation in the risk for addiction. However, unless a variant is functional, considerably more research is required to understand how a particular gene polymorphism actually affects the liability to addiction at a biological level. Thus, genetic epidemiological studies, such as gene association, are only the beginning. GWAS may not be applicable to disorders caused by multiple rare SNPs or small, undetectable copy number variations. Low-frequency variants that are not well tagged by common SNPs, and are not well represented in the current commercial arrays, can be detected by the new high-density sequencing technologies (e.g. Illumina Solexa). The studies will be facilitated by projects such as the "1000 Genomes Project", which is a worldwide collaboration to produce an extensive catalogue of human genetic variation by deep sequencing the genomes of over 1000 individuals from around the world, using next generation sequencing technologies.

Epigenetics

In addition to genetic studies, comparing the outcome of differences in the sequence of the DNA, epigenetic studies have recently been undertaken to determine whether changes in DNA not involving stable changes in code, or changes in chromatin (the DNA/protein scaffold complex making up the chromosomes), are involved in the progression to addiction.[46] For instance, in methadone-maintained heroin addicted persons, differences in methylation of cytosine residues in CpG sites in the promoter of the gene encoding for MOP-r were found in peripheral monocytes. In rodent studies, more comprehensive studies related to changes in chromatin structure of individual genes in individual brain regions in response to drugs of abuse have been performed, with identification of responsible enzymes, potentially leading to novel pharmacotherapeutic targets. The promoter region of the dynorphin gene has been shown to undergo chromatin remodeling in response to stress, a change that may underlie, in part, the close relationship between stress and addiction. Epigenetic studies are still relatively new, and techniques and targets of investigation are rapidly evolving.

QUESTIONS FOR FURTHER RESEARCH

Some key questions need to be considered in the basic neurobiology of addictions:

- How can genetic findings related to vulnerability to specific addictions, or to treatment effectiveness, best be translated, in order to optimize personalized treatment for existing or novel treatments?
- What are the mechanistic underpinnings (genetic, epigenetic, and neurobiological) of the relapsing nature of the addictions? How do relapse mechanisms differ across drug classes, and compared to non-drug reinforcers (e.g. food, in the context of obesity)?
- Can novel mechanistic approaches (focusing on changes in neuropharmacological, mRNA, or epigenetic targets) be developed to reverse the neurobiological changes resulting from chronic drug exposure, which underlie the chronic relapsing nature of specific addictive diseases?

Acknowledgments

The authors gratefully acknowledge funding from the National Institutes of Health: grant P50 DA05130 (MJK), R21 DA031990 (BR), R21 DA036365 (ERB), UL1 TR000043 (Dr Barry Coller, Rockefeller University), RO1 DA018151 (Dr Thomas Prisinzano, University of Kansas, subcontract to ERB), and RO1 DA032928 (Dr Jane Aldrich, University of Kansas, subcontract to MJK). The authors are also grateful to the Adelson Medical Research Foundation.

References

1. Kreek MJ. Molecular and cellular neurobiology and pathophysiology of opiate addiction. In: Davis KL, Charney D, Coyle JT, Nemeroff C, eds. *Neuropsychopharmacology: The Fifth Generation of Progress*. Philadelphia, PA: Lippincott, Williams & Wilkins; 2002:1491–1506.
2. Volkow ND, Wang G-J, Fowler JS, Tomasi D. Addiction circuitry in the human brain. *Annu Rev Pharmacol Toxicol*. 2012;52:321–336.
3. Dole V, Nyswander M, Kreek M. Narcotic blockade. *Arch Intern Med*. 1966;118(4):304–309.
4. O'Brien CP. Drug addiction. In: Brunton L, Chabner B, Knollman B, eds. *Goodman and Gilman's The Pharmacological Basis of Therapeutics*. 12th ed. New York: McGraw-Hill; 2011:649–668.
5. Inturrisi CE, Schultz M, Shin S, Umans JG, Angel L, Simon EJ. Evidence from opiate binding studies that heroin acts through its metabolites. *Life Sci*. 1983;33(suppl 1):773–776.
6. Kreek MJ, Levran O, Reed B, Schlussman SD, Zhou Y, Butelman ER. Opiate addiction and cocaine addiction: underlying molecular neurobiology and genetics. *J Clin Invest*. 2012;122(10):3387–3393.
7. Bond C, LaForge KS, Tian M, et al. Single-nucleotide polymorphism in the human mu opioid receptor gene alters beta-endorphin binding and activity: possible implications for opiate addiction. *Proc Natl Acad Sci U S A*. 1998;95(16):9608–9613.
8. Yaksh TL, Wallace MS. Opioids, analgesia, and pain management. In: Brunton L, Chabner B, Knollman B, eds. *Goodman and Gilman's The Pharmacological Basis of Therapeutics*. 12th ed. New York: McGraw-Hill; 2011:481–526.
9. Butelman ER, Yuferov V, Kreek MJ. Kappa-opioid receptor/dynorphin system: genetic and pharmacotherapeutic implications for addiction. *Trends Neurosci*. 2012;35(10):587–596.
10. Fagergren P, Smith HR, Daunais JB, Nader MA, Porrino LJ, Hurd YL. Temporal upregulation of prodynorphin mRNA in the primate striatum after cocaine self-administration. *Eur J Neurosci*. 2003;17(10):2212–2218.
11. Ahmed SH, Koob GF. Transition from moderate to excessive drug intake: change in hedonic set point. *Science*. 1998;282(5387):298–300.
12. Shaham Y, Rajabi H, Stewart J. Relapse to heroin-seeking in rats under opioid maintenance: the effects of stress, heroin priming, and withdrawal. *J Neurosci*. 1996;16(5):1957–1963.
13. Zernig G, Ahmed SH, Cardinal RN, et al. Explaining the escalation of drug use in substance dependence: models and appropriate animal laboratory tests. *Pharmacology*. 2007;80(2–3):65–119.
14. Chartoff E, Sawyer A, Rachlin A, Potter D, Pliakas A, Carlezon WA. Blockade of kappa opioid receptors attenuates the development of depressive-like behaviors induced by cocaine withdrawal in rats. *Neuropharmacology*. 2012;62:167–176.
15. Breese GR, Chu K, Dayas CV, et al. Stress enhancement of craving during sobriety: a risk for relapse. *Alcohol Clin Exp Res*. 2005;29(2):185–195.
16. Zhou Y, Leri F, Grella SL, Aldrich JV, Kreek MJ. Involvement of dynorphin and kappa opioid receptor in yohimbine-induced reinstatement of heroin seeking in rats. *Synapse*. 2013;67(6):358–361.
17. Zhou Y, Leri F, Cummins E, Hoeschele M, Kreek MJ. Involvement of arginine vasopressin and V1b receptor in heroin withdrawal and heroin seeking precipitated by stress and by heroin. *Neuropsychopharmacology*. 2008;33(2):226–236.
18. Unterwald EM, Kreek MJ, Cuntapay M. The frequency of cocaine administration impacts cocaine-induced receptor alterations. *Brain Res*. 2001;900(1):103–109.
19. Pavlov IP. *Conditioned Reflexes*. Mineola, NY: Dover Publications; 2003.
20. Weeks JR, Collins RJ. Factors affecting voluntary morphine intake in self-maintained addicted rats. *Psychopharmacologia*. 1964;6(4):267–279.
21. Skjoldager P, Winger G, Woods JH. Analysis of fixed-ratio behavior maintained by drug reinforcers. *J Exp Anal Behav*. 1991;56(2):331–343.
22. Ahmed SH. The science of making drug-addicted animals. *Neuroscience*. 2012;211:107–125.
23. Picetti R, Ho A, Butelman ER, Kreek MJ. Dose preference and dose escalation in extended-access cocaine self-administration in Fischer and Lewis rats. *Psychopharmacology (Berl)*. 2010;211(3):313–323.
24. Picetti R, Caccavo JA, Ho A, Kreek MJ. Dose escalation and dose preference in extended-access heroin self-administration in Lewis and Fischer rats. *Psychopharmacology (Berl)*. 2012;220(1):163–172.
25. Vlachou S, Markou A. Intracranial self-stimulation. In: Olmstead MC, ed. *Animal Models of Drug Addiction*. New York: Humana Press; 2010:3–56.
26. Sinha R, Shaham Y, Heilig M. Translational and reverse translational research on the role of stress in drug craving and relapse. *Psychopharmacology (Berl)*. 2011;218(1):69–82.
27. McLellan AT, Luborsky L, Woody GE, O'Brien CP. An improved diagnostic evaluation instrument for substance abuse patients. The Addiction Severity Index. *J Nerv Ment Dis*. 1980;168(1):26–33.
28. Zubieta J-K, Gorelick DA, Stauffer R, Ravert HT, Dannals RF, Frost JJ. Increased mu opioid receptor binding detected by PET in cocaine-dependent men is associated with cocaine craving. *Nat Med*. 1996;2(11):1225–1229.
29. Bart G, Schluger JH, Borg L, Ho A, Bidlack JM, Kreek MJ. Nalmefene induced elevation in serum prolactin in normal human volunteers: partial kappa opioid agonist activity? *Neuropsychopharmacology*. 2005;30(12):2254–2262.
30. Kling MA, Carson RE, Borg L, et al. Opioid receptor imaging with positron emission tomography and [(18)F]cyclofoxy in long-term, methadone-treated former heroin addicts. *J Pharmacol Exp Ther*. 2000;295(3):1070–1076.
31. Weerts EM, McCaul ME, Kuwabara H, et al. Influence of OPRM1 Asn40Asp variant (A118G) on [11C]carfentanil binding potential: preliminary findings in human subjects. *Int J Neuropsychopharmacol*. 2013;16(1):47–53.
32. Lewis DA. The human brain revisited: opportunities and challenges in postmortem studies of psychiatric disorders. *Neuropsychopharmacology*. 2002;26(2):143–514.
33. Yuferov V, Kroslak T, Laforge KS, Zhou Y, Ho A, Kreek MJ. Differential gene expression in the rat caudate putamen after "binge" cocaine administration: advantage of triplicate microarray analysis. *Synapse*. 2003;48(4):157–169.
34. Mash DC, ffrench-Mullen J, Adi N, Qin Y, Buck A, Pablo J. Gene expression in human hippocampus from cocaine abusers identifies genes which regulate extracellular matrix remodeling. *PLoS ONE*. 2007;2(11):e1187.
35. Haile CN, Mahoney III JJ, Newton TF, De La Garza II R. Pharmacotherapeutics directed at deficiencies associated with cocaine dependence: focus on dopamine, norepinephrine and glutamate. *Pharmacol Ther*. 2012;134(2):260–277.
36. Huang P, Kehner GB, Cowan A, Liu-Chen LY. Comparison of pharmacological activities of buprenorphine and norbuprenorphine: norbuprenorphine is a potent opioid agonist. *J Pharmacol Exp Ther*. 2001;297(2):688–695.

37. Yoder KK, Morris ED, Constantinescu CC, et al. When what you see isn't what you get: alcohol cues, alcohol administration, prediction error, and human striatal dopamine. *Alcohol Clin Exp Res.* 2009;33(1):139–149.
38. Spanagel R, Vengeliene V. New pharmacological treatment strategies for relapse prevention. *Curr Top Behav Neurosci.* 2013;13:583–609.
39. Tuesta L, Fowler C, Kenny P. Recent advances in understanding nicotinic receptor signaling mechanisms that regulate drug self-administration behavior. *Biochemical Pharmacology.* 2011;82(8):984–995.
40. Mihalak K, Carroll F, Luetje C. Varenicline is a partial agonist at alpha4beta2 and a full agonist at alpha7 neuronal nicotinic receptors. *Mol Pharmacol.* 2006;70(3):801–805.
41. Joutsa J, Johansson J, Niemelä S, et al. Mesolimbic dopamine release is linked to symptom severity in pathological gambling. *Neuroimage.* 2012;60(4):1992–1999.
42. Grant JE, Odlaug BL, Potenza MN, Hollander E, Kim SW. Nalmefene in the treatment of pathological gambling: multicentre, double-blind, placebo-controlled study. *Br J Psychiatry.* 2010;197(4):330–331.
43. Tomasi D, Volkow ND. Striatocortical pathway dysfunction in addiction and obesity: differences and similarities. *Crit Rev Biochem Mol Biol.* 2013;48(1):1–19.
44. Tsuang MT, Lyons MJ, Meyer JM, et al. Co-occurrence of abuse of different drugs in men: the role of drug-specific and shared vulnerabilities. *Arch Gen Psychiatry.* 1998;55(11):967–972.
45. Nielsen DA, Kreek MJ. Common and specific liability to addiction: approaches to association studies of opioid addiction. *Drug Alcohol Depend.* 2012;123(suppl 1):S33–S41.
46. Robison A, Nestler E. Transcriptional and epigenetic mechanisms of addiction. *Nature Rev Neurosci.* 2011;12(11):623–637.

CHAPTER

36

Sleep Disorders

Birgitte Rahbek Kornum, Emmanuel Mignot*[†]

*Molecular Sleep Laboratory, Department of Diagnostics and Danish Center for Sleep Medicine, Copenhagen University Hospital Glostrup, Glostrup, Denmark; [†]Stanford Center for Sleep Sciences, Stanford University School of Medicine, Palo Alto, California, USA

OUTLINE

Neurobiology of Brain Disorders
http://dx.doi.org/10.1016/B978-0-12-398270-4.00036-7

INTRODUCTION TO SLEEP AND CIRCADIAN NEUROBIOLOGY

Sleep and circadian rhythms are phylogenetically ancient. Plants and animals have clear circadian rhythms while reptiles, insects, and fish have behaviorally defined sleep-like states (typical body posture, physical quiescence, elevated threshold for arousal and reactivity, rapid state reversibility, homeostatic and circadian regulation). In birds and all mammals, including humans, sleep occurs at the behavioral level but also shares stereotyped patterns of changes in cortical activity on electroencephalography (EEG) consistent with the presence of different sleep stages. Almost specific to humans, however, is the unusual distribution of sleep and wakefulness that is restricted to the night and day in single continuous periods, a phenomenon called monophasic sleep.

The Need for Sleep

The importance of sleep is best illustrated by the overpowering effects of total sleep deprivation in humans and the fact that it is recognized as a form of torture. Indeed, sleep deprivation is difficult to sustain for more than a few days, and the longest documented voluntary total sleep deprivation lasted for 11 days.[1] Selective rapid eye movement (REM) sleep deprivation is similarly unsustainable, but it takes a longer time. During complete sleep deprivation, cognition rapidly becomes impaired and mood labile, although large interindividual differences exist. Microsleep episodes, with temporary loss of consciousness, occur with increasing frequency. Endocrine and thermoregulatory changes and hallucinations may follow. These effects are all rapidly reversed by sleep, although rebound sleep following deprivation is not equal in terms of amount of time lost. In humans, sleep deprivation is most often chronic and partial ("sleep restriction"), a phenomenon increasingly recognized as deleterious for human health. Similarly, shift work, an occupation associated with circadian misalignment and sleep deprivation, is associated with well-documented medical problems.

In rats, total sleep deprivation is lethal after 2–3 weeks.[2] Within days, animals become hyperphagic but lose weight, a state associated with increased heart rate and energy expenditure. Body temperature subsequently drops. Animals become increasingly debilitated, develop ulcers on the tail and paws, and eventually die from sepsis. The effects of long-term sleep deprivation have not been documented in any other species, and there is great debate over whether lethality is the result of stress or sleep deprivation, a largely academic debate considering the non-physiological aspect of the experiment.

More physiologically, migrating birds fly at night and forage during the day, thus experiencing substantial sleep loss during migration. In contrast to the above, these show no functional impairment, suggesting that sleep may be temporarily suspended or dramatically decreased in some species without deleterious effects. EEG recordings from migrating birds have shown that they compensate for sleep loss by experiencing multiple short naps during the day. The microsleep can involve the entire brain, or can be unilateral, affecting only one hemicortex.[3] This example is just one among many that demonstrates how sleep can adapt to environmental challenges, but remains vital even under extreme ecological pressure (e.g. in migrating birds and aquatic mammals).

The Day–Night Cycle

The day–night cycle of the natural environment has played a fundamental role in shaping the evolutionary development of sleep homeostatic mechanisms because of the dominating predictability of diurnal changes in illumination, temperature, food availability, and predator activity. Recent evidence suggests that most physiological functions are under the control of the circadian timing system. The mammalian circadian timing system is a hierarchically organized network of molecular oscillators driven by a central pacemaker located in the suprachiasmatic nucleus (SCN) of the hypothalamus. The circadian regulation of sleep and wakefulness will be described in more detail below.

Homeostasis has traditionally been seen as a corrective reaction to a physiological or pathophysiological change. In 1986, Moore-Ede[4] proposed to extend the concept of homeostasis by introducing the term "predictive homeostasis", a response initiated in anticipation of predictably timed challenges. In this model, traditional homeostasis, such as the recovery sleep after sleep deprivation, was called "reactive" homeostasis. The advantage of this concept is that it is easier to explain in the context of natural evolution. It also avoids the artificial dichotomy of circadian versus homeostatic regulation, stressing the fact these two systems are likely to be coordinated and to impinge on similar outputs. Predictive responses have particular value because they enable physiological mechanisms to be utilized immediately, even if they involve a delay of several hours, by activating them at a suitable time in advance of a probable challenge.

Sleep differs from rest during wakefulness because there is reduced responsiveness to stimulation. Although the function of sleep is a highly debated topic, sleep, or at least non-REM sleep, is associated with reduced energy expenditure and may thus have been selected as a way to reduce the need for extra calories, especially at times when the ecology is unfavorable to food seeking (night and day for diurnal and nocturnal animals, respectively).[5] It is also clear that during sleep a large number of physiological processes are returning to sustainable levels in preparation for a new day, suggesting that sleep is a form of coordinated rest. In contrast to this, the exact role of REM sleep has been more complex to explain, as it is energetically unfavorable and more difficult to study.

Unlike other unresponsive states, such as unconsciousness and full anesthesia, sleeping individuals can awaken rapidly following stimulation. The rapid reversibility is made possible through the action of a network of brain areas in the brainstem and hypothalamus that have long projections and far-reaching effects through the entire brain. This network, and how it is regulated, will be described in later sections of this chapter. Following this, the different sleep disorders that can arise when these systems are not functioning optimally will be discussed.

NEUROBIOLOGY OF SLEEP

Sleep States and Sleep Cycles

Sleep consists of two distinct states: REM sleep and non-REM sleep. In humans, these two states are defined using polysomnography, a technique in which electrodes are placed on the head and body, and electric cortical activity patterns (EEG), chin muscle tone [electromyography (EMG)], and eye movements [electro-oculography (EOG)] are recorded. In clinical polysomnographic studies, breathing and electrical activity in legs are also typically monitored, as abnormal respiration can cause pauses in breathing during sleep (sleep apnea) and abnormal repetitive limb movements (periodic leg movements during sleep) can disrupt sleep. These are frequent sleep disorders.

REM sleep is characterized by high levels of desynchronized cortical EEG activity, absence of muscle tone, irregular heart rate and respiratory patterns, and episodic bursts of phasic eye movements; the last feature is why it has been given the name rapid eye movement sleep. Non-REM sleep is characterized by low-frequency, high-voltage EEG activity, low muscle tone, and absence of eye movements. Non-REM sleep is further divided into four stages based on different features. Stage 1 non-REM occurs at transitions of sleep and wakefulness; stage 2 is characterized by frequent bursts of rhythmic EEG activity, called sleep spindles, and high-voltage slow spikes, called K complexes. Stages 3 and 4 are also called slow-wave sleep (SWS), and are characterized by high-voltage activity in the lowest frequency range (<2 Hz).

During the night, human non-REM and REM sleep alternate with a period of about 90 minutes (Fig. 36.1). Non-REM sleep and particularly SWS predominates in first cycles while the proportion of REM sleep increases towards the last part of the night in later cycles. Decreased SWS activity towards the end of the night parallels the dissipation of the sleep debt. In contrast, the increased occurrence of REM sleep in the later part of the night is mostly regulated by the circadian system and occurs when body temperature is the lowest. The mechanism underlying ultradian rhythmicity of non-REM and REM sleep is unknown, although length of the cycle correlates roughly with brain size across mammals.

Homeostatic and Circadian Regulation of Sleep and Wakefulness

It is well known that the need to sleep (and sleepiness) increases the longer you are awake, and, independently of how rested you are, is highest at night (notably during early morning hours), when it is dark (or during the day

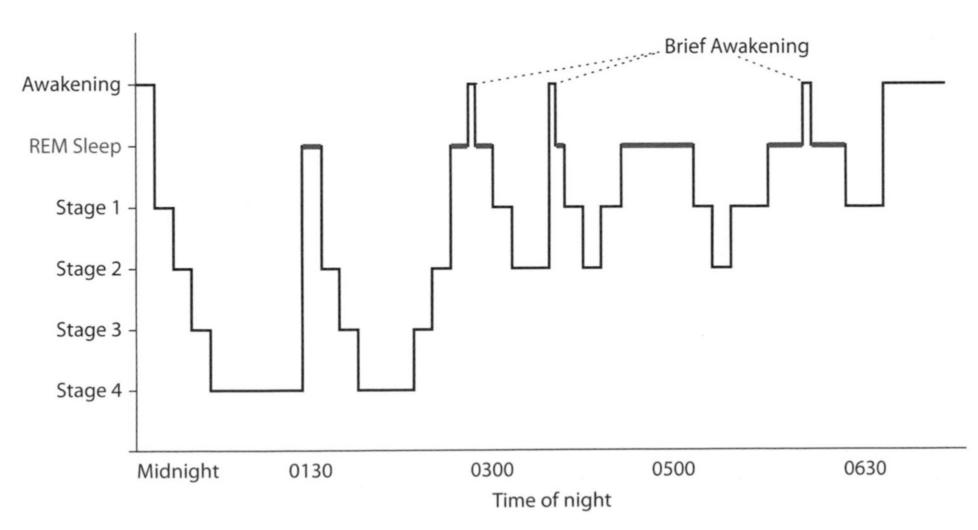

FIGURE 36.1 **Hypnogram illustrating sleep architecture.** The figure shows a simplified hypnogram dividing a night's sleep into different stages and cycles. During the night non-rapid eye movement (non-REM) and REM sleep alternate in cycles with a period of about 90 minutes. Slow-wave sleep (SWS; stages 3 and 4) predominates in the early cycles during the night and the proportion of REM sleep is increased towards the last part of the night. The mechanism underlying this ultradian rhythm is unknown.

for nocturnal animals). This pattern is best modeled by postulating the existence of two independent processes: a circadian sleep process (also called process C) and a homeostatic sleep process (process S) (Fig. 36.2).[6]

In this model, the two processes interact to optimize sleep at night and wakefulness during the day. Humans are awake in the morning because sleep debt is low following a night's rest. Throughout the day, increasingly strong wake-promoting signals, generated by the circadian clock, counteract the mounting sleep debt and keep subjects awake. The increased wake promotion driven by the circadian clock is maximal when body temperature is highest, in the evening. Once the circadian wake drive starts to decrease and the sleep debt is unsustainable, the subject falls asleep. An opposite interaction occurs during the night, helping humans to sleep longer even when their sleep debt is already mostly dissipated. The duality of this interaction explains why sleeping during the night is optimal to restoration in humans, although a small window of opportunity also exists in the middle of the day for a nap, when sleep pressure has increased but the circadian wake signal is not yet too strong.

The circadian sleep process is a clock-like process that is traditionally considered independent of the homeostatic process. This circadian process oscillates with a 24 hour rhythm and is entrained by the light–dark cycle, even though it will persist in complete darkness with a period not much different from 24 hours (the free-running circadian period of an individual). The circadian sleep process is controlled via direct and indirect input, mainly from light but also from other cycling events such as activity patterns and feeding.

The homeostatic sleep process indicates that something increases or accumulates during wakefulness and decreases over the course of a sleep period. This process is the reason why extended wakefulness will result in increased sleep pressure, and it also accounts for the restorative process that occurs during sleep. SWS is used as a marker of the homeostatic sleep process, since EEG slow-wave activity increases as a function of previous wakefulness and decreases during the course of sleep.

Even though circadian processes and sleep homeostasis are generally considered independent, they interact in complex ways to control sleep and wakefulness states and sleep timing.

CIRCADIAN REGULATION OF SLEEP

Circadian rhythms can be produced and sustained by a single mammalian cell in the absence of any external stimuli. These rhythms result from coordinated daily oscillations in the transcription and translation of several clock genes. Some of these important clock proteins feed back to inhibit the transcription of their own genes, a process that takes approximately 24 hours (circadian period), and this is the core mechanism behind the circadian time-keeping mechanism.

All circadian clocks share several fundamental properties: they are entrained each day by external cues, they are self-sustained and produce oscillations that persist even in the absence of any external cues, they are temperature compensated such that temperature changes in the physiological range do not alter their endogenous period, and they are cell autonomous and genetically determined.

The SCN in the hypothalamus is the site of a master circadian oscillator in the brain, which coordinates all rhythms across the entire body. Destruction of the SCN abolishes circadian oscillations in behaviors such as locomotion and drinking, and also plasma concentrations of molecules such as cortisol. Normal circadian rhythms can be restored in animals by transplantation of fetal SCN tissue, and the recipient will then adapt the

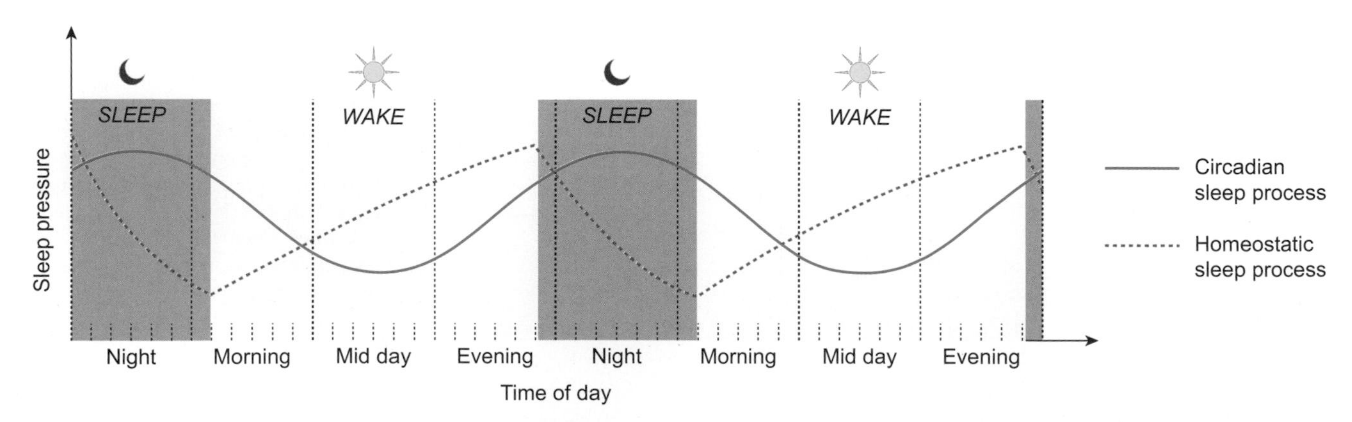

FIGURE 36.2 **Two-process model of sleep–wake regulation.** The two-process model of sleep regulation, including the circadian and homeostatic components, is illustrated. The total sleep pressure an individual experiences is, according to this model, the sum of the sleep pressure from each of the two processes. Traditionally, these two processes were considered independent, but recent evidence points to these processes having molecular links. *Source: Modified from Borbélly and Achermann.* J Biol Rhythms. *1999;14:557–568.*[7]

circadian period of the donor, indicating that the properties of the rhythm are genetically determined and arise in the SCN and not other brain regions.[8] These rhythms are then transmitted to all other cells in the organism, so that activity of various organs and cells may occur at different phases (i.e. peaks at different times of the day) to optimize function at predicted times where it is typically needed (see the predictive homeostasis concept, in "The Day–Night Cycle", above).

Molecular Regulation of the Circadian Clock

The central pacemaker is entrained each day by the environmental light–dark cycle and transmits synchronizing cues to cell-autonomous oscillators in tissues throughout the body. Within cells of the central pacemaker and the peripheral tissues, the underlying molecular mechanism by which oscillations in gene expression occur involves interconnected feedback loops of transcription and translation. Most of our knowledge regarding these loops has come from genetic studies in model organisms, starting with *Drosophila* and being continued in mice and humans.

The core circadian molecular oscillator is cell autonomous. This is also the case for individual neurons of the SCN. In this system, positive and negative regulation of a set of core clock genes forms a rhythmic feedback loop with a time constant of about 24 hours per cycle.

A central element in this regulation is the heterodimer of two transcription factors: CLOCK (circadian locomotor output cycles kaput) and BMAL1 [brain and muscle aryl hydrocarbon receptor nuclear translocator (ARNT)-like]. The CLOCK–BMAL1 heterodimer initiates transcription from genes containing E-box *cis*-regulatory elements, including period genes (*Per1, Per2, Per3*) and cryptochrome genes (*Cry1, Cry2*). PER and CRY proteins heterodimerize as well and suppress the transcription rate of their own genes, by binding to the CLOCK–BMAL1 heterodimer and inhibiting its binding to E-box elements (Fig. 36.3). Following several post-transcriptional and

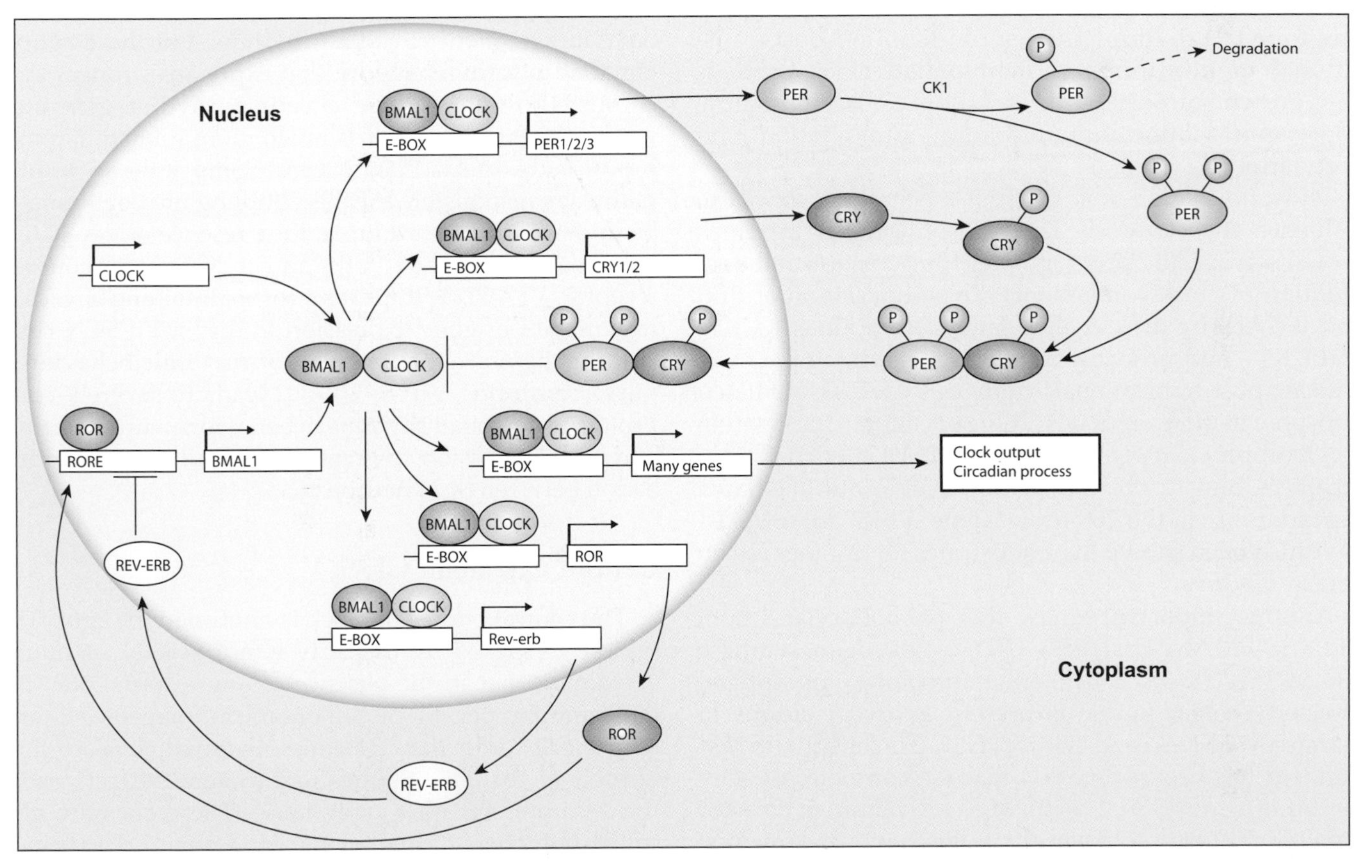

FIGURE 36.3 **Core components of the molecular clock and their interactions.** The molecular mechanism of the circadian clock in mammals is shown. Constituting the core circadian clock is an autoregulatory transcriptional feedback loop involving the activators CLOCK and BMAL1 and their target genes *Per1, Per2, Per3, Cry1,* and *Cry2*, whose gene products form a negative-feedback repressor complex. In addition to this core transcriptional feedback loop, other feedback loops are driven by CLOCK–BMAL1. One feedback loop involves REV-ERBα, which activates *Bmal1* and the RORs (Rorα, Rorβ, Rorγ) that repress *Bmal1* transcription. CLOCK–BMAL1 also regulates many downstream target genes known as clock-controlled genes. At the post-transcriptional level, the stability of the PER and CRY proteins is regulated by phosphorylation. The kinases, casein kinase-1ε/δ (CK1ε/δ) and AMP kinase (AMPK), phosphorylate the PER and CRY proteins, respectively, to promote polyubiquitination by ubiquitin ligase complexes, which in turn tag the PER and CRY proteins for degradation.

post-translational modifications discussed below, PER–CRY is slowly degraded, and when the level becomes adequately low, CLOCK–BMAL1 is released from inhibition, and transcription and translation of *Per* and *Cry* genes can start again.[9]

Several accessory regulatory loops interconnect with the core loop described above, adding robustness and stability to the clock, and providing additional layers of control, linking the clock to a myriad of other pathways within the cell. The best studied of these accessory loops involve members of the large nuclear receptor family, whose transcription is also induced by the CLOCK–BMAL1 heterodimer. These are the orphan nuclear receptors REV-ERBα (reverse orientation c-erbAα) and REV-ERBβ, which inhibits *Bmal1* expression, and RORα (retinoic acid receptor-related orphan receptor-α), RORβ, or RORγ, which activates *Bmal1* expression.

Post-translational modifications of the core clock components play a crucial role in generating the delays necessary to establish the 24 hour rhythm of the mammalian circadian clock. Some of these modifications are essential to clock function, whereas others simply fine-tune the rhythm. The list of identified post-translational modifications of mammalian clock proteins has grown rapidly and includes phosphorylation, dephosphorylation, ubiquitination, sumoylation, and acetylation.

Phosphorylation was the first mechanism shown to influence the clock.[10,11] The phosphorylation of clock proteins at specific sites regulates transcription and degradation of clock components. Phosphorylation of PER and BMAL1 by the closely related casein kinases CK1δ and CK1ε, and probably also CK1α, is among the most critical post-translational changes. CK1δ/ε-mediated phosphorylation regulates PER subcellular localization and its ability to repress CLOCK–BMAL1-mediated transcription, and further promotes its ubiquitin-mediated degradation via the 26S proteasome. PER2 and BMAL1 have also been shown to be substrates for another casein kinase, CK2α.

Another important kinase that phosphorylates both CLOCK–BMAL1 and PER–CRY is glycogen synthase kinase-3 (GSK-3). This serine–threonine, phosphate-directed protein kinase is present as two isoforms in mammals: GSK-3α and GSK-3β. GSK-3 is sensitive to lithium.[12] It has been suggested that Li^+ competes directly for binding to GSK-3 with Mg^{2+}, a required cofactor for GSK-3 function. Several studies have documented the effects of lithium treatment on circadian rhythms in mammals, including a consistent effect of lengthening the free-running period of behavioral rhythms, notably those of locomotor activity and drinking.

Protein phosphatases are fewer in number in the mammalian genome relative to kinases. These also likely play a role in regulating the clock by reversing the effect of kinases and adding another level of flexibility to the clock.

The circadian clock is regulated not only at the protein and transcriptional level, but also at the DNA structural level. Rhythmic changes in chromatin condensation notably participate in the activation and repression of transcription of several clock genes. This regulation involves post-translational modifications at histone N-terminal tail regions. It has been demonstrated in mice that light pulses during the subjective night promote phosphorylation of histone H3, and further that histone H3 is rhythmically acetylated at *Per1* and *Per2* promoters. Finally, CLOCK has been shown to have histone acetyltransferase activity itself. This suggests that CLOCK, while activating transcription with its partner BMAL1, may rhythmically acetylate histones at many clock-controlled genes, thereby participating in far-reaching effects on chromatin remodeling.

As mentioned above, although circadian rhythms occur in many cells, the SCN as a whole is the master pacemaker coordinating all rhythms in the body. In the brain, individual SCN neurons have the capacity to have independent circadian rhythms, but these couple together to form a network that expresses stronger synchronized rhythms. The circadian rhythm generated within the SCN network is much more robust than that produced by individual neurons. Some of the molecular pathways responsible for this coupling are beginning to be understood. For example, the presence of vasoactive intestinal polypeptide (VIP) and its G-protein coupled receptor, VPAC2, is important for maintaining circadian rhythmicity of gene expression in dispersed SCN cells, and for the normal expression of rhythmic behavior in mice. Disrupting VIP or VPAC2 leads to severely compromised circadian rhythms in behavior, neuronal firing, and gene expression, owing to intercellular desynchronization between SCN neurons.[13]

Central Circadian Clock

The central circadian clock is entrained by light. The period generated by the SCN is not exactly 24 hours, explaining that in the absence of any external cue, the free-running period of an organism may be slightly longer or shorter than 24 hours. External cues are thus needed to realign the circadian system perfectly with the environment on a daily basis. These cues are also called zeitgebers ("time givers" in German). The principal and strongest zeitgeber regulating overall behavioral rhythmicity in mammals is light. Light is detected in the eye by rods and cones (also responsible for vision), and by a small subset (≈1%) of intrinsically photosensitive retinal ganglion cells (ipRGCs).[14] ipRGCs respond to light stimulation independently of rods and cones, and use a specific photopigment called melanopsin (Opn4).

Nonetheless, rods, cones, and retinal ganglionic cells all participate in the synchronization of circadian rhythms by light.

Light input to the retina is principally transmitted to the SCN through the ipRGCs via the retinohypothalamic tract. This projection is mediated by glutamate and ultimately activates SCN neurons, leading to the phosphorylation of Ca^{2+}/cAMP-response element binding proteins (CREBs). Within the promoters of many of the core clock genes reside Ca^{2+}/cAMP-response elements (CREs), to which phospho-CREB homodimers bind to activate transcription. The *Per1* and *Per2* genes both contain CREs in their promoters and are rapidly induced in SCN neurons following light exposure.

The effect of light input to the SCN depends on the time of exposure. If light is received at the beginning of one's habitual dark period, it will result in a phase delay (e.g. leading a person to be tired later), whereas if light is received at the end of the dark period, it will result in a phase advance (leading a person to be tired earlier). Besides these longer lasting changes in the timing of the circadian output system, light also has acute, immediate effects. For example, light by itself has acute effects on corticosteroid secretion, heart rate, body temperature, and cognitive performance, and suppresses endogenous melatonin production.

Non-photic zeitgebers such as the sleep–wake cycle itself, locomotor activity, and feeding also influence the phase of the circadian pacemaker, and provide fine-tuning signals to relevant peripheral clocks.

Peripheral Clocks

Peripheral clocks are autonomous and synchronized by signals from the SCN. The CLOCK–BMAL1 oscillator described above controls circadian rhythms in all cells through the entire body in a cell-autonomous fashion. Current estimates indicate that 10–15% of the cellular transcriptome follows a circadian pattern of expression, with only a minority of these changes being driven by humoral or neuronal circadian signals coming from elsewhere in the body. The vast majority of these changes (more than 90%) are dependent on each cell-autonomous local circadian oscillator. This self-autonomous circadian transcription activity allows peripheral tissues to anticipate upcoming circadian environmental challenges such as activity and feeding. However, to be efficient, individual cells of a given tissue need to function in synchrony with the entire body. This synchronization is achieved by signals from the SCN to both other brain centers and the periphery.[15]

The SCN signals to many peripheral tissues including the heart, vasculature, adrenal glands, liver, pancreas, and adipose tissue. The synchronizing signals reach peripheral tissues via autonomic neural connections and the release of hormones such as glucocorticoids. Other factors such as body temperature, rest/activity, and feeding also affect peripheral clocks directly. In the absence of the SCN, circadian rhythms diminish in most peripheral tissues after a few days. The decrement is due to the desynchronization of cells within the tissue, as circadian rhythms persist at the single-cell level. There are, however, oscillators in some mammalian brain regions and tissues that, in the absence of the SCN, can drive local physiological rhythms. One of these is the food-entrainable oscillator that can drive circadian behavioral and endocrine rhythms.[16]

In the brain, the SCN signals to other areas in need of circadian regulation such as sleep regulatory networks. One important output from the SCN is a pathway to the pineal gland, which regulates the endogenous circadian rhythm of melatonin production, a rhythm that peaks during the dark phase. In another important sleep regulatory pathway, the SCN targets the dorsomedial nucleus of the hypothalamus (DMH) via the subparaventricular zone. The DMH then sends projections to many sleep-inducing areas of the brains, coordinating sleep states. These areas will be described in detail below.

CIRCUITRY AND MOLECULAR ASPECTS OF SLEEP

Sleep-state specific changes in neuronal activity can be observed in all parts of the brain. Sleep is thus a distributed process, although it is coordinated by a smaller set of specific neuronal systems. Interactions between sleep- and wake-specific populations of neurons must ensure that the sleep–wake processes occur in synchrony and to the exclusion of each other to create stable states of wakefulness, non-REM, and REM, with limited time spent in transition states. This is regulated by mutual inhibition mechanisms and probably some degree of mutual excitation as well, ensuring that an entire network of neurons specific to a given state is activated at once.

Wakefulness and Cortical Activation

Wakefulness is regulated by multiple waking neurochemical systems located in the upper brainstem, the hypothalamus, and the forebrain. These systems send parallel ascending projections to the entire cortex and also to the thalamus. Overall, this collection of brain nuclei and far-reaching associated brain projections is called the ascending arousal system (Fig. 36.4).[5] This system has two main components. The first are ascending neural pathways that activate the thalamic neurons, a system that conveys sensory input to the cerebral cortex, and also participate in the generation of sleep EEG

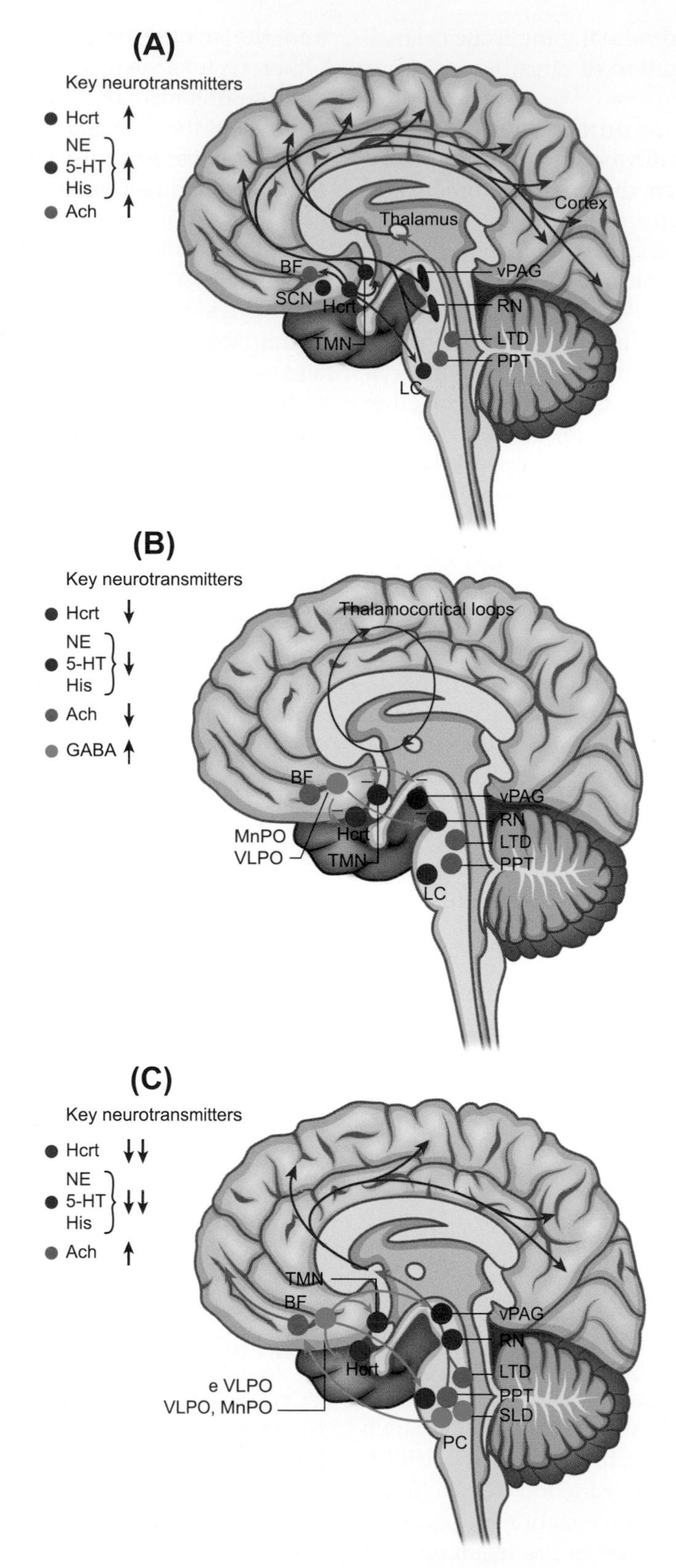

FIGURE 36.4 **Activity of major neural networks during (A) wakefulness, (B) non-rapid eye movement (non-REM) sleep, and (C) REM sleep.** (A) During wakefulness, monoaminergic, hypocretinergic, and cholinergic systems are active and contribute to electroencephalographic (EEG) desynchronization through thalamic and cortical projections. Hypocretin cells excite monoaminergic cells, and possibly cholinergic neurons. Muscle tone on electromyography (EMG) is variable and high, reflecting movements. Dopaminergic cells of the dopaminergic ventral periaqueductal gray matter (vPAG) do not significantly change firing rates across sleep and wakefulness, although the pattern of firing does,

activity patterns. Activating signals to the thalamus arise through projections from two acetylcholine-producing brain nuclei in the upper brainstem: the pedunculopontine and laterodorsal tegmental nuclei (PPT/LDT). The neurons in the PPT/LDT fire most rapidly during wakefulness and REM sleep, which are the states associated with cortical activation.

The second component of the ascending arousal system includes direct and indirect projections stemming from selected midbrain and brainstem nuclei to the cerebral cortex. Indirect projections occur after relays in lateral hypothalamic and basal forebrain areas. The nuclei involved use various neurotransmitters or neuropeptides, serotonin (5-hydroxytryptamine) for the dorsal and median raphe nuclei (DRN and MRN), norepinephrine for the locus ceruleus, dopamine for the ventral periaqueductal gray matter (vPAG), histamine for the tuberomammillary nucleus (TMN), and the neuropeptides hypocretin (also referred to as orexin) and melanin-concentrating hormone (MCH) for the lateral hypothalamus area. In general, these nuclei are most active during wakefulness, less active during non-REM sleep, and inactive during REM sleep. The hypocretin-producing neurons are most active in the waking state, whereas MCH neurons are most active during REM sleep.

The Preoptic Area and Sleep Promotion

The existence of brain regions promoting sleep and wakefulness was first proposed in 1930 by the Austrian neurologist Baron Constantine von Economo. He became particularly interested in the study of encephalitis lethargica, a curious and often fatal condition, which appeared during and after the great influenza pandemic of 1918.[17] Although patients with this condition presented with a diverse set of symptoms, ranging from movement disorders to oculomotor symptoms, almost all had greatly disturbed sleep. In most cases, patients were almost continuously sleeping and nearly comatose, a situation often associated with ophthalmoplegia. However, a few individuals developed severe insomnia, or sleep-cycle inversion, most often in combination with chorea and abnormal movements. Using postmortem samples, van Economo found that patients with sleepiness typically had posterior hypothalamic lesions (probably encompassing the TMN, hypocretin, and MCH neurons) also involving the upper brainstem, while in the more unusual insomnia patients, damage was localized to the basal forebrain and adjacent anterior hypothalamus.[18]

The hypothesis that neurons located within the anterior preoptic hypothalamus could contribute to the promotion of sleep was later confirmed by electrophysiological and c-fos (a marker of activation) studies, indicating sleep active neurons in the area. It was also found that the wake-promoting areas described above do receive inhibitory input from γ-aminobutyric acidergic (GABAergic) sleep-active neurons in the preoptic area.[19] Two key areas in the sleep-promoting network are the ventrolateral preoptic nuclei (VLPO) and the median preoptic nucleus (MnPO), with the VLPO currently being regarded as the most important.[20] VLPO neurons produce the inhibitory neurotransmitter GABA and the inhibitory neuromodulator galanin, and send projections to several systems that regulate arousal, including histaminergic, serotonergic, noradrenergic, and hypocretinergic neurons. MnPO neurons also produce GABA and project to wake-active neurons such as the serotonergic and hypocretinergic neurons (see Fig. 36.4).

Although SWS-active neurons in areas such as the VLPO are considered important in the generation of

contributing to higher dopaminergic release during wakefulness in the prefrontal cortex. The suprachiasmatic nucleus (SCN), the master biological clock, is located close to the optic chiasma, and receives retinal input. Time of the day information is relayed through the ventral subparaventricular zone to the dorsomedial hypothalamus, and other brain areas. (B) During non-REM, slow-wave sleep, γ-aminobutyric acidergic (GABAergic) cells of the basal forebrain (BF), median (MnPO), and ventrolateral preoptic (VLPO) hypothalamic area are highly active. MnPO area GABAergic cells may also be involved in thermoregulation. VLPO and other cells inhibit monoaminergic and cholinergic cells during non-REM and REM sleep. Upon cessation of sensory inputs and sleep onset, thalamocortical loops from the cortex to the thalamic reticular nucleus and relay neurons contribute to the generation of light non-REM sleep. As non-REM sleep deepens, slow-wave oscillations appear on the EEG. Muscle tone is low but not abolished in non-REM sleep. (C) During REM sleep, hypothalamic and basal forebrain sleep active cells fire, but glutamatergic cells in the sublaterodorsal nucleus (SLD REM-on neurons) and preceruleus (PC) also increase activity. These cells trigger REM atonia through caudal projections, while ventral basal forebrain projections contribute to hippocampal theta. Brainstem cholinergic systems are also active, and stimulate thalamocortical loops to generate EEG desynchronization similar to wakefulness. During REM sleep, EMG is low, indicating paralysis through motor neuron inhibition (tonic REM sleep). Twitches (bursting of EMG and small movements) also occur, with intermittent saccades of rapid eye movements (phasic REM sleep). LDT: laterodorsal tegmental cholinergic (ACh) nucleus; PPT: mesopontine pedunculopontine tegmental cholinergic (ACh) nucleus; LC: adrenergic (NE) locus ceruleus; RN: serotoninergic (5-HT) raphe nuclei; TMN: histaminergic (His) tuberomammillary nucleus; Hcrt: hypocretin/orexin-containing cell group. Upward and downward arrows represent increased/decreased activity and release for selected neurotransmitter (e.g. ACh, NE, 5-HT, His, and Hcrt) systems or in selected metabolic pathways (e.g. protein biosynthesis) during the corresponding sleep–wake state. Different colors denote various neurotransmitter systems: red: norepinephrine, histamine, and serotonin; orange: acetylcholine; dark red: hypocretin peptide; pink: dopamine; dark violet: GABAergic and glutamatergic non-REM sleep and REM sleep-on neurons. *Source: Modified from Mignot.* PLoS Biol. *2008;6(4):661–669.*[5]

SWS, sleep-active neurons are present in many other areas of the brain, and these may also play a role as auxiliary or master SWS regulators. As an example, although most neurons in the cortex typically exhibit reduced activity during SWS (which is logical, considering the function of sleep), a subpopulation of GABAergic interneurons, which express the enzyme neuronal nitric oxide synthase (nNOS), has been found to be active during SWS.[21] The extent of activation of these nNOS neurons increases with the length of the preceding period of wakefulness, and these neurons have been suggested to provide a long-sought anatomical link to explain homeostatic sleep regulation. Although these neurons are low in density, their number is far larger than those of the VLPO, considering the size of the cortex. Unlike the VLPO, however, whether or not lesions of these neurons influence sleep has not been tested, and this may be difficult because they are widely distributed throughout the cortex. As for circadian rhythms, it is likely that there is a hierarchy in sleep regulation as well, with sleep being driven by more important coordinating networks, but still including the possibility of sleep-like properties at the local level.

Transitions Between Wakefulness and Sleep

Animals, including humans, are particularly vulnerable during sleep. It is therefore necessary for them to wake up quickly and completely when in danger. Stable sleep–wake transitions are achieved by a finely tuned coordination of networks that enables a complete change in activity across large populations of neurons between wakefulness, SWS, and REM sleep. To achieve state stability, neurons that trigger wakefulness or sleep mutually inhibit each other. Important components of the sleep regulatory network are the preoptic areas and their inhibitory projections to the ascending arousal system (TMN, DRN, locus ceruleus, and lateral hypothalamus). The sleep-active neurons in the VLPO are in turn inhibited by serotonin, norepinephrine (noradrenaline), and acetylcholine from the dorsal and median raphe, locus ceruleus, and PPT/LDT, and by GABA produced with histamine in TNM.[5]

Based on this circuitry, the current hypothesis for transition from wakefulness to sleep is that it is induced by the activation of VLPO neurons resulting in GABA- and/or galanin-mediated inhibition of neurons at the level of arousal systems. MnPO are especially sensitive to sleep deprivation and are likely to be more important for the homeostatic sleep regulation and less for the induction of sleep. This is supported by findings that patterns of neuronal activity across the sleep–wake cycle in the MnPO and VLPO are, for the most part, reciprocal to those observed in the TMN, DRN, locus ceruleus, and lateral hypothalamus.

The mutual inhibition between ascending arousal- and sleep-promoting systems is self-reinforcing and keeps the activity of all neurons that induce wakefulness low during sleep. This system assures the stability of both sleep and wakefulness, and further enables the brain to switch quickly between different states. This minimizes periods of drowsiness, and being half-awake and half-asleep.

Both the VLPO and the MnPO receive input from the SCN, possibly explaining circadian regulation of sleep and wakefulness. Similar inputs, but with opposite effects to wake-promoting systems, are likely. How these networks receive inputs regarding sleep homeostasis, prompting the switching to sleep or wakefulness, is still unknown and probably occurs at multiple levels. One possible mediator of sleep homeostasis may be adenosine, a substance that is produced during neuronal activity. During wakefulness (such as during the day for humans), adenosine activates VLPO neurons via the $A_{2A}R$ receptor (at which caffeine is an antagonist). Past a certain level, however, this does not occur, allowing transitions to sleep. To avoid induction of sleep due to small increases in sleep pressure, there is a need for systems counteracting increased sleep drive during the day. Some of the most important factors stabilizing long periods of wakefulness are the neuropeptides hypocretin-1 and -2.[22] Hypocretin neurons are primarily active during wakefulness and innervate the cerebral cortex and the monoaminergic and cholinergic neurons. When hypocretin neurons are active, this stabilizes the waking state.

Regulation of Rapid Eye Movement Sleep

The regulation of REM sleep mostly occurs at the level of the brainstem, with input from the hypothalamus. The non-REM/REM regulating system also has a mutually inhibitory setup, which contributes to the rapid and complete transition from non-REM to REM sleep. During REM sleep, specific populations of neurons in the sublaterodorsal nucleus (SLD) and the adjacent preceruleus area are active (REM-on neurons). Some of these neurons activate the basal forebrain, hippocampus, and cortex, explaining EEG desynchronization and dreaming; others project caudally directly or indirectly to the spinal cord motor neurons, explaining muscle atonia.

The onset of SLD REM-on neuronal activity is mostly due to the removal of a constant GABAergic tone present during waking and SWS. These GABAergic inputs are from interneurons in the SLD itself and from GABAergic neurons in the tegmentum, including the ventrolateral part of the periaqueductal gray matter (vPAG) and lateral pontine tegmentum (LPT). Lesions in these areas increase REM sleep.[23]

Other REM-on neurons are inhibitory neurons in the SLD that feed back on to non-REM-active GABAergic neurons in vPAG and LPT. The MnPO and VLPO innervate the vPAG, and projections from these sleep-active neurons (notably from the extended VLPO to the vPAG) also suppress neuronal activity in this area during sleep. Suppression of vPAG activity from the VLPO and further from the SLD triggers a cascade of changes including activation of REM-active neurons, resulting in changes in the cortical/hippocampal EEG and induction of atonia during REM sleep.

Several systems act as stabilizers of REM sleep. The cholinergic PPT/LDT excites the REM-active nuclei and inhibits non-REM-active nuclei. MCH neurons are active during REM sleep and play a role in both REM sleep regulation and homeostasis. As MCH is primarily an inhibitory peptide, it has been proposed that MCH neurons might promote REM sleep by inhibiting the REM-sleep suppressing neurons that are active during waking, such as the vPAG neurons and hypocretinergic neurons.

While phasic motor activity such as rapid eye movements and brief and low-amplitude muscle twitches occur as normal phenomena in REM sleep, more elaborate motor activity is suppressed. The absence of motor activity in normal REM sleep is due to the active inhibition of spinal motor neurons plus reduced drive within locomotor generators. Glutamatergic SLD REM-on neurons indirectly inhibit spinal motor neurons. Two different pathways have been proposed to mediate spinal motor inhibition. One is through the activation of inhibitory spinal interneurons, the other is through the activation of glycinergic/GABAergic neurons localized in the medullary ventral magnocellular reticular nucleus (MCRF, where the "F" stands for formation). These two pathways are not mutually exclusive.

CURRENT THEORIES ON WHY WE SLEEP

Studies of sleep and sleep deprivation suggest that the functions of sleep include recovery at the cellular, network, and endocrine system levels, energy conservation and ecological adaptations, and a role in learning and synaptic plasticity. This section will briefly discuss these three main hypotheses.

Sleep Decreases Energy Consumption

As sleep is associated with reduced brain energy expenditure, and given that energy consumed by this organ is a large fraction of total body energy consumption in organisms with large brains (≈30% in humans), sleep may have been selected to reduce energy expenditure at times when food is difficult to access.[5] In this model, performance peaks at specific times when food is available, depending on whether an animal is nocturnal or diurnal. Vision and temperature regulation are also optimized for these times. At other times, reducing energy expenditure ensures survival when food is scarce. Models have shown that small energy savings could have effects on selection. The observation that long-term sleep deprivation in rats is associated with metabolic dysregulation also suggests functional relationships between sleep and energy regulation. Finally, circadian mouse mutants are prone to metabolic abnormalities.

A variant of this hypothesis suggests that sleep allows for the reallocation of energy to the immune system.[24] This is supported by evidence that sleep deprivation reduces the magnitude of the immune response to challenges and also that total sleep length across species correlates with white blood cell count, such that species with longer sleep duration have higher blood cell counts. It has, however, not been shown whether this immune function of sleep is related to the sleep-specific changes in brain activity, or rather the quiescence that occurs during periods of sleep.

Several aspects make the energy-economic model insufficient to fully explain natural selection of sleep. First, if true, sleep would be similar to hibernation, selected to save energy. Against expectation, however, animals coming out of torpor experience a sleep rebound, suggesting sleep deprivation. Furthermore, whereas non-REM sleep may be associated with decreased energy expenditure, REM sleep is most often associated with increased whole-body oxygen consumption.

Overall, although it is likely that energy saving played a role in the selection of non-REM sleep during evolution in specific circumstances (e.g. in mammals with high energy demands, such as mice), it is unlikely to have been the sole and only function. Energy savings may have played a role in the selection of a REM-sleep like state in animals without cortex such as reptiles, but other functions of REM sleep must have taken over to explain its persistence in mammals.[25]

Sleep is Involved in Information Processing and Synaptic Plasticity

Higher cortical functions such as cognition, attention, and memory are rapidly affected by sleep deprivation. Studies have shown that learning and memory are improved by subsequent sleep without repetition of the task, suggesting that information processing occurs during sleep. Imaging studies have shown reduced activity in cortical regions involved in a task learned during prior wakefulness during non-REM and reactivation during REM sleep.

The synaptic homeostasis hypothesis states that a major function of sleep is the scaling down of synapses that have accumulated in size and number during learning, a process that occurs mostly during waking.[26,27] In this model, synaptic potentiation and increase in synaptic strength occur during wakefulness and must be reduced during sleep to return to homeostasis. Synaptic connections accumulate during waking, and this is unsustainable to the brain in terms of both space and energy costs. During SWS, a proportional synaptic downscaling occurs that leaves only the most robust connections intact, reducing energetic and space requirements for the maintenance of the most crucial learned circuits. This synaptic downscaling also increases signal-to-noise ratio for remaining connections, thereby improving performance. The hypothesis implies that without this process, the brain would rapidly lose its capacity to learn new things.

In support of this hypothesis, markers of synaptic potentiation, such as phosphorylation of 2-amino-3-(5-methyl-3-oxo-1,2-oxazol-4-yl)propanoic acid (AMPA) receptors, calmodulin-dependent protein kinase II, and GSK-3β, occur in proportion to sleep debt in the cortex and hippocampus. These effects are independent of light, time of day, and temperature, but track estimated sleep debt. If the process is general, it is likely to involve other metabolic, structural, or electrophysiological processes as well, such as replenishment of neurotransmitter stores in terminals through vesicular trafficking. Indeed, for example, administration of neurotransmitter-depleting agents, such as amphetamine, leads to stronger rebounds in sleep time than administration of substances that preserve dopaminergic storage.

Structural evidence for synaptic homeostasis as a function of wakefulness and sleep comes from studies in *Drosophila*.[28] Protein levels of key structural components of synapses and of the secretory machinery are high after waking and low after sleep, independent of time of day, with a progressive decrease in concentration during sleep. Social experiences that increase the number of synapses between lateral ventral neurons and their partners in the brainstem also increase the need for sleep. Finally, direct structural evidence from imaging of three *Drosophila* neuronal circuits shows that the size and number of synapses increase after a few hours of wakefulness and decreases only if the flies are allowed to sleep. At the functional level, the hypothesis predicts that SWS should increase locally if a particular brain region undergoes a high amount of learning and synaptic potentiation in wakefulness. This has been confirmed by several studies, in both rodents and humans.

The synaptic homeostasis hypothesis reconciles several observations, such as the temporal association between slow-wave activity and sleep debt, local sleep at the cortical level, the importance of sleep for learning and memory, and the possibility that sleep is efficient energetically. It does not, however, consider other forms of learning, such as long-term depression and altered efficacy of inhibitory synapses. Nor does the hypothesis explain the known role of REM sleep in memory consolidation, a process that is widely assumed to depend on strengthening synapses. Finally, the model does not yet explain how plasticity may occur in sleep-active networks.

Sleep Restores Key Cellular Components of Macromolecule Biosynthesis

Similar to the field of circadian regulation discussed above, it is likely that sleep is a complex process that encompasses both the brain and the body, and regulation at the network, humoral, and cellular (cell autonomous) levels. About 10% of genes in the brain change expression with sleep, half of these independently of circadian phase.[29] Similar sleep-dependent changes in gene expression occur in peripheral tissues such as the liver. The reported changes are consistently found across multiple species (including *Drosophila*, rats, mice, and birds) and brain regions (cortex, cerebellum, hypothalamus), and, similarly to circadian regulation, even occur in peripheral organs.

A large fraction of sleep-associated transcripts encodes proteins involved in the synthesis of complex macromolecular components such as cholesterol and protein synthesis, intracellular transport, and endocytosis and exocytosis. This is generally consistent with older studies that showed increased protein synthesis during sleep. In contrast, wakefulness is associated with the upregulation of genes involved in transcription and RNA processing, and at a later stage, with increased expression of molecular chaperones, a finding suggesting cellular stress. These results indicate that a function of sleep may be to restore macromolecules and to replenish or traffic transmitter vesicles that have been used by extended wakefulness.

A strength of the restoration hypothesis is that it is applicable to all organisms and tissues, as it has a cellular basis. A major weakness is that the changes reported have only been correlative, and may thus not be related to the function of sleep. Unlike the circadian field, the field of molecular sleep is still awaiting mutants that have profoundly altered sleep homeostasis in mammals, and possibly point us towards a fundamental core of sleep genes.

INTRODUCTION TO SLEEP DISORDERS

Sleep and circadian disorders are quite common and heterogeneous, affecting 10% of the population. Prevalence increases with age. These disorders affect not only sleep or the circadian clock *per se* but also indirectly other organs and functions, for example breathing and movements. Although a few sleep disorders such as narcolepsy and restless legs syndrome (RLS) have a clear neurological basis, in most other cases pathophysiology is unknown or heterogeneous.

Traditionally, the most common sleep disorders have been subdivided into the following main categories: insomnias, hypersomnias, sleep-related breathing disorders, sleep-related movement disorders, parasomnias, and circadian rhythm disorders (Table 36.1). This section will focus on a few disorders that involve the brain

TABLE 36.1 Selected Sleep Disorders

	Sleep Disorders	Mechanisms Involved
Circadian rhythm disorders	Advanced and delayed sleep phase syndrome	Abnormal endogenous phase relationship with natural light/dark conditions can be due to genetic, anatomic, and/or problematic zeitgeibers; intrinsic circadian period may or may not be abnormal
	Shift work disorder, jet lag	Normal circadian clock subjected to an abnormal, inadaptable light/dark environment
	Free-running circadian disorder	Absence of entrainment of a normal circadian clock due to insufficient zeitgebers, e.g. in some completely blind individuals
	Irregular sleep–wake	Weak circadian regulation, behavioral and psychiatric issues
Hypersomnia	Narcolepsy	Autoimmune destruction of hypocretin neurons leading to sleep state instability and dissociated REM sleep events, such as cataplexy, sleep paralysis, and hypnagogic hallucinations
	Idiopathic hypersomnia	With or without extended sleep amounts over the 24 hours; unknown cause, heterogeneous
	Kleine–Levin syndrome	Unknown
Insomnia	Psychophysiological insomnia	Hyperactive arousal systems (hyperarousal hypothesis), negative conditioning, defective sleep-promoting networks
Sleep-related breathing disorders	Obstructive sleep apnea	Reduced upper airway patency interrupting sleep; may be due to anatomical problems (obesity, craniofacial), reduced airway dilator muscle activity, or instability of brain/carotid body CO_2 and O_2 chemoreception (loop gain)
	Central sleep apnea	Abnormal or delayed CO_2 chemosensitivity due to central abnormalities, genetic factors, environment (altitude), or abnormal circulation time (heart failure)
Sleep-related movement disorders	Restless legs syndrome (RLS)	Urge to move legs in evening associated with periodic leg movements during sleep; associated with developmental gene polymorphisms, dopamine and iron imbalance
Parasomnias	REM behavior disorder (RBD)	REM sleep occurs without atonia; lesions or dysfunction of pontine REM regulatory areas, e.g. due to Lewy body accumulation in these areas in early Parkinson disease
	Sleep walking	Partial arousal in the presence of non-REM sleep; confusional arousal, high sleep pressure

REM: rapid eye movement.

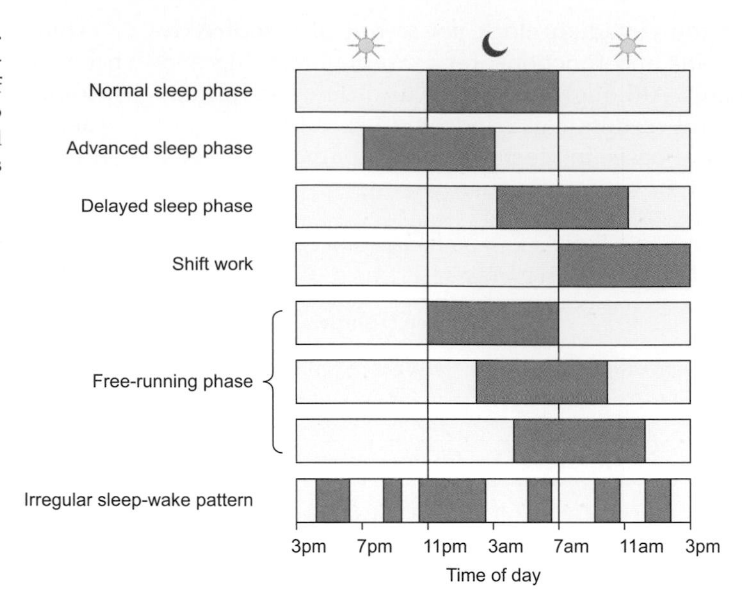

FIGURE 36.5 **Circadian rhythm sleep disorders.** Schematic illustration of sleep–wake patterns in different circadian rhythm sleep disorders and an example of shift work. Patients with free-running disorder will keep advancing their sleep phase in the pattern depicted until it again aligns with the night. The cycle continues like this for their entire life.

and for which the pathophysiology is relatively well understood.

CIRCADIAN RHYTHM SLEEP DISORDERS

Circadian sleep disorders reflect a misalignment of the internal circadian clock with the attempted sleep–wake schedule of a given individual. For example, if the phase of body temperature is misaligned with the expected sleep–wake cycle, individuals may be attempting to fall asleep when body temperature is rising, making sleeping impossible. Circadian misalignment most commonly results from the circadian system being slightly out of phase with the desired sleep–wake cycle, as in advanced or delayed sleep phase syndromes. Alternatively, the internal phase may be completely misaligned due to jet lag or shift work. More rarely, the circadian system can fail to entrain to the natural 24 hour light–dark cycle, and instead just follows its free-running period, its natural period in the absence of any environment clues, for example in completely blind people. Finally, the overall input from the circadian system to the sleep–wake regulatory system can be too weak, causing irregular sleep–wake patterns (Fig. 36.5).

As detailed in earlier sections, single gene mutations that alter the free-running period or the phase angle of entrainment (delay between the onset of activity and the dark period in rodents, equivalent to advanced or delayed sleep phase) or produce complete arrhythmicity have been reported in animal models, illustrating possible mechanisms for these pathologies. As illustrated below, however, only in rare cases have single mutations been found to cause these conditions, suggesting that in most cases the etiology is multifactorial and also involves environmental interactions.

Advanced Sleep Phase Syndromes

Patients with advanced sleep phase syndrome (ASPS, extreme early birds) have sleep times that are much earlier than what would be suitable for a normal light–dark or activity–sleep cycle, and that cannot be entrained by usual light exposure. ASPS is rare and hardly ever needs treatment. Markers of circadian phase, such as melatonin, occur earlier relative to individuals with normal circadian rhythms. In many cases, for example with advancing age, the cycle is self-reinforcing since when the subject goes to sleep early, no light will be received at a time that could shift the circadian phase to later hours. The same patient is likely to turn on a bright light early in the morning, further consolidating the undesired phase advance. In some cases, ASPS can be the result of a shorter free-running period, so that every evening the subject must go to bed earlier. Importantly, however, slightly shorter periods, for example 23.5 hours, can usually be reset to a socially acceptable norm by normal light exposure. In rare cases, the period may be so short that the subject is an abnormally early bird or even may not be able to entrain at all to a 24 hour cycle.

In familial advanced sleep phase syndrome (FASPS), the condition can be caused by mutations in human clock-related genes.[30] In cases where mutations have been identified, transmission is autosomal dominant. Subjects have normal sleep architecture, and a lifelong tendency to wake and sleep at very early times (e.g. wake up 1–3 a.m. and bedtime 6–8 p.m.). In the subjects that have been studied, melatonin and temperature rhythms

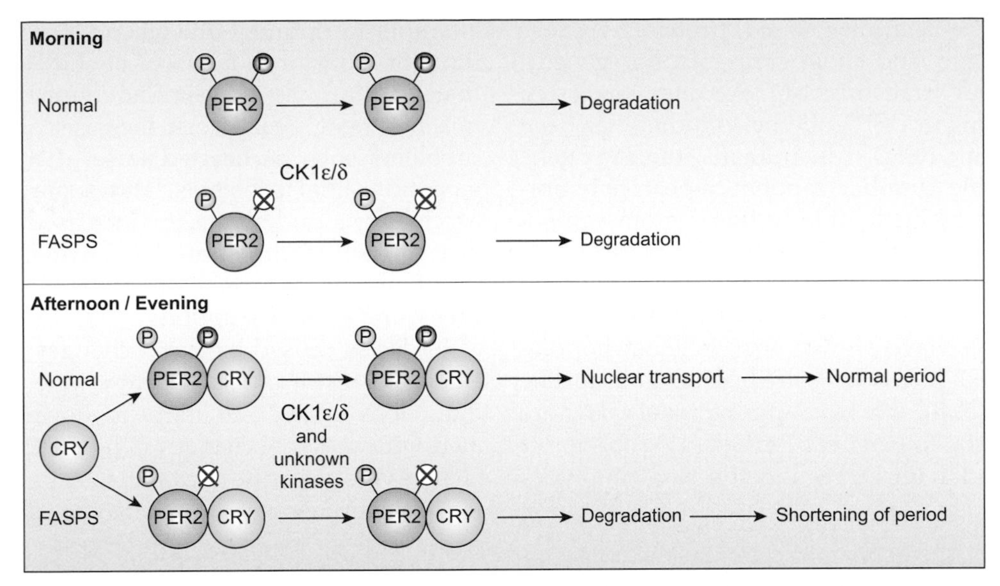

FIGURE 36.6 **Model of the differential effects of PER2 phosphorylation on circadian oscillations.** PER2 contains at least two functionally different sets of phosphorylation sites: one primarily mediating proteasomal degradation (green), the other nuclear retention (purple). In familial advanced sleep phase syndrome (FASPS), when caused by a PER2 mutation, the latter cannot be phosphorylated because Ser-662 is mutated to glycine. At the beginning of the circadian cycle (morning), newly synthesized PER2 protein shuttles between nucleus and cytoplasm, where it is phosphorylated by kinases such as casein kinase-1ε/δ (CKIε/δ) at sites that target it for rapid proteasomal degradation in the cytoplasm. Later (afternoon/evening), complex formation with CRY proteins enhances the nuclear localization of the PER2–CRY complex and probably activates or recruits additional kinases to the PER2–CRY complex. These yet-unknown kinases phosphorylate PER2 at the FASPS site, which serves as a priming site for CKIε/δ phosphorylation at downstream residues. Together, this leads to nuclear accumulation of the PER2–CRY complex and thereby to transcriptional repression of CLOCK–BMAL1 transactivation. At the end of the circadian cycle (late night), the PER2–CRY repression is released because the PER2–CRY complex is degraded in the cytoplasm after nuclear export. In FASPS-PER2, however, the region responsible for nuclear retention cannot be phosphorylated (red crosses), leading to premature nuclear export of the PER2–CRY complex, and thus to an earlier cytosolic degradation and to a faster circadian cycle. *Source: Modified from Vanselow* et al. Genes Dev. *2006;20(19):2660–2672.*[31]

are advanced by 4–6 hours, and the free-running period is 1 hour shorter than in controls (≈23 hours). Genetic studies of different families with FASPS point to defects in phosphorylation of PER2 as a common pathophysiology. The first functional mutation identified was a serine to glycine (S662G) change in the human PER2 gene that removes an important phosphorylation site. This results in decreased phosphorylation of PER2 by CK1δ at nearby serines, causing decreased stability of the PER2 protein. The decreased concentration of PER2 then shortens the period of the circadian oscillator (Fig. 36.6). A mutation responsible for FASPS in another family has been found in the kinase itself, CK1δ. This is a missense mutation (T44A) that causes decreased enzymic activity, leading to hypophosphorylation of PER2 and a similarly more unstable protein. Transgenic mice carrying the same mutation also have a shorter circadian period, but transgenic *Drosophila* carrying the human CKIδ-T44A gene have a lengthened circadian period.

Delayed Sleep Phase Syndromes

Patients with delayed sleep phase syndrome (DSPS, extreme night owls) have sleep times that are much later than what is customary and practical in relation to the normal light–dark cycle. Patients may, for example, be unable to go to bed until 3 or 4 a.m., but will then sleep until noon without any problem. Markers of circadian phase, such as melatonin, occur later relative to individuals with normal circadian rhythms, indicating delayed phase. As described for ASPS, DSPS is also self-reinforcing as light exposure late at night promotes the undesired phase advance. Social and other factors, such as an unusually strong wake-promoting signal coming from the circadian clock in the evening during adolescence, may be mainly responsible. DSPS improves with age, as the phase of the circadian cycle typically advances with aging.

DSPS is often associated with sleep deprivation in adolescence, because of early school and work times. When patients try to sleep earlier, sleep may be fragmented and restless. This results in sleep deprivation at a developmental time where sleep needs are the highest, increasing the risk of sleep attacks and car accidents. Possible mechanisms behind DSPS include dysfunction of systems coordinating circadian and sleep processes, behavior, psychological traits, genetic processes, and combined effects.

Although in animals some clock gene mutations can cause DSPS-like phenotypes with a long free-running period, familial cases of DSPS due to single gene mutations have not been discovered, probably because DSPS is so frequent and most cases are non-genetic, obscuring familial transmission. Among suggested genetic associations in DSPS, the best studied has been an association with PER3.[32] In humans, the *Per3* gene contains a variable number tandem repeat polymorphism, in which a 54-nucleotide coding-region segment is repeated four or five times. The repeated motifs in the PER3 protein contain numerous potential phosphorylation sites, which could affect the stability of the protein. Population studies have shown that in most populations around 10% of individuals are homozygous for the five-repeat PER3(5/5), whereas approximately 50% are homozygous for the four-repeat PER3(4/4). The shorter allele is associated with DSPS and the two genotypes have also been suggested to correlate with a morning versus evening chronotype.

The fact that DSPS is more common among adolescents and young adults is perhaps driven by changes in the mechanisms regulating sleep and wakefulness during pubertal development.[33] Evidence supports hypotheses that circadian mechanisms change across adolescent development, including a change towards evening circadian phase preference and later circadian phase. Furthermore, adolescents tend to have a longer free-running period than adults. Slower accumulation of homeostatic sleep pressure during puberty permits adolescents to stay awake longer and, thus, delay the sleep–wake cycle. In addition to environmental factors, these underlying changes in the circadian and sleep systems are likely to increase the risk of DSPS during puberty.

Shift Work and Jet Lag

During shift work, individuals are forced to sleep during the normal waking period, causing impaired sleep consolidation and shortened sleep duration. In some night shift workers, entrainment may occur, especially if the subject avoids bright light exposure during the day, so that a stable "reverse" cycle sets itself up. Many other subjects, however, remain entrained to the solar day and do not adapt night shifts. People who have rotating shifts or irregular shifts never entrain and are always out of phase at least part of the time. When circadian misalignment is chronic, it reduces productivity and increases the risk of accidents. Shift work also impairs the coordination of peripheral and central circadian rhythms, probably creating hormonal imbalances and explaining increased metabolic and cardiovascular risk in shift workers.

Shift work sleep disorder (SWSD) is defined as insomnia and/or excessive sleepiness associated with working (and sleeping) at non-standard times.[34] It is associated with extreme difficulty in maintaining adequate sleep–wake function, while on a shift work schedule, despite attempts to optimize the sleep environment. SWSD is thought to occur in a subset of shift workers (approximately 10%) who are especially vulnerable to circadian misalignment. Shift work increases the risk of health problems such as heart disease, diabetes, depression, gastrointestinal problems, and some forms of cancer. Mechanisms underlying the increased risk of these disorders in shift workers are unknown, but may be related to inflammatory factors, eating at the wrong circadian times, and sleep disruption.

Jet lag is caused by rapid changes in environmental time as a result of travel across time zones. The circadian clock evolved to adapt to slower changes associated with seasonal changes in light exposure, and rapid changes induce acute circadian misalignment. During jet lag, sleep and wakefulness often occur at abnormal biological times resulting in daytime sleepiness, associated with high melatonin levels, and disturbed sleep, associated with low melatonin levels. Sleepiness in the new time zone can increase the risk of accidents and impair cognitive performance. Jet lag symptoms also include gastrointestinal disruption related to ingestion of meals at abnormal circadian times. Although a few days with the new light schedule are generally sufficient to restore most behavioral rhythms, complete resynchrony of all rhythms takes a few weeks.

Free-Running Circadian Disorder

Free-running circadian disorder occurs when the circadian clock completely fails to entrain to the normal 24 hour light–dark cycle. As a result of this failure to entrain, sleep and wakefulness progressively delay or advance from day to day depending on the free-running rhythm of the patient's internal clock. If a 24 hour sleep–wake schedule is maintained, sleep and wakefulness are disturbed when they occur out of phase with internal circadian time (see Fig. 36.5).

Free-running disorder is most common in totally blind individuals. For this to occur, however, subjects typically need to have not only no light perception using cones and rods (no conscious vision) but also no functional photoreception through ganglion cell melanopsin pigments. An example would be a subject with a complete enucleation of both eyes.

Irregular Sleep–Wake Patterns in Developmental and Neurodegenerative Disorders

In animals, SCN lesions produce a loss of circadian rhythmicity, causing an irregular sleep–wake pattern characterized by sleep times that occur in multiple fragmented and short episodes over the 24 hour cycle. In humans, SCN lesions may be due to tumors or other types of lesion, as exemplified by some cases with encephalitis lethargica and anterior hypothalamic lesions (see "The Preoptic Area and Sleep Promotion", above). Total sleep time is relatively normal or slightly increased. In these cases, there is still some effect of light, so that waking is stimulated by light and sleep by the dark, but the pattern is very fragmented.

The same irregular sleep–wake pattern can also be the result of reduced circadian drive. Some patients will, for instance, have lower melatonin levels in the dark compared with healthy individuals. Such irregular sleep–wake patterns with maintained total sleep times can be caused by both developmental (e.g. autism, mental retardation) and neurodegenerative disorders (e.g. Alzheimer disease).

NARCOLEPSY

Narcolepsy is a disorder of wakefulness and REM sleep. It is characterized by severe, irresistible daytime sleepiness, cataplexy, dissociated REM sleep events, disturbed nocturnal sleep, and abnormally rapid transitions into REM sleep. Sleepiness in narcolepsy is severe, leading to sleep attacks, but brief naps are generally restorative.[35]

Dissociated REM sleep events are occurrences where the patient is neither in full REM sleep nor fully awake, but in between. A typical example is sleep paralysis, a symptom that generally occurs when waking up and where the patient finds himself paralyzed (as in REM sleep) but conscious and awake. Hypnagogic hallucinations, are another example of dissociated REM sleep event, where vivid dreaming or frightening hallucinations (as in REM sleep) occur without the patient actually being asleep, generally when the patient is very tired or falling asleep. Many narcoleptic patients also have disturbed dreams, where excessive motor activity can occur and patients "act out" their dreams because muscle tone is not totally suppressed, a phenomenon called REM sleep behavior disorder (RBD).

About two-thirds of diagnosed narcoleptic patients have cataplexy, a very specific symptom of narcolepsy. Cataplexy is characterized by a sudden loss of muscle tone triggered by emotions, most typically laughing and joking, and occasionally anger. Unlike hypnagogic hallucinations, vivid dreaming, sleep paralysis, and RBD, which can occur in the normal healthy population independently of narcolepsy, cataplexy is almost 100% specific to narcolepsy. It can also be thought of as a REM-sleep atonia phenomenon occurring during wakefulness not unlike sleep paralysis, although here the paralysis is triggered by emotions.

Many patients with narcolepsy also have disturbed nocturnal sleep. In these patients, not only is sleep disturbed by vivid dreaming and abnormal REM sleep events, as described above, but very often, although they fall asleep rapidly, they are also unable to stay asleep for very long, experiencing insomnia. To diagnose narcolepsy, abnormal transitions into REM sleep (sleep-onset REM periods) are documented using EEG sleep tests, such as the multiple sleep latency test.

Narcolepsy with Cataplexy

Narcolepsy with cataplexy is caused by hypocretin (also called orexin) deficiency. In 1999, loss of function mutations in the hypocretin receptor-2 gene were shown to produce narcolepsy–cataplexy in families of Doberman, Labrador, and Dachshund dogs that transmitted the disorder as a single autosomal recessive gene.[36] Hypocretin (orexin) knockout mice were also found to have narcolepsy-like sleep patterns and sudden episodes of behavioral arrests that resembled cataplexy,[37] further establishing a crucial connection between narcolepsy and hypocretin.

Shortly thereafter, it was found that human narcolepsy is not caused by hypocretin receptor mutations (as in dogs), but still involved abnormal hypocretin systems, as documented by low to undetectable levels of hypocretin in the cerebrospinal fluid (CSF) of patients with narcolepsy. Postmortem studies also documented a 90–95% loss of the 70,000 neurons that produce hypocretin in the hypothalamus (Fig. 36.7).[38] Staining of markers colocalizing with hypocretin in this area, such as dynorphin or NARP (NPTX2, neuronal pentraxin II), was also performed and the corresponding signals were found to be lacking, indicating that hypocretin cells have not just stopped producing hypocretin, but rather been lost. Nearby and intermingled MCH-producing neurons are intact. As discussed below, the specific loss of the hypocretin neurons in narcolepsy is now thought to result from an autoimmune attack.

The function of the hypocretin neurons is to sustain wakefulness and suppress REM sleep, which is why the loss of these neurons causes the narcolepsy phenotype. These neurons also project to the spinal cord and may regulate muscle tone independently of sleep. As described in "Regulation of Rapid Eye Movement Sleep" (above), hypocretin excites REM-off neurons. Lesions of these neurons thus result in sleep-onset REM periods, sleep paralysis, and cataplexy-like phenomena in rodents and humans. Besides a primary role in the regulation of the sleep–wake cycle, hypocretin affects several other functions such as feeding, cardiovascular regulation, pain, locomotion, stress, and addiction. In these other functions, however, loss of hypocretin can be compensated, as knockout animals and narcoleptic patients show more variable phenotypes.

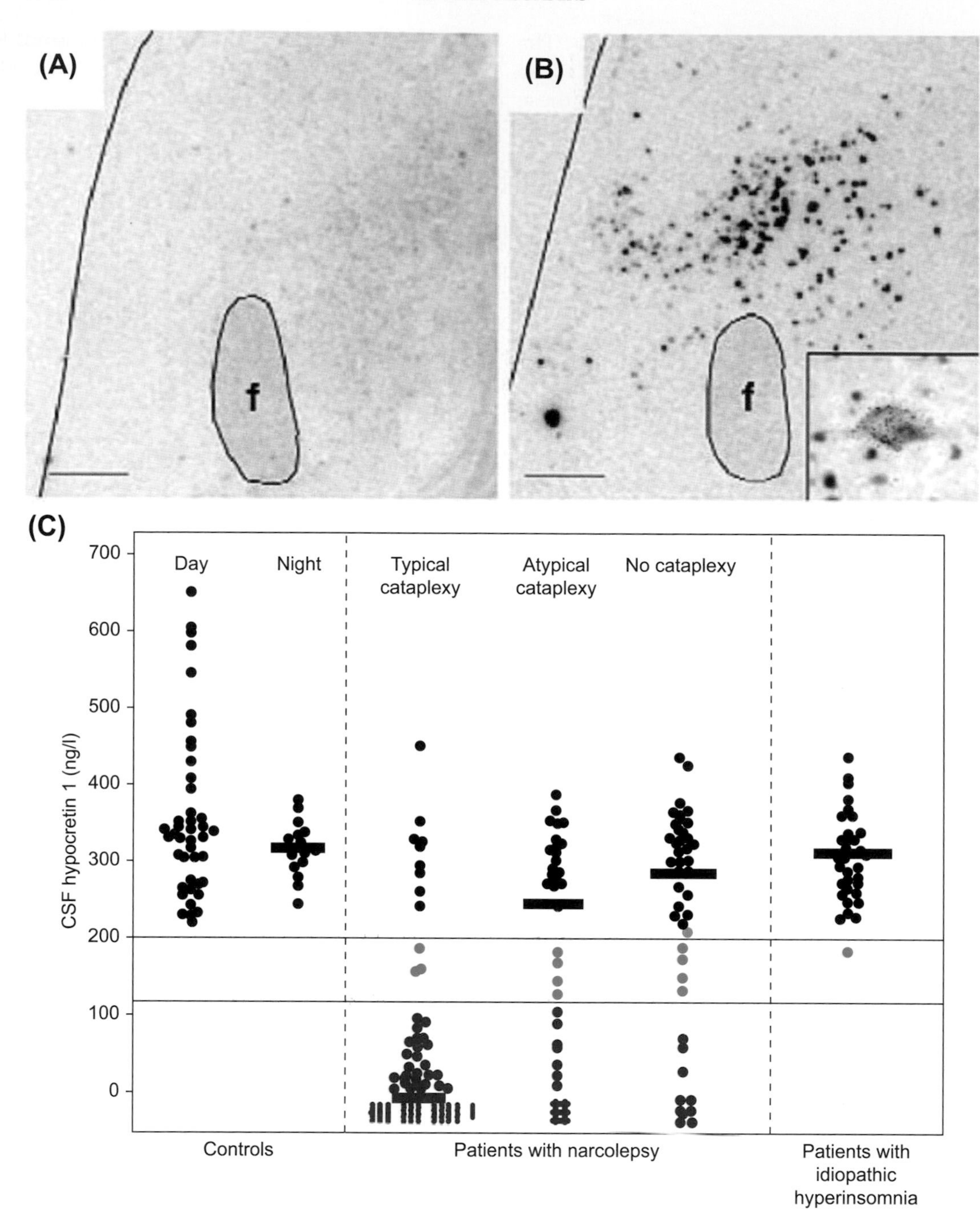

FIGURE 36.7 **Narcolepsy is caused by a loss of hypocretin cells.** (A, B) Distribution of hypocretin-labeled neurons in narcoleptic patients and controls. Each dot is a hypocretin-producing cell and these are seen only in controls. f: fornix. (C) Hypocretin-1 concentrations in cerebrospinal fluid (CSF) in controls, in people with narcolepsy with and without cataplexy, and in those with idiopathic hypersomnia. Cutoffs for normal (>200 pg/ml) and low (<110 pg/ml) hypocretin-1 concentrations are represented by dashed lines. Mean values are represented as a horizontal bar in each group. *Source: (A, B) Reprinted from Peyron* et al. Nat Med. *2000;6(9):991–997*[38]*; (C) adapted from Dauvillers* et al. Lancet. *2007;369(9560): 499–511.*[35]

Autoimmunity Towards Hypocretin Neurons

The specific loss of hypocretin neurons in narcolepsy is believed to be the result of an autoimmune process. This is based on the presence of $CD4^+$ T-cell reactivity towards hypocretin in blood samples from patients with narcolepsy.[39] Several preprohypocretin epitopes with high affinity towards the binding pocket of DQA1*01:02/DQB1*06:02 have been reported, and $CD4^+$ T-cell reactivity towards two of these, $HCRT_{56–68}$ and $HCRT_{87–99}$, is uniquely present in blood samples from narcoleptic patients but not from healthy controls. The two peptides correspond to the C-terminal end of the two mature, secreted functional hypocretin peptides. The first study reporting hypocretin-reactive CD4+ T cells tested 37 patients and found that 33 of these reacted to $HCRT_{56-68}$ or $HCRT_{87-99}$.[39] Three of the four patients without reactivity had long-lasting disease, indicating that the autoimmune process may abate in some subjects over time.

Because the loss of hypocretin causes narcolepsy, and hypocretin-containing neurons are apparently the only cells lost, $CD4^+$ T-cell autoreactivity to DQ0602-HCRT is likely to be a critical contributor to disease pathophysiology, although it is possible that it is only a secondary reaction. The sequence of events through which hypocretin-reactive $CD4^+$ T cells would lead to neuronal death remains obscure. Neurons do not express major histocompatibility complex (MHC) class II, even in the presence of cytokines such as interferon-γ that induce class II on other cell types. It is therefore unlikely that autoreactive $CD4^+$ T cells directly kill hypocretin neurons. Rather, other cells such as $CD8^+$ T cells could be involved, although to date there is no genetic, pathological, or immunological evidence for autoaggressive $CD8^+$ T cells in this disease. Unlike other human leukocyte antigen (HLA) class II-associated diseases there is no evidence for the association of narcolepsy with alleles of HLA class I, the antigen-presenting molecule of $CD8^+$ T cells, although this has not been thoroughly explored. The lack of HLA class I association may suggest the involvement of an epitope with promiscuous binding to many HLA class I alleles, or another mechanism, for example involving microglial cells.

It is still unclear whether autoreactive antibodies contribute to the tissue destruction in narcolepsy. Findings in this area have been either controversial or negative. Older studies have typically searched for serum or CSF autoantibodies directed against preprohypocretin, or against hypocretin cells using immunohistochemistry on brain tissue sections. In general, these studies reported negative results, suggesting that antibody responses and epitope spreading may be limited in diseases such as narcolepsy that selectively implicate neurons.

In 2010, a publication reported autoantibodies against tribbles homologue-2 (TRIB2) in 14% of European patients versus 5% of controls,[40] a finding replicated in other patient populations in the USA and Japan, especially in those with a recent onset of the disease. This finding raised the possibility that an autoimmune marker for narcolepsy had been discovered. Unfortunately, recent-onset narcolepsy patients have all tested negative for TRIB2 antibodies, suggesting that something unique occurred in the previous cases. Indeed, only selected samples collected before 2008 were positive (unpublished analysis by the present authors), a finding that could possibly be explained by the activity of a specific virus strain that has since disappeared. Furthermore, TRIB2 is expressed not only in hypocretin neurons, but also in many other cell populations in both the CNS and the periphery, including immune cells. Thus, TRIB2 autoantibodies are unlikely to be the cause of the specific destruction of hypocretin cells, but may be an upstream or a downstream effect.

Genetic Evidence

The first suggestion that narcolepsy may have an autoimmune basis came from HLA typing studies. Narcolepsy has the strongest HLA association known, with more than 99% of cases carrying a specific pair of alleles at two linked HLA loci, DQA1 and DQB1, namely DQA1*01:02 and DQB1*06:02. As these HLA alleles heterodimerize to function as a unit, the HLA DQA1*01:02/DQB1*06:02 heterodimer (referred to as DQ602) seems to be almost necessary to develop the disorder.

Other HLA alleles also affect predisposition to narcolepsy, but this is only a weak effect. DQB1*06:02/DQB1*03:01 heterozygotes are, for example, at higher risk for narcolepsy than DQB1*06:02 heterozygotes in general. Conversely, carriers of DQB1*06:02/DQB1*06:01, DQB1*06:02/DQB1*05:01, and DQB1*06:02/DQB1*06:03 are at decreased risk. Heterodimerization of DQA1*01:02 and DQB1*06:02 with other DQA1 and DQB1 alleles of the DQ1 group may explain these protective effects, by reducing the abundance of the disease-susceptibility DQA1*01:02/DQB1*06:02 heterodimer (Fig. 36.8). The hetorodimerization hypothesis does not include DQB1*03:01. Instead, DQB1*03:01 has been shown to be associated with earlier disease onset.

Although HLA genes are mostly studied in the context of autoimmune diseases, their natural function is to bind foreign antigens and to present these epitopes to the T-cell receptor (TCR) in the context of the adaptive immune response. Genome-wide association studies have found that the T-cell receptor alpha chain gene (*TRA*) is also an important susceptibility factor for narcolepsy. The TCR is expressed in T cells and initiates an immune response when it interacts with peptide-bound HLA. It is, however, an extremely complex locus, and in this case variation in the peptide–HLA binding domain of the protein is created by random somatic cell recombination (similar to the immunoglobulin loci), followed by negative and positive selection of functional and useful TCRs. This process leads to the generation of a diverse repertoire of unique TCR-bearing T cells that adapt to environmental challenges and keep memories of past events. It is possible that a single peptide unique to hypocretin cells is presented by a HLA-DQB1*0602/DQA1*0102 molecule to specific T cells (Fig. 36.9). This could then trigger an autoimmune reaction leading to the destruction of hypocretin-producing cells.[41]

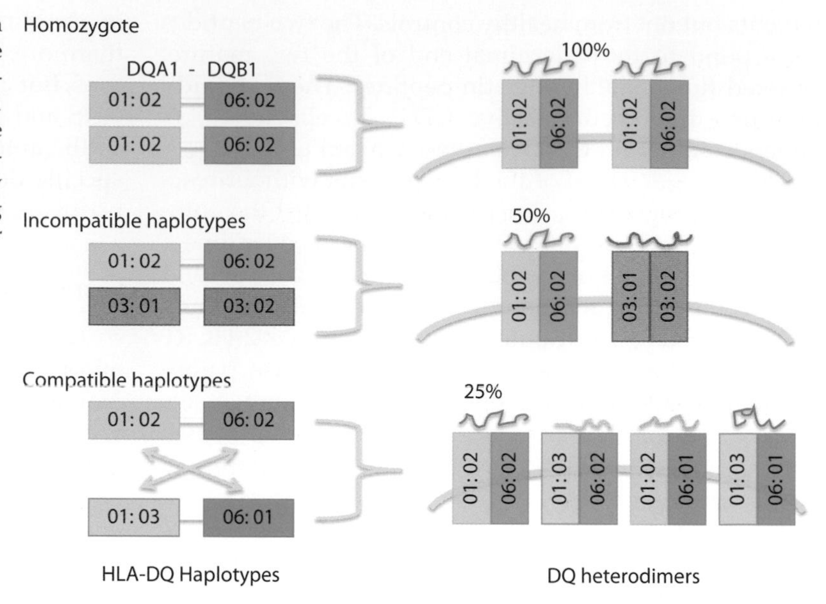

FIGURE 36.8 **Effects of human leukocyte antigen (HLA) haplotypes on susceptibility to narcolepsy.** The figure shows a proposed model of how heterodimerization of DQA1*01:02 and DQB1*06:02 with other DQA1 and DQB1 alleles of the DQ group may explain the protective effects of these alleles. The heterodimerization reduces the abundance of the disease-susceptibility DQA1*01:02/DQB1*06:02 heterodimer, thus lowering the risk of developing narcolepsy. *Source: Courtesy of Dr Juliette Faraco.*

Beside HLA and TCRs, narcolepsy is associated with polymorphisms in other mainly immune-related genes including the purinergic receptor subtype 2Y11 (P2RY11), cathepsin H (*CTSH*), tumor necrosis factor (ligand) superfamily member 4 (*TNFSF4*; also called OX40L or CD252), *ZNF365*, the *IL10RB-IFNAR1* locus, and T-cell receptor beta (*TRB*). Even though the genes associated with narcolepsy are mostly known for their immune-regulatory function, most of them are also expressed in the brain, thus a direct effect in the brain cannot be ruled out. For HLA in particular, whereas MHC class II molecules are not known to be expressed in neurons, MHC class I molecules have an established role in brain development and the regulation of synaptic plasticity. It is thus possible that the cause of narcolepsy will involve novel functions for the growing number of immune-like molecules that have been shown to have functional effects in the CNS.

Environmental Factors

Autoimmune diseases in general result from the interaction of environmental factors on a susceptible genetic background. In the case of narcolepsy, the importance of environmental factors is best illustrated by the fact that most (≈75%) monozygotic twins are discordant for narcolepsy.

In terms of environmental factors, there is now solid evidence that upper airway infections, usually influenza A and β-hemolytic streptococcal infections, can

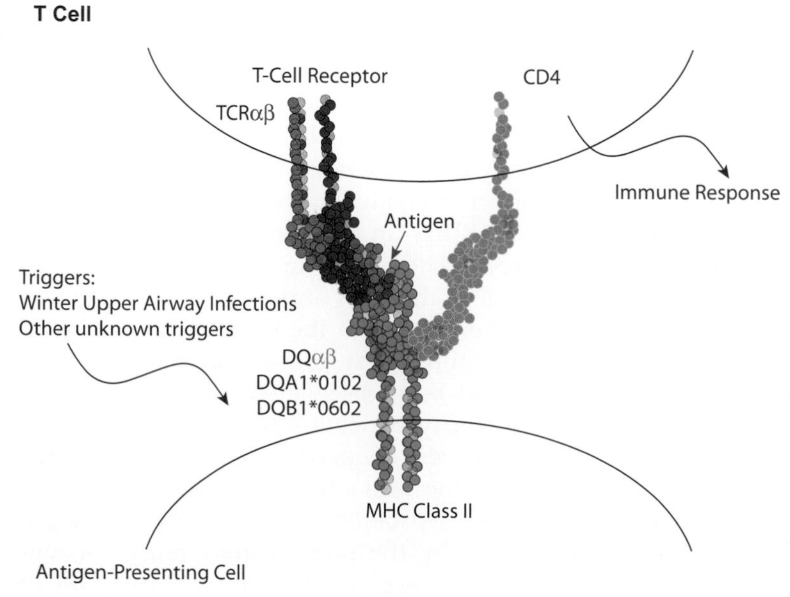

FIGURE 36.9 **Key genetic factors in narcolepsy with cataplexy.** Schematic representation of the most important factors determining risk of narcolepsy with cataplexy. TCR: T-cell receptor; MHC: major histocompatibility complex. *Source: Courtesy of Dr Juliette Faraco.*

precipitate the onset of narcolepsy. In a case–control study, unexplained fevers and influenza infections in the year preceding onset were associated with increased risk of developing narcolepsy. Epidemiological population studies in King County, Washington, USA,[42] showed that the risk of narcolepsy is increased in subjects with a history of a physician-diagnosed streptococcal throat infection. The same group also found that narcolepsy was associated with a history of passive smoking in childhood (also associated with upper airway infections).

Onset of narcolepsy has also been shown to be highly seasonal in children developing the disease, with a six-fold higher incidence in April than in December, suggesting occurrence most typically 5–6 months following the winter. The involvement of streptococcal infections in development of the disease is further supported by the presence of antistreptococcal antibodies in 65% of narcoleptic patients within 1 year of onset compared with 26% in age-matched controls in US studies.

After the 2009–2010 H1N1 influenza pandemic, a significant six- to nine-fold increase in the risk of developing narcolepsy was reported in northern Europe; this followed a campaign of pandemic H1N1 vaccination using Pandemrix™, an AS03-adjuvanted influenza vaccine. In parallel, data from China showed a three-fold increase in the occurrence of childhood cases following the winter of 2009–2010, independent of vaccination, suggesting that H1N1 infections by themselves also increased susceptibility or precipitated earlier onset in susceptible patients. The association between H1N1 vaccination and narcolepsy is still debated and other factors associated with the 2009–2010 pandemic may explain the association. However, data from China suggest that new onset rates in children decreased to prepandemic levels in 2011 and 2012.[41]

Although the association between H1N1 and narcolepsy remains unclear, molecular mimicry has been suggested to be involved. A hemagglutinin (HA) pHA1 epitope specific to the 2009 H1N1 strain, $pHA1_{275-287}$, was found to have homology to $HCRT_{56-68}$ and $HCRT_{87-99}$.[39] Vaccination of narcolepsy patients with a seasonal influenza vaccine containing pH1N1 increased the frequency of T cells reactive not only to $pHA1_{275-287}$ but also to $HCRT_{56-68}$ and $HCRT_{87-99}$. Finally, the study showed that $CD4^+$ T cells from narcolepsy patients stimulated with either the HA epitope or the hypocretin epitopes consequently could be stimulated with the other epitope. These data point towards molecular mimicry between $pHA1_{275-287}$ and $HCRT_{56-68}/HCRT_{87-99}$ as a potential trigger for pH1N1-related narcolepsy. As $pHA1_{275-287}$ seems to be unique to the pH1N1 influenza strain it remains to be studied whether other pathogens also carry epitopes mimicking the hypocretin epitopes, and whether these could be involved in triggering narcolepsy before 2009.

Generalized stimulation of the immune system could also trigger narcolepsy by activating dormant T-cell clones or permeabilizing the blood–brain barrier and thereby facilitating immune cell entry to the brain. The fact that narcolepsy onset is seasonal, and associated with both streptococcus and H1N1 infections, could argue in favor of non-specific immune activation precipitating the disease in predisposed individuals.

KLEINE–LEVIN SYNDROME

Kleine–Levin syndrome (KLS) is a rare sleep disorder that mainly affects adolescents and is characterized by recurrent episodes of severe hypersomnia, cognitive impairment, apathy, derealization, and psychiatric and behavioral disturbances. Boys are more frequently affected than girls. Typical episodes last from a few days to several weeks (median 10 days) and end suddenly, with relapses occurring every 1–12 months for a median of 14 years. Between episodes, patients are completely normal.[43]

The pathophysiology underlying KLS remains unknown. No objective biochemical, neuropathological, or structural imaging alterations have been identified. The symptomatology suggests that the associative cortex and deeper structures involved in sleep regulation are all affected. This hypothesis is supported by functional imaging studies, which have shown hypoactivity in thalamic and hypothalamic regions, and in the frontal and temporal lobes during episodes. During episodes, EEG can show diffuse or local slow activity. Temporal lobe dysfunction may explain altered cognition and derealization, and frontal lobe involvement could explain apathy and disinhibition. Thalamus and hypothalamus involvement could explain the hypersomnia. Basic motor, cerebellar, and sensory functions are intact.

Studies conducted during asymptomatic periods between episodes have shown persistent hypoperfusion of the temporal lobe in 50% of patients and, more rarely, in the frontal lobe and the basal ganglia. These results suggest that functional brain abnormalities are also present between episodes (interictal) in these patients.

Serum biology is normal during and between episodes, including markers of inflammation and white blood cell counts. White blood cell counts and protein concentrations in CSF are also normal in KLS patients. CSF concentrations of hypocretin-1 are also within the normal range, although they are lower during than between episodes when asymptomatic, possibly a passive reflection of sleepiness or brain pathology. An epileptic cause has been ruled out by EEG studies and by a lack of efficacy of antiepileptic medication during episodes. Causal factors may instead be genetic, autoimmune, inflammatory, or metabolic in various combinations.

Birth Difficulties

A four-fold increase in self-reported birth difficulties has been found in KLS patients compared with control subjects.[44] This finding is reminiscent of autism, epilepsy, and schizophrenia, where it is also speculated that birth difficulties could cause brain injury and subsequent symptoms. Alternatively, birth difficulties could reflect an already existing brain abnormality, as birth difficulties are prevalent in children with hypothalamic defects. In light of this, increased birth complications in KLS could signal an underlying brain dysfunction perhaps causing the very unusual response to an external triggering event, such as a stressful situation or an infection.

Infectious Diseases

It has been speculated whether KLS could be caused by an infectious disease or autoimmunity. This is supported by the fluctuating symptomatology, and the presence of mild upper airway infections or fever preceding episodes in 72–96% of cases.[44]

Although infections are known triggers for episode onset, the pathogens involved are highly variable. Onset is too rapid (1–3 days) after infection for an abnormal immune response to have developed, which suggests that infection and immunity are instead precipitating factors. Besides infections and fever, other triggering events include alcohol intake (alone or combined with sleep deprivation) and head trauma. As all these factors can lead to a weakening of the blood–brain barrier, it has also been speculated whether the symptoms are caused by a dysfunctional response to some metabolite from the blood entering the brain.

Some metabolic disorders are associated with hypersomnia, megaphagia, confusion, and disturbed behavior, such as genetic disorders leading to hyperammonemia, and those leading to abnormal serotonin and dopamine pathways (e.g. sepiapterin reductase deficiency). KLS could be caused by similar rare genetic mutations, as familial clustering has been reported; however, the disease most often occurs spontaneously without family history.

RESTLESS LEGS SYNDROME

RLS is a sensorimotor disorder where patients suffer from an urge to move their legs, most often in association with uncomfortable or painful sensations in the lower limbs. RLS mainly affects patients in the evening or at night, culminating between midnight and 2 a.m. The symptoms occur during both relaxation and sleep, and are relieved by moving the affected limb. This can be very disturbing for sleep initiation and maintenance.[45]

Periodic limb movements (PLMs) occur in up to 80% of patients with RLS and can cause sleep disruption. PLMs are repetitive twitching leg movements lasting between 0.5 and 5 seconds and recurring at intervals of 5–90 seconds for minutes to hours during the night. Quantification of PLMs (PLM index) can be used as an objective measure of disease severity in RLS cases, and is often used in clinical trials to assess the therapeutic response to an agent. However, PLMs can occur by themselves in many more healthy subjects than those complaining of RLS. Whether or not having a lot of PLMs at night without RLS can cause disturbed sleep and daytime symptoms is debated. PLMs are also frequently found in other disorders including narcolepsy and neurodegenerative diseases. PLMs can occur in patients with spinal cord transsection, suggesting that the generator for these movements is in the spinal cord.

RLS symptoms are present in up to 10% of the population. The age of onset varies widely, from childhood to over 80 years of age. RLS occurs in nearly 2% of school-age children. Many patients do not develop daily symptoms until 40–60 years of age.

RLS is generally considered to be a CNS-related disorder, although no specific lesion has been found to be associated with the syndrome. Numerous studies have shown that iron deficiency in the brain and reduced dopaminergic neuronal activity are critical pathophysiological factors.[46] RLS can also be secondary to some other conditions such as pregnancy, end-stage renal failure, iron deficiency, and polyneuropathies.

Dopamine Disturbances

Dopamine agonists and levodopa, a precursor of dopamine that can cross the blood–brain barrier and is metabolized in the brain into dopamine, are very effective treatments for RLS.[47] The efficacy of dopaminergic drugs in the treatment of RLS points to disturbed dopaminergic neurotransmission being central in RLS. Long-term use of dopaminergic treatment can, however, worsen RLS symptoms, in a reverse tolerance-like phenomenon called augmentation, where symptoms start to occur earlier and earlier in the day.

Studies in animal models and patients have shown increased concentrations of tyrosine hydroxylase (the rate-limiting enzyme in dopamine synthesis) in the substantia nigra and decreased numbers of D_2 dopamine receptors in the putamen. This suggests that RLS patients have increased dopaminergic signaling compensated by postsynaptic desensitization or *vice versa*. Differences in dopamine-related markers have also been demonstrated in the CSF of individuals with RLS. Since dopaminergic activity shows natural circadian fluctuations, this could offer an explanation for the characteristic circadian pattern of the symptoms. Dopamine agonists are also effective against PLMs, suggesting that dopamine projections to the spinal cord could be involved.

However, despite clear indications of dopamine being implicated in the mode of action of some treatments, a causative role for dopamine in the pathophysiology of RLS remains uncertain.

A role for neuronal nitric oxide synthase-1 (NOS1) in RLS has also been suggested. NOS1 is involved in the production of nitric oxide in the brain, which in turn takes part in the regulation of pain perception, control of sleep–wake regulation, and modulation of dopaminergic activity.

Iron Deficiency

Iron deficiency has been considered to have a causal role in RLS ever since the first modern description of RLS in the middle of the twentieth century and the observation of cases clearly secondary to iron deficiency. Studies have shown that iron substitution can be effective in some RLS cases, especially after intravenous administration. Iron is an essential cofactor for the formation of dopamine, consistent with the hypothesis that disturbances in dopamine function can cause RLS.

Several studies on RLS and iron parameters have been performed. Low iron levels in the substantia nigra and the red nucleus in the brain of RLS patients have been demonstrated. The substantia nigra is a very important brain region for initiating movements, and the red nucleus regulates coordination of limb movements. Differences in iron-related markers such as ferritin and transferrin saturation have also been demonstrated in the CSF of individuals with RLS.

Using blood samples from RLS patients, studies have shown that low serum ferritin, an indicator of the level of body iron storage, is associated with both increased risk of RLS and more severe RLS. However, other studies have not found this. It is possible that peripheral iron parameters do not accurately reflect the iron status in the brain, which would make serum ferritin and related markers less relevant in the context of RLS.

Genetic Predisposition

Approximately six out of every 10 patients with RLS have a family history of the syndrome, which suggests genetic predisposition. Genetic studies of RLS have revealed a role for developmental regulatory factors.[48] The transcription factors involved may affect spinal cord regulation of sensory perception and locomotor pattern generation, and may also interact with brain iron homeostastis. Supraspinal influences may also be very important.

The *Meis1* locus is the most important RLS susceptibility gene.[49] MEIS1 is strongly expressed in dopaminergic neurons of the substantia nigra, in the spinal cord, and in the red nucleus. MEIS1 is part of a transcriptional regulatory network that is very important for motor neurons, suggesting a key link to the pathophysiology of RLS.

Variants in *BTBd9* [BTB (POZ) domain containing-9] have also repeatedly shown association with RLS, and especially PLMs, implying that this gene confers risk specifically for the motor component of RLS. Serum ferritin levels were also found to vary by genotype, potentially underlying the iron deficiency associated with RLS. Little is known of the function of BTBD9 in mammals; however, in *Drosophila*, proteins containing the BTB (POZ) domain have important roles in metamorphosis and limb pattern formation.

Other gene variants have also been found to increase the risk of RLS. These include variants in the region encompassing *Skor1* (SKI family transcriptional corepressor-1) and *MAP2K5* (mitogen-activated protein kinase kinase-5). *Skor1* is expressed selectively in a subset of neurons in the developing spinal cord. These neurons relay pain and touch, and thus may play a role for the sensory components of RLS. *PTPRD* (protein tyrosine phosphatase receptor type delta) is associated with RLS and has established roles in axon guidance and termination of motor neurons during embryonic development in the mouse. *TOX3* (a breast cancer susceptibility locus) and untranslated BC034767 were associated with susceptibility in an extended RLS sample, but a role for these two transcripts in RLS pathogenesis is not yet known.

Taken together, the current understanding of RLS is that the disease is caused by an imbalance in the brain and spinal cord circuitry that integrates sensory input and controls locomotor output. This imbalance is probably developmental and the result of genetic variants in several factors. Disturbances in the level of several substances in the brain, such as iron, dopamine, and nitric oxide, then further destabilize the circuitry, causing the RLS symptoms to appear. Much work remains to be done to determine the exact circuitry involved, but identification of the clear genetic factors described here is likely to aid a rapid increase in understanding of the pathophysiology of RLS.

RAPID EYE MOVEMENT SLEEP BEHAVIOR DISORDER

Normal REM sleep is characterized by an activated brain state with EEG desynchronization in combination with skeletal muscle paralysis, preventing the enactment of dreams. In RBD, the normal atonia is lost and patients therefore move during REM sleep. The spectrum of movement varies from small hand movements to violent activities, such as punching, kicking, or leaping out of bed. These can be dangerous, and may lead to self-injury or trauma to the spouse.[50]

RBD is characterized by two other symptoms besides abnormal motor behavior: abnormal vocalizations and altered dream mentation. Healthy individuals may grunt, speak, laugh, or vocalize in a variety of ways during non-REM and REM sleep, but the vocalizations in RBD tend to be much louder and suggest unpleasant dreams. Shouting, screaming, and swearing are common and are often described as being very unlike the typical soft-spoken nature of the person's tendency to speak during wakefulness. The vocalizations and behaviors that are exhibited are consistent with the content of the dreams later reported by the patient, suggesting that RBD is in fact enactment of the dream content.

RBD frequently occurs as a side effect of antidepressant treatment (loss of muscle atonia during REM sleep due to chronic treatments with adrenergic or serotoninergic reuptake inhibitors), but most patients with primary RBD are male. In primary RBD most develop symptoms in the 40–70 years age range, although onset of symptoms can vary widely. Those with RBD evolving before age 40 typically have narcolepsy, but idiopathic RBD can also develop early in some cases. Idiopathic RBD cases starting around middle age frequently evolve into Parkinson disease or Lewy body dementia, with onset of symptoms occurring up to 20 years after the onset of RBD.

Based on the symptoms, one would predict that the underlying cause of RBD is a brain disorder that affects one of the brain areas in the REM sleep regulatory network. This is indeed the case, and several disorders can lead to this, including other primary sleep disorders (e.g. narcolepsy), structural lesions in the brainstem, neurodegeneration, medications, and head trauma.

Disturbed Rapid Eye Movement Sleep Regulation

The suppression of motor activity during REM sleep is the cumulative result of multiple pathways that terminate at spinal motor neurons. Multiple areas of the brainstem may influence muscle tone during REM sleep. These include the REM active and non-REM active nuclei described above (see "Regulation of Rapid Eye Movement Sleep"), as well as various related brainstem structures. Dysfunction in these structures, as well as their neurotransmitters and pathways, can result in REM sleep without atonia.

Based on animal lesion experiments, several brainstem regions have classically been considered to be core to the RBD pathophysiology (Fig. 36.10). Lesions in the MCRF have been shown to release the tonic inhibition on spinal motor neurons, leading to loss of REM sleep atonia, but since these lesions also destroy passing fibers it has been difficult to draw any conclusions from this. It is also not yet clear whether lesioning or degeneration of the MCRF alone is sufficient to cause RBD in humans. Small lesions limited to the SLD also induce RBD-like symptoms. Importantly, larger lesions of SLD induce a decrease in the total quantities of REM sleep, which is not the case for RBD patients. Lesions in the locus ceruleus also cause loss of REM sleep atonia, and the site and size of the lesion determine whether simple or complex behaviors are exhibited.

The substantia nigra and the dopaminergic system have been implicated in RBD, but this may be due to the fact RBD is associated with Parkinson disease (see below). Neuroimaging reveals coincident and progressive dopaminergic abnormalities in RBD. Investigations have found reduced striatal dopamine transporters and dopaminergic innervation. Transcranial ultrasound has also indicated neuronal dysfunction in the substantia nigra. However, the use of dopaminergic agents usually does not improve RBD.

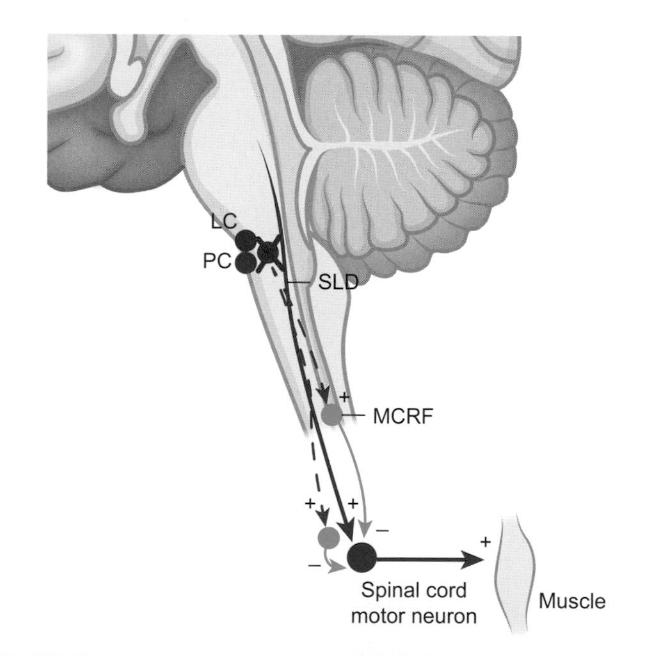

FIGURE 36.10 **Proposed pathophysiology of rapid eye movement (REM) sleep behavior disorder in humans.** The sublaterodorsal nucleus (SLD) projects to spinal interneurons ("direct route") and probably represents the final common pathway that causes active inhibition of skeletal muscle activity in REM sleep. The "indirect route", from the SLD to the medullary ventral magnocellular reticular nucleus/formation (MCRF) to the spinal interneurons, may also contribute to electromyographic (EMG) atonia. Loss of SLD signaling will cause a loss of REM sleep atonia since the inhibition of the spinal cord motor neuron that normally is induced by the active SLD during REM sleep (Fig. 36.4C) is lost. However, it is not yet known whether lesions in structures which project to and from the MCRF, or lesioning the MCRF itself, are critical in affecting EMG atonia during REM sleep. LC: locus ceruleus; PC: preceruleus.

Parkinson Disease

RBD can be an early stage of Parkinson disease. Longitudinal studies estimate that more than 50% of patients with idiopathic RBD will develop neurodegenerative α-synuclein CNS pathologies such as Parkinson disease, multiple system atrophy, and dementia with Lewy bodies.[51] This is not surprising as the brainstem nuclei that control REM sleep are often involved early in the natural history of these synucleinopathies. The time interval between the onset of RBD and the onset of parkinsonian symptoms, such as resting tremor and difficulties in initiating movement, varies from months to decades, with a mean latency from RBD onset to disease diagnosis of 13 years. The hypothesized pathophysiology of RBD in these cases is that the synucleopathy starts in the brainstem (RBD), evolving frontally to affect the substantia nigra (Parkinson disease), and then cortex (Lewy body dementia), recapitulating the Braak staging system for Parkinson disease.

This high conversion rate to neurodegenerative disease provides a unique opportunity to observe directly the development of clinical parkinsonism, and RBD is increasingly being studied with the purpose of increasing our understanding of early-stage Parkinson disease.

Other Brain Disorders

RBD has also been associated with brain disorders such as narcolepsy, limbic encephalitis and Guillain–Barré syndrome. One-third of narcolepsy patients have RBD symptoms, and RBD is much more prevalent in the subgroup of patients who also have cataplexy.[52] The incidence of RBD symptoms in narcolepsy patients is correlated with hypocretin deficiency, which could explain the association between the disorders given the stabilizing role of hypocretin in sleep regulation (see "Transitions Between Wakefulness and Sleep", above). Hypocretin deficiency causes a general instability of sleep–wake regulation and REM sleep motor regulation, which explains not only the sleep attacks and multiple awakenings in narcolepsy patients, but also the coexistence of RBD in narcolepsy. One possibility is that in normal conditions, hypocretin neurons excite the descending pathways from SLD during REM sleep, thus increasing the suppression of the spinal cord motor neurons, in particular during the phasic excitation of the motor neurons in REM sleep. Hypocretin neurons are mainly active during wakefulness, but do display bursts of activity during REM sleep. Application of hypocretin in the SLD region has further been shown to induce REM sleep with atonia.

RBD has been associated (although much less commonly) with neurodegenerative diseases that do not affect brainstem structures, including progressive supranuclear palsy, frontotemporal dementia, and Huntington disease. Drugs, including antidepressants, alcohol, and β-blockers, have also been linked to RBD.

Thus, several diverse pathologies can lead to RBD, including α-synuclein neurodegeneration, non-synuclein neurodegeneration, hypocretin dysfunction, drugs, and direct brain lesions. Common to these is that they affect the REM sleep regulatory network, and thus may result in a loss of behavioral control during REM sleep, resulting in the enactment of dreams.

CONCLUSION AND QUESTIONS FOR FURTHER RESEARCH

Sleep remains one of the big mysteries in biology. As a state that seemingly freezes all productive activity and puts animals in danger of being caught by predators, sleep must serve an important purpose because it has survived many years of evolution. Nevertheless, the function of sleep and the molecular processes that produce the need to sleep remains elusive – even its role in synaptic plasticity remains debatable. Current research is especially focusing on two fronts: the development of model organisms such as zebrafish, *Drosophila*, and worms, for studying the molecular mechanisms of sleep; and attempts to identify genetic underpinnings of human sleep traits and sleep disorders.

Understanding of the circadian regulation and the molecular mechanisms behind the cellular clocks has greatly expanded in the twenty-first century, and these processes are now understood at a highly detailed level. Identification of genetic mutations that cause circadian alterations in different species from humans to *Drosophila* has been particularly helpful in this process. In contrast to the circadian field, however, it has been difficult to find a common core genetic pathway in sleep, and few mutants have been identified. This is one of the reasons why the regulation of sleep architecture and duration is still very poorly understood, and identification of genetic mutations and polymorphisms that affect these traits is needed to move sleep research forward.

One example of a successful genetic study of a sleep trait was the discovery of a mutation in a transcriptional repressor DEC2 (P385R) that is associated with a human short sleep phenotype.[53] The DEC2-P385R mutation also gives a short sleep phenotype in transgenic mice and *Drosophila* models, suggesting that it is possible to find evolutionarily conserved pathways that control the regulation of sleep and wakefulness. Future genetic studies in humans with long or short sleep may lead to a better understanding of the pathways involved.

Some other questions that also need to be addressed in future sleep research include the following.

- *Are homeostatic and circadian mechanisms really independent or are they intimately linked?* Circadian regulation and homeostatic regulation of sleep are usually considered distinct. Although this holds true under various experimental conditions, it would be illogical if these processes, functionally linked by environmental conditions such as light and dark, had not evolved molecular links. The results of recent research indeed suggest such links. Sleep deprivation can impair expression of circadian genes and modulate electrical activity within the SCN, and circadian mutants and polymorphisms have been shown to affect sleep homeostasis. For example, homozygous carriers of the short or long allele of PER3 show differences in several markers of sleep homeostasis including slow-wave activity in non-REM sleep. This suggests that a simple dichotomy between circadian and sleep homeostasis may not be valid.
- *What is the relevance of REM sleep?* Non-REM sleep may have localized restorative effects (in particular SWS in the cortex), and it is tempting to speculate that REM sleep could have a similar role in some non-cortical regions. A cessation of neuronal activity during REM sleep is observed in some key regulatory areas, including those regulating body temperature and various autonomic functions. Homeostasis could thus be specifically restored in these networks during REM sleep. This hypothesis does not explain how REM sleep, a hybrid state with decreased activity in a few networks and increased activity elsewhere, could have evolved. To solve this puzzle, it has been argued that the function of REM sleep has changed across evolution. A primitive sleep state may have emerged first to restore homeostasis in locomotor, sensory, autonomic, and subsequently thermoregulatory networks. Sleep would have then diverged into two states: non-REM sleep to restore metabolic homeostasis in most of the brain, and REM sleep to restore selected primitive networks mentioned above. Activation of forebrain and limbic areas during REM sleep would have finally been selected to optimize learning and creativity, increasing survival and thus mitigating the negative effects of increased energy expenditure. This hypothesis may also explain why long-term REM sleep deprivation is lethal, as it would also perturb primitive networks involved in energy homeostasis and basic functions. These speculations strongly argue for the need to study specific molecular and electrophysiological changes within discrete networks regulating sleep and wakefulness. It may be predicted, for example, that recovery in sleep-promoting networks occurs during wakefulness, with a similar molecular signature as seen in the wake-active systems during sleep (but allowing for neurochemical diversity, as many of the wake-active systems currently reported are glutamatergic or monoaminergic, whereas most known sleep-promoting systems are GABAergic).
- *What controls the molecular and anatomical diversity of sleep regulatory networks across species?* It is becoming easier to conduct molecular studies in exotic species, thanks to genomic sequencing efforts, yet there are almost no data on the functional organization of sleep regulatory networks across species. Studies have shown that hypocretin does not have similar anatomical connections, has only one rather than two receptors, and is not strongly wake-promoting in fish, where light and melatonin have stronger effects. Similarly, birds, which are very sensitive to light and melatonin, also have a single hypocretin receptor. This suggests that the top neural networks orchestrating the occurrence of sleep are more variable across species than are cellular, molecularly based changes; this is analogous to the circadian system, where clock genes are more conserved than SCN organization. Further studies in selected species will be extremely instructive for understanding sleep across evolution, confirming or rejecting some of the hypotheses discussed above.

In conclusion, sleep is as necessary as water and food, yet it is unclear why it is required and maintained by evolution. Recent work suggests multiple roles, a correlation with synaptic plasticity changes in the brain, and widespread changes in gene expression, not unlike what has been discovered in circadian biology. Functional data are, however, still largely lacking, and studies such as functional genomic screens in model organisms, comparative sleep neuroanatomy through phylogeny, and the study of molecular changes within specific wakefulness, REM sleep, and non-REM sleep regulatory systems are needed. The resilience of behavioral sleep in evolution and after experimental manipulations may be due to the fact that it is firmly grounded at molecular, cellular, and network levels across the entire brain.

References

1. Ross JJ. Neurological findings after prolonged sleep deprivation. *Arch Neurol*. 1965;12:399–403.
2. Rechtschaffen A, Bergmann BM, Everson CA, Kushida CA, Gilliland MA. Sleep deprivation in the rat: X. Integration and discussion of the findings. *Sleep*. 1989;12(1):68–87.
3. Fuchs T, Maury D, Moore FR, Bingman VP. Daytime micro-naps in a nocturnal migrant: an EEG analysis. *Biol Lett*. 2009;5(1):77–80.
4. Moore-Ede MC. Physiology of the circadian timing system: predictive versus reactive homeostasis. *Am J Physiol*. 1986;250(5 Pt 2): R737–R752.
5. Mignot E. Why we sleep: the temporal organization of recovery. *PLoS Biol*. 2008;6(4):661–669.
6. Borbély AA. Refining sleep homeostasis in the two-process model. *J Sleep Res*. 2009;18(1):1–2.
7. Borbélly AA, Achermann P. Sleep homeostasis and models of sleep regulation. *J Biol Rhythms*. 1999;14:557–568.

8. Ralph MR, Foster RG, Davis FC, Menaker M. Transplanted suprachiasmatic nucleus determines circadian period. *Science*. 1990;247(4945):975–978.
9. Lowrey PL, Takahashi JS. Genetics of circadian rhythms in mammalian model organisms. *Adv Genet*. 2011;74:175–230.
10. Price JL, Blau J, Rothenfluh A, Abodeely M, Kloss B, Young MW. Double-time is a novel *Drosophila* clock gene that regulates PERIOD protein accumulation. *Cell*. 1998;94(1):83–95.
11. Kloss B, Price JL, Saez L, et al. The *Drosophila* clock gene double-time encodes a protein closely related to human casein kinase Iepsilon. *Cell*. 1998;94(1):97–107.
12. Freland L, Beaulieu J-M. Inhibition of GSK3 by lithium, from single molecules to signaling networks. *Front Mol Neurosci*. 2012;5:14.
13. Aton SJ, Colwell CS, Harmar AJ, Waschek J, Herzog ED. Vasoactive intestinal polypeptide mediates circadian rhythmicity and synchrony in mammalian clock neurons. *Nat Neurosci*. 2005;8(4): 476–483.
14. Schmidt TM, Do MTH, Dacey D, Lucas R, Hattar S, Matynia A. Melanopsin-positive intrinsically photosensitive retinal ganglion cells: from form to function. *J Neurosci*. 2011;31(45): 16094–16101.
15. Mohawk JA, Green CB, Takahashi JS. Central and peripheral circadian clocks in mammals. *Annu Rev Neurosci*. 2012;35:445–462.
16. Hara R, Wan K, Wakamatsu H, et al. Restricted feeding entrains liver clock without participation of the suprachiasmatic nucleus. *Genes Cells*. 2001;6(3):269–278.
17. Von Economo C. *Encephalitis Litargica [Newman KO, Trans.]*. London: Oxford University Press; 1931.
18. Von Economo C. Sleep as a problem of localization. *J Nerv Ment Dis*. 1930;71:249–259.
19. Luppi P-H. Neurochemical aspects of sleep regulation with specific focus on slow-wave sleep. *World J Biol Psychiatry*. 2010;11(suppl 1): 4–8.
20. Saper CB, Fuller PM, Pedersen NP, Lu J, Scammell TE. Sleep state switching. *Neuron*. 2010;68(6):1023–1042.
21. Kilduff TS, Cauli B, Gerashchenko D. Activation of cortical interneurons during sleep: an anatomical link to homeostatic sleep regulation? *Trends Neurosci*. 2011;34(1):10–19.
22. De Lecea L, Kilduff T, Peyron C, et al. The hypocretins: hypothalamus-specific peptides with neuroexcitatory activity. *Proc Natl Acad Sci USA*. 1998;95(1):322.
23. Luppi P-H, Clément O, Sapin E, et al. The neuronal network responsible for paradoxical sleep and its dysfunctions causing narcolepsy and rapid eye movement (REM) behavior disorder. *Sleep Med Rev*. 2011;15(3):153–163.
24. Imeri L, Opp MR. How (and why) the immune system makes us sleep. *Nat Rev Neurosci*. 2009;10(3):199–210.
25. Lesku JA, Roth TC, Rattenborg NC, Amlaner CJ, Lima SL. History and future of comparative analyses in sleep research. *Neurosci Biobehav Rev*. 2009;33(7):1024–1036.
26. Tononi G, Cirelli C. Time to be SHY? Some comments on sleep and synaptic homeostasis. *Neural Plast*. 2012;2012:415250.
27. Frank MG. Erasing synapses in sleep: is it time to be SHY? *Neural Plast*. 2012;2012:264378.
28. Bushey D, Tononi G, Cirelli C. Sleep and synaptic homeostasis: structural evidence in. *Drosophila. Science*. 2011;332(6037): 1576–1581.
29. Cirelli C, Gutierrez CM, Tononi G. Extensive and divergent effects of sleep and wakefulness on brain gene expression. *Neuron*. 2004;41(1):35–43.
30. Sehgal A, Mignot E. Genetics of sleep and sleep disorders. *Cell*. 2011;146(2):194–207.
31. Vanselow K, Vanselow JT, Westermark PO, et al. Differential effects of PER2 phosphorylation: molecular basis for the human familial advanced sleep phase syndrome (FASPS). *Genes Dev*. 2006;20(19):2660–2672.
32. Ebisawa T, Uchiyama M, Kajimura N, et al. Association of structural polymorphisms in the human period3 gene with delayed sleep phase syndrome. *EMBO Rep*. 2001;2(4):342–346.
33. Crowley SJ, Acebo C, Carskadon MA. Sleep, circadian rhythms, and delayed phase in adolescence. *Sleep Med*. 2007;8(6): 602–612.
34. Drake CL, Roehrs T, Richardson G, Walsh JK, Roth T. Shift work sleep disorder: prevalence and consequences beyond that of symptomatic day workers. *Sleep*. 2004;27(8):1453–1462.
35. Dauvilliers Y, Arnulf I, Mignot E. Narcolepsy with cataplexy. *Lancet*. 2007;369(9560):499–511.
36. Lin L, Faraco J, Li R, et al. The sleep disorder canine narcolepsy is caused by a mutation in the hypocretin (orexin) receptor 2 gene. *Cell*. 1999;98(3):365–376.
37. Chemelli RM, Willie JT, Sinton CM, et al. Narcolepsy in orexin knockout mice: molecular genetics of sleep regulation. *Cell*. 1999;98(4):437–451.
38. Peyron C, Faraco J, Rogers W, et al. A mutation in a case of early onset narcolepsy and a generalized absence of hypocretin peptides in human narcoleptic brains. *Nat Med*. 2000;6(9):991–997.
39. De la Herrán-Arita AK, Kornum BR, Mahlios J, et al. CD4+ T-cell autoimmunity to hypocretin/orexin and crossreactivity to a 2009 H1N1 influenza A epitope in narcolepsy. *Sci Transl Med*. 2013;5(216). 216ra176.
40. Cvetkovic-Lopes V, Bayer L, Dorsaz S, et al. Elevated Tribbles homolog 2-specific antibody levels in narcolepsy patients. *J Clin Invest*. 2010;120(3):713–719.
41. Kornum BR, Faraco J, Mignot E. Narcolepsy with hypocretin/orexin deficiency, infections and autoimmunity of the brain. *Curr Opin Neurobiol*. 2011;21(6):897–903.
42. Longstreth WT, Koepsell TD, Ton TG, Hendrickson AF, Van Belle G. The epidemiology of narcolepsy. *Sleep*. 2007;30(1):13–26.
43. Arnulf I, Rico TJ, Mignot E. Diagnosis, disease course, and management of patients with Kleine–Levin syndrome. *Lancet Neurol*. 2012;11(10):918–928.
44. Arnulf I, Lin L, Gadoth N, et al. Kleine–Levin syndrome: a systematic study of 108 patients. *Ann Neurol*. 2008;63(4):482–493.
45. Trenkwalder C, Paulus W. Restless legs syndrome: pathophysiology, clinical presentation and management. *Nat Rev Neurol*. 2010;6(6):337–346.
46. Salas RE, Gamaldo CE, Allen RP. Update in restless legs syndrome. *Curr Opin Neurol*. 2010;23(4):401–406.
47. Scholz H, Trenkwalder C, Kohnen R, Riemann D, Kriston L, Hornyak M. Dopamine agonists for restless legs syndrome. *Cochrane Database Syst Rev*. 2011;(3). CD006009.
48. Trenkwalder C, Högl B, Winkelmann J. Recent advances in the diagnosis, genetics and treatment of restless legs syndrome. *J Neurol*. 2009;256(4):539–553.
49. Winkelmann J, Schormair B, Lichtner P, et al. Genome-wide association study of restless legs syndrome identifies common variants in three genomic regions. *Nat Genet*. 2007;39(8):1000–1006.
50. Boeve BF. REM sleep behavior disorder: updated review of the core features, the REM sleep behavior disorder–neurodegenerative disease association, evolving concepts, controversies, and future directions. *Ann N Y Acad Sci*. 2010;1184:15–54.
51. McCarter SJ, St Louis EK, Boeve BF. REM sleep behavior disorder and REM sleep without atonia as an early manifestation of degenerative neurological disease. *Curr Neurol Neurosci Rep*. 2012;12(2):182–192.
52. Knudsen S, Gammeltoft S, Jennum PJ. Rapid eye movement sleep behaviour disorder in patients with narcolepsy is associated with hypocretin-1 deficiency. *Brain*. 2010;133(Pt 2):568–579.
53. He Y, Jones CR, Fujiki N, et al. The transcriptional repressor DEC2 regulates sleep length in mammals. *Science*. 2009;325(5942): 866–870.

CHAPTER

37

Fear-Related Anxiety Disorders and Post-Traumatic Stress Disorder

Arshya Vahabzadeh, Charles F. Gillespie*, Kerry J. Ressler*, †, ‡*

*Department of Psychiatry and Behavioral Sciences, Emory University School of Medicine, Atlanta, Georgia, USA; †Yerkes National Primate Research Center, Atlanta, Georgia, USA; ‡Howard Hughes Medical Institute, Chevy Chase, Maryland, USA

OUTLINE

INTRODUCTION

Anxiety and fear are distinct, but related, emotional states. Anxiety is experienced as a state of dread or apprehension in anticipation of encountering danger or an aversive experience. In contrast, fear is defined by the subjective experience of intense agitation triggered by the presence of an actual or a perceived threat that in some cases results in the expression of fear-related behaviors such as fleeing, fighting, or freezing. When appropriately triggered, anxiety and fear are useful emotional states that facilitate adaptation to hazardous environmental conditions. Optimal adaptation involves the development of "healthy" fear and anxiety that is neither overly specific nor excessively generalized. When anxiety or fear is inappropriately triggered, or experienced with a sufficient intensity to cause distress and functional impairment, then it is considered an anxiety disorder.[1]

Anxiety disorders are the most common type of psychiatric disorder, with an overall lifetime prevalence of over 28%.[2] As a diagnostic group, anxiety disorders are phenomenologically heterogeneous, with some disorders characterized predominantly by anxiety (generalized anxiety disorder), others by fear-related obsessional thoughts and behavioral compulsions (obsessive–compulsive disorder), and still other disorders characterized predominantly by fear [specific phobias, panic disorder, and post-traumatic stress disorder (PTSD)]. In particular, fear-related anxiety disorders appear to be based, in part, on associative learning that occurs in conjunction with a fear-provoking experience, or the mental representation

Neurobiology of Brain Disorders
http://dx.doi.org/10.1016/B978-0-12-398270-4.00037-9

of such a real or imagined experience, that triggers excessive fear that is behaviorally managed by inappropriate avoidance. This chapter will review the historical classification, clinical manifestations, and psychobiology of anxiety disorders, with an emphasis on fear-related anxiety disorders and PTSD.

CLASSIFICATION OF ANXIETY DISORDERS

The fourth edition of *Diagnostic and Statistical Manual of Mental Disorders* (DSM-IV) and its text revision, DSM-IV-TR (which was current when this chapter was conceived), outlined 12 categories of anxiety disorder.[3] In comparison to the earlier DSM-III, three new anxiety disorders were added to DSM-IV: acute stress disorder, anxiety disorder due to a general medical condition, and substance-induced anxiety disorder. The categorization of anxiety disorders was subsequently divided into anxiety, obsessive–compulsive, and trauma- and stressor-related disorders in the fifth edition, DSM-5, published in 2013 (Table 37.1).[4] The historical aspects of anxiety disorder classification are of interest to clinicians and researchers since they underpin both research methodology and clinical treatment. In DSM-II, anxiety disorders were categorized under "neurosis", a term which was later dropped from DSM-III. Three anxiety disorders were outlined in DSM-II, namely anxiety neurosis, phobic neurosis, and obsessive compulsive neurosis. DSM-III attempted to further subcategorize anxiety disorders, for example incorporating both generalized anxiety disorder and panic disorder. These diagnoses would have both fallen under anxiety neurosis in DSM-II; however, this new categorization reflected increasing knowledge about the disease course and treatment of these conditions. The increase in the number of categories seen in DSM-IV and DSM-5 could be perceived as an increasing recognition and awareness of the importance of anxiety disorders. However, some critics point out that the discriminative validity of these disorders may be undermined as we contend with increasingly significant overlap of symptomatology and comorbidity between anxiety and other psychiatric disorders.

A number of different anxiety disorders are described in DSM-5, including panic disorder, social phobia, PTSD, and generalized anxiety disorder. Several lines of research suggest that a majority of anxiety disorders may be conceptualized as disorders of fear.[5,6] Large-scale population studies such as the NEMESIS study in the Netherlands have highlighted how a "fear" dimension appears to be integral to social phobia, specific phobia, agoraphobia, and panic disorder. Conversely, the NEMESIS study also found that although generalized anxiety disorder shares an "anxiety–misery" dimension similar to that of major depression and dysthymia, it does not share the "fear" dimension common to panic disorder, specific phobia, and PTSD.[6] This reinforced previous findings by Krueger, who investigated patterns of psychiatric comorbidity using a sample of 8098 participants from

TABLE 37.1 Categorization of Anxiety, Obsessive–Compulsive, and Trauma-Related Disorders in DSM-5

ANXIETY DISORDERS
Separation anxiety disorder
Selective mutism
Specific phobia
Social anxiety disorder (social phobia)
Panic disorder
Panic attack (specifier)
Agoraphobia
Generalized anxiety disorder
Substance/medication-induced anxiety disorder
Anxiety disorder due to another medical condition
Other specified anxiety disorder
Unspecified anxiety disorder
OBSESSIVE–COMPULSIVE AND RELATED DISORDERS
Obsessive–compulsive disorder
Body dysmorphic disorder
Hoarding disorder
Trichotillomania (hair-pulling disorder)
Excoriation (skin-picking) disorder
Substance/medication-induced obsessive–compulsive and related disorder
Obsessive–compulsive and related disorder due to another medical condition
Other specified obsessive–compulsive and related disorder
Unspecified obsessive–compulsive and related disorder
TRAUMA- AND STRESSOR-RELATED DISORDERS
Reactive attachment disorder
Disinhibited social engagement disorder
Post-traumatic stress disorder
Acute stress disorder
Adjustment disorders
Other specified trauma- and stressor-related disorder
Unspecified trauma- and stressor-related disorder

Source: American Psychiatric Association. Diagnostic and Statistical Manual of Mental Disorders. *5th ed.* DSM-5. *Arlington, VA: American Psychiatric Publishing; 2013.*[4]

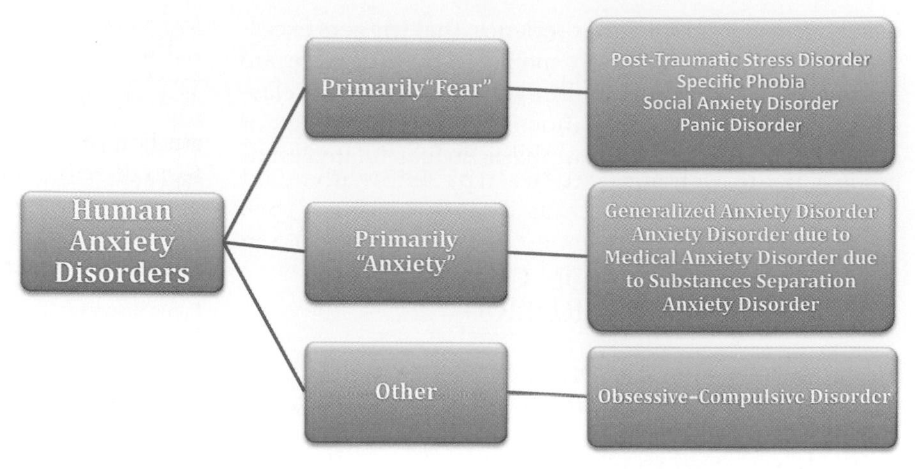

FIGURE 37.1 **Subcategorization of anxiety disorders based on symptoms.**

the National Comorbidity Survey.[5] Collectively, these findings are consistent with a subdivision of the broad class of anxiety disorders into disorders of anxiety and disorders of fear (Fig. 37.1).

NEUROANATOMICAL BASIS OF ANXIETY DISORDERS

The prevailing model of the pathogenesis of fear-related anxiety disorders is based on the acquisition of fear-related memories through associative learning and their maintenance through the reinforcing effects of inappropriate avoidance.[7] Insight into the etiology of anxiety disorders in humans has been acquired, in part, from preclinical research examining fear learning in rodents. In classical fear conditioning paradigms using rodents, an initially neutral conditioned stimulus (CS), such as a tone, is paired with an aversive unconditioned stimulus (US), such as a foot shock. This pairing subsequently elicits a collection of behavioral and physiological responses, known as the conditioned response (CR). After the CS–US pairing is made, the CS is able to elicit the CR in the absence of the US. The environmental context in which the initial conditioning occurred will also be able to elicit the response. This response serves as an observable and quantifiable representation of the emotional memory that was established.

Fear Learning, Fear Expression, and the Amygdala

The amygdala has been recognized as one of the central brain structures involved in fear conditioning, recognition, and resultant fear-related behaviors.[8–10] Within the brain, the amygdala appears to function as one component of a threat-responsive comparator system that acts to differentiate threatening from non-threatening environmental stimuli in real time on the basis of prototype matching to fear memories, and to initiate adaptive behavior to deal with any perceived threat. More specifically, it is a central site for the changes in neuronal plasticity associated with fear conditioning. The name amygdala derives from the Greek for "almond", in reference to its shape. It is a relatively small, evolutionary conserved brain structure that is located bilaterally within the temporal lobe. The amygdala is comprised of different regions composed of clusters of nuclei. Of particular interest in the study of conditioned fear are the lateral (LA), basolateral (BLA), and central (CeA) amygdala regions. The LA is the key input site of the amygdala and receives both cortical and thalamic projections. There is evidence to suggest that the neuroplastic changes responsible for associative learning linking the CS and US occur in the LA.[11] Notably, pharmacological inhibition of the LA results in failure to acquire and express the behavioral manifestations of a conditioned fear response. The LA projects to the CeA both indirectly and directly via the BLA.[12] The CeA is responsible for the physiological, hormonal, and behavioral responses involved in the expression of conditioned fear. The CeA has a plethora of projections to other brain structures such as the hypothalamus, basal forebrain, and central gray area (Fig. 37.2). Projections to the hypothalamus play a key role in the regulation of adaptive autonomic responses to threat, such as heart rate and blood pressure changes, whereas projections to the central gray area from the CeA influence fear-associated behavioral changes such as freezing along with changes in social interaction. Notably, the direct behavioral result of the "fear reflex" following CeA activation directly mimics the panic attack symptoms associated with fear-related disorders. Research has suggested that the CeA not only functions as the central regulatory site for the expression of conditioned fear but may also have a role in the formation and storage of CS–US associations.[13]

Contextual Fear and the Role of the Hippocampus

Owing to its major role in the regulation of both learning and stress, the hippocampus has been a major focus of neuroanatomical studies in patients with mood and anxiety disorders such as depression and PTSD. Functionally, the hippocampus plays a central role in adaptation to stress because of its major role in the explicit/contextual memory systems and regulation of the hypothalamic–pituitary–adrenal (HPA) axis. The hippocampus is a remarkably plastic structure, being a major site of neurogenesis in the adult brain, while concurrently possessing a well-demonstrated vulnerability to stress. Neurons of the hippocampal CA3 region exhibit a loss of dendritic spines, reduced dendritic branching, and impaired neurogenesis in response to stress exposure.[14]

While the amygdala had been noted to be of central importance in classical conditioning, early research suggested a prominent role for the hippocampus in contextual fear conditioning.[15] Early studies highlighted the impairment in contextual fear with damage to the hippocampus.[16] More recent work has also clarified a role for the hippocampus in the extinction of aversive memories. The hippocampus, and more specifically the ventral hippocampus, is heavily connected with both the amygdala and hypothalamus. Through fear conditioning, the pairing of CS–US may result in the presentation of the CS, producing a range of behaviors and physiological responses (i.e. the CR). It has been recognized the context in which the CS is presented may also elicit responses similar to the CR when encountered again by the organism. In this setting, the context refers to the background static environment in the immediate vicinity of the administration of the aversive US.[16] Contextual fear conditioning has also been noted to arise with the administration of an aversive US alone without the presence of a CS.

Fear Extinction and the Prefrontal Cortex

The inhibition of fear acquired by associative learning, a process known as extinction, has been extensively studied in laboratory animals as well as humans. It is a form of active inhibitory learning, and not "unlearning" or forgetting, of a conditioned association (reviewed in Myers and Davis[17]). During extinction training, a neutral CS is repeatedly presented in the absence of the US, leading to inhibition of the previously conditioned fear response to the neutral CS. Considered in operational terms, extinction may thus be defined as "a reduction in the strength or probability of a conditioned fear response as a consequence of repeated presentation of the CS in the absence of the US".[18] Clinically, principles of extinction learning form much of the foundation for the most effective behavioral therapies for fear-related anxiety disorders,[19] although gaps in efficacy remain.[20] In practice, this involves progressively graded exposure to the feared object or event in the absence of any likely actual harm, much like exposure to a fear-related CS without imposition of the aversive US as in rodent extinction training. The environmental context of therapeutic exposure may be imaginal in nature, wherein a narrative is read or listened to by the patient, or *in vivo*, where the patient is directly exposed to the feared stimulus.

A large body of literature derived from preclinical studies using animal models as well as clinical research with human subjects has deepened our present understanding of the systems, cellular, and molecular biology of fear conditioning and extinction.[17] In particular, the excitatory amino acid neurotransmitter glutamate, acting on the *N*-methyl-D-aspartate (NMDA) receptor, has been found to have a central role in the acquisition and extinction of fear learning. Infusion of NMDA receptor antagonists into the rat amygdala blocks the acquisition as well as the extinction[21] of conditioned fear, whereas peripheral or central infusion of the partial NMDA receptor agonist D-cycloserine (DCS) facilitates the extinction

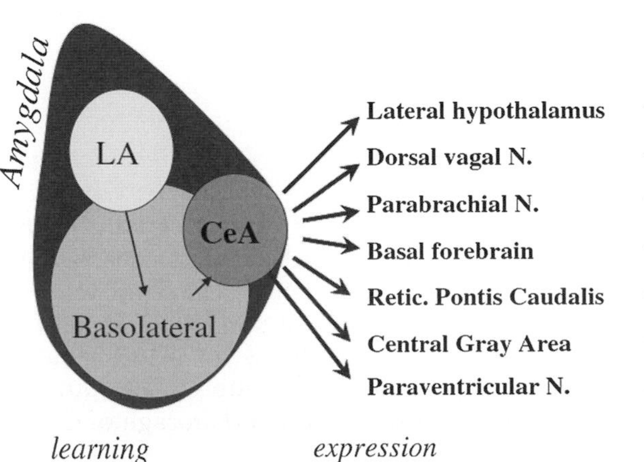

FIGURE 37.2 **Schematic diagram of hardwired amygdala outputs leading to fear and panic symptoms.** LA: lateral amygdala; CeA: central amygdala; N: nucleus.

of conditioned fear in rats but does not influence fear in rats in the absence of extinction training, suggesting that the effects of DCS are not related to anxiolysis but rather are directed towards the facilitation of new learning (replications were described and reviewed by Myers and Davis[17]). In humans, a number of trials have demonstrated the efficacy of DCS in enhancing extinction of fear in specific phobia,[22] as well as in other fear-related anxiety disorders including PTSD, panic disorder, and social phobia, as reviewed in two meta-analytic studies.[23,24]

The site of DCS action is the strychnine-insensitive glycine recognition site of the NMDA receptor complex[25] which, when bound by endogenous glycine or D-serine, or both, facilitates the activity of previously "silent" NMDA synapses. The effects of DCS on NMDA receptor function are complex and contingent on the synaptic concentrations of glycine/D-serine as a consequence of DCS partial agonist properties.[26] DCS administration under conditions of high glycine concentration (in which glycine binding sites are saturated by endogenous glycine) reduces NMDA receptor activity, presumably as a consequence of competitive displacement of glycine by DCS and the lower intrinsic efficacy of DCS compared to glycine with respect to facilitation of NMDA receptor activity. Conversely, administration of DCS under conditions of minimal occupancy of glycine/D-serine binding sites results in facilitation of NMDA receptor activity from baseline by occupancy of glycine binding sites by DCS. The modulating effects of DCS on NMDA receptor activity may thus promote neuroplasticity as well as interfere with the reconsolidation of fear memories, resulting in increased effectiveness of extinction training.[27]

Exploring the Neurobiological Basis of Anxiety Disorders in Humans

Although animal models have been the basis for much of the preclinical research related to anxiety disorders, translational research using human subjects has resulted in enhanced understanding of anxiety disorders in human clinical populations. Human anxiety disorders commonly present with fear and avoidance of fear-related cues. A growing body of data indicates that neural circuits responsible for processing component elements of fear learning, including conditioning and extinction, may function abnormally in individuals with anxiety disorders.[28,29] Neuroimaging data from humans have demonstrated amygdala hyperactivity across a range of anxiety disorders, including PTSD,[30] social anxiety disorder and specific phobia,[31] and panic disorder.[32] A meta-analysis of neuroimaging studies[33] demonstrated a consistent finding of hyperactivation of the amygdala and insula among patients with PTSD, social anxiety disorder, and specific phobia compared with healthy controls.[33] PTSD was also associated with hypoactivation in the ventromedial prefrontal cortex, anterior cingulate cortex, and thalamus. Collectively, these data suggest that there may be a shared neurobiology underpinning several human anxiety disorders.

CLINICAL FEATURES AND PSYCHOBIOLOGY OF ANXIETY DISORDERS

Panic Disorder

Clinical Features of Panic Disorder

Panic attacks are episodes of intense fear or discomfort that typically peak within 10 minutes. They can present with a range of features including unpleasant somatic sensations, physiological changes, and alterations of cognition (Table 37.2). Palpitations, dyspnea, dizziness, and chest discomfort appear to be the most commonly experienced features of panic attacks. Panic attacks can present clinically in any anxiety disorder and also in a wide range of other psychiatric disorders.[34] They also appear to have a strong association with a variety of medical conditions, especially respiratory disorders such as asthma.[35] Clinical research has demonstrated that a history of panic attacks functions as an independent risk factor for cardiovascular mortality and morbidity.[36] Most notably, panic attacks occur in all anxiety disorders of fear regulation, including phobias, panic disorder, and PTSD.

Recurrent and uncued panic attacks resulting in concern about additional attacks, worry about future attacks, and changes in behavior related to attacks suggest a diagnosis of panic disorder (Table 37.3). This diagnosis is only made in the absence of another psychiatric, medical, or substance-related cause for the episodes of panic. Panic disorder is three times more common in women than in men, and typical age of onset is in the third decade of life.[37] The frequency and severity of the panic attacks may be highly variable between individuals. Panic attacks and panic disorder are both widespread conditions, with a lifetime prevalence of 22% and 4%, respectively.[38]

Psychobiology of Panic Disorder

Comprehensive understanding of the neurobiological basis of panic disorder has been somewhat hampered by the range of models used to conceptualize panic disorder combined with a lack of methodologically robust human neuroimaging studies. The discovery of an improvement in clinical symptoms following treatment with tricyclic antidepressants led to the "chemical" hypothesis of panic disorder; this was followed by the dyspnea–hyperventilation and "false suffocation" theories.[39] Increasingly, however, these theories have

TABLE 37.2 Definition of Panic Attack in DSM-5

An abrupt surge of intense fear or intense discomfort that reaches a peak within minutes, and during which time four or more of the following symptoms occur. The abrupt surge can occur from a calm state or an anxious state:

1. Palpitations, pounding heart, or accelerated heart rate
2. Sweating
3. Trembling or shaking
4. Sensations of shortness of breath or smothering
5. Feeling of choking
6. Chest pain or discomfort
7. Nausea or abdominal distress
8. Feeling dizzy, unsteady, lightheaded, or faint
9. Chills or heat sensations
10. Paresthesias (numbness or tingling sensations)
11. Derealization (feelings of unreality) or depersonalization (being detached from oneself)
12. Fear of losing control or going crazy
13. Fear of dying

Source: American Psychiatric Association. Diagnostic and Statistical Manual of Mental Disorders. *5th ed.* DSM-5. *Arlington, VA: American Psychiatric Publishing; 2013.*[4]

TABLE 37.3 Definition of Panic Disorder in DSM-5

Panic disorder (includes previous diagnoses of panic disorder with agoraphobia and panic disorder without agoraphobia)

A. Recurrent unexpected panic attacks
B. At least one of the attacks has been followed by 1 month (or more) of one or both of the following:
   1. Persistent concern or worry about additional panic attacks or their consequences (e.g. losing control, having a heart attack, going crazy).
   2. Significant maladaptive change in behavior related to the attacks (e.g. behaviors designed to avoid having panic attacks, such as avoidance of exercise or unfamiliar situations).
C. The panic attacks are not restricted to the direct physiological effects of a substance (e.g. a drug of abuse, a medication) or a general medical condition (e.g. hyperthyroidism, cardiopulmonary disorders).
D. The panic attacks are not restricted to the symptoms of another mental disorder, such as social phobia (e.g. in response to feared social situations), specific phobia (e.g. in response to a circumscribed phobic object or situation), obsessive–compulsive disorder (e.g. in response to dirt in someone with an obsession about contamination), post-traumatic stress disorder (e.g. in response to stimuli associated with a traumatic event), or separation anxiety disorder (e.g. in response to being away from home or close relatives).

Source: American Psychiatric Association. Diagnostic and Statistical Manual of Mental Disorders. *5th ed.* DSM-5. *Arlington, VA: American Psychiatric Publishing; 2013.*[4]

given way to the fear-conditioning paradigm that has been broadly applied to a range of anxiety disorders. As a consequence, revised psychobiological conceptualizations of panic disorder now incorporate the core neurocircuitry of fear conditioning.

While the amygdala has a central role in the development of conditioned fear, there is significant top–down control by higher processing areas in the prefrontal and anterior cingulate cortices. These areas evaluate sensory stimuli and help to mediate the response of the amygdala.[40] Functional neuroimaging studies suggest that there may be decreased bilateral cerebral blood flow in patients with panic disorder and that this may be a marker of impaired inhibitory capacity of the cortex to modulate the amygdala response. The neural circuit consisting of the amygdala, hippocampus, thalamus, and key brainstem structures has been shown to demonstrate greater glucose utilization in people with panic disorder.[41] The alterations of glucose metabolism in this circuit are consistent with the notion of a dysfunctional fear-processing circuit in individuals with panic disorder. Successful treatment of panic disorder with cognitive behavioral therapy or pharmacotherapy has been associated with normalization of cerebral metabolism in individuals with panic disorder. Further evidence consistent with the presence of a dysfunctional fear circuit in individuals with panic disorder has been generated by translational research with human subjects using doxapram, a respiratory stimulant and panic-evoking compound (panicogen). Experimental subjects with panic disorder treated with doxapram and subsequently imaged using position emission tomography had decreased prefrontal activity and increased cingulate gyrus and amygdala activity compared with controls.[42] These findings support the general hypothesis of reduced cortical inhibitory capacity on the amygdala in individuals affected by panic disorder.

Abnormalities in neurotransmitter systems have also been noted in people with panic disorder. γ-Aminobutyric acid (GABA) is the principal inhibitory neurotransmitter in the brain. GABA is localized predominantly within the synaptic terminals of small, local circuit neurons and exerts its inhibitory effect by actions on $GABA_A$ and $GABA_B$ receptors. It has been postulated that a reduction in GABAergic activity may predispose an individual to increased anxiety.[43] Benzodiazepines are a group of agents that bind to and alter the activity of $GABA_A$ receptors, resulting in increases in GABA-mediated inhibitory activity. Radioligand studies investigating benzodiazepine binding sites have demonstrated decreased binding to $GABA_A$ receptors in the left hippocampus but increased binding in the right lateral frontal gyrus and temporal cortex.[44] In addition, findings from magnetic resonance spectroscopy studies have noted decreased GABA concentrations in the basal ganglia, anterior cingulate cortex and occipital cortex in people with panic disorder.[45]

In addition to a role for altered GABAergic neurotransmission in anxiety disorders, there is recognition of a role for noradrenergic and serotonergic activity in the etiology of panic disorder. Evidence for serotonergic dysfunction has arisen through a variety of findings. Patients with panic disorder may exhibit greater panic and anxiety symptoms compared with controls when exposed to serotonergic agents.[46] In addition, neuroendocrine challenge studies using serotonergic agents consistently demonstrate exaggerated cortisol and prolactin responses in individuals with panic disorder.[46] Finally, radioligand studies have demonstrated decreased serotonin receptor binding (5-HT_{1A}) in several areas of the brain, including the raphe nucleus, orbitofrontal cortex, temporal cortex, and amygdala, in subjects with panic disorder.

Noradrenergic dysfunction in panic disorder has also been observed in both preclinical models of panic disorder and clinical studies using subjects with panic disorder.[47] Challenge studies using the α_2 antagonist yohimbine and the α_2 agonist clonidine have demonstrated abnormal responses in subjects with panic disorder compared with healthy controls. Yohimbine administration results in greater cardiovascular and self-rated anxiety symptoms in people with panic disorder, whereas the administration of clonidine results in an attenuated growth hormone response and marked hypotension.[47]

Specific Phobia

Clinical Features of Specific Phobia

Clinically, specific phobia is characterized by the development of unreasonable and persistent fear in response to exposure to a specific situation or object.[3] Exposure to the situation or object results in an immediate response which may be manifest by a panic attack, or in children by other responses such as crying, tantrums, or freezing. The most common phobias may be generally classified into five groups. These phobic groups are natural environment type (e.g. heights, water, or storms); those that are related to animals (e.g. spiders or snakes); those that are related to blood, injection, and injury; situational type (e.g. airplanes or elevators); and other type (e.g. those that may result in choking, vomiting, or contracting an illness). Specific phobia is a relatively common psychiatric disorder with a lifetime prevalence of approximately 9%.[48] Female gender, younger age, and low income are all associated with higher risk.[48] Different prevalence rates have been noted across the different categories of phobias, with both animal and height specific phobia being particularly prevalent in adults and adolescents.

Post-Traumatic Stress Disorder

Clinical Features of Post-Traumatic Stress Disorder

PTSD is a severely debilitating, stress-related psychiatric illness associated with trauma exposure. The lifetime prevalence of PTSD in the general population has been estimated to be 5–10%,[49] with higher rates of PTSD being observed among combat veterans[50] and individuals living in areas of high violence.[51] Depression,[49] substance abuse,[51] and suicide[52] are commonly observed behavioral comorbidities. The primary symptom clusters include re-experiencing, avoidance/numbing, and hyperarousal symptoms. Patients with PTSD experience intrusive and fearful memories as well as flashbacks and nightmares of the traumatic event (or events) for much of their lives, may avoid other people, and have difficulty in social relationships and retaining jobs. These psychiatric symptoms, in turn, have adverse influences on patient compliance with treatment of concurrent medical illness and substance use, and may independently influence progression and risk for non-psychiatric medical disease.

Psychobiology of Post-Traumatic Stress Disorder

Physiological adaptation to stress is managed primarily by the sympathetic–adrenal–medullary (SAM) system and the HPA axis. Disturbances in the function of the SAM system[53] and the HPA axis[54] are routinely observed clinical features of PTSD. Patients with PTSD are commonly found to have elevated circulating concentrations of norepinephrine and elevated cerebrospinal fluid levels of corticotropin-releasing hormone[55] in conjunction with decreased peripheral concentration of cortisol[56] and altered sensitivity to glucocorticoids,[54] suggesting that endocrine glands downstream of the hypothalamus are downregulated, inhibited by glucocorticoid negative feedback, or insensitive to the increased activity of stress response systems in the brain. Regulation of the sensitivity of glucocorticoid receptors may be significantly influenced by genetic variation in the glucocorticoid receptor chaperone, FKBP5. *FKBP5* codes for FK506 binding protein-51 or FKBP5, which is a cochaperone of hsp90 that regulates glucocorticoid receptor sensitivity.[57] Multiple genetic epidemiological studies have demonstrated that genotypic variation in *FKBP5* alters risk for PTSD[58] and also influences glucocorticoid sensitivity.

CONCLUSION

This chapter has reviewed the clinical manifestations and broad psychobiological mechanisms underlying a number of the most common and debilitating anxiety and fear-related disorders. In particular, panic disorder, specific and social phobia, and PTSD are all characterized by the occurrence of panic attacks. What is particularly striking is that the symptom clusters associated with panic overlap convincingly with the understanding of the basic fear reflex that occurs through chemical or electrical activation of the amygdala, in particular the central nucleus of the amygdala. The hardwired outputs of the amygdala are associated with stereotypical fear responding. It appears that what differentiates these fear-related disorders is the differential responses to cues that activate the fear system. Thus, a neural circuit understanding of the conserved mammalian fear pathway provides hope for the rational design of therapeutic approaches to specifically target, at a mechanistic level, the underlying neurocircuitry of fear.

QUESTIONS FOR FURTHER RESEARCH

There has been much progress in the area of anxiety and fear-related disorders, from neuroimaging to an increased functional understanding of the underlying neural circuitry. However, much remains to be done to further progress and enhance treatment and prevention. In many ways, focusing on disorders of fear regulation allows a bottom–up approach to psychiatric disorders – starting with a well-known circuitry and finding relevant psychiatric disorders that may be explained by this circuitry – instead of the more traditional top–down perspective of starting with psychiatric diagnoses. In what ways is this a useful approach, and which other areas of psychiatric and medical inquiry may benefit from this type of analysis? Furthermore, the mammalian fear circuit is a very good model of the human fear circuit, and thus allows a functional dissection of fear processing and its sequelae, such as the underlying processes of panic attacks. However, other components of anxiety disorders are not as well modeled in such translational approaches: What are these and how might such problems be approached?

From a perspective of heritability, there are numerous examples of genetic variation accounting for up to 30–40% of risk for anxiety disorders, particularly when combined with environmental risk. Why have common gene variants been evolutionarily conserved that increase disease risk? Why, during ancestral periods, might it have been genetically advantageous to quickly learn fear associations? In relation to prevention and treatment, it is hoped that understanding the neural circuitry and neural plasticity components of fear processing will lead to new approaches to prevention and treatment. Examples include using agents that block memory consolidation in the aftermath of trauma to decrease the risk of PTSD development, and using cognitive enhancers to increase synaptic plasticity combined with psychotherapy to enhance the extinction of fear in fear-related disorders such as phobia, panic, and PTSD. In which other ways could understanding the neural circuits, the receptor molecules, or the molecular events underlying behavioral processing lead to rationally designed novel therapeutic approaches to prevention and treatment?

References

1. Etkin A. Functional neuroanatomy of anxiety: a neural circuit perspective. *Curr Top Behav Neurosci*. 2010;2:251–277.
2. Kessler RC, Chiu WT, Demler O, Merikangas KR, Walters EE. Prevalence, severity, and comorbidity of 12-month DSM-IV disorders in the National Comorbidity Survey Replication. *Arch Gen Psychiatry*. 2005;62(6):617–627.
3. American Psychiatric Association. *Diagnostic and Statistical Manual of Mental Disorders*. 4th ed. *Text Revision: DSM-IV-TR*. Washington, DC: American Psychiatric Association; 2000.
4. American Psychiatric Association. *Diagnostic and Statistical Manual of Mental Disorders*. 5th ed. *DSM-5*. Arlington, VA: American Psychiatric Publishing; 2013. http://www.dsm5.org.
5. Krueger RF. The structure of common mental disorders. *Arch Gen Psychiatry*. 1999;56(10):921–926.
6. Vollebergh WA, Iedema J, Bijl RV, de Graaf R, Smit F, Ormel J. The structure and stability of common mental disorders: the NEMESIS study. *Arch Gen Psychiatry*. 2001;58(6):597–603.
7. Mineka S, Oehlberg K. The relevance of recent developments in classical conditioning to understanding the etiology and maintenance of anxiety disorders. *Acta Psychol (Amst)*. 2008;127(3):567–580.
8. Blanchard DC, Blanchard RJ. Innate and conditioned reactions to threat in rats with amygdaloid lesions. *J Comp Physiol Psychol*. 1972;81(2):281–290.
9. Davis M. The role of the amygdala in fear and anxiety. *Annu Rev Neurosci*. 1992;15:353–375.
10. LeDoux JE. Emotion: clues from the brain. *Annu Rev Psychol*. 1995;46:209–235.
11. Maren S, Quirk GJ. Neuronal signalling of fear memory. *Nat Rev Neurosci*. 2004;5(11):844–852.
12. Pitkanen A, Savander V, LeDoux J. Organization of intra-amygdaloid circuitries in the rat: an emerging framework for understanding functions of the amygdala. *Trends Neurosci*. 1997;20(11):517–523.
13. Wilensky AE, Schafe GE, Kristensen MP, LeDoux JE. Rethinking the fear circuit: the central nucleus of the amygdala is required for the acquisition, consolidation, and expression of Pavlovian fear conditioning. *J Neurosci*. 2006;26(48):12387–12396.
14. Duman RS, Monteggia LM. A neurotrophic model for stress-related mood disorders. *Biol Psychiatry*. 2006;59(12):1116–1127.
15. Blanchard RJ, Fial RA. Effects of limbic lesions on passive avoidance and reactivity to shock. *J Comp Physiol Psychol*. 1968;66(3):606–612.
16. Kim JJ, Fanselow MS. Modality-specific retrograde amnesia of fear. *Science*. 1992;256(5057):675–677.
17. Myers KM, Davis M. Behavioral and neural analysis of extinction. *Neuron*. 2002;36(4):567–584.
18. Rothbaum BO, Davis M. Applying learning principles to the treatment of post-trauma reactions. *Ann N Y Acad Sci*. 2003;1008:112–121.
19. Rothbaum BO, Schwartz AC. Exposure therapy for posttraumatic stress disorder. *Am J Psychother*. 2002;56(1):59–75.
20. Brown TA, Barlow DH. Long-term outcome in cognitive–behavioral treatment of panic disorder: clinical predictors and alternative strategies for assessment. *J Consult Clin Psychol*. 1995;63(5):754–765.
21. Falls WA, Miserendino MJ, Davis M. Extinction of fear-potentiated startle: blockade by infusion of an NMDA antagonist into the amygdala. *J Neurosci*. 1992;12(3):854–863.
22. Ressler KJ, Rothbaum BO, Tannenbaum L, et al. Cognitive enhancers as adjuncts to psychotherapy: use of D-cycloserine in phobic individuals to facilitate extinction of fear. *Arch Gen Psychiatry*. 2004;61(11):1136–1144.
23. Norberg MM, Krystal JH, Tolin DF. A meta-analysis of D-cycloserine and the facilitation of fear extinction and exposure therapy. *Biol Psychiatry*. 2008;63(12):1118–1126.

24. Bontempo A, Panza KE, Bloch MH. D-cycloserine augmentation of behavioral therapy for the treatment of anxiety disorders: a meta-analysis. *J Clin Psychiatry*. 2012;73(4):533–537.
25. Monahan JB, Handelmann GE, Hood WF, Cordi AA. D-cycloserine, a positive modulator of the *N*-methyl-D-aspartate receptor, enhances performance of learning tasks in rats. *Pharmacol Biochem Behav*. 1989;34(3):649–653.
26. Hood WF, Compton RP, Monahan JB. D-cycloserine: a ligand for the *N*-methyl-D-aspartate coupled glycine receptor has partial agonist characteristics. *Neurosci Lett*. 1989;98(1):91–95.
27. Krystal JH. Neuroplasticity as a target for the pharmacotherapy of psychiatric disorders: new opportunities for synergy with psychotherapy. *Biol Psychiatry*. 2007;62(8):833–834.
28. Grillon C. Startle reactivity and anxiety disorders: aversive conditioning, context, and neurobiology. *Biol Psychiatry*. 2002;52(10):958–975.
29. Milad MR, Pitman RK, Ellis CB, et al. Neurobiological basis of failure to recall extinction memory in posttraumatic stress disorder. *Biol Psychiatry*. 2009;66(12):1075–1082.
30. Shin LM, Wright CI, Cannistraro PA, et al. A functional magnetic resonance imaging study of amygdala and medial prefrontal cortex responses to overtly presented fearful faces in posttraumatic stress disorder. *Arch Gen Psychiatry*. 2005;62(3):273–281.
31. Straube T, Kolassa IT, Glauer M, Mentzel HJ, Miltner WH. Effect of task conditions on brain responses to threatening faces in social phobics: an event-related functional magnetic resonance imaging study. *Biol Psychiatry*. 2004;56(12):921–930.
32. van den Heuvel OA, Veltman DJ, Groenewegen HJ, et al. Disorder-specific neuroanatomical correlates of attentional bias in obsessive–compulsive disorder, panic disorder, and hypochondriasis. *Arch Gen Psychiatry*. 2005;62(8):922–933.
33. Etkin A, Wager TD. Functional neuroimaging of anxiety: a meta-analysis of emotional processing in PTSD, social anxiety disorder, and specific phobia. *Am J Psychiatry*. 2007;164(10):1476–1488.
34. Barlow DH, Vermilyea J, Blanchard EB, Vermilyea BB, Di Nardo PA, Cerny JA. The phenomenon of panic. *J Abnorm Psychol*. 1985;94(3):320–328.
35. Ortega AN, McQuaid EL, Canino G, Goodwin RD, Fritz GK. Comorbidity of asthma and anxiety and depression in Puerto Rican children. *Psychosomatics*. 2004;45(2):93–99.
36. Smoller JW, Pollack MH, Wassertheil-Smoller S, et al. Panic attacks and risk of incident cardiovascular events among postmenopausal women in the Women's Health Initiative Observational Study. *Arch Gen Psychiatry*. 2007;64(10):1153–1160.
37. Stein DJ, Hollander E, Rothbaum BO. *Textbook of Anxiety Disorders*. Washington, DC: American Psychiatric Publishing; 2010.
38. Kessler RC, Chiu WT, Jin R, Ruscio AM, Shear K, Walters EE. The epidemiology of panic attacks, panic disorder, and agoraphobia in the National Comorbidity Survey Replication. *Arch Gen Psychiatry*. 2006;63(4):415–424.
39. Klein DF. False suffocation alarms, spontaneous panics, and related conditions. An integrative hypothesis. *Arch Gen Psychiatry*. 1993;50(4):306–317.
40. Hariri AR, Mattay VS, Tessitore A, Fera F, Weinberger DR. Neocortical modulation of the amygdala response to fearful stimuli. *Biol Psychiatry*. 2003;53(6):494–501.
41. Sakai Y, Kumano H, Nishikawa M, et al. Cerebral glucose metabolism associated with a fear network in panic disorder. *Neuroreport*. 2005;16(9):927–931.
42. Garakani A, Buchsbaum MS, Newmark RE, et al. The effect of doxapram on brain imaging in patients with panic disorder. *Eur Neuropsychopharmacol*. 2007;17(10):672–686.
43. Nutt DJ, Malizia AL. New insights into the role of the GABA(A)-benzodiazepine receptor in psychiatric disorder. *Br J Psychiatry*. 2001;179:390–396.
44. Malizia AL, Cunningham VJ, Bell CJ, Liddle PF, Jones T, Nutt DJ. Decreased brain GABA(A)-benzodiazepine receptor binding in panic disorder: preliminary results from a quantitative PET study. *Arch Gen Psychiatry*. 1998;55(8):715–720.
45. Goddard AW, Mason GF, Appel M, et al. Impaired GABA neuronal response to acute benzodiazepine administration in panic disorder. *Am J Psychiatry*. 2004;161(12):2186–2193.
46. Targum SD, Marshall LE. Fenfluramine provocation of anxiety in patients with panic disorder. *Psychiatry Res*. 1989;28(3):295–306.
47. Charney DS, Heninger GR, Breier A. Noradrenergic function in panic anxiety. Effects of yohimbine in healthy subjects and patients with agoraphobia and panic disorder. *Arch Gen Psychiatry*. 1984;41(8):751–763.
48. Stinson FS, Dawson DA, Patricia Chou S, et al. The epidemiology of DSM-IV specific phobia in the USA: results from the National Epidemiologic Survey on Alcohol and Related Conditions. *Psychol Med*. 2007;37(7):1047–1059.
49. Kessler RC, Sonnega A, Bromet E, Hughes M, Nelson CB. Posttraumatic stress disorder in the National Comorbidity Survey. *Arch General Psychiatry*. 1995;52(12):1048–1060.
50. Hoge CW, Castro CA, Messer SC, McGurk D, Cotting DI, Koffman RL. Combat duty in Iraq and Afghanistan, mental health problems, and barriers to care. *N Engl J Med*. 2004;351(1):13–22.
51. Alim TN, Graves E, Mellman TA, et al. Trauma exposure, posttraumatic stress disorder and depression in an African-American primary care population. *J Natl Med Assoc*. 2006;98(10):1630–1636.
52. Dube SR, Anda RF, Felitti VJ, Chapman DP, Williamson DF, Giles WH. Childhood abuse, household dysfunction, and the risk of attempted suicide throughout the life span: findings from the Adverse Childhood Experiences Study. *JAMA*. 2001;286(24):3089–3096.
53. Krystal JH, Neumeister A. Noradrenergic and serotonergic mechanisms in the neurobiology of posttraumatic stress disorder and resilience. *Brain Res*. 2009;1293:13–23.
54. Yehuda R. Status of glucocorticoid alterations in post-traumatic stress disorder. *Ann N Y Acad Sci*. 2009;1179:56–69.
55. Bremner JD, Licinio J, Darnell A, et al. Elevated CSF corticotropin-releasing factor concentrations in posttraumatic stress disorder. *Am J Psychiatry*. 1997;154(5):624–629.
56. Heim C, Nemeroff CB. Neurobiology of posttraumatic stress disorder. *CNS Spectr*. 2009;14(1 suppl 1):13–24.
57. Binder EB. The role of FKBP5, a co-chaperone of the glucocorticoid receptor in the pathogenesis and therapy of affective and anxiety disorders. *Psychoneuroendocrinology*. 2009;34(suppl 1):S186–S195.
58. Binder EB, Bradley RG, Liu W, et al. Association of FKBP5 polymorphisms and childhood abuse with risk of posttraumatic stress disorder symptoms in adults. *JAMA*. 2008;299(11):1291–1305.

CHAPTER

38

Obsessive–Compulsive Disorder

Nastassja Koen, Dan J. Stein

Department of Psychiatry, University of Cape Town, Groote Schuur Hospital, Cape Town, South Africa

OUTLINE

Neurobiology of Brain Disorders
http://dx.doi.org/10.1016/B978-0-12-398270-4.00038-0

INTRODUCTION

From a historical perspective, obsessive–compulsive disorder (OCD) was once viewed as a rarity. Today, it is recognized as a common and debilitating disorder. In the *Diagnostic and Statistical Manual of Mental Disorders*, 4th edition (DSM-IV), OCD was categorized as an anxiety disorder.[1] However, it has now been included as part of a new grouping of obsessive–compulsive and related disorders in DSM-5.[2] There is a substantial evidence base to support moving OCD out of the section on anxiety disorders. Most frequently, it is noted that obsessions and compulsions (rather than anxiety) characterize OCD and the related disorders and that this group of disorders differs significantly from the anxiety disorders in terms of pathogenesis and treatment.

OCD is often represented in literary, scientific, and popular culture media, with such varied allusions as Lady Macbeth's compulsive hand washing: "Out, damned spot!" (Shakespeare's *Macbeth*, Act V, Scene 1) and Howard Hughes's exceedingly reclusive and erratic behavior (as depicted in the 2004 Martin Scorsese film *The Aviator*). OCD is widely conceptualized as a neuropsychiatric disorder, in which dysfunctional neural circuitry and neurochemical changes culminate in characteristic symptomatology. This chapter reviews the current understanding of OCD, including epidemiology, disease burden, pathogenesis and natural history, clinical assessment and diagnostic criteria, and available treatment modalities. OCD-related disorders such as trichotillomania, as well as pediatric OCD, will also be discussed briefly.

EPIDEMIOLOGY OF OBSESSIVE–COMPULSIVE DISORDER

Prevalence

In the 1980s, the first large-scale and standardized epidemiological data on OCD were obtained as part of the Epidemiologic Catchment Area (ECA) study.[3] OCD was ranked as the fourth most common psychiatric disorder, with lifetime prevalence reported to be about 2.5% (range 1.9–3.3%), and representative data collected from five US communities and more than 18,500 individuals. Before this study, OCD had been regarded as a rare disorder, affecting only about 0.005% of the population. Thus, the ECA study revolutionized the epidemiological understanding of OCD, with prevalence rates reported as 25–60 times greater than had been estimated in previous clinical samples.[3]

Because of this remarkable diversion from prior findings, the validity of the ECA study has been challenged. A number of small follow-up studies published shortly after the ECA, as well as one larger follow-up investigation, criticized the reliability of the National Institute of Mental Health Diagnostic Interview Schedule (DIS), the psychiatric assessment tool designed for lay interviewers and used in the ECA to establish diagnostic status. It was reported that the DIS showed poor diagnostic validity, as evidenced by the very low temporal stability between those participants interviewed during the initial ECA study (wave 1) and those assessed 1 year later.

However, more recent epidemiological studies employing clinician-administered diagnostic interviews have supported the ECA study findings, with OCD emerging as one of the most common psychiatric disorders in adult populations. For example, in the National Comorbidity Survey Replication (NCS-R), lifetime prevalence of OCD was reported to be 2.3%,[4] a finding comparable to that of the ECA. Large-scale epidemiological data collected in the European Union have also supported the ECA study findings. In their critical review and appraisal of 27 studies of a range of mental disorders, Wittchen and Jacobi found that the 12 month population prevalence of OCD ranged from 0.1 to 2.3%.[5] Taken together, these cross-national findings suggest that, universally, OCD is a prevalent disorder. The earlier underestimation of the prevalence of this disorder may have been attributable to a number of factors, including a reluctance of affected individuals to seek care, a lack of routine screening by health professionals, and misdiagnosis.

Age of Onset

Age of onset of OCD is typically bimodal, with individuals presenting either during childhood/adolescence or later in life. Juvenile-onset OCD is particularly evident in males, while pregnancy-related events (e.g. miscarriage) may precipitate symptoms in women of childbearing age. Symptom onset before age 18 is particularly prevalent in first degree relatives of probands with OCD, suggesting that juvenile-onset OCD may be more highly heritable than later-onset disease.

Demographic Risk Factors

Gender

Overall, disease prevalence seems to be fairly equal in males and females. Although some clinical and epidemiological studies have shown a slight female preponderance,[3,6] this is far less definitive than in the anxiety disorders. However, gender-related clinical differences have been noted in these individuals. For example, symptom onset may be earlier and more insidious in males, with the illness then following a chronic progression with a greater likelihood of comorbid tic disorders.

Culture and Ethnicity

Most studies to date have focused on Caucasian individuals with OCD. While there is evidence of transcultural homogeneity in this disorder, it has also been suggested that its prevalence is lower in certain ethnic groups. For example, findings from the ECA indicated lower prevalence rates among black individuals.[3] However, the paucity and methodological shortcomings of these studies have limited their generalizability.

Socioeconomic Status

Although the ECA reported that unemployed individuals with lower socioeconomic status were more likely to develop OCD,[3] this may well be a reflection of the personal and societal functional impairment experienced by these individuals. Conflicting evidence that adolescents with OCD are more likely to be of middle to upper socioeconomic classes seems to support this notion.

Impact of Disease

OCD is often associated with serious and diverse direct and indirect health costs. Individuals with this disorder may have problems in a number of functional domains, including interpersonal difficulties (including relationship impairment), loss of employment, and poor instrumental role performance (i.e. domestic, academic, or work functioning). Compared with both the general population and those with other psychiatric or biomedical disorders, overall quality of life is also significantly affected in individuals with OCD. On a societal level, loss of productivity and increased use of health-care services may lead to inflated fiscal disease burden.

Unfortunately, individuals with OCD tend to delay their presentation to health-care services,[6] and misdiagnosis remains an obstacle in primary and specialized settings. Thus, there is often a substantial lag between symptom onset and treatment initiation, which worsens the impact of this disorder. Therefore, early diagnosis and effective treatment of OCD are essential to curtail its multidimensional sequelae.

CLINICAL CONSIDERATIONS IN OBSESSIVE–COMPULSIVE DISORDER

Diagnosis

The hallmark clinical characteristics of OCD are obsessions (persistent, recurrent, intrusive, and inappropriate thoughts or images) and compulsions (repetitive behaviors or mental acts aimed at reducing anxiety symptoms) which are not secondary to a general medical condition or substance misuse, and are not better explained by the symptoms of another mental disorder[2] (Table 38.1). The diagnosis of OCD requires comprehensive psychiatric history taking and examination.

Symptomatology

OCD may be subtyped according to the content of the obsessions and compulsions that are present. The Brown Longitudinal Obsessive Compulsive Study (BLOCS), a large-scale, prospective longitudinal clinical study of the natural history of OCD,[6] reported some of the major symptom classes encountered.[7] Non-phenomenological subtyping of OCD has also been the subject of recent work. For example, a three-tiered subclassification comprising abnormal risk assessment, pathological doubt, and incompleteness has been put forward. However, evidence remains inconclusive. Thus, for the purposes of this chapter, the traditional four-factor phenomenological subtyping will be discussed.

Symmetry

The symptoms of OCD may concern the need for symmetry or perfection. Typically, these individuals continue their compulsive ordering and rearranging either out of fear that harm will come to their loved ones (primary magical thinking), or because they take far longer than most to complete tasks of daily living (primary obsessive slowness).[7]

Forbidden Thoughts (Sexual, Aggressive, and Somatic Obsessions; Checking Compulsions)

Individuals may experience intrusive and persistent fears that they may commit, or indeed already have committed, sexually inappropriate, aggressive, or violent acts.[7] They often feel compelled to carry out ritualistic checking, confessional, or reassurance behaviors, such as confessing to imagined crimes. Somatic obsessions may be rooted in irrational fears of body dysmorphisms (defects in physical appearance or form) or of contracting a life-threatening illness. While not disorder specific [individuals with hypochondriasis, body dysmorphic disorder (BDD), and major depression may all experience such fears], OCD-related somatic obsessions are also accompanied by classic fears (e.g. of contamination) and compulsions (e.g. hand washing).[7]

Individuals with OCD may also experience obsessive fears that their own carelessness will directly lead to catastrophic events. For example, fear of a house fire may cause an individual to check and recheck his or her stove before leaving home.[7] This ritualistic checking is often highly time consuming, and the individual may employ strategies to minimize functional impairment, such as counting the number of completed checks. While these individuals are often aware of their misperceptions, the severe anxiety they may experience when checks are not completed arguably perpetuates their compulsive

TABLE 38.1 DSM-5 Diagnostic Criteria for Obsessive–Compulsive Disorder

A. Presence of obsessions, compulsions, or both:
Obsessions are defined by (1) and (2):
(1) Recurrent and persistent thoughts, urges, or images that are experienced, at some time during the disturbance, as intrusive and unwanted and that in most individuals cause marked anxiety or distress.
(2) The individual attempts to ignore or suppress such thoughts, urges, or images, or to neutralize them with some other thought or action (i.e. by performing a compulsion).
Compulsions are defined by (1) and (2):
(1) Repetitive behaviors (e.g. hand washing, ordering, checking) or mental acts (e.g. praying, counting, repeating words silently) that the individual feels driven to perform in response to an obsession or according to rules that must be applied rigidly.
(2) The behaviors or mental acts are aimed at preventing or reducing anxiety or distress, or preventing some dreaded event or situation; however, these behaviors or mental acts either are not connected in a realistic way with what they are designed to neutralize or prevent, or are clearly excessive.
Note: Young children may not be able to articulate the aims of these behaviors or mental acts.

B. The obsessions or compulsions are time consuming (e.g. take more than 1 hour a day) or cause clinically significant distress or impairment in social, occupational, or other important areas of functioning.

C. The obsessive–compulsive symptoms are not attributable to the physiological effects of a substance (e.g. a drug of abuse, a medication) or another medical condition.

D. The disturbance is not better explained by the symptoms of another mental disorder [e.g. excessive worries, as in generalized anxiety disorder; preoccupation with appearance, as in body dysmorphic disorder; difficulty discarding or parting with possessions, as in hoarding disorder; hair pulling, as in trichotillomania (hair-pulling disorder); skin picking, as in excoriation (skin-picking) disorder; stereotypies, as in stereotypic movement disorder; ritualized eating behavior, as in eating disorders; preoccupation with substances or gambling, as in substance-related and addictive disorders; preoccupation with having an illness, as in illness anxiety disorder; sexual urges or fantasies, as in paraphilic disorders; impulses, as in disruptive, impulse-control, and conduct disorders; guilty ruminations, as in major depressive disorder; thought insertion or delusional preoccupations, as in schizophrenia spectrum and other psychotic disorders; or repetitive patterns of behavior, as in autism spectrum disorder].

Specify if:
With good or fair insight: The individual recognizes that obsessive–compulsive disorder beliefs are definitely or probably not true or that they may or may not be true.
With poor insight: The individual thinks obsessive–compulsive beliefs are probably true.
With absent insight/delusional beliefs: The individual is completely convinced that obsessive–compulsive beliefs are true.
Specify if:
Tic-related: The individual has a current or past history of a tic disorder.

Source: American Psychiatric Association. Diagnostic and Statistical Manual of Mental Disorders. *5th ed. Arlington, VA: American Psychiatric Press; 2013 (DSM-5).*[2]

behavior. These cognitive–affective manifestations form the basis of cognitive–behavioral therapeutic techniques for treating OCD (see "Psychological Therapy", below).

Cleaning (Contamination and Washing)

This is the subtype most commonly encountered in individuals with OCD.[6,7] Fearful obsessions centered on dirt or germs often culminate in compulsive and excessive washing (e.g. hand washing) after contact with the feared stimulus. While this type of washing is usually in response to fears of contamination, it may also be a manifestation of obsessions with symmetry (see above). In addition, avoidance behaviors may be employed to avoid perceived contamination.

Hoarding

While hoarding was included in the original four-factor phenomenological symptom structure, it has now been classified as a distinct, OCD-related disorder in DSM-5.[2] Nonetheless, evidence suggests that hoarding is present in almost 20% of individuals with OCD and is characterized by a persistent difficulty discarding one's possessions, driven by the fear of losing items that may be needed in future. This may result in the accumulation of clutter with accompanying distress or functional impairment.

Differential Diagnoses

Obsessions and compulsions may occur in a variety of psychiatric and biomedical conditions. Thus, in the clinical setting, it is important to exclude these differentials before definitively diagnosing OCD. While most anxiety disorders manifest with overwhelming fear, autonomic features, and avoidant behavior, OCD may be distinguished from, for example, panic disorder, by the absence of spontaneously occurring panic episodes (individuals with OCD may experience panic secondary to obsessions). Similarly, while OCD and many psychotic disorders (e.g. schizophrenia) are associated with a chronic course and long-term functional impairment, the delusions of schizophrenia are often not recognized as unreasonable and are thus accepted by the affected individual, whereas individuals with OCD may have varying degrees of insight into their dysfunctional beliefs.[7] Similar specifiers have been included in DSM-5, so that affected individuals are described as having good or fair insight, having poor insight, or having absent insight/delusional beliefs.

OCD-related disorders should also be considered in individuals presenting with obsessions or compulsions. However, unlike in OCD, the recurrent, intrusive thoughts and behaviors of trichotillomania and BDD are not ubiquitous, rather limited to particular contexts. Furthermore, OCD should be carefully distinguished from obsessive–compulsive personality disorder (OCPD). In general, individuals with this disorder do not have obsessions and compulsions meeting the DSM criteria, but rather experience a pervasive preoccupation with orderliness, perfectionism, and interpersonal control, often to the detriment of flexibility and efficiency.[1]

In addition to the range of psychiatric disorders, a number of biomedical conditions should also be excluded before a diagnosis of OCD is made. For example, neurodegenerative disorders of the frontal lobe and basal ganglia (e.g. Huntington disease and Sydenham's chorea), cerebrovascular accidents, or even neurotoxin exposure (e.g. carbon monoxide) may present with signs and symptoms not unlike those of OCD. In these cases, thorough history taking, examination, and investigations where indicated should elucidate the definitive diagnosis.

Symptom Severity

The most widely used assessment and evaluation tool for individuals with OCD is the Yale–Brown Obsessive Compulsive Scale (Y-BOCS). This is a 10-item, clinician-administered tool including five domains for obsessions and five for compulsions. Individuals are instructed to answer each domain on a Likert scale, from 0 (no symptoms) to 4 (extreme symptoms). The Y-BOCS has shown good validity and reliability and has been shown to change with treatment. Thus, it is generally accepted as a standardized outcome measure for individuals with OCD.[7] However, it cannot be used as a diagnostic instrument as the DSM criteria are not specifically assessed. Thus, comprehensive psychiatric history taking and examination remain the bedrock of diagnosis.

Comorbidity

OCD frequently co-occurs with other psychiatric conditions. To date, studies have reported that 50–90% of adults with OCD also meet the diagnostic criteria for another axis I disorder, and at least 40% for an axis II disorder.[6]

Of these comorbidities, major depressive disorder (MDD) has consistently been found to be the most prevalent. Clinical and epidemiological studies have reported that approximately one-third of individuals with OCD experience concurrent MDD, and about two-thirds have a lifetime history of this comorbidity.[8] Most individuals report the onset of depressive symptoms after those of OCD, suggesting secondary MDD (which may manifest in response to OCD-related functional impairment).[3,7] However, a small percentage of patients experiences concurrent onset of obsessive–compulsive and depressive symptoms. Of these individuals, only a minority view their depression as independent of their OCD. In the clinical setting, depressive ruminations (usually centered on a prior event) should be carefully differentiated from pure obsessions (which often focus on a current or future incident and are resisted by the individual).[7]

There has been a relative paucity of literature on comorbid bipolar disorder (BPD) in individuals with OCD. In a comorbidity study of the ECA population, lifetime prevalence rates of bipolar and unipolar disease in individuals with OCD were reported as approximately 20 and 12%, respectively. More recently, it has been estimated that approximately 10% of people with OCD also meet the criteria for BPD, with the majority of these cases being BPD type II. Compared with those individuals with only OCD, those with comorbid OCD and BPD were more likely to be male, to experience hoarding and sexual obsessions, and to have a lifetime history of a comorbid substance use disorder.

Anxiety disorders may also coexist in people with OCD. For example, findings from the BLOCS showed social phobia (28%) and specific phobia (22%) emerging as the most common comorbid anxiety disorders in these individuals.[6]

OCD-related disorders may also be highly comorbid in individuals with OCD. In their study of 315 adult OCD patients, Lochner and colleagues found that 18.1% of their total sample also exhibited hoarding.[9] These high prevalence rates may be due, in part, to shared etiological mechanisms (e.g. genetic and environmental influences).[7–9]

In a substantial number of individuals, OCD also co-occurs with eating disorders. Participants with OCD are more likely to report disturbed eating attitudes and behaviors (e.g. drive for thinness, body dissatisfaction) than healthy controls. Several studies have substantiated these findings, with lifetime rates of comorbid anorexia nervosa and bulimia nervosa in individuals with OCD reported as 10–17% and 15–20%, respectively.[8] Conversely, significant rates of OCD have also been reported in individuals with eating disorders.

The axis II disorders most commonly diagnosed in individuals with OCD are OCPD (24.6%) and avoidant personality disorder (15.3%).[6] While less frequent, comorbid schizotypal personality disorder has also been associated with treatment resistance and poorer prognosis. Attention has been paid to the overlap and discrepancy between OCD and OCPD. Phenomenological, etiological, and treatment response similarities certainly exist between these two disorders. Although very early evidence suggested that all people with OCD may have a premorbid personality that predisposes them to develop the disorder, this theory has since been discounted. Furthermore, owing to evolving definitions and diagnostic criteria for the personality disorders, the literature on comorbid OCD and OCPD remains heterogeneous. Thus, while OCPD may not be as common a comorbidity as had been previously thought, it is widely accepted that this personality disorder occurs more frequently in individuals with OCD than in the general population.

NATURAL HISTORY AND COURSE OF THE DISEASE

Retrospective follow-up studies of individuals with OCD have delineated the following categories of disease course: continuous; waxing and waning; deteriorative (approximately 6–14%); and episodic with full remissions between episodes (approximately 10–15% of individuals).[7] According to these data, as well as clinical lore, most individuals with OCD experience a chronic, progressive course with some symptom fluctuation.[1] Factors associated with decreased remission include male gender, earlier age of symptom onset, older age at intake, and greater symptom severity.

Several prospective studies have also investigated the course of disease of juvenile-onset OCD. Overall, these echo the findings of follow-up studies in adult populations. For example, in a study of 25 individuals who had been diagnosed with severe OCD in childhood or adolescence, the psychiatric assessment conducted 2–7 years later revealed continued psychopathology.[10] Only 28% of this study population (seven individuals) were symptom-free at follow-up. Findings from a 40 year follow-up study conducted by Skoog and Skoog deviated significantly from this natural history.[11] These authors found that 83% of individuals showed improvement in OCD symptoms, with complete recovery in 20% and incomplete recovery with some subthreshold residual symptoms in 28%. This suggests that most individuals with OCD improve over time. However, as this finding has not been widely replicated, it is not yet the general consensus.

PATHOGENESIS OF OBSESSIVE–COMPULSIVE DISORDER

As is the case for many psychiatric disorders, the etiology and pathogenesis of OCD are multifactorial, as evidenced by its broad clinical heterogeneity.

Cognitive–Affective Factors

It has been suggested that cognitive–behavioral factors in the pathogenesis of OCD are rooted in appraisal theory, that is, the idea that emotions are elicited from evaluations (appraisals) of a situation or event. The structural model of appraisal comprises primary and secondary components. In an individual's primary appraisal of a situation, he or she evaluates its motivational relevance (i.e. its importance to his or her own well-being) and its motivational congruence (i.e. how closely it is aligned with his or her goals). Secondary appraisal comprises an evaluation of the individual's capacity to cope with the situation. During primary appraisal of a situation, an individual with OCD experiences intrusive obsessional thoughts. As a result, he or she tends to overestimate the risk of an unfavorable occurrence. Secondary appraisal then leads to dysfunctional coping, in the form of compulsive behavior or avoidance of the perceived threat. Belief domains affected in OCD include amplified sense of responsibility, need for control, overestimation of threat, and intolerance of uncertainty.

Neurobiological Factors

Neuroanatomy

Corticostriatothalamocortical (CSTC) neurocircuitry is thought to be central to the pathogenesis of OCD. In brief, this circuitry comprises a number of parallel, isolated projections. Efferent signals from the prefrontal cortex (PFC) are transmitted to specific regions in the striatum (a subcortical area of the forebrain that is the major input center of the basal ganglia system). In healthy individuals, the PFC mediates a range of higher cognitive processes, such as planning, organization, inhibition, and controlling. Thus, in disorders associated with dysfunctional and dysregulated PFC, symptoms of disorganization, disinhibition, and inflexibility are likely to occur. The functional subterritories of the PFC are: the dorsolateral PFC (involved in learning, memory, and other executive functions); and the ventral PFC, comprising the posteromedial orbitofrontal PFC (motivational and affective functions) and the anterior and lateral orbitofrontal cortex (reverse learning, i.e. behavioral flexibility after negative feedback). The orbitofrontal cortex, in particular, is central to many neurobiological models of OCD, and is often found to show abnormally reduced activation in affected individuals.

When functioning normally, the striatum serves to regulate conscious information processing by refining input and output, and fine-tunes brain efficiency by mediating some non-conscious functions. From the striatum, these signals are then projected via other intermediate basal ganglia to the thalamus. The circuit is then closed via reciprocal projections from the thalamus back to the cortical regions from which the original signals were transmitted.

Structural and functional neuroimaging studies have given credence to the CSTC hypothesis of OCD.

STRUCTURAL NEUROIMAGING

Studies to assess volumetric changes in the brains of individuals with OCD compared with healthy controls have shown abnormalities in four key areas: the caudate (a nucleus located within the basal ganglia), the orbital frontal and anterior cingulate regions (situated in the medial aspect of the cortex), the thalamus, and the cerebral white matter.

Changes in the caudate have been widely investigated, but findings have been mixed. Some studies have reported increased caudate volume in individuals with OCD. For example, in their 2009 meta-analysis of gray matter changes in individuals with OCD, Radua and Mataix-Cols reported increased regional gray matter volumes in bilateral lenticular nuclei, extending to the caudate nuclei.[12] However, other studies have shown decreased caudate volume or indeed no volumetric change. The inconsistency of these findings may be a reflection of the heterogeneity of OCD. In other words, reduction in caudate volume (with enlarged ventricular volume) may be more evident in individuals with OCD and soft neurological signs (e.g. tics). Another explanatory model would be that caudate volume in individuals with OCD changes over time. For example, caudate volume increases in the aftermath of infection in children with streptococcal-associated OCD and/or tics, with shrinkage thereafter.

There are fewer data on the structural changes in the orbital frontal and anterior cingulate regions in individuals with OCD, and findings have again been inconsistent. Some studies have reported increased gray matter volume in posterior orbitofrontal regions, with decreased volume of the left anterior cingulate. However, in their meta-analysis Radua and Mataix-Cols reported decreased volumes in bilateral dorsal medial frontal and anterior cingulate gyri.[12] There is also an evidence base showing reduced bilateral orbital frontal (and amygdala) volumes.

Several studies have shown an increase in thalamic volume in individuals with OCD compared with healthy controls. Thus, an inverse relationship may exist between thalamic and orbitofrontal volumetric changes in OCD. This finding was replicated in a meta-analysis of magnetic resonance imaging (MRI) studies in OCD by Rotge and colleagues.[13] Furthermore, these authors found that the severity of obsessions or compulsions correlated significantly with the effect sizes for the left and right thalamus.

In their MRI study of 16 adults with OCD and 17 healthy controls, Mataix-Cols and colleagues found that both participant groups experienced increased subjective anxiety when exposed to symptom-provoking stimuli (i.e. pictures or imaginal scenarios relating to washing, checking, or hoarding).[14] Furthermore, distinct patterns of regional activation associated with specific symptom dimensions were seen in individuals with OCD, as follows: bilateral ventromedial prefrontal regions and right caudate nucleus (washing); putamen/globus pallidus, thalamus, and dorsal cortical areas (checking); and left precentral gyrus and right orbitofrontal cortex (hoarding). Thus, another explanation for the heterogeneity of neuroimaging findings in individuals with OCD may be that different symptom dimensions are mediated by dysfunctions in distinct components of the frontostriatothalamic circuitry.

While most structural studies have focused on gray matter abnormalities in OCD, there is also evidence of white matter dysfunction in this disorder. For example, MRI studies in the field have demonstrated significantly less total white matter in individuals with OCD, compared with unaffected controls.

FUNCTIONAL NEUROIMAGING

Many neurobiological theories of the pathogenesis of OCD are derived from the results of functional imaging studies. In these studies, techniques are used to measure indirectly the activity levels in specific brain regions. Examples of such techniques are positron emission tomography (PET) and single-photon emission computed tomography (SPECT), which can be used to determine whether certain neuroanatomical areas or structures are more active in individuals with OCD than in healthy controls.

The findings of functional neuroimaging studies have largely echoed those of structural studies. In a number of studies, individuals with OCD have shown increased activity in the orbitofrontal cortex (Fig. 38.1), cingulate, and striatum at rest and during exposure to feared stimuli. A meta-analysis of studies using PET and SPECT to investigate brain activity in OCD partially supported these conclusions, with particular functional differences noted in the orbital gyrus and head of the caudate nucleus.[15] However, no other significant effect sizes were reported (including no differences in the more inclusive regions of the caudate, or in the thalamus, anterior

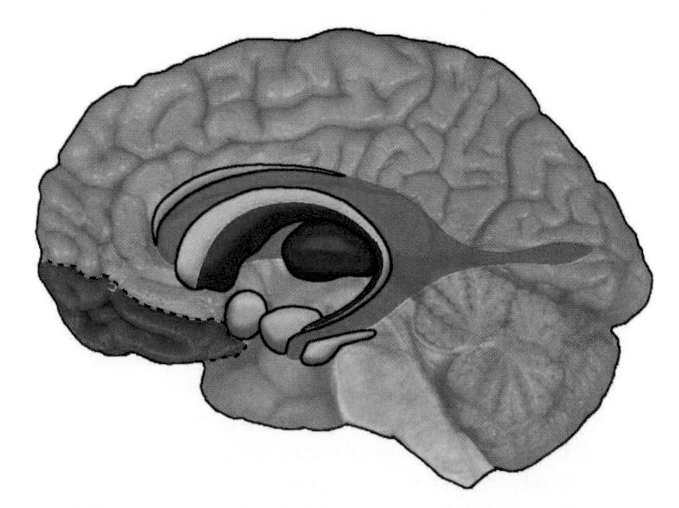

FIGURE 38.1 **Increased activation in ventral corticostriatal–thalamocortical circuitry in obsessive–compulsive disorder.** *Source: MRC Unit on Anxiety and Stress Disorders.*

cingulate, or orbitofrontal cortex). These inconsistencies may again reflect the phenotypic heterogeneity of OCD.

Magnetic resonance spectroscopy (MRS) techniques have also been used to evaluate differences in regional metabolites in the brains of individuals with OCD versus healthy controls. To this end, the concentrations of metabolites such as *N*-acetyl aspartate (NAA), combined glutamate and glutamine, *myo*-inositol, choline, and creatinine in brain tissue are measured. Some small studies have found significantly lower levels of NAA in the striatum of adults with OCD than in matched control participants. This finding has been replicated in pediatric populations with OCD, with decreased NAA levels reported in the thalami of affected children. This reduction was inversely correlated with increased symptom severity. As NAA is only detectable in neurons, decreased concentration would indicate decreased density of healthy neurons in that brain region. In small MRS studies of juvenile-onset OCD, significantly greater glutamate and glutamine concentrations have been found in treatment-naïve children with OCD than in healthy controls. Following serotonergic treatment, these levels decreased significantly to concentrations comparable to those of the control subjects. Reduction in glutamate and glutamine was also associated with decreased symptom severity, suggesting that the mechanism of action of paroxetine may be mediated by a serotonergically modulated reduction in glutamate and glutamine in the caudate.

NEUROANATOMICAL MODELS OF DISEASE

Various models have been posited to relate CSTC dysfunction to the symptoms of OCD. In an early heuristic paradigm, the striatal topography model of OCD and related disorders was articulated. This model suggests that the clinical manifestations of OCD (and, for example, of Tourette syndrome) are governed by specific dysfunctions within the striatum. Building on this original model, it has been hypothesized that the obsessional symptoms of OCD are mediated by the caudate; motor symptoms by the putamen (a component of the dorsal striatum); and affective symptoms (anxiety) by the paralimbic CSTC circuits. Structural imaging studies showing caudate involvement in individuals with OCD (see above) support this striatal topography model of disease.

This model may be extended to include cognitive neuroscience perspectives of OCD. It has been suggested that information processing can occur either explicitly (i.e. consciously) via dorsolateral PFC and medial temporal structures (e.g. hippocampus) or implicitly (i.e. unconsciously) via corticostriatal systems. Thus, as OCD is viewed primarily in the context of CSTC dysfunction, its pathogenesis may be understood as a disorder of implicit processing. In other words, the characteristic intrusive thoughts and behaviors of OCD may result from failure of the thalamus to filter out input information (that is normally efficiently processed via the CSTC), which then erroneously reaches consciousness.

As mentioned above, the striatum is believed to be integral to implicit information processing. However, in OCD, striatal dysfunction necessitates involvement of those structures involved in explicit processing (e.g. medial temporal regions). Compulsive behaviors may arise thus as adaptive mechanisms to recruit these corticostriatothalamic collateral structures. Thus, although relatively inefficient, input filtering at the level of the thalamus is facilitated. This pattern of thalamic deactivation with striatal recruitment has been demonstrated in some functional imaging studies of OCD.

While the striatal topography model focuses on subcortical pathology in OCD, primary cortical dysfunction may also play a role in the pathogenesis of this disorder. According to a cortical excitability model of disease, intracortical neuronal inhibition is defective in individuals with OCD compared with healthy controls. Transcranial magnetic stimulation is a non-invasive technique used to induce regional neuronal activity and thereby assess neuronal inhibition. In humans, a scalp electromagnetic coil is used to transmit magnetic pulses to cortical motor output cells. Commonly used transcranial magnetic stimulation paradigms include the cortical silent period, an interruption of voluntary muscle contraction by transcranial stimulation of the contralateral motor cortex[16]; and intracortical facilitation, a paired pulse paradigm in which conditioning stimuli are applied to the motor cortex before the test stimulus, enabling the investigator to index excitability of the motor cortical excitatory circuits.[17] The cortical silent period is significantly shortened and intracortical facilitation significantly increased in individuals with OCD compared with healthy controls. This suggests that OCD may be associated with dysregulated cortical inhibitory and facilitatory neurotransmission.

Neurochemistry

Several neurotransmitter systems have been implicated in the pathogenesis of OCD. The three most widely investigated are the serotonergic, dopaminergic, and neuropeptidergic systems.

SEROTONIN

Serotonin [5-hydroxytryptamine (5-HT)] is a monoamine neurotransmitter released from neurons with cell bodies in the medium-sized raphe nuclei in the midbrain. A serotonergic hypothesis first arose from the clinical and experimental finding that serotonin reuptake inhibitors (SRIs) were particularly efficacious in alleviating the symptoms of OCD. However, while the idea that the SRIs exert their antiobsessional effect by correcting a fundamental abnormality in the serotonergic system is appealing, it may be that these agents actually modulate an intact system to compensate for an underlying deficit not related to serotonergic function.

Several studies have attempted to elucidate this issue. Indirect measures of serotonergic function include peripheral receptor binding in the blood; measurement of 5-HT metabolites in the cerebrospinal fluid (CSF); and pharmacological challenge studies. Although the literature is vast, studies using these methods have been largely disappointing and inconsistent. More direct methods include functional imaging receptor characterization techniques such as SPECT. Studies using this technique and [^{123}I](β-CIT) (a non-selective dopamine reuptake inhibitor with prominent affinity for the serotonin transporter) found that significantly reduced serotonin transporter (SERT) availability is evident in the thalamus and hypothalamus of individuals with OCD versus healthy controls. SERT availability was also significantly negatively correlated with symptom severity and positively correlated with duration of illness. However, these findings have not been replicated.

Preclinical (animal) studies have also begun to delineate the therapeutic mechanisms of the SRIs in OCD. For example, high and low doses of the selective serotonin reuptake inhibitor (SSRI) fluoxetine significantly reduce spontaneous stereotypic behavior in deer mice. This treatment-response pattern mimics that of individuals with OCD, suggesting predictive validity of stereotypic behavior of deer mice for OCD.

It also appears that SRIs potentiate serotonergic transmission via autoreceptor desensitization. The timing of these receptor changes mimics the delay between SRI initiation and therapeutic response that is often observed. For example, animal studies have shown that SSRI-induced changes may occur more quickly in the lateral frontal cortex than in the medial frontal cortex. This is in line with clinical observations that the SRIs tend to exert their antidepressant effect sooner than their anti-obsessional effect. Thus, high doses of SSRIs over long periods may be required to desensitize the serotonin autoreceptors.

DOPAMINE

Animal studies have suggested a dopaminergic influence in producing stereotypic ("compulsive") behavior. For example, in their study of transgenic mice in which dopamine D_1 receptor-expressing neurons in regional subsets of the cortex and amygdala express a neuropotentiating cholera toxin transgene,[18] Campbell and colleagues suggested that the perseverative motor and behavioral abnormalities observed in these mice may be attributable to chronic potentiation of cortical and limbic D_1 receptor-expressing neurons. This finding has been strengthened by reports that administration of dopamine agonists (e.g. quinpirole) may induce ritualistic (compulsive) behavior in rats.

Human radioligand studies of dopaminergic function in OCD have also been increasing. For example, [^{123}I](β-CIT) SPECT studies of psychotropic-naïve adults with OCD found that these individuals exhibit increased dopamine transporter binding in the basal ganglia, compared with unaffected controls. In addition, dopamine D_2 receptor binding in the basal ganglia is significantly lower, and D_1 receptors are downregulated in the striatum of individuals with OCD compared with healthy controls. Taken together, these findings seem to suggest that there is a higher synaptic concentration of dopamine in the basal ganglia of people with OCD than in the general population.

Several pharmacological studies have also elucidated the role of dopamine in the pathogenesis of OCD. While the response rate of individuals with tic-related OCD to SSRIs (e.g. fluvoxamine) is only about 50–60%, augmentation with dopamine antagonists (e.g. haloperidol, a typical neuroleptic) may yield more promising results. In addition, the atypical neuroleptics (e.g. risperidone), which also function as dopamine antagonists, have shown efficacy in treating refractory OCD without tics. These heterogeneous responses may be attributable to differences between OCD subtypes.

NEUROPEPTIDES

Neuropeptides that may play a role in the pathogenesis of OCD include arginine vasopressin (AVP), oxytocin, adrenocorticotropic hormone (ACTH), corticotropin-releasing factor (CRF), somatostatin, and the opioid system. AVP (also known as antidiuretic hormone) is an endogenous peptide hormone synthesized in the hypothalamus. Most of the synthesized volume is stored in vesicles in the posterior pituitary before secretion into the circulation, where AVP regulates homeostasis via its mediating effects on renal tubular absorption and peripheral vascular resistance. However, some of the hormone is released directly into the brain, where it has been implicated in memory formation and the development of grooming behaviors. In an early clinical study, Altemus and colleagues found that individuals with OCD exhibited significantly higher levels of AVP in the CSF, and significantly increased secretion of AVP into the plasma in response to hypertonic saline administration.[19] However, this finding has not been replicated more recently, as investigators have failed to find significant differences in CSF AVP concentrations between individuals with OCD, those with Tourette syndrome (TS), and healthy controls. Thus, more work is required to delineate this relationship further.

Oxytocin is a major neuromodulator that is synthesized in the hypothalamus and either stored in the posterior pituitary or released directly into the CNS. It is structurally and functionally related to AVP and may play a role in grooming and maternal behavior (as well as sexual arousal). Both animal and human studies have shown an association between oxytocin and the development of obsessive–compulsive symptoms. For example, this neuropeptide has been found to induce maternal and reproductive behaviors in rats primed with gonadal steroids, which may explain the onset of OCD during pregnancy or in the puerperium. Human clinical studies have complemented these findings, reporting higher CSF oxytocin levels in individuals with non-tic-related OCD than in unaffected controls. Oxytocin levels may also be correlated with symptom severity.

CRF is a polypeptide hormone secreted by the paraventricular nucleus (PVN) of the hypothalamus in response to stress. Its main function is to stimulate the anterior pituitary to secrete ACTH. ACTH is an important component of the hypothalamic–pituitary–adrenal axis (HPA) (see "Neuroendocrinology", below) and stimulates corticosteroid production and release. Studies relating ACTH and CRF to OCD in humans have yielded inconsistent results. While there is some evidence that basal plasma and post-CRF-administration CRF levels are lower in individuals with OCD than in healthy controls, this finding has not been widely replicated. One explanation for these inconsistencies may be the non-specific nature of CRF and ACTH. Both are stress hormones, and thus sensitive to alterations during any chronic illness.

Somatostatin is a regulatory (inhibitory) hormone that modulates neurotransmission and cell proliferation via interaction with G-protein coupled somatostatin receptors. It is produced by neuroendocrine neurons of the periventricular nucleus of the hypothalamus and acts to inhibit the release of numerous secondary hormones. Animal studies have shown that somatostatin delays the termination of avoidance behaviors (reminiscent of persistent repetition in people with OCD) and can produce stereotypies. Human studies have supported these findings, with individuals with OCD exhibiting higher levels of CSF somatostatin than healthy controls.

Endogenous opioid peptides include the endorphins, enkephalins, endomorphins, and dynorphins, all of which mediate reward pathways and successful task completion. Thus, dysfunctions of these pathways may account for the self-doubt often experienced by individuals with OCD. Early clinical studies suggested that higher levels of antibodies for prodynorphin (a polypeptide precursor) were present in individuals with OCD than in healthy volunteers and disease controls. However, more recent therapeutic trials of the efficacy of opioid antagonists in OCD have produced disappointing results and further work is required to investigate the role of neuropeptides in the pathogenesis of this disorder.

Neuroendocrinology

Abnormalities in the HPA axis have been well documented in the anxiety disorders. The role of this endocrine system in the development of OCD has also been investigated. For example, multiple studies have shown increased 24 hour urinary cortisol levels as well as increased serum cortisol concentration in individuals with OCD compared with control subjects. However, dexamethasone suppression testing in OCD has yielded inconsistent results, with some studies showing non-suppression and others failing to demonstrate any abnormalities. Thus, the HPA axis may be less central to the pathogenesis of OCD than in anxiety disorders. Further work in this area is warranted to elucidate fully the role of the HPA axis and other endocrine abnormalities in OCD.

Neuroimmunology

In their clinical study of 80 children (aged 5–17) with a diagnosis of tic disorder, Singer and colleagues found that 42 participants described acute symptom onset or exacerbation.[20] Of this subgroup, approximately 35% reported that the exacerbation was historically associated with an infection, most commonly streptococcal.

Pediatric autoimmune neuropsychiatric disorders associated with streptococcal infection (PANDAS) comprise a subset of pediatric OCD cases that meet the following criteria: presence of OCD and/or a tic disorder; prepubertal onset; episodic course of symptom severity; association with group A β-hemolytic streptococcal (GABHS) infection; and association with neurological abnormalities. PANDAS were initially designated following the observation that OCD was common in Sydenham's chorea (a neurological manifestation of acute rheumatic fever), as well as an early suggestion that the basal ganglia damage characteristic of Sydenham's chorea may be mediated by antineuronal antibodies as part of an autoimmune response to GABHS. A number of clinical studies strengthened the case for this subtyping, showing that a subset of children with OCD and TS had antineuronal antibodies and exhibited characteristic clinical features of abrupt symptom onset and discrete, episodic exacerbations, often with demonstrable GABHS infection.

Neuroethology

There is a significant body of work examining animal models of OCD. The symptoms of this disorder are reminiscent of maladaptive, repetitive behavior patterns in animals (including perseverative and stereotyped motor activities). The pathogenesis of these behaviors in animals has also been found to overlap with the neurobiological mechanisms underlying OCD (e.g. excess dopaminergic activity in the basal ganglia). In early animal studies, serotonergic drugs were found to show selective efficacy in treating excessive behaviors, such as the repetitive licking of paws or flank that characterizes canine acral lick dermatitis. These findings mimic the typical pharmacological response pattern of individuals with OCD.

In two noteworthy reviews, Joel and colleagues summarized the five most studied animal models to date: quinpirole-induced compulsive checking; marble burying; signal attenuation; spontaneous stereotypy in deer mice; and decreased alternation induced by 8-hydroxy-2-(di-*n*-propylamino)-tetralin hydrobromide (8-OHDPAT), a serotonin receptor agonist.[21,22] They also evaluated the face validity (derived from the phenomenological similarities between the animal behavior and the corresponding symptoms in humans), predictive validity (derived from similarities in treatment response), and construct validity (derived from similarities in pathogenesis) of these animal models. There is some evidence that optogenetic techniques (i.e. combining genetics and optics to allow precise manipulation of electrical and biochemical neural events using fiberoptic light) may be applied to these and other animal models to identify specific neurocircuits driving repetitive behavior. For example, hyperactivation of the glutamatergic pathway between the orbitofrontal cortex and ventromedial striatum may produce compulsive (grooming) behavior in mice.[23] Taken together, animal models may be of great value in uncovering the pathogenesis of OCD in humans, in elucidating its symptom dimensions, and in detecting novel treatment targets and strategies.

Genetic Factors

There has been much interest in elucidating the genetic factors contributing to OCD. There seems to be an element of heritability in the pathogenesis of this disorder. While a number of methodological approaches has been employed, the four most informative have been family, twin, candidate gene, and genome-wide association studies (GWAS).

Family studies to date have shown that OCD has significant familial aggregation, occurring more frequently in relatives of affected probands (individuals with the disorder) than in those of control subjects. For example, in their meta-analysis of five studies, Hettema and colleagues found that the risk of developing OCD was 8.2% in relatives of probands compared with only 2.0% in relatives of unaffected individuals.[24] Such studies have also delineated a familial relationship between OCD and tic disorders. In two important genetic studies, Pauls and colleagues[25,26] found not only a higher frequency of OCD and subthreshold OCD in relatives of affected probands compared with comparison participants, but also a higher rate of TS and chronic tics in these individuals.[26] Furthermore, the prevalence of OCD without either TS or chronic tics was elevated in first degree relatives of people with both OCD and TS, and those with TS only.[25] These data suggest an autosomal dominant pattern of inheritance, with incomplete penetrance.

The few twin studies of OCD to date have strengthened the case for heritability in this disorder. In their review of these studies,[27] van Grootheest and colleagues found an estimated heritability of 27–47% in adults (and 45–65% in children). Disease concordance among monozygotic twins (approximately 87%) has also been reported to be far higher than in their dizygotic counterparts (approximately 47%). Thus, taken together, family and twin studies suggest that OCD risk is often inherited.

However, in order to identify specific candidate genes that may confer an increased risk of OCD, a candidate gene approach is required. In these studies, discrete genes that may be associated with OCD risk or course of disease are identified and characterized. The choice of candidate gene to be analyzed is often based on underlying neurobiological mechanisms of disease (see above). Thus, in OCD, serotonergic genes (e.g. those coding for the 5-HT receptors), dopaminergic genes (e.g. those related to the dopamine transporter), noradrenergic genes (e.g. the catechol-*O*-methyltransferase gene), and glutamatergic genes (e.g. the glutamate receptor, ionotropic, kainite gene) have been studied. However, owing to the complex phenotype and multifactorial etiology of OCD, it is doubtful that a single causative gene will be identified. Instead, the development of this disorder is likely to be the result of the combined effect of multiple genes. In a meta-analysis of gene association studies published in 2013, Taylor reported a significant association between OCD and two serotonin-related polymorphisms.[28] A secondary meta-analysis identified another 18 polymorphisms with significant odds ratios. This adds credence to a polygenic model of OCD, with most genes having a modest association with the disorder.

Although a candidate approach is useful in identifying specific high-risk genes in OCD, it is also inherently limited. Arbitrary gene selection, insufficient gene coverage, and negligible potential for new gene discovery are characteristically identified as shortcomings of candidate gene studies. A genome-wide (whole exome) association approach addresses these limitations and is increasingly being used in genetic studies on OCD. For example, in their genome-wide linkage analysis of 56 individuals from seven families (identified through pediatric OCD probands), Hanna and colleagues found evidence for linkage (i.e. the tendency of proximally located genes to be inherited together) on chromosome 9.[29] Similarly, in a genome-wide linkage analysis study of 219 families, Shugart and colleagues found suggestive linkage signals on a number of chromosomes, with the strongest evidence for linkage emerging for chromosome 3.[30] Despite being more cost, time, and labor intensive than a candidate gene approach, future GWAS would be of value in further elucidating the genetic etiology of OCD.

TREATMENT OF OBSESSIVE–COMPULSIVE DISORDER

A range of pharmacotherapeutic and psychological interventions has shown efficacy in treating individuals with OCD. SRIs and exposure and response prevention (ERP) are the most widely used therapeutic modalities.

Pharmacotherapy

First Line Treatment

The role of the serotonergic system in the pathogenesis of OCD is well documented and has already been discussed. Thus, it is not surprising that the mainstay of pharmacotherapy for this disorder is the SRIs, primarily clomipramine and the SSRIs.

In the 1960s, clomipramine, a tricyclic antidepressant with marked serotonergic activity, was found to relieve obsessive–compulsive symptoms. Unlike the other tricyclics that did not show such efficacy, this agent was found to block potently serotonin reuptake. Thus, interest in a serotonergic pathogenesis of OCD was sparked. Before the discovery of the SSRIs, clomipramine was widely considered the standard pharmacotherapy for OCD. In an early large multicenter trial of individuals with OCD and no comorbid depression,[31] clomipramine (flexibly dosed to a maximum of 300 mg/day) yielded a statistically and clinically significant improvement in symptoms, when compared to placebo. However, this improvement was not maintained after treatment discontinuation. Furthermore, the adverse side-effect profile reported by individuals with OCD with and without comorbid depression was problematic. Owing to its receptor binding profile, clomipramine is often associated with symptoms typical of anticholinergic blockers (e.g. constipation, dry mouth, blurred vision), antihistaminic (H_1) binding (e.g. sedation, weight gain), and α-adrenergic blockade (e.g. postural hypotension). Like the other tricyclics, clomipramine may also be lethal when taken in overdose, owing to QT interval prolongation. Thus, there was a need for an intervention with comparable efficacy but improved tolerability compared with clomipramine.

In the 1980s, SSRIs were adopted as pharmacotherapy for individuals with OCD. While early meta-analyses suggested that clomipramine may be more efficacious than the SSRIs, these publications had numerous methodological limitations, and direct head-to-head studies have shown equal efficacy between SSRIs and clomipramine. Furthermore, SSRIs are generally better tolerated and are associated with far fewer adverse effects than clomipramine. Fluoxetine, fluvoxamine, sertraline, and paroxetine have all shown superior efficacy in treating individuals with OCD compared with placebo, and all have received US Food and Drug Administration (FDA) approval for clinical use in OCD. A Cochrane review of placebo-controlled randomized controlled trials (RCTs) and quasi-RCTs examining the efficacy of SSRIs in OCD found that these agents were more effective than placebo in achieving clinical response at post-treatment.[32] Pooled effect sizes were similar between individual SSRIs (Fig. 38.2).

In the clinical setting, however, differences in side-effect profile and tolerability between SSRIs may influence treatment decisions. For example, weight gain is

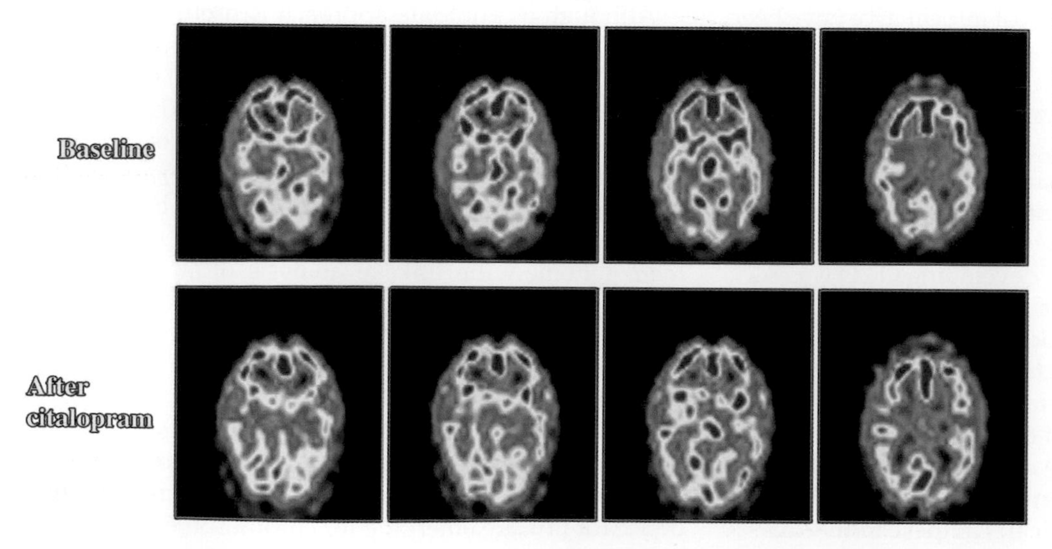

FIGURE 38.2 **Comparison of single-photon emission computed tomography (SPECT) images before and after treatment with the selective serotonin reuptake inhibitor citalopram in individuals with obsessive–compulsive disorder.** *Source: MRC Unit on Anxiety and Stress Disorders.*

more common with paroxetine use, while insomnia and agitation are most often associated with fluoxetine and sertraline use.

Despite some interest in the use of agents other than the SRIs in the acute treatment of individuals with OCD, evidence remains unconvincing. For example, clinical trials of monotherapy with venlafaxine (a serotonin and norepinephrine reuptake inhibitor) and with clonazepam (a benzodiazepine) have yielded inconsistent results. Thus, SSRIs remain the first line choice for individuals with this disorder.

In general, SSRI treatment should be commenced at a low to moderate dose, and then titrated slowly upwards to the maximum tolerated. In their meta-analysis of SSRI dose–response relationships in OCD, Bloch and colleagues found that higher doses of SSRIs were associated with improved treatment efficacy compared with low or medium doses.[33] However, these higher doses were also associated with higher participant dropout rates due to adverse effects. Thus, efficacy and tolerability should be balanced to optimize treatment adherence.

Acute treatment at the maximally tolerated dose should be undertaken for at least 6–12 weeks. Once a therapeutic response is observed, maintenance therapy should be continued for at least 1 year, and then discontinued gradually to minimize the risk of symptom relapse. In the long term, there is evidence that some individuals can be maintained at lower therapeutic doses without worsening of symptoms. Further work is warranted to confirm this preliminary result, as reduced-dose maintenance therapy has clear tolerability and adherence benefits.

Treatment Resistance

Currently, the Y-BOCS is the most widely used standardized rating instrument to measure an individual's response to treatment. Full treatment response has been defined by some as a reduction of at least 35% in Y-BOCS score, partial response as a reduction between 25 and 35%, and no response as a reduction of less than 25%. Unfortunately, up to 40% of individuals treated with an optimal SSRI trial exhibit only partial or no response to treatment.

For patients who experience a partial response to SSRI treatment, second line interventions include augmentation strategies. While a number of agents have been used in the clinical setting, the strongest evidence supports the low-dose second generation (atypical) antipsychotics, including risperidone, olanzapine, and quetiapine. Early data suggested that these agents should be used specifically in individuals with OCD and comorbid tic disorder, but it is now widely accepted that they are efficacious in tic-related and non-tic-related OCD. Haloperidol, a first generation (typical) antipsychotic, showed some promise as a useful augmentation agent, but its adverse side-effect profile (and the availability and efficacy of the better tolerated atypical antipsychotics) has limited its use.

Experimental Treatments

Although various novel and experimental pharmacotherapies for OCD have been investigated, findings have been largely disappointing. Nonetheless, ongoing work in this area is important for continued therapeutic development.

NEUROPEPTIDES

Despite the well-documented role of neuropeptides in the pathogenesis of OCD (see above), clinical efficacy trials of these agents are sparse and inconsistent. For example, in their early RCT, Epperson and colleagues failed to demonstrate any significant difference between intranasal oxytocin and placebo in treating individuals with OCD.[34] The fact that peripherally administered oxytocin does not penetrate the blood–brain barrier well (<0.003% is absorbed) may account for these disappointing findings.

GLUTAMATERGIC AGENTS

There is a growing body of evidence that glutamatergic abnormalities may be involved in the development of OCD. This has ignited interest in the utility of glutamate-modulating agents in treating individuals with this disorder. For example, in an open-label trial of 13 adults with treatment-resistant OCD, seven (54%) of the study participants demonstrated greater than 35% reduction in Y-BOCS scores, with five (39%) being categorized as treatment responders when given riluzole augmentation therapy 50 mg twice daily.[35] This agent was well tolerated, with no serious adverse effects documented.

Similarly, 11 out of 16 SSRI-resistant cases in another study responded favorably to topiramate augmentation (mean dose 253.1 ± 93.9 mg/day).[36] However, this agent (an anticonvulsant with glutamatergic properties) has also been found to induce the symptoms of OCD, and its use warrants further investigation.

Non-Pharmacological Somatic Treatments

Somatic interventions for individuals with OCD are still in the experimental stage, and should be reserved only for the most severe, resistant cases.

Electroconvulsive Therapy

Although there have been reports of isolated success in treating resistant cases of OCD with electroconvulsive therapy, this intervention is likely to be most useful in individuals with severe comorbid depression who are at high risk of suicide.

Repetitive Transcranial Magnetic Stimulation

There is preliminary evidence of the efficacy of repetitive transcranial magnetic stimulation (the use of high-density electromagnetic pulses delivered to the scalp to stimulate or disrupt cortical activity) in the treatment of OCD. However, owing to a relative lack of evidence to date, further work is required before it is widely implemented.

Ablative Neurosurgery

Neurosurgical techniques currently under investigation for the treatment of severe OCD are focused on interrupting CSTC pathways, through cingulotomy and capsulotomy. While there is some evidence of symptom improvement following these procedures, the lack of blinded, comparator-controlled follow-up studies and the irreversible nature of these interventions make them a last resort only.

Deep Brain Stimulation

As a reversible, non-ablative alternative to cingulotomy and capsulotomy, deep brain stimulation has shown some preliminary promise. This technique involves the implantation of electrodes into neuroanatomical structures believed to be involved in the pathogenesis of OCD. These electrodes remain *in situ* and deliver an electric current to the structures, aiming to modulate the neurocircuitry of OCD. Multidisciplinary teams in Europe and the USA have found that this intervention resulted in clinically significant symptom reduction and functional improvement in approximately two-thirds of study participants with severe or highly treatment-resistant OCD. However, as is the case for the other somatic treatments, deep brain stimulation may be associated with a range of serious adverse events including seizure induction, worsening depression, and asymptomatic hemorrhage, and further study on precise indications is needed.

Psychological Therapy

Expert consensus guidelines suggest that ERP should be the first line psychological therapy for individuals with OCD.[37] This intervention comprises controlled, repeated, and prolonged exposure (either *in vivo* or imaginal) to stimuli that are known to provoke obsessional fear, and enforced abstinence from compulsive rituals. For example, an individual suffering from contamination obsessions and compulsive hand washing would be encouraged to touch numerous foreign door handles and handrails, without then resorting to ritualistic hand washing. In the clinical setting, exposure therapy sessions should be hierarchical. In other words, individuals should initially be exposed to less anxiety-provoking situations, which are then escalated gradually. The most anxiety-provoking stimuli should not be left until the end of treatment, but rather included midway through the schedule. This will allow generalization of treatment effects, as individuals can undergo repeated exposure to the most difficult situations in a variety of contexts.

The effectiveness of ERP in treating obsessive–compulsive symptoms was arguably first shown in early animal laboratory trials. In their study of dogs in partially electric-gated shuttle boxes (small rooms divided in two by a hurdle), Solomon and colleagues found that compulsive ritual-like behaviors could be induced in the dogs by pairing a flickering light with an electric shock (that was delivered about 10 seconds after the light).[38] The dogs soon learned that, by jumping across the hurdle into the component of the shuttle box that was not electrified, they could terminate exposure to the shock. Thereafter, they learned that they could avoid this exposure altogether by jumping across the hurdle as soon as the flickering light appeared, thus displaying a conditioned response. Once this was observed, the electric shocks were discontinued, and it was found that the flickering light stimulus alone could induce the conditioning hurdle-jumps; thus, the dogs had acquired obsessive–compulsive-like jumping behavior to reduce the fear and anxiety associated with exposure to the flickering light. Solomon and colleagues then implemented an early form of ERP in their canine subjects. The height of the hurdle in the shuttle box was increased so that the dogs could no longer jump over it. Thus, when the flickering light was shown, the dogs were unable to escape – an analogue of response prevention. While initially exhibiting strong signs of fear and distress, the dogs eventually calmed down after repeated exposures to the flickering light (i.e. fear extinction).

These early animal models were applied to human studies during the 1960s and 1970s. Individuals with contamination obsessions and compulsive hand-washing rituals were required to place their hands into a container with dirt and garbage, and were prevented from washing their hands for some time afterwards. While initially distressed and anxious, these individuals eventually demonstrated fear extinction with reduced urges to wash, thus reflecting the behavior of Solomon's dogs.

RCTs have been undertaken to examine the clinical efficacy of ERP in managing individuals with OCD. Using the Y-BOCS as an outcome measure, these studies showed that ERP may produce clinically significant improvement in obsessive–compulsive symptoms. For example, in their study of 31 outpatients with OCD, Nakatani and colleagues randomly assigned these individuals to one of three treatment groups: behavior therapy (BT) with or without pill placebo; autogenic training (a psychological placebo for OCD) with or without fluvoxamine (FLV); or control group (autogenic training with or without pill placebo).[39] Significant symptom improvement was seen in both the BT and FLV groups compared with the control group. Furthermore, those in the BT group showed significantly more improvement than the FLV group. These findings were supported by a large meta-analytic study, which found substantial post-test treatment effect sizes of 1.16 (self-report measures) and 1.41 (interviewer measures) for ERP in individuals with OCD.[40]

In the clinical setting, evidence suggests that an acute ERP trial of 13–20 weekly sessions, each lasting for 90–120 minutes, should be undertaken. If time and resource constraints limit this implementation, a trial of seven to 12 sessions has also shown efficacy. If a favorable response to treatment is observed, maintenance therapy can then be continued for 3–6 months.[37]

However, not all individuals with OCD respond to ERP. Broadly speaking, predictors of outcome may be grouped into three main categories: treatment-related factors, individual-related factors, and environmental factors. In terms of treatment characteristics, psychoeducation, therapist supervision, combining situational (*in vivo*) and imaginal exposure, and promoting complete response prevention improved treatment outcome. Individual characteristics that may be associated with poorer response to ERP treatment include poor insight, severe comorbid depression, and the presence of prominent hoarding symptoms. Finally, while there is some preliminary evidence that interpersonal hostility towards the individual may be associated with premature dropout and poorer ERP response, further work is required to elucidate environmental influences on treatment outcome.

There is an emerging body of work investigating the efficacy of mindfulness-based therapy in individuals with OCD. Broadly, mindfulness may be conceptualized as self-regulated attention on the immediate (present) moment, followed by orientation towards one's experience in the moment, while maintaining an open, curious, and accepting attitude. Recent controlled pilot data indicate that a mindfulness intervention (meditation) may be associated with a significant reduction in symptoms, compared with waiting-list control groups. However, further work in this area using larger and clinical samples is needed to evaluate systematically the usefulness of mindfulness-based therapy in OCD.

Combined Treatment

Combined pharmacotherapy and ERP is sometimes used in clinical practice, despite a paucity of empirical evidence supporting this approach. Although some studies have reported superior efficacy of simultaneous SRI/ERP therapy compared with SRI monotherapy, these are often limited by methodological biases and low statistical power.

Nonetheless, there is some evidence that adjunctive ERP may be helpful in individuals with SRI-resistant symptoms. Several studies have reported that ERP augmentation of SRI treatment may lead to significant symptom improvement in individuals who have shown minimal response to pharmacotherapy alone. In a large, well-controlled augmentation study, Tenneij and colleagues randomly assigned 96 individuals with OCD who had responded to 3 months of drug treatment either to continue drug treatment alone or to receive additional behavior therapy for 6 months.[41] Those individuals who had received the additional behavior therapy showed a greater symptom improvement than those who had continued with drug treatment alone.

There is also a body of work examining the efficacy of D-cycloserine, a partial agonist at the glycine modulatory site on *N*-methyl-D-aspartate (NMDA) glutamate receptors, in augmenting psychological therapy in treatment-resistant individuals. Enhancing NMDA receptor function is hypothesized to promote synaptic plasticity during learning. For example, 100–125 mg D-cycloserine, administered to patients approximately 1–2 hours before each exposure therapy session, may be associated with significant symptom improvement, decreased number of sessions required to achieve significant clinical improvement, and reduced therapy dropout rate (compared with a placebo control group). Thus, it seems that combination therapy may be efficacious in individuals who have responded to initial pharmacotherapy and in those who have shown treatment resistance.

OBSESSIVE–COMPULSIVE SPECTRUM DISORDERS

It has long been hypothesized that OCD may be phenomenologically and etiologically related to a number of other conditions along the spectrum of disease. While there is no clear consensus about which disorders should be considered along this spectrum, putative inclusions are TS, somatoform disorders such as BDD, and trichotillomania and skin-picking disorder. The DSM-5 chapter on obsessive–compulsive and related disorders now includes several of these conditions.[2]

Phenomenological Correlates

The hallmark feature of OCD and the related disorders is the presence of unwanted repetitive symptoms. Individuals with TS experience motor and vocal tics, those with BDD often experience intrusive thoughts of imagined ugliness and may resort to repetitive behaviors such as mirror checking, and trichotillomania is diagnostically characterized by a compulsive urge to hair pull. In all disorders, obsessive thoughts are distressing and are often only relieved by completion of the corresponding compulsive act. Furthermore, OCD and the related disorders often coexist. For example, tics are more common in individuals with OCD than in the general population, and people with TS often also have OCD.

However, several important phenomenological differences distinguish OCD from these disorders. For example, while the tics associated with TS are often involuntary, individuals with OCD exhibit goal-directed behavior (e.g. compulsive hand washing to relieve contamination-related anxiety). Individuals with BDD are preoccupied solely with their appearance, while those with OCD may experience a wide range of obsessions. Clinical comparisons have also yielded significant differences between OCD and the spectrum disorders. For example, while OCD has been found to be associated with more lifetime disability and comorbidity, treatment response rate among people with trichotillomania is reportedly lower. Furthermore, individuals with OCD have reported more harm avoidance and maladaptive beliefs. These and other differences between OCD and the related disorders may indicate important discrepancies in underlying pathogenesis, which could, in turn, influence response to treatment.

Neurobiological Correlates

Structural and functional imaging studies have shown the involvement of the CSTC in individuals with TS and with BDD, suggesting an overlap in pathogenesis with OCD. Although there is some evidence to suggest a central role of the cerebellum in the pathogenesis of trichotillomania, work by Chamberlain and colleagues supports a CSTC neuroanatomical model.[42] For example, in a morphometric MRI study, individuals with trichotillomania showed increased gray matter density in the striatal, amygdalohippocampal, and cortical regions compared with healthy controls.[42]

Neurogenetic Correlates

Several family and twin studies (as well as some molecular genetic data) support a genetic link between OCD and the related disorders. For example, a segregation analysis study of 30 nuclear families indicated that first degree relatives of probands with TS were more likely to have TS, TS or chronic tics, and OCD compared with relatives of control subjects.[43] This supports an etiological link between OCD, TS, and chronic tics. Similarly, a large study of 2148 female twins (1074 pairs) examining genetic and environmental covariance between BDD and OCD found a significant genetic overlap between these two conditions, whereas environmental factors did not contribute substantially to their covariation.[44]

Candidate genetic studies of trichotillomania have also sought to identify specific genes that may predispose individuals to developing this disorder. While such a gene has not yet been identified, there is evidence that common genetic risk factors may contribute to the development of trichotillomania and OCD. In a case–control association study, Hemmings and colleagues investigated the role of serotonergic and dopaminergic genes in mediating these disorders.[45] In the study sample of 39 individuals with trichotillomania, 250 with OCD, and 152 control subjects, these authors found that the 5-HT receptor 2A (*5-HT2A*) gene may be involved in the molecular etiology of trichotillomania.

Bienvenu and colleagues investigated whether variation in the human *Sapap3* gene was associated with OCD and grooming disorders (pathological nail biting, pathological skin picking, and trichotillomania) in 383 families.[46] This gene codes for the SAP90/PSD95-associated protein (SAPAP) family of proteins: postsynaptic density (PSD) components that interact with other proteins at glutamatergic (excitatory) synapses to form essential scaffolding. Four of the six single-nucleotide polymorphisms (SNPs) of the *Sapap3* gene were associated with at least one grooming disorder, although none of the SNPs was associated with OCD. Thus, while this gene holds promise in elucidating the genetic etiology of grooming disorders and OCD, further work is required to bolster this preliminary evidence.

Pharmacotherapeutic Correlates

The association between OCD and the spectrum disorders is further evidenced by their treatment response pattern. Broadly, individuals with any of these conditions seem to show some symptom improvement when treated with SRIs. However, while SRIs are the first line pharmacotherapeutic option for individuals with OCD and BDD, those with TS have shown preferential response to dopamine antagonists (e.g. risperidone), and clomipramine showed efficacy superior to the SSRIs in individuals with trichotillomania in a systematic review and meta-analysis (while habit-reversal therapy was superior to both pharmacological agents).

Taken together, these data suggest that similarities and associations between OCD and the related disorders exist. However, significant differences have also been demonstrated. Thus, the idea of OCD and related disorders may be clinically useful, provided these important distinctions in diagnostic validity are borne in mind.

OBSESSIVE–COMPULSIVE DISORDER IN PEDIATRIC POPULATIONS

While the focus of this chapter has been adult OCD, pediatric populations also warrant mention, as up to 80% of individuals report symptom onset before age 18. Broadly, the clinical presentation in children and adolescents mimics that in adults, with obsessive–compulsive symptoms concerning contamination/washing and symmetry/ordering emerging as among the most prevalent. However, the DSM-5 stipulates that young children may not be able to articulate the aims of these behaviors or mental acts.[2] Despite having a population prevalence of approximately 1–4%, pediatric OCD remains underdiagnosed and undertreated. Furthermore, early-onset OCD is often associated with greater severity and poorer treatment response than in adult populations. As is the case in adults, SRIs remain the first line pharmacotherapy for children and adolescents with OCD, with the SSRIs preferable to clomipramine owing to their more favorable side-effect profile. Inclusion of CBT in the therapeutic regimen (particularly at the time of treatment initiation) has also shown efficacy in pediatric populations. Indeed, non-pharmacological interventions may be preferable in children and adolescents to avoid the adverse effects of drug treatment. However, much remains to be learned of the optimum treatment of pediatric OCD (including maintenance therapy and treatment-resistant cases), and more work in this area is warranted.

CONCLUSION

While obsessive compulsions have long been documented in historical, literary, and scientific data, OCD has only recently emerged as an archetype of a neuropsychiatric disorder. Much has already been uncovered about the neuroanatomy, neurochemistry, neuroimmunology, neuroendocrinology, and neuroethology of this disorder. Nonetheless, further work is needed to elucidate and integrate the precise factors involved in the pathogenesis of OCD, and may ultimately contribute to expanding current therapeutic strategies.

QUESTIONS FOR FURTHER RESEARCH

Although there are several animal models of OCD, additional research is needed if we are to move from bench to bedside. Do these animal models have face validity at different developmental periods, what is the precise neuroanatomical and molecular basis of compulsive symptoms in these different models, and what is their response to a range of different interventions?

Clinical neuroscience research on OCD, including brain imaging and neurogenetics, has led to a series of neurobiological models of OCD, but much remains unknown. What light will new genetic methods including new-generation sequencing and gene expression studies shed on the neurobiology of OCD? Can new brain imaging methodologies provide new insights? Will the integration of imaging and genetic databases, with longitudinal clinical data, lead to a better understanding of risk and resilience factors in OCD?

Clinical trials in OCD have provided evidence for both first line interventions and some treatments of refractory cases, but much further work is needed. How effective are treatments in real-world settings (rather than in academic contexts)? Will agents with novel mechanisms of action relevant to OCD (e.g. glutamatergic agents) prove efficacious? What is the optimal way to sequence pharmacotherapy, psychotherapy, and their combination in the treatment of OCD?

Epidemiological studies from around the world have shed light on the enormous public health burden of OCD. What is the cross-national prevalence of obsessive–compulsive and related disorders as a whole? What do community studies reveal about their risk and resilience factors? What is the optimal way to improve the recognition, diagnosis, and treatment of OCD in primary care and low- and middle-income settings?

References

1. American Psychiatric Association. *Diagnostic and Statistical Manual of Mental Disorders*. 4th ed. Text Revision. Washington, DC: American Psychiatric Press; 2000.
2. American Psychiatric Association. *Diagnostic and Statistical Manual of Mental Disorders*. 5th ed. Arlington, VA: American Psychiatric Press; 2013.
3. Karno M, Goldin JM, Sorenson SB, Burnam MA. The epidemiology of obsessive compulsive disorder in five US communities. *Arch Gen Psychiatry*. 1988;45(12):1094–1099.
4. Kessler RC, Berglund P, Demler O, Jin R, Merikangas KR, Walters EE. Lifetime prevalence and age-of-onset distributions of DSM-IV disorders in the National Comorbidity Survey Replication. *Arch Gen Psychiatry*. 2005;62(6):593–602.
5. Wittchen HU, Jacobi F. Size and burden of mental disorders in Europe – a critical review and appraisal of 27 studies. *Eur Neuropsychopharmacol*. 2005;15(4):357–376.
6. Pinto A, Mancebo MC, Eisen JL, Pagano ME, Rasmussen SA. The Brown Longitudinal Obsessive Compulsive Study: clinical features and symptoms of the sample at intake. *J Clin Psychiatry*. 2006;67(5):703–711.
7. Eisen JL, Yip AG, Mancebo MC, Pinto A, Rasmussen SA. Phenomenology of obsessive–compulsive disorder. In: Stein DJ, Hollander E, Rothbaum BO, eds. *Textbook of Anxiety Disorders*. 2nd ed. Arlington, VA: American Psychiatric Association; 2010:261–286.
8. Pigott TA, L'Heureux F, Dubbert B, Bernstein S, Murphy DL. Obsessive compulsive disorder: comorbid conditions. *J Clin Psychiatry*. 1994;55 (Suppl):15–27; discussion 28–32.
9. Lochner C, Kinnear CJ, Hemmings SM, et al. Hoarding in obsessive–compulsive disorder: clinical and genetic correlates. *J Clin Psychiatry*. 2005;66(9):1155–1160.
10. Flament MF, Koby E, Rapoport JL, et al. Childhood obsessive–compulsive disorder: a prospective follow-up study. *J Child Psychol Psychiatry*. 1990;31(3):363–380.
11. Skoog G, Skoog IA. 40-year follow-up of patients with obsessive–compulsive disorder. *Arch Gen Psychiatry*. 1999;56(2):121–127.
12. Radua J, Mataix-Cols D. Voxel-wise meta-analysis of grey matter changes in obsessive–compulsive disorder. *Br J Psychiatry*. 2009;195(5):393–402.

13. Rotge JY, Guehl D, Dilharreguy B, et al. Meta-analysis of brain volume changes in obsessive–compulsive disorder. *Biol Psychiatry*. 2009;65(1):75–83.
14. Mataix-Cols D, Wooderson S, Lawrence N, Brammer MJ, Speckens A, Phillips ML. Distinct neural correlates of washing, checking, and hoarding symptom dimensions in obsessive–compulsive disorder. *Arch Gen Psychiatry*. 2004;61(6):564–576.
15. Whiteside SP, Port JD, Abramowitz JS. A meta-analysis of functional neuroimaging in obsessive–compulsive disorder. *Psychiatry Res*. 2004. 15;132(1):69–79.
16. Wolters A, Ziemann U, Benecke R. The cortical silent period. In: *Oxford Handbook of Transcranial Stimulation [online]*. 2012. Available at http://www.oxfordhandbooks.com/view/10.1093/oxfordhb/9780198568926.001.0001/oxfordhb-9780198568926-e-010. Accessed 06.03.13.
17. Radhu N, Ravindran LN, Levinson AJ, Daskalakis ZJ. Inhibition of the cortex using transcranial magnetic stimulation in psychiatric populations: current and future directions. *J Psychiatry Neurosci*. 2012;37(6):369–378.
18. Campbell KM, de Lecea L, Severynse DM, et al. OCD-like behaviors caused by a neuropotentiating transgene targeted to cortical and limbic D1+ neurons. *J Neurosci*. 1999;19(12):5044–5053.
19. Altemus M, Pigott T, Kalogeras KT, et al. Abnormalities in the regulation of vasopressin and corticotropin releasing factor secretion in obsessive–compulsive disorder. *Arch Gen Psychiatry*. 1992;49(1):9–20.
20. Singer HS, Giuliano JD, Zimmerman AM, Walkup JT. Infection: a stimulus for tic disorders. *Pediatr Neurol*. 2000;22(5):380–383.
21. Joel D. Current animal models of obsessive compulsive disorder: a critical review. *Prog Neuropsychopharmacol Biol Psychiatry*. 2006;30(3):374–388.
22. Albelda N, Joel D. Animal models of obsessive–compulsive disorder: exploring pharmacology and neural substrates. *Neurosci Biobehav Rev*. 2012;36(1):47–63.
23. Rauch SAL, Carlezon Jr WA. Illuminating the neural circuitry of compulsive behaviors. *Science*. 2013;340:1174–1175.
24. Hettema JM, Neale MC, Kendler KS. A review and meta-analysis of the genetic epidemiology of anxiety disorders. *Am J Psychiatry*. 2001;158(10):1568–1578.
25. Pauls DL, Towbin KE, Leckman JF, Zahner GE, Cohen DJ. Gilles de la Tourette's syndrome and obsessive–compulsive disorder. Evidence supporting a genetic relationship. *Arch Gen Psychiatry*. 1986;43(12):1180–1182.
26. Pauls DL, Alsobrook II JP, Goodman W, Rasmussen S, Leckman JF. A family study of obsessive–compulsive disorder. *Am J Psychiatry*. 1995;152(1):76–84.
27. van Grootheest DS, Cath DC, Beekman AT, Boomsma DI. Twin studies on obsessive–compulsive disorder: a review. *Twin Res Hum Genet*. 2005;8(5):450–458.
28. Taylor S. Molecular genetics of obsessive–compulsive disorder: a comprehensive meta-analysis of genetic association studies. *Mol Psychiatry*. 2013;18:799–805.
29. Hanna GL, Veenstra-VanderWeele J, Cox NJ, et al. Genome-wide linkage analysis of families with obsessive–compulsive disorder ascertained through pediatric probands. *Am J Med Genet*. 2002;114(5):541–552.
30. Shugart YY, Samuels J, Willour VL, et al. Genomewide linkage scan for obsessive–compulsive disorder: evidence for susceptibility loci on chromosomes 3q, 7p, 1q, 15q, and 6q. *Mol Psychiatry*. 2006;11(8):763–770.
31. Clomipramine Collaborative Study Group. Clomipramine in the treatment of patients with obsessive–compulsive disorder. *Arch Gen Psychiatry*. 1991;48(8):730–738.
32. Soomro GM, Altman D, Rajagopal S, Oakley-Browne M. Selective serotonin re-uptake inhibitors (SSRIs) versus placebo for obsessive compulsive disorder (OCD). *Cochrane Database Syst Rev*. 2008;(1):CD001765. doi: 10.1002/14651858.CD001765.pub3.
33. Bloch MH, McGuire J, Landeros-Weisenberger A, Leckman JF, Pittenger C. Meta-analysis of the dose–response relationship of SSRI in obsessive–compulsive disorder. *Mol Psychiatry*. 2010;15(8):850–855.
34. Epperson CN, McDougle CJ, Price LH. Intranasal oxytocin in obsessive–compulsive disorder. *Biol Psychiatry*. 1996;40(6):547–549.
35. Coric V, Taskiran S, Pittenger C, et al. Riluzole augmentation in treatment-resistant obsessive–compulsive disorder: an open-label trial. *Biol Psychiatry*. 2005;58(5):424–428.
36. Van Ameringen M, Mancini C, Patterson B, Bennett M. Topiramate augmentation in treatment-resistant obsessive–compulsive disorder: a retrospective, open-label case series. *Depress Anxiety*. 2006;23(1):1–5.
37. March JS, Frances A, Carpenter LL, Kahn D. Expert consensus treatment guidelines for obsessive–compulsive disorder: a guide for patients and families. *J Clin Psychiatry*. 1997;58(suppl 4):1–72.
38. Solomon RL, Kamin LJ, Wynne LC. Traumatic avoidance learning: the outcomes of several extinction procedures with dogs. *J Abnorm Psychol*. 1953;48(2):291–302.
39. Nakatani E, Nakagawa A, Nakao T, et al. A randomized controlled trial of Japanese patients with obsessive–compulsive disorder – effectiveness of behavior therapy and fluvoxamine. *Psychother Psychosom*. 2005;74(5):269–276.
40. Abramowitz JS. Variants of exposure and response prevention in the treatment of obsessive–compulsive disorder: a meta-analysis. *Behav Ther*. 1996;27(4):583–600.
41. Tenneij NH, van Megen HJ, Denys DA, Westenberg HG. Behavior therapy augments response of patients with obsessive–compulsive disorder responding to drug treatment. *J Clin Psychiatry*. 2005;66(9):1169–1175.
42. Chamberlain SR, Menzies LA, Fineberg NA, et al. Grey matter abnormalities in trichotillomania: morphometric magnetic resonance imaging study. *Br J Psychiatry*. 2008;193(3):216–221.
43. Pauls DL, Leckman JF. The inheritance of Gilles de la Tourette's syndrome and associated behaviors. Evidence for autosomal dominant transmission. *N Engl J Med*. 1986;315(16):993–997.
44. Monzani B, Rijsdijk F, Iervolino AC, Anson M, Cherkas L, Mataix-Cols D. Evidence for a genetic overlap between body dysmorphic concerns and obsessive–compulsive symptoms in an adult female community twin sample. *Am J Med Genet B Neuropsychiatr Genet*. 2012;159B(4):376–382.
45. Hemmings SM, Kinnear CJ, Lochner C, et al. Genetic correlates in trichotillomania – a case–control association study in the South African Caucasian population. *Isr J Psychiatry Relat Sci*. 2006;43(2):93–101.
46. Bienvenu OJ, Wang Y, Shugart YY, et al. Sapap3 and pathological grooming in humans: results from the OCD collaborative genetics study. *Am J Med Genet B Neuropsychiatr Genet*. 2009;150B(5):710–720.

CHAPTER

39

Schizophrenia

Glenn T. Konopaske, Joseph T. Coyle

Harvard Medical School, McLean Hospital, Belmont, Massachusetts, USA

OUTLINE

CLINICAL ASPECTS OF SCHIZOPHRENIA

Symptomatic Features

Schizophrenia is a chronic, disabling psychiatric condition characterized by psychosis, social withdrawal, amotivation, and cognitive deficits. With a lifetime prevalence of 1% worldwide, schizophrenia generally impairs multiple domains of an individual's life including relationships, self-care, employment, and education.[1] Since schizophrenia often starts in young adulthood and is typically a lifelong disorder, it produces a substantial degree of persistent disability. Schizophrenia has been ranked among the top 10 causes of disability-adjusted life-years and reduces life expectancy by 10 years, often as a result of suicide.[2] Thus, schizophrenia is a relatively common, chronic mental illness associated with substantial morbidity and mortality.

Schizophrenia was conceptualized as a specific disorder at the beginning of the twentieth century. The German psychiatrist Emil Kraepelin (1856–1926) identified a group of psychotic patients characterized by the onset of symptoms in early adulthood with impaired cognition and poor outcomes. Earlier, Benedict Morel (1809–1873), a French psychiatrist, labeled a group of deteriorated patients, whose illness began in adolescence, as having *démence précoce*. Kraepelin adapted Morel's nomenclature and denoted his patients as having *dementia praecox*. The term *schizophrenia* was introduced a decade later by Eugen Bleuler (1857–1939), who characterized the splitting ("schizo") of affect (i.e. emotional tone) from cognition. He further identified four characteristic symptoms: autism, ambivalence, blunted affect, and disturbances of (a)volition. While these early clinicians readily accepted that schizophrenia was a brain disorder with heritable vulnerabilities, this biological conceptualization of the disorder was eclipsed by the rise of psychoanalysis in the 1930s and the dominance for the next half century of its theoretical conceptualization that intrapsychic conflicts are the cause of psychiatric disorders. Thus, until recently, many clinicians viewed schizophrenia as a consequence of pathological maternal–child interactions and deviant family communication.

Schizophrenia is characterized by four, somewhat independent symptom clusters, which are designated as positive symptoms, negative symptoms, affective dysregulation, and cognitive impairments. The *positive symptoms* are the most dramatic manifestations of psychosis.

Neurobiology of Brain Disorders
http://dx.doi.org/10.1016/B978-0-12-398270-4.00039-2

They include hallucinations, in which the patient has sensory experiences in the absence of external stimuli. Hallucinations are most often auditory, but can be visual, tactile, gustatory, or olfactory. A patient also experience delusions, which are fixed, false beliefs that remain intact even in the face of contradictory external evidence. Delusions can range from the mundane (e.g. "my family has placed cameras in my house to watch me") to the bizarre (e.g. "aliens control my thoughts using a chip implanted in my brain"). Mental function can be so disrupted that thoughts become disorganized such that the patient exhibits loosening of associations (e.g. one thought does not logically flow from the previous), disorganized speech, or even quite bizarre behavior (e.g. odd mannerisms, dress, or appearance). Before the emergence of the first positive (psychotic) symptom, patients are said to be in a "prodrome" phase during which the patient can have a host of psychiatric symptoms including depression, anxiety, and substance abuse. Symptoms of *affective dysregulation* include affective flattening, depression, mania, anxiety, impulsivity, and demoralization. The demoralization may account for the 10% lifetime risk of completed suicide in patients with schizophrenia.[2]

The term "negative" refers to a loss or deficit, and includes decreased emotional reactivity, a paucity of speech, amotivation, an inability to feel pleasure (anhedonia), and attentional impairment. Patients with prominent *negative symptoms* tend to be socially isolated and withdrawn, and have difficulties performing activities of daily living (e.g. personal hygiene), often making employment and independent living virtually impossible. Patients with schizophrenia also exhibit a range of *cognitive impairments* affecting multiple domains including selective attention, working memory, learning, and executive function. Cognitive impairment appears to be a core feature of the illness. Among the symptom clusters of schizophrenia, the positive symptoms respond best to antipsychotic medications such as haloperidol or clozapine, whereas negative symptoms and cognitive impairment do not respond well, if at all, to conventional psychopharmacological treatment.[3] Consequently, cognitive impairments and negative symptoms, not the positive symptoms, are the best predictors of functional outcomes.[4]

Neuroimaging

Neuroimaging studies have implicated several cortical regions in the cognitive impairments seen in schizophrenia. Two cognitive tasks relevant to schizophrenia are selective attention and working memory. Selective attention is the ability to remain focused on a particular task in the face of distracting stimuli. Working memory is the ability to hold discrete pieces of information "online" and manipulate that information for a brief duration. In patients with schizophrenia, the anterior cingulate cortex (ACC) fails to activate appropriately during selective attention tasks, and fails to properly regulate the dorsolateral prefrontal cortex (DLPFC) during the performance of working memory. These findings are consistent with the ACC and DLPFC forming an integrated network within the prefrontal cortex (PFC). The DLPFC provides the top–down guidance of task-oriented behaviors and the ACC monitors ongoing behavior and incoming stimuli to ascertain when more control is needed. The functional deficits in PFC are complex such that cortical areas subserving cognitive function are not simply less active in patients with schizophrenia, rather they function inefficiently.[5,6] For example, areas typically activated in healthy control subjects (e.g. ACC and DLPFC) show less activity in individuals with schizophrenia (Fig. 39.1), and other brain areas (e.g. inferior parietal lobule), not typically activated in healthy controls, are engaged. Despite the activation of additional cortical areas, patients with schizophrenia exhibit impaired performance on multiple cognitive tasks. Thus, the activation of accessory cortical areas may serve as unsuccessful compensation by the affected brain.

Although antipsychotic medications are associated with significant side effects, the cognitive impairments in schizophrenia do not appear to be the result of antipsychotics. Indeed, patients with first episode schizophrenia and having minimal antipsychotic exposure exhibit cognitive impairments. In addition, similar cognitive impairments are seen in unaffected biological relatives of individuals with schizophrenia, albeit in an attenuated form.[5] This suggests that the cognitive impairments in schizophrenia may be, at least in part, associated with risk genes and represent an endophenotype, that is, a heritable component of the disorder often observed in first degree relatives.

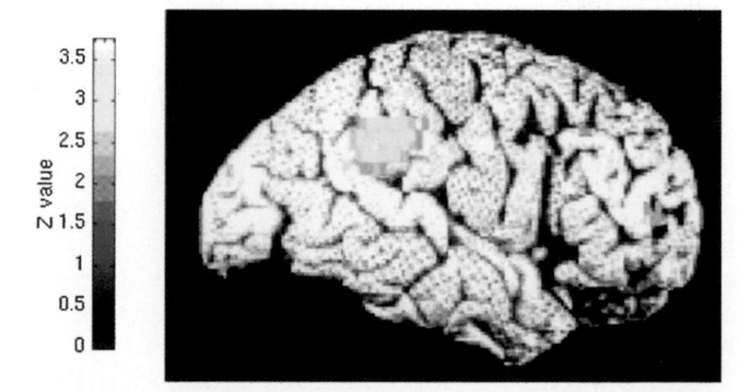

FIGURE 39.1 **Functional magnetic resonance image showing regions with significant differences in activation during a working memory task in patients with schizophrenia relative to controls.** The smaller area on the right represents the right dorsolateral prefrontal cortex and the larger area on the left represents the right posterior parietal cortex. *Source: Carter* et al. Am J Psychiatry. *1998;155(9):1285–1287.*[6]

Schizophrenia and Complex Genetics

Schizophrenia does not follow a traditional Mendelian inheritance pattern and cannot be attributed to a single gene or genetic abnormality. Genetic risk for schizophrenia appears to be the result of allelic variants of many genes, each contributing a small degree of risk. In addition, there is a collection of known environmental risk factors, which may interact with the genetic risk factors to produce the schizophrenia phenotype. The environmental risk factors predominantly involve the perinatal period and include maternal infections during the second and third trimesters, nutritional deficiencies, and birth complications. Numerous twin and adoption studies suggest that the heritability of the disorder is between 65 and 85%.[7] Moreover, the concordance for schizophrenia in monozygotic twins is less than 100%, indicating a significant environmental effect.

Genome-wide association studies (GWAS) using family-based and case–control designs have been conducted to find genetic risk loci. These GWAS have implicated several chromosomal regions that are likely to contain risk genes. More recently, through fine mapping and meta-analyses from 80,000 subjects (≈50% with schizophrenia), nearly 100 risk genes have been identified that reach genome-wide significance.

Copy number variants (CNVs) have been shown to contribute 10% or more to the risk of schizophrenia. A CNV occurs when the number of copies of a DNA region differs in affected individuals (probands) relative to controls. The segment in question may range from one kilobase to several megabases in size, and CNVs may be either inherited or caused by *de novo* mutations. Increased paternal age increases the risk of developing CNVs, consistent with paternal age being a risk factor for developing schizophrenia. Genome-wide scans for CNVs in schizophrenia have identified several regions containing putative risk loci on 1q, 2p, 15q, and 22q. Of these identified regions, 22q is particularly interesting, as it has been implicated in the 22q11.2 deletion syndrome (i.e. velocardiofacial syndrome-1). Patients with the 22q11.2 deletion syndrome often have cardiac abnormalities, abnormal facies, thymic aplasia, cleft palate, hypocalcemia with hypoparathyroidism, and an increased risk of developing schizophrenia.

DYSREGULATED NEUROTRANSMITTER SYSTEMS IN SCHIZOPHRENIA

Dopamine Neurotransmission

Dopamine is an important neurotransmitter involved in multiple functions including movement, emotional regulation, reward, and cognition. In humans, there are five dopaminergic pathways: nigrostriatal, mesolimbic, mesocortical, tuberinfundibular, and thalamic. The mesolimbic and mesocortical dopaminergic pathways have the most relevance to the pathophysiology of schizophrenia. The mesolimbic pathway originates in dopamine-producing cells in the ventral tegmental area (VTA) in the midbrain. These dopaminergic neurons send afferents to the striatum, hippocampus, and amygdala. The mesocortical dopaminergic pathway also originates in the VTA, and projects throughout the cortex including the PFC.[3] Increased dopaminergic activity in the mesolimbic pathway accounts for the production of positive symptoms, whereas decreased dopaminergic activity in the mesocortical pathway is postulated to account for the negative symptoms and cognitive impairments seen in patients with schizophrenia.

Dopamine receptors are G-protein coupled receptors. There are five subtypes of dopamine receptor (D_1–D_5), with D_2 and D_4 having the most relevance to schizophrenia. The primary mechanism of therapeutic action of antipsychotic medications is blockade of D_2 receptors. Most antipsychotics achieve antipsychotic effects at a D_2 receptor occupancy level of 60–80%, whereas occupancy of greater than 80% is associated with neurological side effects of the typical antipsychotics (e.g. chlorpromazine or haloperidol), namely, akathisia (extreme restlessness), and extrapyramidal symptoms, which consist of parkinsonism and dystonias. Chronic use of typical antipsychotics is also associated with tardive dyskinesia, which is manifested by irreversible choreoathetotic movements (e.g. involuntary contractions with twisting and writhing movements). Blockade of D_2 receptors in the mesolimbic pathway produces antipsychotic effects, whereas blockade of D_2 receptors in the nigrostriatal pathway produces extrapyramidal symptoms. In addition, blockade of D_2 receptors in the pituitary causes hypersecretion of prolactin, potentially resulting in galactorrhea and menstrual abnormalities in women, and hypogonadism, infertility, and erectile dysfunction in men.[3]

In response to the neurological side effects of the typical antipsychotics, pharmaceutical companies undertook a search for antipsychotic medications free of these side effects. They developed a class of drugs, termed the atypical antipsychotics, with the prototype being clozapine. The atypical antipsychotics still block D_2 receptors, thereby producing their antipsychotic effect, but they also block other receptors, especially serotonin-2A [5-hydroxytryptamine-2A (5-HT_{2A})] receptors, attenuating the neurological side effects. Antagonism at 5-HT_{2A} receptors enhances striatal dopamine release and may also confer some antipsychotic effects. Nevertheless, atypical antipsychotics are not a panacea since they can result in substantial weight gain and increase the risk for the metabolic syndrome, which includes hyperglycemia, type 2 diabetes mellitus, and dyslipidemia.

Animal studies have revealed that neurotransmission at D_1 receptors in the PFC plays a key role in cognitive functions, especially working memory. Moreover, activity at D_1 has an inverted U relationship with cognitive performance such that too little or too much stimulation at D_1 degrades cognitive performance.[8] In patients with schizophrenia, positron emission tomography (PET) demonstrates decreased D_1 expression in the PFC that correlates with impaired cognitive function. PET and single-photon emission computed tomography (SPECT) imaging reveal increased dopamine neurotransmission in the striatum along with altered neurotransmission at D_2 and D_4 receptors in multiple brain areas of patients with schizophrenia.[9] Together, these data suggest that dopamine neurotransmission is altered in several brain areas including cortical–striatal areas, and contributes to the cognitive impairments in the disorder.

Given the substantial role that dopamine plays in the pathophysiology of schizophrenia, several models of schizophrenia have been developed using animals with dysregulated dopaminergic neurotransmission. After release from presynaptic terminals, dopamine is cleared from the extracellular space by a membrane-spanning dopamine transporter (DAT), which plays a critical role in the regulation of dopamine levels. DAT knockout mice have increased brain dopamine levels, hyperactivity, and deficits in sensory gating (e.g. filtering of unnecessary or redundant external stimuli). A measure of sensory gating is prepulse inhibition (PPI), in which a loud startling tone is preceded by a non-startling stimulus. In wild-type animals, the startle response is attenuated after the warning tone. DAT knockout mice show deficits in PPI, as do individuals with schizophrenia and many of their unaffected first degree relatives.[10]

In addition to genetic animal models, pharmacological approaches to studying the effects of dopamine dysregulation exist. Amphetamine administration to subjects with schizophrenia and genetically susceptible individuals can produce psychosis. In animals, withdrawal states after repeated amphetamine administration produce increased amphetamine sensitivity and dopamine turnover along with alterations in PPI. Animals withdrawing from repeated amphetamine exposure also exhibit deficits in latent inhibition (LI), which is another behavior associated with sensory gating deficits. LI occurs when an irrelevant stimulus inhibits the acquisition of a conditioned response to a given stimulus. Patients with schizophrenia have deficits in LI. Moreover, dopamine agonism via amphetamine administration disrupts LI, and antipsychotic medications restore LI in patients with schizophrenia by antagonism at D_2 receptors. These data suggest that the sensory gating deficits observed in subjects with schizophrenia are related, at least in part, to alterations in dopamine neurotransmission.

Glutamatergic Neurotransmission

Although dopamine-based models of schizophrenia reproduce positive symptoms and some cognitive impairments, they do not replicate negative symptoms. Dissociative anesthetics (e.g. phencyclidine, ketamine) are non-competitive antagonists at the *N*-methyl-D-aspartate (NMDA) subtype of glutamate receptors and can produce psychosis in humans. NMDA receptor antagonists, such as phencyclidine (PCP) and ketamine, exacerbate positive symptoms, negative symptoms, and cognitive impairments in patients with schizophrenia. In addition, they produce psychosis, negative symptoms, and cognitive deficits in healthy volunteers. Based on the clinical findings of patients who abused dissociative anesthetics, a glutamatergic hypothesis of schizophrenia was introduced. A significant amount of evidence has since been gathered suggesting that the pathophysiology of schizophrenia is due, at least in part, to decreased neurotransmission at NMDA receptors.[11]

Glutamate is the major excitatory neurotransmitter in the brain and has two classes of receptors; ionotropic, which are ligand-gated cation channels, and metabotropic (mGluR), which are G-protein coupled receptors. Ionotropic receptors can be further subdivided into N-methyl-D-aspartate (NMDA), α-amino-3-hydroxyl-5-methyl-4-isoxazole-propionate (AMPA), and kainate receptors. AMPA receptors are composed of four subunits ($GRIA_1$–$GRIA_4$) and their stimulation generates fast excitatory postsynaptic potentials (EPSPs). Most AMPA receptors are heterotetrameric, composed of $GRIA_2$ dimers along with dimers of $GRIA_1$, $GRIA_3$, or $GRIA_4$. Compared with AMPA receptors, NMDA receptors mediate slower and longer lasting EPSPs. Activation of NMDA receptors allows entry of calcium and sodium into the cell with potassium efflux. Activation of the NMDA receptor is voltage dependent, a consequence of its ion channel being blocked by a magnesium ion at the resting membrane potential. NMDA receptors are heterotetramers formed by two obligatory $GRIN_1$ subunits and two regionally localized $GRIN_2$ subunits. Kainate receptors have five subunits ($GRIK_1$–$GRIK_5$), which form homotetramers or heterotetramers. The kainate channel is permeable to sodium and potassium, and the receptor is located both presynaptically and postsynaptically. The precise function of kainate receptors is not well defined. mGluRs are involved in diverse functions such as learning, memory, and pain, and have three major subtypes: groups I, II, and III. Groups I ($mGluR_1$ and $mGluR_5$) and III ($mGluR_4$, and $mGluR_6$–$mGluR_8$) are mainly postsynaptic and increase NMDA receptor activity, whereas group II ($mGluR_2$ and $mGluR_3$) are presynaptic and decrease NMDA receptor activity.[3] Thus, glutamate binds to various receptors having diverse downstream effects on cell physiology.

Glutamate metabolism has been investigated *in vivo* in schizophrenia. Proton magnetic resonance spectroscopy (MRS) reveals elevated levels of glutamine in the ACC of antipsychotic-naïve patients with schizophrenia who are experiencing their first episode of psychosis. However, more chronic, antipsychotic-treated, patients with schizophrenia exhibit decreased levels of glutamine and glutamate.[12] Levels of *N*-acetyl aspartate (NAA), a marker of neuronal integrity, are also reduced in the cortex of affected individuals.

Brain abnormalities detected by neuroimaging are influenced by polymorphisms in genes involved in glutamate neurotransmission. A single-nucleotide polymorphism (SNP) in the $mGluR_3$ gene (*GRM3*) is associated with altered activation patterns in the PFC and hippocampus along with PFC NAA levels in patients with schizophrenia. Dystrobrevin binding protein-1 (DTNBP1 or dysbindin) is involved in vesicle trafficking in glutamatergic terminals and dendritic morphology. Evidence implicating dysbindin in the pathophysiology of schizophrenia came from fine mapping of a chromosome 6p risk locus in affected Irish families. Dysbindin regulates glutamate release in cultured cells, and SNPs in the dysbindin gene are associated with reductions in DLPFC and occipital gray matter volume in subjects with schizophrenia. Neuregulin-1 (NRG1) decreases NMDA receptor activation in rats, and is associated with enlarged lateral ventricles in patients early in the course of the disorder. Regulator of G-protein signaling-4 (RGS4) inhibits signal transduction at $mGluR_1$ and $mGluR_5$, and is associated with the regulation of DLPFC volume in the disorder. In addition, RGS4 polymorphisms are associated with the functional coupling of cortical regions during the performance of working memory tasks.[13] Activation of NMDA receptors requires the binding of both glutamate to the $GRIN_2$ subunit and a coagonist to the $GRIN_1$ subunit. D-serine is a coagonist at NMDA receptors in cortical–limbic regions. It is synthesized by serine racemase (SR) and catabolized by D-amino acid oxidase (DAO), the activity of which is modulated by G72 (DAOA). Notably, polymorphisms in SR, DAO, and DAOA are all associated with the development of schizophrenia, and a null mutation of murine SR results in cortical atrophy along with reduced dendritic complexity and reduced dendritic spine density of cortical pyramidal neurons similar to neuropathological findings in schizophrenia.[14] Taken together, these data suggest that polymorphisms in seven glutamate-related genes impact both brain structure and function, and they are in the top 40 risk genes for schizophrenia (see www.szgene.org).

In the brain, pyramidal neurons are the chief glutamatergic neurons, and alterations in pyramidal cell density occur in the DLPFC and ACC from individuals with schizophrenia.[15,16] In the mature brain, most glutamatergic neurotransmission occurs at excitatory synapses on dendritic spines; as a result, considerable emphasis has been placed on characterizing dendritic arbors and spines in subjects with schizophrenia. These findings are supported by reduced dendritic arbors in the PFC and subiculum, and by reductions in dendritic spine density in the DLPFC and temporal cortex in affected subjects (Fig. 39.2).[17] These alterations in pyramidal cell density, and especially reductions in dendritic arborization and spine density, are consistent with attenuated NMDA neurotransmission in schizophrenia.

Molecular studies on postmortem tissue support the histological findings, which indicate a loss of excitatory synapses in people with schizophrenia. Synaptophysin is a marker of synapses, and its messenger RNA (mRNA) expression is reduced in the hippocampus, superior temporal gyrus (STG), and visual cortex from individuals with schizophrenia. Vesicular glutamate transporter-1 (VGLUT1) is a synaptic protein, which concentrates glutamate into synaptic vesicles, and its mRNA expression is reduced in the hippocampus and DLPFC. Together, these data indicate that a loss of glutamatergic synapses occurs in multiple brain areas including the DLPFC, STG, and hippocampus in schizophrenia.

The regulation of the actin cytoskeleton is central to the formation, function, and morphology of dendritic spines. Cell division cycle-42 (CDC42) and kalirin

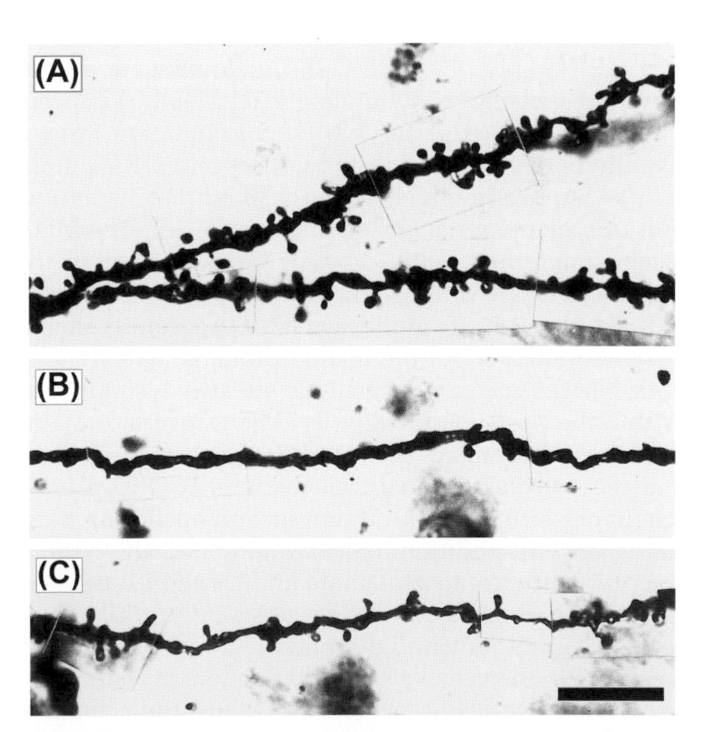

FIGURE 39.2 **Golgi-stained basilar dendrites of deep layer III pyramidal cells from the dorsolateral prefrontal cortex of (A) control and (B,C) schizophrenia subjects.** The small projections off the dendrites are dendritic spines. Note the significantly reduced dendritic spine density in the schizophrenia subjects relative to the control subject. Scale bar = 10 μm. *Source: Glantz and Lewis.* Arch Gen Psychiatry. *2000;57(1):65–73.*[17]

RhoGEF kinase (KALRN) regulate the actin cytoskeleton within dendritic spines. Both CDC42 and KALRN, along with the related proteins CDC42EPS and SEPT7, are significantly altered in the DLPFC in subjects with schizophrenia.[18] Brain-derived neurotrophic factor (BDNF), a member of the nerve growth factor family, promotes the survival of glutamatergic neurons and regulates dendritic growth and synaptogenesis. In the DLPFC, the mRNA and protein expression of BDNF is reduced significantly.[19] Thus, the expression of several genes critical to dendritic spine function and morphology is decreased in the DLPFC of schizophrenia patients.

In addition to the loss of excitatory synapses and dendritic spines, postmortem studies suggest that glutamate metabolism and turnover are dysregulated in the brains of patients with schizophrenia. After release into the synaptic cleft, glutamate is taken up by astrocytes via high-affinity glutamate transporters (excitatory amino acid transporters), the expression of which is altered in the frontal cortex and parahippocampal gyrus in schizophrenia.[20,21] Once inside the astrocyte, a portion of glutamate is converted to glutamine by glutamine synthetase and a portion to α-ketoglutarate and ammonium via glutamate dehydrogenase. In the PFC, the expression of both glutamine synthetase and glutamate dehydrogenase, along with glutamate dehydrogenase enzymic activity, is altered in subjects with schizophrenia.

The NMDA receptor is a heterotetramer of $GRIN_1$ and $GRIN_2$ subunits. The $GRIN_1$ subunits are obligatory and the $GRIN_2$ subunits are regionally localized. Multiple isoforms of the NMDA receptor are formed via alternative splicing of the $GRIN_1$ and $GRIN_2$ transcripts. Supporting the hypothesis of NMDA hypofunction in schizophrenia, differential expression of NMDA receptor subunits $GRIN_1$ and/or $GRIN_2$ occurs in the thalamus, hippocampus, and ACC from affected individuals. Not only are individual NMDA subunits altered in schizophrenia, but the related proteins, which transduce NMDA-mediated signaling, are also dysregulated. Within the postsynaptic density (PSD), several proteins have been identified that modulate receptor function and play a role in signal transduction. PSD-95 plays a role in clustering NMDA receptors and anchoring them to the PSD. In addition, PSD-95 modulates the NMDA receptor response to glutamate and assembles the necessary signaling machinery to affect intracellular processes. Neurofilament-light polypeptide links NMDA receptors to the cytoskeleton and anchors protein phosphatase-1 to the PSD, where it regulates the actions of NMDA subunits and calcium/calmodulin-dependent protein kinase II (CamKII). Reductions in both PSD-95 and neurofilament-light expression occur in multiple brain areas in individuals with schizophrenia. Furthermore, glycogen synthase kinase-3β (GSK3B) plays a role in the trafficking of $GRIN_{2B}$ subunit. GSK3B mRNA, protein, and activity are all decreased in the frontal cortex in subjects with schizophrenia.

D-serine and glycine are required along with glutamate to activate NMDA receptors. D-serine is synthesized from L-serine by SR and degraded by DAO. Extracellular levels of D-serine levels and glycine are regulated by astrocytic and neuronal transporters. The key transporter for D-serine is sodium-independent alanine–serine–cysteine transporter-1 (Asc-1) and for glycine is glycine transporter-1 (GlyT1) or small neutral amino acid transporter-2 (SNAT2). Asc-1 protein and SNAT2 mRNA are decreased in the DLPFC and hippocampus, and DAO activity is increased in the cortex in schizophrenia. Moreover, hippocampal DAO protein levels correlate with illness duration, suggesting that increased DAO expression may be due to antipsychotic treatment. Nevertheless, D-serine levels are reduced in the cerebrospinal fluid (CSF) of schizophrenia patients.[22] Together, these data suggest that dysregulation of D-serine and glycine may contribute to impaired NMDA neurotransmission in patients with schizophrenia.

NAA is synthesized from aspartate by aspartyl *N*-acetyl transferase or produced by the cleavage of *N*-acetylaspartylglutamate (NAAG) by glutamate carboxypeptidase II (also called folate hydrolase-1 or FOLH1). NAAG is the most prevalent and widely distributed neuropeptide in the brain, and inhibits neurotransmission at the NMDA receptor. In agreement with neuroimaging studies, glutamate and NAA levels are decreased while NAAG levels are elevated in postmortem brains from subjects with schizophrenia.[23] Both NAA and NAAG were found to be decreased in the STG in schizophrenia patients. NAA is hydrolyzed to aspartate and acetyl groups by aspartoacylase, the mRNA of which is decreased along with NAA levels in the DLPFC from affected individuals. The mRNA for FOLH1 is increased in the hippocampus, but decreased in the DLPFC and entorhinal cortex. However, overall FOLH1 activity is decreased in brains of patients with schizophrenia, which probably contributes to elevated NAAG levels. Moreover, NAAG also activates the $mGluR_3$ receptor (GRM_3), which inhibits glutamate release, thereby attenuating NMDA receptor function. Thus, in schizophrenia, cortical glutamate levels are decreased and the ability of glutamate to activate NMDA receptors may be further inhibited by elevated NAAG levels.[24]

In addition to changes in NMDA receptor subunits and associated proteins, differential expression of AMPA subunits occurs in schizophrenia. AMPA receptors are composed of four subunits ($GRIA_1$– $GRIA_4$), which combine to form a tetramer. The majority of AMPA receptors are heterotetramers of $GRIA_2$ dimers coupled with dimers of $GRIA_1$, $GRIA_3$, or $GRIA_4$. In the hippocampus, mRNA for $GRIA_1$

is reduced along with $GRIA_2$ and $GRIA_3$ protein levels. $GRIA_2$ mRNA is also decreased in the DLPFC, but $GRIA_1$ mRNA is significantly increased in DLPFC layers II/III and V pyramidal cells in individuals with schizophrenia. Unlike AMPA receptors, the precise function of kainate receptors in the human CNS remains elusive. Nevertheless, there is considerable evidence of differential expression of kainate subunits in schizophrenia. There are five types of kainate receptor subunit ($GRIK_1$–$GRIK_5$), which form tetramers. Kainate receptor binding is increased in the orbitofrontal cortex, and decreased in the DLPFC and hippocampal formation. Reductions in $GRIK_2$ mRNA levels occur in the hippocampus along with reductions in $GRIK_5$ mRNA levels in both the DLPFC and the hippocampus in schizophrenia subjects. In contrast, $GRIK_3$ mRNA is increased in the DLPFC, which may be the result of increased CNVs in patients with schizophrenia.

Metabotropic glutamate receptors regulate glutamate release and its metabolism by astrocytes. $mGluR_1$ and $mGluR_5$ mRNAs are increased in the PFC, whereas $mGluR_3$ mRNA and its dimeric/oligermic protein forms are reduced, possibly as a result of altered NAAG metabolism in patients with schizophrenia. In the cortex, $mGluR_2$ forms functional complexes with a presumably unrelated receptor, the $5\text{-}HT_{2A}$ receptor. In the DLPFC, $mGluR_2$ mRNA and receptor binding are decreased, whereas $5\text{-}HT_{2A}$ receptor binding is increased in subjects with schizophrenia who were not receiving antipsychotics at the time of death. This finding is consistent with increased $mGluR_1$ and $mGluR_{2/3}$ immunoreactivity in the DLPFC. Taken together, these data suggest that alterations occur in AMPA, kainate, and metabotropic glutamate receptors in addition to NMDA receptors in schizophrenia.

Administration of the NMDA antagonists PCP, ketamine, and MK-801 has been used to reproduce some symptoms of schizophrenia in animals and to gain insight into the pathophysiology of the disorder. In animals, PCP causes working memory impairments, social withdrawal, decreased PPI, decreased dendritic spine complexity, and enhanced sensitivity to amphetamines. In addition to pharmacological models, glutamate-related genetic animal models have been developed for schizophrenia. $GRIN_1$ hypomorphic mice exhibit abnormal PPI which is normalized by antipsychotics.[10] Mice lacking $GRIN_{2B}$ in the hippocampus and neocortex exhibit altered NMDA neurotransmission, impaired hippocampal long-term depression and long-term potentiation, decreased dendritic spine density, and impaired spatial working memory. Mice lacking $mGluR_1$ or $mGluR_5$ exhibit decreased PPI. Furthermore, mice lacking SR have altered glutamatergic neurotransmission with hyperactivity and impaired spatial memory.[25] An inability to express dysbindin results in disturbed synaptic morphology in the hippocampus, with altered social behavior and impaired memory. Thus, animal models in which glutamate neurotransmission is disturbed reproduce many of the molecular, cellular, structural, and behavioral disturbances seen in schizophrenia.

Research into the neurobiology of glutamatergic neurotransmission has led to novel psychopharmacological treatment approaches, which hold great promise. LY404039 is a highly selective $mGluR_{2/3}$ agonist, which reverses PCP-induced behavioral changes in rats. LY404039 has been shown in phase II trials to have antipsychotic efficacy similar to that of olanzapine in patients with schizophrenia. Other approaches have capitalized on the NMDA receptor's requirement for coagonists, glycine or D-serine, along with glutamate for activation of the receptor. Several placebo-controlled studies have demonstrated that the addition of agonists at the glycine modulatory site on the NMDA receptor, including glycine, D-serine, and D-cycloserine, in conjunction with conventional antipsychotics produces significant improvements in positive symptoms, negative symptoms, and cognitive functioning in individuals with schizophrenia. Synaptic levels of glycine are regulated by GlyT1, and inhibitors of GlyT1 increase the synaptic concentration of glycine and enhance NMDA receptor function. A GlyT1 inhibitor developed by Roche has been reported to reduce negative symptoms in patients with schizophrenia receiving antipsychotics.[26]

γ-Aminobutyric Acid Neurotransmission

γ-Aminobutyric acid (GABA) is the predominant inhibitory neurotransmitter in the CNS, and increasing evidence has implicated alterations in GABA neurotransmission in the pathophysiology of schizophrenia. GABA plays an especially important role in the cognitive impairments seen in patients with schizophrenia. Oscillatory behavior of cortical neural networks in the range of 30–80Hz, or gamma band, is critical to the performance of working memory tasks. Subjects with schizophrenia exhibit altered gamma-band activity during the performance of working memory tasks and sensory stimulation. Furthermore, gamma-band alterations often correlate with positive and negative symptoms. Gamma-band oscillations are generated when cortical pyramidal neurons stimulate GABAergic interneurons that express the calcium binding protein parvalbumin. Parvalbumin-positive interneurons have two morphological subtypes (basket cells and chandelier cells), and both are affected in schizophrenia. In addition, elevated GABA levels in the ACC and parietal–occipital junction have been detected by MRS in subjects with schizophrenia.

In the CNS, GABA is synthesized from glutamate by the enzyme glutamate decarboxylase (GAD), which occurs in 65kDa and 67kDa isoforms (GAD_{65} and GAD_{67}). There are two broad classes of receptor in the GABA system, $GABA_A$ and $GABA_B$. $GABA_A$ receptors are ionotropic and selectively gate chloride ions to hyperpolarize the neuron. $GABA_B$ receptors are metabotropic and G-protein linked. $GABA_A$ receptors are composed of five subunits, of which there is considerable diversity affecting the function of the receptor. $GABA_B$ receptors form heterodimers and consist of two subtypes, $GABA_{B1}$ and $GABA_{B2}$. After release from the presynaptic terminal, GABA is taken up by GABA transporters (GAT-1, -2, -3) and metabolized by GABA transaminase.

GABAergic alterations have been detected in postmortem schizophrenia studies. In the DLPFC, the density of interneurons immunoreactive to the calcium binding proteins calbindin and parvalbumin is reduced in subjects with schizophrenia (Fig. 39.3).[27] In addition, a decreased density of GABAergic interneurons is seen in the hippocampus. The densities of interneurons that express mRNA for GAD_{67} are decreased in the DLPFC and ACC, and a compensatory upregulation of the $GABA_A$ receptor and its subunits is observed in the DLPFC, ACC, and posterior cingulate cortex from patients with schizophrenia.[28]

Among the multiple morphological subtypes of GABAergic interneurons in the human brain, chandelier cells are unique in that they are the only interneurons that synapse directly on the axon initial segment (AIS) of pyramidal neurons (Fig. 39.4). The AIS is the site of action potential generation, thus placing chandelier cells in a position to exquisitely regulate the action potentials of pyramidal neurons. Chandelier cells get their name from the arrays of axon terminals, termed cartridges, that they form at the AIS. The cartridges can be immunolabeled with antibodies raised against GABA transporter-1 (GAT-1) (Fig. 39.4). In the DLPFC, the density of GAT-1-immunoreactive cartridges is reduced in the disorder. Consistent with decreased innervation, the density of $GABA_A$ receptor α_2-subunit-immunoreactive cartridges in conjunction with $GABA_A$ α_2-receptor mRNA is increased in the PFC from subjects with schizophrenia (Fig. 39.4).[29]

Molecular studies have also detected evidence for alterations in GABAergic neurotransmission in schizophrenia. In the DLPFC, there is reduced expression of mRNA for GAD_{67}, parvalbumin, GAT-1, multiple subunits of the $GABA_A$ receptor, and three neuropeptides also expressed by GABAergic interneurons; namely,

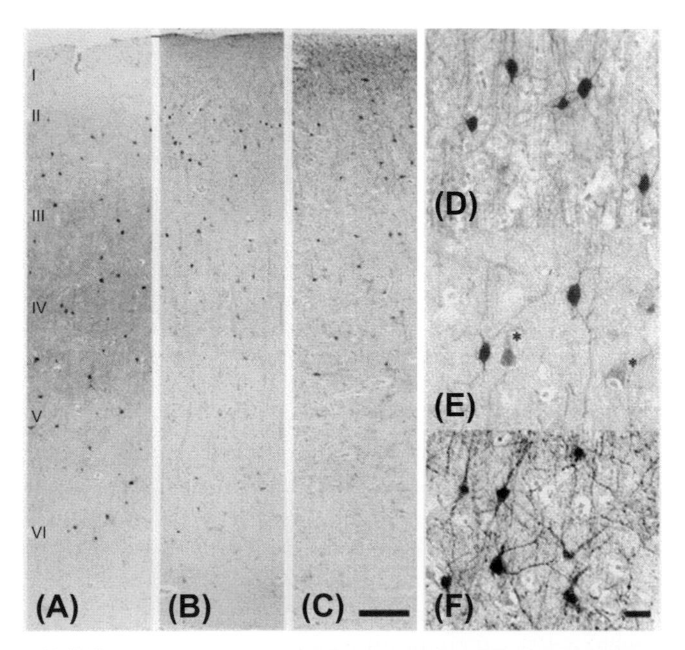

FIGURE 39.3 **Immunoreactivity for calcium binding proteins in the cortex.** (A–C) Low-power photomicrographs from the dorsolateral prefrontal cortex of an unaffected control subject showing interneurons immunoreactive for calcium binding proteins: (A) parvalbumin, (B) calbindin, and (C) calretinin; (D–F) high-power images: (D) parvalbumin-immunoreactive interneurons from layer IV; (E) calbindin-immunoreactive interneurons and pyramidal cells (asterisks) from layer III; (F) calretinin-immunoreactive interneurons from layer III. Scale bars = 200 μm (C), 30 μm (F). *Source: Beasley* et al. Biol Psychiatry. *2002;52(7):708–715.*[27]

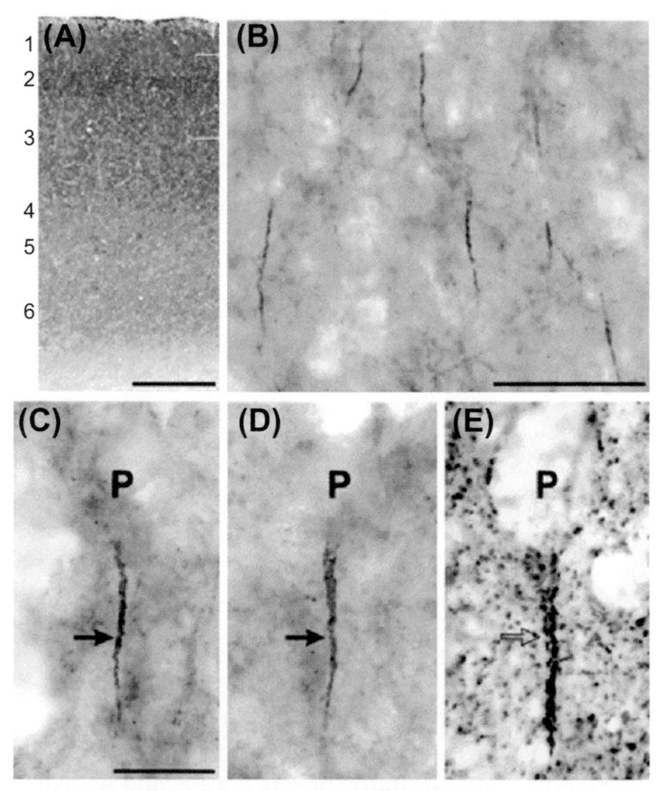

FIGURE 39.4 **γ-Aminobutyric acid-A ($GABA_A$) receptor α_2-subunit immunoreactivity in the human dorsolateral prefrontal cortex.** (A) Low-power photomicrograph depicting $GABA_A$ receptor α_2-subunit immunoreactivity across all six cortical layers; (B–D) higher power photomicrographs showing $GABA_A$ receptor α_2-subunit immunoreactivity in (B,C) control and (D) schizophrenia subjects; (E) high-power photomicrograph of GABA transporter-1 (GAT-1) immunoreactivity in the same region. P: pyramidal cell body; labeled axon initial segments are indicated by solid arrows for $GABA_A$ receptor α_2-subunit immunoreactivity and an open arrow for GAT-1 immunoreactivity. Scale bars = 600 nm (A), 50 mm (B), 20 μm (C–E). *Source: Volk* et al. Cereb Cortex. *2002;12(10):1063–1070.*[29]

somatostatin, neuropeptide Y, and cholecystokinin. Many of these GABAergic changes also occur to a similar degree in the ACC, primary motor cortex, and visual cortex from subjects with schizophrenia. Reductions in GAD_{67} protein also occur in the DLPFC and temporal cortex, indicating that GABA deficits occur throughout the cortex in affected individuals.

Several studies suggest that the GABAergic alterations in schizophrenia are a downstream effect of impaired glutamatergic neurotransmission, especially at NMDA receptors.[11] GABAergic interneurons express several glutamate receptors, which modulate the function of the cell. In the ACC, the density of GAD_{67}^{+} neurons, which coexpress $GRIN_{2A}$ mRNA, is significantly reduced.[28] Likewise, the density of GAD_{67}^{+} neurons, which coexpress $GRIK_1$ mRNA, is reduced in the ACC from individuals with schizophrenia. Nevertheless, alterations in GABA neurotransmission may also involve abnormalities inherent to the GABAergic interneurons. For example, reductions in GAD_{67} mRNA in the DLPFC are associated with a SNP in the GAD1 gene which codes for GAD_{67},[30] and other SNPs in the 5′ region of GAD_{67} are associated with the development of childhood-onset schizophrenia and with gray matter loss over time in schizophrenia patients.[31] GAD1 SNPs are associated with impairments in attention, working memory, and learning, and with increased inefficiency seen in DLPFC activity during the performance of cognitive tasks by individuals with schizophrenia. Thus, it seems likely that deficits in GABA neurotransmission are produced by both downstream effects of hypofunction at the NMDA receptor and alterations inherent to GABAergic interneurons, specifically those expressing parvalbumin.

In addition to genetic effects, environmental risk factors play a significant role in the development of schizophrenia. The use of cannabis is an independent risk factor for the development of psychosis, and patients with schizophrenia have a heightened sensitivity to the propsychotic and cognitive impairing effects of Δ^9-tetrahydrocannabinol (THC), the major psychoactive component in cannabis. The endocannabinoid system consists of neuromodulatory lipids and associated receptors, which modulate GABAergic interneuron function. The chief endogenous ligands are the arachidonic-acid based lipids anandamide and 2-arachidonoylglycerol, which bind to the cannabinoid receptors CB_1 and CB_2. THC exerts its psychoactive effects through interactions with the CB_1 receptor. Axon terminals of cholecystokinin-containing, GABAergic interneurons express CB_1 receptors, which inhibit the release of GABA presynaptically, and the postsynaptic inhibitory potentials mediated by $GABA_A$ receptors on pyramidal neurons. Notably, CB_1 mRNA and protein levels along with the density of CB_1-immunoreactive neurons are reduced in the DLPFC from subjects with schizophrenia.

Animal models of schizophrenia based on impaired GABAergic neurotransmission have been developed. Mice with a point mutation in the $GABA_A$ α_5-subunit in the hippocampus show alterations in learning, reduced PPI, reduced LI, and increased locomotor activity. $GABA_A$ α_3-subunit knockout mice show decreased PPI, which is normalized by haloperidol. Mice lacking GAD_{65} show decreased PPI, as do GAT-1 knockout mice.[10] Thus, several physiological and behavioral features of schizophrenia can be replicated using mice having genetic alterations in the GABA system.

Reelin (RELN), a trophic protein which modulates the migration of interneurons and pyramidal cells during development, has been implicated in the pathophysiology of schizophrenia. Reductions in RELN mRNA expression and density of RELN-immunoreactive neurons have been reported in the DLPFC and the temporal cortex from individuals with schizophrenia. Heterozygous *reeler* mice, which have a RELN haploinsufficiency, show a loss of dendritic spines in the frontal cortex along with decreased GAD_{67} expression and decreased GABA turnover.

Reduced density of parvalbumin-immunoreactive interneurons in the PFC has been repeatedly reported in schizophrenia, and is a common feature of several animal models of the disorder. $ERBB_4$ is a receptor for the schizophrenia risk gene *NRG1*, and knockout mice lacking $ERBB_4$ exhibit reduced density of parvalbumin-immunoreactive interneurons in the hippocampus and reduced density of calbindin-immunoreactive interneurons in the cortex. In addition, $ERBB_4$ knockout mice show reduced power of gamma-band oscillations. Mice expressing a truncated version of the schizophrenia risk gene *DISC1* have reduced parvalbumin immunoreactivity in the PFC and hippocampus, as do rodents subchronically exposed to PCP. However, chronic ketamine and MK-801 produce reduced parvalbumin immunoreactivity in the hippocampus, but not in the PFC.[32] The finding that animals exhibiting impaired NMDA neurotransmission have a loss of parvalbumin-immunoreactive interneurons supports the hypothesis that GABAergic alterations are secondary, at least in part, to impaired NMDA signaling in the brains of patients with schizophrenia.

Since cognitive impairments are one of the best predictors of poor functional outcome in individuals with schizophrenia,[4] considerable research has been conducted to identify cognitive enhancers effective in the disorder. One potential cognitive enhancer is the $GABA_A$ α_2-subunit specific agonist MK-0777, which improves cognitive function related to the DLPFC in subjects with schizophrenia. MK-0777 is also associated with increased power of gamma-band oscillations in the frontal brain areas. Thus, selective modulation of GABAergic neurotransmission may be a strategy for ameliorating the cognitive impairments observed in the disorder.

Cholinergic Neurotransmission

Interest in the role of acetylcholine in the pathophysiology of schizophrenia arose from observations that almost 80% of patients with schizophrenia smoke tobacco, compared with 20–30% of the general population. Acetylcholine is a neurotransmitter in both the peripheral and central nervous systems. In the CNS, acetylcholine is involved in arousal, reward, and plasticity. There are two broad classes of acetylcholine receptor: nicotinic (nAChR) and muscarinic (mAChR). nAChRs are ionotropic, composed of five subunits, and when activated, allow passive transport of sodium into and potassium out of the cell. In addition to acetylcholine, nAChRs are activated by nicotine. mAChRs are metabotropic G-protein linked receptors. There are five subtypes of mAChR, M_1–M_5, each having different downstream effects depending on the G-protein type associated with the receptor.

As stated previously, patients with schizophrenia exhibit impaired selective attention along with behavioral evidence of impaired sensory gating (e.g. deficits in PPI and LI). Polymorphisms in the α_3-subunit of the nAChR ($CHRN_{A3}$) are associated with deficits in PPI in patients with schizophrenia, and are reversed by smoking. Schizophrenia patients also exhibit electrophysiological evidence of impaired sensory gating. An evoked potential is a scalp electrical potential occurring after presentation of a sensory stimulus as detected by electroencephalography (EEG). The P50 is a component of the auditory evoked potential seen on EEG approximately 50 milliseconds after the auditory stimulus. In control subjects, the P50 amplitude following the second stimulus in a pair of auditory stimuli is lower than the P50 following the first stimulus. Subjects with schizophrenia and nearly half of their first degree relatives show less inhibition of the second P50, a deficit briefly normalized by smoking and related to clinical symptoms. Polymorphisms in the promoter of the α_7-subunit of the nAChR ($CHRN_{A7}$) in affected subjects are related to the P50 deficit. Indeed, DMXB-A, a $CHRN_{A7}$ agonist, increases inhibition of P50 in individuals with schizophrenia. Together, these data suggest that alterations in nAChRs may be involved in the impaired sensory gating and selective attention in the disorder.

Acetylcholine-producing neurons in the nucleus basalis of Meynert project to all layers of the cerebral cortex and account for 70–80% of cortical cholinergic input. These afferents play a major role in the modulation of PFC function and cognitive processes. Despite no differences in $CHRN_{A7}$ mRNA in the DLPFC, $CHRN_{A7}$ protein levels are decreased in the frontal cortex from subjects with schizophrenia. Moreover, the density of high-affinity nAChRs containing α_4 and β_2-subunits ($CHRN_{A4}$ and $CHRN_{B2}$) is reduced in the striatum and nAChR binding is reduced in the hippocampus from individuals with schizophrenia. A potential confound of studies assessing nAChR expression in the brain is smoking, which increases nAChR expression in the brains of healthy controls. This increase in nAChR expression is attenuated in subjects with schizophrenia.

In the PFC, M_1 and M_4 mAChRs ($CHRM_1$ and $CHRM_4$) are the major postsynaptic acetylcholine receptors. SPECT imaging reveals decreased mAChR binding in the cortex, basal ganglia, and thalamus. In addition, a polymorphism in the M_1 mAChR is associated with improved performance on the Wisconsin Card Sorting Task in affected individuals. Thus, altered mAChR neurotransmission may also affect cognitive impairments in schizophrenia.

In postmortem samples, mRNA for the $CHRM_1$ is reduced in the PFC, and $CHRM_1/CHRM_4$ protein expression is decreased in the DLPFC, posterior cingulate, and hippocampus from subjects with schizophrenia. Alterations in $CHRM_1/CHRM_4$ protein expression are probably driven by a reduction in $CHRM_4$, at least in the hippocampus. In addition, receptor densities of $CHRM_1/CHRM_4$ and $CHRM_2/CHRM_4$ are decreased in the STG of patients with schizophrenia. Alterations in mAChRs may also influence the function of GABAergic interneurons in the hippocampus, where the vast majority express mAChRs. Therefore, postmortem and neuroimaging studies have found altered nAChR and mAChR neurotransmission in schizophrenia patients.

In the brain, high-affinity nAChRs are composed mainly of α_4 and β_2 ($CHRN_{A4}$ and $CHRN_{B2}$) subunits. Individually, polymorphisms in $CHRN_{A4}$ or $CHRN_{B2}$ do not produce increased risk for the development of schizophrenia. However, an interaction between SNPs in $CHRN_{A4}$ and $CHRN_{B2}$ significantly increases the risk of developing schizophrenia. Polymorphisms in $CHRN_{A7}$ have also been associated with increased risk of developing schizophrenia and tobacco dependence. In addition, there appears to be an interaction between $CHRN_{A7}$ and $CHRM_5$ polymorphisms affecting the risk of developing schizophrenia. Polymorphisms in $CHRN_{A3}$ are also associated with smoking and serve as a marker for disease severity in schizophrenia patients.

Kynurenic acid provides a link between alterations in both acetylcholine and glutamate neurotransmission in schizophrenia. Endogenous kynurenic acid is produced by astrocyte metabolism of the amino acid tryptophan, and is a non-competitive antagonist of $CHRN_{A7}$ as well as the NMDA receptor. In the PFC, $CHRN_{A7}$ regulates glutamate release, and therefore alterations in kynurenic acid may contribute to the deficits in glutamate neurotransmission in the disorder. Indeed, systemically administered kynurenine raises kynurenic acid in the brain and decreases glutamate levels in the PFC. In schizophrenia patients, levels of kynurenine and kynurenic acid are increased in the DLPFC, whereas in the ACC, kynurenine and its precursor, tryptophan, are increased. Moreover, the cytotoxic metabolite of kynurenine, 3-OH-anthranilic acid, is also increased in the ACC in schizophrenia subjects. The first step in the kynurenine pathway is catalyzed by tryptophan 2,3-dioxygenase (TDO_2), and PFC TDO_2 mRNA levels and density of TDO_2-immunoreactive astrocytes are increased in schizophrenia. Thus, the kynurenine pathway may play an important role in the alterations in glutamate neurotransmission and the structural alterations (i.e. loss of dendritic spines) observed in the brains of subjects with schizophrenia.

Given the role of acetylcholine in cognitive function, studies on the effects of EVP-6124, a $CHRN_{A7}$ partial agonist, on cognitive function and event-related potentials have been conducted in schizophrenia patients in a phase IIB study. Promising results in this study and on cognition in Alzheimer disease suggest that such an intervention may be effective for negative and cognitive symptoms in schizophrenia.[33]

GLIAL CELL ALTERATIONS IN SCHIZOPHRENIA

Glial cells provide critical support for neurons in the CNS and are comprised of three major cell types: astrocytes, oligodendrocytes, and microglia. Astrocytes provide metabolic support and play a key role in glutamatergic neurotransmission. Oligodendrocytes also provide metabolic support and produce myelin sheaths, which insulate the axons. Microglia are the chief immune effector cells in the brain. All three cell types have been implicated in the pathophysiology of schizophrenia.

Astrocytes

After release from the presynaptic terminal, the majority of synaptic glutamate is taken up by excitatory amino acid transporters on the cellular membrane of astrocytes. Within the astrocyte, a portion of glutamate is metabolized to glutamine, which is transported back to the presynaptic terminals and converted to glutamate by phosphate-activated glutaminase to replenish the neuronal glutamate pool and tricarboxylic acid (TCA) cycle. Because the neuronal TCA cycle lacks an anaplerotic pathway, it is depleted by glutamate production and critically dependent on metabolic support from astrocytes, which contain the anaplerotic enzyme pyruvate carboxylase.

Glial fibrillary acidic protein (GFAP) is one of the major intermediate filament proteins in mature astrocytes and is used to distinguish astrocytes from other glial cells (e.g. oligodendrocytes and microglia). Reductions in GFAP mRNA expression occur in the ACC with reductions in GFAP protein expression in the PFC and ACC in schizophrenia. Although the density of GFAP-immunoreactive astrocytes is generally unchanged across the six cortical layers in the DLPFC, ACC, and STG, an increase in GFAP-immunoreactive astrocyte density with a concomitant reduction in GFAP-immunoreactive areal fraction has been detected in layer V of the DLPFC from individuals with schizophrenia. Thus, cellular abnormalities in astrocytes occur in the cerebral cortex of schizophrenia patients.

Astrocytes are metabolically quite active and account for 30% of total cortical oxygen consumption. In addition, they consume a large percentage of cortical glucose, the primary oxidative fuel in the brain. Astrocytes have peripheral cellular processes whose end-feet cover over 99% of the vascular surface in the brain and have an intimate relationship with endothelial cells. These astrocytic end-feet possess two astrocyte-specific proteins, aquaporin-4 and the gap junction protein connexin-43, along with glucose transporters. As a result, neurons rely on astrocytes for their supply of nutrients, especially glucose. In response to synaptic activity, astrocytes are able to regulate local cerebral blood flow, which in turn regulates regional oxidative metabolism. Both astrocytes and neurons metabolize glucose using glycolytic, pentose shunt, and oxidative pathways. However, only astrocytes can store glucose as glycogen. In general, brain glycogen levels are relatively stable, but glycogenolysis leading to gluconeogenesis can increase significantly with increased demand.

Alterations in oxidative metabolism have relevance to schizophrenia since the risk of developing the illness is increased in individuals exposed to famine during gestation.[34] In postmortem studies, genes associated with energy metabolism are dysregulated in multiple brain regions in individuals with schizophrenia.[35] In particular, protein expression for enzymes involved in all stages of cellular respiration are decreased in the ACC, suggesting impaired glucose metabolism and increased vulnerability to oxidative stress in the cerebral cortex. Indeed, proteins associated with the oxidative stress response are also dysregulated in the ACC[36] and in the DLPFC from individuals with schizophrenia. In addition, the expression of growth factors and hormones, which regulate glycolytic and other energy pathway enzymes, is altered in the cerebral cortex of patients with schizophrenia, where insulin signaling is significantly increased in the DLPFC. Together, these data suggest that the utilization and regulation of glucose, coupled with increased vulnerability to oxidative stress, occur in the PFC in affected subjects.

Postmortem studies showing that metabolic alterations occur in the brains of subjects with schizophrenia have been bolstered by a genetic study suggesting that increased CNV affects the expression of glutathione *S*-transferase mu-1 (GSTM1) in patients with schizophrenia.[36] Glutathione *S*-transferase occurs in cytosolic and membrane-bound forms and has eight classes: alpha, kappa, mu, omega, pi, sigma, theta, and zeta. GSTM1 encodes a glutathione *S*-transferase in the mu class, which is associated with the detoxification of carcinogens, drugs, toxins, and products of oxidative stress by conjugating these compounds with glutathione. Alterations in GSTM1 could contribute to the increased vulnerability to oxidative stress and structural alterations seen in the cerebral cortex of patients with schizophrenia.

Oligodendrocytes

In the CNS, oligodendrocytes are intimately related to neurons by providing metabolic support and, most importantly, by creating myelin sheaths, which reduce the escape of ions from axons and increase propagation velocities of action potentials. Diffusion tensor imaging is being applied to study the pathophysiology of schizophrenia. This technique provides an *in vivo* assessment of anatomical connectivity in white matter tracts. It measures the fractional anisotropy (i.e. coherence of directionality) of water diffusion to assess white matter integrity. Patients with schizophrenia show lower fractional anisotrophy in white matter, consistent with disruption of the myelin sheath or the directional coherence of axonal fiber tracts. Patients experiencing a first episode of psychosis have less pronounced fractional anisotropic abnormalities than individuals who have a more chronic illness. Subjects with schizophrenia also exhibit white matter deficits in the corpus callosum and temporal and frontal lobes, especially in the left hemisphere. In an exhaustive review of structural MRI studies conducted in schizophrenia subjects, the most common finding was ventricular enlargement, specifically enlargement of the third ventricle, probably reflecting, at least in part, a significant loss of white matter volume.[37] In addition, reductions in white matter volume occur over time in the frontal lobe and correlate with worsening negative symptoms and cognitive impairments in patients with schizophrenia.[38]

Polymorphisms in the schizophrenia risk gene *RGS4* are associated with alterations in regional white matter volume in schizophrenia subjects.[13] Furthermore, ZNF804A, a poorly understood gene expressed throughout the brain, is associated with an increased risk of developing schizophrenia and increased white matter volume in patients with the illness. Together, these data suggest that myelin integrity is disrupted in the brains of schizophrenia subjects and that these disruptions are influenced by genetic abnormalities.

Reductions in oligodendrocyte number occur in the DLPFC, hippocampus, and anterior principal thalamic nucleus in schizophrenia (Fig. 39.5).[39] In addition, reductions in mRNA expression levels of oligodendrocyte–myelin-related (OMR) genes occur in the DLPFC, ACC, and hippocampus.[23,35] Polymorphisms in both 2′,3′-cyclic nucleotide 3′-phosphodiesterase (CNP), an enzyme involved in the production of myelin, and oligodendrocyte lineage transcription factor-2 (OLIG2), a basic helix–loop–helix transcription factor critical to the development of oligodendrocytes, are associated with an increased risk of developing schizophrenia and predict decreased expression of these genes in the PFC. Moreover, OLIG2 polymorphisms are associated with reduced CNP

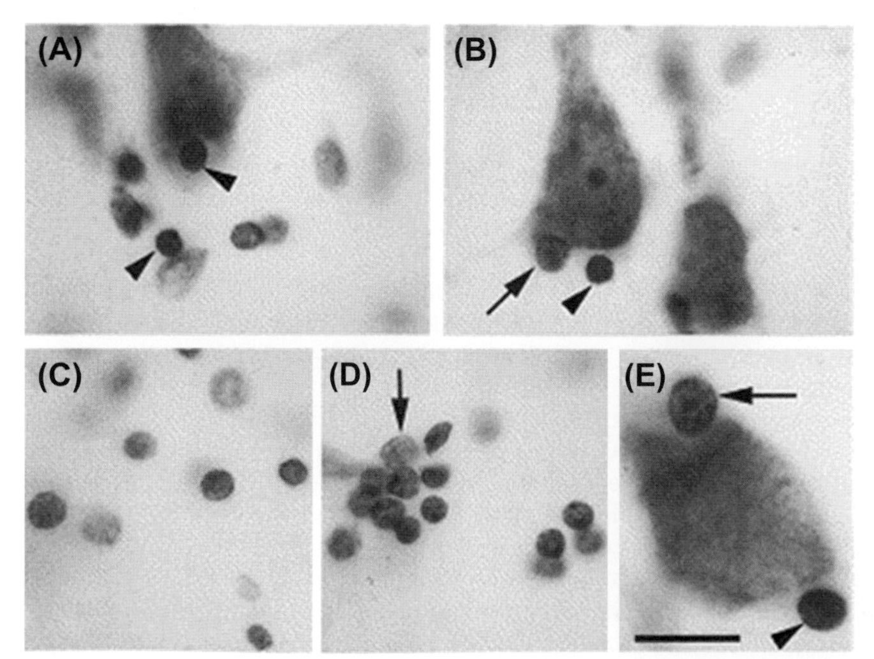

FIGURE 39.5 **High-power photomicrographs of Nissl-stained sections of human dorsolateral prefrontal cortex.** Oligodendrocytes are indicated by arrowheads and astrocytes are indicated by arrows. (A,D,E) Sections from a control subject; (B,C) from a subject with schizophrenia. (A,B,E) Sections from layer III of the gray matter; (C,D) from the white matter. Scale bar = 10 μm. *Source: Hof* et al. Biol Psychiatry. *2003;53(12):1075–1085.*[39]

mRNA levels in the cortex, caudate, and cerebellum from schizophrenia subjects. The cytosine–guanine dinucleotide (CpG) island of sex-determining region Y-box containing gene-10 (*SOX10*), a transcription factor involved in oligdodendrocyte development, is highly methylated in brains of patients with schizophrenia. Hypermethylation of SOX10 is associated with reduced expression of multiple OMR genes including *SOX10*, *CNP*, and *OLIG2* in the PFC in affected individuals. The schizophrenia risk gene *NRG1* promotes oligodendrocyte survival and proliferation of oligodendrocyte precursors. Mice lacking the NRG1 receptor ERBB4 show alterations in oligodendrocyte morphology and decreased myelin in the corpus callosum.[32] Likewise, mRNA for the neuregulin receptor ERBB3 is reduced in the DLPFC in schizophrenia. In addition, reduced expression of ERBB4 in the cortex is associated with OLIG2 polymorphisms. Thus, postmortem studies suggest that the myelin alterations seen in neuroimaging in patients with schizophrenia are likely to be due to a loss of oligodendrocytes and impaired function of the remaining oligodendrocytes.

A loss of oligodendrocytes could also explain the decreased levels of NAA in the brain in schizophrenia. NAA is hydrolyzed by the oligodendrocytic enzyme aspartoacylase, which provides NAA-derived acetyl groups for incorporation into fatty acids and lipids, some of which are incorporated into myelin. Aspartoacylase mRNA levels are decreased in the DLPFC in patients with schizophrenia.[23] Because of the production of myelin, lipid biosynthesis and metabolism are important functions of oligodendrocytes. In the DLPFC of schizophrenia patients, several genes involved in fatty acid and lipid biosynthesis, including those for cholesterol and long-chain fatty acid biosynthesis, are decreased. The carnitine transport system shuttles fatty acids from the cytosol into mitochondria for degradation. Several genes involved in the carnitine transport system are upregulated in the PFC in schizophrenia subjects. However, the protein expression of both 2,4-dienoyl coenzyme A reductase-2 (DECR2), which participates in the degradation of unsaturated fatty acids, and sphingomyelin phosphodiesterase-1 (SMPD1), which degrades the myelin component sphingomyelin, is decreased in the DLPFC from individuals with schizophrenia. Therefore, pathways associated with lipid and fatty acid synthesis and metabolism are altered and may contribute to the myelin abnormalities in the brains of patients with schizophrenia.

Microglia

Microglia are the chief immune effector cells within the CNS and provide immune surveillance within the brain. They begin as myeloid cells in the peripheral hematopoietic system and take up residence in the CNS during development. Alterations in the immune system are detectable in both the CNS and periphery in schizophrenia, which may be due in part to early exposure to pathogenic organisms such viruses, bacteria, or protozoa. Indeed, perinatal infections appear to be associated with an increased risk of developing schizophrenia. The rate of schizophrenia among individuals born 5 months after the peak infection prevalence of the 1957 A2 influenza pandemic was 88% higher than average. In addition, the presence of maternal antibodies against both the protozoal parasite *Toxoplasma gondii* and herpes simplex virus-2 (HSV-2) is associated with an increased risk of developing schizophrenia in the offspring.[34]

Untreated individuals with schizophrenia have altered serum levels of antibodies to cytomegalovirus, *Toxoplasma gondii*, and human herpesvirus type 6, a finding absent in treated patients experiencing recent-onset psychotic symptoms. Individuals with schizophrenia also have elevated antibodies to cytomegalovirus, vaccinia, HSV, and influenza virus in the CSF.[34] Antibodies against HSV-1 are associated with cognitive impairments in schizophrenia, suggesting a link between dysregulated peripheral immune activity and impaired brain function. Heat-shock proteins-70 and -90 (HSP70 and HSP90) are stress response proteins produced during infection. Patients with schizophrenia have elevated antibodies against both HSP70 and HSP90. Moreover, antibody titers against HSP70 in patients positively correlate with more severe clinical symptoms and are decreased following antipsychotic treatment.

Activated immune effector cells release bioactive proteins termed cytokines, which modulate the activity of other immune and non-immune cells. Two important cytokines which activate peripheral lymphocytes are interleukin-2 (IL-2) and IL-6. In schizophrenia, plasma levels of IL-6 are elevated at baseline, associated with treatment resistance, and reduced following antipsychotic treatment.[40] CSF levels of IL-2 in patients with schizophrenia are associated with an increased risk of relapse of psychotic symptoms but are reduced following antipsychotic treatment. In culture, white blood cells from individuals with schizophrenia show altered function. IL-2 production by lymphocytes is elevated, whereas IL-6 production is decreased, and lower IL-2 production is associated with better treatment response.

In the CNS, microglia reside in one of two functional modes: surveillance when they are monitoring for pathogens, and activated when they are triggered by glutamate, cytokines, bacteria-derived lipopolysaccharide, or changes in extracellular potassium concentration. When microglia shift from surveillance mode to activated mode, they alter their morphology, upregulate major histocompatibility complex (MHC) class I and II molecules and cell adhesion molecules, and increase their enzymic activity. As a result, microglia transform from ineffective (surveillance mode) to effective (activated mode) antigen-presenting cells owing to the upregulation of MHC molecules. A surface marker of activated microglia is human leukocyte antigen (HLA)-DR, a MHC class II cell surface receptor, which presents antigens to other immune mediators. An increased density of HLA-DR-immunoreactive microglia occurs in the DLPFC and the STG in schizophrenia. In addition, molecular studies show evidence of altered expression of immune-related genes in the DLPFC and temporal cortex in schizophrenia. Thus, evidence of dysregulated immune activity has been found in the periphery, CSF, and cerebral cortex in the disorder.

Animal models have been developed to explore the effects of immune activation by early pathogen exposure. Injection of double-stranded RNA polyinosinic–polycytidylic acid (poly I:C) recreates the immune response observed during viral infections. Poly I:C injections in pregnant rat dams produce molecular, cellular, and behavioral alterations in offspring reminiscent of those observed in patients with schizophrenia. The offspring display white matter alterations and decreased white matter tract anisotropy in the corpus callosum. Prenatal injection with the bacterial immune activator lipopolysaccharide (LPS) reduces the dendritic arbor and spine density in the PFC and hippocampus.[32] In addition, LPS-exposed rats show increased dopamine levels and turnover in the striatum. Similar dopaminergic alterations are seen in poly I:C-treated offspring. Prenatal poly I:C-exposed offspring also show decreased hippocampal $GRIN_1$ expression, with decreased reelin and parvalbumin expression in the PFC and hippocampus. Poly I:C-treated offspring have increased $GABA_A$ α_2-subunit expression at the AIS of hippocampal pyramidal neurons. Poly I:C-treated offspring also show reduced expression of the oligodendrocyte marker myelin basic protein and decreased myelination in the hippocampus. Prenatal exposure to influenza virus, poly I:C, LPS, and IL-6 results in impaired sensory gating including PPI and LI deficits. In addition, animal studies suggest that prenatal infection in genetically predisposed individuals may significantly increase the risk of developing schizophrenia. Mice overexpressing the cytokine IL-10 show an increased response to prenatal poly I:C. Furthermore, mice with a mutated human version of the schizophrenia risk gene *DISC1* (*hDISC1*) exhibit a synergistic response to prenatal poly I:C treatment that affects social behaviors not observed with *hDISC1* alone.[34]

Many markers of immune dysfunction in patients with schizophrenia (e.g. elevated plasma IL-2 levels) are normalized after treatment with antipsychotic medications. Therefore, in addition to regulating neurotransmitter activity and other cellular processes, antipsychotics may have immunomodulatory functions. The best example of this comes from studies of the effects of the atypical antipsychotic clozapine. Treatment of patients with schizophrenia with clozapine requires frequent assessments of white blood cell counts owing to the increased risk of agranulocytosis, a potentially dangerous reduction in white blood cell counts which puts the patient at risk of serious infection. In cultured mononuclear cells, clozapine inhibits cell proliferation and the release of the growth-promoting cytokine granulocyte–macrophage colony-stimulating factor (GM-CSF). In patients with schizophrenia, treatment with clozapine results in both proinflammatory and immunosuppressant effects. Plasma levels of the proinflammatory cytokine IL-6 increase, but production of an antagonist against the receptor for the proinflammatory cytokine interleukin-1 (IL-1RA) also increases in patients treated with clozapine.

CONCLUSION

Schizophrenia is a relatively common chronic mental illness, which results in substantial disability and has a 10% risk of suicide. It has several symptom clusters including positive symptoms, negative symptoms, cognitive impairments, and affective dysregulation. Positive symptoms respond well to conventional antipsychotic medications. However, negative symptoms and cognitive impairments are the best predictors of poor functional outcome and respond poorly to current treatments. Genetic and environmental factors contribute to the risk of developing schizophrenia. Multiple genetic risk loci for schizophrenia have been identified in GWAS. Modern neurobiological research into the pathophysiology of schizophrenia refutes the psychoanalytic notion that schizophrenia has its origins in pathological early relationships. Rather, schizophrenia is a disorder of molecular, cellular, and structural abnormalities in the brain leading to alterations in emergent behavioral phenomena and observable clinical symptoms.

The neurotransmitter dopamine plays a critical role in several aspects of the pathophysiology of schizophrenia, including psychosis and cognitive impairments. Patients with the disorder show increased sensitivity to drugs that enhance dopamine neurotransmission, and gain relief from positive symptoms with drugs that block dopamine neurotransmission at dopamine D_2 receptors. Molecular, cellular, structural, and behavioral abnormalities in schizophrenia are due, at least in part, to decreased neurotransmission at NMDA glutamate receptors in the brain. Polymorphisms in several glutamate-related genes significantly increase the risk of developing schizophrenia. Parvalbumin-expressing

GABAergic interneurons are also affected in schizophrenia, and these GABAergic alterations probably stem, in part, from impaired NMDA receptor neurotransmission. Both nAChR and mAChR neurotransmission are altered in the brain in schizophrenia. Furthermore, kynurenine pathway abnormalities that disrupt α_7 nAChR function may further impair glutamate neurotransmission.

Deficits in astrocyte function lead to impaired glucose utilization and regulation, coupled with an increased vulnerability to oxidative stress in the cerebral cortex of individuals with schizophrenia. With regard to oligodendrocytes, the integrity of myelin sheaths is disrupted in schizophrenia, and polymorphisms in oligodendrocyte-related genes are likely to contribute to this abnormality. Activation of peripheral and central mediators of inflammation, including microglia, has been described in schizophrenia; when this occurs *in utero* in genetically predisposed individuals, the risk of development of schizophrenia is elevated.

In summary, schizophrenia is a very disabling psychiatric disorder with variable symptomatic manifestations and with widespread pathological alterations in the forebrain that result from the interaction of multiple risk genes with environmental factors. Understanding the final common pathways of risk genes leading to pathology may illuminate targets for more effective pharmacological and behavioral interventions.

References

1. Jablensky A. The 100-year epidemiology of schizophrenia. *Schizophr Res*. 1997;28(2–3):111–125.
2. Siris SG. Suicide and schizophrenia. *J Psychopharmacol*. 2001;15(2):127–135.
3. Coyle JT, Konopaske GT. The neurochemistry of schizophrenia. In: Brady ST, Siegel GJ, Albers RW, Price DL, Benjamins J, eds. *Basic Neurochemistry: Principles of Molecular, Cellular, and Medical Neurobiology*. 8th ed. Amsterdam: Elsevier / Academic Press; 2012: 1000–1011.
4. Green MF. What are the functional consequences of neurocognitive deficits in schizophrenia? *Am J Psychiatry*. 1996;153(3): 321–330.
5. Karlsgodt KH, Glahn DC, van Erp TG, et al. The relationship between performance and fMRI signal during working memory in patients with schizophrenia, unaffected co-twins, and control subjects. *Schizophr Res*. 2007;89(1–3):191–197.
6. Carter CS, Perlstein W, Ganguli R, Brar J, Mintun M, Cohen JD. Functional hypofrontality and working memory dysfunction in schizophrenia. *Am J Psychiatry*. 1998;155(9):1285–1287.
7. Sullivan PF, Kendler KS, Neale MC. Schizophrenia as a complex trait: evidence from a meta-analysis of twin studies. *Arch Gen Psychiatry*. 2003;60(12):1187–1192.
8. Vijayraghavan S, Wang M, Birnbaum SG, Williams GV, Arnsten AF. Inverted-U dopamine D1 receptor actions on prefrontal neurons engaged in working memory. *Nat Neurosci*. 2007;10(3): 376–384.
9. Laruelle M, Abi-Dargham A, van Dyck CH, et al. Single photon emission computerized tomography imaging of amphetamine-induced dopamine release in drug-free schizophrenic subjects. *Proc Natl Acad Sci U S A*. 1996;93(17):9235–9240.
10. Powell SB, Zhou X, Geyer MA. Prepulse inhibition and genetic mouse models of schizophrenia. *Behav Brain Res*. 2009;204(2):282–294.
11. Coyle JT. The GABA–glutamate connection in schizophrenia: which is the proximate cause? *Biochem Pharmacol*. 2004;68(8): 1507–1514.
12. Theberge J, Al-Semaan Y, Williamson PC, et al. Glutamate and glutamine in the anterior cingulate and thalamus of medicated patients with chronic schizophrenia and healthy comparison subjects measured with 4.0-T proton MRS. *Am J Psychiatry*. 2003;160(12):2231–2233.
13. Buckholtz JW, Meyer-Lindenberg A, Honea RA, et al. Allelic variation in RGS4 impacts functional and structural connectivity in the human brain. *J Neurosci*. 2007;27(7):1584–1593.
14. Balu DT, Basu AC, Corradi JP, Cacace AM, Coyle JT. The NMDA receptor co-agonists, D-serine and glycine, regulate neuronal dendritic architecture in the somatosensory cortex. *Neurobiol Dis*. 2012;45(2):671–682.
15. Selemon LD, Rajkowska G, Goldman-Rakic PS. Elevated neuronal density in prefrontal area 46 in brains from schizophrenic patients: application of a three-dimensional, stereologic counting method. *J Comp Neurol*. 1998;392(3):402–412.
16. Todtenkopf MS, Vincent SL, Benes FM. A cross-study meta-analysis and three-dimensional comparison of cell counting in the anterior cingulate cortex of schizophrenic and bipolar brain. *Schizophr Res*. 2005;73(1):79–89.
17. Glantz LA, Lewis DA. Decreased dendritic spine density on prefrontal cortical pyramidal neurons in schizophrenia. *Arch Gen Psychiatry*. 2000;57(1):65–73.
18. Hill JJ, Hashimoto T, Lewis DA. Molecular mechanisms contributing to dendritic spine alterations in the prefrontal cortex of subjects with schizophrenia. *Mol Psychiatry*. 2006;11(6):557–566.
19. Weickert CS, Hyde TM, Lipska BK, Herman MM, Weinberger DR, Kleinman JE. Reduced brain-derived neurotrophic factor in prefrontal cortex of patients with schizophrenia. *Mol Psychiatry*. 2003;8(6):592–610.
20. Ohnuma T, Tessler S, Arai H, Faull RL, McKenna PJ, Emson PC. Gene expression of metabotropic glutamate receptor 5 and excitatory amino acid transporter 2 in the schizophrenic hippocampus. *Brain Res Mol Brain Res*. 2000;85(1–2):24–31.
21. Matute C, Melone M, Vallejo-Illarramendi A, Conti F. Increased expression of the astrocytic glutamate transporter GLT-1 in the prefrontal cortex of schizophrenics. *Glia*. 2005;49(3):451–455.
22. Verrall L, Burnet PW, Betts JF, Harrison PJ. The neurobiology of D-amino acid oxidase and its involvement in schizophrenia. *Mol Psychiatry*. 2010;15(2):122–137.
23. Tkachev D, Mimmack ML, Huffaker SJ, Ryan M, Bahn S. Further evidence for altered myelin biosynthesis and glutamatergic dysfunction in schizophrenia. *Int J Neuropsychopharmacol*. 2007;10(4):557–563.
24. Bergeron R, Coyle JT. NAAG, NMDA receptor and psychosis. *Curr Med Chem*. 2012;19(9):1360–1364.
25. Basu AC, Tsai GE, Ma CL, et al. Targeted disruption of serine racemase affects glutamatergic neurotransmission and behavior. *Mol Psychiatry*. 2009;14(7):719–727.
26. Lin CH, Lane HY, Tsai GE. Glutamate signaling in the pathophysiology and therapy of schizophrenia. *Pharmacol Biochem Behav*. 2012;100(4):665–677.
27. Beasley CL, Zhang ZJ, Patten I, Reynolds GP. Selective deficits in prefrontal cortical GABAergic neurons in schizophrenia defined by the presence of calcium-binding proteins. *Biol Psychiatry*. 2002;52(7):708–715.
28. Woo TU, Walsh JP, Benes FM. Density of glutamic acid decarboxylase 67 messenger RNA-containing neurons that express the *N*-methyl-D-aspartate receptor subunit NR2A in the anterior cingulate cortex in schizophrenia and bipolar disorder. *Arch Gen Psychiatry*. 2004;61(7):649–657.

29. Volk DW, Pierri JN, Fritschy JM, Auh S, Sampson AR, Lewis DA. Reciprocal alterations in pre- and postsynaptic inhibitory markers at chandelier cell inputs to pyramidal neurons in schizophrenia. *Cereb Cortex*. 2002;12(10):1063–1070.
30. Straub RE, Lipska BK, Egan MF, et al. Allelic variation in GAD1 (GAD67) is associated with schizophrenia and influences cortical function and gene expression. *Mol Psychiatry*. 2007;12(9):854–869.
31. Addington AM, Gornick M, Duckworth J, et al. GAD1 (2q31.1), which encodes glutamic acid decarboxylase (GAD67), is associated with childhood-onset schizophrenia and cortical gray matter volume loss. *Mol Psychiatry*. 2005;10(6):581–588.
32. Jaaro-Peled H, Ayhan Y, Pletnikov MV, Sawa A. Review of pathological hallmarks of schizophrenia: comparison of genetic models with patients and nongenetic models. *Schizophr Bull*. 2010;36(2):301–313.
33. EnVivo Pharmaceuticals. *News Release*. February 28, 2013. http://www.forumpharma.com/content/news-events/phase3-trial-evp-6124
34. Brown AS. The environment and susceptibility to schizophrenia. *Prog Neurobiol*. 2011;93(1):23–58.
35. Katsel P, Davis KL, Haroutunian V. Variations in myelin and oligodendrocyte-related gene expression across multiple brain regions in schizophrenia: a gene ontology study. *Schizophr Res*. 2005;79(2–3):157–173.
36. Rodriquez-Santiago B, Brunet A, Sobrinoa B, et al. Association of common copy number variants at the glutathione *S*-transferase genes and rare novel genomic changes in schizophrenia. *Mol Psychiatry*. 2010;15:1023–1055.
37. Shenton ME, Dickey CC, Frumin M, McCarley RW. A review of MRI findings in schizophrenia. *Schizophr Res*. 2001;49(1–2):1–52.
38. Ho BC, Andreasen NC, Nopoulos P, Arndt S, Magnotta V, Flaum M. Progressive structural brain abnormalities and their relationship to clinical outcome: a longitudinal magnetic resonance imaging study early in schizophrenia. *Arch Gen Psychiatry*. 2003;60(6):585–594.
39. Hof PR, Haroutunian V, Friedrich Jr VL, et al. Loss and altered spatial distribution of oligodendrocytes in the superior frontal gyrus in schizophrenia. *Biol Psychiatry*. 2003;53(12):1075–1085.
40. Maes M, Bosmans E, Calabrese J, Smith R, Meltzer HY. Interleukin-2 and interleukin-6 in schizophrenia and mania: effects of neuroleptics and mood stabilizers. *J Psychiatr Res*. 1995;29(2):141–152.

CHAPTER

40

Bipolar Disorder

Heinz Grunze

Institute of Neuroscience, Newcastle University, Newcastle upon Tyne, UK

OUTLINE

INTRODUCTION

Bipolar disorder (BD) is a lifelong and life-threatening illness, with a 12-fold increased risk of suicide compared with healthy people, and high somatic and psychiatric comorbidity rates; for example, 60% of people with BD also abuse alcohol or other substances. With up to 4.5% of the population affected, BD is a common problem, with a direct cost of 342 million pounds to the UK National Health Service in 2009.

The worldwide incidence and prevalence figures for BD are in a similar, narrow range, in contrast to figures for unipolar depression (UPD), which show a much larger regional variation. The similar prevalence rates for BD across countries and cultures suggest that the contribution of environmental factors may be less important than in UPD, and a biological predisposition predominates. Furthermore, the age of symptomatic onset is uniform and significantly earlier than in UPD, with a nearly equivalent age across countries. These epidemiological findings suggest that BD may be an easier disorder than UPD in which to search for a common, underlying neurobiology. The pursuit of a biological understanding of BD dates back to the days of Hippocrates (460–370 BC) and Aretaeus of Cappadocia (AD 81–138). Research methods in neurobiology and genetics have progressed enormously over the past few decades, as have the number of publications, revealing important aspects from genetics to psychosocial stressors. Nevertheless, we are only just beginning to understand the decisive biological mechanisms underlying BD.

Despite attempts to differentiate the neurobiology of BD from UPD, these disorders do not constitute entirely separate biological entities and have many mechanisms in common, and this may be especially true for

Neurobiology of Brain Disorders
http://dx.doi.org/10.1016/B978-0-12-398270-4.00040-9

depressive episodes in BD. Thus, many of the underlying mechanisms in UPD (see Chapter 43, "Depression and Suicide") have their equivalent in BD. This chapter will concentrate on a few principal findings. One focus will be mania, including its overlap with psychosis; another will be the mechanisms leading to increasing vulnerability to episode recurrence, as they are unique to BD.

SPECTRUM OF BIPOLAR DISORDER

The reliability of psychiatric research depends on the homogeneity of the disorder. Thus, categorical clinical diagnostic criteria and symptom clusters have to be defined to guarantee comparability of findings. At the time of writing this chapter, the most widely used diagnostic classification systems are the fourth edition of the *Diagnostic and Statistical Manual of Mental Disorders* (DSM-IV) and the tenth revision of the *International Classification of Diseases* (ICD-10). DSM-5 replaced DSM-IV in 2013, and ICD-11 is scheduled for release in 2017. Although DSM-5 adds dimensional components to the diagnosis of BD, it perpetuates the categorical approach to diagnosis. In clinical reality, however, BD is not a homogeneous disorder; rather, it presents as a spectrum with symptoms ranging from "normal" to "pathological", and with no sharp boundaries with schizophrenia at one extreme and UPD at the other. As supported by a large overlap of candidate risk genes, BD may fit more closely with schizophrenia into a dimensional psychotic cluster than with UPD in an exclusively emotional cluster.[1]

Owing to their severity and impact on life, two subgroups of BD, bipolar I and bipolar II disorders, will be considered here. These subgroups are well defined within categorical systems. Whereas bipolar I disorder is defined by manic or mixed episodes, bipolar II patients show only milder, hypomanic episodes, frequently following a depressive episode. The assumption has been made that these groups are slightly different entities, as far as their prevalence, course, underlying pathology, and stability of diagnosis are concerned.[2] Whereas the lifetime prevalence of bipolar I disorder classified according to DSM-IV has been estimated as 1%,[3] large epidemiological studies have revealed a higher prevalence, probably up to 6%, of bipolar II disorder when using sensitive screening instruments.

Although already well characterized by Wilhelm Weygandt, a medical assistant of Emil Kraepelin, in 1899, mixed states have experienced a revival and the attention that they deserve, given their high frequency and unfavorable prognosis.[4] Whereas DSM-IV only recognized mania combined with depression as "mixed states", DSM-5 corresponds more to the clinical reality, including mixed states as specifiers; that is, depression combined with some features of mania ("mixed depression") or mania or hypomania combined with some features of depression ["mixed (hypo)mania"]. The frequent occurrence of mixed states, with up to 70% of manic patients simultaneously showing symptoms of depression, underlines that BD is not a pure *"folie à double forme"* (dual-form insanity), as coined by the French psychiatrist Jules Baillarger in 1854, with mania and depression being distinct opposite poles. This is important to keep in mind when considering animal models of mania and depression, which are typically based on the assumption of polar opposites and thus may be far from the actual situation.

Another manifestation of BD that is attracting special interest in biological research is "rapid cycling", which has been defined as having four or more mood episodes within 1 year. This definition was made according to a pivotal trial conducted by Dunner and Fiève showing a rapid decline in lithium responsiveness with a cut-off at four episodes. Observations of large patient samples, however, call into question this "magic line" defined by four episodes. Data from the Stanley Foundation Bipolar Network demonstrate that there is a continuous decline in treatment response and worsening of prognosis with increasing number of episodes per year. However, rapid cycling also includes some rare manifestations, probably coupled to aberrations of circadian rhythm and zeitgeber (see Chapter 36), and research conducted in such patients may be of special interest for understanding rhythmicity in BD.

GENETICS OF BIPOLAR DISORDER

The genetic etiology of BD is apparent from its relatively uniform prevalence across cultures,[5] familial transmission, and concordance in twin studies.[6] Relatives of bipolar patients show a significantly higher risk of being affected by psychiatric disorders, not only BD, but also schizoaffective disorder and UPD, as well as alcohol or other drug dependence. A large Swedish study[7] showed a heritability of 59% for BD (compared with 64% for schizophrenia), with the remaining variance being contributed by environmental factors (Fig. 40.1). A large proportion of the genetic effects explaining the variance is similar between these disorders. In addition, the risk of developing schizophrenia is increased two- to four-fold in relatives of bipolar patients.

Several studies estimated the lifetime prevalence of BD in first degree relatives of BD patients as 10% for bipolar I disorder (compared with 1% in the general population) and 15% for a depressive episode (compared with 10%). Studies in monozygotic twins show a concordance rate of 62%, which is about four times higher

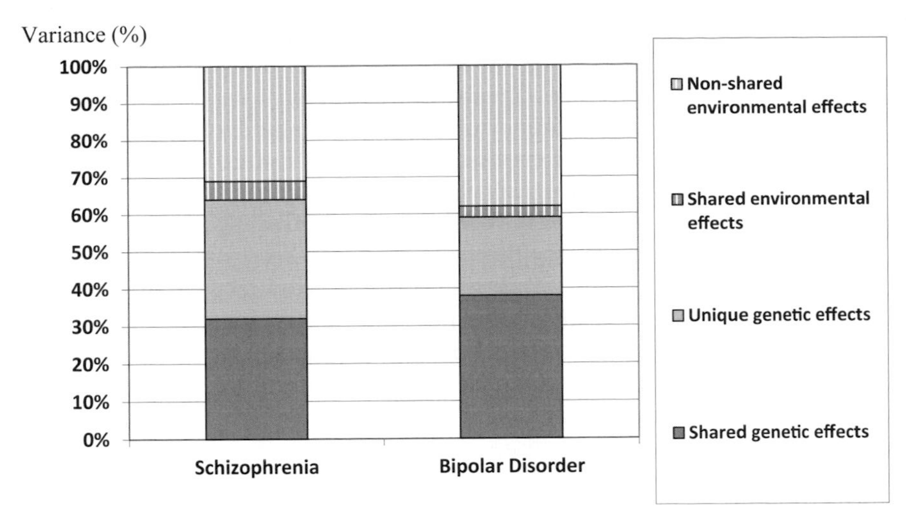

FIGURE 40.1 **Variance accounted for by genetic, shared environmental, and non-shared environmental effects for schizophrenia and bipolar disorder.** *Source: Modified from Lichtenstein* et al. Lancet *2009;373:234–239.*[7]

than in dizygotic twins. The results from adoption studies are also consistent with a genetic vulnerability. Thus, the manifestation of BD appears to be largely genetically determined, but with a low penetration, and modified by environmental factors. In addition, there is accumulating evidence of epigenetic interaction between environment and genes, such as brain-derived neurotrophic factor (BDNF) allelic variants, in BD patients.

BD cannot be explained by a single gene mutation with Mendelian transmission. BD follows an oligogenetic and epistatic mode of inheritance. As no single dominant gene with high penetrance has been identified, several candidate genes located on various autosomes, found by association analyses, have been put forward. However, logarithm of odds (LOD) scores are usually low, and they may explain only a small amount of the variance, if any.

Candidate genes have been identified that connect to assumed or observed neurotransmitter abnormalities in BD, for example to biogenic amines and their G-protein coupled receptors or to the γ-aminobutyric acidergic (GABAergic) system, to genes encoding for resilience factors such as BDNF, or to the action of clinical effective mood stabilizers. These genes encoding for receptors, transporters, metabolic enzymes, or resilience factors were associated with "hot spots" in the search for linkage sites for BD. Association had been found in several studies between BD and single-nucleotide polymorphisms (SNPs) of the genes encoding tryptophan hydroxylase-2 (TPH2), dopamine transporter (SLC6A1), serotonin [5-hydroxytryptamine (5-HT)] transporter (SLC6A3), 5-HT_{2A}, and dopamine D_1 receptor.[8] At least three positive association studies have also been reported for the catabolic enzymes catechol-*O*-methyltransferase (COMT) gene on chromosome 22q11.2 and monoamine oxidase A (MAO-A) gene on chromosome Xp11.

Factors that determine cell survival may play a decisive role in distinguishing the genetic fingerprint of BD from schizophrenia. Resistance to oxidative stress appears to be diminished in BD, and this may, in part, be the result of reduced BDNF brain levels. The BDNF Val66Met polymorphism has been associated with reduced brain volumes in schizophrenia but may be of even greater importance in BD. On the other hand, there are also several candidate genes showing suggestive LOD scores in BD patients, which were thought to be more or less exclusive to schizophrenia (e.g. *DISC1*, *Dysbindin*, and *Neuregulin1*.[9] Some genes, such as *G72* (DAOA), could be associated with common psychopathological features in schizophrenia and BD patients, in this case paranoia or mood-incongruent features in general. The most established findings from association studies are summarized in Table 40.1.

The genetic approach to BD is methodologically difficult and many more studies are needed to identify genes predisposing to this disorder. One major obstacle is the insufficient definition of the phenotype, which is heterogeneous, with the current classification systems providing only very crude diagnostic categories. Therefore, attempts have been made to define subgroups of patients with BD showing high similarities. Such subgroups could be families showing anticipation, paternal or maternal transmission as previously mentioned, or response to a specific treatment, such as lithium.

Response to lithium treatment could be an especially promising approach. Gene associations have been found with dysfunctional intracellular mechanisms as assumed in BD, with lithium acting on those affecting second messengers. A gene for the α-subunit of the G-protein has been localized to the 18p11.2 region, thus making it a hot spot. G-protein activation is the starting point of the major second messenger pathways in the cell. Lithium decreases the binding of this α-subunit to GTP, a requirement for G-protein activation. Thus, a functionally altered G-protein subunit may have a clear impact on the onset and course of BD. However, other

TABLE 40.1 Genetic Associations Found in at Least Three Studies in Patients with Bipolar Disorder

Gene	Chromosome
DISC1	1q42.1
DAT	5p15.3
DRD1	5q35.1
DTNBP1 (*Dysbindin*)	6p21.3
NRG1 (*Neuregulin1*)	8p22–11
BDNF	11p13
TPH2	12q21.1
5-HT2A	13q14–21
DAO-A/G30	13q33.2
5HTT	17q11–12
COMT	22q11.2
MAO-A	Xp11

Source: Modified from Serretti and Mandelli. Mol Psychiatry. *2008;13:742–771.*[8]

investigators could not replicate this finding, and that response to lithium seems not to be restricted to patients with those chromosome 18 markers. More recently, however, the G-protein receptor kinase (*GRK3*) gene on chromosome 22q1 has emerged as a new hot spot associated with BD. In addition, a polymorphism in BD patients has been described for *myo*-inositol monophosphatase-2 (IMPA2), which is also considered as a target of lithium.

Since these earlier hypothesis-driven genome studies could not produce unequivocal results, genome-wide association studies (GWAS) are now the preferred method to identify risk genes for BD and human disease in general. Up to 2010, well over 600 human GWAS had been published yielding genetic associations with stringent statistical significance relevant to the etiology of 92 diseases and 117 other traits. The success of GWAS depends on large samples, both for initial screening and for replication studies. Sample size requirements can easily exceed tens of thousands of cases and controls, which is different from traditional association studies involving usually no more than 300 subjects. It is only more recently that this *a priori* hypothesis-free approach has been adopted by psychiatry.[10]

A GWAS is, in a broad sense, a cross-sectional case–control study. Subjects meet lifetime criteria for a disease (e.g. BD) whereas controls should have never met these criteria and, ideally, have been through the age period of risk. Each individual in the sample is genotyped for a predefined set of a million or more genetic markers spaced across the genome. The genetic markers are SNPs (pronounced "snips"), which are relatively straightforward to discover using commercially available microarrays. Each SNP is then tested for its association with disease. These tests compare the allele frequencies in affected people versus controls, and a large case–control difference suggests an etiological role for a particular SNP or its genomic region after statistical correction for multiple comparisons. However, rare variations and lengthy polymorphisms may not be captured by these arrays.

In more than 40 linkage studies using the GWAS approach, susceptibility loci linked to BD have been identified on chromosomes 3p12–14, 4p16, 4q31–35, 5q31–33, 6q16 and 21–25, 8q21–24, 10q25–26, 11p15.5, 12q23–24, 13q14–32, 18p11 and 18q21–22, 20q13, 21q21–22, 22q11–12, and Xq24–28, often overlapping with findings in schizophrenia.

A genome-wide search in almost 7500 bipolar subjects by the Psychiatric GWAS Consortium Bipolar Disorder Working Group suggested a variation in a calcium channel subunit (CACNA1C, calcium channel, voltage-dependent, L type, α_{1C} subunit) as being highly associated with BD. This GWAS did not include the sex chromosomes in the search. BD, however, has an unequal gender distribution and a greater likelihood of being inherited from the maternal side. For instance, the relative risk (RR) of developing either schizophrenia or BD in half-siblings is much higher when inherited from the maternal side.[7] Baron and colleagues[11] first reported on linkage of X-chromosomal markers and bipolar illness, followed by further studies linking BD with fragile X syndrome or factor IX deficiency. However, these findings could not be replicated, even by the same investigators. Reanalysis of the data linking BD to the Xq27–28 region concluded that abnormalities in the X-chromosome may at least account for an increased vulnerability to BD in some families, although it cannot be reduced to a single locus. Thus, it is difficult to comprehend why that GWAS left out the sex chromosomes.

However, the reason for a gender preference in the inheritance pathway could also be independent of the sex chromosomes. As mitochondrial DNA stems exclusively from the ovum, it may play a role in the inheritance pattern of BD. Another possible explanation for this finding is imprinting, meaning that some alleles of a parent are not transcribed into messenger RNA (mRNA).

Another clinical finding in BD is anticipation, as also observed in highly inherited disorders such as Huntington disease. Manifestation of the disorder starts earlier and is more severe in subsequent generations. The underlying genetic mechanism in Huntington disease is an increase in the length of a trinucleotide repeat with each successive generation. Expansion of trinucleotide repeats has also been found in bipolar patients and may be linked to anticipation; however, a pathognomic triplet repeat has yet to be found for BD.

In which direction will the genetics of BD develop? Although more GWAS in larger samples will be conducted, it will still be difficult to link findings to pathophysiology, and each new SNP will explain only a little of the variance. A fascinating, newly emerging approach is imaging genetics, combining morphometric or functional magnetic resonance imaging (fMRI) with genetics. It may potentially link genetic findings with observable pathology in BD. As an example, gray matter reduction has been noted in relation to aberrations of G72 (DAOA), BDNF, and interleukin-1β (IL-1β), or dysfunctional emotional response in the anterior cingulate as measured with fMRI in relation to a polymorphism of the 5-HT transporter (5-HT-TLPR).

EPIGENETICS OF BIPOLAR DISORDER

Whether a gene becomes activated, leading to functional changes in mRNA expression, protein production, and finally behavioral changes, depends on various epigenetic mechanisms. DNA methylation or histone acetylation massively affect gene regulation and expression. In contrast to the static nature of DNA, epigenetic mechanisms are variable over time and modified by the environment. Thus, they may play an important role in the episodic nature of some disorders, including BD.[12] There is evidence in bipolar patients of a hypomethylation of the COMT promoter, and an aberrant pattern of BDNF gene and X-chromosome methylation. Differences in DNA methylation may mediate different phenotypes of otherwise genetically identical individuals, such as monozygotic twins, as observed in a case–control study,[13] explaining the less than 100% penetration of BD in these cases.

Epigenetic factors modulate gene transcription throughout life, and may influence not only the age of manifestation of an illness but also its progression. If BD is conceptualized as a neurodegenerative disorder, one would expect to see, at some stage, epigenetic changes analogous to those seen in Alzheimer disease. Postmortem BD brains show several epigenetic similarities with postmortem Alzheimer brains, including global DNA hypermethylation and histone H_3 phosphorylation. These changes are associated with hypomethylation and hypermethylation of CpG islands in cyclooxygenase-2 (COX-2) and BDNF promoter regions, respectively.[14] Thus, these epigenetic changes may substantiate the theory of BD as, in part, a disorder modified by inflammation, similar to what is observed in Alzheimer disease. Epigenetics may not only explain the variance in heritability but also foster therapeutic progress. The mood-stabilizing effect of valproate could be, in part, due to epigenetic mechanisms, potentially giving rise to a new generation of histone deacetylase inhibitors as mood stabilizers.

Our understanding of epigenetics is still limited, but this field holds promise for novel treatments that could have an impact on epigenetic risk, including not only pharmacotherapy but also well-tailored psychological and sociotherapeutic interventions.

NEUROMORPHOLOGICAL CHANGES IN BIPOLAR DISORDER

Today, BD is understood mainly as a neurodegenerative or neuroprogressive illness rather than a neurodevelopmental disorder like schizophrenia. This implies that structural and functional changes develop over time. The frequency and severity of episodes, especially the presence or absence of psychotic first rank symptoms, are considered as important modifiers of the disease process. According to a staging model of BD, early intervention favors both better acute response to medication and the maintenance of remission.[15] The underlying molecular mechanisms of neuroprogression include neurotrophins, regulation of neurogenesis and apoptosis, neurotransmitters, inflammatory, oxidative and nitrosative stress, mitochondrial dysfunction, hypothalamic–pituitary–adrenal (HPA) axis dysregulation, and epigenetic influences. Medication appears beneficial for slowing down or even stopping neuroprogression, but it should be kept in mind that some medication that is highly effective in suppressing acute symptoms (e.g. antipsychotics) may have a negative impact on the long-term course, as may be the case in schizophrenia.[16] However, neuronal resilience may be *a priori* different in schizophrenia compared with BD, and thus medication that is potentially harmful in one disease may not be in the other. Solid evidence of progressive neurodegeneration leading to loss of connectivity and cell death in BD comes from *postmortem* examinations, structural and functional neuroimaging, and cognitive testing.

Postmortem Findings

In contrast to schizophrenia, little work has been conducted on the neurohistology of mood disorders. The few studies available suggest that both BD and UPD are associated with circumscribed changes in temporal, prefrontal, and subcortical regions. Early *postmortem* work found a greater brain weight, a thicker parahippocampal cortex, and smaller temporal horns in patients with affective disorder compared with those with schizophrenia, suggesting that temporal lobe structure differs in these disorders. A disturbed morphological integrity of this region has been found in cytoarchitectural studies reporting on malformations in the entorhinal lamination in patients with BD or with major depression, offering a partial explanation for memory deficits in these disorders.

Alteration in the temporal lobe was also reported by Benes and co-workers, with a selective reduction of non-pyramidal neurons in the CA2 region of the hippocampus of patients with BD and schizophrenia.[17] This finding is in line with animal experiments showing developmentally dependent increased sensitivity to *N*-methyl-D-aspartate (NMDA) antagonist neurotoxicity, leading to a selective apoptosis of inhibitory interneurons in this region.[18]

Based on neuroimaging and neuropsychological testing, prefrontal dysfunction has long been implied in mood disorders. Decreased cortical and laminar thickness but unchanged overall neuronal density and laminar density have been demonstrated postmortem in the dorsolateral prefrontal cortex (DLPFC) in both BD and major depressive disorder. In contrast, patients with schizophrenia showed increased neuronal density in this area. Similar to the temporal lobe findings, a layer-specific reduction of interneurons was also observed in the anterior cingulate cortex in subjects with BD.[19]

Basal ganglia, namely the left nucleus accumbens, the right putamen, and right and left palladum externum, are smaller in patients with depression (UPD and BD). This is highly suggestive of an involvement of the basal ganglia in mood dysregulation as part of mood-regulating circuitry. Besides cortical and subcortical structures, brainstem nuclei of the monoaminergic system appear to be affected in BD. Higher neuron numbers have been reported in the locus ceruleus of patients with BD compared with those with UPD. Patients with both BD and UPD show subtle structural deficits in the dorsal raphe. Histochemical data are consistent with a regional reduction in the synthesis of norepinephrine (noradrenaline) and serotonin.

BD has also been hypothesized to be a disorder of astrodendrocytes and oligodendrocytes. This has been supported by neuroimaging findings of a volume reduction and fewer glial (but not neuronal) cells in the subgenual prefrontal cortex in familial mood disorders. Histopathological findings consistently show reductions in glial cell density or glial cell numbers in prefrontal brain regions, such as the (subgenual) anterior cingulate cortex, orbitofrontal cortex (OFC), and DLPFC in association with reduced prefrontal gray matter in patients with mood disorders. Furthermore, alterations have been described histopathologically for astrocytes and oligodendrocytes in these disorders (reviewed by Schroeter *et al.*[20]). Studies have also consistently shown that the glial marker S100B is elevated in mood disorders, and more strongly in UPD than in BD, whereas the neuronal marker protein neuron-specific enolase is unaltered, pointing towards a specific glial dysfunction or cell death. The glial hypothesis is not specific to BD but, given the findings in UPD, may indicate the presence of depressive episodes across disorders.

Beside macrohistological and microhistological examinations, postmortem bipolar brains have also been subjected to GWAS. Increased gene expression has been found for synaptic genes encoding for MAO-A, the α_5-subunit of the $GABA_A$ receptor, and metabotropic glutamate receptor-3. A decrease was observed for glutamate decarboxylase-1 and tyrosine 3-monooxygenase/tryptophan 5-monooxygenase activation protein, zeta polypeptide (reviewed by Lin *et al.*[21]).

A crucial question for clinicians is whether current treatments for BD are capable of changing the expression of genes and consequently protein synthesis. Chen and colleagues compared postmortem brains from individuals with BD who had been exposed to antipsychotics, non-exposed individuals with BD, and controls. They identified 2191 unique genes with significantly altered expression levels in non-antipsychotic-exposed bipolar brains compared with those in the control and exposed groups. The protein products of many of these genes proved critical to the function of synapses, affecting, for example, intracellular trafficking and synaptic vesicle biogenesis, transport, release, and recycling, as well as organization and stabilization of the nodes of Ranvier. These findings support a hypothesis of impaired synaptic and intercellular communication in BD, which may be corrected to some degree by antipsychotic treatment.[22]

Neuroimaging Findings

Morphometric brain changes are not unique to BD, depression, or schizophrenia. They have also been documented in individuals with obsessive–compulsive disorder, post-traumatic stress disorder, and borderline personality disorder, and specificity across disorders is rather low. Morphometric changes may vary with gender and duration of illness, the latter being an important factor, especially in BD.

Numerous imaging studies have investigated the brain structure of patients with BD, using various imaging modalities and analysis techniques. The overall brain volume in bipolar patients seems not to differ significantly from controls. Non-specific increases in lateral and third ventricle volumes appear to be the most robust finding.[23] Meta-analysis has, so far, not established a fully consistent pattern of abnormalities in candidate cortical or subcortical brain regions. The complexity of BD – variable mood states, onset during development, illness progression, medication effects – clearly contributes to the study heterogeneity that hinders effective comparison and conclusions. Whereas earlier studies needed to concentrate on specific regions of interest (ROIs), such as the hippocampus–amygdala complex, voxel-based morphometry techniques now permit the whole brain to be scrutinized without the need to nominate specific ROIs, benefiting further from a high degree of automation and reproducibility. Combining the key findings to date, the corticolimbic structures appear to have important implications in BD.[24]

Besides the well-established ventricular enlargement, BD has been associated with gray matter reduction in the anterior limbic region. Schizophrenic patients show greater dorsomedial and dorsolateral prefrontal cortex gray matter reductions than those with BD.[25] In both schizophrenia and BD, these changes first become evident in the high-risk period just before a patient becomes symptomatic and are expressed by the first episode. This suggests that the underlying disease process in BD either becomes activated at an early stage, or may not only be neurodegenerative but also include neurodevelopmental elements (although not to the same degree as in schizophrenia). Sanches and colleagues[26] reviewed reported changes in neurodevelopment in BD and found strong evidence of premorbid neurobehavioral changes, especially impairment of attention, as well as multiple neuroimaging differences between the children of bipolar parents and age-matched healthy controls.

The prefrontal cortex has been shown to be smaller in people with BD than in healthy subjects, although this is not invariably the case. Considering the prefrontal cortex as a single structural or functional entity is doubtless crude and probably incorrect, such that comparisons of specific regions may be of greater value. Initially reported by Drevets and colleagues, and subsequently replicated by independent groups, gray matter volume appears to be reduced in the subgenual prefrontal cortex (especially on the left).[27] Gray matter volume reductions across the whole left anterior cingulate cortex have also been reported, although negative studies and contrary findings have been published and reviewed. Individual variations in anatomy in this region hinder analysis and as gray matter loss may be associated with rapid cycling, illness characteristics are an important confounder between studies.

As pointed out before, the age of subjects may be crucial. Adolescents with BD demonstrate an acceleration of the volume reduction in the ventral prefrontal cortex that occurs during this stage of development. Adolescents with BD have also consistently been shown to have smaller amygdala volumes than healthy controls. In adults, this difference may be maintained, but normal and increased volumes of the amygdalae have also been reported. In contrast to the reductions seen in UPD, the volume of the hippocampus in BD appears to be preserved.

In brief, corticolimbic dysregulation theories of UPD assert that it is associated with ventral paralimbic overactivity and dorsal cortical inactivity. It has been hypothesized that in BD, a reduction in the modulating capacity of the prefrontal cortex releases limbic structures from inhibition,[28] increasing activity in the amygdala, for example. In proposing neural network dysfunction, subcortical nuclei and their interconnecting tracts may be implicated in addition to cortical regions (Fig. 40.2).[28] This assertion has attracted inconsistent and inconclusive support from structural image analysis studies examining these ROIs: the caudate may be of normal or increased volume, with reductions seen in elderly patients; nucleus accumbens volume may be reduced in young patients; and the putamen may be normal in euthymia but enlarged in mania. Summarizing gray matter involvement in BD, parts of the prefrontal and temporal cortex as well the claustrum seem to have reduced volumes.[30]

A confounding factor when assessing gray matter volume in BD as a potential biomarker is the fact that it may be affected by medication. Whereas antipsychotics have been associated with gray matter loss in schizophrenia,[16] there is substantial evidence that lithium is capable of increasing the gray matter volume of the prefrontal cortex, amygdala, and hippocampus, although the impact of such an increase on the long-term course of BD is unclear.

With regard to white matter, the volume of the corpus callosum is reduced by approximately 7% in bipolar patients; however, this finding is not consistent in all structural studies. Studies using diffusion tension imaging seem to support subtle fiber disruptions between hemispheres,[31] and microstructural abnormalities of the myelinated tracts of the prefrontal cortex have also been observed in BD. Further evidence of disruption to the integrity of interconnecting tracts comes from the consistent findings of an excess of deep white matter hyperintensities, with the frontal regions again most affected.[32] White matter hyperintensities, especially around the third ventricle, are a long-established finding in MRI studies of bipolar patients, and may indicate disrupted connectivity among cortical, subcortical, and limbic areas. However, these findings are not pathognomic for BD; they have also been reported for UPD, dementias, high blood pressure, and normal aging. Linking once more to the progression or consequences of the illness, the presence of white matter hyperintensities has been associated with poor prognosis.

A study from the Netherlands in pairs of twins concordant and discordant for BD found that overall brain volume reduction was, as expected, related to BD. However, white matter reduction was more genetically determined, whereas gray matter loss appeared related to environmental, probably epigenetic factors.[33]

Studies testing ROIs derived from structural MRI with magnetic resonance spectroscopic techniques have found reduced levels of *N*-acetyl aspartate (NAA) in the DLPFC, anterior cingulate, and hippocampus. It remains unclear whether these findings indicate neuronal loss or mitochondrial dysfunction.[34] Some medications

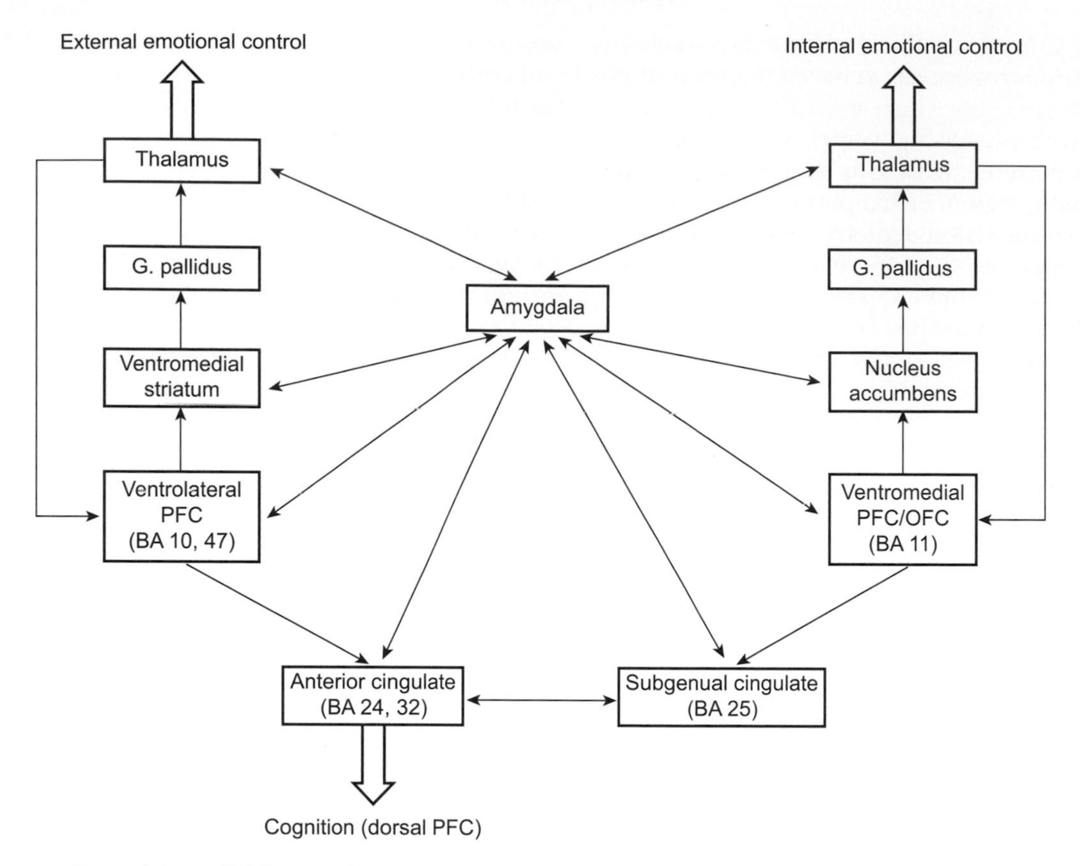

FIGURE 40.2 **A consensus model of neurocircuitry dysfunction in bipolar disorder.** G: globus; PFC: prefrontal cortex; OFC: orbitofrontal cortex; BA: Brodmann area. *Source: Strakowski.* The Bipolar Brain: Integrating Neuroimaging and Genetics. *Oxford University Press; 2012:253–274,*[29] *with permission.*

(lithium, valproate, and quetiapine) appear to normalize NAA levels. Choline levels may also be affected in bipolar patients, with increased concentration in the basal ganglia, suggesting a disturbed phospholipid metabolism.

NEUROBIOLOGICAL CHANGES IN BIPOLAR DISORDER

When considering potential disruptions to signaling on a cellular level, a distinction is made between the extracellular and the intracellular space. The mechanisms of synaptic activation and its potential disruption are not very different from what we have learned from UPD or schizophrenia; therefore, medications correcting for normalizing monoamines or their receptor function are also effective in the acute treatment of bipolar depression or mania. In support of this hypothesis, dysregulated glutamate and dopamine transporters have been reported in postmortem frontal cortex from both bipolar and schizophrenic patients. In addition, neuroendocrine substrates such as cortisol or thyroid hormones modulate neuronal activity. Effective long-term interventions, however, mainly target the intracellular space, namely the signal transduction cascade and mechanisms of cellular resilience. The ways in which these changes affect neuronal networks by changing the response to specific (e.g. emotional) stimuli can be captured by brain imaging techniques, such as positron emission tomography (PET), single-photon emission computed tomography, or fMRI.

The Extracellular Space

Neurotransmission

The catecholamine hypothesis of depression also holds true, in principle, for explaining mechanisms underlying acute bipolar depression. Increased cortical norepinephrine and decreased serotonin (5-HT) and dopamine turnover have been described in bipolar depressed patients.[35] A central serotonergic deficit, in both manic and depressed patients, has also been proposed, which may improve under treatment with valproate. However, other studies reported no change in either 5-HT_{2A} or 5-HT_{1A} receptor density and, regarding noradrenergic action, no change in β-receptor density. Dopaminergic hyperfunction is well accepted as an underlying cause of mania, and genetic aberrations of dopamine receptors and transporters have been demonstrated in bipolar patients. Decreased presynaptic dopamine function in the basal ganglia after successful treatment of mania with valproate has also been demonstrated by PET.[36]

As BD is heterogeneous, more precise results may be expected in a homogeneous group of patients. Therefore, several studies have investigated monoaminergic imbalance in 48 hour rapid cycling patients. Characteristic and reproducible changes in norepinephrine and metanephrine excretion were closely related to mood swings.[37] In general, the urinary excretion of norepinephrine and metanephrine was increased on both manic and depressed days, with higher values during mania, which was generally ameliorated after successful valproate treatment. Swann and colleagues reported on increased urinary excretion of 3-methoxy-4-hydroxyphenylethyleneglycol (MHPG), a metabolite of norepinephrine, in mixed states.[38] Since tricyclic noradrenergic-acting antidepressants, in particular, are capable of inducing a switch from pure depression into mixed states or mania, and may also induce a rapid cycling course, norepinephrine and its metabolites may play at least a strong modulating role in these conditions.

Finally, glutamate and its synaptic receptors appear to be critically involved not only in mediating learning but also in cell death, if overstimulated. NMDA receptors are a subtype of ionotropic glutamate receptor. Inappropriate levels of Ca^{2+} influx through the NMDA receptor can contribute to neuronal loss in acute trauma such as ischemia and traumatic brain injury, certain neurodegenerative disorders such as Huntington disease, and psychiatric disorders such as schizophrenia and probably BD. However, normal physiological patterns of synaptic NMDA receptor activity can promote neuroprotection against apoptotic, oxidative, and excitotoxic insults. NMDA receptor blockade can promote neuronal death outright or render neurons vulnerable to secondary trauma, depending on the nature of the activity. Thus, both overactivation and blockade of NMDA receptors may promote a loss of connectivity or even neuronal death. Under physiological conditions, the glutamatergic system is under inhibitory control of GABAergic interneurons. Disruption of this local circuitry at the glutamatergic synapse connecting with the GABAergic interneuron (e.g. by low-dose NMDA antagonists) can lead to overactivity of glutamatergic neurons, which may cause seizure activity or, perhaps, psychosis in severe mental disorders, including BD.

Neuroendocrinology

The HPA axis plays an important role as the central physiological adaptive mechanism to stress. This axis originates from paraventricular hypothalamic neurons releasing the neuropeptide corticotropin-releasing factor (CRF) into the portal veins. CRF-containing neurons also exist in other limbic and brainstem areas, illustrating the complex circuitry and feedback mechanisms between limbic brain areas and neuroendocrine response organs. They also project to the amygdala and noradrenergic nuclei. CRF acts directly as a neurotransmitter as well as a hormone.

Neuroendocrine dysregulation of the HPA and the hypothalamic–pituitary–thyroid (HPT) axis is a common finding in affective disorders. Increased pituitary volume, more pronounced in females than males, has been described in BD. The functional implication of this finding remains unclear as no correlation with the duration, severity, or pattern of the illness has been established.

Since the early 1950s, the administration of HPA exogenous hormones has been reported to elicit psychiatric symptoms in some patients with no pre-existing psychiatric conditions. In particular, adrenocorticotropic hormone (ACTH) and cortisone have been associated with mood elevation. High-dose (>80 mg/day) glucocorticoids elevate mood in patients with multiple sclerosis, ophthalmological diseases, and asthma, and also in healthy volunteers. Remitted bipolar patients appear to be even more vulnerable to a treatment-emerging mania or rapid cycling when exposed to glucocorticoids. However, constant hypercortisolemia in patients with a current acute mood episode may lead not to mood swings, but rather to persistence of aberrant (manic or depressed) mood. Endocrine dysfunction of the HPA axis in mood disorders is evident from hypersecretion of cortisol in both unipolar and bipolar depression, and from the failure of depressed patients to suppress secretion of cortisol with the dexamethasone suppression test. The more sensitive combined dexamethasone suppression and corticotropin-releasing hormone (DEX/CRH) challenge test has enabled detection of subtle impairments of central glucocorticoid receptor signaling in patients with depression and BD; these findings have been replicated to a lesser extent across a broad range of psychiatric illnesses. The causal chain may thus originate in the limbic area, projecting downstream towards the hypothalamus. HPA axis dysfunction in affective disorders may arise from decreased function or expression of brain glucocorticoid receptors, resulting in glucocorticoid resistance and subsequent elevation of circulating cortisol as a compensatory mechanism.[39]

Healthy first degree relatives of depressed patients display impaired glucocorticoid receptor signaling after DEX/CRH administration, suggesting that glucocorticoid dysfunction may be a familial trait marker for affective disorder. Evidence for state- and trait-related HPA axis dysfunction in BD using the combined DEX/CRH test is shown by the consistently enhanced cortisol response in patients with acute mania or bipolar depression, which persists on a longitudinal follow-up of up to 6 months' remission. Glucocorticoid receptor dysfunction, measured with the dexamethasone suppression test, correlates with declarative and working memory deficits in euthymic BD patients, suggesting that trait-related glucocorticoid receptor dysfunction may be etiologically linked to the cognitive dysfunction frequently reported even in remitted BD.[40]

Combined immunological and endocrinological investigations have also revealed the presence of T-cell steroid resistance as a potential trait marker in BD, linking BD to immune system dysregulation.

Converging evidence from small studies with rapid cyclers or ultrarapid cyclers with BD suggests that HPA hyperactivity is critical for the switch from mania to depression in most of these patients; however, the role of the HPA axis in switching from depression to mania is more controversial. The already cited case report by Juckel and colleagues[37] not only looked into biogenic amines, but also investigated changes in the HPA axis in a 48 hour rapid cycling patient. Dramatic changes in growth hormone and cortisol were seen, with peaks during mania and troughs during depression. Valproate treatment ameliorated this rollercoaster of the HPA axis.

Transgenic mice overexpressing glucocorticoid receptors in the forebrain display enhanced depressive-like behaviors and increased sensitization to cocaine and antidepressants. They also have a wider range of reactivity to stimuli that trigger both negative and positive emotional responses, which may be relevant for the neurobiology of the switching process in BD.

HPT axis dysfunction had been implicated in several studies, especially in rapid cycling patients. Decreased nocturnal thyroid-stimulating hormone secretion, as well as hypothyroid metabolism, has been reported by several authors in rapid cyclers (for a review, see Joffe and Sokolov[41]). Preliminary clinical data suggest that thyroxine may be a helpful augmentation strategy in refractory rapid cycling patients and in women with bipolar depression. The reason for a genuine thyroid dysfunction and its relation to mood states in BD remains unclear. In the largest trial looking for thyroid antibodies in different mood states in 226 bipolar patients (including 28 patients with mixed states), no difference was found in the number of thyroid antibody-positive patients between euthymic, depressed, and mixed patients.

Besides those cited studies, there are several case reports on changes to neuroendocrinological abnormalities in rapid cycling bipolar patients and their amelioration with remission. However, none of these reports or studies solve a general problem; that is, causality. Are hormonal changes the cause or are they simply effects of a different mood state? Or are they completely independent of mood and represent a variable triggered by another unknown, underlying mechanism which also affects the mood state? Not only the evidence but especially the causality of neuroendocrine alterations needs further research.

The Intracellular Compartment

Second Messenger Pathways

Cellular signal transduction describes the process of transmitting information from the extracellular to the intracellular space. This includes short-term and transient functional changes (e.g. phosphorylation of an ion channel), as well as enduring changes (e.g. of gene transcription). Cell-membrane associated G proteins (guanyl nucleotide binding proteins) hold a key position in signal transduction, mediating between extracellular receptors and intracellular effector proteins. G proteins can be subclassified into four main categories (Gs, Gi, Gq, and G1,2) with circumscribed specificity for effector proteins. An estimated 80% of neurotransmitters, neuromodulators, and hormones act via G proteins. G proteins not only are coupled to extracellular receptors, but also play a role in regulating the opening and closing of transmembrane ion channels.

G proteins consist of α- and βγ-subunits. When a membrane receptor is activated, for instance by a neurotransmitter, the G-protein complex dissociates, with the result that the free α- and βγ-subunits activate effector proteins such as adenylate cyclase or phospholipase. Hydrolysis of GTP to GDP by GTPase intrinsic to the α-subunit terminates the process, and the G-protein subunits return to their inactivated heterotrimeric state. This inactivation process is enhanced by RGS (regulators of G-protein mediated signal transduction) proteins.

Second messengers are usually small, water-soluble molecules such as cyclic AMP (cAMP) or inositol phosphate. An exception is diacylglycerine, which is lipophilic and membrane bound. Second messengers activate different protein kinases which, in turn, phosphorylate several target proteins and, by this, change their functional state. Important targets of activated protein kinases include ion channels, cytoskeleton proteins, or transcription-factor regulating genes. The end result is that a weak extracellular stimulus has been potentiated by many times, and both short-term and lasting changes have been induced by the second messenger cascade. Subtle dysregulation of this signal transduction pathway may have deleterious effects on neuronal functioning and cell survival.

Little attention had been paid to the effect of changes to the intracellular proton equilibrium (pH) in bipolar patients. Moderate changes in the intracellular pH have a strong effect on both receptor sensitivity and second messenger systems, and even small fluctuations may have a strong impact on the expression of mood. Thus, changes in the intracellular pH may be potential triggers of mood swings. A decrease in the intracellular pH in the frontal lobe has been described in drug-free bipolar patients, suggesting aberrations of the proton equilibrium in BD.

Several medications used in BD, especially mood stabilizers, affect intracellular signal transduction rather than synaptic receptors. The exponential growth in knowledge in the past few years becomes apparent upon typing the search terms "signal transduction" and "bipolar" in scientific databases. Lithium, in particular, has a multitude of intracellular targets, including cAMP, glycogen synthase kinase-3β (GSK-3β), and phosphatidylinositol phosphatase, and numerous other downstream proteins.[42] Randomized controlled studies with tamoxifen have shown that clinically effective treatment of mania can be achieved by selectively inhibiting protein kinase C. Various effects on signal transduction have also been discovered for valproate and, to a lesser extent, carbamazepine. Since a detailed review of the intracellular signaling cascade and its modulation by medication is beyond the scope of this chapter, the reader is referred to comprehensive reviews on this topic.[43,44]

Probably even more important, at least for the long-term outcome of BD, is the action of mood stabilizers, especially lithium, on neuronal resilience and regeneration, including neurogenesis. As with schizophrenia, neuroprogression occurs in UPD and BD. Cellular resilience mechanisms, especially mitochondrial and endoplasmic reticulum-related responses to stress, seem to be less efficient at later stages of BD. GSK-3β and the extracellular signal-regulated kinase pathway seem to play an important role in resilience to stressors in their widest sense. Stress-induced remodeling of the hippocampus, prefrontal cortex, and amygdala is related to changes in the levels of BDNF, which has been shown to act as a trophic factor facilitating the survival of existing and newly born neurons. BDNF is reduced in unmedicated patients with both depression and BD, and oxidative and nitrosative stress markers are increased.

Depression has been associated with lowered antioxidant status and lowered antioxidant enzyme activity, which may lead to damage to fatty acids, proteins, and DNA. Redox dysregulation is noted in schizophrenia, UPD, and BD, including reduced glutathione defenses and increased oxidative damage, indexed by changes such as lipid peroxidation. Antidepressants and some atypical antipsychotics may restore BDNF transcription, thus reinstalling cellular protection against oxidative stress.

Again, the complexity of these mechanisms and how they are affected in BD exceeds what can be comprehensively dealt with in this chapter, but has been extensively reviewed elsewhere.[45]

Calcium: A Key Player in Bipolar Disorder?

Mobilization of calcium is a key event in presynaptic and postsynaptic signaling, and also in lasting neuronal changes such as long-term potentiation. Subtle dysregulation of Ca^{2+} signals has been linked to major diseases in humans, including cardiac disease, schizophrenia, BD, and Alzheimer disease.[46] Massive disinhibition of calcium fluxes into the cell or release from intracellular stores triggers cell death. Regarding potential dysregulation of signal transduction as well as the mechanisms underlying cell death in BD, the vast majority of these processes is Ca^{2+} dependent.

Mood-stabilizing agents may directly or indirectly regulate intracellular Ca^{2+}. A well-known target of lithium is the inositol phospholipid pathway, which shows increased sensitivity in bipolar patients. Inositol 1,4,5-triphosphate (IP3) synthesis mobilizes intracellular calcium stores in close proximity to their calcium-dependent effector proteins. Lithium inhibits IP3. Valproate is capable of inhibiting the synthesis of inositol monophosphate. This also means a reduction in the activation rate of Ca^{2+} calmodulin kinase II, an enzyme critically involved in long-term synaptic changes, and of myristoylated alanine-rich C kinase substrate (MARCKS), implicated in synaptic neurotransmission and cytoskeletal restructuring, thus inhibiting structural changes to neurons in the course of the disease.

Different mechanisms have been implied in the therapeutic action of carbamazepine. With respect to calcium, carbamazepine exerts strong calcium antagonistic properties *in vitro* by blocking L-type calcium channels, whereas valproate inhibits another voltage-dependent calcium channel, the T channel. Lamotrigine also exerts part of its action by calcium antagonistic effects on L-, N-, and P-type channels.

Besides inhibiting ion channels, valproate acts on the chaperones GRP-78, GRP-94, and calreticulin. These endoplasmic reticulum stress proteins bind calcium and protect against cell death. Chronic treatment with valproate increases the synthesis of all three chaperones in rat C6 glioma cells, thus contributing to calcium homeostasis in the endoplasmic reticulum.

In summary, all currently used effective mood stabilizers have either direct or indirect calcium antagonistic properties. A review by Hollister and Trevino[47] listed 61, partially successful, reports on trials of calcium antagonists in BD. In addition, a randomized study of the L-type calcium channel blocker nimodipine showed efficacy in previously refractory rapid cycling patients.

Bipolar Disorder: An Immunological Disorder?

Immunological aberrations have been implicated in both schizophrenia and BD, based on observations on the neuronal level and by measuring peripheral inflammation markers. Particular viruses or parasites have been thought to have a role in the pathogenesis of schizophrenia or BD but, as most studies were based on serology that essentially detects an "immunological scar" represented by the presence of specific immunoglobulin G antibody, the period of infection is debated and may occur at variable times. GWAS and epigenetic findings linked to inflammation support such hypotheses, although these links seem to be unspecific and not exclusive to BD.

Studies have found increased activity of interleukin 1 (IL-1), IL-6, tumor necrosis factor-α (TNF-α), prostaglandin E_2, chemokine ligand-2, von Willebrand factor, and osteoprotegerin in both BD and schizophrenia; however, the chain of causality remains unclear. Hope and colleagues demonstrated that IL-1 receptor agonist and soluble TNF receptor-1, in particular, correlate with both general disease severity and psychotic features across these disorders.[48]

An immunological and neuroprogressive model of BD would be complementary because a proinflammatory state is one potential cause of excitotoxicity. Peripheral inflammatory signals activate microglia in the brain, inducing an inflammatory cascade of cytokines and free radicals. Cytokines, and reactive oxygen and nitrogen species have a direct toxic apoptotic effect on oligodendrocytes. Through the loss of oligodendrocytes, oxidative stress can lead to demyelination. Such a process may account for the reduction in oligodendroglia found postmortem in the prefrontal cortex in BD.[49]

The therapeutic efficacy of lithium and valproate in BD may be partly attributable to their anti-inflammatory properties.[50] The mood-stabilizing effect of inflammatory intracellular cascades involves proinflammatory eicosanoids, cytokines, and acute-phase proteins, which are elevated during acute mood episodes. In animal models of stroke, lithium suppresses microglial activation and attenuates overexpression of proinflammatory cytokines and chemokines. Similarly, valproate treatment markedly reduces the number of both activated microglia and infiltrating monocytes/macrophages, and suppresses ischemia-induced upregulation of proinflammatory factors, inducible nitric oxide synthase, and COX-2. Therapeutic trials with COX-2 inhibitors indicate moderate benefit in schizophrenia. A trial of a COX-1 inhibitor in BD underway, and due to be completed in late 2014.

In summary, there is reasonable evidence that links BD to inflammatory processes. However, inflammation appears to be more a feature of severe mental illness in general than specific to BD, and the chain of causality needs further evaluation.

Brain Activity Depicted by Functional Neuroimaging

The episodic nature of BD and the ostensibly polar extremes of mania and depression led to the wide acceptance of a functional model of disturbed activity within brain regions postulated to return to normal with time or treatment. It is, however, increasingly recognized that this pattern may not accurately describe the nature of the condition; rather, dysfunction may be detectable during euthymia and many patients experience chronic illness, mixed presentations, rapid cycling, and suboptimal responses to treatment. Underlying the syndromal presentations, derangements in the processes of neural transmission have been proposed, often in a dichotomous manner to mirror the polar nature of the illness; for example, mania has been explored in terms of increased dopaminergic activity and depression discussed in terms of reduced activity. There is merit to such a stance and much evidence to support these assertions, but factors such as medications, psychosis, physical activity, and stress axis activation, all of which can span episodes or emerge during both mania and depression, confound the investigation and interpretation of potential neurobiological abnormalities.[51]

Imaging studies investigating the function of the brain in BD have been reviewed, and differences in imaging modalities, task and rest conditions, mental states, and medication effects have emerged as important confounders. Given these confounders, a synthesis of the findings permits only the broadest of conclusions to be reached; the testing of specific hypotheses remains the remit of individual studies. In general, the areas implicated in structural studies show evidence of dysfunction, although not all reports are in line with this.[52] On the basis of the findings from studies using the blood oxygen level-dependent (BOLD) fMRI technique, it has been argued that during emotional and cognitive tasks, mania is predominantly associated with reduced activation in the ventral prefrontal cortex, whereas in bipolar depression, increased activation is observed. Laterality effects may be of some importance, with abnormalities in mania seen in the right hemisphere and depression in the left. With regard to corticolimbic dysregulation, BOLD fMRI activation of the amygdala at rest and during emotional tasks is typically greater in patients with BD than in controls, and the increase in BOLD signal occurs in both depression and mania.

It would seem reasonable to conclude that the imaging studies support the notion that in mania, frontal inhibition is lost while in depression the reverse occurs, borne out as a disruption to normal emotional, cognitive, and volitional processes. The certainty, as well as the direction of the changes in structure and function observed, requires discussion in light of current theories explaining the neurophysiological basis of MRI. With respect to BD, two areas warrant attention: the effects of medication and the baseline state of the brain during functional studies (BOLD fMRI in particular). Although BOLD fMRI studies seem to be considered the current top standard to gather reliable information on the functional state of the brain, they are not free of potential traps.[53,54]

Taken together, neuroimaging and functional imaging studies point towards state-dependent changes and, with less certainty, neurodegenerative changes in limbic–temporal areas in BD, although most of these data should be considered as preliminary. These observations could be the gross presentation of underlying kindling processes on the cellular level with the consequences of lasting structural changes.

The amygdala has been attributed a key role in mood and anxiety disorders. This evolutionarily ancient structure is responsible for generating fight-or-flight responses to threats. The amygdala is heavily innervated by the prefrontal cortex, which probably serves to modulate the fight-or-flight response into a more complex emotional behavior that defines human interactions. Along with other nearby medial temporal structures (including the paralimbic cortex and hippocampus), the amygdala is responsible for the perception and regulation of emotions.

Disruption in early development within brain networks that modulate emotional behavior (Fig. 40.2) leads to decreased connectivity between ventral prefrontal networks and limbic brain structures, especially the amygdala. This loss of connectivity is associated with abnormal functional responses of emotional networks to various cognitive and emotional tasks in imaging studies, as well as abnormal development of the component brain regions (e.g. failure of the amygdala to mature normally and disruption of prefrontal modulation of limbic structures). Dysregulation of the limbic brain then leads to mood instability. In the absence of healthy prefrontal–striatal–pallidal–thalamic–limbic brain networks that can restore emotional homeostasis, bipolar patients may develop extreme mood states and switching among mood states. During euthymia, recovery of prefrontal function, along with compensation from other brain regions, temporarily restores emotional homeostasis; nonetheless, the underlying functional neuroanatomic abnormalities leave the subject at risk of disruption of this fragile homeostasis under even minor stress in the face of "stably unstable" prefrontal–striatal–thalamic–amygdala mood networks. In brief, developmental failure to establish healthy ventral prefrontal–amygdala networks may underlie the onset of mania and, ultimately, with progressive changes throughout these networks over time, a bipolar course of illness.[55]

Animal Models

In patients, mania is characterized by elated, expansive, or irritable mood together with features such as increased drive, high-risk hedonistic behavior, hyperactivity, pressured speech, flight of ideas, grandiosity, and reduced requirements for sleep; hypomania is less severe according to symptom intensity, duration, and impact on functioning. Depression, in addition to persistently low mood, comprises anhedonia, reduced energy levels, psychomotor retardation, hopelessness, guilt, suicidal ideation, loss of appetite, and disrupted sleep. Animal models can never capture all aspects of the complex nature of the disorder, only very selected symptoms.

Animal models are used to reproduce aspects of the behavioral disorder that can be studied, as a test bed for discovering novel drugs that may treat the disorder, and as procedures through which new molecular targets can be identified for subsequent drug discovery and development. Three aspects are important when judging the suitability of an animal model: the face validity, which refers to the similarity in behavior between the model and the disorder; the construct validity, which evaluates the correlation of the model with molecular changes described for the pathophysiology of the disorder; and the predictive validity, which evaluates the responsiveness of the model to the drugs used to treat the disorder. Given the limited knowledge of the pathophysiology of BD, researchers should be aware that especially the construct validity of animal models for BD may be low.

Behaviors in mania, beyond traditional rating scales, need to be identified that can more reliably be modeled in animals. An example is prepulse inhibition (PPI). Inhibitory abnormalities in mania are observed not only in higher cortical functions but also in other domains such as in sensorimotor gating, quantified as deficits in PPI. PPI is a cross-species measure where a response to a startling stimulus (e.g. a loud noise) is inhibited by the prior presentation of a low-intensity prepulse. When tested, manic patients had significantly lower PPI than did control subjects, as well as less startle habituation.

Traditionally, increased motor activity remains the mainstay of animal models of mania; however, this measure has also been used as a proxy for other neuropsychiatric disorders such as schizophrenia and attention deficit/hyperactivity disorder. Thus, its face validity is low; nevertheless, most models using increased motor activity still have a reasonable predictive validity when drug treatments are tested.

The most frequently used motor activity model in rodents is hyperactivity induced by a psychostimulant, in most instances amphetamine, but also the ATPase inhibitor ouabain or the dopamine agonist quinpirole. A frequently used method to assess activity is by recording the number of entries into the arms of a maze during a given period. After baseline measurement, hyperactivity is induced by subcutaneous injection of a mixture of amphetamine alone or in combination with chlordiazepoxide before retesting. Addition of established antimanic agents, such as lithium, valproate, or carbamazepine, to the injection is able to counteract amphetamine-induced hyperlocomotion. However, this model has several limitations: amphetamine hyperactivity has been interpreted as a model for a number of distinct disorders besides BD (including schizophrenia, drug abuse, and tardive dyskinesia); mania is characterized by a broad set of symptoms that may not always include motor hyperactivity; and most models use acute doses, while mania is a chronic disease with long-term alterations in behavior.[56] Amphetamine-induced behavioral changes cannot reflect a long-lasting, recurrent disorder such as mania. In addition, amphetamine may produce other systemic and pharmacological effects that interfere with measurements or antimanic agents tested. Therefore, models are needed that do not rely on extrinsic injection of stimulants.

Increased locomotion in rodents can also be induced by environmental manipulations, such as sleep deprivation, social isolation of rodents paired with the resident–intruder test, or by surgical brain lesions targeting the hippocampus and amygdala. A "genuine" mania model has also been assessed using knockout or knockdown mice strains [e.g. for the glutamate-6 receptor ($GluR_6$) or the dopamine transporter (DAT) gene].

$GluR_6$ knockout mice exhibit hyperactivity, aggression, and reduced anxiety, which can be reversed with chronic lithium treatment. Similar to $GluR_6$, the DAT has been linked to mania across several genetic linkage studies. Lower levels of DAT have been reported in BD patients. DAT knockout mice can serve as a putative model of mania because these mice exhibit increased motor activity and impaired PPI. Given that the poor physical condition of these mice limits their usefulness in assessing novel therapeutics, a line of DAT knockdown mice was developed that exhibit only 10% DAT compared with littermates. DAT knockdown mice showed increased activity and risk preference as well as hedonia-like behavior in terms of food motivation behavior, and thus mimic several features that are observed in mania in terms of behavior, cognition, and putative neurobiological abnormalities.[56]

Research is also being conducted in knockout mice lacking circadian genes such as d-box binding protein or CLOCK, which exhibit increased activity, especially when paired with stress, sleep deprivation, or other environmental stimuli. This clearly resembles the close interaction of environmental factors and (potential) genetic load in human mania; however, more work is needed to establish these rodent models.

No animal model can be considered specific for BD. All animal models used to explore new drugs for potential use in BD are identical to those used in UPD (see Chapter 43). Similarly to models used in mania, they are based on pharmacological challenges (e.g. reserpine, apomorphine), responses to stress (e.g. forced swimming test, resident–intruder test, early separation) or surgical interventions (e.g. olfactory bulbectomy).

The greatest challenge, however, is to develop an integrative model explaining the repetitive course of BD, with increasing vulnerability to relapse, with free intervals between episodes becoming increasingly shorter, and with shifts in symptomatology over time.

Sensitization and Kindling: Models Explaining Recurrence of Bipolar Disorder?

Kraepelin noticed that a marked psychosocial stressor usually proceeds the first affective episode in BD, whereas subsequent episodes show minor or even absent notable live events. The frequency of episodes tends to increase, in some patients to autonomous rapid cycling, with decreasing efficacy of mood-stabilizing drugs. Neuroprogresssion seems to increase with time and number of episodes, and staging models of BD, with an emphasis on the search for biomarkers associated with illness progression, have been proposed.[57]

Post and colleagues developed the model of cocaine-induced behavioral sensitization (CIBS).[58] Cocaine administration causes hyperlocomotion in rats and hypomanic-like symptoms in humans. Repeated cocaine administration, however, may cause a shift in symptomatology towards signs of dysphoric mania or even paranoid symptoms. Lesion experiments in the amygdala show that CIBS involves different neuromodulatory changes depending on the duration and frequency of cocaine administration. Thus, not only the symptomatology, but also the neuronal pathways involved can shift, becoming independent of a direct action on the amygdala. Cocaine can also influence neuromodulators in a similar fashion to stress, leading to increases in CRF, ACTH, cortisol, cytokines, catecholamines, and indoleamines.

CIBS may be a useful model to study acute events and long-term changes in symptomatology caused by recurrent episodes of affective disorders. However, to explain the aspect of sequential unfolding of episodes with increasing autonomy from stressors, the amygdala-kindled rat appears to be a more suitable model. Although epileptic seizures may have some mechanisms in common with affective disorders, such as increased transmembrane calcium fluxes, they are clearly two different conditions. However, the rough anatomical substrate is similar, as the amygdala plays a key role in both disorders. Repeated electrical stimulation of the basolateral amygdala decreases the threshold for epileptic seizures, often leading to spontaneous epileptic activity. The correlate on the synaptic level is an increase in both NMDA and non-NMDA receptor-mediated glutamatergic transmission with a parallel decrease in inhibitory GABAergic transmission. At the level of expression of early genes and neuropeptides, an increase in c-fos and thyrotropin-releasing hormone mRNA has been observed. With full manifestation of seizures, these changes on the synaptic level and substrate expression also involve the contralateral, non-stimulated amygdala complex. The assumption can be made that, similar to electrical kindling, recurrent affective episodes cause analogous long-term changes in neuronal networks, thus lowering the threshold for any consecutive episode. This hypothesis has been backed up by a clinical study.[59] Despite having previously had an overall comparable total number of episodes, patients who had a pattern of close periodicity of episodes showed an increased risk of relapse during follow-up, interpreted as an indicator of a previous kindling process.

Some mechanisms have been pointed out to explain how mood stabilizers may prevent such a kindling process. Lithium inhibits the activity of MARCKS and calmodulin kinase II, thus inhibiting structural changes in the course of the disease, implying an antikindling potency. Both carbamazepine and valproate also exert antikindling effects, which are probably linked to their calcium antagonistic effects.

BEHAVIORAL MARKERS IN BIPOLAR DISORDER

Emotional Processing

Intact perception and experience of emotion is vital for survival in any social environment, in humans and other animals. Emotional processing involves the identification, appraisal, understanding, and regulation of generated emotions. It requires the ability to discriminate between and make sense of external emotional stimuli, and entails regulation of one's own affective responses. Various tests, such as the emotional Stroop, affective go/no-go, and dot-probe tasks, have been developed to test emotional processing either alone or in combination with fMRI. Areas of emotional processing commonly tested include affective attention, emotional memory, facial emotion recognition, mood induction, and auditory processing.

It has been hypothesized that bipolar patients may have difficulties in emotional processing not only in acute episodes, but also while euthymic. Differences in emotional processing between BD and healthy controls may thus be a trait marker of BD. This view is supported by the fact that not only children with diagnosed BD but also children at risk have impairment in their capacity to recognize emotions in adults' and children's faces.

Mercer and Becerra[60] reviewed the literature on emotional processing in euthymic bipolar patients. Overall, studies indicate that some BD patients do retain their emotional processing deficits during euthymia. Twenty-two out of 34 studies reviewed reported some form of emotional processing deficits in euthymic bipolar patients. Further research is needed to reach clear conclusions. This is of particular importance as the ability to correctly process emotions is directly linked to daily functioning in professional and social life, areas where bipolar patients show deficits.

Subsequent to problems in emotional processing are problems in emotional regulation. Gross and Thompson proposed that "emotion regulation consists of extrinsic and intrinsic processes responsible for monitoring, evaluating, and modifying emotional reactions, especially their intensive and temporal features, to accomplish one's goals". These processes may be "automatic or controlled, conscious or unconscious, and may have their effects at one or more points in the emotion generative process".[61] Phillips and colleagues developed a neural model of emotion regulation that includes neural systems implicated in different voluntary and automatic emotion-regulatory subprocesses.[62]

In their original neural model of emotion processing, Phillips and co-workers highlighted that two systems in the prefrontal cortex may be implicated in emotion regulation: a lateral prefrontal cortical system, including the dorsolateral and ventrolateral prefrontal cortex, which is neocortical in origin and operates by a feedback mechanism; and a medial prefrontal cortical system, including the OFC, subgenual and rostral anterior cingulate gyrus, and mediodorsal prefrontal cortex, which is paleocortical in origin and operates by a feedforward mechanism. Further research suggests that the former neural system may be involved in voluntary subprocesses, whereas the latter neural system may subserve automatic subprocesses. The mediodorsal prefrontal cortex in particular may use feedforward inputs from the OFC, as signals of internal states, to select appropriate behaviors during automatic cognitive change paradigms. These two neural systems may be activated concurrently during regulation of emotional states and behaviors that are initially generated by orienting and emotion-perceptual processes in the amygdala, ventral striatum, and thalamus.

How does this model link to the distinctive changes observed in BD? Whereas findings regarding function in lateral and dorsal prefrontal cortical regions implicated in voluntary emotion regulation in adult BD are inconsistent, studies on automatic emotion regulation in adult BD have produced more reliable findings regarding functional and structural changes in these neural regions.

Relative to healthy controls, studies using automatic attentional control paradigms show reduced activity, predominantly in the left-sided ventromedial prefrontal cortex, in BD. Studies using automatic emotion regulation paradigms also show reduced activity, predominantly in left-sided ventromedial prefrontal cortical regions implicated in automatic emotion regulation, during both remission and mania. These functional neuroimaging findings are paralleled by structural neuroimaging findings in BD showing gray matter structural changes in the left OFC, and abnormal integrity and number of white matter fibers connecting the left OFC and subcortical limbic regions implicated in emotional processing.

Increased recruitment of DLPFC in BD compared with healthy adults has been reported during an automatic attentional control paradigm. This suggests that bipolar patients need to employ more effortful, voluntary, rather than automatic, regulatory control systems during the performance of otherwise automatic control tasks.[62]

The combination of these functional and structural neural abnormalities may underlie mood instability in BD. The significance of the laterality of these findings remains unclear, but may suggest a role for the left hemisphere, previously linked with the perception of positive emotion, in the pathophysiology of BD.

Cognition

More than any other feature of BD, cognitive dysfunction may be most closely linked to functional outcome. Cognitive impairment may affect everyday activities, impair the patient's ability to work, and delay re-employment; therefore, there is a growing need to elucidate not only neurocognitive impairment, but also its effects on functional outcome and therapeutic measurements preventing cognitive decline in BD.

Mania is associated with significant neurocognitive impairment across a broad array of domains.[63] While executive dysfunction appears to be the most prominent aspect of mania, bipolar patients in manic states also exhibit deficits in vigilance, working memory, verbal fluency, inhibitory control, and verbal recall and recognition. Most clinical trials of treatments for mania have focused exclusively on the amelioration of mood symptoms, and relatively little work has been done to examine the efficacy of various therapies on cognitive dysfunction linked with the disorder, despite the close correlation of cognitive disruption with functional outcome. The number of manic episodes predicts poor cognitive performance, suggesting that frequent recurrences of mania have a long-term neuropsychological impact.[64]

Patients with bipolar depression also show deficits in executive functioning, memory, attention and, differently from hypomanic or euthymic patients, fine motor skills. The number and severity of depressive episodes in BD have been linked with poorer long-term cognitive and functional outcome, although the association is not as clear as with mania. Lasting subdepressive symptoms may worsen executive function over time, even in the absence of further full-blown episodes.

Considering the long-term course of cognitive impairment in BD, it may be important to distinguish between potential neurodevelopmental and neuroprogressive contributions, with the latter being an easier target for preventive treatments. Compared with schizophrenia, BD appears to present smaller magnitudes of neurocognitive impairment, even during the first episode of the illness, and may relate to the more neurodevelopmental than neuroprogressive nature of schizophrenia. Some studies have reported differences in premorbid intelligence quotient between patients with schizophrenia and those with BD, with the latter group showing higher scores. In contrast to BD, schizophrenic patients show cognitive impairment even before the onset of illness. However, the overall profile of cognitive impairment as depicted by standard tests is similar between bipolar and schizophrenic patients; the difference appears to be rather quantitative than qualitative.[65]

Executive functioning, inhibition, processing speed, and verbal memory are impaired in euthymic bipolar patients. These cognitive deficits remained stable, on average, throughout a 6-year follow-up, but had enduring negative effects on the psychosocial adaptation of patients. A probable exception is executive functioning. Torrent and colleagues report that cognitive impairment in euthymic bipolar patients remains stable over time on most measures except for a worsening of executive measures, which are associated with the duration of illness and subdepressive symptoms.[66]

Biological Rhythms

Several endogenous rhythms dictate people's daily activity levels and, perhaps, also their mood. These include brief, often 24 hour rhythms, such as the secretion of cortisol or melatonin, or longer cycles, such as the menstrual cycle. External factors can modulate or stop these cycles; for example, daylight and sleep deprivation alter the diurnal secretion of hormones; and pregnancy, postmenopausal age, and contraceptive pills affect the menstrual cycle.

Circadian Rhythmicity

Vulnerability to bipolar mood episodes increases with disruption of diurnal cyclicity, for example lack of sleep or travel across time zones. As a result of the manifestation of a mood episode, both mania and depression, diurnal cycles may then become unresponsive. An example is uncoupled HPA axis activity with hypercortisolemia paired with a blunted response in the DEX/CRH test. Disruption of the sleep–wake cycle also persists as a consequence of dysfunctional cellular circadian clocks. These are located not only in the suprachiasmatic nucleus (SCN), the brain's primary circadian pacemaker, but also throughout the brain and peripheral tissues.

In BD, but also UPD, patients, defects have been found in SCN-dependent rhythms of body temperature and melatonin release.[67] However, the SCN may not be particularly relevant to mood regulation, whereas the lateral habenula, ventral tegmentum, and hippocampus, which also contain cellular clocks, have established roles in this regard. Dysfunction in these non-SCN clocks may contribute directly to the pathophysiology of both unipolar and bipolar mood disorders. Resynchronizing circadian clocks, for example through light or darkness therapy, social rhythm therapy, or administering antidepressant medication with a melatonergic mode of action (e.g. agomelatine), may be indicated as additional therapeutic approaches.

Seasonal Rhythmicity

Recurrent depression, manifesting as seasonal affective disorder, is well recognized. Seasonality of BD, however, has attracted less research, and the results are ambiguous. A multisite Canadian study in more than 400 patients was conducted to test the hypothesis that, on average, depressive symptoms peak in autumn (fall) and winter, and hypomanic and manic symptoms peak in spring and summer. In this prospective study, the largest conducted to date, no evidence of systematic seasonal variation in symptoms was found in the sample as a whole.[68]

Nevertheless, there are compelling case reports on seasonal mania, and a GWAS has identified *NF1A* as a possible susceptibility gene for a seasonal pattern of mania in BD. A seasonal pattern may apply in only a very small subgroup of patients.

Menstrual Cycle

The influence of gonadal hormones on the manifestation of major psychiatric disorders is apparent when it comes to drastic changes, such as with postpartum depression, mania, or psychosis. Shifts in hormone levels are more subtle when following the normal menstrual cycle. Mood variations, often experienced as part of the premenstrual syndrome (PMS), are frequent, and approximately 3% of women develop a more severe form of PMS, named premenstrual dysphoric syndrome (PMDD), which DSM-5 includes under depressive disorders. Women with PMDD are more likely to experience mood disorders. Estrogen and progesterone fluctuation across the menstrual cycle can modulate affective symptoms through their actions in the CNS. A drop in the serotonin concentration in the brain premenstrually and at the time of ovulation has also been demonstrated. Reviewing the literature, Cirillo and colleagues found evidence of increased risks among women with either PMS or PMDD for bipolar I and bipolar II disorder; and, *vice versa*, a greater risk of women with bipolar II having PMS or PMDD. Furthermore, women with BD and PMS reported increased mood lability, anger, irritability, more severe symptoms, more frequent relapses, and a worse response to treatment.[69] Thus, the higher incidence of bipolar II disorder in women than in men may be, in part, due to fluctuations in gonadal hormones.

CONCLUSION

BD is a frequently occurring mental disorder with often devastating consequences, ranking among the top 20 diseases worldwide causing lasting impairment, expressed as years lived with disability. However, compared with other highly disabling disorders such as chronic obstructive pulmonary disease or diabetes mellitus, knowledge of the underlying biological mechanisms of action in BD is lacking and fragmented. This is mainly due to the high complexity of the affected organ, the brain. There are lots of pieces of information, ranging from elementary genetics to behavioral sequelae of BD, but a straightforward chain of causalities has not been formed. In addition, BD is defined by a categorical approach, listing behavioral symptoms that may be the observable result of a multitude of underlying neurobiological aberrations.

GWAS have isolated dozens of candidate genes, each of them contributing only a fractional amount to a genetic template of BD. Since the genes implicated in BD have a large overlap with those implicated in schizophrenia, different severe mental illnesses would perhaps be better understood as a dimensional spectrum rather than as independent categories. Environmental factors contribute at least as much to the variability in BD as an individual's genetic layout does. Genetic and epigenetic factors together may cause an array of subtle changes in signal transmission and intracellular signal transduction, contributing to short-term behavioral changes as mania or depression, as well as to long-term neuroprogression as observed in postmortem brains or on neuroimaging. These changes interact in both directions with the body's endocrine regulatory systems; for example, dysfunction of the HPA or HPT axis may be the consequence of neuronal dysfunction, but may also affect gene transcription and the disease process. As these mechanisms are manifold, drugs used in BD can act at very different points. They may primarily correct an acute imbalance in neurotransmission (e.g. antipsychotics and antidepressant) or have a positive impact on the long-term resilience of the neuron (e.g. lithium). Animal models have been developed to test the effects of medications on behavioral symptoms of BD; however, they cannot be more than a rough proxy of BD in humans. Long-term cognitive impairment leading to loss of functionality, as seen in BD, cannot be recreated in animal models.

Further research is needed so that an understanding of the neurobiology of BD can emerge from isolated laboratory bench observations to a holistic view from gene to behavior.

References

1. Strakowski SM, Fleck DE, Maj M. Broadening the diagnosis of bipolar disorder: benefits vs. risks. *World Psychiatry*. 2011;10:181–186.
2. Goodwin FK, Jamison KR. *Manic–Depressive Illness*. 2nd ed. New York: Oxford University Press; 2007.
3. Merikangas KR, Akiskal HS, Angst J, et al. Lifetime and 12-month prevalence of bipolar spectrum disorder in the National Comorbidity Survey replication. *Arch Gen Psychiatry*. 2007;64:543–552.
4. Grunze H, Walden J. Biological aspects of rapid cycling and mixed states. In: Marneros A, Goodwin F, eds. *Bipolar Disorders. Mixed States, Rapid Cycling and Atypical Forms*. Cambridge: Cambridge University Press; 2005:311–323.
5. Weissman MM, Bland RC, Canino GJ, et al. Cross-national epidemiology of major depression and bipolar disorder. *JAMA*. 1996;276:293–299.
6. Craddock N, Jones I. Genetics of bipolar disorder. *J Med Genet*. 1999;36:585–594.
7. Lichtenstein P, Yip BH, Bjork C, et al. Common genetic determinants of schizophrenia and bipolar disorder in Swedish families: a population-based study. *Lancet*. 2009;373:234–239.
8. Serretti A, Mandelli L. The genetics of bipolar disorder: genome "hot regions", genes, new potential candidates and future directions. *Mol Psychiatry*. 2008;13:742–771.
9. Barnett JH, Smoller JW. The genetics of bipolar disorder. *Neuroscience*. 2009;164:331–343.
10. Sullivan PF. The Psychiatric GWAS Consortium: big science comes to psychiatry. *Neuron*. 2010;68:182–186.
11. Baron M, Risch N, Hamburger R, et al. Genetic linkage between X-chromosome markers and bipolar affective illness. *Nature*. 1987;326:289–292.
12. Connor CM, Akbarian S. DNA methylation changes in schizophrenia and bipolar disorder. *Epigenetics*. 2008;3:55–58.
13. Dempster EL, Pidsley R, Schalkwyk LC, et al. Disease-associated epigenetic changes in monozygotic twins discordant for schizophrenia and bipolar disorder. *Hum Mol Genet*. 2011;20:4786–4796.
14. Rao JS, Keleshian VL, Klein S, Rapoport SI. Epigenetic modifications in frontal cortex from Alzheimer's disease and bipolar disorder patients. *Transl Psychiatry*. 2012;2:e132.
15. Berk M, Brnabic A, Dodd S, et al. Does stage of illness impact treatment response in bipolar disorder? Empirical treatment data and their implication for the staging model and early intervention. *Bipolar Disord*. 2011;13:87–98.
16. Ho BC, Andreasen NC, Ziebell S, Pierson R, Magnotta V. Long-term antipsychotic treatment and brain volumes: a longitudinal study of first-episode schizophrenia. *Arch Gen Psychiatry*. 2011;68:128–137.
17. Benes FM, Kwok EW, Vincent SL, Todtenkopf MS. A reduction of nonpyramidal cells in sector CA2 of schizophrenics and manic depressives. *Biol Psychiatry*. 1998;44:88–97.
18. Rujescu D, Bender A, Keck M, et al. A pharmacological model for psychosis based on *N*-methyl-D-aspartate receptor hypofunction: molecular, cellular, functional and behavioral abnormalities. *Biol Psychiatry*. 2006;59:721–729.
19. Todtenkopf MS, Vincent SL, Benes FM. A cross-study meta-analysis and three-dimensional comparison of cell counting in the anterior cingulate cortex of schizophrenic and bipolar brain. *Schizophr Res*. 2005;73:79–89.
20. Schroeter ML, Abdul-Khaliq H, Sacher J, Steiner J, Blasig IE, Mueller K. Mood disorders are glial disorders: evidence from in vivo studies. *Cardiovasc Psychiatry Neurol*. 2010;2010:780645.
21. Lin CY, Sawa A, Jaaro-Peled H. Better understanding of mechanisms of schizophrenia and bipolar disorder: from human gene expression profiles to mouse models. *Neurobiol Dis*. 2012;45:48–56.
22. Chen H, Wang N, Zhao X, Ross CA, O'Shea KS, McInnis MG. Gene expression alterations in bipolar disorder postmortem brains. *Bipolar Disord*. 2013;15:177–187.
23. Kempton MJ, Geddes JR, Ettinger U, Williams SC, Grasby PM. Meta-analysis, database, and meta-regression of 98 structural imaging studies in bipolar disorder. *Arch Gen Psychiatry*. 2008;65:1017–1032.
24. Womer FY, Kalmar JH, Wang F, Blumberg HP. A ventral prefrontal–amygdala neural system in bipolar disorder: a view from neuroimaging research. *Acta Neuropsychiatr*. 2009;21:228–238.
25. Bora E, Pantelis C. Structural trait markers of bipolar disorder: disruption of white matter integrity and localized gray matter abnormalities in anterior fronto-limbic regions. *Biol Psychiatry*. 2011;69:299–300.
26. Sanches M, Keshavan MS, Brambilla P, Soares JC. Neurodevelopmental basis of bipolar disorder: a critical appraisal. *Prog Neuropsychopharmacol Biol Psychiatry*. 2008;32:1617–1627.
27. Drevets WC, Ongur D, Price JL. Neuroimaging abnormalities in the subgenual prefrontal cortex: implications for the pathophysiology of familial mood disorders. *Mol Psychiatry*. 1998;3:220–221.
28. Adler CM, DelBello MP, Strakowski SM. Brain network dysfunction in bipolar disorder. *CNS Spectr*. 2006;11:312–320.
29. Strakowski SM. Integration and consolidation – a neurophysiological model of bipolar disorder. In: Strakowski SM, ed. *The Bipolar Brain: Integrating Neuroimaging and Genetics*. New York: Oxford University Press; 2012:253–274.
30. Selvaraj S, Arnone D, Job D, et al. Grey matter differences in bipolar disorder: a meta-analysis of voxel-based morphometry studies. *Bipolar Disord*. 2012;14:135–145.
31. Heng S, Song AW, Sim K. White matter abnormalities in bipolar disorder: insights from diffusion tensor imaging studies. *J Neural Transm*. 2010;117:639–654.
32. Macritchie KA, Lloyd AJ, Bastin ME, et al. White matter microstructural abnormalities in euthymic bipolar disorder. *Br J Psychiatry*. 2010;196:52–58.
33. van der Schot AC, Vonk R, Brans RG, et al. Influence of genes and environment on brain volumes in twin pairs concordant and discordant for bipolar disorder. *Arch Gen Psychiatry*. 2009;66:142–151.
34. Langan C, McDonald C. Neurobiological trait abnormalities in bipolar disorder. *Mol Psychiatry*. 2009;14:833–846.
35. Manji HK, Potter WZ. Monoaminergic system. In: Young LT, Joffe RT, eds. *Bipolar Disorder – Biological Models and Their Clinical Application*. New York: Marcel Dekker; 1997:1–40.
36. Yatham LN, Liddle PF, Shiah IS, et al. PET study of [(18)F]6-fluoro-L-dopa uptake in neuroleptic- and mood-stabilizer-naive first-episode nonpsychotic mania: effects of treatment with divalproex sodium. *Am J Psychiatry*. 2002;159:768–774.
37. Juckel G, Hegerl U, Mavrogiorgou P, et al. Clinical and biological findings in a case with 48-hour bipolar ultrarapid cycling before and during valproate treatment. *J Clin Psychiatry*. 2000;61:585–593.
38. Swann AC, Stokes PE, Secunda SK, et al. Depressive mania versus agitated depression: biogenic amine and hypothalamic–pituitary–adrenocortical function. *Biol Psychiatry*. 1994;35:803–813.
39. Pariante CM. The glucocorticoid receptor: part of the solution or part of the problem? *J Psychopharmacol*. 2006;20:79–84.
40. Watson S, Thompson JM, Ritchie JC, Nicol Ferrier I, Young AH. Neuropsychological impairment in bipolar disorder: the relationship with glucocorticoid receptor function. *Bipolar Disord*. 2006;8:85–90.
41. Joffe RT, Sokolov STH. Hormones and bipolar affective disorder. In: Young LT, Joffe RT, eds. *Bipolar Disorder – Biological Models and Their Clinical Application*. New York: Marcel Dekker; 1997:81–92.

42. Pasquali L, Busceti CL, Fulceri F, Paparelli A, Fornai F. Intracellular pathways underlying the effects of lithium. *Behav Pharmacol.* 2010;21:473–492.
43. Bachmann RF, Schloesser RJ, Gould TD, Manji HK. Mood stabilizers target cellular plasticity and resilience cascades: implications for the development of novel therapeutics. *Mol Neurobiol.* 2005;32:173–202.
44. Shaltiel G, Chen G, Manji HK. Neurotrophic signaling cascades in the pathophysiology and treatment of bipolar disorder. *Curr Opin Pharmacol.* 2007;7:22–26.
45. Soeiro-de-Souza MG, Dias VV, Figueira ML, et al. Translating neurotrophic and cellular plasticity: from pathophysiology to improved therapeutics for bipolar disorder. *Acta Psychiatr Scand.* 2012;126:332–341.
46. Berridge MJ. Calcium signalling remodelling and disease. *Biochem Soc Trans.* 2012;40:297–309.
47. Hollister LE, Trevino ES. Calcium channel blockers in psychiatric disorders: a review of the literature. *Can J Psychiatry.* 1999;44: 658–664.
48. Hope S, Ueland T, Steen NE, et al. Interleukin 1 receptor antagonist and soluble tumor necrosis factor receptor 1 are associated with general severity and psychotic symptoms in schizophrenia and bipolar disorder. *Schizophr Res.* 2013;145:36–42.
49. Ongur D, Drevets WC, Price JL. Glial reduction in the subgenual prefrontal cortex in mood disorders. *Proc Natl Acad Sci U S A.* 1998;95:13290–13295.
50. Chiu CT, Wang Z, Hunsberger JG, Chuang DM. Therapeutic potential of mood stabilizers lithium and valproic acid: beyond bipolar disorder. *Pharmacol Rev.* 2013;65:105–142.
51. Cousins DA, Butts K, Young AH. The role of dopamine in bipolar disorder. *Bipolar Disord.* 2009;11:787–806.
52. Moorhead TW, McKirdy J, Sussmann JE, et al. Progressive gray matter loss in patients with bipolar disorder. *Biol Psychiatry.* 2007;62:894–900.
53. Bennett CM, Baird AA, Miller MB, Wolford GL. Neural correlates of interspecies perspective taking in the post-mortem Atlantic salmon: an argument for proper multiple comparisons correction. *J Serendipitous and Unexpected Results.* 2010;1:1–5.
54. Vul E, Harris C, Winkielman P, Pashler H. Puzzlingly high correlations in fMRI studies of emotion, personality, and social cognition. *Perspect Psychol Sci.* 2009;4:274–290.
55. Blond BN, Fredericks CA, Blumberg HP. Functional neuroanatomy of bipolar disorder: structure, function, and connectivity in an amygdala–anterior paralimbic neural system. *Bipolar Disord.* 2012;14:340–355.
56. Young JW, Henry BL, Geyer MA. Predictive animal models of mania: hits, misses and future directions. *Br J Pharmacol.* 2011;164: 1263–1284.
57. Fries GR, Pfaffenseller B, Stertz L, et al. Staging and neuroprogression in bipolar disorder. *Curr Psychiatry Rep.* 2012;14:667–675.
58. Post RM, Contel NR. Cocaine-induced behavioral sensitization: a model for recurrent manic illness. In: Perris C, Struwe G, Janson B, eds. *Biological Psychiatry.* Amsterdam: Elsevier; 1981:746–749.
59. Goldberg JF, Harrow M. Kindling in bipolar disorders: a longitudinal follow-up study. *Biol Psychiatry.* 1994;35:70–72.
60. Mercer L, Becerra R. A unique emotional processing profile of euthymic bipolar disorder? A critical review. *J Affect Disord.* 2013;146:295–309.
61. Gross JJ, Thompson RA. Emotion regulation: conceptual foundations. In: Gross JJ, ed. *Handbook of Emotion Regulation.* New York: Guildford Press; 2007:3–24.
62. Phillips ML, Ladouceur CD, Drevets WC. A neural model of voluntary and automatic emotion regulation: implications for understanding the pathophysiology and neurodevelopment of bipolar disorder. *Mol Psychiatry.* 2008;13:833–857.
63. Goodwin GM, Martinez-Aran A, Glahn DC, Vieta E. Cognitive impairment in bipolar disorder: neurodevelopment or neurodegeneration? An ECNP expert meeting report. *Eur Neuropsychopharmacol.* 2008;18:787–793.
64. Lopez-Jaramillo C, Lopera-Vasquez J, Gallo A, et al. Effects of recurrence on the cognitive performance of patients with bipolar I disorder: implications for relapse prevention and treatment adherence. *Bipolar Disord.* 2010;12:557–567.
65. Schretlen DJ, Cascella NG, Meyer SM, et al. Neuropsychological functioning in bipolar disorder and schizophrenia. *Biol Psychiatry.* 2007;62:179–186.
66. Torrent C, Martinez-Aran A, del Mar BC, et al. Long-term outcome of cognitive impairment in bipolar disorder. *J Clin Psychiatry.* 2012;73:e899–e905.
67. McCarthy MJ, Welsh DK. Cellular circadian clocks in mood disorders. *J Biol Rhythms.* 2012;27:339–352.
68. Murray G, Lam RW, Beaulieu S, et al. Do symptoms of bipolar disorder exhibit seasonal variation? A multisite prospective investigation. *Bipolar Disord.* 2011;13:687–695.
69. Cirillo PC, Passos RB, Bevilaqua MC, Lopez JR, Nardi AE. Bipolar disorder and premenstrual syndrome or premenstrual dysphoric disorder comorbidity: a systematic review. *Rev Bras Psiquiatr.* 2012;34:467–479.

CHAPTER

41

Pain: From Neurobiology to Disease

Michael S. Gold*, †, ‡, §, ¶, Miroslav "Misha" Backonja¶¶, **

‡Pittsburgh Center for Pain Research, Department of Anesthesiology; *Center of Neuroscience; §Departments of Neurobiology and †Medicine; ¶Division of Gastroenterology Hepatology and Nutrition, University of Pittsburgh; ¶¶Departments of Neurology, Anesthesiology, and Rehabilitation Medicine, University of Wisconsin School of Medicine and Public Health, Madison, Wisconsin, USA; **CRILifetree, Salt Lake City, Utah, USA

OUTLINE

INTRODUCTION

Based on its prevalence, potential severity, and repercussions for those who suffer, their loved ones and society, it would be hard to argue that there is any neurological disease with a greater impact than pain. The magnitude of the problem of pain in the USA alone has been summarized in a detailed report from the Institute of Medicine in 2011,[1] where a conservative estimate put the number of people in the USA who experience chronic pain, defined as pain lasting for more than 6 months, at over 100 million, or almost one-third of the population. The financial impact of all this pain was estimated at between 560 and 635 billion USD in medical costs and lost productivity. However, these numbers say nothing about the cost of pain in terms of suffering. Furthermore, the impact of pain is likely to be significantly greater worldwide owing to the increased incidence of chronic pain associated with poverty,[1] limited access to the few effective interventions for the treatment of pain, and medical care needed to treat a variety of associated conditions.

Neurobiology of Brain Disorders
http://dx.doi.org/10.1016/B978-0-12-398270-4.00041-0

To appreciate the significance of the prevalence and impact of pain, one need only consider prevalence rates for some of the more common pain syndromes, such as low back pain, osteoarthritis, and migraine. The incidence of low back pain is 26.4%,[1] while that of osteoarthritis is 13.9% in adults aged 25 and older (according to the Centers for Disease Control and Prevention). Furthermore, over 10% of adults aged 18–65 in the world reported at least one migraine in the past year (according to the World Health Organization). In specific groups of people, the prevalence of pain is even greater; for example, it is estimated that 62% of elderly patients in nursing homes experience chronic pain.[1] The impact of the high incidence and prevalence of pain is compounded by the fact that there are no consistently effective treatments for pain devoid of deleterious consequences. For example, the most commonly used medications for the treatment of mild to moderate pain, the non-steroidal anti-inflammatory drugs (NSAIDs) such as ibuprofen, are associated with a significant risk of gastrointestinal bleeding. The most commonly used medications for the treatment of moderate to severe pain, opioids such as fentanyl and hydromorphone, are associated with significant risk of respiratory depression, somnolence, and altered mentation.[1] Even for the most commonly prescribed medications for the treatment of neuropathic pain, two of every three patients given the medication will experience no relief, and of those who do have a positive response, the average benefit is a reduction in the intensity of pain by two points on a 10-point scale.[2] Multiply the cost of multiple failed attempts to obtain pain relief by the lost productivity of those unable to work because of their pain and it becomes clear why the financial impact of pain is so immense.

TERMINOLOGY OF PAIN

Pain is a private experience for which there are no objective criteria with which to assess its presence, let alone its magnitude. Because it is a term applied to a wide array of experiences, it is necessary to start a discussion of pain by defining terms central to this discussion, particularly in the context of the neurobiology of disease.

The first and potentially most controversial definition, at least in the context of a discussion of pain, is that of disease. The definitions of disease found in popular medical dictionaries are not particularly helpful. For example, disease is defined in *Stedman's Medical Dictionary* as "Morbus, illness, sickness. An interruption or perversion of functions of any of the organs, a morbid change of any of the tissues or an abnormal state of the body as a whole, continuing for a longer or shorter period". The problem with this definition, as pointed out by Wilken,[3] is that it is circular, and therefore not particularly useful from a scientific viewpoint. Consequently, Wilken proposes an alternative definition derived from that found in *Blakiston's New Gould Medical Dictionary*, where disease is operationally defined as a state that "results in a living system whenever the system's stressor adaptability is exceeded by the sum of the stressors acting upon the system". By this definition, everything but the most transient of pain, relieved by escape (with reflex withdrawal) from the inciting stimulus, should be considered a disease.

The discussion of pain in the context of disease is controversial because pain has historically been viewed as a symptom of an underlying disease, rather than a disease itself. This perspective is largely based on the assumption, still reflected in most medical school curricula, that the most effective, if not only way to treat pain is to treat the underlying disease. Indeed, until recently, it was common to withhold analgesics for fear that they would prevent the ability to assess the efficacy of a particular therapeutic course of action. As a result, pain management is still implemented as an afterthought by many specialists treating diseases where pain is a prominent feature as well as a likely consequence of many therapeutic interventions. This historical view probably accounts for why the pain community has been so cautious about the use of the term disease in reference to pain, with some of the strongest language only recently appearing in documents such as the Institute of Medicine Report, with statements such as "chronic pain can be a disease in itself" or "some types of chronic pain are diseases in their own right".[1]

The problems with these more cautious ways of describing pain as a disease are manifold. The use of arbitrarily defined terms such as "chronic" or undefined terms like "some types" are not useful for either patient or clinician. The problem with "some types" is obvious because it begs the question, "which types?" Furthermore, the term "chronic", generally used to define pain present for a duration of 6 months or longer, is problematic because this timeline encompasses everything ranging from the ongoing pain associated with the normal trajectory of recovery from a severe traumatic injury, through the pain of an osteoarthritic knee that has been a problem for years, but which may be completely resolved with a total knee replacement, to the pain associated with a peripheral nerve injury that may be present for a lifetime.

Attempts to address the limited utility of the term "chronic" with the inclusion of contrasting terms such as "acute" or "nociceptive" pain to refer to transient or "physiological" pain, and "persistent" pain to refer to pain lasting longer than acute, but shorter than chronic, are also not particularly illuminating. Terms such as "persistent acute" pain have been proposed to cover even long-lasting pain like that associated with osteoarthritis, which may be completely resolved with the appropriate intervention. Use of these additional terms relegate "chronic" pain to that which we do not understand or cannot treat. Augmenting this unfortunate situation, the use of this terminology ignores the fact that even for the types of pain we think we understand and may be able to effectively treat, there are failures with treatment. For example, as many as 30% of patients undergoing total knee replacement for the treatment of an osteoarthritic knee are still in pain 1 year after surgery.[4] More problematic is that this definition of chronic pain as a "possible disease" ignores the understanding of the array of factors that may serve as a prognosis for recovery and a guide for clinical decisions about treatment (see "Factors Affecting the Emergence, Prognosis, and Severity of Pain", below). There is still vastly more to be learned about pain, its diagnosis, and its treatment. Nevertheless, it may be argued that based on our current understanding of pain and the operational definition of disease, it is appropriate to refer to all but the most transient of pain as a disease, which based on the data described below, should be treated as aggressively as possible.

The next essential term to define is pain. The International Association for the Study of Pain (IASP) defines pain as "an unpleasant sensory and emotional experience associated with actual or potential tissue damage, or described in terms of such damage".[5] The acknowledgement that there is an affective or emotional component to the experience of pain distinguishes it from other sensory experiences afforded by the somatosensory system and the five special senses. This distinction has fundamental implications for the measurement of pain because its emotional valence necessarily makes it a personal experience. It also means that the impact of pain will depend on context and past experience, where both factors influence not only the emotional valence, or suffering associated with pain, but also the perception of intensity via the neural circuitry described below (see "Neurobiology of Pain"). For example, the impact of pain in the abdomen will be very different for someone who has a history of lactose intolerance than for someone just diagnosed with pancreatic cancer. As a result of these unique features, there are no objective measures for pain. Even relatively sensitive measures of acute pain, such as changes in heart rate or galvanic skin response, rapidly lose their predictive validity in the face of ongoing pain, because of what appear to be adaptive changes in the nervous system. This leaves the subjective measure of self-report as the only viable measure of pain, a fact that continues to frustrate those impacted by pain despite many lines of evidence supporting the validity of this measure.

In addition to temporal aspects of pain encompassed by "acute", "persistent", and "chronic", three classification systems for pain are often employed (Table 41.1). One of these is a general reference to the source of the pain, described as primarily inflammatory or neuropathic. Inflammatory pain is classically dominated by inflammatory processes involving the activation of the innate and adaptive immune system and subsequent release of inflammatory mediators. In contrast, the IASP definition of neuropathic pain is "pain caused by lesion or disease of the somatosensory system".[5] While not mutually exclusive, these two general types of pain are commonly associated with distinguishing features. For example, inflammatory pain tends to be more commonly associated with movement-evoked pain as well as the perception that normally painful stimuli are perceived as being even more painful in the presence of inflammation, a phenomenon referred to as hyperalgesia. In contrast, neuropathic pain is commonly associated with ongoing pain, or the perception of pain in the absence of any overt stimulation.[6] Neuropathic pain may also be characterized by the presence of a perception of pain in response to innocuous or light brushing of the affected area, a phenomenon referred to as dynamic mechanical allodynia. As discussed later in this chapter,

TABLE 41.1 Criteria Used in the Classification of Pain

Temporal	Source/cause	Dominant Feature/modality	Symptom Based
Acute	Visceral	Ongoing/spontaneous	TMD
Persistent	Somatic	Evoked/hypersensitivity	IBS
Chronic	Inflammatory	Thermal (hot/cold)	IC
	Neuropathic	Mechanical	Fibromyalgia
		Chemical	Low back pain

TMD: temporomandibular disorder. In addition to the signs and symptoms of TMD, which include clicking and locking of the temporomandibular joint, pain may be a prominent feature, with pain in the temporomandibular joint and/or muscles of mastication. IBS: irritable bowel syndrome, one of the "functional pain disorders" because of the presence of pain in the absence of other pronounced signs of underlying pathology. IC: interstitial cystitis, which is associated with pain and the sensation of urgency with bladder filling.

in the section "Treatment of Pain", these general distinctions have been important from the clinical perspective, because treatment of neuropathic and inflammatory pain may involve very different strategies.

As suggested by terms such as hyperalgesia and dynamic mechanical allodynia, qualities or modalities of pain are also used to classify pain. Modality in this context is used to refer to the nature of the stimulus energy that evokes the sensation of pain. These are generally divided into mechanical, chemical, and thermal, with thermal being further subdivided into heating and cooling. The relevance of this classification system can be immediately appreciated by recalling the changes in perception associated with heat and ultraviolet burns, where normally pleasantly warm stimuli are perceived as painful and previously hot stimuli are unbearable. These are examples of warm allodynia and heat hyperalgesia. The specific properties of the stimulus, at least for chemical and mechanical stimuli, may be context dependent. For example, punctate mechanical stimuli may be very effective for evoking the perception of pain on skin, but torque and flexion are more relevant for joints and distension for hollow visceral organs. These distinctions suggest potentially important differences in the underlying mechanisms of these perceptions that will be discussed below. Finally, these distinctions have several important implications. First, they may be suggestive of underlying pathology and therefore require appropriate therapeutic strategies. For example, in addition to the marked heat sensitivity associated with a burn, pain associated with thalamic stroke is often associated with cool allodynia, where the constellation of sensitivities may be used to identify subpopulations of patients.[7] Second, the nature of the sensitivity may suggest the underlying mechanisms. For example, as discussed in the next section, many proteins have been identified that are specifically activated by various modes of stimuli discussed above. Furthermore, in the presence of or in response to these specific stimuli, these proteins may change in their degree of sensitivity or density of expression, shifting the overall response to the stimulus.[8] Third, the nature of the altered sensitivity combined with knowledge of the underlying proteins suggests novel therapeutic approaches for the treatment of pain.

Finally, and possibly most commonly used, is a symptom-based classification of pain syndromes. For example, pain in the temporomandibular joint and/or muscles of mastication is associated with temporomandibular joint disorder, while pain in the abdomen associated with feelings of bloating and cramping, with or without diarrhea but lacking overt signs of inflammation of the colon, is referred to as irritable bowel syndrome. Some pain, as in the case of headache, is further subdivided based on an array of inclusion and exclusion criteria. Naming a particular compilation of features associated with pain may be useful for both patient and clinician, particularly if it suggests an appropriate course of treatment. However, it is far more often the case that there is tremendous heterogeneity among pain patients grouped within a particular pain syndrome, even those with very well-defined inclusion or exclusion criteria. For example, while common migraine is generally divided into migraine with and without aura, there is evidence that the therapeutic efficacy of botulinum toxin injections depends on the nature of the pain, that is, whether the headache feels like the head is "imploding" or "exploding".[9] The situation is far worse for most pain syndromes, which are little more than a loose aggregate of features generally associated with pain in a particular body region. It has been suggested that poorly defined patient populations have played a significant role in the number of failed clinical trials, prompting a call for detailed phenotyping of patient populations as a way to identify the most appropriate target for a clinical trial, if not a specific therapeutic strategy.[7]

FACTORS AFFECTING THE EMERGENCE, PROGNOSIS, AND SEVERITY OF PAIN

As with most diseases, the impact of pain on those who experience it is widespread. For example, pain may not only be associated with a loss of function, a necessary component to the normal healing process, but also function as a stressor, driving changes in the hypothalamic–pituitary and sympathoadrenal axes that regulate the body's normal response to stress. Pain can disrupt sleep patterns and may also result in anxiety and depression.[10] Importantly, pain can have significant psychological repercussions, as the withdrawal from work and family may lead to social isolation and a sense of loss of control. All of these changes can further exacerbate pain, creating a vicious cycle. Thus, one problem with the current classification schemes for pain is that they are not tied to an integrated approach to the diagnosis and management of the disease. That is, based on our current understanding of pain, like any other disease, there is a wide variety of factors that influence whether a person is likely to develop the disease and, once manifest, the prognosis for recovery with or without treatment (Fig. 41.1).

The factors that influence the emergence and prognosis of a disease may influence strategies for both prevention and treatment. It is therefore important to consider the factors that influence the emergence of pain. In this context, pain can be broken down into two general categories. One category includes pain patients where pain is likely to emerge *de novo*. The clearest examples of these pain syndromes are the relatively rare forms of hereditary pain disorders associated with genetic mutations.

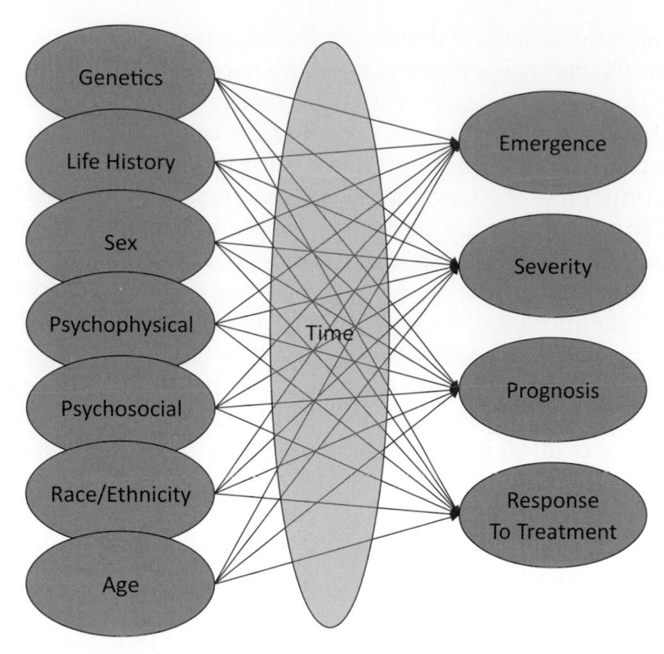

FIGURE 41.1 **Factors that impact pain as a disease.** Available evidence suggests that a variety of factors impacts all aspects of pain from its emergence to the response to treatment. Complicating the relative impact of all of these factors is time, which influences everything from the likelihood that a genetic variant will be manifest in a pain condition to the underlying mechanisms of pain that may serve as viable therapeutic targets.

Examples include severe forms of hereditary migraine [familial hemiplegic migraine (FHM)], primary extreme pain disorder (PEPD), and primary inherited erythermalgia (PIE). The three best characterized forms of FHM are associated with mutations in genes encoding the α-subunit CaV2.1 of the P/Q-type of voltage-gated Ca^{2+} channel, the α_2-subunit of Na^+/K^+-ATPase, and the α-subunit of NaV1.1, a voltage-gated Na^+ channel, CACNA1 (FHM1), ATP1A2 (FHM2), and SCNA1 (FHM3), respectively. PEPD and PIE are associated with mutations in SCN9A, the α-subunit of the voltage-gated Na^+ channel NaV1.7. Mutations in ion channels resulting in a deleterious phenotype are referred to as channelopathies. A discussion of the mechanisms through which the mutations in these genes result in the manifestation of pain syndromes is beyond the scope of this chapter. However, data concerning this issue are covered at length in a number of excellent reviews (e.g. see Dib-Hajj *et al.*[11]). In the context of a discussion of the emergence of pain, what we have learned from these hereditary pain disorders has at least three important implications. First, each of these pain syndromes has a very specific phenotype with pain that is far more localized than the expression pattern of the proteins encoded by the mutant genes. For example, CACNA1 is expressed throughout the central and peripheral nervous system, but the aberrant pain associated with this gene mutation is restricted to the head. Similarly, SCN9A is expressed throughout the peripheral nervous system (both sympathetic postganglionic and sensory neurons) as well as in endocrine tissue, yet PIE is manifest as pain in the hands and feet, whereas PEPD is restricted to the anal sphincter early in life with a possible progression to pain around the jaw and eye.[11] These observations suggest that the context in which the mutant proteins are expressed is critical to the manifestation of the deleterious consequences of the mutation. Second, while PEPD appears to be manifest at birth, there is a considerable delay in the emergence of PIE, and the migraine of FHM appears to be manifest with a time-course comparable to common migraine, with an onset at puberty. These observations suggest that in addition to the local context (e.g. the other proteins present in the cell with the mutant protein), the emergence of a particular pain syndrome is likely to depend on the larger context of the patient, including factors such as the developmental state of the nervous system and hormonal status. Third, the observation that these particular pain syndromes are episodic in nature suggests that there are both specific stimuli or triggers capable of selectively engaging the neural substrates impacted by the mutant protein, and endogenous mechanisms that can normalize the system again after a period of aberrant activity. This is clearly the case for PEPD, where defecation is a trigger for a painful attack. It also appears to be true, however, for PIE and FHM, which may be triggered by stimuli such as heat, exercise, or stress.

While hereditary pain disorders are an extreme example of a situation where pain will emerge *de novo*, there are other situations that are less clear-cut but are nevertheless highly associated with the emergence of a pain disorder. One of the most perplexing of these is an early life experience where trauma in general, and sexual trauma in particular, is highly associated with the subsequent development of pain syndromes in adulthood.[12] Preclinical data suggest that there are critical periods during the development of the nervous system wherein exposure to noxious stimuli can produce persistent changes in the nervous system that impact normal responses to noxious stimuli as well as the response to subsequent injury.[13] Thus, these early life experiences may "prime" the nervous system so that subsequent challenge, such as that associated with a bacterial or viral infection, may result in the development of a pain disorder. It is becoming clear that extreme stress, even in adulthood, may be sufficient for the development of a pain disorder. Traumatic brain injuries (TBIs) have received the most attention as a factor influencing the exceptionally high incidence of headache in postwar veterans. However, comparative datasets in case–control cohorts who were exposed to the same traumatic stress but did not suffer TBI develop headache at only slightly lower rates than those observed in the TBI subjects.[14] These data indicate that some life events significantly increase the likelihood of developing a pain disorder.

The second category of pain patients includes those who experience pain associated with a specific insult or event. In this population, the question is generally not whether they will have pain, but their prognosis for recovery. Several factors influence prognosis, including psychophysical and psychological factors, genetics, ethnicity and race, gender, age, and life history. For example, psychological constructs such as catastrophizing[10] as well as a history of depression and/or anxiety are predictors of the magnitude of postoperative pain. Similarly, the sensitivity to experimentally evoked pain, as assessed using quantitative sensory testing (QST), and the ability to activate endogenous pain inhibitory mechanisms, a phenomenon referred to as conditioned pain modulation, are also predictive of the likelihood that pain will persist for a significant duration after surgery or other types of injury.[15] Genetics contributes to these differences in psychological and psychophysical properties between individuals. However, even when these baseline differences are controlled for there is evidence that genetics contributes to the likelihood of developing a pain disorder.[16] Ethnicity and race also appear to influence the manifestation of pain, with some pain disorders more prevalent in different populations. There is also evidence of race by genetic interactions that influence the emergence of pain disorders. While it is unequivocally true that men and women have different coping strategies and utilize health-care services differently, well-controlled population-based studies clearly indicate that with a few notable exceptions, women experience disproportionately more pain than men.[17] The basis for this difference remains an active area of investigation, with clinical and preclinical data indicating the presence of an interaction between gender and genetics in the response to pain.[18] As noted above, age appears to be a factor in the manifestation of a number of pain syndromes, such as migraine, that appear to be tied to development stages. However, other pain syndromes such as trigeminal neuralgia and postherpetic neuralgia are more common in older people, with the likelihood of developing postherpetic neuralgia following a shingles outbreak increasing dramatically over the age of 60. Finally, life history can influence the manifestation and prognosis of pain. This has been most clearly demonstrated in a series of preclinical studies in which previous injury to a tissue altered the response to subsequent injury: there was a dramatic increase in the duration of the hypersensitivity associated with stimuli that would normally produce a relatively transient response.[19]

Importantly, the same factors that influence pain duration after an insult affect the prognosis for the patient through influences on both the severity of the pain experienced and the efficacy of treatment. For example, catastrophic thinking is a good predictor for postoperative pain.[20] Similarly, patients who effectively engage endogenous pain modulatory circuitry preoperatively report significantly less postoperative pain.[15] Genetics also influences both pain severity and sensitivity to analgesics,[16] and a race by genetic interaction has been shown to impact the analgesic efficacy of morphine. While reported baseline differences in pain sensitivity between men and women are generally small, women report higher levels of clinical pain than men.[17] There is also evidence to suggest that women may be more sensitive to the deleterious side effects of analgesics such as morphine, which then limits the use of what could be otherwise effective therapy in this population. With respect to age, preclinical data highlight the presence of marked developmental changes in opioid receptor expression, which could impact the analgesic efficacy of opioids such as morphine that appear to act primarily through the mu-opioid receptor. Conversely, there are paradoxical changes in elderly people, who are less sensitive to acute somatic stimulation but also report more pain. There are also marked changes in the sensitivity of elderly people to both the therapeutic efficacy and side effects of a number of analgesics. Finally, while opioid tolerance is probably the clearest example of how life history can impact the efficacy of a pain treatment, previous trauma or injury can also influence the severity of the pain following a subsequent injury.

NEUROBIOLOGY OF PAIN

In the absence of tissue injury or disease, the perception of pain generally starts with a noxious or potentially damaging stimulus impinging on the body, and ends with activity in a number of brain regions which include structures such as the primary and secondary somatosensory, cingulate, and insular cortices, now colloquially referred to as the "pain matrix" (Fig. 41.2). In between is a sequence of events and neural circuits that provide information about the site, intensity, and modality of the stimulus, as well as its potential meaning. A considerable amount of neural circuitry engaged well below the level of consciousness influences not only the perceived intensity and meaning of the stimulus, but also the coordinated response to it. The site and nature of the stimulus can influence the resolution of the perception of location and intensity, as well as its meaning based on a number of factors, described below. To demonstrate the complexity of mechanisms underlying pain as a disease, the sequence of events and circuitry associated with pain is presented below.

Stimulus Transduction

A fundamental unit of information transfer in the nervous system is the action potential, which is the rapid, all-or-nothing change in membrane potential. In general, action potentials are initiated from a resting membrane potential of around −65 mV and involve a total change in membrane potential of more than 80 mV (−65 mV to +30 mV and back again) in less than 2 milliseconds. Thus, for a noxious stimulus to generate a response in the nervous system, the "energy" of the stimulus must be converted into a change in membrane potential. This process is referred to as transduction, and several molecules have been identified that are responsible for the process (see Gold[21] for a review). The somatosensory system is responsive to thermal, chemical, and mechanical stimuli. Thus, the transducers responsive to these stimulus modalities are referred to as chemotransducers, thermotransducers, and mechanotransducers, and those responsive to more than one modality are polymodal.

The sensory neurons that are responsive to noxious stimuli are referred to as nociceptors. There is evidence that many, if not all, of the known transducers are expressed in nociceptors,[21] suggesting that this first step in the pain signaling pathway may occur in the primary afferent. Nevertheless, not only are transducers present in the tissue innervated by the nociceptive afferents, but some of these end-organ cell types, such as the epithelial cells that line the urinary bladder (urothelia) or colon, are able to release bioactive substances such as ATP in response to stimulation.[8] Thus, stimulus transduction may also involve a two-step process, whereby the initial stimulus causes the release of a bioactive substance from an end-organ cell type, which then acts directly on the nociceptive afferent.

Spike Initiation

The change in membrane potential associated with the activation of a transducer is referred to as a generator potential. In the somatosensory system, the generator potential is a membrane depolarization that passively spreads through the afferent terminal arbor. The distance of passive spread is determined by the electrical properties of the terminal, one of the most important being the membrane resistance. Membrane resistance is largely determined by the density and distribution of various ion channels that may be active at the resting membrane potential.[8]

The ability of a stimulus to generate an action potential and thereby affect the CNS is dependent on whether the magnitude of the generator potential is sufficient to activate voltage-gated Na^+ channels that are responsible for the depolarizing phase of the action potential. These channels are presumably clustered at one or more points in the terminal arbor at sites referred to spike initiation zones. The opening of voltage-gated Na^+ channels drives further membrane depolarization as Na^+ ions enter the neuron, and this process leads to the activation of additional Na^+ channels: it is a feedforward process that can be exceptionally rapid. The action potential threshold is the membrane potential at which this feedforward process overwhelms any inhibitory currents and proceeds to the generation of the action potential.

Action Potential Propagation

Once an action potential is initiated, the density and distribution of voltage-gated Na^+ channels within a neuron are such that the action potential is actively propagated along the axon. Action potential propagation is faster in larger myelinated axons and slower in unmyelinated axons because of both the diameter of the axon and the presence of myelination. Primary sensory neurons are pseudounipolar neurons, with a cell body located in the dorsal root, trigeminal or nodose ganglion that sits at the end of a small branch of axon formed by the fusion of the peripheral and central segments of the

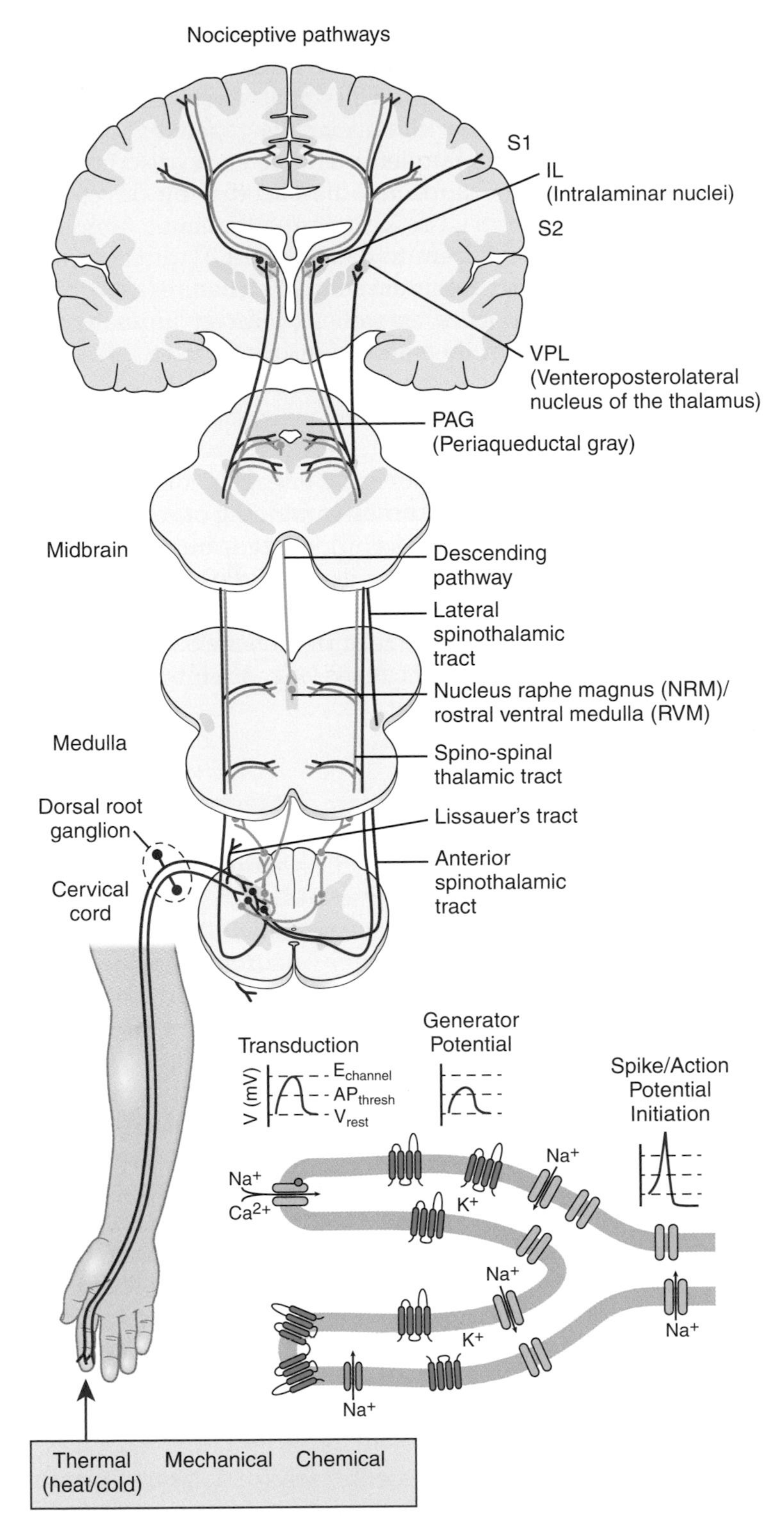

Sites of plasticity

Cortex:
Changes in cortical thickness have been described in association with several chronic pain states.

Thalamus:
Increases in spontaneous and evoked activity following spinal cord injury.

Other cortical/subcortical structures:
Changes synaptic strength and gene expression in structures including the anterior cingulate, amygdala and hippocampus have been observed in several chronic pain states.

PAG:
Loss of descending inhibitory tone from the PAG.

Nucleus raphe magnus (NRM)/ rostral ventral medulla (RVM):
Shift from descending inhibition to facilitation.

Spinal dorsal horn:
Primary site of Central Sensitization that reflects both pre- and post-synaptic facilitation, a decrease in inhibitory (GABA and glycine) signaling, increases in the intrinsic excitability of dorsal horn neurons. Depending on the timing and type of injury, spinal microglia and astrocytes may also contribute to the increase in dorsal horn excitability through the release of a variety of mediators.

Dorsal root ganglion:
Changes within the DRG may enable the emergence of ectopic activity.

Peripheral nerve:
Injury induced changes in action potential conduction velocity and activity-dependent slowing have been described. Nerve-injury may also result in the emergence of ectopic activity along the nerve and neuroma formation which may also become a site of ectopic activity.

Peripheral terminals:
Sensitization of peripheral terminals is due to changes in the density, distribution and/or biophysical properties of channels responsible for sensory transduction (e.g., TRP channels), the spread of the generator potential (e.g., K^+ channels), and action potential initiation (e.g., voltage-gated Na^+ channels)

FIGURE 41.2 **The neurobiology of pain.** While a relatively simply neural circuit involving three neurons may be sufficient for the transmission of pain from the site of the noxious or potentially tissue-damaging stimulation of somatic tissue to perception, the available evidence suggests that nociceptive processing is far more complex. Many of the critical molecular mechanisms underlying the conversion or transduction of a noxious or painful stimulus into a change in membrane potential in nociceptive afferents have been identified. This initial change in membrane potential is referred to as a generator potential, and additional proteins that control the spread of the generator potential within the afferent terminal as well as the initiation and propagation of the action potential (AP) within the afferent axons have been identified. Within the dorsal horn of the spinal cord, the central terminals of nociceptive afferents synapse not only on projection neurons that carry the nociceptive signal to a number of structures in the brainstem and brain, but also on excitatory and inhibitory interneurons that feed back to modulate the incoming afferent activity and feed forward to modulate the ascending signal. Within the brainstem and brain are additional regions that are further integrated, enabling perception of both the sensory discriminative aspects of pain, which include site, intensity, and modality of the stimulus, and the emotional or affective aspects of pain, which include suffering and the perceived meaning of the sensation. Additional levels of modulation occur at each of these sites in the transmission of the nociceptive signal through what is generally referred to as descending nociceptive circuitry, which can both amplify and inhibit the ascending signal. Finally, there is evidence of plasticity at each of these sites that can profoundly influence the response of the nervous system to subsequent stimulation, with evidence that many of these changes may contribute to the perception of ongoing pain.

axon, a branch point referred to as the t-junction. Because of this anatomical structure, propagated action potentials do not need to invade the afferent cell body. Thus, even if the action potential does not invade the afferent cell body, it is propagated to the central terminals of the afferent.

Synaptic Transmission

The membrane depolarization associated with the action potential invading the central terminals of the primary afferent results in the activation of voltage-gated Ca^{2+} channels. The increase in intracellular Ca^{2+} enables the mobilization and release of synaptic vesicles containing glutamate and, in some afferents, peptides that activate receptors on second order spinal/brainstem dorsal horn neurons. The essential role of voltage-gated Ca^{2+} channels in transmitter release from primary afferents makes these channels an attractive target for both endogenous and exogenous regulation of nociceptive signaling.

Spinal/Brainstem Dorsal Horn Circuitry

The complexity of nociceptive signaling increases dramatically in the spinal/brainstem dorsal horn. Not only do afferents synapse on projection neurons that give rise to ascending tracks in the spinal cord, but they terminate on interneurons that both provide inhibitory feedback on to the central terminals of the afferents and inhibit others neurons in the area that may be necessary for proper localization of the afferent input. There are also excitatory interneurons that enable feedforward circuit formation and the activation of projection neurons located in deeper lamina in the spinal cord. While the details of the dorsal horn circuitry are another area of active investigation, the available evidence indicates that the inhibitory and excitatory interneurons are connected in such a way as to facilitate the interaction between types or modalities of afferent input. For example, activity in low-threshold afferents can inhibit input by high-threshold afferent activity, as originally postulated in the gate control theory of Melzack and Wall.[22] Another example of the layers of complexity in the spinal cord is emerging through our understanding of the circuitry underlying itch, which enables a noxious stimulus, such as scratching, to suppress this unpleasant sensation.

Ascending Circuitry

The essential nociceptive circuit is often drawn with three neurons: an afferent terminating on a spinal/brainstem dorsal horn neuron, the dorsal horn neuron projecting to a neuron in the thalamus, and the thalamic neuron projecting to a neuron in the cortex.[23] However, available evidence indicates that sensory/discriminative and affective circuits are engaged by pathways that ascend in parallel.[24] The sensory discriminative aspects of pain appear to be encoded by direct input to somatosensory cortices via the thalamus. Emotional aspects of pain seem to be mediated by direct input to brainstem and forebrain/limbic structures underlying affective responses as well as indirect input to these structures from the somatosensory cortices. These parallel pathways are not equally engaged by all stimuli. Stimuli impinging on somatic structures, particularly those below the neck, are relatively well localized but tend to have a less extensive engagement of emotional circuits. In contrast, stimuli impinging on visceral structures are relatively poorly localized, but engage emotional circuits to a larger degree.[25] Possibly because of the importance of craniofacial structures for survival, stimuli impinging on the structures associated with the head are generally perceived as more intense and with more emotional content.

Descending Circuitry

Finally, there are multiple sites at which ascending sensory information may be modulated. Descending input from brainstem structures such as the periaqueductal gray and rostral ventral medulla (RVM) on to spinal/brainstem dorsal horn neurons and the central terminals of primary afferents have received the most attention in this regard. However, given that there are receptors for both facilitatory and inhibitory transmitters on the peripheral terminals of nociceptive afferents that may be engaged by mediators released systemically, and there is evidence for the interaction between sympathetic efferent and primary afferent terminals in the periphery, modulation of nociceptive signaling starts at the peripheral terminal.[8] Similarly, there is evidence for modulatory processes present at each stage of the ascending nociceptive circuit.

INJURY-INDUCED PLASTICITY

The nervous system is plastic, meaning that it changes in response to stimuli or experience. Tissue injury is a highly effective stimulus for driving plasticity in the nervous system, much of which appears to facilitate pain associated with the injury. The exact nature of the changes depends on both the site and type of injury. Changes in nociceptors in response to both inflammation and nerve injury have been extensively documented and involve processes starting with transduction and ending with transmitter release. With respect to transduction, no molecule has received more attention than the polymodal receptor transient receptor potential vanilloid type 1 (TRPV1). TRPV1 was originally identified as the receptor for capsaicin, the pungent compound of chili peppers.[8] However, this receptor is also activated by heat in the noxious range (>43°C), by protons in relatively highly acidic environments (<pH 5.5), and by a number of endogenous compounds, including the endocannabinoid anadamide, and a class of lipid metabolites, oxidized linoleic acid metabolites, released from heated skin. Several changes in TRPV1 have been described following tissue injury that enable the channel to contribute to the injury-induced enhancement of pain.[8] TRPV1 is a common target for second messenger cascades initiated by inflammatory mediators and is positively modulated by protein kinases A and C, Ca^{2+}-calmodulin dependent kinase II, phosphoinositide 3-kinase, P38, and ceramide. Positive modulation involves an increase in channel open time, a decrease in the threshold for channel activation, and a reduction in channel desensitization. An increase in channel density has also been described as a result of an increase in trafficking of receptors to the membrane. An increase in the translation and transcription of TRPV1 has been described in response to inflammatory mediators, resulting in an increase in channel density. There is also evidence that following nerve injury, there may a shift in the pattern of expression of TRPV1 among subpopulations of sensory neurons, enabling the channel to contribute to nociceptive signaling in afferents that do not normally respond to noxious stimuli.[8]

Because of all the data pointing to TRPV1 as a key contributor to the increase in pain observed following tissue injury, antagonists of this receptor have been pursued as a strategy for the treatment of pain. However, TRPV1 agonists have so far been shown to have the greatest therapeutic efficacy because TRPV1 activation can drive toxic increases in intracellular Ca^{2+}. The result is a generalized inactivation of the nociceptors in which this channel is present, and consequently a dramatic decrease in the sensitivity to noxious stimulation. Since the receptor is also present on the central terminals of nociceptive afferents, this same toxicity can be used to inactivate the central terminals of afferents whose peripheral terminals are not readily accessible. As a result, the intrathecal administration of capsaicin, or the ultrapotent analogue resiniferatoxin, can provide profound relief from bone cancer pain.[26] TRPV1 can also undergo a unique process referred to as pore dilatation, in which prolonged activation is associated with an increase in the channel pore diameter. Because of this unique quality, the channel has also be used to enable the selective delivery of normally membrane-impermeable local anesthetics as a way of producing a long-lasting nociceptor specific block.[27]

In addition to the injury-induced changes in TRPV1 that contribute to an increase in the sensitivity of nociceptive afferents to noxious stimulation, other changes have been described. These include a decrease in leak K^+ currents that, if present in peripheral terminals, would result in an increase in both the amplitude and spread of the generator potential.[8] Increases in voltage-gated Na^+ currents have also been described in association with the sensitization of the channel properties, an increase in trafficking, and an increase in expression.[8] Together, the changes in leak K^+ and voltage-gated Na^+ will result in an increase in the excitability of nociceptive afferents and possibly the emergence of spontaneous activity, a phenomenon referred to as peripheral sensitization. Changes have also been described along afferent axons, the most striking of which have been observed following nerve injury. These include the emergence of sites responsive to thermal and mechanical stimuli along the axon,[8] thought to be due to the aberrant insertion of transducers into the axons of the injured afferents. Aberrant activity also emerges from within the sensory ganglia, possibly as a result of a similar aberrant trafficking of transducers, instability of the neural membrane as a result of changes in the relative density and/or properties of Na^+, K^+, and non-selective cation channels, the activation of satellite glial cells within the ganglia, and/or aberrant coupling between primary afferent and sympathetic postganglionic neurons.[8] Both of these sources of aberrant or ectopic activity, from the axon and from within the ganglia, not only serve as a potential source for spontaneous pain, but also indicate that targeting the primary afferent as a means to treat pain may require blocking the entire neuron. Changes in voltage-gated Ca^{2+} channels have been described, with an increase in the trafficking of channels in the presence of both inflammation and nerve injury. In addition to the changes in ion channels that directly influence neural excitability, there are changes in the expression of receptors that modulate afferent activity secondary to an action on ion channels. Finally, there is evidence for increases in the expression of neurotransmitters that would facilitate afferent input to the CNS.

In the context of dorsal horn circuitry, a distinct change in afferent properties has been described that may have a profound impact on the pain and sensitivity observed following tissue injury. It has long been appreciated that the regulation of intracellular Cl^- in primary afferents is unique compared to neurons in the CNS in that it is maintained at a relatively high level into adulthood.[28] Consequently, activation of ionotropic γ-aminobutyric acid type A ($GABA_A$) receptors, via a feedback inhibitory loop mediated by GABAergic interneuron input on to afferent central terminals, results in a depolarization, termed primary afferent depolarization. The elevated intracellular Cl^- in primary afferents appears to be due to the presence of the $Na^+/K^+/Cl^-$ cotransporter-1 (NKCC1) that carries Cl^- into the neuron down the Na^+ concentration gradient. In the presence of tissue injury, there is first an increase in NKCC1 activity and then a translocation to the membrane, which appears to contribute to an injury-induced shift in GABA signaling from inhibition to excitation.[28] There is evidence that $GABA_A$ receptor activity in the dorsal horn is sufficient to evoke action potentials in the central terminals of nociceptive afferents that are propagated back out to the periphery, a phenomenon referred to as the dorsal root reflex. In the presence of persistent inflammation, other changes may also contribute to the shift in GABA signaling, including an increase in $GABA_A$ receptor density and a decrease in K^+ current.[29] These results suggest that manipulating the intracellular Cl^- concentration in nociceptive afferents may be a novel strategy for treating inflammatory pain.

Plasticity in the Dorsal Horn

Plasticity in the spinal cord/brainstem dorsal horn following tissue injury is also extensive. The overall increase in excitability observed in the dorsal horn has been generically referred to as central sensitization, and involves both intrinsic changes in the excitability of dorsal horn neurons and changes in excitatory and inhibitory synaptic transmission. Increases in excitatory synaptic transmission appear to involve many of the same mechanisms described in the hippocampus that are thought to represent learning, in a process referred to as long-term potentiation.[30] There is evidence for the requisite involvement of the *N*-methyl-D-aspartate (NMDA)-type ionotropic glutamate receptor, which can drive changes in non-NDMA receptors to increase synaptic strength. There is also evidence for multisynaptic or heterosynaptic plasticity, enabling dramatic increases in the peripheral receptive field size of dorsal horn neurons.[31] These changes are due not only to increases in the efficacy of excitatory synaptic transmission, but also to decreases in the efficacy of inhibitory synaptic transmission. Decreases in inhibitory synaptic transmission appear to be due to a number of factors, including a decrease in GABA release, a decrease in the inhibitory receptor activity, and a decrease in the change in membrane potential driven by inhibitory receptor activation. Such changes in dorsal horn circuitry are also thought to contribute to the manifestation of dynamic mechanical allodynia, where low-threshold input is able to drive nociceptive output neurons.

While neurons are ultimately responsible for the transmission of information in the nervous system, there is a growing body of evidence pointing to changes in glial cells as a major contributor to the central sensitization observed following tissue injury.[32] One line of research has focused on microglial cells, the resident immune cells in the CNS. These cells become activated following peripheral injury, releasing a number of cytokines and chemokines that can increase the excitability of dorsal horn neurons. One of the more detailed models to be worked out involves the activation of microglial cells following peripheral nerve injury. While the mechanisms responsible for the upregulation of the ionotropic ATP receptor, P2X4, in microglial cells has yet to be confirmed, once upregulated, ATP-induced activation of P2X4 on microglia results in the release of brain-derived neurotrophic factor (BDNF) from these cells. BDNF then acts on its receptor, TrkB, present on dorsal horn neurons, resulting in a decrease in the activity of the K^+/Cl^- cotransporter KCC2. Because KCC2 is a Cl^- extrusion mechanism, the result is comparable to that associated with an increase in NKCC1 activity in the primary afferent, leading to an increase in intracellular Cl^-. The final result is a dramatic decrease in the efficacy of inhibitory GABA signaling in the dorsal horn.[32] A second line of research has focused on another major class of glial cell in the CNS, the astrocyte. Under normal conditions, this cell facilitates synaptic transmission by helping to remove glutamate and GABA from the synaptic cleft and recycling these amino acids for subsequent use by the neurons. Following tissue injury, however, not only are these transmitter regulatory functions disrupted, resulting in an excess of excitatory neurotransmission, but also these cells may release substances that further increase the excitability of the dorsal horn.

Plasticity at Higher Brain Centers

Injury-induced changes have also been observed at higher brain centers, one of the best described being the changes in the RVM. While there is some debate over whether there is simply a decrease in the descending inhibition from the RVM to the dorsal horn or an increase in descending facilitation, the change in output has a significant influence on the changes in nociceptive threshold observed following tissue injury.[33] One model involves an increase in neuropeptide Y expression in injured primary afferents, resulting in changes in the nucleus gracilis, a structure normally associated with processing low-threshold input to the CNS, that are critical for the changes in the output from the RVM. There is also evidence of changes in glial signaling within the RVM that contribute to the change in output. Additional structures, including the amygdala and prefrontal and cingulate cortices, undergo changes that may contribute to pain associated with tissue injury.[33] Finally, changes in cortical thickness as well as changes in the hippocampus have been applied to explain some of the altered mentation (forgetfulness, trouble concentrating, etc.) observed in the presence of ongoing pain.

LOSS OF HOMEOSTASIS

While mechanistic studies on pain have been focused on factors that contribute to a net increase in nociceptive signaling, several lines of evidence support the suggestion that pain in general, and neuropathic pain in particular, is an emergent property resulting from a loss of balance between nociceptive and non-nociceptive circuitry. In this model, the loss of activity in the nervous system associated with injury or disease may be just as problematic as the increases in excitability that have been the primary focus of mechanistic studies. Evidence in support of this suggestion includes psychophysical data from experiments in which low-threshold afferents were blocked with a pressure cuff where, in the absence of low-threshold afferent input to the CNS, both cooling and low-threshold mechanical stimuli are perceived as painful. These data are consistent with electrophysiological data demonstrating the suppression of nociceptive input to dorsal horn projection neurons by activity in low-threshold afferents.[22] Loss of ascending nociceptive tracks results in the paradoxical hyperexcitability of neurons in the thalamus, a change that correlates with the emergence of hypersensitivity to mechanical and thermal stimuli.[34] These observations take one of the more therapeutically relevant applications of the gate control theory to the next level with the implication that the "normal" level of afferent activity is necessary to restore homeostasis to the nervous system and therefore provide the most effective pain relief.

HOPE FOR THE MAGIC BULLET

With the growing list of changes that appear to contribute to pain associated with tissue injury, it should not be surprising that identification of consistently effective therapeutic interventions has been difficult. This complexity is compounded by a number of problems. First, most proteins serve multiple functions in the body, and selectively targeting these proteins for the treatment of pain is difficult without generating unwanted side effects; second, the underlying mechanisms of pain depend on both the site and type of injury; third, the underlying mechanisms can change with time following an injury; and fourth, the factors that influence the prognosis for pain may influence both the specific mechanisms underlying the pain and the time-course over which the mechanisms are engaged. All of these issues highlight the importance of undertaking a detailed assessment before generating a treatment plan, as well as the need for biomarkers that can be used to mitigate the impact of these complicating features on the development of effective therapeutic strategies. They also stress the need for novel strategies for the treatment of pain that can address these complicating issues.

TREATMENT OF PAIN

Despite considerable progress in the mechanistic understanding of pain, most of the pharmacological approaches introduced over the past decade for the treatment of pain are "improved" formulations of already existing medications (e.g. extended-release opioids). However, mechanistic studies over the past several decades are finally starting to pay off with novel approaches for the treatment of pain. The voltage-gated Ca^{2+} channel toxin ziconotide is one of the best examples of a drug developed specifically for the treatment of pain, although its use is limited by its potential for toxicity and because it can only be delivered to the spinal cord. A major class of new drugs is referred to as "biologicals" because they incorporate biological material such as antibodies.[35] Based on the preclinical data pointing to a role for a number of mediators such as the cytokine tumor necrosis factor-α (TNF-α) in the pain associated with inflammation, drugs that selectively suppress TNF-α signaling are available for the treatment of pain. Similarly, with the understanding that the brain of a migrainuer is hyperexcitable, in combination with the observation that migraine and epilepsy may be comorbid diseases, clinicians have had at least some success using antiepileptics for the prophylactic treatment of migraine. Various neural stimulators have been developed that have therapeutic efficacy under the right conditions. The development of these stimulators was based on data highlighting the interaction between modalities of sensory input, most clearly conceptualized in the gate control theory of pain, highlighting the fact that a loss of neural activity in the nervous system can be just as problematic as the emergence of hyperexcitability. Furthermore, with the understanding that specific populations of afferents may be responsible for ongoing pain, older methodologies such as nerve injections have evolved into a whole new field of interventional pain therapy. Despite these advances, it is becoming increasingly clear that for many who suffer from pain, no therapy is curative. As a result, the best that the medical profession can offer these patients is improvement in pain management and coping strategies for living with pain. From this perspective, quality of life becomes the central focus of treatment, which necessarily involves a team approach with specialists able to maximize the therapeutic utility of the available pharmacotherapy and interventional approaches, combined with experts able to address coping styles and the psychosocial repercussions of pain, as well as physical and occupational rehabilitation.

Pain Assessment

The basis of a successful medical therapy is specific diagnosis and the essential first step in diagnosis is assessment. Unfortunately, this first step is too often ignored in the treatment of pain, with treatments administered based on what is available rather than what is necessary for a particular patient. Despite the proliferation of assessment tools, driven in large part by the complexity of pain, most are not used on a regular basis. The dynamic nature of pain, in terms of both the underlying mechanisms described above (see "Neurobiology of Pain") and its manifestation in those who experience pain, has largely escaped the attention of pain clinicians as something that needs to be assessed systematically. The standard and starting point in pain assessment is a numeric pain rating scale, often referred to as a visual analogue scale, most often 0–10 on a 11-point scale, where 0 is no pain and 10 is the worst pain imaginable. Assessment should also involve a series of questions designed to help determine essential features of the pain such as its localization (to a specific body region), its qualities (ongoing, sharp–shooting, dull aching, or burning), and whether there was an inciting incident. This information can help in the formation of hypotheses about underlying mechanisms for the pain and consequently therapeutic strategies. These initial hypotheses can be further tested with assessment tools designed for more specific types of pain. For example, additional descriptors such as numbness and burning are classic signs of neuropathic pain, whereas a patient with suspected temporomandibular joint disorder might undergo an assessment of range of motion of the jaw, bite force, and tender points. Given the impact of early life experiences on the manifestation of pain, a detailed life history is also essential for a complete patient assessment. Although not widely used because of the time and training required, systematic sensory testing, referred to as QST, can be used to more accurately assess the presence of localized changes in sensory processing, as well as signs of more systemic changes. The results of a detailed QST analysis may also be used to predict the relative amount of pain a patient may experience following a surgical intervention.[20] Similarly, multiple tools are available to assess psychological attributes (e.g. anxiety, depression, catastrophizing) and aspects of daily life (e.g. daily activity, sleep inventory) that can be used to evaluate features that may contribute to pain and be predictive of outcomes for various interventions. Accordingly, the use of currently available assessment tools can guide future research and help clinicians to make informed decisions about patient prognosis and treatment.

Given the premise that the complexity of pain necessitates an integrated approach to management, data on the primary therapeutic modalities used for the treatment of pain are briefly summarized below.

Pharmacotherapy

Administration of analgesics is still the mainstay of pain therapy. NSAIDs are widely used for minor to moderate pain and opioids for moderate to severe pain. While there is disappointment among clinicians and patients, the best we have to offer today are still the essential ingredients of therapies in use for millennia, and both classes of compounds work well in the short term for many patients. Therefore, the most important developments in pharmacotherapy include the more sophisticated use of opioids with the advent of extended-release preparations and patient-controlled delivery devices.

The use of drugs in routine clinical practice is determined by two major factors. The first is the demonstration of efficacy in well-controlled clinical trials. Clinical guidelines are becoming increasingly important in directing clinical care. These are essentially based on data from clinical trials, but are also shaped by the standard of clinical practice, where the balance must be drawn between efficacy and other factors such as side effects and access to care. For example, gabapentin and pregabalin are among the first line therapy for postherpetic neuralgia, while opioids, although effective in clinical trials, are recommended as second or third line therapy because of side effects. In contrast, opioids are first line therapy in cancer and palliative therapy and care.

The following is by no means a comprehensive list. Rather, data on the most commonly used pharmacological approaches are briefly summarized, along with two relatively recent strategies that yielded promising results.

Non-Steroidal Anti-Inflammatory Drugs

NSAIDs remain by far the most commonly used drugs for the treatment of pain.[36] The primary mode of action is blockage of cyclooxygenase (COX), the enzyme responsible for the conversion of arachidonic acid into precursors for a variety of biologically active prostanoids. Two COX isoforms have been identified, COX-1 and COX-2. COX-1 is constitutively expressed throughout the body, but particularly in the gastrointestinal tract, where it contributes to normal upper gastrointestinal function. This distribution and function are responsible for the major deleterious side effects of NSAIDs, which are gastric ulceration and bleeding. COX-2 appears to be upregulated at sites of injury and in the CNS in response to injury. The injury-induced upregulation of COX-2 served as justification for the generation of COX-2-selective antagonists which, despite the revelation of potential cardiac toxicity, were shown to be relatively effective in the treatment of inflammatory pain. Topical NSAID preparations were developed as another solution to the gastrointestinal side effects associated with the systemic administration of NSAIDs. Available evidence suggests that these preparations have therapeutic efficacy and a lower gastrointestinal liability.[37] While oral NSAID preparations are relatively effective for the treatment of mild to moderate pain, intravenous preparations have efficacy for the treatment of more severe pain. One of the most interesting features of NSAIDs is their efficacy in the presence of ongoing inflammation; despite the release of other inflammatory mediators such as interleukin-1β, TNF-α, and nerve growth factor that, like prostanoids, have been shown to directly sensitize nociceptive afferents and therefore should be acting in parallel with the prostanoids, NSAIDs continue to have therapeutic efficacy.

Opioids

The prototypical analgesics, opioids, are among the best studied class of drugs. The major receptor subtypes were pharmacologically characterized approximately four decades ago, with a functional profile and anatomical distribution largely confirmed by subsequent cloning of the three major opioid receptors, mu, delta, and kappa, in the early 1990s. Opioid-induced analgesia is due to a synergistic effect of opioid receptor activity at multiple sites throughout the nervous system, including cortical and subcortical structures, brainstem, spinal cord, and the periphery. Consistent with their widespread distribution, opioid receptors can be therapeutically targeted by delivery of opioids through a variety of routes of administration, including oral, intravenous, transcutaneous, and spinal. Opioids may be used as analgesics in patients of any age group with appropriate adjustment of the dose, which needs to be titrated very carefully to avoid multiple adverse effects, the most serious being respiratory depression. Genetic differences in metabolic enzymes and the opioid receptors themselves may contribute to ethnic, racial, and individual differences in the sensitivity to opioids in general[38] or certain types of opioids such as codeine.

The use of opioids for long-term management of pain has received considerable attention, driven largely by concerns associated with the diversion of these drugs to populations using them for recreational purposes. While the details of this particular issue are beyond the scope of this chapter, several relevant issues should be emphasized. First, the ability to use opioids for long-term management of pain is a relatively new phenomenon, enabled by the development of opioid preparations that provide sustained serum levels of drug. These preparations were a solution to the problem associated with the relatively short half-life of morphine and related compounds, which was difficult to titrate when taken orally. Second, for relatively short-term use (weeks to months), these sustained-release preparations have been shown to be consistently effective for the treatment of a variety of types of pain, including cancer pain as well as a number of neuropathic pain disorders. That said, there have also been a number of negative trials, particularly for the treatment of musculoskeletal pain and headaches. In the latter case, it is generally acknowledged that opioids are contraindicated in treating migraine because of their propensity to drive rebound headache. Third, in the absence of effective ways to mitigate opioid side effects associated with short-term use, these side effects can become particularly problematic with longer term use. Additional problems associated with long-term use include tolerance, functional impairment, and opioid-induced hyperalgesia, a paradoxical condition in which opioids cause pain. Fourth, and probably most importantly, there is little evidence of sustained efficacy for long-term use of opioids for pain management. These data suggest that alternatives to opioid therapy should be pursued in patients not on a trajectory of recovery after an extended period of opioid therapy.

Anticonvulsants and Antiepileptics

Carbamazepine was the first anticonvulsant used for the treatment of pain, and was prescribed for control of paroxysms of trigeminal neuralgia. Carbamazepine also has efficacy in the treatment of PEPD.[11] Despite evidence for the involvement of Na^+ channels in the ongoing pain of nerve injury, this Na^+ channel-blocking drug shows limited efficacy in other pain disorders. Other anticonvulsants such as topiramate and valproate have generated mixed results in a number of trials and, as a class, have been disappointing in the treatment of other types of neuropathic pain. Negative results from some of the older trials should probably be revisited, however, in light of the compelling argument that the pain phenotype may be a far more sensitive way of grouping patients and therefore detecting a therapeutic effect than the pain syndrome *per se*.[7] Finally, anticonvulsants with at least some Na^+ channel-blocking activity appear to have some efficacy as prophylactic treatments for migraine.

The most widely prescribed anticonvulsants, at least for the treatment of neuropathic pain, have been gabapentin and its successor pregabalin. Gabapentin was first reported in open-label trials to be effective for control of complex regional pain syndrome and later in controlled clinical trials for relief of pain in postherpetic neuralgia and diabetic neuropathy. Gabapentin appears to act via binding to the $\alpha_2\delta_1$-subunit of the voltage-gated Ca^{2+} channel, a subunit that appears to be critical for trafficking the α-subunit to neuronal terminals, as well as recycling the α-subunit at the synapse. Gabapentin has little analgesic efficacy in the absence of injury. The observation that it is only in the presence of injury that the $\alpha_2\delta_1$-subunit is dramatically upregulated[39] suggests that the upregulation confers the therapeutic utility and consequently relatively limited efficacy of this drug. Pregabalin is an analogue of gabapentin and has better pharmacokinetic properties than gabapentin; it has been approved for treatment of painful diabetic neuropathy and postherpetic neuralgia as well as fibromyalgia.

Antidepressants

Tricyclic antidepressants (TCAs) were not originally developed for the treatment of major depression, but rather as a class of antipsychotic drug that acted as serotonin reuptake inhibitors. It was quickly realized, however, that TCAs had mood-stabilizing effects that were confirmed in clinical trials, leading to US Food and Drug Administration (FDA) approval of amitriptyline for the treatment of major depression in 1961. TCAs have been used for the treatment of neuropathic pain since the 1960s. This class of compounds has efficacy for the treatment of various pain syndromes including diabetic neuropathy, postherpetic neuropathy, atypical facial pain, and temporomandibular disorder. TCAs also have at least a limited efficacy when used prophylactically for the treatment of migraine. It is still believed that the therapeutic efficacy of TCAs in the treatment of pain involves increasing serotonin levels, possibly engaging the descending pain inhibitory circuitry. However, these compounds have efficacy in the treatment of pain at doses far lower than those needed for the treatment of depression, and with a far more rapid onset of effect. One explanation for these different pharmacological properties is an "off-target effect", given that these compounds also block voltage-gated Na^+ channels and NMDA receptors.[40] However, even in the best of trials, the number of patients needed to treat is close to three. The utility of TCAs is also limited by their side effects, including dizziness, drowsiness, confusion, and hypotension, which can be particularly problematic in elderly people.

While the second generation antidepressants, the selective serotonin reuptake inhibitors (SSRIs), were generally shown to have limited efficacy in the treatment of pain, the third generation of antidepressants that are both serotonin and norepinephrine reuptake inhibitors (SNRIs) appear to have both analgesic efficacy and a more limited side-effect profile than the TCAs. The SNRI duloxetine is approved for the treatment of both diabetic neuropathy and fibromyalgia, while others such as milnacipran are approved only for fibromyalgia. Although the analgesic efficacy of antidepressants may be limited, they have the potential for significantly greater therapeutic utility because they can also help with other symptoms associated with pain, including depression, anxiety, and sleep disorders.

N-Methyl-D-Aspartate Receptor Blockers

The weight of preclinical evidence in support of a role for the NMDA receptor in mediating central sensitization as well as opioid tolerance has had researchers returning repeatedly to this receptor as a potential target for the treatment of pain. The two compounds that have received the most attention are ketamine and dextromethorphan. Ketamine is a relatively potent NMDA receptor antagonist approved for use in general anesthesia. An intraoperative bolus injection of ketamine may have short-term benefits in reducing postoperative pain and opioid consumption. In one study of patients with back pain, opioid consumption at 6 weeks was significantly reduced in the ketamine treatment group, suggesting that ketamine may also have long-term benefits.[41] Very low-dose ketamine has being used for postoperative pain management, with promising results.[42] While more study is needed on the potential long-term benefits of ketamine, the current hypotheses are that it may be able to "reset" the system by reversing the central sensitization that amplifies pain signaling. Unfortunately, the side effects of ketamine, in particular its significant psychotropic actions, continue to limit its more widespread use outside the clinical setting. Dextromethorphan is a relatively weak antagonist of the NMDA receptor, widely used as an antitussive agent, which may have some efficacy for postoperative pain management. Results from trials on the management of ongoing pain have been less impressive, although some of the most compelling data come from trials in the management of cancer pain.

Capsaicin Patch and Intrathecal Resiniferotoxin

As discussed in the section "Injury-Induced Plasticity" (above), it has long been appreciated that the TRPV1 agonist capsaicin can be used to treat pain, as it has been included in rubs and liniments for the treatment of minor aches and pains for centuries. With improved understanding of the mechanisms of analgesic action, more directed use of TRPV1 agonists has been developed. These include the intrathecal resinferatoxin,[26] currently approved for use in veterinary medicine for the treatment of cancer pain and in a clinical trial in humans for the same indication, and a high-concentration capsaicin patch.[43] Clinical trials with the patch were particularly challenging, as the burning sensation of capsaicin makes it particularly problematic for blinding. Nevertheless, using high- and low-concentration patches, both of which cause a burning sensation, it was demonstrated that the high-concentration patch was associated with superior pain relief that lasted significantly longer (up to 3 months) in the treatment of postherpetic neuralgia and HIV neuropathy.

Interventional Pain Therapies

Various manipulations are generally grouped under the heading of "interventional pain medicine". These are the most common procedures performed in pain clinics and usually involve focal injections into sites thought to be responsible for the pain. For example, steroid injections are often used to treat herniated vertebral discs thought to be the source of inflammatory mediators driving back pain. Similarly, local anesthetics are used to transiently block nerves thought to be responsible for pain or hypersensitivity. The site of injection depends on the site of the pain and whether it is necessary to block more than one nerve. If the correct nerve is blocked, this approach can provide rapid and effective pain relief. The duration of pain relief often extends well beyond the duration of the nerve block. The mechanisms underlying these long-lasting effects have yet to be confirmed, but as with NMDA receptor antagonists, the current hypothesis is that block of the offending nerve enables the system to reset such that the re-establishment of central sensitization may take days to weeks. That said, long-term blocks are achieved with the injection of substances such as phenol and ethyl alcohol that damage the axons in the problem nerve, effectively denervating the painful tissue. Special needles may also be used to achieve the same goal by delivery of electric current to produce electrolytic lesions, or radiofrequency pulses to produce thermal lesions. These strategies are more often used to treat visceral pain, where other strategies, such as the topical application of drugs, are not possible. Finally, a variety of nerve stimulators has been developed to enable stimulation of both peripheral nerves and the spinal cord. The interventions for which there are compelling data to support efficacy include both peripheral and central nerve stimulators, catheters for the delivery of local anesthetics for the management of postoperative pain, and intrathecal pumps for the delivery of a variety of compounds to the spinal cord to manage regional pain. Risks associated with the use of invasive stimulators such as those implanted in the spinal cord or brain remain a concern, as does the long-term efficacy of this approach. Furthermore, because of the limitations of the spinal route of delivery, including the risk of infection, intrathecal pumps are most often used in the management of pain at the end of life. Although there are many success stories with all of the interventional approaches currently in use, limited data have been published in the peer-reviewed literature supporting the efficacy, duration, and safety of these interventions.

Psychological Interventions

Patients may be resistant to employing psychology-based approaches to help manage their pain for fear that their very real pain is being dismissed as something "all in their head". The health-care profession has not helped this situation, often dismissing patients who are not responsive to more "mechanism"-based strategies as neurotic or concluding that in the absence of a physical manifestation of a problem, the pain must be "psychogenic". Nevertheless, it is clear that there is a dynamic interplay between pain and psychological constructs that can create a feedforward situation whereby an increase in pain leads to an increase in anxiety and depression that can lead to an increase in pain. Thus, interventions designed to break this cycle, even if they cannot "cure" the pain, may have significant therapeutic utility. Towards that end, three general approaches are employed most often that may be used in combination. These are cognitive behavioral therapy (CBT), stress/relaxation approaches, and biofeedback. CBT is a relatively structured system designed to enable patients to develop strategies to better deal with negative emotions, behaviors, and ways of thinking. There are several stress/relaxation techniques that are designed to enable patients not only to reduce muscle tension and anxiety, but also to develop strategies for more effectively dealing with stressful events in their lives. Finally, biofeedback employs any number of physiological endpoints such as heart rate, galvanic skin conductance, or a functional magnetic resonance imaging (fMRI) signal played back to the patient in real time in the form of a signal such as a tone that the patient can ultimately learn to control. While larger trials will be needed to confirm both efficacy and duration of effect, the limited evidence available suggests that with an fMRI signal based on brain activity in the "pain matrix", the patient can learn to control his or her own pain.[44]

Clinical trials with psychology-based approaches are difficult to run if only because of difficulty in identifying the most appropriate control. Waiting lists are often used for comparisons, but these do not enable researchers to determine whether there are specific aspects to an intervention that are particularly efficacious. Treatment as usual is also used. Nevertheless, the available data suggest that these strategies have weak effects on improving pain, and usually only immediately after treatment. With CBT, longer lasting effects are observed on disability scores as well as mood and catastrophizing outcomes. Relatively short-lasting effects are also observed for stress/relaxation approaches. Finally, as just noted, additional data will be needed to assess the long-term efficacy of biofeedback strategies.

Physical Therapy Modalities

There is a wide range of physical therapy modalities, ranging from passive and active movements to aerobic conditioning. Included under this subheading are chiropractic and massage-based therapies. A number of studies has been published on specific modalities applied to singular areas of the body, and in aggregate each of the modalities provides modest pain relief of limited duration.[45] Importantly, however, as with other therapeutic modalities, demonstrating efficacy with physical therapy modalities may be more of a problem in applying the right approach to the right patient than a limitation of the approaches in general. For example, there is evidence that by appropriately matching a patient to a therapeutic strategy, chiropractic manipulations may be highly effective for the treatment of low back pain.[46] The results of these types of study underscore the need for additional diagnostic criteria that can be used to match patients with therapies.

Complementary and Alternative Medicine

Complementary and alternative medicine is often based on traditional medical practices, and it has received notoriety as a complement to "modern" physiology-based medicine, especially in areas of unmet needs, such as pain. The term is now applied to strategies ranging from the use of diets and food supplements to non-traditional psychological approaches such as hypnosis. To meet the demands of regulatory authorities, there is much data on the efficacy of dietary supplements for the treatment of pain. For example, based on the notion that osteoarthritic joints may be painful because of a loss of the protective lining around joints, glucosamine, chondroitin, and their combination were touted as effective treatments. While negative results may again be due to patient heterogeneity, what looked like a promising non-invasive therapy for a widespread problem has failed to show a significant effect in several large trials.

Of the more traditional CAM approaches, acupuncture has probably received the most attention in both preclinical studies and clinical trials. While specific mechanisms are still debated, there is compelling evidence to support the notion that proper needle placement results in a number of physiological changes. Most relevant to pain control is the release of endogenous opioids and anti-inflammatory mediators in response to acupuncture. Results from well-controlled clinical trials in which sham needling is used as an active control for acupuncture are consistent with preclinical data demonstrating a moderate effect of acupuncture for the treatment of a variety of pain syndromes.[47]

CONCLUSION

Pain continues to be a tremendous problem both for those who suffer and for society. Progress in the development and implementation of consistently effective therapies for the treatment of pain devoid of deleterious consequences has been slow. However, two major lines of investigation appear to be converging. The first of these is the generation of data on the underlying mechanisms of pain. The complexity of the problem has been highlighted by the revelation of details about mechanisms that are influenced by a variety of factors that affect both the prognosis for recovery and the efficacy of available therapeutic strategies. The second is the appreciation of the heterogeneity of pain, with the growing realization that an individualized multidisciplinary approach may be necessary for the development and implementation of the most effective treatment plans. Considerably more work is needed at both the basic science and clinical levels, but there is reason to be hopeful that more effective strategies for the treatment of pain will be available in the near future.

References

1. Committee on Advancing Pain Research, Care, and Education, Pizzo PA, et al. *Relieving Pain in America: A Blueprint for Transforming Prevention, Care, Education, and Research*. Washington, DC: National Academies Press; 2011.
2. Moore RA, Wiffen PJ, Derry S, McQuay HJ. Gabapentin for chronic neuropathic pain and fibromyalgia in adults. *Cochrane Database Syst Rev (Online)*. 2011;(3):CD007938.
3. Wilken T. My search for a definition of disease. SynEARTH.network. vol. 2013. Manila Home Page Archive; 2002. http://www.synearth.net/TheUnifiedStressConcept.pdf. Accessed September 2012.
4. Haddad SL, Coetzee JC, Estok R, Fahrbach K, Banel D, Nalysnyk L. Intermediate and long-term outcomes of total ankle arthroplasty and ankle arthrodesis. A systematic review of the literature. *J Bone Joint Surg Am*. 2007;89(9):1899–1905.
5. Loeser JD, Treede RD. The Kyoto protocol of IASP basic pain terminology. *Pain*. 2008;137(3):473–477.
6. Backonja MM, Stacey B. Neuropathic pain symptoms relative to overall pain rating. *J Pain*. 2004;5(9):491–497.
7. Baron R, Forster M, Binder A. Subgrouping of patients with neuropathic pain according to pain-related sensory abnormalities: a first step to a stratified treatment approach. *Lancet Neurol*. 2012;11(11):999–1005.
8. Gold MS, Gebhart GF. Nociceptor sensitization in pain pathogenesis. *Nat Med*. 2010;16(11):1248–1257.
9. Kim CC, Bogart MM, Wee SA, Burstein R, Arndt KA, Dover JS. Predicting migraine responsiveness to botulinum toxin type A injections. *Arch Dermatol*. 2010;146(2):159–163.
10. Edwards RR, Cahalan C, Mensing G, Smith M, Haythornthwaite JA. Pain, catastrophizing, and depression in the rheumatic diseases. *Nat Rev Rheumatol*. 2011;7(4):216–224.
11. Dib-Hajj SD, Yang Y, Black JA, Waxman SG. The Na(V)1.7 sodium channel: from molecule to man. *Nat Rev Neurosci*. 2013;14(1):49–62.
12. Drossman DA. Abuse, trauma, and GI illness: is there a link? *Am J Gastroenterol*. 2011;106(1):14–25.
13. Ren K, Anseloni V, Zou SP, et al. Characterization of basal and re-inflammation-associated long-term alteration in pain responsivity following short-lasting neonatal local inflammatory insult. *Pain*. 2004;110(3):588–596.
14. Gibson CA. Review of posttraumatic stress disorder and chronic pain: the path to integrated care. *J Rehabil Res Dev*. 2012;49(5):753–776.
15. Yarnitsky D. Conditioned pain modulation (the diffuse noxious inhibitory control-like effect): its relevance for acute and chronic pain states. *Curr Opin Anaesthesiol*. 2010;23(5):611–615.
16. Mogil JS. Pain genetics: past, present and future. *Trends Genet*. Jun 2012;28(6):258–266.
17. Berkley KJ. Sex differences in pain. *Behav Brain Sci*. 1997;20(3):371–380.
18. de Vries B, Frants RR, Ferrari MD, van den Maagdenberg AM. Molecular genetics of migraine. *Hum Genet*. 2009;126(1):115–132.
19. Joseph EK, Levine JD. Hyperalgesic priming is restricted to isolectin B4-positive nociceptors. *Neuroscience*. 2010;169(1):431–435.
20. Nielsen CS, Staud R, Price DD. Individual differences in pain sensitivity: measurement, causation, and consequences. *J Pain*. 2009;10(3):231–237.
21. Gold MS. Molecular biology of sensory transduction. In: McMahon SB, Tracey I, Turk DC, eds. *Wall and Melzack's Textbook of Pain*. 6th ed. London: Elsevier; 2013:31–47.
22. Melzack R, Wall PD. Pain mechanisms: a new theory. *Science*. 1965;150(3699):971–979.
23. Fields HL, Basbaum AI. Brainstem control of spinal pain-transmission neurons. *Annu Rev Physiol*. 1978;40:217–248.
24. Price DD. Central neural mechanisms that interrelate sensory and affective dimensions of pain. *Mol Interv*. 2002;2(6):392–403. 339.
25. Ness TJ, Gebhart GF. Visceral pain: a review of experimental studies. *Pain*. 1990;41:167–234.
26. Brown DC, Iadarola MJ, Perkowski SZ, et al. Physiologic and antinociceptive effects of intrathecal resiniferatoxin in a canine bone cancer model. *Anesthesiology*. 2005;103(5):1052–1059.
27. Binshtok AM, Bean BP, Woolf CJ. Inhibition of nociceptors by TRPV1-mediated entry of impermeant sodium channel blockers. *Nature*. 2007;449(7162):607–610.
28. Price TJ, Cervero F, Gold MS, Hammond DL, Prescott SA. Chloride regulation in the pain pathway. *Brain Res Rev*. 2009;60(1):149–170.
29. Zhu Y, Dua S, Gold MS. Inflammation-induced shift in spinal GABAA signaling is associated with a tyrosine-kinase dependent increase in GABAA current density in nociceptive afferents. *J Neurophysiol*. 2012;108(9):2581–2593.
30. Sandkuhler J, Gruber-Schoffnegger D. Hyperalgesia by synaptic long-term potentiation (LTP): an update. *Curr Opin Pharmacol*. 2012;12(1):18–27.
31. Dubner R. The neurobiology of persistent pain and its clinical implications. *Suppl Clin Neurophysiol*. 2004;57:3–7.
32. Trang T, Beggs S, Salter MW. ATP receptors gate microglia signaling in neuropathic pain. *Exp Neurol*. 2012;234(2):354–361.
33. Hughes JP, Chessell I, Malamut R, et al. Understanding chronic inflammatory and neuropathic pain. *Ann N Y Acad Sci*. 2012;1255:30–44.
34. Wang G, Thompson SM. Maladaptive homeostatic plasticity in a rodent model of central pain syndrome: thalamic hyperexcitability after spinothalamic tract lesions. *J Neurosci*. 2008;28(46):11959–11969.
35. Chessell IP, Dudley A, Billinton A. Biologics: the next generation of analgesic drugs? *Drug Discov Today*. 2012;17(15–16):875–879.
36. Shah S, Mehta V. Controversies and advances in non-steroidal anti-inflammatory drug (NSAID) analgesia in chronic pain management. *Postgrad Med J*. 2012;88(1036):73–78.
37. Derry S, Moore RA, Rabbie R. Topical NSAIDs for chronic musculoskeletal pain in adults. *Cochrane Database Syst Rev (Online)*. 2012;9:CD007400.
38. Hastie BA, Riley III JL, Kaplan L, et al. Ethnicity interacts with the OPRM1 gene in experimental pain sensitivity. *Pain*. 2012;153(8):1610–1619.
39. Luo ZD, Chaplan SR, Higuera ES, et al. Upregulation of dorsal root ganglion (alpha)2(delta) calcium channel subunit and its correlation with allodynia in spinal nerve-injured rats. *J Neurosci*. 2001;21(6):1868–1875.
40. Amir R, Argoff CE, Bennett GJ, et al. The role of sodium channels in chronic inflammatory and neuropathic pain. *J Pain*. 2006;7(5 suppl 3):S1–S29.
41. Loftus RW, Yeager MP, Clark JA, et al. Intraoperative ketamine reduces perioperative opiate consumption in opiate-dependent patients with chronic back pain undergoing back surgery. *Anesthesiology*. 2010;113(3):639–646.
42. Costantini R, Affaitati G, Fabrizio A, Giamberardino MA. Controlling pain in the post-operative setting. *Int J Clin Pharmacol Ther*. 2011;49(2):116–127.
43. Treede RD, Wagner T, Kern U, et al. Mechanism- and experience-based strategies to optimize treatment response to the capsaicin 8% cutaneous patch in patients with localized neuropathic pain. *Curr Med Res Opin*. 2013;29(5):527–538.

44. Chapin H, Bagarinao E, Mackey S. Real-time fMRI applied to pain management. *Neurosci Lett*. 2012;520(2):174–181.
45. Kay TM, Gross A, Goldsmith CH, et al. Exercises for mechanical neck disorders. *Cochrane Database Syst Rev (Online)*. 2012;8:CD004250.
46. Delitto A, George SZ, Van Dillen LR, et al. Low back pain. *J Orthopaed Sports Phys Ther*. 2012;42(4):A1–57.
47. Vickers AJ, Cronin AM, Maschino AC, et al. Acupuncture for chronic pain: individual patient data meta-analysis. *Arch Intern Med*. 2012;172(19):1444–1453.

CHAPTER

42

Migraine

David Borsook, Nasim Maleki*, Rami Burstein*[†]

*PAIN Group, Department of Anesthesia, Boston Children's Hospital and Harvard Medical School, Boston, Massachusetts, USA; [†]Department of Anesthesia, Beth Israel Deaconess Hospital, Harvard Medical School, Boston, Massachusetts, USA

OUTLINE

INTRODUCTION

Migraine is the most common neurological disease and it is ranked 12th among women and 19th in the general population for the degree of handicap it causes. The World Health Organization (WHO) notes migraine to be in the top 20 causes of disability worldwide, affecting about 15% of the population. The condition affects some 45 million Americans.[1] Other estimates are around 11–17% of adults in Western

Neurobiology of Brain Disorders
http://dx.doi.org/10.1016/B978-0-12-398270-4.00042-2

societies, with patients being affected during the formative and most productive periods of their lives, between 25 and 55 years. It frequently starts in childhood, particularly around puberty, and affects women more than men. Estimated annual health-care costs related to migraine are around 1 billion euro in France, and the annual cost to US society is estimated to be around 13 billion USD.[2]

Migraine is a disabling headache disorder characterized by intermittent attacks. The pain associated with migraine usually lasts between 4 and 72 hours (untreated or unsuccessfully treated). Various physiological and emotional stressors (e.g. pain, tiredness, nausea, vomiting, photophobia, phonophobia, or irritability) may be associated with each attack. The headache pain has at least two of the following characteristics: unilateral location, pulsating quality, moderate to severe pain intensity, and aggravation by routine physical activity.

Migraine may be episodic (14 or fewer headache days per month for more than 3 months) or chronic (more than 14 headache days per month for more than 3 months) (see http://ihs-classification.org/en) and occurs with or without aura. Aura refers to focal neurophysiological symptoms that usually precede or sometimes accompany the headache, although migraine aura may occur without headache. The most common type of aura in migraineurs is visual in nature and the most notable of these are scotomas. The divisions of patients with and without aura are somewhat arbitrary in terms of the disorder but reflect increasing severity of the disease when aura is present. Patients with aura are considered to have more damaging effects on their brains than those without aura. As a condition, migraine is complex and may be considered in a continuum, not only in terms of the periheadache changes in episodic migraine but also in the progression to high-frequency and chronic daily headache that takes place in some patients.[3] There are two major processes relating to the latter: adaptive responses to each migraine attack (i.e. the individual migraine attacks), and adaptive responses over time with disease modification (i.e. with disease progression, also known as chronification or transformation).

THE MIGRAINE SPECTRUM: AN OVERVIEW OF CLINICAL MANIFESTATIONS

Fundamental processes going on in the migraine patient may be categorized into peri-ictal attack events (preictal, ictal, and postictal), and interictal migraine events. Related comorbidities are also associated with migraine. The clinical manifestations of migraine within these domains, as summarized in Fig. 42.1, are discussed below.

Peri-ictal Clinical Manifestations

Prodromal Changes

Whereas premigraine events occur immediately before the headache, prodromal events occur hours and sometimes days before the migraine onset. Prodromal events include irritability or other mood changes, appetitive changes, dizziness, confusion, and imbalance. The most common premonitory symptoms described include fatigue (in about 47%), phonophobia (36%), and yawning (36%).[4]

Preictal Changes

The preictal phase is a period immediately before the attack episode that may not occur in all patients. Aura is a pre-eminent clinical manifestation in the preictal phase in some patients. The manifestation of spreading scotoma by Lashley in 1941 is a classic manifestation of visual aura.[5] Other aura symptoms include sensory, motor, or verbal disturbances. Some patients have visual auras without headache. Visual aura symptoms are thought to occur in around 20–30% of patients with migraine and sensory auras in around 30% of patients. Sensory auras include tingling and numbness.

Ictal Changes

Symptoms include pain from the headache itself and exacerbation by normally non-painful sensory stimuli (light, noise, smell), dizziness, and autonomic phenomena (see "Autonomic Features, Including Nausea and Vomiting", below). Photophobia (80%) and phonophobia (76%) are the most common symptoms in migraine patients, followed by nausea (73%) and vomiting (29%).[6]

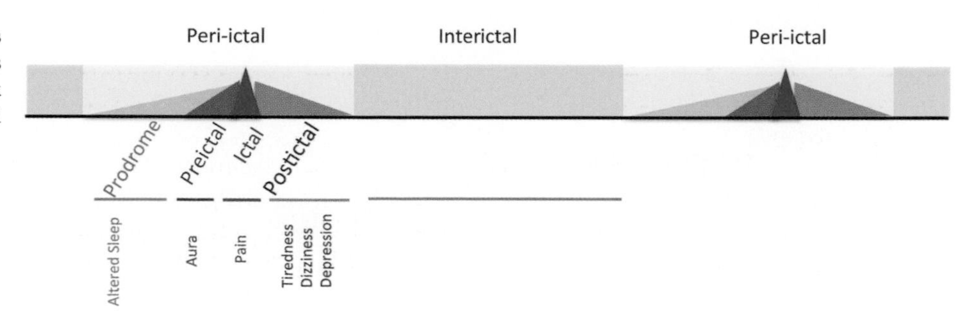

FIGURE 42.1 **Clinical manifestations across the migraine spectrum.** Processes may be categorized into peri-ictal attack (preictal, ictal, and postictal) and interictal migraine events. See text for details.

The pain is usually unilateral but may be bilateral and throbbing in nature. In addition to headache, associated neck pain is common in about 30% of patients.[7] The pain may vary in intensity as well as duration and may last for hours to days. In addition, if the headache persists, allodynia (where non-noxious stimuli evoke pain) may be present in cutaneous areas unaffected by the headache.

Postictal Changes

Symptomatic effects of a migraine may linger for days after an attack. These include tiredness or fatigue, mild headache, dizziness, mood alterations including depression, and diminished appetite. Such changes are suggestive of a major brain insult and presumably indicate recovery processes.

Interictal Migraine Events

The definition of the interictal migraine state is not clear since prodromes and premigraine attack events may be subtle, taking the form of altered behavior or mood changes such as tiredness or irritability.

Comorbidity-Related Issues

Migraine is associated with a number of comorbid processes, including sleep disruption and depression, especially in chronic migraine.[8] Migraine is also comorbid with numerous brain diseases that are associated with stress. However, stress effects may be subtle and not easily detected in migraine patients using standard measures (e.g. cortisol). It may be that the burden of migraine contributes to the psychiatric and neurological changes observed in these patients. Cognitive changes are difficult to determine, and lifetime diagnosis of migraine does not support association with cognitive deficits in middle age.[9] However, migraine is a stressor and as such may affect an individual's disease burden or allostatic load, as will be discussed later (see "Migraine and Allostatic Load".[10]

MIGRAINE MECHANISMS

It is not understood what generates a migraine attack, although several mechanisms based on brain alterations or activation of the trigeminal nerve–vascular system (Fig. 42.2) are clearly implicated. The debate is ongoing as to whether migraine is generated from peripheral or central changes. Some aspects of this relating to the notion that the brainstem may act as a migraine generator were reviewed in a 2012 publication.[11] Other central putative effector regions include the hypothalamus, mainly because of its known role in cyclicity including circadian rhythms, hormones related to stress, and gastrointestinal peptide influences in eating and satiety that also involve the nucleus of the solitary tract. Whatever the initiator, a sequence of events seems to be at play, including: (1) initiator (evoked or spontaneous); (2) clinical or subclinical spreading depression; (3) activation of trigeminovascular nociceptors; (4) peripheral and central sensitization of trigeminovascular pathways; (5) brain hypersensitivity; (6) brain hyperexcitability; and (7) alterations in brain structure. Some of these are noted in Fig. 42.3.

Cortical and Subcortical Spreading Depression

Cortical spreading depression (CSD) is an electrophysiological phenomenon characterized by a wave of excitation followed by inhibition in cortical neurons. The aura phase that precedes migraine headache in about 20–30% of migraineurs may be a direct consequence of the events of CSD. The role of CSD in migraine patients without aura is not well understood. Some studies have shown subcortical spreading depression.[12] Spreading depression may be the basis for neuroanatomical and functional (central hypersensitivity) changes observed in migraine patients. The mechanism of CSD involves complex molecular changes in cortical upregulation of genes involved in inflammatory processing (e.g. for cyclooxygenase-2, tumor necrosis factor-α, interleukin-1β, galanin, or metalloproteinases). Metalloproteinase activation produces leakage of the blood–brain barrier, which leads to the release of a variety of molecules (e.g. potassium, nitric oxide, adenosine) that may then sensitize the dural perivascular trigeminal afferents. CSD is reported to activate dural nociceptors; specifically, migraine with aura is initiated by waves of CSD that lead to delayed activation of the trigeminovascular pathway,[13] thus linking a central event with peripheral nociceptive pathways.

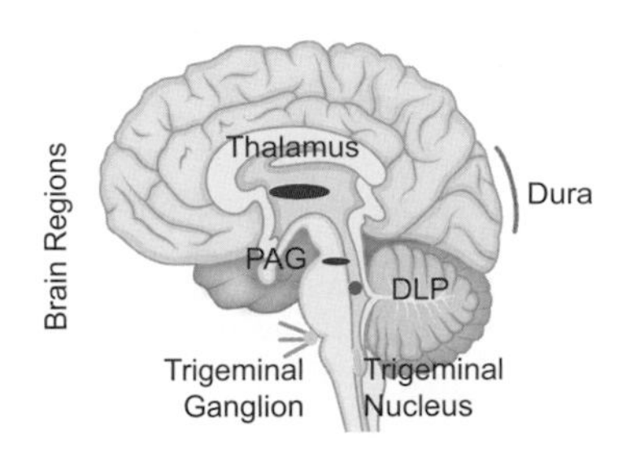

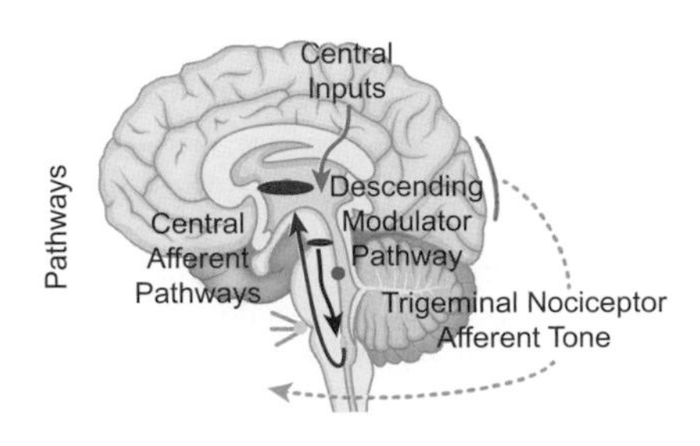

FIGURE 42.2 **Trigeminovascular system.** The figure identifies peripheral and central structures involved in nociceptive drive (dura, trigeminal afferents; trigeminal ganglion and trigeminal nucleus); modulation of nociceptive inputs [dorsolateral pons (DLP), periaqueductal gray (PAG)], and cortical processing (thalamus, S1). *Source: Adapted from Borsook and Burstein.* Cephalalgia. *2012;32(11):803–812.*[11]

Trigeminovascular System

The trigeminovascular system (TVS) is thought to comprise neurons located in the trigeminal ganglion that innervate the cerebral vasculature including the dura mater (Fig. 42.2). Afferent fibers, mainly from the first or ophthalmic division of the trigeminal nerve, project from these latter areas to the ganglion neurons and these neurons have centrally projecting axons to the trigeminal nucleus. Several neurotransmitters and neuropeptides [e.g. calcitonin gene-related peptide (CGRP) and neurokinin-1] are produced in the ganglion neurons that may modulate pain transmission and vascular tone. The TVS in migraine has been the topic of a number of reviews.[14] Following a report of activation in the trigeminal system in migraine subjects,[15] a 2011 study described increased activity during and before migraine attacks in the trigeminal nucleus.[16] Such data support the notion that a presensitized state may exist for the peripheral and central nervous systems in migraine patients.

Genetics

Family history and twin studies have suggested that genetic predisposition plays a major role in the occurrence of migraine. A genetic association with migraine was first observed and defined in patients with familial hemiplegic migraine (FHM).[17] Three FHM genes have been defined, all of which encode ion transporters, and may be the basis for neuronal hyperexcitability observed in migraine. The TT genotype of the methylenetetrahydrofolate reductase (*MTHFR*) gene seems to be associated with migraine and in particular in patients with aura.[18]

Genes associated with hemiplegic migraine have been found on chromosomes 19, 1, and 2 – *ACNA1* (19p13), *ATP1A2* (1q23), and *SCN1A* (2q24) – but these do not describe migraine at a population level. One can

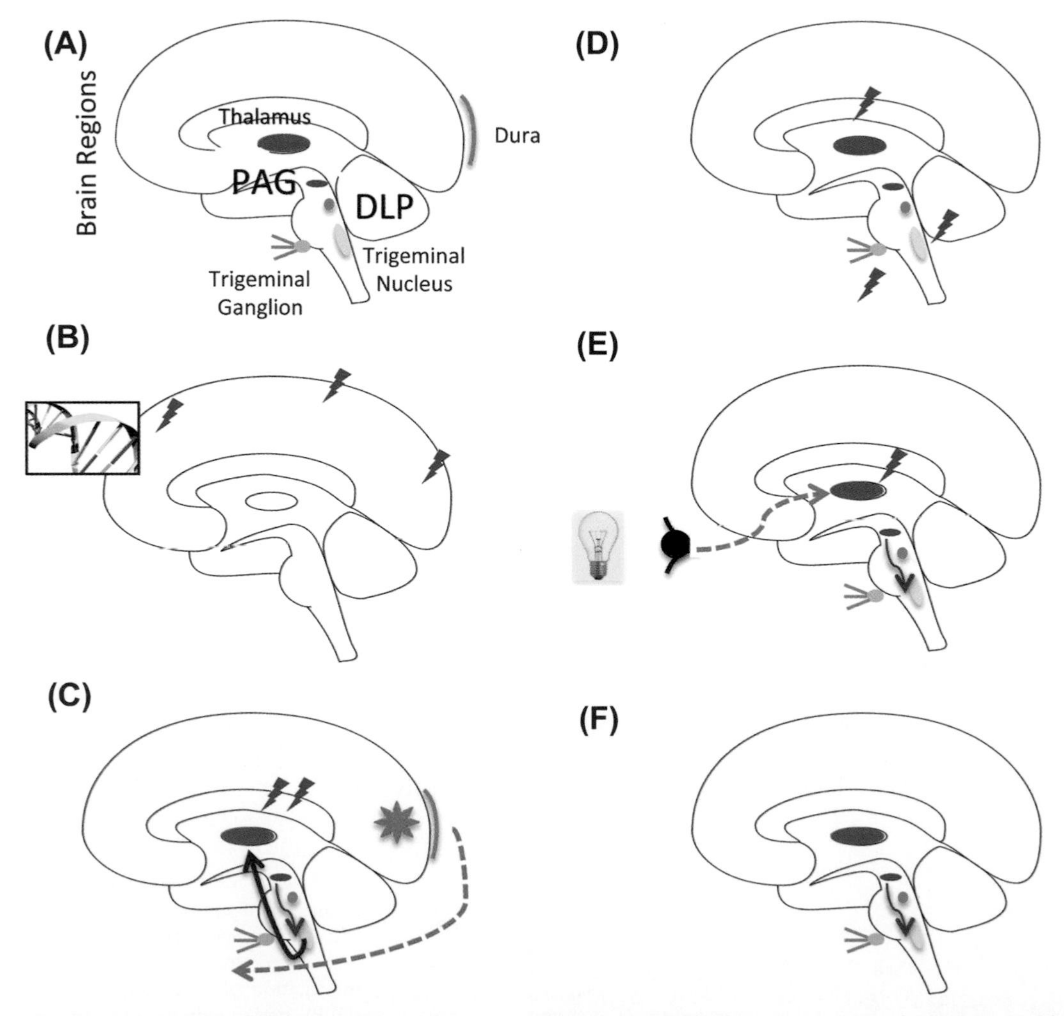

FIGURE 42.3 **Migraine mechanisms.** (A) Trigeminovascular system; (B) genetic contributions to migraine: the hyperexcitable brain; (C) cortical spreading depression: onset of migraine attack; (D) peripheral and central sensitization: progressive cascade leading to localized and extended allodynia; (E) photophobia: exacerbation of migraine pain by light; (F) altered descending modulation: pain facilitation. PAG: periaqueductal gray; DLP: dorsolateral pons.

also look at specific mechanisms (i.e. target genes, e.g. serotonin, dopamine, neurogenic inflammation, hormonal regulation), but most efforts in these domains have failed, except for the TT genotype of *MTHFR6777* (noted above). In a genome-wide association study (GWAS) of genes at a population level, Anttila and colleagues[19] evaluated nearly 3000 migraine cases and found the minor allele of *rs1835740* on chromosome 8q22.1 to be a risk factor for migraine. In another GWAS in nearly 5000 migraineurs and 18,000 non-migraineurs, *Prdm16* (1p36.32), *TRPM8* (2q37.1), and *LRP1* (12q13.3) were found to show the highest association with migraine.[20] Other GWAS have found that 1q32 in *MEF2D* and 3p24 near *TGFBR2* are associated with an increased risk of migraine without aura.[21] Thus, the top five single-nucleotide polymorphisms (SNPs) so far identified at population level that may have effects on migraine are: *Rs18357450*, glutamate transporter; *TRPM8*, sensor for cold and cold-induced pain; *LRP1*, modulates glutamate transmission; *MEF2D*, regulates complex neuronal differentiation; and *TGFBR2*, encodes the growth factor β-receptor-2 (general vascular components that may be important in migraine patients). There are issues with epidemiological studies since differentiation of subgroups, comorbidities, and epigenetic factors is significant in such studies and the phenotype under study is not usually clearly defined. However, no precise mechanism has yet been found that may contribute to predicting migraine in patients or potential treatment approaches. It should also be noted that epigenetic factors may alter gene processing. For example, the prevalence of colic in infants of mothers suffering from migraine is more than twice as high as in infants of healthy mothers (29% vs 11%), and colic may be an early expression of migraine. Colic can thus be considered a manifestation of a migrainous phenotype in childhood, which may evolve to migraine in adulthood as a consequence of alterations over time relating to environmental inputs (stress, hormones, etc.) on an underlying genetic constitution.

Pain Signals from the Dura: Cascade and Crescendo (Central Sensitization)

During a migraine attack, most patients exhibit cutaneous allodynia that may be localized (e.g. to the face) or extended (outside the facial area to include limbs and torso). The evolution of cutaneous allodynia in a patient has been described in detail elsewhere.[22] The initial activation can be considered to be a consequence of peripheral nociceptor activation; that is, sensitization can mediate the symptoms of intracranial hypersensitivity (e.g. throbbing headache). Subsequent sensitization of second order neurons is thought to mediate the development of cutaneous allodynia on the ipsilateral head. The cascade continues with subsequent sensitization of third order neurons that produces the clinical manifestation of cutaneous allodynia on the contralateral head and extracephalic regions. The neurobiology has specific clinical relevance in that the use of antimigraine drugs early in the attack can limit the cascade of sensitization, and specifically central sensitization.[23,24] These sequences are shown in Fig. 42.4.

The Hyperexcitable Brain: Insights from Photophobia, Phonophobia, and Osmophobia

A hyperexcitable brain state exists in both the ictal and interictal periods. The basis for the excitable state is supported by evidence from behavioral and functional imaging studies. The biochemical basis is also supported by alterations in neurotransmitters involved in neurotransmission. Studies have reported increased platelet levels of neuroexcitatory amino acids including aspartate, glutamate, glutamine, and glycine in migraine patients compared with healthy control subjects. Cerebrospinal fluid glutamate, glutamine, and taurine concentrations are elevated in migraineurs, which strongly implicates increased activity in glutamatergic systems in the migraine brain. Increased synaptic concentrations of excitatory amino acid neurotransmitters may lead to

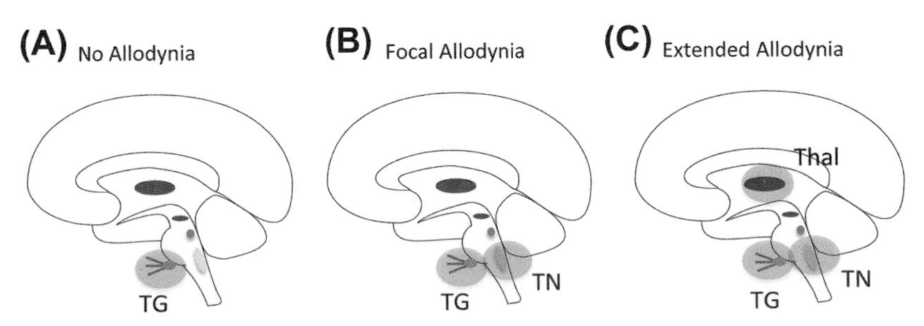

FIGURE 42.4 **Progression from nociception to extended allodynia through progressive sensitization of central pathways.** (A) Initial nociceptive activation in the trigeminal system results in peripheral sensitization of trigeminal afferents from the trigeminal ganglion (TG) associated with localized pain (headache). (B) In patients with localized allodynia affecting the face/head, the model postulates sensitization in the TG and trigeminal nucleus (TN). (C) Generalized allodynia is manifest as a result of sensitization of the whole pathway including the thalamus (Thal) that can result in allodynia in other body areas (limbs and trunk).

excessive activity at the *N*-methyl-D-aspartate (NMDA) glutamate receptor subtype, which in turn may amplify and reinforce pain transmission in migraine and other types of headache. Low-affinity NMDA receptor antagonists, such as memantine, have been shown to reduce the frequency of migraine and tension-type headaches.[25]

Hyperexcitability in the Ictal State

Photophobia, where bright light produces exacerbation of pain, is perhaps the best example of the hyperexcitable state and occurs in most migraine sufferers. How is this possible? Photophobia in blind migraine patients who had no visual perception and blind migraine patients capable of detecting light led to the discovery of the retinothalamocortical pathway. This pathway transmits non-visual forming light signals from the retina to posterior thalamic regions in rats, and putatively to the pulvinar in humans.[26] This region of the thalamus is involved in extracephalic allodynia. An optic tract pathway can also be traced in humans to the pulvinar using diffusion tensor tractography.[26] It is surmised that similar input from auditory or olfactory systems on to these central nociceptive sensitized neurons contributes to the increased sensitivity during the migraine state.

Hyperexcitability in the Interictal State

Evidence for an interictal hyperexcitability state has come from clinical, electrophysiological, and imaging studies that suggest the existence of a hyperexcitable or sensitized brain state. Such evidence has been found in physiological studies reporting a consistent lack of habituation to repetitive stimulation; functional neuroimaging studies showing increased sensitivity to smell, noise, visual stimuli, or diminished modulatory tone; and studies showing altered excitatory neurotransmitters in brain regions in migraine patients. Such data strongly support the hypothesis of central neuronal hyperexcitability playing a key role in the pathogenesis of migraine. Mechanisms of cortical hyperexcitability may result from thalamocortical dysrhythmia that may lead to a lack of habituation in migraine, which could be due to increased neuronal excitability, decreased inhibition, or decreased preactivation levels. Deficient inhibition of these networks involves changes in the basal forebrain, including the basal ganglia and brainstem.

Another common example of migraine-related hyperexcitability is vestibular changes, such as in "vestibular migraine", which is associated with vertigo during the migraine attack. As with the visual hyperexcitability, there may be common or additive processing of vestibular and pain afferent information. The vestibular component is significant because of the purported interictal features of migraine, such as car-sickness.[27] Migraine may affect vestibular pathways in migraine patients with and without vestibular symptoms.[28]

Autonomic Features, Including Nausea and Vomiting

Some headaches manifest with significant autonomic changes (e.g. conjunctival injection, flushing, and tearing), specifically those called trigeminal autonomic cephalalgias. Similar changes are manifest in migraine patients. Such changes may also include other autonomic features such as piloerection, nausea, vomiting, diarrhea, cutaneous vasoconstriction, and diaphoresis, as well as cardiac electrical abnormalities. Some have described migraine as a primary sympathetic disorder[29]; as Peroutka describes this dysfunction: "Migraine is most similar to pure autonomic failure in terms of reduced supine plasma norepinephrine levels, peripheral adrenergic receptor supersensitivity, and clinical symptomatology directly related to sympathetic nervous system dysfunction".[29] Such changes may reflect alterations in hypothalamic nuclei such as the paraventricular nucleus.

A profound feature of migraine attacks is nausea, sometimes accompanied by vomiting. This may occur at any time, but more often during the attack. The mechanism is not well understood, but must involve the nausea and vomiting centers in the brainstem (nucleus ambiguous and nucleus of the solitary tract). Targets for treating nausea include serotonin (5-hydroxytryptamine-3) antagonists that are found in high concentrations in the subnucleus gelatinosa and area postrema of the solitary nucleus, which is the principal site of termination of gastric vagal afferent fibers. Thus, treatments may affect activated regions. Whether nausea and vomiting areas are affected by vagal inputs or secondarily from vestibular alterations is currently unexplored. Motion sickness that is associated with migraine has clear symptoms of nausea and vomiting, suggesting that vestibular imbalance may be the primary process affected.[30] Migraine is also associated with loss of appetite, which may relate in part to those effects on nausea and vomiting. Some patients have increased appetite before the onset of headache, but loss of appetite may relate to pain.

Parasympathetic dysfunction may also be observed. Nitric oxide is thought to be involved in parasympathetic control of the trigeminal vasculature. Nitric oxide may be involved in migraine, tension, and cluster headaches, although the inducible nitric oxide synthase inhibitor has not been effective in the acute prophylaxis of migraine. Nitroglycerine may activate migraine attacks in some but not all patients; it alters the vasculature through vasodilatation and may activate pain fibers through CGRP release.[31]

Mechanisms of Comorbid Depression

Migraine is commonly associated with psychological features including depression. Such changes are perhaps not surprising given that migraine is associated with brain systems involved in stress, such as the hypothalamus, amygdala, and hippocampus. It has been proposed that trigeminovascular projections from the medullary dorsal horn to selective areas in the midbrain, hypothalamus, amygdala, and basal forebrain are functionally positioned to produce migraine symptoms such as irritability, loss of appetite, fatigue, and depression.

Brainstem Mechanisms in Migraine, Including Pain Modulation

Given that subregions of the brainstem, including those within the dorsolateral pons, are involved in more complex integration of pain, baroreceptor, respiratory, and other processes, a number of possibilities may be ongoing individually or in concert. First, the onset of postdrome is considered to be the time-point when the migraine pain stops but other symptoms may persist, including mood change or tiredness. Two symptoms, tiredness and mood, have been linked with locus ceruleus function and may be present in both the prodromal and postdromal periods. Indeed, "somnolence (sleepiness) and asthenia (lack of energy) in patients who respond to treatment suggests that these treatment-emergent neurological symptoms may represent the unmasking of CNS symptoms associated with the natural resolution of a migraine attack, rather than simply representing drug-related side-effects",[32] perhaps in part due to headache relief. Second, parasympathetic systems are involved in vasodilatation in migraine, and triptans produce the opposite effects, suggesting action on sympathetic tone. Sumatriptan induces blood vessel contraction at both cortical and scalp surfaces. This represents a significant sympathetic response. Thus, the observation of persistent activation in the dorsolateral pons in the brainstem may simply be a measure of autonomic drive induced by the drug. Third, persistent use of migraine medication may alter trigeminal systems (medication-induced persistent pronociceptive adaptations). Specifically, repeated use of triptans induces latent sensitization in animals.[33] Afterpain sensations are observed following stimulation of neuropathic pain areas and medications may simply sensitize these effects. Thus, the ongoing allodynia (not spontaneous pain of the migraine) may also persist. Many of the regions noted above are involved in autonomic regulation. Autonomic dysfunction is complex and involves the integration of cortical (e.g. cingulate cortex) and subcortical (e.g. amygdala) structures.

Putative regions involved in descending modulation of pain include the periaqueductal gray and cuneiform nucleus. The cuneiform nucleus may be involved in migraine as assessed in the interictal state.[34] It is thought to be involved in descending modulation of pain, including expectation of pain, and to function together with the periaqueductal gray and superior colliculus to control ventral medullary pain pathways. Studies of pain using functional magnetic resonance imaging (fMRI) and positron emission tomography show this region to be activated. It is also considered to be involved in modulation of locomotor activities and may thus have a role in the inhibition of motor activity associated with pain. Brain systems are clearly involved in the migraine attack, either by triggering attacks (the generator hypothesis) or by producing a predisposition to limiting initiating factors (e.g. cytokines and stressors). Figure 42.5 provides a conceptual approach in which cyclical processes may contribute to an endogenous tone that has a set point related to initiation of a migraine attack.

Migraine, Stress, and Stress Circuits

Migraine as a Stressor

Migraine attacks are stressful events. Migraine is a continuum of processes as outlined above and shown in Fig. 42.1. In addition to the migraine-related stress, stress is considered to be the most frequent cause of migraine attacks. Indeed, stressful experiences related to activities of daily living may trigger migraine in patients and other psychological issues also play a role. Environmental factors, such as weather, are also involved in precipitating migraine. Approximately 50–70% of subjects show a temporal correlation between their daily levels of stress and migraine headaches.[35] While a headache onset may potentially be protective, several behavioral and biological processes are brought into play, including going to sleep to terminate an attack, activating protective responses associated with neurogenic inflammation that may be associated with migraine, decreasing physical activity, or staying away from exacerbators of headache such as bright light. However, stress effects may be subtle and not easily detected in migraine patients using standard measures such as cortisol. The effects of repeated stress (pain, hormonal, CSD, etc.) may result in significant alterations to brain function and structure.

Hypothalamus, Puberty, and Gender

The focus on the hypothalamus relates to the circadian cyclicity of migraines, and some of the symptoms relate to the migraine, including yawning, tiredness, and altered mood. In addition, the hypothalamus is involved in the stress response. Imaging studies have shown clear hypothalamic changes in patients with cluster headache.[36] Hypothalamic volume is increased in hypnic headache, and activation is present in cluster headache[37] but less defined in classical migraine. In patients with cluster headache, deep brain stimulation of the hypothalamus resulted in improvements in 50% of patients.[38] The hypothalamus orchestrates hormonal changes related to the menstrual cycle in addition to appetite and feeding-induced changes. Some important insights relate to hormonal changes: migraine is more frequent in women, the incidence is increased during menarche, treating menstrual migraine decreases overall migraine frequency, and cessation of migraine pain is related to cessation of menstrual cycles. Mechanisms for the cessation of pain are unknown. Insights have come from studies on other pain disorders showing an association between a gene encoding for vasopressin and gender-related sensitivity to pain[39] and alterations in the control of cyclicity through perturbations of hypothalamic suprachiasmatic function.

Other brain regions are involved in autonomic function including the stress response, such as the hippocampus and insula. Both are involved in migraine, and significant changes in these regions have been demonstrated. The hippocampal changes in migraine[40] have also been shown in patients with chronic pain.[41] Studies in rodents suggest that hippocampal spreading depression may be a potential trigger for nociceptive activation of the trigeminal nucleus, perhaps associated with migraine.[42]

Differences in pain activation have been found between male and female migraineurs. Increased thickening in the posterior insula and precuneus in female migraineurs compared with male migraineurs and

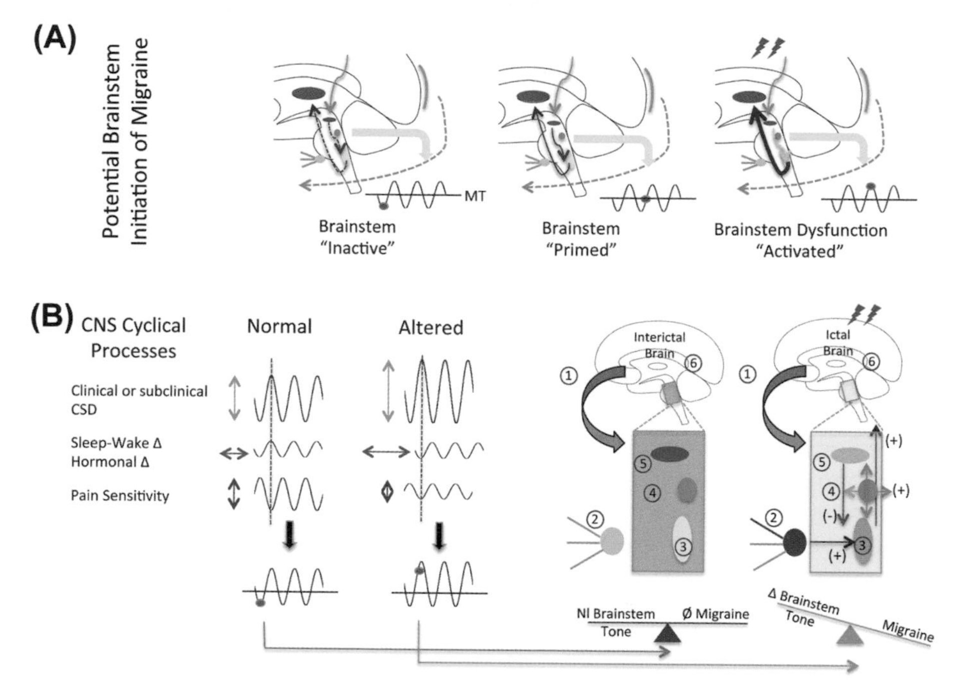

FIGURE 42.5 **Principles that govern brainstem modulation of nociceptive drive during migraine.** Activation of nociceptors, for example by cortical spreading depression (CSD), mild head trauma in migraine patients, or inflammation, triggers activity in central trigeminovascular neurons. The magnitude of activation is then enhanced by insufficient synaptic inhibition due to periaqueductal gray (PAG) deficiency or to abnormally enhanced synaptic strength caused by overactive pain facilitatory neurons in the PAG. (A) Brainstem state of activity can limit afferent nociceptive drive in migraine-susceptible individuals. The robustness of the gate that allows nociceptive signals to drive central trigeminovascular neurons (and thus headache) is dictated by brainstem tone. Thus, when tone is high [(red dot below line of migraine threshold (MT)], signals are inhibited, and when tone is low (red dot above MT), afferent signals are not effectively blocked. Imaging studies showing increased signal in certain dorsolateral pons (DLP) areas/nuclei during migraine support the notion that when cyclical phase brainstem activity is low, the potency of pain facilitation (i.e. enhanced synaptic strength in the dorsal horn) is too low to allow nociceptive signals from the periphery to drive the central neurons into the active state (left). At threshold, the system has reached a primed state that could tip into a functional state that would allow for nociceptive drive from the dura to activate the central trigeminovascular neurons (middle). Migraine state: when cyclical brainstem activity is high, the potency of pain facilitation (i.e. enhanced synaptic strength in the dorsal horn) is sufficient to allow nociceptive signals from the periphery to drive the central neurons into the active state (right). Imaging studies showing decreased signal in certain DLP areas/nuclei during migraine support the notion that when cyclical brainstem tone is low, brainstem inhibition of pain is less ineffective. (B) CNS cyclical process and brainstem thresholds: genetic, physiological, pharmacological, social, and other interactions define migraineurs' susceptibility. When processes are in synchrony (a harmonic or repetitive frequency), the model suggests that the migraine potential is subthreshold (red dot); when these are altered in magnitude, phase, or duration, the system becomes unstable and the migraine threshold is exceeded. These components affect the brainstem tone as shown on the right of the figure for the interictal state: 1, cortical processes affect subcortical processing; 2, normal afferent input through the trigeminal ganglion; 3, trigeminal nucleus function is normal/unchallenged; 4, DLP function is normal or not activated; 5, PAG functioning is not challenged; 6, brain systems acting on the brainstem are in a balanced tone. During migraine, the following cascade of events is postulated: 1, normal cortical inhibition of brainstem pathways is altered; 2, afferent barrage of nociceptive input from trigeminal ganglion neurons to the spinal trigeminal nucleus; 3, increased activation of the DLP; 4, diminished brainstem inhibition of nociceptive signals resulting from altered modulatory/inhibitory tone by the PAG; 5, all of this results in increased afferent inputs to higher brain centers that 6, may be sensitized or affected by brainstem nuclei. *Source: Adapted from Borsook and Burstein.* Cephalalgia. *2012;32(11):803–812.*[11]

healthy controls of both genders is reported, concurrent with differences in the pattern of functional connectivity of these regions from other brain regions involved in pain processing. The results support the notion that gender differences involve greater participation of emotional circuits compared with sensory circuits in females compared with males, and brain morphometry.[43]

Brain Changes Associated with Migraine Attacks

Alterations in brain systems associated with migraine can be categorized into three processes: alterations in brain function; alterations in brain structure; and alterations in brain chemistry (Figs 42.6 and 42.7). Insights into brain systems biology have helped to define migraine as a disease of the brain. Novel findings of brain alterations are described below. Although all of these alterations are associated with migraine, it is unclear whether some of them cause the migraine or are a consequence of the migraine.

Temporal Lobe

fMRI measures of the temporal lobe have been compared in migraine patients and healthy controls in the ictal and interictal state.[45] Increased activation was observed in the anterior temporal pole; notably, activation in the temporal pole was exacerbated during the ictal compared with the interictal phase. In addition, differences in response associated with increased migraine frequency (i.e. between the high- and low-frequency migraine groups) were observed in the temporal lobe.[46] Associated with these observations were differences in cortical thickness in the inferior temporal gyrus. The differences reflect an increased sensitized state in those migraineurs having a higher frequency of attacks.

Basal Ganglia

The basal ganglia consist of the striatum (caudate–putamen and the core of the nucleus accumbens), the external segment of the globus pallidus, the subthalamic nucleus, the internal segment of the globus pallidus, and the substantia nigra. The basal ganglia play an important role in multiple functions including pain processing. These regions are major sites for adaptive plasticity in the brain, affecting a broad range of behaviors in the normal state.The basal ganglia play a role or are affected in patients with increased migraine frequency.[47] Specifically, when comparing low- and high-frequency migraineurs, responses to painful stimuli in high-frequency patients show reductions in the caudate, putamen, and pallidum. In addition, gray matter volume of the caudate nuclei is significantly larger in high-frequency than in low-frequency patients.[47] Alterations in basal ganglia circuits are common in pain imaging studies.[48] Hypometabolism in the caudate has been reported for FHM, and patients with basal ganglia dysfunction (e.g. Tourette syndrome) have an increased incidence of migraine. Repeated migraine attacks are associated with increased iron accumulation in multiple deep nuclei that are involved in central pain processing and migraine pathophysiology, including the caudate. The authors are unaware of any other subcortical regions showing changes in gray matter volume, but given the relative

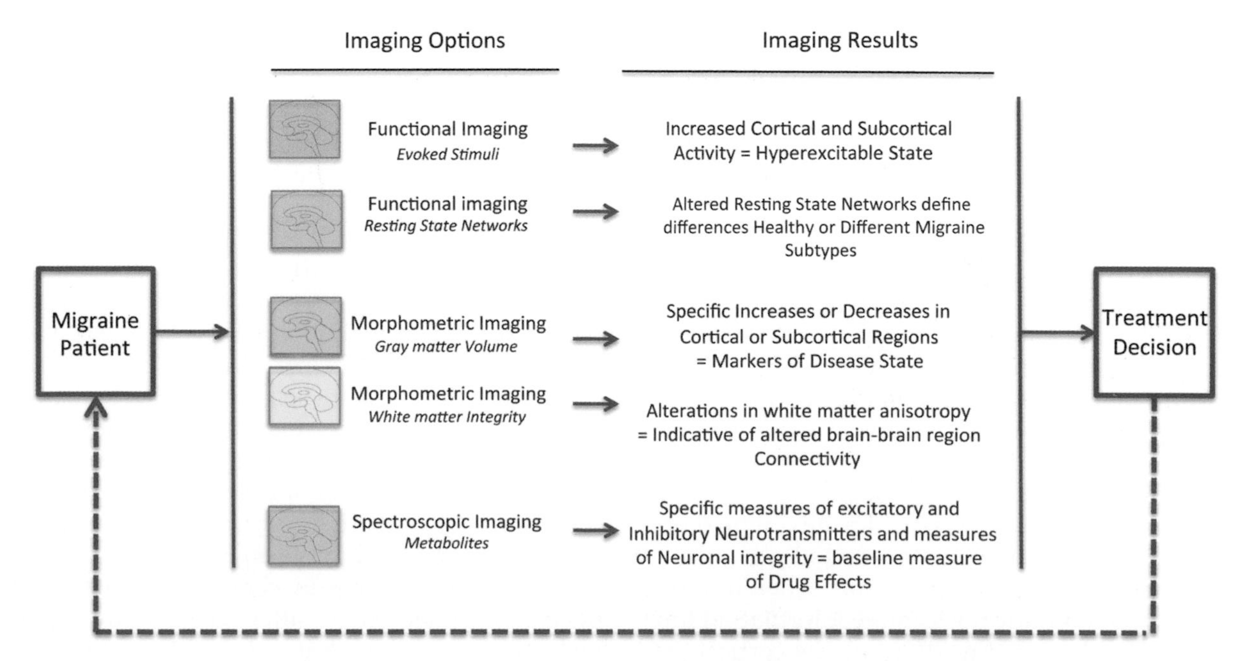

FIGURE 42.6 **Potential use of magnetic resonance imaging in migraine evaluation and treatment.** *Source: Borsook.* The Migraine Brain: Imaging, Structure, and Function. *New York: Oxford University Press; 2012,[44] with permission.*

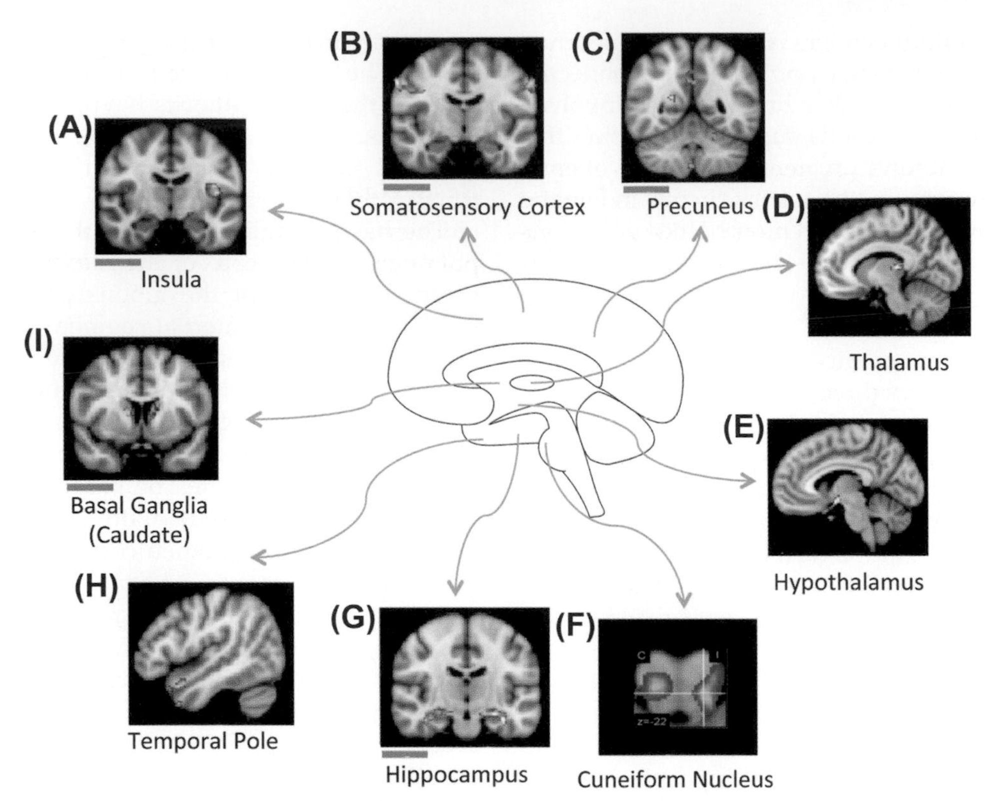

FIGURE 42.7 **Functional magnetic resonance (fMRI) imaging changes in the migraine brain.** The figure summarizes studies reporting fMRI blood oxygen level-dependent (BOLD) activation during the interictal period, which suggests that these regions are functionally different even outside migraine attack. (A) Stronger response to noxious thermal stimulation in anterior insula in high-frequency (HF) versus low-frequency (LF) migraine patients associated with reduction in the cortical volume; (B) concurrent cortical thickening in S1 and stronger response to noxious thermal stimulation in anterior insula in HF migraine patients; the same area is thicker in female migraineurs than in male migraineurs and healthy control subjects of both genders; (C) stronger functional connectivity between posterior insula and precuneus in female migraineurs; stronger response to noxious thermal stimulation in HF versus LF migraineurs in (D) thalamus and (E) hypothalamus; (F) decreased nucleus cuneiformis responses to noxious thermal stimulation relative to controls; (G) bilateral stronger response to noxious thermal stimulation in HF migraineurs associated with bilaterally reduced hippocampal volume compared with LF migraineurs; (H) increased response in the temporal pole in migraineurs versus healthy control subjects in response to noxious thermal stimulus; (I) reduced BOLD response in the caudate associated with a significantly bilateral larger volume in HF versus LF migraineurs. Color bars beneath some of the figures indicate morphometric changes in that same region.

ease of the approach, more studies are expected in this area.

Pulvinar of the Thalamus

The pulvinar is the posterior nuclear complex in the thalamus most frequently associated with processing of visual information. It is considered to be important in two aspects of transmission of information, from layer 5 of one cortical area to another, and in integrating multisensory information.[49] Activation in the pulvinar during migraine corresponds with extended allodynia, presumably as a result of afferent nociceptive neurons converging on to cells that drive secondary responses (i.e. allodynia).[50] In addition, non-visual-forming retinal projections to the pulvinar may contribute to light-induced hypersensitivity in migraine patients.[51] Therefore, the concept is that in migraine nociceptive information travels directly rather than indirectly from one cortical area to another.[52] In this model, diverse responses based on cortical areas that have extensive input to multiple cortical areas are involved in the behavioral phenotypes observed in migraine patients.

Morphometric Cortical Changes

Ever since the landmark reports on alterations in gray matter in patients with cluster headache[53] and chronic pain,[54] evaluation of similar changes in migraine patients has shown a significant loss of gray matter. DaSilva and colleagues reported thickening of the somatosensory cortex in migraineurs relative to healthy controls.[55] Decreased gray matter volume in episodic migraine in superior temporal gyrus, inferior frontal gyrus, and precentral gyrus has also been reported, compared with controls. More changes have been reported in patients with chronic migraine than in controls.[56] While it remains unclear whether gray matter changes are adaptive or maladaptive, reports in the chronic pain literature suggest that when the pain resolves these areas normalize compared with healthy controls. It is an ongoing matter of debate whether such changes are causes or consequence of migraine, but the fact that, at least in many voxel-based morphometry (VBM) studies, changes correlate with disease duration or frequency argues in favor of the latter. Research in osteoarthritis has suggested that gray matter changes observed with VBM are reversible, and the same is likely to be true for migraine. The exact underlying mechanisms leading to alterations in gray matter density remain to be elucidated. Such alterations are reported to correlate with alterations in dendritic complexity or changes in the numbers of synapses. Alterations in neural activity can modify the behavior of neural circuits by eliciting long-lasting changes in the strength or efficacy of synaptic transmission at preexisting excitatory synapses. Drugs or disease may also lead to modification of dendritic spine density and thus synaptic contacts.

White Matter Changes

White matter changes can be measured using fractional anisotropy with diffusion tensor imaging. These changes reflect alterations in the integrity of white matter tracts or fibers that when altered are interpreted as specific changes brought about by the disease state. Alterations in the thalamocortical tract have been reported in migraineurs. Such changes may allow the migraine state to be measured, since they may reflect decreasing white matter integrity, which could be an index of the disorder and its progression and may lead to an effective therapy.

Brainstem Changes

Although some functional data suggest alterations in the brainstem gray matter volume, evidence for alterations in brainstem nuclei has only been reported by a few groups.[57,58]

Hippocampus

The hippocampus, located in the medial temporal pole, is classically involved in memory consolidation and spatial navigation and is well known to be affected by repeated stress. Given that episodic migraine can be viewed as a repeated stressor, alterations in hippocampal function and structure may play an important role in migraine pathophysiology. Increased functional activation to pain stimuli has been reported in high- compared with low-frequency migraineurs.[40] Consistent with this, bilateral increases in volume in the hippocampus were observed in migraineurs compared with healthy controls, with the low-frequency group displaying larger volume changes than the high-frequency group. The plasticity of the hippocampal volume relates to mechanisms that enhance or diminish dendritric complexity, synapse number, dentate gyrus neuronal number, and novel neural connections. A study on the hippocampal correlates of pain in healthy elderly adults[58] found that older adults who report more severe acute or chronic pain have smaller hippocampal volumes and lower levels of hippocampal *N*-acetyl aspartate (NAA)/creatine, a marker of neuronal integrity. Migraine attacks may be considered emotional and physical stressors that alter hippocampal neurogenesis and gene expression with the release of stress hormones with each occurrence.[59]

Precuneus

The precuneus is a brain region involved in a variety of complex functions,[60] which include recollection and memory, integration of information (gestalt) relating to perception of the environment, cue reactivity, mental imagery strategies, episodic memory retrieval, and affective responses to pain. Therefore, the precuneus responds to a wide variety of cognitive processes. Significant changes in the precuneus have been reported in female versus male migraineurs and compared with healthy controls, suggesting a specific gender-disease effect.[43] The precuneus can be divided into regions involved in sensorimotor processing, cognition, and visual processing, and this segregation may provide a model for its involvement in migraine. For example, behavioral changes in migraine patients include disturbances of body image and physical sensations, or alteration of time sense. Such changes may be indicative of alterations in self-perception and issues related to expected pain, as may be the case in a recurrent disease.

Measures of Neurotransmitters Showing Hyperexcitable State

Fundamental changes have been shown to occur in the brain of migraineurs, including abnormal function of key brain areas and networks. In addition, changes in brain gray and white matter structure and changes in brain chemistry have been reported in migraine patients. Changes in multiple brain areas in migraine were summarized by Sprenger and Borsook.[61]

One technique used to assess brain metabolism and energy production is magnetic resonance spectroscopy (MRS). MRS measures brain chemistry in single voxels (i.e. one brain area) or as a composite measure for the entire brain. Changes in excitatory or inhibitory amino acids, alterations in metabolites of energy function (mitochondria), and measures of metabolites that are indicators of neuronal integrity can be measured. Because of sufficiently high concentrations of hydrogen (^{1}H) and phosphorus (^{31}P) in the brain, ^{1}H and ^{31}P spectroscopy can be used to study the level of metabolites in the brain, increases or decreases of which could indicate a disease state or abnormal brain function. For instance, decrease in NAA in the ^{1}H MRS spectrum is associated with neuronal loss. A few reports have measured MRS changes in migraine patients. This technique offers the opportunity not only to measure changes in the disease state, but also to monitor how treatments may alter, reverse, or normalize the chemical milieu in the brain. Migraine patients have higher concentrations of excitatory amino acid neurotransmitters such as aspartate, glutamate, glutamine, and glycine compared with healthy controls, which could be linked to the increased cortical excitability particularly in those with migraine aura. Increased synaptic concentrations of these neurotransmitters may play a role in amplifying pain transmission in migraine. MRS has also been used to assess the metabolic changes in the brain (visual cortex in particular) with aura; patients experiencing visual aura have increased lactate levels,[62] suggestive of a potential underlying mitochondrial dysfunction. The duration of aura in migraine patients could change the metabolite levels in the brain, as evidenced by the decrease in the phosphocreatine/phosphate ratio with increasing aura duration.[63] Reduced NAA and glutamate, and increased *myo*-inositol levels are also reported in FHM, which is a monogenic subtype of migraine.

Migraine Chronification or Transformation

A subset of patients has an increasing frequency of migraine attacks to more than 14 headache days per month, thus transforming from episodic migraine to chronic migraine (Fig. 42.8). Changes in psychometric and imaging studies indicate significant differences in these patients. For example, transformed migraine patients exhibit cutaneous allodynia, as assessed by decreased mechanical and thermal thresholds, during and between migraine attacks.[64]

Aside from a natural progression, other factors could contribute to progression, the most notable of which is medication overuse, specifically of triptans and opioids but also of topiramate. Potential mechanisms of migraine chronification with triptan use include increased frequency of CSD, increased endogenous facilitation in brainstem modulatory networks, and potentially increased sensitivity of nociceptive afferents as a result of increased CGRP expression and release, or increases in nitric oxide synthase.[65] Non-steroidal anti-inflammatory drugs (NSAIDs) seem to be protective when given to patients having fewer than 10 headache days per month.

Placebo and Migraine

When it comes to the placebo effect or response, there are two categories that may relate to underlying biology in the migraine patient. The first is that there are abnormal brain systems including the brainstem modulatory regions but also emotional regions involved in such responses (frontal regions, basal ganglia including the nucleus accumbens). The second is that if there are abnormal neural circuits then, given that migraine is an emotionally laden disease, how do migraineurs respond to placebo? With respect to the latter, even in standardized trials, the placebo response in migraine patients is significant.[66] More interesting, perhaps, is the adverse events in placebo groups in trials of migraine treatment (triptans, NSAIDs, and anticonvulsants).[66] As described by Amanzio and colleagues, "the adverse events in the placebo arms corresponded to those of the anti-migraine

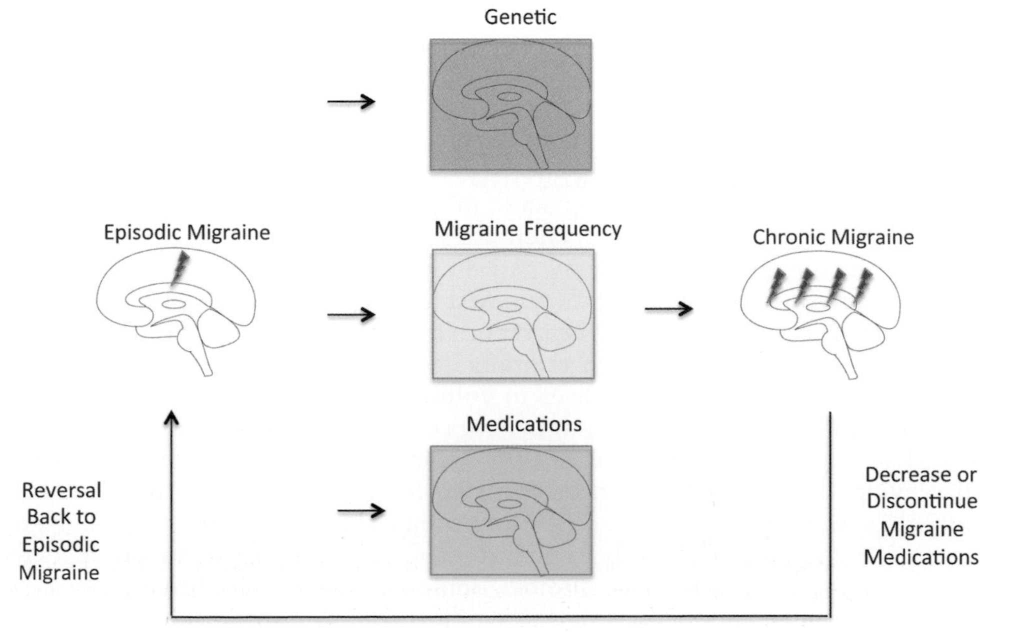

FIGURE 42.8 **Migraine chronification or transformation.** An increasing frequency of migraine attacks, due to genetic factors and/or medication overuse, transforms the disease status from episodic to chronic migraine.

medication against which the placebo was compared".[66] For example, for a particular drug category, the side effects of placebo (i.e. anorexia and memory difficulties) corresponded to those observed in the drug treatment category.

MIGRAINE AND ALLOSTATIC LOAD

The normal response of the brain and body to potential and actual threats (stressors) and changes in the physical and social environment is to activate hormonal and neural mediators as well as behaviors that help the individual to adapt (allostasis) and cope.[10] When these stressors are frequent and/or severe and when the allostatic responses become overused and dysregulated, there is wear and tear on the body and brain (allostatic load). In the case of migraine, allostatic load may alter brain networks both functionally and structurally. As a result, there are altered or abnormal brain responses to stressors. Importantly, this is a failing process since both behavior and systemic physiology are altered in ways that can then add to the allostatic load. Migraine offers a unique model of such a feedforward cascade.

Migraine is a major perturbation of brain and body systems in a number of ways including pain, cardiovascular changes, and immunological changes. Over time these perturbations may lead to an altered brain state characterized by increased cortical excitability, and changes in brain morphology and behavior. Both endogenous (e.g. hormones, migraine attacks) and exogenous (e.g. physical stressors, such as light, and psychological stressors) factors contribute to alterations in the brain that manifest as ictal or interictal alterations in sensory (e.g. allodynia, photophobia, osmophobia) and other neural networks. Such stressors typically occur periodically as part of the nature of the disease in both episodic migraine and chronic daily headache. The role of stress in precipitating migraine is well described and is reviewed elsewhere,[59,67] but migraine itself, in the context of its ability to generate pathophysiological changes in the brain and throughout the body, has not been studied. The cascading pathophysiological changes in brain structure and function with the progression of migraine attacks may contribute to an improved understanding of the full nature and consequences of this condition, which frequently affects the brain (Fig. 42.9).

The effects of migraine attacks relate to frequency, preictal and postictal processes, and the ongoing effects of ictal and peri-ictal challenges. Differences in the temporal nature and magnitude of these challenges drive the allostatic state, probably as a result of overactivity and dysregulation of glutamate, γ-aminobutyric acid (GABA), and glucocorticoids, among other mediators.

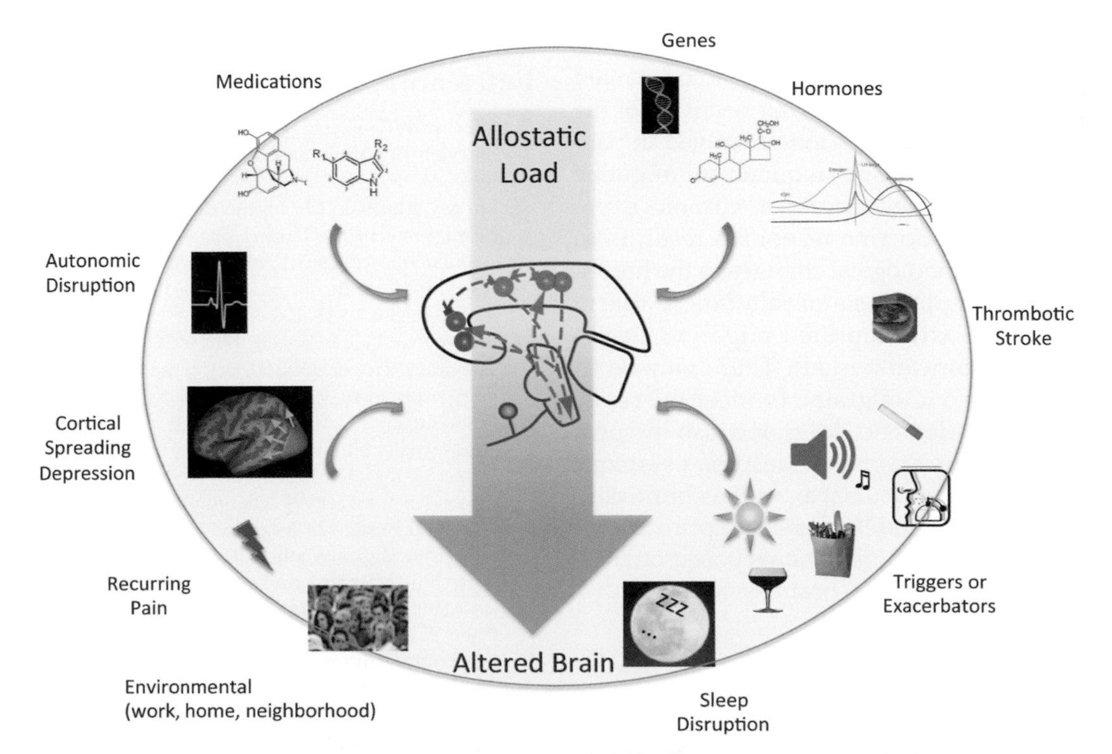

FIGURE 42.9 **Model of migraine onset as a consequence of allostatic load.** The figure summarizes multiple actions on the brain in migraine. The relative contribution of each effector to allostatic load is unknown but may be additive or cumulative over time. Effectors may have a continuous (e.g. genetic contributions), episodic (e.g. pain), or progressive (e.g. white matter lesions and stroke) effect. *Source: Borsook* et al. Neuron. *2012;73(2):219–234,[10] with permission.*

Allostatic challenges may be considered continuous (type 1), intermittent of short duration (type 2), or intermittent of long duration (type 3). Type 1 challenges include ongoing processes such as the increased excitability of cortical and subcortical structures, type 2 include the actual ictal event (pain and associated symptoms), and type 3 include the premonitory symptoms, comprising a wide and heterogeneous collection of cognitive, psychological, and physical changes preceding an attack by a few hours to 2–3 days, as discussed in "Peri-ictal Clinical Manifestations", above.

Four major processes contribute to allostatic load in migraine: (1) migraine attacks produce repeated stress; (2) the brain fails to habituate to stimuli; (3) there is dysregulation of normal adaptive responses where components of the stress response may fail to shut down normally; and (4) compensatory increased responses (e.g. photophobia) may occur during ictal and interictal states. All of these processes act on brain systems to increase the allostatic load. The process is aggravated by the bidirectional effects of migraine on systemic processes that contribute to alterations in brain processing.

The hippocampal volumetric changes as a function of the disease burden (an estimate of the total number of migraine attacks experienced in life) in migraineurs reported by Maleki and colleagues[40] may very well fit with the allostatic load model for migraine. Significantly larger bilateral hippocampal volume was found in low-frequency than in high-frequency migraineurs. This suggests an initial adaptive plasticity that may then become dysfunctional with increased frequency, accompanied by an overall reduction in functional connectivity of the hippocampus with other brain regions involved in pain processing in patients with a high frequency of migraine attacks. Physiological consequences of chronic exposure to neural or neuroendocrine responses result from repeated stress. The morphological changes in the hippocampus may reflect complex dynamic alterations due to neurochemical changes with repeated migraine attacks (and perhaps in the interictal period). There may be an initial adaptive response in migraine to meet the needs of increased processing demand. Migraine also disrupts many systems of the body and has widespread systemic (e.g. metabolic and cardiovascular) as well as brain consequences. The allostatic model offers an approach for evaluating and treating migraine patients based on multiple stressors and their effects on brain systems.

CONCLUSION

The history of quantum leaps in our understanding of the most common neurological disorder, migraine, continues with new developments from genetic discoveries to mechanism-based changes that have a direct impact on patients. Recent examples include the understanding of photophobia and treating migraine early to avoid central sensitization. Migraine as a pain condition holds a notable record of having a specific Food and Drug Administration-approved drug designed on the basis of neurobiological understanding. New, improved treatments should result from new understanding of the disease.

QUESTIONS FOR FURTHER RESEARCH

A great deal that remains to be learned about migraine, a debilitating condition that affects more people than almost any of the other disorders discussed in this textbook. It is hoped that some readers of this chapter will be moved to examine the following research questions: Can the genetic predisposition to migraine be defined? Is it possible to predict which patients will chronify? Can an imaging biomarker for migraine be defined? How does stress contribute to the migraine condition? And how can the interactions of the disease burden of migraine, including comorbidity, be measured?

Acknowledgments

This work was supported by grants from NINDS (R01-NS056195 and R01-NS073997 to DB and R37-NS079678 to RB). The brain outline used in the figures was provided by Eric Moulton PhD.

References

1. Stewart WF, Shechter A, Rasmussen BK. Migraine prevalence. A review of population-based studies. *Neurology*. 1994;44(6 suppl 4):S17–S23.
2. Hu XH, Markson LE, Lipton RB, Stewart WF, Berger ML. Burden of migraine in the United States: disability and economic costs. *Arch Intern Med*. 1999;159(8):813–818.
3. Bigal M. Migraine chronification – concept and risk factors. *Discov Med*. 2009;8(42):145–150.
4. Schoonman GG, Evers DJ, Terwindt GM, van Dijk JG, Ferrari MD. The prevalence of premonitory symptoms in migraine: a questionnaire study in 461 patients. *Cephalalgia*. 2006;26(10):1209–1213.
5. Tfelt-Hansen PC, Koehler PJ. One hundred years of migraine research: major clinical and scientific observations from 1910 to 2010. *Headache*. 2011;51(5):752–778.
6. Lipton RB, Stewart WF, Diamond S, Diamond ML, Reed M. Prevalence and burden of migraine in the United States: data from the American Migraine Study II. *Headache*. 2001;41(7):646–657.
7. Bartsch T. Migraine and the neck: new insights from basic data. *Curr Pain Headache Rep*. 2005;9(3):191–196.
8. Chen YC, Tang CH, Ng K, Wang SJ. Comorbidity profiles of chronic migraine sufferers in a national database in Taiwan. *J Headache Pain*. 2012;13(4):311–319.
9. Gaist D, Pedersen L, Madsen C, et al. Long-term effects of migraine on cognitive function: a population-based study of Danish twins. *Neurology*. 2005;64(4):600–607.
10. Borsook D, Maleki N, Becerra L, McEwen B. Understanding migraine through the lens of maladaptive stress responses: a model disease of allostatic load. *Neuron*. 2012;73(2):219–234.

11. Borsook D, Burstein R. The enigma of the dorsolateral pons as a migraine generator. *Cephalalgia*. 2012;32(11):803–812.
12. Eikermann-Haerter K, Yuzawa I, Qin T, et al. Enhanced subcortical spreading depression in familial hemiplegic migraine type 1 mutant mice. *J Neurosci*. 2011;31(15):5755–5763.
13. Zhang X, Levy D, Noseda R, Kainz V, Jakubowski M, Burstein R. Activation of meningeal nociceptors by cortical spreading depression: implications for migraine with aura. *J Neurosci*. 2010;30(26):8807–8814.
14. May A, Goadsby PJ. The trigeminovascular system in humans: pathophysiologic implications for primary headache syndromes of the neural influences on the cerebral circulation. *J Cereb Blood Flow Metab*. 1999;19(2):115–127.
15. DaSilva AF, Becerra L, Makris N, et al. Somatotopic activation in the human trigeminal pain pathway. *J Neurosci*. 2002;22(18):8183–8192.
16. Stankewitz A, Aderjan D, Eippert F, May A. Trigeminal nociceptive transmission in migraineurs predicts migraine attacks. *J Neurosci*. 2011;31(6):1937–1943.
17. Barrett CF, van den Maagdenberg AM, Frants RR, Ferrari MD. Familial hemiplegic migraine. *Adv Genet*. 2008;63:57–83.
18. Stuart S, Cox HC, Lea RA, Griffiths LR. The role of the MTHFR gene in migraine. *Headache*. 2012;52(3):515–520.
19. Anttila V, Stefansson H, Kallela M, et al. Genome-wide association study of migraine implicates a common susceptibility variant on 8q22.1. *Nat Genet*. 2010;42(10):869–873.
20. Chasman DI, Schurks M, Anttila V, et al. Genome-wide association study reveals three susceptibility loci for common migraine in the general population. *Nat Genet*. 2011;43(7):695–698.
21. Freilinger T, Anttila V, de Vries B, et al. Genome-wide association analysis identifies susceptibility loci for migraine without aura. *Nat Genet*. 2012;44(7):777–782.
22. Burstein R, Cutrer MF, Yarnitsky D. The development of cutaneous allodynia during a migraine attack clinical evidence for the sequential recruitment of spinal and supraspinal nociceptive neurons in migraine. *Brain*. 2000;123(Pt 8):1703–1709.
23. Burstein R, Collins B, Jakubowski M. Defeating migraine pain with triptans: a race against the development of cutaneous allodynia. *Ann Neurol*. 2004;55(1):19–26.
24. Burstein R, Jakubowski M, Rauch SD. The science of migraine. *J Vestib Res*. 2011;21(6):305–314.
25. Krusz JC. Prophylaxis for chronic daily headache and chronic migraine with neuronal stabilizing agents. *Curr Pain Headache Rep*. 2002;6(6):480–485.
26. Maleki N, Becerra L, Upadhyay J, Burstein R, Borsook D. Direct optic nerve pulvinar connections defined by diffusion MR tractography in humans: implications for photophobia. *Hum Brain Mapp*. 2012;33(1):75–88.
27. Furman JM, Marcus DA. Migraine and motion sensitivity. *Continuum (Minneap Minn)*. 2012;18(5 Neuro-otology):1102–1117.
28. Casani AP, Sellari-Franceschini S, Napolitano A, Muscatello L, Dallan I. Otoneurologic dysfunctions in migraine patients with or without vertigo. *Otol Neurotol*. 2009;30(7):961–967.
29. Peroutka SJ. Migraine: a chronic sympathetic nervous system disorder. *Headache*. 2004;44(1):53–64.
30. von Brevern M, Neuhauser H. Epidemiological evidence for a link between vertigo and migraine. *J Vestib Res*. 2011;21(6):299–304.
31. Levy D, Burstein R, Strassman AM. Calcitonin gene-related peptide does not excite or sensitize meningeal nociceptors: implications for the pathophysiology of migraine. *Ann Neurol*. 2005;58(5):698–705.
32. Goadsby PJ, Dodick DW, Almas M, et al. Treatment-emergent CNS symptoms following triptan therapy are part of the attack. *Cephalalgia*. 2007;27(3):254–262.
33. De Felice M, Ossipov MH, Wang R, et al. Triptan-induced latent sensitization: a possible basis for medication overuse headache. *Ann Neurol*. 2010;67(3):325–337.
34. Moulton EA, Burstein R, Tully S, Hargreaves R, Becerra L, Borsook D. Interictal dysfunction of a brainstem descending modulatory center in migraine patients. *PLoS ONE*. 2008;3(11):e3799.
35. Holm JE, Lokken C, Myers TC. Migraine and stress: a daily examination of temporal relationships in women migraineurs. *Headache*. 1997;37(9):553–558.
36. Dunckley P, Wise RG, Fairhurst M, et al. A comparison of visceral and somatic pain processing in the human brainstem using functional magnetic resonance imaging. *J Neurosci*. 2005;25(32):7333–7341.
37. May A, Bahra A, Buchel C, Frackowiak RS, Goadsby PJ. Hypothalamic activation in cluster headache attacks. *Lancet*. 1998;352(9124):275–278.
38. Seijo F, Saiz A, Lozano B, et al. Neuromodulation of the posterolateral hypothalamus for the treatment of chronic refractory cluster headache: experience in five patients with a modified anatomical target. *Cephalalgia*. 2011;31(16):1634–1641.
39. Mogil JS, Sorge RE, LaCroix-Fralish ML, et al. Pain sensitivity and vasopressin analgesia are mediated by a gene–sex–environment interaction. *Nat Neurosci*. 2011;14(12):1569–1573.
40. Maleki N, Becerra L, Brawn J, McEwen B, Burstein R, Borsook D. Common hippocampal structural and functional changes in migraine. *Brain Struct Funct*. 2013;218(4):903–912.
41. Mutso AA, Radzicki D, Baliki MN, et al. Abnormalities in hippocampal functioning with persistent pain. *J Neurosci*. 2012;32(17):5747–5756.
42. Kunkler PE, Kraig RP. Hippocampal spreading depression bilaterally activates the caudal trigeminal nucleus in rodents. *Hippocampus*. 2003;13(7):835–844.
43. Maleki N, Linnman C, Brawn J, Burstein R, Becerra L, Borsook D. Her versus his migraine: multiple sex differences in brain function and structure. *Brain*. 2012;135(Pt 8):2546–2559.
44. Borsook D. *The Migraine Brain: Imaging, Structure, and Function*. New York: Oxford University Press; 2012.
45. Moulton EA, Becerra L, Maleki N, et al. Painful heat reveals hyperexcitability of the temporal pole in interictal and ictal migraine states. *Cereb Cortex*. 2011;21(2):435–448.
46. Maleki N, Becerra L, Brawn J, Bigal M, Burstein R, Borsook D. Concurrent functional and structural cortical alterations in migraine. *Cephalalgia*. 2012;32(8):607–620.
47. Maleki N, Becerra L, Nutile L, et al. Migraine attacks the Basal Ganglia. *Mol Pain*. 2011;7:71.
48. Borsook D, Upadhyay J, Chudler EH, Becerra L. A key role of the basal ganglia in pain and analgesia – insights gained through human functional imaging. *Mol Pain*. 2010;6:27.
49. Sherman SM. The thalamus is more than just a relay. *Curr Opin Neurobiol*. 2007;17(4):417–422.
50. Burstein R, Jakubowski M, Garcia-Nicas E, et al. Thalamic sensitization transforms localized pain into widespread allodynia. *Ann Neurol*. 2010;68(1):81–91.
51. Noseda R, Kainz V, Jakubowski M, et al. A neural mechanism for exacerbation of headache by light. *Nat Neurosci*. 2010;13(2):239–245.
52. Noseda R, Jakubowski M, Kainz V, Borsook D, Burstein R. Cortical projections of functionally identified thalamic trigeminovascular neurons: implications for migraine headache and its associated symptoms. *J Neurosci*. 2011;31(40):14204–14217.
53. May A, Ashburner J, Buchel C, et al. Correlation between structural and functional changes in brain in an idiopathic headache syndrome. *Nat Med*. 1999;5(7):836–838.
54. Apkarian AV, Sosa Y, Sonty S, et al. Chronic back pain is associated with decreased prefrontal and thalamic gray matter density. *J Neurosci*. 2004;24(46):10410–10415.
55. DaSilva AF, Granziera C, Snyder J, Hadjikhani N. Thickening in the somatosensory cortex of patients with migraine. *Neurology*. 2007;69(21):1990–1995.

56. May A. New insights into headache: an update on functional and structural imaging findings. *Nat Rev Neurol*. 2009;5(4):199–209.
57. Schmitz N, Admiraal-Behloul F, Arkink EB, et al. Attack frequency and disease duration as indicators for brain damage in migraine. *Headache*. 2008;48(7):1044–1055.
58. Rocca MA, Ceccarelli A, Falini A, et al. Brain gray matter changes in migraine patients with T2-visible lesions: a 3-T MRI study. *Stroke*. 2006;37(7):1765–1770.
59. Sauro KM, Becker WJ. The stress and migraine interaction. *Headache*. 2009;49(9):1378–1386.
60. Cavanna AE, Trimble MR. The precuneus: a review of its functional anatomy and behavioural correlates. *Brain*. 2006;129(Pt 3):564–583.
61. Sprenger T, Borsook D. Migraine changes the brain: neuroimaging makes its mark. *Curr Opin Neurol*. 2012;25(3):252–262.
62. Sandor PS, Dydak U, Schoenen J, et al. MR-spectroscopic imaging during visual stimulation in subgroups of migraine with aura. *Cephalalgia*. 2005;25(7):507–518.
63. Schulz UG, Blamire AM, Corkill RG, Davies P, Styles P, Rothwell PM. Association between cortical metabolite levels and clinical manifestations of migrainous aura: an MR-spectroscopy study. *Brain*. 2007;130(Pt 12):3102–3110.
64. Burstein R, Yarnitsky D, Goor-Aryeh I, Ransil BJ, Bajwa ZH. An association between migraine and cutaneous allodynia. *Ann Neurol*. 2000;47(5):614–624.
65. De Felice M, Ossipov MH, Porreca F. Update on medication-overuse headache. *Curr Pain Headache Rep*. 2011;15(1):79–83.
66. Amanzio M, Corazzini LL, Vase L, Benedetti F. A systematic review of adverse events in placebo groups of anti-migraine clinical trials. *Pain*. 2009;146(3):261–269.
67. Rains JC. Epidemiology and neurobiology of stress and migraine. *Headache*. 2009;49(9):1391–1394.

CHAPTER

43

Depression and Suicide

Maura Boldrini, J. John Mann

Columbia University, New York State Psychiatric Institute, New York, USA

OUTLINE

INTRODUCTION

Major depressive disorder (MDD) is projected to be the leading cause of disease burden worldwide by 2030. In most Western countries, untreated major depression is the leading cause of suicide. Suicide in its own right is a major global public health problem causing almost half of all violent deaths worldwide, resulting in about 1 million fatalities yearly, and economic costs in the billions of dollars (World Health Organization; see www.who.int). Suicides represent 1.4% of the global burden of disease. In most countries, the number of suicides is higher than the number of deaths due to traffic accidents or homicide. Morbidity due to non-fatal suicidal behavior is also substantial, because non-fatal suicide attempts are about 10 times more frequent than suicide.

The pathogeneses of depression and suicide are described, beginning with epidemiological studies, followed by research into psychological, social, and biological factors. The biology underlies fundamental processes

Neurobiology of Brain Disorders
http://dx.doi.org/10.1016/B978-0-12-398270-4.00043-4

related to mood regulation, reward mechanisms, and stress responses that are related to the onset and recurrence of major depressive episodes.

EPIDEMIOLOGICAL OBSERVATIONS IN DEPRESSION AND SUICIDE

Depression Rates

Major depression is one of the leading causes of disability worldwide, yet epidemiological data are not available for many countries, particularly low- to middle-income countries. Although direct information on the prevalence of depression does not exist for most countries, the available data indicate wide variability in prevalence rates. Data from 1998, consistent with earlier reports, showed that lifetime depression rates range from 1.0% (Czech Republic) to 16.9% (USA), with midrange examples of 8.3% (Canada) and 9.0% (Chile). The 12 month prevalence ranges from 0.3% (Czech Republic) to 10% (USA), with midrange examples of 4.5% (Mexico) and 5.2% (former West Germany). A study conducted in 2011, analyzing data from 18 countries indicated that the average lifetime and 12 month prevalence estimates of DSM-IV major depressive episodes are 14.6% and 5.5% in the 10 high-income countries and 11.1% and 5.9% in the eight low- to middle-income countries.[1] The average age of onset ascertained retrospectively is 25.7 years in high-income and 24.0 years in low- to middle-income countries. In high-income countries, younger age is associated with higher 12 month prevalence; by contrast, in several low- to middle-income countries, older age is associated with greater likelihood of major depressive episodes. The female to male ratio is about 2 to 1 in all countries. Countries with the highest prevalence estimates generally report the lowest impairment associated with depression, which could suggest a different severity threshold for diagnosing depression for inclusion in survey results. Although lifetime prevalence estimates are higher in high- than low- to middle-income countries overall, no significant difference in 12 month prevalence was found between them. These results may reflect a lower lifetime prevalence but higher persistence of depression in low- to middle-income than in high-income countries, but is also plausible that error in recall of previous lifetime episodes is higher in low- to middle-income than in high-income countries. Limitations of these cross-sectional studies are that in some settings, particularly those where treatment is unavailable, the most depressed people were probably unable to participate, and some surveys included mostly metropolitan areas, whereas others involved national samples. In addition, institutionalized patients, people in jails and prisons, people in the military, and people with severe cognitive or physical disabilities or who were receiving intensive treatment at the time of the survey were less likely to be interviewed. The samples also reflect survivor bias, which could be of considerable importance for understanding differences between high-income and low- to middle-income countries, given the gap in life expectancy of 10–15 years between people in developed and developing countries.

In the USA, the Substance Abuse and Mental Health Services Administration (SAMHSA) examines the national prevalence of depression each year through the National Survey on Drug Use and Health (NSDUH). Data from a recent survey show the 12 month prevalence of depression among adults in the USA to be 6.7%, while the lifetime prevalence of depression in the USA is 16.5%.

Depression and Societal Factors

There is an association between lower socioeconomic status and burden of disease, including depression, obesity, and diabetes. The associations between economic hardship and poorer mental and physical health and functioning are evident at a young age and persist across the lifespan. It is difficult to determine the sequential occurrence of these events and which of the elements is cause or consequence. Figure 43.1 summarizes environmental

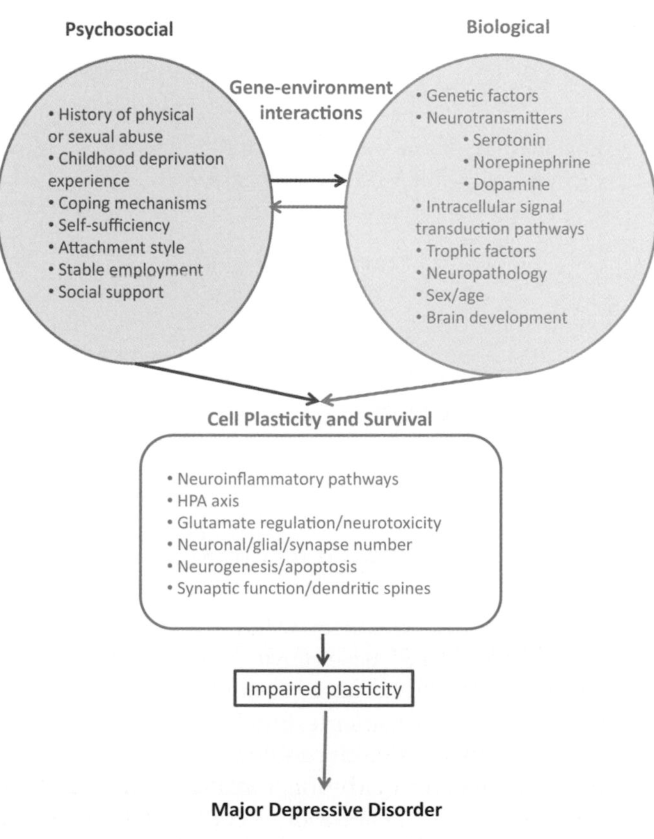

FIGURE 43.1 **Pathophysiological model of depression.**

and genetic factors that mold the phenotype for major depression.

Depression and Childhood Adversity

Relationship status also shows a strong association with depression: the strongest demographic correlate in high-income countries is being separated from a partner, and in low- to middle-income countries, it is being divorced or widowed.[1] Reported childhood abuse experience (physical, emotional, or sexual abuse, or neglect and deprivation) is associated with increased risk of developing depressive disorders later in life, as well as other psychopathology in adulthood, including nonsuicidal self-injury and suicide.[2] Adverse experiences in childhood increase the rate of adult depression and suicidal behavior[3] independently of impulsivity, aggression, and borderline personality disorder. Childhood maltreatment is not only linked to mental disorders, but also correlated with other health outcomes, including diabetes, gastrointestinal problems, obesity, gynecological problems, headaches, arthritis, breast cancer in women, and thyroid disease in men. The rates of substance use disorders, driving while intoxicated, smoking, compulsive overeating, obesity, avoiding exercise, and being involved in risky sexual behavior are all greater in those exposed to childhood adversity. Conversely, social and family supports are protective factors against psychiatric disorders, particularly depression.

Maternal separation in both animals and children has been associated with the development of depressive symptomatology. In primates, separation may induce depression that can be reversed or prevented by antidepressant drugs. A novel social stress paradigm in adult male C57BL/6J mice induces a range of depression- and anxiety-like behaviors both 24 hours and 1 month after the stress. Increased levels of serum corticosterone, part of the stress response, accompany these behavioral deficits. Genome-wide messenger RNA (mRNA) expression patterns in the ventral tegmental area (VTA; a key brain reward region) in stressed mice using RNA-seq show significant overlap between types of stress (active versus passive) in the pattern of altered gene expression, which suggests several potential gene targets for mediating the behavioral responses observed.[4] Witnessing traumatic events is a potent stressor in adult male mice, capable of inducing long-lasting neurobiological perturbations and depressive behavior.

Association of Depression with Medical Illness and Other Psychopathology

Depressed mood is associated with medical illness. The 1 year prevalence of depressive episodes, as defined by the 10th revision of the *International Statistical Classification of Diseases and Related Health Problems* (ICD-10), in participants in the WHO World Health Survey used in 60 countries, was 3.2% in participants without comorbid physical disease and 9.3–23.0% in participants with chronic medical conditions.[1] Depression is an early sign of Parkinson and Alzheimer disease, and is more common than in the general population in acquired immunodeficiency syndrome (AIDS), Huntington disease, temporal lobe epilepsy, diabetes, anemia, chronic obstructive airways disease, hypothyroidism, and multiple sclerosis. Strokes of the left prefrontal cortex and basal ganglia are associated with depression more often than strokes in the right hemisphere. Neoplastic diseases are also associated with depression, especially pancreatic and lung cancers, where there may be circulating mediators released by the tumors.

There is an association between mood and anxiety disorders, with 32% of people with mood disorders also having phobias, 31% panic attacks, and 11% obsessions and compulsions. MDD is found in 8–39% of patients with generalized anxiety disorder, 50–90% of patients with panic disorder, 35–70% of patients with social phobia, and 33% of patients with obsessive–compulsive disorders. Depression is also associated with substance use disorders, in particular alcoholism, and with personality disorders, more often of the cluster B. These associations raise the question of a possible common pathogenic pathway in MDD and these other psychopathologies or psychiatric disorders.

Suicide Rates

The Centers for Disease Control and Prevention report that there were 38,364 suicide deaths in the USA in 2010 (www.cdc.gov/injury/wisqars/index.html) and suicide was the 10th leading cause of death in the USA. Nationally, the annual suicide rate has increased by 3.9% since 2009 to approximately 12.4 suicides per 100,000. In 2005, the WHO sent a questionnaire on suicide mortality to the member states. Data obtained from the study (http://www.who.int/healthinfo/statistics/survey_2005/en/) are patchy because the latest year for which data are available varies from 1978 for Honduras to 2005 for El Salvador; for many countries (Burkina Faso, Eritrea, Haiti, Malawi, Malaysia, Maldives, Qatar, Sierra Leone, Timor-Leste, Turkey, Northern Ireland) there is no known year to which data completeness refers. According to the WHO data, the 15 countries with the highest suicide rates were: Lithuania (88.2 per 100,000), Russian Federation (81.2), Belarus (73.6), Sri Lanka (61.4), Kazakhstan (59.0), Estonia (57.5), Slovenia (57.0), Hungary (56.9), Ukraine (55.1), Latvia (54.7), Japan (48.0), Belgium (42.6), Finland (42.6), Croatia (39.8), and Switzerland (37.1); the US rate was 21.7, while some countries reported a rate of zero (Saint Vincent and the Grenadines, Saint Kitts and Nevis, Jordan, Honduras, Dominican Republic, Antigua and

Barbuda); in these countries the most recent data reported are from 1978 to 1995. Ascertainment modality can also affect the resulting reported suicide rates in different countries. Male suicide rate was highest in the countries with overall highest suicide rates. Female suicide rates were highest in Sri Lanka (16.8), China (in selected rural and urban areas excluding Hong Kong, 14.8), Lithuania (13.9), Japan (13.8), Slovenia, Hungary, Cuba (all 12.0), Russian Federation (11.9), Belgium (11.4), and Republic of Korea (11.2). The lowest male to female ratio was in rural China (0.9) and the highest in Antigua and Barbuda (13.4); in the USA the male to female ratio was 4.3.

These data raise the question of which factors contribute to such dramatically different national suicide rates and gender ratios. One concern is the reliability of collected data, because different definitions of suicide and ascertainment methods may contribute to differences internationally, and even regionally within countries or provinces. Doubt has been cast on the accuracy of suicide verdicts or diagnosis owing to stigma or shame, including designation of suicide as a criminal act in the criminal code of the country (e.g. India), and religion forbidding suicide, all of which contribute to suicide being underreported. There is a need to improve the ascertainment procedures, even within countries, to reduce underreporting of suicide from some areas. Civil, rather than criminal procedures for the ascertainment of suicide might reduce verdicts other than suicide. The lower suicide rate in Scotland compared with England and Wales appears to be due to differences in ascertainment, and ascertainment methods may contribute to the low official suicide rate in Ireland rather than it being entirely attributable to the strength of religious observance acting as a deterrent to suicide. Biases in reporting and classifying deaths may also contribute to gender differences in suicide. Since female suicide in some countries is culturally less acceptable than male suicide, coroners are less likely to record a suicide verdict for female deaths, leading to underreporting. Female suicides may also be underrecorded because females tend to self-poison, and this method is less likely to be identified as suicide.

Suicide and Societal Factors

Brain biology reflects the social context of the individual and the response of the individual to that social context. When social context is adverse, the risk of psychiatric illness or suicidal behavior increases for vulnerable patients. The interplay of society and the individual is reflected in biology associated with the stress response domain and epigenetics, and that modified stress response can in turn become a part of the diathesis that contributes to the triggering of major depression or suicidal behavior by subsequent stressful life events (Fig. 43.2). Societal factors may contribute to differences in suicide rates between countries and between regions within countries. A dramatic change in suicide rates occurred in some Eastern European countries in the transition from communism to democracy, which had more to do with alcohol consumption than with the political systems in place. Between 1990 and 1996, suicide mortality in Lithuania rose by 82.4%. From 1990 to 2002, the increase in suicides in urban areas was around 50% and in rural areas around 72%. When a country experiences a dramatic change in its suicide rates in a short timespan such as a few years, it is improbable that this can be related to genetic factors; instead, one must consider the potential social or environmental factors that are responsible for the change in suicide risk. In the mid-1980s, with the start of political reforms in the Soviet Union, the introduction of perestroika, and the banning of alcohol consumption, Lithuania experienced a sharp decline in suicide rate. In 1991, when alcohol became legally available again, the suicide rate surged, reaching an estimated 44.7 per 100,000 people in 2002, one of the highest suicide rates in the world. Every alcoholic drink per day increases the lifetime mortality due to alcoholism. Other factors have also been proposed to explain the increase in Eastern European suicide rates in the early 1990s, such as the social consequences of the process of democratization, with the loss of personal security associated with the breakdown of the highly structured, paternalistic, social support systems of the communist state.

The media may affect public attitudes to suicide. In countries with high suicide rates (e.g. Lithuania, Hungary) the mass media appear more accepting of suicide than in countries with low suicide rates (e.g. Germany, Austria, Ireland, and Italy). Mass media representations of suicide as an understandable response to stress, or as a glamorous or brave act under difficult circumstances, are thought to increase the risk of imitation suicides. A German television show about a person committing suicide by jumping under a train transiently increased the number of high-speed train suicides in Germany. In the USA, more local suicides after a highly publicized suicide acts have been reported. A similar effect was described following the impact of Goethe's book *Die Leiden des jungen Werther* (*The Sorrows of Young Werther*) and led to the coining of the term the "Werther effect". Television viewing has also been associated with aggressive behavior in adolescents and adults related to the frequency of violent acts (three to five per hour during prime-time television, 20–25 per hour during children's programs). Imitation has been theorized to be the "elementary social phenomenon" shaping learning and behavior.

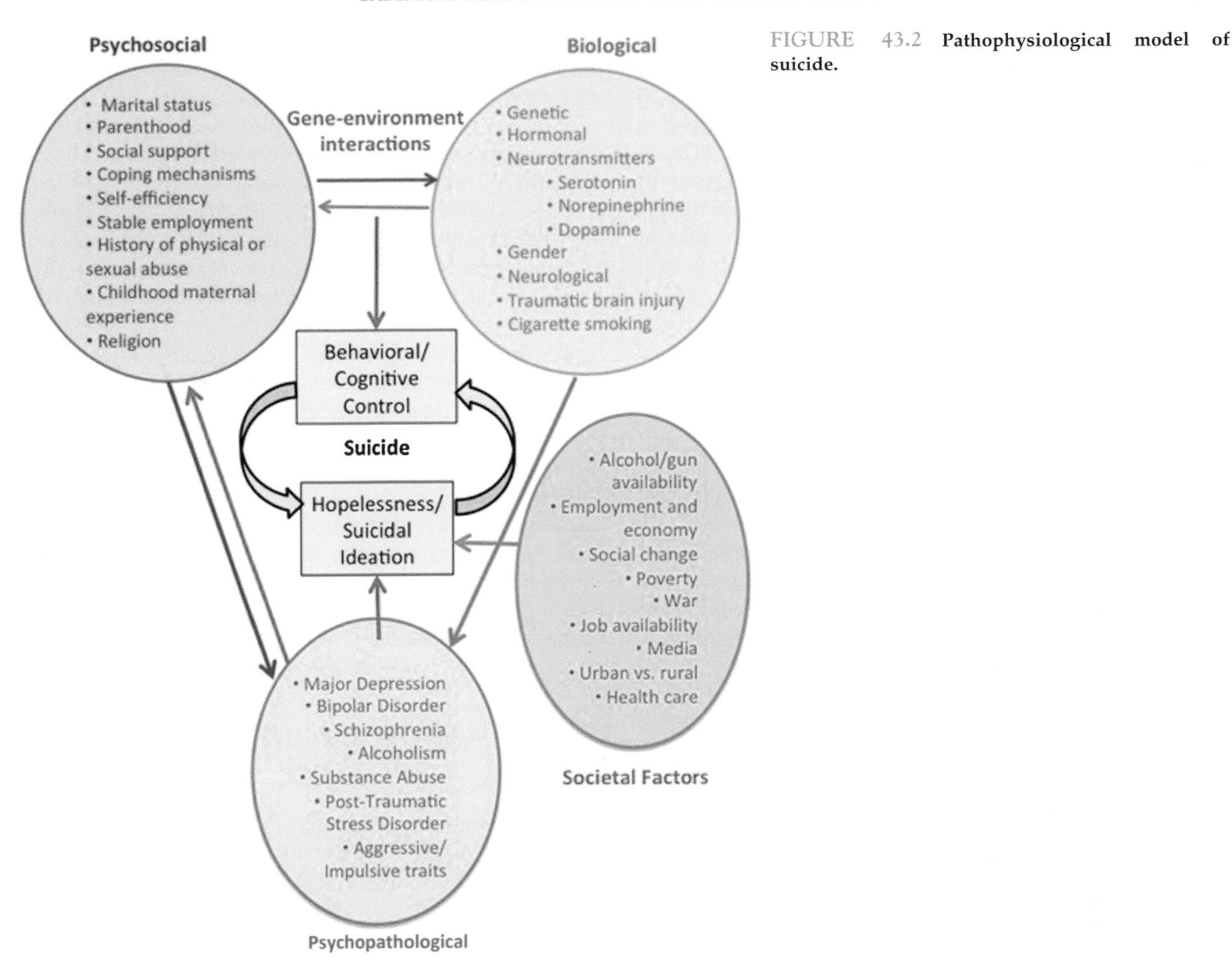

FIGURE 43.2 **Pathophysiological model of suicide.**

Urban and rural areas often report different suicide rates. In Lithuania in 2002 male suicide rates in rural areas were double those in urban areas and 1.4 times higher among rural women. In rural China the rural suicide rate is three or four times higher than the urban rate for both men and women, and for all age groups. Predictors of rural Chinese suicide are psychopathology and poor physical health, followed by poor social support and negative and stressful life events. Other significant correlates include lower education, poverty, religion, and family disputes. Significant predictors of rural suicide were higher depression symptom score, previous suicide attempt, acute stress at time of death, low quality of life, high chronic stress, severe interpersonal conflict in the 2 days before death, and a blood relative or friend or associate with previous suicidal behavior. Specific risk factors for rural suicide in comparison with urban suicide include easy accessibility to lethal suicide methods such as pesticides in the countryside, less accessible medical or emergency care, and negligible psychiatric services in rural China. In the USA, rural counties also have higher suicide rates than urban counties, which seem to be related to poorer medical services, lower income, and lower rates of prescriptions for new-generation antidepressants. Although, suicide in the USA is four times more common in males, non-fatal suicide attempts are more common in females than in males.

Suicide, Adversity, and Psychological Factors

Suicide is associated with a reported history of physical or sexual abuse during childhood. Pessimism, cognitive rigidity, impaired problem solving, and reactive aggression/impulsivity are reported components of the diathesis for suicidal behavior. Childhood adversity is diversely described across studies as maternal neglect, rejection, loss of a primary caregiver, harsh discipline, physical abuse, sexual abuse, interparental violence, parental neglect, and inconsistent discipline. Adverse parenting can involve neglect, but physical abuse and sexual abuse have been more commonly studied for their impact on suicidal behavior. Both neglect and deprivation can coexist with physical or sexual abuse. The two types of adversity are quite different and when they coexist it is hard to determine which is more pathogenic. Occurrence and severity of childhood sexual abuse, but not physical abuse, predict adult suicidal behavior independently of other known risk factors, and the odds of a sexually abused patient attempting suicide in adulthood are 10 times higher than a patient who reports not being sexually abused in childhood. There are also psychosocial factors that protect against suicidal behavior: religious affiliation, parenthood, marital status, reasons for living (such as feelings of responsibility towards family), fear of social disapproval, moral objections to suicide, better survival and coping skills, and greater fear of suicide.[5] Protective factors in economically, educationally, and socially disadvantaged African-American women are hope, spirituality, coping, social support (family and friends), and effectiveness in obtaining resources. Psychosocial factors have a bidirectional relationship with psychiatric illness. Psychiatric disorders can provoke failure in a job, school, or relationship, or prevent the beginning of stable employment or relationships. Conversely, psychosocial adversity can increase the degree of distress, depression, desperation, and hopelessness and, thereby, potentially increase the risk for suicidal behavior. Therefore, it is difficult to separate the impact of psychosocial adversity from that of psychiatric illness.

Association of Suicide with Mental Illness and Other Psychopathology

Over 90% of suicide victims or suicide attempters in Western countries have been found in psychological autopsies to have a diagnosable psychiatric illness, and about 60% of all suicides occur in the context of mood disorders, with the rest mostly related to various other psychiatric conditions, including schizophrenia, alcoholism, substance use disorder, and personality disorders. Lifetime suicide rates are higher in discharged hospital populations with mood disorders (15–20%) and alcoholism (18%) the highest, and lower in patients with borderline and antisocial personality disorders (5–10%). Suicide rates are lower in non-hospitalized psychiatric populations. About 80% of suicides in mood disorders occur during an untreated episode of depression.

Since most psychiatric patients do not commit suicide, it is evident that a psychiatric disorder is usually not a sufficient condition for suicide. Patients with a diathesis or predisposition for suicide are at higher risk, and this diathesis includes more pronounced aggressive/impulsive traits, hopelessness or pessimistic traits, impaired problem solving, and cognitive rigidity. The diathesis is about 50% heritable and contributes to the family history of suicidal behavior. Contributors to or consequences of the same diathesis are comorbidity for substance use disorders and alcoholism, a history of traumatic brain injury or epilepsy, multiple sclerosis, amyotrophic lateral sclerosis, and cigarette smoking.

Most of the reported risk factors are not independent: aggressive/impulsive traits, substance use disorders, MDD, and cigarette smoking are correlated; head injuries occur more frequently in aggressive, impulsive subjects and in people with substance use disorders, and conversely, aggressive behaviors can also follow head injuries. Head injuries in childhood are more common in aggressive subjects and the impact of the head injury on future aggression is greater in subjects who were already more aggressive before the head injury. Moreover, suicide, aggressive behaviors, and alcoholism are more common in males than in females, explaining some of the additional predisposition to suicide that is inherent in males. Early-onset mood disorders are associated with greater risk of suicidal behavior and with more pronounced aggressive/impulsive traits. The tendency for these clinical characteristics to coexist suggests that they may have common genomic and biological features.

PATHOGENIC FACTORS IN DEPRESSION AND SUICIDE

The pathogenesis of depression and suicidal behavior includes factors related to childhood development and genetics that mold the brain and behavior and determine resilience to stress. Child development, genetic and epigenetic factors, stress response systems, neurochemistry, and neuroanatomical changes related to depression and suicide, and their interplay, are reviewed here. A section on the neurobiology of resilience is included as it plays a role in coping and the risk of developing a psychiatric disorder.

Pathophysiological Model of Depression

Major depression has genetics and early life adversity causal factors. The intermediate biological phenotypes of MDD must be distinguished from those of the diathesis for suicidal behavior. Causal factors may lead to the development of MDD, and biological correlates of MDD include alterations of the hypothalamic–pituitary–adrenal (HPA) axis, the noradrenergic stress response system, and altered neural circuitry at rest and under conditions of performance of a variety of tasks, such as recognition of emotion on faces, reappraisal, and the Stroop test. Specific neurotransmitter systems abnormalities are described in MDD, mostly in the serotonergic and γ-aminobutyric acidergic (GABAergic) systems (Fig. 43.1).

Pathophysiological Model of Suicide

The stress diathesis model for suicidal behavior addresses clinical and cognitive factors and social context. Its structure reflects what is known about the interrelationship of the many reported causal factors. Mood regulation, decision making, and aggressive/impulsive traits can be linked to biological intermediate phenotypes and in turn to suicidal behavior. The model thus includes biological, clinical, societal, and psychosocial stress factors (Fig. 43.2). The presence of a psychiatric disorder, most frequently MDD, leads to disproportionately severe distress and suicidal ideation in subjects with suicide diathesis, who are then less able to suppress the urge to act on suicidal thoughts. Religion can help to inhibit the likelihood of acting on such thoughts and feelings. Components of the diathesis may be influenced by familial and genetic factors, resilience in the face of childhood adversity, and other biological underpinnings favoring the risk of suicidal behavior, such as impaired ventromedial prefrontal cortex and anterior cingulate serotonin input, low cholesterol levels, HPA axis hyperactivity, and altered noradrenergic and dopaminergic function. Compromised function of these brain regions can occur as a result of even mild head injuries, especially when repeated. The dorsolateral prefrontal cortex is involved in cognitive regulation of mood and decision making. Inhibition of dorsolateral prefrontal cortex leads to more risky decision making in males. Aggression/impulsivity increases the risk of suicide attempts. This model is heuristically useful and can predict risk of future suicidal behavior. The fact that its components correlate with a past history of suicidal behavior and predict future risk suggests that many elements of the diathesis are traits or relatively stable characteristics of the individual (Fig. 43.2).

GENETIC FACTORS IN DEPRESSION AND SUICIDE

Genetic Factors in Major Depression

In twin studies, 21–45% of the risk for MDD is heritable. Many studies have been undertaken to determine which genes contribute to MDD. Linkage analyses of families and association studies involving either candidate genes selected on the basis of known biological findings, such as abnormal serotonin indices or HPA axis, or the use of genome-wide association studies (GWAS) to search for previously unsuspected genes, form the main corpus of work. A meta-analysis of three large MDD GWAS datasets (with a total of 4346 cases and 4430 controls) found that six genes (*SLC1A4, CACNA1A, GRM8, PARK2, UNC13A,* and *SHC3*) involved in glutamatergic synaptic neurotransmission were associated with MDD.[6] This result is consistent with previous studies that found a role of the glutamatergic system in synaptic plasticity and MDD, and supports the potential utility of targeting glutamatergic neurotransmission in the treatment of MDD. Two genes, *GRM8* and *SHC3*, have previously been implicated in depression-related phenotypes. The metabotropic glutamate receptor *GRM8* inhibits presynaptic glutamate release. Nominal association was reported for *GRM8* with trait depression in a GWAS meta-analysis of European and US samples, and many studies have reported that *GRM8* knockout mice exhibit anxiety-related phenotypes. *SHC3*, a signaling adaptor involved in the signal transduction pathways in neurons, has been implicated in modulating hippocampal synaptic plasticity underlying learning and memory, and associated with nicotine dependence.[6]

Glucocorticoid receptor gene (*NR3C1*) polymorphisms have been shown to have a role in the susceptibility to recurrent MDD. Single-nucleotide polymorphisms (SNPs) were assessed in patients with recurrent MDD and controls, and differences were observed for allele and genotype frequencies of *NR3C1* at three loci: BclI, N363S, and ER22/23EK. The presence of C allele, CC, and GC genotypes of BclI polymorphism, G allele, and GA genotype for N363S and ER22/23EK variants, respectively, was associated with increased risk of recurrent MDD, while haplotype GAG played a protective role.[7] A GWAS revealed a neuron-specific neutral amino acid transporter (SLC6A15) as a susceptibility gene for MDD, with risk allele carrier status in humans and chronic stress in mice associated with downregulation of the expression of this gene in the hippocampus, a brain region implicated in the pathophysiology of MDD, and associations with alterations in hippocampal volume and neuronal integrity.[8] Alterations of adrenocorticotropic hormone (ACTH) and cortisol secretion in the combined dexamethasone/corticotropin-releasing hormone (DEX/CRH) test, as well as memory and attention performance, were found in MDD patients carrying the depression-associated AA genotype of *SLC6A15*. No effects of the *SLC6A15* variant were found in the healthy control group.[9]

A GWAS dataset was used to perform haplotype analysis to test for an association between MDD and the two-SNP tagging haplotype for the serotonin transporter (*HTT*) promoter region, *HTTLPR*, which has two long variants (L_A and L_G) and one short variant, differing in a 44 base pair insertion/deletion. The L_A allele is associated with greater gene expression than the L_G and S alleles. The L_G allele is a minor allele in Caucasian populations (0.09–0.15) and more abundant in African-American populations (0.24) and it is associated with a low rate of transcription, such that the L_G allele is functionally similar to the S allele. The GWAS data analysis found a significant association between the *HTTLPR* S allele and MDD.[10] Genetic variations in other serotonin-related genes, such as the rate-limiting enzymes of serotonin synthesis tryptophan hydroxylase-1 and -2 (*TPH1* and *TPH2*), serotonin receptors 1A, 2A, 2B, and 1B (*HTR1A*, *HTR2A*, *HTR2B*, and *HTR1B*), and monoamine oxidase A (*MAOA*) promoter, have also been studied. There are two *TPH* genes located on different chromosomes: *TPH1*, which is expressed in small intestine and megakaryocytes but is not expressed in adult brain; and *TPH2*, which is expressed in the brain. The less common alleles of *TPH1*, A218 and A779, are associated with a blunted prolactin response to fenfluramine and with low cerebrospinal fluid (CSF) levels of 5-hydroxyindoleacetic acid (5-HIAA) in healthy volunteers, reflecting possible brain serotonergic alterations due to developmental effects related to this gene variant or a gene variant in linkage disequilibrium with this SNP. A truncated form of mRNA for *TPH2* compromises the catalytic site of the enzyme, is found in human brain, and may be related to a SNP variant of the *TPH2* gene, which is associated with MDD but not with suicide. Allele and haplotype frequencies of a promoter polymorphism in *TPH2* did not show an association with mood disorders and did not differ in 83 suicidal schizophrenic patients compared with 170 non-suicidal schizophrenic patients.

Transcription factors regulating expression of the *HTR1A* gene may be relevant for depression and bipolar disorder. Extensive characterization of the transcriptional regulation of *HTR1A* using cell culture systems has revealed a GC-rich "housekeeping" promoter flanked by a series of upstream repressor elements for REST, Freud-1/CC2D1A, and Freud-2/CC2D1B factors that not only inhibit its expression as an autoreceptor on serotonin neurons, but may have different regulatory effects in various subsets of target neurons, where the receptor is also expressed. A separate set of allele-specific factors, including Deaf1, Hes1, and Hes5 repressors around the *HTR1A* C(-1019)G (rs6295) polymorphism, are found in serotonergic neurons in culture, as well as *in vivo*. The G allele does not bind repressors in serotonin neurons and is associated with greater gene expression. Pet1, an obligatory enhancer involved in serotonin neuron *HTR1A* expression and differentiation, has been identified as a potent activator of *HTR1A* autoreceptor expression. A *HTR1A* C(-1019)G allele load relationship with higher binding was confirmed on positron emission tomography (PET) scanning in MDD, bipolar depression, and healthy volunteers. Anxiety-like behavior has been reported in mice lacking *HTR1A*. The C(-1019)G polymorphism in the promoter region of the human *HTR1A* gene is reported to also be associated with schizophrenia and substance use disorder. No associations were found with MDD, bipolar disorder, alcoholism, panic disorder, or suicide attempt, or with personality traits in alcoholic patients. A study investigating 36 SNPs within 10 genes previously associated with MDD and bipolar disorder, as well as with response to treatment, did not confirm such an association with mood disorders or clinical outcomes.[11] The genes studied were *HTR1A*, *CLOCK* (circadian locomotor output cycles kaput protein), *ABCB1* (ATP-binding cassette, subfamily-B member-1), *ABCB4* (ATP-binding cassette, subfamily-B member-4), *TAP2* (transporter-2, ATP-binding cassette, subfamily-B), *CPLX1* (complexin1), *CPLX2* (complexin2), *SYN2* (synapsinII), *NRG1* (neuregulin1), and *GPRIN2* (G-protein regulated inducer of neurite outgrowth-2).

A study of 1255 subjects followed longitudinally for 22 years, from the French-speaking population of Quebec, Canada, investigated the serotonergic diathesis for mood disorders and suicide attempts: *HTR2A*, *TPH1*, *HTR5A*, *HTT*, and *HTR1A*. Of these, the *HTR2A* gene influenced both suicide attempts and mood disorders, although through different variants. In mood disorders, *HTR2A* variants (rs6561333, rs7997012, and rs1885884) were involved through one main effect (rs9316235). Three genes contributed exclusively to mood disorders, *HTR5A* through a main effect (rs1657268).

Variants of the *MAOA* gene include a variable number of tandem repeats in the promoter region. The alleles with two or three tandem repeats and lower expression have been associated with impulsive aggression, and there are lower levels of *MAOA* expression in males. The lower expressing alleles, together with a history of abuse before 15 years of age, are associated with higher impulsivity in males but not in females, and no specific associations with depression or suicide.

Genetic Factors in Suicidal Behavior

Suicide, non-fatal suicide attempts, and even suicidal ideation have a moderate hereditary component. Although the specific genes responsible for this genetic effect are not known, it appears that this genetic effect, is independent of major psychiatric illnesses such as mood disorders which are also partly genetically determined. In other words, the genetic effect related to suicidal behavior must affect the diathesis for suicidal behavior. Suicide and suicide attempts are more frequent in people who have a higher rate of familial suicidal acts in first or second degree relatives. Monozygotic twins have higher concordance rates for suicide and non-fatal suicide attempts than dizygotic twins. Adoption studies report that biological parents of adoptees who commit suicide have a higher rate of suicide than biological relatives of control adoptees, even controlling for psychosis and mood disorders. Parents of adolescent suicides show higher rates of suicidal behavior, independently of the presence of a major psychiatric disorder. Therefore, the heritability of suicide and suicide attempts is comparable to the heritability of other psychiatric disorders, such as bipolar disorder and schizophrenia, and is independent from the heritability of major psychiatric disorders. The specific genes that contribute to suicide risk are unknown.

A GWAS of 2023 depression cases examining a phenotype from suicidal thoughts up to severe suicide attempts or a composite phenotype found no genome-wide significant SNP.[12] Another GWAS of suicide attempts in patients with mood disorders identified one region within the Abl-interactor family member-3 binding protein (*ABI3BP* or *TARSH*) in a group with depression, but did not replicate this finding in a second cohort. A study using 36,048 SNPs compared suicides with sudden death controls and did not find any SNPs that reached the traditional 5×10^{-8} threshold for significance.[13] Another study found linkage results on Xq at DXS1047 (143 cM, 127.8 Mbps, DeltaLOD = 3.87, $p < 0.0001$) in more severe depression that may contribute to the higher rate of suicide attempts among women than men. In a sample of patients with bipolar disorder, an association signal on 2p25 at the threshold of genome-wide significance was identified, although the primary analysis did not detect a signal that reached wide significance. Several GWAS have investigated treatment-emergent suicidal ideation during antidepressant treatment without finding a variant reaching genome-wide significance.

A meta-analysis concluded that the tryptophan hydroxylase *TPH1* intron 7, A218 polymorphism, is associated with suicide-related behavior. This association was replicated in a further meta-analysis that assessed the heterogeneity due to variations in genetic background, which reported that the A allele has a dose-dependent effect on the risk of suicidal behavior, at least in the Caucasian population. *TPH1* genotype may be associated with higher levels of anger and more severe suicidal acts. There is an association with suicide. The T(-473)A and G(-8396)C polymorphisms of the *TPH2* gene were not associated with history of suicide in major psychoses. The *TPH1* variation (rs10488683) is reported to be associated with the diathesis for suicide attempts.

Of six polymorphisms in four genes related to the serotonin system, including *HTTLPR* and *HTTVNTR* in the *HTT* gene, rs6295 in the *HTR1A* gene, rs11568817 and rs130058 in the *HTR1B* gene, and rs6313 in the *HTR2A* gene, an association was found between rs11568817 and suicidal ideation in a Chinese population. The *HTR1A* C(-1019)G functional polymorphism (rs6295) G allele is associated with greater *HTR1A* gene expression in presynaptic raphe neurons, and lower neuronal firing. The frequency of the -1019G allele in the *HTR1A* gene is higher in MDD, but a meta-analysis did not find an association with suicidal behavior. Differences in C(-1019)G genotype frequencies were found in relation to impulsiveness: the GG genotype, more common in MDD, showed higher impulsiveness scores compared with GC or CC carriers.

HTR1B knockout mice are impulsive, aggressive, and more prone to self-administer cocaine and possibly alcohol. In a case–control study, the C129T and G891C polymorphisms of the *HTR1B* were associated with 20% fewer receptors in the postmortem brain, although these polymorphisms were not associated with suicide or other suicidal behavior. *HTR1B* A-161T polymorphism was also not associated with suicide in 160 patients with MDD and 160 healthy controls and in 110 patients with schizophrenic disorders compared with 215 healthy controls. *HTR1B* G861C polymorphism was associated with antisocial alcoholism, and an association was found with non-alcohol-related substance abuse, but not with mood disorders or suicidal acts. *HTR1B* agonists decrease aggressive responses in mice. Therefore, *HTR1B* seems be related to aggression, impulsivity, and drug or alcohol abuse, but not directly to mood disorders or suicidal behavior.

HTR2A variants (rs6561333, rs7997012, and rs1885884) appear to be associated with suicide attempts through interactions with histories of sexual and physical abuse. A meta-analysis failed to find association of the *HTR2A* gene T102C polymorphism with suicidal behavior (H452Y and A1438G). A family-based study on adolescent inpatients with a recent suicide attempt also found no association of *HTR2A* T102C polymorphism with any relevant clinical traits. The T allele was associated with more binding, and more binding is associated with aggressive traits and suicide.

The *HTT* promoter region has a triallelic variant, *HTTLPR*, with two long variants (L_A and L_G) and one short variant. The L_A allele is associated with greater gene expression than the L_G and S alleles. Most studies have not distinguished between the two L alleles and this may contribute to the lack of agreement on results. A meta-analysis of 12 studies investigating *HTT* genotype in suicide found that 5-HTT may play a role in the predisposition to suicide. The short form has been reported to be associated with violent suicide attempts in subjects with mood disorders, alcoholism, and suicide attempts. S carriers (LS and SS genotypes) are reported to make more suicide attempts and more lethal attempts, and to use more violent means compared with LL carriers. In African-Americans with substance abuse, increased risk of a suicide attempt was associated with reports of childhood trauma in those with low expressing alleles (S and L_G) of the *HTTLPR* genotype. Conflicting results showed no association between the *HTTLPR* polymorphism and suicide, and a positive or no association between family suicide history and SS genotype. An association of suicide completion with at least one copy of the 10 repeat allele of the *HTT VNTR* intron-2 variant is reported, while another study found the 10 repeat allele to be protective against suicidal behavior in schizophrenia.

No association between suicide attempt and the *MAOA VNTR* polymorphisms was found in patients with bipolar disorder or schizophrenia. Failure to express the gene in males has been associated with pathological aggressive behavior and lower intelligence.

For the candidate gene approach to be fruitful, more basic behavioral and neurochemical endophenotypes are required, which are likely to have a closer biological relationship to specific genes. MDD and suicidal behavior are complex disorders and the predisposition probably consists of numerous genetic factors, probably of small effect size, which may manifest themselves only when a certain threshold is crossed. Standard SNP-based GWAS analysis typically has limited power to deal with the clinical heterogeneity and polygenic contribution of individually weak genetic effects underlying the pathogenesis of MDD. Depression and suicide may be the result not only of gene effects and gene–gene interactions but also of gene–environmental interactions, and so far insufficient attention has been paid to the last two mechanisms.

GENE BY ENVIRONMENT INTERACTIONS IN STRESS, DEPRESSION, AND SUICIDE

Interactions between genes and environment account for some of the individual variation in consequences of childhood adversity for outcomes in adulthood. Peer-reared monkeys compared with maternally reared young monkeys manifest deficits in parenting style in adulthood and a host of biological differences. Primates or rodents removed from their mothers and deprived of early participation in complex social groups later exhibit aberrations in neurochemical function, neuroendocrine stress axis activity, and many aspects of social behavior and parenting; however, individual variation in these behavioral and brain biology changes also reflects the impact of risk and protective genetic variants.

Maternal deprivation in rodents alters development of the HPA axis and noradrenergic system function and results in a hyperresponsive stress system in adulthood. Both the HPA axis and the noradrenergic responses to stress then remain heightened into adulthood. Abnormalities in terms of excessive HPA axis and noradrenergic stress responses in MDD are also apparently related to childhood adversity, although clinical studies have only been able to examine correlations between reported history of adversity and adult heightened stress response. Data on a possible change in the genome or stress response systems in childhood after a period of adversity have not been studied in detail, and clinical studies are generally retrospective. Major depression and suicide have been associated with HPA axis abnormalities. Epigenetic effects of reported childhood abuse in suicides include greater DNA methylation of CpG sites that are known to be able to suppress expression of the glucocorticoid receptor gene, and less glucocorticoid receptor gene expression in the brain. Suicides without reported childhood abuse do not show elevation of DNA methylation or lower glucocorticoid gene expression. Maternally deprived rodents show the same effect on this gene, and loss of glucocorticoid receptors would mean loss of feedback at higher concentrations of glucocorticoids and contribute to the observed excessive ACTH and corticosterone response to an air-puff on the snout in adulthood. Overactivity of the HPA axis and the alteration in some neuropeptide systems, including neuropeptide Y and CRF, may also be related to the neurobiological consequences of abuse and neglect, as shown by animal studies of maternal separation. Increased noradrenergic and cortisol-dependent stress responses are reported in depression. Cortisol is increased in MDD as part of a hyperactive HPA axis, and treatment with selective serotonin reuptake inhibitors (SSRIs) and recovery from depression are associated with decreases in cortisol concentrations. Prepubertal children with post-traumatic stress disorder (PTSD) secondary to reported maltreatment excrete more urinary norepinephrine, dopamine, and urinary free cortisol over 24 hours than non-traumatized children with overanxious disorder and healthy controls. Levels correlate with the duration and severity of PTSD, and PTSD is associated with increased lifetime suicidal ideation and suicide attempts, suggesting that this altered stress-related biological abnormality may contribute to suicide risk through both MDD and PTSD aside from direct effects of elevated cortisol on the brain.

Familial transmission of suicidal behavior in families of probands with mood disorders appears to be partly related to reported childhood abuse in probands and offspring, together with early onset of mood disorders and aggressive/impulsive traits.[14] This combination of factors may have a common biological component because low serotonin activity is associated with both early-onset recurrent mood disorders and aggressive/impulsive trait severity. The HPA axis has a bidirectional relationship with the serotonergic system. The serotonergic system is involved in cortisol release and corticosteroids suppress *HTR1A* receptor expression in the hippocampus. *HTR1A* and *HTT* genes contribute to mood disorders through gene by environment interactions with reported childhood physical abuse. Dexamethasone and stress increase TPH protein immunoreactivity in the rodent raphe nuclei, and more TPH immunoreactivity[15] and mRNA in brainstem has been reported in suicides. The cause could be a transcription factor enhancing expression of the gene or a stress response, or both. In female rats maternal separation leads to decreased norepinephrine levels in the cingulate cortex and decreased 5-HT, 5-HIAA, dopamine, and norepinephrine levels in the amygdala, but increased levels of both 5-HT and 5-HIAA in dorsal raphe nucleus (DRN) tissue. Peer-reared monkeys have lower serotonergic activity (low CSF 5-HIAA and lower serotonin transporter binding in the brain) than maternally raised monkeys, an alteration that persists into adulthood and is reflected in greater aggression and impulsivity. A gene by environment interaction between the rhesus macaque serotonin transporter gene promoter polymorphism and early rearing experience accounts for some variation in the physiological and behavioral outcomes of group versus peer rearing, whereby abnormalities related to the peer-rearing experience are more pronounced in monkeys with low-expressing *HTTLPR* alleles. Male monkeys raised with or without their mothers, when tested for social competition and *MAOA* gene promoter polymorphism manifested a genotype by environment interaction effect on competitive behavior. In humans, there is a significant interaction between childhood maltreatment and low *MAOA* expressing alleles in modulating the risk for antisocial behavior, aggressiveness, and violence; low *MAOA* expression is a detectable risk factor only in the presence of an adverse childhood environment, and the effect is not trivial. These results have been replicated in a community-based sample of boys from the Virginia Twin Study for Adolescent Behavioral Development (VTSABD). The lower expressing alleles of the *MAOA* gene and a history of abuse before 15 years of age were associated with higher impulsivity in adult male subjects, but not females, suggesting that the genetic variation may sensitize males to the effects of childhood abuse experiences on impulsive traits in adulthood. A history of child abuse appears to be associated with a greater risk of suicidal behavior in adult life, and this effect may be mediated via a history of severe impulsivity and aggression, and more commonly borderline personality disorder.

NEUROTRANSMITTER SYSTEMS IN MAJOR DEPRESSION AND SUICIDAL BEHAVIOR

Direct study of the brain of depressed people and suicides can provide important information about the genome, neurotransmitter systems, morphology, and signal transduction pathways related to trophic and apoptotic pathways, stress responses, and neurotransmission. However, the results are correlative, and causal relationships need to be worked out in animal studies. Postmortem brain studies investigate neurobiological changes in specific brain areas of subjects who died by suicide or from other causes of sudden death without a prolonged agonal period that may result in brain changes unrelated to psychiatric illness. Serotonin, norepinephrine, GABA, and dopamine neurotransmitter systems, signal transduction, and cellular morphology have been studied in mood disorders and suicide. Such studies have the advantage of studying suicides, which are the extreme phenotype for suicidal behavior, and they allow direct study of the brain. At the same time, these studies require accurate information about the subject's diagnosis and history. The use of medical records and a psychological autopsy for clinical data are crucial to separate effects related to mood disorders or schizophrenia and suicide. Regarding antemortem drug treatment, postmortem studies can utilize blood and brain tissue toxicological screens to verify recent drug or medication use but may miss information on more distant medication exposure. Postmortem studies examine the brain only at one point in time, whereas *in vivo* imaging studies can examine the brain at different time-points, evaluate the effect of current clinical or treatment status or specific experimental conditions, and allow prospective re-evaluation. In terms of live patient studies, the subpopulations of suicide attempters or depressed patients who agree to participate in clinical and neurobiological studies, and undergo a medication washout and medication-free period, may limit generalizability to the wider patient population.

The Serotonergic System

Serotonin hypofunction (low CSF 5-HIAA) favors more lethal suicidal behavior and carries an odds ratio for future suicide in MDD of 4.6. Low CSF 5-HIAA is also associated with anxiety, impulsivity, and aggression. The relationship of serotonin hypofunction to suicidal acts or suicide is independent of psychiatric diagnosis. Lower 5-HIAA is reported in the brainstem of suicide victims and in the CSF of suicide attempters regardless of psychiatric diagnosis, and low CSF 5-HIAA can predict future suicide and suicide attempts. In men with mood disorders, followed up (mean 21 years) for causes of death, CSF 5-HIAA was associated with short-term (within 1 year) suicide risk, and HPA dysregulation was a long-term (beyond 1 year) suicide risk predictor.[16] Some studies have linked low CSF 5-HIAA to more lethal suicide attempts and suicide attempts using violent methods. Severity of lifetime aggression is inversely correlated with CSF 5-HIAA, and low CSF 5-HIAA predicts recidivism in murderers and arsonists. These findings further link the diathesis for severe aggression to that for suicidal behavior and may reflect the impaired serotonin input into ventral prefrontal cortex. Postmortem studies have shown that CSF 5-HIAA correlates with levels of 5-HIAA in prefrontal cortex.

The relationship between low serotonergic function and suicidal behavior is further indicated by a blunted prolactin response to serotonin released by fenfluramine in suicide attempters with MDD or personality disorders compared with controls. The more lethal the suicide attempt, the lower the CSF levels of 5-HIAA and the prolactin response to fenfluramine. In parallel, aggressive/impulsive traits are related to low prolactin release after fenfluramine, indicating a serotonin release deficit that could reflect a common biological cause for suicide risk and aggressive/impulsive traits. Depressed high-lethality suicide attempters with MDD had lower regional cerebral metabolic rate for glucose in ventral, medial, and lateral prefrontal cortex compared with low-lethality attempters with MDD, and the size of the brain areas showing this difference doubled after fenfluramine administration. Such correlational studies cannot establish causal relationships, but a relationship is suggested as depletion of serotonin function in animals and humans increases aggressive behavior and impulsivity, whereas serotonergic enhancement decreases such behaviors.

Dysfunctional attitudes about oneself, the world, and the future are reported in a subgroup of depressed patients to be positively correlated with cortical HTR2A binding potential, especially in dorsolateral prefrontal cortex, and with serotonin transporter binding in MDD. More postsynaptic HTR1A and/or HTR2A binding and less HTT binding has been found in prefrontal cortex of suicides, although not all studies confirm these results. Postsynaptic serotonin receptor upregulation may be a compensatory response to less serotonin release at the level of the projection areas in the brain of suicides. In the case of HTR2A, the upregulation involves increased gene expression in at least the adolescent and young adult population of suicide attempters. In suicides, the changes in HTT and HTR1A binding are localized to the ventral prefrontal cortex and anterior cingulate.

In the brainstem, despite postmortem findings of low serotonin and 5-HIAA in several studies, another study found no deficiency in the number of serotonin neurons. Increased TPH immunoreactivity and mRNA has been found in the brainstem in suicides. Another report confirmed that TPH immunoreactivity levels were 46% higher in the DRN in depressed suicides compared with controls. Therefore, serotonin neuron numbers and TPH2 are elevated in suicides, in accordance with some other studies, although fewer HTT mRNA-expressing neurons were found in depressed suicides compared with controls. Studies of inhibitory HTR1A somatodendritic autoreceptors in the DRN report both higher and lower binding in depressed suicides. The binding is higher in depressed suicides in the rostral section of the DRN and lower in the more caudal sections, and this may cause less serotonin firing and release in specific cortical regions, such as the ventral prefrontal cortex, involved in suicide.[17]

The Noradrenergic System

Compared with the serotonergic system, fewer studies have analyzed noradrenergic alterations in suicides postmortem. Conflicting results exist regarding tyrosine hydroxylase (TH), the rate-limiting enzyme in the biosynthesis of norepinephrine (noradrenaline), which has been reported to be higher in suicides by some and lower by other studies. Fewer noradrenergic neurons in the locus ceruleus of suicides with MDD were found in comparison to controls. Norepinephrine levels in locus ceruleus did not did not differ between controls and suicides, but agonist α_2-adrenergic receptor binding was higher in suicides. TH compensatory increases have been reported when increased norepinephrine release leads to transmitter depletion. Therefore, TH immunoreactivity could be a state-dependent, homeostatic response. In the prefrontal cortex, β-adrenergic binding is higher in suicide. Norepinephrine levels in the prefrontal cortex of suicides are higher and α_2-adrenergic receptor antagonist binding is lower, indicating cortical noradrenergic overactivity. In suicide, this may be a stress-related change, since the period before the suicide act is likely to involve high-intensity stress for the subject. Humans exposed to adverse childhood experiences can show exaggerated sympathetic responses in adulthood in response to stress, and prolonged stress may deplete norepinephrine function. *In vivo*, there is no correlation in suicide attempters between the CSF levels of norepinephrine or epinephrine and suicidal behavior. Lower CSF levels of 3-methoxy-4-hydroxyphenylglycol (MHPG) predicts future suicide attempt or suicide and correlates with higher medical lethality of the future suicidal act.[13] Part of the relationship between smoking and suicidal behavior is mediated by low CSF MHPG. The effect of MHPG on suicide risk may be related to its impact on hopelessness and depression. Severe anxiety or agitation may be associated with noradrenergic overactivity in response to stress and overactivity of the HPA axis, and both are related to higher suicide risk. Urinary levels of MHPG were higher in a sample of 111 inpatients after a suicide attempt (axis I diagnosis of adjustment disorder, MDD, schizophrenia, and personality disorder) compared with controls, although no difference in serum MHPG was found between violent and non-violent attempts. These results suggest increased norepinephrine turnover in subjects who attempt suicide, at least within 24 hours after the attempt, and may reflect a stress response. Lower levels of CSF MHPG and fewer noradrenergic neurons may increase the risk of noradrenergic depletion, especially after excessive release when stressed. Noradrenergic depletion may contribute to despair and "giving-up" behavior, as described in rodents exposed to a restraint stressor.

The Dopaminergic System

Mesolimbic dopaminergic reward pathways may be involved in depression, pessimism, hopelessness, and perception of reasons for living. This pathway may also determine how an individual responds to stress and thereby modulate the resilience to depressive and suicidal thoughts. Subcortical structures involved in dopamine signaling include the dopaminergic neurons of the VTA, and the dopaminergic projections to the dorsal striatum, ventral striatum, nucleus accumbens, and amygdala. Higher dopaminergic activity may favor suicidal behavior through greater aggression or behavioral activation and lower dopaminergic activity through anhedonia and depression. Studies indicate a role for lower dopaminergic function in MDD characterized by psychomotor retardation. A dopaminergic system deficit in MDD may be related to defective reward mechanisms and underlie aspects of depression such as anhedonia. Low CSF HVA is found in suicide attempters with MDD by some but not all studies, and in some cases this relationship may be confined to MDD. No alteration of mRNA for the dopamine D_1 and D_2 receptors or dopamine D_4 receptor binding was found in the caudate nucleus of suicides with MDD.

Functional interactions between glutamate, *N*-methyl-D-aspartate (NMDA) receptors, dopamine, and dopamine receptors contribute to reward circuits. Phasic activity of VTA neurons projecting to the nucleus accumbens, but not to the medial prefrontal cortex, induces susceptibility to social defeat stress in mice. Optogenetic inhibition of the VTA–nucleus accumbens projections induces resilience, whereas inhibition of the VTA–medial prefrontal cortex projections promotes susceptibility.[18] Mice susceptible to social defeat stress also show lower disheveled-2 (DVL-2) mRNA in nucleus accumbens, and blockade of DVL function renders mice more susceptible and promotes depression-like behavior; similar depression-like effects were induced by overexpressing glycogen synthase kinase-3β (GSK-3β) in the nucleus accumbens, and downregulation of DVL was also shown in the nucleus accumbens of depressed humans postmortem.[19] These findings reveal a novel role for the DVL–GSK-3β signaling pathway, acting within the brain's reward circuitry and regulating susceptibility to chronic stress. Depression severity correlates inversely with the rewarding effects of amphetamine, with a mechanism possibly related to depletion of synaptic dopamine and upregulation of dopamine receptors. Preclinical studies indicate that there may be a genetically determined mesocortical and mesoaccumbens dopamine response to stress that relates to learned helplessness. Therefore, psychostimulants, dopamine reuptake inhibitors, monoamine oxidase B inhibitors (selegiline), dopamine receptor agonists (pramipexole), and NMDA receptor antagonists (memantine) may be useful for treating anhedonia and hopelessness. The cognitive effect of dopaminergic stimulation is dependent on its activity level. Psychostimulants can decrease or increase impulsiveness in rodents. Acute administration of low doses of amphetamine reduces impulsiveness and enhances alertness or signal discrimination. Thus, the role of dopamine in MDD seems more clearly linked to psychomotor function and reward pathways. Its role in suicidal behavior via mood regulation remains to be determined, although its role in decision making and reward mechanisms seems clearer.

CELL PLASTICITY AND SURVIVAL IN DEPRESSION AND SUICIDE

Intracellular Signal Transduction Pathways

Neurotransmitters, as well as trophic factors, act through G-protein coupled receptors, which in turn act on intracellular kinases such as extracellular signal-regulated kinase-1/2 (ERK1/2), protein kinase B (AKT), and GSK-3β, which regulate cell proliferation and survival. A deficiency of selective G-protein α-subunits is associated with suicide, independently of psychiatric diagnosis. Serotonin-activated intracellular kinases interact with proteins involved in cell viability, including other kinases: the mammalian target of rapamycin (mTOR); transcription factors: cyclic AMP response element binding (CREB); proto-oncogenes: B-cell leukemia/lymphoma-2 (BCL2); polymerases: poly-ADP-ribose-polymerase (PARP); phosphatases: phosphatase and tensin homolog (PTEN); and proteases: cysteine-dependent aspartate-directed protease-3 (Caspase3). Besides its effects as a neurotransmitter, serotonin is a trophic factor in the brain, affecting neuroplastic processes that may remodel neural circuitry as part of the pathogenesis of mood and other psychiatric disorders. *HTR1A* is highly expressed in the hippocampus of rodents and humans. Deletion of *HTR1A* in mice abolishes the effect of fluoxetine on neurogenesis in the dentate gyrus and its behavioral effects in the novelty-suppressed feeding test. This suggests that serotonergic signaling through this receptor is critical for the antidepressant action of SSRIs, which is in turn dependent on an effect related to neurogenesis. If this is so, then serotonin and neurogenesis may play important roles in the pathogenesis of MDD.

Understanding more of the molecular mechanisms of serotonin and neurogenesis may inform us about the action of antidepressants and the pathophysiology of MDD. GTPγS binding to $G\alpha_{i/o}$ protein and HTR1A coupling to adenylyl cyclase, together with phosphatidylinositol 3-kinase (PI3K) and its downstream effector AKT, are lower in the occipital cortex of suicide victims. In addition, the activation of ERK1/2 is attenuated in suicides. HTR1A activates AKT, via PI3K, and PI3K/AKT regulates angiogenesis, growth cone formation, cell proliferation, and survival. It activates BCL2 directly[20] as well as through the activation of nuclear factor-κB. It inhibits apoptosis by inhibition of GSK-3β, which stimulates Caspase3. The activity of protein kinase C, part of the signal transduction pathway for HTR2A, is low in the prefrontal cortex of suicides. Stimulation of HTR2A promotes survival of rat neural progenitor cells and increases neural progenitor cell number and postmitotic neurons.[21] HTR2A increases DNA synthesis, activating ERK1/2 in human choriocarcinoma cells[22] and resulting in proliferation. Subjects with MDD who die by suicide have abnormalities in the activity of the extracellular-signal regulated kinase/mitogen-activated protein kinase (ERK/MAPK) cascade. ERK has a role in learning and memory, and it is a critical player in synaptic and neuronal plasticity. The fluoxetine-induced increase in immature neuronal marker expression in adult hippocampus is attenuated in mice lacking HTR4.[23] Polymorphisms of HTR4 are associated with MDD, and suicides have altered brain expression of HTR4. HTR4 activates CREB, resulting in increased enteric neurogenesis[24] and promoting survival of amyloid-expressing neurons.

Lower expression of mTOR and its downstream protein translation regulator eIF4B is found in MDD, and fluoxetine acts through activity on eIF4B.[25] The multimeric kinase complexes mTOR1 and mTOR2 control metabolism, proliferation, dendritic protein synthesis, and survival downstream of PI3K/AKT signaling.[26] Serotonin can act on mTOR and override rapamycin inhibition of tumor growth,[27] therefore showing potential for a proliferative effect.

CREB is lower in MDD in brain tissue, fibroblasts, and leukocytes. CREB is a leucine zipper transcription factor expressed during neuronal maturation and differentiation, and gradually downregulated when newborn neurons become mature.[28] It facilitates maturation and dendritic development.[29] Antidepressants increase CREB in rat brain. Inhibition of CREB impairs proliferation and differentiation. In rodents and monkeys[30] CREB regulates transcription of the polysialylated neuronal cell adhesion molecule, which is necessary for newborn neurons to extend their dendrites. The levels of CREB, its DNA binding activity, and PKA are low in the hippocampus and prefrontal cortex of suicides. Olanzapine and fluoxetine increase AKT, CREB, and BCL2 levels in prefrontal cortex, hippocampus, and striatum.[31] Those properties may contribute to their antidepressant action or potential prevention of suicide.

Prefrontal cortex mRNA and protein levels of BCL2 are low in bipolar disorder.[32] *BCL2* is a human proto-oncogene that prevents cell apoptosis, and its polymorphism influences gray matter volume in the human striatum.[33] Fluoxetine modulates neural progenitor cell survival and differentiation at least partly through BCL2. Studies of cortical thickness indicate deficiencies in MDD and bipolar disorder, and cortical thinning and lower metabolic levels are reported in future suicide attempters.

Trophic Factors

Brain-derived neurotrophic factor (BDNF) signaling is required for the long-term survival of newborn neurons in mouse dentate gyrus. It is involved in activity-dependent neuronal plasticity, survival, and differentiation of peripheral and central neurons. BDNF is known to increase neuronal survival and axonal length via ERK1/2 and PI3K/AKT pathways. PI3K/AKT protects neuronal cells from apoptosis by phosphorylation and inhibition of GSK-3β, which also blocks cell proliferation and is a target for the mood stabilizer lithium. BDNF binds to its receptor tyrosine kinase receptor B (TrkB) with downstream phosphorylation of CREB, an effect potentiated by chronic administration of fluoxetine and desipramine to rats. Both $TrkB^{+/-}$ and $BDNF^{+/-}$ mice are resistant to antidepressants. Dentate gyrus neurogenesis is reduced in $BDNF^{+/-}$ mice, as well as survival of newly formed cells after chronic imipramine treatment compared with wild-type mice. Suicides have lower BDNF brain expression and BDNF is higher postmortem in those treated with medication. BDNF and TrkB mRNA and protein are also lower in the prefrontal cortex and hippocampus of teenage suicides compared with controls.[34] BDNF mRNA levels in cortex and hippocampus increase with chronic antidepressant administration and electroconvulsive shock in rats. Antidepressants reverse BDNF mRNA downregulation by stress and, when centrally administered to rats, BDNF induces antidepressant-like effects and neurogenesis.

Proliferation of neural stem or progenitor cells in the adult brain is affected by several other growth factors, including epidermal growth factor, stem cell factor, heparin-binding epidermal growth factor, and vascular endothelial growth factor (VEGF). Electroconvulsive seizures increase neurogenesis and angiogenesis in rats via a VEGF-signaling dependent mechanism.[35] VEGF, through the Flk-1 receptor, is also required for antidepressant-induced (fluoxetine, desipramine) cell proliferation. Chronic stress decreases proliferation of vasculature-associated neural progenitor cells to a larger extent than non-vascular-associated cells. Steroids such as estrogen, dexamethasone, and corticosterone are all able to regulate VEGF and/or VEGF receptor mRNA expression. After chronic stress, both VEGF and Flk-1 protein levels are significantly decreased in the hippocampus, and levels recover after three weeks, as well as neural progenitor cell number, suggesting that changes in VEGF are implicated in the decreased adult proliferation found after chronic stress. VEGF could act by directly increasing cell replication, as described above, altering vessel permeability, or increasing BDNF expression, which can be modulated by astrocytes and endothelial cells. Perturbing blood vessel permeability could also affect brain access by other trophic factors, including fibroblast growth factor-2 and insulin-like growth factor.

Thus, there is evidence that trophic pathways may be compromised in MDD, altered by chronic stress, and involved in antidepressant action. The link to suicide is less clear, except for BDNF.

Cell Survival

Indirect evidence suggests that apoptotic processes contribute to MDD. Modest, *in situ* end-labeling for DNA fragmentation was observed in 11 out of 15 MDD cases in one study. However, there are a dearth of data for expression of the prototypical apoptotic initiators Fas, p53, Bax, and downstream caspases in human brain from subjects with MDD. Importantly, animal studies indicated that stress reduces expression of the antiapoptotic factor BCL2 and severe stress exacerbates infarct size through the suppression of BCL2 expression. Sleep deprivation, a treatment for depression, significantly increases cell proliferation and the total number of surviving cells in the hippocampal dentate gyrus. Stress-induced depression in animal models, associated with cell death and dendritic shrinkage in the hippocampus, can be reversed by treatment with antidepressants and neurotrophins. These findings, together with findings that antidepressant treatments increase neurogenesis in the dentate gyrus, indicate that an increase in the number of young neurons may be one of the mechanisms of action of antidepressants, and could explain in part their delayed onset of therapeutic efficacy. However, the role of apoptosis in the pathogenesis of MDD or suicide remains unclear.

Inflammatory Pathways

Chronic inflammation and oxidative stress have been implicated in the pathophysiology of MDD[36] as well as in a number of chronic medical conditions, and it has been hypothesized that the homeostatic buffering mechanisms regulating oxidation and inflammation in healthy individuals become dysregulated in untreated MDD, and may be improved with antidepressant treatment. Subjects with atypical depression had higher levels of inflammatory markers [C-reactive protein, interleukin-6 (IL-6), and tumor necrosis factor-α] compared with those with melancholia and controls, while melancholic subjects had a higher salivary cortisol awakening curve and diurnal slope, suggesting that depressive subtypes differ not only in their symptom presentation, but also in their biological correlates.[37] People with MDD have an altered peripheral immune system, with impaired cellular immunity and increased levels of proinflammatory cytokines. Cytokines can influence neurotransmitter metabolism, neuroendocrine function, and regional brain activity, all of which are relevant to depression. Experimentally induced inflammation can provoke depression-like feelings and behaviors. Interleukin-1 receptor (IL-1R1) is expressed on proliferating neural progenitor cells and IL-1β, and exerts an antiproliferative, antineurogenic, and progliogenic effect on embryonic hippocampal neural progenitor cells.[38] GSK-3β is involved in both IL-1β production and IL-1β-induced effects on embryonic hippocampal neurogenesis *in vitro*.[36] Oxidative stress and inflammatory processes were positively associated with untreated MDD when compared before and after 8 weeks of open-label sertraline treatment, and to healthy non-depressed controls. Suicidal behavior has been linked to allergens and allergic reactions, and it is thought that immune responses involving ventral brain structures may contribute to suicide risk.

Neurotransmitters, growth factors, and inflammation pathways share an effect on cell proliferation and survival. These convergent pathways are altered in depression and in suicide, and these alterations may be reversed by antidepressant treatments including drugs, electric convulsive therapy, and sleep deprivation (see Fig. 43.1).

NEUROANATOMICAL CHANGES IN DEPRESSION AND SUICIDE

Neuroanatomical Correlates of Depression

There is evidence that MDD may lead to neocortical thinning and/or rewiring and hippocampal atrophy. Alternatively, these morphological changes may antedate the onset of the illness clinically and may be responsible for the illness and not a consequence of the illness. Smaller hippocampal volume in MDD has been reported, based on volumetric analysis in some but not all studies.[39] Disease duration and treatment status contribute to the findings in both recurrent and first episode MDD.[39] Seven studies of first episode MDD found smaller hippocampal volume bilaterally, with gender and disease duration effects shown in some but not all studies.[40] The finding of a smaller hippocampus in first episode MDD suggests that smaller hippocampal volume may precede MDD onset or develop shortly thereafter. Thus, it is unclear whether a small hippocampus is a causal factor in MDD or a consequence of depression. Findings that cumulative time spent depressed and untreated correlates with a smaller hippocampus *in vivo*, and the duration of MDD is associated with a smaller hippocampus,[39] suggest that the illness is one cause of a smaller hippocampus. Some magnetic resonance imaging studies have divided the hippocampus into anterior, middle, and posterior regions, but most have not.[41] This matters because mouse studies on the effect of SSRIs and human postmortem brain studies of MDD treated with SSRIs show that effects on neurogenesis are limited to the anterior hippocampus in humans and the ventral hippocampus in mice. Studies differ with regard to their definition of the hippocampus, and some studies[40] included white matter bundles (i.e. alveus and fimbria) in their computation of volume. These methodological differences may have contributed to the heterogeneity of results across *in vivo* studies. One study showed higher density of granule neurons and glia in the dentate gyrus and cornu ammonis regions, and smaller soma of pyramidal neurons in MDD compared with controls, suggesting that a reduction in neuropil may account for smaller hippocampal volume in MDD. Postmortem and *in vivo* data show that a volume increase associated with antidepressant treatment is found selectively in the dentate gyrus[42] and cannot be detected if all the hippocampal subfields are included in the analysis.[43]

Consistent with a stress model of depression, rodents exposed to a chronic stressor regimen, or corticosterone administration, have impaired hippocampal neurogenesis. Altered hippocampal neurogenesis may have a role in the pathogenesis of MDD. Adult neurogenesis has been demonstrated in the hippocampus of mammals including humans. Combined retroviral-based lineage tracing and electrophysiological studies show that newborn neurons in the adult mammalian CNS are functionally and synaptically integrated. Newborn neurons have a physiological role, at least in rodents, because blockade of hippocampal neurogenesis inhibits hippocampal-dependent learning, and the rate of survival of new neurons in the dentate gyrus is activity dependent, as is the determination of the neuronal versus glial fate of neural progenitor cells. Activities such as environmental enrichment and wheel running increase dentate gyrus neurogenesis.

There are more neural progenitor cells (nestin-immunoreactive) and more dividing cells (Ki-67-immunoreactive) in human dentate gyrus of patients with MDD treated with SSRIs (sertraline, fluoxetine) or tricyclic antidepressants (TCAs: nortriptyline, clomipramine) compared with untreated MDD or controls.[44] The number of neural progenitor cells decreases with age, although age does not correlate with the number of dividing cells. Females have more neural progenitor cells than males.[44] Overall, these results are consistent with findings observed in rodents. One study using Ki-67 in human tissue to identify dividing cells did not find an effect of antidepressants on Ki-67-immunoreactive cell number in the human anterior hippocampal formation. All but one of the MDD subjects in that study had been prescribed antidepressant medications, but since no toxicology data were available it is not known whether the medications prescribed were actually taken. Fluoxetine, imipramine, tranylcypromine, and reboxetine increase dentate gyrus cell proliferation in mice and rats. In rodents, fluoxetine administration for 15 days does not affect division of quiescent neural progenitor cells but increases the number and symmetric divisions of amplifying neural progenitor cells. Inhibition of neurogenesis blocks some behavioral effects of antidepressants, suggesting that dentate gyrus cell proliferation may be a mechanism of antidepressant action. Since BDNF expression increases in the granule cell layer and cornu ammonis subfields after chronic antidepressant administration, growth factors may play a role in the increased neurogenesis with antidepressants. Impaired adult hippocampal neurogenesis has been hypothesized to be part of the biological and cellular basis of MDD. Nevertheless, blocking or inhibiting cell proliferation does not induce learned helplessness in rats and does not affect anxiety and depression-related behaviors in mice,[45] raising doubts about the etiological role of cell proliferation *per se* in MDD.

A larger sample of MDD patients treated with SSRIs had more dentate gyrus neural progenitor cells and larger dentate gyrus volume than patients with untreated MDD or those with MDD treated with TCAs. More neural progenitor cells in SSRI-treated cases were found selectively in the anterior and mid-dentate gyrus. In addition, more angiogenesis was found in association with SSRI treatment in the dentate gyrus of adults with MDD. Subjects with MDD treated with SSRIs had larger capillary area in anterior and mid-dentate gyrus and larger dentate gyrus volume than untreated MDDs and subjects with MDD treated with TCAs. The degree of angiogenesis correlated with neural progenitor cell number in the dentate gyrus. Greater neural progenitor cell number and capillary area correlated with larger dentate gyrus volume.[46] Angiogenesis and neurogenesis coregulation in humans could open new ways to enhance cell proliferation and survival, which may be relevant for treating mood disorders and neurodegenerative diseases. Control of angiogenic activity may be key to integration into host neural circuitry of stereotactically transplanted stem cells, which can differentiate into functional excitatory neurons under optogenetic stimulation.

Fewer mature granule neurons in anterior and mid-dentate gyrus were found in untreated MDD compared with controls and treated MDD. Younger age of onset of MDD was associated with fewer granule neurons in anterior dentate gyrus. MDDs treated with SSRIs had granule neuron numbers comparable to controls.[42]

Maturation or survival of neuronal progenitor cells in the dentate gyrus may be impaired in MDD, and these data are consistent with the hypothesis that antidepressant action may reverse this effect. The observed higher neural progenitor and mature granule cell number in subjects with MDD treated with antidepressants could be attributable not only to the increased neurogenesis, but also to enhanced cell survival resulting from antidepressant effects such as increasing *Bcl2*, and BDNF, which regulates cell survival. *In vivo* longitudinal studies of hippocampal structure in patients, using brain imaging, are required to determine how cell number and volume relate to clinical characteristics and severity of depression, and how antidepressant action affects hippocampal structural plasticity.

The increase in neural progenitors and dividing cells in patients with MDD treated with antidepressants was localized to the rostral hippocampus, which, in primates, is interconnected with the prefrontal cortex, amygdala, and nucleus accumbens. The equivalent structure in rodents, the ventral hippocampus, has been shown to be involved in anxiety-related behaviors and to regulate the neuroendocrine responses to psychological stress. A functional dissociation of the anterior and posterior axis of the hippocampus has been demonstrated by lesion studies in rats and functional studies in primates. Moreover, anterograde tracer injections studies show that there is a topographic organization of the intrinsic connections of the dentate gyrus, which is similar in macaque monkey and rat. Changes in neurogenesis are subregion specific, depending on the conditions. Adult animals performing spatial learning tasks have the highest levels of neurogenesis in dorsal hippocampus.[47] Chronic administration of possible antidepressant (agomelatine) increased cell proliferation and neurogenesis, specifically in the ventral dentate gyrus of the rat. The anatomical specificity of findings in mice and humans bears further investigation and may link effects on hippocampal function to neuroendocrine responses to stress.

A larger anterior dentate gyrus volume was found in MDD patients treated with SSRIs[46] and smaller dentate gyrus volume in untreated MDD compared with MDD treated with SSRIs and controls, with no differences between groups in glial numbers in the anterior dentate gyrus.[42] Some imaging studies *in vivo* found smaller hippocampal volume in MDD than in healthy controls, but the effects of treatment have been less well studied. More severe depressive symptoms in MDD subjects are associated with greater hippocampal atrophy, particularly in CA1 subfields and the subiculum.[48] Inescapable foot shock caused loss of spine synapses selectively in CA1, CA3, and dentate gyrus, reversed by desipramine. The increased dentate gyrus volume in subjects with MDD treated with antidepressants may suggest that antidepressant treatment increases the total mature granule neuron number or volume of the neuropil.

Postmortem studies also showed altered numbers of neurons and glial cells in mood disorders, neuronal and glial prefrontal cell pathology in MDD, apoptosis in the hippocampus in MDD, fewer glia in the subgenual prefrontal cortex in mood disorders, and fewer non-pyramidal cells in sector CA2 in schizophrenia and bipolar disorder. Thus, in multiple brain regions, deficits in mature neuron number and glial cell number characterize MDD. The cause is uncertain, although impaired neurogenesis and accelerated apoptosis are possible pathogenic mechanisms. Antidepressant treatment may increase neurogenesis and thereby potentially correct this deficit of mature neurons.

Neuroanatomical Studies in Suicide

The cognitive construct of decision making or willed action is relevant to suicidal behavior. This process involves the ventromedial prefrontal cortex, as well as its connections with the amygdala, anterior cingulate, and somatosensory/insular cortices, and connectivity via the underlying white matter. The anatomical convergence of fewer serotonin transporter sites and upregulation of HTR1A in the ventral prefrontal cortex in the brain of suicides indicates a role for this brain region in suicide. The ventral prefrontal cortex is involved in willed action, and behavioral and cognitive inhibition, and low serotonergic input to this brain region may contribute to impaired inhibition, creating a greater propensity to act on suicidal or aggressive feelings. Ventromedial prefrontal cortex damage results in deficits of social behavior, including a failure to observe social conventions and making poor decisions, while learning, memory, attention, language, and many other cognitive functions are normal. Suicides also have lower neuron density in the dorsal prefrontal cortex, Brodmann area (BA) 9, and ventral prefrontal cortex (BA47).[49] Ventral prefrontal cortex has been shown in functional imaging studies to be activated during response inhibition. The importance of this region for decision making has been shown by lesion studies. In assessing decision-making performance using the Iowa Gambling Task, people with lesions in the ventromedial prefrontal cortex persist in selecting cards from losing decks (incurring a net loss despite occasional bigger payouts) for longer than healthy controls, consistent with impaired decision making and cognitive inflexibility associated with suicidal behavior. The prefrontal cortex is also activated during imagined actions. Therefore, the prefrontal cortex provides insight into suicidal ideation and the perceived consequences of future actions, facilitating the possibility of acting on suicidal thoughts in vulnerable patients. The orbital prefrontal and anterior cingulate cortices are also part of the neural circuitry of emotion and aggression. The activation of these regions may be part of an effort to suppress emotions, and enroll an inhibitory projection to the amygdala.

Imaging studies using PET indicate an inverse relationship between the activity of the prefrontal cortex and the amygdala. Prefrontal cortical activity changes on PET correlate with serotonin release by fenfluramine, and reduced cortical response is associated with suicidal and aggressive behaviors. Lower regional cerebral blood flow in right temporal and prefrontal brain areas is associated with self-harm or aggressive acts towards others. Non-depressed, impulsive females with borderline personality disorder, who had attempted suicide or self-mutilated, showed PET hypometabolism in the medial orbital frontal cortex bilaterally compared with healthy controls. Lower prefrontal and midbrain serotonin transporter binding was observed in impulsive/violent subjects compared with controls, and an inverse correlation was found between HTR1A binding in the orbital frontal cortex and lifetime aggression. Structural and functional abnormalities are related to impulsive homicide. Inferior frontal white matter microstructural defects, as shown by lower fractional anisotropy, have been found in association with aggressive impulsivity in males with schizophrenia and in MDD. Therefore, part of the relationship of the ventral prefrontal cortex to suicide seems as much related to depression as it is to behavioral control. Lower α-[^{11}C]methyl-L-tryptophan trapping in the orbitofrontal cortex is associated with suicidal intent in suicide attempters and with *TPH2* SNPs rs6582071 and rs4641527, respectively, previously associated with suicide.[50] Imaging revealed lower HTR2A binding index in anxious and depressed suicide attempters and higher HTR2A binding in impulsive suicide attempters. These results are in keeping with higher HTR2A binding reported in suicides. Both an increase and a decrease in HTR2A binding index seem to normalize with SSRI treatment. The serotonin system is believed to play a role in modulating impulsivity and violence. Imaging studies have implicated the anterior cingulate and orbitofrontal cortex in impulsive aggression. Regional serotonin transporter distribution in the brain of individuals with impulsive aggression assessed using PET with the serotonin transporter radiotracer [^{11}C] McN 5652 showed that HTT availability was significantly reduced in the anterior cingulate cortex of individuals with impulsive aggression compared with healthy subjects. Less HTT binding was found in depressed suicide attempters compared with depressed non-attempters.

RESILIENCE, DEPRESSION, AND SUICIDE

To understand the pathophysiology of suicide or MDD it is important to consider resilience in the presence of relevant biological or environmental stress or a psychiatric disorder. Studies of children in a variety of settings, including war, family violence, poverty, and natural disasters, have revealed psychological and biological patterns or characteristics associated with successful adaptation to stress.

Psychological Characteristics of Resilience

Resilience is characterized by good intellectual functioning, effective self-regulation of emotions and attachment styles, a positive self-concept, optimism, altruism, a capacity to convert traumatic helplessness into learned helpfulness, and an active coping style in confronting a stressor. The concept of allostasis and allostatic load has been used to explain response to acute and chronic stress and vulnerability to adverse consequences. Much data on adult human resilience come from studies of men in combat and show the importance of the ability to bond with a group with a common mission, placing a high value on altruism, and the capacity to tolerate high levels of fear and still perform effectively. Problem solving, reappraisal, and cognitive flexibility can determine resilience and are lacking in those at risk of suicidal behavior or MDD.

Neurochemistry of Resilience

Part of a resilient response to stress may be related to neurohormones. Psychological and physical stress is associated with a lowering of testosterone levels. Lower plasma testosterone is associated with depression in men, and testosterone may be effective as adjunctive therapy in refractory depression. Aggression is the aspect of human behavior most often linked to testosterone concentrations, and suicide is more frequent in those with more pronounced aggressive traits and in men. Estrogens (as well as progesterone) produce a relative resistance to glucocorticoid feedback, suggesting that estradiol may moderate stress responsiveness. Estrogen may also affect mood and contribute to postpartum and postmenopausal depression. Therefore, sex steroid differences may contribute to different degrees of resilience and potentially rates of depression or suicide in men and women.

Cortisol has important regulatory effects on the hippocampus, amygdala, and prefrontal cortex; facilitates the encoding of emotion-related memory; and has biphasic effects on hippocampal excitability, cognitive function, and memory. It is involved in the stress response by mobilizing energy, potentiating the noradrenergic response, having an anti-inflammatory effect to allow function to be sustained in the short term in the face of injury. Prolonged elevation of glucocorticoids has deleterious effects including muscle atrophy, hippocampal neuron damage, and insulin resistance.

Biological responses to stress include glucocorticoid and dehydroepiandrosterone (DHEA) secretion. DHEA seems to promote psychological resilience. A negative correlation between DHEA and the severity of PTSD has been reported. Several studies have found negative correlations of plasma DHEA levels with depressive symptoms, consistent with a protective effect and possible antidepressant effects of DHEA. Aside from the antiglucocorticoid actions, DHEA has effects on $GABA_A$ receptors and NMDA neurotransmission that may affect the risk of depression and suicidal behavior.

Galanin is a peptide that modulates stress responses. It reduces the firing rate of the locus ceruleus, possibly by stimulating the galanin-1 receptor, which acts as an autoreceptor. Administered centrally, galanin modulates anxiety-related behaviors. Galanin-overexpressing transgenic mice do not exhibit an anxiety-like phenotype when tested under baseline conditions, and these mice are unresponsive to the anxiogenic effects of the α_2-receptor antagonist yohimbine. Galanin administration and galanin overexpression in the hippocampus result in deficits in fear conditioning that could protect against the development of conditions such as PTSD or the impact of childhood adversity on future risk of depression or PTSD.

The upregulation of neuropeptide Y mRNA levels in the amygdala after chronic stress suggests that neuropeptide Y may be involved in the adaptive responses to stress exposure. Neuropeptide Y appears to be involved in the consolidation of fear memories; its injection into the amygdala impairs memory retention in a foot-shock avoidance paradigm. Neuropeptide Y counteracts the anxiogenic effects of CRH, and a CRH antagonist blocks the anxiogenic effects of a neuropeptide Y antagonist. NPY levels in CSF may be higher in response to adversity and major depression as a homeostatic response.

The dopaminergic system may be implicated in mechanisms of resilience. Increasing dopamine function in the nucleus accumbens, the orbital prefrontal cortex, and the VTA, and NMDA receptor blockade in the nucleus accumbens and the medial prefrontal cortex, may enhance sensitivity to reward. VTA dopamine neurons in the brain's reward circuit have a crucial role in mediating stress responses, including determining susceptibility versus resilience to behavioral abnormalities induced by social stress. VTA dopamine neurons show two *in vivo* patterns of firing: low-frequency tonic firing and high-frequency phasic firing. Phasic firing of the neurons, which encodes reward signals, is upregulated by repeated social defeat stress, a mouse model of stress-related depression. Surprisingly, this pathophysiological effect is seen in susceptible mice only, with no apparent change in firing rate in resilient mice. Optogenetic inhibition of the VTA–nucleus accumbens projection induces resilience.[18] Mice susceptible to social defeat stress effects, but not resilient mice, show lower protein levels of DVL-2 mRNA, and an overexpression of *GSK-3β* promotes resilience to social defeat stress.

Neuroanatomy of Resilience

Neurons in the orbital–prefrontal cortex receive dopamine projections from the VTA, and can discriminate different rewards based on their motivational value. This drives behavioral choices between more certain but lower risk choices and less probable, high-risk choices. Low CSF homovanillic acid is found in psychomotor retarded depression, and in some depressed suicide attempters and suicides, indicating a lack of dopaminergic tone and impaired reward circuitry. There is an extensive animal and clinical literature linking reduced dopaminergic function to MDD.

The amygdala also modulates conditioned responses to rewarding stimuli as part of a circuit involving the subiculum, bed nucleus of the stria terminalis, nucleus accumbens, and medial prefrontal cortex. The cyclic AMP pathway and CREB protein in the amygdala affect both aversive and rewarding associations. These neural networks establish the emotional value of a reward memory as well as its strength and persistence. The detailed molecular basis of such plasticity is beginning to be understood. During the expression of fear-related behaviors, the lateral amygdala engages the central nucleus of the amygdala, which, as the principal output nucleus, projects to areas of the hypothalamus and brainstem that mediate the autonomic, endocrine, and behavioral responses associated with fear. Failure to achieve an adequate level of activation of the medial prefrontal cortex after memory extinction may lead to persistent fear responses, as found in PTSD. Individuals with the capacity to function well after experiencing states of high fear may have more potent medial prefrontal cortex inhibition of amygdala responsiveness.

Problem solving and cognitive flexibility are impaired in suicide. Depressed individuals who are at risk of suicidal behavior have a cognitive rigidity, which may result from dysfunctional executive decision making related to prefrontal cortical dysfunction, and impaired connectivity with the anterior cingulate and amygdala. High-lethality suicide attempters show poorer executive functioning than depressed patients with no or low-lethality suicide attempts and non-patients. Assessment of interpersonal problem-solving abilities before and after a mood-induction procedure showed that formerly depressed people with a history of suicidal ideation were less capable of effective problem solving after a mood challenge, potentially accounting for vulnerability to recurrence of suicidal ideation and behavior. Poor problem solving also mediated the effect of family history of suicide in determining the occurrence of multiple compared with single future suicide attempts.

The concept of allostatic load, a cumulative measure of physiological dysregulation over multiple systems, is useful as a predictor of functional decline in elderly men and women. It could be applicable to MDD and suicidal behavior, and help to identify cumulative neurobiological and stress responses with prognostic significance.

CONCLUSION

Suicide and depression share some biological, societal, and psychobiological features, although many alterations in brain function and neural circuitry are specific to the diathesis for suicidal behavior and quite distinct from MDD. Some specific characteristics of the diathesis for suicidal behavior are negative cognition, difficulties in problem solving, and impulsive or reactive behavior in response to emotion such as aggression, which involve changes in specific brain regions such as the ventromedial prefrontal cortex, involved in decision making, and the dorsal lateral prefrontal cortex that is, involved in mood regulation. Behavior, cognition, and emotion are regulated by neurotransmitters, hormones, trophic factors, and genetic and epigenetic changes that reflect the subject's biological predispositions and experience. Brain circuits are regulated and molded by all factors shaping human behavior, emotion, cognition, and coping. The outcomes of MDD and suicidal behavior depend on this complex combination of risk and protective factors and life experiences at key formative stages of development.

References

1. Bromet E, Andrade LH, Hwang I, et al. Cross-national epidemiology of DSM-IV major depressive episode. *BMC Med*. 2011;9:90.
2. Afifi TO, Macmillan H, Cox BJ, Asmundson GJ, Stein MB, Sareen J. Mental health correlates of intimate partner violence in marital relationships in a nationally representative sample of males and females. *J Interpers Violence*. 2009;24(8):1398–1417.
3. Scott KM, McLaughlin KA, Smith DA, Ellis PM. Childhood maltreatment and DSM-IV adult mental disorders: comparison of prospective and retrospective findings. *Br J Psychiatry*. 2012;200(6):469–475.
4. Warren BL, Vialou VF, Iniguez SD, et al. Neurobiological sequelae of witnessing stressful events in adult mice. *Biol Psychiatry*. 2013;73(1):7–14.
5. Oquendo MA, Dragatsi D, Harkavy-Friedman J, et al. Protective factors against suicidal behavior in Latinos. *J Nerv Ment Dis*. 2005;193(7):438–443.
6. Lee PH, Perlis RH, Jung JY, et al. Multi-locus genome-wide association analysis supports the role of glutamatergic synaptic transmission in the etiology of major depressive disorder. *Transl Psychiatry*. 2012;2:e184.
7. Galecka E, Szemraj J, Bienkiewicz M, et al. Single nucleotide polymorphisms of NR3C1 gene and recurrent depressive disorder in population of Poland. *Mol Biol Rep*. 2013;40(2):1693–1699.
8. Kohli MA, Lucae S, Saemann PG, et al. The neuronal transporter gene SLC6A15 confers risk to major depression. *Neuron*. 2011;70(2):252–265.
9. Schuhmacher A, Lennertz L, Wagner M, et al. A variant of the neuronal amino acid transporter SLC6A15 is associated with ACTH and cortisol responses and cognitive performance in unipolar depression. *Int J Neuropsychopharmacol*. 2013;16(1):83–90.
10. Haenisch B, Herms S, Mattheisen M, et al. Genome-wide association data provide further support for an association between 5-HTTLPR and major depressive disorder. *J Affect Disord*. 2013;146(3):438–440.
11. Crisafulli C, Chiesa A, Han C, et al. Case–control association study of 36 single-nucleotide polymorphisms within 10 candidate genes for major depression and bipolar disorder. *Psychiatry Res*. 2013;209(1):121–123.
12. Schosser A, Butler AW, Ising M, et al. Genomewide association scan of suicidal thoughts and behaviour in major depression. *PloS ONE*. 2011;6(7):e20690.
13. Galfalvy H, Huang YY, Oquendo MA, Currier D, Mann JJ. Increased risk of suicide attempt in mood disorders and TPH1 genotype. *J Affect Disord*. 2009;115(3):331–338.
14. Mann JJ, Bortinger J, Oquendo MA, Currier D, Li S, Brent DA. Family history of suicidal behavior and mood disorders in probands with mood disorders. *Am J Psychiatry*. 2005;162(9):1672–1679.
15. Boldrini M, Underwood MD, Mann JJ, Arango V. More tryptophan hydroxylase in the brainstem dorsal raphe nucleus in depressed suicides. *Brain Res*. 2005;1041(1):19–28.
16. Jokinen J, Nordstrom AL, Nordstrom P. CSF 5-HIAA and DST non-suppression – orthogonal biologic risk factors for suicide in male mood disorder inpatients. *Psychiatry Res*. 2009;165(1–2):96–102.
17. Boldrini M, Underwood MD, Mann JJ, Arango V. Serotonin-1A autoreceptor binding in the dorsal raphe nucleus of depressed suicides. *J Psychiatr Res*. 2008;42(6):433–442.
18. Chaudhury D, Walsh JJ, Friedman AK, et al. Rapid regulation of depression-related behaviours by control of midbrain dopamine neurons. *Nature*. 2013;493(7433):532–536.
19. Wilkinson MB, Dias C, Magida J, et al. A novel role of the WNT-dishevelled-GSK3beta signaling cascade in the mouse nucleus accumbens in a social defeat model of depression. *J Neurosci*. 2011;31(25):9084–9092.
20. Crawford N, Chacko AD, Savage KI, et al. Platinum resistant cancer cells conserve sensitivity to BH3 domains and obatoclax induced mitochondrial apoptosis. *Apoptosis*. 2011;16(3):311–320.
21. Klempin F, Babu H, De Pietri TD, Alarcon E, Fabel K, Kempermann G. Oppositional effects of serotonin receptors 5-HT1a, 2, and 2c in the regulation of adult hippocampal neurogenesis. *Front Mol Neurosci*. 2010:3.
22. Oufkir T, Vaillancourt C. Phosphorylation of JAK2 by serotonin 5-HT (2A) receptor activates both STAT3 and ERK1/2 pathways and increases growth of JEG-3 human placental choriocarcinoma cell. *Placenta*. 2011;32(12):1033–1040.
23. Kobayashi K, Ikeda Y, Sakai A, et al. Reversal of hippocampal neuronal maturation by serotonergic antidepressants. *Proc Natl Acad Sci U S A*. 2010;107(18):8434–8439.
24. Liu MT, Kuan YH, Wang J, Hen R, Gershon MD. 5-HT4 receptor-mediated neuroprotection and neurogenesis in the enteric nervous system of adult mice. *J Neurosci*. 2009;29(31):9683–9699.
25. Deschwanden A, Karolewicz B, Feyissa AM, et al. Reduced metabotropic glutamate receptor 5 density in major depression determined by [(11)C]ABP688 PET and postmortem study. *Am J Psychiatry*. 2011;168(7):727–734.
26. Hoeffer CA, Klann E. mTOR signaling: at the crossroads of plasticity, memory and disease. *Trends Neurosci*. 2010;33(2):67–75.
27. Soll C, Jang JH, Riener MO, et al. Serotonin promotes tumor growth in human hepatocellular cancer. *Hepatology*. 2010;51(4):1244–1254.

28. Yamashima T. "PUFA-GPR40-CREB signaling" hypothesis for the adult primate neurogenesis. *Prog Lipid Res*. 2012;51(3):221–231.
29. Jagasia R, Steib K, Englberger E, et al. GABA–cAMP response element–binding protein signaling regulates maturation and survival of newly generated neurons in the adult hippocampus. *J Neurosci*. 2009;29(25):7966–7977.
30. Boneva NB, Yamashima T. New insights into "GPR40–CREB interaction in adult neurogenesis" specific for primates. *Hippocampus*. 2012;22(4):896–905.
31. Reus GZ, Abelaira HM, Agostinho FR, et al. The administration of olanzapine and fluoxetine has synergistic effects on intracellular survival pathways in the rat brain. *J Psychiatr Res*. 2012;46(8):1029–1035.
32. Kim HW, Rapoport SI, Rao JS. Altered expression of apoptotic factors and synaptic markers in postmortem brain from bipolar disorder patients. *Neurobiol Dis*. 2010;37(3):596–603.
33. Salvadore G, Nugent AC, Chen G, et al. Bcl-2 polymorphism influences gray matter volume in the ventral striatum in healthy humans. *Biol Psychiatry*. 2009;66(8):804–807.
34. Pandey GN, Ren X, Rizavi HS, Conley RR, Roberts RC, Dwivedi Y. Brain-derived neurotrophic factor and tyrosine kinase B receptor signalling in post-mortem brain of teenage suicide victims. *Int J Neuropsychopharmacol*. 2008;11(8):1047–1061.
35. Segi-Nishida E, Warner-Schmidt JL, Duman RS. Electroconvulsive seizure and VEGF increase the proliferation of neural stem-like cells in rat hippocampus. *Proc Natl Acad Sci U S A*. 2008;105(32):11352–11357.
36. Zunszain PA, Hepgul N, Pariante CM. Inflammation and depression. *Curr Top Behav Neurosci*. 2013;14:135–151.
37. Lamers F, Vogelzangs N, Merikangas KR, de Jonge P, Beekman AT, Penninx BW. Evidence for a differential role of HPA-axis function, inflammation and metabolic syndrome in melancholic versus atypical depression. *Mol Psychiatry*. 2013;18(6):692–699.
38. Green HF, Nolan YM. Unlocking mechanisms in interleukin-1beta-induced changes in hippocampal neurogenesis – a role for GSK-3beta and TLX. *Transl Psychiatry*. 2012;2:e194.
39. McKinnon MC, Yucel K, Nazarov A, MacQueen GM. A meta-analysis examining clinical predictors of hippocampal volume in patients with major depressive disorder. *J Psychiatry Neurosci*. 2009;34(1):41–54.
40. Cole J, Costafreda SG, McGuffin P, Fu CH. Hippocampal atrophy in first episode depression: a meta-analysis of magnetic resonance imaging studies. *J Affect Disord*. 2011;134(1–3):483–487.
41. Malykhin NV, Lebel RM, Coupland NJ, Wilman AH, Carter R. *In vivo* quantification of hippocampal subfields using 4.7 T fast spin echo imaging. *Neuroimage*. 2010;49(2):1224–1230.
42. Boldrini M, Santiago AN, Hen R, et al. Hippocampal granule neuron number and dentate gyrus volume in antidepressant-treated and untreated major depression. *Neuropsychopharmacology*. 2013;38(6):1068–1077.
43. Cobb JA, Simpson J, Mahajan GJ, et al. Hippocampal volume and total cell numbers in major depressive disorder. *J Psychiatr Res*. 2013;47(3):299–306.
44. Boldrini M, Underwood MD, Hen R, et al. Antidepressants increase neural progenitor cells in the human hippocampus. *Neuropsychopharmacology*. 2009;34:2376–2389.
45. David DJ, Samuels BA, Rainer Q, et al. Neurogenesis-dependent and -independent effects of fluoxetine in an animal model of anxiety/depression. *Neuron*. 2009;62(4):479–493.
46. Boldrini M, Hen R, Underwood MD, et al. Hippocampal angiogenesis and progenitor cell proliferation are increased with antidepressant use in major depression. *Biol Psychiatry*. 2012;72(7):562–571.
47. Snyder JS, Radik R, Wojtowicz JM, Cameron HA. Anatomical gradients of adult neurogenesis and activity: young neurons in the ventral dentate gyrus are activated by water maze training. *Hippocampus*. 2009;19(4):360–370.
48. Bearden CE, Thompson PM, Avedissian C, et al. Altered hippocampal morphology in unmedicated patients with major depressive illness. *ASN Neuro*. 2009;1(4).
49. Underwood MD, Kassir SA, Bakalian MJ, Galfalvy H, Mann JJ, Arango V. Neuron density and serotonin receptor binding in prefrontal cortex in suicide. *Int J Neuropsychopharmacol*. 2012;15(4):435–447.
50. Booij L, Turecki G, Leyton M, et al. Tryptophan hydroxylase(2) gene polymorphisms predict brain serotonin synthesis in the orbitofrontal cortex in humans. *Mol Psychiatry*. 2012;17(8):809–817.

SECTION VI

DISEASES OF THE NERVOUS SYSTEM AND SOCIETY

CHAPTER

44

Introduction

Michael J. Zigmond

Departments of Neurology, Neurobiology, and Psychiatry, University of Pittsburgh, Pittsburgh, Pennsylvania, USA

OUTLINE

In this short, concluding section we attempt to go beyond the consideration of individual disorders and touch on a few of the many important broader issues. In Chapter 45, Judy Illes and Peter Reiner consider the neurobiology of disease from the point of view of neuroethics – the intersection of bioethics and neuroscience. They focus on four ethical considerations that are relevant to many of the previous chapters: (1) the development of clinically relevant animal models; (2) the imperative of sharing data while at the same time protecting privacy; (3) dealing with incidental findings during brain imaging; and (4) research in neuroscience as a public service. In discussing the ethics of using animals in neuroscience research – an often debated issue within the public at large as well as among some neuroscientists – they add to the standard considerations of three Rs, *replacement*, *reduction*, and *refinement*, with two new Rs: *reflection* and *responsiveness*.

These, of course, are not the only ethical issues that must be confronted by neuroscientists, and many of these are of importance to science more generally. Other critical issues include informed consent, particularly within vulnerable populations, conflict of interest, the sharing of reagents, authorship, and the recurring issues of fabrication, falsification, and plagiarism. Many sources exist for those wishing to read further in the area of ethics as it pertains to neuroscience as well as research ethics more generally. Among those are the guidelines for responsible conduct published on the website of the Society for Neuroscience (www.sfn.org), many articles in journals such as the *Journal of Neuroethics* and *Science and Engineering Ethics*, and several excellent monographs.[1–4]

In Chapter 46, Mitchell Wallin and John Kurtzke address the epidemiology of brain disease, focusing on neurological disorders. (General reviews also exist on the epidemiology of psychiatric disorders[5,6] and on the epidemiology of specific psychiatric disorders.) Wallin and Kurtzke begin with an introduction to epidemiology, providing critical background for a subject of vital importance to understanding the literature on who is affected by neurological and psychiatric disorders. The authors then demonstrate how, in attempting to determine the causes of diseases or develop interventions to prevent, slow, or cure them, we must consider who the disease is most likely to affect. In this regard, they discuss gender, age, race/ethnicity, geographical location, and genetic predisposition. In most cases the biological bases for the correlations they present are not yet known, presenting critical challenges for the future.

In the final chapter, Chapter 47, Zinzi Bailey and David Williams address the widespread but often ignored health disparities in the USA and throughout the world. Although stress is easily attributed by many to genetic factors, Bailey and Williams reinforce the discussion of the impact of stress on the brain presented by Bruce McEwen in Chapter 34, and note that stress – whether it derives from malnutrition, environmental pollution, physical danger, racism, or other factors – is not distributed equally among the world's population. Instead, it has a particularly toxic impact on people of color and those in lower socioeconomic groups. The authors cite some data that will astonish many readers: in the USA, African Americans have higher mortality rates than whites for two-thirds of the leading causes of death; immigrants to the USA have better health than native-born

Neurobiology of Brain Disorders
http://dx.doi.org/10.1016/B978-0-12-398270-4.00044-6

Americans, but their health declines over time and with successive generations; and infants born in the USA to African-American mothers who are college graduates have a higher rate of mortality than do children of white women who have not completed high school. Deeply disturbing statistics are also provided for other regions of the world.

These chapters, like the others in this textbook, do more than provide information; they represent a call to those engaged in neuroscience research and teaching to contribute to one or more of the urgent issues of relevance to the field. In nearly all respects, science must proceed dispassionately. However, in selecting the problems on which we work and the manner in which we address them, it is essential that we take into consideration our societal responsibilities.[7–12]

References

1. National Research Council. *On Being a Scientist: A Guide to Responsible Conduct in Research*. 3rd ed. Washington, DC: National Academies Press; 2009.
2. Farah MJ. *Neuroethics: An Introduction with Readings*. Cambridge, MA: MIT Press; 2010.
3. Illes J, Sahakian BJ, eds. *Oxford Handbook of Neuroethics*. Oxford: Oxford University Press; 2011.
4. Macrina FL. Virginia Commonwealth University. 4th ed. *Scientific Integrity*. Washington, DC: ASM Press; 2013.
5. Kessler RC. Psychiatric epidemiology: selected recent advances and future directions. *Bull WHO*. 2000;78:464–474.
6. Kessler RC, Aguilar-Gaxiola S, Alonso J, et al. The global burden of mental disorders: an update from the WHO World Mental Health (WMH) Surveys. *Epidemiol Psichiatr Soc*. 2009;18:23–33.
7. Brunner RD, Ascher W. Science and social responsibility. *Policy Sciences*. 1992;25:295–331.
8. Carlson R, Frankel MS. Reshaping responsible conduct of research education. Professional Ethics Report, 24(1):1–3.
9. Galston AW. Science and social responsibility: a case history. *Ann N Y Acad Sci*. 1972;196:223–235.
10. Grimm A. *Al Gore to Scientists: 'We Need You.' Science*. 2009;323:998. See also: www.youtube.com/watch?v=562uz1cbPlg.
11. *International Council for Science, Freedom, Responsibility, and Universality of Science*. 2008.
12. Lekka-Kowalik A. Why science cannot be value-free: understanding the rationality and responsibility of science. *Sci Eng Ethics*. 2010;16:33–41.

CHAPTER

45

Advances in Ethics for the Neuroscience Agenda

Judy Illes[*, †], *Peter B. Reiner*[*, ‡]

[*]National Core for Neuroethics; [†]Division of Neurology, Department of Medicine; [‡]Department of Psychiatry, The University of British Columbia, Vancouver, British Columbia, Canada

OUTLINE

INTRODUCTION

The primary objective of the basic neuroscientist is to understand the machinery of the brain. The challenges are legion: the brain is considered to be the most complex machinery in the known universe, and unraveling its inner logic is hardly a job for the fainthearted. In addition to the obstacles that routinely arise from investigating neuroscience, occasionally ethical conundrums appear. It is here that neuroethics – the rigorous empirical inquiry that falls squarely at the crossroads of neuroscience and biomedical ethics – can help.

Some neuroscientists will view this claim with suspicion, expecting that neuroethics represents but one more hurdle for them to overcome in achieving their scientific goals. And, while it is true that some ethical dilemmas may place certain experimental manipulations off limits – causing unnecessary pain to an experimental subject is an obvious example – many more can be avoided by proactive consultation with neuroethics colleagues. Indeed, successful collaboration between basic scientists and

Neurobiology of Brain Disorders
http://dx.doi.org/10.1016/B978-0-12-398270-4.00045-8

neuroethicists provides an opportunity to join forces and overcome challenges before they become problematic. In many ways, neuroethics is best considered a discipline of anticipatory ethics, steadfastly asserting that once a problem arises, it is often much more difficult to manage than one for which mitigating strategies were set in advance.

Yet the discipline of neuroethics involves much more than just addressing ethical dilemmas. The field is at the forefront of investigating the ways in which advances in the neurosciences affect society at large; manipulations that affect the function of the brain – be it memory or trust, emotion or motor coordination – are central to the view of who we are as human beings. It is not just funding agencies that request scientists to examine the ethical, legal, and social issues involved in scientific research; empirical data demonstrate that scientists themselves are concerned about the impact that their work may have upon society at large.[1] Many scientists find that the joy of unraveling the inner workings of the brain is made richer by having their work impact society, and increasingly, the public is paying attention: discoveries in the neurosciences are among the fastest growing topics for media portrayals of biological phenomena.[2] As the public's interest grows, the trust that it places in neuroscientists becomes an important element of the social contract between society and the scientific enterprise. By introducing neuroethical analysis early in the development of an experimental paradigm, neuroscientists can not only avoid pitfalls but also reassure a sometimes skittish public that its best interests are being considered.

The current authors believe that neuroethics represents more of an opportunity than a threat to neuroscience. In this chapter, they provide four examples based on their past work by which neuroethical analysis paves the way towards better science.

RESEARCH WITH ANIMALS

Experimental models using animals are a foundation of the biomedical sciences and, as a case in point, for neuroscience research. Over the centuries, results of studies about the brain have yielded important insights into and treatments for diseases such as Alzheimer disease, mental illness, neurodevelopmental disorders including autism and fetal alcoholism syndrome, addiction, multiple sclerosis, and spinal cord injuries. There are few people who do not know at least one person affected by one of these conditions. Countless other studies on brain function have yielded fundamental knowledge about sensory and motor processes, cognition and perception, as well as advanced methods for rehabilitating people with injuries and improving quality of life. Even as these advances in the neurosciences, which are based on animal research, move forward, animal activism is on the rise, not only in terms of the quantity of events but in the quality of the terror they bring to the fore.[3]

Today, four independent paths in neuroscience are converging in unexpected ways. These paths raise the bar for thinking about the nature of work with animals and, in particular, with those that can be considered sentient.[4] They are:

- the steady evolution of sophisticated new technologies that do not use non-human animals, such as in computer science, computational neuroscience, and neuroimaging
- empirical evidence that neuroscientists are keen to engage deeply about the ethical and societal implications of their work
- increasingly extreme activism measured by violence and destruction, especially for higher life-forms such as mammals
- efforts by professional organizations to underscore the importance of and facilitate the ethical use of a full range of research models.

This convergence has created a landscape for research in which the longstanding principles of *replacement*, *reduction*, and *refinement* that are central to considerations of animal research ethics (see NC3Rs at http://www.nc3rs.org.uk/downloaddoc.asp?id=719) may no longer be ethically sufficient. Elsewhere, two new Rs – reflection and responsiveness – have been proposed.[5]

Reflection highlights the explicit consideration by individuals and professional groups of the immediate and downstream ethical and societal implications of complex advances in neuroscience. Engaging with ethicists with specific expertise in neuroscience, creating opportunities for mentored dialogue about the selection of models for research, and integrating principles of biomedical ethics[6] relevant to neuroscience in research planning are examples of ways in which the R of reflection can be advanced.

Responsiveness embodies strategies for both improved education and scientific communication that can be pursued by the wide range of players in the research enterprise: individuals, science groups, institutions, professional societies, policy makers, and sponsors of research. For example, we should pursue innovative and tailored, case-based teaching tools at the intersection of neuroscience and ethics. We should cultivate a growing cadre of neuroscientists with expertise in ethics, promote the visibility of and access to ethicists with expertise in neuroscience, and streamline cross-disciplinary collaborations between biomedical ethics and neuroscience. We must engage neuroscientists in the development and implementation of well-informed neuroscience-relevant courses in law curricula that pertain to animal research.

Improvement in scientific communication will come with a shift in an institutional culture that openly values and rewards education and outreach to the public, rather than a near-exclusive emphasis on research productivity and grant support. Similarly, research sponsors would do well to more actively support conversations about all aspects of the research they support, including animal-based research that can be controversial. Neuroscientists need to appreciate what journalists and the public want to know about, especially with respect to the clinical relevance of research results, but not at the expense of providing false promises or communicating clearly about the importance of basic science.

This is a time for a new kind of action that is both multidimensional and multidisciplinary. The three original Rs have provided the root structure. The two new Rs of reflection and responsiveness extend and strengthen this base. Together with academic openness, these kinds of initiatives will increase the likelihood of continued and enduring commitment to the highest form of ethical standards for all forms of research, including research that will bring immeasurable benefits to promoting brain health and mitigating the ravages of brain disease.

SHARING DATA AND RESOURCES

At the Organisation for Economic Co-operation and Development (OECD) Megascience Forum in 1999, the creation of a neurosciences database was highlighted as a "vital need" and "one of the great challenges for the 21st century".[7] The *Human Brain Project: Phase I Feasibility Studies Report* of the US National Institutes of Health in 1993 was one of the first to describe the practical implications of this effort to the scientific community and signaled the beginning of the initiative. Under the Human Brain Project grant program, first phase studies focused on feasibility and proof of concept; later phase studies were to be devoted to refinements, including further testing of the tools across sites, improvements, models and grids, maintenance, and integration with other related web-based resources. As there is great diversity in the types of data generated by neuroscience research, novel approaches to collecting, manipulating, combining, displaying, retrieving, managing, and disseminating were needed to successfully make these data available for scientific collaboration and electronic use. Neuroscience data repositories developed at a steady pace with repositories for microscopy data, single- and multiple-unit recording data, and structural magnetic resonance imaging (MRI). Others such as the functional Magnetic Resonance Imaging Data Center (fMRIDC) served as repositories for functional imaging data obtained from fMRI, positron emission tomography, electroencephalography, and magnetoencephalography.[8] Beyond statistical power, benefits include the stability, relationships, integration, and distribution of the structure and function of the brain at both the microscopic and macroscopic levels.[9] Cross-modality interoperability, including resources such as genome and protein databases, has been an enduring goal.

Many of the issues that arise in sharing of databases resurface in considering sharing of reagents, competition and intellectual property rights being foremost among them. Journal policies that mandate sharing of reagents have been in place for some time now, but the growth of commercialization of research findings and the need for young scientists to advance their careers serve as bulwarks against free and open exchange of reagents.

Many researchers believe that the development of tools for handling ethical and policy issues in parallel with the development of technical tools is vital to the realization of a truly enabling toolbox.[10,11] Data sharing is both a technical and a human challenge. Unlike ethical responses that may be sought only after difficult issues have surfaced, a proactive, solution-oriented ethical–technical partnership can be a powerful force in nurturing the scientific enterprise. In this context, both the structure of database sharing and the culture of sharing are vital to the success of the enterprise. These are considered next.

Database-Sharing Infrastructure

Content, Access, and Ownership Using Imaging as a Model

Image-based data have been a primary driver for neuroinformatics efforts.[12] These data are rich in content, large, and laborious to maintain. The fMRIDC, for example, was introduced to the neuroscience community in June 2000 by the *Journal of Cognitive Neuroscience*, which began requiring that all authors who publish in the journal submit their data to it. This data center, funded by the US National Science Foundation/National Institutes of Health, the Keck Foundation, and SUN Microsystems, was created to provide an avenue through which neuroscience researchers could share their data from fMRI studies. The goal was to "speed the progress and the understanding of cognitive processes and the neural substrates that underlie them" (www.fmridc.org). The fMRIDC meets these goals as a publicly accessible database of peer-reviewed fMRI studies storing information that may enable others to reuse data, replicate original studies, generate and test new hypotheses, and provide training opportunities.[8] The center's database is fully accessible on the Internet and was one of the first to encourage a multidisciplinary approach to the development of fMRI. As with other repositories that draw on policies for data sequence storage in the genetics community (e.g. GenBank and The Wellcome Trust), anyone has the right to publish findings based on these fMRIDC datasets. Authors whose papers are based on results from datasets obtained from the center are expected to provide meta- (descriptive) information for data use, credit original study authors, and acknowledge the fMRIDC and accession number of the dataset.

The Organization for Human Brain Mapping (OHBM) also favored the concept of brain data sharing for its potential to enable comparison of data across studies, improve reliability and reproducibility, promote meta-analyses, and create access to data for those who cannot afford neuroimaging equipment. The *Journal of Cognitive Neuroscience* data-sharing mandate brought the challenges of data sharing to the foreground, and the OHBM quickly responded with a task force dedicated to the topic. The work of the OHBM Neuroinformatics Subcommittee culminated in a 2001 *Science* publication in which it framed the critical elements necessary for an informed discussion of the issues.[13] Among the most pressing were data content, data access, data ownership, database structure, and interaction with the community. The OHBM also highlighted issues of database structure, including whether hybrid structures should be constructed for the specific purpose of storing and maintaining neuroimaging data. Management of violations and a multitude of issues surrounding interactions with the community were further identified given the multiplicity and diversity of challenges associated with banked neuroscience data.

Consent, Confidentiality, and Commercialization

Internet accessibility of databases has heightened concerns about consent and the confidentiality of research participants' information. US federal human subject protection law mandates that all identifying information be removed from data before submission for sharing,[14] but true deidentification of imaging data may be inherently flawed.[15] New possibilities for reconstructing facial and cranial features from a brain image make old confidentiality rules about identifying information a particularly vexing problem today.[16] Moreover, whereas institutional ethical review, safety, and quality assurances are fundamental, prospective secondary data use expands the horizon of these considerations.

In the mid-1990s, writings by Clayton and others underscored the complexity of the underlying ethical, legal, and social problems surrounding the status, storage, and current and future use of human materials.[17,18] The focus of the work by Clayton and colleagues was on organs, gametes, embryos, tissue, blood and cells, but neuroscience data follow suit. As donors, patients, and research participants everywhere have become "sources", consent, choice, contact, and controls are topics of ongoing interest. Wolf and Lo attended to institutional review board (IRB) issues in the control over future uses of data and disclosure of results to donors in research involving stored biological materials.[19] They found that IRBs address many significant issues but could do more. Best practices within institutions were identified as those that embody a rationale, and examples of protocols provide a checklist to walk investigators through pertinent issues, and highlight particular issues that investigators might not anticipate. They further emphasized the need for scrupulous protection of the rights and welfare of individual subjects, especially those of children and those without decisional capacity to provide informed consent. Novel challenges related to brain banking, such as obtaining consent from groups and protecting groups from harm, have also been addressed.[20,21] Issues regarding confidentiality and consent have resulted in opposition to some publicized projects,[22] and concerns about commercialization of information[23] have led to the rejection of gene banking in at least one population.[24]

Clayton argued for a greater need for more detailed content, scope, and transparency of consent, especially as withdrawal of data or material is a key unresolved area.[25] She argues that general blanket consent to all future research should not be considered sufficient to meet standards of consent; this reality was faced by one of the three partners in the Human Brain Project consortium whose research was held up for several years because the local IRB objected to the blanket consent that subjects were asked to provide.[26] People need to be given adequate information on which to base a decision. What are permissible secondary uses then, if subjects did not expressly consent to them? How can the imperative to align practices of repositories with requirements of ethics committees best be met?[11]

Commercialization raises further ethical issues, and these include preventing exploitation of vulnerable populations, balancing costs and benefits, and avoiding conflicts of interest. [27,28] One example of intellectual property privileges and commercialization is represented by the Brain Resource Company, whose promotional material offers "large quality of controlled databases of normative subjects and a range of clinical disorders" and provides fee-for-service analysis reports to clients. In 2003, the OECD Working Group noted that the short-term impact of proprietary databases on open neuroscience may seem small, but long-term and larger effects should be anticipated.[29] They urged anticipation of issues arising from relationships between public and private contributions to neuroinformatics resources, and the construction of a policy framework.

Data Anonymization, Incidental Findings, and Recontact

Research with identifiable samples involves the risks of discovery of unexpected and potentially unknown clinical significance, missed incidence, violation of the donor's privacy through discovery and disclosure of sensitive information (intrinsic harm), or discrimination by disclosure of information to third parties (consequential harms).[27] Knoppers favors a coded model (double-coded with a third party or "tissue trustee model") for banked samples of biomaterials in that it gives subjects the chance to opt out from the study upfront or from recontact by the researchers downstream.[30] Majumder describes an initiative to create a secure web-based consent mechanism for patients to communicate with researchers in a dynamic and anonymous fashion.[31] But, as Clayton points out, recontact can be a real "wild card".[25] What investigators do when they are faced with undesired information in secondary data analysis from a research participant with whom they have had no prior contact is an open question.

In the 1970s, the National Bioethics Advisory Commission recommended that IRBs should develop general guidelines for disclosure of results from current or future research when: (1) the results are scientifically valid and confirmed; (2) the results have implications for subjects' health concerns; and (3) a course of action to ameliorate or treat the identified health concern is readily available. Although these guidelines provide a strong basis for framing approaches in neuroscience, they do not readily apply to incidental findings from brain imaging today. Discoveries about the frequency and clinical significance of findings, including false positives, are still ongoing, and treatment, especially in the case of certain neurodegenerative diseases, remains elusive. In the case of shared data, the Office of Human Research Protection suggests that the Common Rule, under the Federal Policy for the Protection of Human Subjects, does not apply to investigators who receive coded information as long as they do not have access to the code key. The reasoning is that the research at this point does not involve humans *per se*.[25] Moreover, in light of the dynamic pace of scientific progress, refinement of ethical norms, and changes in public opinion, approaches and protocols may require adjustments that were not foreseeable at the outset.[32]

With increasing demands comes the need for ongoing reform of regulations for protecting human research participants.[33] Inadequate resources for IRBs and costs to academic medical centers for the system of protecting participants that can average nearly 750,000 USD per year per institution in some countries[34] make essential the proactive embodiment of ethical principles that could enhance coherence and efficiency.

Culture of Data Sharing and Contours of an Ongoing Debate

While increased statistical power and cost efficiency are commonly noted as benefits of data sharing, proponents are not without opposition. At the heart of the issues – whether for neuroscience, genomics, DNA, or other data forms – are the entanglement of open science and the proprietization of information.[29] In the genetics literature, researchers have reported intentionally withholding data for reasons related to the sheer workload associated with sharing, as well as to protect publication opportunities for themselves and other faculty, especially junior faculty and fellows.[35] For brain imaging, for example, Toga argued that in order for data to be appropriately understood and used, they must be considered in the context of the sample, methodology, and analysis with which they were collected and generated.[16] Quality control and the relinquishment of personal benefit constitute other central themes in resistance to the principles of brain data sharing.[16] Moreover, in the absence of a standard paradigm for collecting data, comparison across studies may be more difficult than expected. This issue also raises questions about who will be responsible for converting data into a standard format, how this procedure might take place, and the impact that standardizing procedures may have on experimental innovation and the individual creative process.[36]

In studying the issue of trust in data-sharing practices and policies, Beaulieu[26] found that sociological hurdles were profound even though the coupling of sharing and publication was designed to be a trust-building mechanism.[37,38] Even before Beaulieu's work was published, Ari Patrinos, then Director of Biological and Environmental Research at the US Department of Energy, was quoted as saying that: "It would be a mistake to adopt simple rules forcing authors to choose between releasing control of all their data at publication or not publishing".[39]

Lack of clear funding agency policies in the face of competing interests that are "often removed from academic research"[40] also poses problems for scientists, as does perilously unstable funding.[41] Administrative and organizational management and diversity in science may necessitate a variety of institutional data management approaches, and establishing and aligning this infrastructure will require proactive, ongoing, and dedicated budgetary planning. Maximizing effectiveness through the involvement of researchers is critical, since many are unaware of existing policies and opportunities even within their institutions and organizations. Heterogeneity in international policies makes data sharing across borders potentially even more difficult. In the USA, for example, federal government databases are not copyright protected, whereas in the European Union, government databases are eligible for protection under law. Practices may even vary within countries, with major funding agencies subscribing to different principles. Arzberger and colleagues, among others, have called for an empirical analysis of views from researchers, funders, and policy makers, and solutions to barriers through guidelines for best practices.[40,42]

With steadily growing sharing practices in the neurosciences, it is essential to take the attendant ethical issues into consideration. The goal of this consideration is to deliver results and tools that are not prescriptive, but that rather enable a broad approach to systems development[43] and an empowering and streamlining effect on existing and newly evolving database practice standards. Appreciation of methodological considerations is a basic principle and no simple formulae are expected. From the pragmatic perspective, ethical guidance is routinely subject to reconsideration as discoveries are relevant to the time, place, and dynamic purpose of inquiry.

INCIDENTAL FINDINGS

The conversation about incidental findings in brain imaging started in earnest in 2001 when approximately 40 people from a wide range of disciplines spanning neuroscience, ethics, and law across the USA and Canada gathered at a roll-up-your-sleeves workshop in Bethesda, Maryland, to discuss the issue (see http://www.ninds.nih.gov/news_and_events/proceedings/ifexecsummary.htm). The challenge was to define what constitutes incidental findings in brain research, what is clinically significant, who should look, and who should tell what to unsuspecting human subjects. The challenge was so undeveloped at the time, however, that one significant source of contention was whether incidental findings posed an issue for research on human subjects at all. Some neuroscientists argued that science is science: the imaginary line between research and clinical medicine should not be blurred or the entire scientific enterprise will come to a grinding halt. Other scientists in the room, as well as ethicists and legal scholars, were troubled by this position. Responding to actionable, potentially life-saving findings and drawing upon relevant work from the genetics community,[44] they argued that trust and reciprocity, autonomy and transparency are fundamental principles of human subjects research and would be violated by this hard line.

The first deliverable from the meeting represented a compromise: a positive outcome that focused on upfront transparency about incidental findings in the protocol review and consent process.[45,46] This was reiterated in further consensus-based discussions that included incidental findings not only for neuroimaging but also for genetics and cancer screening data.[47]

Incidence of Incidental Findings

In the early 2000s, retrospective reviews were performed of anatomical MRI brain scans obtained from research studies with children[48] and adults[49] presumed to be neurologically healthy. Incidental abnormalities were found in the brain images of 47 children (21%) recruited to studies as healthy controls. Of these 47 abnormalities, 17 (36%) were determined to have required routine referral for further evaluation; a single case (2% of the total abnormalities; 0.5% of the cases studied) was categorized as an urgent referral. In 151 studies on adults, an overall occurrence of incidental findings having required referral of 6.6% was found. By age, there were significantly more findings in the older cohort (60 years and older) than the younger, and in more men than women in the older cohort. Three out of four (75%) findings in the younger cohort were classified in the urgent referral category; 100% of the findings in the older cohort were classified as routine. These trends have since been replicated in other studies.

Research Protocols

With these findings on incidence in hand, different protocols for handling them were examined.[50] The goal was to provide a platform for establishing formal discussions of related ethical and policy procedures. Seventy-four investigators who conduct MRI studies in the USA and six other countries responded to a web-based survey. Eighty-two percent (54/66) reported discovering incidental findings in their studies, such as arteriovenous malformations, brain tumors, and developmental abnormalities. There was substantial variability in the procedures for handling and communicating findings to subjects, neuroradiologist involvement, the academic level of research personnel permitted to operate equipment, and training.

Subject Expectations

What do subjects expect? That was the central question in the next study. Healthy control subjects who had previously participated in brain scans in medical and non-medical settings were surveyed about their expectations and attitudes towards unexpected clinical findings on their research brain scans.[51] It was hypothesized that, although participants consent to a scanning procedure for research purposes alone, they still expect pathology, if present, to be detected and reported to them. Responding to a web-based survey, 54% of 105 participants reported that they expect research scans to detect abnormalities should they exist. Nearly all subjects (over 90%) reported that they would want findings communicated to them by a physician affiliated with the research team. No significant differences were found between participants scanned in medical and non-medical settings.

The Functional Frontier

Let's look ahead now. In a paper published in 2012 in the *Journal of Research on Human Research Ethics*, the possibility of functional incidental findings, particularly those from the resting state, was considered. Investigators have shown that intrinsic connections within the brain are sculpted and trained by learning,[52] substantiating the claim for biological relevance and further lending support to their role in function. But perhaps the greatest potential for understanding brain function lies in the perturbations of these connections. Analyses of functional connectivity in these circuits have revealed changes in synchrony and connection strength correlated with neurological diseases such as Alzheimer disease and stroke, as well as psychiatric diseases such as depression and schizophrenia, attention deficit/hyperactivity disorder, and autism. Accordingly, there is potential for these characterized changes in functional connectivity to be used as clinical biomarkers. To this end, it has been suggested that acquiring resting-state data during the standard clinical workup would be invaluable for diagnosing and prognosticating disease states.[53]

The present authors' approach in this discussion did not anchor the resting state as the *sine qua non* of functional incidental findings, but was intended rather as a case in point to thinking forward to the future. Considering the issues proactively today, within a framework that is maximally flexible and open to modification, is better than responding reactively after the fact and with no framework at all. This is further the case as technologies are increasingly being combined, such as imaging genetics, bringing associated ethical challenges (Fig. 45.1). Overall, there is a duty to consider possible incidental findings despite the ambiguities of data interpretation and increased likelihood of incidence with the increasing power of technologies alone and together, while working hard to prevent unnecessary alarm.

NEUROSCIENCE COMMUNICATION

There is increasing pressure for neuroscientists to communicate their research and the societal implications of their findings to the public. Communicating science to the public is challenging and the transformation of communication by digital and interactive media makes the challenge even greater. To successfully facilitate dialogue with the public in this new media landscape, neuroethical challenges at the interface of neuroscience and communication were studied and three courses of action for the neuroscience community were suggested: a cultural shift so that academic institutions, professional

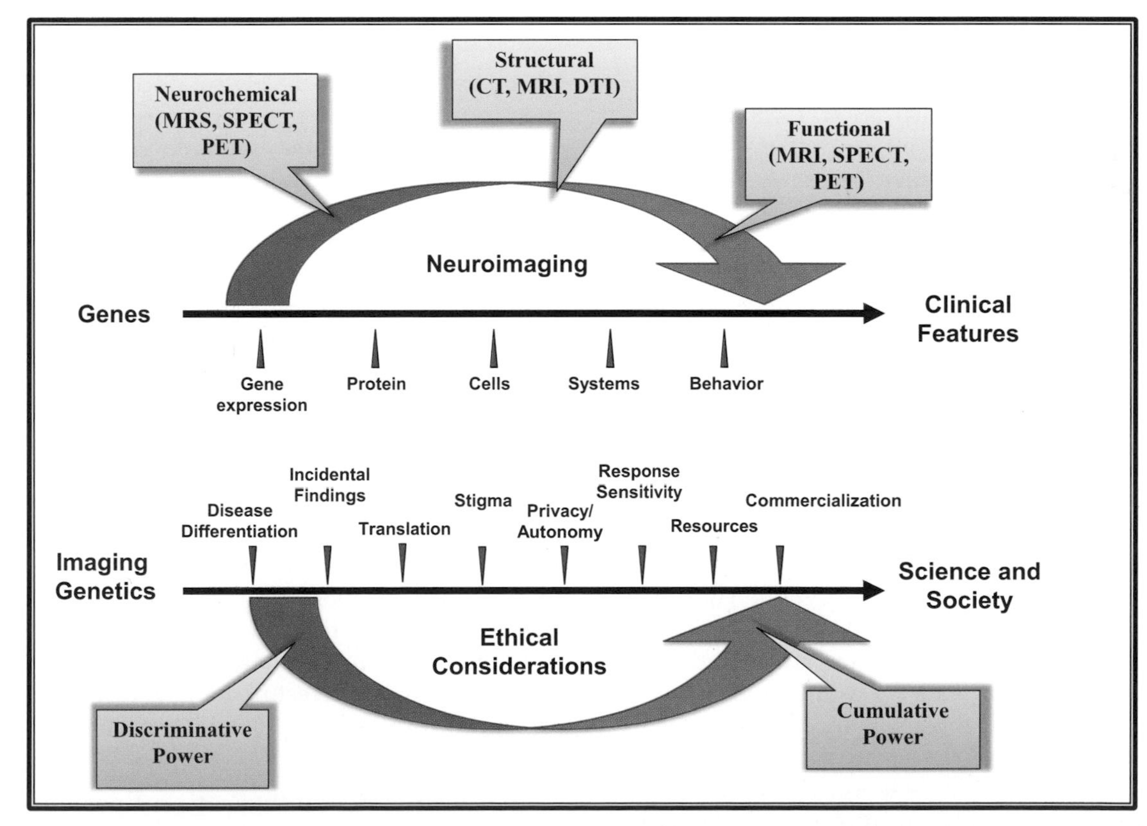

FIGURE 45.1 **Neuroimaging and ethical considerations.** The role of neuroimaging in investigating intermediate phenotypes (top) expanded with ethical features (bottom) to illustrate a logical continuum, but not necessarily a fixed temporal relationship, of considerations for imaging genetics, and the downstream impact of this combined form of neurotechnology on science and society. *Source: Updated from Tairyan and Illes.* Neuroscience. *2009;164(1):7–15.*[54]

societies, and funding agencies explicitly recognize and reward public outreach; the identification and development of neuroscience communication experts; and ongoing empirical research on public communication of neuroscience.[1]

Cultural Shift

Owing to the growing societal relevance of neuroscience, the importance of communication needs to be recognized explicitly and elevated as a priority in the community, akin to protecting the rights of human subjects and ensuring respectful animal care in research. Institutional support is required to advance this goal and that support begins with explicitly valuing the effort. Developing a process for valuing communication will surely be no less complex than the composite metrics used today, for example, for valuing productivity in peer-reviewed publications from a combination of raw numbers, journal impact factor, and individual publication impact. However, the last two do not exist for science communication products. Audience size, evaluations, and local, national and international reach could serve as first proxy measures of impact. These measures must be ultimately factored into the evaluation of junior researchers for promotion and those more senior for advancement. Awards that recognize excellence are important signals of commitment and success. Other long-term rewards should take the form of time off from teaching, research, or administration. Little in this shift will be cost free either in real dollars or in personal effort. Nevertheless, the already skilled must step forward to model these goals with mentorship and action.

Some actions towards the cultural shift can be immediately implemented, such as increasing the professional value of delivering public lectures, media work, and efforts to develop training activities tailored specifically for neuroscientists. Other actions, such as the full integration of communication training into neuroscience curricula and graduate training, will require longer range planning and a more fundamental culture shift to achieve equally full acceptance given already heavily laden schedules. For neuroscientists, the overall continued development of specialized training sessions, online course modules, and boot camps at professional meetings or local institutions will help to realize this culture shift.

Some actions have already been taken and investments made towards this goal. For example, the American Association for the Advancement of Science sponsors a summer internship program that places graduate and postgraduate students in science, engineering, and mathematics at media organizations nationwide; participants "come in knowing the importance of translating their work for the public, but they leave with the tools and the know-how to accomplish this important goal" (see http://www.aaas.org/programs/education/MassMedia). An intensive science communication program for scientists, journalists, and communications professionals takes place each year at the Banff Centre in Alberta, Canada. This immersive residency program pushes midcareer professionals to initiate creative science communication projects, with the goal of fostering a broad, ethical, and more engaging role for science in public culture. Both of these programs cater to all scientific disciplines. These initiatives should be extended directly to neuroscience to create focused communication internships for trainees or midcareer researchers, and immersion opportunities for neuroscience communication experts. Organizations with already existing programs should customize new ones for neuroscience and provide guidance to others who wish to embark on new initiatives building on history and experience.

Some programs aimed specifically at neuroscience led, for example, by the International Brain Research Organization, the Dana Alliance for Brain Initiatives, the Federation of European Neuroscience Societies, and the Society for Neuroscience, already have prominence. The membership of the Society for Neuroscience, for example, has endorsed public education as a key component of its strategic plan and published *Neuroscience Core Concepts*, a document with application to both K-12 educators and the general public that lays out fundamental principles about the brain and nervous system. The neuroscience research community can immediately support the further development, awareness, and uptake of these resources by elevating the visibility of communication in the community and accountability of individuals to the task.

A commitment to a cultural shift will also compel funders of neuroscience research to encourage or even require information on plans for knowledge translation, public engagement, and outreach. The US National Science Foundation, for example, which funds basic research across all disciplines including behavioral and neurobiological sciences, already has a societal impact review requirement. In Canada, many requests for applications and proposals have explicit knowledge translation requirements. Funding agencies that primarily target neuroscience research could follow suit by providing similar societal impact inclusion requirements in submitted proposals and funding opportunities for knowledge translation and public engagement. Even in a difficult economic climate, the prevailing view in science policy is that investment in the future of science and the R&D workforce through education is needed.

Neuroscience Communication Specialists

Specialized training of journalists, editors, and neuroscientists alike is needed to promote increasingly effective communication of newsworthy neuroscience findings and considerations of their ethical, social, and policy impact. People from both the academic and non-academic neuroscience community who can serve as specialists or ambassadors in neuroscience communication should be identified and should identify their interests to their supervisors, faculty heads, and deans. Neuroscientists are not generally trained in communications or in emerging new media and, among those who are, skills are variable. It is not reasonable to assume that all scientists will be able to acquire the specialized skills required to communicate effectively in any medium, despite the suggested heightened level of exposure and activity. Although all neuroscientists need to be aware of the public discussion surrounding neuroscience and the increasingly diverse means by which it is circulated by online, print, television, and radio sources, a cohort of skilled neuroscience ambassadors who are embedded in neuroscience research programs could become experts in new communication tools. These individuals would work with each other, with other science communication experts at institutional press offices, journalists, and their own colleagues and students to foster the communication of accurate and contextualized information. They could become neuroscience "knowledge brokers" by linking the creators of new knowledge with recipients, and could increase the quantity and caliber of communications activity by providing education about and access to new knowledge.[55] They could explore creative uses of new media tools and develop strategic communications for engaging the public using new media platforms. Investment in specialized programs, such as expert workshops in which neuroscientists and journalists exchange knowledge and know-how, will be a further powerful tool in achieving this goal.

The need for such experts is further amplified by the rapid flow of information through continually emerging non-peer-reviewed, non-curated publications and web postings. Organizations and researchers can disseminate their own information directly to the public via blogs and websites. Filtering and discriminating high-quality information in this new landscape is time consuming and will require dedicated and reliable specialists who can provide services for the larger community.

Research on Neuroscience Communication

More empirical data are needed from research on neuroscience communication. It is imperative to understand the receptivity, motivation, and barriers to communication of both neuroscience findings and their social impact. The complexities of commercialization and partnerships between academia and industry, including conflict of interest, intellectual property, and risks to the privacy of brain data, expand this imperative.[56,57] In parallel, the opportunity exists to gather data about activities in which neuroscience engages with the public, to make changes and improve these activities, and to re-engage the communicators. These initiatives will require support from within institutions and funding from research sponsors to support a communication component both for projects that are not specifically focused on communication and for those that are specifically earmarked to meet this objective.

It is also important to understand what different publics know, understand, and value, what is of interest, and how much science non-scientists can absorb, especially in this age when traditional journalistic reporting collides with the worlds of arts, electronic media, and entertainment. Whereas detailed audience profiles can be obtained for print, radio, television, and arts consumers, the same information is not yet available for the conflation of these forms on the Internet. For example, we can gather statistical data on the behavior of visitors to a website but at present need to infer intent. We can tell if someone uses a search engine to find an article on depression, but we do not have an understanding of the motivation or goal for that search. We do not understand how viewers are engaged with the data and how they take it up in everyday life. We do not understand how web-based information shapes public dialogue and participation in events. Empirical research in science communication that draws on quantitative and qualitative data in the Internet age can form the foundation of well-informed strategies. This can include appropriate and rigorous evaluations of current and emerging mechanisms designed to improve public understanding of neuroscience, as well as the effectiveness of public dialogue and engagement activities.

Neuroscience communication requires scientists to explicitly articulate new scientific knowledge and the implications of that knowledge. The community of scientists and scholars with interests in neuroethics – a mixed composition of experts in brain science, social science, law, and philosophy whose multidisciplinary interests lie at the intersection of neuroscience and its impact on people and society – offers a compelling starting point for advancing communication in neuroscience.[58–60]

NEUROETHICS FOR NEUROSCIENCE

The sections above have covered some key areas in which empirical neuroethics work has made a tangible impact on neuroscience. This work has been driven by people such as the authors of this chapter, neuroscientists who have left basic science and turned their attention to the ethics discourse around neuroscience, and others. This last section discusses studies on how still-bench neuroscientists are thinking about neuroethical issues today, and how neuroethicists can provide enabling tools to meet identified needs.

Mentors and Trainees

The first of three studies examined the landscape of ethics training in neuroscience programs, beginning with the Canadian context.[59] Neuroscientists at all training levels were surveyed, and directors of neuroscience programs and training grants were interviewed. From concepts in biomedical ethics prescribing good research conduct, such as informed consent and equipoise, to ethical challenges in application, such as those surrounding new abilities to probe the functional metabolic basis of thought and morality, interest was found to be high. Educational opportunity and time to learn about, understand, and address associated challenges, however, were less than ideal.

Current Approaches

Both survey respondents and interviewees reported having an interest in opportunities to integrate ethics into their professional and academic environment. Some survey respondents reported a general dissatisfaction with current approaches to ethics saying, for example, that "[ethics concerns are] never discussed amongst colleagues – not good. We all should be obliged to take part in neuroethical discussions". Neuroscience research program directors emphasized the importance of understanding and discussing ethics-related topics with other members of the neuroscience community (e.g. in laboratories and conferences, and with members of the public). They emphasized forces external to neuroscience programs, such as the regulatory environment and institutional encouragement, more than internal factors, such as interest from students and faculty, as motivators for including ethics in their curricula. By contrast, training grant directors cited intrinsic interest from faculty and good citizenship, both internal motivators, as the major reasons for enhancing ethics curricula.

New Frontiers

Responding to these data, a first set of recommendations was developed to address the barriers to neuroethics education in neuroscience programs. Resources in the form of case-based materials tailored to neuroscience research are needed for face-to-face ethics training. Modules should complement existing curricular activity and be fully integrated into training programs to maximize receptivity to them. Based on the data here and elsewhere, the first set of modules should cover:

- fundamental principles and contemporary writings in bioethics, biomedical ethics (classic cases with a focus on the nervous system), and neuroethics
- applied societal implications of neuroscience and planning for impact of results upstream during research design, using historical examples and current relevant literature
- translational considerations for clinical trials and other research that moves neuroscience innovation from the bench to the bedside
- communication strategies and innovative approaches to disseminating neuroscience knowledge to the press and public
- commercialization challenges.

To address a second major barrier, lack of expertise, new training and funding opportunities for neuroscience faculty to gain expertise in ethics are suggested. Such mechanisms could support visiting faculty from the humanities to neuroscience programs and neuroscientist stays in ethics programs. Faculty exchanges would foster new dialogue among the groups and in the neuroscience domain, in particular, provide "on-the-ground" support as ethics programs are implemented. By narrowing the existing gap in expertise through interdisciplinary collaborations, neuroscience programs would capitalize on existing tools and have the opportunity to develop well-tailored new ones.

To address the barrier of time, the first step will come from understanding how curricula can be designed or to integrate ethics modules seamlessly. Whether the approach is a series of modules across time or a dedicated unit will depend on the needs and size of the program. There is no reason to think that one size will fit all, nor is this necessary. Feedback on and evaluation of the programs on an ongoing basis, and continuous refinement based on ever-changing needs of programs are, however, unequivocal requirements.

Neuroimagers

This pursuit was taken further by surveying faculty, trainees, and staff whose work involves brain imaging and brain stimulation. A total of 605 respondents completed an online survey about ethics in their research. Factor analysis and linear regression revealed significant effects for professional position, gender, and local presence of bioethics centers. To understand these effects deeply, especially as they apply to the sense of responsibility about integrating ethics into neuroimaging and readiness to adopt new ethics strategies as part of their research, the authors followed up with focus groups and interviews. Here, they learned that safety, trust, and virtue – the duty to do good – were key motivators for incorporating ethics into neuroimaging research. Managing incidental findings emerged as a predominant daily challenge for faculty, while student reports focused on the malleability of neuroimaging data and scientific integrity. The most frequently cited barrier was time, again, and administrative burden associated with the ethics review process. Lack of scholarly training in ethics also emerged as a major barrier. Participants constructively offered remedies to these challenges, including the development and dissemination of best practices and standardized ethics review for protocols. Students, in particular, urged changes to curricula to include early, focused training in ethics.

Researchers in the Domain of Neurodegenerative Disease

In a final phase of work on exploring and defining neuroethics for neuroscientists, the authors turned their attention to researchers working on neurodegenerative disease. The neuroscience in this community has a particular mandate to discover effective treatments, and the ethics landscape surrounding it is in a constant state of flux. The ongoing challenges place ever greater demands on investigators to be accountable to the public and to answer questions about the implications of their work for health care, society, and policy.

Using well-trodden survey methods, US-based investigators involved in neurodegenerative diseases research were asked about how they value ethics-related issues, what motivates them to give consideration to those issues, and the barriers to doing so. Using online databases, researchers with relevant, active grants were identified and invited to complete an online questionnaire. Altogether, 193 responses were received (an 11% response rate). Exploratory factor analysis was used to transform individual survey questions into a smaller set of factors, and linear regression to understand the effect of key variables of interest on the factor scores.

For this cohort, ethics-related issues clustered into two groups: research ethics and external influences. Heads of research groups viewed issues of research ethics to be more important than the other respondents. Concern about external influences was related to overall interest in ethics. Motivators clustered into five groups: ensuring public understanding, external forces, requirements, values, and press and public. Heads of research groups were more motivated to ensure public understanding of research than the other respondents. Barriers clustered into four groups: lack of resources, administrative burden, relevance to the research, and lack of interest. Perceived lack of ethics resources was a particular barrier for investigators working in drug discovery.

These data suggest that senior-level neuroscientists working in the field of neurodegeneration, like their counterparts in other domains of neuroscience, are motivated to consider ethics issues related to their work. The perceived lack of ethics resources again thwarts their efforts. With bioethics centers at more than 50% of the institutions at which these respondents reside, the neuroscience and bioethics communities appear to be disconnected. Dedicated ethical, legal, and social implications programs, such as those fully integrated into genetics and regenerative medicine, provide models for achieving meaningful partnerships that are not yet adequately realized for scholars and trainees in other areas of neuroscience. Strategies for improving communication between neuroscientists and biomedical ethicists, as well as ethics training in graduate neuroscience programs, will go a long way towards realizing mutual goals and interests.

CONCLUSION

This chapter has highlighted some of the opportunities for neuroscientists to participate in the revolution in neuroethics and use it to enrich their work in the field. A summary is provided in Table 45.1.

TABLE 45.1 Example Opportunities for Neuroethics in Neuroscience

Case Studies	Examples of Key Ethics Considerations for Neuroscience
Animal models	Animal safety and care
	Judicious selection of models
	Public trust and understanding
	Transparency
Data sharing	Consent
	Privacy
	Confidentiality
	Responsibility for incidental discoveries of clinical importance
	Conflict of interest
	Intellectual property and commercialization
	Participant recontact
Incidental findings	Transparency
	Consent
	Duty to warn
	Allocation of research resources
	Participant trust
	Reciprocity between investigators and participants
Neuroscience communication	Knowledge sharing
	Public trust
	Neuroscience literacy
	Expanded definition of the duty to care

Whether it be the challenges of research on animals or science communication, managing data sharing in an era of big science, or coping with the intrusion of incidental findings on routine experimental examinations, neuroethics provides an opportunity to proactively meet these challenges head on. It is not that there is no alternative, but rather that engaging with the societal issues deepens our discourse and adds to the value that neuroscience provides to the world at large by investigating the inner workings of the most interesting organ in the body.

Acknowledgments

Judy Illes is supported by the Canada Research Chairs Program, and grants from the Canadian Institutes of Health Research, the National Institutes of Health, the British Columbia Knowledge Development Fund, the Canadian Foundation for Innovation, the Dana Foundation, the Vancouver Foundation, the Stem Cell Network, NeuroDevNet Inc., the North Growth Foundation, and the Foundation for Ethics and Technology. Peter Reiner is supported by the Canadian Institutes of Health Research.

References

1. Illes J, Moser MA, McCormick JB, et al. Neurotalk: improving the communication of neuroscience. *Nat Rev Neurosci*. 2010;11(1): 61–69.
2. Reiner PB. The rise of neuroessentialism. In: Illes J, Sahakian B, eds. *The Oxford Handbook of Neuroethics*. Oxford: Oxford University Press; 2011:161–175.
3. Conn M, Parker J. The animal research war. *FASEB J*. 2008;22: 1294–1295.
4. Clarence WM, Scott JP, Dorris MC, Paré M. Use of enclosures providing vertical dimension by captive animals involved in biomedical research. *J Am Assoc Lab Anim Sci*. 2006;45:31–34.
5. Illes J. Transparency ensures ideals met: UBC's new methods of looking at the ethics of animal research are leading the way to an improved future. *Vancouver Sun*. November 8, 2011.
6. Beauchamp T, Childress J. *Principles of Biomedical Ethics*. 5th ed. Oxford: Oxford University Press; 2001.
7. Organisation for Economic Co-operation and Development. *Final Report of the OECD Megascience Forum Working Group on Biological Informatics*; 1999 (pp. 1–74).
8. Van Horn JD, Grethe JS, Kostelec P, et al. The fMRIDC: The challenges and rewards of large scale databasing of neuroimaging studies. *Philos Trans R Soc Lond B Biol Sci*. 2001;356(1412):1323–1339.
9. Mazziotta J, Toga A, Evans A, et al. A probabilistic atlas and reference system for the human brain: International Consortium for Brain Mapping (ICBM). *Philos Trans R Soc Lond*. 2001;256(1412): 1293–1322.
10. Hyman SE. The millennium of mind, brain, and behavior. *Arch Gen Psychiatry*. 2000;57:88–89.
11. Insel TT, Volkow ND, Landis S, Li TK, Sieveng P. Limits to growth: why neuroscience needs large and small scale science. *Nat Neurosci*. 2004;7(5):426–427.
12. Martone ME, Gupta A, Ellsiman MH. e-Neuroscience: challenges and triumphs in integrating distributed data from molecules to brains. *Nat Neurosci*. 2004;7(5):467–472.
13. OBHM. Neuroimaging databases. *Science*. 2001;292(5522): 1673–1676.
14. Van Horn JD, Gazzaniga MS. Databasing fMRI studies – towards a "discovery science" of brain function. *Nat Rev Neurosci*. 2002;3(4):314–318.
15. Sweeney L. *Maintaining patient confidentiality when sharing medical data requires a symbiotic relationship between technology and policy. Massachusetts Institute of Technology Artificial Intelligence Laboratory, AI Working Paper No. AIWP-WP344b*; 1991.
16. Toga AW. Imaging databases and neuroscience. *Neuroscientist*. 2002;8(5):423–436.
17. Clayton EW, Steinberg KK, Khoury MJ, et al. Informed consent for genetic research on stored tissue samples. *JAMA*. 1995; 33(1):15–21.
18. Sugarman J, Reisner G, Kurtzberg J. Ethical aspects of banking placental blood for bone marrow transplantation. *JAMA*. 1995;274(22):1783–1985.
19. Wolf S, Lo B. Untapped potential: IRB guidance for the ethical research use of stored biological materials. *IRB: Ethics Hum Res*. 2004;26(4):1–8.
20. Malinowski MJ. Technology transfer in biobanking: credits, debuts and population health futures. *J Law Med Ethics*. 2005; 33(1):54–69.
21. Scott NA, Murphy TH, Illes J. Incidental findings in neuroimaging research: a framework for anticipating the next frontier. *J Empir Res Hum Res Ethics*. 2012;7(1):53–57.
22. Austin MA, Harding S, McElroy C. GeneBanks: a comparison of eight proposed international genetic databases. *Commun Genet*. 2003;6:1.
23. Siang S. NIH seeks comment on proposed data sharing policy. *J Natl Cancer Inst*. 2002;94:555.
24. Burton B. Proposed genetic database on Tongans opposed. *BMJ*. 2002;324(7335):443.
25. Clayton EW. Informed consent and biobanks. *J Law Med Ethics*. 2005;33(1):15–21.
26. Beaulieu A. Research woes and new data flows: a case study of data sharing at the fMRI Data Center. In: Wouters P, Schröder P, eds. *The Public Domain of Digital Research Data*. Netherlands Institute for Scientific Information Services; 2003. 2003:65,85.
27. Rothstein M. Expanding the ethical analysis of biobanks. *J Law Med Ethics*. 2005;33(1):41–53.
28. Stein D. *Buying In or Selling Out? The Commercialization of the American University*. Piscataway, NJ: Rutgers University Press; 2004.
29. Amari S, Beltrame F, Bennett R, et al. OECD Working Group on Neuroinformatics. *Neuroinformatics*. 2003;1(2):149–165.
30. Knoppers B. Biobanking: international norms. *J Law Med Ethics*. 2005;33(1):7–14.
31. Majumder MA. Cyberbanks and other virtual research repositories. *J Law Med Ethics*. 2005;33(1):31–39.
32. Deschênes M, Sallée C. Accountability in population biobanking: comparative approaches. *J Law Med Ethics*. 2005;33(1):41–53.
33. Moreno J, Caplan AL, Wolpe PR. Updating protections for human subjects involved in research: policy perspectives. *JAMA*. 1998;280(22):1951–1958.
34. Sugarman J, Getz K, Speckman JL, Byrne MM, Gerson J, Emmanuel EJ. The cost of institutional review boards in academic medical centers. *N Engl J Med*. 2005;352(17):1825–1827.
35. Campbell EG, Clarridge BR, Gokhale M, et al. Data withholding in academic genetics: evidence from a national survey. *JAMA*. 2002;287(4):473–480.
36. Illes J, Racine E. Neuroethics: a dialogue on a continuum from tradition to innovation [Response]. *Am J Bioethics*. 2005;5(2):3–4.
37. Birnholtz J, Bietz M. Data at work: supporting sharing in science and engineering. In: *Proceedings of the 2003 International ACM SIGGROUP Conference on Supporting Group Work*. New York: ACM Press; 2003:339–348.
38. Kotter R. Neuroscience database tools for exploring neuroscience relationships. *Philos Trans R Soc Lond B Biol Sci*. 2001;356(1412): 111–112.
39. Marshall E. Clear-cut publication rules prove elusive. *Science*. 2002;5560(295):1625.

40. Arzberger P, Schroeder P, Beaulieu A, et al. Science and government. An international framework to promote access to data. *Science*. 2004;303(5665):1777–1778.
41. Merali Z, Giles J. Databases in peril. *Nature*. 2005;435(7045):1010–1011.
42. Ascoli GA, Beatty JT, Brinkley JF, et al. Towards effective and rewarding data sharing. *Neuroinformatics*. 2003;1(3):289–296.
43. Jones J, Preston H. Big issues, small systems: managing with information in medical research. *Top Health Inf Manage*. 2000;21(1):45–54.
44. National Bioethics Advisory Commission (1999). Research involving human biological materials: ethical issues and policy guidance. Paper presented at the Report and Recommendations of the National Bioethics Advisory Commission, Rockville, MD.
45. Illes J. Pandora's box of incidental findings in brain imaging research. *Nat Clin Pract Neurol*. 2006;2(2):60–61.
46. Illes J, Kirschen MP, Edwards E, et al. Incidental findings in brain imaging research. *Science*. 2006;311(5762):783–784.
47. Cho MK, Clayton E, Fletcher J, et al. Managing incidental findings in human subjects research. *J Law Med. Ethics*. 2008;36(2):219–248.
48. Kim BS, Illes J, Kaplan RT, Reiss A, Atlas SW. Incidental findings on pediatric MR images of the brain. *Am J Neuroradiol*. 2002;23(10):1674–1677.
49. Illes J, Rosen AC, Huang L, et al. Ethical consideration of incidental findings on adult brain MRI in research. *Neurology*. 2004;62:888–890.
50. Illes J, Kirschen MP, Karetsky K, et al. Discovery and disclosure of incidental findings on brain MRI in research. *J Magn Reson Imaging*. 2004;20:743–747.
51. Kirschen M, Jaworska A, Illes J. Participant expectations of incidental findings in neuroimaging research. *J Magn Reson Imaging*. 2006;23(2):205–209.
52. Scott CT, Caulfield T, Borgelt E, Illes J. Personalized medicine: the next banking crisis. *Nat Biotechnol*. 2012;30:1–7.
53. Dosenbach NUF, Nardos B, Cohen AL, et al. Prediction of individual brain maturity using fMRI. *Science*. 2010;329(5997):1358–1361.
54. Tairyan K, Illes J. Imaging genetics and the power of combined technologies: a perspective from neuroethics. *Neuroscience*. 2009;164(1):7–15.
55. Ward VL, House O, Hamer S. Knowledge brokering: exploring the process of transferring knowledge into action. *BMC Health Serv Res*. 2009;9(12):1–6.
56. Bubela T, Nisbet MC, Borchelt R, et al. Science communication reconsidered. *Nat Biotechnol*. 2009;27(6):514–518.
57. Eaton ML, Illes J. Commercializing cognitive neurotechnology: the ethical terrain. *Nat Biotechnol*. 2007;25(4):1–5.
58. Abi-Rached JM. The implications of the new brain sciences. The "Decade of the Brain" is over but its effects are now becoming visible as neuropolitics and neuroethics, and in the emergence of neuroeconomies. *EMBO Rep. 2008*. 2008;9(12):1158–1162.
59. Lombera S, Illes J. The international dimensions of neuroethics. *Dev World Bioeth*. 2009;9(2):57–64.
60. Illes J, Kirschen MP, Gabrieli JDE. From neuroimaging to neuroethics. *Nat Neurosci*. 2003;6(3):205.

CHAPTER

46

Burden of Neurological Disease

Mitchell T. Wallin, John F. Kurtzke

VA Multiple Sclerosis Center of Excellence–East, Georgetown University School of Medicine, Department of Veterans Affairs Medical Center Neurology Service, Washington, DC, USA

OUTLINE

INTRODUCTION

This chapter provides an overview of the neurological burden of disease. A brief review of epidemiological concepts is provided to assist the reader in interpreting epidemiological studies. The major focus of this chapter is morbidity rates (i.e. incidence and prevalence). Space precludes attention to survey methods, risk factors, treatment comparisons, and statistical methods, which are all intrinsic aspects of epidemiology. For further information on these topics, the reader may consult with one of several excellent sources.[1–3] Regarding neuroepidemiology *per se*, the material presented here is but a sketch of some highlights for a few major diseases chosen to represent the field.

BASIC CONCEPTS IN EPIDEMIOLOGY

Epidemiology is the study of the distribution and determinants of disease. In epidemiology, the unit of study is a person affected by a defined condition. Therefore, diagnosis is the essential prerequisite. Thus, the neurologist must be an essential part of any inquiry into neuroepidemiology, the epidemiology of neurological diseases. The content and uses of epidemiology are described in Fig. 46.1.

After diagnosis, the most important question is the frequency of a disorder. Much of this type of information has been based on case series from clinic and hospital databases. However, whether taken as numerator

Neurobiology of Brain Disorders
http://dx.doi.org/10.1016/B978-0-12-398270-4.00046-X

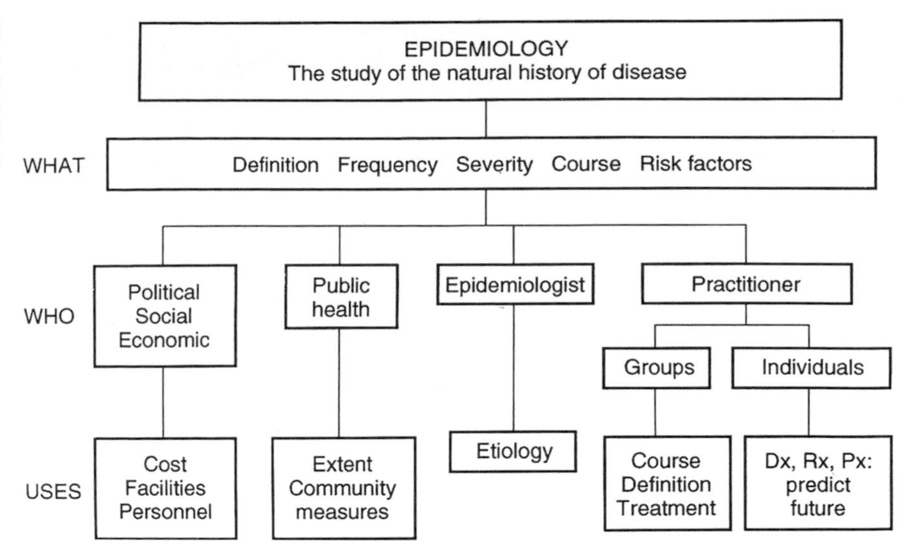

FIGURE 46.1 **Epidemiology: content and uses.** Dx: diagnosis; Rx: treatment; Px: prognosis. *Source: Adapted from Kurtzke JF. Multiple sclerosis from an epidemiological viewpoint. In: Field.* Multiple Sclerosis: A Critical Conspectus. *Lancaster: Medical and Technical Publishing Press; 1977;83–142,[4] with permission.*

alone (case series) or compared with all admissions (relative frequency), the difficulty with such data is that one has little assurance that what has been included is representative of the true distribution. Such case material needs to be referenced to its proper denominator, its true source: the finite population at risk.

Neuroepidemiology is the branch of medical science in which the methods of epidemiology are applied to the problems of clinical neurology. Both of these fields are extensive, requiring a wide base of knowledge and techniques. Clinicians are aware of these necessities for the comprehensive knowledge of neurology, but may not appreciate that epidemiology is equally complex and that a solid grounding in this discipline is also essential for research in and understanding of neuroepidemiology. A good consultant explains and justifies the conclusions, but the physician still needs to interpret them in the context of the individual patient. Similarly, good epidemiological publications provide enough methodology and data so that the critical reader can assess properly the conclusions offered.

Case Ascertainment

To study disease frequency, data based on cases found within a specific population are needed. If the entirety of this population were examined, all the symptomatic cases and those with relevant signs, although without complaints, could be discovered. If an appropriate sample of the population were examined, instead of the total, the numbers expected within the total could still be estimated. This is referred to as a population survey; enumeration is used for census data. Such measures have been taken for common diseases (e.g. hypertension and diabetes), but are largely impractical for rare entities.

One step removed from this true population survey is the ascertainment of all those affected who have come to medical attention, often referred to as a prevalence study rather than a population survey.

Another step still farther removed from complete enumeration is a listing of the deaths that the disease has caused. These cases, however, will refer only to those instances in which the condition was fatal, and this is, for most disorders, only a fraction of those affected. Autopsy series are a subset of hospital data. Even if all autopsies were collected from all the resources of the community, they would still represent only a fragmentary portion of those affected. Moreover, as in most hospital series, there is usually no population denominator.

Population-Based Rates

Ratios of cases to population, together with the period to which they refer, make up the population-based rates. Those commonly measured are the incidence rate, mortality rate, and the so-called prevalence rate, "so-called" because "rate" implies duration as well as frequency. They are usually expressed in unit–population values. For example, 10 cases among a community of 20,000 people represent a rate of 50 per 100,000 population or 0.5 per 1000. Accurate rates require an adequate surveillance system with access to neurological specialists for diagnosis and care.

The incidence or attack rate is the number of new cases of a disease beginning in a unit of time in a population. This is usually given as an annual incidence rate in cases per 100,000 population per year. The date of onset of clinical symptoms ordinarily decides the time of accession, although occasionally the date of first diagnosis is used.

The mortality or death rate is the number of deaths in a population in a period with a particular disease as the underlying cause, such as an annual death rate per 100,000 population. The case–fatality ratio is the proportion of those affected who die from the disease.

The (point) prevalence rate is more properly called a *ratio*, but it refers to the number of those affected at one time within the community, again expressed per unit of population. Prevalence rates include both old and new cases at a specific time-point. If there is no change in case–fatality ratios over time and no change in annual incidence rates (and no migration), then the average annual incidence rate times the average duration of illness in years equals the point prevalence rate.

When numerator and denominator for a rate each refer to an entire community, their quotient is a crude rate for all ages. When both terms of the ratio are delimited by age or gender, these are then age-specific or gender-specific rates. Such rates for consecutive age groups, from birth to the oldest group of each gender, provide the best description of a disease within a community.

In comparing rates between two communities for an age-related disorder (such as stroke or epilepsy), there may be differences in crude rates solely because of differences in the age distributions of the denominator populations. This can be avoided by comparing only the individual age-specific rates between the two, but this rapidly becomes unwieldy. Methods do exist for adjusting the crude rates for all ages to permit such comparisons. One such method involves taking for each community each age-specific rate and multiplying it by the proportion of a "standard" population that the same age group represents. The sum of all such products provides an age-adjusted (to a standard) rate, or a rate for all ages adjusted to a standard population. One common standard in the USA is its population for a given census year.

The denominator for morbidity rates is the resident population of the study site. It is essential that all medical resources serving the site be assessed. Extrapolation from, say, one or two sources will give not only an incomplete count but often also a biased one. This holds whether we are studying a community or a nation.

Mortality rates in most countries are calculated from data recorded on the official death certificates. The appropriate government agency is responsible for coding and calculating mortality rates by cause of death. The code used for mortality rates has been a three- or four-digit number representing a specific diagnosis in the *International Statistical Classification of Diseases, Injuries and Causes of Death* (ICD), which is revised about every 10 years. Major changes were made in the 10th revision of the *International Statistical Classification of Diseases* (ICD-10).[5]

ICD-10 was published in 1989 with the innovation of an alphanumeric coding scheme of one letter followed by three numbers (e.g. I63.1, cerebral infarction due to thrombosis of precerebral arteries). One drawback of the ICD system of classification is that several diseases are frequently subserved under the same primary code and subcodes are often not published. To provide a more refined classification for individual diseases, several disciplines have published specialty-related expansions of the primary ICD structure. ICD-10-NA is the expansion of the codes relating to neurological diseases, so that virtually every known neurological disease or condition has a unique alphanumerical identifier.[6] ICD-11 is in preparation with an expected release in 2017. In the USA, all hospitals and health-care workers currently use the ICD-9 clinical modification (ICD-9 CM), which was replaced by ICD-10 in late 2013. Since 1999, the US National Center for Health Statistics has used ICD-10 coding for vital statistics.

Mortality rates thus arise from administrative files with no diagnostic review, leading to major questions of under- and over-reporting. In recent years there have been similar efforts to provide morbidity rates, incidence or prevalence, from administrative files such as insurance or pharmaceutical records or disability files, all with the same caveats about diagnosis as with death rates.

Nationwide morbidity surveys for neurological disease in the USA are rare. However, for selected diseases and with appropriate precautions and diagnostic review, the military veteran population of the country over time provides a rich potential resource for geographic distributions, in particular. Multiple sclerosis (MS) is perhaps the prime example, with nationwide samples and unbiased preillness controls having been available for nearly a century.

CEREBROVASCULAR DISEASE

The World Health Organization (WHO) defines stroke as a syndrome of rapidly developing clinical signs of focal (or global) disturbance of cerebral function, with symptoms lasting 24 hours or longer or leading to death, with no apparent cause other than of vascular origin.[7] (See Chapter 22 for an in-depth discussion of cerebrovascular disease.) In 2002, 15.3 million strokes occurred worldwide with nearly 6 million deaths.

An estimated 795,000 strokes occur annually in the USA alone, which is the equivalent of one new stroke every 40 seconds.[8] Overall stroke morbidity and mortality have been decreasing in the USA over the past century. In 2008, stroke mortality declined in the USA from the third to the fourth leading cause of death behind heart disease, cancer, and chronic lower respiratory diseases. Despite this decline in mortality in the USA, stroke remains the second leading cause of death worldwide, with over two-thirds of stroke deaths occurring in the developing world. Cerebrovascular disease has been variably classified, particularly in mortality data. The general classification of stroke in morbidity studies has been to subdivide into ischemic stroke, primary intracerebral hemorrhage, subarachnoid hemorrhage, and undefined.

Mortality Rates

International death rates from stroke have varied notably. Data from the WHO from 2002 revealed a 10-fold difference in mortality rates between the highest in Russia (251 per 100,000) and the lowest in Seychelles (24.5 per 100,000). Relatively high rates were seen in Eastern Europe, north Asia, central Africa, and the south Pacific. Looking at stroke mortality studies performed after 2000 in the European Union and the USA, France had the lowest age-standardized stroke mortality rates at 24–38 per 100,000. Italy and the UK had the highest age-standardized rates at 45–63 per 100,000.

In the USA, the annual age-adjusted death rates decreased dramatically during the past century, and the decline continues. From 1998 to 2008, the stroke death rate fell by 35%, with the actual number of stroke deaths declining by 19%. However, US racial and ethnic minorities have a larger stroke mortality burden than non-Hispanic whites. (See Chapter 47 for a general discussion of health disparities.) Stroke mortality among all racial and ethnic minorities compared with non-Hispanic whites is higher in the 35–64-year-old age group and decreases with age. For example, the equivalence point for stroke mortality with non-Hispanic whites is after age 64 years in Hispanics, American Indians (Native Americans)/Alaska Natives, and after age 85 in African Americans.

Early stroke case fatality (21–30 days) is decreasing in both high-income and low-income countries.[9] Between 2000 and 2008, stroke case fatality in high-income countries (USA and Western Europe) was 20%. Middle- and low-income countries (Brazil, Chile, French West Indies, Georgia, India, Nigeria, Mongolia, Sri Lanka, Russia, Ukraine) had a rate of 27%. When evaluating pathological subtypes of stroke, case fatality in ischemic stroke and primary intracerebral hemorrhage was similar for both high- and low- to middle-income countries but the case fatality for subarachnoid hemorrhage was 32% higher in low- to middle-income countries. Improved care and a trend towards more mild strokes are partly responsible for these changes.

Causes of death after stroke are most frequently related to the index stroke, recurrent stroke, or cardiovascular diseases. One of the few studies to evaluate long-term mortality trends after stroke was published by the Copenhagen Stroke Study investigators.[10] Overall 10 year mortality in this population-based cohort after stroke was 81%, which is similar to other recent studies. Stroke was the main cause of death in the first poststroke months, but just the third leading cause of mortality long term, after non-vascular disorders and cardiovascular diseases. Using a competing-risks analysis, stroke mortality was significantly associated with older age, male gender, and greater stroke severity regardless of the cause of death. Possible effects of acute thrombolytic therapy of stroke on long-term mortality were not evaluated in this study. One would expect a significant attenuation in stroke severity with thrombolytics, and therefore a reduced risk of death.

Morbidity Rates

Along with the temporal trends in stroke death rates, stroke incidence has decreased over the past several decades. Since the 1990s, however, the incidence rates have seemed to plateau or decrease only slightly in industrialized countries.[11] Stroke incidence increases logarithmically with increasing age but with a lesser increase beyond age 74. Average annual age-adjusted incidence rates by gender show a modest, but possibly increasing, male excess.

Looking at the past four decades worldwide, stroke incidence rates in high-income countries have decreased by 42%, from an average age-adjusted incidence of 163 per 100,000 person-years in 1970–1979 to 94 per 100,000 in 2000–2008. Significant reductions of over 40% were seen in all age groups during this period. In contrast, incidence rates in low- to middle-income countries more than doubled from 52 per 100,000 person-years in 1970–1979 to 117 in 2000–2008, with their current rates exceeding those in high-income countries. This corresponded to a two-fold increase in rates for individuals younger than 75 years and a four-fold increase for those older than 75 years.

Ischemic stroke is the most frequent subtype but its incidence rates have fallen by 11% over the past four decades in high-income countries. In 2000–2008, the pooled age-adjusted incidence rate for stroke was 70 per 100,000 in high-income and 67 per 100,000 in low- to middle-income countries. The rates for primary intracerebral hemorrhage and subarachnoid hemorrhage in high-income countries were twice those in low-income countries (primary intracerebral hemorrhage 22 *vs* 10 per 100,000 and subarachnoid hemorrhage 7 *vs* 4 per 100,000).

In the USA, nationwide stroke prevalence rates are monitored through the Behavioral Risk Factor Surveillance System.[12] For 2010, the most recent data available at the time of writing, the age-adjusted prevalence of stroke was 2.6%. This was slightly lower than the 2.7% prevalence rate in 2006 (p for trend = 0.05). Among racial groups in 2010, age-adjusted prevalence rates were highest among the categories of American Indians/Alaska Natives and lowest among Asians/Native Hawaiian/Other Pacific Islanders. For individual states, the 2010 age-adjusted prevalence rate ranged between 1.5% in Connecticut and 4.1% in Alabama. Regionally, the southeastern USA continue to have the highest rates (Fig. 46.2).

Whereas much of the research, prevention efforts, and clinical care for stroke occurs in high-income countries, more than 85% of strokes occur in low- to middle-income

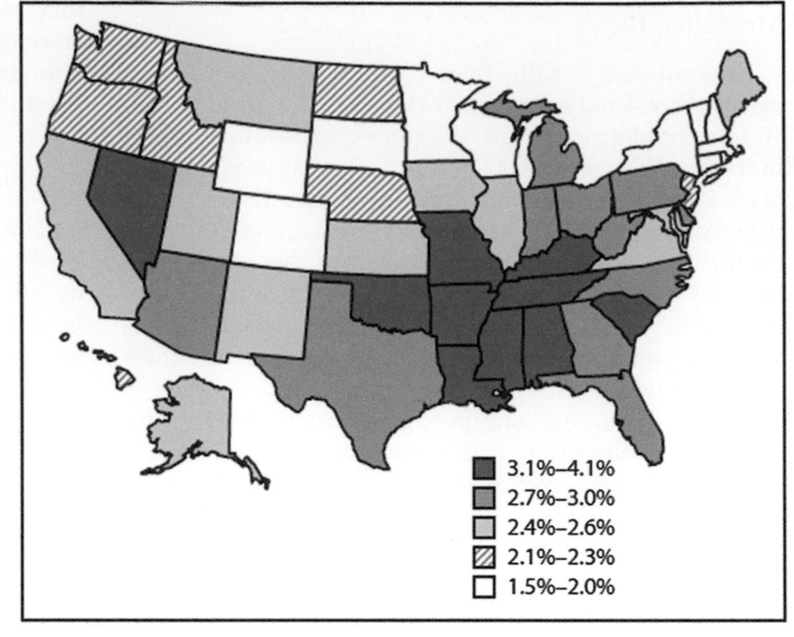

FIGURE 46.2 **Prevalence of stroke.** The figure shows the age-adjusted prevalence of stroke (adjusted to the 2000 US standard population; weighted estimates) among non-institutionalized adults aged 18 years and above, by state (Behavioral Risk Factor Surveillance System, USA, 2010). *Source: Centers for Disease Control and Prevention.* MMWR Morb Mortal Wkly Rep. *2012;61(20):379–382.*[12]

countries. Broad-based stroke prevention and treatment efforts in high-income countries have in part contributed to the fall in incidence and mortality rates for stroke. In a report using data from the WHO, the gap in stroke mortality and disability seen in low- and middle-income countries was surprisingly weakly or not significantly related to population-level risk factors.[13] Poverty, as measured by national income level, was the strongest predictor of stroke mortality and disability. More studies are needed to address this increasing burden of stroke in the developing world.

PRIMARY NEOPLASMS

Primary brain tumors make up less than 2% of all neoplasms in adults and nearly 20% of all malignancies during childhood. In clinical experience, approximately 85% of primary CNS tumors are intracranial and 15% are intraspinal. For the brain, the major groupings are the gliomas (40–50%, of which approximately half are glioblastoma multiforme) and the meningiomas (15–20%). Pituitary adenomas plus neurilemmomas, especially acoustic, comprise another 15–20%. The most common spinal cord tumors are neurofibroma and meningioma, followed by ependymoma and angioma.

Survival

Based on data from the population-based North American Association of Central Cancer Registries (NAACCR), the 5 year survival for all malignant tumors of neuroepithelial tissue (e.g. astrocytic and ependymal tumors) as well as most histological types and age groups has improved over time.[14] For example, 5 year survival for children with malignant neuroepithelial tumors increased from 63% for those diagnosed in 1970–1989 to 75% in 2000–2006. Less impressive changes were seen in older groups diagnosed with neuroepithelial tumors, with those aged 40–64 years having survival rates increasing from 16% to 27% over the same period. Individuals 65 years or older had survival rates under 5% for both 1970–1989 and 2000–2006.

Morbidity Rates

Average annual incidence rates for primary brain tumors in the more complete surveys have ranged mostly between 10 and 19 per 100,000 population overall. A large survey has reported age-standardized (2000 US standard population) incidence rates for adults and children with brain and other nervous system tumors as part of the NAACCR. For the period 2004–2007, benign brain tumors were approximately twice as common as malignant tumors among adults. For children (age 0–19 years), brain tumors were less common than in adults but more likely to be malignant. The overall annual incidence rate for brain and other nervous system tumors was 18.9 per 100,000 people between 2004 and 2007. The adult rate was 24.6, with malignant tumors making up 33.7% of the total. Neuroepithelial tissue tumors were the most common histological group of malignant brain tumors (7.8 per 100,000 population) in adults, with a 1.4 male to female excess. Glioblastoma was the most common malignant tumor, with whites having the highest rates (3.4 per 100,000 people), followed by Hispanics, blacks, American Indians/Alaska Natives and Asian/Pacific Islanders. In contrast, meningiomas made up the most common benign brain tumor (9.21 per 100,000 people), being twice as common among women as men. Within racial groups, black women had the highest incidence rates overall (9.7 per 100,000 people) and black men (4.8 per 100,000) had the highest incidence rate among men.

For children (0–19 years) in the NAACCR survey population, the age-standardized incidence rate for brain and other nervous system tumors was lower than for adults at 4.8 per 100,000 people, but the majority (65.2%) were malignant tumors. The most common childhood brain tumor by histological group was astrocytomas (1.6 per 100,000), the majority of which are histologically benign. Medulloblastoma, a primitive neuroectodermal tumor, was the most common malignant tumor among children. In contrast with adults, germ cell tumors and meningiomas were more likely to malignant in children.

From the late 1960s to the late 1980s, the incidence of primary brain neoplasms in the USA at least doubled. The Connecticut Tumor Registry showed a dramatic increase in incidence for age groups 65–84 years, with relatively stable rates for younger cohorts. For the 65–69-year-old age group, the incidence rate increased from 18.4 per 100,000 in 1965–1969 to 28.3 per 100,000 in 1985–1988. A slowing in brain tumor incidence rates has been documented in recent decades by the Surveillance, Epidemiology, and End Results registry. For malignant neuroepithelial tumors, the overall incidence increased at a rate of 1.9% per year from 1980 to 1987 and decreased at a rate of 0.4% between 1987 and 2007. In effect, this produced no net change between 1980 and 2007. Trends among specific histological groups differed markedly in this period, however. These changes are in part due to the more widespread use of neuroimaging, as well as modifications in coding and histological classification of brain tumors.

Demographic factors may alter the risk of developing a particular type of brain tumor. Genetic syndromes such as neurofibromatosis predispose to the development of brain tumors but they are rare and contribute less than a few percent to incident tumors. High-dose ionizing radiation is a well-established risk factor for brain tumor development, but no other significant environmental risk factors have been clearly established. More work is needed on environmental–genetic interactions and novel environmental exposures in CNS tumors.

EPILEPSY AND SEIZURE DISORDERS

Approximately 65 million people are estimated to have epilepsy worldwide, of whom 80% live in resource-poor countries. Epilepsy is defined as two or more unprovoked seizures occurring at least 24 hours apart (see Chapter 17). Classification of seizures or the epilepsies has varied considerably. The International League Against Epilepsy classification scheme for epidemiological studies divides seizures according to onset (generalized or focal) and predominant ictal features (motor or non-motor).[15] The causes of epilepsy have been divided into three broad categories: genetic/presumed genetic, structural/metabolic, and unknown. Prior epidemiological studies used the term idiopathic or cryptogenic to represent genetic or presumed genetic, and symptomatic to refer to structural or metabolic.

Morbidity Rates

A review of 33 high-quality epilepsy studies published in 2011 found a median incidence of 50.4 per 100,000 per year.[16] In high-income countries the incidence of epilepsy was 45.0 per 100,000 per year and in low-income countries it was 81.7. There was variability across studies in terms of methods, with lower estimates found in hospital-based studies and those with retrospective designs. Uniform standardization of data collection and larger population-based designs, especially in low-income countries, will help to produce more accurate estimates of the epilepsy burden over time.

The elderly population is a growing segment of many Western countries and epilepsy can produce significant physical and social morbidity. Faught and co-workers studied epilepsy among older US Medicare beneficiaries.[17] The average annual incidence rate (age ≥65 years) was 240 per 100,000 population and the average annual prevalence rate (age ≥65 years) was 1080 per 100,000. Epilepsy rates increased with age for all gender and racial groups. Compared with epilepsy incidence for whites (230 per 100,000), blacks had higher rates (410 per 100,000) while Asians (160 per 100,000) and Native Americans (110 per 100,000) had lower rates. These relatively high rates for epilepsy in the elderly population indicate a significant but unrecognized public health issue.

The lifelong incidence for developing epilepsy follows a J-shaped curve, with high rates for infants (100 per 100,000 population), decreasing in early adulthood, and remaining low (20 per 100,000 population) until age 60 when the rates increase dramatically up to 175 per 100,000 at 80 years. A 2011 study from Rochester, Minnesota, USA, evaluated the lifetime risk of developing epilepsy.[18] The traditional approach to understanding longitudinal risk for epilepsy has been cumulative incidence rates. This technique, however, tends to overestimate rates when competing risk for mortality becomes large, as is the case in elderly populations. The lifetime risk approach accounts for competing risk for death. Cumulative incidence was 1.7% to age 50 and 3.4% to age 80. Corresponding lifetime risk estimates were 1.6% to age 50 and 3.0% to age 80. The lifetime risk of epilepsy up to age 87 increased over time from 3.5% in the cohort with onset in 1960–1969 to 4.2% in the cohort with onset in 1970–1979.

Age-specific incidence rates vary by type of seizure (Fig. 46.3). Myoclonic seizures were the major type diagnosed during the first year of life; they were also the most common in the 1–4-year-old age group but rarely occurred after 5 years of age. Absence (petit mal) seizures peaked in the 1–4-year-old age group and did not begin in patients older than 20. Complex partial (psychomotor) and generalized tonic–clonic (grand mal) seizures both had fairly consistent incidence rates of 5–15 per 100,000 in patients aged 5–60 after low maxima at ages 1–4; for patients 70 years old and older, the rates of both were sharply higher.

Point prevalence rates for epilepsy are available from a number of community surveys. Rates from high-income countries range between 300 and 900 per 100,000 population. An overall estimate of the point prevalence of active epilepsy may be taken as approximately 600–700 per 100,000 population. Prevalence rates from medium- and low-income countries have tended to be more variable, with a range between 300 and 5700 per 100,000. Temporally, prevalence rates have tended to increase in recent years in several surveys. Males and blacks have higher rates than females and whites.

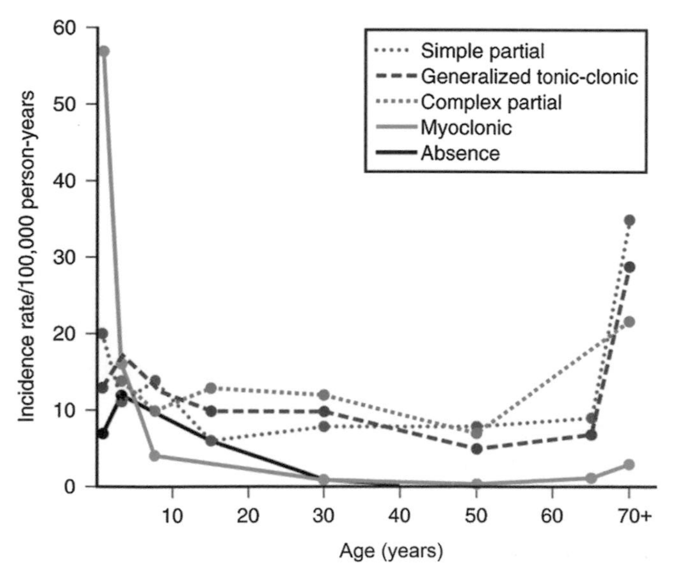

FIGURE 46.3 **Incidence rates of seizure.** Average annual age-specific incidence rates per 100,000 population are shown by clinical type of seizure: absence, myoclonic, generalized, simple, and complex partial. *Source: Reprinted from Kurtzke and Kurland. The epidemiology of nervous system disease. In:* Baker and Joynt's Clinical Neurology 2004. *Baltimore, MD: Lippincott Williams & Wilkins; 2004,*[19] *with permission.*

DEMENTIA

Dementia is an acquired decline in memory and other cognitive functions sufficient to affect daily life in an alert patient. An estimated 35.6 million people were diagnosed with dementia in 2012 worldwide and this figure is expected to almost double to 65.7 million by 2030. Alzheimer disease or dementia of the Alzheimer type is the most common cause of dementia, representing approximately 70% of cases (see Chapter 21). Vascular dementia is the next most common cause, accounting for 10–15% of cases. A small proportion of dementias is reversible, with causes including intoxications, metabolic derangements, infections, deficiency states, and cardiopulmonary disorders.

Morbidity Rates

In general, incidence rates for dementia have been relatively stable in the limited number of studies that have examined them over the past four decades. The Rochester, MN cohort had stable incidence rates between 1975 and 1984 but between 1985 and 1994 there was a 30% decline in rates. The most recent data on the incidence of dementia come from the Rotterdam Study, where a cohort of patients 60–90 years of age was followed from 2000 forward.[20] Age-adjusted dementia incidence rates were 4.9 per 1000 person-years overall and slightly higher in women (5.2 per 1000 person-years) than in men (4.5 per 1000 person-years). These rates were non-significantly lower than those from the 1990 cohort. Whether a true decline in the incidence of dementia is occurring in high-income countries over the past few decades will need to be confirmed by other studies.

Predicting the lifetime risk for dementia is difficult owing to the steep increase in incidence of dementia with increasing age and the competing risk of death due to other causes. Reported cumulative incidence rates have ranged between 21 and 49.6%. Cumulative incidence estimates for dementia overestimate the risk when the likelihood of an alternative cause of mortality is high, as it is with increasing age *per se*. As such, the Framingham cohort was used to estimate the lifetime risk of all-cause dementia.[21] At age 65, the remaining lifetime risk for developing dementia of the Alzheimer type was 6.3% for men and 12% for women. This doubling of risk for women is largely because women live longer and experience a longer period of risk.

For moderate and severe dementia, the estimated prevalence rates from a meta-analysis of studies published between 1945 and 1985 were 2800 per 100,000 for the 70–74-year-old group and 5600 for the 75–79-year-old age group. Rates for dementia doubled every 5.1 years of age. Furthermore, age-specific rates were not different between men and women.

Mild cognitive impairment (MCI) is considered a prodromal stage of dementia, and understanding its frequency and risk factors is critical in mechanistic research and prevention efforts. Roberts and colleagues studied the incidence of MCI in a prospective cohort of adults aged 70–89 years in Rochester, MN.[22] The age- and gender-standardized incidence rate of MCI was 63.6 per 1000 person-years overall. Men had higher rates than women and the rate of amnestic MCI (37.7 per 1000 person-years) was higher than non-amnestic MCI (14.7 per 1000 person-years). Men and those with fewer years of education were at higher risk for both amnestic and non-amnestic MCI. These differential rates by gender and subtype of MCI deserve more intensive study. More population-based prospective studies are required to fully understand the earliest changes that lead to dementia in a given demographic group.

PARKINSON DISEASE

Parkinsonism is a clinical syndrome characterized in part by tremor, bradykinesia, rigidity, and abnormal gait and posture (see Chapter 19). This syndrome has several causes, including drugs, toxins, and head trauma, but the most common variety is Parkinson disease (PD). In the great majority of cases the etiology of PD, the most common type of parkinsonism, is unknown but believed to be related to multiple environmental and genetic influences.

More than a dozen single gene mutations [e.g. *α-synuclein* (*SNCA*)-*PARK1*] have been identified to cause familial PD, but these Mendelian genes account for only 1–2% of cases in Western countries.[23] Studies of families affected by these mutations have produced important details on processes that lead to neuronal death such as mitochondrial dysfunction and aggregation of α-synuclein. Some of these changes can be initiated by environmental toxins, leaving room for both genetic and environmental effects in the pathogenesis of PD.

Demographic variables that have been shown to increase PD risk include advancing age, male gender, and white race. Pesticides, solvents, and heavy metals are environmental factors that have been associated with an increased risk of PD. Protective factors that are associated with a lower risk of PD include consumption of caffeine, smoking, and engaging in physical activity.

The diagnosis of PD is based on clinical criteria, as there are as yet no specific biomarkers. Criteria used include those of Gelb[24] and Calne,[25] which divide the diagnostic confidence into possible, probable, and definite PD. PD is the second most common neurodegenerative disorder after Alzheimer disease.

Mortality Rates

Patients with PD have approximately a two-fold increase in mortality compared with the general population.[26] Studies using a variety of designs have calculated mortality hazard rates between 1.3 and 5.7. Reporting of PD on death certificates is inconsistent. Pneumonia has been the most common cause of death compared to a control group. When evaluating age of onset of PD and its relationship to mortality, some studies have shown later onset of disease to be associated with higher mortality rates but others have shown the reverse to be true. It is unclear how newer medications and deep brain stimulation techniques for PD will affect survival in the coming decades.

Morbidity Rates

Few studies have evaluated the incidence of parkinsonism. For studies that have reported incidence rates for all age groups, the range is between 1.5 and 26 per 100,000 person-years. The estimated median standardized rate in high-income countries based on eight high-quality studies was 14 per 100,000 person-years. When adjusted to a standard population, the incidence rates for PD in North America, Western Europe, and Japan are similar. The incidence rates increase with increasing age. A recent study described the incidence of parkinsonism and its specific types among residents of Rochester, MN, between 1976 and 1990. The average annual incidence of parkinsonism was 26 per 100,000 population for all ages, or 139 per 100,000 person-years in the age group 65–99 years. The incidence increased steeply with age and PD was the most common type of parkinsonism (42%).

Community-based surveys during the past 20 years have reported prevalence rates of PD ranging from 31 per 100,000 in Benghazi, Libya, to 244 per 100,000 in Alberta, Canada. Studies that used door-to-door case ascertainment report higher rates than studies using disease registries. It is critical to evaluate the denominator for a given prevalence or incidence rate as rates for elderly people are dramatically higher than rates for all ages. For example, an overall prevalence rate of 950 per 100,000 person-years for people 65 years or older was calculated from 12 high-quality prevalence studies in the USA and Europe. This would translate into 349,000 cases of PD in the USA.

MULTIPLE SCLEROSIS

MS is an inflammatory neurodegenerative disorder of the CNS (see Chapter 30). It is the most common progressive neurological disease of young adults. In virtually all studies of MS a diagnosis is made clinically with the assistance of magnetic resonance imaging and other paraclinical tests.

Several schemes for diagnostic classification have been proposed. In almost all of these, there are several grades relating to the degree of confidence in the correctness of the label. By limiting attention to the classes considered the better ones and discard "possible MS" and "uncertain MS", the defined groups become quite similar in time and space. The 2010 McDonald criteria are the most current and widely accepted diagnostic classification system for MS.[27] Within these criteria, cases are labeled as "MS", "possible MS", and "not MS". In addition to the McDonald criteria, in its several versions, the major clinical criteria in current use for the diagnosis of MS in epidemiological surveys have been those of Poser and co-workers and Schumacher and co-workers.

The relative influences of genes and environment are the backdrops used to explain the population patterns of MS. The geographic distribution of MS has been the subject of many mortality and morbidity surveys as well as the topic of several symposia. Recent overviews of the epidemiology of MS should be consulted for interpretations that may differ, often drastically, from the views presented here.

While a complete discussion of environmental risk factors for MS is outside the scope of this chapter, some of the major risks identified in the current literature should be discussed. First, infection with the Epstein–Barr virus is associated with an increased risk for developing MS in a number of studies. High titers of the Epstein–Barr virus nuclear antigen have specifically been shown to be associated with a higher risk MS risk. This risk seems minimal for exposure before adolescence. Second, high levels of vitamin D along with outdoor sun exposure have been shown to be associated with a decreased risk of MS. Finally, cigarette smoking has been found to carry a twofold relative risk in the acquisition of MS. Of these three risk factors, smoking and vitamin D intake are relatively easy to modify. More work is required to understand how they may interact with human leukocyte antigen (HLA) genes and other environmental influences.

Prevalence Studies

Well over 500 prevalence surveys for MS have been conducted. Almost all of them have been performed since World War II and most are "spot surveys" of individual communities. To epitomize distributions, prevalence rates of 30 or more per 100,000 were considered high frequency, 5–29 per 100,000 medium frequency, and under 5 per 100,000 low-frequency MS regions. This trichotomy, made in the early 1960s, still provides a valid overview.

Prevalence rates for MS in Europe and the Mediterranean basin as of 1980 were correlated with geographic latitude.[28] This was the last major assessment before the diffusion of the rates in Europe. The distribution then comprised two clusters, one for high prevalence rates and one for medium rates. Taking only the best studies, the high prevalence zone extended from a latitude of 44° to 64° north (N). The medium zone extended from 32° to 47° N, plus two sites from the west coast of southern Norway. The only high rate below 44° N was that for a small survey of Enna, Sicily. However, by 1994 most of Western Europe from northern Scandinavia to Italy was of high frequency (Fig. 46.4). Portugal, Greece, and several countries of the Balkan peninsula (Albania and Turkey) are now also high, with prevalence rates of 30% and above.

A few reports have been published on the distribution of MS in Russia and countries of the former USSR. Much of north-western Russia past Moscow and Kiev (Ukraine) appears to be high prevalence, surrounded to the north, east, and south by medium-prevalence areas. Overall, Ukraine and the Caucasus seem to average in the medium range. In the far east of Russia, medium prevalence again appears, and rates are in the high range in the central and western parts of the Amur region, which abuts the Pacific Ocean north of China and includes Vladivostok. In all these areas, rates are higher for Russian-born people or those of Russian parentage than for the indigenous population.

Prevalence rates from North America as of 2004 were distributed in three risk zones: high frequency from 37° to 52° N, medium frequency from 30° to 33° N, and low frequency from 12° to 19° N and from 63° to 67° N. The prevalence rates for the USA and Canada were similar to the high-frequency rates of Western Europe. More recent works confirm all of the provinces of Canada and the entire coterminous USA as high-frequency regions, with prevalence rates mostly in the range of 60–120 per 100,000, and much less of a gradient in the latter than previous works indicated.

MS estimates for many countries of Latin America indicate that these are medium-frequency areas (5–29 per 100,000), but Colombia and Venezuela have low (<5 per 100,000) and Cuba high rates (≥30 per 100,000). The modest number of studies does not demonstrate any clear latitude gradient in Latin America.

In general, Australia and New Zealand comprise high prevalence areas for 44° to 34° south (S) latitude and medium frequency for 33° to 13° S. This high zone includes all of New Zealand and south-eastern Australia with Tasmania.

Earlier assessments provided low prevalence rates throughout Asia and Africa, except for English-speaking whites of South Africa where prevalence estimates were in the medium range. Rates are still low in Korea, China, and South-East Asia, but the rate for Japan is at least 10 per 100,000.

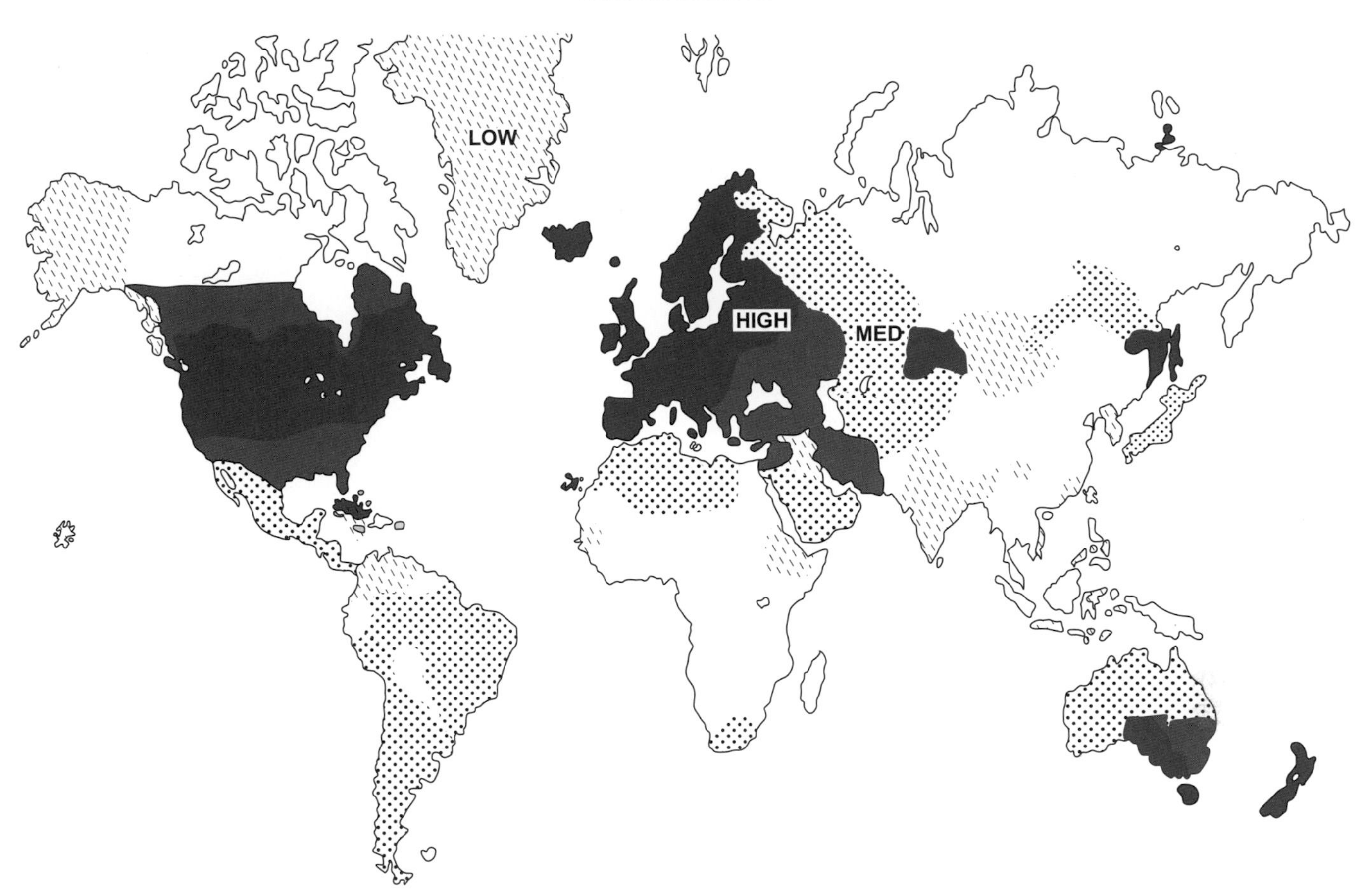

FIGURE 46.4 **Worldwide distribution of multiple sclerosis, 2011.** High (prevalence ≥30, solid), medium (prevalence 5–29, dotted), and low (prevalence 0–4, dashed) regions are defined. Blank areas are regions without data or people.

The northern African shores of the Mediterranean are of medium prevalence, and this extends into the Middle East, with Israel being of high frequency. Recent reports from Kuwait and Jordan also show MS prevalence rates in the high range.

Worldwide Distribution

The general worldwide distribution of MS thus still seems well described by a division into high (≥30 per 100,000), medium (5–29 per 100,000), and low prevalence (<5 per 100,000) regions, as proposed years ago. A "super high" class for prevalence of more than, say, 90 or 100 or so, does not yet seem to be indicated. The overall distribution in 2011 is shown in Fig. 46.4.

Incidence Studies

In North America and Western Europe, annual incidence rates for MS have generally been increasing over the past 50 years, with recent overall rates of 3–10 per 100,000. A review of MS incidence studies between 1966 and 2007 found an age- and gender-specific rate (World Population standard) of 3.6 per 100,000 person-years in women and 2.0 in men.[29] Overall rates at high latitudes were associated with higher MS incidence rates. Temporal trends showed a decreasing latitude gradient after 1980 due largely to greater increased incidence rates at lower latitudes. A steadily increasing female-to-male ratio of MS has been consistently observed across studies over the past 50 years.

Longitudinal trends in risk for MS in three separate cohorts of military veterans across 60 years are shown in Table 46.1 for the major gender and race groups. Risk was calculated as relative risk ratios for the World War II–Korean Conflict series and Vietnam and later series, as opposed to risk ratios from the incidence rates of the Gulf War series. Compared to white males as the reference, there has been a steady rise in risk for MS among black males, other race males, white females, and black females.[32] The rates for blacks exceed those for whites in the Gulf War era cohort.

MS had been predominantly a disease of whites, but incidence is increasing in minority groups once thought not to be susceptible to the disease, including Asians, Native Americans, and Arabs. However, it is clear that, where good data are available, the less susceptible racial groups do share the geographic

TABLE 46.1 Risk of Multiple Sclerosis Compared with White Males, by Race and Gender

	WWII and KC	Vietnam War and Later	Gulf War Era
Group	Adjusted Case–Control Ratio[b]	Adjusted Case–Control Ratio[c]	RR (95% CI)
White males	1.00[a]	1.00[a]	1.00[a]
Black males	0.44	0.67	1.16 (1.03–1.30)
Other males	0.22	0.30	0.77 (0.65–0.92)
White females	1.79	2.99	3.54 (3.20–3.91)
Black females	1.28	2.86	3.62 (3.18–4.11)
Other females	–	3.51	1.98 (1.52–2.58)

WWII: World War II.
[a]*Reference values.*
[b]*KC: Korean Conflict; RR: relative risk; CI: confidence interval.*
[c]*Wallin* et al. Ann Neurol. *2004;55:65–71.*[31]
Source: Data from [b]*Kurtzke* et al. Neurology. *1979;29:1228–1235.*[30]

gradients of the whites, with higher frequencies in high-risk areas than in low.

Mortality

Survival is a fundamental measure of disease severity. Ideally, it is best measured prospectively from disease onset within a defined population. The study of a male World War II army hospital cohort showed a 25 year survival after onset of 69%. The survival curves were similar to those observed in studies from the Mayo Clinic (25 year survival 74%) and Lower Saxony (25 year survival 63%). US veteran cohort studies and other incident studies show median survival by gender and race ranging from 30 to 43 years.

Regarding risk factors for survival, all studies but one showed that later age of onset and male gender were associated with shorter survival. Other risk factors that were in agreement across studies and predicted worse survival were early high initial disability scores and progressive disease course.

Genetics

Family studies in MS have provided a means of assessing environmental factors against a set genetic background, the most convincing of which have been those published by George Ebers and his group based on long-term studies of nationwide MS centers in Canada.[33] Such studies have shown that the risk of MS is 3–4% for primary relatives and 20–30% for monozygotic twins. This is in contrast to the general population prevalence of approximately 0.1%. The increased family frequency may be related to shared environment rather than shared genetic factors because close relatives would be expected to share similar environmental influences. However, the following evidence indicates that MS is under some genetic control:

- Most twin studies show an excess of concordant monozygous twins. The difference in concordance rates between monozygotic and dizygotic twins is primarily attributable to genetic factors.
- Higher rates are found in children than in adoptees of MS patients.
- There is an association between HLA alleles and MS and the higher frequency of HLA sharing in affected sibling pairs.
- Some population groups in high-frequency areas (Native Americans in North America, Lapps in Scandinavia, and Gypsies in Hungary) are relatively resistant to MS.

Support for the fourth observation is limited; many of these groups have not been systematically studied and their true risk for MS is not clear. Moreover, these groups not only differ genetically but also have substantially different lifestyles and environmental exposures from the majority populations in the countries in which they live.

While the HLA DRB1 loci carry the largest contribution to the genetic risk for MS, genome-wide association studies have identified more than 50 non-HLA MS susceptibility variants. These susceptibility loci, however, contribute minimally to the overall genetic risk, with odds ratios for MS between 1.1 and 1.3. Investigations into the interactions between genes, and between genetic and environmental risk factors, should provide more answers on genetic susceptibility in MS. For example, the interplay between major histocompatibility complex alleles via epistasis within racial groups could be a factor, but this has not yet been extensively studied.

Migration

The fate of migrants who move into regions of differing risk for MS is critical to the understanding of the nature of the disease. If migrants retain the risk of their birthplace, then either the disease is innate or it is acquired very early in life. However, if upon moving, their risk changes, then clearly there is a major environmental cause or precipitant active in this disorder. If such altered risk is also dependent on age at migration, then not only external cause but also internal (personal) susceptibility can be defined.

There is considerable evidence that migrants' risk for MS does change. In migrants who move from a country of origin in which MS is common to a country in which it is less common, there is overall a decrease in the rate of disease, but only for those under age 15 or so at migration; the few studies of low to high migration show the reverse effect, with an increase up to about age 45, as exemplified below.

TABLE 46.2 Risk of Multiple Sclerosis expressed as Case Control Ratios of World War II or Korean Conflict Veterans, by Tier of Residence at Birth and at Entry into Active Duty (EAD)[a]

	EAD Tier			
Birth Tier	North	Middle	South	Birth Total
North	1.48	1.27	0.74	1.44
Middle	1.40	1.03	0.73	1.04
South	0.70	0.65	0.56	0.57
EAD total	1.46	1.03	0.58	1.06

[a]*Coterminous USA only.*
Source: Kurtzke et al. Neurology. *1985;35:672–678.*[35]

For the veteran series, three geographic tiers for state of residence at entry into military service are described: a northern tier of states above 41° and 42° N latitude; a middle tier; and a southern tier below 37° N, including California from Fresno south. Migrants are those born in one tier who entered service from another tier.[34] In Table 46.2, the marginal totals provide the ratios for birthplace and residence at service entry for whites of World War II or Korean service. The major diagonal (north–north, middle–middle, and south–south) gives the case–control ratios for non-migrants, and cells off this diagonal define the ratios for the migrants.

All ratios decrease as one goes from north to south. The non-migrant ratios are 1.48 north, 1.03 middle, and 0.56 south. For the migrants, those born in the north and entering service from the northern tier have a ratio of 1.48; if they enter from the south, their ratio is 0.74 – half that of the non-migrants. Birth in the middle tier is marked by an increase in the MS to control ratio for northern entrants to 1.40 and a decrease to 0.73 for the southern ones. Migration after birth in the south increases the ratios to 0.65 (middle) and 0.70 (north), even though the series is truncated by the young age at service entry (mean of 29 years). The migrant risk ratios are intermediate between those characteristic of their birthplace and their residence at service entry compatible with an age at acquisition of some 10–15 years.

In a study of European immigrants to South Africa, the MS prevalence rate, adjusted to a population of all ages, was 13 per 100,000 for immigration younger than age 15, which is the same medium prevalence rate as that for the native-born English-speaking white South Africans.[36] However, for age groups older at immigration, the prevalence was 30–80 per 100,000, the same range as expected from their high-risk homelands. This change was major and occurred exactly at age 15. This indicates that natives of high-risk areas are not very susceptible to MS acquisition before age 15, supporting the findings of the veteran series, and that there is a long interval between acquisition and onset of symptoms.

Inferences regarding the opposite migration, low to high, were afforded by north African migrants to France.[37] Among approximately 7500 respondents with known place of birth who had completed a nationwide questionnaire survey for MS in France in 1986, 260 were born in former French North Africa (Morocco, Algeria, and Tunisia). They had migrated to France between 1923 and 1986, but 66% came between 1956 and 1964. Two-thirds were from Algeria, where virtually the entire European population had emigrated in 1962 at the end of the Algerian war for independence. The 225 migrants with onset more than 1 year after immigration presumably acquired MS in France. They had an age-adjusted (USA 1960) MS prevalence rate 1.54 higher than that for all France. If the latter is taken at 50 per 100,000 population, their estimated adjusted rate is 76.8. The other 27 with presumed acquisition in North Africa had an estimated adjusted prevalence of 16.6 per 100,000. For those migrants with acquisition in France, there was a mean interval of 13 years between immigration or age 11 and clinical onset, with a minimum of 3 years' duration, and with a maximal age of susceptibility near 45 years.

Epidemics

In the past, there has been little reason to consider that MS occurred in the form of epidemics. All known geographic areas that had been surveyed at repeated intervals until 1980 provided either stable or increasing prevalence rates, the latter compatible with better case ascertainment, increasing incidence rates, and perhaps improved survival. Epidemics of MS would serve to define the disease as not only an acquired one but also perhaps a transmittable one. Separate epidemics of MS have been encountered, which may have common precipitants, and which have occurred in the ethnically similar lands of Iceland, Shetland–Orkney, and the Faroe Islands. The experience in the Faroes is the best studied.[38]

The Faroe Islands are a semi-independent unit of the Kingdom of Denmark located in the North Atlantic Ocean between Iceland and Norway. As of 1999, 54 native resident Faroese had onset of MS in the twentieth century. There were none before 1943, but between 1943 and 1949, 17 patients had symptom onset in a populace of fewer than 30,000. All had been at least 11 years old in 1941 and thus had 2 years' "exposure" before first onset. Another four were also at least age 11 then; these 21 cases comprise a type 1 epidemic of clinical MS.

Recall that an epidemic may be defined as disease occurrence clearly in excess of normal expectancy and derived from a common or propagated source. Epidemics are divisible into two types: type 1 epidemics occur in susceptible populations, exposed for the first time to a virulent infectious agent, whereas type 2 epidemics occur in populations within which the virulent organism is already established. If the entire populace is exposed to a type 1 epidemic, the ages of those affected clinically will define the age range of susceptibility to the infection. Type 2 epidemics will tend to have a young age at onset because the effective exposure of the patients will be greatest for those reaching the age of susceptibility.

Since Faroese migrant MS patients required 2 years' exposure in a high-risk area from age 11 to acquire MS, the same criteria were applied to the resident Faroese (the 2 years before 1943 being 1941 and 1942). Epidemic I was followed by three later type 2 epidemics with peaks at 13 year intervals (Fig. 46.5), comprising 10, 10, and 13 cases, respectively.

It was concluded that the disease was introduced to the Faroes by British troops who had occupied the islands for 5 years beginning in April 1940. The authors believe that an infection was introduced that was transmitted during the war to the Faroese population at risk, of which the epidemic I cases of clinical MS were a part. This infection, the primary multiple sclerosis affection (PMSA), was defined as a single, specific, widespread, systemic but unknown infectious disease (that may be totally asymptomatic). PMSA produces clinical neurological multiple sclerosis (CNMS) in only a small proportion of those affected after an incubation period averaging 6 years in virgin populations and perhaps 12 years in endemic areas. Using this hypothesis, transmissibility is limited to part or all of the systemic phase, which ends by the usual age of onset of MS symptoms. The PMSA cases from the first cohort of Faroese transmitted the disease to the next Faroese population cohort, those who reached age 11 in the period in which the first cohort was transmissible. Included in the second Faroese cohort were the epidemic II cases of CNMS, and this cohort similarly transmitted PMSA to the third population cohort with its own (epidemic III) cases, and from there to the fourth cohort with epidemic IV.

Two years of exposure beginning between ages 11 and 45 were required to acquire PMSA in such a virgin but susceptible population. PMSA may be a specific, but unknown, age-limited infection that can be acquired only during these hormonally active years and that only rarely leads to clinical MS. There is evidence that PMSA might be an enteric disease transmitted by the fecal–oral route.

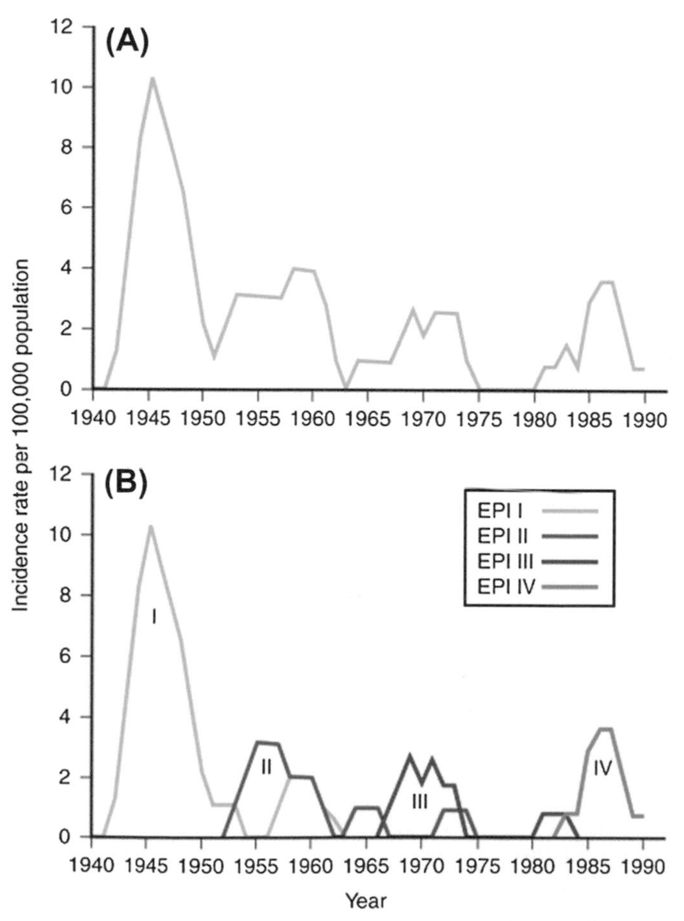

FIGURE 46.5 **Annual incidence rates per 100,000 population for clinical multiple sclerosis in native resident Faroese, 1900–1998, calculated as 3 year centered moving averages.** (A) Total series; (B) rates for each of four epidemics (EPI I–IV). *Source: Reproduced from Kurtzke and Heltberg.* J Clin Epidemiol. *2001;54(1):1–22,*[38] *with permission.*

Summary

In MS there is a female preponderance, which is increasing. There has also been a clear predilection for whites, but the burden for other racial groups is increasing. Moreover, minorities share the geographic distributions of whites but at lower levels. And most recently, in the USA, the rates for blacks have exceeded those for whites. Thus, location, gender, and ethnicity are all independent risk factors for this disease.

The etiological debate in MS has historically been divided between the contributions of genetics and the environment. The distinction between these opposing views has become blurred because many in the field hold that MS is likely to be related to both nature and nurture. The present authors agree, but argue that the environment is the overwhelming driving force in the etiology of MS. First, there is good evidence for the existence of geographic variations in the USA, Australia, and Europe. Diffusion of high-risk areas over one or two generations has also been observed, at a pace too rapid for genetic influences to be effective, and which does not support latitude or sunshine (and thus vitamin D) as being of etiological importance. Second, studies of migration and the epidemics (Faroes, Iceland, Shetland–Orkney) lend further support to an environmental hypothesis.

OVERVIEW OF NEUROLOGICAL DISORDERS

Morbidity and mortality rates for neurological disorders provide data regarding their numerical impact on the community. Tables 46.3 and 46.4 represent the best

TABLE 46.3 Approximate Average Annual Incidence Rates (per 100,000 population) of Neurological Disorders, All Ages

Disorder	Rate
Herpes zoster	400
Migraine	250
Brain trauma	200
Other severe headache	200[a]
Acute cerebrovascular disease	150
Other head injury	150[a]
Transient postconcussive syndrome	150
Lumbosacral herniated nucleus pulposus	150
Lumbosacral pain syndrome	150[a]
Dementia	150
Neurological symptoms (with no defined disease)	75
Epilepsy	50
Febrile fits	50
Ménière disease	50
Mononeuropathies	40
Polyneuropathy	40
Transient ischemic attacks	30
Bell palsy	25
Single seizures	20
Parkinsonism	20
Cervical pain syndrome	20[a]
Persistent postconcussive syndrome	20
Alcoholism	20[a]
Meningitides	15
Encephalitides	15
Sleep disorders[b]	15
Subarachnoid hemorrhage	15
Cervical herniated nucleus pulposus	15
Metastatic brain tumor	15
Peripheral nerve trauma	15
Blindness	15
Benign brain tumor	10
Deafness	10[a]
Cerebral palsy	9.0
Congenital malformations of CNS	7.0
Mental retardation, severe	6.0
Mental retardation, other	6.0[a]
Malignant primary brain tumor	5.0
Metastatic cord tumor	5.0
Tic douloureux	4.0
Multiple sclerosis	3.0[c]
Optic neuritis	3.0[c]
Dorsolateral sclerosis	3.0
Functional psychosis	3.0[a]
Spinal cord injury	3.0
Motor neuron disease	2.0
Down syndrome	2.0
Guillain–Barré syndrome	2.0
Intracranial abscess	1.0
Benign cord tumor	1.0
Cranial nerve trauma	1.0
Acute transverse myelopathy	0.8
All muscular dystrophies	0.7
Chronic progressive myelopathy	0.5
Polymyositis	0.5
Syringomyelia	0.4
Hereditary ataxias	0.4
Huntington disease	0.4
Myasthenia gravis	0.4
Acute disseminated encephalomyelitis	0.2
Charcot–Marie–Tooth disease	0.2
Spinal muscular atrophy	0.2
Familial spastic paraplegia	0.1
Wilson disease	0.1
Malignant primary cord tumor	0.1
Vascular disease cord	0.1

[a]*Rate for those who should be seen by a physician competent in neurology (10% of total).*
[b]*Narcolepsies and hypersomnias (with sleep apnea).*
[c]*Rate for high-risk areas.*
Sources: Modified and updated from Kurtzke. Neurology. *1982;32:1207–1214,*[39] *and Kurtzke and Wallin. The epidemiology of neurologic diseases. In: Joynt and Griggs.* Baker's Clinical Neurology on CD ROM. *Hagerstown, MD: Lippincott Williams and Wilkins; 2000:Chapter 66B.*[40]

estimates for incidence and prevalence available. All rates have been rounded and, unless otherwise specified, refer to the population of all ages, even though the disorder favors one gender or age group. Furthermore, in view of the limited sources of data, the information is largely applicable to sites of economically developed countries, predominantly those of Europe and North

TABLE 46.4 Approximate Average Annual Point Prevalence Rates (per 100,000 population) of Neurological Disorders, All Ages

Disorder	Rate
Migraine	2000[a]
Other severe headache	1500[a]
Brain trauma	800
Epilepsy	650
Acute cerebrovascular disease	600
Lumbosacral pain syndrome	500[a]
Alcoholism	500[a]
Dementia	500
Sleep disorders[b]	300
Ménière disease	300
Lumbosacral herniated nucleus pulposus	300
Cerebral palsy	250
Parkinsonism	200
Transient ischemic attacks	150
Febrile fits	100
Persistent postconcussive syndrome	80
Herpes zoster	80
Congenital malformations of CNS	70
Single seizures	60
Multiple sclerosis	60[c]
Benign brain tumor	60
Cervical pain syndrome	60[a]
Down syndrome	50
Subarachnoid hemorrhage	50
Cervical herniated nucleus pulposus	50
Transient postconcussive syndrome	50
Spinal cord injury	50
Tic douloureux	40
Neurological symptoms without defined disease	40
Mononeuropathies	40
Polyneuropathies	40
Dorsolateral sclerosis	30
Peripheral nerve trauma	30
Other head injury	30
Acute transverse myelopathy	15
Metastatic brain tumor	15
Chronic progressive myelopathy	10
Optic neuritis	10
Encephalitides	10
Vascular disease spinal cord	9
Hereditary ataxias	8
Syringomyelia	7
Motor neuron disease	6
Polymyositis	6
Progressive muscular dystrophy	6
Malignant primary brain tumor	5
Metastatic cord tumor	5
Meningitides	5
Bell palsy	5
Huntington disease	5
Charcot–Marie–Tooth disease	5
Myasthenia gravis	4
Familial spastic paraplegia	3
Intracranial abscess	2
Cranial nerve trauma	2
Myotonic dystrophy	2
Spinal muscular atrophy	2
Guillain–Barré syndrome	1
Wilson disease	1
Acute disseminated encephalomyelitis	0.6
Dystonia musculorum deformans	0.3

[a]*Rate for those who should be seen by a physician competent in neurology (20% of migraine, 10% of all others).*
[b]*Narcolepsies and hypersomnias (with sleep apnea).*
[c]*Rate for high-risk areas.*
Sources: Modified and updated from Kurtzke. Neurology. *1982;32:1207–1214,*[39] *and Kurtzke and Wallin. The epidemiology of neurologic diseases. In: Joynt and Griggs.* Baker's Clinical Neurology on CD ROM. *Hagerstown, MD: Lippincott Williams and Wilkins; 2000:Chapter 66B.*[40]

America. Therefore, these rates should be regarded as a general guide only; they may vary considerably in different populations. The rates are cited as cases per 100,000 population. For cases in a given community or country, these rates are multiplied by the appropriate population factor. For example, in a population of 200 million with a rate of 10 per 100,000, there would be 20,000 cases. Regarding morbidity, for the 63 disorders in Table 46.3, the combined average annual incidence rates equal more than 2600 per 100,000 population or 2.6%.

These disorders includes eight for which only one-tenth of the incident cases were thought to require attention by a physician competent in clinical neurology: the two vertebrogenic pain syndromes, non-migrainous headache, non-brain head injury, alcoholism, psychosis, non-severe mental retardation, and deafness. Total blindness data were taken as an estimate for the proportion of all visually impaired people who should see a neurologist. Overall incidence rates are easily calculated, but the aim here is to provide data likely to be relevant to readers. Even if all headaches, trauma, vertebrogenic pain, vision loss, deafness, and psychosis were excluded, there would be more than 1200 new cases of neurological disease each year in every 100,000 of the population – well over one case for every 100 people.

For the 61 disorders listed in Table 46.4, the point prevalence rates in like manner contain only 10% of the non-migrainous headache, vertebrogenic pain, alcoholism, and non-brain head injury, and 20% of migraine, but they exclude completely all mental retardation, psychosis, deafness, and blindness. The total exceeds 9700 per 100,000 population. Again, if all the disorders previously mentioned were excluded, there would still be a prevalence rate of approximately 3800 per 100,000 population, or 3.8% of the people who at any one time require the care of a physician competent in clinical neurology. In a US population of 308.8 million (US Census Bureau, 2010 Census), this means that 11.7 million people are affected.

Not all neurological disorders are referred to neurologists, even in the most socioeconomically advanced countries; in many countries, the neurologist is primarily a consultant rather than the physician who manages these disorders, and in others there is a mixture. There are also regions, not limited to low-income countries, in which there is a lack of neurologists and other specialists. However, regardless of the type of practice a given country deems appropriate for a neurologist, the patients will still exist. Therefore, the data in Tables 3 and 4 could serve as a basis for at least a rational allocation of available resources in any country for the teaching, research, and patient care of neurological disorders.

CONCLUSION AND FUTURE DIRECTIONS

Disease does not occur randomly, but rather in patterns that reflect the underlying causes or major risk factors of a given condition. For example, stroke, dementia, and PD occur primarily in older individuals, whereas MS and primary headache disorders have their onset at earlier ages. Although the reasons for these age-dependent differences are largely unknown, some possibilities include immunological responses that are initially aggressive in early adulthood and wane with time, altered vascular responses to dietary and environmental stimuli, and a breakdown in the clearance of misfolded proteins. Stroke has emerged as a common disorder across the developed and developing worlds as the lifespan increases and dietary and inactivity patterns become similar to those in the USA. MS has been a geographically spreading disease with its onset in northern Europe, in contrast to primary headache disorders, which appear to be rather homogeneous in their distribution.

Differences in neurological disorders can in some cases be related to the regional gene pool, dietary patterns, or environmental exposures, including infections. In many cases little is known about these issues. This calls for more collaborative work to be done between epidemiologists and basic scientists to uncover how environmental influences may be driving the changing patterns seen in disease frequency. Population patterns should inform the laboratory scientist and focus the directions of study. Such understanding will be helpful in the treatment and prevention of neurological disorders.

References

1. Hoppitt T, Pall H, Calvert M, et al. A systematic review of the incidence and prevalence of long-term neurological conditions in the UK. *Neuroepidemiology*. 2011;36(1):19–28.
2. Sajjad A, Chowdhury R, Felix JF. A systematic evaluation of stroke surveillance studies in low- and middle-income countries. *Neurology*. 2013;80(7):677–684.
3. Wallin MT, Kurtzke JF. Neuroepidemiology. In: Bradley WG, Daroff RB, Fenichel GM, Jankovic J, eds. *Neurology in Clinical Practice*. 6th ed. Philadelphia, PA: Butterworth Heinemann Elsevier; 2012:687–703.
4. Kurtzke JF. Multiple sclerosis from an epidemiological viewpoint. In: Field EJ, ed. *Multiple Sclerosis: A Critical Conspectus*. Lancaster: Medical and Technical Publishing Press; 1977:83–142.
5. Steindel SJ. International classification of diseases, 10th edition, clinical modification and procedure coding system: descriptive overview of the next generation HIPAA code sets. *J Am Med Inform Assoc*. 2010;17(3):274–282.
6. van Drimmelen-Krabbe JJ, Bradley WG, Orgogozo JM, Sartorius N. The application of the International Statistical Classification of Diseases to neurology: ICD-10 NA. *J Neurol Sci*. 1998;161(1):2–9.
7. Stroke – Recommendations on stroke prevention, diagnosis, and therapy. Report of the WHO Task Force on Stroke and other Cerebrovascular Disorders. *Stroke*. 1989;20(10):1407–1431.
8. Roger VL, Go AS, Lloyd-Jones DM, et al. Heart disease and stroke statistics – 2012 update: a report from the American Heart Association. *Circulation*. 2012;125(1):e2–e220.
9. Feigin VL, Lawes CM, Bennett DA, Barker-Collo SL, Parag V. Worldwide stroke incidence and early case fatality reported in 56 population-based studies: a systematic review. *Lancet Neurol*. 2009;8(4):355–369.
10. Mogensen UB, Olsen TS, Andersen KK, Gerds TA. Cause-specific mortality after stroke: relation to age, sex, stroke severity, and risk factors in a 10-year follow-up study. *J Stroke Cerebrovasc Dis*. 2013;22(7):e59–e65.
11. Kleindorfer DO, Khoury J, Moomaw CJ, et al. Stroke incidence is decreasing in whites but not in blacks: a population-based estimate of temporal trends in stroke incidence from the Greater Cincinnati/Northern Kentucky Stroke Study. *Stroke*. 2010;41(7):1326–1331.
12. Centers for Disease Control and Prevention. Prevalence of stroke – United States, 2006–2010. *MMWR Morb Mortal Wkly Rep*. 2012;61(20):379–382.

13. Johnston SC, Mendis S, Mathers CD. Global variation in stroke burden and mortality: estimates from monitoring, surveillance, and modelling. *Lancet Neurol*. 2009;8(4):345–354.
14. Kohler BA, Ward E, McCarthy BJ, et al. Annual report to the nation on the status of cancer, 1975–2007, featuring tumors of the brain and other nervous system. *J Natl Cancer Inst*. 2011;103(9):714–736.
15. Thurman DJ, Beghi E, Begley CE, et al. Standards for epidemiologic studies and surveillance of epilepsy. *Epilepsia*. 2011;52(suppl 7):2–26.
16. Ngugi AK, Kariuki SM, Bottomley C, Kleinschmidt I, Sander JW, Newton CR. Incidence of epilepsy: a systematic review and meta-analysis. *Neurology*. 2011;77(10):1005–1012.
17. Faught E, Richman J, Martin R, et al. Incidence and prevalence of epilepsy among older US Medicare beneficiaries. *Neurology*. 2012;78(7):448–453.
18. Hesdorffer DC, Logroscino G, Benn EK, Katri N, Cascino G, Hauser WA. Estimating risk for developing epilepsy: a population-based study in Rochester, Minnesota. *Neurology*. 2011;76(1):23–27.
19. Kurtzke JF, Kurland LT. The epidemiology of nervous system disease. In: *Baker and Joynt's Clinical Neurology 2004 [CD-ROM]*. Baltimore, MD: Lippincott Williams & Wilkins; 2004.
20. Schrijvers EM, Verhaaren BF, Koudstaal PJ, Hofman A, Ikram MA, Breteler MM. Is dementia incidence declining? Trends in dementia incidence since 1990 in the Rotterdam Study. *Neurology*. 2012;78(19):1456–1463.
21. Seshadri S, Wolf PA, Beiser A, et al. Lifetime risk of dementia and Alzheimer's disease. The impact of mortality on risk estimates in the Framingham Study. *Neurology*. 1997;49(6):1498–1504.
22. Roberts RO, Geda YE, Knopman DS, et al. The incidence of MCI differs by subtype and is higher in men: the Mayo Clinic Study of Aging. *Neurology*. 2012;78(5):342–351.
23. Kieburtz K, Wunderle KB. Parkinson's disease: evidence for environmental risk factors. *Mov Disord*. 2013;28(1):8–13.
24. Gelb DJ, Oliver E, Gilman S. Diagnostic criteria for Parkinson disease. *Arch Neurol*. 1999;56(1):33–39.
25. Calne DB, Snow BJ, Lee C. Criteria for diagnosing Parkinson's disease. *Ann Neurol*. 1992;32(suppl):S125–S127.
26. Wirdefeldt K, Adami HO, Cole P, Trichopoulos D, Mandel J. Epidemiology and etiology of Parkinson's disease: a review of the evidence. *Eur J Epidemiol*. 2011;26(suppl 1):S1–S58.
27. Polman CH, Reingold SC, Banwell B, et al. Diagnostic criteria for multiple sclerosis: 2010 revisions to the McDonald criteria. *Ann Neurol*. 2011;69(2):292–302.
28. Kurtzke JF. Geographic distribution of multiple sclerosis: an update with special reference to Europe and the Mediterranean region. *Acta Neurol Scand*. 1980;62(2):65–80.
29. Alonso A, Hernan MA. Temporal trends in the incidence of multiple sclerosis: a systematic review. *Neurology*. 2008;71(2): 129–135.
30. Kurtzke JF, Beebe GW, Norman Jr JE. Epidemiology of multiple sclerosis in United States veterans: 1. Race, sex, and geographic distribution. *Neurology*. 1979;29:1228–1235.
31. Wallin MT, Page WF, Kurtzke JF. Multiple sclerosis in United States veterans of Vietnam era and later military service. 1. Race, sex and geography. *Ann Neurol*. 2004;55:65–71.
32. Wallin MT, Culpepper WJ, Coffman P, et al. The Gulf War era multiple sclerosis cohort: age and incidence rates by race, sex and service. *Brain*. 2012;135(Pt 6):1778–1785.
33. Giovannoni G, Ebers G. Multiple sclerosis: the environment and causation. *Curr Opin Neurol*. 2007;20(3):261–268.
34. Kurtzke JF, Beebe GW, Norman Jr JE. Epidemiology of multiple sclerosis in US veterans: III. Migration and the risk of MS. *Neurology*. 1985;35(5):672–678.
35. Kurtzke JF, Beebe GW, Norman Jr JE. Epidemiology of multiple sclerosis in United States veterans: III. Migration and the risk of MS. *Neurology*. 1985;35:672–678.
36. Dean G, Kurtzke JF. On the risk of multiple sclerosis according to age at immigration to South Africa. *BMJ*. 1971;3(5777): 725–729.
37. Kurtzke JF, Delasnerie-Laupretre N, Wallin MT. Multiple sclerosis in North African migrants to France. *Acta Neurol Scand*. 1998;98(5):302–309.
38. Kurtzke JF, Heltberg A. Multiple sclerosis in the Faroe Islands: an epitome. *J Clin Epidemiol*. 2001;54(1):1–22.
39. Kurtzke JF. The current neurologic burden of illness and injury in the United States. *Neurology*. 1982;32:1207–1214.
40. Kurtzke JF, Wallin MT. The epidemiology of neurologic diseases. In: Joynt RJ, Griggs RC, eds. *Baker's Clinical Neurology on CD ROM*. Hagerstown, MD: Lippincott Williams and Wilkins; 2000. Chapter 66B.

CHAPTER

47

Stress, Health, and Disparities

Zinzi D. Bailey, David R. Williams

Department of Social and Behavioral Sciences, Harvard School of Public Health, Boston, Massachusetts, USA

OUTLINE

RACIAL DISPARITIES IN HEALTH

National data reveal that African Americans (or blacks) have higher death rates than whites for 10 of the 15 leading causes of death in the USA.[1] Hispanics (or Latinos) and Native Americans (American Indians) have higher death rates than whites for diabetes, liver cirrhosis, and homicide. Native Americans also have elevated mortality rates for accidents and hypertension. Subgroups of Native Americans such as those served by the Indian Health Service (some 60% of that population) have worse health status than Native Americans nationally.[2] Blacks and Native Americans also have higher age-specific mortality rates than whites.[1] This pattern is evident in early childhood and persists until advanced age. In contrast, Hispanics have rates that are roughly equivalent at the youngest ages but at older ages have rates lower than those of whites. Asians, a diverse group with an even higher proportion of immigrants than Hispanics, have mortality rates that are markedly lower than those of whites throughout the life course. Data quality problems with both the numerator and denominator used to calculate death rates lead to mortality statistics being more accurate for blacks and whites than for the other racial populations.[3] Nonetheless, it seems clear that, on average, race and ethnicity play a critical role in determining mortality.

The health profile of Asians and Hispanics is importantly affected by the high proportion of immigrants within these populations. Immigrants to the USA of all racial groups tend to have better health, as captured by infant and adult mortality rates, than their native-born peers, but their health declines with increasing length of stay in the USA and generational status.[4] For example, a study of adults in Chicago, Illinois, found that the health of Hispanics was intermediate between that of blacks and whites.[5] However, when Hispanics were disaggregated by place of birth, the health of foreign-born Hispanics

Neurobiology of Brain Disorders
http://dx.doi.org/10.1016/B978-0-12-398270-4.00047-1

was similar to whites while the health of US-born Hispanics was similar to African Americans. National data from the USA have also documented variations by nativity status in "allostatic load",[6] a term coined by Bruce McEwen and Eliott Stellar to refer to "the cost of chronic exposure to fluctuating or heightened neural or neuroendocrine response resulting from repeated or chronic environmental challenge that an individual reacts to as being particularly stressful"[7] (see Chapter 34, "Stress"). In the Kaestner study, allostatic load was used as a global measure of biological dysregulation that summed 10 indicators of clinical and subclinical status: systolic and diastolic blood pressure, body mass index, glycated hemoglobin, albumin, creatinine clearance, triglycerides, C-reactive protein, homocysteine, and total cholesterol.[6] Mexican adults who had recently immigrated to the USA had a health profile similar to that of whites, while US-born Mexican Americans and Mexican immigrants with more than 20 years of residence in the USA had an allostatic load profile similar to that of African Americans.

The patterns of racial inequities in health in the USA are similar to those in many other countries. In race-conscious societies such as Australia, Brazil, New Zealand, the UK, and South Africa, non-dominant racial groups have worse health than the dominant racial group.[8] For example, analyses of health data for the New Zealand Maori, Australian Aboriginals and Torres Strait Islanders, First Nations Canadians and Native Americans, and Alaskan Natives found that indigenous people had lower life expectancy than non-indigenous people in every country.[8]

Racial Differences in Age at Onset of Disease

One of the striking features of disparities in health is that minorities get sick at younger ages and die sooner than whites. Research has provided many examples of this earlier onset of disease and accelerated aging for minorities across multiple health status indicators. Although white women in the USA have an overall incidence of breast cancer that is higher than that of blacks, the opposite pattern is evident under the age of 40, with young African-American women having a higher incidence of breast cancer than their white peers.[9] Similarly, a 20 year follow-up of American adults in a cardiovascular disease cohort found that incident heart failure before the age of 50 was 20 times more common in blacks than in whites, with the average age of onset being 39 years for blacks.[10] National data also reveal that hypertension occurs earlier in blacks than whites, with 63% of blacks age 60 or younger having hypertension compared with 45% of whites.[11] Using a summary measure of allostatic load, an analysis of national data showed that the early health deterioration of black adults is evident across a range of biological systems.[12] In this study, blacks had poorer health than whites at younger ages and these racial differences cumulated so that the racial gap in health widened with increasing age. Specifically, blacks were more likely than whites to score highly on allostatic load (high on four or more indicators) at all ages. The size of the gap between blacks and whites increased with age; and the average score for blacks, in each age group, was comparable to that of whites who were 10 years older. Blacks continued to have higher allostatic load scores even after adjustment for poverty. In addition, black women consistently had higher allostatic load scores than black men.

Other evidence indicates that the increased risk of disease for minorities is evident early in life. A study of 23-year-old American men found that compared to whites with similar body mass index, body fat, fitness, renal function, blood lipids, and glucose levels, black men had comparable brachial blood pressure, but greater central blood pressure, greater augmentation of central blood pressure from wave reflections, and greater macrovascular and microvascular dysfunction reflected in increased central artery stiffness and reduced peripheral endothelial function.[13] The early deterioration in health of some minority groups is also evident among pregnant women. One national study in the USA documented, as expected, that infant death rates were lower for white and Mexican American women who delayed first births to their twenties compared with those who gave birth in their teens.[14] In striking contrast, the opposite pattern was evident for black and Puerto Rican women, for whom infant mortality for first births was lower for 15–19-year-olds than for women who had their first baby in their twenties. Geronimus argued that this pattern is due to "weathering" – the cumulative wear and tear of multiple social disadvantages.[14] That is, for groups residing in unhealthy social environments, chronological age reflects higher levels of exposure to adverse conditions in social and physical environments and thus greater erosion and dysregulation of physiological systems. A report from a longitudinal birth cohort study of over 8000 children in the USA also illustrated that racial differences in obesity are established early. By age 4, 13% of Asians and 16% of whites were obese, compared to 21% of blacks, 22% of Hispanics, and 31% of Native Americans.[1]

Racial Differences across the Continuum of Disease

Another important characteristic of racial inequities in health is that they are evident not only in the onset but also in the severity and progression of disease. For example, African Americans with chronic kidney disease have faster progression to end-stage renal disease than whites with this condition, with racial differences in the level of chronic kidney disease risk factors not adequately accounting for this difference.[8] Thus, blacks require dialysis or a kidney transplant at younger ages than whites and have a higher incidence of end-stage renal disease at each decade of life. These disparities in the course of disease have been documented even for outcomes that are less prevalent in blacks. Although black women are less likely than whites to be diagnosed with breast cancer each year, they are more likely to die from it because of their higher risk of early-onset, severe types of breast cancer. African-American women are also more likely than their white peers to have tumors that grow quickly, recur more often, are resistant to treatment, and kill more frequently.[8] Similarly, African Americans experience increased risk of stroke compared to whites, and those strokes tend to be more severe and are more likely to lead to death.[15]

With regard to mental health outcomes, the racial differences are more complex. Compared with their white counterparts, African Americans tend to have lower levels of common mental disorders (such as mood, anxiety, and substance use disorders), but higher levels of schizophrenia and other psychotic disorders, and greater likelihood of receiving poorer quality antipsychotic treatment.[16] Greater severity of illness and poorer quality treatment for minorities compared with whites are also evident for other disorders. For example, although African Americans have lower lifetime and current rates of major depression than whites, depressed blacks are more likely than their white counterparts to be chronically depressed, to have higher levels of impairment, to have more severe symptoms, and not to receive any treatment.[17] Likewise, other national data reveal that whites are more likely than blacks (but not Hispanics) to develop alcohol dependence.[18] However, once dependence emerges, both blacks and Latinos/Hispanics are more likely than whites to remain dependent on alcohol.

Research also reveals that risk factors such as alcohol and tobacco have a more adverse impact on blacks than on whites even when their overall levels are lower than or similar to those of whites. For example, the risks of lung cancer do not mirror variations in smoking behavior, with a given level of smoking associated with an elevated risk of lung cancer for African Americans and Native Hawaiians compared to whites, Japanese Americans, and Latinos.[1] Other research indicates that compared to whites, blacks have higher levels of nicotine intake and serum cotinine (a major metabolite of nicotine in the blood) per cigarette smoked.[1] In a similar vein, alcohol-related mortality is more than twice as high for black males than for their white counterparts and almost twice as high for females.[1] Another study found that blacks were more susceptible than whites to liver damage.[19] At every level of alcohol consumption blacks had higher levels of liver damage than whites, with the differences being largest at the highest level of alcohol use.

Similarly, although moderate alcohol consumption is generally associated with better health outcomes, this does not appear to be the case among African Americans. There was no beneficial effect of moderate alcohol consumption on all-cause mortality for black men or women in the USA.[1] In addition, although moderate alcohol consumption tends to be cardioprotective in middle-aged American adults, opposite to the pattern for whites and black females, moderate alcohol use was positively related to incident coronary heart disease and incident hypertension for black men in the Atherosclerosis Risk in Communities (ARIC) study.[1] Similarly, the Coronary Artery Risk Development in Young Adults (CARDIA) study found a positive association between alcohol consumption and the development of coronary calcification, with no beneficial effect of moderate alcohol consumption being clearest for black males.[1] It is not known whether these patterns reflect group differences in the specific types of substance used (blacks are more likely than whites to smoke menthol cigarettes and to use hard liquor), differential exposure to unmeasured physical and chemical exposures in occupational and residential environments, or interactions between health practices and psychosocial stressors that markedly increase health risks. It is also possible that social contextual factors including stressors and socioeconomic status (SES) are contributors. Some limited evidence indicates that when controls are introduced for multiple measures of SES, the inverse association between moderate consumption and mortality is no longer evident.[1] Other research reveals that at least some of the reported protective effects of moderate drinking are likely to be due to residual confounding of moderate alcohol use with high SES and good health practices.[1]

Persistence in Racial Disparities in Health Over Time

These racial disparities in health are also persistent over time. For example, in 1950, blacks had a life expectancy at birth of 60.8 years compared with 69.1 years for whites.[20] Life expectancy has improved over time for both groups and although the racial gap has narrowed, there was still a gap in life expectancy of almost 5 years in 2007 (73.6 *vs* 78.4 years). A 5 year difference in life expectancy is large. If blacks were to improve their life expectancy at the rate at which overall life expectancy increased in the USA between 1980 and 2000 (an average of 0.2 years annually), it would take about 25 years to close a 5 year gap in life expectancy. Instructively, it was not until 1990 – 40 years later – that blacks achieved the life expectancy that whites had in 1950. Trend data for heart disease, which is the leading cause of death in the USA, indicate that blacks and whites had comparable death rates in 1950, but African Americans currently have a higher mortality rate. For example, African-American males and females aged 25–64 years currently die from heart disease at rates two to three times higher than those of their white peers.[21] Long-term trend data are readily available only for blacks and whites. However, trend data exists for the 60% of the Native Americans population served by the Indian Health Service. The Indian Health Service was established in in 1955, and widening disparities in health are evident for Native Americans compared with whites for many causes of death such as diabetes and liver cirrhosis.[1]

STRESS, STRESSORS, AND THEIR ROLE IN HEALTH

Disparities in the incidence, severity, and progression of disease are multifactorial, in that behavior, pathogens, and environment all play a role. Although genetics is unlikely to be the major determinant of the patterns of population health that are so pronounced for racial categories, there may be important interactions between genes and environmental factors that contribute to health disparities. Race is a crude, yet important social category that is likely to reflect residual confounding for both unmeasured genetic factors and unmeasured exposures in physical, social, and chemical environments.[1] A growing body of evidence suggests that differential exposure to stressors and the co-occurrence of stressors may reflect part of the distinctive environmental contexts of racial groups. In turn, those exposures contribute in important ways to the observed patterns of racial and ethnic disparities in health. The following text considers research on the conceptualization and operationalization of stress and how stressors can affect health status and contribute to health disparities.

Psychosocial Stress

Stress is an almost ubiquitous concept that has taken on a variety of lay interpretations. Stress does not reflect an isolated event, but is a dynamic process that involves the interplay between individuals and their social environments. Wheaton and Montazer distinguish between three commonly used and interrelated terms in the stress process: stressors, stress, and distress.[22] *Stressors* take the form of environmental pressure creating "conditions of threat, challenge, demands, or structural constraints that, by the very fact of their occurrence or existence, call into question the operating integrity of the organism".[22] These threats range from distinctive, defined events (e.g. sexual abuse, divorce, or job loss) to chronic stressors in major domains of life (e.g. marital problems or financial difficulties) and chronic, everyday hassles (e.g. rush hour traffic or long workdays). For members of socially stigmatized groups such as racial or ethnic minorities there may be the added element of discrimination, the threat of victimization, and the greater clustering of many acute and chronic stressors.

An individual's subjective experience of stressors is also an important part of the stress process. Depending on the context of a particular stressor, the characteristics of the stressor, and the personal and social environment of the individual, stressors may trigger *stress*, described as "a biological state of the body – a generalized physiological alert – in response to threatening agents".[22] This is often referred to as the "fight or flight" response. That is, a stressor may precipitate hormonal, neurological, and cardiovascular changes in the body, ostensibly to meet the demands of that stressor. There may be differential effects of types of stressors on different systems of the body. The type, severity, and frequency of stressors may influence the perception of those incidents by an individual and affect the biological response to particular stressors. The word "stress" has been used in the fields of physics and engineering to refer to "an external force acting against a resisting body"; however, this force does not become a stressor until the force surpasses the "elastic limit" of the body, at which point the integrity of the body cannot be maintained.[22] Sociologist Leonard Pearlin theorized that psychosocial stress on the human body could be conceptualized in a similar manner. He also distinguished stress from strain. In his view, stress occurs when environmental demands are perceived by the individual to tax or exceed his or her adaptive capacities. The concept of strain is similar to that of allostatic load (see Chapter 34, "Stress"). It occurs when the individual reaches the point, "the ultimate elastic limit", when he or she is no longer able to respond adaptively to the environmental demands. Strain emphasizes the idea that different demands on the "body" can produce different evidence of physiological dysregulation.

The physical, psychological, and social resources that an individual mobilizes to meet the demands of his or her social environment can also alter the physiological response of the body to stressors and determine whether a given stressor triggers a state of *distress*, which is commonly characterized as "stress".[22] While for some, a heightened physiological response could result in directly experienced consequences for the individual (e.g. cardiovascular damage, inflammation, and hyperventilation), for others, psychosocial distress can trigger maladaptation in the neurological system or the development or exacerbation of unhealthy coping behaviors, such as tobacco and other substance use, which can negatively affect the their mental and physical health. Furthermore, an individual whose body has adapted to chronic stressful conditions may still encounter mental health effects of stressors without physiological evidence of them.[22] Regardless of the mechanism by which stressors affect the health of individuals, stress processes reflect one important pathway through which the social environment can "get under the skin" and have biological consequences.

Early Research on Psychosocial Stress and Health

Early research on stress conceptualized it as primarily a biological process. Hungarian endocrinologist Hans Selye, often called the "father of stress research", was one of the key early theorists researching the biology of stress. He conducted experiments with rats and extrapolated their responses to humans.[23] For example, he administered a series of "nocuous agents" to laboratory rats, including exposure to cold, surgical injury, and administrations of sublethal doses of morphine, to induce a stress response in the animals.[23] He observed that the rats all experienced a predictable syndrome, independent of the nature of the agent. He concluded that those nocuous agents represented any non-specific stressor and hypothesized that humans experiencing nocuous agents from their social environments – a viral invader, job strain, chemical stressor, or traumatic life event – would experience similar physiological responses.[23] In Selye's experiments, he controlled and administered the stressors to which the rats were exposed. For humans, the psychosocial environment is a primary driver setting the stress process in motion by providing "a wide variety of events and conditions that represent threat and insult to the organism".[22]

Although Selye did not pay particular attention to distinguishing among types of stressors, he recognized that there were factors, both intrinsic and extrinsic to the individual that may mitigate and exacerbate the impact of the stressors. The average person would be alarmed by close proximity to a lion, activating stress hormones such as epinephrine. However, for a lion tamer, this would be part of a daily routine and he or she could have the psychological and physical resources to draw on that would minimize any potential negative effects. Most humans do not regularly encounter dangerous lions, but we face environmental toxins, rush hour traffic, hard drive failures, and discriminatory experiences. Selye argued that the response in the body is similar regardless of whether the stressor is physical, psychological, or environmental: the body goes into high alert to attack or escape whatever the stressor may be. Although the intensity can vary, daily hassles, stressful life events (e.g. death of a loved one), and the threat of a charging lion can all trigger the stress response.

When an individual is chronically exposed to stressful conditions over a sustained period without sufficient time for repairs, as per Selye's model, the body enters the exhaustion stage of Selye's generalized adaptation syndrome. To meet the demands, the body essentially raises its homeostatic set point for catecholamine levels and certain hormones for as long as the body's systems can continue responding without restocking its stores. This often has undesirable consequences when the body cannot recover and repair itself, much less fight off subsequent infection. The near depletion of the body's resources can lead to enduring impairment of many physiological systems through hormone dysregulation and physical damage to the nervous system. Furthermore, extended exposure to certain stressors (e.g. extreme heat or cold) can be fatal.

Current Views of Stress

Selye can be credited with significant contributions to our knowledge of stress. However, in light of more recent research, there have been changes in how the stress process is conceptualized. The concept of fight or flight becomes insufficient in light of gender differences in physiological stress and, more importantly, the maladaptive nature of sympathetic nervous system activation in response to chronic, often unrelenting stressors.[24] For instance, while activation of the sympathetic nervous system may give an individual avoiding a car accident a physiological advantage that is helpful in the situation, that same activation is unlikely to impart benefits to people who experience chronic discrimination. Stress has been shown to produce changes in the neuroendocrine, autonomic, and immune systems, driving fluctuations in hormones, neurotransmitters, and catecholamines regulating processes in the brain and the rest of the body.[25–27] However, all stressors do not trigger the same pattern of response from the body's systems. The sympathetic–adrenal–medullary (SAM) system and the hypothalamic–pituitary–adrenocortical (HPA) axis are especially responsive to psychosocial stress, but respond differently depending on what type of stressor has prompted the response.[28] Activation of the SAM system stimulates the release of catecholamines, such as epinephrine, norepinephrine, and dopamine, which act together with the autonomic nervous system to apply regulatory effects on the cardiovascular, hepatic, pulmonary, musculoskeletal, and immune systems, in addition to producing mood changes.[28] Activation of the HPA axis stimulates cortisol, which helps to regulate various physiological processes (e.g. anti-inflammatory responses, gluconeogenesis, and carbohydrate and fat metabolism) and plays a role in compulsive overeating.[28]

Research on the biological response to stressors emphasizes the idea of allostasis: "the active processes of adaptation", the overarching process of maintaining stability in the systems that preserve the balance in life-sustaining, homeostatic processes through active release of stress mediators such as cytokines, catecholamines, and the hormones of the HPA axis.[29] Changes in the HPA axis and the SAM system occur to enable other systems to maintain homeostasis. An allostatic state occurs when homeostatic processes exceed the acceptable range of control by changing set points, resulting in the dysregulation of various systems, with some mediators being released at increasingly higher levels and others being inhibited. This dysregulation can damage the regulatory systems in the body, and with sufficient, cumulative wear and tear can result in allostatic load or allostatic overload.[29] Rather than the strict biology of the stress response, the idea of allostatic load includes the idea of "how individuals perceive and have or do not have confidence in their ability to cope with the burdens of life experiences".[30]

Measuring Psychosocial Stressors

Stressors can range from the chronic stressors of daily life to acute life events or even devastating traumatic experiences. Early approaches to measuring stressors emphasized the assessment of acute life events and the change that they produced in the routine of life. These discrete, observable experiences triggered important life changes and could be readily dated because they had a specific onset and end. Typically, "once in motion, there is a relatively well-defined sequence of subevents describing the 'normal' process of the event".[22] The ability to pinpoint a specific date for these life events facilitated the enumeration of this type of stressor. Stress researchers have developed several checklists to capture exposure to stressful life events. For instance, in 1967, Holmes and Rahe developed the Social Readjustment Rating Scale (SRRS), a landmark questionnaire identifying key stressful life events that sparked research that led to more comprehensive and specialized checklists.[22] The SRRS tallied whether an individual had experienced each of 43 life events in the previous 12 months, including death of a spouse, divorce, incarceration, marriage, foreclosure, and Christmas. The scale included both positive and negative events that might lead to change in one's life. Each life event was assigned a value called a life unit that approximates the level of change and stress associated with each event. The sum of life units from each questionnaire was then used to approximate the likelihood that the individual was suffering from the wear and tear of stress. Later scales improved on this iteration by trying to capture additional stressful life events neglected in the SRRS (e.g. sexual assault and other crime victimization).[22] Research has also moved away from a focus on change to an emphasis on negative experiences. It also recognized that checklists like these often assume that each event will be perceived in the same manner by all individuals answering the questionnaire. In addition, they do not take the duration and frequency of each of these life events into account.

Chronic stressors are often harder to measure since they typically develop gradually over time and are generally interwoven into the natural progression of social interactions and the conditions of the social environment. For instance, daily rush hour commutes to and from work may be a vague annoyance when starting a new job; however, over time, those commutes may develop into serious sources of strain. It is difficult to identify a specific onset or resolution of these kinds of stressor, but the effects of the continuous wear and tear of chronic stressors may be cumulative and severe.[28] Over time, like the proverbial straw that broke the camel's back, a morning commute can bear more weight on the body than a more well-defined life event. Research reveals that recurrent or daily stressful experiences can have negative effects on physical and mental health, as large as or larger than those of acute life events.[31]

A key finding in research on psychosocial stressors and health over the past several decades is that failure to measure psychosocial stressors comprehensively will dramatically understate the contribution of stressors to health. For example, in the area of mental health, researchers estimate that 25–40% of the variance in psychological distress and depressive symptoms can be attributed to acute and chronic psychosocial stressors when measured comprehensively, which is a striking improvement over only 1–12% of the explained variance when psychosocial stressors are assessed as negative life events.[31] Table 47.1 provides an example of a comprehensive characterization of stressors from a 2011 study.[5] In addition to life events, it includes chronic stressors in multiple domains of life.

Accounting for the experience of different kinds of stressor over time may play an important role in fully understanding the role of psychosocial stressors. This highlights the necessity of multiple assessments of exposure to psychosocial stressors over time to adequately capture the accumulation and clustering of stressors over the life course. Pearlin, calling for a sociological approach to the stress process, emphasizes the importance of distinguishing between categories and types of stressor because the patterning and sequencing of specific stressors can determine the effect of stress on the body, with the impact of a given stressor shaped by prior and co-occurring stressors.[22] For example, the experience of chronic and acute stressors during the critical first years of life (early life adversity) appears to have

TABLE 47.1 Measuring Stressors Comprehensively: An Example

Domains	Sample Items
ACUTE LIFE EVENTS	
Lifetime traumas	Death of a child; victim of serious physical attack
Past 5 years	Death of someone close; involuntary job loss
EMPLOYMENT STRESSORS	
Job satisfaction	How satisfied with job
No control	Job allows decision making; job requires creativity
Job insecurity	Likelihood of losing main job in next couple of years
Work demands	Asked to do excessive amount of work; conflicting demands
Job–non-job conflicts	Job leaves one too tired/stressed to participate in other activities
Job hazards	Exposure to dangerous chemicals on job; exposure to air pollution
FINANCIAL STRESSORS	
Financial strain	How difficult to meet monthly bill payments
Economic problems	Postponed medical care; borrowed money from friends/relatives
LIFE DISCRIMINATION	
Major life events	Unfairly fired from a job; stopped, threatened, abused by the police
Everyday discrimination	Treated with less courtesy/respect than others
JOB DISCRIMINATION	
Harassment	Supervisor/co-workers make slurs or jokes about racial or ethnic groups
Unfair treatment	Watched more closely than others; unfairly humiliated
RELATIONSHIP STRESSORS	
Marital stressors	Frequently bothered/upset by marriage; spouse is critical
Marital abuse	Spouse drinks too much; spouse pushes, slaps, hits you
Child-related stressors	Children make too many demands; bothered/upset as a parent
Problems of children	Your children currently have problems with finances, health, etc.
Friend criticism	Friends make too many demands; friends critical of you
EARLY LIFE STRESSORS	
Parental stressors	Before you were 12, how much parents made you feel loved
Parental educational involvement	Parents participate in school activities
Hunger	Before age 12, how often went to sleep feeling hungry
COMMUNITY/NEIGHBORHOOD STRESSORS	
Violence	Neighborhood fights; violent argument between neighbors
Total victimization	Home broken into; exterior of home/property damaged
Disorder	Frequency of seeing graffiti; vacant/deserted houses

Source: Sternthal et al. Du Bois Rev. *2011;8(1):95–113.*[5]

particularly deleterious effects on development and subsequent experiences of stressors later in life.[32,33] The neonatal period is a critical period for brain development that will later guide the boundaries of self-regulation and social interactions.[32,33] Thus, understanding the full impact of psychosocial stress will require careful characterization of the complex ways in which stressors unfold and accumulate over time.

A comprehensive conceptualization of psychosocial stress would also take into account the type, frequency, and duration of stressors, the physiological response to stressors, the psychological evaluation of the stressors and the body's response, and the availability and mobilization of resources to deal with stressors over time. However, measurement challenges in fully capturing the dynamic nature of the stress experience over the life course have limited the ability of researchers to fully evaluate and quantify its effects on health. While the effects of psychosocial stress, as currently measured, are likely to be underestimated, a large and growing body of research is linking the experience of psychosocial stress to health.

Psychosocial Stress and Health

For more than seven decades, there has been sustained research into the relationship between psychosocial stress and health. Because exposure to stressors can lead to allostatic load and the dysregulation of various body systems, psychosocial stressors can have potentially wide-ranging effects on health. A review concluded that the "effects of stress on the regulation of immune and inflammatory processes have the potential to influence depression; infectious, autoimmune, and coronary artery disease; and at least some (e.g. virally mediated) cancers".[28] The mechanisms by which psychosocial stress affect health are also diverse, ranging from altered immune function through the stimulation of lymphatic tissue to stress-induced behavioral coping mechanisms, such as increased smoking. Simultaneously, the same, classic neuroendocrine activation has been associated not only with mental health outcomes, but also with cardiovascular outcomes.[28,34] Research in animals and humans suggests that psychosocial stress affects health through two key pathways: early life adversity and cumulative exposure to stressors.

Early Life Adversity

In mammals, the period following birth is a critical period for development, and the social interactions that occur in those weeks, months, and years are decisive in the development of adult behavior.[33] Deprivation of social contact and maternal care constitutes a highly stressful psychosocial environment for young mammals. For example, in studies where female rhesus monkeys were denied maternal contact, the offspring were more aggressive, socially withdrawn, and neglectful of their own offspring.[33] The modified child rearing produced dramatically different social behavior through adulthood and into the next generation. Furthermore, rhesus monkeys with mothers who experienced high levels of anxiety during their early development display chronic anxiety throughout adulthood and indications of metabolic syndrome.[26] Research reveals that through hippocampal dysfunction and hypersensitivity, early life adversity causes enduring changes in cognition, emotionality, and stress responsiveness in animals.[26]

In line with the animal research, studies in humans have found that early life adversity is associated with increased risk of subsequent physical, mental, and cognitive disorders.[30,32] Children who are emotionally neglected or live in environments that are characterized by abuse or severe punishment demonstrate increased risk of anxiety or depression, altered intellectual development, and greater likelihood of mirroring the same parenting with their own children.[33] In addition, those who experience abuse early in life show increased inflammatory markers as older children and young adults, which is correlated with increased cardiovascular risk.[30] When abuse is absent but there is disorder in the home, children tend to develop a decreased ability to regulate their own behavior and are at increased risk of developing obesity.[33]

Cumulative Exposure to Stressors

Exposure to other chronic and acute stressors that occur later in the life course can also have substantial adverse effects on health. However, acute and chronic stressors appear to affect the brain in different ways. For instance, numerous studies have linked acute stressful life events to the onset of depressive symptoms and major depressive disorder.[35,36] For those already diagnosed with depression, stress levels are associated with the clinical course of the disease; they are positively correlated with the duration of major depression, symptom severity, and relapse.[35,36] The experience of an acute stressful life event generally results in delays in the formation of dendritic spines in neurons of the basolateral amygdala, resulting in an increase in anxiety levels.[30] The immediate activation of the sympathetic nervous system could also have adverse effects on physical health, especially for those with pre-existing cardiovascular damage.[29]

The experience of chronic stressors results in evidence of alterations to the hippocampus, amygdala, and medial prefrontal cortex.[30] Brain imaging research shows that chronic stress is associated with neuronal replacement, shrinkage of dendrites, and the loss of glial cells, which produces smaller hippocampal volumes.[30] It would follow that cumulative exposure to chronic stressors may result in impaired learning and memory, since the hippocampus "plays instrumental roles in learning and remembering declarative and spatial information, processing the contextual aspects of emotional events, and regulating visceral functions, including the HPA axis".[27] Impaired learning and memory may reduce academic performance over the life course, reducing levels of educational attainment, which in turn can have independent, deleterious effects on health.[37] Furthermore, since some receptors in the hippocampus are responsive to insulin, ghrelin, insulin-like growth factor-1, and leptin, these changes in hippocampal volume could have an effect on type 2 diabetes and metabolism.[30]

In response to increased exposure to stressors, the dendrites in the amygdala tend to shrink not only in the amygdala, but also in the medial prefrontal cortex.[30] In addition, there is evidence of dendritic spine loss in the medial prefrontal cortex and the increase of dendrites in the orbitofrontal cortex. These changes could lead to several deleterious effects in humans, including poor cognitive flexibility and circadian rhythm disruption, since the prefrontal cortex is involved with higher cognitive functions.[27,30] Sleep deprivation (resulting in circadian rhythm disruption) is correlated with overweight and obesity, cognitive impairment, and increased aggression.[25] Furthermore, research on the health effects of shift work shows that disruption of the circadian rhythm may have effects on cardiovascular health and is clinically related to psychiatric diagnoses and neurodegenerative disease.[38]

The effect of psychosocial stress on health behaviors is another pathway by which the psychosocial environment can affect health. For instance, changes in the hippocampus and hypothalamus are implicated in increased food intake and "emotional eating".[25] Research has found that experiences of stress are associated with compulsive overeating and the consumption of comfort foods.[25] Psychosocial stress also influences tobacco smoking behavior, although the mechanisms are not fully understood.[39] On a basic level, psychosocial stress increases the levels of circulating cortisol, which, in turn, naturally triggers the release of dopamine in the brain.[39] There is evidence that chronic, toxic exposure to stressors over time decreases dopamine transporter binding sites, down-regulating overall dopamine release, and potentially increasing the likelihood of self-medicating (with nicotine or some other substance).[40] In addition, acute psychosocial stress increases the neural salience of cues to smoke.[41] Thus, stressors, ranging from major life events to daily hassles, have been linked to increased odds of smoking initiation among adolescents, tobacco consumption among established smokers, difficulty in smoking cessation, and relapse among those who have quit.[39] Moreover, social environments help to pattern psychosocial stressors, such that people in certain social environments (e.g. low-income neighborhoods, high-strain and monotonous work environments) are more likely to have high stress levels and, therefore, to smoke as a coping mechanism.[39]

UNDERSTANDING RACIAL DIFFERENCES IN HEALTH: A ROLE FOR STRESS?

Research on racial/ethnic differences in health has emphasized that racial differences in SES play a large role.[1] Table 47.2 shows that there are large racial differences in SES (income, education, and wealth). Compared to whites, African Americans and Hispanics have lower levels of education, income, and, especially, wealth. While wealth (the economic value of reserves of accumulated assets) is used less often in research as a measure of SES, racial disparities in wealth are larger than those for income and education, and wealth makes an incremental contribution to health and health disparities over and above the more commonly used indicators of SES.[37] For example, blacks and Hispanics have only one penny of non-home wealth (excluding home equity) for every dollar that whites have. SES, in turn, reflects differential exposure to psychological, social, physical, and chemical factors in occupational, residential, and other societal contexts. Not surprisingly, research finds that racial differences in SES account for a substantial part of the racial differences in health.[1] However, race captures more than socioeconomic inequality: there is an added burden of race, over and above SES being linked to poor health.

Table 47.3 presents national data on life expectancy, illustrating the complex relationship between race and SES in predicting health. At age 25, there are substantial racial differences in life expectancy, with whites living 5 years longer than African Americans.[44] At the same time, for both blacks and whites, differences in life

TABLE 47.2 Socioeconomic Status by Race and Ethnicity, 2007

Indicator	Whites (W) (non-Hispanic)	Blacks (B) (non-Hispanic)	Hispanics (H)	Ratios	
				B/W	H/W
College graduate or more (%)[a]	29.1	18.5	12.7	0.64	0.44
Income[a]	52,115	33,916	38,679	0.65	0.74
Median wealth[b]	143,600	9,300	9,100	0.06	0.06
Non-home wealth[b]	43,600	500	400	0.01	0.01
Home owner (%)[a]	74.8	48.6	49.2	0.65	0.66

[a] *US Census Bureau.* The 2010 Statistical Abstract. *Washington, DC: US Census Bureau; 2010*[42]

[b] *Wolff EN.* Recent Trends in Household Wealth in the United States: Rising Debt and the Middle-Class Squeeze – An Update to 2007. *Annandale-on-Hudson, NY: Levy Economics Institute of Bard College; 2010.*[43]

TABLE 47.3 Life Expectancy at Age 25, USA

Group	White	Black	Difference
All (1998)[a]	53.4	48.4	5.0
By education (1988–1998)[b]			
0–12 years	50.1	47.0	3.1
12 years	54.1	49.9	4.2
Some college	55.2	50.9	4.3
College graduate	56.5	52.3	4.2
Education difference	6.4	5.3	
By income (1988–1998)[b]			
Poor	49.0	45.5	3.5
Near poor	51.4	48.0	3.4
Middle income	53.8	50.7	3.1
High income	55.8	52.6	3.2
Income difference	6.8	7.1	

Poor: below federal poverty level (FPL); near poor: above the FPL but less than twice the FPL; middle income: more than twice but less than four times the FPL; high income: four times the FPL or more.
[a]*National Vital Statistics.*[44]
[b]*National Longitudinal Mortality Study.*[45]

expectancy by income and education are larger than the overall difference between blacks and whites.[45] For example, high-income blacks and whites live 7.1 and 6.8 years longer, respectively, than their counterparts who are poor. For both racial groups, as income and education levels rise, health improves in a stepwise manner, but there are differences between blacks and whites in life expectancy at every level of income and education. For education, the racial differences grow larger as education increases. These national data tell a complex story not only about the centrality of SES in shaping the distribution of health but also about added pathogenic factors linked to race that are not reduced with increasing SES.

Other data provide further examples of this pattern. A striking illustration of large racial differences in health at similar levels of education and occupation comes from a study of the health of black and white physicians who graduated from medical school in the late 1950s and early 1960s.[46] The black physicians completed medical school at Meharry Medical College and the white ones at Johns Hopkins University. When they were assessed 23–25 years later, the black doctors had a higher risk of cardiovascular disease (relative risk = 1.65) and earlier onset of disease, with incidence rates of diabetes and hypertension that were twice as high. The black doctors also had a 40% higher incidence of coronary artery disease than their white peers and strikingly higher case fatality (52% *vs* 9%). National data on infant mortality provide another example.[45] There is an inverse association between mother's education and infant mortality for blacks, whites, and Hispanics. Nonetheless, the infant mortality rate for college-educated African-American women is more than two and a half times as high as that for whites and Hispanics at similar levels of education. Strikingly, black female college graduates (the best off African-American women) have a higher rate of infant mortality than Hispanic and white women who have not completed high school.

Role of Stress in the Added Burden of Race

Research suggests that multiple factors contribute to the residual effect of race after SES is controlled.[47] First, indicators of SES such as income and education are not equivalent across race.[48] For example, compared with whites, blacks and Hispanics have lower earnings at comparable levels of education, less wealth at every level of income, and less purchasing power because the costs of a broad range of goods and services are higher in their communities.[49] Multiple processes linked to the history and current circumstances of racial groups have resulted in racial groups residing in distinctive environments that have important implications for a broad range of health-damaging exposures. For example, residential segregation by race, a policy that ensured that blacks and whites lived in different neighborhoods, has been characterized as one of the most successful domestic policies of the twentieth century in the USA.[50] It has also affected health through multiple pathways.[51] Segregation restricts socioeconomic mobility for racial minorities by limiting access to quality elementary and high school education, preparation for higher education, and employment opportunities. For example, segregated schools have lower teacher quality, fewer educational resources, and lower per-student spending, and are more likely to be located in neighborhoods with elevated exposure to violence, crime, and poverty.[52] The bottom line is that segregation, as it exists in the USA, creates distinct environments of concentrated poverty that help to pattern potential stressors. One study found that the elimination of segregation would erase differences between blacks and whites in earnings, high-school graduation rate, and unemployment, and reduce racial differences in single motherhood by two-thirds.[53] There is nothing inherently negative about living with other members of one's own race; it is the co-occurrence of racial segregation with the concentration of social ills, such as poverty, crime, and public disinvestment, that creates an unhealthy living environment. The challenge for society is create opportunities for educational and occupational advancement for groups trapped in disadvantaged environments. The Moving to Opportunity study, an experimental study randomizing households in public housing to receive vouchers to relocate to low-poverty neighborhoods, has demonstrated that moving people out of areas of concentrated disadvantage can have long-term positive effects on physical and mental health, without any health intervention. Fifteen years after relocation, moving to a more advantaged neighborhood was associated with improved mental well-being and reduced risk of obesity and diabetes in low-income adults.[54]

Second, health is affected not only by one's current SES, but also by exposure to social and economic stressors over the life course. Earlier in this chapter, the importance of early life adversity in health was considered. Racial/ethnic minority populations are more likely than whites to have experienced elevated levels of early life psychosocial and economic adversity that adversely affect health in adulthood.[54] For example, stressors experienced in childhood and adolescence were found to be associated with four of five markers of inflammation for adult African Americans, but not for whites.[55]

Third, psychosocial stressors in adulthood contribute to the patterning of racial disparities in health. Residential segregation creates pathogenic residential conditions, with minorities living in markedly more health-damaging residential environments than whites and facing higher levels of acute and chronic stressors. Because of segregation, racial minorities live in poorer quality housing and in neighborhood environments that are deficient in a broad range of resources that enhance health and well-being, including medical care. Although the majority of poor people in the USA are white, poor white families are not concentrated in contexts of economic and social disadvantage and with the absence of an infrastructure that promotes opportunity, in the ways that poor blacks and Latinos are. Minorities, have higher exposures than whites to a broad range of psychosocial stressors. A 2011 study documented, for example, that compared with whites, blacks and US-born Latinos had both a greater clustering of multiple stressors and higher exposure to psychosocial stressors across most of eight domains of stressors, including a greater clustering of multiple stressors.[5] Specifically, they had higher levels of life events, financial, relationship, discrimination, and neighborhood stressors. For the multiple domains of stressors (listed in Table 47.1) there was a graded association between the number of stressors experienced and poor health, with each additional stressor being predictive of worse health status across multiple indicators of health status. Among the different domains of stressors, financial stressors, relationship stressors, and life events reflected the strongest consistent associations with poor health. Moreover, differential stress exposure by race accounted for some of the residual effect of race on health after income and education were controlled.

Segregation also leads minorities to have a higher risk of exposure to toxic chemicals at the individual, household, and neighborhood level.[56] There is often overlap between psychosocial stressors and physical and chemical stressors among members of the same racial and ethnic groups owing to the social patterning of stressors. These different types of stressor are usually studied separately, but research suggests that chemical and environmental stressors often interact with psychosocial stressors to affect health.[1,31] Residential segregation has played a critical role in the distribution of stressful psychosocial environments in the USA. It has a greater effect on the health of African Americans than that of other groups, because blacks today are more segregated than any other immigrant group in US history.[1] Because of decades of residential segregation, blacks in the USA have been historically relegated to neighborhoods with inferior housing stock, with a higher likelihood of proximity to potentially toxic sites. For instance, black children with asthma living in disadvantaged neighborhoods are likely to experience a higher exposure to noxious fumes from an industrial smokestack than those living elsewhere, which may result in increased asthma attacks, emergency room visits, and vulnerability to the effects of other physical threats.

In addition, residential segregation is highly correlated with concentrated poverty, which leads to the clustering of a broad range of pathogenic exposures in the social and physical environment. This limits access to health-enhancing opportunities for people living in segregated neighborhoods. For example, living in disadvantaged neighborhoods is associated with reduced accessibility to and affordability of healthy food items.[1] Notably, in densely populated urban centers there is often an absence of large supermarkets selling fresh fruit and vegetables. Instead, there are corner stores that sell basic food items, but specialize in unhealthy snack foods. In these corner stores and elsewhere in disadvantaged, segregated communities, there are high levels of targeted marketing of alcohol and tobacco products.[1] Furthermore, the concentrated poverty in disadvantaged neighborhoods allows for stressors associated with poverty to operate simultaneously at the individual, household, and community levels.[1] In sum, the historical legacy of institutional discrimination, in the form of residential segregation, contributes to the differential patterning of stressors by race and promotes elevated exposure to the synergistically ill effects of stressful, unhealthy psychosocial and physical environments.

Experiences of Discrimination as a Pathogenic Stressor

Experiences of discrimination are neglected psychosocial stressors for socially stigmatized minority populations. A large and growing body of research indicates that these personal experiences of discrimination are added pathogenic factors that adversely affect the health of minority group members.[57] Early studies on discrimination and health yielded important but equivocal findings. Most studies were cross-sectional, focusing on mental health outcomes or other self-reported indicators of health. In these mainly US-based studies there was concern about the potential of shared response bias between the independent and dependent variables.[58,59] Recent reviews document important progress in this area of research.[47,60] Several longitudinal studies and other studies have found that the effects of discrimination persist after adjusting for potential psychological confounders such as social desirability bias, neuroticism, self-esteem, negative affect, and hostility. Studies on all major racial and ethnic groups in the USA; non-dominant racial groups in New Zealand, South Africa, and Australia; and immigrants in Canada, Hong Kong, and many countries in Europe show that racial discrimination affects health.

Subjective experiences of discrimination are psychosocial stressors that adversely affect a very broad range of health outcomes and health risk behaviors.[47,60] Discrimination is associated with health conditions ranging from violence, sexual problems, and poor sleep quality to elevated risk of C-reactive protein, high blood pressure and coronary artery calcification, breast cancer incidence, uterine myomas (fibroids), and subclinical carotid artery disease. For example, a review of studies by Tené Lewis and colleagues indicates that chronic everyday discrimination is positively associated with coronary artery calcification, C-reactive protein, blood pressure, giving birth to lower birth weight infants, cognitive impairment, subjective and objective indicators of poor sleep, visceral fat, and mortality.[61] Discrimination has also been associated with delays in seeking treatment, lower adherence to medical regimens, and lower rates of follow-up. Moreover, research in the USA, South Africa, Australia and New Zealand reveals that discrimination makes an incremental contribution over SES in accounting for racial disparities in health.[47] Like stressors more generally, experiences of discrimination can take the form of both acute and chronic stressors. Acute experiences of discrimination include being denied a mortgage, unfairly fired, or denied a promotion, or being unfairly stopped, physically threatened, or abused by the police because of being a member of a stigmatized group. Chronic experiences of discrimination include daily hassles of being treated with less courtesy and respect than others or receiving poorer service than others in restaurants and stores. While acute experiences of discrimination are typically more publicized and recognizable as discrimination, the chronic experiences have a stronger, more consistent positive association with multiple indicators of morbidity.[47,60] If the effects of only major, acute discrete discriminatory events on social status and health problems were considers, only minor associations would be detected, drastically underestimating the true impact of discrimination.[31]

RESEARCH IMPLICATIONS

Based on the existing literature, one of the most important conclusions that can be drawn is that future research needs to take a comprehensive view of stressors to capture the full effect of stress on human physiology, health, and health disparities. This approach requires studies that recognize both variations in the type, frequency, and duration of psychosocial stressors and the biological and psychological responses of the body to those stressors, both chronic and acute. Furthermore, paying attention to how exposure and responses to stressors vary across important social statuses (race, gender, SES, age) can enhance our ability to identify particular vulnerable groups, sources of resilience, and specific mechanisms by which stressors contribute to health disparities.

For example, the way in which race and gender combine to contribute to patterns of disparities is complicated and the contribution of social experiences to these patterns is not well understood. For example, at all ages, black women have higher allostatic load scores than black men.[12] At the same time, racial differences in life expectancy are larger for men than for women, and black women have higher levels of life expectancy than white men, suggesting that factors linked to both race and gender contribute to these patterns.[62] Gender differences in life expectancy are consistently larger for blacks than for whites, and the gap in life expectancy between black men and women is consistently larger, at every age, than the racial gap in life expectancy. The role that stressors play in the complex patterning of health by race and gender is unclear. Research needs to better understand the determinants, levels, and consequences of stressors at the intersection of race and gender.

It has been argued that the fight or flight reaction to stressors is a more apt description of stress responses in males, while the response in females tends to follow a distinct trajectory.[63] Taylor and colleagues propose that, in addition to the traditional fight or flight response to stress, females call on another system during times of acute stress that they call "tend and befriend".[63] According to this theory, females tend either to participate in nurturing activities that promote the well-being and safety of themselves and their offspring, or to create and maintain social networks to promote the same, especially when the stressors are formidable. They argue that this response might have evolved in order for females to take care of their young and respond effectively to exogenous stressors.[63] They also hypothesize that this additional system is mediated through the production of oxytocin and moderated by female hormones. The extent to which differential stress responses by gender occur, the universality of these patterns, and the implications that they may have for racial and gender disparities in health are important questions to be addressed in future research.

Furthermore, there are various mechanisms by which the physical and psychosocial environment is embodied and reflected in racial and socioeconomic disparities in health, including developmental and epigenetic processes.[64] Epigenetics is a term used to describe changes in genetic processes, such as DNA methylation and gene expression, which bring about short- and long-term biological changes. Thus, epigenetic changes beginning *in utero* can have an impact through adulthood and sometimes across generations. These processes are particularly responsive to social and economic change, especially those occurring within childhood and adolescence, critical periods for neurological, tissue, and organ development.[64] Future research delving further into the impact of psychosocial stress on brain development and functioning, from *in utero* into subsequent generations, should further elucidate the mechanisms by which environmental stressors modify physical and mental health across the life course. Research on the epigenetic effects of psychosocial stressors will help to disentangle the relative effects of stressors occurring during childhood and those occurring during adulthood, as well as the cumulative effects of both. This could directly feed into interventions that can reduce health disparities on multiple levels, ranging from the macroeconomic to the epigenetic, and identify potential critical periods in which those interventions can occur.

CLINICAL IMPLICATIONS

In the clinical context, it is essential for health-care providers to consider the role of their patients' psychosocial stress. Finding ways to access the levels of stressors in patients' psychosocial environments and providing resources to enable them to mitigate the physiological and psychological effects of the stressors in their lives may prove essential in enhancing the ability of patients to manage chronic disease and engage in health behaviors that minimize health risks. Depression co-occurs with multiple common diseases such as hypertension and diabetes. Research reveals that depression prevention and intervention programs can enhance protective resources and reduce psychiatric symptoms and new cases of depression.[65] Thus, more systemic efforts to address stressors and symptoms of depression in individuals with chronic illness can reduce disparities in the severity and course of illness.

Research has also shown that positive prostate gene expression among prostate cancer patients is associated with participation in a holistic health behavior improvement program that included stress management.[1] Similarly, a clinical trial that used cognitive behavioral therapy to treat depression had a marked effect on improving smoking cessation outcomes.[66] Therefore, to optimize the likelihood of smoking cessation, a recommendation to stop smoking could be paired with a stress management intervention or therapy, especially in the context of pre-existing mental illness.

Moreover, considering the far-reaching effects of psychosocial stress, a comprehensive understanding of stressors, stress response, and stress response mitigation could be helpful in pharmacological treatment settings. There is evidence that treating mood increases adherence to medication for the treatment of several chronic diseases, including cardiovascular disease and HIV/AIDS, which should improve health outcomes.[34] However, treating one condition does not necessarily take into consideration other illnesses or contributors to illness that could be affected by psychosocial stressors. McEwen notes that, "because of the nonlinearity of the systems of allostasis, the consequences of any drug treatment may be either to inhibit the beneficial effects of the systems in question or to perturb other systems that interact with it in a direction that promotes an unwanted side effect".[25] If neuroendocrine dysregulation from psychosocial stress contributes to the development of hypertension, offering a beta-blocker to a patient with hypertension may help to reduce blood pressure, while neglecting potential damage to other body systems and exacerbating other subclinical conditions. Addressing stress moves the intervention upstream to minimize multisystem dysregulation. Once these complex mechanisms have been more clearly elucidated, it may be possible to optimize pharmacological treatment.

In the meantime, there is promising evidence of amelioration of the effects of allostatic load and possible stress resiliency associated with three behavioral interventions: physical activity, social support, and meditation.[25] Physical activity seems to improve mood, while mitigating some of the remodeling of the cortex and hippocampus due to exposure to stress.[25] In addition, regular exercise is correlated with improved cognitive functioning, verbal learning, and memory.[27] While social support seems to be associated with reduced stress levels and has been linked to maintenance of telomere length of white blood cells, little research has been conducted on its effects on brain circuitry.[25] In addition, Benson and co-workers have demonstrated that meditation results in a relaxation response, with changes in sympathetic nervous system activity and "activation of neural structures involved in attention and control of the autonomic nervous system".[67] While this research is promising, understanding of how well these three behavioral strategies work for individuals in extremely stressful psychosocial environments is an important priority. Further research into the field of resilience to psychosocial stress would not only advance the field, but also identify the preconditions that are necessary for particular interventions to have optimal impact, especially in socially disadvantaged contexts.

CONCLUSION

More attention needs to be paid in both research and policy making to reducing the occurrence of stressors and adversity. Priority needs to be given to population-based interventions that improve living conditions and reduce the prevalence of stressors, and thus improve health. A natural experiment in the Great Smoky Mountain Study revealed that, even without relocating individuals, income supplements due to the opening of a casino led to improved health and reduced health disparities among Native American youth involved in the study. Additional income was associated with reduced psychopathological symptoms, lower occurrence of minor criminal offenses, and increased levels of educational attainment among adolescents.[68] In fact, these income supplements eliminated racial disparities in minor criminal offenses and educational attainment between Native Americans and whites.[68] Thus, although there is much still to learn, there is a scientific base for a broad range of societal interventions that can improve health and reduce racial inequalities in health. One of the greatest challenges is identifying the feasible strategies to raise awareness and build political will for initiatives to reduce a broad range of societal stressors and improve the quality of life and thus the health and well-being of vulnerable populations.[8]

References

1. Williams DR, Mohammed SA, Leavell J, Collins C. Race, socioeconomic status, and health: complexities, ongoing challenges, and research opportunities. *Ann N Y Acad Sci*. 2010;1186(1):69–101.
2. Indian Health Service. *Trends in Indian Health 2002–2003*. Rockville, MD: Indian Health Service, US Department of Health and Human Services; 2009.
3. Williams DR. The health of US racial and ethnic populations. *J Gerontol Ser B*. 2005;60B(Special Issue II):53–62.
4. Singh GK, Miller BA. Health, life expectancy, and mortality patterns among immigrant populations in the United States. *Can J Public Health*. 2004;95(3):I14–I21.
5. Sternthal MJ, Slopen N, Williams DR. Racial disparities in health: how much does stress really matter? *Du Bois Rev*. 2011;8(1):95–113.
6. Kaestner R, Pearson JA, Keene D, Geronimus AT. Stress, allostatic load, and health of Mexican immigrants. *Soc Sci Q*. 2009;90(5):1089–1111.
7. McEwen BS, Stellar E. Stress and the individual: mechanisms leading to disease. *Arch Intern Med*. 1993;153(18):2093.
8. Williams DR. Miles to go before we sleep: racial inequities in health. *J Health Soc Behav*. 2012;53(3):279–295.
9. Anderson WF, Rosenberg PS, Menashe I, Mitani A, Pfeiffer RM. Age-related crossover in breast cancer incidence rates between black and white ethnic groups. *J Natl Cancer Inst*. 2008;100(24):1804–1814.
10. Bibbins-Domingo K, Pletcher MJ, Lin F, et al. Racial differences in incident heart failure among young adults. *N Engl J Med*. 2009;360(12):1179–1190.
11. Hertz RP, Unger AN, Cornell JA, Saunders E. Racial disparities in hypertension prevalence, awareness, and management. *Arch Intern Med*. 2005;165(18):2098–2104.
12. Geronimus AT, Hicken M, Keene D, Bound J. "Weathering" and age patterns of allostatic load scores among blacks and whites in the United States. *Am J Public Health*. 2006;96(5):826–833.
13. Heffernan KS, Jae SY, Wilund KR, Woods JA, Fernhall B. Racial differences in central blood pressure and vascular function in young men. *Am J Physiol Heart Circ Physiol*. 2008;295(6):H2380–H2387.
14. Geronimus AT. The weathering hypothesis and the health of African-American women and infants: evidence and speculations. *Ethn Dis*. 1992;2(3):207–221.
15. Stansbury JP, Jia H, Williams LS, Vogel WB, Duncan PW. Ethnic disparities in stroke: epidemiology, acute care, and postacute outcomes. *Stroke*. 2005;36(2):374–386.
16. Miranda J, McGuire T, Williams D, Wang P. Mental health in the context of health disparities. *Am J Psychiatry*. 2008;165(9):1102–1108.
17. Williams DR, Gonzalez HM, Neighbors H, et al. Prevalence and distribution of major depressive disorder in African Americans, Caribbean Blacks, and Non-Hispanic Whites: results from the national survey of American life. *Arch Gen Psychiatry*. 2007;64(3):305–315.
18. Grant BF. Prevalence and correlates of alcohol use and DSM-IV alcohol dependence in the United States: results of the National Longitudinal Alcohol Epidemiologic Survey. *J Stud Alcohol*. 1997;58(5):464–473.
19. Stranges S, Freudenheim JL, Muti P, et al. Greater hepatic vulnerability after alcohol intake in African Americans compared with Caucasians: a population-based study. *J Natl Med Assoc*. 2004;96(9):1185–1192.
20. National Center for Health Statistics. *Health, United States, 2009: With Special Feature on Death and Dying*. Hyattsville, MD: US Department of Health and Human Services, Centers for Disease Control and Prevention; 2010.
21. Williams DR, Leavell J. The social context of cardiovascular disease: challenges and opportunities for the Jackson Heart Study. *Ethn Dis*. 2012;22(suppl 1):S1–S14.
22. Wheaton B, Montazer S. Stressors, stress and distress. In: Scheid TL, Brown TN, eds. *A Handbook for the Study of Mental Health: Social Contexts, Theories, and Systems*. 2nd ed. Cambridge: Cambridge University Press; 2010:171–199.
23. Selye H. A syndrome produced by diverse nocuous agents. *Nature*. 1936;138:32.
24. Sapolsky RM, Romero LM, Munck AU. How do glucocorticoids influence stress responses? Integrating permissive, suppressive, stimulatory, and preparative actions. *Endocr Rev*. 2000;21(1):55–89.
25. McEwen BS. Physiology and neurobiology of stress and adaptation: central role of the brain. *Physiol Rev*. 2007;87(3):873–904.
26. Eiland L, McEwen BS. Early life stress followed by subsequent adult chronic stress potentiates anxiety and blunts hippocampal structural remodeling. *Hippocampus*. 2012;22(1):82–91.
27. McEwen BS, Gianaros PJ. Stress- and allostasis-induced brain plasticity. *Annu Rev Med*. 2011;62:431–445.
28. Cohen S, Janicki-Deverts D, Miller GE. Psychological stress and disease. *JAMA*. 2007;298(14):1685–1687.
29. McEwen BS. Stressed or stressed out: what is the difference? *J Psychiatry Neurosci*. 2005;30(5):315.
30. McEwen BS. Brain on stress: how the social environment gets under the skin. *Proc Natl Acad Sci U S A*. 2012;109(suppl 2):17180–17185.

31. Thoits PA. Stress and health major findings and policy implications. *J Health Soc Behav.* 2010;51(suppl 1):S41–S53.
32. Shonkoff JP. From neurons to neighborhoods: old and new challenges for developmental and behavioral pediatrics. *J Dev Behav Pediatr*. 2003;24(1):70–76.
33. Cushing BS, Kramer KM. Mechanisms underlying epigenetic effects of early social experience: the role of neuropeptides and steroids. *Neurosci Biobehav Rev*. 2005;29(7):1089–1105.
34. Diener E, Chan MY. Happy people live longer: subjective well-being contributes to health and longevity. *Appl Psychol Health Well Being*. 2011;3(1):1–43.
35. Hammen C. Stress and depression. *Annu Rev Clin Psychol.* 2005;1:293–319.
36. Monroe SM, Simons AD. Diathesis-stress theories in the context of life stress research: implications for the depressive disorders. *Psychol Bull*. 1991;110(3):406.
37. Duncan GJ, Daly MC, McDonough P, Williams DR. Optimal indicators of socioeconomic status for health research. *Am J Public Health*. 2002;92(7):1151–1157.
38. Wulff K, Gatti S, Wettstein JG, Foster RG. Sleep and circadian rhythm disruption in psychiatric and neurodegenerative disease. *Nat Rev Neurosci*. 2010;11(8):589–599.
39. Slopen N, Dutra LM, Williams DR, et al. Psychosocial stressors and cigarette smoking among African American adults in midlife. *Nicotine Tob Res*. 2012;14(10):1161–1169.
40. Isovich E, Mijnster MJ, Flügge G, Fuchs E. Chronic psychosocial stress reduces the density of dopamine transporters. *Eur J Neurosci*. 2000;12(3):1071–1078.
41. Dagher A, Tannenbaum B, Hayashi T, Pruessner JC, McBride D. An acute psychosocial stress enhances the neural response to smoking cues. *Brain Res*. 2009;1293:40–48.
42. US Census Bureau. *The 2010 Statistical Abstract*. Washington, DC: US Census Bureau; 2010.
43. Wolff EN. *Recent Trends in Household Wealth in the United States: Rising Debt and the Middle-Class Squeeze – An Update to 2007*. Annandale-on-Hudson, NY: Levy Economics Institute of Bard College; 2010.
44. Murphy SL. *Deaths: Final Data for 1998*. Hyattsville, MD: Centers for Disease Control and Prevention; 2000.
45. Braveman PA, Cubbin C, Egerter S, Williams DR, Pamuk E. Socioeconomic disparities in health in the United States: what the patterns tell us. *Am J Public Health*. 2010;100:S186–S196.
46. Thomas J, Thomas DJ, Pearson T, Klag M, Mead L. Cardiovascular disease in African American and white physicians: the Meharry cohort and Meharry-Hopkins cohort studies. *J Health Care Poor Underserved*. 1997;8(3):270–284.
47. Williams DR, Mohammed SA. Discrimination and racial disparities in health: evidence and needed research. *J Behav Med*. 2009;32(1):20–47.
48. Kaufman JS, Cooper RS, McGee DL. Socioeconomic status and health in blacks and whites: the problem of residual confounding and the resiliency of race. *Epidemiology*. 1997;8:621–628.
49. Williams DR, Collins C. US socioeconomic and racial differences in health: patterns and explanations. *Annu Rev Sociol*. 1995;21:349–386.
50. Cell JW. *The Highest Stage of White Supremacy: The Origin of Segregation in South Africa and the American South*. New York: Cambridge University Press; 1982.
51. Williams DR, Collins C. Racial residential segregation: a fundamental cause of racial disparities in health. *Public Health Rep*. 2001;116(5):404–416.
52. Orfield G, Frankenberg E, Garces LM. Statement of American social scientists of research on school desegregation to the US Supreme Court in Parents v. Seattle School District and Meredith v. Jefferson County. *Urban Rev*. 2008;40(1):96–136.
53. Cutler DM, Glaeser EL. Are ghettos good or bad? *Q J Econ*. 1997;112:827–872.
54. Ludwig J, Duncan GJ, Gennetian LA, et al. Long-term neighborhood effects on low-income families: evidence from moving to opportunity. *National Bureau Econ. Res. Working Paper Series*. 2013; No. 18772.
55. Slopen N, Lewis TT, Gruenewald TL, et al. Early life adversity and inflammation in African Americans and whites in the midlife in the United States survey. *Psychosom Med*. 2010;72(7):694–701.
56. Morello-Frosch R, Jesdale BM. Separate and unequal: residential segregation and estimated cancer risks associated with ambient air toxics in US metropolitan areas. *Environ Health Perspect*. 2006;114(3):386–393.
57. Williams DR, Mohammed SA. Poverty, migration, and health. In: Lin Ann C, Harris DR, eds. *The Colors of Poverty*. New York: Russell Sage Foundation; 2008:135–169.
58. Krieger N. Embodying inequality: a review of concepts, measures, and methods for studying health consequences of discrimination. *Int J Health Serv*. 1999;29(2):295–352.
59. Williams DR, Neighbors HW, Jackson JS. Racial/ethnic discrimination and health: findings from community studies. *Am J Public Health*. 2003;93(2):200–208.
60. Pascoe EA, Richman LS. Perceived discrimination and health: a meta-analytic review. *Psychol Bull*. 2009;135(4):531–554.
61. Williams DR, Mohammed SA. Racism and health. I: Pathways and scientific evidence. *Am Behav Sci*. 2013;57(8):1152–1173.
62. Arias E. United States life tables, 2004. *Natl Vital Stat Rep*. 2007; 56(9):1–39.
63. Taylor SE, Klein LC, Lewis BP, Gruenewald TL, Gurung RAR, Updegraff JA. Biobehavioral responses to stress in females: tend-and-befriend, not fight-or-flight. *Psychol Rev*. 2000;107:411–429.
64. Kuzawa CW, Sweet E. Epigenetics and the embodiment of race: developmental origins of US racial disparities in cardiovascular health. *Am J Hum Biol*. 2009;21(1):2–15.
65. Muñoz RF, Cuijpers P, Smit F, Barrera AZ, Leykin Y. Prevention of major depression. *Annu Rev Clin Psychol*. 2010;6(1):181–212.
66. Hall SM, Muñoz RF, Reus VI. Cognitive-behavioral intervention increases abstinence rates for depressive-history smokers. *J Consult Clin Psychol*. 1994;62(1):141–146.
67. Lazar SW, Bush G, Gollub RL, Fricchione GL, Khalsa G, Benson H. Functional brain mapping of the relaxation response and meditation. *Neuroreport*. 2000;11(7):1581–1585.
68. Akee R, Copeland W, Keeler G, Angold A, Costello J. Parents' incomes and children's outcomes: a quasi-experiment using transfer payments from casino profits. *Am Econ J Appl Econ*. 2008;2(1):86–115.

Index

Note: Page numbers followed by "b", "f", and "t" indicate boxes, figures, and tables respectively.

B

C

D

E

F

G

N

O

P

Q

R

S